AA002370

# 2006 IEEE International Conference on Power Electronics, Drives and Energy Systems for Industrial Growth

New Delhi, India
12 – 15 December 2006

Volume 1 of 2

IEEE Catalog Number:    06TH8899
ISBN:                   0-7803-9771-1

**Copyright © 2006 by The Institute of Electrical and Electronics Engineers, Inc.**
**All Rights Reserved**

*Copyright and Reprint Permissions:* Abstracting is permitted with credit to the source. Libraries are permitted to photocopy beyond the limit of U.S. copyright law for private use of patrons those articles in this volume that carry a code at the bottom of the first page, provided the per-copy fee indicated in the code is paid through Copyright Clearance Center, 222 Rosewood Drive, Danvers, MA 01923.

For other copying, reprint or republications permission, write to IEEE Copyrights Manager, IEEE Operations Center, 445 Hoes Lane, Piscataway, New Jersey USA 08854. All rights reserved.

IEEE Catalog Number:        06TH8899

ISBN:        0-7803-9771-1

LOC:        2006928206

**Additional Copies of This Publication Are Available from:**

IEEE Service Center
445 Hoes Lane
Piscataway, NJ 08854
IEEE Service Center
445 Hoes Lane
Piscataway, NJ 08854
Phone:        (800) 678-IEEE
        (732) 981-1393
Fax:        (732) 981-9667
E-mail:        customer-service@ieee.org

# 2006 IEEE International Conference on Power Electronics, Drives and Energy Systems for Industrial Growth

New Delhi, India
12-15 December 2006

IEEE Catalog Number: CFP06PED-POD
ISBN: 978-0-78039-771-2

# Table of Contents

**A program for harmonic modeling of distribution network transformers and determination of loss in the transformers and the amount of decrease of their life** ...... 1
*M. Marzband, A. Shaikholeslami*

**Novel Integral Cycle Voltage Controller for Self Excited Induction Generators** ...... 7
*S. S. Murthy, A. J. P. Pinto, A. R. Beig*

**EMI Modeling and Simulation of High Voltage Planar Transformer** ...... 11
*Bai Feng, Niu Zhong-Xia, Shi Yu-Jie, Zhou Dong-Fang*

**Graphical Estimation of Optimum Weights of Iron and Copper of a Transformer** ...... 15
*C. Easwarlal, V. Palanisamy, M.Y. Sanavullah, M.Gopila*

**Nonlinear Behavior of Self-excited Induction Generator Feeding an Inductive Load** ...... 20
*D.D.Ma, B. Zahawi, D. Giaouris, S. Banerjee, V. Pickert*

**Effects of Different Voltage Sags on Three-Phase Transformers** ...... 25
*M. R. Shakarami, A. Jalilian*

**Design and Transient Analysis of Cage Induction Motor Using Finite Element Methods** ...... 30
*Bhoj Raj Singla, Sanjay Marwaha, Anupma Marwaha*

**Methodology for Estimating Performance Characteristics of Three Phase Induction MotorOperating Direct-on-Line or with Six Pulse Inverter** ...... 35
*Slatish Chander Slabharwal*

**Design of Squirrel Cage Induction Motors for Traction Applications** ...... 39
*S. S. Murthy, Bhim Singh, G. Bhuvaneswari, Kiran Naidu, Uddanti Siva*

**Effect of Sequential Phase Energization on the Inrush Current of a Delta Connected Transformer** ...... 46
*K. P. Basu, Ali Asghar, Stella Morris*

**Accurate Performance Prediction of Three-Phase Induction Motor by FEM Using Separate Saturation Curves for Teeth and Yoke** ...... 50
*V. Jaiswal, M. Fazil, A. Hangal, N. Ravi*

**Nonlinear Sliding-Mode Controller for Sensorless Speed control of DC servo Motor Using Adaptive Backstepping Observer** ...... 54
*A. Farrokh Payam, B. Mirzaeian Dehkordi*

**Robust Speed Sensorless Control of Doubly-Fed Induction Machine Based on Input-Output Feedback Linearization Control Using a Sliding-Mode Observer** ...... 59
*A. Farrokh Payam, M. Jalalifar*

**Adaline Based Control of Solid State Voltage Regulator for Isolated Asynchronous Generators** ...... 64
*Bhim Singh, Gaurav Kumar Kasal*

**Development of a Prototype Controller for PMDC Motor Based Portable Telemetry Tracking System for Defense Application** ...... 70
*Parveen Kumar, A K Pradhan, Gautam Sadhukhan*

**Design & Development of a High Performance Electronic Starter for Single- Phase Induction Motors** ...... 75
*T. P. Shenoy, J. S. Nirody*

**Transient Analysis of a Single-Phase Self- Excited Induction Generator using a Three- Phase Machine feeding Dynamic Load** ...... 80
*S. N. Mahato, M. P. Sharma, S. P. Singh*

**Performance Analysis of a Three-Phase Squirrel-Cage Induction Motor under Unbalanced Sinusoidal and Balanced Non-Sinusoidal Supply Voltages** ...... 86
*C. Thanga Raj, Pramod Agarwal, S. P. Srivastava*

**Efficiency Optimization of Induction Motor Using a Fuzzy Logic Based Optimum Flux Search Controller** ...... 90
*L. Ramesh, S. P.Chowdhury, S.Chowdhury, A. K. Saha, Y. H. Song*

**Observer Based Position and Speed Estimation of Interior Permanent Magnet Motor** ...... 96
*Bhim Singh, Prerna Gaur, A.P.Mittal*

**Genetic Algorithm Based Optimal Design of Switching Circuit Parameters for a Switched Reluctance Motor Drive** ...... 101
*Behzad Mirzaeian-Dehkordi, Peyman Moallem*

**Reduction of Cogging Torque in PMBLD Motor with Reduced Stator Tooth Width and Bifurcated Surface Area Using Finite $lement Analysis** ...... 107
*Zx Somanathamv flxVxKxflrasadv and 3x/xZajkumarx*

# Table of Contents

A Novel Phasor Diagram of Interior Permanent Magnet Synchronous Motors based on Spiral Vector Theory.............................. 111
*Bishnu P. Muni*

A Novel DTC Strategy of Torque and Flux Control for Switched Reluctance Motor Drive.................................................... 117
*R. Jeyabharath, P.Veena, M.Rajaram*

Remedial Strategies for the Minimization of Cogging torque in PMBDC Motor possessing Material Saturation ........................... 122
*M. H. Ravichandran, V. T. Sadasivan Achari, C. C. Joseph, Robert Devasahayam*

Fuzzy Pre-compensated PI Controller for PMBLDC Motor Drive..................................................................... 126
*Mukesh Kumar, Bhim Singh, B. P. Singh*

A Slimplified Design M'ethodology for Slwitched Reluctance M'otor using analytical and Finite Llement M'ethod...................... 131
*'0Y0fiavichandranq V0T0fladasivan 4chariq F0F0=osephq fiobert jevasahayam*

Computer Aided Design of Permanent Magnet Brushless DC Motor for Hybrid Electric Vehicle Application ............................... 135
*Bhim Singh, Devendra Goyal*

Design and Analysis of a 3 kVA,28 Permanent Magnet Brushless Alternator for Light Combat Aircraft ............................. 141
*Fhim SinghT , Jally Rlavi*

Estimation of Core Loss in a Switched Reluctance Motor Based on Actual Flux Variations ........................................ 146
*Nimit. K. Sheth, K. R. Rajagopal*

A Novel Hybrid Brushless dc motor/Generator for Hybrid Vehicles Applications ................................................ 151
*E. Afjei, H. Toliyat, H. Moradi*

Computer Aided Design and FE Analysis of a PM BLDC Hub Motor................................................................. 157
*K. R. Rajagopal, Chippa Sathaiah*

Effect of Armature Reaction and Skewing on the Performance of Radial-flux Permanent Magnet Brushless DC Motor .............. 163
*Parag Upadhyay, K. R. Rajagopal*

Torque Ripple Minimization of Interior Permanent Magnet Brushless DC Motor Using Rotor Pole Shaping ..................... 168
*Parag Upadhyay, K. R. Rajagopal*

Design and Development of a In-Wheel Brushless D.C. Motor Drive for an Electric Scooter ................................... 171
*N. Ravi, S. Ekram, D. Mahajan*

Comparative Study of Laminated Core Permanent Magnet Hybrid Stepping Motor with Soft Magnetic Composite
Core Claw Pole Motor........................................................................................................ 175
*E.V. Chandra Sekhara Rao, P.V.N. Prasad, G. Ravindranath*

A Doubly Fed Induction Motor as High Torque Low Speed Drive ............................................................... 179
*Mukhtar Ahmad, M.Rizwan Khan, Atif Iqbal*

Performance of Doubly Salient Permanent Magnet Motors for Parallel and Tapered Rotor Poles............................... 182
*Nimit K. Sheth, K. R. Rajagopal*

Improved Torque Profile of a Doubly Salient Permanent Magnet Motor using Skewed Rotor Teeth and Sinusoidal
Excitation.................................................................................................................. 187
*Nimit. K. Sheth, K. R. Rajagopal*

Dynamic Modeling and Simulation of an Induction Motor with Adaptive Backstepping Design of an Input-Output
Feedback Linearization Controller in Series Hybrid lectric Vehicle ........................................................ 193
*M. Jalalifar, A. Farrokh Payam, B. Mirzaeian, S. M. Saghaeian nezhad*

Prototyping of a Precision Mechanism Using a Hybrid-Driven Piezoelectric Actuator ....................................... 199
*Fu-Shin Lee, Yung-Tsung Lei, Sheng-Feng Chiang, Jyun-Jhong Jhang, Shao-Chun Tseng, Po-Jia Chen*

DSP Based Implementation of Vector Controlled Induction Motor Drive using Fuzzy Pre-compensated Proportional
Integral Speed Controllers................................................................................................. 204
*Bhim Singh, S. Ghatak Choudhuri*

Optimal Controller for High Frequency AC-Link Converter Induction Motor Drive System....................................... 210
*R. A. Gupta, A. K. Wadhwani, R. R. Joshi*

An Adaptive Backstepping Controller for Doubly-Fed Induction Machine Drives................................................ 215
*A. Farrokh Payam*

Application problem of PWM AC drives due to long cable length and high dv/dt .............................................. 219
*B.Basavaraja, D.V.S.S.Siva Sarma*

Adaptive Controller Design for Permanent Magnet Linear Synchronous Motor Control System.................................... 225
*B. Srinivasu, P.V.N. Prasad, M.V. Ramana Rao*

# Table of Contents

**An Overmodulation Scheme for Vector Controlled Induction Motor Drives** ........................................ 231
*S. Venugopal, G. Narayanan*

**Modified Direct Torque Control of Matrix Converter Fed Induction Motor Drive** ........................................ 237
*Bhim Singh, Jally Ravi*

**LMI Based Digital State Feedback Controller for a Wound Rotor Induction Drive with Guaranteed Closed Loop Stability** ........................................ 244
*D. Sivanandakumar, K. Ramakrishnan*

**Open-End Winding Induction Motor Driven With Matrix Converter For Common-Mode Elimination** ........................................ 250
*Krushna K Mohapatra, Ned Mohan*

**Elimination of Common Mode Voltage and Fifth and Seventh Harmonics in a Multilevel Inverter fed IM Drive using 12-Sided Polygonal Voltage Space Phasor** ........................................ 256
*Sanjay Lakshminarayanan, Gopal Mondal, P.N Tekwani, K. Gopakumar*

**A New Space Vector Pulsewidth Modulation for Reduction of Common Mode Voltage in Direct Torque Controlled Induction Motor Drive** ........................................ 262
*Y.V. Siva Reddy, T. Brahmananda Reddy, M. Vijaya Kumar*

**Parallel Power Flow AC/DC Converter with High Input Power Factor and Tight Output Voltage Regulation for Universal Voltage Application** ........................................ 267
*Aman Kumar Jha, K. Hari Babu, B. M. Karan*

**A Generalized Space Vector Modulation with Simple Control technique for Balancing DC-Bus Capacitor Voltages of a Three-Phase, Neutral-Point Clamped Converter** ........................................ 274
*A. H. Bhat, P. Agarwal*

**A Novel Load Compensator for a 12-pulse Diode Converter** ........................................ 280
*Maryclaire Peterson, Brij N. Singh*

**Resonant Operated Buck Converter with Reduced Device Switching Stress with Power Factor Improvement** ........................................ 286
*Vinayak N. Shet*

**A High Power Factor Forward Flyback Converter with Input Current Waveshaping** ........................................ 292
*Vinayak N. Shet*

**A Fuzzy Logic Controller for Direct Power Control of PWM Rectifiers with SVM** ........................................ 298
*R. Skandari, A. Rahmati, A. Abrishamifar, E. Abiri*

**DSP-Based Matrix Converter Operation Under Various Abnormal Conditions with Practicality** ........................................ 303
*Vinod Kumar, R R Joshi*

**Improvement of an input waveform of a Neutral Point Type Step-down Converter** ........................................ 307
*Yoshito KATO, Masaaki NAKAMURA, Nabil M. Hidayat, Nobuo TAKAHASHI*

**Development of Neutral-Point Type Converter and Application to Electronic Ballast** ........................................ 310
*Nabil M Hidayat, Masaaki Nakamura, Yoshito Kato, Nobuo Takahashi, Shun-ichi Adachi, Ichiro Yokozeki*

**Hysteresis-Band Current Control of a Four Quadrant AC -DC Converter giving IEEE 519 compliant performance at any Power Factor** ........................................ 315
*A.N.Arvindan, V.K.Sharma*

**Multiphase Inverter Topology and its Modulation Technique for Optimal Harmonic Output** ........................................ 321
*Ravindra Kumar Singh*

**A PWM Current Source Rectifier with Leading Power Factor** ........................................ 331
*B. Geethalakshmi, P. Sanjeevikumar, P. Dananjayan*

**A Novel Harmonic Mitigation Converter for Variable Frequency Drives** ........................................ 336
*Bhim Singh, Sanjay Gairola*

**Performance Comparison of High Frequency Isolated AC-DC Converters for Power Quality Improvement at Input AC Mains** ........................................ 342
*Bhim Singh, B.P. Singh, Sanjeet Dwivedi*

**Single-Phase Resonant Converter with Active Power Filter** ........................................ 348
*M. A. Chaudhari, H. M. Suryawanshi*

**PV Power Tracking Through Utility Connected Single-Stage Inverter** ........................................ 354
*K. S. Phani Kiranmai, Veerachary. M*

**A Novel Control of Bi-Directional Switches in Matrix Converter** ........................................ 360
*Meharegzi Tewolde, Shyama P. Das*

# Table of Contents

**PWM SHE Switching Algorithm for Voltage Source Vnverter** ........................................................ 366
*Ali. I. Maswood*

**New Fuzzy logic Controller for a Buck Converter** .................................................................. 370
*D. Seshachalam, R. K. Tripathi, D. Chandra, Anil kumar*

**Development of Conventional Control of Parallel Loaded Resonant Converter -Simulation and Experimental Evaluation** ........................................................................................................ 373
*T.S.Sivakumaran, S.P.Natarajan*

**A Novel Technique to Reduce the Switching Losses in a Synchronous Buck Converter** .................... 378
*A. K. Panda, Aroul. K*

**Transformer Core Unbalancing Issue in a Full-Bridge DC-DC Converter with Current Doubler Rectifier** ....... 383
*B.A. Gusev, V.I. Meleshin, D.A. Ovchinnikov*

**Computer Aided Analysis of Fault Tolerant Multilevel DC/DC Converters** ...................................... 389
*K. A. Ambusaidi, V. Pickert, B. Zahawi*

**Auto Voltage Balancing in High Power DC-DC Converter** ......................................................... 395
*S. B. Bodkhe, V. B. Virulkar , S. W. Mohod , M.V. Aware*

**Inrush Current Control of a DC/DC Converter Using MOSFET** .................................................... 401
*Gaddam Mallesham, Keerthi Anand*

**A ZVT Boost Converter using an Auxiliary Resonant Circuit** ...................................................... 407
*M. Phattanasak*

**Adaptive Hysteretic Control of 3rd Order Buck Converter** ......................................................... 413
*Veerachary M, Deepen Sharma*

**A Novel Topology for Multiple Output DC-DC Converters for One Cycle Control** ........................... 417
*Ravindra Kumar Singh*

**New Hybrid SVPWM Methods for Direct Torque Controlled Induction Motor Drive for Reduced Current Ripple** ...................... 424
*T. Brahmananda Reddy, J. Amarnath , D. Subbarayudu*

**Analysis of Experimental Investigation of Various Carrier-based Modulation Schemes for Three Level Neutral Point Clamped Inverter-fed Induction Motor Drive** ........................................................... 430
*Ranjan K. Behera, T. V. Dixit, Shyama P. Das*

**High Frequency SMPS Based Inverter With Improved Power Factor** ............................................. 436
*M. G. Wani, V. K. Sharma, K. M. Soni*

**Comparison of Mode Switched Controllers for a Pseudo Continuous Current Mode Boost Converter** ....... 443
*Sreekumar C, Vivek Agarwal*

**Multi-level inverter for Induction Motor Drive** ...................................................................... 449
*K.Chandra Sekhar, G.Tulasi Ram Das*

**A Unified Model For Auxiliary Switch Commutated DC-DC Converters** ........................................ 455
*N. Lakshminarasamma, V. Ramanarayanan*

**Novel Pulse Power Supply Operating at High Input Power Factor** ................................................ 460
*Vishnu K Sharma, Kishore Chatterjee, Vivek Agarwal*

**System Identification and controller tuning rule for DC-DC converter using ripple voltage waveform** ....... 463
*K. Lavanya, B. Umamaheswari, R. C. Panda*

**Space Vector Modulation with DC-Link Voltage Balancing Control for Three-Level Inverters** .............. 467
*Kalpesh H. Bhalodi, Pramod Agrawal*

**Investigations on Different Multilevel Inverter Control Techniques by Simulation** ............................ 473
*P. K. Chaturvedi, Shailendra K Jain, Pramod Agrawal, P. K. Modi*

**Peak-Current Mode control of Hybrid Switched Capacitor Converter** ........................................... 479
*Veerachary M, Singamaneni Bala Sudhakar*

**Observer based current control of single-phase inverter in DQ rotating frame** ................................. 485
*B.Saritha, and P.A.Jankiraman*

**MATLAB Simulation of current control of PMSM using single sensor technology** ........................... 490
*B. Saritha, P. A. Jankiraman*

**Novel Approach to Develop Behavioral Model Of 12-Pulse Converter** .......................................... 495
*Amit Sanglikar, Vinod John*

# Table of Contents

**Simulation of PMSM VSI Drive for Determination of the Size Limits of the DC-Link Capacitor of Aircraft Control Surface Actuator Drives** ............................................................................................................. 500
*M.Khatre, Alan G. Jack*

**A Novel Soft Switched Improved Power Quality Converter Fed D.C. Motor Drive** ............................................. 506
*M. B. Daigavane, Z. J. Khan, H. M. Suryawanshi*

**Generalized Discontinuous PWM Based Direct Torque Controlled Induction Motor Drive with a Sliding Mode Speed Controller** ................................................................................................................................................. 511
*T. Brahmananda Reddy, J. Amarnath, D. Subbarayudu, Md. Haseeb Khan*

**Hardware-in-Loop Simulation of Direct Torque Controlled Induction Motor** ............................................... 517
*P. K. Gujarathi, M. V. Aware*

**Near-Field Modeling and Prediction of Switched Mode Power Supply** ......................................................... 522
*Bai Feng, Niu Zhong-Xia, Shi Yu-Jie, Zhou Dong-Fang*

**Power Electronic Circuit-oriented Model for the Fuel Cell System** ............................................................. 526
*Veerachary M, Arun Shailendra Kumar*

**A Simplified Space-Vector Modulated Control Scheme for CSI fed IM drive** ............................................... 531
*P.Parthiban, Pramod Agarwal, S.P.Srivastava*

**A Study on Design and Dynamics of Voltage Source Inverter in Current Control Mode to Compensate Unbalanced and Non-linear Loads** ................................................................................................................................ 537
*Mahesh K. Mishra, K. Karthikeyan*

**Optimal Voltage and Reactive Power Control Based on Multi-Objective Genetic Algorithm** ....................... 545
*Behzad Mirzaeian Dehkordi*

**Model Validation Studies in Obtaining Q-V Characteristics of P-Q Loads in Respect of Reactive Power Management and Voltage Stability.** ....................................................................................................................... 550
*G. Govinda Rao, K. V. S. Ramachandra Murthy*

**Simulation Study of a Shunt 5ctive Power Filter Using Nonlinear Least Squares Harmonic Extraction Technique** ...... 555
*RM Bhudamani, JM Vasudevan, BMSM Ramalingam*

**Comparison of Synchronous Detection and I.Cosf Shunt Active Filtering Algorithms** ............................... 560
*G. Bhuvaneswari, Manjula G. Nair, Sathish Kumar Reddy*

**A Nonlinear Control Method for SSSC to Improve Power System Stability** .................................................. 565
*Majid Poshtan, Brij N. Singh, Parviz Rastgoufard*

**An Improved Power Flow Analysis Technique with STATCOM** ................................................................... 572
*Annapurna Bhargava, Vinay Pant, Biswarup Das*

**Design of a Current Hybrid Filter Including Active and Variable Passive Filters** ......................................... 577
*H. Dalvand*

**Grid Connected Photovoltaic Interface with VAR Compensation and Active Filtering Functions** ................ 583
*Aslain Ovono Zué, Ambrish Chandra*

**Design and Implementation of a Current Controlled Parallel Hybrid Power Filter** ..................................... 589
*Bhim Singh, Vishal Verma*

**Active Power Filter Control in Three-Phase four-wire Systems using Space Vector Modulation** .................. 596
*H. Mokhtari, M. Rahimi*

**Operation of a 12-pulse converter in closed loop for controlled P-Q operation** ......................................... 602
*Faisal M. Ahsan, J.K. Chatterjee, Anandarup Das*

**A Novel Structure for Three-Phase Four-Wire Distribution System Utilizing Unified Power Quality Conditioner (UPQC)** ................................................................................................................................................. 608
*V. Khadkikar, A. Chandra*

**Load Compensation for Diesel Generator Based Isolated Generation System Employing DSTATCOM** ........ 614
*Bhim Singh, Jitendra Solanki*

**Automatic Classification of Power Quality Events Using Multiwavelets** ...................................................... 620
*Surender Dahiya, D.K. Jain, Manish Kumar, Ashok Kumar, Rajiv Kapoor*

**Power quality monitoring at the industrial, commercial and educational centers of Mazandaran province and presenting the related solution** ............................................................................................................... 625
*M. Marzband, A. Shaikholeslami*

# Table of Contents

A New Power Quality Enhancement Method for Two-Phase Loads ................................................................ 631
*H. Hojabri, H. Mokhtari*

Three level STATCOM Based Power Quality Solution for a 4 MW Induction Furnace ........................................ 636
*Unnikrishnan A.K, Aby Joseph, Subhash Joshi T G*

Analysis and Simulation of Single Phase Composite Observer for Harmonics Extraction ........................... 641
*K. Selvajyothi, P. A. Janakiraman*

Third Harmonic Current Injection for Power Quality Improvement in Rectifier Loads .............................. 647
*Bhim Singh, Vipin Garg, G.Bhuvaneswari*

Polygon Connected 15-Phase AC-DC Converter for Power Quality Improvement .................................... 652
*Bhim Singh, Vipin Garg, G.Bhuvaneswari*

Power Quality Standards and Their Application to a Granite Factory ................................................... 657
*S. Hasani, F. Donyavi, M. Masoudi, H. Mokhtari*

Minimization of Losses in Radial Distribution System by using HVDS ................................................. 662
*K. Amaresh, S. Sivanagaraju, V. Sankar*

SVPWM Switched DSTATCOM for Power Factor and Voltage Sag Compensation ................................ 667
*Bishnu P. Muni, S. Eswar Rao, JVR Vithal*

Unified Constant frequency Integration Control of Universal Power Quality Conditioner ...................... 673
*Vadirajacharya K, Pramod Agarwal, H.O.Gupta*

Application of a Boundary Model to Assess Power Quality Cost Function ......................................... 678
*J. Ahmadian, A. Jalilian, M.A.S. Masoum*

Active Power Filter Solution without PLL for Fluctuating Industrial Load ......................................... 683
*S. Elangovan*

A Novel Digital Signal Processing Algorithm for On-line Assessment of Power System Frequency .............. 689
*Arghya Sarkar, S. Sengupta*

An Evolutionary Algorithm Approach to Estimate the Parameters of Power Quality Signals ................... 695
*V. Ravikumar Pandi, B. K. Panigrahi*

A 36-Pulse AC-DC Converter for Line Current Harmonic Reduction ............................................... 701
*Bhim Singh, Sanjay Gairola*

A Unified Analysis of CCM Boost PFC for Various Current Control Strategies .................................. 707
*Ranjan K. Gupta, Hariharan Krishnaswami, Ned Mohan*

Minimum Loss Configuration of Power Distribution System ....................................................... 712
*L Jaswant, T. Thakur*

Control of Cascaded H-Bridge Converter based DSTATCOM for High Power Applications ..................... 718
*K. Anuradha, B.P.Muni, A. D. Raj Kumar*

Detection and Classification of Non-stationary Power disturbances in Noisy Conditions ...................... 724
*B. K. Panigrahi, S. K. Sinha*

3-Phase Fault Current Limiter for distribution systems ........................................................... 729
*Vijay K. Sood, Shahabur Alam*

Power Flow Control of a Solid Oxide Fuel-Cell for Grid Connected Operation .................................. 735
*Ankur Goel, S. Mishra, A.N. Jha*

An Universal Interconnection System to Connect Distributed Generation to the Grid ......................... 740
*Vinod John, Eric Benedict, Shazreen*

Transient Fault Response of Grid Connected Wind Electric Generators .......................................... 747
*Vinodh Kumar P, Meera K S, Sasi K Kottayil*

Black Start with DFIG Based Distributed Generation after Major Emergencies ................................. 753
*M. Aktarujjaman, M.A. Kashem, M. Negnevitsky, G. Ledwich*

Fuzzy Logic Based Control of Wind Turbine Driven Squirrel Cage Induction Generator Connected to Grid ......... 759
*CH.Siva Kumar, A.V.R.S.Sarma, P.V.N. Prasad*

Speed Sensor-less Direct Power Control of a Matrix Converter Fed Induction Generator for Variable Speed Wind Turbines ........................................................................................................ 765
*T. Satish, K.K. Mohapatra, Ned Mohan*

*viii*

# Table of Contents

**Stochastic Model for Optimal Selection of DDG by Monte Carlo Simulation** ...................... 771
*N. Vaitheeswaran, R. Balasubramanian*

**Capacitive Self-Excitation in a Six-Phase Induction Generator for Small Hydro Power  An Experimental Investigation** ............ 776
*G. K. Singh, K. B. Yadav, R. P. Saini*

**Grid Power Quality with Variable Speed Wind Energy Conversion** ...................... 782
*S.W. Mohod, M. V. Aware*

**Investigations on Combined Operation of Industrial Distribution System and utility in Distributed Generation Environment** ...................... 787
*K. Manjunatha Sharma, K.P. Vittal, T.K. Nagaraja Rao*

**Rotor Speed Stability Analysis of Constant Speed Wind Turbine Generators** ...................... 792
*M. G. Kanabar, C. V. Dobariya, S. A. Khaparde*

**Performance Evaluation of Indian Electric Power Utilities Based on Data Envelopment Analysis** ...................... 797
*Tripta Thakur*

**Modelling of Hybrid Energy System for Off Grid Electrification of Clusters of Villages** ...................... 801
*Ajai Gupta, R P Saini, M P Sharma*

**PSO-Based Multidisciplinary Design of A Hybrid Power Generation System With Statistical Models of Wind Speed and Solar Insolation** ...................... 806
*Lingfeng Wang, Chanan Singh*

**SVPWM Implementation in dSPACE for Generalized Impedance Controller Used for Self Excited Induction Generation System** ...................... 812
*B.Venkatesa Perumal, J.K. Chatterjee*

**Trajectory Sensitivity Analysis in Distributed Generation Systems** ...................... 818
*Dheeman Chatterjee, Arindam Ghosh, M. A. Pai*

**Steady State Performance Of A Stand-Alone Variable Speed Constant Frequency Generation System Using A New Build Up Algorithm** ...................... 824
*Isha T B, D. Kastha*

**Control Strategy of Distributed Generation for Voltage Support in Distribution Systems** ...................... 830
*M. Negnevitsky, G. Ledwich, An D.T. Le, M. A. Kashem, Seni*

**A Steady State Analysis on Voltage and Frequency Control of Self-Excited Induction Generator in Micro-Hydro System** ...................... 836
*Bhim Singh, S.S. Murthy, Madhusudan, Manish Goel, A. K. Tandon*

**A Novel Digital Control Technique of Electronic Load Controller for SEIG Based Micro Hydel Power Generation** ...................... 842
*S. S. Murthy, Ramrathnam,  M. S. L.Gayathri, Kiran Naidu, U. Siva*

**Analysis and Design of Voltage and Frequency Controllers for Isolated Asynchronous Generators in Constant Power Applications** ...................... 847
*Bhim Singh, Gaurav Kumar Kasal*

**A Simple Controller using Line Commutated Inverter with Maximum Power Tracking for Wind-Driven Grid-Connected Permanent Magnet Synchronous Generators** ...................... 854
*V. Lavanya, N. Ammasai Gounden, Polimera Malleswara Rao*

**A High-power High-frequency and Scalable Multi-megawatt Fuel-cell Inverter for Power Quality and Distributed Generation** ...................... 860
*Sudip K. Mazumder, Rongjun Huang*

**Integrating a Redox Flow Battery System with a Wind-Diesel Power System** ...................... 865
*Shameem Ahmad Lone, Mairaj-ud-Din Mufti*

**Hydrocarbon Fuel Based Micro Battery Power System** ...................... 871
*Surendran Devadoss, Theo Kangsanant, Ian Bates*

**Analysis, Design and Development of Single Switch Forward Buck AC-DC Converter for Low Power Battery Charging Application** ...................... 876
*Bhim Singh, Ganesh Dutt Chaturvedi*

**A Novel Approach for Eco-Friendly and Economic Power Dispatch using MATLAB** ...................... 882
*D.P.Kothari, K.P.Singh Parmar*

**Real Time Based PI-like Fuzzy Controller for DC Servomotor** ...................... 888
*S.G. Kadwane, Swapnil Gupta, B.M. Karan, T Ghose, Amit Kumar*

# Table of Contents

**Neural Network Based DSTATCOM Controller for Three-phase, Three-wire System** ..................... 892
*Bhim Singh, A. Adya, A. P. Mittal, J.R.P Gupta*

**Analysis of the Influence of Control Parameters on Wind Farm Output: a Sensitivity Analysis using ANN Modelling** ................. 898
*E. Fernandez, M. Carolin Mabel*

**An Advanced Control Scheme for Micro Hydro Power Plants** ..................... 902
*M. Hanmandlu, Himani Goyal, D. P. Kothari*

**Application of Fuzzy Logic PSS to Enhance Transient Stability in Large Power Systems** ..................... 909
*P. V. Etingov, N. I. Voropai*

**Neural Approach for Automatic Identification of Induction Motor Load Torque in Real-Time Industrial Applications** ..................... 918
*A. Goedtel, I. N. da Silva, P. J. A. Serni*

**Speed Estimation for Sensorless Technology Using Recurrent Neural Networks and Single Current Sensor** ..................... 926
*A. Goedtel, I. N. da Silva, P. J. A. Serni*

**Electricity Price Forecasting Using Artificial Neural Network** ..................... 931
*M. Ranjbar, S. Soleymani, N. Sadati, A. M. Ranjbar*

**A New Approach for Fault Location Identification in Transmission system using Stability Analysis and SVMs** ..................... 936
*D. Thukaram, H. P. Khincha, B. Ravikumar*

**Fast and Effective Algorithm for Economic Dispatch with Prohibited operating zones** ..................... 942
*T. Adhinarayanan, M. Sydulu*

**Computation & Analysis of End Region EM Force for Electrical Rotating Machines using FEM** ..................... 948
*Manpreet Singh Manna, Sanjay Marwaha, Anupma Marwaha*

**Optimal Reactive Power Dispatch based on Voltage Stability Criteria in a Large Power System with AC/DC and FACTs Devices** ..................... 953
*D.Thukaram, G. Yesuratnam, C. Vyjayanthi*

**Location of Unified Power Flow Controller and its Parameters settin for congestion Management in Pool M arket Model** ..................... 959
*Hassan Barati, Mehdi Ehsan, Mahmud Fotuhi-Firuzabad*

**Security Enhancement of Optimal Power Flow using Genetic Algorithm** ..................... 966
*N.B. Muthuselvan, P. Somasundaram, and Subhransu Sekhar Dash*

**Congestion Management in Nodal Pricing With Genetic Algorithm** ..................... 970
*S.M.H Nabav, Shahram Jadid, M.A.S. Masoum, A. Kazemi*

**Coupled Magneto-Mechanical Field Computations** ..................... 975
*Amogh Kank, G. B. Kumbhar, S. V. Kulkarni*

**Optimizing Voltage Stability Limit and Real Power Loss in a Large Power System using Bacteria Foraging** ..................... 979
*M. Tripathy, S. Mishra*

**Application of Power Flow Sensitivity Analysis and PTDF for Determination of ATC** ..................... 985
*N. D. Ghawghawe, K. L. Thakre*

**Application of Tabu-Search Algorithm for Network Reconfiguration in Radial Distribution System** ..................... 992
*T. Thakur, Jaswanti*

**Comparative Studies of Transient and Steady State Analysis for a Typical 765kV/400kV EHV Transmission System in Indian Power System** ..................... 996
*D. Thukaram, H. P. Khincha, P. Shyamala*

**A Finite Element Modeling and Simulation Method for Time-Varying Field-Circuit Problems** ..................... 1002
*Prem Sagar*

**A Wavelet Based Numerical Technique for Electromagnetic Field Analysis** ..................... 1007
*Kaushik K, S. V. Kulkarni*

**Frequency Linked Pricing as an Instrument for Frequency Regulation Market and ABT Mechanism** ..................... 1013
*K. V. V. Reddy, Ashwani Kumar, Saurabh Chanana*

**Induction Machine Fault Identification using Particle Swarm Algorithms** ..................... 1020
*S. A. Ethny, P. P Acarnley, B. Zahawi, D. Giaouris*

**A Novel Technique for Identification and Condition Monitoring of Nonlinear Loads in Power Systems** ..................... 1024
*Phil Gilreath, Maryclaire Peterson, Brij N. Singh*

# Table of Contents

**Real-Time Identification of Distributed Bearing Faults in Induction Motor** .................... 1031
*Rajesh Patel , S P Gupta, Vinod Kumar*

**Integration of IEDs Using Legacy and IEC61850 Protocol** ........................................ 1036
*Anupama Prakash, Mini S. Thomas, Ashutosh Gautam*

**Ethernet Enabled Fast and Reliable Monitoring, Protection and Control of Electric Power Substation** .................... 1041
*Iqbal Ali, Mini S. Thomas*

**Expert System for Power Transformer Condition Monitoring and Diagnosis** ................... 1047
*M. Ahfaz Khan, A.K. Sharma, Rakesh Saxena*

**Evaluation of Leakage Current Measurement for Site Pollution Severity Assessment** ........ 1053
*S.M.H Nabavi, A. Gholami, A. Kazemi, M.A.S. Masoum*

**Detection of Bearing Failure in Rotating Machine Using Adaptive Neuro-Fuzzy Inference System** ............ 1059
*Sulochana Wadhwani , A.K. Wadhwani, S P Gupta, Vinod Kumar*

**Discrimination between Inrush current and Internal Faults using Pattern Recognition Approach** ........... 1064
*B. K. Panigrahi, S. R. Samantaray, P. K. Dash, G. Panda*

**Stepwise Restoration of Power Distribution Network under Cold Load Pickup** ................ 1069
*Vishal Kumar, Rohith Kumar H.C., I. Gupta, H.O. Gupta*

**Power Sector Reforms in India** .......................................................................... 1074
*Harbans L. Bajaj, Deepak Sharma*

**A New Structure for Electricity Market Scheduling** .............................................. 1079
*S. Soleymani, A. M. Ranjbar, A. R. Shirani*

**Modelling of STATCOM Based Voltage Regulator for Self-Excited Induction Generator with Dynamic Loads** ........... 1084
*Bhim Singh, S. S. Murthy, Sushma Gupta*

**Optimum Design of UPFC Controllers Using GEA: Decoupled Real & Reactive Power Flow and Damping Controllers** ........... 1090
*N. Ray Chaudhuri, M. L. KotharI*

**Application of Static Synchronous Series Compensator to Dam Sub-Synchronous Resonance** ........ 1096
*M. Ehsan, M. Fotuhi-Firuzabad, S. M. T. Bathaee*

**A New 24-Pulse STATCOM for Voltage Regulation** ............................................. 1102
*Bhim Singh, R. Saha*

**A Nonlinear Fuzzy PID Controller for CSI-STATCOM** ........................................ 1107
*A. Kazemi, A. Tofighi, B.Mahdian*

**Distance Relay Tripping Characteristic in Presence of UPFC** ................................... 1114
*S. Jamali, A. Kazemi, H. Shateri*

**Investigations on Boundaries of Controllable Power Flow with Unified Power Flow Controller** .......... 1120
*S. Srividhya, C. Nagamani, A. Karthikeyan*

**VSC Based HVDC System for Passive Network with Fuzzy Controller** ...................... 1127
*A. K. Moharana , Ms. K. Panigrahi, B. K. Panigrahi, P. K. Dash*

**Voltage Regulation and Power Flow Control of VSC Based HVDC System** ................. 1131
*Bhim Singh, B. K. Panigrahi, D. Madhan Mohan*

**Modeling and Simulation of Electromagnetic Conducted Emission Due to Power Electronics Converters** .......... 1137
*A. Farhadi, A. Jalilian*

**Evaluation of Operational Characteristics Of Electronic Ballasts For Metal-Halide HID Lamps** .......... 1143
*Ahteshamul Haque, M. S. Jamil Asghar*

**Active Power Filter Control Algorithm using Wavelets** .......................................... 1150
*Karunesh K Gupta, Rajneesh Kumar, H. V. Manjunath*

**Effects of Power Lines on Performance of Home Control System** ............................... 1154
*V. Chunduru, N. Subramanian*

*xi*

**2006 IEEE International Conference on Power Electronic, Drives and Energy Systems**

# A program for harmonic modeling of distribution network transformers and determination of loss in the transformers and the amount of decrease of their life

M. Marzband, *Student Member, IEEE,* and A. Shaikholeslami, *Student Member, IEEE*

*Abstract*—harmonic modeling of power network transformers is dealt with. The input of the electromagnetic model, voltages and distorting currents and the output of this model are harmonic losses. By using the thermal model and the input of losses temperature distribution is obtained. By using the temperature distribution the amount of transformer's loss of life and the related cost are calculated. For thermal modeling of the transformer existing experimental formulas in the other papers that have been obtained by experimental results have been used. About the presented harmonic model for the transformer's equivalent circuit the following points have been considered: 1) considering the shell and the juxtaposition effect in the modeling of loss. 2) Modeling of the core's saturation effect by adding the current source which is dependant on the voltage. 3) The core's loss as a function of the frequency. 4) Adding leakage and stray capacitors in the model.

*Index Terms*—Transformer, Hot point, Loss of life Transformer, Harmonic loss

## I. INTRODUCTION

THE transformers of the distribution network are the most important sections of this network that have to be considered from the harmonic model's point of view. For considering the harmonic effect over transformers the two following jobs have to be done:
1- Obtaining the electromagnetic model of the transformer.
2- Obtaining the thermal model of the transformer.

The input of the problem here are the distorting voltages on the high pressure side and the distorting currents on the low pressure side. From the electromagnetic model and by the mentioned outputs one has to be able to obtain the harmonic losses in the transformer. These losses that are originally the output of the electromagnetic model and act as thermal

sources are considered as an input for the thermal model. From the thermal model and the loss's input one should be able to obtain the transformer's hot point's temperature that the temperature of these points are used for detaining the transformer's life under harmonic conditions. Also we can calculate the amount of the nominal power's decrease from this point's temperature. Modeling transformers for harmonic frequencies is a very complicated task. The most exact method would be to model the transformer in the ring to ring form and by considering all the capacitance and mutual induction effects. It is clear that this model is very complex and is not proper for use in the computer program [1]-[2]-[4]. For proper modeling of the transformer for the exact examining and simulating of the harmonics the following three points should be considered.
1- Nonlinear properties resulting from over excitation.
2- Phase change resulting from the different type of coil connections
3- Impedance characteristics dependent on the frequency.

In explaining the first case one say that if the transformer works in the saturated condition (because of high voltage) also the magnetic circuit has to come in action and it has to be shown as harmonic sources. In this case the values of the injected harmonics by the transformer to the network will be completely large. Also another important point that should be remembered is that the three phase transformers create various phase differences between the primary and secondary voltages. For example if we consider a $\Delta Y$ Transformer in which there is a $30°$ Phase difference between the primary and secondary voltage by considering the following relationships for the created phase differences at deferent harmonics:

$$pocitive = |h. \ basic \quad basic| = (h \quad 1)| \ basic| \quad (1)$$

This relationship is about positive sequence harmonics and the relationship below is true for negative sequence harmonics:

$$negative = |h. \ basic \ + \ basic| = (h+1)| \ basic| \quad (2)$$

In the above relationships h states the harmonic order and $\varphi basic$ states the phase differences between the primary and secondary transformer voltages at the fundamental frequency.

The transformer's simple model include the series resistance, the series diffusion inductance, the parallel

---

Part of the financial support of this work was supported by the Department of electrical engineering of the Mazandaran Univercity and other part was supported by National Petrochemical Company.

M. Marzband is with National Petrochemical Company, PARS SPECIAL ECONOMIC/ENERGY ZONE, ASSALOUYEH, BUSHEHR, IRAN, PO.BOX NO.75391-115, TEL: +98 7727323250-4, FAX: +98 7727323255 (e-mail: m_marzband2005@yahoo.com)

A. sheikhol-Eslami is with the Department of Electrical Engineering, Mazandaran University, Babol, Iran (e-mail: abdolahinegar@yahoo.com).

0-7803-9771-1/06/$25.00 ©2006 IEEE

magnetizing inductance, and the parallel resistance that are representative of the core's losses. In the general case the series resistance and the diffusion inductance change with the frequency because of the shell effect the shell effect will have more significance at the resonance frequency. In the fig.1 the variations of resistance and the transformer's diffusion inductance has been shown versus the frequency [3].

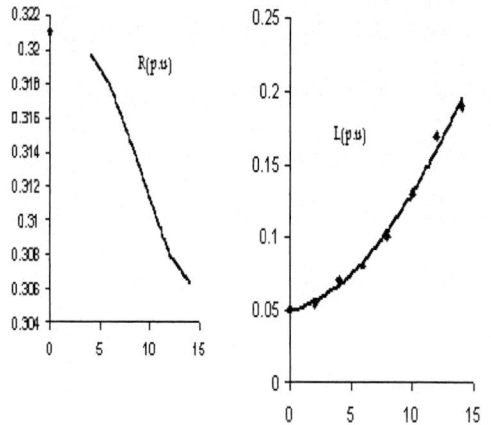

Fig. 1. The Variations of short Circuit Parameters Versus Frequency for a Sample Transformer.

As is inferred from the figure the variations of L with frequency are relatively low so usually L is considered constant in the calculations.

## II. LOSSES RESULTING FROM THE EFFECT OF HARMONICS IN THE TRANSFORMERS

### A- Hysteresis Loss

In references [5]-[6] formula (3) for hysteresis loss under harmonic conditions has been suggested. It has been pointed out in this reference that hysteresis sub loops have been neglected.

$$P_{Hn} = \Sigma \frac{v_n}{nV_1} \cos \varphi_n^{\ S} \qquad (3)$$

In which $P_{Hn}$ is the hysteresis loss according to P.U relative to the fundamental component, n is the harmonic order, $V_n$ is the voltage of the nth harmonic, $V_1$ is the fundamental component's voltage, $\varphi_n$ the phase angle of the voltage of the nth component and s is the coefficient dependent on the core's material.

### B- Eddy current loss

In references [5]-[6] is for eddy current loss which is one of the effects of distorting voltage. Suggests the following formula:

$$P_{Fn} = 1 + \sum_{n=2}^{\infty} \frac{V_n}{V_1}^2 . C_{en} \qquad (4)$$

In the above expression, $C_{en}$ is a function of the depth of penetration of electromagnetic wave at nf frequency. In this definition the flux of the created reaction by eddy current which opposes the main flux has been considered. $C_{en}$ Which is a function of the core is defined as follows:

For $\xi < 3.6$, $C_{en} = 1$   $0.0017\xi^{3.61}$ and for $\xi > 3.6$, $C_{en} = \dfrac{3}{\xi}$

is obtained. Here $\xi = \Delta\sqrt{\pi\mu\gamma nf}$ in which $\Delta$ is the thickness of the magnetic core, $\mu(m)$ is the magnetic permeability coefficient ( $\dfrac{H}{m}$ ), $\gamma$ electric conductivity of the magnetic core ($(\Omega.m)^{\ 1}$) and f is the frequency of the (Hz) main component.

### C- Stray loss

Flux variable with time in the transformer cans create eddy current in the conductive. At [5] the effect of frequency in the two frequency ranges has been modeled this way.

$$R_{AC}^{Lf} = 1.29 \ \frac{fh}{f_1}^{\ 0.8} \ [m\Omega] \qquad (5)$$

$$R_{AC}^{hf} = 9.29 \ 0.58 \ \frac{fh}{f_1}^{\ 0.9} \ [m\Omega] \qquad (6)$$

The Lf and hf indexes show the lower and upper frequency respectively equation (6) can be stated in the from of $C_1 \ \dfrac{fh}{f_1}^{\ \varepsilon_1}$ according to equation (7) which is an inverse dependence on frequency.

$$R_{AC}^{LF} = C_1 \ \frac{fh}{f_1}^{\ E_1} \ [m\Omega] \qquad (7)$$

In which $\varepsilon_1 = 1.87$ and $C_1 = 333.58 \, \text{m}\,\Omega$ .

### D- Coil Losses (loading losses)

These losses include losses in the primary and secondary coils resulting from the passing of harmonic currents. If we consider the DC component also we have the following relation [5]:

$$P_{Jn} = \sum_{n=0, n \neq 1}^{\infty} R_n(P)I_n^2(P) + \sum_{n=0, n \neq 1}^{\infty} R_n(S)I_n^2(S) \qquad (8)$$

$R_n(P)$, $R_n(S)$ are the resistance of the primary and secondary coils belonging to the nth harmonic. Also, $I_n(S)$ $I_n(P)$ are the effective values of the nth harmonics current in the primary and secondary coil.

By separating the two linear and nonlinear load section we have the following relation:

$$I^2 T = I_L^2 + I_{NL}^2 \qquad (9)$$

In which here $I_T$ is the transformer's load current (RMS) [A], $I_L$ linear current (sinusoidal) at nominal frequency [A] and $I_{NL}$ is nonlinear current (RMS) [A]. For considering harmonic losses in the coils (by considering the shell effect and apposition effect), $I_{NL}$ is defined as follows:

$$I_{NL}^2 = \sum_{h=2}^{\infty} \beta_h I_h^2 \qquad (10)$$

Here $I_h$ (RMS) is harmonic current of the hth order [A] and coefficient $\beta_h$ depends on order of harmonic (h), the shape of conductor and the size of conductor; as a result it is related to the nominal power of the transformer. For distribution transformers MV/LV the following formula has been presented [7].

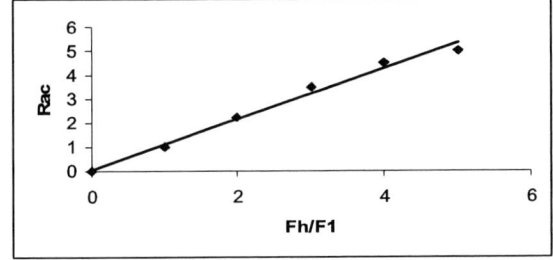

Fig. 2. Dependence of stray loss on frequency in the low frequencies range.

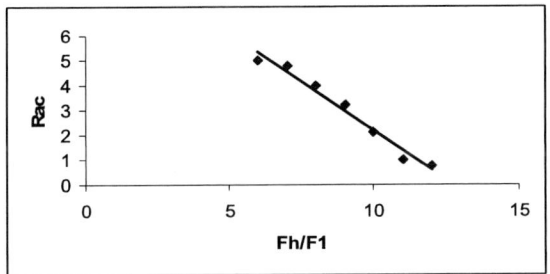

Fig. 3. Dependence of stray loss on frequency in the high frequencies range.

$$\beta_h = \frac{1 + \xi + \xi^3}{1 + \xi^2} \qquad (11)$$

$$\xi^2 \approx \gamma \frac{S_r}{S_{r0}} h \qquad (12)$$

$\frac{S_r}{S_{r0}}$ is the relation of the transformer's nominal power to the base power [P.U]. $\gamma$ is the experimental coefficient ($\gamma \approx \frac{3}{2}$ For the limit 100-1000KVA with $S_{r0} = 1000KVA$) from the above relations we can investigation that $\lim_{h \to 0} B_h = 1$ (DC resistance) and $\lim_{h \to \infty} B_h = constant \times \sqrt{h}$.

The coefficient $\gamma$ has the conductor's next characteristic with it and by the increase of nominal power (at constant current density) it increases continuously [8]-[9].

## III. TRANSFORMER HARMONIC EQUIVALENT CIRCUIT

Noting the stated subjects we can by combining the ideal model of a transformer and the mentioned cases, obtain the equivalent circuit for it. The current source which is seen in the model is a harmonic current source dependant on the middle node's voltage that models the non-liner's effect of saturation; an order such that first the middle node's voltage is calculated then by noting the amount of this node's voltage for simulating the effect of saturation the amount of dependant current's source is determined at that frequency. At harmonic analysis, evaluation of the frequency response of equivalent circuit is important. As we know the frequency response of RL circuit drops as frequency increases. In the case of transformer by increasing frequency primary and secondary impedance increases, and the result of this impedance increase, is the isolation of the primary and secondary coil of transformer from each other.

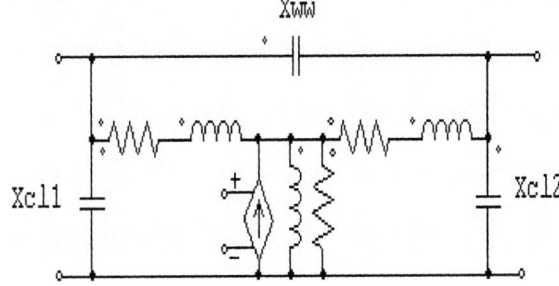

Fig. 4. The equivalent circuit of a transformer for harmonic analysis.

It has been mentioned in most references that by increasing frequency care losses decrease severely (the core's B-H curve travels more revolutions per cycle and as a result causes increase of loss). The presumption of these references is the constant ness of maximum flux at higher frequencies which is an incorrect assumption. At higher frequencies by increasing frequency and with the assumption of constant ness of voltage, flux (with a power greater than one) decreases and as a result core losses decrease by increase of frequency.

On the other hand, the secondary open circuit's voltage at the transformer is equal to $j\omega M$ (M is the mutual inductance between the coils). As a result the voltage of the transformer's open circuit decreases linearly with the decrease of frequency.

Noting the above explanations the transformer's complete equivalent circuit that can analyze its frequency response is shown in Fig.5. $X_{CL1}$ and $X_{CL2}$ capacitors are for modeling capacitance effect between coil rings stray capacitance effects and modeling capacitance effects between coils and the body. These capacitors provide routes with low impedance at high frequencies for leakage currents. The values of capacitance reactants depend mostly on the transformer design. For example the values of capacitance reactants can be in the range of 50 to 200 per unite. $X_{WW}$

states capacitance effects between primary and secondary coil which it's size depends completely on the way secondary coil is added to the primary coil in the transformer coil, geometry and the distance between them. (The importance of this capacitor in modeling autotransformer is mare than $X_{CL}$). It shouldn't be forgotten that the universal model in the presence of high frequencies will be in fact a distributed model of inductive and capacitive reactants. The distributed model is built by the lumped model which is described above. The model of the transformer's distributed parameters is only used for special applications that need a ware from at every point of the transformer.

Transformer's model for zero sequence harmonics (third multiple harmonics) is like models that are used at calculations related to unsymmetrical short circuit and for obtaining zero sequence network. These models are obtained by noting the type of transformer winding and the way they are earthed for two coil and three coil transformers by considering three type of star, triangular and staggered coiling. It is clear that it has been used in the circuit related to the zero sequence harmonics of R and X related to this sequence and also the shell effect is considered.

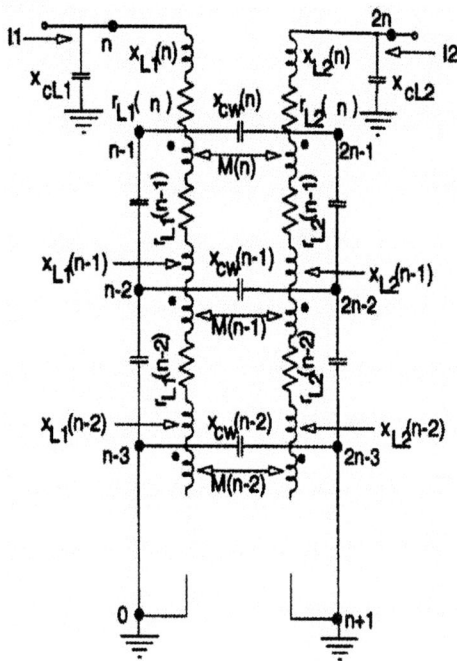

Fig. 5. Model of transformer's distributed parameters.

Transformers that have different coiling at primary and secondary coils don't create any phase difference for zero sequence harmonics. As a result the transformer model that is used in this project will be a model that is used for analysis at power frequency. With this difference that the shell effect creates change in it's resistance in series and by increase of frequency this resistance also increases. We can also consider the L variations according to frequency, but because these variations are little we can consider the dispersal frequency for all the frequencies constant and it's corresponding reactance

will be of the from $X_h = hX_{50}$. for increasing precision the parallel branch related to core losses can also be added to the model and the effects of harmonic generation by the saturation phenomenon by the current source be added to the parallel branch. This source, models the injected harmonic current as a result of saturation. Of course it's possible that internal resonances also come into existence which is a result of existing capacitance vacancies between the coils and the earth; by using the equivalent circuit of a transformer, noting the leakage reactants mentioned, was obtained between 7 to 15 KHz which is very large compared to frequencies under study in harmonic analysis and we can neglect their effect in the calculations.

The equivalent circuit that has been shown in Fig.6 is a circuit which has been suggested for calculating harmonic losses. The parameters of this circuit are all a function of frequency. The precision of the equivalent circuit depends on the precision of each of the components.

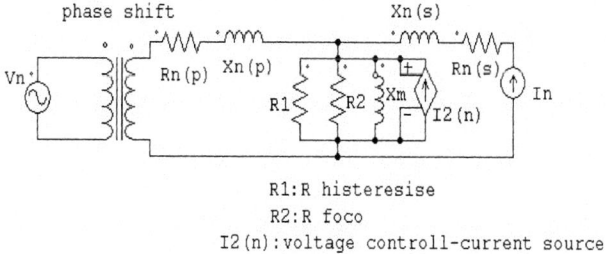

R1:R histeresise
R2:R foco
I2(n):voltage controll-current source

Fig. 6. The transformer's equivalent circuit for analyzing losses.

In Fig.6 two sources are visible. The voltage source from the point of view of the initial bus and the injected current source that is produced by the nonlinear load.

IV. INVESTIGATING THE METHOD OF LOSS VARIATIONS AND THE AMOUNT OF TRANSFORMER DISSIPATIONS IN HARMONIC CONDITIONS

Consider the transformer's harmonic model. In this model by noting the given explanations all the components are a function of the frequency. On the other hand, nonlinear loads are modeled as $I_h$ current source in the secondary coil. As a result the inputs of the problem are the distorting currents on the low voltage side and the distorting voltages on the high voltage side. The amount of voltage distortion on the high voltage in addition to the amount of load nonlinearity depends on the kind of power network which is connected to the high voltage side (strength or weakness of the network). Naturally the amount of voltage distortion of the high voltage side in strong networks is less than weak networks.

*4-1- Node Harmonic Analysis with the Existence of Voltage Sources*

We can solve a system that includes harmonic voltage sources in some of the shins and injection of harmonic currents in the other shins by grouping and inversion of the admittance matrix. This way the shin's unknown voltages and undetermined injective harmonic currents become apparent. If

$V_2$ shows known voltage sources $I_2$ will show unknown variables, the remaining shins are determined as injected harmonic current source $I_1$ (which can be zero and or equal to the harmonic current sources) and the related harmonic voltages vector $V_1$, is unknown.

For separating two kinds of nodes, the matrix equations are grouped and the following relation is obtained:

$$\begin{matrix} Y_{11} & Y_{12} \\ Y_{21} & Y_{22} \end{matrix} \quad \begin{matrix} V_1 \\ V_2 \end{matrix} = \begin{matrix} I_1 \\ I_2 \end{matrix} \qquad (13)$$

The voltage's unknown vector $V_1$ is solved by the following relation:

$$[Y_{11}][V_1] = [I_1] \quad [Y_{12}][V_2] \qquad (14)$$

Then injected harmonic currents by harmonic voltage sources are determined as follows:

$$[Y_{21}][V1] \quad [Y_{22}][V_2] = [I_2] \qquad (15)$$

By this method, with same amount of matrix extra processing injective currents are obtained.

One of the application cases of this problem is in transformer analysis which the turbulent source model is from the low voltage side of the current source and from the high voltage side of the turbulent voltage.

Nothing these explanations and by using the MATLAB software and the presented harmonic model in Fig.6 the way the care loss and transformer coiling is dependant on the frequency spectrum is investigated.

### 4-1-1- The Source Effects $I_h$ $(V_h = 0)$

The simulation done states that if the load model on the low voltage side be in the form of current source $I_h$ the coil loss increase (because of the increase of the shell effect) frequency increase. (By assuming THD constant ness at different frequencies)

The core losses (which are calculated according to formulas (3) & (4) decrease by increase of frequency. The obtained results for core and coil loss have been drawn in the figures below by applying harmonic current with constant amplitude (5%) at different frequencies.

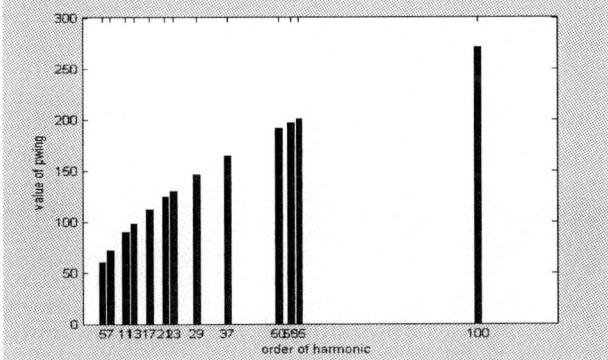

Fig. 7. Transformer's coil losses at different frequencies (constant $I_h$ amplitude).

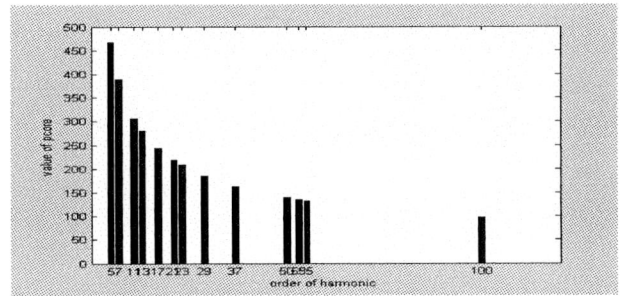

Fig. 8. Transformer's core losses.

Also total loss has been drawn in the figure below.

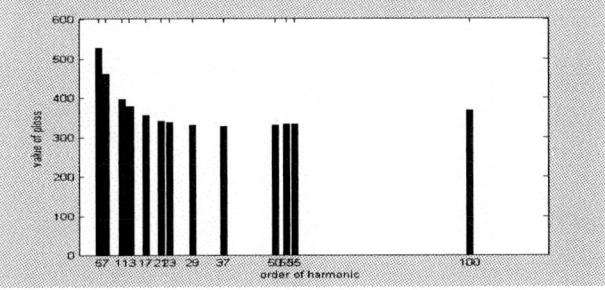

Fig. 9. Transformer's total loss.

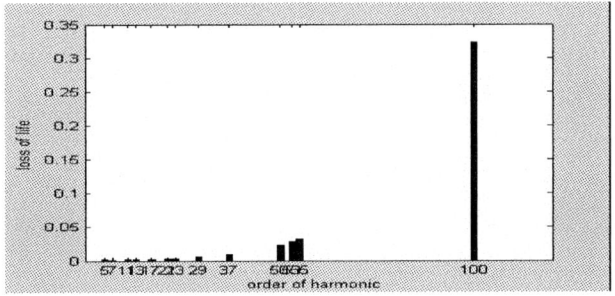

Fig. 10. Life loss in frequencies.

As seen around frequency n=37 the loss reaches its minimum value. The reason for this fact is that by frequency increase the coil losses is increasing and core losses are decreasing. At low frequencies core losses and at high frequencies coil losses are vital. Also the amount of life loss has been drawn according to frequency for the case that there is only distortion on the low voltage side.

### 4-1-2- The Distortion Effects on the High Voltage Side

Distortion on the high voltage side has been modeled as a $V_h$ voltage source. The amount of this distortion is a function of the load and impedance of the network's short circuit which means that it is dependant on the strength or weakness of the network.

In strong networks the high voltage side distortion can be neglected. The way the transformer losses vary with the existence of feed from the voltage source $V_h$ has been drawn below. As seen the core and coil losses decrease with the increase of frequency. (In case the core losses would have increased). Also the total losses have been drawn in Fig.12. These losses have been obtained for constant $V_h$ amplitude. ($V_h = 5\%$) also the amount of life loss has been drawn in Fig.13 as a bar diagram.

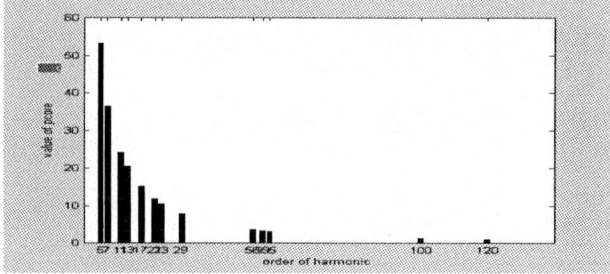

Fig. 11. Transformer core losses ( $V_h$ Amplitude is constant).

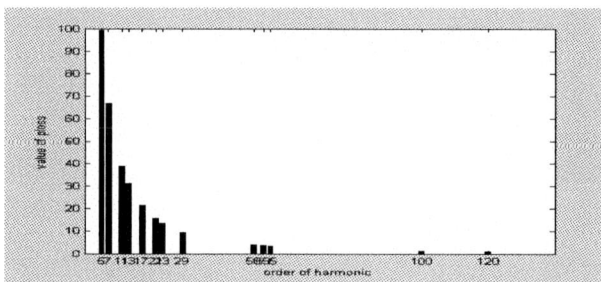

Fig. 12. Transformer's total loss.

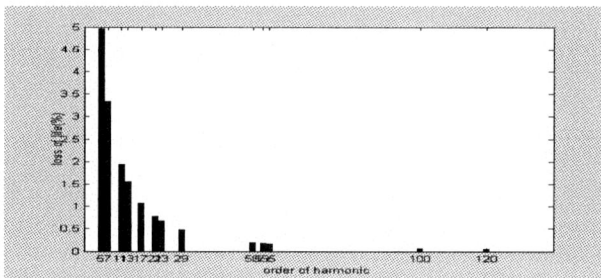

Fig. 13. Life loss in frequencies.

## V. CONCLUSION

In this paper, after presenting the transformers harmonic model, the method of loss variations and also the amount of transformer's useful life loss in the harmonic conditions has been investigated. With the evaluations done we can state the results below briefly:

1- Briefly we can say that distorting currents that pass the transformer cause increase or decrease of loss and increase of temperature. This fact cause the expected life of the transformer's life reduction (or its nominal power decrease), it's necessary that the current's harmonic spectrum ($THD_i$), transformer's electrical characteristics (loss), thermal behavior (thermal model) and the true curves of the ambient temperature and load variations relative to time be at the disposal.

2- The assumption of proportionality of the eddy current loss with the second power of frequency is only correct for small transformers that the dimension of their conductors is less than 3mm. For transformers with thicker conductor using the existing standards doesn't give exact results.

3- In practice Hotpoint is located at the low voltage winding and its end boundary. These conductors are located at a magnetic field which is composed of two Hz components, axial component Hz. Eddy current loss is related of the components. Despite this condition an equivalent harmonic loss coefficient shall be calculated that considers the eddy current loss in both the axial and radial directions. About smaller transformers and with conductor thickness up to $\tau < 3mm$ , the mentioned approximate method is acceptable.

4- With the increase of harmonic order we can't comment about the increase or decrease of core losses, coil losses and also the transformer's useful life.

## VI. REFERENCES

[1] P.Barret, P.Bornard, B.Meyer, "Power System Simulation", book, chapman&hall, 1st edition, 1997.
[2] T.J.Miler, "reactive power control at electrical systems", 1980.
[3] Alexander E.Emanuel, "the effect of harmonic randomness upon temperature rise of electrical equipment", third international conference on harmonics in power system.
[4] Robert L.smith, J.r.Ray, P.strat ford, "Power System Harmonic effects form adjustable speed drivers", IEEE Transactions on industry application, Vol 1A.20,NO.4.july/August 1984, PP.973-977.
[5] A.C.Delaiba, J.C.oliveira, A.L.A.Vilaca, Jose Rose Roberto Cardoso, "The Effect of Harmonics on power Transformers loss of life", IEEE Transactions on power delivery, Vol.5, 1996.
[6] E.Emanuel and xiaoming wang, "Estimation of loss of life of power transformers supplying non-linear load", IEEE Transaction on power Apparatus and systems, Vol.PAS-104, No.3, March 1985, PP.628-636.
[7] L.Pierrat and M.Jose resende and J.Santana, "Power Transformers life Expectancy under Distorting Power Electronic Loads", IEEE Transaction on power Appartus, Vol.PAS-104, No.7, 1996pp578-583.
[8] Deniz Yildirim, Ewald.F.Fuchs, "Measured transformer Derating and comparison with Harmonic Loss factor ($F_{HL}$) Approach", IEEE Transactions on power Delivery, Vol.15 No.1, JANUARY 2000, PP.186-191.
[9] Sergey N.Makarov, Alexander E.Emanuel, "Harmonic Loss Factor for Transformers supplying nonsinusoidal Load Currents", 2000, IEEE.

## VII. BIOGRAPHIES

**Mosa Marzband** received his M.S. the electrical engineering faculty of the school of engineering of Mazandaran University, Iran in 2005. Currently he is working as a senior engineer in the electrical projects of Borzoye Petrochemical Company in the Pars Special Economic Energy Zone. His current interests are harmonics and power quality. He can be connected at m_marzband2005@yahoo.com.

**A.sheikhol-Eslami** was born in Iran on 1956. He received the B.S from Mazandaran University Iran, in 1979 and M.S ad PhD degree in electrical engineering from Strathclyde University U.K. in 1989. Since 1989, he has been Assistant Professor of Mazandaran University Iran. His research interests are power quality, power electronics and reacting power control.

6

**2006 IEEE International Conference on Power Electronic, Drives and Energy Systems**

# Novel Integral Cycle Voltage Controller for Self Excited Induction Generators

S. S. Murthy, Senior Member, *IEEE*, A. J. P. Pinto, *Student Member* and A. R. Beig, Senior Member

*Abstract--* **This paper presents the theory of a new var regulator used for voltage control of self excited induction generators (SEIG). In this approach the excitation capacitors of each phase are switched individually by IGBTs at the zero crossing of respective capacitor currents, thereby eliminating switching losses and harmonics and reducing component count. Simulation results using Matlab–Simulink are presented and compared with experimentally obtained results. The developed integral control scheme presented here overcomes all the shortcomings of existing control schemes and, in addition to being self-starting is also capable of handling capacitive loads, thus making its application in standby generators acceptable with associated reduction in unit costs.**

*Index Terms--***Induction generators, Standby generators.**

## I. NOMENCLATURE

| | |
|---|---|
| $d^s, q^s$ | stator and rotor direct and quadrature axes, |
| $d^r, q^r$ | respectively |
| $k$ | number of pairs of poles |
| $L^s, L^r$ | self inductance of stator and rotor coils, respectively |
| $M^{sr}$ | mutual inductance between any pair of stator and rotor coils with their magnetic axes collinear |
| $p$ | d/dt |
| $R^l, L^l$ | load resistance in Ohms and Inductance in Henry |
| $R^s, R^r$ | resistance of the stator and rotor coils, respectively |
| $v_d^s\ i_d^s$ | stator voltage and current, respectively, associated with the d axis |
| $v_d^r\ i_d^r$ | rotor voltage and current, respectively, associated with the d axis |
| $v_q^s\ i_q^s$ | stator voltage and current, respectively, associated with the q axis. |
| $v_q^r\ i_q^r$ | rotor voltage and current, respectively, associated with the q axis |

---

S. S. Murthy is with the Department of Electrical Engineering, Indian Institute of Technology Delhi, Hous Khas New Delhi-110016, INDIA (e-mail: ssmurthy@ee.iitd.ac.in)

A. J. P. Pinto is with Department of Electrical and Electronics Engineering, National Institute of Technology Karnataka, Surathkal. Mangalore -575025, Karnataka INDIA (e-mail: loypinto@yahoo.com).

A. R. Beig is with Department of Electrical and Electronics Engineering, National Institute of Technology Karnataka, Surathkal. Mangalore -575025, Karnataka, INDIA (e-mail: arbeig@yahoo.com).

## II. INTRODUCTION

THE induction motor operated as an induction generator with terminal capacitors, offers considerable advantages due to its ruggedness, low cost, brush-less squirrel cage rotor, manufacturing simplicity, low maintenance and wide off-the-shelf range as compared to the synchronous machine, but has a major drawback of poor voltage regulation. The rigid frequency and voltage of the grid makes the equivalent circuit model suitable for steady state analyses of the induction motor in the generating mode. However its analysis and operation as a stand-alone power source is complicated, since now both the voltage and frequency are variables, and involve solving non linear equations of higher order [1]. Under these conditions proper selection of equipment and the prediction of the system performance are essential for successful implementation of the scheme. The operation of such machine in any remote or stand alone conditions therefore requires some type of voltage regulator. Considerable literature exists describing various arrangement such as, contactor or thyristor switched capacitors, thyristor controlled inductors, saturable reactors, etc [2,3]. Devices like STATCOM in effect provide a virtual bus for the induction machine to operate. All these systems introduce a lot of harmonics or are expensive to build and complicated to program. The excitation capacitors too need to be sized properly for acceptable operation. The STATCOM based controller overcomes these problems but is not self-starting. The greatest drawback of most of the systems is their inability to handle unintentional capacitive loads as in the case of power factor improvement capacitors left on line, when the motor is disconnected by some fault, resulting in dangerous over voltages. To overcome these shortcomings which prevent the wide acceptance of induction generators in stand-alone engine-driven applications, the present controller was developed. The machine and load were modeled for dynamic analysis and simulation studies were carried out using MATLAB-SIMULINK. The encouraging results of the simulation were instrumental in building the hardware and technology demonstration setup reported here.

## III. DYNAMIC MODELING AND SIMULATION

Connection of excitation capacitors across the terminal can be represented in the d-q model of the machine as two additional equations with the capacitor voltages as state variables. Similarly the connection of the load across the terminal can be represented in two more equation shown here. The complete arrangement is shown in Fig. 1.

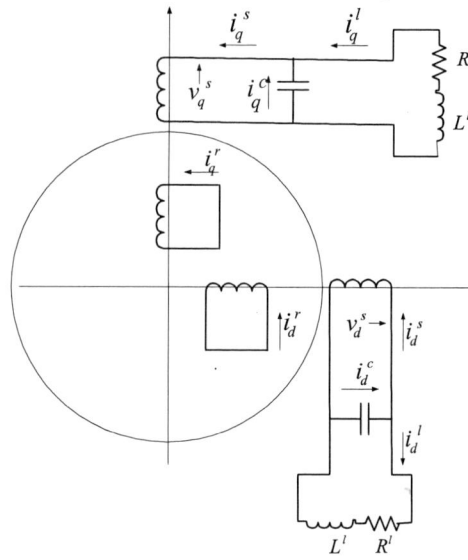

Fig. 1.  d-q model of the induction machine complete with excitation capacitors and load.

Referring to Fig. 1 it is seen that

$$i_q^l = -i_q^s - Cpv_q^s \qquad (1)$$

$$i_d^l = -i_d^s - Cpv_d^s \qquad (2)$$

and

$$v_q^s = i_q^l \left( R^l + L^l p \right) \qquad (3)$$

$$v_d^s = i_d^l \left( R^l + L^l p \right) \qquad (4)$$

combining the above two equations in the standard d-q model[4] of the induction machine we get the dynamic model of a self excited induction generator complete with excitation capacitor and load, giving due representation to the capacitive voltages and inductor currents as state variable is in (5)

$$
\begin{pmatrix} v_q^s \\ v_d^s \\ 0 \\ 0 \\ i_q^l \\ i_d^l \\ 0 \\ 0 \end{pmatrix} =
\begin{pmatrix}
(R+Lp) & 0 & M^rp & 0 & 0 & 0 & 0 & 0 \\
0 & (R+LP) & 0 & M^rp & 0 & 0 & 0 & 0 \\
M^rp & k\omega M^r & (R+Lp) & k\omega L & 0 & 0 & 0 & 0 \\
-k\omega M^r & M^rp & -k\omega L & (R+Lp) & 0 & 0 & 0 & 0 \\
-1 & 0 & 0 & 0 & Cp & 0 & 0 & 0 \\
0 & -1 & 0 & 0 & 0 & Cp & 0 & 0 \\
0 & 0 & 0 & 0 & 1 & 0 & -(R^l+L^lp) & 0 \\
0 & 0 & 0 & 0 & 0 & 1 & 0 & -(R^l+L^lp)
\end{pmatrix}
\begin{pmatrix} i_q^s \\ i_d^s \\ i_q^r \\ i_d^r \\ v_q^c \\ v_d^c \\ i_q^l \\ i_d^l \end{pmatrix}
$$

.......(5)

In order to investigate the acceptability of the proposed integral cycle control under extreme conditions, magnetic saturation is not represented. This approach will then also permit operation in the linear region of the saturation curve. Capacitance value Cp is selected to be greater than that required to achieve self excitation; Loading is controlled by varying $R^l$ and $L^l$. Other variables in (5) being machine parameters. Since the capacitive vars absorbed by the machine

is a function of the loading on the machine too, for low values of capacitance just above minimum required for self excitation, loading of the machine with resistive load only is sufficient to absorb the excess capacitive vars. By monitoring the line voltages and switching  the terminal capacitors- switching operation being carried out only at current zero crossings- the machine model can operate in any average condition, i.e., of excess capacitive vars, deficient capacitive vars or just balanced condition corresponding to voltage build-up, voltage decay or stable voltage operation.

*A.  Simulation Results*

Results of the simulation are shown in Fig 2. It shows the variation in line voltage over time while the capacitive vars are varied in a controlled fashion. The machine is driven at constant speed with no load and minimum required capacitance for self excitation. The exponential build up of terminal voltage on achieving self excitation is seen.

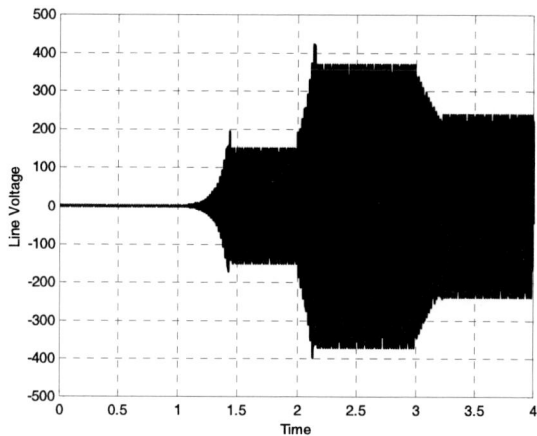

Fig. 2.   Simulation results showing terminal voltage being controlled at desired values.

Referring to Fig. 2: At 1.4s the when the line voltage exceeds the set-point, control system is activated and the capacitors of switched off and on for integral number of cycle in the manner of 'duty cycle'. Upon change of the set-point, at 2s, the capacitors are fully on line and the voltage again rises until the new set-point is reached and the capacitor are again on integral cycle based 'duty cycle'. At 3s, when the line voltage set-point is reduced, the capacitors are switched off at their respective current zeros, until the voltage decays to the new reduced set-point, thus activating the control scheme and stabilizing the voltage.

IV.  EXPERIMENTAL SETUP

The Schematic diagram of the implemented hardware set up in shown in Fig. 3. The capacitors are connected to the terminals through IGBTs. For cold starting, i.e., without the use of an external power source, a normally closed contactor (not shown) bypasses the IGBT's, connecting the capacitors directly across the terminal of the induction machine and providing a low resistance path for the extremely small excitation currents resulting from residual magnetism of the rotor and thus permitting voltage buildup. The contactor opens when the terminal voltage reaches 50% of rated voltage of machine which is sufficient to power up the control circuits.. Hall effect current and voltage sensors are used as isolation devices between power and control circuits.

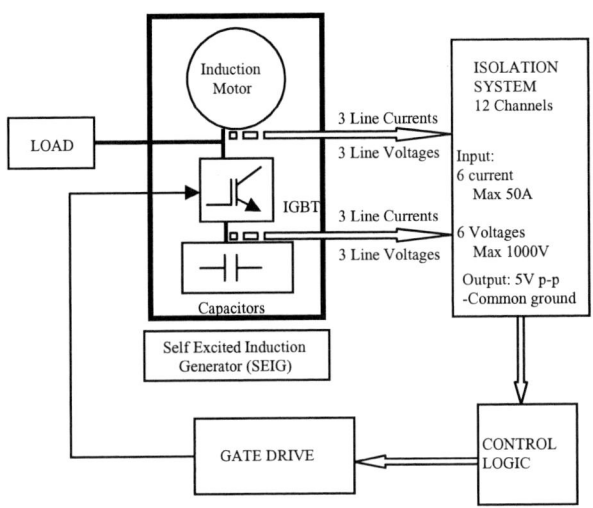

Fig. 3. Schematic diagram of experimental setup.

The Control logic is implemented using CMOS logic ICs. The control circuit monitors and compares machine terminal voltage with the set point. Switching of the capacitors on high or low voltage in each phase is implemented only at the zero crossing of currents, irrespective of the point at which the decision to switch takes place. The dead band between high and low voltage set-point is continually adjustable

Since switching is at zero currents, the device losses are considerably reduced, in addition the integral cycle switching scheme eliminates the need for snubber circuits, and accordingly none have been provided, resulting in reduced component count and complexity.

### A. Experimental Results

The operation of the control scheme may be observed from Fig. 4, 5 and 6 which are Oscilloscope prints taken while the control system was in operation and the machine was loaded by R-L load. The top trace of all the three figures is the line voltage, the middle trace the line current and the bottom trace in the capacitor voltage. Fig. 4 shows the switching in and out of one bank of capacitors in a 'duty cycle' manner. The capacitor bank is on for a integral number of cycles and off for a integral no of cycles. The ratio between number of on cycles to number of off cycles depends on the loading of the machine and the dead-band between the high and low voltage set-points. The rise in line voltage while the capacitors are on line and fall in line voltage when the capacitors are disconnected may be observed in all there prints.

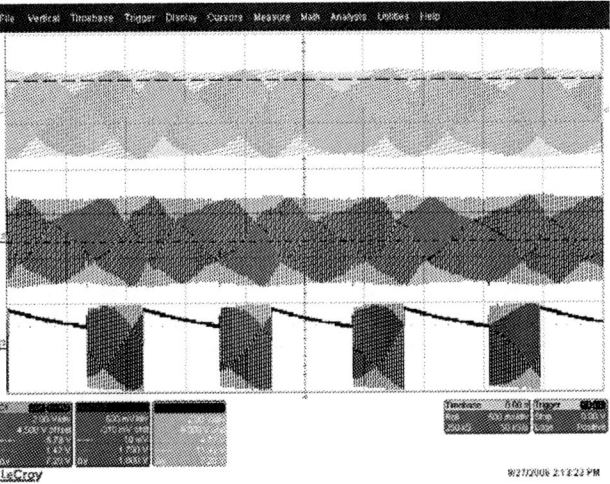

Fig. 4. Oscilloscope print showing 'duty cycle' base integral control action.

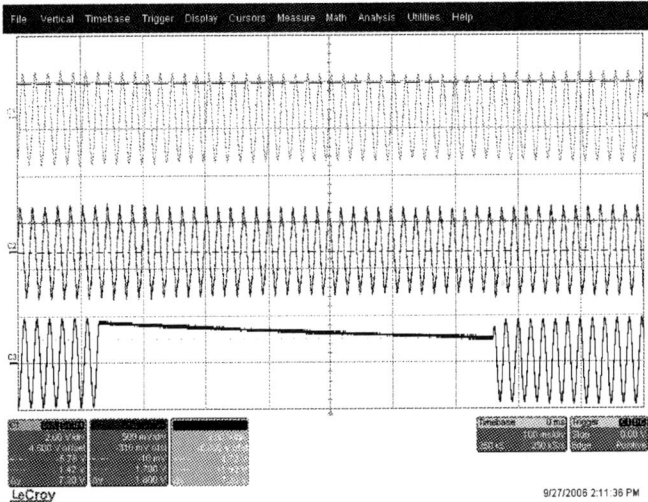

Fig. 5. Oscilloscope print showing Capacitor voltage during off state.

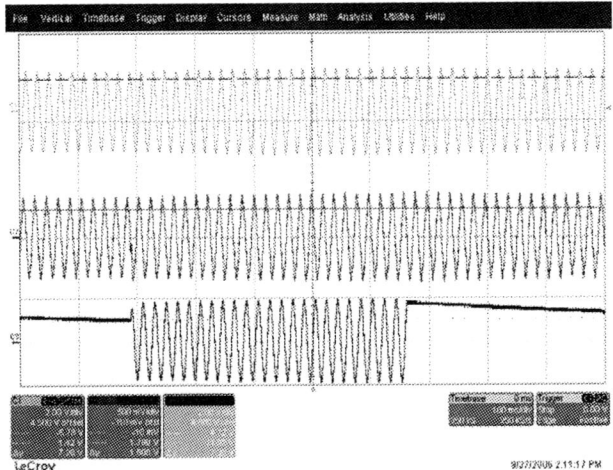

Fig. 6. Oscilloscope print showing capacitor voltage during on state.

Capacitor voltage during the off time is seen in more detail in Fig. 5, the droop in the capacitor voltage during this time is due to the discharge resistors connected across the terminals of the capacitors which should not be removed for safety reasons. The voltage across the capacitor during the on period may be observed in Fig. 6. The absence of switching surges in the traces of the line voltages in the three prints may also be observed.

## V. CONCLUSION

A new control strategy that switches the excitation capacitors into or out of the circuit for integral number of cycles is presented. The effectiveness of this method is studied using the dynamic model and MATLAB-SIMULINK software, which confirm the same. A laboratory technology demonstration model is built and test results are presented. These show the control strategy if very effective in controlling the voltage at any desired value irrespective of the loading conditions. The inherent ability of the control to switch off the capacitors totally, effectively takes care of unintentional capacitive loading of the machine, thus demonstrating its capability to take care of any random loading without restriction, making it a suitable for standby generators.

## VI. REFERENCES

[1]  S.S. Murthy, O.P. Malik and A.K. Tandon, "Analysis of self-excited induction generator", IEE Proc. C. Gener. Transm. Distrib., vol. 129, (6), pp 260-265. 1982.
[2]  S.S. Murthy and B. Singh, "Capacitive VAR controllers for induction generators for autonomous power" proc. IEEE Power Electronics, Drives and Energy Systems Conf., 1996, pp 679-686.
[3]  B. Singh, S.S. Murthy and S. Gupta, "Analysis and design of STATCOM based Voltage Regulator for self-Excited Induction Generators" IEEE Transactions on Energy conversion, Vol 19, No 4, Dec 2004.
[4]  Howard E. Jordan, "Digital Computer Analysis of Induction Machines in Dynamic systems" IEEE Trans. on Power Apparatus and systems, Vol. 86, No 6, pp. 722-728. June 1967.

## VII. BIOGRAPHIES

**S.S. Murthy (SM' 87)** was born in Karnataka, India, in 1946. He received his B.E., M.Tech. and Ph.D.degrees respectively from Bangalore University, Indian Institute of Technology (IIT) Bombay, and IIT Delhi. After serving a year at BITS Pilani, he has been with IIT Delhi as faculty since 1970 and professor since 1983. He has held several visiting Positions in India and Abroad such as: Visiting Fellow Univ. of Newcastle on Tyne (UK), Visiting Professor, Univ. of Calgary (Canada), Visiting Consultant, Kirloskar Elec. Co., Director, ERDA Baroda, Adjunct Prof. IISc Bangalore and Director of NITK, Surathkal He has published over 200 research papers in refereed journals and conferences, guided over 100 theses/dissertations at B. Tech, M. Tech,MS. and Ph. D. levels and completed over 80 sponsored research and industrial consultancy projects dealing with Electrical Machines, Drives and Energy Systems and sponsored by major electrical industries in India. Dr. Murthy has received many awards notable being-ISTE/Maharashtra Govt. Award for outstanding research, .IETE/Bimal Bose Award for contribution in Power Electronics and President of India prize for Best paper published in Journal of Institution of Engineers. In recognition of his outstanding professional contributions the prestigious Indian National Academy of Engineering (INAE) has conferred him as a Fellow of INAE. He is also a Life Senior Life Member of IEEE, Fellow of IEE, Life Fellow of the Institution of Engineers,( India), and Life Member of ISTE.

**Aloysius Jude Pius Pinto (M' 96, Student 05)** was born in Mangalore,

India, on Jan 19, 1955. He graduated from the Karnataka Regional Engineering College, Surathkal in 1976. After working for 21 years went back to college and completed his masters degree in 1999 at the same institution. He is now working as a research/PhD scholar at National Institute of Technology Karnataka, India. His employment experience included Engineering, procurement, installation and commissioning of electrical systems for Power stations, Refineries and petrochemical plants both in India and middle east

**Abdul Rahiman Beig** (M'92) received the B.E. degree in Electrical and Electronics Engineering from National Institute of Technology Karnataka, Surathkal, India, in 1989. M.Tech. and PhD degrees in Electrical Engineering from Indian Institute of Science, Bangalore, in 1998 and 2004 respectively. He is currently an Assistant Professor in the Department of Electrical Engineering, National Institute of Technology Karnataka, Suratkal, India. From 1989 to 1992, he was with Kirloskar Electric Company, Ltd, Mysore, India, as a R&D Engineer in the drives group. His research interests include ac drives and multi level inverters. Dr. Beig received the Innovative Student Project Award for his Ph.D. work from the Indian National Academy of Engineering in 2000 and in 1998 L&T- ISTE National Award for the his M.E. thesis from the Indian Society for Technical Education.

**2006 IEEE International Conference on Power Electronic, Drives and Energy Systems**

# EMI Modeling and Simulation of High Voltage Planar Transformer

Bai Feng, Niu Zhong-Xia, Shi Yu-Jie, and Zhou Dong-Fang

*Abstract* -- **High voltage planar transformer is a new kind of power component using in the special environment. The EMI issue of it acts as an important role in the analysis of product reliability. This paper firstly presents a prototype and a application of the transformer. Then through finite element analysis (FEA) simulation, gives the calculation results of field distributions and current density distributions of the planar transformer windings. The calculation area is divided into two regions, inside the magnetic core and outside the core. The models and simulation methods are of great values to the EMI analysis of the planar transformer.**

*Index Terms* – **Planar transformer, EMI, FEA.**

## I. INTRODUCTION

Compact and high-efficient power converter is critical to the miniaturization of telecommunication, computer, vehicle and other industrial and military systems[1][2]. The creation of any high-efficiency switched mode power supply (SMPS) requires a proper selection of magnetic materials for the transformer and inductor. The answer for miniaturization and material performance is planar magnetic technology. Its benefits include: low profile, high power densities, and highly constant parasitics, etc.

With planar technology, the windings are realized by copper tracks within multi-layer printed circuit boards (PCBs), instead of winding copper wire on a bobbin. Frequently, winding layers with a small number of turns are used. Even layers with just one or two turns are quite common. So it is widely used in the low-voltage and high-current applications. But the advantages of planar transformer are deeply attracted by the high voltage applications.

This paper firstly presents a high voltage application using planar transformer. Then studies the main factors of EMI issues, for instance, the influences of eddy current effects and electromagnetic field distributions due to high frequency and winding arrangement. Through FEA simulation, this paper gives the calculation results of field intensity distributions and current density distributions of the planar transformer

---

Bai Feng is with the department of Communication Engineering, PLA Information Engineering University, Zhengzhou City, Henan 450002, China. (e-mail: thirdplanet@sohu.com)

Niu Zhong-Xia, Shi Yu-Jie and Zhou Dong-Fang are all with the department of Communication Engineering, PLA Information Engineering University, Zhengzhou City, Henan 450002, China.

windings. Finally, points out that the leakage of magnetic field outside the magnetic core is rather stronger.

## II. A HIGH VOLTAGE APPLICATION

Traveling wave tube (TWT) is a kind of microwave amplifier. It is widely used in microwave communication and electromagnetic countermeasure. Comparing with solid-state amplifier, the main advantage of TWT is the high output power, while the disadvantage is the bulk mass and the rather complicated power supply. The power supply usually includes several unequal high voltages, the highest one may up to 5~10KV DC. Thus, it's a great challenge to the design of the power supply and the corresponding transformer.

Fig. 1 gives the transformer section of the TWT power supply of this paper. The main switching converter works at the frequency of 200 KHz. The planar transformer acts as an important role in voltage boosting. It involves four windings, one primary and three secondaries. Although the bulk of the transformer is $66 \times 58 \times 16 mm^3$, it can transform more than 300W power. The magnetic ferrite core is of the *EE* type, whose bulk is $58 \times 38 \times 16 mm^3$. Fig.2 shows the photograph of the prototype planar transformer of this paper. The windings are both manufactured by PCBs. The primary winding has 5 turns ( $W_P=5$ ) and the three secondary windings respectively has 7 turns ( $W_{S1}=W_{S2}=W_{S3}=7$ ), as is shown in Fig. 3. It's a view of *x-y* plane.

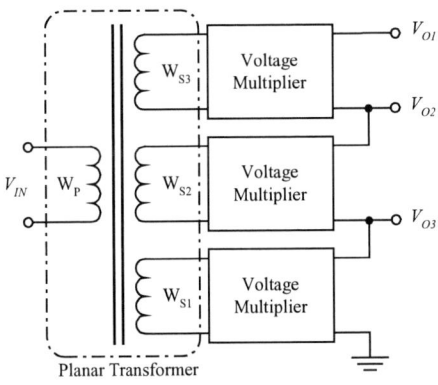

Fig. 1. A high voltage application of planar transformer

0-7803-9771-1/06/$25.00 ©2006 IEEE

Fig. 2. Photograph of the prototype planar transformer

Fig. 3. Windings arrangement of the high voltage planar transformer

TABLE I
MEASURED VOLTAGES AND CURRENTS OF THE WINDINGS OF THE PROTOTYPE
PLANAR TRANSFORMER

| Winding | V (V) | I (A) |
|---------|-------|-------|
| $W_P$ | 300 | 1.50 |
| $W_{S1}$ | 420 | 0.35 |
| $W_{S2}$ | 420 | 0.35 |
| $W_{S3}$ | 420 | 0.35 |

As is shown above, Table I gives the measured voltages and currents of the windings of the prototype planar transformer. The voltage of the primary winding is 300V, while the three secondaries are 420V. And the current respectively are 1.50A and 0.35A. From the measured data we can calculate the efficiency of the planar transformer, which is 98% here.

## III. MODELING OF FEA METHOD

This study is carried out by means of a FEA tool, not only by inspection of the flux and current distribution, but computing the magnetic intensity **H** and the electric-field intensity **E**. The FEA tool used here is Ansoft Maxwell® SV (student version).

The software is powerful and accurate for two-dimensional, electromagnetic, electromechanical, and thermal analysis [3]. Its AC Magnetic capability solves systems that have significant effects from induced eddy currents, skin effect, and proximity effect. Designers may use the AC Magnetic solver in frequency ranges from 0 Hz through several hundred MHz. Applications include bus bars, transformers, coils, and nondestructive evaluation systems. The solver could calculate power loss, core loss, impedance for frequency, force, torque, inductance, and stored energy. Additionally, plots of flux lines,

B and H fields, current distribution, and energy densities over the entire phase cycle are available.

In this case, the calculation window of the planar transformer is shown in Fig. 4. The type of the board is the commonly used FR4_epoxy, and the conduction strip line on it is made of copper. The *EE*-type magnetic core is of ferrite. These settings are coincident with the actual prototype. The dimension of the winding arrangement is shown in Fig. 5, from which we can see that the width of the primary winding is 2.2*mm* while the secondary is 1.2*mm*. And the thickness of the epoxy board is 1*mm*, while the copper strip line is 0.2*mm*.

To analyze the field distribution, we firstly specify the appropriate geometry as above, and then define the material properties and excitations for each copper track according to Table 1. The Maxwell software then does the following: Creates the required finite element mesh. Iteratively calculates the desired electrostatic or magnetostatic field solution and special quantities of interest, including force, torque, inductance, capacitance, and power loss. Here we are interested with the flux and current distribution, the magnetic intensity and the electric-field intensity. The calculated results allow one to analyze, manipulate, and display field solutions.

Fig. 4. Calculation window of the planar transformer

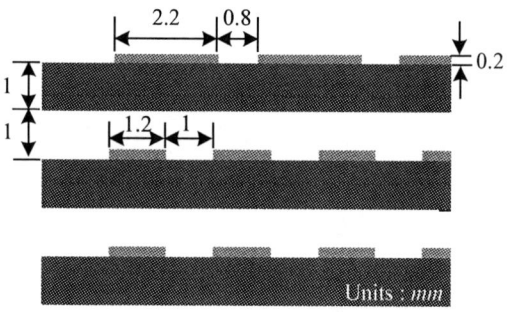

Fig. 5. Winding demensions of the planar transformer

The following Fig. 6 shows the final finite element mesh that was generated for this geometry. It can be refined manually by dividing a big mesh into smaller elements. This refining allows the software to compute the field solution separately in each element. The smaller the elements, the more accurate the final solution will be. In this case, the geometry is divided into 3906 elements.

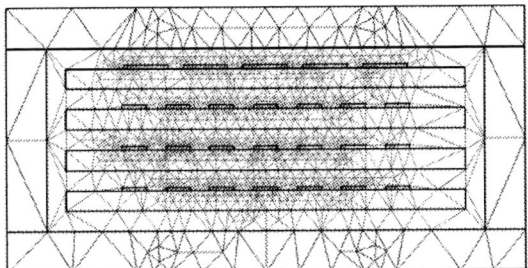

Fig. 6.    Mesh ploting of the calculation area

## IV.    CALCULATION INSIDE THE MAGNETIC CORE

### A. The magnetic intensity

Operating at a frequency of 200 KHz, the magnetic flux distribution of the transformer shows that the magnetic flux is evenly distributed around each primary winding and induces the electromotive force in the secondary winding, as shown in Fig. 7. The problem of unbalance magnetic flux distribution is due to the high frequency and the high current. The ferrite core encloses almost all the magnetic flux inside the transformer, which increases magnetic coupling and input impedance, and reduces EMI as produced by an air core transformer.

Fig. 8 shows the contour line of the magnetic intensity $H$ in details. It characterizes the magnetic intensity clearly. From the Figure, we can see that the magnetic intensity around the primary winding is rather stronger than the secondary windings. And the outer magnetic intensity of the primary winding is fairly stronger than the inner. The magnetic flux distributions of each pair of secondary windings are very similar. This is because the currents flowing in them are as equal to each other. Moreover, most of the contour lines are inside the magnetic core. The EMI issue bought by the leakage of the magnetic field here is the intrasystem type.

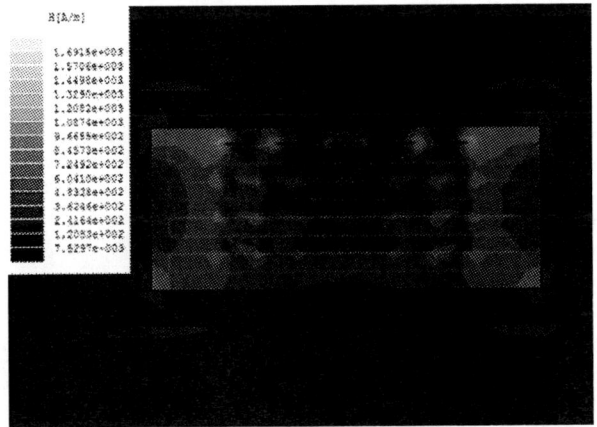

Fig. 7.    Magnetic intensity ( $H$ ) distribution of the windings inside the magnetic core

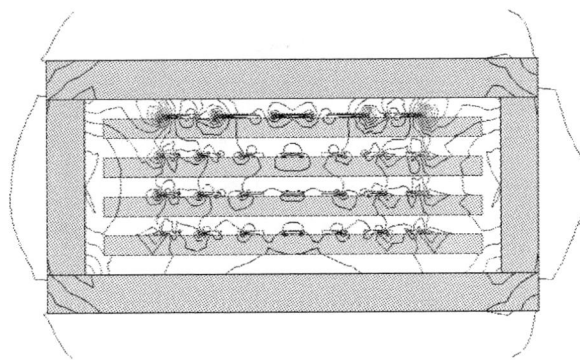

Fig. 8.    Contour line ploting of $H$ inside the core

### B. The eddy current

In the majority of conventional transformer designs, calculating the total power loss was limited to core and copper losses ( $I^2 R_{DC}$ ). In today's high frequency (HF) multilayer planar design, there are two effects: Skin and Proximity, which increase the resistance of a winding above the DC resistance value.

The two effects behave as the eddy current. The eddy current distribution inside the transformer windings will effect the leakage inductance and the overall performance of the HF transformer.

Skin depth or the depth at which the current density decreases to about 37% of the value at the surface can be calculated using the following formula:

$$\delta = \frac{1}{\sqrt{\pi f \sigma \mu}}$$

Where, $\delta$ is the skin depth in meters, $f$ is the frequency of the current, $\sigma$ is the electric conductivity of the metal, $\mu$ is the magnetic conductivity. According to copper, $\sigma = 5.8 \times 10^7$ *Siemens/m*, $\mu = 4 \pi \times 10^{-7}$ *H/m*. When $f$=200 *KHz*, $\delta$=0.1476 *mm*.

The eddy current distribution is shown in Fig. 9. The eddy current is evenly distributed in the whole winding of the HF transformer. It agrees with the flux distribution mentioned in above section. The maximum eddy current flows around the two sides of the strip lines of the winding, because of the magnetic fields generated by the current flowing the strip line.

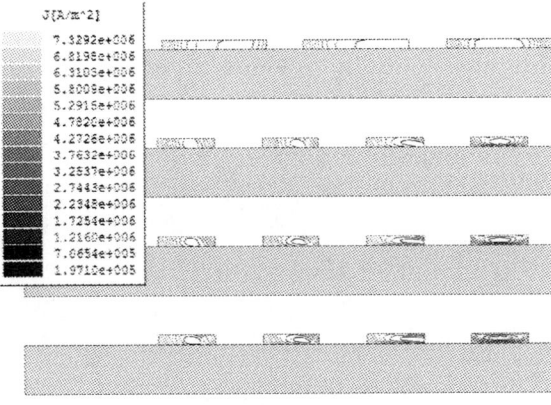

Fig. 9.    Current density distributions of the windings

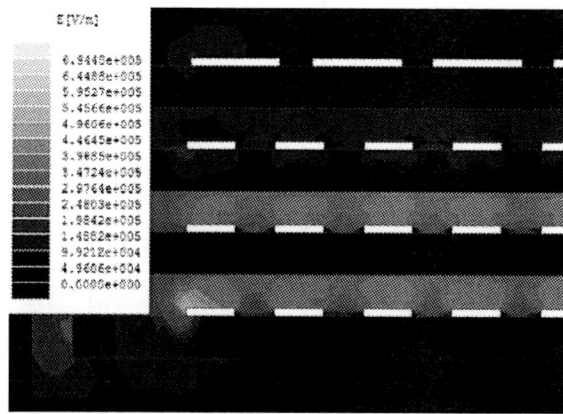

Fig. 10.    Electric-field intensity ( **E** ) distribution of the windings

### C. The electric-field intensity

Since the planar transformer in this paper is used in a high voltage SMPS, the electric potential of the windings are distinctly different. And the potential difference is rather high. In order to keep the insulativity of each winding, we arrange the winding in the order of $W_P$ - $W_{S1}$ - $W_{S2}$ - $W_{S3}$. It's a voltage incremental sequence. The electric-field intensity distribution of the windings is shown in the Fig. 10, from which we can see that: The **E** around $W_{S2}$ and $W_{S3}$ is rather stronger, the **E** around $W_P$ and $W_{S1}$ is fairly faint, and each epoxy board acts as a perfect insulation role.

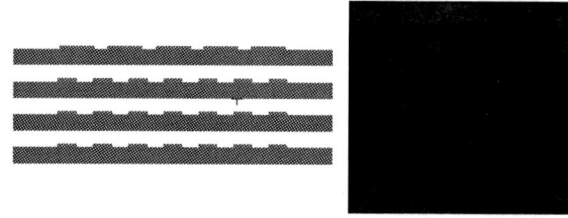

Fig. 11.    Calculation area of the planar transformer outside the magnetic core

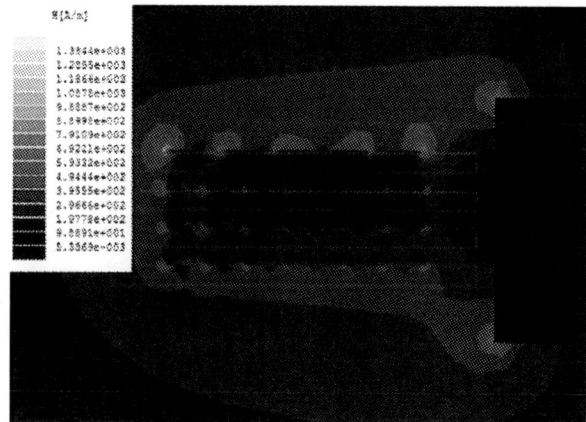

Fig. 12.    Magnetic intensity ( **H** ) distribution of the windings outside the magnetic core

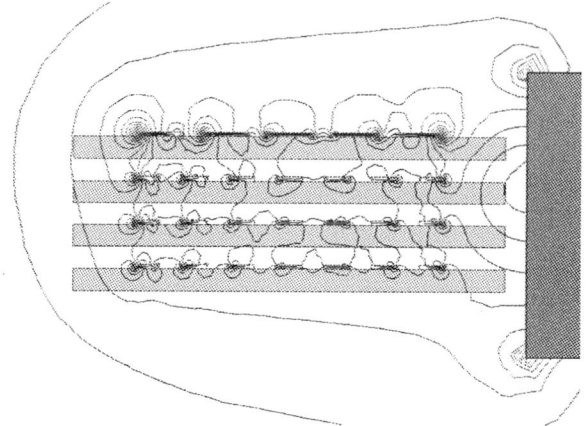

Fig. 13.    Contour line ploting of **H** outside the core

## V.    CALCULATION OUTSIDE THE MAGNETIC CORE

According to planar transformer, the emission extension of magnetic field is generally wider than that of electric field, as Fig. 7 and Fig. 10 showed. The energy of electric field is concentrated among the windings, while the magnetic field is radiated. Fig. 11 gives the calculation area of the planar transformer outside the magnetic core. According to Fig. 2, it's a view of y-z plane. The block filled in black is the magnetic core. Fig. 12 gives the magnetic intensity **H** distribution of the windings outside the magnetic core. According to Fig. 7, the leakage field is rather strong for the reason of the relative position of the magnetic core. Fig. 13 gives the contour line ploting.   From the two figures, we can see that the EMI issue bought by the leakage of the magnetic field here is the intersystem type. In high-density system integration, the sensitive components should keep away from the area.

## VI.    CONCLUSIONS

The benefits of planar magnetics are becoming widely recognised and the diversity of applications is increasing. A high voltage scheme, the TWT power supply, is presented to show the application of planar transformer. An FEA model has been developed for the prediction of the field distribution of the planar transformer. The simulation results show that the emission extension of magnetic field is generally wider than that of electric field. The leakage of magnetic field outside the magnetic core is rather stronger than that inside the magnetic core. The models and simulation methods are of great values to the EMI analysis of the planar transformer and the system integration.

## VII.    REFERENCES

[1] N. Dai and F. C. Lee, "An Algorithm of High- Density Low-Profile Transformers Optimization," in Proc. Of IEEE Applied Power Electronics Conference and Exposition, 1997, pp. 918–924.

[2] H. Chung, S.Y.R. Hui and S.C. Tang, "Development of low profile DC/DC power converter using switched capacitor circuits and coreless PCB gate drive", IEEE PESC'99, South Carolina, USA, 1999, pp48-53.

[3] Ansoft Corporation. Maxwell SV Getting Started – A 2D electrostatic problem. Feb.2002.

**2006 IEEE International Conference on Power Electronic, Drives and Energy Systems**

# Graphical Estimation of Optimum Weights of Iron and Copper of a Transformer

C.Easwarlal , Dr.V.Palanisamy, Dr.M.Y.Sanavullah , and M.Gopila

*Abstract* - Transformer may be designed to make one of the following quantities as minimum. (i) Total volume (ii) Total weight (iii) Total cost (iv) Total losses. In general these requirements are contradictory and it is normally possible to satisfy only one of them in a transformer. However the conditions for all the above quantities end in terms of the ratio of weight of iron and copper. In particular, the ratio is unity for optimum weight design of a transformer. Once we obtain optimum weights, optimum requirement for other quantities can easily be estimated. The design equations are modified so as to have only two variables in the equation. Necessary objective function and constraint equations are formed with two variables. Graphical method is adopted and optimum values are estimated separately and simultaneously. A 5 kVA single phase core type distribution transformer is taken and the condition for optimum weight design is verified and analysed.

*Index Terms*-- Constraint, Graphical optimization, Objective function, Optimum weight, Transformer.

## I. NOMENCLATURE

$\Phi_m$ = Main flux, Wb.
$B_m$ = Maximum flux density, Wb/m$^2$
$\delta$ = Current density, A/m$^2$
$A_i$ = Net core area, m$^2$
$A_c$ = Area of copper in the window, m$^2$
$A_w$ = Window area, m$^2$
$K_w$ = Window space factor
$f$ = Frequency, Hz
AT = mmf
$l_i$ = Mean length of flux path in iron, m
$L_{mt}$ = Length of mean turn of transformer winding, m
$G_i$ = Weight of active iron, kg
$G_c$ = weight of copper, kg
$g_i$ = Weight per m$^3$ of iron, kg
$g_c$ = Weight per m$^3$ of copper, kg

Prof. C. Easwarlal, Prof/EEE, Sona College of Technology, Salem,Tamilnadu State,India Ph.No:+91-0427-2443545 Ext:333 (e-mail: easwarlalc@yahoo.co.in)
Dr.V.Palanisamy,Principal,Government College of Technology, Coimbatore,Tamilnadu State,India. Ph.No:009442141200 (e-mail: vpsamyin@yahoo.co.in)
Dr.M.Y.Sanavullah,Dean/EEE, K.S.R College of Technology, Tiruchengode,Tamilnadu State, India. Ph.No:009442061632 (e-mail: sana-vullah2006@yahoo.co.in)
Ms.M.Gopila, Lect/EEE, Sona College of Technology, Salem, Tamilnadu State, India. Ph.No:009894604809.(e-mail: gopila13@yahoo.com)

$p_i$ = loss in iron per kg, W
$p_c$ = loss in copper per kg, W
$P_i$ = Total iron loss, W
$P_c$ = Toal copper loss, W
$C_i$ = Total Cost of iron, Rs
$C_c$ = Total Cost of copper, Rs
$c_i$ = Specific cost of iron, Rs/kg
$c_c$ = Specific cost of copper, Rs/kg
$x$ = Fraction of full load.

## II. INTRODUCTION

OPTIMIZATION[1] is the act of obtaining the best result under the given circumstances. [3] Mr. Frank F. Judd and D. R. Kressler in their paper emphasized on the ultimate utility of analytical techniques for a comprehensive computer aided design program besides highlighting on graphical design procedure. They suggested two approaches namely the first one is to fix the electrical and magnetic parameters and then choose the transformer geometry parameters to minimize an overall objective such as weight, volume, losses or cost. An alternate approach is to start with assumed core geometry and then find values for the electrical and magnetic parameters, which minimize the design objective. Their objective is to maximize the transformer output VA capability by choosing optimum values for flux density and current density. They preferred the second approach and explained in their paper. In this paper, the transformer problem is stated similar to first method.

While deriving the conditions for the above said four quantities, they end in terms of the ratio of weight of iron and copper. The weight of iron and copper can be found with the help of relevant formulae. The normal method of calculation of weights of iron and copper involves values for some quantities, estimation of overall dimensions of frame and winding design. Moreover, it is not certain to obtain optimum values. But it would be very much useful in the design if we find the values of optimum weights of iron and copper separately. Once the optimum weights are known, the optimum values for the other quantities can be obtained easily. In order to achieve it, objective function and constraint equations are required. The equations are formed with the help of only two variables, which motivated for graphical estimation. Here in this paper, optimum design of total weight is considered and the values are obtained graphically for a given transformer. A 5 kVA single-phase transformer as described below is considered and the data are obtained graphically and analysed.

0-7803-9771-1/06/$25.00 ©2006 IEEE

## III. CONTROLLING FACTOR

[2] The quantities of interest vary with the ratio $r = \phi_m / AT$. If we choose a high value of r, the flux becomes large and consequently a large core cross section is needed which results in higher volume, weight and cost of iron and also gives a higher iron loss. On the other hand owing to decreasing in the value of AT, the volume, weight and cost of copper required decreases and also the $I^2R$ losses decreased. Thus we conclude that the value of r is a controlling factor for the above mentioned quantities.

## IV. DESIGN FOR MINIMUM WEIGHT

Let us consider a single phase transformer. Its kVA output is: $Q = 2.22 f\, B_m\, \delta\, k_w\, A_w\, A_i \times 10^{-3}$

$Q = 2.22 f\, B_m\, \delta\, A_c\, A_i \times 10^{-3}$. Assuming that the flux and current densities are constant, we see that for a transformer of given rating the product $A_c$ and $A_i$ is constant.

Let this product: $\quad A_c A_i = M^2 \qquad (1)$

The optimum design problem is the minimum value of total weight.

Now, $r = \varphi_m / AT$ and

$\phi_m = B_m A_i$ and

$AT = \delta K_w A_w / 2 = \delta A_c / 2$

Therefore, $\quad r = 2 B_m A_i / \delta A_c$

$or \; A_i / A_c = \delta r / 2 B_m$

$$= \beta \qquad (2)$$

where $\beta$ is a function of r only as $B_m$ and $\delta$ are constant.
From (1) and (2), we have:

$$A_i = M\sqrt{\beta} \quad and \quad A_c = M / \sqrt{\beta}$$

Let G be the total weight of transformer active materials (ie., iron and copper):

$G = G_i + G_c$

$\quad = g_i l_i A_i + g_c L_{mt} A_c$

$\quad = g_i l_i M\sqrt{\beta} + g_c L_{mt} M / \sqrt{\beta} \qquad (3)$

Differentiating (3) w.r.to $\beta$,

$$dG/d\beta = \tfrac{1}{2} g_i l_i M\beta^{-\frac{1}{2}} - \tfrac{1}{2} g_c L_{mt} M\beta^{-\frac{3}{2}}$$

For minimum weight,

$dG/d\beta = 0$

$$\tfrac{1}{2} g_i l_i M\beta^{-\frac{1}{2}} = \tfrac{1}{2} g_c L_{mt} M\beta^{-\frac{3}{2}}$$

$$g_i l_i = g_c L_{mt} / \beta$$

$$\beta = g_c L_{mt} / g_i l_i$$

$$\beta = A_i / A_c$$

$$g_i l_i A_i = g_c L_{mt} A_c$$

$$G_i = G_c$$

Weight of iron = Weight of copper
Therefore, the ratio:

$$G_i / G_c = 1 \qquad (4)$$

Similar such conditions are

$G_i / G_c = c_c / c_i \quad$ for minimum cost

$G_i / G_c = g_i / g_c \quad$ for minimum volume

$G_i / G_c = x^2 p_c / p_i \quad$ for minimum losses

It is seen that the conditions for optimization end with the requirements of the ratio of weight of iron and copper. Since $G_i / G_c = 1$ for minimum weight design, it is verified by graphical method for a given transformer.

## V. FORMATION OF OBJECTIVE FUNCTION, CONSTRAINT EQUATION AND OPTIMIZATION

Let us take a single phase distribution transformer. The transformer rating, specifications are known. The objective function for weight to be minimum may be written as

$$G = G_i + G_c \qquad (5)$$

However from the above derivation, we know the condition for minimum weight as $G_i = G_c$. We shall verify the condition using graphical method. For a series of values of G, the points for $G_i$ and $G_c$ are obtained and tabulated. It will be a straight line graph. To form constraint or performance equation, we know that the output of the transformer

$$Q = 2.22 f\, B_m\, \delta\, A_c\, A_i \times 10^{-3}$$

Replacing $A_c$ and $A_i$ in terms of $G_c$ and $G_i$, the above equation becomes:

$$Q = \frac{2.22 f\, B_m\, \delta\, G_c\, G_i \times 10^{-3}}{g_c g_i L_{mt} l_i}$$

Knowing the value for $k = L_{mt} / l_i$, the above equation may be written as:

$$Q = \frac{2.22 f\, B_m\, \delta\, G_c\, G_i \times 10^{-3}}{g_c g_i k l_i^2}$$

Transforming the above equation, the constraint equation can be written as:

$$G_i G_c = k l_i^2 \qquad (6)$$

Where,

$$k = Q g_c g_i k \times 10^3 / 2.22 f B_m \delta$$

Now treating $l_i$ as a parameter, a graph between $G_i$ and $G_c$ is drawn which will be parabolic in nature since equation (6) is a non linear equation. The tangent made by the objective function to the constraint curve gives the optimum value for $G_i$ and $G_c$, which will be equal for minimum weight design. For different values of $l_i$ optimum values can also be easily obtained for the transformer considered.

## VI. DESCRIPTION OF A TRANSFORMER

Design a 5kVA, 11000/400V, 50Hz, single phase core type distribution transformer for optimum weight. The ratio $L_{mt}/l_i = 0.4$ and $l_i = 1m$. Assume a square cross section for the core, a flux density 1Wb/m², a current density 1.4A/mm² and a window space factor 0.2.

## VII. GRAPHICAL OPTIMIZATION

The objective function is $G = G_i + G_c = cons\tan t$, C for the given transformer. For a series of values of C, the points $G_i$ and $G_c$ are obtained and tabulated as shown in table I to V.

TABLE I
C=50

| $G_i$ | 10 | 15 | 20 | 25 | 30 | 35 | 40 |
|-------|----|----|----|----|----|----|----|
| $G_c$ | 40 | 35 | 30 | 25 | 20 | 15 | 10 |

TABLE II
C=60

| $G_i$ | 10 | 15 | 20 | 30 | 35 | 40 | 50 |
|-------|----|----|----|----|----|----|----|
| $G_c$ | 50 | 45 | 40 | 30 | 25 | 20 | 10 |

TABLE III
C=70

| $G_i$ | 10 | 20 | 30 | 40 | 50 | 60 | 70 |
|-------|----|----|----|----|----|----|----|
| $G_c$ | 60 | 50 | 40 | 30 | 20 | 10 | 0 |

TABLE IV
C=72

| $G_i$ | 10 | 20 | 30 | 40 | 50 | 60 | 70 |
|-------|----|----|----|----|----|----|----|
| $G_c$ | 62 | 52 | 42 | 32 | 22 | 12 | 2 |

TABLE V
C=80

| $G_i$ | 10 | 20 | 30 | 40 | 50 | 60 | 70 |
|-------|----|----|----|----|----|----|----|
| $G_c$ | 70 | 60 | 50 | 40 | 30 | 20 | 10 |

The corresponding graphs are drawn which are straight line in nature as shown in Fig. 1.

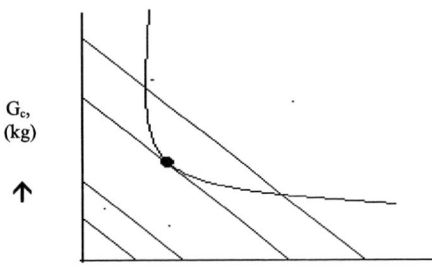

Fig. 1. Graphical optimization of weight.

We know that the transformer is to give an output of 5kVA. Hence the performance or behavior constraint equation is written as

$$5 = 2.22 f B_m \delta \times (G_c / g_c L_{mt}) \times (G_i / g_i l_i) \times 10^3$$

Since, $L_{mt} = 0.4 l_i$ and substituting the other known values in the above equation

$$G_i G_c = \left(5 g_c \times 0.4 l_i^2 g_i\right) / \left(2.22 f B_m \delta \times 10^3\right)$$

$$= \frac{\left(5 \times 8.9 \times 10^3 \times 0.4 \times 1^2 \times 7.8 \times 10^3\right)}{\left(2.22 \times 50 \times 1 \times 1.4 \times 10^3\right)}$$

$$G_i G_c = 893$$

The points $G_i$ and $G_c$ are obtained for the above constraint equation and tabulated as shown in Table VI

TABLE VI
CONSTRAINT EQUATION

| $G_i$ | 10 | 15 | 20 | 25 | 30 | 35 | 40 |
|-------|------|-------|-------|------|------|------|------|
| $G_c$ | 89.3 | 54.53 | 44.65 | 35.7 | 29.8 | 25.5 | 22.3 |

The graph for constraint equation is drawn and it is parabola in shape and indicates that it is a non linear equation. It is as shown in Fig.1. The tangent point is obtained as (30, 30) where the weight of iron is equal to the weight of copper, which is the condition for optimum weight design.

## VIII. ANALYSIS

a) From Fig. 1. the values for $G_i = 30 kg$ and $G_c = 30 kg$ where the condition for optimum weight design of $G_i = G_c$ is verified.

b) For $l_i = 1.18 m$, the new constraint equation (NCE1) is

$$G_i G_c = 893 \times 1.18^2 = 1243$$

The new points are found and shown in Table VII. The constraint equation graph will be shifted and appear as shown in Fig.1. Actually the graph is not shown. The optimum values are obtained as (35.35, 35.35). The output is checked as:

$$Q = \frac{\left(2.22 \times 50 \times 1 \times 1.4 \times 10^6 \times 35.35 \times 10^{-3}\right)}{\left(8.9 \times 7.8 \times 10^6 \times 0.4 \times 1.18^2\right)}$$

$$Q = 5.035 kVA$$

TABLE VII
NCE1

| $G_i$ | 10 | 15 | 20 | 30 | 40 | 50 |
|---|---|---|---|---|---|---|
| $G_c$ | 124.3 | 82.86 | 62.15 | 41.43 | 31.1 | 24.86 |

| $G_i$ | 60 | 70 |
|---|---|---|
| $G_c$ | 20.7 | 17.76 |

c) For $l_i = 1.2m$, the constraint function graph will be further shifted. The new constraint equation (NCE 2) is

$$G_i G_c = 893 \times 1.12^2 = 1286$$

The new points are found and shown in Table VIII

TABLE VIII
NCE2

| $G_i$ | 10 | 15 | 20 | 30 | 40 | 50 |
|---|---|---|---|---|---|---|
| $G_c$ | 128.6 | 85.7 | 64.3 | 42.86 | 32.15 | 25.7 |

| $G_i$ | 60 | 70 | 80 | 60 | 70 | 80 |
|---|---|---|---|---|---|---|
| $G_c$ | 21.43 | 18.37 | 16 | 21.43 | 18.37 | 16 |

The optimum values are obtained as (36, 36). The output is checked as:

$$Q = \frac{\left(2.22 \times 50 \times 1 \times 1.4 \times 10^6 \times 36^2 \times 10^{-3}\right)}{\left(8.9 \times 7.8 \times 10^6 \times 0.4 \times 1.2^2\right)}$$

$$Q = 5.035 kVA$$

d) For $l_i = 2m$, the new constraint equation (NCE 3) is:

$$G_i G_c = 893 \times 2^2 = 3572$$

The new points are found and shown in Table IX.

TABLE IX
NCE3

| Gi | 10 | 20 | 30 | 40 | 50 |
|---|---|---|---|---|---|
| Gc | 357.2 | 178.6 | 190.1 | 89.3 | 71.44 |

| Gi | 60 | 70 | 80 | 90 | 100 |
|---|---|---|---|---|---|
| Gc | 59.53 | 50 | 44.65 | 39.7 | 35.72 |

The optimum values are obtained as (60, 60).
The output is checked as:

$$Q = \frac{\left(2.22 \times 50 \times 1 \times 1.4 \times 10^6 \times 60^2 \times 10^{-3}\right)}{\left(8.9 \times 7.8 \times 10^6 \times 0.4 \times 2^2\right)}$$

$$Q = 5.037 kVA$$

e) Optimum efficiency: Taking resistivity of copper as $2.1 \times 10^{-8} \Omega m$ at $75^0$C and its density as $8.9 \times 10^3 kg/m^3$ the specific copper loss $p_c$ at $75^\circ c = 2.36 \times 10^{-12} \delta^2 W/kg$, where $\delta =$ Current density, $A/m^2$. Substituting the known values, total copper loss $P_c = 138.768 W$. From loss curve for the laminations supplied by the manufacturer and for $B_m = 1 Wb/m^2$, the specific iron loss, $p_i = 1.3 W/kg$. Therefore the total iron loss $P_i = 39 W$. The full load efficiency under UPF, $\eta_{fl} = 96.56\%$. From the condition for optimum efficiency, i.e. $P_i = x^2 P_c$, $x$, the fraction of full load is obtained as $x = 0.53$. Since the transformer considered is small one, the above value is reasonable. The values for other transformer are indicated below for comparison.

| Rating | Voltage V | Type | Pc W | Pi W | η % | x |
|---|---|---|---|---|---|---|
| 500 kvA, Power transformer | 6600/400 | 1 Ph core type | 3865 8138 | 1460 684 | 98.95 98.27 | 0.61 0.29 |

| Rating | Gi, kg | Gc, kg | G, kg | Remarks |
|---|---|---|---|---|
| 500 kVA, Power transformer | 974 456 | 206 456 | 1180 912 | Refer[3] Optimization |

IX. CONCLUSION

In conventional design, estimation of weights involve requirement of values for some quantities and other data. Moreover it is not certain to obtain optimum values. From analysis (a), it is seen that the condition of $G_i = G_c$ for optimum weight design of the given transformer is verified. Further from (b) to (d), it is seen that the weights increase proportionately as shown in Table X with the increase in mean length of flux path, $l_i$, keeping other parameters constant. The optimum weight of iron and copper are found separately and simultaneously. Since it is graphical optimization, the values are more accurate and unique for the given transformer.

TABLE X
INCREASE IN MEAN LENGTH OF FLUX PATH

| $l_i$, m | $G_i$,kg | $G_c$, kg | $G_i/G_c$ |
|---|---|---|---|
| 1.0 | 30 | 30 | 1 |
| 1.18 | 35.35 | 35.35 | 1 |
| 1.2 | 36 | 36 | 1 |
| 2 | 60 | 60 | 1 |

It is stated that optimization of other quantities are dependent upon the weight of $G_i$ and $G_c$. For cost optimization the condition is $G_i/G_c = c_c/c_i$. Taking specific cost as $c_i = 100\,Rs/kg$ and $c_c = 300\,Rs/kg$, the ratio $G_i/G_c = 3$. For 500 kVA transformer with normal design $G_i/G_c = 974/206 = 4.728$ which is not equal to 3. But for the 5kVA transformer with $G_i = G_c$, the cost of iron, $C_i = c_i G_i = Rs.3000$ and $C_c = c_c G_c = Rs.3000$. From the above, it is concluded that optimum weights can be estimated graphically for optimum weight design of a given transformer. The values are useful is estimating other quantities such as cost, efficiency, etc.

## X. REFERENCES

[1] Singiresu S. Rao, "Engineering optimization theory and practice,"3rd Ed, New Age international (P) Ltd., New Delhi – 110002,1999,pp. 11-15

[2] A. K. Sawhney, "A course in electrical machine design," Dhanpat Rai & Co. (P) Ltd., Delhi – 110006, 2001.

[3] Frank F Judd and Durwood R Kressler, "Design optimization of small low frequency power transformers,"*IEEE Trans. on magnetics*, vol. MAG.13, No.4, July 1977

## XI. BIOGRAPHIES

**C. Easwarlal** was born in 1947. He studied his B.E (Electrical and Electronics Engineering) degree course and M.Sc (Engg) (Electrical Machines) degree course at PSG College of Technology, Coimbatore - 641 004 and received the degrees in the years 1971 and 1973 from University of Madras.

He has worked as an Electrical Engineer in a company for two years. Then he took up the teaching assignment. He has put in 28 Years of experience. Currently he is working towards his Ph.D degree in Anna University, Chennai, India. He is a professor of Electrical and Electronics Engineering, Sona College of Technology, Salem – 636 005, India. His area of interests is design of Electrical Machines, optimization in design, power system analysis.

**V. Palanisamy** received the B.E. degree in Electronics and Communication Engineering, from PSG College of Technology, Coimbatore in 1972, the M.E. degree in Communication systems in 1974 from University of Madras and the Ph.D. in Antenna Theory in 1987 from IIT Karagpur, India. Since 1974 he has been working in various capacities of Technical Education in Tamilnadu, and currently he is Principal of Government College of Technology, Coimbatore, India. His interest includes Neural Network, Fuzzy logic and optimization.

**M. Y. Sanavullah** was born in 1942. He received the B.E degree in Electrical & Electronics Engineering and M.Sc degree in Power System Engineering in the years 1965 and 1967 from the University of Madras, Ph.D degree in High Voltage Engineering in 1987 from Anna University, Chennai. He guided a candidate at Ph.D level in Periyar University, Salem during 2001 – 2005. At present he is guiding four candidates leading to Ph.D in Vinayaka Mission's University, Salem. He has 38 Years of teaching experience. He is currently a Professor and Dean with K.S Rangasamy College of Technology, Tiruchengode 637 209, India. He is interested in finite Element Method.

**M. Gopila** received the B.E degree in Electrical and Electronics Engineering from Sona College of Technology, Salem in 2003. She is working as a Lecturer in Sona College of Technology, Salem, India and has got 2 ½ Years of experience in the teaching field. Currently, she is pursuing her Master Degree in Power System Engineering. She is interested in Electrical Machines and Power Systems.

**2006 IEEE International Conference on Power Electronic, Drives and Energy Systems**

# Nonlinear Behavior of Self-excited Induction Generator Feeding an Inductive Load

D.D.Ma, *Student Member, IEEE*, B. Zahawi, *Senior Member, IEEE*, D. Giaouris, *Member, IEEE*,
S. Banerjee, *Senior Member, IEEE, and* V. Pickert, *Member, IEEE*

*Abstract—* The nonlinear behavior of a self-excited, smooth air gap, cage induction generator feeding an inductive load is analyzed in this paper, allowing for the effects of machine saturation. The self autonomous system is shown to exhibit a transition from a periodic orbit to a quasi-periodic orbit through a Neimark bifurcation.

*Index Terms--* Bifurcation theory, induction generator, induction machine, inductive load, nonlinear dynamics

## I. INTRODUCTION

INDUCTION generators are widely used in conjunction with small hydro or wind turbine to produce electric power, mainly due to their low cost, compared with synchronous machines. The generator in such applications is usually connected directly to the ac supply network which also provides the necessary reactive power for the production of the machine rotating magnetic flux. This need for reactive power limits the use of the induction machine as a stand alone generator for remote applications where a supply connection is not available. To overcome this problem the reactive power can be supplied from a capacitor bank connected across the stator terminals, allowing the machine to work as a Self-Excited Induction Generator (SEIG) in the absence of a supply connection.

State space methods [1]-[4] have to be used to model and study the dynamics of these systems. The states of the system may be the machine currents, fluxes or a combination of these [5], [6]. The model must also include components that represent the magnetic nonlinearities (mainly cross-saturation phenomena) of the machine [7] as the machine is normally working with values of magnetic flux density near the saturation level. Hence the overall model of the system will be highly nonlinear and time varying. The dynamical analysis of the system is further complicated by the use of capacitor bank which provides the reactive power to the generator. This paper studies the dynamical behavior of self-excited induction generators and shows that it is possible to have bifurcation phenomena which force the system to change its desired stable

D. D. Ma, B. Zahawi, D. Giaouris and V. Pickert are with the School of Electrical, Electronic and Computer Engineering, Newcastle University, Newcastle upon Tyne, NE1 7RU, UK (email: bashar.zahawi@ncl.ac.uk).

S. Banerjee is with the Indian Institute of Technology, Kharagpur Centre for Theoretical Studies and Department of Electrical Engineering, 721302, India (email: soumitro@iitkgp.ac.in).

response. The bifurcation that causes this loss of stability is shown to be a Neimark bifurcation.

The machine nonlinear model is presented and described and the operation of the self-excited generator on no-load, and when feeding a purely resistive load, is examined to show that the system exhibits a normal period one orbit. When linear inductive components are included in the load the machine undergoes a transition from a stable period one orbit to a quasi-periodic through a Neimark bifurcation.

## II. MODELING OF THE SATURATED INDUCTION MACHINE

The mathematical model of the induction machine uses four states (currents and/or fluxes) and is linear time-varying rotor speed depended. If the chosen states are the stator and rotor currents expressed at a Stationary Reference Frame (SRF) the model is:

$$\mathbf{U} = \mathbf{RI} + \mathbf{L_1}\frac{d\mathbf{I}}{dt} + \omega_r \mathbf{L_2} \qquad (1)$$

where **U** is the vector with the stator and rotor voltages, **I** is the vector with the stator and rotor currents, **R** is the resistive matrix, $\omega_r$ is the rotor speed and $\mathbf{L_1}$, $\mathbf{L_2}$ are inductive matrices.

To model the nonlinearity the last two matrices have to change and to be a function of the magnetizing current instead of being constant. Hence the 5 matrices of the magnetically nonlinear system are:

$$\mathbf{U} = \begin{bmatrix} u_{sD} & u_{sQ} & u_{rd} & u_{rq} \end{bmatrix}^T, \mathbf{I} = \begin{bmatrix} i_{sD} & i_{sQ} & i_{rd} & i_{rq} \end{bmatrix}^T,$$

$$\mathbf{R} = diag(R_s, R_s, R_r, R_r),$$

$$\mathbf{L_1} = \begin{bmatrix} L_{sd} & L_{dq} & L_{md} & L_{dq} \\ L_{dq} & L_{sq} & L_{dq} & L_{mq} \\ L_{md} & L_{dq} & L_{rd} & L_{dq} \\ L_{dq} & L_{mq} & L_{dq} & L_{rq} \end{bmatrix}, \mathbf{L_2} = \begin{bmatrix} 0 & 0 & 0 & 0 \\ 0 & 0 & 0 & 0 \\ 0 & L_m & 0 & L_r \\ -L_m & 0 & -L_r & 0 \end{bmatrix}$$

where $L_m$ is the magnetizing inductance: $L_m = \left|\overline{\psi}_m\right| / \left|\overline{i}_m\right|$

The cross-saturation inductance ($L_{dq}$) is [7]:

$$L_{dq} = \frac{i_{md}i_{mq}}{i_m} \times \frac{dL_m}{d\left|\overline{i}_m\right|} \qquad (2)$$

this equation can be simplified to:

$$L_{dq} = \frac{i_{md}i_{mq}}{i_m} \times \frac{L - L_m}{\left|\overline{i}_m\right|} \qquad (3)$$

where $L = d\left|\overline{\psi}_m\right| / d\left|\overline{i}_m\right|$ is the dynamic inductance.

0-7803-9771-1/06/$25.00 ©2006 IEEE

The nonlinear curves of the magnetizing and dynamic inductance are taken from [7] and are shown in Fig. 1.

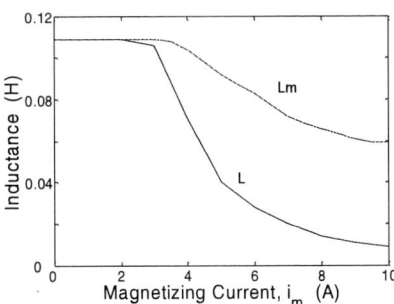

Fig. 1. The saturated magnetizing inductance curve $L_m$ and the dynamic inductance curve $L$.

The direct and quadrature axis saturated inductances are

$$L_{md} = L_m + \frac{i_{md}}{i_{mq}} L_{dq} \; , \quad L_{mq} = L_m + \frac{i_{mq}}{i_{md}} L_{dq} \qquad (4)$$

The stator and rotor dq axis inductance are as following

$$\begin{aligned} L_{sd} &= L_{sl} + L_{md}; \quad L_{sq} = L_{sl} + L_{mq} \\ L_{rd} &= L_{rl} + L_{md}; \quad L_{rq} = L_{rl} + L_{mq} \end{aligned} \qquad (5)$$

where $L_{sl}$ and $L_{rl}$ are the unsaturated stator and rotor leakage inductance, respectively. The mechanical equation between the prime mover and the electrical torque is

$$T_e + T_m = J \frac{d\omega}{dt} \qquad (6)$$

where

$$T_e = \frac{3}{2} \frac{P}{2} \left( \psi_{rq} i_{rd} - \psi_{rd} i_{rq} \right) \qquad (7)$$

### III. THE MACHINE PARAMETERS AND THE PROCEDURES OF RUNNING THE SEIG

By using the equations that were presented in section II a 4-pole start connected IG of 1.5kW, with a capacitor bank (135μF per phase) was simulated. The rated voltage and current of the machine were 220/380V and 7/4A respectively and the rated frequency was 50Hz. The stator and rotor resistances were 0.6Ω and 0.83Ω respectively while the stator and rotor impedances were 1.8Ω/phase and 1.8Ω/phase respectively. The prime mover was represented by a dc machine rotating at 1500rev/min. To represent the effect of the capacitor bank and the various loads that were applied the following dq equivalent circuit was used [8]:

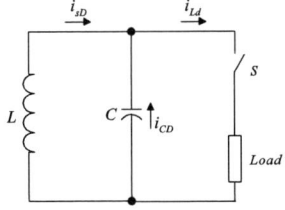

Fig. 2. Stator direct component without load, $i_{CD}$ is the capacitor current and $i_{Ld}$ is the load current.

### A. The Initial Self Excitation of the Induction Machine with No Load

As the machine is working under no load the switch S remains open and hence the d-q voltages are:

$$u_{CD} = -u_{sD} = -\frac{1}{C} \int i_{sD} dt \qquad (8)$$

$$u_{CQ} = -u_{sQ} = -\frac{1}{C} \int i_{sQ} dt \qquad (9)$$

By using the mathematical model which is presented in section II and by using (8) and (9) it is possible to simulate the behavior of the IG which is driven by a dc machine at a constant speed of 1500 rev/min under no load. From that test it can be seen that as the stator voltage increases (entering the saturation area) so does the magnetizing current and hence a big drop of the magnetizing inductance is observed, Fig.4, which agrees with the curves shown in Fig.1.

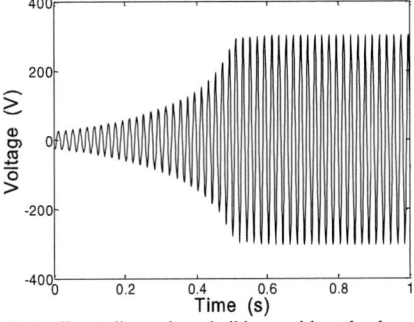

Fig. 3. Stator line to line voltage builds up without load.

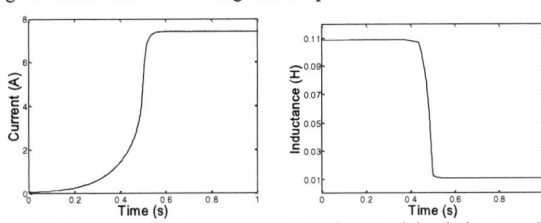

Fig. 4. Variation of magnetizing current (a) and magnetizing inductance (b) with voltage builds up without load.

### B. The SEIG with a Resistive Load

If the contactor S closes and the IG is supplying a resistive load the extra equations needed are:

$$u_{CD} = -u_{Ld} = -R i_{Ld} \qquad (10)$$

$$i_{CD} = -C \frac{du_{Ld}}{dt} = -RC \frac{di_{Ld}}{dt} \qquad (11)$$

$$i_{sD} = -i_{CD} + i_{Ld} \qquad (12)$$

$$i_{sD} = RC \frac{di_{Ld}}{dt} + i_{Ld} \qquad (13)$$

$$u_{CQ} = -u_{Lq} = -R i_{Lq} \qquad (14)$$

$$i_{sQ} = RC \frac{di_{Lq}}{dt} + i_{Lq} \qquad (15)$$

By using these equations the IG was simulated and its response is shown in Figs. 5 & 6. Initially the IG is under no load and at 0.1s a resistive load of 27Ω is applied. It is clear from that figure that there is a drop at the output voltage as the system has to supply the extra load.

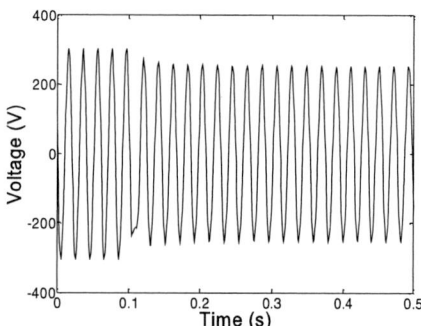

Fig. 5. Stator line to line voltage with a resistive load.

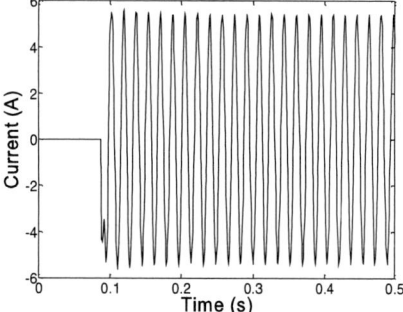

Fig. 6. Load current ($i_{LD}$) with a resistive load.

From these figures it can be seen that (regardless of the voltage drop) the solution curve in the state space follows a closed curve of period 1. At this point it has to be stated that in practice another capacitor is used in series with the resistive load which greatly decreases the voltage drop but from the dynamical point of view the behavior of the system remained qualitatively the same (i.e. the system exhibits a similar stable period one orbit) and hence due to space limitation it is not shown.

## IV. THE NONLINEAR BEHAVIORS OF SEIG FEEDING AN INDUCTIVE LOAD

In this section a balanced three phase inductive load of $30\Omega$ and 15mH per phase has been added to the system when the machine is driven at a constant speed of 1500rev/min. Therefore the equations describing this system are:

$$u_{CD} = -u_{Ld} = -Ri_{Ld} - L\frac{di_{Ld}}{dt} \qquad (16)$$

$$i_{CD} = -C\frac{du_{Ld}}{dt} = -RC\frac{di_{Ld}}{dt} - LC\frac{d^2i_{Ld}}{dt^2} \qquad (17)$$

$$i_{sD} = -i_{CD} + i_{Ld} \qquad (18)$$

$$i_{sD} = RC\frac{di_{Ld}}{dt} + LC\frac{d^2i_{Ld}}{dt^2} + i_{Ld} \qquad (19)$$

$$u_{CQ} = -u_{Lq} = -Ri_{Lq} - L\frac{di_{Lq}}{dt} \qquad (20)$$

$$i_{sQ} = RC\frac{di_{Lq}}{dt} + LC\frac{d^2i_{Lq}}{dt^2} + i_{Lq} \qquad (21)$$

Thus the state equations of capacitor voltages of both axes are obtained by substituting (19) into (16) and (21) into (20).

The response of the system was investigated for various values of the capacitance. For $C$=135µF (see Figs. 7 - 8) the response of the system is a period one closed orbit which indicates that the system operates within the desired specification. As this is a high order system it is not possible to plot all states and therefore only two representative states are shown in Fig. 8. All other combinations gave similar results and hence are not shown here.

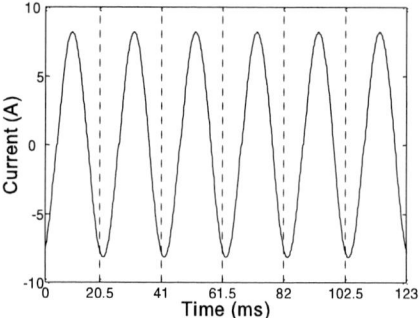

Fig. 7. Stator phase A current for C=135µF

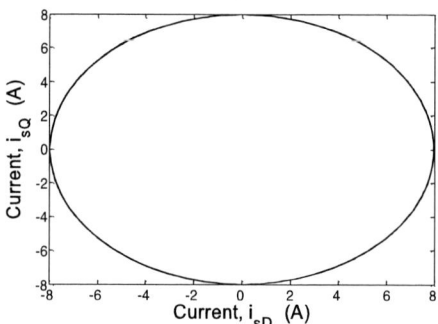

Fig. 8. Phase plane diagram for C=135µF

When the capacitance is increased to 156µF the response of the system changes to what initially appeared to be a period seven limit cycle (Fig. 9). A closer look reveals that the solution does not follow any periodic pattern but is instead a quasi-periodic behavior. By ignoring the initial transients the phase space was plotted using 5000 samples. Fig. 10 shows that the locus of the solution lies on a "toroid typed" manifold (difficult to visualize in such a high order).

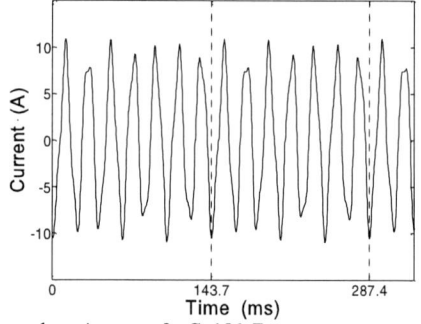

Fig. 9. Stator phase A current for C=156µF

Apart from this to prove that the system exhibits a quasi-periodic behavior it must be shown that the solution is dense on the torus, the Poincaré section is a closed orbit and also to show the bifurcation diagram. Other techniques can also be used like the Lyapunov exponent or the eigenvalues of the monodromy matrix of the period one orbit but in this paper only the first set is presented.

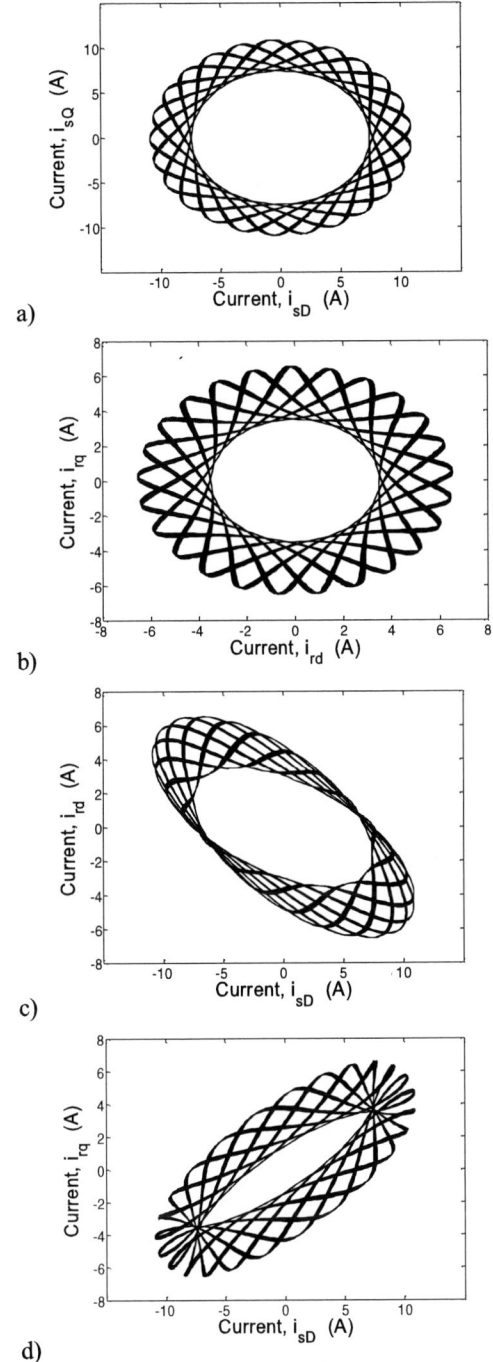

a)

b)

c)

d)

Fig. 10. Phase plane diagram for C=156μF (5000 sample points) (5000 sample points)

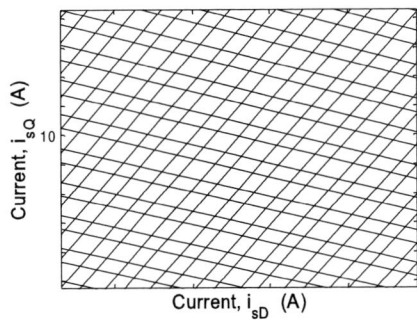

Fig. 11. Dense orbit in the torus

Fig. 11 shows the 20000 samples after the initial transient of one of previous tori and is clearly demonstrated that the orbit is dense on the torus. Furthermore, by sampling the state vector when the current $i_{sD}$ is zero the Poincaré map of the system is derived and as it can be seen by Fig.12 it is a closed orbit which again proves the statement that the orbit is quasi-periodic.

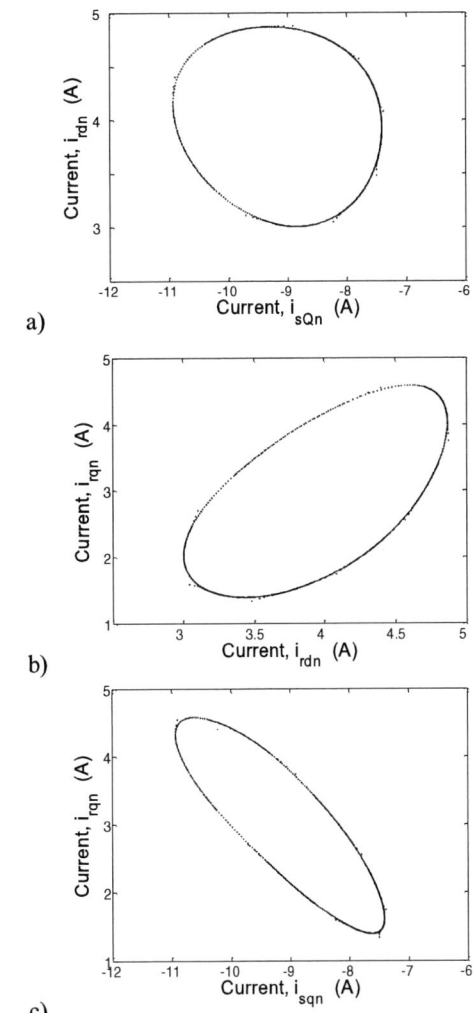

a)

b)

c)

Fig. 12. Poincare section of stator q-axis versus rotor q-axis sample.

The final test is to create the bifurcation diagram of the system which in this cases it was chosen to be the sampled value of the q-axis stator current when the d-axis stator current is zero, the bifurcation variable was the value of the capacitors used in the capacitor bank. Fig. 13 shows that diagram and it can be seen that the system looses its stability through a Neimark bifurcation and hence the system exhibits a quasi-periodic orbit [9].

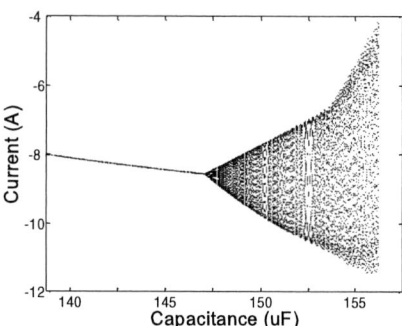

Fig. 13. Bifurcation diagram of stator q-axis current.

## V. CONCLUSIONS

The performance of the self-excited induction generator with no load, resistive load and compensate capacitors is studied. The nonlinear model utilizing currents as state variables is then connected with an inductive load to the stator terminal. The nonlinear behaviors of the induction generator are investigated through a bifurcation diagrams, phase spaces and Poincaré sections while changing a control parameter, the self-excited capacitors. The results show that the autonomous dynamical system loses its stability from period one orbit moving to a quasi-periodic orbit as a result of small changes in the values of system parameters (in this case the self-excited capacitors). The practical experiments of the machine will be examined and compared with the simulation models in the future work.

## VI. REFERENCES

[1] P. K. Shadhu Khan and J. K. Chatterjee, "Three-Phase Induction Generators: A Discussion on Performance," *Taylor & Francis Electric Machines and Power Systems*, vol. 27, pp. 813-832, August 1999.

[2] J. M. Elder, J. T. Boys, and J. L. Woodward," Self-excited induction machine as a small low-cost generator," *IEE Proceedings*, vol. 131, Pt. C, No. 2, pp. 33-41, March 1984.

[3] J. E. Brown, K. P. Kovacs, and P. Vas, "A Method of including the Effects of Main Flux Path Saturation in The Generalized Equations of A. C. Machines," *IEEE Trans. Power Apparatus and Systems*, vol. PAS-102, No. 1, pp. 96-103, January 1983.

[4] P. Vas, K. E. Hallenius, and J. E. Brown, "Cross- Saturation Smooth-Air- Gap Electrical Machines," *IEEE Trans. Energy Conversion*, vol. 1, No. 1, pp. 103-113, March 1986.

[5] P. Vas, *Electrical Machines and Drives — A space-vector theory approach*, Oxford, Clarendon Press, 1992.

[6] F. A. Farret, B. Palle, and M. G. Simões, "State Space Modeling Parallel Self-excited Induction Generators for Wind Farm Simulation," *Industry Applications Conference, 2004. 39th IAS Annual Meeting. Conference Record of the 2004 IEEE*, vol. 4, pp. 2801-2807, October 2004.

[7] K. E. Hallenius, P. Vas, and J. E. Brown, "The Analysis of a Saturated Self-Excited Asynchronous Generator," *IEEE Trans. Energy Conversion*, vol. 6, No. 2, pp. 336-341, June 1991.

[8] M. G. Simões and F. A. Farret, *Renewable Energy Systems: Design and Analysis with Induction Generators*, CRC Press, 2004.

[9] S. Banerjee and G. C. Verghese, *Nonlinear Phenomena in Power Electronics*, New York, IEEE Press, 2001.

## VII. BIOGRAPHIES

**Dandan Ma** was born in 1980, at Shanghai, China. She received the BEng degree in Electrical Engineering and Automation from Shanghai Teacher's University in 2002, the MSc in Automation and Control from Newcastle University in 2004. She is currently a PhD student at School of Electrical and Electronic Engineering at Newcastle University.

**Bashar Zahawi** received his BSc and PhD degrees in electrical and electronic engineering from the University of Newcastle, England, in 1983 and 1988, respectively. From 1988 to 1993 he was a design engineer at Cortina Electric Company Ltd, a UK manufacturer of large ac variable speed drives and other power conversion equipment. In 1994, he was appointed as a Lecturer in Electrical Engineering at the University of Manchester and in 2003 he joined the School of Electrical, Electronic & Computer Engineering at Newcastle University, where he is currently the Director of Postgraduate Studies. His research interests include power conversion, variable speed drives and the application of nonlinear dynamical methods to transformer and power electronic circuits.

**Soumitro Banerjee** did his B.E. from the Bengal Engineeringn College (Calcutta University) in 1981, M.Tech. from IIT Delhi in 1983, and Ph.D. from the same Institute in 1987. He has been in the faculty of the Department of Electrical Engineering, IIT Kharagpur, since 1986. Dr. Banerjee's areas of interest are the nonlinear dynamics of power electronic circuits and systems, and border collision bifurcations. He has published three books: "Nonlinear Phenomena in Power Electronics" (IEEE Press, 2001), "Dynamics for Engineers" (Wiley, 2005), and "Wind Electrical Systems" (Oxford University Press, 2005). He served as Associate Editor of the IEEE Transactions on Circuits & Systems II (2003-05), and is currently serving as Associate Editor of the IEEE Transactions on Circuits & Systems I. He is a recipient of the S. S. Bhatnagar Prize (2003), and the Citation Laureate Award (2004). He is a Fellow of the Indian Acad. of Sci. and of the Indian National Academy of Engineering.

**Damian Giaouris** received the diploma of Automation Engineering from the Automation Department, Technological Educational Institute of Thessaloniki, Greece, in 2000, the MSc degree in Automation and Control with distinction from Newcastle University in 2001 and the PhD degree in the area of control and stability of Induction Machine drives in 2004. His research interests include advanced nonlinear control & estimation of electromagnetic devices, and nonlinear phenomena in power electronic converters. He is currently a lecturer in Control Systems at Newcastle University, UK.

**Volker Pickert** received his Dipl.-Ing. in Electrical and Electronic Engineering from the RWTH Aachen, Germany and the University of Cambridge, UK (1994). He received his PhD from the Newcastle University in 1998. From 1998 to 2003 he worked first for Semikron International as an application engineer and then for Volkswagen as project manager responsible for power electronic systems and electric drives for electric-, hybrid- and fuel cell vehicles. Since 2003 he is Senior Lecturer at Newcastle University. His research interests are power electronics for automotive applications, thermal management, fault tolerant converters and non-linear controllers.

**2006 IEEE International Conference on Power Electronic, Drives and Energy Systems**

# Effects of Different Voltage Sags on Three-Phase Transformers

M. R. Shakarami, and A. Jalilian

*Abstract*--**This paper studies the effects of different types of voltage sags on three-phase transformers by means of computer simulation. The study shows that sags can produce transformer saturation during voltage recovery process. The saturation produces transformer inrush current which might be harmful in some cases. The influences of sag parameters (initial point–on-wave, duration and depth) have been analyzed in this paper. The simulation results show that current peak has a periodical dependence on sag duration and a linear dependence on sag depth. The highest peak currents are obtained for specific initial point-on-wave when the transformer supplied by different types of voltage sag.**

*Index Terms*-- **Inrush current, transformer saturation, voltage sag.**

## I. INTRODUCTION

POWER quality has a significant influence on different industries including high-technology equipment related to the communication systems, advanced control, automation, precise manufacturing techniques and on-line services. As an example, voltage sag can have a serious influence on the products of semiconductor fabrication with considerable financial losses. Power quality problems consist of transients, sags, interruption and other disturbances to the sinusoidal waveform [1]. One of the most important power quality problems is known as voltage sag (dip) which is a short duration (from 0.5 cycle to 1 minute) reduction between 10% and 90% in rms voltage [1].

Voltage sag can be either balanced or unbalanced, depending on its causes. If the three-phase voltages are equal, the sag is balanced (symmetrical) and if the phase voltages are different or the phase relationship is other than 120 degrees, sags are unbalanced (unsymmetrical) [2]. A three-phase short circuit or a large motor starting can produce symmetrical voltage sags. Single-line-to-ground, phase-to-phase or two phase-to-ground faults can cause unsymmetrical sags. Different categories of industrial equipment have different sensitivities to voltage sag [3]. The main categories of sensitive loads are: motors, adjustable-speed drives, some power electronic equipment, discharge lamps and control devices. Reference [4] describes a new phenomenon produced by voltage sag: according to measurements at different voltage levels.

Transformer saturation can occur when the voltage is recovered after a voltage sag event. The transformer saturation is caused precisely by the sudden variations in voltage after the sag, introducing high inrush currents.

The transformer behavior is very different to that of the three phase induction machine [5]. In induction machines,

there are current peaks after the initial and the final instants of sag while the initial point-on-wave has a minimal influence on the performance of machine.

In this paper, the performance of a transformer under balanced and unbalanced voltage sags are studies and influence of depth, duration and initial point-on-wave are analyzed.

A 60 kVA, 380/220V, three phase no-load transformer is implemented and simulated using MATLAB/SIMULINK software. The equivalent circuit parameters and magnetic curve in [6] is used to study the transformer performance.

## II. VOLTAGE SAG CLASSICICATION

Measured voltage at the customer bus is transient during a fault and after fault clearing (which means voltage is non-sinusoidal). Voltage sag may also involve a phase angle shift (phase jump) during the fault. In this study it is assumed that supply voltages before, during and after the sag are purely sinusoidal.

The phase to ground voltages during sag are given as:

$$V_{ah}(t) = \sqrt{2}V_{ah}\sin(\omega t + \alpha_{ah})$$
$$V_{bh}(t) = \sqrt{2}V_{bh}\sin(\omega t + \alpha_{bh}) \qquad (1)$$
$$v_{ch}(t) = \sqrt{2}V_{ch}\sin(\omega t + \alpha_{ch})$$

where $V_{ah}$, $V_{bh}$, $V_{ch}$, $\alpha_{ah}$, $\alpha_{bh}$ and $\alpha_{ch}$ are the magnitude and phase angle of the corresponding phases.

Depending on magnitude and angle of phases, voltage sags are defined in seven types, denoted as A, B, C, D, E. F and G [7]. In this paper, the initial instant $t_i$ -or fault instant- is the instant when the sag begins (drop voltage point), and the final instant $t_f$ -or fault clearing instant- is the instant when the sag ends (recovery voltage point). In the literature, it is usual to characterize voltage sags by only depth h and duration $\Delta t$. In the analyzed cases it has been observed that the voltage point-on-wave when the sag begins (called initial point-on-wave, $\psi_i = \omega t_i + \alpha$) is a parameter that also influences the inrush current peak.

To clearly show the voltage sag dependence on depth, duration and initial point-on-wave, the voltage sag will be represented by notation $v(h, \Delta t, \psi_i)$ and the inrush current will be represented by $i(h, \Delta t, \psi_i)$, where h, $\Delta t$, and $\psi$ are depth, duration and initial point-on-wave respectively. Duration is also shown as $\Delta t = t_f - t_i$ where $t_f$ and $t_i$ are final instant and initial instant of fault respectively. The current peak is also labeled as $i_{peak}(h, \Delta t, \psi_i)$ and it is defined as:

$i_{peak}(h, \Delta t, \psi_i)=\max\{| i_a(h, \Delta t, \psi_i) |, | i_b(h, \Delta t, \psi_i) |, | i_c(h, \Delta t, \psi_i) |\}$

It must also be considered that no phase jump during the fault has been assumed. Otherwise, the phase jump would be another parameter to characterize the sags.

---

M. R. Shakarami is a PhD student and A. Jalilian is an academic member of the Department of Electrical Engineering, Iran University of Science and Technology, Tehran, Iran. (email: jalilian@iust.ac.ir).

## III. TRANSFORMER MODEL

In this paper a transformer bank model has been implemented in MATLAB/SIMULINK for simulation purposes. Although, there are many different methods for representing core nonlinear behavior (power series, piecewise linear or arctangent functions) a function relationship presented in [6] has been employed as:

$$\phi = \left( k_1 \left( 1 + \left( \frac{|F|}{F_o} \right)^p \right)^{\frac{-1}{p}} + k_2 \right) F \qquad (2)$$

where $k_1$, $k_2$, p and $F_0$ are experimental parameters which allows this single-value function to fitted to the $(\Phi$-f$)$ transformer saturation curve (Fig. 1). These four parameters have a clear physical interpretations as:
- $k_1$ and $k_2$ are defined by slope in the linear and nonlinear zones of the$(\Phi$-f$)$ curve.
- p influences the shape of curve.
- $F_0$ is the magnetic potential where saturation begins.

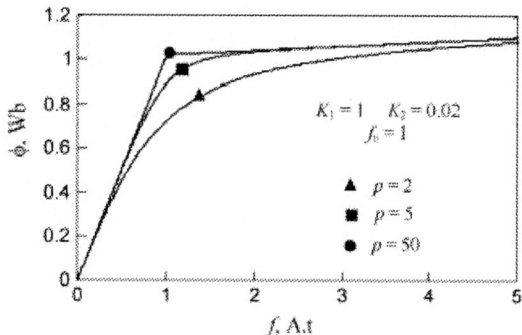

Fig. 1. $\left( \phi - f \right)$ Characteristic of a transformer core [6].

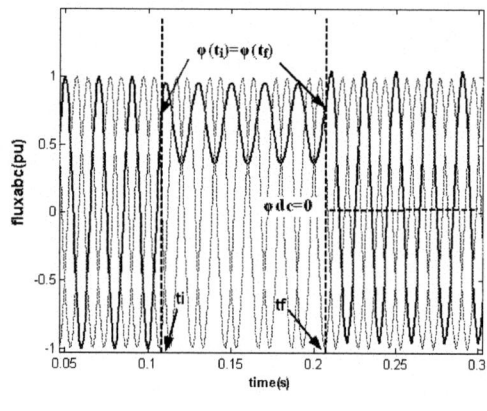

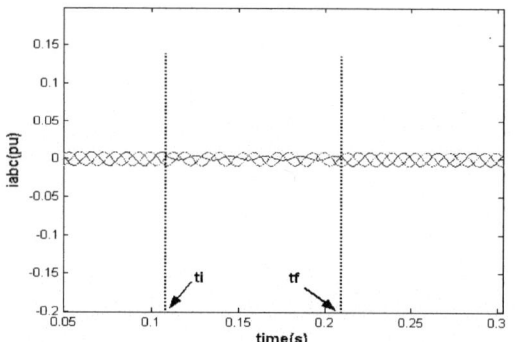

Fig. 2. (a) Flux and current for voltage sag type B with h=0.3, ψ=150, Δt=5T.

## IV. SIMULATION RESULTS

In this section the influence of different sag parameters (duration, initial point-on-wave and depth) are investigated on the performance of the transformer under different voltage sags. Fig. 2(a) shows the primary currents and fluxes in a voltage sag type B with the following characteristics: depth h=0.3, duration Δt=5T, initial point-on-wave $\psi_i$=150°, ie v(0.3, 5T, 150°). In this example no inrush current is obtained and the dc component of flux after voltage sag is null.

Fig. 2(b) shows the primary currents and fluxes in a case of voltage sag type B with characteristics v(0.3, 5.5T, 0°) where a sever peak inrush current is obtained. The reason for such a high current is found in the dc component of the magnetic flux after the sag. A detailed analytical explanation of this phenomenon is given in Appendix A. Fig. 2 shows that the flux is identical in both cases when the voltage drops ($\Phi_i$) and when the voltage recovers ($\Phi_f$) and no inrush current is experienced.

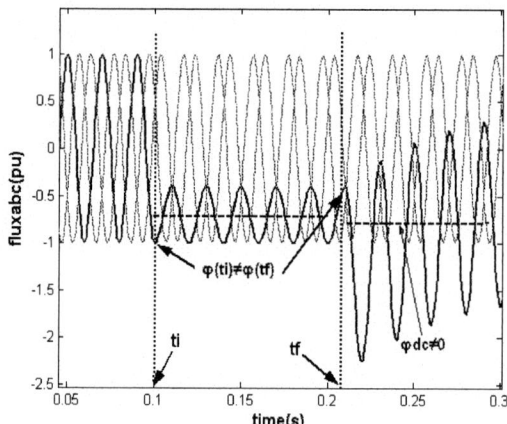

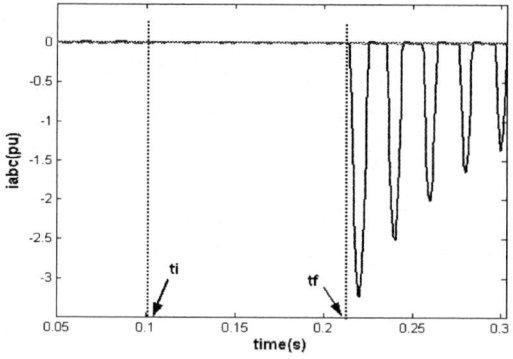

Fig. 2. (b) Flux and current for voltage sag type B with h=0.3, ψ=0, Δt=5.5T.

### A. Effect of Duration

Fig. 3 shows primary current for voltage sag type B with long duration Δt=15.5T and h=0.3, ψ=0°, this example shows that sag duration influence on inrush current peak is periodical.

$$i_{Peak}(h, \Delta t, \psi_i) \approx i_{Peak}(h, \Delta t + nT, \psi_i), \, n=integer \qquad (3)$$

To show details of the effect of voltage sag duration on peak current and maximum dc flux, different types of voltage sag have been studied with:

$v(h=0.3, \Delta t, \psi_i),\quad \psi_{i=0^\circ, 90^\circ, 180^\circ, 270^\circ}$

$\Delta t=[5T\ldots6T]$ (4)

Fig. 4 shows the results for sag duration ranging from 5T to 6T. Other conclusions that obtain from Fig. 4 and Fig. 2 are:

1) No inrush current peak obtained when duration is a multiple of full period $\Delta t=nT$, n=1, 2, 3,…

2) The sever current peak is obtained when sag duration is multiple of full period plus half period and initial point-on-wave 0° or 180° for voltage sags type A, B, D ,F and 90° or 270° for type C, E, G.

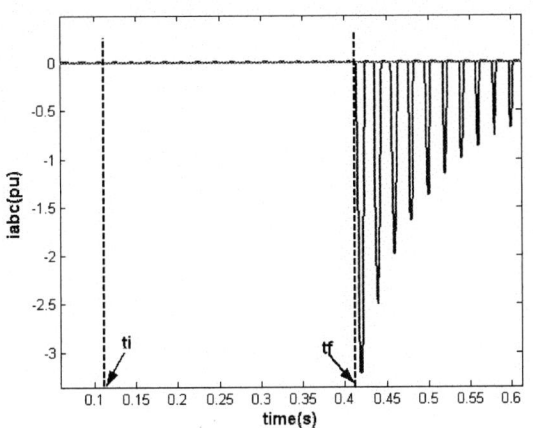

Fig. 3. Flux and current for voltage sag type B with h=0.3, ψ=0, Δt=15.5T.

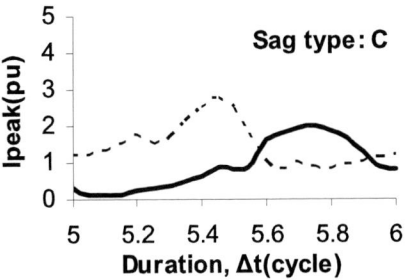

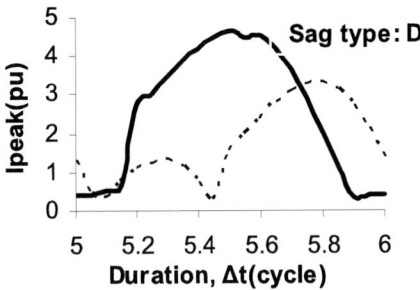

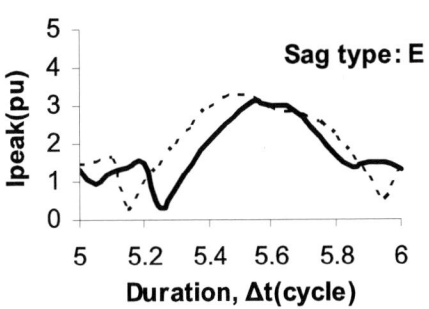

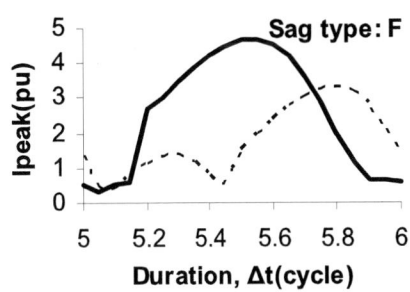

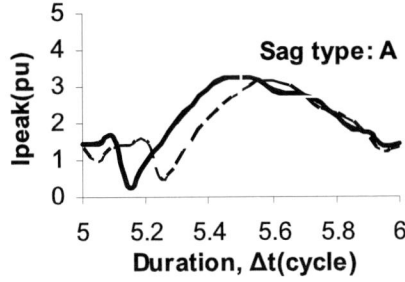

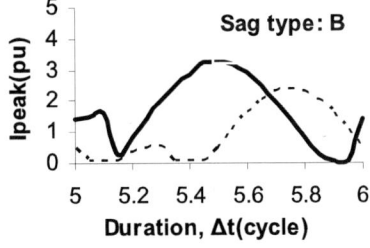

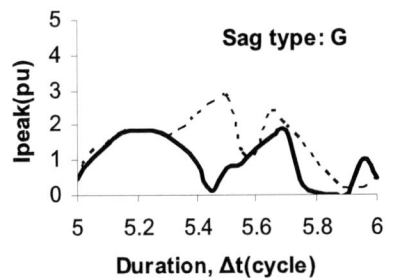

Fig. 4. Effect of duration with h=0.3 (solid line: ψ_i=0°, 180°, dashed line: ψ_i=90°, 270°).

## B. Effect of Initial Point-on-Wave

Various simulations are conducted in order to study the effect of initial point-on-wave on peak inrush current and dc flux under different types of voltage sags. With v*(h=0.3, Δt=5.5T, ψᵢ)*, different initial point-on-wave from 0° to 180° have been applied. The simulation results in this case are summarized in Table I and shown in Fig. 5. It can be seen from Table I that the peak inrush current can be occurred when dc flux component is high. If this component is very low inrush current is not significant. Generally, the value of inrush current is related to the value

of dc flux after voltage sag. Also TableI shows the maximum peak of inrush current is occurred for specific initial point-on-wave. These points are very important for transformer protection. The simulation results reveal that voltage sags type D and F are among the most dangerous types of voltage sag for transformer.

The most unfavorable initial points obtained are as:
0°, 60° and 120° for sag type A,
0° and 180° for sags type B, D and F,
90° and 270° for sags type C and G, and
60° and 120° for sag type E.

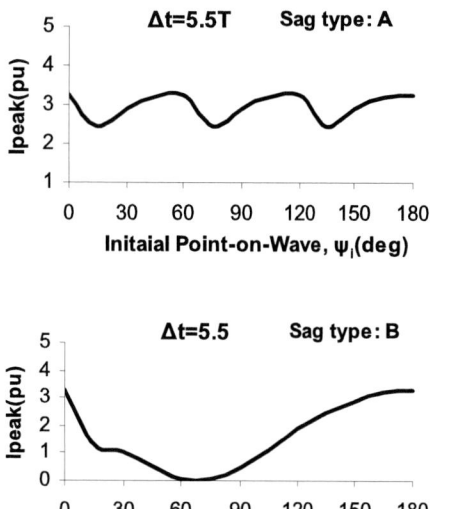

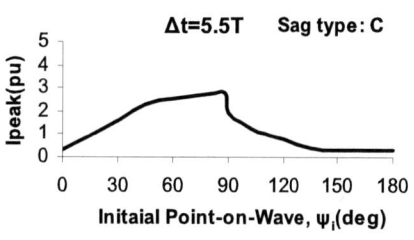

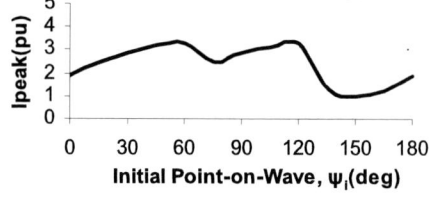

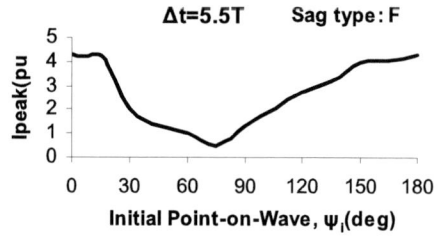

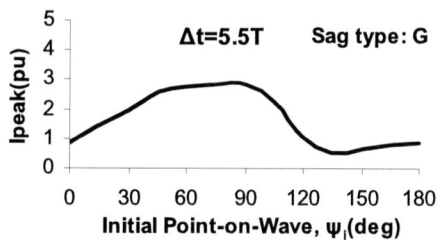

Fig. 5. Effect of initial point-on-wave of peak current with h=0.3.

TABLE I
EFFECT OF INITIAL POINT-ON-WAVE ON DC FLUX AND PEAK CURRENT FOR DIFFERENT VOLTAGE Sags

| Sag Type | $\psi_i$ | 0° | 15° | 30° | 45° | 60° | 75° | 90° | 105° | 120° | 135° | 150° | 165° | 180° |
|---|---|---|---|---|---|---|---|---|---|---|---|---|---|---|
| A | $\Phi dc_{max}$ (pu) | 1.55 | 1 | 1.2 | 1.3 | 1.4 | 1 | 1.2 | 1.3 | 1.38 | 1 | 1.2 | 1.4 | 1.55 |
|  | Ipeak(pu) | 3.25 | 2.46 | 2.89 | 3.19 | 3.25 | 2.47 | 2.89 | 3.19 | 3.25 | 2.47 | 2.9 | 3.2 | 3.25 |
| B | $\Phi dc_{max}$ (pu) | 1.5 | 0.7 | 0.65 | 0.4 | 0.1 | 0.1 | 0.35 | 0.63 | 0.32 | 1.18 | 1.25 | 1.35 | 1.5 |
|  | Ipeak(pu) | 3.25 | 1.32 | 1.03 | 0.56 | 0.1 | 0.1 | 0.45 | 1.12 | 1.9 | 2.5 | 2.9 | 3.2 | 3.25 |
| C | $\Phi dc_{max}$ (pu) | 0.35 | 0.55 | 0.8 | 0.95 | 1.1 | 1.15 | 0.9 | 0.62 | 0.5 | 0.2 | 0.19 | 0.3 | 0.35 |
|  | Ipeak(pu) | 0.32 | 0.97 | 1.58 | 2.31 | 2.56 | 2.69 | 1.9 | 1.08 | 0.82 | 0.41 | 0.35 | 0.37 | 0.32 |
| D | $\Phi dc_{max}$ (pu) | 1.8 | 1.82 | 0.95 | 0.88 | 0.7 | 0.2 | 0.89 | 0.98 | 1.3 | 1.5 | 1.7 | 1.65 | 1.8 |
|  | Ipeak(pu) | 4.6 | 4.72 | 2.23 | 1.84 | 1.2 | 0.3 | 1.54 | 2.26 | 2.95 | 3.6 | 4.2 | 4.4 | 4.6 |
| E | $\Phi dc_{max}$ (pu) | 0.85 | 1.18 | 1.3 | 1.5 | 1.5 | 1 | 1.3 | 1.4 | 1.45 | 0.7 | 0.58 | 0.7 | 0.85 |
|  | Ipeak(pu) | 1.9 | 2.5 | 2.9 | 3.2 | 3.3 | 2.5 | 2.9 | 3.2 | 3.3 | 1.3 | 1 | 1.2 | 1.9 |
| F | $\Phi dc_{max}$ (pu) | 1.8 | 1.68 | 0.9 | 0.75 | 0.58 | 0.2 | 0.8 | 0.98 | 1.3 | 1.5 | 1.7 | 1.75 | 1.8 |
|  | Ipeak(pu) | 4.32 | 4.21 | 2 | 1.3 | 1 | 0.43 | 1.32 | 2 | 2.7 | 3.2 | 4 | 4.1 | 4.32 |
| G | $\Phi dc_{max}$ (pu) | 0.52 | 0.8 | 0.9 | 1.1 | 1.17 | 1.2 | o.95 | 0.62 | 0.45 | 0.49 | 0.5 | 0.51 | 0.52 |
|  | Ipeak(pu) | 0.83 | 1.46 | 2 | 2.56 | 2.82 | 2.8 | 2.2 | 1.07 | 0.52 | 0.67 | 0.78 | 0.82 | 0.83 |

## C. Effect of Depth

Depth influence for different types of voltage sag is analyzed considering $v(h, \Delta t=5.5T, \psi_i=0°)$ where voltage sag depth (h) is varied from 0 to 0.9. The simulation results of this study are shown in the Fig. 6. It can be observed that the effect of depth on the inrush current peak is almost linear for most types of voltage sag.

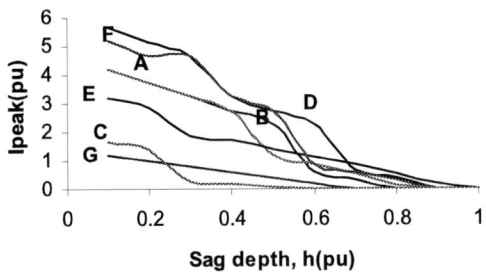

Fig. 6. Effect of depth on the current peak.

## V. CONCLUSIONS

Effects of balance and unbalance voltage sags on three-phase transformers have been studied in this paper with the aim of computer simulation. When the voltage sag recovers, a dc magnetic flux is produced which can cause transformer saturation and an inrush current is produced. Duration has a periodical and depth has a linear impact on inrush current. When duration equals to half of a full period plus any full periods of the waveform, severe peaks are obtained. Impact of each type of voltage sag on transformer performance is investigated in this paper. For different voltage sag types the most unfavorable initial points are given in the paper.

### Appendix: Analytical Study [6]

This study is made in a Wye-G-Wye-G three phase transformer bank where the voltage drops in the primary winding resistances, and the leakage reactances are neglected. Under this assumption for example the magnetic flux for phase a is

$$N_p \frac{d\phi(t)}{dt} = ep_a(t) \approx up_a(t) = v_a(t) = \sqrt{2}\sin(\omega t + \alpha)$$

$$\Rightarrow \phi_a(t) \approx \frac{1}{N_p}\int v_a(t)dt = \phi_a(0) + \frac{1}{N_p}\int_0^t v_a(t)dt$$

$$= \frac{\sqrt{2}V}{N_p\omega}\sin(\omega t + \alpha - \frac{\pi}{2}) + \phi_{a,DC}$$

where this component is null in steady state conditions. Here, we assume that the transformer is in the steady state condition. When the voltage sag is produced in the supply, the rms voltage changes from V to hV at instant $t_i$ and is recovered at instant $t_f$. Therefore, the magnetic flux during the steady state conditions is purely sinusoidal as:

$$\phi_a(t) \approx \frac{\sqrt{2}V}{N_p\omega}\sin(\omega t + \alpha - \frac{\pi}{2}) \quad (t \le t_i)$$

The magnetic flux during the sag is

$$\phi_a(t) \approx \phi_a(t_i) + \frac{1}{N_p}\int_{ti}^t hv_a(t)dt$$

$$= \phi_a^{i},_{DC} + \frac{\sqrt{2}hV}{N_p\omega}\sin(\omega t + \alpha - \frac{\pi}{2}) \qquad (t_i \le t \le t_f)$$

where

$$\phi_a^{i},_{DC} = \phi_a(t_i) - \frac{\sqrt{2}hV}{N_p\omega}\sin(\omega t + \alpha - \frac{\pi}{2})$$

The magnetic flux when voltage sag recovered is

$$\phi_a(t) \approx \phi_a(t_f) + \frac{1}{N_p}\int_{t_f}^t v_a(t)dt$$

$$= \phi_a^{f}{}_{DC} + \frac{\sqrt{2}hV}{N_p\omega}\sin(\omega t + \alpha - \frac{\pi}{2}) \qquad (t_f \le t)$$

If the total flux of any phase is high (because of the constant flux), the magnetic circuit of that phase will be saturated and so it's current will be high.

## VI. REFRENCES

[1] B. H. Chowdhury, " Power Quality", IEEE Potential, Vol. 20, No. 2, pp.5-11, 2001.

[2] L. Guash, F. Coroles and J. Pedra, "Effect of Unsymmetrical Voltage Sags on Induction Motors", IEEE SDEMPED, Spain, Vol. 1, pp. 265-272, Sept. 1999.

[3] M. F. Mcgranaghan, D. R. Mueller and M. J. Samotyj, "Voltage in Industrial Systems" , IEEE Trans. on Ind. Appl., Vol. 29, No. 2, pp.397-403, Apr. 1993.

[4] E. Styvaktakis, M.H. J. Bollen and I.Y.H. Gu, "Transformer Saturation after a Voltage Dip", IEEE Power Eng. Rev. , Vol.20, pp. 62-64, Apr. 2000.

[5] L. Guash, F. Corcles and J. Pedra, "Effects of Symmetrical and Unsymmetrical Voltage Sags on induction machines" IEEE Trans. Power Del. , Vol. 19, No. 2, pp. 774-782, Apr. 2004.

[6] L. Guash, F. Corcles, J. Pedra and L. Sainz, "Effects of Symmetrical Voltage Sags on Three-phase Three-legged Transformers," IEEE Trans. On Power Del. , Vol. 19, No. 2, pp. 875-883, Apr. 2004.

[7] M. H. J. Bollen, "Voltage Recovery after Unbalance and Balance Voltage Dips in Three-phase Systems," IEEE Trans. On Power Del., Vol. 18, pp.1376-1381, Oct. 2003.

## V. BIOGRAPHIES

**M. R. Shakarami** is a PhD student at the power group of the Department of Electrical Engineering, Iran University of Science and Technology, Tehran, Iran. He is also an academic member of Lorestan University in Iran.

**Alireza Jalilian** was horn in Yazd, Iran in 1961. He received his BSc degree in Electrical Engineering from Mazandran University, Iran in 1989 and his ME (Hons) and PhD degree in Electrical Engineering from University of Wollongong, Australia in 1992 and 1997 respectively. Dr Jalilian joined the power engineering group of the Department of Electrical Engineering of Iran University of Science and Technology (IUST) in 1998 as an academic member. Dr Jalilian's research interests are Power Quality causes, effects and mitigations.

**2006 IEEE International Conference on Power Electronic, Drives and Energy Systems**

# Design and Transient Analysis of Cage Induction Motor Using Finite Element Methods

Bhoj Raj Singla, Sanjay Marwaha, and Anupma Marwaha

*Abstract*--During the past few decades, the numerical computation of magnetic fields has gradually become a standard in electrical machine design. Finite element method (FEM) has the advantage over previous design calculation procedures in that geometric variations and irregularities, saturation and eddy current effects can be modeled with a high degree of accuracy. Besides these advantages, the method does allow the designer to obtain a clearer understanding of the flux distribution in the machine and so indicates directly the corrective action to be taken if necessary. The paper describes a method for the design and transient analysis of induction motor using two-dimensional finite-elements. It relies on the decomposition of the rotor currents into harmonic distributions and the use of auxiliary circuit equations to take the time stepping into account.

*Index Terms*-- Cage Induction Motor, Electromagnetic Torque FEM, MFSS, Transient Response.

## I. INTRODUCTION

AMONG electromagnetic devices, the induction motor consumes a great deal of electric power energy. The suitability of the finite-element method for the analysis of electrical machines has long been recognized. In the early days, the method was restricted to being a research tool or used for troubleshooting purposes due to the high cost of computing power and the amount of work force required to define the input data. However through the development and implementation of special Engineering techniques that increased the modeling capability of the method without increasing the solution time, the technique is now considered part of the designers ' tool-kit ' for everyday design. Recently the induction motor has received scant attention, despite its pre-eminence in industrial drives. It is not feasible to construct a three-dimensional field model for an induction motor, so that two-dimensional field models must be used [2]. Chief

---

This work is supported in part by the Electrical & Electronics Department of SLIET research Laboratories.

Bhoj Raj Singla is with the Department of Electrical Engineering, Guru Teg Bahadar Khalsa Institute of Engineering & Technology, Chhapianwali, Malout, Punjab 152107, India (e-mail: brsingla@rediffmail.com).

Sanjay Marwaha is with the Department of Electrical Engineering, Sant Longowal Institute of Engineering & Technology, Longowal, Sangrur, Punjab 148106, India (e-mail: marwaha_sanjay@yahoo.co.in).

Anupma Marwaha is with the Department of Electronics and Communication Engineering, BHSB Institute of Engineering & Technology, Lehragaga, Sangrur, Punjab, India (e-mail: marwaha_anupma@yahoo.co.in).

amongst these is the effect of the rotor end-rings and the stator end-windings. These present impedance to their respective current flows and yet lie outside the jurisdiction of a two-dimensional field solution, which can only cope with currents flowing in the axial direction. Equally, some means must be found for dealing with the effects of rotor skew. This axial twist is put on the rotor stack so that the orientation of the rotor laminations with respect to the stator varies in the axial direction. Skew can have a profound effect on machine performance. Circuit equations are used to relate the emf induced in each stator winding to the current flowing in it and to the applied terminal voltage. The stator circuit equations are then combined with the field equations, which include an eddy-current formulation for the rotor currents, and the whole is time-stepped. Stator end-winding effects are dealt with simply by adding appropriate impedance in the stator circuit equations. Rotor end-ring effects require consideration that is more careful. A possible approach is to represent the skewed rotor using several un-skewed axial segments, which have a combined length equal to that of the real machine. The stators of all such segments are aligned while the rotors are progressively indexed round to give a piece-wise representation of the real skewed rotor. The difficulty now is to ensure continuity of current flow along the rotor bars in the model. This will probably require simultaneous solution of all segment models, so that the size of the set of equations will expand rapidly as the number of segments increases. In essence, to obtain a transient solution the circuit equations are time-stepped to determine the currents, and the field solution is used periodically to update the circuit parameters. In this method, both stator end-windings and rotor end-rings are dealt with in the same manner, by adding appropriate impedance to the relevant circuit model. Rotor skew may also be dealt with using circuit techniques. That is, the mutual inductance between a stator winding and the rotor may be determined using the two-dimensional model and then simply multiplied by the appropriate skew-factor, as calculated using classical expressions. Alternatively, if it is thought that the skew may produce significant differences in saturation along the length of the machine, the same type of segmented model outlined above may be used. The difference here is that because all currents are specified for each field solution, the models for the individual segments are solved separately, rather than simultaneously, representing a significant saving in computer resource [3]-[4]-[7].

0-7803-9771-1/06/$25.00 ©2006 IEEE

## II. EQUATION OF CIRCUIT

The stator currents flow in windings, which define unique current flow paths. The rotor current paths though well defined are multiply connected. The most convenient form is the harmonic distribution. If the *pth* rotor bar is located at $\theta = \theta_p$ in the rotor reference frame, the current flowing in that bar may be expressed in the form

$$i_p = \sum \left[ i_d^k \cos k\theta_p + i_q^k \sin k\theta_p \right] \text{------------- (1)}$$

Where $k$ is a harmonic number, and $i_d^k$ and $i_q^k$ are the *kth* harmonic direct-axis and quadrature-axis bar currents. The circuit equation governing the current flow in the *nth* stator winding may be written in the form

$$\frac{d\psi_n}{dt} = V_n - i_n R_n \text{----------------------------------- (2)}$$

Where $\psi_n$ is the flux linkage of the *nth* winding, V is the impressed voltage and $R_n$ is the winding resistance. $\psi_n$ is related to the currents flowing in the stator windings and to the rotor current distributions by

$$\psi_n = \sum_m M_{nm} i_m + \sum_k M_{nd}^k i_d^k + \sum_k M_{nq}^k i_q^k \text{---------------------- (3)}$$

$M_{nm}$ is the mutual inductance between the *nth* and *mth* stator windings, and the summation index covers all stator windings. $M_{nd}^k$ and $M_{nq}^k$ are the mutual inductances between the *nth* stator winding and the *kth* harmonic rotor current distributions and the summation index now covers all harmonics.

Voltage balance equations for the rotor in a form similar to "equation (2)", except that the impressed voltage term is now zero. For the *kth* harmonic sine distribution, for example, we may write

$$\frac{d\psi_q^k}{dt} = i_q^k R^k \text{--------------------------------------------- (4)}$$

Where $R^k$ is the resistance of the cage to a *kth* harmonic distribution of rotor currents, and the flux linkage $\psi_q$ has the form

$$\psi_q^k = \sum_m M_{qm}^k + \sum_i M_{qq}^{ki} i_q^i + \sum_i M_{qd}^{ki} i_d^i \text{----------------------- (5)}$$

$M_{qq}^{ki}$ is the mutual inductance between the *kth* harmonic sine distribution and the *ith* harmonic sine distribution of rotor currents. $M^{ki}$ is the mutual interface between the *kth* harmonic sine distribution and the *ith* harmonic cosine distribution of rotor currents. "Equations (2) and (4) may be written for convenience of notation in matrix form"

$$\frac{d\psi}{dt} = V - Ri \text{----------------------------------------- (6)}$$

and "equations (3) and (5)" similarly expressed as

$$\psi = Mi \text{------------------------------------------------------- (7)}$$

From which

$$i = M^{-1}\psi \text{----------------------------------------------- (8)}$$

## III. FIELD MODEL

The finite-element model is a two dimensional representation of the machine. For simulation the action of the rotor, it is necessary to allow continuous movement of the rotor with respect to the stator. This is accomplished by means of a limited series of meshes that depict the rotor being progressively rotated through one rotor slot pitch [7]. If the rotor has $N_b$ bars, and the number of meshes generated is $N_m$, the angular displacement of the rotor between successive meshes is $2\pi/N_m N_b$ when it is necessary to carry out a field solution with the rotor in a position that does not corresponds to one of these stored meshes. The mesh that is closest is selected and the air-gap elements distorted to allow the rotor to adopt the required position. When the rotor has rotated through one full slot pitch, the rotor is rocked back one slot pitch to its start position and the bar currents indexed forward one slot pitch, to simulate continuous movement. Both rotor and stator currents are source currents in the field model, so that a magneto static formulation is used. The bar currents are assumed to be evenly distributed over the bar cross-sectional area. This assumption affects that part of the rotor self inductance that is due to flux which actually passes through the rotor bar itself, but this component is amenable to pre-calculation by either classical means or by specialist finite-element models [1]-[10]. The rotor resistance is also obtained by the same method and augmented to take account of the end-ring resistance using the classical expressions. It is worthwhile noting that these expressions give a different effective end-ring resistance for each harmonic rotor current distribution that is present, as would be expected from physical considerations.

## IV. CALCULATION OF COUPLING INDUCTANCES

The coupling inductances used in "equations (7) and (8)" are obtained from the field model. The general method is to excite each winding or current distribution in turn and then calculate the flux linkages with each other. For example, suppose that a field solution is obtained with $i_m$ amps flowing in the *mth* stator winding, all other currents having been set to zero. The mutual inductance $M_{nm}$ is obtained from the flux linkages with winding n.

$$M_{nm} = \frac{w}{i_m} \sum_{j=1}^{N_s} C_{nj} \sum_{i=1}^{N_e} \Delta_i A_{oi} \text{------------------------------------- (9)}$$

In which N, is the number of stator slots, and $C_{nj}$ is the density of conductors of the *nth* winding in the *jth* slot (signed to take sense of connection into account) and $w$ is the axial length of the machine. $\Delta_i$ is the cross sectional area of the *ith* element and $A_{oi}$ is the average magnetic vector potential over that element, with the summation for $i$ extending over all elements in the cross sectional area of winding $n$ in slot $j$. The inductance calculated in accordance with "equation (9)" must be augmented by the mutual inductance between the two windings due to end-winding leakage flux, calculated by any convenient means. The mutual inductance between the *mth* stator winding and the *kth* harmonic rotor current distributions may be obtained from the same field solution.

$$M_{qm}^{k} = \frac{w}{S_b i_m} \sum_{p=1}^{N_b} \sin k\theta_p \sum_{i=1}^{N_e} \Delta_i A_{oi} \text{-------------------- (10)}$$

$$M_{dm}^{k} = \frac{w}{S_b i_m} \sum_{p=1}^{N_b} \cos k\theta_p \sum_{i=1}^{N_e} \Delta_i A_{oi} \text{-------------------- (11)}$$

Where $S_b$ is the cross-sectional area of the rotor bar, $N_b$ is the number of bars, and $\theta_p$ is the angular co-ordinate of the *pth* bar (measured in the rotor reference frame). The index $i$ for the finite element summation is now taken to range over all elements in the cross-section of the *pth* rotor bar.

The mutual inductances between harmonic current distributions are similarly evaluated by solving for the field produced with say just $i_d^k$ non-zero see "equation (1)" and then using "equations (9) to (11)" with $i_m$ replaced by $i_d^k$. "Equation (9) will then give $M_{nd}^k$" whilst "equations (10) and (11)" give $M_{qd}^{kk}$ and $M_{dd}^{kk}$ respectively. $M_{dd}^{kk}$ is the self-inductance of the *kth* harmonic cosine distribution of rotor currents. This will include a slot leakage term that must be modified to consider skin effect. This may be done most conveniently for regular-shaped bars by subtracting out the component due to slot leakage, modifying it by the appropriate skin-effect correction factor then adding it back in. $M_{dd}^{kk}$ must also be augmented by the rotor end-ring leakage inductance, so that finally

$$M_{dd}^{kk} \rightarrow M_{dd}^{kk} + (\psi - 1)\frac{N_b}{2}L_b + \frac{L_e}{4\sin^2\left(\frac{k\pi}{N_b}\right)} \text{------------- (12)}$$

$\psi$ is the skin-effect correction factor [9] and $L_b$ is the bar inductance due to flux passes through the bar itself. $L_e$ is the self-inductance of one complete rotor end-ring. The same corrections must be made to $M_{qq}^{kk}$. If the bar shape is irregular, so that the usual skin effect correction factors become suspect, an alternative method for calculating $\psi$ using a finite-element model can be derived from references [11].

## V. PROCEDURE OF TIME STEPPING

The time-stepping procedure is described in terms of the following sequence of steps, which define a single time step. At the start of this time step (i.e. at $t = t_0$) we will know all the currents $(i_0)$ and will be able to estimate the time derivatives $(i_0')$ by backward differences. In addition, we will know the instantaneous torque $T_0$, the orientation of the rotor with respect to the stator, and the rotor speed.

### A. Step1

"Equation (6) is marched forward in time until either a preset maximum interval $(\Delta t_{\max.})$ has elapsed, or one of the components of $\psi$ has changed by more than a specified amount". During this procedure $V$ varies in accordance with the impressed supply voltages, whilst the variation of $i$ is estimated using $i_0$. This step fixes the duration of $\Delta t$ of the current time step, and enables both $\psi$ and $I$ to be estimated at $t = t_0 + \Delta t$.

### B. Step2

The angle through which the rotor rotates in $\Delta t$ seconds is calculated using the torque $T_0$ and the mechanical properties of the load and rotor. The rotor in the finite element model is rotated though this angle.

### C. Step3

With the rotor in this new position, and the iron reluctivities frozen at their recently calculated values, the finite-element model is used to calculate the elements of the matrix $M$.

### D. Step4

"Equation (8) is used to solve $i$ at $t = t_0 + \Delta t$", given the inductance matrix determined in Step 3 and the flux linkages obtained at the end of Step 1.

### E. Step5

If the new estimate for $i$ (calculated at Step 4) agrees with the old estimate (obtained at Step 1) then the torque can be calculated and the solution moved on to the next time step. If it does not, then this new value for $i$ is used in conjunction with $i_0$ to determine an improved estimate of the variation of the currents through the time step, and "equation (6)" re-solved to obtain an up-dated estimate of $\psi$ at $t = t_0 + \Delta t$. In addition, the new current set is used to excite the finite-element model, which is then solved in a non-linear sense to reset the elemental reluctivities with the rotor in its new position, corresponding to the end of the time step. Steps 3-5 are repeated until successive estimates of $i$ at $t = t_0 + \Delta t$ are in close agreement.

Torque calculations involve multiplying the current flowing in each slot by the average flux density over the corresponding slot pitch. The philosophy behind choosing a time step based on a maximum permitted change in flux linkages when the magnetic field changes, so that a non-linear field solution is carried out only when the magnetic field has changed. The authors have found that 40 time-steps are required per supply cycle using this technique, whilst other authors have reported 100-200 steps per cycle [11].

## VI. EXPERIMENTAL VERIFICATION

The procedure outlined above is verified by means of tests carried out on a 400V, 4-pole cage motor. The stator and rotor laminations for this motor are shown in "Fig. 1 and 2" and other details are summarized in Table 1. Performance characteristics are shown in "Fig. 3(a) and 3(b)". "Fig. 4(a) and 4(b)" show the computed and measured variations of red and blue phase current during a direct-on-line start with the rotor coupled to an inertial load. "Fig. 5" shows the computed variation of electromagnetic torque during the same test. The simulation runs on a MFSS to produce "Fig. 3 to 5". The results are produced using only the fundamental rotor currents and taking rotor skew into account using equivalent-circuit techniques (i.e. skew-factors). The mesh used has rotor 3191 nodes, 4796 elements and stator 4512 nodes, 7440 elements and the maximum permissible change in any component of $\psi$ (i.e. $\Delta\psi_{max.}$) was set to 22% of the corresponding peak value [5]-[6].

### TABLE I
### DETAILS OF EXPERIMENTAL MACHINE

| | |
|---|---|
| Connection | Star |
| Stator outer diameter | 160 mm |
| Rotor diameter | 83.6 mm |
| Gross core length | 80 mm |
| No. of stator slots | 24 |
| No. of turns per coil | 30 |
| Winding type | Standard |
| Stator resistance per phase | 1.437 Ohm |
| No. of rotor slots | 22 |
| Outer cage conductivity | 30.00e6 1/(ohm-m) |
| Inner cage conductivity | 37.71e6 1/(ohm-m) |
| Total Moment of Inertia | 0.1065 kg-m |
| End-turn inductance per phase | 2.069e-3 Henry |
| Arm winding copper weight | 2.413 kg |
| Arm coil wire length per turn | 361.2 mm |
| Type of rotor cage | M19-0.50mm |
| Rotor skew | 1 rotor slot pitch |

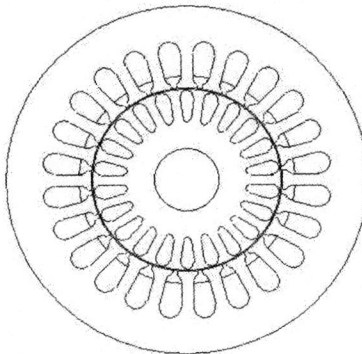

Fig. 1. Test Induction Motor.

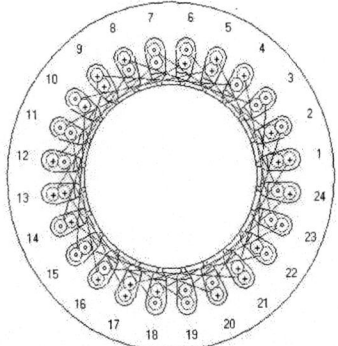

Fig. 2. Test Induction Motor with Stator windings.

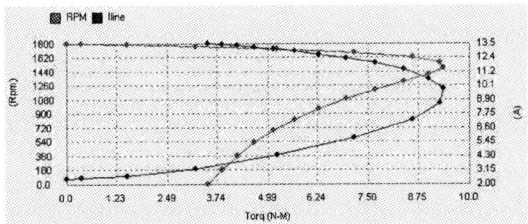

Fig. 3 (a). Performance Characteristics of Test Induction Motor.

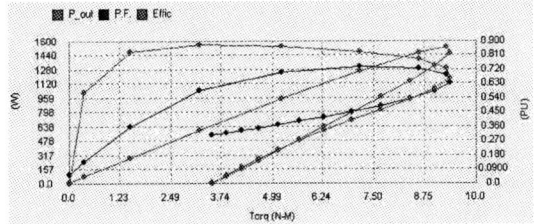

Fig. 3 (b). Performance Characteristics of Test Induction Motor.

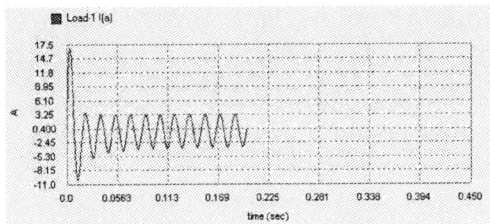

Fig. 4 (a). Red Phase current.

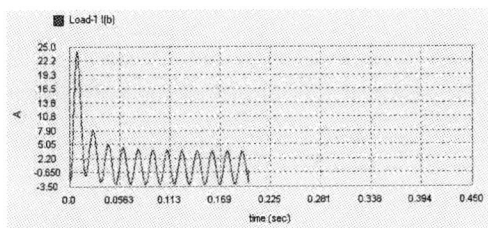

Fig. 4 (b). Blue Phase current.

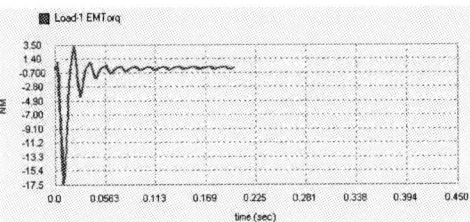

Fig. 5. Torque.

It is well known that lumped parameters determined from field solutions are relatively insensitive to the accuracy of the solution. To determine a suitable convergence criterion numerical experiments are carried out for the Newton-Raphson procedure for the non-linear field solution. The solution is assumed to have converged when

$$\frac{\sum_i \left|\Delta A_i\right|^{r+1}}{\sum_i \left|A_i\right|^r} \le \varepsilon \text{-------------------------------------------------- (13)}$$

Where the index $i$ range over all nodes, and the superscript $r$ indicates the iteration number. It is found that no noticeable difference is observed in the terminal characteristics for values of $\varepsilon$ smaller that 0.01, and so this value is adopted. Inclusion of the third and fifth harmonic rotor current distributions is found to produce a negligible effect on the predicted starting performance whilst boosting the time required by approximately 50%. The inclusion of harmonic currents is also found to necessitate a reduction in $\Delta\psi_{max}$ from 20% to 15%. Indeed, it is the reduction in maximum flux linkage change that was primarily responsible for increasing the computing time [8]-[9].

## VII. CONCLUSION

This paper has described a new method for the analysis of transient phenomena in cage induction machines. The procedure described is implemented directly into design routines by automating the complete finite element calculation. Its main point of interest is the way in which the circuit and field models are developed. The adoption of this technique yields some distinct advantages. Time stepping is limited to the circuit model alone, thereby allowing an optimum step size to be selected at all times. The use of a magneto static finite element formulation permits non-linear solutions of comparatively low accuracy to be used, thus bringing substantial savings in computer resources. The extensive use of circuit techniques also allows the effects of axial skew and spatially distributed air gap flux harmonics to be incorporated.

## VIII. REFERENCES

[1] Tenhunen A., Arkkio A., "Modeling of Induction Machines With Skewed Rotor Slots", *IEE Procedures*, vol. 148. No 1, January 2001.

[2] S.L.Ho and W.N.Fu, "Review an future application of finite element method in the induction motor," Electrical Machines and Power Systems, Taylor and Francis, vol.26, No.2, pp.111-125, 1998.

[3] S.J.Salon, "Finite element analysis of electrical machines", Kluwer Academic Publishers, 1995.

[4] "Finite element methods in electrical power engineering" oxford university press, oxford, 2000.

[5] P. Dziwniel, B. Boualem, F. Piriou, J. P. Ducreux, and P. Thomas, "Comparison between two approaches to model induction machines with skewed slots", *IEEE Trans. Magn.*, vol. 36, no. 4, pp. 1453–1457, Jul. 2000.

[6] Hamid A Toliyat, Gerald B Kliman Handbook of electrical motors, 2004.

[7] "Finite element analysis of electrical machines" *Kluwer academic publication*, Boston, 1995.

[8] A. M. Oliveira, R. Antunes, P. Kuo-Peng, N. Sadowski, and P. Dular, "Electrical machine analysis considering filed-circuit-movement and skewing effect", COMPEL vol. 23, no. 4, pp. 357–360, 2004.

[9] G. H. Jang and S. J. Park, "Simulation of the electromechanical faults in a single phase squirrel cage induction motor," *IEEE Trans. Magn.*, vol. 39, pp. 2618–2620, Sept. 2003.

[10] D. Meeker, "Finite Element Method Magnetics FEMM", User's Manual, Ver. 4.0, *Foster-Miller*, MA, USA, 2004.

[11] Williamson, S., Flack, T.J. and Volschenk, A.F. (1995), "Representation of skew in time-stepped two-dimensional finite-element models of electrical machines", *IEEE Transactions on Industrial Applications*, Vol. 31 No. 5, pp. 1009-15.

## IX. BIOGRAPHIES

**Bhoj Raj Singla** born on July 1972 at Bathinda. Presently he is working as Asstt. Prof. in the Department of Electrical Engineering at Guru Teg Bahadar Khalsa Institute of Engineering & Technology, Chhapianwali Malout, India. He did his BE (Electrical Engg.) from Nagpur University, Nagpur in 1993, M. Tech. (Electrical Engineering) from Guru Nanak Dev College of Engineering & Technology, Ludhiana in 2005. He is life member of ISTE. His area of research interests includes Finite Element applications on various Electromagnetic devices.

**Sanjay Marwaha** born on April 1966 at Nahan. Dr. Marwaha is Professor and Head in the Department of Electrical and Instrumentation Engineering at SLIET, Longowal. He did his BE (Electrical Engg.) from Gorakhpur University, Gorakhpur in 1988, ME (Power Systems) from Punjab University, Chandigarh in 1990 and Ph.D. from GNDU, Amritsar in 2000. He is life member of ISTE and Member, Institution of Engineers (India) and has published around 60 research papers in National and International journals/conferences of repute. His area of interest includes Design and Analysis of Electromagnetic Devices, Power Systems and HV Engg. Electrical and Electronic Measurements and Instrumentations, Industrial Electronics and Microwave Engineering.

**Anupma Marwaha** born on April 1969 at Chandigarh. Presently she is working as Asstt. Prof. and Head in the Department of Electronics and Communication Engg. at BHSB Institute of Engineering and Technology, Lehragaga. Dr. (Mrs.) Marwaha did her BE (Electronics Engg.) from Punjab University, Chandigarh in 1990 and M. Tech. (Electronics and Communication Engg.) from Kurukshetra University in 1992 and Ph.D. from GNDU, Amritsar in 2003. She is life member of ISTE and Member, Institution of Engineers (India) and has published around 50 research papers in National and International journals/conferences of repute. Her area of interest includes Communication Systems, Microwave and Antennas and application of Finite Difference Time Domain and Finite Element tools in design of various electromagnetic structures.

**2006 IEEE International Conference on Power Electronic, Drives and Energy Systems**

# Methodology for Estimating Performance Characteristics of Three Phase Induction Motor Operating Direct-on-Line or with Six Pulse Inverter

Satish Chander Sabharwal, *Senior Member, IEEE*

*Abstract*-- The paper presents a methodology to derive parameters of an equivalent circuit representation of a three phase induction motor from the catalogue data provided by the electric motor manufacturer. The performance of the motor in terms of torque, efficiency and power factor is then calculated from the equivalent circuit. The paper presents validity of methodology with results of analysis with data of a motor from a catalogue and its comparison with the results of manufacturer's tests in the laboratory .The paper then also attempts to further analyze motor performance if it is driven by a six pulse inverter.

*Index Terms*-- Equivalent Circuit, Induction Motor Performance, Methodology, Six Pulse Inverter.

## I. NOMENCLATURE

Following nomenclature is used in this paper:

| | | |
|---|---|---|
| $n$ | Number of harmonics | |
| $I_{in}$ | input Current | (ampere) |
| $I_m$ | Current in magnetizing -branch in Fig.1 | (ampere) |
| $I_{out}$ | Rotor Current | (ampere) |
| $N$ | Motor shaft speed | (rpm) |
| $Ns$ | synchronous speed | (rpm) |
| $Pf$ | Power factor | |
| $P_{in}$ | Input motor Power | (kW) |
| $R_1$ | Stator Resistance | (ohm) |
| $R_2$ | Rotor Resistance | (ohm) |
| $R_c$ | Core loss Resistance | (ohm) |
| $s$ | Motor slip (Ns-N)/Ns | |
| $T$ | Motor Output torque | (Nm) |
| $V$ | Stator phase voltage | (Volt) |
| $V_m$ | Back Emf of motor | (Volt) |
| $X_1$ | Stator leakage reactance | (ohm) |
| $X_2$ | Rotor leakage reactance | (ohm) |
| $X_m$ | Reactance due to Magnetizing inductance | (ohm) |
| $Z_{in}$ | input reactance of motor | (ohm) |

---

Satish Chander Sabharwal is an Energy Economist with Bureau of Energy Efficiency, Core IV,NBCC Tower, Bhikaji Cama Place ,New Delhi ,India , 110066 E-mail (satishsabharwal@ieee.org)

## II. INTRODUCTION

THREE phase AC Induction motors are ideal for most industrial and agricultural applications because of construction, low maintenance and robustness in field conditions. With varying load requirements the performance of the motor varies in terms of torque, efficiency and power factor [1]. The induction motor's performance characteristics provides important information to the motor designer and also to the actual users for evaluating operating costs and monitoring motor performance under actual load conditions.

Traditionally, the slip method is used for quick calculations. However, it is not accurate and does not provide efficiency and power factor. As an alternative method, field measurements by decoupling the motor have been proposed [3] that require decoupling of the motor shaft and is time consuming and have practical difficulties. Some methods to estimate efficiency based on field measurements at two load conditions have also been proposed in the past which has difficulties due to maintaining constant power supply and load conditions and need for calibrated and accurate measuring instruments. Also all the performance characteristics cannot be produced by this method. This paper presents an alternative methodology for a desk study which does not require any field testing and is developed based on the catalogue data.

The motor manufacturer generally provide following data to its customers: rated output in kilo-Watt (kW), supply voltage in Volts (V), full load speed in revolutions per seconds (rpm), Full load current in Ampere (A), Locked rotor current (A), full load torque in Newton-meters (Nm), locked rotor torque (Nm), breakdown torque (Nm), efficiency in per cent at full load, three quarter and half load and temperature rise in degree Centigrade. Please see Appendix I.

## III. PROPOSED METHODOLOGY

The three phase induction motor connected directly to supply is represented by an equivalent circuit comprising of resistances and inductances on per phase basis as shown in Fig. 1

The losses in the motor are mainly copper loss, core loss, windage loss and stray loss. The core loss is relatively fixed

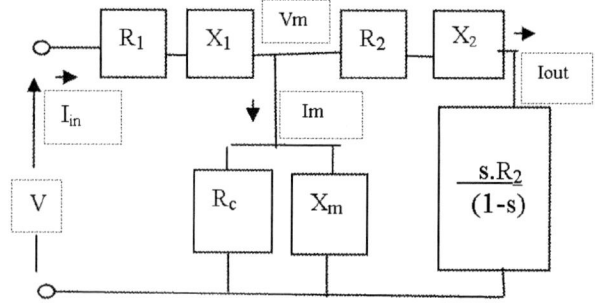

Fig. 1. Equivalent Diagram of Induction motor for one phase.

with respect to output. However, the copper loss in stator and rotor are proportional to square of input current and square of rotor current respectively. It is to be noted that windage and stray loss is small percentage and have been omitted in the analysis.

The remaining losses can be calculated from following equations.

Stator copper loss = $3 I_{in}^2 R_1$      (1)
Rotor copper loss = $3 I_{out}^2 R_2$     (2)
Core loss       = $3 V_m^2 R_c$     (3)

By ignoring Stray and Windage losses as they are small as compared to above remaining losses, we get

$$\text{Total loss} = (1) + (2) + (3)$$

The losses consist of relatively fixed loss comprising of core losses due to hystersis and eddy currents in the core material, and variable loss due to copper losses that vary as square to output power. This implies that the losses at rated and part load can be represented by the following:

$$L_{or} = P_{out} a^2 + b \qquad (4)$$
$$L_{op} = P_{outp} a^2 + b \qquad (5)$$

where, $P_{out}$ and $P_{outp}$ are output shaft power at rated and part load respectively and a and b are constants.

Also, from the efficiency values ($E_{ffr}$) & ($E_{ffp}$) of motor at 100% and 50 % of rated load we can calculate the losses in motor at the rated load and part load at 50 % from the following equations.

$$\text{Lor} = kW [1/E_{ffr} - 1] \qquad (6)$$
$$\text{Lop} = kW [1/Effp - 1] \qquad (7)$$

From the above equations (4), (5), (6) and (7) we can calculate the value for constants a and b by substituting catalogue values of the motor. Solving equations for a 15 KW motor with specifications at Appendix I, the values of b and a are 1130 and 0.075 respectively. It can be easily seen that at no load, the three phase fixed loss as given by (5) with shaft power output equal to zero would give core loss equal to b .We can calculate core loss/phase from b by dividing it by a factor of three. Hence the value of Rc can be easily calculated from no load loss and supply voltage. In the example considered, Rc is 376 ohm.

At the time of motor starting from rest almost all the starting current flows through rotor circuit producing torque at synchronous speed .The value of R2 can be calculated by considering core loss as given in the following equation

$$3 R_2 I_2^2 = T_s w - 3 V^2 / R_c \qquad (8)$$

In the motor under study R2 is 0.3.

We know that torque is proportional to the square of rotor current and we calculate R2/X2 from the equations at rated slip and on locked rotor conditions. We can easily get X2 equal to 1.39 by substituting value for R2 .

The input current and its power factor is calculated from the analysis of equivalent circuit in Fig.1. The magnitude and phase angle of Iout is calculated from value of output Power and R2.

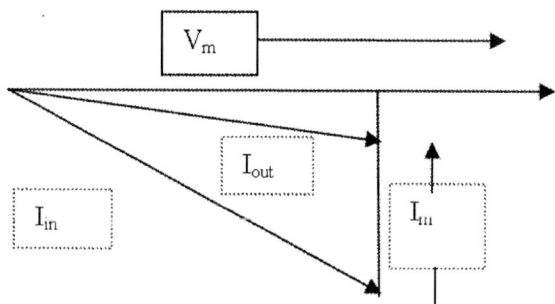

Fig. 2. Phasor Diagram for different Currents in motor circuit.

The magnitude of $X_m$ can be the estimated by drawing Iin and $I_{out}$ phasors as shown in Fig. 2 and measuring $I_m$.

At rated speed the total loss is given by sum of stator copper loss, core loss and rotor copper loss. The rotor copper loss is given by slip times the input power. By substituting the values, we get the value for R1.

## IV. CIRCUIT ANALYSIS

In order to produce the motor performance characteristics following analysis is performed. From Fig.1 the input impedance (Zin) of the motor is represented in terms of stator, rotor and core impedances and slip. The input phase current (Iin) can easily calculate from the following equation

Iin = Vin/Zin

The power factor can be calculated from phase angle from the cosine of phase angle of input current phasor as given in (9).

Thevenin`s voltage and Thevenin`s impedances are calculated at the magnetizing branch in the direction of input power supply, in terms of equivalent circuit parameters. Rotor resistance is found by solving the current produced by Thevenin's Voltage in the rotor circuit and the output power. The shaft power output is calculated from the power dissipated across resistance R2 (1/s-1) and is given by

$$P_{out} = 3 I_{out}^2 R_2 (1/s - 1) \qquad (10)$$

Also, the torque output is calculated from

$$T = 60 P /2 (3.1416 N) \qquad (11)$$

The motor efficiency at any load can be either calculated

from input voltage, current and power factor, or by calculating individual losses and using output power. In this paper both approaches were followed .These two efficiency values help to fine tune the equivalent circuit parameters.

## V. COMPUTER SIMULATION AND RESULTS

A computer simulation was performed on EXCEL worksheet and tabulated below in Table I  First section requires information on motor from catalogue and is shown below:

Following procedure is adopted for obtaining results of performance characteristics of motors shown in TableII below.

Step1: Data entry for equivalent circuit parameters is made in the respective fields of worksheet along with motor and supply data.

Step2 :Once all the input values are entered in the worksheet, the resulting efficiency values and losses in the worksheet are studied with respect to catalogue data.

Step3: In case of slight deviation of results from catalogue value the equivalent circuit parameters are fine tuned to get accurate results.

Step4: Run the Program for different values of parameter under study, for example to study efficiency and power factor for different values of rpm change only rpm values and get desired output values.

Step5: A chart for different parameters can be easily generated with chart option in the EXCEL. For example the following Fig. 3 shows the efficiency and power factor The values are found to be within 5 % of  manufacturer test result data supplied by the manufacturer

## VI. HARMONIC ANALYSIS

The use of six pulse solid state inverters for variable speed application with these motors have increased for varying speed applications which generates harmonic currents in the motor. The harmonic voltages induce harmonic currents in the motor resulting in harmonic torques and cause heating in copper windings and the core. Harmonics are present due to inverter applications.

A preliminary analysis to estimate the effect of $5^{th}$ and $7^{th}$ harmonics has been attempted below. The losses can be modeled as the sum of the additional stator copper loss, rotor copper losses and iron losses for a given harmonic current (In). For simplicity we do not take the skin effect for low order harmonics. From literature we find that for $5^{th}$ and $7^{th}$ harmonic voltage may be assumed as 20% and 14.3% of the fundamental voltage respectively.

In the computer simulation on worksheet the fields for % harmonic voltage and harmonic number are used for feeding harmonic data. The results for torque and losses are produced in the respective fields. The results obtained are tabulated in Table III below.

Following observations are made from the study done:
1. With increase of harmonic number the core loss becomes more  prominent than the copper loss.

2. The losses due to $5^{th}$ and $7^{th}$ harmonic are about 5 % of total loss with fundamental harmonic.

3. From the results obtained the torques generated by harmonic currents are negligible.

TABLE I
EXCEL WORKSHEET FOR MOTOR PROGRAM

| Input Data | |
|---|---|
| Rated Voltage | 415 |
| Rated Output kW | 15 |
| Rated Current | 27.5 |
| Operating Voltage | 415 |
| Rated frequency | 50 |
| Torque ratio B | 6 |
| Rated Power factor | 0.83 |
| Number of poles | 4 |
| Rated syn. Speed | 1500 |
| Operating frequency | 50 |
| Operating rpm | 1465 |
| Operating slip | 0.02333 |
| Rated Slip | 0.03333 |
| Rated Current | 27.5 |
| **Equivalent Circuit Parameters** | |
| $R_1$ | 0.1 |
| $R_2$ | 0.27 |
| $R_m$ | 320 |
| $X_1$ | 2.18 |
| $X_2$ | 1.35 |
| $X_m$ | 45 |
| **Input/ Output Values** | |
| Input $I_1$ | 20.57248 |
| Input PF | 0.866381 |
| Input W | 12811.25 |
| Effi | 93.28528 |
| $I_2$ | 18.77478 |
| $P_o$ | 11951.01 |
| Torque | 102 |
| **Losses** | |
| Core | 489.3433 |
| Copper | 412.4872 |
| Total Loss | 901.8305 |
| Efficiency | 92.98341 |

## VII. CONCLUSIONS

A new methodology is presented for estimating the induction motor performance without performing any field measurements. The results are validated with catalogue data of an actual motor. The method can help the designers and users to study and optimize the motor performance for any specific load requirement. Further preliminary analysis for including the effect of harmonics due to six pulse inverter application on motor performance have been presented for torque and relative increase in losses for  a typical 15 kW motor for estimating the de-rating of the motor.

## VIII. ACKNOWLEDGEMENT

The author gratefully acknowledges Dr. Ajay Mathur, Director General, Bureau of Energy Efficiency for enabling the author to present this paper and my wife, Dr. Nikki Sabharwal for her encouragement.

TABLE II
THE PERFORMANCE CHARACTERISTICS OF MOTOR FOR DIFFERENT RPM

| Rpm | Input $I_{in}$(A) | PF | Input Power (W) | Eff. (%) | Eff. | Pout (W) | Pout/Pout(rated) |
|------|------|------|------|------|------|------|------|
| 1495 | 6.21 | 0.54 | 2426 | 79 | 0.79 | 1918 | 0.13 |
| 1490 | 8.2 | 0.73 | 4323 | 87.6 | 0.88 | 3793 | 0.25 |
| 1485 | 10.66 | 0.8 | 6161 | 90.6 | 0.91 | 5608 | 0.37 |
| 1480 | 13.1 | 0.85 | 7930 | 92 | 0.92 | 7343 | 0.49 |
| 1475 | 15.55 | 0.86 | 8986 | 92.6 | 0.93 | 8986 | 0.6 |
| 1470 | 18.1 | 0.87 | 11515 | 93 | 0.93 | 10525 | 0.7 |
| 1465 | 20 | 0.87 | 13177 | 93.1 | 0.93 | 11951 | 0.8 |
| 1460 | 22.96 | 0.86 | 14218 | 93 | 0.93 | 13258 | 0.88 |
| 1455 | 25.27 | 0.85 | 15509 | 92.8 | 0.93 | 14444 | 0.96 |
| 1450 | 27.49 | 0.84 | 16758 | 92.5 | 0.93 | 15507 | 1.03 |
| 1445 | 29.61 | 0.83 | 17828 | 92.3 | 0.92 | 16450 | 1.1 |
| 1440 | 31.61 | 0.82 | 18689 | 92 | 0.92 | 17275 | 1.15 |
| 1435 | 33.56 | 0.81 | 19523 | 91.6 | 0.92 | 17989 | 1.2 |

TABLE III
RESULTS OF HARMONIC LOSSES IN MOTOR

| | Fundamental | N= 5 | N=7 |
|------|------|------|------|
| % of Fund. Voltage | 100 % | 20 % | 14.3 % |
| Core Loss (W) | 489 | 19.6 | 10 |
| Copper Loss (W) | 761 | 20.5 | 5.4 |
| Total Loss (W) | 1250 | 40.1 | 15.4 |

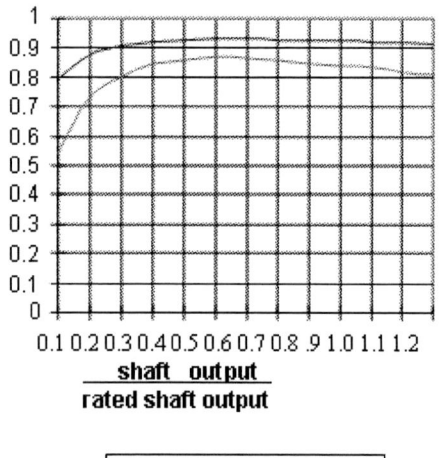

shaft output / rated shaft output

----- Efficiency ----- Pf

Fig. 3. Load Performance Curves for 15 kW motor.

## IX. REFERENCES

[1] A.E. Fitzgerald, Charles Kingsley Jr., Stephen D. Umans, Electric Machinery, New York, NY: McGraw Hill Book Company 1983.
[2] IEEE Standard 112 – 1996
[3] Tata Energy Research Institute, " Survey of Industrial Motors" interim and final report ,1987

## X. BIOGRAPHIES

**Satish Chander Sabharwal** (SM'2006) was born in New Delhi, India on 22nd May 1957.He graduated in Electrical Engineering with B. Tech. degree with Distinction from Indian Institute of Technology, Delhi and Masters from University of Manitoba, Winnipeg, Canada in 1983. He was selected for BOYSCAST fellowship by Ministry of Science and Technology. His employment experience includes the Bharat Heavy Electrical Limited, Tata Energy Research Institute, Energy Management Centre and is presently with Bureau of Energy Efficiency, Ministry of Power. His special fields of interest includes Energy Policy ,Energy Conservation, Power Electronics , Lighting, Motors and Drives.

**2006 IEEE International Conference on Power Electronic, Drives and Energy Systems**

# Design of Squirrel Cage Induction Motors for Traction Applications

S. S. Murthy, *Senior member, IEEE*, Bhim Singh, *Senior Member*, IEEE, G. Bhuvaneswari, *Senior Member, IEEE*, Kiran Naidu, and Uddanti Siva

*Abstract*-**This paper highlights the design issues of high power induction motors for traction applications. The circuit modeling and analysis of an induction machine is carried out with the parameters calculated from physical dimensions. The modification of the parameters to incorporate the time harmonic effects of inverter PWM supply, flux and frequency variation effect during closed loop V/F control is investigated in detail. The design program of the machine is developed using MATLAB.**

*Index Terms*--**VVF, Squirrel cage Induction motor (IM), Finite element.**

## I. INTRODUCTION

MAJOR part of electrical energy is utilized by squirrel cage induction motors (SCIM) to convert into mechanical work in large number of applications. Due to the present advancements in power electronics SCIM is being widely used in variable speed applications and advancements in computer hardware and software made design/ analysis of induction motor less complicated than ever before. None of the prevalent softwares such as SPEED, MAGNET, FLUX2D, RM-Expert etc, do provide direct optimum design as they do not do the engineer's job. These are simply specialized calculating tools to assist the design engineer with initial sizing and preliminary design of motors by providing a simple intuitive interface and simulation.

Design of the motor for adjustable-speed Variable Voltage Variable Frequency (VVVF) drives is complicated as the supply waveforms are 'switch mode' chopped waveforms and motor parameters like coreloss resistance ($R_c$), rotor resistance ($R_r$) and rotor inductance ($L_r$) vary considerably with the change in frequency of supply [1-3]. So an accurate analysis is required for calculating magnetic as well as electrical performance. To get an optimum design suitable for VVVF applications commercial softwares need many simulation iterations and time for implementation of numerical methods on complicated design equations [22-24]. Moreover due to the modern control strategies for VVVF drives, motor design cannot be isolated from the controller

S. S. Murthy, Bhim Singh, G. Bhuwaneshwari, Kiran Naidu and Uddanti Siva are with Department of Electrical Engineering, Indian Institute of Technology, Delhi, New Delhi -110016. (email : ssmurthy@ee.iitd.ac.in, bsingh@ee.iitd.ac.in, bhuvan@ee.iitd.ac.in, kirannaiduv@gmail.com, uddantisiva.iitd@gmail.com)

design; both the designs should be integrated in one module, where no commercial software serves as single destination. Therefore, either simple tailor made design tools have to be developed or the existing softwares need to be modified by integrating inverter circuit configurations along with load patterns and control strategies.

This paper deals with the design analysis of high power induction motors for traction application. A software program has been developed in MATLAB to perform the design analysis. Design can be done either from classical methods or from FE numerical methods, but later is always time consuming, however one can use the FE tools to validate the results of classical methods and this approach is used in this paper. Analysis of machines under VVVF application is quite involved as conventional equivalent/d-q models are limited, to completely incorporate the non sinusoidal effects of supply voltage. The effects of non sinusoidal voltage supply on the induction motor behavior have been examined extensively in the literature [14-18]. Many modified equivalent circuits [10-13] and dynamic models [3-9] of the induction motor are proposed, taking into account the frequency-dependent nature of the motor parameters due to various phenomena like skin effect [3-6], saturation [5-6] and various losses like iron loss [7-9] and stray load loss [19-20]. In this paper a modified equivalent circuit and dynamic model [17] is used to incorporate strayload loss and ironloss variation in steadystate and transient operation. The developed software program performs the electro magnetic calculations and circuit modeling of induction machine from the physical dimensions of the machine as input and thus calculates the steady state and transient performance by incorporating nonlinearities due nonsinusoidal excitation. The design analysis is carried out for two practical induction motors and the results are discussed in detail.

## II. DESIGN ANALYSIS OF INDUCTION MOTOR

The design steps have been discussed for performance evolution of the motor with the given physical dimensions. Initially from the given physical dimensions and electrical and magnetic specifications, the equivalent circuit parameters are calculated by the conventional methods. Average flux/pole in the air gap is calculated from equation (1). $K_\phi$ is stator drop constant, can vary from 0.95 to 0.97.

$$E_{rms/ph} = (K_\phi)4.44 f\phi NK_w \qquad (1)$$

$\Phi$: Average fundamental flux/pole

0-7803-9771-1/06/$25.00 ©2006 IEEE

N: Number of series turns/phase

$K_w$ : Stator winding factor

Average flux density in air gap is given by (2). The effective air gap length is calculated using carter's coefficient $K_g$ given by (3).

$$G = K_g g \qquad (2)$$

$$K_g = \frac{t+s}{t+s-g*f(\alpha)} \qquad (3)$$

$$f(\alpha) = f(\frac{s}{2g}) = \alpha \tan(\alpha) - \log \sec(\alpha) \qquad (4)$$

$$B_g = \frac{P\phi}{\pi DL} K_g \qquad (5)$$

t: Tooth width

s: Slot width

P : number of poles

D: stator inner diameter in m

L: stack length in m

Similarly the flux densities in core and tooth are given by following equations.

$$B_c = \frac{\phi}{2d_c L_i} \qquad (6)$$

$$B_t = \frac{P\phi}{S L_i W_t} \qquad (7)$$

$d_c$: core depth in m

$L_i$: effective core length in m

S: number of slots

$W_t$ average tooth width in m

Magnetizing current is given by

$$I_m = \frac{\sqrt{2} P^2 G \phi K_i}{32 q N K_w DL} 10^7 \qquad (8)$$

$K_i$ is constant factor >1, which includes tooth magnetization ($M_t$) and core magnetization ($M_c$) and is given as

$$K_i = 1 + \frac{M_t + M_c}{M_g} \qquad (9)$$

All the above calculations are given for sinusoidal flux densities. However for calculating extra magnetizing current the flux density is considered at $53.2^0$ in waveform. Further considerations regarding core and stray load losses are discussed in the following sections. In calculating leakage reactance of stator and rotor the following reactances are considered primary slot leakage, secondary slot leakage, zigzag leakage, belt leakage, coil end leakage, incremental reactance, peripheral leakage. The design analysis is carried out for two typical high power motors of rating 297KW and 850 kW traction motors, whose specifications are given in Table. I and Table-II. The design results are compared with simulated values in MAGNET and flux density distribution is given in Fig .1. The peak flux density is observed in rotor

tooth to be 1.95 T, which is nearer to calculated value, 1.9 T. current densities in stator winding are around 5 to 7.

TABLE I
MOTOR SPECIFICATIONS

| Rating | Motor I | Motor II |
|---|---|---|
| Stator frequency | 65-132 Hz | 64-132 Hz |
| Motor voltage | 2180 V | 1716 V |
| Stator current | 270-393 A | 136-195 A |
| Power factor | 0.88 | 0.82 |
| Torque | 6330 Nm | 2256 Nm |
| Power | 850 KW | 297 KW |
| Motor speed | 1283 rpm | 1259 rpm |

TABLE II
EQUIVALENT CIRCUIT PARAMETERS

| Power | 850KW | 297KW |
|---|---|---|
| $R_s$ | 0.0762 Ω | 0.16 Ω |
| $R_r$ | 0.0639 Ω | 0.122 Ω |
| $R_m$ | 267.492 Ω | 304.6 Ω |
| $L_s$ | 0.001223 H | 0.00249 |
| $L_r$ | 0.001226 H | 0.00217 II |
| $L_m$ | 0.038 H | 0.041 H |

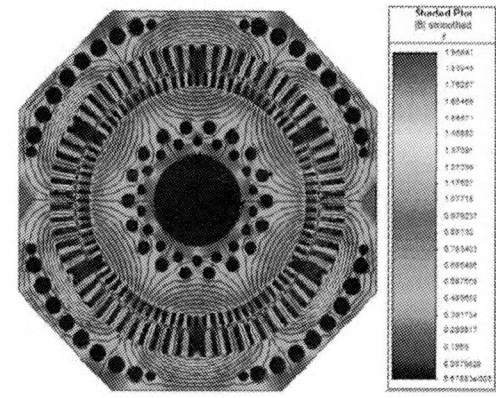

Fig. 1. Magnetic flux density distribution – MAGNET.

## III. PERFORMANCE ANALYSIS OF SCIM

### A. Skin effect

In the equation used for steadystate analysis, the motor parameters are normally considered as constant values. If an induction motor is supplied by an inverter, beside the fundamental wave, the rotor currents contain harmonics. The frequency of these harmonics could have very high values. The high frequency harmonic currents make the rotor parameters change, due to skin effect and deep bar effect. Further, the induction machine is highly affected by the main magnetizing flux paths saturation. On the other hand, for high currents, the leakage flux path saturation occurs [4, 5]. For these reasons and in order to have a more precise and more faithful to reality model, these phenomena must be taken into account.

Skin effect is related to the flux and current density distribution in a conductor (or a group of conductor) [5]. Skin effect effectively increases rotor resistance and decreases rotor leakage inductance [5]. The variation of parameters regarding to slip can be determined using correction coefficient factors for resistance and leakage inductance. To calculate these coefficients, one has to know the rotor slot geometry. These coefficients vary with slip. For rectangular slots, $K_r$, $K_l$ correction coefficients for resistance and reactance are given in (10) and (11) respectively.

$$K_r = \xi \frac{\left(\sinh 2\xi + \sin 2\xi\right)}{\left(\cosh 2\xi - \cos 2\xi\right)} \tag{10}$$

$$K_l = \frac{3}{2}\xi \frac{\left(\sinh 2\xi - \sin 2\xi\right)}{\left(\cosh 2\xi - \cos 2\xi\right)} \tag{11}$$

Where

$$\xi = h_s \sqrt{\frac{s f \mu_0 \sigma_{al} b_c}{b_s}}$$

s: p.u. slip at frequency, f
$\mu_0$: Magnetizing space constant
$\sigma_{al}$: Electrical conductivity
$b_c$: Width of conductor
$b_s$: Width of slot
$h_s$: height of conductor

$$R_r = K_r R_{rdc} \tag{12}$$
$$X_r = K_l X_{rdc} \tag{13}$$

The main losses of induction machines are the following.

*B. Copper loss*

The copper loss is due to electric current flow through the stator and rotor windings and is given by

$$P_{copper} = r_s i_s^2 + r_r' i_r'^2 \tag{14}$$

$r_s$ and $r_r$ are stator and rotor resistance respectively, $i_s$ and $i_r$ are stator and rotor current respectively.

*C. Iron loss*

It is composed of the hysteresis and eddy current loss in stator and rotor core. Iron loss is frequency and flux dependent and is expressed by

$$
\begin{aligned}
P_{core} &= P_{cs} + P_{cr} \\
&= \{k_e \omega_e^2 \phi_m^2 + k_h \omega_e \phi_m^n\} \\
&\quad + \{k_e (s\omega_e)^2 \phi_m^2 + k_h (s\omega_e)\phi_m^n\}
\end{aligned}
$$

$$P_{core} = k_e(1+s^2)\omega_e^2\phi_m^2 + k_h(1+s)\omega_e\phi_m^2 \tag{15}$$

Rotor iron loss is quite lower than stator iron loss, because stator flux varies at the supply frequency $\omega_e$ and the rotor flux varies at very lower slip frequency $s\omega_e$ ($s \ll 1$). In general if the supply voltage contains lower order harmonics $m^{th}$ switching harmonic iron loss can be considered by (16). Total iron loss of the squirrel cage rotor type motor used in this paper is observed to be 2% of rated output power.

$$R_{mk} = \frac{1}{\dfrac{K_h}{k\omega} + K_e} \tag{16}$$

*D. Stray load loss*

This loss is mainly torque dependent. When the load torque and rotor current increase, a percentage of the air-gap flux diverts from its axial path and the leakage flux is magnified. In addition to the above loss, the harmonic loss that is due to the non sinusoidal motor excitation inflicted by the pulse-width modulation (PWM) inverter causes an increase in copper, iron and strayload loss [18]. Two major contributions for stray load losses are no-load losses in rotor teeth because of stator slot opening modulation of fundamental flux density and load losses in the rotor teeth because of stator zigzag mmf. They primarily cause a drag on the rotor, so one way to account them is to subtract their power from the mechanical output of the machine, however here a modified equivalent circuit shown in fig.2 to account for stray load losses. In the equivalent circuit shown in Fig.2, rotor inductance $L^1_{lr}$ and resistance $R^1_r$ are functions of frequency due to skin effect, magnetizing inductance $L_m$ vary due to main flux saturation. The core loss resistance $R_c$ varies due to frequency and stray load resistance $R_{str}$ varies with loading.

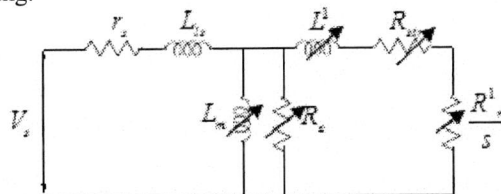

Fig. 2. Equivalent circuit of IM with iron and stray losses.

*E. Steady state model*

In the steady state condition from a given operating point to other operating point all the parameters shown in Fig 2 change considerably due to reasons discussed so far. In the developed program these variables are calculated at each operating point and steady state analysis is carried out using these parameters.

*F. Dynamic model*

The inclusion of iron loss in space vector and *d-q* axis models of induction machines is possible only in an approximate way. Stator iron loss, the dominant part of the overall iron loss in induction machines, may be modeled with an equivalent iron loss resistance placed in parallel to the magnetizing branch in the dynamic equivalent circuit. The equivalent inductance is a fictitious parameter that attempts to model the rate of change of eddy current losses during transients, and it has not much impact on steady-state operation. The equivalent iron loss resistance models the

41

power dissipated in the stator core. The PWM voltage source is known to cause a significant increase in total iron losses, compared with purely sinusoidal supply. Indeed, in any steady-state *d-q* axis stator current commands are constant dc quantities, so that stator phase current or voltage commands, created by coordinate transformation, are pure sine waves of required fundamental frequency. Thus iron loss produced by higher order voltage harmonics of the inverter supply has no impact on accuracy of control schemes generally that one uses. The same applies, for the additional copper losses produced by higher order current harmonics. The first harmonic iron loss with PWM supply is essentially equal to the iron loss with sinusoidal supply. Therefore the core loss branch models the fundamental stator iron loss in the machine of dynamic model unlike steady state model that we considered earlier. $R_c = f(W_e)$ and $R_{str} = g(W_e^2)$.

Steady state equivalent circuit given in Fig 2 is modified eliminating rotor inductance according to (17), (18), (19) and corresponding dynamic models are given in Fig 3 and Fig 4.

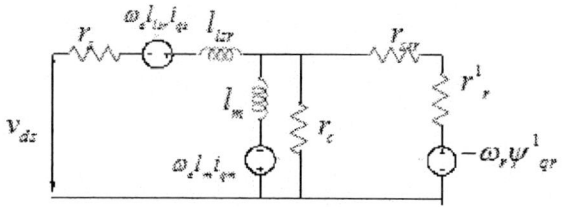

Fig. 3. d-axis modified dynamic model of IM.

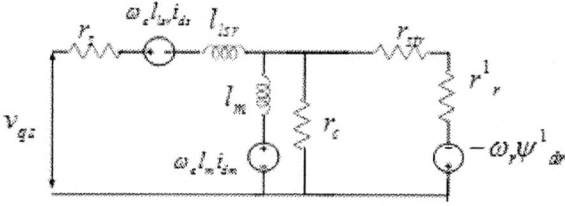

Fig. 4. q-axis modified dynamic model of IM.

$$l_m = \frac{L_m^2}{L_m + L_{lr}^1} \tag{17}$$

$$l_{lsr} = L_{ls} + \frac{L_m L_{lr}^1}{L_m + L_{lr}^1} \tag{18}$$

$$r_r^1 = R_r^1 \left(\frac{L_m}{L_m + L_{lr}^1}\right)^2 \tag{19}$$

d-q modeling equations in synchronously rotating reference frame are given from (20) to (30) using normal conventions.

$$V_{qs} = r_s i_{qs} + \omega_e \psi_{ds} + \frac{p}{\omega_b} \psi_{qs} \tag{20}$$

$$V_{ds} = r_s i_{ds} - \omega_e \psi_{ds} + \frac{p}{\omega_b} \psi_{ds} \tag{21}$$

$$0 = (r_r^1 + r_{str})i_{qr}^1 + (\omega_e - \omega_r)\psi_{dr}^1 + \frac{p}{\omega_b}\psi_{qr}^1 \tag{22}$$

$$0 = (r_r^1 + r_{str})i_{dr}^1 - (\omega_e - \omega_r)\psi_{qr}^1 + \frac{p}{\omega_b}\psi_{dr}^1 \tag{23}$$

$$\psi_{ds} = l_{lsr} i_{ds} + \psi_{dr}^1 \tag{24}$$

$$\psi_{qs} = l_{lsr} i_{qs} + \psi_{qr}^1$$

$$\psi_{qr}^1 = l_m i_{qm} \tag{25}$$

$$\psi_{dr}^1 = l_m i_{dm} \tag{26}$$

$$x_m \frac{p}{\omega_b} i_{dm} - r_c i_{dc} = \omega_e x_m i_{qm} \tag{27}$$

$$x_m \frac{p}{\omega_b} i_{qm} - r_c i_{qc} = -\omega_e x_m i_{dm} \tag{28}$$

$$i_{ds} + i_{dr}^1 = i_{dm} + i_{dc} \tag{29}$$

$$T_e = \psi_{dr}^1 (i_{qs} - i_{qc}) - \psi_{qr}^1 (i_{ds} - i_{dc}) \tag{30}$$

*where* $p = \dfrac{d}{dt}$ *is time differential operator*

The complete design and analysis procedure is given in the form of flow chart below.

## IV. FLOWCHART OF DESIGN ANALYSIS PROGRAM

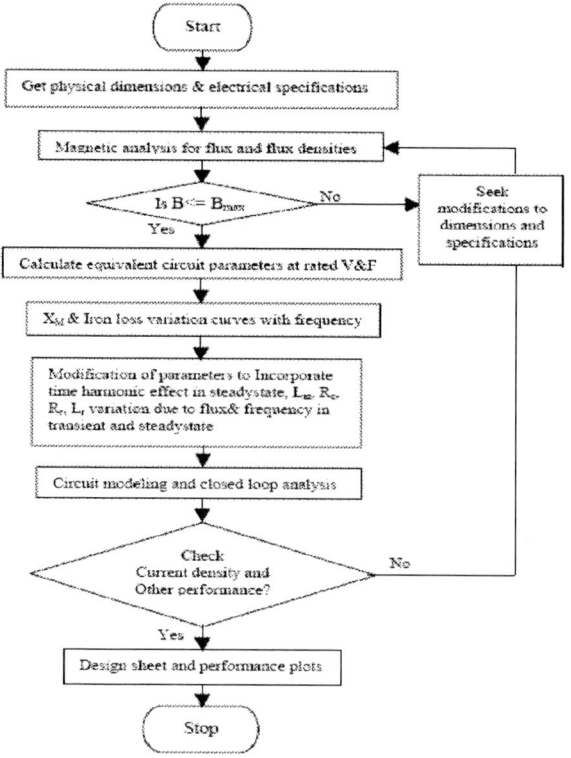

## V. RESULTS AND DISCUSSION

The output of software program is compared with experimental results in steady state the modified equivalent circuit model give results comparable to practical results. Test results therefore at variable inverter supply of 16-132Hz are compared with simulated results in Fig. 1 to Fig. 16. Inverter employs SPWM in the low frequency range and quasi-square wave operation in high frequency range in a typical traction scenario. The carrier frequency is in the range of 1000-2000 Hz and DC link voltage is 2800 V.

The parameter variation of the designed motor under constant V/F control is given. The designed motor are supposed to operate under constant rated torque for speed below rated speed and in field weakening region for speeds above base speeds. Therefore the parameter variations are given for these conditions in Fig.17 to Fig. 22. In simulation the parameters at each operating point can be taken directly form the graphs obtained.

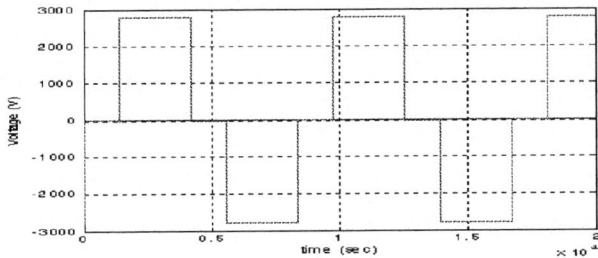

Fig. 5. Supply voltage(line) at 60 Hz frequency.

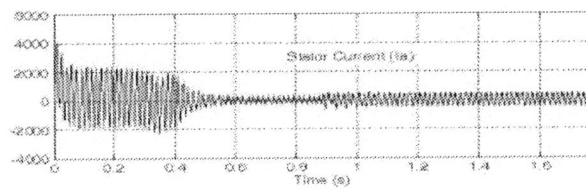

Fig. 6. Stator current and Harmonics Spectrum.

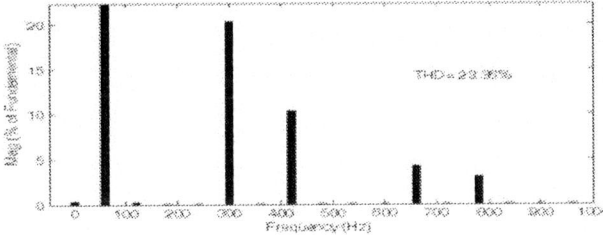

Fig. 7. Motor under rated torque of 9200 N-m.

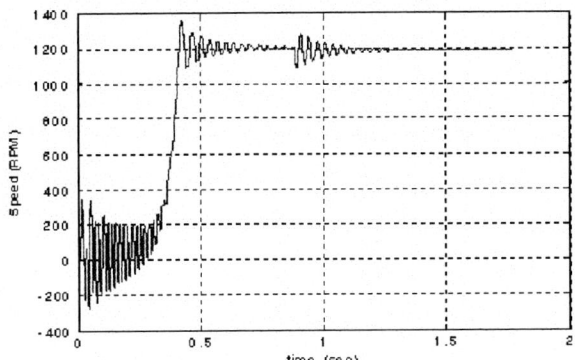

Fig. 8. Speed of the motor for 60 HZ supply.

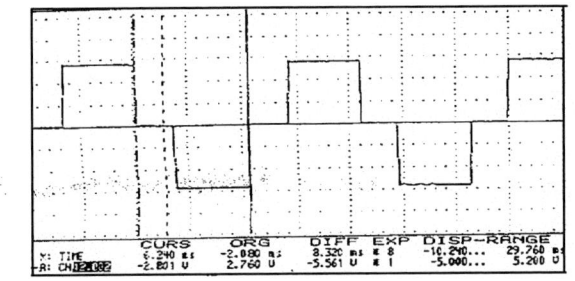

Fig. 9. Experimental stator voltage wave form at 60 Hz.

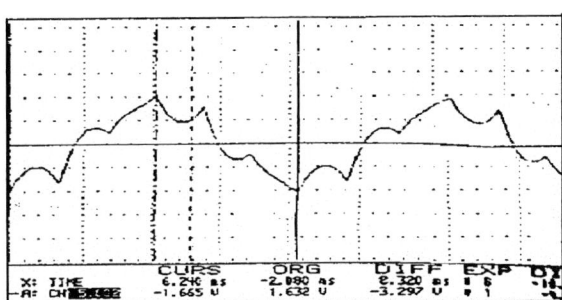

Fig. 10. Experimental stator current wave form at 60 Hz.

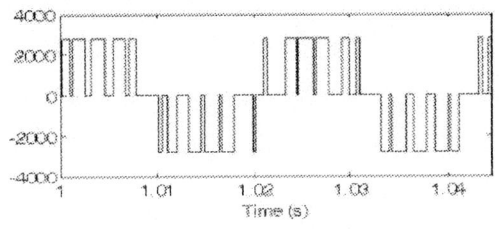

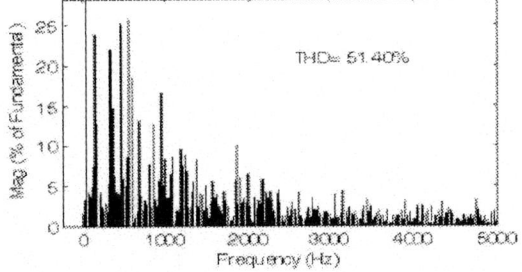

Fig. 11. Supply voltage(line) at 16 Hz frequency.

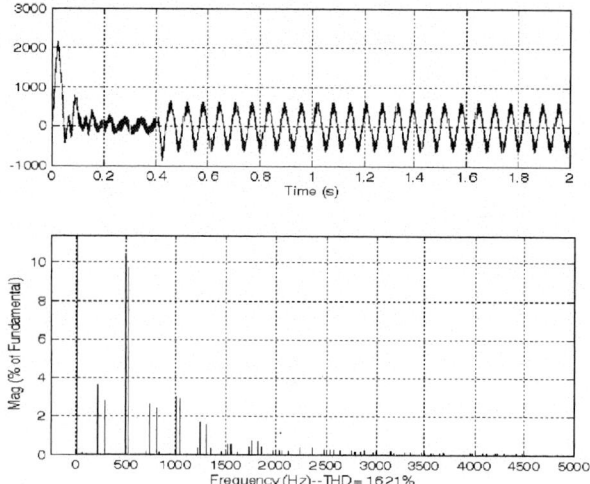

Fig. 12. Stator current and Harmonics Spectrum at 16 Hz.

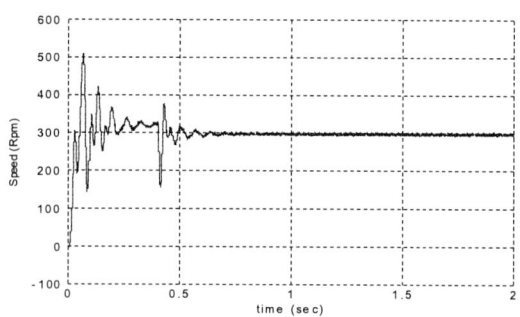

Fig. 13. Speed of the motor for 16 HZ supply.

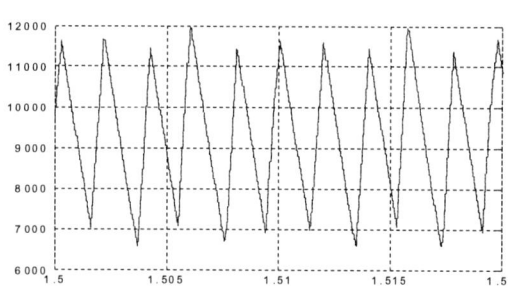

Fig. 14. Motor steady state torque ripple under rated torque of 9200 N-m.

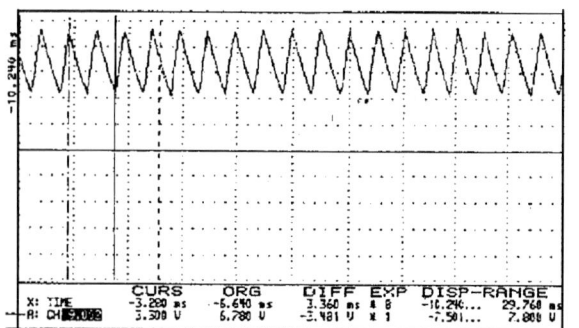

Fig. 15. Experimental Motor steady state torque ripple under rated torque of 9200 N-m.

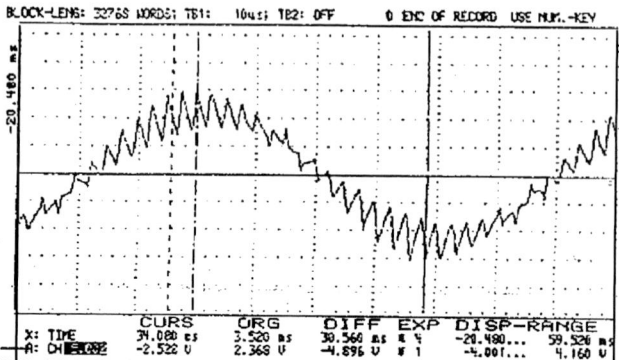

Fig. 16. Experimental Stator current wave form at 16 Hz.

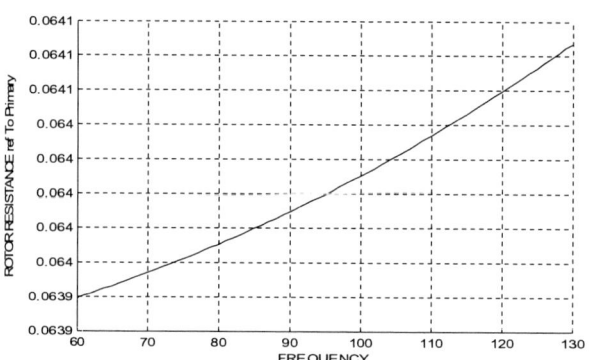

Fig. 17. Rotor resistance variation with frequency.

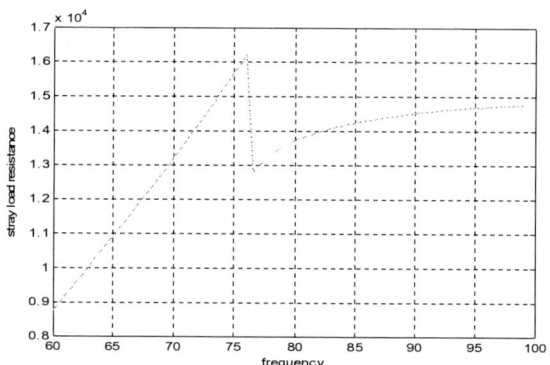

Fig. 18. Stray load resistance variation with frequency.

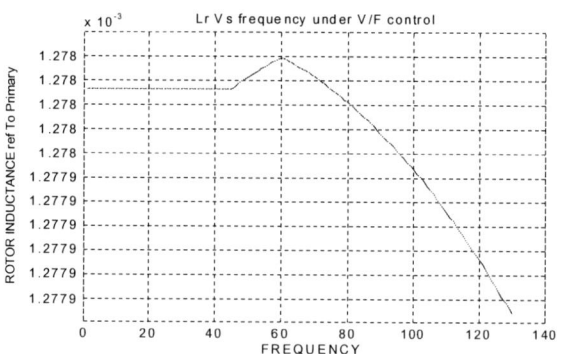

Fig. 19. Rotor inductance variation with frequency under V/F control.

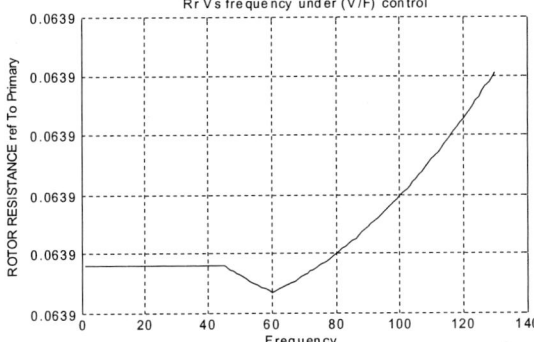

Fig. 20. Rotor resistance variation with frequency under V/F control.

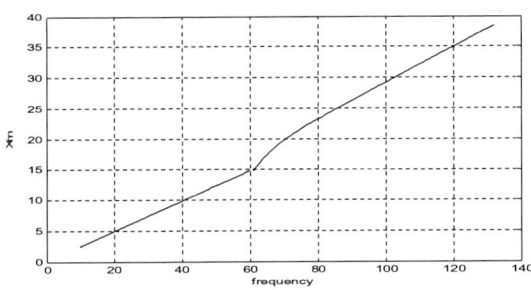

Fig . 21. Magnetizing reactance variation with frequency.

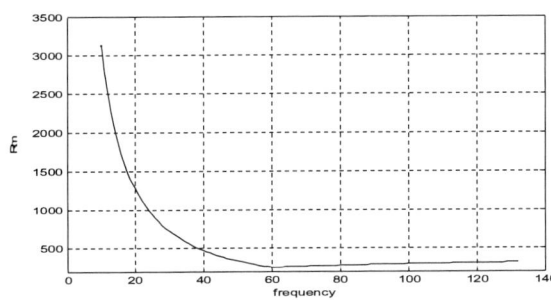

Fig. 22. Core loss resistance variation with frequency.

## VI. CONCLUSION

This paper has dealt the issues involved in the design of induction motors for VVVF drives. A design and analysis software is developed considering input frequency and operating flux variation effect on parameter variation. The developed program can perform both transient and steady state analysis of induction machines integrating the design of the machine with inverter circuit configurations and control strategies.

## VII. REFERENCES

[1] S J Loddick , "Modelling Frequency Dependency Of Induction Machine Equivalent Circuit Parameters" *In Proc. Of IEE* 1996 Power Electronics And Variable Speed Drives *Conf.,* Publication No. 429, pp. 30-35

[2] Sanchez, I.; Pillay, P. "Sensitivity Analysis Of Induction Motor Parameters", In Proc. *of IEEE 1994 Southeastcon Creative Technology Transfer - A Global Affair.* pp. 50 – 54.

[3] Erdogan, N.; Assaf, T.; Grisel, R.; Aubourg, M.; "An Accurate 3-Phase Induction Machine Model Including Skin Effect And Saturations For Transient Studies" *In Proc. Of 2003 ICEMS Electrical Machines And Systems International Conf,. 2003. Sixth International Conf. pp.* 646 – 649.

[4] J. Langheim, "Modeling of ratorbars with skin effect for dynamic simulation of induction machines", *In Proc of 1989 IEEE industry Applications Sociery Anniial Meeting record. .* vol. 1.

[5] J.Li, L. Xu, "lnvestigotion of Cross-Saturation and Deep Bar Efleclecrs of lnducrion Motors by Augmented d-q Modeling Method" IEEE Tansactions 2001.

[6] A.C. Smith, R.C. Healey, S . Williamson, "A Transient Induction Motor Model Including Saturation And Deep Bar Effect", *IEEE Trons. On Energv Conversion*, Vol. 11, No. I, March 1996.

[7] Z. Papazacharopoulos, K. Tatis, A. Kladas And S. Manias, "Dynamic Induction Motor Model For Non-Sinusoidal Supply", *In Proc. Of IEEE Power Electronics Specialists Conference*, 2002. Vol 2, 23-27 June 2002 pp. 845 – 850.

[8] A. Kandianis, A. Kladas, S. Manias And J. Tegopoulos, "Electrical Vehicle Drive Control Based On Finite Element Induction Motor Model", *IEEE Transactions On Magnetics*, Vol. 33, No. 2, March 1997, Pp. 2109-2112.

[9] A. Vamvakari, A. Kandianis, A. Kladas And S. Manias, "High Fidelity Equivalent Circuit Representation Of Induction Motor Determined By Finite Elements For Electrical Vehicle Drive Applications", *IEEE Transactions On Magnetics*, Vol. 35, No 3, May 1999, Pp. 1857-1860.

[10] T. J. White and J. C. Hinton, "Improved dynamic performance of the 3-phase induction motor using equivalent circuit parameter correction," *In Proc. Of 1994 International Conference on Control,* vol. 2, 1994, pp. 1210–1214.

[11] Z. Papazacharopoulos, K. Tatis, A. Kladas, S. Manias And J. Tegopoulos, "Inverter-Fed Induction Motor Simulation Under Overloading For Regenerative Braking Management In Electric Vehicle", *In Proc of 2000 ICEM. p*p. 417- 42 1.

[12] G. Champenois, R. Perret, and D. S. Zhu, "Losses and torques in inverter-fed induction motor drives," in *IEEE Industry Applications Society Annual Meeting*, vol. 1, 1988 Record, pp. 175–180.

[13] Vamvakari, A. Kandianis, A. Kladas, S. Manias And J. Tegopoulos "Analysis Of Supply Voltage Distortion Effects On Induction Motor Operation", *IEEE Transactions On Energy Conversion*, Vol. 16 Pp. 209 – 213, September 2001.

[14] M. Amar And R. Kaczmarek, "A General Formula For Prediction Of Iron Losses Under Nonsinusoidal Voltage Waveform," *IEEE Trans. Magnetics*, Vol. 31, No. 5, Pp. 2504–2508, Sept. 1995.

[15] D. Lin, T. Batan, E. F. Fuchs, And W. M. Grady, "Harmonic Losses Of Single-Phase Induction Motors Under Nonsinusoidal Voltages," *IEEE Trans. Energy Conversion,* Vol. 11, No. 2, Pp. 273–286, June 1996.

[16] V. Kinnares, P. Jaruwanchai, D. Suksawat, And S. Pothivejkul, "Effect Of Motor Parameter Changes On Harmonic Power Loss In Pwm Fed Induction Machines," *In Proc. Of 1999 IEEE International Conference On Power Electronics And Drive Systems*, Vol. 2, Pp. 1061–1066.

[17] G.K. Singh, "A research survey of induction motor operation with non-sinusoidal supply wave forms Electric Power Systems Research 75, Elsevier (2005) 200–213.

[18] P. K. Sen and H. A. Landa, "Derating of induction motors due to waveform distortion," *IEEE Trans. Ind. Applicat*ions., vol. 26, no. 6, pp. 1102–1107, Nov./Dec. 1990.

[19] Emil Levi, Adoum Lamine, and Andrea Cavagnino, "Impact of Stray Load Losses on Vector Control Accuracy in Current-Fed Induction Motor Drives", *IEEE Transactions on energy conversion,* vol. 21, no. 2, june 2006.

[20] Constantine Mastorocostas, Iordanis Kioskeridis, Member, IEEE, and Nikos Margaris, Member, IEEE "Thermal and Slip Effects on Rotor Time Constant in Vector Controlled Induction Motor Drives", *IEEE Transactions on power electronics*, vol. 21, no. 2, march 2006

[21] P.L. Alger, Induction Machines: Their Behaviour and Uses, second edition., Gordon and Breach Publishers, 1975.

[22] C. G. Veinott, "Theory and Design of Small Induction Motors". New. York: McGraw-Hill, 1959

[23] "Speed' Electric Motors Reference Manual", Tje Miller, University Of Glasgow, 2002-2004.

[24] T. J. E. Miller And M. I. Mcgilp, Pc-Fea 5.0 For Windows—Software. Glasgow, U.K.: Speed Laboratory, Univ. Glasgow, 2002.

[25] "Infolytica's Magnet Low Frequency Electromagnetics Simulation Software Manuals- 2006" Infolytica Corporation.

**2006 IEEE International Conference on Power Electronic, Drives and Energy Systems**

# Effect of Sequential Phase Energization on the Inrush Current of a Delta Connected Transformer

K. P. Basu, *Senior Member, IEEE*, Ali Asghar, *Member, IEEE*, and Stella Morris, *Senior Member, IEEE*

*Abstract–* **Magnetizing inrush current in power transformer creates severe power quality problems. Energization of the transformer from the delta side of a delta-star transformer does not allow the control of neutral resistor at the time of switching the supply for reducing the inrush current. Controlled switching at peak value of the voltage may reduce inrush current in one winding of the transformer, but the line currents are not reduced. Delayed and controlled energization of the third phase may reduce inrush winding currents in two windings and only one line current is reduced, but the inrush currents in the other two lines become very high. Simulation studies of delta side energization of a delta-star transformer depict the results under various switching conditions.**

*Index Terms–***Magnetizing inrush current, delta side energization, sequential and controlled phase energization.**

## I. INTRODUCTION

**P**RODUCTION of high inrush current in a big transformer during switching in the supply always creates a big power quality problem. In a deregulated market with disperse generation having transformers located at different points in a system makes the problem more acute. The transient voltage dip may cause motor tripping, protective relay mal-function etc.

Over the past many decades, a large number of simulation and experimental studies have been reported on the production and control of inrush currents in single and three-phase transformers [1-8]. Commonly used techniques to reduce inrush current are insertion of series resistor, point-on-wave switching etc. Energization of the transformer from the star side allows the control of neutral resistor utilizing the advantage of unbalancing in the 3-phase inrush current and sequential switching of phases to achieve the reduction of inrush current [9-10]. No extensive research on inrush current is reported for the delta side energization of the transformer, which may be necessary due to various practical reasons.

This paper reports the simulation studies on the inrush current produced by the delta side energization of delta-star transformer with and without sequential phase switching.

## II. DELTA SIDE ENERGIZATION

Fig.1 shows the delta-side energization of a delta-star transformer. Unlike star-side energization, series resistance insertion to reduce inrush current is possible only in the line and not in the delta winding of the transformer.

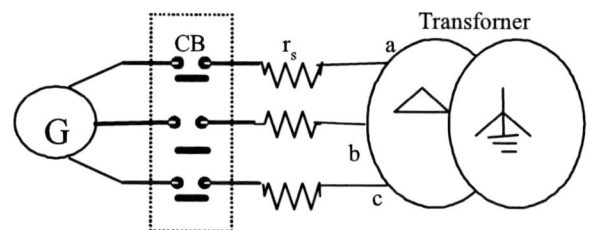

Fig.1. Delta-side switching of transformer.

The equations governing currents and voltages of the delta winding with the star side open are presented below. Leakage reactance and remnant flux in the transformer core are neglected.

$$E_{ab} = E_{max}\sin(\omega t+\delta) = n_1(d\varphi_{ab}/dt) + r_1 i_{ab} + r_s(i_a + i_b) \quad -----(1)$$
$$E_{bc} = E_{max}\sin(\omega t+\delta-2\pi/3) = n_1(d\varphi_{bc}/dt) + r_1 i_{bc} + r_s(i_b + i_c) \quad -- (2)$$
$$E_{ca} = E_{max}\sin(\omega t+\delta+2\pi/3) = n_1(d\varphi_{ca}/dt) + r_1 i_{ca} + r_s(i_c + i_a) \quad --(3)$$

Where, $E_{ab}$, $E_{bc}$, $E_{ca}$ – instantaneous voltages across windings

$i_{ab}, i_{bc}, i_{ca}$- instantaneous winding currents

$i_a$, $i_b$, $i_c$- instantaneous line currents

$\varphi_{ab}$, $\varphi_{bc}$, $\varphi_{ca}$- instantaneous fluxes linking each winding

$n_1, r_1$- number of turns/phase, resistance/phase of winding

$r_s$ – series resistance in each line; $\delta$- switching angle

The first peak of inrush current depends upon the (i) instant of switching ($\delta$) or the voltage magnitude at the instant of energization, (ii) winding resistance and external resistance connected in series with the winding or with the line, and (iii) remnant flux in the core. As the no-load impedance angle of the transformer is almost 90°, controlled switching at $\delta = 90°$ causes maximum reduction in inrush current. Balanced or unbalanced external resistance may only be inserted in the lines of a delta connected transformer. Though high value of resistance at the time of switching may help to reduce the initial peak and quick damping of dc offset current and thereby, core saturation, it causes extra loss and voltage drop during normal operation. Therefore, series resistance insertion

---

K. P. Basu (kartik.basu@mmu.edu.my), Ali Asghar (ali-asghar@mmu.edu.my) and Stella Morris (stella.morris@mmu.edu.my) are with the Faculty of Engineering, Multimedia University, 63100 Cyberjaya, Malaysia.

0-7803-9771-1/06/$25.00 ©2006 IEEE

is generally avoided for the reduction of inrush current.

A bank of 3 single phase transformers, each having a rating of 4 kVA, 230 V/115 V, 50 Hz, connected in delta-star is used for the simulation study. $n_1 = 135$ turns and $r_1 = 0.37\ \Omega$. No external series resistance $r_s$ is inserted. Simulation models [11-12] are generally used to obtain magnetic and electric circuit parameters of 3-phase, 3-limb or other type of transformers. However, magnetization curve with deep saturation data of a stalloy core [7] is used for the simulation. A MATLAB program was prepared for the simulation study. Simulation results are shown in figures 2-5.

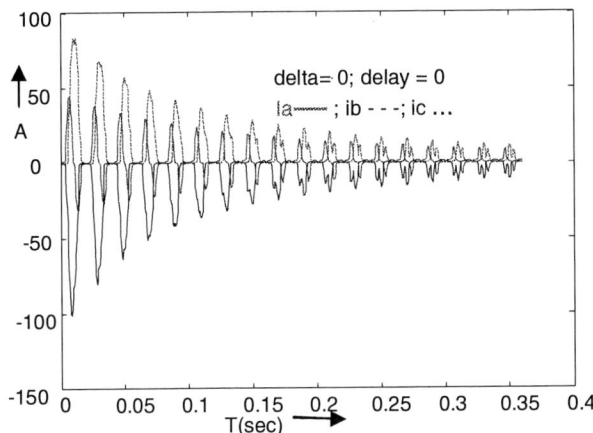

Fig. 2. Line currents – simultaneous switching of all phases; $\delta=0$.

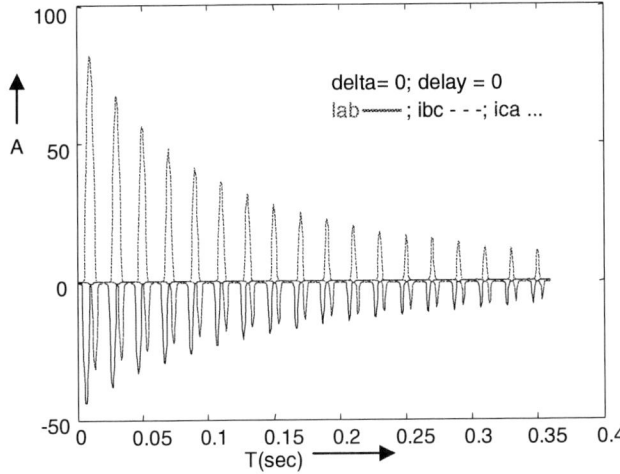

Fig. 3. Winding currents – simultaneous switching of all phases; $\delta=0$.

*Sequential Switching* - Due to backlash and wear and tear of Circuit Breaker contacts, all the contacts of the CB may not close simultaneously. Delayed and controlled closing of the CB poles, which may be termed as sequential switching, may also be used to reduce the inrush current. It is well known that modern high voltage circuit breakers have a built-in mechanical delay among three phases or independent operating mechanism for each phase. So delayed switching of phases is not difficult but controlled switching at any angle $\delta$ needs the use of point-on-wave switching device.

Closing of any 2 CB poles (say a, b) applies line voltage $E_{ab}$ across the winding ab and across windings bc and ca in series. Due to symmetrical construction of the transformer, voltages across bc and ca are half of $E_{ab}$ until the third pole of CB is closed. $E_{ab}/2$ produce winding currents $i_{bco}$ and $i_{cao}$, which can not saturate the cores bc and ca. Controlled switching of ab winding with $E_{ab}$ at $\delta = \pi/2$ produces no transient inrush current $i_{ab}$ in winding ab.

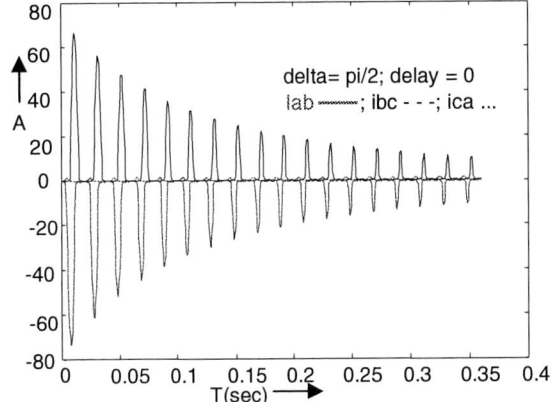

Fig. 4. Winding currents–simultaneous switching of phases; $\delta=\pi/2$.

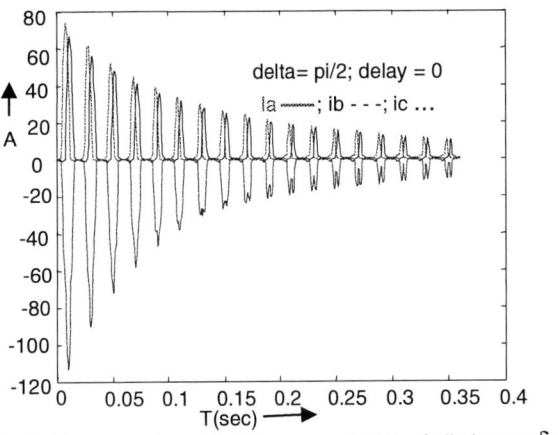

Fig. 5. Line currents – simultaneous switching of all phases; $\delta=\pi/2$.

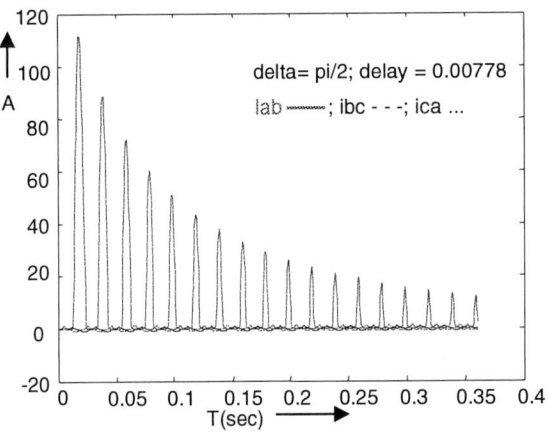

Fig. 6. Winding currents – $\delta=\pi/2$; $t_d = 0.00778$ sec.

So, $i_{ab} \approx E_{max} \sin(\omega t)/X_m$; $i_{bco} = i_{cao} \approx -E_{max} \sin(\omega t)/2X_m$ —(4)
$X_m$ – magnetizing reactance of each winding.
If the 3$^{rd}$ pole c of the CB is closed after a time delay $t_d$ the steady state magnetizing currents in bc and ca become;
$i_{bc} \approx E_{max} \sin[\omega(t-t_d) - 7\pi/6]/X_m$ ---(5) for $t \ge t_d$
$i_{ca} \approx E_{max} \sin[\omega(t-t_d) + \pi/6]/X_m$ ---(6) for $t \ge t_d$

For no transient inrush currents in bc and ca at the instant of switching of the pole c, $t_d$ should be such that $i_{bco} = i_{bc} = i_{ca}$, which can not be achieved simultaneously.

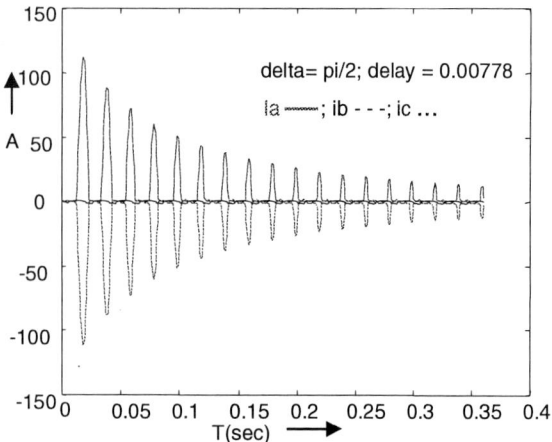

Fig. 7. Line currents – $\delta$=pi/2; $t_d$ = 0.00778 sec.

For $t_d \approx 0.00778$ sec inrush current in winding bc becomes negligible but the inrush in winding ca becomes very high, producing high values of $i_a$ and $i_c$ with negligible $i_b$. Similarly, for $t_d \approx 0.0225$ sec inrush current in winding ca becomes negligible but the inrush in winding bc becomes very high, producing high values of $i_b$ and $i_c$ with negligible $i_a$ (refer figures 6-9).

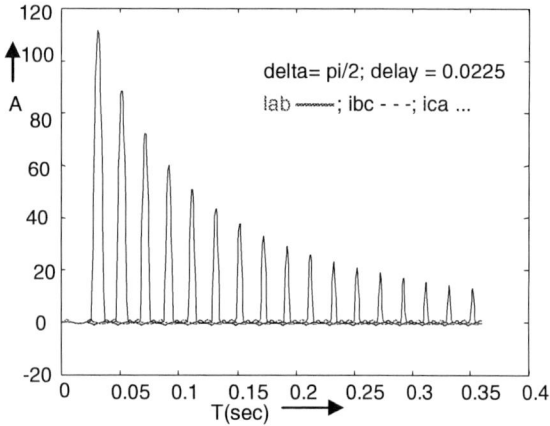

Fig. 8. Winding currents – $\delta$=pi/2; $t_d$ = 0.0225 sec.

MATLAB simulation results of a 50MVA, 33kV/132kV, delta-star power transformer also show the nature of inrush current similar to those depicted in figures 2-9.

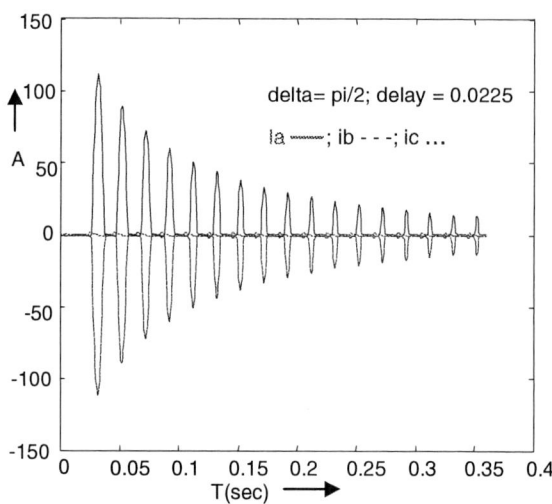

Fig. 9. Line currents – $\delta$=pi/2; $t_d$ = 0.0225 sec.

## III. CONCLUSION

Simulation results of the production of magnetizing inrush currents due to delta-side energization of a delta-star transformer have been presented. Delayed and controlled switching of the third pole of the CB may produce negligible inrush current in one line only. Currents in the other two lines become very high.

## IV. REFERENCES

[1]  T. Specht, " Transformer inrush and rectifier transient currents," *IEEE Trans. Power App. Syst.*, vol. 88, no. 4, pp. 269-276, Apr. 1969
[2]  K. Smith, L. Ran, and B. Leyman, "Analysis of transformer inrush transients in offshore electrical systems," *Proc. Inst. Elect. Eng. Gen. Trans. Distrib.*, vol. 146, no. 1, pp. 89-95, Jun. 1999.
[3]  CIGRE working group task force 13.07, Controlled switching of HVAC circuit breakers, in 1st Part Elektra, no. 183, pp. 43-73, Apr. 1999.
[4]  Westinghouse, *Electrical Transmission and distribution Reference Book.* Chicago: R. R. Donnelley & Sons co., 1944, pp. 411-417.
[5]  B. Holmgrem, R. S. Jenkins, and J. Rinbrugent, " Transformer inrush current", in *CIGRE Proc. 22nd session*, vol. 1, 1968, pp. 1-13.
[6]  R. Yacamini and A. Abu-Nasser, "The calculation of inrush current in three-phase transformer," *IEE. Proc. Elect. Power Appl.*, vol. 133, no.1, pp.31-40, Jan. 1986.
[7]  Mohibullah and Basu K. P., "Computerised Evaluation of the Magnetising Inrush Current in Transformers," *Electric Power System Research*, vol.2, pp. 179-182, 1979.
[8]  M. A. Rahman and A. Gangopadhayay, "Digital Simulation of magnetizing inrush currents in three-phase transformers," *IEEE Trans. Power Del.*, vol. PWRD-1, no. 4, pp. 235-242, Oct. 1986
[9]  Yu Cui, S. G. Abdulsalam, S. Chen and Wilsun Xu, " A Sequential Phase Energization Technique for Transformer Inrush Current Reduction-Part I: Simulation and Experimental Results," *IEEE Trans. On Power Delivery*, Vol. 20, No.2, pp. 943-949, Apr. 2005.
[10] Wilson Xu, S. G. Abdulsalam, Yu Cui, and X. Liu, " A Sequential Phase Energization Technique for Transformer Inrush Current Reduction-Part II: Theoretical Analysis and Design Guide," *IEEE Trans. On Power Delivery*, Vol. 20, No.2, pp. 950-957, Apr. 2005.
[11] M. Elleuch and M. Polujadoff, "A contribution to the modeling of three phase transformers using reluctance," *IEEE Trans. Magn.*, vol. 32, no. 4, pp. 1199-1204, Oct. 2000.
[12] S. G. Abdulsalam, Wilson Xu, W. L. A. Neves, and X. Liu, " Esyimation of Transformer saturation characteristics from inrush currents waveforms," *IEEE Trans. On Power Delivery*, Vol. 21, No.1, pp. 170-177, Jan. 2006.

## V. BIOGRAPHIES

**Dr. K. P. Basu** (M'99 and SM'02) received his B.E.E., M.E.E. and Ph.D.(Eng.) degrees from the Jadavpur University, Kolkata, India in 1961, 1967 and 1974 respectively. His research interest is power system operation, control, protection, and electric drives. He has authored and co-authored more than 70 journal and conference papers.

**Mr. Ali Asghar** received his B.Sc. Degree in Electrical Engineering from the Azad University in Tehran, Iran in 1993. After many years of working and research experience, he continued his studying and received his MSc in Electrical Engineering from University Putra in Malaysia in 2001. His research interest is Power system protection, Electrical Machines and Electric drives, and Power Electronic. He has authored and co-authored some journal and conference papers.

**Dr. Stella Morris** received her B.E degree from Madurai Kamaraj University, India, her M.E degree from Anna University, India and her Ph.D degree from Multimedia University, Malaysia. Her research area includes power system reliability, power system dynamics and control, soft computing and FACTS devices. She has published and presented over 15 papers in refereed journals and conferences.

**2006 IEEE International Conference on Power Electronic, Drives and Energy Systems**

# Accurate Performance Prediction of Three-Phase Induction Motor by FEM Using Separate Saturation Curves for Teeth and Yoke

V. Jaiswal, M. Fazil, A. Hangal, and N. Ravi

*Abstract*--**In this paper, the performance analyses of the three phase induction motors have been done by Finite Element Method (FEM) using time harmonic and transient with motion solvers. If there is no saturation in the ferrous portion of the mutual-flux path, the flux wave in the motor approaches a sine-wave shape, but some amount of saturation always exists in a typical three phase induction motor that has the effect of reducing the peak value of the flux wave. To study this phenomena, initially, 60 Hz saturation curves were used for the teeth as well as core yoke and performance analyses of the 2 motors (5hp 4P, and 15hp 4P) were done by FEM and then in turn, separate AC saturation curves for the teeth and the core yoke derived from the normal DC saturation curve were used, and FEM analyses were done to calculate all the electrical parameters of the motors. In both the cases, computational results are compared with the tested results, and it has been found that the percentage error between computational and tested results is less in case of separate saturation curves for teeth and core yoke compared to the other case.**

*Index Terms*--**Core yoke, FEM analysis, Induction motor, saturation curve, teeth.**

## I. NOMENCLATURE

N1 = number of stator slots
p = number of poles
B = flux density
H = magnetic field strength

## II. INTRODUCTION

INDUCTION motor performance can be predicted by FEM analysis. Theoretical models that compute the electrical parameters of the induction motor are useful for two main purposes: (i) to better understand the physical processes during the motor operation, and (ii) to predict the electrical performance of the induction motor over a wide range of load conditions, if the model can be shown to be well correlated with tested results. Such a model could help to reduce the number of laboratory experiments needed for the design optimizations of motors.

The general solution for the low frequency electromagnetic fields in motors can be obtained by solving Maxwell's equations. To apply the finite element method, a functional based on the system energy need to be defined that can be applied for the stated problem. Minimization of the system energy leads to a system of algebraic equations, whose solution, under the application of corresponding boundary conditions (Dirichlet, Neuman or mixed), supplies the required nodal variables.

## III. PERFORMANCE ANALYSIS

In this work, electromagnetic analyses of the two three phase induction motors were done by a commercially available program based on FEM. The motor model was assumed to be two dimensional; the resistance of end ring was neglected, but all other three dimensional components- overhang winding resistance and reactance, the end ring leakage reactance were taken into consideration by including them into the circuit [1]. In time harmonic analysis, the slip in motor was simulated by varying the conductivity of rotor bar material, so the results of simulations greatly depend upon the accuracy with which conductivity is assigned to the bar materials. The polynomial order and Conjugate Gradient tolerance were set to be 2 and 0.001%, respectively.

Due to the presence of slots, the wave shape of the magnetomotive-force wave is reasonably close to that of sine wave, and this similarity increases as the ratio of N1/p increases. If there is no saturation in the ferrous portion of the mutual-flux path, the flux wave in the motor approaches a sine-wave shape, but some amount of saturation (around 10% to 20% of the total magnetizing ampere-turns) always exists in a typical three phase induction motor. This saturation has the effect of reducing the peak value of the flux wave.

To incorporate the saturation phenomena in a three phase induction motor, AC saturation curves have been derived from the normal DC saturation curve. For tooth magnetizing curve, a value maximum flux density B is selected and this value is then multiplied by 0.8. This value corresponds to the value of the actual flux wave and its fundamental component. Then the value of the magnetic field strength H which corresponds to (0.80)(B) is read from the normal dc saturation curve for the grade of steel being considered. This value of H is divided by 0.80 to give the average value of H that corresponds to the selected value B. This value of H is then multiplied by $\sqrt{2}$ to

---

Authors are with the Crompton Greaves Ltd., Corporate R & D and Quality, Kanjur Marg (East), Mumbai, India, Phone: 022-6755 8825, Fax: 022-6755 8867, E-mail: vinay.jaiswal@cgl.co.in

0-7803-9771-1/06/$25.00 ©2006 IEEE

give the equivalent maximum value of ampere-turns/meter for use with ac current. This effective value of H is then plotted against the selected value of B to give one point on the ac saturation curve. This process is repeated until the curve is completed.

To determine the core-yoke saturation curve, half of a sine wave (i.e., a wave from 0 to 90 degrees) is divided into increments of equal angle, such as 15, 30, 45, 60, 75 and 90 degrees, and the corresponding values of sin $\theta$ calculated. The tooth magnetizing curve is then used to construct the core magnetizing curve. A given value of flux density B is selected, which corresponds to the maximum value of flux density at the inter-polar point. Using the choice of increments above, the six values of incremental flux densities are B sin(15), B sin(30), B sin(45), B sin(60), B sin(75), and B sin(90). The corresponding values of H for each of the increments are then read from the tooth magnetizing curve. The six values o H are then averaged to give the value of H that corresponds to the selected value of B. Corresponding pairs of B and H, so determined, are then plotted, to give the core magnetizing curve, as described in [2].

Fig. 1 shows AC saturation curves for the teeth and the core yoke derived from the normal DC saturation curve.

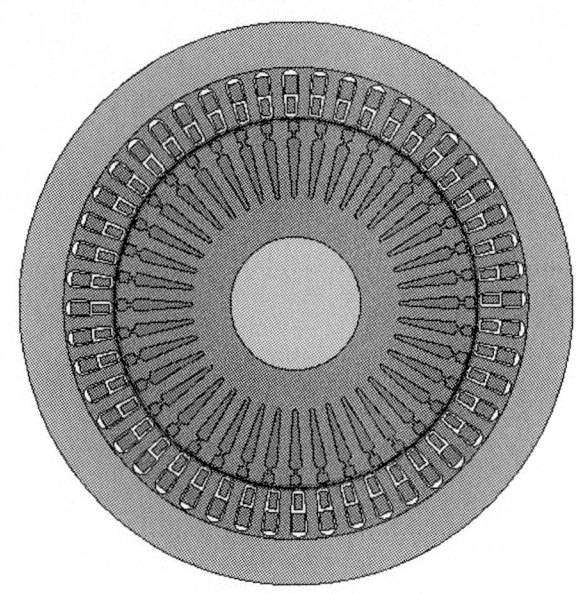

Fig. 2. Separate components for the core yoke and teeth in a 15hp, 4P, three phase induction motor.

Fig. 3 shows the 60Hz B-H curve for the material EBG450 used in the simulations.

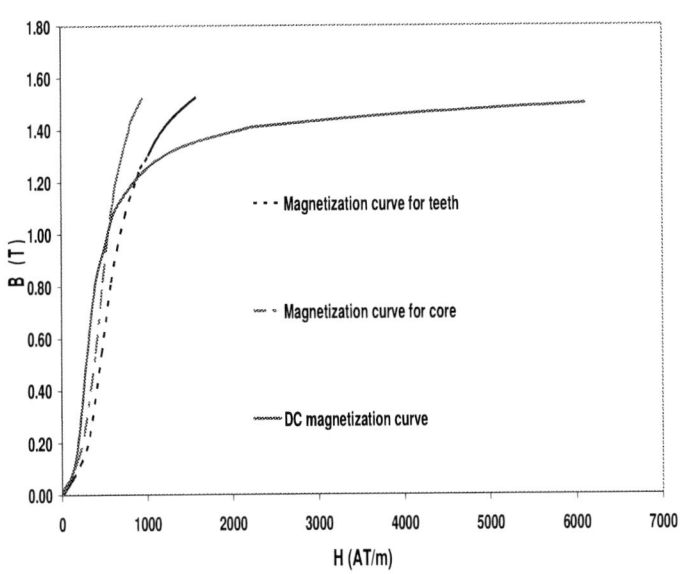

Fig. 1. AC magnetization curves for the teeth and core yoke and the DC magnetization curve.

Two components were created in the stator core, first one was for the core yoke and material property was set to be as core yoke saturation cure, and for the second one material property was set to be as teeth saturation curve [3] as shown in Fig. 2.

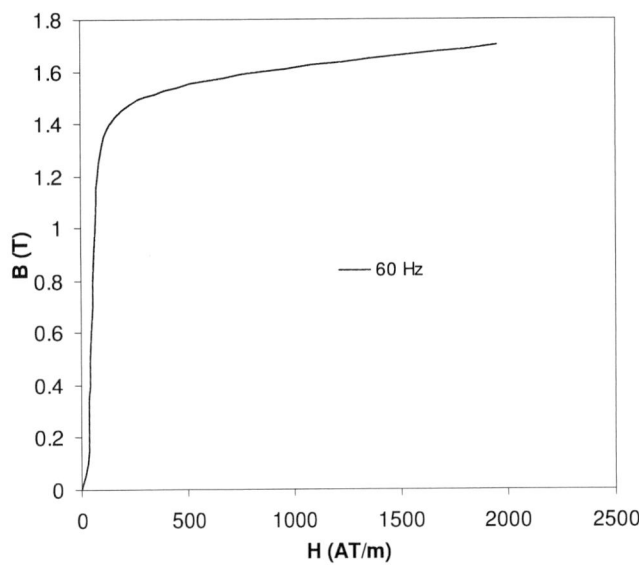

Fig. 3. B-H curve for EBG450 at 60 Hz.

51

## IV. RESULTS AND DISCUSSION

FEM analyses have been done for the two three phase, induction motors. Specifications of the analyzed motors are given in Table I and Table II.

TABLE I

SPECIFICATIONS OF THE FIRST MOTOR ANALYZEZ

| Rated Power | 5 hp |
|---|---|
| Voltage | 230 Volts, 60Hz |
| Phase, Pole | 3Phase, 4pole |
| Connection | STAR |
| Number of stator and rotor slots | 36, 28 |
| Stamping material | EBG450 |
| Full load torque | 20.34 N-M |
| Rated Current | 12.8 A |
| Rated speed | 1740 rpm |

TABLE II

SPECIFICATIONS OF THE SECOND MOTOR ANALYZEZ

| Rated Power | 15 hp |
|---|---|
| Voltage | 460 Volts, 60Hz |
| Phase, Pole | 3Phase, 4pole |
| Connection | DELTA |
| Number of stator and rotor slots | 48, 40 |
| Stamping material | EBG450 |
| Full load torque | 61.28 N-M |
| Rated Current | 18 A |
| Rated speed | 1752 rpm |

Fig. 4 shows the flux distribution in the mentioned 15 hp motor.

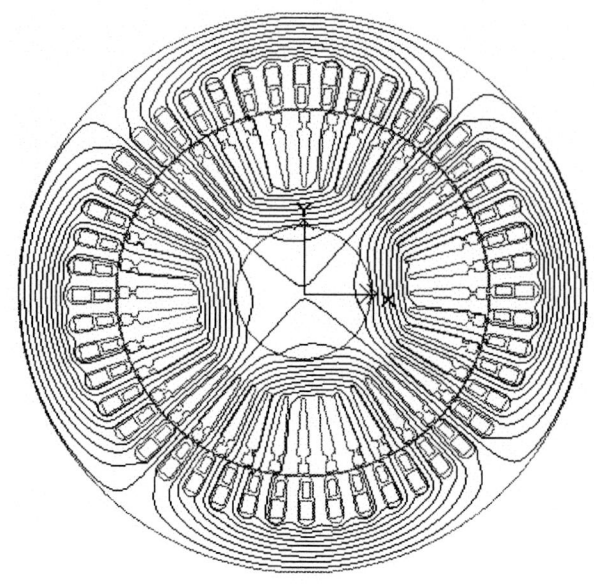

Fig. 4. The flux distribution in a 15 hp motor.

Table III shows the computational and tested line currents (A) for 5hp and 15hp 4P three phase induction motors at no-load. It can be seen from the table that the percentage error between the computational and tested results reduces considerably if separate saturation curves for teeth and core yoke are used instead of one ac saturation curve for both the teeth and core yoke.

TABLE III

TESTED AND COMPUTATIONAL LINE CURRENTS (A) AT NO-LOAD

| SN | Rati-ng | Tested result | 60Hz saturation curve | | Separate saturation curves for teeth and core yoke | |
|---|---|---|---|---|---|---|
| | | | Comput-ational | % error w.r.t. tested result | Compu-tational | % error w.r.t. tested result |
| 1 | 5hp | 5.56A | 4.55 A | 18.17 % | 5.53 A | 0.54 % |
| 2 | 15hp | 6.36 A | 5.98 A | 5.97 % | 6.60 A | 3.77 % |

Table IV shows the computational and tested line currents (A) for 5hp and 15hp 4P three phase induction motors at full-load. It is observed from the table that the percentage error between the computational and tested results reduces considerably in this case also if separate saturation curves for teeth and core yoke are used instead of one ac saturation curve for both the teeth and core yoke.

TABLE IV

TESTED AND COMPUTATIONAL LINE CURRENTS (A) AT FULL-LOAD

| SN | Rating | Tested result | 60Hz saturation curve | | Separate saturation curves for teeth and core yoke | |
|---|---|---|---|---|---|---|
| | | | Compu-tational | % error w.r.t. tested result | Computa-tional | % error w.r.t. tested result |
| 1 | 5hp | 12.83A | 12.07 A | 5.95 % | 12.19 A | 5.01 % |
| 2 | 15hp | 18.81A | 17.77 A | 5.53 % | 18.08 A | 3.88 % |

## V. CONCLUSION

The percentage error between the computational and tested results reduces, when separate saturation curves for teeth and core yoke derived from the normal dc saturation curve are used, instead of one ac saturation curve for both the teeth and core yoke.

## VI. ACKNOWLEDGMENT

The authors would like to thank the authorities of Crompton Greaves Limited for their permission to publish this work.

## VII. REFERENCES

[1] K. Yamazaki, "induction motor analysis considering both harmonics and end effects using combination of 2D and 3D finite element method," IEEE Trans. Energy Conversion, vol. 14, Sept. 1999, pp. 698-703.

[2] Paul L. Cochran, "Polyphase Induction Motors," Marcel Dekker, 1989, ch. 14.

[3] E. Vassent, G. Meunier and J. C. Sabonnadiere, "Simulation of Induction machine Operation using Complex Magnetodynamic Finite Elements," IEEE Tran. on Magnetics, vol. 25, no.4, July 1989, pp. 3064-3066.

**2006 IEEE International Conference on Power Electronic, Drives and Energy Systems**

# Nonlinear Sliding-Mode Controller for Sensorless Speed control of DC servo Motor Using Adaptive Backstepping Observer

A. Farrokh Payam, and B. Mirzaeian Dehkordi

*Abstract–* In this paper, the robust nonlinear sliding-mode speed sensorless speed controller for a dc servo motor is proposed. Based on state-space model representing the speed and current dynamics, the nonlinear sliding mode control is designed to track a linear reference model. An observer based on adaptive backstepping approach is used to estimate speed and uncertain parameters. With the proposed speed controller, the controlled dc servo motor possesses the advantages of good transient performance and robustness for parameters deviations, and transient dynamics of dc servo motor can be regulated through the design of a linear reference model which has the desired dynamic behaviors for the drive system. Finally, simulation results are demonstrated to validate the proposed controller.

*Index Terms–* Sliding-Mode Control, Dc Servo Motor, Sensorless Speed Control, Adaptive Backstepping.

## I. INTRODUCTION

NONLINEAR sliding mode control has been extensively studied in recent years. Many publications consider and study the theory of nonlinear sliding-mode control [1]-[6]. Since the sliding mode control has many advantages such as invariance condition, insensivity and robustness and fast dynamic response over the controllers therefore, the application of sliding mode control to electrical drives are recently dedicated much attenuation [7], [8]. DC drives are widely used in applications requiring adjustable speed regulation and frequent starting, braking and reversing. Some important applications are rolling mills, paper mills, mine winders, machine tools, etc.  In recent years nonlinear control has many interested to application in electrical drives. Many nonlinear control methods are proposed for dc drives such as [9]-[14]. In this paper by using a nonlinear sliding-mode controller combined with adaptive backstepping observer design a sensorless speed controller for DC servo motor.

Using the model of DC motor [12]-[14], in order to follow a reference model and robust speed tracking a sliding-mode controller is presented. In addition for position/speed sensorless operation a nonlinear observer structure based on the backstepping principle is presented. A full proof for

A. Farrokh Payam is with Isfahan University of Technology, Isfahan, Iran (e-mail: amir_farrokh@yahoo.com ).

B. Mirzaeian Dehkordi is with University of Isfahan, Isfahan, Iran (e-mail: Mirzaeian@eng.ui.ac.ir ).

convergence of states is presented together with a stability proof for the parameter estimates and controller. The combination of sliding-mode controller and adaptive backstepping observer makes the DC servo motor robust and controller stable against to the parameters uncertainties. Computer simulation results obtained, confirm the effectiveness and validity of the proposed control approach.

## II. DC SERVO MOTOR MODEL

The model of a dc servo motor [12]-[14], is given by:

$$
\begin{bmatrix} \dot{\omega}_r \\ \dot{i}_a \end{bmatrix} = \begin{bmatrix} -\dfrac{F_d}{J} & \dfrac{K_t}{J} \\ -\dfrac{K_t}{L_a} & -\dfrac{R_a}{L_a} \end{bmatrix} \begin{bmatrix} \omega_r \\ i_a \end{bmatrix} + \begin{bmatrix} -\dfrac{T_L}{J} \\ 0 \end{bmatrix} + \begin{bmatrix} 0 \\ \dfrac{1}{L_a} \end{bmatrix} V_a \tag{1}
$$

Where $\omega_r$, $i_a$ and $V_a$ are rotor speed, armature current and armature voltage respectively. $T_L$ is load torque, $R_a$ and $L_a$ are armature resistance and inductance, $K_t$ is torque constant which is equal to Back-EMF constant, and $F_d$, $J$ are dynamic friction constant and inertia respectively.

## III. NONLINEAR SLIDING-MODE CONTROL

Sliding mode theory is one of the prospective control methodologies for a DC servo drive speed controls because of its disturbance rejection, strong robustness and particularly simplicity of implementation. In this section a model-following nonlinear sliding-mode control for DC servo motors is proposed. The most important advantage of this model is a good transient performance and robustness to parametric deviations. Let ,

$$
\begin{aligned} z_1 &= \omega_r \\ z_2 &= i_a \end{aligned} \tag{2}
$$

and

$$
\begin{aligned}
f_1(z_1) &= -\frac{F_d}{J}\omega_r - \frac{T_L}{J} \\
f_2(z_1, z_2) &= -\frac{K_t}{L_a}\omega_r - \frac{R_a}{L_a}i_a \\
a_n &= \frac{K_t}{J} \\
\hat{u} &= \frac{V_a}{L_a}
\end{aligned} \tag{3}
$$

0-7803-9771-1/06/$25.00 ©2006 IEEE            54

The dynamical model of system (1) become as follows:

$$\dot{z}_1 = f_1(z_1) + a_n z_2$$
$$\dot{z}_2 = f_2(z_1, z_2) + \hat{u} \qquad (4)$$

The reference model is designed in a linear form as

$$\begin{bmatrix} \dot{z}_{m1} \\ \dot{z}_{m2} \end{bmatrix} = \begin{bmatrix} a_{m1} & a_{m2} \\ -a_{m3} & -a_{m4} \end{bmatrix} \begin{bmatrix} \omega_r \\ i_a \end{bmatrix} + \begin{bmatrix} a_{m5} & 0 \\ 0 & a_{m3} \end{bmatrix} \begin{bmatrix} T_L \\ \omega_{ef} \end{bmatrix} \qquad (5)$$

where $a_{m1}$, $a_{m3}, a_{m4}$ are the designed positive constants and $a_{m2} = a_n, a_{m5} = -\dfrac{1}{J}$.

Furthermore define the tracking errors between plant and reference model as:

$$e_z = [z_1 - z_{m1} \quad z_2 - z_{m2}]^T = [e_{z1} \quad e_{z2}]^T \qquad (6)$$

and its error dynamics are derived as follows

$$\dot{e}_z = A(x) + B(x)\bar{u} \qquad (7)$$

Where

$$A(x) = \begin{bmatrix} f_1(z_1) + a_n e_{z2} \\ f_2(z_1, z_2) \end{bmatrix} \quad B(x) = \begin{bmatrix} 0 \\ 1 \end{bmatrix} \qquad (8)$$

and

$$\bar{u} = \hat{u} + a_{m3} z_{m1} + a_{m4} z_{m2} - a_{m3} \omega_{ref} \qquad (9)$$

According to the system (7), the sliding surface is designed as:

$$\sigma(e_z) = S e_z(x) \qquad (10)$$

Where $S \in \Re^{1 \times 2}$ is a constant linear matrix and inverse of $SB(x)$ must exist follows i.e. $\det(SB(x)) \neq 0$ for all $x$. Here the dynamics of the sliding surface as a specified form is designed as:

$$\dot{\sigma} = S\dot{e}_z = SA(x) + SB\bar{u} = -Q\,\text{sgn}(\sigma) - K\sigma \qquad (11)$$

where $Q > 0$ and $K > 0$.

And sgn(.) is the sign function defined as

$$\text{sgn}(\sigma) = \begin{cases} 1 & as \ \sigma > 0 \\ -1 & as \ \sigma < 0 \end{cases} \qquad (12)$$

Then, the controller in terms of $\bar{u}$ can be described as

$$\bar{u} = -(SB)^{-1}[SA(x) + Q\,\text{sgn}(\sigma) + K\sigma] \qquad (13)$$

Using the proposed controller (13), the reachibility of sliding-mode control system (7) is guaranteed.

**Theorem**: With the developed nonlinear sliding surface (10), the reaching condition $\sigma^T \dot{\sigma} < 0$ is satisfied, and the controlled system (7) will be stabilized.

**Proof**: From the designed dynamic sliding surface (11), the following equation can be derived:

$$\dot{\sigma} = -Q\,\text{sgn}(\sigma) - K\sigma \qquad (14)$$

Multiplying $\sigma$ on the both side of the above equation yields

$$\sigma\dot{\sigma} = -Q\sigma\,\text{sgn}(\sigma) - K\sigma^2$$
$$= -Q|\sigma| - K\sigma^2 < 0 \qquad (15)$$

From the above analysis, it is evident that reaching condition is guaranteed.

## IV. ADAPTIVE BACKSTEPPING OBSERVER

For the adaptive backstepping design of the observer the notation is simplified by the definitions:

$$x_1 = \omega_r, x_2 = i_a, y = i_a$$
$$a_1 = \frac{F_d}{J}, a_2 = \frac{K_t}{J}, a_3 = \frac{1}{J}$$
$$a_4 = \frac{K_t}{L_a}, a_5 = \text{Ra/La}, \alpha_6 = 1/\text{La} \qquad (16)$$

giving the motor equations

$$\begin{bmatrix} \dot{x}_1 \\ \dot{x}_2 \end{bmatrix} = \begin{bmatrix} -a_1 & a_2 \\ -a_4 & -a_5 \end{bmatrix} \begin{bmatrix} x_1 \\ x_2 \end{bmatrix} + \begin{bmatrix} -a_3 T_L \\ 0 \end{bmatrix} + \begin{bmatrix} 0 \\ a_6 \end{bmatrix} V_a$$
$$y = i_a \qquad (17)$$

Because only the output $y$ is measurable an observer has to be designed. The prediction model for the backstepping observer design is chosen to be

$$\begin{bmatrix} \dot{\hat{x}}_1 \\ \dot{\hat{x}}_2 \end{bmatrix} = \begin{bmatrix} -\hat{a}_1 & \hat{a}_2 \\ -\hat{a}_4 & -\hat{a}_5 \end{bmatrix} \begin{bmatrix} \hat{x}_1 \\ y \end{bmatrix} + \begin{bmatrix} -\hat{a}_3 T_L \\ 0 \end{bmatrix} + \begin{bmatrix} 0 \\ \hat{a}_6 \end{bmatrix} V_a + \begin{bmatrix} 0 \\ v \end{bmatrix}$$
$$\hat{y} = \hat{i}_a \qquad (18)$$

Where $v$ is the control input. This input is to be designed by the backstepping method.

The dynamical equations for the prediction errors

$$\tilde{x}_1 = \hat{x}_1 - x_1 \ \tilde{x}_2 = \hat{x}_2 - x_2 \ \tilde{a}_1 = \hat{a}_1 - a_1 \ \tilde{a}_2 = \hat{a}_2 - a_2$$
$$\tilde{a}_3 = \hat{a}_3 - a_3 \ \tilde{a}_4 = \hat{a}_4 - a_4 \ \tilde{a}_5 = \hat{a}_5 - a_5 \ \tilde{a}_6 = \hat{a}_6 - a_6$$

are then given by

$$\begin{bmatrix} \dot{\tilde{x}}_1 \\ \dot{\tilde{x}}_2 \end{bmatrix} = \begin{bmatrix} -\tilde{a}_1 & \tilde{a}_2 \\ -\tilde{a}_4 & -\tilde{a}_5 \end{bmatrix} \begin{bmatrix} x_1 \\ y \end{bmatrix} + \begin{bmatrix} -\hat{a}_1 \\ -\hat{a}_4 \end{bmatrix} \tilde{x}_{1+}$$
$$\begin{bmatrix} -\hat{a}_3 T_L \\ 0 \end{bmatrix} + \begin{bmatrix} 0 \\ \hat{a}_6 \end{bmatrix} V_a + \begin{bmatrix} 0 \\ v \end{bmatrix}$$
$$\tilde{y} = \tilde{x}_2 = \hat{y} - y \qquad (19)$$

The first step in the backstepping strategy is to design a stable controller for the integral of the prediction errors $\tilde{y}$ using $\tilde{x}_2$ as a virtual control variable with stabilizing function $\phi$ (the reference for the virtual variable). The integral of the prediction error $\tilde{x}_3$ is given by

$$\dot{\tilde{x}}_3 = \tilde{x}_2 \qquad (20)$$

Adding and subtracting $\phi$ to the equation

$$\dot{\tilde{x}}_3 = (\tilde{x}_2 - \phi) + \phi = z - c_1 \tilde{x}_3 \qquad (21)$$

gives

$$z = \tilde{x}_2 - \phi$$
$$\phi = -c_1 \tilde{x}_3 \qquad (22)$$

The second step in the backstepping strategy is the control of $z$, by proper choice of the control input $v$. This parameter is defined as follow:

$$z = \tilde{x}_2 + c_1 \tilde{x}_3 \qquad (23)$$

Taking the derivative of $z$ gives

$$\dot{z} = c_1\tilde{x}_2 - \tilde{a}_4 x_1 - \tilde{a}_5 y - \hat{a}_4 \tilde{x}_1 + \tilde{a}_6 V_a + v =$$
$$-\tilde{x}_3 - c_2 z - \tilde{a}_4 x_1 - \tilde{a}_5 y + \tilde{a}_6 V_a \qquad (24)$$

where the second equation is obtained for

$$v = -(c_1 + c_2)\tilde{x}_2 + \hat{a}_4 \tilde{x}_1 - (1 + c_1 c_2)\tilde{x}_3 \qquad (25)$$

In this equation, $c_1$ and $c_2$ are the parameters that stability analysis is the proper method for choice of these parameters.

## V. STABILITY ANALYSIS

The equations obtained in step 1 and step 2 of the backstepping method give the error dynamics for the field observer

$$\dot{\tilde{x}}_3 = z - c_1 \tilde{x}_3$$
$$\dot{z} = -\tilde{x}_3 - c_2 z - \tilde{a}_4 x_1 - \tilde{a}_5 y + \tilde{a}_6 V_a$$
$$\dot{\tilde{x}}_1 = -\tilde{a}_1 x_1 - \hat{a}_1 \tilde{x}_1 + \tilde{a}_2 y - \tilde{a}_3 T_L \qquad (26)$$

This system has equilibrium at $z = \tilde{x}_1 = \tilde{x}_3 = 0$

Furthermore the derivative of the Lyapanouv function candidate as follow:

$$V = \frac{1}{2}\Big\{ \|\tilde{x}_3\|^2 + \|z\|^2 + \|\tilde{x}_1\|^2 + \frac{1}{\gamma_1}\tilde{a}_1^2 + \frac{1}{\gamma_2}\tilde{a}_2^2 + \frac{1}{\gamma_3}\tilde{a}_3^2 +$$
$$\frac{1}{\gamma_4}\tilde{a}_4^2 + \frac{1}{\gamma_5}\tilde{a}_5^2 + \frac{1}{\gamma_6}\tilde{a}_6^2 \Big\} \qquad (27)$$

along the solution of (27) is after some calculations reduced to

$$\dot{V} = -\operatorname{Re} c_1 \|\tilde{x}_3\|^2 - \operatorname{Re} c_2 \|z\|^2 - \hat{a}_1 \|\tilde{x}_1\|^2$$
$$+ \tilde{a}_1 \Big\{ \frac{1}{\gamma_1}\frac{d\tilde{a}_1}{dt} - real(x_1 \tilde{x}_1^*) \Big\}$$
$$+ \tilde{a}_2 \Big\{ \frac{1}{\gamma_2}\frac{d\tilde{a}_2}{dt} + real(y \tilde{x}_1^*) \Big\}$$
$$+ \tilde{a}_3 \Big\{ \frac{1}{\gamma_3}\frac{d\tilde{a}_3}{dt} - real(T_L \tilde{x}_1^*) \Big\}$$
$$+ \tilde{a}_4 \Big\{ \frac{1}{\gamma_4}\frac{d\tilde{a}_4}{dt} - real(x_1 z^*) \Big\}$$
$$+ \tilde{a}_5 \Big\{ \frac{1}{\gamma_5}\frac{d\tilde{a}_5}{dt} - real(y z^*) \Big\}$$
$$+ \tilde{a}_6 \Big\{ \frac{1}{\gamma_6}\frac{d\tilde{a}_6}{dt} + real(V_a z^*) \Big\} \qquad (28)$$

For $\operatorname{Re} c_1 > 0, \operatorname{Re} c_2 > 0, \hat{a}_1 > 0$ and

$$\dot{\tilde{a}}_1 = \gamma_1 real(x_1 \tilde{x}_1^*)$$
$$\dot{\tilde{a}}_2 = -\gamma_2 real(y \tilde{x}_1^*)$$
$$\dot{\tilde{a}}_3 = \gamma_3 real(T_L \tilde{x}_1^*)$$
$$\dot{\tilde{a}}_4 = \gamma_4 real(x_1 z^*)$$
$$\dot{\tilde{a}}_5 = \gamma_5 real(y z^*)$$
$$\dot{\tilde{a}}_6 = -\gamma_6 real(V_a z^*) \qquad (29)$$

## VI. SIMULATION RESULTS

A $C^{++}$ computer program was developed to model this system on P.C.

In this program, a static Runge-Kutta fourth order method is used to solve the system equations. The effectiveness and validity of the proposed approach is tested for a DC servo motor with these parameters:

$R_a = 3.2\Omega$ , $L_a = 8.6 mH$ , $K_t = .0319 Nm/A$

$F_d = .000012 Nm/rad$ and $J = 3 \times 10^{-5} Kgm^2$

In the simulation tests the applied load torque is

$T_L = .01$ $Nm$ and changes to $.02$ $Nm$ at t=4.5 sec.

Test One:

The results of the first test simulation of speed control for $\omega_{ref}$ that shows in Fig. 1, is shown in Figs. 2-3. Note that in this test the value of all parameters of dc servo motor equal to nominal value. The validity and performance of proposed sliding-mode controller is considered from these results.

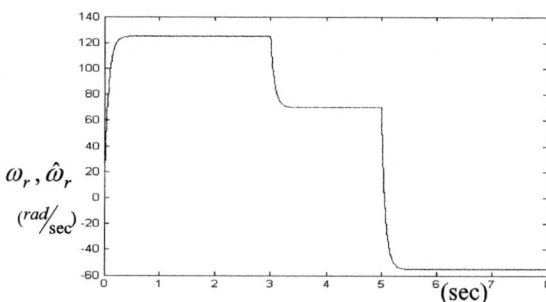

Fig. 1. Estimated and Actual Speed of the dc servo drive.

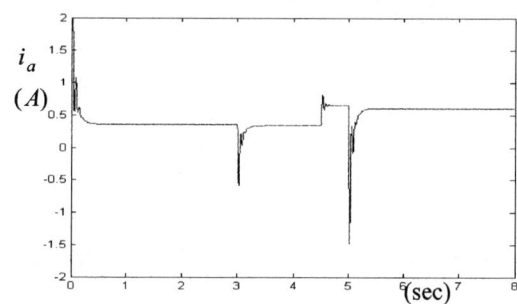

Fig. 2. Armature current of the dc servo drive.

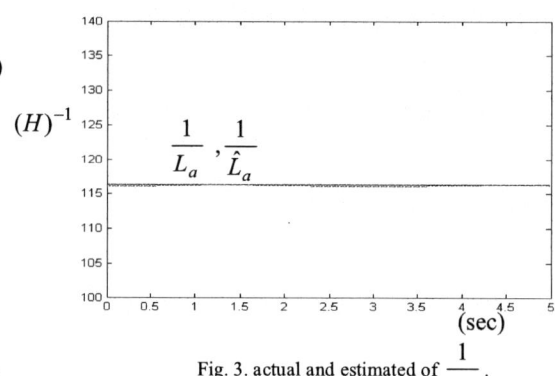

Fig. 3. actual and estimated of $\frac{1}{L_a}$ .

Test Two:

In the second test a non-ideal case is considered. In this case the armature resistance is third of the nominal value and inductance is twice of the nominal value and inertia is twice of the nominal value. The speed reference is shown in Fig. 1 and in Figs. 4-5 simulation results are reported. These results are shown the proposed controller is robust to parameter deviations.

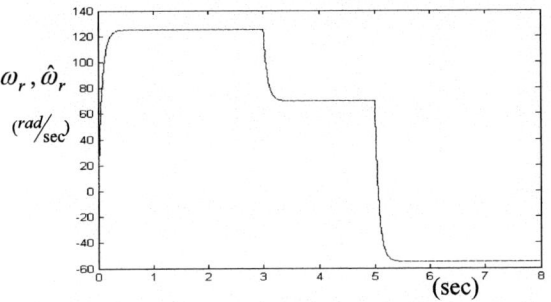

Fig. 4. Estimated and Actual Speed of the dc servo drive.

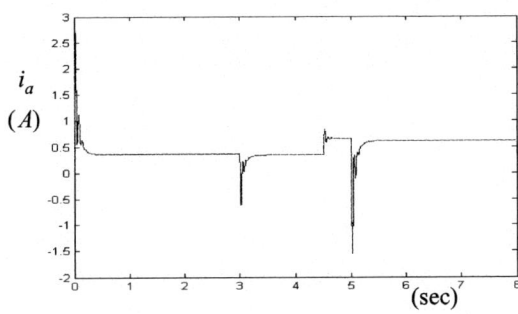

Fig. 5. Armature current of the dc servo drive.

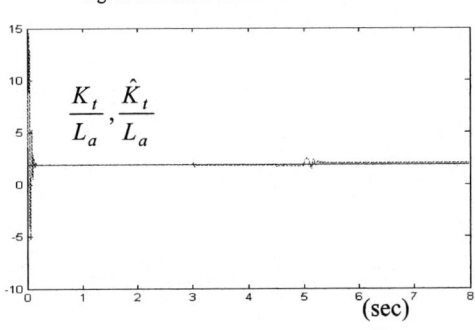

Fig. 6. actual and estimated of $\dfrac{K_t}{L_a}$ .

**Test Three:**

The results of another test simulation of speed control for $\omega_{ref}$ that shows in Fig. 7, is shown in Figs. 8-11. Note that in this test the value of all parameters of DC servo motor equal to nominal value. The validity and performance of proposed sliding-mode controller is considered from these results.

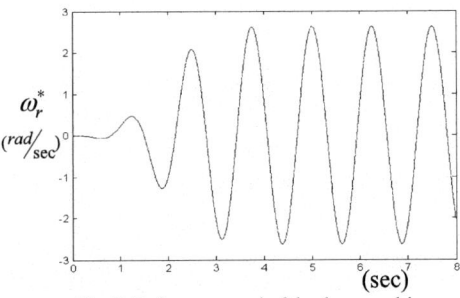

Fig. 7. Reference speed of the dc servo drive.

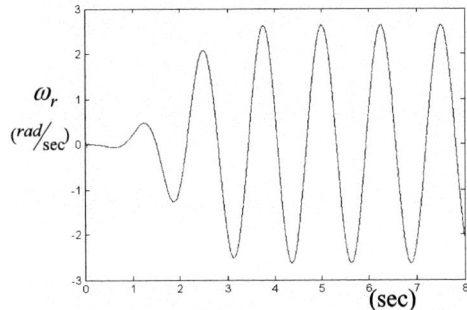

Fig. 8. Speed of the dc servo drive.

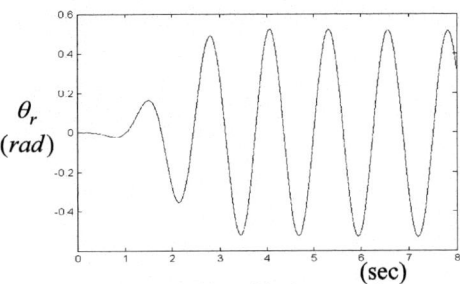

Fig. 9. Position of the dc servo drive.

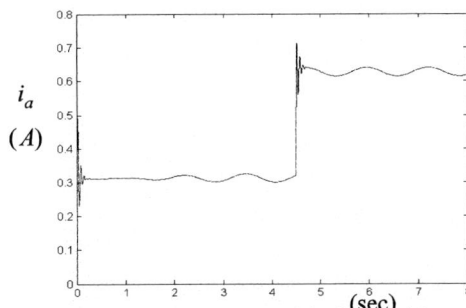

Fig. 10. Armature current of the dc servo drive.

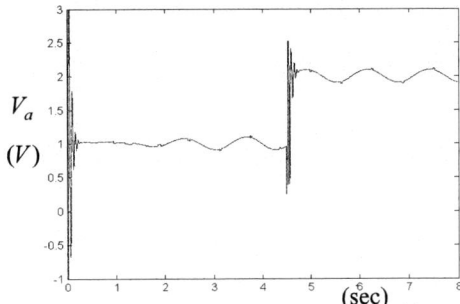

Fig. 11. Armature voltage of the dc servo drive.

**Test Four:**

In fourth test a non-ideal case is considered. In this case the armature resistance is third of the nominal value and inductance is twice of the nominal value and inertia is twice of the nominal value. The speed reference is shown in Fig. 7 and in Figs. 12-14 simulation results are reported. As shown from these results the proposed controller is robust to parameter deviations.

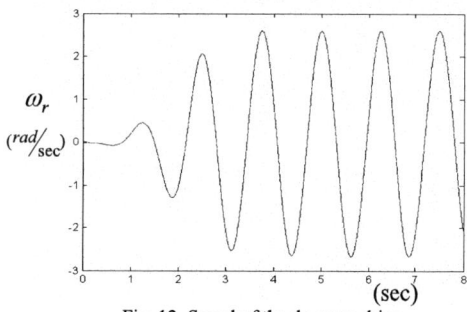

Fig. 12. Speed of the dc servo drive.

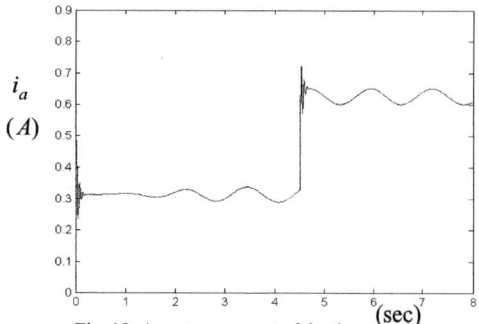

Fig. 13. Armature current of the dc servo drive.

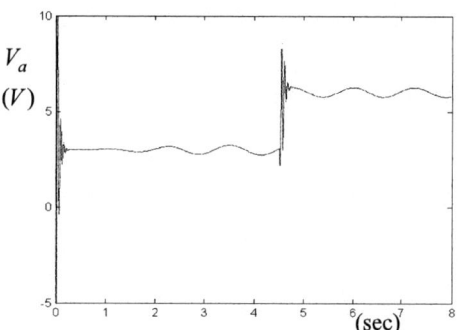

Fig. 14. Armature voltage of the dc servo drive.

## VII. CONCLUSION

In this paper a novel nonlinear sliding-mode speed control of DC servo motor is proposed. The transient dynamics of the speed can be precisely regulated by the design of the linear reference model, since the tracking errors between the state-space model system and reference model converge to zero asymptotically. In addition by using an adaptive backstepping observer speed and electrical uncertainties are estimated, therefore the controller is robust against uncertainties and parameters deviation. Simulation results obtained evident the validity of proposed controller.

## VIII. REFERENCES

[1] M.Vidysagar, *Nonlinear Systems Analysis*, Englewood Cliff's, NJ:Prentice-Hall, 1993.

[2] R.A.Decarlo, S.H.Zak, and G.P.Matthews, "Variable structure control of nonlinear multivariable systems:A tutorial," in *Proc. 1988 IEEE Conf.*, vol.76, pp. 212-231.

[3] W.Gao and J.C.Hung, "Variable structure control of nonlinear systems:A new approach," *IEEE Trans. Ind. Electron.*, vol. 40, pp. 45-55, Feb.1993.

[4] H.S.Ramirez, "Nonlinear variable structure systems in sliding mode:The general case ," *IEEE Trans. Automat. Contr.*, vol. 34, pp. 1186-1188, Nov.1989.

[5] V.I.Utkin, , *Sliding Modes in control and optimization*, Berlin, Germany: Springer-Verlag, 1992.

[6] Christopher Edvards and Sarah K.Spurgeon, *Sliding Mode Control*, Taylor & Francis, 1998.

[7] E.Y.Y.Ho and P.C. SEN, "Control dynamics of speed drive systems using sliding mode controllers with integral compensation," *IEEE Trans. Ind Appllicat.*, vol. 27, pp. 883-892, Sept./Oct. 1991.

[8] H.J. Shieh and K.K.Shyu, "Nonlinear Sliding-Mode torque control with adaptive backstepping approach for induction motor drive," *IEEE Trans, Ind. Electron.*, vol. 46, pp. 380-388, April.1999.

[9] Dubey, G.K, *Power semiconductor control drives*, Prentice-Hall, 1989.

[10] D.M.Dawson, J.Hu, T.C.Burg, *Nonlinear control of electrical machinery*, Marcel-Dekker,Inc

[11] M.R.Bolivar, H.S.Ramirez and A.S.I.Zinober, "Output tracking control via adaptive input-output linearization:A backstepping approach," in *proc. LA-December 1995 Decision & Control conf.*, New Orleans, pp. 1579-1584.

[12] F. G¨urb¨uz and E. Akp_nar, Stability Analysis of a Closed-Loop Control for a Pulse-Width Modulated DC Motor Drive, *Turk J Elec Engin.*, vol. 10, No. 3, pp. 427-438, 2002.

[13] A. Sevnic, "A full adaptive observer for DC servo motors," *Turk J Elec Engin.*, vol. 11, no. 2, pp. 117-130, 2003.

[14] A. Farrokh Payam, B. Mirzaeian, R. Yazdanpanah, "Nonlinear sliding-mode speed control of the DC servo motor," in *Proc. 2006 AIESP Conf.*

## IX. BIOGRAPHIES

**Amir Farrokh Payam** was born in Tehran on June 1979. He received the B.Sc. degree of Electrical Engineering from K.N.Toosi University of Technology, Tehran, Iran in 2003 and the M.Sc. degree of Power Engineering from Isfahan University of Technology in 2006. His special fields of interest are Electrical Drives and Machines, Power Quality, Power Electronics and Applied Nonlinear Control.

**2006 IEEE International Conference on Power Electronic, Drives and Energy Systems**

# Robust Speed Sensorless Control of Doubly-Fed Induction Machine Based on Input-Output Feedback Linearization Control Using a Sliding-Mode Observer

A. Farrokh Payam, and M. Jalalifar

*Abstract*—**In this paper a nonlinear controller is presented for Doubly-Fed Induction Machine (DFIM) drives. The nonlinear controller is designed based on input-output feedback linearization control technique, combined with Sliding-Mode observer. Using the fifth order model of induction machine in fixed stator d ,q axis reference frames with stator current and rotor flux components as state variables. The nonlinear controller can perfectly track the torque and flux reference signals inspite of stator and rotor resistance variations. In order to make the drive system capable of operating in the motoring and generating modes below and above synchronous speed, two level SVM-PWM back-to-back voltage source inverters are employed in the rotor circuit. Computer simulation results obtained, confirm the effectiveness and validity of the proposed control approach.**

*Index Terms*-- **Input-output Feedback Linearization, Sliding-Mode Observer, Rotor dc Link Voltage Nomenclature.**

## I. INTRODUCTION

THE principle theory of a vector controlled DFIM drive has been described in [1]. The structure of the system controller design in [1] is based on neglecting the voltage drop across the stator leakage impedance. Having mode this assumption, a steady-state error is always expected. To solve the above problems, a few researchers have tried to apply the nonlinear control approaches to DFIM drive system [2], [3]. The nonlinear control technique reported so far for DFIM drive are almost parameters dependent or in other words are seriously affected by the machine parameters deviation.

Our purpose in this paper is to introduce a combined adaptive nonlinear controller for DFIM drive system for generating and motoring modes of operation below and above the base speed. According to this method, based on input-output feedback linearization and Sliding-Mode observer, an ideal linear torque controller is first designed for DFIM. The ideal input-output controller is capable of tracking a desired

---

A. Farrokh Payam is with Isfahan University of Technology, Isfahan, Iran (e-mail: farokhpayam@alumni.iut.ac.ir ).

M. Jalalifar is with Isfahan University of Technology, Isfahan, Iran (e-mail: mehran_j1356@yahoo.com).

---

second order reference model. In this control system, the active and reactive power reference signals, (injected to the stator circuit) are perfectly tracked in generating mode of operation. One may note that motoring mode of operation the tracking signals are chosen to be the rotor speed and reactive power reference signal, injected to the stator circuit. The ideal nonlinear controller is designed on the basis of DFIM fifth order model in a stator d &q axis reference frame, with rotor fluxes ($\psi_{dr}, \psi_{qr}$) and stator currents ($i_{ds}, i_{qs}$). Considering a special synchronous reference frame with d axis in coincide with stator voltage space vector, the active and reactive power reference commands are converted into the norm of rotor flux and torque reference signals. However in motoring mode of operation, the torque reference signal is generated by a conventional speed PI controller.

Moreover to preserve to the drive system robustness and also in order to remove the rotor speed sensor, a sliding-mode observer is designed that online detects the rotor resistance and rotor speed. In this observer the estimation laws are derived based on Lyapanouv stability theory and the rotor fluxes ($\psi_{dr}, \psi_{qr}$) are calculated by using the stator and rotor real currents. Moreover, using the two level back-to-back SVM-PWM voltage source inverters in the rotor circuit, the proposed control method can be applied for motoring and generating modes of operation below and above synchronous speed. In addition the rotor dc link voltage is maintain constant also based on input-output linearizing, using a rotating reference frame with d axis is coincide with the space voltage vector of the main ac supply.

## II. DOUBLY-FED INDUCTION MACHINE MODEL

Under assumption of linear magnetic circuits and balanced operating condition, the equivalent two-phase model of the symmetrical DFIM with stator connected to line, represented in fixed stator d-q reference frame is

$$\frac{di_{ds}}{dt} = -(\frac{R_s}{L_\sigma} + \frac{R_r L_m^2}{L_r^2 L_\sigma})i_{ds} + \frac{R_r L_m}{L_r^2 L_\sigma}\psi_{dr} + \frac{\omega_r L_m}{L_r L_\sigma}\psi_{qr} +$$

$$\frac{u_{ds}}{L_\sigma} - \frac{L_m}{L_r L_\sigma}u_{dr}$$

---

0-7803-9771-1/06/$25.00 ©2006 IEEE

$$\frac{di_{qs}}{dt} = -\left(\frac{R_s}{L_\sigma} + \frac{R_r L_m^2}{L_r^2 L_\sigma}\right)i_{qs} + \frac{R_r L_m}{L_r^2 L_\sigma}\psi_{qr} - \frac{\omega_r L_m}{L_r L_\sigma}\psi_{dr} +$$

$$\frac{u_{qs}}{L_\sigma} - \frac{L_m}{L_r L_\sigma}u_{qr}$$

$$\frac{d\psi_{dr}}{dt} = \frac{R_r L_m}{L_r}i_{ds} - \frac{R_r}{L_r}\psi_{dr} - \omega_r\psi_{qr} + u_{dr}$$

$$\frac{d\psi_{qr}}{dt} = \frac{R_r L_m}{L_r}i_{qs} - \frac{R_r}{L_r}\psi_{qr} + \omega_r\psi_{dr} + u_{qr}$$

(1)

where $i_s$, $\psi_r$, $u_s$, $u_r$, $R$ and $L$ denote stator currents, rotor flux linkage, stator terminal voltage, rotor terminal voltage, resistance and inductance, respectively. The subscripts s and r stand for stator and rotor while subscripts d and q stand for vector component with respect to a fixed stator reference frame. $\omega_r$ denote the rotor electrical speed and $L_m$ is the mutual inductance. $L_\sigma = L_s\left(1 - \left(\frac{L_m^2}{L_r L_s}\right)\right)$ is the redefined leakage inductance.

The generated torque of DFIM can be expressed in terms of stator currents and rotor flux linkage as

$$T_e = \frac{3P}{2}\frac{L_m}{L_r}(\psi_{dr}i_{qs} - \psi_{qr}i_{ds})$$

(2)

where $P$ is the number of poles. The mechanical dynamic equation is given by

$$J\frac{d\omega_m}{dt} + B\omega_m + T_L = T_e$$

(3)

where $J$ and $B$ denote the moment of inertia of the motor and viscous friction coefficient, respectively, $T_L$ is the external load and $\omega_m$ is the rotor mechanical speed ($\omega_r = \left(\frac{P}{2}\right)\omega_m$).

Let

$$x = \begin{bmatrix} i_{ds} & i_{qs} & \psi_{dr} & \psi_{qr} \end{bmatrix}^T$$

(4)

be the state vector and let the generated torque $T_e$ be the output $y$ of the dynamic system (1), that is

$$y = T_e = \frac{3P}{2}\frac{L_m}{L_r}(\psi_{dr}i_{qs} - \psi_{qr}i_{ds})$$

(5)

It is well known that the torque control is very important for high-performance motion control. However from (1) and (5), we can view that the generated torque $T_e$ of DFIM is a nonlinear output with respect to state variables $x$ of dynamic model(1), $i_{ds}$, $i_{qs}$, $\psi_{dr}$ and $\psi_{qr}$. Therefore, it is difficult to evaluate the torque response from (5) by the control input $u_{dr}$ and $u_{qr}$ designed for the model (1). So based on the dynamic model (1), it is a task for the torque control of induction machine to industrial and practical applications.

## III. Adaptive Input-Output Feedback Control

For the proposed nonlinear sliding mode controller, the state coordinate transformation is applied. Therefore the state-coordinates transformed model from (1) can be rewritten in a compact form as

$$\dot{x} = f(x) + g_1 u_{dr} + g_2 u_{qr}$$

(6)

where $x$ is defined in (4) and

$$f(x) = \begin{bmatrix} -\left(\frac{R_s}{L_\sigma} + \frac{R_r L_m^2}{L_r^2 L_\sigma}\right)i_{ds} + \frac{R_r L_m}{L_r^2 L_\sigma}\psi_{dr} + \frac{\omega L_m}{L_r L_\sigma}\psi_{qr} + \frac{u_{ds}}{L_\sigma} \\ -\left(\frac{R_s}{L_\sigma} + \frac{R_r L_m^2}{L_r^2 L_\sigma}\right)i_{qs} + \frac{R_r L_m}{L_r^2 L_\sigma}\psi_{qr} - \frac{\omega L_m}{L_r L_\sigma}\psi_{dr} + \frac{u_{qs}}{L_\sigma} \\ \frac{R_r L_m}{L_r}i_{ds} - \frac{R_r}{L_r}\psi_{dr} - \omega_r\psi_{qr} \\ \frac{R_r L_m}{L_r}i_{qs} - \frac{R_r}{L_r}\psi_{qr} + \omega_r\psi_{dr} \end{bmatrix}$$

(7)

and

$$g_1 = \begin{bmatrix} -\frac{L_m}{L_r L_\sigma} & 0 & 1 & 0 \end{bmatrix}^T$$

$$g_2 = \begin{bmatrix} 0 & -\frac{L_m}{L_r L_\sigma} & 0 & 1 \end{bmatrix}^T$$

(8)

At this stage the generated torque $T_e$ and the squared modules of the rotor flux linkage, $|\psi_r|^2 = \psi_{dr}^2 + \psi_{qr}^2$, are requested to be the controlled output. Therefore, by considering

$$h_1(x) = \frac{3P}{2}\frac{L_m}{L_r}(\psi_{dr}i_{qs} - \psi_{qr}i_{ds})$$

$$h_2(x) = \psi_{dr}^2 + \psi_{qr}^2$$

(9)

The following notation is used for the lie derivative of a function $h(x): \Re^n \to \Re$ along a vector field $f(x) = (f_1(x), ..., f_n(x))$

$$L_f h(x) = \sum_{i=1}^{n}\frac{\partial h}{\partial x_i}f_i(x)$$

(10)

Iteratively, we define $L_f^i h = L_f(L_f^{i-1}h)$

Define the change of coordinates as

$$z_1 = h_2(x)$$

$$z_2 = h_1(x)$$

(11)

Then, the dynamic model of DFIM is given in new coordinates by

$$\begin{bmatrix} \dot{z}_1 \\ \dot{z}_2 \end{bmatrix} = \begin{bmatrix} L_f h_2 \\ L_f h_1 \end{bmatrix} + \begin{bmatrix} L_{g_1}h_2 & L_{g_2}h_2 \\ L_{g_1}h_1 & L_{g_2}h_1 \end{bmatrix}\begin{bmatrix} u_{dr} \\ u_{qr} \end{bmatrix}$$

(12)

where

$$L_f h_2 = 2\frac{R_r L_m}{L_r}(i_{ds}\psi_{qr} - i_{qs}\psi_{dr}) - 2\frac{R_r}{L_r}(\psi_{dr}^2 + \psi_{qr}^2)$$

$$L_f h_1 = \frac{3P}{2}\frac{L_m}{L_r} \times$$

$$\begin{pmatrix} -(\frac{R_s}{L_\sigma} + \frac{R_r L_m^2}{L_r^2 L_\sigma} + \frac{R_r}{L_r})(i_{qs}\psi_{dr} - i_{ds}\psi_{qr}) - \\ \frac{\omega_r L_m}{L_r L_\sigma}(\psi_{dr}^2 + \psi_{qr}^2) - \omega_r(\psi_{qr}i_{qs} + \psi_{dr}i_{ds}) + \frac{u_{qs}\psi_{dr}}{L_\sigma} - \frac{u_{ds}\psi_{qr}}{L_\sigma}) \end{pmatrix}$$

$$L_{g1}h_2 = 2\psi_{dr}$$

$$L_{g2}h_2 = 2\psi_{qr}$$

$$L_{g1}h_1 = \frac{3P}{2}\frac{L_m}{L_r}(i_{qs} + \frac{L_m}{L_r L_\sigma}\psi_{qr})$$

$$L_{g2}h_1 = -\frac{3P}{2}\frac{L_m}{L_r}(i_{ds} + \frac{L_m}{L_r L_\sigma}\psi_{dr}) \tag{13}$$

Furthermore, a nonlinear state feedback decoupling the control inputs method is employed. We construct the new control inputs as follows

$$\begin{bmatrix} \hat{u}_{dr} \\ \hat{u}_{qr} \end{bmatrix} = \begin{bmatrix} L_{g1}h_2(x)u_{dr} + L_{g2}h_2(x)u_{qr} \\ L_{g1}h_1(x)u_{dr} + L_{g2}h_2(x)u_{qr} \end{bmatrix} \tag{14}$$

Then, the system (12) becomes

$$\begin{bmatrix} \dot{z}_1 \\ \dot{z}_2 \end{bmatrix} = \begin{bmatrix} L_f h_2(x) \\ L_f h_1(x) \end{bmatrix} + \begin{bmatrix} 1 & 0 \\ 0 & 1 \end{bmatrix}\begin{bmatrix} \hat{u}_{dr} \\ \hat{u}_{qr} \end{bmatrix} \tag{15}$$

By defining errors

$$e_{z1} = z_1 - \psi_r^{*2}$$

$$e_{z2} = z_2 - T_e^* \tag{16}$$

derivating error model (16) gives

$$\dot{e}_{z1} = L_f h_2(x) + \bar{u}_{dr} + \phi_1 d_1(x)$$

$$\dot{e}_{z2} = L_f h_1(x) + \bar{u}_{qr} + \phi_2 d_2(x)$$

$$\dot{e}_z = [A(x) + \Delta A(x)] + B(x)\bar{U} \tag{17}$$

It is obvious that the controllers $\bar{u}_{dr}$ and $\bar{u}_{qr}$ are decoupling with respect to two dynamic models, $[e_{z1}, e_{z2}]$. According to the equations of (17), the adaptive input-output control for first equation of (23) is designed as

$$\bar{u}_{dr} = -L_f h_2(x) - \phi_1 d_1(x) - k_1 e_{z1} \tag{18}$$

where $k_1 > 0$. Similarly, the adaptive input-output feedback control for the second dynamic equation of (17) is designed as follows

$$\bar{u}_{qr} = -L_f h_1(x) - \hat{\phi}_2 d_2(x) - k_2 e_{z2} \tag{19}$$

where $k_2$ is positive constant feedback gain.

**Theorem 2:** Using the controller described by (18)-(19), the controlled torque and flux amplitude of the DFIM is stable and robust to the mismatched uncertainties due to parameter variations.

**Proof:** The proof is obtained by choosing the following Lyapanov function

$$V_1 = \frac{1}{2}\left[e_{z1}^2 + e_{z2}^2\right] \tag{20}$$

Taking the derivative of (20) with respect to time and then substituting (17) into this derivative, we can obtain

$$\dot{V}_1 = e_{z1}\left[L_{f(x)}h_2(x) + \bar{u}_{dr} + \phi_1 d_1(x)\right] + e_{z2}\left[L_{f(x)}h_1(x) + \bar{u}_{qr} + \phi_2 d_2(x)\right] \tag{21}$$

Substituting (18), (19) into equation (29), gives

$$\dot{V}_1 = -k_1 e_{z1}^2 - k_2 e_{z2}^2 \le 0 \tag{22}$$

Define the following equation

$$M(t) = k_1 e_{z1}^2 + k_2 e_{z2}^2 \ge 0 \tag{23}$$

also, defining the following function

$$V_1(t) = V_1(e(0),\phi(0)) + \int_0^t \dot{V}_1(\tau)d\tau = V_1(e(0),\phi(0)) - \int_0^t M(\tau)d\tau \tag{24}$$

where $e = [e_{z1}, e_{z2}]^T$. From the definition of the Lyapanouv function $V_1(t) \ge 0$ and the above equation, the following result can be deduced

$$\lim_{t\to\infty}\int_0^t M(\tau)d\tau \le V_1(e(0),\phi(0)) < \infty \tag{25}$$

Based on the Barbalat's Lemma [5], we can obtain

$$M(t) \to 0 \text{ as } t \to \infty \tag{26}$$

That is, $e_{z1}$ and $e_{z2}$ will converge to zero as $t \to \infty$.

Therefore, the proposed controller is stable and robust, even if parametric uncertainties exist.

## IV. DESIGN OF ROTOR FLUX OBSERVER

Using the stator measured voltage and currents, the space vector model of DFIM in the stator reference frame can be expressed by:

$$\frac{di_s}{dt} = -(\gamma_1 + j\omega_0)i_s + \beta(\alpha - j\omega)\psi_r + \frac{u_s}{\sigma} - \beta u_r$$

$$\frac{d\psi_r}{dt} = \alpha L_m i_s - (\alpha - j(\omega_0 - \omega))\psi_r + u_r \tag{27}$$

the adaptive rotor flux observer is introduced as:

$$\frac{d\hat{i}_s}{dt} = -(\hat{\gamma}_1 + j\omega_0)i_s + \beta(\hat{\alpha} - j\hat{\omega})\hat{\psi}_r + \frac{u_s}{\sigma} - \beta u_r - K\,\text{sgn}(\tilde{i}_s)$$

$$\frac{d\hat{\psi}_r}{dt} = \hat{\alpha}L_m \hat{i}_s - (\hat{\alpha} - j(\omega_0 - \hat{\omega}))\hat{\psi}_r + u_r \tag{28}$$

where :

$$\hat{\alpha} = \alpha_n + \Delta\hat{\alpha} \quad \text{з} \quad \hat{\gamma}_1 = (\frac{R_s}{\sigma} + \hat{\alpha}L_m\beta)$$

In order to develop the adaptation laws, one assumes that the motor speed, rotor and stator resistances are unknown constant parameters. Therefore from (27) and (28), the observer error dynamic is expressed by:

$$\frac{d\tilde{i}_s}{dt} = -(\tilde{\gamma}_1)i_s + \beta(\tilde{\alpha} - j\tilde{\omega})\hat{\psi}_r + \beta(\alpha - j\omega)\tilde{\psi}_r - K\,\text{sgn}(\tilde{i}_s)$$

$$\frac{d\tilde{\psi}_r}{dt} = \tilde{\alpha}L_m i_s - (\tilde{\alpha} + j\tilde{\omega})\hat{\psi}_r - (\alpha - j(\omega_0 - \omega))\tilde{\psi}_r \tag{29}$$

where:
$$\tilde{i}_s = \hat{i}_s - i_s, \tilde{\psi}_r = \hat{\psi}_r - \psi_r, \tilde{\alpha} = \hat{\alpha} - \alpha, \tilde{\gamma}_1 = \hat{\gamma}_1 - \gamma_1, \tilde{\omega} = \hat{\omega} - \omega$$

Candidate the following Lyapunov function

$$V = \frac{1}{2}\left(\tilde{\psi}_r^T \tilde{\psi}_r + \frac{1}{c_\omega}\tilde{\omega}^2 + \frac{1}{c_\alpha}\tilde{\alpha}^2\right) \qquad (30)$$

Derivating $V$ with respect to time gives

$$\dot{V} = -\alpha\tilde{\psi}_r^T \tilde{\psi}_r + \tilde{\omega}\left(\frac{1}{c_\omega}\dot{\tilde{\omega}} - (j\hat{\psi}_r)^T \tilde{\psi}_r\right)$$

$$+ \tilde{\alpha}\left(\frac{1}{c_\alpha}\dot{\tilde{\alpha}} - (\hat{\psi}_r - L_m i_s)^T \tilde{\psi}_r\right) \qquad (31)$$

As a result from (31), the adaptation laws are obtained as:

$$\dot{\tilde{\omega}} = c_w (j\hat{\psi}_r)^T \tilde{\psi}_r$$
$$\dot{\tilde{\alpha}} = c_\alpha (\hat{\psi}_r - L_m i_s)^T \tilde{\psi}_r \qquad (32)$$

## V. ACTIVE AND REACTIVE STATOR POWER CONTROL

For both the motoring and generation modes of operation of DFIM, it is desired to regulate the stator active-reactive power, whose references are $P_s^*$ and $Q_s^*$ respectively. Considering a synchronous d & q axis rotating reference frame with the d axis coincide with space voltage vector for the main ac supply, reference [4] shows that:

$$i_d^* = \frac{3}{2}\frac{P_s^*}{U}$$

$$i_q^* = \frac{3}{2}\frac{Q_s^*}{U} \qquad (33)$$

in addition the rotor flux references can be obtained as:

$$\psi_d^* = \frac{1}{\beta\omega_0}\left(-\frac{R_s}{\sigma}i_q^* - \omega_0 i_d^*\right)$$

$$\psi_q^* = \frac{1}{\beta\omega_0}\left(\frac{R_s}{\sigma}i_d^* - \omega_0 i_q^* - \frac{1}{\sigma}U\right) \qquad (34)$$

so rotor flux reference and torque reference are calculated as below

$$\psi_r^* = \sqrt{\left(\psi_d^{*2} + \psi_q^{*2}\right)}$$
$$T_e^* = \mu\left(\psi_d^* i_q^* - \psi_q^* i_d^*\right) \qquad (35)$$

## VI. STABILIZATION OF ROTOR DC LINK VOLTAGE

In this paper the rotor dc link voltage is maintained constant on the basis of input-output feedback linearization technique. The control strategy is shown in Fig. 1. [6].

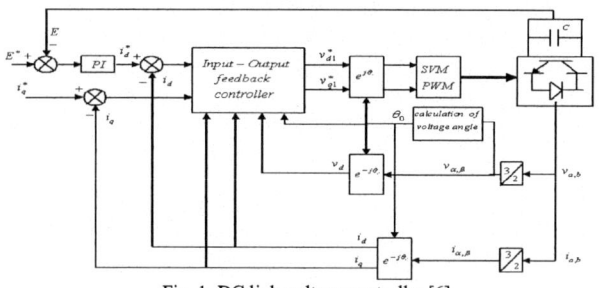

Fig. 1. DC link voltage controller [6].

## VII. SYSTEM SIMULATION

A C$^{++}$ computer program was developed to model this system on P.C.

The overall block diagram of the proposed controller is shown in Fig. 2.

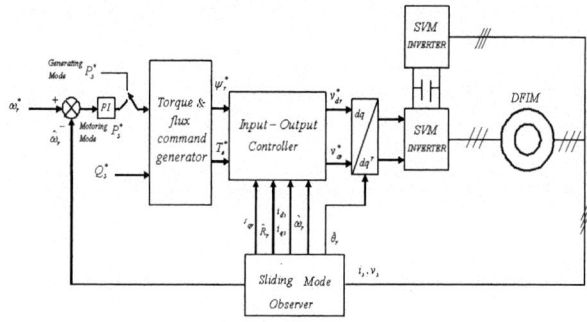

Fig. 2. Block diagram of proposed controller.

In this program, a static runge-kutta fourth order method is used to solve the system equations. The effectiveness and validity of the proposed approach is tested for a three-phase 5 KW, 380 V, six poles,50 Hz DFIM drive [4] by simulation.

Fig. 3 shows the drive system performance in generating mode of operation above the synchronous speed. These results are obtained for the nominal condition and active and reactive power references and torque and flux references are shown in Fig. 3. These results obtained with the rotor speed $\omega_r = 375\left(\frac{rad}{\sec}\right)$.

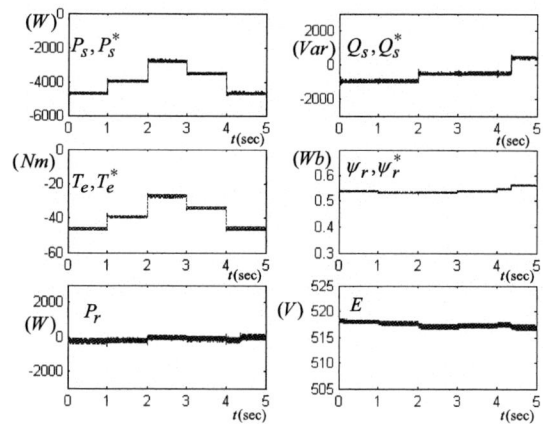

Fig. 3. Simulation results for generating mode above synchronous speed.

Fig. 4 shows the system performance in the motoring mode of operation below and above synchronous speed. These results are obtained for $R_r = 2R_{rn}$. Where $R_{rn}$ is the nominal rotor resistance.

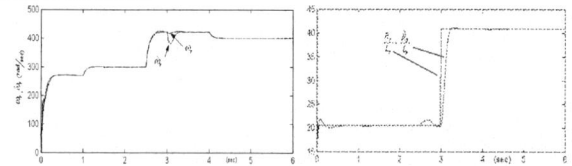

Fig. 4. Simulation results for motoring mode of operation.

62

Also in the all mode of operations, the torque and rotor reference flux signals are obtained based on desired active and reactive power reference profiles which are injected to the stator from main ac supply.

## VIII. CONCLUSION

In this paper an adaptive nonlinear controller has been introduced for DFIM drives. The proposed controller is designed based on input-output feedback linearization combined with Sliding-Mode observer. The proposed control approach has been tested for both the motoring and generating modes of operation bellow and above synchronous speed, using the two level SVM-PWM back-to-back voltage sources inverters in the rotor circuit. Computer simulation results obtained, confirm the validity and effectiveness of the proposed control approach.

## IX. REFERENCES

[1] B.Hopfensperger, J.Atkinson, and R.A.Lakin, "Stator-flux-oriented control of a doubly-fed induction machine with and without position encoder," *in Proc. July 2000 IEE Electr. Power Appl Conf.,* vol. 147, no. 4, pp. 241-250.

[2] S.Peresada, A.Tilli, and A.Tonielli, "Robust active-reactive power control of a doubly-fed induction generator," in *Proc. Sept. 1998 IEEE-IECON'98 Conf., Aachen, Germany,* pp. 1621-1625.

[3] S.Peresada, A.Tilli, and A.Tonielli, "Indirect stator flux-oriented output feedback control of a doubly fed induction machine," *IEEE Trans. Contr Sys,* vol. 11, no. 6, pp. 875-888, November 2003.

[4] S.Peresada, A.Tilli,and A.Tonielli, "Robust Output Feedback Control of a Doubly Fed Induction Machine," in *Proc. 1999 IEEE Conf.,* pp. 1348-1354.

[5] Jeffery T.Spooner, Manfred Maggiore, Raul Ordonez and Kevin M.Passino, *Stable Adaptive Control And Estimation For Nonlinear Systems,* John Wiley and Sons, Inc, NewYork.

[6] J.Soltani, A.Farrokh Payam, "A Robust Adaptive Sliding-Mode Controller for Slip Power Recovery Induction Machine Drives," in *Proc. 2006 IEEE/IPEMC Conf.*

**2006 IEEE International Conference on Power Electronic, Drives and Energy Systems**

# Adaline Based Control of Solid State Voltage Regulator for Isolated Asynchronous Generators

Bhim Singh, *Senior Member, IEEE*, and Gaurav Kumar Kasal

*Abstract*-- This paper deals with the adaline based control of a solid state voltage regulator (SSVR) for isolated asynchronous generator (IAG). With indirect current control, proposed control scheme uses a fast adaptive linear element (adaline) based neural network reference current extractor, which extracts the real positive sequence current component without any phase shift. The estimation of the reference current through adaline utilizes LMS (Least Means Squire) algorithm to maintain online calculation of weights. A closed loop PI (Proportional-Integral) controller is used for regulating the DC bus of SSVR. The control scheme is verified for regulating the voltage of an isolated asynchronous generator in constant speed variable power application through simulation in MATLAB using Simulink and PSB (power system block) toolboxes.

*Index Terms*-- Asynchronous generator, SSVR, Adaline, Neural network.

## I. INTRODUCTION

RECENT emphasis on distributed power generation has given an impetus to research which is evolving suitable generating systems especially for remote locations. Use of locally available renewable energy sources such as bio fuel, small hydro potential and wind for such a purpose is naturally attractive. Asynchronous generator driven by an engine or a turbine is a source of electric power to such a scheme. However such systems must be simple, reliable and user friendly. Isolated asynchronous generator with excitation capacitor is considered as a viable option due to its specific advantages such as low cost, less weight, less maintenance compared to other isolated electric generators. The fundamental problems for its commercialization is its inability to control the terminal voltage and frequency under varying load conditions [1, 2]. Hence the limitations of asynchronous generator system with capacitor excitation are poor voltage and frequency regulation result in under utilization of the machine. In order to regulate its terminal voltage with the load and utilize the machine to its rated capacity an external source of reactive current is required for its excitation and meets reactive power requirements of the loads. Attempts have been made to regulate its voltage for 3-phase [3, 4] as well as single phase loads [5] in constant speed, variable power application but here control strategy to regulate the voltage of isolated asynchronous generator is based on the neural network. The

control of SSVR for isolated generation using asynchronous generator with capability of reactive power, harmonics compensation and unbalanced load compensation is achieved by LMS algorithm [6, 7] based adaline. The adaline is used to extract the positive sequence fundamental frequency real component of load current. The extraction of reference currents using adaline involves calculation of the weights, these weight are a measure of peak of fundamental frequency current component of the load current. Control technique is comparatively fast and results in less drop in DC bus voltage during transients.

## II. SYSTEM CONFIGURATION

Fig 1 shows the schematic diagram of an asynchronous generator (AG) with excitation capacitor, SSVR and consumer load. Excitation capacitors are selected such that AG generates rated voltage at no load. The additional demand of reactive power is met using the SSVR under varying loads. The SSVR acts as a source of lagging or leading current to maintain the constant terminal voltage with variation in loads with harmonic elimination and load balancing. The SSVR consists of IGBT based current controlled voltage source converter (VSC), DC bus capacitor and AC inductors. The output of the VSC is connected through the AC filtering inductor to the AG terminals. The DC bus capacitor is used as an energy storage device and provides self-supporting DC bus of SSVR.

Fig 2 shows the control technique to regulate the terminal voltage of the AG which is based on the generation of reference source currents (have two components, in-phase and quadrature, with AC voltage). The in-phase unit vectors ($u_a$, $u_b$ and $u_c$) are three-phase sinusoidal functions, computed by dividing the AC voltages $v_a$, $v_b$ and $v_c$ by their amplitude $v_t$. Another set of quadrature unit vectors ($w_a$, $w_b$ and $w_c$) are sinusoidal function obtained from in-phase vectors ($u_a$, $u_b$ and $u_c$). To regulate AC terminal voltage ($v_t$), it is sensed and compared with the reference voltage. The voltage error is processed in the PI (Proportional-Integral) controller. The output of the PI controller ($I_{smq}^*$) for AC voltage control loop decides the amplitude of reactive current to be generated by the SSVR. Multiplication of quadrature unit vectors ($w_a$, $w_b$ and $w_c$) with the output of PI based AC voltage controller ($I_{smq}^*$) yields the quadrature component of the reference source currents ($i_{saq}^*$, $i_{sbq}^*$ and $i_{scq}^*$). The active or in-phase component of source current ($i_{sad}^*$, $i_{sbd}^*$, $i_{scd}^*$) is estimated using adaline control. The real power supplied by the source has two parts: one part of real power is needed by consumer loads while other one to meet the switching and conduction losses of the SSVR to maintain the constant dc link voltage. Therefore

---

Bhim Singh and Gaurav Kumar Kasal are with the Dept. of Electrical Engineering, Indian Institute of Technology Delhi, 110016 New Delhi India (e-mail: bhimsinghr@gmail.com , gauravkasal@gmail.com)

0-7803-9771-1/06/$25.00 ©2006 IEEE

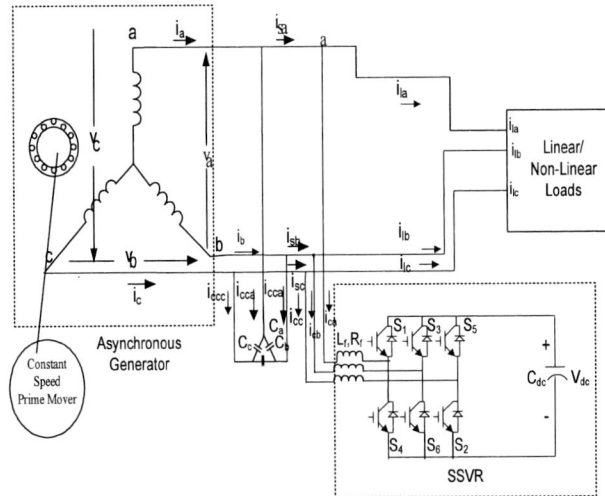

Fig. 1. Schematic diagram of SSVR-AG system.

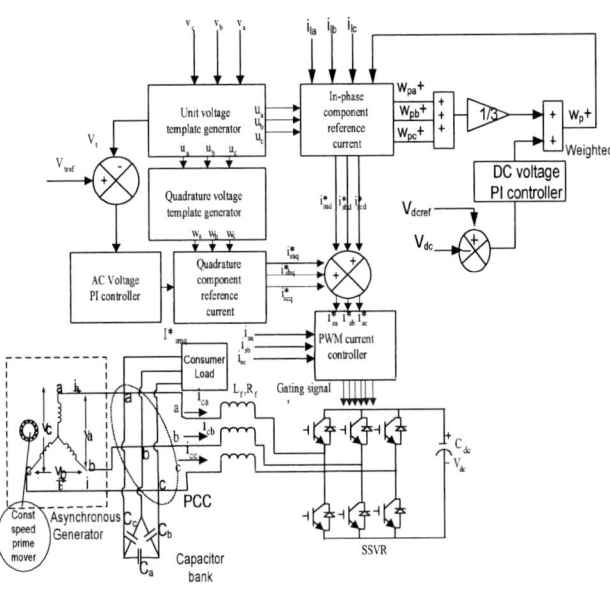

Fig. 2. Schematic diagram of control scheme for SSVR-AG system.

the reference current corresponding to the fundamental frequency real current has two components one that is computed by neural network based current decomposer of the load current and other is to be computed by the PI controller at the DC bus of VSC. To provide self-supporting DC bus of SSVR, its DC bus voltage is sensed and compared with DC reference voltage. The error voltage is processed in another PI controller. The output of the PI controller ($I_{smd}^*$) is added to the weight corresponding to the fundamental frequency positive sequence real current component, such that total real current or in phase component ($i_{sad}^*$, $i_{sbd}^*$ and $i_{scd}^*$) gives corresponding to the load and component to meet the losses of SSVR. The instantaneous sum of quadrature and in-phase components gives the total reference source currents ($i_{sa}^*$, $i_{sb}^*$ and $i_{sc}^*$), which are fed to PWM hysteresis current controller to

switch the VSC of SSVR through forcing the source currents to follow these reference three phase currents.

### III. CONTROL ALGORITHM

Modeling of control scheme of AG-SSVR system is given as follows.

Three-phase voltages at the AG terminals ($v_a$, $v_b$ and $v_c$) are considered sinusoidal and hence their amplitude is computed as:

$$V_t = \sqrt{ \{(2/3) (v_a^2 + v_b^2 + v_c^2)\} } \qquad (1)$$

The unit vector in phase with $v_a$, $v_b$ and $v_c$ are derived as:

$$u_a = v_a/V_t; \quad u_b = v_b/V_t; \quad u_c = v_c/V_t \qquad (2)$$

The unit vectors in quadrature with $v_a$, $v_b$ and $v_c$ may be derived using a quadrature transformation of the in-phase unit vectors $u_a$, $u_b$ and $u_c$ as:

$$w_a = -u_b / \sqrt{3} + u_c / \sqrt{3} \qquad (3)$$
$$w_b = \sqrt{3}\, u_a / 2 + (u_b - u_c) / 2\sqrt{3} \qquad (4)$$
$$w_c = -\sqrt{3}\, u_a / 2 + (u_b - u_c) / 2\sqrt{3} \qquad (5)$$

### A. Quadrature Component of Reference Source Currents

The AC voltage error $V_{er(n)}$ at the $n^{th}$ sampling instant is:

$$V_{er(n)} = V_{tref(n)} - V_{t(n)} \qquad (6)$$

Where $V_{tref(n)}$ is the amplitude of reference AC terminal voltage and $V_{t(n)}$ is the amplitude of the sensed three-phase AC voltage at the AG terminals at $n^{th}$ instant. The output of the PI controller ($I_{smq(n)}^*$) for maintaining AC terminal voltage constant at the $n^{th}$ sampling instant is expressed as:

$$I_{smq(n)}^* = I_{smq(n-1)}^* + K_{pa} \{ V_{er(n)} - V_{er(n-1)} \} + K_{ia} V_{er(n)} \qquad (7)$$

Where $K_{pa}$ and $K_{ia}$ are the proportional and integral gain constants of the proportional integral (PI) controller. $V_{er(n)}$ and $V_{er(n-1)}$ are the voltage errors in $n^{th}$ and $(n-1)^{th}$ instant and $I_{smq(n-1)}^*$ is the amplitude of quadrature component of the reference source current at $(n-1)^{th}$ instant. The quadrature components of the reference source currents are computed as:

$$i_{saq}^* = I_{smq}^* w_a; \quad i_{sbq}^* = I_{smq}^* w_b; \quad i_{scq}^* = I_{smq}^* w_c \qquad (8)$$

### B. In-Phase Component of Reference Source Currents

In-phase component or real component of source current estimation is based upon Least Mean Square (LMS) algorithm and its training is through Adaline. Load current is made up of active current ($i_d$), reactive current ($i_q$) and negative sequence current ($i^-$) can be decomposed in part as:

$$i_L(t) = i_d^+(t) + i_q^+(t) + i^-(t) \qquad (9)$$

The control algorithm estimation is based on the extraction of current component in phase with the unit voltage template. To estimate fundamental frequency positive sequence real component of current, the unit voltage template should be in phase with the system voltage and having unit amplitude given in eq. (2).

The initial estimate of 'first' part of active current represent on single phase basis as:

$$i_d(t) = W_p u_p(t) \qquad (10)$$

Where weight ($W_p$) is estimated using Adaline. The weight is variable and changes as per the load current and magnitude of phase voltage. The scheme for estimating weights corresponding to fundamental frequency real component of current based on LMS algorithm tuned adaline tracks the unit vector templates to maintain minimum error. The weight is estimated by following iterations:

$$W_p(k+1) = W_p(k) + \mu\{i_L(k)\text{-}W_p(k)u_p(k)\}u_p(k) \qquad (11)$$

The value of $\mu$ which is called as convergence coefficient decides the rate of convergence and accuracy of estimation. The practical range of convergence coefficient lies from 0.1 to 1.0. The average value of weight is represented as:

$$W_{pavg}^{+} = (W_{pa}^{+} + W_{pb}^{+} + W_{pc}^{+})/3 \qquad (12)$$

For proper estimation of reference signals, the weights are averaged to compute the equivalent for positive sequence and negative sequence current component in the decomposed form. The averaging of weights helps in removing the unbalance from the current components.

'Second' part of active component of reference source current is estimated using DC bus voltage PI controller. The error in DC bus voltage of SSVR ($V_{dcer(n)}$) at $n^{th}$ sampling instant is:

$$V_{dcer(n)} = V_{dcref(n)} - V_{dc(n)} \qquad (13)$$

Where $V_{dcref(n)}$ is the reference DC voltage and $V_{dc(n)}$ is the sensed DC link voltage of the SSVR. The output of the PI controller for maintaining DC bus voltage of the SSVR at the $n^{th}$ sampling instant is expressed as:

$$I^*_{smd(n)} = I^*_{smd(n-1)} + K_{pd}\{V_{dcer(n)} - V_{dcer(n-1)}\} + K_{id} V_{dcer(n)} \qquad (14)$$

$I^*_{smd(n)}$ is considered as the amplitude of active source current. $K_{pd}$ and $K_{id}$ are the proportional and integral gain constants of the DC bus PI voltage controller. Therefore final weighted average for calculating active component of source current is:

$$W_p^{+} = I^*_{smd(n)} + W_{pavg}^{+} \qquad (15)$$

Total active component of source current is represented as:

$$i^*_{sad}(t) = W_p^{+} u_a, \; i^*_{sbd}(t) = W_p^{+} u_b, \; i^*_{scd}(t) = W_p^{+} u_c \qquad (16)$$

### C. Reference Source Currents

Total reference source currents are sum of in-phase and quadrature components of the reference source currents as:

$$i^*_{sa} = i^*_{saq} + i^*_{sad} \qquad (17)$$
$$i^*_{sb} = i^*_{sbq} + i^*_{sbd} \qquad (18)$$
$$i^*_{sc} = i^*_{scq} + i^*_{scd} \qquad (19)$$

### D. PWM Current Controller

The total reference currents ($i^*_{sa}$, $i^*_{sb}$ and $i^*_{sc}$) are compared with the sensed source currents ($i_{sa}$, $i_{sb}$ and $i_{sc}$). The ON/OFF switching patterns of the gate drive signals to the IGBTs are generated from the PWM hysteresis controller. The current errors are computed as:

$$i_{saerr} = i^*_{sa} - i_{sa} \qquad (20)$$
$$i_{sberr} = i^*_{sb} - i_{sb} \qquad (21)$$
$$i_{scerr} = i^*_{sc} - i_{sc} \qquad (22)$$

These current error signals are amplified and are fed to PWM hysteresis controller to generating gating signals for IGBTs.

## IV. MATLAB BASED MODELING

Figs.3-5 shows the MATLAB model of the AG-SSVR generating system driven by constant speed variable prime mover. Main system consists of asynchronous machine with capacitor bank and SSVR. The model of SSVR based voltage regulator is shown in Fig. 3 while corresponding subsystems for demonstrating control scheme are shown in Figs. 4-5. The modeling of AG is carried out using 7.5 kW, 415V, 50Hz, Y-connected induction machine. The controller is realized with voltage source converter and capacitor at DC bus for realizing self supporting DC bus. Both linear and non-linear loads are considered here to demonstrate the capability of voltage

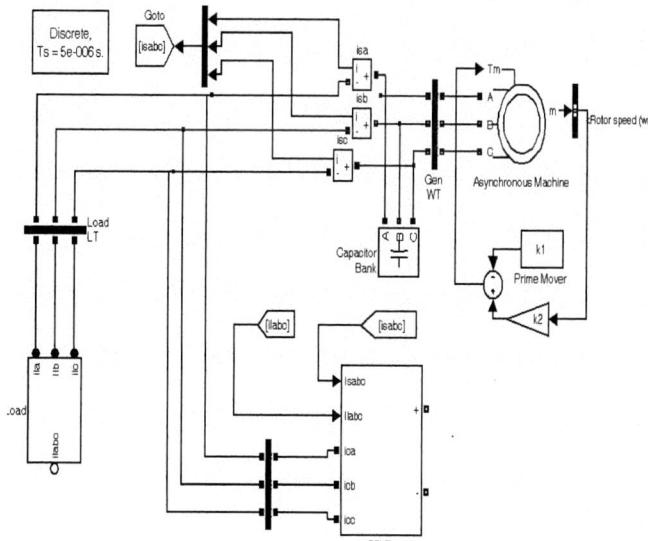

Fig. 3. MATLAB based simulation model of SSVR-AG system.

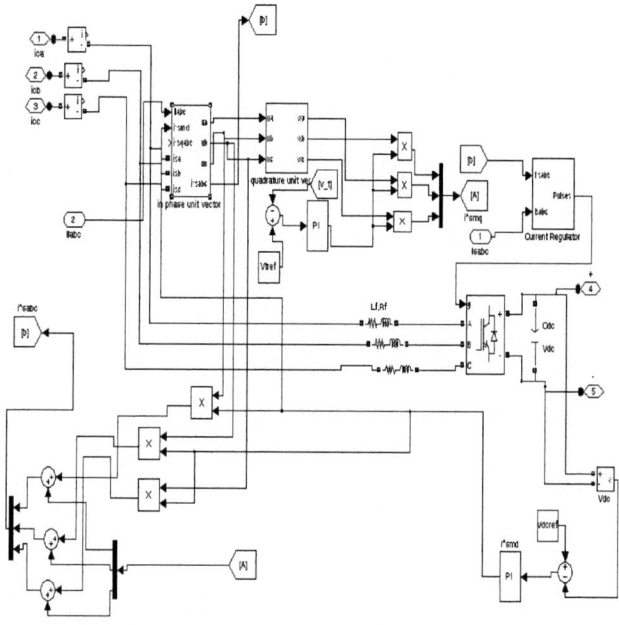

Fig. 4. Subsystem of SSVR control.

controller. Simulation is carried out in discrete mode at 5e-6 step size with ode23tb (stiff/ TR-BDF-2) solver.

## V. RESULTS AND DISCUSSION

The performance of AG-SSVR system using adaline control scheme is demonstrated for feeding linear/ non-linear, balanced and unbalanced loads in constant speed variable power applications and waveforms of the generator voltage ($v_{abc}$) and current ($i_{abc}$), capacitor current ($i_{cca}$), load current ($i_{labc}$), controller currents ($i_{cabc}$), terminal voltage ($v_t$) DC link voltage ($v_{dc}$) and speed ($\omega$) etc are shown in Figs. 6 and 8 for linear resistive and reactive load respectively and Fig 9 for non-linear load. For the simulation, a 7.5 kW, 415V, 14.8A, 4 pole induction machine has been used as an asynchronous generator and its detailed parameters are given in Appendix.

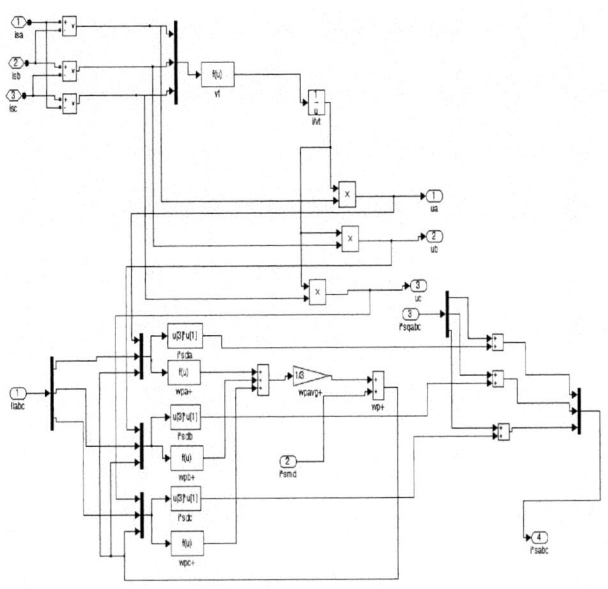

Fig. 5. Subsystem of reference current generation.

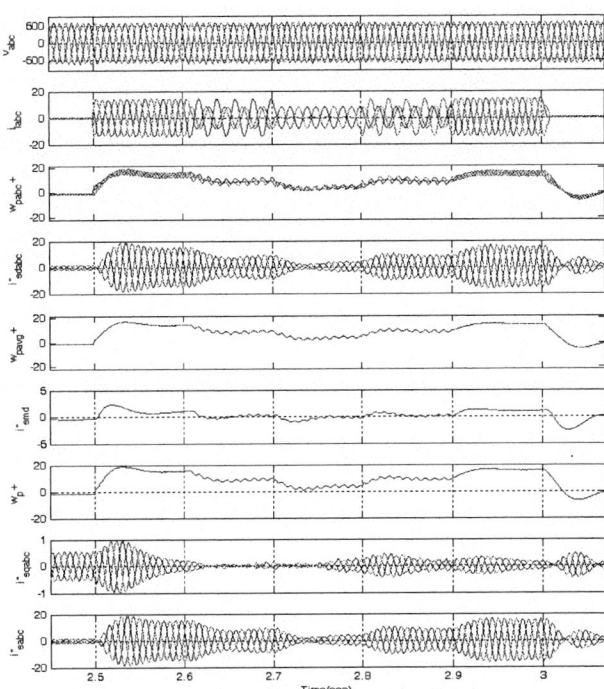

Fig. 7. Transient waveforms of load current decomposition.

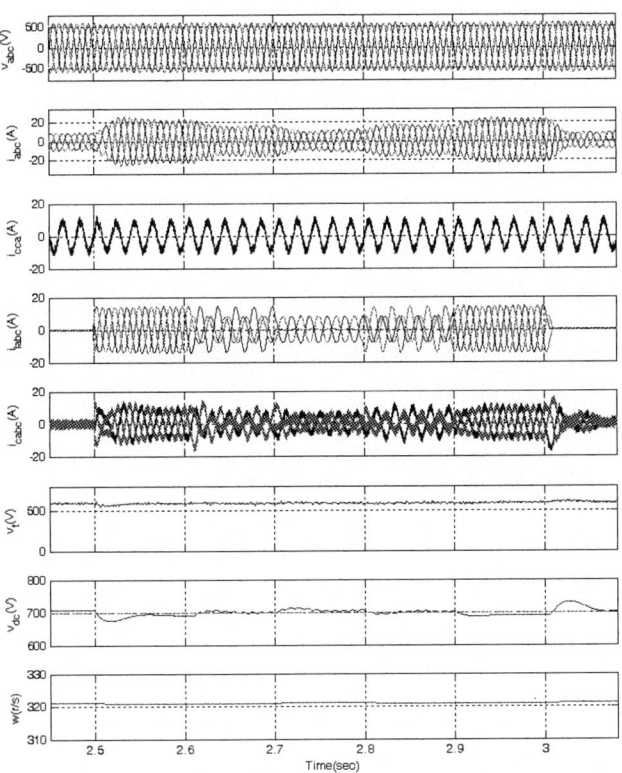

Fig. 6. Transient waveforms of 7.5 kW AG with SSVR for feeding resistive load.

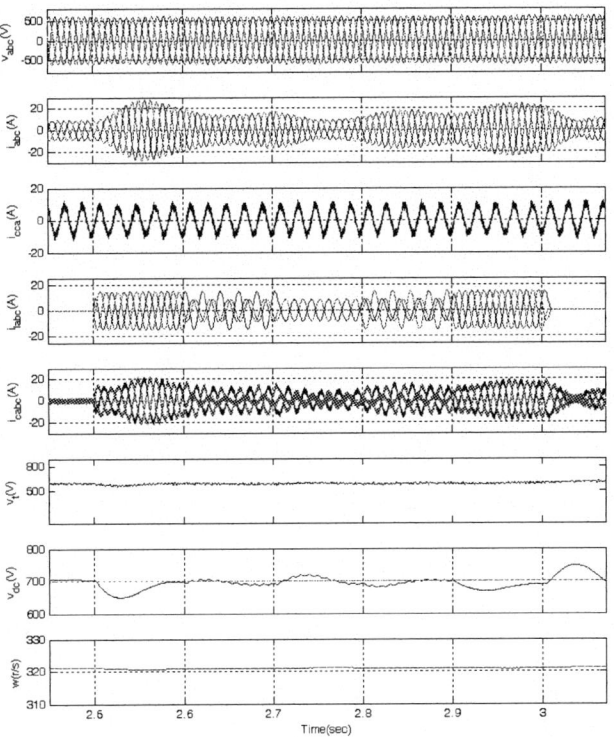

Fig. 8. Transient waveforms of 7.5 kW AG with SSVR for feeding 0.8 PF lagging reactive load.

### A. Performance of AG-SSVR System Feeding 3-Phase Linear Loads

Fig. 6 shows the performance of AG-SSVR system with balanced/unbalance resistive loads. Three, single phase load each of 2.5 kW is applied between phase to phase at 2.5 sec, and it is observed that magnitude of terminal voltage ($v_t$) remains constant and DC link voltage ($v_{dc}$) settles to at reference value. At 2.6 sec when opening of one phase and

later on at 2.7 sec opening of second phase of load are performed the load becomes unbalanced, which shows the load balancing aspect of SSVR based voltage regulator system. Fig 7 demonstrate the adaline based decomposition waveforms of load currents ($i_{labc}$) for determining the reference source currents these waveforms represent weight of each phase ($w_{pa}^+$, $w_{pb}^+$, $w_{pc}^+$), average of weights ($w_{pavg}^+$), output of

67

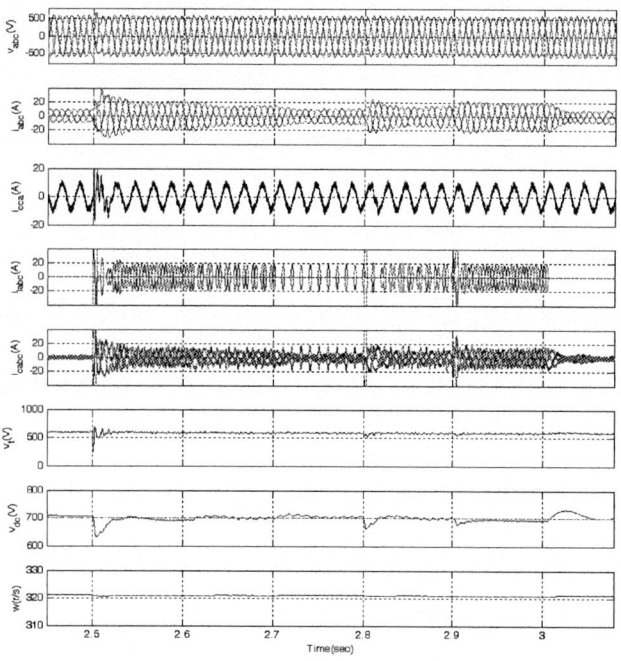

Fig. 9. Transient waveforms of 7.5 kW AG with SSVR for feeding nonlinear load.

DC bus PI controller ($i^*_{smd}$), in-phase and quadrature component of reference source current ($i^*_{sdabc}$ and $i^*_{sqabc}$) and reference source current ($i^*_{sabc}$). Similarly, Fig. 8 shows the performance of the system with 0.8 PF lagging balanced/unbalanced reactive loads. When this load is applied then reactive currents flowing through controller ($i_{cabc}$) are increased for regulating the system voltage which demonstrates the voltage regulating feature of AG-SSVR system.

### B. Performance of AG-SSVR System Feeding 3-Phase Non-linear Loads

Fig.9 shows the performance of AG-SSVR system with balanced/unbalanced non-linear loads using single phase diode rectifier with resistive load and L-C filter at its DC side. At 2.5 sec a balanced non- linear load is applied then controller regulates the voltage and controller currents ($i_{cabc}$) become non-linear for eliminating harmonic currents. During load unbalancing at 2.6 sec DC link capacitor charging and discharging are observed similar to the case of linear load. Fig. 10 and Fig. 11 show the harmonic spectrum of source voltage ($v_a$) and current ($i_a$) with un-balanced and balanced non-linear load current ($i_{la}$) respectively. It can be observed from these figures that total harmonic distortion (THD) of terminal voltage and generator current is less than 5%, the limit imposed by IEEE – 519 Standard.

## VI. Conclusions

The capability of AG-SSVR system has been observed with the proposed control technique to regulate the terminal voltage under load change and unbalance conditions in constant speed variable power applications. The scheme has envisaged nearly zero phase shifts to extract the reference current with simplicity. MATLAB based simulated results have shown quite satisfactory operation and demonstrated that the proposed controller is suitable for voltage regulation,

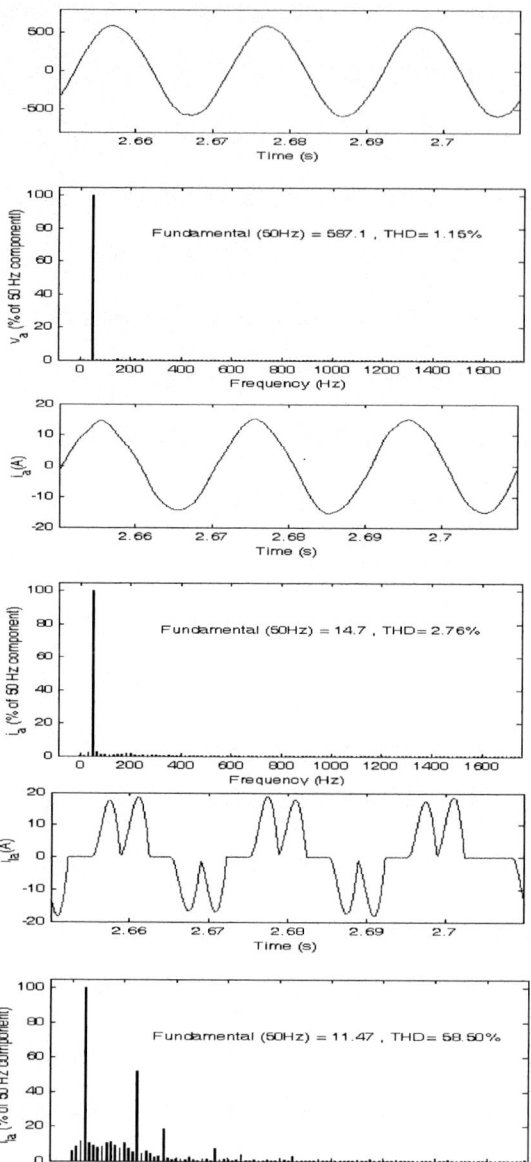

Fig. 10. Harmonic spectrum of generator voltage, current and load current with un-balanced nonlinear load.

harmonic elimination and load balancing in isolated AG system.

## VII. Appendix

### A. The parameters of 7.5kW, 415V, 5□Hz, Y-connected, 4-pole Asynchronous machine are given below.

$R_s = 1\Omega$, $R_r = 0.77\Omega$, $X_{lr} = X_{ls} = 1.5$ $\Omega$, J = 0.1384 kg-m$^2$
$L_m = 0.134$H ($I_m$<3.16)
$L_m = 9e\text{-}5I_m^2 - 0.0087I_m + 0.1643$ (3.16<$I_m$<12.72)
$L_m = 0.068$H ($I_m$>12.72)

### B. Controller parameters

$L_f = 4$mH, $R_f = 0.1\Omega$, and $C_{dc} = 4000\mu$F.
$K_{pa} = 0.04$, $K_{ia} = 0.0015$,
$K_{pd} = 0.02$, $K_{id} = 0.001$.

### C. Consumer Loads

Resistive load        2.5kW single phase loads.

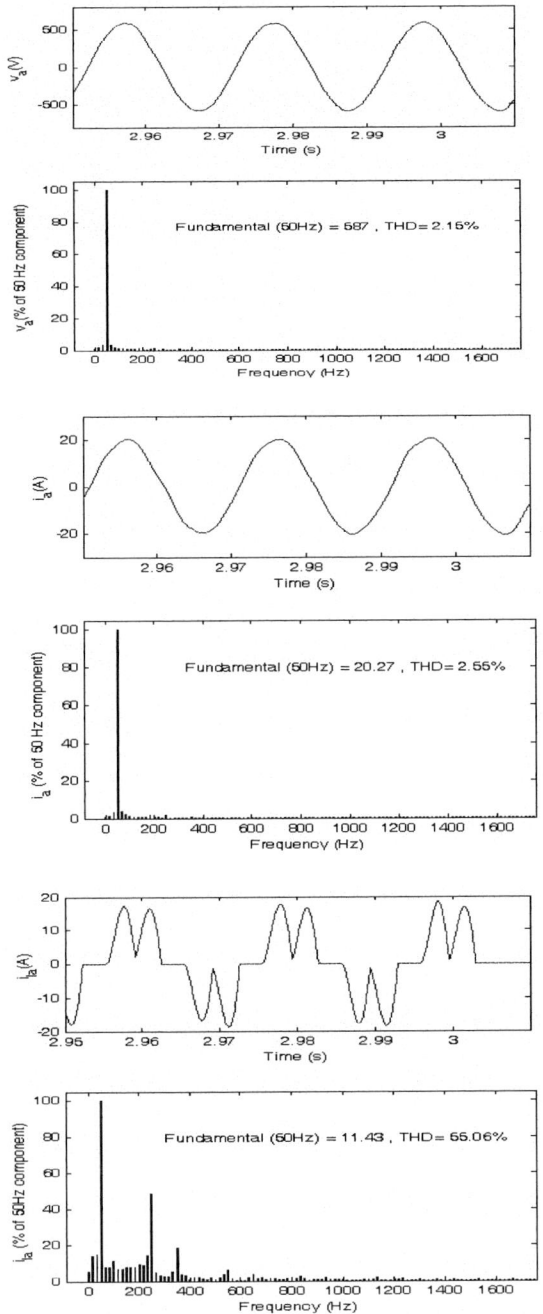

Fig. 11. Harmonic spectrum of generator voltage, current and load current with balanced nonlinear load.

| | |
|---|---|
| Reactive load | 2.5kW, 1.875 kVAR 0.8PF lagging single phase loads. |
| Non-linear load | 2.5kW with 200μF capacitor and 1mH inductor at DC end of single phase diode rectifier |

*D Prime Mover Characteristics*

$T_{sh} = K_1 - K_2 \omega_r$
$K_1 = 33000, K_2 = 100$

## VIII. REFERENCES

[1] E.D. Basset and E.M. Potter, "Capacitive excitation for induction generator," *AIEE Trans. on Electrical Engineering*, vol.54, pp. 540-545, May 1935.

[2] G.K.Singh, "Self-excited induction generator research- a survey" *Electric Power Systems Research*, vol 69, no. 2-3, pp 107-114, May 2004.

[3] Bhim Singh and L. B. Shilpakar, "Analysis of a novel solid state voltage regulator for a self-excited induction generator," *IEE Proc.-Gener. Transm. Distrib.*, vol. 145, no. 6, pp. 647-655, November 1998.

[4] Bhim Singh, S.S. Murthy and Sushma Gupta, "Modelling and analysis of SSVR based voltage regulator for self-excited induction generator with unbalanced loads", In *Proc. of IEEE Conf TENCON 2□3*, Bangalore, pp. 1109-1114.

[5] T. Ahmed, K. Nishida and M. Nakaoka, "Static Var compensator based voltage regulation implementation of 1-phase self excited induction generator" in *Proc. of IEEE/IAS Industry Applications Conf.*, vol. 3, 3-7 Oct 2004, pp. 2069-2076.

[6] N. Pechanranin, H Uitsui and M. Sone, "Harmonic detection by using neural network," in *Proc. of IEEE conf. on Neural Network*, vol 2, 1995, pp. 923-926.

[7] B. Singh, V.Verma, J. Solanki, "Active power filter selective compensation of current using neural network" in *Proc. of IEEE ISIE'□4*, May 2004.

**Bhim Singh** (SM'99) was born in Rahampur, India, in 1956. He received the B.E (Electrical) degree from the University of Roorkee, Roorkee, India, in 1977 and the M.Tech and Ph.D. degree from the Indian Institute of Technology (IIT) Delhi, New Delhi, India, In 1979 and 1983, respectively.

In 1983, he joined the Department of Electrical Engineering, University of Roorkee, as a Lecturer, and in 1988 became a Reader. In December 1990, he joined the Department of Electrical Engineering , IIT Delhi, as an Assistant Professor. He became an Associate Professor in 1994 and Professor in 1997. His area of interest includes power electronics, electrical machines and drives, active filters, and analysis and digital control of electrical machines.

Dr. Singh is a fellow of Indian National Academy of Engineering (INAE), the Institution of Engineers (India) (IE(1)), and the Institution of Electronics and Telecommunication Engineers (IETE), a life Member of the Indian Society for Technical Education(ISTE), the System Society of India (SSI), and the National Institution of Quality and Reliability (NIQR) and Senior Member of Institute of Electrical and Electronics Engineers (IEEE).

**Gaurav Kumar Kasal** was born in Bhopal, India, in Nov, 1978. He received the B.E (Electrical) and M.Tech degree from the National Institute of Technology (NIT) Allahabad and National Institute of Technology (NIT) Bhopal, India respectively in 2002 and 2004. Since Dec 2004, he has been pursuing the Ph. D. degree with the Department of Electrical Engineering, Indian Institute of Technology (IIT) Delhi, New Delhi, India. His field of interest includes power electronics and drives, renewable energy generation and applications, FACTS and electrical machines.

**2006 IEEE International Conference on Power Electronic, Drives and Energy Systems**

# Development of a Prototype Controller for PMDC Motor Based Portable Telemetry Tracking System for Defense Application

Parveen Kumar, A K Pradhan, and Gautam Sadhukhan

*Abstract-* **In the Range, Block House Telemetry plays an important role in providing quick look telemetry data of the airborne vehicle, particularly for initial phase of the trajectory to the range experts for their decision-making and control. Presently blockhouse telemetry front-end helical antenna is manually steered at the blockhouse top. This is a concern of safety and does not give any guarantee of acquiring data due to the poor visibility when the airborne vehicle goes far off. The paper describes prototype model of portable tracking antenna system that facilitates the user to control antenna both in Manual and Auto mode remotely from the Block House. An array of four helical antennas were used to generate the dc tracking error in both azimuth and elevation plane. It also explains a control mechanism by which antenna systems are steered in both azimuth as well as elevation plane using power transistors and permanent magnet direct current motors. This method uses AGC comparison, which is relatively simpler and economical than conventional Single Channel Amplitude Comparison Mono pulse or Electronic Scanning tracking method. The prototype was developed, tested and successfully driven the antenna in both manual as well as auto mode for real time telemetry data acquisition for airborne vehicle at blockhouse.**

*Index Terms--* **AGC, ESCAN, LNA, PMDC, RF, RIC, SCACM.**

## I. INTRODUCTION

IN any mission planning and configuration, minimum two auto track telemetry systems are required for real time tracking and data acquisition from on-board airborne vehicle but a portable blockhouse telemetry checkout system is configured to carry out Phase III, Phase IV as well as range integration check (RIC) and other launch pad checks for data validation and initial health parameter monitoring. Also this portable telemetry checkout system is very useful for initial data capture when airborne vehicle is at launch pad. Some of the quick look telemetry parameters are also required to be

---

Parveen Kumar is with Telemetry Group, Integrated Test Range(ITR), Chandipur, Orissa, Defence R & D Organisation(DRDO), Ministry of Defence, India (e-mail: kumarparveen_kaushik@rediffmail.com).

A. K. Pradhan is with Telemetry Group, Integrated Test Range(ITR), Chandipur, Orissa, Defence R & D Organisation(DRDO), Ministry of Defence, India (e-mail: akp68@hotmail.com).

Gautam Sadhukhan is with Telemetry Group, Integrated Test Range(ITR), Chandipur, Orissa, Defence R & D Organisation(DRDO), Ministry of Defence, India (e-mail: gsadhukhan@hotmail.com).

displayed at blockhouse decision-making consoles for mission director (MD), vehicle director (VD) and range safety officer (RSO). Experts sitting at the consoles analyze telemetry data in real time and take decisions regarding state of the ongoing launch sequence. Apart from that, blockhouse telemetry plays an important role in capturing full telemetry data for short-range airborne vehicle and initial phase of trajectory of long-range air vehicle. But presently, blockhouse telemetry front-end helical antenna is manually steered at the block house top. This is one of the safety concerns and does not give any guarantee of acquiring data due to the poor visibility when the airborne vehicle goes far off. In this case, telemetry data quality is entirely dependent on the human perception and capability. So, there was a need for developing a portable tracking antenna system, which can be steered both manually and automatically depending on the on-board vehicle position. An array of four helical antennas is used for RF reception and error generation, two in azimuth plane and other two in elevation plane. Each axis consists of a permanent magnet direct current (PMDC) motor [1] and two potentiometers coupled with the motors for position sensing and error generation. RF signal is fed to the receivers via Low noise amplifiers (LNA). Four receivers are tuned to incoming RF carrier and used to demodulate the RF. Demodulated Signal is used for the telemetry data extraction and recording of the same. Receivers generate the Automatic Gain Control (AGC) voltage, which is proportional to the incoming signal strength. In Auto mode, position error is measured by comparing the AGC of the receivers and fed to the power amplifier stage. Power amplifier supplies the proportionate voltage to motor and the antenna is steered accordingly. In manual mode, command is given by a potentiometer from controller which compare it's signal with the reference generated by potentiometer fixed on pedestal. Then, this error is amplified and fed to the power transistor amplifier stage. Power transistor stage applies the proportional voltage to the DC motor to rotate the antenna system. This forms a complete closed loop system with feedback. The system block diagram is shown in Fig. 1. Now, at each plane, radiation pattern of antenna element 1 and 2 are plotted as shown in Fig. 2. It shows that if target is at some angular offset from the antenna boresight, then target axis will cross both the pattern at different points, thus producing a voltage difference at the reference voltage in the auto mode. At the next stage, the same

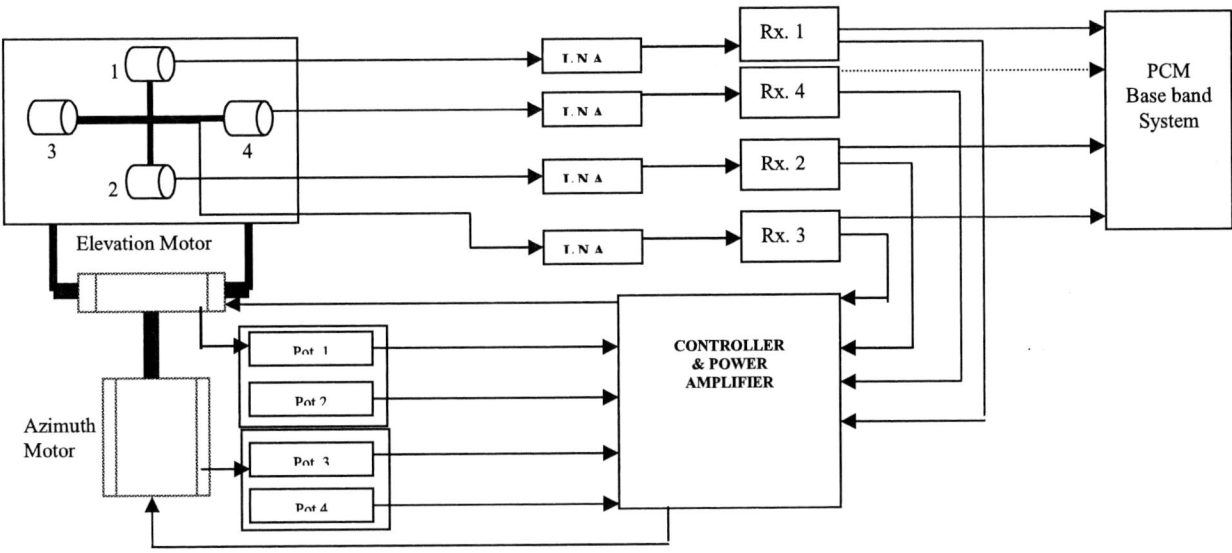

Fig. 1. Block diagram of the system.

receiver end. If the target is aligned along the antenna boresight then these voltages are exactly equal [2]. This is the key idea taken into consideration for our prototype model.

## II. SYSTEM DESCRIPTION

Controller consists of five major stages. Adder stage calculates an average of the AGC signals received from two receivers say 1 and 2 in elevation plane and another for azimuth plane. There is a switch to change over the operational mode between Auto and Manual. In auto mode, the reference voltage is generated by adding these two received AGC voltages and dividing by 2. This reference voltage is subtracted in the subtractor stage with the input AGC voltages. The outcome of this subtractor is ultimately the error in reverse or forward direction for the power amplifier stage. If the polarity of the added signal is positive, antenna will rotate in the clockwise direction and if the polarity is negative, then antenna will rotate in the counterclockwise direction. The Power amplifier stage supplies amplified voltage to the pedestal motors, which is proportional to this error. When mode switch is in manual mode, potentiometer 2, which is mechanically coupled with pedestal motor shaft, senses its position in each plane. Position sensor sends a feedback to the power amplifier stage and then compared with the command given by another Potentiometer 1. As a result, error is generated which is further amplified and fed to power transistor stage. Then, the power amplifier feeds a proportionate voltage to the motor, which steers the antenna in desired direction [3]. Similar logic also holds good for another plane. The block diagram of the system is shown in the Fig. 3.

### A. Auto Mode:

Received AGC voltages say AGC1 and AGC2 are added at the first stage then divided by 2, which is taken as the

reference is added algebraically with the AGC 1. If the transmitting antenna is in the center of the two receiving antennas mounted at pedestal, then the signal received by both the antennas will be equal and it will generate same AGC voltages. So, the output of the subtractor stage will be zero. As there is no error voltage, none of the power transistors is switched on. As a result, pedestal will remain stationery. In case, transmitting antenna is in some angular offset from the center, then the AGC voltages developed by two receivers will not be equal.

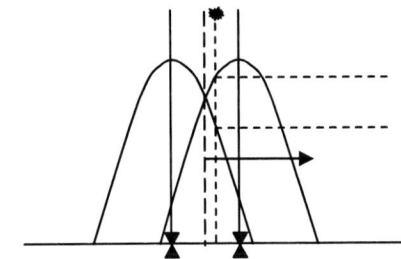

Fig. 2. Design philosophy.

Antenna closer to the transmitting antenna will produce more AGC voltage than that farthest one. So, the developed error will have either positive polarity or negative polarity and corresponding transistor will be switched on and the antenna will move in the direction of transmitting antenna to nullify the error voltage. When the transmitting antenna will be exactly in the center of pedestal, antennas movement will stop.

### B. Manual Mode:

Two potentiometers are used for the manual mode. One potentiometer is mounted at the controller side for the command given from user and the other one is coupled with the motor shaft at the pedestal, which senses the pedestal position. The command given and the position of the antenna, both are compared and if there is a difference, an actuating command (error) is fed to the second stage of the circuit. Further, this error is amplified and the corresponding transistor drives the motor and rotates the pedestal in a desired direction. It stops when the signal of both the voltages at the

potentiometers are same. The polarity of the error generated decides the direction of rotation of the antenna pedestal.

## III. CIRCUIT DESIGN

TL074 OPAMP is used in adder and amplification stages. This OPAMP typically have low noise, high input impedance,

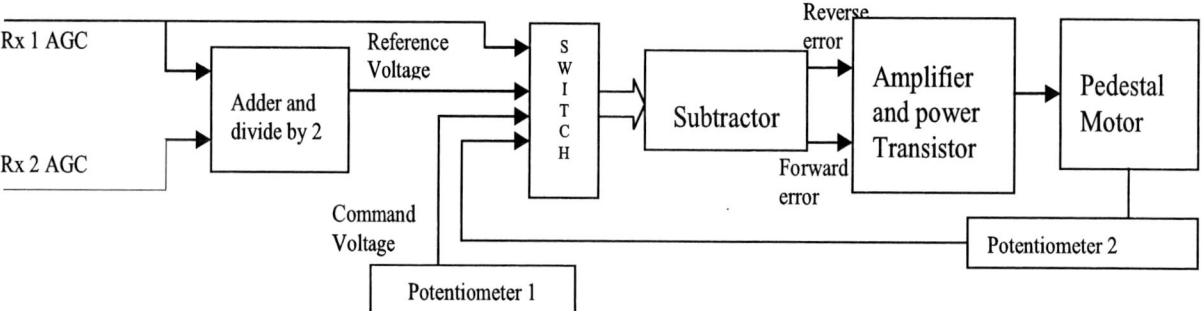

Fig. 3. Block diagram of controller for each plane.

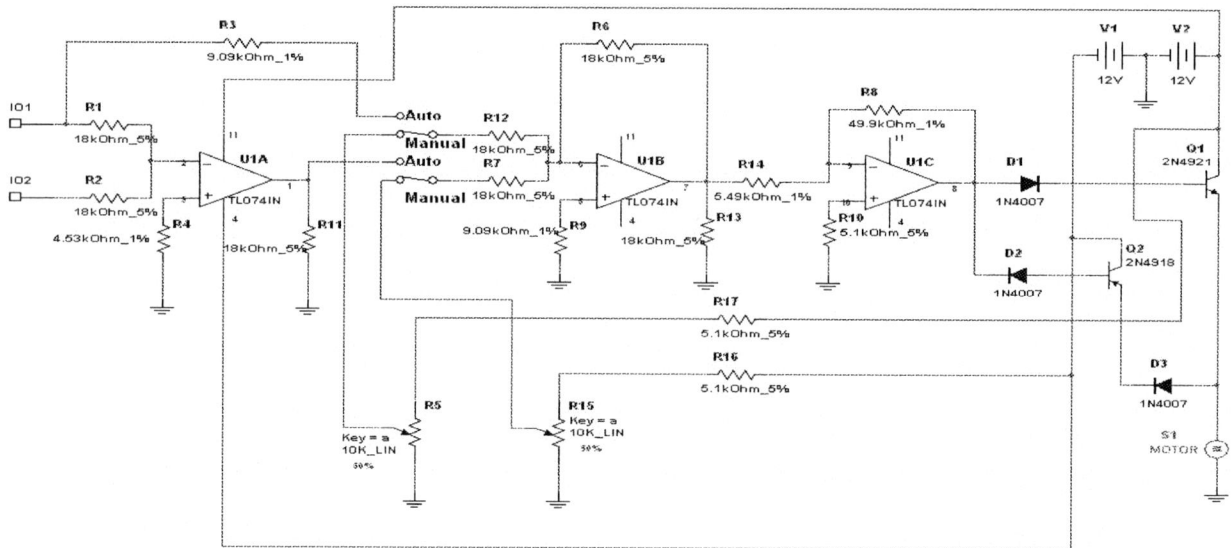

Fig. 4. Circuit diagram of the controller for each plane.

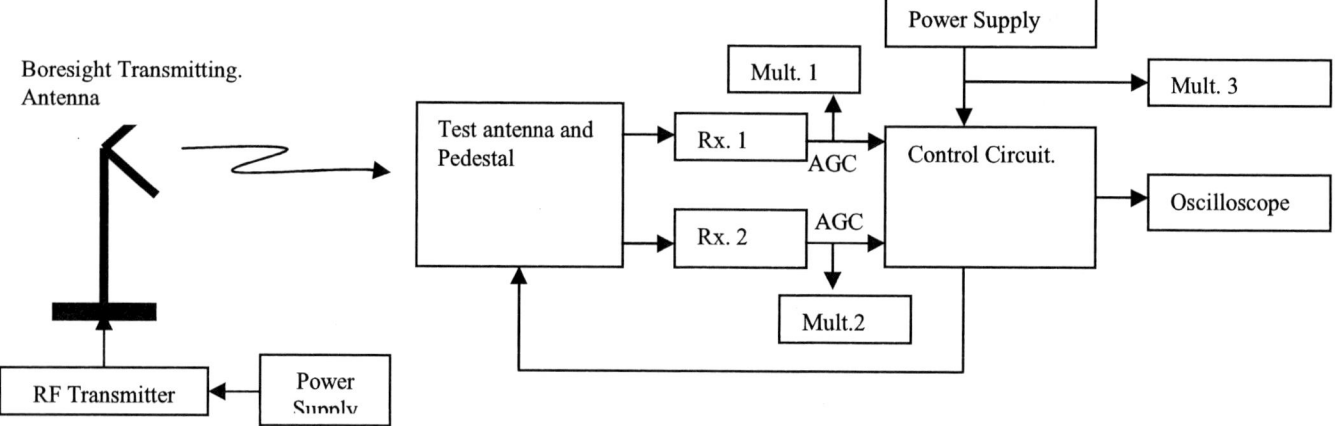

Fig. 5. Block diagram of test setup.

72

internal frequency compensation, latch up free operation, high slew rate (13V/µ sec.) and low THD features. First two stages are used as the adder and third stage is used as the amplifier. Two transistors, 2N4921 (NPN), Q1 and 2N4918 (PNP), Q2 are used at the power stage. If error is positive, then transistor Q1 conducts and motor rotates in the clockwise direction. If the error is negative then transistor Q2 conducts and motor rotates in the anticlockwise direction. A permanent magnet direct current (PMDC) motor S1 is used for steering the antenna in each plane. Voltage fed to the motor terminal by the transistors is directly proportional to the error. Diodes are used in the circuit so that OPAMP stage will not sink the current from the power stage.

## IV. TEST SETUP

Experimental studies were carried out at S band (2200-2300 MHz). The prototype system was kept on a tower. RF signal was transmitted from a bore sight antenna system. Bore sight consists of RF transmitter, a power supply and test antenna.

### A. Auto Mode

AGC output of the receivers was processed at the control Circuitry and a corrective voltage is fed to the antenna pedestal. AGC, Errors and degree of rotation were recorded for both the axes.

### B. Manual Mode

Manual command was given with the help of potentiometer and output voltage with respect to degree of rotation was recorded for both the axes. Test set up block diagram for one axis is shown in Fig.5. The photographs of the test set up and prototype are shown in Fig. 6 and Fig. 7.

Fig. 6. Prototype and its test setup.

Fig.7. Prototype and its test setup.

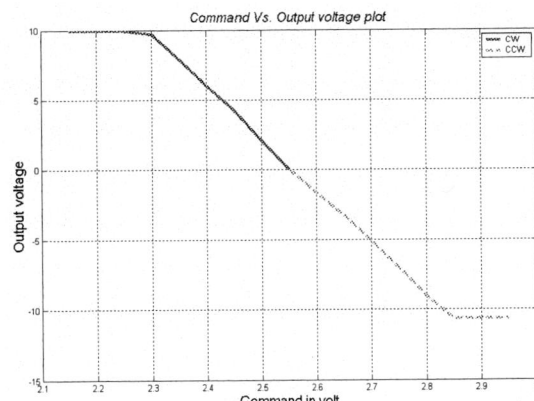

Fig. 8. Command V/s. output voltage Plot.

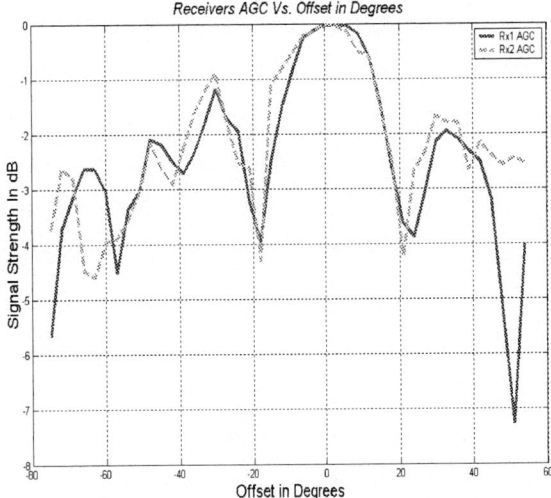

Fig. 9. Antenna pattern.

## V. EXPERIMENTAL RESULTS

### A. Manual Mode

#### 1) Input error vs. output voltage:

The antenna pedestal is driven manually with potentiometer provided to user. Initially both reference and command potentiometers are set at 2.5 volts. Then Command voltage is slowly increased and output voltage to motor is monitored up to its saturation voltage of ± 10V. The Fig. 8 shows a perfect linearity between input command voltage and output voltage from control circuit to the motor from 2.3 volts to 2.8 volts both in clockwise and anti-clockwise direction.

### B. Auto Mode

#### 1) Pattern Measurement and Error Slope:

Keeping the test antenna fixed, antenna under test was rotated in azimuth plane. The AGC voltages generated by the two receivers were plotted together with respect to the angle in degree. The angular offset between two antenna elements was found 3° approximately with beam width of ~30°. The error slope was measured over ±3° and found linear over this range. The combined radiation pattern and the error slope are shown in Fig. 9 and Fig. 10 [4].

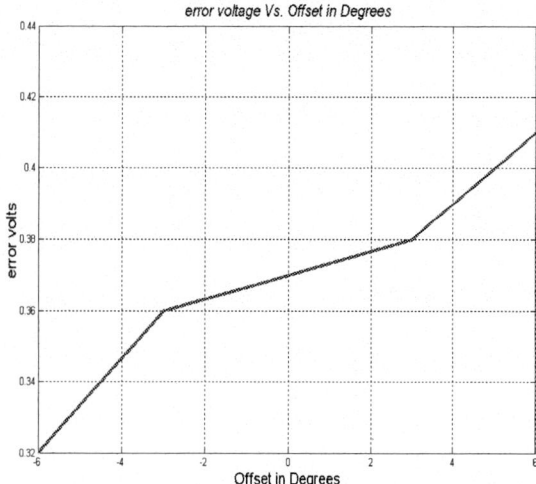

Fig. 10. Error slope.

## VI. CONCLUSION

The proposed system has shown an improved performance in the tracking and range data acquisition system over the manual tracking. The designed prototype demonstrates a better signal strength, automation of antenna steering both in manual as well as in auto mode. The control circuit shows perfect linearity in error slope in both forward and reverse direction. Further, improvement and ruggedness in the potentiometric arrangement on the antenna pedestal will enhance the system's capability and stability. The prototype of controller and set-up of BHTM facilitates control of the antenna remotely in auto and manual mode and proves to be more worthy and reliable as far as reception of data in the initial phase for long range missiles and almost full for the short range missiles is concerned.

## VII. ACKNOWLEDGEMENT

The authors express their sincere thanks to Shri A. K. Checker, Sc'G', Director, Integrated Test Range Chandipur for valuable suggestions and facilities provided for completing the studies and realization of the prototype. Also Technical discussions held with Dr. S.N. Banerjee, Sc. 'F', JD(Instrumentation), Mrs. B. Sucharita, Sc. 'E', Head Telemetry, are gratefully acknowledged. The authors would like to thank Dr. B. K. Das, Sc. 'F', JD(Optronics), members of the Technical Expert Committee and HRD group, Integrated Test Range, Chandipur for their valuable suggestions.

## VIII. REFERENCES

[1] Richard Vlentine, '*Motor control electronics hand book*', Mc Graw Hill.
[2] Horen, S., 'Introduction to PCM Telemetering Systems', CRC Press, New York, 2002.
[3] Bill Drury , *The Control Techniques Drives and Controls Handbook*
[4] O.J. Stroke, '*Telemetry Computer Systems*', Instrument Society of America.

## IX. BIOGRAPHIES

**Parveen Kumar** received B.E. degree in Electrical Engineering from National Institute of Technology Kurukshetra, Kurukshetra, Haryana, India in 2000. In 2000, he joined S. M. Creative Electronics Ltd., Gorgon as an R & D Executive in the Power Electronics field. He worked as an R & D Engineer in Autometer alliance Ltd., Noida. in the area of power electronics. Then from 2002 onward, he is serving as Scientist, Integrated Test Range, Defence R & D Organisation, Chandipur, Ministry of Defence India & working in the area of Ground Telemetry instrumentation systems. His research activities were in the control and Power electronics based design and Development of Range Telemetry Instrumentation system.

**Asit Kumar Pradhan** received Diploma in Electronics and Telecommunication Engineering from U.C.P. Engineering School Berhampur, Orissa, India in 1988. He joined O. S. E. B., Orissa as Junior Engineer Telecom in 1989 and area of work was Power line Carrier communication. Then from, 1997 onwards he is working as Senior Technical Assistant in Integrated Test Range, Chandipur. His field of Activities is in the area of Telemetry instrumentation and R. F. He has also completed AMIE in 2004. His research activities were in the control and Power electronics based design and Development of Range Telemetry Instrumentation system.

**Gautam Sadhukhan** born in 1968 in India. He received B.Sc. (physics) degree from University of Calcutta in 1989. Then he received B.Tech degree in Radio Physics & Electronics, University of Calcutta in 1992 and received M.Tech in Electronics and Electrical Communication from Indian Institute of Technology (IIT), Kharagpur in 2002. In 1994, he joined Defence Electronics Research Laboratory (DLRL), Defence R & D Organization (DRDO), Hyderabad, India. He was involved in design and development of broadband antenna and their characterization. From 1996 onwards he is serving in Integrated Test Range, Chandipur, DRDO, India as Scientist in the area of ground telemetry instrumentation systems. His research activities were in the area of VLSI based design and development of Range Instrumentation System and Computer networking.

**2006 IEEE International Conference on Power Electronic, Drives and Energy Systems**

# Design & Development of a High Performance Electronic Starter for Single – Phase Induction Motors

T. P. Shenoy, and J. S. Nirody

*Abstract*--Single Phase Induction Motors are not self-starting unless they are started as quasi-two phase motors. This necessitates a starting winding, with or with out starting capacitors, which is switched out once motor is accelerated. A centrifugal gear having flying weights and a mechanical contact type switch is most commonly used to do this function. This paper describes a non-contact type; non – mechanical type, electronic switch that eliminates contact sparking and mechanical assembly criticality, thereby improving the reliability, motor life and servicing cost. The various types of electronic starters available in market operate on different principles like time delay or capacitor voltage or winding turns ratio or phase difference between start and run windings.

The operation of the electronic starter, described in this paper, is based on actual shaft speed, similar to mechanical switch, but switching the capacitor / winding electronically. The starting capacitor and/or the winding is switched out based on the simple speed feed back data obtained. The switching instant is decided by a micro-processor based logic circuit using speed feedback and the switching is done using single/multiple semi-conductor switches. Electronic starter reduces the inventory cost due to its flexibility.

*Index Terms*— Drives, Electrolytic Capacitors, Electronic Switching System, Feedback, Induction Motors, Life Estimation, Microprocessors, Semiconductor Switches, Rotating machines, Squirrel Cage Motors.

## I. INTRODUCTION

SINGLE phase induction motors are extensively used as drives for industrial as well as domestic applications and they form a large segment of total motor market. The main reason for this is the economics involved along with motor ruggedness. Basic single phase induction motor will be having a single winding construction that is excited by a single phase supply. As the magnetic flux created in the motor air gap by such a winding is pulsating type, one will have a pulsating

This work was carried out at the Advance Motor Design & Technology Center operating under the Corporate Research and Development Center of Crompton Greaves Ltd. India.

torque generated by motor, which cannot make the rotor to rotate. To achieve a unidirectional rotation, single phase induction motors are to be operated as quasi two phase motors. This is achieved by having a run winding connected directly to the single phase supply and a start winding is connected through a series connected capacitors to the single phase supply, in a capacitor motor. To achieve the application related need of developing starting torques of the order of 220% full load torque and having a good full load performance, one has to have large value capacitors during starting and same are replaced by low value capacitors on running. The change over from the start capacitors to run capacitors is done at a shaft speed of approximately 80% of the motor synchronous speed, through electromechanical system like a mechanical switch operated by a centrifugal gear. This type of construction will have sparking contacts and the operation of which is dependent on positional accuracy of the centrifugal gear. The sparking contacts are the area of concern when the motor is to be used in explosive atmospheres. Replacing the mechanical switch by an electronic switch takes care of such problems and can also address other pain areas of the mechanical switch system
World over people have tried to make use of different principles as basis for switch operation point determination, like giving time delay or tracking voltage across capacitor etc., each resulting in to a new patent. This paper describes a new method of mimicking the operations of the mechanical switch with centrifugal electronically. The switch operation is totally dependent on shaft speed and operates with better accuracy and reliability as compared to mechanical switch centrifugal gear. The new starter is already been made, tested and found functioning satisfactorily along with meeting the stringent switch life requirements.

## II. CONVETIONAL METHODS

### A. Mechanical Starter

Till the advancement of semi-conductor switches the most popular means of switching off start capacitor and switching on run capacitor was done by mechanical switch operated by a centrifugal gear. A typical construction of such a starter has two parts; viz. a centrifugal gear (CFG) which is press fitted on the rotor shaft (refer Fig. 1) rotating along with it and the other is a mechanical switch popularly known as OC (open circuiting) switch (refer Fig. 2) mounted inside motor that connects the start winding to supply through a start capacitor. During the motor start up the centrifugal gear weights fly out due to the centrifugal force acting on them against the retaining force of a set of springs. This leads in to an axial movement of a lever away from the OC Switch to open the contacts thereby disconnecting the start capacitor. Under standstill condition of motor the CFG lever presses the OC Switch contact to close it so that the start capacitor connected to start winding. Thus depending on the shaft speed, at cut – in speed, contact opens fully to cut out the start capacitor and cut – in speed, contact re-closes to reconnect the start capacitor.

Fig. 1.    Centrifugal gear (shaft mounted component).

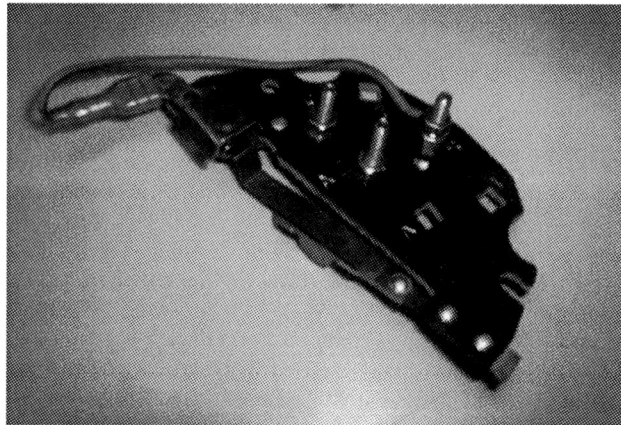

Fig. 2.    Open circuiting switch (end – shield mounted).

With the advent of power electronics and semi-conductor switches like Triacs and Thyristors etc. single phase motor manufacturer are looking in to possibilities of using them in place of the mechanical switch as indicated by various patent applications on electronic starting switches for single phase induction motors.

The electronic starter given in this paper is also exploring one more of such solutions.

*B. Electronic Starter*

Unlike mechanical starter, there is no single electronic starter solution available for single phase motors. To make the electronic starter cost economic to its mechanical counterpart various methods have been tried out of which major are –

- Based on Time delay Principle, which have their logic circuit designed to open the start capacitor

circuit after the lapse of a pre-set delay from the instant of motor switch on.
- Based on Start Capacitor Voltage Pattern Principle, which track the voltage across start capacitor from the instant of motor switch on to open the start capacitor circuit when the voltage changes its pattern from decreasing to increasing mode
- Based on Current through winding principle like operation of a relay when the current in specially wound additional stator winding exceed pre-set value or operation of the electronic switch, based on sensing run winding current using NTC resistor
- Based on Phase Angle Relationship Between start and run winding voltages during acceleration that operates the switch when the phase angle relationship changes from decreasing value to increasing value.

All these ideas, which have been patented, have their own advantages and disadvantages in mimicking the mechanical switch behavior and very few are commercially exploited due to cost economics or inherent weakness of operating principle. The paper describes an attempt made to give a robust design / operation of electronic switch based actual speed feedback as that of the mechanical CFG.

### III. NEW ELECTRONIC STARTER – DETAILS

The new method proposed here with works on the principle of continuously checking the motor shaft speed during starting as well as during normal running conditions and taking a logical decision regarding semi-conductor switch state to be ON or OFF. The decision / logic block is designed to open the switch at 80% of synchronous speed during startup and close the switch back at 50% of the synchronous speed, there by mimicking the mechanical switch driven by centrifugal gear.

The basic block model of the new electronic starter is as shown below in Fig. 3.

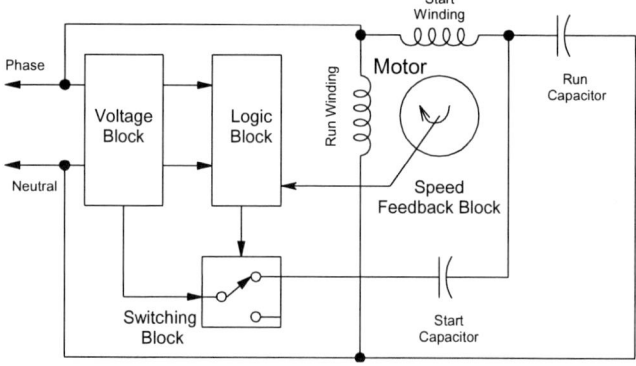

Fig. 3.    Basic model of the electronic starter.

The electronic starter is made up of following blocks –

Voltage Block which generates a stable DC supply for operation of the electronic circuits from single phase AC supply lines

A Logic Block that has micro-processor and its peripherals to make logical decisions regarding state of switch to be at any given instant

A Speed feedback Block made of a simple encoder mounted on rotor shaft and a receiver on end-shield of the motor that gives shaft speed related information

A switching block consisting of semiconductor switch which is switched ON or OFF based on the speed feedback received from speed feedback block, by the logic block and a decision taken thereby

A component level diagram of same is as shown below

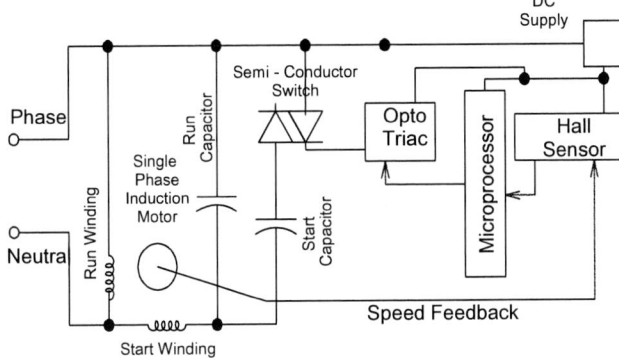

Fig. 4. Component level model of the electronic starter.

The operation the electronic switch is as described below – On application of single phase supply to the electronic switch, its low voltage logic circuits get energized first and the microprocessor sends the gate turn on signal for the semiconductor switch, through a opto-triac, so that the start capacitor is connected with start winding to start and accelerate the motor from its standstill condition. As the motor picks up speed, the hall sensor based speed feedback system output tracking is initiated by the microprocessor. When the shaft speed reaches cut out value, the microprocessor sends the gate turn off signal to the semiconductor switch and the start capacitor is disconnected from start winding circuit. The microprocessor actively tracks the shaft speed to take care of any eventuality of speed falling below cut-in speed. The semiconductor switch gate signal is turned on to reconnect the start capacitor, so that any spurious overload conditions are taken care of.

## IV. PROTOTYPE TEST DETAILS

A prototype of the starting switch suitable for operating single phase induction motors up to 1HP rating was fabricated and tested for its operating point accuracy as well repeated on-off of hot motor after the temperature rise test of the motor. The following oscilloscope recording gives the operating point accuracy evaluated by checking the voltage across the switch vis-à-vis the hall sensor signal generated. The hall sensor output variation is dependent on the magnetic flux polarity reversal it senses during acceleration or on steady

state running of motor shaft and is shaft speed dependent.

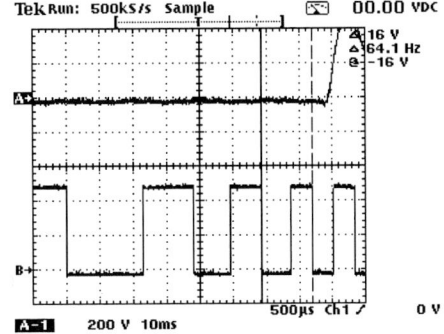

Fig. 5. Oscilloscope waveform record.

The level change waveform of the voltage across the electronic switch terminals and the hall sensor output shown here indicates the hall signal frequency at which the switch operates. In the above given result, for single revolution of motor shaft hall sensor gives three pulse output so that there are six level changes happening and the linking of the speed information to hall output pulse frequency is arrived at using a simple relation of

$$\text{Frequency of Hall Output Signal} = \frac{\text{Shaft Speed} * \text{No. of level changes}}{120}$$

To make the semiconductor switch to operate at a shaft of 1280 rpm and having speed feedback that has with 6 level changes per RPM, as described above the frequency to be set, for the Hall output waveform, at which the switch will be operating is calculated as

$$f = (1280 * 6) / 120 = 64 \text{ Hz.}$$

The result as obtained from oscilloscope recording is tabulated below

TABLE I

COMPARISON OF PREDICTED TO TEST OPERATION OF SWITCH

| Frequency Set for Switch Operation (Prediction) | Frequency at which Switch Operation happens (Test) |
|---|---|
| 64.0 Hz | 64.1 Hz |

In the above given result the actual changeover point for switch was considered at the point where the voltage across the switching device begins to rise and corresponding hall output frequency was measured which was found to be 64.1Hz (refer Fig. 5).

## V. RELIABILITY ANALYSIS

During the design of the new electronic starter an attempt was made to inculcate sufficient reliability in to the new system being developed by carrying out a reliability analysis on the intended circuit vis-à-vis the popular mechanical switch. The following graphs indicate the stage-wise improvements done on electronic starter circuit.

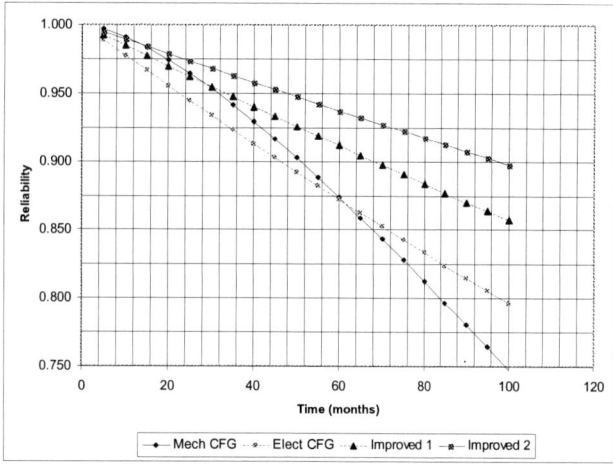

Fig. 6. Reliability prediction curves.

The prediction was based on failure rates derived from MIL-217 standard. The two improvement details are as follows,

Improvement 1: The electrolytic capacitors used in the starter circuit are made to have a higher voltage rating (from 25V to 50V) and higher temperature grade (i.e. from 85°C to 105°C). Rest is same as the original circuit.

Improvement 2: Improvements in stage 1 will remain. In addition, instead of one capacitor of voltage block, two capacitors will be used and the stressing will be reduced to 0.45 instead of 0.9 in the original circuit.

TABLE II

RELIABILITY COMPARISONS

| Time (Months) | Mechanical CFG | Electronic CFG | | |
|---|---|---|---|---|
| | | First design | Improved 1 | Improved 2 |
| 5 | 0.997 | 0.987 | 0.991 | 0.994 |
| 15 | 0.983 | 0.963 | 0.975 | 0.982 |
| 25 | 0.964 | 0.939 | 0.958 | 0.970 |
| 35 | 0.942 | 0.915 | 0.942 | 0.959 |
| 45 | 0.916 | 0.893 | 0.926 | 0.947 |
| 55 | 0.888 | 0.870 | 0.910 | 0.936 |
| 65 | 0.859 | 0.849 | 0.894 | 0.925 |
| 75 | 0.828 | 0.827 | 0.879 | 0.914 |
| 85 | 0.796 | 0.807 | 0.864 | 0.903 |
| 95 | 0.764 | 0.787 | 0.850 | 0.892 |
| 100 | 0.748 | 0.777 | 0.842 | 0.887 |

The reliability index calculations described above clearly indicate a higher reliability for the electronic starter as compared to the mechanical starter, which is also confirmed by the life cycle testing carried out on the electronic starter.

The electronic starter was evaluated for its life cycle also, on a specially developed test jig, by subjecting it to continuous ON-OFF operation so that both cut-in and cut-out settings are utilized and switch experiences the starting current on each switch on. It was observed that the electronic

starter was still functioning satisfactorily even after 2Lakh operations of the switch.

## VI. ELECTRONIC STARTER CONSTRUCTIONS

The electronic starter can be given in various constructions depending on the space available inside the motor, space available in the terminal box, motor enclosure type, winding absolute temperature on continuous duty etc. A few constructions possible are given below

### A. Two PCB Construction

This is suitable for motor having their internal working ambient temperature well above 55 – 60°C. It will have one PCB inside the motor having components that are less sensitive to temperature and the other PCB will be located in motor terminal box carrying all temperature sensitive components.

### B. Single PCB Construction

This is suitable when motor having their internal working ambient temperature below 55 – 60°C. In such cases the PCB can be mounted inside the motor itself.

## VII. CONCLUSIONS

A new concept in electronic starter for single phase induction motors that mimics the conventionally used mechanical starting switch with centrifugal gear for its functionality having operational reliability better than mechanical or other electronic starters is shown to be a practical reality. The major advantages observed of this new electronic starter are –

- Switch Operation dependent on the actual motor speed
- Accurate single point of operation when compared to the large speed band for mechanical switch operation like 900rpm – 1200rpm
- Increased operational life, in excess of 2lakh switching operations
- Higher Reliability Index as compared to mechanical starter
- Mimics mechanical starter for switch cut-in and cut-out operations more accurately.

Patent application for the design described above is already done and field trials are being conducted to establish the lab findings on the actual applications also.

## VIII. REFERENCES

[1] C.G. Vienott, *Theory and Design of Small Induction* Motors, Fractional *Horsepower Single Phase Motors,* E. Clarke, *Circuit Analysis of AC Power Systems*, vol. I. New York: Wiley, 1950, p. 81.

[2] Military Handbook, MIL-HDBK-217F.

[3] James F Gordon, "Motor Starting Circuit", U.S. Patent 3,071,717, Jan. 1, 1963.

[4] Ricky L. Bunch, "Positive Temperature Coefficient Start Winding Protection", U.S. Patent 5,345,126, Sept. 6, 1994.

[5] Richard M. Burkhart, "Starting Device for Single Phase Induction Motor having a Start Capacitor", U.S. Patent 5,559,418, Sept. 24, 1996.

[6] Umakant Dipa Dubhashi, "Motor Starting Circuit", U.S. Patent 6,570,359 B2, May 27, 2003.

[7] Eric A. Depner, James A Butcher, "Motor Start Switch", U.S. Patent 6,204,628 B1, Mar. 20, 2001.

[8] "SINPAC" Switch catalogue – Reliance Motors Start Switch Catalogue

[9] GE Industrial Systems, "Current Type Motor Starting Relays", 3ARR2, 3ARR12, 3ARR12P, 3ARR18, Product Catalogue

## IX. BIOGRAPHIES

**T.Purushotham Shenoy** born in Karnataka, India on Feb' 1957 and did his B.Tech in Electrical Engineering from Karnataka Regional Engineering College, Surathkal, Karnataka. He joined Crompton Greaves Ltd in 1979 and has worked in the fields of conventional and unconventional electric machines and electronic controllers for them. His special fields of interest include induction motors, DC/BLDC Motors, Fan motors and electronic control of single phase induction motors. Presently he is working as Senior Technology Manager, in Advance Motor Design and Technology Center, Corporate R&D and Quality of Crompton Greaves Ltd.

**Jaishankar Sidhanand Nirody** was born in Mumbai in April 1956 and did his Diploma in Electrical Engineering from Somaiya Polytechnic in Mumbai. He joined Crompton Greaves Ltd. In 1977 and worked through various divisions of Crompton Greaves in the fields of AC motors, AC generators, Television, Lighting Electronics and Fan Controllers. Presently he is an independent consultant for development of Consumer Electronics related products.

**2006 IEEE International Conference on Power Electronic, Drives and Energy Systems**

# Transient Analysis of a Single-Phase Self-Excited Induction Generator using a Three-Phase Machine feeding Dynamic Load

S. N. Mahato, M. P. Sharma, and S. P. Singh

*Abstract--* This paper presents the transient behavior of a single-phase self-excited induction generator (SEIG) supplying a dynamic load i.e., induction motor. The generator consists of a three-phase star connected induction machine with three capacitors and a single-phase induction motor load. The dynamic models of the SEIG and the motor load have been developed based on stationary reference frame d-q axes theory and the equations of excitation capacitors are described by three-phase *abc* model. Heavy transients occur during the switching of induction motor and the system becomes unstable. The use of damping resistors across series capacitors is proposed to damp out the transients for the stable operation. Using the damping resistors, the motor can be started up successfully. Simulated results have been compared with the experimental results for unsuccessful starting without damping resistances and successful starting with damping resistances to validate the developed model.

*Index Terms--* Damping resistance, Dynamic load, Dynamic model, Self-excited induction generator.

## I. NOMENCLATURE

$C_p$, $C_s$ : capacitances.
$C_m$ : start/run capacitor of the motor.
$R_s$, $R_r$ : per phase stator and rotor (referred to stator) resistances of the SEIG.
$R_{dsm}$, $R_{qsm}$ : resistance of stator main and auxiliary winding of the motor.
$R_{drm}$, $R_{qrm}$ : rotor main and auxiliary winding resistance referred to stator winding.
$X_{ls}$, $X_{lr}$ : per phase stator and rotor (referred to stator) leakage reactances of the SEIG.
$L_{ls}$, $L_{lr}$ : per phase stator and rotor (referred to stator) leakage inductances of the SEIG.
$L_{ldsm}$, $L_{lqsm}$ : stator main and auxiliary winding leakage inductances of the motor.
$L_{ldrm}$, $L_{lqrm}$ : rotor main and auxiliary winding leakage inductances referred to stator of the motor.

$L_m$ : magnetizing inductance of the SEIG.
$L_{mdm}$, $L_{mqm}$ : d, q components of the magnetizing inductance of the motor.
$L_{dqm}$, $L_{qdm}$ : cross-saturation inductances of the motor.
$I_m$, $I_{Lm}$ : magnetizing current of the SEIG and the motor respectively.
$i_{ds}$, $i_{qs}$ : d-axis and q-axis components of stator currents of the SEIG.
$i_{dr}$, $i_{qr}$ : d-axis and q-axis components of rotor currents (referred to stator) of the SEIG.
$i_{dsm}$, $i_{qsm}$ : main and auxiliary winding currents of the motor.
$i_{drm}$, $i_{qrm}$ : d, q components of rotor currents (referred to stator) of the motor.
$i_{sa}$, $i_{sb}$, $i_{sc}$ : stator phase currents in a, b, c phases of the SEIG.
$i_{cp}$ : current through capacitance $C_p$.
$v_{sa}$, $v_{sb}$, $v_{sc}$ : stator voltages of a, b, c phases of the SEIG.
$v_{ds}$, $v_{qs}$ : d-axis and q-axis components of stator voltages of the SEIG.
$v_{dsm}$, $v_{qsm}$ : voltage across main and auxiliary winding of the motor.
$v_{mc}$ : voltage across the start/run capacitor of the motor.
$V_L$ : output voltage of the generator.
$\omega_r$, $\omega_{rm}$ : rotor speed of the SEIG and the motor respectively.
$T_{shaftm}$ : shaft load torque.

## II. INTRODUCTION

THE squirrel cage induction generators are being increasingly used for power generation from renewable energy sources, such as, wind, biomass, biogas and mini-hydro plants. Since, in remote and rural areas, the electric loads are usually of single-phase types, the single-phase power supply is preferred over three-phase one in order to make the distribution system simple and cost effective. Three-phase induction motors being available over a wide power rating can be used as generators to supply single-phase loads to give better efficiency [1]. The single-phase induction motors used for different home appliances, constitute a major part of the domestic loads. However, a very little information is available on the operation of the SEIG feeding single-phase dynamic load.

---

S. N. Mahato is with Alternate Hydro Energy Centre, Indian Institute of Technology, Roorkee- 247 667, India (e-mail: snmrec@yahoo.co.in).

M. P. Sharma is with Alternate Hydro Energy Centre, Indian Institute of Technology, Roorkee- 247 667, India (e-mail: mpshafah@iitr.ernet.in ).

S. P. Singh is with Department of Electrical Engineering, Indian Institute of Technology, Roorkee- 247 667, India (e-mail: spseefee@iitr.ernet.in).

0-7803-9771-1/06/$25.00 ©2006 IEEE

The steady-state performance of three-phase SEIG with static load has been reported in [2-4]. Chan and Lai [3] has presented the steady-state performance of a single-phase self-regulated self-excited induction using a three-phase machine. Kumaresan [4] has developed genetic algorithm based approach for the analysis of single-phase operation of three-phase SEIG. But very few literatures are available on transients of three-phase SEIG feeding single-phase static load [5-7]. The transient performance of three-phase SEIG supplying single-phase load with C-2C configuration of excitation has been analyzed by Shilpakar et al. [5]. Jain et al. [6] have presented the generalized dynamic model of a delta connected three-phase SEIG under unbalanced operating conditions. Wang and Deng [7] have specified the transient performance of the induction generator under unbalanced excitation capacitor.

Singh et al. [8] shown the unstable behavior of short-shunt SEIG-IM configuration during the starting of the induction motor load and proposed damping resistors across series capacitors to obtain the stable operation. The transient analysis of an SEIG supplying dynamic load has been given in [9]. Considering the importance of single-phase induction motor in domestic use, further investigation is very much required for the single-phase SEIG feeding dynamic load.

This paper therefore deals with the transient behavior of a single-phase SEIG using a three-phase induction machine feeding induction motor load. The dynamic models of the SEIG and the motor have been developed in d-q axes stationary reference frame and the equations of the excitation capacitors are described by three-phase *abc* model. The machine performances are computed by solving the non-linear first order differential equations of the dynamic model using fourth-order Runge-Kutta numerical technique of integration.

### III. DYNAMIC MODEL

The schematic diagram of the single-phase SEIG feeding dynamic load is shown in Fig. 1. The generator consists of a three-phase star connected induction machine with three capacitors, $C_p$ and two $C_s$'s, and a single-phase induction motor. Two capacitors ($C_s$'s) are connected in series with the two phases of the stator winding and one ($C_p$) is connected in parallel with the motor load.

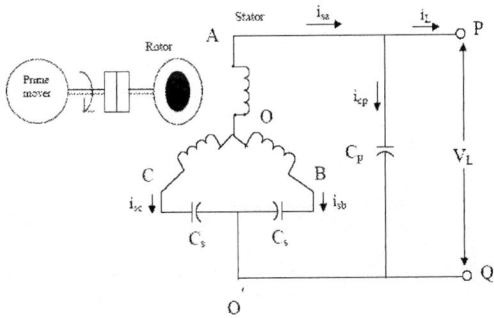

Fig. 1. Connection diagram of the single-phase self-excited induction generator.

### A. Model of the SEIG

The dynamic model of the three-phase squirrel cage induction generator is developed by using stationary d-q axes reference frame and the volt-ampere equations are as:

$$[v] = [R_l][i] + [L_l] p[i] + \omega_r [G_l][i] \tag{1}$$

The current derivative can be expressed as:

$$p[i] = [L_l]^{-1} \{ [v] - [R_l][i] - \omega_r [G_l][i] \} \tag{2}$$

where, $[v]$, $[i]$, $[R_l]$, $[L_l]$ and $[G_l]$ are defined in Appendix A. From Fig.1, the following equations can be written:

$$pv_{cp} = \frac{1}{C_p} i_{cp} = \frac{1}{C_p} (i_{sa} - i_L) \tag{3}$$

$$pv_{bcs} = \frac{1}{C_s} i_{sb} \tag{4}$$

$$pv_{ccs} = \frac{1}{C_s} i_{sc} \tag{5}$$

$$v_{sb} + v_{bcs} - v_{ccs} - v_{sc} = 0 \tag{6}$$

$$\text{and,} \quad v_{sa} + v_{cp} - v_{bcs} - v_{sb} = 0 \tag{7}$$

where $v_{cp}$, $v_{bcs}$ and $v_{ccs}$ are voltages across capacitances $C_p$, $C_s$ of phase B and $C_s$ of phase C respectively and $p = (d/dt)$ is a time derivative operator.

Expressing the stator phase variables of equations (6) and (7) in terms of d-q variables, we get,

$$v_{ds} = \frac{v_{bcs} + v_{ccs} - 2v_{cp}}{3} \tag{8}$$

$$\text{and,} \quad v_{qs} = \frac{v_{bcs} - v_{ccs}}{\sqrt{3}} \tag{9}$$

where $v_{ds}$ and $v_{qs}$ are d-axis and q-axis components of stator voltages respectively.

The magnetizing inductance depends on the instantaneous value of the magnetizing current. The magnetizing current must be calculated at each step of the integration as:

$$i_m = \sqrt{i_{md}^2 + i_{mq}^2} / \sqrt{2},$$

where $i_{md} = i_{ds} + i_{dr}$ and $i_{mq} = i_{qs} + i_{qr}$ are direct and quadrature axes components of the magnetizing current.

The torque balance equation of SEIG is defined as:

$$T_{\text{shaft}} = T_e + J \left( \frac{2}{P} \right) p\omega_r \tag{10}$$

where the developed electromagnetic torque of the SEIG is expressed as:

$$T_e = \left( \frac{3P}{4} \right) L_m(i_{dr} i_{qs} - i_{qr} i_{ds}) \tag{11}$$

The shaft torque, $T_{\text{shaft}}$ of prime-mover and speed is represented by a linear curve given as:

$$T_{\text{shaft}} = k_1 - k_2 \omega_r$$

where $k_1$ and $k_2$ are constants. $J$ is the moment of inertia of the induction machine including the machine (prime-mover) coupled on its shaft.

From equation (10), we get,

$$p\,\omega_r = \frac{P}{2J}(T_{\text{shaft}} - T_e) \qquad (12)$$

The differential equations (2), (3), (4), (5) and (12) describe the dynamic model of the three-phase SEIG for single-phase power generation.

### B. Model of the single-phase induction motor

The dynamic model of the single-phase induction motor (shown in Fig. 2) has been developed in d-q axes stationary reference frame. The voltage-current equations can be written as:

$$[v_m] = [R_2][i_L] + [L_2]\,p[i_L] + \omega_{rm}[G_2][i_L] \qquad (13)$$

The current derivative can be expressed as:

$$p[i_L] = [L_2]^{-1}\{\,[v_m] - [R_2][i_L] - \omega_{rm}[G_2][i_L]\,\} \qquad (14)$$

where $[v_m]$, $[i_L]$, $[R_2]$, $[L_2]$ and $[G_2]$ are defined in Appendix A. The magnetizing current space vector is evaluated as:

$$i_{Lm} = \sqrt{i_{Lmd}^2 + i_{Lmq}^2}\,/\,\sqrt{2}$$

where $i_{Lmd} = i_{dsm} + i_{drm}$ and $i_{Lmq} = i_{qsm} + i_{qrm}$ are direct and quadrature axes components of the magnetizing current space vector.

The derivative of the voltage across the capacitor connected in series with the auxiliary winding is defined as:

$$pv_{mc} = \frac{1}{C_m} i_{qsm} \qquad (15)$$

The equation of the mechanical motion is given by,

$$T_{em} = T_{\text{shaftm}} + J_m\left(\frac{2}{P_m}\right)p\,\omega_{rm} \qquad (16)$$

From (16), the speed derivative may be defined as:

$$p\,\omega_{rm} = \frac{P_m}{2J_m}(T_{em} - T_{\text{shaftm}}) \qquad (17)$$

Hence, the equations (14), (15) and (17) give the dynamic model of the single-phase induction motor.

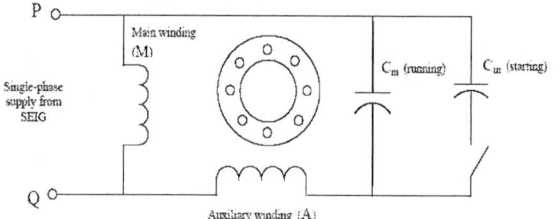

Fig. 2. Circuit diagram of the single-phase induction motor.

## IV. RESULTS AND DISCUSSION

The simulated and experimental results are presented for starting of the induction motor load with different values of capacitors and also with and without damping resistances. In the laboratory test rig, a three-phase squirrel cage induction machine is coupled to a separately excited DC motor, which operates as the prime-mover. One single-phase induction motor is used as dynamic load. The parameters and the magnetization characteristics of both the induction machines obtained by DC resistance test, block rotor test and

synchronous speed test are given in Appendix B. From the steady-state model of the generator, it has been found that to obtain a voltage of 240 V at no-load, the required capacitances are $C_p = C_s = 44\ \mu F$. The performance of the SEIG has been studied during switching of the induction motor load for different cases as follows:

### A. Without damping resistance

The generator is initially excited at no-load with $C_p = C_s = 44\ \mu F$. Keeping the values of the capacitors unchanged when the single-phase motor load is switched on to the SEIG terminals, the inrush transients are obtained in both the generator and motor currents. The motor cannot be started-up successfully and highly oscillatory behavior of sustained nature, known as subsynchronous resonance (SSR) is observed. Figures 3 and 4 show such oscillatory waveforms of the voltage and current of the phase A of the generator, and main winding of the motor respectively.

The starting of the single-phase motor without any load has been attempted by changing the capacitances $C_p$ and $C_s$. The waveforms of the voltage and current of main winding of the motor when the value of $C_p$ is changed to 80 $\mu F$ at the instant of switching of the motor load keeping $C_s$ constant, are shown in Fig. 5. It is found that sustained oscillations are present in this case also and the system becomes unstable. In the similar way, attempts have been made to start the motor successfully by increasing or decreasing the values of $C_p$ and/or $C_s$ at the instant of switching of the motor, but without result. It is found that sustained oscillations are present in all the cases. Hence, simply by changing the values of the capacitors at the instant of switching of the motor, it is not possible to start it successfully.

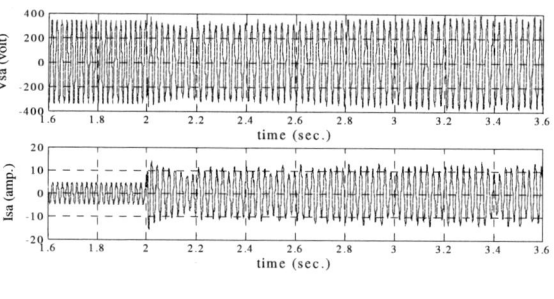

(a) Simulated

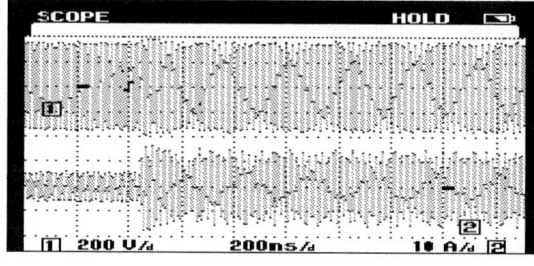

(b) Experimental

Fig. 3. Voltage and current waveforms of phase A of the SEIG during switching of the motor load with constant $C_p$ and $C_s$.

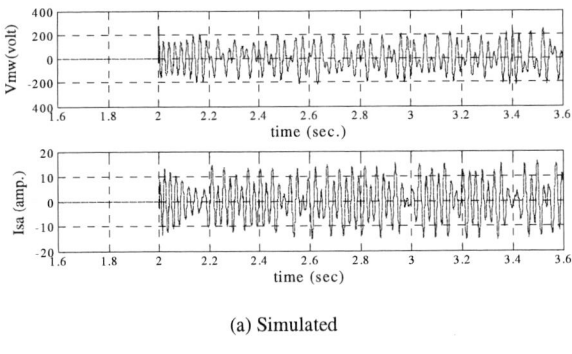

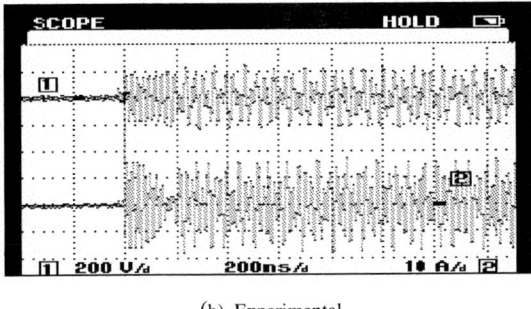

(a) Simulated

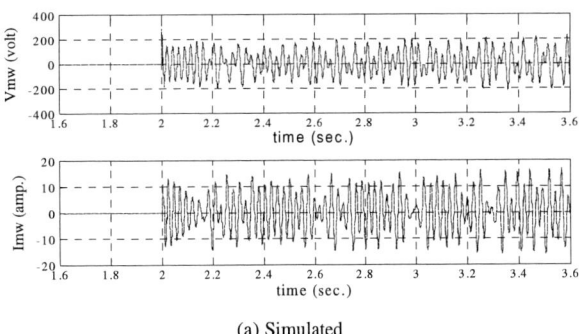

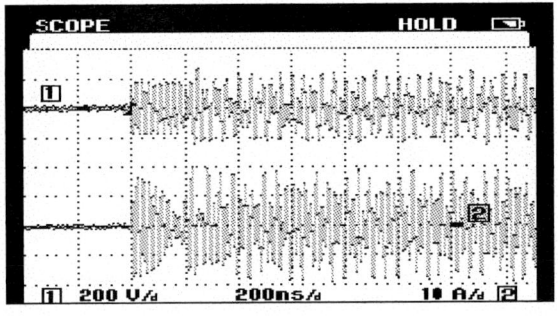

(a) Simulated

(b) Experimental

Fig. 4. Voltage and current waveforms of main winding of the single-phase motor during switching on with constant $C_p$ and $C_s$.

(b) Experimental

Fig. 5. Voltage and current waveforms of the main winding of the motor when the value of $C_p$ is increased at the time of switching of the motor load.

## B. With damping resistance

The use of damping resistances across the capacitors connected in series with two phases of the stator winding of the SEIG is proposed for successful starting of the motor. The generator is initially excited at no-load with $C_p = C_s = 44$ μF. After successful build-up of voltage, the motor is switched on to the terminals of the SEIG. Simultaneously the suitable values of damping resistances are connected across the capacitors of phases B and C of the generator to damp out the starting oscillations. Fig. 6 shows the experimentally obtained variations of damping resistances with the capacitors ($C_p = C_s$) required at the time of switching of the motor for its successful starting. It is found that successful starting of the motor is possible for different values of capacitances ranging from 44 μF to 80 μF with suitable values of damping resistances. The values of capacitances cannot be taken beyond 80 μF, since it is found that the stator current of phase B exceeds the rated current of the generator for the values of the capacitances above 80 μF. If the values of the capacitances are kept unchanged (i.e., $C_p = C_s = 44$ μF) after the switching of the motor, the damping resistances required for successful starting are $R_b = 28$ Ω and $R_c = 400$ Ω, as given in Fig. 6.

The voltage and current waveforms of the main winding of the motor during starting of the motor with capacitances $C_p = C_s = 44$ μF and damping resistances $R_b = 28$ Ω and $R_c = 400$ Ω which are connected across phases B and C respectively of the generator at the time of switching of the motor are given in Fig. 7. It is found that the motor is started up successfully using the damping resistances.

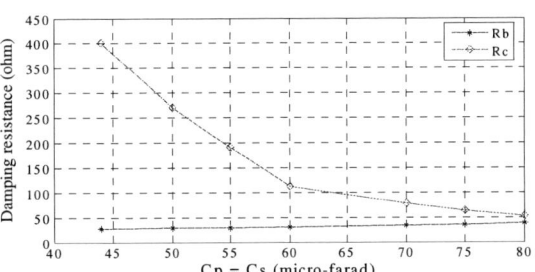

Fig. 6. Variation of damping resistances with capacitances for successful starting of the motor.

Table-I indicates the effect of damping resistances on starting of motor load. It is found that successful starting is possible for $C_p = C_s = 44$ μF with damping resistances $R_b = 28$ Ω and $R_c = 400$ Ω . If the higher values of damping resistances (i.e., $R_b = 50$ Ω and $R_c = 500$ Ω ) are used keeping the values of the capacitors unchanged, there is sustained oscillation and with the lower values of the damping resistances (i.e., $R_b = 10$ Ω and $R_c = 300$ Ω ), the voltage of the generator collapses. It has been observed that after successful starting of the motor without any load, if the damping resistances are increased, the oscillation occurs. Hence, the damping resistances should be kept in the circuit while the motor is at no-load.

After successful starting of the motor with $C_p = C_s = 44$ μF and damping resistances $R_b = 28$ Ω and $R_c = 400$ Ω, the output voltage of the generator becomes 220 V. Now if the load on the motor is gradually increased, the output voltage

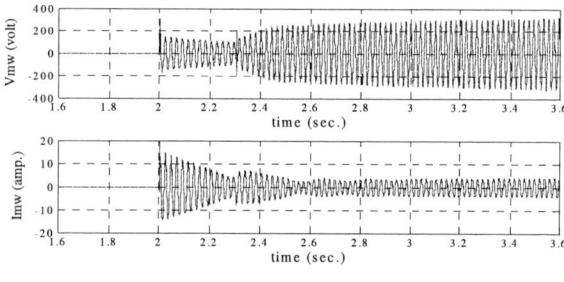

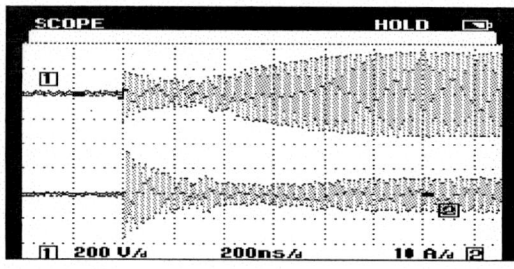

(a) Simulated

(b) Experimental

Fig. 7. Voltage and current waveforms of the main winding of the motor during switching on with damping resistance.

reduces. To keep the output voltage constant at this value, either the capacitances have to be increased if damping resistances are kept unchanged, or the damping resistances have to be changed keeping the capacitors' values fixed. The variations of damping resistances up to a motor load of 700 W have been given in Fig. 8 which shows that for successful starting of the motor at no-load, $R_b = 28\ \Omega$ and $R_c = 400\ \Omega$. With increase of load on the motor up to 580 W (i.e., 26.4 % of the rating of the three-phase induction machine), $R_b$ is kept constant at 28 Ω and $R_c$ is increased gradually. At 580 W, $R_b = 28\ \Omega$ and $R_c$ is removed. Above this load, $R_b$ is reduced gradually to maintain the terminal voltage constant. Also, it is found that the motor draws 200 W from the generator at no-load.

## V. CONCLUSION

The transient behavior of a three-phase SEIG feeding single-phase dynamic load is presented. It is found that the motor cannot be started up successfully while switched on to the SEIG terminals directly, and sustained oscillation, named as SSR is observed. Attempts have been made for successful starting of the motor by changing the values of the capacitors, but without result. For successful start-up of the motor, use of damping resistances in parallel with the series capacitors is proposed. It is seen that with suitable value of damping resistances, the motor can be started-up successfully. During starting, the lower value of damping resistance results into voltage    collapse, while the higher value results in sustained oscillations. With the increase in motor load, either the capacitances may be increased or the damping resistances may be changed (i.e., $R_b$ is reduced or $R_c$ is increased) to maintain the constant terminal voltage. Up to motor load of 26.4 % of the rating of the three-phase induction machine, $R_c$ may be

increased keeping $R_b$ fixed and beyond this load, $R_c$ is removed and $R_b$ is decreased gradually to maintain constant terminal voltage and hence only one damping resistance is sufficient to damp out the oscillations. The experimental results show close agreement with the simulated results and thus the proposed model is validated.

TABLE I
EFFECT OF DAMPING RESISTANCE ON STARTING OF THE MOTOR

| Sl No. | Values of $C_p$ and $C_s$ before switching on the motor | | Values of $C_p$ and $C_s$ after switching on the motor | | Values of damping resistances | | Remarks |
|---|---|---|---|---|---|---|---|
| | $C_p$ (μF) | $C_s$ (μF) | $C_p$ (μF) | $C_s$ (μF) | $R_b$ (Ω) | $R_c$ (Ω) | |
| 1 | 44 | 44 | 44 | 44 | 50 | 500 | Sustained oscillation |
| 2 | 44 | 44 | 44 | 44 | 28 | 400 | Successful starting |
| 3 | 44 | 44 | 44 | 44 | 10 | 300 | Voltage collapse |

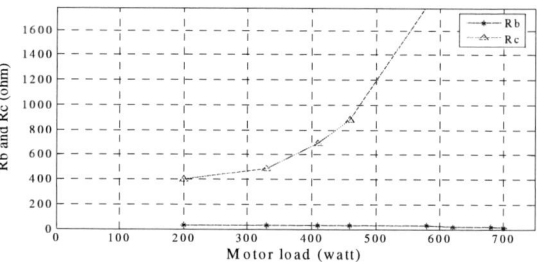

Fig. 8. Variation of $R_b$ and Rc with increase in load on the motor.

## VI. APPENDIX A

The matrices of equation (1) are defined as:
$$[v] = [v_{ds}\ v_{qs}\ v_{dr}\ v_{qr}]^T, \qquad [i] = [i_{ds}\ i_{qs}\ i_{dr}\ i_{qr}]^T$$
$$[R_1] = \text{diag}\ [R_s\ R_s\ R_r\ R_r]$$

$$[L_1] = \begin{bmatrix} L_{ls} + L_m & 0 & L_m & 0 \\ 0 & L_{ls} + L_m & 0 & L_m \\ L_m & 0 & L_{lr} + L_m & 0 \\ 0 & L_m & 0 & L_{lr} + L_m \end{bmatrix}$$

$$[G_1] = \begin{bmatrix} 0 & 0 & 0 & 0 \\ 0 & 0 & 0 & 0 \\ 0 & -L_m & 0 & -L_{lr} - L_m \\ L_m & 0 & L_{lr} + L_m & 0 \end{bmatrix}$$

Here, suffixes d, q refer to d and q axis (in stationary reference frame), and s and r refer to stator and rotor.
The matrices of equation (13) are defined as:
$$[v_m] = [v_{dsm}\ v_{qsm}\ v_{drm}\ v_{qrm}]^T, \qquad [i_L] = [i_{dsm}\ i_{qsm}\ i_{drm}\ i_{qrm}]^T$$
$$[R_2] = \text{diag}\ [R_{dsm}\ R_{qsm}\ R_{drm}\ R_{qrm}]$$

$$[L_2] = \begin{bmatrix} L_{ldsm}+L_{mdm} & L_{dqm} & L_{mdm} & L_{dqm} \\ L_{qdm} & L_{lqsm}+L_{mqm} & L_{qdm} & L_{mqm} \\ L_{mdm} & L_{dqm} & L_{ldrm}+L_{mdm} & L_{dqm} \\ L_{qdm} & L_{mqm} & L_{qdm} & L_{lqrm}+L_{mqm} \end{bmatrix}$$

$$[G_2] = \begin{bmatrix} 0 & 0 & 0 & 0 \\ 0 & 0 & 0 & 0 \\ 0 & \dfrac{L_{mqm}}{a} & 0 & \dfrac{(L_{lqrm}+L_{mqm})}{a} \\ aL_{mdm} & 0 & a(L_{ldrm}+L_{mdm}) & 0 \end{bmatrix}$$

The transformation of stator phase variables to stationary reference frame d-q variables is carried out by the transformation as follows:

$$[V_{dq0}] = [K_s][V_{abc}] \qquad (A\text{-}1)$$

where, the transformation matrix $[K_s]$ is given by

$$[K_s] = \frac{2}{3} \begin{bmatrix} \cos\theta & \cos(\theta-\phi) & \cos(\theta+\phi) \\ \sin\theta & \sin(\theta-\phi) & \sin(\theta+\phi) \\ \dfrac{1}{2} & \dfrac{1}{2} & \dfrac{1}{2} \end{bmatrix}$$

where $\phi = 120°$ and $\theta$ is the angle between phase-$a$ axis and $d$-axis. Selecting the $d$-axis to align phase-$a$ axis, in stationary reference frame, $\theta = 0°$.

## VII. APPENDIX B

The specifications of the induction machines are given below:

### A. Induction generator

2.2 kW, 3-phase, 4-pole, 50 Hz, 415 V, 4.5 A, star connected, 1440 rpm, $R_s = 3.735\ \Omega$, $R_r = 2.91\ \Omega$, $X_{ls} = 4.727$ $\Omega$, $X_{lr} = 4.727\ \Omega$. Base impedance = 53.24 $\Omega$. Base speed = 1500 rpm.

The magnetizing inductance $L_m$ is related to the magnetizing current in the following manner:

$$\begin{aligned} L_m &= 0.3177 & &\text{for } I_m \le 0.75 \\ &= 0.3502 - 0.0349\,I_m - 0.0017 I_m^2 & &\text{for } 0.75 < I_m \le 4.25 \\ &= 0.17667 & &\text{for } I_m > 4.25 \end{aligned}$$

### B. Induction motor

1 hp, 230 V, 6 A, 4-poles, capacitor-start capacitor-run single-phase induction motor.

$R_{dsm} = 3.41\ \Omega$, $R_{drm} = 4.37\ \Omega$, $X_{ldsm} = X_{ldrm} = 3.99\ \Omega$, $R_{qsm} = 11.22\ \Omega$, $R_{qrm} = 8.01\ \Omega$, $X_{lqsm} = X_{lqrm} = 6.433\ \Omega$, Turns ratio, $a = N_q/N_d = 1.4$, and $C_m = 100\ \mu F$ for starting and 6.3 $\mu F$ for running.

The magnetization curves are modeled as given under based on test data:

Main winding :

$$\begin{aligned} L_{mdm} &= 0.3204 & &\text{for } I_{Lm} \le 1.22 \\ &= -0.6692 + 1.695\,I_{Lm} - 0.9292 I_{Lm}^2 + 0.168\,I_{Lm}^3 \\ & & &\text{for } 1.22 < I_{Lm} \le 2.24 \end{aligned}$$

$$\begin{aligned} &= 0.4523 - 0.0476\,I_{Lm} + 0.00164\,I_{Lm}^2 \\ & \qquad\qquad \text{for } 2.24 < I_{Lm} \le 5.32 \\ &= 0.244 \qquad\qquad\quad \text{for } I_{Lm} > 5.32 \end{aligned}$$

Auxiliary winding :

$$\begin{aligned} L_{mqm} &= 0.662 & &\text{for } I_{Lm} \le 0.93 \\ &= -2.074 + 7.066\,I_{Lm} - 6.03\,I_{Lm}^2 + 1.713\,I_{Lm}^3 \\ & & &\text{for } 0.93 < I_{Lm} \le 1.32 \\ &= 0.6856 + 0.0427\,I_{Lm} - 0.0317\,I_{Lm}^2 \\ & & &\text{for } 1.32 < I_{Lm} \le 2.29 \\ &= 0.6158 & &\text{for } I_{Lm} > 2.29 \end{aligned}$$

## VIII. REFERENCES

[1] T. F. Chan, "Performance analysis of a three-phase induction generator connected to a single-phase power system", IEEE Trans. on Energy Conversion, vol. 13, no. 3, pp. 205-213, September1998.

[2] T. Fukami, Y. Kaburaki, S. Kawahara and T. Miyamoto, "Performance analysis of a self-regulated self-excited single-phase induction generator using a three-phase machine", IEEE Trans. on Energy Conversion, Vol. 14, no. 3, pp. 622-627, Sept. 1999.

[3] T. F. Chan and L. L. Lai, "Steady-state analysis and performance of a single-phase self-regulated self-excited induction generator", IEE Proceedings, Generation, Transmission and Distribution, Vol. 149, Issue 2, pp. 233-241, 2002.

[4] N. Kumaresan, "Analysis and control of three-phase self-excited induction generators supplying single-phase AC and DC loads", IEE Proceedings, Electric Power Applications, Vol. 152, no. 3, pp. 739-747, May 2005.

[5] L. B. Shilpakar, B. Singh and B. P. Singh, "Dynamic behavior of three-phase self-excited induction generator for single-phase power generation", Electric Power Systems Research, Vol. 48, pp. 37-44, 1998.

[6] S. K. Jain, J. D. Sharma and S. P. Singh, "Transient performance of three-phase self-excited induction generator during balanced and unbalanced faults", IEE Proceedings, Generation, Transmission and Distribution, Vol. 149, no.1, pp. 50-57, 2002.

[7] L. Wang and R. Y. Deng, "Transient performance of an isolated induction generator under unbalanced excitation capacitor", IEEE Trans. on Energy Conversion, Vol. 14, no. 4, pp. 887-893, 1999.

[8] S. P. Singh, S. K. Jain and J. Sharma, "Voltagr regulation optimization of compensated self-excited induction generator with dynamic load", IEEE Trans. on Energy Conversion, Vol. 19, no. 4, pp. 724-732, 2004.

[9] B. Singh, L. Sridhar and C. S. Jha, "Transient analysis of self-excited induction generator supplying dynamic load", Electric Machines and Power Systems, Vol. 27, pp. 941-954, 1999.

## IX. BIOGRAPHIES

**S. N. Mahato** obtained B.E. in Electrical Engineering in 1992 and M.Tech. in 2002 from National Institute of Technology, Durgapur. He is lecturer in the Department of Electrical Engineering, National Institute of Technology, Durgapur. He is currently a research scholar at Alternate Hydro Energy Centre, Indian Institute of Technology, Roorkee. His research interests are non-conventional energy sources.

**Dr. M. P. Sharma** has been working as Senior Scientific Officer at Alternate Hydro Energy Centre, Indian Institute of Technology, Roorkee since the last 20 years. His area of research are Renewable Energy Sources with special reference to Modelling of IRES, Hybrid Energy Systems, Modelling of Induction Generators, Energy Conservation and Environment Impact Assessment of renewables and other projects.

**Dr. S. P. Singh** received B.Sc. in Electrical Engineering from Aligarh Muslim University, Aligarh in 1978, ME in 1980 and PhD in 1993 from Univeristy of Roorkee, India. He is serving as Associate Professor at Indian Institute of Technology, Roorkee. His area of research includes power plant operation and control, electric machine analysis, self and line excited induction generators.

**2006 IEEE International Conference on Power Electronic, Drives and Energy Systems**

# Performance Analysis of a Three-Phase Squirrel-Cage Induction Motor under Unbalanced Sinusoidal and Balanced Non-Sinusoidal Supply Voltages

C. Thanga Raj, *Student Member, IEEE*, Pramod Agarwal, *Member, IEEE,* and S. P. Srivastava

*Abstract*--As a result of the increasing use of electronic devices and other non-linear loads, the waveforms of the electricity supply voltage are being distorted and inequalities are appearing between the phases. This deterioration is associated with problems of electromagnetic incompatibility and reductions in the efficiency of loads such as motors. This paper investigates the negative effects of unbalanced sinusoidal voltage (0.96% unbalance and 2.9% THD) which always present in the power supply over balanced (inverter supply) non-sinusoidal voltage (23.7% THD) on the performance of induction motor in terms of line currents, power factor and efficiency. Under both supply conditions, the performance of a 5 HP three-phase squirrel cage induction motor was measured through a real load test. According to the test results and analysis, the negative effects of unbalanced sinusoidal voltage are more than the balanced non-sinusoidal voltage on the motor's performance. The financial losses caused by unbalanced voltage of the same motor are determined.

*Index Terms*--Economics, induction motors, non-sinusoidal voltage, power quality, unbalanced voltage.

## I. INTRODUCTION

UNTIL around twenty years ago, electrical systems were formed principally by linear loads. The technological developments of the last twenty years have led to an ever-increasing use of electronic devices, constituting non-linear loads which require non-sinusoidal current. The propagation of these currents through the supply network means that the supply voltage is being distorted and is taking on non-sinusoidal forms, suffering unbalances and asymmetries between the phases.

In practice, induction machines experience overvoltages and undervoltages, depending on the location of the motor and the length of the feeder used. During peak hours, some

---

C. Thanga Raj is with Department of Electrical Engineering, Indian Institute of Technology Roorkee, India (e-mail: optimalraj@yahoo.com).

P. Agarwal is with Department of Electrical Engineering, Indian Institute of Technology Roorkee, India (e-mail: pramgfee@iitr.ernet.in).

S.P. Srivastava is with Kumaon College of Engineering, Nainital, India, on leave from Indian Institute of Technology Roorkee, India (e-mail: satyafee@iitr.ernet.in).

customers with three phase motor could experience minimum voltage guaranteed by the supply utility. Furthermore, the supply voltage is not always balanced. Therefore, the motor will experience a combination of over- or undervoltages with unbalance voltages [1]. Induction motors, even under normal operating conditions, involving perfectly sinusoidal voltage supply produce a relatively limited amount of current harmonics due to the winding arrangement and the iron core non linear behavior. In case of considerably distorted voltage supply, in centre parts, the induction motor performance is heavily affected and derating must be applied. The mentioned derating in motor characteristics is strongly dependent on the harmonic content of the voltage waveform, both in magnitude and other, and its analysis is mainly associated with the involved iron losses [5]. Non-sinusoidal voltages applied to electric machines may cause overheating, pulsating torques, or noise. In addition to across the line applications, adjustable speed drive motors are fed from inverters that can produce significant voltage distortion [7].

Due to voltage unbalance, induction motors face four kinds of problems. First, the machine cannot produce its full torque as the inversely rotating magnetic field of the negative-sequence system causes a negative braking torque that has to be subtracted from the base torque linked to the normal rotating magnetic field. Secondly, the bearings may suffer mechanical damage because of induced torque components at double system frequency. Third, reduction in efficiency and finally, the stator and, especially, the rotor are heated excessively, possibly leading to faster thermal ageing. Chapter IV and V analyzed the operating performance of a three-phase squirrel-cage induction motor under unbalanced sinusoidal (0.96% unbalance, 2.9% Total Harmonic Distortion (THD)) and balanced non-sinusoidal (inverter) supply voltages (23.7% THD) through a real load test and the economic analysis included in the chapter VI.

## II. TEST PLAN, INSTRUMENTATION

The testing of the 4 pole, 400V, 5 HP squirrel-cage induction motor covered stator and rotor copper losses, slip, power factor and iron losses were measured. The experimental set up for testing induction motor is shown in Fig. 1. This experimental set-up is capable to handle non-sinusoidal and unbalanced quantities, again with the aim of treating inverter supplied drives. The input power, input voltages and currents are measured using voltage and non-contacting current probes, and the signals are digitally processed by the power

---

0-7803-9771-1/06/$25.00 ©2006 IEEE

quality (PQ) analyzer (Fluke 434). A DC generator is used for applying variable load to the motor.

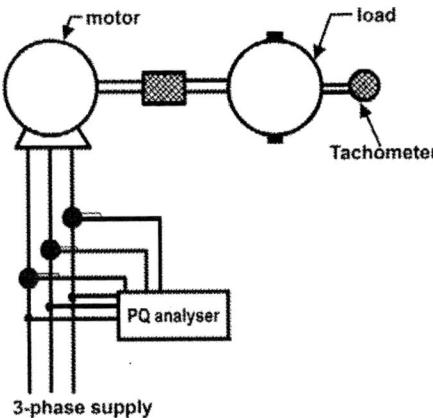

Fig. 1. Test bench layout.

## III. DEFINITIONS OF VOLTAGE UNBALANCE

Voltage imbalance or unbalance is the difference in phase (or phase to phase) voltage magnitudes of the three phase system. There are many possible causes of voltage unbalance, including the malfunction of automatic power factor correction equipment and voltage regulators in the utility distribution lines, unevenly distributed single-phase loads in a facility, high-impedance connections, and an unbalanced transformer bank. An unbalanced three-phase voltage causes three-phase motors to draw unbalanced current, which can cause the rotor of a motor to overheat. In fact, the temperature rise caused by unbalanced current is much greater than the rise caused by balanced current [8]. There are three definitions of voltage unbalance:

(1) Line Voltage Unbalance Rate (LVUR) as defined by National Electrical Manufactures Association (NEMA) [2], the ratio of maximum voltage deviation from the average line voltage magnitude to the average line voltage magnitude:

$$LVUR = \frac{\max\left[|V_{ab} - V_{avg}||V_{bc} - V_{avg}||V_{ca} - V_{avg}|\right]}{V_{avg}} * 100 \quad (1)$$

where $V_{avg} = \dfrac{V_{ab} + V_{bc} + V_{ca}}{3}$

$V_{ab}, V_{bc}, V_{ca}$ are line-to-line voltages

(2) Phase Voltage Unbalance Rate (PVUR) as defined in IEEE std 141, the ratio of maximum voltage deviation from average phase voltage magnitude to the average phase voltage magnitude:

$$PVUR = \frac{\max\left[|V_a - V_{avg}||V_b - V_{avg}||V_c - V_{avg}|\right]}{V_{avg}} * 100 \quad (2)$$

where $V_{avg} = \dfrac{V_a + V_b + V_c}{3}$

$V_a, V_b, V_c$ are phase voltages.

(3) True Voltage Unbalance Factor (VUF), the positive and negative sequence components of a three-phase, sinusoidal and unbalanced voltage system can be calculated exactly without the application of the Fortescue transformation in the complex plane, which only the RMS line-to-line voltages are required [3]:

$$VUF = \frac{V_-}{V_+} \quad (3)$$

where $V_+$ and $V_-$ represent the voltages of the positive and negative sequence components, respectively. Kennelly provided analytical expressions (using only real mathematics) for the results of the graphical procedure, yielding $V_+$ and $V_-$:

$$V_+ = \frac{\sqrt{A_m^2 + \dfrac{4A_s^2}{\sqrt{3}}}}{2} \quad ; \quad V_- = \frac{\sqrt{A_m^2 - \dfrac{4A_s^2}{\sqrt{3}}}}{2}$$

where $A_m^2 = (V_{ab}^2 + V_{bc}^2 + V_{ca}^2)/3$

$$A_s^2 = \sqrt{p(p - V_{ab})(p - V_{bc})(p - V_{ca})}$$

$$P = \frac{V_{ab} + V_{bc} + V_{ca}}{2}, \text{ the voltage unbalance}$$

factor is then calculated as in (3) [3].

## IV. TESTING

At the time of testing, the authors use a 5.5 kW voltage source inverter for setting balanced non-sinusoidal voltage. In order to match the frequency of both supplies (sinusoidal and non-sinusoidal), the authors set the voltages nearly 425 V which is somewhat higher than rated voltage of the motor. If the three-phase voltages are over-compensated to different degrees, then all these three phase voltages will be higher than the rated value and not equal. This type of unbalanced condition is called three-phase over-voltage unbalance (3ΦOV) [10]. Also the authors use the definition in (1) for calculating voltage and current unbalance since only the magnitude of the line voltages was collected at the time of testing. Percent current unbalance is defined in a similar manner to the voltage unbalance [9]. IEC 34-2, summation of losses method used for efficiency determination of the motor. Additional load losses are assumed to be equal to an estimated 0.5% of the power input of the motor and to vary as the square of the stator current. The rotor Joule losses are evaluated as the product of the rotor slip for the air gap transmitted power [4].

### A. No-Load Test Results

The motor was first operated in the unbalanced sinusoidal voltage no load condition to establish the baseline for normal performance. Tests were then performed at no load for the non-sinusoidal balanced condition. Table I gives a summary of the performance of the motor characteristics during no load testing for the 0.96% unbalanced sinusoidal voltage condition and balanced non sinusoidal voltage. In table I, the voltage unbalance is 0.96% which creates the current unbalance is 6.58%. The no-load losses have increased at unbalanced voltage by 12.82% from 0.195 kW to 0.22 kW.

### B. Load Test Results:

The next test to be undertaken was the load tests by varying applied load to the motor up to full load. Since the motor coupled with a DC generator, lamp load applied with the generator so that the motor is indirectly loaded. Measurements were taken for the entire range of loading i.e. from no-load to full load. The summary of the load tests is tabulated in table II and III.

TABLE I
NO- LOAD PERFORMANCE

| Parameter | 0.96% Unbalanced sinusoidal voltage | Balanced non-sinusoidal voltage |
|---|---|---|
| Voltage, Volts | 425.5, 425.4, 419.3 | 425 |
| Current, Amps | 2.2, 2.7, 2.5 | 2.1 |
| Power, kW | 0.22 | 0.195 |
| Power factor | 0.13 | 0.14 |

TABLE II
LOAD TEST UNDER UNBALANCED SINUSOIDAL VOLTAGE

| Power Input, (KW) | Line currents, (Amps) | Total losses, (kW) | Efficiency (%) | Power factor (%) |
|---|---|---|---|---|
| 0.72 | 2.8, 3, 2.6 | 0.27 | 62.5 | 35 |
| 1.46 | 3.5, 3.6, 3.2 | 0.31 | 78.8 | 58 |
| 2.1 | 4.2, 4.3, 3.8 | 0.35 | 83.2 | 70 |
| 3.09 | 5.3, 5.6, 5.1 | 0.49 | 84.3 | 80 |
| 3.9 | 6.5, 6.7, 6.1 | 0.61 | 84.5 | 84 |
| 4.66 | 7.5, 7.8, 7.3 | 0.74 | 84.0 | 85 |

TABLE III
LOAD TEST UNDER BALANCED NON-SINUSOIDAL VOLTAGE

| Power Input, (kW) | Line current, (Amps) | Total losses, (kW) | Efficiency (%) | Power Factor (%) |
|---|---|---|---|---|
| 0.6 | 2.3 | 0.23 | 59.4 | 33 |
| 1.26 | 3.0 | 0.26 | 79.4 | 58 |
| 2.25 | 4.3 | 0.35 | 84.6 | 72 |
| 2.76 | 5.1 | 0.40 | 85.4 | 75 |
| 3.74 | 6.5 | 0.57 | 84.7 | 78 |
| 4.39 | 7.5 | 0.68 | 84.5 | 79 |

## V. EXPERIMENTAL RESULTS AND DISCUSSION

Fig 2, 3, 4 shows the experimental results for a 5HP, 400V, three-phase squirrel-cage induction motor supplied by both unbalanced sinusoidal and balanced non-sinusoidal supply voltages. Efficiency of the tested motor under unbalanced sinusoidal voltage is slightly smaller than the motor operated with balanced non-sinusoidal (inverter) voltage shown in Fig. 2. Line currents drawn by the motor under both operating conditions is shown in Fig. 3. This figure shows that the current drawn by the motor under first condition is higher than the current under second condition at low loads and this

situation reversed at high loads. Also, improved power factor is obtained by use of non-sinusoidal balanced supply at low loads and unbalanced sinusoidal voltage at high loads. Lower power factor is obtained by use of unbalanced sinusoidal voltage at low loads and non-sinusoidal voltage at high loads shown in Fig. 4.

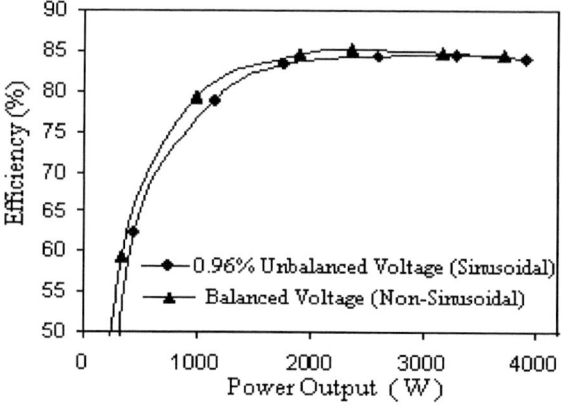

Fig. 2. Efficiency of the motor at different load.

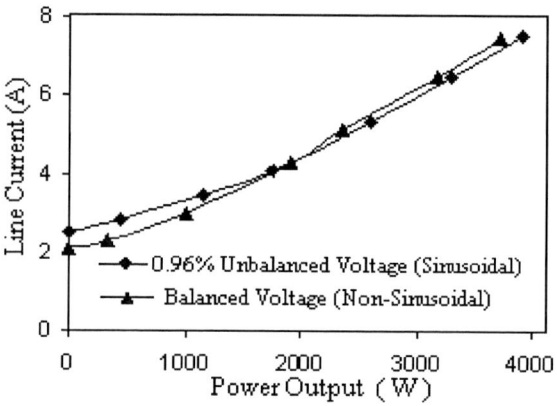

Fig. 3. Line current of the motor at different load.

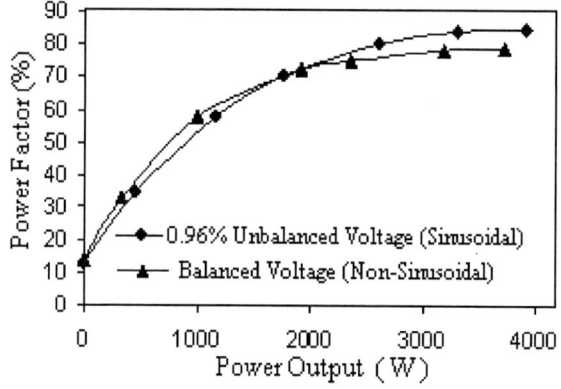

Fig. 4. Power factor of the motor at different load.

## VI. ECONOMIC ANALYSIS

From the consumer's point of view, efficiency reduction means paying more for energy as a result of unbalanced voltage supplies. From the power system point of view, operation of motor with lower efficiency implies an increase of the system load and a reduction of the power plant reserves [6]. An analysis of a 5 HP motor operating under small unbalanced sinusoidal voltage compared with balanced non-sinusoidal voltage at the following electricity tariff (Tamil Nadu Electricity Board, HT tariff I for the industries situated in Non-Metropolitan localities) and assuming 8000 hours of operation/year is summarized in table IV. Extra KW consumption by the motor due to unbalanced voltage = 4.66 - 4.39= 0.27 KW. US $ 188 additionally paid per year to the electricity supplier due to 0.96% voltage unbalance in one motor.

Maximum demand (KVA) charges: US $ 6.66/month
Energy (kWh) charges: US $ 0.077/kWh
(1 US $= IRS 45 approximately)

TABLE IV
ECONOMIC LOSSES DUE TO 0.96% VOLTAGE UNBALANCE IN A 5 HP MOTOR

| Extra KW consumption at full load | Extra consumed energy (KWh / year) | Extra demand charge (US $/year) | Extra payment (US $/year) |
|---|---|---|---|
| 0.27 | 2160 | 21.5 | 187.8 |

## VII. CONCLUSION

This study has investigated the negative effects of three-phase unbalanced (0.96%) sinusoidal voltage over balanced non-sinusoidal voltage on the performance of 5HP, squirrel cage induction motor, focusing on the efficiency, line currents and power factor. The main conclusions from the study are (i) efficiency of the motor under three-phase unbalanced voltage is lower than the efficiency at balanced non-sinusoidal voltage, (ii) current drawn by the motor under unbalanced supply is higher at low loads, (iii) lower power factor is obtained by use of unbalanced supply at low loads, and (iv) US $ 188 additionally paid per year to the electricity supplier due to 0.96% voltage unbalance in one motor.

## VIII. REFERENCES

[1] P.Pillay, M. Manyage, "Loss of life in induction machines operating with unbalanced supplies," *IEEE Trans. Energy Conversion*, to be published.

[2] Ching-Yin Lee, "The effects of unbalanced voltage on the performance of a 3 phase induction motor," *IEEE Trans. Energy conversion*, Vol. 14, No. 2, pp. 202-208, Jun. 1999.

[3] Jozef A. L Ghijselen, Alex P. M Van den Bossche, "Exact voltage unbalance assessment without phase measurements," *IEEE Trans. Power Systems*, Vol. 20, No. 1, pp. 519-520, Feb. 2005.

[4] Auinger H, "Determination and designation of the efficiency of electrical machines," *Power Engineering journal*, pp. 15-23, Feb. 1999.

[5] G.K. Singh, "A research survey of induction motor operation with non-

sinusoidal supply wave forms," *Electric Power Systems Research*, Vol. 75, pg. 200–213, May 2005.

[6] Jawad Faiz, H. Ebrahimpour, P. Pillay, "Influence of unbalanced voltage supply on efficiency of three phase squirrel cage induction motor and economic analysis," *Energy Conversion and Management*, 47 (2006) pg. 289-302.

[7] W.E. Wagner, "Report of the IEEE task force on the effects of harmonics on equipment," 1992.

[8] California Energy Commission, "power quality solutions for industrial customers," Final Report, Public Interest Energy Research Program (PIER), August 2000.

[9] James H Dymond, Nick Stranges P, "Operation on unbalanced voltage: one motor's experience and more," *IEEE PCIC Conference* - 2005-35, pp. 311-319.

[10] Arfat Siddique, G S Yadava, Bhim Singh, "Effects of Voltage unbalance on induction motors" *IEEE Int. Conf. Electrical Insulation*, USA, pp. 26-29, Sep. 2004.

## IX. BIOGRAPHIES

**Thanga Raj Chelliah** was born in Kanyakumari, India, on June 20, 1978. He received the diploma in Electrical and Electronics Engineering from the Government Polytechnic College, Nagercoil, India in 1996, Bachelor's degree in Electrical and Electronics Engineering from Bharathiar University, Coimbatore, India in 2002 and the Master's degree in Power Electronics and Drives from Anna University, Chennai, India in 2005. He is currently working towards the Ph. D degree at Indian Institute of Technology Roorkee, India.

From 1996 to 2002, he was with Haitima Textiles Limited, Coimbatore, as an Assistant Electrical Engineer. While there, he was involved in energy conservation activities in the electrical equipments. From 2002 to 2003, he was with PSN College of Engineering and Technology, Tirunelveli, as a Lecturer. His research interests include power quality and induction machine efficiency improvement.

**Pramod Agarwal** received the Bachelor's, Master's and Ph. D degrees in Electrical Engineering from the University of Roorkee (now, Indian Institute of Technology Roorkee), India in 1983, 1985, and 1995 respectively.

Currently he is with Indian Institute of Technology Roorkee, India, where he is a Professor in the Department of Electrical Engineering. His special fields of interests include electrical machines, power electronics, power quality, microprocessors and microprocessor-controlled drives, active power filters, high power factor converters, multilevel inverters, and dSPACE-controlled converters.

**S. P. Srivastava** received the Bachelor's and Master's degrees in Electrical Technology from I.T. Banarus Hindu University, Varanasi, India in 1976, 1979 respectively and the Ph. D degree in Electrical Engineering from the University of Roorkee, India in 1983.

Currently he is with Indian Institute of Technology (IIT) Roorkee, India, where he is an Associate Professor in the Department of Electrical Engineering. He is also a principal at the Kumaon Engineering College, Nainital, India, on taking leave from the IIT Roorkee. His research interests include power apparatus and electric drives.

# Efficiency Optimization of Induction Motor Using a Fuzzy Logic Based Optimum Flux Search Controller

L. Ramesh, S. P.Chowdhury, *IEEE*, S.Chowdhury, *IEEE*, A. K. Saha, and Y. H. Song

*Abstract--* Induction motors are, without any doubt, the most used in industry. Motor drive efficiency optimization is important for two reasons: economic saving and reduction of environmental pollution. In this paper, advantages of using fuzzy logic in steady-state efficiency optimization for induction motor drives are described. Experimental results of a fuzzy logic based optimum flux search controller are presented. There are so many speed control techniques available, like Scalar control, Vector control, sensor less control etc. Due to coupling effect, the Scalar control has inferior performance and Vector control is a method of speed control in which both magnitude and phase angle of current can be controlled. In Vector control, the presence of speed sensor at shaft decreases the reliability and ruggedness. For transient states, a new original idea is introduced: a fuzzy logic based controller, actuating as a supervisor, is proposed to work with reduced flux levels during transients to optimize efficiency also in dynamic mode. Two different rule tables are designed, for torque transitions and for reference speed changes. With this controller, efficiency can be improved in transients, and also search controller convergence speed is increased. In my work efficiency optimization of induction motor using Fuzzy logic controller was carried out using Matlab.

*Index Terms--* Adjustable speed drives, fuzzy logic control, speed control.

## I. INTRODUCTION

ELECTRIC motors consume more than 50% of the electrical power generated in the world. More than 80% of the electricity used by motors is consumed by less than 1% of the motor population [motors greater than 20 hp (142 kW)]. Each 1% improvement in motor efficiency could result in savings of over $1Billion per year in energy

---

L. Ramesh is with Electrical Engineering Department, Dr.M.G.R University,Chennai,India(raameshl@rediffmail.com)

S.P.Chowdhury is with Electrical Engineering Department,JadavpurUniversity,Kolkota, India spchowdhury@yahoo.com

S. Chowdhury is with Electrical Engineering Department, Women's polytechnic,Kolkota,India.( sunetra69@yahoo.com)

A. A. Natarajan is principal with Sree Sasta Institute of Engineering and Technology,Chennai,India.(ssiet@eth.net)

A. K. Saha is with Electrical Engineering Department,Jadavpur University,Kolkota,India

Y. H. Song is Director with Brunel University,,UK

costs, 6-10 million tons (5.4-9.1 million tones) less per year of combusted coal and approximately 15-20 million tons (13.6-18.1 million tones) less carbon dioxide released into the atmosphere. To minimize power losses, it is necessary to control motor speed and thereby 2 match motor speeds to load requirements. The most recent, universal, and successful approach is the ASD. ASDs use semiconductors and switching circuits to vary the voltage and frequency of a motor's power supply. ASDs essentially operate on the principle of rectifying ac input (line) voltage, filtering the signal, and then switching the dc power on and off in an inverter section to form a variable frequency ac power output to the motor. A microprocessor control block modifies the inverter switching characteristics so that the Connected motor speed may be controlled to satisfy the Process requirements. The rectifier voltage or current is also directly controlled in conjunction with frequency. FLMC is being developed as the core of the ASD control block, to analyze system feedback and select frequency/ voltage/current combinations to optimize energy efficiency. An embedded fuzzy logic controller can be added to conventional ASDs. ASDs have now been applied to over 1% of the motors larger than 7.5 hp (5.6 kW) in India and over 10% of motors larger than 100 hp. At least 85% of the energy that could be saved by better motor control would be associated with motors larger than 5 hp.The complexity of the system model and uncertain parametric relationships mean that an accurate system model for simulation, performance prediction, and control is very difficult to achieve. This is especially true considering the precision and accuracy needed to obtain the last few percent of efficiency from a motor system.

Adjustable speed drives (ASDs) [1] are power electric devices, which allow control of the speed of rotation of electric motors. ASDs can provide a significant savings in energy for motors, which spend a portion of their duty cycle operating at less than their rated speed and torque. Prior to the introduction of ASDs, control of motor driven devices such as fans and pumps were always controlled by valves, vanes, dampers, and other mechanical devices, which are inherently inefficient. Conventional ASDs do not optimally minimize motor input power at any given motor speed and load torque. The objective of the research described in this paper had been to improve ASDs by adding controls, which optimize the ASD on the basis of energy efficiency.

## A. Need for Speed Control

Volts/hertz (v/f) control and vector control are the most generally used control strategies of induction motor. In general v/f control method is used in fans, conveyors, centrifugal pumps, etc. where high performance and fast response is not needed. The v/f principle adjusts a constant Volts-per-Hertz ratio of the stator voltage by feed forward control. It serves to maintain the magnetic flux in the machine at desired level. The absences of closed loop c make v/f controlled drives very robust. Scalar control is the control and the restriction to low dynamic performance technique in which the control action is obtained by the variation of only magnitude of control variables and disregards to control the coupling effect in the machine. The voltage of the machine can be controlled to control the flux and frequency, or slip can be controlled to control the torque. The control is provided by frequency and voltage reference generator with constant volt per hertz ratio. Scalar control technique is somewhat simple to implement, but the inherent coupling effect results sluggish response and the system is easily prone to instability because of higher order system effect. The particular attraction of v/f-controlled drives is their extremely simple control structure, which favors an implementation by a few highly integrated electronic components. There is no direct or indirect control of torque and flux. The status of the rotor is ignored, i.e. no speed or position signal is feedback. These cost-saving aspects are especially important for applications at low power below 5 kW. The cost advantage makes v/f control very attractive for low power applications, while their robustness favors its use at high power when a fast response is not required.

## II. FUZZY LOGIC CONTROL

It is estimated [3] that more than 50% of the world electric energy generated is consumed by electric machines improving efficiency in electric drives is important, mainly, for two reasons: economic saving and reduction of environmental pollution .Induction motors have a high efficiency at rated speed and torque. However, at light loads, the iron losses increase dramatically; reducing considerably the efficiency. The main induction motor losses are usually split into 5 components: stator copper losses, rotor copper losses, and iron losses, mechanical and stray losses. To improve the motor efficiency, the flux must be reduced, obtaining a balance between copper and iron losses. Basically, there exist two different approaches [11] to improve the efficiency.

## A. Loss Model Based Approach

If a motor loss model is available the loss minimization optimum flux is computed analytically. The main advantage is the simplicity of this method, not requiring extra hardware. However, it is mandatory an accurate knowledge of motor parameters, which change considerably with temperature, saturation, skin effect, etc.

## B. Power Measure Based Approach

An optimum flux search algorithm is used, and the drive power consumption is measured. Obviously, this approach does not require the knowledge of motor parameters. However, it is only efficient at steady-state condition and with high powers, the method has convergence problems. Although there exist many search algorithms, their main limitations are speed of convergence and undesirable oscillations around the minimum loss point.

In this work, advantages of using a fuzzy logic based search controller to optimize efficiency at steady state are described. For transients, a fuzzy logic based controller, actuating as a supervisor, is proposed to optimize efficiency with reduced flux levels in dynamic mode. Experimental results with an indirect field oriented control are obtained using a standard 1.5 kW induction motor in laboratory.

## C. Fuzzy Logic Applied to Bearch Controllers

There exist many search algorithms to optimize efficiency for induction motor drives using the power measure based approach. Their main limitations are speed of convergence, perturbations in the electromagnetic torque and oscillations around the optimum flux. Fuzzy logic can help avoid these problems, improving the performance of the search controller. This idea of using fuzzy logic for search controllers was recently reported in. The FL controller antecedents are the power change $\Delta Pn$) and the last flux current change $\Delta i*sd$) n-1 and the consequent is the new flux change $\Delta i*sd$) n

TABLE I
FUZZY RULES TO FIND THE OPTIMUM FLUX (OUTPUT $\Delta I*_{SD}$) N)

| ($\Delta Pn$)/ $\Delta i*sd$) n-1 | N | P | Description |
|---|---|---|---|
| PB | PM | NM | PB - Positive Big |
| PM | PS | NS | PS  - Positive Small |
| PS | PS | NS | PM - Positive Medium |
| ZE | ZE | ZE | ZE - Zero |
| NS | NS | PS | NM - Negative Medium |
| NM | NM | PM | NL - Negative Large |
| NB | NB | PB | NS -Negative Small |

With this mentioned (Table I) controller, some improvements may be obtained. First, the speed of convergence increases due to the adaptive size of the flux steps imposed to the machine. Moreover, when we are approaching to the optimum flux, the step size is reduced dramatically, avoiding unnecessary movements around the minimum loss point. Finally, the fuzzy logic based search is infinitum, i.e., the algorithm is always working, avoiding deviations of the optimum flux due to motor parameter changes by temperature effect, and so on. It can be demonstrated the fuzzy logic based search method consists of a gradient based search with adaptive flux steps and infinitum search with non-null flux step around the minimum loss point. The only disadvantage of this approach is that the search controller requires adjustable gains (speed and torque dependent), which depend on the motor power.

### D. Fuzzy Logic Optimization in Transients

So far, we have been only considering optimization [6] at steady-state condition. The above fuzzy logic based search controller is only effective with constant speed and torque. When a torque perturbation or command speed change is produced, the search process stops. All the works in recent literature propose to establish the rated flux during a dynamic transition. This permits to have the maximum torque capability in order to get the optimum transient response. However, the transient torque and speed response may not be so important for some practical applications (it is the case of an elevator or a crane). For example, if the process has many small load torque perturbations. In these cases, efficiency optimization can take priority. In this paper, we propose to work with reduced flux levels during dynamic transitions, optimizing power in the dynamic mode and accelerating the search process at steady state. For this, a fuzzy logic controller, actuating as a supervisor, is designed. During transients, the controller increases the flux, depending on the speed error and its derivative. If the transient is very large or the speed error is very large, the flux is established at rated level, in order to let the drive track the reference command. The proposed controller is very simple, containing two tables of heuristic rules, distinguishing a torque transition and a reference speed change. The controller only provides positive flux increments, actuating as a supervisor, since it supervises the right command tracking, incrementing the flux if necessary. The antecedents are the speed error $e_\omega$ and its derivative $\dot{e}_\omega$ and the consequent is the flux current increment $\Delta i*sd$. Figure1 shows the proposed fuzzy logic supervisor to increment the flux current. Table II and III shows the rules for a positive torque transition and for a positive reference speed change, respectively.

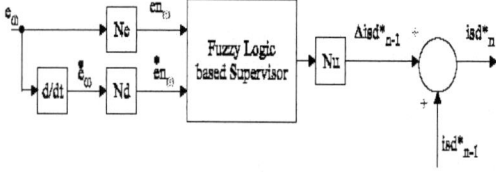

Fig. 1. Diagram of the proposed fuzzy logic supervisor to increment the flux current.

The product-sum was used as inference method and the center of gravity was implemented as defuzzification method to get a fast algorithm in real time.

TABLE II
RULES FOR POSITIVE TORQUE TRANSITION

| $e_\omega / \dot{e}_\omega$ | ZE | PS | PM | PB |
|---|---|---|---|---|
| ZE | ZE | ZE | PS | PM |
| PS | PS | PS | PM | PB |
| PM | PM | PM | PS | PB |
| PB | PB | PB | PM | PB |

TABLE III
RULES FOR A POSITIVE REFERENCE SPEED CHANGE

| $e_\omega / \dot{e}_\omega$ | NB | NM | NS | ZE |
|---|---|---|---|---|
| ZE | ZE | ZE | ZE | ZE |
| PS | ZE | ZE | ZE | PB |
| PM | PS | PS | PS | PB |
| PB | PS | PM | PM | PB |

### III. FUZZY LOGIC BLOCK DIAGRAMS

Efficiency optimization of induction motor using fuzzy logic controller is simulated on MATLAB/FUZZY LOGIC-platform to study the various aspects of the controller. It provides a user interactive platform and a wide variety of numerical algorithms. Figure 2. Shows the root block diagram for simulation.

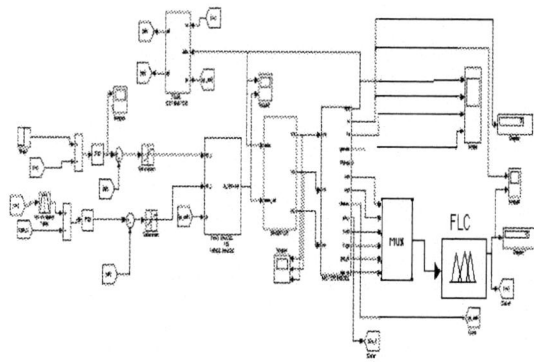

Fig. 2. Simulink root block diagram of Efficiency Optimization of induction motor using Fuzzy logic controller.

### A. Induction MotorMode

The motor is modeled in stator reference frame. By using equations we can develop the induction motor model in stator reference frame. Figure 3. Shows the simulink block diagram for motor model. Inputs to this block are direct and quadrature axes voltages and load torque. The outputs are direct axis stator and rotor fluxes, quadrature axis stator and rotor fluxes, direct and quadrature axes stator currents, electrical torque developed and rotor speed.

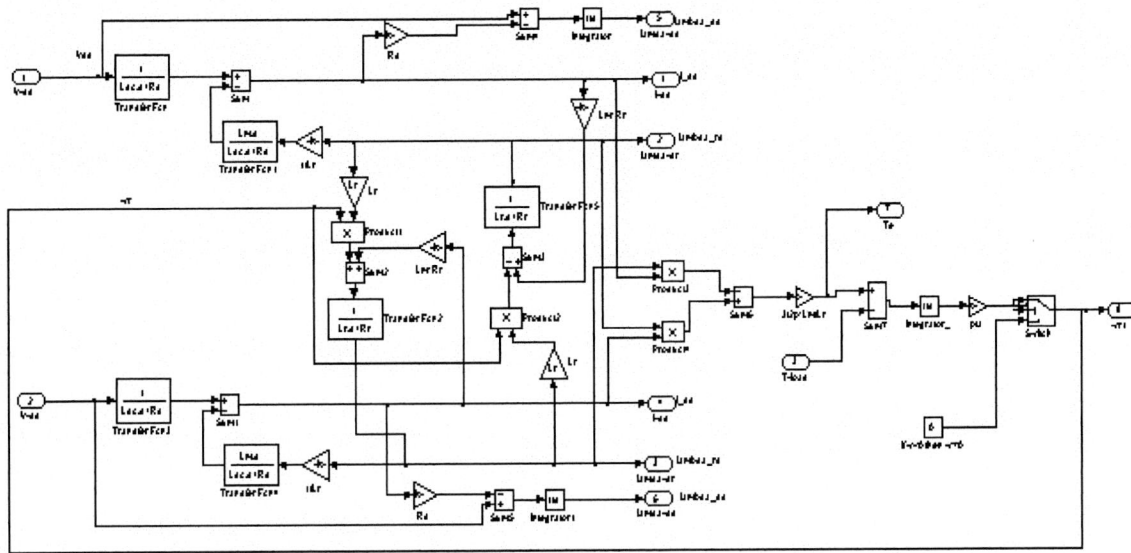

Fig. 3. Simulink block diagram for induction motor model.

## B. 2-φ to 3-φ Transformation

The feedback quantities are in rotating frame. First these quantities are transferred to stationary frame and then from 2-φ to 3-φ. The simulink block diagram of 2-φ to3-φ transformation in Fig.4.

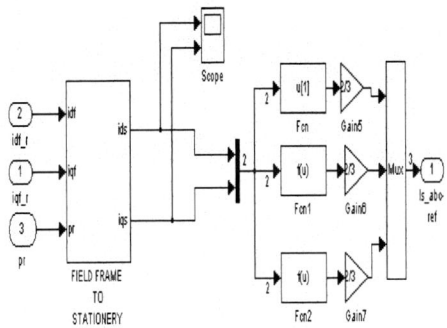

Fig. 4. The simulink block diagram of 2-φ to3-φ transformation.

## C. Optimal Switching Logic

The function of the optimal switching logic is to select the appropriate stator voltage vector that will satisfy both the torque status output and the flux status output. Processing of the torque status output and the flux status output is handled by the optimal switching logic.

## IV. SIMULATION

### A. Parameters for Simulation

Parameters of the induction motor:
The parameters for 1.5 hp, 4-pole, and 50Hz induction motor are given below
Stator circuit resistance = 4.495 ohms
Rotor circuit resistance = 5.365 ohms
Inductance of stator circuit = 0.165H
Inductance of rotor circuit = 0.162H
Mutual inductance = 0.149H
Moment of inertia = 0.095 Kg.m$^2$
Simulation parameters:
DC link voltage = 250V
Simulation solver is fixed step, Euler method
Step size = 25 microseconds

### B. Simulation Results

The simulation of Efficiency optimization of induction motor using fuzzy logic controller is done by using MATLAB-fuzzy toolbox. The results for different cases are given below
*Case 1-Without load change*

### C. Comments

Figure 10 shows the power consumed by the induction motor when a fuzzy controller is used

From the above figures 5, 6,7,8,9 and10 it is clear that the power consumed by the machine in both the cases of rated & reduced flux is less if a fuzzy logic supervisor is used.

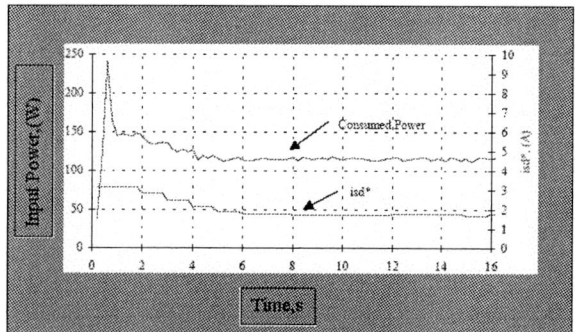

Fig. 5. Consumed input power and reference flux current isd*.

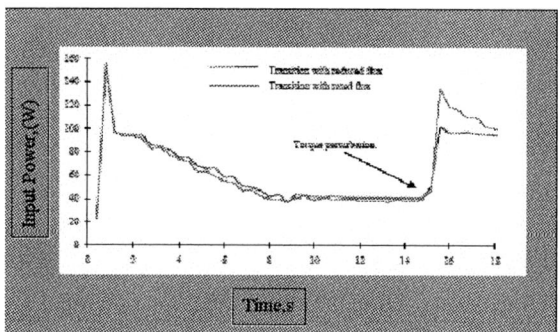

Fig. 6. Comparison of consumed power with rated and reduced flux during the torque transition.

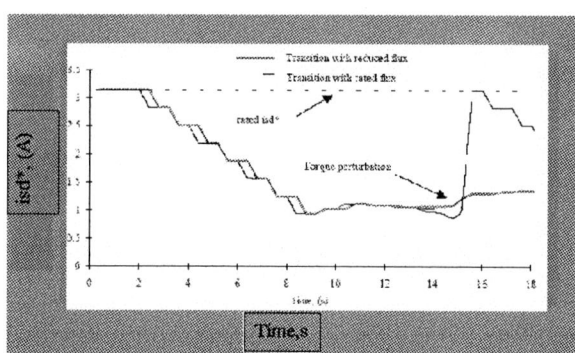

Fig. 7. Reference flux currents for the torque transition.

*CASE – 2: Change in load*

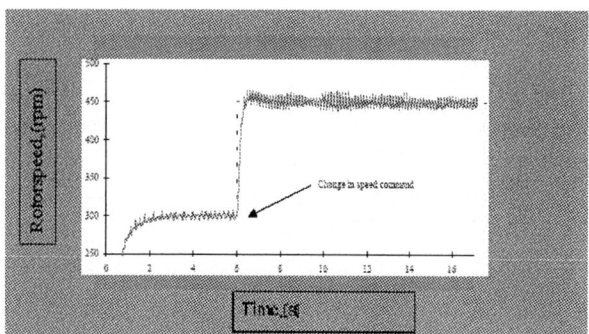

Fig. 8. Speed evolution with transition at rated flux.

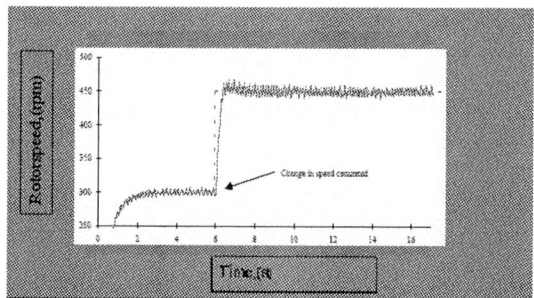

Fig. 9. Speed evolution with transition at reduced flux.

*CASE – 3: Fuzzy logic supervisor output*

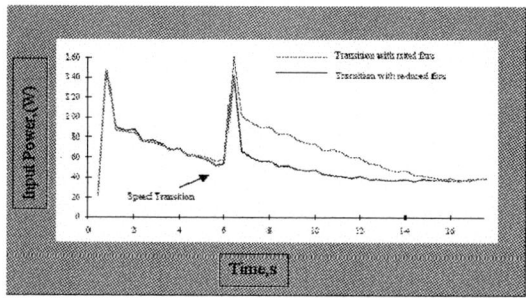

Fig. 10. Comparison of consumed power for the speed transition With rated and reduced flux.

## V. CONCLUSIONS

In this work, advantages of using fuzzy logic in efficiency optimization for induction motor drive have been described. A fuzzy logic based search controller improves the speed of convergence and reduces oscillations around the optimum flux using a simple rule table. However, the controller requires adjustable speed and torque gains which depend on the motor power. A new fuzzy logic based controller, actuating as a supervisor has been proposed to optimize efficiency during transients. Two different rule tables have been designed for torque transitions and for reference speed changes. Experimental results of the proposed controllers with 1.5 kW induction motor drive demonstrate the validity of the described methods. From simulation results it can be observed that in steady state there are ripples in torque wave and starting current is high. The main results obtained in this are efficiency is improved considerably, speed evolution with transition is same in both rated flux and reduced flux and using fuzzy logic controller we are estimating speed, which is same as that of actual induction motor speed.

## VI. REFERENCES

[1] R. Krishnan, "Electric Motor Drives: Modeling, Analysis, and Control", Prentice Hall of India, 2002.

[2] Bimal K. Bose, "Modern Power Electronics and AC Drives", Pearson Education Singapore, 2003.

[3] G.C.D. Sousa, B.K. Bose, J.G. Cleland, "Fuzzy logic based on-line efficiency optimization control of an indirect vector-controlled induction motor drive*", *IEEE Transactions on Industrial Electronics*, April 1995.

[4] Joachin Holtz,"Sensorless Control of Induction Motor Drives", Proceedings of the IEEE, vol.90, No.8, Aug-2002, P-P1359-1394.

[5] Joachin Holtz,"Sensorless speed and position control Of Induction Motors", 27th annual Conference of the *IEEE Industrial Electronics Society*, Nov, 28.

[6] R.D. Lorenz, and S. Yang, "Efficiency-optimized flux trajectories for closed-cycle operation of field-orientation induction machine drives,"IEEE Transactions on Industry Applications, vol. 28, no. 3, pp. 574-580, May/June 1992.

[7] J.C. Moreira, T.A. Lipo, and V. Blasko, "Simple efficiency maximizer for an adjustable frequency induction motor drive," *IEEE Transactions on Industry Applications*, vol. 27, no. 5, pp. 940-945,Sep/October 1991.

[8] I.Kioskeridis, and N. Margaris, "Loss minimization in scalar controlled induction motor drives with search controllers," *IEEE Transactions on Power Electronics*, vol. 11, no. 2, pp. 213-220, March 1996.

[9] G. Kim, I. Ha, and M. Ko, "Control of induction motors for both high dynamic performance and high power efficiency," *IEEE Transactions on Industrial Electronics*, vol. 39, no. 4, pp. 323-333, August 1992.

[10] M. Cipolla, J. M. Moreno-Eguilaz, and J. Peracaula "Fuzzy control of an induction motor with compensation of system dead-time," IEEE Power Electronics Specialist Conference, PESC'96, Baveno, Italy, June 24-27, 1996.

[11] J. M. Moreno-Eguilaz, M. Cipolla, P. Branco, and J. Peracaula, "Fuzzy logic based improvements in efficiency optimization of induction motor drives," IEEE International Conference on Fuzzy Systems 1997, FUZZ-IEEE'97, Barcelona, Spain, July 1-5, 1997, in press.

## VII. BIOGRAPHIES

**L.Ramesh**, born in 1977, is presently Assistant professor of Electrical and Electronics Engineering Department, Dr.M.G.R.University,Chennai,India and Research Scholar of Jadavpur University,Kolkota,India.He obtained B.E in M.S.Universitu and M.Tech in Kerala University, India. He is a member of IET (UK). raameshl@yahoo.co.in

**Dr.S.P.Chowdhury**, born in 1963, is presently Reader of Electrical Engineering Department and In-Charge of the Power System Section, Electrical Engg. Dept., Jadavpur University, Kolkata, India. He obtained BEE (Hons.), MEE and Ph.D. (Engg.) Degrees from the Jadavpur University in 1987, 1989 and 1992 respectively. He has published more than 100 research papers in different International and National Journals and Conference Proceedings. He has co-authored two books entitled "Electrical Engineering Computations by Computer Graphics Aided Basic Programming", published by M/S B.P.B. Publications in 1997 and "Microprocessor and peripherals" published by Scitech publications Pvt. Ltd.,Chennai and Hydrabad. Dr.S.P.Chowdhury has been awarded several Medals/Prizes/Certificates of merit, Memento for his contributions in technical papers in IE (India) and IEEE (USA). He made several technical visits to UK, Canada, Sri Lanka and Singapore. He has successfully served as the Principal Investigator in a microprocessor –based SCADA Project sponsored by the AICTE. Currently, he is an active member of the professional bodies IE (India), the IET (UK) and IEEE (USA). He holds the position of Member, the IET (UK) Membership and Regions Board (MRB), the IET (UK) MRB Finance Committee and the IET (UK) Council. He is also acting as counselor of the IET(UK), Kolkata Students Chapter. He acted as the Organizing Secretary of PEITSICON-2005. (Mail : spchowdhury@yahoo.com)

**Dr. S.Chowdhury**, born in 1969, is presently Lecturer in Electrical Engineering at the Women's Polytechnic, Jodhpur Park, and Kolkata, India. She obtained her BEE (Hons) and Ph.D. (Engg.) Degrees from the Jadavpur University in 1991 and 1998 respectively .She has published more than 45 research papers in different International and National Journals and Conference Proceedings and has co-authored two books entitled "Electrical Engineering Computations by Computer Graphics Aided Basic Programming" published by M/S B.P.B. Publications in 1997 and "Microprocessor and peripherals" published by SciTech publications Pvt. Ltd., Chennai and Hyderabad. Dr.S.Chowdhury has been awarded several Medals/Prizes/Certificates of merit, Memento for her contributions in technical papers in IE (India). Presently she is a member of the professional bodies IE (India), IET (UK) and IEEE (USA). (Mail: sunetra69@yahoo.com)

**Dr.A.A.Natarajan**, born in 1946,is presently Principal, Sree Sastha Institute of Engineering and Technology, Chennai, India. He is a member of MPSS, LMISTE, MISLE, and MCDA.He has published 100 papers and produces 65 R&D reports.

**A.K.Saha** born in 1970 is a research fellow in J.U. Kolkata and associate member of IE (I) and IET (UK).

**Yong-Hua Song** was born in 1964 in China and received his BEng, MSc and PhD in 1984, 1987 and 1989 respectively. In 1991, he joined Bristol University, and then held various positions at Liverpool John Moores University and Bath University before he joined Brunel University in 1997 as Professor of Network Systems at the Department of Electronic and Computer Engineering. Currently he is Director of Brunel Advanced Institute of Network Systems and Pro-Vice-Chancellor of the University. He has published four books and over 300 papers mainly in power systems. He was awarded the Higher Doctorate of Science (DSc) in 2002 by Brunel University for his significant research contributions. He is a fellow of the IET and the Royal Academy of Engineering.

**2006 IEEE International Conference on Power Electronic, Drives and Energy Systems**

# Observer Based Position and Speed Estimation of Interior Permanent Magnet Motor

Bhim Singh, *Senior Member IEEE*, Prerna Gaur, *Senior Member IEEE*, and A.P.Mittal, *Senior Member IEEE*

*Abstract--***This paper presents Position sensorless Interior Permanent magnet synchronous motor drive using a discretized Extended Kalman Filter algorithm (EKF) . An observer based speed estimator which can be used for the state estimation of a non linear dynamic system in real time is presented here. Speed and position estimation of IPM is simulated using MATLAB and results of step variation in speed, load perturbation and flux weakening are presented to substantiate the proposed estimation of the speed.**

*Index Terms--* **Sensorless control, Extended Kalman Filter algorithm (EKF). Interior Permanent Magnet Synchronous Motor (IPM).**

## I. INTRODUCTION

AMONG AC motors, the interior permanent magnet synchronous motor (IPM) has high power density and torque to inertia ratio that makes it the most popular candidate for replacing DC motors for servo applications[1]-[12]. The IPM with sinusoidal flux distribution is preferred over the one with trapezoidal flux distribution due to lower torque ripple. In IPM based adjustable speed drives, two current sensors and an absolute rotor position sensor are normally taken as its integral part. An absolute rotor position is required for closed loop speed and position control in both vector and scalar controlled drives of IPM. For an IPM the rotor position information is fed back to the controller which generates the stator voltages according to the rotor position. Hence the rotor is always in synchronism with the rotating magnetic field, the rotor and stator fluxes are always at right angles under normal operating conditions, thus giving maximum torque. A speed signal is also required in indirect vector control in whole speed range, and in direct vector control for the low speed range, including zero speed start up operation. However, position sensors increase the size and are costly as compared to the cost of low power motor, and also require regular

---

B. Singh is with Electrical Engineering department, IIT New Delhi-110016, India. (e-mail: bhimsinghr@gmail.com).

P. Gaur is with Instrumentation and control engineering Division, Netaji Subhas Institute of Technology. Dwarka, New Delhi-110075, India. (e-mail:prernagaur@yahoo.com)

A.P.Mittal is with Instrumentation and control engineering Division, Netaji Subhas Institute of Technology. , Dwarka, New Delhi-110075, India. (e-mail: mittalap@yahoo.com)

maintenance. The use of speed sensors also compromised on ruggedness and reliability of the drive. Hence, the control and operation of IPM drive without a rotor position sensor would enhance its applicability to many sensitive applications and provide a backup control in sensor based drives during sensor failures [1]-[5]. It is possible to estimate the speed signal from machine terminal voltages and currents using DSP. An observer based speed estimator which can be used for the state estimation of a non linear dynamic system. The improved method of speed estimation uses a plant model and a feedback loop with measured plant variables.

## II. MODEL OF IPM

The dynamic model of PMSM motor can be written in state space derivation form as [6].

$$pi_d = (v_d - Ri_d + \omega_r L_q i_q)/L_d \tag{1}$$

$$pi_q = (v_q - Ri_q - \omega_r L_d i_d - \omega_r \lambda_f)/L_q \tag{2}$$

$$p\omega_r = (T_e - T_l - B\omega_r)/J \tag{3}$$

$$p\theta_r = \omega_r \tag{4}$$

The electromagnetic torque $(T_e)$ is expressed as:

$$T_e = \frac{3}{2}\frac{p}{2}[\lambda_f i_q + (L_d - L_q)i_d i_q] \tag{5}$$

### A. Sensorless Control Based on EKF

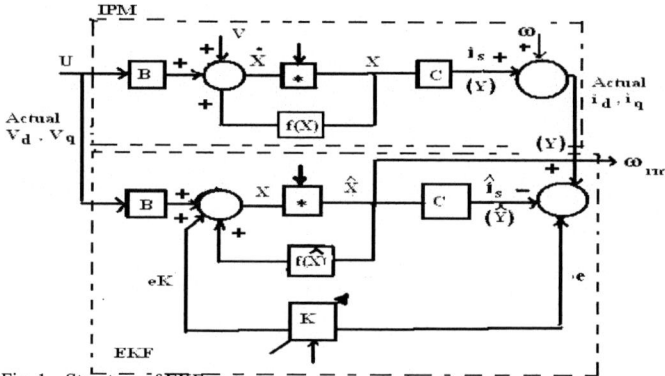

Fig. 1. Structure of EKF.

The block diagram of Kalman filter is shown in Fig.1. A kalman filter provides an optimum observation from noisy sensed signals (actual $i_d$ and $i_q$ in case of IPM) and processes that are disturbed by random noise. This assumes that measurement noise and disturbance noise are uncorrelated. The Kaman filter approach is viable and computationally efficient candidate for online estimation of the speed and rotor

---

0-7803-9771-1/06/$25.00 ©2006 IEEE

position. This is possible because a mathematical model describing the PM motor dynamics is sufficiently well known.

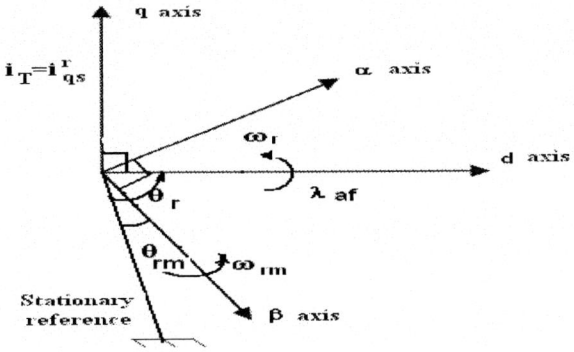

Fig. 2. Phasor diagram corresponding to an error between the actual and assumed rotor position.

The rotor position can be determined based on measured voltages and currents. The Consider that motor is running at a speed $\omega_r$ whereas the model starts with an assumed rotor speed $\omega_{rm}$. The assumed rotor position $\theta_{rm}$ lags behind the actual rotor position $\theta_r$ by $\delta\theta$ radians. As shown in Fig.2. they are related to the actual and assumed or model speed as follows:

$$\theta_r = \int \omega_r dt \qquad (6)$$

$$\theta_{rm} = \int \omega_{rm} dt \qquad (7)$$

$$\delta\theta = \theta_r - \theta_{rm} = \int (\omega_r - \omega_{rm}) dt \qquad (8)$$

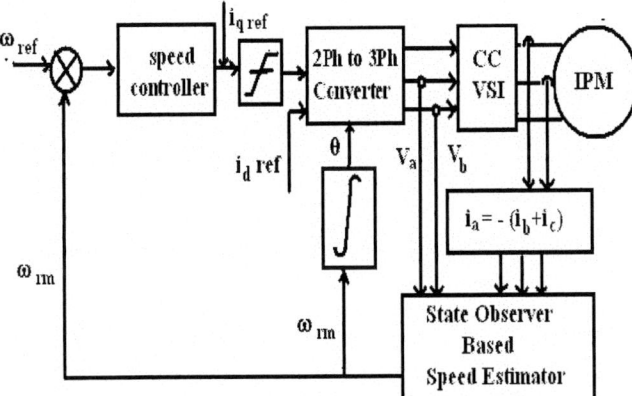

Fig. 3. Block diagram for sensorless control of IPM.

The line diagram in Fig.3 shows the observer based speed estimation block. The machine model is utilized to compute the stator currents. Substituting value of $T_e$ in (3) the state equations can be expressed purely in terms of state variables. Thus one can achieve the state equations as given below:

$$pi_d = -\frac{R}{L_d} i_d + \frac{L_q i_q}{L_d} \omega_{rm} Sin\,\theta_{rm} + \frac{1}{L_d} V_d \qquad (9)$$

$$pi_q = -\omega_r \frac{L_d}{L_q} i_d + \frac{-R}{L_q} i_q + \frac{\lambda_f}{L_q} \omega_r Cos\,\theta_{rm} + \frac{1}{L_q} V_q \qquad (10)$$

$$p\omega_r = 3*p((L_d - L_q)i_d * i_q + \lambda_f * i_q)/(4*J) - (B*\omega_r + T_l)/J \qquad (11)$$

### B. State Observer

The above equations contain the states $i_d, i_q, \omega_{rm}$ and $\theta_{rm}$, where the later two variables are estimated. The dynamic model of the machine in state variable form can be expressed as

$$\frac{dX}{dt} = f(X) + BU \qquad and \qquad (12)$$

$$Y = CX \qquad (13)$$

Where $X = [i_d \quad i_q \quad \omega_{mr} \quad \theta_{mr}]$ is the state vector, $U = [v_d \quad v_q]^T$ is the input vector, Y= $[i_d \quad i_q]^T$ is the output vector, and the matrices $f(X)$ and $B$ are parameter matrices and $C$ is a constant matrix. The stator phase voltages in dq rotor frame can be expressed in terms of their abc-reference frame values. If one can measure the motor phase current $i_a, i_b, i_c$ then those can also be transformed to rotor reference dq frame, in the similar way as for voltages. The state equation of the motor is given in (12). This state equation can be used for the design of state observer. Now if one wants to estimate the only some of the state variables say $[i_d \quad i_q]^T$ then the standard form of the state observer equation is given by:

$$\frac{d(X)}{dt} = f(X,V) + G(\hat{I} - I) \qquad (14)$$

V is the estimated input vector and is equal to $[v_d \quad v_q]$. It is obtained by transforming the phase voltage ($v_a$, $v_b$, $v_c$) into rotor reference frame using the estimated rotor position, not the actual rotor position. The phase voltages are measured from the terminals of the motor. $I = [i_d \quad i_q]$, obtained by transforming the measured value of phase current ( $i_a, i_b, i_c$) into rotor reference frame using the estimated rotor position. $\hat{I} = [\hat{i}_d \quad \hat{i}_q]$, is the estimated value obtained by the solution of (14). In (14), G is the observer gain matrix, which is the result of tuning the system in such a way, that (14) becomes stable. G matrix can be given as:

$$G = \begin{pmatrix} g_{11} & g_{12} \\ g_{21} & g_{22} \\ g_{31} & g_{32} \end{pmatrix} \qquad (15)$$

Using this matrix, the matrix equation (14), can be expressed in detail as:

$$pi_d = -\frac{R}{L_d} i_d + \frac{\omega_r L_q}{L_d} i_q + \frac{1}{L_d} v_d + g_{11}(\hat{i}_d - i_d) \qquad (16)$$
$$+ g_{12}(\hat{i}_q - i_q)$$

$$pi_q = -\frac{\omega_r L_d}{L_q} i_d + \frac{(-R)}{L_q} i_q + \frac{-\lambda_f}{L_q} \omega_r + \frac{1}{L_q} v_q \qquad (17)$$
$$+ g_{21}(\hat{i}_q - i_q)$$

$$p\omega_r = 3 * P((L_d - L_q)i_d * i_q + \lambda_f * i_q)/(4 * J)$$
$$- (B * \omega_r + T_l)/J + g_{31}(\hat{i}_d - i_d) + g_{32}(\hat{i}_q - i_q) \qquad (18)$$
$$+ g_{32}(\hat{i}_q - i_q)$$

In any motor the electrical time constant is much smaller than the mechanical time constant, that is electrical sub dynamics are much as faster as compared to that of mechanical. So any error in electrical quantities can be used as feedback to rectify the estimated values of both electrical and mechanical quantities. This is the basic concept used in the observer model defined by (16), (17) and (18).

### III. MATLAB MODEL

An observer block for sensorless control of IPM is developed in MATLAB. Three phase voltages and currents are taken as input to the observer. Using Kalman filter and (14) to (18) the observer block is simulated. The flux weakening block is also added in the model to run the IPM motor above the base speed [7], [8].

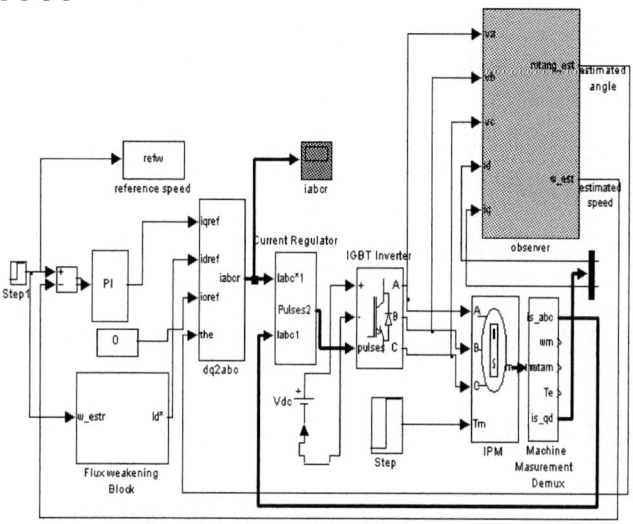

Fig. 4. MATLAB model of the observer based IPM drive.

In order to get desired output voltage and current the IGBT based inverter of Fig.4 is triggered with the help of pulses that are generated using PWM techniques at 16 kHz. Estimated angle and estimated speed are calculated using gain matrix of (15) and are the output of the observer block which in turn are fed back to the PI speed controller of the model in order to minimize the speed error and to run the motor at the desired reference speed. The value of $i_d$ is zero on or below the base speed. The flux weakening block is developed to get different values of $i_d$ in order to achieve the speed higher than the rated speed. There are speed estimation and position estimation block inside the observer block of the model.

#### A. Three Phase to Two Phase Converter block.

The input to the observer are three phase voltages and currents $i_d$ and $i_q$ .The three phase voltages $V_a$ ,$V_b$ and $V_c$ are first converted in $V_d$ an $V_q$ to be utilized using (16),(17) and (18)in simulation.

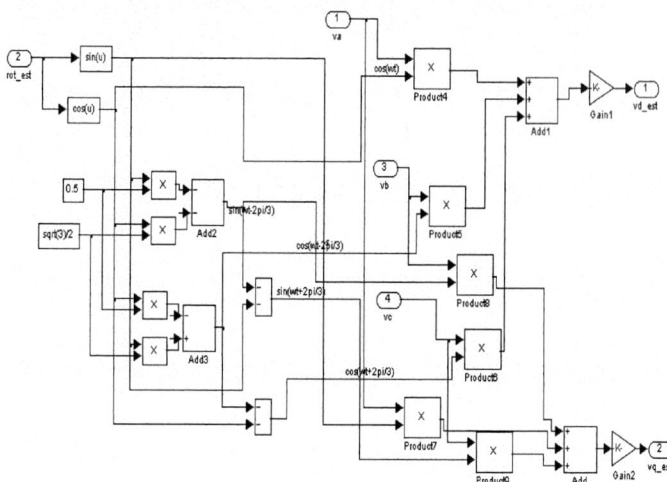

Fig. 5. Three phase to two phase converter block.

The Three phase to two phase converter block as shown in Fig.5.is used for the above purpose.

#### B. Speed Estimation Block block.

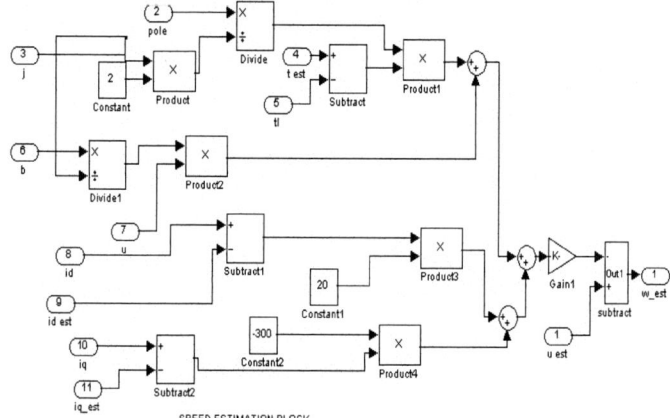

Fig. 6. Speed estimation block.

The speed estimation block as shown in Fig.6 is developed using (18) in which $g_{31}$ and $g_{32}$ are the gains which need to be determined by hit and trial method to achieve the desired estimation of speed and the angle.

### IV. RESULTS AND DISCUSSION

Simulation tests have been carried out in MATLAB in order to evaluate the theoretical behavior of the proposed sensorless control scheme such as starting performance, speed reversal , step change in load torque , step change in speed and flux weakening etc[11],[12]. The simulation of observer based sensorless drive system has been developed and implemented including a 2kW, 3 phases, 4 pole IPM synchronous motor with a rotor inertia of 0.0001278 kgm² damping of 9.4 × 10 $^{-5}$ Nm/rad/sec and a 16 kHz current regulated PWM inverter[9],[10]. To see the response of the drive for wide range of speed, the motor was made to run at half speed, at rated speed and above the rated speed

## A. Response of Speed Reversal

The motor was run at 340 rad/sec i.e. approximately at half of the base speed. The response of the reversal of speed from 340 rad/sec to -340 rad/sec in 0.0125 sec is satisfactory and load perturbation of 1Nm to 2 Nm at 0.06 sec is achieved

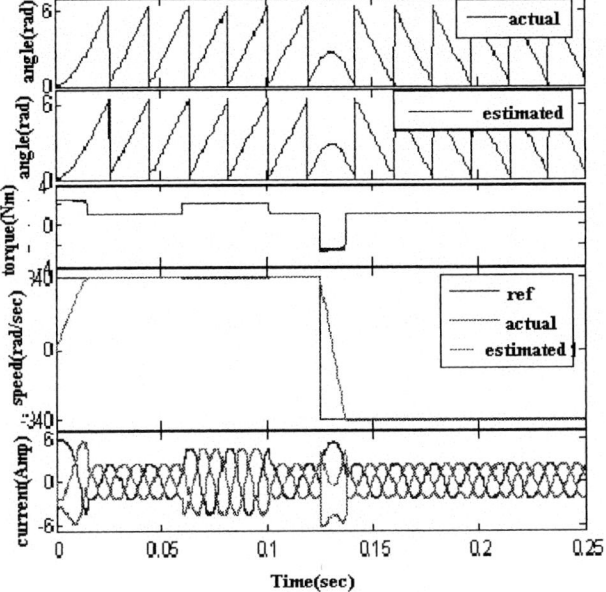

Fig.7. Simulated response of speed reversal and step variation in load.

at constant speed of 340 rad/sec The fig. also show the torque, estimated and actual speeds, estimated and actual angles. There is negligible error between actual and estimated speeds and angle. The motor attains the speed of 340 rad/sec.in 0.016 sec. and it takes 0.014 sec in reversing the speed. A small dip in speed at load perturbation at 0.06 to 0.1 sec. is seen in the Fig.7.

## B. Response of step change in speed

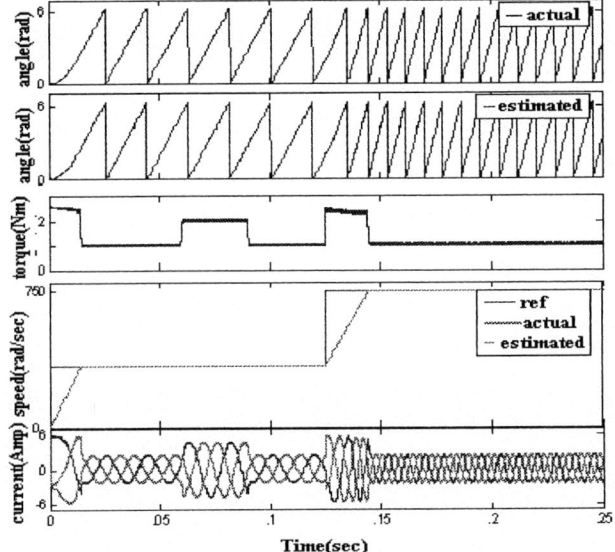

Fig. 8. Simulated response of step variation in speed and load.

To see the performance of the motor from half of the base speed to the base speed, the step variation in speed from 340rad/sec to the speed of 750 rad/sec.is applied. The settling time from the speed of 340 rad/sec. to 750 rad/sec. is 0.02 sec and the performance in this duration is satisfactory. The load perturbation of 1 Nm to 2 Nm at 0.06 sec to 1.0 sec is achieved at the constant speed of 340 rad/sec according to Fig.8. There is negligible error between actual and estimated angle .

## C. Response of Flux Weakening

To see the response of the motor up to the base speed in starting and then in flux weakening i.e. above the base speed, the reference speed was applied in step of 700 rad/sec(near base speed) to 1400 rad / sec. (near double the base speed), at starting and at 0.08 sec respectively.

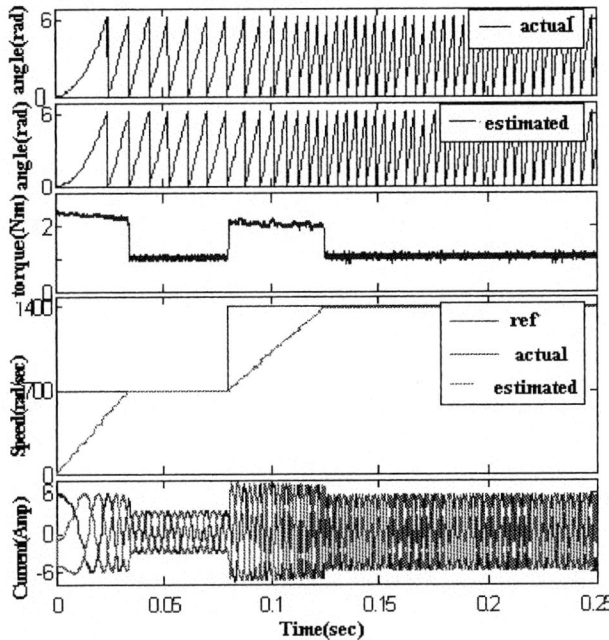

Fig.9. Simulated response of step variation in speed with flux weakening.

The motor reaches the base speed in 0.04 sec. hence a smooth start up to the base speed is seen according to the Fig.9. The response of the flux weakening with constant speed of 1400 rad./sec is also achieved in 0.04 sec. as shown in Fig.9.The frequency increases as the speed is increasing. The error between actual angle and the estimated angle is negligible.

## V. CONCLUSION

The developed sensorless algorithm for IPM drive has been found quite suitable to provide satisfactory performance for a wide speed range during steady state and dynamic operating conditions such as step response, load perturbation, starting performance and reversal of speed. The proposed sensorless technique can also be used successfully in flux weakening mode for speeds above the base speeds of the IPM synchronous motor. Hence the proposed scheme can be utilized successfully for eliminating the speed sensor.

## VI. Appendix

### MOTOR PARAMETERS

| | |
|---|---|
| Rated speed | 780(rad/sec) |

Permanent magnet synchronous motor parameters

| | |
|---|---|
| Type | IPMSM |
| Motor rating | 2 HP |
| Number of phases | 3 |
| Number of poles | 4 |
| Base current | 8 Amp |
| Rated voltage | 240 V |
| Stator resistance per phase | 1.22 ohm |
| Stator flux linkages per phase due to rotor magnet | |
| | 0.1432V/ (rad/sec) |
| Moment of inertia | 0.001278Kg/m$^2$ |
| Viscous coefficient | 0.000094Nm/ |
| (rad.sec) | |
| d-axis inductance | 11.5 mH |
| q-axis inductance | 9.5 mH |

## VII. References

[1] M. A. Rahman and T. A. Little and G. R. Slemon, "Analytical Models for Interior Permanent Magnet Synchronous Motors", *IEEE Transactions on Magnetics*, Vol. Mag-21, No.5, pp 1741-1743. Sept. 1985.

[2] Eike Ritcher, T. J. E. Miller and T. W. Neuman and T. L. Hudson , "The Ferrite Permanent Magnet AC Motors – A Technical and Economical Assesment ", *IEEE Transactions on Industry Applications*, VOL. 1A-21, NO.4, PP 640-650,May/June 1985,

[3] Peter Vas, *"Sensorless Vector and Direct Torque Control"*, Oxford University Press, 1998.

[4] Bimal K. Bose, *"Modern Power Electronics and AC-Drives"*, Pearson Education Asia, Low Price Edition (LPE), 2003.

[5] A.Consoli, G.Scarcella and A.Testa, "Industry Application of Zero-Speed Sensorless Control Technique for PMSM," *IEEE Trans. Ind. App.* Vol. 37, No. 2, pp 513-159 March/April 2001

[6] M. A. Rahman and T. A. Little, "Dynamic Performance Analysis of Permanent Magnet Motors", *IEEE Transactions on Power Apparatus and Systems*, Vol. PAS-103, No.6, June 1984, pp 1277-1282.

[7] P. K. HO. AND C. K. LEE, "Modeling and Simulation of a Permanent Magnet Synchronous Motor Under the Flux-weakening Control", *IEEE Transactions on Industry Applications*, VOL.34,no.4, pp 462-467,july/august,1998

[8] Thomas M. Jahns, "Flux Weakening Regime Operation of an Interior Permanent Magnet Synchronous Motor Drive", *IEEE Transactions on Industry Applications*, Vol. 1A-23, No.4, pp 681-689 July/Aug. 1987.

[9] B. K. Bose, "A High Performance Inverter Fed Drive System of an Interior Permanent Magnet Synchronous Motor", *IEEE Transactions on Industry Applications*, Vol. 24, No.6, , pp 987-997, Nov./Dec. 1988.

[10] J. P. Verster and J. H. R. Enslin, "Practical Design Approach for PWM Technique Controllers in the Application of Permanent Magnet Synchronous Machine (PMSM) Drives", *CD-324, Fourth International Conference on Power and Variable Speed Drives*,17-19 July 1990, London, UK. , pp 40-45.

[11] R. Krishnan and R. Gosh, "starting algorithm for permanent magnet brush less DC -motor drive with no position sensor", in *Proc. of IEEE Power Electronics Specialist Conference*, 1989.

[12] Jainhua Quain and M. A. Rahman, "Analysis of Field Oriented Control for Permanent Magnet Hystersis Synchronous Motors", *IEEE Transactions on Industry Applications*, Vol. 29, No.6, pp 1156-1163, Nov./Dec. 1993.

### TABLE I

#### LIST OF SYMBOLS

| Symbol | Quantity | Unit |
|---|---|---|
| $V_a, V_b, V_c$ | Phase voltages of stator winding | Volt |
| $\lambda_a, \lambda_b, \lambda_c$ | Flux linkages of phases a, b, c | Volt/(rad/sec) |
| $i_a, i_b, i_c$ | Phase current of stator winding | Ampere |
| $\lambda_f$ | Stator flux linkage due to permanent magnet | Volt/(rad/sec) |
| $R$ | Stator resistance per phase | Ohm |
| P | Number of poles | |
| $\omega_r$ | Rotor speed | radian/second |
| $T_e$ | Electromagnetic torque | Newton-meter |
| P | Differentia operator d/dt | |
| $V_d, V_q$ | d-axis and q-axis voltage | Volt |
| $i_d, i_q$ | d-axis and q-axis current | Amperes |
| $\lambda_d, \lambda_q$ | d-axis and q-axis flux linkages | Volt/(rad/sec) |
| $L_d, L_q$ | d-axis and q-axis inductances | Henry |
| $V_{dc}$ | DC-link voltage | Volt |
| $J$ | moment of inertia | Kg/meter$^2$ |

**2006 IEEE International Conference on Power Electronic, Drives and Energy Systems**

# Genetic Algorithm Based Optimal Design of Switching Circuit Parameters for a Switched Reluctance Motor Drive

Behzad Mirzaeian-Dehkordi, and Peyman Moallem

*Abstract*--In this paper, an optimization method based on Genetic Algorithms (GA) is applied to find the best design parameters of the switching power circuit for a Switched Reluctance Motor (SRM). The optimal parameters are found by GA with two objective functions, i.e. efficiency and torque ripple. A fuzzy expert system for predicting the performance of a switched reluctance motor has been developed. The design vector consists of design parameters, and output performance variables are efficiency and torque ripple. An accurate analysis program based on Improved Magnetic Equivalent Circuit (IMEC) method has been used to generate the input-output data. These input-output data are used to produce the optimal fuzzy rules for predicting the performance of SRM. Table look-up scheme and gradient decent training are used for optimal fuzzy prediction designed. The results of the optimal switching power circuit design for a 8/6, four phase, 4 kW, 250V, 1500rpm SR motor

*Index Terms*--**Fuzzy Prediction, Genetic Algorithms, SRM Drive.**

## I. INTRODUCTION

GENETIC algorithms (GA's) have been successfully applied to various optimization problems. The application of GA's to multi-objective optimization has been reported in several research works, for example see [1], [2].

In this paper, a new method based on genetic algorithm for multi-objective optimization is proposed. GA is applied to find the best design parameters of the switching power circuit for a SRM.

SR motor and it's switching parameters design are relatively new in the field of electrical machines and there is not as much experience and publication in this area as for the classical machines. Because of the highly saturated operating condition and the doubly salient structure of the motor, accurate analysis of SR motor is difficult and time consuming. Optimal design process of switching parameters, need the information of motor performance. The information can be

---

Behzad Mirzaeian Dehkordi is with the Department of Electronic Engineering,Faculty of Engineering, Isfahan University, Isfahan, Iran (e-mail: mirzaeian@eng.ui.ac.ir).

Peyman Moallem is with the Department of Electronic Engineering, Faculty of Engineering, Isfahan University, Isfahan, Iran (e-mail: P_moallem@eng.ui.ac.ir).

obtained from an expert designer or from sensitivity analysis of the switching parameters. The information can be formulated into a fuzzy expert system. The response of fuzzy expert system is fast and relatively accurate. Hence, it can be used for optimal design of switching parameters for SR motor in optimizing processes, which are based on genetic algorithms, immune systems, and or neural networks and need numerous performance evaluations of motor performance during the optimization process.

## II. GENETIC ALGORITHM

The general multi-objective optimization problem can be stated as to find an n dimensional vector, X , such that:

$$Maximize \quad [\,f_1(x),\ f_2(x),.....,\ f_M(x)\,]$$
$$S.T. \quad x_{io} \le x_i \le x_{if} \quad i = 1,..., n$$

Suppose that the number of population, N, is fixed and each solution of the multi-objective problem defined as a chromosome, $S_L$, with the length L.

In GA, a probability function has been defined to select the best-fitted chromosomes for existing population. Mutation and recombination operators applied to create the new chromosomes for the new population. In the probability function, the objective functions are combined by crisp weights as follow

$$f(x) = K_1 f_1(x) + K_2 f_2(x) \tag{1}$$

So, that the chromosomes with best performances for all objective functions have more chances to be chosen for participation in the next generation. Meanwhile, chances for participation of other chromosomes are not null. The probability function is defined as:

$$P_{L,i} = \frac{\big[C(S_{L,i})\big]}{\sum\limits_{L=1}^{N}\big[C(S_{L,i})\big]} \tag{2}$$

---

0-7803-9771-1/06/$25.00 ©2006 IEEE

In this equation, $C(S_{L,i})$ is the fitness number of the objective function for the L'th chromosome, N is the number of the chromosomes in population.

The fitness number of the objective function for L'th chromosome defined as:

$$C(S_{L,i}) = \begin{cases} \dfrac{C^o(S_{L,i})}{\gamma} & if \quad C^o(S_{L,i}) \succ 0 \\ 0 & if \quad C^o(S_{L,i}) \leq 0 \end{cases} \qquad (3)$$
$$L = 1,2,...,N$$

In this equation $C^o(S_{L,i})$ is the fitness value for the objective function and $\gamma$ is the summation of the positive fitness values. Fitness value for objective function is defined as:

$$C^o(S_{L,i}) = f(S_{L,i})$$
$$L = 1,2,...,N \qquad (4)$$

Another way of multi-objective optimization is to defined the probability function as follows:

$$P_{L,i} = \frac{\prod\limits_{m=1}^{M} \left[ C_m(S_{L,i}) \right]^{K_m}}{\sum\limits_{L=1}^{N} \left[ \prod\limits_{m=1}^{M} \left[ C_m(S_{L,i}) \right]^{K_m} \right]} \qquad (5)$$

In this equation, $C_m(S_{L,i})$ is the fitness number of the m'th objective function for the L'th chromosome, M is number of the objective functions, N is the number of the chromosomes in population and, $K_m$, is the weight which shows the importance degree of the m'th objective function. The fitness number of the m'th objective function for L'th chromosome defined as:

$$C_m(S_{L,i}) = \begin{cases} \dfrac{C_m^o(S_{L,i})}{\gamma_m} & if \quad C_m^o(S_{L,i}) \succ 0 \\ 0 & if \quad C_m^o(S_{L,i}) \leq 0 \end{cases} \qquad (6)$$
$$L = 1,2,...,N$$
$$m = 1,2,...,M$$

In this equation $C_m^o(S_{L,i})$ is the fitness value for the m'th objective function and $\gamma_m$ is the summation of the positive fitness values. Fitness value for each objective function is defined as:

$$C_m^o(S_{L,i}) = f_{objm}(S_{L,i}) - \mu_m$$
$$m = 1,2,...,M \qquad (8)$$
$$L = 1,2,...,N$$

Where, $\mu_m$ is the mean value of the m'th objective function in population sequences. In equations(1) and (5), $K_m$ is the weight, which shows the importance degree of the m'th objective function. In this paper for two objective functions, i.e., efficiency and torque ripple the impotence degrees are defined as :

$$k1 = k2 = 1 \qquad (9)$$

## III. DESIGN OF FUZZY SYSTEM

A number of methods for constructing fuzzy systems from input-output pairs exist such as:
- Table look-up scheme. [3], [4]
- Gradient Descent training. [3], [4]

In this paper, a set of fuzzy rules has been devised for a SR motor by using a table look-up scheme using the information obtained from an accurate analysis program. Then, using gradient descent training optimizes the fuzzy rules. This fuzzy expert system is used to predict the performance of the SR motor.

## IV. DESIGN THE FUZZY SYSTEM USING A LOOK-UP TABLE SCHEME [11]

Suppose that the following input-output pairs are given from IMEC analysis program. This information has been obtained for the range of design parameter variations within the feasible limits.

where:

$$(x_o^p ; y_o^p) \qquad p=1,2,...,N \qquad (10)$$

The objective is to design a fuzzy system, F(x), based on the N input-output pairs.

F(x) is defined as follows:

$$F(x) = \frac{\sum\limits_{L=1}^{M} \bar{y}^L [\prod\limits_{i=1}^{n} \mu_{A_i^L}(x_i)]}{\sum\limits_{L=1}^{M} [\prod\limits_{i=1}^{n} \mu_{A_i^L}(x_i)]} \qquad (11)$$

where, $\mu_{A_i^L}(x_i)$ is the i'th membership value of the i'th fuzzy set in the precedence predicate and $\bar{y}^L$ is the center of fuzzy set in the consequence predicate for i'th rule.

## V. DESIGN OF FUZZY SYSTEM USING GRADIENT DESCENT TRAINING [11]

In this section, a procedure is presented for a fuzzy system in which the membership functions are chosen to optimize a certain criterion.

First, it is assumed that the designed fuzzy system is of the following form [9]:

$$F(x) = \frac{\sum_{L=1}^{M} \bar{y}^L [\Pi_{i=1}^n \exp(-(\frac{x_i - \bar{x}_i^L}{\sigma_i^L})^2)]}{\sum_{L=1}^{M} [\Pi_{i=1}^n \exp(-(\frac{x_i - \bar{x}_i^L}{\sigma_i^L})^2)]} \quad (12)$$

Where M is the fixed number of fuzzy rules, and $\bar{y}^L, \bar{x}_i^L$ and $\sigma_i^L$ are free parameters.

The fuzzy system can be represented by F(x) given in equation (11) as a feed forward network.

For a given input-output pair :

$$(x_o^p ; y_o^p) \qquad p=1,2,\dots,N$$

and at the q'th stage of training , q=0,1,2,.. present $x_0^p$ to the input layer of the fuzzy system and compute the output. Then we must update the free parameters by using the matching error as follows:

$$e^p = \frac{1}{2}[f(x_0^p) - y_0^p]^2$$

$$\bar{y}^L(q+1) = \bar{y}^L(q) - \alpha \frac{\partial e}{\partial \bar{y}^L}\Big|_q \quad (13)$$

L=1,2, …, M

$$\bar{x}_i(q+1) = \bar{x}_i^L(q) - \alpha \frac{\partial e}{\partial \bar{x}_i}\Big|_q \quad (14)$$

i=1,2,…n      L=1,2,…M

$$\sigma_i^L(q+1) = \sigma_i^L(q) - \alpha \frac{\partial e}{\partial \sigma_i^L}\Big|_q \quad (15)$$

i=1,2,…n      L=1,2,…M

In (13), (14) and (15) $\alpha$ is the constant step size.

## VI. SWITCHED RELUCTANCE MOTOR

A SR motor is a variable reluctance motor that is designed to convert energy efficiently. The motor is double salient, and it is essential for the machine operation that the number of rotor and stator poles be different. Torque is produced by the tendency of the rotor poles to align with the poles of the excited stator phase. Torque is independent of the direction of phase current, giving rise to the possibility of unipolar phase current in which only one main switching device may be required per phase. A SRM is of very simple structure: its rotor is brushless and has no winding of any kind. The motor is singly excited from stator windings, which are concentric coils wound in series on diagonally opposite stator poles. Both rotor and stator are made of laminated iron.

In spite of the simple structure of the motor, because of the highly saturated operating condition and the doubly salient structure of the motor, the accurate analysis of SR motor is very difficult and conserving. Several methods are reported for the analysis of SR motors, such as: Finite Element Method (FEM), [5],[6], Magnetic Equivalent Circuit (MEC),[7],and piecewise linear model [8]. FEM is applied for accurate prediction of the machine parameters and performances. This method requires complicated modeling and large computational time, which is prohibitive for optimal design of SR motors. Improved MEC is the other method that is used for modeling and analysis of SR motors, which is based on the magnetic field distribution in the several parts of the motor. The IMEC modeling [8], results show good agreement with the results obtained from FEM, however, the modeling is less complicated and the computational time in IMEC is much lower. For optimal design of SR motor a large number of performance evaluation are required and even with IMEC method, the computational time would be very large. Thus, in this study a fuzzy expert system for performance prediction of the SR motor has been developed. The fuzzy expert system can be used as a very fast performance prediction tool in optimal design programs.

## VII. SENSITIVITY ANALYSIS OF SRM

To design the fuzzy expert performance predictor for SR motor which is shown in   figure 1 as cross section form and with parameters are given in Table III, a set of input-output information pairs is needed. These information pairs can be obtained from sensitivity analysis of the motor parameter controllers in switching circuits. Sensitivity curves of efficiency and torque ripple with respect to switching parameters are shown in Figs. 2 to 5. In this paper the switching circuit which is shown in Fig. 6-a for a 6/4 SR motor   is extended for a 8/6, four phase, 4 kW, 250V, 1500rpm SR motor and is used for simulation.

In this section, for each set of design parameters, the performance characteristics of SR motor are obtained using IMEC method, which is a fast and relatively accurate analysis method.   The design vector includes three parameters and output performance variables are efficiency and torque ripple. The design vector is defined as: electrical turn-on angle, electrical turn-off angle,$(\theta_{0n}, \theta_{off})$ and limited current in switching elements (Ip).

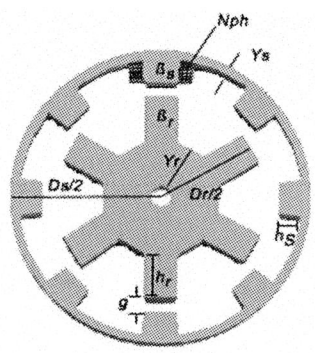

Fig.1. Design parameters of SR motor.

## VIII. CONSTRUCTING THE OPTIMAL FUZZY RULES FOR PREDICTING THE SR MOTOR PERFORMANCES

Using the information from sensitivity analysis, a set of fuzzy rules will be devised. The i'th rule of this fuzzy system is as follows:

Ri: If x1 is A1 and x2 is A2 and x3 is A3 Then y1 is B1 and y2 is B2
  for  i=1:M                    (14)

Where, $x1(\theta_{on})$, $x2(\theta_{off})$ and $x3(Ip)$ are the controller parameters of SR motor in switching circuits and y1 and y2 are the efficiency and torque ripple . Based on the high or low sensitivity of performance variable to design controller parameters and output performances, the number of fuzzy membership functions defined for each parameter could be high or low. The fuzzy rules with triangular shape membership functions are designed by table-look up scheme described in section (IV). Setting the rules for all parameters, we have a fuzzy rule base with a set  of          IF-THEN rules. Using the gradient descent, the fuzzy rules with triangular membership function can be optimized to Gaussian membership functions based on the method described in section (V). By using this method, ten rules are obtained for predicting the performance of SR motor.

The results of performance prediction of optimized fuzzy rule base for the 4 Kw, 8/6 motor are given in Table II based on probability function (1) and in Table III based on probability function (5), which is compared with the results obtained by fuzzy expert system with triangular shape membership function, IMEC method and FEM analysis methods. The results shows good accuracy compared with results obtained by accurate methods; however, the computational time is very low compared to FE and IMEC method. The parameters of the SR motor are given in Table IV.

## IX. CONCLUSION

In this paper, a new method based on genetic algorithms for multi-objective optimization process is proposed. The method is successfully applied for optimal design of a SR motor switching parameters.      A fuzzy expert system has been devised which is used for fast performance prediction during the optimization process. The new GA method has shown promising success in the design optimization of SR motor switching parameters in power electronic circuit, which consist of many parameters and contradicting objective functions.

TABLE I
THE NUMBER OF FUZZY SETS FOR EACH PARAMETER IN OPTIMIZED FUZZY EXPERT SYSTEM

| Parameter | Minimum | Maximum | Number of fuzzy sets with Gaussian membership function |
|---|---|---|---|
| $\theta_{on}$ | 6 (elec.deg.) | 16 (elec.deg.) | 6 |
| $\theta_{off}$ | 21 (elec.deg.) | 28 (elec.deg.) | 8 |
| Ip | 26(Amp.) | 30(Amp.) | 9 |

TABLE II
PERFORMANCE RESULTS OF SR MOTOR USING PROBABILITY FUNCTION (1)

| Method | % Efficiency | % Torque Ripple ( %T.R) | % Error (w.r.t. FEM) | |
|---|---|---|---|---|
| | | | % Eff | % T.R |
| Fuzzy predictor with triangular membership function | 92.5 | 17 | 1.65 | 41.6 |
| Optimized Fuzzy predictor with Gaussian membership function | 91.7 | 13 | 0.77 | 8.33 |
| IMEC Method | 91.8 | 14 | 0.88 | 16.67 |
| IMEC Method | 91 | 12 | | |

### TABLE III
### PERFORMANCE RESULTS OF SR MOTOR USING PROBABILITY FUNCTION (5)[10]

| Method | % Efficiency | % Torque Ripple ( %T.R) | % Error (w.r.t. FEM) | |
|---|---|---|---|---|
| | | | % Eff | % T.R |
| Fuzzy predictor with triangular membership function | 91.9 | 15 | 1.29 | 44.23 |
| Optimized Fuzzy redictor with Gaussian membership function | 93 | 11.2 | 0.11 | 7.69 |
| IMEC Method | 92.7 | 14.7 | 041 | 41.35 |
| IMEC Method | 93.1 | 10.4 | | |

### TABLE IV
### SR MOTOR PARAMETRS

| | |
|---|---|
| $\beta_s{}^\circ$ | 21 |
| $\beta_r{}^\circ$ | 23 |
| $h_r(mm)$ | 17.5 |
| $y_r(mm)$ | 12.1 |
| $h_s(mm)$ | 29 |
| $y_s(mm)$ | 13.9 |
| $D_s(mm)$ | 177.5 |
| $D_r(mm)$ | 91.1 |
| $L_s(mm)$ | 150 |
| $N_{ph}/2$ | 20 |
| g(mm) | 0.3 |

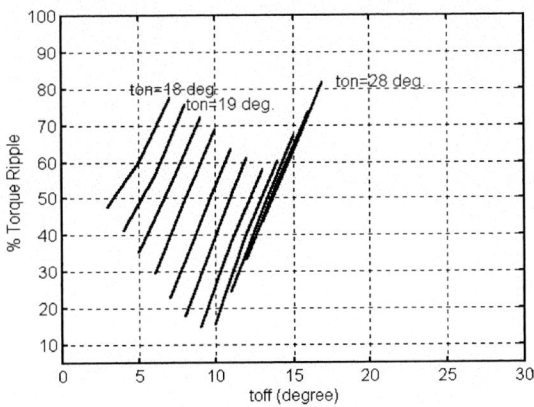

Fig. 2. Sensitivity curve of torque ripple w.r.t turn- off and turn-on electrical switched angle ( limited current=31amp.).

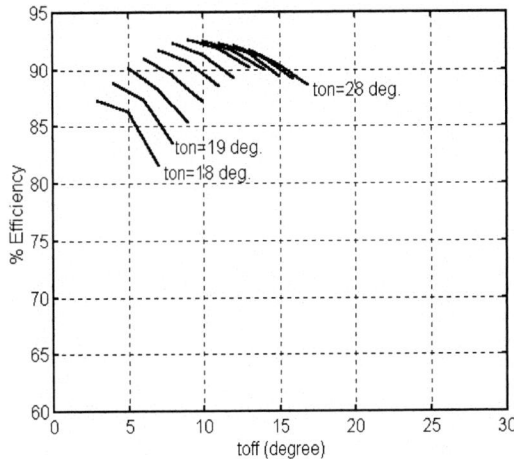

Fig. 3. Sensitivity curve of efficiency w.r.t turn-off and turn-on electrical switched angle ( limited current=31amp.).

## X. References

[1] A.Trebi-Ollennu, B.A. White," Multi-objective genetic algorithm optimization approach to nonlinear control system design," *IEE. Proc. Control Theory App1.*, vol. 144, no. 2, pp. 137-142, March 1997.

[2] C.M. Fonseca and P.J. Fleming," Multi-objective genetic algorithms made easy, selection, sharing and making restriction,"*in Proc.1995 1st IEE/IEEE Int. conf. GA's in Eng. Systems, Innovations and Applications, Sheffield. UK.*,pp. 42-52.

[3] Li-Xin Wang, *A Course in Fuzzy Systems and Control*, Prentice-Hall Int., Inc.1997.

[4] Wang,L.X.,and J.M.Mendel, "Generating fuzzy rules by learning from examples", *IEEE Trans. on system , Man , and Cyber.*, 22.no.6.,pp.1414-1427,1992.

[5] Moallem ,M., " *Performance characteristic of switched reluctance motor drive,* Purdue University, Ph.d. Thesis, USA,1989.

[6] Arkadan, A.A. and Kielagas, B.W. ," Switched reluctance motor drive system dynamic performance prediction and experimental verification," *IEEE Trans. Energy Conv.*, vol.9, no. 1, pp.36-44, 1994.

[7] M.Moallem and G.E.Dawson , "An improved magnetic equivalent circuit method for predicting the characteristic of highly saturated electromagnet devices", *IEEE Trans.on Magnetic*, vol.34, no.5, pp.337-347, 1990.

[8] T.J. Miller *Switched Reluctance Motor and Their Control*, Oxford 1993.

[9] Li-Xin Wang, *A Course in Fuzzy Systems and Control*, Prentice-Hall Int., Inc.2000.

[10] B.Mirzaeian Dehdordi,A.Kiyoumarsi,P.Moallem, M.Moallem, "Optimal design of switching-circuit parameters for switched reluctance motor drive based on genetic algorithm," *Electro motion,* vol.13,no.3,pp. 213-220,July-Sept. 2006

## XI. BIOGRAPHY

**Behzad Mirzaeian Dehkordi** was born in Shahr-e-kord, in the year 1966. He received the B.Sc. of Electronics Engineering from Shiraaz University, Iran in 1985 and M.Sc. and Ph.D. of Electrical Engineering form Isfahan University of Technology in the years 1994 and 2000 respectively. He is currently an assistant professor at the department of electrical and electronics engineering, the university of Isfahan.

His field of interests includes power electronics and drives and power quality problems.

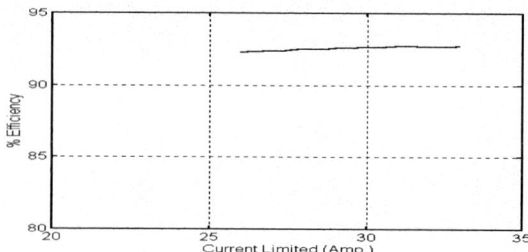

Fig. 4. Sensitivity curve efficiency w.r.t Limited current (turn-on current= 8 degrees and turn-off current =23 degrees w.r.t aline axis.

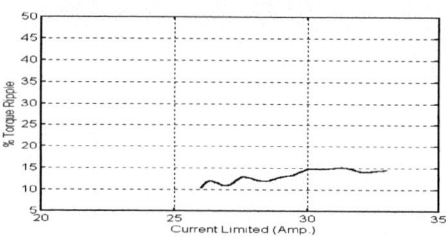

Fig. 5. Sensitivity curve of torque ripple w.r.t Limited current (turn-on current= 8 degrees and turn-off current =23 degrees w.r.t aline axis.

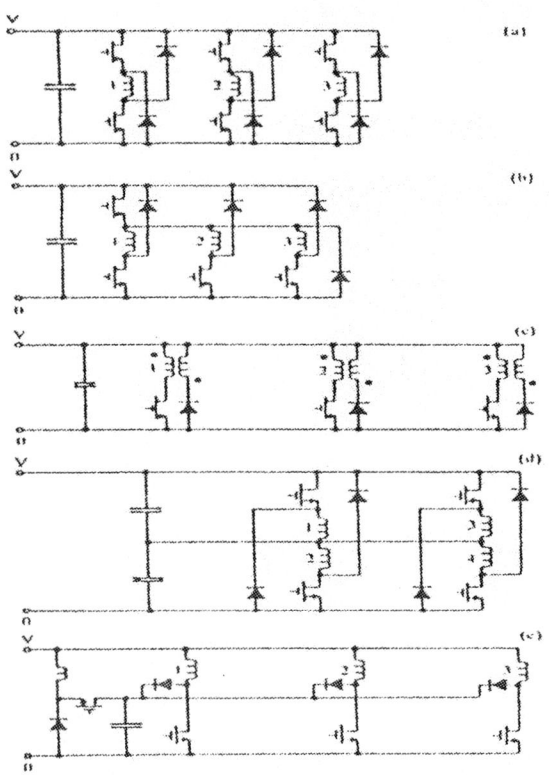

Fig. 6. Typical switching power electronic circuits for 6/4 SR motor.

**2006 IEEE International Conference on Power Electronic, Drives and Energy Systems**

# Reduction of Cogging Torque in PMBLDC Motor with Reduced Stator Tooth Width and Bifurcated Surface Area Using Finite Element Analysis

R. Somanatham, P.V.N.Prasad, and A.D.Rajkumar.

*Abstract --* **This paper proposes a method for reduction of cogging torque with reduced stator tooth width and bifurcated active surface area using Finite Element analysis. For the proposed changes in the motor design the variation in the flux density distribution in the air gap, cogging torque and reluctance torque ripple are determined. The variations in the torque developed by the motor, cogging torque, output power, losses and efficiency for different stator tooth shapes are recorded. From the present study it is observed that there is a significant reduction of cogging torque with reduced stator tooth width & bifurcated active surface area compared with a specified motor.**

*Index Terms--***Cogging torque, Finite element analysis, Permanent magnets, PMBLDC motor, Torque ripple.**

## I. NOMENCLATURE

| | |
|---|---|
| $B_{mg}$ | flat topped value of the trapezoidal magnetic flux density wave form |
| $b_{14}$ | slot opening |
| $b_{sk}$ | skew of stator slots |
| $D_{1in}$ | stator inner diameter |
| g | equivalent air gap |
| g' | air gap(mechanical clearance) |
| $h_m$ | height of PM |
| $k_c$ | Carter's coefficient |
| $k_{skk}$ | stator slot skew factor refer to slot pitch $t_1$ |
| $L_i$ | armature stack effective length |
| p | number of pole pairs |
| $T_c$ | Cogging Torque |
| $t_1$ | slot pitch |
| $\mu_0$ | magnetic permeability of free space |
| $\mu_{rec}$ | recoil permeability |
| $\tau$ | pole pitch |
| $X$ | rotor position |

## II. INTRODUCTION

ALTHOUGH PM machines are high performance devices, there are two torque components that affect their output produced from the harmonic content of the current and voltage

waveforms in the machine. The second component called cogging torque is due to the physical structure of the machine [1]. The cogging torque, is due to the physical structure of the machine and is produced by the magnetic attraction between the rotor mounted permanent magnets and the stator teeth. It is the circumferential component of attractive force that attempts to maintain the alignment between the stator teeth and the permanent magnets. Cogging torque produces both vibration and noise, both of which may be amplified in variable frequency drives when the torque frequency coincides with a mechanical resonant frequency of the stator or rotor. Cogging torque is also detrimental to the performance of position control systems such as robots and to the performance of speed control systems particularly at low speed [2].

Reduction of cogging torque in permanent magnet motors is gaining increasing importance along with the demand for high- performance PM brushless motors. It is a vital design consideration in power steering, robotics, machine spindle, high-precision position control and any application where minimizing torque ripple, vibration and noise is an essential requirement. In order to be able to develop alternate and improved techniques of reducing cogging torque, it is important to predict it accurately for any given motor geometry and configuration. The torque ripple can be minimized both by the proper motor design and motor control. There are various methods available in design perspective to reduce cogging torque in brushless machines. They are slot less windings, skewing of stator slots, shaping of stator slots, adjusting slot opening, selection of the number of stator slots, shaping of PMs, skewing of PMs, shifting PM segments, selection of PM width, magnetic strength of PMs and creating magnetic circuit asymmetry [3].

The electromagnetic torque can be calculated analytically or numerically in a variety of ways such as by the Maxwell stress and co-energy methods. They need very accurate global and local field solutions particularly for the determination of cogging torque. In other words, a high level of mesh discretization is required in finite element calculations, whilst a reliable physical model is essential to an analytical prediction. The analytical model used for predicting the cogging torque is also capable of quantifying the effects of the various design parameters [4].

Cogging torque can be determined analytically as well as by Finite element methods. The FEM is a powerful and economical approach to characterize the torque ripple of a given design without hardware proto type. The main objective of the present chapter is to demonstrate that cogging torque can be reduced to generally acceptable levels by appropriate

---

The authors express their deep sense of gratitude to the Principal, University College of Engineering, Osmania University, Hyderabad for the financial support rendered under TEQIP programme.

R.Somanatham (email: rsm2006@rediffmail.com), P.V.N. Prasad (email: polaki@rediffmail.com), and A.D.Rajkumar (email: adrajkumar@yahoo.com) are with the Department of Electrical Engineering, University College of Engineering, Osmania University, Hyderabad - 500 007, India.

---

0-7803-9771-1/06/$25.00 ©2006 IEEE

selection of motor design. For certain standard stator slot (teeth) shapes the cogging torque can be determined.

A 36 slot, 4pole PMBLDC motor with certain dimensions [5] is considered as the specified motor for which the cogging torque is determined analytically. But, for some special designs the cogging torque can be determined by FE method only. In the present work the cogging torque is determined analytically and by FEA for the specified motor. Further by FE analysis a comparative study has been made for the effect of (a) reduced tooth width compared to active surface area (b) reduced tooth width with bifurcated active surface area, on flux density distribution in air gap, cogging torque and reluctance torque ripple. Three different tooth shapes of stator under consideration are shown in Fig. 1.

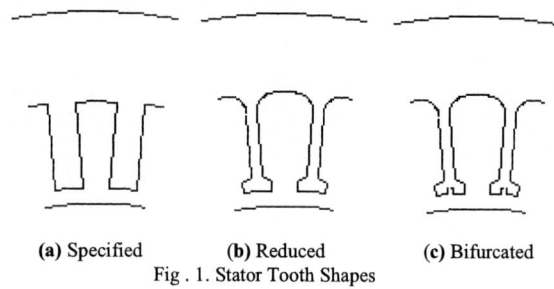

**(a)** Specified     **(b)** Reduced     **(c)** Bifurcated

Fig . 1. Stator Tooth Shapes

## III.  DETERMINATION OF COGGING TORQUE

The cogging torque for the specified machine is determined analytically using MATLAB. The analytical expression for the cogging torque is,

$$T_c(X) = -\frac{g'L_i}{\mu_0}\frac{D_{lin}}{2}A_T\frac{B_{mg}}{k_C}\sum_{k=1,2,3,\ldots}^{\infty}(-1)^k\zeta_k\sin\left(k\frac{\pi}{t_1}c_t\right)\sin\left(\frac{2k\pi}{t_1}X\right) \quad (1)$$

where

$$k_C = \frac{t_1}{t_1 - \gamma g'} \quad (2)$$

$$g' = g + h_m/\mu_{rec} \quad (3)$$

$$\zeta_k = k_{ok}k_{skk} \quad (4)$$

$$A_T = -2\gamma\frac{g'}{t_1}B_{mg} \quad (5)$$

$$\sigma = \frac{b14}{g'} \quad (6)$$

$$\gamma = \frac{4}{\pi}[0.5\sigma\arctan(0.5\sigma) - \ln\sqrt{1+(0.5\sigma)^2}] \quad (7)$$

$$\rho = \left(\frac{\sigma}{5+\sigma}\right)\frac{2\sqrt{1+\sigma^2}}{\sqrt{1+\sigma^2}-1} \quad (8)$$

$$k_{ok} = \frac{\sin[k\pi\rho b_{14}/(2t_1)]}{k\pi\rho b_{14}/(2t_1)} \quad (9)$$

$$k_{skk} = \frac{\sin(kb_{sk}\pi/t_1)}{kb_{sk}\pi/t_1} \quad (10)$$

$$c_t = t_1 - b_{14} \quad (11)$$

The analytical expression for cogging torque is useful only for standard design of the motor. But, for the proposed design of reduced stator tooth width, with and without bifurcation of active surface area, it is not possible to determine the cogging torque analytically and thus FE analysis is the only alternative. The cogging torque for the specified motor [2] is determined analytically and by FE analysis and both the results are found to be very close to each other as shown in  Fig. 2.  A comparative study has been made for cogging torque and flux density distribution in the air gap, for three different shapes of the stator teeth. The FEA software tool MotorPro [6] is used in the present simulation study. The pre & post processing and calculations are carried out by exciting the stator winding with 3 phase $120^0$ Square wave Inverter.

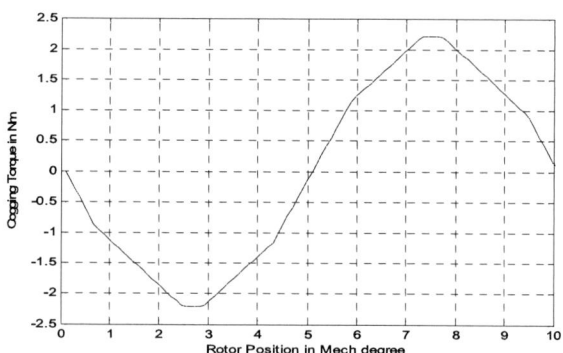

Fig. 2(a). Cogging Torque (Analytical)

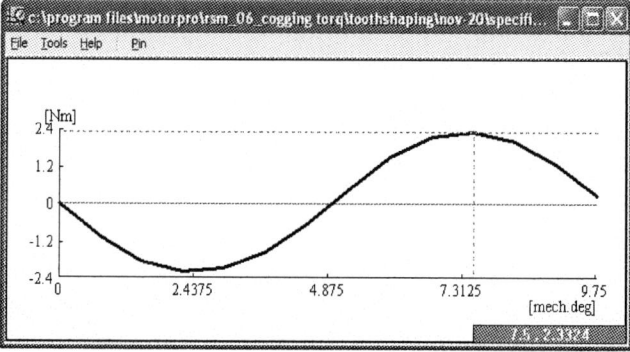

Fig. 2(b).  Cogging Torque (FE Analysis)

## IV. RESULTS & CONCLUSIONS

Few important results of the simulation study are incorporated here. The mesh distribution, equipotential lines of the proposed design and flux density distribution diagrams for bifurcated stator tooth are shown in Fig. 3. The FEM profiles of flux density distribution in the air gap, cogging torque, and reluctance torque ripple are recorded and are shown in Fig. 4. For the three motor designs torque developed by the motor, peak to peak values of the cogging torque output power, total losses, the efficiency are recorded and they are shown in Table I.

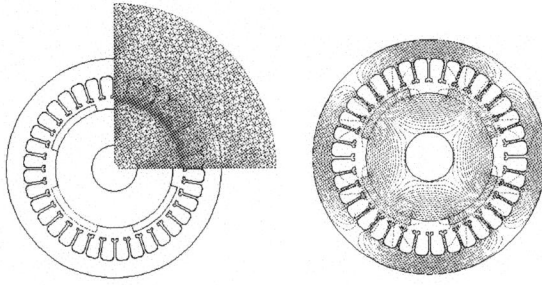

Fig. 3(a). Element Mesh Distribution    Fig. 3 (b) Equipotential Line

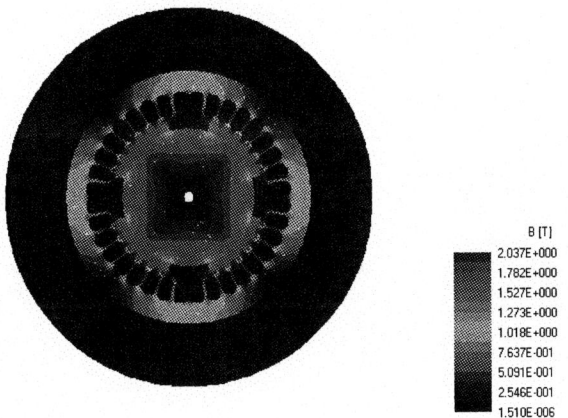

Fig. 3(c). Flux Density Distribution

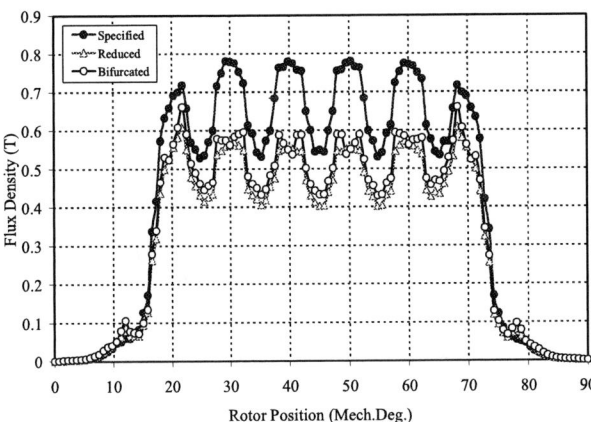

Fig. 4(a). Flux Density Distribution in the Air Gap

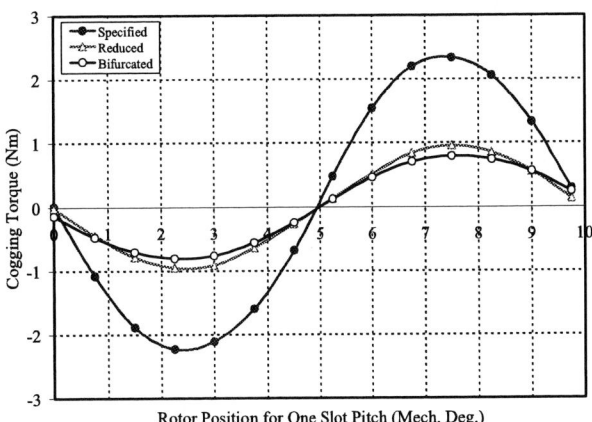

Fig. 4(b). Cogging Torque

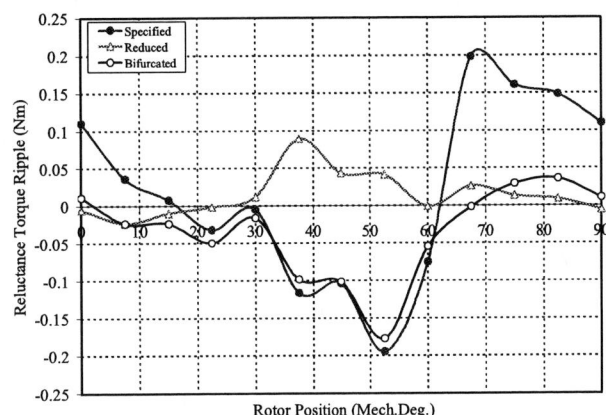

Fig. 4 (c). Reluctance Torque Ripple

TABLE I
PERFORMANCE OF MOTOR FOR THREE STATOR TOOTH SHAPES

| S.No | Performance | Specified Slot opening (5.3mm) | Reduced Slot opening (2mm) | Bifurcated Slot opening (2mm) |
|------|-------------|-------------------------------|----------------------------|-------------------------------|
| 1. | Torque | 11.85 Nm | 4.71 Nm | 8.914 Nm |
| 2. | Cogging Torque (pk-pk) | 4.55 Nm | 1.97 Nm | 1.596 Nm |
| 3. | Output Power | 3639 W | 1478 W | 2860 W |
| 4. | Total losses | 443 W | 277 W | 387 W |
| 5. | Efficiency | 89.13 % | 84.26 % | 87.86 % |

From the observations it is evident that there is a significant reduction of more than 60 % peak to peak cogging torque with reduced stator tooth width with bifurcated active surface area compared with the specified motor.

## V. MOTOR SPECIFICATIONS

The specifications of the motor are:

| | |
|---|---|
| $V_{DC\ link}$ | 310V |
| $I_{Peak}$ | 16.5 A |
| Connection | Star |
| Speed | 3000 rpm |
| Number of phases | 3 |
| Yoke depth | 17.4 (mm) |
| Number of poles | 4 |
| Tooth depth | 17.2 (mm) |
| Number of slots | 36 |
| Tooth width | 5.3 (mm) |
| Winding pitch | 6/9 |
| Magnet depth | 6.3 (mm) |
| Stator length | 76 (mm) |
| Magnet width | 5.05 (slot pitch) |
| Outer radius of stator | 95 (mm) |
| Air gap radius | 58.5 (mm) |
| Surface-magnet of rotor | Neodymium |

## VI. REFERENCES

[1] C. Studer, A. Keyhani, T. Sebastian and S. K. Murthy, "Study of Cogging Torque in Permanent Magnet Machines," *IEEE IAS Annual Meeting, USA*, pp 42-49 (1997).

[2] Touzbu Li and Gordon Slemon, "Reduction of cogging torque in permanent magnet motors," *IEEE Transactions on Magnetics*, Vol. No.6, November 1998, pp 2901-2903.

[3] Jacek F. Gieras and Mitchell Wing, "Permanent Magnet Motor Technology, Design and Applications," Second Edition, Marcel Dekker, Inc., 2002.

[4] Z.Q. Zhu and David Howe, "Influence of design parameters on cogging Torque in Permanent Magnet Machines," *IEEE Transactions on Energy Conversion*, Vol. 15, No. 4, December 2000.

[5] Takeo Ishikawa, Gordon R.Slemon, "A Method of Reducing Ripple Torque in Permanent Magnet Motors without Skewing," *IEEE Transactions On Magnetics*, pp 2028-2031, 1993.

[6] "MotorPro Manual," Komotek Co., Ltd., Korea, 2003.

[7] "MATLAB 7.0.4 Version," The Math Works Inc., USA.

## VII. BIOGRAPHIES

**R.Somanatham** was born on 4.12.1950 in Secunderabad, India. He received BE. and M.Tech. degrees from Osmania University, Hyderabad, India in 1973 and 1977 respectively. His field of specialization is Industrial Drives and Control. Since 1987 he is working as Associate professor in the Department of Electrical Engineering, Osmania University, Hyderabad. His main field of interest is Machine modeling, Power Electronics, Solid state ac and dc drive systems. He is a Fellow of Institute of Engineers (India).

**P.V.N.Prasad** was born in Hyderabad, India in January 1960. He graduated in Electrical & Electronics Engineering from Jawaharlal Nehru Technological University, Hyderabad in 1983 and received M.E in Industrial Drives & Control from Osmania University, Hyderabad in 1986. He served as faculty member in Kothagudem School of Mines during 1987 - 95 and at presently serving as Associate Professor in the Department of Electrical Engineering, Osmania University. He received his Ph.D in Electrical Engineering in 2002. His areas of interest are Simulation of Electrical Machines & Power Electronic Drives and Reliability Engineering. He is a member of Institution of Engineers (India) and Indian Society for Technical Education. He is recipient of Dr.Rajendra Prasad Memorial Prize, Institution of Engineers (India), 1993 - 94 for best paper. He has published over 30 papers in National and International Journals, Conferences & Symposia in the fields of Reliability Engg. and Electric drives. He has visited Thailand and Italy to present technical papers.

A.D.Rajkumar was born in Hyderabad, India in December, 1947. He received B.Sc., B.E and Ph.D degrees from Osmania University, Hyderabad, India in 1965,1968 and 1983 respectively. In 1971 he received M.Tech degree from the IIT-Madras, Chennai, India. Since 1987 he has been Professor in the Department of Electrical Engineering, Osmania University, Hyderabad , India. He is engaged in the teaching, research and development of Power Electronics, Microprocessor applications and Frequency response analysis of power transformers in which he has published papers. He has visited Singapore and China to present papers. He is a Fellow of the Institution of Engineers (India)

**2006 IEEE International Conference on Power Electronic, Drives and Energy Systems**

# A Novel Phasor Diagram of Interior Permanent Magnet Synchronous Motors based on Spiral Vector Theory

Bishnu P. Muni

*Abstract*— The paper presents a novel phasor diagram of interior permanent magnet synchronous (IPMS) motor for steady state analysis. Starting from first principle, the phase decoupled spiral vector model of the IPMS motor has been obtained. A phase decoupled voltage equation in phasor can be obtained from the spiral vector IPMS motor at steady state. In the phasor diagram, the effect of reluctance variation is represented as reluctance voltage. The shaft power output of the motor is the sum of power due to field voltage and reluctance voltage. The steady state performance of the IPMS motor can be predicted with the use of this phasor diagram. The novel phasor diagram can also be used for steady state performance analysis of salient pole synchronous and reluctance machine.

*Index Terms*-- Interior permanent magnet synchronous machine, Phasor diagram, Salient pole synchronous machine, and Spiral vector.

## I. INTRODUCTION

PHASOR diagrams are widely used for steady-state analysis of electrical circuits and machines. Lumped parameter based sinusoidal equivalent circuits are being used for predicting steady-state performance of all types of electrical machines. The phasor diagrams can be considered vectorial representation of voltages in AC circuit at any particular instant of time. In case of salient pole synchronous motor, the air-gap is non-uniform. The inductance of the machine in direct and quadrature axes are different. Blondel's two reaction theory is being applied for drawing steady-state phasor diagram.

In synchronous motor, the field mmf can be generated by a DC excited field winding mounted on rotor or permanent magnet poles mounted on rotor or buried inside rotor iron. In interior permanent magnet synchronous (IPMS) motor, the reluctance offered in direct axis is higher than reluctance offered in quadrate axis as the permanent materials has recoil permeability of approximately unity. The IPMS motor

behaves like a salient pole synchronous machine with direct axis reactance less than quadrature axis reactance.

Generally, the steady-state analysis of the IPMS motor is carried out using the phasor diagram based on Blondel's two reaction theory and transient analysis is carried out by two phase models. The spiral vector [1-4] method of analysis can be used for steady-state and transient analysis of ac circuits and machines in a unified way. The application of spiral vector results in representing a three-phase AC machine by a phase decoupled single phase model. The steady-state and transient behaviour of the ac motors can be analysed by this phase decoupled model.

In IPMS motor, since the reluctances offered in direct-axis and quadrature-axis are different, the corresponding inductances are also different. Hence, IPMS motor behaves like a salient pole synchronous motor with inverted saliency. Generally, the steady-state analysis of the IPMS motor is carried out using the phasor diagram based on Blondel's two reaction theory and transient analysis is carried out by two phase model. The spiral vector method of analysis has given us a new method for analyzing the steady-state and transient behaviour of three-phase motors with the help of phase decoupled voltage equation [1-4]. Yamamura has presented the spiral vector model of synchronous motor model along with closed form solution [4]. This paper presents derivation of phase decoupled spiral vector model of IPMS motor. Under balanced operation, it is possible to represent the three-phase motor by a single phase decoupled equivalent circuit. A new phasor diagram based on this equivalent circuit is also presented. The transient behaviour of the motor can be studied either by analytical or numerical methods.

This paper is organized in five sections. In section II, the phase decoupled analysis of IPMS motor has been derived from first principle. The steady state analysis of IPM motor is presented in section III. Section IV presents the novel phasor diagram. The power and torque equations of the IPMS motor have been derived in Section V. Finally conclusions are given in Section VI.

## II. PHASE DECOUPLED ANALYSIS OF IPMS MOTOR

The phase decoupled spiral vector model of IPMS motor can be obtained by representing the ac quantities by spiral vectors. This section gives a systematic derivation of the phase decoupled spiral vector model of IPMS motor starting from first principles.

---

Bishnu P. Muni is with BHEL, Corporate R&D, Hyderabad – 500093, India (e-mail:bpmuni@bhelrnd.co.in).

0-7803-9771-1/06/$25.00 ©2006 IEEE

The schematic of an IPMS motor is shown in Fig. 1. The following assumptions have been made while deriving the mathematical model.

(a) The air-gap flux density distribution due to the rotor magnet is sinusoidal in nature, so that it induces sinusoidal voltage in the stator winding.
(b) The stator mmf distribution is sinusoidal in nature.
(c) There is no damper winding on the rotor.
(d) The flux linkage per pole due to permanent magnet is constant.
(e) The saturation of the machine has been neglected.

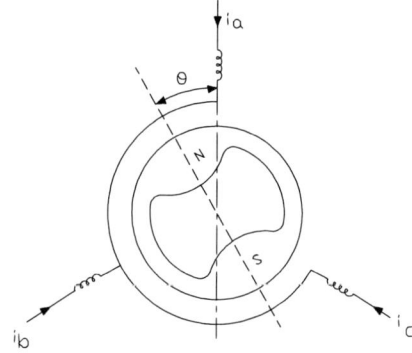

Fig. 1. Schematic of IPMS Motor.

With above assumptions, the spiral vector voltage equation of the IPMS motor in matrix form can be written as,

$$\begin{bmatrix} v_a \\ v_b \\ v_c \end{bmatrix} = \begin{bmatrix} R & 0 & 0 \\ 0 & R & 0 \\ 0 & 0 & R \end{bmatrix} \begin{bmatrix} i_a \\ i_b \\ i_c \end{bmatrix} + p \begin{bmatrix} \lambda_{ga} \\ \lambda_{gb} \\ \lambda_{gc} \end{bmatrix} \quad (1)$$

The flux linkage spiral vector matrix can be given as,

$$\begin{bmatrix} \lambda_{ga} \\ \lambda_{gb} \\ \lambda_{gc} \end{bmatrix} = \begin{bmatrix} L_{aa} & L_{ab} & L_{ac} \\ L_{ba} & L_{bb} & L_{bc} \\ L_{ca} & L_{cb} & L_{cc} \end{bmatrix} \begin{bmatrix} \underline{i_a} \\ \underline{i_b} \\ \underline{i_c} \end{bmatrix} + \underline{\lambda_f} \begin{bmatrix} e^{j(\theta)} \\ e^{j(\theta-2\pi/3)} \\ e^{j(\theta-4\pi/3)} \end{bmatrix} \quad (2)$$

Where,

$$\theta = \omega_e t + \theta_o$$

Since the air-gap is non-uniform in IPMS motor, the self and mutual inductances are dependent on the rotor position. The self inductances and mutual inductances between phases are as given below:

$$L_{aa} = L_l + L_k + L_{l\theta}\cos(2\theta)$$
$$L_{bb} = L_l + L_k + L_{l\theta}\cos(2(\theta - 2\pi/3))$$

$$L_{cc} = L_l + L_k + L_{l\theta}\cos(2(\theta - 4\pi/3))$$
$$L_{ab} = L_{ba} = -L_m + L_{m\theta}\cos(2(\theta - \pi/3)) \quad (3)$$
$$L_{bc} = L_{cb} = -L_m + L_{m\theta}\cos(2\theta)$$
$$L_{ca} = L_{ac} = -L_m + L_{m\theta}\cos(2(\theta + \pi/3))$$

Substituting the expressions for $L_{aa}$, $L_{ab}$ and $L_{ac}$ in the flux linkage equations, the air-gap flux linkage spiral vector for phase 'a' can be written as:

$$\underline{\lambda_{ga}} = L_l\underline{i_a} + (L_k\underline{i_a} - L_m\underline{i_b} - L_m\underline{i_c}) + [(L_{l\theta}\cos(2\theta)\underline{i_a}) + (L_{m\theta}\cos(2(\theta-\pi/3))\underline{i_b}) + (L_{m\theta}\cos(2(\theta+\pi/3)\underline{i_c})] + \underline{\lambda_f}e^{j\theta} \quad (4)$$

The fixed and variable part of the self and mutual inductances are related as given below,

$$L_m = (1/2)L_k$$
$$L_{l\theta} = L_{m\theta} \quad (5)$$

The currents in three phases under symmetrical operation can be written as,

$$\underline{i_a} = I e^{-\lambda t + j(\phi_1(t))}$$
$$\underline{i_b} = I e^{-\lambda t + j(\phi_1(t) - 2\pi/3)} \quad (6)$$
$$\underline{i_c} = I e^{-\lambda t + j(\phi_1(t) - 4\pi/3)}$$

The instantaneous phase angle of current spiral vector can be given as,

$$\phi_1(t) = \omega_{ei} t + \phi_1$$

Under symmetrical operation,

$$\underline{i_a} + \underline{i_b} + \underline{i_c} = 0 \quad (7)$$

Substituting eqs. (5), (6) and (7) in eq. (4), and simplifying [3], we get,

$$\underline{\lambda_{ga}} = (L_l + 3/2 L_k)\underline{i_a} + 3/2 L_{l\theta}e^{j2\theta}\underline{i_a}^* + \underline{\lambda_f}e^{j\theta} \quad (8)$$

The flux-linkage spiral vector can also written as [6]: (Appendix - I)

$$\underline{\lambda_{ga}} = (L_l + 3/2 L_k)\underline{i_a} + 3/2 L_{l\theta}e^{-j2\theta}\underline{i_a} + \underline{\lambda_f}e^{j\theta} \quad (9)$$

Substituting eq. (8) in eq. (1), the phase 'a' voltage equation can be written as,

$$\underline{v_a} = R\underline{i_a} + (L_l + 3/2 L_k)p(\underline{i_a}) + p(3/2 L_{l\theta}e^{j2\theta}\underline{i_a}^*) + p(\underline{\lambda_f}e^{j\theta}) \quad (10)$$

Eq. (10) shows that, the voltage equation is independent of the variables of the other two phases. Hence, the phase decoupled voltage equation of the IPMS motor can be written as:

$$v = R\underline{i} + (L_l + 3/2\,L_k)\,p(\underline{i}) + p(3/2\,L_{l\theta}\,e^{j2\theta}\,\underline{i}^*) + p(\lambda_f\,e^{j\theta}) \quad (11)$$

By using the flux linkage expression given in eq. (9), the phase decoupled voltage equation of the IPMS motor can be written as:

$$v = R\underline{i} + (L_l + 3/2\,L_k)\,p(\underline{i}) + p(3/2\,L_{l\theta}\,e^{-j2\theta}\,\underline{i}) + p(\lambda_f\,e^{j\theta}) \quad (12)$$

The steady-state and transient behavior of the motor can be predicted from the phase decoupled spiral vector voltage equation (11) or (12).

The third term in right hand side of (11) represents the voltage induced in the stator winding due to change in flux-linkage because of reluctance variation. Using the expression for current spiral vector and simplifying (derivation given in Appendix - II), the reluctance voltage spiral vector can be given as,

$$\underline{e}_r = -3/2\,L_{l\theta}\,p(e^{j\beta(t)}\,\underline{i}) \quad (13)$$

Where,

$$\beta(t) = 2(\omega_e - \omega_{ei})t + 2(\phi_o - \phi_1)$$
$$= 2(\omega_e - \omega_{ei})t + \beta_o \quad (14)$$
$$\beta_o = 2(\phi_o - \phi_1)$$

The last term in the phase decoupled voltage equation (eq. (11)) represents the voltage induced in stator winding because of rotation of permanent magnet excited rotor. The field induced voltage spiral vector can be given as,

$$\underline{e}_f = p(\lambda_f\,e^{j\theta}) = \omega_e\,\lambda_f\,e^{j(\theta+\pi/2)} = \omega_e\,\lambda_f\,e^{j\phi_o(t)} \quad (15)$$

The instantaneous phase angle of the field induced voltage is given below,

$$\phi_o(t) = \theta + \pi/2 = \omega_e t + \theta_o + \pi/2 = \omega_e t + \phi_o \quad (16)$$

The angle $\beta$ (t) can also be expressed in terms of instantaneous IPF angle, which has been defined as the instantaneous phase angle between the fields induced and phase current spiral vectors.

$$\beta(t) = 2[(\omega_e t + \phi_o) - (\omega_{ei} t + \phi_1)]$$
$$= 2(\phi_o(t) - \phi_1(t)) = 2\phi_i(t) \quad (17)$$

The phase decoupled spiral vector voltage equation of the IPMS motor can be written as,

$$v = R\underline{i} + L_s\,p(\underline{i}) + (-3/2 L_{l\theta}\,p(e^{j\beta(t)}\underline{i})) + \omega_e\lambda_f\,e^{j\phi_o(t)}$$
$$= R\underline{i} + L_s\,p(\underline{i}) + \underline{e}_r + \underline{e}_f \quad (18)$$

Where,

$L_s$ - synchronous inductance excluding the saliency effect

$(= L_1 + 3/2\,L_k)$.

$e_r$ - reluctance voltage; this voltage is being generated because of reluctance variation as the rotor revolves in the air gap $(= -3/2\,L_{l\theta}p(e^{j\beta(t)}i))$.

$e_f$ - field voltage; it represents the voltage generated in the phase winding due to the rotation of the rotor magnet

Under balanced condition, the voltage equation of a phase is independent of variables of the other phases. Hence, the three-phase IPMS motor can be represented by a phase decoupled equivalent circuit as given in Fig. 2.

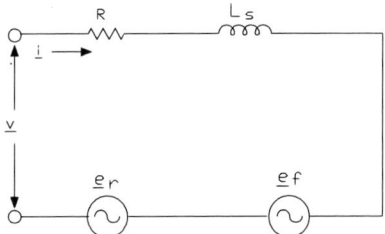

Fig. 2. Phase Decoupled equivalent Circuit of IPMS Motor.

### III. STEADY-STATE ANALYSIS OF IPMS MOTOR

The steady-state performance of the IPMS motor can be studied by using the phase decoupled voltage equation as given in eq.18. Under steady-state operation, the rotor speed is equal to the supply angular frequency.

Substituting $\omega_{ei} = \omega_e$ in eq.(18), we get,

$$v = R\underline{i} + L_s\,p(\underline{i}) - (3/2)L_{l\theta}\,e^{j\beta_o}\,p(\underline{i}) + p(\lambda_f\,e^{j\theta}) \quad (19)$$

By substituting $p = j\,\omega_e$ in eq. (19) and taking the phase angles with reference to an arbitrary reference frame, the steady-state voltage equation of the motor with circular vector notation can be written as,

$$Ve^{j(\omega t + \phi_v)} = RIe^{j(\omega t + \phi_1)} + j\omega L_s I e^{j(\omega t + \phi_1)} + 3/2 L_{l\theta}\omega_e e^{j(\beta_o - \pi/2)}I e^{j(\omega t + \phi_1)} + \omega\lambda_f e^{j(\omega t + \phi_o)} \quad (20)$$

The field voltage and reluctance voltage are ac voltages of supply frequency and constant amplitude. The expressions for $E_f$ and $E_r$ are given below.

$$E_f = \omega_e\,\lambda_f\,e^{j(\omega_e t + \phi_o)}$$
$$E_r = 3/2\,\omega_e\,L_{l\theta}\,e^{j(\beta_o - \pi/2)}\,I\,e^{j(\omega_e t + \phi_1)} \quad (21)$$

### IV. NOVEL PHASOR DIAGRAM OF IPMS MOTOR

Phasor diagram is a very important tool in steady-state analysis of ac circuits and machines. A new type of phasor diagram shown in Fig. 3 for the IPMS motor can be drawn based on eq. 20. The phasor diagram has been drawn to the scale for the parameters of IPMS motor given in Appendix – III. This

113

phasor diagram when used for steady-state analysis gives exactly same results as obtained from a conventional phasor diagram. This has been clearly demonstrated by drawing both the diagrams together as shown in Fig. 4 and by deriving the following relations trigonometrically from the conventional and the novel phasor diagrams shown in Fig. 4.

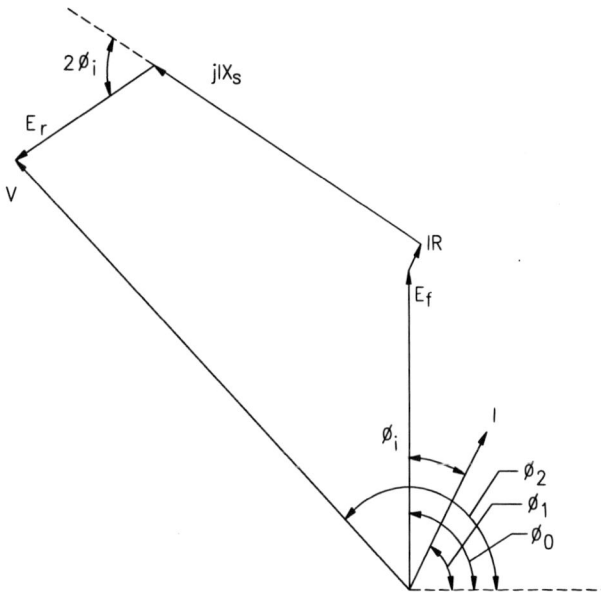

Fig. 3. Novel Phasor Diagram of IPMS Motor based on Spiral Vector Theory.

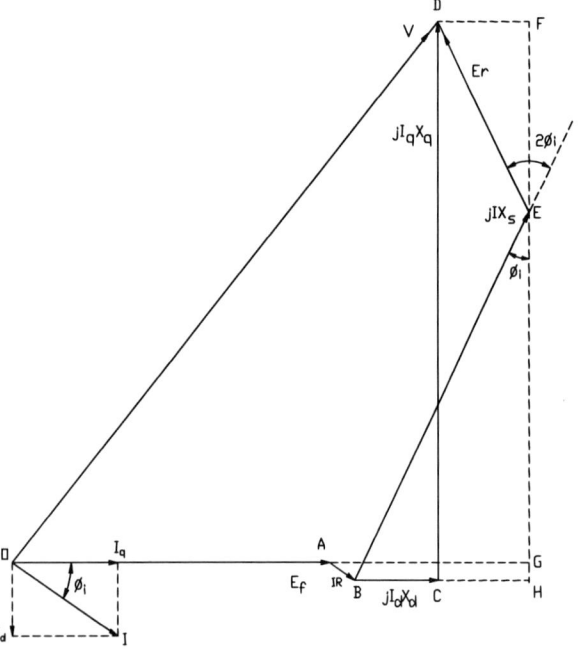

Fig. 4. Novel and Conventional Phasor Diagram of IPMS Motor.

$$I_q X_q = CD = I X_s \cos(\phi_i) + E_r \cos(\phi_i) \qquad (22a)$$

$$I_d X_d = BC = I X_s \sin(\phi_i) - E_r \sin(\phi_i) \qquad (22b)$$

$$V = OD = \sqrt{[(E_f + I R \cos(\phi_i) + (I X_s \sin(\phi_i) - E_r \sin(\phi_i)))^2} \qquad (22c)$$
$$+ ((I X_s \cos(\phi_i) + E_r \cos(\phi_i)) - I R \sin(\phi_i))^2]$$

## V. POWER AND TORQUE EQATIONS OF IPMS MOTOR

In the phase decoupled voltage equation of IPMS motor as given in eq.18, if the voltage and currents are expressed in spiral vectors, the phase decoupled spiral vector voltage equation can be written as,

$$\underline{v} = R\underline{i} + L_s \, p(\underline{i}) + \underline{e}_r + \underline{e}_f \qquad (23)$$

The instantaneous shaft power per phase is the sum of the instantaneous power due to the field i.e. field power and instantaneous power due to the reluctance variation i.e. reluctance power. The total power output is the sum of instantaneous field power of each phase. Similarly, the total reluctance power is the sum of instantaneous reluctance power of each phase. The instantaneous total field power ($P_f$) and total reluctance power ($P_r$) of the motor are given below.

$$P_f = Re(\underline{e}_{fa})Re(\underline{i}_a) + Re(\underline{e}_{fb})Re(\underline{i}_b) + Re(\underline{e}_{fc})Re(\underline{i}_c)$$
$$(24)$$
$$P_r = Re(\underline{e}_{ra})Re(\underline{i}_a) + Re(\underline{e}_{rb})Re(\underline{i}_b) + Re(\underline{e}_{rc})Re(\underline{i}_c)$$

The total field power and reluctance power can also be written as,

$$P_f = (3/2)Re[\underline{e}_f\underline{i}^*] + (3/2)Re[\underline{e}_r\underline{i}^*] \qquad (25)$$

Since, the field-induced voltage is proportional to the flux-linkage due to the magnet, which is constant, the field torque is proportional to the phase current and cosine of the instantaneous angle between the field induced voltage and phase current spiral vectors i.e. the instantaneous IPF angle. The reluctance voltage being proportional to current, the instantaneous reluctance torque is directly proportional to the square of current and cosine of the angle between reluctance voltage and phase current spiral vectors. It can be shown that the angle between reluctance voltage and phase current spiral vector is ($\pi/2 - \varphi(t)$). (Refer Appendix - II).

The steady-state power of the motor can also be obtained from the phasor diagram. Power due to the field excitation can be obtained by multiplying the component of current phasor along the field voltage phasor.

$$P_f = 3 E_{frms} I_{rms} \cos(\phi_o - \phi_1) \qquad (26)$$

Similarly, power due to reluctance variation can be obtained by multiplying the component of current phasor along reluctance voltage phasor.

$$P_r = 3 E_{rrms} I_{rms} \sin 2(\phi_o - \phi_1) \qquad (27)$$

For a three-phase motor with P poles, the power developed is given below:

$$P_t = (3/2)[\lambda_f \omega I \cos(\phi_o - \phi_1) + (3/2)\omega L_{I0} I^2 \sin 2(\phi_o - \phi_1)] \qquad (28)$$

The torque developed by the motor can be given as,

114

$$T_e = (3/2)(P/2)[\lambda_f I \cos(\phi_o - \phi_l) + (3/2)L_{l\theta}I^2 \sin 2(\phi_o - \phi_l)] \quad (29)$$

In this expression, $(\varphi_0 - \varphi_1)$ represents the IPF angle. The field torque versus IPF angle, the reluctance torque versus IPF angle and total torque versus internal PF angle graphs of an IPMS motor are shown in Figs. 5 (a), 5(b) and 5(c), respectively.

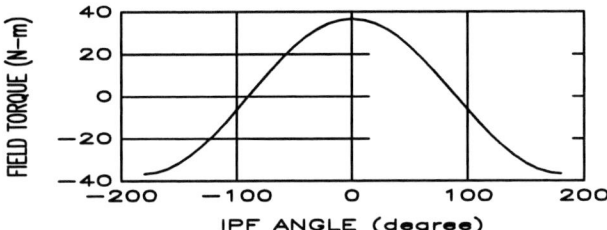

Fig. 5 (a). Field Torque versus IPF Angle.

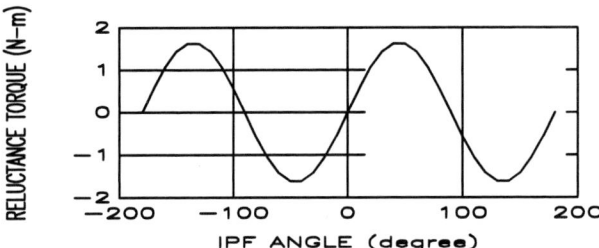

Fig.. 5 (b). Reluctance Torque versus IPF Angle.

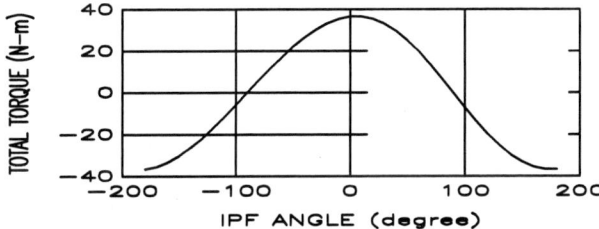

Fig. 5 (c). Total Torque versus IPF Angle.

From (29), it is observed that the reluctance torque vanishes and the total torque is directly proportional to current, if the instantaneous IPF angle is maintained at zero degree. It must be mentioned that in the case of IPMS motor, the reluctance torque vanishes when the IPF angle is kept at zero. Hence, if the instantaneous IPF angle is maintained at zero both during transient and steady-state, the torque of the motor will be directly proportional to the motor current and thus separately excited dc motor like characteristics can be obtained from IPMS motor [6]. For any other IPF angle, the torque has two components: the field component, directly proportional to current, and reluctance component, proportional to the square of the current. For applications where the power output per ampere is important, the IPF angle can be maintained at a value for which the torque developed is maximum.

## VI. CONCLUSIONS

An IPMS motor, which behaves like a salient pole synchronous motor with inverted saliency, can be represented by a phase decoupled voltage equation with the help of spiral vector method of analysis. A novel phasor diagram for IPMS motor can be drawn from the phase decoupled spiral vector model of IPMS motor. The equivalence of the proposed phasor diagram and the conventional pahsor diagram has been shown trigonometrically. Further, the power and torque equations of IPMS motor have been derived form the phase decoupled model and the novel phasor diagram. The torque developed by the IPMS motor has two components: field torque and reluctance torque. When the IPF angle is maintained at zero degree, the reluctance torque vanishes and the torque is directly proportional to current. This mode of operation of the drive is suitable for high dynamic performance applications.

## VII. APPENDIXES

### APPENDIX – I

### IPMS MOTOR AIR-GAP FLUX LINKAGE

This appendix gives the derivation of IPMS motor flux-linkage expression eq.9 starting from eq. 4.

The air-gap flux linkage expression of the IPMS motor (eq.4)) is given below.

$$\lambda_{ga} = L_l \underline{i_a} + (L_k \underline{i_a} - L_m \underline{i_b} - L_m \underline{i_c}) + [(L_{l\theta} \cos(2\theta) \underline{i_a}) +$$
$$(L_{m\theta} \cos(2(\theta - \pi/3)) \underline{i_b}) + (L_{m\theta} \cos(2(\theta + \pi/3)) \underline{i_c})] + \lambda_f e^{j\theta}$$
$$(I.1)$$

By using eq. (6) and eq.(.8), the second term in right hand side of eq.(I.1) can be simplified as,

$$(L_k \underline{i_a} - L_m \underline{i_b} - L_m \underline{i_c}) = (L_k \underline{i_a} - L_m(\underline{i_b} + \underline{i_c})) = (L_k \underline{i_a} - L_m(-\underline{i_a}))$$
$$= 3/2 L_k \underline{i_a}$$
$$(I.2)$$

By using (6), (7) and (8), the third term in right hand side of (I.1) can be simplified as,

$$(L_{l\theta} \cos(2\theta) \underline{i_a}) + (L_{m\theta} \cos(2(\theta - \pi/3)) \underline{i_b}) + (L_{m\theta} \cos(2(\theta + \pi/3)) \underline{i_c})$$
$$= (L_{l\theta} \cos(2\theta) \underline{i_a}) + (L_{l\theta} \cos(2\theta)(-1/2)(\underline{i_b} + \underline{i_c}))$$
$$+ (L_{l\theta} \sin(2\theta)(\sqrt{3}/2)(\underline{i_b} - \underline{i_c}))$$
$$= 3/2 L_{l\theta} \cos(2\theta) I e^{-\lambda t} e^{j(\omega_{el}t + \phi_l)} +$$
$$\sqrt{3}/2 L_{l\theta} \sin(2\theta) I e^{-\lambda t} e^{j(\omega_{el}t + \phi_l)} (e^{-j2\pi/3} - e^{j2\pi/3})$$
$$= L_{l\theta} I e^{-\lambda t} e^{j(\omega_{el}t + \phi_l)} (3/2 \cos(2\theta) + \sqrt{3}/2 \sin(2\theta)(-j)2 \sin(2\pi/3))$$
$$= 3/2 L_{l\theta} \underline{i_a} e^{-j2\theta} \quad (I.3)$$

By substituting (I.2) and (I.3) in (I.1), the IPMS motor air-gap flux-linkage equation can be written as,

$$\lambda_{ga} = (L_l + 3/2 L_k) \underline{i_a} + 3/2 L_{l\theta} e^{-j2\theta} \underline{i_a} + \lambda_f e^{j\theta} \quad (I.4)$$

APPENDIX - II

IPMS MOTOR RELUCTANCE VOLTAGE

This appendix gives the derivation of the reluctance voltage spiral vector ($\underline{e}_r$) expression given in (13).

The phase decoupled voltage equation of IPMS motor given in eq.(11) is given here again for the sake of clarity.

$$\underline{v} = R\underline{i} + (L_l + \frac{3}{2}L_k)p(\underline{i}) + p(\frac{3}{2}L_{l\theta}e^{j2\theta}\underline{i}^*) + p(\lambda_f e^{j\theta}) \quad (II.1)$$

The fourth term on the right hand side of .(V.1) represents the field induced voltage spiral vector. It can be simplified as given below,

$$\underline{e}_f = p(\lambda_f e^{j\theta}) = w_e \lambda_f e^{j(\theta+\pi/2)} = \omega_e \lambda_f e^{j\phi_o(t)} \quad (II.2)$$

Where,

$$\phi_o(t) = \theta + \pi/2 = \omega_e t + \theta_o + \pi/2 = \omega_e t + \phi_o \quad (II.3)$$

The third term on the right hand side of eq.(V.1) represents the reluctance voltage. It can be simplified as given below,

$$\underline{e}_r = p(\frac{3}{2}L_{l\theta}e^{j2\theta}\underline{i}^*)$$

$$= \frac{3}{2}L_{l\theta} p(e^{j(2\omega_e t+2\theta_0)} I e^{-\lambda t} e^{-j(\omega_{ei}t+\phi_1)})$$

$$= \frac{3}{2}L_{l\theta} p(e^{j2(\omega_e t+\theta_0-\omega_{ei}t-\phi_1)} I e^{-\lambda t} e^{j(\omega_{ei}t+\phi_1)})$$

$$= \frac{3}{2}L_{l\theta} p(e^{j2(\omega_e t-\omega_{ei}t+\phi_0-\phi_1-\frac{\pi}{2})} I e^{-\lambda t} e^{j(\omega_{ei}t+\phi_1)}) \quad (II.4)$$

$$= \frac{3}{2}L_{l\theta} p(e^{j(\beta(t)-\pi)}\underline{i})$$

$$= -\frac{3}{2}L_{l\theta} p(e^{j\beta(t)}\underline{i})$$

where,

$$\phi_o = \theta_o - \pi/2$$
$$\beta(t) = 2(\omega_e - \omega_{ei})t + 2(\phi_o - \phi_1) \quad (II.5)$$

APPENDIX - III

IPMS MOTOR PARAMETERS

| | |
|---|---|
| No. of Poles | : 6 |
| Rated torque | : 25.5 Nm |
| Rated current | : 5.2 A |
| Resistance per phase | : 4.3 ohm |
| Leakage inductance per phase | : 4.25 mH |

| | |
|---|---|
| Self inductance (constant part) per phase | : 42.47 mH |
| Self inductance (Amplitude of variable part) per phase | : -3.0 mH |
| Direct-axis inductance | : 72.45 mH |
| Quadrature-axis inductance | : 81.45 mH |
| Maximum flux linkage due to magnet | : 0.7834 V.s/rad |
| Rated speed | : 1000 RPM |
| Moment of inertia (only motor) | : 0.012 kg. m$^2$ |
| Moment of inertia (With loading generator) | : 0.3721 kg. m$^2$ |
| Permanent magnet material used in rotor | : Nd-Fe-B |

VIII. ACKNOWLEDGMENT

The authors gratefully acknowledge the support of BHEL, Corporate R&D for providing facilities to carry out work and permission to present the paper in PEDES 2006.

IX. REFERENCES

[1] S. Yamamura, *AC Motors for High-Performance Applications*, Marcel Dekker Inc., New York, 1986.

[2] S. Yamamura, "Spiral Vector Theory of AC motor Analysis and Control", Conf. record, IEEE, 5th. Applied Power Electronics Conference 1990, pp 77 - 82.

[3] S. Yamamura, "Spiral Vector Theory of Salient-Pole Synchronous Machine", Conf. record, IEEE, IAS Annual Meeting 1992, pp 204 - 211.

[3] S. Yamamura, "Spiral Vector Theory of AC motor Analysis and Control", Conf. record, IEEE, 5th. Applied Power Electronics Conference 1990, pp. 77 - 82.

[4] B.P. Muni, S.K. Pillai and S.N. Saxena, "Digital Simulation of Internal Power Factor Angle Controlled Surface Mounted Permanent Magnet Synchronous Motor", Conf. record, IEEE, International Conference on Power Electronics, Drives and Energy Systems (PEDES - 96), 1996, pp 900 - 906.

[5] B.P. Muni, S.K. Pillai and S.N. Saxena, "A PC Based Internal Power Factor Angle Controlled Interior Permanent Magnet Synchronous Motor Drive", Conf. record, IEEE, Power Electronics Specialist Conference (PESC - 96), 1996, pp. 931 - 937.

[6] Bishnu P. Muni, "Internal Power Factor Angle Controlled Permanent Magnet Synchronous Motor Drive," Ph.D. dissertation, Dept. Electrical Eng., IIT, Bombay, 1997.

X. BIOGRAPHY

Bishnu P. Muni was born in India on November 16, 1961. He obtained B. Sc.(Engg.) in Electrical Engineering from Regional Engineering College, Rourkela in the year 1983. Subsequently, he obtained Ph. D from IIT, Bombay in 1997. Since 1983, he is working with BHEL, Corporate R&D in the area of power electronics for industrial, traction, power system and distributed generation applications.

**2006 IEEE International Conference on Power Electronic, Drives and Energy Systems**

# A Novel DTC Strategy of Torque and Flux Control for Switched Reluctance Motor Drive

R. Jeyabharath, P.Veena, and M.Rajaram

**Abstract** - In this paper, a novel methodology for control of torque and flux of switched reluctance motor (SRM) drive in a hystersis manner is described. The philosophy of control is the direct torque control (DTC) by analysis the non-uniform torque characteristics of the motor. The scheme directly controls the torque and the torque ripple by controlling the magnitude of flux linkage and change in speed of stator flux vectors. Further importance is given to overcome the difficulties of conventional linear and non-linear controls of SRM drives. This new technique can be implemented in real time with low cost microcontrollers and higher simplicity.

**Keywords** - Direct torque control, Flux control, Switched Reluctance Motor.

## I. INTRODUCTION

SWITCHED reluctance motor,the doubly salient,singly excited motor has simple and robust construction. Although, the induction motor is still the workhorse of the industries, the promising feature of the high torque to mass ratio, high torque to inertia ratio, low maintenance, high specific output and excellent overall performance of SRM make it an efficient competitor for ac drives. The simplified converter topology and switching algorithm due to the unipolar operation avoiding shoot through faults makes SRM advantageous in applications of aerospace, which require high reliability. Also it finds wide application in automotive industries, direct drive machine tools etc.

The main disadvantages of SRM are the highly nonlinear and discrete nature of torque production mechanism. The total torque in SRM is the sum of torques generated by each of the stator phase, which are independently controlled. When torque production mechanism is transferred from one active phase to another, pulsations are produced leading to vibrations and acoustic noise. The nonlinear magnetization characteristics make the control of motor really complex.

The control of SR motor is the recent trend of research as there are complications implemented due to mutual coupling of the motor phase and parameter variation of inductance characteristics [1],[2]. Previous control schemes involve using of linear or nonlinear models. An adaptive feedback controller assuming linear magnetic circuits was proposed in [3]. In another scheme, an analytical solution was developed for production motor voltages to provide a smooth torque [4].

R.Jeyabharath, and P.Veena Department of EEE, KSRCT, Tiruchengode, Tamil Nadu, India.
Email: jeya_psg@rediffmail.com, veena_gce@yahoo.co.in.
M.Rajaram, Department of EEE, Thanthai Periyar Govt. Institute of Technology, Vellore, India.
Email: rajaramgct@rediffmail.com

Though linear systems were simple, they were highly inaccurate as torque and flux are both nonlinear functions. So some schemes developed non linear characteristics of SR motor. In [5] feedback linearization provides compensation for the magnetic nonlinearity. Also ripples were reduced to provide smooth torque in [6]. Some nonlinear adaptive schemes were also developed.

The nonlinear model implementations were complex and could not be implemented in real time. They were expensive and affected by variations in saturation. To overcome all these problems, the Direct Torque Control (DTC) was proposed which provided simple solution to control the motor torque and speed and minimized torque ripple [7]. Early scheme used concept of short flux pattern that links two separated poles of the SRM stator [8] [9]. However, this needed a new winding configuration which is expensive and inconvenient. This scheme can only be theoretically achieved as they required bipolar currents in opposition to unipolar currents SRM.

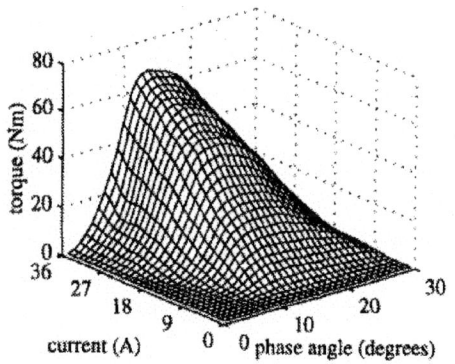

Fig. 1. Measured nonlinear magnetization characteristics of SR motor.

In this paper, a new strategy of control is proposed from the analysis of the non uniform torque characteristics of the motor. This scheme does not involve short flux pattern, a change of winding configuration and can be implemented on any normal type SR motor drive. This proposed scheme provides the advantage of directly controlling the torque by controlling the magnitude of flux linkage and the change in speed of the stator flux vector. Calculations involving nonlinear magnetization characteristics and model parameters are not required. Problems associated with torque ripple control are eliminated as the scheme directly regulates the torque output with in a hystersis band. Further, the scheme is simple and can be implemented with low cost microprocessor hardware.

0-7803-9771-1/06/$25.00 ©2006 IEEE 117

## II. CONVENTIONAL DTC METHOD FOR AC MACHINES

DTC is based on theories of field oriented (FOC) control and torque vector control. Field Oriented Control uses space vector theory to optimally control magnetic field orientation. The DTC principle is to select stator voltage vectors according to the differences between the reference torque and stator flux linkage with exact value.

Voltage vector are so chosen to limit the torque and flux errors within hysteresis bands. The required optimal voltage vectors are obtained from the position of the stator flux linkage space vector, the available switching vectors and the required torque and flux linkage [9].

The principle of DTC in AC machines can be derived from motor equations.

$$\vec{e} = \vec{v} - \vec{i}\vec{R} \tag{1}$$

$$\vec{\psi_s} = \int \left( \vec{v} - \vec{i}\vec{R} \right) dt \tag{2}$$

Where $\vec{\psi_s}$ =stator flux vector component, $\vec{v}$ =stator voltage vector component, $\vec{i}$ = stator current vector component, R= stator resistance.

The torque expression for ac machine is given by

$$T = \frac{3}{2} p \frac{L_m}{\sigma L_s L_r} \vec{\psi_s} \times \vec{\psi_r'} \tag{3}$$

Where $L_m$ =mutual inductance, $L_s$ =stator self indctancence, $L_r$ =rotor self inductance, σ =leakage coefficient of the motor, $\vec{\psi_r'}$ =rotor flux linkage vector in the stationary reference frame.

The space vector of rotor flux is related to stator flux in s domain as

$$\vec{\psi_r'} = \frac{\dfrac{L_m}{L_s}}{1 + s\,\sigma\,\tau_r} \vec{\psi_s} \tag{4}$$

The rotor flux has a first order relationship with stator flux. The control scheme is based on controlling the amplitude and phase variation of the stator flux assuming that the rotor flux is constant. So torque controlled by varying the angle between the two fluxes. Hence torque can be controlled by accelerating, decelerating or stopping the rotation of the stator flux vector relative to rotor flux vector.

## III. THE NEW PROPOSED SCHEME FOR SRM

SRM has nonlinear and non sinusoidal currents and each phase is independently excited. Thus, conventional ac machine rotating field theory cannot be directly applied to the switched reluctance motor. This combined with the motor's highly nonlinear magnetic and torque characteristics has been one of the major problems in the widespread adoption of switched reluctance motor drives in industry. However, by re-examining the torque equation of the switched reluctance motor, a new control technique can be found which uses a similar philosophy as the conventional DTC of ac machines. As in conventional DTC, the control scheme directly controls the amplitude of the flux and torque within hysteresis bands. However, the motor phase switching strategy is based instead on the non-uniform torque characteristics of the SR motor. To drive the new control scheme for the SR motor, the non-uniform torque characteristics will firstly be examined. The motor torque output can be found using the motors electromagnetic equation. (1)

$$v = Ri + \frac{d\psi(\theta, i)}{dt} \tag{5}$$

The energy equation is

$$dW_e = dW_m + dW_f \tag{6}$$

Where $dW_m$ = differential mechanical energy, $dW_f$ =differential field energy. But

$$dW_e = eidt \tag{7}$$

$$dW_f = \frac{\partial W_f}{\partial i} di \Big|_{\theta=const} + \frac{\partial W_f}{\partial \theta} d\theta \Big|_{i=const} \tag{8}$$

The instantaneous torque expression is

$$T = \frac{dW_m}{d\theta} \tag{9}$$

Hence by substitutions torque expression is derived considering the variation of magnetic co-energy and is given by

$$T \approx i \frac{d\psi(\theta, i)}{d\theta} \tag{10}$$

This approximate equation is sufficient for control purpose as it controls the general characteristics of torque production and not the magnitude of torque. The current is always positive as SRM is a unipolar drive. Hence, the sign of torque is directly related to the sign of $\partial\psi/\partial\theta$. The increase of stator flux amplitude with respect to rotor position (positive value of $\partial\psi/\partial\theta$) produce a positive torque and is called "flux acceleration". Whereas a negative value of $\partial\psi/\partial\theta$ called "flux deceleration" produce a negative torque. As this is held for both directions of rotation a four quadrant operation is achieved using unipolar currents.

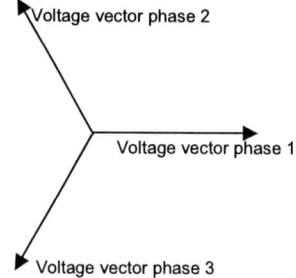

Fig. 2. Definition of phase voltage vectors in three phase SR motor.

The new proposed technique can be defined as follows.

a) The stator flux linkage vector of the motor is kept at constant amplitude.

b) Torque is controlled by accelerating or decelerating the stator flux vector.

This is similar to conventional DTC except that it is dependent on stator current. The stator current is found to have a first order delay relative to the change of stator flux $\partial\psi/\partial\theta$.

The stator current equation in time domain is

$$\frac{di}{dt} = \frac{e - \dfrac{\partial\psi}{\partial\theta}\dfrac{d\theta}{dt}}{\dfrac{\partial\psi}{\partial i}} \qquad (11)$$

and in $s$ domain

$$i = \frac{e - \dfrac{\partial\psi}{\partial\theta}\omega}{sl} \qquad (12)$$

as $\dfrac{d\theta}{dt} = \omega$ -speed of the rotor and

$\dfrac{d\psi}{di} = l$ - incremental inductance.

It is seen that in the new proposed scheme the stator current has first order delay relative to the change in applied voltage $e$ and change in flux with respect to rotor position $\partial\psi/\partial\theta$.

## IV. VOLTAGE VECTORS FOR SRM

Similar to the AC drives, equivalent space vectors can be defined for SRM. The voltage space vector for each phase is defined as lying on the center axis of the stator pole because the flux linkage for a current and voltage applied to the motor phase will have phasor direction in line with the centre of the pole axis. This does not need any change in physical winding topology.

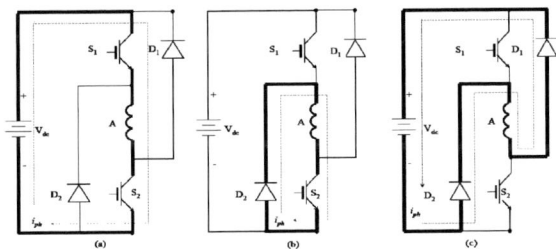

Fig. 3. SR motor phase voltage states.

In SRM, each motor phase can have three possible voltage states ($S_q$) for a unidirectional current.

i   When both devices are ON and positive voltage is applied $S_q=1$.

ii.  For $S_q=0$, one device is turned OFF and a zero voltage loop occurs.

ii.  For negative state $S_q=-1$, both devices are OFF. The freewheeling current flows through the diodes.

So with each phase having three possible states (0, 1,-1) unlike conventional DTC for ac drives with two states, a total of 27 possible configuration is possible.

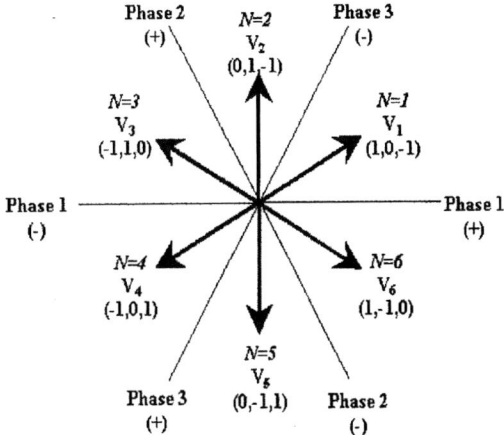

Fig. 4. Definition of SR motor voltage vectors for DTC drive.

Only six equal magnitude voltage vectors separated by $\pi/6$ radians (Fig.4.) is considered as DTC allows no other states to be chosen by the controller. One out of the six states is chosen to keep torque and flux within the hystersis bands. Let the stator flux vector be located in the $K^{th}$ sector (K=1,2,3,4,5,6). In order to increase the amplitude of the stator flux, the voltage vector $V_K, V_{K+1}, V_{K-1}$ can be applied and $V_{K+2}, V_{K+3}, V_{K-2}$, can be applied to decrease the flux. $V_K$ and $V_{K+3}$ are zero space vectors.

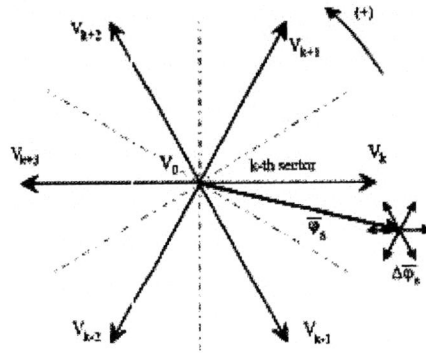

Fig. 5. Inverter voltages and corresponding stator flux variations.

As in conventional DTC, the proposed control scheme of SRM is based on the results as follows.

a)   The motor is solicited only through the converter component of voltage space vectors along the same flux.

b)   The motor torque is affected by the component of the voltage space vector orthogonal to the stator flux.

c)   The zero space vectors do not affect the space vector of the stator flux.

119

So the stator flux when increased by $V_{K+1}$ and $V_{K-1}$ vectors and decreased by $V_{K+2}$ and $V_{K-2}$ affect the torque. As $V_{K+1}$ and $V_{K+2}$ vector advance the stator flux linkage in the direction of rotation they tend to increase the torque. But $V_{K-1}$ and $V_{K-2}$ decelerate the flux in opposite direction and decrease the torque. So the switching table becomes as TABLE.I

TABLE I
STATOR FLUX AND TORQUE VARIATIONS DUE TO APPLIED INVERTER VOLTAGE SPACE VECTOR

| | $V_{k-2}$ | $V_{k-1}$ | $V_k$ | $V_{k+1}$ | $V_{k+2}$ | $V_{k+3}$ |
|---|---|---|---|---|---|---|
| $\varphi_s$ | ↓ | ↑ | ↑↑ | ↑ | ↓ | ↓↓ |
| T ($\omega_m>0$) | ↓↓ | ↓↓ | ↓ | ↑ | ↑ | ↓ |
| T ($\omega_m<0$) | ↓ | ↓ | ↑ | ↑↑ | ↑↑ | ↑ |

The flux control using hystersis is achieved by vectors $V_{K+2}$ and $V_{K+1}$ that increase the torque as shown in Fig.6. The instantaneous torque and flux magnitudes are to be known as they are controlled in the hystersis manner. In conventional DTC, the stator flux vector $\vec{\psi}_S$ is expressed as

$$\vec{\psi}_s = \int_0^t \left( \vec{V}_s - \vec{R}_s \vec{i}_s \right) + \vec{\psi}_{s0} \tag{13}$$

Where $\vec{\psi}_{s0}$ the initial value of stator flux vector.

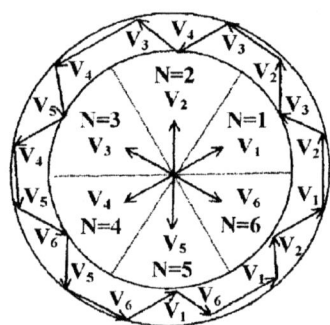

Fig. 6. Illustration of flux control using hysteresis.

In SRM, as the motor is robust to problems of integration offset the $\vec{\psi}_{s0}$ is assumed to be zero and the individual flux linkages are calculated using equation (13). The magnitude of individual phase flux linkage varies with time but the direction is always along the stator pole axis.

To resolve these phase flux vectors the three phases of SRM are transformed onto a stationary orthogonal two axis α-β reference frame as shown in Fig.7.

The orthogonal flux vectors are expressed as

$$\psi_\alpha = \psi_1 - \psi_2 Cos60° - \psi_3 Cos60° \tag{14}$$

$$\psi_\beta = \psi_2 Sin60° - \psi_3 Sin60° \tag{15}$$

The magnitude of $\vec{\psi}_s$ and angle δ of the equivalent flux vector is defined as

$$\psi_s = \sqrt{{\psi_\alpha}^2 + {\psi_\beta}^2} \tag{16}$$

$$\delta = \arctan\left( \frac{\psi_\beta}{\psi_\alpha} \right) \tag{17}$$

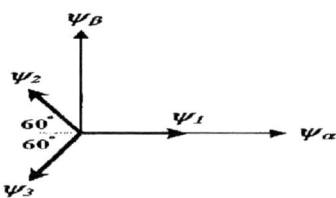

Fig. 7. Definition of two frame reference axis for motor voltages.

## V. SIMULATION RESULT

To simulate the system a Matlab/ Simulink model was constructed for the SR motor and the control system Fig.8. In this simulation test, the motor command flux was maintained a constant 0.3wb and a motor torque reference of 5Nm was used. The speed of the motor in this test was a constant 2800rpm.The hysteresis bands were defined to be of ±0.01Wb and ±0.1Nm for the flux linkages and torque respectively. The result of the torque and flux control can be seen in Fig.9 and Fig.10.

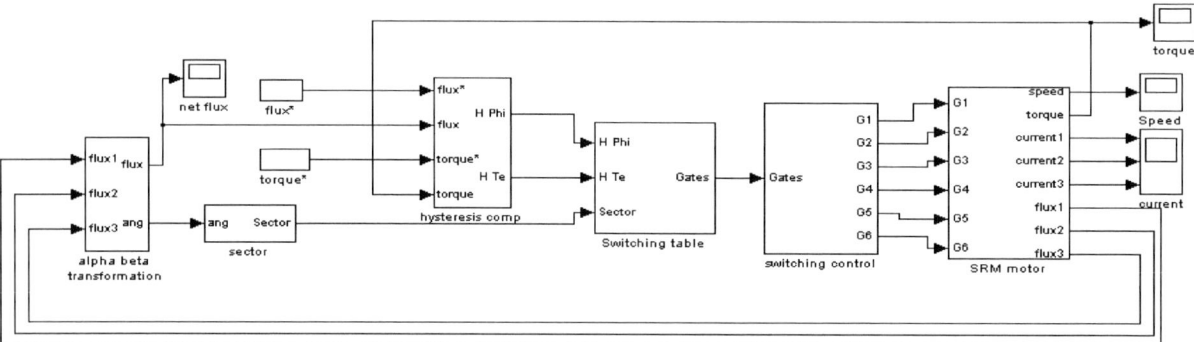

Fig. 8. Block diagram of DTC algorithm.

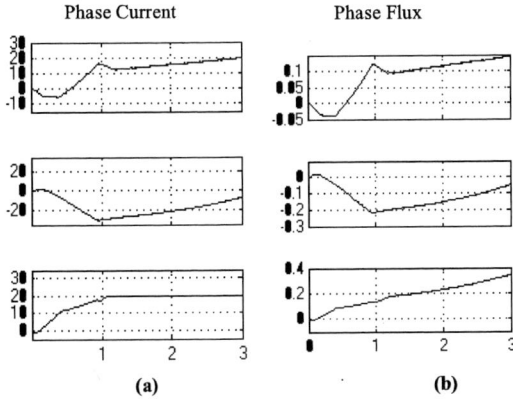

Fig. 9. Simulation results for (a) Phase current (b) Phase flux.

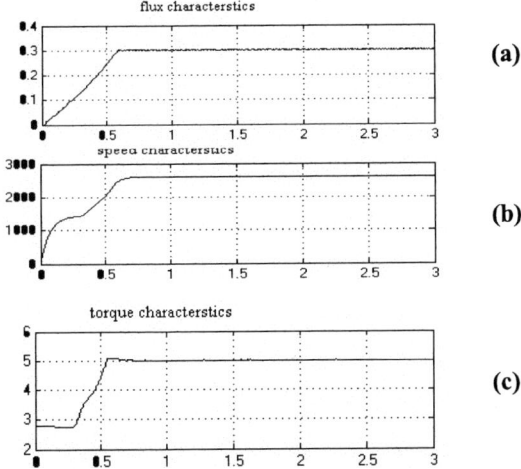

Fig. 10. Simulation results for (a) flux (b) speed (c) torque characteristics.

## VI. CONCLUSION

In this paper, a novel control methodology for the SR motor was derived from analysis of the non uniform torque characteristics of the motor. In the method, torque and torque ripple is directly controlled through the control of the magnitude of the flux linkage and the change in speed of the stator flux vector. Although the method is based on the nonlinear characteristics of the motor torque output during operation, the method does not require the nonlinear magnetization characteristics to be used in the real-time torque control. Furthermore, the scheme is not dependent on the accuracy of the estimated model parameters, as no model calculation is required during operation.

Hence, this overcomes the disadvantages and difficulties faced by conventional linear or nonlinear controllers of the SR motor mentioned above, which is due to the inaccuracy of linear control schemes, or the computational complexity of nonlinear model based schemes.

## VII. REFERENCES

[1] L. Xu and E. Ruckstadter, "Direct modeling of switched reluctance machine by coupled field-circuit method," *IEEE Trans. Energy Conv.*, vol. 10, pp. 446–454, Sept. 1995.

[2] N. J. Nagel and R. D. Lorenz, "Modeling of a saturated switched reluctance motor using an operating point analysis and the unsaturated torque equation," *IEEE Trans. Ind pplicat.*, vol. 36, pp. 714–722, May/June2000.

[3] D. G. Taylor, "Adaptive control design for a class of doubly-salient motors,"in *Proc. 30th IEEE Conf. Dec. Contr.*, vol. 3, 1991, pp. 2903–2908.

[4] N. J. Nagel and R. D. Lorenz, "Complex rotating vector method for smooth torque control of a saturated switched reluctance motor,"in *Proc. 34th Annu. Meeting IEEE Ind. Applicat.*, vol. 4, 1999, pp.2591–2598.

[5] M. Ilic'-Spong, R. Marino, S. M. Peresada, and D. G. Taylor, "Feedbacklinearizing control of switched reluctance motors," *IEEE Trans.Automat. Contr.*, vol. AC-32, pp. 371–379, May 1987.

[6] M. Stankovic, G. Tadmor, Z. J. Coric, and I. Agirman, "On torque ripple reduction in current-fed switched eluctance motors," *IEEE Trans.Ind. Electron.*, vol. 46, pp. 177–183, Feb. 1999.

[7] P. Jinupun and P. C.-K. Luk, "Direct torque control for sensorless switched reluctance motor drives," in *Proc. 7th Int. Conf. PowerElectron. Variable Speed Drives*, 1998, pp. 329–334

[8] M. Michaelides and C. Pollock, "Modeling of a newwinding arrangement to improve performance in the switched reluctance motor," in *Proc.6th Int. Conf. Elect. Mach. Drives*, 1993, pp. 213–218.

[9] Takahashi and T. Noguchi, "Take a look back upon the past decade of direct torque control," in *Proc. IECON '97, 23rd Int. Conf. Ind. Electron.,Contr. Instrum.*, vol. 2, 1997, pp. 546–551.

## VIII. BIOGRAPHIES

**Mr. R.Jeyabharath** received his bachelor degree from Thiagarajar College of Engineering, Madurai and got his master degree from PSG College of Technology, Coimbatore. He is doing Ph.D in the field of AC Drives and Vector control.

**Ms.P.Veena,** received her bachelor degree, and master degree from Government College of Engineering, Salem. She is doing her Ph.D in the field of intelligent control for AC Drives.

**Dr.M.Rajaram** received his bachelor degree from Alagappa Chettiar College of Engineering and Technology, Karaikudi and got his master degree from GCT, Coimbatore. He received his PhD from PSG College of technology, Coimbatore. His field of interest is control systems and special machines.

**2006 IEEE International Conference on Power Electronic, Drives and Energy Systems**

# Remedial Strategies for the Minimization of Cogging torque in PMBDC Motor possessing Material Saturation

M. H. Ravichandran, V. T. Sadasivan Achari, C. C. Joseph, and Robert Devasahayam

*Abstract*--This paper investigates the various strategies for the minimization of cogging torque produced in a Permanent Magnet Brushless DC (PMBDC) motor for a low speed, speed stability stringent spacecraft application with cold coil redundancy. The effect of the material saturation on cogging torque is analyzed using both FE software simulation and hardware simulation. Based on the simulation, suitable method has been derived for the minimization of the cogging torque for this particular application and the machine is realized. Test results shows that a decrease of 90% of the cogging component with 20% of alignment component in the developed torque pattern once saturation is removed.

*Index Terms*--Alignment torque, Cogging torque, Finite Element, Saturation, Skewing.

## I. Introduction

PERMANANT Magnet Brushless DC motor offers significant advantage over conventional motors being lighter, higher efficiency and power density. However, Torque pulsations are the inherent problem results in poor speed stability. The removal of torque pulsations is an essential requirement for high performance application [1]. Cogging torque plays a major role in these torque pulsations. It is caused by the airgap permeance variation due to the slotting effect, because of the attraction of PM with stator a slot where the permeance is maximum.

This paper reports the various strategies for the minimization of the cogging torque and the effect of magnetic saturation on it. The analysis has been carried out using both FE software simulation and hardware simulation. The standard methods for the reduction of cogging torque are studied and the suitable method for this particular machine is identified. The effect of the magnetic saturation on cogging torque is studied and implemented. Fig. 1. shows the cross section of the experimental outer rotor PMBDC motor. The

---

This work was supported by the Indian Government, Department of Space, Indian Space Research Organization.

M.H.Ravichandran, V.T.Sadasivan Achari, C.C.Joseph, Robert Devasahayam are with Spacecraft Inertial Systems Group of ISRO Inertial Systems Unit, Indian Space Research Organization, Trivandrum, India (e-mail: mh_ravichandran@vssc.gov.in).

dimensions and the properties of the experimental machines were:

| | | |
|---|---|---|
| Number of phases | : | 3 |
| Number of poles | : | 16 |
| Number of slots | : | 48 |
| Stator length | : | 15 (mm) |
| Outer Diameter of the rotor | : | 101.5 (mm) |
| Inner Diameter of the stator | : | 34.5 (mm) |
| Airgap thickness | : | 0.5 (mm) |
| Magnet thickness | : | 4 (mm) |
| Stator and Back iron material | : | SS 410 |
| Permanent Magnet material | : | $Sm_2Co_{17}$ |
| Redundancy | : | coil (slot redundancy) |

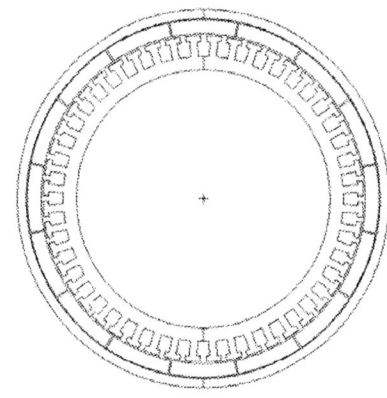

Fig. 1. Cross-section of the experimental motor.

## II. Cogging Torque

The torque produced in a PMBDC motor is broadly divided in to two major categories, Alignment torque (useful torque) and Cogging torque. Alignment torque is produced by the interaction of permanent magnet with the current carrying conductors and the cogging torque is produced by the interaction of salient poles, permanent magnet and the slotted stator. The alignment torque is proportional to the airgap flux density, $Bg$ and the cogging torque us proportional to the square of the airgap flux density, $Bg^2$. This torque component is typically independent of stator current unless saturation is present [2], [3]. The theoretical basics of the cogging torque are studied by several authors, [2], [3]. Its well known that this parasitic torque is a magnetostatic effect due to the

0-7803-9771-1/06/$25.00 ©2006 IEEE

permeance variation encountered by the rotor magnets during their rotation. If $Wa$, is the coenergy, the cogging torque given as (1).

$$Tc = \left( \frac{dWa}{d\theta} \right)_{i=const} \quad (1)$$

where,

$$Wa = \frac{1}{2\mu_0} \int_V B^2 dV \quad (2)$$

where, B is the flux density.

In a simple way the motor cogging torque may be considered as the interaction of each edge of the rotor permanent magnet with the slot opening. Thus the study of the cogging torque is reduced to the analysis of these interactions, considering each of them independent of others. If $Ns$ and $Np$ are the number of slots and poles respectively, the number of periods, $Nperiods$, of cogging waveform in one slot pitch is given in (3)

$$Nperiods = \frac{Ns}{HCF(Ns, Np)} \quad (3)$$

### III. COGGING TORQUE MINIMIZATION METHODS

#### A. Skewing

The well-known method for the minimization of the cogging torque is skewing. Skewing of stator slots or permanent magnet gives the same result. In eliminate the cogging torque, skew angle, $\theta sk$ should be of one cogging period as give in (4). [4] points out that either underskew or overskew yields a better result.

$$\vartheta sk = \frac{2\pi}{Ns.Np} \quad (4)$$

The Skew angle for this 16-pole, 48-slot experimental machine is $7.5^0$. Based on this, the stator is skewed for one slot pitch. The test results of the machine show an appreciable amount of cogging torque even with skewing. This problem is dealt in the next section.

#### B. Change in pole arc to pole pitch ratio

The pole arc width can be rearranged to reduce or eliminate some of the cogging torque harmonics. The pole arc optimization generally requires Finite Element analysis of the motor. Normally permanent magnet spans almost 'n' times that of the slot pitch are chosen. where, 'n' is an integer. [5] outlines that the pole pitch of (n+0.17) or (n+0.14) yields the optimized results. The machine is modeled in FE and the effect of variation of pole pitch on cogging torque is studied and tabulated in Table I.

TABLE I
VARIATION OF COGGING TORQUE WITH POLE ARC

| Pole Arc | 14 | 15 | 16 | 18 | 20 | 22.5 |
|---|---|---|---|---|---|---|
| Cogging Torque (Normalized Value) | 0.39 | 0.29 | 0.69 | 1.0 | 0.82 | 0.58 |

Perfect value of the permanent magnet pole arc width depends on the magnetization direction (radial or parallel) and the airgap thickness. It is found from these results, that a small variation of pole arc variation causes a drastic variation of the peak value of the cogging torque. This requires perfectly machined permanent magnet and stringent assembly procedure. This doesn't suit for mass production and hence not implemented.

#### C. Providing Notches in stator teeth

The cogging harmonic can be reduced by introducing dummy slots in the stator teeth. With the introduction of the dummy slots, the frequency of the interaction between the salient poles increases, thus reducing the amplitude of the cogging torque, for this particular model, FE analysis shows material saturation in the stator teeth and stem. Introduction of the dummy slots requires the removal of material, which additionally introduce material saturation and hence this method is not implemented.

#### D. Shifting permanent magnet pole pairs.

The order of the fundamental frequency of the cogging torque is closely related to the symmetry of the motor. Any operation that removes the symmetry eliminates the cogging torque. [6] investigates a method for the removal of the symmetry, shifting the permanent magnet poles. From Table I, it is found that the pole arc for minimum cogging torque is $15^0$. With this pole pitch, the deviation angle for each magnet is $1.875^0$. When the magnets arranged with this deviation angle, it eliminates the cogging torque, on the other hand, it results in the unsymmetrical back emf pattern and hence this method is not implemented.

#### E. Fractional Slot pitch

The number of cogging cycles in one complete mechanical rotation is given by the least common multiple of the Number of stator slots, $Ns$ and Number of poles, $Np$. When the frequency of the cycle increases, the peak amplitude of the cogging torque comes down. Based on this strategy, the 16-pole motor is simulated with stator of different slot numbers. The variation of cogging torque with the slot number is tabulated in Table II.

TABLE II
VARIATION OF COGGING TORQUE WITH SLOT NUMBER-FE RESULTS

| Slot Number | 48 | 49 | 50 | 51 |
|---|---|---|---|---|
| Cogging Torque (Normalized Value) | 1 | 0.06 | 0.01 | 0.01 |

It shows, that the cogging torque reduces drastically once the slot number is changed from 48. Taking coil redundancy in consideration, number of slots is chosen as 50. Fig. 2 shows hardware experimental results of the cogging torque for 48 and 50 slot configurations.

## IV. EFFECT OF MATERIAL SATURATION

With the modification of slots from 48 to 50, cogging torque reduces drastically, but still a reasonable amount of cogging torque is seen. As pointed out in the earlier section, the cogging torque doesn't reduce with the introduction of skewing in 48-slot configuration. The factor other than the slot-pole configuration, which plays a role in the development of cogging torque, is material saturation. The material chosen for the back iron and stator are SS 410, having nonlinear BH characteristics. The machine is analyzed using FE method. The FE results show that, at some portion at stem and tooth, the maximum flux density crosses 2.0Tesla. The maximum saturation flux density for this particular material is 1.8Tesla. Saturation can be eliminated either by adding extra material or reducing the operating flux density. As the envelope of the machine is fixed, it is not possible to add extra material, and hence, operating flux density is to be reduced.

The permanent magnet used in this particular motor is Samarium Cobalt, because of its higher energy product thermal stability. Once this PM is magnetized to saturation, it attains the open-air flux density of 3.3Kgauss, and the magnetic circuit decides the operating point of the PM. The operating point can be changed by changing the magnetization level of the PM. With the aid of the magnetizer assembly, three types of PM of different open-air flux density are obtained and they are tabulated in Table III.

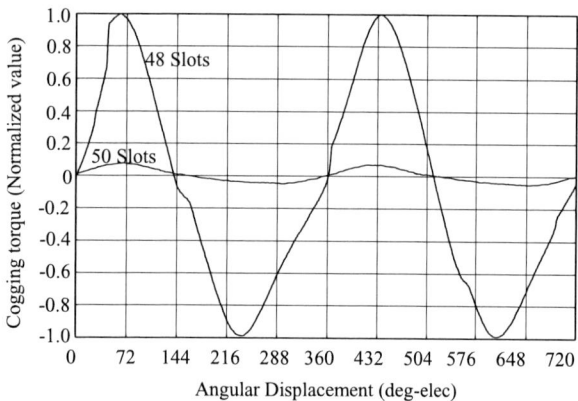

Fig. 2. Cogging torque variation with 48 slots and 50 slots. (hardware experimental results – normalized value).

TABLE III
PM OF DIFFERENT OPEN AIR FLUX DENSITY LEVELS

| P.M Type | Open air Flux Density (Kgauss) |
|---|---|
| Type-1 (High Power) | 3.3 |
| Type-2 (Medium Power) | 2.1 |
| Type-3 (Low Power) | 1.5 |

The machine is realized and tested with these three types of permanent magnets. The cogging torque pattern is plotted in Fig. 3. The corresponding developed torque is plotted in Fig. 4. It is observed that there is considerable amount of reduction of cogging component with only marginal reduction in alignment component. Table IV gives the harmonic analysis of cogging and alignment component with the different slot configuration and with the change in permanent magnet open-air flux density.

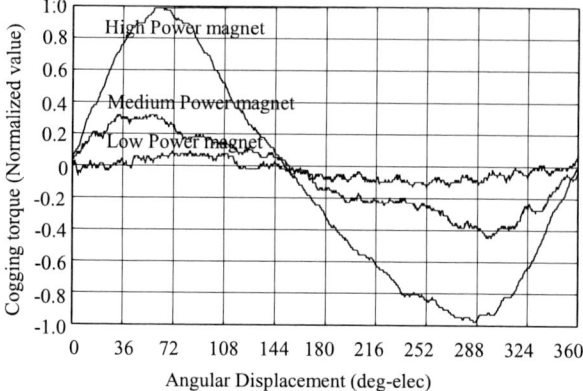

Fig. 3. Cogging torque variation for 50-slot configuration with change in permanent magnet open-air flux density.

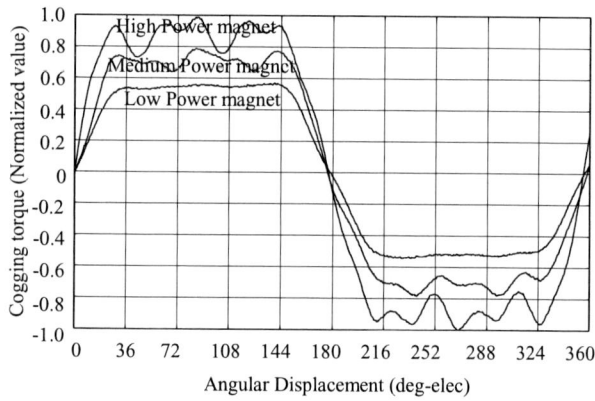

Fig. 4. Developed torque variation for 50-slot configuration with change in permanent magnet open-air flux density.

TABLE IV
HARMONIC CONTENT OF DIFFERENT CONFIGURATION

| Configuration | Alignment Component | Cogging Component |
|---|---|---|
| 48 slots, High power magnet | 1.00 | 1.00 |
| 50 slots, High power magnet | 0.94 | 0.10 |
| 50 slots, Low power magnet | 0.78 | 0.01 |

## V. CONCLUSION

Various methodologies for the minimization of cogging torque is studied using both FE and hardware simulations. The strategy for the minimization of cogging torque depends on particular application and its requirements like redundancy, speed of operation and speed stability. The number of slots is chosen as 50 taking redundancy in consideration. With the introduction of dummy slots, the cogging component reduces by 90% with 6% reduction in alignment component. In addition to the standard technique for the minimization of cogging torque, removal of material saturation plays a major role. As the envelope of the machine is fixed, saturation removal is carried out by changing the open-air flux density level of the permanent magnet, With this change, there is again 90% reduction in cogging component with 22% reduction in alignment component.

## VI. Acknowledgment

The authors gratefully thank the support and encouragement given by Dr.V.Krishnan, Associate Director and Mr. P.S.Veeraghavan, Director ISRO Inertial Systems Unit.

## VII. References

[1] T.M.Jahns, "Pulsating torque minimization techniques for permanent magnet AC motor drives-A review". *IEEE Trans. Ind. Electronics,* pp. 321-330, 1996.

[2] B.Ackermann, "New technique for the reduction of cogging torque in a class of brushless motors". *IEE Proc. B,* pp. 315-320, 1992.

[3] D.C.Hanselman, "Effect of skew, pole count and slot count on brushless radial force, cogging torque and back EMF". *IEE Proc. B,* pp. 325-330, 1997.

[4] Min Dai, Ali Keyhani, Tomy Sebastin, "Torque Ripple Analysis of a PM Brushless DC Motor using finite Element Method", *IEEE Trans. Energy Conversion,* Vol 19. No. 1. pp. 40-45, March 2004

[5] Touzhu Li, Gordon Slemon, "Reduction of Cogging torque in permanent magnet motors ", *IEEE Trans. Magnetics,* Vol. 24, No. 6, pp. 2901-2903, November 1988..

[6] C.Breton, j.Bartolome, "Influence of Machine Symmetry on reduction of Cogging Torque in Permanent Magnet Brushless Motors", *IEEE Trans. Magnetics,* Vol. 36, No. 5, pp. 3819-3823, September 2000.

**2006 IEEE International Conference on Power Electronic, Drives and Energy Systems**

# Fuzzy Pre-compensated PI Controller for PMBLDC Motor Drive

Mukesh Kumar, Bhim Singh, *Senior Member, IEEE*, and B. P. Singh, *Senior Member, IEEE*

*Abstract*--This paper presents the fuzzy pre-compensated PI (FPPI) speed controller for Permanent Magnet Brushless DC (PMBLDC) Motor drive. Various aspects of DSP based implementation of PMBLDC motor drive system such as inverter, position and current sensing, reference current generation and PWM current control are discussed in detail. The fuzzy logic algorithm used in this work particularly suitable for digital control because of the simplified way of handling the rule base and considerable reduction of computation during defuzzification process. Transient performance of the drive system obtained using FPPI speed controller is compared with the conventional PI speed controller.

*Index Terms*-- DSP based control, Fuzzy pre-compensated, PI controller, PMBLDC motor

## I. INTRODUCTION

PERMANENT Magnet Brushless DC (PMBLDC) motors are becoming popular in electric drives [1-3]. High efficiency, small size and inertia, ease of speed control and suitability for explosive environment are some of the main advantages of these motors. The use of these motors in high performance applications requires advance control method for their speed control. The conventional PI speed controller is simple to implement and gives good dynamic performance with zero steady state error. But, as shown by many studies [7], the PI speed controllers are very sensitive to perturbations and variations of system parameters. The PI controller requires accurate mathematical model of the drive system for tuning the PI gains. The rectifier and inverter combination used to feed power to the motor makes the drive system more complex and difficult to model. The other control schemes such as sliding mode control suffer from chattering. The fuzzy logic based controllers are useful when precise mathematical modeling of the system is not feasible. Moreover, fuzzy logic based controller often yield better results as compared to conventional controllers [5-6].

The inverter fed PMBLDCM drive system is a non-linear system and the PI controller parameters for one operating condition does not work satisfactorily for another operating condition. The variation of the operating point of the system (such as variation of reference speed, load and the DC link voltage) leads to oscillations in the system.

These difficulties are resolved by employing fuzzy logic based speed controllers

In this paper a fuzzy logic based pre-compensation PI speed controller for PMBLDCM drive system is presented. The performance of the drive system obtained with this speed controller is compared with that obtained with conventional PI speed controller.

## II. THE DRIVE SYSTEM

Fig.1 shows the PMBLDC motor drive system. It consists of the following parts:

### A. The Power Circuit

The power circuit of the PMBLDC motor drive consists of a rectifier, filter, and a three-phase Current Controlled Voltage Source Inverter (CC-VSI). Three-phase diode bridge rectifier is used to obtain DC voltage by rectification of line frequency AC supply. The rectified DC voltage is smoothened by a LC filter. The CC-VSI uses IGBT's (Insulated gate bipolar transistor) operating at high switching frequencies thereby achieving nearly ideal current waveform in the motor.

### B. The Controller

The closed loop speed control of PMBLDC motor is implemented using TMS 320F240 digital signal processor. The details of DSP based implementation are discussed below.

*1) Reference Speed*

The speed control of the drive system needs the reference speed signal. A reference signal (0 to 5 V) for speed control of drive is obtained through a multi turn pot. The output of the pot is connected to the ADC channel of the DSP.

*2) Rotor Speed Monitoring*

The speed of the rotor is determined from the two-encoder pulses (HS4, HS5). The encoder gives 30 pulses per revolution. These pulses are connected to the capture unit of the DSP. The rotor speed is determined by measuring the width of these pulses.

*3) The Speed Controller*

The closed loop control of the drive system requires a speed controller, which takes appropriate action to control the drive system at reference speed. The speed controller determines the reference torque for the motor, which is then used to find the reference values of winding currents.

*4) Rotor Position Sensing*

The rotor position information is essential for proper synchronization of motor currents with the rotor position. Three-position signal (HS1, HS2, HS3) are obtained from the encoder mounted on the motor shaft. These signals are connected to the DSP through its high-speed digital I/O port.

---

Mukesh Kumar is in Electrical Engineering Department, National Institute of Technology Hamirpur, H.P (mukesh@recham.ernet.in)
Bhim Singh and B. P. Singh are in Electrical Engineering Department, Indian Institute of Technology Delhi (bsingh@ee.iitd.ac.in, bpsingh@ee.iitd.ac.in)

0-7803-9771-1/06/$25.00 ©2006 IEEE

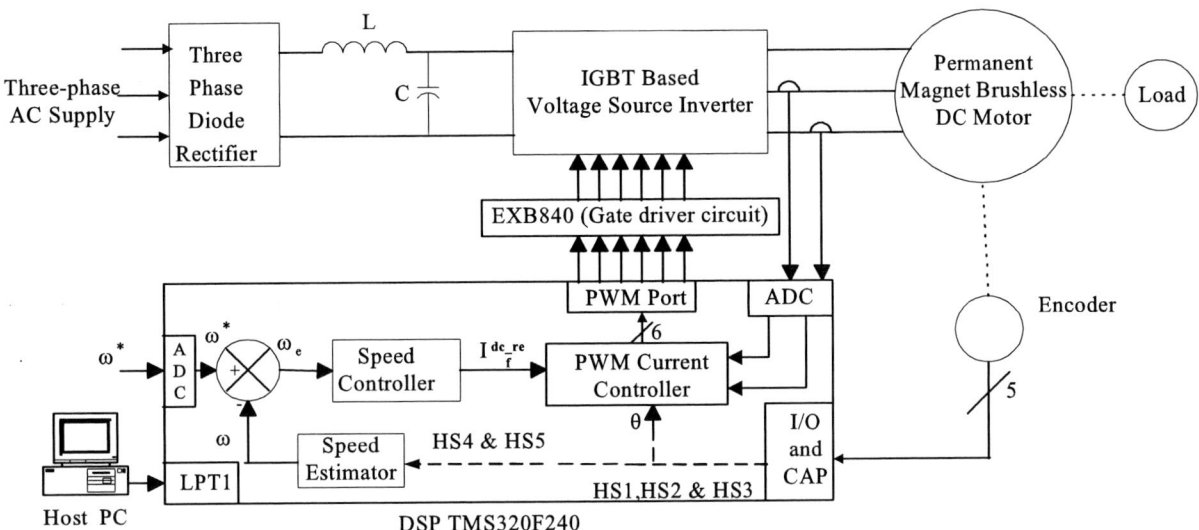

Fig. 1. DSP based closed loop speed controlled PMBLDC motor drive system.

### 5) Reference Current Generation

The reference currents are generated ($i_a^*$, $i_b^*$, $i_c^*$) in software using rotor position and the reference torque is computed from the speed controller. The relationship between the rotor position and the reference currents is shown in Table I.

### 6) Winding Current Sensing

An inner control loop consists of PWM current controller, which controls the motor current and torque. The winding current sensing is essential for current control. Since the three-phase stator windings of the PMBLDC motor are star connected with isolated neutral, measurement of two line currents is sufficient. Two Hall effect current sensors along with a signal conditioning circuit (to add an offset of 2.5 V to the signal so that it becomes unipolar) are used to sense any two-line currents. These signals are interfaced to the DSP through its ADC unit.

TABLE I

ROTOR POSITION AND REFERENCE CURRENTS

| Rotor Position | | | Reference Phase Current | | |
|---|---|---|---|---|---|
| HS1 | HS2 | HS3 | $i_a^*$ | $i_b^*$ | $i_c^*$ |
| 1 | 0 | 0 | + $I_{ref}$ | - $I_{ref}$ | 0 |
| 1 | 1 | 0 | + $I_{ref}$ | 0 | - $I_{ref}$ |
| 0 | 1 | 0 | 0 | + $I_{ref}$ | - $I_{ref}$ |
| 0 | 1 | 1 | - $I_{ref}$ | + $I_{ref}$ | 0 |
| 0 | 0 | 1 | - $I_{ref}$ | 0 | + $I_{ref}$ |
| 1 | 0 | 1 | 0 | - $I_{ref}$ | + $I_{ref}$ |

### 7) The Current Controller

The pulse width modulation (PWM) is used to control the winding currents. In PWM current control the PWM duty cycle (ratio of ON time to PWM cycle time) is determined for each phase leg. For example, when the current error is the maximum positive in a phase, the current controller generates PWM signal to the upper device of the phase leg with 100% duty cycle and that for the lower device with 0% duty cycle. Similarly, when the current error is the maximum negative in a phase, the current controller generates PWM signal to the upper device of the phase leg with 0% duty cycle and that for the lower device with 100% duty cycle. For any other value of current error in a phase the duty cycle for upper and lower devices of that phase will vary between the two extremes (0 to 100%), decided by the controller.

### 8) Gate Drive Control Unit

The signals generated by current controller through PWM unit of the DSP cannot be directly used to drive the gate of the IGBT's of the inverter, because of different voltage needed for the gate of the IGBT and the need for the isolation. Hence, the signals obtained from the PWM unit of DSP are fed to Gate Drive Control Unit, where they are isolated and transformed to –5 to +15 V (from 0 to 5V).

## III. FUZZY PRE- COMPENSATED PI SPEED CONTROLLER

### A. The Speed Controller

The basic structure of Fuzzy Pre-compensated PI speed controller is shown in Fig.2. It consists of a pre-compensator block followed by a conventional PI speed controller. The pre-compensator block uses reference speed ($\omega^*$) and the actual speed ($\omega$) as inputs. It generates the output signal ($\omega^{*'}$) on the basis of these inputs and fuzzy compensation term as follows:

$$e(n) = \omega^*(n) - \omega(n) \qquad (1)$$

$$\Delta e(n) = e(n) - e(n-1) \qquad (2)$$
$$\gamma(n) = f[e(n), \Delta e(n)] \qquad (3)$$
$$\omega'^* = \omega^* + \gamma(n) \qquad (4)$$

Where $e(n)$ is speed error, $\Delta e(n)$ is change in error and $\gamma(n)$ is a compensation term (determined on the basis of fuzzy logic mapping function) at an instant n.

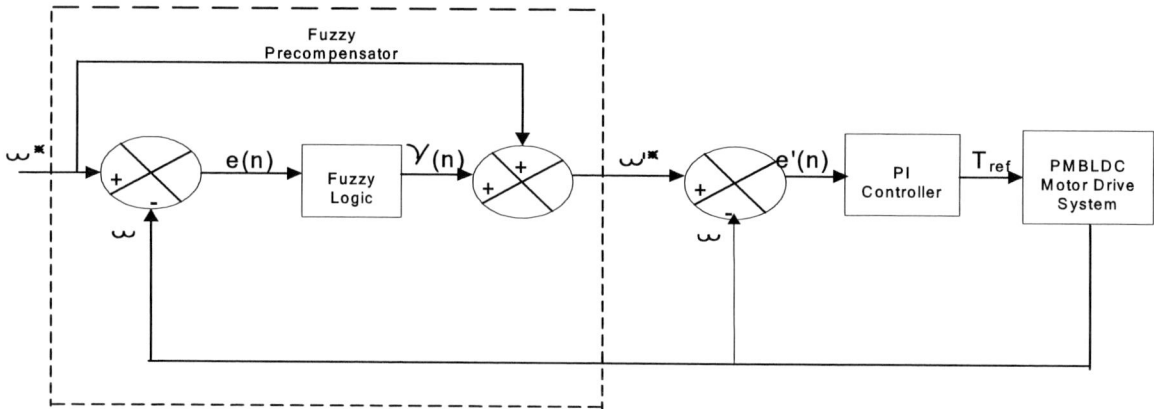

Fig. 2. Fuzzy precompensated PI speed controller.

The compensation term $f[e(n), \Delta e(n)]$ is used to change the reference speed so that the transients are avoided. This term is computed using fuzzy logic control as discussed in the next section.

The compensated reference signal $(\omega'^*)$ is then applied to the conventional PI controller to obtain the torque reference $(T_{ref})$ at that instant as follows:

$$e'(n) = \omega'^* - \omega \qquad (5)$$
$$\Delta e'(n) = e'(n) - e'(n-1) \qquad (6)$$
$$T_{ref}(n) = T_{ref}(n-1) + K_p \Delta e'(n) + K_i e'(n) \qquad (7)$$

*B. Fuzzy Pre-compensator*

As discussed in the previous section the compensation term $\gamma(n) = f[e(n), \Delta e(n)]$ is computed using fuzzy logic. The error $e(n)$ and change in error $\Delta e(n)$ are the inputs to the fuzzy mapping function $f$, and $\gamma(n)$ is the output. The intricate details of the fuzzy logic control used in this work is given below:

*1) Structure of Fuzzy Logic Control*

The basic structure of fuzzy logic speed controller is given in Fig.3. The input to the fuzzy logic speed controller is speed error and change in error (which is calculated from error and previous error). These two inputs are normalized to obtain error $(\varepsilon)$ and change in error $(\Delta\varepsilon)$ in desired range (-3 to 3 is the normalized range taken for these variables in this study). The compensation term $\gamma(n)$ is then determined by using fuzzy logic control, which consists fuzzification, application of control rules, decision table and defuzzification. Each of these steps are discussed as follows:

*Fuzzification:* The process of fuzzification is basically assigning membership functions for various fuzzy subsets (seven fuzzy subsets namely PB, PM, PS, ZE, NS, NM, NB

are considered in this study) on the basis of normalized inputs (error and change in error). As shown in Fig. 4 for any value of universe of discourse of error or change in error the corresponding membership function has non-zero value for any two fuzzy subsets (at the most).

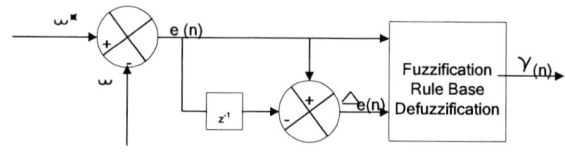

Fig. 3. Basic structure of fuzzy logic control.

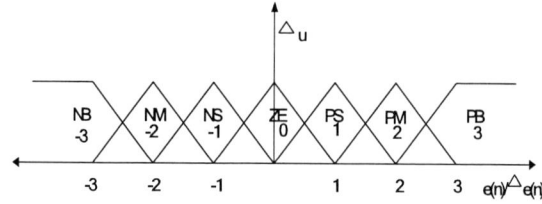

Fig. 4. Fuzzy membership function.

*Control Rules:* The decision-making in fuzzy logic controller is based on rules written using linguistic terms as follows:
IF $e(n)$ is "negative big (NB)" AND $\Delta e(n-1)$ is "positive big (PB)" THEN $\Delta u$ is "Zero (ZE)".
..
..

128

IF ε is "negative big (NB)" AND Δε is " Zero (ZE)" THEN Δu is "negative medium (NM)".

The seven fuzzy sub sets are then assigned numerical values (+3 to –3).

*Decision Table*: With the seven fuzzy subsets forty-nine rules are written. These rules are given in TableII. For example:

IF ε = +1   AND   Δε = -2   THEN   Δu = -1.

### TABLE II
### DECISION TABLE

| e(n) \ Δe(n) | -3 | -2 | -1 | 0 | +1 | +2 | +3 |
|---|---|---|---|---|---|---|---|
| -3 | -3 | -3 | -2 | -2 | -1 | -1 | 0 |
| -2 | -3 | -2 | -2 | -1 | -1 | 0 | +1 |
| -1 | -2 | -2 | -1 | -1 | 0 | +1 | +1 |
| 0 | -2 | -1 | -1 | 0 | +1 | +1 | +2 |
| +1 | -1 | -1 | 0 | +1 | +1 | +2 | +2 |
| +2 | -1 | 0 | +1 | +1 | +2 | +2 | +3 |
| +3 | 0 | +1 | +1 | +2 | +2 | +3 | +3 |

*Inference:* The inference is the process of determining the degree of belonging of the output (Δu) to the various fuzzy subsets, for given values of ε and Δε.  Larsen's's inference method [5] is used to determine the degree of belonging of output. This method is explained with the following example.

Consider  ε = -2.6 and  Δε= 0.7

The degree of belonging of error (ε) and change in error (Δε) to fuzzy subsets is shown in Table III. The degree of belonging for output (Δu )is also calculated in this table.

### TABLE III
### INFERENCE

| ε | | Δε | | Δu | |
|---|---|---|---|---|---|
| Fuzzy Subset | Degree of Belonging | Fuzzy Subset | Degree of Belonging | Fuzzy Subset | Degree of Belonging |
| -3 | 0.6 | 0 | 0.3 | -2 | 0.18 |
| -3 | 0.6 | 1 | 0.7 | -1 | 0.42 |
| -2 | 0.4 | 0 | 0.3 | -1 | 0.12 |
| -2 | 0.4 | 1 | 0.7 | -1 | 0.28 |

*Defuzzification:* The output of the inference process gives degree of belonging of Δu to different fuzzy subsets. Defuzzification is the process of obtaining crisp output from these fuzzy values. To obtain Δu, the method of center of gravity is used.

$$\Delta u = \frac{\sum_i A_i \times \alpha_i}{\sum_i \alpha_i} \qquad (8)$$

Where $A_i$ is output fuzzy subset and $\alpha_i$ is the degree of belonging.

The value of Δu for the example taken in previous section is:

$$\Delta u = \frac{-2 \times 0.18 - 1 \times 0.42 - 1 \times 0.12 - 1 \times 0.28}{0.18 + 0.42 + 0.12 + 0.28} = 1.18$$

*DSP Based Implementation of the Drive System:* The assembly language software is developed for implementing the PMBLDC motor drive system with PI speed controller and fuzzy pre-compensated PI speed controller. The assembly language software is developed for closed loop speed control of the PMBLDC drive system with Fuzzy Pre-compensated PI and PI speed controllers.

## IV. RESULTS AND DISCUSSION

The experimental set up used to implement the proposed scheme, consists of a 2-h.p. PMBLDC motor coupled to a DC generator and an IGBT based Inverter (parameters given in Appendix). The performance of the PMBLDC motor drive system with PI and fuzzy pre-compensated PI speed controller is compared for starting, and reversal.

Fig.5 shows the starting response of the drive speed at a reference speed of 500rpm. It is seen that with pre-compensated PI speed controller the starting time is 20 msec less than the one obtained with PI speed controller.

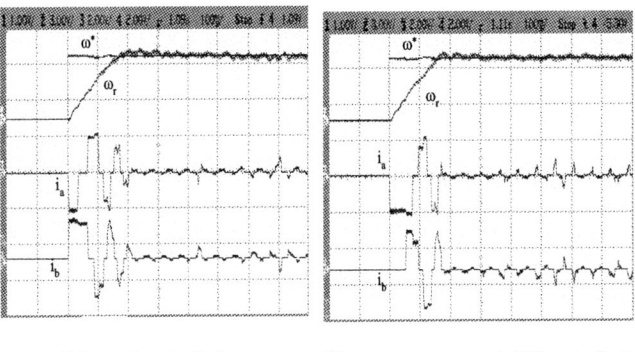

(PI speed controller)          (Fuzzy pre-compensated PI controller)

Scale: 300 rpm /div (channel 1 and 2), 4 A/div (channel 3 and 4)
100-ms/ div. (time scale)

Fig. 5. Starting response of the PMBLDC motor drive.

The response of the drive system during reversal with these speed controllers is shown in Fig.6. It is seen in this figure that the reversal time is 700 ms with fuzzy pre-compensated PI and 900 ms with PI speed controllers. Hence, there is reduction in reversal time by 200 ms.  In both the cases the motors speed changes from 500 rpm to –500 rpm.

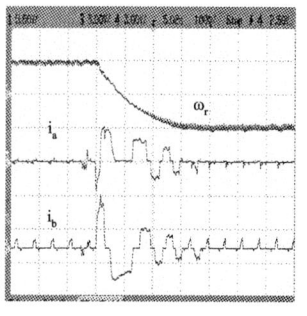 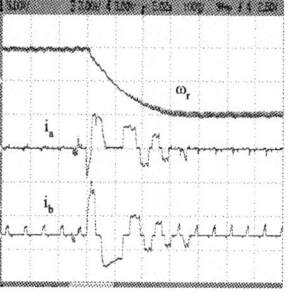

(PI speed controller)　　　　(Fuzzy pre-compensated PI controller)

Scale: 300 rpm /div (channel 1 and 2), 4 A/div (channel 3 and 4)
100-ms/ div. (time scale)

Fig. 6.  Speed reversal response of the PMBLDC motor drive.

## V. CONCLUSION

The DSP based implementation of PMBLDC drive system is presented. A fuzzy logic algorithm suitable for fixed point DSP implementation is discussed in detail. The response of the drive system obtained with a conventional PI speed controller is compared with the one obtained from a fuzzy pre-compensated PI speed controller. It is seen that response of the PMBLDC drive system is superior with the fuzzy pre-compensated PI speed controller. The experimental result of the drive system in terms of the winding currents and speed response has been presented.

## VI. REFERENCES

[1]  T.Kennnjo and S. Nagamori, "Permanent Magnets Brushless DC Motors," Clarendon Press, Oxford, 1985.

[2]  T.J.E. Miller, "Brushless Magnets and Reluctance Motor Drive," Clarendon Press, Oxford, 1989.

[3]  P. Pillay and P. Freere, "Literature survey of permanent magnet AC motors and drives," in *Proc. IEEE IAS*, 1989, vol. 1, pp. 74-84.

[4]  P. Pillay and R. Krishnan, "Modeling, Simulation, and Analysis of Permanent Magnets Motor Drives, Part II: The Brushless DC Motor Drive," IEEE Trans. on Industry Applications, Vol. 25, No. 2, March/April 1989, pp. 274-279.

[5]  J-H. Kim, K-C. Kim and E.K.P. Chong, "Fuzzy Precompensated PID Controllers," IEEE Trans. on Control Systems Technology, Vol. 2, No. 4, December 1994, pp. 406-410.

[6]  Th. Lubin, E. Mendes and C.Marchamd, "Fuzzy Controller in A.C. servo Motor Drive," Electrical Machines and Drives Conference, No. 412,11-13 September 1995, pp. 320-324.

[7]  Gilberto C. D. Sousa and Bimal K. Bose, "A Fuzzy Set Theory Based Control of a Phase-Controlled Converter DC Machine Drive," IEEE Trans. on Industry Applications, Vol. 30, No. 1, Jan./ Feb. 1994, pp. 34-44.

[8]  TMS320C24x DSP Controllers Reference Set, Vol.1 and 2, Texas Instruments Incorporated, Dallas, 1997.

## VII. APPENDIX

Motor Parameters: 2 h.p, 4 pole, 1500 rpm
Current Sensor: Hall Effect Current sensor ABB (EL 50 P1)
Inverter:  IGBT module 6MBI50L –120 (FUJI)

## VIII. BIOGRAPHIES

**Mukesh Kumar** was born in Hamairpur , H. P., India in 1965. He received B. E. (Electrical) degree from Gujarat University Ahmedabad, India in 1987 and M.Tech. and Ph. D. degrees from Indian Institute of technology (IIT), New Delhi, in 1995 and 2006, respectively. He is presently working as an Assistant Professor in Electrical Engineering Department, NIT Hamirpur, H.P, INDIA, where he joined as a lecturer in 1989. His field of interest includes power electronics, electrical machines and drives, and DSP based control of electrical machines. He is a Life Member of Indian Society for Technical Education (ISTE).

**Bhim Singh** (SM'99) was born in Rahamapur, U. P., India in 1956. He received B. E. (Electrical) degree from University of Roorkee, India in 1977 and M.Tech. and Ph. D. degrees from Indian Institute of technology (IIT), New Delhi, in 1979 and 1983, respectively. In 1983, he joined as a Lecturer and in 1988 became a Reader in the Department of Electrical Engineering, University of Roorkee. In December 1990, he joined as an Assistant Professor, became an Associate Professor in 1994 and Professor in 1997 at the Department of Electrical Engineering, IIT Delhi. His field of interest includes power electronics, electrical machines and drives, active filters, static VAR compensator, analysis and digital control of electrical machines. Prof. Singh is a Fellow of Indian National Academy of Engineering (INAE), Institution of Engineers (India) (IE (I)) and Institution of Electronics and Telecommunication Engineers (IETE), a Life Member of Indian Society for Technical Education (ISTE), System Society of India (SSI) and National Institution of Quality and Reliability (NIQR) and Senior Member of IEEE (Institute of Electrical and Electronics Engineers).

**B. P. Singh** (SM'96) was born in Singhia, Bihar, India, in 1940. He received B.Sc.(Engg.) degree from the Bihar Institute of Technology Sindri(Bihar), India, in 1963, M.E. from Bengal Engineering College Howrah(W.B.)and Ph.D. degrees from Indian Institute of Technology(IIT), New Delhi, India, in 1966 and 1974, respectively. In 1966, he joined the Department of Electrical Engineering, M.I.T. Muzaffarpur, as an Assistant Professor. In 1978, he joined Department of Electrical Engineering, IIT, Delhi, India, as an Assistant Professor. He became full Professor in 1985. He was visiting professor at California State University Long Beach USA from 1988 to 1990. Presently he is an Emeritus Fellow in the Department of Electrical Engineering at IIT New Delhi. His fields of interest include Energy conservation in electrical machines and drives, and analysis and control of electrical machines. Prof. Singh is a Fellow of the Institution of Engineers (India) and a Life Member of the Indian Society for Technical Education.

**2006 IEEE International Conference on Power Electronic, Drives and Energy Systems**

# A Simplified Design Methodology for Switched Reluctance Motor using Analytical and Finite Element Method

M.H.Ravichandran, V.T.Sadasivan Achari, C.C.Joseph, Robert Devasahayam

*Abstract--***Switched Reluctance Motors (SRM) is gaining wider popularity for variable speed drives. These motors have been developed for electric propulsion, fan, pump and even as starter/alternator for aerospace applications. This paper gives the step-by-step design procedure of a 16/12 Switched Reluctance Motor for a spacecraft actuator. Machine design has been started with analytical method to fix the major parameters and optimization is carried out using Finite element technique. Inductance and Torque developed has been computed using both the methods and a good comparison seem between them. The motor has been realized and tested for its capability.**

*Index Terms--* **Analytical method, FEM, Flux tubes, Inductance, Ripple, Torque.**

## I. INTRODUCTION

THE concept of SRM is extremely old dated back 1838 when a locomotive was propelled by this motor[1]. Savings in manufacturing cost of the motor because of the simplicity in construction, high torque to weight ratio and use of minimum number of switching devices in the drive circuit are the important advantages compared to other motors. The period of 70s is the time of development of SRM. Since then there have been massive developments in both the design and control of SRM. The torque production in the machine is based on the reluctance principle. Here the developed torque is the function of current and position. The inductance variation from aligned to unaligned position makes the torque profile highly non-linear. The direction of the torque does not depend on the polarity flux-linkage and current, but only on the polarity of the rate of change of inductance with rotor position. The controller must supply unipolar current pulses, precisely phased relative to the rotor position. It requires an accurate shaft position sensor for getting the position feedback signal

implying the need for a servo-quality encoder/Resolver or alternatively, a sophisticated sensor less controller that will be, in general, specific to particular application.

The fundamental analytical design procedure is outlined in [2] and the finite element based design is first briefed in [3]. The procedure to calculate the developed torque is explained in detail in [4], [5]. There are number of papers that rely heavily in the process of design solely based on the Finite Element Method. The major disadvantage of FEM is, it gives microscopic details where macroscopic details are required. Here in this paper the design starts with the analytical method to fix the major design parameters and dimension of the machine and optimization is carried out using FEM. The aligned and unaligned inductance, developed torque is calculated using both Analytical and Finite Element method.

## II. NOMENCLATURE

| | | |
|---|---|---|
| m | Number of Phases | |
| Nr | Number of Rotor Poles | |
| Ns | Number of Stator Poles | |
| $\beta r$ | Rotor pole Angle | (deg) |
| $\beta s$ | Stator pole Angle | (deg) |
| $\varepsilon$ | Stroke Angle | (deg) |
| C | Back Iron thickness | (mm) |
| g | Airgap thickness | (mm) |
| hr | Rotor Pole height | (mm) |
| hs | Stator Pole Height | (mm) |
| Bmax | Stator pole Flux Density | (T) |
| Brc | Rotor Core Flux Density | (T) |
| By | Stator Yoke Flux density | (T) |
| La | Aligned Inductance | (H) |
| Lu | Unaligned Inductance | (H) |

## III. DESIGN PROCEDURE

This section deals with the selection of the major dimensions of SRM using analytical method and optimization of the same using Finite Element Method.

*A. Specifications of the machine*

| | | |
|---|---|---|
| Speed | : | $\pm 1500$ RPM |
| Torque | : | 2 Nm |

---

This work was supported by the Indian Government, Department of Space, Indian Space Research Organization.

M.H.Ravichandran, V.T.Sadasivan Achari, C.C.Joseph, Robert Devasahayam are with Spacecraft Inertial Systems Group of ISRO Inertial Systems Unit, Indian Space Research Organization, Trivandrum, India (e-mail: mh_ravichandran@vssc.gov.in).

0-7803-9771-1/06/$25.00 ©2006 IEEE

| Supply Voltage | : 42 V |
| Motor Current | : 1.5 A |
| Dimensions | : $\phi$ 148mmX30mm |
| Weight | : 3.0 kg |

### B. Selection of m:

The design process starts with the selection of Number of phases, m. The increase in phase number decreases the torque ripple and vice versa. The problem of torque dips can be reduced by increasing the number of strokes per rotation, which leads to smaller stroke angle, ε. It can be achieved by increasing either rotor pole number or phase number. Increase in Nr reduces the inductance ratio, the ratio of aligned to unaligned inductance. Low inductance ratio increases the controller volt-ampere requirement and decreases the specific output. The increase of m will complicate the controller circuit and increase the switching losses. With a tradeoff between this data, the number of Phases is chosen as 4

### C. Selection of Nr and Ns:

The rotor pole number, $N_r$ is selected such that Nr=Ns ± 2. Ns = Nr is not recommended, since all positions will be aligned positions and hence creates starting problem. The number of strokes per revolution is higher for machines with Nr>Ns. whereas, inductance ratio is higher for machines with Nr<Ns. It is due to the fact that the stator pole arc is smaller for a motor with Nr>Ns and this decreases the aligned inductance and hence lowers inductance ratio, even though it increases the slot area. So the reduction in available conversion energy cancels the effect of increase in number of strokes per revolution. Core losses will be higher because of higher switching frequency. Hence lesser value of Nr is preferable. A configuration with repetition of the basic one is also possible. With repetitive configuration, the inductance ratio is not reduced, but, the number of strokes per revolution increases and hence improves torque ripple. If 'n' is the multiplication factor, the number of poles per phase becomes 'n' times that of the basic form. Thus there is an added advantage of inclusion of short flux paths because of '2n' pole magnetic field configuration unlike the two-pole configuration. Based on the above factors, Number of phases and envelope of the machine, Ns and Nr is selected as 16 and 12 respectively.

### D. Selection of βs and βs:

The three important factors that governs the selection of βs and βr are,

$$\beta s < \beta r \qquad (1)$$
$$\beta s < \varepsilon \qquad (2)$$
$$\frac{2\pi}{Nr} > \beta r + \beta s \qquad (3)$$

Equation (1) to be satisfied to maximize the aligned inductance, (2) to be satisfied to achieve starting torque at all rotor positions and (3) to be satisfied to avoid overlap between poles in unaligned conditions. Based on the above factors a feasible triangle is drawn and shown in Fig. 1. The value of βs and βs is selected such that, it lies in the hatched area. Ripple Torque is also low for optimum values of βs and βs. The variation of torque pattern for different values of βs and βs is modeled in FE software (FE method is outlined in section III). Based on simulation results βs and βs is optimized to as $15.2^0$ and $13.2^0$ in order to minimize the ripple.

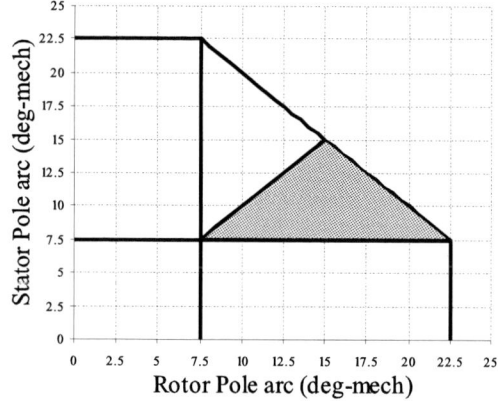

Fig. 1. Feasible Triangle of 16/12 SRM

### E. Selection of g

The airgap should be of minimal thickness and concentric to maximize specific torque, maintain the balanced phase current and to minimize the acoustic noise. The variation of developed torque with change in excitation and airgap is carried out using FE method and plotted in Fig. 2. Based on the torque developed and manufacturing feasibility, airgap thickness, g, is chosen as 0.1 mm.

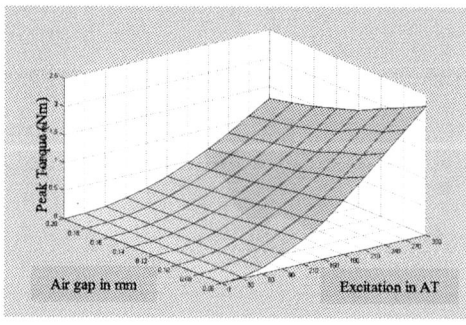

Fig. 2. Variation of Developed torque with change in airgap and excitation

The values of C, hs, hr are computed taking By and Brc as 50% and 80% of Bs and the same is verified using FE method. Thus the major design parameters and dimensions are initially fixed using the analytical method fine optimized using FE method.

## IV. ESTIMATION OF TORQUE

The inductance and developed torque are estimated both by analytical and Finite Element method and is described in this section.

### A. Analytical Method:

The procedure for the calculation of average torque is outlined in [5]. The average torque, *Tav*, is given as (4).

$$Tav = \frac{WNrNs}{4\pi} \qquad (4)$$

where, W is the difference between aligned and unaligned energy.

The unaligned inductance, *Lu,* is calculated by dividing the total flux path in to definite flux tubes. The magnetic circuit equation is derived for all the flux tubes. The minimum flux density, $Bs_{min}$' is calculated for all the flux tubes, and corresponding Inductance is calculated using (5).

$$L_u = \tau \frac{(Bs_{min}.A)}{i^2} \qquad (5)$$

where $\tau$ is the excitation in AT, A is area of the flux tube in m³ and i is the phase current in A

The total unaligned inductance is calculated by summing up the unaligned inductance of all the flux tubes. Equation (6) gives the total unaligned energy.

$$Wu = \frac{1}{2}i^2 Lu \qquad (6)$$

The magnetic circuit of the one fourth of the machine in aligned condition is shown in Fig. 3. The magnetic circuit is similar to all four branches. $R_s$, $R_y$, $R_r$, $R_{rc}$ and $R_g$ represent the reluctance of the stator pole, yoke, rotor pole, rotor core and air-gap respectively.

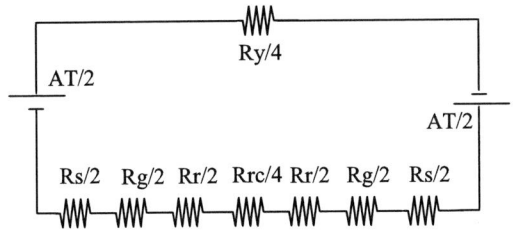

Fig. 3. Magnetic Circuit model of SRM

The magnetic circuit equation for aligned condition is given in (7),

$$\tau = 4(HsIs + HrIr) + (HrcIrc + HyIy) + \frac{BgAg}{Pa} \qquad (7)$$

Where, $\tau$ is the excitation in AT, H is the magnetic field intensity and l is the length of various sections. Pa is the permeance of the airgap. For the given excitation, the airgap flux density, Bg, is calculated by iterating (7). The value of Bg is obtained from the above expression and the aligned inductance is calculated using (8).

$$La = \tau \frac{BsAs}{i^2} \qquad (8)$$

The Variation of aligned and unaligned inductances with change in excitation is computed by analytical method and plotted in Fig. 4

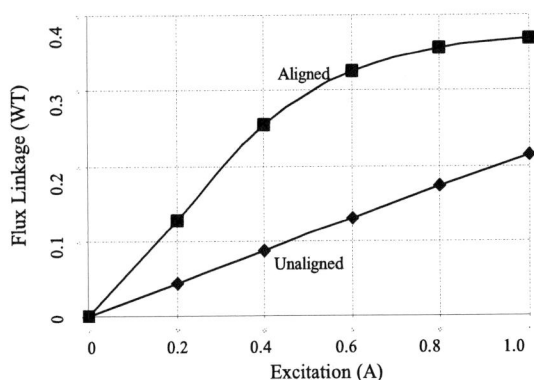

Fig. 4. Variation of aligned and unaligned inductances with excitation using analytical method

### B. Finite Element Method

Finite Element modeling of 16/12 SRM is done using Maxwell 2D electromagnetic software. As the machine is axis symmetric the deviation between 2D and 3D model is negligible. The flux path during aligned and unaligned condition is shown in Fig. 5. The Variation Torque Pattern for different excitation with rotor position and the $T\ I\ \theta$ characteristics of the machine is plotted in Fig. 6 and Fig. 7 respectively

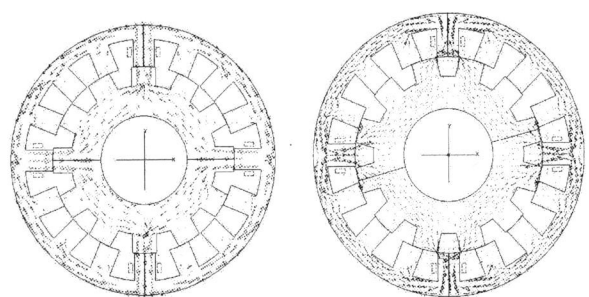

Fig. 5. Flux pattern in aligned and unaligned condition

## V. RESULTS AND DISCUSSIONS

Table I give the comparison of the analytical and FE simulation results and are fairly matching. The leakage flux is not perfectly modeled in analytical method. This

133

accounts for the dispersion between these two results.

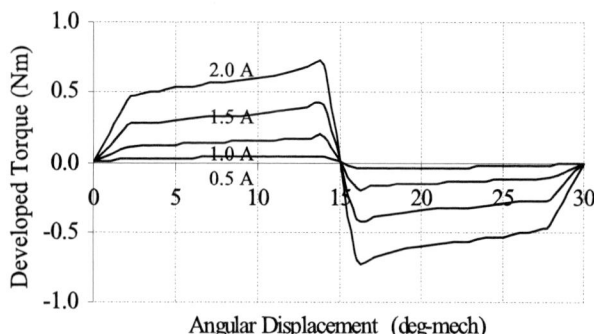

Fig. 6. Torque pattern of SRM (FE simulation results)

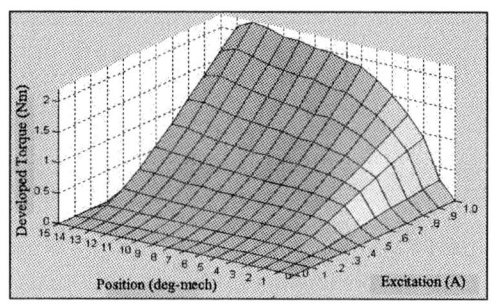

Fig. 7. *T I θ* Characteristics of 16/12 SRM

TABLE I
COMPARISON OF ANALYTICAL AND FE SIMULATION RESUKTS

| Parameters | Analytical Method | FE Method |
|---|---|---|
| La | 318 mH | 290 mH |
| Lu | 215 mH | 183 mH |
| Tav @ 2 A | 0.72 Nm | 0.68 Nm |

Based on the above analytical and FE method the machine is realized and tested. The block diagram of the drive scheme for driving SRM is shown in Fig 8. The continuous torque pattern is plotted in Fig. 9. The torque pattern and the average torque fairly match with that of the FE simulation results. The ripple content is in actual machine is comparatively higher than that of the FE results.

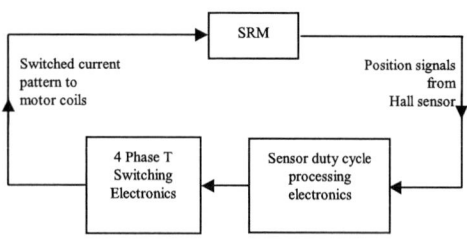

Fig. 8. Drive scheme for SRM

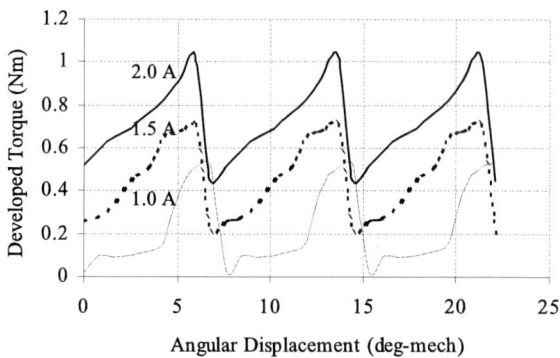

Fig. 9. Continuous torque pattern of SRM (Hardware test results)

## VI. CONCLUSION

A Simplified Design Methodology has been formulated for the design of the Switched Reluctance Motor. The Basic design has been started with Analytical method to fix up the major parameters and optimization of dimensions has been carried out using FE method. Developed torque has been computed by FE method and Analytical method using Flux tube technique. The machine is designed and fabricated based on this method and tested for its performance.

## VII. ACKNOWLEDGMENT

The authors gratefully thank the support and encouragement given by Dr.V.Krishnan, Associate Director and Mr. P.S.Veeraghavan, Director ISRO Inertial Systems Unit.

## VIII. REFERENCES

[1] Lawrenson, P.J., Stephenson, J.M., Blenkinsop, P.T., Corda, J. and Fulton, N.N., "Variable-speed Switched Reluctance Motors", *Proceedings of IEE*, Vol. 127, Pt B, No. 4, pp. 253-265, July 1980.

[2] R.Krishnan, and J.F.Lindsay, "Design Procedure for Switched Reluctance Motors", *IEEE Trans. Industry Applications*, Vol. 24, No. 3, pp. 456-461, May/June 1988.

[3] R.Arumugam, D.A.Lowther, R.Krishnan and J.F.Lindsay, "Magnetic Field Analysis of Switched Reluctance Motor using Two Dimensional Finite Element Model", *IEEE Trans. Magnetics,* vol. 21, No. 5, pp. 1883-1885, September 1985.

[4] M.N.Anwarm Iqbal Hussain and Arthur V Radun, "A Comprehensive Design Methodology of Switched Reluctance Motor", *IEEE Trans. Industry Applications*, Vol. 37, No. 6, pp. 1684-1692, Nov/Dec 2001.

[5] A.Radun, "Analytically Computing the flux linked by the Switched Reluctance Motor when stator and rotor poles overlap", *IEEE Trans. Magnetics*, Vol. MAG-36, No. 4, pt 2, pp. 1996-2003, July 2000.

**2006 IEEE International Conference on Power Electronic, Drives and Energy Systems**

# Computer Aided Design of Permanent Magnet Brushless DC Motor for Hybrid Electric Vehicle Application

Bhim Singh, *Senior Member, IEEE,* and Devendra Goyal

*Abstract—***This paper deals with a method of design of permanent magnet brushless dc machine (BDCM), primarily aimed for hybrid electric vehicle applications. The design variables such as airgap flux density, slot electric loading, stacking factor, coil fill factor, end turn coil factor, magnet fraction, slot fraction, flux density in the stator back iron, etc., are used in design process. The simplified design equations for trapezoidal back emf motor of rating 16 hp (12kW), 1100 rpm, 72 V radial flux surface mounted BDCM are obtained and used for CAD algorithm which gives the calculated performance of the motor. Since the motor is for high current and low voltage application so water cooling is used. Sequence Quadratic Programming Technique is used as a constraint optimization tool to achieve feasible and acceptable design. Finite element analysis are then carried out to obtain the electromagnetic characteristics of the motor for modification and verification of the obtained computer aided design.**

*Index Terms—***Computer-Aided Design (CAD), Design Equations, Finite Element (FE) Analysis, Permanent Magnet Brushless DC Motor (PMBLDCM), Sequential Quadratic Programming (SQP) Technique.**

## I. INTRODUCTION

DUE to their high efficiency and power density, high speed permanent magnet brushless machines are emerging as a key technology for applications as spindle drives, compressors, pumps, gas turbine micro-generators, and electrical hybrid vehicle traction systems. Permanent magnet synchronous machine is a good candidate for most of the future industrial applications due to their distinct advantages over the induction machine. They are usually more efficient because of the fact that field excitation losses are eliminated. In addition, copper losses in general are reduced in PM machines compared to conventional machines. In other words, due to lower losses, heating of the PM machines will be less, which can result either run the machine at low temperatures or to increase the shaft power so that the maximum allowable temperature has been reached.

The design of permanent magnet motors requires a series of iterative computations based on the selection of different configurations. These include the choice of geometrical dimensions, materials, parameter calculations, etc. The designer needs to consider certain dimensions and materials

and then calculate the performance of the designed motor. Then performance is compared with desired specification. If specification is not according to desired one, then designer has to modify the design to improve performance of the motor. Initially basic design equations are used to obtain the motor design and then use finite element analysis for performance evaluation. Performance equations for torque, power, inductance, magnetic operating point, etc. of the PM BLDC motor are available in the literature [1-5]. In this paper, development of a computer-aided design (CAD) program for achieving the design data and performance evaluation of radial-flux surface mounted PMBLDC motor is achieved and the details of the steps involved along with a flow chart is shown. Design optimization is carried out on nonlinear constraint objective function which is the minimization of the volume as an objective function subject to constraints on the possible values of independent variables using sequential quadratic programming technique [6]. Obtained design is analyzed using two-dimensional (2-D) FE technique which confirms the validity of the developed CAD program.

## II. BASIC DESIGN CONSIDERATIONS

### A. Air-Gap Flux Density Distribution

The shape and the magnetization pattern of the PMs dictate the air-gap flux density distribution. Furthermore, they have substantial influence on cogging torque, harmonic contents, and magnetic saturation so trapezoidal flux distribution is considered here.

### B. Magnet Material

Normally, a magnet is a integral part of the motor so mechanical and electrical properties have to be considered. Nd-Fe-B and samarium cobalt permanent magnets are preferred because of their high-energy product ($BH_{max}$) and retentivity ($B_r$). The selection depends on the cost and the thermal stability. SmCo offers the resistance to temperature effects and several grades are suitable upto $350^{\circ}C$ so samarium cobalt magnet with 26H grade is chosen here.

### C. Specific Loadings

Specific magnetic loading depends on the type of configuration and permanent magnet properties. Higher values of specific slot electric loading leads to increase in copper loss; but because of the reduction in the permanent magnet requirement, reduces the overall cost.

---

Bhim Singh and Devendra Goyal are with the Department of Electrical Engineering, Indian Institute Technology Delhi, Hauz Khas, New Delhi-110016, India (e-mail: bsingh@ee.iitd.ac.in and goyal_devin@yahoo.co.in).

0-7803-9771-1/06/$25.00 ©2006 IEEE

## D. Winding Configuration

Both amplitude and shape of the back EMF and the stator MMFs in these machines are determined by the winding arrangements and general machine geometry. These configurations in turn are dictated by optimum use of space and materials in the machine. Here double layer winding configuration is proposed which also minimizes the end turns length.

## E. Length of Airgap

Larger length of airgap results in reduced phase-inductance, armature reaction effects, and also the cogging torque, but requires bigger magnets, thereby increased cost so 1mm gap is selected for design of PMBLDCM.

## F. Slot Shape

It is usually desirable to have deep, narrow slots. Ratio of slot depth to width would be typically about 4. In applications where space is limited this factor must be optimized to obtain the highest possible force density. Chosen stator slots are semi closed and rectangular in shape.

## G. PM Pole Arc Width

The PM pole arc width can be arranged to reduce or eliminate some harmonics and to reduce the cogging torque. It is expected that the PM spans almost an integer number of slot pitches. Optimized value of magnetic fraction of order of 0.65 is considered here.

## III. FORMULATION OF DESIGN OPTIMIZATION PROBLEM

The heart of the motor design is a link between a fairly conventional design process and optimization procedure implemented via the design environment. The choice of appropriate optimization technique is rather important. As it is often the case that there is no single method which would be best in all circumstances and some judgment has to be applied when making the choice. The electrical and mechanical performances of the motor depend on its geometry and properties of magnetic materials. In the past, basic electromagnetic theories and industrial experience are used for preliminary motor design, and the finite element analysis contributes to the detailed modification and verification of the final shape of the motor. However, performance of a motor can be enhanced by the optimal design in terms of efficiency, weight, torque, frequency response, volume and so on. Optimizing the design for lowest volume is one of the important criteria for achieving optimized design. Because of the complicated nature of the objective function describing the electrical motor performance and constraints, the optimization problem is a nonlinear, nongradient constrained minimization problem, which can be formulated as follows:

$$P = \min F(\overline{X}) \tag{1}$$

Subject to the constraints:

$$g_i(\overline{X}) \leq 0, \quad i = 1, 2, \ldots k \tag{2}$$

where P is the objective function, describing the volume which is to be optimized, $\overline{x}$ are the design vectors, whose elements are the optimum design principle variables and $g_i$ are constraint functions, describing the limitations of the performance and the dimensions in the design.

## A. Design Variables

Design vector X which affects constraint and/or low volume objective function is listed below

(a) stator slot height (mm),      $10 \leq X_1 \leq 20$

(b) average air gap flux density (T),      $0.3 \leq X_2 \leq 0.7$

(c) ampere-conductor per meter (ac/m),      $60000 \leq X_3 \leq 80000$

(d) magnet fraction,      $0.6 \leq X_4 \leq 0.7$

(e) rotor outer diameter (mm),      $80 \leq X_5 \leq 100$

## B. Design Constraints

The following constraints are imposed to make the design practically feasible and acceptable:

(a) maximum stator tooth flux density (T)    < 2.0

(b) full load efficiency    > 85%

(c) conductor current density (A/mm$^2$)    < 20.00

(d) permeance coefficient    > 4 and < 16

(e) slot opening over air gap    > 1.9

(f) slot opening over wire diameter    > 2.0

(g) full load stator temperature rise ($^o$C)    < 75 $^o$C

(h) magnet operating point (T)    < 0.7 and > 0.25

## C. Optimization Technique

Design optimization of PMBLDC motor is a non-linear and multivariable constrained optimization problem. In constrained optimization, the general aim is to transform the problem into an easier sub-problem that can then be solved and used as the basis of an iterative process. A characteristic of a large class of early methods is the translation of the constrained problem to a basic unconstrained problem by using a penalty function for constraints that are near or beyond the constraint boundary. In this way the constrained problem is solved using a sequence of parameterized unconstrained optimizations, which in the limit (of the sequence) converge to the constrained problem. The general constrained optimization problem is to minimize a nonlinear function subject to nonlinear constraints. Two equivalent formulations of this problem are useful for describing algorithms. They are

$$\min\left\{ f(x), \quad c_i(x) \leq 0, i \in \tau, \quad c_i(x) = 0, i \in \varepsilon \right\}, \tag{3}$$

where $c_i(x)$ are the constraint functions and $\tau$ and $\varepsilon$ are index sets for inequality and equality constraints, respectively; and

$$\min\left\{ f(x) \quad c(x) = 0, \quad lb \leq x \leq ub \right\}, \tag{4}$$

where $l_b$ and $u_b$ are the lower and upper bound conditions for design variables. The main technique that is implemented for solving constrained optimization problem is sequential quadratic programming (SQP) method. Fundamental to the understanding of this algorithm is the Lagrangian function, for which formulation as given below

$$L(x,\lambda) = f(x) + \sum_{i \in \tau \cup \varepsilon} \lambda_i c_i(x) \tag{5}$$

The sequential quadratic programming algorithm is a generalization of Newton's method for unconstrained optimization in that it finds a step away from the current point by minimizing a quadratic model of the problem. The principal idea is the formulation of a QP sub-problem based on a quadratic approximation of the Lagrangian function. At each major iteration, an approximation is made of the Hessian of the Lagrangian function using a quasi-Newton updating method. This is then used to generate a QP sub-problem whose solution is used to form a search direction for a line search procedure.

## IV. DESIGN AND PERFORMANCE EQUATIONS

Surface-mounted structures are relatively simple to manufacture and assemble. The first step is to organize the design input data and basic equations in a logical way so that classical design approach can be used to achieve initial accepted design which can be further iterated and validated using FE analysis. In this process, the designer needs to assume certain parameters which are fixed throughout the design process and materials and then calculate the performance of the designed motor. All design inputs i.e. full load power, rated and no load speeds, DC link voltage, saturation flux density of steel, electrical loading, magnet remanence, magnet fraction etc. are formulated to get output design data consisting of all mechanical dimensions i.e. stator back iron width, tooth width, slot tip and wedge width, slot opening, shaft diameter, active axial length, end turn length, etc., electrical parameters i.e. number of conductors per slot, winding inductance and resistance, etc., thermal analysis include temperature rise on winding and total temperature rise of the motor and other performance parameters i.e. motor inertia, weight, magnet operating point, conductor current density, efficiency, etc., in terms of input design parameters. Some of the design parameters are non-linear function of input data. The cross section of the designed motor is shown in Fig.1.

### A. Derivation of Output Design Data and Performance Parameters

Performance parameters describe the machine's operation. Machine efficiency, weight, inertia are the important and interest. Therefore, output design data and performance parameters are calculated in terms of input parameters and considered in detail. Input design parameters, output design data and performance parameters are tabulated as shown in Tables I and II respectively. The necessary relations used in design are as follows:

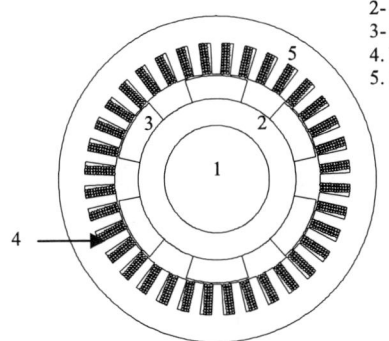

1- Shaft
2- Rotor back iron
3- Magnet pole
4- Winding
5- Stator back iron

Fig. 1. Cross-section of designed PMBLDC motor.

Rotor outer diameter ($D_{ro}$) is as:

$$D_{ro} = \sqrt{\frac{2 T_e}{\pi B_g K_w K_s (ac) L_i}} \qquad \text{(m)} \qquad (6)$$

Carter's coefficient ($K_c$) is as:

$$K_c = [1 - \alpha_s + \frac{4(g + \frac{l_m}{\mu_r})}{\pi \tau_s} \ln(1 + \frac{\pi \alpha_s \tau_s}{4(g + \frac{l_m}{\mu_r})})]^{-1} \qquad (7)$$

Stator slot area ($A_s$) is as:

$$A_s = (d_s - d_1 - d_2)[\frac{2\pi}{N_s}(\frac{D_{so}}{2} - W_{sb} - \frac{(d_s - d_1 - d_2)}{2}) - W_{tb}] \qquad \text{(m}^2\text{)} \qquad (8)$$

Stator slot tip and wedge width ($d_1$ and $d_2$) is as:

$$d_1 = d_2 = \frac{W_{tb} \alpha_{sd}}{2} \qquad \text{(m)} \qquad (9)$$

Number of turns per phase ($N_{ph}$) is as:

$$N_{ph} = \frac{V_{dc}}{4 N_p K_w \tau_p L_i B_g N_{nl}} \qquad (10)$$

Mean length of turn ( MLT) is as:

$$MLT = \frac{2 L_i + [\frac{\pi(D_{sb} + D_{si})}{N_p}] + 2 W_{tb}}{K_{et}} \qquad \text{(m)} \qquad (11)$$

Stator iron weight ($W_{st}$) is as:

$$W_{st} = L_i \rho_s K_{st} [\frac{\pi}{4}(D_{so}^2 - D_{si}^2) - N_s A_s] \qquad \text{(kg)} \qquad (12)$$

Wire weight ($W_{cu}$) is as

$$W_{cu} = 3 \times 0.0254 \times MLT \times \rho_{cu} N_{ph} N_c A_w a \qquad \text{(kg)} \qquad (13)$$

Magnet weight ($W_{mag}$) is as:

$$W_{mag} = \frac{\pi}{4} L_i \rho_{mag} \alpha_m [D_{ro} - (D_{ro}^2 - 2g)^2] \qquad \text{(kg)} \qquad (14)$$

Rotor weight ($W_{rt}$) is as:

$$W_{rt} = \frac{\pi}{4} L_i \rho_r [(D_{ro} - 2l_m)^2 - D_{sh}^2] \qquad \text{(kg)} \qquad (15)$$

Total active material weight ($W_T$) is as:

$$W_T = W_{mag} + W_{rt} + W_{st} + W_{cu} + W_{sh} \qquad \text{(kg)} \qquad (16)$$

Moment inertia of the motor ($J_T$) is as:

$$J_T = \frac{1}{8}\{W_{rt}[(D_{ro} - 2l_m)^2 - D_{sh}^2] + W_{mag}[D_{ro}^2 - (D_{ro} - 2l_m)^2] + W_{sh}D_{sh}^2\}$$
$$\text{(kg.m}^2\text{)} \qquad (17)$$

Full load efficiency at rated speed ($\eta$) is as:

$$\eta = \frac{P_o}{P_o + W_i \times W_{st} + 3 I_a^2 R_{ph} + P_w} \times 100\% \qquad (18)$$

where $P_w$ is friction and windage losses in Watts which are assumed to be maximum 3% of input power and all other dimensions are in MKS unit.

## V. COMPUTER AIDED DESIGN OF PMBLDC MOTOR

The optimization flow chart of developed CAD program for surface mounted PM motor is shown in Fig. 2 and developed on MATLAB platform. Objective function which is volume and constraints are represented in terms of variables. Input motor specification consists of nominal DC voltage, power rating, rated speed etc. Proper materials for stator and rotor back iron, shaft, permanent magnet, etc. are chosen which are grain oriented laminated steel, stainless steel and samarium cobalt, respectively and their properties i.e. specific

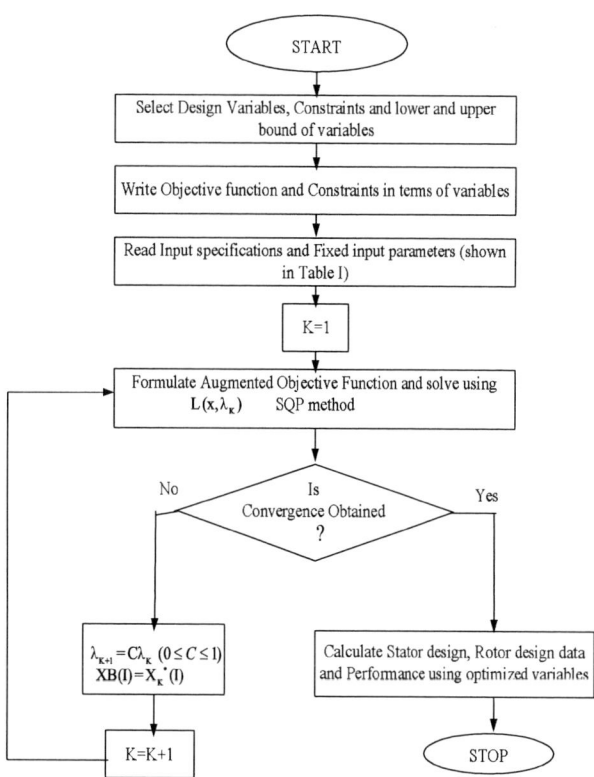

Fig. 2. Optimization flow chart for CAD of PMBLDC Motor.

### TABLE I
### DESIGN INPUT DATA

| Parameter | Symbol | Normal Design | Low Volume Optimized |
|---|---|---|---|
| **1. Ratings and Configuration** | | | |
| Full Load Power (kW) | $P_o$ | 12 | 12 |
| Full load speed (rps) | $N_{fl}$ | 30 | 30 |
| No load speed (rps) | $N_{nl}$ | 18.33 | 18.33 |
| DC link voltage (Volts) | $V_{dc}$ | 72 | 72 |
| Stator winding configuration | Star/Delta | | |
| No. of phase | 3 | | |
| **2. Other data:** | | | |
| No. of stator slots | $N_s$ | 36 | 36 |
| No. of slots per pole per phase | $N_{spp}$ | 2 | 2 |
| No. of poles | $N_p$ | 6 | 6 |
| Winding arrangement | Double layer | | |
| No. of parallel circuit/phase | a | 6 | 6 |
| Saturation flux density of steel (T) | $B_m$ | 2.0 | 2.0 |
| Slot fill factor | $K_f$ | 0.41 | 0.41 |
| Stack factor | $K_{st}$ | 0.7 | 0.7 |
| End turn coil factor | $K_{et}$ | 0.95 | 0.95 |
| Slot insulation thickness (mm) | $t_{si}$ | 0.07 | 0.07 |
| Slot fraction | $\alpha_s$ | 0.5 | 0.5 |
| Stator slot wedge angle (Degree) | $\alpha$ | 40 | 40 |
| Magnet remanence (T) | $B_r$ | 1.09 | 1.09 |
| Magnet recoil permeability | $\mu_r$ | 1.05 | 1.05 |
| Air gap (mm) | g | 1 | 1 |
| Stator lamination thickness (mm) | $t_s$ | 0.2 | 0.2 |
| Rotor steel thickness (mm) | $t_{rs}$ | 0.0022 | 0.0022 |
| Rotor steel density (Kg/m$^3$) | $\rho_r$ | 8150 | 8150 |
| Stator steel density (Kg/m$^3$) | $\rho_s$ | 8150 | 8150 |
| Magnet material density (Kg/m$^3$) | $\rho_m$ | 8850 | 8850 |
| Copper density (Kg/m$^3$) | $\rho_{cu}$ | 8900 | 8900 |
| Shaft material density (Kg/m$^3$) | $\rho_{sh}$ | 7850 | 7850 |
| Iron loss/kg (W/kg) | $W_i$ | 5 | 5 |
| Specific heat coefficient of copper (Joule/kg) | $S_c$ | 0.385 | 0.385 |
| Specific heat co-efficient of stator steel (Joule/kg) | $S_{st}$ | 0.1 | 0.1 |
| Specific heat co efficient of rotor steel (Joule/kg) | $S_r$ | 0.1 | 0.1 |
| Time for temperature rise | t | 90 | 90 |

heat coefficient, material density, etc., are provided as input in design. Stacking factor, slot fill factor, magnet fraction, slot fraction, end turn coil factor, saturation flux density in stator back iron, etc., are considered as input fixed parameters. The phase current is decided by the power requirement and induced EMF decides the number of conductors per slot. Magnet operating point at load is decided by armature reaction and electrical time constant is dependent on winding inductance. Design variables have to be varied within lower and upper bound. Now sequential quadratic programming technique is used to achieve optimum value of variables for which volume function is minimum. Using these optimum values of variables motor design output data and performance parameters are calculated.

The PMBLDC motor is designed using above mentioned optimized CAD program. Important input parameters and optimal output design data and performance parameters are shown in Table I and II respectively.

## VI. VALIDATION OF THE DESIGN BY FINITE ELEMENT MAGNETIC ANALYSIS

It is difficult to predict the precise performance of the designed motor using magnetic circuit analysis due to non linear characteristics of magnetic material which causes saturation during overload when vehicle is accelerating. The finite element analysis on the preliminary design prototype become necessary to provide detailed information on the magnetic flux and torque distribution, steady-state temperature distribution and modal dynamics. As opposed to the magnetic

circuit analysis, the 2-D finite element tool (MAGNET) numerically calculates the magnetic field of the 3D motor configuration and accuracy of the developed CAD program is validated. Flux density distribution for designed motor at full load is shown in Fig. 3. Fig. 4 shows the magnetic flux line distribution in the motor at full load where flux per pole in air gap is 0.005 Wb which is approximately equal to the theoretical calculated value of flux. Difference in designed motor parameters through CAD and FE magnetic analysis is shown in Table III.

## VII. RESULTS AND DISCUSSION

The result having all output design data and performance parameters pertaining to the low volume optimized design of 16 hp (12kW), 1100 rpm and 72 V surface mounted permanent magnet brushless motor is shown in Table II.

138

## TABLE II
### DESIGN OUTPUT DATA AND PERFORMANCE PARAMETERS

| Design output data/ Performance parameters | Symbol | Normal Design | Low Volume Optimized |
|---|---|---|---|
| 1. Objective function | | | |
| **Volume (mm³)** | $V_T$ | **291.39 x10⁴** | **244.164x10⁴** |
| 2. Design variables | | | |
| Slot depth (mm) | $d_s$ | 19.31 | 14 |
| Average air gap flux density (T) | $B_g$ | 0.533 | 0.525 |
| Electric loading (A.cd/m) | ac | 75000 | 72000 |
| Magnet fraction | $\alpha_m$ | 0.67 | 0.65 |
| 3. Design constraints | | | |
| Stator tooth flux density (T) | $B_{st}$ | 1.78 | 1.47 |
| Full load efficiency | $\eta$ | 95.06 | 95.11 |
| Conductor current density (A/mm²) | $J_s$ | 11.83 | 15.3 |
| Permeance coefficient | PC | 12.46 | 12.69 |
| Slot opening over air gap | $W_o/g$ | 3.05 | 4.1 |
| Slot opening over wire diameter | $W_o/D_w$ | 1.9 | 2.56 |
| Temperature rise at winding (°C) | $\Delta T_w$ | 44.3 | 70 |
| Magnet operating point (T) | $B_{mg}$ | 0.68 | 0.667 |
| 4. Other data: | | | |
| Stator inner diameter (mm) | $D_{si}$ | 93.78 | 96 |
| Stator slot tip depth (mm) | $d_1$ | 0.4 | 0.5 |
| Stator slot wedge depth (mm) | $d_2$ | 0.4 | 0.5 |
| Back iron width at inner radius (mm) | $W_{bi}$ | 13.8 | 12 |
| Rotor outer diameter (mm) | $D_{ro}$ | 91.78 | 94 |
| Slot area (mm²) | $A_s$ | 98.4 | 76.2 |
| Stator axial length (mm) | $L_i$ | 145 | 142 |
| Tooth width at inner radius(mm) | $W_{tb}$ | 4.6 | 3.9 |
| Magnet spacer thickness (mm) | $\zeta_{ms}$ | 16.20 | 17.7 |
| Stator outer diameter (mm) | $D_{so}$ | 160 | 148 |
| No. of series coil/phase | $N_c$ | 1 | 1 |
| No. of conductors/slot | $Z_s$ | 20 | 20 |
| No. of turns per phase | $N_{ph}$ | 20 | 20 |
| Winding distribution factor | $K_d$ | 0.966 | 0.966 |
| Skew factor | $K_s$ | 0.917 | 0.917 |
| Winding factor | $K_w$ | 0.886 | 0.886 |
| End turns length (from AUTOCAD) (mm) | $L_{te}$ | 0.015 | 0.015 |
| Magnet radial thickness (mm) | $l_m$ | 10 | 10 |
| Armature reaction flux (T) | $B_a$ | 0.029 | 0.028 |
| Shaft diameter (mm) | $D_{sh}$ | 44.1 | 50.4 |
| Total shaft length (mm) | $L_{sh}$ | 212 | 206 |
| Carter's coefficient | $K_c$ | 1.06 | 1.07 |
| Effective air gap length (mm) | g' | 1.06 | 1.07 |
| Area of magnet (mm²) | $A_m$ | 4772.7 | 4673 |
| Flux per pole (Wb) | $\Phi_p$ | 0.0047 | 0.00467 |
| Skin depth of conductor | X | 0.0089 | 0.0089 |
| Mean length per turn (mm) | MLT | 440.4 | 429.26 |
| Phase resistance (mΩ) | $R_{ph}$ | 12.9 | 13.4 |
| Phase inductance (mH) | $L_{ph}$ | 0.105 | 0.0978 |
| Stator iron volume (mm³) | $V_{st}$ | 980.35 | 708.38 |
| Volume of copper wire (mm³) | $V_{cu}$ | 321.10 | 241.61 |
| Weight of stator (kg) | $W_{st}$ | 7.99 | 5.77 |
| Weight of magnet (kg) | $W_{mag}$ | 2.035 | 2.00 |
| Weight of rotor (kg) | $W_{rt}$ | 2.97 | 2.70 |
| Weight of wire (kg) | $W_{wr}$ | 2.85 | 2.15 |
| Weight of shaft (kg) | $W_{sh}$ | 2.83 | 3.73 |
| Total active material weight (kg) | $W_T$ | 18.68 | 16.35 |
| Net temperature rise of the motor (°C) | $\Delta T_m$ | 20.37 | 29.68 |
| Electrical time constant (msec) | $\zeta$ | 8.1 | 7.29 |
| Total inertia of motor (Kg.m²) | $J_t$ | 0.0027 | 0.0031 |
| Copper losses at rated current (W) | $P_{cu}$ | 528.77 | 550.56 |
| Core losses (W) | $P_i$ | 42.95 | 28.86 |
| Friction and windage losses at rated seed (W) | $P_w$ | 37.92 | 37.92 |

## TABLE III
### COMPARISON OF CAD AND FE-BASED RESULTS OF THE DESIGNED MOTOR

| Performance Parameter | Computed by CAD | Computed by FE analysis |
|---|---|---|
| Developed average torque (N.m.) | 104.22 | 102.10 |
| Inductance (mH) | 0.0978 | 0.0982 |
| Average air gap flux density (T) | 0.525 | 0.532 |
| Flux density in stator teeth (T) | 1.47 | 1.50 |

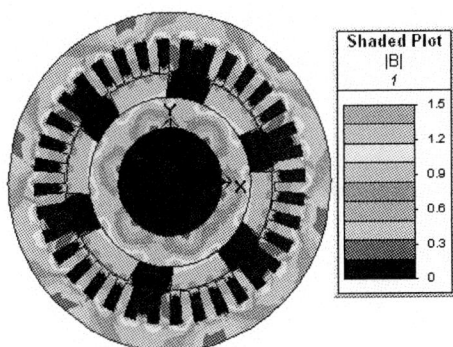

Fig. 3. Magnetic flux density distribution in designed PMBLDC motor at full load.

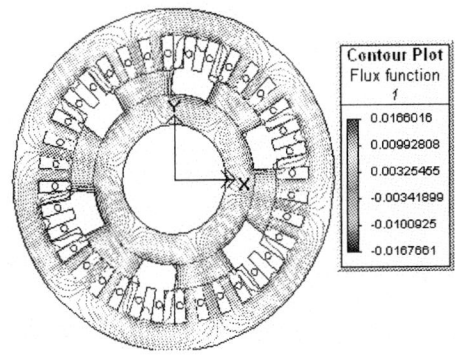

Fig. 4. Magnetic flux distribution in designed PMBLDC motor at full load.

Sequential quadratic programming algorithm has been found to be suitable to achieve feasible and optimized design which minimize the volume objective function to 244.164x10⁴ mm³ having a major constraint for hybrid vehicle application and also leads to achieve low core losses which helps to increase efficiency and low weight of the motor. Motor inertia and electrical time constant of the winding is found to be good which are considered as critical parameters while designing controller of the motor. Advantage of this method is that much time is saved in developing the optimized program. Moreover, the motor design engineer does not have to be an expert in optimization theory in order to obtain a superior design with a very short time.

Finite element magnetic analysis is performed on optimal design motor for refining the analytical design and leads to validate the design procedure. Developed average torque, inductance of the stator winding and average air gap flux density are the few important parameters decides the performance of the designed motor which are verified on MAGNET toolbox and gives acceptable change between CAD and FE computed results.

# VIII. CONCLUSIONS

A procedure for the design of a surface-mounted PM brushless DC motor has been developed in tightly combined with the thermal and magnetic analysis. A steady-state analysis of a brushless motor supports the design procedure. It gives the expressions of the main electrical, magnetical, mechanical and thermal quantities as a function of the machine dimensions and working conditions. Thus the motor design is reached by solving a system of non-linear equations, derived from the aforementioned working analysis, where the motor performance, the material stress limits, and other constraints are imposed. Taking advantage of the completely analytical design procedure, an optimization procedure has been implemented for achieving best design. The above CAD procedure has been specifically developed to support design of brushless permanent magnet DC motor under above constraint and finally FE magnetic analysis has validated the motor design.

# IX. APPENDIX

## A. Expression for Objective Function

Volume as the objective function has been chosen here which can be expressed in terms of design variables as given below

$$V_T = \frac{\pi}{4}[x_5 + 2x_1 + 2g + \frac{\pi x_2(1+x_4)(x_5+2g)}{(4B_m K_{ST} P)}]^2 \times [\frac{2T_e/K_w K_S \pi}{x_2 \times x_3 \times x_5^2}] \quad (19a)$$

## B. Expression for Constraint Functions

The considered few constraint functions are formulated as follows:

Tooth flux density is maximum at near the stator surface and decreases up to the depth of tooth so flux density at $1/3^{rd}$ of the distance may be assumed for whole tooth which is

$$2B_m K_{ST} - 2 - 2x_4 \leq 0 \quad (19b)$$

Slot opening and air gap relation is represented as

$$1.9g - \frac{\pi}{N_s}(x_5+2g) - \frac{\pi}{2N_s B_m K_{ST}}(x_2)(1+x_4)(x_5+2g) - 0.001 \leq 0 \quad (19c)$$

Permeance coefficient has a range between 4 to 16 and represented as

$$4x_4 - \frac{l_m}{2g} - \frac{l_m}{2g}x_4 \leq 0 \quad (19d)$$

$$\frac{l_m}{2g} + \frac{l_m}{2g}x_4 - 16x_4 \leq 0 \quad (19e)$$

Magnetic operating point checking limits the cause of demagnetization and has a limit between 0.25 to 0.7 T and represented as

$$0.25[(\frac{l_m}{2g})(1+x_4+\mu_{rec}x_4)] - B_r(\frac{l_m}{2g})(1+x_4) \leq 0 \quad (19f)$$

$$B_r(\frac{l_m}{2g})(1+x_4) - 0.7[(\frac{l_m}{2g})(1+x_4+\mu_{rec}x_4)] \leq 0 \quad (19g)$$

Slot opening should be at least twice of wire diameter and represented as

$$2 \times \sqrt{(\frac{4Kf}{\pi})[\frac{(x_1-0.001)(\frac{\pi}{N_S}(x_5+x_1+2g+0.001) - \frac{\pi}{2N_S B_m K_{ST}}(x_2)(1+x_4)(x_5+2g))}{[\frac{x_3 x_5^2}{(x_5+2g)}] \bowtie [\frac{V}{8T_e N_{noload}}]}]}$$
$$- \frac{\pi}{N_S}(x_5+2g) - \frac{\pi}{2N_S B_m K_{ST}}(x_2)(1+x_4)(x_5+2g) - 0.001 \leq 0 \quad (19h)$$

Safe limit of the current density at full load for water cooling provided design should be less than 20 A/mm$^2$ and represented as

$$20 - (\frac{I_{rms}}{a \times 10^6})[\frac{(x_1-0.001)(\frac{\pi}{N_S}(x_5+x_1+2g+0.001) - \frac{\pi}{2N_S B_m K_{ST}}(x_2)(1+x_4)(x_5+2g))}{[\frac{x_3 x_5^2}{(x_5+2g)}] \bowtie [\frac{V}{8T_e N_{noload}}]}] \leq 0 \quad (19i)$$

# X. REFERENCES

[1] Duane C. Hanselman, *Brushless Permanent-Magnet Motor Design*, New York: McGraw-Hill Inc., 1994.

[2] T. J. E. Miller, *Brushless Permanent-Magnet and* Reluctance *Motor Drives*, Clarendon Press, Oxford, 1989.

[3] A. H. Wijenayake, J. M. Bailey, P. J. McCleer, "Design Optimization of an Axial Gap Permanent Magnet Brushless DC Motor for Electric Vehicle Applications", in *Conf. Rec. 1995 13th IAS Ann. Mtg. IEEE Industry Applications Soc.*, vol. 1, pp. 685-692.

[4] Gieras and Wing, *Permanent Magnet Motor Technology*, Marcel Dakker Inc., New York, 1997.

[5] J. R. Handershot Jr. and T. J. E. Miller, *Design of Permanent Magnet Motors*, Oxford, U.K.: Oxford Univ. Press, 1994.

[6] R. Fletcher, *Practical Methods of Optimization*, John Wiley and Sons, 1987.

# XI. BIOGRAPHIES

**Bhim Singh** (SM'99) was born in Rahamapur, U. P., India in 1956. He received B. E. (Electrical) degree from University of Roorkee, India in 1977 and M. Tech. and Ph. D. degrees from Indian Institute of technology (IIT), New Delhi, in 1979 and 1983, respectively. In 1983, he joined as a Lecturer and in 1988 became a Reader in the Department of Electrical Engineering, University of Roorkee. In December1990, he joined as an Assistant Professor, became an Associate Professor in 1994 and Professor in 1997 at the Department of Electrical Engineering, IIT Delhi. His field of interest includes power electronics, electrical machines and drives, active filters, static VAR compensator, analysis and digital control of electrical machines. Prof. Singh is a Fellow of Indian National Academy of Engineering (INAE), Institution of Engineers (India) (IE (I)) and Institution of Electronics and Telecommunication Engineers (IETE), a Life Member of Indian Society for Technical Education (ISTE), System Society of India (SSI) and National Institution of Quality and Reliability (NIQR) and Senior Member of IEEE (Institute of Electrical and Electronics Engineers).

**Devendra Goyal** was born in Bharatpur, Rajasthan, India in 1982. He received B. Tech. (Electrical) degree from National Institute of Technology, Kurukshetra, India in 2004. In 2005, he worked as a GET in Automobile Industry, Subros Limited, Noida. Presently he is a pursuing M.Tech degree in the Department of Electrical Engineering, IIT Delhi. His field of interest includes power electronics, electrical machines and drives.

**2006 IEEE International Conference on Power Electronic, Drives and Energy Systems**

# Design and Analysis of a 3 kVA, 28 V Permanent Magnet Brushless Alternator for Light Combat Aircraft

Bhim Singh, *Senior Member, IEEE,* and Jally Ravi

*Abstract*--**Permanent magnet (PM) brushless machines have been increasingly used in aircrafts and automobiles due to their light weight, small size, high efficiency, high reliability, variable speed operation and good dynamic performance. The Permanent Magnet Brush less Alternator has a permanent-magnet rotor, and the stator windings are wound such that the induced electromotive force (EMF) is trapezoidal. This paper presents the optimized design of a 3kVA, 28V permanent magnet brushless (PMBL) alternator for light combat aircraft application (LCA). The proposed Alternator has two poles made of NdFeB and 12 stator slots. An analytical algorithm is developed for the design of PMBL alternator. The finite element analysis is carried in MAGNET 2D FEA-package for refining the design and performance evaluation of a three-phase permanent magnet brushless alternator.**

*Index Terms*--**PM Alternator, Light Combat Aircraft, Electrical Power Generating System (EPGS).**

*List of symbol*

| | |
|---|---|
| p | number of poles |
| n | machine speed; |
| $D_i$ | inner stator diameter; |
| $\gamma$ | angle between two conscecutive slots; |
| $\alpha$ | short pitching angle; |
| $\sigma$ | skewing angle; |
| $B_{sat}$ | saturation flux density; |
| $B_r$ | remanent flux density; |
| $B_g$ | air gap flux density; |
| $B_m$ | magnet flux density; |
| $\mu_{rec}$ | recoil permeability of the magnet; |
| $C_\Phi$ | flux concentration factor; |
| $\tau_p$ | rotor pole pitch; |
| $\tau_s$ | stator slot pitch; |
| $\omega_s$ | slot width; |
| $t_t$ | tooth width; |
| $k_{st}$ | lamination stacking factor; |
| $k_{sf}$ | slot filling factor; |
| $N_S$ | number of slots; |
| $A_{slot}$ | slot area; |
| $\alpha_m$ | pole pitch coverage co-efficient; |

Bhim Singh and Jally Ravi are with the Department of Electrical Engineering, Indian Institute Technology Delhi, Hauz Khas, New Delhi-□□00□6, India. (E-mail: bhimsinghr@gmail.com and jallyravi@gmail.com).

## I.INTRODUCTION

PERMANENT magnet brushless machines (PMBL) present certain advantages over other electrical motors such as dc motors, induction motors, synchronous motors and switched reluctance machines. Due to absence of the field current and field winding, permanent magnet generators exhibit high efficiency in operation, simple and robust structure in construction and high power to weight ratio. The attractiveness of the permanent magnet generators is further enhanced by the availability of high-energy permanent magnet materials like NdFeB. As a result, PMBL machines have been increasingly used in small motor drives for automobiles, aircrafts particularly in high-end vehicle models. Automotive applications require highly competitive costs, low acoustic noise and high efficiency. The stringent requirements and unique operating conditions demand effective methodologies for the design of PMBL machines.

The Permanent-magnet machines allow a great deal of flexibility in their geometry. The permanent magnets of radial-flux machines are radially oriented. Radial-flux permanent-magnet machines can be divided mainly into two types, surface-magnet and buried-magnet machines. The simple way of constructing the rotor with high number of poles is by gluing the permanent magnets on the rotor surface of the machine. The Permanent Magnet Brush less Alternator has a permanent-magnet rotor, and the stator windings are wound such that the induced electromotive force (EMF) is trapezoidal.

Many generators have been proposed in the literature as radial-flux generators. In [5], have designed and constructed two small multi-pole radial-flux permanent-magnet test machines for use as a directly coupled generator in wind turbines. In [6], has investigated how a direct driven wind turbine generator should be designed and how small and efficient such a generator will be. In [7], has described the arrangement of multi-pole radial-flux permanent-magnet synchronous machine. In [8], have designed gearless radial-flux buried and surface mounted permanent magnet wind energy converters. In [9], have presented the design of outer-rotor (the positions of the rotor and stator are exchanged) radial-flux permanent-magnet multi-pole low-speed directly coupled wind power converter for standalone applications. In [□0], has proposed a dual rotor, radial-flux, toroidally wound permanent-magnet machines to substantially improve machine torque density and efficiency. In [4], has designed and

0-7803-9771-1/06/$25.00 ©2006 IEEE

analyzed a 42V permanent magnet generator for automotive application [4].

The electrical power generating system (EPGS) for the light combat aircraft consists of one 30 kVA integrated drive generator (IDG) and its generator control unit (GCU). In the absence of the main generator a 3 kVA alternator and its GCU are required to take of the critical loads. There is a requirement to indigenously design, development and testing of a 3 kVA brushless alternator for operating the electrical systems of LCA. The PMBL alternator with control unit is required to produce 28 Vdc, 2.5 kW power. It has to operate efficiently over a high-speed range of 6200 rpm to ☐2500 rpm and provide 2.5 kW at a constant output voltage of 28 Vdc. When the main generator 30 KVA fails or Transformer Rectifier unit fails, the 3KVA brushless alternator shall operate DC emergency bus for power supply to: Power management relays, System management relays, Radio altimeters, ECS controller, FCS channel, Artificial horizon, Head up display.

The main objective of this work is the design of a minimum volume and good efficient three-phase PMBL alternator supplying 2.5 KW and 28 V to the dc loads. The speed of the alternator varies in the range of 6200 – ☐2500 rpm. The main requirement is that the PMBL alternator should provide the 2.5 kW power and 28 V in the given speed range. The converter topology of Generator control unit consists of an uncontrolled three-phase bridge rectifier followed by a buck converter shown in fig.☐ This paper presents the design algorithm of the three phase PMBL alternator. The Finite element analysis has done using Magnet 2D FEA-package. The iron saturation is validated at no load and full load. The input data and design results are given in the tables I and II.

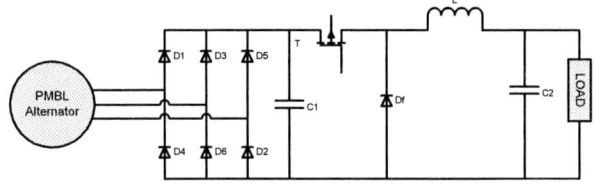

Fig. ☐ Un-controlled diode bridge rectifier with buck converter.

## II. ALTERNATOR DESIGN

### A. Topology of alternator

The surface mounted rotor structure is selected for the proposed machine as shown in Fig. 2. The main advantage of this topology is that all of the magnetic flux produced by the magnets links the stator, and therefore, takes part in energy conversion. The type of the winding selected is double layer distributed winding; the air gap is designed such that the output voltage of the machine is trapezoidal wave. The advantage of designing the machine for trapezoidal output voltage wave is the weight and size of the LC filter reduces. The choice of the number of poles depends upon many factors, some of which are as follows: Magnetic material and grade, Mechanical assembly of the rotor and magnets, Speed of rotation, Inertia requirements. The number of poles should

be inversely proportional to the maximum speed of rotation. The reason, of course, is to limit the commutation frequency to avoid excessive switching losses in the transistors and iron losses in the stator. For very high speeds two or four pole motors are preferred. Because of very high speed range specified (6200 to ☐2500 rpm), number of poles selected as 2 for the proposed permanent magnet blush less alternator.

### B. Methodology of the design algorithm

The objective of this design is to maximize the efficiency and minimize the volume of the alternator. The design starts with a set of input data that includes apparent power, phase voltage, speed range, stacking factor, and slot filling factor, remanent flux density of Magnet and saturation flux density. Based the practical limitations the number of poles p and the no. of slots/pole/phase m are predefined.

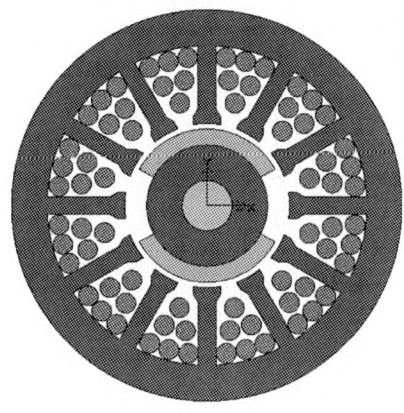

Fig. 2. Structure of the designed machine.

From the expression of the air gap power (☐), the minimum volume of the machine is obtained by iterating the diameter of the rotor $D_r$ and slot depth $d_s$ between its minimum and maximum limits specified. The minimum volume of the rotor is obtained when the product of the electric and magnetic

$$P_{gap} = 2k_{sh}k_d k_s \omega_S B_g A\pi R_{or}^2 L_i \qquad (☐)$$

where $k_{sh}$, $k_d$, and $k_s$ and shorting, distribution and skewing factors can be computed from the following equations (2), (3)

$$k_{sh} = \frac{sin\left(\frac{m\gamma}{2}\right)}{m.sin\left(\frac{\gamma}{2}\right)} \qquad (2)$$

$$k_d = cos\left(\frac{\alpha}{2}\right) \qquad (3)$$

$$k_s = \frac{sin\left(\frac{\alpha}{2}\right)}{\left(\frac{\alpha}{2}\right)} \qquad (4)$$

The next task is to determine the pole pitch coverage coefficient of the magnet $\alpha_m$. According to [3] the $\alpha_m$ is given as (5).

$$\alpha_m = \frac{(n+0.14)}{\left(N_{spp}N_{ph}\right)} \tag{5}$$

where n is any integer satisfying $\alpha_m < \square$

The areas of the magnet and air gap per pole are found by using (6) and (7).

$$A_m = \frac{(\alpha_m \pi D_{ro}L)}{p} \tag{6}$$

$$A_g = \left[\frac{\alpha_m \pi (D_{ro}+g)}{p}+2g\right](L+2g) \tag{7}$$

The length of the air gap is usually determined by mechanical constraints. The flux concentration factor and length of the magnet are computed by using (8) and (9).

$$C_\phi = \frac{A_m}{A_g} \tag{8}$$

$$l_m = g.C_\phi.PC \tag{9}$$

The permeance co-efficient represents the slope of the air gap line in the second quadrant of the B-H plane. This is the measure of the capacity of the magnet to withstand demagnetization. According to [$\square$], PC is chosen as 5 in order to minimize the magnet cost.

The values of air gap and magnet flux densities are computed from ($\square$0) and ($\square$).

$$B_g = \frac{C_\phi.B_r}{1+\frac{C_\phi k_c g \mu_{rec}(1+p_{r1})}{l_m}} \tag{$\square$0}$$

$$B_m = \frac{\left(1+\frac{p_{r1}C_\phi k_c g \mu_{rec}}{l_m}\right)B_r}{1+\frac{C_\phi k_c g \mu_{rec}(1+p_{r1})}{l_m}} \tag{$\square$}$$

According to [2], Carter's factor, $k_c$ can found by using ($\square$2).

$$k_c = \left[\square - \frac{w_s^2}{5\tau_s(g_c+w_s)}\right]^{-\square} \tag{$\square$2}$$

The value of the n is iterated between minimum and maximum values and the value of $\alpha_m$ is computed by using (5) such that the satisfied air gap and magnet flux densities are obtained.

The shape of the slots chosen is semi open type. Once the values of $\alpha_m$ and $B_g$ are set, the dimensions of slots are computed according [2]. The dimensions of Stator back iron and rotor back iron are computed by using

$$w_{st} = \frac{\pi D_i \alpha_m B_g}{2k_{st}pB_{sat}} \tag{$\square$3}$$

$$w_{rt} = \frac{\pi(D_{or}-2l_m)\alpha_m B_g}{2k_{st}pB_{sat}} \tag{$\square$4}$$

where

$$D_i = D_{or} + 2\left(g+d_s\right) \tag{$\square$5}$$

The outer diameter of the machine OD is

$$OD = D_{or} + 2\left(g+d_s+w_{st}\right) \tag{$\square$6}$$

The ampere loading, current density, diameter of the

conductor and total number of the conductors are computed from ($\square$7), ($\square$8), ($\square$9) and (20) respectively.

$$A = \frac{Z.I_{ph}}{a.\pi.D_{si}} \tag{$\square$7}$$

$$J = \frac{I_{ph}.Z}{a.A_{slot}.K_{sf}.N_S} \tag{$\square$8}$$

$$d_c = \sqrt{\frac{4A_{slot}.K_{sf}.N_S}{Z.\pi}} \tag{$\square$9}$$

$$Z = 2N_{ph}T_{ph} \tag{20}$$

### C. Calculation of efficiency

The main Losses in permanent magnet machines include copper losses in the stator windings, iron losses in the stator. Of these, the core losses is the most difficult to compute accurately. The magnet and rotor back iron experience little variation in flux and therefore the iron losses in them can be neglected. According to [4], the iron losses in the stator and copper losses in the stator winding are computed from (2$\square$) and (22)

$$P_{iron} = \rho_{iron}.k_{st}.\left(N_S V_t \Gamma_t + V_y.\Gamma_y\right) \tag{2$\square$}$$

$$P_{copper} = \frac{\rho_{cu} Z I_{rms}^2 \left(L+\left(\alpha_{oc}p(D_{si}+2.d_S)\right)\right)}{2.a.A_c} \tag{22}$$

And the friction and windage losses have taken according to [4]. The output power and the efficiency of the alternator can be found from (25) and (26).

$$P_{out} = S.pf \tag{25}$$

$$\eta\% = \frac{P_{out} \times 100}{P_{gap}+P_{copper}+P_{iron}+P_f+P_w} \tag{26}$$

### D. Armature Reaction

The proposed design considered the effect of the armature reaction. Current flowing in the stator tends to distort the magnetic field set up by the permanent magnet. The flux density due to armature reaction, according to [$\square$], is found by using.

$$B_a = \frac{(T_{ph}I_{ph}\mu_o \mu_{rec})}{2(l_m+\mu_{rec}g)} \tag{27}$$

In permanent magnet machines where the magnets are of surface type, the effect of armature reaction is weak. Because the low recoil permeability of the magnets and their long relative length make the stator field as increased air gap. As long as the stator teeth and shoes are not highly saturated due to the permanent magnets acting alone, armature reaction is not a problem. The greatest concern with respect to armature reaction normally occurs under faulty conditions, where machine currents exceeds their normal range, the armature reaction field can become large enough to demagnetize the rotor magnets[3].

*E. Input Data and design Results*

The input data and design results are given in the Tables I and II.

TABLE I
DESIGN INPUT DATA

| Apparent Power, S | 3 kVA |
|---|---|
| phase voltage, $V_{rms}$ | 25 V |
| phase current,, $I_{rms}$ | 40 A |
| power factor, pf | 0.85 |
| Speed, n | 6200 rpm |
| $B_{sat}$ (cobalt iron alloy) | 2.2 T |
| Remnant flux density $B_r$ | □□7 T |
| stacking factor, $k_{sk}$ | 0.94 |
| slot filling factor, $k_{sf}$ | 0.6 |
| short pitching, | □slot |

## III. RESULTS OF FINITE ELEMENT ANALYSIS

The magnets are NdFeB with a remanent flux density $B_r$ of about □□7T, coercivity $H_c$ of about –700kA/m and maximum energy product $(BH)_{max}$ of about 370kJ/m$^3$ at room temperature. The cobalt iron alloy is chosen as the stator and rotor core material with a saturation flux density $B_{sat}$ of about 2.2T. The finite element analysis has been carried out by using Magnet 2D FEA-package. The iron saturation is verified for alternator for both no load and full load. The distribution of flux lines in the machine for no-load condition at two different rotor positions are shown in Fig.2. The distribution of the magnitude of the flux density for both no-load and full load conditions is shown in Fig.3 and 4. The maximum values of air gap flux densities at no load and full load are of about 0.55T and 0.68T respectively. The maximum values of iron flux densities at no load and full load are of about □45T and □96T respectively, while the saturation value of cobalt iron had been set at 2.2T. The value of the flux density was verified in various points of the teeth, stator yoke, and rotor core of the machine shown in Table III. The results obtained shows that the iron sections of the machine are not saturated.

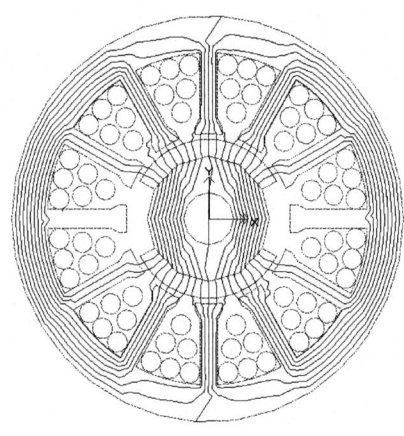

(a)

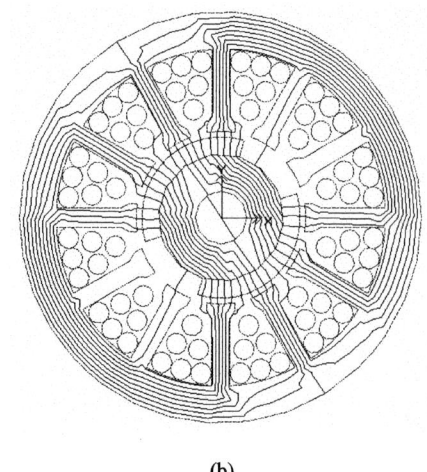

(b)

Fig. 3. Flux distribution at no-load (open-circuit condition) at two different positions of the rotor.

TABLE II
DESIGN RESULTS

| Number of poles, p | 2 |
|---|---|
| number of slots per pole per phase, q | 2 |
| Rotor outer diameter $D_r$ | 35.3 mm |
| Motor diameter, OD | □0 mm |
| Length of the stator core, L | □03 mm |
| Air gap flux density, $B_g$ | 0.7□3 T |
| Magnet flux density, $B_m$ | 0.828 T |
| Current density, J | 4.224 A/mm$^2$ |
| Ampere loading, A | 44376.5 A/m |
| Stator yoke thickness, $w_{st}$ | 8.□8mm |
| Number of coils per phase | 4 |
| Number of turns per coil | 8 |
| Number of parallel paths | 2 |
| Pole coverage coefficient, $\alpha_m$ | 0.69 |
| Slot depth, $d_s$ | □7.88mm |
| Tooth width at the slot opening | 6.□5mm |
| Slot width at the slot opening | 3.76mm |
| Tooth width just below the stator yoke | 4.09mm |
| Slot width at just below the stator yoke | □5.□9mm |
| Air gap power, $P_{gap}$ | 2730.□9W |
| Copper losses, $P_{copper}$ | 83.86W |
| Core losses, $P_{iron}$ | 40.□5 W |
| Output power, $P_{out}$ | 2550W |
| Efficiency, η | 93.4□% |

TABLE III
MAGNITUDE OF THE FLUX DENSITY AT DIFFERENT POINTS OF THE MACHINE

| B (Tesla) | No load | Full load |
|---|---|---|
| Teeth | □448 | □954 |
| Rotor core | □432 | □903 |
| Stator Yoke | □4□5 | □932 |

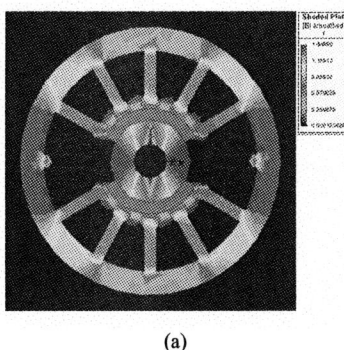

(a)

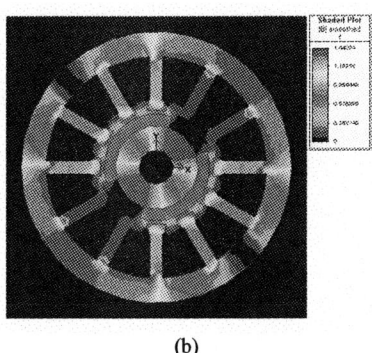

(b)

Fig. 3. Magnitude of flux density at no load at two different positions of rotor.

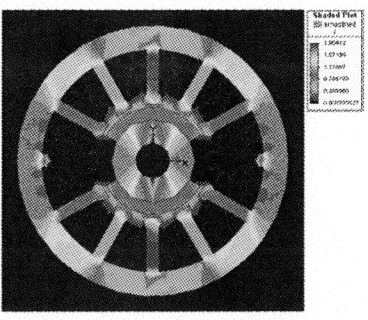

(a)

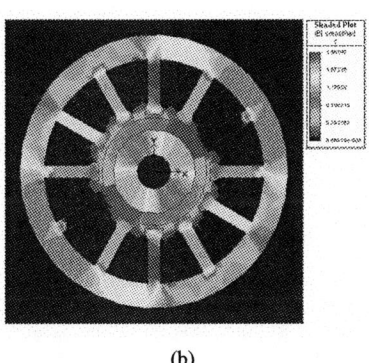

(b)

Fig. 4. Magnitude of flux density at full load at two different positions of rotor.

## IV. CONCLUSION

A methodology of design algorithm of a PMBL alternator for light combat aircraft is presented. The design input data and design results are given. In the proposed design the effect of the armature reaction is considered. The magnitude of the armature reaction flux density in the air gap is of about 0.06□9T. The design results of the Permanent Magnet Brushless alternator are verified using MAGNET2D, FEA-package. The distribution of flux lines and the magnitude of the flux density in the machine are verified at no load and full load conditions. The value of the flux density was verified in various points of the teeth, stator yoke, and rotor core of the machine. The results obtained shows that the iron sections of the machine are not saturated.

## V. REFERENCES

[□] J. R. Hendershot and T. J. E. Miller, *Design of Brushless Permanent Magnet Motors*. Oxford, U.K.: Magna Physics Publishing and Clarendon Press, □994.

[2] D. C. Hanselmann, *Brushless Permanent Magnet Motor Design*. New York: McGraw-Hill, □994.

[3] D. C. Hanselmann, *Brushless Permanent Magnet Motor Design*. Published by the writers collective, rhode island, 2003.

[4] M. Comanescu, A. Keyhani, and Min Dai, "Design and Analysis of 42-V Permanent-Magnet Generator for Automotive Applications," *IEEE Transaction on Energy Conversion*, vol. □8, no.□, March 2003.

[5] E. Spooner and A. Williamson, "Direct-coupled, permanent magnet generators for wind turbine applications," *IEE Proceeding of Electric Power Applications,* □996, vol. □43, no. □, pp. □+8.

[6] Grauers, O. Carlson, E. Högberg, P. Lundmark, M. Johnsson and S. Svenning, "Test and evaluation of a 20 kW direct-driven permanent-magnet generator withfrequency converter," *Proceedings of the European Wind Energy Conference (EWEC'97)*, Dublin, Ireland, □997, pp. 686-689.

[7] E. Spooner, A. Williamson and G. Catto, "Modular design of permanent-magnet generators for wind turbine," *IEE proceedings of Electric Power Applications*, vol. □43, No. 5, September □996.

[8] S. A. Papathanassiou, A. G. Kladas and M. P. Papadopoulos, "Direct-coupled permanent magnet wind turbine design considerations," *Proceedings of theEuropean Wind Energy Conference (EWEC'99)*, Nice, France, □999

[9] J. Chen, C. Nayar and L. Xu, "Design and finite-element analysis of an outerrotor permanent-magnet generator for directly-coupled wind turbine applications," *Proceedings of the IEEE Trans. on Magnetics,* vol. 36, no. 5, September 2000, pp. 3802-3809.

[□0] Q. Ronghai and A. Thomas, "Dual-rotor radial-flux toroidally wound permanentmagnet machines," *IEEE Trans. on Industrial Applications*, vol. 39, No. 6, November/December 2003, pp.□665-□673.

**Bhim Singh** (SM'99) was born in Rahamapur, U. P., India in □956. He received B. E. (Electrical) degree from University of Roorkee, India in □977 and M. Tech. and Ph. D. degrees from Indian Institute of technology (IIT), New Delhi, in □979 and □983, respectively. In □983, he joined as a Lecturer and in □988 became a Reader in the Department of Electrical Engineering, University of Roorkee. In December□990, he joined as an Assistant Professor, became an Associate Professor in □994 and Professor in □997 at the Department of Electrical Engineering, IIT Delhi. His field of interest includes power electronics, electrical machines and drives, active filters, static VAR compensator, analysis and digital control of electrical machines. Prof. Singh is a Fellow of Indian National Academy of Engineering (INAE), Institution of Engineers (India) (IE(I)) and Institution of Electronics and Telecommunication Engineers (IETE), a Life Member of Indian Society for Technical Education (ISTE), System Society of India (SSI) and National Institution of Quality and Reliability (NIQR) and Senior Member of IEEE ( Institute of Electrical and Electronics Engineers).

**Jally Ravi** was born in Thanakalan, Nizamabad, A.P., India in □984. He received Diploma (Electrical) from Govt. Polytechnic collage, Nizamabad and B.Tech (Electrical) degree from Vignan Institute of Technology & Science, Hyderabad and presently he is pursuing M.Tech degree from Department of Electrical Engineering, IIT Delhi. His field of interest includes power electronics, electrical machines and drives.

**2006 IEEE International Conference on Power Electronic, Drives and Energy Systems**

# Estimation of Core Loss in a Switched Reluctance Motor Based on Actual Flux Variations

Nimit. K. Sheth, *Student Member, IEEE*, and K. R. Rajagopal, *Senior Member, IEEE*

*Abstract--* **Accurate calculation of the core loss in a motor is essential for computing the actual efficiency. In a switched reluctance motor (SRM), the non-sinusoidal flux waveform with different frequencies of flux reversals in various parts of the motor makes the core loss calculation difficult. The conventional method for calculation of the core loss is based on the stator pole flux density variation in the triangular fashion with the peak flux density at the rotor position corresponding to the conduction angle from the unaligned position, which will not give accurate results. In this paper, an accurate method for the calculation of the core loss based on actual flux waveforms in various parts of the SRM is presented. The proposed method is validated using the FE analysis. The difference in the core loss estimated using the proposed and the conventional methods with the one obtained using the FE analysis is 4.89 % and 17.23 % respectively.**

*Index Terms--* **Core loss, FE analysis, Losses, Motor, SRM.**

## I. INTRODUCTION

SWITCHED reluctance motor (SRM) is a doubly salient motor having concentrated winding on the stator and without any winding or permanent magnet on the rotor. It has advantages like simplicity, controllability and high efficiency, which make it popular as an alternative to the conventional drive in many applications. To work out the actual efficiency of the SRM, it is very much essential to accurately estimate the copper and core losses. For the SRM, saturation of various magnetic parts and the non-sinusoidal flux waveform with different frequencies of flux reversals in various parts makes the core loss calculation difficult. The conventional method for the calculation of core loss is based on the triangular variation of the stator pole flux density with the peak at the conduction angle from the unaligned rotor position using an idealized current waveform in which the saturation is neglected [1]. This will lead to erratic results. The other

---

N. K. Sheth is with Electrical Engineering Department, Institute of Technology, Nirma University of Science and Technology, Ahmedabad 382481, Gujarat, India (e-mail: nimit75@yahoo.com) and pursuing PhD at Indian Institute of Technology Delhi, New Delhi 110016, India.

K. R. Rajagopa is with the Department of Electrical Engineering, Indian Institute of Technology Delhi, New Delhi 110016 (e-mail: rgopal@ee.iitd.ac.in).

approach is based on the measured flux waveform in one part of the motor [2], which needs a sensor to measure flux leading to additional cost. In this paper, an accurate method of core loss estimation based on the actual flux waveforms in various parts of the SRM is presented and the results are validated using the Finite element (FE) analysis.

## II. PROCEDURE FOR THE CALCULATION OF CORE LOSS

In the SRM, the actual phase current is dependent on the rotor position and phase inductance, which in turn is dependent on the rotor position and phase excitation. Considering this aspect, it is very much essential to precisely estimate the variations in the actual phase current. Using the flux tube method for inductance estimation [3] and the flow chart shown in Fig. 1, for a 5 hp 8/6 SRM the actual phase current based on the excitation for best step angle region [4] is obtained. Table I gives the major dimensions of the motor.

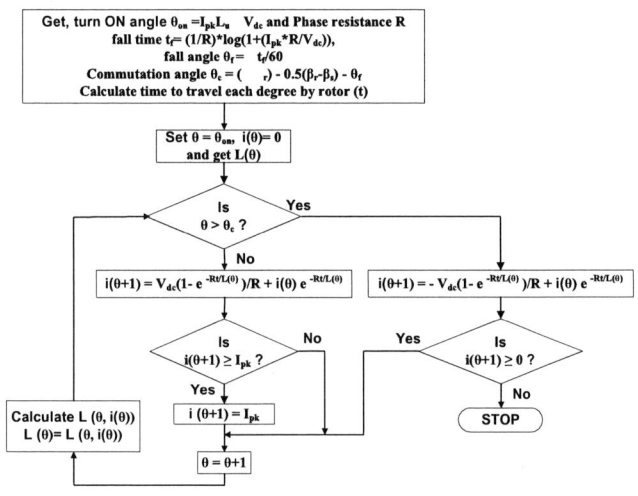

Fig. 1. Flow chart to calculate the actual phase current.

Fig. 2 shows the calculated phase current profile using which the flux density in the stator pole is estimated. The peak flux density obtained for the stator pole is 1.56 T. Fig. 3 shows the normalized stator pole flux density. Once the stator pole flux density for one pole is known the flux density in other stator poles have been calculated using the Fig. 3 and matrix equation (1) as proposed by Hayasahi and Miller [5].

TABLE I
MAJOR DIMENSION AND MATERIAL DETAILS OF THE MOTOR

| | |
|---|---|
| No. of stator poles $N_s$ = 8 | No. of rotor poles $N_r$ = 6 |
| Stator bore diameter $D$ = 99.4 mm | Stack length $L$ = 119.28 mm |
| Stator pole arc $\beta_s$ = 26° | Rotor pole arc $\beta_r$ = 22 ° |
| Overall diameter $D_0$ = 200.61 mm | Shaft diameter $D_{sh}$ = 33.33 mm |
| Stator back iron width $b_{sy}$ = 11.93 mm | Stator pole height $h_s$ = 38.68 mm |
| Rotor pole height $h_r$ = 20.36 mm | Airgap length $g$ = 0.75 mm |
| Material for stator and rotor stack = M43 | |

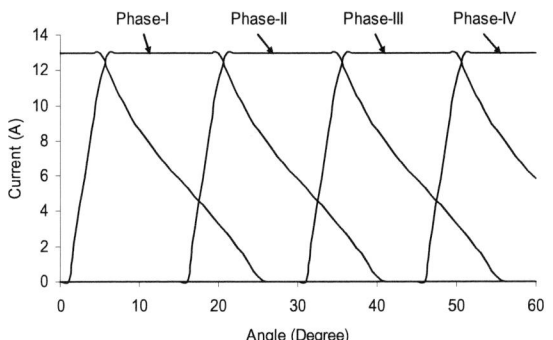

Fig. 2. Actual phase current waveforms of the motor.

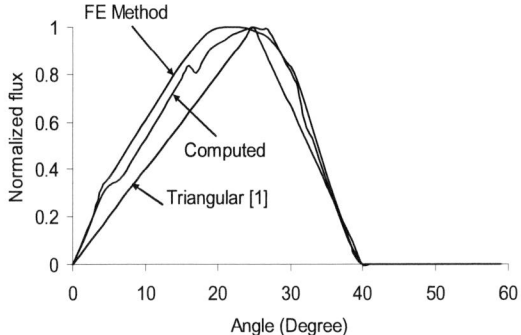

Fig. 3. Normalized flux density plot in the stator pole ($B_{sp3}$).

$$\vec{B}_{sp} = \begin{bmatrix} B_{sp1}(\theta) \\ B_{sp2}(\theta) \\ B_{sp3}(\theta) \\ B_{sp4}(\theta) \end{bmatrix} = \begin{bmatrix} B_{sp1} & 0 & 0 & 0 \\ 0 & B_{sp1} & 0 & 0 \\ 0 & 0 & B_{sp1} & 0 \\ 0 & 0 & 0 & B_{sp1} \end{bmatrix} \begin{bmatrix} \theta \\ \left(\theta - \dfrac{2\pi}{mN_r}\right) \\ \left(\theta - \dfrac{4\pi}{mN_r}\right) \\ \left(\theta - \dfrac{6\pi}{mN_r}\right) \end{bmatrix} \quad (1)$$

where, $m$ is the number of phase and $N_r$ is representing the number of rotor poles.

From the known flux density variation in one of the stator poles now the flux density in various parts of the motor as

indicated in Fig. 4 is worked out. For 8/6 SRM there will be two groups of four stator yoke sections respectively having the same flux density variation. The flux density variation in the various stator yoke portion is obtained from the flux density variation in the stator poles and using (2).

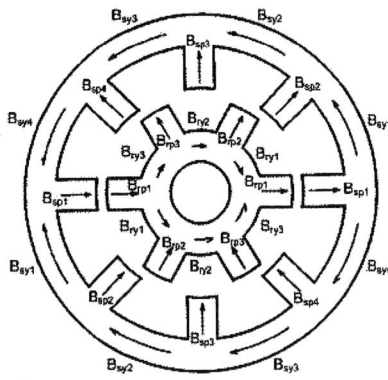

Fig. 4. Motor cross section indicating flux density in various parts of the motor.

$$\vec{B}_{sy} = \begin{bmatrix} B_{sy1}(\theta) \\ B_{sy2}(\theta) \\ B_{sy3}(\theta) \\ B_{sy4}(\theta) \end{bmatrix} = \frac{A_{sp}}{2A_{sc}} \begin{bmatrix} B_{sp1} & -B_{sp2} & -B_{sp3} & -B_{sp4} \\ B_{sp1} & B_{sp2} & -B_{sp3} & -B_{sp4} \\ B_{sp1} & B_{sp2} & B_{sp3} & -B_{sp4} \\ B_{sp1} & B_{sp2} & B_{sp3} & B_{sp4} \end{bmatrix} \begin{bmatrix} 1 \\ 1 \\ 1 \\ 1 \end{bmatrix} \quad (2)$$

Fig. 5 to Fig. 8 shows the variation of normalized flux in various sections of the stator yoke. From these figures it is observed that in three of the stator yoke sections the flux is bidirectional while in one section it is unidirectional.

The flux variation in the rotor pole is obtained using the stator pole flux variation of Fig. 3 and (3). Fig. 9 shows the normalized flux variation in the rotor pole. In other rotor poles the flux variation waveform will be similar to the one shown in Fig. 9 and will be obtained by (4).

$$B_{rp1} = \begin{bmatrix} B_{sp1} & -B_{sp2} & -B_{sp3} \cdots & -B_{spl} & -B_{sp1} & B_{sp2} & B_{sp3} & B_{spl} \end{bmatrix}_{1 \times 2l}$$
$$\begin{bmatrix} \theta & \left(\theta - \dfrac{2\pi}{N_s}\right) & \left(\theta - \dfrac{4\pi}{N_s}\right) & .. & \left(\theta - \dfrac{2(2l-1)\pi}{N_s}\right) \end{bmatrix}^T_{2l \times 1} \quad (3)$$

$$\vec{B}_{rp} = \begin{bmatrix} B_{rp1} \\ B_{rp2} \\ B_{rp3} \end{bmatrix} = \begin{bmatrix} B_{rp1} & 0 & 0 \\ 0 & B_{rp1} & 0 \\ 0 & 0 & B_{rp1} \end{bmatrix} \begin{bmatrix} \theta \\ \left(\theta - \dfrac{2\pi}{N_r}\right) \\ \left(\theta - \dfrac{4\pi}{N_r}\right) \end{bmatrix} \quad (4)$$

The flux variation in the rotor yoke is obtained using the flux density variation in the rotor pole as shown in Fig. 9 and the matrix equation (5). Fig. 10 shows the flux density variation in one of the rotor yoke portion.

$$\vec{B}_{ry} = \begin{bmatrix} B_{ry1} \\ B_{ry2} \\ B_{ry3} \end{bmatrix} = \begin{bmatrix} B_{rp1} & B_{rp2} & -B_{rp3} \\ B_{rp1} & -B_{rp2} & -B_{rp3} \\ B_{rp1} & B_{rp2} & B_{rp3} \end{bmatrix} \begin{bmatrix} 1 \\ 1 \\ 1 \end{bmatrix} \quad (5)$$

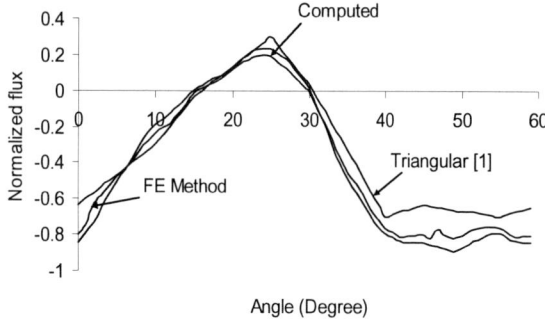

Fig. 5. Normalized flux density plot in the stator yoke part I ($B_{sy1}$).

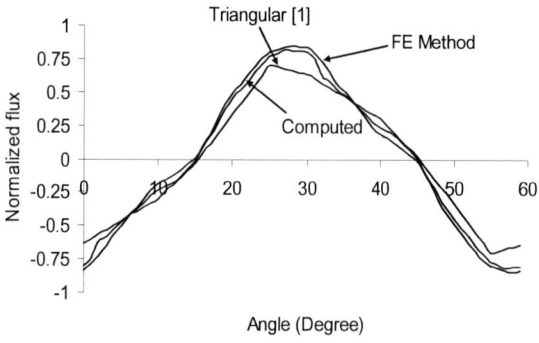

Fig. 6. Normalized flux density plot in the stator yoke part II ($B_{sy2}$).

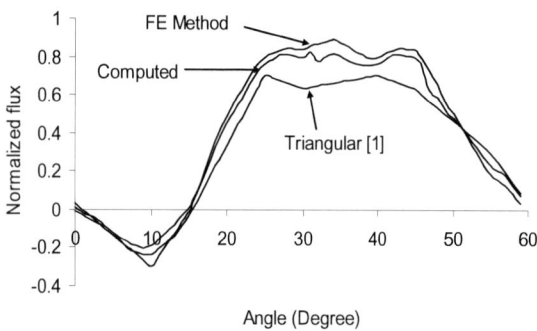

Fig. 7. Normalized flux density plot in the stator yoke part III ($B_{sy3}$).

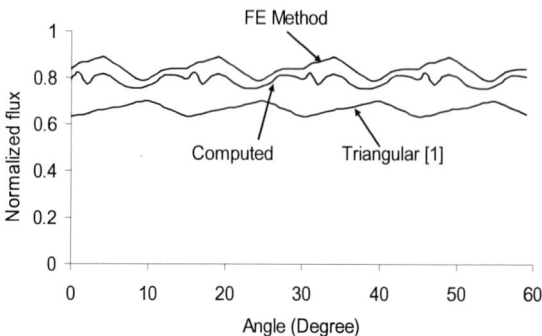

Fig. 8. Normalized flux density plot in the stator yoke part IV ($B_{sy4}$).

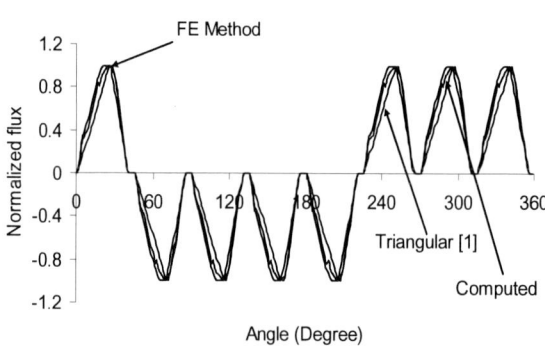

Fig. 9. Normalized flux density plot in the rotor pole ($B_{rp1}$).

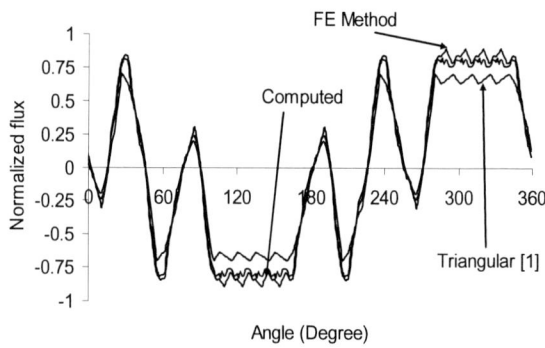

Fig. 10. Normalized flux density plot in the rotor yoke ($B_{ry1}$).

Table II shows the maximum flux density for various sections of the motor and the corresponding fundamental frequency, where $A_{sp}$, $A_{rp}$, $A_y$ and $A_{rc}$ are the area of the stator pole, rotor pole, stator yoke and rotor core respectively, and $\omega_m$ is the angular speed of the rotor. Loss characteristics of the material are represented as a third order polynomial. Various harmonic components of the flux waveforms of various sections are obtained using a FFT and corresponding loss per unit weight is obtained for first six harmonics from the loss characteristics and is summed up. It is observed from the harmonic analysis that the fundamental flux component in the stator and rotor poles has peak magnitude more than half of the overall peak. It is also observed that for the rotor pole and rotor yoke flux the odd harmonics are predominant.

Core loss for various parts of the motor is calculated using (6) to (9) and the total core loss of the motor is obtained by adding the core loss of the individual section, where $b_{ry}$ is the rotor core thickness, $D_{sh}$ is the shaft diameter and $\rho_1$ is the density of the material for stator and the rotor stack.

TABLE II
MAXIMUM FLUX DENSITY AND FUNDAMENTAL FREQUENCY OF FLUX FOR VARIOUS SECTIONS OF THE MOTOR

| Section of the Motor | Maximum Flux Density | Fundamental Frequency |
|---|---|---|
| Stator pole | $B_{spm}$ | $f_{sp} = (\omega_m \times N_r)/(2\pi)$ |
| Rotor pole | $B_{rpm} = B_{spm} \times A_{sp}/A_{rp}$ | $f_{rp} = f_{sp}/N_r$ |
| Stator yoke | $B_{scm} = B_{spm} \times A_{sp}/(2 \times A_{sy})$ | $f_{sc} = k \times f_{sp}$, $1 \leq k \leq m$ |
| Rotor yoke | $B_{rcm} = B_{spm} \times A_{sp}/(2 \times A_{rc})$ | $f_{rc} = f_{sp}/N_r$ |

Stator pole core loss = Loss / kg for a stator pole
$$\times N_r N_s \times 0.5 D \beta_s h_s L \rho_1 \quad (6)$$

Rotor pole core loss = Loss / kg for a rotor pole
$$\times N_r N_s \times (0.5D - g) \beta_r h_r L \rho_1 \quad (7)$$

Stator yoke core loss $= \sum_{i=1}^{N_s}$ Loss/kg for a yoke portion$_i$
$$\times 2 N_r (0.5 D_0 - b_{sy}) \left(\frac{2\pi}{N_s}\right) b_{sy} L \rho_1 \quad (8)$$

Rotoe core loss $= \sum_{i=1}^{N_r}$ Loss / kg for a core portion$_i$
$$\times b_{ry} L \times \pi (D - 2g - 2h_r + D_{sh}) \quad (9)$$

The total core loss calculated based on the actual flux density waveform is 253.07 W. To compare the core loss computed based on the actual flux density waveform with that of the flux density waveform obtained using the conventional method of considering the stator pole flux variation as triangular one with peak flux density occurring at commutation angle, normalized flux density waveform for various sections are obtained and shown in Fig. 3 and Fig. 5 to Fig. 10. Using the harmonic analysis of these waveforms obtained based on the conventional method the total core loss obtained is computed as 199. 7 W.

III. VALIDATION OF THE PROPOSED METHOD BY FE ANALYSIS

To validate the method presented here, two dimensional (2-D) model of the same motor is made and analyzed using the finite element (FE) method. Fig. 11 shows the flux density plot for the fully aligned position of the excited phase. Flux density waveforms in various parts of the motor have been obtained for the excitation of the stator phase based on the excitation shown in Fig. 2.

Fig. 3 and Fig. 5 to Fig. 10 show the normalized flux density waveforms for various sections of the motor. In Fig. 3 and Fig. 5 to Fig. 10, it is observed that the flux density plots obtained from the approach presented here is fairly matching to the ones obtained from the FE analysis in comparison to the ones obtained based on the conventional method. The peak stator pole flux density based on the FE analysis is 1.48 T. Based on the harmonic analysis of the flux density waveforms of the FE analysis the total core loss calculated is 241.27 W. Table III shows the comparison of the core loss computed by various methods and its difference from the one obtained from the FE analysis. It is observed that the core loss obtained from the proposed method is having 4.89% difference from the FE based core loss in comparison to 17.23 % when the conventional method of triangular stator flux variation is used hence validates the approach that core loss estimation based on the actual flux variation in various parts of the motor is more realistic and accurate compared to the core loss estimation based on the conventional approach.

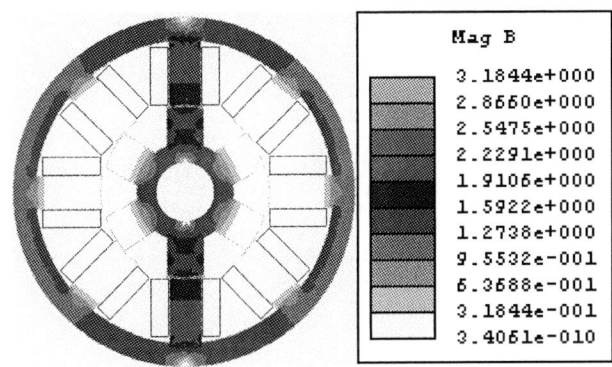

| Mag B |
|---|
| 3.1844e+000 |
| 2.8660e+000 |
| 2.5475e+000 |
| 2.2291e+000 |
| 1.9106e+000 |
| 1.5922e+000 |
| 1.2738e+000 |
| 9.5532e-001 |
| 6.3688e-001 |
| 3.1844e-001 |
| 3.4061e-010 |

Fig. 11. Normalized flux density plot in the rotor yoke ($B_{ry1}$).

TABLE III
COMPARISON OF CORE LOSS CALCULATED BASED ON VARIOUS METHODS

| Method | Proposed | FE Analysis | Triangular |
|---|---|---|---|
| Loss (W) | 253.07 | 241.27 | 199.70 |
| Change from the FEA method | + 4.89 % | - | - 17.23 % |

IV. CONCLUSION

Core loss calculated based on the actual waveform of flux in various parts of the motor gives more accurate results than the core loss calculated based on the assumption of triangular flux in stator pole. The flux waveforms obtained for various parts of the motor are in good agreement with the one obtained from the FE method. In the motor analyzed, the difference in the core loss with the proposed and conventional methods with that obtained from FE method based core loss is 4.5% and 17.23 % respectively.

## V. REFERENCES

[1] P. N. Materu and R. Krishnan, "Estimation of switched reluctance motor losses", *IEEE Trans. Ind. Appln.*, vol. 28, pp. 668-679, May/June 1992.

[2] J. Faiz and M. B. B. Sharifian, "Core loss estimation in a multiple teeth per stator pole switched reluctance motor", *IEEE Trans. Magn.* vol. 30, pp. 189-195, Mar. 2004.

[3] N. K. Sheth, and K. R. Rajagopal, "Calculation of the flux-linkage characteristics of a switched reluctance motor by a flux tube method", *IEEE Trans. Magn.* vol. 41, pp. 4069-4071, Oct. 2005.

[4] N. K. Sheth, and K. R. Rajagopal, "Optimum pole arcs for a switched reluctance motor for higher torque with reduced ripple", *IEEE Trans. Magn.* vol. 39, pp. 3214-3216, Sept. 2003.

[5] Y. Hayashiu, and T. J. E. Miller, "A new approach to calculating the core losses in the SRM", *IEEE Trans. Ind. Appln.*, vol. 31, pp. 1039-1046, Sept./ Oct. 1995.

## VI. BIOGRAPHIES

**Nimit K. Sheth** (S'2003) was born in Nadia, Gujarat, India in 1975. He received B. E. Degree in Electrical Engineering from the Gujarat University, Ahmedabad, India in 1996, M. Tech. Degree in Power Electronics, Electrical Machines and Drives in Electrical Engineering from the Indian Institute of Technology Delhi, New Delhi, India in 2002. Since, 2003 he is working towards his PhD degree at Indian Institute of Technology Delhi, New Delhi, India.

From 1997 to 1998, he was with Ahmedabad Electricity Company Ltd. Ahmedabad, India, as a Trainee Engineer. Since 1998, he is with the Institute of Technology, Nirma University of Science and Technology, Ahmedabad, India, where currently he is an Assistant Professor in Electrical Engineering Department.

He has published more than 20 papers in International Journals and Conference proceedings. He received Prof. A. K. Sinha Award for securing highest CGPA among all the M. Tech. graduating students from the Electrical Engineering Department of Indian Institute of Technology Delhi, New Delhi, India during the year 2002. His fields of interest include Electrical Machines and Drives, Special Electrical Machines (Switched Reluctance Motors, DSPM Motors, Flux Reversal Motors, Stepper Motors, etc.,), Finite Element Analysis and CAD of Electrical Machines.

**K. R. Rajagopal** (M'1998, SM'2000) was born in Alappuzha, Kerala, India in 1961. He received Diploma in Electrical Engineering from Carmel Polytechnic, Alappuzha, India in 1979, B. Tech. Degree in Electrical Engineering from the College of Engineering, Trivandrum, India in 1988, M. Tech. Degree in Power Electronics, Electrical Machines and Drives and Ph. D. Degree in Electrical Engineering from the Indian Institute of Technology Delhi, New Delhi, India during 1991 and 1998 respectively.

From 1980 to 1983, he was with Aluminum Industries Ltd. (ALIND), Trivandrum, India, as an Application Engineer (Relays), from 1983 to 1999, he was with the Indian Space Research Organization (ISRO), Trivundrum, India, where he was engaged in Analysis, Design, Development and Testing of Special Electrical Machines/Devices used in space applications. Since 1999, he is with the Indian Institute of Technology Delhi, New Delhi, India, where currently he is a Professor in Electrical Engineering Department.

He has published more than 30 papers in International Journals and more than 60 papers in International conference proceedings. He received Indian National Academy of Engineering (INAE) award for most Innovative Potential Project in Engineering during the year 1998. His fields of interest include Electrical Machines and Drives, Special Electrical Machines (Stepper Motors, Switched Reluctance Motors, PM BLDC Motors, Hysterisis Motors, etc.,), Magnetic Devices, Finite Element Analysis and CAD of Electrical Machines and Design of Energy Efficient Motors for Home Appliances.

**2006 IEEE International Conference on Power Electronic, Drives and Energy Systems**

# A Novel Hybrid Brushless dc motor/Generator for Hybrid Vehicles Applications

E. Afjei, H. Toliyat, *Senior Member, IEEE,* and H. Moradi

*Abstract--***The Brushless dc motor is a simple and robust machine, which has found application over a wide power and speed ranges in different shapes and geometries. This paper presents a new configuration for an integrated starter-generator system based on brushless dc motor without permanent magnet technology. The proposed novel motor consists of two magnetically dependent stator and rotor sets (layers), where each stator set includes nine salient poles with windings wrapped around them while, the rotor comprises of six salient poles with no windings. The magnetic field passes through a guide to the rotor then the stator and finally completes its path via the motor housing. To evaluate the motor performance, two types of analysis, namely numerical technique and experimental study have been utilized. In the numerical analysis the finite element analysis is employed, where as in the experimental study, a proto-type motor has been built and tested. The calculated results compare favorably with the test results. Due to the ruggedness of the proposed motor in comparison with the conventional and brushless dc motors it looks very promising for hybrid vehicle.**

*Index Terms--* **Brushless dc motor, dc motor, hybrid vehicle.**

## I. INTRODUCTION

DUE to the increasing demand for higher power and less fuel consumption in cars, the concept of starter-generator integrated into the flywheel has been considered over the past several years. It is intended to provide the starter for the thermal engine and the generator for charging the car battery and supplying the on board equipment. Significant enhancement of vehicle driving performance and improvements in fuel economy and exhaust emissions has been demonstrated by the introduction of more-electric drive concepts for road transportation. Although the application of electrical machines and drives systems in all-electric and hybrid-electric vehicles has been widely reported in recent years [1-4], there has been a relatively slow progress in these technologies due to the cost of major vehicle technological change. However, mild hybrid solutions have been recognized

as a next solution, since they are viable within the existing automotive infrastructure. Switched Reluctance Generator (SRG) is an attractive solution for worldwide increasing demand of electrical energy. It is low cost, fault tolerant with a rugged structure and operates with high efficiency over a wide speed range. Merits of using SRG have been proved for some applications like starter/generator for gas turbine of aircrafts [5, 6], windmill generator [7, 8] and as an alternator for automotive applications [9]. In [10], principle of operation of SRG has been presented and the necessity for closed loop control is proved. In [11], the control of excitation of SRG for maximum efficiency at single pulse mode of operation has been presented. Turn on and turns off angles are defined as control variables, turn on angle is set based on the output power and the turn off angle is selected to achieve optimal efficiency at each power level and speed [12]. Fig. 1 shows the general block diagram of a hybrid vehicle.

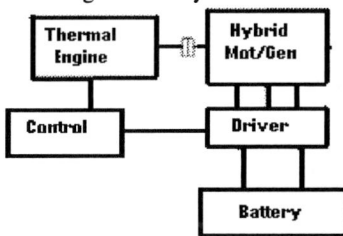

Fig. 1. The general block diagram of a hybrid vehicle.

Block diagrams contains the combustion engine, the electric motor [integrated starter–generator (ISG)], the battery pack and the power electronics system. The traction system also contains the gearbox. The main advantage of an ISG system over a conventional starter motor can be found in the fast engine stop/start behavior and the potential of reduced noise, vibration, and harshness during actively controlled engine startup and shutdown procedures [13].

This paper presents construction, numerical analysis, experimental results of a new type of hybrid dc motor/generator.

## II. MOTOR/GENERATOR DESCRIPTION

The proposed novel motor/generator consists of two magnetically dependent stator and rotor sets (layers), where each stator set includes nine salient poles with windings wrapped around them while, the rotor comprises of six salient poles without any windings. Every stator and rotor pole arcs is about $30^{\circ}$. The two layers are exactly symmetrical with respect to a plane perpendicular to the middle of the motor shaft. This

---

E. Afjei is with the Department of Electrical and Computer Engineering, Shahid Beheshti University , Tehran, Iran (e-mail: e-afjei@sbu.ac.ir).

H A. Toliyat is with the Department of Electrical Engineering, Texas A&M University, TX. , USA. (e-mail: toliyat@ee.tamu.edu).

H. Moradi is with the Shahid Beheshti University.(e-mail: hmch1580@yahoo.com).

0-7803-9771-1/06/$25.00 ©2006 IEEE    151

is a three phase motor/generator, therefore, three coil windings from one layer is connected in series with the other three coil windings in the other layer. Fig. 2 shows the shape of the stator and the rotor laminations.

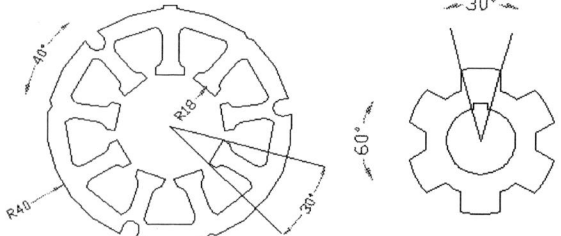

Fig. 2. (a) Stator lamination.          (b) Rotor lamination.

There is a stationary reel, which has the field coils wrapped around it and is placed between the two-stator sets. This reel has a rotating cylindrical core, which guides the magnetic field. The magnetic flux produced by the coils travels through the guide and shaft to the rotor and then to the stator poles, and finally closes itself through the motor housing. Therefore, one set of rotor poles is magnetically north and the other set is magnetically south. In this motor, the magnetic field has been induced to the rotor without using any brushes. A cut view of the motor/generator is shown in Fig. 3.

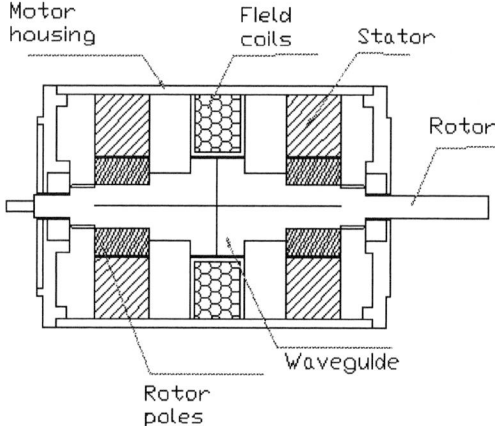

Fig. 3. The cut view of the motor/generator.

In order to get a better view of the motor/generator configuration, the complete motor /generator assembly is shown in Fig. 4.

Fig. 4. The complete motor/generator assembly.

There are two stators and rotors sections placed on both sides of the field coil assembly which has the rotor shaft as its main core and two front / end caps plus the motor housing. A set of photo interrupters are also place in the back of the motor for the detection of rotor position. One of the most widely used methods for analysis of any types motor or generator is the finite element technique [7].

## III. NUMERICAL ANALYSIS

One side or layer of the motor/generator cross section is shown in Fig. 5.

Fig. 5. One layer of the motor/generator cross section.

As seen from Fig. 5, this motor has nine stator poles as well as six rotor poles, which will be engaged in the torque production mechanism. The design of the motor becomes complicated due to complex geometry and material saturation. The reluctance variation of the motor has an important role on the performance; hence an accurate knowledge of the flux distribution inside the motor for different excitation currents and rotor positions is essential for the prediction of motor performance. The motor can be highly saturated under normal operating conditions. To evaluate properly the motor design and performance a reliable model is required. The finite-element technique can be conveniently used to obtain the magnetic vector potential values throughout the motor in the presence of complex magnetic circuit geometry and nonlinear properties of the magnetic materials. These vector potential values can be processed to obtain the field distribution, torque, and flux leakage.

In order to be able to analyze this motor in 2-D case, only one side of the motor which is symmetric to the field coil is considered, normal boundary conditions are applied to the outer and inner borders of the motor, and a magnet producing the same magnetic field density is considered to act as the field coil. It is worth mentioning that the other options for the field coils could have been using small coil windings on the rotor poles.

The field analysis has been performed using a Magnet CAD package [8], which is based on the variational energy minimization technique to solve for the magnetic vector potential. The partial differential equation for the magnetic vector potential is given by [9-10].

$$-\frac{\partial}{\partial x}\left(\gamma \frac{\partial A}{\partial x}\right) - \frac{\partial}{\partial y}\left(\gamma \frac{\partial A}{\partial y}\right) = J \qquad (1)$$

Where, A is the magnetic vector potential.

In the variational method (Ritz) the solution to (1) obtained by minimizing the following functional

$$F(A) = \frac{1}{2} \iint_{\Omega} [\gamma(\frac{\partial A}{\partial x})^2 + \gamma(\frac{\partial A}{\partial y})^2] d\Omega - \iint_{\Omega} JAd\Omega \qquad (2)$$

Where $\Omega$ is the problem region of integration.

In order to be able to analyze the motor in two dimensions, the field normal boundary conditions are used over the inner and outer boarders of the motor. In the finite element analysis second order triangular elements with dense meshes at places where the variation of fields are greater have been used.

Figs. 6 and 7 show the magnetic flux and the magnetic field density for aligned and non-aligned cases when the machine acting as a motor.

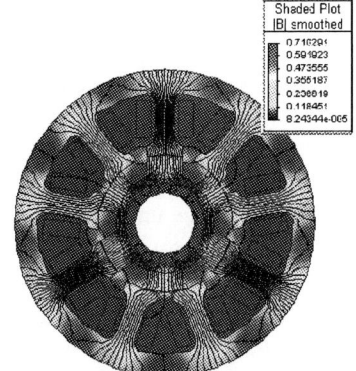

Fig. 6. Plots of magnetic field density and magnetic flux for aligned case.

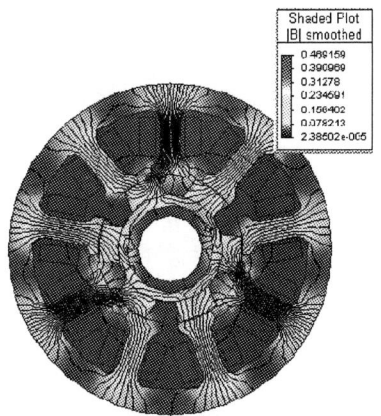

Fig. 7. Plots of magnetic field density and magnetic flux for non-aligned case.

The plot of static torque versus rotor positions developed by the hybrid brushless dc motor is shown in Fig. 8.

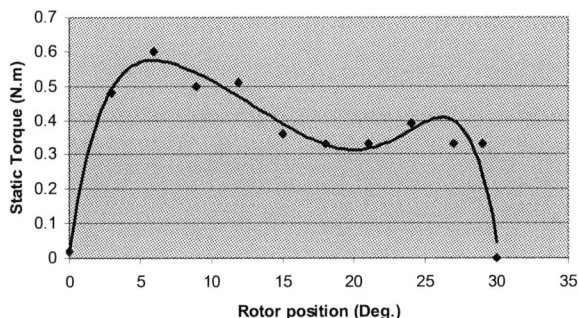

Fig. 8. Static torque versus rotor angle.

The Torque versus angle characteristics of the motor obtained from using the finite element method by giving constant current in two phases of the motor in appropriate switching cycle.

There is a good discussion on the static torque vs. position for BLDC motor in [11] which models the effects of skewing in BLCD motor on its performance. It is worth mentioning here that, the stator and rotor cores are made of a non-oriented silicon steel lamination. The magnetization curve is taken from manufacturer's data sheet for M-27 steel. In the Fig. 8 zero degree is considered as unaligned case.

In the generator mode, the plots of magnetic field density and magnetic flux for only field coil considered to be turned on are shown in Fig. 9.

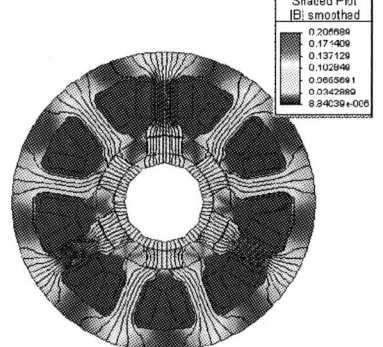

Fig. 9. Plots of magnetic field density and magnetic flux.

## IV. Experimental Results

The motor has been fabricated and tested for performance and functionality in the laboratory. Fig. 10 illustrates the novel brushless dc motor fabricated in the laboratory.

Fig. 10. The actual brushless dc motor.

The static torque of the motor was obtained by blocking the motor at different angle. The average static torque for a rated current of 3A was measured to be about 46 N.cm over the stator pole arc (0 to $30^0$). It suddenly went to zero at the start of stator to rotor complete overlap. It was observed that the static torque shows lower value than computed which is expected, since, the silicon sheet steel material used to build the motor is not quite what is used for the numerical analysis.

Using a motor generator assembly, the dynamic torque for the motor versus speed has been measured by loading the motor. The torque speed characteristics of the motor for two different field currents is shown in Fig. 11.

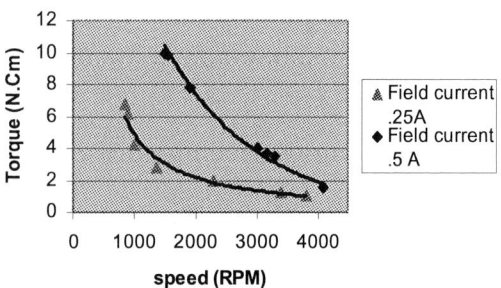

Fig. 11. Dynamic torque of the motor versus speed.

The power curve fitting has been used for the data points. The torque speed characteristics of the motor behave like a series dc motor and switched reluctance motor.

Fig. 12 shows the plot of the motor torque versus current under different loads for the motor.

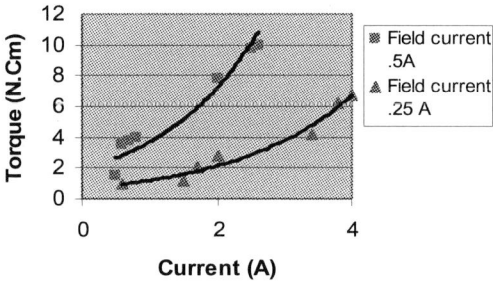

Fig. 12. Plot of the motor torque versus current.

As seen from the Fig. 12 the torque is proportional to the square of motor current, which resembles the switched reluctance motor. The static torque versus rotor position is also obtained using a torque meter which in general agrees with the one found numerically. There is a good discussion on the static torque vs. position for BLDC motor in [11] which explains the torque curvature produced under different skewing in this type of motor.

In BLCD motors, each individual phase excitation must be synchronized with the rotor position, which necessitates the need for a position sensing scheme. In general there are two types of rotor position sensing method namely, direct and indirect. In the direct position sensing method usually, a mechanical shaft position transducer, such as opto-couplers with a slotted disk, Hall-effect sensors and embedding permanent magnets within the teeth of the slotted disk, or a high precision encoder is mounted on the motor housing to produce the necessary and accurate rotor position information for the proper motor operation. Fig. 13 shows opto-couplers with slotted disk together used in detecting the rotor position for the new brushless dc motor.

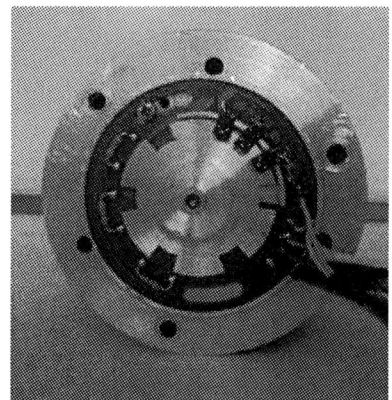

Fig. 13. Opto- interrupters with Slotted Disk.

The output signals come from the photo-interrupters mounted on the back of the motor. There are three 30° pulses produced by the motor shaft position sensors and each pulse appears 6 times in one rotation. Fig. 14a shows the resulting pulses produced by the sensing unit for 30° duration while 14b shows two consecutive photo-interrupters.

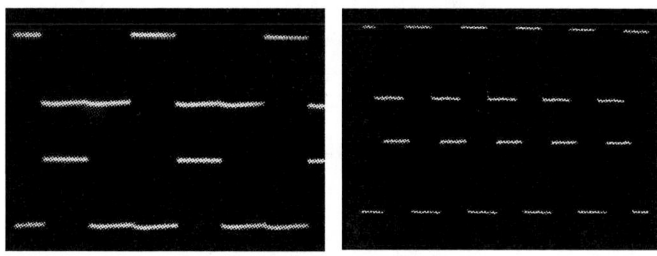

Fig. 14a. The output signals from the photo-interrupters.
Fig. 14b. Resulting pulses from two consecutive photo-interrupters.

In Fig. 14a there is no overlapping between the phases and each photo-interrupter works one third of the period. In Fig. 14b the overlapping of two photo-interrupters output signals which corresponds to two of the motor phases to be on at one time is clearly shown. It is possible to adjust the unit for any overlapping needed.

The shaft of the motor/generator machine is connected to a motor to act as prim mover. The speed of the motor kept constant first at 1000 and then at 2000 rpm under different loads for various field current. The results of these tests are shown in Figs. 15 and 16.

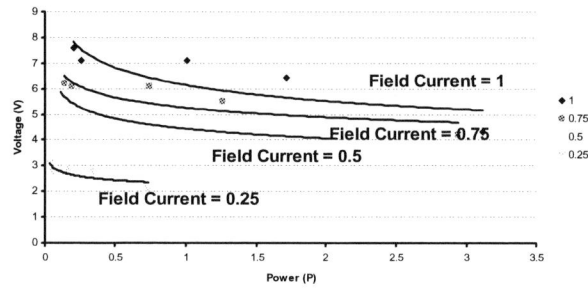

Fig. 15. Output voltage vs. power (1000 rpm).

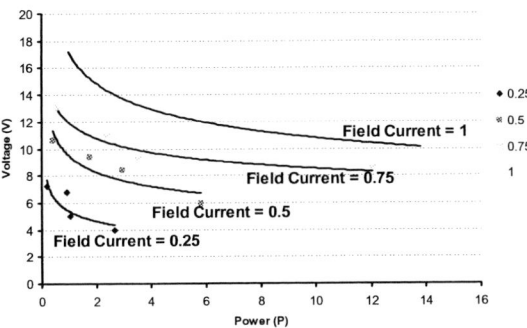

Fig. 16. Output voltage vs. power (2000 rpm).

In these Figs. curve fitting (power) has been used for better presentation of the data points. The actual output voltages for two consecutive phases are also shown in Fig. 17.

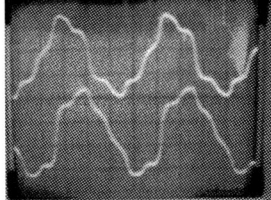

Fig. 17. output voltages for two consecutive phases.

Figs. 18a and 18b show the output voltages from one phase of the generator for a field current of 0.25A and 0.5A, respectively

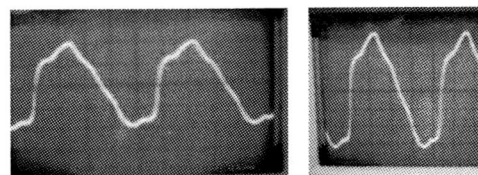

Fig. 18. Output voltage from one phase of the generator
A) Field current of .25 A          B) Field current of .5 A.

The voltages have harmonics which are due to the shape of the stator and rotor poles. These voltages are then rectified by a three phase full bridge configuration.

## V. CONCLUSION

In this paper a novel motor/generator was fabricated in the laboratory. Some of the motor parameters numerically computed and experimentally measured and tested. The main objective of this paper namely, introduction of a new motor/generator configuration with out using any permanent magnet to increase its durability has been achieved. The experimental analysis shows the functionality of the motor/generator in its new configuration, meaning, it has the ability and the potential of becoming a motor potentially be used in hybrid vehicle.

## VI. REFERENCES

[1] K. M. Rahman, B. Fahimi, G. Suresh, A. V. Rajarathnam, and M. Ehsani, "Advantages of Switched Reluctance Motor Applications to EV and HEV: design and control issues", IEEE Trans. Industry. Applications, vol. 36, pp. 111-121, Jan./Feb. 2000.

[2] Emadi, " Low-Voltage Switched Reluctance Machine Based Traction Systems for lightly Hybridized vehicles", Society of Automotive Engineers, pp. 1-7, 2001.

[3] Koch, J. Lehmann, G. Probst, and H. Schäfer, "The Integrated Starter Generator as Part of the Power Train Management", Aachener Kollequium Fahrzeug- und Motorentechnik, pp. 11-11, 2000.

[4] S. A. Long, N. Schofield, D. Howe, M. Piron and M. McClelland, "Design of a switched reluctance machine for extended speed operation," Proceedings of IEEE International Electrical Machines and Drives Conf. (IEMDC), Wisconsin, pp. 235-240, 1-4 June 2003.

[5] S. R. MacMinn and J. W. Sember, "Control of a Switched-Reluctance Aircraft Starter-Generator Over a very wide Speed Range," in Proc. Intersociety Energy Conversion Engineering Conf., 1989, pp. 631–638.

[6] C.A Ferreira., S.R Jones, W.S. Heglund, W.D. Jones, "Detailed Design of a 30-kW Switched Reluctance Starter/Generator System for a Gas Tturbine Engine Application", IEEE Transactions on Industry Applications, Volume: 31 , Issue: 3 , May-June 1995.

[7] M.A Mueller, "Design of Low Speed Switched Reluctance Machines for Wind Energy Converters", Electrical Machines and Drives, 1999. Ninth International Conference on (Conf. Publ. No. 468), 1-3 Sept. 1999, pp. 60 – 64.

[8] R. Cardenas, W. F. Ray, and G. M. Asher, "Switched Reluctance Generators for Wind Energy Applications," in Proc. IEEE PESC'95, 1995, pp. 559–564.

[9] B. Fahimi, A. R.B. Emadi, Jr. Sepe, "A Switched Reluctance Machine Based Starter/Alternator for More Electric Cars" Energy Conversion, IEEE Transactions on Energy Conversion, Volume: 19 , Issue: 1 , March 2004, pp.116 – 124.

[10] A. Radun, "Generating With the Switched-Reluctance Motor," in Proc. IEEE APEC'94, 1994, pp. 41–47.

[11] Y. Sozer, D.A. Torrey, "Closed Loop Control of Excitation Parameters for High Speed Switched-Reluctance Generators", IEEE Transactions on Power Electronics, Volume: 19, Issue: 2 , March 2004, pp. 355 – 362.

[12] B. Fahimi, S. Dixon, "Enhancement of Output Electric Power in Switched Reluctance Generators", . IEEE International on Electric Machines and Drives Conference, 2003, IEMDC'03, Volume: 2 , 1-4 June 2003,pp.849 -856.

[13] Ion Boldea,, Lucian Tutelea, and Cristian I. Pitic, "PM-Assisted Reluctance Synchronous Motor/Generator (PM-RSM) for Mild Hybrid Vehicles: Electromagnetic Design", IEEE Transactions on Industry Applications, Vol. 40, No. 2, March, 2004, pp. 492-498.

## VII. BIOGRAPHIES

**Ebrahim S. Afjei** received the B.S. and M. S. degrees in electrical engineering from the University of Texas in 1984 and 1986, and the Ph.D. degree from New Mexico State University, Las Cruces, in 1991. He is currently a professor in the Department of Electrical & Computer Engineering, Shahid Beheshti University, Tehran, Iran. His research interest is in switched reluctance motor drives.

**Hamid A. Toliyat** (S'87–M'91–SM'96) received the B.S, degree from Sharif University of Technology, Tehran, Iran, in 1982, the M.S. degree from West Virginia University, Morgantown, in 1986, and the Ph.D. degree from the University of Wisconsin, Madison, in 1991, all in electrical engineering. Following receipt of the Ph.D. degree, he joined the Faculty of Ferdowsi University of Mashhad, Mashhad, Iran, as an Assistant Professor of electrical engineering. In March 1994, he joined the Department of Electrical Engineering, Texas A&M University, where he is currently a Professor. His main research interests and experience include multi-phase variable speed drives for traction and propulsion applications, fault diagnosis of electric machinery, analysis and design of electrical machines, and sensorless variable speed drives. He has published over 190 technical papers in these fields.

Dr. Toliyat received the Texas A&M Select Young Investigator Award in 1999, the Eugene Webb Faculty Fellow Award in 2000, the Space Act Award from NASA in 1999, the Schlumberger Foundation Technical Awards, in 2000 and 2001, and the 1996 IEEE Power Engineering Society Prize Paper Award. He is a member of Sigma Xi and an Editor of the IEEE TRANSACTIONS ON ENERGY CONVERSION. He is a member of the Editorial Board of the Electric Machines and Power Systems Journal. He is also Vice-Chairman of IEEE-IAS Electric Machines Committee.

**H. Moradi** received the B.S. and M. S. degrees in electrical engineering from the University of Esfahanin 2002 and 2006, respectively. His research interest is in motor drives and power electronics.

**2006 IEEE International Conference on Power Electronic, Drives and Energy Systems**

# Computer Aided Design and FE Analysis of a PM BLDC Hub Motor

K. R. Rajagopal, *Senior Member, IEEE,* and Chippa Sathaiah

*Abstract*--This paper presents the computer aided design (CAD) of a PM BLDC hub motor. Using the developed CAD program, a 30W, 48V, 310rpm PM BLDC hub motor meant for a ceiling fan application is designed. The design variables such as flux density in air gap and iron, slot space factor, stack length of the motor, air gap length, number of magnet poles, etc., are assumed. Basic output equation is used for the design algorithm. Output of the developed CAD program gives the design data and the same is validated by FE analysis. This paper also presents the parametric analysis of the designed PM BLDC hub motor and the results of parametric analysis are submitted.

*Index terms*--Brushless motor, ceiling fan, computer aided design (CAD) of motor, finite element (FE) analysis, hub motor, motor, parametric analysis, permanent magnet brushless dc (PM BLDC) motor, permanent magnet (PM) motor.

## I. INTRODUCTION

Due to the worldwide growing concern over the energy conservation, the development and commercialization of high efficiency electric motor drives becomes a major necessity. The high efficiency electric motor drives systems can replace the conventional drive systems in home appliances and in industrial applications. This of course will reduce the consumption of the electric energy. As most of the electric power is consumed by electric motors, any improvement in the motor efficiency, even by one percent, can save a lot of energy.

PM BLDC motors have various advantages like high efficiency, compact volume and high torque-to-power ratios. Reduction in the cost of permanent magnet materials and power electronic devices have collectively brought the attention of motor designers to look for cost effective and high performance applications for these motors in various domestic and industrial applications. The choice of motor type is the most fundamental design decision, because of the relatively high cost of magnets, together with issues related to packaging, magnet retention, and winding. There are several different configurations of brushless motors which use rotating permanent magnets and stationary phase coils. The main reason for so many different variations has to do with the utilization of different magnet grades in addition to the

---

K. R. Rajagopal is with the Electrical Engineering Department, Indian Institute of Technology Delhi, New Delhi-110016, India (e-mail: rgopal@ee.iitd.ac.in).
Chippa Sathaiah is an M. Tech student in Power Electronics, Electrical Machines and Drives of Electrical Engineering Department at Indian Institute of Technology Delhi, New Delhi-110016, India (email: sathaiah_c@rediffmail.com).

wide range of applications. The most cost effective use of permanent magnets in brushless DC motors requires a configuration with the rotor outside the stator. As ceiling fan application requires constant speed at low to medium speed it may make more sense to use an exterior rotor configuration with rotating member on the outside of the wound stator.

This paper presents a computer aided design (CAD) program for the design of a PM BLDC hub motor. Flow chart for the development of the CAD program for the design of PM BLDC hub motor and its description is given in this paper. Details of the steps involved for the development of the computer aided design (CAD) of a PM BLDC hub motor are also given. Using the developed CAD program, a 30W PM BLDC motor meant for ceiling fan application is designed. Design outputs from the CAD program are used as inputs for 2-D FE analysis for validation of design. Results obtained from the FE analysis, like flux density plot and torque angle characteristics are validating the design. The design data of the motor obtained from the CAD program is used for parametric analysis. Design outputs from parametric analysis are also validated using FE analysis. The output from the parametric analysis can be used as the input for further optimization of the motor design.

## II. COMPUTER AIDED DESIGN

The flow chart of the developed CAD program for design of a PM BLDC hub motor is shown in Fig.1. Motor specifications, type of configuration, material types and other assumed data for the design are provided as inputs.

The input motor specifications are:

> Stack length of the motor = 15.5mm
> Air gap length of the motor = 0.25mm
> Remanent flux density of the permanent magnet = 1.12T
> Relative permeability of the permanent magnet = 1.1
> Slot fill factor, $S_f = 0.3$
> Flux density in rotor = 2T
> Flux density in stator = 1.5T
> Teeth flux density = 2T
> Number of poles = 8
> Lamination thickness of stator iron = 0.35 mm
> Friction and windage loss = 1W
> Number of slots = 9
> Magnet fraction=0.89
> Number of phases=3

The developed CAD program contains four loops. The first loop is used to find the length of the magnet to get the required flux density in the air gap; second loop is used to find the air gap diameter to get the required shaft diameter;

---

0-7803-9771-1/06/$25.00 ©2006 IEEE  157

third loop is used to find phase current and number turns per phase based on the calculation resistance of the winding and the fourth loop is used to find the mechanical efficiency of the motor. The computer aided design of a PM BLDC hub motor is divided into three major parts. They are:

### A. Magnetic circuit design

#### 1) Number of poles:

Number of poles will be selected based on the speed of rotation, commutation frequency and cogging torque [3]. Increase in the number of poles for the same speed results in increased commutation frequency, hence increase in switching losses in transistors and iron loss in the stator. And for the fixed main dimensions, with increase in the number of poles, either the width of magnet spacer or the magnet fraction will decrease, which will result in decrease in specific magnetic loading and thereby the developed torque. Also end turn becomes shorter and the leakage inductance becomes lower for higher pole number. But the reduction in pole number increases the cogging torque and end copper requirement.

#### 2) Number of slots:

Selection number of slots mainly depends on the cogging torque and end copper requirement. If the ratio of slot number to pole number is even, then every edge of every pole lines up with every slot, causing cogging. If a fractional slot combination is used, fewer pole edges line up with the slots and hence reduced cogging torque. And a fractional slot configuration minimizes the need for skewing of either the poles or the lamination stack to reduce cogging. A final point to be made about the slot and pole relationship concerns the winding pitch. Since the coils can be wound only over an integral number of slots, the winding pitch is determined by dividing the number of slots by the number of poles and rounding off to the next lower number, or in the case of the 0.75 slots/pole series, the next larger whole number. It should be obvious that the end turns are shortest when the pitch is one or two slot pitches [3].

#### 3) Length of air gap:

Of the three components of the phase-inductance, the air gap inductance is the predominant. Larger air gap will result in reduced phase-inductance, armature reaction effects, and also the cogging torque, but will necessitate bigger magnets, thereby increased cost. And as the air gap increases, the efficiency of the motor reduces.

#### 4) Magnetic materials:

Choice of soft magnetic materials for stator and rotor core depends on the frequency and operating flux density. As negligible flux levels occur at rotor back iron, the choice can be on the flux density alone. And the selection of Permanent Magnet depends on the energy product and commercial availability. Nd-Fe-B and samarium cobalt permanent magnets are preferred because of their high-energy product and retentivity. Nd-Fe-B can be selected for the design of the PM BLDC hub motor because of its maximum energy product, $(BH)_{max} = 30$ MGOe and higher residual flux density, $(Br) = 1.12$T.

#### 5) Magnet fraction:

Magnet fraction is particularly an important parameter in regard to the level of cogging torque, and it has been found that when magnet fringing is neglected, the optimum ratio of pole-arc to pole-pitch, $\alpha_p$, for minimizing the fundamental component of cogging torque, for any combination of slot and pole number, is:

$$\alpha_p = \frac{(N - k_1)}{N}, \qquad k_1 = 1, 2, 3, \ldots\ldots, N-1 \qquad (1)$$

Where $N = N_c/P$
$N_c$ is the smallest common multiple of slot number and pole number
and P is the number of poles.

In practice, however, due to fringing of the magnet flux into the slots, the optimum value of $\alpha_p$ should be increased by adding a small factor typically ranges from 0.01 to 0.03 depending on the air gap length [5]. Clearly, in order to maximize the air gap flux, and thereby the excitation torque, the optimal ratio of pole-arc to pole-pitch should be as high as possible. Hence, in practice $k_1=1$, is usually the preferred value.

#### 6) Magnet length (thickness):

Based on the magnetic circuit equations [1], the length of the permanent magnet can be found by reducing the difference between the assumed and actual flux densities in the air gap with the variation of permanent magnet length. In order to reduce the requirement of the ampere conductors to improve efficiency, the magnet length can be increased and hence specific magnetic loading. But the increase in magnet length requires more magnet material and hence increased cost of the motor.

### B. Electrical circuit design

#### 1) Number of phases:

In case of 2-phase configuration there is a problem of stalling and uncertainty in direction of rotation. Three phase motors are the most common choice for all but the lowest power levels. Although the utilization can theoretically be argued to be higher in motors of higher phase number, the gains would be offset by the increased number of leads and transistors, which increases cost and may severely compromise reliability. Three phase motors have the flexibility afforded by star or delta connected windings, or even unipolar windings. They can operate with only three connecting leads with no loss of control flexibility. They have excellent starting characteristics, with smooth rotation in either direction, or low torque ripple. They can work with a wide range of magnet configurations and an enormous range of winding configurations, and can take advantage of the coil winding technology that has been developed for both AC induction and DC brush type motors. They can operate with either square wave drive of\r sine wave drive, and are well adapted to the development of sensor less controllers that requires no physical shaft position sensor [3].

*2)   Conductor current density:*

Conductor current density is decided based on the recommendations given for permanent magnet motor windings and the cooling method employed [3]. Lower values of conductor current density are advisable for small and totally enclosed. Also lower values of current density requires more quantity of permanent magnet material resulting in increased cost and still smaller current densities can be used for higher efficiency but with higher size and cost for the motor.

Where as higher values of current density gives rise to more copper loss and reduces the efficiency of the motor. But higher values of current density reduce permanent magnet material required and hence overall cost of the motor.

*3)   Winding design:*

Based on the winding tables for different pole and slot combination for PM BLDC motor [2], winding connections for the designed motor can be made and the winding factor which depends on the pitch factor and distribution factor can be easily calculated. Slot space factor ranges from 0.3-0.35 for double layer winding and 0.65-0.7 for single layer winding [3].

*4)   Number of turns per phase and Phase current:*

From the output equation of the PM BLDC motor [1], the expression for product of number of turns per phase and phase current can be derived as follows:

During any $120^0$ interval of phase current the instantaneous power being converted from electrical to mechanical is given by:

$$P_d = \frac{P_o}{\eta_m} = mB_gLD_{go}I_{ph}W_mN_{ph} \qquad (2)$$

Where,

$P_o$ is the rated output power
$\eta_m$ is the mechanical efficiency of the motor
m is the number of phases conducting simultaneously
$B_g$ is the air gap flux density
L is the stack length of the motor
$D_{go}$ is the air gap diameter
$I_{ph}$ is the phase current
$w_m$ is the speed in rad/sec
$N_{ph}$ is the number turns per phase

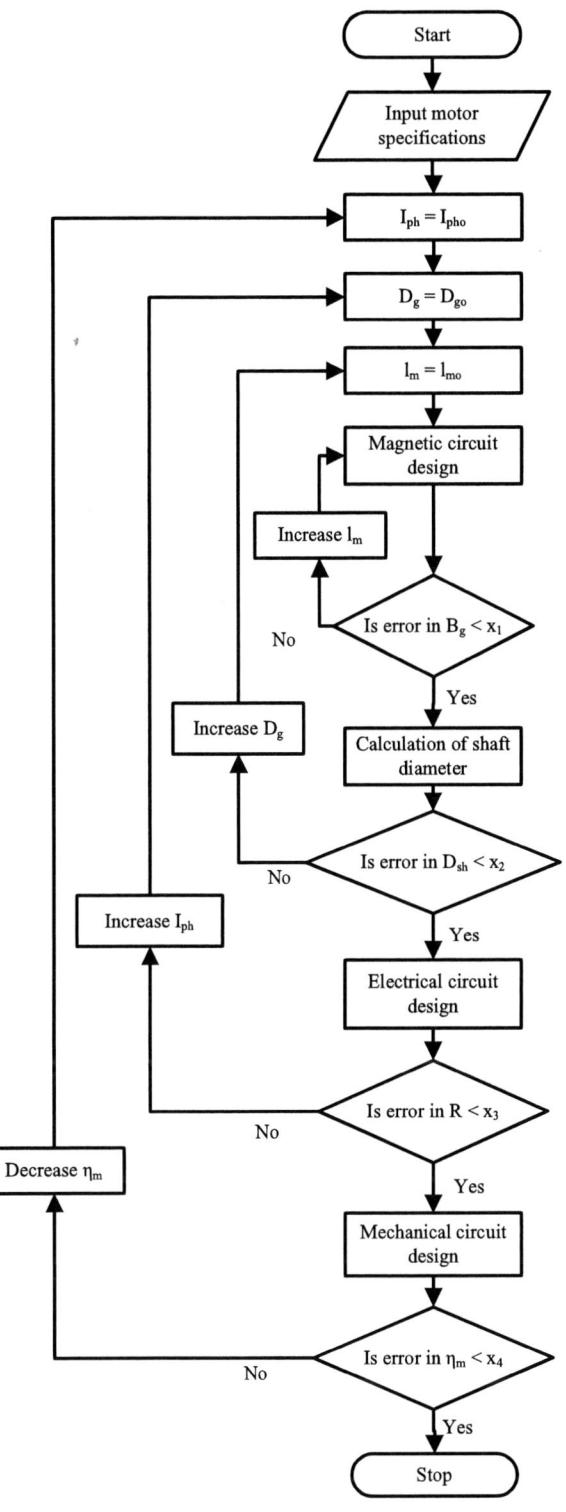

Fig. 1. Flow chart of the CAD program for the design of PM BLDC hub motor.

From the output equation, the product of number of turns per phase and phase current can be derived and is given by:

$$N_{ph}I_{ph} = \frac{P_d}{mB_gLD_{go}W_m} \qquad (3)$$

The two parameters, one is number of turns per phase and the other is phase current can be calculated by varying one parameter and calculating the other, until the two parameters satisfies the two equations for the resistance of the winding given by:

$$R = \frac{V - E_b}{I_{ph}} \qquad (4)$$

Where,

V is the supply voltage

$E_b$ is the back emf across the number of phases conducting simultaneously and

$I_{ph}$ is the supply current

$$R = \frac{\rho l}{a} \qquad (5)$$

Where,

$\rho$ is the resistivity of copper

l is the total conductor length of numbers of phases conducting simultaneously and

$\alpha$ is the area of cross section of conductor

C. *Mechanical design*

*1) Air gap diameter:*

Air gap diameter is given by the sum of shaft diameter, twice the stator iron thickness and twice the slot height. The shaft diameter can be assumed based on the application, where the motor is intended to be used. Next the thickness of the stator back iron can be easily calculated by assuming the flux density in the iron less than the saturation level of the ferromagnetic material used. And finally the slot height can be calculated from the number of turns per slot, conductor cross sectional area and the slot space factor.

*2) Weight of the motor:*

Weight of the designed PM BLDC hub motor can be easily calculated from the motor dimensions and mass density of different materials used. From the weight of the motor, the value of torque to weight ratio of the designed PM BLDC hub motor can be calculated.

*3) Mechanical efficiency:*

Mechanical efficiency is calculated by reducing the assumed mechanical efficiency in small steps until the error between the assumed and calculated mechanical efficiencies of the motor is less than the predetermined value. Hence the overall efficiency of the designed PM BLDC hub motor can be calculated easily.

III.    VALIDATION OF THE DESIGN USING FE ANALYSIS

Finite Element Analysis (FEA) is a computer simulation technique used in engineering analysis. It uses a numerical technique called the finite element method (FEM). In general,

there are three phases in any computer-aided engineering task: They are preprocessing, analysis and visualization. Design outputs of the designed PM BLDC hub motor obtained using the developed CAD program are given in Table I. Accuracy of the developed CAD program is established by conducting 2-D FE analysis in MAGNET software (student version) of the designed motor with two of the three phases excited at a time with rated current fed to the phase windings. Fig.2 gives the flux density plot of designed 30W PM BLDC hub motor obtained from FE analysis and this figure also helps in visualizing the geometry of the designed motor. The comparison of the CAD results with FE analysis results of the designed PM BLDC hub motor are given in Table II. From the comparison, it can be concluded that the FE analysis is validating the design output of CAD program for 30W PM BLDC hub motor meant for ceiling fan application.

Further validation of CAD results can be done with respect to average torque developed by the designed motor using FE analysis. The average torque developed by a motor using FE analysis can be calculated by finding the instantaneous torque developed by the motor at different rotor positions. Torque versus angle characteristics of the designed 30W PM BLDC hub motor for one cycle of the fundamental frequency, obtained using 2-D FE analysis by supplying currents in two of the three phases depending on the position of the rotor of the motor is given in fig.3. The average torque computed using the FE analysis is 1.02Nm against the designed torque of 0.998Nm.

TABLE I
DESIGN OUTPUTS OF THE 30W PM BLDC HUB MOTOR OBTAINED USING THE DEVELOPED CAD PROGRAM

| | |
|---|---|
| Full load efficiency (%) | 78.72 |
| Rotor outer diameter (mm) | 98.62 |
| Stack length of the motor (mm) | 15.5 |
| Total weight of the motor (gms) | 861.06 |
| Average torque developed (Nm) | 0.998 |
| Inductance per phase (H) | 0.0162 |

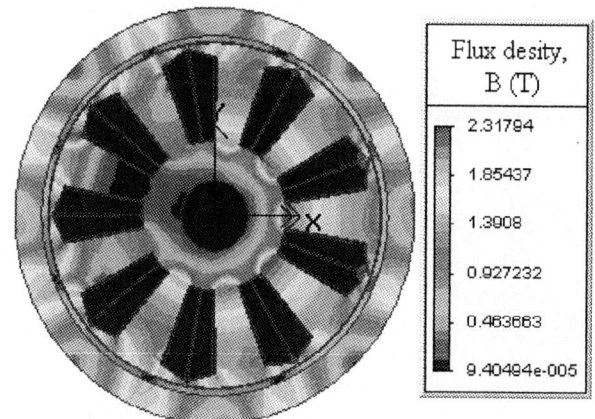

Fig. 2. Flux density plot of the designed 30W PM BLDC hub motor.

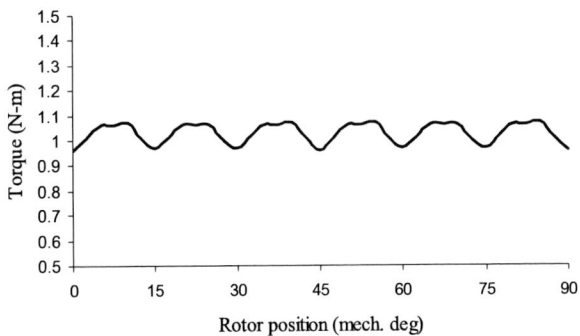

Fig. 3. Torque versus angle characteristics of the designed 30W PM BLDC hub motor.

TABLE II
COMPARISON OF DESIGN OUTPUTS OF 30W PM BLDC HUB MOTOR

| Parameter | Using the developed CAD program | Using FE analysis |
|---|---|---|
| Average flux density in air gap (T) | 0.6 | 0.596 |
| Flux density in teeth (T) | 2.0 | 1.96 |
| Flux density in stator iron (T) | 1.5 | 1.46 |
| Flux density in rotor iron (T) | 2.0 | 1.92 |
| Average torque (Nm) | 0.998 | 1.02 |

## IV. PARAMETRIC ANALYSIS OF THE DESIGNED PM BLDC MOTOR

Parametric analysis is carried out on the designed PM BLDC hub motor to get improved design in terms of efficiency and size. Stack length of the motor, number of poles and slot number are the three parameters considered for parametric analysis of the motor. Variation of efficiency with slot number is given in fig.4. As the slot number is increased, efficiency is increasing linearly up to slot number around 27. For further increase in slot number, there is no appreciable increase in efficiency of the motor.

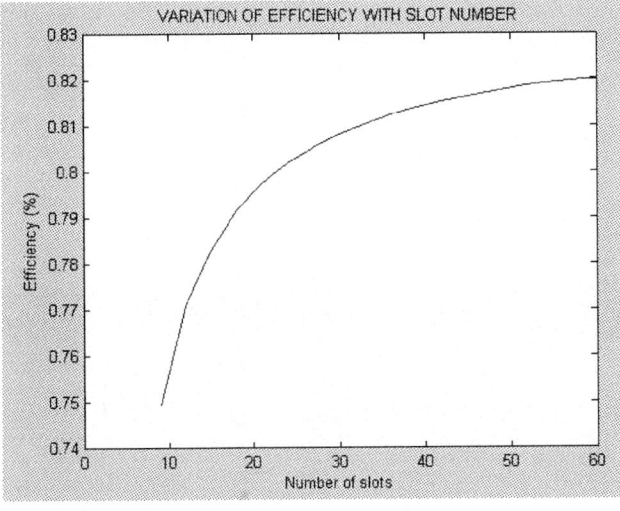

Fig. 4. Variation of efficiency with slot number of the designed 30W PM BLDC hub motor.

Variation of efficiency with stack length and pole number of the motor is given in fig.5 and fig.6 respectively. Parametric analysis results of the designed motor gives increase efficiency and reduced weight and size. The design outputs after parametric analysis are given in table III.

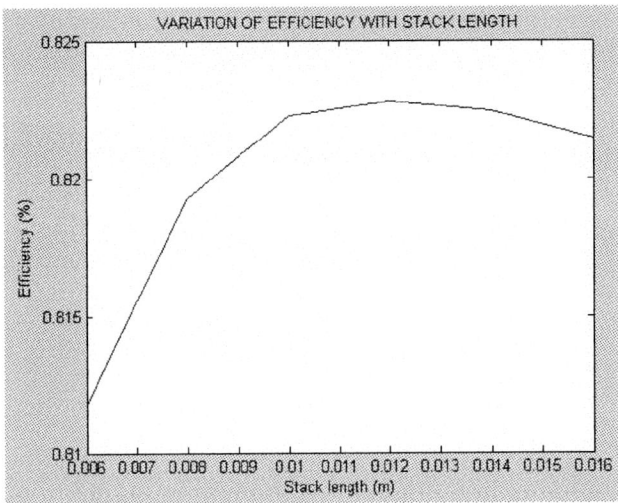

Fig. 5. Variation of efficiency with stack length of the designed 30W PM BLDC hub motor.

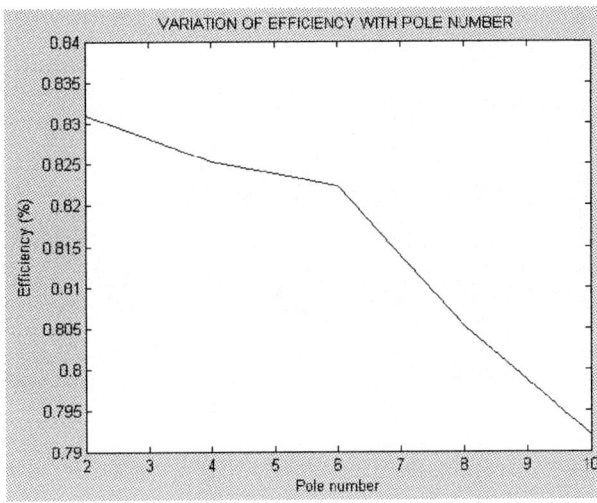

Fig. 6. Variation of efficiency with pole number of the designed 30W PM BLDC hub motor.

TABLE III
FINAL DESIGN OUTPUTS OF THE 30W PM BLDC HUB MOTOR AFTER INCORPORATING THE RESULTS OBTAINED FROM THE PARAMETRIC ANALYSIS

| | |
|---|---|
| Full load efficiency (%) | 82.65 |
| Rotor outer diameter (mm) | 88.62 |
| Stack length of the motor (mm) | 14 |
| Total weight of the motor (gms) | 464.46 |
| Average torque developed (Nm) | 0.978 |
| Inductance per phase (H) | 0.0579 |

Comparison between the motor parameters before and after parametric analysis is given in table IV. The design outputs from the parametric analysis are also validated using FE analysis. Fig.7. shows the flux density plot of the designed 30W PM BLDC hub motor using FE analysis after parametric analysis. The torque angle characteristics of the designed motor after parametric analysis can be obtained by the similar procedure given for the motor before parametric analysis.

TABLE IV
COMPARISON OF MOTOR PARAMETERS BEFORE AND AFTER PARAMETRIC ANALYSIS OF THE DESIGNED 30W PM BLDC MOTOR

| Parameter | Original design (before parametric analysis) | Final design (after parametric analysis) |
|---|---|---|
| Number of poles | 8 | 6 |
| Number of slots | 9 | 27 |
| Stack length (mm) | 15.5 | 14 |
| Magnet fraction | 0.89 | 0.67 |
| Average torque (Nm) | 0.998 | 0.978 |

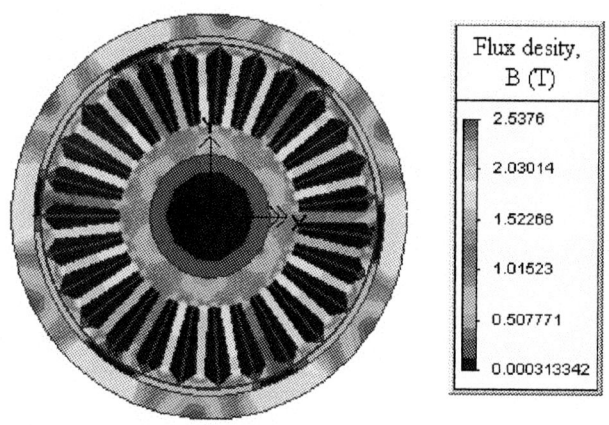

Fig. 7. Flux density plot of the final design of the 30W PM BLDC hub motor after parametric analysis.

## V. CONCLUSION

Computer aided design of a PM BLDC hub motor is presented in this paper. Using the developed CAD program a 30W PM BLDC hub motor meant for ceiling fan application is designed. The design procedure is divided into three major parts and is clearly described. The design validation with FE analysis is also given in this paper. There is design improvement in terms of efficiency and size is obtained using parametric analysis. The design outputs of the CAD program and parametric analysis for 30W PM BLDC hub motor are submitted. The design outputs from CAD program and parametric analysis are validated using FE analysis. The torque angle characteristics obtained using FE analysis of the designed motor also given in this paper.

## VI. REFERENCES

[1] T. J .E. Miller, Brushless Permanent-Magnet and Reluctance motor drives. Claredon press Oxford, 1989.
[2] D. C. Hanselman, Brushless Permanent Magnet Motor Design. New York: McGraw-Hill, 1994.
[3] I. R. Handershot and T. J. E. Miller, Design of Brushless Permanent Magnet Motors, Oxford, 1994.
[4] Parag R. Upadyay and K. R. Rajagopal, "FE analysis and CAD of Radial-Flux Surface Mounted Permanent Magnet Brushless DC Motors" IEEE Transactions on magnetic, Vol. 41 No. 10, October 2005.
[5] Z. Q. Zhu and David Howe, "Influence of design parameters on cogging torque in Permanent Magnet Machines" IEEE Transactions on energy conversion, Vol. 15, No. 4, December 2000.
[6] Min Dai, Ali Keyhani and Tomy Sebastian, "Torque ripple analysis of a PM Brushless DC motor using Finite Element Method" IEEE Transactions on energy conversion, Vol. 19, No. 1, March 2004.
[7] Jinyun, K. T. Chau, C. C. Chan and J. Z. Jiang, "Design and analysis of a new Permanent Magnet Brushless DC Machine" IEEE Transactions on magnetics, Vol. 36, No. 5, September 2000.
[8] P. R. Upadyay, K. R. Rajagopal and B. P. Singh, "Computer aided design of an Axial Field Permanent Magnet Brushless DC Motor for an electric vehicle" Journal of applied physics, Vol. 93, No. 10, 15 May 2003.

## VII. BIOGRAPHIES

**K. R. Rajagopal** (M'1998, SM'2000) was born in Alappuzha, Kerala, India in 1961. He received Diploma in Electrical Engineering from Carmel Polytechnic, Alappuzha, India in 1979, B. Tech. Degree in Electrical Engineering from the College of Engineering, Trivandrum, India in 1988, M. Tech. Degree in Power Electronics, Electrical Machines and Drives and Ph. D. Degree in Electrical Engineering from the Indian Institute of Technology Delhi, New Delhi, India during 1991 and 1998 respectively.

From 1980 to 1983, he was with Aluminum Industries Ltd. (ALIND), Trivandrum, India, as an Application Engineer (Relays), from 1983 to 1999, he was with the Indian Space Research Organization (ISRO), Trivundrum, India, where he was engaged in Analysis, Design, Development and Testing of Special Electrical Machines/Devices used in space applications. Since 1999, he is with the Indian Institute of Technology Delhi, New Delhi, India, where currently he is a Professor in Electrical Engineering Department.

He has published more than 30 papers in International Journals and more than 60 papers in International conference proceedings. He received Indian National Academy of Engineering (INAE) award for most Innovative Potential Project in Engineering during the year 1998. His fields of interest include Electrical Machines and Drives, Special Electrical Machines (Stepper Motors, Switched Reluctance Motors, PM BLDC Motors, Hysterisis Motors, etc.,), Magnetic Devices, Finite Element Analysis and CAD of Electrical Machines and Design of Energy Efficient Motors for Home Appliances.

**Chippa Sathaiah** was born in Chendoli, Karimnagar district, Andhra Pradesh, India in 1979. He received Diploma in Electrical and Electronics Engineering from the Jawaharlal Nehru Government Polytechnic, Hyderabad, Andhra Pradesh, India in 2000. He graduated from Osmania University with B. E. (Electrical and Electronics Engineering) Degree, Hyderabad, Andhra Pradesh, India in 2005.

He joined the Andhra Pradesh Transmission Corporation, Hyderabad, Andhra Pradesh, India as Sub Engineer Electrical during the year 2002. Currently he is pursuing his M. Tech. in Power Electronics, Electrical Machines and Drives at the Indian Institute of Technology Delhi, New Delhi, India. His fields of interest include Power Electronics, Electrical Machines and Drives, Finite Element Analysis and Computer Aided Design of Electrical Machines.

**2006 IEEE International Conference on Power Electronic, Drives and Energy Systems**

# Effect of Armature Reaction and Skewing on the Performance of Radial-flux Permanent Magnet Brushless DC Motor

Parag Upadhyay, and K. R. Rajagopal, *Senior Member, IEEE*

*Abstract* –In this paper the effect of armature reaction and skewing on the performance of a 70 W, 24 V, 350-rpm surface mounted radial-flux permanent magnet brushless dc (PM BLDC) motor is presented. The net reduction in peak torque of the motor due the effect of armature reaction is only 2.8%. The rate of reduction in airgap flux density because of the armature reaction is 5.59 mT/A from no-load to full-load. The 2D Finite element (FE) results are exploited for the analysis of skewing. The percentage torque ripple is reduced from 24% to 13.63% and 7.07% for the half and full slot pitch skewing respectively.

*Index Terms* — Motor, PM BLDC Motor, Armature reaction, Skewing, Permanent Magnet Motor, FE Analysis, Brushless Motor.

## I. INTRODUCTION

PERMANENT magnet brushless dc (PM BLDC) motors have advantages such as high efficiency, high torque density, high power density, and high reliability[1],[2]. These motors are inherently maintenance free because of the absence of a mechanical commutator. These advantages, combined with the ease of control had made them very attractive candidates increasingly being used in various domestic and industrial applications.

As per definition, the armature reaction refers to the magnetic field produced by currents in the stator coils and their interaction with the magnetic field produced by the permanent magnets[3]. In some locations in the airgap, the armature reaction may be adding the PM field and at some other locations in the airgap, it may be opposing the PM field. The presence of saturation phenomenon in soft magnetic materials in the stator, obviously lead to nonlinearity; therefore, addition or deletion of fields for the same armature MMF need not be the same at all locations in the airgap, which leads to variation in performance of the

motor for various armature MMF. The saturation in the magnetic circuit will increase the reluctance of the circuit and drive the PM to operate at a lower permeance coefficient. That means, the armature reaction determines the movement of the PM operating point under dynamic conditions.

## II. COMPUTER AIDED DESIGN

The computer aided design (CAD) program of PM BLDC motor is a two-loop MATLAB-program with different function call. The outer loop is to set and correct the assumed efficiency. Initially, the efficiency of the motor is assumed. The CAD program designs the motor and calculates the actual efficiency. The correction loop reduces the assumed efficiency and it is active till the error between the assumed efficiency and the actual efficiency is within the given limit. The inner loop is for reducing the difference between the assumed and actual airgap flux densities by changing the length of the magnet in a similar way. The magnet length is increased till the error between the two is less than given limit.

The airgap flux density, slot electric loading, winding factor, stacking factor, stator current density, slot space factor, magnet fraction, slot fraction, flux density in the stator back iron, etc. are assumed as fixed input parameters in the design. The phase current is decided by the power requirement and induced EMF decides the number conductors per slot. Motor specifications, type of configuration, material types, and other assumed data for the design are provided as the input.

Selection of standard wire gauge (SWG), material data for selected material number, specific iron loss data for a given material flux density and the frequency, etc. are also part of the developed program. Number of magnet poles, Number of slots/pole/phase, type of permanent magnet material and its grade, type of soft magnetic material, airgap, current density, stator flux density, airgap flux density, slot electric loading, magnet fraction, slot fraction, stacking factor, slot space factor, winding factor, flux density in the rotor core are the parameters required for the design.

---

Prof. P. R. Upadhyay is with the Electrical Engineering Department, Nirma University of Science and Technology, Ahmedabad, INDIA. (e-mail: pru_nirma@yahoo.com).

Prof. K.R.Rajagopal is with the Electrical Engineering Department, Indian Institute of Technology Delhi, New Delhi, INDIA. (e-mail: rgopal@ee.iitd.ac.in).

0-7803-9771-1/06/$25.00 ©2006 IEEE

## III. ARMATURE REACTION EFFECTS

In the PM BLDC motor, when there is no armature current, magnetic neutral axis (MNA) coincides with the geometrical neutral axis (GNA) of the PM poles; that means the interpolar axis. But when the armature field is also present, MNA will shift from the GNA in a direction and by an angle, decided respectively by the polarity and strength of the armature field. This deviation depends on the amount of magnetic loading and the slot electrical loading.

The torque vs. rotor position characteristics as well as the developed average torque will change with the MNA. The net effect of the armature reaction is to reduce the torque developed by the motor.

Considering current to be constant through the coils, in the two-phase excited condition, so as to have maximum torque, the armature field vector is always considered to be perpendicular to the PM field vector as shown in Fig. 1. It can also be seen that the resultant airgap MMF can be more or less in magnitude than the PM field MMF depending on the angle at which the armature MMF acts.

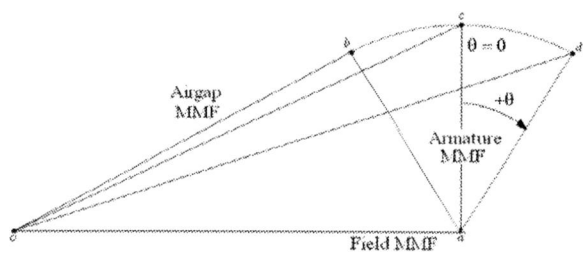

Fig. 1. Vector representation of the airgap MMF of the radial-flux surface mounted PM BLDC motor.

A typical case of a 3-phase, bipolar PM BLDC motor, which will be excited in 2-phase in series configuration changing the phase combinations at every 60° electrical can be well explained using Fig. 1. Ideally, we would like to have the armature MMF as perpendicular to the PM field MMF. But the phase combinations will have the currents in them for 60° electrical. That means, the phases must be excited 30° electrical prior to the perpendicular position and up to 30° electrical beyond the perpendicular position. Therefore, as shown in Fig.1, the resultant airgap MMF ($MMF_{AG}$) vector can vary in magnitude and also in phase from $ob$ to $od$. The mathematical expression for $MMF_{AG}$ is given below:

$$MMF_{AG} = \sqrt{\left(MMF_{PM} + MMF_A \sin \theta\right)^2 + \left(MMF_A \cos \theta\right)^2}$$

where, $MMF_{PM}$ is the PM field MMF/pole in the airgap, $MMF_A$ is the armature MMF/pole in the airgap and $\theta$ is the angle between the ideal armature MMF vector $ac$ and the actual armature MMF vector (which can vary from $ab$ to $ad$). The counter clockwise (CCW) direction of rotation of the motor is taken as positive. However, in Fig.1, the CW direction appears to be positive, which may be clearly understood as the actual rotation of the motor to the CCW direction.

For example, in the 70 W, 24 V, 350 rpm surface mounted radial-flux PM BLDC motor designed using the developed CAD program with the desired airgap flux density of 0.8 T, and with the Nd-Fe-B 35 PM for which the retentivity, coercive force, and recoil permeability are 1.23 T, 890 kA/m and 1.01 respectively, the no-load PM operating point is given by the CAD program as $Hm = -579$ A/mm and $Bm = 0.93$ T, based on which the PM MMF/pole ($H_c l_m$) is worked out to be 2895 A. The armature MMF/pole is also worked out based on the slot electric loading and its value is 221 A. Knowing these two MMF values, resultant airgap MMF for various rotor positions from $-30°$ to $30°$ electrical can be worked out based on which corresponding developed torques can be calculated.

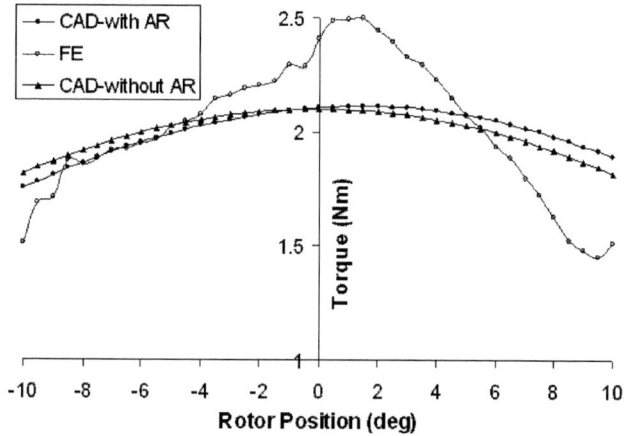

Fig. 2. Torque profiles of the designed 70 W radial-flux surface mounted PM BLDC motor.

The calculated torque profiles with and without considering for the armature reaction are given in Fig. 2. The rotor position in mechanical degrees is taken in the x-axis of this figure. It may be noted that $-10°$ to $10°$ mechanical shown in this figure corresponds to $-30°$ to $30°$ electrical in the 6-pole motor. It can be observed that the armature reaction reduces the developed torque and also shifts the torque profile symmetry away from the polar axis towards the direction of rotation of the motor, CCW. That means, there is a forward shift of the MNA; in this case it is observed that the shift is 1.9° mechanical. Effectively, the shifting of the MNA depends on the PM length (or the magnetic loading) and the slot electric loading. Thus, the important inference is that in motor if the slot electric loading is more, then the shift of MNA is more, which will result in reduced average developed torque and increased torque ripples. Instead, if we provide more magnetic loading, then the shift in MNA will be less, but will result in more cogging torques and increased cost.

Figure 3 shows the torque profiles for the 70 W PM

BLDC motor at full-load and also at half load conditions from which it can be observed that the armature reaction effects are more predominant at higher loads.

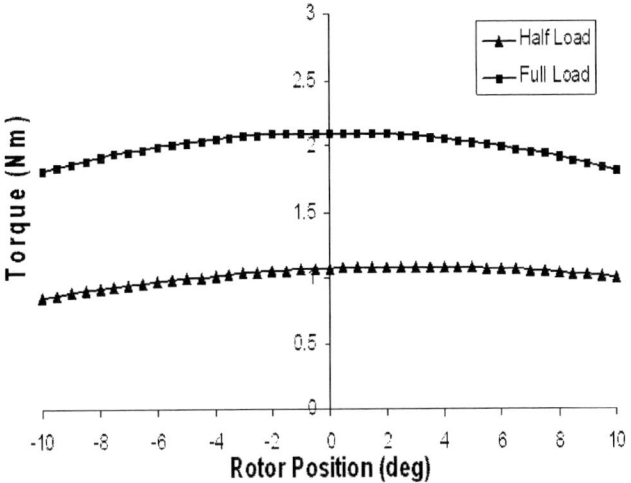

Fig. 3. Torque profiles of radial-flux surface mounted PM BLDC motor obtained using the CAD program at full-load and half-load.

## IV. VERIFICATION BY FE ANALYSIS

FE analysis helps in visualizing the effects of armature reaction very clearly. Analysis is carried out on the designed 70 W, 24 V, 350 rpm surface mounted radial-flux PM BLDC motor for the no-load and the full-load conditions. The actual torque profile with the armature reaction obtained from the FE analysis is also given in Fig.2. Even though the FE based and CAD based torque profiles are not that close, two observations are very clear; (i) the MNA shifts towards the direction of rotation, and (ii), the average torque decreases with the armature reaction. The average torque worked out by the CAD program and by the FE analysis are 1.90 Nm and 2.02 Nm, which are fairly matching, but with the deviations caused by the slotting, which will very effectively be considered in the FE analysis.

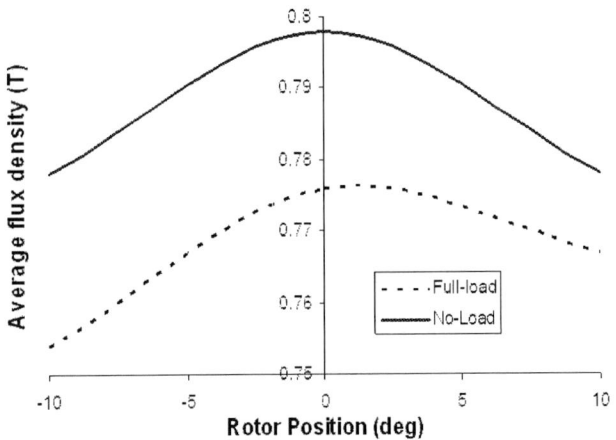

Fig. 4. Airgap flux density vs. rotor position plots of the designed 70 W radial-flux surface mounted PM BLDC motor.

Figure 4 gives the average airgap flux density at no-load as well as at full-load for the 70W PM BLDC motor obtained from the FE analysis. It clearly indicates the reduction in the airgap flux density and also the shifting of the MNA. The average airgap flux density (effectively, this is the average of the average flux densities obtained for the rotor positions varying from –30° to 30° electrical) in the airgap changes from the no-load value of 0.785 T to 0.766 T at full-load; the decrease of about 2.4% is obviously due to the armature reaction effects.

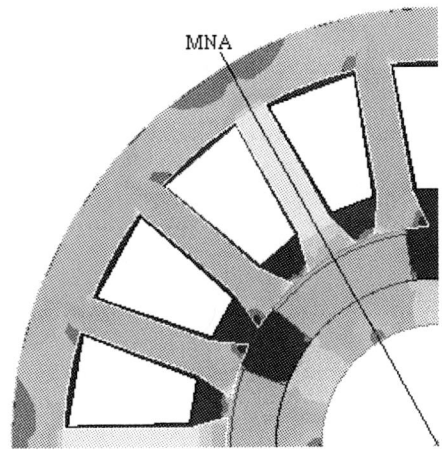

Fig. 5. Flux density plot of the designed 70 W radial-flux surface mounted PM BLDC motor at no-load.

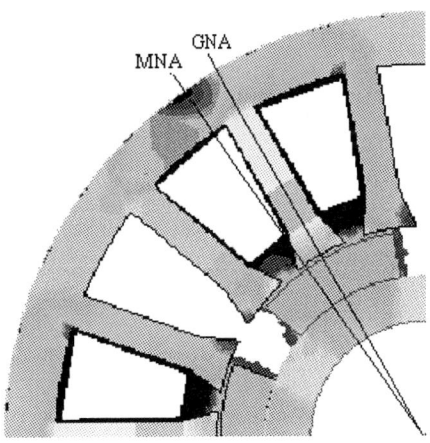

Fig. 6. Flux density plot of the designed 70 W radial-flux surface mounted PM BLDC motor at full-load.

The flux density plots of the motor for no-load and full-load conditions obtained from the FE analysis are shown in figures 5 and 6 respectively. From the Fig. 6, the actual shift of the MNA is observed as 1.8° as against the CAD based calculated value of 1.9°. Even though, it is not linear, the rate of reduction in the average airgap flux density because of armature reaction from the no lad to full-load condition of the motor is accounted to be 5.59 mT/A; this is an acceptable rate in majority of the motors used in domestic and industrial applications – thanks to the large airgap (contributed by the physical airgap and the PM length and also the availability of high energy of the PM materials) which keeps the adverse effects of armature reaction to very low limits.

## V. SKEWING OF SURFACE MOUNTED MOTOR

The same 70 W, 24 Volt, 350 rpm, radial-flux surface mounted PM BLDC motor is used in this analysis. In case of the radial-flux surface mounted PM BLDC motors, skewing of stator slots is a viable and easy method for the torque ripple minimization by design. Here, rather than doing a 3-D FE analysis, 2-D FE parametric analysis is carried out by rotating the stator to the required skew angle in small steps by maintaining the rotor to the initial position itself and obtaining the torque profile for each step and then finally adding the torque profiles of all the steps to get the torque profile of the motor with the stator slots skewing.

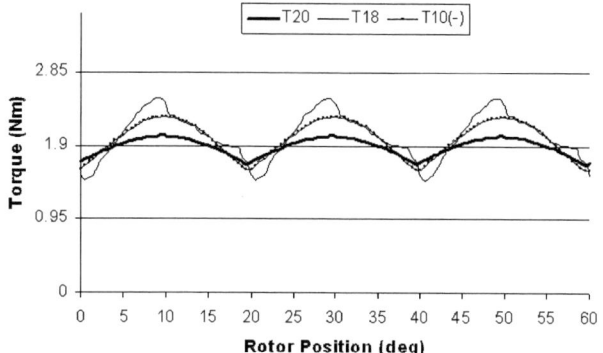

Fig. 7. Torque profiles of the designed 70 W radial-flux surface mounted PM BLDC motor with skewing in stator slots.

TABLE I
AVERAGE TORQUE AND TORQUE RIPPLE FOR DIFFERENT SKEW ANGLES IN A 70 W RADIAL-FLUX SURFACE MOUNTED PM BLDC MOTOR

| Skew angle (° mechanical) | 20 | 0 | 10 |
|---|---|---|---|
| Average torque (Nm) | 1.89 | 2.03 | 1.99 |
| Percentage torque ripple | 7.07 | 24.18 | 13.63 |

The torque profiles of the motor with two different skew angles in stator slots, (i) with skew angle equal to a slot pitch (20°) and (ii) with the skew angle equal to a half of the slot pitch (10°) along with the torque profile of the motor with no skewing are given in Fig. 7. The average torque and the ripple toque in each case is worked and given in table-1.

It can be observed that the skewing reduces the torque ripple, but with a penalty of reduction in average torque. For a full slot pitch skewing, the torque ripple is the least, but with a 6.7% reduction in the average torque. With half a slot pitch skewing, the torque ripple is 13.63%, but the reduction in the average torque is only 1.97%.

## VI. CONCLUSIONS

The effect of armature reaction reduces the flux

density in the airgap. Larger airgap and high-energy permanent magnet material with low permeability will reduce the adverse effects of the armature reaction. In the typical 70 W radial-flux surface mounted PM BLDC motor designed in this paper, the net reduction in peak torque of the motor due the effect of armature reaction is only 2.8%. The rate of reduction in airgap flux density because of the armature reaction is 5.59 mT/A from no-load to full-load. From the analysis it is observed that the torque ripple is reduced from 24% to 13.63% and 7.07% for the half and full slot pitch skewing respectively. The effect of skewing can be observed from the 2D FE analysis.

## VII. REFERENCES

[1] D. C. Hanselman, Brushless permanent magnet design McGraw Hill Inc. Publ., 1994.
[2] Handershot J.R., Miller TJE, "Design of brushless Permanent magnet motors", Oxford Sc. Publ., 1994.
[3] Parag R. Upadhyay, K. R. Rajagopal, "Effect of Armature Reaction on the Performance of an Axial-Field Permanent Magnet Brushless DC Motor Using FE Method" IEEE Trans. Magnetics Vol. 40, NO. 10, pp-2023-2025, July 2004.

## VIII. BIOGRAPHIES

**K. R. Rajagopal** (M'1998, SM'2000) was born in Alappuzha, Kerala, India in 1961. He received Diploma in Electrical Engineering from Carmel Polytechnic, Alappuzha, India in 1979, B. Tech. Degree in Electrical Engineering from the College of Engineering, Trivandrum, India in 1988, M. Tech. Degree in Power Electronics, Electrical Machines and Drives and Ph. D. Degree in Electrical Engineering from the Indian Institute of Technology Delhi, New Delhi, India during 1991 and 1998 respectively.

From 1980 to 1983, he was with Aluminum Industries Ltd. (ALIND), Trivandrum, India, as an Application Engineer (Relays), from 1983 to 1999, he was with the Indian Space Research Organization (ISRO), Trivandrum, India, where he was engaged in Analysis, Design, Development and Testing of Special Electrical Machines/Devices used in space applications. Since 1999, he is with the Indian Institute of Technology Delhi, New Delhi, India, where currently he is a Professor in Electrical Engineering Department.

He has published more than 30 papers in International Journals and more than 60 papers in International conference proceedings. He received Indian National Academy of Engineering (INAE) award for most Innovative Potential Project in Engineering during the year 1998. His fields of interest include Electrical Machines and Drives, Special Electrical Machines (Stepper Motors, Switched Reluctance Motors, PM BLDC Motors, Hysterisis Motors, etc.,), Magnetic Devices, Finite Element Analysis and CAD of Electrical Machines and Design of Energy Efficient Motors for Home Appliances.

**Parag Upadhyay** was born in Bhavnagar, Gujarat, India in 1971. He received B.E. Degree in Electrical Engineering from the L.E. College Morbi, Gujarat, India in 1992, M. Tech. Degree in Power Electronics, Electrical Machines and Drives in Electrical Engineering from the Indian Institute of Technology Delhi, New Delhi, India during 2000 and he is pursuing Ph.D. from the Indian Institute of Technology Delhi, New Delhi, India since 2003.

In the year 1993, he worked in Menpara Pumps (P) Ltd. as testing and quality engineer and Coretech Int (P) Ltd. as Project Engieer. From 1993 to 1996, he was with L. D. Engineering Colege, Ahmedabad, Gujarat, India as a lecturer. Since 1996, he is with the Institute of Technology, Nirma University of Science and Technology, Ahmedabad, Gujarat, India, where currently he is an Assistant Professor in Electrical Engineering Department.

He has published about 9 papers in International Journals and about 12 papers in International conference proceedings. He received Prof. A. K. Sinha award for securing the highest CGPA in the electrical engineering department of IIT Delhi for the year 2000 and also received L&T, ISTE, Best M.Tech thesis award in 2001. His fields of interest include Electrical Machines and Drives, Special Electrical Machines (PM BLDC Motors, Axial Flux Motors etc.), Finite Element Analysis and CAD of Electrical Machines.

**2006 IEEE International Conference on Power Electronic, Drives and Energy Systems**

# Torque Ripple Minimization of Interior Permanent Magnet Brushless DC Motor Using Rotor Pole Shaping

Parag Upadhyay, and K. R. Rajagopal, *Senior Member, IEEE*

*Abstract –* **In this paper effort is made for the torque ripple minimization of a 70 W, 24 Volt, 350 rpm, interior permanent magnet brushless dc (IPM BLDC) motor. The rotor pole shaping along with the pole shifting is analyzed. The results obtained from FE analyses have been discussed. The proper design and geometry of the motor reduces the cogging torque and the proper excitation reduces the mutual torque ripple. The torque ripple minimization using only the stator excitation makes the controller inefficient as well as costlier. Using the methods used in this paper, the torque ripple of the IPM BLDC motor is reduces and also the average torque is improved. In this method concept of magnet shifting is utilized for the torque improvement and rotor saliency is provided to reduce the cogging torque. Detailed analysis and explanation is given in this paper. The wide range of torque control by changing the switching intervals is an additional advantage of this design.**

*Index Terms —* **Motor, PM BLDC Motor, Torque Ripple, Magnet shifting, Pole shaping**

## I. INTRODUCTION

$\mathbf{P}$ERMANENT magnet brushless dc (PM BLDC) motors have advantages such as high efficiency, high torque density, high power density, and high reliability. These motors are inherently maintenance free because of the absence of a mechanical commutator. These advantages, combined with the ease of control had made them very attractive candidates increasingly being used in various domestic and industrial applications.

The interior permanent magnet (IPM) motors are widely used for the home as well as industrial applications. The interior PM BLDC motor is a special case of radial-flux PM BLDC motor, in which the permanent magnets are embedded inside the rotor, thereby making this motor suitable for high-speed applications. The stator of the motor is similar to that of the radial-flux surface mounted PM BLDC motor. The geometry of a typical 6-pole interior PM BLDC motor is shown in Fig. 1.

Prof. P. R. Upadhyay is with the Electrical Engineering Department, Nirma University of Science and Technology, Ahmedabad, INDIA. (e-mail: pru_nirma@yahoo.com).

Prof. K.R.Rajagopal is with the Electrical Engineering Department, Indian Institute of Technology Delhi, New Delhi, INDIA. (e-mail: rgopal@ee.iitd.ac.in

A CAD program for the interior PM BLDC motor is developed. This program has two loops for the correction of efficiency and the airgap flux density. The outer loop is to set and correct the assumed efficiency. Initially, the efficiency of the motor is assumed. The CAD program designs the motor and calculates the actual efficiency. The correction loop reduces the assumed efficiency and it is active till the error between the assumed efficiency and the actual efficiency is within the given limit. The inner loop is for reducing the difference between the assumed and actual airgap flux densities by changing the length of the magnet in a similar way. The magnet length is increased till the error between the two is less than given limit.

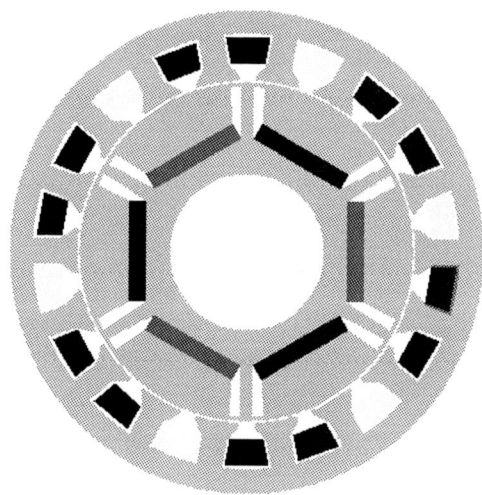

Fig. 1. Cross-sectional view of a 6-pole interior PM BLDC motor.

In case of interior the PM BLDC motor, the magnet length is increased till it reaches to the maximum available length that can be accommodated in the rotor. If the required magnet length is more than the available length in the rotor, then the assumed airgap flux density is reduced and the output equation is solved again to obtain the main dimensions; the entire design is repeated and obviously bigger mean diameter is selected for the motor. Even though, such a phenomenon can happen in a radial-flux surface mounted PM BLDC motor also, it is not that critical as in the interior PM BLDC motor because the magnet

0-7803-9771-1/06/$25.00 ©2006 IEEE

leakage in the surface mounted case will be very low as compared to that of the interior PM motor.

The phase current is decided by the power requirement and induced EMF decides the number conductors per slot. Motor specifications, type of configuration, material types, and other assumed data for the design are provided as the input. Selection of standard wire gauge (SWG), material data for selected material number, specific iron loss data for a given material flux density and the frequency, etc. are also part of the developed program.

Basically, the developed CAD program consists of four stages, namely (i) the main dimensions, (ii) the stator design, (iii) the rotor design, and (iv) the performance calculations.

The permanent magnet rotor design executes the inner loop of the CAD program. The program, as a first approximation, assumes the length of the PM as equal to that of the airgap, and proceeds with any of the following two approaches, which has to be selected by the designer.

The first approach is to achieve the maximum available area for the magnet by placing the magnet pole towards the outer surface of the rotor, such that the corner points of the rectangular magnets are on the external periphery of the rotor core and makes the selected fraction of the pole pitch; any further increase in length required for the magnet will be considered in subsequent runs, radially inwards. The difference in actual flux density and the assumed flux density is reduced by the corrective loop where the length of the magnet is increased such that the value of this difference is obtained within the tolerable value. When the required magnet length exceeds the limit of proximity with the neighboring magnet, the program reverts back and restarts by assuming a decreased airgap flux density. When the assumed flux density is reduced, the rotor outer diameter will increase, hence for reduced flux density there will be more scope for increase in the magnet length.

In the second approach, the designer can decide at what fraction of the rotor outer diameter, the rectangular magnets having selected width based on the fraction of the pole pitch, must initially be placed, and moved outwards for further increase in magnet lengths. Here also, the difference in actual flux density and the assumed flux density is reduced by the corrective loop where the length of the magnet is increased such that the value of this difference is obtained within the tolerable value. When the required magnet length exceeds the available space in the designed rotor, the program reverts back and restarts by assuming a decreased airgap flux density. Due to this reason, normally, the outer diameters of the stator as well as the rotor will be more in the interior PM BLDC motors compared to the surface mounted PM BLDC motors.

## II. ROTOR POLE SHAPING

Torque ripple minimization is also a requirement for improved performance of the PM BLDC motors. The torque ripple in a trapezoidal PM BLDC motor is higher when compared to the sinusoidal PM BLDC motor or the PM synchronous motor (PMSM). There are basically two

sources for the torque ripple namely, (i) cogging torque created by the stator slots when interacted with the rotor magnetic field and (ii) mutual torque ripple created by the mismatch between excitation currents to cancel cogging torque [1]. The torque ripple can be minimized by reducing the cogging torque or by reducing the mutual torque ripple. The proper design and geometry of the motor reduces the cogging torque and the proper excitation reduces the mutual torque ripple. The torque ripple minimization using only the stator excitation makes the controller inefficient as well as costlier.

In this work, an effort is made to have both the improvements in the torque, namely, the torque enhancement and the torque ripple minimization for the designed 70 W, 24 Volt, 350 rpm, interior PM BLDC motor. Magnet shifting is utilized for the torque improvement and shaping of the rotor pole outer surface is done to reduce the cogging torque.

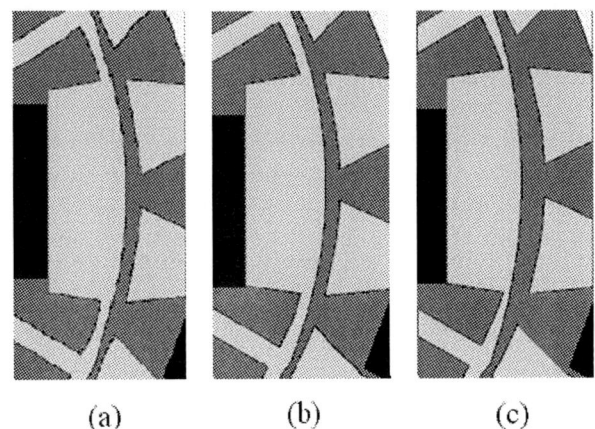

Fig. 2. Magnet shifting and rotor pole shaping in 70 W radial-flux interior PM BLDC motor
  (a) Rotor pole shaping – airgap increased at the rotor poles and decreased at the interpolar axes.
  (b) Rotor pole shaping – airgap increased at the rotor poles and decreased at the interpolar axes along with the backward magnet shifting.
  (c) Case (b) but with more airgap.

Figure 2 (a) shows the case with only the rotor pole shaping in which airgap increased at the rotor poles and decreased at the interpolar axes. The airgap variation is kept between 0.5 mm to 1.0 mm, whereas the original airgap was 0.5 mm. Figures 2 (b) and 2 (c) have both the techniques, backward magnet shifting and also the rotor shaping are used in them; only difference being larger airgap (0.5 mm to 1.5 mm) in Fig. 2 (c).

The torque profiles for all these geometries obtained from the 2-D FE analyses are shown in Fig. 3. It can be seen that with the rotor pole shaping, the torque profile in the forward angle segment of –20° to 0° is better for the phase excitation than the segment of –10° to 10° in the sense that it gives higher average torque and also reduce torque ripple. Also, with the rotor pole shaping, the torque profile in the backward angle segment of 0° to 20° is better for the phase excitation than the segment of –10° to 10° in

169

the sense that it gives less average torque and reduce torque ripple. The reduced ripple in this case is because of the fact that with the new rotor pole shape, the airgap flux density becomes more trapezoidal and with the rectangular current fed to the motor phases, the developed torque shall move more towards a rectangular shape.

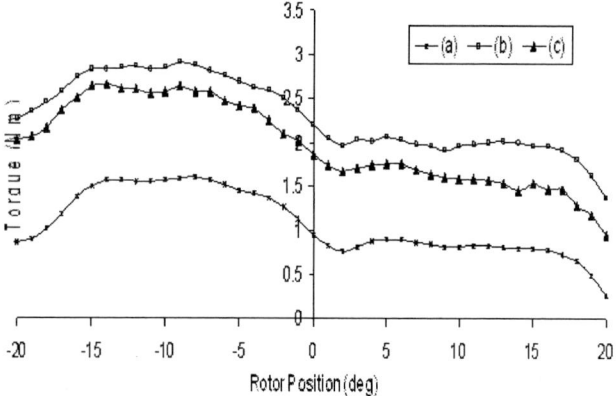

Fig. 3. Torque profile for different saliencies of PM BLDC motor.

## III. CONCLUSIONS

The proper design and geometry of the motor reduces the cogging torque and the proper excitation reduces the torque ripple. Using the rotor pole shaping, the torque ripple of the IPM BLDC motor is reduces and also the average torque is improved. Magnet shifting is also helps in improving the torque of IPM BLDC motor. The torque profile obtained has two ranges of operation in which the torque remains constant; one with the higher torque and the other with the reduced torque. Hence, the wide range of torque control by changing the switching intervals is obtained.

## IV. REFERENCES

[1]  Hanselman D. C., "Effect of skew, pole count and slot count on brushless motor radial force, cogging torque and back EMF" IEE Proceedings on EPA, Vol. 144, Issue: 5, pp 325 – 330, Sept. 1997.

[2]  Zeroug H.; Boukais B. and Sahraoui, H., "Analysis Of Torque Ripple In A Brushless DC Motor", IEEE Transactions on Magnetics, Vol. 38, pp-1293 – 1296, March 2002.

[3]  Parsa L, Hao L and Toliyat H.A, "Optimization of average and cogging torque in 3-phase IPM motor drives", Industry Applications Conference, 2002. 37th IAS Annual Meeting. Vol. 1, pp-417 – 424.

## V. BIOGRAPHIES

**K. R. Rajagopal** (M'1998, SM'2000) was born in Alappuzha, Kerala, India in 1961. He received Diploma in Electrical Engineering from Carmel Polytechnic, Alappuzha, India in 1979, B. Tech. Degree in Electrical Engineering from the College of Engineering, Trivandrum, India in 1988, M. Tech. Degree in Power Electronics, Electrical Machines and Drives and Ph. D. Degree in Electrical Engineering from the Indian Institute of Technology Delhi, New Delhi, India during 1991 and 1998 respectively.

From 1980 to 1983, he was with Aluminum Industries Ltd. (ALIND), Trivandrum, India, as an Application Engineer (Relays), from 1983 to 1999, he was with the Indian Space Research Organization (ISRO), Trivundram, India, where he was engaged in Analysis, Design, Development and Testing of Special Electrical Machines/Devices used in space applications. Since 1999, he is with the Indian Institute of Technology Delhi, New Delhi, India, where currently he is a Professor in Electrical Engineering Department.

He has published more than 30 papers in International Journals and more than 60 papers in International conference proceedings. He received Indian National Academy of Engineering (INAE) award for most Innovative Potential Project in Engineering during the year 1998. His fields of interest include Electrical Machines and Drives, Special Electrical Machines (Stepper Motors, Switched Reluctance Motors, PM BLDC Motors, Hysterisis Motors, etc.,), Magnetic Devices, Finite Element Analysis and CAD of Electrical Machines and Design of Energy Efficient Motors for Home Appliances.

**Parag Upadhyay** was born in Bhavnagar, Gujarat, India in 1971. He received B.E. Degree in Electrical Engineering from the L.E. College Morbi, Gujarat, India in 1992, M. Tech. Degree in Power Electronics, Electrical Machines and Drives in Electrical Engineering from the Indian Institute of Technology Delhi, New Delhi, India during 2000 and he is pursuing Ph.D. from the Indian Institute of Technology Delhi, New Delhi, India since 2003.

In the year 1993, he worked in Menpara Pumps (P) Ltd. as testing and quality engineer and Coretech Int (P) Ltd. as Project Engieer. From 1993 to 1996, he was with L. D. Engineering Colege, Ahmedabad, Gujarat, India as a lecturer. Since 1996, he is with the Institute of Technology, Nirma University of Science and Technology, Ahmedabad, Gujarat, India, where currently he is an Assistant Professor in Electrical Engineering Department.

He has published about 9 papers in International Journals and about 12 papers in International conference proceedings. He received Prof. A. K. Sinha award for securing the highest CGPA in the electrical engineering department of IIT Delhi for the year 2000 and also received L&T, ISTE, Best M.Tech thesis award in 2001. His fields of interest include Electrical Machines and Drives, Special Electrical Machines (PM BLDC Motors, Axial Flux Motors etc.), Finite Element Analysis and CAD of Electrical Machines.

**2006 IEEE International Conference on Power Electronic, Drives and Energy Systems**

# Design and Development of a In-Wheel Brushless D.C. Motor Drive for an Electric Scooter

N. Ravi, S. Ekram, and D. Mahajan

*Abstract*--The gloomy answers to the oil crises compel to strive for sustainable road transportation. One such effort in this regard is reported in this paper. Due to the advent of power electronics; Brush-less DC motor is considered as potential drive for automotive applications. Design of a direct drive, Brush less DC Motor is presented. Many constraints have been considered, such as the size of the motor, maximum driving current and maximum output power. The diameter and the stack length are limited by using modern analytical tools of the prototype vehicle. Nobel analytical tools are used to obtain static and dynamic characteristics of the motor and geometrical verifications for low cost efficient design. A low cost electronic controller compatible for present application has been developed and discussed in this paper. The prototype is fabricated and tested on a lightweight electric vehicle. The performance results are presented and discussed in this paper.

*Index Terms*-- Brushless machines, Microcontrollers, Road vehicle electric propulsion.

## I. INTRODUCTION

INDIA possesses sufficient conditions for developing electric scooters, with the highest scooter per capita density. It is also the major producer of motor scooters around the world. Many configurations that are suitable for the electric vehicles (EV) exist and as a consequence, the required characteristics for the electric machine vary significantly. Vehicles have typical power requirements ranging from a few kW's to several tens of kW and the speed varies from a few thousand rev/min to several tens of thousand rev/min. the EV targeted under this project is a mid range two wheeler. Most electric scooter currently developed in China employ high-speed (3500 r/min) dc commutators motors (some using rare-earth permanent magnets), the speed being reduced by mechanical gears. The presence of the commutators and brush-gear gives rise to sparking and brush wear problems, requiring frequent maintenance. The use of speed-reduction gears also lowers the drive efficiency and increases the weight of the drive train. To overcome these problems, a brushless dc motor drive may be employed. Brushless motors provide less maintenance, long life, low EMI, and quiet operation. They produce more output power per frame size than PM or shunt

---

This work was carried in AMDTC Lab, Corporate R&D and Quality, Crompton Greaves Ltd, Kanjurmarg (East), Mumbai-400042, India.

wound motors and gear motors. Low rotor inertia improves acceleration and deceleration times while shortening operating cycles and their linear speed/torque characteristics produce predictable speed regulation. With brushless motors, brush inspection is eliminated making them ideal for limited access areas and applications where servicing is difficult. Low voltage models are ideal for battery operation and making it ideal for electric scooter. In this paper, the design and construction of a direct-drive, brushless dc motor for use on an electric scooter will be described. Certain design features pertinent to the specific application are highlighted. Experimental results obtained to demonstrate the feasibility of the proposed motor design.

## II. MOTOR DRIVE REQUIREMENTS

THE electric motor for use on an electric scooter should be small in size, light in weight, and should have a high efficiency. Due to space limitations, the motor must have a short axial length. The motor must provide sufficient torque to overcome the road load which consists of the aerodynamic drag, the rolling resistance, the tractive force against the gravitational pull when climbing up a slope, and the tractive force for acceleration. Rolling friction depends on both tyre and road surface characteristics, namely the interaction between these two. However, since the tires in focus of this material are conventional types, it can be said that it depends on the road type. Many constraints have to be considered, such as the size of the motor maximum driving current and maximum output power. The diameter and the stack length are limited by the tire size of the prototype vehicle.

TABLE I
SCOOTER PARAMETERS

| Sr. No | Parameter | Values | Unit |
|---|---|---|---|
| 1 | Total mass of vehicle, people, and cargo (m) | 200 | kg. |
| 2 | Maximum acceleration from IDC | 0.65 | $m/s^2$ |
| 3 | Coefficient of normal tyre rolling resistance (Crr) | 0.013 | |
| 4 | Density of air, approximately air approximately ($\rho_a$) | 1.23 | $Kg/m^3$ |
| 5 | Drag coefficient (Cd) | 0.6 | |
| 6 | Frontal area (Af) | 0.8 | $m^2$ |
| 7 | Angle of slope ($\theta$) | 12 | degree |

Using these values together with some appropriate wheel dimensions, it is quite straightforward to calculate necessary torque and power as a function of road surface conditions and vehicle speed.

In order to study the feasibility of the motor design, a prototype motor was designed and constructed. The required technical specifications are given in Table II.

TABLE II
DESIGN SPECIFICATION

| Sl No | Item | Description |
|---|---|---|
| 1 | Volts | 48 V, DC (Voltage range from 43V to 55V) |
| 2 | RPM | 520 RPM (40 KMph at max load) |
| 4 | Output power | 500 watt |
| 5 | Current (amps – max) | 12.5 amps |
| 6 | Efficiency | >80% |
| 7 | Insulation class | B |
|  | Cooling | Natural |
| 8 | Winding temp. rise | 45° C |
| 9 | Max body temp | 55° C |
| 10 | Protection | IP 65 |
| 11 | Bearings | Sealed Bearings |
| 12 | Body type & material | Aluminum /cast iron body |
| 14 | Mounting | Hub mounted |
| 16 | Ambient ° C | -15° C to 55° C |
| 17 | Relative humidity | 98% |
| 19 | Rated torque | 9.18 Nm |
| 20 | Duty | S1 (Continues) |

III. OPTIMAL DESIGN USING FEA

The classical design gives the basic dimensions of the product. To get optimal design the analytical tools requires. Maxwell equations and the virtual work method are the theory behind this. 2D finite element method is used to analyze the electromagnetic field distribution in the machines. The simulated results of the present design are shown in Fig.1. It is the equipotential contour plot of magnetic flux. The results are coherent with the theory, and show that torque production is possible. There exist some flux paths not obeying the conventional. Leakage is one of the factors which degrade the efficiency of the motor. It can be diminished, but never can totally be snuffed out.

For computation of the motor performance, two dimensional finite element method (2-D FEM) is used and back emf is computed using the FEM solver. The respective results are shown in Fig. 2. The back emf obtained experimentally is also shown in Fig.3. The Fig.2 and Fig.3 represents the symmetry in simulated back emf and the experimental back emf. The developed proto is tested in AMDTC Lab and the test results are shown in Table III. The table shows the matching in designed value and experimental value.

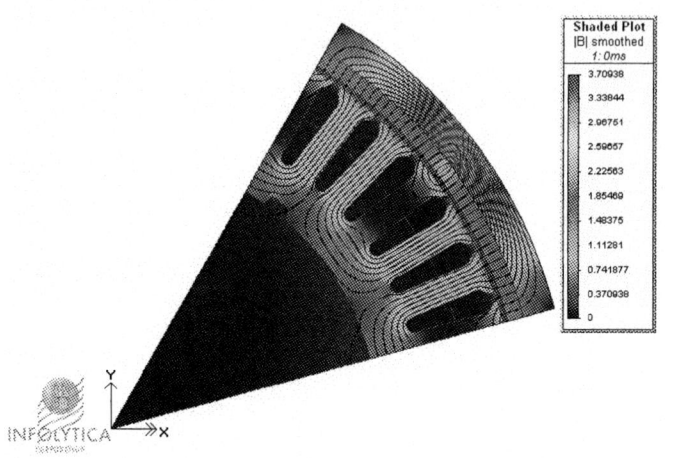

Fig. 1. Flux path of the brushless dc motor.

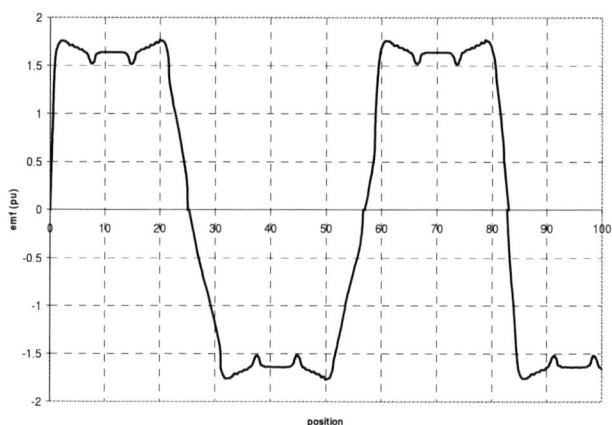

Fig. 2. Computed back emf using the FEM solver.

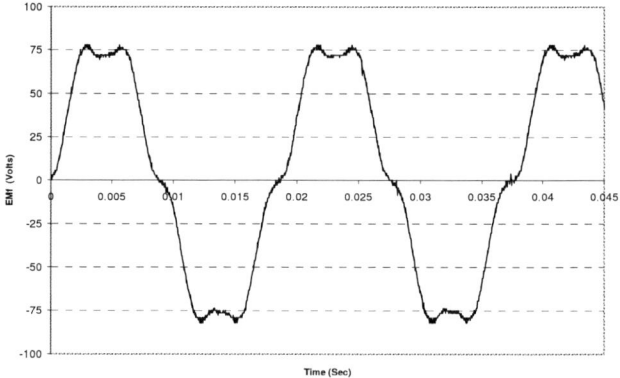

Fig. 3. Experimental back emf.

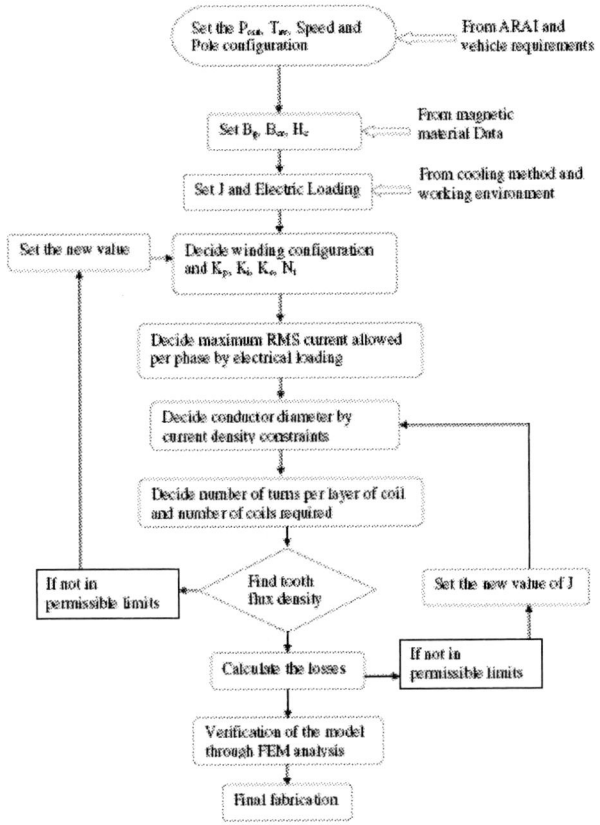

Chart 1. Design flow chart.

The design flow chart is presented in Chart 1. The required critical factors essential for the present design are considered as shown in the chart. In the flow chart, the design considered as per the norms of standards specifies by ARAI. The required cooling and the environmental effects like humidity, rains etc. are also considered in the present proto development.

TABLE III
DESIGN AND EXPERIMENTAL PERFORMANCE COMPARISON

| Sr No | Parameter | Designed Value | Experimental Value |
|-------|-----------|----------------|--------------------|
| 1 | Full Load Output (Watt) | 500 | 504 |
| 2 | Torque (N-m) | 9.18 | 9.2 |
| 3 | Full Load Speed (RPM) | 520 | 562 |
| 4 | Full Load Current (Amp) | 12.5 | 12 |
| 5 | Full Load Efficiency (%) | 86.89 | 87.88 |

## IV. RESULTS AND DISCUSSION

A prototype scooter containing fabricated motor and controller are developed. Fig.4 shows the developed controller compatible for the present 500W scooter. The present development is compatible to carry load of 200Kg (80Kg scooter weight and 120 kg of two men weight). The developed proto is also compatible to support load at 12° slope. The required heat-sink is calculated and accordingly the heat-sink cum cover box is developed. The top side of the box contains the three wires as red, yellow and blue for motor supply. The bottom side of the box contains the DC supply wires, brake wires, hall-sensor wires and the speed controlled pot wires. An electronics brake is provided in the present developed controller. The electronic brake will cutoff the motor supply, when mechanical brake implements by the user. To improve the system reliability the normal resistive pot are replaced by the linear hall-sensor. The required load test on the developed system is done. Fig.5 and Fig.6 show the current characteristics of the PM brushless dc motor during starting and running condition respectively. During start duty cycle is kept at maximum so that the required starting torque shall get exerted. Fig.6 shows the current envelop at the rated motor speed. A torque of 9.55 N-m could be developed at a speed of 520 RPM and an armature current of 12.12 Amp. Fig.7 and Fig.8 shows that the torque and power capability of the prototype motor. The present requirement requires open loop control. The torque speed characteristic as shown in Fig.7 indicates the speed regulation of the present system. The efficiency at different load is shown in Fig.8.

Fig. 4. The developed controller.

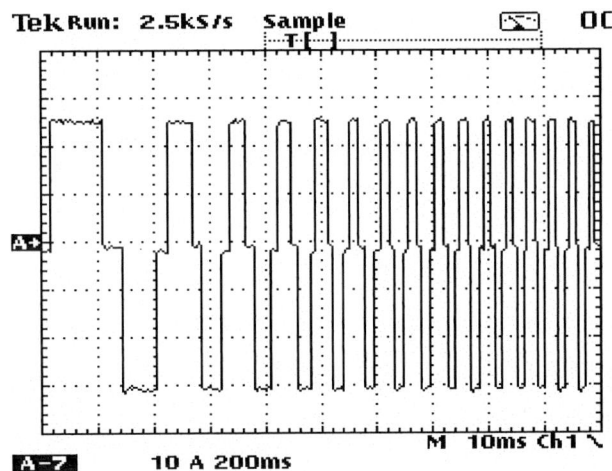

Fig. 5. Current waveform during starting condition.

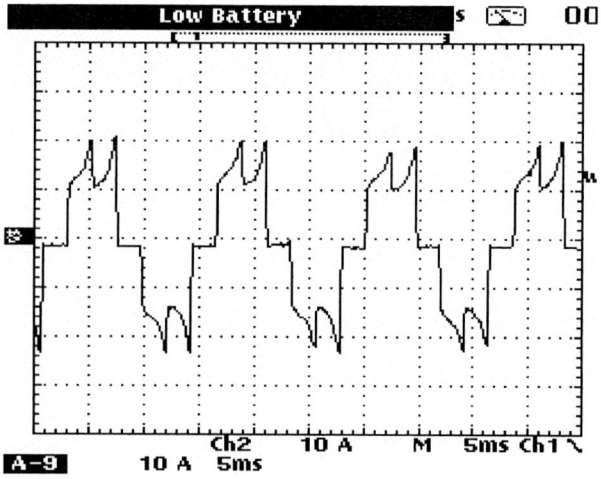

Fig. 6. Current waveform during running condition.

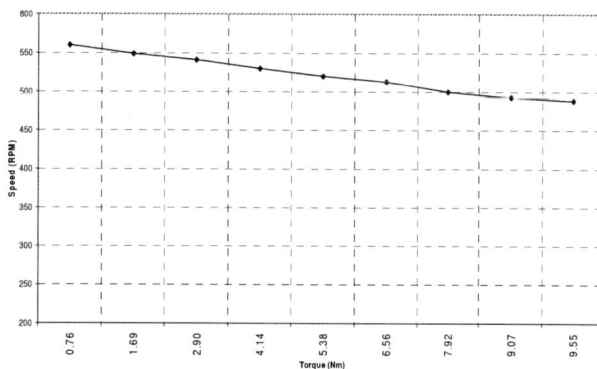

Fig. 7. Speed torque characteristics of BLDCM for scooter.

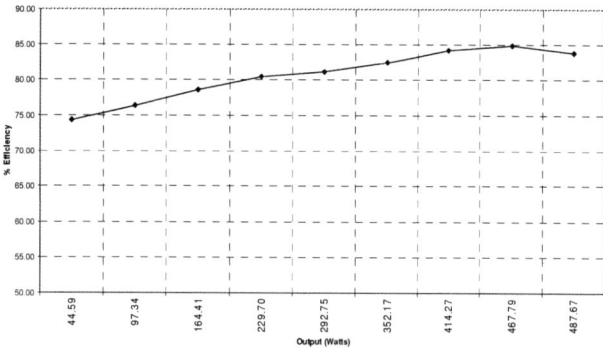

Fig. 8. Efficiency Vs output plot of BLDCM for scooter.

## V. CONCLUSION

In this paper, the design of an in-wheel, permanent-magnet BLDC motor drive for powering an Electric scooter is described. Two dimensional finite element methods are used for computing the magnetic field distribution, from which the motor performance can be determined. A prototype motor has been built, together with the electronic converter circuit. Preliminary tests on the prototype motor drive demonstrate the feasibility of the proposed design for electric scooter applications.

## VI. REFERENCE

[1] M. Ehsani, *et. al*, "Modern Electric, Hybrid Electric, And Fuel Cell Vehicles: Fundamentals, Theory, And Design," CRC Press, 2005.
[2] Hendershot, J. R., and T. J. E. Miller, Design of Brushless Permanent-Magnet Motors, Clarendon Press, Oxford, 1994.
[3] Miller, T. J. E., Brushless Permanent-Magnet and Reluctance Motor Drives, Clarendon Press, Oxford, 1989.
[4] K.T. Chau, Zheng Wang, "Overview of power electronic drives for electric vehicles," HAIT Journal of Science and Engineering A, Volume 2, Holon Academic Institute of Technology, 2005.
[5] US Patent No. 6,802,385, "Electrically powered vehicles having motor and power supply contained within wheels," filed by Wavecrest laboratories LLC. Dulles (VA), US.
[6] The Society of Automotive Engineers (SAE), http://www.sae.org/
[7] The Automotive Research Association of India, http://www.araiindia.com/

## VII. BIOGRAPHIES

**N. Ravi** was born in 1958 in Mysore, Karnatak. He graduated from University of Mysore post graduated from university of Madras.

He has been with Crompton Greaves Ltd. since 1983. Presently he is heading the Advanced Motor Design and Technology Center of R & D division in Crompton Greaves Ltd, Mumbai. His areas of interest are special machines and FEM analysis. E-mail:- ravi.nagaraj@cgl.co.in

**S. Ekram** born in Malda, West-Bengal, India on Dec 25, 1970. He graduated from the B. E. College, Shivpur, Howrah, and Post-graduated form SGSITS, Indore, India.

His employment experience included the SIGCOE & TEC (during post graduation) and Crompton Greaves Ltd. His special fields of interest included design and development of special motor (SRM, BLDC) and IM drives. E-mail:- samsul.ekram@cgl.co.in

**D. Mahajan** born in Bhusawal, Maharashtra State, India, on May 20, 1981. He graduated in Electrical Engineering from University of Mumbai, India. He did his masters in power electronics and power systems from Indian Institute of Technology Bombay, India.

His employment experience includes the Reliance Energy Ltd. Crompton Greaves Ltd. He is currently working in Corporate R & D division of Crompton Greaves Ltd. His special fields of interest include BLDC drives, SRM Drives for Electric Vehicles. E-mail:- deepak.mahajan@cgl.co.in

**2006 IEEE International Conference on Power Electronic, Drives and Energy Systems**

# Comparative Study of Laminated Core Permanent Magnet Hybrid Stepping Motor with Soft Magnetic Composite Core Claw Pole Motor

E.V. Chandra Sekhara Rao, P.V.N. Prasad, and G. Ravindranath

*Abstract* - **This paper presents the comparison and analysis of cogging torque and core losses of Permanent Magnet hybrid (PMH) stepping motor made up of laminated steel core with claw pole (PM) motor made up of soft magnetic composite core. The material used as Permanent Magnet in both the machines is NdFeB. Basic analysis was done using equivalent magnetic circuit for both machines and refined analysis is carried on Claw Pole (PM) machine using finite element magnetic field analysis. Cogging torque is found to be double in claw pole machine than that of PMH machine. Core losses are more at normal frequencies and reduced to one-third at higher frequencies for soft magnetic core PM.**

*Index Terms* – **Claw pole motor, Cogging torque, Core losses, Permanent magnet hybrid stepping motor.**

## I. INTRODUCTION

PERMANENT magnet hybrid stepping motor (PMH) and claw pole motor (PM) are both permanent magnet stepper motors having permanent magnet as rotor. PMH motor is the most popular motor available as monofilar, bifilar and enhanced types. Stator construction is simple with laminated steel core. The main advantage of claw pole machine is its low manufacturing cost, but it is difficult to construct claw poles using steel laminations. Soft magnetic core (SMC) material offers an opportunity to overcome this problem. SMC material is in general magnetically isotopic due to its powdered nature and has crucial design benefits.

Besides the favorable properties mentioned above the SMC materials have some disadvantages and that should be carefully considered in the design, manufacturing and application of electrical machines. The permeability of SMC material is significantly lower than that of electrical steels because it has lesser flux density. Therefore, it is expected that this material would be appropriate for construction of permanent magnet motors for which the magnetic reluctance of the magnet dominates the magnetic circuit, making such motors less sensitive to the permeability of the core than armature magnetized machines, such as inductive and reluctance machines.

------------------------------------------------------------------------

Comparative Study of Laminated Core Permanent Magnet Hybrid Stepping Motor with Soft Magnetic Composite Core Claw Pole Motor
E. V. C. Sekhararao, Asst. Professor, M.V.S.R.Engineering College, Hyderabad, India (email: chandrasekharev@yahoo.co.in).
Dr. P. V. N. Prasad, Assoc. Professor, O.U. College of Engineerimg, Hyderabad, India (email: polaki@rediffmail.com).
Dr.G. Ravindranath, Professor, M.V.S.R.Engineering College, Hyderabad, India(email: g_ravindranath@hatmail.com).

Because of the significant difference in magnetic, thermal and mechanical properties, simply replacing the existing laminated steel core in an electrical machine with an SMC material will result in a loss of performance with very small compensating benefits. To fully take the advantage of the SMC material and overcome its disadvantages, a great amount of research is required on material properties, novel motor topologies, advanced field analysis design and optimization techniques, and appropriate power electronic drive system.

This paper reports cogging torque calculations of laminated steel core permanent magnet hybrid stepping motor and soft magnetic core claw pole motor. The material NdFeB is used as permanent magnet for both the machines. No load core losses are calculated for PMH motors using classical method and finite element method for PM motor.

## II. CLAW POLE MOTOR PROTOTYPE

Monofilar permanent magnet hybrid stepper motor with a sheet of Kawasaki 65RM800 and 1 mm air gap is considered for analysis.

Electrical machines with claw pole rotors or stators have been manufactured in mass production for many years. These machines have quite simple excitation coil and pole system producing the excitation magnetic fields. They are capable of producing power densities up to three times greater than conventional machines because the topology allows the pole number to be increased without reducing the magnetomotive force per pole. The excessive eddy currents in the commonly used solid steel core, limits the motors to very small size and/or low speeds which results in low efficiency.

Because of the complex structure, it is very difficult to construct the claw poles using electrical steel laminations, SMC materials offer an opportunity to overcome these problems, Fig, 1 illustrates the magnetically relevant parts of the rotor and the stator of the claw pole SMC motor prototype [1]. The three phases of the motor are stacked axially with an angular shift of 120° electrical from each other. Each stator phase has a single coil around an SMC core, which is molded in two halves. The outer rotor comprises a tube of mild steel with an array of magnets for each phase mounted on the inner surface. Mild steel is used for the rotor because the flux density in the yoke is almost constant.

The motor data is given as 300 Hz. 3 phase, 500 W, 64 V line to neutral voltage, 4.1 A phase current, 1800 rpm, 20

0-7803-9771-1/06/$25.00 ©2006 IEEE            175

poles, 1mm air gap length, 60 magnets and SMC stator core, Permanent magnet for both the machines is NdFeB

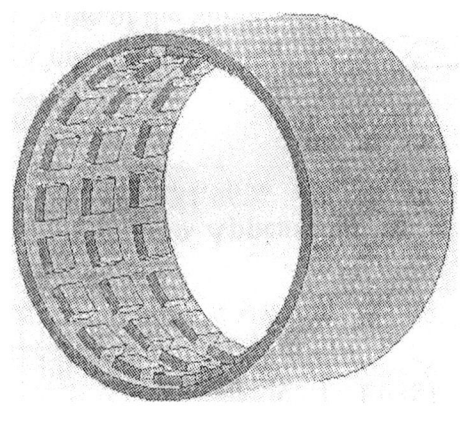

(a)

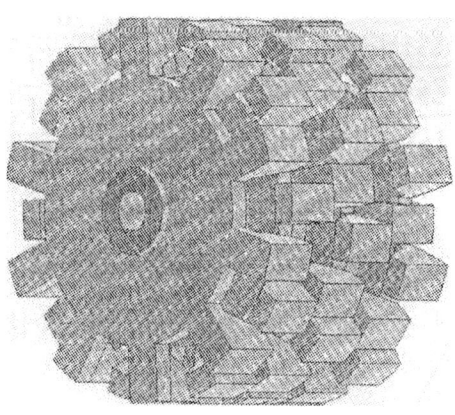

(b)

Fig. 1 The magnetically relevant parts of claw pole motor:
(a) Outer rotor, (b) SMC stator.

## II. NUMARICAL FIELD ANLYSIS

### A. PMH Motor

Flux in laminated PMH motor is given by:

$$\Phi_i = P_i (F_i + F_0) \qquad (2)$$

where,

$\Phi_i$ is Flux in $i^{th}$ winding in Webers

$P_i$ is Gap permeance in $\mu$ H

$F_i$ is MMF due to $i^{th}$ winding in ampere-turn

$F_0$ is MMF drop between rotor and stator in ampere-turn

and

$$P_i = 4\rho_o \qquad (3)$$

where,

$\rho_o$ is Fourier constant in $\mu$H

$$F_i = NI_i \qquad (4)$$

where,

N is Number of turns of the $i^{th}$ pole and $I_i$ is Current in the $i^{th}$ winding in ampere

$$F_0 = (4 \rho_o F_j + P_m F_m) / (4\rho_o + P_m) \qquad (5)$$

where,

$P_m$ is permeance of permanent magnet and $F_m$ is MMF of permanent magnet

### B. Claw Pole Motor

It is evident that the complicated shape of a claw pole machine leads to a truly 3D magnetic flux. Therefore, it is necessary that 3D finite element analysis be conducted for accurate determination of the parameters and performance of the electrical machine. The magnetic circuits of three stacks (or phases) of the motor are basically independent. For each stack, because of the symmetrical structure, it is only required to analyze the magnetic field in one pole pitch, as shown in fig.2.

Fig. 2. Region for Field Solution.

At the two radial boundary planes of one pole pitch, the magnetic scalar potential obeys the so-called half periodical boundary conditions:

$$\Phi_m(r,\Delta\theta/2,z) = -\Phi_m(r,-\Delta\theta/2,z) \qquad (6)$$

where,

$\Delta\theta = 18"$ is the angle of one pole pitch. The original point of the cylindrical coordinate is located at the center of the stack.

The stator coil flux curve and the corresponding electromotive force with respect to time can be calculated from the no-load flux density distribution at various rotor positions. The flux is calculated by rotating the rotor for one pole pitch in 12 steps. This flux waveform versus the rotor position is almost perfectly sinusoidal. In the design, the motor structure and dimensions have been adjusted such that the peak flux linkage of the stator winding is maximum. The flux distribution is a function of angle in claw pole motor and it is perfectly sinusoidal and distribution is smooth. Whereas in PMH motor it depends on air-gap permeability and permanent magnet permeability, so the distribution is not perfect sinusoidal and smooth.

## III. COGGING TORQUE

For PMH motor, the cogging torque is because of 4th harmonic component of the gap permeance and has a four cycle-variation over one pitch. Thus multiples of 4th harmonic exists, but their magnitudes are negligible. The other harmonics drop out because of the phase cancellation. The cogging torque PMH motor is given by:

$$T_{Cog(PMH)} = 0.5 N_R (P_m F_m / \rho_0)^2 \rho_4 \sin 4\theta_e \qquad (7)$$

where,

$T_{Cog}$ is Cogging torque in N-m
$N_R$ is No of rotor teeth
$P_m$ is Permeance of Permanent magnet in $\mu H$
$F_m$ is MMF of Permanent Magnet in Ampere-turns
$\rho_o$, $\rho_4$ are Fourier constants in $\mu H$
$\theta_e$ is Electrical step angle in degrees

For three-stack claw pole motor, cogging torque has only 6th harmonic and it's multiples because the three stacks are shifted by 120° electrical and resultant cogging torque is given by:

$$T_{Cog(PM)} = -0.381 \sin 6\theta_e + 0.003 \sin 12\theta_e \qquad (8)$$

In some applications like paper advance in printers, positive cogging torque is an attractive feature. But in microstepping applications, cogging torque introduces a positive error. It is minimized by minimizing 4th harmonic component of gap permeance for PMH motor and 6th harmonic for PM motor by proper choice of the valley-to-tooth ratio.

The cogging torques of both the machines as a function of stepping angle of $1.5^0$ are shown in Fig.3.

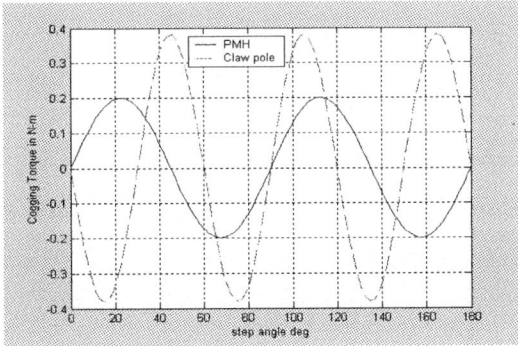

Fig. 3. Cogging Torques of PMH and Claw pole motors.

Peak cogging torque of PMH motor is nearly half of that of claw pole motor and frequency of PMH motor is 2/3rd of the claw pole motor.

## IV. CORE LOSS CALCULATIONS

### A. PMH Motor:

To measure the core loss of the PMH motor, a calibrated dc motor is used as a prime mover. The power fed into the dc motor is measured under two conditions: Stand-alone and when PMH motor is coupled. The difference in measured power gives the total of the core loss and mechanical loss of the tested PMH motor.

Replacing the laminated steel core with a wooden tube and then repeating the previous test pocedure measure the mechanical loss of the PMH motor. The core losses are obtained by subtracting mechanical losses from total losses.

### B. Claw Pole Motor

Due to the complex shape of the claw pole machine, the flux density locus at one position can be altering with or without harmonics, two dimensional or even 3D rotating with purely circular or elliptical patterns. Fig.4 illustrates the calculated 3D flux density locus in a typical element of claw pole, showing that the flux density in the claw pole is truly 3D and rotating elliptically [2]. The variables Bz, Br and $B_\theta$ are axial, radial and circumferential components of flux density respectively.

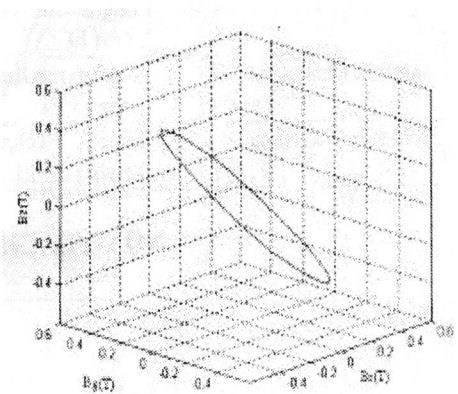

Fig. 4. Flux Density Locus at a Typical point of Claw pole.

An improved method is applied for predicting core losses of 3D flux SMC machines [2]. The following formula is used for core loss predictions.

$$P_a = C_{ha} f B^h + C_{ca}(fB)^2 + C_{aa}(fB)^{1.5} \qquad (9)$$

where,

B is the peak flux density
f is the frequency
$C_{ha}$, $C_{ca}$, and $C_{aa}$ and are loss coefficients

A series of 3D finite element analysis is conducted to determine the flux density locus in each element when the rotor rotates [3].

The calculated no load core loss varied approximately linearly with frequency and at 50 Hz. for PMH motor it is 7 W and for SMC claw pole motor it is 14 W. At 300 Hz, no load core loss for PMH motor is 168 W and 58 W for SMC claw pole motor.

## IV. CONCLUSION

The peak cogging torque of PMH motor is nearly half of that of claw pole motor. The calculated no load core loss varied approximately linearly with frequency for both the machines and core loss of PMH motor is half of that of claw pole machine at lower frequencies but thrice at higher frequencies. The frequency of cogging torque of PMH motor is 2/3$^{rd}$ the frequency of claw pole motor.

## V. REFERENCES

[1] Y.G. Guo, J.G. Zhu, J.J. Zhong and W. Wu, "Improved Design and Performance Analysis of a Claw Pole Permanent Magnet SMC Motor with Sensorless Brushless DC Drives ", *The 5th IEEE Conf. on Power Electronics and Drives, Singapore, 17-20 Nov.2003, pp. 704-709*

[2] Y.G. Guo, J.G. Zhu, J.J. Zhong P.A. Watson and W. Wu, "Comparative Study of 3D Flux Electrical Machines with Soft Magnetic Composite Core", *IEEE Transactions on Magnetics*, Vol. 39, No. 6, Nov. 2003, pp.1696-1703

[3] Y.G. Guo, J.G. Zhu, J.J. Zhong and W. Wu, "Core Losses in Claw Pole Permanent Magnet Machine with Soft Magnetic Composite Stators", *IEEE Tran. on Magnetics*, Vol. 39, No. 5, Sep. 2003, pp.3199-3201

## VI. BIOGRAPHIES

Mr. E.V.Chandra.Sekhar Rao did his graduation in Electrical & Electronics Engg. from Jawaharlal Nehru Technological University, Hyderabad in the year 1999, and post-graduation from Osmania University, Hyderabad in 2005. He has 12 years of industrial experience and six years of academic experience. His areas of interest are in Electrical machine design and Special machines. At present he is member of faculty in Electrical & Electronics Engineering department in M.V.S.R.Engg. College, Hyderabad, India.

Dr. P.V.N.Prasad graduated in Electrical & Electronics Engineering from Jawaharlal Nehru Technological University, Hyderabad in 1983 and received M.E in Industrial Drives & Control from Osmania University, Hyderabad in 1986. He served as faculty member in Kothagudem School of Mines during 1987 - 95 and at presently serving as Associate Professor in the Department of Electrical Engineering, Osmania University, Hyderabad. He received his Ph.D in Electrical Engineering in 2002. His areas of interest are Reliability Engineering and Computer Simulation of Electrical Machines & Power Electronic Drives. He is recipient of Dr.Rajendra Prasad Memorial Prize, Institution of Engineers (India), 1993 - 94 for best paper. He has got over 30 publications in National Journals & Magazines and International Conferences & Symposia and presented technical papers in Thailand and Italy.

Dr. G. Ravindranath did his graduation in Electrical & Electronics Engineering from Jawaharlal Nehru Technological University, Hyderabad in 1981 and received M.Tech in Power Systems from R.E.C.Warangal in 1983. He has 18 yeaars of academic experience and 4 years of experience in software. His areas of interests are Electrical machines, Power Electronics drives, power Systems and Special Motors.

**2006 IEEE International Conference on Power Electronic, Drives and Energy Systems**

# A Doubly Fed Induction Motor as High Torque Low Speed Drive

Mukhtar Ahmad, *Senior Member, IEEE*, M.Rizwan Khan, and Atif Iqbal

*Abstract*--**For high torque low speed applications in pulp and paper and cement industries a dc motor or a cage type motor with reducer has been used. In this paper use of a doubly fed induction motor as a high torque very low speed drive is presented. It is shown that such a motor works as constant speed motor without any problem of stability.**

*Index Terms*-- **AC motors, low speed drives, sensorless control**.

## I. INTRODUCTION

BEFORE 1980, DC motor with reduction gear was the main machine used in high torque low speed drives. Later, with the availability of power semiconductor devices for variable frequency applications cage type induction motor was also used. These motors were connected to the drive equipment through a gear box. The higher speed motor and gear box provided a cost effective solution for low speed drives. However, the maintenance and reduced efficiency of such drives makes them unsuitable for high torque drives. In this paper a doubly fed induction motor is proposed for high torque and very low speed operation.

The recent interest in the study of doubly fed induction machines is mainly as variable speed wind generator or for driving fan and pump loads [1]-[2]. However, if in the doubly fed induction machine the stator is supplied from the 50 Hz Grid supply and the rotor is connected to a power converter with 49.5 Hz frequency., the motor will run at very low speed. For a 4 pole motor the speed is 15 rev. per minute.

Presently, active research is also going on to eliminate position / speed sensors from most of high performance drives [3]-[5]. In doubly fed machine, the rotor field is explicitly defined by the external source which requires synchronization with respect to stator field. The vector control therefore, without a position sensor is more complicated in a doubly fed machine compared to a single fed machine.

In this paper a scheme for low speed drive of a doubly fed induction motor is presented. It is shown that the machine runs as synchronous motor at very low speeds and the vector control for the speed is possible with feed back from rotor

---

M.Ahmad (e-mail:mukhtar@ieee.org) M.R. Khan (km_rizwan@ rediffmail.com) and A. Iqbal (atif_iqbal@refiffmail.com) are with the Department of Electrical Engineering , Aligarh Muslim University , Aligarh India.

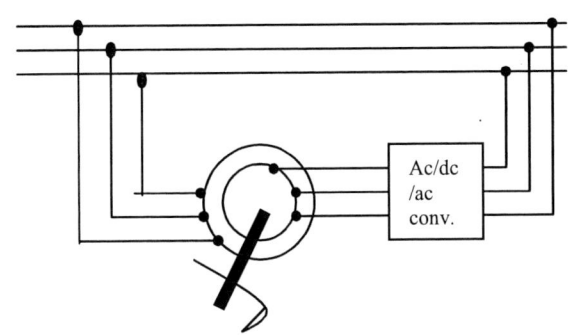

Fig. 1. Doubly fed machine configuration.

voltages and currents only.

## II. MACHINE FOR LOW SPEED OPERATION

A doubly fed induction machine is shown in Figure1. As shown, the stator of the machine is connected to 50 Hz ac source directly. The rotor winding is connected to an ac/dc/ac converter with bi-directional power flow. Considering the power balance equation of the system, the electrical power from the stator and rotor must be equal to the mechanical power output plus losses. Since the motor is working at very low speed, the slip is high and the voltage at the rotor circuit is very nearly equal to the standstill voltage. The steady state behavior of the machine can be easily understood from the equivalent circuit as shown in Fig. 2. Unlike doubly fed induction motors working with low value of slip, the rating of the converter here will be comparable to the motor rating. When an induction motor is supplied from both sides, there will be fluxes produced by both stator and rotor winding. In case of motor running at near synchronous speed the rotor voltage is very small and the torque developed by the motor is due to induction only. However, if the rotor supply frequency is very nearly equal to the supply frequency of the stator, the rotor voltage is also comparable to stator supply voltage. In such cases the induction torque is very small as s is nearly equal to 1.

$$torque = I_r^2 r_r (1-s)/s\omega_r \qquad (1)$$

As the motor now behaves as a synchronous machine, the torque is produced by virtue of the tendency of stator and rotor air gap fluxes to align their axes. Now if the motor is running on no load with rotor connected to 49.5 Hz supply, the speed of the motor is 15 rpm and slip s =0.99, the power required to run the motor is low. In this condition the power is supplied from rotor and stator sides almost equally. The

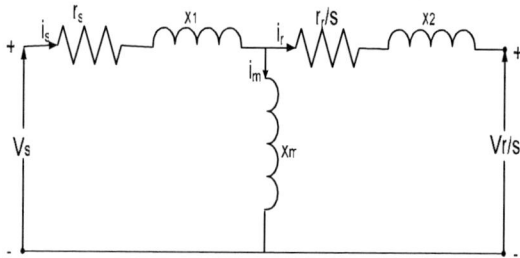

Fig. 2. Steady state equivalent circuit.

motor runs exactly like asynchronous motor without any problem of stability.

## III. DYNAMIC PERFORMANCE OF DOUBLY FED INDUCTION MACHINE

The dynamic performance of the doubly fed motor can be conveniently understood from the well-known d-q-0 transformation. The voltage equations in terms of flux linkages for the stator and rotor can be written in matrix form as:

$$\mathbf{v_{abcs}} = r_s \mathbf{i_{abc\,s}} + \frac{d\lambda_{\mathbf{abcs}}}{dt} \tag{2}$$

$$\mathbf{v_{abcr}} = r_r \mathbf{i_{abc\,r}} + \frac{d\lambda_{\mathbf{abcr}}}{dt} \tag{3}$$

$$\lambda_{\mathbf{abcs}} = \int \mathbf{v_{abcs}} - r_s \mathbf{i_{abcs}} \tag{4}$$

$$\lambda_{\mathbf{abcr}} = \int \mathbf{v_{abcr}} - r_r \mathbf{i_{abcr}} \tag{5}$$

Since $r_s$ is very small and the applied voltage to the stator is kept constant, the stator flux linkage remains constant. When a doubly fed motor is running at near synchronous speed, the rotor voltage and therefore the rotor flux linkage is very small. However, if the motor runs at low speed the rotor voltage is comparable to stator voltage and the flux in the rotor circuit is also comparable to stator flux.

If d-q-0 transformation is used, the equations can be written as-

$$\mathbf{v_{dqs}} = r_s \mathbf{i_{dqs}} + \frac{d\lambda_{\mathbf{dqs}}}{dt} + \omega_e \times \lambda_{\mathbf{dqs}} \tag{6}$$

$$\mathbf{v_{dqr}} = r_r \mathbf{i_{dqr}} + \frac{d\lambda_{\mathbf{dqr}}}{dt} + (\omega_e - \omega_r) \times \lambda_{\mathbf{dqr}} \tag{7}$$

where $\omega_e$ is the synchronous speed and $\omega_r$ is the speed of rotor. Normally the term $(\omega_e - \omega_r)$ is very small, but in this case it will have a high value. The electromagnetic torque developed by the motor is-

$$T_e = \frac{3P}{4} |\lambda_{qdm}||i_{qdr}|\sin\delta \tag{8}$$

where $\delta$ is the angle between air gap flux and rotor current vectors. If the air gap flux is aligned with the d- axis of the synchronously revolving reference vector

$$\lambda_{qdm} = \lambda_{dm}, \text{ and } \lambda_{qm} = 0 \tag{9}$$

The torque equation therefore reduces to

$$T_e = \frac{3P}{4} |\lambda_{dm}||i_{qdr}| \tag{10}$$

Also the relationship between the rotor speed, rotor current frequency, and stator current frequency is given by

$$\omega_s = \omega_e - \omega_r \tag{11}$$

For vector control $\delta$ must be maintained at 90 degrees or the current vector is to be maintained orthogonal to the air gap flux.

## IV. VECTOR DIAGRAM

The vector diagram of doubly fed induction machine is shown in Fig 3. The rotor current can be changed by adjusting the rotor voltage. Since the induction motor works at very low speed the rotor voltage vector and current vector have opposite direction. The power is therefore supplied to the mains by the converter connected to the rotor.

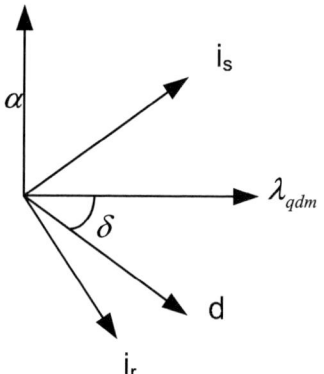

Fig. 3. Vector Diagram.

## V. CURRENT CONTROL METHOD

The inverter voltage vector supplying power to the rotor has eight states; out of which only six are active the other two are zero. The space distribution of the voltage vectors is shown in Fig.4. The complete region of 360 degrees is divided into six regions I,II, III.........VI and each region lies between the two adjacent voltage vectors. In order to change the current $i_r$ due to change in load, the phase angle and magnitude of rotor current can be changed by controlling the adjacent voltage vectors in the region where $i_r$ is placed. For example if $i_r$ in region II, its phase angle and magnitude can be controlled by $v_5$ and$v_6$ . as is normally done with space vector modulated PWM inverter . These two voltages can be applied one at a time for a fraction of the switching period depending on the magnitude required. For remaining switching period $v_0$ or $v_7$ is applied.

In order to generate the required voltage vectors , the torque difference $\Delta T$ , and the angle between rotor current and flux vector is required. These quantities are calculated depending on the position of rotor current. The proposed control system block diagram is shown in Fig.5. The block diagram has two control loops. The outer loop is for the speed and the inner loop is for torque control. The speed of the motor is measured using a measuring system described in (6). However, the scheme is being developed to have sensorless system .The desired value of torque $T_e^*$ is determined based on speed error. The machine torque is calculated using equation (10). The angle $\alpha$ between rotor flux and rotor current can be obtained as

$$\alpha = arctan \frac{\left|\lambda_{dr}\right|}{\left|\lambda_{qr}\right|} + arctan \frac{\left|i_{dr}\right|}{\left|i_{qr}\right|} \qquad (12)$$

The controller keeps the rotor flux linkage constant and the angle $\alpha$ is forced to have 90 degrees value. Thus the torque can be controlled linearly by controlling the rotor current.

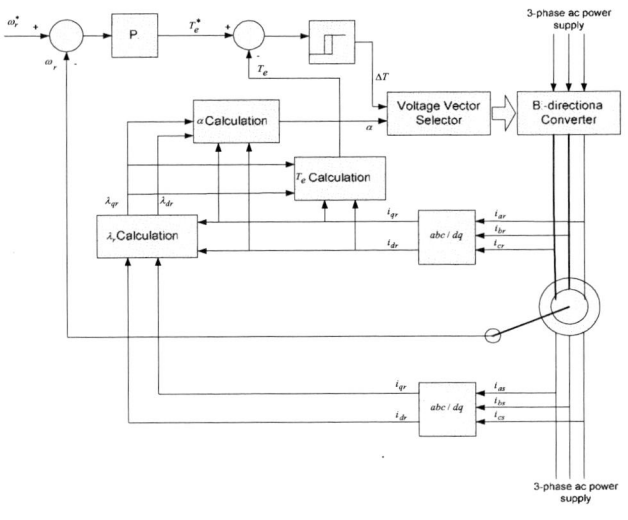

Fig. 5. Block diagram of control circuit.

## VII. REFERENCES

[1] M. G. Jovanovic, R.E.Betz, and Yu Jian, "The use of doubly fed reluctance machines for large pump and wind turbines" *IEEE Trans. Ind. Appl.* Vol.38.no.6, pp. 1508-1516, Nov.-Dec.2002.

[2] C. Abbey, and G. Joos, **"**Integration of energy storage with a doubly-fed induction machine for wind power applications" *in Proc. IEEE 35th Power Electronics Specialist conference2004, vol.3* pp1964-1968.

[3] Longya Xu, and Wei Cheng, "Torque and Reactive Power Control of a doubly fed Induction Machine by Position sensorless Scheme" *IEEE trans. Ind Appl.* Vol.31 no. 3 pp636-642 May/June 1996.

[4] O.A. Mohammed, Z.Liu, and S. Liu, "A novel sensorless control strategy of doubly fed induction motor and its examination with the physical modeling of machines" *IEEE Trans. Magnetics vol.*41 no.5 pp.1852-1855 May 2005.

[5] Z.Wang, F. Wang, M. Zong, and F. zhang, "A new control strategy by combining direct torque control with vector control for doubly fed machine" *in Proc. International Conference on Power System Technology*, 2004

[6] Mukhtar Ahmad, "A digital tachometer for measurement of low speeds" *Proceedings IEEE*, pp. 1096, 1984

## VIII. BIOGRAPHIES

**Mukhtar Ahmad** received his B.Sc , M. Sc. And Ph. D from Aligarh Muslim University Aligarh in 1969, 1972 , and 1991department of electrical engineering A.M.U. Aligarh. He is currently working as professor in the Department of electrical engineering AMU Aligarh. He has also worked in Multimedia University Malaysia, and University Putra Mal;aysia as associate professor . He is a senior member of IEEE since 2002. His research interests are in power electronics; electric drives and power system control.

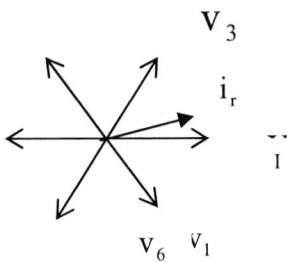

Fig. 4. Space distribution of voltage vector.

## VI. CONCLUSION

A method has been proposed to obtain very low speed drive using doubly fed induction motor. Since the problem of estimating the value of flux accurately is very difficult in single fed induction machines, this method will be very useful for such drives. Now with the availability of fast switching devices it is easy to control the rotor current quickly thereby getting high performance torque control.

**2006 IEEE International Conference on Power Electronic, Drives and Energy Systems**

# Performance of Doubly Salient Permanent Magnet Motors for Parallel and Tapered Rotor Poles

Nimit K. Sheth, *Student Member, IEEE*, and K. R. Rajagopal, *Senior Member, IEEE*

*Abstract--* This paper presents the results of a finite element (FE) analysis carried out to study the effects of variation of rotor pole arc and rotor pole shapes on the performances of a doubly salient permanent magnet (DSPM) motor; based on which a clear comparison can be done on two types of rotor pole geometries namely, the parallel and tapered. Two-dimensional (2-D) FE models of a 8/6 DSPM motor for parallel sided and tapered rotor poles with rotor outer pole arcs varying from of 15°to 38° have been analyzed with appropriate phase excitations. It is observed that in all the cases, the tapered rotor pole gives higher average torque. The maximum average torques in both the cases, occur for an outer pole arc of 22°, where the maximum torque in tapered case is more by 1.8%. The average torques in both the cases varies similarly for rotor outer pole arcs above 22° and up to 30°. This is also true in case of torque ripples. For outer rotor pole arcs of less than the optimum 22°, it is observed that the tapered case is better; at 15° pole arc, the average torque developed by the motor is 14.8% higher than the corresponding value of parallel-sided case. The magnet utilization in case of the tapered poles is found to be better for rotor pole arcs up to 26°.

*Index Terms--* Doubly Salient Motor, DSPM, FE Analysis, Motor, Permanent Magnet Motor.

## I. INTRODUCTION

DOUBLY permanent magnet (DSPM) motor inherits the merits of both the permanent magnet brushless dc (PM BLDC) motor and the switched reluctance motor (SRM). In a DSPM motor, the permanent magnets are located in the stator eliminating the problems of irreversible demagnetization and mechanical instability while retaining the merits of high efficiency and high power density. The corresponding rotor is same as that of the SRM; hence the advantage of mechanical robustness and simple construction is also available [1, 2]. The windings of the DSPM motor are of concentric type which gives an advantage of extended constant power region

---

N. K. Sheth is with Electrical Engineering Department, Institute of Technology, Nirma University of Science and Technology, Ahmedabad 382481, Gujarat, India (e-mail: nimit75@yahoo.com) and pursuing PhD at Indian Institute of Technology Delhi, New Delhi 110016, India.

K. R. Rajagopa is with the Department of Electrical Engineering, Indian Institute of Technology Delhi, New Delhi 110016 (e-mail: rgopal@ee.iitd.ac.in).

by splitting the phase winding [3]. Effects of stator and rotor pole arcs [4] and the pole face shapes [5] on the performance of SRM are reported. The effects of slanted and saw-toothed stator poles on the performance of a DSPM motor are also reported [6], but there is no publication dealing with the effect of rotor pole arc variation and rotor pole shape on the performance of DSPM motor. This paper presents the results of a finite element (FE) analysis carried out to study the effects of variation of rotor pole arc and rotor pole shapes on the performances of the DSPM motor; specifically for two rotor pole shapes, namely the parallel sided and tapered.

## II. FINITE ELEMENT ANALYSIS OF THE DSPM MOTORS WITH PARALLEL SIDED AND TAPERED ROTOR POLES

Two-dimensional (2-D) FE model of an existing 1 hp, 8/6 DSPM motor with parallel sided rotor poles and with the stator pole arc of 22° and the rotor pole arc of 26° is prepared and analyzed in detail with and without the rated excitation given to the phases. Table I gives the major dimensions and material details of the motor analyzed. For the analyzed motor the flux-linkage, inductance, detent torque and the developed torque profiles are computed. Apart form this, the magnet leakage factor, which is the ratio of the total flux produced by the magnet and the sum of the total airgap flux of all the phases, is also calculated in each case. By a spectrum analysis, harmonic torque components of the both the detent and the developed torques are also obtained. This analysis is repeated 60 times by changing the rotor position in steps so as to cover a full rotor pole pitch. The entire analysis is repeated with rotor outer pole arcs varying from 15° to 38° such that rotor poles become parallel sided poles and by keeping the stator pole arc to its original value of 22°. Fig. 1 shows the combination of the stator and rotor pole, for the original motor and motor with rotor outer pole arc of 28° for parallel rotor poles. Fig. 2 shows the flux density plot of the motor when stator excitation is absent, and when the rotor outer pole arc is 28°.

The entire 2-D FE analyses are repeated by keeping the stator pole arc and the rotor base arc to the original value, and applying tapering to the rotor poles by changing the outer rotor pole arcs from 15° to 38° in steps. Fig. 3 shows the combination of the stator and rotor pole, for the original motor and motor with rotor outer pole arc of 28° for tapered rotor poles. Fig. 4 shows the flux density plot of this motor when stator excitation is absent in a typical case, when the rotor outer pole arc is 28°.

TABLE I
MAJOR DIMENSION AND MATERIAL DETAILS OF THE MOTOR

| | |
|---|---|
| Stator outer diameter = 128 mm | Stack length = 75 mm |
| Stator pole height = 13 mm | Rotor outer diameter =74.1 mm |
| Length of airgap = 0.45 mm | Height of the permanent magnet = 6 mm |
| Rotor pole height = 10 mm | Width of the permanent magnet = 37 mm |
| Shaft diameter = 22 mm | Length of the permanent magnet = 75 mm |
| Stator pole arc =22° | Rotor pole arc =26° |
| Stator and rotor core material : M19 | |
| Permanent magnet material : NdFeB 35 | |

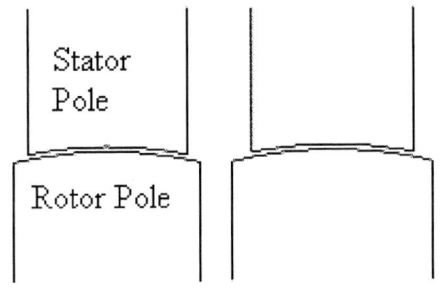

Fig. 1. Stator and rotor pole combination for original motor and motor having parallel rotor pole with rotor outer pole arc of 28°.

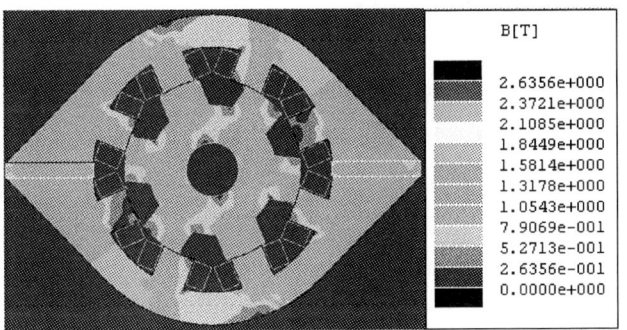

Fig. 2. Flux density plot o the 8/6 DSPM motor having parallel sided rotor poles of rotor outer pole arc of 28° with the stator excitation absent.

## III. COMPARISON OF THE PERFORMANCE WITH PARALLEL SIDED AND TAPERED ROTOR POLES

Fig. 5 shows the flux-linkage versus rotor position characteristics of phase of the motor when the stator excitation is absent, from which it is observed that with reduction in rotor pole arc, the flux-linkage increases. It is also observed that that the flux-linkage of a phase reaches its maximum or

minimum when the stator and rotor poles are fully aligned or unaligned respectively.

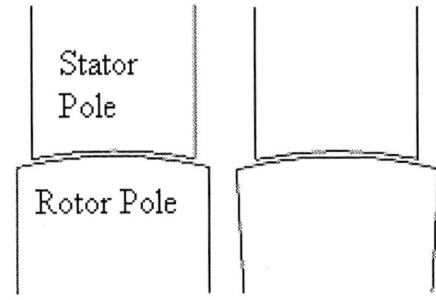

Fig. 3. Stator and rotor pole combination for original motor and motor having tapered rotor pole with rotor outer pole arc of 28°.

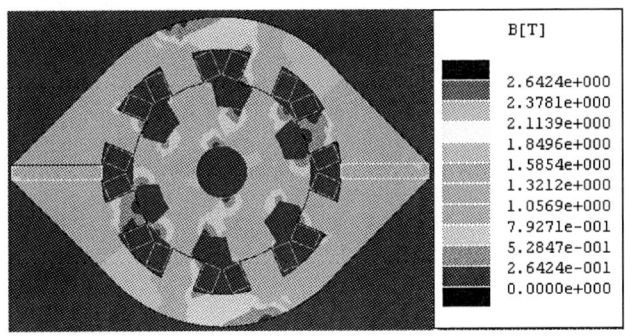

Fig. 4. Flux density plot of the 8/6 DSPM motor having tapered rotor poles of rotor outer pole arc of 28° with the stator excitation absent.

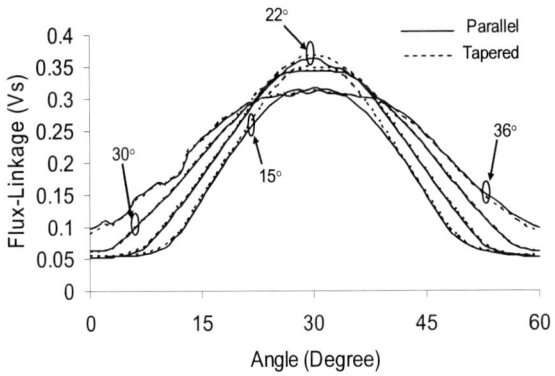

Fig. 5. Flux-linkage vs. rotor position characteristics of the 8/6 DSPM motor when stator phase excitation is absent.

In case of DSPM motor to get the phase inductance at any rotor position , flux-linkage values at the same rotor position needs to be calculated when phase excitation is present and absent. Then after using (1) phase inductance is calculated, where, $\psi_{(PM \pm i_{phase})}$ is the phase flux-linkage, when phase is excited with current $i_{phase}$, and $\psi_{(PM)}$ is the phase flux-linkage when the stator excitation is absent.

$$L_{self} = \left| \frac{\psi_{(PM \pm i_{phase})} - \psi_{(PM)}}{i_{phase}} \right| \qquad (1)$$

Fig. 6 shows the variation of phase-A inductance for an excitation of 540 AT. From Fig. 6 it is observed that with the increase in the rotor pole arcs the aligned and unaligned inductances increase and the phase inductance found in case of tapered pole will be higher when the rotor outer pole arc is less than the stator pole arc; and lower when the rotor outer pole arc is greater than the stator pole arc. It is observed from the Fig. 6 that the phase inductance will have maximum phase inductance when stator and rotor poles are nearer to half aligned position and compared to this maximum inductance the inductance at the fully unaligned and aligned position will be less, because the existence of the PM constitutes a very high reluctance path for the armature reaction flux and thus forces the bulk of the armature reaction flux to circulate through another overlapped pair. It is also observed that the inductance profile in both halves considering fully aligned condition at the center point (i.e. 30° in Fig. 6) is not symmetrical because in one half the winding excitation supports the PM excitation resulting in to higher saturation, while in the other half it is opposing the PM excitation resulting in to lower saturation.

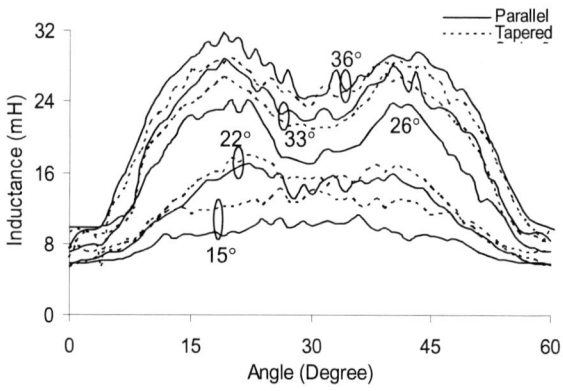

Fig. 6. Phase inductance variation at the excitation of 540 AT/phase.

Figure 7 shows the variation of the detent torque for various rotor outer pole arcs, from which it is observed that the peak detent torque for tapered rotor pole is more in comparison to the parallel sided rotor pole with the same outer rotor pole arc. It is also observed from Fig. 7 that the detent torque is anti-symmetric about the center point and detent torque cycle is $2\pi / S$ degree mechanical, where $S$ is the least common multiplier of the stator and the rotor pole numbers.

Harmonic torque components of the detent torque waveforms are obtained using the Fast Fourier Transform (FFT). Fig. 8 shows the predominant harmonic torques, from which it is observed that the fourth and its multiple harmonic

torques are the predominant ones and the fourth harmonic torque is the most predominant one. It is also observed that the rotor pole arc of 28° is giving the minimum harmonic torques for both type of rotor poles and for rotor outer pole arcs less than the stator pole arc or more than 34°, the tapered pole will have higher fourth order harmonic torque compared to parallel sided rotor poles of same outer pole arcs.

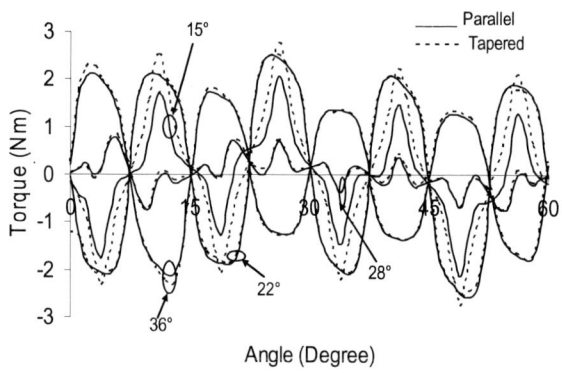

Fig. 7. Detent torque characteristics of the 8/6 DSPM motor.

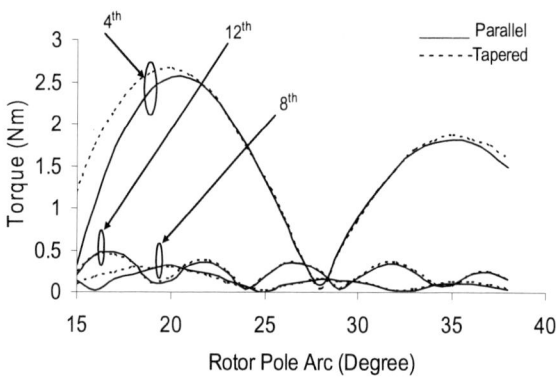

Fig. 8. Variation of harmonic torque components of the detent torques.

Fig. 9 shows the developed torque profiles for both types of rotor poles for various values of rotor pole arc for stator phase excitation of 540 AT/pole. Using the developed torque profile the average developed torque values have been calculated which are shown in Fig. 10. The torque ripple values for these developed torque profile is calculated using (2). Fig. 11 shows the variation of torque ripple for different rotor pole arcs when both type of rotor poles are considered.

$$Torque\ ripple = \frac{Maximum\ torque - Minimum\ torque}{Average\ torque} \qquad (2)$$

184

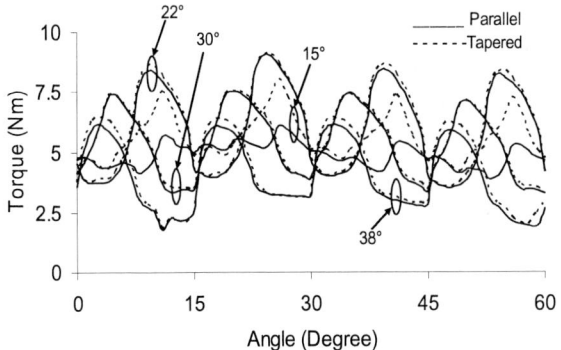

Fig. 9. Developed torque characteristics of the motor at the stator excitation of 540 AT.

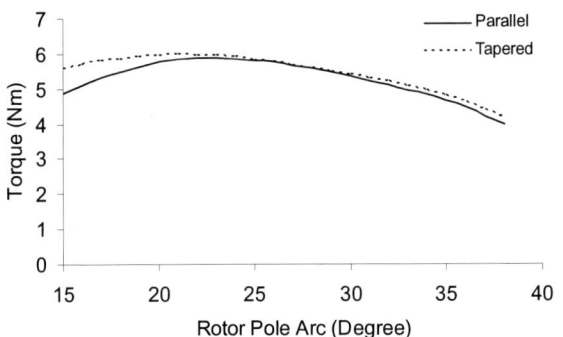

Fig. 10. Comparison of the average torques of the motor at the stator excitation of 540 AT/phase.

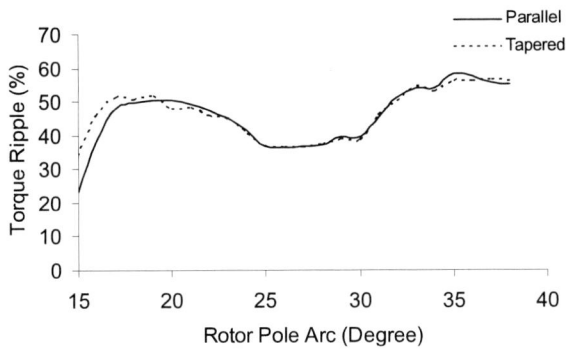

Fig. 11. Comparison of the average torques of the motor at the stator excitation of 540 AT/phase.

Table II gives these average developed torque and torque ripple values for various rotor outer pole arcs for both the parallel sided and tapered rotor pole cases. From the Table II, it is observed that for both types of rotor poles, the maximum average torque occurs when the stator and rotor pole arcs are equal, but with higher value of the torque ripple. AT the rotor outer pole arc of 15°, the torque ripple is the minimum. It is also observed that the tapered rotor pole gives more average torque compared to the parallel rotor pole of the same pole

arc, but with more torque ripple if the rotor outer pole arc is less than 19° or greater than 33°.

TABLE II
AVERAGE TORQUE, TORQUE RIPPLE AND MAGNET LEAKAGE FACTOR FOR THE
8/6 DSPM MOTOR WITH PARALLEL SIDED AND TAPERED ROTOR POLES

| Rotor outer pole arc (°) | Parallel Rotor Pole | | | Tapered Rotor Pole | | |
|---|---|---|---|---|---|---|
| | Average torque (Nm) | Torque ripple (%) | Magnet leakage factor | Average torque (Nm) | Torque ripple (%) | Magnet leakage factor |
| 15 | 4.86 | 23.64 | 1.336 | 5.58 | 34.21 | 1.305 |
| 18 | 5.49 | 49.72 | 1.29 | 5.86 | 50.60 | 1.275 |
| 20 | 5.76 | 50.37 | 1.268 | 5.94 | 47.82 | 1.26 |
| 22 | 5.88 | 47.13 | 1.251 | 5.97 | 45.88 | 1.247 |
| 26 | 5.76 | 36.36 | 1.226 | 5.76 | 36.36 | 1.226 |
| 30 | 5.36 | 39.51 | 1.206 | 5.41 | 38.36 | 1.208 |
| 34 | 4.83 | 54.01 | 1.191 | 4.95 | 52.87 | 1.194 |
| 36 | 4.45 | 57.48 | 1.184 | 4.62 | 55.89 | 1.188 |
| 38 | 3.98 | 55.23 | 1.176 | 4.18 | 55.97 | 1.181 |

Harmonic torque components of the developed torques are calculated using FFT and given in Fig. 12, from which, it can be observed that the fourth and its multiple harmonic torques are predominant. Fig. 13 shows the magnetic leakage factor, from which it is observed that higher the rotor pole arc the lesser or better the magnet leakage factor is. It is also observed that for the rotor outer pole arcs of less that 24°, the magnet leakage factor is less for tapered rotor pole; for outer rotor pole arcs of 24° to 28°, it is the same for both types of rotor poles and for rotor outer pole arc greater than 26° the magnet leakage factor is better in case of parallel rotor poles.

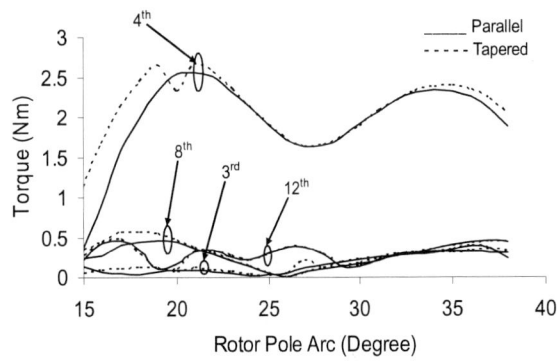

Fig. 12. Variation of harmonic torque components of the developed torques of the 8/6 DSPM motor at the stator excitation of 540 AT/phase.

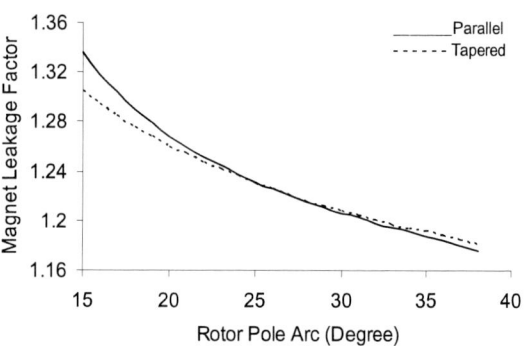

Fig. 13. Variation of the magnet leakage factor with rotor outer pole arcs.

## IV. CONCLUSION

The tapered rotor pole gives higher average torque than the parallel sided rotor poles. The maximum average torques in both the cases, occur for an outer pole arc of 22°, where the maximum torque in tapered case is more by 1.8%. The torque ripples for this particular pole arc in case of tapered is slightly less compared to parallel sided rotor pole. The average torques in both cases varies similarly for rotor outer pole arcs above 22° and up to 30°. This is also true in case of torque ripples. For outer rotor pole arcs of less than the optimum 22°, it is observed that the tapered case is better; at 15° pole arc, the average torque developed by the motor is 14.8% higher than the corresponding value of parallel-sided case. The magnet utilization in case of the tapered poles is found to be better for rotor pole arcs up to 26°. The analysis has also revealed that the DSPM motor is immune to the type of restrictions of certain stator and rotor pole arc combinations prevalent in switched reluctance motors for having non-zero torque zones.

## V. REFERENCES

[1] Yuefeng Liao, Feng Liang and Thomas A Lipo, "A novel permanent magnet motor with doubly salient structure", *IEEE Trans. Ind. Applicat*, vol. 31, pp. 1069-1078, Sept./Oct. 1995.

[2] M. Cheng, K. T. Chau, C. C. Chan, and Q. Sun, "Control and operation of a new 8/6-pole doubly salient permanent-magnet motor drive", *IEEE Trans. Ind. Applicat*, vol. 39, pp. 1363-1371, Sept./Oct. 2003.

[3] Ming Cheng, K. T. Chau and C. C. Chan, "Performance analysis of split-winding doubly salinet permanent magnet motor for wide speed operation", *Electrical Machines and Power Systems*, vol. 28, pp. 277-288, 2000.

[4] N. K. Sheth, and K. R. Rajagopal, "Optimum pole arcs for a switched reluctance motor for higher torque with reduced ripple", *IEEE Trans. on Magnetics*, vol. 39, pp. 3214-3216, Sept. 2003.

[5] N. K. Sheth and K. R. Rajagopal, "Torque profiles of a switched reluctance motor having special pole face shapes and asymmetric stator poles", *IEEE Trans. on Magnetics*, vol. 40, pp. 2035-2037, July. 2004.

[6] A. R. C. Sekhar Babu, and K. R. Rajagopal, "Slanted and saw-toothed stator poles for improved performance of doubly salient permanent magnet motors", *J. Appl. Phys.*, vol. 97, pp. 10Q514-1-3, May 2005.

## VI. BIOGRAPHIES

**Nimit K. Sheth** (S'2003) was born in Nadia, Gujarat, India in 1975. He received B. E. Degree in Electrical Engineering from the Gujarat University, Ahmedabad, India in 1996, M. Tech. Degree in Power Electronics, Electrical Machines and Drives in Electrical Engineering from the Indian Institute of Technology Delhi, New Delhi, India in 2002. Since, 2003 he is working towards his PhD degree at Indian Institute of Technology Delhi, New Delhi, India.

From 1997 to 1998, he was with Ahmedabad Electricity Company Ltd. Ahmedabad, India, as a Trainee Engineer. Since 1998, he is with the Institute of Technology, Nirma University of Science and Technology, Ahmedabad, India, where currently he is an Assistant Professor in Electrical Engineering Department.

He has published more than 20 papers in International Journals and Conference proceedings. He received Prof. A. K. Sinha Award for securing highest CGPA among all the M. Tech. graduating students from the Electrical Engineering Department of Indian Institute of Technology Delhi, New Delhi, India during the year 2002. His fields of interest include Electrical Machines and Drives, Special Electrical Machines (Switched Reluctance Motors, DSPM Motors, Flux Reversal Motors, Stepper Motors, etc.,), Finite Element Analysis and CAD of Electrical Machines.

**K. R. Rajagopal** (M'1998, SM'2000) was born in Alappuzha, Kerala, India in 1961. He received Diploma in Electrical Engineering from Carmel Polytechnic, Alappuzha, India in 1979, B. Tech. Degree in Electrical Engineering from the College of Engineering, Trivandrum, India in 1988, M. Tech. Degree in Power Electronics, Electrical Machines and Drives and Ph. D. Degree in Electrical Engineering from the Indian Institute of Technology Delhi, New Delhi, India during 1991 and 1998 respectively.

From 1980 to 1983, he was with Aluminum Industries Ltd. (ALIND), Trivandrum, India, as an Application Engineer (Relays), from 1983 to 1999, he was with the Indian Space Research Organization (ISRO), Trivundrum, India, where he was engaged in Analysis, Design, Development and Testing of Special Electrical Machines/Devices used in space applications. Since 1999, he is with the Indian Institute of Technology Delhi, New Delhi, India, where currently he is a Professor in Electrical Engineering Department.

He has published more than 30 papers in International Journals and more than 60 papers in International conference proceedings. He received Indian National Academy of Engineering (INAE) award for most Innovative Potential Project in Engineering during the year 1998. His fields of interest include Electrical Machines and Drives, Special Electrical Machines (Stepper Motors, Switched Reluctance Motors, PM BLDC Motors, Hysterisis Motors, etc.,), Magnetic Devices, Finite Element Analysis and CAD of Electrical Machines and Design of Energy Efficient Motors for Home Appliances.

**2006 IEEE International Conference on Power Electronic, Drives and Energy Systems**

# Improved Torque Profile of a Doubly Salient Permanent Magnet Motor using Skewed Rotor Teeth and Sinusoidal Excitation

Nimit. K. Sheth, *Student Member, IEEE*, and K. R. Rajagopal, *Senior Member, IEEE*

*Abstract--* **This paper presents the results of a three dimensional (3-D) finite element (FE) analysis carried out to get the improved torque profile of doubly salient permanent magnet (DSPM) motor by optimum skewing of rotor teeth and type of excitation from two types of excitations namely, rectangular and sinusoidal. Effect of skewing the rotor teeth on various performance characteristics like phase flux-linkage, self and mutual inductances, detent torque, average developed torque and torque ripple have been presented. It is observed that with the increase in the skew angle of the rotor pole the phase flux-linkage and the difference between the phase inductance for phase excitation supporting or opposing the permanent magnet excitation reduces. Result of harmonic analysis of the detent torque and developed torque for both type of excitation at various skew angles are presented from which it is observed that the 4th and its multiple torque harmonics are the predominant ones for both type of torque. It is also observed that for all skew angles the sinusoidal excitation of motor with two phase excitation mode is giving more average torque in comparison to the rectangular current excitation this is also true for torque ripple if the skew angles less than 24º. It is also observed that skewing the rotor teeth by 12º to 15º with sinusoidal excitation will give higher average torque with reduced ripple, so becomes the best choice.**

*Index Terms--* **Doubly Salient Motor, DSPM, FE Analysis, Motor, Permanent Magnet Motor, Skewing, Torque ripple.**

## I. INTRODUCTION

DOUBLY salient permanent magnet (DSPM) motor inherits the merits of both the permanent magnet brushless dc (PM BLDC) motor and the switched reluctance motor (SRM). In a DSPM motor, the permanent magnets are located in the stator eliminating the problems of irreversible demagnetization and mechanical instability while retaining the merits of high efficiency and high power density. The corresponding rotor is same as that of the SRM; hence the

N. K. Sheth is with Electrical Engineering Department, Institute of Technology, Nirma University of Science and Technology, Ahmedabad 382481, Gujarat, India (e-mail: nimit75@yahoo.com) and pursuing PhD at Indian Institute of Technology Delhi 110016, India.

K. R. Rajagopa is with the Department of Electrical Engineering, Indian Institute of Technology Delhi, New Delhi 110016 (e-mail: rgopal@ee.iitd.ac.in).

advantage of mechanical robustness and simple construction is also available [1, 2]. Similar to SRM, the DSPM motor suffers from severe torque ripple because of the nature of the salient poles in both the stator and the rotor. Motor saturation and finite mutual inductance will also lead to the torque ripple [3]. Literature on the reduction of cogging torque and torque ripple indicates that shape of the torque waveform for the PMBLDC motor can be changed by shifting the stator pole pair by half a slot pitch and by choosing the appropriate value of the magnet width [4] or by skewing either the stator slot or by rotor magnets or rotor magnetization patter for an optimum skew angle [5] but at the cost of the reduction in average torque [6]. Proper skewing of the rotor will lead to reduced self inductance, mutual inductance and cogging torque for flux reversal machines and makes it suitable for servo applications [7]. For 8/6 DSPM motor by operating motor in the two phase mode where two groups of two phases connected back to back having sinusoidal back-emf and supplied with a sinusoidal current under ideal condition of no net reluctance torque [2, 8] or conduction angle control will help in reducing the torque ripple [3]. There is no literature giving the comprehensive results on the performance of DSPM motor with skewed rotor excited with sinusoidal current. This paper presents the results of a three dimensional (3-D) finite element (FE) analysis carried out to study the effects of skewing the rotor teeth on various performance parameters of the DSPM motor for three type of excitations namely; (i) only permanent magnet excitation, (ii) permanent magnet excitation with appropriate polarity of rectangular current excitation for windings and (iii) permanent magnet excitation with appropriate polarity of sinusoidal current excitation for windings and a comparison has been made for motor performance with skewed rotor pole for both type of combined excitations and type of excitation has been arrived which gives improved torque profile in terms of higher average torque and reduced torque ripple.

## II. PERFORMANCE OF DSPM MOTORS WITH SKEWED ROTOR TEETH FOR ONLY PERMANENT MAGNET EXCITATION

For a normal induction motor to improve the torque profile the rotor is generally skewed by one stator slot pitch [9], considering this three-dimensional (3-D) finite element (FE) model of an 1 hp, 8/6 DSPM motor with rotor poles having no

0-7803-9771-1/06/$25.00 ©2006 IEEE

skewing to skewing of one stator slot pitch (45° in case of 8/6 DSPM) have been made and analyzed for three type of excitations namely; (i) only permanent magnet (PM) excitation, (ii) combined PM excitation with appropriate polarity of rectangular current excitation for windings and (iii) combined PM excitation with appropriate polarity of sinusoidal current excitation for windings. Table I gives the major dimensions and material details of the motor analyzed.

TABLE I
MAJOR DIMENSION AND MATERIAL DETAILS OF THE MOTOR

| | |
|---|---|
| Stack length = 75 mm | Stator outer diameter = 128 mm |
| Stator pole height = 13 mm | Rotor outer diameter =74.1 mm |
| Length of airgap = 0.45 mm | Height of the permanent magnet = 6 mm |
| Rotor pole height = 10 mm | Width of the permanent magnet = 37 mm |
| Shaft diameter = 22 mm | Length of the permanent magnet = 75 mm |
| Stator pole arc =22° | Rotor pole arc =26° |
| Stator and rotor core material : M19 | |
| Permanent magnet material : NdFeB 35 | |

For the analyzed motor flux-linkage characteristics for various phases have been obtained for only PM excitation for various rotor pole skew angles. Fig. 1 shows the phase-A flux-linkage characteristics for various skew angles, from which it is observed that with the increase in the skew angle the peak flux-linkage reduces. At skew angle of 30° the variation in flux-linkage is the minimum and above that the shape of the waveform is inverted. In the case of the motor analyzed the peak flux-linkage becomes 0.087 Vs for skew angle of 30° from 0.3961 Vs for no skewing of rotor poles and again increases to 0.104 Vs for 42°. It is also observed from the Fig. 1 that with the increase in the skew angle the flux-linkage characteristics is getting shifted towards one side this is because the unstable equilibrium point for a particular phase is shifting by a half a skew angle in the opposite direction of the skew and this must be considered while designing the control circuit.

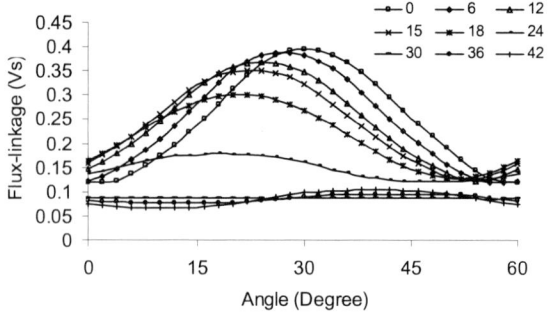

Fig. 1. Flux-linkage profiles for phase A at various skew angles for only PM excitation.

Fig. 2 shows the variation of flux-linkages for various phases for the conditions of no skewing of rotor pole and rotor pole skewed by 15° from which it is observed that phase A and C

and phase B and D can be connected back to back to form two phase system. Fig. 3 shows the variation of the detent torque at various skew angles for the rotor poles, from which it is observed that the detent torque is anti-symmetric about the center point and detent torque cycle is $2\pi /S$, where S is the least common multiple of Stator and rotor pole numbers. By a spectrum analysis, harmonic torque components of the detent torques are obtained and shown in Fig. 4 from which it is observed that the predominant harmonic torque is the 4th and multiple of it. It is also observed that the fourth harmonic torque is the minimum for the skew angle of 12°, other harmonic torques are also small at the same skew angle, so one can expect to have very less torque ripple at skew angle of 12° or nearer to it.

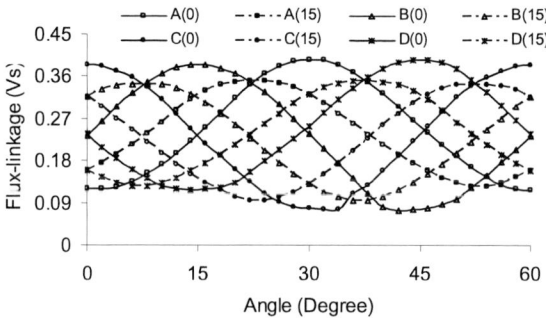

Fig. 2. Flux-linkage profiles for various phases when rotor pole are not skewed and skewed by 15° for only PM excitation.

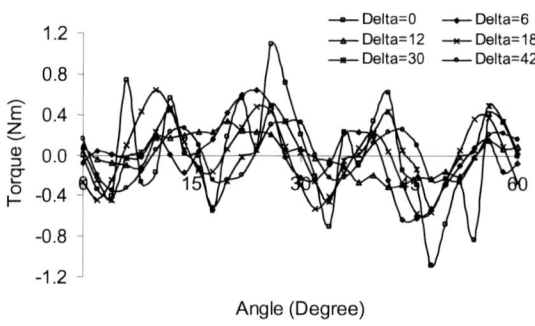

Fig. 3. Detent torque profiles at various skew angles of rotor poles.

Fig. 5 shows the variation of the magnet operating point at various skew angles from which it is observed that there no much variation in the magnet operating point with rotor position for the same skew angle but as the skew angle is reduced the magnet operating flux density is reduced, this is because of the increase in effective reluctance due to skewing. For the analyzed motor the operating flux density of the magnet reduces to 0.253 T for a skew angle of 42° from 0.786 T when there is no skewing. It is also observed that the reduction in the magnet operating flux density is not mush for the skew angle up to 15° and then after there is drastic reduction in the operating flux density.

188

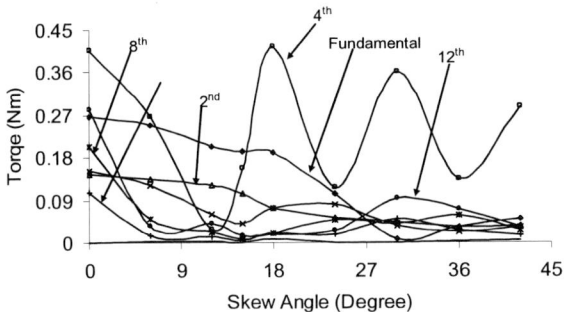

Fig. 4. Variation of harmonic torques of the detent torque profiles at various skew angles of rotor poles.

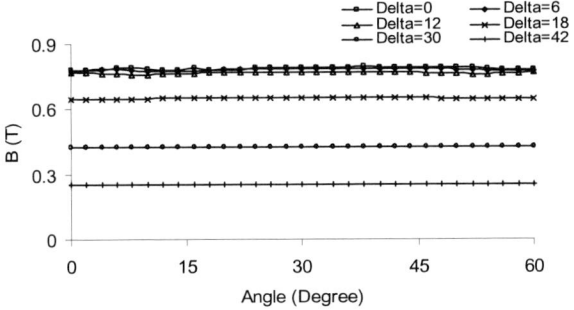

Fig. 5. Magnet operating flux density at different rotor positions for various skew angles.

## III. PERFORMANCE OF DSPM MOTORS FOR SKEWED ROTOR TEETH AND COMBINED PERMANENT MAGNET AND WINDING EXCITATION

Now motors with rotor poles having particular skew angle have been analyzed for two type of combined excitation as mentioned previously and various performance characteristics like flux-linkage, back-emf, self inductance of various phases, mutual inductance between various phases, and the developed torque profiles when only one phase is excited and when all the phases are excited in a particular phase sequence are computed. By a spectrum analysis, harmonic torque components of the developed torques are also obtained.

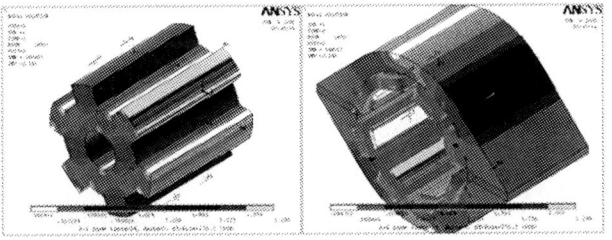

Fig. 6. Flux density plot for rotor and stator when rotor poles are not skewed and 34° away from unaligned position for phase-A with excitation for phase A is 270.5 AT/pole.

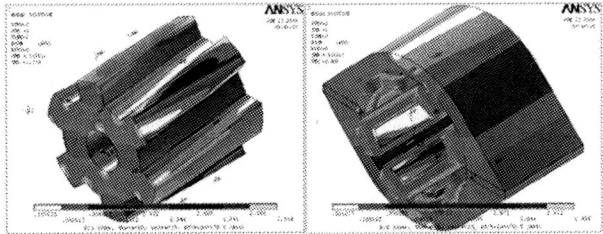

Fig. 7. Flux density plot for rotor and stator when rotor poles are skewed by 15° and rotor is at unaligned position for phase-A with excitation for phase A is -270.5 AT/pole.

Fig. 6 shows the flux density in the rotor and stator of the motor when the rotor poles are not skewed and rotor is at 34° away from phase-A unaligned position with phase-A excited with 270.5 AT/pole. Fig. 7 shows the flux density in the rotor and stator of the motor when rotor poles are skewed by 15° and rotor is at unaligned position with phase-A when phase-A is excited by -270.5 AT/pole. Here negative sign indicates that the flux produced by the winding excitation opposes the flux produced by the magnet excitation for the excited phase. Fig. 8 shows the phase-A flux-linkage characteristics for an excitation of 270.5 AT/pole for phase-A. Here positive and negative sign indicates that the flux produced by the winding excitation supports or opposes the magnet excitation respectively. It is observed from the Fig. 8 that the with the increase in the skew angle the peak flux-linkage reduces and with the increase in the skew angle the flux-linkage characteristics is getting shifted towards one side this is because the unstable equilibrium point for a particular phase is shifting by a half a skew angle in the opposite direction of the skew and this must be considered while designing the control circuit. For the skew angle of 30° the variation in the flux-linkage is almost negligible and for skew angle above 30° the direction of shift in the flux-linkage characteristics is reversed.

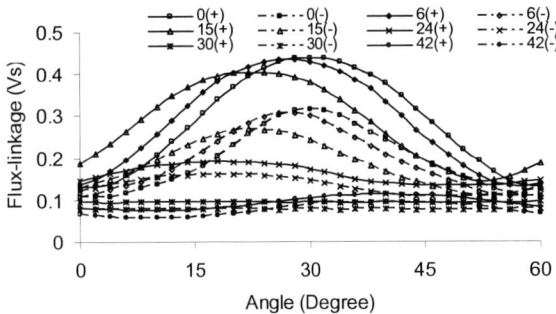

Fig. 8. Variation of phase-A flux-linkage at various skew angles when phase-A is excited with 270.5 AT/pole.

The phase inductance for DSPM motor is calculated using (1), where $\psi_{(PM \pm i)}$ and $\psi_{(PM)}$ is the phase flux-linkage for the combined PM and winding excitation of $i$ A/ph and for only PM excitation and $L$ is the self inductance of the winding.

$$L = \frac{\psi_{(PM \pm i)} - \psi_{(PM)}}{i} \qquad (1)$$

Fig. 9 shows the variation of phase-A inductance for various skew angles. Here positive and negative signs indicate that the winding excitation is additive or subtractive to the PM excitation. It is observed from the Fig. 9 that the winding self inductance is more when the flux due to winding excitation opposes the flux produced by the PM excitation; this is because of the reduction in the saturation. It is also observed that up to the skew angle of 15° there is not much change in the aligned inductance but the unaligned inductance increases with the increase in the skew angle. In DSPM motor each phase will have additive and subtractive winding excitation to PM excitation for half a rotor pole pitch resulting in to different self inductance profiles for the winding, this difference will lead to different reluctance profiles in the both halves of rotor movement and in turn induces the torque ripple. Skewing the rotor teeth reduces this difference and helps in reducing the net reluctance torque and torque ripple.

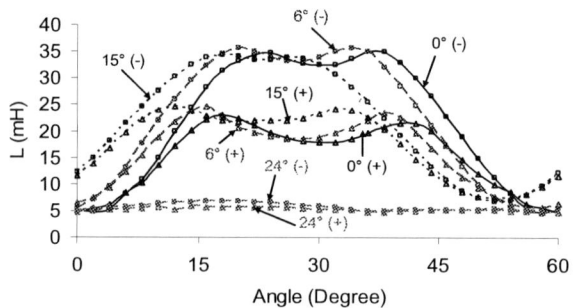

Fig. 9. Variation of phase-A self inductance at various skew angles when phase-A is excited with 270.5 AT/pole.

The mutual inductance between the phases $x$ and $y$ has been calculated using (2). Here, $M_{xy}$ is the mutual inductance between the phases $x$ and $y$, $\psi_{y(PM \pm i)}$ is the phase-y flux-linkage when phase-x is excited with current $i_x$ A/phase and $\psi_{y(PM)}$ is the phase-y flux-linkage when only PM excitation is there.

$$M_{xy} = \left. \frac{\psi_{y(PM \pm i_x)} - \psi_{y(PM)}}{i_x} \right|_{i_y = \text{constant}} \qquad (2)$$

Fig. 10 shows the variation of the mutual inductance between phase-A and phase-B ($M_{AB}$), when phase-A is excited with 270.5 AT/pole. It is observed from the Fig. 10 that mutual inductance for both additive and subtractive winding excitation to the PM excitation is different where for the subtractive excitation it is more, this is because of the non-linear B-H characteristics of the material used for the stator and rotor stacks. It is also observed that $M_{AB}$ is about one third of the self-inductance of phase-A, and for the skew angle of

30° the magnitude as well as variation of $M_{AB}$ is negligible. Similarly mutual inductances $M_{AC}$ and $M_{AD}$ have been calculated and shown in Fig. 11 and Fig. 12 respectively. It is observed from the analysis on mutual inductance that skewing will reduce the mutual inductances between the phases and thus will help in reducing the torque ripple produced by the non-ideality of the motor characteristics [3].

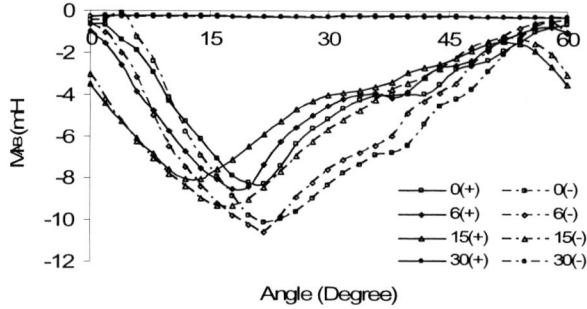

Fig. 10. Variation of the mutual inductances between phase-A and B at different skew angles.

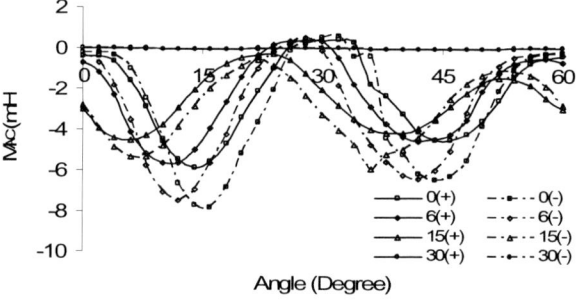

Fig. 11. Variation of the mutual inductances between phase-A and C at different skew angles.

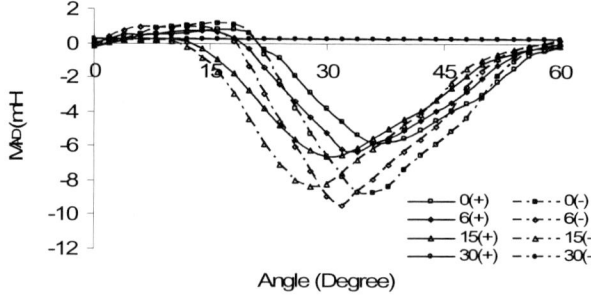

Fig. 12. Variation of the mutual inductances between phase-A and D at different skew angles.

Fig. 13 shows the phase-A torque for an excitation of 270.5 AT/pole additive to the PM excitation respectively. It is observed from this figure that the cogging torque acts as a ripple torque on top of the torque produced due to interaction of PM flux and winding excitation. Harmonic analysis of these torque waveforms is done and various harmonic torques are achieved. Fig. 14 shows the variation of the predominant harmonic torques at various skew angles from which it is observed that besides fundamental, fourth and its multiple harmonic torques are the predominant ones and the fundamental torque reduces with the increase in the skew angle. It is also observed that for a skew angle up to 15° the reduction in the fundamental torque is less but beyond 15° the reduction in fundamental torque is drastic, this restricts the further analysis of finding the actual torque profile, harmonic torques and torque ripple analysis of the motor for an appropriate excitation of various phases including the effect of mutual inductance for the combined PM and winding excitation when the windings are excited either with a rectangular pulse or sinusoidal pulse of appropriate polarity.

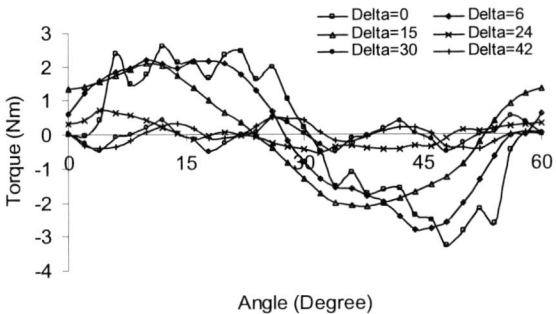

Fig. 13 . Torque profile for various skew angles when phase-A is excited with 270.5 AT/pole.

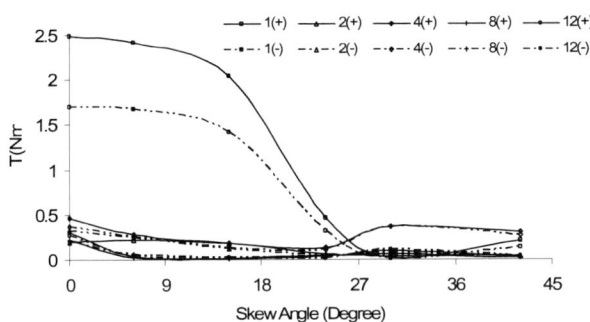

Fig. 14. Variation of harmonic torque for phase torque at various skew angles.

The overall torque profile is now achieved for the motor with skew angle up to 24°. Fig, 15 shows the overall torque profile for various skew angles when combined excitation due to permanent magnet and winding is considered. Here during the winding excitation two phase mode in which all the four phases are excited simultaneously with appropriate polarity of current have been considered. Torque ripple for various excitations have been calculated using (3).

$$T_{ripple}(\%) = \left( \frac{T_{max} - T_{min}}{T_{avg}} \right) \times 100 \qquad (3)$$

Various harmonic torque component of the developed

torque profile has been done and various torque harmonics are achieved. Fig. 16 gives the significant harmonic for developed torque profile of Fig. 15, from which it is observed that on the average torque there is the fourth and its multiple harmonic components superimposed giving the torque ripple.

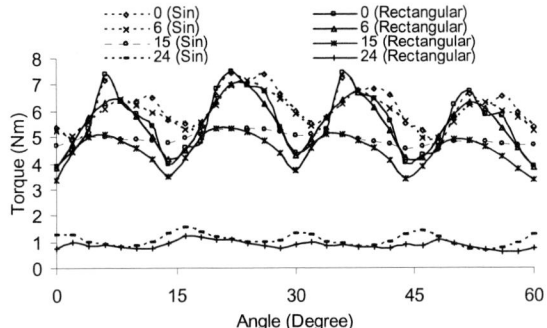

Fig. 15. Developed torque profiles for various skew angles when all the phases are excited with rectangular and sinusoidal excitation.

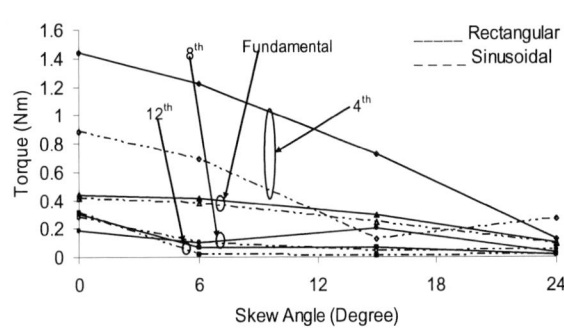

Fig. 16. Variation of harmonic torques for developed torque for both type of excitations.

Table II gives the values of the average torque and torque ripple for motor with different skew angles for both rectangular and sinusoidal excitations. Fig. 16 shows the variation of average torque and torque ripple for different skew angles of the rotor poles for both types of excitations. It is observed from Fig. 16 and Table II that for all the skew angles the sinusoidal current is giving more average torque compared to the rectangular excitation. It is also observed that the torque ripple is less for all skew angles except 24° when DSPM motor is excited with sinusoidal current in comparison to rectangular current. From Table II it is observed that with the increase in the skew angle of rotor teeth the average torque reduces. The reduction in average torque at the skew angle of 15° is 19.2%, and 17.4% respectively for both sinusoidal and rectangular excitation and the corresponding reduction in torque ripple is 67.4% and 24.9% which is significant for sinusoidal excitation this is due to at the skew angle of 15° the fourth harmonic reduces significantly as shown in Fig. 16.

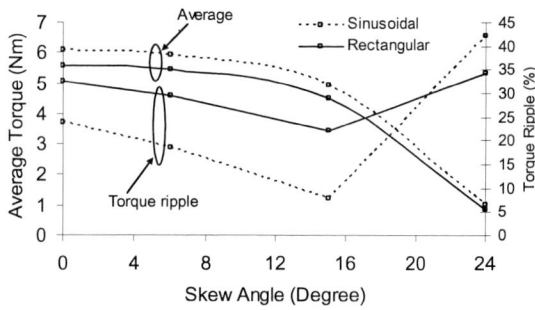

Fig. 17. Variation of average torque and torque ripple with different skew angles for both rectangular and sinusoidal excitations.

TABLE II
MAJOR DIMENSION AND MATERIAL DETAILS OF THE MOTOR

| Excitation | Sinusoidal Excitation | | | | Rectangular Excitation | | | |
|---|---|---|---|---|---|---|---|---|
| Skew angle (°) | 0 | 6 | 15 | 24 | 0 | 6 | 15 | 24 |
| $T_{avg}$ (Nm) | 6.1 | 5.9 | 4.9 | 1.1 | 5.6 | 5.4 | 4.5 | 0.9 |
| $T_{ripple}$ (%) | 23.9 | 18.5 | 7.81 | 42.3 | 32.5 | 29.5 | 22.1 | 34.3 |

## IV. CONCLUSION

This paper had presented the effect of skewing the rotor teeth on various performance parameters of the DSPM motor for only permanent magnet excitation and permanent magnet and stator winding combined excitations. It is observed that skewing the rotor teeth reduces the phase flux-linkage and for the skew angle more than half of rotor pole pitch the amount of flux-linkage is very small with negligible variation. Skewing also reduces the difference between the phase inductance when phase inductance is supporting or opposing the permanent magnet excitation. Skewing the rotor teeth by 12° to 15° will result in to minimum detent torque. It is also observed that for all skew angles the sinusoidal excitation of motor with two phase excitation mode is giving more average torque in comparison to the rectangular current excitation this is also true for torque ripple if the skew angles less than 24°. It is also observed that skewing the rotor teeth by 12° to 15° will give us much reduced torque ripple without much reduction in the torque capability of the motor, thus a combination of sinusoidal excitation and skew angle of 12° to 15° is an ideal combination for an 8/6 DSPM motor to improve torque profile.

## V. REFERENCES

[1] Yuefeng Liao, Feng Liang, and Thomas A Lipo, "A novel permanent magnet motor with doubly salient structure", *IEEE Trans. Ind. Applicat*, vol. 31, pp. 1069-1078, Sept./Oct. 1995.

[2] M. Cheng, K. T. Chau, C. C. Chan, and Q. Sun, "Control and operation of a new 8/6-pole doubly salient permanent-magnet motor drive", *IEEE Trans. Ind. Applicat*, vol. 39, pp. 1363-1371, Sept./Oct. 2003.

[3] K. T. Chau, Qiang Sun, Ying Fan, and Ming Cheng, "Torque ripple minimization of doubly salient permanent-magnet motors", *IEEE Trans. on Energy Conversion*, vol. 20, pp. 352-358, June 2005.

[4] Takeo Ishikawa, and Gordon R. Slemon, "A method of reducing ripple torque in permanent magnet motors without skewing", *IEEE Trans. on Magnetics*, vol. 29, pp. 2028-2031, March 1993.

[5] Min Dai, Ali Keyhani, and Tomy Sebastain, "Torque ripple analysis of a PM brushless dc motor using finite element method", *IEEE Trans. on Energy Conversion*, vol. 19, pp. 40-45, March 2005.

[6] R. P. Deodhar, D. A. Staton, and T. J. E. Miller, "Modelling of skew using the flux-mmf diagram", *IEEE Trans. Ind. Applicat*, vol. 31, pp. 1069-1078, Sept./Oct. 1995.

[7] C. Wang, S. A. Nasar and I. Boldea, "Three-phase flux reversal machine (FRM)", *IEE Proc.Electric Power Application*, vol. 146, pp. 139-146, August 1999.

[8] N. K. Sheth, A. R. C. Sekharbabu, and K. R. Rajagopal, "Effect of skewing the rotor teeth on the performance of doubly salient permanent magnet motors", *J. Appl. Phys.*, vol. 99, pp. 08R320-1-08R320-3, May 2006.

[9] A. K. Sawhney, *A course in electrical machine design*, 5th Edition, Dhanpat Rai & Co. Ltd., New Delhi, 1991.

## VI. BIOGRAPHIES

**Nimit K. Sheth** (S'2003) was born in Nadia, Gujarat, India in 1975. He received B. E. Degree in Electrical Engineering from the Gujarat University, Ahmedabad, India in 1996, M. Tech. Degree in Power Electronics, Electrical Machines and Drives in Electrical Engineering from the Indian Institute of Technology Delhi, New Delhi, India in 2002. Since, 2003 he is working towards his PhD degree at Indian Institute of Technology Delhi, New Delhi, India.

From 1997 to 1998, he was with Ahmedabad Electricity Company Ltd. Ahmedabad, India, as a Trainee Engineer. Since 1998, he is with the Institute of Technology, Nirma University of Science and Technology, Ahmedabad, India, where currently he is an Assistant Professor in Electrical Engineering Department.

He has published more than 20 papers in International Journals and Conference proceedings. He received Prof. A. K. Sinha Award for securing highest CGPA among all the M. Tech. graduating students from the Electrical Engineering Department of Indian Institute of Technology Delhi, New Delhi, India during the year 2002. His fields of interest include Electrical Machines and Drives, Special Electrical Machines (Switched Reluctance Motors, DSPM Motors, Flux Reversal Motors, Stepper Motors, etc.,), Finite Element Analysis and CAD of Electrical Machines.

**K. R. Rajagopal** (M'1998, SM'2000) was born in Alappuzha, Kerala, India in 1961. He received Diploma in Electrical Engineering from Carmel Polytechnic, Alappuzha, India in 1979, B. Tech. Degree in Electrical Engineering from the College of Engineering, Trivandrum, India in 1988, M. Tech. Degree in Power Electronics, Electrical Machines and Drives and Ph. D. Degree in Electrical Engineering from the Indian Institute of Technology Delhi, New Delhi, India during 1991 and 1998 respectively.

From 1980 to 1983, he was with Aluminum Industries Ltd. (ALIND), Trivandrum, India, as an Application Engineer (Relays), from 1983 to 1999, he was with the Indian Space Research Organization (ISRO), Trivundrum, India, where he was engaged in Analysis, Design, Development and Testing of Special Electrical Machines/Devices used in space applications. Since 1999, he is with the Indian Institute of Technology Delhi, New Delhi, India, where currently he is a Professor in Electrical Engineering Department.

He has published more than 30 papers in International Journals and more than 60 papers in International conference proceedings. He received Indian National Academy of Engineering (INAE) award for most Innovative Potential Project in Engineering during the year 1998. His fields of interest include Electrical Machines and Drives, Special Electrical Machines (Stepper Motors, Switched Reluctance Motors, PM BLDC Motors, Hysterisis Motors, etc.,), Magnetic Devices, Finite Element Analysis and CAD of Electrical Machines and Design of Energy Efficient Motors for Home Appliances.

# 2006 IEEE International Conference on Power Electronic, Drives and Energy Systems

# Dynamic Modeling and Simulation of an Induction Motor with Adaptive Backstepping Design of an Input-Output Feedback Linearization Controller in Series Hybrid Electric Vehicle

M. Jalalifar, A. Farrokh Payam, B. Mirzaeian, and S. M. Saghaeian nezhad

*Abstract*— In this paper using Adaptive Backstepping approach an adaptive rotor flux observer which provides stator and rotor resistances estimation simultaneously for induction motor used in series hybrid electric vehicle is proposed. The controller of induction motor (IM) is designed based on input-output feedback linearization technique. Combining this controller with adaptive backstepping observer the system is robust against rotor and stator resistances uncertainties. In additional, mechanical components of a hybrid electric vehicle are called from the Advanced Vehicle Simulator Software Library and then linked with the electric motor. Finally, a typical series hybrid electric vehicle is modeled and investigated. Various tests, such as acceleration traversing ramp, maximum speed, fuel consumption and emission are performed on the proposed model of a series hybrid vehicle. Computer simulation results obtained, confirm the validity and performance of the proposed IM control approach using for series hybrid electric vehicle.

*Index Terms*— Adaptive Backstepping Observer, Series Electric Vehicle.

## I. INTRODUCTION

NOWADAYS the air pollution and economical issues are the major driving forces in developing electric vehicles (EVs).

In recent years EVs and hybrid electric vehicles (HEVs) are the only alternatives for a clean, efficient and environmentally friendly urban transportation system [1]. HEVs meet both consumer needs as well as car manufacturer needs. They give the consumer the ability to use the car for long periods of time without recharging. HEVs also take a giant step forward in meeting low emission standards set by the Partnership for a New Generation of Vehicles.

Because of simple and rugged construction, low cast and maintenance, high performance and sufficient starting torque and good ability of acceleration, squirrel cage induction motor is a good candidate for EVs [2].

In this paper by using an input-output feedback linearization technique combined with an adaptive backstepping observer in stator reference frame the induction motor [3] using in series hybrid electric vehicle is controlled. One of the best advantages of this control method is eliminating the flux sensor and decreases the cost of controller, in addition the control system is robust respect to resistances variations and external load torque.

Advanced vehicle simulator (ADVISOR) provides the vehicle engineering community with an easy-to-use, flexible, yet robust and supported analysis package for advanced vehicle modeling. It is primarily used to quantify fuel economy, the performance, and the emissions of vehicles that use alternative technologies including fuel cells, batteries, electric motors, and ICE in hybrid configurations. But the components in ADVISOR have been modeled simply and only with static model to decrease the simulation time [4].

In this paper using MATLAB/SIMULINK software, dynamic modeling of an induction motor that is used in series hybrid electric vehicle and controlled by input-output feedback linearization method combined with adaptive backstepping observer is investigated and then simulated separately by linking the mechanical components for a series hybrid electric vehicle from the ADVISOR software library. At the end, a typical HEV is modeled and investigated. Simulation results obtained show the IM and other components performances for a typical city drive cycle.

---

M. Jalalifar is with Isfahan University of Technology, Isfahan, Iran (e-mail: mehran_j1356@yahoo.com).

A. Farrokh Payam is with Isfahan University of Technology, Isfahan, Iran (e-mail: amir_farrokh@yahoo.com , farokhpayam@alumni.iut.ac.ir ).

B. Mirzaeian Dehkordi is with University of Isfahan, Isfahan, Iran (e-mail: Mirzaeian@eng.ui.ac.ir ).

S. M. Saghaeian Nezhad is with Isfahan University of Technology, Isfahan, Iran

0-7803-9771-1/06/$25.00 ©2006 IEEE

## II. THE PERFORMANCE OF AN ELECTRIC VEHICLE

The first step in vehicle performance modeling is to write an equation for the electric force. This is the force transmitted to the ground through the drive wheels, and propelling the vehicle forward. This force must overcome the road load and accelerate the vehicle as shown in Fig. 1 [5].

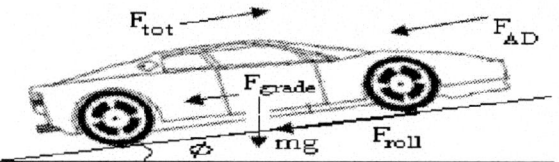

Fig. 1. A summary of forces on a vehicle.

The rolling resistance is primarily due to the friction of the vehicle tires on the road and can be written as:

$$f_{roll} = f_r . M . g \qquad (1)$$

where, $M$ is the vehicle mass, $f_r$ is the rolling resistance coefficient and $g$ is gravity acceleration.

The aerodynamic drag is due to the friction of the body of vehicle moving through the air. The formula for this component is as in the following:

$$f_{AD} = \frac{1}{2} . \xi . C_D . A . V^2 \qquad (2)$$

where, $\xi$ is the air mass density, and $V, C_D$, and $A$ are the speed, the aerodynamic coefficient, and the frontal area of the vehicle, respectively.

The gravity force due to the slope of the road can be expressed by:

$$f_{grade} = M . g . \sin \alpha \qquad (3)$$

where, $\alpha$ is the grade angle.

In addition to the forces shown in Fig. 3, another one is needed to provide the linear acceleration of the vehicle given by:

$$f_{acc} = M . \alpha = M . \frac{dV}{dt} \qquad (4)$$

The propulsion system must now overcome the road loads and accelerate the vehicle by the tractive force, $F_{tot}$, as follows:

$$F_{tot} = f_{roll} + f_{AD} + f_{grade} + f_{acc} \qquad (5)$$

A typical road load characteristic as a function of the speed and mass of a vehicle is shown in Fig. 2.

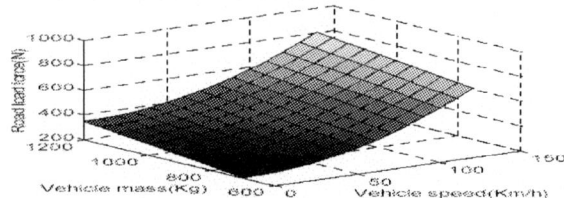

Fig. 2. The road profile as a function of speed and mass of a vehicle. ( $\alpha$ =0 deg.)

Wheels and axels convert $F_{tot}$ and the speed of vehicle to torque and angular speed requirements for the differential as follows [4]:

$$T_{wheel} = F_{tot} . r_{wheel}$$

$$\omega_{wheel} = \frac{V}{r_{wheel}} \qquad (6)$$

where, $T_{wheel}$, $r_{wheel}$, and $\omega_{wheel}$ are the tractive torque, the radius, and the angular velocity at the wheels, respectively.

The angular velocity and torque of the wheels are converted to motor rpm and motor torque requirements using the gears ratio at differential and gearbox as follows:

$$\omega_m = G_{fd} G_{gb} . \omega_{wheel}$$

$$T_m = \frac{T_{wheel}}{G_{fd} G_{gb}} \qquad (7)$$

where, $G_{fd}$ and $G_{gb}$ are respectively differential and gear box gears ratios.

A series HEV consists of two major group components as shown in Fig. 3:

- mechanical components ( engine, wheels, axels and transmission box).
- Electrical components ( batteries and electric motor).

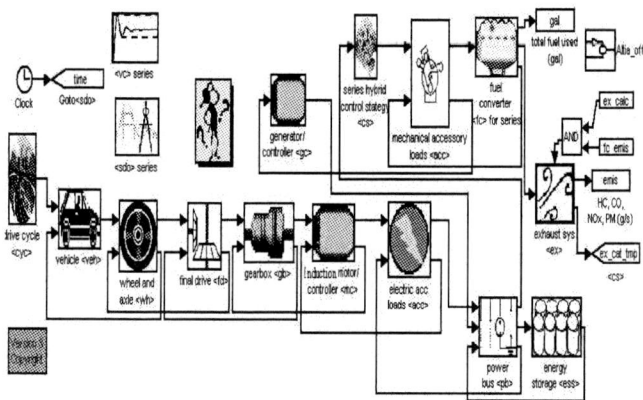

Fig. 3. A series hybrid electric vehicle.

## III. ELECTRIC MOTOR

Due to the simple and rugged construction, low cast and maintenance, high performance and sufficient starting torque and good ability of acceleration, squirrel cage induction motor is one of the well suited motors for the electric propulsion systems [6]. In this section first modeling of the induction motor in stator fixed reference frame and then designing an input-output controller combined with adaptive backstepping observer for IM [3] is investigated.

### A. Input-Output Feedback linearization Controller Design

The square of rotor flux amplitude is taken as the first output of the controlled system

$$y_1 = \psi_{ra}^2 + \psi_{rb}^2 = \psi_r^2$$

$$e_1 = y_1 - y_{1d} = \psi_r^2 - \psi_r^{r2} \qquad (8)$$

where $\psi_r^r$ is the reference signal for rotor flux amplitude.

Differentiating $y_1$ so much inputs $(u_{sa}, u_{sb})$ appear in our equations

$$\dot{y}_1 = 2\psi_{ra}(-\frac{R_r}{L_r}\psi_{ra} + \frac{R_r}{L_r}Mi_{sa}) +$$

$$2\psi_{rb}(-\frac{R_r}{L_r}\psi_{rb} + \frac{R_r}{L_r}Mi_{sb}) \qquad (9)$$

Then

$$\ddot{y}_1 = -4\frac{R_r}{L_r}\psi_{ra}(-\frac{R_r}{L_r}\psi_{ra} + \frac{R_r}{L_r}Mi_{sa})$$

$$+2\frac{R_r}{L_r}M\psi_{ra}(\frac{MR_r}{\sigma L_s L_r^2}\psi_{ra} - \frac{M^2 R_r + L_r^2 R_s}{\sigma L_s L_r^2}i_{sa} + \frac{u_{sa}}{\sigma L_s})$$

$$-4\frac{R_r}{L_r}\psi_{rb}(\frac{R_r}{L_r}\psi_{rb} + \frac{R_r}{L_r}Mi_{sb}) + 2\frac{R_r}{L_r}M\psi_{rb} \times \qquad (10)$$

If $e_1$ dynamic force to be $k_{11}\ddot{e}_1 + k_{12}\dot{e}_1 + k_{13}e_1$, we have

$$k_{11}(\ddot{y}_1 - \ddot{y}_{1d}) + k_{12}(\dot{y}_1 - \dot{y}_{1d}) + k_{13}(y_1 - y_{1d}) = 0$$

$$\rightarrow \ddot{y}_1 = \frac{k_{12}}{k_{11}}(\dot{y}_{1d} - \dot{y}_1) + \frac{k_{13}}{k_{11}}(y_{1d} - y_1) + \ddot{y}_{1d} = \frac{k_{13}}{k_{11}}(\psi_r^{*2} - \psi_r^2)$$

$$-\frac{k_{12}}{k_{11}}\{2\psi_{ra}(-\frac{R_r}{L_r}\psi_{ra} + \frac{R_r}{L_r}Mi_{sa}) + 2\psi_{rb}(-\frac{R_r}{L_r}\psi_{rb} + \frac{R_r}{L_r}Mi_{sb})\}$$

$$-2\psi_r^*\dot{\psi}_r^* + 2\left(\dot{\psi}_r^{*2} + \psi_r^*\ddot{\psi}_r^*\right) \qquad (11)$$

Now the motor developed electromagnetic torque is considered as the second output

$$y_2 = te = \frac{n_p M}{2L_r}(\psi_{ra}i_{sb} - \psi_{rb}i_{sa})$$

$$e_2 = y_2 - y_{2d} = te - te^r \qquad (12)$$

where $te^r$ is the reference signal for developed electromagnetic torque.

Time derivative of $y_2$ is

By setting $e_2$ dynamic as $k_{21}\dot{e}_2 + k_{22}e_e$, we have

$$\dot{y}_2 = \frac{n_p M}{2L_r}\{\psi_{ra}(-\frac{n_p M}{\sigma L_s L_r}\omega\psi_{ra} - \frac{M^2 R_r + L_r^2 R_s}{\sigma L_s L_r^2}i_{sb} + \frac{u_{sb}}{\sigma L_s})$$

$$+i_{sb}(-\frac{R_r}{L_r}\psi_{ra} - n_p\omega\psi_{rb}) - \psi_{rb}(-\frac{n_p M}{\sigma L_s L_r}\omega\psi_{rb} - \frac{M^2 R_r + L_r^2 R_s}{\sigma L_s L_r^2}i_{sa} + \frac{u_{sa}}{\sigma L_s})$$

$$-i_{sa}(-\frac{R_r}{L_r}\psi_{rb} + n_p\omega\psi_{ra})\} \qquad (13)$$

By setting $e_2$ dynamic as $k_{21}\dot{e}_2 + k_{22}e_2 = 0$, we have

$$k_{21}(\dot{y}_2 - \dot{y}_{2d}) + k_{22}(y_2 - y_{2d}) = 0$$

$$\rightarrow \dot{y}_2 = \frac{k_{22}}{k_{21}}(y_{2d} - y_2) + \dot{y}_{2d} = \frac{k_{22}}{k_{21}}(te^r - te) + \dot{te}^r \qquad (14)$$

Setting right side of (13) equal to right side of (14), another equation for inputs is achieved

$$\frac{n_p M}{2L_r}\{\psi_{ra}(-\frac{n_p M}{\sigma L_s L_r}\omega\psi_{ra} - \frac{M^2 R_r + L_r^2 R_s}{\sigma L_s L_r^2}i_{sb} + \frac{u_{sb}}{\sigma L_s})$$

$$+i_{sb}(-\frac{R_r}{L_r}\psi_{ra} - n_p\omega\psi_{rb}) - \psi_{rb}(-\frac{n_p M}{\sigma L_s L_r}\omega\psi_{rb} - \frac{M^2 R_r + L_r^2 R_s}{\sigma L_s L_r^2}i_{sa} + \frac{u_{sa}}{\sigma L_s})$$

$$-i_{sa}(-\frac{R_r}{L_r}\psi_{rb} + n_p\omega\psi_{ra})\} = \frac{k_{22}}{k_{21}}(te^r - te) + \dot{te}^r \qquad (15)$$

Now, from system of two equations, (6,10), inputs $(u_{sa}, u_{sb})$ are found.

To proof the stability of proposed controller, note that according to dynamics of $e_1, e_2$, if iniyially $e_1(0) = \dot{e}_1(0) = 0$ and also $e_2(0) = 0$, then $e_1(t), e_2(t) \equiv 0, \forall t \geq 0$, perfect tracking is achieved; otherwise, $e_1(t), e_2(t)$ converges to zero exponentially [7].

### B. Adaptive Backstepping Observer Design

Stator current is measurable and is taken as output:

$$y_a = i_{sa}, y_b = i_{sb} \qquad (16)$$

The prediction model for the backstepping observer is chosen to be

$$p\hat{\psi}_{ra} = -\frac{\hat{R}_r}{L_r}\hat{\psi}_{ra} - n_p\omega\hat{\psi}_{rb} + \frac{\hat{R}_r}{L_r}My_a$$

$$p\hat{\psi}_{rb} = -\frac{\hat{R}_r}{L_r}\hat{\psi}_{rb} + n_p\omega\hat{\psi}_{ra} + \frac{\hat{R}_r}{L_r}My_b$$

$$p\hat{i}_{sa} = \frac{M\hat{R}_r}{\sigma L_s L_r^2}\hat{\psi}_{ra} + \frac{n_p M}{\sigma L_s L_r}\omega\hat{\psi}_{rb} - \frac{M^2\hat{R}_r + L_r^2\hat{R}_s}{\sigma L_s L_r^2}y_a + \frac{1}{\sigma L_s}u_{sa} + v_a$$

$$p\hat{i}_{sb} = \frac{M\hat{R}_r}{\sigma L_s L_r^2}\hat{\psi}_{rb} - \frac{n_p M}{\sigma L_s L_r}\omega\hat{\psi}_{ra} - \frac{M^2\hat{R}_r + L_r^2\hat{R}_s}{\sigma L_s L_r^2}y_b + \frac{1}{\sigma L_s}u_{sb} + v_b \qquad (17)$$

where $p$ is the differential operator and $v_a, v_b$ are the control input to be designed by the backstepping method.

The dynamical equations for the prediction errors are

$$p\tilde{\psi}_{ra} = -\frac{\tilde{R}_r}{L_r}\psi_{ra} - \frac{\hat{R}_r}{L_r}\tilde{\psi}_{ra} - n_p\omega\tilde{\psi}_{rb} + \frac{\tilde{R}_r}{L_r}My_a$$

$$p\tilde{\psi}_{rb} = -\frac{\tilde{R}_r}{L_r}\psi_{rb} - \frac{\hat{R}_r}{L_r}\tilde{\psi}_{rb} + n_p\omega\tilde{\psi}_{ra} + \frac{\tilde{R}_r}{L_r}My_b$$

$$p\tilde{i}_{sa} = \frac{M\tilde{R}_r}{\sigma L_s L_r^2}\psi_{ra} + \frac{M\hat{R}_r}{\sigma L_s L_r^2}\tilde{\psi}_{ra} + \frac{n_p M}{\sigma L_s L_r}\omega\tilde{\psi}_{rb} - \frac{M^2\tilde{R}_r + L_r^2\tilde{R}_s}{\sigma L_s L_r^2}y_a + v_a$$

$$p\tilde{i}_{sb} = \frac{M\tilde{R}_r}{\sigma L_s L_r^2}\psi_{rb} + \frac{M\hat{R}_r}{\sigma L_s L_r^2}\tilde{\psi}_{rb} - \frac{n_p M}{\sigma L_s L_r}\omega\tilde{\psi}_{ra} - \frac{M^2\tilde{R}_r + L_r^2\tilde{R}_s}{\sigma L_s L_r^2}y_b + v_b$$

$$\tilde{y}_a = \tilde{i}_{sa}$$

$$\tilde{y}_b = \tilde{i}_{sb} \qquad (18)$$

where

$$\tilde{\psi}_{ra} = \hat{\psi}_{ra} - \psi_{ra}, \tilde{\psi}_{rb} = \hat{\psi}_{rb} - \psi_{rb}, \tilde{i}_{sa} = \hat{i}_{sa} - i_{sa}$$

$$\tilde{i}_{sb} = \hat{i}_{sb} - i_{sb}, \tilde{R}_s = \hat{R}_s - R_s, \tilde{R}_r = \hat{R}_r - R_r \qquad (19)$$

The first step in the backstepping strategy is to design a stable controller for the integral of the prediction errors $\tilde{y}_a, \tilde{y}_b$ using $\tilde{i}_{sa}, \tilde{i}_{sb}$ as virtual control variables with stabilizing functions $\phi_a, \phi_b$ which is reference for virtual variables. The integral of the prediction errors $\tilde{x}_a, \tilde{x}_b$ are

$$p\tilde{x}_a = \tilde{i}_{sa} \quad , p\tilde{x}_b = \tilde{i}_{sb} \qquad (20)$$

Adding and subtracting $\phi_a, \phi_b$ to above equations

$$p\tilde{x}_a = z_a - c_1\tilde{x}_a \quad , p\tilde{x}_b = z_b - c_1\tilde{x}_b$$
$$z_a = \tilde{i}_{sa} - \phi_a \quad , z_b = \tilde{i}_{sb} - \phi_b$$
$$\phi_a = -c_1\tilde{x}_a \quad , \phi_b = -c_1\tilde{x}_b \tag{21}$$

The second step in the backstepping strategy is the control of

$$z_a = \tilde{i}_{sa} + c_1\tilde{x}_a \quad , z_b = \tilde{i}_{sb} + c_1\tilde{x}_b \tag{22}$$

By proper choice of the control inputs $v_a, v_b$. Taking the derivative of $z_a, z_b$

$$\dot{z}_a = \frac{M\tilde{R}_r}{d_sL_r^2}\psi_{ra} + \frac{M\hat{R}_r}{d_sL_r^2}\tilde{\psi}_{ra} + \frac{n_pM}{d_sL_r}\omega\tilde{\psi}_{rb} - \frac{M^2\tilde{R}_r + L_r^2\tilde{R}_s}{d_sL_r^2}y_a + v_a + c_1\tilde{i}_{sa}$$

$$\dot{z}_b = \frac{M\tilde{R}_r}{d_sL_r^2}\psi_{rb} + \frac{M\hat{R}_r}{d_sL_r^2}\tilde{\psi}_{rb} + \frac{n_pM}{d_sL_r}\omega\tilde{\psi}_{ra} - \frac{M^2\tilde{R}_r + L_r^2\tilde{R}_s}{d_sL_r^2}y_b + v_b + c_1\tilde{i}_{sb} \tag{23}$$

and selecting following control inputs

$$v_a = -\frac{M\hat{R}_r}{\sigma L_sL_r^2}\tilde{\psi}_{ra} - \frac{n_pM}{\sigma L_sL_r}\omega\tilde{\psi}_{rb} - c_1\tilde{i}_{sa} - c_2z_a - \tilde{x}_a$$

$$v_b = -\frac{M\hat{R}_r}{\sigma L_sL_r^2}\tilde{\psi}_{rb} + \frac{n_pM}{\sigma L_sL_r}\omega\tilde{\psi}_{ra} - c_1\tilde{i}_{sb} - c_2z_b - \tilde{x}_b \tag{24}$$

Yields

$$\dot{z}_a = \frac{M\tilde{R}_r}{\sigma L_sL_r^2}\psi_{ra} - \frac{M^2\tilde{R}_r + L_r^2\tilde{R}_s}{\sigma L_sL_r^2}y_a - c_2z_a - \tilde{x}_a$$

$$\dot{z}_b = \frac{M\tilde{R}_r}{\sigma L_sL_r^2}\psi_{rb} - \frac{M^2\tilde{R}_r + L_r^2\tilde{R}_s}{\sigma L_sL_r^2}y_b - c_2z_b - \tilde{x}_b \tag{25}$$

$c_1, c_2$ are positive constant design parameters.
Stability analysis of observer is done by the following Lyapunov candidate

$$V = \frac{1}{2}\left\{\tilde{x}_a^2 + \tilde{x}_b^2 + z_a^2 + z_b^2 + \tilde{\psi}_{ra}^2 + \tilde{\psi}_{rb}^2 + \frac{1}{\gamma_s}\tilde{R}_s^2 + \frac{1}{\gamma_r}\tilde{R}_r^2\right\} \tag{26}$$

Derivating $V$ along the dynamics of (18,21,25) yields

$$\dot{V} = -c_1\tilde{x}_a^2 - c_1\tilde{x}_b^2 - c_2z_a^2 - c_2z_b^2 - \frac{\hat{R}_r}{L_r}\tilde{\psi}_{ra}^2 - \frac{\hat{R}_r}{L_r}\tilde{\psi}_{rb}^2$$

$$+\tilde{R}_s\{-\frac{z_aL_r^2}{d_sL_r^2}y_a - \frac{z_bL_r^2}{d_sL_r^2}y_b + \frac{1}{\gamma_s}\frac{d\tilde{R}_s}{dt}\}$$

$$+\tilde{R}_r\{\frac{z_aM}{d_sL_r^2}\psi_{ra} - \frac{M^2z_a}{d_sL_r^2}y_a + \frac{z_bM}{d_sL_r^2}\psi_{rb} - \frac{M^2z_b}{d_sL_r^2}y_b - \frac{\tilde{\psi}_{ra}\psi_{ra}}{L_r}$$

$$+\frac{M}{L_r}y_a\tilde{\psi}_{ra} - \frac{\tilde{\psi}_{rb}\psi_{rb}}{L_r} + \frac{M}{L_r}y_b\tilde{\psi}_{rb} + \frac{1}{\gamma_r}\frac{d\tilde{R}_r}{dt}\} \tag{27}$$

By selecting following adaptation laws

$$\frac{d\tilde{R}_s}{dt} = \gamma_s\{\frac{z_aL_r^2}{d_sL_r^2}y_a + \frac{z_bL_r^2}{d_sL_r^2}y_b\}$$

$$\frac{d\tilde{R}_r}{dt} = \gamma_r\{\frac{z_aM}{d_sL_r^2}\psi_{ra} + \frac{M^2z_a}{d_sL_r^2}y_a - \frac{z_bM}{d_sL_r^2}\psi_{rb} + \frac{M^2z_b}{d_sL_r^2}y_b$$

$$+\frac{\tilde{\psi}_{ra}\psi_{ra}}{L_r} - \frac{M}{L_r}y_a\tilde{\psi}_{ra} + \frac{\tilde{\psi}_{rb}\psi_{rb}}{L_r} - \frac{M}{L_r}y_b\tilde{\psi}_{rb}\} \tag{28}$$

we have $\dot{V} < 0$ outside the equilibrium point $(\tilde{x}_a, \tilde{x}_b, z_a, z_b, \tilde{\psi}_{ra}, \tilde{\psi}_{rb}) = (0,0,0,0,0,0)$. Based on the Barbalat's Lemma, we can obtain $\tilde{x}_a, \tilde{x}_b, z_a, z_b, \tilde{\psi}_{ra}, \tilde{\psi}_{rb}$ will converge to zero as $t \to \infty$. Therefore, the proposed observer is stable, even if parametric uncertainties exist [8].
The block diagram of the proposed controller is given in Fig. 4.

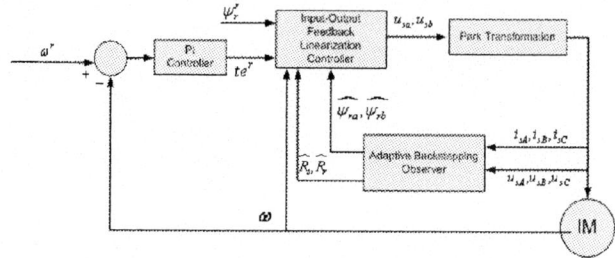

Fig. 4. Block diagram of proposed controller.

## IV. BATTERY MODELING

The battery considered in this paper is of the NiMH type for which a simple model is assumed. Therefore, a simplified version of the complex battery model reported in [9] is used.

## V. TRANSMISSION BOX

In electric vehicles the differential box can be eliminated if one uses two electric motors. Also, due to wide speed range operation of electric motors enabled by power electronics control, it is possible for an electric vehicles to use a single gear ratio for instantaneous matching of the available motor torque ($T_m$) with the desired tractive torque ($T_{wheel}$). The ratio of this single gear box and its size depends on the maximum speed of the motor and the vehicle as well as radius of the wheel as in the following [3]:

$$GR = \frac{(\omega_m)_{max}}{V_{max}}.r_{wheel}$$
$$GR = G_{fd}.G_{gb} \tag{29}$$

where $GR$ is the single gear ratio.

## VI. SIMULATION RESULTS

Simulation results presented in this section, focus on the dynamic behavior of IM and the battery of the vehicle. First the system is simulated for ECE+EUDC test cycle. This cycle is used for emission certification of light duty vehicles in Europe. Due to the electric motor has been modeled dynamically in SIMULINK. The data for IM and EV is in the appendix.
Fig. 5 shows the simulation block diagram.

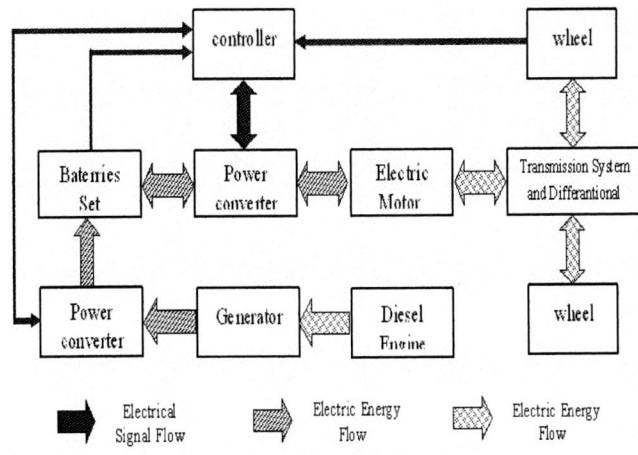

Fig. 5. simulation block diagram.

The drive cycle gives the required vehicle speed then the torque and speed requested from the electric motor. The current drawn from IM power supply shows the battery performance. The dynamic behavior of the IM in the ECE+EUDC drive cycle is shown in Figs. 6(a-d). Fig. 6(a) shows the ECE+EUDC drive cycle. Figs. 6(c,d) show the IM torque and average torque.

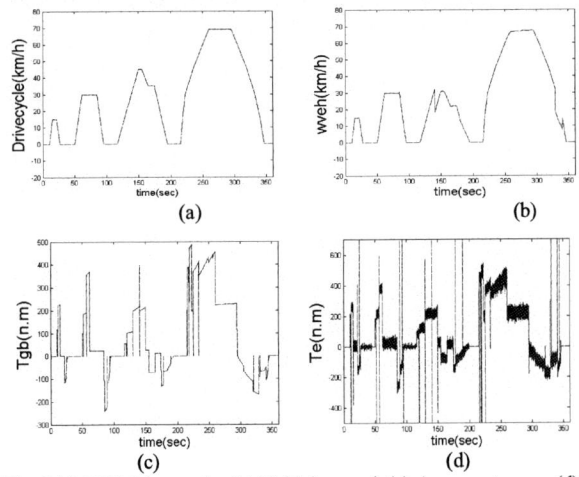

Fig. 6.(a) ECE drive cycle, (b) Vehicle speed, (c) Average torque, (d) IM torque.

Finally, the system is simulated for fuel consumption and emissions test. The results obtained are shown in Fig. 7 and Fig. 8. Fig. 7 shows the fuel consumption and emissions of diesel engine when the motor is hot at the start of test, and Fig. 8 shows the results when the engine is cool at the start of test. It is obvious that when the engine is cool, fuel consumption and emissions is increases respect to the engine is hot.

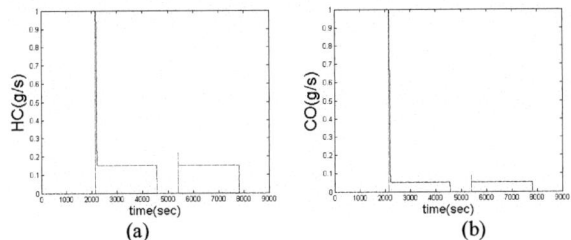

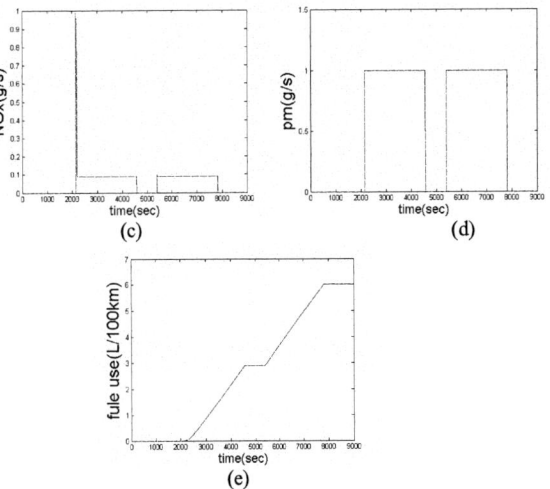

Fig. 7.(a) HC, (b) CO, (c) NOX, (d) PM, (e) Fuel consumption.

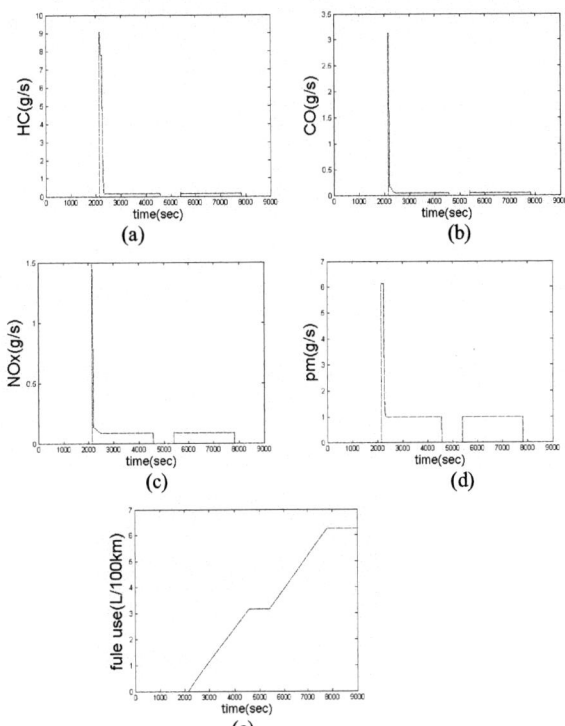

Fig. 8.(a) HC, (b) CO, (c) NOX, (d) PM, (e) Fuel consumption.

## VII. CONCLUSION

Steady-state simulation tools such as ADVISOR have been developed in recent years for the design and analysis of electric and hybrid electric vehicles. In the past, dynamic simulation models have focused mainly on the analysis of control strategies. In this paper, the dynamic behavior of the electric motors with input-output state feedback controller combined with adaptive backstepping observer and batteries of a typical series hybrid EV is investigated and simulated by Matlab/Simulink, has been presented and the performance and ability of control strategy is investigated. Simulation results have also been shown the IM dynamic behavior and batteries for various tests such as ECE+EUDC drive cycle, maximum speed, traversing ramp and fuel consumption and emissions.

## VIII. APPENDIX

*A. EV Data:*

Vehicle total mass of 8200 kg; Air drag coefficient of .79; Rolling resistance coefficient of .008; Wheel radius of .41 m; Level ground; Zero head wind.

*B. The IM Parameters:*

| | | |
|---|---|---|
| $P_n$ | 75 | KW |
| $P_{max}$ | 189 | KW |
| $f$ | 60 | HZ |
| $T_n$ | 209 | Nm |
| $T_{max}$ | 520 | Nm |
| $R_s$ | .02 | $\Omega$ |
| $R'_r$ | .01 | $\Omega$ |
| $X_m$ | 1.95 | $\Omega$ |
| $X_{ls}$ | .06 | $\Omega$ |
| $X'_{lr}$ | .06 | $\Omega$ |
| $n_s$ | 3600 | r.p.m |
| $J$ | 1.2 | Kg.m$^2$ |
| $P$ | 2 | |

## IX. REFERENCES

[1] M. Ehsani, K.M. Rahman, H.A. Toliyat, "Propulsion System Design of Electric and Hybrid Vehicle", *IEEE Tran. On industrial Electronics*, Vol. 44, No.1, February 1997.

[2] http://www.ott.doe.gov/hev/

[3] R. Yazdanpanah, A. Farrokh Payam, "Direct Torque Control of An Induction Motor Drive Based on Input-Output Feedback Linearization Using Adaptive Backstepping Flux Observer", in *Proc. 2006 AIESP Conf.*, Madeira, Portugal.

[4] T. Markel, A. Brooker, T. Hendricks, V. Johnson, K. Kelly, B. Kramer, M. O' Keefe, S. Sprik, K. Wipke, "ADVISOR: a systems analysis tool for advanced vehicle modeling", *ELSEVIER Journal of Power Sources 110,* (2002) 255-266.

[5] James Larminie, and John Lowry, *Electric Vehicle Technology Explained*, John Wiley, England, 2003.

[6] A. Rahide, "Vector Control of Induction Motor Using Neural Network in Electric Vehicle", Scientific Report, IUST, 2000.

[7] R. Marino, S. Peresada, and P. Valigi, "Adaptive Input-Output Linearization Control of Induction Motors", *IEEE Trans. Automatic Cont.*, Vol. 38, No. 2, PP. 208-220, Feb. 1993.

[8] J. E. Slotine, *Applied Nonlinear Control*, Prentice-Hall International Inc.

[9] S. Sadeghi, J. Milimonfared, M. Mirsalim, M. Jalalifar, "Dynamic Modeling and Simulation of a Switched Reluctance Motor in Electric Vehicle", in *Proc., 2006 ICIEA Conf.*

**2006 IEEE International Conference on Power Electronic, Drives and Energy Systems**

# Prototyping of a Precision Mechanism Using a Hybrid-Driven Piezoelectric Actuator

Fu-Shin Lee, Yung-Tsung Lei, Sheng-Feng Chiang, Jyun-Jhong Jhang, Shao-Chun Tseng, and Po-Jia Chen

*Abstract*--**The object of this research is to prototype a precision system, which mainly consists of a piezoelectrically actuated cantilever beam structure as powered by an AC + DC hybrid driving circuit. Pulse width modulation (PWM) control algorithm as well as nonlinear tabulation are developed to capture characteristics of the specific hybrid driver and actuator, while dedicated driving techniques are elaborately explored for achieving fine motions of the beam. Enhanced apparatus such as a Doppler interferometer, a laser interferometer, and a conventional accelerometer are employed to quantify the dynamic as well as static deflections of the piezoactuator–driven structure. The experimental results validate the efficacy of the hybrid driver circuit as it is powering the piezoactuator-driven structure, and finite element method (FEM) approach also confirms the resonance frequency of the assembled mechanism.**

*Index Terms*—**Cantilever beam, Doppler interferometer, hybrid driver, laser interferometer, piezoelectric actuator.**

## I. INTRODUCTION

NANO-technology has been a major research topic among various industries for the last decade. Especially in the semiconductor and manufacturing industries. Certainly, demanding of ultimate size, speed, and weight in all kinds of fabrication equipments, entertainment gadgets, or daily facilities push apparatus resolution asked for in manufacturing down to the limit. Among all the available motion mechanisms, piezoelectric actuators are usually first selected to achieve ultra-fine displacements or strokes for most manufacturing machines.

---

The authors are grateful to the National Science Council for partial financial support under contract No. NSC 95-2221-E-211-013.

F. S. Lee is with the Mechatronical Engineering Department, Huafan University, No. 1 Huafan Road, Shihtin, Taipei, Taiwan, R.O.C. (e-mail: fslee@huafan.hfu.edu.tw).

Y. T. Lei is with the Mechatronical Engineering Department, Huafan University, No. 1 Huafan Road, Shihtin, Taipei, Taiwan, R.O.C. (e-mail: d9362003@cat.hfu.edu.tw).

S. F. Chiang is is with the Mechatronical Engineering Department, Huafan University, No. 1 Huafan Road, Shihtin, Taipei, Taiwan, R.O.C. (e-mail: d9462003@cat.hfu.edu.tw).

S. C. Tseng is with the Mechatronical Engineering Department, Huafan University, No. 1 Huafan Road, Shihtin, Taipei, Taiwan, R.O.C. (e-mail: m9022010@cat.hfu.edu.tw).

J. J. Jhang is with the Mechatronical Engineering Department, Huafan University, No. 1 Huafan Road, Shihtin, Taipei, Taiwan, R.O.C. (e-mail: m9422009@cat.hfu.edu.tw).

P. J. Chen is with the Mechatronical Engineering Department, Huafan University, No. 1 Huafan Road, Shihtin, Taipei, Taiwan, R.O.C. (e-mail: d9062003@cat.hfu.edu.tw).

Piezoelectric actuators can be adequately designed in various systems, such as applications require AC, DC, or AC+DC driving sources.

As for researches emphasized on precision mechanisms using piezoactuators, Fu [1] developed a nanostage implementing the concept of integrating course motion platform accompanied with a fine motion mechanism. Su [2] adopted real-time close-loop PI-Fuzzy control approach to the precision mechanism as to accomplish a fine resolution of a step tracking error within 6 nm for the two-stage precision system. Wu [3] prototyped a parallel-connected mechanism for gaining 6-DOF fine motions. Inside the structure the mechanism seats plural flexible joints for subtle movements controlled by a computer, and dynamic characteristics, motion resolutions, as well as resonance frequency of the mechanism were thoroughly studied.

In the aspect of controlling nanostages, Xie [4] adopted PID control basics and N-time feed-forward compensation approach to maneuver a 2-axis 3DOF flexible stage, and the attained positioning errors were reduced to 40 nm. Ge [5] established an asymmetric hysteretic model based upon the Preisach representation, and essential PID technique accompanied with feed-forward control algorithm were exercised on the model for precision positioning purposes.

Hence, it is motivated to prototype a precision mechanism in this research using an AC + DC hybrid driver to power the commonly used piezoelectric actuators. The precision mechanism consists of a cantilever beam driven by a piezostack actuator. Dynamic as well as static deflections of the piezoelectrically actuated cantilever beam were closely examined. A Doppler interferometer, a laser interferometer, and an accelerometer were utilized to reckon the beam deflections upon AC power resonating or DC power biasing. Simulations based upon finite element method (FEM) as well as frequency spectrum scanning were both performed for locating resonance frequency of the assembled structure. Experiment as well as simulation results conclude the efficacy of the piezoelectrically powered mechanism.

The outcome of this research can be efficiently exercised in applications such as nano-manipulations, biotech optics, and semiconductor fabrications, which all demand nano-scale precisions on related actuation mechanisms.

## II. DESIGN STRATEGIES FOR THE HYBRID DRIVER

The hybrid driver circuit for piezoelectric actuators is composed of two sets of independent power suppliers,

---

0-7803-9771-1/06/$25.00 ©2006 IEEE

including a DC voltage source supposed with an AC voltage source. The hybrid driver circuit design and its functional behaviors were well discussed in a referred literature [6].

### A. Performance Design of the Hybrid Driver Circuit

The DC-to-DC converter and DC-to-AC converter circuits composing the hybrid driver for piezoelectric actuators are designed and simulated in this section. A modified DC-to-DC converter with zero-voltage switching and snubber circuits designed in is shown in Fig. 1.

The designed DC-to-AC driving circuit for piezoelectric materials is shown in Fig. 2, which is composed of a series of a PWM adjustable DC-to-DC converter and a DC-to-AC converter, and the circuit features its adjustable AC amplitude and variable frequency.

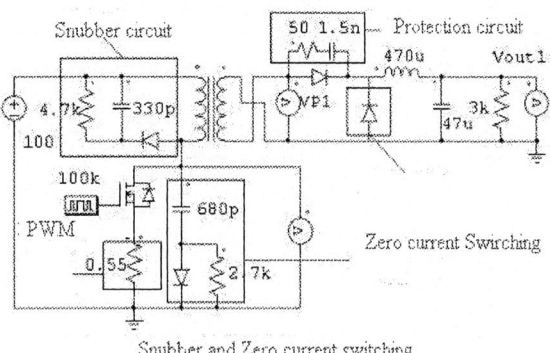

Fig. 1. Modified DC-to-DC forward converter portion of the hybrid driver.

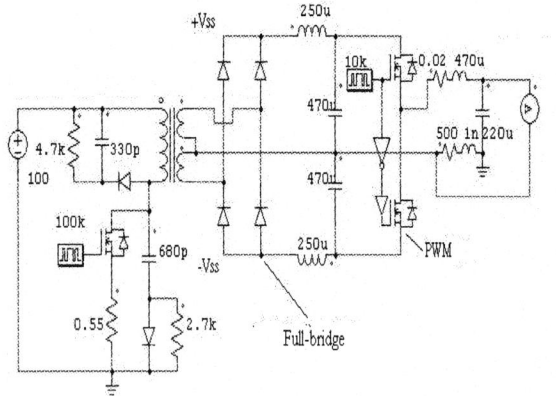

Fig. 2. DC-to-AC converter portion of the hybrid driver.

From above, the overall hybrid AC+DC circuit to drive piezoelectric materials is finally concluded in Fig. 3, in which the PZT-model stands for the equivalent circuit of the driven piezoelectric actuator, and the DC-to-AC converter accompanied with the DC-to-DC converter are depicted in the upper and lower portions, respectively. PWM modulations are practiced to control DC output level as well as AC voltage output amplitude from the hybrid driver, and varying clocking frequencies are used to investigate

influences of the resonant frequencies onto the piezoelectric actuator.

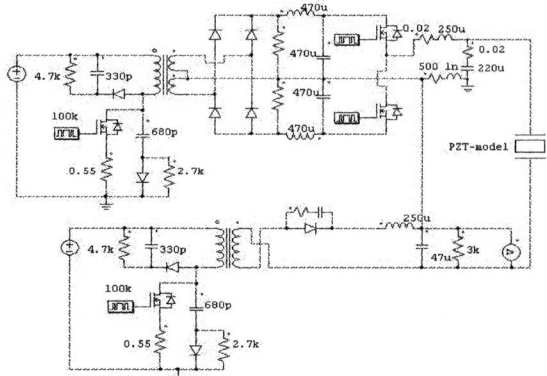

Fig. 3. AC+DC hybrid driver for piezoelectric actuators.

### B. Digital Driving Circuit Design for Piezoelectric Actuators

A CPLD IC is used to implement digital schemes as triggering the AC-to-DC driver circuit for the piezoelectric actuators, and a personal computer is used through USB protocols to perform graphic user interface functions for monitoring/controlling purposes. The codes embedded in the CPLD chip contains encoding, decoding, triggering, and latching functions. A/D signals sent to the PC through the CPLD mechanism are converted first and encoded using logic gates within the digital chip. Signals received from the PC are also converted first and then decoded to calculate triggering commands. The output commands are latched and utilized to trigger three sets of D/A converter in sequence. Also, an USB chip (Cypress cy7c63101) is employed between the CPLD chip and the PC terminal to perform series/parallel signal conversions.

### III. TESTING OF THE DRIVER PERFORMANCE

The output voltage across the piezoelectric actuator using the prototyped AC+DC hybrid driver circuit is shown in Fig. 4. Since the capacitive piezoelectric element is series connected with an inductor to form a filter in the circuit, the filtered output load voltage is similar to a sinusoidal waveform, with its working frequency as 15 kHz, peak-to-peak voltage as 60.4V, and DC bias as 38V.

### IV. DISPLACEMENTS FOR A CANTILEVER BEAM

The designed hybrid circuit is used to drive a piezoelectric actuator attached to a cantilever beam, and the experiment setup is illustrated in Fig. 5. The setup is served to measure the beam deflections caused by the AC periodical as well as DC bias sourcing onto the piezoelectric actuator. A laser interferometer is used to measure low-frequency large-magnitude displacements of the beam with, and a Doppler interferometer is served to quantify fine deflections of the beam. Both apparatus are endeavored to evaluate the performance of the hybrid driver upon actuating the cantilever beam structure. Furthermore, an accelerometer is exercised to verify the resonant behavior of the assembled structure, and frequency responses of the system among various driving frequencies are studied for further references in designing the hybrid driver.

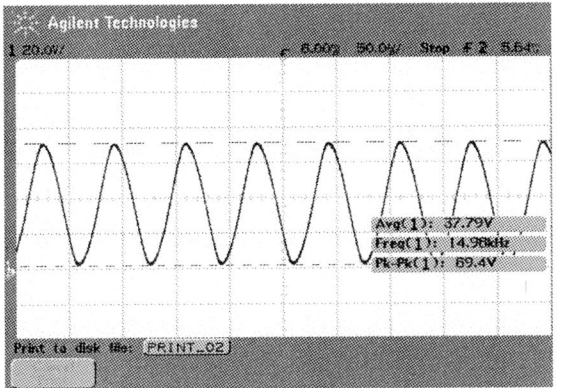

Fig. 4. Measured output voltage waveform from the hybrid driver (inductor=220uH, frequency =15kHz, peak-to-peak voltage=69.4V).

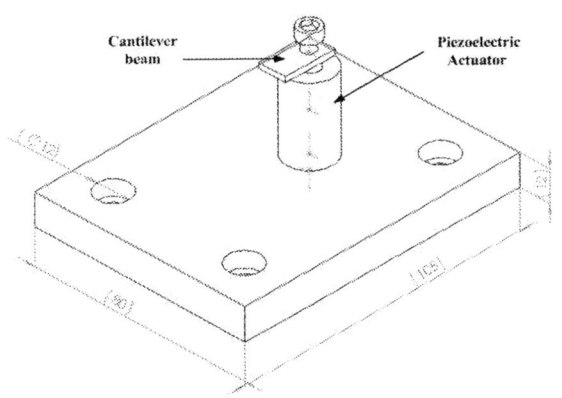

Fig. 5. Experiment setup with a cantilever beam driven by a piezoelectric actuator.

### A. Doppler Interferometer Measurement for the Piezoelectrically Actuated Cantilever Beam

The hybrid driver is commanded to pulsate the piezoelectric actuator with a 21 kHz 0.5 V AC voltage, and a Doppler interferometer, which characterizes its extreme resolution in ±40 nm range, is used to measure the free end displacements of the cantilever beam.

As a result, a high frequency displacement profile of a vertically-sectioned line on top of the resonated cantilever beam is shown in Fig. 6.

Also, the experiment substantiated that the free end vibrating amplitude would roughly amplify 25 nm more for every 0.1 V increased AC voltage, in the range of 0 V ~ 0.5 V.

Furthermore, Doppler interferometer measured maximum displacement of a horizontal surface on top of the piezoelectrically resonated cantilever beam is shown in Fig. 7, Accordingly, average deflection of the free end is

about 35 nm with a 0.5 V AC voltage of delivered through the hybrid driver.

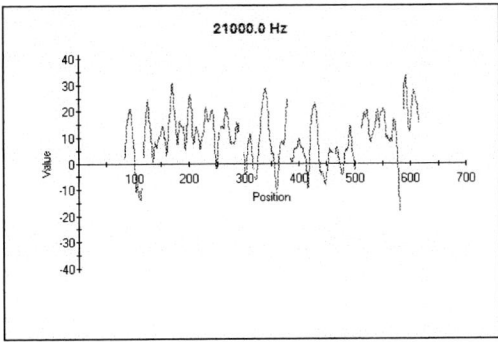

Fig. 6. Doppler interferometer measured displacement profile of a vertically-sectioned line on top of the resonated cantilever beam (V= ± 0.5V).

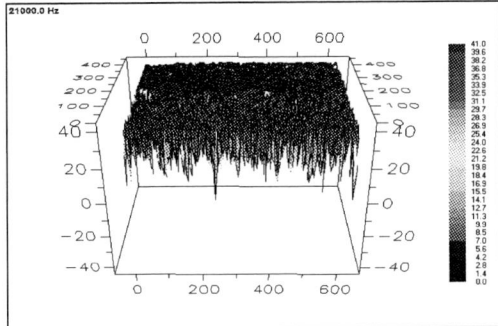

Fig. 7. Doppler interferometer measured maximum displacement of a horizontal surface on top of the resonated cantilever beam (V= ± 0.5V).

### B. Laser Interferometer Measurement for the Piezoelectrically Resonated Cantilever Beam

Moreover, the precision mechanism is driven with 1 Hz piezoelectric actuations to investigate its low- frequency behaviors, and a laser interferometer, which features in larger working range ($0.1\ \mu m \sim 10\ \mu m$), is employed to evaluate the overall system response upon low-frequency driving to the cantilever beam. Fig. 8 shows acquired movements of a target point on the free end surface of the cantilever beam at 1Hz, and Fig. 9 demonstrates the same target point motions as the same system is resonated in 10 Hz.

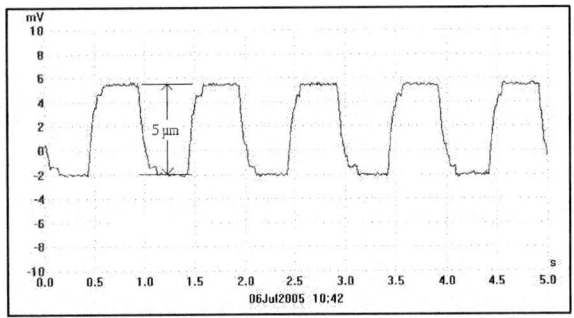

Fig. 8. Laser interferometer measured movement of a target point on the free end surface of the resonated cantilever beam (1Hz).

201

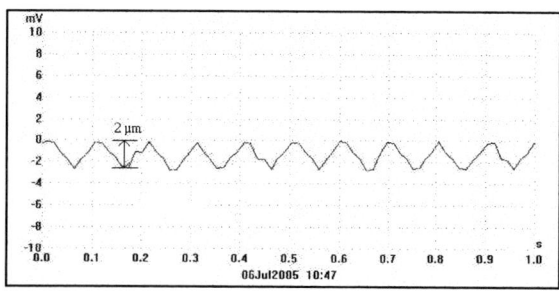

Fig. 9. Laser interferometer measured movement of a target point on the free end surface of the resonated cantilever beam (10Hz).

The experimental results verify a fact that the higher actuating frequency the hybrid driver sends off the square waveform voltage with, the more distorted free end displacement waveform obtained. In other words, the waveform would be more sinusoidal alike, due to the low pass filtering nature of the mechatronical system.

## V. MEASUREMENTS USING AN ACCELEROMETER

This research applied an accelerometer to examine the resonance characteristics of the piezoelectrically actuated cantilever beam as well, and the responses were monitored using a spectrum analyzer.

Since the accelerometer has a compatible mass as attached to the cantilever beam assembly, a finite element method (FEM) model, as shown in Fig. 10, was built in a commercial development environment (ANSYS) with the aim of locating resonance frequency of the overall system.

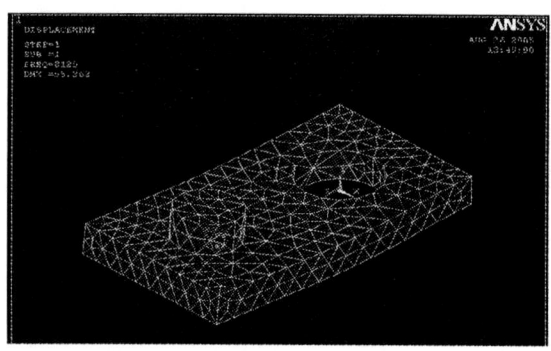

Fig. 10. FEM model (ANSYS) of the assembly of a cantilever beam and an accelerometer.

The resonance frequency of the assembly is about 8 kHz after several FEM simulations, and it is close to the experimental result as 8.3 kHz. The experiment was conducted through a frequency spectrum scanning process using the spectrum analyzer.

During the experiment, an AC voltage of 80 V amplitude was delivered with varying frequencies to the hybrid driver, and the acquired frequency response using the accelerometer is shown in Fig. 11.

## VI. EXPERIMENTS FOR HYSTERETIC BEHAVIOR

The hysteretic behavior of the assembly due to the employed piezoelectric actuator is also studied through examining its DC biased deformations. The DC bias voltage delivered to the hybrid driver is monotonously increased first and then decreased to the initial level.

In the meantime, the laser interferometer is employed again to measure the displacement of a designated point, and the recorded history is shown in Fig. 12.

It is apparent that the hysteresis phenomenon goes by with the piezoelectric actuator, and the average displacement of the measured point is about 24.8 nm/V, with the maximum deviation between the displacements upon increasing and decreasing driving voltages as much as 600 nm. The averaged amplitudes acquired for the actuated cantilever beam using the Doppler interferometer and laser interferometer are registered in TABLE I.

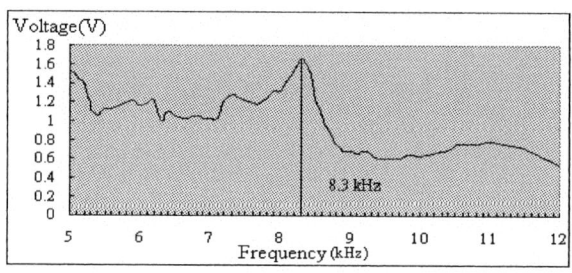

Fig. 11. Accelerometer measured frequency response of the piezoelectrically resonated cantilever system.

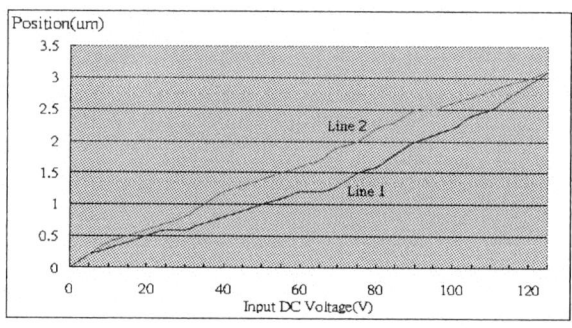

Fig. 12. Laser interferometer measured target point displacements versus biased DC voltage lever on the piezoelectrically cantilever beam system (1: monotonously increasing DC bias voltage, 2: monotonously decreasing DC bias voltage).

TABLE I
DOPPLER INTERFEROMETER, LASER INTERFEROMETER MEASURED CANTILEVER BEAM DISPLACEMENT AVERAGE, AND PIEZOELECTRIC ACTUATOR RATED DISPLACEMENT

|  | piezoelectric actuator rated displacement | laser interferometer measured displacement average | Doppler interferometer measured displacement average |
|---|---|---|---|
| nm/V | 25 | 24.8 | 25 |

Also in the same table, the rated stroke of the piezoelectric actuator is listed, and it is useful to verify the efficacy of the prototyped precision mechanism using piezoelectric actuators in this research.

## VII. CONCLUSIONS

A precision mechanism implementing a piezoelectric actuator is successfully prototyped in this research. An AC + DC hybrid driver is employed to feed the piezoelectric actuator, and high resolution AC resonant movements as well as DC bias deformations are obtained for the actuated cantilever beam structure. The precision control of the piezoelectric actuator motion is accomplished through dedicated triggering on the hybrid driver, and the PWM control algorithm as well as nonlinear tabulation techniques are developed to capture characteristics of the specific hybrid driver. Also, triggering patterns to hybrid driver are dedicatedly schemed for attainable displacement resolutions of beam through the well-designed digital controller.

Excellent apparatus such as a Doppler interferometer, a laser interferometer, and a conventional accelerometer are exploited to study the high-frequency small-amplitude resonances as well as low-frequency large-amplitude vibrations of the piezoelectrically actuated cantilever beam structure. The hysteretic phenomenon of the assembled structure is also investigated through monotonously increasing and monotonously decreasing the DC bias voltage levels in turns as delivered from the hybrid driver.

Experiment results as well as simulation data all confirm that the prototyped system exhibits its expected AC as well as DC performances, as evaluated in the aspects of its high resolution vibrating amplitude, designed resonance frequency, and statically deformed characteristics.

## VIII. ACKNOWLEDGMENT

The authors are grateful to the National Science Council for partial financial support under contract No. NSC 95-2221-E-211-013.

## IX. REFERENCES

[1] S. T. Fu, *"Optimal Design and Characterization of a Nanometer Positioning Stage"*, master thesis, Mechanical Eng. Dept., National Chung Hsing Univ., Taiwan, R.O.C., 2001.

[2] Y. Y. Su, *"Nano-positioning Control of a Dual-stage and Long-range System"*, master thesis, Mechanical Eng. Dept., National Chung Hsing Univ., Taiwan, R.O.C., 2002.

[3] T. L. Wu, *"Design of a Six Degrees-of-freedom Nanometer Resolution Parallel Micro-Positioning Stage"*, master thesis, Mechanical Eng. Dept., National Taiwan Univ., Taiwan, R.O.C., 2002.

[4] M. C. Xie, *"A Study on the Control of Dual-axis 3DOF Precision Positioning System"*, master thesis, Mechanical Eng. Dept., National Chung Hsing Univ., Taiwan, R.O.C., 2003.

[5] P. Ge, and M. Jouaneh, "Tracking control of a piezoceramic actuator," IEEE *Transactions on Control System Technology*, Vol. 4, pp. 209-216, 1996.

[6] J. J. Jang, *"Prototyping of a Hybrid Driving Circuit for Nano-Manipulation"*, master thesis, Electromechanical Eng. Dept., Hua Fan Univ., Taiwan, R.O.C., 2005.

## X. BIOGRAPHIES

**Fu-Shin Lee** received the B.S. degree and the M.S. degree in Mechanical Engineering from the National Chiao-Tung University at Taiwan in 1983 and 1985, respectively, and the Ph. D. in Mechanical Engineering from the University of Texas at Austin in 1993. Since 1993, he has been Associate Professor of the Mechatronical Engineering Department, Huafan University at Taipei, Taiwan. His research team is currently working on nano-manipulation technologies, piezoelectric actuator design, embedded controller prototyping, and MEMS system developments.

**Yung-Tsung Lei** was born in Taipei, Taiwan in 1980. He received the M.S. degree in the Department of Mechatronic Engineering from the Huafan University, Taipei, Taiwan, in 2004. He is currently working toward the Ph.D. degree in the Department of Mechatronic Engineering, Huafan University, Taipei, Taiwan. His research interests include piezoelectric actuators, electronic circuits, and power electronics.

**Sheng-Feng Chiang** was born in Kaoshung, Taiwan in 1981. He received the B.S. and M.S. degrees in the Department of Mechatronic Engineering from the Huafan University, Taipei, Taiwan, in 2003 and 2005, respectively. He is currently working toward the Ph.D. degree in the Department of Mechatronic Engineering, Huafan University, Taipei, Taiwan. His research interests include system modeling, power electronics, automatic control technology, and embedded controller prototyping.

**Jyun-Jhong Jhang** was born in Taipei, Taiwan in 1981. He received the B.S. and M.S. degrees in the Department of Mechatronic Engineering from the Huafan University, Taipei, Taiwan, in 2003 and 2005, respectively. He is currently working toward the Ph.D. degree in the Department of Mechatronic Engineering, Huafan University, Taipei, Taiwan. His research interests include electronic circuits, power electronics, and automatic control technology.

**Shao-Chun Tseng** was born in Taipei, Taiwan in 1978. He received the B.S. degree in 2001 from the Department of Mechatronic Engineering of the Huafan University, Taipei, Taiwan. He is currently a Ph.D. candidate in the Department of Mechatronic Engineering, Huafan University, Taipei, Taiwan. His research interests include MEMS devices and micro mechatronics.

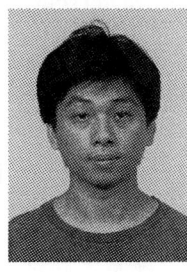

**Po-Jia Chen** was born in Taipei, Taiwan in 1977. He received the B.S. and M.S. degrees in the Department of Mechatronic Engineering, Huafan University, Taipei, Taiwan in 1998 and 2000, respectively. He is currently a Ph.D. candidate in the Department of Mechatronic Engineering, Huafan University, Taipei, Taiwan. His research interests include piezoelectric actuators, ultrasonic motors, power electronics, and mechatronics.

**2006 IEEE International Conference on Power Electronic, Drives and Energy Systems**

# DSP Based Implementation of Vector Controlled Induction Motor Drive using Fuzzy Pre-compensated Proportional Integral Speed Controllers

Bhim Singh, *Senior Member, IEEE*, and S. Ghatak Choudhuri, *Member, IEEE*

*Abstract--* **This paper deals with implementation of vector controlled induction motor drive (VCIMD) using DSP TMS320F240 and a comparison between proportional integral (PI) and fuzzy pre-compensated proportional integral (FPPI) speed controllers. The power circuit consists of a three-phase IGBT based voltage source inverter (VSI) feeding a three-phase squirrel cage induction motor. The control circuits consist of sensing circuits□interfacing circuits and control software. The sensed speed (ωᵣ) and the sensed winding currents in two of the three phases (iₐₛ and i_{bs}) are used as feedback signals for the closed loop control structure. Test results are presented for drive starting□speed reversal and load perturbation modes using PI and FPPI speed controllers respectively and the advantage of pre-compensation is demonstrated in detail.**

*Index Terms--* **Vector Control (VC)□Indirect Field Oriented Control (IFOC)□Fuzzy pre-compensated proportional integral (FPPI).**

## I. INTRODUCTION

DC motors are simple in control and offer better dynamic response inherently. Numerous economical reasons, for instance high initial cost, high maintenance cost for commutaters, brushes and brush holders of DC motors call for a substitute which is capable of eliminating the persisting problems in dc motors and has all the advantages present in dc motors. Freedom from regular maintenance and a brushless robust structure of the three phase squirrel cage induction motor are among the prime reasons, which brings the motor forward as a good substitute. Vector Control (VC) or field oriented control (FOC) is defined as a control technique by virtue of which two control signals are produced to control torque and flux respectively in an independent decoupled manner. When the three-phase squirrel cage induction motor is operated in VC mode, its response improves considerably and it acts as a better substitute for the separately excited DC motor [1-18].

In the present investigation, an Indirect Field Oriented Control (IFOC) of induction motor is implemented with FPPI closed loop speed controller. The three-phase squirrel cage induction motor is fed from a current controlled (CC) voltage source inverter (VSI). The current and speed signals are fed back to the closed loop control structure. The control algorithm is processed in real time using TMS32□F24□digital signal processor (DSP). Digital control eliminates drifts and by using programmable processors, upgradation is achieved with the developed software. High accuracy and fast speed of DSPs such as TMS32□F24□(used for implementation here) enables high-resolution control. DSP TMS32□F24□ has built in features like timers, 16 channels of ten bits analog to digital converter (ADC), four channels of digital to analog converter (DAC), digital input and output (I/O) units and twelve pulse width modulated (PWM) outputs along with central processing unit (CPU) [1□22]. Therefore, using such DSPs the hardware required for realizing a real time controller is reduced leading to improvement in reliability. In this implementation, the external hardware electronics consists of the IGBT based VSI bridge, current sensing circuit, speed sensing circuit, and the gate driver circuits

## II. SYSTEM CONFIGURATION AND CONTROL SCHEME

Fig.1 shows the basic building block diagram of the proposed drive system. VCIMD system comprises of: (a) three phase IGBT based VSI, (b) three phase squirrel cage induction motor, (c) the feedback control circuits receiving sensed signals, namely, the speed signal (ωᵣ) and current signals (iₐₛ and i_{bs}) respectively. The induction motor is star connected with an isolated neutral. Sensing currents corresponding to two phases is sufficient to yield feedback signals for all three-phase currents (iₐₛ, i_{bs} and i_{cs}).

The speed of the motor (ωᵣ) is compared with its reference value (ωᵣ*) and the error is processed in the speed controller. A limit is kept on the output of the speed controller depending on the maximum permissible winding current. The output of the speed controller after the limiter, is considered as the reference torque (T*) signal and similarly the output of the field weakening block is considered as the reference flux signal

---

Bhim Singh is with the Dept. of Electrical Engineering, Indian Institute of Technology Delhi, 11□□16 New Delhi India and S. Ghatak Choudhuri with the Dept. of Electrical Engineering, Indian Institute of Technology Roorkee, U.A. India (e-mail: bhimsinghr@gmail.com , sgceefeel@iitr.ernet.in)

0-7803-9771-1/06/$25.00 ©2006 IEEE

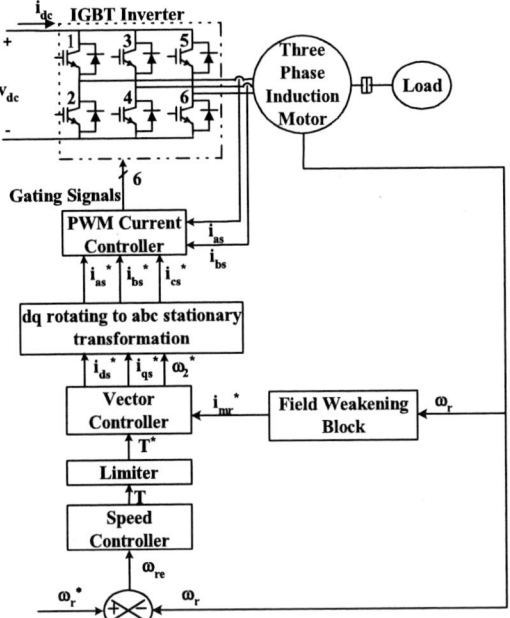

Fig. 1. Block diagram of vector controlled induction motor drive (VCIMD).

($i_{mr}^{*}$) respectively. These two signals are the main command signals for the vector controller. The controller calculates the d component ($i_{ds}^{*}$) and the q component ($i_{qs}^{*}$) of the reference current signal (which are responsible for the torque control and field control respectively) and also the slip

frequency reference signal ($\omega_2^{*}$) in the synchronously rotating reference frame (SRRF) aligned with the rotor field. These signals are then converted into three-phase reference currents ($i_{as}^{*}$, $i_{bs}^{*}$ and $i_{cs}^{*}$) in the stationary reference frame (SRF). For the induction motor to be operated in the VC mode, it is necessary that the VSI is controlled in CC mode so that the winding currents are in the same pattern as the reference currents ($i_{as}^{*}$, $i_{bs}^{*}$ and $i_{cs}^{*}$) computed by the vector controller in the SRF. Therefore, these reference currents are fed into the PWM current controller which simultaneously receives the sensed winding currents also. For current sensing, two of the three-phase currents are sensed using Hall Effect current sensors and the third current is computed from the two sensed currents (sum of all the three currents for a star connected balanced motor is zero). The current error signals are amplified and used as modulating signals for the Pulse Width Modulated (PWM) current controller. A Virtual Triangular Carrier Waveform (VTCW) is generated at the required switching frequency. The point of intersection of the VTCW and the modulating signals, acts as the point of state hange over for the resulting PWM driver signals. Six driver PWM signals emanate from the output of the PWM current controller and these are then fed to the respective gate driver circuits of the devices forming the VSI. Because of the closed loop current controlled operation of VSI, the winding currents follow the same pattern as the reference currents computed by the vector controller.

## III. HARDWARE IMPLEMENTATION

Some of the major parts of the drive system, namely, the speed controller, limiter, field weakening block, vector controller, and PWM current controller are implemented through the control software. The remaining parts are implemented using the developed hardware. Both the hardware and the software implementations are explained in this section.

### A.. System Hardware

The drive hardware consists of three-phase IGBT based VSI, gate driver circuit, sensed speed and reference speed scaling circuits and the sensed current (using Hall Effect sensors) scaling circuits. The gate driver circuit comprises of EXB84☐ driver chip which is shown in Fig.2. Four separate DC power supplies are developed which are isolated from the mains and from each other to supply to the driver circuits. For designing the driver power supplies, a 23☐V, single phase, 5☐Hz supply is stepped down into four-isolated 24 V, single phase, 5☐Hz signals. These signals are rectified and regulated to make necessary four DC supplies for six gate driver circuits. Each gate driver has an optical isolation in its initial stage to isolate the control circuit from the power circuit. Six gating signals achieved at the output of driver chips are applied between gate and emitter of the respective IGBTs.

Sensed speed signal, reference speed signal and sensed current signals are scaled by their respective signal conditioning circuits (shown in Fig.3 (a) and Fig.3 (b)) and are brought in the acceptable range of ☐5V to be fed to the DSP respective ADC channels. Both sensed speed and reference speed signals are adjusted in range from +3☐☐☐rpm to −3☐☐☐ rpm. Fig.4 shows and the connection of the DSP with the remaining system.

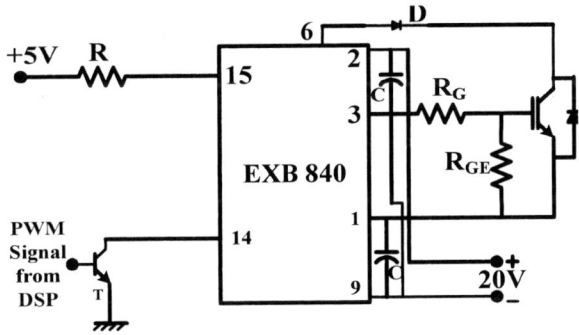

Fig. 2. Gate driver circuit.

### B. Control Software

The modules implemented through software are the speed controller and limiter, field weakening block, vector controller, d-q rotating to abc stationary transformation and PWM current controller. The software has been developed in assembly language. The main equations used in the software modules have been described here.

#### I. Speed controller with limiter

The speed controller and the limiter routines compute the reference torque ($T^{*}$) using the error between the reference speed ($\omega_r^{*}$) and the rotor speed ($\omega_r$). Speed controller

205

algorithm calculates the output torque (T) and limits it to the reference value (T*) decided by the permissible winding current. In this implementation of the VCIMD, the speed controllers considered are namely, proportional integral (PI) controller and fuzzy pre-compensated proportional integral (FPPI) controller respectively.

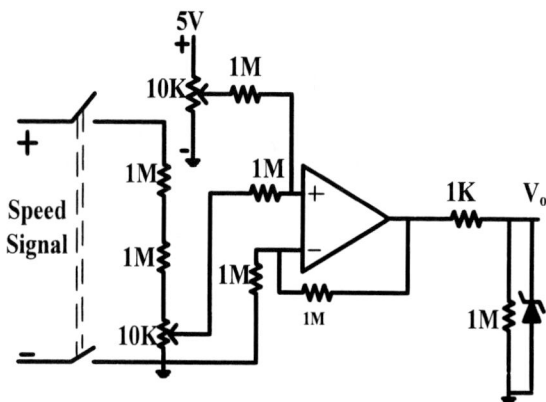

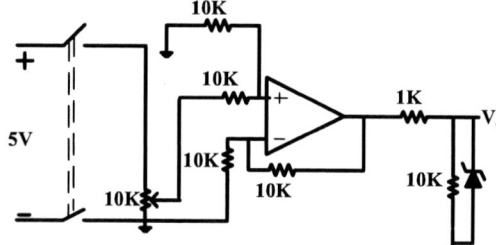

Fig. 3(a). Speed scaling circuit.

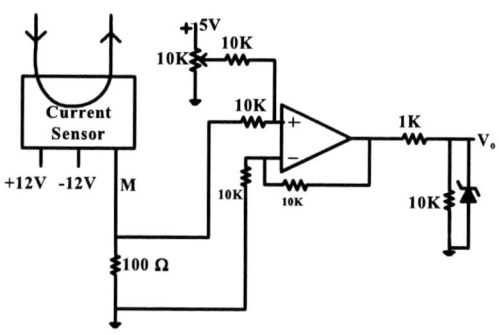

Fig. 3(b) Current scaling circuit.

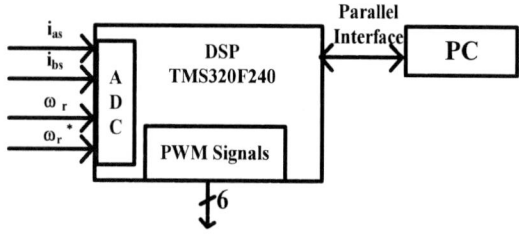

**To gate driver circuit**

Fig. 4. DSP block diagram.

The general block diagram of the PI controller has been shown in Fig.5. The working logic of the PI speed controller routine in discrete format may be mathematically stated as:

$$T_{(n)} = T^*_{(n-1)} + K_P \{ \omega_{e(n)} - \omega_{e(n-1)} \} + K_i \omega_{e(n)} \qquad (1)$$

where $T_{(n)}$ is controller output at the $n^{th}$ instant, $T^*_{(n-1)}$ is controller output after limit application at the $(n-1)^{th}$ instant, $K_P$ is proportional gain constant, $K_i$ is integral gain constant, $\omega_{e(n)}$ is speed error at the $n^{th}$ instant, $\omega_{e(n-1)}$ is speed error at the $(n-1)^{th}$ instant.

The block diagram for the fuzzy pre-compensated proportional integral (FPPI) speed controller is shown in Fig.6. Fuzzy pre-compensation refers to the advance alteration of the reference speed signal $(\omega_r^*)$ using fuzzy logic $(\omega_1^*)$ in accordance with the rotor speed $(\omega_r)$ to eliminate specific features such as overshoot and undershoot in the speed response which are visible when the PI speed controller is used alone. The fuzzy rules are stated in Table I. According to the general block diagram shown in Fig. 6, the fuzzy logic (FL) controller accepts the speed error $(\omega_{e(n)})$ and the change in speed error $(\Delta\omega_{e(n)})$ as its inputs at the $n^{th}$ instant. The output of the FL controller $(\gamma)$ is added to the reference speed $(\omega_r^*)$ and generates the pre-compensated reference speed $(\omega_1^*)$ which is to be used as the reference signal by the PI controller. The fuzzy pre-compensator can be mathematically stated as:

$$\omega_{e(n)} = \omega_{r(n)}^* - \omega_{r(n)} \qquad (2)$$

$$\Delta\omega_{e(n)} = \omega_{e(n)} - \omega_{e(n-1)} \qquad (3)$$

$$\gamma_{(n)} = F[\omega_{e(n)}, \Delta\omega_{e(n)}] \qquad (4)$$

$$\omega_{1(n)}^* = \gamma_{(n)} + \omega_{r(n)}^* \qquad (5)$$

where $F$ is fuzzy logic mapping.

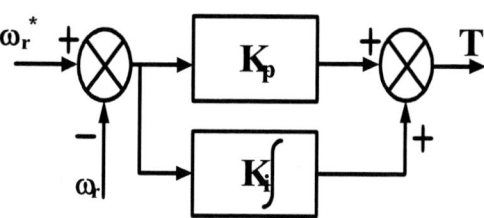

Fig. 5. Block schematic of proportional integral (PI) speed controller.

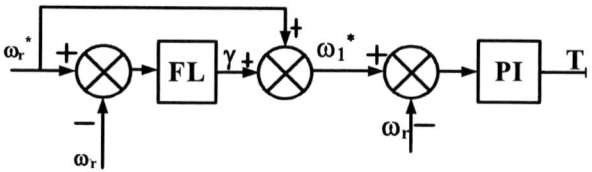

Fig. 6. Block schematic of fuzzy pre compensated proportional integral (FPPI) speed controller.

### II. Field Weakening Controller

The reference value of the exciting current $(i_{mr}^*)$ is a function of the rotor speed. Mathematically, the logic for computation of the exciting current $(i_{mr}^*)$ may be stated as follows:

$$i_{mr(n)}^* = I_m \qquad \text{if } \omega_{r(n)} < \text{base speed of the motor} \qquad (6)$$

$$i_{mr(n)}^* = K_f I_m / \omega_{r(n)} \qquad \text{if } \omega_r \geq \text{base speed of the motor} \qquad (\square)$$

where $K_f$ is flux constant and $I_m$ is rms value of the magnetising current. The value of $K_f$ can be calculated by

206

### TABLE I
#### FUZZY LOGIC DECISION TABLE

| ΔE | NB | NM | NS | ZE | PS | PM | PB |
|----|----|----|----|----|----|----|----|
| NB | NB | NB | NM | NM | NS | NS | ZE |
| NM | NB | NM | NM | NS | NS | ZE | PS |
| NS | NM | NM | NS | NS | ZE | PS | PS |
| ZE | NM | NS | NS | ZE | PS | PS | PM |
| PS | NS | NS | ZE | PS | PS | PM | PM |
| PM | NS | ZE | PS | PS | PM | PM | PB |
| PB | ZE | PS | PS | PM | PM | PB | PB |

substitution of the base speed and magnetising current magnitudes in eqn (□).

### III. Vector Controller

a) This routine calculates the direct axis and the quadrature axis stator current components ($i_{ds}^*$ and $i_{qs}^*$) in the synchronously rotating reference frame (SRRF) inclined at the flux angle (Ψ) w.r.t. stationary frame. A complete control is achieved in wide operating range, above and below the base speed. Using the excitation current ($i_{mr}^*$) and the torque component of the stator reference current vector ($i_{qs}^*$), the slip frequency ($\omega_2^*$) is computed. Mathematically, the equations in the discretised form are stated as follows:

$$i_{ds}^*{}_{(n)} = i_{mr}^*{}_{(n)} + \tau_r \, di_{mr}^*{}_{(n)} / dt \tag{8}$$

$$i_{qs}^*{}_{(n)} = T^* / \{K_m \, i_{mr}^*{}_{(n)}\} \tag{□}$$

$$\omega_2^*{}_{(n)} = i_{qs}^*{}_{(n)} / \{\tau_r \, i_{mr}^*{}_{(n)}\} \tag{1□}$$

The rotor time constant is defined as:

$$\tau_r = L_{rr} / R_r \tag{11}$$

where $K_m$ is $(3/2)(P/2)[M/(1+ \sigma_r)]$, $i_s^*$ is stator reference current space vector, $i_{ds}^*{}_{(n)}$ is flux component of $i_s^*$ at the $n^{th}$ instant, $i_{qs}^*{}_{(n)}$ is torque component of $i_s^*$ at the $n^{th}$ instant, $\omega_2^*{}_{(n)}$ is reference slip speed of rotor at the $n^{th}$ instant, P is the number of poles, M is the mutual inductance, $\sigma_r$ is rotor leakage factor, $\tau_r$ is rotor time constant, $L_{rr}$ is rotor self inductance, Rr rotor resistance.

The reference slip frequency of the rotor ($\omega_2^*{}_{(n)}$) is added to the sensed rotor speed ($\omega_{r(n)}$) and the discrete integration is carried out to calculate the flux angle at the $n^{th}$ instant ($\psi_{(n)}$) as:

$$\psi_{(n)} = \psi_{(n-1)} + \Delta\psi \tag{12}$$

The small incremental value of the flux angle (Δψ) can be computed as:

$$\Delta\psi = \{\omega_2^*{}_{(n)} + \omega_{r(n)}\} \Delta t \tag{13}$$

where $\psi_{(n-1)}$ is flux angle of the SRRF at the $(n-1)^{th}$ instant, ΔT is sampling interval.

### IV. d-q Rotating to abc Stationary Transformation

This routine is used for the transformation of the calculated d-q components of the reference current ($i_s^*$) in the SRRF to the three phase reference currents ($i_{as}^*$, $i_{bs}^*$, $i_{cs}^*$) in the stationary reference frame (SRF). The transformations can be stated mathematically as follows:

$$i_{as}^* = i_{ds}^* \sin\psi + i_{qs}^* \cos\psi \tag{14}$$

$$i_{bs}^* = i_{ds}^* (1/2)\{-\sin\psi + \sqrt{3}\cos\psi\} + i_{qs}^* (1/2)\{-\cos\psi + (\sqrt{3})\sin\psi\} \tag{15}$$

$$i_{cs}^* = i_{ds}^* (1/2)\{-\sin\psi - \sqrt{3}\cos\psi\} + i_{qs}^* (1/2)\{-\cos\psi + \sqrt{3}\sin\psi\} \tag{16}$$

### V. Pulse Width Modulated (PWM) Current Controller

The PWM current controller generates six PWM signals to the respective gate driver circuits driving IGBTs of the three-phase VSI bridge. The controller operation takes place as follows:

#### Current error calculation

Current error for a given phase is defined as the difference between the reference current (computed in the DSP software) and the sensed winding current for the respective phase. Mathematically, it may be stated in a discrete form as follows:

$$i_{ek\,(n)} = i_{k(n)}^* - i_{k(n)} \tag{1□}$$

where k is as, bs, cs, $i_{ek\,(n)}$ is current error for the $k^{th}$ phase at the $n^{th}$ instant, $i_{k(n)}^*$ is reference current for the $k^{th}$ phase at the $n^{th}$ instant, $i_{k(n)}$ is sensed winding current for the $k^{th}$ phase at the $n^{th}$ instant.

#### Modulating signal generation

For a given phase, processing the current error through a proportional (P) controller generates the modulating signal. A limit is then put on the amplified error to prevent the current controller operation to avoid overmodulation. Mathematically, the process is defined in a discretised manner as follows:

$$v_{ck(n)} = K_c \, i_{ek(n)} \tag{18}$$

where $K_c$ is gain parameter of the P controller and $v_{ck(n)}$ is the modulating signal for the $k^{th}$ phase at the $n^{th}$ instant

#### Switching signal generation

Switching signal generation requires comparison of modulating signal with a triangular carrier waveform. The frequency of the modulating signal is the fundamental frequency of the inverter output voltage and the frequency of triangular carrier waveform is the switching frequency ($f_s$) of the VSI. The instant, at which the two signals equate, is the instant when the switching signal changes its logic state.

The triangular carrier waveform required in the switching signal generation is achieved by means of digital counting in up-down counting mode. The maximum count is initialized in accordance to the required switching frequency ($f_s$). Such a triangular carrier waveform is referred to as virtual triangular carrier wave (VTCW).

The modulation signal is compared with the VTCW to generate the PWM switching signals. Six PWM switching signals are generated, wherein two are meant for each leg of the bridge inverter namely, leg A, leg B and leg C respectively. Two paired complementary signals for a respective leg are used for switching the upper and the lower devices of that leg of the VSI with proper dead band to avoid shoot through fault in the VSI

### IV. IMPLEMENTATION PROCEDURE

The DSP control software execution is based upon two modules: the initialization module and the run module [1□21]. The initialization module executes once. The second module namely, the Run Module (RM) is based upon a waiting loop interrupted by the PWM underflow. When the interrupt flag is set, acknowledgement signal is generated and the corresponding Interrupt Service Routine (ISR) is served. The complete vector control algorithm is computed within the PWM ISR and thus runs at the same frequency as switching frequency.

The compare unit of DSP is used for the PWM signal generation. Timer1 operation is set at the continuous Up-Down counting mode. The Timer1 [2□] operation is responsible for the internal generation of the virtual triangular carrier wave. Programming has been done to generate complementary PWM signals at a switching frequency of 1□ kHz. Therefore, the maximum value of count corresponding to the sampling interval is calculated as follows:

$$(1.\square / f_s) = (2.\square)(COUNT)(Z) \qquad (1\square)$$

Where $f_s$ is switching frequency and COUNT is maximum value of numerical count for Up-Down Counting, and Z is the operating time period of the processor corresponding to the processor frequency of 2□MHz.

Therefore, whenever the up-down counting process undergoes an underflow (indicating one cycle of counting is complete), an interrupt is received. As a result, the control jumps to the PWM Interrupt Service Routine (ISR). At the end of the PWM ISR execution, the count content of the compare registers is upgraded and the control returns back to the waiting loop. With continuous counting as soon as underflow occurs again, interrupt occurs and same procedure is repeated.

## V. PERFORMANCE OF VECTOR CONTROLLED INDUCTION MOTOR DRIVE MPLEMENTATION PROCEDURE

The developed prototype of the VCIMD is tested for a □□5kW, 2-pole, 5□Hz, three-phase, squirrel cage induction motor (rating is given in the Appendix). Test results for starting, speed reversal and load perturbation are shown in Figs.□+□ The controller parameters have been stated in the Appendix. The recorded test results are discussed as follows:

### A .Starting Dynamics

The three-phase induction motor is fed from a controlled voltage and frequency source using CC-VSI, which operates the three-phase induction motor in vector control (VC) mode. The motor starts at a low frequency decided by the controller and finally runs at the reference speed in the steady state condition. The reference speed is set to 1□□□ rpm (1□4.□2 electrical rad/sec) with a limit of torque at twice the rated value.

Fig.□ shows set of currents and speed response for the VCIMD when PI and FPPI speed controllers are used during starting of the motor. The steady state motor winding current at no load is observed to be □65A rms with a starting peak current of about 2.□6A. Fuzzy pre-compensation in the existing PI control loop helps to eliminate the demerits of PI controller such as occurrence of overshoot, undershoot etc in the speed response of VCIMD. When the VCIMD reaches the reference speed (1□□□rpm), the motor currents are reduced to the no load value and the VCIMD runs under steady state condition.

### B. Speed Reversal

The response of the VCIMD system is recorded in terms of sensed currents from two of the phases ($i_{as}$ and $i_{bs}$) and rotor speed ($\omega_r$) for the speed reversal phenomenon employing PI and FPPI speed controllers respectively and are shown in Fig. 8. When the three-phase induction motor is running at a steady state speed of +1□□□ rpm (1□4.□2 electrical rad/secs), the reference speed is suddenly changed to -1□□□ rpm (-1□4.□2 electrical rad/secs), the control structure is activated and a

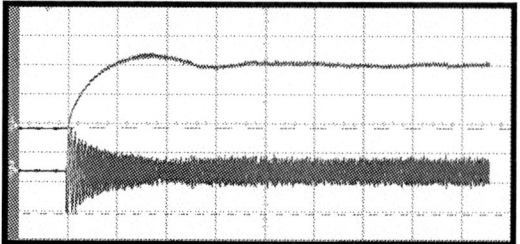

Speed Y : 1□□rpm/div Current: Y: 1.838 A/div Time: X: 1.□ secs/div
Fig. □(a). Starting response of VCIMD using PI speed controller.

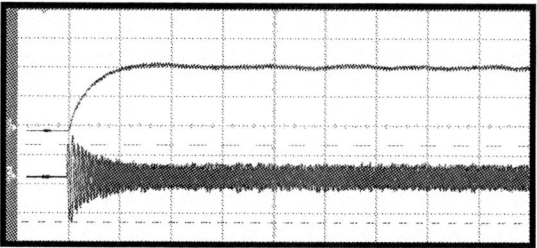

Speed Y : 1□□rpm/div Current: Y: 1.838 A/div Time: X: 1.□ secs/div
Fig. □(b). Starting response of VCIMD using FPPI speed controller.

large value of speed error saturates the output of the speed controller. As a result, regenerative braking followed by reverse motoring takes place. Therefore, the controller first reduces the frequency of the stator currents having regenerative braking and then reverses phase sequence of the currents for reversed motoring operation. The steady state current in either direction of operation of the three-phase induction motor is observed to be □65A. Since just before and after the reversal phenomenon, the drive is in the same dynamic state and the steady state values of the inverter currents are found to be the same, both in magnitude and frequency in either direction of rotation. However, the phase

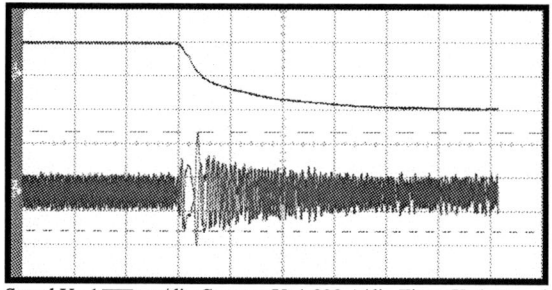

Speed Y : 1□□rpm/div Current: Y: 1.838 A/div Time: X: 1.□secs/div
Fig. 8(a). Reversal response of VCIMD using PI speed controller.

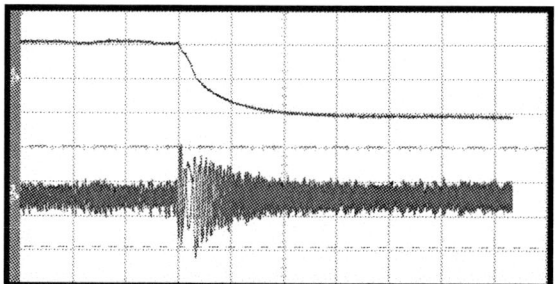

Speed Y : 1□□rpm/div Current: Y: 1.838 A/div Time: X: 1.□secs/div
Fig. 8(b). Reversal response of VCIMD using FPPI speed controller.

sequences of the currents in the two directions are different.

## C. Load Perturbation

Load perturbation response of the VCIMD, as shown in Fig. □, is carried out through a coupled DC generator for the sudden application and removal of load on the motor shaft when the motor is running under steady state condition at a speed of 1□□□rpm (1□4.□ electrical rad/secs). A load torque of about 8□% of the rated value is suddenly applied to the motor shaft. The closed loop action of the speed controller causes the motor speed to recover back to the defined reference value. The sensed currents in two of the phases (i$_{as}$ and i$_{bs}$) are observed to recover back to the no load value on load removal.

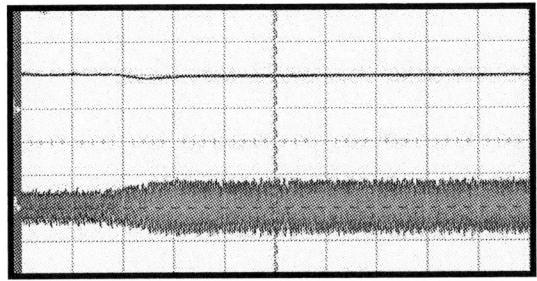

Speed: Y: 1□□□rpm/div  Current: Y: 2.263A/div  Time: X: 1.□secs/div
Fig. □(a).  Load Application Response of VCIMD using FPPI speed Controller

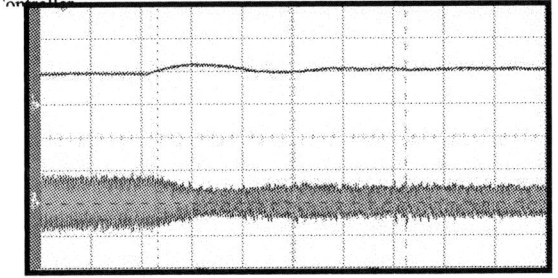

Speed: Y: 1□□□rpm/div  Current: Y: 2.263A/div  Time: X: 1.□secs/div
Fig. □(b).  Load Removal Response of VCIMD using FPPI speed Controller

## VI. CONCLUSION

A developed VCIMD has been realised using DSP TMS32□F24□. The use of such a dedicated DSP for motor control has reduced the hardware complexity and it has provided all functions of speed control. Operation of PI speed controller results in occurrence of specific characteristics such as overshoot, undershoot, etc., during starting response which gets eliminated on use of Fuzzy logic pre-compensation. It has also been observed that there is a fast change in the stator currents in accordance with the change in speed or change in load on the motor shaft. The variation in frequency of the stator currents in the desired manner, results in a quick acceleration torque. The control structure implements regenerative braking as well as changes in phase sequence, during speed reversal of the drive resulting in a fast response of the VCIMD.

## VII. APPENDIX

*Specifications of motor*
1HP, □□5kW, 3-Phase, 2-Pole, Y-Connected, 415V, 1.□A, 5□Hz, squirrel cage induction motor.

$R_s$ = □45Ω, $R_r$ = 11.□2Ω, $X_{ls}$ =11.□4Ω, $X_{lr}$ =11.□4Ω, $X_m$=212.□□□Ω, J=□.□□18Kgm$^2$

Proportional Integral (PI) speed controller parameters:
$K_p$=□3□453125, $K_i$=□.□□1□53125

## VIII. REFERENCES

[1] W. Leonard, "*Control of Electric Drives*", Narosa Publication, New Delhi, 1□85.

[2] S. Yamamura, "*AC Motors for Performance Applications, Analysis and Control*", Marcel Dekker, New York, 1□86.

[3] B. K. Bose, "*Power Electronics and AC Drives*", Prentice-Hall, New Jersey, 1□86.

[4] J. M. D. Murphy and F. G. Turnbull, "*Power Electronic Control of AC Motors*", Pergamon Press, Oxford, 1□88.

[5] P. Vas, "*Vector Control of AC Machines*", Oxford University Press, Oxford, 1□□□

[6] A. Kelemen and M. Imecs, "*Vector Control of AC Drives*", Vol. 1, Omikk Publisher, Budapest, 1□□1.

[□] I. Boldea and S. A. Nasar, "*Vector Control of AC Drives*", CRC Press, Florida, 1□□2.

[8] A. Kelemen and M. Imecs, "*Vector Control of AC Drives*", Vol. 2, Omikk Publisher, Budapest, 1□□1.

[□] A. M. Trzynadlowski, "*The Field Orientation Principle in Control of Induction Motors*", Kluwer Academic Publishers, Netherlands, 1□□4.

[1□] D. W. Novotny and T. A. Lipo, "*Vector Control and Dynamics of AC Drives*", Oxford University Press, New York, 1□□6.

[11] P. Vas, "*Sensorless Vector and Direct Torque Control*", Oxford University Press, New York, 1□□8.

[12] B.N. Singh, "*Investigations on Vector Controlled Induction Motor Drive*", Ph.D. Thesis, Indian Institute of Technology, Delhi, India, 1□□5.

[13] V.R.Stefanovic and R.M. Nelms, "*Microprocessor control of Motor Drives and Power Converters*", Tutorial course of *IEEE Industry Applications Society Annual Meeting*, Michigan, 1□□1.

[14] S. Sathikumar and Joseph Vithayathil, "Digital Simulation of Field-Oriented Control of Induction Moto*r*", *IEEE Trans. on IE*, Vol.IE-31, No.2, pp.141-148, May 1□84.

[15] Thomas R. Doll and Andrew H. Kaiser, "Vector Controls", *in Proc. 199□ IEEE Electrical Engineering Problems in Rubber and Plastics Industries Conf.*, pp.5□-61.

[16] James A. Norris, "Vector Control of A.C. Motors", *in Proc. 199□ IEEE Textile, Fibre and Film Industry Technical Conf.*, pp. 3/1-3/8.

[1□] B. N. Singh, Bhim Singh and B.P.Singh, "Performance Analysis of a closed-loop field oriented cage induction motor drive", *Journal of Electrical Power Systems Research*, Vol.2□, No.2, pp.6□-81, February 1□□4.

[18] E. S. Tez, "A Simple Understanding of Field- Orientation for AC Motor Control", *in Proc. 199□ IEE Vector Control and Direct Torque Control of Induction Motors Colloquim*, pp. 3/1-3/4.

[1□] Texas Instruments, "TMS32□C24X DSP Controllers Evaluation Module Technical Reference", Manual, Literature Number: SPRU248A, 1□□□

[2□] Texas Instruments, "TMS32□C24X DSP Controllers Peripheral Library and Specific Devices", Preliminary, Digital Signal Processing Solutions, 1□□□

[21] Texas Instruments, "TMS32□C24X DSP Controllers CPU, System and Instruction Set", Reference Set, Volume 1, Digital Signal processing Solutions,1□□□

[22] Texas Instruments, "Implementation of a Speed Field Oriented Control of Three Phase AC Induction Motor using TMS32□F24□", Application Report, Literature Number: BPRA□6, March 1□□8.

**2006 IEEE International Conference on Power Electronic, Drives and Energy Systems**

# Optimal Controller for High Frequency AC-Link Converter Induction Motor Drive System

R. A. Gupta, A. K. Wadhwani, and R. R. Joshi

*Abstract*--This paper proposes a new control algorithm for a high frequency ac-link converter (HFAC) induction motor drive system. A two-degree-of-freedom controller is proposed to improve the system performance. The controller design algorithm can be applied in an adjustable speed control system and a position control system to obtain good transient responses and good load disturbance rejection abilities. Several experimental results are shown to validate the theoretical analysis.

*Index Terms*– Fuzzy logic, high frequency ac-link converter, induction motor, optimal controller, transient response.

## I. INTRODUCTION

THE matrix converter (MC) has received considerable attention in recent years. Several methods have been proposed to improve the performance of the matrix converter. Many researchers have studied the matrix converters that drive AC motors. For example, a matrix converter was developed to drive an induction motor [1] and [2]. These studies focused on the new type of matrix converter and the field-oriented control in adjustable-speed induction drive systems. In addition, a 30-hp matrix converter was applied in an induction motor to adjust its speed. The current harmonics, however, were large because only the high bus voltage was selected [3]. AC drive has been investigated. A two-degree-of-freedom optimal control algorithm is applied. By using this optimization algorithm, a satisfactory Servo drive can be achieved. The forward controller determines the transient response. On the other hand, the load compensator determines the load disturbance response. The forward controller and the load compensator can be individually designed. As a result, a two-degree-of-freedom controller can be obtained.

## II. HIGH-FREQUENCY LINK MATRIX CONVERTER

Fig. 1 (a) shows the circuit diagram of the high frequency link matrix converter topology and it has two power conversion stages, i.e., the variable speed source 3f/1f matrix converter part (primary side converter) for balanced three-phase variable-frequency ac to high frequency ac (HFAC) and the utility interactive 1f/3f matrix converter part (secondary side converter) for HFAC to fixed-frequency utility ac, respectively.

---

R. A. Gupta is with Department of Electrical Engineering, MNIT, Jaipur Raj., India (e-mail: rag_mnit@rediffmail.com).
A. K. Wadhwani is with Department of Electrical Engineering, MITS, Gwalior, India (e-mail: arun_wadhwani@rediffmail.com).
R.R.Joshi is with Department of Electrical Engineering, CTAE, MPUAT, Udaipur (Raj.), India (e-mail: rrjoshi_iitd@yahoo.com).

Fig. 2 shows an actual implementation of the proposed matrix converter.

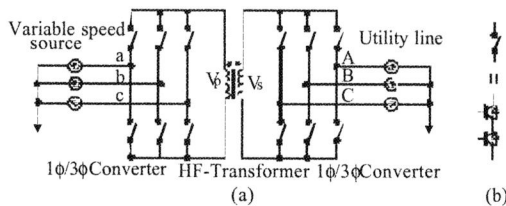

Fig. 1. (a) Proposed 3f/3f high frequency link matrix converter (b) bi-directional switch configuration.

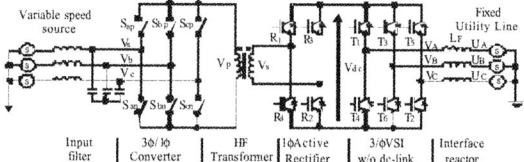

Fig. 2. Real implementation of the proposed matrix converter.

## III. TORQUE PULSATION IMPROVEMENT AND FUZZY CONTROLLER

Unfortunately, in the real world, the torque pulsation can be produced in many different ways. First, the rotor magnetic flux flows through a minimum reluctance path to the stator, which produces cogging torque. Second, the analog-to-digital *(A/D)* converter and Hall-effect current sensors have offset biases and thus produce torque pulsation. Finally, the phase currents and motor back emf are affected by the undesired high-order harmonics [4]-[8]. In order to describe the real torque, a simple mathematical model, which is related to the rotor position of the motor and the current command, is used as [9]

$$T_e(\theta_r, I^*) = a_0(\theta_r) + K_T I^*(1 + a_1(\theta_r)) \quad (1)$$

where $a_0$ and $a_1$ are the parameters relative to torque pulsation, $K_T$ is the torque constant which is equal to $L_{md} I_{fd}$ and $I^*$ is the ideal current command. The parameter $a_0(\theta_r)$ includes the torque pulsation due to the cogging torque and the current offset in the drives. It can be measured by setting the $I^*$ to zero, and then measuring the relative torque in different shaft angles. The parameter $a_1(\theta_r)$ includes the torque pulsation related to the harmonic contents, and the phase misalignment of the current and the back emf profiles. It can be calculated by using the following equation :

$$a_1(\theta_r) = \frac{T_e(\theta_r, I^*) - a_0(\theta_r) - K_T I^*}{K_T I^*} \quad (2)$$

A compensation current command is designed to produce an ideal torque. The relationship between the compensation command and the ideal torque is shown as follows

$$T_e^* = K_{TI}^* = a_0(\theta_r) + K_T I_{ref}^e (1 + a_1(\theta_r)) \quad (3)$$

where $T_e^*$ is the ideal torque command, and $\hat{I}_{ref}$ is the required compensating current command. Now, it is easy to obtain the compensation current command as

$$I_{ref}^c = \frac{I^*}{1 + a_1(\theta_r)} - \frac{a_0(\theta_r)}{K_T(1 + a_1(\theta_r))} \quad (4)$$

It is possible to tune the PI controller by using a self-tuning algorithm [10]. However, this method is very complicated and is difficult to implement. A fuzzy algorithm is therefore proposed to adjust the parameters of the PI controller. Both a good transient response and a good load disturbance rejection capability can be achieved by using this method. First, the state variables of this system are defined as

$$\Delta\omega_r(k) = \omega_r^* - \omega_r(k)$$

$$\alpha(k) = [\Delta\omega_r(k) - \Delta\omega_r(k-1)]/T \quad (5)$$

where $\Delta\omega_r(k)$ is the speed error of the motor at sampling interval k, $\alpha(k)$ is the change of the speed error at sampling interval $k$, and $T$ is the sampling interval of the drive system.

The acceptable tuning range of the PI controller can be determined as follows. A saturation-type current limiter exists in the forward loop. The reason is that the matrix converter has its allowable maximum current to avoid damage to the solid-state power devices. The current limiter is a nonlinear element which may cause limit cycles of the speed response [11].

In order to avoid producing limit cycles, the parameters of the PI controller should be limited within a reasonable range. The describing function technique is used here to determine the range of the parameters of the PI controller. In the discrete-time domain, the transfer function in the forward-loop is

$$Gf(z) = Gc(z)\, Gp(z) = \frac{K_T T}{J_m} \frac{K_p(z-1) + K_1 z}{(z-1)^2} \quad (6)$$

where $G_c(z)$ is the transfer function of the PI controller, $G_p(z)$ is the transfer function of the simplified plant, and $J_m$ is the equivalent inertia of the system. Then, we can use the bilinear transformation, $z = (1 + \omega)/(1 - \omega)$, to obtain

$$G_f(\omega) = \frac{K_T T[-(K_1 + 2K_p)\omega^2 + 2K_p\omega + K_1]}{4 J_m \omega^2} \quad (7)$$

The $G_f(w)$ can be transformed into the frequency domain. The current limiter can be approximated by its describing function

$$N(I^*) = \frac{2}{\pi}\left[\frac{I^*\max}{I^*}\cos\left(\sin^{-1}\frac{I^*\max}{I^*}\right) + \sin^{-1}\frac{I^*\max}{I^*}\right] \quad (8)$$

where $N(I^*)$ is the describing function of the current limiter, $I^*_{max}$ is the maximum allowable current of the inverter. The closed-loop system exhibits limit cycles if $G_f(j\omega)$ and $-1/N$ intersect. It is easy to observe that the magnitude of $N(I^*)$ is always positive and varies from 0 to 1. As a result, the phase of $-1/N$ is always $-180$ degrees and the magnitude of $-1/N$ varies from 1 to infinity. Assume $\omega_c$ is the phase crossover frequency for $G_f(j\omega)$. Then, the $G_f(j\omega_c)$ is approximated by

$$|G_f(j\omega_c)| \cong K_T T (2K_p + K_I)/(4J_m) \quad (9)$$

The condition for the closed-loop system to be stable, is one in which $G_f(j\omega_c)$ cannot intersect with $-1/N$. This, then implies that the $|G_f(j\omega_c)|$ satisfies the following inequality

$$0 < |G_f(j\omega_c)| < 1 \quad (10)$$

Finally, the tuning range of the PI controller can be obtained as

$$0 < K_T T (2K_p + K_I) < 4J_m \quad (11)$$

### A. Membership Functions and Fuzzy Control Law:

The membership functions are used to describe the compatibility or degree of truth. The speed error is divided into three categories: the large error, the small error, and the zero error and is shown in Fig. 3(a). Then, any $\Delta w_r(k)$ may have different membership functions which are related to the three different categories. In Fig. 3(b), the change of the speed error is $\alpha(k)$, which can be divided into three categories as well. The maximum value of the change of the speed error is set as $400$ rad/s$^2$ due to the current limitation and inertia of the induction motor drive system.

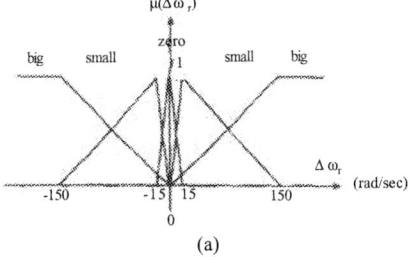

(a)

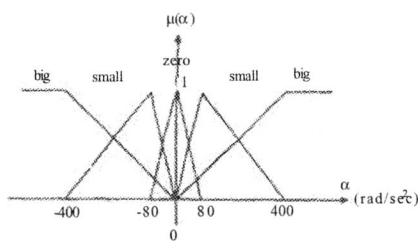

(b)

Fig. 3. Membership functions (a) Speed error. (b) Change of speed error.

The fuzzy-reasoning algorithm is based on the characteristics of the induction motor drive system and the designer's experience. For example, when the motor is starting, the PI controller is adjusted to achieve a fast response. However, when the motor is in a close to steady-state condition, the PI controller is tuned to avoid oscillations of the speed response. The details are as follows:

IF $|\Delta\omega_r(k)|$ is big THEN $K_p = K_{p1}$, $K_I = K_{I1}$.

IF $|\Delta\omega_r(k)|$ is small and $|a(k)|$ is big THEN $K_p = K_{p2}$, $K_I = K_{I2}$.

IF $|\Delta\omega_r(k)|$ is small and $|\alpha(k)|$ is zero THEN $K_p = K_{p4}$, $K_I = K_{I4}$.

IF $|\Delta\omega_r(k)|$ is zero THEN $K_p = K_{p5}$, $K_I = K_{I5}$.

The parameters $K_{p1} - K_{p5}$ and $K_{I1} - K_{I5}$ should satisfy tuning range of PI controller, keeping in mind, these parameters are determined by the designer's experience.

### B. Defuzzification:

The center of area (COA) method is used to obtain the PI parameters. In order to consider the effect of both the state variables, $\Delta w_r$ and $\alpha$ the intersection of the grade of the membership function with $\Delta \omega_r$ and $\alpha$ is calculated as

$$\mu_n = \mu_{\Delta\omega r}\Delta_a = \min((\mu(\Delta\omega_r), \mu(\alpha))$$

$$K_p = \sum_{n=1}^{5} \mu_n K_{pn} / \sum_{n=1}^{5} \mu_n \qquad (12)$$

$$K_I = \sum_{n=1}^{5} \mu_n K_{In} / \sum_{n=1}^{5} \mu_n \qquad (13)$$

where $\mu_n$ is the grade of the membership function in the different categories. Finally the control input u is set as

$$u = K_p \Delta\omega_r(k) + K_I \sum_{i=0}^{k} \Delta\omega_r(i) \qquad (14)$$

## IV. EXPERIMENTAL RESULTS

An experimental prototype matrix converter drive system based on the proposed control method has been constructed. The motor line currents are measured by two Hall-effect current sensors. The other feedback quantities are the absolute rotor shaft angle and motor speed, which can be obtained by the revolver mounted on the motor shaft [12].

The parameters of the induction motor are shown as follows. The stator resistance is 0.40 $\Omega$/phase, the d-axis inductance is 6.7 mH, the q-axis inductance is 6.7 mH, the inertia of the motor and dynamometer is 0.0233 N*m*s$^2$/rad, and the torque constant is 0.643 N*m/A. The current limitation of the matrix converter is $\pm$ 14 A.

Fig. 4 shows the measured speed responses of the matrix converter drive system with different control methods. Fig. 4(a) is the transient responses with a speed command of 500 r/min. Fig. 4(b) is the load disturbance responses when a 2 N*m external load is added at a speed of 500 r/min.

According to Fig. 4(a) and Fig. 4(b), the proposed control algorithm, which uses fuzzy logic to adjust the PI controller parameters, performs better than the fixed PI controllers.

Fig. 5 shows the measured a-phase currents in steady-state. Fig. 5(a) shows the a-phase current at 100 r/min and 2 N*m by using the traditional method. Fig. 5(b) shows the a-phase current in the same condition by using the proposed method. The proposed method performs better. Fig. 5(c) shows the a-phase current at 500 r/min and 2 N*m by using the traditional method proposed in reference paper [5].

Fig. 5(d) shows the a-phase current in the same situation by using the proposed method. The proposed method performs better again. Fig. 5(e) shows the a-phase current at 1500 r/min and 2 N*m by using the traditional method. Fig. 5(f) shows the a-phase current in the same condition by using the proposed method. The proposed method has similar performance to the traditional one when the motor is operated at a rated speed. The major reason for this is that almost the highest bus voltage switching patterns are selected as the motor is operated at a rated speed. According to Fig. 5, we can observe that the proposed method can effectively reduce the current deviations in a low or middle speed range.

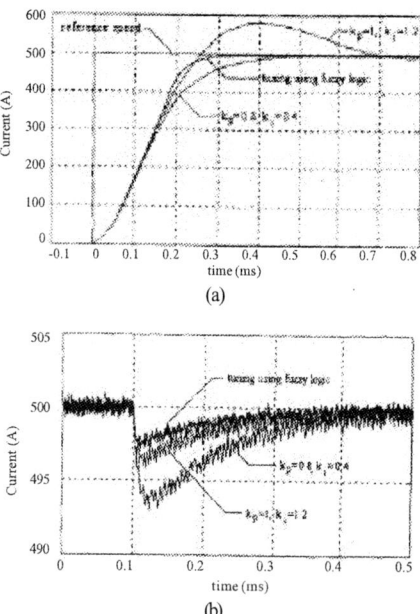

Fig. 4. Measured speed response at 500 r/min (a) transient (b) load disturbance.

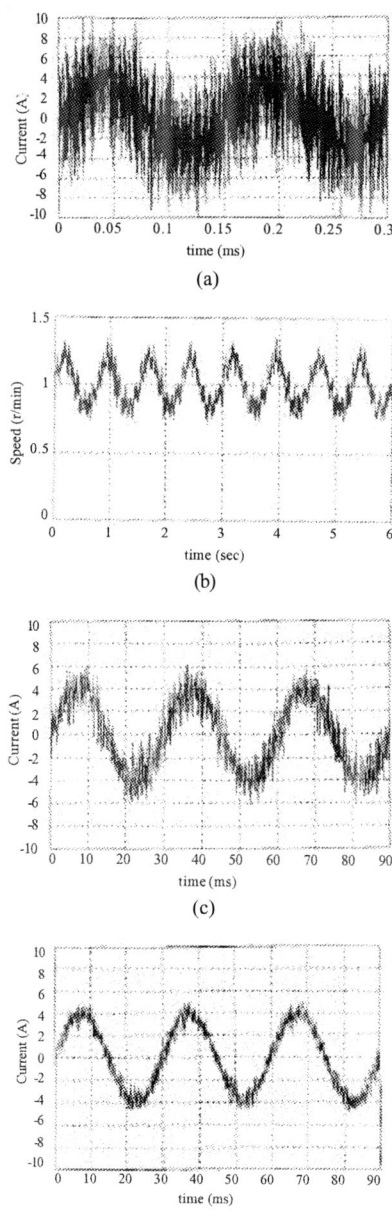

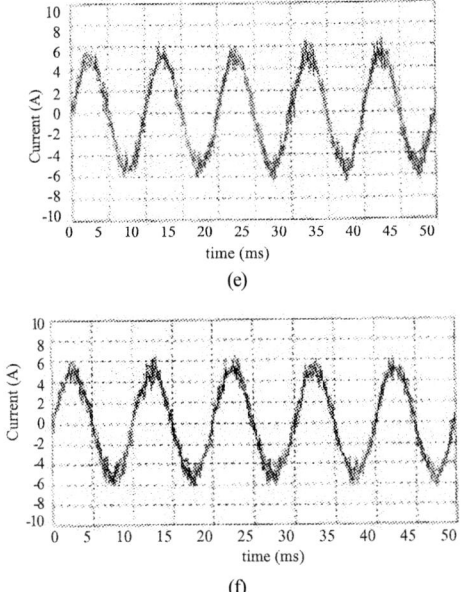

(e)

(f)

Fig. 5. Measured steady-state stator currents. (a) Traditional at 100 r/min. (b) Proposed at 100 r/min. (c) Traditional at 500 r/min. (d) Proposed at 500 r/min. (e) Traditional at 1500 r/min. (f) Proposed at 1500 r/min.

Fig. 6 shows the low speed responses of 1 r / min. Fig. 6(a) is the speed response of the traditional method proposed in [7]. Fig. 6(b) is the speed response obtained by using the neural-network learning technique without using torque compensation. Fig. 6(c) is the speed response when using the neural network learning technique and torque compensation. This method can obtain a smooth speed with small ripples.

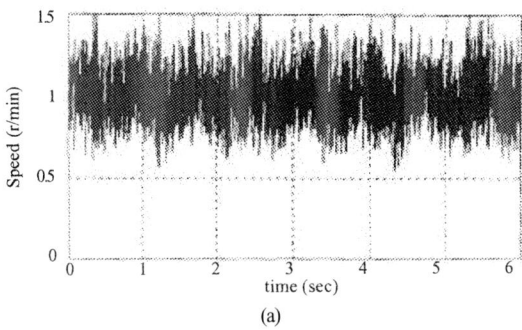

(a)

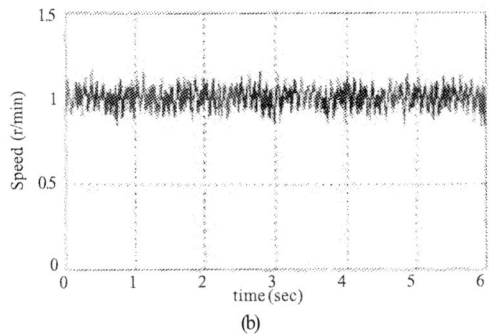

(b)

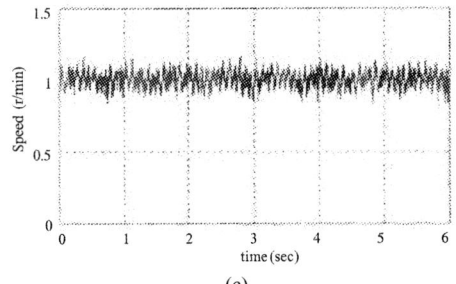

(c)

Fig. 6. Low speed responses. (a) Traditional method. (b) Proposed method without torque compensation.(c) Proposed method with torque compensation.

## V. CONCLUSIONS

This paper has provided a systematic and analytic approach for designing a high performance matrix converter induction motor drive system. Moreover, this design approach can be applied in both the speed and the position control system with an added advantage that the implemented system is flexible. As a result, it is likely to play major role in future converter designs both as a motor drive at low and high powers and as a power converter linking two electrical power systems having different voltages and frequencies.

## VI. REFERENCES

[1]  L. Huber and D. Borojevic, "Space vector modulated three-phase to three-phase matrix converter with input power factor correction," IEEE Trans. Industrial Application, vol. 31, no. 1, pp. 1234–1246, Nov. / Dec. 1995.

[2]  M. Milanovic and B. Dobaj, "Unity input displacement factor correction principle for direct AC to AC matrix converters based on modulation strategy," IEEE Trans. Circuits System vol. 47, no. 1, pp. 221–230, Feb. 2000.

[3]  A. Alesina and M. G. B. Venturini, "Analysis and design of optimum amplitude nine-switch direct ac -ac converters," IEEE Trans. Power Electronics, vol. 4, pp. 101–112, Jan. 1989.

[4]  D. Casadei, G. Serra, and A. Tani, "Reduction of the input current harmonic content in matrix converters under input/output unbalance," IEEE Trans. Ind. Electronics, vol. 45, pp. 401–411, June 1998.

[5]  J. H.Youm and B. H.Kwon, "Switching technique for current-controlled AC -to-AC converters," IEEE Trans. Industrial Electronics, vol. 46, pp. 309–318, April 1999.

[6]  S. Kim, S. K. Sul, and T. A. Lipo, "AC/AC power conversion based on matrix converter topology with unidirectional switches," IEEE Trans. Industrial Application, vol. 36, pp.139–145, Jan./Feb. 2000.

[7]  T. Matsuo, S. Bernet, R. S. Colby, and T. A. Lipo, "Application of the matrix converter to induction motor drives," in Proc. IEEE - IAS'96, San Diego, CA, , pp. 60–67, Oct. 1996.

[8]  C. L. Neft and C. D. Schauder, "Theory and design of a 30-hp matrix converter," IEEE Trans. Industrial Application, vol. 28,pp. 546–551, May/June 1992.

[9]  G. J. Wang, C. T. Fong, and K. J. Chang, "Neural -network-based self tuning PI controller for precise motion control of PMAC motors," IEEE Trans. Industrial Electronics, vol. 48, pp. 408 –415, April 2001.

[10]  C.Lascu, I. Boldea and F. Blaabjerg, "Variable structure direct torque control – A class of fast and robust controllers for induction machine drives," IEEE Trans. On Industrial electronics, vol. 51, no. 4, Aug. 2004.

[11]  C. Lascu and A.M. Trzynadlowski, "Combining the principle of sliding mode, DTC, and SVM in high performance sensor less AC drive," IEEE Transaction on Industrial Application, vol.40, no.1, Jan. / Feb. 2004.

[12] M. Matsui, M. Nagai, M. Mochizuki, and A. Nabae, "High-frequency link DC/AC converter with suppressed voltage clamp circuits-naturally commutated phase angle control with self turn-off devices", IEEE Trans. on Industry Applications, vol. 32, pp. 293-300, March/April 1996.

## VII. BIOGRAPHIES

**R. A. Gupta** born in India. He received bachelor degree in Electrical Engineering from MBM Engineering College, Jodhpur, India. He received Ph.D. degree in Electrical Engineering from IIT, Roorkee, India. He is faculty in the Department of Electrical Engineering, MNIT, Jaipur, India. His research interests are in the area of Power Electronics Drives.

**A. K. Wadhwani** born in India. He received Ph.D. degree in Electrical Engineering from IIT, Roorkee, India. He is faculty in the Department of Electrical Engineering, MITS, Gwalior, India. His research interests are in the area of Intelligent Control.

**R. R. Joshi** born in India. He received bachelor degree in Electrical Engineering from MBM Engineering College, Jodhpur, India and M. Tech. degree from IIT, Delhi, India. He is working towards the Ph.D. Degree in Electrical Engineering from MITS, Gwalior. He is faculty in the Department of Electrical Engineering, MPUAT, Udaipur, India. His current research interests are in the area of Neural Networks, Fuzzy Logic, Genetic Algorithms, Matrix Converter Technology and DSP based Implementation of DTC controlled Drive .

**2006 IEEE International Conference on Power Electronic, Drives and Energy Systems**

# An Adaptive Backstepping Controller for Doubly-Fed Induction Machine Drives

A. Farrokh Payam

*Abstract*—In this paper a nonlinear controller is presented for Doubly-Fed Induction Machine (DFIM) drives. The nonlinear controller is designed based on adaptive Backstepping control technique, using the fifth order model of induction machine in a synchronous d & q axis rotating reference frame with the d axis coincide with space voltage vector for the main ac supply with rotor current and stator flux components as state variables. The nonlinear controller can perfectly track the torque signal in the condition of unity power factor regulation, measured in the stator terminals, inspite of stator and rotor resistance variations. In order to make the drive system capable of operating in the motoring and generating modes below and above synchronous speed, two level SVM-PWM back-to-back voltage source inverters are employed in the rotor circuit. Computer simulation results obtained, confirm the effectiveness and validity of the proposed control approach.

*Index Terms*— Adaptive Backstepping, Doubly-Fed Induction Machine, Unity Power Factor.

## I. INTRODUCTION

IN field oriented methods applied to DFIM drives, the voltage drop across the stator leakage impedance is neglected. Such an assumption, forces a steady-state error both in motoring and generating modes of operation [1]. In References [2], [3] a backstepping torque tracking controller has been presented for DFIM drive on the basis of unity power factor operation (measured in the stator terminals). In [2], [3], the drive system performance has been obtained only for generating mode of operation above the synchronous speed and with nominal parameters. The main purpose of this paper is to continue the research work described in [2] with main objective of presenting the DFIM performance in both motoring and generating modes of operation below and above the synchronous speed and the proposed controller robust against rotor and stator resistances uncertainties. A full proof for convergence of states is verified together with stability for the quantity estimates. It will be shown that the proposed controller in this paper is capable of perfectly tracking control of torque reference commands with unity power factor regulation inspite of machine resistance variations and external load torque disturbance.

Moreover, in order to DFIM drive system capable of operating in motoring and generating modes of operation,

A. Farrokh Payam is with Isfahan University of Technology, Isfahan, Iran (e-mail: farokhpayam@alumni.iut.ac.ir ).

below and above the synchronous speed, two level SVM-PWM back-to-back Voltage-Source inverters are employed in the rotor circuit. In this control scheme the rotor dc link voltage is maintained constant on the basis of input-output feedback linearization technique. Finally, the drive system performance is demonstrated by some simulation results.

## II. DOUBLY-FED INDUCTION MACHINE MODEL

Under assumption of linear magnetic circuits and balanced operating condition, the equivalent two-phase model of the symmetrical DFIM with stator connected to line, represented in stator voltage d-q reference frame is

$$\frac{di_{dr}}{dt} = -\gamma_2 i_{dr} + \omega_2 i_{qr} + \beta\alpha_1\psi_{ds} - \beta\omega\psi_{qs} - \beta U + \frac{1}{\sigma_2}u_{dr}$$

$$\frac{di_{qr}}{dt} = -\gamma_2 i_{qr} - \omega_2 i_{dr} + \beta\alpha_1\psi_{qs} + \beta\omega\psi_{ds} + \frac{1}{\sigma_2}u_{qr}$$

$$\frac{d\psi_{ds}}{dt} = \alpha_1 L_m i_{dr} - \alpha_1\psi_{ds} + \omega_0\psi_{qs} + U$$

$$\frac{d\psi_{qs}}{dt} = \alpha_1 L_m i_{qr} - \alpha_1\psi_{qs} - \omega_0\psi_{ds}$$

$$\frac{d\omega}{dt} = \frac{1}{J}[\mu(\psi_{qs}i_{dr} - \psi_{ds}i_{qr}) - T] \qquad (1)$$

where $i_r$, $\psi_s$, $U$, $u_r$, $R$ and $L$ denote rotor currents, stator flux linkage, stator terminal voltage, rotor terminal voltage, resistance and inductance, respectively. The subscripts s and r stand for stator and rotor while subscripts d and q stand for vector component with respect to a stator voltage reference frame. $\omega$ denotes the rotor electrical speed and $L_m$ is the mutual inductance. $\omega_2 = \omega_0 - \omega$ is the slip speed and $L_\sigma = L_r(1 - (\frac{L_m^2}{L_r L_s}))$ is the redefined leakage inductance. And

$$\alpha_1 = \frac{R_s}{L_s}, \beta = \frac{L_m}{L_\sigma L_s}, \gamma_2 = (\alpha_1\beta L_m + \frac{R_r}{L_\sigma}), \mu = \frac{3PL_m}{2L_s} \qquad (2)$$

where $P$ is the number of poles.

The generated torque of DFIM can be expressed in terms of rotor currents and stator flux linkage as

$$T_e = \mu(\psi_{qs}i_{dr} - \psi_{ds}i_{qr}) \qquad (3)$$

The mechanical dynamic equation is given by

$$J\frac{d\omega_m}{dt} + B\omega_m + T_L = T_e \qquad (4)$$

0-7803-9771-1/06/$25.00 ©2006 IEEE

where $J$ and $B$ denote the moment of inertia of the motor and viscous friction coefficient, respectively, $T_L$ is the external load and $\omega_m$ is the rotor mechanical speed ($\omega = (\frac{P}{2})\omega_m$).

As said in [2], [3], for achieving steady-state unity power factor on the stator side of the machine

$$\lim_{t \to \infty} \psi_{ds} = 0 \tag{5}$$

Perfect torque tracking is the other considered specification, inparticular $T^*$ is the torque reference, which is assumed to be bounded together with its first and second derivative.

At this point, the two outputs of the DFIM to be controlled can be defined as:

$$y = \begin{pmatrix} T_e \\ \psi_{ds} \end{pmatrix} \tag{6}$$

Control input is the two-dimensional rotor voltage vector $u_r = [u_{dr}, u_{qr}]^T$. In the motor operation of this drive for the speed control objective, a PI controller is used as the speed controller. The following control task is then formulated:

Using the measured variables, design rotor voltages in order to perform asymptotic torque tracking, achieve a unity power factor during steady state, while guaranteeing internal stability, this implies:

$$\lim_{t \to \infty} T_e = 0 \,,\, \lim_{t \to \infty} \psi_{ds} = 0 \tag{7}$$

where the tracking error is defined as: $\tilde{T} = T_e - T^*$.

### III. ADAPTIVE BACKSTEPPING TORQUE-FLUX CONTROLLER DESIGN

As shown in in this section design a controller for tracking torque and reactive-power regulation based on adaptive backstepping approach and Lyapanov like method.

The design procedure is performed in two steps: first flux-torque control algorithm is developed to designed rotor current references. Second based on rotor currents references designed rotor voltages in order to perform asymptotic torque tracking, achieve a unity power factor during steady state. By considering

$$T_e = \mu(\psi_{qs}i_{dr} - \psi_{ds}i_{qr})$$

$$\frac{d\psi_{ds}}{dt} = \alpha_1 L_m i_{dr} - \alpha_1 \psi_{ds} + \omega_0 \psi_{qs} + U$$

$$\frac{d\psi_{qs}}{dt} = \alpha_1 L_m i_{qr} - \alpha_1 \psi_{qs} - \omega_0 \psi_{ds} \tag{8}$$

Defining the stator flux errors as:

$$\tilde{\psi}_{ds} = \psi_{ds} \,,\, \tilde{\psi}_{qs} = \psi_{qs} - \psi^* \,,\, \tilde{\alpha}_1 = \hat{\alpha}_1 - \alpha \tag{9}$$

where $\psi^* \neq 0$ is the reference for the modulus of flux vector.

Using error definitions the equation (8) can be rewritten as:

$$\dot{\tilde{T}} = \mu\left((\tilde{\psi}_{qs} + \psi^*)i_{dr} - \tilde{\psi}_{ds}i_{qr}\right)$$

$$\dot{\tilde{\psi}}_{ds} = -\alpha_1\tilde{\psi}_{ds} + \omega_0(\tilde{\psi}_{qs} + \psi^*) + \alpha_1 L_m i_{dr} + U$$

$$\dot{\tilde{\psi}}_{qs} = -\omega_0\tilde{\psi}_{ds} - \alpha_1(\tilde{\psi}_{qs} + \psi^*) + \alpha_1 L_m i_{qr} - \dot{\psi}^* \tag{10}$$

Defining:

-torque control algorithm:

$$i_{dr} = \frac{T^*}{\mu\psi^*} \tag{11}$$

-flux reference desired trajectory given by:

$$\omega_0\psi^* + \hat{\alpha}_1 L_m \frac{T^*}{\mu\psi^*} + U = 0 \tag{12}$$

-flux control algorithm:

$$i_{qr} = \frac{1}{\hat{\alpha}_1 L_m}(\hat{\alpha}_1\psi^* + \dot{\psi}^*) \tag{13}$$

error dynamics (10) becomes:

$$\dot{\tilde{T}} = \mu\left(\tilde{\psi}_{qs}i_{dr} - \tilde{\psi}_{ds}i_{qr}\right)$$

$$\dot{\tilde{\psi}}_{ds} = -\alpha_1\tilde{\psi}_{ds} + \omega_0\tilde{\psi}_{qs} - \tilde{\alpha}_1 L_m i_{dr}$$

$$\dot{\tilde{\psi}}_{qs} = -\omega_0\tilde{\psi}_{ds} - \alpha_1\tilde{\psi}_{qs} - \frac{\tilde{\alpha}_1}{\hat{\alpha}_1}\dot{\psi}^* \tag{14}$$

Define the rotor current tracking error as:

$$\tilde{i}_{dr} = i_{dr} - i_{dr}^* \,,\, \tilde{i}_{qr} = i_{qr} - i_{qr}^* \tag{15}$$

the stator flux, rotor current error dynamics can be computed from (14) and (1) as:

$$\dot{\tilde{\psi}}_{ds} = -\alpha_1\tilde{\psi}_{ds} + \omega_0\tilde{\psi}_{qs} + \alpha_1 L_m \tilde{i}_{dr} - \tilde{\alpha}_1 L_m i_{dr}^*$$

$$\dot{\tilde{\psi}}_{qs} = -\alpha_1\tilde{\psi}_{qs} - \omega_0\tilde{\psi}_{ds} + \alpha_1 L_m \tilde{i}_{qr} - \frac{\tilde{\alpha}_1}{\hat{\alpha}_1}\dot{\psi}^*$$

$$\dot{\tilde{i}}_{dr} = -\gamma_2\tilde{i}_{dr} + \omega_2\tilde{i}_{qr} + \beta\alpha_1\tilde{\psi}_{ds} - \beta\omega\tilde{\psi}_{qs} - \gamma_2 i_{dr}^* + \omega_2 i_{qr}^* - \beta\omega\psi^* - \beta U + \frac{1}{L_\sigma}u_{dr} - \dot{i}_{dr}^*$$

$$\dot{\tilde{i}}_{qr} = -\gamma_2\tilde{i}_{qr} - \omega_2\tilde{i}_{dr} + \beta\alpha_1\tilde{\psi}_{qs} + \beta\omega\tilde{\psi}_{ds} - \gamma_2 i_{qr}^* - \omega_2 i_{dr}^* + \beta\alpha_1\psi^* + \frac{1}{L_\sigma}u_{qr} - \dot{i}_{qr}^* \tag{16}$$

The rotor voltages should be designed to achieve asymptotic rotor current tracking and to guarantee global asymptotic stability for the full system. The current control is selected as:

$$u_{dr} = L_\sigma[\hat{\gamma}_2 i_{dr}^* - \omega_2 i_{qr}^* + \beta\omega\psi^* + \beta U + \dot{i}_{dr}^* - k_i\tilde{i}_{dr} - k_{ii}x_d]$$

$$u_{qr} = L_\sigma[\hat{\gamma}_2 i_{qr}^* + \omega_2 i_{dr}^* - \beta\hat{\alpha}_1\psi^* + \dot{i}_{qr}^* - k_i\tilde{i}_{qr} - k_{ii}x_q]$$

$$\dot{x}_d = \tilde{i}_{dr}$$

$$\dot{x}_q = \tilde{i}_{qr} \tag{17}$$

Where $k_i > 0, k_{ii} > 0$ are current controllers proportional and integral gains. Substituting (16) into (17) resulting system error dynamic is:

$$\dot{\tilde{\psi}}_{ds} = -\alpha_1\tilde{\psi}_{ds} + \omega_0\tilde{\psi}_{qs} + \alpha_1 L_m \tilde{i}_{dr} - \tilde{\alpha}_1 L_m i_{dr}^*$$

$$\dot{\tilde{\psi}}_{qs} = -\alpha_1 \tilde{\psi}_{qs} - \omega_0 \tilde{\psi}_{ds} + \alpha_1 L_m \tilde{i}_{qr} - \frac{\tilde{\alpha}_1}{\hat{\alpha}_1}\psi^*$$

$$\dot{\tilde{i}}_{dr} = -(\gamma_2 + k_i)\tilde{i}_{dr} + \tilde{\gamma}_2 i_{dr}^* + \omega_2 \tilde{i}_{qr} - k_{ii}x_d + \beta\alpha_1\tilde{\psi}_{ds} - \beta\omega\tilde{\psi}_{qs}$$

$$\dot{\tilde{i}}_{qr} = -(\gamma_2 + k_i)\tilde{i}_{qr} + \tilde{\gamma}_2 i_{qr}^* - \omega_2 \tilde{i}_{dr} - k_{ii}x_q + \beta\alpha_1\tilde{\psi}_{qs} - \beta\tilde{\alpha}_1\psi^* + \beta\omega\tilde{\psi}_{ds}$$

$$\dot{x}_d = \tilde{i}_{dr}$$

$$\dot{x}_q = \tilde{i}_{qr} \tag{18}$$

Control gains $k_i > 0, k_{ii} > 0$ must be selected to stabilize system (18). For this extent consider positive-definite function:

$$V = \frac{1}{2}\left( (\tilde{\psi}_{ds}^2 + \tilde{\psi}_{qs}^2) + \gamma_i(\tilde{i}_{dr}^2 + \tilde{i}_{qr}^2) + \gamma_{ii}(x_d^2 + x_q^2) + \frac{1}{\gamma_\alpha}\tilde{\alpha}_1^2 + \frac{1}{\gamma_\beta}\tilde{\gamma}_2^2 \right) \tag{19}$$

where $\gamma_i, \gamma_{ii}$ are arbitrary parameters.

When $\gamma_{ii} = \gamma_i k_{ii}$ is selected, the time-derivative of $V$, along the trajectories of (18) is:

$$\dot{V} = -\alpha_1(\tilde{\psi}_{ds}^2 + \tilde{\psi}_{qs}^2) + \alpha_1(\gamma_i\beta + L_m)(\tilde{\psi}_{ds}\tilde{i}_{dr} + \tilde{\psi}_{qs}\tilde{i}_{qr}) - $$
$$\gamma_i(k_i + \gamma_2)(\tilde{i}_{dr}^2 + \tilde{i}_{qr}^2) + \gamma_i\beta\omega(\tilde{\psi}_{ds}\tilde{i}_{qr} - \tilde{\psi}_{qs}\tilde{i}_{dr}) + $$
$$\tilde{\alpha}_1(-L_m i_{dr}^* \tilde{\psi}_{ds} - \frac{\tilde{\psi}_{qs}\psi^*}{\hat{\alpha}_1} - \gamma_i\beta\psi^*\tilde{i}_{qr} + \frac{1}{\gamma_\alpha}\dot{\tilde{\alpha}}_1) + $$
$$\tilde{\gamma}_2(\frac{1}{\gamma_\beta}\dot{\tilde{\gamma}}_2 + \gamma_i(\tilde{i}_{dr}i_{dr}^* + \tilde{i}_{qr}i_{qr}^*)) \tag{20}$$

function (20) is negative semi-definite if the following condition for the current controller proportional gain is satisfied:

$$k_i > \frac{1}{2\alpha_1}\left( \gamma_i\beta^2(\omega_{max}^2 + \alpha_1^2) + \frac{\alpha_1^2 L_m^2 + 4k_p}{\gamma_i} + 2\alpha_1^2\beta L_m \right) \tag{21}$$

and adaptation lows:

$$\dot{\hat{\alpha}}_1 = \gamma_\alpha\left( L_m i_{dr}^* \tilde{\psi}_{ds} + \frac{\tilde{\psi}_{qs}\psi^*}{\hat{\alpha}_1} + \gamma_i\beta\psi^*\tilde{i}_{qr} \right)$$

$$\dot{\tilde{\gamma}}_2 = -\gamma_\beta\gamma_i\left( \tilde{i}_{dr}i_{dr}^* + \tilde{i}_{qr}i_{qr}^* \right) \tag{22}$$

where $k_p$ is arbitrary, but greater than 0, and $\omega_{max}$ is the maximum value for the system. From (19), (20) and (21) and (22) it can be concluded that $x = [\tilde{i}_{dr}, \tilde{i}_{qr}, \tilde{\psi}_{ds}, \tilde{\psi}_{qs}, x_d, x_q]^T$ is bounded. From boundness of torque, of its first and second derivatives and from direct application of Barbalat's lemma [5] it follows that:

$$\lim_{t\to\infty} x = 0 \tag{23}$$

and the stability of control system is proved.

## IV. STABILIZATION OF ROTOR DC-LINK VOLTAGE

In this paper the rotor dc link voltage is maintained constant on the basis of input-output feedback linearization technique. The control strategy is shown in Fig. 1. [6].

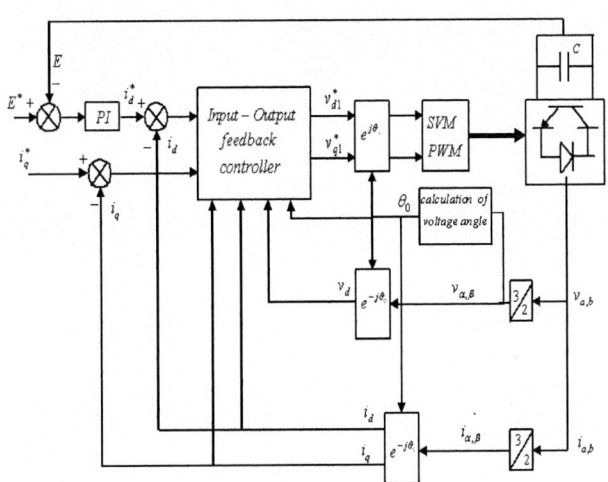

Fig. 1. DC link voltage controller [6].

## V. SIMULATION RESULTS

The overall block diagram of the proposed control approach is shown in Fig. 2. A C$^{++}$ computer program was developed to model this system on P.C.

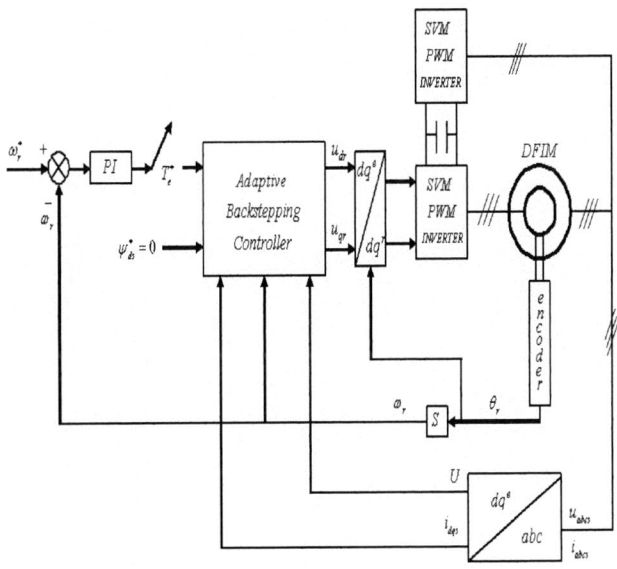

Fig. 2. Block diagram of proposed controller.

In this program, a static runge-kutta fourth order method is used to solve the system equations. The effectiveness and validity of the proposed approach is tested for a three-phase 5 KW, 380 V, six poles, 50 Hz DFIM drive [4] by simulation.

Fig. 3 shows the drive system performance in motoring mode of operation below and above the synchronous speed.

These results are obtained with $R_r = 2R_{rn}, R_s = 2R_{sn}$.

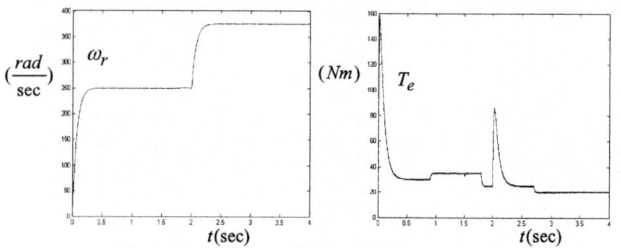

Fig. 3. Simulation results for motoring mode of operation.

Fig. 4 shows the drive system performance in the generating mode of operation above the synchronous speed. These results are obtained with $R_r = 2R_{rn}, R_s = 2R_{sn}$.

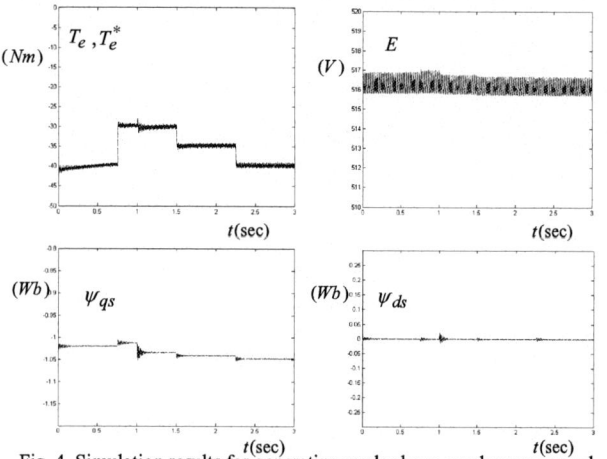

Fig. 4. Simulation results for generating mode above synchronous speed.

Fig. 5 shows the drive system performance in the generating mode of operation below the synchronous speed. These results are obtained with $R_r = 2R_{rn}, R_s = 2R_{sn}$.

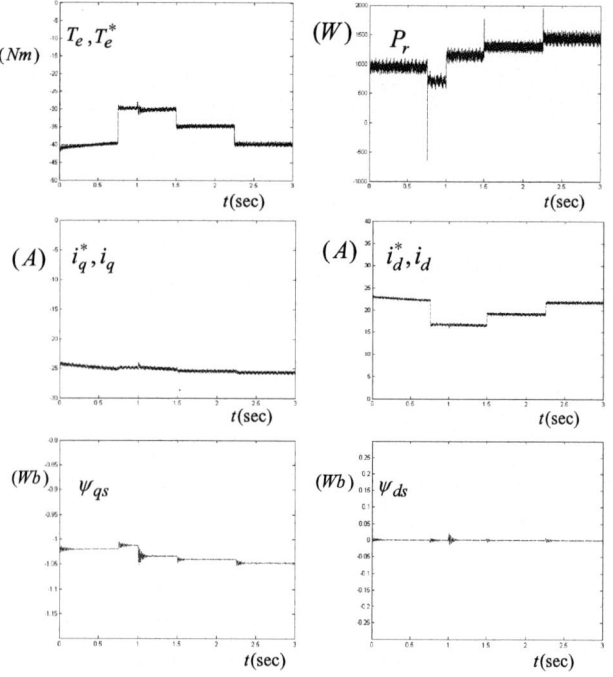

Fig. 5. Simulation results for generating mode below synchronous speed.

As it is shown of these results the proposed controller is robust against rotor and stator resistances uncertainties.

## VI. CONCLUSION

In this paper an adaptive nonlinear controller has been introduced for DFIM drives. The proposed controller is designed based on adaptive backstepping control approach and is capable of making the system states trajectories follow the torque reference signal with unity power factor condition inspite of stator and rotor resistance uncertainties and external load torque disturbance. The proposed control approach has been tested for both the motoring and generating modes of operation bellow and above synchronous speed; using the two level SVM-PWM back-to-back voltage sources inverters in the rotor circuit. Furthermore the rotor dc link voltage is maintained constant also based on input-output control method, using a rotating synchronous reference frame with d axis coincide with the direction of space voltage vector of the main ac supply. Computer simulation results obtained, confirm the validity and effectiveness of the proposed control approach.

## VII. REFERENCES

[1] B.Hopfensperger, J.Atkinson, and R.A.Lakin, "Stator-flux-oriented control of a doubly-fed induction machine with and without position encoder," in *Proc. July 2000 IEE Electr. Power Appl Conf.*, vol. 147, no. 4, pp. 241-250.

[2] S.Peresada, A.Tilli, and A.Tonielli, "Robust active-reactive power control of a doubly-fed induction generator," *in Proc. Sept. 1998 IEEE-IECON'98, Aachen, Germany*, pp. 1621-1625.

[3] S.Peresada, A.Tilli, and A.Tonielli, "Indirect stator flux-oriented output feedback control of a doubly fed induction machine," *IEEE Trans. Contr Sys*, vol. 11, no. 6, pp.875-888, November 2003.

[4] S.Peresada, A.Tilli,and A.Tonielli, "Robust Output Feedback Control of a Doubly Fed Induction Machine," in *Proc. 1999 IEEE Conf.*, pp. 1348-1354.

[5] Jeffery T.Spooner, Manfred Maggiore, Raul Ordonez and Kevin M.Passino, *Stable Adaptive Control And Estimation For Nonlinear Systems*, John Wiley and Sons, Inc, NewYork.

[6] J.Soltani, A.Farrokh Payam, "A Robust Adaptive Sliding-Mode Controller for Slip Power Recovery Induction Machine Drives," in *Proc. 2006 IEEE/IPEMC Conf.*

**2006 IEEE International Conference on Power Electronic, Drives and Energy Systems**

# Application problem of PWM AC drives due to long cable length and high dv/dt

B.Basavaraja,*Member,IEEE*, and D.V.S.S.Siva Sarma,*Member,IEEE*

*Abstract--*Advances in power electronics technology have improved the performance and output waveforms of PWM voltage source inverters. Switching frequencies of 2 to 15 kHz with 0.1μs rise times are common with the current IGBT technology while allowing for power levels over 200 kW. While the high switching speeds and zero switching loss schemes drastically improve the performance of the PWM inverters, the high rate of voltage rise (dv/dt) of 0 to 600V in less than 0.1μs has adverse effects on the motor insulation and bearings and deteriorates the waveform quality.

Long cables contribute to a damped high frequency ringing due to the distributed nature of the cable leakage inductance and coupling capacitance (L-C) at the motor terminals resulting in over voltages which further stress the motor insulation.

Voltage reflection is a function of inverter output pulse rise time and the length of the motor cables, which behave as a transmission line for the inverter output pulses.

## I. EFFECT OF USING LONG CABLES BETWEEN THE INVERTER AND THE MOTOR

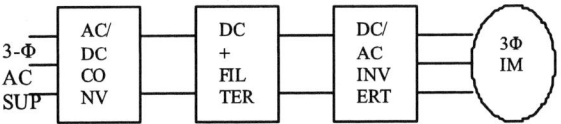

Fig. 1. PWM inverter driving an induction motor using long cable leads

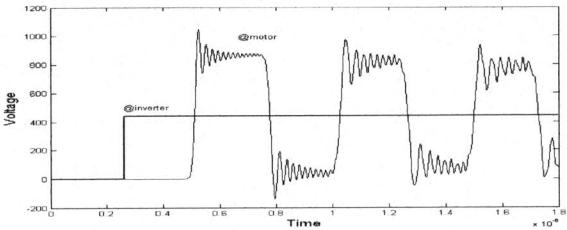

Fig. 2.. Inverter Dc-Link voltage and motor over voltage

Fig.1.shows schematic of the PWM inverter driving an induction motor using long cable. Fig.2 shows the simulated Inverter DC-Link voltage and over voltage waveform at motor terminal due to Reflected Wave Phenomenon.

Difference in impedances of the cable and motor leads to voltage reflection and hence over voltage appears at motor terminal.

---

Basavaraja Member,IEEE',Research Scholar,NIT Warangal/Associate Professor SREC,Warangal,AP,India( banakara_36@rediffmail.com )

D.V.S.S.Sivasarma Member, IEEE' Assistant Professor, EED, NITWarangal Andhra Pradesh, India (Sivasarma@gmail.com )

## II. REFLECTED WAVE PHENOMENON

Cause of Reflected Wave at the Motor Terminals is mismatch between cable surge impedance and motor surge impedance.

Cable Surge Impedance

$$z_o = \sqrt{\frac{L}{C}} \quad \text{--------}1(a)$$

$L$ = inductance per unit length

$C$ = capacitance per unit length

$Z_o$ – Varies with wire gauge and cable construction

$Z_o$ – Range is 80 to 180 ohms

| Motor Size | Z-load Impedance Range |
|---|---|
| < 5 HP | 2000– 5000 ohms |
| 125 HP | 800 ohms |
| 500 HP | 400 ohms |

Difference in impedances of the cable and motor leads to voltage reflection and hence over voltage appears at motor terminal.Fig3 shows the Voltage Reflection Analysis due to long motor leads. To better understand the repeated reflections on a finite length of cable with infinite *dv/dt*, one reflection of an incident wave will be considered

Fig. 3 shows  the different steps involved in the voltage reflection analysis of over voltages at the motor terminal. Fig. 3(a) shows the PWM pulse input at the sending end of the cable. The motor terminal is considered as open circuit for high frequencies. Fig3 (b) shows the incident voltage wave and the associated current waves when the voltage input is switched on. Fig (c) shows the reflected voltage pulse and associated current pulse at the motor end of the cable. As the motor terminals are open circuited, current must be zero. Hence the reflected component of the current will have same amplitude but with opposite sign, where as the incident voltage wave will be reflected with positive coefficient towards the sending end. The voltage at motor terminals which is the sum of incident and reflected voltage will be nearly double that of incident voltage. The reflected component at the motor terminals travels towards inverter. However, at the sending end inverter output voltage is E and hence a traveling wave of –E reflects and travel towards the motor. This voltage is associated with a current waveform. This second incident wave soon reaches the receiving end and is reflected again.Fig3 (d) and (e) shows these reflected waveforms.

0-7803-9771-1/06/$25.00 ©2006 IEEE

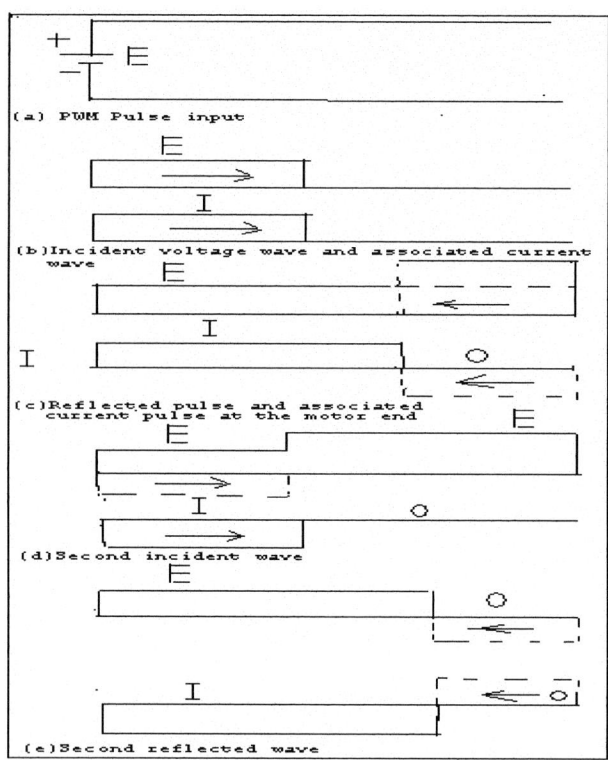

Fig. 3. Voltage Reflection analysis

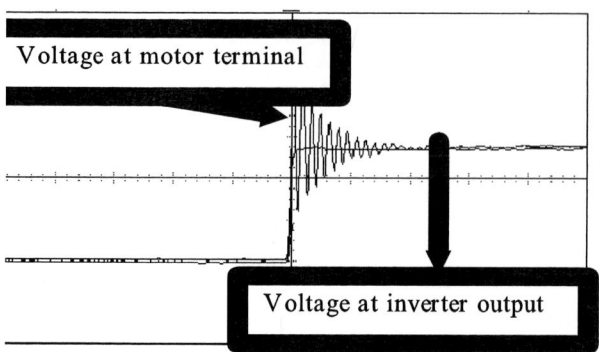

Fig. 4(a). Expanded Waveform

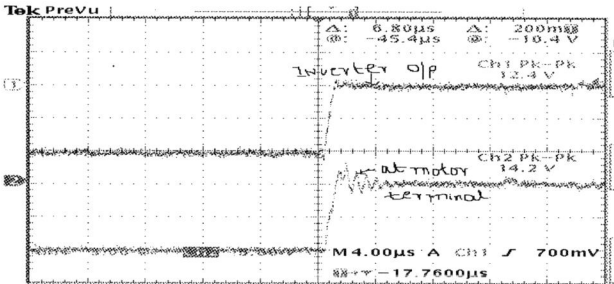

Fig. 4(b). Experimental result

During the second reflection at the motor terminal, voltage waveform will be reflected with negative pulse and the current pulse be reflected with positive pulse so that the current at the motor terminals remains zero. These second reflected current and voltages will travel towards the inverter and the cycle repeats.

If $Z_m$ is the motor impedance and $Z_c$ is the cable surge impedance, reflection coefficient at the motor terminals

$$\tau_m = (Z_m - Z_c)/(Z_m + Z_c) \text{------1(b)}$$

$$Z_c = SQRT (Lc/C_c) \text{-----------} (2)$$

Where Lc and Cc are cable inductance and capacitance per unit length.

Reflected coefficient at the inverter terminal

$$\tau_s = (Rs - Z_c)/(Rs + Z_c) \text{------} (3)$$

Where Rs is the source resistance.

At the inverter, the reflected forward-traveling wave has the same shape as the incoming backward-traveling wave but with corresponding points reduced by $\tau_s$. Due to the dominating winding inductance, the characteristic impedance of the motor can be ten to one hundred times that of the characteristic impedance of the cable connecting the drive to the motor. Therefore, the incident wave voltage will be reflected back towards the inverter as a function of eqn. (1)(b), and the voltage amplitude at the terminals of the motor will approximately double [2][3], as shown in Fig. 4.

## III. EFFECT OF PWM RISE TIME (DV/DT)

Peak voltage at the motor terminal can be determined by using wave propagation theory and voltage reflection analysis. The traveling time ($t_t$ in μs) for the inverter output pulse to travel from the inverter terminals to the motor terminals can be expressed as:

$$t_t = \frac{l_c}{v} \text{------------} (4)$$

where $v$ is the pulse velocity and is given by

$$v = \frac{1}{\sqrt{L_c C_c}} \text{--------} (5)$$

$l_c$ = cable length in feet
$L_c$ = inductance per foot
$C_c$ = capacitance per foot
$t_t$ = time for pulse to transit the length of the cable once

Forward traveling inverter output pulse will be reflected at the motor terminals after $t_t$ time and the resulting backward traveling wave, moving towards the inverter, will have an magnitude of:

$$E_t(t_t) = \frac{t_t * E_{dc} * \tau_m}{t_r} \quad \text{for } t_t < t_r \text{ (6)}$$

and

$$E_t(t_t) = E_{dc} * \tau_m \quad \text{for } t_t \geq t_r \quad (7)$$

where
$E_{dc}$ = dc bus voltage

$\tau_m$ = reflection coefficient at the motor (typically 0.9 for motors less than 20hp)

$t_r$ = inverter output pulse rise time (in μs)

Hence when $t_t \geq t_r$ equation (7) applies and no reflection and the peak motor terminal voltage will be reached after the pulse travels the length of the cable. Usually cable length of 15mt or less will result in $t_t < t_r$, and therefore eqn. (6) would apply.

The backward traveling wave will then be reflected at the inverter terminals in the same manner, however, now as a function of the reflection coefficient of the inverter (or source), $\tau_s$. From eqn. (3), it can be seen that for a typical low impedance source, $\tau_s$ will approach -1, and therefore, the resulting reflected wave traveling back towards the motor will be negative in amplitude.

Therefore, after three transitions of the cable, the increasing motor terminal voltage will be reduced by this negative reflected wave, after it has traveled back and reached the motor. Therefore, the peak voltage can be found by determining the total voltage due to reflections at the terminals of the motor, from eqns. (4-7), after three transitions of the cable, and adding this to the incident wave voltage magnitude, $E_{dc}$, as shown in eqns. (8-9).

$$E_{LL,p} = \frac{3 * l_c * E_{dc} * \tau_m}{v * t_r} + E_{dc} \quad \text{for } t_t < t_r/3 \quad ----(8)$$

$$E_{LL,p} = E_{dc} * \tau_m + E_{dc} \quad \text{for } t_t \geq t_r/3 ---------(9)$$

The normalized peak motor terminal voltage for longer pulse rise times, i.e. $t_r/3 > t_t$, can be written as a function of rise time as:

$$\frac{E_{LL,P}}{E_{dc}} = \frac{3 * l_c * \tau_m}{v * t_r} + 1 \quad ---------- (10)$$

$$\text{Therefore} \quad \frac{3 * l_c * \tau_m}{v * t_r} \ll 1 \quad --- (11)$$

for minimum or no overvoltage to occur.

The critical rise time can then be computed by substituting the maximum desired voltage overshoot as shown below,

$$\text{Set} \quad \frac{3 * l_c * \tau_m}{v * t_r} \quad 0.2 \quad ----(12)$$

$$\Rightarrow t_r = \frac{3 * l_c * \tau_m}{v * 0.2} \quad -------(13)$$

$$\Rightarrow l_c = \frac{t_r * v * 0.2}{3 * \tau_m} \quad ----------(14)$$

i.e. for a 440V ac system with 594V dc bus, the allowable peak voltage would be 1.2*594V = 712.8V. From the cable parameters $v$ = 160m/μs is obtained and the critical rise time ($t_r$) for 10m of cable and $\tau_m$= 0.9 would be 2.566μs. Therefore, a rise time of less than 2.566μs (higher *dv/dt*) will result in an over-voltage at the terminals of the motor greater than 20%.

Table I shows the minimum cable length and rise time after which virtual voltage doubling occurs at the terminals of the motor. The cable measurement carried by using a LCR meter

TABLE I
MINIMUM CABLE LENGTH FOR VOLTAGE DOUBLING

| Rise time (μs) | Cable length(mt) |
|---|---|
| 0.2566 | 3.6 |
| 0.5 | 18 |
| 2.566 | 36 |
| 5.132 | 54 |
| 2.566 | 92.32 |
| 3.0 | 106.44 |
| 3.5 | 126 |
| 4.0 | 144 |
| 5.0 | 180 |

Fig.4 (b) shows the experimental waveforms of motor terminal voltages for 14m cable, 200V input systems and reflection coefficients of 0.9.

## IV. ANALYSIS OF THE OVER VOLTAGE

Using the high frequency models [4] of Induction motor and cable as shown in fig5 and fig6, a simulation program has been developed in Matlab to calculate the over voltage in a range of voltage pulse rise times (350ns-1.2μs,) driving two different induction motor of power ratings 3hp and 15hp. Experimental results for 3HP, 10mt and 14mt cables give the validity for the simulation results [4].The simulation and experimental waveforms are shown in fig 7 and fig8.

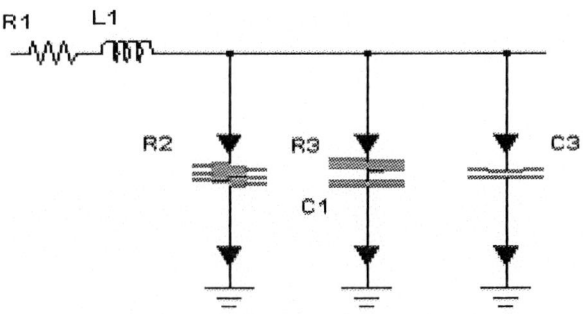

Fig. 5. High Frequency Model of the Power Cable

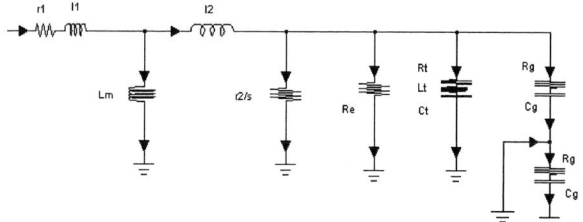

Fig. 6. High Frequency Model of the Induction motor

### A. Cable Length Vs Peak Voltage

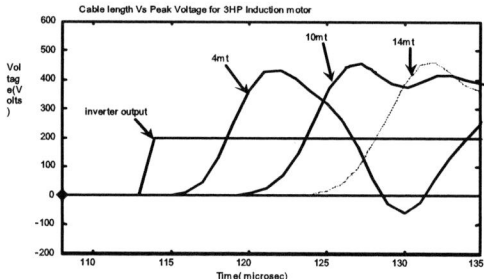

Fig. 7(a). Simulation results for 3HP, 4mt, 10mt and 14mt cable respectively.

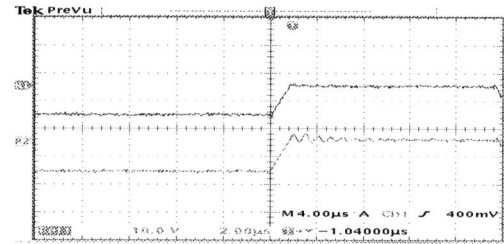

Fig. 7(b). Experimental results for 3HP, 10mt cable

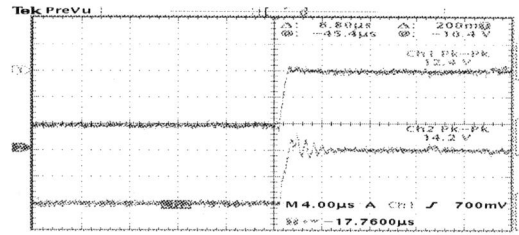

Fig. 7(c). Experimental results for 3HP, 14mt cable

Fig8 shows simulation results of voltage vs rise time for 3HP motor with other cable lengths.

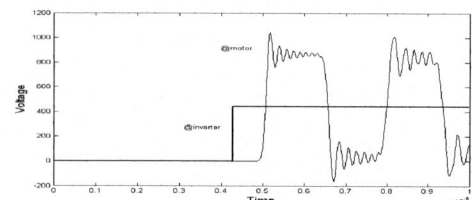

Fig. 8(a). Line to Line Voltage Peak Vs Time for Rise time=350ns; cable=60m;

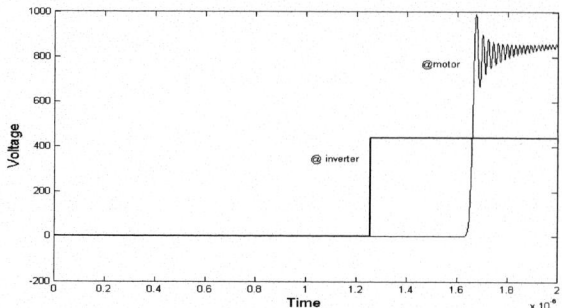

Fig. 8(b). Line Voltage Peak Vs Time for Rise time=1.2μs; cable=60m.

TABLE II
EFFECT OF CABLE LENGTH FOR 3 HP MOTOR: RISE TIME 350 NSEC AND
INVERTER OUTPUT VOLTAGE OF 400VOLTS

| Cable Length (in meters) | Over Voltage (in Volts) | Output Pulse Rise Time (μs) |
|---|---|---|
| 10 | 1050 | 0.325 |
| 20 | 1060 | 0.375 |
| 30 | 1075 | 0.4 |
| 60 | 1100 | 0.5 |

From the above waveforms and table.II, it can be noted that the voltage at the motor terminals increases as the length of the cable increases. As the rise time increases the over voltage reduces for the same length of cable. Fast rise time causes high dv/dt which stresses the motor insulation [5].

### Cable length Vs Propagation time

Fig9.shows the effect of cable length on Propagation time.

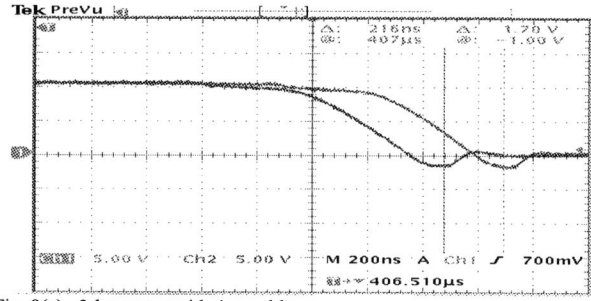

Fig. 9(a). 3-hp motor with 4mt cable

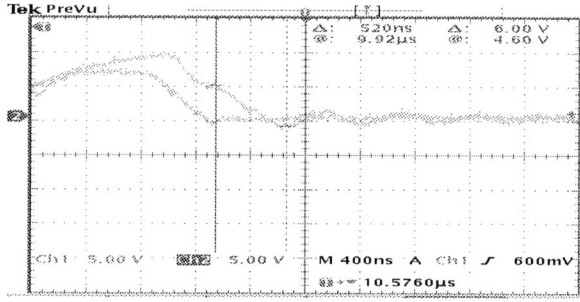

Fig. 9(b). 3-hp motor with 10mt cable

TABLE III
EFFECT OF CABLE LENGTH ON PROPAGATION TIME FOR 3 HP MOTOR

| Cable length (mt) | Over Voltage (in Volts) | Propagation Time (in ns) |
|---|---|---|
| 4 | 250 | 216 |
| 10 | 280 | 520 |
| 14 | 300 | 640 |

By observing the experimental waveform and table III, as cable length increases propagation time increases and peak voltage will shift to the higher rise time. So this will happen up to critical cable length, for which voltage doubling occurs as shown in table1

### C. Effect of High dv/dt

High switching speeds and zero switching loss schemes drastically improve the performance of the PWM inverter, but high rate of voltage rise (dv/dt) of 0 to 400v in less than 0.1μs has adverse effects on the motor insulation and bearings and deteriorates waveform quality. This leads to overvoltage problems and hence reduces the motor life.

Fig 10 shows the dv/dt for supply voltage of 440v, 10mt cable length and different frequencies. Usually PWM Inverters operates at higher switching frequencies and for these values of frequencies the dv/dt is more as shown in fig10.Hence shorter rise time will have more dv/dt.

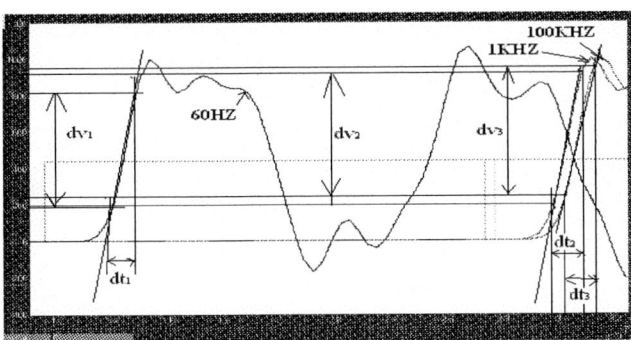

Fig. 10. dv/dt for supply voltage of 440v, 10mt cable length and different frequencies

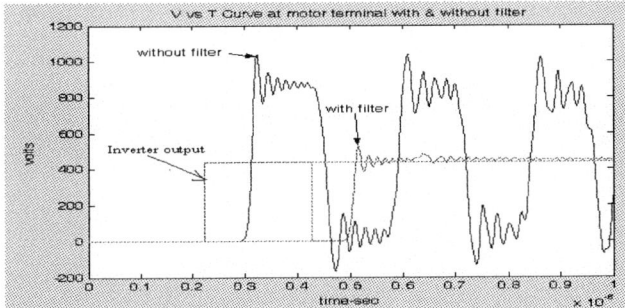

Fig. 11. Voltage and time curve at inverter output and at motor terminal with filter and without filter.

Over voltage can reduce by using filters at motor terminal or at output of inverter. One of the filter design method to reduce the overvoltage is RC filter at motor terminal. In this method cable is terminated by a first order filter consisting of a capacitor in series with the resistor to match with the cable and provide the proper level of damping to control the voltage overshoot.

By observing fig11 over voltages are more at motor terminal without filter and less over voltages with filter. Hence dv/dt can be reduced by adopting filter at motor terminal or at output of the inverter terminal.

## V. CONCLUSION

This paper discusses the analysis of the over-voltage phenomena in long cable PWM drives due to voltage reflection theory. Experimental and simulation results are compared. Effect of cable length and rise time on the over voltages are analysed. As the cable length is increased over voltage magnitude increases. Also the effect of rise time on dv/dt is presented. Shorter rise time results in high dv/dt which stresses the motor insulation. Also simulation results of mitigating high dv/dt is explained by adopting filter.

## VI. REFERENCES

[1] G.Suresh, Hamid A Tollyat,A Rendusara, and Prasad N.Enjeti, "Predicting the transient Effects of PWM Voltage wave form on the stator windings of random wound Induction motors " IEEE Transaction on power electronics Vol 14 No.1,January 1999.

[2] Melhorn,Christopher J.,Le Tang, "Effects of PWM ASDs on Standard Squirrel Cage Induction Motors" 1994 PCIM Conference Proceedings, Dallas Texas,September 1994, pp.356- 364.

[3] J.A Pomilio,C R de Souza, L,Matias, P.L.D.Peres and I.Sbonatti "Driving AC Motors through a Long Cable: The Inverter Switching Strategy", IEEE Transactions on Energy Conversion, Vol.14,No.4, December 1999.

[4] A. F. Moreira, T. A. Lipo, G. Venkataramanan, and S. Bernet, "High Frequency Modeling for Cable and Induction Motor Over-Voltage Studies in Long Cable Drives" IEEE Industrial Application Society 36th Annual Meeting Chicago, Illinois, USA, September 30 October 5, 2001.

[5] A.von Jouanne and P.N. Enjeti, "Design Considerations for an Inverter Output Filter to Mitigate the Effects of Long Motor Leads in ASD Applications," IEEE Transactions on Industry Applications, vol. 33, no. 5, pp. 1138-1145, Sep/Oct 1997.

[6] S. Evon, D. Kempke, L. Saunders, and G. Skibinski, "Riding the reflected wave—IGBT drive technology demands new motor and cable considerations," in Proc. IEEE Petroleum and Chemical Industry Conf., Philadelphia, PA, Sept. 23–26, 1996, pp. 75–84.

[7] E. Persson, "Transient effects in application of PWM inverters to induction motors," IEEE Trans. Ind. Applicat., vol. 28, pp. 1095–1101,Sept./Oct. 1992.

[8] P. Van Paucke, R. Belmans, W. Geysen, and E. Ternier, "Overvoltages in inverter fed induction machines using high frequency power electronic components," in Proc. IEEE APEC'94, Mar. 1994, pp. 536–541.

[9] T. Takahashi, M. Termeyer, T. Lowery, and H. Tsai, "Motor lead length issues for IGBT drives," in Proc. IEEE Pulp and Paper Conf., 1995, pp.21–27.

[10] R. Kerkman, D. Leggate, and G. Skibinski, "Interaction of drivemodulation and cable parameters on ac motor transients," IEEE Trans.Ind. Applicat., vol. 33, pp. 722–731, May/June 1997.

[11] G. Skibinski, R. Kerkman, D. Leggate, J. Pankau, and D. Schlegel,"Reflected wave modeling techniques for PWM ac motor rives," in Proc. IEEE APEC'98, Anaheim, CA, Feb. 15–19, 1998, pp. 1021–1029.

[12] K. A. Corzine, J. T. Tichenor, J. L. Drewniak, and S. D. Sudhoff, "High frequency characterization of a 3-phase induction motor," in *Proc.1997 Naval Symp. Electric Machines*, Newport, RI, July 28–31,1997,pp.251–256.

[13] G. Grandi, D. Casadei, and A. Massarini, "High frequency lumped parameter model for ac motor windings," in *Proc. EPE'97*, 1997, pp.2.578–2.583.

## VII. BIBLIOGRAPHIES

**Mr.Bsavaraja Banakar** was born in 1970.He is IEEE Member since2005. He obtained his B.Tech(EEE) degree from Gulbarga University and M.Tech from Karantaka Unversity, India. He worked as a Lecturer in VEC Bellary & Associate Professor at SSJ Engineering College Mahaboobnagar & Presently he is pursuing a Doctoral program at National Institute of Technology, Warangal, India & working as a Asso.Prof. at SR Engineering College Ananthasagar, Waranagal, India. His areas of interest include power electronics and drives, High voltage Engineering and EMTP applications.

**Dr.D.V.S.S.Siva Sarma** was born in 1964. He is IEEE Member since2004. He obtained his B.Tech (EEE) and M.Tech(Power Systems) from JNTU College of Engineering, Anantapur in 1986 and 1988 respectively. He obtained his Doctorate degree from Indian Institute of Technology, Chennai in 1993. Since 1992, he is working as Faculty member of Department of Electrical Engineering at National Institute of Technology, Warangal, Andhra Pradesh (Formerly Regional Engineering College, Warangal). His areas of interest include Power System Transients, Fault diagnostics, Protection and Condition Monitoring of Power Apparatus, High Voltage Engineering and EMTP applications. Presently he is a chairman of EMTP Indian user group and counselor for IEEE student branch of NIT Warangal.

**2006 IEEE International Conference on Power Electronic, Drives and Energy Systems**

# Adaptive Controller Design for Permanent Magnet Linear Synchronous Motor Control System

### B. Srinivasu, P.V.N. Prasad, and M.V.Ramana Rao

*Abstract* - **The different adaptive controllers for a permanent magnet linear synchronous motor (PMSLM) position control system are proposed. The proposed controllers include a back-stepping adaptive controller and a self-tuning adaptive controller. The detailed controller design procedures are discussed. The transient responses of position and speed due to load disturbances are discussed and analyzed. The tracking response of the system for different input position commands like step, square, triangular and sinusoidal time-varying signals are studied and compared. The system is simulated using Simulink.**

*Index Terms* – **Permanent magnet linear synchronous motor (PMLSM), Integral square of position error (ISE)**

## I. INTRODUCTION

MODERN control systems, such as manufacturing equipment, transportation, and robots, usually require high-speed/high-accuracy linear motions. Generally, these linear motions are traditionally implemented using rotary motors with mechanical transmission mechanisms such as reduction gears and lead screws. Such mechanical transmissions, not only seriously reduce linear motion speed and dynamic response, but also introduce backlash, large frictional and inertial loads. Recently, more and more industrial products, such as: semiconductor manufacturing equipment, X-Y driving devices, robots, and artificial hearts require high-speed/high-accuracy motions. The linear motors, which do not use mechanical transmissions, show promise for widespread use in these high-speed/high-accuracy systems. Without using the conventional gears or ball screws, the permanent magnet linear synchronous motor (PMLSM), however, is more easily affected by load disturbance, torque ripple, and parameter variations. As a result, the design of a controller and load disturbance compensator is very important in PMLSM drives. The linearization method has been successfully used for a PMLSM. This method requires accurate parameters of the PMLSM and complex control procedures. Recently, some researchers have used fuzzy rules or neural networks to control PMLSM drives. The idea is new and good. However, it requires on-line training or designer

The authors express their deep sense of gratitude to the Principal, University College of Engineering, Osmania University, Hyderabad for the financial support rendered under TEQIP programme.

B.Srinivasu is presently working as a junior telecom officer in Bharath Sanchar Nigam Limited, Andhra Pradesh, INDIA (e-mail: sreenivasubpd@ yahoo.co.in )

M.V.Ramana Rao (email: ramanarao2@yahoo.co.in ) , P.V.N. Prasad (email: polaki@rediffmail.com), are with the Department of Electrical Engineering, University College of Engineering, Osmania University, Hyderabad - 500 007, India.

experience. In addition, the execution of the fuzzy rules requires a lot of selections and that of the neural networks a lot of computations. Some researchers proposed robust controllers for PMLSM drives. The robust controller design is based on the parameters and model of the PMLSM. In the real world, unfortunately, the measurement of the parameters of the PMLSM is very difficult. To solve the difficulty, adaptive controllers have been proposed for the PMLSM. Adaptive control does not require the accurate parameters of the motor. As a result, it can be a feasible control algorithm for a PMLSM. So for controlling the position or speed of Permanent Magnet Linear Synchronous Motor(PMLSM) two different control algorithms, which include back-stepping adaptive control, and self-tuning adaptive control algorithms are discussed and compared. The superiority of this control algorithms of PMLSM system are compared with PID (Proportional, Integral and Derivative) controller.

## II. MATHEMATICAL MODEL

The mathematical model of a permanent magnet synchronous motor can be described, in the two-axis d-q synchronously rotating frame, by the following differential equations:

$$ pi_d = \frac{1}{L_d}(V_d - R_s i_d + \frac{\pi}{\tau} v_e L_q i_q) \tag{1} $$

$$ pi_q = \frac{1}{L_q} V_q - R_s i_q + \frac{\pi}{\tau} v_e L_d i_d - \sqrt{\frac{3}{2}} \frac{\pi}{\tau} \lambda_{max} v_e \tag{2} $$

Where $p$ is the differential operator $d/dt$, $i_d$ and $i_q$ are the d-axis and q-axis currents, $L_d$ and $L_q$ are the $d$ and $q$-axis inductances, $V_d$ and $V_q$ are the d-axis and q-axis stator voltages, $R_s$ is the stator resistance, $\tau$ is the pole pitch, $v_e$ is the velocity, and $\lambda_{max}$ is the maximum flux linkage due to permanent magnet in each phase. The electro-mechanical equation of the PMLSM is

$$ F_e = \frac{\pi}{\tau} \left[ \left( L_d - L_q \right) i_d i_q + \sqrt{\frac{3}{2}} \lambda_{max} i_q \right] \tag{3} $$

Consider that PMLSM is a surface-mounted PMLSM. As a result, $L_d$ is equal to $L_q$. Then, (3) can be simplified as

$$ F_e = \frac{\pi}{t} \sqrt{\frac{3}{2}} \lambda_{max} i_q \tag{4} $$

$$ = k_t i_q \tag{5} $$

where $K_t$ is the torque constant of the motor. The dynamic mechanical speed-movement equation is

0-7803-9771-1/06/$25.00 ©2006 IEEE      225

$$pv_m = \frac{1}{M}\left(F_e - Bv_m - F_L\right) \qquad (6)$$

and
$$v_m = \frac{2}{P_o}v_e \qquad (7)$$

Where $v_m$ is the mechanical speed, $M$ is the mass, $B$ is the friction coefficient, $F_L$ is the external force, and $P_o$ is the pole number of the motor. The dynamic mechanical position-movement equation is

$$py_m = v_m \qquad (8)$$

Where $y_m$ is the mechanical position of the motor. Substituting equation (5) in equation (6) then

$$pv_m = \frac{1}{M}\left(K_t i_q - Bv_m - F_L\right) \qquad (9)$$

Apply Laplace transform on both sides of equation (9)

$$Msv_m = K_t i_q - Bv_m - F_L \qquad (10)$$

$$(Ms + B)v_m = K_t i_q - F_L \qquad (11)$$

## III. CONTROLLER DESIGN

The proportional-integral (PI) controller has been widely used in industry for a long time due to its simplicity and reliability. Unfortunately, using a fixed PI controller, it is difficult to obtain both a good transient response and a good load disturbance rejection capability. To solve the difficulty, an adaptive controller has been proposed and discussed. The details are discussed as follows.

### A. Backstepping Adaptive Controller Design

From equation (9)

$$\frac{d}{dt}v_m = \dot{v}_m = A_1 i_q + A_2 F_L + A_3 v_m \qquad (12)$$

Where $A_1 = K_t/M$, $A_2 = -1/M$ $A_3 = -B/M$ are the parameters of the system. In the real world, the motor parameters cannot be precisely measured. In addition, these parameters are varied while the motor is saturated or its temperature is increased. As a result, the real parameters of the system include the nominal parameters and their variations. Then, eq (12) can be rewritten as $\dot{v}_m = A_1 i_q + A_2 F_L + A_3 v_m + \Delta A_1 i_q + \Delta A_2 F_L + \Delta A_3 v_m$

$$= A_1 i_q + (A_2 F_L + \Delta A_1 i_q + \Delta A_2 F_L + \Delta A_3 v_m) + A_3 v_m$$

$$= A_1 i_q + d_1 + A_3 v_m \qquad (13)$$

Where $d_1$ is the amount of the parameter variations and external load, and it can be expressed as

$$d_1 = A_2 F_L + \Delta A_1 i_q + \Delta A_2 F_L + \Delta A_3 v_m \qquad (14)$$

The position error, the difference between the position command and the real position, is expressed as

$$e_1 = y_m^* - y_m \qquad (15)$$

From eq (15), we can take the derivation of the position error and express the result as

$$\dot{e}_1 = \dot{y}_m^* - \dot{y}_m \qquad (16)$$

We define the velocity command of the PMLSM as

$$v_m^* = \dot{y}_m^* + De_1 + F\int e_1 dt$$

$$= \dot{y}_m^* + De_1 + Fx_1 \qquad (17)$$

Where $e_1$ is the position error. $D$ is the proportional controller gain, $F$ is the integral gain, and $x_1$ is the integral result of $e_1$. By taking the differential of eq (17), we can obtain

$$\dot{v}_m^* = \ddot{y}_m^* + D\dot{e}_1 + Fe_1 \qquad (18)$$

After that, we can define the speed error of the motor as

$$e_2 = v_m^* - v_m \qquad (19)$$

By taking the differential of eq (19) and substituting eq (15) and eq (13) into it, we can easily obtain

$$\dot{e}_2 = \ddot{y}_m^* + D\dot{e}_1 + Fe_1 - A_1 i_q - d_1 - A_3 v_m \qquad (20)$$

Combining $py_m = v_m$, eq (16), and eq (19), we can derive

$$\dot{e}_1 = \dot{y}_m^* - \dot{y}_m$$

$$= \dot{y}_m^* - v_m$$

$$= \dot{y}_m^* - v_m^* + e_2 \qquad (21)$$

By substituting eq (15) into eq (21), we can easily obtain

$$\dot{e}_1 = -De_1 - Fx_1 + e_2 \qquad (22)$$

Then, we can define a Lyapunov function as

$$V = \frac{1}{2}e_1^2 + \frac{1}{2}e_2^2 + \frac{1}{2}\frac{1}{\gamma}\tilde{d}1^2 + \frac{1}{2}Fx_1^2 \qquad (23)$$

Where $V$ is the Lyapunov function, $e_1$ is the position error, $e_2$ is the speed error, $\gamma$ is the positive weighting factor, and $\tilde{d}_1$ is the error between the real uncertainty($d_1$) and the estimated uncertainty($\hat{d}_1$). The estimated error $\tilde{d}_1$ is defined as

$$\tilde{d} = d_1 - \hat{d}_1 \qquad (24)$$

From eq (23), we can obtain the derivation of the Lyapunov function as

$$\dot{V} = e_1\dot{e}_1 + e_1\dot{e}_2 + \frac{1}{\gamma}\tilde{d}_1\dot{\tilde{d}}_1 + Fx_1e_1 \qquad (25)$$

By combining eq (20), eq (22), eq (24), and eq (25), we can obtain

$$\dot{V} = -De_1^2 + e_2[(1+F)e_1 + D\dot{e}_1 + \ddot{y}_m^* - A_1 i_q - \hat{d}_1 - \tilde{d}_1 - A_3 V_m] - \frac{1}{\gamma}\tilde{d}_1\dot{\hat{d}}_1 \qquad (26)$$

In this work, to simply the control algorithm, we assume the real current can track the current command well. Then, we set the current command as the control input, and it is expressed as $u = i_q^* = i_q$

$$= \frac{1}{A_1}[(1 + F)e_1 + D\dot{e}_1 + \ddot{y}_m^* - \hat{d}_1 - A_3 v_m + Ge_2] \quad (27)$$

Where $i_q^*$ is the current command amplitude, and $G$ is a positive gain. Substituting eq (6.18) into eq (6.17), we obtain

$$\dot{V} = -De_1^2 - Ge_2^2 - \frac{1}{\gamma}\tilde{d}_1 \dot{\hat{d}}_1 - \tilde{d}_1 e_2 \quad (28)$$

By designing an adaptive law $\dot{\hat{d}}_1 = -\gamma e_2$ we an remove the last two terms. Equation eq (28), therefore, can be reduced as

$$\dot{V} = -De_1^2 - Ge_2^2 \le 0. \quad (29)$$

In order to obtain the convergence of the drive system, the Barbalat's lemma (10) is applied here. First, the function $Z(t)$ is defined as follows

$$Z(t) = -De_1^2 - Ge_2^2. \quad (30)$$

By integrating both sides of eq (6.21), we can obtain

$$\int_0^\infty Z(\tau)d\tau = V(e_1(0),(0),\tilde{d}_1(0)) - V(e_1(\infty),e_2(\infty)\tilde{d}_1(\infty)). \quad (31)$$

The values of $e_1(0)$, $e_2(0)$, and $\tilde{d}_1(0)$ are bounded. In addition, the $\dot{V} \le 0$, therefore, the values of $e_1(\infty)$, $e_2(\infty)$, and $\tilde{d}_1(\infty)$ are also bounded. Then, from eq (31), we can observe that the integration of $Z(t)$ is bounded. From eq (32), it is easy to understand that the integration of the position error square $e_1^2(t)$ and the integration of the speed error square $e_2^2(t)$ are bounded as well. After deriving that the $e_1^2(t)$ and $e_2^2(t)$ are uniformly continuous, one can use Barbalat's lemma[5] and obtain that

$$\lim_{t \to \infty} Z(t) = 0. \quad (32)$$

Equation eq (32) means while the time approaches infinite, the position error and the speed error converge to zero. As a result, the proposed PMLSM drive system is stable.

*B. Self-Tuning Adaptive Controller Design*
Equation (12) can be easily rewritten as

$$\overline{M}\dot{v}_m = -\overline{B}v_m + i_q - \overline{F}_L \quad (33)$$

The parameters of eq (33) can be defined as follows

$$\overline{M} = \frac{M}{K_t} \quad , \overline{B} = \frac{B}{K_t} \quad , \overline{F}_L = \frac{F_L}{K_t}$$

Then, by substituting $py_m = v_m$ into eq (33), we can obtain

$$\overline{M}\ddot{y}_m = -\overline{B}\dot{y}_m + i_q - \overline{F}_L \quad (34)$$

In this work, we select the variable W as

$$W = \lambda_1 e_1 + e_2 = \lambda_1 e_1 + \dot{e}_1 \quad (35)$$

By taking the differential of eq (35), we can obtain

$$\dot{W} = \lambda_1 \dot{e}_1 + \dot{e}_2 \quad (36)$$

Then, by combining eq (16), eq (19), eq (34), and eq (36), we can obtain

$$\overline{M}\dot{W} = \overline{M}[\lambda_1(\dot{y}_m^* - \dot{y}_m) + (\dot{v}_m^* - \dot{v}_m)]$$
$$= -i_q + \overline{M}(\lambda_1 e_2 + \dot{v}_m^*) + \overline{B}v_m + \overline{F}_L \quad (37)$$

In the real world, however, the $M$, $B$, and $F_L$ cannot be precisely measured. The estimated parameters to replace the real parameters of the motor are used here. We select the estimating parameter vector as

$$\hat{\theta}^T = [\hat{\overline{M}} \ \hat{\overline{B}} \ \hat{\overline{F}}_L]. \quad (38)$$

Then, we define the state vector $\overline{Y}$ as

$$\overline{Y} = [\lambda_1 e_2 + \dot{v}_m^* \ v_m \ 1]^T \quad (39)$$

By substituting eq (38) and eq (39) into eq (37), we can obtain

$$\overline{M}\dot{W} = -i_q + \theta^T \overline{Y} \quad (40)$$

We select the control input $i_q$ as

$$i_q = \hat{\theta}^T \overline{Y} + \lambda_2 W \quad (41)$$

Substituting eq (41) into eq (40), it is not difficult to obtain

$$\overline{M}\dot{W} = \phi^T \overline{Y} - \lambda_2 W \quad (42)$$

And

$$\phi^T = \theta^T - \hat{\theta}^T$$
$$= [\overline{M} - \hat{\overline{M}} \ \overline{B} - \hat{\overline{B}} \ \overline{F}_L - \hat{\overline{F}}_L] \quad (43)$$

We select the Lyapunov function V as

$$V = \frac{1}{2}\gamma_1 \overline{M}W^2 + \frac{1}{2}\phi^T \phi. \quad (44)$$

By taking the differential of equation eq (44), we can obtain

$$\dot{V} = \gamma_1 W \overline{M}\dot{W} + \phi^T \dot{\phi}. \quad (45)$$

Substituting eq (42) into eq (45), we can obtain

$$\dot{V} = \phi^T(\gamma_1 W\overline{Y} + \dot{\phi}) - \gamma_1 \lambda_2 W^2 \quad (46)$$

We select the adaptive law a

$$\dot{\phi} = -\gamma_1 W\overline{Y} \quad (47)$$

Substituting eq (47) into eq (46), we can obtain

$$\dot{V} = -\gamma_1 \lambda_2 W^2 \quad (48)$$

From (42), we can derive

$$\int_0^\infty W^2 d\tau = -\frac{1}{\gamma_1 \lambda_2}\int_0^\infty \dot{V}d\tau$$
$$= \frac{1}{\gamma_1 \lambda_2}[V(0) - V(\infty)]. \quad (49)$$

The $W$ is bounded because $V(0)$ and $V(\infty)$ are bounded. In addition, the $W$ is uniformly continuous. According to Barbalat's lemma[10] we can obtain

227

$$\lim W(t) = 0.$$
$$t \to \infty$$
(50)

Then, from eq (6.27), we can obtain
$$\lim e_1(t) = 0.$$
$$t \to \infty$$
(51)

Equation (51) shows that using a self-tuning controller and in steady-state condition, the position error of the PMLSM is equal to zero.

## IV. RESULT

Some measured waveforms are provided here to validate the theoretical analysis. Fig. 1(a), 6(a) & 11(a) shows the position response and Fig.1(b) 6(b) & (b) relative speed response of system with PID controller, with backstepping controller and with self-tuning controller respectively to square wave(10cm) input position command at 10 Nt load. Fig. 2(a), 7(a) &12(a) shows the position response and 2(b), 7(b) &12(b) relative speed response with PID controller, with backstepping controller, with self-tuning controller respectively to square wave(10cm) input position command, when the load is changed to 20 Nt from 10 Nt load. Fig. 3(a), 8(a) &13(a) shows the position response and 3(b), 8(b) &13(b) relative speed response with PID controller, with backstepping controller and with self-tuning controller respectively to sinewave(5cm) input position command at 10 Nt load.Fig.4(a), 9(a) &14(a) shows the position response and 4(b), 9(b)&14(b) relative speed respons with PID controller, with backstepping controller and with self-tuning controller respectively to triangular wave(10cm) input position command at 10 Nt load. Fig. 5(a), 10(a)&15(a) shows the position response and 5(b), 10(b)&15(b) relative speed respone with PID controller, with backstepping controller and with self-tuning controller respectively to step wave(1cm) input position command at 10 Nt load. The proposed controllers have good load rejection capability. Self-tuning controller and backstepping controller perform better than PID controller. In order to evaluate the dynamic performance, the different performance indices, which are the integration of the square of the position error, are shown in Table I

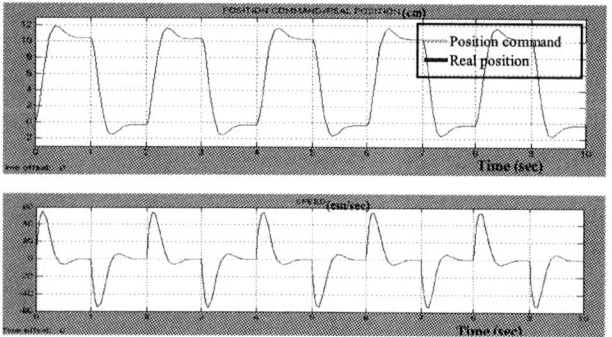

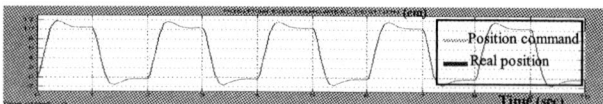

Fig. 1. Measured square-wave (10cm) response with 10Nt load
(a) Position command/Real position (b) Speed.

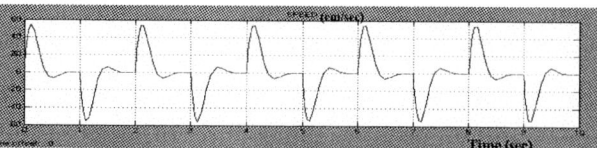

Fig. 2. Measured square-wave (10cm) response with 20Nt load
(a) Position command/Real position (b) Speed.

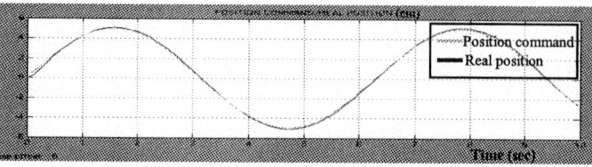

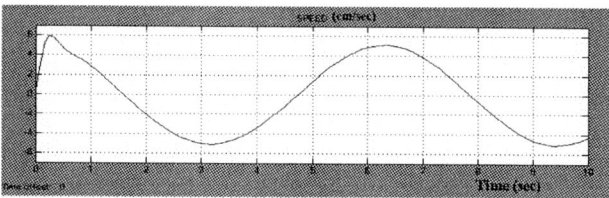

Fig. 3. Measured sine wave (5cm) response with a 10Nt load
(a) Position command/Real position (b) Speed.

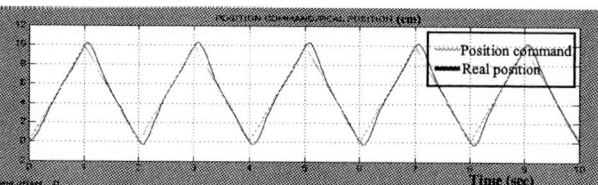

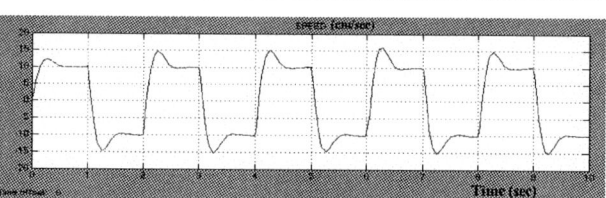

Fig. 4. Measured triangular-wave (10cm) response with a 10Nt load
(a) Position command/Real position (b) Speed.

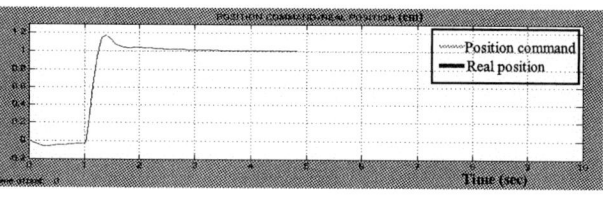

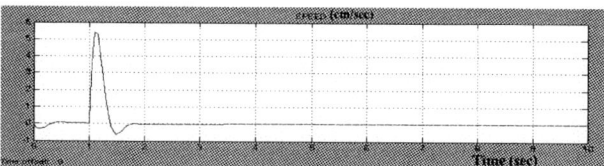

Fig. 5. Measured step (1cm) response with a 10Nt load
(a) Position command/Real position (b) Speed.

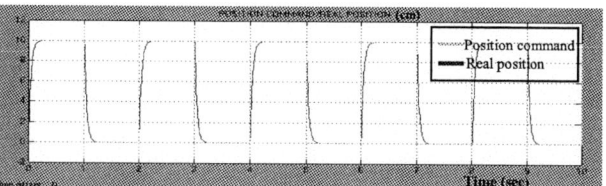

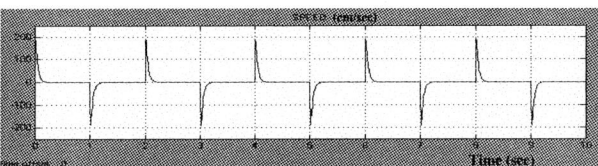

Fig. 6. Measured square-wave (10cm) response with 10Nt load
(a) Position command/Real position (b) Speed.

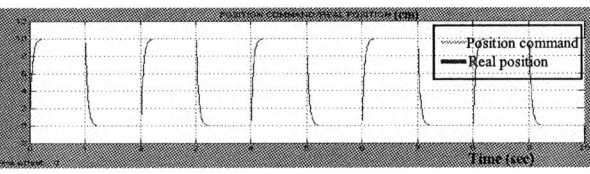

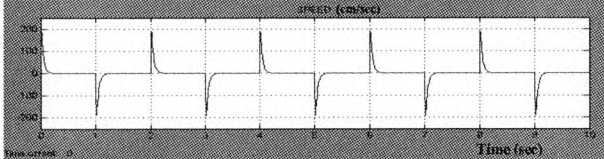

Fig. 7. Measured square-wave (10cm) response with a 20Nt load
(a) Position command/Real position (b) Speed.

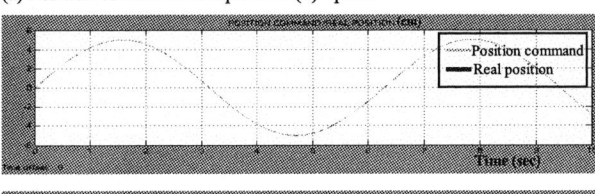

Fig. 8. Measured sine wave (5cm) response with a 10Nt load
(a) Position command/Real position (b) Speed.

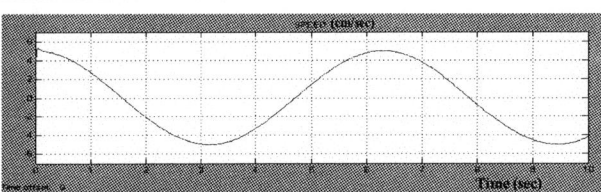

Fig. 9. Measured triangular-wave (10cm) response with a 10Nt load
(a) Position command/Real position (b) Speed.

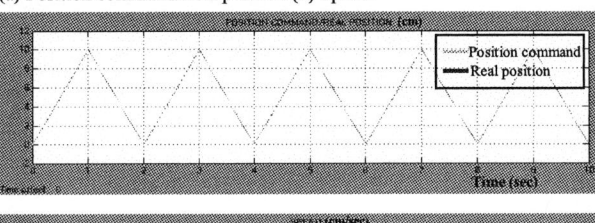

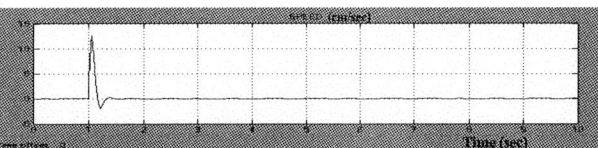

Fig. 10. Measured step (1cm) response with a 10Nt load
(a) Position command/Real position (b) Speed.

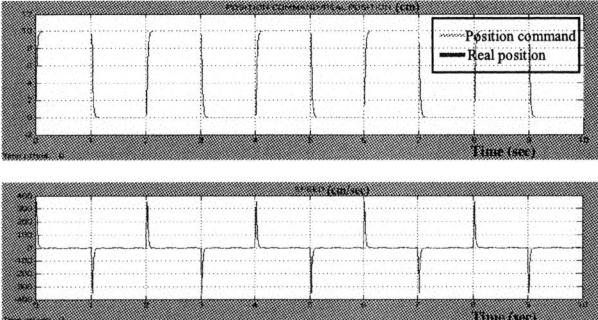

Fig. 11. Measured square-wave (10cm) response with 10Nt load
(a) Position command/Real position (b) Speed.

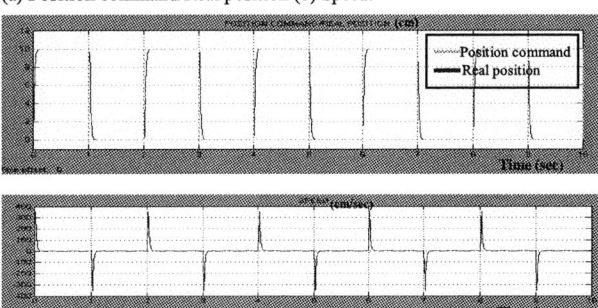

Fig. 12. Measured square-wave (10cm) response with a 20Nt load
(a) Position command/Real position (b) Speed.

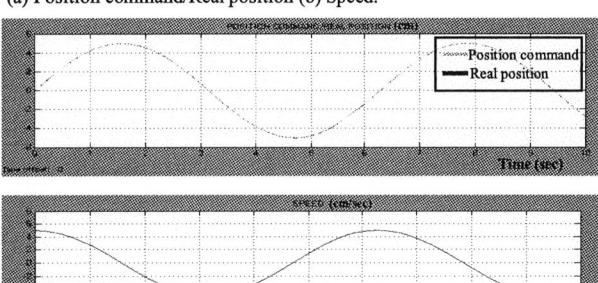

Fig. 13. Measured sine wave (5cm) response with a 10Nt load
(a) Position command/Real position (b) Speed.

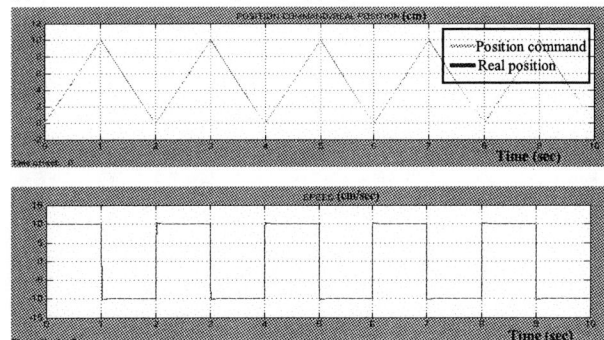

Fig. 14. Measured triangular-wave (10cm) response with a 10Nt load
(a) Position command/Real position (b) Speed.

229

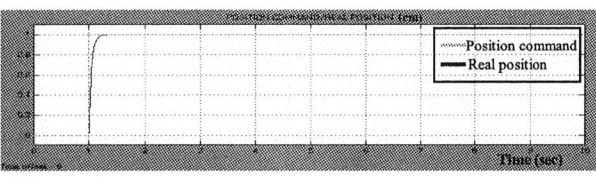

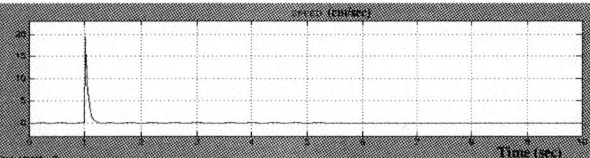

Fig. 15. Measured step (1c.m) response with a 10Nt load
(a) Position command/Real position (b) Speed.

TABLE I

INTEGRATION OF SQUARE OF POSITION ERROR (CM² .SEC)
FOR DIFFERENT CONTROLLERS TO DIFFERENT INPUT POSITION
COMMAND

| Position Command | PID Controller | Back Stepping | Self Tuning |
|---|---|---|---|
| Square 10 cm at 10Nt load | 0.0629 | 0.00406 | 0.001976 |
| Sinusoidal 5cm at 10Nt load | 0.004289 | 0.0009544 | 0.00079 |
| Triangular 10cm at 10Nt load | 0.001645 | 0.0003955 | 0.0002567 |
| Square 10cm at 20Nt load | 0.4041 | 0.01526 | 0.005383 |
| Step 1cm at 10Nt load | 0.1188 | 0.01377 | 0.01041 |

## V.
## VI. CONCLUSIONS

In this thesis, two different adaptive control techniques have been applied to a PMLSM. The systematic design procedures have been discussed. By using the proposed controllers, the PMLSM control system has satisfactory performance in transient response, load disturbance rejection capability, and tracking ability compared to conventional PID controller. Several varying commands like square, step, sinusoidal and triangular signals are used to evaluate the proposed PMLSM system.

## VII.   REFERENCES

[1]  Gerco Otten, Theo J. A. de Vries, , Job van erongen, Adrian M. Rankers, and Erik W. Gaal, Linear Motor Motion Control Using a Learning Feed forward Controller IEEE/ASME transactions on mechatronics, vol. 2, no. 3, September 1997.

[2]  Haojian  Xu and Petros A. Ioannou, Robust Adaptive Control for a Class of MIMO Nonlinear Systems with Guaranteed Error Bounds. IEEE transactions on automatic control, vol. 48, no. 5, may 2003.

[3]  K. K. Tan, S. N. Huang, and T. H.  LeeRobust Adaptive Numerical Compensation for Friction and Force Ripple in Permanent-Magnet Linear Motors.IEEE transactions on magnetics, vol. 38, no. 1, january 2002 .

[4]  Kinjiro Yoshida, Hiroshi Takami, Xiaoming Kong, and Akihiro Sonoda. Mass Reduction and Propulsion Control for a Permanent-Magnet Linear Synchronous Motor Vehicle. IEEE transactions on industry applications, vol. 37, no. 1, january/february 2001.

[5]  Miroslav Krstic, Ashish Krupadanam, and Clas Jacobson, Self-Tuning Control of a Nonlinear Model of Combustion Instabilities .IEEE transactions on control systems technology, vol. 7, no. 4, july 1999.

[6]  Li Xu and Bin Yao, Adaptive Robust Precision Motion Control of Linear Motors With Negligible Electrical Dynamics: Theory and Experiments IEEE/ASME transactions on mechatronics, vol. 6, no. 4, December 2001.

[7]  Karl Johan Astrom and Bjorn Wittenmark, "Adaptive Control", *Second Edition Lund Institute of Technology,* Pearson Education,

[8]  Petros A.Ioannou and Jing Sun, "Robust Adaptive Control", Prentice Hall, Englewood Cliffs, Elsevier Science, 2002

[9]   Digital control and State Variable Methods by *M.Gopal, Tata McGraw-Hill Publishing Company Limited*

[10]  I.J.Nagrath, M.Gopal, *Control Systems Engineering,Third Edition, New Age International Publishers.*

[11]  M.V. Shutov, E.E.  Sandoz , D.L. Howard , T.C. Hsia , R.L. Smith , S.D. Collins. A micro fabricated electromagnetic linear synchronous motor.Avilable on line at  www.sciencedirect.com.

[12]  Comparison of Linear Synchronous and Induction Motors, Urban Maglev Technology Development Program

[13]  Colorado Maglev Project. Report Number: FTA-DC-26-7002.2004.01 SAND2004-2734P.

## VIII.   BIOGRAPHIES

**B.Srinivasu** received B.E. degree from Andhra University, Visakhapatnam in 2002 and received M.E. from Osmania University, Hyderabad in 2006. During 2003-2004, he was section engineer in Indian railways. During 2004-2005 he was trainer in ACE Engineering Acadamy, Hyderabad. Presently he is working as Telecom Officer in BSNL, Inida. His area of interests are computer simulation of power electronic drives and electric machines.

**P.V.N.Prasad** was born in Hyderabad, India in January 1960. He graduated in Electrical & Electronics Engineering from Jawaharlal Nehru Technological University, Hyderabad in 1983 and received M.E in Industrial Drives & Control from Osmania University, Hyderabad in 1986.

He served as faculty member in Kothagudem School of Mines during 1987 - 95 and at presently serving as Associate Professor in the Department of Electrical Engineering, Osmania University. He received his Ph.D in Electrical Engineering in 2002. His areas of interest are Computer Simulation of Electrical Machines & Power Electronic Drives and Reliability Engineering.

He is a member of Institution of Engineers (India) and Indian Society for Technical Education. He is recipient of Dr.Rajendra Prasad Memorial Prize, Institution of Engineers (India), 1993 - 94 for best paper. He has got over 30 publications in National and International Journals, Magazines, Conferences & Symposia and presented technical papers in Thailand and Italy.

**M.V. Ramana Rao** graduated in Electrical Engineering from National Institute of Technology (Formerly Regional Engineering College), Warangal in 1997 and received M.Tech in Power Electronics from Jawaharlal Nehru Technological University, Hyderabad in 2003.  He presently serving as Assistant Professor in the Department of Electrical Engineering, Osmania University, Hyderabad. His areas of interest are Computer Simulation of Electrical Machines & Power Electronic Drives.

**2006 IEEE International Conference on Power Electronic, Drives and Energy Systems**

# An Overmodulation Scheme for Vector Controlled Induction Motor Drives

S. Venugopal, and G. Narayanan

*Abstract*— This paper presents an algorithm for control of line side voltage of a voltage source inverter upto six-step mode. This is a modified version of an existing overmodulation algorithm. The modified algorithm maintains proportionality between the reference voltage and the output fundamental voltage, and also reduces the computational effort required for implementation, while resulting in a marginally higher harmonic distortion. An estimation method is proposed for calculation of lower order ripple current. This estimation method is applied to a sensorless vector controlled induction motor drive to improve the performance of the drive during overmodulation.

## I. INTRODUCTION

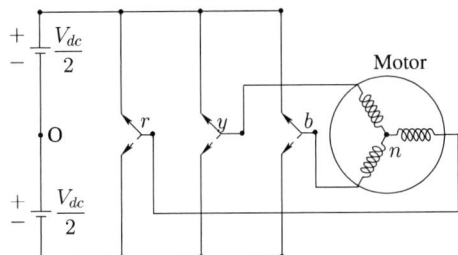

Fig. 1.  Two-level voltage source inverter (VSI).

**A** Two level voltage source inverter (VSI) shown in Fig. 1 is commonly used to feed an induction motor in a variable speed application. The pulse width modulated (PWM) inverter is usually operated in linear modulation zone, where the line side fundamental voltage is proportional to the reference voltage. The operation of the PWM inverter can be extended up to the six-step mode to obtain the highest possible line side voltage for a given dc bus voltage ($V_{dc}$) [1] [2]. This paper presents a modified version of an existing overmodulation algorithm [1] to control the inverter output voltage in the extended region or the overmodulation region. The modified algorithm maintains linearity between the reference voltage and the line side fundamental voltage. Also it reduces the computational effort required at the expense of a moderate increase in the harmonic distortion.

This work was supported by the Department of Science and Technology, Govt. of India, under the SERC Fast Track Scheme for Young Scientists.

First author S. Venugopal completed M.Sc(Engg) in the Department of Electrical Engineering, Indian Institute of Science and presently working in Meher Capacitors Private Limited, Bangalore, Email id: venugopals@gmail.com.

Second author G. Narayanan is currently an Assistant Professor in the Department of Electrical Engineering, Indian Institute of Science, Bangalore, Email id: gnar@ee.iisc.ernet.in

Overmodulation results in lower order ripple current in the output, which is also fed back to the current controller in vector controlled drives. These low frequency current components get amplified by the current error amplifier, resulting in oscillations in the drive.

The present work proposes an algorithm for real-time estimation of lower order current ripple. The measured motor current with the ripple subtracted is fed back to the current controller in a sensorless vector controlled drive. This is shown to reduce the oscillations and performance degradation in the drive during overmodulation. Simulation and experimental results are presented.

## II. EXISTING OVERMODULATION ALGORITHM

In space vector based PWM, the three-phase voltage reference is provided as a voltage reference vector $\mathbf{V_{ref}}$. During linear modulation, the average applied voltage vector over a subcycle $T_s$ equals the reference vector provided for the given subcycle. If $\mathbf{V_{ref}}$ falls in sector-I as shown in Fig. 2, the active vector $\mathbf{V_1}$, active vector $\mathbf{V_2}$ and the zero vector $\mathbf{V_z}$ of the inverter are applied for durations $T_1$, $T_2$ and $T_z$ respectively as shown in (1). The average voltage vectors applied during consecutive subcycles in sector-I are as shown in Fig. 3(a).

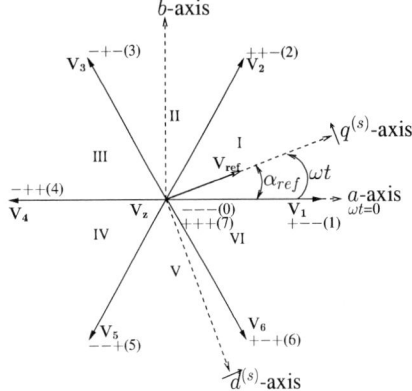

Fig. 2.  Voltage vectors produced by a two-level VSI and I, II, III, IV, V, VI are sectors.

$$\frac{T_1}{T_s} = \left[\frac{|\boldsymbol{V_{ref}}|}{V_{dc}}\right]\left[\frac{\sin\left(\frac{\pi}{3} - \alpha_{ref}\right)}{\sin\frac{\pi}{3}}\right]$$

$$\frac{T_2}{T_s} = \left[\frac{|\boldsymbol{V_{ref}}|}{V_{dc}}\right]\left[\frac{\sin\alpha_{ref}}{\sin\frac{\pi}{3}}\right]$$

$$\frac{T_z}{T_s} = 1 - \frac{T_1}{T_s} - \frac{T_2}{T_s} \qquad (1)$$

0-7803-9771-1/06/$25.00 ©2006 IEEE

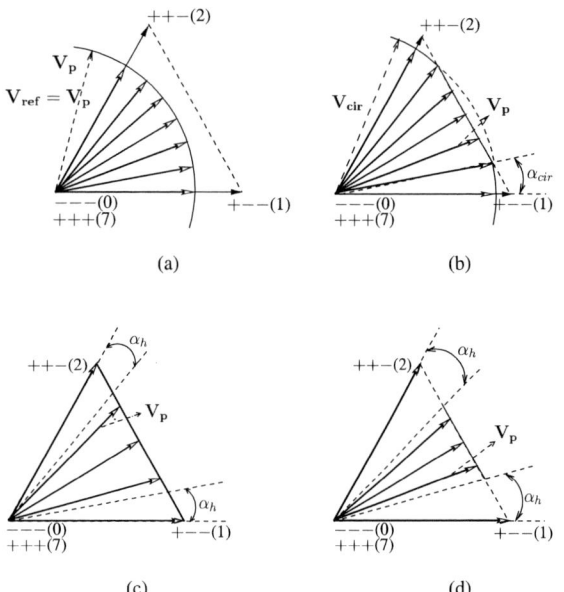

(a)　　　　　　　　　　(b)

(c)　　　　　　　　　　(d)

Fig. 3. Average voltage vector applied during different subcycles in sector-I in (a) linear modulation zone, (b) overmodulation zone-I, (c) overmodulation zone-II with existing algorithm and (d) overmodulation zone-II with proposed algorithm.

When the magnitude of the reference vector is greater than $0.866V_{dc}$, the inverter is said to be operating in the overmodulation zone. The overmodulation zone can be divided into zone-I and zone-II. In overmodulation zone-I, the average voltage vector applied in different subcycles in a sector is illustrated in Fig. 3(b). The trajectory of the tip of the average voltage vector is partly circular and partly hexagonal as shown. The radius of the circular trajectory is as given in (2).

$$V_{cir} = \left[ \frac{0.866V_{dc}}{\cos\left(\frac{\pi}{6} - \alpha_{cir}\right)} \right] \qquad (2)$$

The fundamental voltage can be increased by increasing $V_{cir}$ (or equivalently, decreasing $\alpha_{cir}$) till the trajectory becomes fully hexagonal. The angle and magnitude of the average voltage vectors applied are given by (3) and (4) respectively.

$$\alpha_p = \alpha_{ref} \quad \text{if } 0 \leq \alpha_{ref} \leq \left(\frac{\pi}{3}\right) \qquad (3)$$

$$|V_p| = \begin{cases} \dfrac{0.866V_{dc}}{\cos\left(\frac{\pi}{6} - \alpha_p\right)} & \text{if } \alpha_{cir} < \alpha_{ref} < \left(\frac{\pi}{3} - \alpha_{cir}\right) \\ V_{cir} & \text{otherwise} \end{cases}$$
$$\qquad (4)$$

In overmodulation zone-II, the average voltage vector in different subcycles in a sector are as shown in Fig. 3(c). The hold angle $\alpha_h$ is used to control the fundamental voltage. The angle and magnitude of the average voltage vector applied are as given in (5) and (6), respectively. The trajectory of the tip

of the average voltage vector is fully hexagonal in this zone.

$$\alpha_p = \begin{cases} 0 & \text{if } 0 \leq \alpha_{ref} \leq \alpha_h \\ \dfrac{\pi}{6}\left(\dfrac{\alpha_{ref} - \alpha_h}{\frac{\pi}{6} - \alpha_h}\right) & \text{if } \alpha_h < \alpha_{ref} < \left(\frac{\pi}{3} - \alpha_h\right) \\ \left(\dfrac{\pi}{3}\right) & \text{if } \left(\frac{\pi}{3} - \alpha_h\right) \leq \alpha_{ref} \leq \left(\frac{\pi}{3}\right) \end{cases}$$
$$\qquad (5)$$

$$|V_p| = \frac{0.866V_{dc}}{\cos\left(\frac{\pi}{6} - \alpha_p\right)} \quad \text{if } 0 \leq \alpha_{ref} \leq \left(\frac{\pi}{3}\right) \qquad (6)$$

## III. PROPOSED OVERMODULATION ALGORITHM

Calculation of $|V_p|$ as per (4) and (6) and that of $\alpha_p$ as given in (5) require division. The magnitude $|V_p|$ is a function of only one variable namely $\alpha_p$. The division operation required can be avoided by using a secant lookup table. However, $\alpha_p$ in (5) is a function of two variables, namely $\alpha_{ref}$ and $\alpha_h$. Hence evaluation of $\alpha_p$ is computationally involved.

The modified overmodulation algorithm is similar to the existing one in zone-I. However, in zone-II, the average voltage vectors applied over the different subcycles are as illustrated in Fig. 3(d). The angle $\alpha_p$ of the average vector is given in (7). The magnitude $|V_p|$ of the average vector is as in (4).

$$\alpha_p = \begin{cases} 0 & \text{if } 0 \leq \alpha_{ref} \leq \alpha_h \\ \alpha_{ref} & \text{if } \alpha_h < \alpha_{ref} < \left(\frac{\pi}{3} - \alpha_h\right) \\ \left(\dfrac{\pi}{3}\right) & \text{if } \left(\frac{\pi}{3} - \alpha_h\right) \leq \alpha_{ref} \leq \left(\frac{\pi}{3}\right) \end{cases}$$
$$\qquad (7)$$

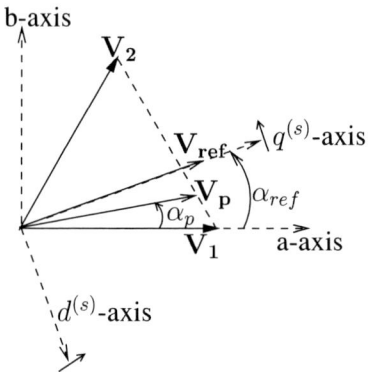

Fig. 4. Rotating $d^{(s)}$-$q^{(s)}$ frame of reference.

The average applied voltage vector $\mathbf{V_p}$ can be resolved along the $d^{(s)}$ and $q^{(s)}$ axes, which are the reference axes of a synchronously revolving reference frame illustrated in Fig. 4. The $q^{(s)}$-axis is aligned with $\mathbf{V_{ref}}$ as shown. The average $d^{(s)}$-axis and $q^{(s)}$-axis applied voltages are given by (8).

$$\begin{aligned} V_{p,qs} &= |V_p| \cos\left(\alpha_{ref} - \alpha_p\right) \\ V_{p,ds} &= |V_p| \sin\left(\alpha_{ref} - \alpha_p\right) \end{aligned} \qquad (8)$$

Fig. 5 illustrates the variation in $V_{p,qs}$ and $V_{p,ds}$ over a sector. In overmodulation zone-I, $V_{p,qs}$ and $V_{p,ds}$ are as shown in Fig. 5. With the proposed algorithm, both $V_{p,qs}$ and $V_{p,ds}$ have discontinuities at $\alpha_{ref} = \alpha_h$ and $\left(\frac{\pi}{3} - \alpha_h\right)$ in zone-II

232

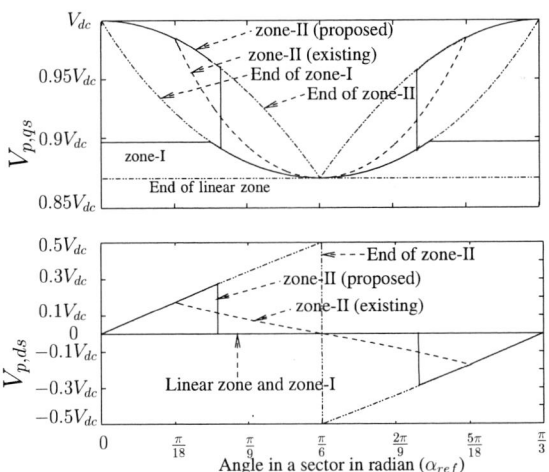

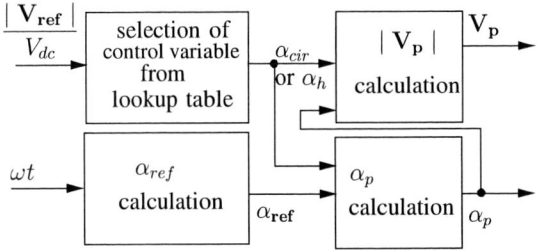

Fig. 7. Premodulation of reference vector.

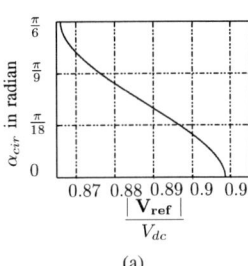 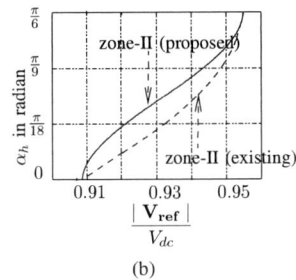

Fig. 5. Variation of average $q^{(s)}$-axis and $d^{(s)}$ axis voltages in overmodulation zone-I for $|\boldsymbol{V_{ref}}| = 0.885V_{dc}$ and in zone-II for $|\boldsymbol{V_{ref}}| = 0.933V_{dc}$.

as shown in the figure. The variations in $V_{p,qs}$ and $V_{p,ds}$ for the existing algorithm are as shown in dashed lines.

$$V_{p,qs(AV)} = \frac{3}{\pi}\left[\int_0^{\frac{\pi}{3}}(V_{qs})d\alpha_{ref}\right] = |\boldsymbol{V_{ref}}|$$

$$V_{p,ds(AV)} = \frac{3}{\pi}\left[\int_0^{\frac{\pi}{3}}(V_{ds})d\alpha_{ref}\right] = 0 \qquad (9)$$

Now, to maintain linearity, the average $q^{(s)}$-axis voltage $(V_{p,qs(AV)})$ should be equal to the magnitude of the reference voltage vector as given by (9). In zone-I of overmodulation, for given $|\boldsymbol{V_{ref}}|$ appropriate value of $\alpha_{cir}$ is selected from Fig. 6(a) to achieve linearity. In zone-II of overmodulation, appropriate value of $\alpha_h$ is chosen from Fig. 6(b). The average $d^{(s)}$ voltage is always equal to zero as shown by (9).

Fig. 6. (a) Relationship between $\alpha_{cir}$ and fundamental line side voltage in zone-I, (b) Relationship between $\alpha_{cir}$ and fundamental line side voltage in zone-II.

The premodulation algorithm is illustrated in Fig. 7. From $|\boldsymbol{V_{ref}}|$ the value of control variable $\alpha_{cir}$ or $\alpha_h$ is chosen. The angle $\alpha_p$ and the magnitude $|\boldsymbol{V_p}|$ are then calculated as shown. Instead of $|\boldsymbol{V_{ref}}|$ and $\alpha_{ref}$, $|\boldsymbol{V_p}|$ and $\alpha_p$ are used in (1) to calculate $T_1$, $T_2$ and $T_z$.

## IV. LOWER ORDER RMS VOLTAGE RIPPLE

In overmodulation zone, the average voltage vector realized using switching states is not the same as the reference voltage vector in an arbitrary subcycle. The average error voltage vector in a subcycle is given by (10). This is of zero magnitude during linear modulation.

$$\tilde{\mathbf{V}} = \mathbf{V_p} - \mathbf{V_{ref}} \qquad (10)$$

The average error voltage vector can be resolved along $d^{(s)}$ and $q^{(s)}$-axes as shown in (11).

$$\begin{aligned} \tilde{V}_{qs} &= V_{p,qs} - |\mathbf{V_{ref}}| \\ \tilde{V}_{ds} &= V_{p,ds} \end{aligned} \qquad (11)$$

The expressions for the RMS value of the $q^{(s)}$-axis voltage ripple and $d^{(s)}$-axis voltage ripple are given by (12). The total RMS ripple is defined as in (13).

$$\tilde{V}_{qs,RMS} = \sqrt{\frac{3}{\pi}\left[\int_0^{\frac{\pi}{3}}\left(\tilde{V}_{qs}\right)^2 d\alpha_{ref}\right]}$$

$$\tilde{V}_{ds,RMS} = \sqrt{\frac{3}{\pi}\left[\int_0^{\frac{\pi}{3}}\left(\tilde{V}_{ds}\right)^2 d\alpha_{ref}\right]} \qquad (12)$$

$$\tilde{V}_{RMS} = \sqrt{\tilde{V}_{qs,RMS}^2 + \tilde{V}_{ds,RMS}^2} \qquad (13)$$

The RMS value of lower order voltage ripple for both the algorithms in zone-II are obtained and compared in Fig. 8. The proposed overmodulation algorithm leads to a marginal increase in $\tilde{V}_{RMS}$ in zone-II. The worst case increase is 6.5% of the $\tilde{V}_{RMS}$ corresponding to existing algorithm.

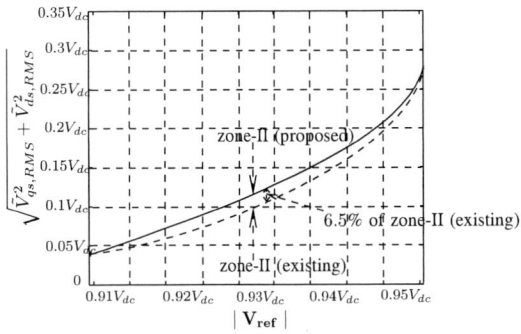

Fig. 8. RMS value of the lower order voltage ripple in overmodulation zone-II.

## V. ESTIMATION OF LOWER ORDER CURRENT RIPPLE

The lower order current ripple during overmodulation is estimated in rotating $d^{(s)}$-$q^{(s)}$ reference frame in [3]. In the present work, the same is estimated in stationary $a$-$b$ frame of reference. The average error voltages along $a$-axis and $b$-axis, namely $\tilde{V}_{sa}$ and $\tilde{V}_{sb}$, can be obtained using $\tilde{V}_{qs}$ and $\tilde{V}_{ds}$ as given in (14).

$$
\begin{aligned}
\tilde{V}_{sa} &= \tilde{V}_{ds}\sin\omega t + \tilde{V}_{qs}\cos\omega t \\
\tilde{V}_{sb} &= -\tilde{V}_{ds}\cos\omega t + \tilde{V}_{qs}\sin\omega t
\end{aligned}
\tag{14}
$$

The lower order ripple current in the motor due to the average error voltages $\tilde{V}_{sa}$ and $\tilde{V}_{sb}$ are as given by (15), where $\sigma_s L_o$ and $\sigma_r L_o$ are the stator and rotor leakage inductances, $R_s$ and $R_r$ are stator and rotor resistances (all quantities are referred to stator).

$$
\begin{aligned}
(\sigma_s + \sigma_r)L_o\frac{d\tilde{i}_{sa}}{dt} + (R_s + R_r)\tilde{i}_{sa} &= \tilde{V}_{sa} \\
(\sigma_s + \sigma_r)L_o\frac{d\tilde{i}_{sb}}{dt} + (R_s + R_r)\tilde{i}_{sb} &= \tilde{V}_{sb}
\end{aligned}
\tag{15}
$$

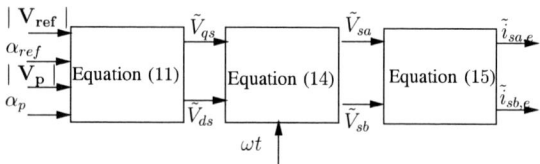

Fig. 9. Lower order current ripple estimation algorithm.

Now, given magnitude and angle of the $\mathbf{V_{ref}}$ and $\mathbf{V_p}$, $\tilde{V}_{qs}$ and $\tilde{V}_{ds}$ can be obtained using (11). Similarly $\tilde{V}_{sa}$ and $\tilde{V}_{sb}$ can be computed using (14). Further $\tilde{i}_{sa,e}$ and $\tilde{i}_{sb,e}$ are obtained by solving the differential equation (15). The letter 'e' in the subscript is to indicate that these quantities are estimated ones using the motor model. The estimation algorithm is illustrated in Fig. 9.

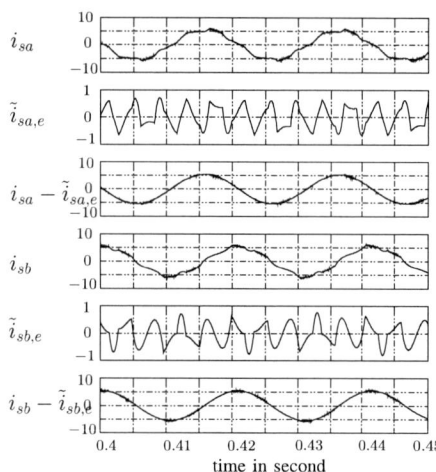

Fig. 10. Estimation of lower order current ripple for $\mid\mathbf{V_{ref}}\mid = 0.93V_{dc}$ and ($f = 48.5Hz$) with load torque equal to $3Nm$- simulation results.

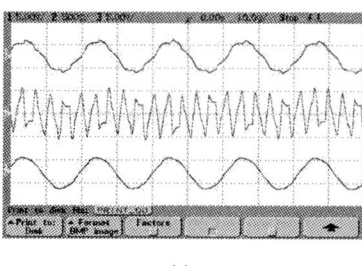

(a)

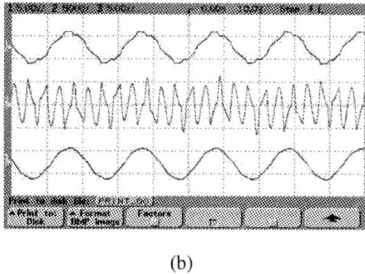

(b)

Fig. 11. Experimental results corresponding to Fig. 10. Scale: X-axis, 10ms/div Y-axis, channel 1: 7.35A/div, channel 2: 0.735A/div, channel 3: 7.35A/div.

The proposed ripple estimation algorithm with the modified overmodulation algorithm is simulated using MAT-LAB/SIMULINK. Fig. 10 presents the simulated waveforms of stator current and estimated current ripple along a-axis and b-axis in the stationary reference frame. Fig. 11 presents the corresponding experimental results on a 550W, 220V, 50Hz, three-phase induction motor fed from an IGBT inverter. Subtracting the estimated ripple from the stator current leads to a near sinusoidal waveform as shown.

## VI. VECTOR CONTROL OF INDUCTION MACHINE IN OVERMODULATION ZONE

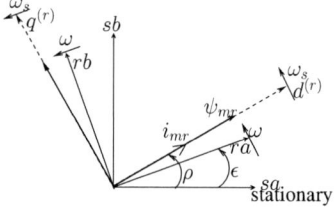

Fig. 12. Rotating $d^{(r)}$-$q^{(r)}$ reference frame.

Vector control [4] of the induction motor enables decoupled control of torque and flux in the machine. Conventionally, vector control of squirrel cage induction machine is done in the rotor flux reference frame, illustrated in Fig. 12.

234

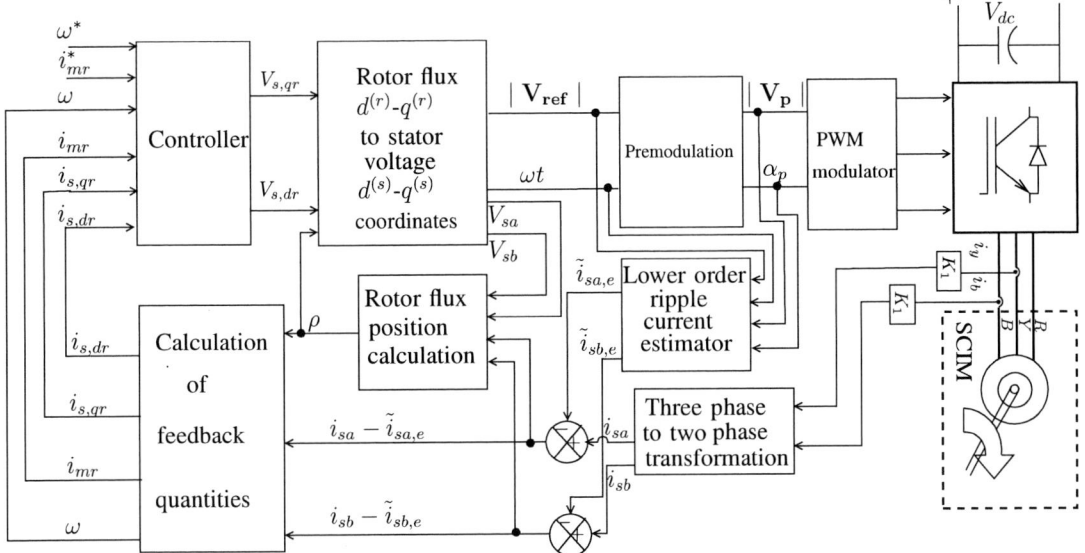

(a) Closed loop control of the induction machine with PWM modulator operating in both linear region and overmodulation region.

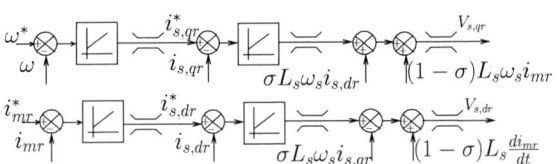

(b) Controller used in closed loop control of induction machine.

Fig. 13. Vector control scheme including overmodulation zone.

The $d^{(r)}$ and $q^{(r)}$-axes revolve at synchronous speed $\omega_s$. The $d^{(r)}$-axis is aligned with the rotor flux $\psi_{mr}$, which is related to $i_{mr}$ as shown in (16), where $L_o$ is the mutual inductance of the motor. The rotor axes $ra$ and $rb$ revolve at rotor speed $\omega$.

$$\psi_{mr} = L_o i_{mr} \qquad (16)$$

The equations governing the induction machine in rotor flux coordinates are given by (17) [4].

$$
\begin{aligned}
V_{s,dr} &= R_s i_{s,dr} + \sigma L_s \frac{di_{s,dr}}{dt} - \sigma L_s \omega_s i_{s,qs} + \\
&\quad (1-\sigma) L_s \frac{di_{mr}}{dt} \\
V_{s,qr} &= R_s i_{s,qr} + \sigma L_s \frac{di_{s,qr}}{dt} + \sigma L_s \omega_s i_{s,ds} + \\
&\quad (1-\sigma) L_s \omega_s i_{mr} \\
i_{s,dr} &= T_r \frac{di_{mr}}{dt} + i_{mr} \\
m_d &= \frac{p}{3} \frac{L_o}{(1+\sigma_r)} i_{s,qr} i_{mr} \\
J \frac{d\omega}{dt} &= m_d - m_l \\
\omega_s &= \frac{d\rho}{dt} = \omega + \frac{i_{s,qr}}{T_r i_{mr}}
\end{aligned}
\qquad (17)
$$

where $L_s$ is total inductance of the stator; $L_r$ is total inductance of the rotor; $\sigma = 1 - \frac{L_o^2}{L_s L_r}$, total leakage coefficient; $T_r = \frac{L_r}{R_r}$, rotor time constant; $J$ is the moment of inertia of the system; $V_{s,dr}, V_{s,qr}$ are stator voltages in the rotor flux coordinates; $i_{s,dr}, i_{s,qr}$ are stator currents in the rotor flux coordinates; $m_d$ is developed torque in the machine; $m_l$ is load torque on the machine.

The overall vector control scheme with the PWM modulator is shown in Fig. 13(a). The controller in this figure includes the controller for $\omega$, $i_{mr}$, $i_{s,dr}$ and $i_{s,qr}$. The details of the controller are as shown in Fig. 13(b) [4]. PI controllers are used for control of $i_{s,dr}$ and $i_{s,qr}$ as shown in Fig. 13(b). The feedback quantities are obtained through measurement of $V_{dc}$, $i_y$ and $i_b$ as shown in Fig. 13(a) and through transformations.

If the PWM modulator is allowed to operate in overmodulation region, the line voltage consists of low frequency components in addition to fundamental component and switching frequency harmonic components. These low frequency voltage harmonics cause corresponding low frequency harmonic currents like $5^{th}$, $7^{th}$ $11^{th}$, $13^{th}$ etc., to flow in the motor. The frequencies of these components are close to the bandwidth of the controller and some are even less than the bandwidth. In addition to the fundamental component, these low frequency components are also amplified by the current error amplifier. Hence the voltage reference produced by the current controller has considerable low frequency component, degrading the performance of the drive. This is illustrated by the simulation results presented in Fig. 14.

If a low pass filter is used to filter these low frequency components in the current, then the bandwidth of the current loop is reduced even during operation in the linear modulation zone. Instead, the lower order current ripple can be estimated as discussed in the previous section. The estimated ripple current is subtracted from the measured current before the latter is fed back to the current controller as shown in Fig. 13(a). The corresponding simulation results are shown in Fig.

235

15. Oscillation in $|V_{ref}|$ is now reduced. The distortion in motor current is also decreased considerably.

Simulation and experimental results demonstrating the dynamic performance of the drive are presented in Fig. 16 and Fig. 17 respectively.

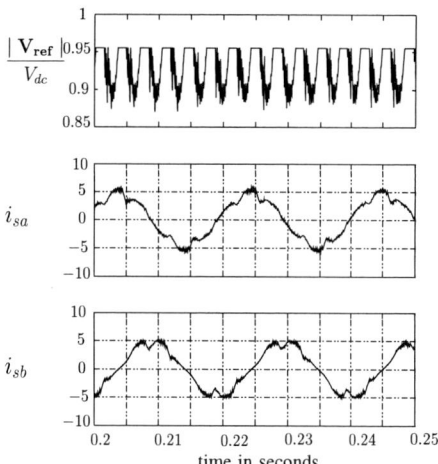

Fig. 14. Closed loop current control in overmodulation zone without ripple estimation with speed reference of $f = 48.5Hz$- simulation results.

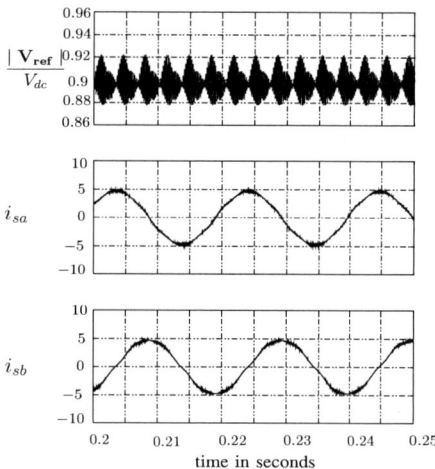

Fig. 15. Closed loop current control in overmodulation zone with ripple estimation with speed reference of $f = 48.5Hz$- simulation results.

## VII. CONCLUSION

A modified overmodulation algorithm is proposed to reduce the computational complexity in zone-II of the existing algorithm. With the modified overmodulation algorithm, the worst case increase in harmonic distortion is 6.5% over the existing algorithm. Further, an estimation method is proposed to estimate the lower order current ripple in overmodulation. The estimation algorithm is verified both through simulations and experiments. The estimation method developed is applied to a sensorless vector controlled drive and the performance of the drive is found to be improved. The experimental and simulation results of the same are presented.

Fig. 16. Dynamics in vector controlled drive operating in overmodulation zone- simulation results.

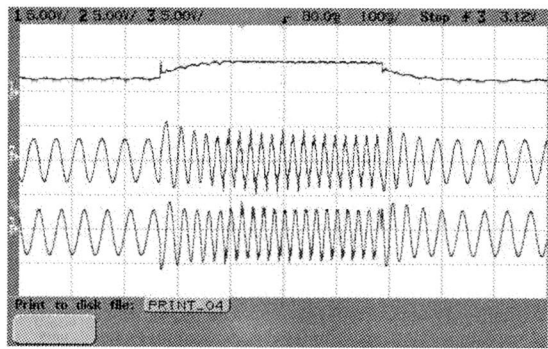

Fig. 17. Dynamics in vector controlled drive operating in overmodulation zone- Experimental results.

## VIII. ACKNOWLEDGMENT

This work was supported by the Department of Science and Technology, Govt. of India, under the SERC Fast Track Scheme for Young Scientists.

## REFERENCES

[1] J. Holtz, Wolfgang Lotzkat, and Ashwin M. Khambadkone, "On continuous control of PWM inverters in the overmodulation range including the six-step mode", *IEEE Transactions on Power Electronics*, Volume 8, 1993, Pages 546- 553.
[2] G. Narayanan and V. T. Ranganathan, "Extension of operation of space vector pwm strategies with low switching frequencies using different overmodulation algorithms", *IEEE Transactions on Power Electronics*, Volume 17, Issue 5, Sep 2002, Pages 788- 798.
[3] Ashwin M. Khambadkone and J. Holtz, "Compensated synchronous PI current controller in overmodulation range and six-step operation of spacevector modulation based vector controlled drives", *IEEE Transactions on Industrial Electronics*, Volume 49, June 2002, Pages 574- 580.
[4] Werner Leonhard, "Control of Electrical Drives", Springer International Edition.

**2006 IEEE International Conference on Power Electronic, Drives and Energy Systems**

# Modified Direct Torque Control of Matrix Converter Fed Induction Motor Drive

Bhim Singh, *Senior Member, IEEE,* and Jally Ravi

*Abstract*--**In this paper, a torque ripple minimization technique with constant switching frequency is proposed for Direct Torque Control (DTC) of a matrix converter fed squirrel cage induction motor drive. Some drawbacks of conventional DTC are the generation of relatively large torque ripple in a low speed region and a variation of switching frequency according to the amplitudes of torque and flux hysteresis bands. In this proposed strategy, an optimal switching instant during one switching cycle is calculated for torque ripple minimization, which is derived from torque ripple based on an instantaneous torque. The proposed strategy improves the performance of the DTC-IM drive by combining a low torque ripple characteristic in the steady state and the conventional fast torque dynamic characteristic. Simulation results prove the feasibility of proposed strategy compared with conventional method.**

*Index Terms*--**Matrix converter, MATLAB, direct torque controlled induction motor drives (DTCIMD).**

## I. INTRODUCTION

THE matrix converter offers an all silicon solution for AC-AC power conversion[1-5]. The circuit consists of an array of bi-directional switches arranged so that any of the output lines of the converter can be connected to any of the input lines. Fig. 1 shows a typical three-phase to three phase matrix converter, with nine bi-directional switches. The switches allow any input phase to be connected to any output phase. The output voltage waveform is then created using a suitable PWM modulation pattern similar to a normal inverter, except that the input is a three-phase supply instead of a fixed DC voltage. This approach removes the need for the large reactive energy storage components used in conventional inverter based converters. An input line filter is included to circulate the high frequency switching harmonics.

The direct torque control (DTC) technique for induction motors has been initially proposed as direct self-control, then the method is generalized to current- source-inverter (CSI) fed induction motors and to voltage source inverter (VSI)-fed and CSI-fed synchronous machines. The main advantages of DTC are robust and fast torque response, no requirements for coordinate transformation, no requirements for PWM pulse generation and current regulators.

Bhim Singh and Jally Ravi are with the Department of Electrical Engineering, Indian Institute Technology Delhi, Hauz Khas, New Delhi-110016, India. (E-mail: bhimsinghr@gmail.com and jallyravi@gmail.com).

In [6], a control scheme for matrix converter fed induction motor drives with DTC has been analyzed, but the rotor flux is assumed as reference, instead of stator flux, in order to achieve the highest pullout torque. Using a VSI, different vector selection criteria can be employed to control the torque and the flux leading to different switching strategies. Each strategy affects the drive behavior in terms of torque and current ripple, switching frequency, and two- or four-quadrant operation capability [7-8]. In [9], a speed-dependent switching strategy has been proposed in order to achieve fast torque response in a wide speed range.

For such reasons, the combination of the advantages of the matrix converter with those of the direct torque control method is effectively possible according to reference [1]. However, the drawbacks exist in the DTC method because the torque and the flux ripples always exceed their bandwidths. Many researchers have paid much attention to solve these problems in DTC methods applied to voltage source converter fed induction motors.

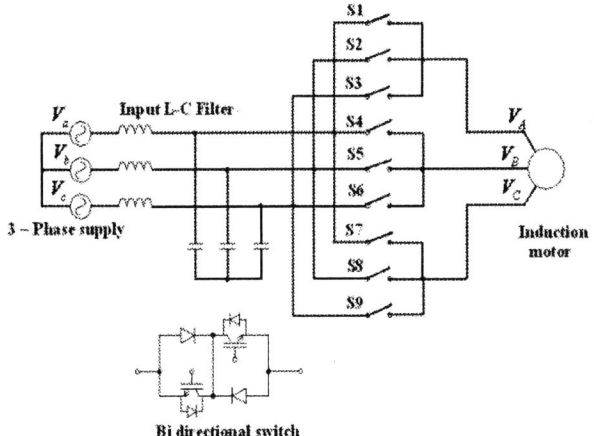

Fig.1. Matrix Converter switch layout.

In this paper, a modified DTC technique is used for matrix converter fed induction motor drive, which minimizes torque ripple with keeping constant switching frequency. Output voltage vector, $V_{sk}$ is selected using conventional DTC switching table, but pulse duration of $V_{sk}$ is derived from ripple minimization. The rms torque ripple during one switching period is obtained using instantaneous torque variation by means of the stator voltage vector, $V_{sk}$. Moreover, the optimal switching instant, $t_s$, is determined using rms torque ripples. Basic principle of the proposed method is

0-7803-9771-1/06/$25.00 ©2006 IEEE            237

demonstrated, control block diagram is illustrated, and simulation results are presented to verify the effectiveness of the algorithm.

## II. BASIC DIRECT TORQUE CONTROL THEORY

The basic principle in conventional DTC for induction motors is to directly select stator voltage vectors by means of a hysteresis stator flux and torque control. As it is shown in Fig.2, stator flux $\psi_s^*$ and torque $T_e^*$ reference values are compared with the corresponding estimated values. Both stator flux and torque errors, $E_\psi$ and $E_{Te}$, are processed by means of hysteresis band comparators. In particular, stator flux is controlled by a two-level hysteresis comparator, whereas the torque is controlled by a three-level comparator. On the basis of the hysteresis comparators and stator flux sector a proper VSI voltage vector is selected by means of the switching table given in Table I.

The reference flux vector and the hysteresis band tracks a circular trajectory as shown in Fig. 3 (a). Thus, the actual flux follows its reference within the hysteresis band in a zigzag path. Fig. 3 (b) shows the voltages vectors a VSI can generate and the corresponding flux variation in time $\Delta t$.

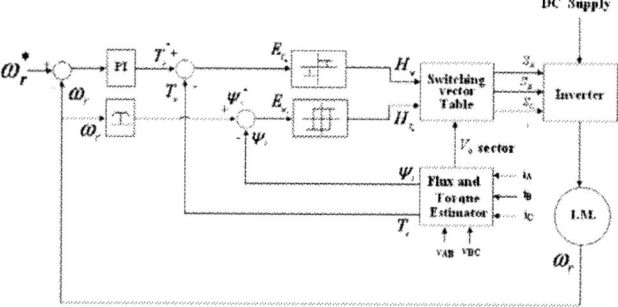

Fig.2. Block diagram of basic DTC.

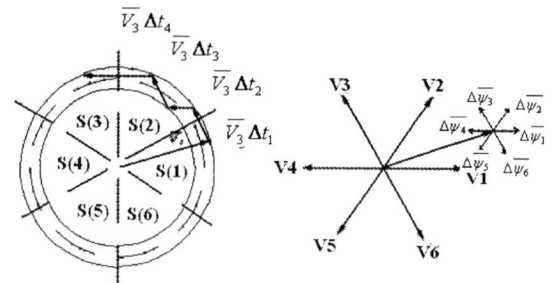

Fig.3. Flux trajectory and VSI voltage vectors and the corresponding flux increment.

The electromagnetic torque in the three-phase induction motor can be expressed as follows:

$$T_e = \frac{3}{2} P \overline{\psi}_s \times I_s \tag{1}$$

where $\psi_s$ is the stator flux, $I_s$ is the stator current and p is

the number of pair poles. Both, stator and rotor fluxes can be expressed in terms of stator and rotor currents as follows:

$$\overline{\psi}_s = L_s \overline{I}_s + L_m \overline{I}_r \tag{2}$$
$$\overline{\psi}_r = L_r \overline{I}_r + L_m \overline{I}_s \tag{3}$$

TABLE I
BASIC DTC SWITCHING TABLE

| $H_\psi$ | $H_{T_g}$ | S(1) | S(2) | S(3) | S(4) | S(5) | S(6) |
|---|---|---|---|---|---|---|---|
| 1 | 1 | V2 | V3 | V4 | V5 | V6 | V1 |
| | 0 | V0 | V7 | V0 | V7 | V0 | V7 |
| | -1 | V1 | V2 | V2 | V3 | V4 | V5 |
| -1 | 1 | V3 | V4 | V5 | V6 | V1 | V2 |
| | 0 | V7 | V0 | V7 | V0 | V7 | V0 |
| | -1 | V5 | V6 | V1 | V2 | V3 | V4 |

Eliminating $\overline{I}_r$ from (2), the following equation can be derived:

$$\overline{\psi}_s = \frac{L_m}{L_r} \overline{\psi}_r + L_s' \overline{I}_s \tag{4}$$

where $L_s' = L_s L_r - L_m^2$  Thus, stator current can be expressed as

$$\overline{I}_s = \frac{1}{L_s'} \left( \overline{\psi}_s - \frac{L_m}{L_r} \overline{\psi}_r \right)_s \tag{5}$$

By substituting (5) into (1), the following expressions for the torque can be derived:

$$\overline{T}_e = \frac{3P}{2} \frac{L_m}{L_r L_s'} \overline{\psi}_r \times \overline{\psi}_s \tag{6}$$

Thus, the torque modulus is:

$$\overline{T}_e = \frac{3P}{2} \frac{L_m}{L_r L_s'} |\overline{\psi}_r| |\overline{\psi}_s| \sin \gamma \tag{7}$$

where $\gamma$ is the angle between stator and rotor fluxes, Fig. 4 shows the phase diagram corresponding to (7).

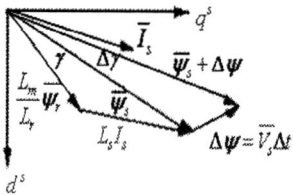

Fig. 4. Phasor diagram of stator flux, rotor flux and stator currents.

If the rotor flux remains constant and the stator one is changed incrementally by means of the stator voltage $V_s$, and the corresponding change of $\gamma$ is $\Delta\gamma$, the incremental torque $\Delta T_e$ can be expressed as follows:

$$\Delta T_e = \frac{3}{2} P \frac{L_m}{L_r L_s'} |\overline{\psi}_r| . |\overline{\psi}_s + \Delta \overline{\psi}_s| \sin \gamma \tag{8}$$

The estimator block calculates stator flux and torque feedback signals. This block uses the motor speed $\omega_m$, $i_a$ and

$i_b$ quantities and the motor dynamic model, in the stationary reference frame, to estimate the stator flux and torque values. The estimator block also calculates the sector $s(n)$ in which the flux vector $\psi_s$ lies. From the dynamic motor model in the stationary reference frame the following equations for both $d^s-q^s$ components of rotor flux are obtained:

$$\psi^s{}_{dr}(k)=\psi^s{}_{dr}(k-1).e^{-\frac{R_r}{L_r}T_z}+\frac{1-e^{-\frac{R_r}{L_r}T_z}}{R_r}$$
$$.\left(L_mR_ri^s{}_{ds}(k-1)-L_rp\psi^s{}_{qr}(k-1).\omega_m(k-1)\right) \tag{9}$$

$$\psi^s{}_{qr}(k)=\psi^s{}_{qr}(k-1).e^{-\frac{R_r}{L_r}T_z}+\frac{1-e^{-\frac{R_r}{L_r}T_z}}{R_r}$$
$$.\left(L_mR_ri^s{}_{qs}(k-1)-L_rp\psi^s{}_{dr}(k-1).\omega_m(k-1)\right) \tag{10}$$

where $R_r$ is the rotor resistance, $L_r$ is the rotor leakage inductance, $L_m$ is the magnetizing inductance, $i_{ds}$ and $i_{qs}$ are the d and q components of the stator currents respectively. The superscript s denotes quantities fixed to the stationary reference frame. $\omega_m$ is the motor speed and k is the actual sample instant.

The $d^s-q^s$ components of the stator flux can be obtained as follows:

$$\psi^s{}_{ds}(k)=i^s{}_{ds}(k).\frac{L_s}{L_m}+\psi^s{}_{dr}(k).\frac{L_m}{L_r} \tag{11}$$

$$\psi^s{}_{qs}(k)=i^s{}_{qs}(k).\frac{L_s}{L_m}+\psi^s{}_{qr}(k).\frac{L_m}{L_r} \tag{12}$$

Resolving the variables in (1) into $d^s-q^s$ components, the following expression for the torque is obtained:

$$T_e=\frac{3}{2}.P.\left(\psi^s{}_{ds}.i^s{}_{qs}-\psi^s{}_{qs}.i^s{}_{ds}\right) \tag{13}$$

### III. DTC USING MATRIX CONVERTER

Since a matrix converter MC generates a higher number of output voltage vectors with respect to a VSI, the introduction of a third variable, such as the average value of the sine of the displacement angle $\phi_i$ between the input line current vector and the input line to- neutral voltage vector, can be used to control the input power factor. This requirement is accomplished if this third variable is kept close to zero. Furthermore a new hysteresis controller, shown in Fig. 5, is introduced, which controls this variable. The flux and torque, and the average value of $\sin\phi_i$ estimation require the knowledge of voltages and currents at the input and output side of the MC. Nevertheless, only the input voltages and output currents are measured. The other quantities are calculated on the basis of the switching states of the matrix converter. The new control algorithm will select the switching state of the matrix converter that generates a voltage vector similar to that selected by the basic DTC control algorithm. Assuming that $V_l$ is the output voltage vector selected by the basic DTC control algorithm, from Fig.5 it is obvious that one

of the switching states $\pm 1$, $\pm 2$ and $\pm 3$ must be selected. Since the magnitude and direction of these voltage vectors depends on the input line-to-neutral voltage vector, only those having the same direction of $V_l$ and the maximum magnitude is taken into consideration. If the input line-to-neutral voltage vectors lies in sector 1, the switching states +1 and –3 can be chosen. From Fig. 5, it can be seen that this two switching states corresponds to input current vectors lying on the sector 1 boundary directions. If an increase in power factor is needed, that is, the average value of $\sin\phi_i$ has to be decreased, the switching state –3 has to be selected. On the other hand, if decrease in power factor needed, that is, the average value $\sin\phi_i$ of has to be increased, the switching state +1 has to be selected. Table II shows the switching table for matrix converter. In the first column, the voltage vectors selected by the basic DTC are present. The top row contains the sector in which the input line-to-neutral voltage vector lies. Depending on the power factor control needs, the output value of the hysteresis comparator $H_\phi$, one of the two sub-columns may be selected.

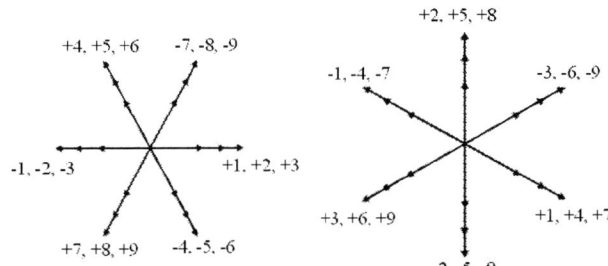

Fig.5. Space-vector hexagons for the active switch combinations.

TABLE II
DTC SWITCHING TABLE USING MATRIX CONVERTER

| $H_\phi$ | 1 | | 2 | | 3 | | 4 | | 5 | | 6 | |
|---|---|---|---|---|---|---|---|---|---|---|---|---|
| | +1 | -1 | +1 | -1 | +1 | -1 | +1 | -1 | +1 | -1 | +1 | -1 |
| V1 | -3 | 1 | 2 | -3 | -1 | 2 | 3 | -1 | -2 | 3 | 1 | -2 |
| V2 | 9 | -7 | -8 | 9 | 7 | -8 | -9 | 7 | 8 | -9 | -7 | 8 |
| V3 | -6 | 4 | 5 | -6 | -4 | 5 | 6 | -4 | -5 | 6 | 4 | -5 |
| V4 | 3 | -1 | -2 | 3 | 1 | -2 | -3 | 1 | 2 | -3 | -1 | 2 |
| V5 | -9 | 7 | 8 | -9 | -7 | 8 | 9 | -7 | -8 | 9 | 7 | -8 |
| V6 | 6 | -4 | -5 | 6 | 4 | -5 | -6 | 4 | 5 | -6 | -4 | 5 |

### IV. MODIFIED TORQUE RIPPLE MINIMIZATION STRATEGY

The induction motor can be modeled with stator and rotor fluxes as the state variables by the following equation.

$$\frac{d}{dt}\begin{bmatrix}\psi_s\\\psi_r\end{bmatrix}=\begin{bmatrix}\dfrac{R_s}{\sigma L_s}I & \dfrac{R_sL_m}{\sigma L_sL_r}I\\[2mm]\dfrac{R_sL_m}{\sigma L_sL_r}I & \left(j\omega_m-\dfrac{R_s}{\sigma L_s}\right)I\end{bmatrix}\begin{bmatrix}\psi_s\\\psi_r\end{bmatrix}+\begin{bmatrix}I\\0\end{bmatrix}V_S \tag{14}$$

where

$\psi_{s} = \begin{bmatrix} \psi_{ds} & \psi_{qs} \end{bmatrix}^{T}$  d and q-axis stator flux matrix

$\psi_{r} = \begin{bmatrix} \psi_{dr} & \psi_{qr} \end{bmatrix}^{T}$  d and q-axis rotor flux matrix

$V_{s} = \begin{bmatrix} V_{ds} & V_{qs} \end{bmatrix}^{T}$  d and q-axis stator voltage flux matrix

$\sigma = 1 - L_{m}^{2} / L_{s}L_{r}$  leakage coefficient

$I = \begin{bmatrix} 1 & 0 \\ 0 & 1 \end{bmatrix}$  identity matrix

The electromagnetic torque can be written in terms of stator and rotor flux as

$$T_{e} = \left(\frac{3}{2}\right)\left(\frac{p}{2}\right)\frac{L_{m}}{\sigma L_{s}L_{r}} \text{Im}\left[\psi_{s}.\psi_{r}^{*}\right] \quad (15)$$

With a small control sampling time, $t_{sp}$, stator and rotor fluxes at sampling instant $(k+1)t_{sp}$ can be written as

$$\psi_{sk+1} = \psi_{sk} + \left(-\frac{R_{s}}{\sigma L_{s}}\psi_{sk} + \frac{R_{s}L_{m}}{\sigma L_{s}L_{r}}\psi_{rk} + V_{sk}\right)t_{sp} \quad (16)$$

$$\psi_{rk+1} = \psi_{rk} + \left(\frac{R_{s}L_{m}}{\sigma L_{s}L_{r}}\psi_{rk} + \left(j\omega_{m} - \frac{R_{r}}{\sigma L_{r}}\right)\psi_{rk}\right)t_{sp} \quad (17)$$

By substituting (16) and (17) into discrete form of (15) and neglecting the square of $t_{sp}$, the torque variation $\Delta T_{e}$ during $t_{sp}$, at $(k+1)^{th}$ sampling instant can be obtained as

$$\frac{+\Delta T_{ek+1}}{t_{sp}} = -T_{ek}\left(\frac{R_{s}}{\sigma L_{s}} + \frac{R_{r}}{\sigma L_{r}}\right) + \left(\frac{3}{2}\right)\left(\frac{p}{2}\right)\frac{L_{m}}{\sigma L_{s}L_{r}}.$$

$$\text{Im}\left\{\left[V_{sk}.\psi_{rk}^{\dagger}\right] - j\omega_{m}\left[\psi_{sk}.\psi_{rk}^{\dagger}\right]\right\} = f_{1} \quad (18)$$

$$\frac{-\Delta T_{ek+1}}{t_{sp}} = -T_{ek}\left(\frac{R_{s}}{\sigma L_{s}} + \frac{R_{r}}{\sigma L_{r}}\right) - \left(\frac{3}{2}\right)\left(\frac{p}{2}\right)\frac{L_{m}}{\sigma L_{s}L_{r}}.$$

$$\text{Im}\left\{j\omega_{m}\left[\psi_{sk}.\psi_{rk}^{\dagger}\right]\right\} = f_{2} \quad (19)$$

In (18), "$f_{1}$" represents the slope of torque variation by active stator voltage vector, $V_{sk}$. And "$f_{2}$" in (19) is the slope of the torque by zero voltage vectors. If voltage command $V_{s}^{*}$ is determined by the DTC algorithm, the torque slope can be determined by DTC algorithm, the torque slope can be predicted as

$$f_{1} = -\frac{T_{e}}{\sigma\tau_{sr}} + \frac{3p}{4}\frac{R_{s}}{\sigma L_{s}L_{r}}\{-V_{ds}^{*}\psi_{qr} + V_{qs}^{*}\psi_{dr} - \omega_{m}.$$

$$(\psi_{ds}\psi_{dr} - \psi_{qs}\psi_{qr})\} \quad (20)$$

$$f_{2} = -\frac{T_{e}}{\sigma\tau_{sr}} - \frac{3p}{4}\frac{R_{s}L_{m}}{\sigma L_{s}L_{r}}.\omega_{m}.\left(\psi_{ds}\psi_{dr} - \psi_{qs}\psi_{qr}\right) \quad (21)$$

where $T_{e}$ is estimated torque and $\sigma\tau_{sr} = \left(\frac{R_{s}}{\sigma L_{s}}\frac{R_{r}}{\sigma L_{r}}\right)^{-1}$ For the small value of $t_{sp}$, the slope $f_{1}$ in (20) and $f_{2}$ in (21) can be considered as constant values during sampling time and then the torque increase and decrease can be approximately as a straight line. Thus if active vector and zero vectors are combined during $t_{sp}$, rms torque ripple $T_{e}$ during one sampling time can be expressed as

$$T_{e\_ripp}^{2} = \frac{1}{t_{sp}}\int_{0}^{t_{s}}\left(f_{1}t + T_{e} - T_{e}^{*}\right)^{2} dt +$$

$$\frac{1}{t_{sp}}\int_{t_{s}}^{t_{sp}}\left(f_{2}t - f_{2}t_{s} + f_{1}t_{s} + T_{e} - T_{e}^{*}\right)^{2} dt \quad (22)$$

where $t_{s}$ is the switching instant at which zero voltage vector is applied.

Then the optimal switching instant $t_{s}$ which minimizes the torque ripple must (23)

$$\frac{\partial T_{e\_ripp}^{2}}{\partial t_{s}} = (2f_{1} - f_{2})t_{s}^{2} + 2\left[T_{e} - T_{e}^{*} - (f_{1} - f_{2})t_{sp}\right]t_{s}$$

$$-\left[2\left(T_{e} - T_{e}^{*}\right)t_{sp} + f_{2}t_{sp}^{2}\right] = 0 \quad (23)$$

By neglecting the square of $t_{s}$ and $t_{sp}$, and solving (23), optimal switching instant $t_{s}$ is obtained as

$$t_{s} = \frac{2\left(T_{e} - T_{e}^{*}\right)t_{sp}}{2\left[T_{e} - T_{e}^{*} - (f_{1} - f_{2})t_{sp}\right]} \quad (24)$$

In the proposed strategy as shown in Fig. 6, the switching instant $t_{s}$ is calculated from (24), stator voltage vector, $V_{s}$ is applied at $t=0$ and is switched to zero voltage vector $V_{s}$ at $t = t_{s}$, then torque decreases with a slope $f_{2}$ during $(t_{sp}-t_{s})$, If $t_{s}>t_{sp}$, upload voltage vector $V_{s}$ is fully turned on during whole sampling period $t_{sp}$. For every switching cycle, time $t$ is reset to zero and new voltage vector is selected and the above sequence is repeated.

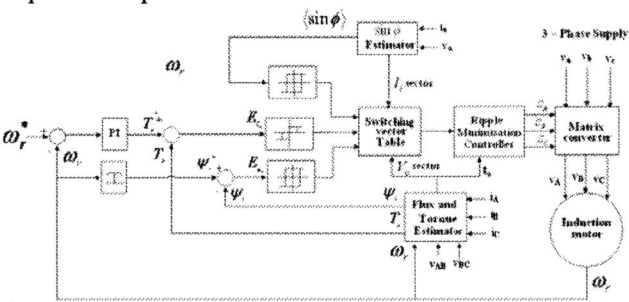

Fig. 6. Block diagram of ripple minimization DTC for Matrix converter.

## V. MODELING AND SIMULATION

To validate the performance of modified DTC, a three phase matrix converter fed induction motor drive is simulated in MATLAB environment using Simulink and power system blockset toolboxes as shown in Fig.7. The matrix converter model has been developed by using ideal switches and switching table has been developed by using direct look up tables. The switching frequency $f_{s}$ is taken as 20 kHz. The simulation is carried out by assuming a sampling period of 20 μs.

240

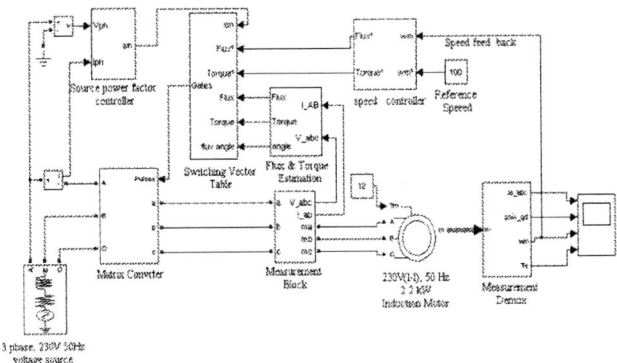

Fig. 7. Simulation model of direct torque control of matrix converter fed induction motor drive in MATLAB.

## VI. RESULTS AND DISCUSSION

The matrix converter fed modified DTCIMD is simulated in MATLAB environment along with Simulink and Power System Blockset toolboxes. The simulated results of the modified DTCIMD are compared with the conventional DTCIMD. The test machine is a 2.2 kW, four pole 230-V 50 Hz cage induction motor having the following parameters $R_s$ = 1.325$\Omega$, $R_r$ = 1.0302$\Omega$, $L_{ls}$ = 4.57mH, $L_{lr}$ = 4.57mH, $L_m$=80.7mH.

### A. Performance of Matrix Converter Fed Conventional DTCIMD

The steady state performance of matrix converter fed conventional DTCIMD has been tested in low and high speed ranges. The simulated results are shown in Fig. 8 at rated load torque and 1000 rpm speed, in terms of source voltage $v_s$, source current $i_s$, speed N, electromagnetic torque T, three phase motor currents $i_{abc}$ and stator dq-axes fluxes $\psi_{sq}$, $\psi_{sd}$. Conventional DTCIMD has also been tested at very low speed of 100 rpm, rated load torque and results are shown in Fig. 9. Fig 8 show high ripples in load torque, which need to improve. The wave form of filtered AC source current along with its harmonic spectrum is shown in Fig. 14 showing the THD of AC main current at rated torque as 18%. The dynamic performance of matrix converter fed conventional DTCIMD is shown in Fig.12.

### B. Performance of Matrix Converter Fed Modified DTCIMD

The steady state performance of matrix converter fed modified DTCIMD has been tested in low and high speed ranges. The simulated results are shown in Fig.10 at rated load torque and 1000 rpm speed. The simulated results at rated torque and 100 rpm speed are also shown in Fig. 11. Figs. 10-11 emphasize good performance of modified DTCIMD in terms reduction of torque ripple, as compared to the conventional DTC shown in Figs 8-9. The wave form of filtered AC source current along with its harmonic spectrum is shown in Fig. 15 showing the THD of AC source current at rated torque as 15.17%. The transient performance of the modified DTCIMD method has been tested with reference to

both step speed command from 1000 to -1000 rpm and step torque command from +12 to -12 N-m. Fig 13 shows the results obtained.

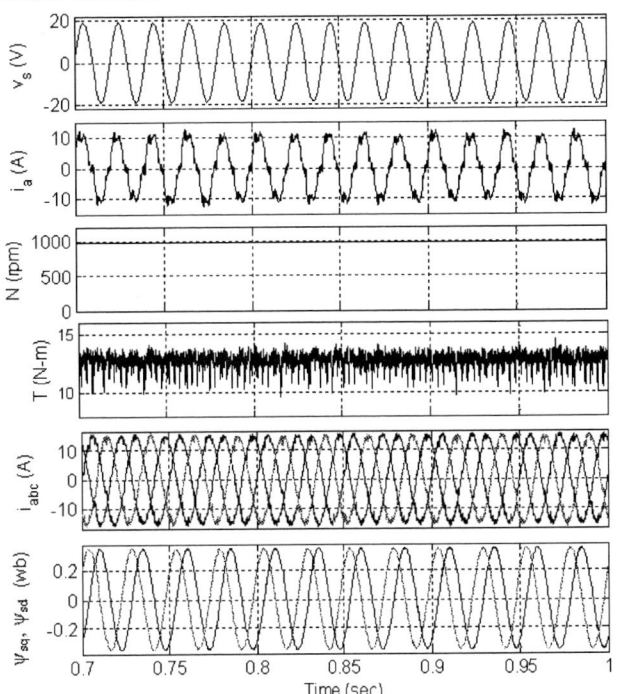

Fig. 8. Steady state response of matrix converter fed conventional DTCIMD at 12 N-m load torque and 1000 rpm speed.

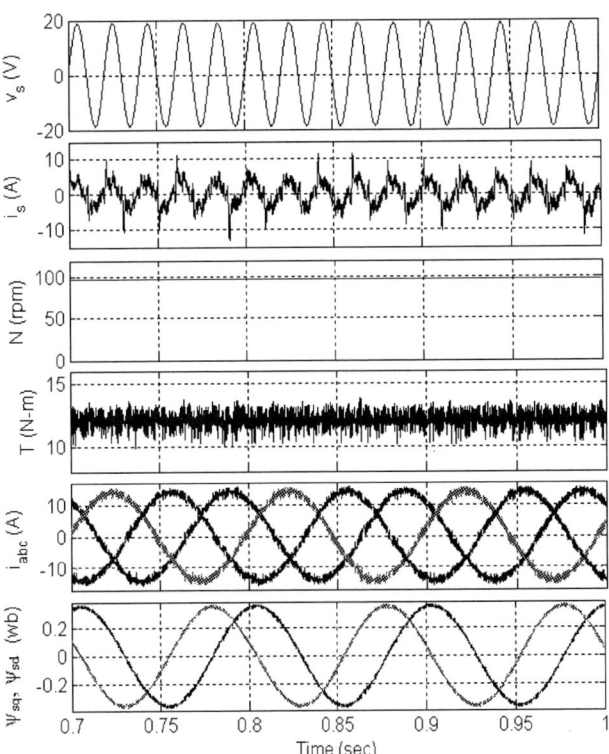

Fig. 9. Steady state response of matrix converter fed conventional DTCIMD at 12 N-m load torque and 100 rpm speed.

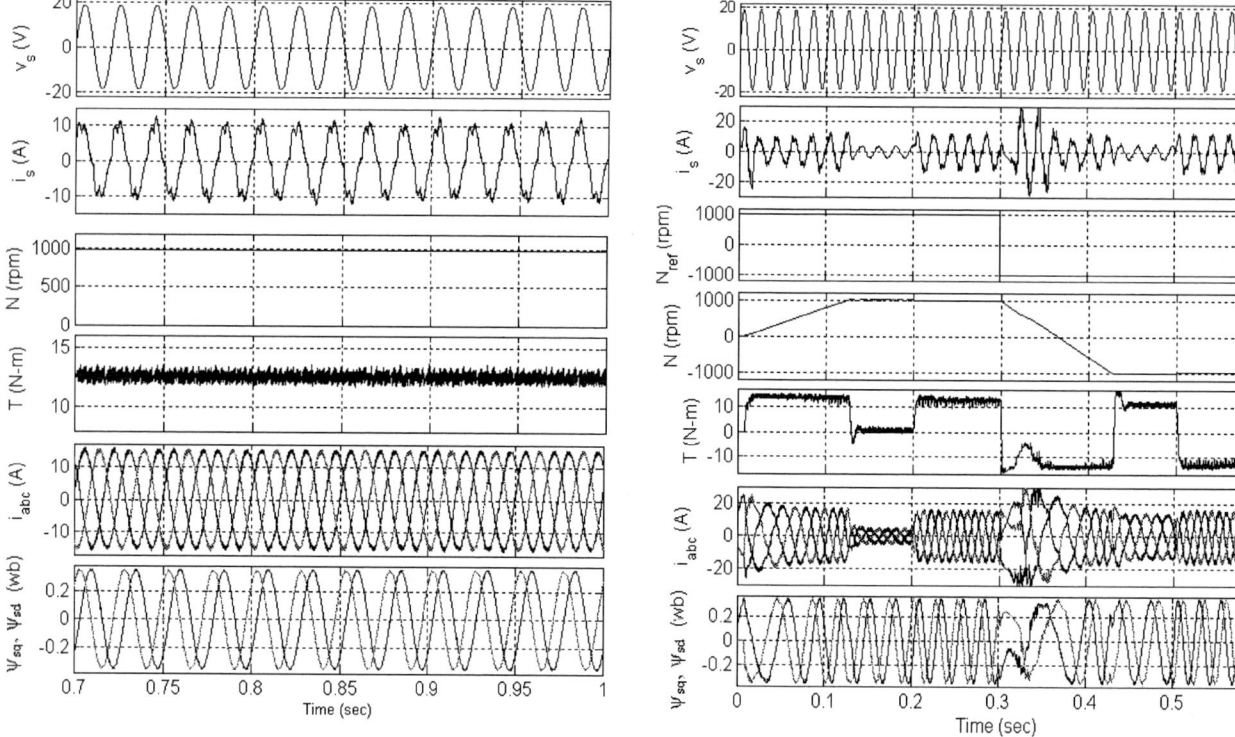

Fig. 10. Steady state response of matrix converter fed modified DTCIMD at 12 N-m load torque and 1000 rpm speed.

Fig. 12. Transient response of matrix converter fed conventional DTCIMD.

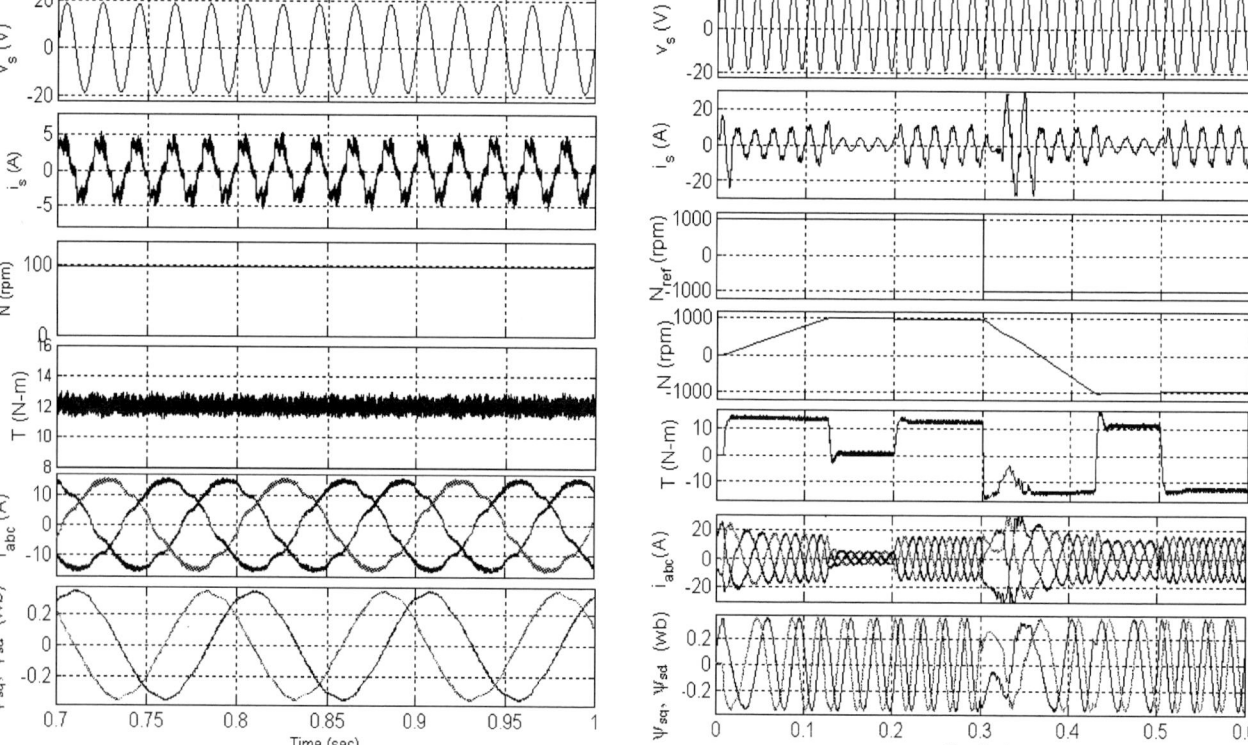

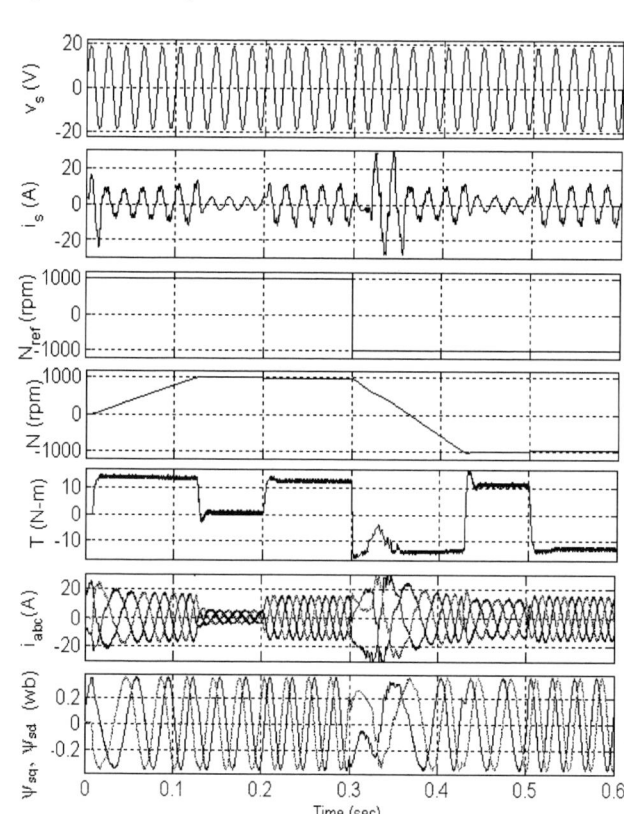

Fig. 11. Steady state response of matrix converter fed modified DTCIMD at 12 N-m load torque and 100 rpm speed.

Fig. 13. Transient response of matrix converter fed modified DTCIMD.

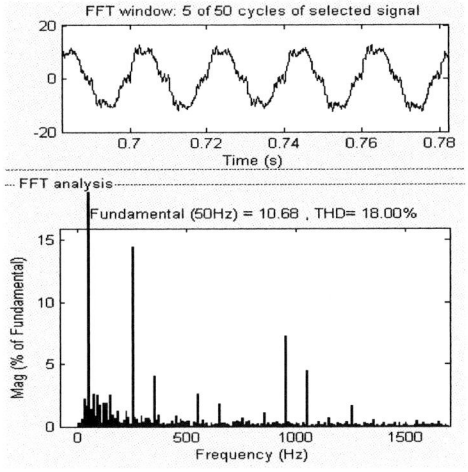

Fig. 14. filtered AC source current of matrix converter fed induction motor drive (conventional DTC) with its harmonic spectrum at rated torque and 1000 rpm speed.

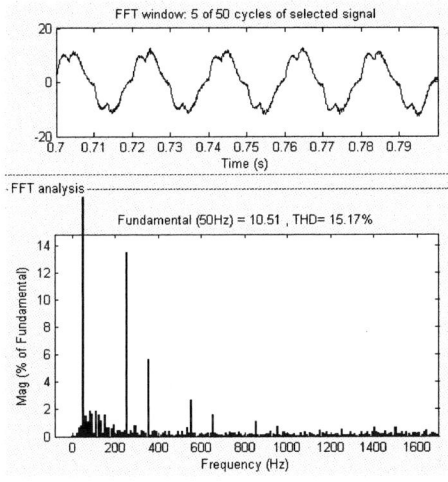

Fig. 15. filtered AC source current of matrix converter fed induction motor drive (modified DTC) with its harmonic spectrum at rated torque and 1000 rpm speed.

It can be observed from the Fig. 13, that modified DTCIMD provides reduced torque ripple, improved THD of AC main current and unity displacement factor.

## VII. CONCLUSION

A modified direct torque control method has been used to improve the performance of a matrix converter fed induction motor in terms reduction of the torque ripple within specified boundaries. The performance of matrix converter fed modified DTCIMD has been tested in steady-state and transient conditions in the low- and high-speed ranges and compared with the matrix converter fed conventional DTCIMD. The torque waveforms have demonstrated the effectiveness of the modified control scheme. The dynamic behavior has been tested with both step change in torque and step change in speed from the motor in regenerative breaking operating condition. The simulation results verify the feasibility of the

modified DTC method for matrix converter and also show that the Modified DTC improves the torque control characteristic with the THD of the AC source current.

## VIII. REFERENCES

[1] D. Casadei, G. Serra, and A. Tani, "The Use of Matrix Converters in Direct Torque Control of Induction Machines" *IEEE Trans. on Ind. Elect.*, vol. 48, No. 6, Dec 2001, p. 1057.

[2] L, Huber; D, Borojevic. "Space Vector Modulated Three-Phase to Three-phase Matrix Converter with Input Power Factor Correction", *IEEE Trans. Industry Applications*, Vol. 31, No. 6, November/December 1995.

[3] M. Venturini, "A new sine wave in sine wave out, conversion technique which eliminates reactive elements," in *Proc. POWERCON 7*, 1980, pp. E3_1–E3_15.

[4] M. Venturini and A. Alesina, "The generalized transformer: A new bidirectional sinusoidal waveform frequency converter with continuously adjustable input power factor," in *Proc. IEEE PESC'80*, 1980, pp. 242–252.

[5] P, Wheeler; J, Rodriguez; J, Clare; L, Emphringham; and A , Weinstein, "Matrix Converters: A Technology Review", *IEEE Trans. On Industrial Electronics*, Vol. 49, No. 2, April 2002, pp 276-288.

[6] D. Casadei, G. Grandi, and G. Serra, "Study and implementation of a simplified and efficient digital vector controller for induction motors," in *Proc. EMD*, Oxford, U.K., Sept. 8–10, 1993, pp. 196–201.

[7] D. Casadei, G. Grandi, G. Serra, and A. Tani, "Effects of flux and torque hysteresis band amplitude in direct torque control of induction machines," in *Proc. IEEE IECON'94*, Bologna, Italy, Sept. 5–9, 1994, pp. 299–304.

[8] Ch. Lochot, X. Roboam, and P. Maussion, "A new direct torque control strategy for an induction motor with constant switching frequency operation," in *Proc. EPE'95*, vol. 2, Seville, Spain, Sept. 18–21, 1995, pp. 431–436.

[9] D. Casadei, G. Grandi, G. Serra, and A. Tani, "Switching strategies in direct torque control of induction machines," in *Proc. ICEM'94*, Paris, France, Sept. 5–8, 1994, pp. 204–209.

**Bhim Singh** (SM'99) was born in Rahamapur, U. P., India in 1956. He received B. E. (Electrical) degree from University of Roorkee, India in 1977 and M. Tech. and Ph. D. degrees from Indian Institute of technology (IIT), New Delhi, in 1979 and 1983, respectively. In 1983, he joined as a Lecturer and in 1988 became a Reader in the Department of Electrical Engineering, University of Roorkee. In December1990, he joined as an Assistant Professor, became an Associate Professor in 1994 and Professor in 1997 at the Department of Electrical Engineering, IIT Delhi. His field of interest includes power electronics, electrical machines and drives, active filters, static VAR compensator, analysis and digital control of electrical machines. Prof. Singh is a Fellow of Indian National Academy of Engineering (INAE), Institution of Engineers (India) (IE(I)) and Institution of Electronics and Telecommunication Engineers (IETE), a Life Member of Indian Society for Technical Education (ISTE), System Society of India (SSI) and National Institution of Quality and Reliability (NIQR) and Senior Member of IEEE ( Institute of Electrical and Electronics Engineers).

**Jally Ravi** was born in Thanakalan, Nizamabad, A.P., India in 1984. He received Diploma (Electrical) from Govt. Polytechnic collage, Nizamabad and B.Tech (Electrical) degree from Vignan Institute of Technology & Science, Hyderabad and presently he is pursuing M.Tech degree from Department of Electrical Engineering, IIT Delhi. His field of interest includes power electronics, electrical machines and drives.

**2006 IEEE International Conference on Power Electronic, Drives and Energy Systems**

# LMI Based Digital State Feedback Controller for a Wound Rotor Induction Drive with Guaranteed Closed Loop Stability

D. Sivanandakumar, *IEEE*, and K. Ramakrishnan, *IEEE*

*Abstract* --In this write-up, a novel approach for the design of a digital state feedback controller for the speed control of slip energy recovery system employing a wound rotor induction machine is proposed using digital redesign approach. The state feedback controllers for digital implementation are suitably redesigned in such a way that the discrete-time-system response and continuous-time response match at all sampling instants. As the digital implementation of the continuous time controller is very desirable when the designed continuous-time controller uses some recent and advanced control algorithms, the redesign problem boils down to an optimization problem minimizing the difference between the continuous-time system and discrete-time system subjected to some constraints. In this paper, the redesign problem is formulated as a generalized eigenvalue problem with linear matrix inequality constraints. The results obtained are compared with other methods. The proposed algorithm uses LQR approaches for the design of continuous-time controller and the approach is illustrated for slip energy recovery system with encouraging results.

*Index Terms* --Digital redesign, Digital control, State feedback controller, Slip Energy Recovery System.

## I. INTRODUCTION

MOST complex dynamical system is described by continuous-time models. It is, therefore common practice and infact, advantages to design a controller in the continuous time frame work. The continuous-time controller, on the contrary is preferable to be implemented by using a digital device for better performance, flexibility and lower cost. A digital implementation of the continuous-time controller is indeed highly desirable when the continuous-time controller uses some advanced control algorithms. There are three digital design approaches for digital control system. The first approach or conventional approach is to first discretize the continuous-time plant (ZOH equivalent) and then design a controller to meet control specifications that are suitable transformed from continuous-time framework to discrete time frame work. The draw back with this approach is that there is nothing physical called the *discretized plant* for a real-time system and designing a controller directly in digital domain by making use of this so called discretize model may not work

satisfactorily at all times when real time implementation is attempted. This argument is supplemented by the fact that for certain sampling frequencies, the discretized system may become unstable in spite of the original continuous-time system being stable. The second approach called direct sampled data approach is to directly design a digital controller for analogue plant. The third approach, called the digital redesign approach encompasses the digital design in two steps. In the first step a suitable analogue controller is designed to meet control specification described in continuous-time domain; the second step is to convert the obtained analogue controller to equivalent digital controller retaining the properties of the original analogously controlled system by which benefits of both continuous time controller and the advanced digital technology can be obtained. The digital system is equivalent to continuous time system if the response of the system is closely matched at all sampling instant for the same input and initial condition. Hence without *discretizing* the plant and without designing a controller explicitly in the discrete domain, the process of converting a continuous time controller in to an equivalent digital controller to facilitate ease of implementation using digital device is known as digital redesign.

Digital redesign technique was first considered by Kuo. B.C[1]. There are many redesign techniques that are reported in the literature like Taylor series approximation method [1], bilinear transformation method [2]-[5], the block pulse approximation method [3]-[4], optimization method [6] and the frequency domain method [7]. In this paper the redesign problem is proposed as a generalized eigen value problem with Linear Matrix Inequality (LMI) constraints and by solving the optimization problem the digital state feedback controller gains are extracted.

The test system dealt in this paper namely the Slip energy recovery system is an energy saving system associated with wound rotor induction drive saving the rotor energy and feeding it at suitable level to the stator supply. The success of the slip energy recovery scheme, apart from loss less switches and static transformer, depends on the speed control scheme employed.

In section II, a detailed system description of the slip energy recovery system is discussed .The state space modeling of the system is dealt in section III. Section IV discusses the design of a linear quadratic regulator for the slip energy recovery scheme. In section V, LMI based digital redesign of continuous-time LQR controller is proposed. The

---

D. Sivanandakumar is with the department of Electrical and Electronics Engineering, Pondicherry Engineering College, Pondicherry 605 014, India (e-mail: svkmreee@yahoo.co.in).

K. Ramakrishnan is with the department of Electrical and Electronics Engineering, Pondicherry Engineering College, Pondicherry 605 014, India (e-mail: ramss_k@yahoo.co.in).

0-7803-9771-1/06/$25.00 ©2006 IEEE

experimental results are presented in section VI. The conclusions are drawn in section VII.

## II. SYSTEM DESCRIPTION

The block schematic of the slip energy recovery system using twelve-pulse converter and wound rotor induction machine is shown in Fig. 1. Reduction in output harmonics is accomplished by employing two six-pulse bridge circuits. One of the bridges is supplied through a Y-Y transformer and the other through a Y-Δ transformer. The purpose of the Y-Δ transformer is to introduce a phase shift of 30° between the source and the bridge. This results in two bridges having similar output but shifted by 30°. The overall output voltage is the sum of the two bridge outputs. Controlling the firing pulses of the twelve-bridge line commutated converter controls the speed of the motor. The digital state feedback controller is designed for this purpose.

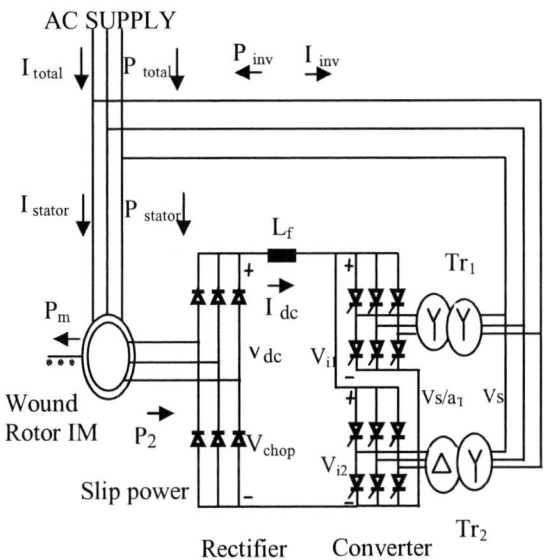

Fig. 1. Block diagram of twelve-pulse converter for slip energy recovery system.

The overall output voltage of the circuit is the sum of two bridge output and the delay angles for the bridges are same. The dc output of the overall circuit $V_i$ is the sum of dc output of the individual bridges, $V_{i1}, V_{i2}$ and it is given by

$$V_i = \frac{V_{i1} + V_{i2}}{2} = \frac{1}{2} \left[ \frac{3\sqrt{2}\ V_{LL}}{\pi a_T} \cos\alpha + \frac{3\sqrt{2}V_{LL}}{\pi a_T} \cos\alpha \right] \quad (1)$$

Since the transition between conducting switches occurs every 30°, there a total of 12 such transition for each period of the ac source. Hence the output has harmonics frequencies, which are multiple of 12 times the source frequency. So, filtering to obtain a relatively a pure dc output less costly than that required for 6-pulse rectifier. Further more another advantage of using twelve-pulse converter instead of a six-pulse converter is the reduced harmonics that occur in the ac

system. This phenomenon is explained in mathematical way as follows. The current in the ac lines supplying Y-Δ Transformer is represented by the Fourier series as

$$i_\Delta(t) = \frac{2\sqrt{3}}{\pi} i_0 [\cos\omega_0 t + \frac{1}{5}\cos 5\omega_0 t - \frac{1}{7}\cos 7\omega_0 t - \frac{1}{11}\cos 11\omega_0 t + \frac{1}{13}\cos 13\omega_0 t - ....](2)$$

where $\omega_0$ is the fundamental frequency.

Similarly the current in the ac lines supplying Y-Y transformers is represented by the Fourier series as,

$$i_y = \frac{2\sqrt{3}}{\pi} i_0 [\cos\omega_0 t - \frac{1}{5}\cos 5\omega_0 t + \frac{1}{7}\cos 7\omega_0 t - \frac{1}{11}\cos 11\omega_0 t + \frac{1}{13}\cos 13\omega_0 t - ....](2.1)$$

The ac system current $i_{ac}(t)$ is given by,

$$i_{ac}(t) = i_\Delta(t) + i_y(t)$$

$$= \frac{4\sqrt{3}}{\pi} i_0 [\cos\omega_0 t - \frac{1}{11}\cos 11\omega_0 t + \frac{1}{13}\cos 13\omega_0 t - ....] \quad (2.2)$$

Thus, some of the harmonics on the ac side are cancelled by using the twelve-pulse scheme rather than the six-pulse scheme. The harmonics that remain in the ac system are of order $12k \pm 1$ and the cancellation of harmonics $6(2n-1) \pm 1$ has resulted from the transformer and the converter configuration.

## III. STATE SPACE MODELING OF THE SYSTEM

In this section we develop the state space model equation for the speed control of the slip recovery scheme. The equivalent circuit diagram of dc link is in Fig. 2.

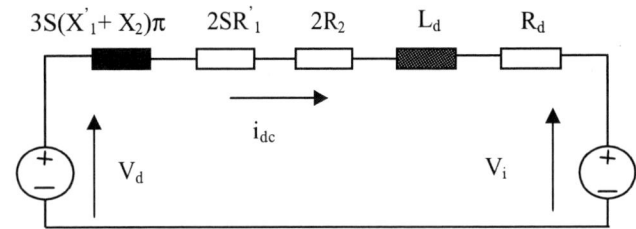

Fig. 2. Equivalent circuit diagram of dc link.

In this equivalent circuit the following relationship is satisfied.

$$V_d + V_i = 0 \quad (3)$$

where, $V_d$ is the output voltage of the rectifier and $V_i$ is input voltage of the inverter. These voltages are given as,

$$V_d = \frac{1.35\,S\,V_s}{a_m} \quad \text{and}\, V_i = \frac{1.35\,V_s}{a_T} \cos\alpha \quad (4)$$

245

where, $V_s$ is the stator voltage, $a_T$ is the turn ratio of the transformer, $a_m$ is the turn ratio of motor, α is the Firing angle of the converter and S is the slip of the motor. From (3) we get,

$$\frac{1.35\,S\,V_s}{a_m} + \frac{1.35\,V_s}{a_T}\cos\alpha = 0 \tag{5}$$

Since S is given by,

$$S = -\frac{a_m}{a_T}\cos\alpha \ \text{ or } \ S = a\cos\alpha$$

where a is the effective turn ratio of the motor. The speed of the motor is controlled by varying the firing angle α. The total resistance in DC link circuit are given by,

$$R_A = 2R'_1 + \frac{3}{\pi}\left(X'_1 + X_2\right)$$
$$R_B = 2R_2 + R_d$$

The dc current $i_d$ in the dc link is given by,

$$L_d\frac{di_d}{dt} = -R_B\,i_d - SR_A\,i_d + \frac{1.35}{a_m}S\,V_s - \frac{1.35}{a_T}V_s\cos\alpha \tag{6}$$

Substituting $S = \frac{\omega_s - \omega_m}{\omega_s}$ in (6) and simplifying we get,

$$\frac{L_d}{R_a + R_B}\frac{di_d}{dt} = -i_d + \frac{\omega_m}{\omega_s}\frac{R_A}{R_A + R_B}i_d + \frac{1.35\omega_m}{a_m(R_A + R_B)\omega_s}V_s - \frac{1.35}{a_T}\frac{V}{(R_A + R_B)}\cos\alpha \tag{7}$$

Following assumptions (a-c) are made to simplify (7)

$$a)\ \left(\frac{\omega_m}{\omega_s}\right)\left(\frac{R_A}{R_A + R_B}\right) \ll 1$$

$$b)\ T_d = \frac{L_d}{(R_A + R_B)}\ (Electrical\ Time\ Constant)$$

$$c)\ a_m = a_T = a$$

Equation (7) becomes

$$\frac{di_d}{dt} = -\frac{1.35V_s}{L_d\omega_s a}\omega_r - \frac{1}{T_d}i_d + \frac{1.35V_s}{aL_d}(1 - \cos\alpha) \tag{8}$$

Electrical developed power output, $P_2$ for the developed torque is given by

$$P_2 = \frac{1.35\,S\,V_s}{a_m}i_d - (S\,R_A + R_B)\ i_d^2 \tag{9}$$

Mechanical torque, $M_d$ is given by

$$M_d = \frac{P_2}{S\,\omega_S} = \frac{1.35V_s}{a\,\omega_s}i_d \tag{10}$$

Now,

$$T_m\frac{d\omega_r}{dt} = M_d \tag{11}$$

where, $T_m$ is the mechanical time constant and $\omega_r$ is the rotor speed

Substituting (10) and (11), we get

$$\frac{d\omega_r}{dt} = \frac{1.35V_S}{a\omega_s T_m}i_d \tag{12}$$

Let us define the state variable as follows $x_1(t) = \omega_r$ (rotor angular speed) and $x_2(t) = i_d$ (dc link current) and adopt the following notations (for $U_a$, $K_T$ and $K_a$) in the equations (8) and (12) to get the state model in its standard form (13).

$$U_a = (1 - \cos\alpha),\ K_T = \frac{1.35V_s}{a\,\omega_s}\ and\ K_a = \frac{\omega_s K_T}{L_d}$$

$$\begin{bmatrix} \overset{\circ}{x}_1(t) \\ \overset{\circ}{x}_2(t) \end{bmatrix} = \begin{bmatrix} 0 & \dfrac{K_T}{T_m} \\ -\dfrac{aK_T}{L_d} & -\dfrac{1}{T_d} \end{bmatrix}\begin{bmatrix} x_1(t) \\ x_2(t) \end{bmatrix} + \begin{bmatrix} 0 \\ K_a \end{bmatrix}U_a \tag{13}$$

$$y = \begin{bmatrix} 1 & 0 \end{bmatrix}\begin{bmatrix} x_1(t) \\ x_2(t) \end{bmatrix}$$

## IV. LQR CONTROLLER DESIGN USING LMIS

The LQR controller is designed to optimize the performance index, J of the system specified in the continuous-time framework. The controller is designed using the Linear Matrix Inequality (LMI) approach. The **lmilqr** problem is an Eigen value problem [8]-[9] and is stated as follows

$$\min_{Y,S,X}\ Tr(QS) + Tr(X) + Tr(YN) + Tr(N^T Y^T)$$

subject to

$$AS\text{-}BY + SA^T - Y^T B^T + EE^T < 0 \text{ and }\begin{bmatrix} X & R^{\frac{1}{2}}Y \\ Y^T R^{\frac{1}{2}} & S \end{bmatrix} > 0 \tag{14}$$

The system being described by

$$\overset{\circ}{X}(t) = AX(t) + Bu(t) + Ew(t) \tag{15}$$

and the performance index given by

$$J = \int (X^T Q X + u^T R u + 2 X^T N u) dt \qquad (16)$$

X, Y and S in the problem equations are the lmi variables and the optimal value of the state feedback gain matrix is given by

$$K_{copt} = Y_{opt}\, S_{opt}^{-1} \qquad (17)$$

## V. DIGITAL REDESIGN OF STATE FEEDBACK CONTROLLER

There are several approaches for the digital redesign of static state feedback controllers. In this paper, following four approaches are employed for digital redesign: Taylor series approximation approach, Bilinear transformation approach, Improved block-pulse approximation approach and compared with the proposed LMI based optimization approach.

### A. Optimization approach using linear matrix inequality

In this section we propose an optimization based digital redesign technique involving linear matrix inequality Constraints [13].

Consider the linear time-invariant continuous-time control system,

$$\overset{\circ}{X}_c(t) = A_c X_c(t) + B_c u_c(t),\ X_c(0) = X_0 \qquad (18)$$
$$y_c(t) = C_c X_c(t)$$

where $X_c(t) \in \Re^n$ is the state vector, $u_c(t) \in \Re^m$ is the control vector, $y_c(t) \in \Re^p$ is the Output vector, and $A_c \in \Re^{nxn}$ is the system matrix, $B_c \in \Re^{nxm}$ is the input matrix, $C_c \in \Re^{pxn}$ is the output matrix. The control vector $u_c(t)$ is given by

$$u_c(t) = -K_c X_c(t) + E_c r(t) \qquad (19)$$

where, $K_c \in \Re^{mxn}$ is the state feedback gain matrix, $E_c \in \Re^{mxp}$ is the feed forward gain, and $r \in \Re^p$ is the constant reference vector.

The resulting closed-loop system is given by,

$$\overset{\circ}{X}_c(t) = \left(A_c - B_c K_c\right) X_c(t) + B_c E_c r,\ X_c(0) = X_0 \qquad (20)$$

The discrete model of the closed-Loop system with T as sampling period is given by

$$X_c(KT + T) = G_c X_c(KT) + H_c E_c r(KT)$$
$$y_c(KT) = C X_c(KT)$$
$$\text{where,} \qquad\qquad\qquad\qquad\qquad\qquad\qquad (21)$$
$$G_c = e^{((A_c - B_c K_c)T)}$$

$$H_c = \int_{KT}^{KT+T} e^{((A_c - B_c K_c)T)(KT+T-\tau)} B d\lambda$$

Consider the same continuous time system with digital input $u_d(t)$. Now, the state equation becomes

$$\overset{\circ}{X}_d(t) = A_d X_d(t) + B u_d(t),\ X_d(0) = X_o$$
$$y_d = C X_d(t) \qquad\qquad\qquad\qquad (22)$$

where $u_d(t) = u_d(KT) = -K_d X_d(KT) + E_d r(KT)$

The resulting closed loop system is given by

$$X_d(KT + T) = \left(G - HK_d\right) X_d(KT) + HE_d r(KT)$$
$$y_d(KT) = C X_d(KT) \qquad\qquad\qquad (23)$$
$$\text{where } G = e^{(AT)} \text{ and } H = \int_{KT}^{KT+T} e^{(A(KT+T-\tau))} B d\tau$$

By Lyaponov stability criterion, the redesigned closed loop system (23) is stable in the sense of Lyapunov if following inequality is satisfied.

$$\left(G - HK_d\right)^T P \left(G - HK_d\right) - P < 0 \qquad (24)$$

where $P = P^T > 0$

### B. Redesign problem formulation

The objective is to find the digital gains $K_d$ and $E_d$ from the analogue gains $K_c$ and $E_c$ so that the outputs of digitally controlled system (23) match those of continuous-time system (21) at every sampling period as closely as possible.

### C. Main Result

To solve the redesign problem, we state the following lemma. Let O, N and L be constant matrices of appropriate dimension. Then following two statements are equivalent.

1) Lemma 1; *Schur complement formula* [10]

a) $O > 0,\ N + L^T O L < 0$

b) $\begin{bmatrix} N & L^T \\ L & -O^{-1} \end{bmatrix} < 0 \qquad (25)$

By applying the Lemma 1 the redesign problem is stated as a generalized eigen value problem (GEVP) in which an optimization is done to reduce the difference between the two systems in (21) and (23) as close as possible. The Lemma 1 is also applied to (24).The complete GEVP is stated as follows:

247

$$\min_{\Gamma,F} \alpha$$

subject to

$$\begin{bmatrix} -\alpha\Gamma & (G_c\Gamma - G\Gamma + HF)^T \\ (G_c\Gamma - G\Gamma + HF) & -\alpha I \end{bmatrix} < 0 \qquad (26)$$

$$\begin{bmatrix} -\Gamma & (G\Gamma - HF)^T \\ (G\Gamma - HF) & -\Gamma \end{bmatrix} < 0$$

## VI EXPERIMENTAL RESULTS

The parameters of the experimental system [11] are as follows: 3 Phase Wound Rotor Induction Machine: 0.22kW, 230/440V, 1.12/0.6 A, $V_2 = 100V$, 1.5A, 50Hz, Pf = 0.78lag, 1410 rpm, $V_s = 380V$, turn ratio = 4.4, $\omega_s = 157$rad/sec, $L_d = 25$mH, $R_a = 3$ Ohm and $R_b = 8$ Ohm, Mechanical time Constant Tm = 120 ms. The state model of the slip recovery system in accordance with the equation (13) is given by

$$A = \begin{bmatrix} 0 & 6.18 \\ -29.7 & -440 \end{bmatrix}$$

$$B = \begin{bmatrix} 0 \\ 4663 \end{bmatrix}$$

$$C = \begin{bmatrix} 1 & 0 \end{bmatrix}$$
and
$$u = U_a$$
$$y = \omega_m$$

The continuous time LMI_LQR controller and the digital controller gains obtained using four approaches are given in the Table I, I I and I I I. The step response of the digital closed loop system and the associated control effort $U_a$ (t) are given in Fig. 3. From the Fig. 3, we see that the closed loop performance of continuous-time system and discrete-time system are closely matched at all sampling instant all the methods. The control effort required for the LMI based digital control technique is lesser compared to the other approaches. The proposed method uses less amount of control energy to implement the objective.

TABLE I
STATE FEEDBACK GAINS IN CONTINOUS TIME DOMAIN

| METHOD | State feedback gain $K_c$ | Feed forward gain $E_c$ |
|---|---|---|
| LMI_LQR METHOD | [0.9921  1.3160] | [1.0558] |

TABLE I I
REDESIGN BASED STATE FEEDBACK GAINS IN DISCRETE TIME DOMAIN

| METHOD | Ts | State feedback gain $K_d$ | Feed forward gain $E_d$ |
|---|---|---|---|
| KUO's Taylor Series Approximation | 0.003 | [0.0209 -0.7539] | [0.0846] |
| TSAI's Bilinear Transformation Method | 0.05 | [0.5109 0.3615] | [0.5747] |
| Improved Block Pulse Approximation Method | 0.05 | [0.3624 0.0271] | [0.4228] |
| LMI Method | 0.05 | [0.3305 0.0036] | [0.3941] |

TABLE I I I
INTEGRAL ERROR INDEX

| Digital redesign method | Ts = 0.003 | Ts = 0.05 |
|---|---|---|
| Taylor series method | 5.081 | Unstable |
| Bilinear Transformation Method | 0.0414 | 01724 |
| Improved Block Pulse Approximation Method | 0.0000 | 0.0000 |
| LMI Method | 0.2137 | 0.4257 |

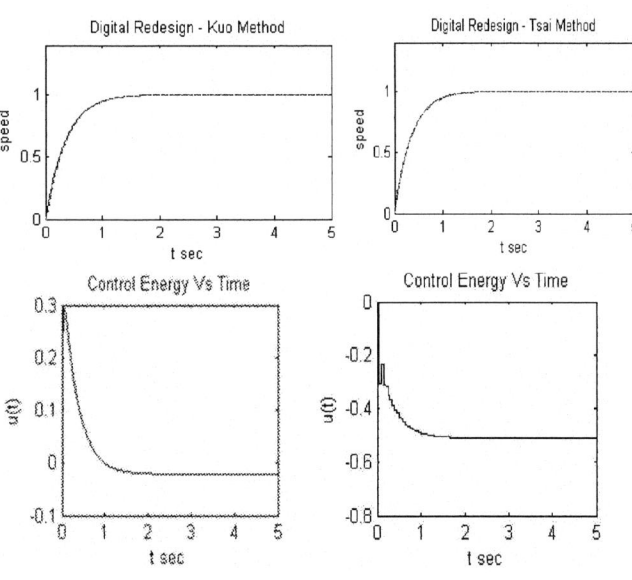

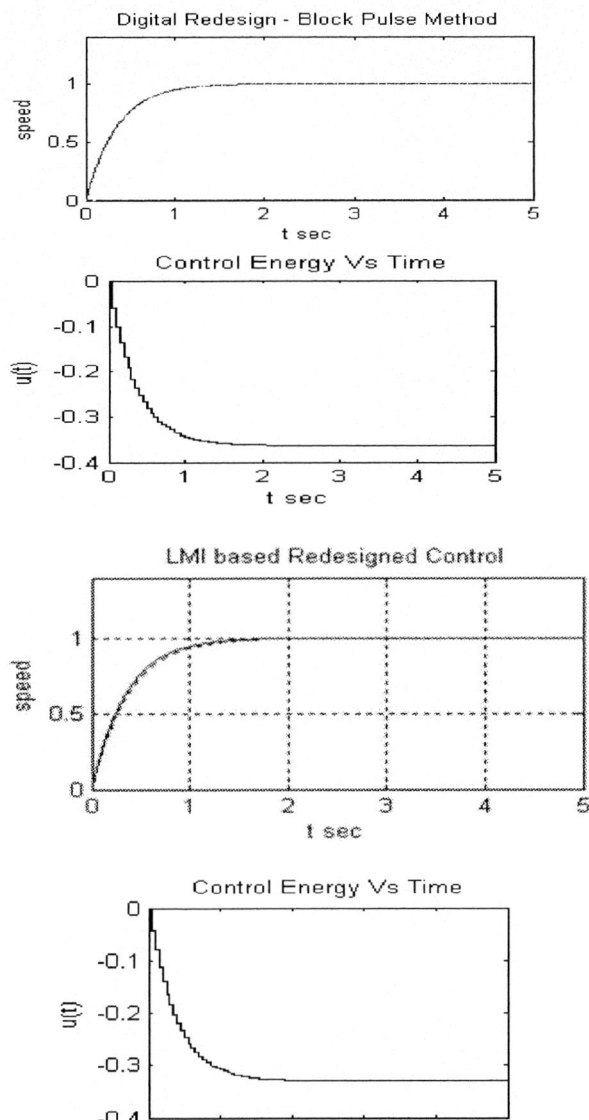

PROPOSED LMI METHOD

Fig. 3. System Response Curves.

## VII. CONCLUSION

The design of the digital state feedback speed controller using redesign concept for slip energy recovery scheme employing wound rotor induction machine is presented in this paper. The redesign problem is configured as a generalized eigen value problem with matrix inequality constraints. By solving the LMI problem the digital state feedback gains are extracted. The closed loop performance of the continuous and discrete-time systems matches accurately for the at all sampling periods. The proposed approach is applied to the Slip Energy Recovery System to effect the lmi redesign of an optimal digital controller to improve the speed control response thereby leading to the saving of control energy in the

overall system. The future work is at present oriented towards incorporating robustness issue into the redesign problem.

## VIII. REFERENCES

[1] B. C. Kuo, "Digital Control Systems," Saunders College Publishing, 2nd Edn,1992.

[2] J. S. H. Tsai, L. S. Shieh, J. L. Zhang and N. P. Coleman, "Digital redesign of pseudo-continuous-time suboptimal regulators for large-scale discrete systems," *Control* Theory *Adv. Tech.*, Vol. 5, No. 1, pp. 37-65, 1989.

[3] L. S. Shieh, J. L. Zhang and S. Ganesan, "Pseudo-continuous-time quadratic regulators with pole placement in a specific region," *Proc. IEE-D*, Vol.137, No. 5, pp. 338-346, 1990.

[4] J. S. H. Tsai, L. S. Shieh and J. L. Zhang, "An Improvement on the Digital Redesign Method based on the Block-Pulse Function Approximation," Circuits. Syst. Signal Process, Vol. 12, No. 1, pp. 37-49, 1993.

[5] L. S. Shieh, J. L. Zhang and X. M. Zhao, "Locally Optimal-Digital Redesign of Continuous-Time Systems," *IEEE Trans. Ind. Electron.*, vol. 36, No. 4, pp. 511- 515, 1989.

[6] N. Raffee, T. Chen, and O. P. Malik, "A technique for optimal digital redesign of analog controllers," *IEEE. Trans. Control Syst. Technol.*, 5, pp. 89-99, 1997.

[7] Y. N. Rosenvasser, K.Y. Polyakov and B. P. Lampe, "Application of Laplace transformation for digital redesign of continuous control system," *IEEE .Trans. Autom. Control*, 44, 4, pp. 883-886, 1999.

[8] B. Erkus and Y. J. Lee, "Linear Matrix inequalities and MATLAB LMI toolbox,"University of Southern California, Los Angeles, CA, 2004.

[9] Erik. A. Johnson and Baris Erkus 'Dissipative and performance analysis of smart dampers VIA LMI synthesis' University of southern California, L.os Angeles, C.A, 2004.

[10] G. E. Dullend and F. Paganini (2000). A course in robust control theory: A convex approach, springer, Newyork.

[11] S. Tunyasrirut, "Implementation of a dSPACE-based Digital State Feedback Controller for a Speed Control of Wound Rotor Induction Motor," *IEEE International Conference, ICIT*, Industrial Technology, pp. 1198-1203, 2005.

[12] P. Gahinet, A. Nemirovshi, A. J. Laub and M. Chilali, LMI control Toolbox-For use with Mat lab. The Math Works Inc., 1995.

[13] W. Chang, J. B. Park, H. J. Lee and Y. h. Joo, "Lmi approach to digital redesign of linear time-invariant systems, "*IEE. Proc-Control Theory Appl.* Vol.149.No.4, 2002.

## IX.BIOGRAPHIES

**Durairajan Sivanandakumar** completed bachelor's degree in Electrical and Electronics Engineering from Mailam Engineering College, Mailam, Tamilnadu, India. In due he is presently doing Master degree in Electric drives and controls in Pondicherry Engineering College, Pondicherry, India. His area of interest includes Electric machine, State feedback controller and Power electronics.

**Krishnan Ramakrishnan** completed bachelor's degree in Electrical and Electronics Engineering from Government College of Technology, Coimbatore, Tamilnadu and Postgraduate degree in Control Systems Engineering from P.S.G college of Technology, Coimbatore, Tamilnadu, India. He is presently employed as a lecturer faculty in the dept. of Electrical and Electronics Engineering, Pondicherry Engineering College, Pondicherry, India. His areas of interest include Robust controller synthesis and digital controller design techniques and fault diagnosis of analog and discrete electronic circuits.

**2006 IEEE International Conference on Power Electronic, Drives and Energy Systems**

# Open-End Winding Induction Motor Driven With Matrix Converter For Common-Mode Elimination

Krushna K Mohapatra *Member, IEEE,* and  Ned Mohan, *Fellow, IEEE*

*Abstract--*In this article a novel scheme using matrix converter is proposed for control of three phase machines. The analysis shows that the scheme has several benefits. By eliminating the common mode voltage from the three phases of the machine the motor bearing current is entirely removed. The maximum insulation stress in the motor winding is limited to the peak value of input phase voltage. The power factor at the utility end is controllable. By using matrix converters for power conversion bulky capacitors are eliminated from the drive.

*IndexTerms--Bearing Current, Common-mode Voltage,* Matrix Converter, Open-end, Three-Phase Motor.

## I. INTRODUCTION

IN the conventional drive systems   two three phase two-level inverters are connected to a common DC-link.  One of the two inverters is connected to the utility end and the other is connected to the motor end.  The utility end converter can be simply a diode bridge rectifier or a three phase PWM rectifier. It  is controlled such that the DC-link voltage is equal to the peak line-to-line voltage of utility. In this type of power conversion system the peak motor phase voltage is equal to the peak value of the utility phase voltage.

Matrix converters are well known for many years for direct frequency and voltage conversion from a set of three phase input voltages to a set of three phase output voltages [1-7]. Generally they consist of nine four-quadrant switches, which enable any of the three output phases to be connected to any of the three input phases. A four-quadrant switch can conduct in either direction when it is on, and can block voltages in either direction when it is off. In the  matrix converter driven direct link power conversion system the peak motor phase voltage is equal to the 86.6% of peak value of the utility phase voltage at unity power factor at the utility. The input power factor is controllable at the expense of output voltage capability.

The open ended three phase drive systems are obtained by opening the stator winding neutral of the three phase motor and driving the motor from both sides of the stator winding using two inverters [8] [9].  In the proposed system as shown in Fig.1 the motor is driven from two matrix converters which are connected to the opposite sides of the open-end winding. The open-end winding is obtained by opening the neutral of the motor winding. The topology shown in Fig.1 consists of 18 four-quadrant switches. Four quadrant switches have

Krushna K. Mohapatra is with University of Minnesota, Minneapolis-55455 USA (e-mail: mohap002@umn.edu).

Ned Mohan is with University of Minnesota, Minneapolis-55455 USA (e-mail: mohan@umn.edu).

capacity to block voltage in either direction and can carry current in either direction. Each terminal of the open-ended motor is connected to each phase of the utility by a four quadrant switches.

As shown in Fig.1 the topologies and control are such that a first set of multi-phase voltages are connected to two sets of multi-phase voltages, at the terminals of a motor/generator, by wires and solid-state switches (without energy storage capacitors and inductors) such that all currents at the two sets of multi-phase terminals always have paths to flow to the first set of multi-phase voltages [10]. The maximum motor voltage capability is 1.5 per unit (pu), that is, $V_{ph,m} = 1.5 V_{ph,in}$, while eliminating the common-mode voltages at both ends to eliminate the motor-bearing currents due to dv/dt of common-mode voltages. At the same time, the insulation stress is only $V_{ins} = \hat{V}_{ph,in}$.

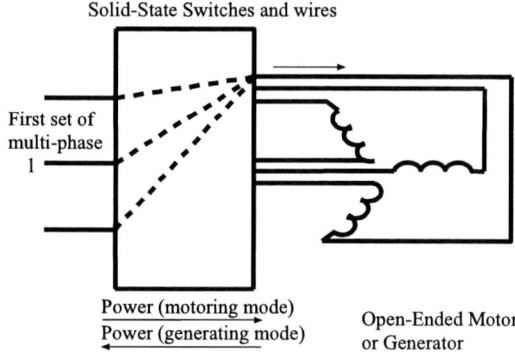

Fig. 1.  The proposed (non-isolated) system .

## II. OUTPUT VOLTAGE SYNTHESIS IN MATRIX CONVERTER WITHOUT COMMON MODE VOLTAGES

In conventional matrix converter as shown in Fig.2 there are total 27 possible switching combinations in order to realize three-phase output from the three-phase input. All possible switching and the corresponding common mode voltages are listed in Table-1. As shown in Table-1 there are three groups of switching combinations.

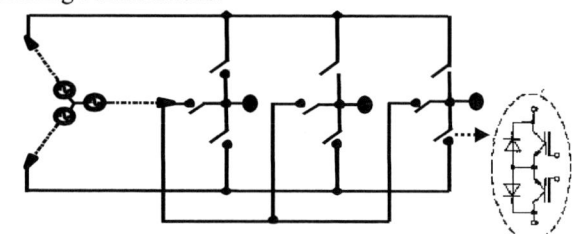

Fig. 2.  Matrix Converter.

0-7803-9771-1/06/$25.00 ©2006 IEEE        250

In conventional matrix converter modulation, the second group of switching combinations (II-A, II-B, and II-C) are used. The second group of switching combination generates maximum possible output voltage vector; however this switching combination also inherently generates significant amount of common mode potential at output. The third group of switching combinations has maximum common mode voltage and produces zero output voltage. This combination is used for controlling the output voltage amplitude. Using these switching voltage vectors, for a given input voltage, the synthesized output voltage is limited in the maximum amplitude of $\sqrt{3}/2 (= 0.867)$ times the input voltage [4-7].

Based on space vectors, the first group (Group-I) of switching combinations shown in TABLE-1 generates null common mode voltage and finite amounts of output voltage vector. Thus output voltage generation from the first group of

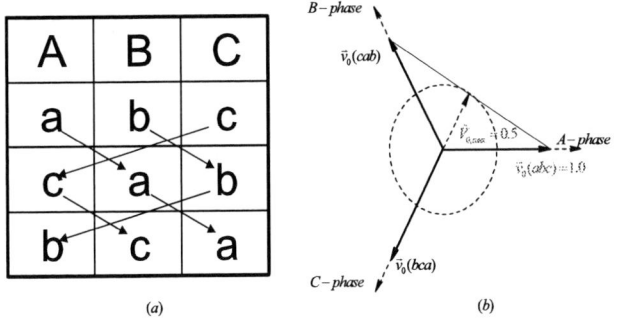

*(a)*      *(b)*

Fig. 3. Output synthesis from counterclockwise (CCW) rotating voltage vectors.

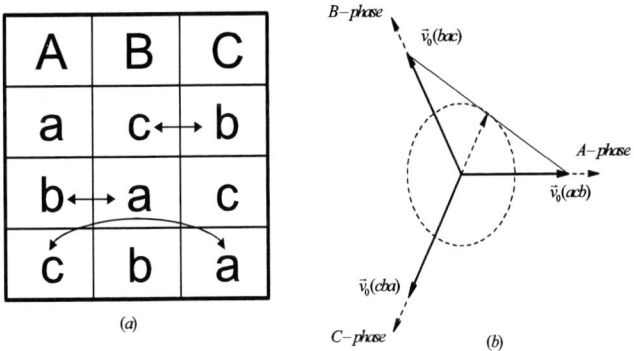

*(a)*      *(b)*

Fig. 4. Output synthesis from clockwise (CW) rotating voltage vectors.

switching combinations is a viable option for PWM generation with zero common mode voltages [11]. There are six possible combinations and they can be divided into two sub-groups. The first sub-group as shown in Fig. 3a results in counterclockwise (CCW) rotating vectors.

These vectors, as shown in Fig. 3b, can synthesize the output voltage, which is limited in the maximum amplitude of 0.5 times the input voltage Similarly, the second sub-group, as shown in Fig. 4a, results in clockwise (CW) rotating vectors. These vectors, as shown in Fig. 4b, can synthesize the output voltage, which is also limited in the maximum amplitude of 0.5 times the input voltage [12].

TABLE I
THE SWITCHING TABLE AND CORRESPONDING OUTPUT LINE VOLTAGE AND COMMON MODE VOLTAGE.

| Group | A B C | $V_{AB}$ | $V_{BC}$ | $V_{CA}$ | $V_{COM}$ |
|---|---|---|---|---|---|
| | $a\ b\ c$ | $v_{ab}$ | $v_{bc}$ | $v_{ca}$ | $0$ |
| | $a\ c\ b$ | $-v_{ca}$ | $-v_{bc}$ | $-v_{ab}$ | $0$ |
| | $b\ c\ a$ | $v_{bc}$ | $v_{ca}$ | $v_{ab}$ | $0$ |
| I | $b\ a\ c$ | $-v_{ab}$ | $-v_{ca}$ | $-v_{bc}$ | $0$ |
| | $c\ a\ b$ | $v_{ca}$ | $v_{ab}$ | $v_{bc}$ | $0$ |
| | $c\ b\ a$ | $-v_{bc}$ | $-v_{ab}$ | $-v_{ca}$ | $0$ |
| | $a\ c\ c$ | $-v_{ca}$ | $0$ | $v_{ca}$ | $(v_c-v_b)/3$ |
| | $b\ c\ c$ | $v_{bc}$ | $0$ | $-v_{bc}$ | $(v_c-v_a)/3$ |
| II-A | $b\ a\ a$ | $-v_{ab}$ | $0$ | $v_{ab}$ | $(v_a-v_c)/3$ |
| | $c\ a\ a$ | $v_{ca}$ | $0$ | $-v_{ca}$ | $(v_a-v_b)/3$ |
| | $c\ b\ b$ | $-v_{bc}$ | $0$ | $v_{bc}$ | $(v_b-v_a)/3$ |
| | $a\ b\ b$ | $v_{ab}$ | $0$ | $-v_{ab}$ | $(v_b-v_c)/3$ |
| | $c\ a\ c$ | $v_{ca}$ | $-v_{ca}$ | $0$ | $(v_c-v_b)/3$ |
| | $c\ b\ c$ | $-v_{bc}$ | $v_{bc}$ | $0$ | $(v_c-v_a)/3$ |
| II-B | $a\ b\ a$ | $v_{ab}$ | $-v_{ab}$ | $0$ | $(v_a-v_c)/3$ |
| | $a\ c\ a$ | $-v_{ca}$ | $v_{ca}$ | $0$ | $(v_a-v_b)/3$ |
| | $b\ c\ b$ | $v_{bc}$ | $-v_{bc}$ | $0$ | $(v_b-v_a)/3$ |
| | $b\ a\ b$ | $-v_{ab}$ | $v_{ab}$ | $0$ | $(v_b-v_c)/3$ |
| | $c\ a\ c$ | $0$ | $v_{ca}$ | $-v_{ca}$ | $(v_c-v_b)/3$ |
| | $c\ b\ c$ | $0$ | $-v_{bc}$ | $v_{bc}$ | $(v_c-v_a)/3$ |
| II-C | $a\ b\ a$ | $0$ | $v_{ab}$ | $-v_{ab}$ | $(v_a-v_c)/3$ |
| | $a\ c\ a$ | $0$ | $-v_{ca}$ | $v_{ca}$ | $(v_a-v_b)/3$ |
| | $b\ c\ b$ | $0$ | $v_{bc}$ | $-v_{bc}$ | $(v_b-v_a)/3$ |
| | $b\ a\ b$ | $0$ | $-v_{ab}$ | $v_{ab}$ | $(v_b-v_c)/3$ |
| | $a\ a\ a$ | $0$ | $0$ | $0$ | $v_a$ |
| III | $b\ b\ b$ | $0$ | $0$ | $0$ | $v_b$ |
| | $c\ c\ c$ | $0$ | $0$ | $0$ | $v_c$ |

## III. MATRIX CONVERTERS ON EACH SIDE OPEN-ENDED MACHINE FOR ELIMINATING COMMON-MODE VOLTAGE

As shown in Fig. 5, if the machine is supplied from opposite ends of the stator winding through two matrix converters, then the net voltage vector across the stator winding is a linear combination of voltage vectors generated from the individual matrix converters. The common mode potential generated from each individual matrix converter must be zero. Fig. 5 illustrates voltages and currents, where the lowercase letters (a, b, and c) are used to designate voltages and currents at the ac side, while capital letters (A, B, and C) are used to designate output voltages from each of theconverters (unprimed and primed) and currents flowing in the phase windings of the ac machine. Using CCW rotating vectors, each of the set of voltage vectors from each of the matrix converters is illustrated in Fig. 6. It can be observed

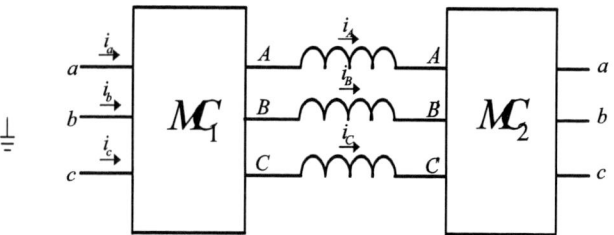

Fig. 5. Matrix converters on each side of the ac machine

from Fig. 6 that the set of voltage vectors generated from switching combinations abc, cab and bca of the second matrix converter, because they are applied to the opposite ends of the motor windings, are 180° phase displaced from the set of voltage vectors generated from switching combinations abc, cab and bca of first matrix converter. Fig. 6 shows that the two sets of voltage vectors, from the two matrix converters, combine vectorially to result in an output vector of motor voltages as shown, across the windings of the AC machine. Six such motor voltage can be obtained as shown. In addition, three zero-vectors without common mode voltages can be generated for controlling the motor voltage amplitude. Each of these voltage vectors rotates at angular speed of $\omega$ counterclockwise around the origin with respect to the stator reference frame of the AC machine.

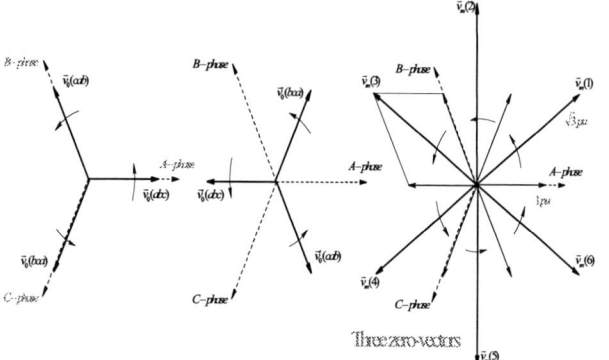

Fig. 6. Applying counterclockwise (CCW) rotating vectors from opposite sides.

In a manner similar to that described above and illustrated in Fig. 7, the three switching combinations acb, bac and cba of the first matrix converter form another set of voltage vectors. The generated vectors from any of these switching combinations always rotate in clockwise direction with an angular frequency of $\omega$. Fig. 7 further illustrates that by combining any two voltage vectors picked separately from the two set of clockwise rotating voltage vectors a set consisting of nine vectors are obtained across the windings of the AC machine. Out of them three are zero vectors. It should be noted that any one of the two output voltage sets (rotating

counterclockwise or clockwise) across the windings of the AC machine is capable of producing the desirable output.

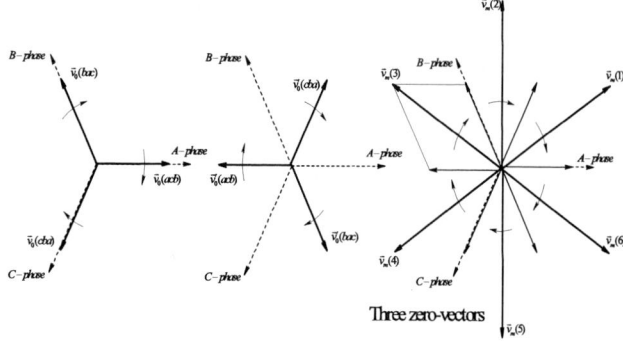

Fig. 7. Applying clockwise (CW) rotating vectors from opposite sides.

The amplitude of the vectors belonging to the switching combinations (abc, cab and bca) and (acb, bac and cba) is same as the input voltage vector amplitude which is equal to $3/2V$, where $V$ is the amplitude of the input phase voltage. Thus, the amplitude of the resultant voltage vector generated by combining the voltage vectors of both converters (Fig. 6 and Fig. 7) equals ($\sqrt{3}$ x 3/2V).

## IV. UTILITY-SIDE POWER FACTOR CONTROL

If the direction of rotation of the resultant output voltage vector coincides with the direction of rotation of the set of vectors generating it, then the input power factor is equal to the motor power factor, otherwise the input power factor is opposite of motor power factor. By using the counterclockwise (CCW) rotating vectors and clockwise (CW) rotating vectors in appropriate ratio for generation of resultant output voltage vector, the input power factor can be controlled in any range. If the output vectors generated from both CCW and CW rotating vectors are same in phase and amplitude during a switching period the input power factor is equal to unity. If controlling the input power factor is not the objective and it is sufficient to operate at either equal or opposite (negative) of the machine power factor, then this property can be used to reduce switching and hence the switching losses to one-half.

If the utility side angular frequency is equal to $\omega$, at any instant the angular displacement of utility side voltage vector is equal to $\theta = \omega t$. In the output side the counterclockwise (CCW) rotating voltage vectors of Fig.6 rotate with an angular frequency equal to $\omega$ in CCW direction and the clockwise (CW) rotating voltage vectors of Fig.7 rotate in opposite direction inside the machine. Similarly if the output side angular frequency is equal to $\omega_o$, at any instant the angular displacement of the machine side current vector is equal to $\theta_o = \omega_o t - \rho$, where **t** is the time and $\rho$ is the output power factor. Correspondingly the switching configurations generating the rotating voltage vectors of Fig.6 will produce a set of current vectors rotating with angular speed of $\omega_o$ and that of Fig.7 will produce current vectors rotating with angular speed of $-\omega_o$ at the utility side.

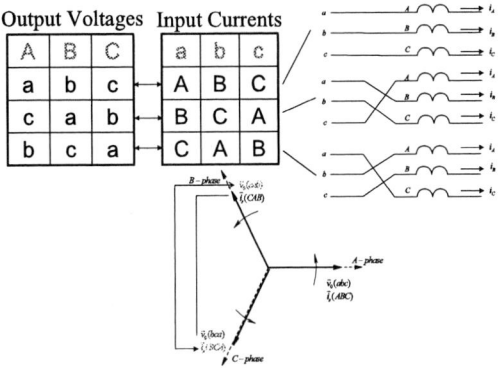

Fig. 8. Utility-side power factor applying CCW vectors.

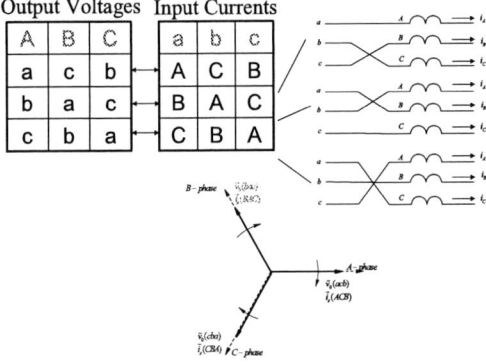

Fig. 9. Utility-side power factor applying CW vectors.

Fig.8 illustrates the connectivity matrix showing output phase voltages in terms of the input phase voltages, and input phase currents in terms of the output phase currents when CCW rotating vectors of Fig.6 are used. Fig.9 shows the relationship between input and output when CW rotating vectors of Fig.7 are used.      In Fig.10a the relationship between output voltage and output current is shown. The output $V_{o\_ccw}$, generated from CCW vectors lags by a phase of $\angle K$ from the resultant output voltage $V_o$ in counter clockwise direction and the output $V_{o\_cw}$, generated from CW rotating vectors lags by a phase of $\angle K$ from $V_o$ in clockwise

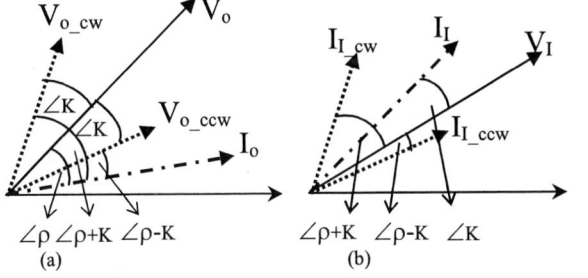

Fig. 10. Relationship between Input side power factor and output side power factor.

direction. Both $V_{o\_ccw}$ and $V_{o\_cw}$ have equal amplitude. The output current $I_o$ lags the resultant output $V_o$ by a phase $\angle\rho$.

Thus     $V_o = V_{o\_ccw} \cos(K) = V_{o\_cw} \cos(K)$ ---------- (1)

Therefore Io lags $V_{o\_ccw}$ by $\angle\rho$-K   in counter-clockwise direction and lags $V_{o\_cw}$ by $\angle\rho$+K in clockwise direction.

The corresponding input side voltage and current are shown in Fig.10b. The input voltage vector $V_I$ rotates in CCW direction. The reflected input current $I_{I\_ccw}$ due to $V_{o\_ccw}$ lags the input voltage vector $V_I$ by $\angle\rho$-K in the CCW direction. The reflected input current $I_{I\_cw}$ due to $V_{o\_cw}$ would lag the input voltage vector $V_I$ by $\angle\rho$+K in the CW direction which implies $I_{I\_cw}$ leads $V_I$ by $\angle\rho$+K in CCW direction.

Therefore the resultant input current $I_I$ leads the input voltage $V_I$ by $(\angle\rho$+K - $\angle\rho$-K$)/2$ i.e. $\angle K$ . Thus the input power factor and output power factor are independent of each other and by maintaining power balance
$V_I\, I_I \cos(K) = V_o\, I_o \cos(\rho)$          ---------- (2)

By varying $\angle K$ the input power factor is controllable in any range. But by operating at non unity power factor the output amplitude is limited to $\cos(K)$ of maximum possible output. By making $\angle K$ = zero unity power factor opertion is achieved.

There is yet another way to control the input power factor without reducing the output voltage capability, provided the input power factor is to be controlled within a range that spans the motor power factor and negative of the motor power factor. In this range, without introducing the phase shift, the sets of CCW and CW vectors can be used in an appropriate ratio to yield the desired input power factor.

Switching between CCW and CW voltage vectors results in switching losses. In controlling the input power factor, switching losses can be minimized, without affecting the output voltage ripple, by transitioning between CCW and CW at a lower frequency, irrespective of the desired input power factor and the method of achieving it. To reduce switching losses where there are a collection of drives, one-half the drives may be operated using CCW rotating vectors and the other one-half using CW rotating vectors, thus resulting in a combined power factor that is essentially unity. Of course, it is possible to obtain the combined power factor to be leading or lagging by unequal distribution of drives using CCW and CW rotating vectors.

V. SIMULATION RESULTS

A open-ended three phase R-L load driven from two matrix converters is simulated in matlab and simulink. The PWM output is generated for unity input power factor operation. The PWM output, the averaged output voltage, output current, input voltage and input current are plotted.

253

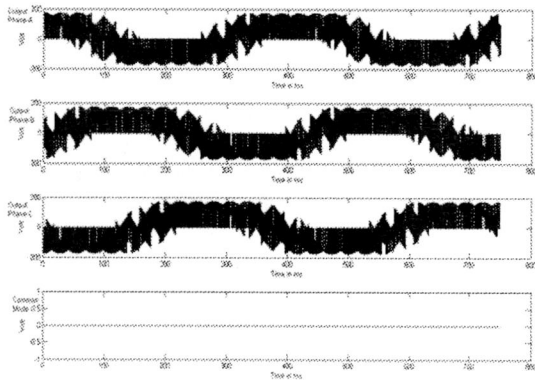

Fig.11a. The PWM output phase voltage and corresponding common mode voltage.

A 60 Hz 100 V per phase three phase source is used in simulation. The PWM output phase voltage and corresponding common mode voltage are plotted in Fig.11a. It may be noted that the common mode voltage is absolutely zero for whole fundamental period of the output.

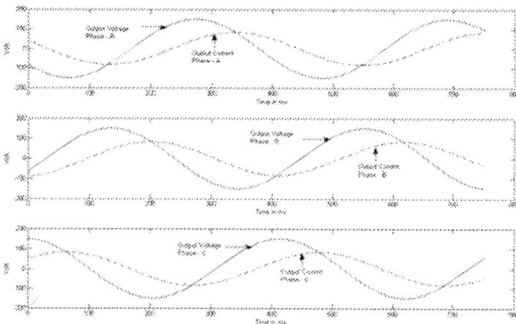

Fig. 11b. The output phase voltage averaged over every switch cycle and the output current.

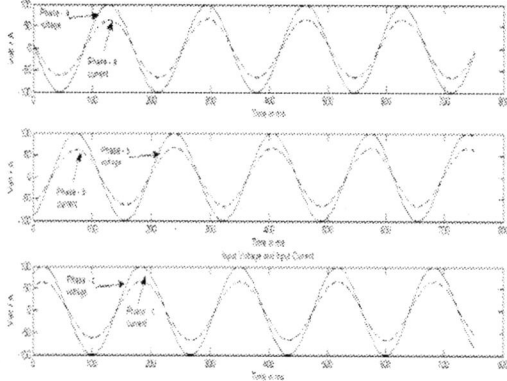

Fig. 11c. The input phase voltage and the input phase current.

The PWM voltage is averaged over every switching cycle and the averaged output phase voltages are plotted in Fig.11b.The lagging output phase currents are also plotted along with the output voltage. It can be observed that the

output phase voltage amplitude is 150V, which is 3/2 times the amplitude of the input phase voltage amplitude of 100V.

The reflected input side phase current along with the input phase voltage plotted in Fig.11c. It can be observed that though the output current is lagging (Fig.11b) the input current, it is in phase with the input voltage(Fig.11c).

## VI. CONCLUSIONS

This paper describes novel topologies and control to extract more power throughput in ac machines, including induction, synchronous, and synchronous-reluctance machines. operating in their motoring or generating modes. Several simultaneous benefits are illustrated, including the elimination of common-mode voltages, switching of which causes bearing currents. These topologies can be used with existing solid-state devices and are ideally suited for SiC devices where elimination of the dc-link capacitor and the common-mode voltages are highly desirable. Some additional topologies with reduced common mode elimination with matrix converter is given in Fig.12a and Fig.12b. The scheme shown in Fig.12a shows a cuircuit arrangement where reduced amount of common mode voltage is generated with 24 switches. Each of the two inverters use six unidirectional switches and the controlled rectifier uses six four-quadrant switches. Thus 12 switches are eliminated from the scheme described in the paper. The controlled rectifier in Fig.12a can be replaced by a uncontrolled rectifier as shown in Fig.12b. The uncontrolled rectifier uses only six unidirectional switches in place of six four-quadrant switches. Thus total number switches can be reduced to 18. In the scheme of 12.b a common mode voltage with harmonics of tripplen order will be available. But switching transition of common mode voltage is eliminated and the input side current is distorted.

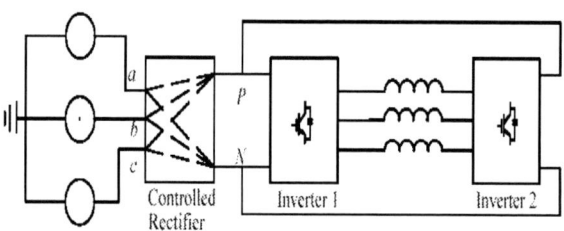

Fig. 12a. The reduced common mode scheme for open_ended drive with controlled rectification.

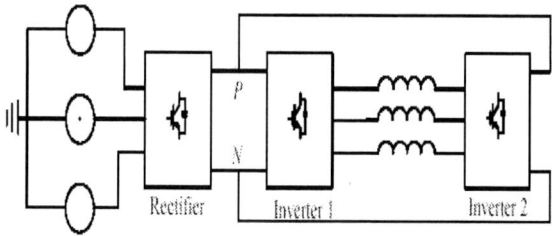

Fig. 12b. The reduced common mode scheme for open_ended drive with uncontrolled rectification.

## VII. REFERENCES

[1] N. Mohan, *First Course in Power Electronics*, year 2005 edition, ISBN 0-9715292-4-8, published by MNPERE (www.mnpere.com).

[2] Gyugyi, L. G. and B. R. Pelly, *Static Power Frequency Changers, Theory, Performance, and Applications*, Wiley, 1976. 442 pages.

[3] Peter Wood, <u>*Switching Power Converters*</u>, New York, Van Nostrand Reinhold Co., c 1981. 446 pages, ISBN: 0442243332.

[4] A. Alesina, M. Venturini, "Analysis and design of optimum amplitude nine-switch direct AC-AC converters," *IEEE Transactions on Power Electronics*, vol: 4, Issue: 1, pp.101 – 112, Jan. 1989.

[5] P.W. Wheeler, J. Rodriguez, J.C. Clare,and L. Empringham, "Matrix converter,A technology review," *IEEE Trans. Ind. Electron*, vol. 49, no. 2, Apr. 2002.

[6] Intellectual Property Protection Application by the University of Minnesota on highly simplified control of Matrix Converters.

[7] N. Mohan, K.K. Mohapatra, P. Jose, A. Drolia, G. Aggarwal, T. Satish, "A Novel carrier based PWM scheme for matrix converters that is easy to implement," *Conf. record of PESC'05*, pp. 2410-2414.

[8] Shivakumar, E.G., Gopakumar, K., and Ranganathan, V.T.: 'Space vector PWM control of dual inverter fed open-end winding induction motor drive'. *IEEE-APEC-2000*, pp. 394-404.

[9] Stemmler and P. Guggenbach, "Configurations of high power voltage source inverter drives," in *Proc. EPE'93 Conference*, Brighton, U.K., 1993, pp. 7–12.

[10] Intellectual Property Protection Application by the University of Minnesota on open-ended drives through matrix converters.

[11] Baiju, M.R.; Mohapatra, K.K.; Kanchan, R.S.; Gopakumar, K.; "A dual two-level inverter scheme with common mode voltage elimination for an induction motor drive" *IEEE Transaction on Power Electronics*, vol 19, Issue 3, May 2004 Page(s):794 - 805

[12] Rzasa, J.; " Control of a matrix converter with reduction of a common mode voltage," *Conf. record of Compatibility in Power Electronics*, 2005 June 1,2005 Page: 213-217.

Krushna K. Mohapatra (SM'00-M'04) received the M.Tech degree in electrical engineering from the Indian Institute of Technology, Kharaghpur, India, in 1996, Ph.D. degree from the Indian Institute of Science, Bangalore, India, in 2004, and is currently a Post Doctorate in the Electrical Engineering Department at the University of Minnesota, Minneapolis, USA. He was a Design and Development Engineer in National Radio and Electronics Company Ltd., from 1995 to 2000. His research interests are in the area of power converters, PWM strategies, and motor drives.

Ned Mohan (M'73-SM'91-F'96) received the Ph.D. degree in electrical engineering from the university of Wisconsin, Madison, in 1973. He is the Oscar A. Schott professor of power electronics at the University of Minnesota, Minneapolis, where he has been teaching since 1976. He has numerous patents and publications in the field of power electronics, electric drive and power systems. He has written five textbooks. Prof. Mohan is a recipient of the Distinguished Teaching Award presented by the Institute of Technology, University of Minnesota.

**2006 IEEE International Conference on Power Electronic, Drives and Energy Systems**

# Elimination of Common Mode Voltage and Fifth and Seventh Harmonics in a Multilevel Inverter fed IM Drive using 12-Sided Polygonal Voltage Space Phasor

Sanjay Lakshminarayanan, *Student Member, IEEE*, Gopal Mondal, *Student Member, IEEE*,
P.N Tekwani, *Student Member, IEEE,* and K. Gopakumar, *Senior Member, IEEE*

*Abstract* - A multilevel inverter with 12-sided polygonal voltage space vector structure is proposed in this paper. The present scheme provides elimination of common mode voltage variation and $5^{th}$ and $7^{th}$ order harmonics in the entire operating range of the drive. The proposed multi level structure is achieved by cascading only the conventional two-level inverters with asymmetrical DC link voltages. The bandwidths problems associated with conventional hexagonal voltage space vector structure current controllers, due to the presence of $5^{th}$ and $7^{th}$ harmonics, in the over modulation region, is absent in the present 12-sided structure. So a linear voltage control up to 12-step operation is possible, from the present twelve sided scheme, with less current control complexity. An open-end winding structure is used for the induction motor drive.

*Index Terms* – **Common Mode Voltage Elimination, Multilevel Inverters.**

## I. INTRODUCTION

VARIOUS multilevel inverter structures have been proposed for high power applications. The neutral point clamped three level inverters, the cascaded H- Bridge and the flying capacitor multi level structures are some of the popular schemes used for high power applications [1]-[7]. Normally the number of switching devices and complexity goes up with the number of levels [3]. All these multi level inverters produce a hexagonal structure for the voltage space phasor from the inverter out put. In the extreme modulation range, these hexagonal voltage space phasor based PWM will have substantial $5^{th}$ and $7^{th}$ harmonics in the motor phase voltage. This requires a different approach for the current controller design taking care of the over modulation region [8].

In inverter fed motor drives, the presence of common mode voltage variations at the pole voltages can lead to leakage currents between motor and the inverter, through motor bearings [9]. Different schemes and modulation techniques have been reported to eliminate common-mode voltage

Sanjay Lakshminarayana ,Gopal Mondal, P.N. Tekwani and K.Gopakumar are with Centre for Electronics Design and Technology, Indian Institute of Science Bangalore-560012, INDIA.
E-mail: kgopa@cedt.iisc.ernet.in

variation at the inverter poles [10][11]. A 12-sided polygonal space vector based multilevel inverter is an improvement over the conventional hexagonal space vector based inverter [12]. For 12-sided voltage space phasor scheme, the linear range of modulation is shown to be 0.64Vdc when compared to 0.577Vdc for a hexagonal space vector scheme (Vdc is the radii of the voltage space vector polygon). The maximum value of the fundamental component that can be generated is 0.658Vdc (12-step operation) compared to 0.637Vdc (six-step operation) in the conventional hexagonal space vector method. Since a 12-sided polygon is used for the space vector PWM control, all the $6n \pm 1$(n=1,3,5..) harmonics will be absent through the modulation range extending up to the 12-step operation. This will enable a simple voltage and current control for the entire modulation range, when compared to the complicated hexagonal space vector based PWM control, especially in the over modulation region [8]. The disadvantage of the 12-sided voltage phasor generation, proposed in [12] is that large common-mode voltage is present at the motor pole voltages. In this paper a method is proposed for achieving the 12-sided polygonal space vectors, for an induction motor drive, with common mode voltage elimination, for the entire modulation range.

## II. POWER CIRCUIT OF THE PROPOSED DRIVE

A schematic of the power circuit of the proposed induction motor drive (open-end winding) is shown in Fig.1. The overall configuration consists of two three-level inverters INV-A and INV-B. One end of each phase is connected to INV- A, and the other end to INV-B. The inverters INV-A and INV-B are further realized by cascading two conventional two-level inverters, fed from two asymmetrical DC link voltages of $1kV_{dc}$ and $0.366kV_{dc}$ (Fig.1) [10]. The factor 'k' is chosen such a way that the voltage space phasor magnitude is made equal to that of hexagonal voltage space phasor amplitude, from a conventional two-level inverter. Each pole of an inverter (INV-A or INV-B) can attain three different voltage levels depending on the inverter switching state. Switches on the same leg such as $S_{21}$ and $S_{24}$ are operated complementary to each other. When switch $S_{24}$ is on, pole A assumes a voltage level of zero, when $S_{14}$ and $S_{21}$ are on the inverter pole A attains 0.366kV$_{dc}$. When S$_{11}$ and S$_{21}$ are on, inverter pole A attains 1.366kV$_{dc}$ (Table 1). A '0' represents the off state and '1" the on state of a switch. A '0/1' implies that the switch can be on or off ('don't care')

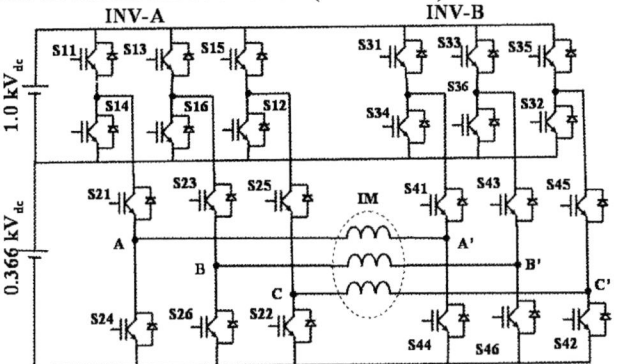

INV-A    INV-B

Fig. 1.   Power circuit of the 12-side polygonal space vector based multilevel.

TABLE I
SWITCHING STATES TO REALIZE VARIOUS VOLTAGE LEVELS IN ONE LEG.

| A phase | | | |
|---|---|---|---|
| Pole voltage | Level | S$_{11}$ | S$_{21}$ |
| 1.366kV$_{dc}$ | 2 | 1 | 1 |
| .366kV$_{dc}$ | 1 | 0 | 1 |
| 0V$_{dc}$ | 0 | 0/1 | 0 |

## III. GENERATION OF 12-SIDED POLYGONAL VOLTAGE VECTORS

Any of the poles of the two inverters (Fig. 1) can take on one of three levels, independently of the other. A 'φ' represents any arbitrary level and a carries a 'don't care' like meaning. Fig. 2 shows twelve pole voltage space vectors from INV-A and INV-B. The voltage space vectors at location P,Q,E,F,I,J ( Fig.2) are realized from the pole voltages of INV-A and the Voltage space vectors at location R,S,G,H,K,L are realized from the pole voltages of INV-B.  Consider the space vector OQ of INV-A shown as (210) (φφφ) in Fig.-2. The Pole- A is at level 2 producing 1.366kV$_{dc}$ ( Table-1) shown by OA along A phase axis, pole B is at level 1 producing 0.366kV$_{dc}$ along B axis shown as AQ, pole C is at level 0,  and the resultant is OQ. The resultant voltage space vector OQ will have a magnitude equal to 1.225 V$_{dc}$. This can be easily verified from the geometry of the Fig.-2[12]. The voltage space vector 'OQ' is generated from INV-A alone and is independent of the switching state of INV-B. So for the voltage space vector generation OQ, the poles of INV-B can be in any state, and is represented by (φφφ) in Fig. 2.  In a similar way, vector OP is a voltage space vector generated form the pole voltages of INV-A, represented by (201)(φφφ). Voltage space vectors OR, OS are close to the  negative axis-c, and are generated form the pole voltages of INV-B. For the voltage space vector 'OR', the pole voltage levels of INV-B are given by (φφφ)(0'1'2'). Here also the INV-A can be in any state. The voltage space vector 'OB' is along the negative axis-c with a magnitude of 1.366kV$_{dc}$, the magnitude of 'BR'

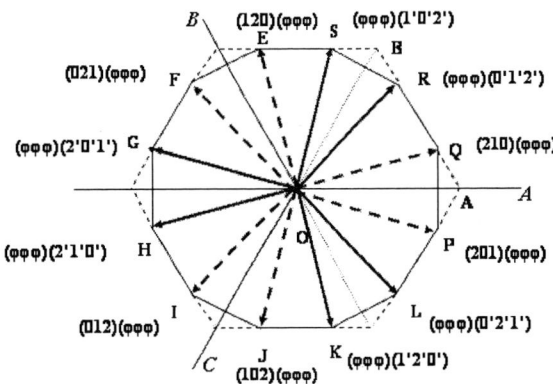

Fig. 2.   Pole voltage space vectors of INV-A and INV-B.

is 0.366kV$_{dc}$ along the negative axis-b, and the vector addition results in the voltage space vector 'OR. The inverter switching states for the other locations are shown in Fig.2. The phase winding of the open-end induction machine is connected across the poles of the two inverters INV-A and INV-B. If we select voltage space vector combinations from the two inverters of INV-A and INV-B, which are 60$^0$ apart, it can be seen that the pole voltages at the opposite end of the motor terminals have identical common mode voltages[10].   The common mode voltage generated at the INV-A pole voltages, for the vector generation 'OP'(Fig.2) is

$$V_{cm-A}(201) = (1.366k\ V_{dc} + 0 + 0.366k\ V_{dc})/3 \qquad (1)$$

Similarly the common voltage generated at the pole voltages of INV-B for the vector generation 'OR' (Fig.3) is

$$V_{cm-B}(021) = (0 + 1.366k\ V_{dc} + 0.366k\ V_{dc})/3 \qquad (2).$$

The combination of these two vectors will have the resultant location at '1' shown in Fig.3. Similar voltage space vector combinations from individual inverters (INV-A and INV-B) can be used to get a resultant 12-sided polygonal voltage space vector structure with zero common mode voltage variation at the inverter poles (Table-2). The resultant 12-sided voltage space vector combinations from INV-A and INV-B are shown in Fig.3. The 12-sided voltage space vector structure of Fig.3 is achieved by the vector addition of voltage space vectors (with a magnitude of 1.225 V$_{dc}$ each) from INV-A and INV-B which are separated by 60$^0$. The resultant radii of the 12-sided polygon generated from the combined inverters is (Fig.3) Magnitude of the vector OR is

$$= 2 \times 1.225 kV_{dc} \times \frac{\sqrt{3}}{2} \qquad (3)$$

The value of 'k' can be selected to be 0.471 ( $1/1.225 \times \sqrt{3}$ ), so that the magnitude of the radii of the 12-sided voltage space vector ( Fig.3) will be 'V$_{dc}$', where V$_{dc}$ is the magnitude of the radii of the hexagonal voltage space vector structure from the conventional 2-level inverter.

Thus by properly selecting the voltage space vectors from INV-A and INV-B of Fig.2, a resultant 12-sided  polygonal voltage space vector structure with zero common mode voltage variation can be achieved for the induction motor drive. The switching state combinations from INV-A and INV-B for the generation of the 12-sided polygonal voltage

257

space vectors with zero common mode voltage are shown in Table-2. All the combinations have identical common mode voltage ((1) and (2)) at the poles of INV-A and INV-B. Now for a

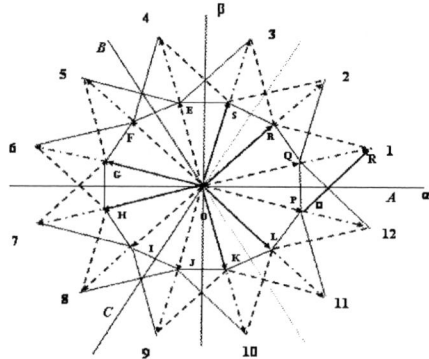

Fig. 3. The twelve resultant voltage space vectors from INV-A and INV-B.

resultant 12-sided polygonal radii of $V_{dc}$ (same as that of a hexagonal structure from the conventional 2-level inverter), The lower DC link requirement of the inverter is ( k=0.338) Lower DC Link voltage (Fig.1)

$$0.366k\ V_{dc} = 0.366 \times 0.471 \times V_{dc} = 0.172\ V_{dc} \qquad (4).$$

And the Upper DC link Voltage (Fig.1)

$$1.0\ k\ V_{dc} = 1.0 \times 0.471 \times V_{dc} = 0.471 V_{dc} \qquad (5)$$

This shows that the proposed multi level inverter with 12-sided polygonal voltage space vector structure can be achieved, with low voltage devices (17% and 47% of the DC link voltage ($V_{dc}$) of a conventional two–level inverter) and the inverter structure can be realized by cascading conventional two-level inverters.

TABLE II
COMBINATIONS GIVING 12-SIDED POLYGONAL SPACE PHASOR.

| space vector location (Fig.3) | INV-A Levels | INV-B Levels | Zero vectors in INV-A | Zero vectors in INV-B |
|---|---|---|---|---|
| 1 | 201 | 0'1'2' | 201 | 2'0'1' |
| 2 | 210 | 1'0'2' | 210 | 2'1'0' |
| 3 | 120 | 0'1'2' | 012 | 0'1'2' |
| 4 | 021 | 1'0'2' | 102 | 1'0'2' |
| 5 | 120 | 2'0'1' | 120 | 1'2'0' |
| 6 | 021 | 2'1'0' | 021 | 0'2'1' |
| 7 | 012 | 2'0'1' | 201 | 2'0'1' |
| 8 | 102 | 2'1'0' | 210 | 2'1'0' |
| 9 | 012 | 1'2'0' | 012 | 0'1'2' |
| 10 | 102 | 0'2'1' | 102 | 1'0'2' |
| 11 | 201 | 1'2'0' | 120 | 1'2'0' |
| 12 | 210 | 0'2'1' | 021 | 0'2'1' |

## IV. PWM SIGNAL GENERATION

Consider the case of a reference space vector $V_r$ moving inside the 12-sided polygon formed by space vectors from INV-A and INV-B, in Figure. 3. The rotating voltage space vector $V_r$ is sampled with a time period 'T$_S$'. The reference vector can be generated by time averaging the two nearest voltage space vectors, forming a sector (12- sectors), within which the reference voltage falls. For example, for the

reference vector $V_r$ lying in the sector between the space vector locations '2' and '3' in Figure. 3, the vector location at '2' is switched on for time period $T_1$ and the vector '3' for time $T_2$. A zero voltage is maintained for a time $T_0 = T_S - T_1 - T_2$. In the present scheme the zero voltage is maintained for time period $T_0/2$ at the beginning of the sampling period $T_S$ and at the end of the sampling period. Note that a zero vector is chosen as in Table-2, such that the there is no variation in the common-mode voltage in INV-A and INV-B.

### A. $T_1$ and $T_2$ Computation from Sampled Reference Phase Amplitudes:

Let V be the magnitude of the space vectors 1-12 in figure. 3, and 'm' be the sector number of the sector in which $V_r$ lies.

In Fig. 3 the α- axis is not coinciding with the sector side (Fig.3)).If the twelve voltage space vectors were shifted by $15^0$ clockwise, the α- axis will coincide with sector side ( Fig.4).

$$T_1.V \angle (m-1)30^0 + T_2.V \angle m30^0 = T_S.(V_\alpha + jV_\beta)$$
(6)

where '*m*'is the sector (*m* varies from 1 to 12*)*

On equating real and imaginary parts and simplifying we get:

$$\begin{pmatrix} T_1 \\ T_2 \end{pmatrix} = \frac{2T_S}{V} \begin{pmatrix} \sin(m30^0) & -\cos(m30^0) \\ -\sin((m-1)30^0) & \cos((m-1)30^0) \end{pmatrix} \times \begin{pmatrix} V_\alpha \\ V_\beta \end{pmatrix} \quad (7)$$

$V_\alpha$ and $V_\beta$ can be substituted in terms of the sampled reference phase amplitudes, $v_A$, $v_B$ and , $v_C$ using the following relationships:

$$V_\alpha = \frac{3}{2} v_A = -\frac{3}{2}(v_B + v_C) \ , \quad V_\beta = \frac{\sqrt{3}}{2}(v_B - v_C) \quad (8)$$

In this way the switching periods $T_1$ and $T_2$ can be evaluated in each sector, in terms the sampled reference phase amplitudes, in that sector, and it does not require any look up tables and complicated computations.

### B. Sector Identification

Fig.4 shows the 12-sided polygonal vectors with the sectors numbered in clock wise direction. The combined voltage space vector structure of Fig.4 is also divided into four quadrants. The quadrant in which the reference voltage lies is found out from the phase voltages as below,

'$v_A$' positive, and '$(v_B - v_C)$' positive: 1st quadrant.
'$v_A$' negative, and '$(v_B - v_C)$' positive: 2nd quadrant.
'$v_A$' negative, '$(v_B - v_C)$' negative: 3rd quadrant.
'$v_A$' positive, '$(v_B - v_C)$' negative: 4th quadrant.

Once the quadrant is found, the sector in which the reference vector lies can be found using the following method.

If in quadrant 1: If $|v_B - v_C| \le |v_A|$ then sector 1 else If $|v_B - v_C| \le 3.|v_A|$ then sector 2 else sector 3. In quadrant 2: If $|v_B - v_C| \le |v_A|$ then sector 6 else If $|v_B - v_C| \le 3.|v_A|$

258

then sector 5 else sector 4. In quadrant 3:

If $\left|v_B - v_C\right| \le \left|v_A\right|$ then sector 7 else If $\left|v_B - v_C\right| \le 3.\left|v_A\right|$

then sector 8 else sector 9. In quadrant 4:

If $\left|v_B - v_C\right| \le \left|v_A\right|$ then sector 12 else If $\left|v_B - v_C\right| \le 3.\left|v_A\right|$

then sector 11 else sector 10.

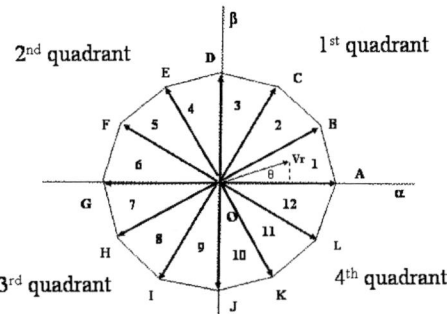

Fig. 4. Sectors and quadrants of 12-side polygonal space vector.

Here also it can be seen that the sector identification is also very fast and only requires sampled reference phase amplitudes, and it can be easily implemented in a drive. scheme with DSP based controllers. Once the inverter switching times are computed for various sectors, the voltage space vector structure of Fig.3 can be realized by using the inverter gating sequence from Table-2., for the corresponding sector.

## V. SIMULATION AND EXPERIMENTAL VERIFICATION

A simple V/f scheme is used for the present study. The drive scheme is first simulated using Simulink and later experimentally verified by using a digital processor based control circuit along with IGBT inverters and an open-end winding induction machine.

### A. Number of samples in a sector for different speed ranges

The number of samples in a sector is so chosen so that the overall switching frequency is limited (<1000) in order to keep switching losses low, while maintaining sufficient resolution.

If 'f' be the operating frequency, the following sampling scheme is used:

1) 0<f<=15Hz: 4 samples per sector
2) 15<f<=30Hz: 3 samples per sector
3) 30<f<=45Hz: 2 samples per sector
4) 45<f<=50Hz: 1 sample per sector.

A total DC-link voltage of 124V is used for the simulation study. The upper DC-link voltage is 91V and the lower one 33V. The pole voltages of INV-A and INV-A' and the corresponding motor phase voltage are plotted, for different speed ranges in Fig.5. The pole voltages are measured with respect to the negative terminal of the lower DC link. The pole voltage shows the three level structures (asymmetric) of the inverters INV-A and INV-B. It can be noted from the pole voltages that the high voltage inverter (Upper inverter of the cascaded structure) is less switched in a cycle of operation compared to the low voltage inverter. In an inverter pole, the

high voltage inverter leg is switched only for the 50% duration, in a cycle of operation. This is very clear from the 50

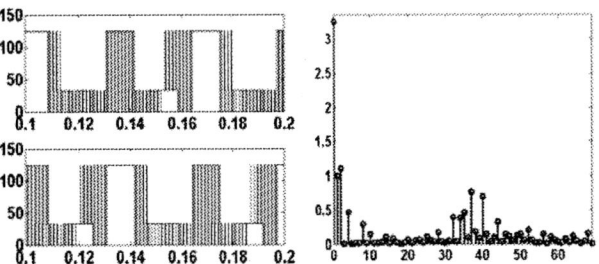

Fig. 5a. Pole voltages of INV-A and INV-B and harmonics in the pole voltage, at a operating frequency of 15Hz.

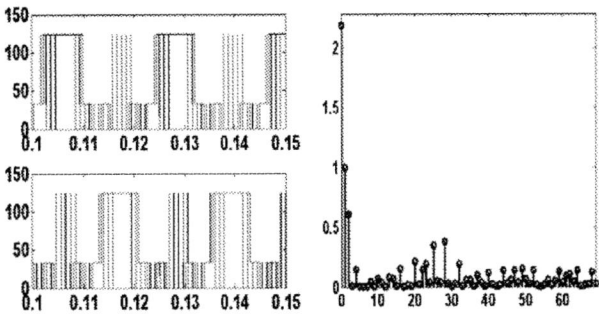

Fig. 5b. Top and bottom waveform: pole voltages at 45 Hz. Also the harmonic spectrum of the pole voltage.

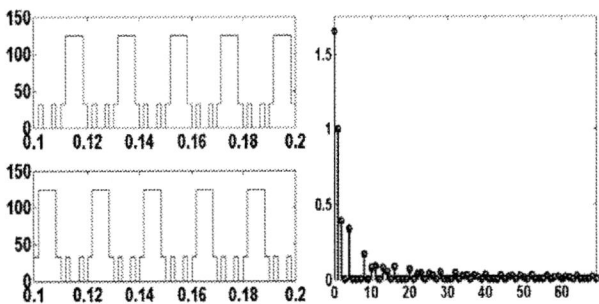

Fig. 5c. Top and bottom waveform: pole voltages at 50 Hz. The harmonic components of the pole voltage is shown.

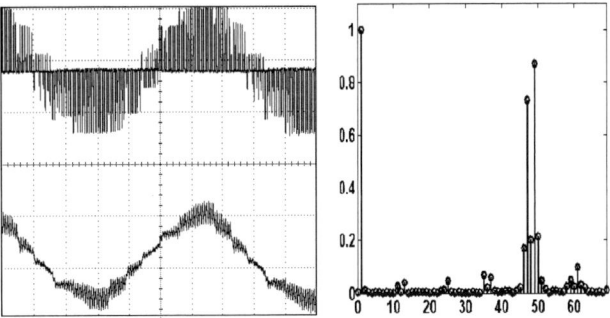

Fig. 6a. Phase voltage, current and relative harmonic spectrum at 15Hz. x-axis: 1div=10ms, y-axis: upper trace: 1div=100V, lower trace: 1div=1A.

Hz operation of Fig.5c.The motor phase voltage and current waveforms from the experimental set up is shown in Fig.6. The proposed scheme is experimentally verified for a 2kW induction motor drive .A simple V/f scheme is implemented using a TMS 320 F2407 platform. This will also considerably reduce the inverter switching losses, for the proposed

259

multilevel inverter scheme. The fundamental and the harmonic amplitudes of the motor phase voltage is computed for different speeds of operation up to 50 Hz operation (12-step mode), for the above mentioned number of samples in a sector.

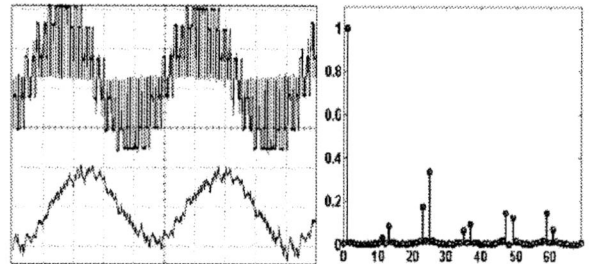

Fig. 6b. Phase voltage current and the relative harmonic spectrum nt at 45Hz. X-axis: 1div =5ms, y-axis: 1div=70V, lower trace : 1div=1 A.

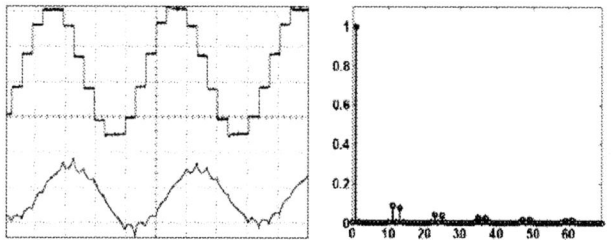

Fig. 6c. Phase voltage and current at 50Hz. X-axis: 1div =5ms, y-axis: upper trace: 1div=70V, lower trace: 1div=1A.

The harmonic spectrum of the pole voltages and the phase voltages show the absence of triplen order (zero common mode voltage). It can also be seen that the lower order harmonic amplitudes are highly suppressed with the total absence of $5^{th}$, $7^{th}$, $17^{th}$ and $19^{th}$ etc., harmonics.

This will enable a simple current control operation and special compensated synchronous reference frame controllers [8] are not needed, for the proposed scheme, when used in high dynamic performance applications such as vector control. Also, the proposed scheme gives a very simple linear control through the modulated range. The experimental results and the harmonic analysis shows that the proposed 12-sided polygonal voltage space phasor based PWM drive is capable of eliminating the common mode voltage variation at the poles and in the motor phase for the entire modulation range.

## VI. CONCLUSION

A 12-sided polygonal voltage space phasor generation for an induction motor drive with common mode elimination is proposed in this paper, for an open-end winding IM drive. The proposed multi level structure is achieved by using only the conventional two-level inverters and by cascading them. The inverters are fed from two asymmetrical DC link voltages of the ratio 1:0.366.When compared to a conventional two level inverter( with a DC link of $V_{dc}$ ), the voltage requirement for the high DC link, for the present scheme is only 0.471 $V_{dc}$ and for the low DC link the voltage requirement is only 0.172 $V_{dc}$ . The proposed PWM voltage control is based on a 12-

sided polygon and it will give an increased modulation range, with the absence of 6n ± 1(n=1,3,5..) order harmonics for the entire modulation range. In the extreme modulation range, the phase peak maximum amplitude is 0.658 $V_{dc}$. Complicated compensated synchronous reference frame PI controllers [8] are not needed in the over modulation region, for operation in high dynamic performance applications, such as vector control scheme. Also, the proposed scheme gives motor operation with zero common mode voltage. So the proposed scheme can be considered for inverter fed induction motor drive for low and medium voltage range high power applications.

## VII. ACKNOWLEDGEMENT

The authors would like to thank Mr. Bhartendu Sinha of Texas Instruments, for providing the DSP tools for the experimental work

## VIII. REFERENCES

[1] A. Nabae, I. Takahashi, and H. Akagi, "A new neutral point clamped PWM inverter," *IEEE Transactions on Industry Applications*, vol 1A-17, No.5, Sept/ pp.518-523, Oct. 1981.

[2] P. M. Bhagawat, and V. R. Stefanovic, "Generalised structure of a multi-level PWM inverter," *IEEE Transactions on Industry Applications*, Vol.1A-19, No.6, pp.1057-1069, Nov/Dec 1983.

[3] J-S. Lai and F. Z. Peng, "Multi-level converters-a new breed of power converters", *IEEE Transactions on Industry Applications*, Vol. 32, No. 3, pp: 509 – 517, May-June 1996.

[4] M. D. Manjrekar, P. K. Steimer, and T. A. Lipo, "Hybrid multi level power conversion systems A competitive solution for high- power applications", *IEEE Transactions on Industry Applications*, vol.IA-36, pp. 834-841, May-June 2000.

[5] J. Rodriguez, J. S. Lai, and F. Z. Peng, "Multi-level inverters: A survey of topologies, controls, and Applications", *IEEE Transactions on Industrial Electronics*, Vol. 49, No. 4, pp. 724 – 738,Aug.2002,

[6] K. Corzine, and Y. Familiant, "A new cascaded multilevel H-bridge drive", *IEEE Transactions on power electronics*. Vol.17, NO.1, pp. 125-131, January 2002.

[7] A. Rufer, M. Veenstra, and K. Gopakumar, "Asymmetrical multilevel converters for high resolution voltage phasor generation", *Conf. Proc. EPE'99*, Lausanne. pp. 1-10,

[8] A. M. Khambadkone, J. Holtz, "Compensated synchronous PI Current controller in overmodulation range and six-step operation of space-vector modulation- based vector- controlled drives," *IEEE Trans. on Industrial electronics*, Vol.49, No.3, pp.574-580, 2002.

[9] S. Chen, T. A. Lipo, and D. Fitzgerald, "Source of induction motor bearing currents caused by PWM inverters", *IEEE Trans. Energy Converse.*, pp. 25-32, 1996.

[10] R. S. Kanchan, P. N. Tekwani, M. R. Baiju, K. Gopakumar and A. Pittet, "Three-level inverter configuration with common-mode voltage elimination for induction motor drive", *IEE Proc.-Electr. Power Appl.*, Vol. 152, No. 2, April 2005, pp. 261-270.

[11] P. N. Tekwani, R. S. Kanchan, K. Gopakumar, and A. Vezzini, "A five-level inverter topology with common-mode voltage elimination for induction motor drives," *11th European Conference on Power Electronics and Applications, EPE-2005*, 11-14, Dresden, Germany, no. 003, pp. 1-10, September 2005.

[12] K. K. Mohapatra , K.Gopakumar, V.T. Somasakhar and L.Umanand: "A Harmonic Elimination and Suppression Scheme for an Open-End Winding Induction Motor Drive." *IEEE transactions on Industrial Electronics*, Vol. 50, No.6, pp. 1187-1198,December 2003.

## IX. Biographies

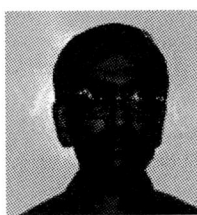

**Sanjay Lakshminarayanan** did his B.Tech from IIT Kharagpur, West Bengal, India and his Masters from EE dept. of Indian Institute of Science, Bangalore. He has nearly 10 years R&D experience with HICAL Magnetics, GE Medical systems India and Electrohms. At present he is pursuing his Doctoral programme in Power Electronics at CEDT, Indian Institute of Science, Bangalore, India.

**Gopal Mondal** received his BE from College of Engineering and management, Kolaghat West Bengal, INDIA. He did his ME in Control systems from Jadavpur University, Calcutta, West Bengal. At present he is pursuing his PhD programme in Power electronics, at CEDT, Indian Institute of Science, Bangalore.

**P. N. Tekwani** received his B.E. degree (Uni. First, Gold Medallist) in Power Electronics from the LEC (Saurashta University), Morbi, India, in 1995, the M.E. degree in Electrical Engineering (Industrial Electronics, First Rank) from the M.S. University, Vadodara, India, in 2000, and is currently pursuing the Ph.D. degree at CEDT, Indian Institute of Science, Bangalore, India. He was with Amtech Electronics Pvt. Ltd., Gandhinagar, INDIA from 1995 to 1996. From 1996 to 2001, he was with Electrical Research and Development Association (ERDA), Vadodara, India, and since 2001, he has been a Member of the Faculty at the Nirma Institute of Technology (Nirma University of Science and Technology), Ahmedabad, India.

**K. Gopakumar** received his B.E., M.Sc.(Engg.) and Ph.D. degrees from Indian Institute of Science in 1980, 1984 and 1994 respectively. He was with the Indian Space Research Organization from 1984 to 1987. He is currently Associate Professor at CEDT, Indian Institute of Science, Bangalore, INDIA. He is an associate editor for IEEE transactions on Industrial Electronics. His fields of interest are Power Converters, PWM Techniques and AC Drives.

**2006 IEEE International Conference on Power Electronic, Drives and Energy Systems**

# A New Space Vector Pulsewidth Modulation for Reduction of Common Mode Voltage in Direct Torque Controlled Induction Motor Drive

Y.V. Siva Reddy, T. Brahmananda Reddy, and M. Vijaya Kumar

*Abstract--* This paper presents a new space vector pulsewidth modulation (SVPWM) algorithm for reduction of common mode voltage in direct torque controlled induction motor drives. The proposed PWM technique does not use any zero voltage vectors for inverter control; hence it can restrict common mode voltage better than conventional SVPWM algorithm. The main advantage of the proposed algorithm is that the number of switchings required in one sector and from one sector to the next sector remains the same as in conventional SVPWM technique. To validate the proposed method, the simulation results are presented and compared to those obtained with conventional SVPWM based direct torque control algorithm.

*Index Terms--* Common mode voltage, direct torque control, SVPWM.

## I. NOMENCLATURE

| | |
|---|---|
| $v_{qs}, v_{ds}$ | d and q axis stator voltages, V |
| $v_{dr}, v_{qr}$ | d and q axis rotor voltages, V |
| $i_{ds}, i_{qs}$ | d and q axis stator currents, A |
| $i_{dr}, i_{qr}$ | d and q axis rotor currents, A |
| $\omega_r$ | Electrical rotor speed, rad/sec |
| $T_e$ | Electromagnetic torque, N-m |
| $\lambda_{ds}, \lambda_{qs}$ | Stator flux linkages in d and q axes, V-sec |
| $\lambda_{dr}, \lambda_{qr}$ | Rotor flux linkages in d and q axes, V-sec |
| $V_{com}$ or $V_{sn}$ | Common mode voltage |
| $V_{an}, V_{bn}, V_{cn}$ | Pole voltages |

---

Y.V. Siva Reddy and T. Brahmananda Reddy are with the Department of Electrical and Electronics Engineering, G. Pulla Reddy Engineering College, Kurnool, A.P, INDIA-518002. (e-mail: yvsreddy_123@rediffmail.com, tbnr@rediffmail.com)
Dr. M. Vijaya Kumar is with the Department of Electrical and Electronics Engineering, Jawaharlal Nehru Technological University, Anantapur, AP, INDIA. (e-mail:mvk_2004@rediffmail.com)

## II. INTRODUCTION

IN the recent years, direct torque control (DTC) [1]-[2] has proven to be a powerful method for controlling induction motors. Despite being simple, DTC is able to produce very fast torque and flux control and also robust to parameter variations. However its inherent fast switching frequency generates high level common mode voltage variations, thus causing the drive itself to be less reliable. Also the common mode voltage results in undesired electromagnetic interference [3], fault actuation of detection circuits [4] and damage to motor bearings [5]. To mitigate the problems of common mode voltage, a new DTC algorithm is presented in [6], which is based on the application of only odd or only even voltage vectors in each sector in which the stator flux lies. Though, it reduces the common mode voltage variations, but it does not overcome the problems like variable switching frequency. To investigate the effects of PWM methods on common mode voltage and also to reduce the common mode voltage, various space vector based PWM algorithms have been considered in [7]. To reduce the common mode voltage, an asymmetrical PWM algorithm and a space vector based PWM technique which does not use any zero voltage vectors have been proposed in [8]. In case of asymmetrical PWM technique the switching number inside the sector is not the same and switching pattern is not symmetrical as in conventional method and hence increases current disturbance. In case of new space vector based PWM technique though the switching pattern is symmetrical and number of switchings with in the sector remains the same as in conventional SVPWM, but the number of switchings from one sector to the next are more and hence switching losses increases.

In this paper, a novel SVPWM technique has been proposed which reduces the common mode voltage. In the proposed SVPWM technique instead of zero voltage vectors, non-zero nonadjacent vectors are used in composing the reference voltage vector. Also the proposed technique preserves the best of conventional SVPWM in two ways 1) The vector times associated with the proposed technique are same as that of conventional SVPWM technique and 2) The number of commutations in a sector and from one sector to the next remains the same as in

0-7803-9771-1/06/$25.00 ©2006 IEEE

conventional SVPWM technique.

## III. MACHINE MODELING

The induction motor model can be developed from its fundamental electrical and mechanical equations. In stationary reference frame the voltage equations are given by

$$
\begin{aligned}
v_{ds} &= R_s i_{ds} + p\lambda_{ds} \\
v_{qs} &= R_s i_{qs} + p\lambda_{qs} \\
0 &= R_r i_{dr} + \omega_r \lambda_{qr} + p\lambda_{dr} \\
0 &= R_r i_{qr} - \omega_r \lambda_{dr} + p\lambda_{qr}
\end{aligned}
\tag{1}
$$

Where $p$ indicates the differential operator (d/dt). The stator and rotor flux linkages are defined using their respective self leakage inductances and mutual inductance as given below

$$
\begin{aligned}
\lambda_{ds} &= L_s i_{ds} + L_m i_{dr} \\
\lambda_{qs} &= L_s i_{qs} + L_m i_{qr} \\
\lambda_{dr} &= L_r i_{dr} + L_m i_{ds} \\
\lambda_{qr} &= L_r i_{qr} + L_m i_{qs}
\end{aligned}
\tag{2}
$$

The electromagnetic torque in the stationary reference frame is given as

$$
T_e = \frac{3}{2}\frac{P}{2}\left(\lambda_{ds} i_{qs} - \lambda_{qs} i_{ds}\right)
\tag{3}
$$

## IV. COMMON MODE VOLTAGE

The common mode voltage generated by a Voltage source inverter controlled drive as shown in Fig.1 is given by

$$
V_{com} = \frac{1}{3}(V_{an} + V_{bn} + V_{cn})
\tag{4}
$$

Where $V_{an}, V_{bn}, V_{cn}$ are inverter output phase voltages

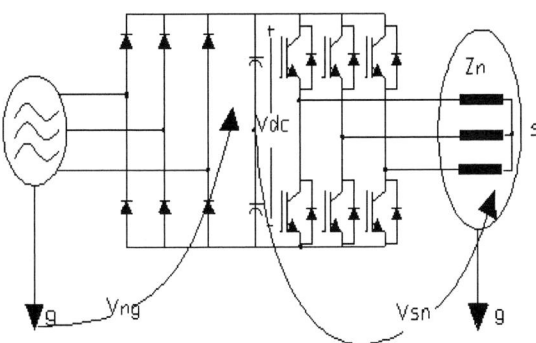

Fig. 1. Three phase voltage source inverter.

If the drive is fed by balanced three phase supply, the common mode voltage is zero. But, the common mode voltage exists inevitably when the drive is fed from an inverter employing PWM technique. It can be shown that the switching state and DC bus voltage decides the common mode voltage. There are eight available output voltage vectors in accordance with the eight different switching states of the inverter as depicted in Fig. 2.

The common mode voltage changes by $V_{dc}/3$ for every switching state of the inverter and the variation of $V_{com}$, the common mode voltages according to output voltage vectors of the inverter are summarized in Table. 1.

According to the switching states of the inverter the common mode voltage can be expressed as

$$
V_{com} = \frac{V_{dc}}{6}(S_a + S_b + S_c)
\tag{5}
$$

Where $S_a, S_b, S_c$ denotes the switching states of each phase. The motor pole voltages can be derived as

$$
\begin{aligned}
V_{an} &= \frac{V_{dc}}{6}(2S_a - S_b - S_c) \\
V_{bn} &= \frac{V_{dc}}{6}(2S_b - S_c - S_a) \\
V_{cn} &= \frac{V_{dc}}{6}(2S_c - S_a - S_b)
\end{aligned}
\tag{6}
$$

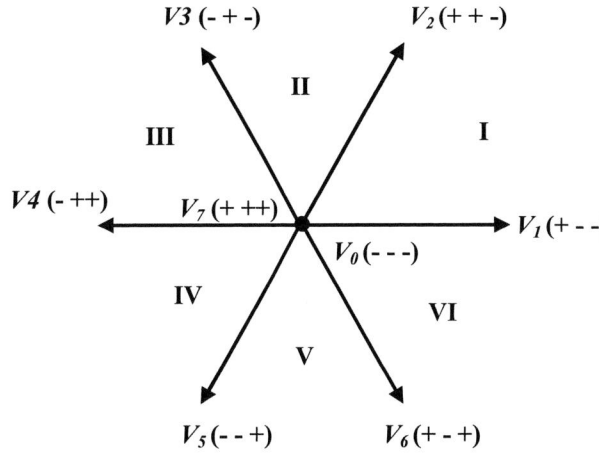

Fig. 2. Voltage space vectors of a three-phase voltage source inverter,

TABLE I
COMMON MODE VOLTAGE AND OUTPUT VOLTAGE GENERATED BY A SWITCHING STATE

| Switching state | Inverter output voltage | | | $V_{com}$ |
|---|---|---|---|---|
| | $V_{an}$ | $V_{bn}$ | $V_{cn}$ | |
| (-1,-1,-1) | $-V_{dc}/2$ | $-V_{dc}/2$ | $-V_{dc}/2$ | $-V_{dc}/2$ |
| (-1,-1, 1) | $-V_{dc}/2$ | $-V_{dc}/2$ | $V_{dc}/2$ | $-V_{dc}/6$ |
| (-1, 1,-1) | $-V_{dc}/2$ | $V_{dc}/2$ | $-V_{dc}/2$ | $-V_{dc}/6$ |
| (-1, 1, 1) | $-V_{dc}/2$ | $V_{dc}/2$ | $V_{dc}/2$ | $V_{dc}/6$ |
| (1,-1,-1) | $V_{dc}/2$ | $-V_{dc}/2$ | $-V_{dc}/2$ | $-V_{dc}/6$ |
| (1,-1, 1) | $V_{dc}/2$ | $-V_{dc}/2$ | $V_{dc}/2$ | $V_{dc}/6$ |
| (1, 1,-1) | $V_{dc}/2$ | $V_{dc}/2$ | $-V_{dc}/2$ | $V_{dc}/6$ |
| (1, 1, 1) | $V_{dc}/2$ | $V_{dc}/2$ | $V_{dc}/2$ | $V_{dc}/2$ |

## V. CONVENTIONAL SVPWM TECHNIQUE

If a constant reference voltage vector $V_{ref}$ at an angle $\alpha$ in any given sector is to be generated then it can be done by using two zero vectors $V_0$ and $V_7$ in combination with two adjacent non zero vectors $V_n$ and $V_{n+1}$ as given in Fig. 3.

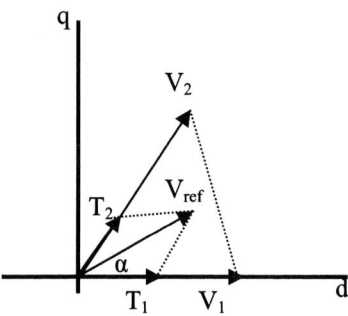

Fig. 3. Vector composition of reference vector for conventional SVPWM technique in first sector.

The times $T_1$, $T_2$ and $T_0$ for which $V_n$, $V_{n+1}$ and $V_o$ or $V_7$ (the duration of zero vector) act respectively can be expressed as

$$T_1 = T.M. \frac{\sin(60^o - \alpha)}{\sin 60^o}$$

$$T_2 = T.M. \frac{\sin \alpha}{\sin 60^o} \qquad (7)$$

$$T_0 = T - T_1 - T_2$$

Where T=sampling period, $T_1$=the duration of vector $V_n$, $T_2$= the duration of vector $V_{n+1}$ and $T_0$=the duration of zero vector $V_0$ or $V_7$ and M is the modulation index, given by $\frac{3V_{ref}}{2V_{dc}}$. The zero vectors $V_0$ and $V_7$ are applied for the same duration. $V_0$ is equally applied at the beginning and at the end of the switching period, where as $V_7$ is applied at the middle and the switching sequence in Sector I are 0127-7210.

## VI. Novel svpwm based DTC

Block diagram of Fig. 4 depicts the direct torque control of induction motor using the proposed SVPWM algorithm. In this approach the actual stator flux vector is determined by using the stator currents and voltage measurements. The calculated one is then compared with a reference and an error signal in flux is generated which when divided by the sampling time period gives a reference voltage vector used for direct control of torque and flux. In the proposed SVPWM technique for direct torque controlled induction motor drive there is no usage of zero vectors in the composition of reference vector. The vector composition of the reference vector in first sector is as shown in Fig. 5. The switching sequences in each sector for the proposed SVPWM technique are given in Table 2.

The switching sequences illustrates that the number of switchings are same as in conventional SVPWM technique. Also, the no of switchings from one sector to the other are same as in conventional SVPWM. Hence the novel SVPWM method preserves the best of conventional SVPWM method in addition decreses the common mode voltage.

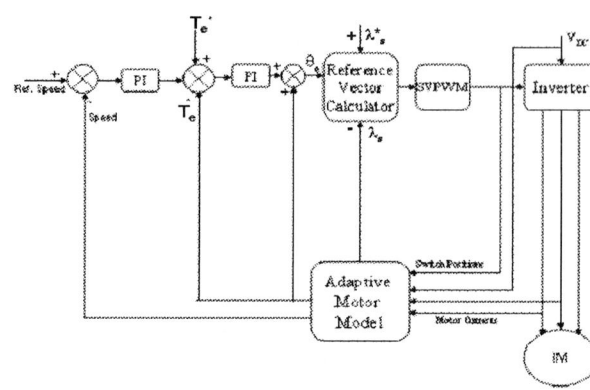

Fig. 4. Block diagram of proposed SVPWM based DTC.

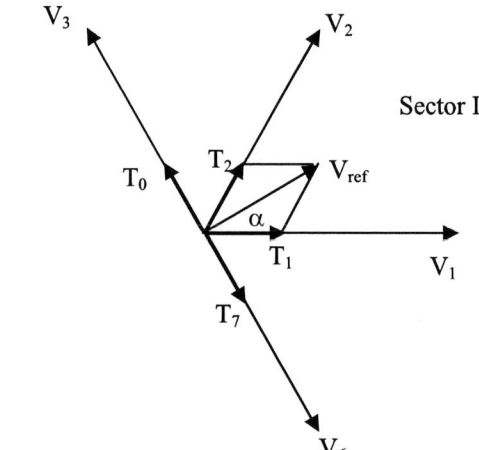

Fig. 5. Vector composition of the reference vector for the proposed method in first sector.

TABLE II
SWITCHING SEQUENCES OF PROPOSED PWM TECHNIQUES

| Sector number | Proposed SVPWM technique |
|---|---|
| I | 3216-6123 |
| II | 4321-1234 |
| III | 5432-2345 |
| IV | 6543-3456 |
| V | 1654-4561 |
| VI | 2165-5612 |

## VII. SImulation results and discussion

To verify the proposed scheme, a numerical simulation has been carried out by using Matlab/Simulink. Sampling time of 125µs and ode 4 (Runge-Kutta) methods are used for a fixed step size of 10µs. For the simulation, reference stator flux is taken as 1wb and starting torque is limited to 40 N-m. The induction motor used in this case study is a 4 KW, 400V, 30 N-m, 1470 rpm, 4-pole, 50 Hz, 3-phase induction motor having the following parameters:

$R_s = 1.57\Omega$  $R_r = 1.21\Omega$, $L_s = 0.17H$, $L_r = 0.17H$,
$L_m = 0.165$ H , $J = 0.089$ Kg.m$^2$

Various conditions of the drive system such as starting, steady state, step change in load and speed reversal are simulated. The simulation results for conventional SVPWM are given in Fig. 6 – Fig. 7.

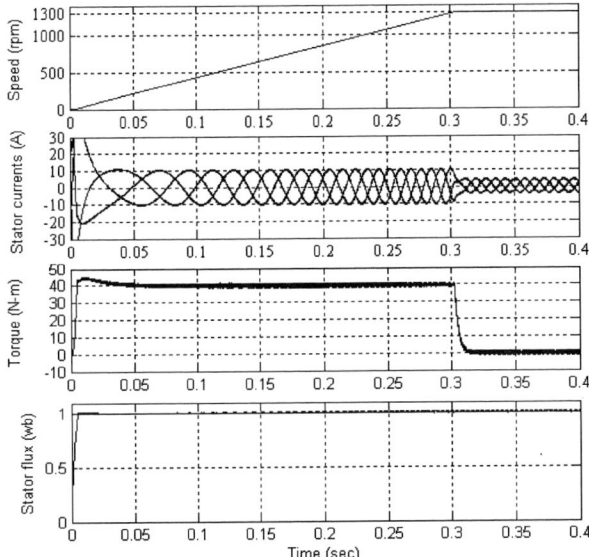

Fig. 6. Simulation results of conventional SVPWM based DTC: starting transients.

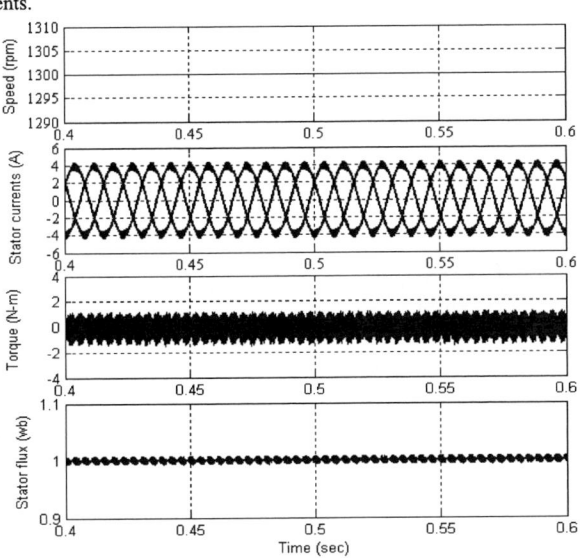

Fig. 7. Simulation results of conventional SVPWM based DTC: steady state plots at a speed of 1300rpm.

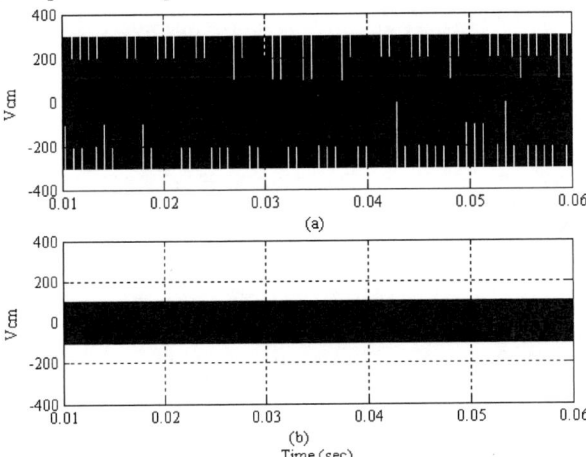

Fig. 8. Common mode voltages with (a) conventional SVPWM and (b) proposed SVPWM.

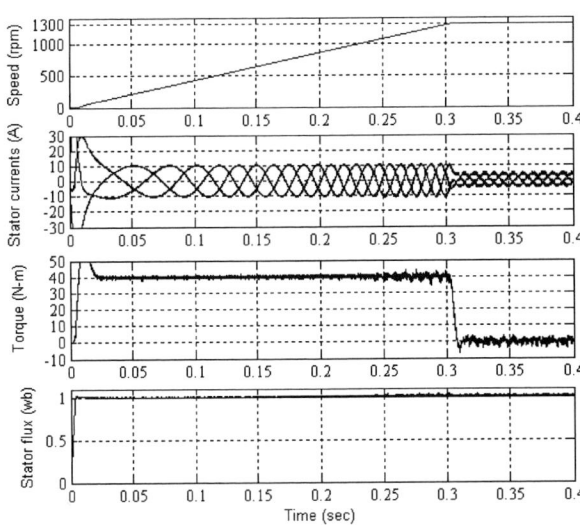

Fig. 9. Simulation results of proposed SVPWM based DTC: starting transients.

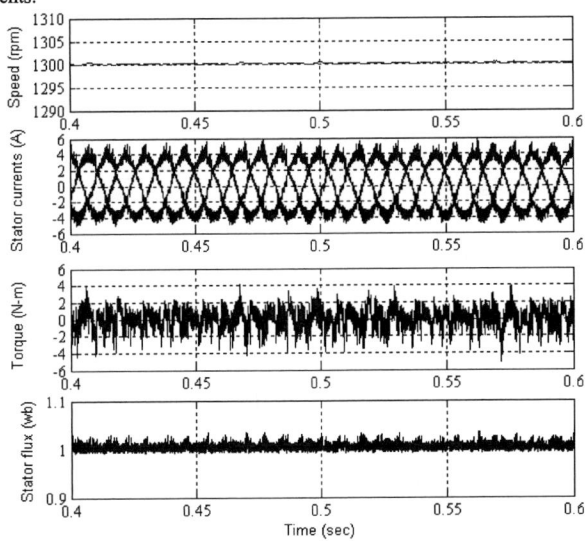

Fig. 10. Simulation results of proposed SVPWM based DTC: steady state plots at a speed of 1300rpm.

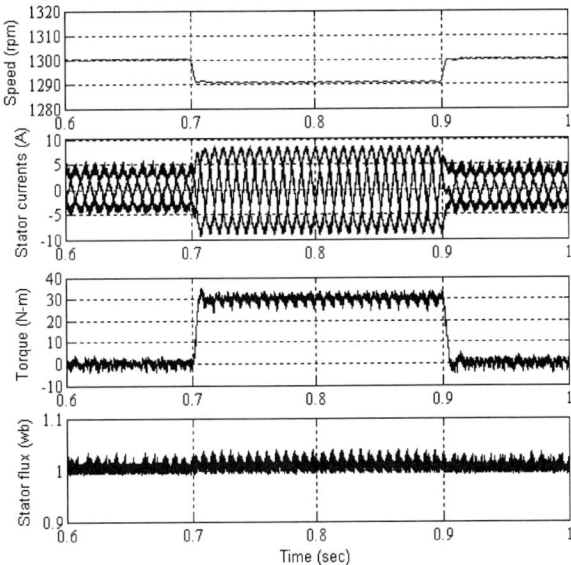

Fig. 11. Simulation results of proposed SVPWM based DTC: a 30N-m load is applied at 0.7sec and removed at 0.9sec.

265

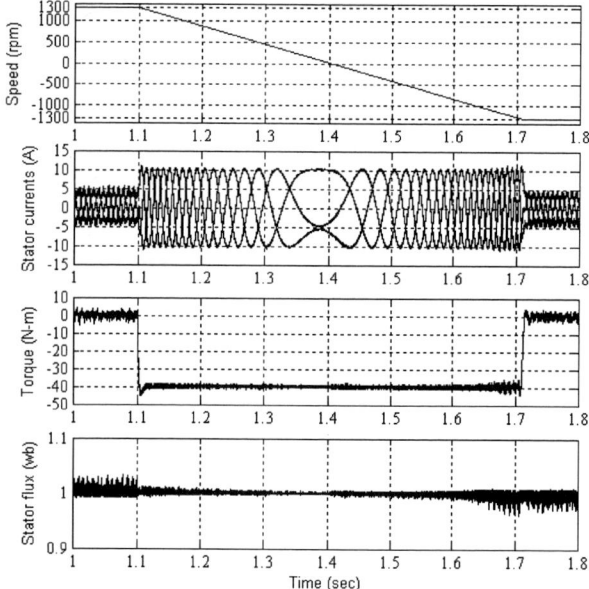

Fig. 12. Transient responses during speed reversal for proposed SVPWM based DTC; Speed is changed from +1300rpm to -1300rpm.

Fig. 8 gives the simulation results for common mode voltage with conventional SVPWM and with the proposed SVPWM technique. The simulation results for the DTC with proposed SVPWM algorithm are given in Fig. 9 – Fig. 12. From the simulation results it is observed that there is a reduction in common mode voltage at the expense of a slight increase of ripple in flux, current and torque.

## VIII. CONCLUSIONS

In this paper a new SVPWM algorithm for direct torque controlled induction motor drives has been proposed, which reduces common mode emissions of the drive. In the proposed SVPWM technique instead of zero voltage vectors, non-zero nonadjacent vectors are used in composing the reference voltage vector. It preserves the best of Conventional SVPWM technique in two ways 1) The vector times associated with the proposed technique are same as that of conventional SVPWM technique and 2) The number of commutations in a sector and from one sector to the next remains the same as in conventional SVPWM technique.

## IX. REFERENCES

[1]  Isao Takahashi, Toshihiko Noguchi, "A new quick-response and high-efficiency control strategy of an induction motor", *IEEE Trans Ind Appl*, Vol.IA-22, No.5, pp. 820-827, Sep/Oct, 1986.

[2]  Thomas G. Habetler, Francesco Profumo Michele Pastorelli and Leon M. Tolbert , "Direct Torque Control of Induction Machines Using Space Vector Modulation", *IEEE Trans. Ind. Appl.*, Vol. 28, No.5, pp. 1045-1053, Sep/Oct, 1992.

[3]  S.Ogasawara and H.Akagi, "Modelling of high frequency leakage currents in PWM inverter- fed Ac motor drive systems" *IEEE Trans. Ind. Appl.*, Vol. 32, No.4, pp. 1105-1114, Sep/Oct, 1996.

[4]  Y.Murai,T.Kobota and Y.Kawase "Leakage current reduction for a high frequency carrier inverter feeding an induction motor", *IEEE Trans. Ind. Appl.*, Vol. 28, No.4, pp. 858-863, July/August, 1992.

[5]  J.M.Erdman, et al, "Effect of PWM inverters on AC motorsbearing currents and shaft voltages" *IEEE Trans. Ind. Appl.*, Vol. 32, No.2, pp. 250-259, March/April, 1996.

[6]  M.Cirrincione,M.Pucci,G.VitaleandG.Cirrincione " A new direct torque control strategy for the minimization of common ode emissions", *IEEE Trans. Ind. Appl.*, Vol.42, No.2, pp. 504-517, Mar/April, 2006.

[7]  Yen-Shin Lai "Investigations into the effects of PWM techniques on Common mode voltage for inverter controlled induction motor drives",*IEEE PES winter meeting* 1999, pp 35-40, 1999.

[8]  Lee-Hun Kim et al "A new PWM method for conducted EMI reduction in inverter fed Motor dives", *IEEE proc. APEC'2005*, Vol.3, pp. 1871-1876, March 2005.

## X. BIOGRAPHIES

**Y.V. Siva Reddy** received B.Tech and M.Tech degrees from JNT University, Anantapur in the year 1995 and 2000 respectively. He is currently pursuing Ph.D at J.N.T.University, Hyderabad. He is presently Associate Professor in the Electrical and Electronics Engineering Department, G. Pulla Reddy Engineering College, Kurnool, Andhra Pradesh, India.
His research areas include Power systems and Electrical Drives.

**T. Brahmananda Reddy** graduated from Sri Krishna Devaraya University, Anantapur in the year 2001, M.E from Osmania University, Hyderabad in the year 2003. He is currently pursuing Ph.D at J.N.T.University, Hyderabad. He is presently Assistant Professor in the Electrical and Electronics Engineering Department, G. Pulla Reddy Engineering College, Kurnool, A.P, India.
His research areas include PWM techniques, and Electrical Drives.

**Dr. M. Vijaya Kumar** graduated from NBKR Institute of Science and Technology, Vidyanagar, A.P, India in 1988. He obtained M.Tech degree from Regional Engineering College, Warangal, India in 1990. He received Doctoral degree from JNT University, Hyderabad, India in 2000.
Currently he is working as Professor and head of Electrical and Electronics Engineering Department, JNT University, Anantapur, A.P, India. He is a member of Board of studies of S.V. University, Tirupathi and JNT University, Hyderabad, India. He has published over 40 research papers. He received two research awards from the Institution of Engineers (India). His areas of interests include Electrical Machines, Electrical Drives, Microprocessors and Instrumentation.

**2006 IEEE International Conference on Power Electronic, Drives and Energy Systems**

# Parallel Power Flow AC/DC Converter with High Input Power Factor and Tight Output Voltage Regulation for Universal Voltage Application

Aman Kumar Jha, K. Hari Babu, and B. M. Karan, *Member, IEEE*

*Abstract*--In this paper, a new parallel-connected single phase power factor correction (PFC) topology using flyback converter in parallel with forward converter is proposed to improve the input power factor with simultaneously output voltage regulation taking consideration of current harmonic norms. Paralleling of converter modules is a well-known technique that is often used in medium-power applications to achieve the desired output power by using smaller size of high frequency transformers and inductors. The proposed approach offers cost effective, compact and efficient AC-DC converter by the use of parallel power processing. Forward converter primarily regulates output voltage with fast dynamic response and it acts as master which processes 60% of the power. Flyback converter with AC/DC PFC stage regulates input current shaping and PFC, and processes the remaining 40% of the power as a slave. This paper presents a design example and circuit analysis for 300 W power supply. A parallel-connected interleaved structure offers smaller passive components, less loss even in continuous conduction inductor current mode, and reduced volt-ampere rating of DC/DC stage converter. MATLAB/SIMULINK is used for implementation and simulation results show the performance improvement.

*Index Terms*-- Circuit analysis, PFC, Power Conversion.

## I. INTRODUCTION

A number of power factor correction circuits have been developed recently [1]–[5]. Normally a boost converter is employed for PFC with DC/DC stage to improve performance or a flyback converter is used to reduce the cost. Although both boost converter and flyback converter are capable for PFC applications [6], [7], the main difficulty in two-stage scheme employing a PFC boost and a DC/DC converter is the high cost and lower efficiency. However, single-stage method using the simplest flyback converter is not able to tightly regulate the output voltage.

Paralleling of converter power modules [8-9] is a well-known technique that is often used in high-power applications

Aman Kumar Jha is with the Electrical and Electronics Engineering Department, Birla Institute of Technology, Ranchi, Jharkhand, India. (Email: aman_art@yahoo.co.in)
Hari Babu Kothurthi is doing M.E. in Electrical & Electronics Engineering, Birla Institute of Technology, Ranchi, Jharkhand, India.
    Dr. B.M.Karan is with the Head of Deptt. EEE, BIT, Mesra.

to achieve the desired output power with smaller size power transformers and inductors [10]. Since magnetics are critical components in power converters because generally they are the size-limiting factors in achieving high-density and/or low-profile power supplies, the design of magnetics becomes even more challenging for high-power applications that call for high power-density and low-profile packaging. Instead of designing large-size centralized magnetics that handle the entire power, low-power distributed high density low-profile magnetics can be utilized to handle the high processing power, while only partial load power flow through each individual magnetics [10, 11].

In addition to physically distributing the magnetics and their power losses and thermal stresses, paralleling also distributes power losses and thermal stresses of the semiconductors due to a smaller power processed through the individual paralleled power stages. As a result, paralleling is a popular approach to eliminating "hot spots" in power supplies. In addition, the switching frequencies of paralleled, lower-power power stages may be higher than the switching frequencies of the corresponding single, high-power processing stages because lower-power, faster semiconductor switches can be used in implementing the paralleled power stages. Consequently, paralleling offers an opportunity to reduce the size of the magnetic components and to achieve a low-profile design for high power applications.

Without increasing the number of power stages and control-circuit components, the transformer magnetics can be distributed by direct transformer paralleling. Not only that transformer paralleling distributes the processed power in each magnetics components, but also their power losses and thermal stresses are distributed at the same time. However, current sharing among the paralleled transformers needs to be maintained to ensure power balance.

In its basic form, the interleaving technique can be viewed as a variation of the paralleling technique, where the switching instants are phase-shifted within a switching period [12]. By introducing an equal phase shift between the paralleled power stages, the total inductor current ripple of the power stage seen by the output filter capacitor is lowered due to the ripple cancellation effect [12].

This chapter discusses the paralleling techniques to achieve high-density, low-profile designs for relative high

0-7803-9771-1/06/$25.00 ©2006 IEEE

power applications. It analyzes and compares the current sharing in various implementations of transformer paralleling. The goal of the proposed PFC scheme is to reduce the passive component size, to employ lower rated semiconductor, and to improve total efficiency. Simulation results show that the proposed topology is capable of offering good power factor correction and fast dynamic response.

## II. PFC CELLS

### A. Two Stage PFC Approach

A two-stage scheme shown in Fig. 1 is mainly employed for the switching power supplies since the boost stage can offer good input power factor with low total harmonic distortion (THD) and regulate the dc-link voltage and the DC/DC stage is able to obtain fast output regulation without low frequency ripple due to the regulated dc-link voltage [13]. These two power conversion stages are controlled separately. However, two-stage scheme suffers from higher cost, complicated control, low-power density, and lower efficiency.

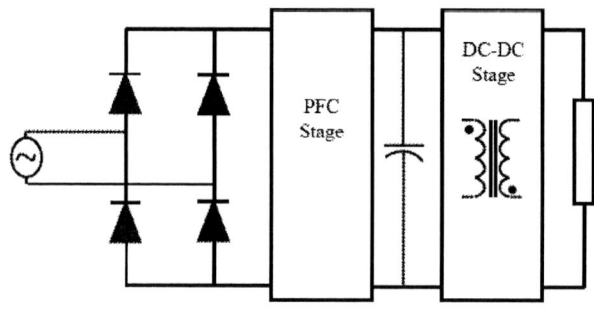

Fig. 1. Two-stage PFC.

### B. Single Stage Approach

For low power applications, where cost is a dominant issue, a single-stage scheme using the flyback converter is more attractive than a two-stage scheme [14]. A single-stage scheme Fig.2, cannot provide good performance in terms of ride-through or hold-up time since it mainly regulates input current with rectified voltage input and also output voltage is normally too small to provide hold-up time. Therefore, most of flyback converters need a large electrolytic capacitor at the output terminal to reduce the second harmonic ripple. But its transient response is still poor. The limitation of the flyback PFC is the output power level and the high breakdown voltage is required for the switch. When the output power increases, both voltage and current stress increase. Due to the high ripple currents the flyback is less efficient than other designs. That is why the two-stage scheme is more attractive for higher power rating.

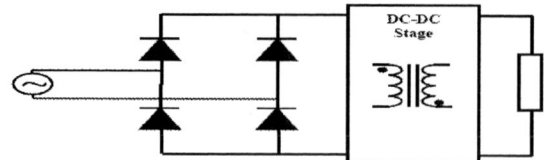

Fig. 2. Single – stage PFC.

### C. Parallel PFC Approach

At higher power levels, since it may be beneficial to parallel two or more DC/DC converters rather than using a single higher power unit, a parallel-connected scheme is proposed as shown in Fig. 3. This approach can offer fast output voltage regulation and high efficiency. The forward converter with DC/DC stage can offer good output voltage regulation due to the pretty dc input voltage and the flyback converter with AC/DC PFC stage fulfills input current regulation to obtain highly efficient power factor. The advantages of the proposed approach are as follows.

1) This scheme offers good input power factor and output regulation.
2) Input inductor and dc-link capacitor can be smaller.
3) The power rating of flyback converter-I is lower than that of two-stage structure due to low dc-link voltage and lower current rating.
4) The diode reverse recovery losses can be minimized due to the tailed operating mode in diode current

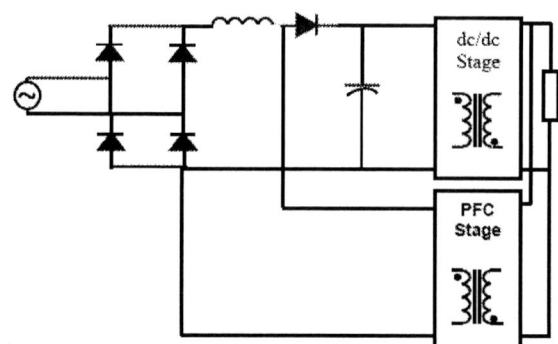

Fig. 3. Proposed parallel – connected single- stage PFC scheme.

## III. PROPOSED PARALLEL PFC SCHEME

Fig. 4 shows the proposed parallel-connected PFC scheme which employs a diode rectifier, dc-link capacitor, forward converter and flyback converter. The function of a forward converter with an electrolytic capacitor is to support output voltage regulation.

A flyback converter fulfills the function of power factor correction by making input current sinusoidal and regulating dc-link voltage. The operation of the flyback converter is given in this paragraph considering that the forward converter operates ideally. The PFC Cell (forward converter) operates with continuous conduction mode in both an input inductor and a flyback transformer. The dc-link voltage in this scheme can be lower than other schemes as

268

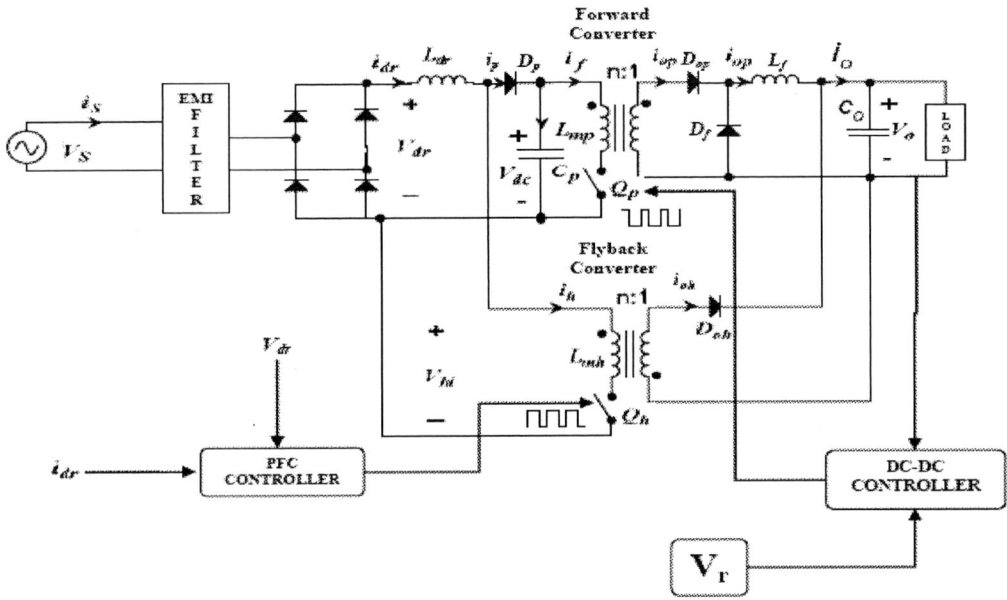

Fig. 4. Proposed parallel PFC converter.

$$V_{dc} = \sqrt{2}V_s \qquad (1)$$

The transfer function of the flyback converter is expressed by defining a conversion ratio as the ratio of the dc output voltage to the input voltage

$$M = \frac{V_o}{V_{dr}} = \frac{D}{D(1-n)} \qquad (2)$$

Where, D is the duty ratio of the switch $Q_h$, n ($=N_p/N_s$), is defined as the ratio of $N_p$ to $N_s$, and $N_p$ and $N_s$ denote the number of turns of primary and secondary side, respectively. The operational waveforms are shown in Fig. 6. To analyze the circuit parameters, basic equations for voltages and currents are given by

$$i_{dr} = i_p + i_h, \qquad (3)$$

$$V_{Ldr} = L_{dr}\frac{di_r}{dt}, \qquad (4)$$

$$V_{hi} = V_{dr} - V_{Ldr}, \qquad (5)$$

$$V_{h1} = V_{hi} - V_{Qh}. \qquad (6)$$

Where $i_{dr}$, $i_p$, and $i_h$ are the rectified, DC/DC Cell, and PFC Cell input currents on dc side. $V_{Ldr}$, $V_{hi}$, $V_{hi}$, $V_{dr}$, $V_{h1}$, $V_{Qh}$, and $V_o$ and are the input inductor, flyback converter input, rectified input, transformer primary winding, switch, and output voltages, respectively. Since the two input currents, $i_p$ and $i_h$, are interleaved, input current, $i_{dr}$, ripple can be significantly reduced. The operational sequences are as follows.

$t_0$-$t_1$: As shown in Fig. 5(a) The current of flyback transformer does not flow simultaneously in both windings.     When the switch $Q_h$ is turned ON at $t_o$, $V_{Qh}$ becomes zero and diode $D_{oh}$ is turned OFF with a reverse bias. The voltage across the diode $D_{oh}$ equals to $V_o + V_{hi}/n$. Energy, $L_{mh}I^2$, is charged in the magnetic field in the primary winding of the flyback transformer. Primary current, $i_h$, ramps up from the remaining magnetizing current and reaches $I_{dr}$ with the slope, ($V_{hi}/L_{mh}$), $i_p$ decreases with a slow current tail, and slowly decreases until $i_p$ reaches zero. At the same time the forward converter switch $Q_p$ is OFF because, as the switch $Q_h$ is ON the potential at he junction of diode $D_p$ and input inductor $L_{dr}$ i.e. $V_{Ldr} > V_{hi}$, the diode $D_p$ is reverse biased. The diode $D_{op}$ is also reverse biassed due to the polarity of the forward transformer and a negative voltage of $-nV_o$. The voltage across the output inductor is $V_L = -V_o$ and the inductor current $i_{Lf}$ decreases and $i_{Lf}$ along with $i_{oh}$, circulates through diode $D_f$, and supplied to load.

$t_1$-$t_2$: The primary current of flyback converter increases by $V_{dr}/(L_{mh} + L_{dr})$. The voltage across switch $Q_p$ decreases from $2V_{dr}$ to $V_{dr}$. The input inductor current ramps up to till switch $Q_h$ is OFF.

$t_2$-$t_3$: As shown if Fig. 5(b), when the switch $Q_h$ is turned OFF, $D_{oh}$ is turned ON with forward bias. The current in the primary winding ceases to flow. The stored energy is transferred to the secondary winding. At this time, the switch voltage, $V_{Qh}$, becomes $V_{hi} + nV_o$, $i_p$ becomes $i_{dr}$ and decreases depending on input voltage, and the secondary current decreases with the slope ($n^2 V_o/L_{mh}$). When the switch $Q_p$ is ON, the diode $D_p$ is forward biassed because the     potential at junction between the diode and inductor is $V_{Ldr} < V_{hi} + nV_o$. The primary current of forward transformer ramps up and the energy stored in the primary winding is instantaneously transferred to secondary, because of the same polarity of the forward transformer. The diode $D_{op}$ is forward biassed and diode $D_f$ is reversed biassed. The output inductor current $i_{Lf}$ increases along with $i_{oh}$ which is delivered to load.

The current slope through the magnetizing inductor when the switch $Q_h$ is turned off is given as

$$\Delta i_{mh} = -\frac{nV_o}{L_{mh}} T_{off} \tag{7}$$

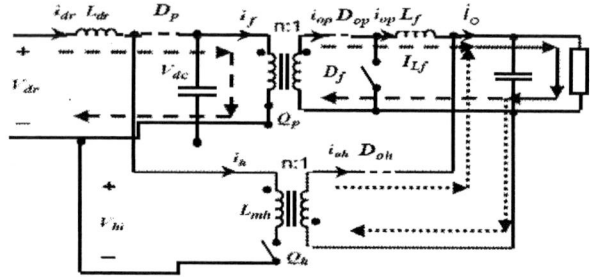

Fig. 5(a). Forward converter switch $Q_p$ is *OFF* and flyback converter switch $Q_h$ is *ON*.

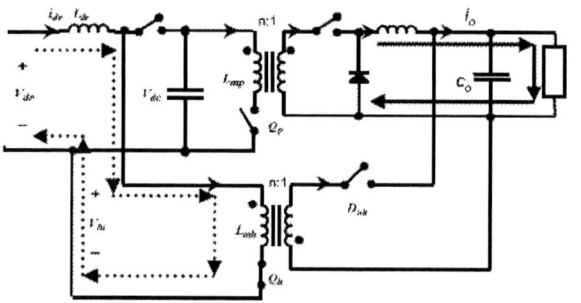

Fig. 5(b). Forward converter switch $Q_p$ is *ON* and flyback converter switch $Q_h$ is *OFF*.

Where, $T_{off}$ is the turn-off time. Similarly, the change of the flyback converter input current $i_p$ through a diode is

$$\Delta i_p = -\frac{Vdc - Vdr}{L_{dr}} T_{off} \tag{8}$$

Based on two slopes of $i_{mh}$ and $i_p$, the tailed diode current mode in which the diode current has current tail is defined as shown Fig. 7 when the slope of $i_{mh}$ is greater than that of $i_p$

$$v_{dr} > \frac{L_{mh}Vdc - nV_oL_{dr}}{L_{mh}} \tag{9}$$

In continuous conduction input inductor current mode, when the MOSFET is switched on, the diode $D_p$ is forced into reverse recovery at a high rate of change in the diode current ip. In this tailed mode operation, however, the diode current slowly decreases so that the reverse recovery effect can be minimized.

To analyze the flyback converter operation, an open loop duty ratio is obtained from (2) as

$$D_{open,h} = \frac{nV_o}{v_{dr} + nV_o} \tag{10}$$

Where, input voltage $v_{dr} = \sqrt{2}V_s|Sin\,\omega t|$. Assuming two input currents of each converter have the waveforms shown in Fig. 8, two currents depend on the duty ratio from (10)

$$i_h = \frac{i_{dr}T_{on}}{T_s} = D.i_{dr} \tag{11}$$

$$i_p = (1-D).i_{dr} \tag{12}$$

Where, input current $i_{dr} = \sqrt{2}I_s|Sin\,\omega t|$.

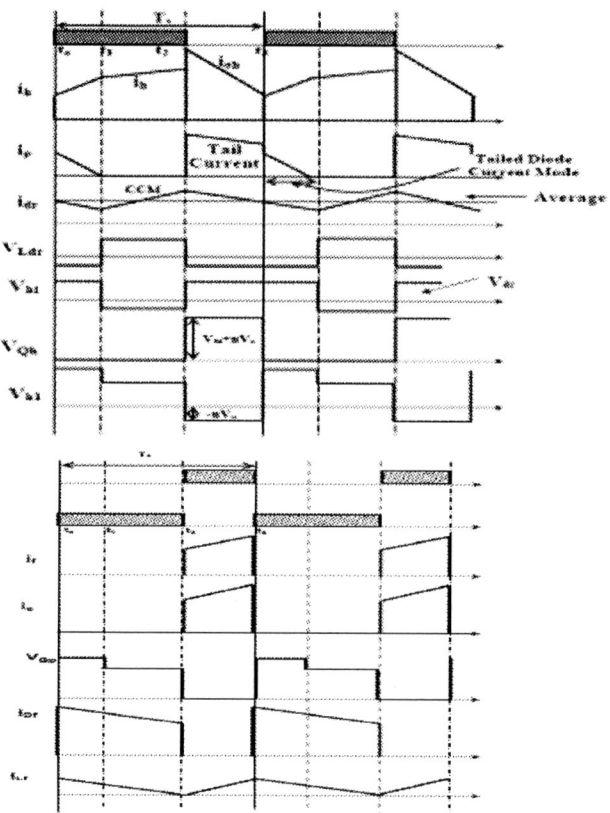

Fig. 6. Operational waveform of proposed parallel PFC topology.

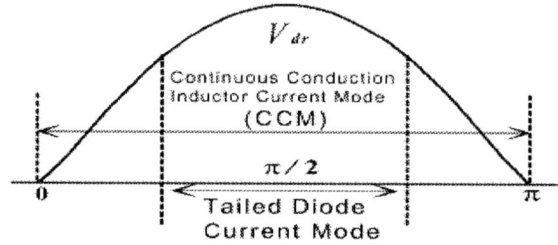

Fig. 7. Operation modes in input inductor and diode current.

Therefore, two currents can be obtained

$$i_h = \frac{\sqrt{2}\,nV_o I_s|Sin\,\omega t|}{\sqrt{2}V_s|Sin\,\omega t| + nV_o} \tag{13}$$

270

$$i_p = \sqrt{2}I_s \left|Sin\omega t\right| \left\{ 1 - \frac{nV_o}{\sqrt{2}V_s \left|Sin\omega t\right| + nV_o} \right\} \quad (14)$$

$$= \sqrt{2}I_s \left|Sin\omega t\right| - i_h.$$

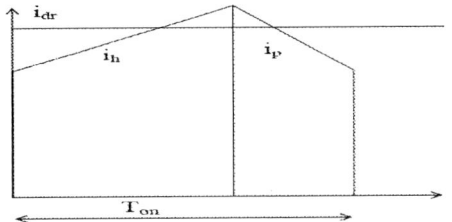

Fig. 8. Current waveforms in switching period.

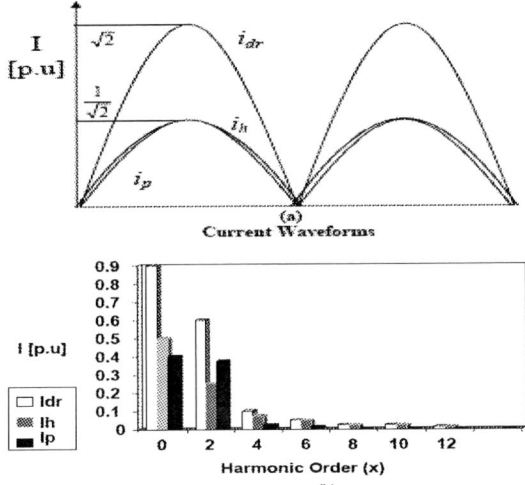

Fig. 9. Input current analysis ($I_{dr,rms}$ = 1 [ p.u], $I_{p,rms}$ = 0.6 [ p.u], $I_{h,rms}$ = O.40 [ p.u]).

Fig.9 shows those current waveforms and harmonic components. Now, instantaneous powers through the diode $D_p$ and the transformer T2 are calculated by using the input inductance and the magnetizing inductance

$$P_{p,pu} = \frac{L_{mh} + L_{dr}}{2L_{mh} + L_{dr}}[p.u] \quad (15)$$

$$P_{h,pu} = \frac{L_{mh}}{2L_{mh} + L_{dr}}[p.u] \quad (16)$$

Where, $L_{mh}$ and $L_{dr}$ denote the magnetizing inductance of T2 and input inductance, respectively, and the input total power $P_{in,pu} = P_{p,pu} + P_{h,pu} = 1[p.u]$. On the other hand, by employing the open loop duty ratio $D_{open,\,h}$, two instantaneous powers can be derived by

$$P_p = V_{dc}.i_p = \sqrt{2}V_s\left[\sqrt{2}I_s\left|Sin\omega t\right| - i_h\right] \quad (17)$$

$$P_h = P_{in} - P_p \quad (18)$$

Where, $P_{in} = V_sI_s\{1 - \cos 2\omega t\}$. The relations between two inductances and two input average powers of two converters are expressed as

$$\frac{L_{mh} + L_{dr}}{L_{mh}} = \frac{P_{p,ave}}{P_{h,ave}} \quad (19)$$

$$L_{dr} = L_{mh}\left\{\frac{P_{p,ave}}{P_{h,ave}} - 1\right\}, \quad (20)$$

The output currents of the two cells are given by turns ratio

$$i_{op} = n.i_f \quad (21)$$

$$i_{oh} = n.i_h$$

Since the output load current $i_o$ may contain only dc and switching frequency components, the harmonic contents for the primary current of the flyback converter-I is expressed as

$$i_o = i_{op} + i_{oh} = \frac{V_sI_s}{V_o}, \quad (22)$$

$$i_{f,x} = i_{h,x} \quad (23)$$

Where, x(= 2 , 4, 6, etc.) is harmonic order. From (23), the dc-link capacitor current can be estimated as a second harmonic

$$i_{dc,2} = -(i_{p,2} + i_{h,2})Cos2\omega t. \quad (24)$$

Therefore, the voltage ripple of the dc-link capacitor is obtained as

$$V_{dc,ripple} = -\frac{i_{p,2} + i_{h,2}}{2\omega C_p}Sin2\omega t \quad (25)$$

Where, $C_p$ is the capacitance of the dc-link capacitor.

## IV. CONVERTER CONTROLS

To control the proposed approach, two control stages are required for PFC and output voltage regulation as shown in Fig. 10. Flyback converter is regulated by a conventional PFC controller which consists of inner input current loop and outer dc-link voltage [16] to obtain high power factor Fig. 11. DC-link voltage is 1.414*$V_s$, which is better to reduce the voltage across drain-source of MOSFET $Q_p$ [16]. Based on the PFC controller, a feed-forward control block is added to improve input current shape. Since the open loop duty ratio $D_{open,h}$ of the PFC cell is calculated from (10), the final duty ratio for the switch gate input is obtained as

$$D_h = D_{open,h} + D_{pi} \quad (26)$$

where $D_{pi}$, is the closed loop duty ratio obtained from S-R flip flop current controller. The output $D_{pi}$ of the S-R flip flop current regulator containing a small amount of variations provides the correction to the final duty ratio. On the other hand, output voltage $V_o$ control is achieved by forward converter. Fig. 10 shows a simple PI voltage controller with a open loop duty ratio $D_{open,h}$ which is calculated similarly to $D_{open,p}$ in terms of power ratings of each converter

$$D_{open,p} = \frac{nV_oP_{p.pu}}{v_{dr} + nV_oP_{p.pu}} \quad (27)$$

271

Final duty ratio $D_p$ is obtained by adding the duty ratio $D_{pi}$ from controller with $D_{open,p}$. The output voltage control response is much faster than single stage scheme since two converters are employed for separate control function

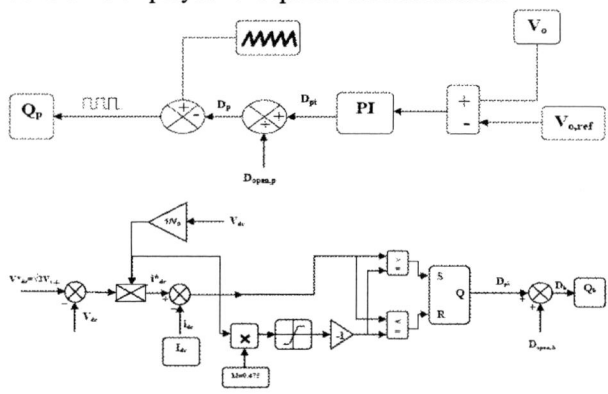

Fig. 10. Converter Controls.

## V. DESIGN EXAMPLE

The proposed PFC circuit is designed according to the following parameters.

Total output power (Po)          = 300 W
Input voltage (Vs)               = 230 V
Output voltage (Vo)              = 48 V
Line frequency                   = 50Hz
Switching frequency              = 50000Hz
Output dc capacitance (Co)       = 1500 μF
Transformer turns ratio (n)      = 6.77:1
**Forward Converter**
Power rating                     = 180 W
Magnetizing inductance (Lmp)     = 0.5 mH
DC capacitance                   = 660uF
DC-link voltage (Vdc)            = 325 Vdc
**Flyback Converter**
Power rating                     = 120 W
Magnetizing inductance (Lmh)     = 1.2 mH

The proposed scheme provides small input inductor since the inductor current depends on, dc-link voltage is smaller so that the voltage stress on the switch of the forward converter is less, the power rating of DC/DC stage is a bit higher than average power due to lower harmonic components, and the diode reverse recovery loss is minimized because of the tailed diode conduction mode.

The simulation results of the proposed topology are shown in Fig.11. Unity power factor and tight output voltage (48 V) regulation can be achieved. The dc-link voltage is $1.414*V_s=325V$, and the voltage ripple of the dc-link is 3.4 V which mainly depends on the dc-link capacitance. Fig. 11 shows the analysis of the circuit currents. The primary side current of forward converter has 2nd and 4th harmonics due to the harmonics on the flyback converter. Two control systems are implemented to prove the proposed scheme. In the diode current, current tail is appeared when the diode is turned off. 5.2% input current THD and 86% efficiency are obtained.

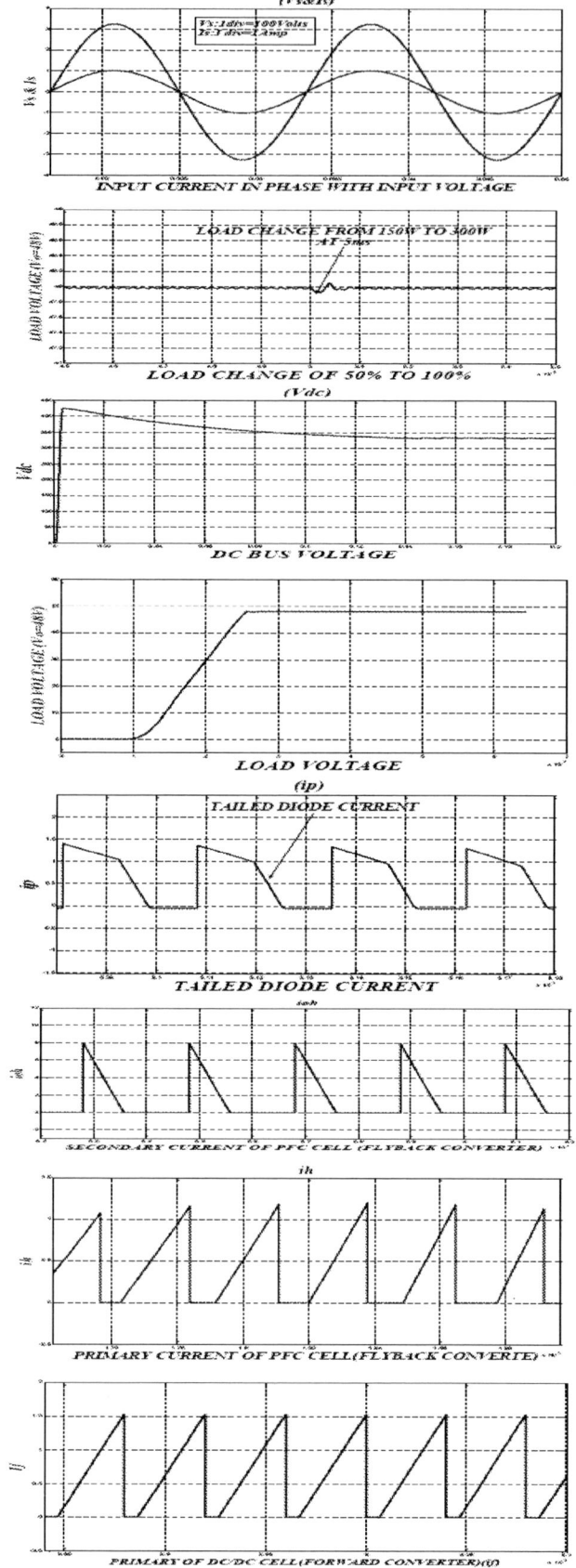

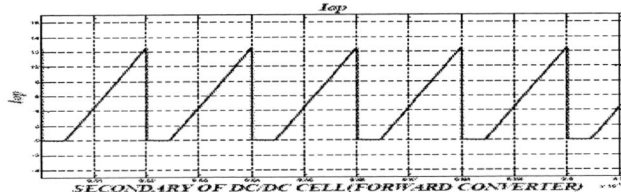

Fig. 11. Simulation results of the proposed approach.

## VI. CONCLUSION

A parallel-connected single phase power factor correction (PFC) topology using forward converter and a flyback converter has been proposed. It has been shown that output voltage regulation is, achieved by DC/DC cell and the input power factor correction is achieved by AC/DC PFC cell. These two power stages have 60% and 40% power sharing, respectively. The proposed approach offers the following advantages: smaller size passive components, lower voltage-ampere rating of DC/DC stage, and higher efficiency. Simulation results demonstrate the capability of the proposed scheme.

## VII. REFERENCES

[1] W. F. Ray and R. M. Davis, *"The definition and importance of power factor for power electronic converters,"* Proc. European conference on Power Electronics and Applications (EPE), 1989, pp. 799-805.

[2] L. Huber, J. Zhang, M. M. Jovanovic, and F. C. Lee, "Generalized topologies of "Single-stage input-current-shaping circuits," IEEE Trans. Power Electron., vol. 16, pp. 508–513, July 2001.

[3] R. Redl, L. Balogh, and N. O. Sokal, *"A new family of single-stage isolated power-factor correctors with fast regulation of the output voltage,"* in *Proc. PESC'94*, 1994, pp. 1137–1144.

[4] R. Erickson, M. Madigan and S. Singer, *"Design of a simple high power factor rectifier based on the flyback converter,"* IEEE Applied Power Electronics Conference, 1990, pp. 792-801.

[5] G. Choe and M. Park, *"Analysis and control of active power filter with optimized injection,"* IEEE Power Electronics Specialists Conference, 1986, pp. 401-409.

[6] K. K. Sen, A. E. Emanuel, *"Unity power factor single phase power conditioning,"* IEEE Power Electronics Specialists Conference, 1987, pp. 516-524.

[7] W. Tang, Y. Jiang, G. C. Hua, F. C. Lee, and I. Cohen, *"Power factor correction with flyback converter employing charge control,"* in *Proc. PEC'93*, 1993, pp. 293–298.

[8] R. Srinivasan and R. Oruganti, *"Single phase parallel power processing scheme with power factor control," Power Electron. Drive Syst.*, pp. 40–47, 1995.

[9] Y. Jiang and F. C. Lee, "Single-stage single-phase parallel power factor correction scheme," in Proc. PESC'94, 1994, pp. 1145–1151.

[10] W. A. Tabisz, M. M. Jovanovi}, and F. C. Lee, *"Present and future of distributed power systems,"* Proc. IEEE Appl. Power Electron. Conf., 1992, pp. 11-18.

[11] G. Suranyi, *"The value of distributed power,"* Proc. IEEE Appl. Power Electron. Conf., 1996, pp. 104-110.

[12] B.A. Miwa, D.M. Otten, and M.F. Schlecht, "High efficiency power factor correction using interleaving techniques," Proc. IEEE Appl. Power Electron. Conf., 1992, pp. 557- 568.

[13] J. Zhang, M. M. Jovanovic, and F. C. Lee, *"Comparison between CCM single-stage and two-stage boost converter,"* in *Proc. APEC'99*, 1999, pp. 335–341.

[14] M. Daniele, P. K. Jain, and G. Joos, *"A single-stage power-factor-corrected AC/DC converter," IEEE Trans. Power Electron.*, vol. 14, pp. 1046–1055, Nov. 1999.

[15] Unitrode Application Note U-140, *"Average Current Mode Control of Switching Power Supplies"*

[16] W. G. Dawes and A. Lyne, *"Improved efficiency constant output power rectifier,"* in Proc. INTELEC'00, 2000, pp. 24–27.

## VIII. BIOGRAPHIES

**Aman Kumar Jha**, born in 1977 in India, he received his BE in electrical engineering from B. R. A. Bihar University, Muzaffarpur in 2002 and M.Tech in Energy Systems Engineering from Indian Institute of Technology, Bombay in 2005. Currently working as a faculty at Birla Institute of Technology, Mesra, Ranchi. His field of interest is power electronics, electric drives, energy efficiency.

**B. M. Karan**, born in 1944 in India. He has received his BE and ME in electrical engineering from Ranchi University, India and PhD from Awadh University, India. He has 38 years of teaching and research experience. His area of research is control system, neural network, fuzzy logic, system biology and digital signal processing. He has 39 research papers in national and international journals and conferences. He is currently professor and head of department of Electrical and Electronics Engineering, BIT, Mesra , Ranchi.

**2006 IEEE International Conference on Power Electronic, Drives and Energy Systems**

# A Generalized Space Vector Modulation with Simple Control technique for Balancing DC-Bus Capacitor Voltages of a Three-Phase, Neutral-Point Clamped Converter

A. H. Bhat, and P. Agarwal, *Member, IEEE*

*Abstract*— In this paper, a generalized space vector modulation technique is used for the control of a three-phase, neutral-point clamped converter in both rectification as well as inversion modes. The SVPWM can be extended to higher-level converters. A simple control algorithm is used for the balance of dc-bus capacitor voltages by identifying the unbalance generated by different conduction states and using a very simple method to achieve balanced capacitor voltages by using the redundant switching vectors. The analytical study is validated by the simulation results of the converter.

*Index Terms*— Control strategy, Harmonic pollution, Multilevel Converters, Power converters, Power factor correction, Supply power factor, Space vector modulation

## I. INTRODUCTION

POWER Converters are extensively used in various applications like power supplies, dc motor drives, front-end converters in adjustable-speed ac drives, and so on. The conventional ac/dc converters used for various applications act as highly non-linear loads on the power system and introduce the power quality problems like harmonic pollution and reactive power flow alongwith other related problems in the system. Active power factor correction circuits called High Power Factor Converters perform the task of power conversion efficiently and at improved power quality both at the input as well as the output sides of the converter. Recently for medium and high power applications at high voltages, multilevel converters have been found a viable solution. They are gaining widespread popularity among the power factor correction converters because of their excellent performance like sinusoidal input currents with negligible THD, high supply power factor, reduced-rippled regulated dc output voltage, reduced voltage stress, reduced dv/dt stresses and hence low EMI emissions[1],[4],[5],[6],[7]. This is achieved at reduced switching frequencies in comparison with

A. H. Bhat is with the Department of Electrical Engineering, National Institute of Technology, Kashmir, India (e-mail: bhat_68@rediffmail.com).

P. Agarwal is with the Department of Electrical Engineering, Indian Institute of Technology Roorkee, India (e-mail: pramgfee@iitr.ernet.in).

their two-level counterparts. With higher number of levels, high voltage, high power applications are possible. Moreover since an MLC itself consists of series connection of switching power devices and each device is clamped to the dc-link capacitor voltage through the clamping diodes, it does not require special consideration to balance the voltages of the power devices. Nevertheless, the neutral point of the neutral point clamped (NPC) converter is prone to fluctuations due to the irregular charging and discharging of the output capacitors. In this paper, a generalized SVPWM technique which can be used for higher-level converters also has been discussed and used. The calculating difficulty in SVPWM is the same independent of the number of converter levels in comparison to the conventional SVM techniques where the computational complexity increases with the number of levels. The coordinates system translation method is used here. For the control of dc-bus capacitor voltages, a simple control algorithm is used which identifies the unbalance generated by different conduction states and uses the various redundant states to balance the capacitor voltages.

To verify the validity of the proposed control scheme, the computer simulations are provided and discussed.

## II. SYSTEM DESCRIPTION AND MATHEMATICAL MODELING

The adopted three-phase power factor correction converter based on neutral point clamped scheme is shown in Fig. 1. The input terminals of the rectifier a, b and c are connected to the terminals of three-phase source A, B and C through the filter inductances $L_s$. Each power switch has a voltage stress of half the dc bus voltage instead of full dc bus voltage as in the two-level PFCs. The Power switches are commutated with a high switching frequency to generate the PWM voltages Va, Vb, and Vc.

Using the definitions of space vectors:

$$v = \tfrac{2}{3}\left(v_A + a.v_B + a^2.v_C\right) \tag{1}$$

$$v^{'} = \tfrac{2}{3}\left(v_a + a.v_b + a^2.v_c\right) \tag{2}$$

$$i_L = \tfrac{2}{3}\left(i_{LA} + a.i_{LB} + a^2.i_{LC}\right) \tag{3}$$

The following vector equation can be written.

$$v = L.\tfrac{di_L}{dt} + v^{'} \tag{4}$$

0-7803-9771-1/06/$25.00 ©2006 IEEE

This equation can be expressed in a rotating reference frame (d-q), with the d-axis oriented in the direction of source voltage vector $v$ as depicted in Fig. 2. Thus (4) can be written as,

$$v_d = L.\frac{di_{Ld}}{dt} - \omega.L.i_{Lq} + v_{d'} \qquad (5)$$

$$v_q = 0 = L.\frac{di_{Lq}}{dt} + \omega.L.i_{Ld} + v_{q'} \qquad (6)$$

where $\omega$ is the angular frequency of three-phase voltage and $v_d, v_{d'}, i_{Ld}$ and $v_q, v_{q'}, i_{Lq}$ are the components of $v, v'$ and $i_L$ in the d and q-axis respectively.

Equations (5) and (6) show that the behaviour of currents $i_{Ld}$ and $i_{Lq}$ can be controlled by using the voltages $v_{d'}$ and $v_{q'}$ generated by the rectifier. In this way, the active as well as the reactive powers delivered by the mains to rectifier can be controlled.

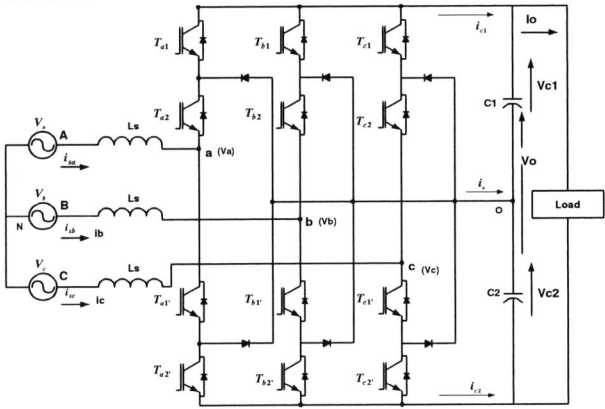

Fig. 1. Adopted three-phase Neutral Point Clamped Converter.

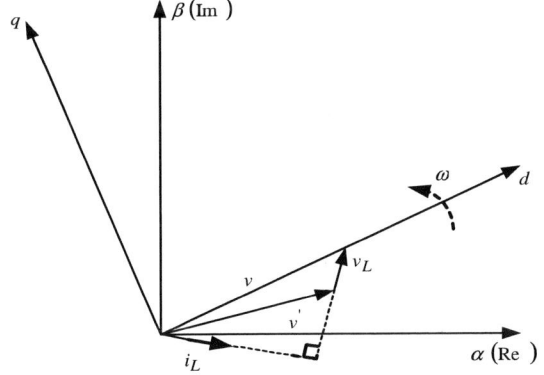

Fig. 2. Vector diagram of the converter.

III. CONTROL STRATEGY AND SPACE VECTOR MODULATION

The control strategy of this rectifier is the same as employed in two-level PWM rectifiers using SVM [2]. Fig. 3 depicts the block diagram of closed-loop control of the converter. The controller is supposed to fulfill the above mentioned control objectives. A PI controller is used to control the converter

output voltage $V_c$. The output of this controller, $i_{Ld^*}$ is used as reference for an inner closed-loop used to control the direct current $i_{Ld}$. The current in the q-axis, $i_{Lq}$ is controlled by a similar loop with reference, $i_{Lq^*} = 0$, to obtain operation with unity power factor. It is to be emphasized that this method only controls the total dc-bus voltage $V_o$ and does not ensure the balance of capacitor voltages $V_{C1}$ and $V_{C2}$. For proper converter operation, $V_{C1} = V_{C2}$. The current controllers deliver the reference values for the voltages in the d and q-axis, $V_{d^*}$ and $V_{q^*}$ respectively. By using coordinates transformation, we obtain $V_{\alpha^*}$ and $V_{\beta^*}$ in the stationary reference frame $(\alpha, \beta)$. Voltages $V_{\alpha^*}$ and $V_{\beta^*}$ are delivered as inputs to the space vector modulator which generates the control pulses for the converter switches using the Nearest Three Vector (NTV) approach of SVM.

Applying the definition of (2) to all the 27 possible conduction states of power semiconductors, the converter generates 19 different space vectors as shown in Fig. 4. This figure also depicts the commutation states used to generate each space vector. The complex plane is divided into 24 triangles.

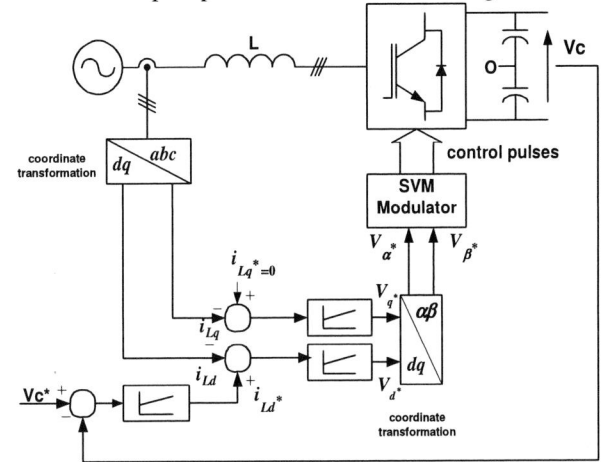

Fig. 3. Block diagram of controller of SVPWM Rectifier.

Four types of space vectors can be identified in Fig. 4 as:
- Zero Vector $(V_o)$: It is generated by three different conduction states connecting the terminals a, b and c simultaneously to the same bus bar. Therefore it is said that zero space vector has double redundancy.
- Internal Vectors $(V_1 to V_6)$: These vectors are generated by connecting three terminals a, b and c to only one capacitor, $C_1$ or $C_2$. They have simple redundancy.
- Medium Vectors $(V_8, V_{10}, V_{12}, V_{14}, V_{16} and V_{18})$: These vectors are generated by connecting the terminals a, b and c to different bus bars (+, 0 and -). Each vector can be generated by a unique commutation state which means that they have no redundancy.

- External Vectors $(V_7, V_9, V_{11}, V_{13}, V_{15} and V_{17})$: They are generated when the terminals a, b and c are connected to the positive and negative bus-bars. These vectors have no redundancy.

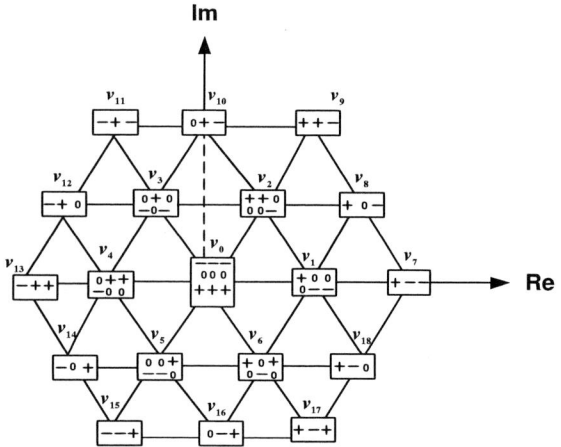

Fig. 4. Space Vectors generated by SVPWM Converter.

For the generalized SVM, a particular scaling factor is used for all magnitudes which yields a scaled modulation index as,

$$M_s = M \frac{\sqrt{3}}{2}(n-1) \qquad (7)$$

where $n$ is the number of levels and $M$ is modulation index.
For the sake of simplification of calculations, the reference vector is translated from its original sector to the first sector as follows,

$$S_n = \text{int}\left(\frac{\theta}{60}\right) \qquad (8)$$

$$\theta_1 = \theta - S_n.60 \qquad (9)$$

Where $S_n$ is the number of the sector the reference vector originally belongs to [0,....,5] and $\theta_1$ is the equivalent angle referred to the first sector.
In classic $\alpha$-$\beta$ co-ordinate system, the reference vector coordinates are given by,

$$V_x = M_s \cos(\theta_1)$$
$$V_Y = M_S \sin(\theta_1) \qquad (10)$$

Fig. 5 shows an example of the reference vector placed in a vector map of a 5-level converter [6]. The components for the reference vector in a system with axes separated 60° and centred in the central origin $A$ are given by,

$$V_{AX} = V_X - 0.57735.V_Y$$
$$V_{AY} = 1.1547.V_Y \qquad (11)$$

Considering the new origin $B$, the components of this new origin with respect to the centered origin $A$ are calculated as follows:

$$B_{AX} = \text{int}(V_{AX}) \qquad B_{AY} = \text{int}(V_{AY}) \qquad (12)$$

The reference vector components with respect to the new origin $B$ are calculated as,

$$V_{BX} = V_{AX} - B_{AX} \qquad V_{BY} = V_{AY} - B_{AY} \qquad (13)$$

After the first co-ordinates translation to origin $B$, the reference vector can belong to two possible regions. $V_{BX} + V_{BY}$

allows knowing if $V_B$ belongs to the lower region (type 1) or the upper region (type 2). We can use the translation factor $T$ to do this as,

$$T = \text{int}(V_{BX} + V_{BY}) \qquad (14)$$

For $T=0$, the reference vector belongs to a region of type 1 and for $T=1$, it belongs to a region of type 2.
In order to avoid a negative value for reference vector components, when it belongs to a region of type 2, its co-ordinate system requires an axes rotation given by the rotation factor $R$ as,

$$R = 1 - 2T \qquad (15)$$

Now if the reference vector belongs to the upper region, it is necessary to make a new co-ordinate system translation from origin $B$ to point $C$, and rotate the axes. For this, the components of new origin $C$ are calculated as,

$$C_{AX} = B_{AX} + T \qquad C_{AY} = B_{AY} + T \qquad (16)$$

As shown in Fig. 6, the new reference vector components are calculated with the help of factor $R$ as given by,

$$V_{CX} = (V_{AX} - C_{AX})R \qquad V_{CY} = (V_{AY} - C_{AY})R \qquad (17)$$

The three adjacent vectors of Fig. 6 are as:

$\overrightarrow{V_{1C}}$ is always the vector located at the origin $C$.

$\overrightarrow{V_{2C}}$ is always the vector located on the axis $x$ at a distance of 1 from the corresponding origin.

$\overrightarrow{V_{3C}}$ is always the vector located on axis $y$ at a distance of 1 from the corresponding origin.

Hence the components of $\overrightarrow{V_{1C}}$ are always $\overrightarrow{V_{1C}} = (0,0)$ and following expression is obtained,

$$\overrightarrow{V_{2C}}.T_2 + \overrightarrow{V_{3C}}.T_3 = \overrightarrow{V_C}.T_S \qquad (18)$$

The components of the vectors $\overrightarrow{V_{2C}}$ and $\overrightarrow{V_{3C}}$ will always have the same value:

$$\overrightarrow{V_{2C}} = (1,0), \qquad \overrightarrow{V_{3C}} = (0,1) \qquad (19)$$

Therefore from (18), we get

$$T_2 = T_S.V_{CX} \qquad T_3 = T_S.V_{CY} \qquad T_1 = T_S - T_2 - T_3 \qquad (20)$$

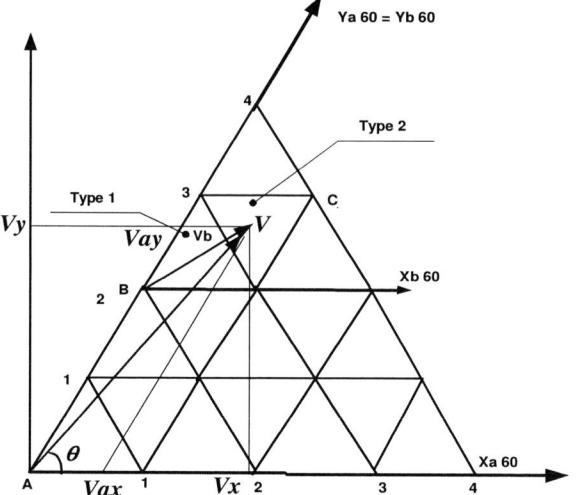

Fig. 5. Coordinates translation technique.

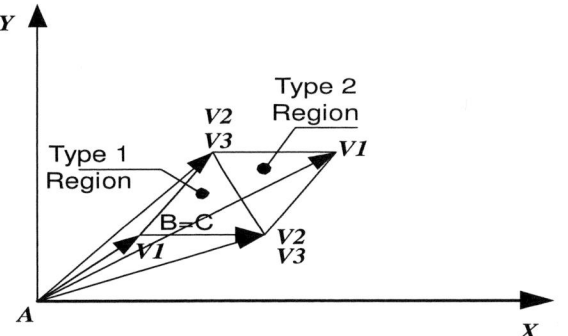

Fig. 6. Reference vector with respect to origins B and C.

This generalized scheme of SVPWM technique is applicable to higher levels of the converter also without increasing the calculating difficulties and it offers a somewhat easy implementation.

## IV. PROPOSED SVM CONTROL STRATEGY

### A. Voltage unbalance generation

Assuming that $C_1 = C_2$ and that initially $V_{C1} = V_{C2}$, the neutral current $i_o$ is defined by:

$$i_o = i_{C1} - i_{C2} \qquad (21)$$

When the neutral current is zero, $i_{C1} = i_{C2}$, the capacitor voltages are balanced. The generation of an internal space vector $V_1$ by two different switching states is depicted in Fig. 7. One of these states uses only capacitor $C_1$, connecting the three-phase source to terminals + and 0. This commutation state can be called positive commutation state $V_1^+$. On the other hand, the negative commutation state $V_1^-$ is obtained by connecting the three-phase source to terminals – and 0. As shown in Fig. 7, the commutation state $V_1^+$ produces:

$$i_o = i_{LR} \qquad (22)$$

And the commutation state $V_1^-$ produces:

$$i_o = -i_{LR} \qquad (23)$$

This means that these commutation states produce neutral current with different polarity and, consequently, generate dc-bus capacitor voltage unbalance with different polarity. This fact forms the basis of the voltage unbalance compensation.

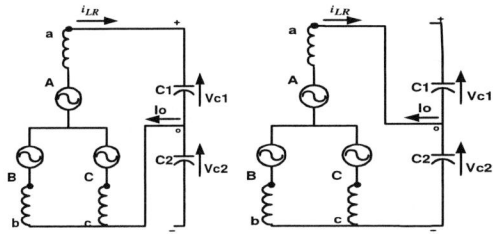

Fig. 7. Switching states of $V_1$: a) Positive switching state $V_1^+$ (+,0,0);

b) Negative switching state $V_1^-$ (0,-,-).

### B. Influence of direction of power flow

The direction of power flow between source and converter changes the polarity of the capacitor voltage unbalance as clearly depicted in Fig. 7. If the three-phase source delivers active power to the converter in commutation state $V_1^+$, this power is received by the capacitor $C_1$, causing an increase in $V_{C1}$. The action of voltage controller is to keep the total dc-bus voltage $V_o$ constant, implying that the capacitor $C_2$ voltage $V_{C2}$ will be reduced by the closed loop voltage control action. On the other hand, if the converter in conduction state $V_1^+$ regenerates active power to three-phase source, this power will be delivered by the capacitor $C_1$, thus causing a reduction in voltage $V_{C1}$ and an increase in voltage $V_{C2}$.

If the three-phase source delivers active power to the converter in commutation state $V_1^-$, this power is received by the capacitor $C_2$, thus causing an increase in $V_{C2}$ and a reduction in $V_{C1}$. On the other hand, if the converter in the state $V_1^-$ delivers active power to three-phase source, this power comes from capacitor $C_2$, thus causing a reduction in $V_{C2}$ and an increase in $V_{C1}$. This analysis is valid for all the internal space vectors.

### C. Capacitor voltage unbalance control strategy

The dc-bus capacitor voltage difference is given by,

$$\Delta V = V_{C1} - V_{C2} \qquad (24)$$

The polarity of the current $i_{Ld}$ determines the direction of power flow between the converter and source. Since the actual current $i_{Ld}$ has more ripples than the reference current $i_{Ld^*}$, the direction of power flow can be determined advantageously by $i_{Ld^*}$.

Thus according to the polarity of $\Delta V$ and $i_{Ld^*}$ in each modulation period $T$, the reference vector $v_{ref}$ is generated by using the redundant states of internal vectors to compensate the voltage unbalance.

As shown in Fig. 7, if $\Delta V > 0$, $V_{C1} > V_{C2}$ and consequently $V_{C2}$ must increase and $V_{C1}$ must reduce to compensate the dc-bus capacitor voltage unbalance. In this case, if power flows from source to converter $\left( i_{Ld^*} > 0 \right)$, a negative redundant state like $V_1^-$, as shown in Fig. 7(b), must be chosen for internal space vectors. On the other hand, if the power flows from converter to the source $\left( i_{Ld^*} < 0 \right)$, the power must be delivered by the capacitor $C_1$ to cause a reduction of $V_{C1}$ and an increase in $V_{C2}$. Therefore a positive redundant state, like $V_1^+$ as shown in Fig. 7(a), must be chosen for internal space vectors.

The strategy to select a redundant switching state is implemented using the following control logic:

If $\Delta V$ is positive & $i_{Ld}{}^*$ is positive, select a negative redundant state, otherwise select a positive redundant state for negative $i_{Ld}{}^*$ ;

If $\Delta V$ is negative & $i_{Ld}{}^*$ is positive, select a positive redundant state, otherwise select a negative redundant state for negative $i_{Ld}{}^*$ .

## V. SIMULATION RESULTS

To validate the proposed control scheme of the adopted three-phase neutral point clamped rectifier, computer simulations using MATLAB/Simulink and SimPowerSystems software are provided.

Following system parameters are chosen for the simulation of converter:

The mains phase voltage is chosen as 110V rms with a frequency of 50Hz. The capacitance of two dc bus capacitors is 5000µF each and the boost inductance is 5mH for each line. A sampling frequency of 5 kHz is chosen. A passive load of 15Ω and 5mH is used for the simulation. The desired dc-link voltage of the proposed rectifier is set at 500V.

The simulated waveforms for phase a voltage, line current and ac side line-to-line voltage of the converter in the rectification mode are shown in Fig. 8 and the harmonic spectrum of the phase a line current is shown in Fig 9. The line current is a sinusoidal waveform with a nearly unity power factor. It has a THD of only 1.23%. This is a highly desirable feature of a high power factor converter. Fig. 10 depicts the simulated waveforms for load voltage and load current in the rectification mode. It can be seen that the load voltage and load current are almost level waveforms with negligible ripple content. This is a highly desirable feature required in some applications like telecommunications. Fig. 11 shows the simulated waveforms of the phase a source voltage and line current when the operation of converter is changed from rectification to inversion mode. The converter still draws the sinusoidal currents with negligible THD at nearly unity input power factor and the transition from rectification to inversion takes place within three cycles only which shows a fast dynamic response of the converter. The proposed converter has the capability of regulating the load voltage even when it is subjected to the changes in the load as shown in Fig. 12. Fig. 13 depicts the two dc-bus capacitor voltages under unbalanced capacitor voltage conditions and under unbalance control strategy. The amplitude of the capacitor voltage unbalance $\Delta V_C$ is about 15 volts before the application of control strategy and it becomes negligible (less than one volt) after the application of unbalance control algorithm as shown. Thus the utility of control algorithm is validated by the simulation results. This solves the burning problem of dc-bus capacitor voltage balancing in case of neutral-point clamped converters. It is to be emphasized that a high performance is achieved in the converter at a lower switching frequency as compared to sinusoidal PWM technique. Fig. 14 depicts the phase a voltage

and line current under source voltage variations. It is observed that the source current decreases with an increase in source voltage and increases with a decrease in source voltage in order to satisfy the power balance condition on the input side of the converter. Fig. 15 depicts the source voltage and line current waveforms under distorted mains condition. The source voltage has a predominant third harmonic component. It can be seen that even under distorted mains condition, the control algorithm yields a sinusoidal source current though at a somewhat increased THD which is still under prescribed harmonic standards [3]. This is a desirable feature as it makes the converter maintain its power quality even when it is fed from a distorted supply voltage.

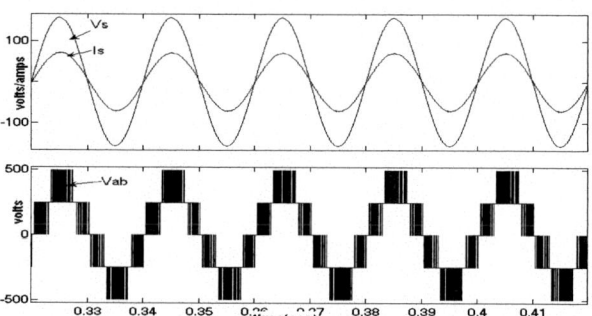

Fig. 8. Phase a voltage, line current, and line-to-line voltage waveforms in rectification mode.

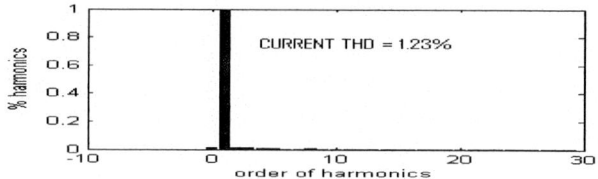

Fig. 9. Harmonic spectrum of phase a line current.

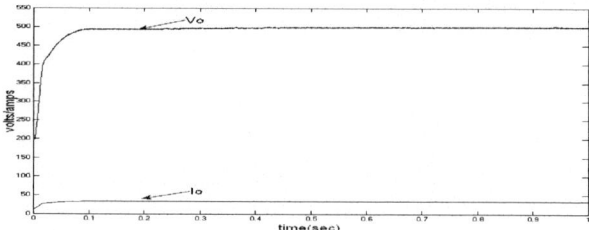

Fig. 10. Load voltage and Load current in rectification mode.

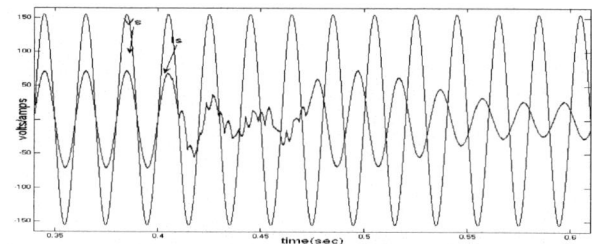

Fig. 11. Phase voltage and line current in the Rectification and Inversion modes (Rectification –to-Inversion takes place at $t = 0.32\,\text{sec}.$ ).

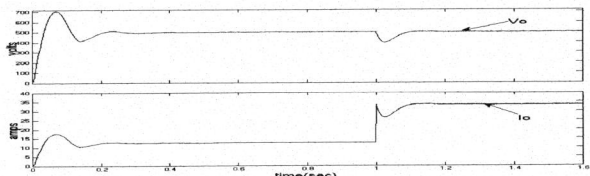

Fig. 12. Load voltage and load current waveforms for a step change in the load current (step change in load current takes place at $t = 1\sec$.).

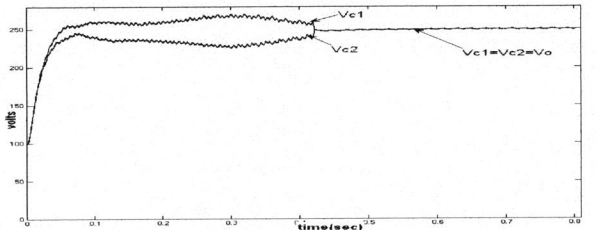

Fig. 13. DC-bus capacitor voltages under unbalanced and balanced conditions.

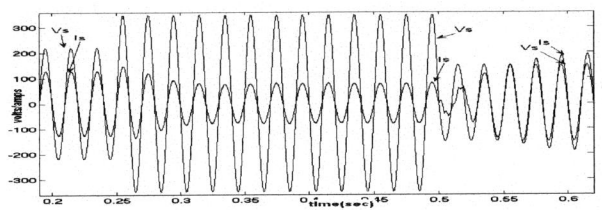

Fig. 14. Phase voltage and line current for variations in phase voltage.

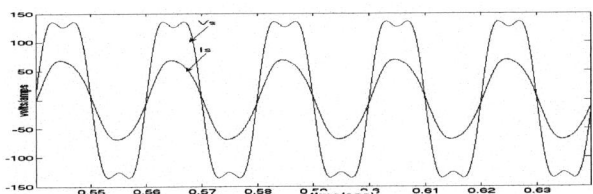

Fig. 15. Phase voltage and line current waveforms under distorted mains.

VI CONCLUSION

A space-vector modulated three-phase power factor correction bidirectional converter based on NPC configuration is proposed to achieve sinusoidal supply currents with nearly unity input power factor, low voltage and dv/dt stresses, low EMI emissions, well-regulated dc output voltage with almost level waveform and balanced neutral point voltage. The proposed control scheme is based on capacitor voltage balancing using space vector modulation. In the proposed control algorithm, the influence of each space vector on capacitor voltages has been studied and a relationship between the direction of power flow and voltage unbalance has been clearly established. Based on this, a non-linear control technique to balance the dc-bus capacitor voltages, by using the redundant switching states, has been developed. The excellent performance of converter is achieved at a reduced switching frequency as compared to three-phase two-level

PWM converter. Hence the proposed converter involves reduced switching power losses and increased efficiency of conversion. The converter also exhibits the properties of drawing sinusoidal line currents and regulating dc output voltage even under the conditions of variations in source voltage and load. An important feature of the converter is its ability to draw nearly sinusoidal currents at unity power factor even under the distorted mains conditions. These characteristics make the converter highly suitable for medium voltage and medium power industrial applications where it acts as almost linear load on the distribution network, thereby causing no harmonic pollution and reactive power flow in the power network. This causes no ill effects on the operation of other equipments including the sensitive equipments being fed from the same feeder. Thus the proposed converter is a useful power quality improvement ac/dc converter suitable for various types of applications including domestic, commercial and industrial applications.

VII REFERENCES

[1] P. M. Bagwat and V. R. Stefanovic, "Generalized structure of a Multilevel PWM Inverter," *IEEE Trans. Ind. Applicat.*, vol. IA-19, No. 6, pp. 1057-1069, Nov./Dec. 1983.

[2] J. W. Dixon, "Boost type PWM rectifiers for high power applications," Ph.D dissertation, Dept. Elect. Comput. Eng., McGill Univ., Montreal, QC, Canada, Jun. 1988.

[3] IEEE Recommended Practices and Requirements for Harmonics Control in Electric Power Systems, IEEE std. 519, 1992.

[4] J. S. Lai and F. Z. Peng, "Multilevel Converters: A New Breed of Power Converters," *IEEE Trans. Ind. Applicat.*, vol. IA-32, No. 3, May/June 1996, pp. 509-517.

[5] S. Fukuda and Y. Matsumoto, "Neutral point potential and unity power factor control of NPC boost converters." in *Proc. Power Conversion Conf. (PCC-Nagaoka)*, vol. 1, 1997, pp. 231-236.

[6] O. Alonso, L. Marroyo, and P. Sanchis, "A Generalized Methodology to Calculate Switching Times and Regions in SVPWM Modulation of Multilevel Converters," *European Power Electronics,* 2001, pp. P.1-P.8.

[7] L. Yacoubi, Kamal Al-Haddad, F. Fnaiech, and L. A. Dessaint, "A DSP-Based Implementation of a New Nonlinear Control for a Three-Phase Neutral Point Clamped Boost Rectifier Prototype," *IEEE Trans. Ind. Electr.*, vol. 52, No. 1, Feb. 2005, pp. 197-205.

VIII BIOGRAPHIES

**Abdul Hamid Bhat** received his Bachelor's degree in Electrical Engineering from the National Institute of Technology Srinagar, Kashmir and received his PG from Indian Institute of Technology Roorkee, India. He is presently pursuing his Ph.D. in Electrical Engineering from Indian Institute of Technology Roorkee, India. He is a faculty member in the department of Electrical Engineering, National Institute of Technology Srinagar. His fields of interest include Power quality problems, Power Electronics, High power factor AC/DC Converters, Multilevel Rectifiers, and applications of dSPACE for the control of power converters.

**Pramod Agarwal** obtained his Bachelor's degree in Electrical Engineering from University of Roorkee and now Indian Institute of Technology Roorkee, India. He received his PG and completed his Ph.D in Electrical Engineering from the same institute in 1985 and 1995 respectively. He is currently Professor in the department of Electrical Engineering, Indian Institute of Technology Roorkee, India. His special fields of interest include Electircal Machines, Power Electronics, Power quality, Microprocessors and microprocessor-controlled Drives, Active power filters, High power factor Converters, Multilevel Converters, and application of dSPACE for the control of Power Converters.

**2006 IEEE International Conference on Power Electronic, Drives and Energy Systems**

# A Novel Load Compensator for a 12-pulse Diode Converter

Maryclaire Peterson, *Student Member, IEEE*, and Brij N. Singh, *Member, IEEE*

*Abstract*– **This paper presents a novel active current shaper, called hereinafter a load compensator (LSTATCOM), for a 12-pulse diode converter. The LSTATCOM injects compensation current through a split winding interphase reactor connected on dc bus of the 12-pulse converter. It is desired that the LSTATCOM provides sinusoidal shaping of the source currents for an output load varying over a wide range. The proposed LSTATCOM should be capable of overcoming the problems of voltage unbalance and transformer design defects. This paper develops a novel technique to make the LSTATCOM inherently capable of addressing these power quality issues. To simulate the dynamic and steady state performance of the converter with the proposed LSTATCOM, a detailed mathematical model of the system is developed. A thorough investigation of the simulation results reveals that the proposed LSTATCOM provides sinusoidal shaped source currents while keeping the system power-factor pegged at unity for a wide range of conditions.**

*Index Terms*--**Multipulse converter, hybrid current shaping, harmonic reduction, current injection**

## I. INTRODUCTION

A variety of active and passive harmonic filtering techniques have been proposed to satisfy the power quality standard [1],[2] when nonlinear loads are deployed in different applications. Passive filters are attractive solutions due to their low cost, simplicity, and control-less operation [3]-[5]. These filters are tuned to attenuate certain harmonics present in the line current. However, passive filters suffer from the problem of mistuning when the load deviates from the designed value [6]. Active filters are controlled power converters connected either in series or in parallel with the nonlinear load. Active filters generate voltage/current waveforms which cancel the harmonics produced by nonlinear loads. Unlike their passive counterparts, the active filters do not suffer from mistuning and can provide simultaneous or selective compensation for all harmonics [7]-[9]. However, cost, complexity, control system design, and reliability are considered some of the major drawbacks of active filters. Consequently, in place of the standard six-pulse rectifier, the

Maryclaire Peterson and Brij N. Singh are with the Department of Electrical Engineering and Computer Science, Tulane University, New Orleans, LA 70118, USA (e-mails: mpeters2@tulane.edu, singh@eecs.tulane.edu). Maryclaire Peterson thanks National Science Foundation for financial support through Graduate Research Fellowship Program.

12-pulse diode converters are gaining an increased acceptance due to their simplicity, robustness, reliability, reduced cost, and control-less operation. As compared to a standard six-pulse rectifier, a 12-pulse converter exhibits a negligible level of $5^{th}$ and $7^{th}$ harmonic distortion and also a reduced level of ripple in the output voltage [10]. In general, a 12-pulse converter consists of two six-pulse rectifiers connected in parallel via an interphase reactor. The 12-pulse converter uses an isolation transformer with a single primary and two secondary windings connected in a $\Delta$-Y and $\Delta$-$\Delta$ configuration to provide the necessary 30° phase-shift between the two secondary voltages [10]. The source current for a 12-pulse converter is the phasor sum of the input currents to the two six-pulse rectifiers [11]. In general, a multipulse converter provides a multi-step source current on the primary winding of the isolation transformer [10]-[13]. This is due to time-displaced currents flowing through multiple secondary windings of the isolation transformer. The lowest order of a multipulse converter is the 12-pulse topology with a typical TDD value in the range of 12-15% and 1-2% ripples in dc bus voltage. However, for further reduction in the TDD and dc bus voltage ripples, the pulse number of the converter can be raised with the addition of more rectifiers. The higher pulse converters provide additional steps in the transformer primary current making it appear closer to sinusoidal. An increase in the pulse number to reduce the TDD and dc bus voltage ripple compromises the converter simplicity and cost because additional six-pulse rectifiers require complex transformer winding arrangements, component packaging, etc. Nevertheless, multipulse diode converters are gaining an increased acceptance due to their robustness, reliability, and control-less operation.

This paper proposes a new topology for a multipulse converter, which consists of a standard 12-pulse diode converter for ac-dc conversion and a novel LSTATCOM on dc bus for sinusoidal shaping of the source currents locked in-phase with the source voltages. The proposed LSTATCOM injects compensation current in series with the dc load current shared by each rectifier. It is desired that the LSTATCOM provides sinusoidal shaping of the source currents for a wide variation in the output load changing from 10% to 110% of the rated value. Also, the LSTATCOM compensates a limited amount of sag and swell in the source voltages and keeps the source current close to sinusoidal for design flaws in the isolation transformer. A detailed mathematical model of the proposed multipulse converter is developed to simulate its dynamic and steady state performance.

0-7803-9771-1/06/$25.00 ©2006 IEEE                    280

## II. BACKGROUND AND STRUCTURE OF PROPOSED CONVERTER

A 12-pulse diode converter exhibits a simple power electronics structure and transformer winding arrangement. However, it suffers from numerous problems, such as an increased value of TDD in the source current and no control over unexpected sags and swells in the source voltage. Moreover, due to the absence of any control system, a 12-pulse diode converter does not offer any remedy for a rise in the source current TDD when the output load drops below its rated value. Also, a 12-pulse diode converter does not address the problems arising from any design mishaps that occur in the isolation transformer. To overcome the problems of higher TDD in the source current, Choi, etal [12] have proposed a low power interphase reactor based triangular current injector, which provides a sinusoidal shaping of the source currents of a 12-pulse diode converter. A current injector scheme, similar to that proposed by Choi, etal [12], is depicted in Fig. 1.

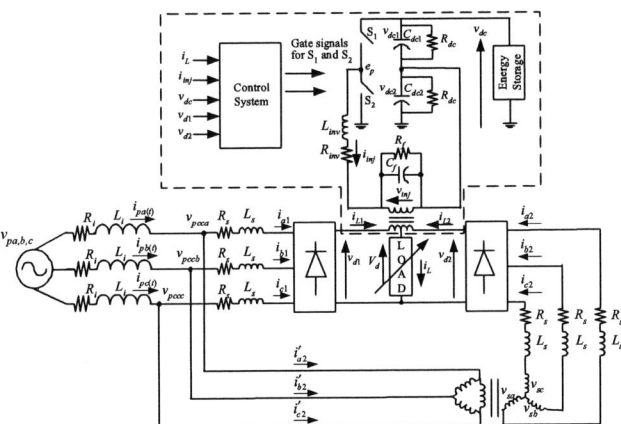

Fig. 1. Current injector for a 12-pulse diode converter [12].

Although the simple structure of the scheme depicted in Fig. 1 is desirable, it requires voltage control of both capacitors of the current source connected on the secondary winding of the interphase reactor. The single-phase dc-ac inverter based current source is required to produce a triangular waveshaped current to be injected in series with dc load currents, $i_{L1}$ and $i_{L2}$, shared by each rectifier unit. However, any mismatch in the average values of the voltages ($v_{dc1}$ and $v_{dc2}$) across the current source capacitors will produce an unsymmetrical triangular waveshaped current resulting is unwanted distortion in the source currents ($i_{pa,b,c}$). This topology also fails to address the problem of voltage unbalance at the input of both rectifiers. The occurrence of any voltage unbalance may result in unsymmetrical and dissimilar primary voltages across both halves of the interphase reactor. Therefore, the desired compensation for voltage unbalance requires equal splitting of the secondary winding of the interphase reactor and the addition of an extra leg to the single-phase dc-ac based current source, resulting in the scheme shown in Fig. 2. Although the additional leg in the single-phase dc-ac inverter increases the control system complexity of the LSTATCOM, the system hardware allows the control system to gain flexibility to address an array of

power quality problems such as single/three-phase voltage swells, sags, voltage unbalance, wide range of load variations, etc.

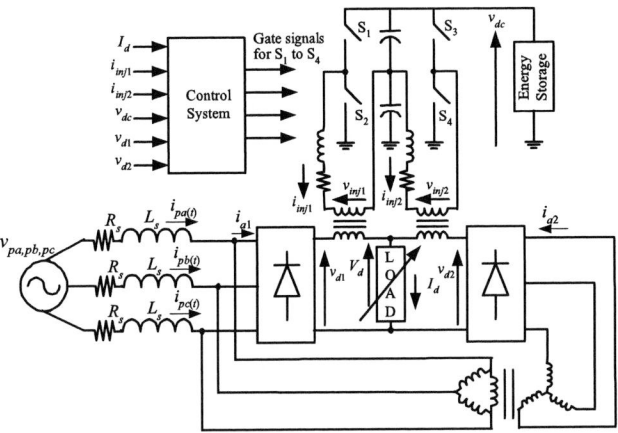

Fig. 2. Proposed LSTATCOM based 12-pulse diode converter.

## III. MATHEMATICAL MODEL OF PROPOSED LSTATCOM-BASED 12-PULSE CONVERTER

Fig. 1 shows a standard 12-pulse diode converter with the current source scheme and interphase reactor enclosed in a discrete block. The 12-pulse diode converter consists of two six-pulse rectifiers connected in parallel. In Fig. 1, rectifier #1 is directly fed from the ac mains, and rectifier #2 is fed from a Δ-Y transformer to provide the required 30° phase-shift between the supply voltages, ($v_{pcca,b,c}$) and ($v_{sa,b,c}$), for both rectifiers. This phase-shift is essential for substantial elimination of the 5th and 7th harmonics in the source current [10] of the 12-pulse converter. The converter source currents ($i_{pa,b,c}$) are phasor sums of the individual rectifier input currents, ($i_{a1,b1,c1}$) for rectifier #1 and ($i'_{a2,b2,c2}$) for rectifier #2. The equivalent circuit for the 12-pulse converter during the 30°-60° period of the source voltage is depicted in Fig. 3. Equation (1) represents the dynamics of rectifier #2 during this period.

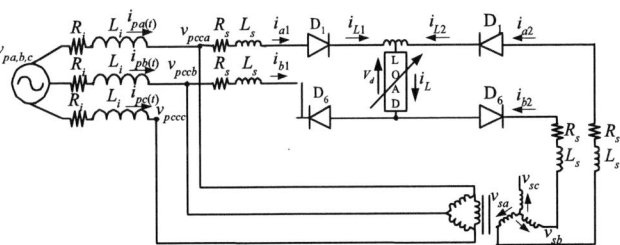

Fig. 3. Equivalent circuit of the 12-pulse converter during the 30°-60° period of the source voltage.

$$v_{sa} - R_s i_{a2} - L_s \frac{di_{a2}}{dt} - R_L i_L - L_L \frac{di_L}{dt} + R_s i_{b2} + L_s \frac{di_{b2}}{dt} - v_{sb} = 0 \quad (1)$$

In Equation (1), $i_{a2}$, $i_{b2}$, and $i_{c2}$ are the input phase currents to rectifier #2 and $i_L$ is the converter load current. From the circuit shown in Fig. 3, Equation (1) can be simplified by realizing the equality $i_{a2} = -i_{b2} = i_{L2}$. Also, the load current ($i_L$)

281

of the converter is the phasor sum of the rectifier load currents, $i_{L1}$ and $i_{L2}$. In addition, the difference between the supply voltages $(v_{sa}-v_{sb})$ can be expressed as the line-to-line supply voltage to rectifier #2, $v_{LLs2}$. The rearranged simplified equation, which represents the dynamic response of rectifier #2, is given below.

$$\frac{di_{L2}}{dt} = \frac{v_{LLs2} - 2R_s i_{L2} - R_L\left(i_{L1}+i_{L2}\right) - L_L\dfrac{di_{L1}}{dt}}{\left(2L_s+L_L\right)} \quad (2)$$

Derived from Fig. 3, Equation (3) represents the dynamic response of rectifier #1 during the 30°-60° period of the source voltage.

$$\left.\begin{aligned} v_{pa} - R_i i_{pa} - L_i\frac{di_{pa}}{dt} - R_s i_{a1} - L_s\frac{di_{a1}}{dt} - R_L i_L - L_L\frac{di_L}{dt}\\ + R_s i_{b1} + L_s\frac{di_{b1}}{dt} + R_i i_{pb} + L_i\frac{di_{pb}}{dt} - v_{pb} = 0 \end{aligned}\right\} \quad (3)$$

The difference between the source voltages $(v_{pa}-v_{pb})$ can be expressed as the line-to-line source voltage to the converter, $v_{LLs1}$. The source currents $(i_{pa,b,c})$ to the converter, which are phasor sums of the individual rectifier currents drawn at the PCC, are expressed in Equation (4).

$$\begin{aligned} i_{pa} &= i_{a1} + i'_{a2} = i_{a1} + \tfrac{1}{\sqrt{3}}\left(i_{a2}-i_{c2}\right)\\ i_{pb} &= i_{b1} + i'_{b2} = i_{b1} + \tfrac{1}{\sqrt{3}}\left(i_{b2}-i_{a2}\right)\\ i_{pc} &= i_{c1} + i'_{c2} = i_{c1} + \tfrac{1}{\sqrt{3}}\left(i_{c2}-i_{b2}\right) \end{aligned} \quad (4)$$

In Equation (4), $i_{a1}$, $i_{b1}$, and $i_{c1}$ are the input phase currents to rectifier #1, and $i_{a2}$, $i_{b2}$, and $i_{c2}$ are the input phase currents to rectifier #2. The input phase currents to rectifier #1 can be written in terms of the load current for rectifier #1 $(i_{L1})$ by substituting $i_{a1} = -i_{b1} = i_{L1}$. The input phase currents to the rectifier #2 can be expressed as functions of the load current $(i_{L2})$ since the input phase currents $(i_{a2}, i_{b2}, i_{c2})$ can only acquire a discrete value from the set, $i_{L2}$, 0, and $-i_{L2}$. This inherent operating characteristic of rectifier #2 allows for further simplification of Equation (3). Consequently, the terms $\left(\tfrac{1}{\sqrt{3}}i_{a2}-\tfrac{1}{\sqrt{3}}i_{c2}\right)$ and $\left(\tfrac{1}{\sqrt{3}}i_{b2}-\tfrac{1}{\sqrt{3}}i_{a2}\right)$ can be written as $k_1\cdot i_{L2}$ and $k_2\cdot i_{L2}$, respectively. Equation (3) can be rewritten strictly in terms of the rectifier load currents $(i_{L1}$ and $i_{L2})$.

$$\left\{\begin{aligned} v_{LLs1} - 2R_i i_{L1} - R_i i_{L2}\left(k_1-k_2\right) - 2L_i\frac{di_{L1}}{dt} - L_i\left(k_1-k_2\right)\frac{di_{L2}}{dt}\\ - 2R_s i_{L1} - 2L_s\frac{di_{L1}}{dt} - R_L\left(i_{L1}+i_{L2}\right) - L_L\frac{d}{dt}\left(i_{L1}+i_{L2}\right) = 0 \end{aligned}\right. \quad (5)$$

After a thorough investigation, it is realized that $(k_1 - k_2)$ is a constant equal to $\sqrt{3}$. Consequently, Equation (5) can be further simplified and made free from terms $k_1$ and $k_2$. Therefore, the final form of the mathematical equation which

represents the dynamic response of rectifier #1 is given below.

$$\frac{di_{L1}}{dt} = \frac{v_{LLs1} - 2R_i i_{L1} - \sqrt{3}R_i i_{L2} - \sqrt{3}L_i\dfrac{di_{L2}}{dt} - 2R_s i_{L1} - R_L\left(i_{L1}+i_{L2}\right) - L_L\dfrac{di_{L2}}{dt}}{\left(2L_i+2L_s+L_L\right)} \quad (6)$$

As seen from Figs. 1 and 2, the high frequency switching of the current controlled PWM inverter will produce the desired injected voltage $(v_{inj})$ across the interphase reactor, which is proportional to the injected current $(i_{inj})$. To include the dynamics of the LSTATCOM, Equation (2) for rectifier #2 and Equation (6) for rectifier #1 are modified to include the effect of the injected voltage $(v_{inj})$. The modified equations given below will be used to represent the converter dynamics in the proposed compensation scheme.

$$\frac{di_{L1}}{dt} = \frac{v_{LLs1} - \dfrac{T_r v_{inj}}{2} - 2R_i i_{L1} - \sqrt{3}R_i i_{L2} - \sqrt{3}L_i\dfrac{di_{L2}}{dt} - 2R_s i_{L1} - R_L\left(i_{L1}+i_{L2}\right) - L_L\dfrac{di_{L2}}{dt}}{\left(2L_i+2L_s+L_L\right)} \quad (7)$$

$$\frac{di_{L2}}{dt} = \frac{v_{LLs2} + \dfrac{T_r v_{inj}}{2} - 2R_s i_{L2} - R_L\left(i_{L1}+i_{L2}\right) - L_L\dfrac{di_{L1}}{dt}}{\left(2L_s+L_t\right)}$$

The quantity $T_r$ in Equation (7) is the turn ratio of the interphase reactor. For sinusoidal shaping of the 12-pulse converter source current, the injected voltage $(v_{inj})$ appearing in Equation (7) can be produced by the proposed LSTATCOM shown in Fig. 2. For this, an injected current with an appropriate magnitude and shape is obtained using the PWM current control scheme (Fig. 4) modeled below.

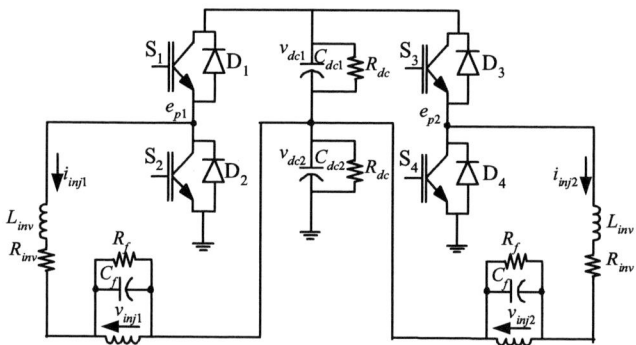

Fig. 4. Two single-phase semi-controlled PWM inverters (Proposed LSTATCOM Scheme).

As shown in Fig. 4, each half of the interphase reactor is used to operate a dedicated current controlled voltage source inverter (CC-VSI) for an appropriate current shaping of both rectifiers under a variety of power quality problems, such as voltage swells, voltage sags, and voltage unbalance. The dynamic response of the proposed LSTATCOM scheme shown in Figs. 2 and 4 can be represented by the following mathematical equations.

282

$$\frac{di_{inj\,1}}{dt} = \frac{\left(e_{p1} - R_{inv}\,i_{inj\,1} - R_t\,i_{inj\,1} - v_{inj\,1}\right)}{L_{inv} + L_t}$$
$$\frac{di_{inj\,2}}{dt} = \frac{\left(e_{p2} - R_{inv}\,i_{inj\,2} - R_t\,i_{inj\,2} - v_{inj\,2}\right)}{L_{inv} + L_t} \qquad (8)$$

$$\frac{dv_{inj\,1}}{dt} = \frac{\left(i_{inj\,1} - \dfrac{v_{inj\,1}}{R_f}\right)}{C_f}$$
$$\frac{dv_{inj\,2}}{dt} = \frac{\left(i_{inj\,2} - \dfrac{v_{inj\,2}}{R_f}\right)}{C_f} \qquad (9)$$

where $R_{inv}$ and $L_{inv}$ are the inverter resistance and inductance, $R_t$ and $L_t$ are the transformer resistance and inductance, and $R_f$ and $C_f$ are the parameters of the low-pass filter. In Equation (8), quantities $e_{p1}$ and $e_{p2}$ are the pole voltages of both inverters (Fig. 4), and $v_{inj1}$ and $v_{inj2}$ are the voltages produced across both halves of the interphase reactor winding. In Equation (9), the term $v_{inj}/R_f$ accounts for the dielectric loss in the filter capacitor $C_f$ as well as the ohmic loss in the low pass filter. The quantities, $i_{inj1}$ and $i_{inj2}$, are the injected currents produced by the PWM inverters.

## IV. CONTROL STRATEGY OF LSTATCOM

The switching states of the CC-VSI devices ($S_1$, $S_2$, $S_3$, $S_4$, $D_1$, $D_2$, $D_3$, and $D_4$) shown in Fig. 4 can be obtained using the PWM current control scheme of Fig. 5. For this, the injected current ($i_{inj}$) is compared with its reference counterpart ($i_{inj}^*$).

The resulted error is processed in a PI controller to ensure that the injected current ($i_{inj}$) follows a trajectory around the reference current ($i_{inj}^*$). The output of the PI controller is subjected to a limiter to avoid the problem of under modulation. The limiter output is compared with a high frequency carrier signal to obtain the control signals for the switching devices used in the LSTATCOM inverter circuit (Figs. 2 and 4). For the desired control of the LSTATCOM current ($i_{inj}$), it is essential to obtain an injected reference current ($i_{inj}^*$) with an appropriate shape. Since the proposed scheme (Fig. 2) is aimed to provide sinusoidal shaping of the 12-pulse converter source current for a variety of operating conditions, such as a wide variation in the load, voltage unbalance, voltage sag, and voltage swell, the reference current ($i_{inj}^*$) needs to be a function of the system quantities. In this investigation, it has been determined that the reference current ($i_{inj}^*$) is expressed as a function of voltages, $v_{d1}$ (output voltage of rectifier #1), $v_{d2}$ (output voltage of rectifier #2), and $i_L$ (converter load current).

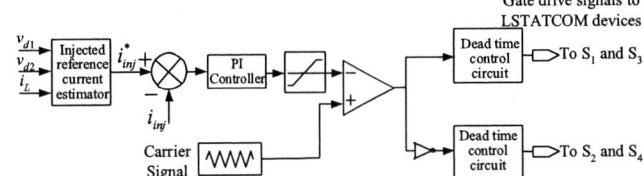

Fig. 5. Control scheme of the LSTATCOM shown in Fig. 4.

## V. PERFORMANCE OF PROPOSED CONVERTER

The derived mathematical model of the ac-dc converter with the proposed LSTATCOM is used to develop the simulation model to obtain the dynamic and steady state response of the 12-pulse converter. The simulation results reveal that the proposed LSTATCOM not only provides sinusoidal-shaped source currents but also keeps the 12-pulse ac-dc converter power-factor pegged at unity under a variety of power quality conditions. To validate the viability of the developed simulation model of the proposed LSTATCOM, the 12-pulse converter has been tested for a variety of operating conditions.

The proposed LSTATCOM is capable of providing sinusoidal shaped source currents over a wide range of output load variations. Fig. 6 shows the current shaping capability of the LSTATCOM for transitions of the converter load from a low value (0.22 pu) to a high value (0.72 pu) to a medium value (0.47 pu).

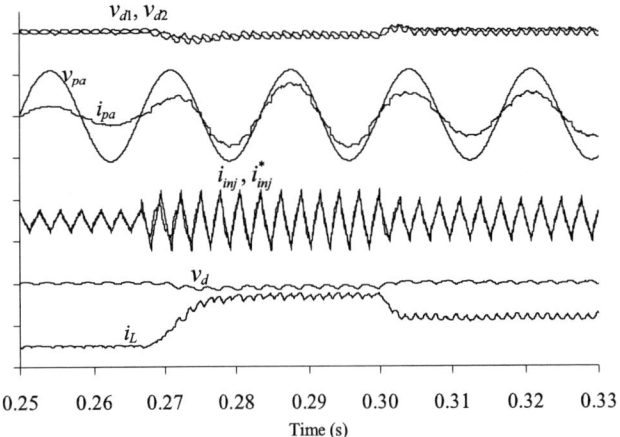

Fig. 6. Transient response of converter for a load change from 0.22 pu to 0.72 pu to 0.47 pu.

It is observed that the converter system needs a finite amount of time of approximately a quarter cycle to a half cycle to restore proper energy balance. This is due to the unidirectional power flow in the system. It is also observed from Fig. 6 that regardless of any loading condition in the 12-pulse converter, the LSTATCOM maintains a sinusoidal-shaped source current by appropriately changing the reference current ($i_{inj}^*$), which is expressed as a function of the load current ($i_L$). From this figure, it is observed that for a wide variation in the load current ($i_L$), the load voltage ($v_d$) remains well within the desired range.

It is demonstrated in Figs. 7 and 8 that the LSTATCOM

control system compensates a 30% sag in the source voltage. Fig. 7 pertains to a 30% sag in all three-phase source voltages ($v_{pa,b,c}$), and Fig. 8 depicts the LSTATCOM performance for a 30% sag in one ($v_{pb}$) of the three-phase source voltages ($v_{pa,b,c}$). Since the reference current ($i_{inj}^{*}$) of the current controlled LSTATCOM is expressed as a function of $i_L$, $v_{d1}$, and $v_{d2}$, a drop in $v_{d1}$ and $v_{d2}$ during the sag forces the LSTATCOM control system to compensate the unwanted sag by injecting a higher value of current through the interphase reactor. It is also essential to evaluate the effectiveness of the LSTATCOM for voltage sag compensation by turning-off its devices. It is found that an absence of the LSTATCOM switching during the voltage sag results in a substantial drop in the load voltage ($v_d$). The difference between the voltage sag in all three-phases or in any of the three-phases is also highlighted. In the presence of a voltage sag in any of the three-phases, the rectifier output voltages ($v_{d1}$ and $v_{d2}$) in Fig. 8 are plagued with a dominant second harmonic, which also affects the waveshape of load voltage ($v_d$). However, for sag in all three-phases or in any of the three-phases, an increased magnitude of the injected current keeps the output load voltage in the desired range. Any distortion in the source current during the transition from the normal condition to the voltage sag condition and vice versa is due to a sudden energy imbalance in the system. In Figs. 7 and 8, the distortion in the source current lasts approximately one cycle, which is the time needed by the 12-pulse converter to restore an appropriate energy balance.

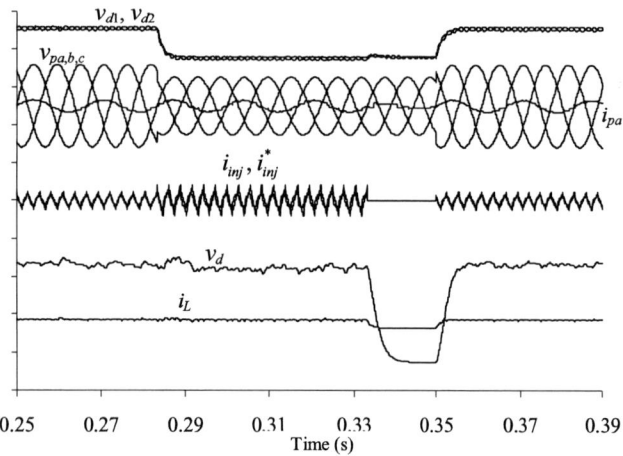

Fig. 7. Transient response of the converter for a 30% sag in the source voltages ($v_{pa,b,c}$).

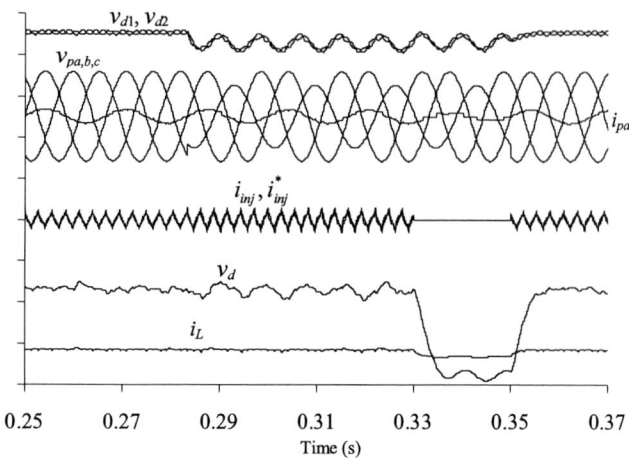

Fig. 8. Transient response of the converter for a 30% sag in phase $b$ ($v_{pb}$) of the source voltages ($v_{pa,b,c}$).

There could also be some design defects in the transformer winding, resulting in a non-optimal phase difference ($\neq 30°$) between the supply voltages for both rectifiers, voltages ($v_{pcca,b,c}$) for rectifier #1 and voltages ($v_{sa,b,c}$) for rectifier #2. Fig. 9 provides the steady state performance of the converter followed by a switching-off of the LSTATCOM devices for the case of a 30° and 34° phase-shift between the supply voltages ($v_{pcca,b,c}$) for rectifier #1 and supply voltages ($v_{sa,b,c}$) for rectifier #2. This figure demonstrates that the LSTATCOM can address non-optimal pulse placement of currents ($i_{a2,b2,c2}$) drawn by rectifier #2. It is observed from Fig. 9 that the LSTATCOM is found capable of maintaining a near sinusoidal source current despite this design flaw in the isolation transformer. It is also observed from Fig. 9 that to provide a sinusoidal current shaping of the source currents ($i_{pa,b,c}$) for a 34° degree phase-shift between the supply voltages ($v_{pcca,b,c}$) for rectifier #1 and supply voltages ($v_{sa,b,c}$) for rectifier #2, the LSTATCOM injects unequal currents ($i_{inj1} \neq i_{inj2}$) through both halves of the interphase reactor. Therefore, for a decoupled operation of the LSTATCOM, two different PWM current controllers are needed to obtain switching signals for device pairs ($S_1S_2$ and $S_3S_4$) of the LSTATCOM, which is essential to meet the condition of unequal injected currents ($i_{inj1} \neq i_{inj2}$).

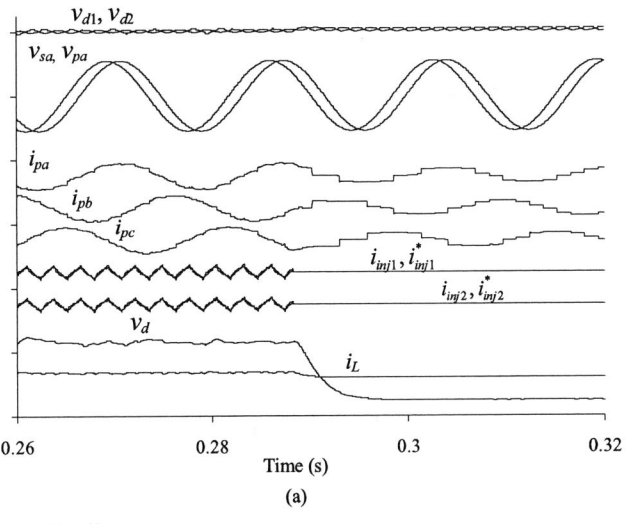

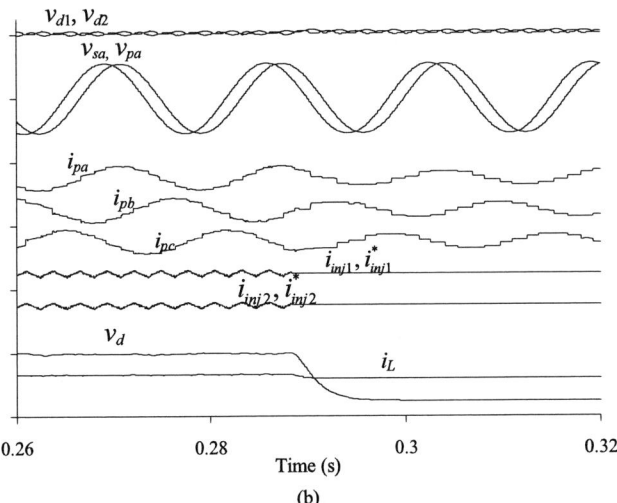

Fig. 9. Performance results of the LSTATCOM for optimal and non-optimal phase-shift between the supply voltages of rectifiers with (a) 30° degree phase-shift and (b) 34° degree phase-shift between the primary and secondary winding voltages.

## VI. CONCLUSIONS

This paper has proposed a novel load compensator (LSTATCOM) scheme for a 12-pulse diode converter. The mathematical model of the proposed converter system has been segmented into two parts, the first part for the controlled component (LSTATCOM) and the second part for the individual rectifiers in the converter. The developed mathematical model has been converted into a simulation model using c-codes. The simulation model has been tested for different operating conditions, such as a wide range of load perturbation and a significant amount of source voltage sag, swell, and unbalance. It has been demonstrated that the decoupled operation of the inverters in the LSTATCOM

system makes the load connected at output of the 12-pulse converter system free from expected power quality problems, resulting in sinusoidal and unity power-factor source current and a nearly constant output voltage. It is expected that the proposed converter system, which is free from power quality problems, could become an enabling technique for the high power dc supplies used in a variety of engineering applications.

## VII. REFERENCES

[1] Recommended Practices and Requirements for Harmonic Control in Power Systems, IEEE Standard 519-1992.
[2] Electromagnetic Compatibility (EMC), IEC Standard 1000-3-2, 1995.
[3] C. Kawann and A. E. Emanuel, "Passive shunt harmonic filters for low and medium voltage: a cost comparison study," *IEEE Trans. on Power Systems*, vol. 11, no. 4, pp. 1825-1831, Nov. 1996.
[4] S. Bhattacharya, D. M. Divan, and B. B. Banerjee, "Control and reduction of terminal voltage total harmonic distortion (THD) in a hybrid series active and parallel passive filter system", in *Proc. IEEE PESC'93*, 1993, pp. 779-786.
[5] J. Arrillaga, D. A. Bradley, and P. S. Bodger, *Power System Harmonics.* New York: Wiley, 1985.
[6] S. Bhattacharya, P. Cheng, and D. M. Divan, "Hybrid solutions for improving passive filter performance in high power applications," *IEEE Trans. on Industry Applications*, vol. 33, no. 3, pp. 732-747, May/June 1997.
[7] H. Akagi, "New trends in active filters for power conditioning," *IEEE Trans. on Industry Applications*, vol. 32, no. 6, pp. 1312-1322, Nov./Dec. 1996.
[8] H. Fujita and H. Akagi, "The unified power quality conditioner: the integration of series- and shunt-active filters," *IEEE Trans. on Power Electronics*, vol. 13, no. 2, pp. 315-322, Mar. 1998.
[9] M. Aredes, J. Hafner, and K. Heumann, "Three-phase four-wire shunt active filter control strategies," *IEEE Trans. on Power Electronics*, vol. 12, no. 2, pp. 311-318, Mar. 1997.
[10] D. A. Rendusara, A. von Jouanne, P. N. Enjeti, and D. A. Paice, "Design considerations for 12-pulse diode rectifier systems operating under voltage unbalance and pre-existing voltage distortion with some corrective measures," *IEEE Trans. on Industry Applications*, vol. 32, no. 6, pp. 1293-1303, Nov./Dec. 1996.
[11] M. Peterson and B. N. Singh, "Modeling and analysis of multipulse uncontrolled/controlled ac-dc converters," in *Proc. IEEE ISIE'06*, 2006, pp. 1400-1407.
[12] S. Choi, P. Enjeti, H. Lee, and I. Pitel, "A new active interphase reactor for 12-pulse rectifiers provides clean power utility interface," *IEEE Trans. Ind. Applicat.*, vol. 32, no. 6, pp. 1304-1311, Nov./Dec. 1996.
[13] S. Choi, B. S. Lee, and P. N. Enjeti, "New 24-pulse diode rectifier systems for utility interface of high-power ac motor drives," *IEEE Trans. on Industry Applications*, vol. 33, no. 2, pp. 531-541, Mar./Apr. 1997.
[14] S. Bhattacharya, A. Veltman, D. M. Divan, and R. D. Lorenz, "Flux-based active filter controller," *IEEE Trans. on Industry Applications*, vol. 32, no. 3, pp. 491-502, May/June 1996.
[15] L. A. Moran, L. Fernandez, J. W. Dixon, and R. Wallace, "A simple and low-cost control strategy for active power filters connected in cascade," *IEEE Trans. on Industrial Electronics*, vol. 44, no. 5, pp. 621-629, Oct. 1997.
[16] B. T. Ooi, J. W. Dixon, A. B. Kulkarni, and M. Nishimoto, "An integrated ac drive system using a controlled-current PWM rectifier/inverter link," *IEEE Trans. on Power Electronics*, vol. 3, no. 1, pp. 64-71, January 1988.
[17] D. A. Paice, Power Electronic Converter Harmonics: Multipulse Methods for Clean Power. New York: IEEE Press, 1996.

**2006 IEEE International Conference on Power Electronic, Drives and Energy Systems**

# Resonant Operated Buck Converter with Reduced Device Switching Stress with Power Factor Improvement

Vinayak N. Shet

*Abstract*- This paper presents a new resonant operated method for an ac to dc converter having reduced device stress. The converter is operated in buck converter mode. An additional switch is used to operate the devices in such a manner, that the charging and discharging take place in a resonant fashion. A higher efficiency can be achieved by turning off the switch under zero current switching (ZCS). The converter is further modified to operate with two switches in a diode bridge converter instead of three switches. Complete theoretical analysis, simulation results and experimental data on a 500W converter are presented to demonstrate the superiority of the new control strategy with reduced device stress. The input power factor is found to be over 0.996.

*Index Terms*— Buck converter, current wave shaping, power factor, resonant, switching stress, zero switching current.

## I. INTRODUCTION

IN designing conventional switching converters, efforts to increase the operating frequency for reducing the weight and size of the magnetics and the filter elements is constantly hampered by the higher switching stress and switching losses. To overcome these obstacles, the concept of resonant switching is used [1]. This provides a general zero current switching technique, applicable to all conventional converters. By employing an additional inductor and capacitor to shape the switching device's current and voltage waveforms, the resonant-switching technique can eliminate switching stresses and losses, thus boosting the switching frequency into the low megahertz range. In fact, the operation of a quasi-resonant switch is independent of the converter topology and the state trajectories all have the same shape. However, the input voltage and output loading to the switch depend on the type of converter.

The power factor correction for the buck converter, in general, can be achieved with a single-stage or two-stage operation. Intensive research is aimed at combining the two stages into one stage. The converter usually employed for single-stage, single-phase power factor correction consists of a front-end diode rectifier followed by a boost converter. This converter, however, presents conduction and commutation losses, which will contribute to the reduction in the efficiency of the converter. The commutation losses occur due to the hard switching of the power semiconductors. The reduction of

the commutation losses can be achieved by operating them under zero-voltage switching (ZVS) or zero current switching (ZCS). With these modifications, the efficiency can be improved. In order to improve the efficiency even more, power factor correction rectifiers with soft commutation and reduced conduction losses are proposed [3-5]. Some of these converters [3] employ continuous conduction mode to achieve high power factor and due to the complexity and cost, they are suitable for high power applications.

Buck boost converter operating in Discontinuous input current mode (DICM) has a very good inherent Power factor correction (PFC) properties. Single-stage PFC converters were developed by combining a dc-dc converter operating in DICM, for PFC, with a second converter operating in DICM or continuous capacitor voltage mode (CICM), for output voltage regulation [6]. However, DICM operation has disadvantages. Boost converter PFC in DICM has discontinuous input current, which poses tougher requirements for the input filter, while in the CICM the input current is continuous. Moreover, the peak current in the switch is larger in DICM operation when compared to CICM operation, for the same level. Hence, a converter with continuous input current, but with inherent PFC properties, is attractive for single stage converters having high P.F. The strategy proposed in [7] to operate the buck converter with LC input filter operating in DCVM still suffers from the high switching stress on the active switch and the diode. This hard switching condition, when turning on the active switch with a high voltage across it, has a negative impact on the efficiency. And, also, a variable-switching frequency is needed for compensating the load variations and the operation at light load is not possible. The main disadvantage of the buck PFC converter (in DICM operation) is the relatively large peak voltage stress on the semiconductor devices. This capacitor peak voltage $V_{1p}$ is given as [7]

$$V_{1p} = \frac{2\,V_i}{1 - D + D_1}$$

Where $V_i$ = Input dc voltage

$D$ = duty cycle

$D_1$ = off period

Because of the large voltage stress, the semiconductor devices used need to have a higher voltage rating, resulting in higher internal resistance and cost [8].

In this section, a buck converter is proposed to operate at high frequency so that the size of the components is reduced. Further, since the converter is operated under resonance condition, the voltage stress across the switching devices can be reduced and in addition, if it is operated under discontinuous input current conduction, the turn off of the switch takes place at zero current switching (ZCS) thereby,

---

Dr. V. N. Shet, Professor and Head of Electrical & Electronics Department, Goa College of Engineering, Farmagudi, Ponda, Goa, India.
Email: vns@gec.ac.in

0-7803-9771-1/06/$25.00 ©2006 IEEE

achieving inherent power factor correction along with high efficiency. The proposed scheme of high frequency resonant operated converter is shown in Fig. 1.

## II. WORKING AND DESIGN CONSIDERATION

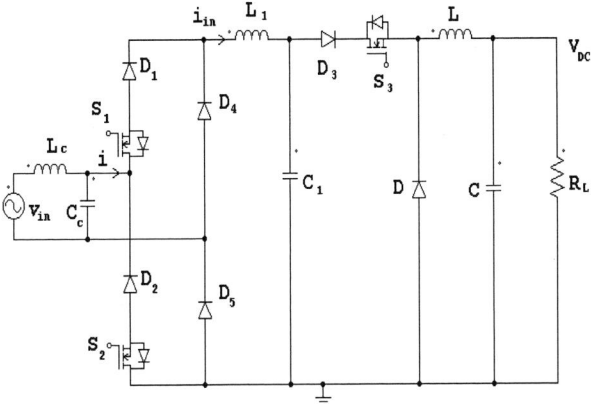

Fig. 1. High frequency resonant operated converter.

The modified scheme is shown in Fig.1, and will function as a resonant operated buck converter. The converter consists of a bridge made up of two switches and two diodes, with one arm consisting of two diodes and other arm having two switches (MOSFET APT6030) in series with diodes to maintain unidirectionality. An additional switch $S_3$ alongwith a diode $D_3$ is connected in series with the bridge configuration. The switches are switched at about 25kHz. Because of this high frequency, the input sinusoidal voltage can be considered to remain constant during each period of the switching frequency. The output voltage is a D.C voltage smaller than the input rms voltage. The operation is explained as follows. During the first positive half cycle of the input ac waveform the switch $S_1$ is turned on and off at the switching frequency. When the switch $S_1$ is switched ON, the switch $S_3$ remains OFF, and vice versa. Similarly, during the next half cycle, switch $S_2$ and $S_3$ are operated. During the turn-on period of $S_1$ the capacitor $C_1$ gets charged through inductor $L_1$. When the capacitor $C_1$ gets charged above the instantaneous value of the input voltage, diode $D_1$ gets reverse biased. In the next half period of the switching cycle, the switch $S_1$ is turned-off and switch $S_3$ is turned ON. The energy stored in the capacitor $C_1$ is discharged into the load through the inductor L. The time required for discharge is determined by L and $C_1$ and it must be less than 20µsec i.e. half the period of the switching cycle. This discharge period remains constant throughout the load as well as the input supply variations. The charging duration of $C_1$ is a function of the load and the instantaneous voltage of the ac main supply. The turn-on and turn-off of the switch $S_1$ and $S_3$ will continue for the remaining half cycle as described above. During the negative half cycle, the switch $S_2$ gets turned on and off with $S_3$ in a complementary fashion. The advantage of the use of resonant principle is to allow the capacitor to charge slowly and thereby reduce the stress on the switching devices. Further, since $i_{in}$ becomes zero during the on period of $S_1$ at full load there is zero current switch off for $S_1$.

The operation of the proposed converter can be represented in terms of the various modes shown in Fig. 2.

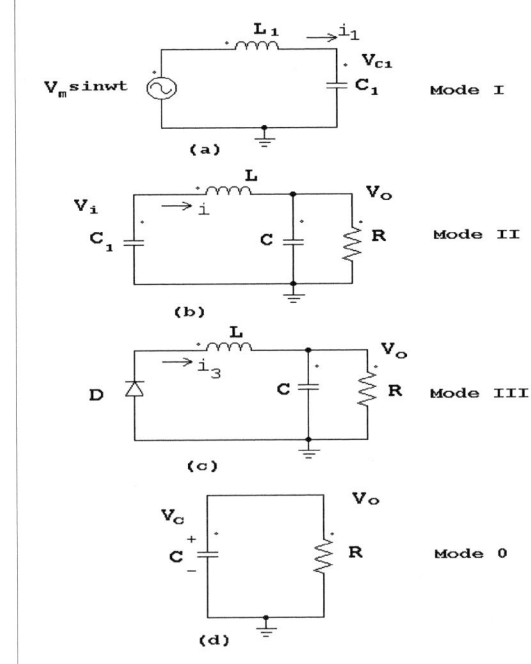

Fig. 2. Operational mode of diagram of the proposed converter.

## III. ANALYSIS OF THE CONVERTER

In the first positive half cycle of the ac mains, switch $S_1$ is turned on at $t = t_1$ and the resonant capacitor $C_1$ gets charged through the resonant inductor $L_1$. The switch remains on for a duty period of $\delta T_{on}$ where $T_{on}$ is the period of the switching frequency. It is to be ensured that the duty cycle is 0.5 for the full load at the rated voltage. The mode ends at t = $t_1 = t_0 + \delta T_s$. At the end of the period of the switching cycle, the switch $S_1$ is turned off and switch $S_3$ is turned on.

*Mode I:* The equation representing this mode for t = 0 is:

$$V_m \sin\omega t_1 = L_1 \frac{di_1}{dt} + V_{C1} \tag{1}$$

Since, $i_1 = C_1 \frac{dV_{C1}}{dt}$ (2)

substituting in equation (1)

$$V_m \sin\omega t_1 = L_1 C_1 \frac{d^2 V_{C1}}{dt^2} + V_{C1} \tag{3}$$

The solution of equation (3) for the initial condition $V_{C1} = 0$ is $V_{C1} =$

$$\frac{V_m \omega}{L_1 C_1 (\omega^2 - 1/L_1 C_1)} \left\{ \frac{-\sin \omega t_1}{\omega} + \sqrt{L_1 C_1} \sin\left( \frac{t}{\sqrt{L_1 C_1}} \right) \right\} \tag{4}$$

The voltage developed across $C_1$ can be determined from the above relationship. The current in the inductor $L_1$ goes to zero at $t_2$ when $V_{C1}$ reaches its maximum value at the end of the half resonance period. Equation (2) is differentiated and equated to zero to determine $t_2$ at which the resonant current $i_1$ goes to zero.

$$i_1 = C_1 \frac{dV_{c1}}{dt} =$$

287

$$C_1 \frac{V_m \omega}{L_1 C_1(\omega^2 - 1/L_1 C_1)} \left\{ \frac{-\omega\cos\omega t}{\omega} + \frac{1}{\sqrt{L_1 C_1}} \sqrt{L_1 C_1} \cos(\frac{t}{\sqrt{L_1 C_1}}) \right\}$$

(5)

At $t = t_2$, $i_1 = 0$

$$0 = C_1 \frac{V_m \omega}{L_1 C_1(\omega^2 - 1/L_1 C_1)} \left\{ -\cos\omega t_2 + \cos(\frac{t_2}{\sqrt{L_1 C_1}}) \right\}$$ (6)

$$\cos\omega t_2 = \cos(\frac{t_2}{\sqrt{L_1 C_1}})$$ (7)

Hence $t_2$ is determined by trial after substituting for $\omega$ ($2\pi50$), $L_1$ and $C_1$. The voltage developed across $C_1$ can be determined from equation (4) by substituting $t = t_2$ in equation (3). This voltage is $V_i$. For lesser loads, $V_{C1}$ must be smaller to satisfy the energy requirement.

*Mode II:* This mode starts at time $t = t_2$ as soon as the switch $S_3$ is turned on. The energy stored in the capacitor $C_1$ is transferred to the Load and capacitor through the inductor L. The current carried by the switch is for a duration of 20μsec. The equivalent circuit during the discharge of the capacitor $C_1$ through the switch is shown in Fig. 2 (b). Assuming that the output voltage across the capacitor C acts as a constant voltage source, the control equation under steady state conditions is given by

$$V_{C1} = L \frac{di}{dT} + V_o$$ (8)

In equation (8) $T = 0$ corresponding to the beginning of mode II

The boundary conditions for equation (8) are

   $i = 0$  at  $T = 0$  $V_{C1} = V_i$  which is the voltage to which the capacitor is charged at the end of mode I.

The solution to the above equation for an initial condition of $V_{c1}=V_i$ is given as

$$i = (V_i - V_o)\sqrt{\frac{C_1}{L}} \sin(T/\sqrt{LC_1})$$ (9)

we can determine $V_{C1}$ as follows,

$$V_{C1} = -\frac{1}{C_1} \int_0^T i\, dT$$ (10)

$$V_{C1} = -\frac{1}{C_1} \int_0^T (V_i - V_o)\sqrt{\frac{C_1}{L}} \sin(\frac{T}{\sqrt{LC_1}}) dT$$ (11)

$$= \frac{(V_i - V_o)}{C_1} \sqrt{\frac{C_1}{L}} \sqrt{LC_1} \cos(\frac{T}{\sqrt{LC_1}})$$ (12)

At $T = T_1$, $V_{C1} = 0$

Hence, $0 = \cos(\frac{T_1}{\sqrt{LC_1}})$ (13)

$$\pi/2 = \frac{T_1}{\sqrt{LC_1}}$$ (14)

$$T_1 = \sqrt{LC_1} \frac{\pi}{2}$$ (15)

Further, from equation (9) the peak current in the switch is given by

$$I_p = (V_i - V_o)\sqrt{\frac{C_1}{L}}$$ (16)

This mode ends at $T = T_1$ when $V_{C1} = 0$. Hence, from the equation (15)

$T_1 = \sqrt{LC_1} \frac{\pi}{2}$ which is the duration of mode II.

The mode ends at $t = t_2 + T_1 = t_3$.
The current at the end of the mode is

$$i_2 = (V_i - V_o)\sqrt{\frac{C_1}{L}} \sin(T_1/\sqrt{LC_1})$$ (17)

*Mode III:* This mode starts at $t = t_3$ when the switch $S_3$ gets turned-off and remains off till it is turned on in the next half cycle. During this period, the freewheeling diode D comes into conduction. This mode can operate either in a continuous current mode or a discontinuous current mode.

*Continuous current conduction mode:*
The equivalent circuit for this mode is shown in Fig. 2(c). The current at the beginning of the mode (T = 0) is given by $i_3 = i_2$. The equation governing the mode is

$$0 = L \frac{di_3}{dT} + V_o$$ (18)

$$i = -\frac{V_o}{L} T + K$$ (19)

at $t = 0$   $i = i_2$

The solution of equation (11) gives the freewheeling current $i_f$

$$i_f = i_2 - \frac{V_o}{L} T$$ (20)

This mode ends at $T = \frac{1}{f_s}$ when the switch $S_3$ is again triggered on after a full cycle of the switching frequency. Hence, the freewheeling current $i_f$ at the end of the mode is

$$i_f = i_2 - \frac{V_o}{L} \frac{1}{f_s}$$ (21)

Where $f_s$ is the switching frequency of the controller. The duration of this mode is $\frac{1}{f_s}$ sec. The mode ends at $t = t_3 + \frac{1}{f_s}$.

*Discontinuous current conduction mode:*
   The equation for this condition also remains the same as equation (20).

$$i = i_2 - \frac{V_o}{L} T$$

Thus,   $i = 0$ at  $T = T_2$

or $T_2 = \frac{i_3}{V_o} L$ (22)

The duration of this mode is given as $T_2$ and ends at $t = t_3 + T_2 = t_4$. If the current is discontinuous and becomes zero at time $t = t_4$, there exists an additional mode-0 as explained below.

*Mode 0:*
   Now, the output capacitor C discharges through the load resistor R. The equivalent circuit is shown in Fig. 2(d). The equation describing this mode is

$$V_C = V_o e^{-T/RC}$$ (23)

This mode ends at $t = t_3 + (\frac{1}{f_s} - T_2)$. The above equation can be solved throughout the cycle to determine the operation.

## IV. SELECTION OF COMPONENTS

The selection of the components begins with the selection of $C_1$. Since, the maximum voltage of operation is fixed at 130V, the peak voltage to which the capacitor should charge should be atleast greater than $130\sqrt{2}$ and as high as possible so that the capacitor value is less. In this case, it is fixed at 300V. Hence, from the energy consideration:

The energy stored in the capacitor $= \frac{1}{2} C_1 V_i^2$

Since the capacitor is fully discharged to the output, the power supplied by the capacitor is

$$= \frac{1}{2} C_1 V_i^2 \times 2 f_s \quad (24)$$

Output power $P_o = V_o I_o$
Therefore for energy balance

$$\frac{1}{2} C_1 V_i^2 \times 2 f_s = P_o \quad (25)$$

Where $V_i$ is the peak value of the capacitor voltage, $V_o$ is the dc output voltage of the converter and $I_o$ is the output load current. The multiplication factor 2 on the left-hand side is because of full wave rectification. Hence,

$$C_1 = \frac{V_o I_L}{f_s V_i^2} \quad (26)$$

For $V_o = 24V$, $I_o = 20A$, $f_s = 25kHz$, $V_i = 300V$. Substituting, the value $C_1 = 0.6\mu F$.
The capacitor used for the experiment is $0.68\mu F$ to take care of the losses.

Next $L_1$ is selected from mode I. For this, it is assumed that Mode-2 exists for about 20μsec. At this time $V_{C1}$ = 300V and I = 0. Substituting in equation (4) $L_1$ is determined. The value $L_1$ calculated is 72μH. Since $V_{C1}$ = 300V and this must be discharged into C in 20μsec through L, it's value is calculated using equation (9) of Mode-2. This gives L = 72μH. To keep the ripple low, C is selected to be 8000μF from the usual ripple consideration.

## V. SIMULATION OF THE CONVERTER

The proposed high frequency resonant converter scheme was simulated using SABER. The trigger signal for the switch $S_1$ and $S_3$ are complementary to each other and operated during the positive half cycle. Similarly, during the negative half cycle, the switch $S_2$ and $S_3$ are switched on and off in a complementary fashion. Due to the presence of diode $D_1$ and $D_2$ in series with switch $S_1$ and $S_2$ respectively, the unidirectionality of the switch is maintained. Hence, the switch $S_1$ will be having conduction path during the positive half cycle and switch $S_2$ in the negative half cycle. In the simulation, the switches are turned on and off at a switching frequency of 25kHz in the open loop condition. To maintain a constant output voltage at varying load and varying input

voltage, the duty period of the controller is adjusted to stabilize the output voltage. The simulation is carriedout for different loads and line voltage conditions to determine its performance.

The component values calculated using the equations (4), (9) and (26) are, $L_1 = 72\mu H$, $C_1 = 0.68\mu F$, $L = 72\mu H$, $C = 8000\mu F$ Load R = 2.5Ω, switching frequency $f_s$ =25kHz, Input voltage = 110V rms Using the above set of values for $C_1$, $L_1$ and L, simulation is carried out.

The simulation waveforms obtained are shown in Fig. 3. Fig. 3, shows the voltage from the input ac mains and the current drawn from the mains. It can be seen from the waveform that the input current drawn from the mains is near sinusoidal with a small cross over distortion. The current follows the voltage without any displacement. Fig. 4, shows the rectified voltage at the output of the bridge and the line current without the input filter. From this waveform, the peak current of the devices can be decided.

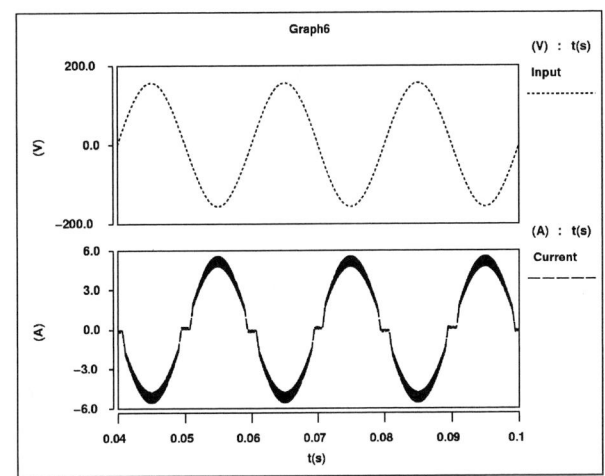

Fig. 3. Input voltage and current drawn from the ac mains.

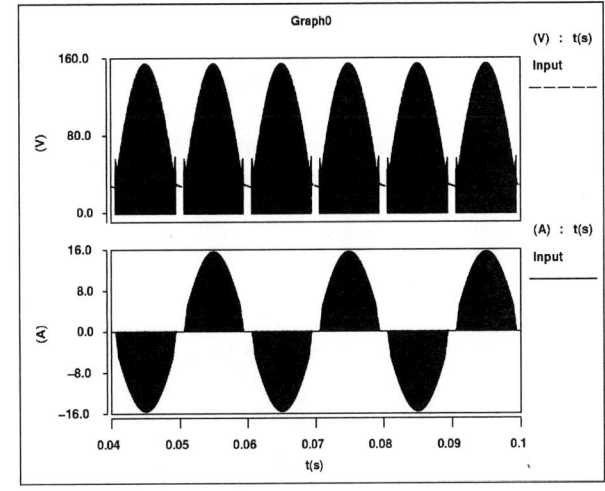

Fig. 4. Rectified voltage at the output of bridge and the line current without input filter.

## VI. WORKING OF THE CONTROLLER

The high frequency resonant operated buck converter discussed above for achieving high power factor throughout its range is designed and built to investigate its operation. A suitable controller was designed using SG 3525. The complimentary control signal is generated by inverting the pwm signal and then given to the driver stage through a buffer. The driver stage consists of optocoupler and a complementary connected driver transistor.

## VII. EXPERIMENTAL RESULTS

The various values of components selected for the experimental verification are as follows:
Mosfet $S_1$ to $S_3$ are APT6030BVR, Diode $D_1$, $D_2$, $D_4$ to $D_5$ are BYX 61-400 D and $D_3$ are BYX 64-600 $L_1$ = 72$\mu$H, $C_1$ = 0.68$\mu$F, L = 72$\mu$H and C = 8000$\mu$F/100V.

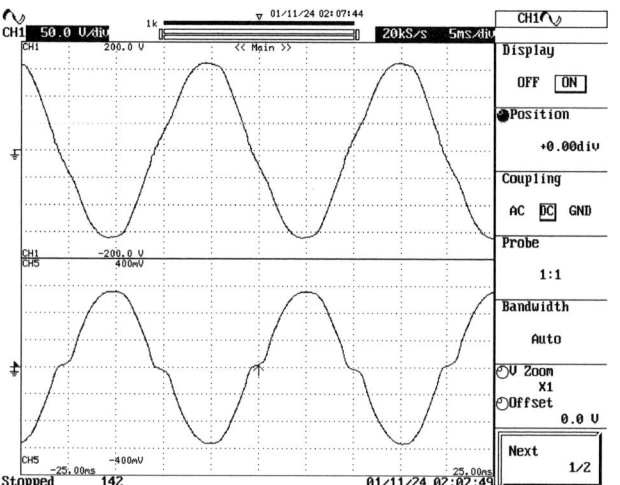

Fig. 5. Voltage and ac current drawn from the mains.(scale: upper 50V/div, lower 1.33A/div).

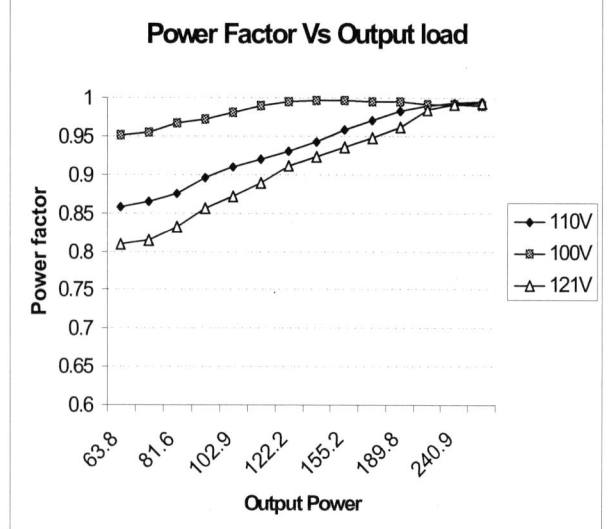

Fig. 6. Experimental Power factor for different load at 110V input mains supply.

Finally, the input voltage from the mains and the input current drawn from the mains are shown in Fig. 5. It can be seen from the waveform that the current drawn from the mains is near sinusoidal with no displacement between the voltage and current thereby achieving a high power factor of 0.994 at full load. The input active power, power factor and input current harmonics content were measured using a Voltech make power Analyser PM100. Which measures the true rms current and voltage and the distortion in the current and voltage. Fig. 6, shows the graphical representation of the experimental power factor for different load conditions under different voltages.

## VIII. MODIFICATION SUGGESTED

The proposed configuration of the half-controlled converter with three switches shown in Fig.1 can be modified so that only two switches as shown in Fig. 7, are used. In this configuration, the additional two switches are used to operate in a resonant fashion and to improve the power factor. The advantage of this converter is the reduction of one switch as compared to the earlier configuration. This scheme is also simulated and the simulation result obtained for the extended scheme matches with the results obtained for the earlier proposed
scheme.

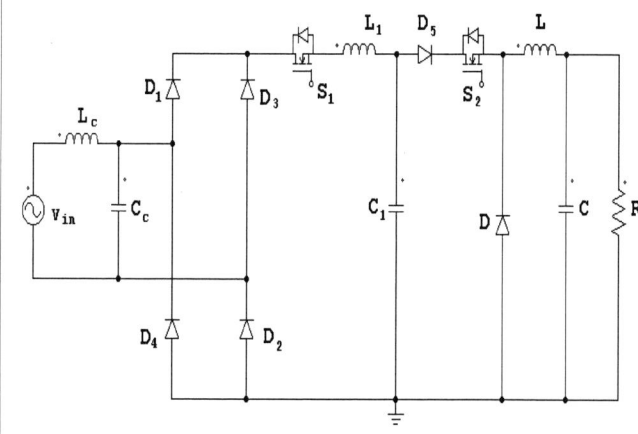

Fig. 7. Resonant converter proposed for diode rectifier with additional switches.

## IX. CONCLUSIONS

A resonant operated buck converter is proposed having an additional switch when compared with the conventional controlled converter. This converter is capable of drawing a higher quality input-current at nearly unity power factor from the mains supply. Moreover, the converter has a wide load range and creates low stress on the semiconductor devices. The scheme was simulated on saber, designed and implemented to deliver a load of 500W at an output voltage of 24V DC from the single–phase ac mains. The controller was designed to operate from 90V to 130V at various loads and the total harmonic distortion (THD) of the line current is less than 8%, and the system power factor is about 0.996 at full load.

## X. REFERENCES

[1] M. M. Jovanvic, K. Liu, R. Oruganti and F.C.Y.Lee, "State plane analysis of quasi resonant converters," *IEEE Trans. Power Electronics.* pp. 36-44, 1987.

[2] V. Grigore, and J. Kyyra, "High power factor rectifier based on buck converter operating in discontinuous capacitor voltage mode," *IEEE Trans. Power Electronics.*, Vol. 15, pp. 1241-1249, 2000.

[3] A. F. de Souza, and I. Barbi, "A new ZVS semiresonant high power factor rectifier with reduced conduction losses," *IEEE Trans. Industrial Electronics.*, Vol. 46, pp. 82-90, 1999.

[4] A. F. de Souza, and I. Barbi, "A new ZVS quasiresonant unity power factor rectifier with reduced conduction losses," *in Proc. 1995 IEEE PESC Conf.*, pp.1172-1176.

[5] K. H. Liu, and Y. L. Lin, "Current wave form distortion in power factor correction circuits employing discontinuous mode boost converter," *in Proc. 1989 IEEE PESC Conf.,in Conf.*, pp.825-829.

[6] H. Pinheiro, P. Jian, and G. Joos, "Self oscillating resonant ac/dc converter topology for input power factor correction," *IEEE Trans. Industrial Electronics.*, Vol. 46, pp. 692-702, 2000.

[7] Y. S. Lee, S. J. Wang, and S. Y. R. Huli, "Modeling, analysis, and application of buck converters in discontinuous input voltage mode operation," *IEEE Trans. Power Electronics.*, Vol. 12, pp. 350-360, 1997.

[8] M. C. Ghanem, K. Al-Haddad, and G. Roy, " A New control strategy to achieve sinusoidal line current in a cascade buck boost converter," *IEEE Trans. Industrial Electronics.*, Vol. 43, pp. 441-449, 1996.

## XI. BIOGRAPHIES

**Vinayak N. Shet** was born in Honnavar at Karnataka in India on 3rd December 1963, and obtained the B. E degree in Electrical Engineering from Karnataka University, Dharwad, India in July 1984, M. Tech degree in Electrical Engineering with specialization in Power electronics from IIT Bombay, Mumbai in January 1995 and Ph.D. degree in Electrical Engineering on the topic of "Power Factor Correction in Converters" from IIT Bombay in December 2002. His fields of interest are power factor correction, resonant converter and energy saving ballast and microcontroller application to power electronic systems. He is presently working as Professor and Head Electrical & Electronics Department in Goa Engineering College, Goa, India.

# A High Power Factor Forward Flyback Converter with Input Current Waveshaping

Vinayak N. Shet

*Abstract*--A novel power factor correction circuit is proposed as an improvement over a one stage forward converter. This PFC circuit will achieve high power factor by making the circuit operate in a discontinuous conduction mode and changing the duty period after the cross over instant. The switching technique is such that the off time of the pwm generated signal shall always remains equal to or higher than the on-time period by maintaining a minimum off duration related to the peak voltage magnitude. The near sinusoidal input current is achieved by modifying the switching technique so that the duty period is changed over to a new one after the reflected output voltage in the primary of the transformer is smaller than the input voltage. The current drawn is fully sinusoidal without and initial delay as in the case of conventional equal width pwm techniques. A high power factor could be achieved with this simple controller.

*Index Terms*—Power factor, Forward, flyback converter, buck-boost, current wave shaping,

## I. INTRODUCTION

SINGLE-PHASE converters were developed to simplify the two cascade-stage systems by incorporating power factor correction and the regulator into one [1]. However, these systems need an inductor in the input side to operate as a boost-type converter and do not have higher efficiency than the two-cascade-stage circuits because the single-stage systems are constructed using cascade connection in principle.

The use of a buck type power preregulator allows one to overcome a few of these drawbacks of a boost topology at the expense of greater input current distortion and higher magnitude of switching frequency in the line [2]. The low frequency distortion in the line current arises from its unavoidable notches around the zero crossing of the line voltage, caused by the inability of the buck converter to draw current when the instantaneous input voltage is lower than the output one. As a consequence, the power that a buck-type preregulator can handle while still complying with the IEC 1000-3-2 standard is limited. Application of forward converter alone result in partial correction, because the power transfer becomes difficult under a low input voltage threshold and the forward converter cannot follow the shape of the rectified line voltage. In order to achieve total correction, a forward converter is associated with a flyback converter such that both topologies [3] share a common transformer and power switches. The configuration by H. E. Tacca [4] uses a combination of flyback and forward converter in such a way that when the input voltage is between zero and output voltage referred to primary, only the flyback subconverter will transfer

Dr. V. N. Shet, Professor and Head of Electrical & Electronics Department, Goa College of Engineering, Farmagudi, Ponda, Goa, India.
Email: vns@gec.ac.in

central energy and the forward converter will operate in the region of the ac cycle. However, the controller becomes more complex and the input wave shaping achieved is far from that expected and maintains a peaky current at the center of the input supply. Further, a quasiconstant duty-cycle operations had to be carriedout, i.e. the duty cycle also gradually increases until it reaches a maximum duty cycle required, for improving the waveshape.

Single-stage forward converter proposed by M. Nagao [5] uses the over peak compensation. The method proposed in this article, is a forward converter to improve the power factor by operating it in the discontinuous conduction mode and the output voltage ripple was reduced by controlling the exciting energy of the transformer using an extra active switch. The duty ratio is smaller than 0.1, under this circumstance and the power factor becomes smaller due to the effect of the input capacitor. It can be presumed that the power factor varies with the duty ratio leading to a poor power factor at light load. In addition, an additional switch along with its controller is required to control the exiting energy from the winding and hence, becomes more costly and complicated.

To overcome the above disadvantages, a modified control strategy and configuration is proposed in this article, which has good operating characteristics, from light load to full load. The proposed configuration is basically a forward converter with modified connection. In this converter, the need for an additional switch as in reference [5] has been eliminated and the number of components is reduced. Hence, the controller becomes simpler. The switch is controlled with a variable duty cycle technique such that at any instant of duty cycle operation, the OFF duration shall be equal to or greater than the turn-on duration. The power factor is improved by making the PFC circuit to operate in a discontinuous conduction mode (DCM) and ripple is controlled because power is feed throughout the cycle. An additional configuration is also proposed, where the combined converter configuration can be achieved by using a conventional two winding transformer instead of a transformer with an additional tertiary winding, as conventionally used for the forward converter. In both the cases, the same controller is capable of transferring the power to the output throughout the input ac cycle, thereby achieving a sinusoidal input current waveform.

The operation and the execution of this converter is discussed below.

## II. CIRCUIT CONFIGURATION AND OPERATING PRINCIPLE

Fig. 1, shows the configuration where a modified controller is used to improve the input power factor. This proposed configuration consists of a forward converter along with a flyback converter interlaced to achieve optimum power factor. In this circuit, the converter will operate at two different duty cycles at a constant frequency. The operation of this scheme is explained using the following modes.

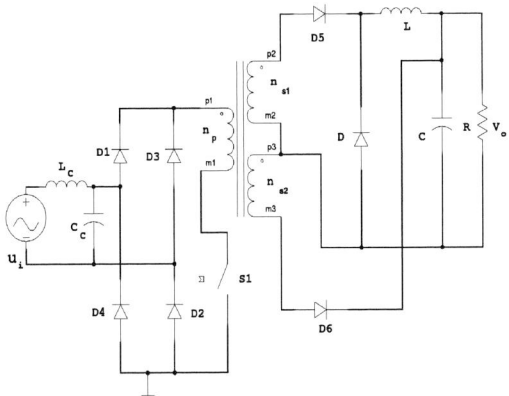

Fig. 1. Forward converter to achieve a near sinusoidal input current.

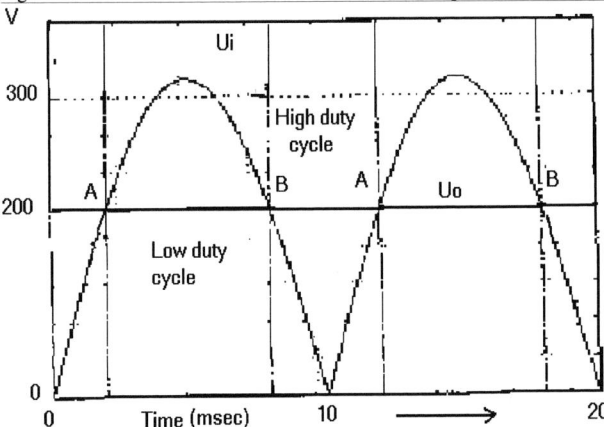

Fig. 2. Input voltage and output voltage with change of duty cycle instant.

*Operating modes:*

The converter is basically operated under a discontinuous mode, and the maximum duty cycle remains $\leq$ 50%. The two operation regions are indicated in Fig. 2. For $|u_i| \leq U_o$, the converter operates in a constant high duty cycle mode until point A is reached. For $|u_i| \geq U_o$, the operation falls between A and B and the converter operates in a constant low duty cycle mode when compared to the earlier case.

The average current drawn by the converter over a one cycle period can be calculated as follows.

Since, the input voltage source is given by $V_m \sin \omega t$, the current from this source for unity power factor should be

$i = I_m \sin \omega t$

When $(V_m \sin \omega t) = v_o \dfrac{n_p}{n_{S1}}$, (1)

Let $t = t_1$

Hence i at $t_1$ is given by

$i = I_m \sin \omega t_1$

Let $L_2 = \left(\dfrac{n_p}{n_{S1}}\right)^2 L$ and (2)

$L_P$ be the primary inductance of the transformer $X_T$.

For $t < t_1$ and under discontinuous conduction, the peak current during the on time of switch $S_1$ is due to the flyback converter alone and is given as

$$i_{p1} = \frac{V_m \sin \omega t}{L_p} \delta_1 \, T \qquad (3)$$

In equation (3), $\delta_1$ is the duty period during the flyback mode

Hence, the average current during the one switching period T due to the flyback action is given by

$i_1 = $ peak current $(i_{p1})$ x $\dfrac{\text{dutyperiod } (\delta_1)}{2}$

$$i_1 = \frac{V_m \sin \omega t}{2L_p} \delta_1^{\,2} T \qquad (4)$$

For $t > t_1$ and under discontinuous conduction

The peak current during one switching period T is due to both the flyback and the forward converter operation and is given by

Hence,

$$i_{p2} = \frac{V_m \sin \omega t}{L_p} \delta_2 T + \frac{(V_m \sin \omega t - v_o \frac{n_p}{n_{S1}})}{L_2} \delta_2 T \qquad (5)$$

In equation (5), $\delta_2$ is the new duty period, $n_p$ is the number of turns in the primary and $n_{S1}$ and $n_{S2}$ are the number of turns in the secondary of the transformer.

Hence

$$i_{p2} = V_m \sin \omega t \, \delta_2 \, T \left(\frac{1}{L_2} + \frac{1}{L_P}\right) - \frac{v_o}{L_2} \frac{n_p}{n_{S1}} \delta_2 T \qquad (6)$$

Hence, the average current during one switching period T due to both the flyback and the forward converter operation and is given by

$i_2 = $ peak current $(i_{p2})$ x $\dfrac{\text{dutyperiod } (\delta_2)}{2}$

or

$$i_2 = V_m \sin \omega t \, \delta_2^{\,2} \frac{T}{2} \left(\frac{1}{L_2} + \frac{1}{L_P}\right) - \left(\frac{v_o}{L_2}\right) \frac{n_p}{n_{S1}} \delta_2^{\,2} \frac{T}{2} \qquad (7)$$

$$i_2 = \frac{V_m \sin \omega t}{2L_p} \delta_2^{\,2} T + \frac{(V_m \sin \omega t - V_o \frac{n_p}{n_s})}{2L_2} \delta_2^{\,2} T \qquad (8)$$

where $L_2$ is the reflected inductance of L in the primary given by equation (2).

For perfect current matching at the cross over is at $t = t_1$,

$i_1 = i_2 \dfrac{V_m \sin \omega t_1}{2L_p} \delta_1^{\,2} T$

$$= V_m \sin \omega t_1 \, \delta_2^{\,2} \frac{T}{2} \left(\frac{1}{L_2} + \frac{1}{L_P}\right) - \left(\frac{v_o}{L_2}\right) \frac{n_p}{n_{S1}} \delta_2^{\,2} \frac{T}{2} \qquad (9)$$

Hence, $\dfrac{V_m \sin \omega t_1}{L_p} \delta_1^{\,2}$

$$= V_m \sin\omega t_1 \, \delta_2^2 \left( \frac{1}{L_2} + \frac{1}{L_P} \right) - \left( \frac{v_o}{L_2} \right) \frac{n_p}{n_{S1}} \delta_2^2 \qquad (10)$$

Hence, the ratio between the flyback duty period ($\delta_2$) and the forward duty period ($\delta_1$) during the operation can be determined as

$$\delta_1 = \frac{\delta_2}{\sqrt{1 + \frac{L_p}{L_2}\left[1 - \left(\frac{v_o}{V_m}\right)\left(\frac{n_p}{n_{S1}}\right)\left(\frac{1}{\sin\omega t_1}\right)\right]}} \qquad (11)$$

## III. SIMULATION OF THE CONVERTER

The proposed scheme was simulated using SABER from Analogy. In the simulation, the change over of duty ratio of the switching pulses is decided by the instant at which the input instantaneous voltage referred to secondary is greater than the output voltage. At full load, the duty ratio will remain maximum from the ac zero crossing instant to the cross over instant of the voltage. From the cross over point to the maximum input instantaneous voltage and upto the lower cross over point the duty ratio will be smaller than the earlier period.

The cross over instant is decided by

$$\theta_c = \arcsin\left(\frac{n_p}{n_{S1}}\right)\frac{v_o}{V_m} \qquad (12)$$

For the simulation, the instant of crossover angle is determined from equation (12). During the simulation, to derive the required switching instants, a constant magnitude, constant frequency triangular voltage is compared with a piece wise linear (pwl) source shown in Fig. 3b. The magnitude of the pwl source is selected as per the duration at which the cross over shall occur. The magnitude of the pwl source is a constant dc voltage from lower crossover point (A) to upper cross over point (B) as shown in Fig. 2. At the end of the upper cross over point (B), the magnitude of the pwl source changes instantly, and remains at the same value upto the next cycle's lower cross over point (A). The ratio between the high duty cycle to low duty cycle is maintained as per the equation (11). Accordingly, the magnitude of the pwl source is adjusted to achieve near sinusoidal input current from the mains at all loads and for variations in the input voltage. The triangular wave and the pwl are applied to a comparator and the output of this comparator is a square wave where the duty period gets changed at the cross over instant. The advantage of this scheme is that the duty period remains constant in between the cross over instant and there is no gradual increase of duty period after the cross over instant as proposed in [4]. The trigger pulse generated for the switch is shown in Fig. 3(d).

The inductance for the primary of the transformer is selected by trial so as to develop a rated output voltage under the minimum input voltage for the full load rated current. Similarly, the selection of the other passive components like output filter inductor L, is also carried out by trial to satisfy the discontinuous conduction under all loads. Simulation was carried out for an output voltage of 24V from a 110V ac

supply to deliver a load of 500W. Finally, the component values selected in the simulation are $L_p$ (primary inductance) = 400µH, $L_2$ = 0.5 mH, C = 5000 µF $f_s$ = 20-25 kHz, $L_c$ = 0.2mH and $C_c$ = 1µF.

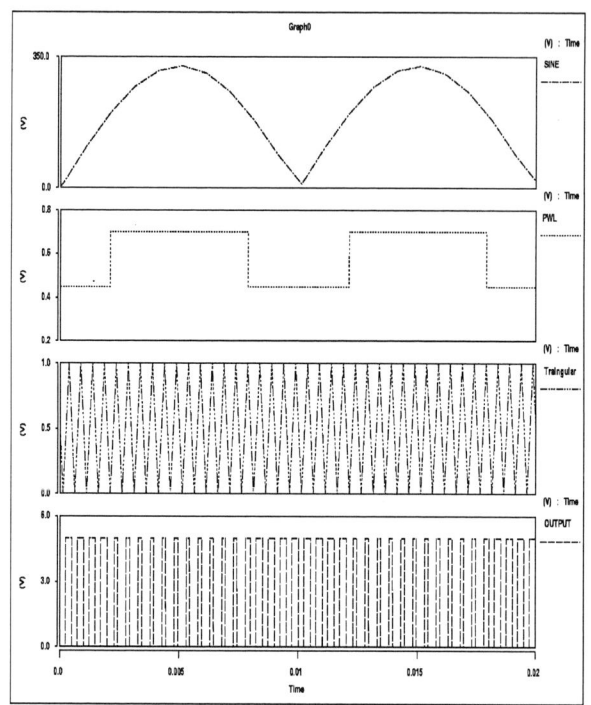

Fig. 3. The controller signal derived during the simulation (a) rectified ac (b) voltage from the pwl source (c) triangular voltage (d) output of the comparator.

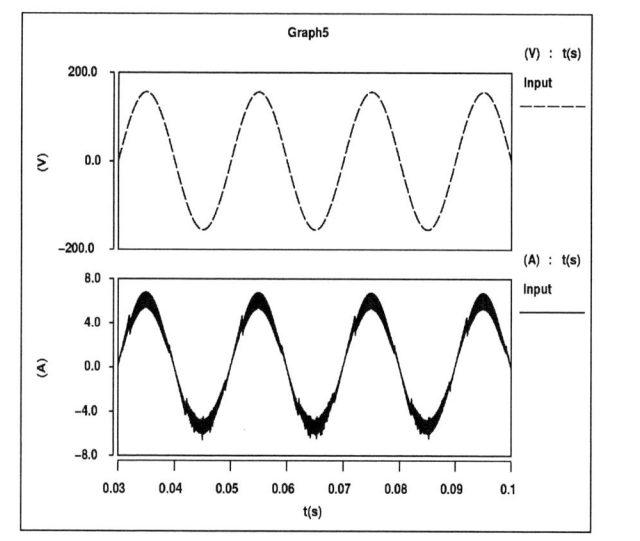

Fig. 4. Input voltage and current drawn from the ac mains.

Fig. 4, shows the input voltage and current drawn from the ac mains. It can be seen from the waveform that the input current drawn from the mains is a nearly sinusoidal in shape with minimal distortion. Fig. 5, shows the input current drawn from the ac mains at different loads such as quarter,

294

half and full loads. In all these cases, we can observe that the input current drawn from the mains remains nearly sinusoidal, thereby the converter can be used from light load to full load without creating higher distortion in the input current.

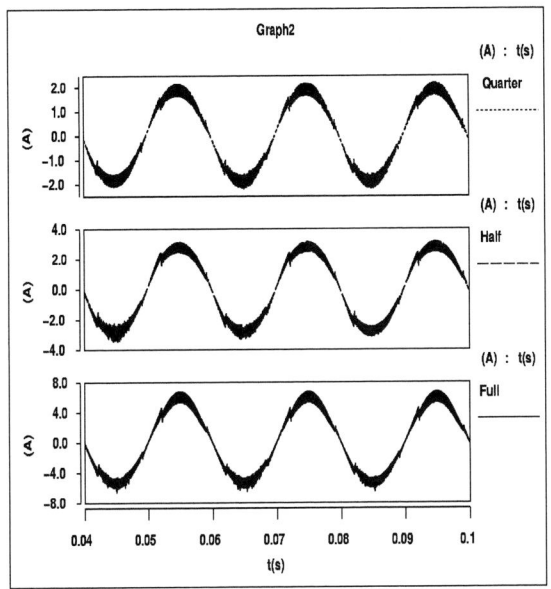

Fig. 5. The input current drawn from the ac mains at different load such as quarter, half and full loads.

## IV. MODIFIED CONFIGURATION

The controller scheme proposed in Fig. 1, can be adopted for a two winding transformer and is shown in Fig. 6. The input section of the forward converter remains the same, however in the secondary side a rearranged diode bridge is used for rectification and for the transfer of magnetising energy from the core to the output. The above proposed control technique is also used in this configuration and simulation was carriedout. The simulation results matches with the earlier configuration shown in Fig. 1.

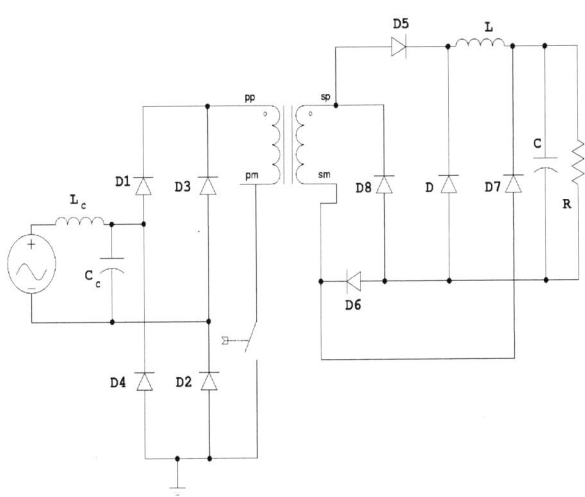

Fig. 6. New configuration for forward converter using two winding transformer.

## V. EXPERIMENTAL INVESTIGATION

The various values of components selected for experimental investigation are as follows:
Switch $S_1$ is IRF 450 (2 nos. in parallel), Diode $D_1$ to $D_4$ are BYW88-600 D, $D_5$ and $D_6$ are BYX64-600, $L_c = 0.2$ mH, $C_c = \mu F$, $L = 0.5$ mH and $C = 5000 \mu F/100V$.

The block diagram of the controller used to stabilize the output, operate the switch and to achieve waveshaping is shown in Fig. 7, and it operates under the following condition:

(a) The control switch $S_1$ is turned on and off at a constant switching frequency of 20-25khz.

(b) The converter is operated in a discontinuous conduction mode throughout the switching cycle.

(c) The maximum duty period at any instant will not exceed 50%.

(d) The duty period is changed after the crossover instant.

Thus, under these conditions when the mains input voltage

$$V_m \sin \omega t < V_o \frac{n_2}{n_1}$$ current is drawn from the mains. Hence, the

current supplied from the mains during this period varies in a sine pattern without any dead band. When the input voltage

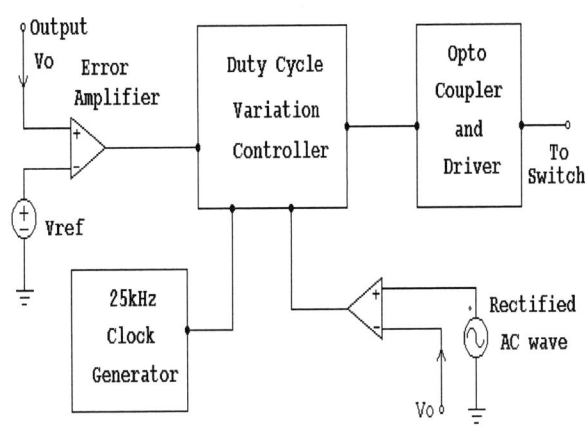

Fig. 7. Block diagram representation of controller.

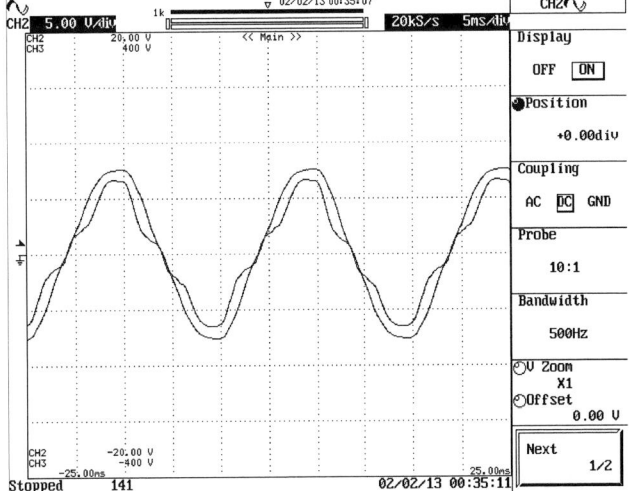

Fig. 8. Voltage from the ac mains and the current drawn from the mains without the compensation.

295

$V_m\sin\omega t > V_o\dfrac{n_2}{n_1}$, the duty period was changed to a lower value and continuation of the sine current occurs. The actual duty cycle during the above two conditions is decided by the load requirement and the input voltage at that instant. Hence, the value of the duty period is also a function of the load.

Fig. 8, to Fig. 12, shows the various experimental results obtained. Fig. 8, shows the voltage from the ac mains and the current drawn from the mains without the compensation incorporated in the proposed configuration. It can be seen from the waveform that the current drawn using these controller draws a distorted current. Fig. 9, show the input ac mains voltage and the input current drawn from the mains after compensation. It can be seen from this waveforms that the current drawn from the mains is near sinusoidal and in phase with the input voltage and thereby achieving high power factor of 0.995 at full load. The converter was tested for various loads with varying input voltages. The various results were recorded for comparison and verification. The input

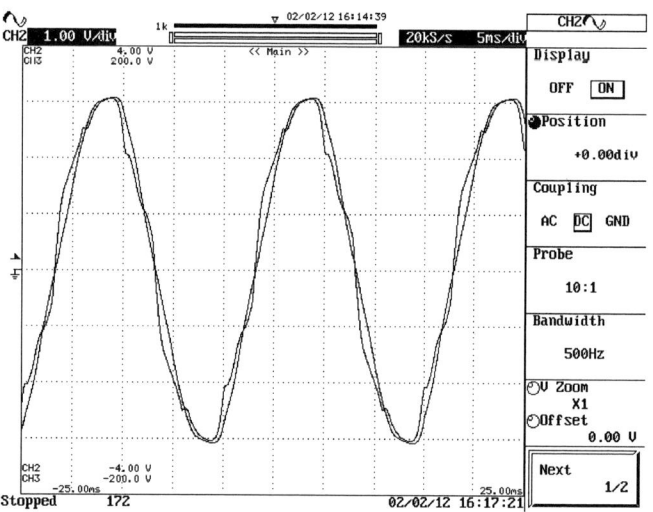

Fig. 9. Input ac mains voltage and the input current drawn from the mains.

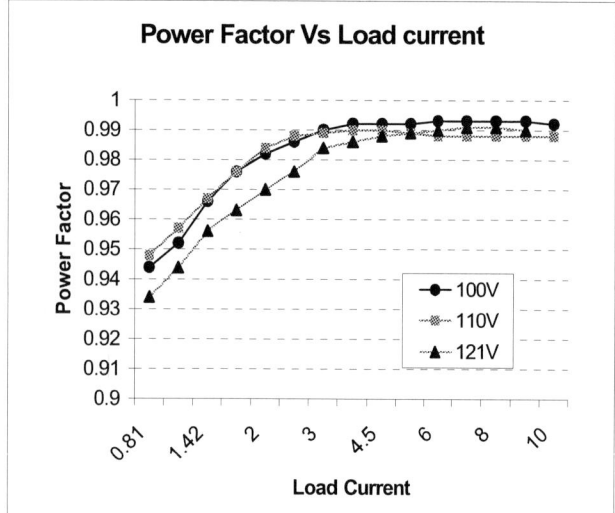

Fig. 10. Experimental power factor for different load conditions under different voltages.

active power, power factor and input current were measured using a Voltac PM 100 Power Analyzer. Fig.10, shows the graphical representation of the experimental power factor for different load conditions under different voltages. Fig.11, shows graphically the experimental efficiency variation for different load conditions and under different voltages. Fig.12, shows the harmonic distortion measured and the magnitude of the harmonic current at 4A current output.

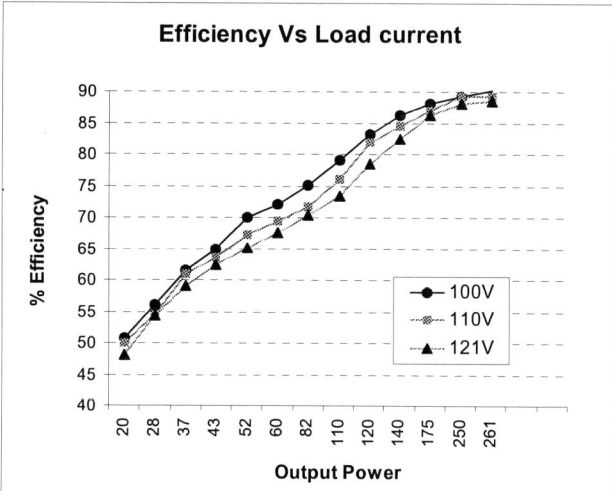

Fig. 11. Experimental efficiency for different load conditions under different voltages.

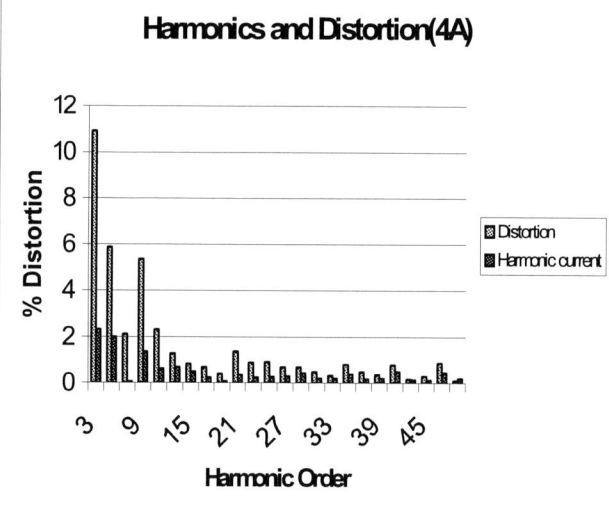

Fig. 12. THD measured for lower harmonics and the magnitude of harmonic current.

## VI. CONCLUSIONS

A modified simple control for a power factor correction scheme is proposed as an improvement to the single stage forward converter. This PFC circuit will achieve high power factor and is achieved by making the circuit operate in a discontinuous conduction mode having varying duty period in the two region. The switching technique is such that the off time of the pulse generated signal will always remains equal to or higher than the on-time period by maintaining a constant off duration. The duty cycle is changed when the input voltage become greater than the reflected output voltage and this resulted in a sinusoidal current without an initial dead period. A high power factor could also be achieved with this simple controller. Finally, power circuit is also modified to obtain forward conversion with a two winding transformer instead of the three winding transformer.

## VII. REFERENCES

[1] M. Daniele, P.K.Jain, and G. Joos, "A single-stage power factor corrected AC/DC converter" *IEEE Trans. Power Electronics.*, vol.14, no.6, pp.1046-1055, Nov. 1999.

[2] D. Divan, G. Venkataramanan, and C. Chen, "A unity power factor forward converter," *in Proc. 1992 IEEE APEC Conf.,* pp. 666-6722.

[3] G. Spiazzi, and S. Buso, "Power factor preregulator based on combined buck flyback topologies", *IEEE Trans. Power Electronics.*, vol.15, no.2, pp.197-204, March 2000.

[4] Hernan Emilio Tacca, "Single-switch two output flyback-forward converter operation", *IEEE Trans. Power Electronics.,* vol.13, no.5, pp.903-911, Sept.1998.

[5] M. Nagao, "A novel one-stage forward type power factor correction circuit", *IEEE Trans. Power Electronics.*, vol.15, no.1, pp.103-110, Jan. 2000.

[6] Y. Jang and R.W. Erickson, "New single switch three-phase high power factor rectifiers using multiresonant zero-current switching," *in IEEE Trans. Power Electronics.*, vol. 13, No.1, pp.194-201, January 1998.

## VIII. BIOGRAPHIES

**Vinayak N. Shet** was born in Honnavar at Karnataka in India on 3$^{rd}$ December 1963, and obtained the B. E degree in Electrical Engineering from Karnataka University, Dharwad, India in July 1984, M. Tech degree in Electrical Engineering with specialization in Power electronics from IIT Bombay, Mumbai in January 1995 and Ph.D. degree in Electrical Engineering on the topic of "Power Factor Correction in Converters" from IIT Bombay in December 2002. His fields of interest are power factor correction, resonant converter and energy saving ballast and microcontroller application to power electronic systems. He is presently working as Professor and Head Electrical & Electronics Department in Goa Engineering College, Goa, India.

**2006 IEEE International Conference on Power Electronic, Drives and Energy Systems**

# A Fuzzy Logic Controller for Direct Power Control of PWM Rectifiers with SVM

R. Skandari, A. Rahmati, A. Abrishamifar, and E. Abiri

*Abstract--A novel direct power control of three-phase PWM rectifiers with constant switching frequency using Fuzzy logic controller for space-vector modulation (Fuzzy DPC-SVM) is proposed in this paper. The errors of active and reactive powers are the inputs of fuzzy controller and switching times are the outputs of controller. Line voltage sensors are replaced by a virtual flux estimator. Fuzzy DPC-SVM has several features, such as a simple algorithm, good dynamic response, constant switching frequency, and provides sinusoidal line current (THD of line current 2.2%) when supply voltage is not ideal like as DPC-SVM in ideal conditions (THD of line current 2.01%).*

*Index Terms-- Fuzzy Logic Controller, DPC, SVM*

## I. INTRODUCTION

FOR wide industrial applications of controlled rectifiers, many efforts are done to control of these converters. For extended request of industry to more and more precise controlled rectifiers, there are several alternatives that must be considered by researchers. Fig.1 presents a three-phase voltage source PWM converter [1].

The rectifiers have nonlinear nature and they generate harmonic currents in to the AC line power. The high harmonic contents of the line current and resulted low power factor of the load cause a lot of problems in the power distribution system.

Various control strategies have been reported to control of this type of PWM rectifier [2], [3], [4]. A well-known method of indirect active and reactive power control is based on current vector orientation with respect to the line voltage vector [voltage-oriented control (VOC)] [2] and [3]. VOC guarantees high dynamics and static performance because of its internal current control loops. Another method based on instantaneous direct active and reactive power control is called direct power control (DPC) [4].

Both of these strategies do not have sinusoidal current when the line voltage is distorted. Only a DPC strategy based on virtual flux instead of the line voltage vector orientation, called VF-DPC, provides sinusoidal line current and lower harmonic distortion [1], [5]. The most important disadvantage of the VF-DPC method is high sampling frequency needed for digital implementation of hysteresis comparators. Therefore, it is difficult to implement VF-DPC in industry.

This work was supported in part by the Electrical Department of Iran University of Science and Technology.

All the above drawbacks can be eliminated by a PWM voltage modulator. In [6] a simple method of line voltage sensorless DPC has been presented with constant switching frequency using space-vector modulation (DPC-SVM).

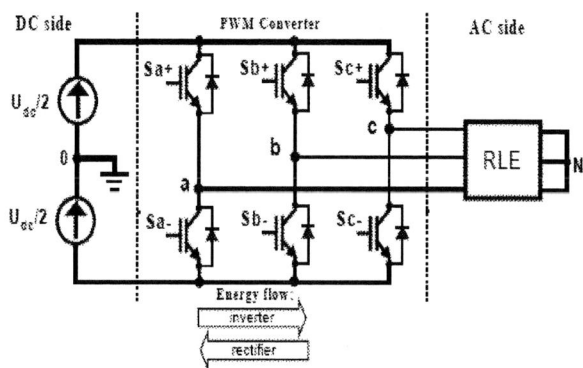

Fig. 1. A three-phase voltage source PWM converter.

THD of line current is reduced in this method in order of constant switching frequency. It has lower sampling frequency and good dynamics too. But we can reduce the line current THD with other strategies. For nonlinear behavior of rectifiers, nonlinear controllers like fuzzy controllers, neural network controllers and ... can help us to solve these problems. Especially systems such as Fuzzy logic controllers can be used for more reduction of THD and power ripple [7-10].

In this paper a fuzzy logic controller has been designed to implement the Space Vector Modulation (SVM) for direct power control of three phase rectifier. Design and simulation of this controller for more reduction of THD is the main purpose of this paper.

## II. DIRECT POWER CONTROL (DPC)

DPC is based on the instantaneous active and reactive power control loops [4]. In DPC there are no internal current control loops and no PWM modulator block, because the converter switching states are selected by a switching table based on the instantaneous errors between the commanded and estimated values of active and reactive power.

Therefore, the key point of the DPC implementation is a correct and fast estimation of the active and reactive line power.

It is possible to replace the ac-line voltage sensors with a virtual flux estimator, which gives technical and economical advantages to the system such as simplification, isolation between the power circuit and control system, reliability, and cost effectiveness.

The voltages imposed by the line power in combination with the ac-side inductors are assumed to be quantities related to a virtual ac motor.

Thus, $R$ and $L$ represent the stator resistance and the stator leakage inductance of the virtual motor and line-to-line voltages: $Uab$, $Ubc$, $Uca$ would be induced by a virtual air-gap flux. These voltages can be estimated as:

$$u_{s\alpha} = \sqrt{\frac{2}{3}} Udc(S_A - \frac{1}{2}(S_B + S_C)) \tag{1}$$

$$u_{s\beta} = \frac{1}{\sqrt{2}} Udc(S_B - S_C) \tag{2}$$

In other words the integration of the voltages leads to a virtual flux (VF) vector $\underline{\psi}_L$, in stationary $\alpha - \beta$ coordinates [1].

$$\psi_{L\alpha(est)} = \int (u_{s\alpha} + L\frac{di_\alpha}{dt})dt \tag{3}$$

$$\psi_{L\beta(est)} = \int (u_{s\beta} + L\frac{di_\beta}{dt})dt \tag{4}$$

The measured line currents, and the estimated virtual flux components, are used to the power estimation [5].

$$p = \omega\psi_{Ld}i_{Lq} = \omega(\psi_{L\alpha}i_{L\beta} - \psi_{L\beta}i_{L\alpha}) \tag{5}$$

$$q = \omega\psi_{Lq}i_{Ld} = \omega(\psi_{L\alpha}i_{L\alpha} + \psi_{L\beta}i_{L\beta}) \tag{6}$$

## III. DPC USING SPACE-VECTOR MODULATION (DPC-SVM)

The SVM strategy, based on space vector representation (Fig. 2) became very popular due to its simplicity [1].

A three phase two-level converter provides eight possible switching states, made up of six active and two zero switching states. Active vectors divide plane for six sectors, where a reference vector $U^*$ is obtained by switching on (for proper time) two adjacent vectors. It can be seen that vector $U^*$ (Fig. 2) is possible to implement by the different switch on/off sequence of $U1$ and $U2$, and that zero vectors decrease modulation index.

Reference vector $U^*$ is sampled with fixed clock frequency $2fs = 1/Ts$, and next $U^*(Ts)$ is used to solve equations which describe times $t_1$, $t_2$, $t_0$ and $t_7$ (Fig. 2). These times can be calculated for the first sector as:

$$t_1 = \frac{2\sqrt{3}}{\pi} MT_s \sin(\pi/3 - \alpha) \tag{7}$$

$$t_2 = \frac{2\sqrt{3}}{\pi} MT_s \sin\alpha \tag{8}$$

After $t_1$ and $t_2$ calculation, the residual sampling time is reserved for zero vectors $U_0$, $U_7$ with condition $t_1 + t_2 \leq Ts$. The next equation is identical for all variants of $SVM$. The only difference is in different placement of zero vectors $U_0$ $(000)$ and $U_7(111)$.

It gives different equations defining $t_0$ and $t_7$ for each of method, but total duration time of zero vectors must fulfill conditions:

$$t_{0,7} = T_s - t_1 - t_2 = t_0 + t_7 \tag{9}$$

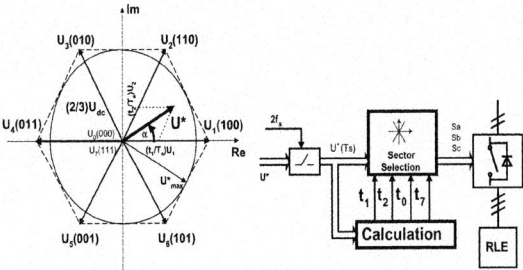

Fig. 2. (a) Space vector representation of three-phase converter, (b) Block scheme of SVM.

The concept of DPC and VF can also be applied to new control scheme. The DPC-SVM with constant switching frequency uses closed-loop power control, as shown in Fig. 3. The commanded reactive power (set to zero for unity power factor operation) and (delivered from the outer PI dc voltage controller) active power (power flow between the supply and the dc link) values are compared with the estimated and values (5 and 6), respectively. The errors are delivered to PI controllers, where the variables are dc quantities, which eliminate steady-state error.

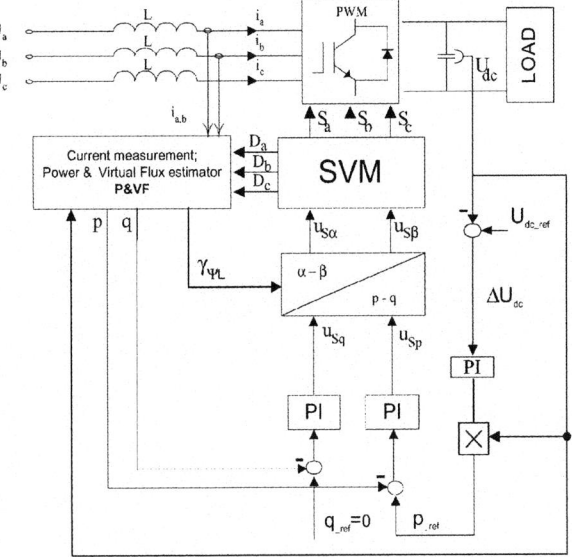

Fig. 3. Block scheme of DPC-SVM.

## IV. SVM USING FUZZY LOGIC CONTROLLER

The proposed fuzzy controller is a three-input two- output controller. The first input variable is the angle *Theta* determining the position of the reference vector $U^*$. The second and third variables are the errors in demanded active $p$ and reactive $q$ powers, which should be withdrawn from the mains to compensate effects of a nonlinear load:

$$dp = p_{ref} - p \tag{10}$$
$$dq = q_{ref} - q \tag{11}$$

The outputs of the fuzzy controller are the switching times: $t_1$ and $t_2$. The membership function of inputs and outputs are shown in fig 4.

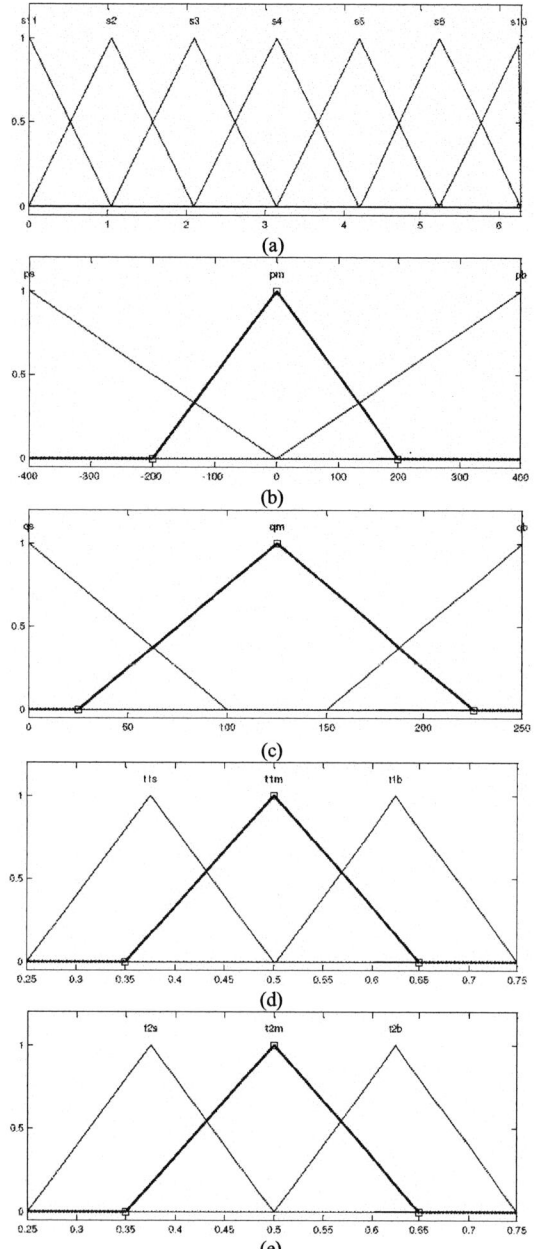

Fig. 4. Membership functions of: (a) *Theta*, (b) *dp*, (c) *dq*, (d) $t_1$ and (e) $t_2$.

## V. SIMULATION

To study the operation of the fuzzy DPC-SVM system under different line conditions, the PWM rectifier with the whole control scheme has been simulated using MATLAB Simulink software. The main electrical parameters of the power circuit and control data are given in TABLE II.

The simulation study has been performed with three main objectives in mind:

• explaining and presenting the steady-state operation of the DPC-SVM with a purely sinusoidal and distorted supply line voltage;

• explaining and presenting the steady-state operation of the Fuzzy DPC-SVM with a distorted supply line voltage;

• presenting the dynamic performance of Fuzzy DPC-SVM.

The simulated waveforms for the DPC-SVM are shown in Fig. 5. These results were obtained for purely sinusoidal supply line voltage.

Similarly, Fig. 6 shows results for distorted (5% of fifth harmonic and 4.5% unbalanced) line voltages. DPC-SVM provides sinusoidal line current (Figs. 5 and 6) and low total harmonic distortion (THD) for ideal (THD of line current 2.01%) as well as for distorted (THD of line current 2.36%) line voltage.

This is thanks to low-pass filter behavior of the integrators used in (3 and 4). Proposed Fuzzy DPC-SVM provides sinusoidal line current (Figs. 7) and low total harmonic distortion (THD) for distorted (THD of line current 2.2%) line voltage. This is thanks to high speed calculation of times in fuzzy logic controller compared with traditional DPC-SVM.

The excellent dynamic behavior under a step change of the load is presented in Fig.8 for proposed Fuzzy DPC-SVM.

TABLE II
PARAMETER USED IN SIMULATION OF FUZZY DPC-SVM.

| Sampling frequency | 10kHz |
|---|---|
| Resistance of reactors (R) | 100m• |
| Inductance of reactors (L) | 10mH |
| DC-link capacitor | 1mF |
| Load resistance ($R_l$) | 100• |
| Switching frequency (f) | 10kHz |
| Phase voltage (V) | 230RMS |
| Source voltage frequency | 50Hz |
| DC-link voltage | 600V |

*Theta* is in the range of *S10, S2, ... S11* and *dp* in range *ps* (small *p*), *pm* (middle *p*) and *pb* (big *p*) and etc.

A fuzzy inference system consists of a number of *IF-THEN* rules.

The example of fuzzy base rules is shown in TABLE I. In this table we assumed that *dq* is *qs* and it informs us the output of $t_1$.

TABLE I
FUZZY BASE RULE FOR $t1$ WITH ASSUMPTION OF: dq=qs.

| $\frac{Theta}{dp}$ | $S_{10}$ | $S_2$ | $S_3$ | $S_4$ | $S_5$ | $S_6$ | $S_{11}$ |
|---|---|---|---|---|---|---|---|
| $p_s$ | $t_{1b}$ | $t_{1s}$ | $t_{1s}$ | $t_{1m}$ | $t_{1s}$ | $t_{1s}$ | $t_{1b}$ |
| $p_m$ | $t_{1b}$ | $t_{1b}$ | $t_{1s}$ | $t_{1m}$ | $t_{1s}$ | $t_{1b}$ | $t_{1b}$ |
| $p_b$ | $t_{1b}$ | $t_{1b}$ | $t_{1b}$ | $t_{1m}$ | $t_{1b}$ | $t_{1b}$ | $t_{1b}$ |

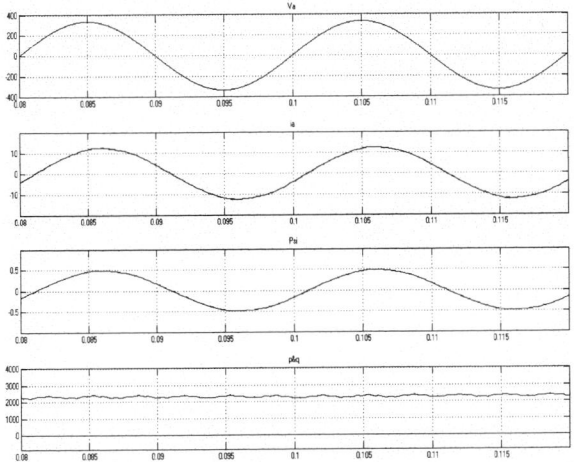

Fig. 5. Simulated basic signal waveforms under purely sinusoidal line voltage for DPC-SVM. From the top: line voltage (V), line current (A), virtual flux (V), instantaneous active (W) and reactive (Var) power. (THD = 2.01%).

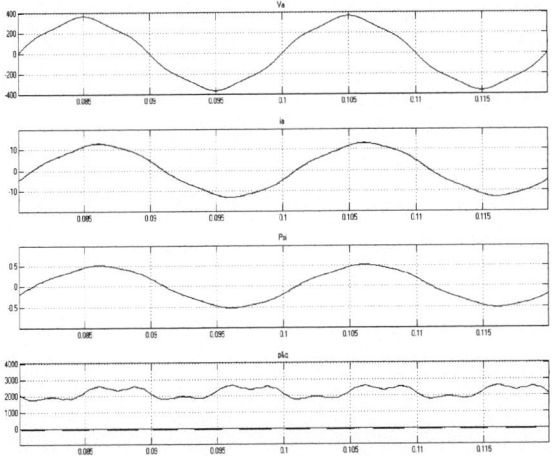

Fig. 6. Simulated basic signal waveforms under distorted (5% of fifth harmonic and 4.5% unbalanced) line voltages for DPC-SVM. From the top: line voltage (V), line current (A), virtual flux (V), instantaneous active (W) and reactive (Var) power. (THD = 2.36%).

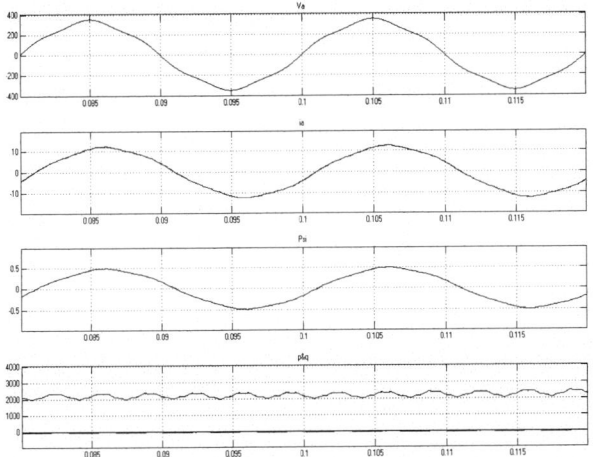

Fig. 7. Simulated basic signal waveforms under distorted (5% of fifth harmonic and 4.5% unbalanced) line voltages for proposed Fuzzy DPC-SVM. From the top: line voltage (V), line current (A), virtual flux (V), instantaneous active (W) and reactive (Var) power. (THD = 2.2%).

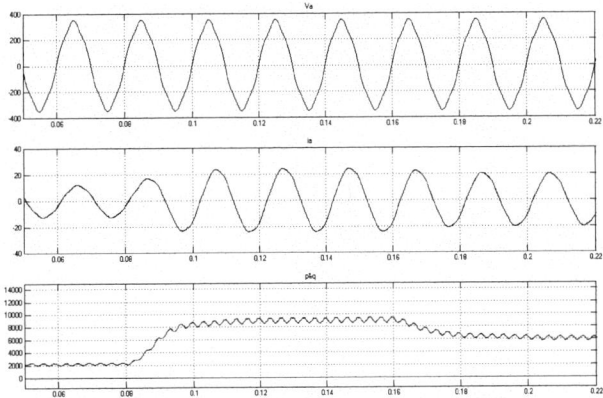

Fig. 8. Simulated basic signal waveforms under step change of the load. From the top: line voltage (V), line current (A), instantaneous active (W) and reactive (Var) power.

## VI. CONCLUSION

A line voltage sensorless DPC with constant switching frequency (DPC-SVM with fuzzy calculation of switching times) for a three-phase PWM boost type rectifier has been presented.

The Fuzzy DPC-SVM system constitutes a variable alternative to the conventional control strategies and it has many features and advantages like: Sensorless control, easy implementation of power estimation algorithm in a DSP or Fuzzytech, lower sampling frequency, good dynamics, sinusoidal line currents (THD=2.2%) for distorted line voltage, constant switching frequency and so on.

## VII. REFERENCES

[1] Mariusz Malinowski, "Sensorless control strategies for three-phase PWM rectifiers," *Ph.D. dissertation, Inst. Control Ind. Electron., Warsaw Univ. Technol., Warsaw, Poland*, 2001.

[2] S. Hansen, M. Malinowski, F. Blaabjerg, and M. P. Kazmierkowski, "Control strategies for PWM rectifiers without line voltage sensors," *in Proc. IEEE APEC*, vol. 2, pp. 832–839, 2000.

[3] B. H. Kwon, J. H. Youm, and J.W. Lim, "A line-voltage-sensorless synchronous rectifier," *IEEE Trans. Power Electron.*, vol. 14, pp. 966–972, Sept. 1999.

[4] T. Noguchi, H. Tomiki, S. Kondo, and I. Takahashi, "Direct power control of PWM converter without power-source voltage sensors," *IEEE Trans. Ind. Applicat.*, vol. 34, pp. 473–479, May/June 1998.

[5] M. Malinowski, M. P. Kaz´mierkowski, S. Hansen, F Blaabjerg, and G. D Marques, "Virtual flux based direct power control of three-phas PWM rectifiers," *IEEE Trans. Ind. Applicat.*, vol. 37, pp. 1019–1027, July/Aug. 2001.

[6] Mariusz Malinowski, P.Kazmierkoweski, " Simple Direct Power Control of Three-Phase PWM rectifier using Space-Vector Modulation (DPC-SVM)," *IEEE Transaction on Indus. Electronics,* Vol.51, No.2, pp.447-454, April 2004.

[7] Carlo Cecati, Antonio Dell'Aquila, Marco Liserre, , and Antonio Ometto, "A Fuzzy-Logic-Based Controller for Active Rectifier," *IEEE Trans. on Ind. App.*, Vol. 39, No. 1,January/February 2003.

[8] M.Rukonuzzaman, M.Nakaoka, "Fuzzy Logic Current Controller for Three-phase Voltage Source PWM-Inverters," *IEEE Ind. App. Conf.,* Volume 2, pp.1163 – 1169, 8-12 Oct. 2000.

[9] Josk A. Torrico , Edson Bim, *"Fuzzy Logic Space Vector Current Control of Three-phase Inverters,"* IEEE Power Electronics Conf., Vol. 1, pp.147 - 152, 18-23 June 2000.

[10] Roland0 P. Burgos, Eduardo P. Wiechmann, Jose R. Rodriguez, "A Simple Adaptive Fuzzy Logic Controller for Three-phase PWM Boost Rectifiers," *in proc. IEEE-Ind. Elec. Conf.,* pp.321-326, 1998.

**2006 IEEE International Conference on Power Electronic, Drives and Energy Systems**

# DSP-Based Matrix Converter Operation Under Various Abnormal Conditions with Practicality

Vinod Kumar, and R. R. Joshi

*Abstract* - The matrix converter connects the three phase power supply with the three phase load directly through a switching matrix composed of four- quadrant switches. The operation of matrix converter with unbalanced power supply has been analyzed.

For this, a 230V, 250VA three phase to three phase matrix converter prototype is implemented using DSP based controller and tests have been carried out to evaluate and improve the stability of system under typical abnormal conditions. Digital storage oscilloscope & power quality analyzer are used for experimental observations.

*Index Terms--* Matrix converter, DSP, Harmonics, Non sinusoidal supply.

## I. INTRODUCTION

THREE phase matrix converters have received considerable attention in recent years In fact; the matrix converter provides bidirectional power flow, sinusoidal input/output waveforms, and controllable input power factor. Furthermore, the matrix converter allows a compact design due to the lack of dc-link capacitors for energy storage.

Till date, much analysis is based on the assumption that the input voltages are well balanced sinusoidal and which results in the ideal output waveform. But it should be noted that harmonics would be always introduced while non-sinusoidal or unbalanced conditions are practically unavoidable. Due to lack of internal energy storage, matrix converter is highly sensitive to the disturbances in the input voltages. Furthermore, input voltage sag and short time blackout of the mains not only causes distortion of the output but also brings some more serious problems. Therefore, it's essential to make harmonic analysis under these conditions. Estimation of harmonics in motor currents is necessary when the input voltages are non-sinusoidal. It enables identifications of the limits on the operating conditions, if any, of the drive. Matrix Converter (MC) has low input current THD characteristics but in some application areas much lower THD level is required [1]. It is known that the supply voltage waveforms may often

Vinod Kumar is with Department of Electrical Engineering, CTAE, MPUAT, Udaipur (Raj.), India (e-mail: vinodcte@yahoo.co.in).

R.R.Joshi is with Department of Electrical Engineering, CTAE, MPUAT, Udaipur (Raj.), India (e-mail: rrjoshi_iitd@yahoo.com).

show typical distortion due to the presence of nonlinear load connected to the grid.

Some useful converter harmonic analysis methods under abnormal power supply conditions have been revised here. In [2] a novel analysis method for calculating harmonic components in unbalanced input conditions is proposed. The method gives a clear expression and a better understanding of the converter.

In this paper, a 230V, 250VA three phase to three phase matrix converter prototype is implemented which consists of a DSP board using TMS320C671, a FPGA board and Analog board for 4 step commutation and several functional digital logics programmed in Altera EPM7128S, a Gate driver & 6 isolated power supply board and a power board containing IGBT Matrix module, voltage and current sensors and snubbers. On the basis of experimental tests the operation of matrix converter in abnormal conditions has been analyzed. The influence of unbalanced input voltage on input/output performance of matrix converter has also been discussed.

The accordance of the experimental results with the theoretical investigation validates the credibility of the analysis.

## II. ANALYSIS OF MATRIX CONVERTER IN ABNORMAL CONDITIONS

The basic structure of the matrix converter is shown in fig.1

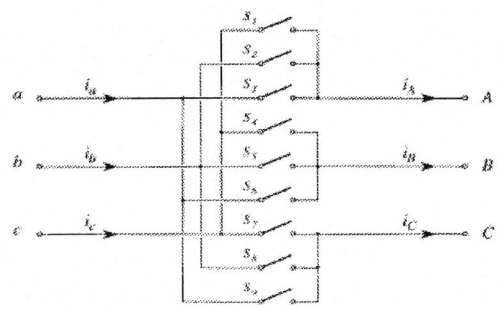

Fig. 1. Basic structure of matrix converter.

### A. Operation with Balanced Power Supply

Reference [3] has proposed space vector modulation algorithm (SVM) for matrix converters. It is considered as two steps of input voltage processing. First, the input supply voltages are multiplied with a PWM switching function of the rectifier to yield a dc link. Second, dc link is multiplied by a PWM switching function of inverter to yield the final sine

0-7803-9771-1/06/$25.00 ©2006 IEEE

output voltages. The process can be expressed as the following, taking 3×1 matrix converter as an example.

$$V_0(\omega_0 t) = [V(\omega_i t)]^T \cdot [T_R(\omega_i t)] \cdot [T_1(\omega_0 t)] \qquad (1)$$

where

$[V_i(\omega_i t)]$ is three phase input voltages;

$V_0(\omega_0 t)$ is one phase output ac voltage;

$[T_R(\omega_i t)]$ is switching function of "rectifier",

$[T_1(\omega_0 t)]$ is switching function of "inverter",

The input currents are derived by

$$\begin{bmatrix} i_a \\ i_b \\ i_c \end{bmatrix} = [T_R(\omega_i t)] \cdot [T_1(\omega_0 t)]^T \cdot i_0(\omega_0 t) \qquad (2)$$

where $i_0(\omega_0 t)$ is the output current.

The switching functions for rectifier part can be written in Fourier series as:

$$[T_R(\omega_i t)] = \begin{bmatrix} A_1 \sin(\omega_i t) + \sum\limits_{n=n_1 n_2 \ldots}^{\alpha} A_n \sin(n\omega_i t) \\ A_1 \sin(\omega_i t - 120^\circ) + \sum\limits_{n=n_1 n_2 \ldots}^{\alpha} A_n \sin n(n\omega_i t - 120^\circ) \\ A_1 \sin(\omega_i t + 120^\circ) + \sum\limits_{n=n_1 n_2 \ldots}^{\alpha} A_n \sin n(n\omega_i t + 120^\circ) \end{bmatrix}$$

where $\sum\limits_{n=n_1 n_2 \ldots}^{\alpha} A_n \sin(n\omega_i t)$ is the high order harmonics generated PWM.

Also, the switching function for inverter part can be written in Fourier series as:

$$[T_1(\omega_i t)] = [B_1 \sin \omega_0 t + \sum\limits_{m=m_1, m_2 \ldots}^{\alpha} B_k \sin(k w_0 t)] \qquad (3)$$

where $\sum\limits_{m=m_1, m_2 \ldots}^{\alpha} B_k \sin(k w_0 t)]$ is the high order harmonics generated by PWM.

If the input three voltages are balanced and sinusoidal

$$V_i = \begin{bmatrix} V \sin(w_i t + \alpha) \\ V \sin(w_i t - 120^\circ + \alpha) \\ V \sin(w_i t + 120^\circ + \alpha) \end{bmatrix} \qquad (4)$$

Substituting (3) and (4) in (1) and ignoring the high order harmonics yields,

$$V_0 = \frac{3 A_1 B_1 V \cos \alpha}{2} \sin(w_0 t) \qquad (5)$$

So, it can be concluded that in the balanced conditions, the output voltages and input currents do not contain low order harmonics.

### B. Operation with Unbalanced Power Supply

For unbalanced sinusoidal power supply, the three-phase unbalanced input voltages can be decomposed into two balanced sets of positive and negative components as equation(6),[4].

$$V_i = V_P + V_N \qquad (6a)$$

where

$$V_P = \begin{bmatrix} V_P \sin(w_i t + \alpha) \\ V_P \sin(w_i t - 120^\circ + \alpha) \\ V_P \sin(w_i t + 120^\circ + \alpha) \end{bmatrix} \qquad (6b)$$

$$V_N = \begin{bmatrix} V_n \sin(w_i t + \beta) \\ V_n \sin(w_i t - 120^\circ + \beta) \\ V_n \sin(w_i t + 120^\circ + \beta) \end{bmatrix} \qquad (6c)$$

According to (1),

$$V_0 = (V_P^T + V_N^T)[T_R(w_i t)] T_i(w_0 t) \qquad (7)$$

$$= V_P^T [T_R(w_i t)] T_1(w_0 t) + V_N^T [T_R(w_i t)] T_1(w_0 t)$$

For the positive component, it is the same as in the balanced condition where no harmonics is generated. For the negative component, it can be derived that,

$$V_N^T [T_R(\omega_i t)] T_1(\omega_i t) =$$

$$\frac{3 A_1 B_1 V_n}{4} \{\sin[2\omega_i + \omega_0)t + \beta] - \sin[2\omega_i - \omega_0)t + \beta]\} + M_1$$

(8)

where $M_1$ is the high harmonic component generated by PWM.

So, equation (8) shows that harmonic components will be introduced by the negative components with the frequency of $(2\omega_i + \omega_0)$ and $(2\omega_i - \omega_0)$. Thus, it is concluded that unbalance in input supply results in abnormal harmonics in the matrix converter output.

### C. Operation under Input Voltage Sag and Short time Blackout of Power Supply Conditions

Voltage sag is a reduction of voltage magnitude with the duration between one cycle and a few seconds which is usually caused by the fault conditions in the plant or within the utility system. It normally does not cause equipment damage but can disrupt the operation of sensitive loads.

Unlike the VSI (Voltage Source Inverter), matrix converter has no dc-link capacitor, so power interruption and deep voltage sags will be fatal to the ASD fed matrix converter.

### D. Operation With Balanced non-sinusoidal Power Supply

The three-phase voltages of balanced non-sinusoidal power supply have the same waveforms delayed by $2\pi/3$. Similar to the unbalanced condition, the input voltages can be represented as the sum of two parts:

$$V_i = V_i + V_k \qquad (9)$$

where $V_k$ is the harmonics with the order of k.

Similar to the analysis of unbalanced condition, it can be derived for this case also, that if the input power supply voltages contains the harmonics with the order of k, the harmonic components with the frequency of $(k-1) \omega_i \pm \omega_0$ and

$(k+1) \omega_i \pm \omega_0$ will be introduced in output voltages.

## III. EXPERIMENTAL RESULTS

The block diagram of matrix converter prototype and the main circuit as well as Power-control-isolation module of prototype are shown in fig.2, fig.3.and fig.4.

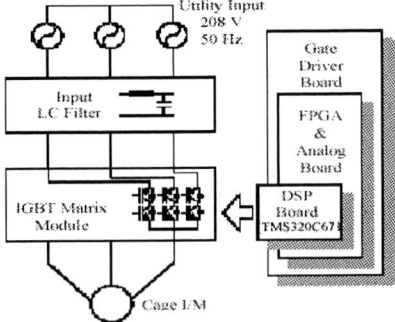

Fig. 2. Block diagram of 250 VA matrix converter prototype.

For the purpose of generating various AC voltages, an AC source equipment-Elgar SW5253A is used which generates three phase output voltages with various amplitudes, various frequency and various waveforms. A three-phase squirrel cage induction motor has been employed as the load of the matrix converter as shown in fig.2.

Fig. 3. 230-V, 250-VA laboratory prototype matrix converter.

Fig. 4. Power-control-isolation module of matrix converter.

The equivalent circuit parameters of the test motor are obtained through light running and blocked rotor tests. Effect of saturation on magnetizing reactance has been found from

zero slip tests .The test motor's name plate data is as given in Table I.

TABLE I
TEST MOTOR DATA

|  | In actual unit |
|---|---|
| Nominal line voltage | 2 30 V |
| Nominal line current | 0.8 Amps |
| Nominal output power at 50 Hz | 186.5 W |
| Nominal speed | 1460 r/min |

Fig.5 and Fig.6 shows the output/input current & voltages, when system is operated with unbalanced power supply. The rms values of the input three-phase voltages are 72.6V, 109.8V and 128.2V respectively. Fig.7 shows the input voltage recorded with power quality analyzer, when operated with balanced power supply.

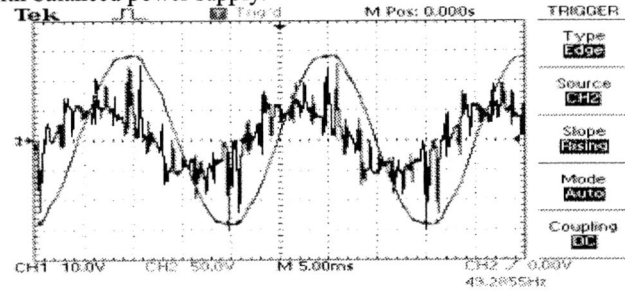

Fig. 5. Input & output phase voltage at 30 Hz.

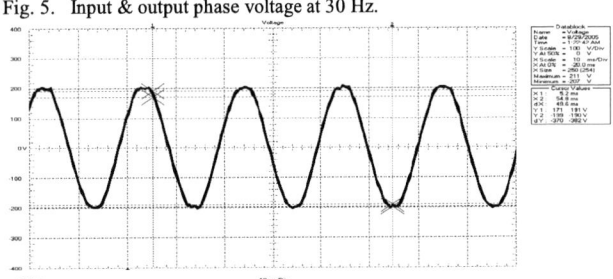

Fig. 6. Iinput voltage with power quality analyzer.

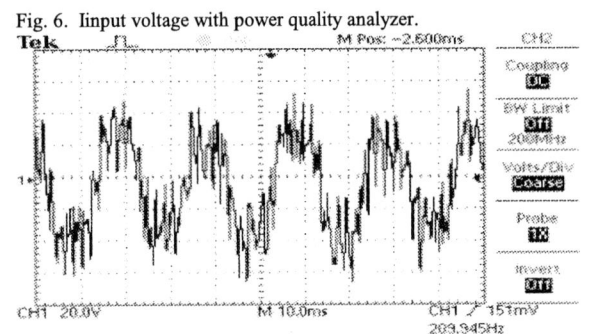

Fig. 7. Output line current at 30 Hz.

When operated with balanced non-sinusoidal power supply, a 3rd order harmonics is added into the input three phase voltages. The input phase voltage and output line voltage is shown in Fig.8, Fig.9 and Fig. 10.

AC source provides 450mS of blackout of the three phase voltages. Fig.11 shows the matrix converter output line voltage. The rapid re-starting capability of matrix converter is

305

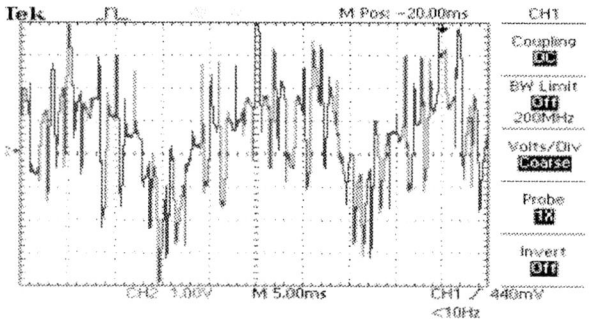

Fig. 8. Input phase voltage with 3$^{rd}$ order harmonics.

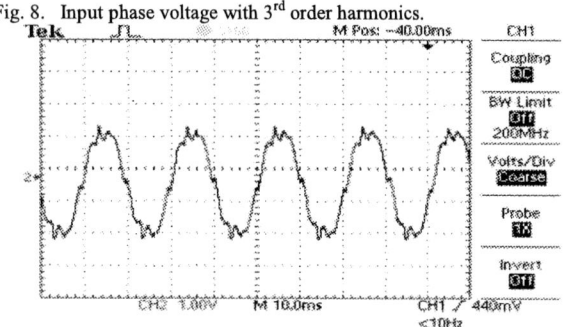

Fig. 9. Output line voltage.

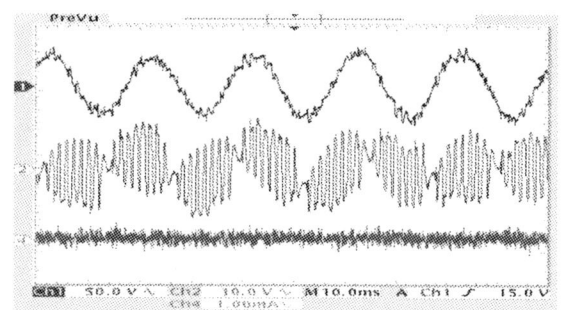

Fig. 10. Input phase voltage with 3rd order harmonics (ch: 1), output line voltage ( ch : 2) and output line current ( ch : 4).

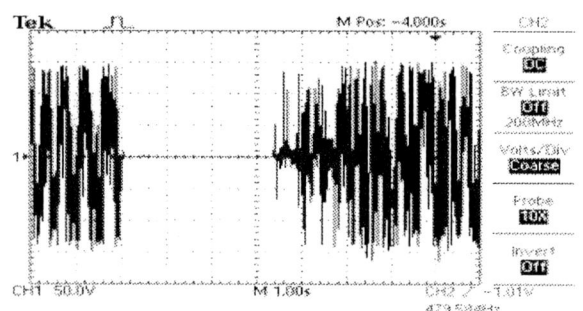

Fig. 11. Output line voltage.

demonstrated in Fig.11. During the power interruption the power delivered to load is also interrupted. However, following the line voltage restoration, the matrix converter rapidly resumes operation and delivers power quickly as compared to PWM converters.

## IV. CONCLUSION

In this paper, the performance of matrix converter in abnormal conditions is analyzed. These abnormal conditions of input power supply have a great influence on the input/output performance of matrix converters.

However, high performance of matrix converter could be obtained over a wide operating range, if some compensation techniques are used in abnormal conditions. So, research on compensation strategy of matrix converter is one of the further extensions of present work.

## V. REFERENCES

[1] Patrick W. Wheeler, Jon C Clare, Lee Empringham, "Matrix converters: a technology review," IEEE Trans. Ind.Electron,Vol.49,No.2,April 2002
[2] A. Alcsina and M.G.B. Venturini, "Analysis and design of optimum-amplitude nine-switch direct ac-ac converters," *IEEE Trans. Power Electron.,* vol.4, pp. 101-112, Jan. 1989.
[3] D.G. Holmes and T.A. Lipo, "Implementation of a controlled rectifier using ac-ac matrix converter theory," *IEEE Trans. Power Electron.,* vol. 7, pp. 240-250, Jan. 1992.
[4] C.L. Neft and C.D. Schauder, "Theory and design of a 30-Hp matrix," *IEEE Trans. Ind. Applicat.,* vol. 28, pp. 546-551, May/June 1992.
[5] L. Huber and D. Borojevic, "Space vector modulated three-phase to three-phase matrix converter with input power factor correction," *IEEE Trans. Ind. Applicat.,* vol. 31, pp. 1234-1246, Nov./Dec. 1995.
[6] M. Kazerani and B.T. Oai, "Feasibility of both vector control and displacement factor correction by voltage source type ac-ac matrix, converter," *IEEE Trans. Ind. Electron.,* vol. 42, pp. 524-530, Oct. 1995.
[7] I. Takahashi and T. Noguchi, "A new quick-response and high-efficiency control strategy of an induction motor," *IEEE Trans. Ind. Applicat.,* vol. IA-22, pp. 820-827, Sept./Oct. 1986.
[8] D. Casadei, G. Serra, and A. Tani, "Improvement of direct torque control performance by using a discrete SVM technique," *IEEE Trans. Power Electron.,* vol. 15, pp. 769-777, July, 2000.

## VI. BIOGRAPHIES

**Vinod Kumar** born in India. He received bachelor degree in Electrical Engineering from College of Tech. & Engg., Udaipur, India and M. Tech. degree from VIT, Vellore, India. He is faculty in the Department of Electrical Engineering, MPUAT, Udaipur, India.

**R. R. Joshi** born in India. He received bachelor degree in Electrical Engineering from MBM Engineering College, Jodhpur, India and M. Tech. degree from IIT, Delhi, India. He is working towards the Ph.D. Degree in Electrical Engineering from MITS, Gwalior. He is faculty in the Department of Electrical Engineering, MPUAT, Udaipur, India

**2006 IEEE International Conference on Power Electronic, Drives and Energy Systems**

# Improvement of an input waveform of a Neutral Point Type Step-down Converter

Yoshito KATO, Masaaki NAKAMURA, Nabil M. Hidayat, and Nobuo TAKAHASHI

*Abstract--***This paper presents one of the methods to improve a input current waveform of a neutral point type step-down converter circuit which we developed before. A neutral point type step-down converter circuit shares switching devices with half bridge inverter, and it is the circuit which can constitute an electronic ballast of a one converter method. A neutral point type step-down converter circuit has the problem that an input current does not flow through in the vicinity of zero cross point because It has some characteristic as step-down chopper circuit. This time, we solved this problem by adding an inductor to a circuit.**

*Index Terms--* **Converters, Electronic ballast, Harmonics, Inverters.**

## I. INTRODUCTION

WE study the electronic ballast for fluorescent lamps and develop an inverter circuit performing high frequency lighting. The inverter circuit performing this AC-AC conversion is realized by a cascade arrangement of a high frequency DC-AC inverter and an AC-DC converter with an active filter to reduce harmonics of the input current what gives a bad effect to the power supply. The development of a circuit called the one converter method aiming at reducing circuit devices, by sharing switching devices of inverter circuit with converter circuit is popular now days.[1]

If an electronic ballast of a one converter method has half bridge inverter as a DC-AC inverter part, an AC-DC converter part can have two switching devices allowing as half bridge inverter and switching devices. Therefore I thought about a circuit including a neutral point with two switching devices as a converter circuit. And I developed the neutral point type step-down converter circuit that put two half-wave rectification step-down chopper circuits together to polarity of an AC power supply. [2]-[8]

---

Yoshito Kato is with the Department of Electrical and Electronic Engineering, Tottori University, Tottori, JAPAN (e-mail: kato@ele.tottori-u.ac.jp)

Masaaki Nakamura is a graduate student of Tottori University. Tottori, JAPAN (e-mail: b01t3040@faraday.ele.tottori-u.ac.jp).

Nabil M. Hidayat is a graduate student of Tottori University. Tottori, JAPAN (e-mail: b02t3076@faraday.ele.tottori-u.ac.jp).

Nobuo Takahashi is with the Department of Control Engineering, Matsue National College of Technology, Shimane, JAPAN

## II. NEUTRAL POINT TYPE STEP-DOWN CONVERTER

A neutral point type step-down converter circuit is shown in Fig. 1. In addition, Cf and Lf are LPF, and C3 is a smoothing capacitor. I was able to constitute an electronic ballast of a one converter method by using a neutral point step-down converter circuit (Fig. 2), but a neutral point type step-down converter circuit has a problem that is the input current waveform doesn't become sinusoidal wave, and harmonics of input current increases (Fig. 3). This problem is generated due to the step-down chopper circuit. This is because the condenser is connected in series to a power supply and input current cannot flow when the input voltage is lower than the voltage of the condenser.

Solution to this problem is important as the main purpose of the neutral point type step-down converter circuit is to reduce harmonics in the power supply. I used a condenser at a point as a charge pump on my first proposition and made the input current continuance (Fig. 4). This circuit has capacitor C4 as a charge pump and found the course that an electric current flowed through for the period when the power supply voltage was low. Fig. 5 shows an input current waveform of this circuit. In this circuit, C4 having big capacitance is connected to series to a power supply. Therefore L2 which phase prevents from advancing is necessary.

On this paper further improvement is reported when inductance is used.

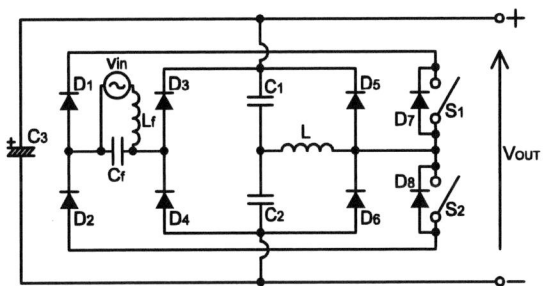

Fig. 1. The neutral point type step-down converter circuit.

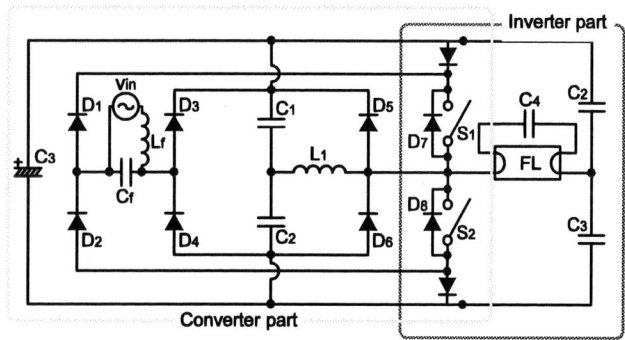

Fig. 2. The electronic ballast circuit.

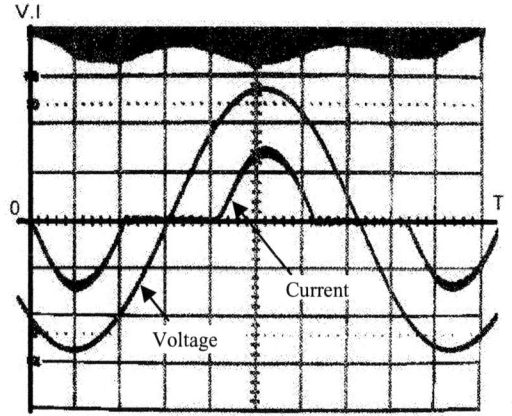

Voltage,V:50(V/div)  Current,I:236(mA/div)  Time,T:2(ms/div)
Fig. 3. The input wave-form of neutral point type step-down converter circuit.

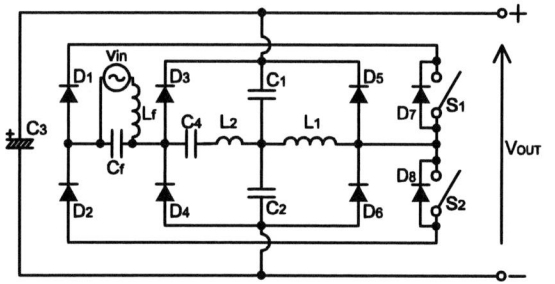

Fig. 4. The conventional modified circuit.

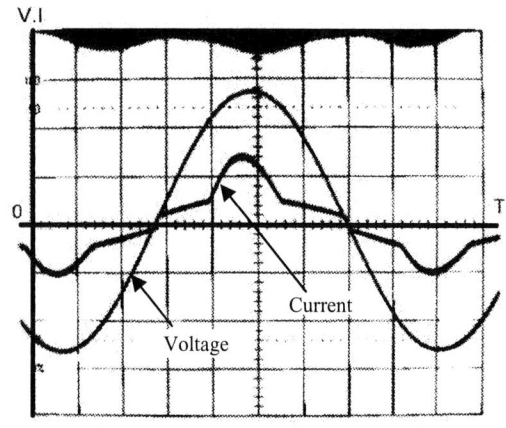

Voltage,V:50(V/div)  Current,I:236(mA/div)  Time,T:2(ms/div)
Fig. 5. The input wave-form of conventional modified circuit.

## III. PROPOSED CIRCUIT

The proposed circuit is shown in Fig. 6. The proposed circuit is made of two pairs of a step-down chopper circuit comprising D1,S1,L1,C2,D4,D6 and a step-down chopper circuit comprising D3,C1,L1,S2,D2,D5. Switching devices are shared with half bridge inverter's switches. I added L2 to a basic circuit and made a closing circuit of power switch, L1 and L2, and it flowed through this loop for the period when the power supply voltage was lower than the voltage of a condenser.

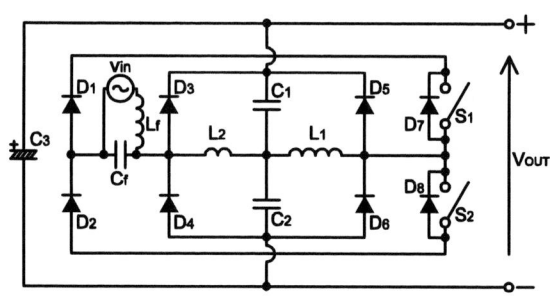

Fig. 6. The proposed circuit.

## IV. EXPERIMENTAL RESULT

I tested it about the proposed circuit. Table I shows the circuit parameters and Table II shows the measured results. Switching frequency of the circuit is 38.4 kHz. The observed waveforms of input current and input voltage by the experiment are shown in Fig. 7. As you can see, it is improved in the shape of consecutive sine waves by the proposed circuit whereas an input electric current is discontinuous by a conventional circuit. The relative harmonic content of input current is shown in Fig. 8.

TABLE I
CIRCUIT PARAMETERS

| Part | Specification |
|---|---|
| Capacitor Cf | 98.9nF |
| Inductor Lf | 8.03mH |
| Capacitor C1,C2 | 3.3nF |
| Capacitor C3 | 22uF |
| Inductor L1 | 1.23mH |
| Inductor L2 | 123.4uH |
| Diode D1-D6 | 5GLZ47 |
| MOS-FET S1:D7,S2:D8 | IRFIB6N60A |
| Load | 20W light bulb x3 |

TABLE II
MEASURED RESULTS

| Input voltage | 100.0V | Output voltage | 152.0V |
|---|---|---|---|
| Input current | 0.237A | Output current | 0.132A |
| Input power | 23.2W | Output power | 20.1W |
| Power factor | 97.89% | Efficiency | 86.48% |

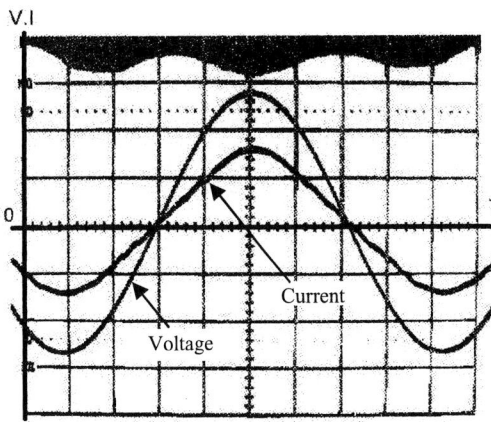

Voltage,V:50(V/div) Current,I:236(mA/div) Time,T:2(ms/div)
Fig. 7. The input wave-form of proposed circuit.

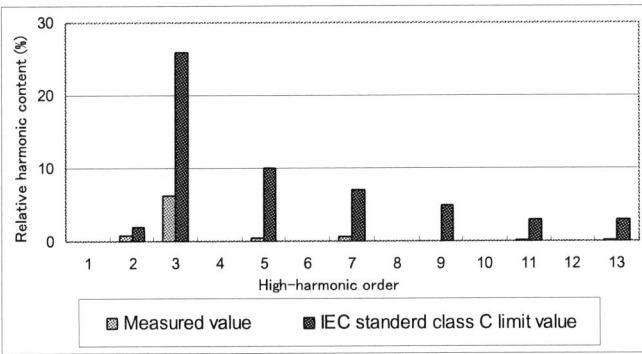

Fig. 8. The relative harmonic content of input current.

## V. CONCLUSION

This paper showed a method to improve an input current waveform of a neutral point type step-down converter circuit with an inductor. a neutral point type step-down converter circuit became the circuit which satisfied harmonics restrictions of an input electric current by this.

## VI. REFERENCES

[1] H.Matsuo, K.Shimizu, F.Kurokawa and L.Tu: Performance characteristics of a novel modified Half-bridge inverter as an electronic ballast for lighting,Proceeding of PESC98, pp.2028-2034, June 1998.

[2] M.Maehara, K.Sato and H.Nishimura: A Current Source Type Charge Pump High Power Factor Electronic Ballast Combined with Buck Converter, Proceeding of PESC98, pp.2016-2020 June 1998.

[3] Y.Kato, H.Terasaka, N.Takahashi and M.Nakaoka: Development of an electronic ballast in decreasing the harmonics current, Journal of Light & Visual Enviroment, Vol.21, No.1, pp.18-21 (1997).

[4] N.Takahashi, Y.Kato, M.Ohkita, K.Okutsu, M.Matsuyama and M.Nakaoka: An Electronic Ballast Using Modified Two-Switch Boost Converter, Journal of Light & Visual Environment, Vol.24, No.2, pp.8-14 (2000).

[5] S.Adachi, Y.Kato, N.Takahashi, I.Yokozeki and M.Nakaoka: Development of back-boost inverter with two switches and the application to the electronic ballast, Proc. IEEE Conf. on Power Electronics and Drive Systems REDS2003, LS5.3-04, pp.1146-1149, Nov. 2003.

[6] Y.Kato, T.Kuratani, Y.Okamura, S.Adachi, N.Takahashi and I.Yokozeki: "Development of the electronic ballast transformed the neutral point type back boost inverter", International Conference on Power Electronics Systems and Applications 9-12 November 2004.

[7] Y.Kato, Y.Okamura, T.Kuratani, S.Adachi, N.Takahashi and I.Yokozeki: "Development of the back-boost inverter suitable for

compact lamp", International Conference on Power Electronics Systems and Applications 9-12 November 2004.

[8] Y.kato, M.Nakamura, T.Nobuo, Y.Ichiro, "Electronic Ballast Using Neutral Point Type Step-down Converter", PEDS2005, pp.672-675, Norvenver 2005

## VII. BIOGRAPHIES

**Yoshito Kato** was born in Tottori in Japan, on February 24, 1944. He graduated from Hiroshima Institute of Technology in 1967.He was been working as a associate professor of Tottori University, Tottori. He received the doctor degree from Osaka Prefecture University in 1995. He has been engaged in the research of Power Electronics, in particular, the development of electronic ballast for improvement of Power Factor and for reduction of input current harmonics.

**Msaaki Nakamura** was born in Tottori in Japan, on 1982. He graduated from Tottori University in 2006. Now, he is graduate student of power electronics administration at Tottori University.

**Muhamad Nabil Hidayat** was born in Selangor, Malaysia on January 18, 1982. He graduated from Tottori University in March 2006 and now a graduate student of Power Electronics in Tottori University

**Nobuo Takahashi** was born in Shimane in Japan, on July 12, 1952. He graduated from Osaka University in 1975. He was been working as a teacher of Matsue National College of Technology, Shimane. He received the doctor degree from Tottori University in 2001. He has been engaged in the research of Power Electronics, in particular, the development of electronic ballast for improvement of Power Factor and for reduction of input current harmonics.

**2006 IEEE International Conference on Power Electronic, Drives and Energy Systems**

# Development of Neutral-Point Type Converter and Application to Electronic Ballast

Nabil M Hidayat, Masaaki Nakamura, Yoshito Kato
Nobuo Takahashi, Shun-ichi Adachi, and Ichiro Yokozeki

*Abstract*-- **This paper presents a neutral-point type buck-boost converter that can decrease input current harmonics and reduce the generation of inrush current at start when turning on. It is possible to utilize this circuit as a converter for the electronic ballast, and the switches in the converter are shared with the inverter for the lighting. It is confirmed by experiments that the proposed electronic ballast can satisfy IEC regulation Class C and can achieve high Power Factor 99.6□ .**

*Index Terms*-- **converters, electronic ballast, harmonics, inverters**

## I. INTRODUCTION

IN order to reduce harmonics of input current, many types of electronic ballast have been proposed [1]-[7], which combines a converter circuit with an inverter circuit. Authors of this paper have proposed the neutral-point type boost converter. It was shown that the converter could be utilized for the electronic ballast. In this paper, after analyses of operating modes of the newly developed neutral-point type buck-boost converter, the relation between the output voltage and the inductance of the coil used in the buck-boost converter is analyzed. And the value of inductance suitable for the converter is obtained from experiments.

The feature of this circuit is that the circuit can reduce the generation of inrush current at start when turning on and also can depress the voltage-stress for circuit elements. It is also confirmed from experiments that the electronic ballast based on the proposed converter satisfies IEC regulation Class C and achieved high Power Factor.

---

Nabil M Hidayat is a graduate student of Tottori University. Tottori, JAPAN (e-mail: b02t3076@faraday.ele.tottori-u.ac.jp).

Masaaki Nakamura is a graduate student of Tottori University. Tottori, JAPAN (e-mail: b02t3076@faraday.ele.tottori-u.ac.jp).

Yoshito Kato is with the Department of Electrical and Electronic Engineering, Tottori University, Tottori, JAPAN. (e-mail: kat@ele.tottori-u.ac.jp).

Nobuo Takahashi is with the Department of Control Engineering, Matsue National College of Technology, Shimane, JAPAN.

Shun-ichi Adachi is with Nihonkai Telecasting Corp., Tottori, JAPAN.

Ichiro Yokozeki is with Harison Toshiba Lighting Corp., Kanagawa, JAPAN

## II. BASIC CONVERTER CIRCUIT

The configuration of the proposed neutral-point type buck-boost converter circuit is shown in Fig.1 [8]. The feature of this circuit is that there is no short-circuit between the power supply (Vin) and the smoothing capacitor(C). Therefore, this circuit can reduce the generation of inrush current at start when turning on and also can depress the voltage-stress for circuit elements.

### A. Circuit Operations

The power supply (Vin) and the inductor (L) are being connected in series. Therefore the inductor L repeats accumulation and release of electric energy following to operations of two switches (S1 or S2). This converter consists of a main circuit and an auxiliary circuit. The main circuit works as a half-wave rectifier that is with two diodes (D1, D2), two switches (S1 and S2) and an inductor (L). When the power supply is in positive half cycle, current flows through D1, S1 and L. When the power supply is in negative half cycle, current flows through L, S2 and D2. In both cases, the inductor accumulates electric energy while the corresponding switch is on and releases electric energy while the corresponding switch is off.

The auxiliary circuit consists of four diodes (D3, D4, D5 and D6), a smoothing-capacitor (C) and an inductor (L). Four diodes are connected in order to constitute the full-bridge for L. DC voltage (Vout) can be obtained at the smoothing-capacitor C due to the counter-electromotive force of L.

### B. Operation Mode

Analyses of inductor current are described under following assumptions.

(1) Switching frequency is higher than the frequency of input power supply. Therefore, it can be considered that the voltage of the input power supply is constant during a switching cycle.

(2) Circuit elements have ideal properties.

(3) Terminal voltage (Vout) of smoothing-capacitor is constant.

(4) An impedance of the load is a pure resistance (R).

---

0-7803-9771-1/06/$25.00 ©2006 IEEE

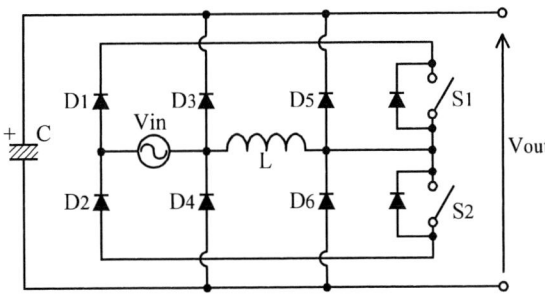

Fig. 1. Basic converter circuit.

The circuit operations are explained here during power supply is in positive half cycle. Behaviors of the inductor currents are different depending on the state of switches and input power voltage.

[Mode 1-a] S1: On, S2: Off, V1 (input voltage) < Vc

The equivalent circuit of Mode 1-a is shown in Fig. 2. In this case, current (I1) flows to the inductor through diode (D1) and switch (S1) from input power supply. The inductor accumulates electric energy. Simultaneously, there is another current flow to the load from the smoothing-capacitor (C). During this mode, the load received energy only from the smoothing-capacitor (C).

[Mode 1-b] S1: On, S2: Off, V1 (input voltage) > Vc

The equivalent circuit of Mode 1-b is shown in Fig. 3. In this case, three current flows are considered. First is the current (I1) that flows to the inductor through diode (D1) and switch (S1) from input power supply. Second is the current that flows to the load directly from input power supply through D1, S1, D5 and D4. Third is the current that flows to the smoothing-capacitor (C) from input power supply through D1, S1, D5 and D4.

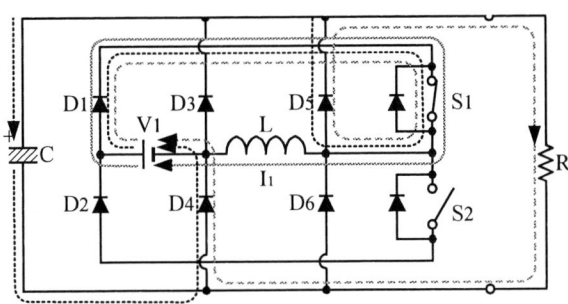

Fig. 2. The equivalent circuit model of 1-a.

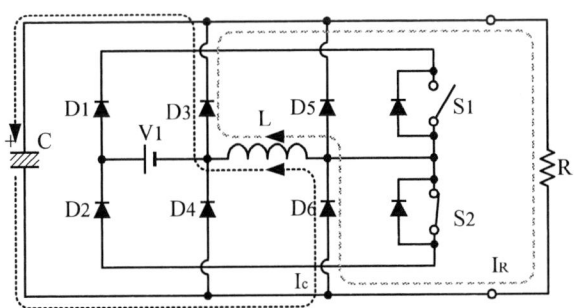

Fig. 3. The equivalent circuit model of 1-b.

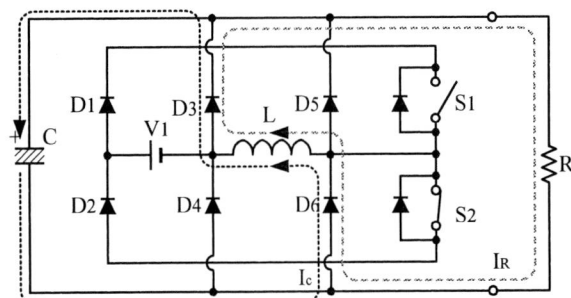

Fig. 3. The equivalent circuit model of 2 and 3.

In both Modes, the inductor accumulates electric energy (Pon) in the inductor (L).

[Mode 2] S1: Off, S2: Off
[Mode 3] S1: Off, S2: On

Both of the switches are "Off" during very short time after Mode 1, and then switch S2 turns on. The behaviors of the current flows are also the same in Mode-2 and Mode-3. In this case, there are two current flows as shown in Fig.4. First is the current that charges the smoothing-capacitor (C) through D3, C, D6 and L. Second is the current that supply the energy to the load through D3, R, D6 and L. Both flows are caused by the counter electromotive force of inductor (L) due to the energy it accumulated in Mode 1. This operation continues until all of the electric energy of inductor (L) is used.

### C. Analysis of Circuit Current

The terminal voltage of the inductor, the voltage of capacitor and terminal voltage of the load are expressed as $V_L$, $V_c$ and $V_R$, respectively.

$$V_L = V_C = V_R \tag{1}$$

The inductor current, the capacitor current and the load current are expressed as $I_L$, $i_c$ and $i_R$, respectively.

$$I_L = i_c + i_R \tag{2}$$

From these two equations, the terminal voltage of the inductor is expressed as follows.

$$V_L = A \exp\{(-1/RC + \sqrt{1/R^2C^2 + 4/LC})t/2\} + B \exp\{(-1/RC - \sqrt{1/R^2C^2 + 4/LC})t/2\} \tag{3}$$

where A and B are constants, $R$ is the resistance of the load , $C$ is the capacitance of the smoothing-capacitor.

The inductor current ($I_L$) at mode 2 can be calculated using following equation.

$$I_L = A\{1/R + C(\alpha + \beta)\}\exp\{(\alpha + \beta)(t - t_1)\} + B\{1/R + C(\alpha - \beta)\}\exp\{(\alpha - \beta)(t - t_1)\} \tag{4}$$

where $t_1$ is time differences from mode 1 to mode 2, and symbols $\alpha$ and $\beta$ are defined as follows.

$$\alpha = -1/2RC, \quad \beta = \sqrt{1/R^2C^2 + 4/LC}/2 \tag{5}$$

In mode 2, if the time period during the inductor releases all energies is assumed to be t, A and B in (1) will be expressed as follows.

$$A = \frac{V_{in}t_1 R}{L\{1 + RC(\alpha + \beta)\}\{1 - \exp(2\beta(t_2 - t_1))\}} \tag{6}$$

311

$$B = \frac{V_{in}t_1 R}{L\{1 + RC(\alpha - \beta)\}\{1 - \exp(-2\beta(t_2 - t_1))\}} \quad (7)$$

Equation (4) shows that the current flow in the inductor decreases as the value of inductance increases. Moreover, the current flow in the inductor decreases as the switching frequency becomes high, and output voltage is influenced by switching frequency.

### D. Experimental Results of Neutral-Point Ttype Buck-Boost Converter Circuit

The relation between the value of inductance and output voltage is experimented using an experimental circuit based on basic neutral-point type buck-boost converter circuit. In this case, incandescent lamps are used as loads. The amount of power consumption is adjusted in the experiments so that it might become the same as a (FLR36T/FLR42T) fluorescence lamp. Experimental results are shown in Fig.5.

When an input current waveform is mostly sinusoidal, the current that flows in the inductor behaves in intermittence mode. The output voltage at this time is near the maximum value of a commercial AC line. An observed waveform of input current is shown in Fig.6. In this experimental circuit, a L.P.F. (Low Pass Filter) that cuts off frequencies of 15 kHz or more is utilized. An observed waveform of output voltage under the condition that the duty factor of switches is 50% is shown in Fig.7. The result proves that the output voltage reaches the maximum value of a commercial AC line due to boost action of the proposed converter.

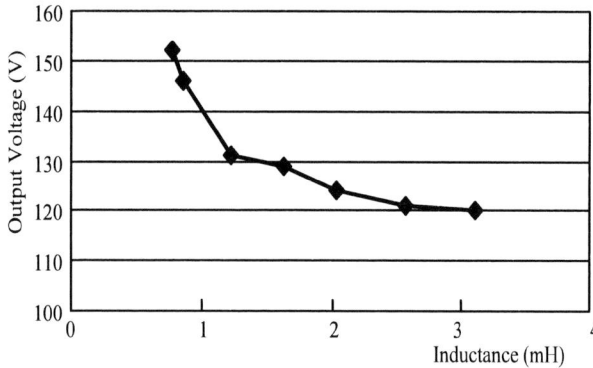

Fig. 5. The relation between inductance and output voltage.

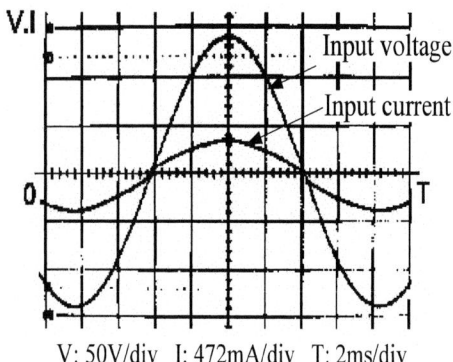

V: 50V/div   I: 472mA/div   T: 2ms/div

Fig. 6. Observed waveforms of input voltage and current.

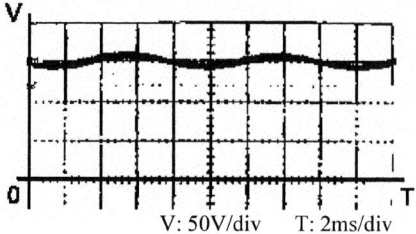

V: 50V/div   T: 2ms/div

Fig. 7. Observed waveforms of input voltage and current.

## III. APPLICATION TO ELECTRONIC BALLAST

In this section the proposed neutral-point type buck-boost converter is applied to the electronic ballast for fluorescent lamp.

### A. Basic Circuit of Proposed Electronic Ballast

Basic configuration of electronic ballast using the proposed neutral-point type buck-boost converter is described in Fig.8. The electric power charged in the smoothing-capacitor (C) is supplied to the inverter for lighting through diodes D7 and D8. In this circuit, the converter and inverter shares the two switches (S1 and S2). The circuit for lighting shows a typical form of half-bridge inverter.

A capacitor (Cr) is utilized to warm up filaments of a lamp, and in a steady state, this capacitor and the secondary side coil of a transformer (T) perform resonance operation. The equivalent circuit of electronic ballast is shown in Fig.9.

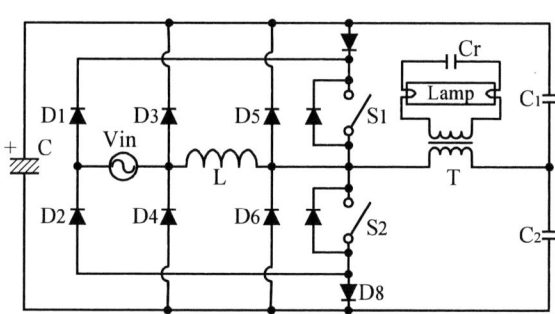

Fig. 8. Basic configuration of proposed electronic ballast.

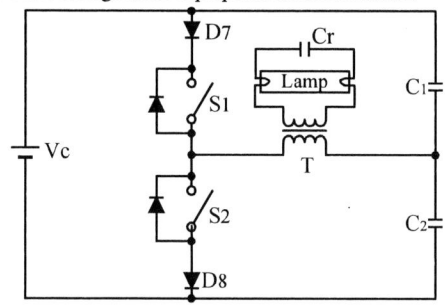

Fig. 9. Equivalent circuit of proposed electronic ballast.

### B. Experimental Results of Proposed Electronic Ballast

Based on the basic circuit shown in Fig.8, a circuit is constructed and experimented, and the results are shown as follows. Table 1 shows specifications of circuit elements. Conditions and results of lighting experiment are shown in Table 2. Two power MOSFET's are used as switching elements and they are controlled by a Driving IC for half-bridge inverter.

TABLE 1
SPECIFICATIONS OF CIRCUIT ELEMENTS

| Part | Specification |
|------|---------------|
| FET    S1, S2 | IRF840L |
| Diode    D1-D8 | 5GLZ47 |
| Electrolytic Capacitor C | 500V22uF |
| Capacitor   Cr | 1.6kV8.2nF |
| Capacitor   C1, C2 | 630V100nF |
| Inductor   L | 0.905mH |
| Transformer   T | primary    0.946mH secondary   3.64mH |
| Driver for Half-bridge | IR2153 |
| Fluorescent lamp | FLR36T6W |

TABLE 2
CONDITIONS AND MEASURED RESULTS

| | | | |
|------|------|------|------|
| Input Voltage | 100.0V | Lamp Voltage | 133.5V |
| Input current | 0.239A | Lamp current | 0.340A |
| Input power | 24.0W | Lamp power | 18.2W |
| Power factor | 99.6% | Oscillation Frequency | 49.5kHz |

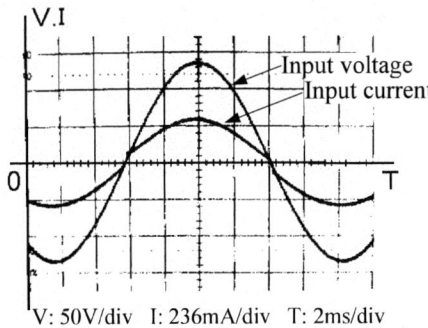

V: 50V/div   I: 236mA/div   T: 2ms/div

Fig. 10. Observed waveforms of input voltage and current.

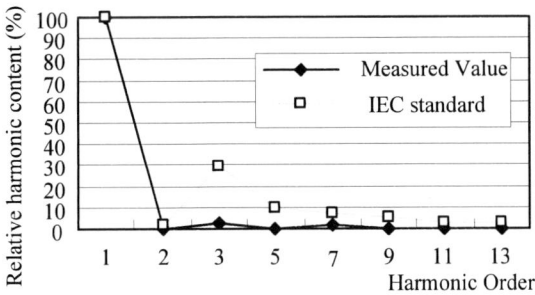

Fig. 11. The relative harmonic component of input current.

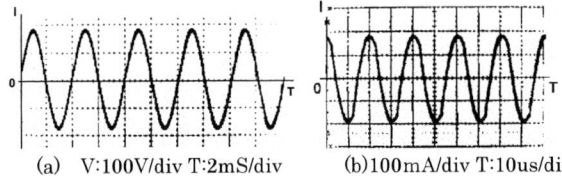

(a)   V:100V/div T:2mS/div       (b)100mA/div T:10us/div

Fig. 12. Observed waveforms of input voltage and current.

Observed input current waveform and input voltage waveform to the proposed electronic ballast are described in Fig.10. The figure shows that the input waveform is almost similar to a sinusoidal wave. The relative harmonic content of this input current and the ICE regulation value are shown in Fig.11. It is proven that relative harmonic contents of the proposed electronic ballast are lower than the IEC regulation values in any order.

Observed lamp current and observed lamp voltage are shown in Fig.12. Both waveforms are also similar to sinusoidal waves that are suitable for the lighting of fluorescent lamp.

## IV. CONCLUSIONS

In this paper, a newly developed neutral-point type buck-boost converter was proposed. The newly proposed converter is an improvement of the previous circuit that is the neutral point type boost converter. The proposed circuit can reduce the generation of inrush current at start when turning on and can depress the voltage-stress for circuit elements.

The feature of this converter was confirmed by experiments and it was proven that the electronic ballast could be designed using this converter. From lighting experiments, the proposed electronic ballast satisfies IEC regulation Class C and achieved high Power Factor. And it was confirmed that waveforms of lamp current and lamp voltage are similar to sinusoidal waves.

## V. REFERENCES

[1] Investigation report by the Illuminating Engineering Institute of Japan," Research of the input harmonics problem in the electronic ballast for lighting", May, 1999.
[2] K. Simizu, Y. Takahashi, N. Kitamura, "Electronic Ballast Circuit for Fluorescent Lamps that Reduce Current Harmonics" , J. Light & Visual Environment, Vol.21, No.2, pp.26-31, 1997.
[3] M. Maehare, K. Sato, H. Nishimura, "A Current Source Type Charge Pump High Power Factor Electronic Ballast Combined with Buck Converter", Record of PESC'98, pp.2016-2020, 1998.
[4] H. Matsuo, K. Shimizu, F. Kurokawa, "Performance Characteristics of a Novel Modified Half-bridge Inverter as an Electronic Ballast For Lighting", Record of PESC'98, pp.2028-2034, 1998.
[5] N. Takahashi, Y. Kato, M. Ohkita, K. Okutsu, M. Matsuyama, M. Nakaoka, "An Electronic Ballast using Modified Two-Switch Boost Converter", J. Light & Visual Environment, Vol.20, No.4, pp.265-271, 2000.
[6] N. Takahashi, Y. Kato, M. Ohkita, K. Okutsu, M. Matsuyama, M. Nakaoka, "An Electronic Ballast Using Modified Two-Switch Boost Converter", J. Light & Visual Environment, Vol.26, No.2, pp.8-13, 2000.
[7] K. Takahashi, M. Aono, M. Kamino, "Electronic Ballast with Combined Half-bridge Inverter and Buck-boost Converter", J. Illuminating Engineering Institute of Japan, Vol.86, No.2, pp.92-96, 2002.
[8] S. Adachi1, Y. Kato, N. Takahashi, I. Yokozeki, M. Nakaoka, "Development of the Buck-Boost Inverter with Two Switches and the Application to the Electronic Ballast", Proceedings of PEDS'03, pp.1146-1149, 2003.

## VI. BIOGRAPHIES

**Nabil M Hidayat** was born in Selangor, Malaysia on January 18, 1982. He graduated from Tottori University in March 2006 and now a graduate student of Power Electronics in Tottori University

 **Msaaki Nakamura** was born in Tottori in Japan, on 1982. He graduated from Tottori University in 2006. Now, he is graduate student of power electronics administration at Tottori University.

 **Yoshito Kato** was born in Tottori in Japan, on February 24, 1944. He graduated from Hiroshima Institute of Technology in 1967.He was been working as a associate professor of Tottori University, Tottori. He received the doctor degree from Osaka Prefecture University in 1995. He has been engaged in the research of Power Electronics, in particular, the development of electronic ballast for improvement of Power Factor and for reduction of input current harmonics.

 **Nobuo Takahashi** was born in Shimane in Japan, on July 12 1952. He graduated from Osaka University in 1975. He has been working as a teacher of Matsue National College of Technology, Shimane, Japan. He received the doctor degree from Tottori University in 2001. He has been engaged in the research of Power Electronics, in particular, the development of electronic ballast for improvement of Power Factor and for reduction of input current harmonics.

 **Shun-ichi Adachi** graduated with masters from Tottori University on March 2004. He now works with Nihonkai Telecasting Corp.

 **Ichiro Yokozeki** was born in Tokushima Japan on May 22 1963. He graduated with masters from Tottori University on March 1990 and received a doctorate from Tottori University on September 2005. He now works at Harison Toshiba Lighting Corp as a system engineer.

**2006 IEEE International Conference on Power Electronic, Drives and Energy Systems**

# Hysteresis-Band Current Control of a Four Quadrant AC-DC Converter giving IEEE 519 compliant performance at any Power Factor

A.N.Arvindan, and V.K.Sharma

*Abstract--*This paper presents unity, lagging and leading power factor operations, with near sinusoidal line currents, of the bi-directional buck-boost type improved power quality ac-dc converter (IPQC) that employs power MOSFET embedded four - quadrant switches (4QSWs). The topology of the IPQC itself ensures variable bi-directional dc voltage and reversible current in both buck and boost modes, however, application of the hysteresis-band current control based closed loop technique renders operation of the converter in either mode at any power factor, in any quadrant, with a harmonic profile at the utility interface conforming to the IEEE 519 stipulations by appropriately modifying the switching pattern of the 4QSWs. The control strategy is evaluated for the IPQC by simulations implemented in the single and three phase topologies of the converter. Results confirm that the technique confers IEEE 519 compliance on the converter with regard to the Total Harmonic Distortion (THD), and, are presented for unity, leading and lagging power factor operations in both modes for the four quadrants.

*Index Terms--* Four-Quadrant Switch, Hysteresis-Band Current Control (HCC), Improved Power Quality AC-DC converter, Power Quality, Pulse Width Modulation (PWM).

## I. INTRODUCTION

TRADITIONALLY , ac-dc converters are developed using diodes and thyristors to provide controlled and uncontrolled unidirectional and bi-directional dc power. They have the problems of poor quality in terms of injected current harmonics, resultant voltage distortion and poor power factor at input ac mains and slowly varying rippled dc output at loan end, low efficiency, and large size of ac and dc filters. To alleviate some of these problems and meet the contemporary stringent power quality standards [1] a new breed of ac-dc converters referred to as improved power quality ac-dc

converters (IPQCs) are increasingly being used for various applications. These converters employ self-commutating devices and have been classified [2] as unidirectional and bidirectional buck, boost, buck-boost etc. The bi-directional buck-boost type IPQC is used in certain applications, which require output dc voltage widely varying from low voltage to high voltage with bi-directional dc current as four-quadrant operation and bi-directional power flow. These converters can be implemented in many ways, such as cascading the buck and boost converters, but the simplest way of realizing them is by using a matrix converter. With high frequency switching, the size of the input ac filter and output dc filter is reduced which allows the fast response of this converter. It is capable of working as bi-directional buck and bi-directional boost converters and is an ideal solution for ac-dc conversion. It is derived from matrix converters normally used for ac-dc conversion for a wide frequency range at input as well as output.

This paper presents unity, lagging and leading power factor operations, with near sinusoidal line currents, of the bi-directional buck-boost type improved power quality ac-dc converter that employs power MOSFET embedded four-quadrant switches (4QSWs) [3]. The topology of the IPQC itself ensures variable bi-directional dc voltage and reversible current in both buck [4] and boost [5] modes, however, application of the hysteresis -band current control (HCC) based closed loop pulse width modulation (PWM) technique renders operation of the converter in either mode at any power factor, in any quadrant, with an improved power quality at the ac interface by appropriately modifying the switching pattern of the 4QSWs. The control strategy is evaluated by simulations implemented in the single and three phase topologies of the converter. The simulations confirm that adoption of the control technique in the converter renders the harmonic profile at the ac utility interface compliant with the revised 1992, IEEE 519, stipulations with regard to the total harmonic distortion (THD).

## II. TOPOLOGICAL CONSIDERATIONS

The bi-directional buck-boost converters are the IPQC version of the conventional thyristor dual converters. Their topology is derived from ac-ac matrix converters using four

---

A. N. Arvindan and V. K. Sharma are with the Electrical Engineering Department, Faculty of Engineering and Technology, Jamia Millia Islamia ( A Central University ), New Delhi, India (e-mail: lkana0@yahoo.com and viren_krec@yahoo.com).

0-7803-9771-1/06/$25.00 ©2006 IEEE

quadrant switches (4QSWs). Since no four-quadrant switch is currently commercially available these are realized by embedding a transistor inside a diode bridge or by inverse parallel connections of transistors as shown in Fig.1. Power MOSFETs are employed because they have the following advantages: the PWM frequency can be higher, and the minimum pulse and notch widths are smaller.

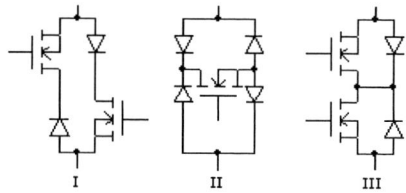

Fig. 1.  Four-Quadrant Switch (4QSW) realizations.

Topologies of single-phase and three-phase versions of the bi-directional buck-boost converter using type I 4QSWs are shown in Fig.2 and Fig.3 respectively.

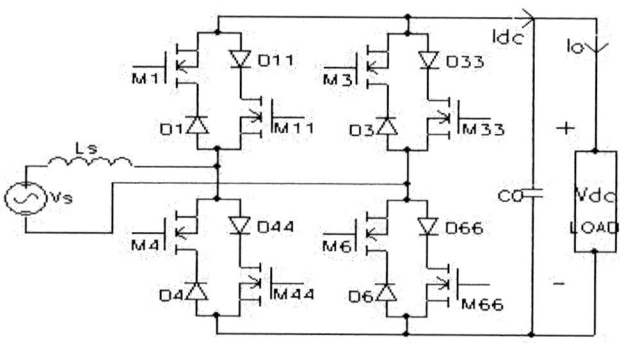

Fig. 2.  Single-phase bi-directional buck-boost IPQC.

In the circuit shown in Fig. 2, there are four 4QSWs, two in each limb. Each 4QSW comprises two 2QSWs (two quadrant switches), each two-quadrant switch consisting of a MOSFET with series diode, connected in inverse-parallel. The three-phase bi-directional buck-boost converter, shown in Fig.3, comprises three limbs, each comprising two 4QSWs.The operation of the bi-directional buck-boost converter in buck or boost mode and in a particular quadrant in the V-I plane shown

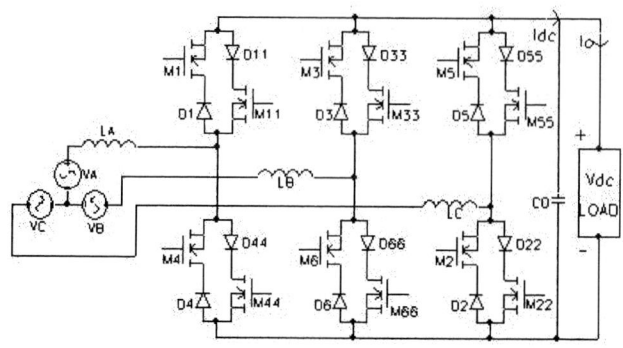

Fig. 3.  Three-phase bi-directional buck-boost IPQC.

in Fig. 4 is determined by the conditioning of the switching states of two sets (I and II) of devices. In the single-phase version each set comprises four MOSFETs; set I MOSFETs - M11, M66, M33, M44 and set II MOSFETs - M1, M6, M3, M4; while in the three-phase version each set comprises six MOSFETs; set I MOSFETS - M11, M33, M55, M44, M66, M22, and set II MOSFETs - M1, M3, M5, M4, M6, M2.

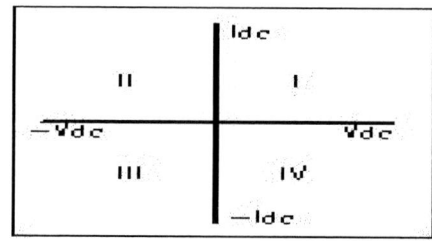

Fig. 4.  V-I plane with the quadrants demarcated.

Table I shows the conditioning of the switching states of the two sets of devices in the converter corresponding to the four quadrants in the buck and boost modes pertaining to the rectification and inversion operations. For the successful operation of the converter the prerequisite conditions must be satisfied. For instance, the boost mode is effective only when the dc link voltage is greater than the amplitude of the line voltage. The inversion operation in the buck mode is possible only if an external dc source of a magnitude greater than the peak line voltage is connected across the dc link.

TABLE I
CONDITIONING OF SWITCHING STATES OF THE TWO SETS OF DEVICES IN THE BIDIRECTIONAL BUCK-BOOST CONVERTER

| Mode | Operation | Quadrant | Set I Devices | Set II Devices |
|---|---|---|---|---|
| Buck | Rectification | I | OFF | Switching pattern 1 |
| | Inversion | II | OFF | Switching pattern 1 displaced by 180° |
| | Rectification | III | Switching pattern 1 | OFF |
| | Inversion | IV | Switching pattern 1 displaced by 180° | OFF |
| Boost | Rectification | I | Switching pattern A | ON |
| | Inversion | II | ON | Switching pattern A displaced by 180° |
| | Rectification | III | ON | Switching pattern A |
| | Inversion | IV | Switching pattern A displaced by 180° | ON |

## III. IMPLEMENTATION OF HCC PWM TECHNIQUE IN BI-DIRECTIONAL BUCK-BOOST CONVERTERS

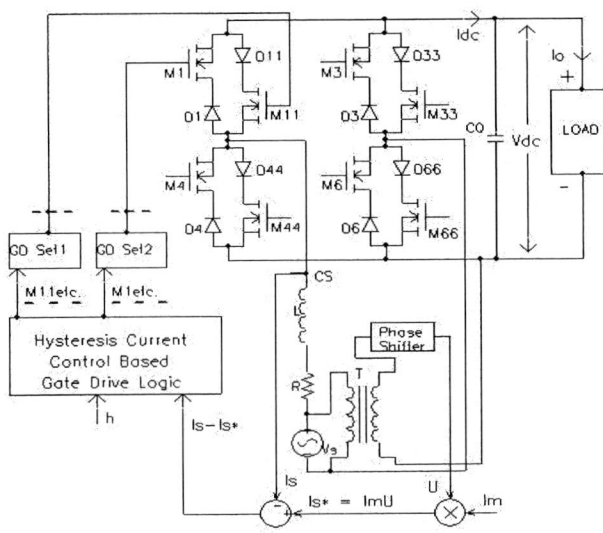

Fig. 5. Schematic of HCC PWM control in the single-phase bi-directional buck-boost type IPQC with type I 4QSWs.

Fig.5 shows the implementation of the current feedback loop based hysteresis-band current control (HCC) in the single-phase bi-directional buck-boost converter. It is essentially a control that tracks a current reference, which is derived from a waveform template. From Fig.5 it is evident that

the current magnitude control $I_m$ is considered to be an open loop controller. This is generally not the practice with closed loop control associated with IPQCs because though primacy is given to power quality parameters at the ac interface of the converter the voltage regulation of the dc link is also of importance. For instance, in the boost mode of operation, the dc link voltage has to be maintained well above the peak value of the line voltage to overcome the constraints imposed by the *loss of control* and *current distortion limits* [5] in addition to the power matching requirement on the ac and dc sides. Usually $I_m$ is derived from the regulated voltage feedback wherein the error between the reference and dc link voltages is fed to a PID controller and then a current limiter. The feed-forward technique is another technique of $I_m$ control. In this paper, the $I_m$ value has been carefully selected after theoretical calculations and conducting several trials, therefore, the dc link voltage magnitude is not a problem. This also aids convergence of the simulations. The reference waveform U is derived via the voltage transformer T and multiplied by $I_m$ to obtain the template $I_s^*$ for the current $i_s$. The switching patterns corresponding to operations pertaining to the two modes and four quadrants are generated by the hysteresis current controller based gate drive logic block shown in Fig.5 that comprises logic gates supplemented by pulse generators apart from the comparators and relays. The block provides apt signals to the gate drive circuits (GD set1 and GD set2) of the two MOSFET sets- set1 and set2. The switching patterns of the MOSFETs in a set (all those of the other set will be always ON or always OFF depending on the mode) make the current

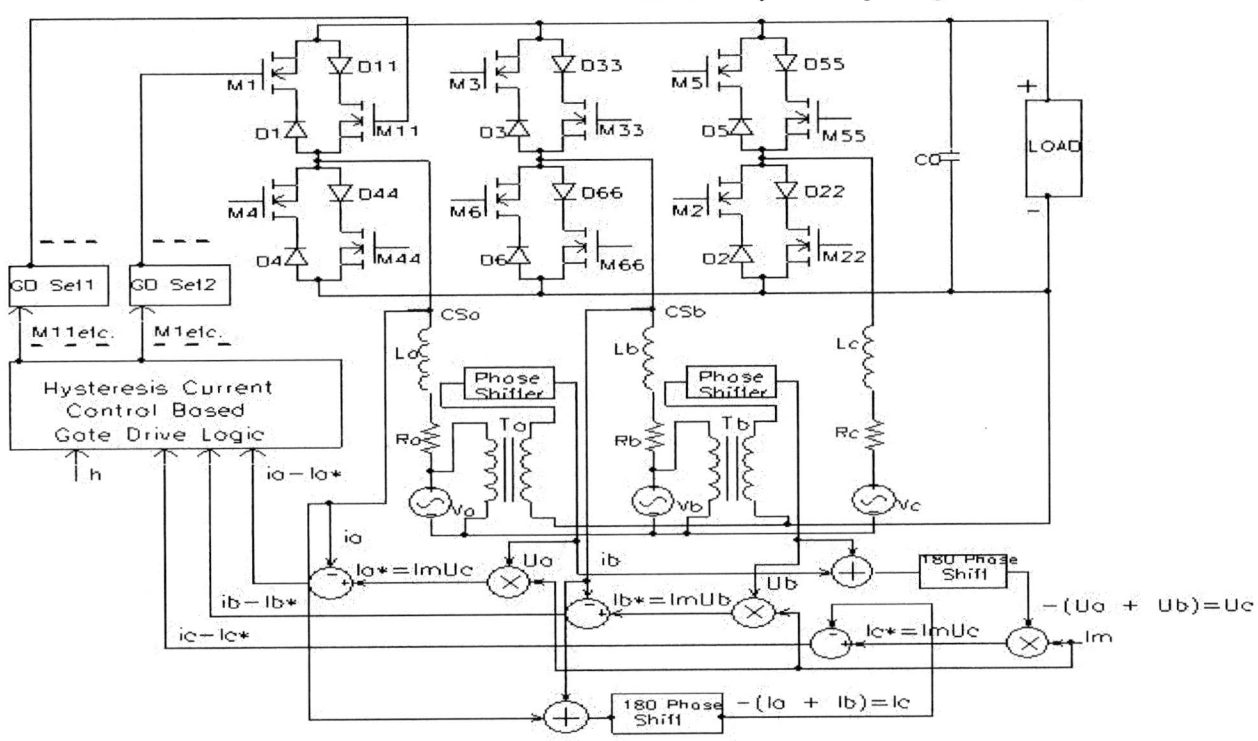

Fig. 6. Schematic of HCC PWM technique controlled three-phase bi-directional buck-boost type, improved power quality ac-dc converter.

317

$i_s$ which is measured by the current sensor CS, follow the template $I_s^*$ within a hysteresis band.

The controller for the three-phase topology is shown in Fig. 6. From Figs. 5 and 6 it is evident that the three-phase control is not realized by duplicating the single-phase control scheme in the three phases. The control scheme is based on the three-phase three-wire phase relationship of the phase voltages and currents. The following equations are valid for the Y-connected three-phase source fed converter in Fig. 6.

$$Va + Vb + Vc = 0 \qquad (1)$$

$$Ia + Ib + Ic = 0 \qquad (2)$$

From (1) and (2) it is obvious that if two phase voltages and currents are known the corresponding third voltage and current respectively, can be determined precisely. This is true for both balanced and unbalanced conditions of the three-phase three-wire system and is put to use in the HCC PWM based control scheme, shown in Fig. 6, for the bi-directional buck-boost converter. The adoption of the simple phasor theory results in requirement of two source voltage ($T_a$ and $T_b$) and line current ($CS_a$ and $CS_b$) sensors rather than three in each type that would be required if duplication of the single-phase control scheme is undertaken in the three phases of the three-phase topology of the converter. This reduces the requirement of high precision hardware and thus lowers the cost in addition to making the control scheme compact.

## IV. CONVERTER SIMULATION

All simulations are conducted with the objective of obtaining bilateral power flow i.e. rectification and inversion operations in the buck and boost modes, with sinusoidal line currents, at unity, leading and lagging power factors. The simulations are conducted for single and three phase versions of the converter, with the same parameters using the three types of 4QSWs, in both the modes for the four quadrants. The simulations are conducted with ideal a.c. sources; and batteries as dc voltage sources rather than charged capacitors, to facilitate simulation convergence.

In the simulation models, the line capacitors usually provided for filtering and improving the power factor, are not included as, primacy is given to the investigation of the efficacy of the control technique alone to provide a high power quality interface with the utility. A hysteresis band (h) of 1A width has been used in all the simulations, except those of the 1-phase buck and boost modes wherein; 0.25/0.5A and 0.5/1A widths have been used respectively.

Phase peak of the 1-phase and balanced 3-phase supplies is 150V.

THD = Total harmonic distortion in the phase current(s).
Ls = Line inductance for single-phase supply.
La = Lb = Lc = Ll ; Ra = Rb = Rc = R1
DCLV = Initial d.c link voltage (for boost mode)
R = Load resistance; L = Load inductance
Vdc = External dc voltage magnitude (for buck lead / inversion)

The simulation data are given in Table II. The first column (OP Code) provides the operation information in alphanumeric code. It uses 1and 3, bk and bst, iu and ru, d and g, and, 90 and 45 to indicate single and three phase configurations, buck and boost modes, rectification and inversion operations at unity power factor (UPF), leading and lagging power factors, and, 90° and 45° phase angles respectively.

TABLE II
SIMULATION DATA: UNITY, LEADING AND LAGGING POWER FACTORS

| OP Code | THD % | Ls or Ll mH | R1 Ohm | DCLV Volt | R Ohm | L mH | Vdc Volt |
|---|---|---|---|---|---|---|---|
| 1bkiu | 4.52 | - | - | - | 8.75 | 0.1 | 350 |
| 1bstru | 3.49 | 2 | 8.75 | 300 | 15 | - | - |
| 3bkru | 4.45 | - | - | - | 25 | $10^4$ | - |
| 3bstiu | 3.31 | 2 | 8.75 | 1200 | 30 | - | - |
| 1bkd90 | 2.89 | - | - | - | 8.75 | 0.1 | 450 |
| 1bstg45 | 2.95 | 2 | 8.75 | 600 | 15 | - | - |
| 3bkg45 | 4.87 | - | - | - | 25 | $10^4$ | - |
| 3bstd90 | 4.20 | 2 | 8.75 | 900 | 30 | - | - |

## V. RESULTS AND DISCUSSION

From Table II it is clear that the HCC PWM technique is able to limit the THD for operations in any quadrant, at any power factor, in both buck and boost modes, to the permissible limit of 5% stipulated by [1].

Figs. 7 and 8 show the magnitudes of the even and odd harmonics respectively of the ac side current $I_s$, obtained for the IV-quadrant inversion operation at unity power factor (UPF) of the single-phase bi-directional buck-boost converter in the buck mode. The waveforms pertaining to the operation are shown in Fig. 9. It is evident that the even harmonics <17 and the 18th are much less than the limits, stipulated by IEEE 519,1992, i.e. 1% and 0.375% respectively. The odd ones<17 are restricted to 4%, however, those of higher order i.e.17<order<23 exceed the 1.5% limit.

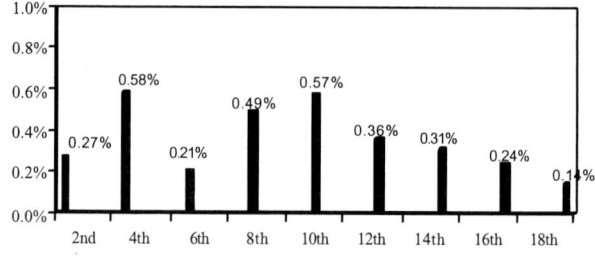

Fig. 7. Even harmonics by FFT analysis of ac side current $I_s$ in Fig.9.

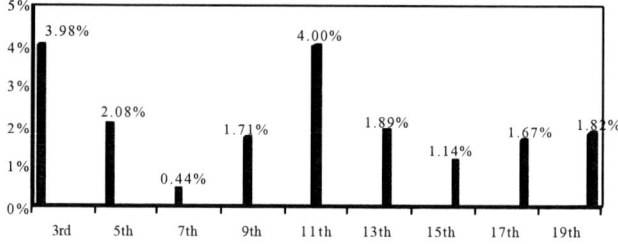

Fig. 8. Odd harmonics by FFT analysis of ac side current $I_s$ in Fig. 9.

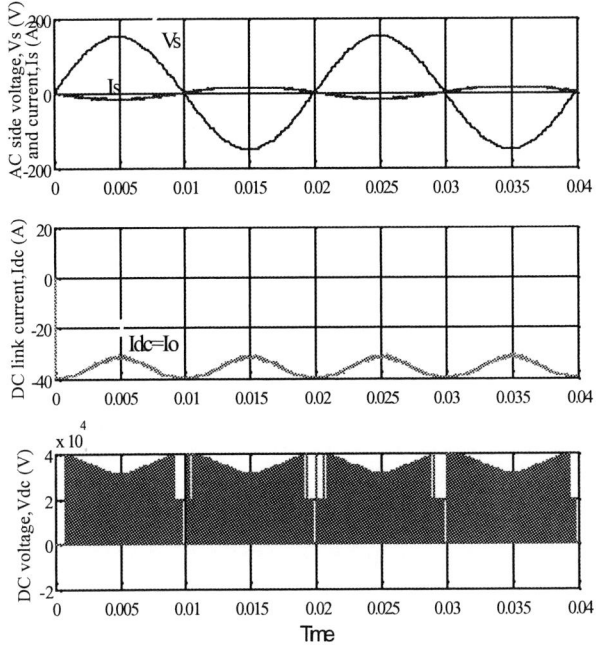

Fig. 9. Waveforms for single-phase IV-quadrant buck inversion at UPF.

The THD of the ac side current $I_s$ in Fig. 9 is determined by the fast fourier transform (FFT) analysis to be 4.52%.

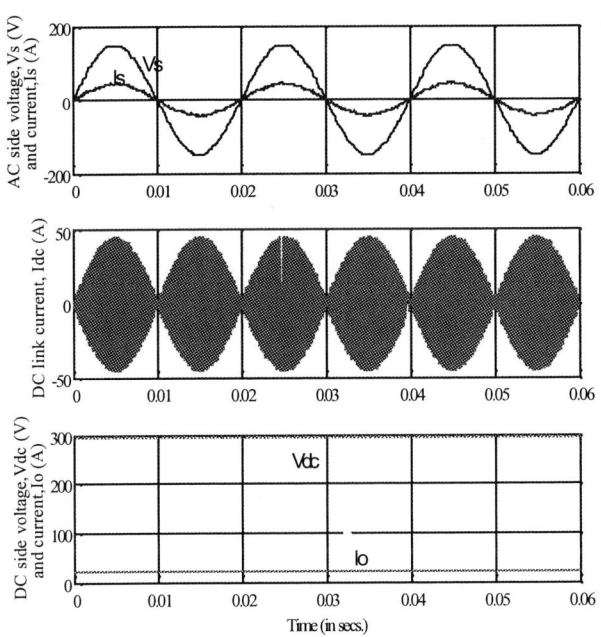

Fig. 10. Waveforms for 1-phase I-quadrant boost rectification at UPF.

Fig. 10 depicts the waveforms associated with the converter for the I-quadrant based rectification operation at UPF in the boost mode. The THD of the ac side current $I_s$ in this case is 0.0349. The II quadrant inversion operation at UPF in the boost mode wherein the three phase currents are displaced from their corresponding phase voltages by $180°$ is shown in Fig. 11.

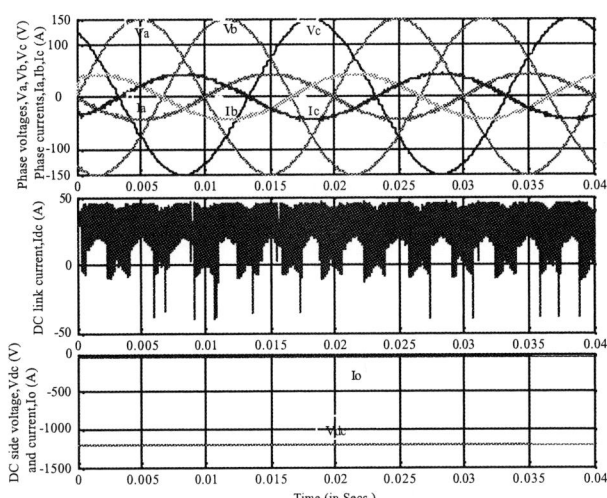

Fig. 11. Waveforms for 3-phase II-quadrant boost inversion at UPF.

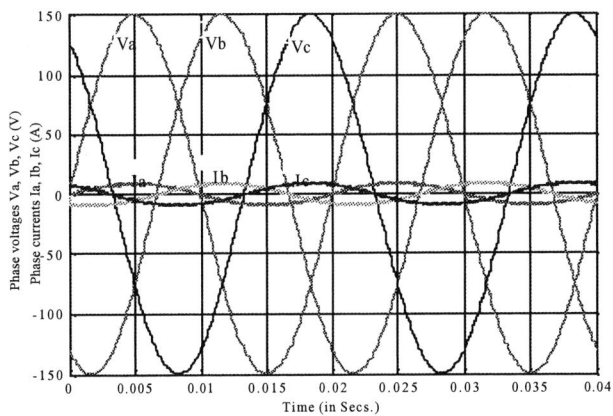

Fig. 12. AC utility waveforms for 3-phase buck rectification at UPF.

The three-phase voltages and currents in the buck mode corresponding to rectification at UPF (in phase) and lagging power factor ($45°$ phase lag) are shown in Figs. 12 and 13 respectively.

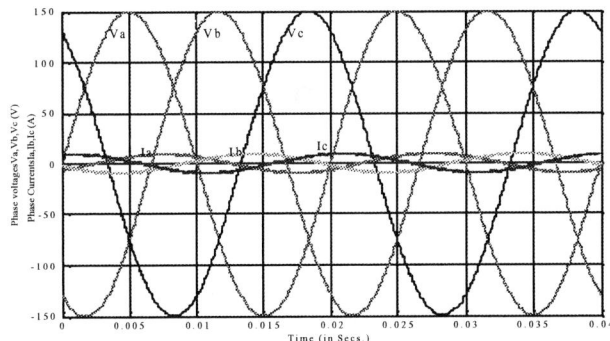

Fig. 13. AC utility waveforms for 3-phase $45°$-lag buck operation.

Figs. 14 and 15 show the waveforms at the ac utility for the single-phase converter corresponding to zero power factor (ZPF) ($90°$ phase lead) in buck mode and lagging power ($45°$ phase lag) in boost mode respectively.

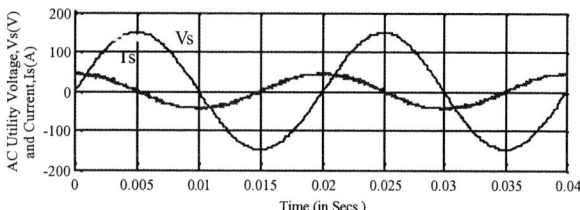

Fig. 14. AC utility waveforms for 1-phase ZPF (lead) buck operation.

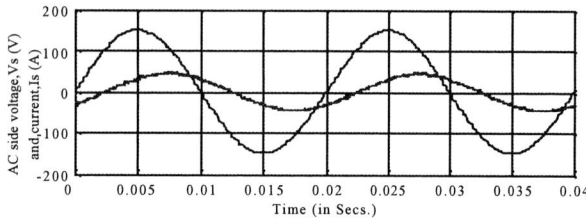

Fig. 15. AC utility waveforms for 1-phase 45°-lag boost operation.

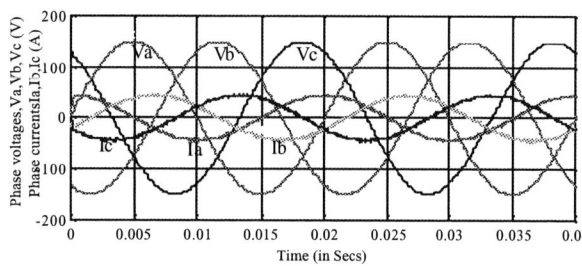

Fig. 16. AC utility waveforms for 3-phase ZPF (lead) boost operation.

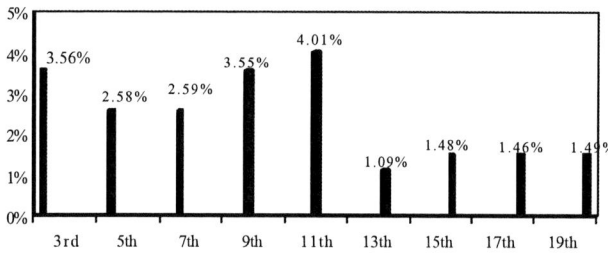

Fig. 17. Odd harmonics of phase currents in Fig.15 by FFT analysis.

In Fig. 16, the three-phase currents lead the voltages by 90°, corresponding to ZPF (lead) operation, in the boost mode. The converter presents itself as a three-phase capacitor bank in this operation. The FFT analysis of the phase currents pertaining to the operation for the odd harmonic currents is shown in Fig. 17. It is seen that the 11th harmonic exceeds the 4% limit and imposed by [1]. The 15th, 17th and 19th harmonics are also very close to the 1.5% limit. The even harmonics are within the acceptable limits but their magnitudes are not depicted because of space constraints.

## VI. CONCLUSIONS

The HCC PWM technique is capable of conferring IEEE 519,1992 compliant performance on the power MOSFET based single- and three-phase versions of the bi-directional buck-boost converter with regard to the THD. The technique renders individual even harmonic currents within the stipulations of the standard; however, further refinement is required in it for limiting the higher order odd harmonics.

## VII. REFERENCES

[1] IEEE Recommended Practices and Requirements for Harmonics Control in Electric Power Systems, IEEE Std. 519,1992.
[2] Bhim Singh, B. N. Singh, A. Chandra, Kamal Al-Haddad, Ashish Pandey, and D. P. Kothari, "A Review of Three-Phase Improved Power Quality AC-DC Converters," IEEE Trans. Ind. Electron., vol. 51, No. 3, pp. 641-660, June 2004.
[3] Burany, N., "Safe Control of Four-Quadrant Switches," in Conf. Rec. IEEE- IAS, 1989, Part I, pp. 1190-1194.
[4] R. Oruganti and M. Palaniapan,"Extension of inductor voltage control to three-phase buck-type ac-dc converter," IEEE Trans. Power Electron., vol. 15, No.2, pp.295-302, Mar. 2000.
[5] B. T. Ooi, J. C. Salmon, J. W. Dixon, and A. B. Kulkarni, "A 3-phase controlled current PWM converter with leading power factor," IEEE Trans. Ind. Applicat., vol. IA-23, No.1, Jan./Feb. 1987, 78-84.

## VIII. BIOGRAPHIES

**A. N. Arvindan** graduated in Electrical and Electronics Engineering from College of Engineering, Guindy, Anna University, in 1988. He obtained the masters degree in Power Systems from Regional Engineering College (now NIT), Tiruchirapalli, in 1995.

He has over sixteen years of professional experience including ten years in industry and six years in academics. His industrial experience includes a stint at Asea Brown Boveri, where he designed high current rectifiers in the power electronics division. He has wide field exposure and has completed several projects including electrification of large continuous process plants and O.H. transmission. In 1999, he took to academics by joining St.Joseph's Engineering College, Anna University, Chennai, as Assistant Professor and was conferred 'The Best Teacher Award' for three successive academic years. He is currently pursuing the PhD programme as a full-time research scholar in the Electrical Engineering Department of Jamia Millia Islamia, New Delhi.

**V. K. Sharma** received his B.E. degree from Karnataka Regional Engineering College, Surathkal, in 1984, M.Tech and PhD degrees from IIT, Delhi, in 1993, and 2000, respectively. He was a Post Doc Fellow at Ecole de technologie supereiure, at Montreal, Canada.

He started his academic career in College of Engineering, Pravaranagar, Maharashtra. He is the recipient of Railway Board Medal for his research paper published in Institution of Engineers (India). He has received merit scholarships including QIP research fellowship for higher studies. He has completed major R&D and Thrust area projects funded by AICTE. He has been visiting research associate at IIT, Delhi. He has a few papers published in IEEE transaction, journal of IE (I), IETE and NISCAIR. He has visited USA and Germany for presenting his research findings, and undertook one month training on real-time microprocessor applications at ICTP, Trieste, Italy. He is a member of IE (I), IETE, IEE and IEEE. Currently, he is a professor in electrical engineering at Jamia Millia Islamia, New Delhi.

**2006 IEEE International Conference on Power Electronic, Drives and Energy Systems**

# Multiphase Inverter Topology and its Modulation Technique for Optimal Harmonic Output

### Ravindra Kumar Singh

*Abstract--* **Space vector pulse width modulation (SVPWM) technique is being increasingly used because of its easy digital implementation. Multiphase load like Six-phase motor drive is an active area of research and industrial applications, but very less work has been done on multiphase inverters. In this paper a six-phase inverters circuit has been taken as a representative problem of multiphase inverters to investigate the best converter topology and its modulation technique such that output has minimum or optimal. This paper presents implementation of SVPWM technique to a six-phase inverter using unified approach. Simulation results are provided to validate the proposed theory. It has been found that a three-phase space vector based inverter topology has better THD than a six phase topology.**

*Index Terms*—**Six-phase inverter, Multi level Inverter, Space Vector Modulation.**

## I. INTRODUCTION

Amultiphase load/ source is a load/source, which has higher phase order than conventional three termed as the three-phase load/source. High phase order or Multi-phase (phase order more than three) machine drives are gaining growing attention in recent years, due to their several inherent benefits such as reduced torque pulsation, harmonic content, current per phase without increasing the voltage per phase, higher reliability and increased power in the same frame as compared to their three phase counterpart. Multi-phase inverter fed induction motor drives has been found to be quite promising for high power ratings and other specialized applications.

These multiphase load requires a multiphase inverter as source for these load. An inverter circuit topology uses two switches connected in series as one inverter arm. The number of inverter arms depends on number of phases. For example, a three-phase inverter will have three inverter arms whereas a six-phase inverter circuit will have six inverter arms. All the inverter circuits used in literature for multiphase inverters employ the same topology. A six-phase inverter therefore has six arms and twelve devices. An alternative way is to use a three-phase inverter and use a 3-phase/6-phase transformer to realize a six-phase source. If both 3-phase and 6-phase inverters are space vector modulated, the question is which topology will give a better harmonic spectrum? The other choice is to use a three level inverter which has approximately same number of silicon switches, why not to use a multilevel

three-phase inverter and use a transformer to convert it into six phase inverter.

In this paper, all these possibilities has been compared in light of the harmonic contents in the output of the inverter. The comparative harmonic based analysis has been used to optimize the inverter topology and its modulation technique.

### A. Space Vector Modulation

Space vector modulation technique is a type of modulation technique used for the changing the pulse width according to given reference vector [1]. In this technique all the possible switching states are represented as vectors in a two dimensional voltage space, which are obtained by transforming six phase dependent vectors to three phase independent vectors. Once the vectors are decided and the reference vector is known the best possible vectors for production of the reference voltage are found and are used in a predefined sequence for their corresponding timings, calculated for each sampling time. The reference vector is assumed to be constant for the corresponding sampling time. Ideally the sampling frequency should be infinity but the sampling frequency is limited by the on and off times the device used, therefore the maximum possible sampling frequency is chosen for minimum total harmonic distortion. The sampling frequency should be multiple of the output frequency and number of sectors for perfect operation. A unified approach for space vector modulation is presented in [2] which involves the following steps.

- Definition of switching vectors.
- Identification of separation planes.
- Identification of boundary planes.
- Obtaining the decomposition matrices.
- Definition of switching sequence.

## II. THREE PHASE INVERTER

The circuit of a three-phase inverter is shown in Fig. 1. The phasor diagram and the corresponding equations of the three phase voltages are as shown in Fig 2.

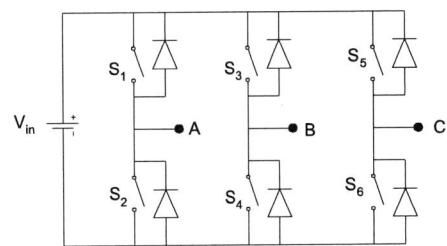

Fig. 1. Circuit topology of a three phase inverter.

---

Ravindra Kumar Singh is with Motilal Nehru National Institute of Technology, Allahabad, India (e-mail: rksingh @mnnit.ac.in).

0-7803-9771-1/06/$25.00 ©2006 IEEE     321

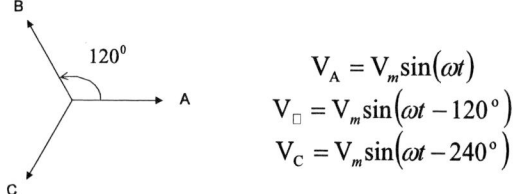

$$V_A = V_m \sin(\omega t)$$
$$V_\square = V_m \sin(\omega t - 120°)$$
$$V_C = V_m \sin(\omega t - 240°)$$

Fig. 2. Phasor diagram and output voltage equations.

### A. Definition of Switching Vectors

The switches in the same leg are assumed to be complimentarily commutated, i.e. if switch S1 is ON then S2 is OFF and vice versa. Thus eight switching vectors are possible, which are shown in Table 1. A '0' indicates that the corresponding switch is off and '1' indicates that the corresponding switch is on.

TABLE I
SWITCHING VECTORS OF THREE PHASE INVERTER

| S.no | S1 | S3 | S5 | Vector |
|------|----|----|----|--------|
| 1 | 0 | 0 | 0 | V0 |
| 2 | 0 | 0 | 1 | V1 |
| 3 | 0 | 1 | 0 | V2 |
| 4 | 0 | 1 | 1 | V3 |
| 5 | 1 | 0 | 0 | V4 |
| 6 | 1 | 0 | 1 | V5 |
| 7 | 1 | 1 | 0 | V6 |
| 8 | 1 | 1 | 1 | V7 |

□y transforming these three phase dependent vectors to three phase independent vectors using the transformation matrix given in equation (1).

$$T_{dq} = \sqrt{\frac{2}{3}} \begin{bmatrix} 1 & -\frac{1}{2} & -\frac{1}{2} \\ 0 & \frac{\sqrt{3}}{2} & -\frac{\sqrt{3}}{2} \end{bmatrix} \quad (1)$$

Thus the distribution of these vectors in two dimensional voltage space, i.e. in d-q plane can be obtained. The distribution of vectors can be seen in Fig. 3.

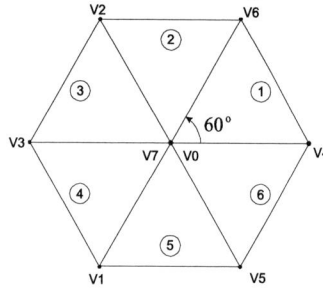

Fig. 3. Space vector distribution of three phase inverter.

### □. Identification of Separation Planes

From the vector distribution in voltage space it can be seen that the voltage space can be divided in to six sectors. Using the planes, which separate these sectors, one can identify the six sectors. These planes are called separation planes. The equations of the separation planes are given in equation (2).

$$u_\beta = 0, \quad \sqrt{3}u_\alpha - u_\beta = 0, \quad \sqrt{3}u_\alpha + u_\beta = 0 \quad (2)$$

### C. Identification of Boundary planes

□y the use of two vectors in each sector the maximum output that can be obtained is fixed and if the command vector is above this value, over-modulation occurs. The plane, which defines the maximum value of the command vector, is called the boundary plane. The boundary planes for different sectors are given in Table II.

TABLE II
□OUNDARY PLANES OF DIFFERENT SECTORS

| Sectors | □oundary Planes |
|---------|-----------------|
| 1 | $\sqrt{3}u_\alpha + u_\beta - 2\sqrt{3} = 0$ |
| 2 | $u_\beta - \sqrt{3} = 0$ |
| 3 | $\sqrt{3}u_\alpha - u_\beta + 2\sqrt{3} = 0$ |
| 4 | $\sqrt{3}u_\alpha + u_\beta + 2\sqrt{3} = 0$ |
| 5 | $u_\beta + \sqrt{3} = 0$ |
| 6 | $\sqrt{3}u_\alpha - u_\beta - 2\sqrt{3} = 0$ |

### D. Obtaining the decomposition matrices

Assuming that the command vector is in first quadrant and the implemented switching sequence as $V^0, V^4, V^6, V^7$. The average output voltage produced by these vectors must be equal to the average voltage produced by the command vector for total sample time $T_s$.

$$V^4 t_1 + V^6 t_2 + V^0 t_0 = u_{cmd} T_s \quad (3)$$

□ut we know that the total time must be equal to the sampling time.

$$t_1 + t_2 + t_0 = T_s \quad (4)$$

Adding these two equations and writing in matrix form.

$$\begin{bmatrix} V^4 & V^6 & V^0 \\ 1 & 1 & 1 \end{bmatrix} \begin{bmatrix} t_1 \\ t_2 \\ t_0 \end{bmatrix} = u_{cmd} T_s \quad (5)$$

$$\begin{bmatrix} t_1 \\ t_2 \\ t_0 \end{bmatrix} = \begin{bmatrix} V^4 & V^6 & V^0 \\ 1 & 1 & 1 \end{bmatrix}^{-1} u_{cmd} T_s \quad (6)$$

This Equation can be written as

$$\begin{bmatrix} t_1 \\ t_2 \\ t_0 \end{bmatrix} = M_1 T_s \quad (7)$$

Where

$$M_1 = \begin{bmatrix} V^4 & V^6 & V^0 \\ 1 & 1 & 1 \end{bmatrix}^{-1} u_{cmd} \qquad (8)$$

Where $M_1$ is called the decomposition matrix. Like this, decomposition matrices can be calculated for different sectors.

### E. Definition of Switching Sequence

After calculation of switching times the vectors are operated for their particular times in a predefined sequence. This sequence is decided for minimum switching loss and minimum Total Harmonic Distortion. The switching sequence for each sector is shown in Table III. The switching sequence is inverted for odd and even sample times so as to have minimum number of switching in each sample time.

TABLE III
SWITCHING SEQUENCES FOR DIFFERENT SECTORS

| Sector | Switching Sequence |
|---|---|
| S1 | $V^0, V^4, V^6, V^7$ |
| S2 | $V^0, V^2, V^6, V^7$ |
| S3 | $V^0, V^2, V^3, V^7$ |
| S4 | $V^0, V^1, V^3, V^7$ |
| S5 | $V^0, V^1, V^5, V^7$ |
| S6 | $V^0, V^4, V^5, V^7$ |

The Output waveforms of phase voltage and line voltage, taken for a symmetric R load, and their corresponding Fourier spectrum are shown in Fig. 4 and Fig. 5.

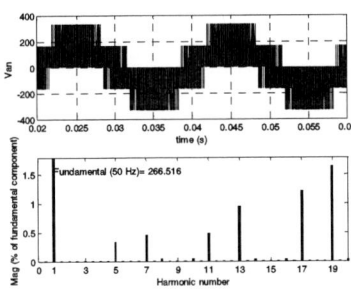

Fig. 4. Output phase voltage and corresponding Fourier spectrum.

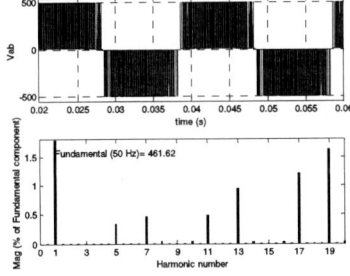

Fig. 5. Output line voltage and corresponding Fourier spectrum.

In case of high power applications Total Harmonic Distortion cannot be reduced by increase in the switching frequency. Thus multi-level inverters are employed for such applications. The normal three-phase inverter has only two levels i.e. either '1' or '0', but multi level inverter has more than two levels i.e. '1/2', '0' and '–1/2'. A three level inverter is simulated using space vector modulation technique in this paper. Circuit of a Diode clamped three level inverter is shown in Fig. 6

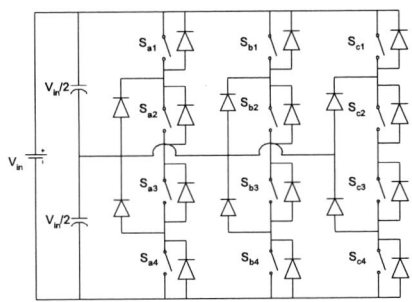

Fig. 6. Circuit topology of a three level Diode-clamped multi-level inverter.

The phasor diagram and the output voltage equations are same as that of a simple three-phase inverter.

### F. Definition of Switching Vectors

The three levels possible in a three-level inverter are '+1/2', '0' and '-1/2'. The different switch states for different output levels of three phases are shown in Table IV.

TABLE IV.
DIFFERENT SWITCH POSITIONS FOR DIFFERENT LEVELS.

| Sa | Sa1 | Sa2 | Sa3 | Sa4 |
|---|---|---|---|---|
| -½ | 0 | 0 | 1 | 1 |
| 0 | 0 | 1 | 1 | 0 |
| ½ | 1 | 1 | 0 | 0 |

Therefore number of switching vector possible is equal to $3^3$=27 vectors. Different vectors possible are tabulated in Table V.

TABLE V
SWITCHING VECTORS FOR A THREE LEVEL INVERTER

| S.No | Sa | Sb | Sc | Vector |
|---|---|---|---|---|
| 1 | -½ | -½ | -½ | $V^1$ |
| 2 | -½ | -½ | 0 | $V^2$ |
| 3 | -½ | -½ | ½ | $V^3$ |
| 4 | -½ | 0 | -½ | $V^4$ |
| 5 | -½ | 0 | 0 | $V^5$ |
| 6 | -½ | 0 | ½ | $V^6$ |
| 7 | -½ | ½ | -½ | $V^7$ |
| 8 | -½ | ½ | 0 | $V^8$ |
| 9 | -½ | ½ | ½ | $V^9$ |
| 10 | 0 | -½ | -½ | $V^{10}$ |
| 11 | 0 | -½ | 0 | $V^{11}$ |
| 12 | 0 | -½ | ½ | $V^{12}$ |
| 13 | 0 | 0 | -½ | $V^{13}$ |
| 14 | 0 | 0 | 0 | $V^{14}$ |
| 15 | 0 | 0 | ½ | $V^{15}$ |
| 16 | 0 | ½ | -½ | $V^{16}$ |
| 17 | 0 | ½ | 0 | $V^{17}$ |
| 18 | 0 | ½ | ½ | $V^{18}$ |
| 19 | ½ | -½ | -½ | $V^{19}$ |
| 20 | ½ | -½ | 0 | $V^{20}$ |
| 21 | ½ | -½ | ½ | $V^{21}$ |
| 22 | ½ | 0 | -½ | $V^{22}$ |
| 23 | ½ | 0 | 0 | $V^{23}$ |
| 24 | ½ | 0 | ½ | $V^{24}$ |
| 25 | ½ | ½ | -½ | $V^{25}$ |
| 26 | ½ | ½ | 0 | $V^{26}$ |
| 27 | ½ | ½ | ½ | $V^{27}$ |

The distribution of these vectors in dq plane can be obtained by multiplication of these vectors with the transformation matrix provide in equation (1) The distribution of vectors is shown in Fig. 7.

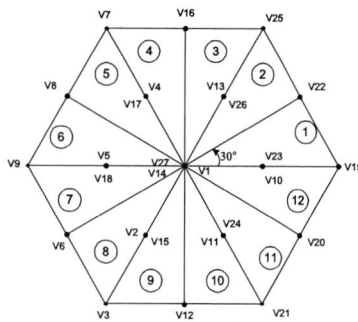

Fig. 7. Distribution of switching vectors in voltage space.

### G. Identification of Separation Planes

The six separation planes, which separate the twelve sectors, are given in equation (9).

$$u_\beta = 0, \; u_\alpha - \sqrt{3}u_\beta = 0, \; \sqrt{3}u_\alpha - u_\beta = 0, \; u_\alpha = 0,$$
$$\sqrt{3}u_\alpha + u_\beta = 0, \; u_\alpha + \sqrt{3}u_\beta = 0 \qquad (9)$$

### H. Identification of Boundary Planes

The ☐oundary planes for the twelve sectors are shown in Table VI.

TA☐LE VI
. ☐OUNDARY PLANES FOR DIFFERENT SECTORS

| Sectors | ☐oundary Planes |
|---|---|
| 1 and 2 | $\sqrt{3}u_\alpha + u_\beta - 2\sqrt{3} = 0$ |
| 3 and 4 | $u_\beta - \sqrt{3} = 0$ |
| 5 and 6 | $\sqrt{3}u_\alpha - u_\beta + 2\sqrt{3} = 0$ |
| 7 and 8 | $\sqrt{3}u_\alpha + u_\beta + 2\sqrt{3} = 0$ |
| 9 and 10 | $u_\beta + \sqrt{3} = 0$ |
| 11 and 12 | $\sqrt{3}u_\alpha - u_\beta - 2\sqrt{3} = 0$ |

### I. Obtaining the Decomposition Matrices

Assuming that the command vector is in sector 1 and the switching sequence is $V^1, V^{10}, V^{19}, V^{22}, V^{23}, V^{14}$. As $V^1$ and $V^7$ are zero vector let their switching time be $t_0$. For $V^{10}$ and $V^{23}$ switching time be $t_2$. The average output voltage produced by these vectors must be equal to the average voltage produced by the command vector for total sample time $T_s$.

$$V^{19}t_1 + V^{10}t_2 + V^{22}t_3 + V^1 t_0 = T_s \qquad (10)$$

We know that $V^{10} = \dfrac{V^{19}}{2}$ there fore

$$V^{19}\left(t_1 + \frac{t_2}{2}\right) + V^{22}t_3 + V^0 t_0 = u_{cmd}T_s \qquad (11)$$

Let $t_1' = t_1 + \dfrac{t_2}{2}$

$$V^{19}t_1' + V^{22}t_3 + V^0 t_0 = u_{cmd}T_s \qquad (12)$$

☐ut

$$t_1' + t_3 + t_0 = T_s$$

Adding these two equations and writing in matrix form.

$$\begin{bmatrix} V^{19} & V^{22} & V^0 \\ 1 & 1 & 1 \end{bmatrix} \begin{bmatrix} t_1' \\ t_3 \\ t_0 \end{bmatrix} = u_{cmd}T_s \qquad (13)$$

$$\begin{bmatrix} t_1' \\ t_3 \\ t_0 \end{bmatrix} = \begin{bmatrix} V^{19} & V^{22} & V^0 \\ 1 & 1 & 1 \end{bmatrix}^{-1} u_{cmd}T_s \qquad (14)$$

This Equation can be written as

$$\begin{bmatrix} t_1' \\ t_3 \\ t_0 \end{bmatrix} = M_1 T_s \qquad (15)$$

Where

$$M_1 = \begin{bmatrix} V^{19} & V^{22} & V^0 \\ 1 & 1 & 1 \end{bmatrix}^{-1} u_{cmd} \qquad (16)$$

$M_1$ is called the decomposition matrix. Like this decomposition matrices are calculated for different sectors. For calculation of $t_1$ and $t_2$ different methods can be used. ☐y linking these times with modulation index $mi$ the output will be good for all modulation indexes.

$$t_1 + \frac{t_2}{2} = mi \qquad (17)$$

### J. Definition of Switching Sequence

The switching sequence is defined for obtaining the minimum switching loss and minimum harmonic distortion. The switching sequences for different sectors are shown in Table VII.

TABLE VII
SWITCHING SEQUENCES FOR DIFFERENT SECTORS

| Sector | Switching Sequence |
|--------|--------------------|
| S1 | $V^1, V^{10}, V^{19}, V^{22}, V^{23}, V^{14}$ |
| S2 | $V^{14}, V^{13}, V^{25}, V^{22}, V^{26}, V^{27}$ |
| S3 | $V^{27}, V^{26}, V^{25}, V^{16}, V^{13}, V^{14}$ |
| S4 | $V^{14}, V^4, V^7, V^{16}, V^{17}, V^1$ |
| S5 | $V^1, V^{17}, V^7, V^8, V^4, V^{14}$ |
| S6 | $V^{14}, V^5, V^9, V^8, V^{18}, V^{27}$ |
| S7 | $V^{27}, V^{18}, V^9, V^6, V^5, V^{14}$ |
| S8 | $V^{14}, V^{15}, V^3, V^6, V^2, V^1$ |
| S9 | $V^1, V^2, V^3, V^{12}, V^{15}, V^{14}$ |
| S10 | $V^{14}, V^{11}, V^{21}, V^{12}, V^{24}, V^{27}$ |
| S11 | $V^{27}, V^{24}, V^{21}, V^{20}, V^{10}, V^{14}$ |
| S12 | $V^{14}, V^{10}, V^{19}, V^{20}, V^{23}, V^1$ |

The Output waveforms of phase voltage and line voltage, taken for a symmetric R load, and their corresponding Fourier spectrum are shown in Fig. 8 and Fig. 9.

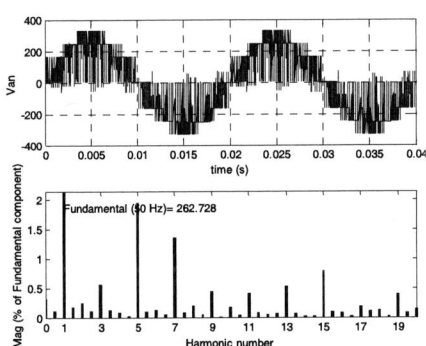

Fig. 8. Output phase voltage and corresponding Fourier spectrum.

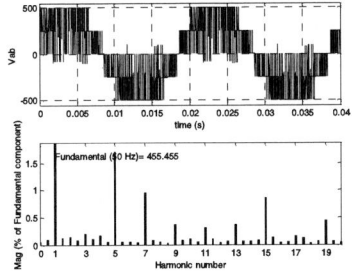

Fig. 9. Output line voltage and corresponding Fourier spectrum.

## III. SIX PHASE INVERTER

Six-phase supply can be of two types symmetrical supply and asymmetrical supply. In symmetric supply the output phases are separated by an equal angle i.e. they are separated by 60 degrees each. Fig.10 shows the symmetrical six phase voltages and their corresponding equations. Asymmetrical

supply is a combination of two three-phase supplies (A⬜C and DEF) separated by an angle of 30 degrees. Fig. 11 shows asymmetrical phase voltages and their corresponding equations.

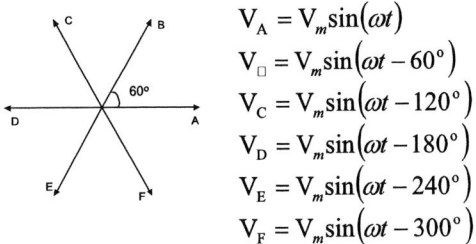

$$V_A = V_m \sin(\omega t)$$
$$V_\square = V_m \sin(\omega t - 60°)$$
$$V_C = V_m \sin(\omega t - 120°)$$
$$V_D = V_m \sin(\omega t - 180°)$$
$$V_E = V_m \sin(\omega t - 240°)$$
$$V_F = V_m \sin(\omega t - 300°)$$

Fig. 10. Symmetrical six-phase voltages.

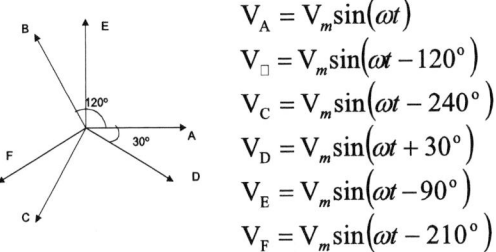

$$V_A = V_m \sin(\omega t)$$
$$V_\square = V_m \sin(\omega t - 120°)$$
$$V_C = V_m \sin(\omega t - 240°)$$
$$V_D = V_m \sin(\omega t + 30°)$$
$$V_E = V_m \sin(\omega t - 90°)$$
$$V_F = V_m \sin(\omega t - 210°)$$

Fig. 11. Asymmetrical six phase voltages.

Six-phase inverter is an inverter, which gives six phase supply from a dc source. Increase in number of phases increases the reliability of operation. With the increase in number of phases the number of devices used increases thus for the same rating of the device high power output can be obtained. Fig.12 shows the circuit topology of a six phase six leg voltage source inverter.

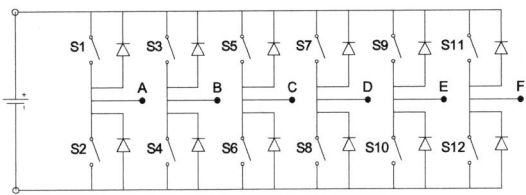

Fig. 12. Six-phase inverter circuit topology.

### A. Symmetric Type Output

#### 1) Definition of switching vectors

The switch state can be either ON or OFF, let us assume "1" represents ON-state and "0" represents OFF-state. To reduce complexity switches in the same leg are complimentarily commutated i.e. if S1 is ON then S2 is OFF and vice versa. Therefore the possible options will be S1, S3, S5, S7, S9, and S11. There are six possible options thus the number of switching vectors will be $2^6 = 64$. The sixty-four switching vectors with their corresponding switch positions are

325

shown in Table VIII.

### TABLE VII
#### SWITCHING VECTORS OF A SIX-PHASE INVERTER

| S.no | S1 | S3 | S5 | S7 | S9 | S11 | Vector |
|---|---|---|---|---|---|---|---|
| 1 | 0 | 0 | 0 | 0 | 0 | 0 | $V^0$ |
| 2 | 0 | 0 | 0 | 0 | 0 | 1 | $V^1$ |
| 3 | 0 | 0 | 0 | 0 | 1 | 0 | $V^2$ |
| 4 | 0 | 0 | 0 | 0 | 1 | 1 | $V^3$ |
| 5 | 0 | 0 | 0 | 1 | 0 | 0 | $V^4$ |
| 6 | 0 | 0 | 0 | 1 | 0 | 1 | $V^5$ |
| 7 | 0 | 0 | 0 | 1 | 1 | 0 | $V^6$ |
| 8 | 0 | 0 | 0 | 1 | 1 | 1 | $V^7$ |
| 9 | 0 | 0 | 1 | 0 | 0 | 0 | $V^8$ |
| 10 | 0 | 0 | 1 | 0 | 0 | 1 | $V^9$ |
| 11 | 0 | 0 | 1 | 0 | 1 | 0 | $V^{10}$ |
| 12 | 0 | 0 | 1 | 0 | 1 | 1 | $V^{11}$ |
| 13 | 0 | 0 | 1 | 1 | 0 | 0 | $V^{12}$ |
| 14 | 0 | 0 | 1 | 1 | 0 | 1 | $V^{13}$ |
| 15 | 0 | 0 | 1 | 1 | 1 | 0 | $V^{14}$ |
| 16 | 0 | 0 | 1 | 1 | 1 | 1 | $V^{15}$ |
| 17 | 0 | 1 | 0 | 0 | 0 | 0 | $V^{16}$ |
| 18 | 0 | 1 | 0 | 0 | 0 | 1 | $V^{17}$ |
| 19 | 0 | 1 | 0 | 0 | 1 | 0 | $V^{18}$ |
| 20 | 0 | 1 | 0 | 0 | 1 | 1 | $V^{19}$ |
| 21 | 0 | 1 | 0 | 1 | 0 | 0 | $V^{20}$ |
| 22 | 0 | 1 | 0 | 1 | 0 | 1 | $V^{21}$ |
| 23 | 0 | 1 | 0 | 1 | 1 | 0 | $V^{22}$ |
| 24 | 0 | 1 | 0 | 1 | 1 | 1 | $V^{23}$ |
| 25 | 0 | 1 | 1 | 0 | 0 | 0 | $V^{24}$ |
| 26 | 0 | 1 | 1 | 0 | 0 | 1 | $V^{25}$ |
| 27 | 0 | 1 | 1 | 0 | 1 | 0 | $V^{26}$ |
| 28 | 0 | 1 | 1 | 0 | 1 | 1 | $V^{27}$ |
| 29 | 0 | 1 | 1 | 1 | 0 | 0 | $V^{28}$ |
| 30 | 0 | 1 | 1 | 1 | 0 | 1 | $V^{29}$ |
| 31 | 0 | 1 | 1 | 1 | 1 | 0 | $V^{30}$ |
| 32 | 0 | 1 | 1 | 1 | 1 | 1 | $V^{31}$ |
| 33 | 1 | 0 | 0 | 0 | 0 | 0 | $V^{32}$ |
| 34 | 1 | 0 | 0 | 0 | 0 | 1 | $V^{33}$ |
| 35 | 1 | 0 | 0 | 0 | 1 | 0 | $V^{34}$ |
| 36 | 1 | 0 | 0 | 0 | 1 | 1 | $V^{35}$ |
| 37 | 1 | 0 | 0 | 1 | 0 | 0 | $V^{36}$ |
| 38 | 1 | 0 | 0 | 1 | 0 | 1 | $V^{37}$ |
| 39 | 1 | 0 | 0 | 1 | 1 | 0 | $V^{38}$ |
| 40 | 1 | 0 | 0 | 1 | 1 | 1 | $V^{39}$ |
| 41 | 1 | 0 | 1 | 0 | 0 | 0 | $V^{40}$ |
| 42 | 1 | 0 | 1 | 0 | 0 | 1 | $V^{41}$ |
| 43 | 1 | 0 | 1 | 0 | 1 | 0 | $V^{42}$ |
| 44 | 1 | 0 | 1 | 0 | 1 | 1 | $V^{43}$ |
| 45 | 1 | 0 | 1 | 1 | 0 | 0 | $V^{44}$ |
| 46 | 1 | 0 | 1 | 1 | 0 | 1 | $V^{45}$ |
| 47 | 1 | 0 | 1 | 1 | 1 | 0 | $V^{46}$ |
| 48 | 1 | 0 | 1 | 1 | 1 | 1 | $V^{47}$ |
| 49 | 1 | 1 | 0 | 0 | 0 | 0 | $V^{48}$ |
| 50 | 1 | 1 | 0 | 0 | 0 | 1 | $V^{49}$ |
| 51 | 1 | 1 | 0 | 0 | 1 | 0 | $V^{50}$ |
| 52 | 1 | 1 | 0 | 0 | 1 | 1 | $V^{51}$ |
| 53 | 1 | 1 | 0 | 1 | 0 | 0 | $V^{52}$ |
| 54 | 1 | 1 | 0 | 1 | 0 | 1 | $V^{53}$ |
| 55 | 1 | 1 | 0 | 1 | 1 | 0 | $V^{54}$ |
| 56 | 1 | 1 | 0 | 1 | 1 | 1 | $V^{55}$ |
| 57 | 1 | 1 | 1 | 0 | 0 | 0 | $V^{56}$ |
| 58 | 1 | 1 | 1 | 0 | 0 | 1 | $V^{57}$ |
| 59 | 1 | 1 | 1 | 0 | 1 | 0 | $V^{58}$ |
| 60 | 1 | 1 | 1 | 0 | 1 | 1 | $V^{59}$ |
| 61 | 1 | 1 | 1 | 1 | 0 | 0 | $V^{60}$ |
| 62 | 1 | 1 | 1 | 1 | 0 | 1 | $V^{61}$ |
| 63 | 1 | 1 | 1 | 1 | 1 | 0 | $V^{62}$ |
| 64 | 1 | 1 | 1 | 1 | 1 | 1 | $V^{63}$ |

The distribution of voltage vectors in two-dimensional space can be obtained by using six phase to dq0 transformation. The transformation matrix is shown in equation.

$$T_{dq} = \begin{bmatrix} 1 & \frac{1}{2} & \frac{-1}{2} & -1 & \frac{-1}{2} & \frac{1}{2} \\ 0 & \frac{\sqrt{3}}{2} & \frac{\sqrt{3}}{2} & 0 & \frac{-\sqrt{3}}{2} & \frac{-\sqrt{3}}{2} \end{bmatrix} \quad (18)$$

The distribution of vectors in $dq$ plane is shown below. The other vectors, which produce zero output, are $V^9, V^{18}, V^{21}, V^{27}, V^{36}, V^{42}, V^{45}, V^{54}$.

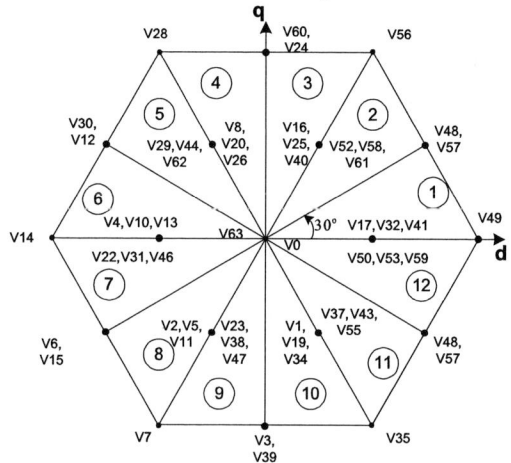

Fig. 13. Vector distribution in voltage space.

#### 2) Separation Planes

As shown the output voltage space can be divided in to twelve sectors. The equations (19) show the six separation planes

$$u_\beta = 0, \quad u_\alpha - \sqrt{3}u_\beta = 0, \quad \sqrt{3}u_\alpha - u_\beta = 0, \quad u_\alpha = 0,$$
$$\sqrt{3}u_\alpha + u_\beta = 0, \quad u_\alpha + \sqrt{3}u_\beta = 0. \quad (19)$$

#### 3) Boundary Planes
The boundary planes for different sectors are shown in Table 9

### TABLE IX
. Boundary planes of different sectors

| Sectors | Boundary Planes |
|---|---|
| 1 and 2 | $\sqrt{3}u_\alpha + u_\beta - 2\sqrt{3} = 0$ |
| 3 and 4 | $u_\beta - \sqrt{3} = 0$ |
| 5 and 6 | $\sqrt{3}u_\alpha - u_\beta + 2\sqrt{3} = 0$ |
| 7 and 8 | $\sqrt{3}u_\alpha + u_\beta + 2\sqrt{3} = 0$ |
| 9 and 10 | $u_\beta + \sqrt{3} = 0$ |
| 11 and 12 | $\sqrt{3}u_\alpha - u_\beta - 2\sqrt{3} = 0$ |

### 4) Decomposition Matrices

Let us assume that the command vector is in first quadrant and the implemented switching sequence as $V^0, V^{32}, V^{48}, V^{49}, V^{57}, V^{59}, V^{63}$. The output voltage state is same for $V^0$ and $V^{63}$ let the time for this state be $t_0$. The output voltage is same for $V^{32}$ and $V^{59}$ let the time for this state be $t_2$. The output voltage is same for $V^{48}$ and $V^{57}$ let the time for this state be $t_3$. Therefore the average output voltage produced by these vectors must be equal to the average voltage produced by the command vector for total time.

$$V^{49}t_1 + V^{59}t_2 + V^{48}t_3 + V^0 t_0 = u_{cmd}T_s \qquad (20)$$

We know that $V^{59} = \dfrac{V^{49}}{2}$ there fore

$$V^{49}\left(t_1 + \frac{t_2}{2}\right) + V^{48}t_3 + V^0 t_0 = u_{cmd}T_s \qquad (21)$$

Let $t_1^{'} = t_1 + \dfrac{t_2}{2}$

$$V^{49}t_1^{'} + V^{48}t_3 + V^0 t_0 = u_{cmd}T_s \qquad (22)$$

But

$$t_1^{'} + t_3 + t_0 = T_s$$

Adding these two equations and writing in matrix form.

$$\begin{bmatrix} V^{49} & V^{48} & V^0 \\ 1 & 1 & 1 \end{bmatrix}\begin{bmatrix} t_1^{'} \\ t_3 \\ t_0 \end{bmatrix} = u_{cmd}T_s \qquad (23)$$

$$\begin{bmatrix} t_1^{'} \\ t_3 \\ t_0 \end{bmatrix} = \begin{bmatrix} V^{49} & V^{48} & V^0 \\ 1 & 1 & 1 \end{bmatrix}^{-1} u_{cmd}T_s \qquad (24)$$

This Equation can be written as

$$\begin{bmatrix} t_1^{'} \\ t_3 \\ t_0 \end{bmatrix} = M_1 T_s$$

Where

$$M_1 = \begin{bmatrix} V^{49} & V^{48} & V^0 \\ 1 & 1 & 1 \end{bmatrix}^{-1} u_{cmd} \qquad (25)$$

$M_1$ is called the decomposition matrix. Like this decomposition matrices are calculated for different sectors. For calculation of $t_1$ and $t_2$ different methods can be used. By linking these times with modulation index mi the output will

good for all modulation indexes.

$$t_1 + \frac{t_2}{2} = mi$$

### 5) Switching Sequences

Many switching sequences can be obtained resulting in different total harmonic distortion values and switching frequencies. Table 10 shows the switching sequence with low THD. The switching frequency of one leg can be reduced by using $V^0$ and $V^{63}$ in alternate sectors

TABLE X
SWITCHING SEQUENCE FOR DIFFERENT SECTORS

| Sector | Switching Sequence |
|--------|--------------------|
| S1 | $V^0, V^{32}, V^{48}, V^{49}, V^{57}, V^{59}, V^{63}$ |
| S2 | $V^0, V^{16}, V^{48}, V^{56}, V^{57}, V^{61}, V^{63}$ |
| S3 | $V^0, V^{16}, V^{24}, V^{56}, V^{60}, V^{61}, V^{63}$ |
| S4 | $V^0, V^8, V^{24}, V^{28}, V^{60}, V^{62}, V^{63}$ |
| S5 | $V^0, V^8, V^{12}, V^{28}, V^{30}, V^{62}, V^{63}$ |
| S6 | $V^0, V^4, V^{12}, V^{14}, V^{30}, V^{31}, V^{63}$ |
| S7 | $V^0, V^4, V^6, V^{14}, V^{15}, V^{31}, V^{63}$ |
| S8 | $V^0, V^2, V^6, V^7, V^{15}, V^{47}, V^{63}$ |
| S9 | $V^0, V^2, V^3, V^7, V^{39}, V^{47}, V^{63}$ |
| S10 | $V^0, V^1, V^3, V^{35}, V^{39}, V^{55}, V^{63}$ |
| S11 | $V^0, V^1, V^{33}, V^{35}, V^{51}, V^{55}, V^{63}$ |
| S12 | $V^0, V^{32}, V^{33}, V^{49}, V^{51}, V^{59}, V^{63}$ |

The Output waveforms of phase voltage and line voltage, taken for a symmetric R load, and their corresponding Fourier spectrum are shown in Fig. 14 and Fig 15.

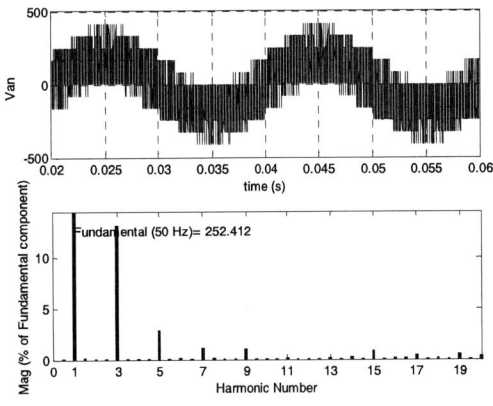

Fig. 14. Output phase voltage and corresponding Fourier spectrum.

327

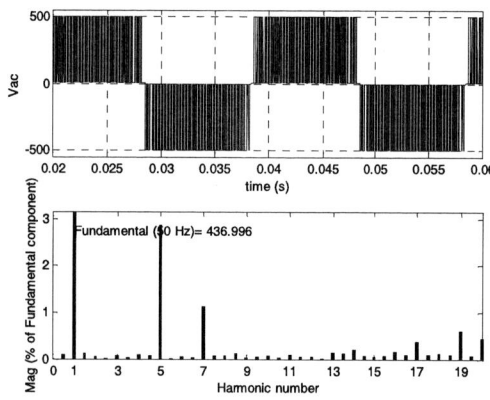

Fig. 15. Output line voltage and corresponding Fourier spectrum.

## B. Asymmetric Type Output

### 1) Identification of Switching Vectors

The number of switching vectors is same as that of symmetric type output. But the space vector distribution will be obtained from a different transformation matrix provided in equation (26). Thus the voltage space distribution will be different and it is as shown in the Fig. 16.

$$T_{dq} = \begin{bmatrix} 1 & -\frac{1}{2} & -\frac{1}{2} & \frac{\sqrt{3}}{2} & 0 & -\frac{\sqrt{3}}{2} \\ 0 & \frac{\sqrt{3}}{2} & -\frac{\sqrt{3}}{2} & -\frac{1}{2} & 1 & -\frac{1}{2} \end{bmatrix} \quad (26)$$

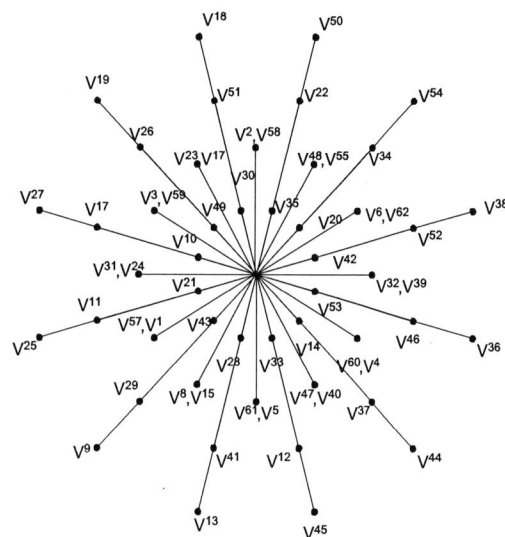

Fig. 16. Vector distribution in voltage space.

Considering all the vectors will make the calculations complex. Considering only the outermost vectors the voltage space can be divided into twelve sectors as shown in Fig. 17.

### 2) Identification of Separation Planes

The six separation planes are shown in equation (27).

$$(2+\sqrt{3})u_\alpha - u_\beta = 0,$$

$$u_\alpha - u_\beta = 0, u_\alpha - (2+\sqrt{3})u_\beta = 0,$$

$$u_\alpha + (2+\sqrt{3})u_\beta = 0,$$

$$u_\alpha + u_\beta = 0, (2+\sqrt{3})u_\alpha + u_\beta = 0 \quad (27)$$

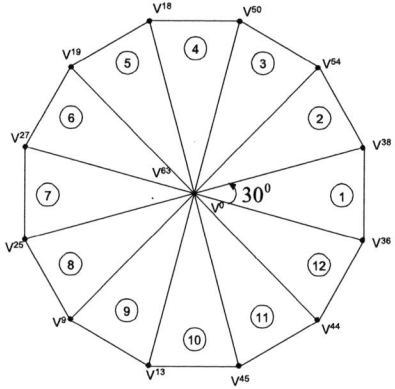

Fig. 17. Sector distribution of voltage space.

### 3) Identification of Boundary Planes

The boundary planes of the twelve sectors are shown in Table 11.

TABLE XI
BOUNDARY PLANES FOR DIFFERENT SECTORS

| Sectors | Boundary Planes |
|---------|-----------------|
| 1 | $2u_\alpha = 2+\sqrt{3}$ |
| 2 | $\sqrt{3}u_\alpha + u_\beta = 2+\sqrt{3}$ |
| 3 | $u_\alpha + \sqrt{3}u_\beta = 2+\sqrt{3}$ |
| 4 | $2u_\beta = 2+\sqrt{3}$ |
| 5 | $u_\alpha - \sqrt{3}u_\beta + 2+\sqrt{3} = 0$ |
| 6 | $\sqrt{3}u_\alpha - u_\beta + 2+\sqrt{3} = 0$ |
| 7 | $2u_\alpha + 2+\sqrt{3} = 0$ |
| 8 | $\sqrt{3}u_\alpha + u_\beta - 2+\sqrt{3} = 0$ |
| 9 | $u_\alpha + \sqrt{3}u_\beta + 2+\sqrt{3} = 0$ |
| 10 | $2u_\beta + 2+\sqrt{3} = 0$ |
| 11 | $u_\alpha - \sqrt{3}u_\beta = 2+\sqrt{3}$ |
| 12 | $\sqrt{3}u_\alpha - u_\beta = 2+\sqrt{3}$ |

### 4) Obtaining the Decomposition Matrices

Assuming that the command vector is in first quadrant and the implemented switching sequence as

$V^0, V^{36}, V^{38}, V^{63}$. The average output voltage produced by these vectors must be equal to the average voltage produced by the command vector for total sample time $T_s$.

$$V^{36}t_1 + V^{38}t_2 + V^0 t_0 = u_{cmd}T_s \quad (28)$$

But we know that the total time must be equal to the sampling time.

$$t_1 + t_2 + t_0 = T_s$$

Adding these two equations and writing in matrix form.

$$\begin{bmatrix} V^{36} & V^{38} & V^0 \\ 1 & 1 & 1 \end{bmatrix} \begin{bmatrix} t_1 \\ t_2 \\ t_0 \end{bmatrix} = u_{cmd} T_s \qquad (29)$$

$$\begin{bmatrix} t_1 \\ t_2 \\ t_0 \end{bmatrix} = \begin{bmatrix} V^{36} & V^{38} & V^0 \\ 1 & 1 & 1 \end{bmatrix}^{-1} u_{cmd} T_s \qquad (30)$$

This Equation can be written as

$$\begin{bmatrix} t_1 \\ t_2 \\ t_0 \end{bmatrix} = M_1 T_s$$

Where

$$M_1 = \begin{bmatrix} V^{36} & V^{38} & V^0 \\ 1 & 1 & 1 \end{bmatrix}^{-1} u_{cmd} \qquad (31)$$

Where $M_1$ is called the decomposition matrix. Like this, decomposition matrices can be calculated for different sectors.

*5) Definition of Switching Sequence*
The switching sequence for each sector is defined as shown in the Table 12.

TABLE XII
SWITCHING SEQUENCE FOR DIFFERENT SECTORS

| Sector | Switching Sequence |
|--------|--------------------|
| S1 | $V^0, V^{36}, V^{38}, V^{63}$ |
| S2 | $V^0, V^{54}, V^{38}, V^{63}$ |
| S3 | $V^0, V^{54}, V^{50}, V^{63}$ |
| S4 | $V^0, V^{18}, V^{50}, V^{63}$ |
| S5 | $V^0, V^{18}, V^{19}, V^{63}$ |
| S6 | $V^0, V^{27}, V^{19}, V^{63}$ |
| S7 | $V^0, V^{27}, V^{25}, V^{63}$ |
| S8 | $V^0, V^9, V^{25}, V^{63}$ |
| S9 | $V^0, V^9, V^{13}, V^{63}$ |
| S10 | $V^0, V^{45}, V^{13}, V^{63}$ |
| S11 | $V^0, V^{45}, V^{44}, V^{63}$ |
| S12 | $V^0, V^{36}, V^{44}, V^{63}$ |

The Output waveforms of phase voltage and line voltage, taken for a symmetric R load, and their corresponding Fourier spectrum are shown in Fig. 18 and 19.

## IV. RESULTS AND DISCUSSION

The output voltage waveforms and their corresponding Fourier spectrums are shown in Fig.4, 5, 8, 9, 14, 15,18 and 19. Table 13 shows the comparison of different inverters in terms of Total Harmonic Distortion (THD) for different modulation indices. The terms used in table are as follows:

MI : Modulation Index
TPI : Three Phase Inverter
TTI : Three level Three Phase Inverter
SPSI: Six Phase Symmetrical Inverter
SPAI: Six Phase Asymmetrical Inverter

TABLE XIII
. COMPARISON OF DIFFERENT INVERTERS BY THD.

| MI | TPI | TTI | SPSI | SPAI |
|------|-------|-------|-------|-------|
| 0.25 | 1.829 | 1.43 | 2.215 | 2.051 |
| 0.5 | 1.1 | 0.816 | 1.311 | 1.095 |
| 0.75 | 0.676 | 0.46 | 0.815 | 0.705 |
| 0.866 | 0.538 | 0.305 | 0.635 | 0.545 |

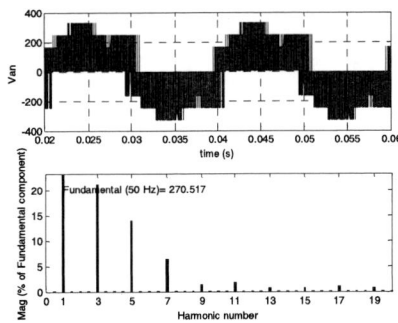

Fig. 18. Output phase voltage and corresponding Fourier spectrum.

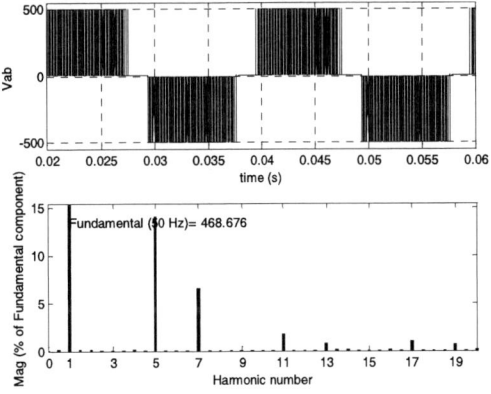

Fig. 19. Output line voltage and corresponding Fourier spectrum.

## V. CONCLUSION

It is evident from the Table 13 that three level three phase inverter based topology has minimum THD at all modulation index and therefore TTI based six phase inverter with a 3-phase/6-phase transformer is better solution for a six phase inverter source rather than a six phase inverter.

## VI. REFERENCES

[1] V.T. Ranganathan, "Space Vector Pulse width Modulation – A Status Review", *Sadhana*, V.22, pp.675-688, Dec. 1997.

[2] H. Pinheiro, F. Dotteron, C. Rech, L. Schuch, R. F. Camargo, H. L. Hey, H. A. Grundling and J. R. Pinheiro, "Space vector modulation for voltage source inverters: A unified approach", *28th Annual Conference of the Industrial Electronics Society*, Volume: 1 , 5-8 Nov. 2002

[3] Jae Hyeong Seo, Chang Ho Choi, Dong Seok Hyun, "A New Simplified Space-Vector PWM Method for Three-Level Inverters", IEEE Transactions on power electronics, vol 16, No. 4, July 2001.

[4] M.□.R. Correa, C.□. Jacobina, C.R. Da Silva, A.M.N. Lima and E.R.C. Da Silva, "Vector and Scalar Modulation for Six-Phase Voltage Source Inverters", *34th Annual Conference on Power Electronics Specialist*, Volume: 2 , 15-19 June 2003

[5] K. Gopakumar, V.T. Ranganathan, and S.R. □hat, "Split-phase induction motor operation from pwm voltage source inverter", *IEEE Trans. Ind. Applicat.*, 29(5):927-932, Sept./Oct. 1993.

[6] Y. Zhao and T.A. Lipo, "Space vector pwm control of dual three-phase induction machine using vector space decomposition", *IEEE Trans. Ind. Applicat*, 31(5):110G1109, Set./Oct.1995.

[7] Ned Mohan Tore and M. Undeland, "Power Electronics: Converters, Applications and Design," *John Wiley & Sons, inc.*

## VII. □IOGRAPHIES

**Ravindra Kumar Singh** was born on December 30, 1965 in Chapra, □ihar, India. He graduated from the □hagalpur University, □ihar. He did his post graduation from IT, □HU and Ph. D. fro IIT Kanpur, India

He is presently serving in M.N.N.I.T. Allahabad, India. He is also Fellow of Institution of Engineers (India).

**2006 IEEE International Conference on Power Electronic, Drives and Energy Systems**

# A PWM Current Source Rectifier with Leading Power Factor

B. Geethalakshmi, P. Sanjeevikumar, and P. Dananjayan

*Abstract--* An unidirectional three-phase AC/DC converter is proposed to obtain leading power factor, sinusoidal line current and to keep DC-bus voltage constant without DC link capacitor. Two active switches and two power diodes are used in each converter leg which is different from conventional AC/DC and three level converters. The scheme is adopted with simple commutation procedure to reduce the complexity of control to ensure effective power transformation. Conventional AC/DC converter and three level converters usually have large value of capacitor at the dc side with neutral balancing which will increase the structural complexity. This paper encloses a converter topology with simple commutation procedure to have good voltage transfer ratio capacity, leading power factor and pure sine wave with only higher order harmonics in the line side. By providing small LC filter on the line side even this harmonics can be eliminated. Switches in each leg of same phase will turn on and turn off at zero current. Theoretical analyses and simulation results are provided to verify its performance.

*Keywords--*AC/DC converter, Current Source Rectifier, Triangular-Ramp PWM, three level converters.

## I. INTRODUCTION

DIODE or phase-controlled rectifiers are widely used in the frond-end converter for the uncontrollable or controllable DC-bus voltage in industrial and commercial applications. However, low power factor and non-sinusoidal line currents are drawn from the AC source owing to a large electrolytic capacitor used on the DC link. Power pollutants such as reactive power and current harmonics result in line-voltage distortion, heating of the transformer core and electrical machines, and increased losses in the transmission and distribution line. To meet the relevant standards in Europe and America, several current wave-shaping solution [1-4] have been proposed to achieve power factor correction and current-harmonic reduction. In [1] conventional single phase rectifiers with one, two or four switches were used to

achieve power factor correction based on two-level (unipolar or bipolar) pulse-width modulation (PWM).

In [2] the single-phase voltage-doubler boost rectifier with one, two, three or four switches was used to achieve power factor correction and DC-bus voltage regulation. Hence the DC-bus voltage is twice the peak main voltage. Switched mode rectifiers with three or four rectifier legs can achieve high power factor and low current harmonics in three-phase three-wire or four-wire systems. Six or eight power switches are used in the three-leg or four leg-converters [5-9] to generate bipolar PWM waveforms on the AC terminal. If the bidirectional power flow is not necessary in the application system, switched-mode rectifiers are not a good choice for the large number of power switches. Multilevel rectifier and inverters have been proposed for high power and medium-voltage applications because they provide advantages such as the low voltage harmonics. However disadvantages of the multilevel rectifiers are the large number of power semiconductors in the circuit, a complex control scheme and the neutral-point voltage balance problem.

In industrial application with a unidirectional power flow, conventional multilevel converters are too expensive and complicated to implement. Moreover classical proportional-integral voltage controller and the hystersis current controller scheme are to be incorporated under closed loop with neutral balancing to achieve fixed DC voltage at the output. However open loop controls of this converter are not effective. For real time implementation these conventional converters are very difficult to implement and not economical for high power ratings. Hence they are not found much application in industry due to its complex control problem.

In this paper, an AC/DC converter topology is analyzed without DC link capacitor which is shown Fig. 1. The performance of this AC/DC converter is superior to conventional AC/DC and Multilevel converters and has the following advantages:

- It performs similar to conventional AC/DC converter such as good voltage transfer ratio capacity, pure sine waveforms with only higher order harmonics in input current.

- No large energy storage elements like inductor or capacitor except a relatively small size ac filter at the input side are required making this filter more easily to be integrated into a system package.

---

P. Sanjeevi Kumar, is with EEE Department in IFET College of Engg., Tamilnadu, India (e-mail: sanjeevi_12@yahoo.co.in).

B. Geethalakshmi, is with EEE Department in Pondicherry Engg., Pondicherry, India (e-mail: geethalakshmi_pec@yahoo.co.in).

P. Dananjayan, is with ECE Department in Pondicherry Engg., Pondicherry, India (e-mail: pdananjayan@hotmail.com).

This project is supported by the Management of IFET college of Engg. for utilizing the sources with in the campus.

0-7803-9771-1/06/$25.00 ©2006 IEEE

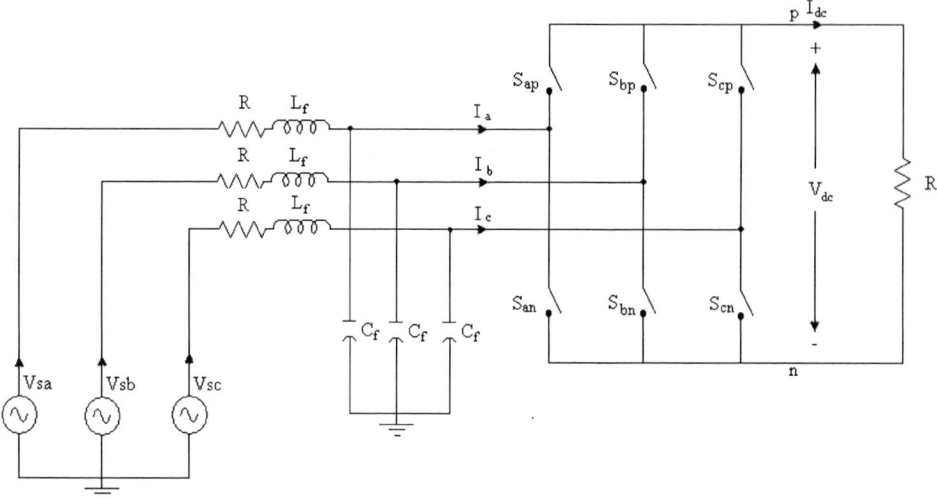

Fig. 1. Basic Topology of Proposed AC-DC Converter.

- Moreover leading power factor can be achieved which makes this AC/DC power converter superior to other topologies.

In this paper, the basic operation of this topology is first described. A PWM algorithm is developed on the rectifier side to ensure proper operation and to improve power factor on the line side. This PWM switching scheme on the rectifier side maintains a fixed dc voltage on the DC link and leading power factor on the input line side. Finally system level simulation study is conducted using MATLAB/SIMULINK to verify its sinusoidal input and output voltage performance.

## II. PROPOSED TOPOLOGY

Fig.1. illustrates the structural view of the proposed converter which is similar to a conventional AC/DC converter but without DC link capacitor. The line side converter is a rectifier similar to a traditional rectifier with two unidirectional switches and two power diodes in each leg of the converter. Here the rectifier is assumed to be a current source rectifier, which is also different from conventional AC/DC converter where line side converter is a voltage source rectifier. The objective of the proposed converter on the line side rectifier is to maintain a fixed DC voltage on the DC link without capacitor and improve the power factor on the input supply side. Now the capacitor in the DC link is replaced by an AC filter on the line side with a much small value to reduce higher order harmonics. Since converter has no large energy storage elements like a capacitor or an inductor it can be designed for higher capacity utilization.

Conventional AC/DC power converters are normally equipped with an electrolytic capacitor in their DC link which has a short life compared with an AC capacitor (metalised polyester film, etc.) [12] and also they occupy a considerably greater space. This supports the proposed topology for neglecting the DC link capacitor.

For the purpose of analysis purpose switching frequency on the rectifier side is assumed to be far greater than the fundamental frequency of input voltage source. The filter components C, L and R are assumed to be zero.

i.e. $L_s = 0$; $C_f = 0$; $V_{sx} = V_x$; $I_{sx} = I_x$ where, $V_x$ is the line side voltage at converter side in phase x.

x = a, b, c phase.

$I_x$ is the phase current at converter side,

$L_s$ is the line side filter inductance,

$C_f$ is the line side filter capacitance

The input side voltages are assumed as:

$$V_{sa} = V_m \cos\theta_{av} = V_m \cos(\omega_i t) \tag{1}$$

$$V_{sb} = V_m \cos\theta_{bv} = V_m \cos(\omega_i t - \frac{2\pi}{3}) \tag{2}$$

$$V_{sc} = V_m \cos\theta_{cv} = V_m \cos(\omega_i t + \frac{2\pi}{3}) \tag{3}$$

$\omega_i$ are the input and output angular frequencies respectively,

$\psi_{in} = \theta_{av} - \theta_a$ is the line side power factor angle,

$V_m$ and $I_0$ is the amplitude of the input phase voltage.

Moreover, the expected line side currents are described as

$$I_a = I_m \cos\theta_a = I_m \cos(\omega_i t - \psi_{in}) \tag{4}$$

$$I_b = I_m \cos\theta_b = I_m \cos(\omega_i t - \frac{2\pi}{3} - \psi_{in}) \tag{5}$$

$$I_c = I_m \cos\theta_c = I_m \cos(\omega_i t + \frac{2\pi}{3} - \psi_{in}) \tag{6}$$

DC side voltage is essentially decided by the switching function of the rectifier and the input voltage. Analyzing the six interval of a 3Φ sinusoidal voltage of a cycle, during each interval one of the line or phase voltages will have the maximum absolute value as shown in Fig.2. For example during the interval 1, $V_{sa}$ has the largest absolute voltage and during the interval 2, $V_{sc}$ has the largest absolute voltage and so forth in all the intervals of the three phase sinusoidal voltage of a cycle.

Each of the switching cycle is split into two portions. In one portion, the switching state of line side switch is fixed and the DC side voltage $V_{dc}$ equals to one of the two highest positive line voltages.

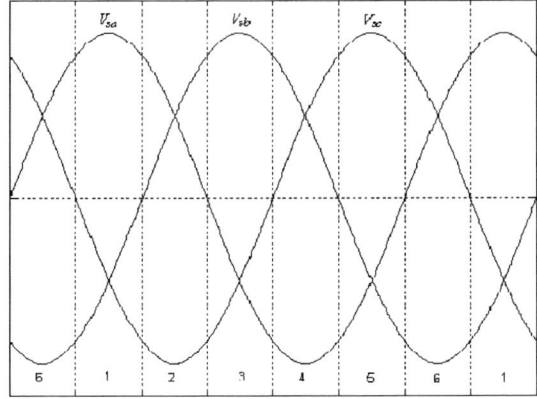

Fig. 2. Six intervals of a switching cycle.

PWM sequence on the rectifier side is developed by splitting each interval of conduction period into two portions, in which the largest absolute voltage corresponding to a particular switch will conduct for both the portions and one of the switches of opposite arm, adjacent to the leg will conduct in one portion and other switch will conduct in the second portion. For example during interval 1, $V_{sa}$ has the largest absolute voltage, with the corresponding line voltages $V_{sa}$-$V_{sb}$ and $V_{sa}$-$V_{sc}$. The line side switching state in each portion can be determined by the following sequence:

In portion 1, for the first 30° conduction period $S_{ap}$, $S_{bn}$ remain turned on while all other line side switches are turned off. The DC side voltage $V_{dc}$ is equal to $V_{sa}$-$V_{sb}$. In portion 2, for the next 30° conduction period $S_{ap}$, $S_{cn}$ remain turned on and all other line side switches are turned off. The DC side voltage $V_{dc}$ is equal to $V_{sa}$-$V_{sc}$ for 30° conduction period. The above sequence is applicable for all other intervals. By providing this switching sequence, DC voltage at the DC link can be maintained with a fixed value. To overcome commutation problem the rectifier switches available on the same leg are provided with sufficient time to turn on and turn off for achieving zero current turn on and turn off condition. This greatly simplifies commutation problem that are associated with conventional AC/DC converter.

## III. SIMULATION RESULTS

The proposed AC-DC converter is analyzed using MATLAB/SIMULINK software for its performance and all the switches utilized are ideal switches. Simulation parameters taken for analysis are:

Input line Voltage = 220V (rms)
Input Frequency = 60Hz
Modulation Index = 0.8
Carrier Frequency =12 KHz
Filter Inductance = 200μH
Filter Capacitance = 800μF
Load Resistance = 45 ohm
Input filter resistor = 0.2 ohm

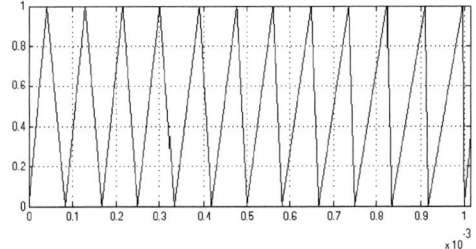

Fig. 3. PWM Carrier for Rectifier.

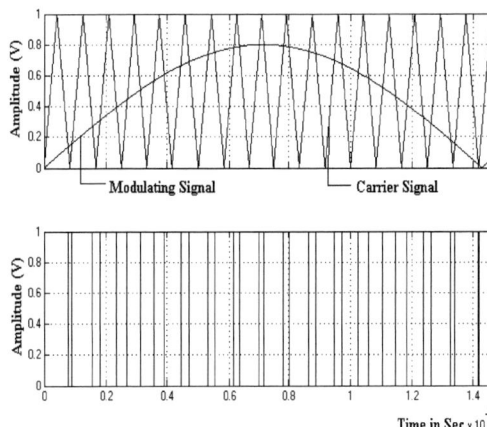

Fig. 4. PWM scheme for Rectifier.

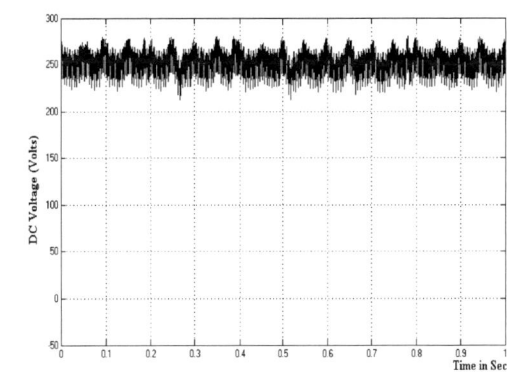

Fig. 5. Fixed DC side voltage $V_{dc}$.

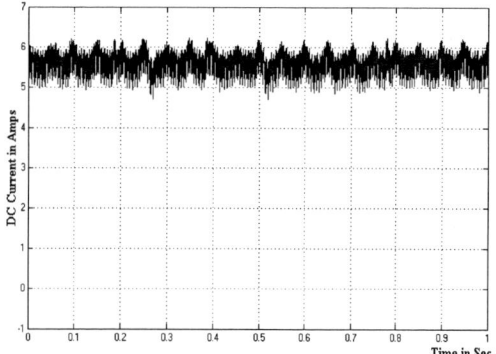

Fig. 6. Fixed DC side current $I_{dc}$.

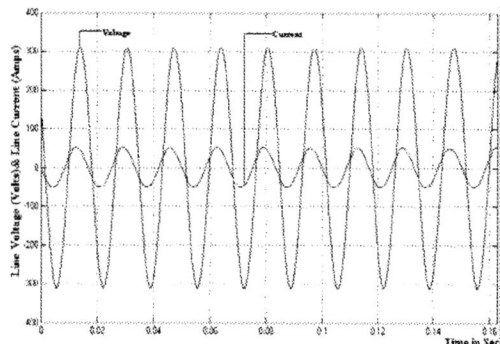

Fig. 7. Line Voltage & Current (Leading Power Factor).

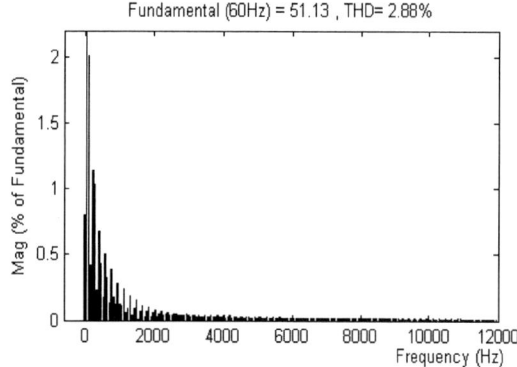

Fig. 8. THD of line current at switching frequency of 12KHz.

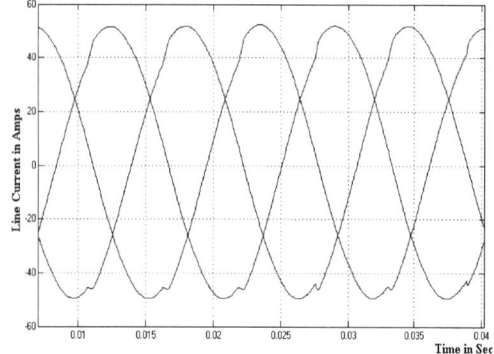

Fig. 9. Three Phase Line current.

Figs. 3 - 4 PWM signal applied to the input converter (rectifier) using sinusoidal modulation technique is given where the carrier is initially triangular in nature, which slowly changes into ramp. With this type of carrier line side THD is significantly reduced. Figs. 5 and 6 depict the DC voltage $V_{dc}$ and current $I_{dc}$.

Fig. 7 shows the waveform of line voltage and line current at the input side providing leading power factor. This result shows that at line side there are no lower order harmonics and are suppressed by the filter.

Fig. 8 depicts that total harmonics distortion is about 2.88%. This ensures feasibility of the performance of the converter under higher switching frequency.

Fig. 9 shows line current is essentially sinusoidal providing leading power factor.

## IV. CONCLUSIONS

This paper presents AC-DC power converter with simplified commutation strategy which reduces complexity of control when compared with the conventional converter. The input current can be pure sine waveforms with only harmonics around or above the switching frequency. The proposed converter provides leading power factor at the line side. DC link capacitor is eliminated hence large capacity; compact converter system can be designed. It has a good voltage transfer ratio of $V_{dc}=3V_m/2$ from AC to DC than conventional AC/DC converter. THD is reduced on line side and is about 2.88% which provides better performance. The same THD is maintained even for ten times of switching frequency. Proposed AC/DC converter scheme can be used in Dual Converters operating non-circulating current mode, traction and various drives applications.

## V. REFERENCES

[1] Salmon. J.C, "Techniques for minimizing the input current distortion of current-controlled single-phase boost rectifier", IEEE Trans. Power Electron., 1993. 8, (4), pp. 509-520.

[2] Salmon. J.C, "Circuit topologies for single-phase voltage doubler boost rectifier", IEEE Trans. Power Electron., 1993. 8, (4), pp. 521-529.

[3] Manias. S., "Novel full-bridge semi controlled switch mode rectifier", IEEE Trans. Power Electron., B, 1991. 1`38, (5), pp. 252-256.

[4] Martinez. R. and Enjeti. P.N.: "A high-performance single phase rectifier with input power factor correction", IEEE Trans. Power Electron., 1996, 11, (2), pp. 311-317.

[5] Wong, c., Mohan. N., and He, J.: "Adaptive phase controls for three phase PWM AC to DC converters with constant switching frequency". Proc. Conf. PCC-Yokohama, Yokohama. Japan. 1993, pp. 73-78.

[6] Kwon, B.H., and Min. B.D., "A fully software-controlled PWM rectifier with current link", IEEE Trans. Ind. Electron., 1993, 40, (3), pp. 353-363.

[7] Dawarde. M.S., Kanetkar. V.R and Dubey. G, "Three phase switch mode rectifier with hystersis current control", IEEE Trans. Power Electron., 1996, 11, (3), pp. 466-471.

[8] Itoh. R., Ishizaka. K., and Goromaru. T., " Three-Phase voltage source converter with controlled DC for the minimization of filter capacitance", IEEE Proc. B, 1990, 137, (5), pp. 327-333.

[9] Wu. R., Dewan. S.B. and Slemon. G.R., "A PWM AC-DC converter with fixed switching frequency", IEEE Trans. Ind. Appl, 1990, 26, (5), pp. 880-886.

[10] B.R. Lin and T. Y. Yang "Three-Phase high power factor AC/DC converter" IEEE Proc.-Electr. Power Appl. Vol. 152, No. 3, may 2005.

[11] L.Wei, and T.A. Lipo, "A Novel Matrix converter topology with simple commutation", in Proceedings of 36th IEEE Industry applications society conf. vol.3 pp.1749-1754, Chicago, 2001.

[12] Kenichi Limori, Katsuji Shinohara, Kichiro Yamamoto, " A Study of Dead-time of PWM Rectifier of Voltage Source Inverter without DC Link Components and Its operating Characteristics of Induction Motor", IEEE proc. 0-7803-8486-5/04/2004 pp. 1638-1645.

[13] Kenichi Limori, Katsuji Shinohara, Mitsuhiro Muroya, Yoichi Matsushita, " Zero-Switching- Loss PWM Rectifier of Converter without DC-Link Components for Induction Motor Drive" IEEE proc. 0-780307156-9/02/2002 pp. 409-414.

[14] K. Mino, Y. Okuma and K. Kuroki, "Direct-linked-type frequency changer based on DC clamped bilateral switching circuit topology" IEEE Trans. On Industry Applications, vol.34, No.6, 1998, pp.1309 1317.

## VI. BIOGRAPHIES

**P. Sanjeevikumar** received Bachelor of Engineering in Electrical & Electronics from the University of Madras in 2002 and Master of Technology (Electrical Drives & Control) in Pondicherry Engineering College, Pondicherry University in 2006. He is working as a Lecturer in the Department of Electrical & Electronics Engineering in IFET College of Engineering, Tamilnadu, India. He is currently working in the area of alternate topology for matrix converter, soft switching PWM schemes.

**B. GeethaLakshmi** received Bachelor of Engineering (ECE) in 1996 and Master of Engineering (Power Electronics & Drives) in 1999 from Bharathidasan University. She is working as Senior Lecturer in Department of Electrical & Electronics in Pondicherry Engineering College and teaching in the area of Power Electronics and Power Systems. She is currently doing her research work in power electronics application in power systems and also working in the area of power converters such as ac-dc-ac, matrix converter & power factor correction techniques & drives.

**P. Dananjayan** received Bachelor of Engineering in 1982 and Master of Technology in 1984 from the Madras Institute of Technology, Chennai and Ph.D degree from Anna University, Chennai in 1998. He is working as a Professor and Head of the Department of Electronics and Communication Engineering, Pondicherry Engineering College. He has more than 30 publications in National and International Journals. He is currently guiding seven Ph.D students. His areas of interest include power electronics application in power system, ATM Networks, Wireless Communication and Spread spectrum Techniques.

**2006 IEEE International Conference on Power Electronic, Drives and Energy Systems**

# A Novel Harmonic Mitigation Converter for Variable Frequency Drives

Bhim Singh, *Senior Member, IEEE,* and Sanjay Gairola

*Abstract*--In this paper, a novel autotransformer for 56-pulse AC-DC converter configuration is designed, modeled and simulated to feed vector controlled induction motor drive (VCIMD). The proposed autotransformer configuration consists of two paralleled 14-pulse AC-DC converters involving seven-phase shifted uncontrolled diode bridges with pulse doubling circuit. It improves power quality at AC mains and it meets IEEE-519 standard requirements at varying loads.

*Index Terms*--28-pulse, 56-pulse, AC-DC converter, power quality, vector controlled induction motor drive.

## I. INTRODUCTION

THE current harmonics are problems in AC-DC converters because they cause increased losses in the customer and utility components. Induction motors are widely used in industry (70-80% of motors are induction motors) and are also known as main work-horse. The evolution of solid state converters has led to their use in variable frequency drives (VFD) employing in various control techniques such as vector control (VC) and direct-torque-control (DTC) due to their superior performance. These VFD's are generally fed by 6-pulse diode bridge rectifiers which results in injection of harmonic currents into AC mains and does not meet IEEE standard 519 [1] requirements. Such a six-pulse diode bridge fed vector controlled induction motor drive (VCIMD) and the controller is shown in Fig. 1.

Six-pulse diode bridge rectifiers with delta and star connected autotransformers are described in the literature [2]. These diode bridges may be connected in series or parallel to form higher pulse AC-DC converter using six-pulse converters depending on output DC voltage requirement in a particular application. The diode bridges connected in series for high voltage multilevel converter fed AC drives are reported by Wu [3] up to 36-pulse AC-DC converters. The VFDs using different controllers are explained in the literature [4]. The harmonic pollution created by these non-linear loads must be checked to meet the standard regarding limiting harmonics outlined by IEEE-519 [1].

A number of isolated and non-isolated AC-DC converters for improved power quality are reported by Paice [5]. Chen

and Horng [6] have also described a 28-step current wave shaper for harmonic reduction but it has four diodes in the path of load current and needs more interphase reactors. A 38-pulse converter is reported by Johnson et. al [7] for meeting US navy requirement of input THD less than 3%. A fork-connection based 24-pulse AC-DC converter is described by Singh et al [8] for power quality improvement in vector controlled induction motor drives.

The harmonics in input current and output voltage of conventional 12-pulse uncontrolled rectifiers can be reduced by using tuned passive filters. But these filters are bulky and lossy. Moreover, some applications have stringent power quality specifications and it is inevitable to use 24-pulse or higher pulse AC-DC converter system. Therefore, it is recommended that higher pulse AC-DC converter must be used so that AC-DC conversion meets IEEE-519 standard requirements. With this in view, a 56-pulse AC-DC converter is designed using polygon connected auto-transformer.

In this paper, a novel 56-pulse rectifier is fed from novel polygon autotransformer. As the power to the load is transferred at low voltage levels, the use of parallel bridge configuration is justified. The input autotransformer secondary is asymmetric polygon extending two sets of seven-phases for the two diode bridge converters. The design is based on the fact that for a rectifier having prime number of input phases (excluding 3) have rectified output waveform that repeats itself only after one complete cycle of transformation [9]. This is unlike the conventional rectifiers where the cyclical repetition in the rectifier output can occur within one cycle of transformation. Hammond et. al. [9] have described

---

Bhim Singh and Sanjay Gairola are with Department of Electrical Engineering, Indian Institute of Technology, Delhi, Hauz-Khas, New Delhi-110016, India (e-mail: bhimsinghr@gmail.com, sanjaygairola@gmail.com).

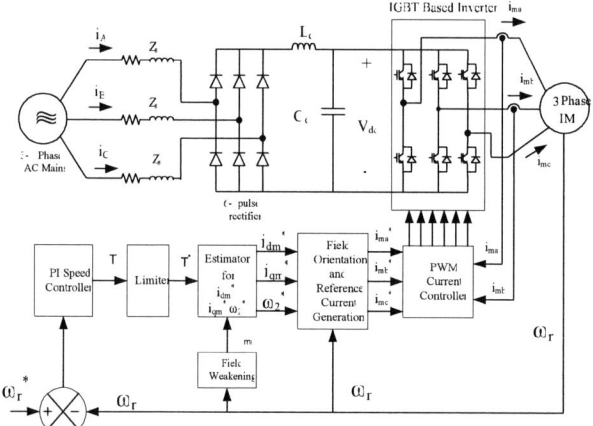

Fig. 1. A six pulse diode bridge fed vector controlled induction motor drive (VCIMD) and the controller.

0-7803-9771-1/06/$25.00 ©2006 IEEE

autowound polygon transformer design for prime number of phases which can feed a single multi-leg diode bridge converters, however, it is stated that the even number of phases are not effective.

The proposed transformer is capable of feeding two diode bridges (each having seven legs) that can be connected in parallel. This parallel connection produces 28-pulse configuration for AC-DC conversion. To further increase the pulse number of the 28-pulse rectifier, the pulse multiplication is achieved using a small rating diode-tapped interphase reactor with two additional diodes. The taps on the interphase transformer are chosen such that 56-pulse AC-DC converter characteristic appears in the input AC line currents.

The pulse multiplication [10] is based on ripple re-injection where the power of the circulating ripple frequency is fed back to the DC system via the interphase reactor, which acts as an autotransformer. The DC voltage ripple acts as the frequency source for the derivation of appropriate voltage and current waveforms capable of modifying AC current and DC voltage to eliminate 28–pulse converter related harmonics. Detailed design of the tapped interphase reactor and resulting 56-pulse diode rectifier system is carried out to study the behaviour of 56-pulse AC-DC converter. The designed converter system is modeled and simulated in MAT☐AB to demonstrate its power quality improvement at AC mains. A prototype is made in the laboratory to validate the design and model.

## II. PROPOSED 56-PU☐SE AC-DC CONVERTER

Fig. 1 shows 6-pulse AC-DC converter systems generally used. In basic 28-puse AC-DC converter configuration an autotransformer is used which generates two sets of 7-phase supply for each bridge. The 28-pulse AC-DC converter system

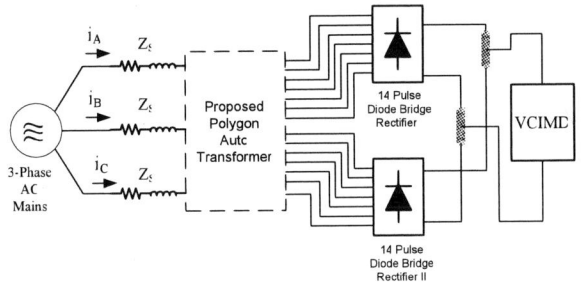

Fig. 2. A 28-pulse AC-DC converter using polygon autotransformer.

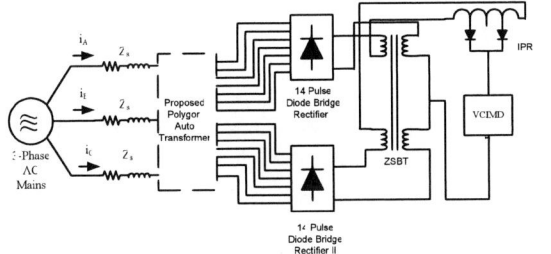

Fig. 3. A 56-pulse AC-DC conversion employing polygon autotransformer, ZSBT and tapped IPR.

is shown in Fig. 2. The 56-pulse AC-DC converter is depicted in Fig. 3, which additionally employs a tapped interphase reactor (IPR) and a zero sequence blocking transformer (ZSBT). The transformer winding arrangement and its connection to two diode bridges is shown in Fig. 4.

### A. Design of auto transformer for 56-pulse AC-DC converter

Fig. 4 shows the schematic of the proposed polygon autotransformer arrangement and its graphical representation depicting angular position of various phasors. The phase angles of various phasors shown are:

$\theta_1 = 6.43°, \theta_2 = \theta_4 = \theta_9 = 38.57°,$
$\theta_3 = \theta_5 = \theta_8 = 12.86°, \theta_6 = 10.71°,$
$\theta_7 = 27.86°.$

Two seven-leg diode bridge converters I and II are connected to two sets of seven-phase output voltages at DB11, DB12, DB13, DB14, DB15, DB16, DB17 and DB21, DB22, DB23, DB24, DB25, DB26, DB27. These two sets are displaced by 12.86° from each other and at -6.43° and +6.43° respectively from input voltage of phase A. The number of turns for every winding is determined as a function of the supply phase voltage, $V_A$. These winding voltages, as marked in Fig. 4b, are expressed by following relationships.

Consider that the input phase voltage is $V_A$ (=$V_{AC}/\sqrt{3}$) and two set of seven phase voltages fed to each bridge be $V_s$ ($V_{A1}, V_{A2}, V_{A3}, V_{A4}, V_{A5}, V_{A6}, V_{A7}$ to the seven leg bridge converter I and $V_{B1}, V_{B2}, V_{B3}, V_{B4}, V_{B5}, V_{B6}, V_{B7}$ to the converter II).

The output voltage phasor $V_S$ is related to input phase voltage $V_A$ as:

$$| V_S | = 0.8419 \ | V_A | \tag{1}$$

Two set of required voltages for the converters I and II are:

$$V_{A1} = V_s \angle 6.43°, V_{A2} = V_s \angle -45°, V_{A3} = V_s \angle -96.43°,$$
$$V_{A4} = V_s \angle -147.86°, V_{A5} = V_s \angle 199.28°, \tag{2}$$
$$V_{A6} = V_s \angle -250.71°, V_{A7} = V_s \angle 302.14°$$

$$V_{B1} = V_s \angle -6.43°, V_{B2} = V_s \angle -57.86°,$$
$$V_{B3} = V_s \angle -109.28°, V_{B4} = V_s \angle -160.71°,$$
$$V_{B5} = V_s \angle 212.14°, V_{B6} = V_s \angle -263.57°, \tag{3}$$
$$V_{B7} = V_s \angle -315°$$

These voltages for the converter I are related as:

$$V_{A1} = V_A + K_1 V_{CA} \tag{4}$$
$$V_{A2} = V_A - (K_1 + K_2) V_{AB} + K_3 V_{BC} \tag{5}$$
$$V_{A3} = V_{A2} + K_5 V_{CA} - K_6 V_{AB} - K_7 V_{BC} - K_8 V_{AB} \tag{6}$$
$$V_{A4} = V_B - K_{12} V_{BC} + K_{13} V_{CA} \tag{7}$$
$$V_{A5} = V_C + (K_{12} + K_{14}) V_{BC} - (K_{13} + K_{15}) V_{AB} \tag{8}$$
$$V_{A6} = V_C + K_{11} V_{CA} + K_{10} V_{AB} \tag{9}$$
$$V_{A7} = V_{A1} + (K_2 + K_4) V_{CA} - (K_3 + K_5) V_{BC} \tag{10}$$

The voltages for the converter II are related as:

$$V_{B1} = V_A - K_1 V_{AB} \tag{11}$$
$$V_{B2} = V_{A2} + K_4 V_{AB} - K_5 V_{BC} \tag{12}$$

337

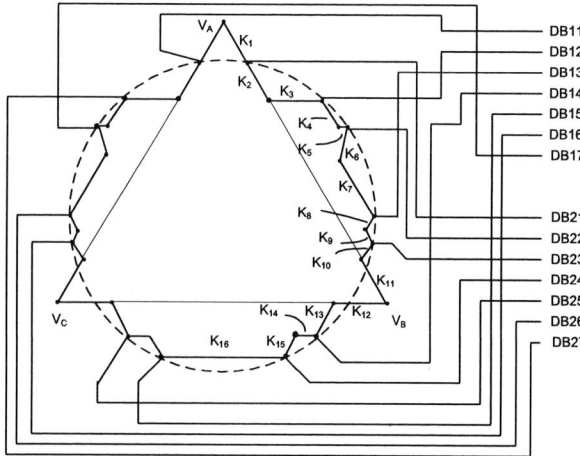

Fig. 4a. The proposed autotransformer winding arrangement.

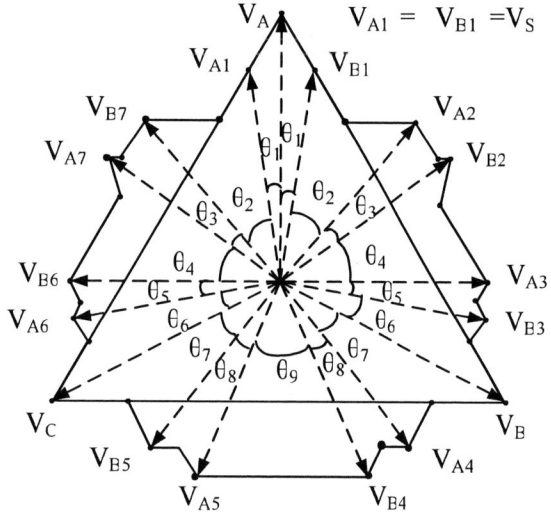

Fig. 4b. Single polygon secondary connection of the input transformer and its phasor representation.

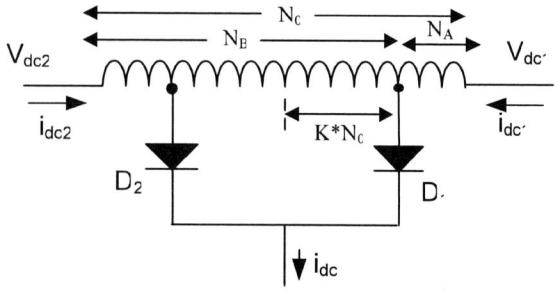

Fig. 4c. Tapped interphase reactor circuit.

$$V_{B3} = V_B + K_{11}V_{AB} - K_{10}V_{CA} \qquad (13)$$

$$V_{B4} = V_{A4} - K_{14}V_{BC} + K_{15}V_{CA} \qquad (14)$$

$$V_{B5} = V_C + K_{12}V_{BC} - K_{13}V_{AB} \qquad (15)$$

$$V_{B6} = V_{A6} + K_9V_{CA} - K_{13}V_{AB} \qquad (16)$$

$$V_{B7} = V_{A1} + K_2V_{CA} - K_3V_{BC} \qquad (17)$$

where,

$$V_{AB} = \sqrt{3}V_A \angle 30^0, \ V_{BC} = \sqrt{3}V_B \angle 30^0, \ V_{CA} = \sqrt{3}V_C \angle 30^0 \qquad (18)$$

Eqns. (4-17) give the values of constants K1 to K16 for desired phase shift as

$$\begin{aligned} &K_1 = 0.2239, \quad K_2 = 0.3309, \quad K_3 = 0.4297, \quad K_4 = 0.2022, \\ &K_5 = 0.0386, \quad K_6 = 0.2248, \quad K_7 = 0.5188, \quad K_8 = 0.1759, \\ &K_9 = 0.0762, \quad K_{10} = 0.0675, \quad K_{11} = 0.2368, \quad K_{12} = 0.3505, \\ &K_{13} = 0.2920, \quad K_{14} = 0.1457, \quad K_{15} = 0.1122, \quad K_{16} = 0.6606. \end{aligned} \qquad (19)$$

The values of these constants $K_1$ to $K_{16}$ determine the winding turns as a fraction of input phase voltage. These values are used for the simulation and developing the prototype of this converter.

### B. Design of Interphase Transformer (IPT)

A required condition to achieve the pulse doubling is to ensure that the average output voltages two converters, bridge converter I and II, are same and displaced by an angle of 6.428°. It is already known that an IPT or interphase reactor (IPR) with suitably tapped diodes [10] can effectively double the pulses in 12-pulse converters where the two converters are fed from 30° phase shifted voltages. The details of pulse multiplication arrangement for diode bridge rectifiers have been given in the literature [10]. The same concept is used here for achieving the pulse multiplication for line current harmonic reduction. The interphase reactor and tapped diodes are arranged in the fashion shown in Fig. 4c. The voltage appearing across the interphase transformer $V_m$ is an AC voltage ripple of fourteen times the source frequency, resulting in smaller size, weight and volume of transformer. Depending upon the polarity of the impressed voltage $V_m$ across the interphase transformer diodes $D_1$ and $D_2$ conduct and result in pulse multiplication. The MMF relationship of the IPR when diode $D_1$ is conducting is

$$i_{dc1}.N_A = i_{dc2}.N_B \qquad (20)$$

where $N_A$ and $N_B$ are the number of turns as shown for the IPR.

Moreover,

$$i_{dc1} + i_{dc2} = i_{dc} \qquad (21)$$

From eqn. (20) and eqn. (21) the output currents of the two diode converters I and II are given by

$$i_{dc1} = (0.5 + K)i_{dc} \qquad (22)$$

$$i_{dc2} = (0.5 - K)i_{dc} \qquad (23)$$

Where $K = (N_B - 0.5N_0)/N_0$ and $N_0$ is total turns ($= N_A + N_B$).

Similarly, the MMF relationships can be written for the case when diode $D_2$ is conducting. Therefore, depending on the polarity of voltage $V_m$, the magnitudes of the converter output currents are modulated and this changes the shape of rectifier input currents and thereby doubles the pulses. The turn ratio of the interphase transformer for suppressing 27th and 29th harmonics is given as: K = 0.2457.

### C. Design of Zero Sequence Blocking Transformer (ZSBT)

The ZSBT provides an independent operation of the two

rectifier bridges, thus eliminating the unwanted conducting sequence of the rectifier diodes. ZSBT offers very high impedance for zero sequence current components, thereby eliminating unwanted conduction sequence of the rectifier diodes [10]. The voltage waveform across ZSBT contains components of order seven times the fundamental frequency resulting in smaller size, weight and volume of the transformer. Its design is carried out to impede the flow of these components of order fourteen times the fundamental frequency of current.

## III. MATLAB BASED SIMULATION

Six-pulse, 28–pulse and 56-pulse AC-DC converters are modeled and simulated in MATLAB environment along with Simulink and Power System Blockset (PSB) toolboxes. The AC-DC converter feeds the VCIMD load (7.5kW) with an input of 415V, 50Hz AC supply and detailed data are given in Appendix.

The MATLAB model is developed for 56-pulse AC-DC converter and results obtained are shown in Figs. 5-6. The value of source impedance is considered to be 3% in these simulations. The results obtained from the simulations for proposed 28-pulse and 56-pulse AC-DC converters are given in Tables I. The resulting waveforms for the 6-pulse and 56-pulse AC-DC converters at full-load along with VCIMD transient response for load perturbation are shown in Fig. 5. Variation of THD of supply current with load in 28-pulse and 56-pulse AC-DC converters is shown in Fig. 7. Table II compares the power quality parameters of 6-pulse, 28-pulse and 56-pulse AC-DC converters at light-load (20% of full-load) and full-load. The magnetic rating involved in the two topologies is given in Table III.

## IV. HARDWARE IMPLEMENTATION

To validate the proposed 56-pulse AC-DC configuration, a prototype has been developed in the laboratory for 10kW output power in three-phase 230V, 50Hz system. The developed fifty-six pulse AC-DC converter is tested with equivalent resistive load and exhaustive tests are carried out. The results are recorded using Fluke 43B Power Analyzer. The recorded waveforms showing power quality parameters at full-load are shown in Fig. 8. It can be seen that the current waveform is nearly sinusoidal at full load and THD variation with the load (as shown in Table IV) show similar trends as simulated results. The test results showing effect of load variation on various power quality parameters is given in Table IV. The experimental results show that the THD of AC mains current with the 56-pulse AC-DC converter vary in the range of 2.6% to 3.8% on load variation while DPF is above 0.98 and power factor is almost unity. The details of developed autotransformer for 56-pulse AC-DC converter are given in Appendix.

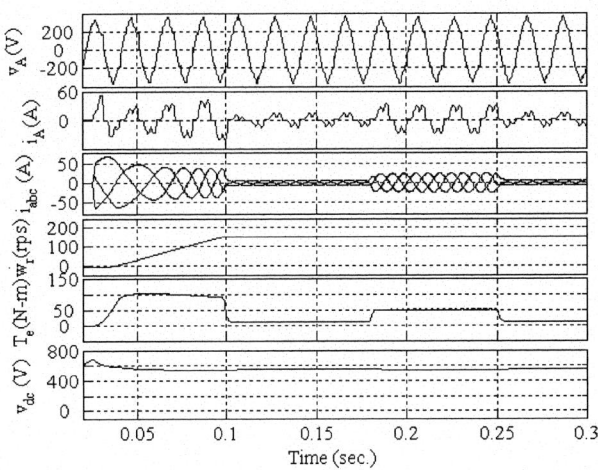

Fig. 5a. Dynamic response of 6-pulse diode rectifier fed VCIMD with load perturbation--supply phase voltage $V_A$, source current $i_{sA}$, motor currents $i_{abc}$, speed $w_r$, developed electromagnetic torque $T_e$ and DC link voltage $V_{dc}$.

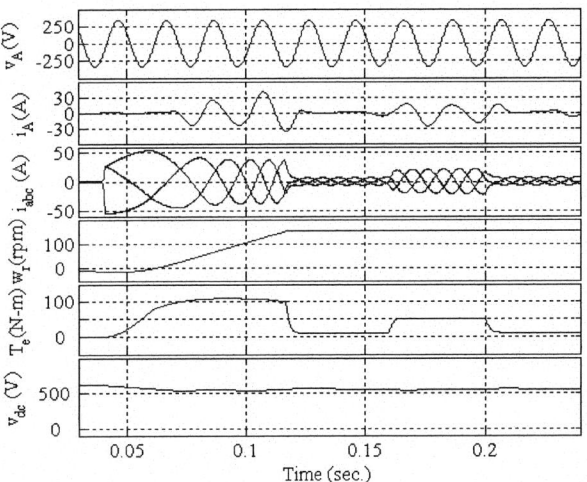

Fig. 5b. Dynamic response of 56-pulse diode rectifier fed VCIMD with load perturbation.

## V. RESULTS AND DISCUSSION

The power quality indices of 28-pulse and 56-pulse AC-DC converters are given in Table I. The waveform of AC mains current has significantly improved in 56-pulse AC-DC converter compared to the 6-pulse AC-DC converters. The simulation results show that THD$_i$ (total harmonic distortion of AC mains current) at full-load in 28-pulse AC-DC converter is 3.652%. These results clearly show that harmonics upto 26th are suppressed in 28-pulse AC-DC converters.

The power quality indices THD$_i$, THD$_v$, distortion factor (DF), displacement factor (DPF) and power factor (PF) are also obtained at varying loads in 56-pulse AC-DC converters. The comparison of power quality indices obtained for proposed 28-pulse and 56-pulse AC-DC converter topologies is also made with 6-pulse AC-DC converter and these are given in Table II. It can be seen that the performance of 56-pulse AC-DC converter is much superior compared to six-pulse AC-DC converter. The input voltage THD, current THD

TABLE I
COMPARISON OF POWER QUALITY PARAMETERS OF THE LOAD FED FROM PROPOSED AC-DC CONVERTERS

| Sr. No. | Topology | Load | THD $V_A$ (%) | AC Mains Current $I_A$ (A) | THD of $I_A$ (%) | DF | DPF | PF | DC Voltage $V_{dc}$ (V) | Load Current $I_{dc}$ (A) | RF (%) |
|---------|----------|------|------|------|------|------|------|------|------|------|------|
| 1 | 28-pulse | 20% | 1.627 | 3.083 | 5.448 | 0.9984 | 1.000 | 0.9984 | 549.4 | 3.981 | 0.0045 |
|  |  | 40% | 2.414 | 5.422 | 4.932 | 0.9984 | 0.997 | 0.9982 | 548.7 | 7.033 | 0.0168 |
|  |  | 60% | 3.084 | 7.91 | 4.526 | 0.9984 | 0.9995 | 0.9980 | 548.0 | 10.3 | 0.0331 |
|  |  | 80% | 3.624 | 10.48 | 4.009 | 0.9984 | 0.9993 | 0.9978 | 547.2 | 13.65 | 0.0285 |
|  |  | 100% | 4.097 | 13.07 | 3.652 | 0.9984 | 0.9991 | 0.9976 | 546.4 | 17.02 | 0.0334 |
| 2 | 56-pulse | 20% | 1.269 | 3.60 | 2.411 | 0.9996 | 1.000 | 0.9996 | 550.6 | 4.645 | 0.0029 |
|  |  | 40% | 1.705 | 5.968 | 2.174 | 0.9996 | 1.000 | 0.9996 | 550.1 | 7.724 | 0.0058 |
|  |  | 60% | 1.978 | 8.496 | 1.978 | 0.9999 | 0.9999 | 0.9995 | 549.5 | 11.0 | 0.0045 |
|  |  | 80% | 2.148 | 11.08 | 1.764 | 0.9995 | 0.9999 | 0.9995 | 549.0 | 14.37 | 0.0058 |
|  |  | 100% | 2.258 | 13.72 | 1.626 | 0.9995 | 0.9999 | 0.9995 | 548.4 | 17.82 | 0.0092 |

TABLE II
COMPARISON OF POWER QUALITY PARAMETERS OF DIFFERENT AC-DC CONVERTERS

| Sr. No. | Topology | % THD of $V_A$ | AC Mains Current $I_A$ (A) | | % THD of $I_A$ at | | Distortion Factor DF | | Displacement Factor DPF | | Power Factor PF | | DC Voltage ($V_{dc}$) | |
|---------|----------|------|------|------|------|------|------|------|------|------|------|------|------|------|
|  |  |  | Light Load | Full Load | Light Load | Full Load | Light Load | Full Load | Light Load | Full Load | Light Load | Full Load | Light Load | Full Load |
| 1 | 6-pulse | 10.58 | 8.701 | 19.12 | 74.68 | 31.24 | 0.9110 | 0.9491 | 0.9798 | 0.9768 | 0.8926 | 0.9271 | 552.9 | 542.8 |
| 2 | 28-pulse | 4.097 | 3.083 | 13.07 | 5.448 | 3.652 | 0.9984 | 0.9984 | 1.0 | 0.9991 | 0.9984 | 0.9976 | 549.4 | 546.4 |
| 3 | 56-pulse | 2.258 | 3.60 | 13.72 | 2.411 | 1.626 | 0.9996 | 0.9995 | 1.0 | 0.999 | 0.9996 | 0.9995 | 550.6 | 548.4 |

and power factor have improved and the waveforms and harmonic spectrum of AC mains current can also be compared in Fig. 6. It can be observed that the proposed 56-pulse AC-

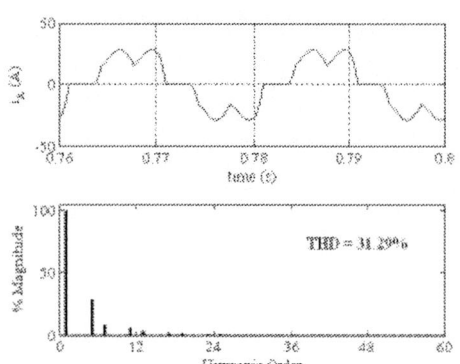

Fig. 6a. Input current waveform and harmonic spectrum of 6-pulse AC-DC converter at full-load.

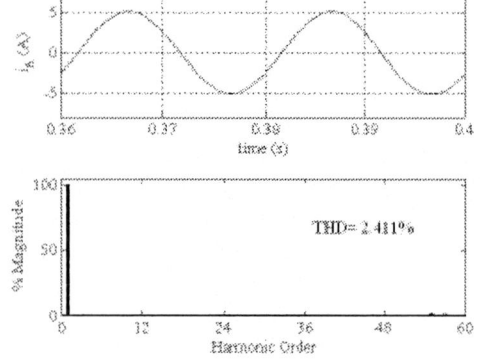

Fig. 6b. Input current waveform and harmonic spectrum of 56-pulse AC-DC converter at light-load.

DC converter has input current THD in range of 2.41% to 1.63% with varying loads. The total magnetic rating is estimated for 28-pulse and 56-pulse AC-DC converters as 45.94% and 45.24% respectively. Fig. 7 depicting variation of $THD_i$ in 28 and 56-pulse AC-DC converters clearly shows that the $THD_i$ in 56-pulse AC-DC converter is almost half that of 28-pulse AC-DC converter at all loads. The hardware results of 56-pulse AC-DC converter is shown in Fig. 8 at full-load which validates the simulation model and design of the proposed converters.

## V. CONCLUSIONS

Based on design, simulation and test results, it is observed that power quality can be improved significantly by employing the proposed polygon-connected autotransformer based 56-pulse AC-DC converter. The resulting 56-pulse converter has exhibited high level of performance with clean power characteristics required for diode based front end rectifiers. The results have shown that the total harmonic distortion of input current remains below 5% and power factor remains above 0.99 at varying loads and meets the requirements of IEEE-519 Standard for loads with short-circuit current ratio in range of 20-50.

## VI. APPENDIX

### A. Motor and controller specifications

Three-phase squirrel cage induction motor −10 hp (7.5 kW), 3-phase, 4-pole, Y-connected, 415 V, 50 Hz, $R_s$ = 0.74 ohms, $R_r$ = 0.74 ohms, $X_{ls}$ = 0.956 ohms, $X_{lr}$ = 0.956 ohms, $X_m$ = 38.99 ohms, J = 0.0343 kg-m$^2$.

PI controller: $K_p$ = 25, $K_i$ = 4.5

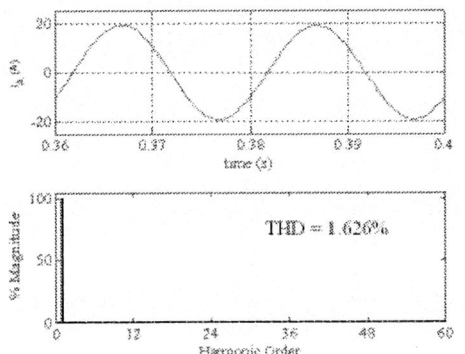

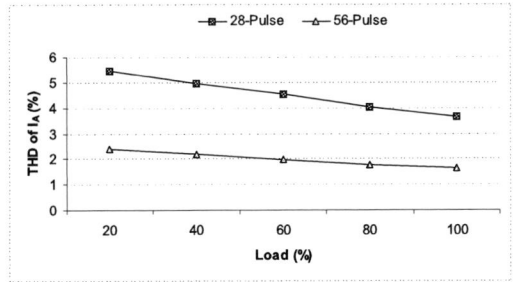

Fig. 6c. Input current waveform and harmonic spectrum of 56-pulse AC-DC converter at full-load.

Fig. 7. Variation of THD of supply current with load in 28-pulse and 56-pulse AC-DC converters.

DC link parameters: $L_d$ = 0.2mH, $C_d$ = 4700 µF.

### B. Design of Interphase Transformer:

The flux density is taken as 0.8 Tesla and the current density is considered as 2.3 A/m². The interphase transformer is wound using core of size No. 3 with E-I laminations of size (76 mm × 127 mm) and (127 mm × 19 mm), respectively. Based on the voltage across different windings, the number of turns are calculated and based on the current flowing through different windings, the gauge of wire is calculated and these are: $N_A$-35T-11SWG, $N_B$-67T-18SWG

### C. Design of Autotransformers:

Autotransformer rating:    3.18kVA (=1060 VA x 3)

Transformer design details:

TABLE III

COMPARISON OF ACTIVE POWER RATING MAGNETICS IN DIFFERENT AC-DC CONVERTERS

| Topology | Autotransformer rating (% of load) | IPT (% of load) | ZSBT (% of load) | Total rating , (% of load) |
|---|---|---|---|---|
| 28-pulse | 44.6 | 1.34 | - | 45.94 |
| 56-pulse | 43.68 | 0.32 | 1.24 | 45.24 |

TABLE IV

POWER QUALITY PARAMETERS OF THE TEST RESULTS OBTAINED FROM 56-PULSE AC-DC CONVERTER

| Load, kW | THD $V_A$ (%) | Current $I_A$ (A) | THD of $I_A$(%) | PF | Load Voltage, V | Load Current, A |
|---|---|---|---|---|---|---|
| 2.11 | 1.2 | 5.307 | 3.8 | 0.976 | 298.3 | 6.370 |
| 4.13 | 1.6 | 10.44 | 3.0 | 0.990 | 289.7 | 13.02 |
| 6.18 | 1.9 | 15.56 | 2.8 | 0.993 | 288.0 | 19.45 |
| 8.10 | 2.1 | 20.43 | 2.6 | 0.994 | 281.2 | 25.89 |
| 10.5 | 2.0 | 26.49 | 2.6 | 1.00 | 276.4 | 31.80 |

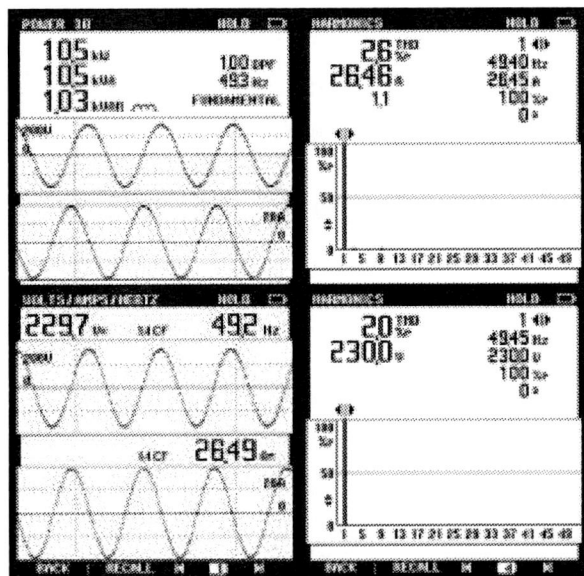

Fig. 8. Test result showing input voltage and current waveform with their harmonic spectrum for 56-pulse AC-DC conversion at full-load.

Flux Density: 0.8Tesla, Current Density: 2.3A/mm², Core size

E-Laminations: Length= 23.5cm, Width= 16cm

I-Laminations: Length= 23.5cm, Width= 4cm

Area of cross-section of core= 58 cm² (7.6 cm X 8.6 cm)

Autotransformer winding details (winding-no. of turns-gauge of wire): $K_1$-21.5T-16SWG, $K_2$-31.5T-16SWG, $K_3$-41T-16SWG, $K_4$-19.5T-16SWG, $K_5$-3.5T-16SWG, $K_6$-21.5T-16SWG, $K_7$-49T-16SWG, $K_8$-17T-16SWG, $K_9$-7.5T-16SWG, $K_{10}$-6.5T-16SWG, $K_{11}$-23T-16SWG, $K_{12}$-33.5T-16SWG, $K_{13}$-28T-16SWG, $K_{14}$-14T-16SWG, $K_{15}$-63T-16SWG.

## VI. REFERENCES

[1] IEEE Standard 519-1992, IEEE Recommended Practices and Requirements for Harmonic Control in Electrical Power Systems, *IEEE Inc.*, New York, 1992.

[2] G. Seguier, *"Power Electronic Converters: AC-DC Conversion"*, McGraw Hill Book Company, New York, 1987.

[3] Bin Wu, High-Power Converters and AC Drives, IEEE Press, Wiley-Interscience, 2006.

[4] B. K. Bose, "Modern Power Electronics and AC Drives", Pearson Education, New Delhi, 2001.

[5] Derek A. Paice, Power Electronic Converter Harmonics: Multipulse Methods for Clean Power, *IEEE Press*, New York, 1996.

[6] C. J. Chen and G. K. Horng, "A new passive 28-step current shaper for three-phase rectification," IEEE Trans. on Industrial Electronics, Vol.47, No.6, Dec.2000, pp. 1212-1219.

[7] J. Johnson and R. E. Hammond, "Main and Auxiliary Transformer Rectifier System for Minimizing Line Harmonics", US Patent 5,063,487, Nov. 5, 1991.

[8] B. Singh, G. Bhuvaneshwari and Vipin Garg, "Power-Quality Improvements in Vector-Controlled Induction Motor Drive Employing Pulse Multiplication in AC–DC Converters", *IEEE Trans. on Power Delivery*, vol. 21, no. 3, pp. 1578-1586, July 2006.

[9] R Hammond, J. Johnson, A.Shimp and D. Harder, "Magnetic solutions to line current Harmonic reduction", *in Proc. of Conf. Power Con.- 1994*, pp.354-364, Sept., 1994.

[10] S. Choi, B. S. Lee and P. N. Enjeti, "New 24-pulse Diode Rectifier Systems for Utility Interface of High Power AC Motor Drives", *IEEE Trans. on Ind. Appl.*, vol.33, no. 2, pp. 531-541, March/April 1997.

**2006 IEEE International Conference on Power Electronic, Drives and Energy Systems**

# Performance Comparison of High Frequency Isolated AC-DC Converters for Power Quality Improvement at Input AC Mains

Bhim Singh, *Senior Member, IEEE*, B.P. Singh, *Senior Member, IEEE*,
and Sanjeet Dwivedi, *Student Member, IEEE*

*Abstract--* **This paper deals with the analysis, design and DSP based hardware implementation of four different topologies of high frequency isolated AC-DC converters in discontinuous current mode (DCM) of operation used for power quality improvement at input AC mains. The design equations, modeling ,simulated and experimental results of these AC-DC converters are presented for their performance evaluation. Modeling and simulation is carried out in a standard PSIM software environment. The obtained results demonstrate the effectiveness in improving the power quality at AC mains.**

*Index Terms--* **Power Quality, High Frequency Isolation, Total Harmonic Distortion, Digital Signal Processor, Cuk, SEPIC, Flyback, Zeta Converter.**

## I. INTRODUCTION

THE single phase Diode Bridge Rectifier (DBR) with DC link capacitive filter is generally used for AC-DC power conversion. This DBR draws non sinusoidal input AC current leading to low input power factor and injection of harmonic into the AC mains. The stringent power quality requirement imposed by the regulating authorities like IEEE and IEC enforce the strict limit on the power quality standard which includes IEEE 519-1992 and IEC 61000-3-2 [1-2].

Normally AC-DC conversion is carried out by simply rectifying the AC input voltage and the rectifier output is filtered by means of large valued capacitance to get nearly constant DC voltage output. In this conversion the input AC supply current is drawn in narrow pulses since the capacitor voltage is nearly constant [3].This narrow pulse current of large peak, results in the power quality problems to nearby consumers, which include higher value of Total Harmonic Distortion (THD) of supply current, higher THD of input supply voltage, lower value of power factor(PF) and displacement factor(DPF), and poor distortion factor(DF). These large harmonic currents are undesirable because they not only produce distortion of AC line voltage but also result in conducted and radiated electromagnetic interference (EMI). The problem becomes more serious particularly when several such loads are connected to single-phase supply where the input power pulsates at twice the supply frequency. Recent

Bhim Singh and B.P. Singh are with the Department of Electrical Engineering, Indian Institute of Technology, New Delhi-110016, India. (e-mail : bsingh@ee.iitd.ac.in,bpsingh@ee.iitd.ac.in).

Sanjeet Dwivedi is with the Department of Electrical Engineering, I.G. Engineering College Sagar (M.P.) India (e-mail: sanjeetkd@gmail.com.).

international regulations governing the power quality and harmonic currents pollutions limit at the utility, have placed an increased emphasis on the application of improved power quality AC-DC converters. For ideal sine wave line voltage, harmonic currents do not contribute to active power, this result in increased value of rms current and therefore produce higher losses in the utility line. It is prime concern to consider these power quality issues to design an AC-DC power converter, which provides unity power factor at input AC mains and also results in close regulation of output DC voltage [4-8]. Moreover, isolation is also desirable in number of small power ratings applications in appliance sector such as computer peripherals, small refrigerators, washing machines, machine tools etc.

This paper presents four different topologies of high frequency transformer isolated AC-DC converters for providing regulated DC voltage to DC loads [9-15]. The proposed converters provide improved power quality in terms of low total harmonic distortion (THD), reduced crest factor (CF) of supply current, high power factor of AC mains and regulated output DC voltage. The complete scheme of these power factor corrected (PFC) AC-DC converters are designed, modeled, simulated in standard PSIM software [16] and implemented on Analog Devices Digital Signal Processor (DSP) ADMC401 [17] to demonstrate their effectiveness in improving the power quality at AC mains. These converters topology are also attractive choice for low cost variable speed drive applications, which are connected to single phase AC supply of rating less than 2kW.

## II. HIGH FREQUENCY ISOLATED AC-DC CONVERTERS

There is a need to improve the quality of the current drawn from the AC mains used in applications such as compressor drive of the residential air-conditioner and refrigerator units. These appliances are operated in unidirectional power flow mode and employing rectifier-inverter combination of order of 2kW. These appliances are fed from single phase AC supply and AC-DC conversion takes place in these drives units with the single phase full bridge diode rectifier and large value of capacitive filter is used to reduce DC voltage ripple, which produces increased THD of input mains current and excessive peak input currents leading to poor power factor. Due to application of boost type of power factor preregulators (PFP) these problems of poor power quality of input AC mains and low power factor have been overcome but their use is restricted to the applications where the output voltage required always more than input peak voltage and also the isolation is

0-7803-9771-1/06/$25.00 ©2006 IEEE

not required between output and input supply by these PFP circuits. To provide the buck-boost control and isolation between the input and output the Cuk, SEPIC Flyback and Zeta converters have advantages of proper placement of the inductor, inherent startup capability, over current control and no restriction on obtained output voltage [10-15]. These buck-boost type PFPs have utilized two approaches for their control, the multiplier approach and voltage follower approach. Due to inherent advantages of voltage follower approach like elimination of input current sensor, simple control scheme, with only one control loop, these AC-DC converter topologies with high frequency transformer isolation in discontinuous current mode (DCM) are employed in many applications such as switched mode power supplies electronic ballasts, small rating fans and compressor drive, with high efficiency permanent magnet brushless motors at low voltages.

The high frequency transformer isolated active waveshaping AC-DC converters considered here are classified into four categories as:
(i) AC-DC Cuk converter
(ii) AC-DC SEPIC converter
(iii) AC-DC Flyback converter
(iv) AC-DC Zeta converter
The circuit diagrams of these topologies are shown in Figs.1-4.

The parameters of these AC-DC converters are obtained from design equations. These parameters are used to model the converter in DCM of operation while feeding load. The complete modeling is performed in standard PSIM software. Design equations are derived, and obtained parameters of converter are tested in simulation model and experimental setup. The dynamic performances of these converters and their steady state responses are obtained. The power quality performance indices of these converters at different load are compared to demonstrate their effectiveness.

### III. DESIGN OF HIGH FREQUENCY ISOLATED AC-DC CONVERTERS

The design of the active waveshaping isolated buck-boost converters topologies are described in this section. The design equations are derived for active waveshaping converters used to feed the small rating appliances. The parameters of these buck-boost PFC converters are selected based on design considerations for power quality improvement at AC mains, allowable ripple content in source current and in DC-link voltage, and the switching frequency of active devices.

#### A. Design AC-DC Cuk Converter

Fig.1 shows the circuit diagram of AC-DC Cuk converter. The design of components is presented to operate the Cuk converter in the discontinuous conduction mode (DCM) of current operation. In DCM of current operation, converter acts as a voltage follower and provides inherent power factor correction at the input AC mains. The output stage is referred to the primary side of the. The DC link voltage $v'_{dc}$ and equivalent load R' are referred values to the primary side of the transformer. The derived design equations are given in Table I and converter design data are given in Table II.

#### B. Design AC-DC SEPIC Converter

The design of components of AC-DC SEPIC (Single Ended Primary Inductance Converter) converter are given to operate the SEPIC converter in the discontinuous conduction mode (DCM) of current operation. In DCM, this converter acts as a voltage follower for providing inherent power factor correction at the input AC mains. The output stage is referred to the primary side of the transformer. The dc link voltage $v'_{dc}$ and the equivalent load R' are referred values to the primary side of the transformer.

#### C. Design AC-DC Flyback Converter

The design of AC-DC flyback converter components are given for the operation of flyback converter in the discontinuous conduction mode (DCM) of current operation. In DCM, the converter acts as a voltage follower with inherent power factor correction at the input ac mains. The input voltage, input current and output current waveforms are obtained. All these values are referred to the secondary side of the flyback transformer.

#### D. Design AC-DC Zeta Converter

The design of AC-DC Zeta converter components is given for the operation of Zeta converter in discontinuous conduction mode (DCM) of current operation. The Zeta converter is designed to operate in DCM as voltage follower for providing inherent power factor correction at the input AC mains. The output stage is referred to the primary side of the transformer. The equivalent DC link voltage $v'_{dc}$ and the

TABLE I
DESIGN EQUATION OF AC-DC CONVERTERS

| Sr. No. | Cuk Converter | SEPIC Converter | Flyback Converter | Zeta Converter |
|---|---|---|---|---|
| 1. | Voltage ratio $M=v_{dc}/v_{in(pea}$ | Voltage Ratio $M=V_{DC}/V_{IN(PEK)}$ | Voltage Ratio $M=V_{DC}/V_s$ | Voltage Ratio $M=V_{DC}/V_s$ |
| 2. | Turn Ratio $n=N_2/N_1$ | Turn Ratio $n=N_2/N_1$ | Turn Ratio $n=N_1/N_2$ | Turn Ratio $n=N_2/N_1$ |
| 3. | Conduction Parameter $K_e=1/(2(M+n)^2)$ | Conduction Parameter $K_e=1/(2(M+n)^2)_e$ | Critical Conduction Parameter $K_{crit}=1/[2\{1+(n\ v_{dc}/V_s)\}^2]$ | Duty Ratio of Switch $d=M/(M+n)$ |
| 4. | Duty Ratio of Switch $d=\sqrt{2}\ M\ \sqrt{K_e}$ | Duty Ratio of Switch $d=\sqrt{2}\ M\ \sqrt{K}$ | Inductance $L_{mcrit}=R_{min}$ $T_s/[4\{1+(nv_{dc}/V_s)\}^2]$ | Ripple Current $I_{Ripple}=r_i(2P_{in}/V_s)$ |
| 5. | Ripple Current $I_{Ripple}=r_i\{2P_{in}/V_{in(peak)}$ | Ripple Current $I_{Ripple}=r_i\{2P_{in}/V_{in(peak)}\}$ | Duty Ratio of Switch $d=(nv_{dc}/V_s)\sqrt{(2K)}$ | Inductance $L_{eq}=RT_s (1-d)^2/2$ |
| 6. | Inductance $L_{eq}=R\ T_s\ K_e/2$ | Inductance $L_{eq}=R\ T_s\ K_e/2$ | Conduction Parameter $K=(2L_m/RT_s)$ | Inductance $L'_o=V_{1r}\ d$ $T_s/I_{Ripple}$ |
| 7. | Inductance $L_1=V_{in(peak)}d$ $T_s/I_{Ripple\ 1}$ | Inductance $L_1=V_{in(peak)}\ d$ $T_s/I_{Rippl}$ | Capacitance $C_{dc}=1/(2\omega_L r_v\ R)$ | Inductance $L_m =L'_o$ $L_{eq}/(L'_o- L_{eq})$ |
| 8. | Inductance $L'_2=L_1\quad L_{eq}/(L_1-L_{eq})$ | Inductance $L'_2=L_1\ L_{eq}/(L_1-L_{eq})$ | ---------- | Capacitance $C'_1=1/\omega^2_{ar}(L_m+L'_o)$ |
| 9. | Capacitance $C_1=2/\omega^2_{ar}(L_1+L'_2)$ | Capacitance $C_1=2/\omega^2_{ar}(L_1+L'_2)$ | ---------- | Capacitance $C'_{dc}=n^2/(2\omega_r r_v R)$ |
| 10 | Capacitance $C_2=2/\omega^2_{ar}(L_1+L'_2)$ | Capacitance $C_2=2/\omega^2_{ar}(L_1+L'_2)$ | ---------- | |
| 11 | Capacitance $C_{dc}=1/(2\omega_L r_v R)$ | Capacitance $C_{dc}=1/(2\omega_L r_v R)$ | ---------- | |

TABLE II
DESIGN PARAMETERS OF AC-DC CONVERTERS

| Sr. No. | Cuk Converter | SEPIC Converter | Flyback Converter | Zeta Converter |
|---|---|---|---|---|
| 1. | Voltage Ratio 'M'= 0.707 | Voltage Ratio 'M'= 0.707 | Voltage Ratio 'M'= 0.707 | Voltage Ratio 'M'= 0.707 |
| 2. | Turn Ratio of Transformer 'n'=1 | Turn Ratio of Transformer 'n'=1 | Turn Ratio of Transformer 'n'=1 | Turn Ratio of Transformer 'n'=1 |
| 3. | Duty Ratio 'd'=0.1714 | Duty Ratio 'd'=0.3161 | Duty Ratio 'd'=0.406 | Duty Ratio 'd'=0.151 |
| 4. | Conduction Parameter '$K_a$'=0.1 | Conduction Parameter '$K_a$'=0.1714 | Conduction Parameter '$K_a$'=0.165 | Inductance '$L_m$'=2.196mH |
| 5. | Critical Value of Conduction Parameter of Converter '$K_{acrit}$'=0.1715 | Critical Value of Conduction Parameter of Converter '$K_{acrit}$'=0.125 | Critical Value of Conduction Parameter of Converter '$K_{acrit}$'=0.1715 | Inductance '$L_o$'=4.5316mH |
| 6. | Inductance '$L_1$'=3.02mH | Inductance '$L_1$'=4.78mH | Inductance '$L_1$'=0.25mH | Capacitance '$C_1$'= 10µF |
| 7. | Inductance '$L_2$'=283µH | Inductance '$L_2$'=196µH | Filter Capacitance '$C_f$'=200µF | Filter Capacitance '$C_f$'=200µF |
| 8. | Capacitance '$C_1$'=0.616 µF | Capacitance '$C_1$'=0.616 µF | Output Capacitance $C_{dc}$=1000 µF | Output Capacitance $C_{dc}$=1000 µF |
| 9. | Capacitance '$C_2$'=0.616 µF | Output Capacitance $C_{dc}$=1000 µF | Ripple current '$I_{Ripple}$'=0.257 A | Ripple current '$I_{Ripple}$'=0.257A |
| 10 | Output Capacitance $C_{dc}$=1000 µF | Ripple current '$I_{Ripple}$'=0.257 A | PI Voltage Controller Gain '$K_{pdc}$, $K_{idc}$'=0.3, 30.15 | PI Voltage Controller Gain '$K_{pdc}$, $K_{idc}$'=0.3, 30.15 |
| 11 | Ripple current '$I_{Ripple}$'=0.257A | PI Voltage Controller Gain '$K_{PDC}$, $K_{IDC}$'=0.28, 33.21 | ----------- | ---------- |
| 12 | PI Voltage Controller Gain '$K_{pdc}$, $K_{idc}$'=0.3, 25.23 | --- | ----------- | ---------- |
| 13 | $P_{in}$=250W, $V_{in}$=110V, $V_{dc}$=110V, | $P_{in}$=250W, $V_{in}$=110V, $V_{dc}$=110V, | $P_{IN}$=250W, $V_{IN}$=110V, $V_{DC}$=110V, | $P_{IN}$=250W, $V_{IN}$=110V, $V_{DC}$=110V, |
| 14 | $f_s$=20kHz | $f_s$=20kHz | $f_s$=20KHZ | $f_s$=20KHZ |

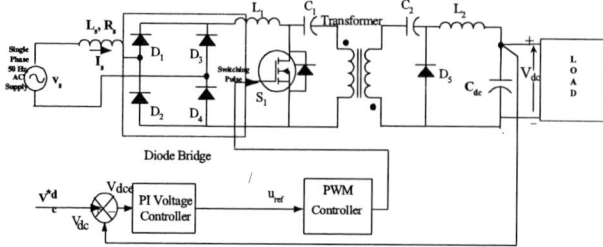

Fig.1. High frequency isolated AC-DC cuk converter.

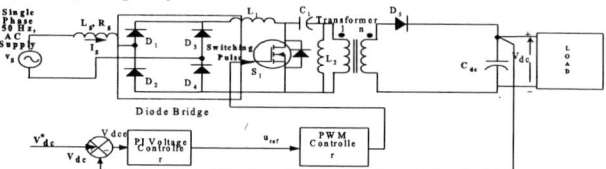

Fig. 2. High frequency isolated AC-DC SEPIC converter.

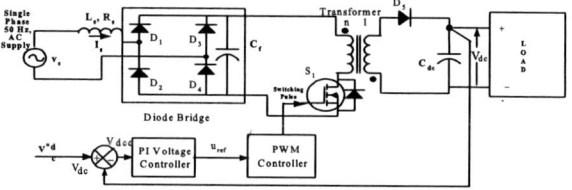

Fig. 3. High frequency isolated AC-DC flyback converter.

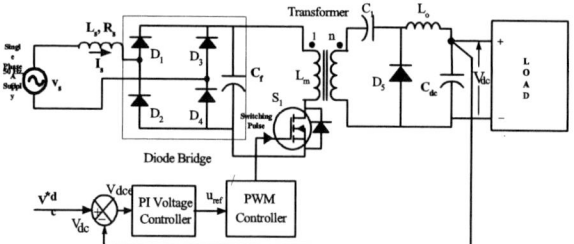

Fig. 4. High frequency isolated AC-DC zeta converter.

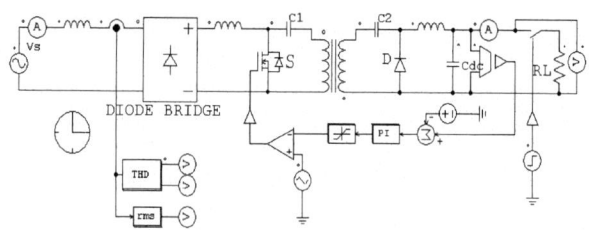

Fig. 5. PSIM based model of high frequency isolated AC-DC cuk converter.

equivalent load 'R' are the values referred to primary side of high frequency transformer.

## IV. PSIM BASED SIMULATION OF HIGH FREQUENCY ISOLATED AC-DC CONVERTERS

The standard PSIM software is used to model the high frequency transformer isolated AC-DC converters. The PI (Proportional-Integral) voltage controller block, PWM duty ratio generator block, circuit components and high frequency transformer isolation transformer with MOSFET as an active switching device are used to form the different AC-DC converters topologies. After designing the components of AC-DC converters, the appropriate components are selected from the PSIM software library to model the particular topology of high frequency transformer isolated AC-DC converters. The PSIM based model developed for AC-DC Cuk converter is shown in Fig.5. The equivalent resistive load is connected with these different front-end AC-DC converters through a switch so that after stabilization of DC link voltage, the load can be energized. The power quality is analyzed through FFT of input AC mains current and % ripple content in the DC link voltage of the converter.

## V. DSP BASED HARDWARE IMPLEMENTATION OF HIGH FREQUENCY ISOLATED AC-DC CONVERTERS

The proposed high frequency transformer isolated AC-DC converters are implemented using the Analog Devices (AD) make Digital Signal Processor (DSP) ADMC401. The Cuk, SEPIC, Flyback and Zeta converters are realized in the discontinuous conduction mode (DCM) of operation. The DC link voltage is the feedback signal for the closed loop structure of these AC-DC converters in the voltage follower mode of operation. The control algorithm is implemented in real time using DSP ADMC401.

Fig.6 shows the block diagram of the developed hardware setup for implementation of AC-DC converters. The power circuit consists of single-phase 50Hz AC supply, single-phase AC-DC diode bridge converter, input and output inductors, high frequency transformer for isolation, capacitors connected in the primary and secondary windings of transformer for blocking the DC components in the windings, DC link capacitor for filtering out the ripple contents in output DC voltage, high frequency diode. The control circuit hardware consists of reference voltage scaling circuits, sensed DC link voltage scaling circuit using the Hall effect voltage sensor, gate driver circuit consists of Toshiba make TLP250 driver. The gate driver has an optical isolation in its initial stage to isolate the control circuit from the power circuit. The gating signal achieved at the output of driver chips is applied between gate and emitter of the insulated gate bipolar transistor (IGBT), which is used as an active device in the implementation of different converter topologies.

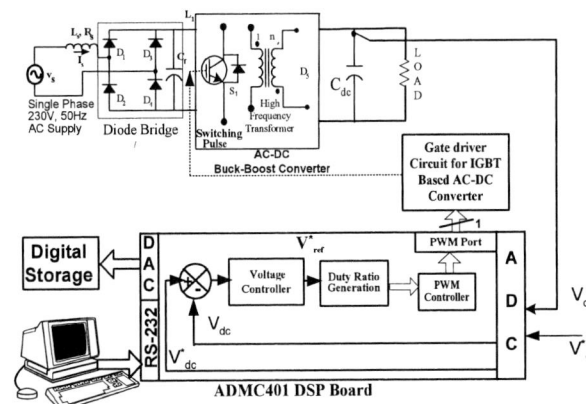

Fig. 6. Experimental setup for DSP based implementation of high frequency isolated AC-DC converters.

## VI. RESULTS AND DISCUSSION

The parameters of AC-DC converters are obtained from design equations. These parameters are used to model the converter in DCM of operation while feeding DC load. The complete modeling is performed in standard PSIM software. Design equations are shown in Table I, and obtained parameters of converter are given in Table II. The obtained dynamic performance of Cuk converter is shown in Fig.7 and the steady state responses is given in Fig.8. The harmonic spectrums of AC mains current are given in Fig.9. The power quality performance indices of these converters at different load are tabulated in Table III. From these results, the following observations are made.

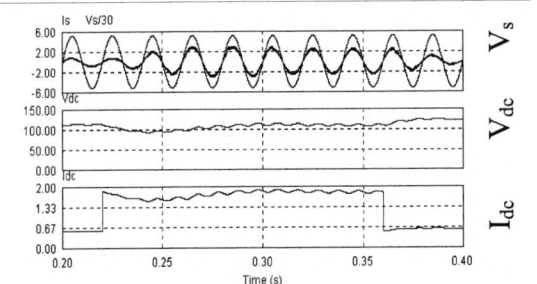

Fig. 7. Simulated and Experimental responses of AC mains voltage, AC mains current, output DC voltage and output DC current waveform of AC-DC Cuk converter for load perturbation response on resistive load (60W to 200W to 60W).

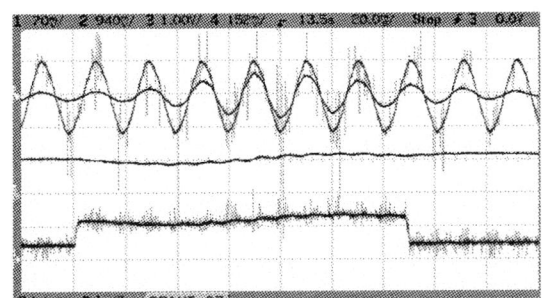

Fig. 8. Simulated and Experimental responses of AC mains voltage, AC mains current, output DC voltage and output DC current waveform of AC-DC Cuk converter for steady state response on resistive load (60W).

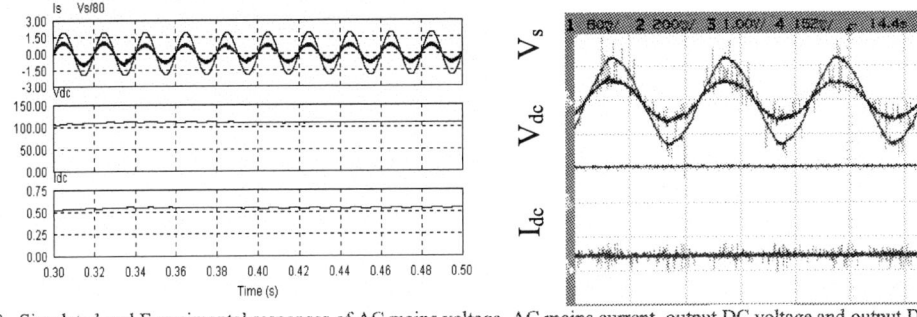

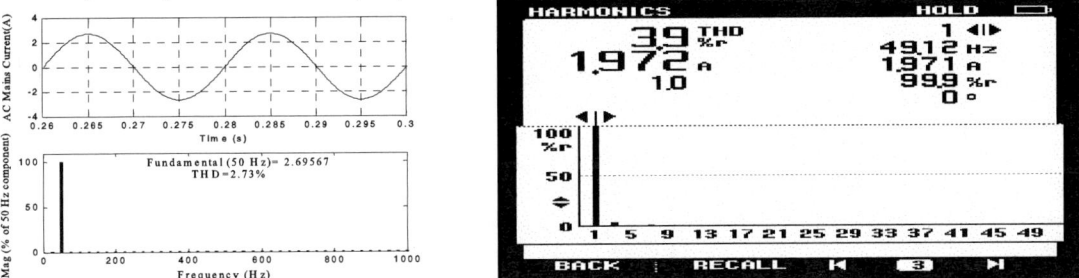

Fig. 9. Simulated and Experimental harmonic spectrum of AC mains current for AC-DC Cuk converter feeding the resistive load (200W).

345

TABLE III
POWER QUALITY PERFORMANCE INDICES OF AC-DC CONVERTERS

| Sr. No | Type of Converter | % Load | % THD of $I_s$ | DF | DPF | PF | CF of $I_s$ | $I_{srms}$ | $P_{in}$ Watts | RF % | %p-p Ripple | $V_{dc}$ Volts |
|--------|-------------------|--------|---------------|------|------|------|-------------|-----------|----------------|------|-------------|----------------|
| 1. | Cuk Converter | 20 | 5.93 | .998 | .999 | .997 | 1.60 | .686 | 75.3 | .427 | 1.14 | 110 |
| | | 40 | 5.60 | .998 | .999 | .997 | 1.56 | .919 | 100.8 | .935 | 2.55 | 110 |
| | | 60 | 4.78 | .999 | .999 | .998 | 1.51 | 1.321 | 145.1 | 1.22 | 3.43 | 110 |
| | | 80 | 4.65 | .999 | .999 | .998 | 1.47 | 1.736 | 190.6 | 1.76 | 4.37 | 110 |
| | | 100 | 4.12 | .999 | .999 | .998 | 1.42 | 2.099 | 230.5 | 2.12 | 5.35 | 110 |
| 2. | SEPIC Converter | 20 | 6.27 | .996 | .999 | .996 | 1.91 | .697 | 76.4 | .621 | 1.36 | 110 |
| | | 40 | 5.73 | .997 | .999 | .997 | 1.84 | .932 | 102.3 | .953 | 2.72 | 110 |
| | | 60 | 5.25 | .997 | .999 | .997 | 1.63 | 1.342 | 147.2 | 1.34 | 3.86 | 110 |
| | | 80 | 4.70 | .998 | .999 | .997 | 1.56 | 1.719 | 188.6 | 2.04 | 5.12 | 110 |
| | | 100 | 4.61 | .999 | .999 | .998 | 1.42 | 2.188 | 240.3 | 2.66 | 5.43 | 110 |
| 3. | Flyback Converter | 20 | 5.24 | .997 | .999 | .996 | 1.58 | .687 | 75.3 | .511 | 1.36 | 110 |
| | | 40 | 4.92 | .998 | .999 | .997 | 1.52 | .931 | 102.2 | .924 | 2.56 | 110 |
| | | 60 | 3.80 | .999 | .999 | .998 | 1.49 | 1.358 | 149.1 | 1.28 | 3.63 | 110 |
| | | 80 | 3.71 | .999 | .999 | .998 | 1.45 | 1.742 | 191.3 | 2.11 | 4.91 | 110 |
| | | 100 | 3.56 | .999 | .999 | .998 | 1.42 | 2.231 | 245.2 | 2.61 | 5.87 | 110 |
| 4. | Zeta Converter | 20 | 7.23 | .997 | .999 | .996 | 1.71 | 0.735 | 80.54 | .415 | 1.21 | 110 |
| | | 40 | 6.82 | .997 | .999 | .996 | 1.65 | 1.101 | 120.6 | .827 | 2.32 | 110 |
| | | 60 | 6.52 | .998 | .999 | .997 | 1.51 | 1.471 | 161.2 | 1.31 | 3.15 | 110 |
| | | 80 | 6.31 | .998 | .999 | .997 | 1.45 | 1.833 | 201.1 | 1.91 | 4.36 | 110 |
| | | 100 | 6.00 | .999 | .999 | .998 | 1.42 | 2.075 | 228.0 | 2.34 | 5.81 | 110 |

The performance of the high frequency transformer isolated Cuk, SEPIC, flyback and Zeta AC-DC converters feeding equivalent resistive load are simulated using the standard PSIM software under different dynamic conditions such as load perturbation (load application and load removal) and steady-state operating condition. A set of response is obtained showing the variations of AC mains voltage, AC mains current DC link voltage and DC link current. The control algorithm for these converters is realized in discrete time domain with a sampling interval of 50μsecs.

The simulated and test results are compared for validating the design of these converters for feeding the equivalent resistive load and are shown in Figs.7-9. The simulated and test results of Cuk AC-DC converter for load perturbation response with load change from 60W to 200W and again back to 60W are shown in Fig.7. The simulated and test results of AC-DC SEPIC, Flyback and Zeta converters for load perturbation from 60W to 200W to 60W resistive load are also obtained. It is observed from these results that AC mains current is always in phase with the source voltage and results in unity power factor operation of converter both during load perturbation and under steady-state operating conditions. The simulated and test results for the harmonic spectrum of AC mains current for these converters are obtained for evaluating the power quality of AC mains in terms of THD of source current. The results have revealed that the THD always remains less than 8% and thus these topologies fulfill the requirements of the standard IEEE-519 1992.

## VII. CONCLUSIONS

The design, modeling and development of high frequency transformer isolated AC-DC converters have been carried out in discontinuous conduction mode of operation with high frequency transformer isolation. The modeling and simulation of these AC-DC converters have been carried out in standard PSIM software with the parameters of converters obtained from the design. The high frequency transformer isolated AC-DC converters namely Cuk, SEPIC, Flyback and Zeta have

been developed in buck-boost mode in the laboratory environments. The simulation and test results on the developed converters have shown improved performance of these proposed converters in terms of low THD of supply current, and improved power factor of AC mains. It has been observed that these AC-DC converters in discontinues mode of conduction provide an improved power quality and have performed as PFP with reduced sensors and higher reliability of these converters configurations. From the careful comparison among these proposed AC-DC high frequency transformer isolated converter topologies, the flyback converter topology is most suitable among all because it gives lowest THD of AC mains current and reduced number of circuit components are needed to realize this topology.

## VIII. REFERENCES

[1] IEEE Recommended Practices and Requirements for Harmonics Control in Electric Power System, IEEE Standard 519, 1992.

[2] Electromagnetic compatibility (EMC)-Part-3: Limits- Section 2: Limits for Harmonic Current Emissions (equipment input current<16A per phase), IEC 61000-3-2 Document, Second Edition, 2000.

[3] M.H. Rashid, Power Electronics Circuits, Devices, and Applications, 3rd Edition, Prentice-Hall, New Delhi, 2005.

[4] I. Pressman, Switching Power Supply Design, McGraw Hill, New York, 1991.

[5] A. Qiao and K.M. Smedley, "A Topology Survey of Single-Stage Power Factor Corrector with a Boost Type Input-Current Shaper", IEEE Trans. on Power Electron., Vol.16, No.3, May 2001, pp.360-367.

[6] Bhim Singh, B.N. Singh, A. Chandra, K.A. Haddad, A.Pandey and D.P. Kothari, "A Review of Single-Phase Improved Power Quality Converters", IEEE Trans on Indus. Electron., Vol. 50, No. 5, pp.962-981, Oct.2003.

[7] T. Suntio, "Unified Derivation and Analysis of Duty Ratio constraints for Peak-Current Mode Control in Continuous and Discontinuous Modes," in Proc. IEEE IECON'02, 2002, pp. 1398-1403.

[8] J. Sebastian, J.A. Cobos, J. M. Lopera and J. Uceda, "The Determination of the Boundaries Between Continuous and Discontinuous Conduction Modes in PWM DC-to-DC Converters Used as Power Factor Preregulators", IEEE Trans on Power. Electron., Vol. 10, No. 5, pp.574-582, Sept. 1995.

[9] M. Kajerani, P.D. Ziogas and G. Joos, "A Novel Active Current Waveshaping Technique for Solid-State Input Power Factor Conditioners", IEEE Trans. Indus. Electron., Vol.38, No.1, pp.72-78, Feb. 1991.

[10] D.S.L. Simonetti, J. Sebastain and J. Uceda, "The Discontinuous Conduction Mode Sepic and Cuk Power Factor Preregulators: Analysis and Design", *IEEE Trans. on Indus. Electron*, vol.44, no.5, pp.630-637, Oct. 1997.

[11] M. Brkovic and S. Cuk, "Input Current Shaper using Cuk converter", in *Proc. IEEE INTELEC'92*, 1992, pp 532-539.

[12] R. Erickson, M. Madigan and S. Singer, "Design of a Simple High-Power-Factor Rectifier Based on the Flyback Converter", in *Proc. IEEE APEC'90*, 1990, pp.792-801.

[13] S. Howimanporn, C. Bunlaksananusorn, "Performance Comparison of Continuous Conduction Mode (CCM) and Discontinuous Conduction Mode (DCM) Flyback Converter", in *Proc. IEEE PEDS'03*, 2003, pp.1434-1438.

[14] A. Peres, D. C. Martins and I. Barbi, "Zeta Converter Applied in Power Factor Correction", in *Proc. IEEE PESC'94*, 1994, pp.1152-1157.

[15] J. L. Lin, S.P. Yang and P.W. Lin, "Small Signal Analysis and Controller Design for an Isolated Zeta Converter with High Power Factor Correction," *Electric Power Systems Research*, Vol. 76, pp. 67-76, July 2005.

[16] "PSIM User's Guide, Copyright©" Powersim Inc, 2003.

[17] "User's Guide of ADMC401 Motion Control DSP" of Analog Devices, 1999.

## IX. BIOGRAPHIES

**Bhim Singh** (SM'99) was born in Rahamapur (U.P.), India, in 1956. He received B.E.(Electrical) degree from the University of Roorkee, Roorkee, India, in 1977 and M.Tech. and Ph.D. degrees from Indian Institute of Technology(IIT), New Delhi, India, in 1979 and 1983, respectively.

In 1983, he joined Department of Electrical Engineering, University of Roorkee, as a lecturer. In 1988, he became a reader. In December 1990, he joined Department of Electrical Engineering, IIT, New Delhi, India, as an Assistant Professor. He became an Associate Professor in 1994 and a Professor in 1997. His fields of interest include power electronics, electrical machines and drives, active filters, static compensators, and analysis and digital control of electrical machines. Prof. Singh is a Fellow of the Indian National Academy of Institution of Engineers (India), and Institution of Electronics.

**B. P. Singh** (SM)was born in Singhia, Bihar, India, in 1940. He received B.Sc.(Engg.) degree from the Bihar Institute of Technology Sindri(Bihar), India, in 1963, M.E. from Bengal Engineering College Howrah(W.B.)and Ph.D. degrees from Indian Institute of Technology(IIT), New Delhi, India, in 1966 and 1974, respectively.

In 1966, he joined the Department of Electrical Engineering, M.I.T. Muzaffarpur, as an Assistant Professor. In 1978, he joined Department of Electrical Engineering, IIT, Delhi, India, as an Assistant Professor. He became full Professor in 1985. He was visiting professor at California State University Long Beach USA from 1988 to 1990. Presently he is an Emeritus Fellow in the Department of Electrical Engineering at IIT New Delhi. His fields of interest include Energy conservation in electrical machines and drives, and analysis and control of electrical machines. Prof. Singh is a Fellow of the Institution of Engineers (India), and a Life Member of the Indian Society for Technical Education.

**Sanjeet Dwivedi** (Student Member) was born in Chhatarpur (M.P.) India, in 1968. He received B.E. (Electrical) degree from the Government Engineering College Jabalpur, India, in 1991 and the M.E. degree (with Gold Medal) from the University of Roorkee, Roorkee, India, in 1999. and Ph.D. Degree from Electrical Engineering Department of IIT Delhi in 2006 .

In 1991, he joined Larson and Toubro Ltd as Graduate Engineer Trainee. In November 1993 he joined as Lecturer at Department of Electrical Engineering, Indira Gandhi Engineering College Sagar(M.P.), whrere he became Senior Lecturer in 1999 and Reader in 2004. He is presently working in Indira Gandhi Engineering College Sagar (M.P.) India. His research interests are in area of, digital control of Brushless Motors, sensor reduction techniques in ac drives and power quality improvement aspects of ac drives.

**2006 IEEE International Conference on Power Electronic, Drives and Energy Systems**

# Single-Phase Resonant Converter with Active Power Filter

M. A. Chaudhari, and H. M. Suryawanshi, *Member, IEEE*

*Abstract--* **A single-phase ac-to-dc converter using modified series-parallel resonant converter (MSPRC) in conjunction with active power filter (APF) is presented for improvement of power factor on input ac line. The pulsating DC voltage obtained from the single phase uncontrolled diode bridge rectifier is applied to high frequency full bridge configuration of the MSPRC. For high power factor operation small active power filter is used. Analysis and design of converter for low current harmonic distortion and high power factor operation is given. An APF consists of pulse width modulated (PWM) voltage source inverter that is operated in variable switching frequency mode. The operation of the converter with and without active control is presented. The output voltage of the converter is regulated independently by using variable frequency control and duty ratio control. The MSPRC operates on lagging power factor mode for the entire load range to maintain the zero-voltage switching. The design procedure of the converter is outline in detail. The converter is simulated using MATLAB (Simulink). To validate the simulated performance an experimental (10 kW) prototype of single-phase ac-to-dc resonant converter is built using high frequency devices, MOSFETs. Experimental results show the high power factor operation of the converter through out the load range.**

*Index Terms*—**Ac-to-dc converter, Active Power Filter, power factor, PWM, Resonance frequency, zero-voltage switching (ZVS) etc.**

## I. INTRODUCTION

SINGLE ended converters like Buck, Boost, Buck-Boost converters presented in the literature [1]-[2], use PWM hard switching technique. These converters have high switching stresses and high electromagnetic interference (EMI). The EMI is produced due to high $di/dt$ and high $dv/dt$ caused by PWM process. High frequency (HF) transformer gets saturated due to the unidirectional core excitation. The transformer core loss increases with the increase in switching frequency. They are not suitable for high power applications.

Resonant converters offer a novel solution to the problems faced by the PWM hard-switched converters. A number of high frequency resonant converters have been proposed and analyzed in the literature [3]-[9]. The series resonant converter [4] provides high efficiency at all load conditions but light load voltage regulation is not possible. It does not provide the voltage-boosting capability. The parallel resonant converter

Mrs. M. A. Chaudhari is with Department of Electrical Engineering, Y. C. C. E., Wanadongri, Nagpur, (M. S.), India. (e-mail: macavc@yahoo.com).

H. M. Suryawanshi is with the Department of Electrical Engineering, V. N. I. T., Nagpur, (M. S.), India. (e-mail: hms_1963@rediffmail.com).

[5] can provide considerable voltage boosting that is needed near the valleys of the ac line voltage but efficiency decreases with the decrease in load. The series-parallel resonant converter [6]-[7] combines the desirable characteristics of the series and parallel resonant converter. The modified series-parallel resonant converter has lower component stresses, high power factor performance and high efficiency compared to the series-parallel resonant converter [8]. But at light load conditions the power factor of MSPRC decreases with duty ratio control. Hence for improvement of power factor at any desired load active power filter (APF) is used. A simple P-I based APF are presented in [10], [11] to provide reactive power and harmonic compensation for variety of single-phase loads. An APF can be operated in variable frequency [12]-[13] or fixed switching frequency [14] mode. A fixed frequency switching is achieved by comparing current error signal with a triangular reference waveform.

In this paper, an improvement of power factor on the ac line side of the single-phase ac-to-dc converter using modified series-parallel resonant converter in conjunction with active power filter is proposed. The high power factor operation of the converter with reduced line current harmonic distortion is presented. The output voltage of the converter is regulated independently by using variable frequency control and duty ratio control. The power factor of the input line of the converter decreases with the decrease in load. However to avoid this, a small APF is used in conjunction with MSPRC to maintain high power factor at any desired load.

The proposed scheme consists of an uncontrolled bridge rectifier followed by small dc link capacitor $C_{in}$, connected to MSPRC. The single-phase APF is used to compensate the reactive power, to eliminate the current harmonic content and to get high input power factor of the ac-to-dc converter. It consists of voltage source inverter (VSI) with a dc capacitor as energy storage device. A simple PI (Proportional-Integral) controller is used to control the dc bus capacitor voltage. The APF eliminates harmonics and provides reactive power requirement of load, so the ac source feeds only fundamental frequency sinusoidal component of current at unity power factor. Since the APF is connected in shunt with the ac-to-dc converter, it improves the system efficiency because it does not process the power delivered to the converter. The main features of the ac-to-dc converter are high frequency (HF) transformer isolation, use of leakage inductance of HF transformer as part of the resonant tank circuit, low input line current harmonic distortion, high power factor through out the loading conditions and high efficiency. This ac-to-dc converter with its active control is best suited for high voltage

0-7803-9771-1/06/$25.00 ©2006 IEEE

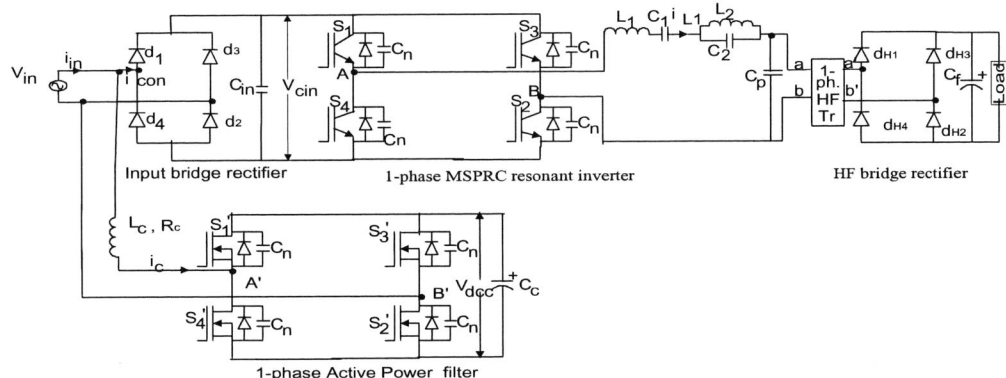

Fig. 1. Single-phase ac-to-dc resonant converter with active power filter.

## II. OPERATING PRINCIPLE

The HF transformer isolated single-phase ac-to-dc converter using modified series-parallel resonant converter in conjunction with the APF is shown in Fig. 1. The proposed converter consists of an uncontrolled bridge rectifier followed by a small dc link capacitor $C_{in}$ (high frequency by pass), connected to the modified series-parallel resonant converter. The input voltage to MSPRC is the rectified line voltage. The MSPRC is operated above the resonant frequency to maintain the zero-voltage switching.

Fig. 2 shows the idealized waveforms of the MSPRC, which includes voltage and current waveforms of reactive elements. The main section of the APF shown in Fig. 1 is a voltage source inverter (VSI) connected to the dc bus capacitor. The current harmonic compensation is achieved by injecting the equal but opposite current harmonic at the point of connection, thereby canceling the original distortion and improving the power quality on connected power system. Fig. 3 shows the control scheme of APF. Instantaneous dc bus voltage ($V_{dcc}$), supply voltage ($V_{in}$) and converter current ($i_{con}$) are sensed to obtain the switching signals to control the switching devices of APF. The ac source supplies the fundamental active power component of the converter current and the fundamental component of current to maintain average dc bus voltage of the APF to a constant value. The later component of current is to supply losses in VSI, such as switching losses, capacitor leakage current etc. The sensed dc bus voltage ($V_{dcc}$) is compared with the dc reference voltage ($V_{dcref}$). The output of the comparator is error signal $e(t)$. This error signal is then processed in a P-I controller and the peak value of reference supply current ($I_{sm}^{*}$) is obtained. The unit vector $u(t)$ of supply voltage is derived from its sensed value. The peak value of reference supply current ($I_{sm}^{*}$) is multiplied with the unit vector to generate reference sinusoidal unity power factor current ($i_s^{*}$). The reference supply current

($i_s^{*}$) is compared with the actual converter current ($i_{con}$) to give reference APF current ($i_c^{*}$). The actual APF current ($i_c$) and the reference APF current ($i_c^{*}$) are processed in a hysteresis current controller to derive gating signals of the devices (*MOSFETs*) of the APF. The corrective action is taken in each half cycle of the ac source, resulting in fast dynamic response of the APF. The output voltage of the converter is regulated by varying the switching frequency or by controlling the duty ratio.

## III. MODELING OF MODIFIED SERIES-PARALLEL RESONANT CONVERTER

The MSPRC with output capacitor filter has been analyzed using an approximate complex ac circuit analysis. The sine wave approximations are quite good [8]. Due to the filtering action of the resonating tank circuit, the fundamental current is much larger than the harmonics. Under these assumptions the simplified equivalent circuit at output terminal AB of HF inverter is shown in Fig. 4. The ac equivalent circuit resistance $R_{ac}$ replaces the HF transformer, the rectifier, capacitive filter and the load. Dynamic equations of resonant tank circuit are;

$$dV_{c1}/dt = i_{L1}/C_1 \tag{1}$$

$$dV_{c2}/dt = (i_{L1} - i_{L2})/C_2 \tag{2}$$

$$dV_{cp}/dt = i_{Cp}/C_p \tag{3}$$

$$di_{L2}/dt = V_{C2}/L_2 \tag{4}$$

$$di_{L1}/dt = (V_{AB} - V_{C1} - V_{C2} - V_{Cp})/L_1 \tag{5}$$

$$V_{Cp} = (I_{L1} - I_{Cp}) \cdot R_{ac} \tag{6}$$

$$V_{Cp}' = V_{Cp}/n \tag{7}$$

$$n = N_1/N_2 \tag{8}$$

where, $n$ is the turn ratio of the HF transformer.

$$R_{ac} = k \cdot R_L{}'$$ (9)

where, $R_L{}' = n^2 \cdot R_L$ and $k$ is a constant. For capacitive inductive output filter it is given by,

$k = 8/\pi^2$ and $k = \pi^2/8$ respectively.

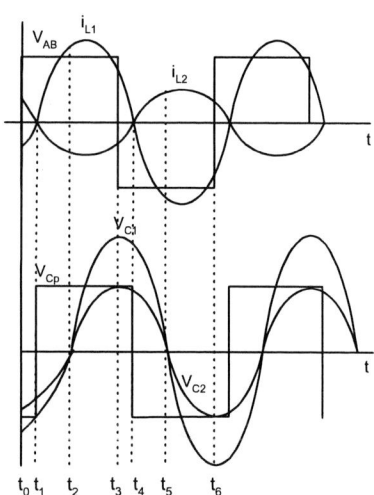

Fig. 2. Idealised waveforms of modified series-parallel resonant tank circuit.

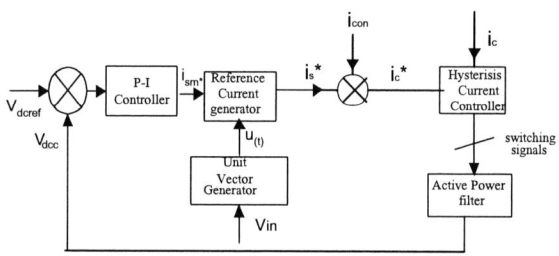

Fig. 3. Control scheme of APF.

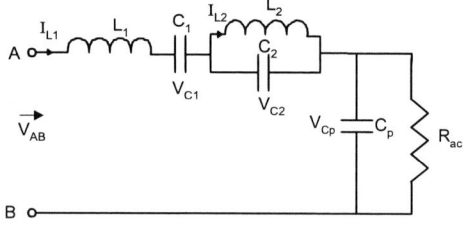

Fig. 4. Equivalent circuit of MSPRC.

## V. DESIGN OF RESONANT COMPONENTS

The single-phase ac-to-dc converter of 10 kW is simulated at switching frequency of 10 kHz. Simulation is carried out at lower switching to avoid large memory and high simulation time requirement problem. The experimental prototype of 10 kW is designed using *IRFP*450 MOSFETs. The MSPRC is operated on 200 kHz switching frequency. The specifications of the converter are as,

Minimum input ac voltage = $V_{in}$ = 230 V (r.m.s.), 50 Hz
Output dc voltage = $V_o$ = 1000 volt
Output power of the converter = $P_o$ = 10 kW
Minimum switching frequency = $f_s$ = 200 kHz

$$\omega_0 = \sqrt{\left[\left(\frac{1}{2L_1 C_1}\right) \cdot a \cdot \left(b + \sqrt{b^2 - 4}\right)\right]}$$ (10)

$$a = \sqrt{\frac{L_1}{L_2} \frac{C_1}{C_2}},$$

$$b = \sqrt{\frac{L_1}{L_2} \frac{C_1}{C_2}} + \sqrt{\frac{L_2}{L_1} \frac{C_1}{C_2}} + \sqrt{\frac{L_2}{L_1} \frac{C_2}{C_1}},$$

$$Q_s = \omega_o L_1 / R_L{}'$$ (11)

$$Y_s = \omega_s / \omega_o$$ (12)

where, $\omega_s$ is switching frequency, $\omega_0$ is resonance frequency, $Q_s$ is the quality factor of MSPRC, and $Y_s$ is the ratio of switching frequency to resonance frequency. Following the design procedure of the converter [8], near optimum design the ratios are selected as,

$L_1/L_2$ =0.1, $C_1/C_2$ = 0.1, $C_1/C_p$ = 10, $Q_s$ = 1.5.
$Y_s$ = 1.03 is set to achieve ZVS operation of the switches. Using above ratios and equations (9) to (12),
$L_1$ = 689 µH, $L_2$ = 6890 µH, $C_1$ = 0.4047 µf, $C_2$= 4.047 µf, $C_p$ = 0.0 4047 µf, $R_L{}'$ = 5.29Ω.

## V. MODELING OF SINGLE-PHASE APF

### A. P-I controller:

$$e(t) = V_{dcref} - V_{dcc}$$ (13)

$$I_{sm}{}^* = e(t) \cdot K_p + K_i / T_i \cdot \int e_{(t)} dt$$ (14)

where, $K_p$ and $K_i$ are the proportionality and integral gain constants of the P-I controller. The $I_{sm}{}^*$ is the peak value of reference supply current.

### B. Estimation of reference supply current:

$$i_s{}^* = u(t) \cdot I_{sm}{}^*$$ (15)

where, $u(t)$ is the unit vector for input voltage $V_{in.}$

### C. Estimation of reference APF current:

$$i_c{}^* = i_s{}^* + i_{con}$$ (16)

where, $i_{con}$ is the input current of the ac-to-dc converter without APF.

### D. Hysteresis current controller:

If $\left(i_c > i_c{}^* + h_b\right)$ $S_1'$ and $S_2'$ ON, $S_3'$ and $S_4'$ OFF

If $\left(i_c < i_c{}^* - h_b\right)$ $S_1'$ and $S_2'$ OFF, $S_3'$ and $S_4'$ ON

350

$S_1'$, $S_2'$, $S_3'$, $S_4'$ are the switching devices of the APF and $h_b$ is the hysteresis bandwidth in ampere.

### E.  APF analysis:

The $V_s$ is the ac PWM voltage reflected on the ac input side of the APF. $V_s$ can be expressed in terms of switching functions as;

$$V_s = V_{dcc} \cdot (S_A - S_B) \tag{17}$$

where, $V_{dcc} = 1 / C_c \cdot \int i_{dc} dt$

$S_A$ and $S_B$ are the switching functions. These functions generate gating signals for the switches of the APF.

$$S_A = 1, \text{ if } S_1' \text{ is ON and } S_A = 0, \text{ if } S_4' \text{ is ON} \tag{18}$$

$$S_B = 1, \text{ if } S_3' \text{ is ON and } S_B = 0, \text{ if } S_2' \text{ is ON} \tag{19}$$

$$Rc \cdot ic + Lc \cdot (dic / dt) + Vs = Vin \tag{20}$$

$$i_{dc} = (S_A - S_B) \cdot i_C \tag{21}$$

$i_c$ is the actual current generated by the APF. $R_c$ and $L_c$ are the resistance and inductance of the APF inductor respectively. The parameters of APF are;

$$L_c = 0.051 mH, C_c = 8000 \mu f, R_c = 0.01 \Omega, h_b = 0.1,$$
$$V_{dcref} = 600 V, K_p = 0.32, K_i = 0.3.$$

## VI.  SIMULATION RESULTS

MATLAB (SIMULINK) simulation results are given for the single-phase ac-to-dc converter (230 V, 10 kW). The switching frequency of the MSPRC was 10 kHz. The output voltage is regulated at 985 volt using variable frequency control and duty ratio control independently. The performance of the converter from full load to 25% of the rated load with and without active control of the line current is studied.

Fig. 5(a) and Fig. 5(b) show the waveforms at full load (5.29 $\Omega$) for MSPRC. High frequency (10 kHz) square-wave output voltage of the inverter ($V_{AB}$), input current of the resonant tank circuit ($i_{L1}$), voltage across the parallel capacitor ($V_{c2}$) and current through the parallel inductor ($i_{L2}$) are plotted here. These waveforms clearly demonstrate that all the switches turned on under zero voltage switching (ZVS). Hence; the turn on losses are eliminated. It is also observed that the switch peak current decreases with the decrease in load current. Thus high part load efficiency is obtained.

Fig. 6(a) shows the waveforms for input voltage ($V_{in}$), input current ($i_{in}$) with active power filter on 50% of full load. Fig. 6(b) shows the actual current drawn by converter ($i_{con}$), compensated current drawn by the APF ($i_c$), and the instantaneous dc bus voltage across the dc capacitor of the APF ($V_{dcc}$). In order to avoid time and memory requirement problem at higher switching frequency, simulation is carried out at lower switching frequency (10 kHz). Increase in switching frequency may increase the input power factor of the converter current.

The input supply voltage ($V_{in}$) and the input line current ($i_{in}$) of the ac-to-dc converter with active power filter are in phase. The switching frequency required to regulate the output voltage on 50% and 25% of the full load was 15 kHz and 22 kHz respectively.

The performance of the ac-to-dc converter with and without active power filter for duty ratio control is also studied. The waveforms for input voltage ($V_{in}$), input line current ($i_{in}$), actual current drawn by the converter ($i_{con}$), the compensated current drawn by the APF ($i_c$) and instantaneous dc bus voltage across the dc capacitor ($V_{dcc}$) of the APF are shown in Fig. 7(a, b) on 50% of full load.

The power factor of the converter decreases with the decrease in load under duty ratio control. Thus, the proposed converter operates on high input power factor through out the loading conditions. The dc bus voltage across the capacitor overshoots to maximum value and then settles to final value. Initial charging current of the capacitor is very high. This can be controlled by initially placing a resistor in series with the capacitor, and shunting it after it is charged or by increasing the supply voltage from zero value to the required value in a controlled way. The inner close loop of the APF generates switching signals for APF devices. Current harmonic compensation is achieved by injecting equal but opposite current harmonic ($i_c$) at the point of connection there by canceling the original distortion in converter current ($i_{con}$) and improving input power factor.

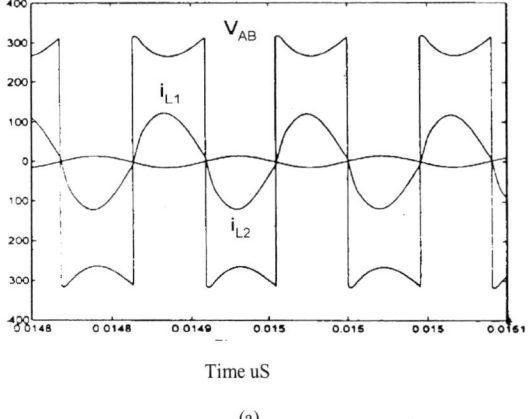

Time uS

(a)

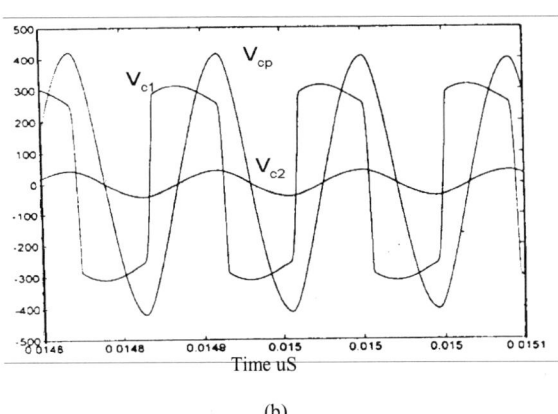

Time uS

(b)

Fig. 5. Simulated waveforms for MSPRC (a) Output voltage of the HF full bridge inverter ($V_{AB}$), Current through series inductance ($i_{L1}$), Current through parallel inductance ($i_{L2}$). (b) Voltage across parallel capacitor ($V_{cp}$), Voltage across capacitor C$_1$ ($V_{c1}$), Voltage across parallel capacitor C$_2$ ($V_{c2}$). Time: µsec.

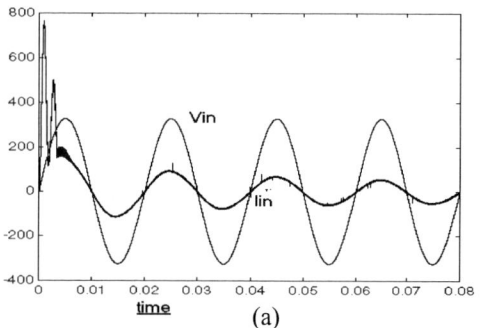

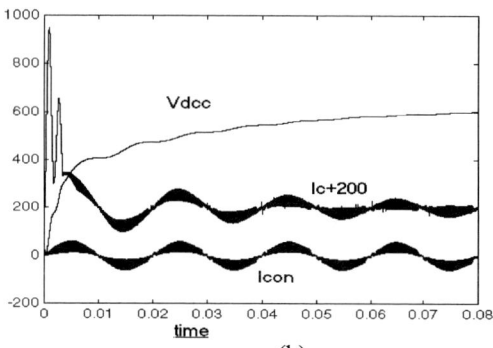

Fig. 6. Simulated waveforms for the ac-to-dc resonant converter with the APF on 50% load with the variable frequency control. (a) Input Voltage ($V_{in}$), Input current ($I_{in}$). (b) Voltage across capacitor of APF ($V_{dcc}$), Compensation current drawn by APF ($I_c+200$), Current drawn by ac-to-dc resonant converter ($I_{con}$).

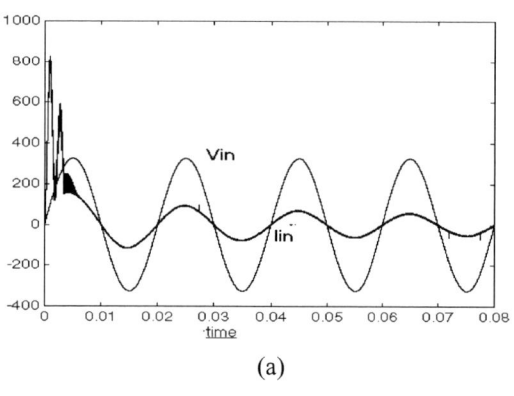

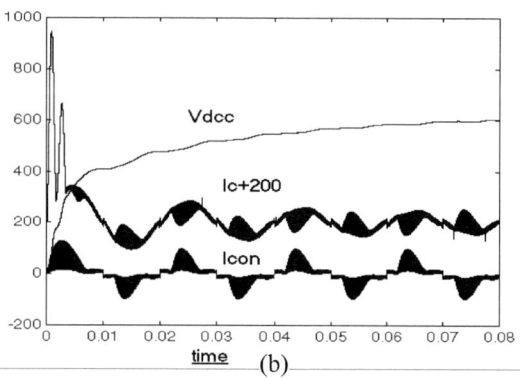

Fig. 7 Simulated waveforms for the ac-to-dc resonant converter with the APF on 50% load with the duty ratio control. (a) Input Voltage ($V_{in}$), Input current ($I_{in}$). (b) Voltage across capacitor of APF ($V_{dcc}$), Compensation current drawn by APF ($I_c+200$), Current drawn by ac-to-dc resonant converter ($I_{con}$).

## VII. EXPERIMENTAL RESULTS

A 10 kW experimental prototype of the ac-to-dc resonant converter using MOSFETs was implemented and successfully tested. The MSPRC was operated on 200 kHz switching frequency and on lagging power factor mode (above resonance frequency operation). The steady state and the transient results with and without active power filter are depicted here. The output voltage is regulated at 982 Volt using variable frequency and duty ratio control independently.

Fig. 8 shows the waveforms for input voltage ($V_{in}$), input current ($i_{in}$) on full load, 50% load and 25 % load. The output voltage is controlled using variable frequency control. The Fig. 9(a, b) shows the waveforms for input voltage ($V_{in}$), input current ($i_{in}$) with and without active power filter on 25 % of full load. The output voltage is controlled using duty ratio control. The input supply current without APF is highly non-sinusoidal and the power factor is low. If the APF is connected across the ac-to-dc resonant converter the input line current is almost sinusoidal and in phase with the input voltage. It is observed that the line current remains less than actual converter because the source feeds only active power component and reactive power component is supplied by the APF.

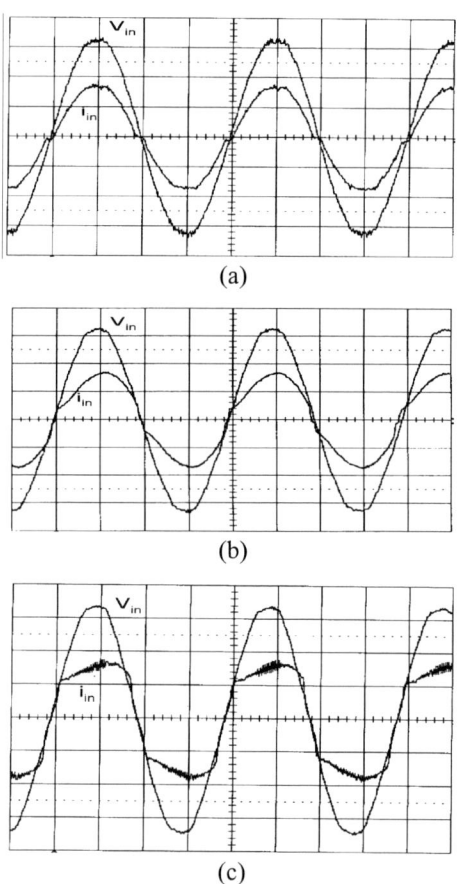

Fig. 8. Experimental waveforms for AC-to-DC resonant converter for input voltage $V_{in}$, input current $i_{in}$, with variable frequency control. (a) on full load $V_{in}$ (100V/div), $i_{in}$ (4 A/div) (b) on 50 % of full load $V_{in}$ (100V/div), $i_{in}$ (2A/div) and (c) on 25 % of full load.. $V_{in}$ (100V/div), $i_{in}$ (1A/div), Time scale: 5 ms/divison.

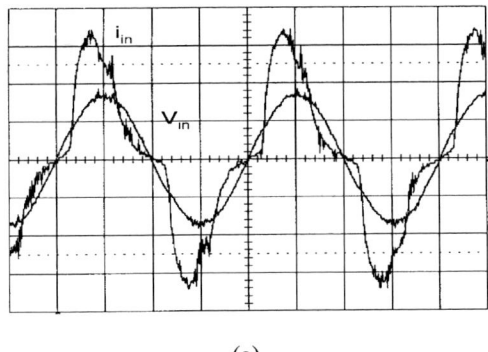

(a)

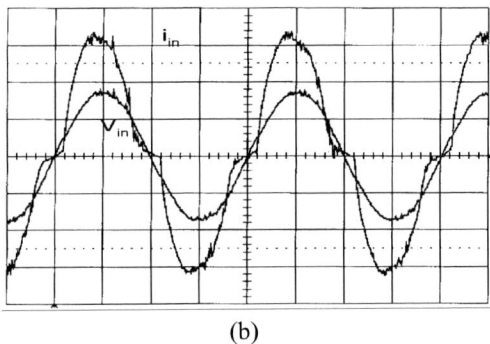

(b)

Fig. 9. Experimental waveforms for the ac-to-dc converter for input voltage $V_{in}$, input current $i_{in}$. (a) on 25 % of full load, without APF using duty ratio control $V_{in}$ (200V/div), $i_{in}$ (0.5A/div), (b) on 25 % of full load, with APF using duty ratio control $V_{in}$ (200V/div), $i_{in}$ (0.5A/div), Time scale: 5 ms/divison.

It is also noted from these waveforms that the power factor decreases with decrease in load under duty ratio control. This is due to the fact that the power factor depends on ratio of switching frequency to resonant frequency. The APF eliminates harmonic components effectively and is able to compensate the reactive power required.

## VIII. CONCLUSIONS

In this paper, a single-phase ac-to-dc converter using modified series-parallel resonant converter in conjunction with the APF is presented. 10 kW experimental prototype converter using MOSFETs was implemented and successfully tested. MATLAB (Simulink) results are substantiated by the experimental observations. It was noted that the input line current and the switch peak current decreases with the decrease in load thus the converter has high part load efficiency. The efficiency of the converter on full load is 94%.

Active control of the line current is used to minimize the current harmonics generated by the converter particularly when it is not operated on full load. APF improves the power factor of the input line current through out the loading conditions. The APF eliminates frequency harmonic components effectively and is able to compensate the reactive power required by the ac-to-dc converter. Hysteresis current controller is used to obtain the gate signals for switching devices of APF. Since all quantities such as dc voltage are symmetric and periodic corresponding to half cycle of ac source, corrective action is taken in each half-cycle result in fast dynamic response of APF. It is noted from the waveforms that the converter enter from continuous current mode to discontinuous current mode with decrease in load. But as active control of input line is used the power factor of the converter is high at light load also. Hence, the proposed converter operates on high input power factor through out the loading conditions.

## REFERENCES

[1] M. C. Ghanem, K.A. Haddad, G. Roy, "A New control strategy to active sinusoidal line current in a cascade Buck-Boost converter," *IEEE Trans. Industrial Electronics*, Vol. 43, No.3, pp 441-449, 1996

[2] Y. Toshiya, S. Osamu, M. Osamu , "An improvement technique for the efficiency of high frequency switch mode rectifier," *IEEE Trans. on Power Electronics*, Vol. 15, No. 6, PP 1118-1123, 2000.

[3] H. M. Suryawanshi, S. G. Tarnekar, "Resonant converter in high power factor, high- voltage dc applications," *IEE Proc.-Electr. Power App.*, Vol 145, No. 4, pp.307-314, 1998.

[4] Quero J. M., Carrasco J.M., Franquelo L.G., "Implementation of a neural controller for the series resonant converter", *IEEE Transactions on Industrial Electronics*, vol. 49, no.3, pp. 628-639, June 2002.

[5] Wong S.C., Brown A.D., "Parallel resonant converter as a circuit simulation primitive", *IEE Proceedings on Circuits, Devices and Systems*, vol. 142, no. 6, pp. 379-386, Dec. 1995.

[6] Belaguli V., Bhat A.K.S., "Series-parallel resonant converter operating in discontinuous current mode. Analysis, design, simulation, and experimental results", *IEEE Transactions on Circuits and Systems I: Fundamental Theory and Applications*, Vol. 47, no. 4, pp. 433-442, April 2000.

[7] A J. Forsyth, G. A.Ward and S. V. Mollov, "Extended fundamental frequency analysis of the LCC resonant converter," *IEEE trans. Power Electronics*, Vol. 18, No. 6, pp 1286-1292, November 2003.

[8] H. M. Suryawanshi, K. L. Thakre, S. G. Tarnekar, D. P. Kothari, A. G. Kothari, "Power factor improvement and closed loop control of an ac-to-dc resonant converter", *IEE Proceedings - Electric Power Applications*, vol. 149, no. 2, pp. 101-110, March 2002.

[9] S. O. Demercil Jr., B. Ivo, " A Three-phase ZVS PWM DC/DC converter with Asymmetrical duty cycle for high power applications," *IEEE Trans. Power Electronics*, Vol. 20 , No. 2, PP 370-377, March 2005.

[10] B. Singh, K. A. Haddad, A Chandra, "Universal Active Power Filter for single-phase reactive power and harmonic compensation," *International conference on IEEE Power Quality*, pp.81 –87, 1998.

[11] A. Mauricio and et al, "An universal active power line conditioner," *IEEE Trans. On Power delivery*, Vol. 13, No. 2, pp.969-977, 1998.

[12] *Q. Chongming, "Three-phase Bipolar Mode Active Power Filters,"* IEEE Trans. Industry Appl., *Vol. 38, No. 1, pp 149-158, 2002.*

[13] P. Jintakosonwit, H. Fujita, H. Akagi, "Control and performance of a fully- digital controlled shunt active filter for installation on a power distribution system," *IEEE Trans. Power Electronics*, Vol. 17 , No. 1, PP 132-140, January 2003.

[14] L. A. Moran., J. W. Dixon and R. R. Wallace, "A three-phase active power filter operating with fixed switching for reactive power and current harmonics compensation," *IEEE Trans. on Industrial Electronics*, Vol.42, No.4, pp.402-408, Aug.1995.

**2006 IEEE International Conference on Power Electronic, Drives and Energy Systems**

# PV Power Tracking Through Utility Connected Single-Stage Inverter

K. S. Phani Kiranmai, *Student Member, IEEE*, and Veerachary. M, *Senior Member, IEEE*

*Abstract*— A three-loop control scheme for the PV supplied single stage inverter is proposed in this paper. Integrating the tracking algorithm together with hysteresis control technique evolves this scheme. The feed-forward loop performs the maximum power extraction, while the remaining two inner-loops maintains unity power factor operation of the inverter. To study the effectiveness of the proposed control scheme simulation models are developed in PSIM platform, which includes photovoltaic source realization, control scheme and utility system. The concept is demonstrated through simulation results.

*Index Terms*— Hysteresis control, Maximum power point tracking, Single-stage inverter, Three-loop control.

## I. INTRODUCTION

THE photovoltaic (PV) source is a non-linear DC source and it needs a maximum power point tracking (MPPT) scheme to improve the overall efficiency. The most important design constraints of a PV system are low mass and size, high efficiency, and reliability, etc. In remote areas where enough sunlight is present then the PV power is the best solution for providing electricity. These PV power systems are broadly classified into: (i) stand-alone systems, (ii) grid/ utility connected systems. Each of these systems has their own advantage and limitations. Despite their disadvantages the stand-alone systems are still in use when there is no possibility of extending the grid-power. Grid-connected PV systems (GCPV) are popular in hybrid power systems wherein the PV supplies full/ partial load during daytime, while the rest of the load demand is compensated from other power sources. In any of these systems the efficiency of power conversion is of at most important issue. Depending on the available options, the PV system designer employs either a Single-stage (SS) or Multi-stage (MS) approach for the PV power conversion schemes. Although the MS approach has

salient features, its main limitations are: (i) overall efficiency is low, (ii) requires more number of intermediate power converters, (iii) complexity in the associated control circuitry, (iv) poor utilization of the switching devices at lower solar insolations, etc. In view of above disadvantages the existing MS PV systems are slowly replacing with SS systems. Important benefits obtained with these SS systems are: (i) eliminates additional intermediate dc-dc conversion stage, (ii) possibility of embedding maximum power point tracking within the dc-ac conversion systcm, (iii) better overall efficiency, etc. As the conversion efficiency of the PV array is about 20 to 30% adding more number of power processing stages reduces the overall efficiency. The only way to realize high efficiency PV system is by having single-stage power conversion scheme (SPCS). A SPCS uses one inverter and its controller must be designed such that maximum power should be extracted from the PV source side, while maintaining the sinusoidal current on the load side.

Several inverter topologies suitable for SPCS's have been reported in literature [1]-[8]. In all these schemes MPPT loop is decoupled from the inverter load side control loops and hence these schemes falls under decentralized schemes. However, the centralized scheme is most suitable for the SPSC's, as it avoids the problems arising due to decoupled loop structure, and hence one such scheme, three loop control scheme (TLCS), is proposed in this paper.

## II. SINGLE-STAGE PV CONVERSION SYSTEM

The block diagram of the proposed single-phase single-stage GCPV is shown in Fig. 1. It essentially consists of: (i) PV source, (ii) an H-bridge inverter connected to utility, (iii) three-loop control scheme. Detailed description of these components is given in the following lines.

### A. PV Generator Model

The PV generator is formed by the combination of many PV cells connected in series and parallel fashion to provide desired value of output voltage and current. This PV generator exhibits a non-linear insolation dependent *v-i* characteristic, mathematically expressed for the PV array [10]-[11] consisting '$N_s$' cells in series and '$N_p$' cells in parallel as

---

This work was supported in part by the MHRD, Govt. of India under R & D Project Grant.

K. S. Phani Kiranmai is with Dept. of Electrical Engineering, Indian Institute of Technology Delhi, New Delhi, India (e-mail: ksphanikiran@ee.iitdac.in).

Veerachary. M is with the Department of Electrical Engineering, Indian Institute of Technology Delhi, New Delhi, India (e-mail: mvchary@ee.iitdac.in).

0-7803-9771-1/06/$25.00 ©2006 IEEE

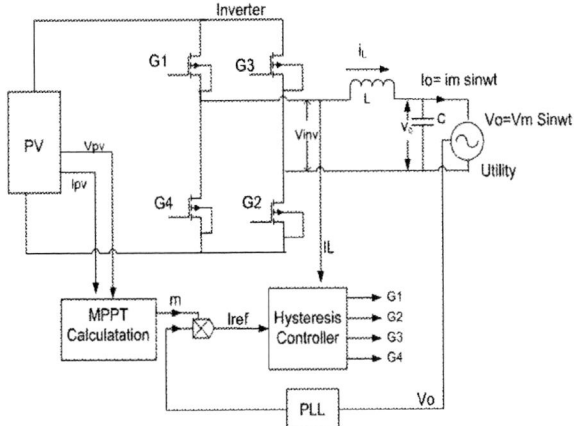

Fig. 1. Block diagram of the proposed single-stage inverter with TLCS.

$$I_{pv} = I_{Ph} - I_O (e^{q*V_{pv}/A*k*T_C*N_S} - 1) * N_P \qquad (1)$$

where q-electric charge; A- Completion factor; K- Boltzman's constant; T- Absolute temperature; $I_{ph}$- photo current; $I_o$- cell reverse saturation current; $I_{pv}$, $V_{pv}$ are the PV array current and voltage respectively. Using eqn. 1 the PV source model was developed for use in PSIM simulator [13].

The eqn. 1 forms the basis for PV source representation within the simulator and it is used to study the proposed SPCS performance. To extract maximum power from the PV array, it is required to adjust the load impedance, connected across the PV terminals, which may be composed of a dc-dc converter or an inverter. Since the proposed studies are focusing on the inverter-based schemes, we will restrict ourselves to inverter matching methodologies and their mathematical analysis.

### B. Inverter modeling and controlling

The proposed SPCS consists of an H-bridge inverter with necessary control circuitry in addition to the PV source. This inverter is simple in construction and uses only four switching devices, $G_1 - G_4$. However, the inverter performance mainly depends on: (i) type of switching, i.e. unipolar or bipolar, (ii) fixed frequency or variable frequency operation, (iii) number of control loops used in the switching signals generation etc. Unipolar switching together with fixed frequency operation extensively studied and some investigations were also reported [12]. In this paper the authors have made an attempt to combine the inverter bipolar operation with variable switching frequency for the PV applications. Although there may be several possibilities in the control schemes for realizing the variable frequencies, but we will be concentrating on the hysteresis control technique.

Here the inverter is controlled by means of hysteresis technique wherein it produces variable switching frequency within the given ac line voltage cycle. At the outset modeling of such inverter systems appears to be difficult due to variable switching frequency operation. Although the switching frequency of the inverter is variable but the ac line frequency

is constant, which is equal to utility frequency 50 or 60 Hz, and if we observe closely each half line cycle consists of switching pulses which are repetitive in the subsequent half-line cycles. Due to this fact the inverter ac voltage waveform can be decomposed into Fourier series/ components (FC). However, while finding the FC of the inverter ac voltage waveforms, it is required to decompose the ac waveform into "$k_n$" number of different waveforms, each having periodicity of "$T_L$", where $T_L$ is the time period of ac voltage. In order to estimate the FC of the resulting waveform it is better to evaluate the FC's of individual waveforms, "k" different waveforms, and then combine the respective FC's. Adopting above-mentioned procedure the inverter voltage is expressed in terms of Fourier Series as

$$V_{t(n)} = \left( A_n^2 + B_n^2 \right)^{1/2} = \frac{4V_{pv}}{n\pi} \sum_{j=1}^{N_p} \sin \frac{n}{2} (\delta_j - \delta_{j-1}) \qquad (2)$$

where $A_n = \frac{2V_{pv}}{n\pi} \sum_{i=1}^{N_p} (-1)^{i+1} [\sin n(\delta_i) - \sin n(\delta_{i-1})]$,

$B_n = \frac{2V_{pv}}{n\pi} \sum_{j=1}^{N_p} (-1)^{j+1} [\cos n(\delta_{j-1}) - \cos n(\delta_j)]$,

$\delta_j = \omega t_j$ , where

$$t_j = \frac{2h + St_{j-1}}{S} + \frac{k\left[\sin(\omega t_{j-1}) - \sin(\omega t_j)\right]}{(-1)^j S}, \quad k = \frac{V_m}{V_j},$$

$$S = \frac{(V_j - V_o)}{L} \quad \text{when } G_1, G_2 \text{ ON.}$$

$$= -\frac{(V_j + V_o)}{L} \quad \text{when } G_3, G_4 \text{ ON.}$$

Using simple mathematical operations it is easy to establish the relationship between the inverter load and PV source voltage as

$$\frac{V_t}{V_{pv}} = \frac{4}{\pi} \sum_{k=1}^{n} \left[ \frac{1}{k} \sum_{j=1}^{N_p} \sin \frac{n}{2} (\delta_j - \delta_{j-1}) \right] \qquad (3)$$

Neglecting the losses in the inverter switches, the relationship between input and load current of the inverter is

$$\frac{I_{pv}}{I_t} = \frac{4}{\pi} \sum_{k=1}^{n} \left[ \frac{1}{k} \sum_{j=1}^{N_p} \sin \frac{n}{2} (\delta_j - \delta_{j-1}) \right] \qquad (4)$$

From eqns. 3 and 4 the effective impedance on the PV array $Z_i$ is given by

$$Z_i = \frac{V_t}{I_t \left( \frac{4}{\pi} \sum_{k=1}^{n} \left[ \frac{1}{k} \sum_{j=1}^{N_p} \sin \frac{n}{2} (\delta_j - \delta_{j-1}) \right] \right)^2} \qquad (5)$$

From the above equation it can be seen that the effective impedance appearing at the PV terminals depends on the time durations of inverter switching devices. Since the hysteresis

355

band is of sinusoidal nature, the time durations variations across the line cycle, during which the switching devices conducting, are also of similar kind. In other words above equation gives important information about the control strategy, which ultimately controlling injected current into the utility or supplied to the stand-alone ac load as well as the PV power extraction.

## III. CONTROL STRATEGY

The block diagram of the proposed SPCS with TLCS is shown in Fig. 1. Here the inverter is controlled by means of hysteresis strategy wherein at any given point of time either $(G_1 - G_2)$ or $(G_3 - G_4)$ switching pair will be conducting and thus load circuit is impressed with either $+V_{pv}$ or $-V_{pv}$. The gate switching signals for these devices are obtained by the three-loop control structure proposed in this paper. Various different loops involved in this control scheme are classified into: (i) feed-forward loop: which controls the operating point of the PV array close to maximum power point (MPP) by changing the turn ON and OFF periods of the inverter switching devices, (ii) feedback loops: there are two feedback loops one to sense the current and the other for load/grid voltage. The function of these two control loops is to generate an appropriate turn ON and OFF signals for the inverter switching devices in such a manner that the current injected by the inverter follows the inverter voltage or grid voltage leading to unity power factor operation (UPF). The inner current loop provides the instantaneous injected current information, while the outer voltage loop instantaneous grid voltage information. The outer voltage loop can't be combined directly with the feed-forward MPPT loop, because its function is to identify the instantaneous phase information (IPI). Simple voltage sensors will provide the IPI but synchronization of the utility voltage with the injected current will not be possible. In such cases the phase locked loop (PLL) [9] is need to be included in the outer voltage loop. This PLL gives a reference voltage/ synchronization signal, which will be combined with the feed-forward MPPT loop in order to generate the current reference signal to the inner current loop. Although we can use several possible structure for the PLL, but a proportional plus integral type is used in this analysis.

In both the loops, feed-forward and feedback, the operation mainly depends on the switch ON/OFF durations and hence demands different ON/OFF sequence if at all loops are decoupled. As a result, with decoupled loop control, it may not be possible to realize MPPT with UPF operation for all operating conditions. The proposed TLCS integrates the two different loops into a single scheme and also avoids problems mentioned above. Here the feed-forward loop, MPPT loop, generates the desired modulation index, while the feedback loop fixes the wave shape of the inverter current. Effectively, the product of the feed-forward and feedback loop quantities, as shown in Fig. 2, generates the current reference to the hysteresis controller. The hysteresis band, h, need to be chosen by the designer, which ultimately defines the maximum switching frequency for the inverter. Attention

must be paid while choosing the "h" magnitude, particularly in the PV systems operating at lower solar insolations.

### A. Hysteresis control

The hysteresis scheme used here for the single-phase full-bridge inverter is the conventional hysteresis current controller and its control logic schematic representation is shown in Fig. 2. Inverter switching devices $G_1$-$G_2$ and $G_3$-$G_4$ are switched whenever the injected current is out of the hysteresis band and it is mathematical representation is:

$$\text{if } i_0 > (i_{ref} + h) \text{ then } G_1 \text{ and } G_2 \text{ are ON}$$

$$\text{if } i_0 < (i_{ref} - h) \text{ then } G_3 \text{ and } G_4 \text{ are ON}$$

where '$i_0$' and '$i_{ref}$' are the actual and reference currents, respectively, and '$h$' is the hysteresis band. Since there are only two structural changes the load voltage waveform is bipolar nature, switching from plus to minus of dc voltage source and vice versa, and keeps the load current is within the hysteresis band.

The hysteresis current-controlled PWM scheme is very simple and also provides good current amplitude control. Due to variable frequency operation this control scheme shows robustness against parameter variation. However, its main limitation is the variable switching frequency, which depends largely on the variation of load parameters and source voltage. This presents little bit difficulty in the filter design. The advantage of hysteresis controller over sinusoidal PWM (SPWM) scheme is the controller is insensitive to filter parameter variations.

### B. Three-loop control strategy

The proposed TLCS basically contains (i) the feed-forward loop, (ii) inner current loop, (iii) outer voltage-loop. These three loops are interdependent and hence the abnormalities,

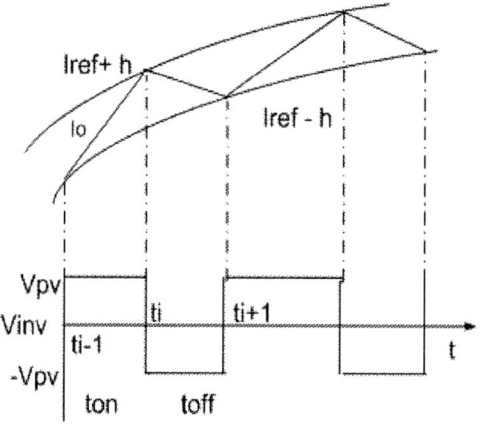

Fig. 2. Hysteresis control scheme for the proposed TLCS inverter.

either on the PV source side or on the load side, are immediately reflected in the inverter action. The feed-forward loop continuously measures the source voltage, current and generates an appropriate control signal, which will force the inverter operation such that the power extracted from the PV source is optimum. Although several possibilities exist for MPPT realization, a simple perturb and observation (P & O) method is employed in this paper. Detailed explanation of this algorithm was reported in [10]. Since the proposed STPC is designed for both MPPT operations as well as for sinusoidal current injection, the MPPT loop must be combined with the feedback-loop. In such cases the control signal produced by the MPPT loop will be useful to generate an appropriate reference signal for the inner loop. Depending on the way in which this signal is utilized in the subsequent loops gives different control options to the inverter, i.e fixed or variable frequency of operation, sinusoidal current injection, load voltage harmonic elimination/ minimization, etc. One such option, multiplying the feed-forward/MPPT loop control signal with grid voltage reference signal and generating and appropriate modulating signal, is reported for fixed frequency operation [12]. However, this two-loop control will not ensure the UPF operation on the inverter downstream side, and in order to inject sinusoidal current into the grid the inner current loop must be included. The detailed simulation diagram of the TLCS is shown in Fig. 3, wherein various loops are clearly identified.

## C. Inverter Dynamic Model

For the purpose of dynamic analysis the inverter internal behavior is neglected and it is assumed that inverter dynamics is mainly due to filter elements and down stream loads.

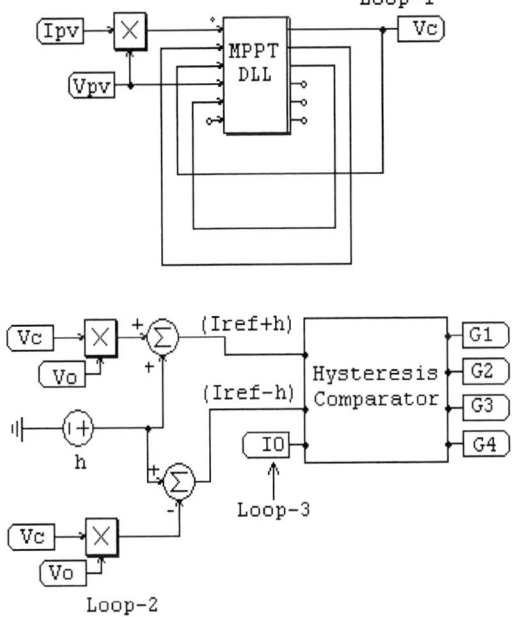

Fig. 3. Block diagram of proposed TLCS.

The inverter dynamic model needs to be developed in order to investigate the inverter stability aspects. To this direction state-space model of the single-stage inverter (SSI) is developed based on the following assumptions: (i) switching devices are ideal, (ii) non-idealities of the inverter are not going affect its dynamics, (iii) filter corner frequency is much smaller than the switching frequency of the inverter. For the purpose of analysis the inverter is assumed to be composed of two-buck converters one operates in the positive half line cycle (HLC) and the other comes into action during negative HLC. In bipolar switching, or wherever the in inverters using hysteresis control, the load circuit is impressed with either positive or negative dc voltage resulting two different linear circuits within one switching cycle. On the other hand in uni-polar constant switching frequency applications the load circuit is impressed either with a dc voltage or it is shorted through complementary switching device pairs. In any case there will not be much difference in the state-matrices except in the input or excitation matrix. This minor difference can easily be taken care and its generalized model can easily be developed. Assuming the inverter supplying an R-L load a state-space model is developed and it is given by

$$
\begin{bmatrix} \dot{i}_L \\ \dot{v}_c \\ \dot{i}_0 \end{bmatrix} =
\begin{bmatrix} -\dfrac{r_f}{L_f} & -\dfrac{1}{L_f} & 0 \\ \dfrac{1}{C_f} & 0 & -\dfrac{1}{C} \\ 0 & \dfrac{1}{L_0} & -\dfrac{R_0}{L_0} \end{bmatrix}
\begin{bmatrix} i_L \\ v_c \\ i_0 \end{bmatrix} +
\begin{bmatrix} \dfrac{d(-1)^{k+1}}{L_f} \\ 0 \\ 0 \end{bmatrix} V_{pv} \quad (6)
$$

where k= 1 for positive half cycle and 2 for negative half cycle. This equation is non-linear because of duty factor. To study the system stability aspects we have to linearize this model about the operating point and then a linear system tools can be utilized. In these studies the inverter stability issues have been studied only with filter parameters, which will be discussed in the next section.

## IV. SIMULATION RESULTS AND DISCUSSIONS

To verify the proposed single-stage inverter suitability for the PV applications extensive simulation studies have been carried out in the power electronic simulator PSIM. The PV system parameters used in the simulation are given in Table I. For demonstration few sample simulation results are presented here for the cases: (i) starting tracking capability, (ii) solar insolation change and, (iii) UPF operation on the inverter load side. Firstly, the simulation model for the PV source is developed using eqn. 1 and then model is verified. For illustration the v-i and v-p characteristics for two different solar insolations, ($\psi_1$, $\psi_2$), are shown in Fig. 4. Power tracking capability was also verified for different solar insolations and sample result is shown in Fig. 5 for the starting and steady-state power tracking characteristics when the solar insolation change from 1.0 → 0.7 p.u. From this figure it is clear that the maximum power extraction is possible at different solar insolations. To verify the UPF operation on the load side, while the inverter is tracking maximum power on source side, the load voltage and current

357

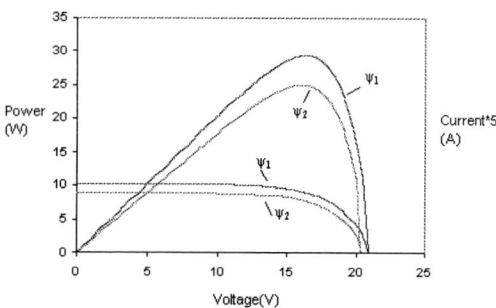

Fig. 4. Simulated (*v-i*)/ (*v-p*) characteristics of the PV source.

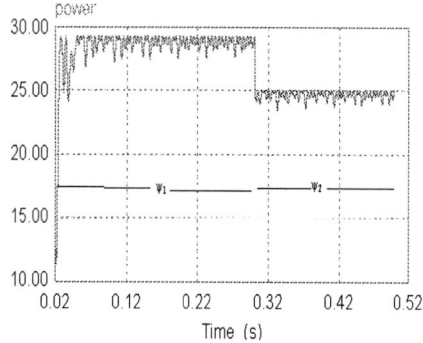

Fig. 5. Tracking characteristics of the PV-array.

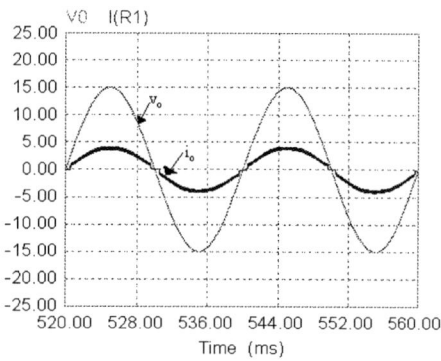

Fig. 6. Inverter load voltage and current waveforms.

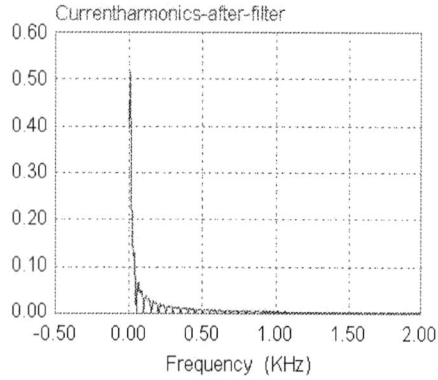

Fig. 7. THD profile of the load current.

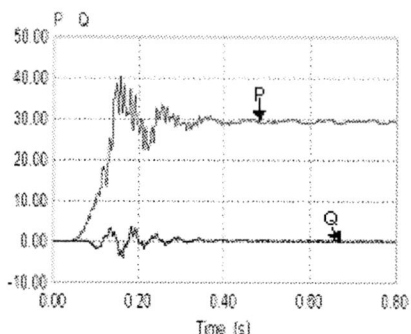

Fig. 8. Inverter active and reactive power.

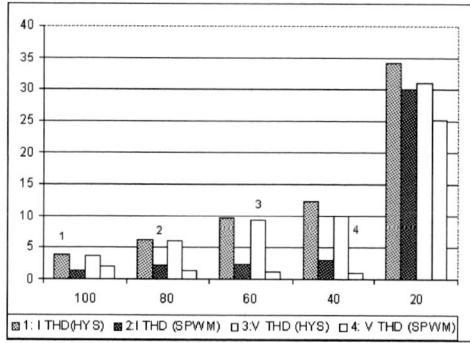

Fig. 9. Comparison of Load voltage and current THD.

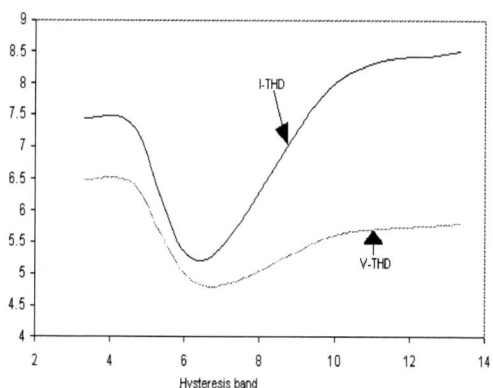

Fig. 10. THD variation with hysteresis band.

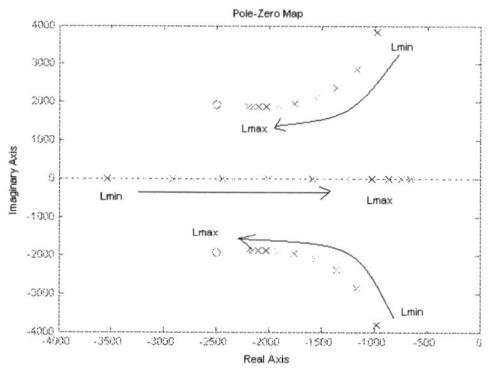

Fig. 11. Pole zero movement against filter inductor variation ($L_{min}$ = 1 mH, $L_{max}$=10 mH).

358

patterns are drawn in Fig. 6. Here the load current, '$i_o$' and grid voltage are almost in phase and thus resulting in UPF operation. Further, it is also observed, through simulations, that the load current and voltage are in phase even when the solar insolation is changing. Total harmonic distortion is plotted in Fig. 7 to verify the quality of load voltage and current harmonic profile. This shows that the harmonic content is very less which is mainly due to the modulation technique employed in this scheme. Active and reactive power supplied by the inverter to the load is also measured and these are shown in Fig. 8.

The performance of the proposed TLCS is compared with that of the simple two-loop sinusoidal PWM constant frequency control scheme. The load voltage and current total harmonic distortion bar chart is plotted in Fig. 9 for these schemes. This clearly shows that the proposed TLCS is resulting little higher THD content as compared to constant frequency control. The other reason, for higher THD, is the bipolar switching. However, as already mentioned that with the proposed scheme it is possible to get UPF operation at the expense of little bit higher THD. Proposed control scheme suitability and THD variation with hysteresis band is observed as shown in Fig. 10. To study the inverter stability aspects the pole-zero movement is plotted for different values of filter inductance, $L_{nominal} \pm \Delta L$ %, in Fig. 11. For obtaining these poles and zero's eqn. 6 was used. It is clear that the complex conjugate pair is moving away from the $j\omega$, while the other pole is moving toward the right half of s-plane. In any case for the expected inductance variation, the load current or solar insolation may influence, the inverter is within the stability limits.

## V. CONCLUSIONS

A SPCS suitable to achieve MPPT and UPF operation was proposed and demonstrated. To achieve the MPPT together with UPF we have used three different loops. To track maximum power the conventional perturb and observation method was used while hysteresis control was used to achieve UPF operation. Hysteresis current control reference was derived from the combination of the feed-forward and load voltage loops. Although the proposed scheme resulting little bit higher THD, as compared to constant frequency control, but capable of injecting sinusoidal grid current.

## VI. REFERENCES

[1] Enslin. J. H. R, Snyman. D. B, "Combined low-cost, high-efficient inverter, peak power tracker and regulator for PV applications," IEEE Trans. on Power Electronics, Vol. 6(1), 1991

[2] Hussein. K. H, Muta. I, Hoshino. T, Osakada. M, "Maximum photovoltaic power tracking: an algorithm for rapidly changing atmospheric conditions," in IEE Proc. Generation, Transmission and Distribution, Vol. 142(1), pp. 59 – 64, 1995.

[3] Yeong-Chau Kuo, Tsorng-Juu Liang, Jiann-Fuh Chen, "A High-Efficiency Single-Phase Three-Wire Photovoltaic Energy Conversion System," IEEE Trans. on Industrial Electronics, 2003, Vol.50(1), pp. 116-122.

[4] Yang Chen, Keyue Ma Smedley, "A Cost-Effective Single-Stage Inverter With Maximum Power Point Tracking, " IEEE Trans. on Power Electronics, 2004, Vol.19(5), pp. 1289-1294.

[5] Soeren Baekhoej Kjaer, John K. Pedersen, Frede Blaabjerg," A Review of Single-Phase Grid-Connected Inverters for Photovoltaic Modules ", IEEE Trans. On Industry Applications, 2005, Vol. 41(5), pp. 1292-1306.

[6] Billy M. T. Ho,Henry Shu-Hung Chung, "An Integrated Inverter With Maximum Power Tracking for Grid-Connected PV Systems," IEEE Trans.on Power Electronics, 2005, Vol.20(4), pp. 953-961.

[7] Tsai-Fu Wu, Hung-Shou Nien, Chih-Lung Shen, Tsung-Ming Chen, "A Single-Phase Inverter System for PV Power Injection and active Power Filtering With Nonlinear Inductor Consideration," IEEE Trans. on Industry Applications, Vol. 41(4), pp. 1075-1083, 2005.

[8] Naser Abdel Rahim, J. E. Quaicoe, "A single-phase voltage source utility interface system for weak ac network applications," IEEE Trans. on Power Electronics, 1996, Vol. 11(4) pp.532 – 541.

[9] Silva, S. M, Lopes, B.M, Filho, B. J. C. Campana, R. P. Bosventura, W.C, "Performance evaluation of PLL algorithms for single-phase grid-connected systems," IEEE Proc. On Industry applications, 2004, Vol. 4, pp. 2259-2263.

[10] Veerachary. M, Senjyu. T, Uezato. K, "Maximum power point tracking control of IDB converter supplied PV system," IEE Proc. Electr. Power Appl., 2001, Vol.148(6), pp. 494-502.

[11] Phani Kiranmai. K. S, Veerachary. M. , "Maximum Power Point Tracking: A PSPICE Circuit Simulator Approach," Power Electronics and Drive Systems, 2005.

[12] Phani Kiranmai.K. S, Veerachary. M, "A single-stage for PV MPPT applications," IEEE Int. Conference, ICIT2006 (to be presented).

[13] PSIM user manual, 2004.

TABLE I. PV SYSTEM PARAMETERS

| Parameter | Value |
|-----------|-------|
| $v_{oc}$ | 21 V |
| $i_{sc}$ | 2.1 A |
| $V_{mp}$ | 16.2 V |
| $I_{mp}$ | 1.8 A |
| $P_{max}$ | 29.3 W |
| $N_s$ | 36 |

## VII. BIOGRAPHIES

**K. S. Phani Kiranmai** was born in Guntur, A.P, India. She received the Bachelor degree from the Srikrishnadevaraya University, AP, India, in 2001, and Masters degree from Osmania University, Hyderabad, India, in 2003. Currently she is research scholar in the Dept. of Electrical Engineering, IIT Delhi and working towards Ph. D degree. Her fields of interests are power electronic converters applications to renewable energy systems and grid-connected systems.

**M.Veerachary** was born in India in 1968. Mr. Chary completed his Dr. Eng from University of the Ryukyus, Okinawa, Japan in 2002. Since then he is with Dept. of Electrical Engineering, Indian Institute of Technology Delhi, India and currently he is an Associate Professor. His research interests are High frequency dc-dc conversion, design and analysis of satellite/spacecraft power conditioning systems, modeling and simulation of large power electronic systems, control theory application to power electronic systems, and intelligent controller applications to power supplies.

**2006 IEEE International Conference on Power Electronic, Drives and Energy Systems**

# A Novel Control of Bi-Directional Switches in Matrix Converter

Meharegzi Tewolde, and Shyama P. Das, *Senior Member IEEE*

*Abstract-*This paper presents a novel control of bi-directional switches (BDS's) used in matrix converter. An extensive literature review and experimental study have been carried out to obtain a modern solution of commutation. An improved technique and control algorithm is used for commutation of bi-directional switch. A simple voltage sensor has been developed to control the matrix converter switches. This technique is simple and effective in comparison to the existing methods. Finally, this paper includes the detailed practical issues of freewheeling path and integration of over voltage protection of BDS in practical implementation. Experimental results are demonstrated to validate the proposed control strategy of freewheeling method.

*Keywords-*Matrix converter, ac-ac converter, Bi-directional switch (BDS), Freewheeling path

## I. INTRODUCTION

A Matrix converter (MC) consists of N-input supply terminals and M-output terminals with N x M-number of bi-directional Switches (BDS's). Since no single device is available as a bidirectional self commutated switch, a BDS is synthesized from the combination of commonly used solid state devices. In order to form a composite BDS with the capability of conduction in both directions, it is necessary to connect two discrete devices in "inverse parallel" with one another (Common Emitter BDS and Common Collector BDS) or to surround a single unidirectional switching device with rectifiers (Diode Bridge BDS), as shown in Fig. 1, in such a way that external bi-directional current flow is routed unidirectionally through the switching device itself. Possible configurations have been made to realize BDS from discrete devices, and they are now available in modular form in the market. Recently BDS has also been realized using the newly developed IGBT with reverse blocking capability (RB-IGBT) [1].

The authors are gratefully acknowledge financial support provided by THE MINISTRY OF EDUCATION of ETHIOPIA to carry out this research work.

Meharegzi Tewolde is with Department of Electrical Engineering, Indian Institute of Technology, Kanpur-208016, INDIA. ( e-mail: meharitw@iitk.ac.in)

Shyama P. Das is with Department of Electrical Engineering, Indian Institute of Technology, Kanpur-208016, INDIA. ( e-mail: spdas@iitk.ac.in)

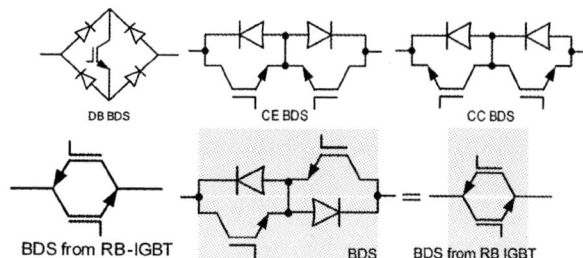

Fig. 1. Bidirectional switches.

To produce the mean average output voltage at desired frequency from MC we are forced to switch the BDS's at different instances [2] according to Venturini algorithm or space vector modulation. For example, in Fig. 2, any of the input supply phases can be connected to the output. In this case, if the applied voltage is greater than the incoming voltage and current is flowing to the load (positive), hard switching-off of the outgoing switch takes place. For the same condition of voltage if the current is negative, soft switching-off of the outgoing switch takes place. In case, the applied voltage is less than the incoming voltage, and at the same time if the current is positive, switching-off of the outgoing switch takes place. For the same condition of voltage if the current is negative, hard switching-off of the outgoing switch takes place. Commutation between supply phase A and supply phase B is shown in Table-I

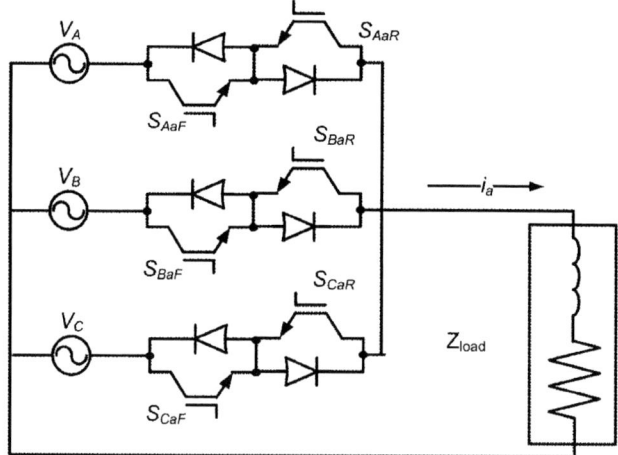

Fig. 2. Three phase to single phase ac-ac converter.

0-7803-9771-1/06/$25.00 ©2006 IEEE

TABLE – I

| Switching Condition | | During $S_{Aa}$ turn-off & $S_{Ba}$ turn-on Switching for $S_{Aa}$ is | During $S_{Ba}$ turn-off & $S_{Aa}$ turn-on Switching for $S_{Ba}$ is |
|---|---|---|---|
| $i_a$ | Voltage | | |
| +Ve | $V_A > V_B$ | Hard | Soft |
| -Ve | $V_A < V_B$ | Hard | soft |
| +Ve | $V_A < V_B$ | soft | Hard |
| -Ve | $V_A > V_B$ | Soft | Hard |

Depending on the power level and type of load as well as protection and control strategy of the converter, the BDS topology can be selected from (Fig. 1).

Beside the BDS topology, the switching algorithms and protection strategies are also important factors during current commutation [5]. Freewheeling path and varistor over voltage protection scheme have been adapted to overcome the commutation problem. In this paper, a novel control scheme of the BDS's in the MC is proposed based on the current freewheeling path method.

## II. FEATURES AND APPLICATIONS OF MATRIX CONVERTER

For years, drives experts have been talking about a technology known as matrix conversion as being the next generation power modulator for drives. In essence, a matrix converter uses an array of semiconductor power switches, operated in a precisely controlled sequence, to connect the three phase lines directly to a motor.

The matrix converter uses all three phases of the input voltage to control the output voltage. MC not only absorbs current disturbances, but also provides a clean voltage to the load. The input current of a matrix converter is maintained sinusoidal and for any type of load current, this input current can be controlled to be in phase with the input voltage.

The matrix converter can operate in regenerative mode without any special converter for regeneration of energy. Hence it has inherent four-quadrant capability. Another advantage of MC is that, it does not need a DC link, so it should operate reliably for a long time. Elimination of the dc link electrolytic capacitor makes the converter compact, lightweight and the converter can work in extreme ambient temperature.

The drive system using matrix converter is expected to be used mainly on applications that exploit its advantages, such as a regenerative capability, compactness, and high ambient temperature operation. Examples include lifts, escalators, cranes, centrifuges and eccentric machines such as presses and cutters running in continuous motoring and regenerative cycles.

Among the most desirable features in ac to ac converters are given as follows [3], [4].

- simple and compact power circuit;
- generation of load voltage with variable amplitude and frequency;
- sinusoidal output and input currents;
- operation with unity input power factor for any load;
- regenerative capability.

These ideal characteristics can be obtained by using matrix converters. Polyphase ac-ac conversion can be realized by suitable arrangement of the semiconductor switches. An M-phase to N-phase matrix converter consists of an array of M x N bi-directional switches arranged so that any of the output lines of the converter can be connected to any of the input lines [5] (Fig. 3).

## III. BI-DIRECTIONAL SWITCHES

The Diode Bridge BDS (DB BDS) Cell Configuration, as shown in Fig.1, conducts the current through two fast recovery diodes (FRD's) and one IGBT. This will increase the conduction loss and the switch always works under forced commutation that leads to switching loss as well. In addition, advanced switching techniques for safe commutation can not be applied.

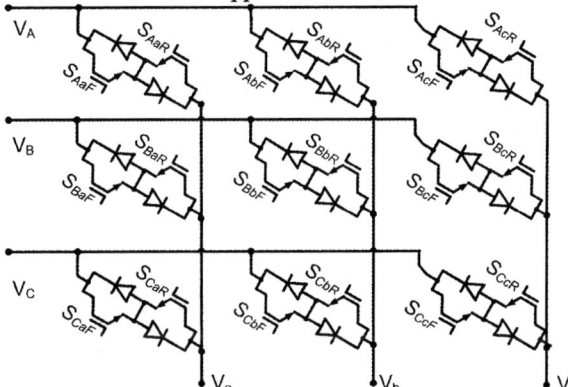

Fig. 3. Three phase to three phase AC/AC converter.

Common Emitter BDS (CE BDS) cell configuration offers a very easy way to build the BDS in a modular structure. The current conducts through one Diode and one IGBT. It reduces the conduction loss compared to DB BDS. It also helps minimize stray inductance for high power application as it is available as a single module. CE-BDS needs one isolated power supply per each BDS, therefore for 9 CE BDS, 9 isolated power supplies are necessary to drive a three-phase MC [6]. The CE BDS requires 3 control terminals per each BDS; thus we need a total of 27 terminals if M=N=3 [7]. CE BDS configuration is convenient to apply the advanced switching techniques for safe commutation and to reduce the switching loss, because it is possible to operate under semi-soft switching condition (Table I & II).

Common Collector BDS (CC BDS) is the most favored combination for low power industrial applications. It needs a minimum of 6 isolated power supplies to build up the three phase to three phase ac-ac MC (3x3 MC), i.e. M + N = 6 [6]. The CC BDS requires N x (M + 1) + M x (N +1) =24 control terminal ends for 3x3 MC. CC-BDS configuration is convenient to apply the advanced switching techniques for safe commutation and for reduction of switching loss like the CE BDS.

Reverse Blocking IGBT based BDS (RB IGBT BDS) needs a minimum of 6 isolated power supplies to build up 3x3 MC, i.e., M + N = 6, like CC BDS. Less conduction loss, less switching loss inside the RB-IGBT [1] are some of the advantages. But, it does not permit to use the step switching technique, as we cannot detect the current direction precisely like the CC BDS or CE BDS.

Using common collector and common emitter BDS's, it is possible to control the direction of the current through the BDS independently. Conduction and switching losses are also reduced by employing Semi soft switching (Table II) or freewheeling method of operation (Table III) (Fig.11).

## IV. OVERVOLTAGE AND OVERCURRENT PROTECTION OF BDS

Bi-directional switches are subjected to over voltage during turn off and due to input side originated line perturbations [4]. To overcome this problem, overvoltage protection becomes necessary. Various voltage clamping methods are discussed as follows.

A. Two 3-φ Bridge rectifiers with one electrolytic capacitor that can clamp the input and output voltage to the safe value Fig. 4. This method has a drawback that it increases the number of devices [7].

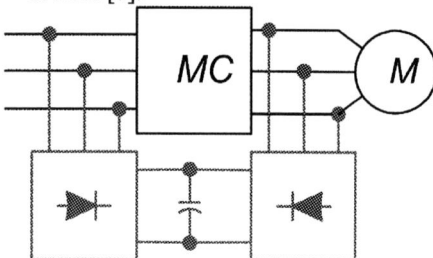

Fig. 4. Voltage clamping methods using two three phase Bridge rectifiers with one capacitor.

B. Individual level of BDS voltage clamping method by using four extra fast recovery diodes and one more capacitor (Fig. 5). This configuration is advisable for discrete IGBTs as it helps to remove the effect of leakage inductance.

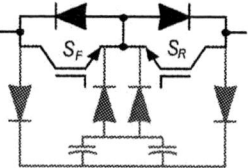

Fig. 5. Voltage clamping methods for discrete IGBTs.

C. Voltage clamping is also possible using Star or Delta connected metal oxide varistors (MOV's) (Fig. 6) with voltage suppressor for each IGBTs. Since the varistors response time is very fast (< 0.5ns) and they have sufficient surge current handling capability, this configuration works well. It is also possible to select and use appropriate voltage clamping varistor. Besides all its advantage, the MOV's are cheaper and affordable to use them as additional protection along with other protection strategy.

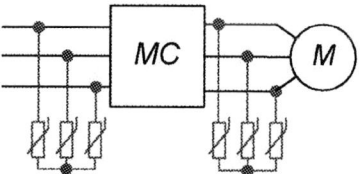

Fig. 6. Voltage clamping methods using *metal oxide varistors.*

D. Current-Direction-Based Commutation method avoids rise of induced emf (Ldi/dt) in the commutating circuit and assures safe commutation. And it is called Step-by-step switching (semi-soft switching) method [4]. This can be practiced by detecting the current direction. Detecting the current direction is not so easy using the current sensors as the current sensors are not accurate nearby the zero values. Therefore, current direction detection has been made by measuring the voltage across the switching IGBTs ($V_{SAF}$ & $V_{SAR}$), as shown in Fig. 7. If $V_{SAF}$ =$V_{CE} \approx 2.5V$ and $V_{SAR} \approx -0.6V$ the current ($i_A$) is in positive direction passing through $S_{AF}$ and $D_F$. If $V_{SAF}$ =$V_{DR} \approx -0.6V$ and $V_{SAR} \approx 2.5V$ the current ($i_A$) is in Negative direction passing through $S_{AR}$ and $D_R$.

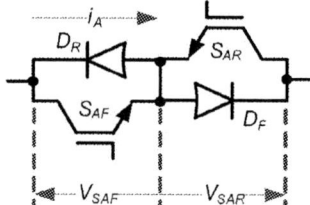

Fig. 7. Current direction detection method.

Step-by-step current direction-based commutation method from phase A to phase B is shown in Table II. Four step ('overlap') commutation method for the circuit in Fig.

362

2 [8] is discussed here. If the current direction ($i_a$) is positive, we can switch off the $S_{AaR}$ at t=$t_1$ because no current is passing in reverse direction. At t=$t_2$ we can switch on $S_{BaF}$ and make overlap commutation of phase A & B using $S_{AaF}$ & $S_{BaF}$. At t= $t_3$ we can switch off $S_{AaF}$ and by now the current will be shifted completely to pass through $S_{BaF}$. At last to allow bi-directional flow of current $S_{BaR}$ will be switched at t= $t_4$. By using 4-step commutation safe and uninterrupted load current commutation can be achieved. However, since the number of commutation steps and the time lag for switching are unnecessarily increased, three-step commutation is often preferred.

In three-step current-direction-based commutation method [10], the understanding is that for most switching devices, turn-off time is longer than turn on time. Thus, if we turn-off $S_{AaF}$ & turn-on $S_{BaF}$ at the same time of $t_2$, $S_{AaF}$ will be turned off after $S_{BAF}$ is turned on. So the overlap commutation without load current interruption can be done using three steps with reduction of switching time.

In case of two-step commutation method [4], we can avoid unused switch early or switch over when the current is changing direction inside each BDS, so that we can have only one switch $S_{AaF}$ on. At a time of t=$t_1$ we can switch on $S_{BaF}$ and switch off at t = $t_2$ $S_{AaF}$. In this case we can reduce time of switching off $S_{AaR}$.

In one-step current-direction-based commutation method, the overlap time is reduced. So by reducing the overlap time from the two-step method we can have one-step current-direction-based commutation method [9]. $S_{BaF}$ is gated on the basis of the current direction information passed from $S_{Aa}$, and $S_{AaF}$ is subsequently turned off. After a short interval the current direction information is taken from the detection circuit in $S_{Ba}$ rather than cell $S_{Aa}$.

To provide for both directions of load current between commutations, the active BDS drive circuit automatically transfers the drive between the devices within the BDS, if the detection circuit determines that the current has fallen to zero. A potential difficulty occurs if the load current changes direction when a commutation between BDSs is required. The Solution is to give dead time during near Zero current value.

TABLE –II

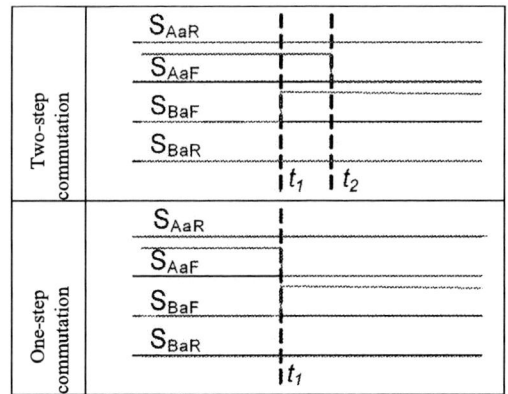

This method of commutation needs absolute current direction detector with power isolation circuit and complex intelligent logic controller.

E.  The proposed voltage claming method for MC is done by integrating the freewheeling path strategy [10] with the metal oxide varistors voltage clamping method to the MC. This can be accomplished by detecting the real time maximum and minimum input voltages. The real time detection has been done using hardware setup, as shown in Fig. 8. It is shown that, the output of voltage sensor will be compared and using logic devices, the maximum and minimum of input voltage is detected.

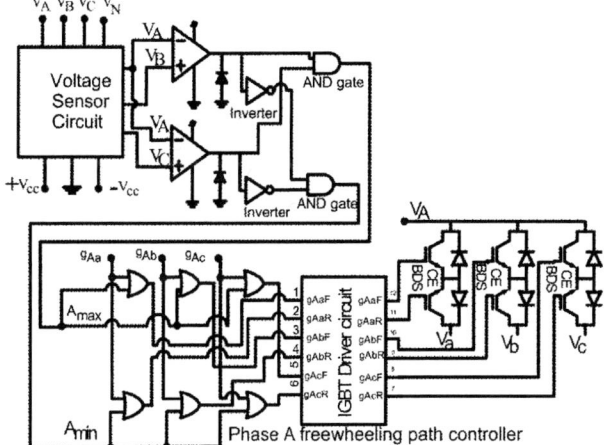

Fig. 8. Freewheeling path controller.

Then the freewheeling path controller circuit as shown in Fig. 8, provides triggering signal during all the times of maximum voltage detection of input phase M, and the respective phase reverse biased IGBT's ($S_{MaR}$ $S_{MbR}$, $S_{McR}$) will be switched on for freewheeling. Besides, it also provides triggering signal during the minimum voltage detection of input phase M, the respective phase reverse biased IGBT's ($S_{MaF}$ $S_{MbF}$, $S_{McF}$). This protection strategy provides almost the same as the classical power electronics converters freewheeling path Fig. 10 and Fig. 11.

This method uses freewheeling paths for the load current like the conventional inverters (which have anti-parallel

363

diodes). The insertion of freewheeling path in MC was not possible earlier with such a simple and low-cost hardware.

Some of the freewheeling switches are mentioned in Table - III. To analyze the working of these switches, we can see from Fig. 9 that for the interval of $30^0$ to $90^0$ of input voltage period, the phase A has maximum amplitude and phase B has minimum amplitude. So for this case, switches $S_{AaR}$, $S_{AbR}$, $S_{AcR}$, $S_{BaF}$, $S_{BbF}$ and $S_{BcF}$ of Fig. 10 are reverse biased. The simplified circuit diagram of Fig. 10 is shown in Fig. 11.

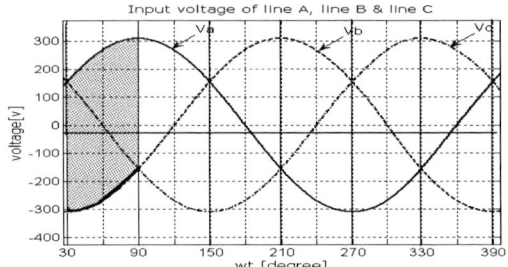

Fig. 9. Input voltage wave form.

TABLE – III

| Interval | $30^0$ up to $90^0$ | | $90^0$ up to $150^0$ | | $150^0$ up to $210^0$ | |
|---|---|---|---|---|---|---|
| Free wheeling switches | $S_{AaR}$ | $S_{BaF}$ | $S_{AaR}$ | $S_{CaF}$ | $S_{BaR}$ | $S_{CaF}$ |
| | $S_{AbR}$ | $S_{BbF}$ | $S_{AbR}$ | $S_{CbF}$ | $S_{BbR}$ | $S_{CbF}$ |
| | $S_{AcR}$ | $S_{BcF}$ | $S_{AcR}$ | $S_{CcF}$ | $S_{BcR}$ | $S_{CcF}$ |

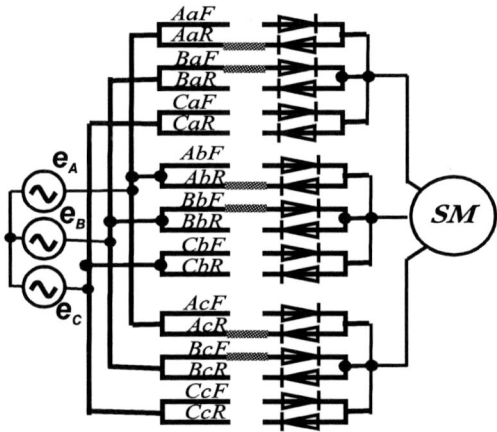

Fig. 10. Three phase Matrix converter and its closed switches for free wheeling, during the interval of $30^0$ to $90^0$ of wt.

The above strategies can ensure *di/dt* protection of the input side of the inverter. In addition, we need to add Voltage clamp using 6 Metal Oxide Varistors (MOV's)[11] for the sake of critical zone (near equal voltage zone), which is not supported by the above strategy.

The described integrated switching strategy makes possible safe commutation of four-quadrant MC switches. Triggering signals of the nine BDS's ($g_{Aa}$,$g_{BA}$,$g_{CA}$ --- $g_{CC}$) for commutation comes from the circuit which defines the control algorithm of the MC.

BDS over current protection is possible using current limiter of control circuit under normal operation condition

Short circuit protection circuit to block the IGBT during abnormal condition is embedded in EXB841 driver circuit and in most of the driver circuits available in market.

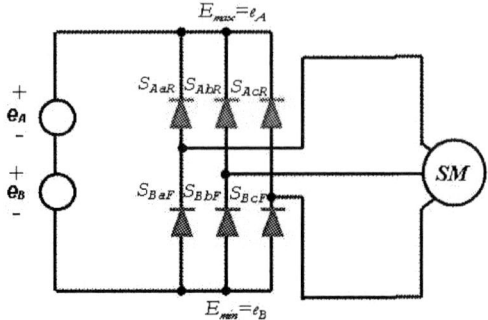

Fig. 11. Simplified circuit diagram of Fig.10.

## V. SIMULATION & HARDWARE RESULT

A single-phase voltage regulator simulation has been performed using MATLAB/SIMULINK. The performance of freewheeling supported power controlling BDS ($S_p$) is being observed by switching the power BDS at 1kHz and freewheeling BDS at line frequency.

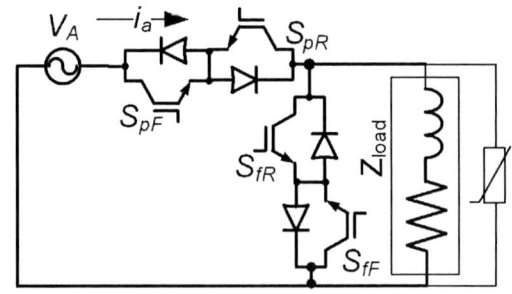

Fig. 12. single phase voltage regulator using BDS supported by free wheeling technique.

The simulation result shows that, the chopped sinusoidal regulated voltage is applied on the load (Figs. 13-14) and continuous current flows through the load. Load current is not interrupted throughout the voltage regulation process. The experimental results are shown from Fig. 15 – Fig. 18.

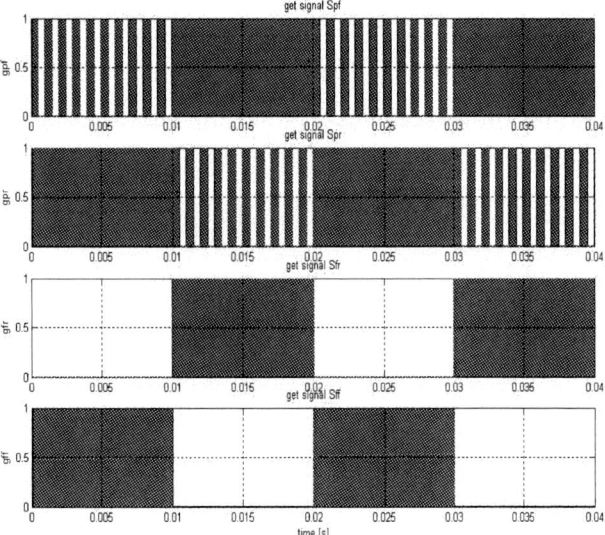

Fig. 13 Simulation of triggering pulses of power switching BDS($S_P$) and free wheeling BDS ($S_F$).

364

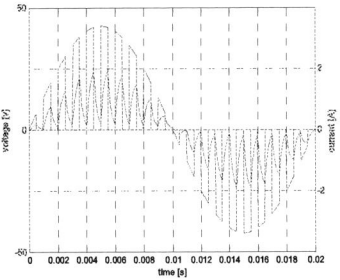

Fig. 14. Simulation output voltage across load and load current.

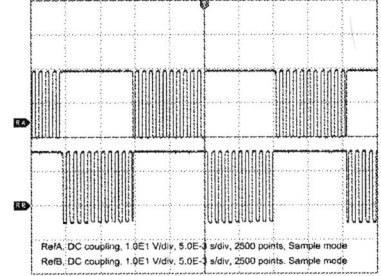

Fig. 15. Experimental result of triggering pulse of power switching BDS ($S_{PF}$ & $S_{PR}$).

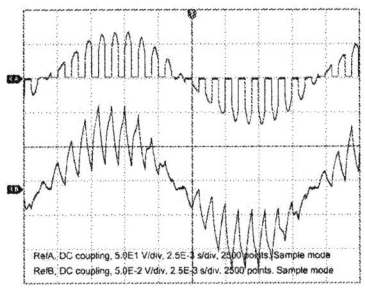

Fig. 16. Experimental result: Voltage across power controlling BDS and load current.

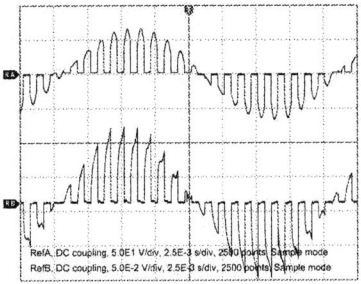

Fig. 17. Experimental result: Voltage across power controlling BDS and line current $i_a$.

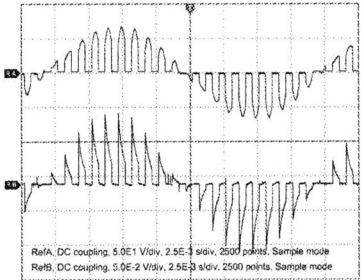

Fig. 18. Experimental result: Voltage across power controlling BDS and freewheeling current.

## VI. CONCLUSION

The proposed BDS switching and protection scheme, is quite practical and easy to implement. It removes the commutation problem during the switching of BDS's in Matrix Converter (MC). The results of simulation work and experimental test show that the circuit is simple and effective in overcoming the over voltage problem during turn off and turn on. Thus, this new strategy makes the matrix converter more reliable and cost effective compared to the existing methods. All the control algorithms for switching freewheeling BDS's have been done using hardware only. This paper has provided the detailed circuit topology and control logic, which have been used to realize the novel control of Bi-directional Switches.

## VII. ACKNOWLEDGMENTS

The authors are gratefully acknowledge financial support provided by THE MINISTRY OF EDUCATION of ETHIOPIA to carry out this research work.

## VIII. REFERENCES

[1] M.Takei, T.Naito, and K.Ueno, *"The Reverse Blocking IGBT for Matrix Converter With Ultra-Thin Wafer Technology,"* Fuji Electric Corporate Research and Development, Ltd. 4-18-1 *Tsukama, Matsumoto, Nagano, 390-0821, Japan.*

[2] P. W. Wheeler, *"Gate Drive Level Intelligence and Current Sensing for Matrix converter Current Commutation,"* IEEE Trans. on Industrial Electronics, Vol. 49, No. 2, April 2002.

[3] A. Alesina and M. Venturini, *"Analysis and design of optimum-amplitude nine-switch direct ac-ac converters,"* IEEE Trans. on Power Electronics, vol. 4, no. 1, pp. 101-112, 1989.

[4] P. W. Wheeler, J. Rodríguez, J. C. Clare, L. Empringham, and A. Weinstein *"Matrix converter: A technology review,"* IEEE Trans. on Industrial Electronics, vol. 49, no. 2, pp. 276-288, Apr. 2002.

[5] P.W. Wheeler, J. C. Clare, L. Empringham, and M. Bland, *"Matrix Converters: The technology and potential for exploitation,"* The Drives and Controls Power Electronics Conference, London, Section 5, March 2001.

[6] P. Nielsen, and F. Blaabjerg *"New Protection Issues of a Matrix Converter: Design Considerations for Adjustable-Speed Drives,"* IEEE Trans. on Industry Applications, Vol. 35, No. 5, 1999

[7] C. Klumpner, P. Nielsen, I. Boldea, and F. Blaabjerg, *"New solutions for a low-cost power electronic building block for matrix converters"*,IEEE Trans. on Industrial Electronics ,Vol.49, 2002,

[8] P. W. Wheeler, J. C. Clare, and L. Empringham *"A Vector Controlled MCT Matrix Converter Induction Motor Drive with Minimized Commutation Times and Enhanced Waveform Quality"*, School of Electrical and Electronic Engineering, University of Nottingham,Nottingham, NG7 2RD, UK

[9] L. Empringham, P. W. Wheeler, and J. C Clare, *"Intelligent Commutation of Matrix Converter Bi-directional Switch Cells using Novel Gate Drive Techniques,"* PESC, Japan, May 1998

[10] M. H. Rashid, Power Electronics Hand Book, *Academic Press, San Diego, California 2001.*

[11] J. Mahlein, *"Passive Protection Strategy for a Drive System with a Matrix Converter and an Induction Machine,"* IEEE Trans. on Industrial Electronics, Vol. 49, No. 2, April 2002.

**2006 IEEE International Conference on Power Electronic, Drives and Energy Systems**

# PWM SHE Switching Algorithm for Voltage Source Inverter

Ali. I. Maswood, *Senior Member, IEEE*

*Abstract--* **This work presents a novel technique for developing a near-optimal on-line real-time PWM-SHE switching pattern in which the piece-wise linearization of a fully optimal PWM-SHE switching strategy is applied. As investigated in this paper, this new linearization technique is applicable not only to PWM-SHE switching, but also to other optimal modulation strategies in general. Implementation of the optimal PWM-SHE switching pattern to VSI have brought the switching losses substantially down.**

*Index Terms -* **PWM, SHE, Harmonics, Switching Angles, On-line, Linear Equation.**

## I. INTRODUCTION

THE on-line real-time application of the PWM-SHE technique has recently become a major research topic in the field of PWM switching. However, the difficulty of computing optimal switching strategies on-line in real-time is widely recognized. The key problem in the iteration process is the computing time for a nonlinear transcendental equation. This paper provides a simple yet a very effective method of real-time on-line solution for the PWM-SHE technique that readily accommodates a researcher's choice of a particular set of harmonic. The piece-wise linear representations of PWM-SHE switching angles are utilized to formulate the on-line linear equations.

A detailed analysis of linearization and on-line capability of an optimized PWM specific harmonic elimination technique is described. The waveforms are characterized by the angles α, which define the pulses in the first 60° interval. A set of equations have been derived to calculate the amplitudes of the fundamental component and the non-zero 3n±□(n=2, 4,··· N) harmonics. A set of these equations may be solved for the $N$ angles of α which realize a given fundamental amplitude and eliminate N-□low-order odd, non-triplen harmonics.

The PWM-SHE harmonic elimination switching angles can be represented by the following set of equations □ This equation is solved with the help of a computer program using the conjugate gradient descent method. Generally for any given modulation index (governing fundamental component),

the output of the optimal PWM-SHE switching angles can be presented as a set of curves as shown in Fig.□ The linearization process presented in this work proceeds by replacing these non-linear curves with piece-wise linear representations, with each curve being made up of one or more straight-line segments. These straight-line segments are then used in a simple on-line real-time computation to determine the placement and period of the inverter switching pulses for each cycle of fundamental output frequency. The principal question to be resolved with such a linearisation technique are:

How many linear segments are required to adequately represent the optimal PWM-SHE non-linear switching curves?

Where should the break-points between linear segments be located?

$$
\begin{cases}
f\left(a_\square\right) = \sum_{k=\square}^{2N} (-\square)^{k+\square} \cos \alpha_k - \dfrac{\pi a_\square}{4} = 0 \\
\cdots \vdots \\
f\left(a_x\right) = \sum_{k=\square}^{2N} (-\square)^{k+\square} \cos x\alpha_k = 0 \quad x = 3N-\square
\end{cases}
\quad (\square)
$$

Where: $a_\square, a_x$ - Harmonic orders to be eliminated, and
$\alpha_k$ - Required Kth switching angle.

The approach presented here is to use the difference analysis and the harmonic variation analysis to resolve these issues. It also determines the accuracy of the pulse timing that is required in a practical system to maintain a satisfactory minimal harmonic standard.

The linearization of the PWM-SHE switching angle proceeds by creating a linear set of equations which explicitly define each switching angle, of the form:

$$[\alpha_i] = V_i[K_{i,j}] + [C_{i,j}], \qquad i = \square, 2, 3, 4, 5 \qquad j = \square, 2, ....(2)$$

The accuracy of the resultant modulation pattern depends on the number of piece-wise linear segments that are used to approximate each of the exact switching angle patterns. But increasing on the number of the piece-wise will also cause the increasing of the complication of the linearization equation and computing time [2].

$K_{i,j}$ and $C_{i,j}$ can be found in terms of the exact solution of the switching angles $\alpha_i$ by the following procedure:

Selecting a series of points for various switching angles $\alpha_i$ and modulation index $M$ along each of the switching angle curves, and allocating more points where the curves deviate more from a straight line.

---

Centre for Advanced power Electronics (CAPE), School of Electrical & Electronics Engineering, Nanyang Technological University, Nanyang Ave. Singapore-639798.

0-7803-9771-1/06/$25.00 ©2006 IEEE

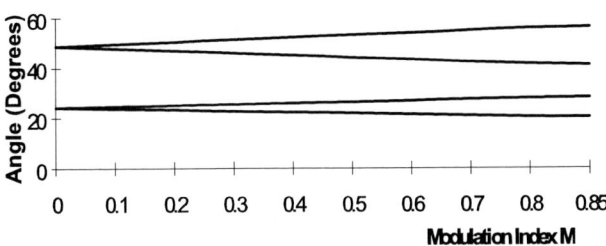

Fig. □ Solution Trajectories for the PWM-SHE Switching Angles Using the Conjugate Gradient Descent Method (Number of pulses per quarter cycle, N = 4).

The slope and offset constants for the approximate linear section defined by any two adjacent points( $\alpha_{i,\,j\text{-}\square}$ , $M_{j\text{-}\square}$ ) and ( $\alpha_{i,j}$ , $M_j$ ) can be expressed as:

$$K_{i,j} = \frac{\alpha_{i,j} - \alpha_{i,j\text{-}\square}}{M_j - M_{j\text{-}\square}} \tag{3}$$

$$C_{i,j} = \alpha_{i,j\text{-}\square} - K_{i,j} M_{j\text{-}\square} \tag{4}$$

Any two adjacent sections identified by points ($\alpha_{i,\,j\text{-}1}$ , $M_{j\text{-}1}$), ($\alpha_{i,\,j}$ , $M_j$) and ($\alpha_{i,\,j+1}$ , $M_{j+1}$) can be combined, provided their slope and offset constants meet the following conditions:

$$\left| ( K_{i,j} - K_{i,j\text{-}\square}) \cdot M + (C_{i,j} - C_{i,j\text{-}\square}) \right| \leq \Delta\alpha \tag{5}$$

for; $\quad M_{j\text{-}\square} < M < M_j$, or $M_j < M < M_{j+\square}$

## II. ONE-SEGMENT ON-□INE EQUATION DERIVATION

From the solution trajectories of α shown in Fig.□, the variation of α with the modulation index M seems to have approximately linear dependence relation. As *N* increases, the linearity is better. This criterion makes the PWM-SHE technique to be superior in linearization of switching angles and further in on-line application over other harmonic elimination techniques. Fig.2 shows the relationship between the new one-segment linear approximate equation and the exact switching angle trajectory calculated by conjugate gradient descent method. It is obvious that some error exists between two trajectories. Based on the principle of linearization of PWM-SHE switching angles discussed in last section, we use one-segment linear equation to approximate PWM-SHE switching angle curves. Its results also can be used as the criteria for the purpose of selecting the break-points and construction of multi-segments linear equation. Two points, $M_0$ and $M_{0.85}$ , are used to set up one-segment linear equation

The next step is the utilization of the solution of the nonlinear transcendental equation (□) to construct the linear algebraic equation. Regarding these solution as $\alpha^0$, based on the formulae (2) to (5), in any *N* (the number of pulses within first 60° interval) for an odd *i*, following one-segment general linear equation can be obtained

$$\alpha_i = \frac{\alpha_{i+\square}^0 + \alpha_i^0}{2} - \frac{V_i}{0.85}\left(\frac{\alpha_{i+\square}^0 + \alpha_i^0}{2} - \alpha_i^0\right) \tag{6}$$

and for an even *i:*

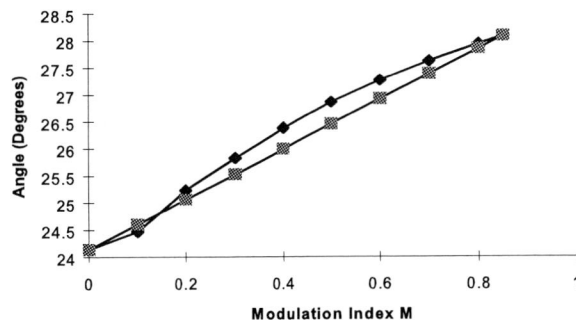

Fig. 2. The relationship between the one-segment linear switching angle equation and the exact switching angle curve ( N=4, α2 ).

$$\alpha_i = \frac{\alpha_i^0 + \alpha_{i\text{-}\square}^0}{2} + \frac{V_i}{0.85}\left(\alpha_i^0 - \frac{\alpha_i^0 + \alpha_{i\text{-}\square}^0}{2}\right) \tag{7}$$

Equations (6) and (7) are the simple first-order linear equation, which are solved using the conjugate gradient descent method. This is expected to take very short time for a microprocessor based computation and generation. The following equation (8) is the one-segment linear equivalent equation for N=6,

$$\begin{cases} \alpha_\square = \square7.05276 - \dfrac{2.03265V_L}{0.85} & \alpha_2 = \square7.05276 + \dfrac{2.03265V_L}{0.85} \\[2mm] \alpha_3 = 34.28269 - \dfrac{3.94366V_L}{0.85} & \alpha_4 = 34.28269 + \dfrac{3.94366V_L}{0.85} \\[2mm] \alpha_5 = 5\square774595 - \dfrac{5.605355V_L}{0.85} & \alpha_6 = 5\square774595 + \dfrac{5.605355V_L}{0.85} \end{cases} \tag{8}$$

where $V_i$ (fundamental amplitude) is the only independent variable changed continuously, switching angles $\alpha_\square$ to $\alpha_6$ can be calculated easily by simple and explicit relationship with $V_i$ , and then the pulse width can be defined, this is reverse of the solution process of the nonlinear transcendental equation.

## III. EVA□UATION OF ON-□INE A□GORITHM

To evaluate the one-segment linearization equation of PWM-SHE switching angles. A comparison, between the solution ($\alpha'$) of using one-segment linear equation (6) and (7) and the solution ($\alpha$) using conjugate gradient method, is conducted.

$$\Delta = \frac{\square}{N}\sum_{i=\square}^N |\alpha_i - \alpha'_i| \tag{9}$$

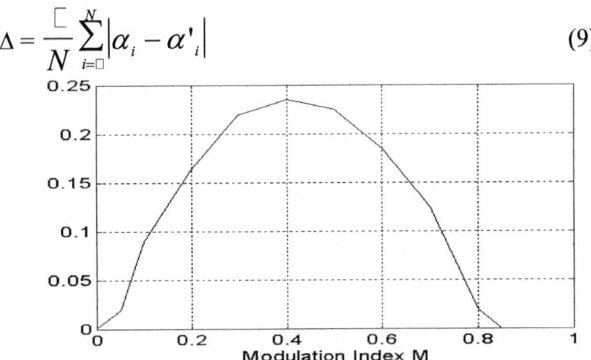

Fig.3. Deviation of α from its exact value.

The variation of Δ with the modulation index *M* is shown in Fig.3 (for *N*=4). As can be seen, the largest difference is only about 0.23 degree from its exact value. The deviation of

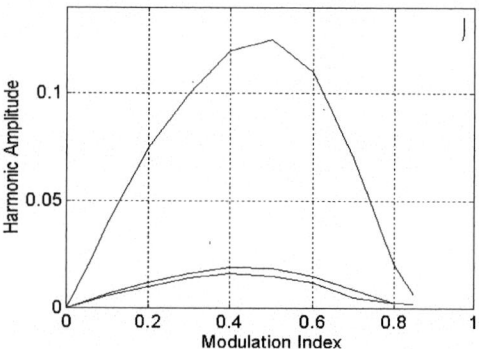

Fig. 4. The variation of the harmonic amplitude with modulation index by using on-line equation (N=4).

solution of one-segment linear equation from the exact solution trajectory will influence the harmonic content. The effect of the difference between the solution of (6) and (7) and that of (□) can be analyzed from Fig.4.

From Fig.4, it can be seen that only □□th harmonic amplitude (for $N = 4$ in PWM-SHE strategy, 5th, 7th and □□th harmonics should be eliminated from the output voltage) vary with the change of $\Delta$ (below □5%), and 5th and 7th harmonic amplitudes have little variations (below 2%). In practical applications, such as AC motor drive systems, the lower-order harmonics particularly 5th and 7th harmonics are to be avoided. So, one-segment linear equation can be used in some applications where harmonic elimination demands are not very high. The validity of harmonic elimination by the one-segment linear switching angle equation (6) and (7) is proved by this analysis, so they can be defined as allowable approximations for equation(□). A good way to make two solution trajectories fit better is by applying the multi-segment linear switching angle equation to approximate the exact switching angles trajectories.

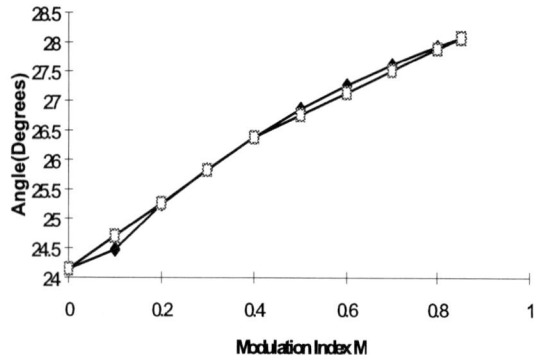

Fig. 5. The relationship of $\alpha_2$ calculated by two segment linear equation and its exact values.

## IV. DERIVATION AND EVALUATION OF THE TWO-SEGMENT ON-□INE EQUATION

As mentioned before, increasing the number of the linear segments will make the linear equivalent equation better coincide with the solution trajectories of nonlinear

transcendental equations governing PWM-SHE switching strategy. For a two-segment linear equation, the key problem is to decide the location of the break-point which connects the two linear segments [6]. From Fig.3, we can see that the largest switching angle error $\Delta\alpha$ occur at the point $M = 0.4$, which is the result of mean analysis. So, this point is selected as break-point for the two-segment linear approximate switching angle equation. We also utilize the solution results of nonlinear transcendental equation. Among them, the values of $\alpha$ when $M = 0.4$ are highlighted. The two-segment linear approximate equation with $M = 0.4$ as the break-point can be expressed as follows:

for $\qquad\qquad\qquad\qquad\qquad\qquad 0 < M \leq 0.4$

$$
\begin{cases}
\alpha_i = \alpha_{0,i} - \dfrac{V_i}{0.4}\left(\alpha_{0,i} - \alpha_{0.4,i}\right) & \text{for an odd } i; \\[2mm]
\alpha_i = \alpha_{0,i} + \dfrac{V_i}{0.4}\left(\alpha_{0.4,i} - \alpha_{0,i}\right) & \text{for an even } i.
\end{cases}
\tag{□0}
$$

for $\quad 0.4 < M \leq 0.85$

$$
\begin{cases}
\alpha_i = \alpha_{0.4,i} - \dfrac{\alpha_{0.4,i} - \alpha_{0.85,i}}{0.85 - 0.4}\left(V_i - 0.4\right) & \text{for an odd } i; \\[2mm]
\alpha_i = \alpha_{0.4,i} + \dfrac{\alpha_{0.85,i} - \alpha_{0.4,i}}{0.85 - 0.4}\left(V_i - 0.4\right) & \text{for an even } i.
\end{cases}
\tag{□□}
$$

For a microprocessor, it is very straight forward to realize the change from (□0) to (□□) or vice versa. A mark will be set to check $V_i$ from the input port whether it is larger or smaller than 0.4.

The values of $\alpha$ at $M = 0$, $M = 0.4$ and $M = 0.85$ are utilized to construct the two-segment linear equation, then it can be expressed as follows:

in the first segment ($0 < M \leq 0.4$):

$\alpha_□ = 24.□4239 - 4.688975\, V_i$

$\alpha_2 = 24.□4239 + 5.605□25\, V_i$

$\alpha_3 = 48.67643 - 8.0□505\, V_i$

$\alpha_4 = 48.67643 + 9.466525\, V_i \qquad\qquad (□2)$

Fig.5 shows the relationship between the $\alpha_2$ calculated by using above two sets of linear equation and its exact solution curve.

Comparing Fig.5 with Fig.2, it is obvious that the error between the two algorithm decreases drastically. Fig.6 shows the variation of $\Delta$ with the modulation index M. Comparing it with Fig.3, one can note that the biggest $\Delta$max is only about 0.□8 degree. It is smaller than that of the one-segment case(about 0.25 degree). The error in most part is less than 0.08 degree that will have little effect on the harmonic content.

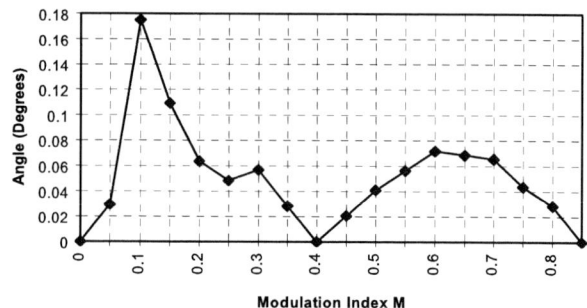

Fig. 6. The variation of $\Delta$ with the modulation index M, for N = 4.

From Fig.7, one can observe the variation of each harmonic with the modulation index $M$ (for $N$ = 4th, 5th, 7th, and □th harmonics should be eliminated). The □th harmonic amplitude varies with the Δ, its amplitude reaches 7.5% at M=0.65. However, it is 5% or below along rest of M. Since it is a relatively high order harmonic and inverters efficiently operate in the range of M=0.75 to 0.85, this may hardly cause any problem. The amplitudes of 5th and 7th harmonic change little and always are below 0.03. Comparing Fig.7 with Fig.4, it can be concluded that the harmonic contents caused by the linearization of the two-segment linear equation are much smaller than that in Fig.4 of one-segment linear equation. So, two-segment linear equation can be defined as an ideal equivalent equation of the nonlinear transcendental equation governing the PWM-SHE switching angles.

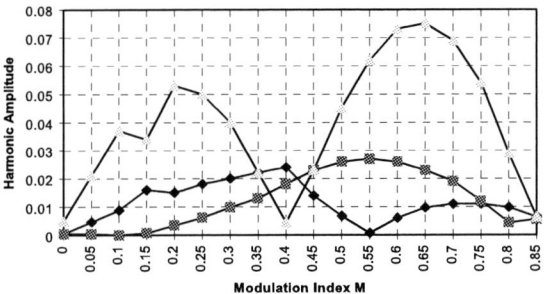

Fig. 7. The variation of each harmonic (5th, 7th, □th)with the modulation index M in the application of the two-segment linear equation.

From the above analysis, one can see the obvious trend that, with the increase of the piece of the segment in the linear approximate equation, the error of α will be reduced drastically.

## V. APPLICATIONS

To validate the simulation results, □ □kW NPC voltage source inverter with Mosfet switches is especially built. The calculated optimal switching angles are stored in an EPROM-27□6 memory device. EPROM-27□6 has a memory space of 2048 bits and possess a clock speed of □02.4kHz and a operating frequency of 50Hz. The Mosfet gate drivers get their gating signal from the EPROM. The work is currently under progress.

## VI. CONCLUSIONS

A novel procedure for generating the initial set of α is proposed. This is the key issue in the conjugate gradient descent method that guarantees convergence of the numerical solution of the nonlinear transcendental equation governing the PWM-SHE switching pattern. The generalized method used in solving the equations makes it simpler to obtain solution for eliminating as many harmonics as required. The solution angles of α vary linearly with respect to the required fundamental amplitude. This property simplifies the nonlinear transcendental equation solution and makes the on-line calculation of the PWM-SHE switching angles possible. The linearization technique uses piece-wise linear representation of

the optimal PWM-SHE switching angles which allow these angles to be computed with one multiplication and addition per switching angle at any modulation depth. This technique has been verified by the error analysis.

Implementation of the optimal PWM-SHE switching pattern to voltage source inverters have brought the switching losses down to about 50% compared to the currently used SPWM technique. This is an important gain specially in high power conversion. PWM-SHE offers the possibility of using forced commutated switches in relatively high power conversion which was traditionally avoided due high switching losses.

## VII. REFERENCES

[□] Boost, M. A., and Ziogas, P. D.: 'State-of-the-art carrier PWM techniques: A critical evaluation', IEEE Trans. Industry Applications, vol. 24, no. 2, Mar./Apr. □988, pp. 27□280.

[2] Maswood, A.I., and □iu, J.: 'Optimal online algorithm derivation for PWM-SHE switching', Electron. □ett., □998, **34**, (8), pp. 82□823.

[3] Enjeti, P. N., Ziogas, P. D., and □indsay, J. F.: 'Programmed PWM techniques to eliminate harmonics: A critical evaluation', IEEE Trans. Industry Applications, vol. 26, no. 2, Mar./Apr. □990, pp. 302-3□6.

[4] Sapre, U. S., Ashar, N. K., and Joshi, S. A.: 'A new approach to programmed PWM implementation', Proceedings of the □993 Applied Power Electronics Conference and Exposition, pp. 793-798.

**2006 IEEE International Conference on Power Electronic, Drives and Energy Systems**

# New Fuzzy logic Controller for a Buck Converter

D. Seshachalam, R. K. Tripathi, *Member, IEEE*, D. Chandra, and Anil kumar

*Abstract*--A Fuzzy logic controller based on sliding mode control is developed for a buck converter operating in CCM. It is an offline fuzzy controller. An important claim of this paper is a drastic reduction in rule base as compared to a conventional fuzzy logic controller, where the number of rules increases exponentially with increase in input membership functions. Simulation studies are carried out using MATLAB SIMULINK, fuzzy logic toolbox. Obtained result is compared with a conventional 7x7 rule based fuzzy controller and also a PID controller for the same buck converter.

*Index Terms*--Continues conduction mode (CCM), Fuzzy logic control (FLC), PID control, Reduced rule base (RRB), Sliding mode control (SMC).

## I. INTRODUCTION

SWITCHED mode power supplies are nonlinear and time varying systems. Design of high performance controllers for such converters is always a challenging issue. In SMC, control action is switched, when the system states cross the switching surface. The main drawback of the SMC control is chattering phenomenon [1] which is nonlinear in nature. From last decade research in the area of fuzzy logic and SMC applied to nonlinear systems has reduced the effect of chattering [2] [3]. Applying fuzzy logic along with SMC is used to reduce the chattering. In this paper a fuzzy logic controller based on sliding mode concept with reduced rule base, applied to a buck converter has been discussed. Simulation studies are carried out using MATLAB-SIMULINK, Fuzzy logic tool box Also the same results are compared with a PID and a conventional 7x7 rule based fuzzy controller. This control results in a better dynamic response and robustness. Because of the improvements in the field of both semiconductor and power devices technologies SMC is being made practically possible in many application areas.

D. Seshachalam is with Department of Electronics and communication, Dr AIT, Bangalore, INDIA-560056 (e-mail: dschalam@mnnit.ac.in ).
R. K. Tripathi is with Department of Electrical Engineering, MNNIT, Allahabad, INDIA-211004 (e-mail: rktripathi@mnnit.ac.in )
D. Chandra, is with Department of Electrical Engineering, MNNIT, Allahabad, INDIA-211004 (e-mail: dinesh@mnnit.ac.in).
Anil kumar, Wing commander, Indian air force, New Delhi, INDIA, (e-mail: anil_1tiwari@hotmail.com).

## II. SIMULINK MODEL OF DC-DC SWITCHING CONVERTERS

A System modeling is an important phase in any controller design work. The choice of a system model depends upon the objectives of the simulation. If the goal is to predict the behavior of a circuit before it is built. A buck converter is modeled using the following state equations [5].

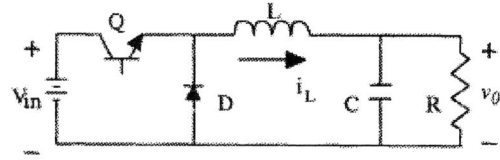

Fig. 1. Buck converter circuit diagram.

When the switch is ON circuit can be represented by

$$\begin{cases} \dfrac{di_L}{dt} = \dfrac{1}{L}(V_{in} - v_o) \\ \dfrac{dv_o}{dt} = \dfrac{1}{C}(i_L - \dfrac{v_o}{R}) \end{cases}, \quad 0 < t < dT, \quad Q:ON \quad (1)$$

When the switch is OFF circuit can be represented by

$$\begin{cases} \dfrac{di_L}{dt} = \dfrac{1}{L}(-v_o) \\ \dfrac{dv_o}{dt} = \dfrac{1}{C}(i_L - \dfrac{v_o}{R}) \end{cases}, \quad dT < t < T, \quad Q:OFF \quad (2)$$

TABLE I
CONVERTER DESIGN PARAMETERS

| Sl No | Circuit Element | Value |
|---|---|---|
| 1 | Inductor | 10mH |
| 2 | Capacitor | 470µF |
| 3 | Input Supply Voltage | 25V(nominal) |
| 4 | Load resistance | 8Ω -4Ω |
| 5 | Output Power | 60W (approx) |

## III. CONVENTIONAL FUZZY CONTROLLER

A conventional fuzzy controller [4] is designed for buck converter with output voltage error and rate of change of error used as fuzzy controller input and duty cycle as output variable. Input and output variables are expressed using triangular membership functions with NL, NM, NS, ZE, PS, PM, PB. A general rule base with 49 rules is framed with the help of IF AND THEN rules. Mamdani type Fuzzy Inference system (FIS) and defuzzification of Smallest of maximum (SOM) is used.

## IV. NEW FUZZY LOGIC CONTROLLER

Let the system be represented by an equation

$$x^{(n)} = f(x,t) + u + d \qquad (3)$$

Where $x$, $d$, $u$ are state vector, disturbance and control variable respectively.

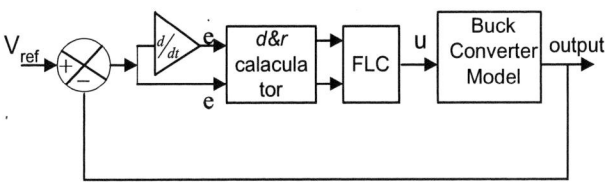

Fig. 2. New Fuzzy Logic Control Block diagram.

With tracking error $e = x - x_d$;

$x_d$ is the desired output voltage. Approach for implementation of the scheme is as shown in the block diagram of Fig.2 .For a second order system a sliding surface happens to be a switching line given by

$$s(x,t) = 0$$

and

$$s(x,t) = \left(\frac{d}{dt} + \lambda\right)^{n-1} e \qquad \lambda \geq 0; \qquad (4)$$

The fuzzy control can be written as

$$U = -K_{fuzz}(e\,\dot{e}\,\lambda)sign(\sigma) \qquad (5)$$

where $\sigma = sx$

Modified fuzzy input variables are 'd' and 'r' and output variable is $K_{fuzz}$ whose input and output membership functions are as shown in Fig. 3. Value of 'r' and 'd' is calculated using equation (6) and (7). The sign of control action is decided by sign$\sigma$.

The value of $K_{fuzz}$ is evaluated based on the radial distance of state from origin 'r' and 'd' the magnitude of Euclidean distance from switching line. This approach leads to appreciable reduction in rule base and much better performance of fuzzy controller than the previously reported algorithms.

$$r = \sqrt{e^2 + \dot{e}^2} \qquad (6)$$

$$d = \frac{|s|}{\sqrt{\lambda^2 + 1}} \qquad (7)$$

If we incorporate a boundary layer above expression can be rewritten as

$$U = -K_{fuzz} sat(\sigma) \qquad (8)$$

where sat function is defined as

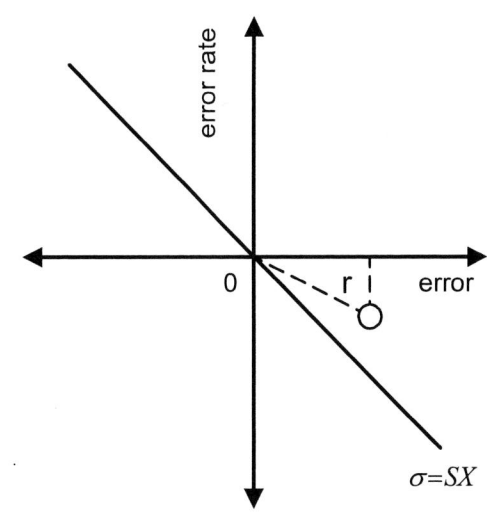

Fig. 3. Graph of error VS error rate with sigma plane.

$$
\begin{aligned}
sat(\sigma) &= sgn(\sigma) & |\sigma| &> \varepsilon & \varepsilon &> 0 \\
&= \frac{\sigma}{\varepsilon} & |\sigma| &\leq \varepsilon
\end{aligned}
\qquad (9)
$$

here $\varepsilon$ is the width of the boundary layer and $K_{fuzz}$ is the fuzzy gain which is evaluated through linguistic rule base. Assigning a fuzzy set to each of the variables we obtain fuzzy rule base as stated below.

Rule 1 IF r is small AND d is small Then $K_{fuzz}$ is small
Rule 2 IF r is large AND d is small Then $K_{fuzz}$ is medium
Rule 3 IF r is large AND d is large Then $K_{fuzz}$ is large

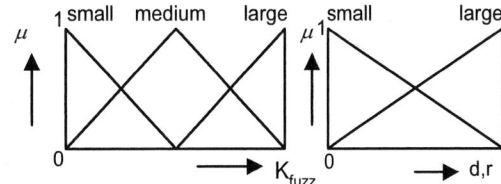

Fig. 4. Input and output membership functions for a RRB fuzzy.

The above rule base is applied to the buck converter using SIMULINK and fuzzy logic tool box. Results are discussed below.

## V. RESULTS AND DISCUSSIONS

The performance of buck converter with three rule base is studied with Fuzzy Logic Toolbox in Simulink. Three rule base is giving faster transient response with no overshoot. The results are tabulated in table II. A comparison is also made with a PID controlled converter. With P= 0.1, I=10, D=0.1 PID control also gives a good transient response. Comparing between conventional fuzzy and a RRB fuzzy, later one gives a better response in terms of peak overshoot and oscillatory response.

371

TABLE II
COMPARATIVE RESULTS

| Parameter | 3 Rule | 7x7 Rule | PID |
|---|---|---|---|
| Rise time | 4.5 m Sec | 4m Sec | 2msec |
| Peak Overshoot | Negligible | 10% | 2% |
| Settling Time | 8 m Sec | 7 m Sec | 2msec |

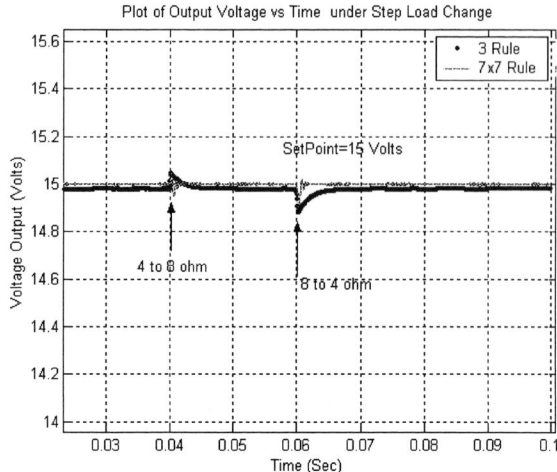

Fig. 6. Plot of output voltage with time for a step change.

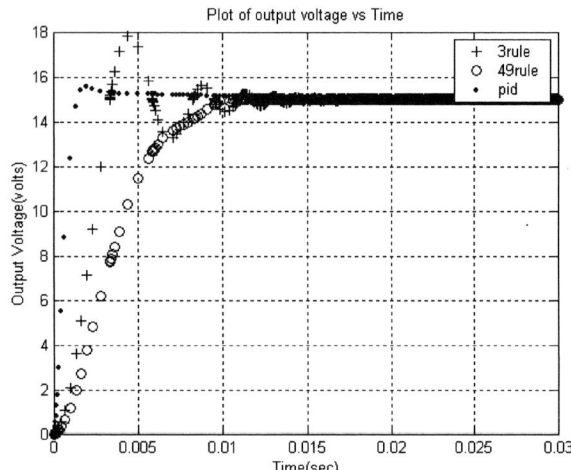

Fig . 5. Plot of output voltage with time. ($V_{ref}$=15V, $R_L$=4Ω) [1].

## VI. REFERENCES

[1] R Palm, Sliding mode Fuzzy control, *IEEE Trans.on fuzzy systems,* pp 519-526, 1992

[2] Xinghuo yu, zhihong Mau ,"Design of fuzzy sliding mode control systems", *Trans on Fuzzy sets and system* 95(1998) 295-306, Elsevier publisher

[3] V. Utkin , "Sliding mode control in electromechanical systems" Taylor and Francis 1998, ISBN: 0748401164

[4] D. Drainkov, H. Hellendoorn, M. Reinfrank, "An Introduction to fuzzy control," Narosa Publishing house. ISBN 81-7319-069-0

[5] J. Mahdavi, A. Emadi, H. A. Toliyat, "Application of State Space Averaging Method to Sliding Mode Control of PWM DC/DC Converters", presented at conf on *IEEE Industry Applications Society*, Oct. 1997.

**2006 IEEE International Conference on Power Electronic, Drives and Energy Systems**

# Development of Conventional Control of Parallel Loaded Resonant Converter -Simulation and Experimental Evaluation

### T.S.Sivakumaran, and S.P.Natarajan

*Abstract*-- **This paper presents simulation and implementation of PI controller for a high frequency soft-switched DC-DC converter for distributed energy generation systems like solar systems, vertical axis aero-generator and fuel-cell power systems. This converter is widely used to achieve reduction in size of the passive components of the converter such as inductors, capacitors and transformers. Control algorithm is developed in this work to ensure tracking of the reference voltage and rejection of system disturbances by successive measurement of the converter output voltage at certain time instants within a conduction period. The main objectives are (i) development of conventional control of parallel loaded resonant converter using MATLAB (version 7.01) software (ii) TMS320F2407 DSP based hardware implementation of above controller and (iii) evaluation of the controller's performance. The controller developed provides fast response and good performance under supply and load disturbances. The results validate the effectiveness of proposed scheme.**

*Index Terms*—**Parallel loaded Resonant Converter (PRC), Proportional plus Integral (PI) controller.**

## I. INTRODUCTION

Parallel loaded Resonant Converter (PRC) which is a subset of DC-DC converter can be operated with either zero-voltage turn–on (above resonant frequency) or zero-current turn-off (below resonant frequency) to eliminate the turn-on or turn- off losses of the semi-conductor switches. Due to the reduced switching losses, this converter is particularly suited for high power and high frequency operation. The operating frequency of a PRC usually varies over a wide range to regulate the output. This results in a penalty in filter design and poor utilization of magnetic components. Instead of frequency modulation control, the resonant converters [1-7] can also be regulated by phase shift control where the duty ratio D is modulated and the switching frequency is kept constant. Power electronics systems like PRC have non-linear characteristics and conventional control is found to provide satisfactory control for these converters. Such control is used

---

T.S.Sivakumaran, Assistant Professor in Electrical and Electronics Engg. Department, Mailam Engineering College, Mailam-604 304, India (email:praveen_tss@rediffmail.com).
S.P.Natarajan, Professor in Instrumentation Engg. Department, Annamalai University, Annamalai Nagar (email:spn_annamalai@rediffmail.com).

in this work to vary the duty cycle for regulating the output voltage of PRC. Tests for load and line regulation are carried out to evaluate the controller's performances. The results are presented.

## II. PARALLEL LOADED RESONANT CONVERTER

A schematic diagram of full-bridge parallel loaded resonant converter is shown in Fig.1. The resonant circuit consists of series inductance Ls and parallel capacitor Cp.Q1-Q4 are switching devices having base/gate turn-on and turn-off capability. D1 to D4 are anti-parallel diodes across the switching devices. The MOSFET (say Q1) and its anti parallel diode (D1) act as a bidirectional switch.

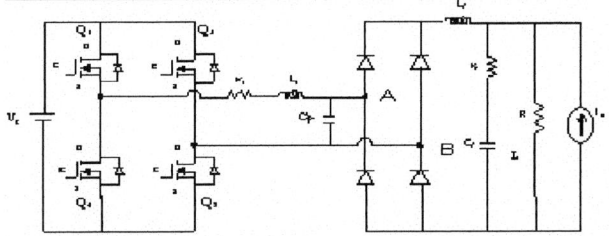

Fig. 1. Schematic diagram of full-bridge parallel loaded resonant converter.

The gate pulses for Q1 and Q2 are in phase but 180 degree out of phase with the gate pulses for Q3 and Q4. The positive portion of switch current flows through the MOSFET and negative portion flows through the anti-parallel diode. The load resistance R is connected across bridge rectifier via low pass filter Cf. The voltage across the point AB is rectified and fed to load R through Cf. io is a current source representing the load disturbance. In the analysis that follows, it is assumed that the converter operates in the continuous conduction mode and the semiconductors have ideal characteristics.

## III. OPEN LOOP CONVERTER DYNAMICS

The open loop converter system (Fig.2) comprises the power stage modeled in the above section. The inputs to the power stage are supply voltage Vs and duty ratio D and the output is Vo. Generally, Vs is maintained at a constant value. Fig.3 shows the open loop output voltage of PRC. Fig.4 shows the open loop response of PRC under sudden supply change from 30V to 27V.

---

0-7803-9771-1/06/$25.00 ©2006 IEEE

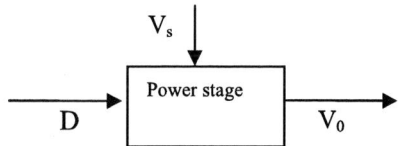

Fig. 2. Block diagram of open loop PRC.

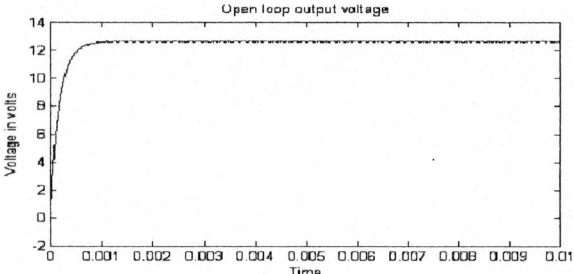

Fig. 3. Open loop output voltage of PRC.

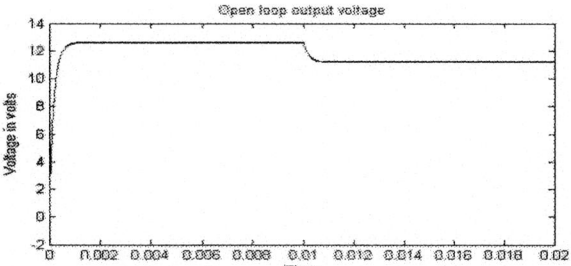

Fig. 4. Open loop output voltage of PRC for sudden supply change from 30V to 27V at t=0.01sec.

## IV. DESIGN OF CONVENTIONAL CONTROLLER

Many industrial processes are non-linear and are thus complicated to be described mathematically. However, it is known that many non-linear processes can satisfactorily be controlled using PID controllers provided the controller parameters are tuned well. Practical experience shows that this type of control has a lot of sense since it is simple and based on three basic behavior types or modes: proportional (P), integrative (I) and derivative (D). Instead of using a small number of complex controllers, a larger number of simple PID controllers are used to control complex processes in an industrial assembly in order to automate the process. Controllers of different types such as P, PI and PD are today basic building blocks in control of various processes. In spite of simplicity, they can be used to solve even a very complex control problem, especially when combined with different functional blocks, filters (compensators or correction blocks) etc. A continuous development of new control algorithms insure that the PID controller has not become obsolete and that this basic control algorithm will have its part to play in process control in foreseeable future. It can be expected that it will be a backbone of many complex control systems. While proportional and integrative modes are also used as single control modes, derivative mode is rarely used in control systems.

PI controller forms the control signal u(t) in the following way with e(t) being the error between set point and actual output:

$$U(t) = K_p \left[ e(t) + \frac{1}{T_i} \int_0^t e(\tau) d\tau \right],$$

From the state transition matrices, the transfer function of the converter is obtained. Using the transfer function, the tuning of the controller is done by the reaction curve method. Controller tuning involves the selection of the best values of $K_p$ and $T_i$. This is often a subjective procedure and is certainly process dependent. In this work $K_p$=2.5 and $T_i$=5 are the values of the controller settings tuned to provide satisfactory response of the converter.

## V. SIMULATION RESULTS

The servo and regulatory responses obtained by simulation using MATLAB software with conventional control under supply and load disturbances for continuous mode of operation of the converter are presented in this section

Figs. 5 and 6 show the start up transient of output voltage and current of PRC with PI controller for 13.5V set point. Figs. 7 and 8 show the output voltage and current with 10% line disturbance at t=0.01sec with set point 13.5V. The output voltage and current with PI control and 10% load disturbance are shown in Figs.9 and 10. The next two Figs.11 and 12 displays the corresponding responses under sequential supply and load disturbances with PI control. The above simulation results show that the proposed PI control regulates the output voltage satisfactorily under disturbances.

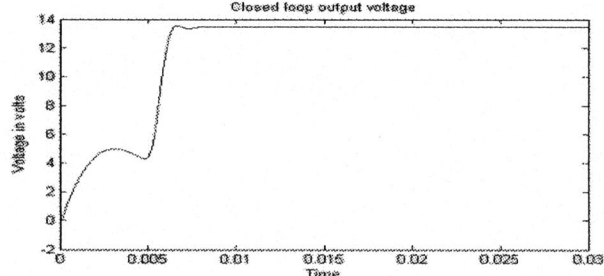

Fig. 5. Output voltage with set point 13.5V.

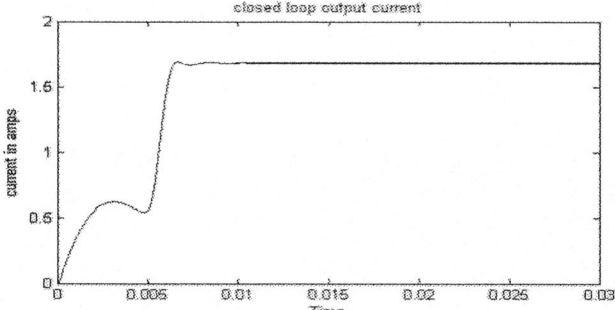

Fig. 6. Output current with set point 13.5V and 8 ohm load.

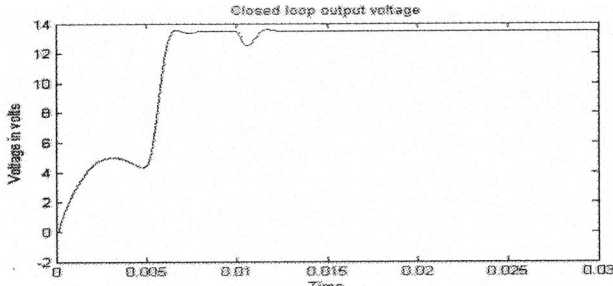

Fig. 7. Output voltage with 10% line disturbance at t=0.01sec with set point 13.5V.

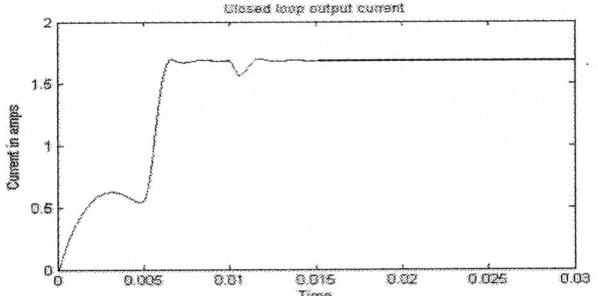

Fig. 8. Output current with 10% line disturbance at t=0.01sec with set point 13.5V.

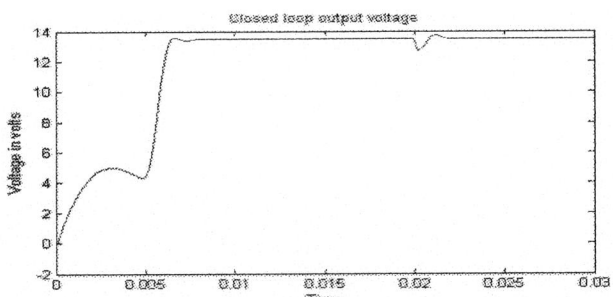

Fig. 9. Output voltage with 10% load disturbance (from nominal load 8 ohm) at t=0.02sec with set point 13.5V.

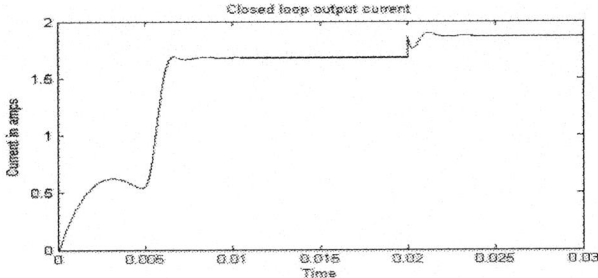

Fig. 10. Output current with 10% load disturbance (from nominal load 8 ohm) at t=0.02sec with set point 13.5V.

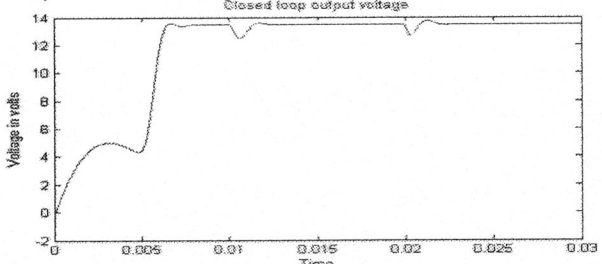

Fig. 11. Output voltage with 10% of line and load disturbances at t=0.01sec and t=0.02sec respectively with set point 13.5V.

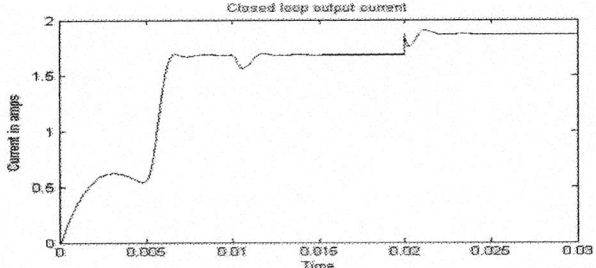

Fig. 12. Output current with 10% of line and load disturbances at t=0.01sec and t=0.02sec respectively with set point 13.5V.

## VI. HARDWARE IMPLEMENTATION

The cost of digital controllers decreased continuously during the last decade while the computing power of such systems increased simultaneously. Today Digital Signal Processors (DSPs) with ten to hundred Million Instructions Per Second (MIPS) are available. This results in effective use of the digital control for power electronics. For testing the proposed control algorithm for a resonant converter, a DSP from Texas Instruments has been used. This fast DSP has been developed for power electronics control. It is thus well suited for the present work.

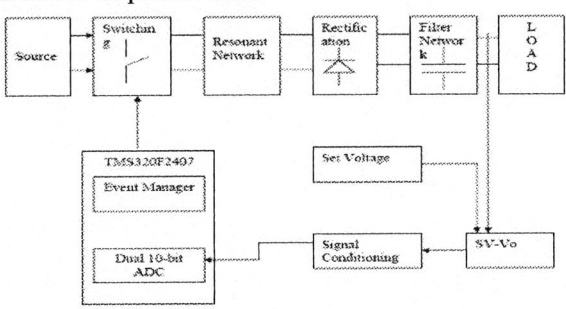

Fig. 13. Block diagram of DSP based PI controller for PRC.

Hardware implementation of the conventional control for PRC is carried out in this work using TMS320F2407 Digital Signal Processor (DSP). TMS320F2407 is a 16-bit fixed point DSP that combines the flexibility of a high-speed controller with numerical capability of an array processor thereby offering an inexpensive alternative to multi-chip bit-slice processors. The chosen DSP comprises a 40 MIPS CPU, 500 ns A/D converter with 10-bit resolution, on-chip data and program memory, a double event manager, a watchdog timer and several communication interfaces. Hence this DSP is used in this work.

The converter output after suitable signal conditioning is fed to the on-chip ADC of DSP. The signal conditioning circuit is tuned to convert the converter output voltage (0-30v) into (0-3V) to be fed as input to ADC (Fig.13). The controller output is sampled every 5μs. In order to provide isolation between the power converter circuit and the ADC of DSP, a high impedance differential amplifier is provided. The error between the required and actual output voltages are manipulated by the DSP based conventional controller employing position form PI algorithm to provide an appropriate change in duty ratio of the firing pulses to the

MOSFET so as to maintain the average output voltage constant in spite of line and load variations. The event manager module of the DSP generates the firing pulses. Optocouplers HCPL 4506 provides isolation between the event manager module of DSP and gate of MOSFET. The PWM signal from the DSP is not capable of driving MOSFET. In order to strengthen the pulses, IR 2110 driver is used.

A drop in the output voltage level triggers the conventional controller to increase the output voltage of the converter by modifying the duty cycle of the converter. Fig.14 shows the PWM pulses for Q1 and Q2 of PRC. Fig.15 shows the start up transient of the output voltage of parallel loaded resonant converter. Fig.16 shows the current through the series resonant inductor (Ls). Fig.17 shows the voltages across the parallel resonant capacitor (Cp). Fig.18 shows the voltage across the terminals A, B of PRC. Fig.19 shows the output voltage ripple for 100 ohm load. Fig.20 shows the transient response of the output voltage of converter for step changes in supply voltage from 30V to 17V. Fig.21 shows the transient response of the output current of converter for step load changes from 500 ohm to 400 ohm and vice versa. The outputs are obtained using 100 MHz digital storage oscilloscope through software [SW205-2 ver 1.4] with 5V/div setting in y-axis and 0.2 ms/div setting in x-axis. These experimental results are satisfactory for power supply applications.

TABLE I
CIRCUIT PARAMETERS OF THE TEST CONVERTER

| PARAMETER | VALUE |
|---|---|
| Inductor Ls | 101.7µH |
| Capacitor Cp | 470nF |
| Inductor $L_f$ | 3.3mH |
| Capacitor $C_f$ | 32µF |
| Load resistance R | 100ohm |
| Input voltage $V_s$ | 30V |
| Switching frequency $f_s$ | 100kHz |
| Duty ratio D | 0.25-0.99 |
| MOSFETs | IRFP9240 |
| DIODEs | UF5042 |

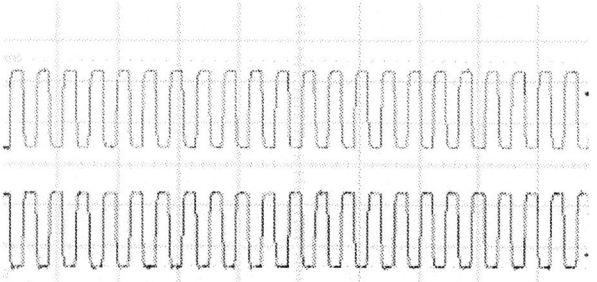

Fig. 14.   PWM pulses for Q1 and Q2 of parallel loaded resonant converter.

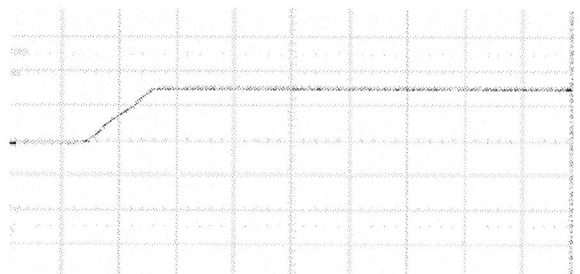

Fig. 15.   Start up transient of the output voltage of parallel loaded resonant converter.

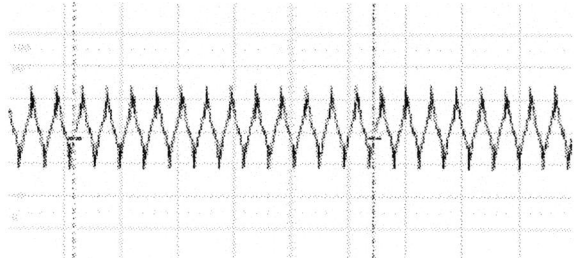

Fig. 16.   Current through the series resonant inductor (Ls).

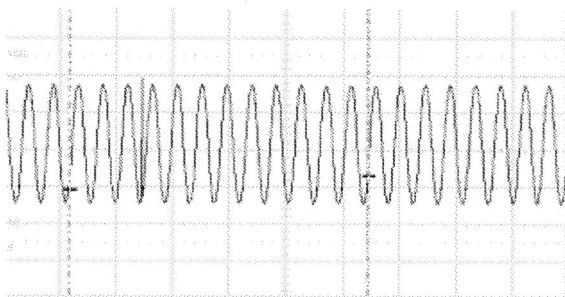

Fig. 17.   Voltages across the parallel resonant capacitor (Cp).

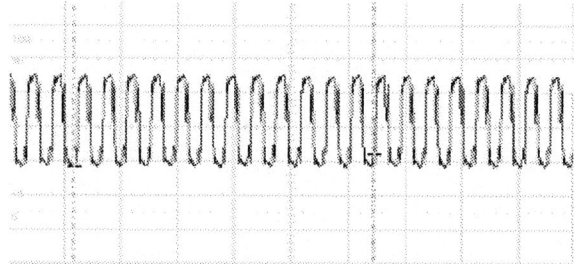

Fig. 18.   Voltage across the terminals A, B of PRC.

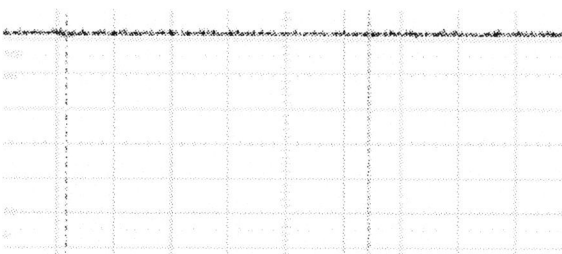

Fig. 19.   Output voltage ripple for 100 ohm load.

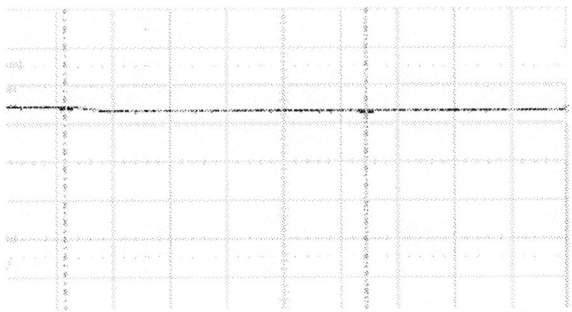

Fig. 20. Transient response of the output voltage of converter for step changes in supply voltage from 30V to 17V.

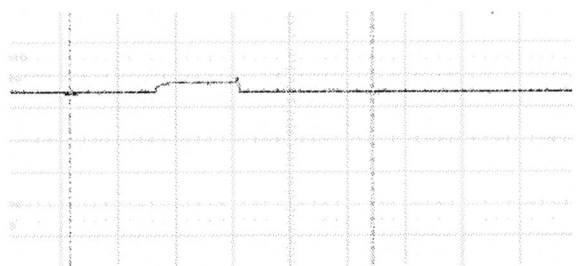

Fig. 21. Transient response of the output current of converter for step load changes from 500ohm to 400ohm and vice versa.

## VII. CONCLUSION

The simulation and experimental results show that the proposed conventional controller regulates satisfactorily the output voltage of PRC irrespective of line and load disturbances. This establishes the validity of the proposed conventional controller that effectively rejects changes in supply and load while achieving quick regulation of the converter output voltage.

## VIII. REFERENCES

[1] Eric X. Yang, Fred C. Lee and Milan M. Jovanovic "Small ignal modeling of series and parallel resonant converters", in *IEEE Record 0- 7803-0485-3/92*, pp.785-792

[2] Christian Hattrup, and Heinz W. Van der Broeck, "Fast estimation techniques for digital control of resonant converters", *IEEE Trans. Power Electronics*, vol. 18, no.1, pp.365-372, January 2003.

[3] Chan Ee Lyn, N.A. Rahim and S. Mekhilef, "DSP-based fuzzy logic controller for a battery charger",in *IEEE Conf. Proceedings TENCON'02*,2002, pp.1512-1515.

[4] O.Ojo, "Robust control of series parallel resonant converters", *IEE Proceedings on Control theory Application,* vol.142, no.5, pp 401-410, September 1995.

[5] V. Vorperian and S. Cuk, "Small signal analysis of resonant converters " in IEEE *Power Electronics Specialist Conf. Rec., 1983,* pp.269-282.

[6] S.Johnson and R.Erickson, "Steady- state analysis and design of the parallel resonant converter ",in *IEEE PowerElectronics Specialists conf. Rec., 1986*, pp.154-156

[7] F.S. Tsai, Y.Chin and F.C.Lee, "State-plane analysis of a constant-frequency clamped-mode parallel-Resonant converter", *IEEE Trans. on Power Electronics,* vol.3, no.3, pp.364-378, 1988.

## IX. BIOGRAPHIES

**S.P.Natarajan** was born in 1955 in Chidambaram. He has obtained B.E (Electrical and Electronics) and M.E. (Power System) in 1978 and 1984 respectively from Annamalai University securing distinction and then Ph.D in Power Electronics from Anna University, Chennai in 2003. He is currently a Professor in Instrumentation Engineering Department at Annamalai University where he has put in 25 years of service. He is presently guiding nine Ph.D scholars and so far guided fifty M.E students. His research papers (15) have been presented at IEEE / International Conferences in Mexico, Virginia, Hong Kong, India, Singapore, Japan, Malaysia and Korea. He has 2 publications in national journals. His research interests are in modeling and control of DC-DC converters and multiple connected power electronic converters, control of Permanent Magnet Brushless DC motor, embedded control for multi level inverters and matrix converters etc. He is a life member of Instrument Society of India and Indian Society for Technical Education. He has completed an AICTE R & D project on "Investigations on Controllers for Permanent Magnet Brushless DC motor.

**T.S.Sivakumaran** was born in Panruti, India, on December 12,1969. He has obtained B.E (Electrical and Electronics) and M.Tech (Power Electronics) in 1998 and 2002 respectively from Annamalai University and Vellore Institute of Technology (Deemed University) Vellore. He is currently Assistant Professor in Department of Electrical and Electronics Engineering, Mailam Engineering College, Mailam, India. He is presently pursuing Ph.D in the Department of Instrumentation Engineering, Annamalai Univesity His areas of interest are: modeling, simulation and implementation of intelligent control strategies for power electronic converters. He has two national and two international publications. He is a life member of Institution of Engineers (India) and Indian Society for Technical Education.

**2006 IEEE International Conference on Power Electronic, Drives and Energy Systems**

# A Novel Technique to Reduce the Switching Losses in a Synchronous Buck Converter

A. K. Panda, and Aroul. K

*Abstract*—This paper proposes a zero-voltage transition (ZVT) PWM synchronous buck converter, which is designed to operate at low output voltage and high efficiency typically required for portable systems. To make the DC-DC converter efficient at lower voltage, synchronous converter is an obvious choice because of lower conduction loss in the diode. The high-side MOSFET is dominated by the switching losses and it is eliminated by the soft switching technique. Additionally, the resonant auxiliary circuit designed is also devoid of the switching losses. The suggested procedure ensures an efficient converter. Using an example design, the converter operation is explained and analyzed.

*Index Terms*-- DC-DC Converter, Switching loss, Synchronous Buck, ZVT

## I. INTRODUCTION

THE next generation of portable products, such as personal communicators and digital assistants, will have to provide long hours of operation between battery charges. A key element in this task, especially at low output voltages that future microprocessor and memory chips will need, is the synchronous rectifier. A synchronous rectifier is an electronic switch that improves power-conversion efficiency by placing a low-resistance conduction path across the diode rectifier in a switch-mode regulator. MOSFETs usually serve this purpose [1], [2].

However, higher input voltages and lower output voltages have brought about very low duty cycles, increasing switching losses and decreasing conversion efficiency. So in this paper, we have optimized the efficiency of the synchronous buck converter by eliminating switching losses using soft switching technique [3].

The voltage-mode soft-switching method that has attracted most interest in recent years is the zero voltage transition. This is because of its low additional conduction losses and because its operation is closest to the PWM converters. The auxiliary circuit of the ZVT converters is activated just before the main switch is turned on and ceases just after it is accomplished.

---

The authors are with the Department of Electrical Engineering, National Institute of Technology, Rourkela, India (e-mail: anuppanda@rediffmail.com; k.aroul@yahoo.com).

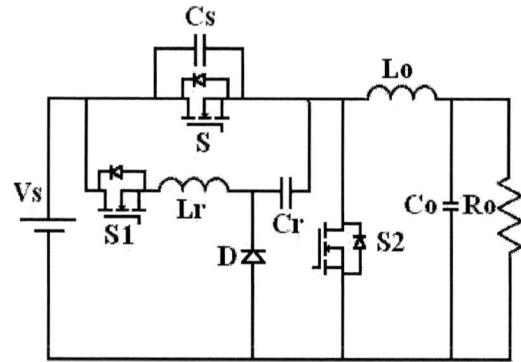

Fig. 1. The proposed converter.

The additional conduction losses are therefore substantially reduced. Moreover, it has little effect on the converter operation characteristics. Many techniques to reduce switching losses in high power have been proposed using both active and passive snubbers [4] – [7]. Reducing switching losses for low power circuit such as synchronous buck is not known to be present in the literatures. The converter shown in Fig.1 is designed for a low voltage, high current circuit and found to be highly efficient.

Hence, this paper presents a new class of ZVT synchronous buck converter. By using a resonant auxiliary network in parallel with the main switch, the proposed converters achieve zero-voltage switching for the main switch and zero-current switching for the auxiliary switch without increasing their voltage and current stresses.

The paper is organized as follows. The next section gives a short description of the proposed circuit followed by review of the various modes of operation with their key waveforms and the representation of their equivalent operation modes. Section III presents the design considerations. Section IV gives the simulation results to illustrate the features of the proposed converter scheme. Section V includes some conclusions.

## II. MODES OF OPERATION

The circuit scheme of the proposed new ZVT synchronous buck converter is shown in Fig.1. The auxiliary circuit consists of switch $S_2$, resonant capacitor Cr, Resonant inductor Lr. The auxiliary circuit operates only during a short switching-transition time to create ZVS condition for the main switch.

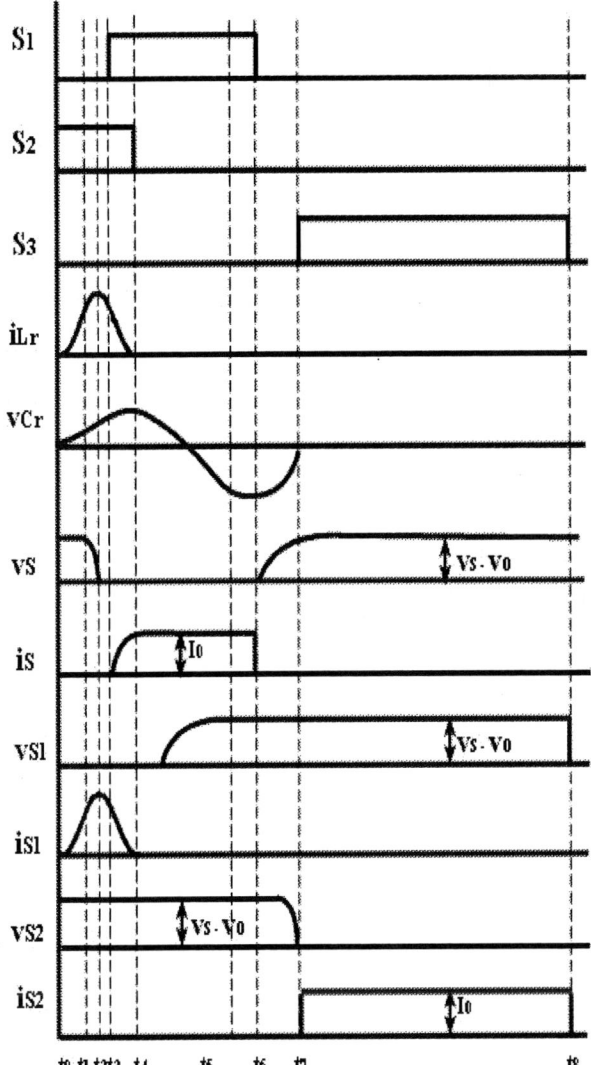

Fig. 2. ZVT synchronous buck waveforms.

A high frequency schottky diode D is used for discharging the capacitor voltage to the output, which happens before the turn on of the synchronous switch. The various modes of steady-state converter operation that goes through during a switching cycle are explained in this section of our paper. Typical waveforms of the converter are shown in Fig. 2, and the equivalent circuit for each mode of operation is shown in Fig. 3. For the sake of simplicity, output inductor, Lo, can be considered to be large enough to be a constant current source, Io. The description of each mode is presented as follows:

*Mode 1 ($t_0$, $t_1$):* At $t_0$, the switch $S_1$ is turned on. $S_1$ realizes zero-current turn-on as it is in series with the resonant inductor Lr. The current through Lr and Cr increases. The resonant network consists of $Lr_1$, Cr and $S_2$. The inductor current $i_{Lr}$ increases linearly from $i_{Lr}$ ($t_0$) = 0 and the capacitor voltage $v_{Cr}$ increases from $v_{Cr}$ ($t_0$) = 0. The mode ends at t = t1, when the capacitor across the main switch Cs begins to conduct.

*Mode 2 ($t_1$, $t_2$):* At $t_1$, the capacitor of the main switch Cs starts discharging through Lr, Cr, Cs and $S_1$. Again here, the current through Lr and Cr increases. $V_S$ is totally discharged at t = $t_2$, when the main switch body-diode begins to conduct and $i_{Lr1}$ reaches its maximum value.

*Mode 3 ($t_2$, $t_3$):* $V_S$ is fully discharged and the body diode of the main switch begins to conduct at $t_2$ as $i_{Lr}$ is greater than the output current. To achieve ZVS, the main switch S should turn on before turn off of the body diode, which occurs when $i_{Lr}$ equals the output current. The resonant process continues in this mode too and the current through $i_{Lr}$ starts to decrease.

*Mode 4 ($t_3$, $t_4$):* At $t_3$, the main switch S is turned on by ZVS. During this stage the growth rate of $i_s$ is determined by the resonance between Lr and Cr. The resonant process continues in this mode also and the current $i_{Lr}$ becomes zero at the end of this mode. The voltage $v_{Cr}$ continues to increase.

*Mode 5 ($t_4$, $t_5$):* At $t_4$, the auxiliary switch $S_1$ can be turned off with ZCS as current through the inductor, $i_{Lr}$ is zero. The source current fully flows to the main switch S.

*Mode 6 ($t_5$, $t_6$):* Since $S_1$ has already turned off, there is no resonance in this mode. Circuit operation is identical to the conventional PWM buck converter.

*Mode 7 ($t_6$, $t_7$):* At t = $t_6$, the main switch S turns off with ZVS. The capacitor Cs is charged by the input current. The resonant energy stored in the capacitor Cr starts discharging through the high frequency schottky diode D for a very short period of time, hence body – diode conduction losses and drop in output voltage is too low. This mode finishes when $v_{Cs}$ is charged to input voltage Vs and $v_{Cr}$ is fully discharged.

*Mode 8 ($t_7$, $t_8$):* At t = $t_7$, switch $S_2$ is turned on as soon as $v_{Cr}$ is fully discharged and $v_{Cs}$ is charged to Vs. Dead time loss is negligibly small compared to the conventional synchronous buck converter. During this mode, the converter operates like a conventional PWM buck converter until the switch $S_2$ is turned on in the next switching cycle.

III. DESIGN CONSIDERATIONS

*A. Inductor selection*

The lower inductor values are best for high frequency converters, since the peak-to-peak current increases linearly with switching frequency. A good rule of thumb is to select an inductor that produces a ripple current of 10 to 30% of full load DC current. Too large an inductance leads to poor loop response, and too small an inductance leads to high AC losses.

*B. MOSFET Selection*

A method to choose the MOSFETs for the converter is to compare the power dissipation values for a number of different MOSFET types. Usually, a low on-state drain resistance MOSFET is chosen for the synchronous Rectifier, and a MOSFET with a low gate charge is chosen for the switches.

*C. Resonant capacitor selection*

Since the current in the switching MOSFET is pulsating, a large resonant capacitor is used.

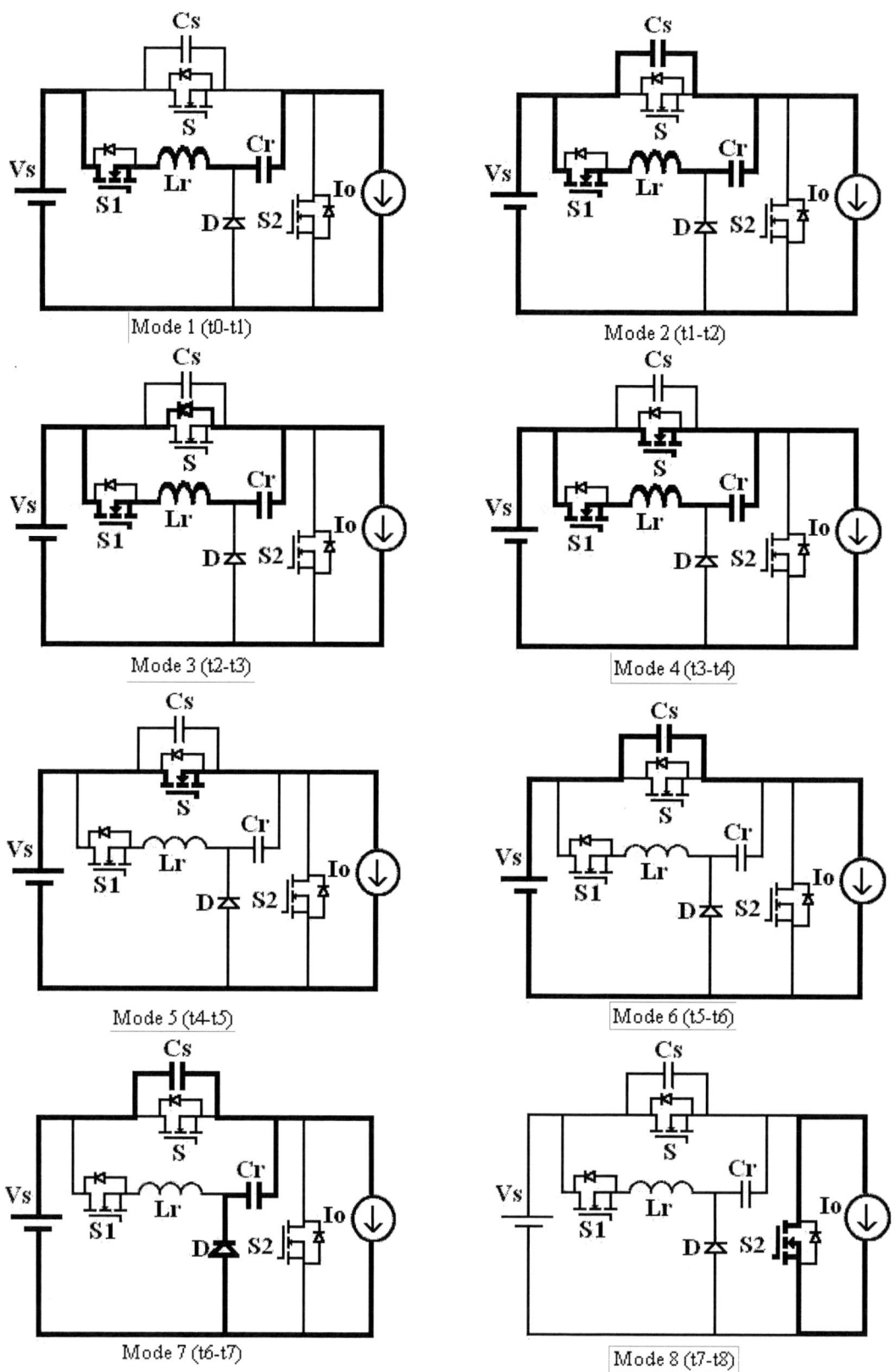

Fig. 3. Converter operation modes.

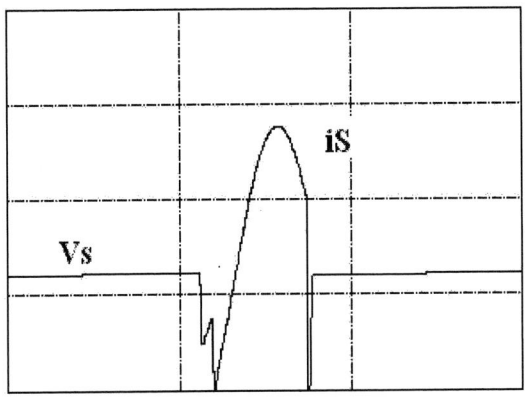

Main switch S1: vS1, iS1 – (5V, 5A)/div

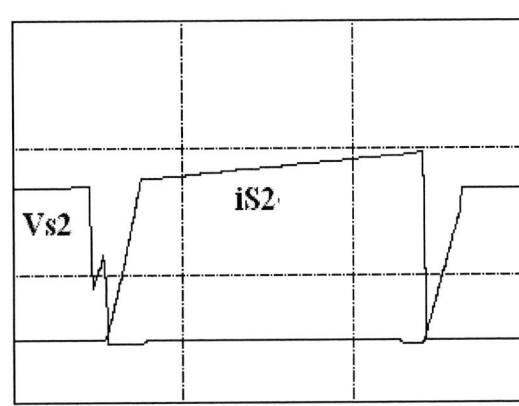

Synchronous switch S3: vS3, iS3 – (2V, 2A)/div

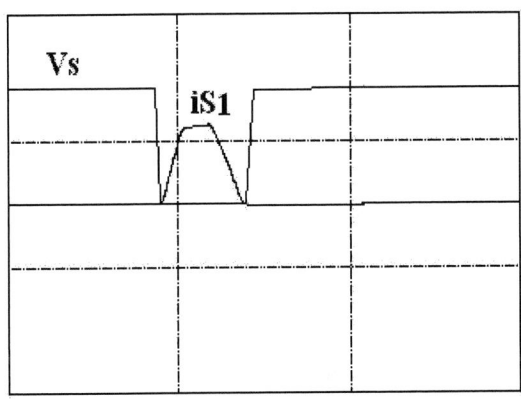

Auxiliary switch S2: vS2, iS2 – (5V, 5A)/div

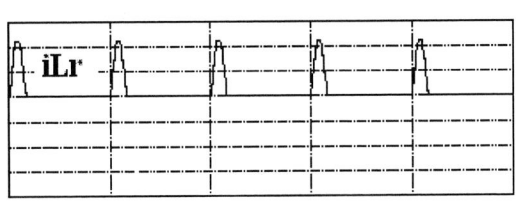

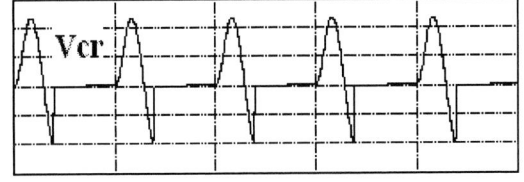

iLr1, vCr - (3V, 3A)/div

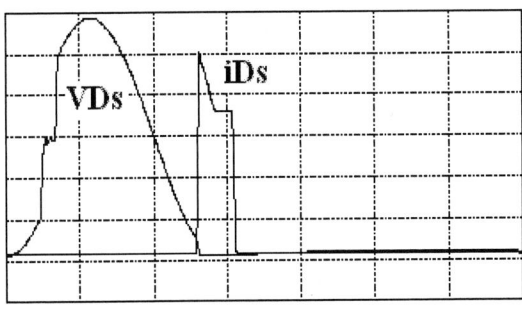

Schottky diode Ds: vDs, iDs – (6V, 6A)/div

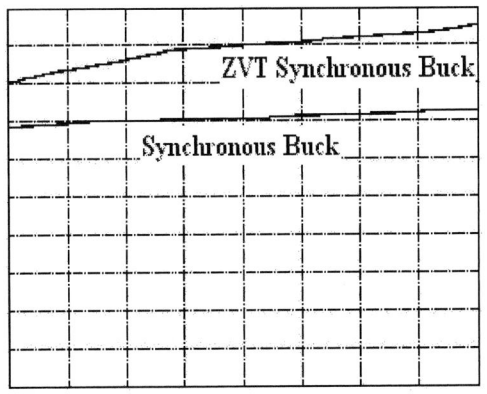

X-axis : Output power - 5W/div
Y-axis : Efficiency - 80 - 100%

Fig.4. Simulation waveforms

### D. Output capacitor selection

The output capacitor is chosen to minimize output noise voltage and to guarantee regulation during transient loads. For a low value of inductance, bulk storage may not be necessary, which is designed using ceramic capacitors.

### E. Feedback loop design

The converter power stage frequency response frequency response characteristics are determined, and then based on the resulting curves; a compensation network is designed to give the desired result. The L-C filter gives the output voltage a double pole response to the output of a compensation network. The compensation network can be designed to have a crossover frequency either above or below the L-C filter's double pole frequency.

## IV. RESULTS

The newly proposed converter operates with an input voltage Vs = 12V, output voltage Vo = 3.3V, load current of 12A and a switching frequency of 1MHz and the circuit parameters are: output inductance Lo = 1μH, output capacitance Co = 30μF, resonant inductors Lr1, Lr2 = 60nH, 90nH, resonant capacitor Cr = 0.1μH, capacitor Cs =0.5nF.

The switching loss of a synchronous buck converter without the soft-switching technique for the above parameters accounted for 50 % of the total losses. High side switching losses alone accounted for 45% of the total losses. So eliminating switching losses on high side becomes prime importance.

Fig. 4 shows the simulation waveforms of this converter. All the waveforms except the efficiency curve represents a time period of one switching cycle, which is 1μs in this case. The amplitudes are denoted below each of their waveforms respectively.

### A. Main Switch S

It is noted from the figure that the main switch operates with the soft switching technique. The converter has not exceeded the voltage and current limits in the main switch. The conduction time of the switch can also be noticed, which is very low appropriate to the design parameters.

### B. Auxiliary Switch S_1

It is noted from the figure that auxiliary switch also operates with the soft switching. The shape of the figure is identified to confine much with the theoretical waveform. The auxiliary switch is active only for a short period of time, which is verified by its conduction period and it's too small. Also, this switch's current and voltage are well within the operating limits.

### C. Schottky Diode D

The schottky diode works for a very short period to discharge the resonant capacitor Cr. A high-frequency schottky diode which is available at high-current, low voltages can be used. The conduction of schottky diodes may cause a considerable drop in output voltage for low power circuits but due to the advancement in semiconductor techniques, schottky diodes are also now available with a low forward voltage drop for high frequency circuits.

### D. Synchronous switch S_2

The synchronous switch also has characteristics similar to the switches S, S_1. They operate within the safe limits and it can be noted here, the conduction period of S_3 is more confining to the design values and it operates at a low power when compared to the other switches.

### E. Efficiency curve

The ZVT synchronous buck converter is found to be more efficient when compared to the conventional synchronous buck converter. The efficiency value is found for different values of output power. The high efficiency concludes the correctness of the design values.

## V. CONCLUSION

The concepts of ZVT used in high power were implemented in synchronous buck converter and it was shown that the switching losses in synchronous buck were eliminated. Hence the newly proposed ZVT synchronous buck is highly efficient than the conventional converter. The additional voltage and current stresses on the main devices do not take place, and the auxiliary devices are subjected to allowable voltage and current values. Moreover, the converter has a simple structure, low cost and ease of control.

## VI. REFERENCES

[1] O.Djekic, M.Brkovic, "Synchronous rectifiers vs. schottky diodes in a buck topology for low voltage applications," Power Electronics Specialists Conference, 1997. PESC '97 Record., 28th Annual IEEE , Volume: 2 , 22- 27 June 1997 Pages:1374 - 1380 vol.2.

[2] O.Djekic, M. Brkovic, A. Roy "High frequency synchronous buck converter for low voltage applications." IEEE PESC'98 Record, vol.2, pp. 1248 – 1254.

[3] M.L.Martins, J.L.Russi, H.L.Hey, "Zero-voltage transition PWM converters: a classification methodology,"IEEE proceedings on electric power applications, Volume 152, Issue 2, 4 March 2005 Page(s):323 – 334.

[4] Ching-Jung Tseng, Chern-Lin Chen "Novel ZVT-PWM converters with active snubbers", IEEE transaction power electronics, sep.1998, pp. 861-869.

[5] Elasser and D. A. Torrey, "Soft switching active snubbers for dc/dc converters," IEEE Trans. Power Electron., vol. 11, no. 5, pp. 710–722, 1996.

[6] M.L. Martins, J.L. Russi, H. Pinheiro, H.A. Grundling, H.L. Hey, " Unified design for ZVT PWM converters with resonant auxiliary circuit," Electric power applications, IEE proceedings, vol.151, issue 3, 8 May 2004, pp. 303-312.

[7] S. Kaewarsa, C. Prapanavarat, U. Yangyuen, "An improved zero-voltage-transition Technique in a single-phase power factor correction circuit," International conference on power system technology – Powercon 2004, Volume 1, 21-24 Nov. 2004 Page:678 – 683.

**2006 IEEE International Conference on Power Electronic, Drives and Energy Systems**

# Transformer Core Unbalancing Issue in a Full-Bridge DC-DC Converter with Current Doubler Rectifier

B.A. Gusev, V.I. Meleshin, and D.A. Ovchinnikov

*Abstract*—Transformer core saturation in a full-bridge, phase-shift DC-DC converter with current doubler rectifier has been considered. Equations allowing to calculate maximum value of the flux density in the core have been obtained. The theoretical analysis and calculations were verified by simulation and experiments of DC-DC converter with 2500 $W$ output power. Main causes of asymmetrical regimes have been revealed and suggested methods preventing a phenomena.

*Index terms*—current doubler rectifier, full-bridge DC-DC converter, transformer core, unbalancing.

## I. INTRODUCTION

**P**OSSIBILITY of one side transformer core saturation caused by flux unbalancing was always an issue for isolated DC-DC converters operating in a symmetrical mode, like bridge, half-bridge and push-pull configurations.

At the time, when mostly bipolar transistors were used as the power switches, the main reason of one side saturation was an unequal turn-off delay time caused by the differences in base storage charges. Another reason was unequal duration control pulses because of using control circuits built on discrete components. As the result, this asymmetrical operation of the transformer led to serious problems with power mesh, such as increased current stresses of switches, increased losses and even failure of the converter [1]. Different technique to solve this problem has been suggested along with the use of more accurate integrated PWM-controllers.

Currently, for the output power more than 1000 $W$, a phase-shifted, full-bridge converter is a very popular solution. It has several advantages including zero-voltage switching of power transistors and moderate voltage and current stresses of transistors and diodes. The secondary side of such converter can be implemented as a full wave rectifier with low pass output LC filter (Fig. 1a) or as a current doubler rectifier (Fig. 1b) [2, 3].

Modern PWM controllers, specifically designed for phase-shifted control applications, provide accurate symmetrical control pulses.

---

Boris A. Gusev is with ZAO "Svyaz engineering", 6-ya Radialnaya str., 9, 115492, Moscow, Russia (e-mail: boris@sving.ru ).

Valery I. Meleshin is with ZAO "Svyaz engineering", 6-ya Radialnaya str., 9, 115492, Moscow, Russia (e-mail: meleshin@sving.ru ).

Denis A. Ovchinnikov is with ZAO "Svyaz engineering", 6-ya Radialnaya str., 9, 115492, Moscow, Russia (e-mail: denis@sving.ru ).

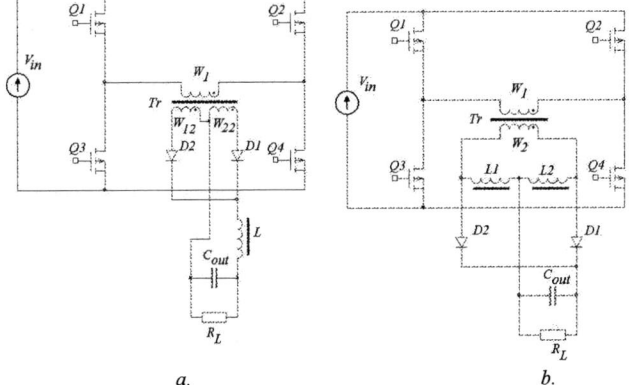

Fig. 1. Full –bridge DC-DC converters. *a* - Secondary side with a low-pass output LC-filter. *b* - Secondary side with a current doubler rectifier.

Still unbalancing of the transformer might happen caused by the inductor current differences on the secondary side of DC-DC converter during each half switching cycles of operation. Obviously, at nominal load current the most critical condition is the one when an input voltage applied to the diagonal on the primary side of full-bridge power stage. Output filter chokes might be considered as the current sources for a steady-state regime. Therefore, if the low pass output LC-filter is used in the full-bridge converter, the occurrence of asymmetrical current in the secondary winding is practically unlikely because the inductor maintains the same average current in each half switching cycle.

The use of current doubler rectifier where in each time interval two chokes operate simultaneously, can cause unequal average currents through the each diode and, as a result, appearance of DC current through the secondary winding of the transformer. At this condition, the transformer with a high permeability magnetic material and ungapped core is prone to the one-sided saturation resulting in significant current stresses on the primary side transistors.

## II. THEORETICAL ANALYSIS

The major factors causing an unbalancing are considered and analyzed in theoretical part. To derive equations defining a bias current through the secondary winding, an equivalent circuit of the converter shown in Fig. 2 has been used.

---

0-7803-9771-1/06/$25.00 ©2006 IEEE          383

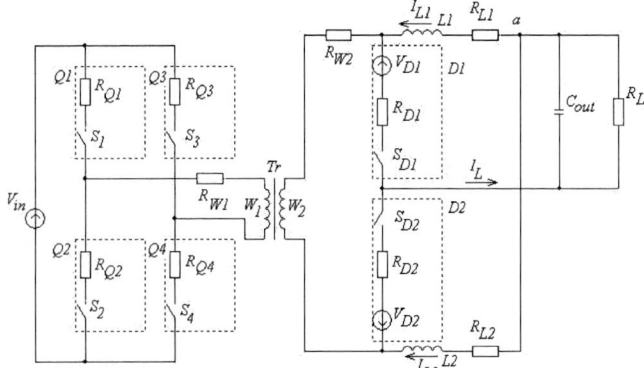

Fig. 2. Equivalent circuit of the converter.

Each transistor is shown as a series connection of a resistor and an ideal switch; $R_{Q1}$ ... $R_{Q4}$ – resistances of switches in ON-state; $S_1$ ... $S_4$ – ideal switches. The diodes are represented by the connection of an ideal switch ($Sw_{D1}$, $Sw_{D2}$) with the differential resistance ($R_{D1}$, $R_{D2}$) and threshold voltage ($V_{D1}$, $V_{D2}$) of each diode. Parameters $R_D$ and $V_D$ present an approximation of the forward biased part of a diode characteristic. Other notations are: $R_{W1}$, $R_{W2}$ – resistances of the primary and secondary windings accordingly; $R_{L1}$, $R_{L2}$ – resistances of choke windings and resistances of printed board traces in related branches.

It is assumed for simplicity that the flux density $B$ of the transformer core is a linear function of magnetizing force $H$ and there is no hysteresis.

The DC portion of the current flowing through the magnetizing inductance referred to the secondary winding $W_2$ indicates an unbalancing of the transformer core Assuming that the resistances of the primary side switches are equal for each half switching cycle intervals ( $R_{Q1} = R_{Q2} = R_{Q3} = R_{Q4}$), an equivalent circuit shown in Fig. 3 can be used. In this figure, $R'_{W1}$ is the primary side resistances referred to the secondary side:

$$R'_{W1} = (\frac{W_2}{W_1})^2 (2R_Q + R_{W1}) \qquad (1)$$

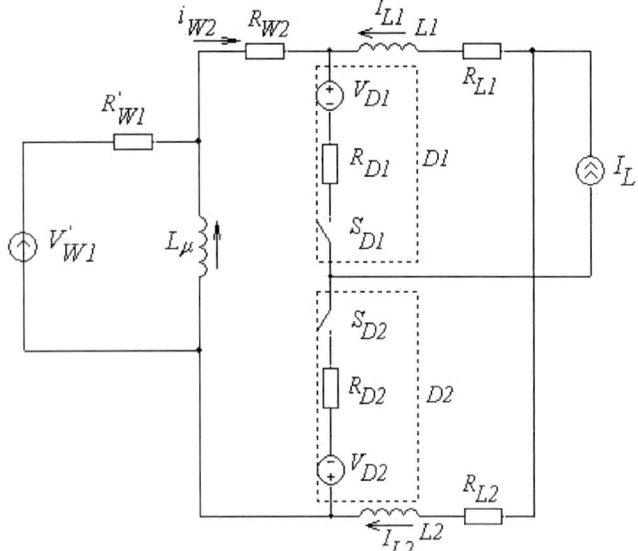

Fig. 3. Equivalent circuit of the secondary side with referred primary one.

Assuming, that an output ripple is quite low, a load with an output capacitor can be represented as the current source $I_L$. The ac voltage waveform on primary side is symmetrical and there is no a bias component in it. It is assumed that transistor resistances are equal at all time intervals. With these assumptions, an average current in the branch containing resistance $R'_{W1}$ should be equal to zero. Therefore, the transformer magnetizing current equals to an average current of the secondary winding.

The secondary winding waveforms at different time intervals of operation are shown in Fig. 4.

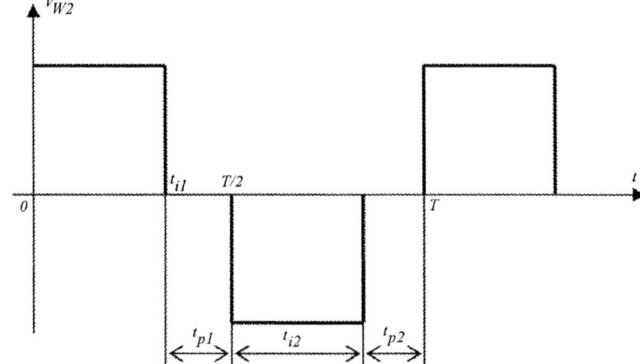

Fig. 4. The secondary winding wave.

The circuit operation is divided at four intervals (two impulses and two pauses):
$t_{i1}$: Q1, Q4 – ON, Q2, Q3 – OFF, D1 – ON;
$t_{p1}$: Q1, Q2 – ON, Q3, Q4 – OFF, D1, D2 – ON;
$t_{i2}$: Q2, Q3 – ON, Q1, Q4 – OFF, D2 – ON;
$t_{p2}$: Q3, Q4 – ON, Q1, Q2 – OFF, D1, D2 – ON.

The currents through winding $W_2$ can be determined for each time intervals:

$t_{i1}$: $\qquad I_{W2i1} = I_{L2} \qquad (2)$

$t_{p1}$: $\qquad I_{W2p1} = \dfrac{V_{D2} - V_{D1} + I_{L2}R_{D2} - I_{L1}R_{D1} + I_{W2}R'_{W1}}{R_{W2} + R_{D1} + R_{D2} + R'_{W1}} \qquad (3)$

$t_{i2}$: $\qquad I_{W2i2} = I_{L1} \qquad (4)$

$t_{p2}$: $\qquad I_{W2p1} = \dfrac{V_{D2} - V_{D1} + I_{L2}R_{D2} - I_{L1}R_{D1} + I_{W2}R'_{W1}}{R_{W2} + R_{D1} + R_{D2} + R'_{W1}} \qquad (5)$

Determine duty ratio $D$ as following:
$$D = 2t_{i1}/T = 2t_{i2}/T$$
where
$T$ – a switching cycle.
After substitution the following symbol,
$$R_\Sigma = R_{W2} + R_{D1} + R_{D2} + R'_{W1}$$
the average secondary winding current can be derived from (2) ... (5):

$$I_{W2} = \frac{D}{2}(I_{L2} - I_{L1}) + (1-D)(\frac{V_{D2} - V_{D1} + I_{L2}R_{D2} - I_{L1}R_{D1}}{R_\Sigma} + \qquad (6)$$

$$+ \frac{R_{W2} + R_{D1} + R_{D2} + DR'_{W1}}{R_\Sigma}I_{W2})$$

Denote:

$$k_I = \frac{DR'_{W1} + R_{W2} + R_{D1} + R_{D2}}{R_\Sigma}$$

From (6) obtain equation in three unknowns: $I_{W1}$, $I_{L1}$ and $I_{L2}$:

$$k_I I_{W2} + (\frac{D}{2} + \frac{(1-D)R_{D1}}{R_\Sigma})I_{L1} - (\frac{D}{2} + \frac{(1-D)R_{D2}}{R_\Sigma})I_{L2} = 0 \quad (7)$$

Consider a node "a" shown in Fig. 3:

$$I_{L1} + I_{L2} = I_L \quad (8)$$

Now consider a loop shown in Fig. 3. Applying Kirchhoff's voltage law for the average values within switching cycle yields:

$$-I_{L1}R_{L1} + I_{L2}R_{L2} + I_{W2}R_{W2} = 0 \quad (9)$$

Solving a system of equations (7), (8) and (9), obtain required currents:

$$I_{W2} = \frac{\frac{D}{2}(R_{L1} - R_{L2})I_L + (1-D)\frac{R_{D2}R_{L1} - R_{D1}R_{L2}}{R_\Sigma}I_L + (1-D)\frac{(V_{D2} - V_{D1})(R_{L2} + R_{L1})}{R_\Sigma}}{DR_{W2} + (1-D)\frac{(R_{D2} + R_{D1})R_{W2}}{R_\Sigma} + k_I(R_{L1} + R_{L2})} \quad (10)$$

$$I_{L1} = \frac{(R_{L2}k_I + \frac{D}{2}R_{W2} + (1-D)\frac{R_{W2}R_{D2}}{R_\Sigma})I_L + (1-D)\frac{R_{W2}(V_{D2} - V_{D1})}{R_\Sigma}}{DR_{W2} + (1-D)\frac{(R_{D2} + R_{D1})R_{W2}}{R_\Sigma} + k_I(R_{L1} + R_{L2})} \quad (11)$$

$$I_{L2} = \frac{(R_{L1}k_I + \frac{D}{2}R_{W2} + (1-D)\frac{R_{W2}R_{D1}}{R_\Sigma})I_L + (1-D)\frac{R_{W2}(V_{D1} - V_{D2})}{R_\Sigma}}{DR_{W2} + (1-D)\frac{(R_{D2} + R_{D1})R_{W2}}{R_\Sigma} + k_I(R_{L1} + R_{L2})} \quad (12)$$

If $I_{w2}$ according to (10) is not equal to zero, it means that DC bias of the magnetic flux density appears. Simultaneously currents $I_{L1}$ and $I_{L2}$ become unequal to each other. After that, if parameters of core and number of turns of the transformer windings are known, then the maximum value of the flux density in the core can be calculated.

Carry on a calculation using (10) and taking as example the following data: D = 0.8, $I_L$ = 50 A, $R_Q$ = 110 *mOhm*, $R_{W1}$ = 1 *mOhm*. As a required function accept the bias flux density:

$$B_= = 4\pi 10^{-7} \mu \frac{W_2}{lm} I_{w2} \quad (13)$$

where

$l_m$ – mean magnetic length.

We use a ring core R63 (63x38x25 *mm*), high frequency ferrite N87 with permeability $\mu$ = 2200; $W_1$ = 18, $W_2$ = 7; $l_m$ = 0.152 *m*.

A function of the flux density values against spread of the resistances in the choke branches is plotted in Fig. 5. Initial values of the resistance $R_{L1}$ were accepted 3.5, 4 and 4.5 *mOhm*. As an independent variable was assumed a magnitude

of spread $\Delta R_{L2}$: $\Delta R_{L2} = 100\left(\frac{R_{L2}}{R_{L1}} - 1\right)$, for $R_{L2} > R_{L1}$. The

rest data were accepted as following: $V_{D1} = V_{D2} = 0.52$ *V*, $R_{D1}$ = $R_{D2}$ = 21.5 *mOhm*.

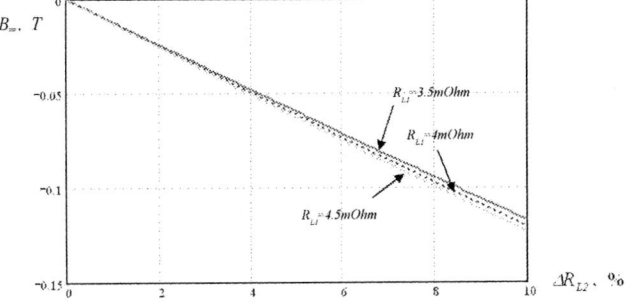

Fig. 5. Bias flux density as a function of resistance spread of chokes.

Fig. 5 shows that an increase of resistances spread of choke leads to a significant gain of the bias flux density. Note, that the higher absolute values of choke branches resistances results in the higher bias flux density for the same spread of the resistances $R_{L1}$ and $R_{L2}$. The analysis was performed for $R_{L2} > R_{L1}$ and for $R_{L2} < R_{L1}$ the bias flux density reverse its sign.

The plot of $B_=$ against spread of output diode voltage threshold and spread of differential resistances of these diodes is shown in Fig, 6. The voltage threshold and differential resistance of $D1$ was secured and as variables used values:

$$\Delta V_{D2} = 100\left(\frac{V_{D2}}{V_{D1}} - 1\right) \text{ for } V_{D2} > V_{D1},$$

$$\Delta R_{D2} = 100\left(\frac{R_{D2}}{R_{D1}} - 1\right) \text{ for } R_{D2} > R_{D1}.$$

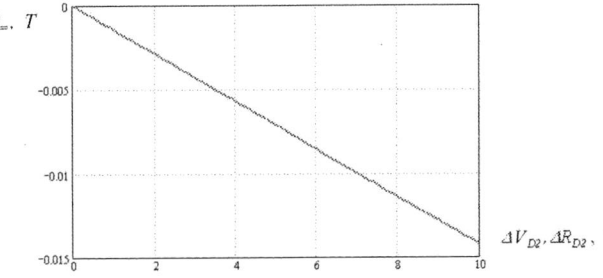

Fig. 6. Bias flux density as a function of spread voltage thresholds and differential resistances of diodes.

Fig.6 shows that dependence $B_=$ against the spread of diode threshold voltages or differential resistances has the same behavior as the spread of choke branches. But in this case the bias flux density is by an order of magnitude less than in the previous one. General conclusion concerning of the influence of power mesh parameters could be made as following: the major effect on the DC bias flux density ($B_=$) inflicted by the choke branch resistances – spread of their copper and traces of a printed board.

## III. SIMULATION RESULTS

To verify theoretical analysis, a full-bridge DC-DC converter with current doubler rectifier has been simulated using Spice. The simulation circuit is shown in Fig. 7.

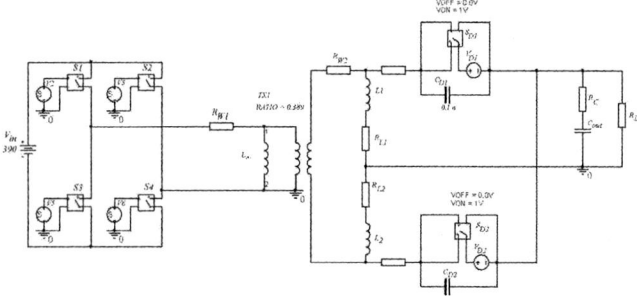

Fig. 7. A circuit of modeling of DC-DC converter.

In this circuit ideal models of switches and diodes have been used: MOSFET – switch with resistance in ON state; diode – switch with voltage source and the differential resistance.

To compare the analysis and simulation results the bias flux density for different variants of a test breadboard with parameters: $P_{out} = 2500\ W$, $V_{out} = 48\ V$ have been calculated. The parameters of real chokes and printed boards with different resistances in the circuit branches have been used for calculations and simulations. The results are presented in Table I.

TABLE I
RESULT OF CALCULATION AND MODELING.

| $R_{VD1}$, mOhm | $R_{VD2}$, mOhm | $R_{L1}$, mOhm | $R_{L2}$, mOhm | $B_=$, T, calc. | $B_=$, T, model |
|---|---|---|---|---|---|
| 21,5 | 21,5 | 3,288 | 3,448 | -0,071 | -0,067 |
| -\|\|- | -\|\|- | **3,288** | **3,574** | **-0,115** | **-0,109** |
| -\|\|- | -\|\|- | 3,524 | 3,522 | -0,013 | -0,011 |
| -\|\|- | -\|\|- | 3,789 | 3,792 | -0,015 | -0.014 |
| -\|\|- | -\|\|- | 4,968 | 4,964 | -0,014 | -0.013 |
| -\|\|- | -\|\|- | **5,469** | **5,182** | **0,056** | **0.055** |
| -\|\|- | -\|\|- | 5,538 | 4,858 | 0,158 | 0.158 |
| -\|\|- | -\|\|- | **5,339** | **3,632** | **0,474** | **0.465** |
| -\|\|- | -\|\|- | 5,794 | 4,932 | 0,198 | 0,199 |
| 23,65 | 19,35 | 3,524 | 3,522 | -0.042 | -0.039 |
| -\|\|- | -\|\|- | 5,794 | 4,932 | 0.166 | 0.165 |
| 19,35 | 23,65 | 3,524 | 3,522 | 0.016 | 0.015 |
| -\|\|- | -\|\|- | 5,794 | 4,932 | 0.230 | 0.225 |

In addition to the presented results some variants of DC-DC converters have been simulated using models of real transistors instead of switches shown in Fig. 7. The results of such simulations are also practically very close to the ones shown in Table 1.

## IV. EXPERIMENTAL RESULTS

Some DC-DC converters with parameters shown in Table 1 by bold fonts have been tested. The output power is 2500 W and switching frequency – 100 $kHz$. The values of maximum core flux density assessed quantitatively by the expression:

$$B_m = B_= + B_\sim /2$$

where

$$B_\sim = \frac{V_{in}D}{2\,fS_CW_1}$$

In this case $V_{in} = 400\ V$, $D = 0.8$, $S_C = 305.9\ mm^2$, $W_1 = 18$ t, which gives $B_\sim = 0.29\ T$. Flux density saturation for ferrite N87 and $T = 25\ ^oC$ according [4] accounts 0.44 ... 0.49 T (H = 200 ... 1200 A/m) and for $T = 100\ ^oC – 0.37 ... 0.39\ T$ (H = 200 ... 1200 A/m). Instantaneous primary winding current was taken as a monitoring parameter. The results are shown in Fig. 8, 9, 10 and 11. Analysis of the primary current waveforms shows very close repeatability of calculations, simulation and experimental results.

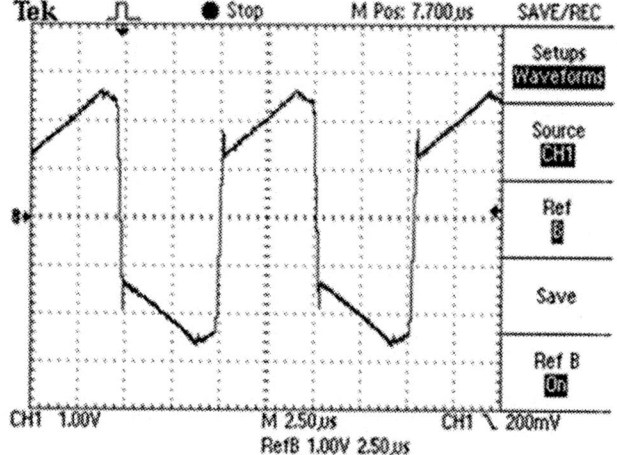

Fig. 8. A primary current waveform (line 2 of table 1).

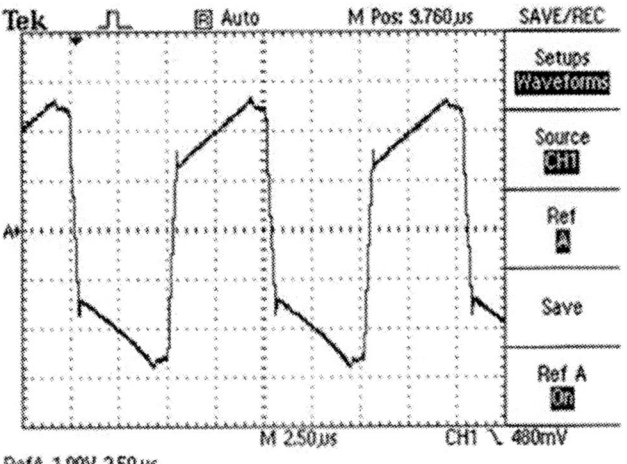

Fig. 9. A primary current waveform (line 6 of table 1). Temperature 25 $^oC$ of the both chokes.

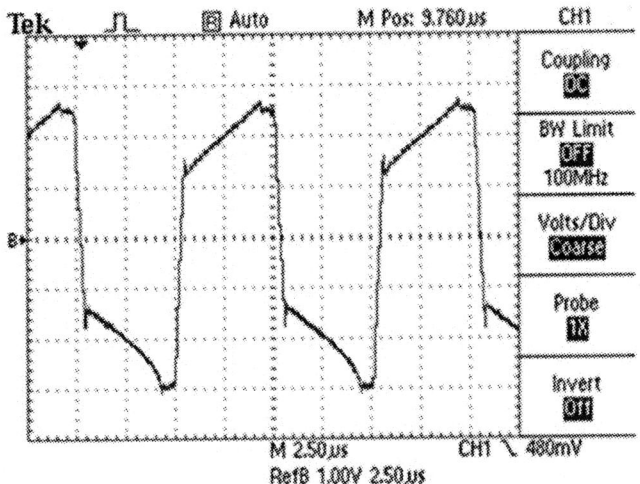

Fig.10. A primary current waveform (line 6 of table 1). The same design, temperature of the choke L1 is higher than one of the choke L2.

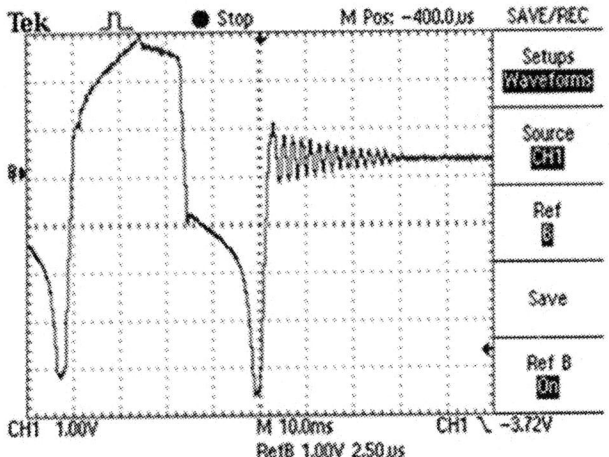

Fig.11. A primary current waveform (line 8 of table 1, the second comparator of PWM controller actuates).

It is necessary to point out that different forced air cooling conditions for the output chokes results in increased differences in the winding resistances even if initially the resistances were almost equal. In this case, after some time the choke windings' spread will be able to inflict significant asymmetry into the transformer operation though by electrical design estimation supposed to be quite safe. For example, the results in Fig. 9 are for the equal initial temperatures *25 °C*, while the waveforms in Fig. 10 are for the same design but the temperature of the choke L1 is 85 *°C* and it is 69 *°C* for the choke L2. This difference is because of different cooling conditions.

The question arises if it is possible to avoid the asymmetrical behavior using means of a cycle-by-cycle over current protection that is usually available in modern PWM controllers?

There are few possible ways how the control and protection circuit can be used to avoid unbalancing:

1. Addition of a ramp with a voltage proportional to instantaneous primary transformer current provides correction of control pulse durations in each half switching cycle of operation.

2. Current limiting comparator in the PWM controller will be tripped if the over current occurs and this limits related control pulse duration.

3. The second over current protection comparator with higher current limit threshold can be used to turn off the PWM controller and then activate its again.

If an asymmetry of the choke branches is very significant, the primary current increases quickly with a high slew rate. This can lead to a situation when only limiting of control pulse duration in each half switching cycle does not manage to improve an asymmetry. In this case, the second comparator operates, PWM comparator operates, PWM controller actuates once again and normal operation of DC-DC controller is broken. That regime is shown in Fig. 11 and 12.

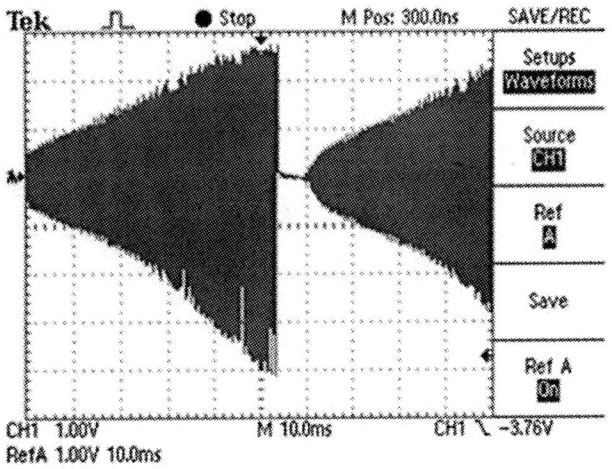

Fig. 12. A primary current waveform (line 8 of table 1, an other scale.

Sometimes, when the asymmetry is too large and occurs because of poor quality soldering of choke multicable winding or a skewness of the printed board resistances the second protection stage won't be able to manage to work on time. As a result the DC-DC converter will be damaged in consequence of prohibitive primary side currents.

Based on this the following ways to eliminate asymmetrical operation of the transformer are suggested:

1. Use of transformer cores with a gap. However, the following considerations can make this solution unacceptable:

   ➤ gapped transformers require special bobbins thus increasing their cost;
   ➤ gapped transformers have higher leakage inductance which affects negatively on a converter transfer ratio and increases the reverse recovery voltage spikes across the output rectifier diodes;
   ➤ usually the cost of gapped cores higher;
   ➤ transformer with the gapped core can block air flow to the other components because of increased height.

2. Ensure during design and production that the choke winding resistances and traces of choke branches on the printed board are within specified tolerances.

3. Use of instantaneous primary current control circuit to compensate asymmetry. This method requires a high gain current loop and the minimum time delay between controller output and gate of the power MOSFET.

## V. CONCLUSION

1. The presented analysis, simulation and experimental results of asymmetrical operation in case of DC-DC phase-shifted converter with current doubler rectifier have shown that the main causes of unbalancing are unequal resistances of choke windings and traces on PCB board that are part of choke current flow branches. Hence, it is necessary to choose a type of choke winding and assembling technology of chokes very carefully; also it is necessary to ensure the equal resistances of windings and traces on printed board.

2. It is shown that parameter tolerances of MOSFETs and output diodes practically do not affect on DC-DC converter unbalancing.

3. The current protection capabilities of modern PWM controllers are not sufficient to protect from unbalancing if the output chokes design or their assembly has been made incorrectly and if the PCB layout of power stage does not guarantee the required symmetry.

## VI. REFERENCES

[1] C.Wm.T. McLyman "Transformer and Inductor Design Handbook, 2nd ed.", revised and expanded; New York and Basel: M.Deccer Inc., 1988.

[2] D. Sable and F.C. Lee, "The operation of a full-bridge zero-voltage-switced PWM converter, in Proc. *Virginia Power Electron. Center Seminar*, 1989, pp.120-126.

[3] N.H. Kutkut, D.M. Divan and R.W. Gascoigne, "An improved full-bridge zero-voltage switching PWM converter using a two-inductor rectifier", *in IEEE Transactions on industry applications*, Jun./Feb. 1995, Vol. 31, No.1, pp. 119-126.

[4] *Ferrite and Accessories*. Data Book 2001. Company EPCOS.

## VII. BIOGRAPHIES

**Boris Gusev** received the M.S. degree in electrical engineering from the Moscow Aviation Institute (Technical University - MAI), Russian Federation. He taught undergraduate course in area of energy conversion. Since 2001 to 2002 he had been working with MAI as an engineer. Since 2002 he has been working as an electrical engineer in JSC "Svyaz Engineering", Moscow. His research focused on various power converters controlled by microprocessors. Mr Gusev published some papers in Russian magazines and has RF patents.

**Valerie Meleshin** received the M.S. degree in electrical engineering from Moscow Aviation Institute (Technical University – MAI), USSR, the P.D. degree in Electrical Engineering from MAI in 1968 and P.D. degree in Power Conversion from Power Energy Institute, Moscow, in 1988. Since 1963 to 2001 he had been working with MAI as a lecturer and then as a Professor. He was a Research Professor and chief of R&D CENTER working with US company AT&T, later Lucent Technologies in the field of power conversion since 1993 to 2000. Since 2001 he has been working as a Leading Research Officer in JSC "Svyaz Engineering", Moscow. Mr. Meleshin published papers in Proceedings of APEC'95, 97, 98, INTELEC'99, TELESCON'97 and PEDS'05. Four US patents.

**Denis Ovchinnikov** received the M.S degree in Power Conversion from Moscow Aviation Institute (Technical University-MAI), Moscow, RF, in 2000, and Ph.D. degree in Power Conversion from Power Energy Institute, Moscow, Russia in 2004. Since 2002 he has been working in JSC "Svyaz Engineering", Moscow. Since 2002 to 2005 his responsibility focused on power supply systems for telecommunications. Nowadays he is a deputy director of the R&D department. Mr. Ovchinnikov published paper in Proceedings of PEDS'05, more twenty publications in Russian magazines, has three RF patents.

**2006 IEEE International Conference on Power Electronic, Drives and Energy Systems**

# Computer Aided Analysis of Fault Tolerant Multilevel DC/DC Converters

K. A. Ambusaidi, *Student Member, IEEE,* V. Pickert, *Member, IEEE,* and B. Zahawi,
*Senior Member, IEEE*

*Abstract*—**The paper presents a comparison of five fault tolerant multilevel dc/dc converters. Initially the paper describes the transitions needed to change each standard multilevel dc/dc converter into a fault tolerant circuit. Then an assessment is carried out to identify the most cost-effective fault tolerant multilevel dc/dc converter topology.**

*Index Terms*—**dc/dc converter, fault tolerant, multilevel converter,**

## I. INTRODUCTION

FAULT tolerant power conversion systems become more and more important in future aerospace and automotive applications. Up to now research focused mainly on fault tolerant drives, machines and control algorithms [1]-[3]. Little research has been done on fault tolerant dc/dc converters [4]-[7].

Usually, in fault tolerant dc/dc converter systems an identical back up dc/dc converter with the same VA rating is implemented, increasing cost, volume and weight. Reference [4] proposed a fault tolerant dc/dc converter named the generalized multilevel dc/dc converter. The converter showed fault tolerant operations for various input voltages except for the maximum and the minimum input voltage level. Reference [5] demonstrated a modified version of [4] allowing also fault tolerant operation at maximum and minimum input voltage. Both multilevel dc/dc converters, however, demonstrated fault tolerant ability for faults in the switches but not for faults in diodes or capacitors.

This paper shows a comprehensive comparison of fault tolerant multilevel dc/dc converters, which find its application in 12V/42V electrical automotives systems for example [4]. The modified converter type described in [5] and four additional multilevel dc/dc converter topologies which are derived from the generalized multilevel dc/dc converter proposed in [4]: the diode clamped, the diode and capacitor clamped, the swapped diode clamped and the flying capacitor multilevel dc/dc converter. All four are inherently not fault tolerant.

The first part of the paper described the changes that are needed (in terms of switches, passive components and

---

K. A. Ambusaidi, V. Pickert and B. Zahawi are with the School of Electrical, Electronics and Computer Engineering, Newcastle University, Newcastle Upon Tyne, NE1 7RU, UK (email: k.a.k.ambusaidi@ncl.ac.uk).

sensors) to convert all five circuit to a fault tolerant design with the help of PSPICE simulation. Each individual component was added with ideal switches emulating open circuit and closed circuit faults. The impact of these faults was monitored and measures were taken accordingly to generate a fault tolerant design. The second part compares all the multilevel dc/dc converters in terms of total component count, total installed power rating, sensor requirements and design complexity to identify the most cost effective fault tolerant multilevel dc/dc converter circuit. For the comparison each multilevel dc/dc converter is fed by four 10V idealized voltages allowing five different output voltage levels (0V, 10V, 20V, 30V, 40V) depending on the switching states used. The average output voltage is controlled by fixed voltage levels with constant time periods rather then a constant output voltage with varying duty cycle. The design of a fault tolerant circuit is greatly dependent on the control strategy. Therefore, this paper describes fault tolerant multilevel dc/dc circuits applying the former control method. In the future, a second paper will be presented showing circuits that are operated with the latter control method.

Each converter is connected to a highly resistive load (1Ω). The maximum output power is 1.6kW for all five converters. Fault tolerant investigations were done for both unidirectional and bidirectional power flow.

## II. MODIFIED GENERALIZED MULTILEVEL DC/DC CONVERTER

The 5-level modified generalized multilevel dc/dc converter presented in [5] consists of twenty two switches, twenty two diodes and three capacitors. In [5] the switches ST4, ST3, SB4 and SB3 are single devices which would conduct continuously when operating the output of the converter at 40 V or 0V (in the reverse flow of operation). As a result, modifications to the switches (ST4 and ST3) and (SB4 and SB3) must be introduced to limit high conduction losses. A parallel switch to each has been added to overcome the problem (Fig.1).

### A. Operating Principles

The switching rules of the modified generalized multilevel dc/dc converter are identical to the multilevel converter proposed in [5] and can be summarized as follows: Any two adjacent switching devices on each pole are complementary, i.e., if one is ON the other is OFF and vice versa. According to [4] there are four switching states to produce 10V and four other switching states to produce 30V. Six switching states are

---

0-7803-9771-1/06/$25.00 ©2006 IEEE     389

possible to produce 20V. Using two additional switches and diodes [5] provide more redundancy to achieve 0V and 40V. For example, the voltage level 40V can be generated through paths (ST4, ST3, S3, S4, D1, and ST1) or (ST4, ST3, ST2, S1, S2 and DB1). The voltage level 0V can be achieved through the paths (DB4, DB3, DB2, D2, D1 and ST1) or (DB4, DB3, D6, D5, S2 and DB1). Other operating principles of the topology can be found in [4] and [5].

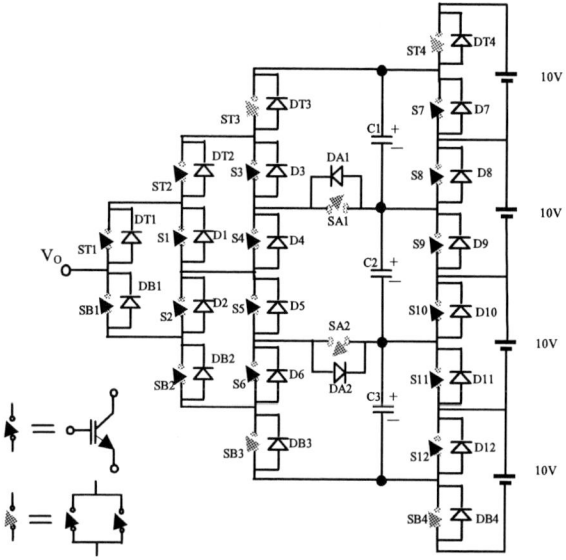

Fig.1. Modified 5-level generalized multilevel converter.

### B. Fault Tolerant Investigations

The following assumptions have been made for the fault tolerant investigation: occurrence of only one fault, the input voltage is always constant, no loss of sensor signals and electric connections and no short circuit between electric connections. Each converter must demonstrate to detect a short circuit or open circuit component and must change the switching states to recover any output voltage level.

### 1) Open Circuit Fault

Table I shows some of possible switching states for each voltage level and its corresponding current paths. If an open circuit occurs in any component of the faulty path, then the switching state must change because it can not deliver the demanded output voltage. No additional components are needed to achieve all the voltage levels because of the large number of switching states. At least two states are needed after a fault to allow fault tolerant operation of the switching devices. Fig.2 shows an example of open circuit fault occurs in S8. In case of 40V one additional switch must be connected in parallel to ST4 and ST3 as these two switches are always involved to generate 40V. Also DB4 and DB3 need to have one parallel diode to guaranty 0V under fault conditions (Fig. 4). When the additional switches (SA1, SA2) or diodes (DA1, DA2) fail open, their corresponding switching states become invalidated. As a result, each diode should be provided with a backup diode connected in parallel.

### 2) Short Circuit Fault

The investigations show that most of the voltage levels can be recovered in case of short circuit in switches. This is because of the variety of the switching states for each voltage level. 0V cannot be recovered if a short circuit occurs in S12 or D12. Also, 40V cannot be recovered if a short circuit occurs in S7 or D7. As a result, each of the mentioned switches and diodes must have one additional component that is connected in series (Fig. 4). The additional switches (SA1, SA2) or diodes (DA1, DA2) will not impact the circuit operation when they fail short because they are ON in normal operating conditions.

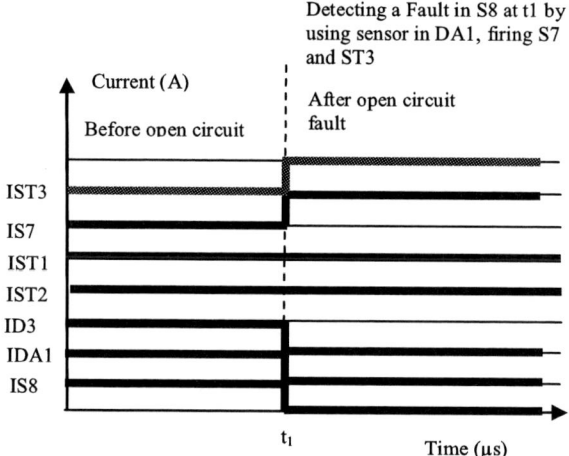

Fig.2. Open circuit fault occurs in S8.

The self balancing voltage through some of the clamping switches and the capacitors cannot be obtained if a short circuit occurs in the capacitors. Therefore, additional capacitors connected in series should be added to the existing one (Fig. 4).

### C. Sensor Requirements

Sensors are needed in order to find where the fault occurs. It is very important to put sensors in the right place to keep cost down. For open circuit faults, current sensors are required and for short circuit faults, voltage sensors are required. The additional switches (SA1, SA2) and diodes (DA1, DA2) are involved in most of the current paths for both power flow directions. As a result all four additional devices require a current sensor. The following flow chart (Fig.3) explains the procedure applied to identify sensor requirements for all other switches and diodes. As an example diode D11 in Fig. 4 is open circuit. Table I shows three current paths which contains D11. Each of the three switching pattern generates a 10V output voltage. A comparison of the output voltages of all three switching patterns leads to the information of the faulty diode D11.

Applying this procedure to all components, the total number of current sensors needed for the modified generalized multilevel converter for bi-directional power flow is 16 sensors. A short circuit in the switch is detected by the driver circuit.

390

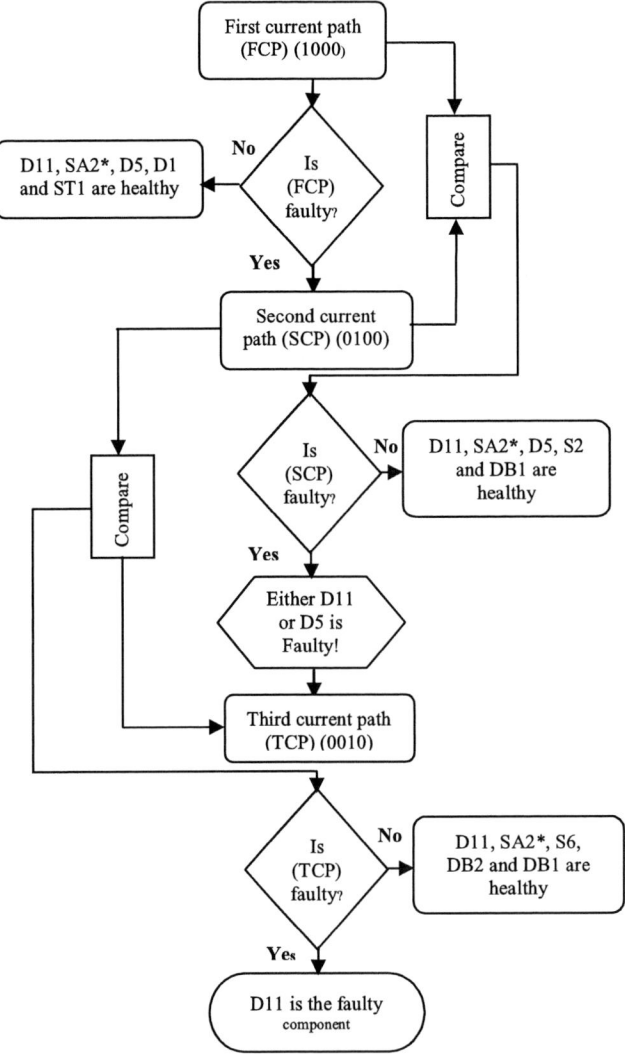

Fig.3. Flow chart to for sensor requirement (e.g circuit fault in D1).

### D. Power rating

The component power rating (VA) is calculated by multiplying the maximum current (when the device is conducting) with the maximum voltage (when the device is not conducting). For this each current path for each voltage level was traced separately. The power rating for the capacitors can be calculated by finding the maximum inrush current and the maximum voltage across the capacitor. The power ratings for all installed components are shown in Table II.

### E. Fault Tolerant Modified Generalized Multilevel Converter

Fig. 4 shows a fault tolerant 5-level modified generalized multilevel converter after adding the additional switches, diodes, capacitors and sensors. The circuit in Fig.4 is a complete fault tolerant system for both unidirectional and bidirectional power flow. The total number of components is 72 (36 switches, 30 diodes and 6 capacitors).

## III. DIODE CLAMPED MULTILEVEL DC/DC CONVERTER

A 5-level diode clamped multilevel dc/dc converter

consists of sixteen switches and twenty diodes. The lower number of components compared to the modified generalized multilevel dc/dc converter is the main advantage of the circuit.

The switching rules of the diode clamped multilevel dc/dc converter are the same for the generalized one. Four switching states are possible to produce 10V and two switching states are possible to produce 20V. There is only one switching state to produce 0V, 30V and 40V.

Table III shows an example of possible switching states for each voltage level and its corresponding current path in unidirectional power operation. There is only one current path for each voltage level even if the switching states are changing. If open circuit fault occurs in one component of that path, the output voltage level will be lost.

TABLE I
OPEN CIRCUIT FAULT INVESTIGATIONS

| Voltage levels | Switching states | | | | Current path |
|---|---|---|---|---|---|
| | ST1 | ST2 | ST3 | ST4 | |
| 0 | 0 | 0 | 0 | 0 | DB4, DB3, DB2,DB1 |
| 10 | 1 | 0 | 0 | 0 | D11, SA2, D5, D1, ST1 |
| | 0 | 1 | 0 | 0 | D11, SA2, D5, S2, DB1 |
| | 0 | 0 | 1 | 0 | D11, SA2, S6, DB2,DB1 |
| | 0 | 0 | 0 | 1 | S12, DB3, DB2, DB1 |
| 20 | 1 | 1 | 0 | 0 | D9, DA1, D3, ST2, ST1 |
| | 1 | 0 | 1 | 0 | D9, DA1, S4, D1, ST1 |
| | 1 | 0 | 0 | 1 | S10, SA2, D5, D1, ST1 |
| | 0 | 1 | 1 | 0 | D9, DA1, S4, S2, DB1 |
| | 0 | 1 | 0 | 1 | S10, SA2, D5, S2, DB1 |
| | 0 | 0 | 1 | 1 | S10, SA2, S6, DB2, DB1 |
| 30 | 1 | 1 | 1 | 0 | D7, ST3, ST2, ST1 |
| | 1 | 1 | 0 | 1 | S8, DA1, D3, ST2, ST1 |
| | 1 | 0 | 1 | 1 | S8, DA1, S4, D1, ST1 |
| | 0 | 1 | 1 | 1 | S8, DA1, S4, S2, DB1 |
| 40 | 1 | 1 | 1 | 1 | ST1,ST2,ST3,ST4 |

TABLE II
POWER RATING

| Circuit components | Power ratings (VA) | | | | Total |
|---|---|---|---|---|---|
| | 400 | 300 | 200 | 100 | (VA) |
| Switches | 8 | 11 | 8 | 9 | |
| Diodes | 6 | 10 | 7 | 7 | |
| Capacitors | - | 2 | 2 | 2 | |
| Total (VA) | 5,600 | 6,900 | 3,400 | 1,800 | 17,700 |

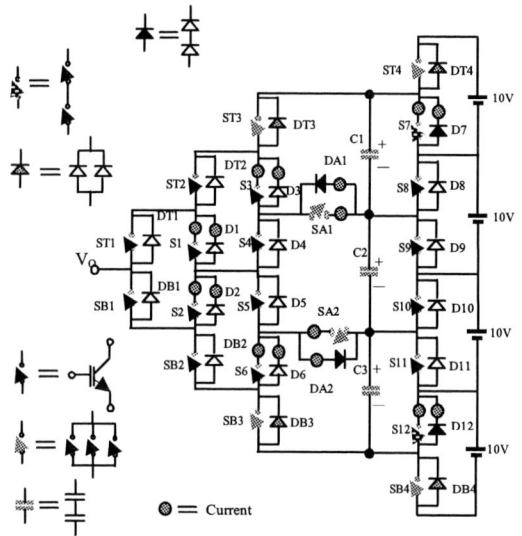

Fig.4. Fault tolerant 5-level modified generalized multilevel dc/dc converter.

TABLE III
OPEN CIRCUIT FAULT INVESTIGATIONS

| Voltage levels | Switching states | | | | Current path |
|---|---|---|---|---|---|
| | ST1 | ST2 | ST3 | ST4 | |
| 0 | 0 | 0 | 0 | 0 | DB4, DB3, DB2, DB1 |
| 10 | 1 | 0 | 0 | 0 | D11, D5, D1, ST1 |
| 20 | 1 | 1 | 0 | 0 | D9, D3, ST2, ST1 |
| 30 | 1 | 1 | 1 | 0 | D7, ST3, ST2, ST1 |
| 40 | 1 | 1 | 1 | 1 | ST4, ST3, ST2, ST1 |

The short circuit investigations show that if a short circuit occurs in components ST2, D4 or D10, the voltage level 10V will be lost. If the fault occurs in ST3 or D8, the voltage level 20V will be lost. The voltage level 30V will be lost if a short circuit fault occurs in ST4. Also, 0V will be lost if a short circuit occurs in ST1, D2, D6 or D12. The 40V voltage level will be lost if a short circuit occurs in D7 or D3, D1 or DB1. As a result an identical component must be connected in series of each component indicated above.

There is no sensor required for the diode clamped multilevel dc/dc converter because an in series connected switch or diode will conduct current in case the other device is short circuit or will not have any effect in case the other device is open circuit.

Power ratings of all components of the 5-level diode clamped multilevel dc/dc converter in both unidirectional and bidirectional power flow is maintained with the same procedure in section II. The overall VA rating is 31,400VA.

Fig. 5 shows a fault tolerant 5-level diode clamped multilevel converter after adding the additional switches and diodes. The total number of the components needed to achieve a fault tolerant system is 128 (48 switches, 80 diodes).

## IV. DIODE AND CAPACITOR CLAMPED MULTILEVEL DC/DC CONVERTER

A 5-level diode and capacitor clamped multilevel dc/dc converter has the same structure as the diode clamped multilevel dc/dc converter except that six capacitors are added in order to balance the voltage levels which gives an

advantage over the diode clamped multilevel converter for voltage transition modes (e.g. 10V to 20-V). However, the capacitor voltage levels are not self-balanced. Both converters have the same operating principles.

The open circuit fault investigations have shown that both converters have the same current paths for each voltage level (Table III) and require the same number of additional components (switches and diodes) in order to achieve fault tolerant operation. The short circuit fault investigations result in exactly the same outcome found for the diode clamped multilevel dc/dc converter (Fig. 6).

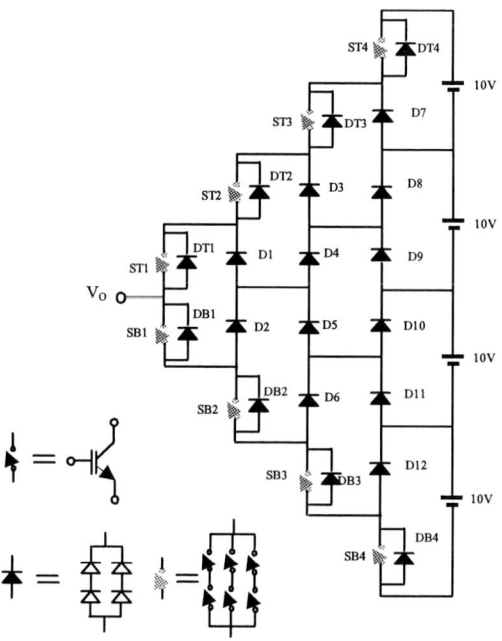

Fig.5. Fault tolerant 5-level diode clamped multilevel dc/dc converter.

There are no sensors required for the diode and capacitor clamped multilevel dc/dc converter for the same reason stated for diode clamped multilevel dc/dc converter.

The overall power rating for the diode and capacitor clamped multilevel dc/dc converter is 32,600VA.

Fig. 6 shows a fault tolerant 5-level diode and capacitor clamped multilevel dc/dc converter after adding the additional switches, diodes and capacitors. The total number of the components needed to achieve fault tolerant system is 140 (48 switches, 80 diodes and 12 capacitors).

## V. SWAPPED DIODE CLAMPED MULTILEVEL DC/DC CONVERTER

The swapped diode clamped multilevel dc/dc converter performs the same like the diode clamped multilevel dc/dc converter and has the same output voltage levels. Keeping the same number of switches, the swapped diode clamped multilevel dc/dc converter has less diodes than the diode clamped multilevel dc/dc converter

Table IV shows an example of possible switching states for each voltage level and the corresponding current paths in unidirectional power operation. There is only one current path for each voltage level even if the switching states are changing. If open circuit fault occurs in one component of that path, its corresponding voltage level will be lost.

The short circuit investigations show that if a short circuit occurs in components ST2 or SB1 the voltage level 10V will be lost. If the fault occurs in ST3, D2 or SB1 the voltage level 20V will be lost. The voltage level 30V will be lost if a short circuit fault occurs in ST4, D1, D2 or SB1. Also, 0V will be lost if a short circuit occurs in ST1, D3, D4 or D6. The 40V voltage level will be lost if a short circuit occurs in D5, D1, D2 or SB1. As a result an identical component must be connected in series of each component indicated above.

TABLE IV
OPEN CIRCUIT FAULT INVESTIGATIONS

| Voltage levels | Switching states | | | | Current path |
|---|---|---|---|---|---|
| | ST1 | ST2 | ST3 | ST4 | |
| 0 | 0 | 0 | 0 | 0 | DB4, DB3, DB2, DB1 |
| 10 | 1 | 0 | 0 | 0 | D2, ST1 |
| 20 | 1 | 1 | 0 | 0 | D1, ST2, ST1 |
| 30 | 1 | 1 | 1 | 0 | D5, ST3, ST2, ST1 |
| 40 | 1 | 1 | 1 | 1 | ST4, ST3, ST2, ST1 |

There are no sensors required for the swapped diode clamped multilevel dc/dc converter because for the same reason given in section IV.

The overall power ratings of all components of the 5-level swapped diode clamped multilevel dc/dc converter in both unidirectional and bidirectional power of operation is 26,600VA.

Fig. 7 shows a fault tolerant 5-level swapped diode clamped multilevel dc/dc converter. The total number of the components is 104 (48 switches, 56 diodes).

## VI. CAPACITOR CLAMPED MULTILEVEL CONVERTER

A 5-level capacitor clamped (flying capacitor) multilevel dc/dc converter consists of sixteen switches, eight diodes and six capacitors. This converter has the lowest component count.

The capacitor clamped multilevel converter has the same switching states as the generalized multilevel converter but the operating principles are different. In the modified generalized multilevel dc/dc converter the voltage across the capacitor is more stable and it can be considered constant. In the other hand, the capacitor clamped multilevel dc/dc converter has unstable and load dependent capacitor voltage.

Even though the converter has redundant switching states for 10V, 20V and 30V it cannot be considered as fault tolerant. Therefore, additional parallel components are needed for all the switches, diodes and capacitors.

There are no sensors required for the capacitor clamped multilevel dc/dc converter. The overall Power rating of all components is 35,200VA.

Fig. 8 shows a fault tolerant 5-level capacitor clamped multilevel converter. The total number of the components is 104 (48 switches, 32 diodes and 24 capacitors).

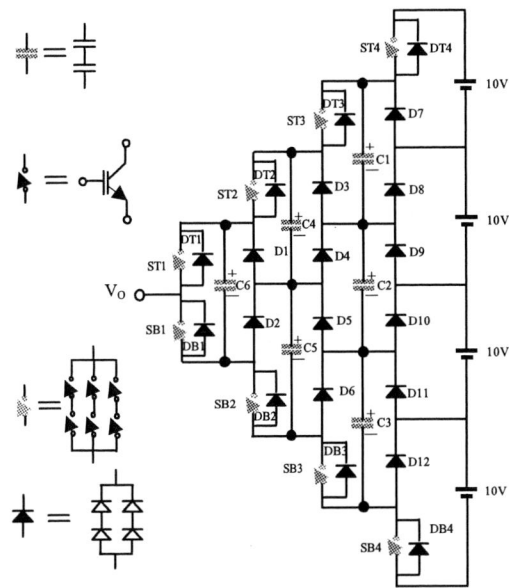

Fig.6. Fault tolerant 5-level diode and capacitor clamped multilevel dc/dc.

## VII. COMPARISON OF FAULT TOLERANT MULTILEVEL DC/DC CONVERTERS

Table V gives a comparison between all five multilevel dc/dc converters. The converter which has the highest number of components is the diode and capacitor clamped multilevel dc/dc converter while the modified generalized multilevel dc/dc converter has the lowest component count. The capacitor clamped multilevel dc/dc converter has the highest total installed power rating while the modified generalized multilevel dc/dc converter shows the lowest total installed VA rating. All multilevel converters do not require sensors except the modified generalized multilevel dc/dc converter which needs 16 sensors. The diode and capacitor clamped multilevel dc/dc converter has a very complex converter structure while the modified generalized multilevel dc/dc converter has the lowest design complexity. In terms of control, the modified generalized multilevel dc/dc converter is difficult because of the variety of the redundant switching schemes. The other topologies are less difficult to control because of the smaller number of switches. The modified generalized multilevel dc/dc converter has the advantage of self balancing of the voltage levels compared to diode and capacitor clamped and flying capacitor multilevel dc/dc converter.

## VIII. CONCLUSION

A comparison of fault tolerant multilevel dc/dc converter shows that the most cost effective fault tolerant multilevel dc/dc converter is the modified generalized multilevel dc/dc converter. It has the lowest total installed power rating and the lowest total components count. Its design complexity is low compared to diode and capacitor clamped multilevel dc/dc converter. Its voltage levels are automatically balanced without the need of external circuits.

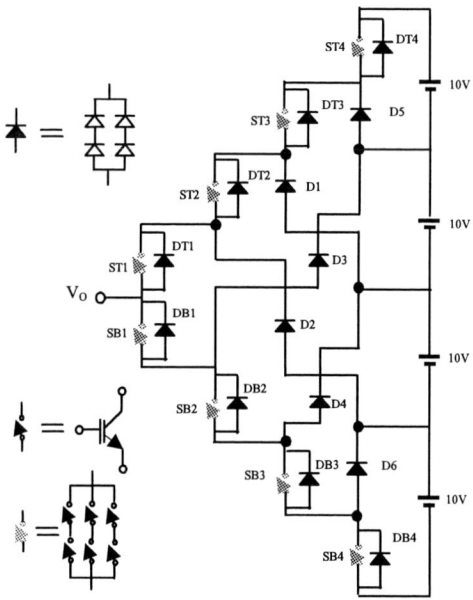

Fig.7. Fault tolerant 5-level swapped diode clamped multilevel dc/dc converter.

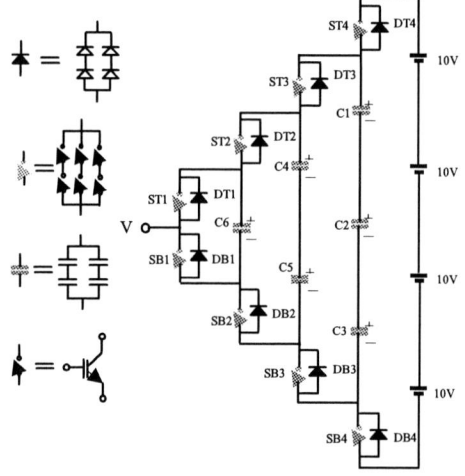

Fig. 8. Fault tolerant 5-level capacitor clamped multilevel dc/dc converter.

TABLE V
COMPARISON BETWEEN CONVERTERS

| Converters | Total components | Total installed rating (VA) | Sensors | Design complexity |
|---|---|---|---|---|
| Modified Generalized | 72 | 17,700 | 16 | Low |
| Diode clamped | 128 | 31,400 | Nil | Complex |
| Diode and capacitor clamped | 140 | 32,600 | Nil | Very complex |
| Swapped diode clamped | 104 | 26,600 | Nil | Moderate |
| Capacitor clamped | 104 | 35,200 | Nil | Moderate |

## IX. REFERENCES

[1] J. Haylock, B.C Mecrow, A.G Jack, and D.J, Atkinson., "Operation of fault tolerant machines with winding failures", *IEEE Trans. on Energy Conversion*, vol. 14, no. 4, pp. 1490-1495, December 1999.

[2] J. Haylock, B.C Mecrow, A.G Jack, and Atkinson, D.J., "Operation of a fault tolerant PM drive for an aerospace fuel pump application" in Proc. *IEE in Electrical Power Applications*, Sep 1998 ,vol. 145, no. 5, pp. 441-448.

[3] B.C Mecrow, A.G Jack, D.J. Atkinson, S. Green, G.J. Atkinson, "Design and testing of a four-phase fault-tolerant permanent-magnet machine for an engine fuel pump", *IEEE Trans. on Energy conversion*, vol. 19, No. 4, pp. 671- 678, December 2004.

[4] F. Z. Peng, "A generalized multilevel inverter topology with self voltage balancing," *IEEE Trans. on Industrial Application.*, vol. 37, no. 2, pp. 611–618, March/April 2001.

[5] A Chen, L. Hu, L. Chen, Y. Deng, X. He, "A multilevel Converter Topology with Fault-tolerant ability", *IEEE Trans. on Power Electronics*, vol. 20, no. 2, pp. 405-415 March 2005.

[6] C. Turpin, P. Baudesson, F. Richardeau, F. Forest, and T. A. Meynard, "Fault management of multicell converters," *IEEE Trans. Ind. Electron.*, vol. 49, no. 5, pp. 988–997, Oct. 2002.

[7] X. Kou, K. A. Corzine, and Y. Familiant, "A unique fault-tolerant design for flying capacitor multilevel inverters," in *Proc. IEEE IEMDC'03Conf.*, Jun. 1–4, 2003, pp. 531–538.

## X. BIOGRAPHIES

**Khalid Ambusaidi** received bachelor degree in Electrical and Electronics Engineering from Sultan Qaboos University, Oman, in 2000 and the Master of Science in electrical engineering from Newcastle University, UK in 2005. He is working towards the Ph.D. Degree in power electronics system from Newcastle University, UK. His research interests are in the area of fault tolerant DC/DC converters.

**Volker Pickert** received his Dipl.-Ing. in Electrical and Electronic Engineering from the RWTH Aachen, Germany and the University of Cambridge, UK (1994). He received his PhD from the Newcastle University in 1998. From 1998 to 2003 he worked first for Semikron International as an application engineer and then for Volkswagen as project manager responsible for power electronic systems and electric drives for electric-, hybrid- and fuel cell vehicles. Since 2003 he is Senior Lecturer at Newcastle University. His research interests are power electronics for automotive applications, thermal management, fault tolerant converters and non-linear controllers.

**Bashar Zahawi** received his BSc and PhD degrees in electrical and electronic engineering from the University of Newcastle, England, in 1983 and 1988, respectively. From 1988 to 1993 he was a design engineer at a UK manufacturer of large ac variable speed drives and other power conversion equipment. In 1994, he was appointed as a Lecturer at the University of Manchester and in 2003 he joined the School of Electrical, Electronic & Computer Engineering at Newcastle University, where he is currently the Director of Postgraduate Studies. His research interests include power conversion and the application of nonlinear dynamical methods to transformer and power electronic circuits. Dr. Zahawi is a senior member of the IEEE and a chartered electrical engineer.

**2006 IEEE International Conference on Power Electronic, Drives and Energy Systems**

# Auto Voltage Balancing in High Power DC-DC Converter

S. B. Bodkhe ,V. B. Virulkar , S. W. Mohod , and M.V. Aware

*Abstract--* This paper develops a systematic and mathematical study of the balancing theory of a three level DC-DC converter. It is propose to make dc link voltages balanced across the capacitor stack. This voltage distribution gets equally distributed by selecting the appropriate switching frequency. This work is to investigate the new control scheme to incorporate the current adaptive switching in DC-DC converters. The major advantage of this novel scheme is to develop the bi-directional high voltage multilevel converters required in many high power industrial applications. Simulation validates of this proposed scheme.

*Index Terms--* Adaptive current control, DC-DC bi-directional Converter, High power converters, Nomenclature

## I. INTRODUCTION

MULTILEVEL DC-DC converters are mostly used in high power applications in many industrial applications. They offer advantage of operating at relatively high DC bus voltages with reduced harmonic content, low EMI and low voltage stress on the devices. Multilevel converters can achieve smoother and less distorted AC-DC, DC-AC, and DC-DC power conversion. These technologies are used in the utility and large motor drives applications. These are presented in many of the literature [1].

In multilevel converters, DC bus capacitor banks are stacked in series. These converters will meet more voltage balance problem than that in AC-DC or DC-AC converter. Because of the asymmetric DC output voltage, fewer redundant switching states will be available in DC-DC multilevel converter. Also the difference of characteristics for each individual component, either on semiconductors or on passive components, will cause the voltage unbalance [2]-[4]. In most cases, a dynamic balance control strategy is necessary to balance the capacitor voltages, which requires enough redundant switching states

---

V.B.Virulkar is with the Department of Electrical Engineering, Govt. College of Engineering, Amravati. (e-mail: vbvirulkar@yahoo.com)

S.W Mohod is with Electronics Engineering Department, Ram Meghe Institute of Technology & Research, Badnera, Amravati Maharashtra (E-mail: shardmohod@rediffmail.com).

S. B. Bodkhe is with G.H.Raisoni College of Engineering, Nagpur (e-mail:s_b_bodkhe@yahoo.co.in).

M.V. Aware is with the Department of Electrical Engineering, Visveshvaraya National Institute of Technology, Nagpur, Maharashtra, INDIA(E-mail: mva_win @ yhaoo.com).

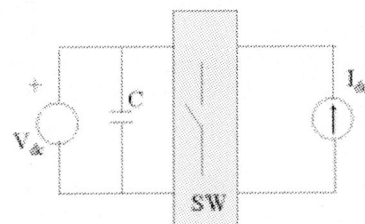

Fig. 1. Generalized Converter.

The transfer of energy from one end to another in DC – DC converter is application dependent and also decided by the type of load. The most convenient structure is to control the load current while maintaining the source side voltage constant across the supply side. This simple configuration is presented in fig. 1. The energy balance theory with basic circuit laws can be used for analysis this circuit. However, the controlled power conversion process involves the switching structure of the DC-DC converters. One such converter structure is analyzed in this paper. In this paper, analysis of DC-DC converter is presented with the voltage balancing across the capacitor. The typical design considerations are also presented. In fig. 2, three level topology of a proposed DC-DC converter is shown.

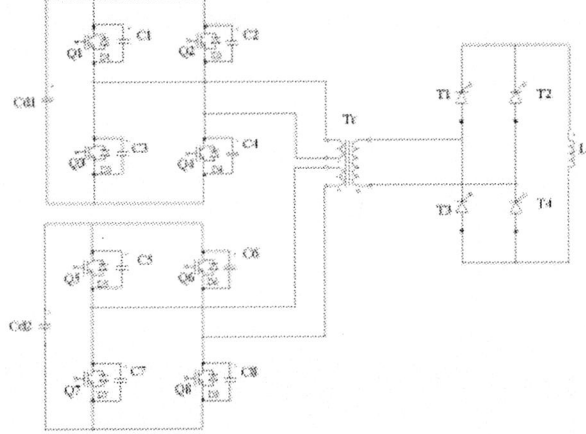

Fig. 2. Multilevel Converter.

The operating principal of the three level DC-DC converter and current adaptive switching control is presented. The benefits of the proposed circuit with high power

---

0-7803-9771-1/06/$25.00 ©2006 IEEE

capability are as under.
- Voltage balancing at DC bus
- Soft-switching is realized
- High power applications
- Lower cost due to high frequency transformer
- Use of multilevel structure can reduce the Total Harmonic Distortion (THD).

## II. CONVERTER OPERATION

The circuit topology is shown in fig. 2 and is analyzed for its operation. The one complete cycle is presented . The circuit is built with the switches Q1-Q8 and C1-C8 capacitors connected across them. The secondary of the transformer ($T_r$) is having full bridge converter connected to the inductor coil (L). The switching signals to H bridge on primary side and for the full bridge thyristors on secondary are shown in fig. 3 (a) and fig.3 (b) respectively. The voltages and currents across the each primary winding of the transformer is shown in fig. 4. The secondary side currents ($I_{T1}$, $I_{T2}$) and voltages are shown in fig. 5.

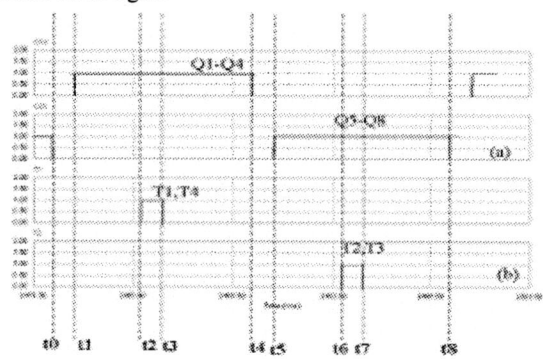

Fig. 3. (a) Switching Pulses for H bridge on Primary
(b) Full bridge on Secondary Side of Transformer.

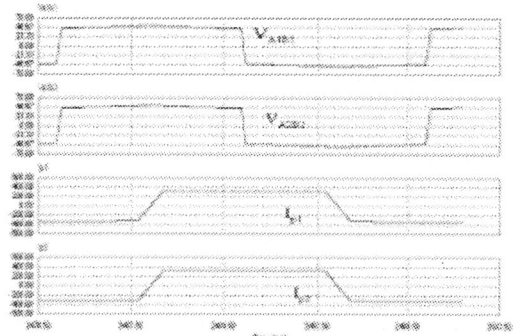

Fig. 4. Voltages $V_{A1B1}$ $V_{A2B2}$ and Current $I_{p1}$ and $I_{p2}$.

The switch Q2 and Q3, Q6 and Q7 at the $t_o$ instant are off and circuit is shown in fig. 6.The time $t_o$-$t_1$ indicates the blanking time and its operating condition is indicated in fig. 7.The operation of the switching, Q1 and Q4, Q5 and Q8 indicates the discharging operation during the time $t_1$-$t_2$. This is shown in fig. 8. The change over of the switching on

secondary side by triggering the thyristors is shown in the fig. 9.This is causing current ($I_{T23}$) to transfer to another pair of thyristor. This current transfer is due to the voltage ($V_L$) applied across the coil. The average current through the coil ($I_o$) is constant and maintains its direction. After this half cycle, the charging cycle starts as the voltage across coil is positive. The subsequent operation is shown in fig. 10, 11 and 12.

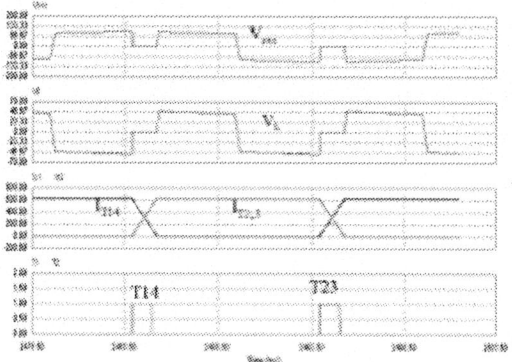

Fig. 5. Output Voltages- $V_{scc}$ and and Coil Voltage $V_L$ with Thyristor Currents $I_{T14}$ and $I_{T23}$ with respect to switching pulses.

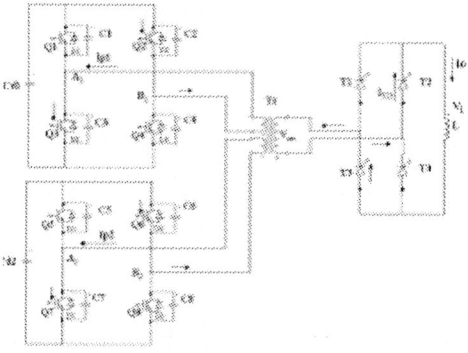

Fig. 6. Operation at $t_0$.

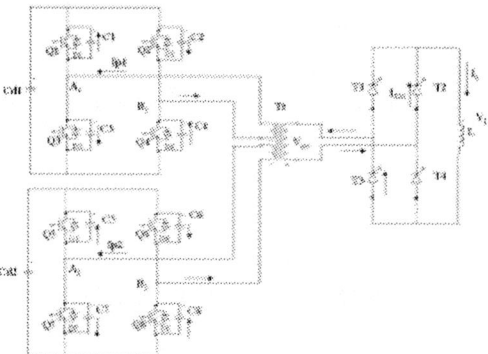

Fig. 7. Operation during $t_0$-$t_1$.

396

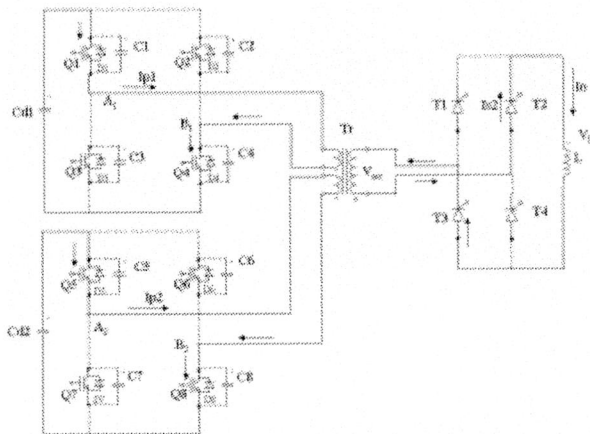

Fig. 8. Operation during $t_1$-$t_2$.

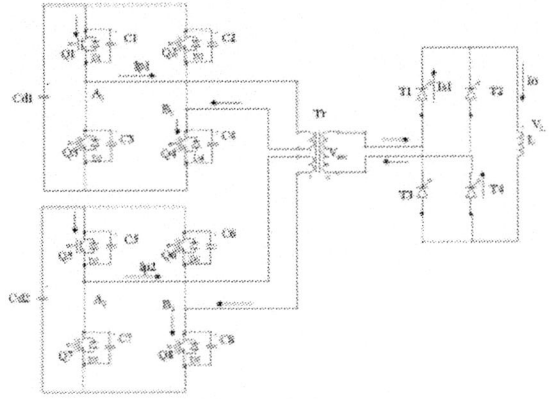

Fig. 9. Operation during $t_2$ - $t_3$.

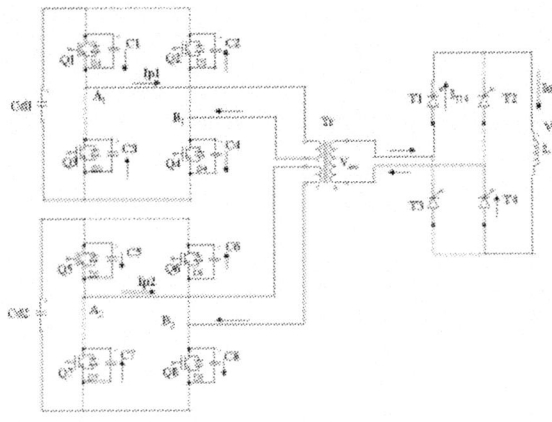

Fig. 10. Operation during $t_3 - t_4$.

Fig. 11. Operation during $t_4 - t_5$.

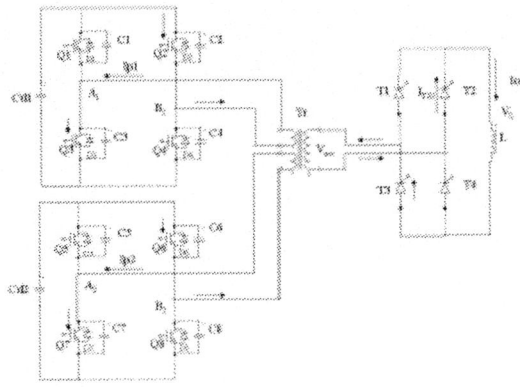

Fig. 12. Operation during $t_5 - t_6$.

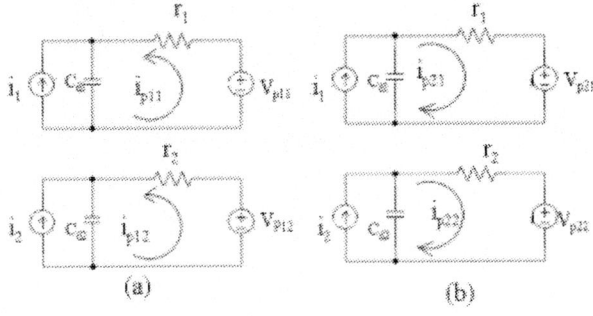

Fig. 13. Circuit Equivalent Stages under first and third stage.

## III. BALANCING THEORY OF MULTI-LEVEL CONVERTERS

The equivalent circuit for H- bridge under energy transferred across the transformer, that is from capacitor to inductive coil and from coil to capacitor is shown in figure 13 (a) and (b) respectively.

The energy transfer across the transformer is simply represented from current source ($i_1$) and voltage source ($Vp_{11}$) with one stage converter having equivalent resistances $r_1$, representing the drop across converter and transformer winding. Similarly another winding is also represented.

The mathematical analysis is carried out with these circuits. The ampere turns across the transformer winding is written as

$$N_p * (i_{p11} + i_{p12}) = N_s * I_0 \qquad (1)$$

The $N_p$ and $N_s$ represents the primary and secondary number of turns respectively. The current through coil is $I_o$.

$$i_{p11} + i_{p12} = \frac{I_0}{K} \qquad (2)$$

Where, $\square$ is the turns ratio.

Assuming that the sum of the voltage across $C_{d1}$ and $C_{d2}$ are 2V, the voltage across $C_{d1}$ is $V + \Delta V$ and the voltage across $C_{d2}$ is $V - \Delta V$, then form figure 3, current equations can be written as

$$i_{p11} = \frac{V_{p1} - (V + \Delta V)}{r_1}, i_{p21} = \frac{V_{p11} - (V + \Delta V)}{r_1} \qquad (3)$$

$$i_{p12} = \frac{V_{p1} - (V + \Delta V)}{r_2}, i_{p22} = \frac{V_{p12} - (V + \Delta V)}{r_2} \qquad (4)$$

The average current on the primary can be written as

$$i_{p1} = \frac{\frac{V_{p1} - (V + \Delta V)}{r_1}T_1 + \frac{V_{p2} - (V + \Delta V)}{r_1}T_2}{T} \qquad (5)$$

$$= \frac{V_{p1} - (V + \Delta V)}{r_1}d + \frac{V_{p2} - (V + \Delta V)}{r_1}(1 - d)$$

$$i_{p2} = \frac{V_{p1} - (V + \Delta V)}{r_2}d + \frac{V_{p2} - (V + \Delta V)}{r_2}(1 - d) \qquad (6)$$

The period $T_1 = t_2 - t_o$ and $T_2 = t_4 - t_3$ can be represented in terms of duty cycle $d = \frac{T_1}{T}$, and $\frac{T_2}{T} = 1 - d$.

The difference of current is $\Delta i_p$ on the primary side is

$$\Delta i_p = i_{p1} - i_{p2}$$

$$= \frac{[V_{p1} - (V + \Delta V)r_2] - [V_{p1} - (V - \Delta V)]r_1}{r_1 r_2}d$$

$$+ \frac{V_{p2} - (V + \Delta V)r_2 - [V_{p2}(V - \Delta V)]r_1}{r_1 r_2}(1 - d) \qquad (7)$$

The unbalance due to equivalent resistance of these two stages can be represented with the reference value of resistance R. Which is represented as $r_1 = R$ and $r_2 = R + \Delta R$.

From above equations and after simplification it is represented as

$$\Delta i_p = \frac{-2\Delta VR - \Delta R[(V_{p2} - V_{p1})d + V - V_{p2}]\Delta R\Delta V}{R(R + \Delta R)} \qquad (8)$$

Neglecting the term $\Delta R\Delta V$ and considering the balance of the charging and discharging mode, the currents are represented as $(i_{p11} + i_{p12}) = -(i_{p21} + i_{p22})$. By substituting the values of currents and simplifications

$$[4V - 2(V_{p1} + V_{p2})]R + [2V - (V_{p1} + V_{p2})]\Delta R + \Delta R = 0 \qquad (9)$$

It can be written as

$$2V - (V_{p1} + V_{p2}) = 0 \qquad (10)$$

This gives the value of voltage V as

$$V = \frac{V_{p1} + V_{p2}}{2} \qquad (11)$$

From equation 2, it can be written as

$$V_{p1} - V_{p2} = 2\frac{I_0}{K} * \frac{R(R + \Delta R)}{(2R + \Delta R)} \qquad (12)$$

Substituting these values of voltages in the equation 8,

$$\Delta i_p = \frac{-2\Delta VR - \Delta R\left(\frac{1}{2} - d\right)2\frac{I_0}{K} * \frac{R(R + \Delta R)}{(2R + \Delta R)}}{R(R + \Delta R)} \qquad (13)$$

The current flow in capacitors are

$$i_{cd1} = (i_1 + i_{p1}), i_{cd2} = (i_2 + i_{p2}) \qquad (14)$$

$$\Delta i_{cd} = i_1 + i_{p1} - (i_2 + i_{p2}) = \Delta i + \Delta i_p \qquad (15)$$

combining (13) and (15), the effective current difference is written as

$$\Delta i_{cd} = i_1 - i_2 + \frac{-2\Delta UR - \Delta R\left(\frac{1}{2} - d\right)2\frac{I_0}{K} * \frac{R(R + \Delta R)}{(2R + \Delta R)}}{R(R + \Delta R)} \qquad (16)$$

From the above equation, if the current $\Delta i_{cd}$ is positive then current flowing in to the $C_{d1}$ is larger than current flowing into the $C_{d2}$, and the voltage across $C_{d1}$ raises relatively more than the $C_{d2}$. It will be reverse if the $\Delta i_{cd}$ is negative. If $i_1 = i_2$ then $\Delta i_{cd}$ become

$$\Delta i_{cd} = \frac{2\Delta V}{R} \qquad (17)$$

If $\Delta V$ is positive and $\Delta i_{cd}$ becomes negative, a very small voltage difference will result in a very large $\Delta i_{cd}$, which will cause the voltage across the $C_{d1}$ diminish relatively very quickly to the voltage across $C_{d2}$. The voltage $\Delta V$ will decrease quickly to zero. The voltage balance will be achieved across the $C_{d1}$ and $C_{d2}$.

The difference of voltage $\Delta V$ is more sensitive to the change in converter current and the equivalent resistance of the switches and circuit components. The design parameters of the converters are selected to maintain the voltage drop near to zero to operate this bidirectional converter in stable condition. One of the control factor to achieve the voltage balance is to operate the converter with selected switching frequency zone. This is one of the most important parameters to make these converters operations more robust and economical. The concept of variable switching frequency control through adaptive current control is introduce and presented in next section.

## IV. ADAPTIVE CONTROL OF DC-DC CONVERTER

The control scheme is incorporated to maintain the voltage balance across the dc link capacitors by sensing the dc voltages and controlling the switching frequency. The dc link capacitor and front end high frequency transformer are the two components, which selects the switching frequency. By keeping the duty ratio constant, frequency is varied through the current balance sensing circuit. The limit is decided upon the maximum allowable safe operating voltages required for the DC converters. Analysis indicates the balances are maintained from no load to full load operating conditions with 15-20 % variation in switching frequency.

The inductive coil current regulation is carried out by regulating the switching pulses of thyristors. This regulates the energy storage in the inductor. There are three stages of operation. In the first case, the current is maintained constant, in the second, coil current is increased and in third case, current is reduced. The pulse position is shown in the figure 14. The thyristor pulse width is $20^\circ$ and positioning of the pulses for both the thyristors are placed $180^\circ$ apart. The location of the pulse in the half cycle decides the charging or discharging. To maintain the current through the coil constant, these pulses are to be located at $90^\circ$ in each half cycle. The average coil current is constant. If the charging is required then the pulses are to be moved above $90^\circ$ and for discharging the pulses are to be placed below the $90^\circ$.

The control of current in coil is one variable while the dc voltage across the capacitors is another.

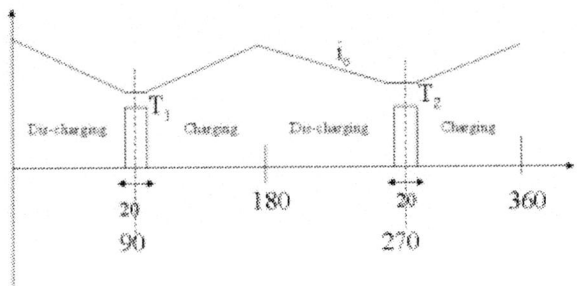

Fig. 14. Charging and Discharging pulse position.

The control scheme is implemented as shown in figure 15. The reference current set for the coil is generally a maximum current to which the coil can be charged. This can be made to follow the load demand if the system is interfaced with the power system with the balance voltage source inverters. The switching of DC-DC multilevel converter is set to the frequency from 50 Hz to 5 kHz. This is depending on the power handling capacity of the converter. The switching frequency will be low as the power rating of the converter is more. The frequency is regulated by using voltage controlled oscillator (VCO). The required control on position ($\alpha$) of the pulses for the half bridge converter is generator through the synchronizing circuit. The charging or discharging mode is regulated by comparing the actual coil current with the set reference value of the current. This control algorithm can be implemented on digital signal processor.

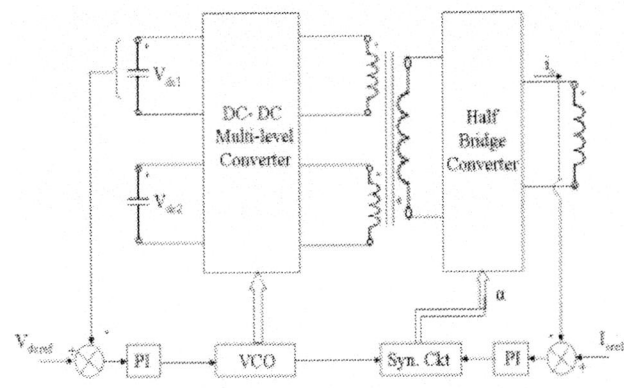

Fig. 15. Control Scheme for DC-DC Converter.

## V. PERFORMANCE OF THE CONVERTER

This control with the circuit considering the IGBT switches is simulated in the PSIM. The design parameters are chosen to make it more realistic so that the prototype could be built. The simulation parameters are given in the Table-I.

TABLE I
SIMULATION PARAMETERS

| $C_{d1}$ , $C_{d2}$ | $20e^{-2}$ F | $V_{d1}$- Volts | 300 |
|---|---|---|---|
| L | 0.1 H | $V_{d2}$ -Volts | 500 |
| $C_1$-$C_8$ | $10 e^{-4}$ | $I_o$ -Amps | 0-1000 |
| Sw. Freq. | 50 Hz | Thy. pulse width | $20^\circ$ |
| Tr-leakage inductance (Primary) | 0.0001 | Tr-leakage inductance (Secondary) | 0.00001 |

The switching frequency of the DC-DC converter is set to visualize the operating of the converter switches on both the sides of the transformer. The voltages across the switch (Q1) are shown with the capacitor voltage (C1) in the figure 16. The switching of these devices takes place exactly at the zero voltages. This confirms the ZVS operation. Similarly the secondary side converter is operating at the zero current (ZCS). This is observed from the current and operating pulses as shown in the figure 17.The results of simulation indicating the charging operation of the coil is presented in figure 18.This indicates the pulse positioning of the thyristors on secondary side of the transformer governs the control of coil current. The capacitor voltages across each of the stages ($V_{d1}$ and $V_{d2}$) and current $I_o$ is shown in figure 19. The effective charging of the current with the balancing of the voltage are taking the place. The voltages on these capacitors are set to a value of 500 volts and 300 volts and charging starts from the zero current in the coil. The final steady state is achieved where both the capacitors are having same voltages. This voltage is 40 volt. The current increases in the coil up to the value of 550 amps.

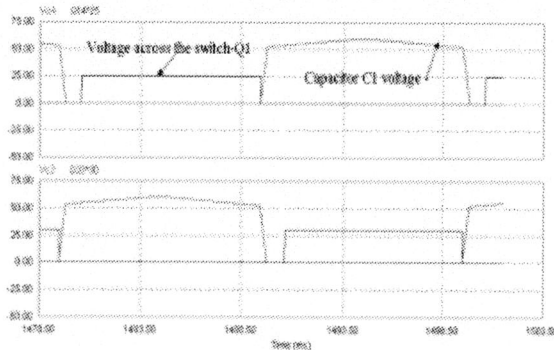

Fig. 16.  Voltages across the switch Q1 and Capacitor C1  A -ZVS Operation.

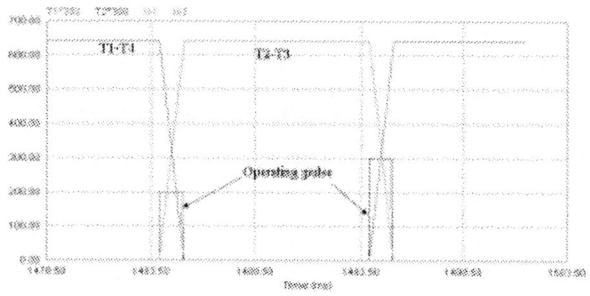

Fig. 17.  Current through the Thyristor – T1 and T4 and T2-T3 ZCS Operation.

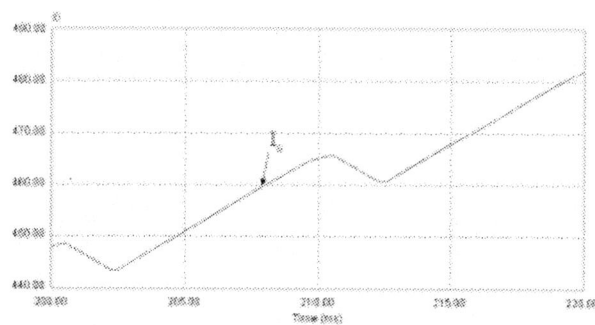

Fig. 18.  Coil Charging – Charging period is more than discharging.

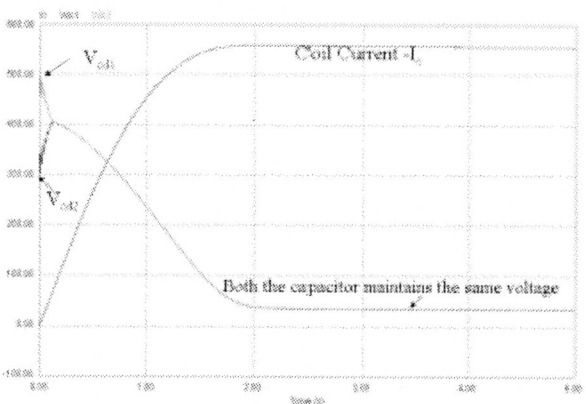

Fig. 19.  Capacitor Voltages are balancing with the charging coil current.

## VI. CONCLUSION

This paper presents the analysis of the voltage balancing at the dc bus of multi- level converters. The high power converter design with selection of circuit parameters is also presented. The operation of the circuit indicates the multi-level dc-dc converters gets balance with the appropriate switching of the H-bridge. The control scheme proposed can regulate the coil current. The new adaptive current controlled auto balancing is implemented to get stable operation of multilevel DC-DC converter. This converter works in soft switching mode hence over all efficiency and operating stability is ensured with more economical design. This has applications in interfacing of superconducting magnetic energy storages to be interfaced with the power system.

## VII. REFERENCES

[1]  J. S. Lai and F. Z. Peng, "Multilevel converters-A new breed of power converters," IEEE Trans. Ind. Applications, Vol.32, pp.509-517, May/June 1996.

[2]  F. Zhang, F. Z. Peng, and Z. Qian, "Study of the multilevel converters in DC-DC Applications," IEEE Conf. PESC 04, pp. 1702-1706, 2004.

[3]  J. □ung, and B.T.Ooi, "Series connected voltage source converter modules for forced commuted SVC and DC transmission", IEEE Trans. Power Deliv., Vol.9, No.2, pp.977-983, April 1994.

[4]  F.Z.Peng, "A generalize multilevel inverter topology with self voltage balancing". IEE Trans. on Industry Appl.,Vol.37,No.2,pp.611-618, March/April 2001.

[5]  A. Nabae, I. Takahashi, and H. Akagi, " A new neutral-point-clamped PWM inverter," IEEE Trans. Ind. App., vol. IA-17,no.5,pp.518-523,sept/oct 1981.

[6]  A. M. Trzynadlowski, Introduction to Modern Power Electronics, John Wiely, 1998.

[7]  N. Mohan, Power Electronics, converter, applications and Design, John Wiely, 1995.

**2006 IEEE International Conference on Power Electronic, Drives and Energy Systems**

# Inrush Current Control of a DC/DC Converter Using MOSFET

Gaddam Mallesham, *Member IEEE,* and Keerthi Anand

*Abstract* -- A DC / DC power converter is designed which boosts the 5V DC obtained from the Universal Serial Bus (USB) port of the computer to the meet the load of the programming circuitry of Electronically Controlled Motor (ECM). When a DC/DC converter is turned on with loads consisting of large input capacitors it causes a large inrush current. To eliminate this inrush current the MOSFET, which is used to switch the load in this case can be utilized to limit this inrush current by designing its gate drive circuit so as to take it into saturation region only when the drain current is below designed limit. The paper will present in detail the design of power converter with inrush current protection.

*Index Terms*--Boost Converter, Electronically Commutated Motor (ECM), Inrush current, Soft Universal Serial Bus (USB).

## I. INTRODUCTION

ELECTRONICALLY Commutated Motor (ECM) is a Brushless DC motor with all of its speed and torque controls built in. Their wide speed range, high efficiency and programmability give them a virtually unlimited range of performance characteristics, which results in a highly reliable, field-proven, convenient usage. Programming options for the ECM include rotation direction, start/stop ramp rates, on/off blower delays and many other functions--all stored in the motor's memory.

Programming of the onboard micro-controller in Electronically Controlled Motor (ECM) control circuit is performed using Programming Module with computer interface through the serial port. The DC power required by the Programming Module and the ECM control circuit in present setup is provided from the 110VAC or 220VDC adaptor.

This arrangement has many disadvantages, such as bulky setup, requires different power supply kits for US and European markets (50Hz and 60Hz), and requires 110VAC or 220VAC power points in the field where motors are installed. To overcome all these problems a boost converter is designed which takes the 5VDC input from the USB [2] port and

This work was supported in part by the Department of Electrical Engineering, Osmania University.

Gaddam Mallesham is with Department of Electrical Engineering, University College of Engineering (A) Osmania University, Hyderabad, Andhra Pradesh, India (e-mail: malleshiitd@yahoo.com).

Keerthi Anand is with HBLNIFE Power Systems Ltd., Hyderabad, Andhra Pradesh, India (e-mail: anand0201@yahoo.com).

supplies to the Programming Module.

A DC/DC boost converter [1] was designed to supply the control circuit of BLDC motor. USB port can supply 5V, this 5V is boosted to 30V using the DC/DC boost converter. This boosted voltage is supplied to the control circuit of the BLDC motor for performing its programming using the computer. Earlier the power required for the programming circuit of the BLDC motor was obtained from the step-down transformer and rectifying network and programming is done using the serial port of the computer but this arrangement is expensive and it requires an additional power source in the field where the motors are installed. Instead of this the newly designed boost converter draws power from the USB port, thus the USB port enables both the programming and the power.

The schematic block diagram of the USB power converter with all the required features is shown in Fig. 1.

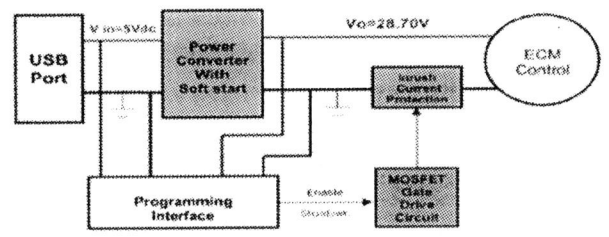

Fig. 1. ECM Programming Power Converter Block Diagram.

The control circuit has 1000uF capacitor in its input side. When this load is switched using a MOSFET [4] it is taking a high inrush current of 1.2A, but the USB port can deliver a maximum of 500mA. This calls for the design of an inrush current protection circuit to limit this current below 500mA.

The main functional blocks in the proposed setup can be broadly classified as below.
- Boost Converter [3]
- Soft Start Circuit.
- Inrush Current Protection circuit
- MOSFET Gate Drive Circuit. [4]

## II. POWER CONVERTER DESIGN

Different stages involved in the design of the power converter for the desired output are presented below.

- Finding the maximum current requirements of the Programming module.
- Finding the minimum operating DC voltage of the ECM.

0-7803-9771-1/06/$25.00 ©2006 IEEE

- Finding the maximum operating DC current of the ECM at the minimum operating voltage.
- Inrush Current Protection circuit.
- Selecting a suitable boost converter IC.
- Determining the required voltage to boost.
- Designing the boost circuit for the desired voltage.

## A. Finding the Maximum Current requirement for the Programming Module

Programming Module (PM) is an interface between the computer and the ECM motor using which programming of the ECM motor for the desired operation can be accomplished. At present, it requires 5VDC for its microcontroller circuitry and 24VAC from the adaptor is given to it. This AC voltage is rectified using a bridge rectifier present onboard and the rectified output is supplied to a 20V regulator, which is supplied to the Op-amp circuitry in the programming module. For determining the maximum current requirement of the Programming Module the setup shown in the Fig. 2, is used. 21VDC is directly supplied to the 20V regulator and 5VDC is given to the microcontroller circuitry. The current consumed by each of the input is recorded using a Current Probe Amplifier and digital Oscilloscope.

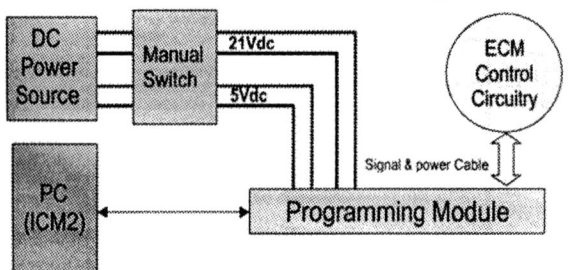

Fig. 2. Setup to determine the minimum operating voltage and maximum current requirement for programming of ECM.

At this operating point the load current on external power supply is measured for obtaining the Maximum current drawn from supply. The maximum current drawn by the regulator is 25mA. The micro controller circuit draws 30mA current from the 5Vdc input source.

The current waveforms of both the 20V and 5V input can be seen as below.

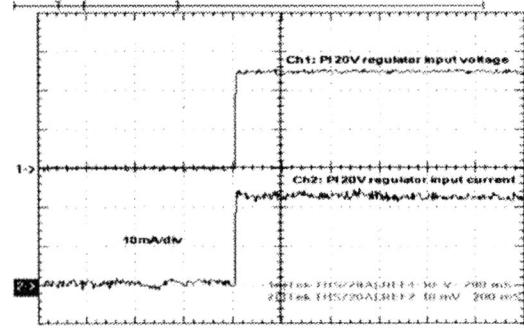

Fig. 3. Programming interface 20V regulator input.

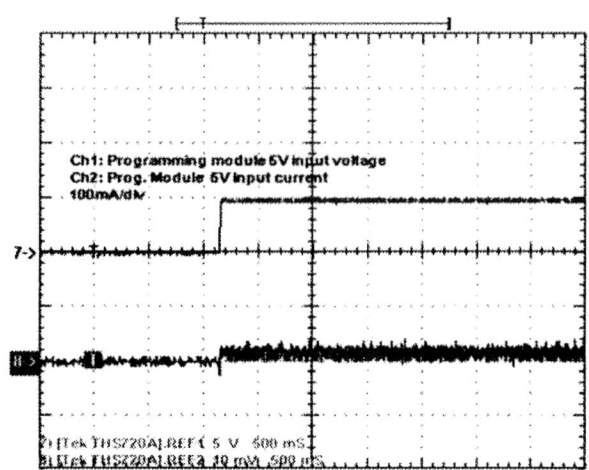

Fig. 4. Programming interface 5V input.

## B. Finding the Minimum operating voltage & Maximum Current for the ECM Control

The minimum operating voltage of the ECM for obtaining programming should be determined, it is the minimum voltage required to power up the electronics of the ECM Control, so that the programming is possible. This is the minimum voltage, which the boost converter should be capable of boosting.

There are four different power rated ECM motors are available with 1/3, 1/2, 34 and 1 HP. In order to design the power converter to work with all rated motors, the minimum voltage required for programming these motors have to be determined.

The setup shown below is used for measuring the maximum voltage requirement of the ECM control circuit for all HP. A DC voltage source is connected to provide the required voltage for the ECM circuit for obtaining programming. The DC voltage slowly increased from 0VDC until the programming operation is obtained. These readings are taking for 12 different ECM control of different HP rating.

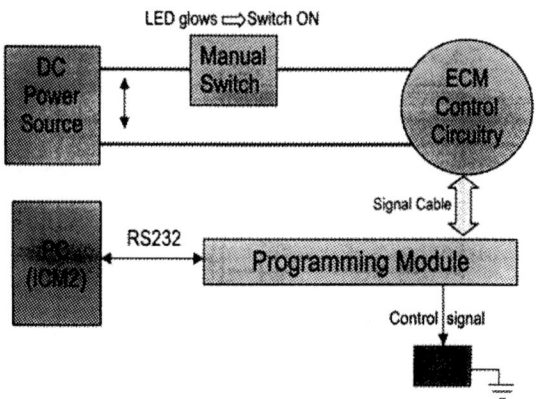

Fig. 5. Setup to determine the maximum voltage requirement of the ECM control.

From the above observations, the maximum voltage requirement for the ECM control is found to be 27.20VDC. This should be the minimum output voltage of the power converter to supply the required voltage for programming for all controls irrespective of their ratings.

402

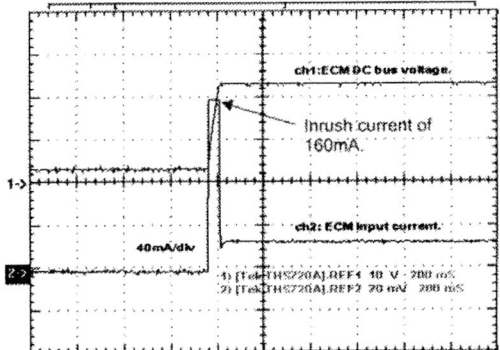

Fig. 6. ECM control input voltage & current waveforms.

During the voltage measurements using the above setup the input current for the ECM control circuit is captured. As it can be seen from Fig. 6, due to the presence of l000uF Capacitor on the DC bus of ECM control it is drawing sudden high current. An inrush current control circuit has to be designed at the output side of the power converter to limit this current below safe operating limit (500mA).

From the above current measurements the total load current on the USB power converter is calculated below.
Load to be met by the power converter:
• Programming module: 21Vdc and 25mA
• ECM Control Circuit: 27.20Vdc and 30mA

The total load current is 55mA on the power converter. If we consider a 75% efficiency converter then the input current drawn from the USB port can be calculated as below.

$$I_{IN} = \frac{V_0 * I_0}{V_{IN} * \eta} = \frac{27.20 * 55}{5 * 0.75} = 398.933mA \qquad (1)$$

The programming module also requires 5Vdc and 25mA, which can be directly supplied from the USB port. The total load on the USB port will be addition of this 25mA.

Total Curent drawn from USB port = 398.933+25 = 423.933mA

According to its specifications USB port can supply a maximum current of 500mA. As the requirement is 424mA we can design a boost converter to boost the 5Vdc available from the USB port to 27.20Vdc without overloading the USB port.

## C. Selecting Boost Converter IC

The USB voltage variation is: 4.5Vdc to 5.5Vdc
The tolerable voltage at ECM DC Bus: >=27.20Vdc
The boost converter IC should have the capability of boosting from an input voltage range of 4.5Vdc-5.5Vdc to an output voltage of up to 30Vdc.

The following are the desirable features of booster IC
• Good transient response.
• Better load regulation.
• Adjustable output voltage from 5.0Vdc to 40.0Vdc
• Few external components (Around 10 components).
• Current mode operation for improved current limiting

The step-up boost regulator UC2577 [3] IC of Texas Instruments has most of the desirable features. This boost IC requires very few external components, it can boost up to 60Vdc from an input voltage of 3Vdc.

## III. PROBLEM FORMULATION

DC/DC converters [1] have an input filter to reduce magnitude of the ripple current reflected back to the input voltage source. When a DC/DC converter is connected to the input voltage source, it causes a large inrush current as a result of the application of the high dv/dt to the filter capacitance. These filter capacitors act like a short circuit, producing an immediate inrush surge current with a fast rise time. A similar phenomenon occurs during load switching i.e., applying loads with large input capacitors to the converter already connected to the voltage source (ECM Control has 1000uF capacitors on the input bus).

The peak inrush current will be significantly greater than the steady state current. If the inrush current is not limited, it may cause dip in the input voltage, and generate high di/dt and dv/dt. Therefore, the peak current and current ramp must be controlled. Fig. 7. shows the inrush current of the USB power converter when connected to input voltage source.

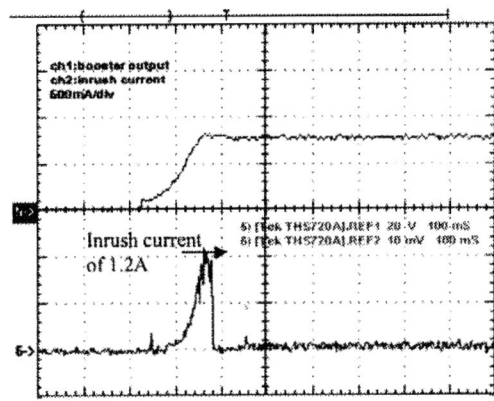

Fig. 7. Inrush current without soft start on No-Load.

Considering that maximum current limit that the USB port can deliver is 500mA, this inrush current has to below 400mA for the proper functioning. This high inrush current has to be totally eliminated or else it should be limited to a value less then 400mA for proper operation. The magnitude of the in rush current depends on many factors, such as the input voltage, the source impedance, capacitance and Effective Series Resistance (ESR) of the input filter of the load. The most accurate way to determine the inrush current is to measure it in the application

USB port can enable a maximum of 500mA to be drawn from its port, because of the input filter capacitors of the converter an inrush current of 1.2A is occurring when the converter is connected to the input DC supply source as seen in Fig. 7.

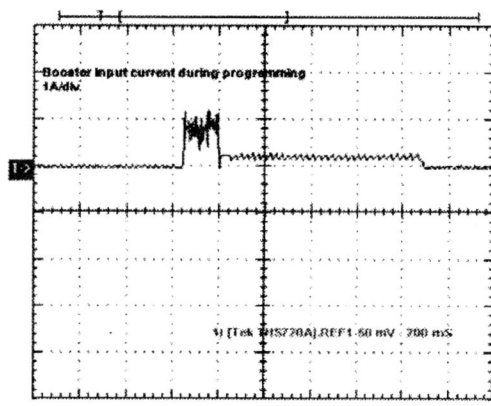

Fig. 8. Inrush current of converter without Soft start.

Due to this surge current the input supply voltage to the boost converter dips as seen in Fig. 9. This results in reduced output voltage for the control circuit during programming, which causes programming failure.

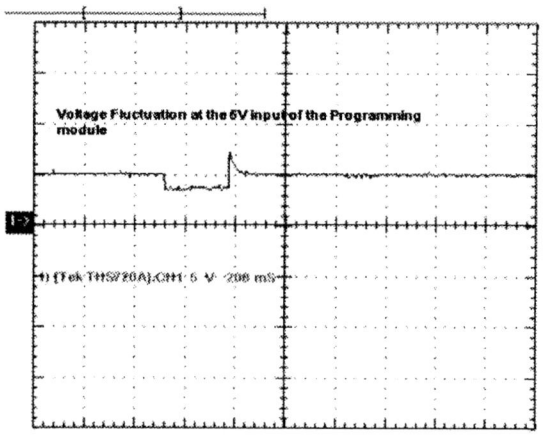

Fig. 9. Dip in the input voltage due to inrush current.

## IV. PROBLEM SOLUTION

### A. Inrush Current control

There are different methods available to control this inrush current. Inrush current control can be achieved by incorporating a resistor or thermistor in series with the load.

#### 1) Inrush current control by resistor

The schematic in Fig.10 shows the resistor R1 that limits the inrush current during plugging boards in the live system. The short pins short the resistor out when the board is fully inserted. Initially when the resistor is in series the input capacitor of the power converter will charge slowly. When it is sufficiently charged the series resistor can be shorted out.

The problem with this method is that the insertion time is not controlled. A very fast insertion may not have the input capacitor of power converter fully charged before the resistor is shorted out, this will result in inrush currents which will compel the USB port to shut down its power as a safety feature.

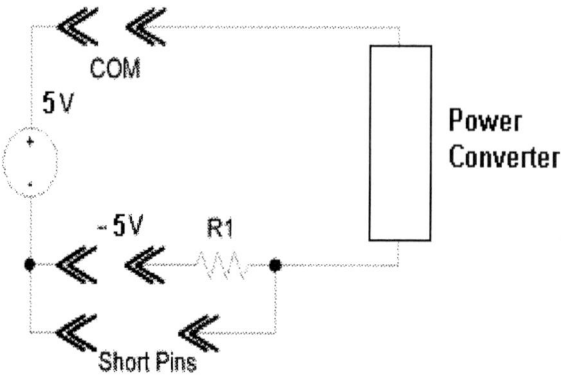

Fig. 10. Inrush current limiting by series resistor.

#### 2) Inrush current control by thermistor

The resistor can be replaced with a thermistor, which does not need to be shorted out when the board is fully inserted. A thermistor is a thermally sensitive resistor with a resistance that changes significantly and predictably as a result of temperature changes. The resistance of the current limiting thermistor decreases as its temperature increases due to the current flowing through the device. As the thermistor self-heats, its resistance begins to drop and relatively small current charges filter capacitors. After the capacitors become charged, the self-heated thermistor introduces very low resistance in the circuit.

Because current limiting thermistors are hot after they suppress inrush currents, these devices require a cool-down time after power is removed. This cool-down or "recovery" time allows the resistance of the thermistor to increase sufficiently to provide the required inrush current suppression the next time it is needed. A cool-down time varies according to the particular device, its mounting method and the ambient temperature. The typical cool-down time is roughly one to two minutes. It is unacceptable in systems requiring high availability.

#### 3) Inrush current control by Active Current Control

The limitations of passive methods can be overcome by utilizing semiconductor devices for inrush current control. The active inrush current control can be accomplished with a single semiconductor switch and a few external passive components.

MOSFET which is used for switching the load can be utilized for inrush current control. The MOSFET Gate drive circuit has to be designed such that it allows the MOSFET to go into saturation region [4] only when the input current of the converter is below 500mA, this demands a current sensing circuit to sense the current passing through the MOSFET. If a current of more than 500mA is flowing then it should not allow the MOSFET to go into saturation region, hence the MOSFET stays in the active region of its operation, which will present high impedance to the inrush current. When the drain current is below a specified limit the MOSFET goes into saturation region presenting only ON- resistance of the MOSFET to the load current.

Steady state load current of the BLDC control circuit is 35mA. The minimum Gate to Source voltage required for a Drain current ID = 50mA has to be calculated and this voltage should be applied only when the drain current is less than 50mA if it is more than this value then the Gate to Source voltage should fall below this minimum level so as prevent it from going into saturation region.

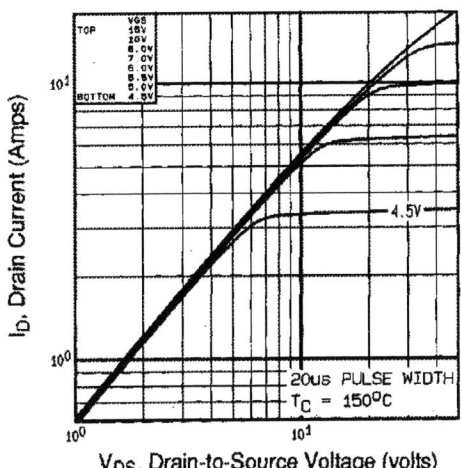

Fig. 11. Typical output characteristics of MOSFET.

From the output characteristics and forward trans-conductance (gTS) of the MOSFET, minimum gate to source voltage (VGS) required to enter saturation region has to be calculated using (1). This voltage has to be supplied to the MOSFET only when the drain current is below 50mA.

$$g_{TS} = \Delta I_d / \Delta V_{GS} \qquad (2)$$

Where $g_{TS}$ is Forward Trans-conductance of the MOSFET.

Forward Trans-conductance,

$g_{TS} = \Delta I_d / \Delta V_{GS}$
    $= (6 - 3.5) / (5 - 4.5)$
    $= 5$
At $I_D$ = 50mA,
$\Delta V_{GS} = \Delta I_d / g_{TS}$
$4.5 - V_{GS} = (3.5 - 0.05) / 5 = 0.69V$

$V_{GS} = 4.5 - 0.69 = 3.81V$

*B. Drain resistance calculation*

Gate Voltage, $V_G$ = 5.0V
Minimum Gate-to-Source Voltage = 3.81V
Voltage to be dropped in the Resistor
    = 5.0 – 3.81 = 1.19V
Maximum Drain Current allowed = 50mA
Current Limiting Resistance,
$R_{SL}$ = 1.19/ 0.05 = 23.8Ω.

A resistance is incorporated in series with the MOSFET

drain; this resistance acts as the current limiting resistance. When the current passing through this resistance is more than 50mA the voltage drop across the drain resistance will be more resulting in a gate to source voltage for the MOSFET less than required to enter saturation region. This causes the MOSFET to offer high impedance to the initial surge current. When the current falls below 50mA the voltage drop across the drain resistance will be minimal, this results in a gate to source voltage for the FET to enter into saturation region. Now the MOSFET offers only its ON- state resistance to the load current.

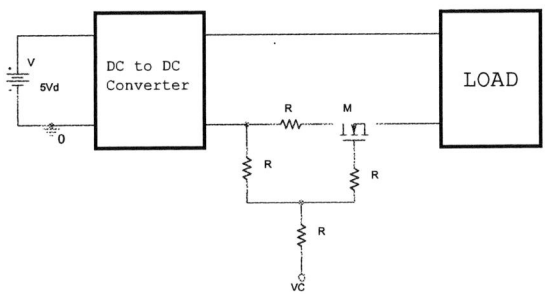

Fig. 12. Inrush current limiting circuit-using MOSFET.

## V. RESULTS

In this work USB Power Converter Design and the total system with the existing and proposed setup are presented. Different aspects in the designing of the USB Power Converter are presented in detail. The design of the circuit is tested for its required operation i.e., supplying power to the programming interface to enable programming through the USB port and the results are found to be fitting the desired requirement. In this work, the idea of providing power to the Programming Interface through the USB port instead of an external power source is implemented by designing a USB Power Converter.

With the design of the inrush current protection circuit using the MOSFET, the inrush current is considerably controlled as seen in Fig. 13

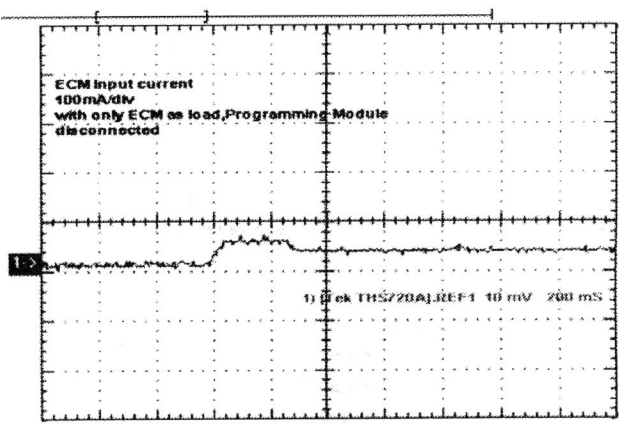

Fig. 13. Inrush current of converter with Soft start.

From Fig. 13, the flat part in the load current indicates the 50mA limit.

The input current waveform shown below clearly depicts the current limiting feature, the flat part of the wave in the initial period when the ECM circuit is switched is the maximum current allowed. The current is limited when it tries to go above this value. The MOSFET will not turn ON totally when the current is above 50mA presenting high impedance to the current.

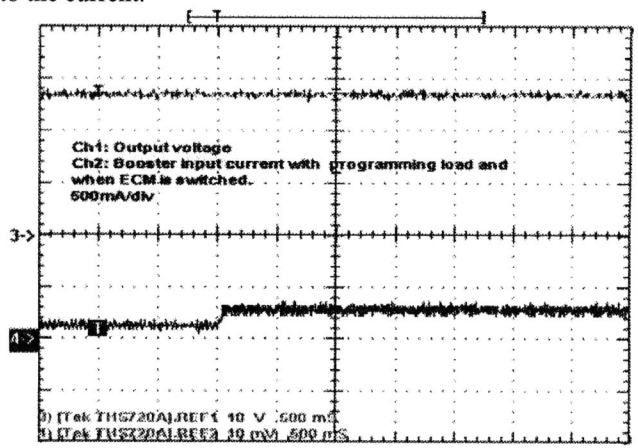

Fig. 14. Smooth switching of the load with inrush current protection circuit.

## VI. REFERENCES

[1] Ned Mohan. Tore M. Undeland, William P. Robbins *"Power Electronics Converters, Applications and Design,"* Wiley, John.

[2] Compaq, Intel, Microsoft, NEC *"Universal Serial Bus Specifications"* Revision1.1, September 1998.

[3] Unitrode. "Step- up Voltage regulator UC-2577 ADJ," Product datasheet.

[4] Abhijit D. Pathak *"MOSFET/IGBT driver's theory and applications,"* IXYS.

## VII. BIOGRAPHIES

**Gaddam Mallesham** was born in Bollepally (Vil), Bhongir (Mon), Nalgonda (Dist.) on 20th August, 1977. He received the B.E degree in Electrical and Electronics Engineering from University College of Engineering (A), Osmania University, Hyderabad, India in 2000. He received his Masters degree in Control Engineering and Instrumentation from Indian Institute of Technology, Delhi, India in 2002. Since 2002, he has been an Assistant Professor in the Department of Electrical Engineering, University College of Engineering (A), Osmania University, Hyderabad, India. His main interest includes Artificial Intelligence Techniques applied to Electrical Engineering. He is a Secretary of IEEE joint chapter of PES/IAS Societies, Hyderabad Section. He has visited Greece and Canada to present technical papers.

**Keerthi Anand** was born in Hyderabad in India, in 1981. He received the B.Tech degree in Electrical and Electronics Engineering from the Jawaharlal Nehru Technological University, Hyderabad, India in 2002. He received his Masters degree in Industrial Drives and Controls specialization from Osmania University, Hyderabad, India. His employment experience included working as a Design Engineer at HBLNIFE Power Systems Ltd., Hyderabad. Special fields of interest include Motor Drives Control.

**2006 IEEE International Conference on Power Electronic, Drives and Energy Systems**

# A ZVT Boost Converter using an Auxiliary Resonant Circuit

## M. Phattanasak

*Abstract*--This paper presents a ZVT Boost converter using an auxiliary resonant circuit. The main switch of the converter can be completely tuned on with zero voltage switching (ZVS) while the auxiliary switch is turned on and off with zero current switching (ZCS). The proposed circuit is analyzed, simulated, designed, and implemented. Both simulated and experimental results show that the soft switch functions are achieved and the maximum efficiency of the converter is approximately 91.5%.

*Index Terms*--Boost converter, zero voltage transition.

## I. INTRODUCTION

NOWADAYS, power electronics trends are aimed at reducing the weight, size, electromagnetic interference (EMI), and also alleviated the losses happening from power switches' actions [1]-[3]. Zero voltage transition (ZVT) is one of the techniques commonly used to get rid off loss from the power switch. The ZVT uses an auxiliary circuit comprising a switch, which is used to force the voltage across the main power switch to be zero before it is turned on. This process is called zero voltage switching (ZVS). In [1]-[3] they used a resonant circuit allowing soft-switching operation in the resonant switch. In [2] they have been proposed new topologies based on boost converter that allow all diode commutated under soft condition. However, one of those used two diodes in a resonant path to clamp the voltage across the resonant capacitor to be zero. Furthermore, they present the converter only operated at medium and high power output. In this paper, another topology in [2] operating at low power output range is proposed, but in another control scheme.

This paper proposes an overview of the proposed circuit and its operation classifying into nine modes in Section II. Besides the analyzed state equations are also given and the converter is simulated with computer software, Pspice. Later, the conventional with an optimum snubber and the proposed

---

This work was supported by the Science and Technology Research Center (STRC), King Mongkut's Institute of Technology North Bangkok (KMITNB).

M. Phattanasak is with Teacher training in electrical engineering department, Faculty of Technical Education, King Mongkut's Institute of Technology North Bangkok, Bangkok 18000 Thailand (e-mail: matheepotp@kmitnb.ac.th).

circuit rated at 400W, input voltage of 120V, output voltage of 400V operating at 100kHz was built and the efficiencies are measured.

## II. ZVT BOOST CONVERTER

The proposed circuit comprises a boost converter using a power mosfet as the main power switch $S_m$ and two sub-auxiliary circuits; Circuit A and Circuit B. Both circuits share an auxiliary switch, $S_a$ together. The auxiliary switch $S_a$ is implemented by an IGBT and a diode, as shown in Fig. 1.

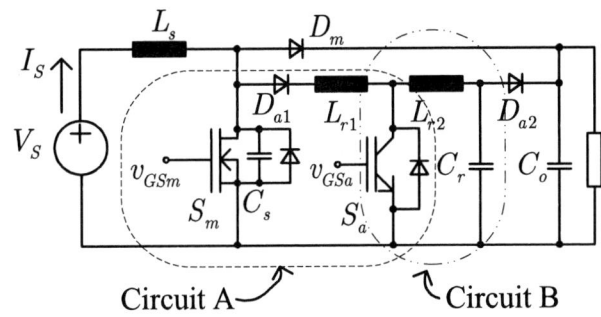

Fig. 1. The proposed converter.

Circuit A consists of a power mosfet including its output capacitor and diodes $C_s$, $D_{a1}$, $L_{r1}$, and $S_a$. Circuit B is made of $L_{a2}$, $C_r$, and $S_a$. The key waveforms of the proposed converter are depicted in Fig. 2.

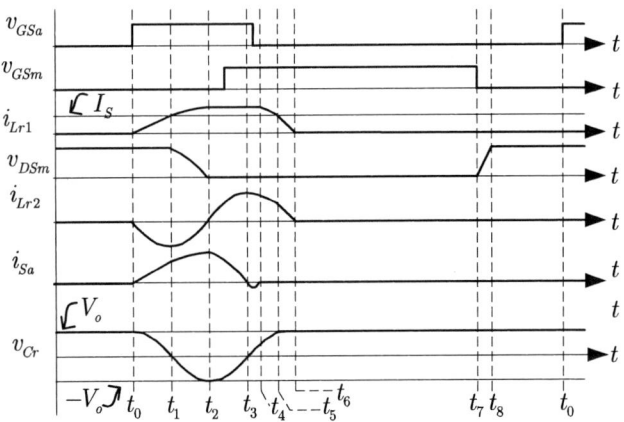

Fig. 2. The key waveforms of the proposed circuit.

---

0-7803-9771-1/06/$25.00 ©2006 IEEE

## A. Circuit Operation

In order to analyze easily, some hypotheses are made:

- All devices are ideal
- $L_1$ is so large enough, thus, can be considered an input voltage source as a current source, $I_s$, which is constant.
- $C_o$ is so large that output voltage $V_o$ is constant.
- $V_S$ is constant
- The circuit is working in steady state.

The operation of the converter can be divided into nine modes in each switching cycle relating with the relevant waveforms in Fig.2. Each circuit mode is illustrated in Fig. 3. Moreover, the operation of each mode is clearly described with equations of relevant resonant elements.

### 1) Mode 1 $[t_0 < t < t_1]$:

In this mode, the auxiliary switch $S_a$ is turned on and Circuit B is started to work in a resonant mode, while the inductor current $i_{Lr1}$ is increased linearly. The steady state equations for this mode are given by

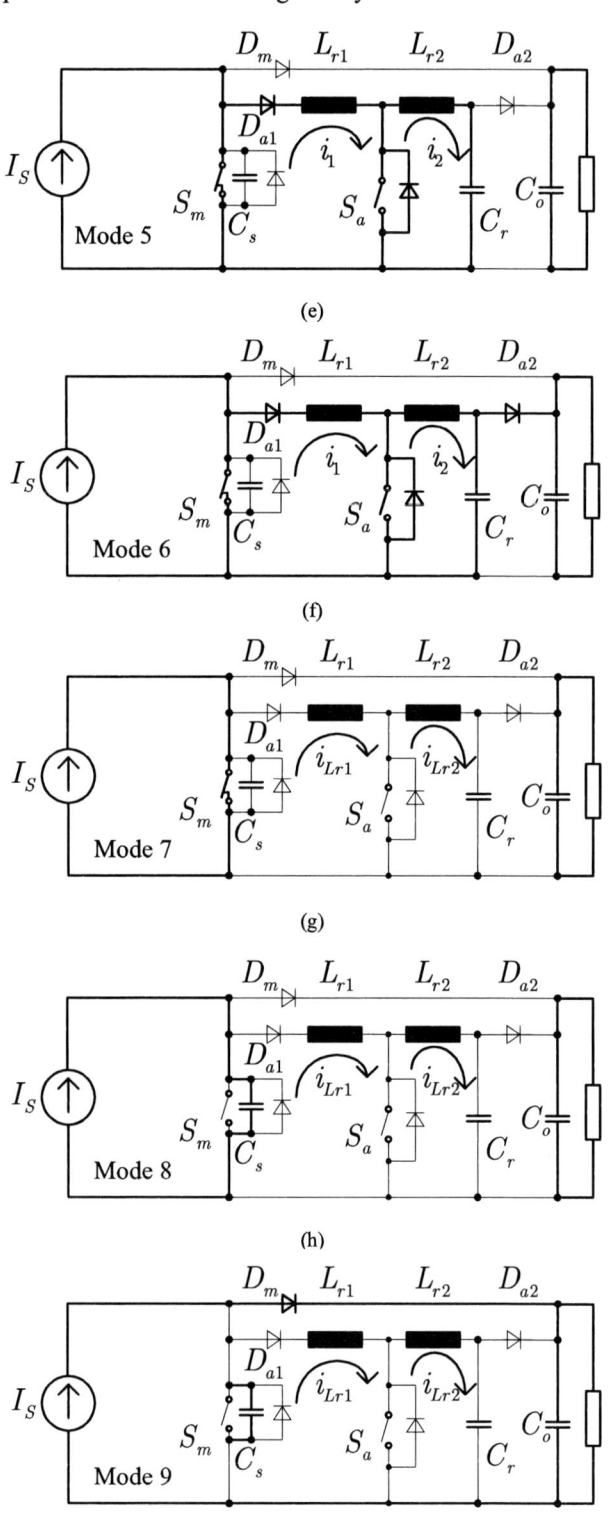

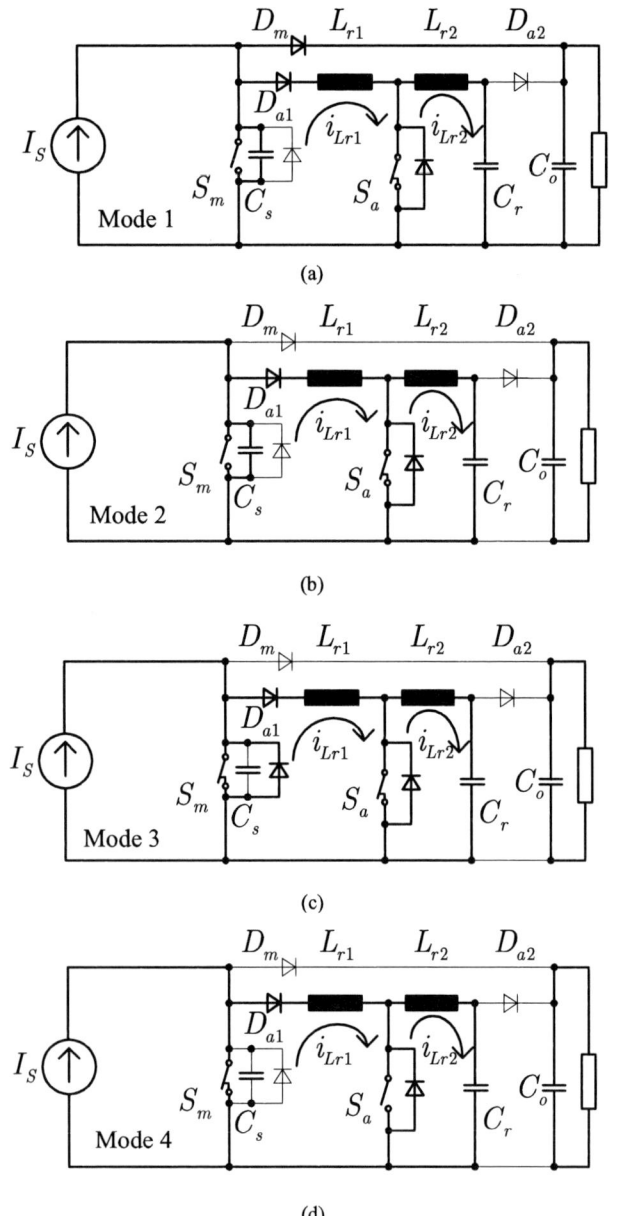

Fig. 3.   The topological state of the proposed circuit.

$$i_{Lr1}(t) = \frac{V_o}{L_{r1}}(t - t_0) \tag{1}$$

$$v_{cs}(t) = V_o \tag{2}$$

$$i_{Lr2}(t) = -\frac{V_{Cr(0)}}{Z_2}\sin\left(\omega_2\left(t - t_0\right)\right) \tag{3}$$

$$v_{cr}(t) = V_{Cr(0)}\cos\left(\omega_2\left(t - t_0\right)\right) \tag{4}$$

where $\omega_2 = \dfrac{1}{\sqrt{L_{r2}C_r}}, V_{Cr(0)} = V_o$ and $Z_2 = \sqrt{\dfrac{L_{r2}}{C_r}}$.

Mode 1 ends when the inductor current $i_{Lr1}$ reaches the input current $I_s$ at $t_1$ and the diode $D_m$ turns off.

2) Mode 2 $[t_1 < t < t_2]$:

In this mode, Circuit A starts operation as a resonant circuit, while Circuit B is continuously resonant from mode 1. The resonant equations are

$$i_{Lr1}(t) = I_s + \frac{V_{Cs(0)}}{Z_1}\sin\left(\omega_1\left(t - t_1\right)\right) \tag{5}$$

$$v_{cs}(t) = V_{Cs(0)}\cos\left(\omega_1\left(t - t_1\right)\right) \tag{6}$$

$$i_{Lr2}(t) = -V_{Cr(0)}\sin\left(\omega_2\left(t - t_0\right)\right) \tag{7}$$

$$v_{cr}(t) = V_{Cr(0)}\cos\left(\omega_2\left(t - t_0\right)\right) \tag{8}$$

where $\omega_1 = \dfrac{1}{\sqrt{L_{r1}C_s}}, V_{Cs(0)} = V_o$ and $Z_1 = \sqrt{\dfrac{L_{r1}}{C_s}}$.

This mode stops when the voltage of the output capacitor of the main switch $v_{cs}$ equals zero.

3) Mode 3 $[t_2 < t < t_3]$:

In this mode, the inductor current $i_{Lr1}$ is constant. The voltage $V_{Cs}$ is zero and the anti-parallel diode of the main switch $S_m$ is turned on. Thus, the main switch $S_m$ can be turned on under the ZVS manner. The equations of the resonant elements are

$$i_{Lr1}(t) = I_s + \frac{V_{Cs(0)}}{Z_1} \tag{9}$$

$$v_{cs}(t) = 0 \tag{10}$$

$$i_{Lr2}(t) = -V_{Cr(0)}\sin\left(\omega_2\left(t - t_0\right)\right) \tag{11}$$

$$v_{cr}(t) = V_{Cr(0)}\cos\left(\omega_2\left(t - t_0\right)\right) \tag{12}$$

where $V_{Cs(0)} = V_{Cs(t2)}$ and $V_{Cr(0)} = V_o$.

This mode ends when the auxiliary switch current $i_{Sa}$ reaches zero.

4) Mode 4 $[t_3 < t < t_4]$:

In this mode, the operation starts when the anti-parallel diode of the auxiliary switch $D_{Sa}$ turns on and Circuit B is continuously resonant as shown in (13)-(16). Thereby, the auxiliary switch $S_a$ can be turned off under zero current switching (ZCS) condition.

$$i_{Lr1}(t) = I_s + \frac{V_{Cs(0)}}{Z_1} \tag{13}$$

$$v_{cs}(t) = 0 \tag{14}$$

$$i_{Lr2}(t) = -V_{Cr(0)}\sin\left(\omega_2\left(t - t_0\right)\right) \tag{15}$$

$$v_{cr}(t) = V_{Cr(0)}\cos\left(\omega_2\left(t - t_0\right)\right) \tag{16}$$

where $V_{Cs(0)} = V_{Cs(t3)}$ and $V_{Cr(0)} = V_o$.

5) Mode 5 $[t_4 < t < t_5]$:

This mode starts when the anti-parallel diode of the auxiliary switch $D_{Sa}$ turns off. The resonant inductor $L_{r1}, L_{r2}$ and resonant capacitor $C_r$ are formed as a series resonant circuit. The process can be explained as follows

$$\begin{aligned} i_{Lr1}(t) = I_x\cos(\omega_3\left(t - t_4\right)) \\ -\frac{v_{Cr(t4)}}{Z_3}\sin(\omega_3\left(t - t_4\right)) \end{aligned} \tag{17}$$

$$v_{cs}(t) = 0 \tag{18}$$

$$i_{Lr2}(t) = i_{Lr1}(t) \tag{19}$$

$$\begin{aligned} v_{cr}(t) = Z_3 I_x\sin\left(\omega_3\left(t - t_4\right)\right) \\ + v_{cr}\left(t_4\right)\cos\left(\omega_3\left(t - t_4\right)\right) \end{aligned} \tag{20}$$

where $i_x = \left\{\dfrac{L_{r1}}{L_{r1} + L_{r2}}i_{Lr1(t4)} + \dfrac{L_{r2}}{L_{r1} + L_{r2}}i_{Lr2(t4)}\right\}$ and $Z_3 = \sqrt{\dfrac{L_{r1} + L_{r2}}{C_r}}$.

This mode ends when the inductor currents $i_{Lr1}$ and $i_{Lr2}$ equal to the input current $I_S$.

6) Mode 6 $[t_5 < t < t_6]$:

At $t_5$, the voltage across resonant capacitor $V_{Cr}$ is equal to the output voltage $V_o$. Hence, the energy in both inductor $L_{r1}$ and $L_{r2}$ is released into the output capacitor $C_o$ and resonant capacitor $C_r$. So, the inductor current is decreased linearly as shown in (21). The system is described as follows

$$i_{Lr1}(t) = \frac{V_o}{L_{r1} + L_{r2}}(t - t_5) \tag{21}$$

$$v_{cs}(t) = 0 \tag{22}$$

$$i_{Lr2}(t) = i_{Lr1}(t) \tag{23}$$

$$v_{cr}(t) = V_o \tag{24}$$

7) Mode 7 $[t_6 < t < t_7]$:

During this interval, the main switch $S_m$ is still turned on by the PWM command. This mode ends when the PWM command is zero. The state equations are shown in (25)-(28).

$$i_{Lr1}(t) = 0 \tag{25}$$

$$v_{cs}(t) = 0 \tag{26}$$

$$i_{Lr2}(t) = 0 \tag{27}$$

$$v_{cr}(t) = V_o \tag{28}$$

*8) Mode 8 $[t_7 < t < t_8]$:*

At instant $t_7$, the main switch's capacitor $C_s$ is charged linearly by the input current $I_s$, until the voltage of the main switch's capacitor is equal to the output voltage, as shown in (32)-(32).

$$i_{Lr1}(t) = 0 \qquad (29)$$

$$v_{cs}(t) = \frac{I_s}{C_s}(t - t_7) v_{cs}(t) = \frac{I_s}{C_s}(t - t_7) \qquad (30)$$

$$i_{Lr2}(t) = 0 \qquad (31)$$

$$v_{cr}(t) = V_o \qquad (32)$$

*9) Mode 9 $[t_8 < t < t_9]$:*

In this mode, all switches are turned off, but the diode $D_m$ turns on to deliver energy absorbed in the inductor $L_S$ to the output capacitor. It makes the voltage across the main switch equal to the output voltage as shown in (33)-(36).

$$i_{Lr1}(t) = 0 \qquad (33)$$

$$v_{cs}(t) = V_o \qquad (34)$$

$$i_{Lr2}(t) = 0 \qquad (35)$$

$$v_{cr}(t) = V_o \qquad (36)$$

After time instant $t_9$, the converter is in the same state as in the mode 1 and the whole process is repeated for the next cycle.

### B. Characteristic curves

To design conveniently, the characteristic curves of the auxiliary switch $S_a$ are provided. Due to the operations of the proposed converter, the current inductor $i_{Lr2}$ must has the amplitude more than that of the $i_{Lr1}$ in order to meet the ZCS condition, $i_{Sa} = 0$, at the auxiliary switch $S_a$ before it is turned off. The relationships between the auxiliary switch current $i_{Sa}$ and the impedance of Circuit B are illustrated in Fig. 4. It shown that the per unit current $i_{Sa} / I_S$ is zero when the impedance is a value that makes the auxiliary switch current is zero.

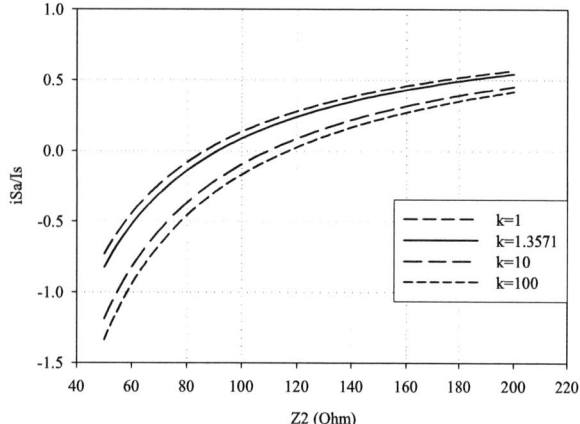

Fig. 4. The relationships between the $i_{Sa} / I_S$ and Circuit B's impedance.

The curves are plotted under the different ratio of the resonant inductor $L_{r1}$ and $L_{r2}$, namely $k$, as defined in (37). In addition, to plot these graphs the ratio between the power mosfet capacitor $C_s$ and the resonant capacitor $C_r$ was chosen equal to 0.0788.

$$k = \frac{L_{r1}}{L_{r2}} \qquad (37)$$

According to graphs in Fig. 4, we can use a value of $Z_2$ to calculate the value of $L_{r2}$ by choosing $C_r$ and $C_s$ and finally employ (37) to find a value of resonant inductor $L_{r1}$.

## III. SIMULATED AND EXPERIMENTAL RESULTS

The proposed circuit prototype is simulated on computer simulation program, Pspice, and is implemented with specifications as shown in Table I. The inductor $L_S$ is chosen equal to 15mH to make the input current $I_S$ constant.

TABLE I
PROPOSED CONVERTER SPECIFICATION

| | |
|---|---|
| Output voltage, $V_o$ | 400 V |
| Output power, $P_o$ | 400 W |
| Input voltage, $V_s$ | 120 V |
| Switching frequency, $f_s$ | 100kHz |

### A. Design Procedure

The boost converter is designed to operate with continuous inductor current mode (CICM). Thus, the equations of a standard boost converter can be applied [6]. The design procedure of the proposed converter is explained as follows.

*1) Calculation of the maximum input current:*

The maximum input current $I_S$ can be calculated as

$$I_{S,\max} = \frac{P_o}{V_S} = 3.33A.$$

*2) Calculation of Circuit A parameters:*

Using power mosfet data, IRFP460, we know $C_s$ is 130pF. In order to reduce losses in the main power diode $D_m$, the resonant inductor $L_{r1}$ should be selected as [3]:

$$L_{r1} \geq \frac{V_o - V_i}{di / dt}$$ where $di / dt$ is the rate of recovery of the diode $D_m$. We chose $L_{r1}$ equal to $19\mu H$. The resonant frequency of Circuit A is $f_1 = \dfrac{1}{2\pi\sqrt{L_{r1}C_s}} = 3.2\text{MHz}$ and the resonant impedance of Circuit A is $Z_1 = \sqrt{\dfrac{L_{r1}}{C_s}} = 382.3\Omega$.

*3) Calculation of Circuit B parameters:*

In order to make the auxiliary circuit operating under ZCS condition, as mentioned before, the maximum current of $i_{Lr2}$ must larger than that of $i_{Lr1}$. Using data from Circuit A, the maximum current of $i_{Lr1}$ is 3.33A. Thus, the maximum

410

current of $i_{Lr2}$ is $I_S + i_{Lr1,max} = 4.3A$ and the impedance of Circuit B is $Z_2 = V_o / i_{Lr1,max} = 93\,\Omega$. We have the resonant capacitor $C_r = 1.65nF$, hence, $L_{r2} = C_r Z^2_{rB} = 14.27\mu H$. The resonant frequency of Circuit B is $f_2 = \dfrac{1}{2\pi\sqrt{L_{r2}C_r}} = 1\text{MHz}$. However, we can use the graph in Fig. 4, to determine the value of inductance as mentioned above.

To verify the operations of the proposed circuit, we used Pspice software with the parameter setting as shown in Table II to simulate. Then, the prototype was built and tested.

TABLE II
PROPOSED CONVERTER DEVICES

| Parameters | Simulation | Experiment set |
|---|---|---|
| $S_m$ | IRFP460 | IRFP460 |
| $D_m$ | Ideal | MUR3060 |
| $D_{a1}$ | Ideal | HFA15TB60 |
| $D_{a2}$ | Ideal | RHRP8120 |
| $S_a$ | Ideal | G12N60 |
| $D_{Sa}$ | Ideal | RHRP8120 |
| $C_r$ | 1.65nF | 1.65nF |
| $L_{r1}$ | 19$\mu$H | 19$\mu$H |
| $L_{r2}$ | 14$\mu$H | 14$\mu$H |

Fig. 5-6, show the simulation results. As can be seen, the main switch $S_m$ turns on under ZVS condition and the auxiliary switch $S_a$ turns on and turns off under ZCS condition. Thereby, the switching losses of $S_a$ and $S_m$ are reduced.

The experimental results of the proposed circuit are depicted in Fig. 7-9. In Fig. 7, the main switch $S_m$ is operated under ZVS condition and in Fig. 8, the voltage and current of the auxiliary switch which are working under ZCS condition are demonstrated.

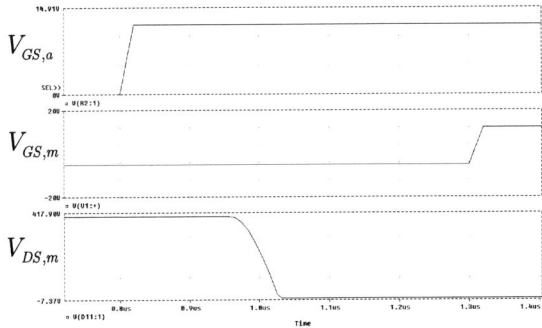

Fig. 5. Simulation waveforms of the auxiliary switch's gate signal, the main switch's gate signal, and the main switch voltage.

In Fig. 9, both resonant inductor currents and resonant capacitor $C_r$ voltage of the auxiliary circuit are shown. It confirms that the circuit can be operated successfully.

To compare the efficiency of the proposed circuit, the conventional PWM Boost converter with a snubber circuit [7] is built. Then, the efficiencies of the proposed circuit that has the efficiency of 91.5% and the conventional circuit are measured as shown in Fig. 10.

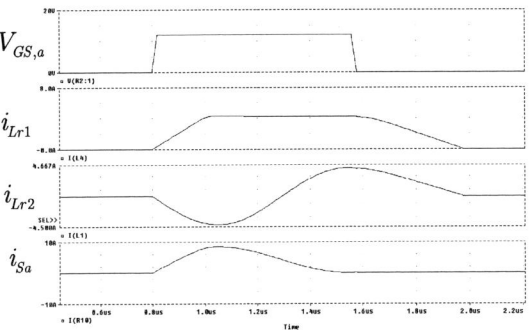

Fig. 6. Simultion waveforms of the auxiliary gate drive signal and currents in the auxiliary circuits.

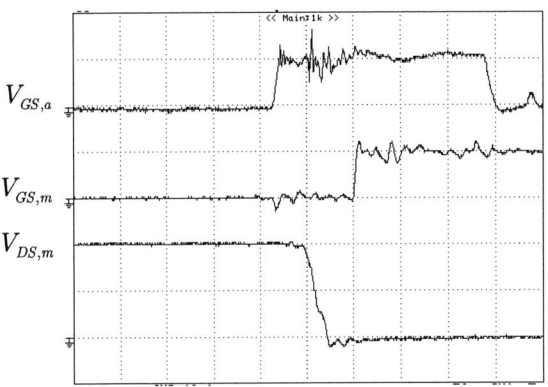

Fig. 7. Experimental waveforms of gate drive signals $V_{GS,a}$ and $V_{GS,m}$ (10V/div) and the voltage across the main switch $V_{DS,m}$ (200V/div), time 200ns/div.

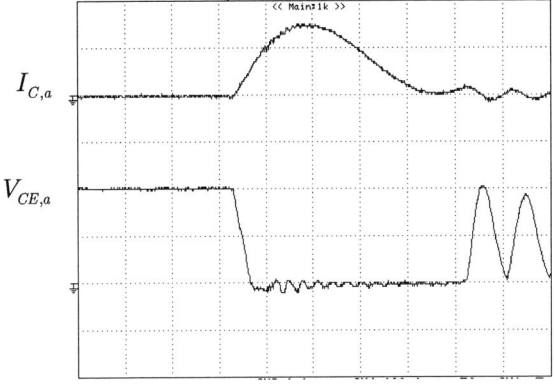

Fig. 8. Experimental waveforms of the auxiliary switch current $I_{C,a}$ (5A/div) and the auxiliary switch voltage $V_{CE,a}$ (200V/div) of the auxiliary switch, time 200ns/div.

## IV. CONCLUSION

In this paper, a ZVT Boost converter, featuring zero voltage switching at the main switch when it turns on and turns off is proposed. With the soft-switching operations, losses are reduced. Resonant circuits are used to complete such the operations presented into nine modes. Simulated and experimental results are confirmed clearly that the converter is

suitable for using at high frequency operation and it can be used as a high efficient power supply at low and medium output power range. The prototype of a ZVT boost converter rated at 400W, input voltage of 120V, output voltage of 400V, and operating at 100kHz was implemented to verify the system performance.

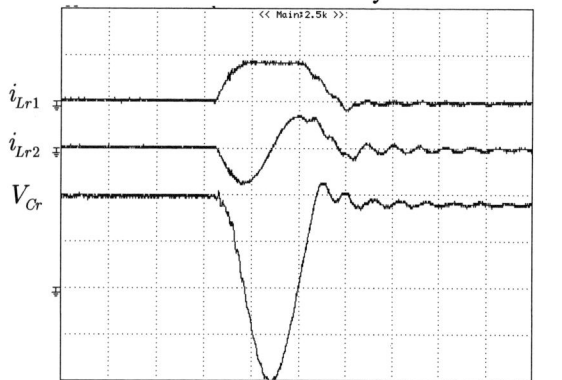

Fig. 9. Currents $i_{Lr1}$, $i_{Lr2}$ (5A/div) and voltage $V_{Cr}$ (200V/div) of the auxiliary circuit, time 500ns/div.

## V. REFERENCES

[1] H. Guichao, L. Ching-Shan, J. Yimin, F.C.Y. Lee, "Novel zero-voltage-transition PWM converters," *IEEE Trans. Power Electron.*, Vol. 9, Issue 2, pp. 213–219, March, 1994.

[2] M.L. Martins, H.L. Hey, "Self-commutated auxiliary circuit ZVT PWM converters," *IEEE Trans. Power Electron.*, Vol. 19, Issue 6, pp. 1435–1445, 2004.

[3] C. M. Stein, H.L. Hey, "A True ZCTZVT Commutation Cell for PWM Converters," *IEEE Trans. Power Electron.*, Vol. 15, Issue 1, pp. 185–193, January, 2000.

[4] M.L. Martins, H. Pinheiro, J.R. Pinheiro, H.A. Grundling, H. L. Hey, "Family of improved ZVT PWM converters using a self-commutated auxiliary network," *IEE Proc. Electr. Power Appl.*, Vol. 150, No 6, pp. 680–688, November, 2003.

[5] M.L. Martins, J. L. Russi, H. L. Hey, "Zero-voltage transition PWM converters: a classification methodology," *IEE Proc.-Electr. Power Appl.*, Vol. 152, No 2, pp. 323–334, March, 2005.

[6] S. S. Ang, Power Switching Converters, New York: Marcel Dekker, 1995, pp. 27-32.

[7] T. Krein, Power ElectronicsElements, New York: Oxford press, 1998, pp. 499-503.

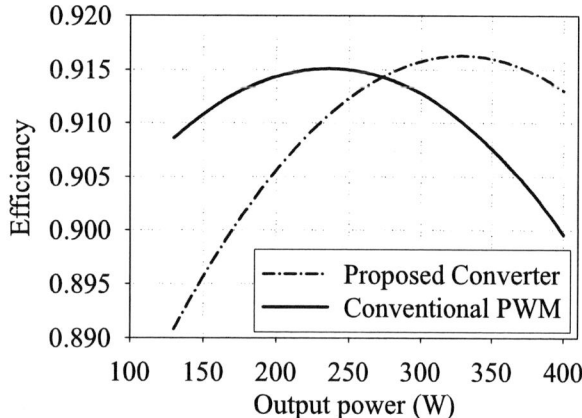

Fig. 10. The efficiencies of the proposed circuit and the conventional circuit.

**Matheepot Phattanasak** was born in Ranong, Thailand, in 1974. He received the B.Sc. in Tech. Ed. and the M.Eng. degree in electrical engineering from King Mongkut's Institute of Technology North Bangkok (KMITNB), Bangkok, Thailand, in 1997 and 2004, respectively. He has been with KMITNB as a lecturer since 2004. His interests include power converters, soft switching and its applications.

**2006 IEEE International Conference on Power Electronic, Drives and Energy Systems**

# Adaptive Hysteretic Control of 3$^{rd}$ Order Buck Converter

Veerachary M, *Senior Member, IEEE*, and Deepen Sharma

*Abstract—* **In this paper analysis and hysteretic controller design of a 3$^{rd}$ order step-down buck converter is presented. Various small-signal models valid up-to half of the switching frequencies, using state-space averaging method, are developed and then step-by-step controller design procedure is described. The results of controller design and closed-loop analysis are illustrated through computer simulations. To validate the controller design and analysis, simulation observations are provided for a 50 W 24/ 15 V point of load application.**

*Index Terms—***Hysteresis control, Step-down buck converter, Variable frequency control.**

## I. INTRODUCTION

IN recent years, the high frequency switching converters application in low power compact electronic circuits is increasing. As the power conversion system is becoming miniaturized, increasing the power density is one of challenging issue for the power supply (PS) designers. Lightweight, small size and high power densities with faster dynamic response time are some of the requirements for the power supplies. Different types of topologies have been developed, both in the non-isolated and isolated topologies, to meet the load requirements. For primary as well as secondary bus applications the isolated converters are most suitable, while non-isolated converters give best performance for the point of load (POL) applications. In this paper one such converter, 3$^{rd}$ order buck converter (TOBC) suitable to the POLA, is proposed.

Several control strategies, including voltage-mode and current-mode, have been reported in literature. Each of these control strategies has their own limitations. However, controller robustness against parameters is one of the important concern in the filed of PS technologies. Traditionally the hysteretic control [1]-[3] is the robust

---

This work was supported in part by the MHRD, Govt. of India under R & D Grant.

Veerachary. M is with Dept. of Electrical Engineering, Indian Institute of Technology Delhi, New Delhi, India (e-mail: mvchary@ee.iitdac.in).

Deepen Sharma is with the Department of Electrical Engineering, Indian Institute of Technology Delhi, New Delhi, India (e-mail: mvchary@ee.iitdac.in).

control. However, robustness can be increased, further, by using adaptive hysteretic band (AHB) instead of using fixed band hysteretic control. In this paper author's have made an attempt to use adaptive hysteresis band for the TOBC topologies.

## II. THIRD ORDER BUCK CONVERTER

Buck topologies are widely used in point of load applications. However, simple buck topologies pose the electromagnetic interference problems and to eliminate/reduce its severity input filters must be used. However, addition of filters results in higher order converter and it's interfacing capability mainly depends on the output impedance ($Z_0$) of the front-end converter as well as input impedance ($Z_{in}$) of POL. Sometimes addition of input filter may destabilize its front-end converter topology due to violation of the constraint $|Z_{in}| \geq |Z_0|$. To avoid some of these problems a simple buck topology with lower source ripple is proposed and it is shown in Fig. 1. This topology has the following advantages: (i) lower source current ripple, (ii) low EMI, (iii) simpler dynamics than the buck converter with input filter, (iv) simplicity in control due to single switch topology.

The proposed TOBC can be operated in several operating modes depending on the load, supply voltage and load side capacitor equivalent series resistance (ESR). However, due to the presence of inductor on source side, for most of the loads, the converter input current exhibits lower ripple. Only at very light loadings it exhibits higher ripple content, and for such cases the proposed converter will not provide any benefit of lower EMI on the source side. In the following lines the converter operation for continuous inductor current mode (CICM) is discussed and then mathematical models are formulated. The assumptions used in the analysis are: (i) switching devices are assumed to be ideal, (ii) the ripple voltage in the capacitor $C_1$ is very small, (iii) converter parameters are linear time-invariant. Various waveforms of the TOBC for the CICM case are shown in Fig. 2. Due to the converter structure the capacitor $C_2$ current, $i_{C2}$, is alternating but not symmetrical about the time axis. Since load voltage is the sum of the capacitor voltage and voltage drop in the ESR, the load voltage ripple waveform also is identical with that of the $i_{C2}$.

---

0-7803-9771-1/06/$25.00 ©2006 IEEE    413

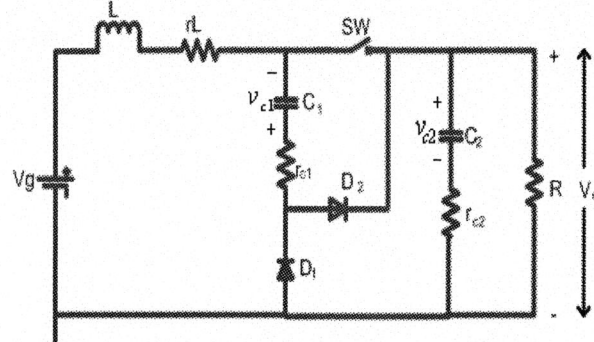

Fig. 1. Circuit diagram of the 3rd order buck converter.

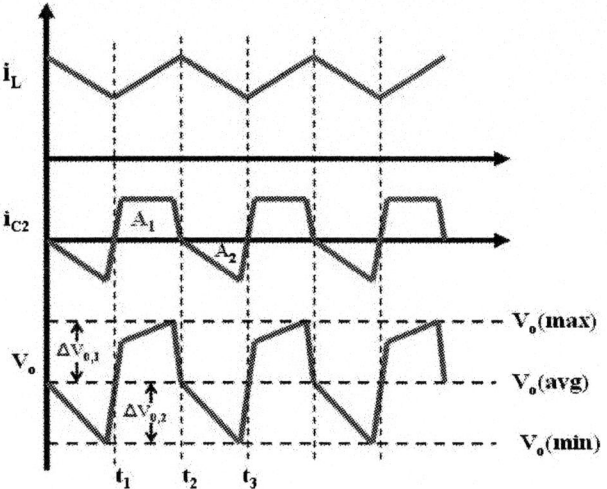

Fig. 2. Steady-state converter waveforms for CICM operation.

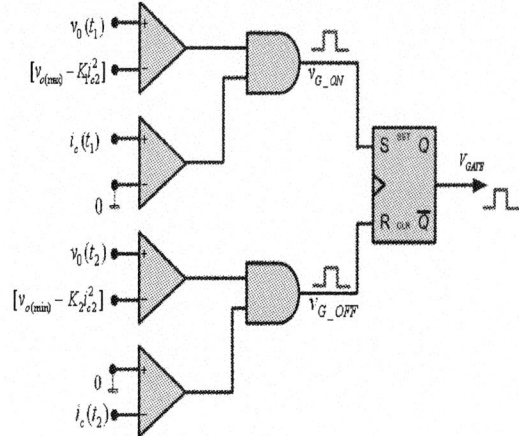

Fig. 3. Logic realization of the proposed control scheme.

The proposed hysteretic control switching strategy is formulated based on the understanding of the capacitor current, $i_{C2}$, and load voltage instantaneous variation. By performing a time-domain analysis of $i_{C2}$ and $V_0$ it is possible to establish the following ON and OFF-state switching constraints.

Switch-ON case:

$$i_{C2}(t_1) \geq 0 \qquad (1)$$

$$v_0(t_1) \geq v_{0(max)} - k_1 i_{C2}^2(t_1) \qquad (2)$$

Switch-OFF case:

$$i_{C2}(t_1) \leq 0 \qquad (3)$$

$$v_0(t_1) \geq v_{0(min)} - k_2 i_{C2}^2(t_2) \qquad (4)$$

where $k_1$, $k_2$ are dependent on the converter parameters, source and load voltages. Although above defined control strategy equations appears to be purely mathematical, but they can easily be realized by means of analog IC's. The control logic diagram is shown in Fig. 3.

The proposed topology can be operated in several operating modes depending on the load, switching frequency and supply voltage. The mathematical model is developed for the CICM operation and in this case the circuit has two operating modes; Mode-1: S-ON ($0<t<dT$); Mode-2: S-OFF ($dT<t<T$). In each mode of operation the circuit is linear and its behaviour can easily be described by the state-space model [4] given by $\dot{x} = A_K x + B_K u$; $v_0 = C_k x$; where $x = \begin{bmatrix} i_L & v_{c1} & v_{c2} \end{bmatrix}^T$, $u = \begin{bmatrix} v_g \end{bmatrix}$, $k=1,2$.

$$[A_1] = \begin{bmatrix} \dfrac{r_l}{L} & \dfrac{1}{L} & 0 \\ \dfrac{1}{C_1} & -\dfrac{(R+r_c)}{Rr_cC_1} & \dfrac{1}{r_cC_1} \\ 0 & \dfrac{1}{r_cC_2} & -\dfrac{1}{r_cC_2} \end{bmatrix};$$

$$[A_2] = \begin{bmatrix} -\dfrac{1}{L}\left[r_l + \dfrac{Rr_c}{(R+r_c)}\right] & -\dfrac{1}{L} & -\dfrac{R}{L(R+r_c)} \\ \dfrac{1}{C_1} & 0 & 0 \\ \dfrac{R}{C_2(R+r_c)} & 0 & -\dfrac{1}{C_2(R+r_c)} \end{bmatrix};$$

$$[B_1] = [B_2] = \begin{bmatrix} \dfrac{1}{L} \\ 0 \\ 0 \end{bmatrix}$$

Above modeling equations are useful for constructing simulation diagram in SIMULINK and controller design. For time-domain analysis purpose, here, we have used PSIM [5] simulator. The final simulation diagram is shown in Fig. 4.

414

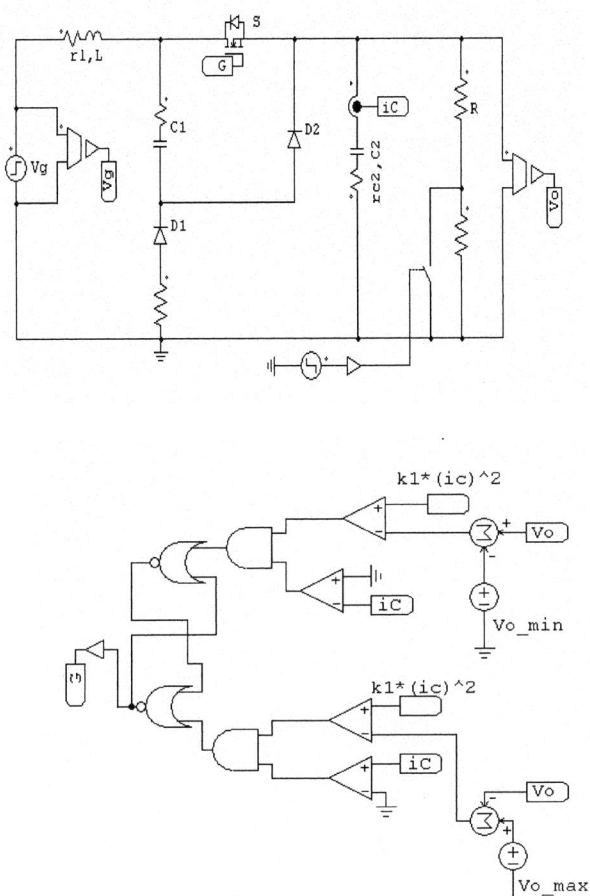

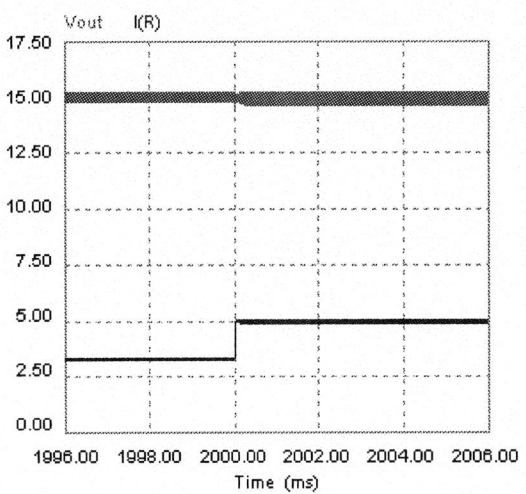

Fig. 5. Simulated dynamic response of load voltage (R: 4.5 → 3 ).

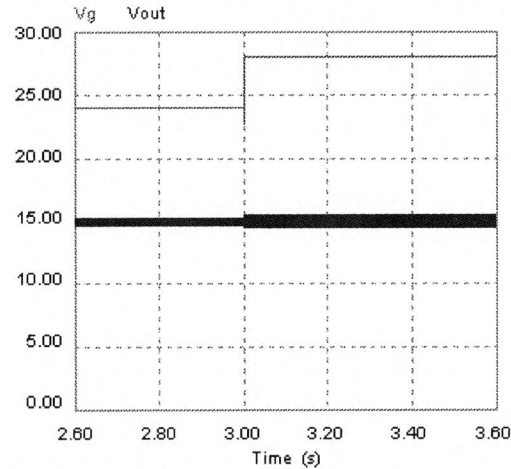

Fig. 6. Simulated dynamic response of load voltage (Vg: 24 → 28 V).

Fig. 4. Simulation diagram of the proposed control strategy.

## III. RESULTS AND DISCUSSIONS

For demonstration the concept is applied to TOBC, as shown in Fig. 4, and its suitability is investigated. In case of simple hysteresis control either the inductor current or load voltage ripple band will be the band limit and the converter operating frequency depends on this hysteresis band. In the proposed AHB control scheme the capacitor current together with load voltage is used. Adaptive nature is due to use of "$k_1$, $k_2$", which are dependent on load, source voltage and converter parameters also. To realize the adaptive hysteresis band the capacitor current information is utilized in the inner loop, while to achieve the voltage regulation the load voltage is used in the outer loop. As a result this scheme has all the features of two-loop cascade control in addition to the controller robustness. The justification for the use of capacitor current in place of inductor current is as follows. If the inductor current is used in the hysteresis control the average inductor current will not have any role in finding the actual switching sequence, but the inductor ripple current ultimately defines the control switching sequence. As the load capacitor current is a function of inductor current, the inductor current ripple information can easily be captured from the capacitor current waveform. Further, employing the adaptive band of capacitor current can enhance the control performance.

The proposed control scheme is shown in Fig. 4. To verify the developed modeling and controller design, a 50 W 24 V/ 15 V POLA converter is chosen for the investigations. The power stage parameters are: $L_1$= 0.1 mH, $C_1$=100 uF, $C_2$= 220 uF, R= 4.5 Ω. Controller coefficients $k_1$ and $k_2$ are obtained based on the load voltage ripple and capacitor current, complete design equations will be given in the final paper. With these parameters the closed-loop converter system regulation capability is tested for (i) supply voltage change 24 V ±15%, (ii) load disturbance of 4.5 Ω ±50%. For simulation PSIM power electronics simulator was used. For demonstration load voltage regulation against load and source voltage disturbances are shown in Fig. 5 and 6, respectively. Note that in all the cases the controller is regulating the load voltage, which is within the hysteresis band and the response time is also very low. But in other type of control strategies the intermediate dynamics exists for considerable amount of time. In the proposed scheme, this time is almost near to zero. The effectiveness of the control scheme for variable reference set point voltages is under investigation.

## IV. CONCLUSIONS

An adaptive hysteresis band control was proposed for the TOBC. The controller realization in terms of capacitor current and load voltage has been developed. This controller contains variable gains, $k_1$ and $k_2$, which are dependent on the load, source voltage and gives the adaptive property to the AHB scheme. Simulation results demonstrated the proposed scheme suitability. Although the load voltage has definite amount of ripple, which is less than the allowable limits, but the dynamic response of the proposed converter is much better than other types of converters.

## V. ACKNOWLEDGMENT

The authors would like to thank the MHRD, Govt. of India for supporting this research through R & D scheme.

## VI. REFERENCES

[1] Kelvin Ka Sing Leung & Henry Shu-Hung Chung, "Dynamic hysteresis band control of buck converter with fast transient response", *IEEE Trans. Power Electron,* Nov.2004.

[2] Kelvin K.S. Leung & Henry S.H. Chung, "State Trajectory Prediction Control of Boost Converters", *IEEE Trans. Power Electron,* Nov.2004.

[3] Song, C., and Nilles, J. "Accuracy Analysis of Hysteretic Current-Mode Voltage Regulator," *Proc. IEEE APEC, 2005, pp. 276-280.*

[4] Robert W. Erickson & Dragan Maksimovic, "Fundamentals of Power Electronics", Springer, 2001.

[5] PSIM user manual, 2004.

## VII. BIOGRAPHIES

**M.Veerachary** was born in India in 1968. Mr. Chary completed his Dr. Eng from University of the Ryukyus, Okinawa, Japan in 2002. Since then he is with Dept. of Electrical Engineering, Indian Institute of Technology Delhi, India and currently he is an Associate Professor. His research interests are High frequency dc-dc conversion, design and analysis of satellite/spacecraft power conditioning systems, modeling and simulation of large power electronic systems, control theory application to power electronic systems and intelligent controller applications to power supplies.

**Deepen Sharma** born in Bhutan. He received the Bachelor degree from the NIT Allahabad, India, in 2003, and currently he is working towards Master of Technology degree in the Dept. of Electrical Engineering, IIT Delhi. His fields of interest are power electronic converters applications to small power supplies and adaptive controllers design for the switch-mode converters.

**2006 IEEE International Conference on Power Electronic, Drives and Energy Systems**

# A Novel Topology for Multiple Output DC-DC Converters for One Cycle Control

Ravindra Kumar Singh

*Abstract*-- **Multiple output DC-DC converters are needed by many utilities/ industry applications. The Control of multiple output converters is challenging as the outputs derive the input power form same source and thereby have a strong coupling. A novel Multiple Output converters topology has been proposed to have decoupling of outputs from individual control. A closed loop control has been suggested for the proposed converter such that it has one cycle control of all the outputs individually as well as collectively. The proposed converter has been verified by simulation and experimentation.**

*Index Terms*-- DC-DC Converter.

## I. INTRODUCTION

A multiple output DC-DC converter consists of a DC power supply and desired number of regulated outputs [1]. The power circuit of such converters normally employs either a fly back or a forward topology. The control design of such converters is aimed at providing constant and stable output voltages at individual outputs for admissible load and input voltage changes. Regulation against load changes of individual outputs by different methods reported in literature has a minimum error in all outputs. In [2-5] the circuit topology has been modified to minimize the errors in outputs by magnetic coupling of secondary windings and other techniques. For example, in [3] one output is operated in Continuous Conduction Mode (CCM) and the other output in Discontinuous Conduction Mode (DCM) to regulate load perturbations in a dual converter. However the uncontrolled charging of the capacitors does not allow independent regulation of the outputs and the response time is a strong function of the system parameters and the switching frequency. The problem of uncontrolled charging has been solved in [6] for a single output DC-DC converter. The conventional fly back converter has a diode for each output. A switch to obtain better control has replaced the diode in this topology.

The same principle has been applied to multiple-output converter to obtain to obtain a new converter topology. In the proposed converter shown in Fig. 1, each of the two outputs can be individually controlled between specified minimum and maximum instantaneous values. An energy recovery winding has been used to facilitate opening of all switches whenever

---

Ravindra kumar singh is working as a senior Lecturer with the department of Electrical Engineering, Motilal Nehru National Institute of Technology, Allahabad, U.P, India.

required. If we call load switch along with capacitor and load as load sub-circuit, additional load sub-circuits will realize a converter topology with increased numbers of outputs. Fig.1 (a) can be called non-isolated topology as the fly back inductor is common to input and output, whereas Fig.1 (b) is isolated topologies for two output converter.

The proposed converter has been analyzed in steady state operating condition to obtain input/output voltage relationship. Small signal analysis has been carried out to characterize the dynamic behavior of the converter. It has been found that the dynamic model is to complex for designing a physically realizable controller. Moreover, the proposed converter can be used for achieving faster dynamic response. Therefore, a hysteresis based closed loop controller has been designed. The proposed converter has response time of utmost one cycle of the hysteresis controlled closed loop converter frequency.

## II. STEADY STATE ANALYSIS

The steady state operation of the converter in constant frequency has been shown by switching diagram in Fig. 2. The switching diagram is drawn for two outputs. The main switch builds flux in the core and is used in sequence by two load circuits in intervals $d_1T$ and $d_2T$ respectively to charge the output capacitors. This may be followed by an energy recovery interval $d_{er}T$ if required. However, the energy recovery period can be positioned in any place in a switching period $T$. The average output voltages are not a function of the position of the energy recovery position as can be seen from (1). The isolated version of the converter topology also has almost the same switching diagram. The output voltage governs the turn ratio of the main transformer. The equivalent circuit diagrams for different switching intervals are shown in Fig. 3.

The inductor-charging interval $DT$ starts with closing of main switch and opening of load switches $Sw_1$ & $Sw_2$. The equivalent circuit is shown in Fig. 3(a). This interval follows by opening of the main switch $Sw_m$ and closing of load switch $Sw_1$ (equivalent circuit given in Fig. 3(b).

This interval is followed by closing of load switch $Sw_2$ and opening of $Sw_1$ (equivalent circuit given in Fig. 3(c). Fig. 3(d) shows the equivalent circuit of the optional energy recovery interval, where all the switches are open.

The core mmf

is defined as $\Im = N_1 i_{L1} + N_2 i_{er}$. Fig. 2 shows the waveform of

---

0-7803-9771-1/06/$25.00 ©2006 IEEE      417

the converter operating at constant frequency ($1/T$) in CCM.

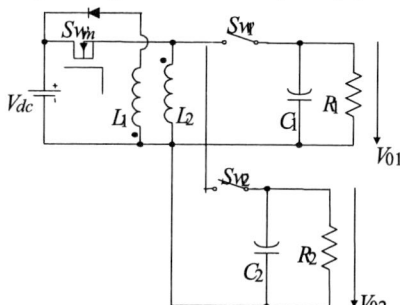

(a) Non-Isolated multiple output fly back converters

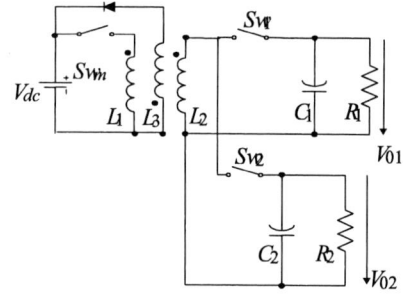

(b) Isolated multiple output fly back converters

Fig.1. Proposed converter in two-topology.

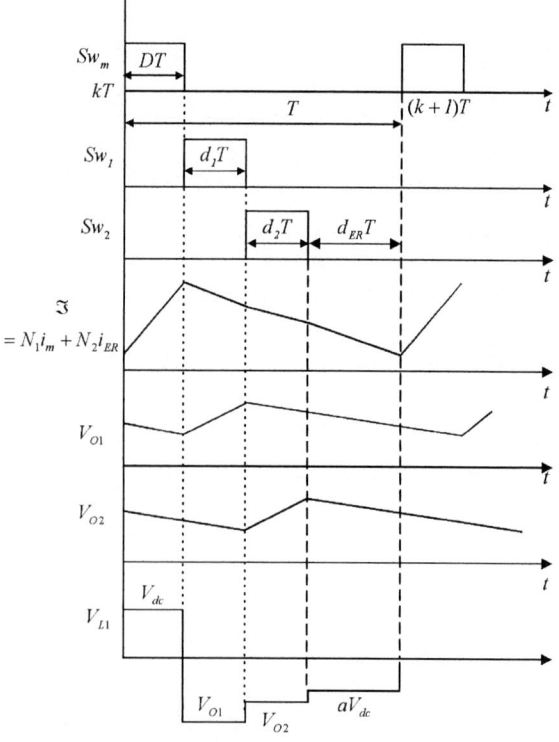

Fig. 2. Switching diagram for multiple output converter.

The main switch $Sw_m$ is closed at the beginning of each cycle, for a time $DT$. Note that $D$ in this chapter refers to duty ratio of the main switch $Sw_m$.

For a converter operating in steady state the voltage-second product across the inductor, which is equal to net change in core flux over a cycle, must be zero. Similarly, net increment in capacitor charge must be zero over a cycle in steady state.

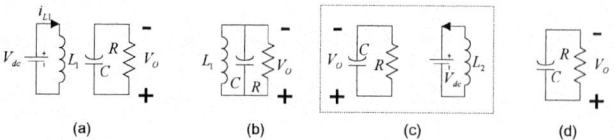

Fig. 3. Equivalent circuits of multiple output converter for (a) main Switch conducting (b) load Switch 1 conducting (c) load switch 2 conducting and (d) Energy recovery operating.

The increment in flux takes place when the inductor is connected to dc source and it reduces when the stored. magnetic energy of the inductor is used for charging capacitor connected across the load and during the energy recovery interval. The voltage waveform across the inductor represented by $v_{L1}$ is shown in Fig. 2. The magnitudes of the average inductor voltage in different duty intervals are marked in the figure. Applying volt-sec balance across the inductor gives

$$V_o dT + aV_{dc} d_{ER} T = V_{dc} DT$$

Where $a = N_1 / N_2$ is the turn ratio of the coupled inductors as stated above. Inductors $L_1$ and $L_2$ are assumed to be tightly coupled and there is no leakage.

The expression for output voltages are obtained by volt-sec balance of the core and charge balance across capacitors. The resulting equation are given by

$$\frac{V_{O1}}{V_{dc}} = \frac{D - a\, d_{er}}{d_1 \left[ 1 + \left( \frac{d_2}{d_1} \right)^2 \frac{R_2}{R_1} \right]} \quad , \quad \frac{V_{O2}}{V_{dc}} = \frac{D - a\, d_{er}}{d_2 \left[ 1 + \left( \frac{d_1}{d_2} \right)^2 \frac{R_1}{R_2} \right]} \qquad (1)$$

Fig. 4. Timing diagram of the converter.

The single output modified buck-boost converter has three independent subintervals, namely main switch ON interval, load switch ON subinterval and energy recovery interval. The main and load duty ratio is independently controllable. The state space equations for the converter in different intervals of the converter operation can be written from equivalent circuits shown in Fig. 3. The small signal model for a multiple sub-interval has been derived in section 2.5 of chapter 2 of [ ] and it is given by

$$\tilde{x}(t_n) = \prod_{i=1}^{n} [\Phi_i'] \tilde{x}(t_0) + \left[ \prod_{i=1}^{n} \Phi_i' - \prod_{i=1}^{n} \Phi_i \right] x(t_0) + \sum_{j=1}^{n} \left( \prod_{j=1}^{n-j} \Phi_j' \Theta_j' - \prod_{j=1}^{n-j} \Phi_i \Theta_j \right) V_{dc} \quad (2)$$

III. SMALL SIGNAL MODEL

The timing diagram of the converter is drawn in Fig. 4 on the basis of the switching diagram given in Fig. 2. The switching instants and interval lengths are given along with the state matrices in these subintervals. The timing diagram of the converter is drawn in Fig. 4 on the basis of the switching diagram given in Fig. 2. The switching instants and interval lengths are given along with the state matrices in these subintervals.

Let the state transition matrices in the intervals $DT$, $dT$ and $d_{er}T$ be $\Phi_1, \Phi_2$ and $\Phi_3$ respectively. Substituting these in (2), we get

$$\tilde{x}(t_3) = \prod_{i=1}^{3}\left[\Phi_i'\right]\tilde{x}(t_0) + \left[\prod_{i=1}^{3}\Phi_i' - \prod_{i=1}^{3}\Phi_i\right]x(t_0)$$
$$+ \sum_{j=1}^{3}\left(\prod_{j=1}^{3-j}\Phi_i'\Theta_j' - \prod_{j=1}^{3-j}\Phi_i\Theta_j\right)V_{dc} \qquad (3)$$

### A. Linear Model of the Converter

The expanded expression for the small signal perturbation model has been given in (4).

The higher order terms of the expansion of the series has not been considered. Equation 4 is linear perturbation model of the converter. It is a MISO (Multi-Input Single Output) system.

$$\tilde{x}(t_3) = \Phi_3\Phi_2\Phi_1\tilde{x}(t_0)$$
$$+ \begin{bmatrix}(\Phi_3\Phi_2 A_1\Phi_1 - A_3\Phi_3\Phi_2\Phi_1)x(t_0)T_s \\ + (\Phi_3\Phi_2\Phi_1 B_1 - A_3\Phi_3\Phi_2\Theta_1)V_{dc}T_s - \Phi_3 B_3 V_{dc}T_s\end{bmatrix}\tilde{D} \quad (4)$$
$$+ \begin{bmatrix}(\Phi_3 A_2\Phi_2\Phi_1 - A_3\Phi_3\Phi_2\Phi_1)x(t_0)T_s + \\ (\Phi_3 A_2\Phi_2\Theta_1 - A_3\Phi_3\Phi_2\Theta_1)V_{dc}T_s - \Phi_3 B_3 V_{dc}T_s\end{bmatrix}\tilde{d}$$

The model developed is validated through extensive simulation study. The simulation results are given in Fig. 5 and

Fig.6. The converter in Fig. 5 is running at duty ratio of main and load switches at 0.5 and 0.2 respectively. 0.2, which is a considerably large perturbation, perturbs the main duty, while the load duty ratio remains constant. The prediction and converter state trajectories overlap.

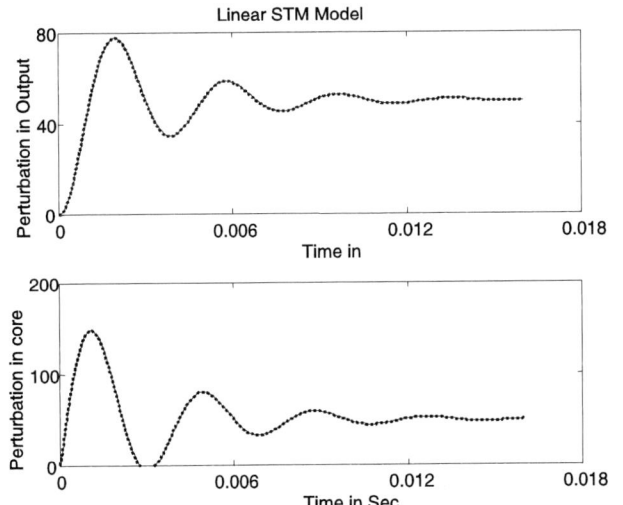

Fig. 5. Linear STM based prediction of perturbation of states of the converter.

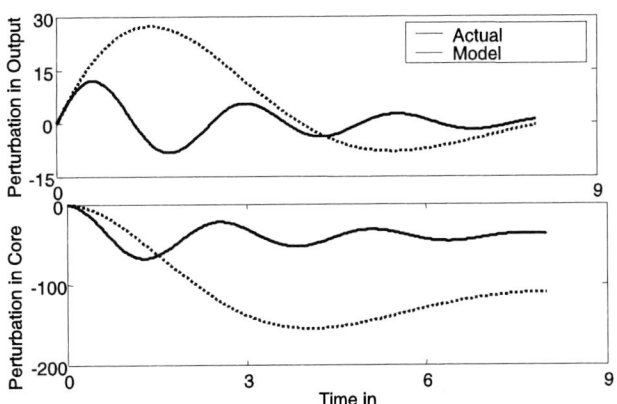

Fig. 6. Linear STM based prediction of perturbation of states of the converter.

However, the prediction model has been tested at different operating point did not succeed as shown in Fig. 6. The converter is assumed to be in steady state at D=06 and d=0.1. The main duty remains unchanged while load duty is changed by 0.2. The state dynamics and prediction dynamics don't match at all. Thus linear prediction model is not sufficient to model the converter for all operating modes.

### B. Bilinear Modeling and Controller Design

The linear prediction model of the modified buck-boost converter is not capable of predicting the perturbation in states for all possible modes. Hence a bilinear model of the converter has been developed in this section. The bilinear model has been derived from (3) by retaining product term of state and duty ratio along with the linear terms of the expansion. The model thus obtained is given in (5).

$$\tilde{x}(t_3) = \Phi_3\Phi_2\Phi_1\tilde{x}(t_0)$$
$$+ \begin{bmatrix}(\Phi_3\Phi_2 A_1\Phi_1 - A_3\Phi_3\Phi_2\Phi_1)x(t_0)T_s + \\ (\Phi_3\Phi_2\Phi_1 B_1 - A_3\Phi_3\Phi_2\Theta_1)V_{dc}T_s - \Phi_3 B_3 V_{dc}T_s\end{bmatrix}\tilde{D}$$
$$+ \begin{bmatrix}(\Phi_3 A_2\Phi_2\Phi_1 - A_3\Phi_3\Phi_2\Phi_1)x(t_0)T_s + \\ (\Phi_3 A_2\Phi_2\Theta_1 - A_3\Phi_3\Phi_2\Theta_1)V_{dc}T_s - \Phi_3 B_3 V_{dc}T_s\end{bmatrix}\tilde{d} \quad (5)$$
$$+ \left[(\Phi_3\Phi_2 A_1\Phi_1 - A_3\Phi_3\Phi_2\Phi_1)T_s\right]\tilde{x}(t_0)\tilde{D}$$
$$+ \left[(\Phi_3 A_2\Phi_2\Phi_1 - A_3\Phi_3\Phi_2\Phi_1)T_s\right]\tilde{x}(t_0)\tilde{d}$$

### C. Model Verification

In order to verify the model, a number of simulations have been carried out. The simulation results are given in Figs. □ to □ and are summarized in Table 1.

It can be seen that the predicted curve almost matches the simulated curve for fairly large perturbations. The failure case Fig. □ has been reported to show that bilinear model too has a limitation.

### D. Controller Design Based on Small Signal Model

Let the instant $t_0$ be the reference instant k that coincides with the closing of the switch. The final switching instant $t_3$ then can be defined as (k+1), where k = 0,1,2,.......... is the discrete time index. We can therefore express (4) in discrete

419

### TABLE I
### SUMMARY OF BILINEAR PREDICTION RESULTS

| Steady state main duty ($D_0$) | Steady state load duty ($d_0$) | ($\tilde{D}$) | ($\tilde{d}$) | Model Prediction (Figure) |
|---|---|---|---|---|
| 0.6 | 0.1 | 0 | 0.2 | Fig. □ |
| 0.2 | 0.□5 | 0.6 | - 0.65 | Fig. □ |

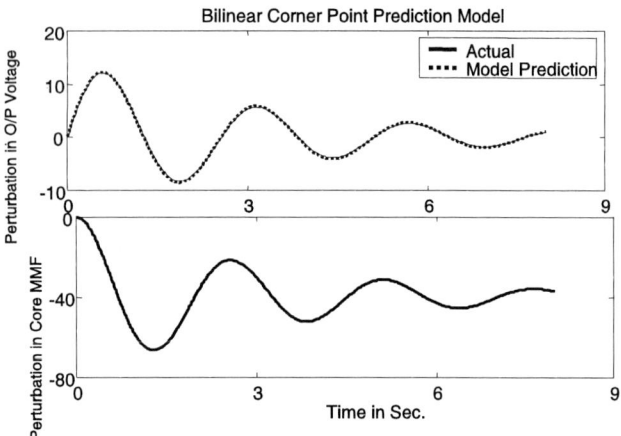

Fig.□ Bilinear prediction for $D_0 = 0.6$, $d_0 = 0.1$ and $\tilde{D} = 0$, $\tilde{d} = 0.2$.

$$\tilde{x}(k+1) = F\tilde{x}(k) + G\tilde{d}(k) \qquad (6)$$

and the output equation can be expressed as (□)

$$\tilde{y}(k) = C\tilde{x}(k) \qquad (\square)$$

It is desired that the controller be able to track any step change in the reference voltage $\tilde{y}_{ref}$. Note that $\tilde{y}_{ref}$ is a change/-desired perturbation in the output from the steady state value. For example, if the converter output is operating at 20V in steady state and we want an output of 30V, then the $\tilde{y}_{ref}$ will be 10V. Similarly, if it is desired to have an output voltage of 10V, $\tilde{y}_{ref}$ will be –10V.

To facilitate the tracking function; we include a pseudo-integrator action on the error function $e(k)$ in the feedback loop. This is given by

$$e(k) = \tilde{y}(k) - \tilde{y}_{ref}(k) \qquad (\square)$$

$$z(k+1) = (1-\varepsilon)z(k) + e(k) \qquad (\square)$$

where $\varepsilon$ is a very small positive number. We now define an extended state vector as

$$x_e(k) = \begin{bmatrix} \tilde{x}(k) \\ z(k) \end{bmatrix} \qquad (10)$$

Then combining (6) and (□) we get an extended state space representation as

$$\tilde{x}_e(k+1) = \begin{bmatrix} F & 0 \\ C & 1-\varepsilon \end{bmatrix} \tilde{x}_e(k) + \begin{bmatrix} G \\ 0 \end{bmatrix} \tilde{u}(k) + \begin{bmatrix} 0 \\ 0 \\ -1 \end{bmatrix} \tilde{y}_{ref}(k) \qquad (11)$$

$$= F_e\,\tilde{x}_e(k) + G_e\tilde{u}(k) + J_e\,\tilde{y}_{ref}(k)$$

The output equation thus becomes

$$\tilde{y}_{ref}(k) = \begin{bmatrix} C & 0 \end{bmatrix} \tilde{x}_e(k) \qquad (12)$$
$$= C_1\tilde{x}_e(k)$$

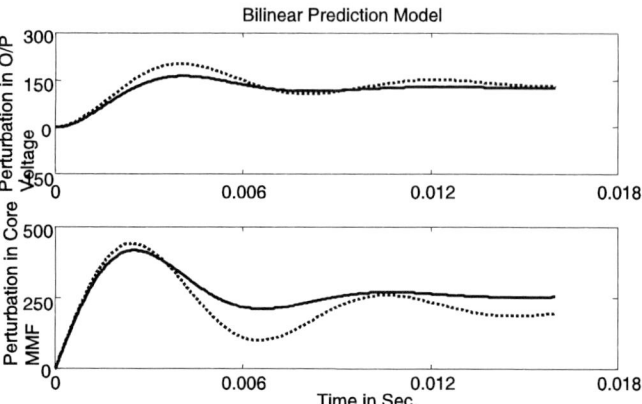

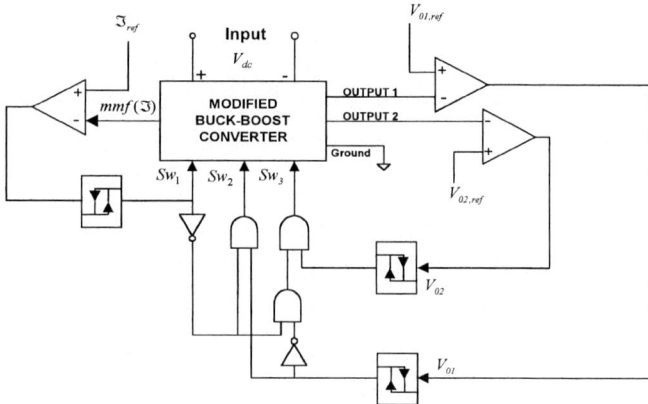

Fig. □ Bilinear prediction for $D_0 = 0.2$, $d_0 = 0.\square5$ and $\tilde{D} = 0.6$, $\tilde{d} = -0.65$.

Based on this state space formulation of the converter, a controller can be designed by control techniques such as pole placement, LQR or LQG controller. In these methods, a control law will be derived in terms of the states of the converter. The control law in this case happens to be the expression of the duty ratio. This duty ratio can be computed from DSP based processors if the states are measured and fed as input to the DSP controller. The complexity of the controller therefore results in smart algorithm development on DSP processors.

Fig. □ Schematic diagram of hysteresis controller for two outputs modified Buck-Boost converter.

The model proposed for controller design is however based on linear model of the converter. As discussed above that the linear model of the converter has a poor open loop prediction, i.e. the model is valid for small perturbation only. Therefore a nonlinear (bilinear) model based controller needs to be designed. The derivation of the control law may be mathematically complex in algebraic expression but can always be derived. The derived model however will become practically impossible or cost ineffective. Therefore a simple control law has been proposed in next section.

The prime objective of a closed loop controller for a DC-

DC converter is to reject input and load disturbances without losing stability. Improvements in response time, extension of zone of validation, etc. are some of other possible secondary objectives of closed loop controllers. But a better dynamic performance is essential for critical loads.

Changing the duty ratio controls the output of a DC-DC converter. Other methods are derived methods of duty ratio control to achieve a better control than direct control of duty. One such method is known as current programmed converter. The current programmed closed loop control has been widely studied in the literature. Inclusion of feed-forward loop in closed loop controller has been suggested to improve dynamic response. It is possible to compensate the supply disturbance in one cycle as reported in [□].

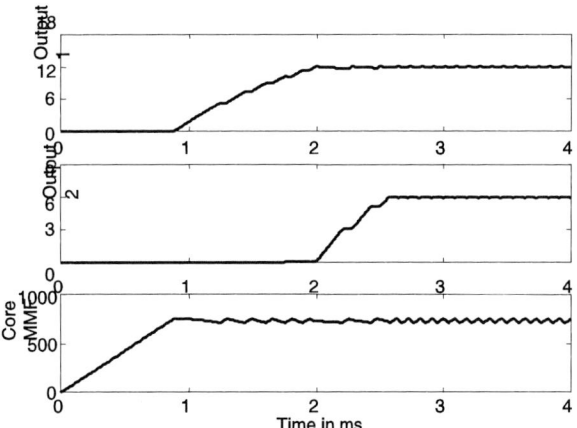

Fig. 10. Cold start behavior of the converter.

TABLE II
SIMULATION RESULTS OF THE CONVERTER

| Test Conditions | $\Delta V_{dc}$ | $\Delta R_1$ | $\Delta R_2$ | Result Shown in |
|---|---|---|---|---|
| Cold Start Behavior | --- | --- | --- | Fig. 10 |
| Supply Disturbance | 10 | --- | --- | Fig. 11 |
| Load & supply Disturbance on Output 1 & 2 | -10 | -10 | -10 | Fig. 12 |
| Load & supply Disturbance on Output 1 & 2 | 10 | 25 | 25 | Fig. 13 |

Although the theory of hysteresis control is well established, this is repeated here for sake of continuity. A variable(s) to be controlled is identified in a given circuit. The desired closed loop state trajectory is described for the variable. The feasibility study and stability requirements of the closed loop trajectory are made by suitable control theorems. This trajectory is now set as the reference value for the variable. We form a band called hysteresis band around this reference. The circuit is switched in such a manner that the actual state trajectory never goes out of this band. The average states of the converter therefore follow the set reference with high frequency component due to switching. This method of closed loop is called hysteresis control.

The operation of the converter in closed loop control can be described as follows. The core mmf builds when the main switch of the converter is closed. The stored magnetic energy is then used for charging load circuit. If the core energy is more than the energy required for load circuit in a switching cycle, the excess energy is fed back to the input by energy recovery circuit. In case the load rejection takes place in the circuit, the amount of the excess energy goes up and thus the energy recovery interval is proportionately increased. However, if the energy demand of the load goes up, there is no

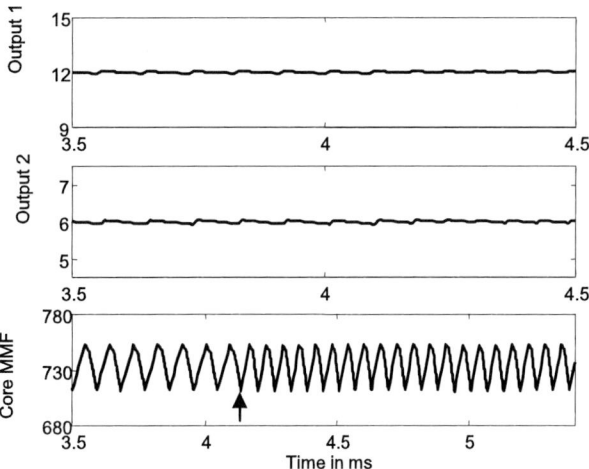

Fig. 11. Effect of supply disturbance, supply going up by 10V.

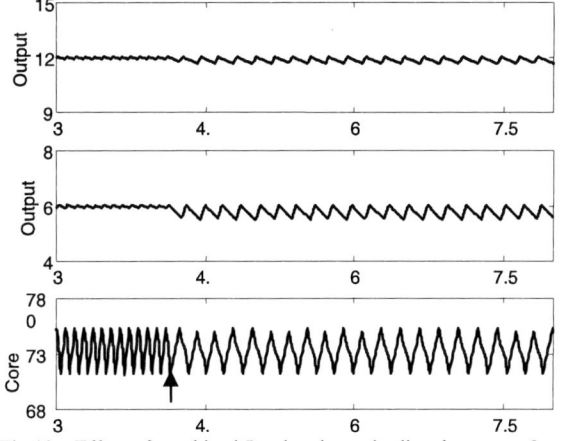

Fig.12. Effect of combined Load and supply disturbance on Output 1 & 2, both resistance decreased by 10 Ω and supply dips by 10V.

This will make the dynamic response slow. Had there been some excess energy available when required by the load, the compensation could have been done instantaneously. Keeping this in view, the closed loop controller has been designed. The reference core energy is chosen to account for maximum loading condition. If the loading is less, the energy recovery is more and if the loading is at its peak, there is no energy

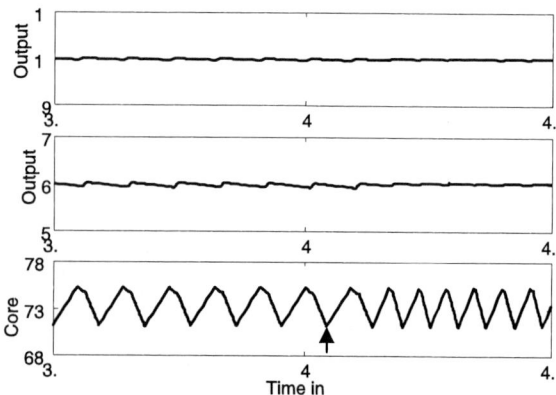

Fig. 13. Effect of combined Load and supply disturbance on Output 1 & 2; both the Output resistances increased by 25 Ω and supply rises by 10V.

recovery. The output voltage starts dropping for loads more then the designed maximum load. This is also known as folding characteristics of the converter.

The block diagram of the hysteresis controller for this converter is shown in Fig. □ The performance of the closed lop controller has been tested by extensive simulation and experiments.

The closed controller uses three feedback loops. The first feedback loop is for core mmf, second for output 1 and third for the output 2. The core mmf controller is given the top priority followed by controller for the output 1 and the output 2 is assigned last in the priority. The choice of the priority among the voltage loops is not strict and can be changed if required by the logic circuit. In Fig. □, the voltage loop for output 1 is given higher priority compared to output 2. The switching ON of the load switch 1 is done when the switching state of the main switch is in OFF condition and similarly the switching ON of the load switch 2 is carried out when the main switch and load switch is simultaneously OFF.

### A. Analysis of Hysteresis Controlled Closed Loop System

The reference voltages for different outputs are set equal to the desired outputs and the ripple requirements decide the hysteresis bandwidth. It is necessary to get a criterion to choose a reference value of core MMF for given outputs and possible variations. It is assumed here that the ripple content of the core MMF and output voltages are very small. Let the core mmf is operated with hysteresis band of $\Delta \Im$ around an average core mmf of $\Im_{av}$ and let the ON for a period of main switch, load switch 1 and load switch 2 be denoted by $t_{onM}$, $t_{on1}$ and $t_{on2}$ respectively. These intervals can be calculated from the averaging technique and are given by,

$$t_{onM} = \frac{\Delta \Im}{N_1 V_{dc}} L_1 \quad \& \quad t_{on1} \approx \frac{\Delta \Im}{N_1 V_{o1}} L_1 \quad t_{on2} \approx \frac{\Delta \Im}{N_1 V_{o2}} L_1 \quad (13)$$

The average core MMF calculated over any subinterval in a cycle must be equal to the average value over a cycle since the ripple in the MMF is small. Thus,

$$F_{av} = N_1 \left( \frac{T}{t_{on1}} \right) \left( \frac{V_{01}}{R} \right) \tag{14}$$

where $T = t_{onM} + t_{on1} + t_{on2}$ is the cycle period if there is no energy recovery. Solving (13) and (14) we obtain (15).

$$F_{av} = \left( \frac{N_1 V_{01}}{R_1} \right) \left( 1 + \frac{V_{01}}{V_{dc}} + \frac{V_{01}}{V_{02}} \right) \tag{15}$$

In order to calculate the reference MMF for this experiment, we calculate the MMF from (15) and perform the simulation study starting with this value. The actual reference MMF is 50% to 60% of the calculated average MMF as calculated from (15). The critical resistance for outputs can be calculated in the same line as done for the single output [6].

### B. Simulation Results of Two-Output Converter

The summary of the simulation study for converter operation in Buck mode with possible disturbances is given in Table II The output voltages of the output 1 and output 2 are 12 V and 6 V respectively. The nominal load resistances are 25 Ω and 15 Ω for output 1 and output 2 respectively. Simulation study has been carried out for cold start behavior, load and source disturbances when acting alone and when they are acting together. The disturbance in supply or load (if any) has been marked by an arrow in the simulation results presented below.

## IV. EXPERIMENTAL VERIFICATION

The parameters of the hardware components with their nominal loads are given in Appendix. The experiment has been carried out to verify the some of the simulation results. The scaling factors for voltages at output 1 and 2 are respectively 4.□2 and 2.□5 in the oscillograms i.e. waveforms of the variables shown on the oscilloscope.

The core MMF has got a scaling factor of 150. The experiments are carried out at low frequency as the energy recovery transformer has got a considerable leakage inductance and hence coupling noise among the windings are high. The frequency of operation can be made high if a better energy recovery winding is used

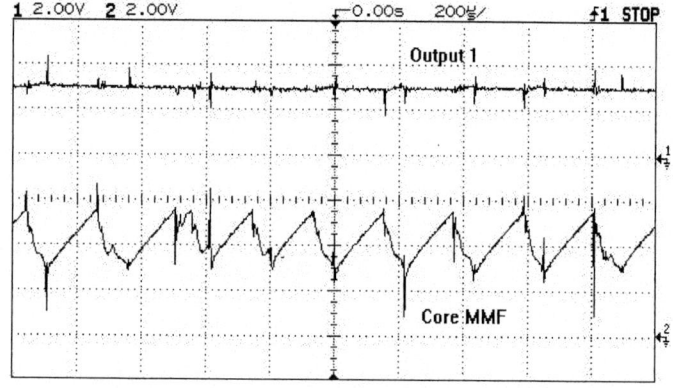

Fig. 14. Experimental waveforms of Output voltage 1 and core MMF in steady state.

The experiments have been carried out for different condition to verify the simulation study for closed loop change

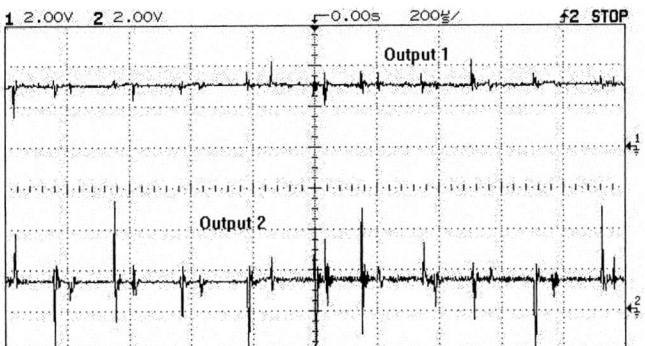

Fig. 15. Experimental waveforms of Outputs 1 & 2 when load is rejected on output 1.

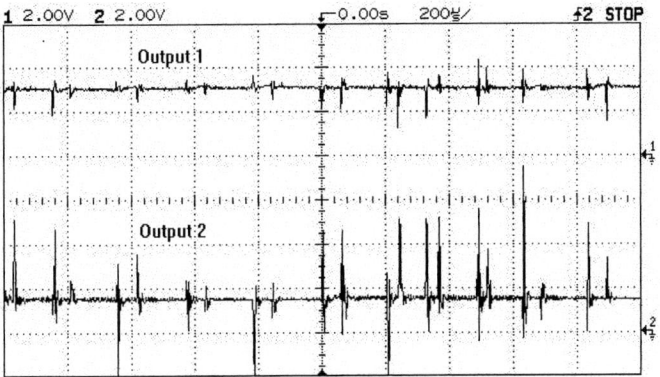

Fig. 16. Experimental waveforms of Outputs 1 & 2 when load on output 1 is increased.

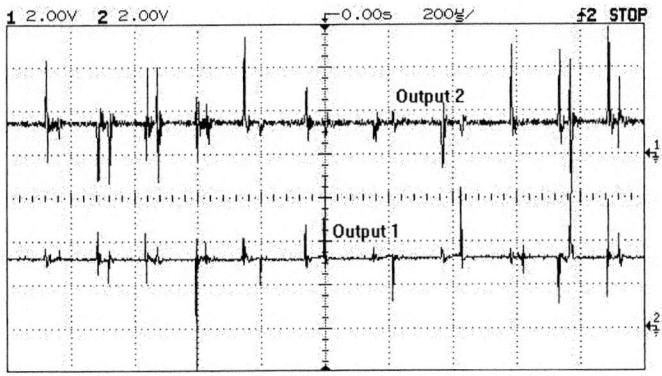

Fig. 1□ Experimental waveforms of Outputs 1 & 2 when load is rejected on output 2.

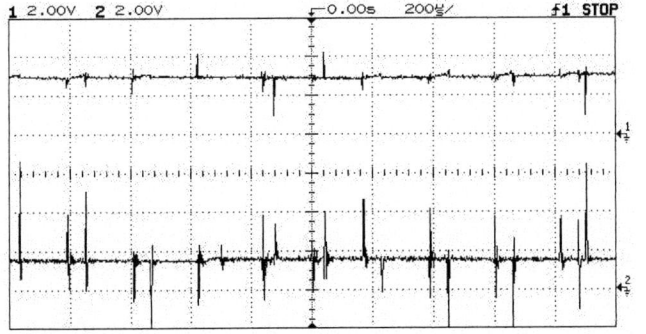

Fig. 1□ Experimental waveforms of Outputs 1 & 2 when load on output 2 is increased.

hysteresis control of converter. It can be seen from the above figures that the dynamics of the converter in closed loop converter has minor frequency deviation corresponding to supply is rejected in same cycle.

## V. CONLUSION

A new topology for multiple output DC-DC converter has been proposed which is capable of rejecting the disturbance if operated with hysteresis band closed loop controller in utmost one switching cycle.

## VI. APPANDIX

The parameters for simulation and experiments are given below.

$$L_1 = 3.5\,\text{mH}, \ C_1 = C_2 = 330\,\mu\text{F}, \ R_1 = R_2 = 15\,\Omega, \ V_{dc} = 20\,\text{V}, \ a = 3/2$$

$$\mathfrak{I}_{av} = \Box 50\,\text{AT}, \quad \Delta\mathfrak{I} = 0.1 * \mathfrak{I}_{av}, \Delta V_0 = 0.05 V_0 \ N_1 = 150$$

It is to be noted that parameter 'a' is the turn ratio of the main winding to energy recovery winding. The value of the inductance of the energy recovery winding can be obtained from turn ratio as $L_2 = L_1/a^2$. Please note that ESR of the capacitor and resistance of the inductance have been calculated indirectly from open loop experiment of the converter.

## VII. ACKNOWLEDGMENT

The author gratefully acknowledges contributions of Arindham Ghosh and Avinash Joshi for the supervision of Ph. D. thesis of the author. This work has been carried out as Ph. D. work of the author.

## VIII. REFERENCES

[1] N. Mohan, T. M. Undelend, and W. P. Robbins: *Power Electronics: Converters, Applications and Design*, John Wiley & Sons, Singapore, 1□□□.

[2] Y. Chen, D. Y. Chen and Y. Wu, "Control Loop Modeling of Multiple Output Feedback of Forward Converter," *IEEE Transactions on Power Electronics*, Vol. □, No. 3, pp. 320-32□, July 1□□3.

[3] J. sebastián and J. Uceda, "The Double Converter: A fully Regulated Two Output Dc-DC Converter," *IEEE Transactions on Power Electronics*, Vol. PE-2, No. 3, pp. 23□-246, July 1□□□

[4] F. Kurokawa and H. Matsuo, "A Multiple –Output Hybrid Power Supply," *IEEE Transactions on Power Electronics*, Vol. 3, No. 4, pp. 23□-246, October 1□□□

[5] T. Charanasomboon, M. J. Devaney and R. G. Hoft, " Single Switch Dual Output DC-DC Converter Performance," *IEEE Transactions on Power Electronics*, Vol. 5, No. 2, pp. 241-244, April 1□□0.

[6] R. K. Singh, A. Joshi, A. Ghosh, "A modified flyback DC-DC converter," in *Proc. 1998 IEEE Applied Power Electronics Conf.*, pp. 33□-343.

[□] R. K. Singh, "Modeling and Control of Conventional and Modified Buck-Boost DC-DC Converter," Ph.D. dissertation, Dept. Elec. Eng., I.I.T. Kanpur, India, 2001.

[□] K. M. Smedley and S. Cûk, "One-cycle control of switching converter," *IEEE Trans. Power Electronics*, Vol. 10, No. 6, pp. 625-633, 1□□5.

## IX. BIOGRAPHIES

**Ravindra Kumar Singh** was born on December 30, 1□65 in Chapra, Bihar, India. He graduated from the Bhagalpur University, Bihar. He did his post graduation from IT, BHU and Ph. D. fro IIT Kanpur, India

He is presently serving in M.N.N.I.T. Allahabad, India. He is also Fellow of Institution of Engineers (India).

**2006 IEEE International Conference on Power Electronic, Drives and Energy Systems**

# New Hybrid SVPWM Methods for Direct Torque Controlled Induction Motor Drive for Reduced Current Ripple

T. Brahmananda Reddy, J. Amarnath, and D. Subbarayudu

*Abstract*--This paper presents novel hybrid SVPWM (HSVPWM) methods for direct torque controlled induction motor drives without angle estimation to reduce steady state ripple in current, torque and flux. The proposed PWM technique is designed based on the notion of stator flux ripple, which is a measure of line current ripple. Expressions for RMS ripple, over a sub-cycle are derived for each switching sequence in terms of reference voltage vector, imaginary switching times and sub-cycle duration. This analysis together with the THD performance is used to design new HSVPWM techniques, which result in reduction in THD. The proposed PWM methods simplify the control algorithm and also reduce the execution time. Also to improve the speed performance under uncertainties caused by load torque, a robust speed controller with an integral sliding switching surface is used. To validate the proposed method, the results are presented.

*Index Terms*—Direct torque control, flux ripple, hybrid SVPWM, sliding mode controller.

## I. NOMENCLATURE

$R_s$, $R_r$      stator and rotor resistances

$L_s$, $L_r$, $L_m$      self and mutual inductances

$P$      number of poles

$\overline{v}_s$      stator voltage phasor

$\overline{i}_s$, $\overline{i}_r$      stator and rotor currents phasors

$\overline{\psi}_s$, $\overline{\psi}_r$      stator and rotor flux linkage phasors

$\omega_r$      rotor electrical speed in radians

$T_e$      electromagnetic torque

$B$      friction coefficient

$J$      inertia constant of the induction motor

$T_L$      load torque.

$\omega_m$      rotor mechanical speed in radians

---

T. Brahmananda Reddy and Dr. D. Subbarayudu are with Department of Electrical and Electronics Engineering, G. Pulla Reddy Engineering College, Kurnool, AP 518002, INDIA (e-mail: tbnr@rediffmail.com).

Dr. J. Amarnath is with Jawaharlal Nehru Technological University, Hyderabad, AP, INDIA (e-mail: amarnathjinka@yahoo.com).

## II. INTRODUCTION

VARAIBALE speed induction motor drive, based on direct torque control (DTC), is receiving wide attention in the literature [1]-[5], [9], [11]. The conventional DTC (CDTC) [1] consists of adaptive motor model, two hysteresis comparators, a switching table and a voltage source inverter (VSI). In DTC the flux and torque are controlled independently by selecting one of the voltage space vectors of the VSI, in order to keep the stator flux and torque with in the limits of the hysteresis bands. Despite its simplicity, CDTC is able to produce quick torque response and is robust with respect to motor parameter variations. Hence this control algorithm is increasingly being used in the industry and is considered to be the next generation motor control method [2]. However, the presence of torque and flux hysteresis controllers leads to variable switching frequency. Also CDTC has considerable ripple in torque, flux and current during steady state, which results in harmonics, power loss and incorrect speed estimation. In order to improve the performance of CDTC in terms of torque, flux and current ripple, discrete space vector modulation (DSVM) is proposed in [3]. DSVM can generate a higher number of voltage vectors than that used in CDTC, which allow a sensible reduction of torque and current ripple in all speed ranges. Only few schemes can produce the constant switching frequency operation with DTC. In [4-5], a voltage reference is generated based upon the errors of torque and flux and reference voltage is realized by using the principle of space vector pulsewidth modulation (SVPWM) to achieve the constant switching frequency operation. Also a substantial reduction in torque and current ripple can be obtained by using SVPWM-DTC methodology. In [6-8] HSVPWM techniques are developed for reduction of current ripple and switching losses. In [9] DTC is developed based on the 7-zone HSVPWM method to reduce the ripple. To reduce the computational burden involved in CSVPWM, a novel voltage modulation technique is proposed in [10] using the concept of effective time. To avoid the requirement of reference voltage vector, sector identification and angle determination, the effective time is determined using the concept of imaginary switching times in [11-12] and then this concept is used for different switching sequences. Industrial applications exhibit

0-7803-9771-1/06/$25.00 ©2006 IEEE

significant uncertainties, so that performance may deteriorate, if conventional controller such as PI controller is used. For this reasons it is worth to develop controllers that have capabilities of to handle uncertainties caused by parameter variations. The sliding mode control can offer good performance against insensitivities to parameter variation and load disturbance [13]. Hence, to improve the speed performance under uncertainties, a sliding mode speed controller is used for field oriented control in [14].

The main objective of this paper is to develop few HSVPWM algorithms for direct torque controlled induction motor drive based on imaginary switching times to reduce the steady state ripples at all modulation indices. Also to improve the speed performance, under uncertainties, an integral switching surface sliding mode speed controller is developed, which is robust under uncertainties caused by load torque disturbances.

### III. PRINCIPLE OF PROPOSED DTC

In this paper, DTC is developed based on HSVPWM methods; the block diagram of the proposed method is as shown in Fig. 1.

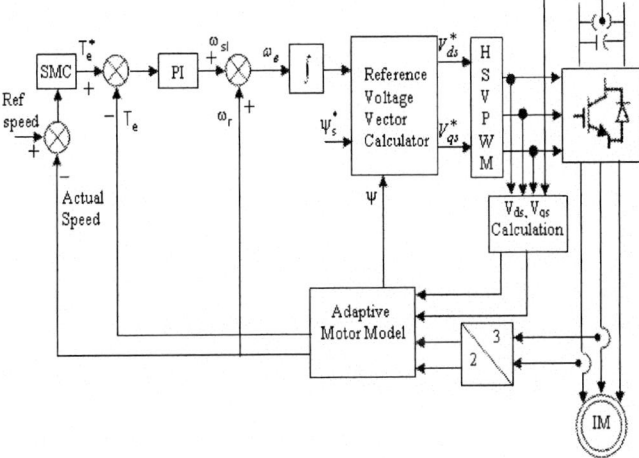

Fig. 1. Block diagram of proposed method.

In the proposed method, speed of the reference stator flux vector $\overline{\psi}_s{}^*$ is derived by the addition of slip speed and actual rotor speed. The actual synchronous speed of the stator flux vector $\overline{\psi}_s$ is calculated from the adaptive motor model. After each sampling interval, actual stator flux vector $\overline{\psi}_s$ is corrected by the error and it tries to attain the reference flux space vector $\overline{\psi}_s{}^*$. Thus the flux error is minimized in each sampling interval. The d-axis and q-axis components of the reference voltage vector can be obtained as follows:

Reference value of the d-axis and q- axis stator fluxes and actual value of the d-axis and q-axis stator fluxes are compared in the reference voltage vector calculator block and hence the error in the d and q-axes stator flux vectors is obtained as in (1).

$$\Delta\psi_{ds} = \psi_{ds}{}^* - \psi_{ds} \qquad (1.1)$$

$$\Delta\psi_{qs} = \psi_{qs}{}^* - \psi_{qs} \qquad (1.2)$$

The knowledge of flux error and stator ohmic drop allows the determination of appropriate reference voltage space vectors as given in (2).

$$V_{ds}^* = R_s\, i_{ds} + \frac{\Delta\psi_{ds}}{T_s} \qquad (2.1)$$

$$V_{qs}^* = R_s\, i_{qs} + \frac{\Delta\psi_{qs}}{T_s} \qquad (2.2)$$

These d-q components of the reference voltage vector are transformed to three phase reference voltages $V_a^*$, $V_b^*$ and $V_c^*$ from which, the actual switching times for each inverter leg are calculated. Finally, to improve the speed performance against uncertainties caused by load variations, an integral switching surface sliding mode speed controller is used and this is shown by SMC block in the block diagram.

#### A. Switching Sequences

The conventional approach to obtain seven possible switching sequences is given in [6-9]. The conventional sequences reported in [6-9], use the reference frame transformation, which increases the complexity of control algorithm. In the proposed method the switching sequences can be generated as follows:

The d-q components of reference voltage vectors are transformed to three phase reference voltages $V_{as}$, $V_{bs}$ and $V_{cs}$. Then the imaginary switching time periods proportional to the instantaneous values of the reference phase voltages are defined as

$$T_{as} \equiv \left(\frac{T_s}{V_{dc}}\right)v_{as}^* ; \quad T_{bs} \equiv \left(\frac{T_s}{V_{dc}}\right)v_{bs}^* ; \quad T_{cs} \equiv \left(\frac{T_s}{V_{dc}}\right)v_{cs}^* \quad (3)$$

Where $T_s$ is the sampling time and $V_{dc}$ is dc link voltage.

Then in every sampling time, the maximum and minimum values of imaginary switching times are calculated as given in (4).

$$T_{\max} = Max(T_{as}, T_{bs}, T_{cs}) \qquad (4.1)$$
$$T_{\min} = Min(T_{as}, T_{bs}, T_{cs}) \qquad (4.2)$$

The active vector switching times $T_1$ and $T_2$ may be expressed as

$$T_1 = T_{\max} - T_X ; T_2 = T_X - T_{\min} \qquad (5)$$

Where $T_X \in (T_{as}, T_{bs}, T_{cs})$ and is neither minimum nor maximum switching time.

The zero voltage vectors switching time is calculated as

$$T_Z = T_s - T_1 - T_2 \qquad (6)$$

The CSVPWM uses 0127-7210 (CS) and employs equal division of zero state times, which has proven to result in least harmonic distortion in comparison to its unequal division. There are three switchings within a sampling period. By using only one zero state, either 0 or 7, the bus clamping sequences 012 (S1) and 721 (S2) can be obtained. Utilizing the freedom of active state division, the remaining bus clamping switching sequences 0121 (S3), 7212 (S4), 1012 (S5) and 2721 (S6) can be generated. These novel sequences involve a double

425

switching of a phase while another phase is clamped. These sequences can be generated using the additional degrees of freedom offered by the space vector approach and cannot be generated using the per-phase approach [8]. In order to obtain minimum switching frequency for each of the power devices, the switching sequences are arranged such that the transmission from one state to the next is performed by switching only one inverter leg. These sequences are illustrated in Fig. 2, which shows the switching sequences corresponding to sector-I (first sixty degrees) only, since all the six sectors are symmetric; the discussion is limited to first sector alone.

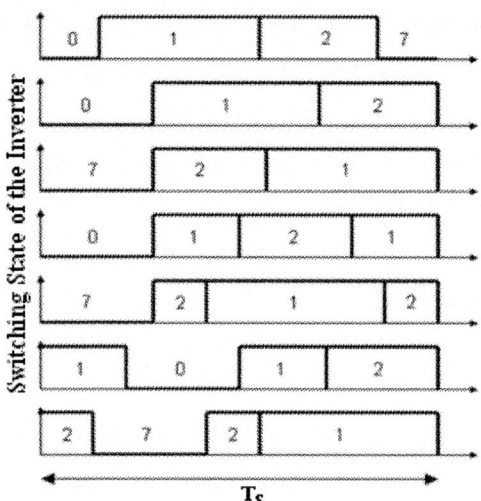

Fig. 2. Various switching sequences in first sector.

### B. Analysis Based on Stator Flux Ripple

The states of the inverter are switched at appropriate instants to generate the required fundamental voltage sample in an average sense and not in an instantaneous fashion. The difference between the instantaneous applied voltage vector and the instantaneous fundamental voltage vector is the instantaneous voltage ripple vector. The time integral of the voltage ripple vector gives the 'stator flux ripple vector', which is a measure of the ripple in line current [6]-[9], [12]. The stator flux ripple vector over a subcycle for above sequences for a given reference vector and the corresponding q-axis ripple and d-axis ripple are also shown in [8].

The values of q-axis and d-axis ripples are given in terms of $Q_Z$, $Q_1$, $Q_2$ and D, which are defined as in (7).

$$Q_Z = -V_{REF} T_Z \qquad (7.1)$$

$$Q_1 = (\frac{2}{3}V_{dc}Cos\alpha - V_{REF})T_1 \qquad (7.2)$$

$$Q_2 = (\frac{2}{3}V_{dc}Cos(60^0 - \alpha) - V_{REF})T_2 \qquad (7.3)$$

$$D = \frac{2}{3}V_{dc}Sin\alpha T_1 \qquad (7.4)$$

The expressions for RMS ripple over a subcycle are also given in [8] as a functions of $Q_Z$, $Q_1$, $Q_2$, D, $T_1$, $T_2$ and $T_s$. In the expressions the angle can be eliminated as follows:

In the conventional approach, the switching times can be defined as

$$T_1 = M * \frac{Sin(60^o - \alpha)}{Sin60^o} * T_s \qquad (8.1)$$

$$T_2 = M * \frac{Sin\alpha}{Sin60^o} * T_s \qquad (8.2)$$

$$T_Z = T_s - T_1 - T_2 \qquad (8.3)$$

From (8) and (5), the values of cos(α), cos(60°-α) and sin(α) can be obtained as given in (9), which are required for the estimation of stator flux ripple.

$$cos(\alpha) = \frac{(T_1 + 0.5 * T_2)}{M * T_s} \qquad (9.1)$$

$$cos(60^\circ - \alpha) = \frac{(0.5 * T_1 + T_2)}{M * T_s} \qquad (9.2)$$

$$sin(\alpha) = \frac{\sqrt{3} * T_2}{2 * M * T_s} \qquad (9.3)$$

Thus, by using the imaginary switching times concept, the angle can be eliminated in the mean square stator flux ripple expressions, which are given in [8].

### C. Proposed Hybrid SVPWM Method

The stator flux mean square ripple trajectory over a sixty degrees period for every sequence at different modulation indices are plotted and are given in Fig. 3.

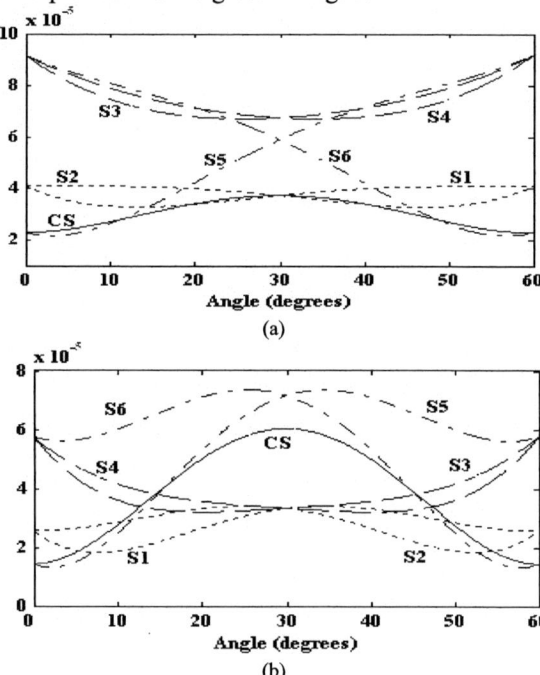

Fig. 3. Mean square flux ripple within a sector (a) for M =0.534 (b) for M=0.73.

By comparing the sequences 0127, 012, 721, 0121, 7212, 1012 and 2721 with respective to each other, the zones of superior performance of each sequence can be identified. The development of HSVPWM techniques for reduced current ripple involves determination of superior performance for every sequence. The zone of superior performance for a given sequence is the spatial zone within a sector where the given sequence results in less RMS ripple than other sequences considered. The superior performance of sequences in the space vector plane is shown in [6]-[9], [12]. Among the various HSVPWM methods, 7-zone HSVPWM method gives better performance [7]. The 7-zone HSVPWM method is shown in Fig. 4, and the proper selection of sequences within a sector gives reduced current ripple.

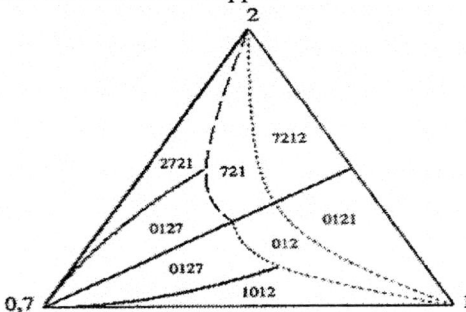

Fig. 4. 7-zone HSVPWM method.

### D. Sliding mode speed controller

The electromechanical equation of an induction motor is described as

$$J\frac{d\omega_m}{dt} + B\omega_m + T_L = T_e \qquad (10)$$

The electromechanical equation can be modified as

$$\dot{\omega}_m + a\omega_m + d = bT_e \qquad (11)$$

Where $a = \dfrac{B}{J}$, $b = \dfrac{1}{J}$ and $d = \dfrac{T_L}{J}$.

Now consider (11) with uncertainties as

$$\dot{\omega}_m = -(a + \Delta a)\omega_m - (d + \Delta d) + (b + \Delta b)T_e \qquad (12)$$

$\Delta a$, $\Delta b$ and $\Delta d$ represents the uncertainties of the terms $a$, $b$ and $d$ introduced by system parameters $J$ and $B$. Now let us define the tracking speed error further as

$$e(t) = \omega_m(t) - \omega_m^*(t) \qquad (13)$$

where $\omega_m^*$ is the rotor reference speed command in angular frequency. Taking derivative of (13) with respect to time yields

$$\dot{e}(t) = \dot{\omega}_m(t) - \dot{\omega}_m^*(t) = -ae(t) + f(t) + x(t) \qquad (14)$$

where the following terms have been collected in the signal $f(t)$,

$$f(t) = bT_e(t) - a\omega_m^* - d(t) - \dot{\omega}_m^*(t) \qquad (15)$$

and the $x(t)$, lumped uncertainty, defined as

$$x(t) = -\Delta a\omega_m(t) - \Delta d(t) + \Delta bT_e(t) \qquad (16)$$

Now, the sliding variable with integral component, is defined as

$$S(t) = e(t) - \int_0^t (h - a)e(\tau)d\tau \qquad (17)$$

where $h$ is a constant gain. Also in order to obtain the speed trajectory tracking, the following assumptions are made.

**Assumtion-1:** The $h$ must be chosen so that the term $(h-a)$ is strictly negative and hence $h<0$.

Then the sliding surface is defined as follows:

$$S(t) = e(t) - \int_0^t (h - a)e(\tau)d\tau = 0 \qquad (18)$$

Based on the developed switching surface, a switching control that guarantees the existence of sliding mode, a speed controller is defined as

$$f(t) = h\,e(t) - \beta sgn(S(t)) \qquad (19)$$

Where $\beta$ is the switching gain and $sgn(\cdot)$ is the sign function defined as

$$sgn(S(t)) = \begin{cases} +1, & \text{if } S(t) > 0 \\ -1, & \text{if } S(t) < 0 \end{cases} \qquad (20)$$

**Assumtion-2:** The gain $\beta$ must be chosen so that $\beta \geq |x(t)|$ for all time.

When the sliding mode occurs on the sliding surface (18), then, $S(t) = \dot{S}(t) = 0$ and the tracking error $e(t)$ converges to zero exponentially. Finally, the reference torque command $T_e^*$ can be obtained by substituting (19) in (15) as follows:

$$T_e^*(t) = \frac{1}{b}\Big[(h.e) - \beta\,sgn(S) + a\omega_m^* + \dot{\omega}_m^* + d\Big] \qquad (21)$$

## IV. SIMULATION RESULTS AND DISCUSSION

To verify the proposed scheme, a numerical simulation has been carried out by using Matlab/Simulink. Sampling time of 125μs and ode 4 (Runge-Kutta) methods are used for a fixed step size of 10μs. For the simulation, the reference flux is taken as 1wb and starting torque is limited to 15 N-m. The induction motor used in this case study is a 1.5 KW, 1440 rpm, 4-pole, 3-phase induction motor having the following parameters:

$R_s = 7.83\Omega$, $R_r = 7.55\Omega$, $L_s = 0.4751$H, $L_r = 0.4751$H, $L_m = 0.4535$ H, $J = 0.06$ Kg.m$^2$ and $B = 0.01$ N-m.sec/rad

The values, which have chosen for integral switching surface variable structure speed controller, are $h = -200$ and $\beta = 10$. Various conditions such as starting, steady state, step change in load and speed reversal are simulated. The results for CDTC are shown in Fig. 5 to Fig. 7.

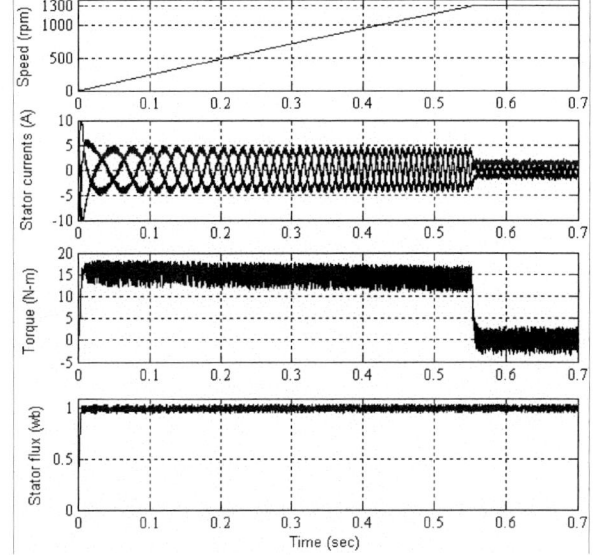

Fig. 5. Simulation results of CDTC: No-load Starting transients.

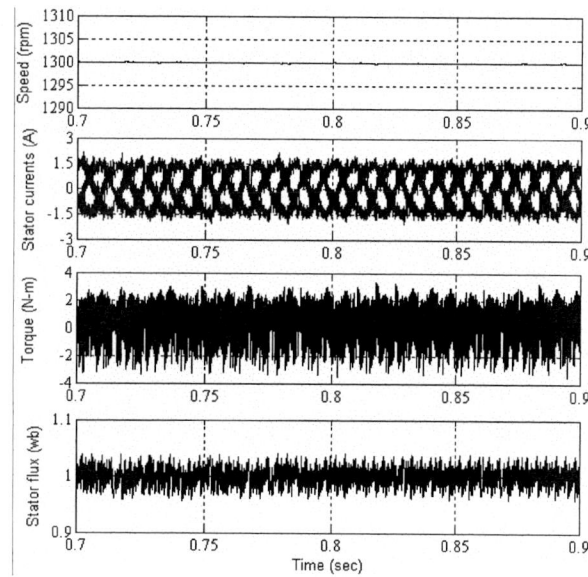

Fig. 6. Simulation results of CDTC: Steady state plots.

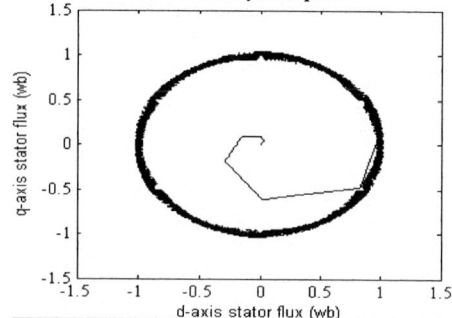

Fig. 7. Stator flux locus in CDTC.

From the simulation results of CDTC, it can be observed that, the ripple in torque, flux and current is more in steady state. To reduce these ripples and to get constant switching frequency, few HSVPWM methods are proposed in [6-9]. Among the proposed HSVPWM methods, 7-zone HSVPWM method gives less harmonic distortion in comparison with the remaining methods. Hence in this paper the simulation results are given for 7-zone HSVPWM based DTC only. These are given in Fig. 8 to Fig. 12.

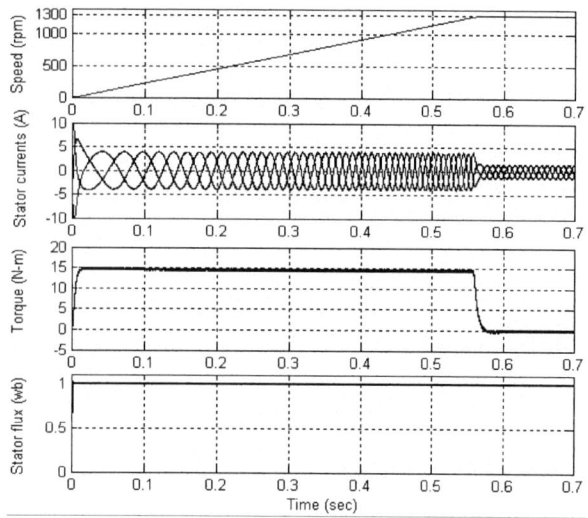

Fig. 8. Proposed DTC: No-load starting transients.

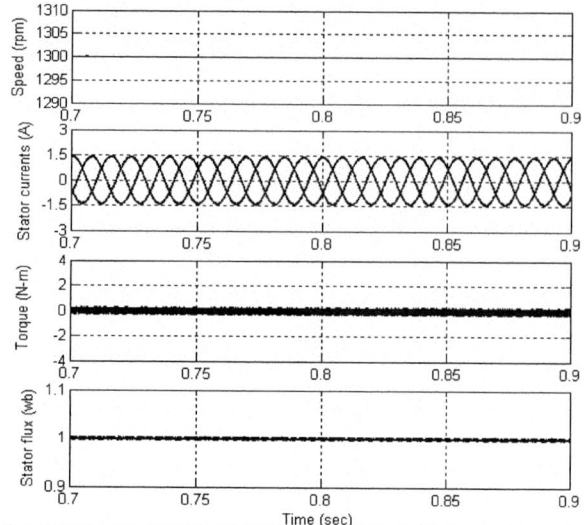

Fig. 9. Results of proposed DTC: No-load steady state plots.

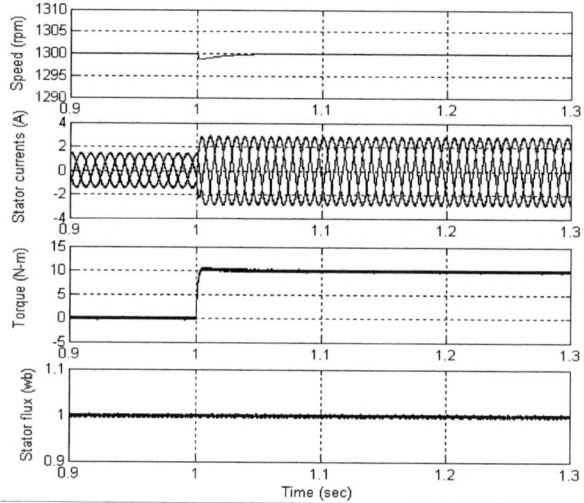

Fig. 10. Proposed DTC: Load torque of 10 N-m is applied at t=1 sec.

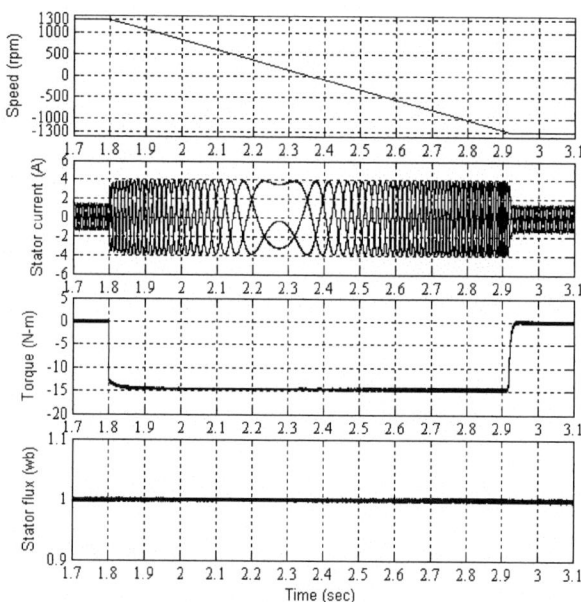

Fig. 11. Proposed DTC: Speed is changed from +1300 rpm to -1300 rpm.

428

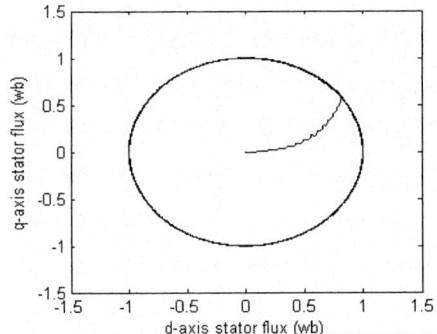

Fig. 12. Locus of stator flux in proposed DTC.

From Fig. 9, it can be observed that, the steady state ripple in torque, flux and current can be reduced with the proposed method compared to that of CDTC and hence acoustical noise can be reduced. Also to improve the speed performance of DTC under uncertainties caused by load torque changes, an integral switching surface variable structure speed controller is used. From Fig. 10 it can be observed that, though the load disturbance is added, the speed response is almost the same with the proposed speed controller. Thus, the speed tracking is not affected by the load torque. Hence, the proposed controller provides the robustness for the system.

## V. CONCLUSIONS

The CDTC gives a considerable ripple in steady state. In this paper, 7-zone HSVPWM method has been proposed by using the imaginary switching times to reduce the steady state ripple. From the simulation results, it can be observed that, with the proposed method, the ripple in steady state can be reduced with compared to that of CDTC. Also, the simulation results show that, the proposed speed controller is robust to the variations in load torque. As the bus-clamping sequences are being used in proposed strategies, the switching losses of the inverter can also be reduced. Hence with the proposed method, THD can be reduced. Also the execution time and memory required can be reduced.

## VI. REFERENCES

[1] Isao Takahashi, Toshihiko Noguchi, "A new quick-response and high-efficiency control strategy of an induction motor", *IEEE Trans Industrial Applications*, Vol.IA-22, No.5, pp. 820-827, Sep/Oct, 1986.

[2] P.Tiitinen, M.Surandra, "the next generation motor control method, DTC Direct torque control", *IEEE Proc on Power Electronics, Drives and Energy Systems for industrial growth*, Vol. 1, pp. 37-43, 1996.

[3] Domenico Casadei, Giovanni Serra, Angelo Tani, "Implementation of direct torque control algorithm for induction motors based on discrete space vector modulation" *IEEE Trans. Power Electronics*, Vol.15, No.4, pp. 769-777, July 2000.

[4] Thomas G. Habetler, et,al, "Direct Torque Control of Induction Machines Using Space Vector Modulation", *IEEE Trans. Industrial Applications.*, Vol. 28, No.5, pp. 1045-1053, September/October 1992.

[5] L.Tang, L.Zhong, M.F.Rahman, Y.Hu, "An investigation of a modified direct torque control strategy for flux and torque ripple reduction for induction machine drive system with fixed switching frequency" *IEEE proc. Ind. Applications society*, pp. 837-844, 2002.

[6] H. Krishnamurthy, G. Narayanan, V.T.Ranganathan, R. Ayyar , "Design of spacevector-based hybrid PWM techniques for reduced current ripple" *Proc. IEEE Applied Power Electronics Conference (APEC'2003)*, Vol.1, pp. 583-588, 2003.

[7] Di Zhao, G.Narayanan, Raja Ayyanar, "Switching loss characteristics of sequences involving active state division in space vector based PWM" *Proc. IEEE Applied Power Electronics Conference (APEC'2004)*, pp 479-485, 2004.

[8] Raja Ayyanar, D. Zhao, H. Krishna Murthy and G. Narayanan, "Space Vector Methods for AC drives to achieve high efficiency and superior waveform quality" Final technical report submitted to Office of Novel research, Arlington, VA 22217-5660, 2004.

[9] T. Brahmananda Reddy, et al, "Sensorless Direct Torque Control of Induction Motor based on Hybrid Space Vector Pulsewidth Modulation to Reduce Ripples and Switching Losses – A Variable Structure Controller Approach" *IEEE proc. Power India Conference, 2006*, paper no. 208, April, 2006.

[10] Dae-Woong Chung, Joohn-Sheok Kim, Seung-Ki Sul, "Unified Voltage Modulation Technique for Real-Time Three-Phase Power Conversion" *IEEE Trans. Ind Applic*, vol.34, No.2 pp 374-380, March/April, 1998.

[11] Arbind Kumar, "Direct Torque Control of Induction Motor Using Imaginary Switching Times with 0-1-2-7 & 0-1-2 Switching Sequences: A Comparative Study" the *30th Annual Conference of the IEEE IES*, pp 1492-1497, November 2-6, 2004.

[12] T. Brahmananda Reddy et.al, "New space vector based hybrid PWM techniques for AC drives without angle estimation to reduce current ripple", *PCEA-IFToMM International conference*, PICA 2006, Nagpur, India, paper no. TG-328, July 11-14, 2006.

[13] Vadim I. Utkin, "Sliding Mode Control Design Principles and Applications to Electric Drives" *IEEE Trans. on Industrial. Electronics*, vol. 40, no. 1, Feb 1993.

[14] Oscar Barambones, Aitor J. Garrido, Francisco J. Maseda, "A Robust Field Oriented Control of Induction Motor with Flux Observer and Speed Adaptation" *Proc. IEEE-ETFA*, pp. 245-252, 2003.

## VII. BIOGRAPHIES

**T. Brahmananda Reddy** graduated from Sri Krishna Devaraya University, Anantapur in the year 2001, M.E from Osmania University, Hyderabad in the year 2003. He is currently pursuing Ph.D at Jawaharlal Nehru Technological University, Hyderabad, India. He is presently working as Assistant Professor in the Electrical and Electronics Engineering Department at G. Pulla Reddy Engineering College, Kurnool, Andhra Pradesh, India.

His research areas include PWM techniques, and control of electrical drives.

**Dr. J. Amarnath** graduated from Osmania University in the year 1982, M.E from Andhra University in the year 1984 and Ph.D from J.N.T.University, Hyderabad in the year 2001.

He is presently Professor and Head of the Electrical Engineering Department, JNTU College of Engineering, Hyderabad. He presented more than 60 research papers in various national and international conferences and journals. His research areas include Gas Insulated Substations, High Voltage Engineering, Power Systems and Electrical Drives.

**Dr. D. Subba Rayudu** received B.E degree in Electrical Engineering from S.V. University, Tirupati, India in 1960, M.Sc (Engg) degree from Madras University in 1962 and Ph.D degree from Indian Institute of Technology, Madras, India in 1977.

He is at present working as professor in Department of Electrical Engineering at G. Pulla Reddy Engineering College, Kurnool, India. His research interests include Electrical machines, Power Electronics and electrical drives.

**2006 IEEE International Conference on Power Electronic, Drives and Energy Systems**

# Analysis of Experimental Investigation of Various Carrier-based Modulation Schemes for Three Level Neutral Point Clamped Inverter-fed Induction Motor Drive

Ranjan K. Behera, *Student Member, IEEE*, T. V. Dixit,and Shyama P. Das, *Senior Member, IEEE*

*Abstract--*The harmonic analysis of the stepped waveform of the three-level inverter, based on number of triangular carriers and a sinusoidal modulating signal has been described. Analytical solutions of different multilevel Pulse Width Modulation (PWM) strategies for three-level Neutral Point Clamped (NPC) has been presented. Analysis of the Total Harmonic Distortion (THD) of voltage and stator current has been carried out by simulation study. These solutions show that In-Phase Sinusoidal Pulse Width Modulation (IPSPWM) strategy shows better spectral performance compared to the Phase Opposite Sinusoidal Pulse Width Modulation (POSPWM) and dipolar modulation techniques. This is because of the fact that it places significant harmonic energy in to main carrier component and relies on the cancellation of this component between phase legs when the line-to-line voltages are formed. All these techniques are applied to an induction motor drive for comparison. Torque ripple is also estimated for different techniques. A laboratory model is developed to verify validity of different modulation schemes.

*Index Terms--*Carrier based pulse width modulation, diode-clamped inverter, multilevel converter, multilevel inverter, and THD.

## I. INTRODUCTION

MULTILEVEL converters are being used for high voltage and high power industrial applications. In high power drives application, the most popular topology used is the three level three phase diode clamped converter (also known as neutral point clamped (NPC) converter) topology, due to serial power switch connection that reduces the device voltage stress [1]. This topology produces three level step line voltages; hence the input current harmonics are reduced.

Theoretically diode clamped topology with any number of levels can be visualized. But some of the problems like unbalanced voltages across capacitors, voltage clamping

Ranjan K. Behera is with Department of Electrical Engineering, Indian Institute of Technology, Kanpur-208016, INDIA (e-mail: ranjanee@iitk.ac.in).

T. V. Dixit is with Department of Electrical Engineering, Indian Institute of Technology, Kanpur-208016, INDIA (e-mail: tvdixit@iitk.ac.in).

Shyama P. Das is with Department of Electrical Engineering, Indian Institute of Technology, Kanpur-208016, INDIA (e-mail:-spdas@iitk.ac.in)

requirements, complexity of switching algorithm, circuit layouts and package constraints have limited the level in practical multilevel converter to seven [2]. In this paper we have considered a three level diode clamped topology for the convenience of fabrication and experimentation in the laboratory.

Multilevel PWM inverters have been developed to overcome shortcomings in solid-state switching device ratings, so that large motors can be controlled by high-power adjustable-frequency drives. Digital FFT analysis uses a sampled representation of the switched waveform, and this requires very large number of sampled points to maintain acceptable accuracy, if the PWM carrier frequency is above several kHz. Therefore, there is always the possibility of a subtle simulation programming error, which slightly degrade the results [3]. Clearly, it would be very useful to be able to determine the exact theoretical harmonic performance of the various PWM strategies that are now regarded as standard without relying on simulation results.

Three phase three level NPC converters controlled by PWM have a wide range of applications for dc-to-ac power supplies and ac machine drives. Important quantities to be considered with machine loads are the distribution of current densities and flux linkages in ac machine windings. Performance criteria will be then introduced to enable the evaluation and comparison of different PWM technique [4]. This paper examines commonly used Sinusoidal Pulse Width Modulation (SPWM) techniques for three-level inverter fed induction motor drive. The comparison is on the basis of THD in output voltage and current, as this strongly affects torque ripple, current ripple and stator flux ripple in an ac drive. These techniques are attractive for industrial applications; they are easily implemented and produce good results for moderate switch frequencies. The conventional switching scheme for two-level can be easily extended to multi-level converters. In particular, the limitation and estimation of harmonics of these techniques are examined for three-level NPC inverter fed induction motor.

The three level diode clamped converter has two capacitors ($C_1$ and $C_2$) in series with a center tap at O as shown in Fig. 1a. Each phase of the three level converter has two pair of switching device ($S_{i1}$, $S_{i2}$) and ($S_{i3}$, $S_{i4}$) in series, where i =a, b, c. The center of each pair is clamped to the neutral of the

0-7803-9771-1/06/$25.00 ©2006 IEEE

center-tapped capacitor through clamping diodes ($D_1$, $D_2$, $D_3$, $D_4$, $D_5$ and $D_6$). Fig. 1a and Fig. 1b show a three level inverter with an induction motor load and the switching states respectively.

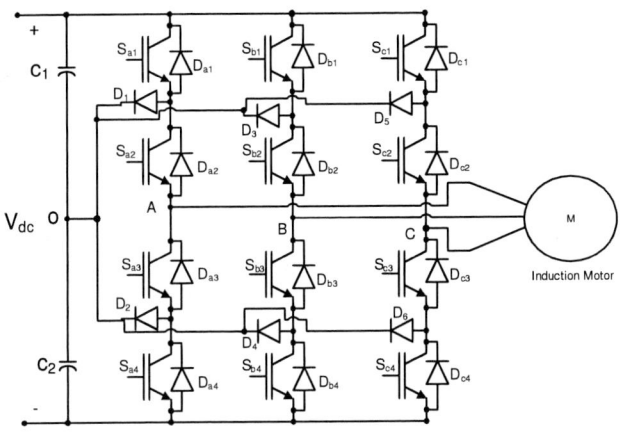

Fig. 1a.  Three phase three-level NPC inverter.

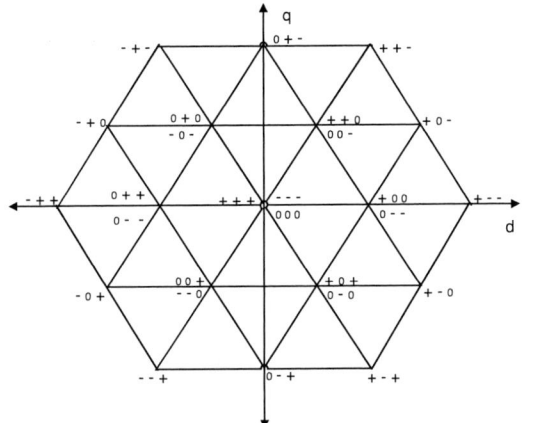

Fig. 1b.  Switching states of three-level inverter.

## II. MODULATION TECHNIQUES

In naturally sampled PWM, the switching instants are generated by the intersection of a triangle carrier wave with the input reference waveform. The input waveform is naturally sampled by the carrier wave, the sampling instant occurring at the same instant as the output edge. SPWM occurs when the carrier or switch frequency is a multiple of the input signal's fundamental frequency and is the usual practice for a low-pulse number $m_f = f_c/f_r$. For natural SPWM, the spectrum consists of the input signal $f_r$, the carrier $f_c$ and its harmonics $mf_c$, and their associated sidebands $mf_c \pm nf_r$. The amplitudes of these spectral terms are given by Bessel functions and are functions of modulation depth $m_a$ (the amplitude of the modulating signal $f_r$) as well as frequency [5]. A limit is, however, imposed on the input signal's slew rate if multiple switching edges per switch cycle are to be avoided. The slope of the input signal should not exceed the slope of the triangular carrier wave by remaining within the boundary $m_a f_r < 2/\pi\ f_c$ [6].

### A. Subharmonic multilevel PWM methods

For N-level inverter, N-1 carrier with same frequency $f_c$ and same peak-to-peak amplitude $A_c$ are disposed such that the bands are contiguous. The reference or modulation waveform has peak-to-peak amplitude $A_m$ and frequency $f_r$, and it is centered in the middle of carrier set. The reference is continuously compared with each of the carrier signals. If the reference is grater then the carrier signal, then the active device corresponding to that carrier is switched on; and if reference is less than the carrier signal, then the active device corresponding to that carrier is switched off.

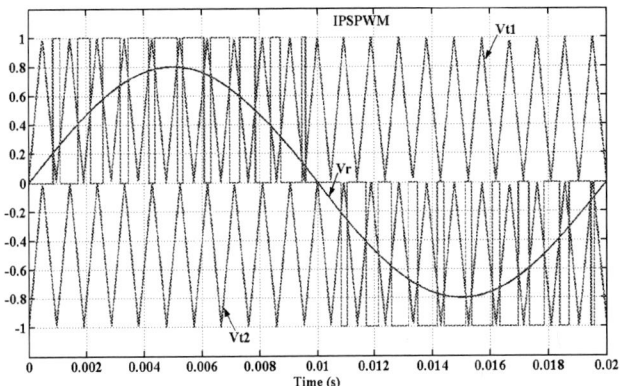

Fig. 2a.  IPSPWM technique.

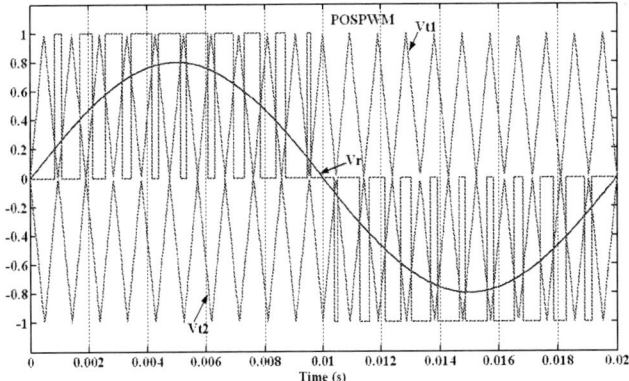

Fig. 2a.  POSPWM technique.

$$m_a = \frac{A_m}{(N\_-1) * A_c} \quad \text{and} \quad m_f = \frac{f_c}{f_r}$$

In this method, two-carrier signal ($V_{t1}$ and $V_{t2}$) are compared with a single reference signal ($V_r$). The carrier signal can be IPSPWM or POSPWM as shown in Fig. 2a and Fig. 2b respectively. In both approach the switching logic is decided as follows. In this type of SPWM switching will take place between state +1 and state 0 during positive half cycle of the fundamental and between -1 and state 0 in the negative half cycle of the fundamental. A set of carriers ($m_f = 21$) with all carriers are in phase, phase opposition and dipolar modulation for 3-level inverter are shown in Figs. 2a-2b and 2c respectively.

## B. Dipolar modulation

This method employs two modulation or reference waveforms (upper waveform $V_{r1}$ and lower waveform $V_{r2}$) and a single carrier signal as shown in Fig. 2c. The two modulating signals are obtained by adding positive or negative offset to the conventional modulating signals. The offset may be uniform [7] or variable [8] in magnitude.

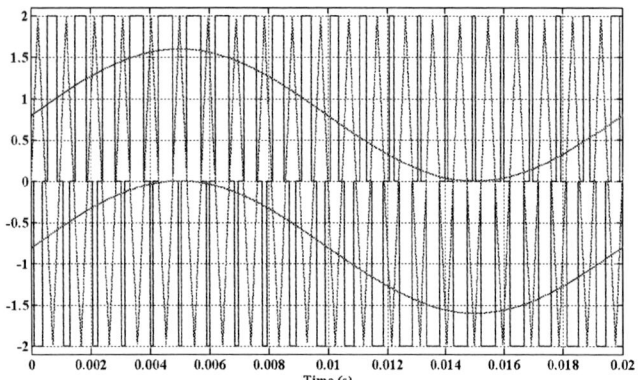

Fig. 2c. Dipolar modulation technique.

## III. SPECTRAL ANALYSIS OF CARRIER BASED PWM MODULATION STRATEGIES

In principle, any time varying waveform can be described by an infinite series of harmonic components, but in practice the non-periodic nature of PWM switched waveform makes the determination of these components difficult. The function is periodic in both dimensions even if the switched waveform is not periodic, and hence can be expressed in general harmonic form as a double Fourier series in (2). The frequency components are obtained by evaluating the double integration of (1) for a particular PWM strategy [9].

$$F(t) = \frac{A_{00}}{2} + \sum_{n=1}^{\infty} \{A_{0n}\cos(n\omega_r t) + B_{0n}\sin(n\omega_r t)\} + \sum_{m=1}^{\infty} \{A_{m0}\cos(n\omega_c t) + B_{m0}\sin(n\omega_c t)\} \quad (1)$$
$$+ \sum_{m=1}^{\infty} \sum_{\substack{n=-\infty \\ n \neq 0}}^{\infty} \{A_{mn}\cos(m\omega_c t + n\omega_r t) + B_{mn}\sin(m\omega_c t + n\omega_r t)\}$$

$$C_{mn} = A_{mn} + jB_{mn} = \frac{1}{2\pi^2}\int_{-\pi}^{\pi}\int_{-\pi}^{\pi} F(x,y)e^{j(mx+ny)}\,dxdy, \quad x = \omega_c t, \ y = \omega_r t \quad (2)$$

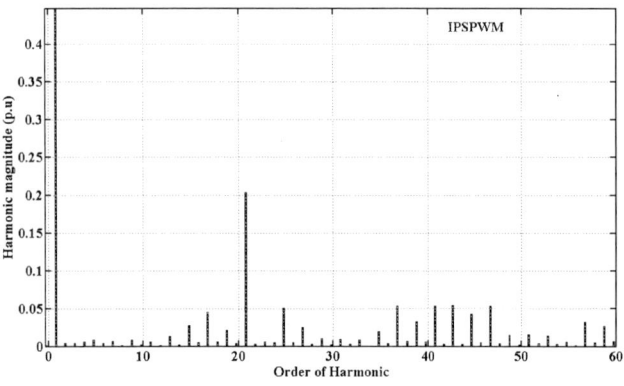

Fig. 3a. Spectrum of phase voltage for IPSPWM.

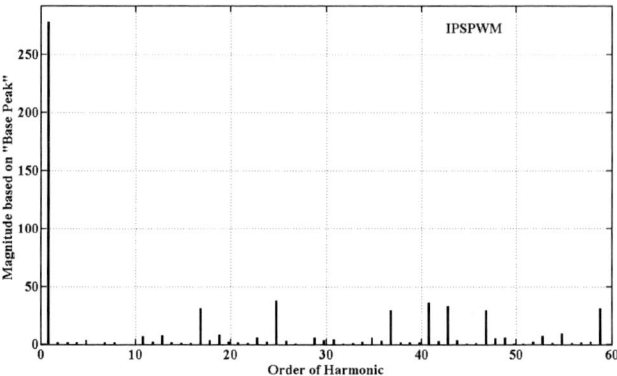

Fig. 3b. Spectrum of line voltage for IPSPWM.

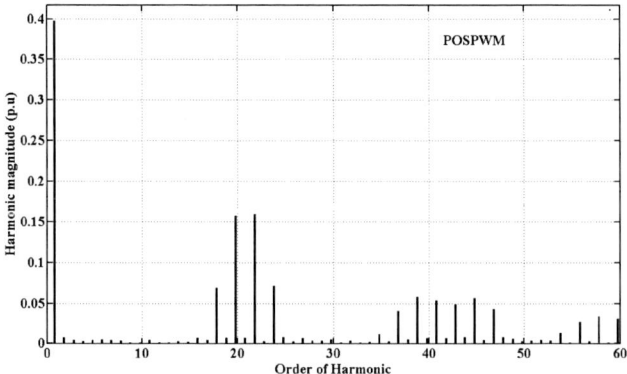

Fig. 4a. Spectrum of phase voltage for POSPWM.

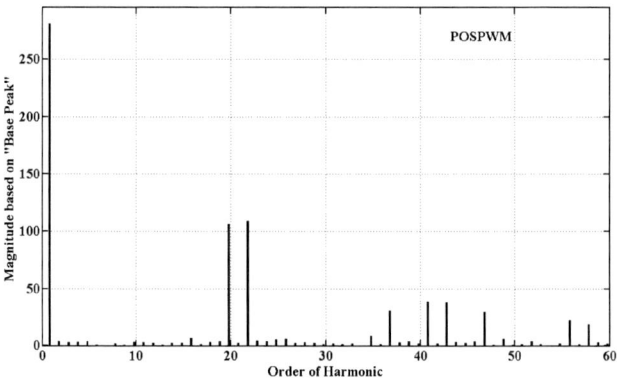

Fig. 4b. Spectrum of line voltage for POSPWM.

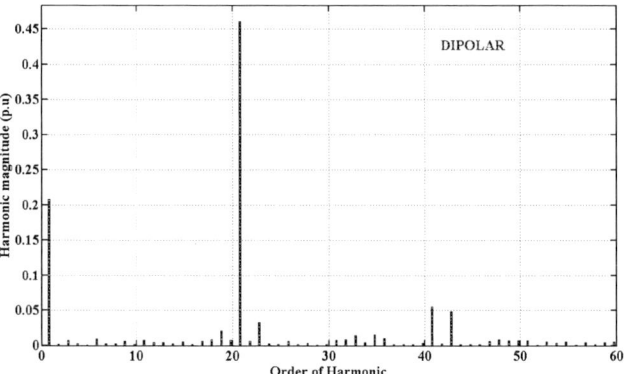

Fig. 5a. Spectrum of phase voltage for dipolar modulation.

432

To evaluate (2), the function f(x,y) must be properly defined in the x,y plane. In a three-level inverter the values that the function f(x,y) can take is determined by the number of levels (+$V_{dc}$, 0, -$V_{dc}$) of the converter. The complete harmonic solution for thee level IPSPWM of phase leg-a is given in (3). The time-varying switched phase leg voltage $v_{ao}(t)$ can be expressed in term of its harmonic components as [10] shown here.

$$V_{ao}(t) = V_{dc} M \cos(\omega_r t) + \frac{8V_{dc}}{\pi^2} \sum_{m=1}^{\infty} \frac{1}{2m-1} \sum_{k=1}^{\infty} \frac{J_{2k-1}([2m-1]\pi M)}{[2k-1]} \cos([2m-1]\omega_c t) \quad (3)$$

$$+ \frac{2V_{dc}}{\pi} \sum_{m=1}^{\infty} \frac{1}{2m} \sum_{n=-\infty}^{\infty} J_{2n+1}(2m\pi M) \cos n\pi \cos(2m\omega_c t + [2n+1]\omega_r t)$$

$$+ \frac{8V_{dc}}{\pi^2} \sum_{m=1}^{\infty} \frac{1}{2m-1} \sum_{\substack{n=-\infty \\ (n \neq 0)}}^{\infty} \sum_{k=1}^{\infty} \frac{J_{2k-1}([2m-1]\pi M)[2k-1]}{[2k-1+2n][2k-1-2n]} \cos([2m-1]\omega_c t + 2n\omega_r t)$$

Phase and line-to-line voltage spectrum plots of different multilevel PWM method for three level inverter for conditions of M=0.8, $f_c/f_r = 21$ are shown in Figs. 3a-5b.

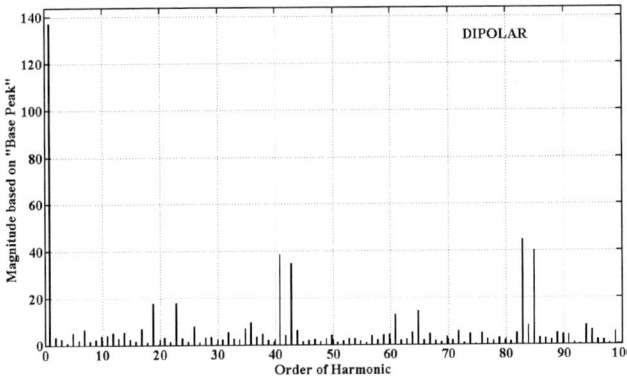

Fig. 5b. Spectrum of line voltage for dipolar modulation.

From the above spectral analysis, Fig. 3a shows the harmonic spectra for one phase leg of NPC inverter under IPSPWM, where the most significant harmonic is the first carrier component. This is in sharp contrast to POSPWM which generates carrier sideband components, as shown in Fig. 4a. In dipolar modulation strategy it is observed that first carrier component is very significant and generates more voltage harmonics as compare to IPSPWM and POSPWM. Clearly, IPSPWM places significant harmonic energy in to a carrier component for each phase leg, and relies on common mode cancellation between the inverter phase legs to eliminate this carrier from the final line-to-line output voltage, as shown in Figs. 3b and 4b. Consequently, the harmonic sidebands have less energy. So IPSPWM shows improve performance as compare to other two modulation strategy. The performance of total harmonic distortion of line-to-line voltage is shown in Fig.6 for varying modulation depth. This shows that as modulation depth increases the THD reduces.

## IV. DISCUSSION ON SIMULATION AND EXPERIMENTAL RESULTS WITH INDUCTION MOTOR LOAD

Optimal utilization of the machine and the power converter, and requirement of high power density with good efficiency are important factor for large industrial and traction drive. The revolving field of induction machine should be free from other components other than the fundamental. This necessitates the reduction of lower order harmonics of the stator current [11]. Different PWM techniques play an important role in satisfying above demand. The switching harmonics are suppressed to a large extent by the low pass characteristics of machine inductances, and by the inertia of the mechanical system.

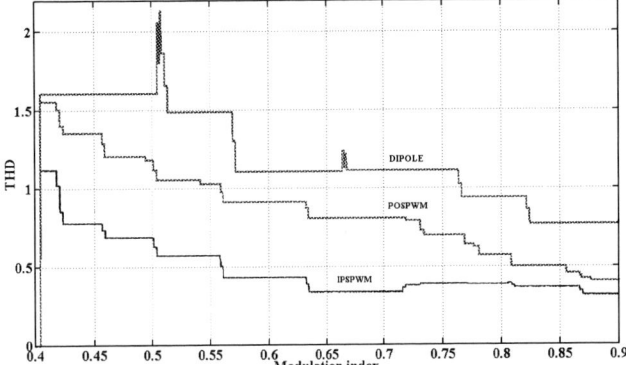

Fig. 6. Voltage THD versus modulation depth.

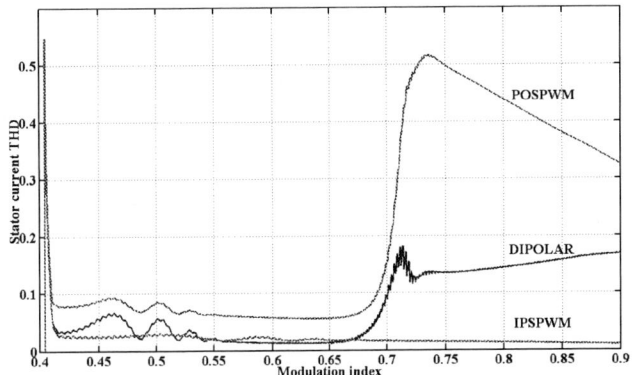

Fig. 7. Stator current THD versus modulation depth.

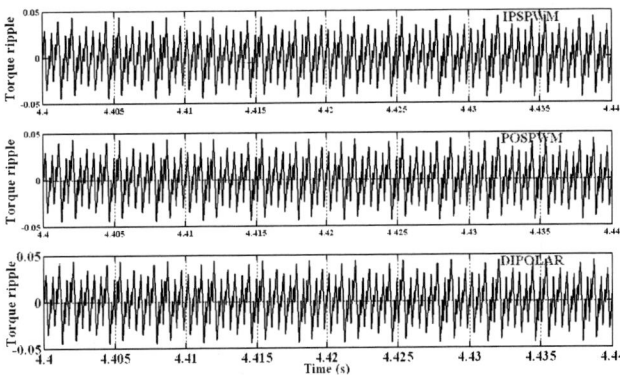

Fig. 8. Torque ripple of the induction motor load.

In the induction machine load, harmonic currents basically determine the cupper losses of the machine, which account for a major portion of the machine losses. Main objective of modulation scheme is to reduce current harmonic. In addition, the performance also depends on machine parameters. To validate the performances of the proposed switching technique for IPSPWM, POSPWM and dipolar modulation, extensive

433

computer simulations are conducted using Matlab/simulink. All the modulating techniques are simulated for $m_f$ =21 and $m_a$ =0.8. It is observed that the total harmonic distortion of stator current is minimal in case of IPSPWM technique. Fig. 7 shows stator current THD performance for different modulation depth. It is observed that during low modulation depth IPSPWM shows more THD as compared to other modulating techniques. However as modulation depth increases, current THD reduces and shows minimum current THD compared to other two techniques. Dipolar modulating technique shows moderate current THD and more voltage THD, but switching frequency is double compared to other two techniques. Although torque harmonics are produced by the harmonics current, there is no stringent relationship between them. Lower torque ripple can go along with higher current harmonics, and vice versa [4]. It is observed that when switching frequency increases torque ripple reduces, however dynamic performance becomes sluggish due to open loop structure. But a closed loop scheme can be incorporated to improve dynamic performance [4]. Fig. 8 shows the torque ripple for different modulating techniques. It is observed that for all scheme percentage of torque ripple is approximately same. It is also observed that when the switching frequency and bandwidth of three-level inverter increase the torque ripple reduces [13], although the torque ripple reduction is not very significant.

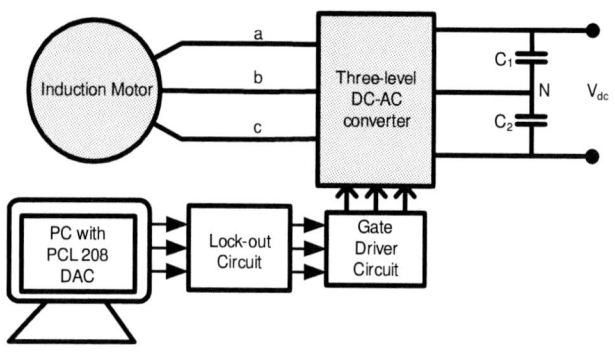

Fig. 9. Block diagram of PC based hybrid system implementation.

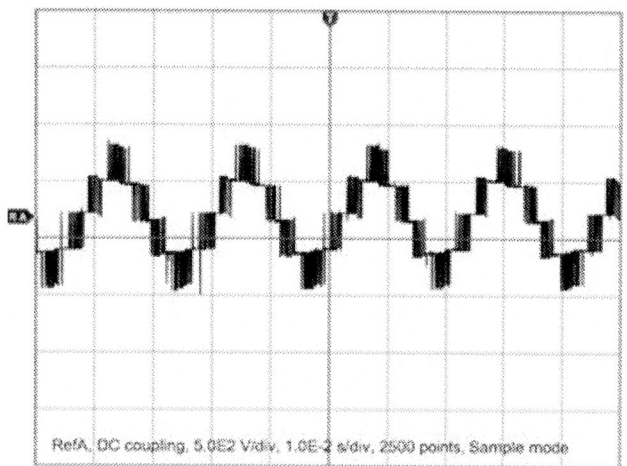

Fig. 10. Line-to-line voltage with induction motor load.
Ref: A: Line to line voltage: 100 V/Div.

The experimental setup for three-phase three-level inverter fed 1.1 kW induction motor drive is shown in Fig. 9. The three-level converter is comprised of six fast recovery power diodes (RM25HG-24S, 1200V, 25A) and six dual IGBT modules (CM50DU-24F, 1200V, and 50A).

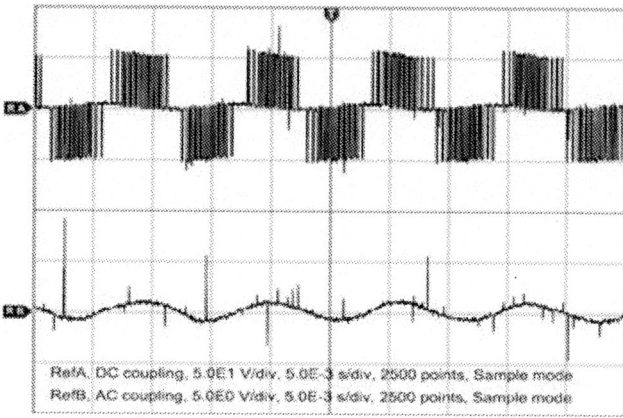

Fig. 11. Line-to-line voltage and current during low modulation depth.
Ref: A: Line to line voltage: 100 V/Div.
Ref: B: Stator phase current: 5.0 A/Div.

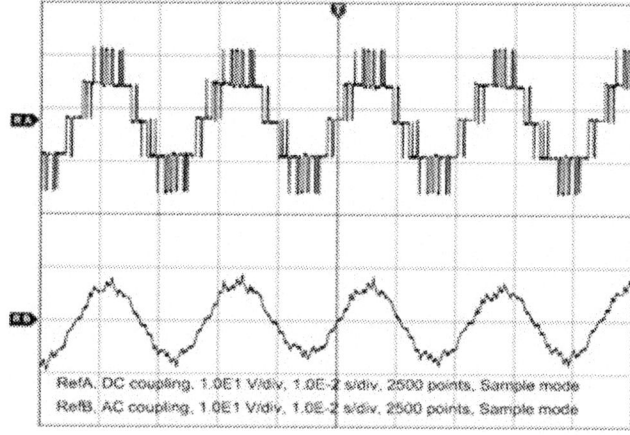

Fig. 12. Line-to-line voltage and current of IPSPWM.
Ref: A: Line to line voltage: 100 V/Div.
Ref: B: Stator phase current: 3.0 A/Div.

The induction motor parameters are given in Table-1. The input capacitor bank has a rating of 2200pF, 400V DC. Selected experimental results for IPSPWM are given in Figs. 10-12 for the three-phase three-level inverter of Fig. 1. Experimentation has been carried out for $m_f$ =21 and $m_a$ =0.4 and 0.8 respectively. Fig. 10 gives line-to-line voltage of dc-ac converter system. Fig. 11 shows the line-to-line voltage during low modulation index and phase current of the induction motor drive. It is observed that the voltages across each capacitor are balanced and less ripple in the stator current. Fig. 12 shows the line-to-line voltage of IPSPWM techniques for $f_c$ =11 and phase current of the induction motor drive. As the switching frequency increases the current ripple decreases. Fig. 13 shows the line-to-line voltage of dipolar PWM techniques for $f_c$ =11 and phase current of the induction motor drive.

434

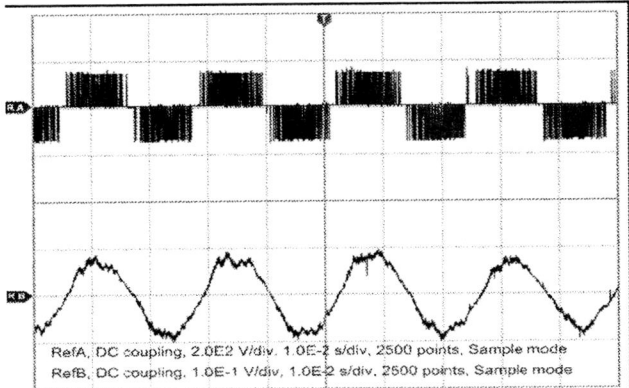

Fig. 13. Line-to-line voltage and current of dipolar PWM.
Ref: A: Line to line voltage: 200 V/Div.
Ref: B: Stator phase current: 3.0 A/Div.

TABLE I

INDUCTION MOTOR RATINGS AND PARAMETERS

(REFERRED TO STATOR)

| DC link capacitor $C_1=C_2$ | 2200 μF, 400V |
|---|---|
| Power | 1.1kW |
| Voltage (*L-L*) | 415 V |
| Current | 2.6 A |
| Frequency | 50 Hz |
| Rated Torque | 4.5 N-m |
| Speed | 1440 rpm |
| Stator Resistance ($R_s$) | 1.00 Ω |
| Rotor Resistance ($R_r$) | 2.256 Ω |
| Stator Leakage Inductance ($L_{ls}$) | 0.03195 H |
| Rotor Leakage Inductance ($L_{ls}$) | 0.03195 H |
| Mutual Inductance between stator and rotor ($L_m$) | 0.57192 H |

## V. CONCLUSION

Generalized analytical solutions are presented for fixed carrier frequency PWM methods for a three-level neutral point clamped (NPC) inverter. Analytical solutions are provided for three major PWM strategies, and have been verified against FFT analysis of simulated switched waveforms for selected conditions. Using the analytical solutions for different PWM method of three-level NPC inverter, it has been identified that IPSPWM has a superior spectral performance compared to other two modulation schemes because it places significant harmonic energy into main carrier component and relies on the cancellation of this component between phase legs when the line-to-line voltages are formed. Analytical study for torque ripple is also carried out for an induction motor load. A laboratory model is developed to validate the superior performance of IPSPWM modulation scheme.

## VI. REFERENCE

[1] A. Nabae, I. Takahashi, and H. Akagi, "A new neutral-point-clamped PWM inverter," *IEEE Trans. Ind. App.,* vol. IA-17, pp. 518–523, Sept./Oct. 1981.

[2] S. Chattopadhyay, A. R. Beig and V. Ramanarayanan, "An Input voltage Sensorless Input Current Shaping Method for Three Level Three Phase High Power Factor Boost Rectifier," in *Proc. of APEC'02,* pp. 2110-2116, 2002.

[3] D. G. Holmes, "A general analytical method for determining the theoritical harmonic component of carrier based PWM strategies,"in *Conf. Rec. IEEE-IAS Annu. Meeting,* 1998, pp. 1207-1214.

[4] J. Holtz, " Pulsewidth Modulation for Electronic Power conversion," *Proceedings of the IEEE,* Vol. 82, No. 8, Aug. 1994, pp. 1194-1214.

[5] H. S. Black, *Modulation Theory,* D. Van Nostrand Company, Inc, Princeton, 1960.

[6] L. M. Tolbetr, F. Z. Peng, and T. G. Habetler, "Multilevel PWM methods at low Modulation Indices," in *APEC'99,* Dallas, Texas, March 14-18, pp. 1032-1039.

[7] B. Velaerts, P. Mathys and E. Tatakis, "A Novel Approach to the Generation and Optimization of three level PWM waveforms," in *Proc. IEEE PESC'88. Conf.,* 1988, pp. 1255-1262.

[8] B. Velaerts, and P. Mathys, "New Developments of three level PWM Strategies,"in *Proc. EPE'89. Conf.,* 1989, pp. 411-416.

[9] B.P.McGrath and D.G. Holmes, " A comparision of multicarrier PWM starategies for cascaded and neutral point clamped Multilevel Inverter," in *Proc. IEEE PESC2000,* 2000, pp. 674-679.

[10] D.G. Holmes and T. A. Lipo, *"Pulse Width Modulation For Power Converters Principles and Practice,* IEEE press series on Power Engineering, IEEE Press, Wiley-Interscience, Piscataway, 2002.

[11] G. K. Dubey and C R. Kasarabada, *Power Electronics and Drives, IETE Book Series.* Vol. I. Tata-Mc Graw Hill, 1993.

[12] Y. Liu, X. Wu, and L. Huang, "Implementation of three-level inverter using a novel space vector modulation algorithm," in *IEEE Conference Proc. on Power System Technology,* vol. 1, pp.606-610, Oct. 2002.

[13] G. Walker and G. Ledwich, "Bandwidth Considerations for Multilevel Converters," *IEEE Trans. Power Electron.,* vol. 14, no. 1, pp. 74-81 Jan.1999.

**2006 IEEE International Conference on Power Electronic, Drives and Energy Systems**

# High Frequency SMPS Based Inverter With Improved Power Factor

### M. G. Wani, V. K. Sharma, and K. M. Soni

*Abstract:-- This paper presents a novel 12 V dc to 230 V, 50 Hz, 200 watt system offering improved power factor at utility lines. The details of electronic system along with ferrite core design are provided. Experiences of experimental details are discussed. At higher power, however, much more precautions and care is needed for circuit and component layout, PCB design, shielding etc.*

*Index Terms— Improved power factor, Inverter, SMPS*

## I. INTRODUCTION

MAJOR chunk of the power requirement is met using storage batteries and dc to ac converters. Presently 5@Hz/23@V power generation is met using magnetic core transformer based on metals and metal alloys. However these are heavy in weight, bulky in volume and give rise to weighty boxes and overall much higher volume. Additionally metal transformer has inherent loss factor of about 1@%.

To overcome these problems, use of high frequency ferrite based transformer less system has become the standard norm of industrial power supply system. This leads to weight reduction by one fifth of conventional inverters, one third of volumetric reduction, 5 to 1@% improved efficiency (and hence equally longer back up), and overall cost reduction by almost 25 to 35 %. This will lead to technology boost in continual reduction in ferrite and ferrite prices, which will provide very good scope for further price reduction in future.

This paper presents a novel 12 V dc to 23@V 5@Hz, 2@@ watt system offering improved power factor at utility lines. The same design with some modification can easily be adopted up to 2KW power levels. At higher power, however, much more precautions and care is needed for circuit and component layout, PCB design, shielding etc.

An advance and compact power factor correction system [1] using international rectifier make PF correction chip IR 115@ is also incorporated in the proposed scheme. This simple eight-pin IC with fewer components effectively counters low power factor due to inductive loads, which is a commercial requirement in 9@to 95% applications. It uses step up conversion topology and operates at 1@@ kHz

---

M. G. Wani is with Maharashtra Academy of Engineering, Alandi, Pune, India (e-mail: mgwanisir@rediffmail.com ).
Currently V.K. Sharma is with the Maharashtra Academy of Engineering, Alandi, Pune, India (e-mail: viren_krec@yahoo.com ).
K.M. Soni is with Galgotia College of Engineering & Technology, Greater Noida, India (email: kmsoni@yahoo.com).

operating frequency [2], which further reduces the system size, and effective corrected power factor comes to almost @998.

The schematic block diagram of the proposed unity power factor SMPS inverter is shown in Fig. 1. Apart from the battery source, it has the push-pull converter operating at 5@khz, a rectifier with voltage sensor & current sensor for over load protection. The next stage has the power factor correction circuit, which receives the feedback from the ac supply generated. The next stage has a full bridge converter with 5@Hz PWM control. Output of this converter is given to the active wave filter, which is a low pass filter with CMR. With this configuration, the PWM pulses are [3] generated such that the switching of the full bridge converter leads to near sinusoidal supply, which is in phase with the supply voltage, thereby leading to almost unity power factor operation.

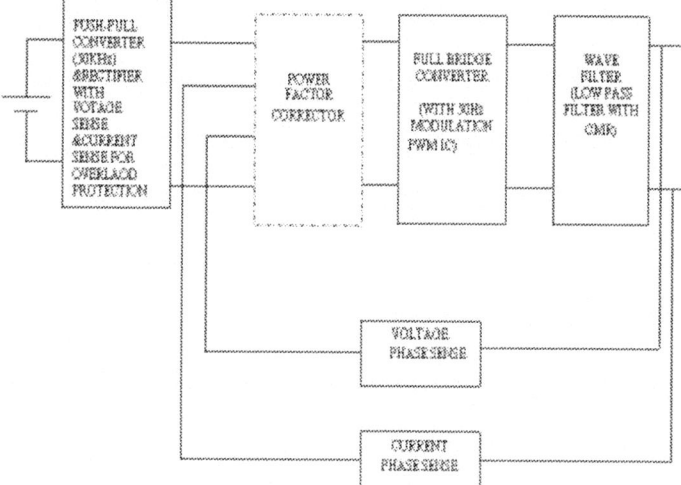

Fig. 1 Schematic of proposed SMPS with power factor correction.

## II. SYSTEM SPECIFICATIONS

The specifications of the high frequency SMPS system developed is given below.

### A. Electrical Specifications

The inverter should work on an input of 12V battery with voltage varying from varies from 1@13.6V. The output available should be 21@22@V, 5@Hz, Quasi- Sine Wave. The output current limit is 1A and wattage is 2@@W. The system should have an efficiency of 95% with an improved power factor up to 92%.

---

0-7803-9771-1/06/$25.00 ©2006 IEEE

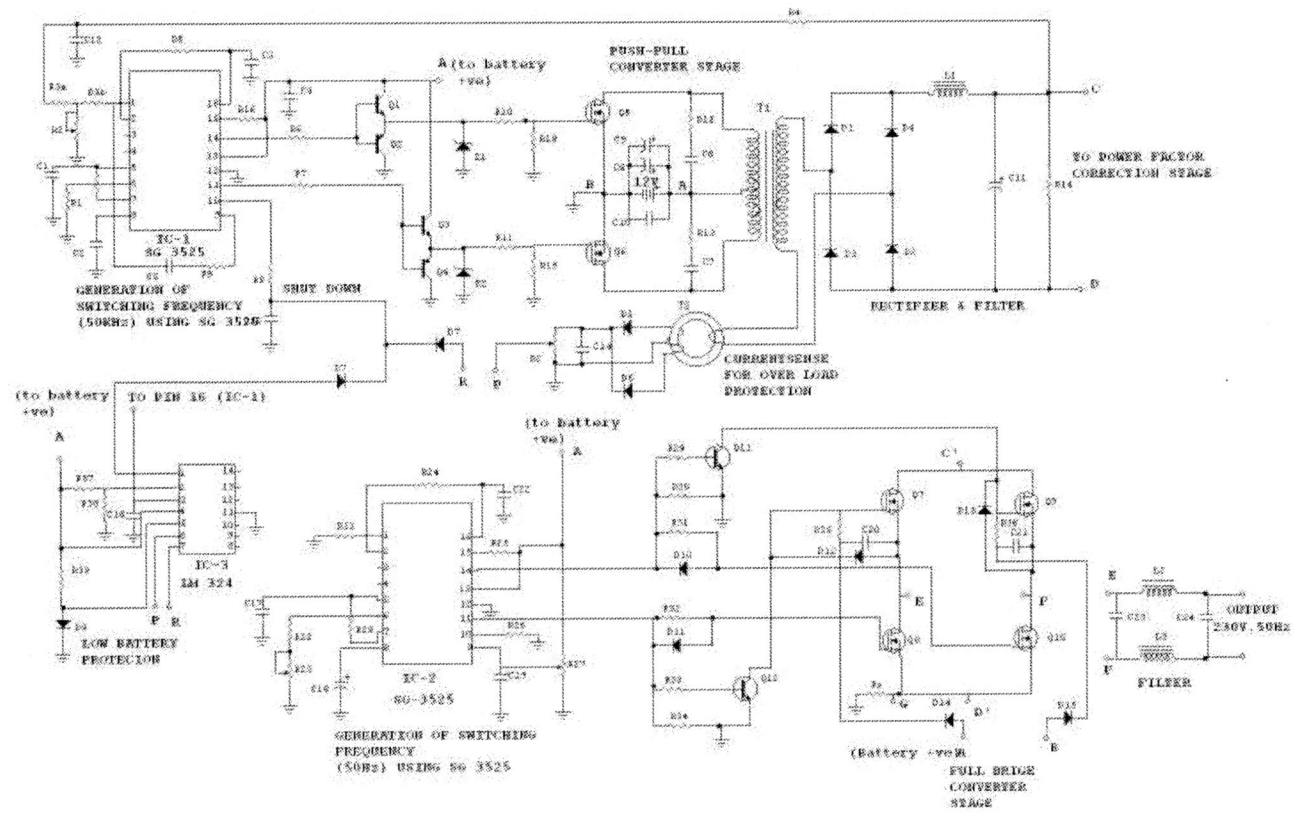

Fig. 2 SMPS based high frequency inverter.

## B. Salient Features

The salient features of the system are that it is extremely low weight than conventional, compact, better and efficient, economical by 3@% of reasonable scale production, power factor ranging between @95 to @998, short circuit, overload protection, and low battery protection

## III. SPECIFICATIONS OF MAJOR COMPONENTS

The specifications of other major components used in the design are described in this section.

Switching is carried out through SG- 3525 Pulse width modulator which operates on supply voltage of 8 to 35 V, oscillator frequency range @1 to 4@@KHz, output sink or source current of ± 1@@mA, and pulse-by-pulse Shutdown.

For power factor control, the scheme makes use of IR make 115@S PFC which is a Continuous Conduction Mode (CCM), one cycle control PFC IC. The switching frequency is in the range of 5@KHz to 2@@KHz with peak gate drive of 1.5 A.

The amplification is carried out through LM 324 Single Supply Quad Operational Amplifier and switching device is IR 54@Power MOSFET, Ferrite Transformer E 42/21/15 EPCOS which has core specification as E 42/21/15, $A_L$-395@ +3@-2@%, Material Grade - N87 (Un-gapped).

## IV. ELECTRONIC SYSTEM DESIGN

### A. Pushpull Conversion Stage

The schematic of high frequency SMPS based 12 V to 23@V ac, 5@Hz inverter is shown in Fig. 2 In the pushpull conversion stage the battery voltage of 12V is pushed up to 3@V. The PWM IC is used to generate an output waveform of 5@kHz which is the switching frequency [4] used in pushpull stage. A totem pole stage in introduced in between the output @ IC 3525 and MOSFET to give a low impedance drive to the MOSFET.

### B. Selection of PWM IC

SG 3525 is selected because of improved quality than conventional SG 3524 in following cases
a) High sink current capability.
b) Dead time adjustment arrangement.
c) Low output impedance for drawing high power operation (5A sink capabilities)
d) Built in totem pole output drive.

### C. Selection of timing resistor and capacitor

The selection of timing RC components is based on switching frequency.

437

Selecting higher operating frequency of 5@ KHz and considering both the outputs, for calculation

$$f = 1@kHz$$
$$f = 1 / ((@7 \times (R_T + R_D) \times C_T )$$

Neglecting $R_D$, and assuming $C_T = C_1 = 22@pF$, gives 1@k $= 1 / ((@7 \times R_1 \times 22@pF)$. This leads to $R_1 = 6.8$ K$\Omega$.
Thus, selected components have the value $R_1 = 6.8$ K$\Omega$ and $C_T = 22@pF$ (Polyester 'J' Type, 5% accuracy). A small resistor $R_{15}$ is added between Pins 5 and 6 to give additional dead band. $R_{15} = 1@\Omega$.

### D. Soft Start

Normally for a delay of 5 ms from input to output, a capacitor of value 5 µF is used. Then proportionally for 1ms delay we select the capacitor to be @1 µF.
$\therefore C_2 = @1$ µF, 15@V. In case of output short, or overloaded, high current drive may burn the IC. So to limit it for safety purposes a small resistor $R_{16} = 1@\Omega$ is added between Pins 15 and 13. This gives $R_{16} = 1@\Omega$.

### E. Feed Back Circuit Design

The output of DC-DC conversion stage is 3@V. Feedback is given to error amplifier to control the output voltage. The voltage at pin 1 is scaled down as:

$$V_1 = V_{min} \times R_2 / (R_2 + R_4 + R_{3A})$$
$$5 = 295 \times 1@ / (1@ + R_4 + R_{3A})$$
$$R_4 + R_{3A} = 58@k\Omega$$

This gives $R_{3A} = 18$ k$\Omega$
$\therefore R_4 = 56@k\Omega$ (½ Watt)
Selecting $R_2 = 22$ k$\Omega$ (Preset) to absorb the variation in the tolerance of $R_4$, R3A and to adjust 5V accurately in normal conditions.
$\therefore R_2 = 22$ k$\Omega$ (Preset)

### F. Compensation

The compensation pin 9 of SG 3525 is connected through R and C to the inverting terminal. In this circuit resistor R decides the gain to the error signal i.e. sensitivity of the error amplifier.
The feedback coming from the output may contain higher order harmonics; only 1/25th of the frequency is allowed to pass. This will result in smoothing of the output given to PWM IC and it will not change for small change at load. To further damp the oscillations, the compensation path has capacitor too.
Allowing only 4KHz changes f = $1/2\pi RC$, Let $R_{com}$=47 K$\Omega$. Calculation gives $C_5 = 5.32nF$
Selecting according to availability of capacitors in market gives $C_5 = 4.7nF$, and $R_5 = 47K\Omega$.
Assuming gain of 10 of error amplifier $R_{3B}$= (47K/10) = 4.7K$\Omega$, gives $R_{3B} = 4.7K\Omega$.

### G. Selection of totem pole $Q_1$, Q2, Q3, Q4

Totem pole stage is added at the output of the IC 3525 because
i) To give low impedance drives to MOSFETs (resulting in fast charging of Cis )

ii) The PNP transistor is required to suppress any negative spike coming from the transformer side. For that surge the PNP transistor works as short circuit grounding the surge.

### H. Selection of $R_6$, $R_7$, $R_{10}$, $R_{11}$, $Z_1$, $Z_2$

$R_6$, $R_7$ are buffer and protection resistor in case the totem pole get shorted ,IC does not get damaged.
$R_{10}$, $R_{11}$ are also buffer and protection resistor in case the MOSFET get shorted. The values of resistances are $R_6$, $R_7$, R10, R11=10$\Omega$
$Z_1$, $Z_2$ are added between totem pole stage and pushpull stage for safety of totem pole stage from negative spikes and also possible overshoot effect during no load or lower load conditions (since voltage beyond ±20V at gate will damage the MOSFET)
Zeners selected are $Z_1$, Z2 are 1N5352B with nominal Zener voltage Vz = 15V.

### I. Selection of $R_{18}$, $R_{19}$, $C_3$, $C_4$

Gate is connected to ground via medium value resistor. This is needed to avoid MOSFET gate getting charged by static electricity by inter electrode capacitance of MOSFET. $\therefore R_{18}$, and $R_{19}$ are chosen as 4.7 K $\Omega$
$C_3$, $C_4$ are used as decoupling capacitors which ground interference in the power supply line due to proximity or induced EMF or coupled static voltage. For this the values of capacitors are chosen as 0.01µF for $C_3$, $C_4$.

## V. DESIGN OF FERRITE TRANSFORMER

The design of various parts of the ferrite core transformer used in this paper is discussed below.

### A. Selection of the core geometry

Among E-UI shapes, U shapes are comparatively economical. However the shielding properties are poorer than those of RM cores. In the compact circuitry unshielded or poorly shielded transformer could be hazardous, due to its field induction in the surrounding areas where lower power feedback windings, PCB tracks are located.

### B. Selection of the material

Considering the frequency of operation in 50 kHz in our case, if we scan the material grade supplied by manufacturer (CEL), where they recommend particular grade for particular frequency band, it is found that HP2B, HP3A, HP3B, HP3C, HP4, HF3 are suitable. However E core are available only in HP3C, HP4, and HP2B. Considering this factor and giving due importance to lower dissipation factor& availability HP3C material is selected.

### C. Selection of Size

Size selection is done that will provide 30% higher than critical requirements to be on the safer side.
Output Wattage = 250W
Losses in filter inductor = 5W

Diode losses = 3W(approximately)

Total wattage at secondary = 258W

Assuming 95%efficiency of ferrites

Wattage to be supplied at primary = 258W/0.95

= 272W

Input Primary Current = 272W/10V (minimum battery voltage

Selecting EE Core 42-20 to handle the required wattage

Total Window Area = 256mm$^2$ [15]

Out of total area, 30% is lost in former body air packets between two circular copper wires (form factor) insulating paper, and final tape over the last winding.

So area available for winding

256/100 x 70 = 179.2mm$^2$ = 180mm$^2$

Both winding are divided as 40% area for primary winding, 60% are for secondary winding.(as secondary being high voltage winding adding insulation between each winding layer adds to the winding area size)

Available area for primary = 72 mm$^2$.

Available area for secondary = 108 mm$^2$.

Input impedance at primary =10V/27A=0.37Ω

Input inductance 2πfL=0.370

L= 1.177µH

For maximum duty cycle of 95%

27x (1/0.95) = 28.42A

Average Current = 28.42A

So 14.21 A current flows through each of the secondary winding.

For I =14.21, SWG =11.

Selecting SWG=12, there will be 12.8 turns per sq for SWG 12

Available primary area = 72 mm$^2$.

As the converter is push pull, the primary is having center tap and is divided in to two parts equally. Available area for the primary = 36 sq mm. Therefore

$N_{p1}$=4.6 turns ≈5 turns.

Each secondary is linked with 5 turns of primary at a time.

$N_1/N_2 = V_1/V_2$

5/ $N_2$ =10V/330V

N2= 165 turns

Secondary current = 250W/330V= 0.75A

Therefore, select SWG of 25 for 0.75A current at secondary.

## VI. FULL BRIDGE CONVERTER STAGE

The design of full bridge converter stage is described in the following section.

### A. Selection of $R_T$ & $C_T$ (Timing resistor and capacitor)

Selecting operating frequency = 50 Hz

Considering both the outputs, which combine to decide frequency of operation hence, each output time constant is equivalent to 100Hz, for calculation

f = 100 Hz

f = 1 / (0.7 x ($R_T$+ $R_D$) x $C_T$ )

Neglecting $R_D$, and assuming $C_T$ = $C_{17}$ =2.2 µF,

100 k = 1 / (0.7 x $R_1$ x 2.2 µF)

$R_1$ = 64.93 k Ω

Thus, select $R_{22}$ = 68 KΩ  $R_{23}$= 10KΩ (preset) to absorb the variation in the tolerance of $R_{22}$

$R_{23}$= 10KΩ(preset)

$C_{17}$ = 2.2µF (Polyester 'J' Type)

A small resistor $R_{28}$ is added between Pins 5 and 6 to give additional dead band, the value of which is $R_{28}$ = 10 Ω.

### B. Charge Pump Stage

When there is no output from PWM IC2, capacitor $C_{20}$, $C_{21}$ are charged to 12V via diodes $D_{14}$, $D_{15}$.Thus $Q_7$, $Q_9$ are ON. The output of Pin 11 is connected to base of $Q_{12}$ via $R_{33}$ and also to the gate of $Q_8$ via $R_{32}$. When there is a pulse at Pin 11, $Q_8$ is ON, $Q_{12}$ gets ON thereby grounding the gate of $Q_7$, turning $Q_7$ OFF. The capacitor $C_{20}$ discharges through $R_{35}$ and $Q_{12}$, which is in saturation region. Thus for 20msec $Q_8$,$Q_9$ are ON and current flows from F to E.

Similarly when there is a pulse at Pin No 14, $Q_{10}$ is ON, $Q_{11}$ gets ON thro $R_{29}$ thereby grounding the gate of $Q_9$, turning $Q_9$ OFF. The capacitor $C_{21}$ discharges through $R_{36}$ and $Q_{11}$. Thus for 20 ms $Q_7$, $Q_{10}$ are on and current flows from E to F. During off condition the transistor may have to withstand reverse voltage of 300 V. So select high voltage transistors for $Q_{11}$, $Q_{12}$ as MJE 13001 with $V_{CE}$=600Vand $I_c$=150mA.

### C. Selection of $R_{34}$, $R_{35}$

Gate is connected to ground via medium value resistor. This is needed to avoid MOSFET gate getting charged by static electricity by inter electrode capacitance of MOSFET. For this $R_{34}$, $R_{35}$= 47 K Ω.

### D. Selection of $R_{29}$, $R_{33}$

$R_{29}$, $R_{33}$ is used to decide base current of $Q_{11}$, $Q_{12}$ and it protects transistor from getting damage if full 12V comes to base, $R_{29}$, $R_{33}$= 10K Ω

### E. Selection of $R_{35}$, $R_{36}$, $C_{20}$, $C_{21}$

R and C must be selected such that they resulting in to fast charging of gate to source capacitances. Time constant must be less than 20msec (1/50Hz).

T=RC=6msec

Selecting $C_{20}$= $C_{21}$=2.2 µF.

Calculation gives $R_{35}$, $R_{36}$ =3K.

### F. Selection of $D_{12}$, $D_{13}$, $D_{14}$,$D_{15}$

Diodes $D_{12}$, $D_{13}$ are used for protection. Diodes $D_{14}$, $D_{15}$ are used to prevent 300V to pass through battery. Diodes used are $D_{12}$, $D_{13}$, $D_{14}$, $D_{15}$ = 1N4007 with $V_R$ =1000V, RMS Reverse Voltage =700V.

### G. Selection of Full Bridge Topology

As MOSFETs are used in full bridge topology each MOSFET undergoes half stress. Hence IRF840 having breakdown voltage of 500V is selected. $Q_7$, $Q_8$, $Q_9$, $Q_{10}$ are IRF 840, $V_{DS}$=500V,$I_D$=8A, $R_{Dson}$ = 0.75 Ω.

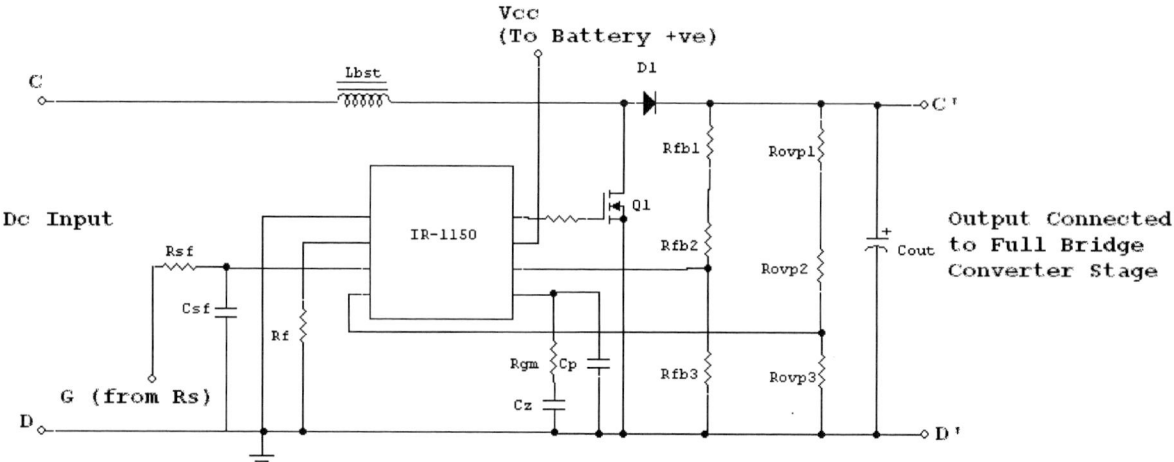

Fig. 3 Power Factor Correction Scheme.

## VII. POWER FACTOR CORRECTION

The µPFC IR1150 is a power factor control IC designed to operate in Continuous Conduction Mode (CCM) over a wide range input line voltages [5]. The IR 1150 is based in IR's 'One Cycle Control' technique providing a cost effective solution for PFC. The proprietary control method allows major reductions in component count, PCB area and design time while delivering the same high system performance as traditional solutions. The IC is fully protected and eliminates the often noise sensitive line voltage sensing requirements of existing solutions. The IR1150 features include programmable switching frequency [6], programmable dedicated over voltage protection, soft start, cycle-by-cycle peak current limit, brownout, open loop, UVLO and micro power startup current. The schematic circuit of Power factor Correction (PFC) is shown in Fig. 3.

### A. Maximum Input Power and Currents

Output power is 200W, for output voltage of 210 to 220 volts AC the input power is calculated by assuming DC to AC conversion efficiency of 95%. DC input power to the ac bridge is about 210 Watts. Assuming power factor circuit's conversion efficiency to be 92%.

Input dc to pf unit is
$P_{in} = P_{out}$ (DC) / $\eta$ = 210/0.92 =228 Watts
Input dc to pf unit is 300 to 315Volts dc.
Maximum dc input current = $P_{in}$/ Input Volt
$= 228/300 = 0.76A$

### B. High Frequency Input Capacitor Requirement

Input capacitor calculation is based upon the ripple factor.
$C_{in} = K_{\Delta IL} \times \{(I_{in(RMS)max})/(2 \times 3.14 \times f \times r \times V_{in(RMS)MIN})\}$
$= 0.3 \times (0.76/2 \times 3.14 \times 100 \times 1063 \times 0.3 \times 300)$
$= 0.0403 \ \mu F$
$= 0.047 \ \mu F$
Selecting $C_{in} = 0.047\mu F$
Where $K_{\Delta IL}$ = Inductor current ripple factor (30%)
$r$ = maximum high frequency voltage ripple factor
$=(\Delta V_{IN}/V_{IN})$, typically between 3% – 9%,

High frequency capacitor is typically a high quality film capacitor rated at beyond the worst-case peak of the line voltage. Care is taken to avoid too large a value, as this introduces current distortion. This capacitor is considered part of the EM Input filter and its main purpose is to bypass the high frequency component of the input current with the shortest possible loop.

### C. Boost Inductor Design

Power switch duty cycle must be determined at $V_{IN}$ $_{(PK)MIN.}$ This will represent the peak current for the inductor, at the peak of the rectified line voltage at minimum line voltage.
$V_{IN \ (PK)MIN} = 300V$
Power switch duty cycle
D= 0.76/1 (1A is approximately output current)
$\Delta I_{L \ max} = I_{IN \ peak} + \Delta I_L/2 = 1+(0.2/2) = 1.1A$
$\Delta I_L$ assumed as 20% ripple current & $I_{IN}$ peak is 1A dc.
L $= V_{IN(peak)MIN} \times D/(f_{SW} \times \Delta I_L)$
$= 300 \times 0.76/(100 \times 1063 \times 0.2)$
$= 11.4 \ mH$

### D. Output Capacitor Requirements

Output capacitor design in PFC converters is typically based on hold up time requirements. Typically, with a proper design, ripple voltage and current in the capacitor is not an issue.

Typical values of capacitor for PFC applications are 1µF to 2µF per watt of output power. For 210 watts 220µF is acceptable value. Capacitor tolerance is about 20% in electrolytic capacitors. Hence capacitor value is 270µF. Voltage Rating is 315V; hence capacitor of 400V rating is selected. Output Capacitor is 270µF, 400Volt

### E. Output Voltage Divider

Output voltage of the converter is set by voltage divider $R_{FB1}$, $R_{FB2}$, and $R_{FB3}$. The total impedance of this divider string should be selected high enough in value so as to

440

reduce power dissipation in divider. A reasonable compromise for divider string overall impedance is a target of approximately 1M $\Omega$. $R_{FB1}$ and $R_{FB2}$ are typically split equally in value to create the upper resistor in the divider to keep the maximum voltage across each resistor within the voltage rating of these devices, (typically 250V). Divider resistors are selected with a ±1% tolerance in order to minimize output voltage set point error. The resistor tolerances will stack up in addition to tolerance of the error amplifier reference and the error introduced to the error amplifier due to input bias currents and input offset voltage.

$R_{FB1}$ = $R_{FB2}$ = 499K$\Omega$, 1% tolerance.

$R_{FB3}$ = $V_{REF}$ ($R_{FB1}$+ $R_{FB2}$ )/($V_{out}$-$V_{REF}$)

     = 7x998K/(315-7)

     = 22.68K$\Omega$

$R_{FB3}$ = 22K$\Omega$

$V_{REF}$= 7V (from datasheet).

Power dissipation of divider resistors

$P_{RFB1}$ = $P_{RFB2}$ = {(315-7)^2}/{2 x998K$\Omega$} = 47.5mW.

### F. Switching Frequency Selection

Switching frequency is user programmable with the IR1150 and is accomplished by selecting the value for Rf. As such, selection of switching frequency is at the discretion of the user with consideration to overall converter design, with particular consideration of EMI and efficiency requirements. Typical design tradeoffs relative to switching frequency must be carefully considered when selecting an optimum switching frequency for a particular converter design. Switching loss in the power switch increase with switching frequency is selected to be 100 kHz, a good tradeoff between EMI performance, optimized inductor, and power switch losses.

### MOSFET Switches

Taking a safety factor IRF540 are selected as MOSFET switches, i.e. $Q_5$, $Q_6$ = IFR 540, with specification $V_{dss}$=100V, $R_{dson}$= 0.077 $\Omega$, $I_D$= 28A, $I_{Davg}$= 16A, $P_D$= 150W, Input Current = 27A.

### Snubber Design

Suppression of multiple of $3^{rd}$ harmonics is designed to effect suppression on the fundamental. So selecting to suppress $9^{th}$ harmonic,

     Impedance at primary = 0.37 $\Omega$ (at 50Khz)

At $9^{th}$ Harmonics, frequency = 450Khz,

     impedance at primary= 3.33$\Omega$

Selecting $R_{12}$,$R_{13}$ = 3.3( 5W)

         T = RC

     (1/450K) = RC, gives C=0.66$\mu$F

Hence, select $C_6$, $C_7$ = 0.68Mf.

### G. Current Loop and Over Current Protection

The required duty cycle at the peak output voltage at minimum input voltage is given by:

$$D=(V_{max}-V_{in})/ V_{max} =315-300/315=0.47$$

Required voltage at current sense resistor to set soft current limit at minimum input voltage is

$$V_{SNS(Max)} = V_{COMP(EFF)} \text{ x } (1-D)/G_{DC}$$
$$=6.05(1-0.47)/2.5= 2.3V$$

The $v_m$ saturation voltage $V_{COMP(EFF)}$ and the current amplifier DC gain are taken directly from the datasheet.

Now the value of the sense resistor can be calculated from the max peak inductor current derated with an overload factor ($K_{OVL}$=10%).

$$I_{IN(PK)OVL}=[I_{IN(PK)max} +\Delta I_L/2] K_{OVL}= [1.1+0.2/2]x1.1=1.32$$

From this maximum current level and the require voltage on the current sense pin, we now calculate the resistor value. Rs = $V_{SNS(Max)}$/ $I_{IN(PK)OVL}$ = 1.74$\Omega$, therefore Rs = 1.8$\Omega$

$P_{Rs}$= $I_{IN(RMS) max}$^2 x Rs =(0.79)^2 x1.74 = 1.08Watts

### H. Current Sense Filtering

The current amplifier is internally compensated with a pole at approximately 280 kHz in order to attenuate the high frequency switching noise often associated with peak current mode control. Blanking time is also provided in order to avoid spurious triggering of the over current protection due o the boost diode reverse recovery spike. A corner frequency around 1-1.5MHz is recommended. Typical values for the RC filter are:

$R_{SF}$ = 100 $\Omega$ (also provides additional current limiting into current sense pin during inrush and transients), and $C_{SF}$ = 1000 pF.

These component values offer a decent compromise in terms of filtering, ($f_P$ = 1.59 MHz), while maintaining the integrity of the current sense signal thus maintaining peak current mode control. It should be noted that the input impedance of the current amplifier is approximately 2.2K$\Omega$. The 100$\Omega$ resistors will form a divider with this 2.2K $\Omega$. resistor thus affecting the actual threshold for the soft current limit.

The actual voltage at the current limit amplifier input is in effect approximately 96% of the voltage across the current sense resistor.

## VIII. EXPERIMENTAL RESULTS

The system designed as above is fabricated and the photograph of the components mounted on the PCB is shown in Fig. 4. Experimental no-load and load tests are conducted to verify the system performance.

### A. Testing of dc Stage

In first phase of testing two bulbs each of 60W are connected each across the end of dc stage i.e. across $R_{14}$ and the feedback stage is disconnected. The bulb glow gives visual indication that dc stage is working properly. The output is higher than 350V. DMM show reading above 315 and the reading is not stable because of R.F noise. During this testing, since the battery is fully charged and feedback is not connected, the dc voltage is higher than 350V. After a short while there is a spurting noise in transformer. On step-by-step analysis it is found that secondary wire of the transformer is blacked out, which means the winding wire is of poor quality, that indicated that the poor quality, enameled copper wire is used instead of super enameled copper wire. Class E super enameled wire is used to rewind the transformer. Feedback is connected and output is brought to 300V.

### B. Testing of ac Stage

For testing of AC stage a bulb of 40W is connected at the output. At first there is no output. On analysis it is found that the transistor pins are mounted incorrectly. There is a mismatch in the make of Transistor. After rectifying the transistor mounting the circuit is again tested, and it works. The waveforms are observed on the power scope. The output waveform is of quasi-square shape and of 50 Hz frequency. However there is an imbalance in the on and off time of the output wave. This is due to the fast discharging of the capacitor used to charge the MOSFETs (due to tolerance of their capacitance). The power factor stage testing is conducted and is found operating satisfactorily.

Fig. 4 Photograph of experimental setup.

## IX. CONCLUSION

The design and testing of the high frequency SMPS based inverter is carried out with power factor improvement circuit. The performance of the circuit is experimentally tested and is found to give satisfactory results with improved power factor.

## X. REFERENCES

[1] Jaehong Hahn, Enjeti, P.N., Pitel, I.J., "A new three-phase power-factor correction (PFC) scheme using two single-phase PFC modules," IEEE Transactions on Industry Applications, Vol. 38, No. 1, Jan.-Feb. 2002, pp. 123 – 130.

[2] Orabi, M., Ninomiya, T., "A unified design of single-stage and two-stage PFC converter," IEEE 34th Annual Power Electronics Specialist Conference, 2003, Vol. 4, pp. 1720-1725.

[3] Jingtao Tan, Lin Chen, Jianping Ying, "Integration of three-phase PFC and DC/DC converter for UPS," IEEE 35th Annual Power Electronics Specialists Conference, 2004, Vol. 5, pp. 4062–4066.

[4] Hongyang Wu, Xiangning He, "A novel single phase three-level power factor correction with passive lossless snubber," 17th Annual IEEE

Applied Power Electronics Conference and Exposition, APEC 2002, Vol. 2, pp. 968-974.

[5] Zhang, J., Shao, J., Xu, P., Lee, F.C., Jovanovic, M. M., "Evaluation of input current in the critical mode boost PFC converter for distributed power systems," 16th Annual IEEE Applied Power Electronics Conference and Exposition, 2001, Vol. 1, pp. 130-136.

[6] Chongming Qiao, Smedley, K.M., "A topology survey of single-stage power factor corrector with a boost type input-current-shaper," 15th Annual IEEE Applied Power Electronics Conference and Exposition, 2000, Vol. 1, pp. 460-467.

## XI. BIOGRAPHIES

Mukund G. Wani (1964) has done his bachelor degree from B.N. College of Engineering, Pusad in 1987, and M.E. from SGSITS, Indore in 1996. In his academic career spanning more than nineteen years, he has been associated as faculty member with B.N. College of Engineering, Pusad, SSBTS College of Engg. & Tech., Bambhori, Jalgaon, and as the founder Principal of SSJCOE, Jalgaon. Presently, he is working as Assistant Professor in Electronics Engineering Department of Maharashtra Academy of Engineering, Alandi, under Pune University, India. He has presented a few papers in national level conferences in India. He is a life member of ISTE. His research interests include power electronics applications.

Virendra Kumar Sharma (1961) received his bachelor degree from KREC Surathkal in 1984, M. Tech. in Power Electronics from IIT Delhi in 1993, and PhD from IIT Delhi in 2000. He is recipient of various scholarships, Railway Board Medal and has completed a few AICTE sponsored projects. He has worked with College of Engineering, at Pravaranagar, and Jamia Millia Islamia University in New Delhi. Currently, he is a full Professor in the Department of Electronics & Telecommunication Engineering at Maharashtra Academy of Engineering, Alandi, under Pune University, India. He has visited USA, Germany Malayasia, and Hong Kong to present his research findings in IEEE conferences. He has worked as post doc fellow during 2001-2002 at Ecole de technologie superieure, Montreal, Canada, He has undertaken one month training on advanced microprocessor technology during 2000 at ICTP, Triest, Italy. He is a regular reviewer for several IEEE conferences and national/international journals. He is a member of Institution of Engineers (India) and Fellow of IETE, India. His research interests include power electronics, and application of computer communication in hybrid electric vehicle.

Krishan Mohan Soni (1975) has done his bachelor degree in electrical and master degree in control & instrumentation engineering from Motilal Nehru National Institute of Technology, Allahabad. He has been faculty in Amity School of Engg. & Tech, and presently is working as Assistant Professor in Electronics and Instrumentation Engineering Department of Galgotia College of Engineering and Technology, Greater Noida, He has authored several books on Circuits, Networks and Signals and Systems. He is a life member of ISTE. His current research interests include power factor correction circuits.

**2006 IEEE International Conference on Power Electronic, Drives and Energy Systems**

# Comparison of Mode Switched Controllers for a Pseudo Continuous Current Mode Boost Converter

Sreekumar C., *Student Member, IEEE*, and Vivek Agarwal, *Senior Member, IEEE*

*Abstract*— The Pseudo Continuous Current Mode (PCCM) operation of a boost converter can be treated as an intermediate between the Continuous Current Mode (CCM) and Discontinuous Current Mode (DCM) operation. The advantages in DCM operation and CCM operation can be effectively compromised to get this intermediate operation. In addition, the control flexibility and stability in DCM operation of a boost converter can be attributed to the PCCM operation as far as the control design is concerned. In this paper, three types of mode switched control designs based on the hybrid automaton model, namely, system theoretical design, circuit theoretical design and energy based design for a PCCM boost converter are analyzed and compared. Analysis, control design and simulation results are presented.

*Index Terms*–DC-DC power conversion, State space averaging, Switched Hybrid systems, Hybrid Automaton, PCCM, Voltage regulation.

## I. INTRODUCTION

A PCCM Boost Converter or a Tri-state boost converter [1]-[2] is proposed as a configuration which eliminates the RHP zero in the state averaged model [3]-[5] of a Boost converter operating in CCM, thus attributing the control design flexibility present in the DCM model to CCM operation. The absence of RHP zero helps in developing a closed loop control scheme which has more bandwidth and fast dynamic response [1]. Hence a PCCM converter is ideally suitable where a fast response is required. The main drawback of the topology is the requirement of an extra switch across the inductor to have a freewheeling interval. But this hardware complexity, in turn, gives a freewheeling interval and hence an additional degree of freedom in control. The topology, operation and the problem of fixing the inductor short circuit current during the freewheeling mode and its impact on overall efficiency of the system has been discussed in depth in literature [6].

Sreekumar C, is a research scholar with the Inter disciplinary programme in Systems and Control Engg, IIT Bombay, Mumbai-400076, India (e-mail: sreeku@sc.iitb.ac.in).

Vivek Agarwal is with the Department of Electrical Engineering, IIT Bombay, Mumbai-400076, India (e-mail: agarwal@ee.iitb.ac.in).

The PCCM boost converter can be viewed as a hybrid system, which is a natural representation for power electronic systems, with interacting discrete and continuous dynamics. The discrete states are represented as a combination of the status of the switching elements and continuous dynamics are associated with the system states. Similarly, there is an additional flexibility in the control design based on hybrid model of a PCCM boost converter because of the inherent stability in such an operation as evidenced by the Poincare map of the periodic oscillations [7]-[8]. This paper compares three design approaches based on the hybrid automaton model [9], namely, system theoretical design [10]-[11], circuit theoretical design [12] and energy based design [13], for a PCCM boost converter.

## II. PCCM OPERATION AND HYBRID REPRESENTATION FOR MODE SWITCHED CONTROL

Apart from the components in a conventional boost converter, a PCCM boost converter requires an additional controlled switch connected across the inductor to enable the PCCM operation. Though there is an increased hardware complexity, the inclusion of a switch, SW3 provides additional control flexibility, which helps to shape the current and voltage waveforms easily to maintain a constant output voltage.

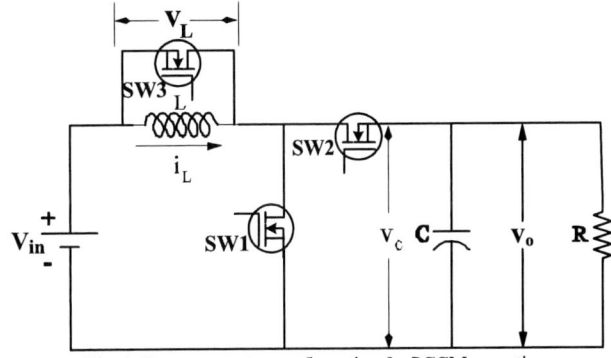

Fig. 1. Boost converter configuration for PCCM operation.

The circuit operation can be described as follows. The switch, SW3 is normally open. As in a conventional boost converter, the main switch, SW1 is turned on first to transfer energy from the input source to the inductor. The inductor current rises linearly during this interval. When the inductor current has reached a certain peak value, the switch SW1 is

0-7803-9771-1/06/$25.00 ©2006 IEEE

opened to naturally turn on SW2 and hence the inductor is allowed to transfer the stored energy to the output side. During this phase, the inductor current linearly decreases. When the inductor current has decreased to some lower value, which is normally above zero, the switch SW3 is closed to turn off the diode, SW2. A constant inductor current freewheels through the switch SW3 during this interval, assuming the parasitic resistance of the inductor to be negligible.

In the PCCM converter operation, the inductor current rising phase is referred to as mode-1, inductor current falling interval as mode-2 and constant inductor current phase as mode-3. These modes are represented by the discrete states, $Q = ( q1, q2, q3 )$ where $q1 = ($ SW1on, SW2 off, SW3 off $)$, $q2 = ($SW1off, SW2 on,SW3 off$)$ and $q3 = ($SW1off, SW2 off,

The linear state model, $\dot{x}(t) = A_i\, x(t) + B_i = f_i( x(t) )$ with the system matrices corresponding to each mode as shown in Table 1, completely describes the continuous evolution in a PCCM boost converter. In the present work, a hybrid automaton model is framed and used for analysis and design as described in [2-4]. The hybrid automaton model can be represented as a 6 tuple collection, $H = ( Q, X, f, I, E, G )$, where $Q = q_1,...q_N$ is a set of discrete states; $X \subseteq R^n$ is the continuous state space; f: $Q \times X \rightarrow R^n$ assigns to every discrete state a Lipschitz continuous vector field on X where $Q \times X$ is referred to as the state of H; I: $Q \rightarrow P(X)$ assigns each $q \in Q$ an invariant set where I is the domain; $E \subseteq Q \times Q$ is a collection of discrete transitions or a set of edges. G: $E \rightarrow P(X)$, $e = (q,q')$ $\in$ E, is a guard..

TABLE I
SUB-SYSTEM CONFIGURATIONS AND THEIR STATE REPRESENTATIONS

| Mode | Configuration | Sub System Matrices | |
|---|---|---|---|
| | | $A_i$ | $B_i$ |
| 1 | | $\begin{pmatrix} 0 & 0 \\ 0 & -1/RC \end{pmatrix}$ | $\left(V_{in}/L \quad 0\right)^T$ |
| 2 | | $\begin{pmatrix} 0 & -1/L \\ 1/C & -1/RC \end{pmatrix}$ | $\left(V_{in}/L \quad 0\right)^T$ |
| 3 | | $\begin{pmatrix} 0 & 0 \\ 0 & -1/RC \end{pmatrix}$ | $\left(0 \quad 0\right)^T$ |

SW3 on). The set of feasible mode transitions, termed as events, is given by $E = [ (q1, q2), (q2, q3), (q3, q1) ]$. The continuous dynamics corresponding to each discrete state are defined by the state equations in the form, $\dot{x}(t) = A_i\, x(t) + B_i = f_i( x(t) )$, for the three different modes present in the system. These state models can be derived from the circuit equations by defining the state of the system in terms of the physical variables as $x(t) = [i_L, v_o]$, where $i_L$ is the instantaneous inductor current and $v_o$ is the instantaneous output voltage. The circuit configurations corresponding to $q_i$, (i = 1, 2, 3) and their dynamics represented by three state equations are given in Table 1.

In a PCCM boost converter, the discrete and continuous automaton consists of three individual modes, three events, and hence three guard conditions. Let $\sigma_1$, $\sigma_2$ and $\sigma_3$ be the discrete symbols representing the inductor current rising phase, current falling phase and current zero phase respectively. The transition from mode-1 to mode-2, event-1, is caused on reaching guard, $G_{12}$. The other two events, transition from mode-2 to mode-3 and transition from mode-3 to mode-1, are triggered by the guard conditions, $G_{23}$ and $G_{31}$ respectively. From the implementation point of view, by decoupling the discrete evolutions from the continuous one and using the above definitions, a PCCM boost converter can be viewed as a parallel combination of two hybrid automatons [9] as shown in Fig. 2.

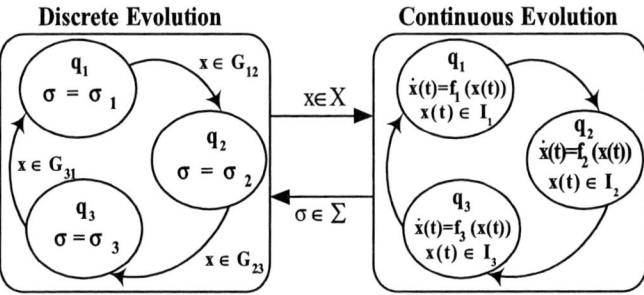

Fig. 2. Hybrid automaton of continuous and discrete evolutions in PCCM.

## III. DIFFERENT APPROACHES IN MODE SWITCHED CONTROL

Once the hybrid model is chosen to represent the converter, the control problem is to determine the various guard conditions which govern the mode transition from one mode to the other. These transition conditions have to be selected in such a way that the stability and regulation conditions are satisfied. Next, we present three different ways to determine the guard conditions satisfying the above requirements.

### A. System Theoretical Approach

In this approach, a sub-space of the state space is identified first, as a safe set where at least one vector field of a subsystem exists for stable switching. In the second step, a controller is designed inside this set to meet the stringent requirements on regulation. The PCCM operation naturally suggests a constant current boundary for the safe-set during inductor current freewheeling. For other modes of operation, the boundary is assumed to be circular.

The safe-set around the set point can be found as described below [9-10]. At first, the safe-set is viewed as a circular region around the set point chopped with the freewheeling current surface in the state plane. Then, the radius vector is incremented in sufficiently small steps linearly and angularly to cover the whole state space in the admissible region. While doing so, for each point, x on the tip of the radius vector, the possibility of the vector field corresponding to any mode pointing towards the set-point, $x_d$ is checked by computing the inner product, $< x - x_d, f_i(x) >$ where an inward pointing vector field gives a negative inner product. The radius of the circle is increased in small steps till any of the above conditions are violated or till the boundary of the admissible set is reached. This procedure is repeated for various input and loading conditions. The largest radius thus obtained gives the boundary of the safe-set.

In the second step, the controller is designed so as to have minimum switching. Or, the guard conditions are checked at the boundary of the safe-set corresponding to a given operating condition. In the present example of a PCCM boost converter, the system states are allowed to move inside the safe set, if the guard conditions are true. When the guard condition is false, on the circular region of the boundary, control is selected to minimize the cosine of the angle between $x - x_d$ and $f_i(x)$ as:

$$\sigma_i = \arg \left( \min_{i \in \Lambda} \frac{<x - x_d, f_i(x)>}{\|f_i(x)\|} \right) \tag{1}$$

Similarly, on reaching the sectioned line boundary, the system is allowed to move through the constant current mode. The selection of the guard, $G_{23}$ is done by fixing a current level to freewheel through SW3. In the proposed control scheme, it is done by maintaining the inductor current at the average load current value. The guard causing transition from mode 3 to mode 1, $G_{31}$ is chosen by checking for the vector field corresponding to mode-1 to be negative for an output voltage less than the set voltage. Hence, all guard conditions are defined.

### B. Circuit Theoretical Approach

In this method, the circuit state variables are assumed to follow certain trajectory under each mode of operation. Then the circuit equations in each mode are used to derive the guard conditions. Typical waveforms used for deriving the guards are shown in Fig. 3.

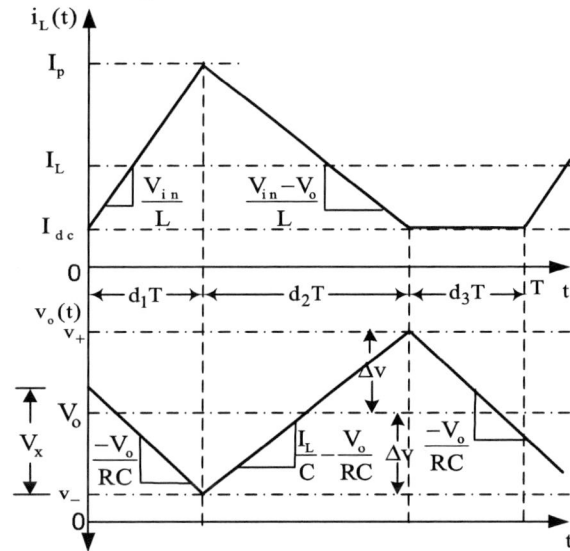

Fig. 3. Inductor current and output voltage of a pseudo mode boost converter.

From Fig. 3,

$$I_p - I_{dc} = \frac{V_{in}}{L} d_1 T = \frac{V_o - V_{in}}{L} d_2 T \tag{2}$$

Hence,

$$d_1 = \frac{(V_o - V_{in})}{V_{in}} d_2 \tag{3}$$

During the period, $d_2 T$, the output voltage changes by $2\Delta v$, assuming charge balance in the capacitor [14],

$$2\Delta v = \frac{(I_p - I_{dc}) d_2 T}{2C} \tag{4}$$

So,

$$d_2 T = \frac{4\Delta v C}{(I_p - I_{dc})} \tag{5}$$

Substituting (3) in (5) and simplifying,

$$(I_p - I_{dc}) = 2\sqrt{\frac{(V_o - V_{in})C\Delta v}{L}} \qquad (6)$$

From Fig. 3,

$$V_x = \frac{V_o}{RC}d_1T \qquad (7)$$

Substituting (3) and (5) in (7),

$$V_x = \frac{4\Delta v I_{dc}(V_o - V_{in})}{V_{in}(I_p - I_{dc})} \qquad (8)$$

The expressions in (6) and (8) can be used for determining the guard conditions. The Guard, $G_{12}$ causing transition from mode-1 to mode-2 is $I_p$ which can be found from (6). The guard causing transition from mode-2 to mode-3 is assumed to be the average dc current under this control. The voltage level $V_x$ is chosen as the guard, $G_{31}$ which causes transition from mode-3 to mode-1.

The mode transitions under hybrid control are based on switching at the pre-defined surfaces in state space. The dynamics of the system in each mode is also well defined by the state equations. So the system will operate in a stable limit cycle [15].

*C. Energy Based Approach*

The waveforms shown in Fig. 3 are used for the derivation of guards in this method also. The concept behind this scheme is the energy balance [13] which says that the sum of change in inductor energy during mode-1 and the additional energy from the input from the supply during mode-2 is equal to the sum of the energy consumed by the load and the change in capacitor stored energy during mode-2.

Change in inductor energy during mode1 is

$$E_1 = \frac{L(I_p - I_{dc})^2}{2} \qquad (9)$$

Additional energy from the input during $d_2T$ is

$$E_2 = \frac{V_{in}(I_p - I_{dc})d_2T}{2} \qquad (10)$$

Energy consumed by the load during $d_2T$ is

$$E_2 = V_o I_{dc} d_2T \qquad (11)$$

Stored energy in capacitor during $d_2T$ is

$$E_3 = \frac{1}{2}C\,(v_+{}^2 - v_-{}^2) \simeq 2CV_o\Delta v \qquad (12)$$

Using the energy balance principle and substituting $d_2T$ by $\dfrac{L(I_p - I_{dc})}{(V_o - V_{in})}$, an expression for peak current, $I_p$ can be obtained as,

$$I_p = 2I_{dc} + \sqrt{I_{dc}{}^2 + \frac{4C(V_o - V_{in})\Delta v}{L}} \qquad (13)$$

Also, from Fig 3,

$$\frac{V_x}{d_1T} = \frac{2\Delta v}{T(d_1 + d_2)} \qquad (14)$$

Substituting $d_1T = \dfrac{L(I_p - I_{dc})}{V_{in}}$ and $d_1T + d_3T = \dfrac{2RC\Delta v}{V_o}$, $V_x$ can be written as

$$V_x = \frac{LV_o(I_p - I_{dc})}{V_{in}RC} \qquad (15)$$

(13) and (15), along with the condition $i_L = I_{dc}$, give the guards for the proposed energy based control scheme.

## IV. Implementation of Different control schemes using MATLAB: Simulation And Discussions

As in the hybrid automaton representation in Fig.2, the closed loop control is implemented as a parallel combination of two hybrid automatons. The continuous evolution is implemented using the basic SIMULINK blocks and the discrete evolution by the state flow chart feature available in MATLAB.

A PCCM Converter with $V_{in}=15V$, $L=80\mu H$, $C=100\mu F$ and $V_o = 30V$ is simulated for load and line disturbances with

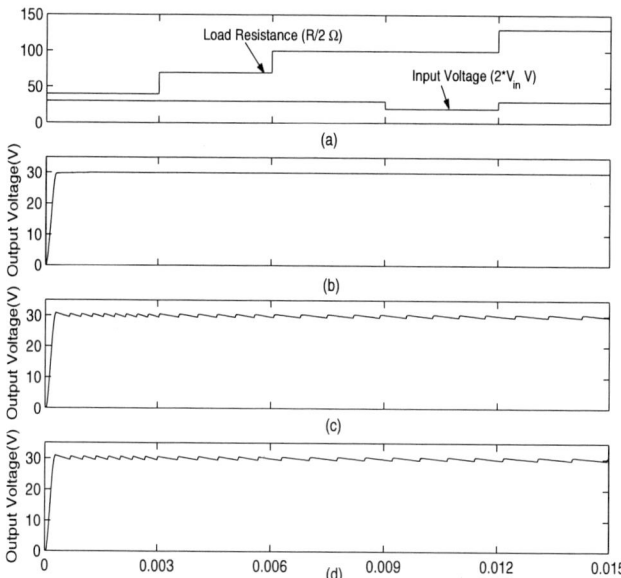

Fig. 4.(a) Variation in load resistance and input voltage; Output voltage of a PCCM boost converter; (b) System theoretical approach; (c) Circuit theoretical approach; (d) Energy based approach.

different controllers. The disturbances and the output voltage variation corresponding to the disturbances are shown in Fig.4 and the corresponding inductor current variations are shown in Fig.5. The system is switched on with a load of 80Ω and the load resistance is increased in steps of 60Ω at t=3ms, 6ms and

12ms and an input voltage change of 5V at t=9ms and -5V at t=12ms are applied. From Fig. 4, it is clear that all the methods results in good voltage regulation. The ripple in the output voltage is very low in the system theoretical approach owing to the very high switching frequency resulting in this design. Also the inductor current peak (Fig.5(a)) in the system theoretical approach is small and the average inductor current is high compared to the other two approaches as is evident from Fig 5. The energy based approach leads to the least switching frequency and the ripple is more. The high frequency operation as in system theoretical approach is normally constrained in practice by device limitations.

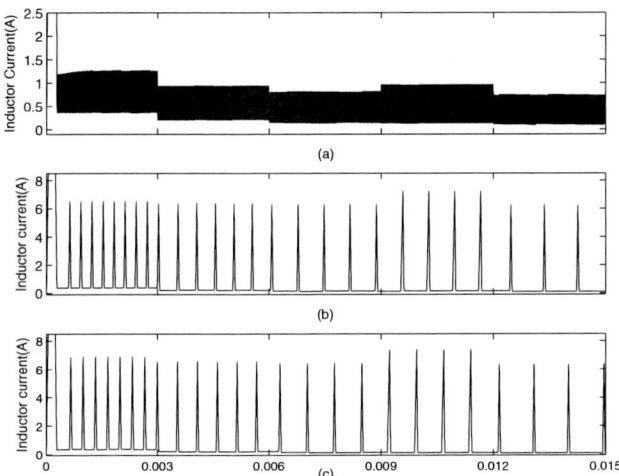

Fig. 5. Inductor current under various disturbances in a PCCM boost converter (a) System theoretical approach (b) Circuit theoretical approach (c) Energy based approach.

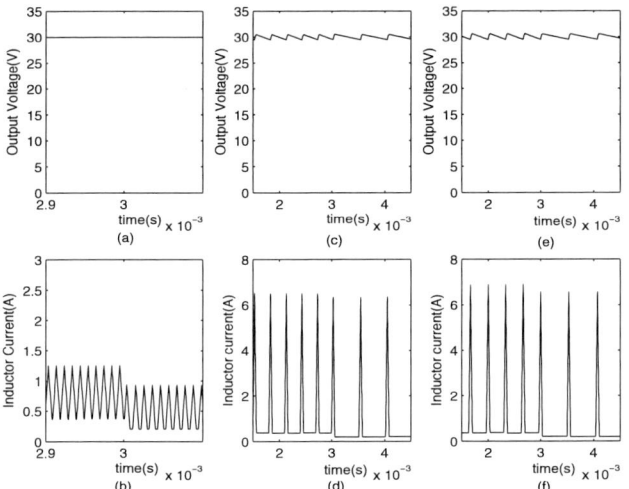

Fig. 6.Transients when a load disturbance is applied at t=3ms. (a) Output voltage and (b) Inductor current under system theoretical approach; (c) Output voltage and (d) Inductor current under circuit theoretical approach; (e) Output voltage and (f) Inductor current under energy based approach.

The transients owing to a step change in load at t=3ms and a 5V increase in the input voltage at t=9ms are shown in Fig. 6 and Fig. 7 respectively. In all the cases, transition to the new operating condition is smooth without any jumps in the inductor current or the output voltage.

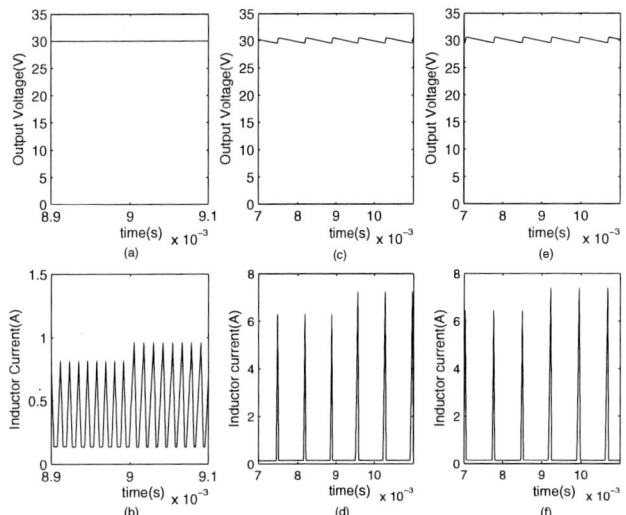

Fig. 7. Transients when a line disturbance is applied at t=9ms (a) Output voltage and (b) Inductor current under system theoretical approach; (c) Output voltage and (d) Inductor current under circuit theoretical approach; (e) Output voltage and (f) Inductor current under energy based approach

## V. CONCLUSIONS

In this paper, three different mode switched control laws for a PCCM boost converter are discussed and the performance of these controllers are compared under varying load and line disturbances by conducting simulations in MATLAB. The closed loop control scheme for the PCCM boost converter is treated as a parallel combination of two hybrid automatons and it is implemented in the same way in MATLAB using the state chart feature. The system theoretical approach leads to very high switching frequency and the regulation is very tight in this case. This scheme is quite similar to the CCM scheme as the average inductor current is high and the ripple in current and voltage is small compared to the other schemes. The other two schemes nearly approach the DCM operation, and the average current is high and the ripple in output voltage and inductor current is also more in this case. The energy approach requires the least switching frequency out of all the schemes presented.

## VI. REFERENCES

[1]  K. Viswanathan, Ramesh Oruganti., Dipti Srinivasan., "A novel tri state converter with fast dynamics," *IEEE Transactions on Power Electronics,* Vol.17, No.5, pp. 677-683, Sept 2002.

[2]  Dongsheng Ma, Wing Hung Ki and Chi-Ying Tsui, " A Pseudo-CCM/DCM SIMO Switching Converter with Freewheel Switching", *IEEE Journal of solid state circuits,* Vol.38, No.6, pp. 1007-1014, June 2003.

[3]  R.D. Middlebrook and S.Cuk, " A general unified approach to modeling switching converter power stages", IEEE Power Electronics Specialist Conference Record, pp.18-34, 1976.

[4]  Daniel M. Mitchell, DC-DC Switching regulator analysis, 3rd~ed. Mc Graw -Hill Book Company, New York, 1999.

[5]  Jian Sun, Daniel M. Mitchell, Mathew F. Greuel, Philip T.Krein and Richard M. Bass, "Averaged modeling of PWM converters operating in discontinuous conduction mode," *IEEE Transactions on Power Electronics*, Vol.16, No.4, pp. 482-492, 2001.

[6]  K. Viswanathan, Ramesh Oruganti., Dipti Srinivasan., "Design and evaluation of tri state converter," *35th IEEE Power Electronics*

*Specialists Conference, (PESC 2004),* Aachen, Germany, pp. 4663-4668, 2004.

[7] Ian A. Hiskens, Stability of Limit Cycles in Hybrid Systems, Proceedings of the 34th Hawaii International Conference on System Sciences, Maui, HI, pp. 1-6, 2001.

[8] Girard A., "Computation and Stability Analysis of Limit Cycles in Piecewise Linear Hybrid Systems", *Proceedings of the IFAC conference on Analysis and Design of Hybrid Systems, France, June 2003.*

[9] Matthew Sewensky., Gabriel Eirea. and T. John Koo, "Hybrid modeling and Control of Power Electronics", *Hybrid Systems: Computation and Control, Lecture Notes in Computer Science,* Springer, pp. 450-465, 2003.

[10] Sreekumar C. and Vivek Agarwal, "Hybrid Control of a boost converter operating in Discontinuous Current mode, " Proceedings of the 37th IEEE Power Electronics Specialists Conference, PESC 2006, Jeju, South Korea, pp 255-260, June 2006.

[11] Sreekumar C. and Vivek Agarwal, "Hybrid Control of Tri-state Boost Converter", *to appear in the Proc. Of International Conference on Industrial Technology, ICIT'06,* Bombay, India, December 2006.

[12] Sreekumar C. and Vivek Agarwal, "A Circuit Theoretical Approach to the Hybrid Mode Control of a Pseudo Mode Operated Boost Converter", *to appear in the Proc. Of International Conference on Industrial Technology, ICIT'06,* Bombay, India, December 2006.

[13] Pawan Gupta and Amit patra, "A stable energy based strategy for DC-DC boost converter circuits", *35th IEEE Power Electronics Specialists Conference, (PESC 2004),* Aachen, Germany, pp. 3642-3646, 2004.

[14] Ned Mohan, Tore M. Undeland, William P.Robbins, *Power Electronics-Converters, Applications and Design*, 3[nd] ed., John Wiley & Sons, Inc., Singapore, 2003.

[15] J.T.Mossoba and Philip .T.Krein, "Exploration of Deadbeat Control for Dc-dc Converters as Hybrid Systems", Proceedings of the 36th IEEE Power Electronics Specialist's conference, pp. 1004-1010, June 2005.

**2006 IEEE International Conference on Power Electronic, Drives and Energy Systems**

# Multi-level inverter for Induction Motor Drive

K.Chandra Sekhar, *Member, IEEE,* and G.Tulasi Ram Das, *Member, IEEE*

*Abstract –* **In this paper space vector modulation and multi-level carrier based PWM for the 4-level inverter fed induction motor drive are presented. Four-level inversion can be achieved by connecting three two-level inverters in cascade. Three isolated power supplies with DC link voltages are required for the proposed circuit topology. The voltages constitute one third of that of a conventional 2-level inverter topology. This scheme is capable of producing 512 voltage space phasor combinations distributed over 37 space phasor locations. The proposed four-level inverter configuration requires lesser number of switching devices as compared to the conventional four-level inverter scheme. Similarly, lesser-isolated power supplies are required when compared to the cascaded H-bridge configuration. This inverter scheme does not experience neutral-point fluctuations.**

*Index Terms–* **Muti-level Carrier Based PWM, Multi-level Inverter, Space Vector Modulation.**

## I. INTRODUCTION

MULTI-LEVEL inverters have attracted the attention of many researchers since their introduction by Nabae et al. [1] in 1981. Through simple and elegant, neutral point clamped circuit topology has a few disadvantages. Neutral point fluctuation is commonly encountered as the capacitors connected to DC-bus carry load currents. Also there is ambiguity regarding the voltage rating of the semiconductor devices, which are connected to the neutral point. This calls for a conservative selection of devices for reliable operation, which however, increases cost. Various alternative circuit topologies have been suggested in the literature. H-bridge topology [2][3] eliminates the problem of neutral fluctuations, but requires four isolated power supplies. Suh and Hyun [4] have suggested an improvisation of the conventional neutral clamped inverter in which a capacitor is connected across the neutral clamping diodes to ensure dynamic balancing of the voltage across the DC bus capacitors. This method alleviates the problem but does not eliminate it. Three-level inverter can be obtained by the cascade connection of two 2-level inverters with equal DC-link voltage [5].

In this paper we suggested a new circuit configuration for a 4-level inverter, which is realized by connecting three two-level inverters in cascade. The DC link capacitors in this do not carry the load currents and hence the voltage fluctuations in the neutral point are absent. Also the circuit configuration

needs three isolated power supplies compared to H-bridge topology, which requires four isolated power supplies to achieve 4-level inversion. However the power semiconductor switches in one bank (three in number) in one of the inverters of this circuit have to be rated for the full DC link voltage.

The performance of the multi-level inverters very much depends on its PWM modulator. There have been many multi-level modulation techniques developed during the past decades. Both space vector modulation [6] and multi-level carrier based PWM [7] schemes have been used in many multi-level inverter topologies. These two techniques applied to the proposed topology and the advantages and disadvantages of these techniques are presented.

## II. PROPOSED 4-LEVEL INVERTER CONFIGURATION

In the proposed 4-level inverter circuit topology, the cascade connection of three 2-level inverters accomplishes 4-level inversion (Fig.1). The output phases of inverter-1 are connected to the DC input points of the corresponding phases in inverter-2 and the output phases of inverter-2 are connected to the DC input points of the corresponding phases in inverter-3. Each inverter is powered with an isolated DC power supply, with a voltage of $V_{dc}/3$ (Fig.1). where $V_{dc}$ is the DC-link voltage of an equivalent conventional single 2-level inverter drive.

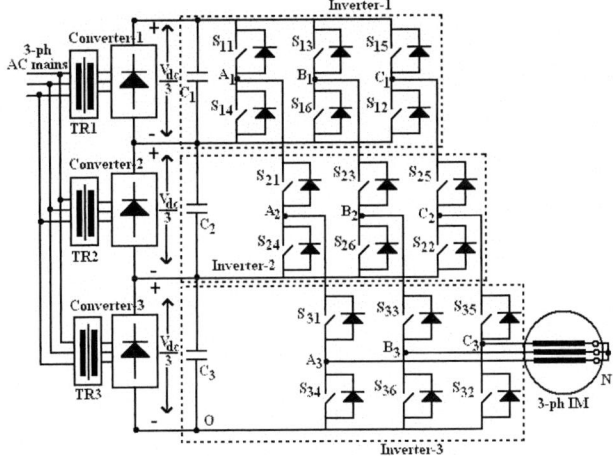

Fig. 1. The power circuit configuration of the Four-level inverter.

For inverter-3, let $V_{A3o}$, $V_{B3o}$ and $V_{C3o}$ represent the pole voltages of A, B and C phases respectively, referred to the point 'O' (Fig.1). The pole voltage, of any phase for inverter-3 for example $V_{A3o}$ (Fig.1) attains a voltage of $(1/3)V_{dc}$, if the following conditions are satisfied:

(i)     The top switch of that leg in inverter-3, in this case $S_{31}$, is turned on (Fig.1).

---

K.Chandra Sekhar is with the Department of Electrical & Electronics Engineering, R.V.R & J.C College of Engineering, Chowdavaram, Guntur-520 019, India (e-mail: cskoritala@hotmail.com).

G.Tulasi Ram Das is with the Department of Electrical & Electronics Engineering, J.N.T.U College of Engineering, Kukatpally, Hyderabad-500 072, India (e-mail: das_tulasiram@yahoo.co.in).

---

0-7803-9771-1/06/$25.00 ©2006 IEEE

(ii)  The bottom switch of the corresponding leg in inverter-2 in this case $S_{24}$, is turned on (Fig.1)

(iii)  The bottom switch of the corresponding leg in inverter-1 in this case $S_{14}$, is turned on (Fig.1)

Similarly the pole voltage of any phase in inverter-3, for example $V_{A3o}$ attains a voltage of $(2/3)V_{dc}$, if the following conditions are satisfied:

(i)  The top switch of that leg in inverter-3, in this case $S_{31}$, is turned on (Fig.1).

(ii)  The top switch of the corresponding leg in inverter-2, in this case $S_{21}$, is turned on (Fig.1).

(iii)  The top switch of the corresponding leg in inverter-1, in this case $S_{14}$ is turned on (Fig.1).

Similarly the pole voltage of any phase in inverter-3, for example $V_{A3o}$ attains a voltage of $(3/3)V_{dc}$, if the following conditions are satisfied:

(i)  The top switch of that leg in inverter-3, in this case $S_{31}$, is turned on (Fig.1).

(ii)  The top switch of the corresponding leg in inverter-2, in this case $S_{21}$, is turned on (Fig.1).

(iv)  The top switch of the corresponding leg in inverter-1, in this case $S_{11}$ is turned on (Fig.1).

Thus, the DC-input points of individual phases of inverter-3 may be connected to a DC-link voltage of either $(3/3) V_{dc}$ or $(2/3) V_{dc}$ or $(1/3) V_{dc}$ by turning on the top switch or the bottom switch of the corresponding phase leg in inverter-2 and inverter-1. Additionally, the pole voltage of a given phase in inverter-3 attains a voltage of zero, if the bottom switch of the corresponding leg in inverter-3 is turned on. Thus, the pole voltage of a given phase for inverter-3 is capable of assuming one of the four possible values 0, $(1/3)V_{dc}$, $(2/3)V_{dc}$ and $(3/3)V_{dc}$, which is the characteristic of a four level inverter.

The triplen harmonic currents are absent in this case, for the lack of a return path, as the neutral point 'N' of the motor is not connected to the point 'O'. Hence all the triplen harmonic voltages appear across the points 'O' and 'N' (Fig.1). Table I depicts individual inverter states and the switches turned on to realize that state. In the Table I, a '+' and a '-' indicate, respectively, that the top and bottom switches in an inverter leg are turned on.

TABLE I
INVERTER STATES FOR INDIVIDUAL INVERTERS

| State of Inverters- 1,2,3 | Switches turned on for inverter –1 | Switches turned on for inverter –2 | Switches turned on for inverter –3 |
|---|---|---|---|
| 1 (+ - -) | $S_{16}, S_{11}, S_{12}$ | $S_{26}, S_{21}, S_{22}$ | $S_{36}, S_{31}, S_{32}$ |
| 2 (+ + -) | $S_{11}, S_{12}, S_{13}$ | $S_{21}, S_{22}, S_{23}$ | $S_{31}, S_{32}, S_{33}$ |
| 3 (- + -) | $S_{12}, S_{13}, S_{14}$ | $S_{22}, S_{23}, S_{24}$ | $S_{32}, S_{33}, S_{34}$ |
| 4 (- + +) | $S_{13}, S_{14}, S_{15}$ | $S_{23}, S_{24}, S_{25}$ | $S_{33}, S_{34}, S_{35}$ |
| 5 (- - +) | $S_{14}, S_{15}, S_{16}$ | $S_{24}, S_{25}, S_{26}$ | $S_{34}, S_{35}, S_{36}$ |
| 6 (+ - +) | $S_{15}, S_{16}, S_{11}$ | $S_{25}, S_{26}, S_{21}$ | $S_{35}, S_{36}, S_{31}$ |
| 7 (+ + +) | $S_{11}, S_{13}, S_{15}$ | $S_{21}, S_{23}, S_{25}$ | $S_{31}, S_{33}, S_{35}$ |
| 8 (- - -) | $S_{12}, S_{14}, S_{16}$ | $S_{22}, S_{24}, S_{26}$ | $S_{32}, S_{34}, S_{36}$ |

An example is presented to evaluate the space vector location for a given space vector combination from individual inverters. The combination '126' means, inverter-1 is switched with a state of '1'(+--), inverter-2 is switched with a

state of '2'(+ + -) and inverter-3 is switched with a state of '6' (+ - +) as depicted in Table-I.

The space vector $V_S$ constituted by the pole voltages $V_{A3O}$, $V_{B3O}$ and $V_{C3O}$ is defined as:

$$V_S = V_{A3O} + V_{B3O}. \, e^{j(2\pi/3)} + V_{C3O}. \, e^{j(4\pi/3)} \qquad (1)$$

If inverter-1 assumes a state of ' 1' (+ - -) , Inverter-2 '2' (+ + -) and inverter-3 '6' (+ - +), it follows the earlier discussion that

$$V_{A3O} = (3/3)V_{dc} \; ; \; V_{B3O} = 0 \; ; \; V_{C3O} = (1/3)V_{dc.}$$

Consequently, the space-vector location for the above set of pole voltages is given by:

$$V_S = (3/3)V_{dc} + 0. \, e^{j(2\pi/3)} + (1/3)V_{dc}. \, e^{j(4\pi/3)} \qquad (2)$$

It may be verified that the tip of the space-vector corresponding to the above phase voltages is located at the point 'C$_2$' in Fig.2. The space-vector locations corresponding to the rest of the combinations may similarly be determined.

This configuration of four-level inverter eliminates the neutral point fluctuations associated with the conventional neutral clamped four-level inverter [1] as the capacitors $C_1$ and $C_2$ do not carry the load current but only the ripple currents. Also, the fast recovery neutral clamping diodes are eliminated in this topology of four-level inverter. This four-level inverter can be synthesized by reconnecting three existing two-level inverters as a retrofit. When these inverters drive the induction motor, each phase of the induction motor can attain four different levels.

## III. MODULATION SCHEME FOR THE PROPOSED INVERTER

### A. Space Vector Modulation

The 37 voltage space vector locations form the vertices of 54 equilateral triangles, which are referred to as sectors (Fig.2). These sectors are distributed into three layers (Fig.2). The equilateral triangles numbered '1' through '6' belonging to the inner most layer are referred to as 'inner sectors' (Fig.2). Layer-2 consists of the sectors numbered - '7' through '24' and layer-3 consists of the sectors numbered - '25' through '54' (Fig.2).

Six adjacent sectors constitute a sub-hexagon. Eighteen such sub-hexagons can be identified with the present scheme (Fig.2). In addition, there is one inner sub-hexagon with its center at O (Fig.2). Each outer sector can be mapped to the inner sector by shifting the outer sub- hexagonal center to the inner hexagonal center-O.

In this paper, the method employed to determine the timing periods $T_0$, $T_1$ and $T_2$ to realize the reference voltage space phasor $v_{sr}$, involves the following steps:

(i)  Finding the sector in which the tip of the reference space phasor is situated;

(ii)  Finding the outer sub-hexagon to which the sector belongs;

(iii)  Shifting the outer sub-hexagonal center to the inner most hexagonal center using an appropriate

450

coordinate transformation so that the reference voltage space phasor is mapped to the corresponding sector in the inner most sub-hexagon.

(iv) Determining the time periods $T_0$, $T_1$ and $T_2$ to realize the mapped reference voltage space phasor in the inner most hexagon [8];

(v) Employing these time periods to switch the space vector combinations available at the vertices forming the sector in which the tip of the reference space phasor is situated [8] [6].

Thus, this procedure is conceptually equivalent to realize the mapped reference space phasor in the inner hexagon and applying a vectored offset to realize the actual reference space phasor in the outer sector. It may be noted that this procedure ensures that the reference space phasor is realized by switching amongst the three vertices, which are situated in the closest proximity to the tip of the actual reference space phasor. Consequently, the switching ripple in the output voltage waveform is minimized.

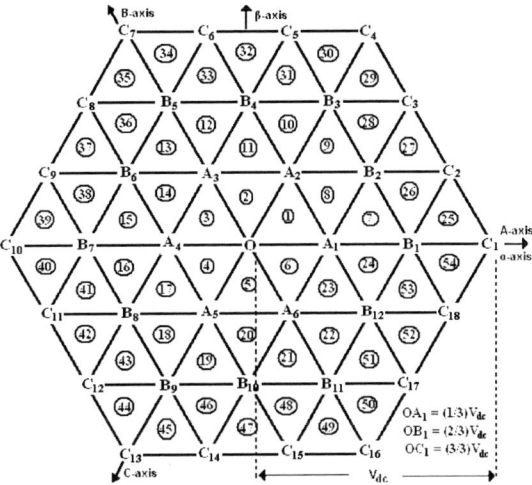

Fig. 2. Space phasor locations in the proposed circuit topology.

### B. Multi-level Carrier Based PWM

Multilevel carrier based PWM is used for the proposed inverter scheme. The multilevel carrier based PWM for an N-level inverter uses a set of N-1 adjacent level shifted triangular carrier waves with the amplitude and frequency [6]. If reference wave has peak amplitude $V_m^*$ and frequency $f_m$, the modulation index is defined with reference to a triangular wave of peak-to –peak amplitude of $V_C$ (N-1) as

$$M_a = 2V_m^* / V_C \ (N-1). \tag{3}$$

For the 4-level inverter drive structure, three triangular waves C1 to C3, with peak-to-peak amplitude of $V_C$ are used, as shown in Fig.3a. The peak-to-peak amplitude of each carrier is $V_C = (1/3)V_{max}$, where $V_{max}$ is the maximum value possible for the modulating signal. These three carriers divide the entire range of modulating signal to four regions R1 to R4, R1 being the region below the lowest carrier C1 and R4 the

region above the highest carrier C3. The regions between these two adjacent carriers are referred as R2 to R3. When the modulating signal is in a particular region a corresponding voltage level is applied across the motor phase winding as assigned below:

$$R1 => 0; \ R2 =>(1/3)V_{dc}; \ R3 => (2/3)V_{dc};$$
$$R4 => (3/3)V_{dc}; \tag{4}$$

Three 120 degree phase shifted sinusoids with 20% third harmonics content are used as the reference waves for the proposed carrier based PWM as the addition of third harmonic will result in a higher modulation index. These reference waves are continuously compared with the carrier set to determine the region (R1, R2...R4) in which the instantaneous value of the reference wave exists. This comparison is performed simultaneously for all the three phases. Control signals for the three inverters then can be generated such that the appropriate devices are switched to realize the particular level in a particular phase depending upon the region. It may be noted that the proposed inverter can realize the even numbered levels also, and can start with the 2-level operation and progressively move to the 3-level, and to the 4-level operation as the modulation index increases. For low modulation index such that $V_m^* \leq V_C/2$ where $V_m^*$ is the peak value of the modulating signal. If the reference wave is placed at the middle of the lowest carrier C1 as Fig.3b, the modulating signal exists only in two regions R1 or R2 and it will result in only two levels L1 (0) and L2 ((1/3) $V_{dc}$). In this case the switching losses are only due to two level inverter. When the modulating index increases such that $V_C/2 \leq V_m^* \leq V_C$, an additional DC bias of $V_C/2$ is given to the reference wave such that it is at the middle of the two lower carriers C1 and C2 and results in 3-level operation (Fig.3c). A similar progressive DC shift in steps $V_C/2$ of is gives such that the inverter progressively moves through the 3-level and to 4-level operation (Fig.3d ). When the V/f control is used, these 4 ranges of voltage amplitudes correspond to 4 ranges in frequency. Therefore the range (denoted by n = 1,2,..4) in which the frequency command falls can be used to determine the DC shift to be given to the reference waves and the reference waveforms can be represented by,

$$V_a^* = V_m \ sinwt + 0.2 \ V_m \ sin3wt + n \ V_c/2, \tag{5}$$

$$V_b^* = V_m \ sin(wt - 2\pi/3) + 0.2 \ V_m \ sin3wt + n \ V_c/2, \tag{6}$$

$$V_c^* = V_m \ sin(wt - 4\pi/3) + 0.2 \ V_m \ sin3wt + n \ V_c/2. \tag{7}$$

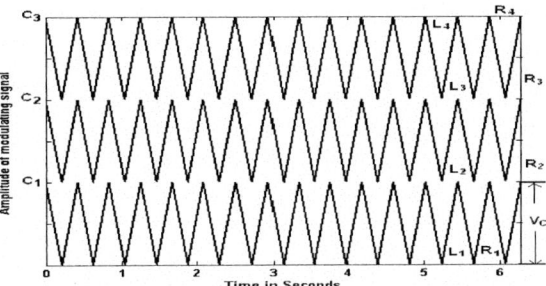

Fig. 3 a. The carrier and the different regions in the multi carrier PWM.

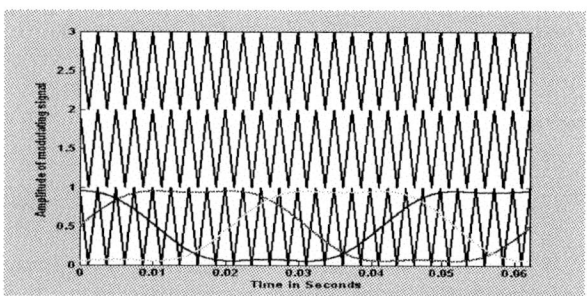

Fig. 3 b. The reference wave set for 2-level operation in the proposed multi-level carrier based PWM.

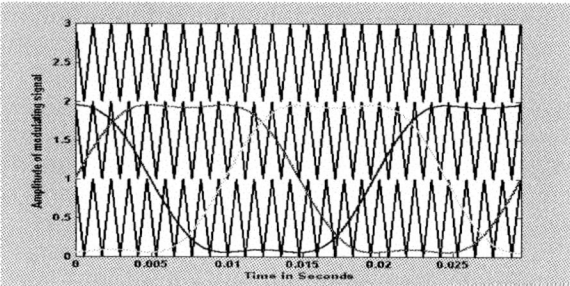

Fig. 3 c. The reference wave set for 3-level operation in the proposed multi-level carrier based PWM.

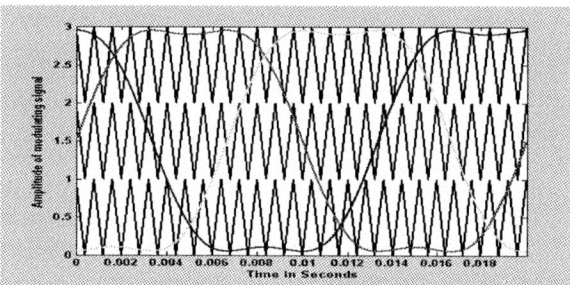

Fig. 3 d. The reference wave set for 4-level operation in the proposed multi-level carrier based PWM.

## IV. SIMULATION RESULTS AND DISCUSSION

Simulation studies have been carried out for the proposed inverter scheme in open-loop v/f control with space vector modulation and multilevel carrier based PWM using SIMULINK software in MATLAB environment. Figure 4 shows the block diagram of the proposed inverter scheme. The respective DC-bus voltages are $(1/3)V_{dc}$, $(1/3)V_{dc}$ and $(1/3)V_{dc}$ for the inverter-1, inverter-2 and inverter-3. This means that the DC-bus voltage of an equivalent conventional 2-level inverter is $V_{dc}$. In case of space vector modulation, look-up tables are employed for the generation of PWM signals in each layer. In case of multi-level carrier based PWM, the speed reference is translated to the frequency and voltage commands maintaining V/f. Depending upon the range in which the frequency command falls the reference waves are generated according to Eqn. (5), (6) and (7). The three reference waves are simultaneously compared with the carrier set and the level at which the instantaneous value of the reference wave exists is determined.

A DC-bus voltage ($V_{dc}$) of 300 is assumed for simulation studies. The motor phase voltage waveform and the normalized harmonic spectrum of the motor phase voltage for $|v_{sr}| = 0.2V_{dc}$ are presented in Fig.5a and Fig.5b. In this case, the tip of the reference voltage space phasor $v_{sr}$ is confined to the inner sectors i.e. sectors '1' through '6' (Fig.2). The motor phase voltage shows the familiar 2-level waveform as the switching is confined to the inner hexagon. The corresponding motor phase voltage waveform and its normalized harmonic spectrum in the proposed multi-level carrier based PWM are presented in the Fig.8a and Fig.8b. Only inverter-3 is switched in this case. Similar simulation results are presented for $|v_{sr}| = 0.5V_{dc}$. In this case, the tip of the reference voltage vector is confined to the layer-2, which consists of sectors numbered '7' through '24'. In this operating region, inverter-2 and inverter-3 are switched while inverter-1 is clamped to a state of '8'(---). Fig.6a and Fig.6b shows the waveforms of the motor phase voltage and normalized harmonic spectrum of the motor phase voltage. In this case, the motor phase voltage shows a 3-level waveform. Fig.9a and Fig.9b show the corresponding motor phase voltage waveform and the normalized harmonic spectrum of the motor phase voltage in the proposed multi-level carrier based PWM. Similar conclusions can be drawn from the simulation result shown in Fig.7a and Fig.7b corresponding to the case $|v_{sr}| = 0.8V_{dc}$. In this operating condition, the tip of $v_{sr}$ is situated exclusively in the sectors of layer-3 (sectors numbered '25' through '54', Fig.2). Unlike the two previous cases, all the inverters are switched in this operating condition. In this case, the motor phase voltage shows a 4-level waveform. Fig.10a and Fig.10b show the corresponding motor phase voltage waveform and the normalized harmonic spectrum of the motor phase voltage in the proposed multi-level carrier based PWM.

From the above simulation results, it may be concluded that the proposed 4-level inverter is capable of rendering a good performance with the proposed PWM techniques. Space vector modulation provides a more efficient use of the dc bus as well as smaller torque ripple, lower switching loss and lower total harmonic distortion in an ac motor drive application. The multi-level carrier based PWM does not require the look-up tables to realize the switching sequences as in case of space vector modulation. As number of levels increases the space vector modulation becomes formidable and cumbersome proposition to switch appropriate space vector combinations using look-up table approach.

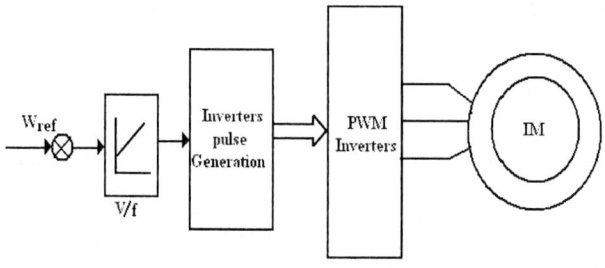

Fig. 4. Block diagram of the proposed multi-level-Inverter for an Induction motor drive system in open loop v/f control.

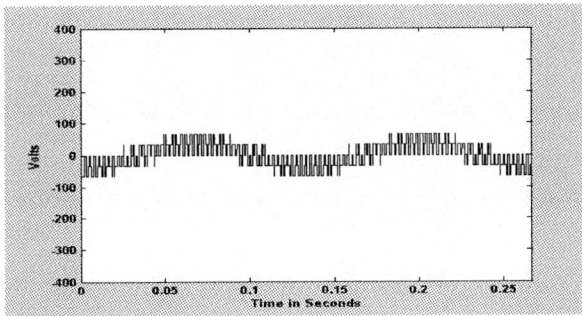

Fig. 5a. The motor phase voltage when $|v_{sr}| = 0.2V_{dc}$.

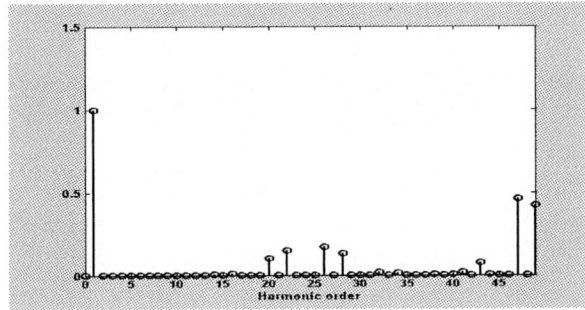

Fig. 5b. Normalized harmonic spectrum of the motor phase voltage when $|v_{sr}| = 0.2V_{dc}$.

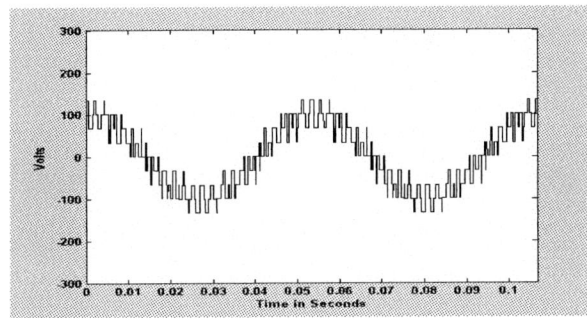

Fig. 6a. The motor phase voltage when $|v_{sr}| = 0.5V_{dc}$.

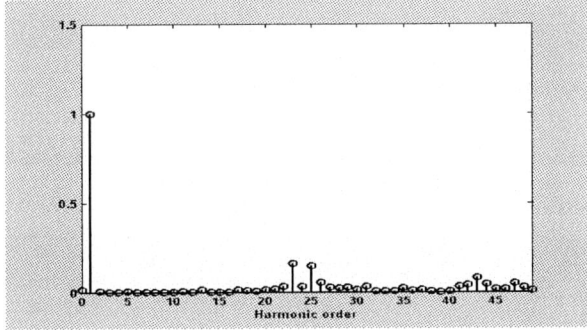

Fig. 6b. Normalized harmonic spectrum of the motor phase voltage when $|v_{sr}| = 0.5V_{dc}$.

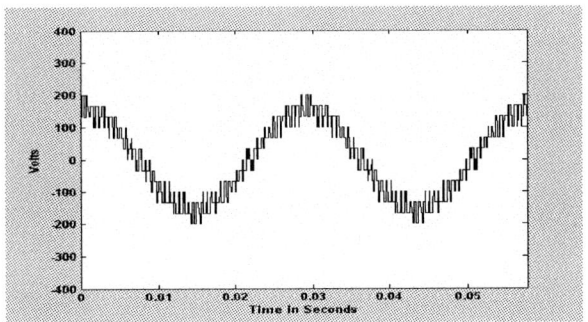

Fig. 7a. The motor phase voltage when $|v_{sr}| = 0.8V_{dc}$.

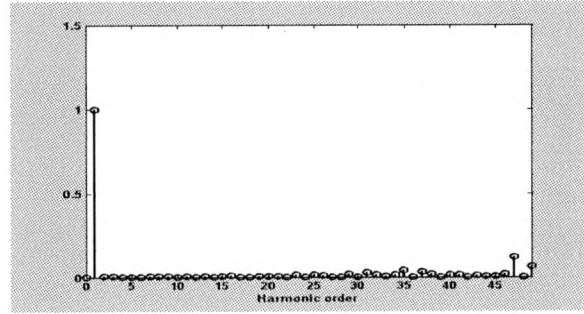

Fig. 7b. Normalized harmonic spectrum of the motor phase voltage when $|v_{sr}| = 0.8V_{dc}$.

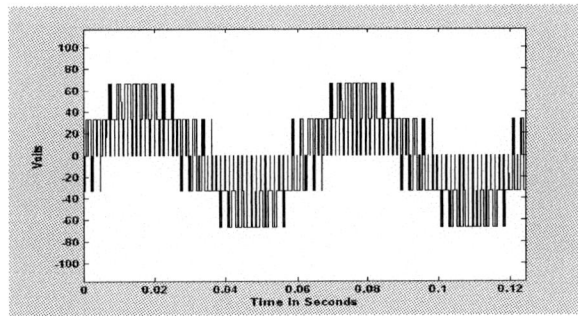

Fig. 8a. Motor phase voltages during 2-level operation in the multi-level carrier based PWM.

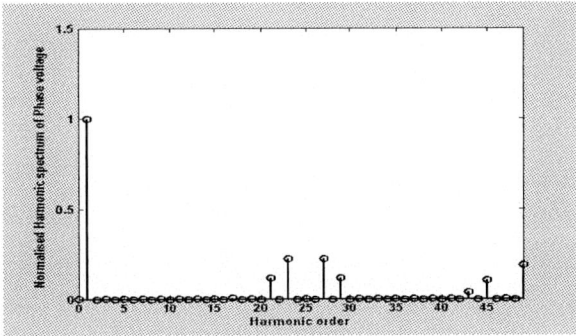

Fig. 8b. Normalized harmonic spectrum of the motor phase voltage during 2-level operation in the multi-level carrier based PWM.

453

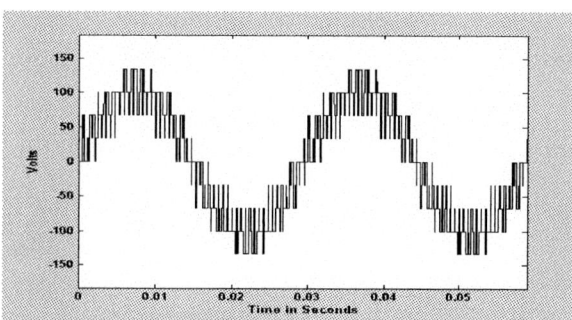

Fig. 9a. Motor phase voltages during 3 -level operation in the multi-level carrier based PWM.

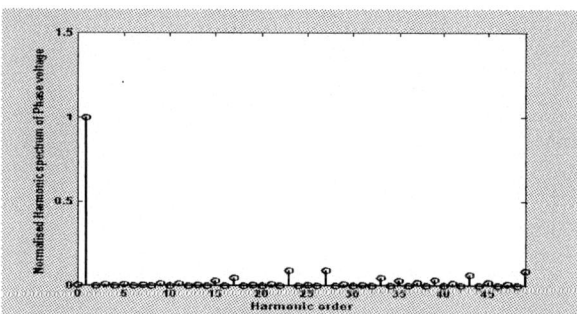

Fig. 9b. Normalized harmonic spectrum of the motor phase voltage during 3-level operation in the multi-level carrier based PWM.

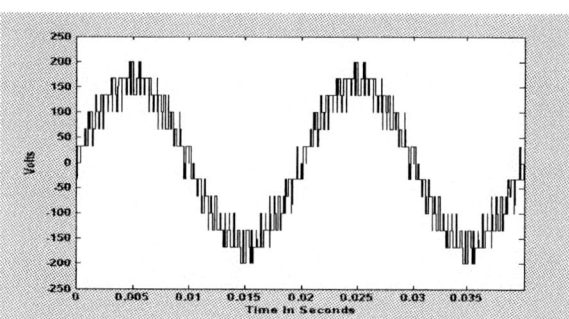

Fig. 10a. Motor phase voltages during 4-level operation in the multi-level carrier based PWM.

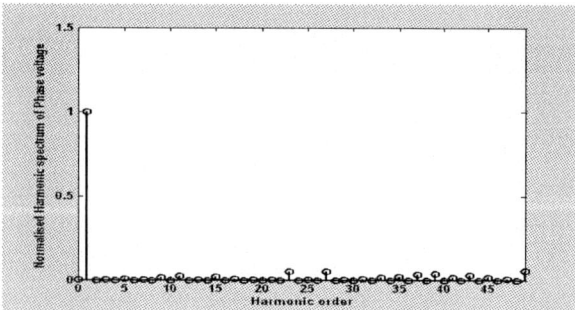

Fig. 10b. Normalized harmonic spectrum of the motor phase voltage during 4-level operation in the multi-level carrier based PWM.

## V. CONCLUSION

A four-level inverter topology cascading three 2-level inverters is presented. Each inverter is operated with an isolated DC-link voltage of $V_{dc}/3$, where $V_{dc}$ is the DC-link voltage of an equivalent conventional 2-level inverter drive. This scheme is capable of producing 512 voltage space phasor combinations distributed over 37 space phasor locations. The proposed four-level inverter configuration requires lesser number of switching devices as compared to the conventional four-level inverter scheme. Similarly, lesser-isolated power supplies are required when compared to the cascaded H-bridge configuration. This inverter scheme does not experience neutral-point fluctuations. Space vector modulation provides a more efficient use of the dc bus as well as smaller torque ripple, lower switching loss and lower total harmonic distortion in an ac motor drive application. The multi-level carrier based PWM does not require the look-up tables to realize the switching sequences as in case of space vector modulation.

## VI. REFERENCES

[1]  A.Nabae, I.Takahashi, and H.Agaki,"A New Neutral- Point-Clamped PWM Inverter", *IEEE Transactions on Industry Applications, vol. IA-17, Sept./Oct. 1981*, pp 518- 523.

[2]  Madhav D. Manjrekar and Thomas A. Lipo,  "A Hybrid Multilevel Inverter Topology for Drive Applications", in *Proceedings of the 1998 IEEE – APEC Conference*, pp.523-529.

[3]  A.Rufer, M.Veenstra and K.Gopakumar, "Asymmetric Multilevel Converter for High Resolution Voltage Phasor Generation", in *Proceedings of the 1999 EPE Conference*, pp. P1-P10.

[4]  B.S. Suh and D.S. Hyun, A new N-level high voltage inversion system, IEEE Trans., IE-44, 107-115 (1997).

[5]  V.T Somasekhar, K.Gopakumar, " Three-level inverter configuration cascading two-level inverter" IEE *Proc.Electr.Power Appl.Vol.150, No.3 may 2003*,pp.245-254.

[6]  Joohn-Sheok Kim and Seung-Ki Sul, "A Novel Voltage Modulation Technique of the Space Vector PWM", in *Proceedings of the 1995 IPEC Conference*, pp.742-747.

[7]  G.Carrara, S.G.Garedella and M.Marchesoni, R.Salutary and G.Sciutto, "A new multilevel PWM method: a  theoretical analysis", IEEE Trans. Power Electronics, Vol.7, No 3, July 1992. pp.497-505.

[8]  E.G.Shivakumar, K.Gopakumar and V.T.Ranganathan, "Space vector PWM control of Dual Inverter fed Open-end winding Induction Motor drive", in *Proceedings of the 2001 IEEE – APEC Conference*,  pp.394 – 404.

## VII. BIOGRAPHIES

**K.Chandra Sekhar** born in 1968 in India.  He received B.Tech degree in Electrical & Electronics Engineering from V.R.Siddartha Engineering College, Vijayawada, India in 1991 and M.Tech in Electrical Machines & Industrial Drives from Regional Engineering College, Warangal, India in 1994. Currently, he is faculty in the Department of Electrical & Electronics Engineering, R.V.R & J.C.College of engineering Guntur, India. He is working towards the Ph.D. Degree in power electronics and drives from J.N.T.U College of Engineering, Hyderabad- 500072, India. His research interests are in the areas of power electronics, industrial drives and pwm techniques.

**G.Tulasi Ram Das** born in 1960 in India. He received B.Tech degree in Electrical & Electronics Engineering from J.N.T.U college of Engineering, Hyderabad, India in 1983 and M.E in Industrial Drives & Control from O.U College of Engineering, Hyderabad, India in 1986. He Received PhD, degree from the Indian Institute of Technology, Madras, India in 1996.  Currently he is faculty in the Department of Electrical & Electronics Engineering, J.N.T.U College of Engineering, Hyderabad- 500 072, India. His Research interests are in the areas of Power Electronics, Industrial   Drives & FACTS Devices.

**2006 IEEE International Conference on Power Electronic, Drives and Energy Systems**

# A Unified Model For Auxiliary Switch Commutated DC-DC Converters

### N. Lakshminarasamma, and V. Ramanarayanan

*Abstract*— A novel ZVS auxiliary switch commutated variation for all DC-DC converter topologies has been proposed in 2006. With proper designation of the circuit variables (throw current I and the pole voltage V), all these converters are seen to be governed by an identical set of equations. With idealized switches, the steady-state performance is obtainable in an analytical form. The conversion ratio of the converter topologies is obtained. A generalized equivalent circuit emerges for all these converters from the steady-state conversion ratio. It also provides a dynamic model as well. With these generalized steady-state equivalent circuits, small signal analysis of these converters may be carried out readily. It enables one to use the familiar state space averaged results of the standard PWM DC-DC converters for the resonant counterparts. Th dc and ac models reveals that dc and low frequency behaviour of the proposed family of converters is similar to that of its PWM parent.

*Index Terms*— Auxiliary switch commutation, Dynamic model, Steady-state model, Zero voltage switching.

## I. INTRODUCTION

THE demand for smaller and lighter power DC-DC converters is pushing switching frequencies well into mega Hertz range. Such high-frequency switching is possible by resonant topologies. In contrast to the sharp-edged switching waveforms of PWM converters, the resonant converter topologies feature smooth waveforms resulting in reduced switching losses and lesser interference.

Steve Freeland and R.D. Middlebrook has brought out the unified model of all quasi-resonant converters elegantly in a paper in 1987 [2]. Unified steady state model for the family of active clamp converters was reported in [3]. The present paper identifies such a unified performance results for family of auxiliary switch commutated converters, proposed in [4]. This is done by defining a normalized current $I_N$, through the pole voltage V, throw current I, and the switching period $T_s$ of the switch. With such a definition, the circuit intervals and the defining equations become identical in all the auxiliary switch commutated converters. In the sections that follow, the method of analysis is outlined through the example of auxiliary switch commutated buck converter. The circuit-topologies in the sub-interval and the solution for the same under idealized operation are derived. Equivalent circuits valid for steady state as well

This work was supported by the research grant from Indian space research organization.

First author N. Lakshminarasamma is a research student in the Department of Electrical Engineering, Indian Institute of Science Email id: lakshmin@ee.iisc.ernet.in

Second author V.Ramanarayanan is currently Professor in the Department of Electrical Engineering, Indian Institute of Science, Bangalore, Email id: vram@ee.iisc.ernet.in

as dynamic performance are proposed. The paper is organized as follows: Section II presents the generic requirements of ZVT PWM Converters with auxiliary switch. Section III presents the mathematical analysis of the performance of a sample buck converter with auxiliary switch circuit. Steady-state modeling of auxiliary switch commutated converters are presented in section IV and section V presents the small signal circuit averaged model of auxiliary switch commutated buck converter. Simulation and experimental results - steady-state results and small signal model verification of 33 Watt, 400 KHz converter are presented in section VI. Section VII gives the conclusion and the references.

## II. ZVT PWM CONVERTERS MECHANISM

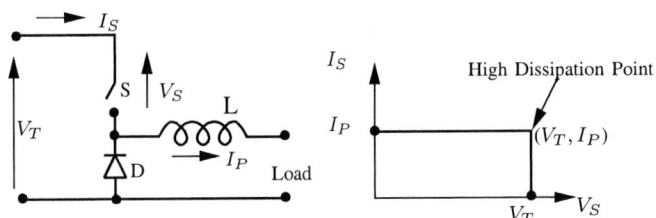

Fig. 1. A typical switching pole in a power converter.

Fig. 1 shows the basic switching element common to all switching power converters [1]. The throw voltage $V_T$, and the pole current $I_P$ are defined as shown in Fig. 1. The active and the passive switches are S and D respectively. The switch voltage $V_S$ and the switch current $I_S$ trajectories are shown in Fig. 1. Every turn-on and turn-off process transits through the high dissipation point of $(V_T, I_P)$. This results in high switching losses which is proportional to the switching frequency. The proposed scheme introduces an auxiliary circuit connected in parallel to the active switch. The auxiliary circuit consists of auxiliary switch $S_a$, a series diode $D_a$, a set of resonant elements $L_a$ and $C_a$ and a dependant voltage source $V_a$ as shown in Fig. 2. The auxiliary circuit when switched properly, ensures lossless switching.

## III. STEADY-STATE ANALYSIS OF AUXLIARY SWITCH COMMUTATED BUCK CONVERTER

Fig. 3 shows the primitive auxiliary circuit for a buck converter employing this method. The commutation process and the mathematical analysis is explained for the buck converter with auxiliary switch. To simplify the analysis, it is considered that, the converter is operating in steady state and the following assumptions are made.

1) All components and devices are ideal.

0-7803-9771-1/06/$25.00 ©2006 IEEE

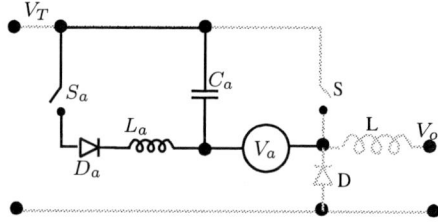

Fig. 2. Auxiliary circuit cell.

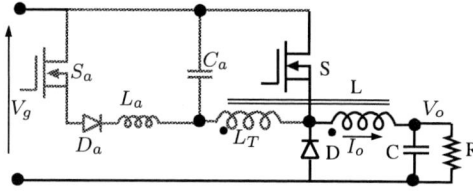

Fig. 3. Buck converter with primitive auxiliary switch commutation circuit.

2) The output filter inductor L is large enough to assume that the output current $I_o$ is constant.
3) The output capacitor C is large enough to assume that the output voltage is constant and ripple free.

The turns ratio between L and $L_T$ may be chosen conveniently. The winding $L_T$ has to carry the commutation current and reset current only. Therefore the RMS value of this coupled winding will be a small fraction of the current flowing in the main inductor L. Accordingly, this will not demand a higher size of inductor. The complete commutation process and the mathematical analysis are explained below for a buck converter. To simplify the analysis, turns ratio for the coupled inductor (L and $L_T$) is taken to be 1. Switching sequences are as shown in Fig. **??**.

**Interval 0** $(t < t_o)$: Prior to time $t = t_o$, the main switch S and the auxiliary switch $S_a$ is in OFF state. The load current is freewheeling through the diode D. The resonant capacitor is charged to voltage $(V_g + V_0)$.

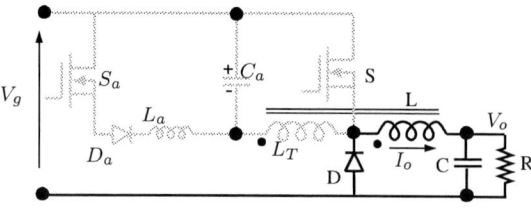

Fig. 4. ZVS buck converter: Interval 0: $(t < t_o)$.

**Interval 1** $(t_o < t < t_1)$: This interval begins when the auxiliary switch $S_a$ is turned-on with ZCS at $t = t_o$. The equivalent circuit is shown in Fig. 5 with $I_{La}(t_o) = 0$ as the initial current of resonant inductor $L_a$. The load current is freewheeling through the passive switch D. The current in the auxiliary switch will increase linearly as in Eq. 1.

$$i_{sa}(t) = \frac{(V_g + V_o)}{L_a}(t - t_o) = \frac{V_g(1 + D)}{L_a}(t - t_o) \quad (1)$$

At t = $t_1$, when $i_{sa}(t)$ reaches $\frac{I_o}{2}$, the passive switch D turns-

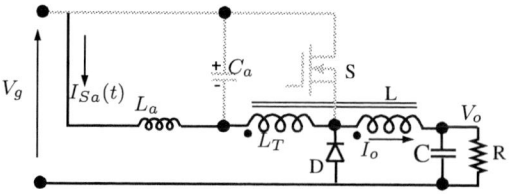

Fig. 5. ZVS buck converter: Interval 1: $(t_o < t < t_1)$. off.

$$T_1 = (t_1 - t_o) = \frac{I_o L_a}{2V_g(1 + D)} \quad (2)$$

**Interval 2** $(t_1 < t < t_2)$: The turn-off of the passive switch D is followed by the resonant interval. The resonant elements $L_a$ and $C_a$ resonate during this interval. This interval ends, when the voltage across the resonant capacitor $v_{Ca}(t)$ reaches $(V_o - V_g)$. This forward biases the body diode of the main switch S.

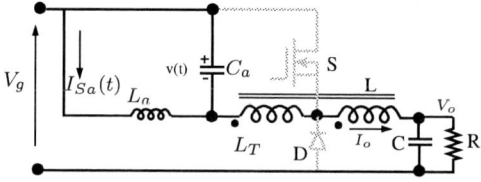

Fig. 6. ZVS buck converter: Interval 2: $(t_1 < t < t_2)$.

$$i_{Sa}(t) = \frac{I_o}{2} + V_g(1 + D)\sqrt{\frac{C_a}{L_a}}sin(\omega(t - t_1)) \quad (3)$$

$$v_{Ca}(t) = V_g(1 + D)cos(\omega(t - t_1)) \; ; \; \omega = \frac{1}{\sqrt{L_a C_a}} \quad (4)$$

$$V_{Ca}(t_2) = V_o - V_g = -V_g(1 - D) \; ; \; where \; t_1 < t < t_2 \quad (5)$$

$$\omega T_2 = \omega((t_2 - t_1)) = cos^{-1}\left[\frac{-(1 - D)}{(1 + D)}\right] \quad (6)$$

$$I_{Sa}(t_2) = 2V_g\sqrt{\frac{C_a}{L_a}}\sqrt{D} + \frac{I_P}{2} \quad (7)$$

At time $t = t_2$, body diode of the main switch S is ON and main switch S can now be turned-on with ZVS. **Interval 3** $(t_2 < t < t_3)$: The resonant inductor current flows through the main switch S, auxiliary switch $S_a$ and the auxiliary diode $D_a$. The trapped energy in the auxiliary circuit inductor $L_a$ is recovered into the coupled inductor $L_T$. The voltage across the coupled winding is $(V_g - V_o)$. The negative voltage across the resonant inductor $L_a$ will reset $i_{La}(t)$ linearly to zero as given by Eq. 8. Turn-off of the auxiliary switch at t = $t_3$, ensures ZCS.

$$i_{Sa}(t) = \frac{-V_g(1 - D)}{L_a}(t - t_2) + I_{Sa}(t_2) \quad (8)$$

End of interval $T_3$ is when $i_{Sa}(T_3) = 0$

$$T_3 = t_3 - t_2 = \sqrt{L_a C_a}\frac{\sqrt{D}}{(1 - D)} + \frac{I_P L_a}{V_g(1 - D)} \quad (9)$$

At the end of $t_3$, turn-off of the auxiliary switch $S_a$ is therefore lossless (ZCS). Following the interval $T_3$, the gate drive to the auxiliary switch may be turned-off.
**Interval 4** $(t_4 < t < t_5)$: The main switch S is switched-off at $t = t_4$ i.e at the end of $DT_S$. The turn-off of the main switch

456

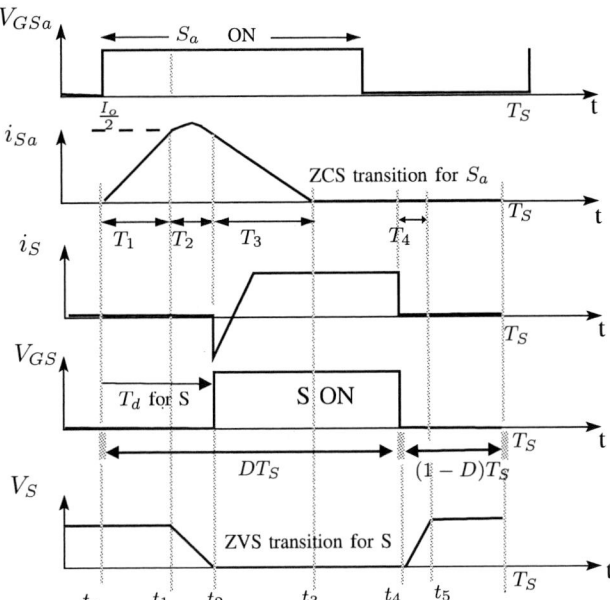

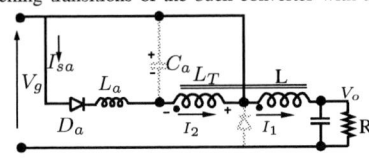

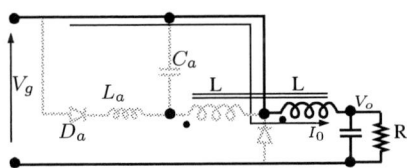

Fig. 7. Switching transitions of the buck converter with auxiliary switch.

Fig. 8. ZVS buck converter: Interval 3: ($t_2 < t < t_3$).

is at zero voltage, on account of capacitor across the main switch. The voltage across the switch raises slowly thereby reducing the turn-off transition losses.

Fig. 9. ZVS buck converter: End of commutation.

It is observed that turn-on and turn-off transitions of main switch and auxiliary switch are loss-less.

## IV. STEADY-STATE MODEL OF AUXILIARY SWITCH COMMUTATED CONVERTER

### A. Steady state conversion ratios

The conversion ratio $M = V_o/V_g$ evaluated by averaging the pole voltage over a full cycle, is given in table. I Under the assumption that the resonant frequency is much higher than the switching frequency, the conversion ratio for the buck, boost and buckboost converter is given in table. I.

### B. Equivalent circuits of Resonant commutated converters

The conversion ratio ($V_o/V_g$) shown for the buck converter in the Eq. IV-B may be simplified as follows:

$$V_o = DV_g - \frac{L_r I}{2T_S(1+D)} \qquad (10)$$

## TABLE I
### DEFINITIONS OF POLE VOLTAGE, CURRENT FOR CONVERTERS

|  | Buck converter | Boost converter | Buck-Boost |
|---|---|---|---|
| V | $V_g$ | $V_o$ | $V_g + V_o$ |
| I | $I_o$ | $I_g$ | $I_L$ |
| $R_d$ | $\dfrac{L_r}{2T_s(1+D)}$ | $\dfrac{L_r}{2T_s(1+D)}$ | $\dfrac{L_r}{2T_s(1+D)}$ |
| M | $D - \dfrac{I_n}{2(1+D)}$ | $\dfrac{1}{1-(D-\dfrac{I_n}{2(1+D)})}$ | $\dfrac{D - \dfrac{I_n}{2(1+D)}}{1-(D-\dfrac{I_n}{2(1+D)})}$ |

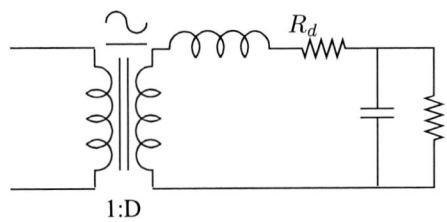

Fig. 10. Eqvt. circuit of resonant switch commutated buck converter.

This may be shown by the following equivalent circuit Fig. 10. Steady-state equivalent cirucits of boost and buck boost converters are shown in Fig. 11 The interesting result

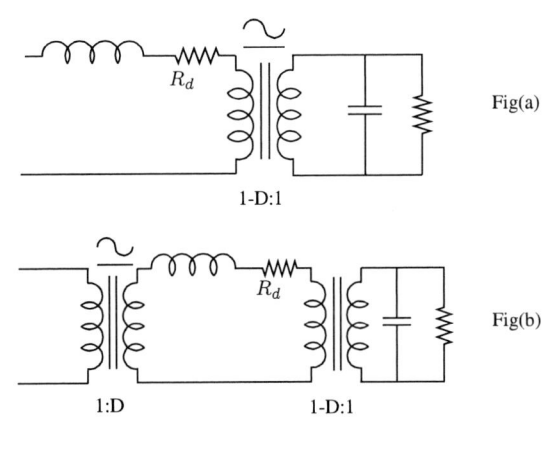

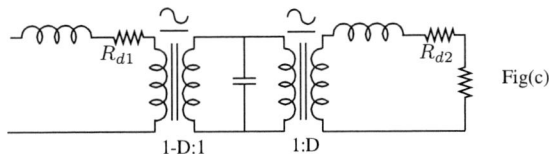

Fig. 11. Equivalent Circuits of the Auxiliary switch commutated Boost, buck-boost and cuk converters.

is that the resonant switch commutated converters retain the qualitative nature of the hard-switched counterparts with additional lossless damping introduced in the equivalent circuit as shown in Fig. 10. This simplification process may be done for all types of converters.

## V. SMALL SIGNAL MODEL

The signal model is obtained by the basic assumption that the natural time constants of the converter network are much longer than the switching period. This assumption coincides with the requirement of small switching ripple. Hence, with this basic assumption, the resulting averaged model is obtained in the fig. 8 by averaging the converter waveforms over the switching period T.

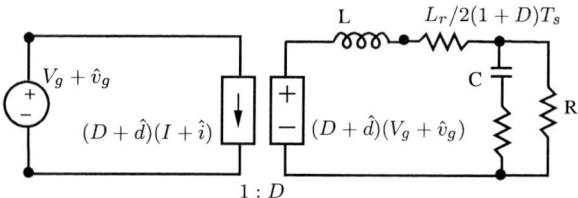

Fig. 12. Perturbation of the nonlinear circuit averaged model about a quiescent operating point.

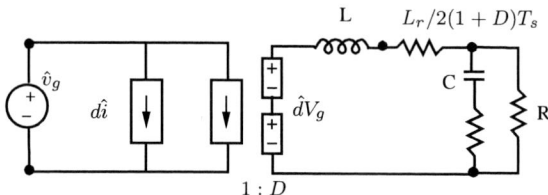

Fig. 13. Lineralized circuit averaged model.

The dependant linear sources are replaced by an equivalent ideal transformer, yielding the final small signal ac circuit averaged model.

Fig. 14 shows the averaged model of a auxiliary switch commutated ZVS buck converter obtained following the linear circuit reduction technique.

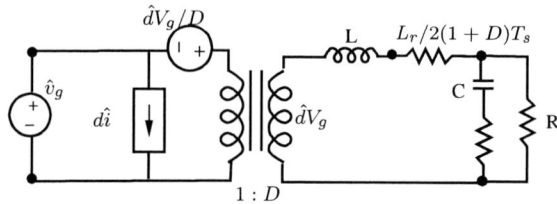

Fig. 14. Small signal AC model of an auxiliary switch commutated buck converter.

## VI. Experimental Measurements

### A. Steady state experimental waveforms

The circuit diagram is shown in Fig. 15. Following is the specifications of the prototype: Input voltage $V_g = 18 - 25$ V, Output voltage $V_o = 30$ V, Output power $P_o = 33$ W, Switching frequency $F_S = 400$ KHz. Simulation results and experimental results are shown in Fig. 16 and Fig. 17 respectively for the boost converter with coupled inductor.

The 33 Watt 400KHz boost converter with auxiliary switch is simulated in sequel and the results of the same are presented Fig. 16.

1) Fig. 17(a) shows the gate pulses of frequency 400 KHz driving the auxiliary switch $S_a$ and the main switch S.

2) Fig. 17(b) shows the pole voltage and the resonant inductor current waveforms. As seen in theoretical and simulated results [Fig. 7 and Fig. 16d], there is a linear resetting of the resonant inductor current. In the experimental results, high frequency oscillations are observed

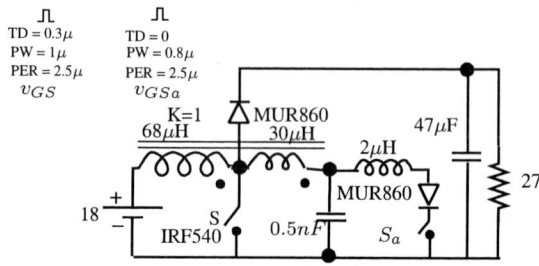

Fig. 15. Simulated boost converter with coupled inductor.

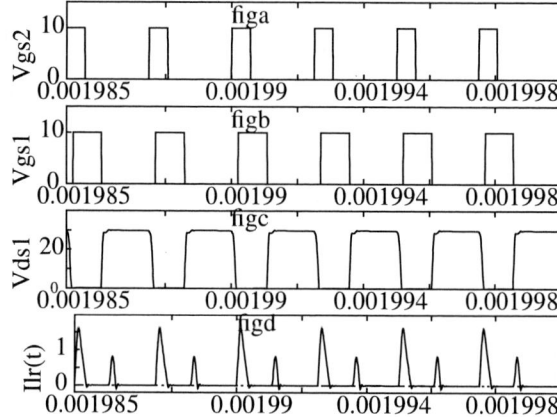

Fig. 16. Simulated waveforms of 33 Watt auxiliary switch boost converter with coupled inductor.

during the resetting interval. These are due to parasitics of the resonant inductor.

3) Fig. 17(c) indicates the ZVS turn-on of the main switch. The gate drive is turned-on after the drain-to-source voltage $V_{ds}$ of main switch has reached zero.

4) Fig. 17(d) indicates the Zero current switching transitions of the auxiliary switch $S_a$. The auxiliary switch is turned-on ($V_{dsa} = 0$) at zero current. After the linear resetting of $i_{La}(t)$ to zero, the auxiliary switch is turned-off at ZCS.

### B. Small signal Model of Auxiliary switch commutated boost converter

As a means of verifying the small signal model developed in the section 4, output impedance was measured using the netqork analyzer. The control gain measurement is done with both the auxiliary switch circuit disabled (hard switched boost converter) and the auxiliary switch circuit enabled for the circuit shown in Fig. 15. Fig. 18 shows the magnitude and phase plot of control gain of the hard switched boost converter and the auxiliary switch commutated boost converter prototype: 18-25 V input voltage, 30 V output voltage, 33 W, 400 kHz.

The control gain plot of the hard-switched boost converter shows Q effect (seen by the peak in the magnitude as well as a sharp change in the phase angle at the natural frequency of the filter). The Q for the hard-switched buck converter is seen to be 2 dB. The damping exhibited by the auxiliary

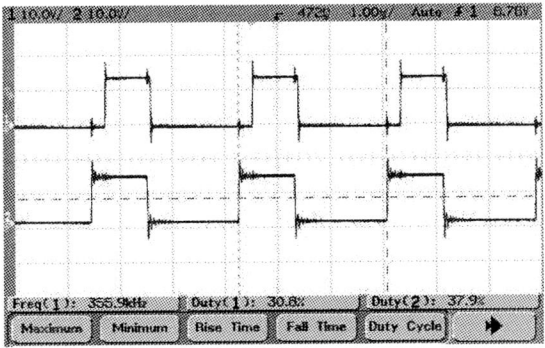

(a) Ch1: 10 volts/div ; Ch2: 10 volts/div :$V_{gs}$ and $V_{gsa}$ of auxiliary switch $S_a$ and active switch S.

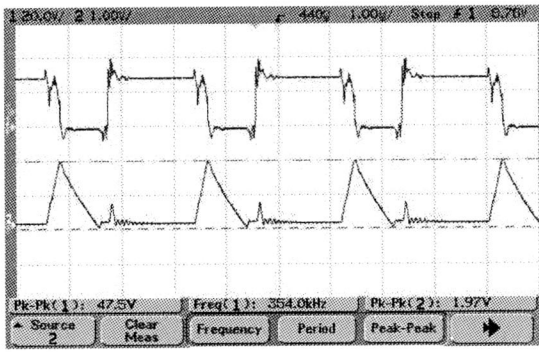

(b) Ch1: 20 volts/div ; Ch2: **1 Amp**/div :pole voltage $V_p$ and resonant inductor current $i_{La}(t)$.

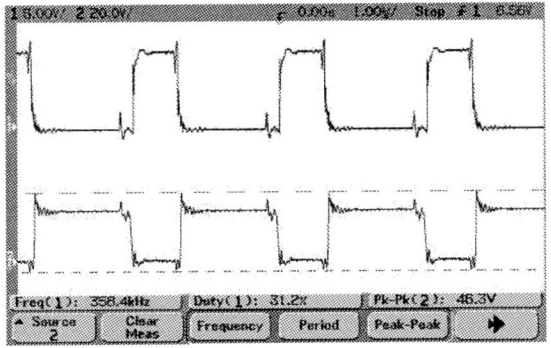

(c) Ch2: 20 volts/div ; Ch1: 5 volts/div :$V_{ds}$ and $V_{gs}$ of active switch S indicating ZVS turn-on.

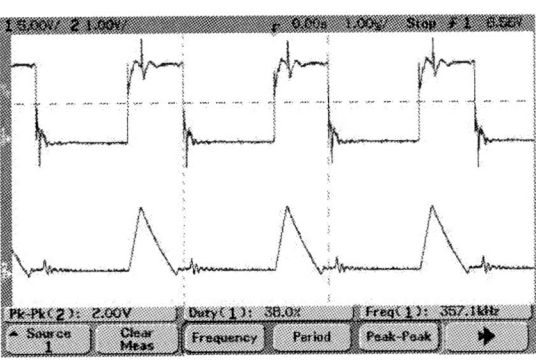

(d) Ch1: 5 volts/div ; Ch2: **1 Amp**/div :$V_{gsa}$ and $i_{Sa}(t)$ of auxiliary switch indicating ZCS.

Fig. 17. Experimental waveforms of 33 Watt, 400 kHz Auxiliary switch commutated boost converter with tapped-coupled inductor.

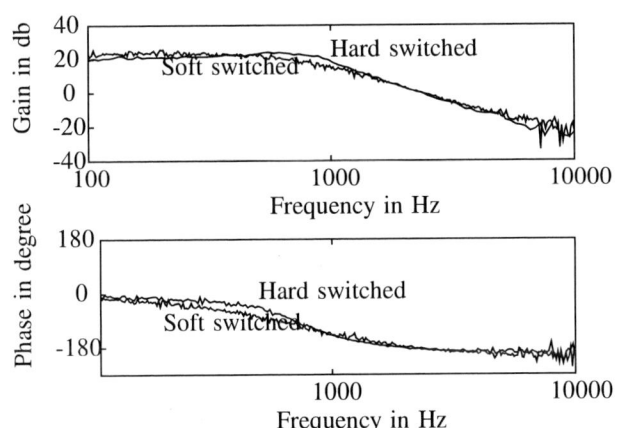

Fig. 18. Measured Control gain frequency responses of Hard-switched and Auxiliary switch commutated boost converter prototype: 33 watt, 500 KHz.

switch commutated converters is approximately 1/6 of the active clamp converters [3].

## VII. Conclusions

This family of auxiliary switch commutated converters are derivable from the hard-switched dc-dc converters by addition of resonant circuit elements following simple rules. Though there are a large number of variations present, all these converters are seen to have the same 4 sub-intervals per cycle. Further the circuit equations governing these sub-intervals are identical when expressed in terms of pole current; throw voltage and freewheeling resonant circuit voltage (I, V). It is seen that the steady state and dynamic equivalent circuits can be obtained from this idealized analysis. Apart from reducing the switching losses in the converter, the resonant sub-interval introduces lossless damping in the converter dynamics. The most striking result is the simple elegance of the equivalent circuit of this family of converters. Simulation and experimental results are presented for a 33 watt, 400 kHz boost converter.

## VIII. Acknowledgments

The work, reported in this paper was supported by a research grant from Indian space research organization.

## References

[1] Middlebrook, R.D., and Sloobodan Cuk, 'Advances in Switched-Mode Power Conversion', Volumes I and II, 2nd Edition, TESLOco, 1983'

[2] 'Freeland, S., Middlebrook, R.D., 'A Unified Analysis of converters with Resonant Switches', *PESC*, pp. 20-30, 1987.

[3] 'Lakshminarasamma N, Swaminathan B, Ramanarayanan V, 'A unified Model for ZVS DC to DC Converters with Active Clamp', *PESC*, pp. 2441-2447, 2004.

[4] 'N.Lakshminarasamma, V.Ramanarayanan, 'A Family of Auxiliary Switch ZVS-PWM DC-DC Converters with Coupled Inductor', *IECON*, 2006.

**2006 IEEE International Conference on Power Electronic, Drives and Energy Systems**

# Novel Pulse Power Supply Operating at High Input Power Factor

Vishnu K Sharma, Kishore Chatterjee, and Vivek Agarwal

*Abstract-* **Pulse Power Supplies (PPS) are used in several applications like food processing, plasma operation, water sterilization etc. Most of the existing topologies of PPS suffer from drawbacks such as requirement of high voltage DC power supply, series operation of switching devices and use of a pulse transformer. In this paper, an improved PPS configuration, based on the combination of an array of boost converters and an input boost rectifier stage is proposed. Not only does this topology overcomes the above mentioned limitations, but also results in high power factor at the AC mains input and allows independent control of the output pulses' amplitude and duration. Results of computer simulations and laboratory experiments are presented**

## I. INTRODUCTION

MARX generator [1]-[4] is a conventional topology for PPS, and does not require any pulse transformer, series connection of switches and high voltage DC power supply. But in this topology spark gaps are used as switches, which causes of demerits like low frequency operation, and short life time. The efficiency of system is also low as capacitors are charged through resistors. A boost converter array based PPS [5] is a modification of Marx generator, which overcomes the limitations of conventional Marx generator. In this topology, semiconductor switches and inductors respectively replace spark gaps and resistances of Marx generator. Fig. 1 shows the circuit diagram of this topology. Modified circuit is nothing but the cascaded connection of boost converter and capacitor of previous stage works as a DC input for next stage.

Vishnu. K. Sharma is with Bhabha Atomic Research Cente, Mumbai-400085, India (e-mail: vishnuiitb@yahoo.com).

K. Chaterjee is with the Department of Electrical Engineering, Indian Institute of Technology Bombay, Mumbai-400076, India (e-mail: kishore@ee.iitb.ac.in).

V. Agarwal is with the Department of Electrical Engineering, Indian Institute of Technology Bombay, Mumbai-400076, India (e-mail: agarwal@ee.iitb.ac.in).

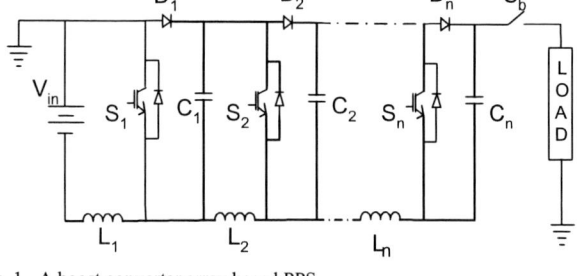

Fig. 1. A boost converter array based PPS.

All switches $S_1$, $S_2$,…etc. are turned on simultaneously, and all capacitors come in series with the load during turn on as shown in fig. 2.

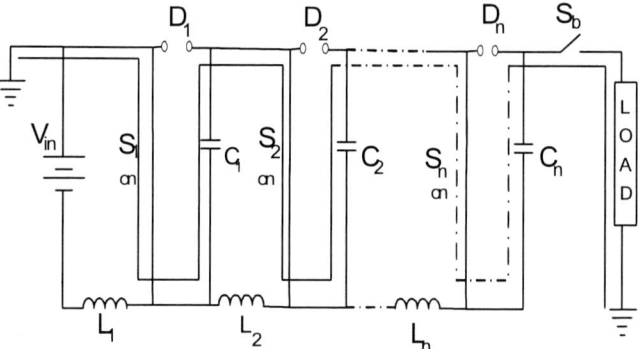

Fig. 2. Equivalent circuit during turn on time.

Magnitude of output voltage pulse is given by:
$$V_0(p) = V_{C1} + V_{C2} + \ldots V_{CN} \quad (1)$$
The width of the output pulses is determined by the load requirements.

If the current is continuous, the capacitor voltage of $n^{th}$ stage is given by;
$$V_{cn} = V_{cn-1}\frac{1}{1-D}, \; D = \frac{T_{on}}{T_s} \quad (2)$$

where D is the duty ratio, $T_{on}$ is pulse width and $T_s$ is the period of pulse. Therefore, the output voltage of the $n^{th}$ stage is given by;
$$V_{cn} = V_{c1}\frac{1}{(1-D)^n} \quad (3)$$

where D is the duty ratio, $T_{on}$ is pulse width and $T_S$ is pulse period. If inductor current is discontinuous, capacitor voltage is given by [5];

$$V_{cn} = \left( \frac{T_{on} + T_\Delta}{T_\Delta} \right) V_{c1} \qquad (4)$$

where $T_\Delta$ is the time until the current becomes zero after the switch turns off. It is clear from equation 1-4 that the voltage gain of a boost converter depends on its duty cycle. Fig. 3 shows the simulated output voltage pulse for 5% and 10% duty ratio, for a three stage 2.5kHz PPS, keeping input voltage constant. It is clear form fig. 3 that duty ratio and output voltage can't be controlled independently for constant input DC voltage.

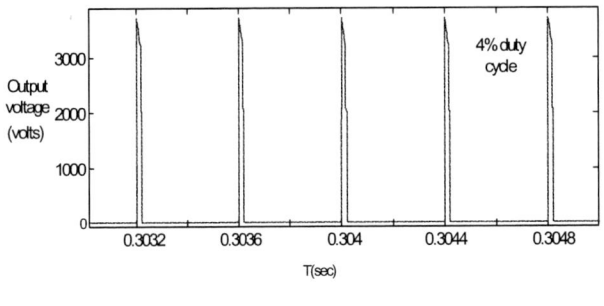

(a)

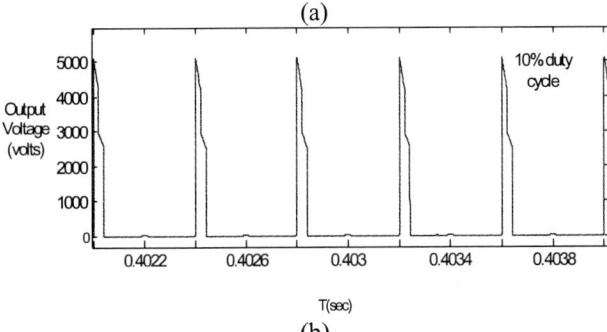

(b)

Fig. 3. Output voltage pulse (a) for 5% duty ratio (b) for 10% duty ratio.

This paper proposes a new PPS configuration, which allows independent control of output pulse width and amplitude of voltage simultaneously. Moreover, it also operates at near unity power factor.

## II. PROPOSED PPS

Fig. 4 shows the circuit diagram of the proposed PPS. In proposed PPS, Constant DC supply is replaced by boost rectifier. Boost rectifier provides variable DC input to boost converter array for regulating the output voltage. There are two switching pulses in the circuit, one for boost rectifier and another for remaining switches of boost converter array.

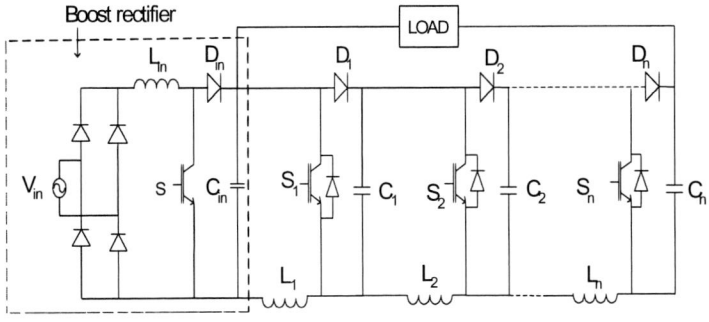

Fig . 4. Circuit diagram of the proposed PPS.

Control scheme of proposed PPS is shown in fig. 5.

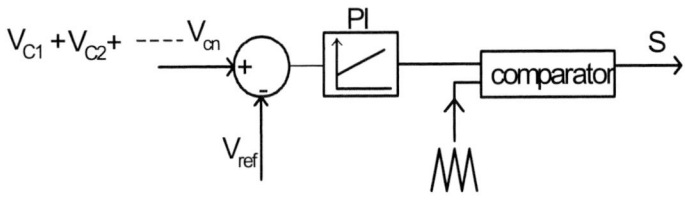

Fig. 5. Output voltage control scheme.

As amplitude of output pulse $V_0(p)$ is sum of voltages of $C_1$, $C_2...C_n$, capacitors of boost converter array, it is compared with a reference voltage, which is processed through the PI controller to generate the control command for the converter to reduce the error. For variable duty ratio, PI scheme automatically control the output voltage. Thus proposed topology allows independent control of duty ratio and output voltage.

Figs. 6 and 7 show the simulation results of a three stage 10kV PPS. Fig.4 shows the output voltage waveform, while individual filter capacitor's voltages have been shown in Fig.5. All the Converters are operated in DCM.

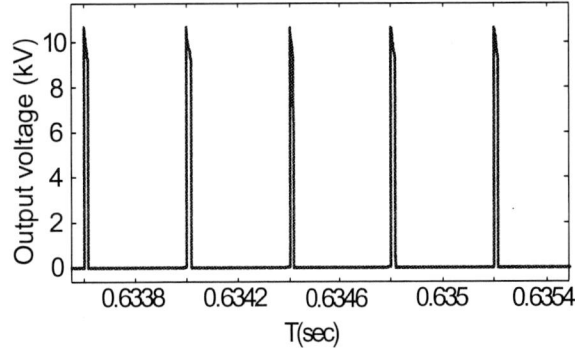

Fig. 6. Output voltage.

461

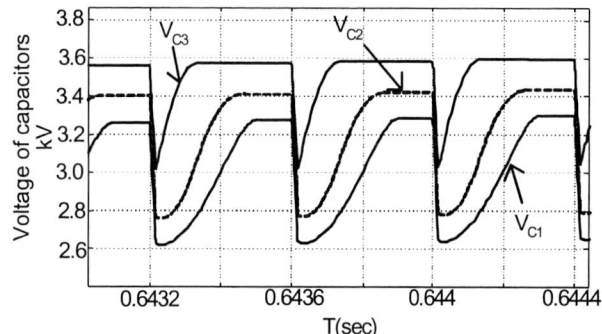

Fig. 7. Voltage of capacitors.

## III. DESIGN CONSIDERATIONS

The designing of circuit is based on characteristics of output pulse, which are described by following parameters

1.  The pulse width of output voltage
2.  The energy transferred in each pulse duration
3.  The magnitude of pulse voltage
4.  The frequency of the voltage pulse

The capacitor's value of boost converter array is given by [5]

$$C_{1.....n} = \frac{\tau \times V_{out}}{\Delta V \times R_{load}} \qquad (5)$$

where $\tau$ is the pulse width and $\Delta V$ is voltage drop during pulse. $\Delta V$ is decided by energy packets, which are to be transferred in each pulse duration. Energy transferred into a single pulse is given as follows;

$$E = \frac{1}{2} C_{eq}(V_1^2 - V_2^2) \qquad (6)$$

where $V_1$ and $V_2$ is the output voltage at the initial and final instant during the pulse and $C_{eq}$ is the equivalent capacitance, when all capacitance of boost converters are in series during switching.

Inductors of boost converter array are preferably operated in discontinuous current mode for reducing reverse recovery problem of diodes, so inductor value is given by [5];

$$L \leq \frac{T_s V_{out}}{2 I_{out}} D(1-D)^2 \qquad (7)$$

Filter capacitor and inductor of boost rectifier should satisfy following inequality; [6]

$$C_{in} \geq \frac{2 I_0}{V_{r1}} |f(t)|_{peak} \qquad (8)$$

where $V_{r1}$ is the voltage ripples across the capacitor $C_{in}$. $f(t)$ is a function of time, and is given as follows

$$f(t) = \{\frac{\pi}{\omega\lambda} \int_0^{\omega t} \frac{\sin^2 \omega t}{a - \sin \omega t} d\omega t - t\} \qquad (9)$$

where

$$a = \frac{V_{Cin}}{V_m} \qquad (10)$$

where $V_{Cin}$, $V_m$ are the capacitor voltage across $C_{in}$ and peak value of AC side voltage of boost rectifier. The inductor in the rectifier circuit is designed by following inequality.

$$\text{And } L_{in} \leq \frac{V_m}{I_P} D_m T_S \qquad (11)$$

## IV. CONCLUSION AND FUTURE SCOPE

A boost converter array topology has been proposed. A 10kV, 20A, 5 kHz PPS has been designed and simulated. Prototype hardware is implemented in open loop with 500V, 4.5 kHz, $40\mu s$ output pulse. For controlling the voltage of output pulse, boost rectifier is used as a DC input for boost converter array. Proposed PPS can be used in applications such as food processing and water sterilization. Most of the problems of conventional topologies like high DC input voltage, high voltage pulse transformer, series operation of switching devices are absent in proposed topology.

The prototype hardware, that is implemented is a crude model, and has some possibilities of improvements. The capacitors, which are discharged in load periodically, have some parasitic inductance, which distort the shape of output pulse, due to rise time and fall time. This phenomenon comes in picture more significantly at high voltage, and high frequency. The stray impedance of connecting wires also affects the voltage swing in switching period. The load is modeled in simulation and experimental work. A sophisticated model with least connecting wires and stray impedance of capacitors has to be made with actual load.

## V. REFRENCES

[1] J. H. Kim, M. H. Ryu, B.D. Min, S. V. Shenderey, "High voltage pulse power supply using Marx generator and solid state switches," in *Conf. Rec. IEEE Industrial Electronics Soc. Annu. Meet.*, Nov. 2005, pp. 1244-1247.

[2] S. G. E. Pronko, M. T. Ngo, and R. F. K. Germer, "A Solid-state Marx-type Trigger Generator," *8th Power Modulator Symposium*, 1988, pp. 211 – 214.

[3] W. J. Carey, and, J. R. Mayes, "Marx generator design and Performance," *24th Power Modulator Symposium and High-Voltage Workshop*, 2002, pp. 625 – 628.

[4] S. E. Calico and M. C. Scott, "Development of a Compact Marx Generator for High-power Microwave Applications," *11th IEEE International Pulse Power Conf.*, 1997, pp. 1536 – 1541.

[5] J.W. Baek, M.H. Ryu, D. W. Yoo, H.G. Kim, "High voltage pulse generator using boost converter array," in *Conf. Rec. IEEE Industrial Electron. soc. Annu. Meet.*, Nov. 2002, pp. 395-399.

[6] T.C. Chen, C.T. Pan, "Modelling and design of a single phase AC to DC converter," in *IEE Proc.B.*, Sep. 1992, pp. 465-470.

[7] M. H. Rashid, *Power Electronics Hand Book*, Academic Press California, 2001.

# 2006 IEEE International Conference on Power Electronic, Drives and Energy Systems

# System Identification and controller tuning rule for DC-DC converter using ripple voltage waveform

K. Lavanya, *Student Member, IEEE*, B. Umamaheswari, and R. C. Panda

*Abstract*—The Ripple voltage waveform of DC-DC buck converter is used for system identification. Based on the parameters identified and using the user specification of gain margin and phase margin, PI controller parameter tuning rule is proposed. The closed loop performance with line and load variation of the Buck converter is simulated using MATLAB/SIMULINK and analyzed. The robustness measure of the PI controller on the performance is also studied.

*Index Terms*—System identification, PI controller, DC-DC converter

## I. Introduction

IN most switched mode power supplies DC-DC converter output voltage is controlled using PI controller. Normally in designing PI parameters, Root locus technique is used if the system model is known. If not, any system identification technique is used to find the system parameters and then a suitable controller is designed. Relay feedback technique is used for system identification for autotuning of controllers. In many cases the system is assumed as a first order with dead time or second order with dead time. The relay feedback experiment gives the parameters for the chosen system model. [1] discusses on the PI tuning and analysis for different types of systems. [2] discusses on identifying the system parameters using the waveshape that results form the relay feedback experiment. The idea of using the waveshape for system identification is used in this work.[3] describes the use of phase margin and gain margin to tune the controllers. Mainly the focus of such tuning rules is for chemical processes. [4] describes the ripple voltage regulation of Buck converter.

When the DC-DC conveter is controlled by means of a switching device, output has a ripple. This work explore the

---

K. Lavanya Research Scholar is with the Department of Electrical and Electronics Engineering, Anna University, Chennai-25 (e-mail: lavanyakrishnasamy@yahoo.com).

B. Umamaheswari, Professor and Head of the Department of Electrical and Electronics Engineering, Anna University, Chennai-25 (e-mail: umamahesb@annauniv.edu).

R. C. Panda , is with the Department of Chemical Engineering, Curtin university, Perth , Australia (e-mail: rcpanda@yahoo.com)

possibility of using this ripple information for identification of the model.

Buck converter model is subjected to relay feedback and the parameters are identified assuming a second order system with dead time. Then using the ripple information with out the relay feedback, the system is identified. Using gain and phase margins and the identified parameters, tuning rule for PI controller is proposed.

Section II gives the details about the relay feedback method of identifying the system parameters of Buck converter and the proposed identification of system parameters with ripple information. Section III describes the simulation results and the robustness measure of the PI controller for the system model. Concluding remarks are drawn at the end.

## II. System Identification and Controller Design

If a relay with amplification '*h*' is inserted in a feedback loop, the input *u(t)* becomes '±*h*' as shown in Fig 2 [3]. The output y(t) oscillates with period '*Pu*' and a limit cycle of amplitude '*a*' for a dead time '*D*'. From the principle harmonic approximation of the oscillations, the ultimate gain ($K_u$) can be approximated as $K_u = 4h/\pi a$, and ultimate frequency ($\omega_u$) thus becomes $\omega u = 2\pi/P_u$ (where $P_u$ is the period of oscillation).

### A. Identification of Transfer Function

Consider an SOPDT model defined by (1).

$$G_P(s) = \frac{K_p e^{-Ds}}{(\tau^2 s^2 + 2\xi\tau s + 1)} \tag{1}$$

where ξ is damping coefficient, D is the time delay and τ is the time constant. A priori guess is made about the unknown process from the shape of the relay response. It is required to find the parameters like $K_p$, D, τ and ξ, in order to find the model. From the relay response, $K_p$, D, and the period of oscillation Pu, are estimated and a priori guess of ξ and τ is made. The analytical expression for the relay feedback response of this kind of processes is given in [2].

$$y(t) = k_p h \left\{ 1 - 2\frac{e^{\frac{-\xi t}{\tau}}}{\beta} \sin\left(\frac{\beta t}{\tau} + \alpha\right) \right\} \tag{2}$$

Using Pu, ξ and τ, the analytical expression is superimposed on the relay response as shown in Fig.2. The landmark values are the starting (A, at t=0) and ending (B, at t=Pu/2 ), $t_p$ and zero-crossing of response. Using the landmark points, the boundary conditions (3) are used to estimate the parameters [2].

$$(y)_{t=P_u/2} = a$$

$$\left(\frac{dy}{dt}\right)_{t=t_p} = 0$$

$$(y)_{t=0} + (y)_{t=pu/2} = 0 \qquad (3)$$

The following are assumed to formulate the boundary conditions.

    (i)    The relay responses are symmetric about zero base line.

    (ii)    The response starts from bottom most point and ends at next top point

    *(iii)*    If more than one peak exists, consider the last peak

D, Pu and $t_p$ are easily measured from the relay response curve. Height of point B in Fig.2 (where $t=0.5P_u$) from zero base line gives an estimate of $K_P$. Let $t_P$ is given by (4)

$$t_p = D^* - D + Pu/2 \qquad (4a)$$

$$\tan\left(\frac{\beta t_p}{\tau} + \alpha\right) = \frac{\beta}{\xi} \qquad (4b)$$

$$k_p h\left(1 - \frac{\sin\alpha}{\beta}\right) = a \qquad (4c)$$

Equation (4) is solved simultaneously to find τ and ξ. Estimate of τ has no errors, whereas estimation of ξ is found to vary with D. Hence an empirical formula (5) is arrived at for correcting ξ as a function of D.

$$\xi_{corrected} = 0.26\exp\left\{\left(\frac{D+1}{D}\right)\xi\right\} \qquad (5)$$

### B. Tuning Rule

The controller, H(s) in Fig 3 is replaced by a relay with amplitude 'h' resulting in limit cycle operation with oscillation condition at the process critical frequency is given by (6).

$$G(j\omega) = -\frac{\pi a}{4h} \qquad (6)$$

As the same PI tuning formula is used for various systems the relay amplitude and the corresponding output magnitude of the relay test varies. Also the limit cycle occurrence of the relay test varies for all systems. Hence with all these under consideration, the estimated gain margin given by (7).

$$|GH| = \frac{K}{\sqrt{1 + \omega^2\lambda^2}} \qquad (7)$$

The desired transfer function is given by (8)

$$G(s)H(s) = \frac{1}{1 + s\lambda}e^{-i\phi} \qquad (8)$$

where $\phi = \pi - \frac{PM * \pi}{180} - \tan^{-1}(\omega\lambda)$. The loop transfer function is given by (9).

$$G(s)H(s) = \frac{K_p K_c e^{-Ds}}{1 + s\tau_p} \qquad (9)$$

using (8) and (9) the PI parameters are obtained.

$$T_i = \frac{1}{\omega}\tan(-D\omega - \theta) \quad where \quad \theta = \frac{\pi}{180}PM - \pi$$

$$K_c = \frac{\sqrt{1 + \omega^2\tau_p^2}}{K_p(GM)} \quad where \quad K = \begin{cases} 0.03, D > 1 \\ \dfrac{D}{\tau_p}, D < 1 \end{cases} \qquad (10)$$

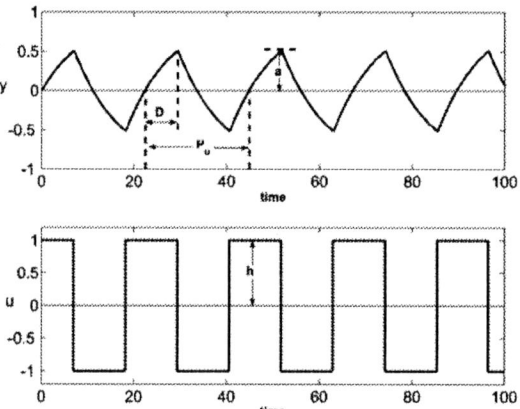

Fig. 1. Typical relay feedback input and output.

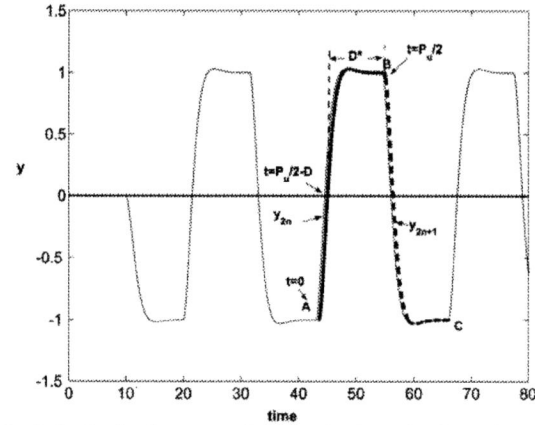

Fig. 2. Relay feedback response for second order under damped system and corresponding analytical expression (thick lines) and landmark values.

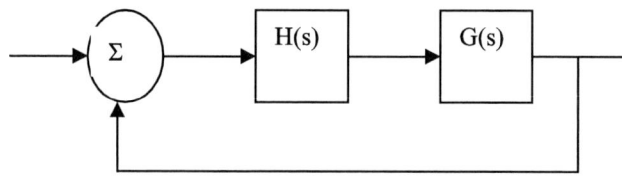

Fig. 3. Schematic Block diagram of closed loop system.

## C. Example System

The Buck converter with two stage filter is considered for the analysis. The parameters and the specification is given in table 1. Using the relay feedback and the identification technique, the parameters are identified.

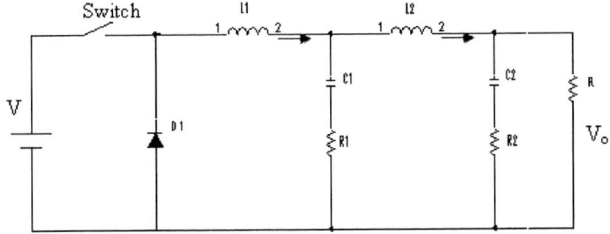

Fig. 4. Buck converter with two stage filter.

### TABLE I
BUCK CONVERTER PARAMETERS AND SPECIFICATION

| Inductor, L1 | 368μH |
|---|---|
| Inductor,L2 | 65.3 μH |
| Capacitor,C1 | 99 μF |
| Capacitor,C2 | 10 μF |
| Resistor,R1 | 0.01ohms |
| Resistor,R2 | 0.01 ohms |
| Output voltage | 5V |
| Switching frequency | 50Khz |
| Input voltage (nominal) | 15V |
| Load | 1A |

## III. SIMULATION RESULTS

### A. Relay feedback and Ripple voltage responses

Fig. 5. shows the relay feedback response for Buck converter system and the identified system model which is obtained from the analytical expression using (2). Equation (4) is used to find the parameters.

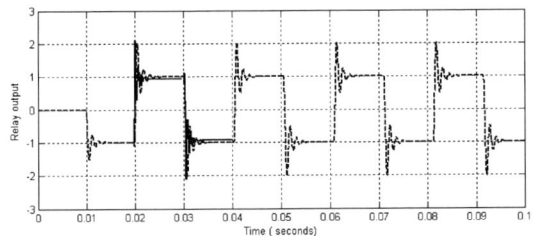

Fig. 5. Relay feedback response for DC-DC Buck Converter system and corresponding analytical expression (thick lines).

PI controller parameters are obtained using (10) and the corresponding step response is shown in Fig. 6. Instead of relay being replaced by the controller in Fig. 3, the ripple information is used to find out the parameters. It is seen that when the ripple information is used, the period of oscillation obtained from the relay feedback experiment is replaced by the switching frequency of the Buck converter. For making

the identification simple, the delay obtained from the relay feedback is kept as such and the identification is done. The identified parameters and the PI controller parameters are shown in table 2.

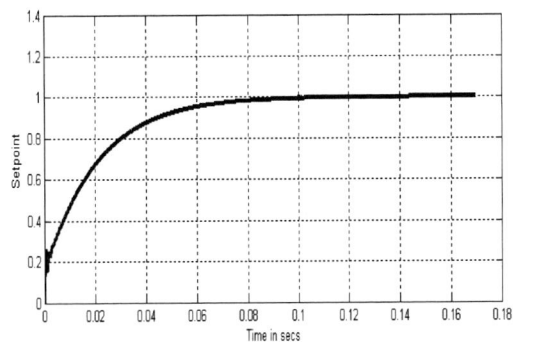

Fig. 6. Step Response of the system when subjected to relay feedback.

### TABLE II
IDENTIFIED AND TUNED CONTROLLER PARAMETERS

| Method of identification | Identified model parameters | | PI controller parameters | |
|---|---|---|---|---|
| | $T_p$ | D | $K_c$ | $T_i$ |
| Ripple | 0.0003 | 0.01 | 35.463 | 0.098 |
| Relay feedback | 0.0001 | 0.01 | 0.18 | 0.003 |

The Buck converter is subjected to step response using the Pi parameters. The results are shown in Fig. 7.

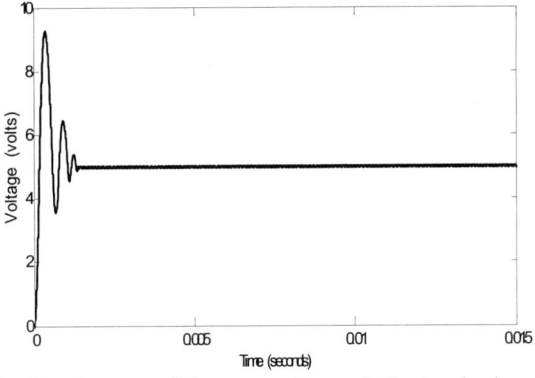

Fig. 7. Step Response of the system when controller tuned using ripple voltage.

It is seen from Fig. 7 that settling time is 1millsecond. The same system with the controller is subjected to line and load variation. The results are shown in Fig.8 and Fig.9 respectively. Line variation of 20% is applied above and below the nominal voltage. Also the variation is subjected for about 2milliseconds. Similarly the system is subjected to load variation for about 2milliseconds. It is seen that there exists overshoot. This could be reduced by choosing the PI parameters which in turn depend on the parameters identified.

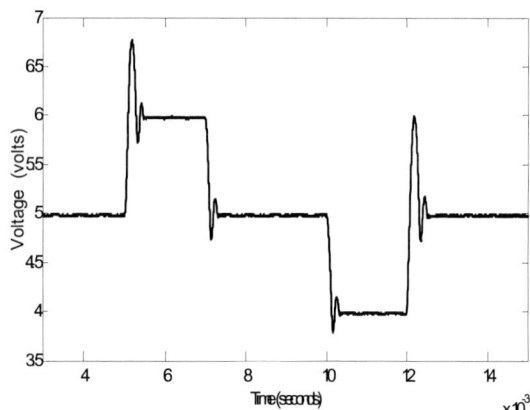

Fig. 8. Response due to line variation.

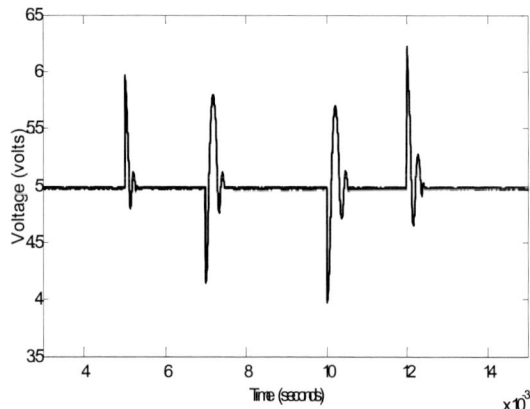

Fig. 9. Response due to load variation.

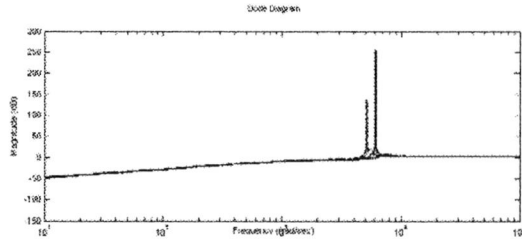

Fig. 8. Robustness analysis.

## IV. REFERENCES

[1]  C. C. Yu, "Autotuning of PID controllers," Springer verlag, London 1999.

[2]  R. C. Panda, C. C. Yu, "Analytical expressions for relay feedback responses," Journal on process control., vol 13(6) 2003.

[3]  K. J. Astrom and T. Hagglund, "Automatic tuning of simple regulators with specifications on phase and amplitude margins,"Automatica, vol. 20, 1984,p. 645-655.

[4]  Chung-Hsien Tso and Jiin Chuan Wu, "A ripple control Buck regulator with fixed output frequency," IEEE power Electronics letters, vol.1, no.3, sep 2003.

### B. Robustness study

The present PI tuning rule has some distinct advantages. Firstly, user can specify desired PM to have a desired closed loop response. Secondly, with the help of tuning parameter, $\lambda$, closed-loop trajectory can be made faster or sluggish.    Fig. 10. explains the robust performance of the proposed scheme. A nominal process of general structure  is considered for this study. Changes on process parameters are given and robust performance is evaluated with synthesized controller. Using MATLAB command 'robuststab', a performance margin with, upper bound: 0.5715, lower bound: 0.5715 & destabilizing frequency: 0.0021 was obtained. Thus, from the observations it can be concluded that the scheme is robust under ±33% with dead-time model–uncertainty as well as other parametric uncertainty.

### C. Conclusion

This work describes a method to identify the parameters of the model chosen using the ripple voltage. Using the identified parameters, and the user information of phase margin and gain margin, the controller is tuned. It is found from the responses that the PI parameters obtained results in more overshoot. This could be reduced if more precise model is obtained using the ripple information.

**2006 IEEE International Conference on Power Electronic, Drives and Energy Systems**

# Space Vector Modulation with DC-Link Voltage Balancing Control for Three-Level Inverters

Kalpesh H. Bhalodi, *Member, IEEE, and* Pramod Agrawal, *Member, IEEE*

*Abstract*--The DC-link voltage balancing scheme for three-level inverter is proposed in this paper. Dependence of the DC-link capacitor voltage deviation on DC-link current and inverter switching states is established for proposed three-level inverter. Pulse pattern rearrangements for space vector PWM (SVPWM) using degree of freedom available in choice of redundant space vectors, sequencing of vectors, and splitting of duty cycles of vector are best exploited. Self neutral-point voltage deviation control in feed forward and simplified closed loop scheme are proposed in this paper. The effectiveness of proposed scheme is verified by computer simulations.

*Index Terms*—Diode-clamped inverter, Multilevel inverter, Space-vector modulation, Voltage source inverter.

## I. INTRODUCTION

**M**ULTILEVEL inverters have gained much attention for the next generation medium voltage and high power applications. Three-level diode-clamped inverter also known as neutral point clamped (NPC) inverter is most favorable among various multilevel configurations explored in the literature. The three-level NPC inverter is used in this paper. Problems due to neutral-point voltage-unbalance and its various balance control methods are discussed at length [1], [2]. DC-link unbalance may overstress the capacitors and devices during a sudden regenerative load increase, and it can also cause nuisance over voltage or under voltage trips. Active front-end converter with coordinated control from grid end and load end for DC-link balancing control is proposed [1].

The effect of capacitor voltage unbalance during transient and steady state condition are analyzed it this paper. In the worst case of unbalance one capacitor is fully charged to full DC-link voltage that results in double stress on the capacitor

This work was supported in part by All India Council of Technical Education, New Delhi, A statutory body of Government of India under National Doctoral Fellowship 1-10/FD/NDF-PG/(IIT-(22))/2005-06.

Kalpesh H. Bhalodi is with Department of Electrical Engineering, Indian Institute of Technolgy Roorekee, Roorkee - 247667 INDIA (e-mail: kalpeshbhalodi @yahoo.co.in).

Pramod Agrawal is with Department of Electrical Engineering, Indian Institute of Technolgy Roorekee, Roorkee - 247667 INDIA (e-mail: pramgfee @iitr.ernet.in).

and the switching devices, reducing output waveforms to two-level from normal three-level. The effect of the zero sequence voltage on the neutral point variation and the dependence of DC-link voltage unbalance on the system parameters like load currents, load power factor, value of capacitance of capacitor, and modulation index have been extensively analyzed for three-level NPC inverter [3], [4]. The neutral point balancing schemes, for the three-level neutral point clamped inverter, are based on the effective use of the redundant switching states of the inverter voltage vectors. The redundant switching states are used alternately such that the neutral point voltage unbalance caused by the first switching state combination is compensated by another state; thus, bringing the total unbalance in one switching cycle to zero [5]-[7]. Detailed study of NPC inverter, space vectors, dwell timings, and pulse pattern arrangement with division of middle regions for neutral point balance and even harmonic elimination scheme are addressed [6]. Neutral point voltage control is achieved by utilizing the phase current polarity and distribution of the redundant voltage vectors. A control strategy is proposed to maintain average current drawn from neutral-point to the minimum [8], [9]. Hysteresis control for DC-link variation control and common mode voltage elimination in an open end winding induction motor fed from two three-level inverters from either side is investigated [10]. ANN based neutral-point self-voltage balancing SVPWM is discussed for NPC inverter [11]. Mathematical modeling and neutral-point control with charge balance is proposed for four-level voltage source inverter [12].

## II. THREE-LEVEL INVERTER

Fig. 1 shows the simplified circuit diagram of a popular three-level neutral point clamped (NPC) inverter. The inverter leg 'a' is composed of four IGBT switches $S_1$ to $S_4$ with four antiparallel diodes $D_1$ to $D_4$. On the DC side of the inverter, the DC bus capacitor is split into two, providing a neutral point 'n'. When switches $S_2$ and $S_3$ are turned on, the inverter output terminal a is connected to the neutral point through one of the clamping diodes $Dn_1$ and $Dn_2$. Ideally, the voltage across each of the DC capacitors is Vdc/2, which is half of the total DC-link voltage $V_{dc}$. With a finite value for $C_1$ and $C_2$, the capacitors can be charged or discharged by neutral current $i_2$, causing neutral-point voltage deviation. The important problem of voltage unbalance between upper and lower capacitors in three-level NPC inverter is discussed further.

0-7803-9771-1/06/$25.00 ©2006 IEEE

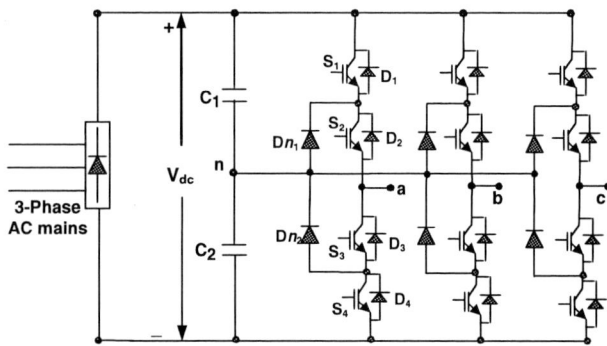

Fig. 1. Three-level neutral-point clamped inverter topology.

As indicated earlier, the neutral-point voltage $V_n$ varies with the operating condition of the NPC inverter. If the neutral-point voltage deviates too far, an uneven voltage distribution takes place, which may lead to premature failure of the switching devices and cause an increase in the harmonic of the inverter output voltage [1].

The operating status of the switches in the NPC inverter can be represented by the switching states shown in Table I. Switching state 'P' denotes that the upper two switches in leg 'a' are on and the inverter pole voltage $V_a$, which is ideally $+V_{dc}/2$, whereas 'N' indicates that the lower two switches conduct, leading to $V_a = -V_{dc}/2$. Switching state 'O' signifies that the inner two switches $S_2$ and $S_3$ are on and $V_a$ is clamped to zero through the clamping diodes. Depending on the direction of the load current $i_a$, one of the two claming diodes is turned on. For instance, a positive load current ($i_a > 0$) forces $Dn_1$ to turn on, and the terminal 'a' is connected to the neutral point 'n' through the conduction of $Dn_1$ and $S_2$. The switches $S_1$ and $S_3$ operate in a complementary manner similar to switches $S_2$ and $S_4$.

TABLE I
DEFINITION OF SWITCHING STATES

| Switching State | Device Switching Status (Phase a) | | | | Pole Voltage $V_a$ |
|---|---|---|---|---|---|
| | $S_1$ | $S_2$ | $S_3$ | $S_4$ | |
| P | On | On | Off | Off | $+ V_{dc}/2$ |
| O | Off | On | On | Off | 0 |
| N | Off | Off | On | On | $- V_{dc}/2$ |

As indicated earlier, the operation of each inverter phase lag can be represented by three switching states P, O, and N. Taking all three phases in account, the inverter has a total of 27 possible combinations of switching states. Fig. 2 shows space vector diagram of total 27 switching states corresponding to 19 voltage vectors for three-level NPC inverter. Fig. 3 shows vector placement in a sector A. Based on magnitude, the voltage vectors can be divided into four groups: Zero vector ($\underline{u}_O$), Small vector ($\underline{u}_S$), Medium vector ($\underline{u}_M$), and Large vectors ($\underline{u}_L$). All zero vectors have zero magnitude, small vectors have a magnitude of $V_{dc}/3$, medium vectors have magnitude of $V_{dc}/\sqrt{3}$ and large vectors have

magnitude of $2V_{dc}/3$. Each small vector has two switching states, one containing P and other containing N, and therefore can be further classified into a P-type or N-type vector.

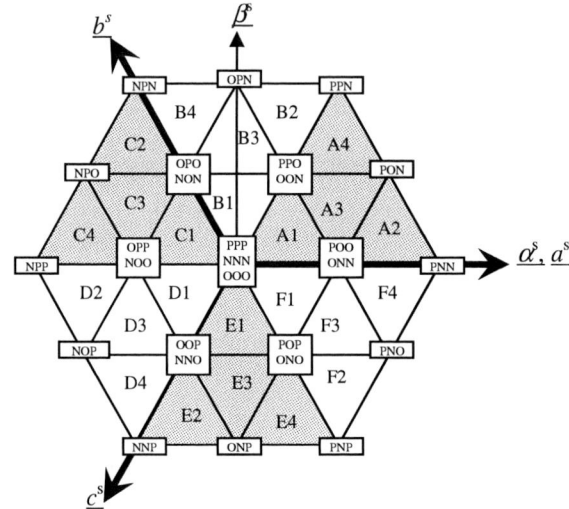

Fig. 2. Space-vector diagram showing switching states.

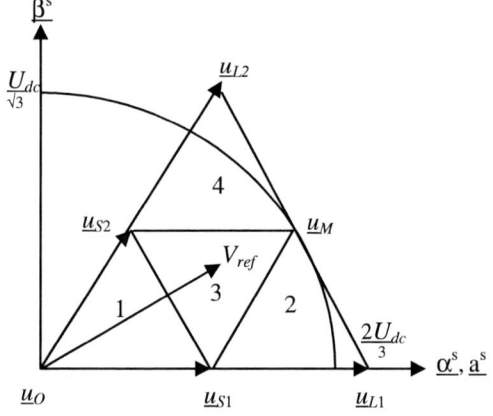

Fig. 3. Vector placement for SVPWM in sector A.

## II. EFFECT OF SWITCHING STATES ON NEUTRAL-POINT VOLTAGE DEVIATION

The effect of switching states on neutral voltage deviation is illustrated in fig. 4. When the inverter is operated with switching state [PPP] of zero vector $\underline{u}_O$, the upper two switches in each of the three inverter legs are turned on, connecting the inverter terminals a, b, and c to the positive DC bus as shown in fig. 4(a). Since the neutral point 'n' is left unconnected, this switching state does not affect $V_n$. Similarly, the other zero switching states [OOO] and [NNN] do not cause $V_n$ to shift either. Fig 4(b) shows the inverter operation with P-type switching state [POO] of small vector $\underline{u}_{Sp}$. Since the three-phase load is connected between the positive DC bus and neutral point 'n', the neutral current $i_n$ flows in 'n', causing $V_n$ to increase. On the contrary, the N-type switching state [ONN] of small vector $\underline{u}_{Sn}$ makes $V_n$ to decrease as shown in fig. 4(c). For medium vector $\underline{u}_M$ with switching state [PON] in fig. 4(d),

468

load terminals a, b, and c are connected to the positive bus, the neutral point, and the negative bus, respectively. Depending on the inverter operating conditions, the neutral-point voltage $V_n$ may rise or drop. Considering a large vector $\underline{u}_L$ with switching state [PNN] as shown in fig. 4(e), the load terminals are connected between the positive and negative DC buses. The neutral point 'n' is left unconnected and thus the neutral voltage is not affected.

It is summarized that zero and large vectors do not affect the neutral point voltage. Medium vectors affect $V_n$, but the direction of voltage deviation is undefined so, redundant small vectors having dominant influence on $V_n$ are used for neutral point voltage control. Above discussion is made under the assumption that the inverter is in motoring mode. In addition to the influence of switching states, the neutral-point voltage may also be affected by a number of other factors like unbalanced DC capacitors due to manufacturing tolerances, inconsistency in switching device characteristics, unbalanced three-phase operation, motoring/regenerative mode of operation etc. As compared to motoring mode, an opposite capacitor voltage charging-discharging action takes place in the regenerative mode due to the reversal of current.

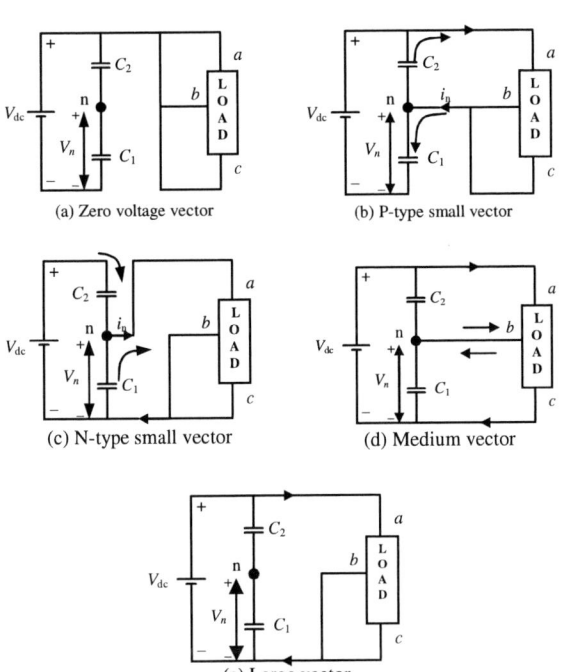

Fig. 4. Effect of switching states on Neutral point voltage deviation.

## IV. MATHEMATICAL MODELLING OF THREE-LEVEL INVERTER

The terminology in this paper is explained with reference to fig. 5, in which DC bus structure and each inverter phases is modeled as 3-pole switch. The inverter switching functions $S_a$, $S_b$, and $S_c$ assume value equal to 1, 2 or 3, which means that the pole of the switch in connected to bottom, middle or top DC-link respectively. The load currents are denoted as $i_a$, $i_b$, and $i_c$ while the currents drawn by the inverter from bottom,

middle and top rails of the DC-link are $i_1$, $i_2$, and $i_3$ respectively.

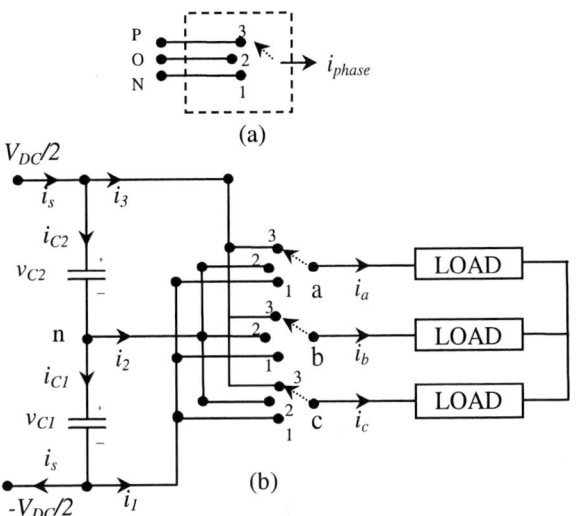

Fig. 5. Three-level inverter and load system model.

In general the inverter pole voltage with respect to the negative DC bus can be written in terms of capacitor voltages and the switching functions as

$$
\begin{aligned}
v_{a1}(S_a) &= \delta(S_a-1)v_{C1} + \delta(S_a-2)(v_{C1}+v_{C2}) \\
v_{b1}(S_b) &= \delta(S_b-1)v_{C1} + \delta(S_b-2)(v_{C1}+v_{C2}) \\
v_{c1}(S_c) &= \delta(S_c-1)v_{C1} + \delta(S_c-2)(v_{C1}+v_{C2})
\end{aligned} \tag{1}
$$

In (1) $\delta(.)$ is the Dirac delta function, $v_{C1}$ and $v_{C2}$ are the voltage across lower and upper capacitors. The currents drawn from the DC-link nodes can be represented in terms of the motor currents as shown in (2)

$$
\begin{bmatrix} i_1 \\ i_2 \\ i_3 \end{bmatrix} = \begin{bmatrix} \delta(S_a-1) & \delta(S_b-1) & \delta(S_c-1) \\ \delta(S_a-2) & \delta(S_b-2) & \delta(S_c-2) \\ \delta(S_a-3) & \delta(S_b-3) & \delta(S_c-3) \end{bmatrix} \begin{bmatrix} i_a \\ i_b \\ i_c \end{bmatrix} \tag{2}
$$

For the three wire load,

$$
\begin{aligned}
i_a + i_b + i_c &= 0 \\
i_1 + i_2 + i_3 &= 0
\end{aligned} \tag{3}
$$

Substituting $i_c = -(i_a + i_b)$ and removing the dependent variable $i_3$, (2) gets reduced to

$$
\begin{bmatrix} i_1 \\ i_2 \end{bmatrix} = \begin{bmatrix} [\delta(S_a-1)-\delta(S_c-1)] \\ [\delta(S_a-2)-\delta(S_c-2)] \end{bmatrix} \begin{bmatrix} i_a \\ i_b \end{bmatrix} \tag{4}
$$

The current flowing through capacitor is given by

$$
\begin{aligned}
i_{C2} &= i_s - i_3 = i_s - (-i_2 - i_1) = i_s + i_2 + i_1 \\
i_{C1} &= i_s + i_1
\end{aligned} \tag{5}
$$

$$
\therefore \begin{bmatrix} i_{C2} \\ i_{C1} \end{bmatrix} = \begin{bmatrix} 1 & 1 & 1 \\ 1 & 0 & 1 \end{bmatrix} \begin{bmatrix} i_s \\ i_2 \\ i_1 \end{bmatrix} \tag{6}
$$

469

Thus, the currents flowing through capacitors is directly related to the voltage across the capacitors and the relationship is given by

$$\begin{bmatrix} v_{C2} \\ v_{C1} \end{bmatrix} = \frac{1}{C} \int \left\{ \begin{bmatrix} 1 & 0 \\ 0 & 1 \end{bmatrix} \begin{bmatrix} i_{C2} \\ i_{C1} \end{bmatrix} dt \right\} = \frac{1}{C} \int \left\{ \begin{bmatrix} 1 & 1 & 1 \\ 1 & 0 & 1 \end{bmatrix} \begin{bmatrix} i_s \\ i_2 \\ i_1 \end{bmatrix} dt \right\} \quad (7)$$

Let $\Delta v_C$ be the change in the capacitor voltage, Substituting value from (4) into (6), results in

$$\Delta v_C = v_{C2} - v_{C1}$$

$$\therefore \Delta v_C = \frac{1}{C} \int i_2 dt \quad (8)$$

Thus the load current drawn from the middle node of the DC-link is responsible for the variation in the capacitor voltages. Whenever switching functions $S_a$, $S_b$, and $S_c$ assume value equal to 2, there exists a tendency of capacitor voltage unbalancing.

## V. DC-LINK CAPACITOR VOLTAGE BALANCING SCHEME

Various space vector modulation (SVM) schemes have been proposed for the three-level NPC inverter using either open loop scheme or closed loop scheme [6].

### A. Open loop capacitor voltage self balancing scheme

This section proposes modified SVM scheme for better neutral point stabilization. To reduce neutral-point voltage deviation, the dwell time of a given small vector can be equally distributed between the P-type and N-type switching states over a sampling period. For nearest three vectors (NTV) selection SVM either one small vector or two small vectors among the three selected vectors are available according to the triangular regions in which the reference vector $V_{ref}$ lies. When the reference vector $V_{ref}$ is in region 2 or 4, only one small vector is in NTV where as in region 1 or 3 two small vectors are in NTV as shown in fig. 3. In conventional SVM seven segment pulse pattern is chosen for all regions. Its pulse pattern arrangement for region A1 is shown in fig. 6. Seven segment SVM divides dwell time of only one small vector in P-type and N-type out of two small sectors available in region 1 and 3. So, neutral point deviation is not minimized in these regions. To reduce neutral-point voltage deviation according to location of region, optimized pulse pattern arrangement is proposed here. Negative sequences of modified SVM pulse pattern arrangement for regions A1, A2, A3, and A4 of sector A are shown in fig 7. Here, two small vectors are used for neutral point voltage control for regions 1 and 3. As shown in fig. 8 negative sequence (NEG_SEQ) pulse patterns are arranged in exact reverse order of positive sequence (POS_SEQ) pulse pattern and vice versa. Positive sequence and negative sequence are switched alternatively. Switching sequence in opposite sectors (A-D, B-E and C-F) is selected to be of a complimentary nature for neutral point balancing.

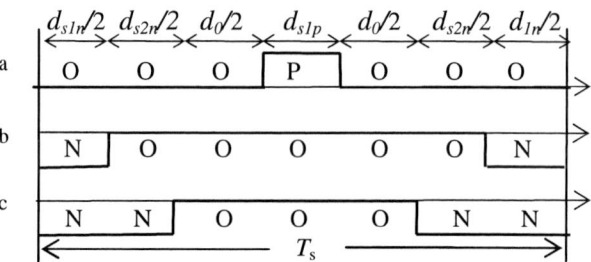

Fig. 6. Pulse pattern arrangement for conventional seven-segment SVPWM in region A1.

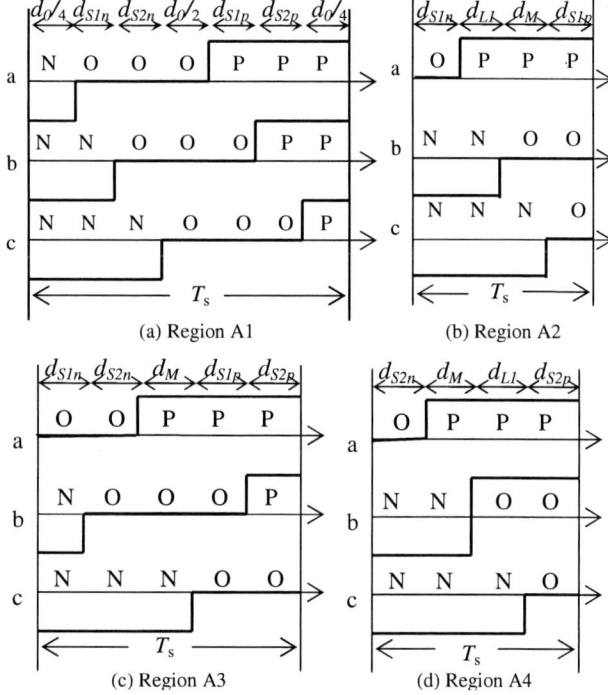

(a) Region A1      (b) Region A2

(c) Region A3      (d) Region A4

Fig. 7. Pulse pattern arrangement of modified SVPWM in sector A.

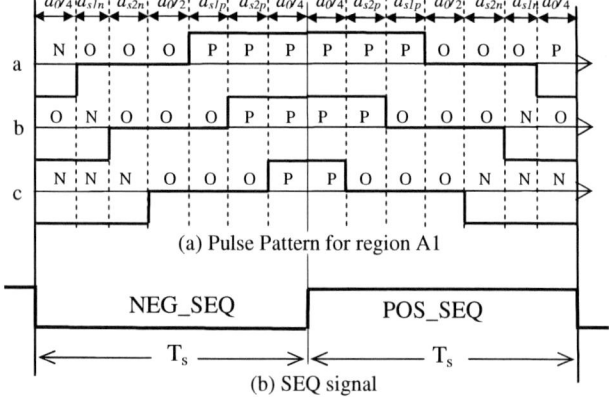

(a) Pulse Pattern for region A1

(b) SEQ signal

Fig. 8. The sequence of various switching combinations.

Number of switching per phase in sampling period $T_s$ for modified SVM are one in region 2 or 4, two in region 1 and one or two in region 3. In conventional SVM as shown in fig.

470

6 two switching per phase are required in sampling period $T_s$. Here, $T_s$ is sum of dwell times of NTV in a sequence.

### B. Linear closed loop capacitor voltage balancing scheme

There always exists a small-voltage vector in each switching sequence, whose dwell time is divided into subperiods, one for its P-type and the other for its N-type switching state. For instance, the dwell time $ds_1$ for $\underline{u}_{S1p}$ and $s_{1n}$ for $\underline{u}_{S1n}$, which is half/half split normally, can be distributed as

$ds_1 = d_{S1p} + d_{S1n}$
where $d_{S1p}$ and $d_{S1n}$ are given by
$d_{S1p} = ds_1 /2(1+\Delta t)$ and
$d_{S1n} = ds_1 /2(1-\Delta t)$ where $-1 <= \Delta t <= 1$     (9)

The deviation of the neutral point voltage can be further reduced by adjusting the incremental time interval $\Delta t$ in (9) according to the detected DC capacitor voltages $v_{C1}$ and $v_{C2}$. The input to the voltage balancing scheme is the difference in capacitor voltages $\Delta v_C$ where, $\Delta v_C = v_{C2} - v_{C1}$. For instance, if $\Delta v_C$ is greater than the maximum allowed DC voltage deviation $\Delta V_m$ for some reasons, we can increase $d_{S1p}$ and decrease $d_{S1n}$ by $\Delta t$ ($\Delta t>0$) simultaneously for the drive in a motoring mode. A reverse action ($\Delta t<0$) should be taken when the drive is in a regenerative mode. The relationship between the capacitor voltages and the incremental time interval $\Delta t$ is summarized in Table II.

#### TABLE II
RELATIONSHIP BETWEEN CAPACITOR VOLTAGES AND INCREMENTAL TIME INTERVAL $\Delta t$

| Neutral-point deviation level | Motoring mode | Regenerating node |
|---|---|---|
| $(v_{C2} - v_{C1}) > \Delta Vm$ | $\Delta t > 0$ | $\Delta t < 0$ |
| $(v_{C2} - v_{C1}) > \Delta Vm$ | $\Delta t < 0$ | $\Delta t > 0$ |
| $\lvert v_{C2} - v_{C1} \rvert < \Delta Vm$ | $\Delta t = 0$ | $\Delta t = 0$ |

$\Delta Vm$ - maximum allowed DC voltage deviation ( $\Delta V_C > 0$ )

## VI.   RESULTS AND DISCUSSIONS

Here, DC-link voltage value of 200 V and inductive load with power factor value of 0.9 is used in all schemes. Fig. 9 shows the variation in the DC-link capacitor voltages, phase voltage, and phase current when open loop DC-link balancing control is on with modulation index value of 0.8. Fig. 10 shows the variation in the DC-link capacitor voltages, phase voltage, and phase current when open loop DC-link balancing control is off with modulation index value of 0.8. As soon as DC-link voltages become unbalanced, the quality of output phase voltage, and current waveform deteriorates. The variation in the DC-link capacitor voltages, phase voltage, and phase current when open loop and closed loop DC-link balancing control is on are shown in fig. 11 with modulation index value of 0.55. The average value of DC-link capacitor

voltages settles towards half the DC-link voltage, hence output voltage and current waveform quality is increased even after slight unsymmetrical pulse pattern due to slight unequal dwell time of P-type and N-type small vectors. Control of proposed scheme is more effective in region 1 and 3, where two small vectors are utilized for pulse pattern arrangement. Linear close loop control accelerates the neutral-point voltage control and improves average DC-link capacitor voltages.

The robustness and effectiveness of the both open loop and close loop control is illustrated in fig. 12 with modulation index value 0.8. In fig. 12, from time t = 6.76 to 13.21 (rad) feed forward and feed back controls are off, in rest of the time controls are made on. As soon as the controls are made effective the deviation in capacitor voltages sharply reduces. Thus, the proposed balancing scheme is capable of fundamentally redistributing charges amongst the DC capacitors.

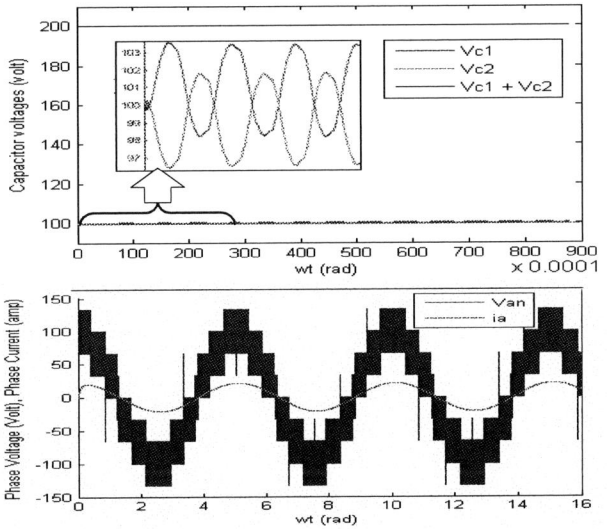

Fig. 9.   The variation in the capacitor voltages, phase voltage, and phase current when open loop DC-link balancing control is on.

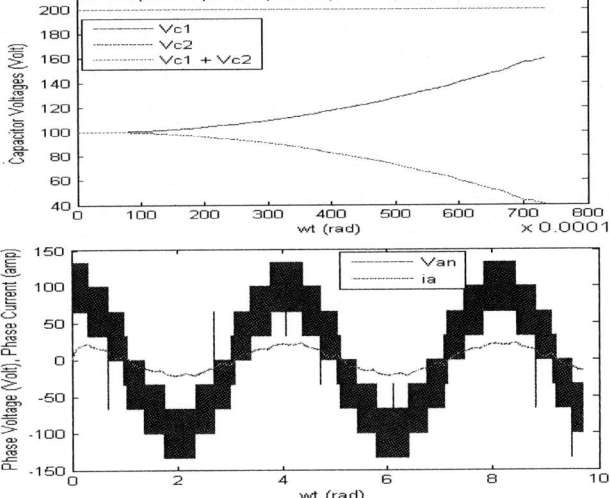

Fig. 10.   The variation in the capacitor voltages, phase voltage, and phase current when open loop DC-link balancing control is off.

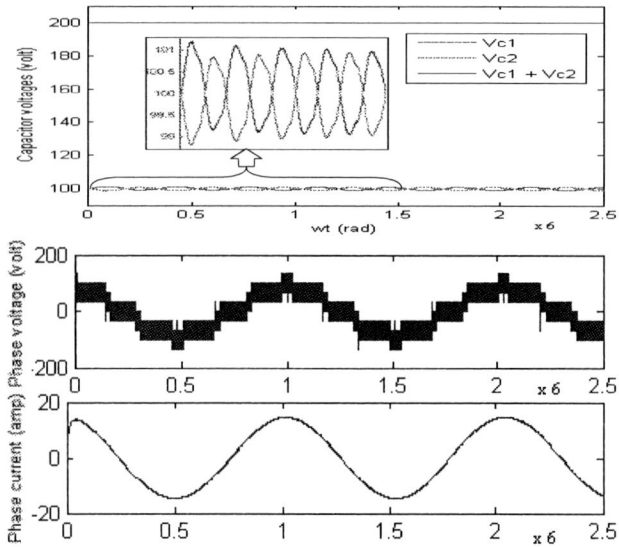

Fig. 11. The variation in the capacitor voltages, phase voltage, and phase current when open loop and closed loop DC-link balancing control is on.

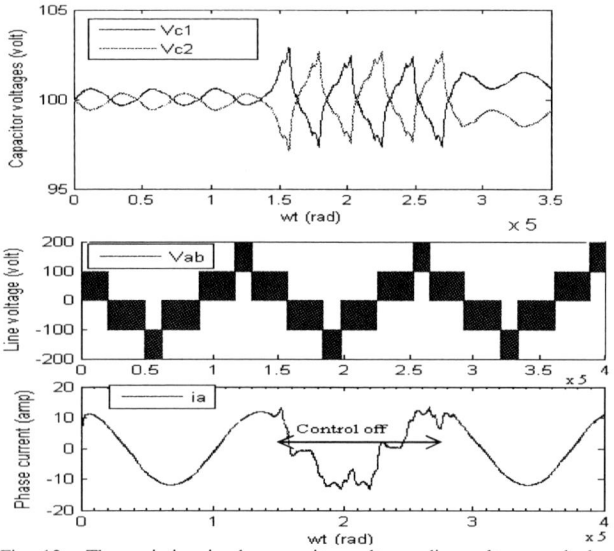

Fig. 12. The variation in the capacitor voltages, line voltage, and phase current when open loop and close loop DC-link balancing control is on or off.

## VII. CONCLUSION

The DC-link voltage balancing scheme for three-level diode clamping inverter is investigated in this paper. The detailed analysis investigates the DC-link voltage control behavior of proposed scheme and popular SVPWM scheme. The closed loop scheme is used with simple control technique. Redundant space vectors, its sequencing, and splitting of its duty cycle are used for control. Thus, single front-end rectifier can be used with reduced rating DC-link capacitors. Large unbalance in the load can be addressed further.

## VIII. REFERENCES

[1] F. Wang, "Multilevel PWM VSIs: Coordinated control of regenerative three-level neutral point clamped pulsewidth-modulated voltage source inverters", *IEEE Ind. App. Mag.*, pp. 51-58, July- Aug. 2004.

[2] C. Newton and M. Summer, "Neutral point control for multi-level inverters: Theory, design and operation limitations," in *Proc. 1997 IEEE IAS Conf. Rec.*, pp. 1336-1343.

[3] S. Ogasawara and H. Akagi, "Analysis of variation of neutral point potential in neutral-point-clamped voltage source PWM Inverters", in *Proc. 1993 IEEE Industrial Applications Society Conf.*, pp. 965-970.

[4] K. R. M. N. Ratnayake, Y. Murai, T. Watanabe, "Novel PWM scheme to control neutral point voltage variation in three-level voltage source inverter", in *proc. 1999 of IEEE Ind. App. 34th IAS Annual Meeting Conf.*, vol. 3, pp.1950 – 1955.

[5] R. Rojas, T. Ohnishi and T. Suzuki, "An improved voltage vector control method for neutral point clamped inverters", *IEEE Trans. Power Electronics*, vol. 10(6), pp. 666- 672, Nov. 1995.

[6] B. Wu, *High-Power Converters and AC Drives*, IEEE Press and Wiley, 2006, pp 143-176.

[7] N. Celanovic and D. Boroyevich, "A comprehensive study of neutral-point voltage balancing problem in three-level neutral-point-clamped voltage source PWM inverters," *IEEE Trans. on Power Electronics*, vol. 15(2), pp. 242-249, Mar. 2000.

[8] C. Newton and M. Summer, "Neutral point control for multi-level inverters: Theory, design and operation limitations," in *Proc. 1997 IEEE IAS Conf. Rec.*, pp. 1336-1343.

[9] K. Yamanaka, A. M. Hava, H. Kirino, Y. Tanaka, N. Koga and T. J. Kume, "A novel neutral potential stabilization technique using the information of output current polarities and voltage vector," *IEEE Trans. on Ind. App.*, vol. 38(6), pp. 1572-1580, Nov/Dec. 2002.

[10] R. S. Kanchan, P. N. Tekwani and K. Gopakumar, "Three-Level Inverter Scheme With Common Mode Voltage Elimination and DC-link Capacitor Voltage Balancing for an Open-End Winding Induction Motor Drive", *IEEE Trans. Power Electronics,* vol. 21(6), pp.1676-1683, Nov. 2006.

[11] J. O. P. Pinto, B. K. Bose, L. E. B. Da Silva and M. P. Kazmierkowski, "Neural-network-based space-vector PWM controller for voltage-fed inverter induction motor drive", *IEEE Trans. Ind. App.*, vol. 36(6), pp. 1628-1636, Nov/Dec 2000.

[12] G. Sinha, T. A. Lipo, "A four-level inverter based drive with a passive front end", *IEEE Trans. Power Electronics*, vol. 15(2), pp. 285-294, Mar. 2000.

## IX. BIOGRAPHIES

**Kalpesh H. Bhalodi** received his Bachelor's and Master's degree in Electrical Engineering Department from L. D. College of Engineering, Ahmedabad, India. He is presently pursuing his Ph.D. in Electrical Engineering Department from Indian Institute of Technology Roorkee, India. He is a faculty member in the department of Electrical Engineering, S. P. College of Engineering, Visnagar, India. His fields of interest include Power Electronics, Electric Drives, Multilevel Inverters, and Microcomputer based Control.

**Pramod Agarwal** obtained his Bachelor's degree in Electrical Engineering from University of Roorkee now, Indian Institute of Technology Roorkee, India. He received his PG and completed his Ph.D in Electrical Engineering from the same institute in 1985 and 1995 respectively. He is currently Professor in the department of Electrical Engineering, Indian Institute of Technology Roorkee, India. His special fields of interest include Electircal Machines, Power Electronics, Electric Drives, Power quality, Microcomputer Controlled Electric Drives, and Multilevel Converters.

**2006 IEEE International Conference on Power Electronic, Drives and Energy Systems**

# Investigations on Different Multilevel Inverter Control Techniques by Simulation

P. K. Chaturvedi, Shailendra K Jain, Pramod Agrawal, *Member IEEE*, and P. K. Modi

*Abstract--* **This paper presents simulation studies on different control techniques of 3- and 5- level diode clamped multilevel inverters i.e. Sinusoidal PWM, Selective Harmonic Elimination, and Optimized Harmonic PWM techniques. Paper contributes towards the detailed simulation and analysis of these techniques which has not been given in previous work. The performance of each technique has been investigated based upon reduction in Total Harmonic Distortion. For simulation, Matlab/Power System Blockset/PowerGUI has been used.**

*Index Terms—* **Matlab, Multilevel Control Techniques, Multilevel Inverter, Total Harmonic Distortion.**

## I. INTRODUCTION

CURRENT and voltage harmonics have attracted growing interest with the increase in use of static power converters. These converters produce distorted current and voltage waveforms. The result is harmonic pollution that degrades the power quality. One of the biggest problems in power quality aspects is the harmonic contents in the electrical system. Generally, harmonics may be divided into two types: 1) voltage harmonics, and 2) current harmonics. Current harmonics is usually generated by harmonics contained in voltage supply and depends on the type of load such as resistive load, capacitive load, and inductive load. Both harmonics can be generated by either the source or the load side. Harmonics generated by load are caused by nonlinear operation of devices, including power converters, arc-furnaces, gas discharge lighting devices, etc [1]-[5]. Load harmonics can cause the overheating of the magnetic cores of transformer and motors. On the other hand, source harmonics are mainly generated by power supply with non-sinusoidal voltage waveform. Voltage and current source harmonics imply power losses, Electromagnetic Interference (EMI) and pulsating torque in AC motor drives. There are two practical ways by which the harmonic content can be brought down to

---

P. K. Chaturvedi is Research Scholar, MANIT Bhopal & Lecturer, Electrical Engg Deptt, SATI, Vidisha (MP), India (e-mail: pradyumnc74@rediffmail.com).

Shailendra K Jain is with the Deptt. of Electrical Engg, Maulana Azad National Institute of Technology, Bhopal (MP), India, e-mail: shailjain02@rediffmail.com.

Pramod Agrawal is with the Electrical Engineering Department, Indian Institute of Technology, Roorkee, UP, India.

P. K. Modi is with the Electrical Engineering Department, UCE, Burla, India.

---

a low value or at least within acceptable limits. One method is to use a filter circuit on the output side of the inverter. The filter circuit will, of course, have many disadvantages like, it should handle the large power output from the inverter, and it has to carry high currents so the cost and size will be more. The second scheme employs a pulse width modulation strategy that will change the harmonic content in the output voltage in such a way that the filtering needed will be minimal or zero depending on the type of application. This technique based on a suitable designed switching strategy within inverter [6]-[8].

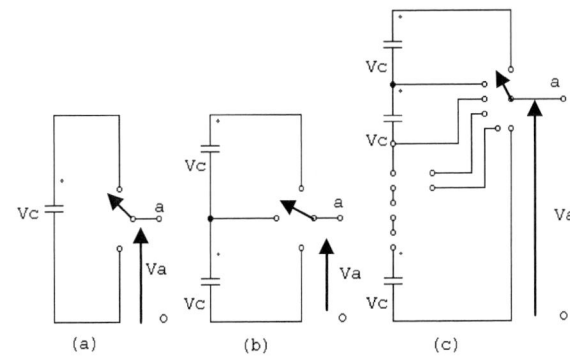

Fig. 1. One phase leg of an inverter with (a) two levels, (b) three levels and (c) m levels.

Traditional 2-level high-frequency pulse width modulation (PWM) inverters for motor drives have several problems associated with high frequency switching, which produce common-mode voltages and high voltage change (dV/dt) rates to the motor windings. Some other noticeable drawbacks of conventional 2-level pwm inverter are:

1. Switching losses will be more,
2. Switches must have very low turn-on and turn-off times,
3. Large dv/dt rating,
4. Problem in voltage sharing in series connected devices, and
5. Higher order harmonics will be introduced.

Because of these shortcomings of conventional 2-level inverter, there is need to develop more efficient and high power inverter. Multilevel inverters have found better counterpart to the conventional 2-level pulse width modulated converter. Main features of multilevel inverters are [8]-[11]:

1. less switching stress on devices,
2. high voltage & high power capability,

---

0-7803-9771-1/06/$25.00 ©2006 IEEE

3. reduced harmonic contents without increasing switching frequency or decreasing the inverter power output,
4. no need of extending the device rating,
5. reduced switching losses,
6. reduced dv/dt,
7. reduced (or even eliminated) common mode voltages,
8. good electromagnetic compatibility (EMC),
9. elimination of the problem of unequal device ratings,
10. capacitor voltage balancing along with significant reduction in Device Count.

These Inverters recently have found many applications in the medium and high power applications such as ac power supplies, static reactive power compensators, adjustable speed drive systems and many more. The main disadvantages of this technique are that a larger number of switching semiconductors are required for lower-voltage systems and the small voltage steps must be supplied on the dc side either by a capacitor bank or isolated voltage sources. The output voltage waveform of a multilevel inverter is composed of number of levels of voltages, typically obtained from capacitor voltage sources. Fig. 1, shows a schematic diagram of one phase leg of inverters with different number of levels, for which the action of the power semiconductors is represented by an ideal switch with several positions. There are 2, 3 and 5-levels of output voltage across one pole of inverter (Va across a-o in Fig. 1) as shown in Fig 2.

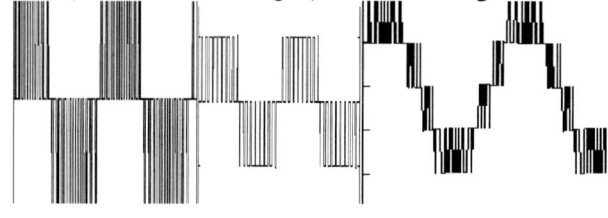

(a)                (b)                (c)

Fig 2. Output voltage profile of (a) 2-level, (b) 3-level, and (c) 5-level inverter.

Ideally, as the number of dc sources approaches infinity, the staircase waveform will approach the desired sinusoidal waveform with zero output THD approaches. The number of achievable voltage levels, however, is limited by voltage unbalance problems, voltage clamping requirement, circuit layout and packaging constraints.

There is a large number of control algorithm developed so far to control the operation of multilevel inverters. In this paper commonly used control techniques, Sinusoidal PWM (SPWM), Selective Harmonic Elimination (SHEPWM), and Optimized Harmonic PWM (OHPWM) techniques have been investigated via simulation and compared based upon THD. SHEPWM can eliminate a particular lower order harmonics but at the cost of increased higher order harmonics. In Optimized harmonic technique, switching angles may be optimized to eliminate or reduce THD and specific harmonics [8].

This paper presents the phase voltages, line voltages, line currents and their Total Harmonic Distortions in 3-level & 5-level inverters with 3-phase highly inductive load with SPWM, SHEPWM and OHPWM techniques. Obtained results indicate significant improvement in inverter performance

when switching from 3-level to 5-level. In Optimized Harmonic PWM technique, THD in phase voltage have been found 27.83% for 3-level inverter and 17.61% for 5-level inverter with 3rd and 5th harmonics almost eliminated. Current THD is 20.11% and 11.65% for 3-level & 5-level inverters respectively with 3rd harmonic almost eliminated. In SPWM technique, the switching frequency used is 1 kHz to maintain low switching losses. Paper also reviews different types of multilevel SPWM techniques according to carrier and modulating signals.

## II. MULTILEVEL INVERTER TOPOLOGY FOR INVESTIGATION

The most widely used multilevel inverter topologies are Diode Clamped type, Flying Capacitor type and Cascaded H-bridge inverter with separate dc sources. Diode clamped multi-level inverter is a very general and widely used topology for real power flow control and is considered for investigation purpose in this paper. The three-level and five-level diode clamped inverters are shown in Fig 3 and Fig 4. The switching tables and switching states for one phase leg are given in Table I and Table II.

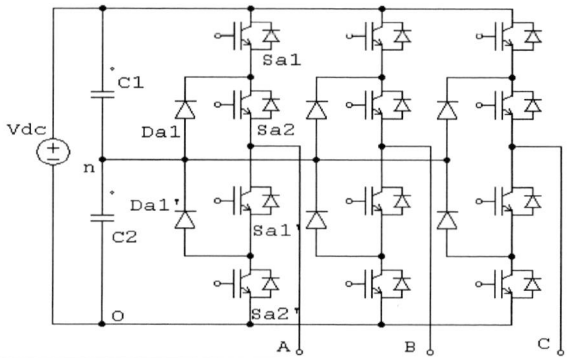

Fig. 3. Three-level, 3-phase diode clamped inverter.

TABLE I
SWITCHING STATES FOR 3-LEVEL INVERTER

| Vab | Output Pole Voltage (Vao) | Switch States | | | |
|-----|---------------------------|--------|--------|---------|---------|
|     |                           | $S_{a1}$ | $S_{a2}$ | $S_{a1}'$ | $S_{a2}'$ |
| $-V_{dc}/2$ | 0 | 0 | 0 | 1 | 1 |
| 0 | $V_{dc}/2$ | 0 | 1 | 1 | 0 |
| $V_{dc}/2$ | $V_{dc}$ | 1 | 1 | 0 | 0 |

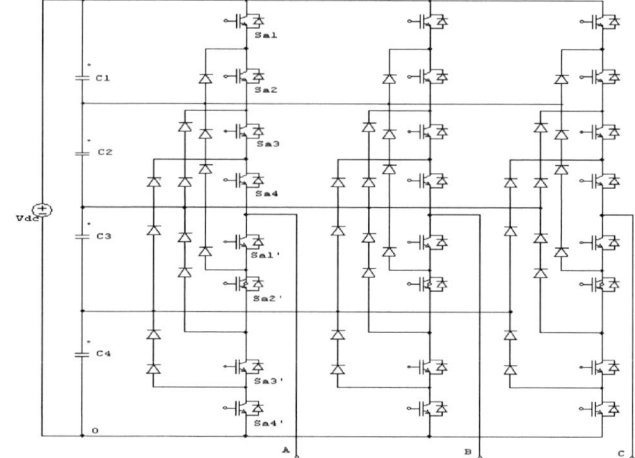

Fig.4. Five-level, 3-phase diode clamped inverter.

TABLE II
SWITCHING STATES FOR 5-LEVEL INVERTER

| Output $V_{AO}$ | Switch States | | | | | | | |
|---|---|---|---|---|---|---|---|---|
| | Sa1 | Sa2 | Sa3 | Sa4 | Sa1' | Sa2' | Sa3' | Sa4' |
| V5=Vdc | 1 | 1 | 1 | 1 | 0 | 0 | 0 | 0 |
| V4=3Vdc/4 | 0 | 1 | 1 | 1 | 1 | 0 | 0 | 0 |
| V3=Vdc/2 | 0 | 0 | 1 | 1 | 1 | 1 | 0 | 0 |
| V2=Vdc/4 | 0 | 0 | 0 | 1 | 1 | 1 | 1 | 0 |
| V1=0 | 0 | 0 | 0 | 0 | 1 | 1 | 1 | 1 |

According to the Table I & Table II, pulses can be generated for all the three phases as explained later.

## III. MULTILEVEL INVERTER CONTROL TECHNIQUES

Pulse width modulation (PWM) techniques are employed to achieve high quality output voltage waveforms of desired amplitude and frequency which are as close as possible to sinusoidal wave. Any deviation from the sinusoidal wave shape will results in harmonic currents in the load which result in electromagnetic interference (EMI), harmonic losses and torque pulsation in case of motor drives. The quality of the output waveform will improve with increase in switching frequency. Higher switching frequency can be employed only for low power levels, as managing the switching losses at high power levels will be a difficult task.

It is generally accepted that the performance of an inverter, with any switching strategies, can be related to the harmonic contents of its output voltage. Power electronics researchers have always studied many control techniques to reduce harmonics in such waveforms. In multilevel inverter technology, there are several well-known modulation topologies out of which presented work focuses on:

1. Sinusoidal Pulse Width Modulation (SPWM),
2. Optimized Harmonic Stepped-Waveform Technique (OHPWM), and
3. Selective Harmonic Eliminated Pulse Width Modulation (SHEPWM).

### A. Sinusoidal Pulse Width Modulation(SPWM)

The control principle of the SPWM is to use several triangular carrier signals keeping only one modulating sinusoidal signal. For the five level inverter, four triangular carriers are needed (Generally speaking, if a m-level inverter is employed, (m-1) carriers will be needed). The carriers have the same frequency $f_c$ and the same peak-to-peak amplitude $A_c$. The zero reference is placed in the middle of the carrier set. The modulating signal is a sinusoid of frequency $f_m$ and amplitude $A_m$. At every instant, each carrier is compared with the modulating signal. Each comparison switches the switch "on" if the modulating signal is greater than the triangular carrier assigned to that switch. Obviously, the actual driving signals for the power devices can be derived from the results of the modulating-carrier comparison by means of a logic circuit. SPWM technique can be classified according to carrier and modulating signals as given in [8]. The main parameters of the modulation process are:

1. The frequency ratio $k=f_c/f_m$, where $f_c$ is the frequency of the carriers, and $f_m$ is the frequency of the modulating signal.

2. The modulation index $M=A_m / (m'^* A_c)$, where $A_m$ is the amplitude of the modulating signal, $A_c$ is the peak-to-peak amplitude of the carriers, and $m'= (m-1)/2$, where m is the number of level (which is odd).

Fig 5. shows generation of pulses for 3-level DCMI.

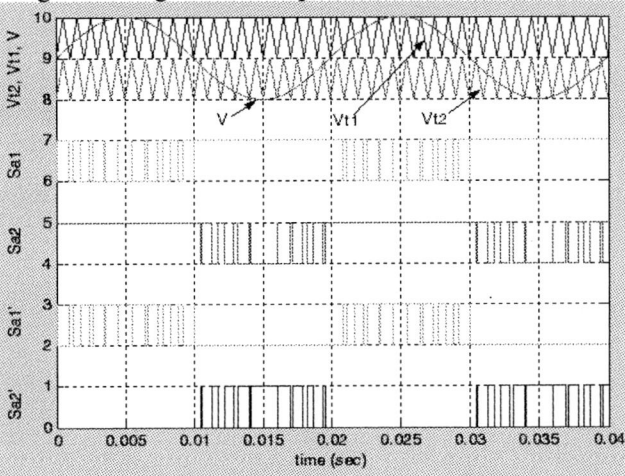

Fig. 5. Generation of firing pulses for three-level, 3-phase diode clamped inverter.

TABLE III
SWITCHING LOGIC FOR 3-LEVEL INVERTER

| Condition | Switch Status for A-phase | State |
|---|---|---|
| $V_r > (V_{t1}, V_{t2})$ | $S_{a1}$ = ON, $S_{a2}$ = ON, $S'_{a1}$ = OFF, $S'_{a2}$ = OFF, | $S_a$ = +1 |
| $V_{t1} > V_r > V_{t2}$ | $S_{a1}$ = OFF, $S_{a2}$ = ON, $S'_{a1}$ = ON, $S'_{a2}$ = OFF, | $S_a$ = 0 |
| $V_r < (V_{t1}, V_{t2})$ | $S_{a1}$ = OFF, $S_{a2}$ = OFF, $S'_{a1}$ = ON, $S'_{a2}$ = ON, | $S_a$ = -1 |

A reference modulating signal is compared with two triangular carrier signals which are in the same phase but phase disposed. Comparison logic is given in Table III. Based upon this logic, control pulses $S_{a1}$, $S_{a2}$, $S_{a1}'$ and $S_{a2}'$ are generated as in Fig 5. Similar logic is applied for other phases [5], [6], [7], [9].

The switching logic for generating the sequence of pulses in five-level inverter is shown in the Table IV. The output line to line voltage waveform contains (2m-1) voltage levels, where $m$ is the number of levels in output phase voltage.

TABLE IV
SWITCHING LOGIC FOR 5-LEVEL INVERTER

| Condition | Switch States for R-Phase | Output Voltage |
|---|---|---|
| $V_r > (V_{t1}, V_{t2}, V_{t3}, V_{t4})$ | $S_{a1} = S_{a2} = S_{a3} = S_{a4}$ = ON | $V_{dc}/2$ |
| $V_{t1} > V_r > (V_{t2}, V_{t3}, V_{t4})$ | $S_{a2} = S_{a3} = S_{a4} = S'_{a1}$ = ON | $V_{dc}/4$ |
| $(V_{t1}, V_{t2}) > V_r > (V_{t3}, V_{t4})$ | $S_{a3} = S_{a4} = S'_{a1} = S'_{a2}$ = ON | 0 |
| $(V_{t1}, V_{t2}, V_{t3}) > V_r > V_{t4}$ | $S_{a4} = S'_{a1} = S'_{a2} = S'_{a3}$ = ON | $- V_{dc}/4$ |
| $V_r < (V_{t1}, V_{t2}, V_{t3}, V_{t4})$ | $S'_{a1} = S'_{a2} = S'_{a3} = S'_{a4}$ = ON | $- V_{dc}/2$ |

### B. Optimized Harmonic Stepped Waveform Technique

The objective of the optimized harmonic stepped-waveform technique is to reduce, as much as possible, the harmonic distortion in the load voltage, working with a reduced switching frequency. The concept of this technique is to combine the idea of the selective harmonic elimination PWM with the quarter-wave symmetric idea concept. The concept of the harmonic reduction is to eliminate the specific lower order harmonics. There are three possible optimization techniques to reduce the voltage THD: (1) step heights are optimized with equally spaced steps, (2) step spaces are optimized with the steps of equal height, and 3) optimizing both heights and spaces.

This paper focuses on the second method, which the switching angles. To achieve these optimized angles, the numerical calculation will be applied. The optimized harmonic stepped waveform is assumed to be the quarter-wave symmetric. The relationship among the switching angles of the waveform are shown in Fig 6.

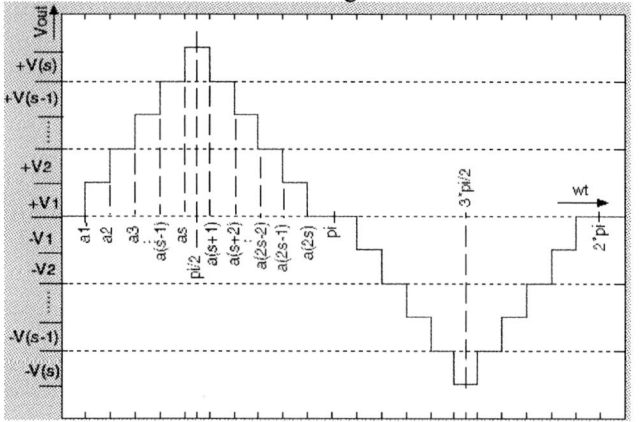

Fig. 6. The relationship among the switching angles of the waveform of m-level DCMLI.

In the second quarter of the waveform,

$\alpha_{s+1} = \pi - \alpha_s$

$\alpha_{2s-1} = \pi - \alpha_2$

$\alpha_{2s} = \pi - \alpha_1$

In the third quarter of the waveform,

$\alpha_{2s+1} = \pi - \alpha_1$

$\alpha_{3s-1} = \pi + \alpha_{s-1}$

$\alpha_{3s} = \pi + \alpha_s$

In the fourth quarter of the waveform,

$\alpha_{3s+1} = 2\pi - \alpha_s$

$\alpha_{4s-1} = 2\pi - \alpha_2$

$\alpha_{4s} = 2\pi - \alpha_1$

The output voltage level is zero from $\omega t = 0$ to $\omega t = \alpha_1$. At $\omega t = \alpha_1$, the output voltage level is changed from zero to $+V_1$, and from $+V_1$ to $+(V_1+V_2)$ at $\omega t = \alpha_2$. The process will be repeated until $\omega t = \pi/2$, and the output voltage level becomes $(+V_1+V_2+\ldots+V_{(s-1)}+V_s)$. Then, in the second quarter, the level of output voltage will be decreased to $(+V_1+V_2+\ldots+V_{(s-1)})$ at $\omega t = \pi - \alpha_s$. The process will be repeated until $\omega t = (\pi - \alpha_1)$ and output voltage becomes zero again. In the second half of the waveform, the process will repeat all previous steps except the amplitude of the dc sources change from positive to negative. The next period will then repeat the same cycle.

The Fourier series of a periodic waveform is expressed as,

$f(\alpha t) = a_0 + \sum a_n \cos n\alpha t - \sum b_n \sin n\alpha t$

Quarter and half wave symmetry ensures that no even harmonics will exist in the output spectrum. The final equation after solving the above fourier analysis is given as,

$$V_{out}(\omega t) = \sum_{n=1}^{\infty} \left[ \frac{4E}{n\pi} \sum_{k=1}^{S} \cos(n\alpha_k) \right] \sin(n\omega t)$$

where $(\alpha_1, \alpha_2 \ldots \alpha_s) < \pi/2$

E = amplitude of dc voltage

n = odd harmonic order

S = (number of levels-1)/2 = (m-1)/2 ..

In the three-level inverter to eliminate the fifth harmonic equate the term

$\cos 5\alpha_{1=} 0$ .The angle obtained is $18^0$.

In the five- level inverter for example, to eliminate first two harmonics ($3^{rd}$ and $5^{th}$) then, the coefficient of $3^{rd}$ and $5^{th}$ harmonics should be equal to zero.

i.e.  $b_3 = \cos 3\alpha_1 + \cos 3\alpha_2 = 0$

$b_5 = \cos 5\alpha_1 + \cos 5\alpha_2 = 0$

By solving these two non-linear equations, $\alpha_1$ and $\alpha_2$ values will be obtained The angles obtained after solving the equations by using the C program are $12^0$ and $48^0$. These values are used for simulation investigation later.

*C. Selective Harmonic Elimination Technique SHEPWM)*

Selective Harmonic Elimination (SHE) is an off-line (pre-calculated) non carrier based PWM technique. In this method the basic square-wave output is "chopped" a number of times, which are obtained by proper off-line calculations. The SHEPWM technique is used to synthesize an output waveform of 3-level and 5-level diode clamped multilevel inverter.

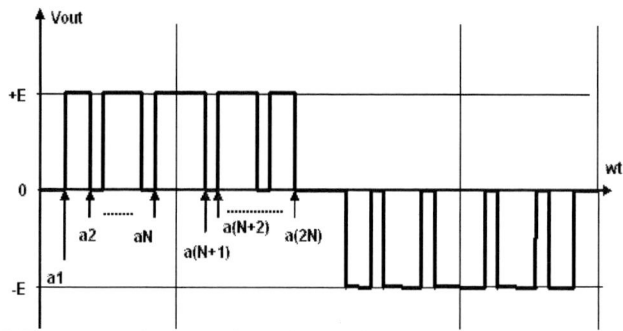

Fig. 7. The output voltage waveform of a three-level DCMI with SHEPWM technique.

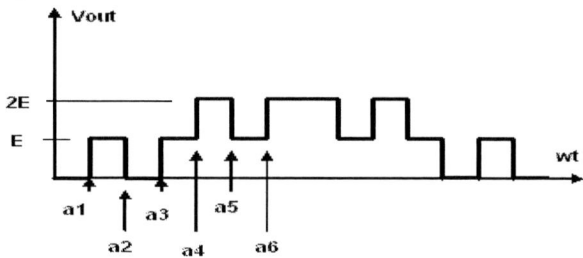

Fig. 8. The first half of output voltage waveform of a 5-level six-angle SHEPWM.

The output voltage waveform of a five level multilevel inverter is shown in the Fig. 8. Now fourier series is being developed for the above waveform. Let N be the number of switching angles per quarter-cycle. The output waveform is assumed to be odd quarter wave symmetry, whose amplitude equals E. Because of odd quarter-wave symmetry, the dc component and the even harmonics are equal to zero. Fourier series of the SHEPWM is as follows [11]:

$$V_{OUT}(\alpha) = \sum_{n=1}^{\infty} a_n \sin(n\alpha t) \text{ where,}$$

$$a_n = \frac{4E}{n\pi} \sum_{K=1}^{N} (-1)^{K+1} \cos(n\alpha_K) ; \quad \text{for odd 'n'}$$

$\alpha_k$ is the switching angles, which must satisfy the following condition:

$$\alpha_1 < \alpha_2 < \alpha_3 < \dots < \alpha_N < \pi/2$$

$E$ is the amplitude of the dc source, and $n$ is the harmonic order. The nonlinear equation system of SHEPWM waveform can be written as follows:

$$\cos(\alpha_1) - \cos(\alpha_2) + \dots \pm \cos(\alpha_N) = (\pi/4)*M$$
$$\cos(3\alpha_1) - \cos(3\alpha_2) + \dots \pm \cos(3\alpha_N) = (3\pi/4)*(h_3/E)$$
$$\cos(5\alpha_1) - \cos(5\alpha_2) + \dots \pm \cos(5\alpha_N) = (5\pi/4)*(h_5/E)$$

where,

$M$ is the modulation index, and $M=h_1/E$
$E = V_{dc} / (m-1)$, where $V_{dc}$ is supply voltage and 'm' is number of levels. In this paper, equations considered are, for three-level inverter,

$$b_3 = \cos 3\alpha_1 - \cos 3\alpha_2 + \cos 3\alpha_3 = 0$$
$$b_5 = \cos 5\alpha_1 - \cos 5\alpha_2 + \cos 5\alpha_3 = 0$$
$$b_7 = \cos 7\alpha_1 - \cos 7\alpha_2 + \cos 7\alpha_3 = 0$$

The angles obtained after solving the equations by using the C program are $22^0$, $38^0$, and $47^0$. For five level inverter, equations are,

$$b_3 = \cos 3\alpha_1 - \cos 3\alpha_2 + \cos 3\alpha_3 - \cos 3\alpha_4 + \cos 3\alpha_5 - \cos 3\alpha_6 = 0$$
$$b_5 = \cos 5\alpha_1 - \cos 5\alpha_2 + \cos 5\alpha_3 - \cos 5\alpha_4 + \cos 5\alpha_5 - \cos 5\alpha_6 = 0$$
$$b_7 = \cos 7\alpha_1 - \cos 7\alpha_2 + \cos 7\alpha_3 - \cos 7\alpha_4 + \cos 7\alpha_5 - \cos 7\alpha_6 = 0$$
$$b_9 = \cos 9\alpha_1 - \cos 9\alpha_2 + \cos 9\alpha_3 - os\ 9\alpha_4 + \cos 9\alpha_5 - \cos 9\alpha_6 = 0$$
$$b_{11} = \cos 11\alpha_1 - \cos 11\alpha_2 + \cos 11\alpha_3 - \cos 11\alpha_4 + \cos 11\alpha_5 - \cos 11\alpha_6 = 0$$
$$b_{13} = \cos 13\alpha_1 - \cos 13\alpha_2 + \cos 13\alpha_3 + \cos 13\alpha_4 - \cos 13\alpha_5 - \cos 13\alpha_6 = 0$$

The equations are solved by using the C program and the angles obtained are $5^0$, $11^0$, $20^0$, $49^0$, $58^0$, $64^0$.

## IV. RESULTS

Simulation has been carried out in MATLAB. Fig. 9 gives the phase voltage, line voltage and line current of 3-level DCMI with SPWM. All the comparative results have been tabulated for easy reference. Fig. 10 to Fig. 12 show output phase and line voltage and current THD for 3-level inverter with SPWM technique. Fig. 13 gives pulse generation for 5-level inverter in SPWM technique. Table V gives harmonic performance of 3-level & 5-level DCMI with SPWM technique for all the three phase & line. Switching frequency used is 1kHz.

TABLE V
THD IN OUTPUT VOLTAGES AND CURRENT WITH SPWM TECHNIQUE

| Phase/ | % THD | | | | | |
|---|---|---|---|---|---|---|
| Line | Phase Voltage | | Line Voltage | | Current | |
| | 3-level | 5-level | 3-level | 5-level | 3-level | 5-level |
| An/AB | 16.27 | 11.65 | 15.05 | 7.35 | 2.62 | 5.87 |
| Bn/BC | 15.92 | 10.47 | 14.98 | 7.93 | 2.66 | 13.10 |
| Cn/CA | 16.07 | 10.32 | 15.08 | 7.75 | 4.68 | 15.53 |

Output voltage THD has been significantly reduced as we switch from 3-level to 5-level. Table VI and Table VII give harmonic performance with optimized harmonic stepped PWM & selective harmonic elimination techniques for three phases. In both these techniques, lower order harmonics (3rd, 5th, 7th etc.) have been significantly reduced as seen from Fig. 14 & Fig. 15 for example. Overall THD performance is given in Table VIII. For simulation, 3-phase load of active power (8kW) & inductive reactive power (6 kVar) has been used.

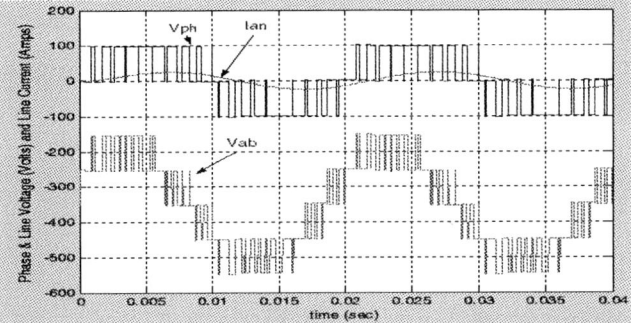

Fig. 9. Phase voltage, line voltage and line current of 3-level DCMI with SPWM.

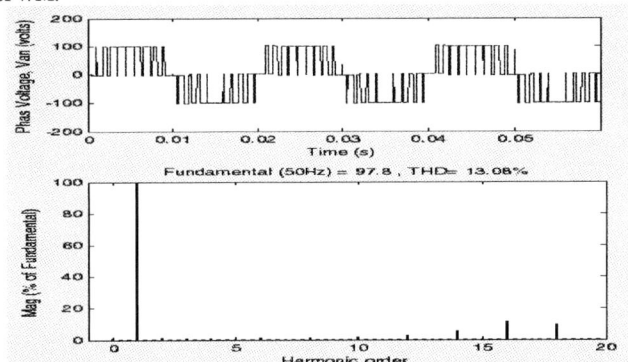

Fig. 10. Phase voltage & its harmonic spectrum of 3-level DCMI with SPWM.

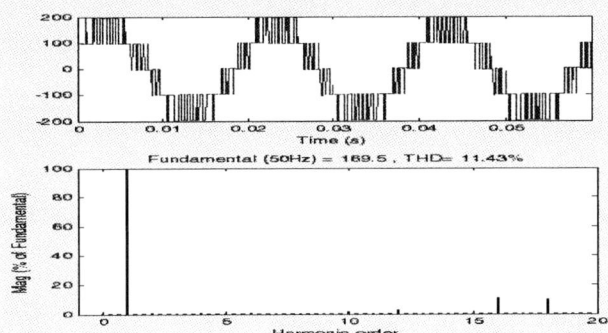

Fig. 11. Line voltage & its harmonic spectrum of 3-level DCMI with SPWM.

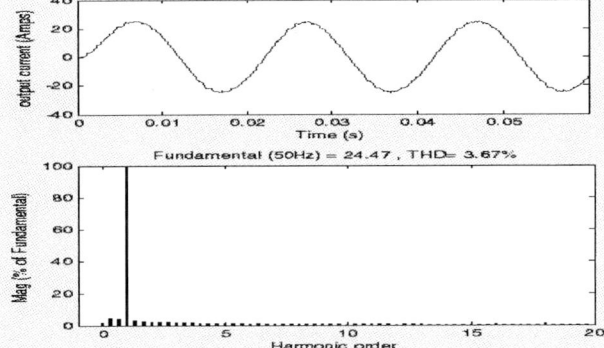

Fig. 12. Line current & its harmonic spectrum of 3-level DCMI with SPWM.

### TABLE VI
#### THD IN OUTPUT VOLTAGES AND CURRENT WITH OHPWM TECHNIQUE

| Phase/ | % THD | | | | | |
|---|---|---|---|---|---|---|
| Line | Phase Voltage | | Line Voltage | | Current | |
| | 3-level | 5-level | 3-level | 5-level | 3-level | 5-level |
| An/AB | 27.83 | 18.69 | 14.97 | 16.52 | 20.11 | 11.75 |
| Bn/BC | 27.81 | 16.39 | 14.98 | 16.28 | 20.53 | 11.96 |
| Cn/CA | 27.81 | 17.61 | 14.96 | 16.52 | 20.34 | 11.65 |

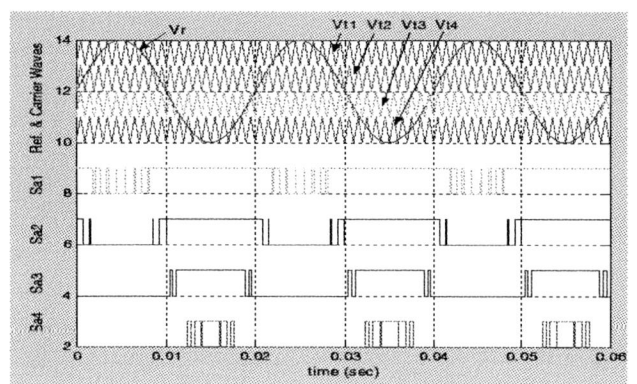

Fig. 13. Generation of firing pulses for five-level, 3-phase diode clamped inverter with SPWM.

### TABLE VII
#### THD IN OUTPUT VOLTAGES AND CURRENT WITH SHEPWM TECHNIQUE

| Phase/ | % THD | | | | | |
|---|---|---|---|---|---|---|
| Line | Phase Voltage | | Line Voltage | | Current | |
| | 3-level | 5-level | 3-level | 5-level | 3-level | 5-level |
| An/AB | 39.51 | 19.01 | 25.55 | 23.46 | 8.35 | 14.38 |
| Bn/BC | 39.26 | 35.69 | 25.47 | 16.15 | 7.92 | 12.30 |
| Cn/CA | 39.43 | 22.57 | 25.48 | 17.30 | 7.95 | 6.38 |

### TABLE VIII
#### COMPARISON OF THD IN OUTPUT VOLTAGES AND CURRENT

| Control | % THD | | | | | |
|---|---|---|---|---|---|---|
| Technique | Phase Voltage | | Line Voltage | | Current | |
| | 3-level | 5-level | 3-level | 5-level | 3-level | 5-level |
| SPWM | 15.92 | 10.32 | 14.98 | 7.35 | 2.62 | 3.87 |
| OHPWM | 27.81 | 16.39 | 14.96 | 14.28 | 20.11 | 11.75 |
| SHEPWM | 39.26 | 19.01 | 25.47 | 16.15 | 7.92 | 6.38 |

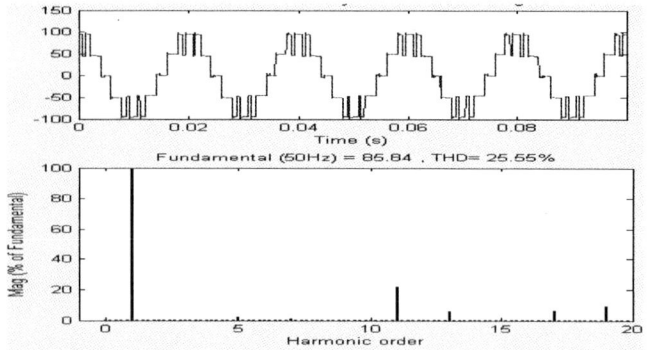

Fig. 14. Line voltage & its harmonic spectrum of 3-level DCMI with SHEPWM technique.

## V. CONCLUSION

The concepts of sinusoidal pulse width modulation, optimized harmonic stepped waveform, and selective harmonic elimination techniques have been investigated. Three-phase three-level and five-level diode clamped inverters have been simulated. The lower order harmonics were considerably reduced in all the techniques. As seen from Table VIII, output line voltage THD has been greatly reduced at 1kHz of switching frequency in SPWM technique and $3^{rd}$, $5^{th}$, $7^{th}$ harmonics have been almost eliminated in OHPWM & SHEPWM techniques, though it shifts the harmonic pollution towards higher order.

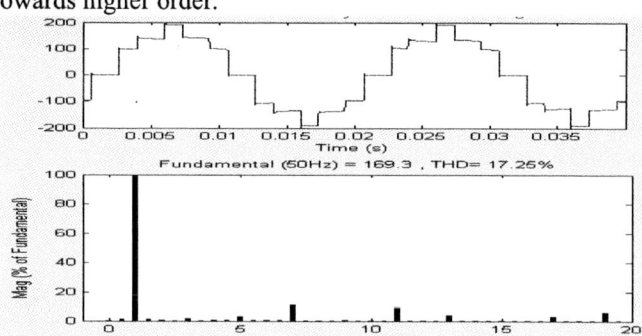

Fig. 15. Line voltage & its harmonic spectrum of 5-level DCMI with OHPWM.

## VI. REFERENCES

[1] Pradeep M. Bhagwat, V. R. Stefanovic, "Generalized Structure of a Multilevel Inverter", IEEE Trans. on Industrial Applications, Vol 19, No. 6, Nov/Dec 1983, pp 1057-1069.

[2] T. Salzmann, G. Kratz, C. Daubler, "High Power Drive System with Advanced Power Circuitry and Improved Digital Control", IEEE Trans. on Industrial Applications, Vol 29, No. 1, Jan/Feb 1993.

[3] F. Zeng Peng, "A Generalized Multilevel Inverter Topology with Self Voltage Balancing", IEEE Trans on Industry Applications, Vol. 37, No. 2, Mar/Apr 2001, pp 611-618.

[4] H. Zhang, A.V. Jouanne, S. Dai, A.K. Wallace, F. Wang, "Multilevel Inverter Modulation Schemes to Eliminate Common-Mode Voltages", IEEE Trans on Industry Applications, Vol. 36, No. 6, Nov/Dec 2000, pp 1645-1653.

[5] Leon M. Tolbert, Fang Z. Peng, T.G. Habetler, "Multilevel Converters for Large Electric Drives", IEEE Trans on Industry Applications, Vol. 35, No. 1, Jan/Feb 1999, pp 36-44.

[6] J.S. Lai, F.Z. Peng, "Multilevel Converters – A New Breed of Power Converters", IEEE. Trans. on Industry Applications, Vol. 32, No. 3, May/Jun 1996, pp 509-517.

[7] T.A. Meynard and H. Foch, "Multilevel Conversions: High Voltage Choppers and Voltage Source Inverters", IEEE-PESC, 1992, pp. 397-403.

[8] J. Rodriguez, J.S. Lai, Fang Z. Peng, "Multilevel Inverters: A Survey of Topologies, Controls and Applications", IEEE Trans on Industrial Electronics, Vol 49, No 4, Aug 2002, pp 724-738.

[9] Leon M. Tolbert, T.G. Habetler, "Novel Multilevel Inverter Carrier-Based PWM Methods", IEEE IAS Annual Meeting, Oct. 10-15, 1998, pp 1424-1431.

[10] Surjeet Ghani Lalla, "Simulation and Performance Investigation of Multilevel Inverter Fed Induction Motor", M .Tech Thesis, Indian Institute of Technology, Roorkee, Uttaranchal, India, Feb 2003.

[11] Li Li, Dariusz Czarkowski, Yaguang Liu, and Pragasen Pillay "Multilevel Selective Harmonic Elimination PWM Technique in Series-Connected Voltage Inverters", *IEEE Trans. Industrial Applications*, vol. 36, no. 1, January/February 2000.

**2006 IEEE International Conference on Power Electronic, Drives and Energy Systems**

# Peak-Current Mode control of Hybrid Switched Capacitor Converter

Veerachary M, *Senior Member, IEEE*, and Singamaneni Bala Sudhakar

*Abstract*– In this paper analysis and controller design of a Peak-Current Mode Controlled higher order buck converter is presented. Various small-signal models valid up-to half of the switching frequencies, are developed using state-space averaging method, and then a step-by-step controller design procedure is described. The results of controller design and closed-loop analysis are illustrated through computer simulations. To validate the controller design and analysis, experimental observations of a prototype hybrid switched-capacitor converter are provided for 28 V applications.

*Index Terms*– Hybrid Switched Capacitor Converter, Peak current-mode control, Point-of-load converters, Type-II Compensator.

## I. INTRODUCTION

IN recent years, the high frequency switching converters application in low power compact electronic circuits is increasing. As the power conversion system is becoming miniaturized, increasing the power density is one of challenging issue for the power supply (PS) designers. Light weight, small size and high power densities are possible using minimum number of inductors and using switched capacitor type of topologies. Although complete elimination of inductors is the desirable condition to realize the compact PS, but its elimination degrades the system performance. 48 V bus structure is common in telecommunication and automotive applications and there several other loads, which requires wide range of voltages such as 28, 24, 15 and 12 V. For these applications it is necessary to employ buck topologies in between the dc bus and the load. While selecting converters for such applications the following important points need to be considered: (i) simple from the topology point of view, (ii) load voltage regulation capability, (iii) compact in size or higher power density, (iv) better reliability. It is possible to use simple switched capacitor based topology (SCBT) solutions for the above said point-of-load (POL) applications. However,

the simple SCBT's are most suitable only for fixed conversion ratio's and they are not having voltage regulation capability. Hence, there is a need to develop hybrid converters by combining the salient features of switched capacitor converter with conventional inductor based topologies. Here, the switched capacitor converters results in compact size, while the inductor based solutions results in load voltage dependence on the duty ratio.

## II. HYBRID SWITCHED CAPACITOR CONVERTER

Buck topology is most widely used for step-down applications. However, use of simple buck topology presents high pulsating current stress on the source and in some cases, such as in battery driven applications, it will affect the source reliability also. The buck converter with input filter (BCIF) will be the alternative solution, but, however, it increases the number of inductive energy storage elements and ultimately the analysis of converter dynamics becomes a complex task. It is possible to reduce number of inductive energy storage elements while retaining the buck conversion property, just by employing one additional diode in place of inductor and its final circuit connection is shown in Fig. 1a. This topology has the following advantages: (i) lower source current ripple, (ii) low EMI, (iii) simpler dynamics than the BCIF, (iv) simplicity in control due to single switch topology.

This hybrid switched-capacitor converter (HSCC) can be operated in several operating modes depending on the load, switching frequency and supply voltage. However, due to the presence of inductor on source side, for most of the loads, the converter input current exhibits lower ripple. Only for very light loadings it exhibits higher ripple content and for such light load applications the proposed converter will not provide any benefit as compared to BCIF topology. In the following lines the converter operation for continuous inductor current mode (CICM) is discussed and then mathematical models, suitable for controller design, are formulated.

The converter acts as a time-invariant system both in switch-ON and OFF cases. But, however, over one full switching cycle the circuit switches between two different linear circuits and thus it become non-linear. In order to study the converter behavior, both steady-state and dynamic, a linear model is developed in the following lines using the state-space

---

This work was supported in part by the MHRD, Govt. of India under R & D Project Grant.

Veerachary. M is with Dept. of Electrical Engineering, Indian Institute of Technology Delhi, New Delhi, India (e-mail: mvchary@ee.iitd.ac.in).

S. Bala Sudhakar is with the Department of Electrical Engineering, Indian Institute of Technology Delhi, New Delhi, India (e-mail: singamaneni2u@gmail.com).

0-7803-9771-1/06/$25.00 ©2006 IEEE

averaging method [12]. This averaged model is nonlinear and time-invariant and its linear model can easily be constructed by using small-signal approximations. State model of the converter for ON-state condition is

$$\dot{X} = [A_1][X] + [B_1][u] \qquad (1)$$

while state model for OFF-state condition is

$$\dot{X} = [A_2][X] + [B_2][u] \qquad (2)$$

The average state-space model is

$$\dot{X} = [A][X] + [B][u] \qquad (3)$$

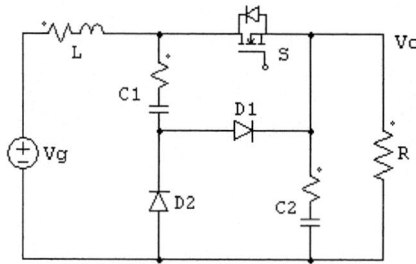

(a) Circuit diagram of HSC converter.

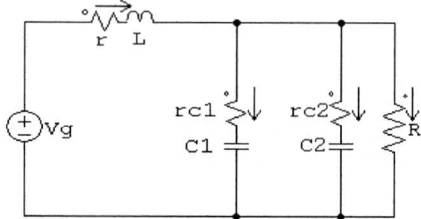

(b) Equivalent circuit during Mode-1 operation.

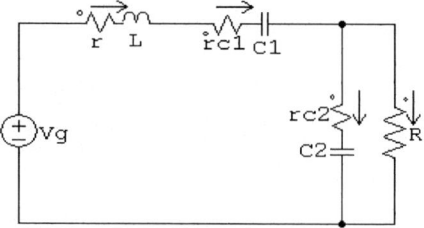

(C) Equivalent circuit during Mode-2 operation.

Fig. 1. Hybrid switched capacitor converter.

where $A = A_1 D + A_2 (1 - D)$ ; $B = B_1 D + B_2 (1 - D)$.

$$A_1 = \begin{pmatrix} -\frac{1}{L}\left(\frac{Rr_{c1}r_{c2}C_2}{k}+r\right) & -\frac{1}{L}\left(\frac{Rr_{c2}C_2}{k}\right) & -\frac{1}{L}\left(1-\frac{r_{c2}C_2(R+r_{c1})}{k}\right) \\ \left(\frac{Rr_{c2}C_2}{C_1 k}\right) & \left(\frac{Rr_{c2}C_2}{r_{c1}C_1 k}-\frac{1}{r_{c1}C_1}\right) & \left(\frac{1}{r_{c1}C_1}-\frac{r_{c2}C_2(R+r_{c1})}{r_{c1}C_1 k}\right) \\ \frac{Rr_{c1}}{k} & \frac{R}{k} & \left(-\frac{R+r_{c1}}{k}\right) \end{pmatrix}$$

$$A_2 = \begin{pmatrix} -\frac{1}{L}\left(\frac{Rr_{c2}}{R+r_{c2}}+r_{c1}+r\right) & \frac{-1}{L} & -\frac{1}{L}\left(\frac{R}{R+r_{c2}}\right) \\ \frac{1}{C_1} & 0 & 0 \\ \frac{1}{C_2}\left(\frac{R}{R+r_{c2}}\right) & 0 & -\frac{1}{C_2}\left(\frac{1}{R+r_{c2}}\right) \end{pmatrix}$$

$$[B_1] = [B_2] = \begin{bmatrix} \frac{1}{L} \\ 0 \\ 0 \end{bmatrix}$$

$$k = C_2[R(r_{c1}+r_{c2})+r_{c1}r_{c2}]$$

### III. PEAK CURRENT-MODE CONTROL

It is well known that the two-loop control strategy, inner current-loop and outer voltage-loop, results in faster dynamic response as compared to single-loop voltage-mode control scheme. In two-loop control strategies there are two possibilities (i) average current-mode control, (ii) peak current-mode control. However, the peak current-mode control (PCC) has the following advantages:

- faster dynamic response due to the presence of inner current loop
- pulse-by-pulse current limiting
- automatic feed-forward compensation
- less circuit complexity compared to average current-mode control

#### A. Loop gain transfer functions

In the following lines the controller design is discussed. The small-signal model of the converter with inner current-loop and outer voltage-loop is shown in Fig. 2. From this block diagram the inner current-loop gain, ($T_i(s)$, is defined as

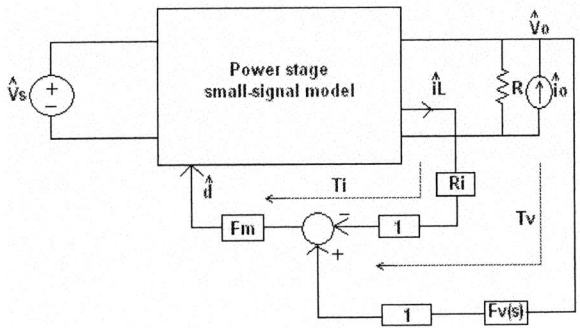

Fig. 2. Small-signal block diagram for peak current-mode control.

$$T_i(s) = G_{id}(s)R_iF_m \tag{4}$$

where '$F_m$' PWM transfer function. The outer voltage loop gain, $T_v(s)$, is

$$T_v(s) = G_{vd}(s)F_V(s)F_m \tag{5}$$

where '$F_v$' compensator, shown in Fig. 3, transfer function. The overall loop gain is

$$T_1(s) = T_i(s) + T_v(s). \tag{6}$$

The outer loop gain is

$$T_2(s) = \frac{T_V(s)}{1 + T_I(s)}. \tag{7}$$

### B. Compensator selection

A closed loop system can be implemented with different types of compensation network. The Type −I compensation network can give good phase margin, but bandwidth is usually too low and hence results in poor transient response. The Type −II compensation network can improve the transient response but phase boost is limited to less than 90°. However, the Type −III compensation network can give the fastest transient response and sufficient phase margin to ensure system stability among three types of compensation networks. Here selection of compensator also depends on the type of the converter to be controlled. In most of the cases the Type −II compensator is sufficient for getting faster dynamic response and also simple from the design point of view. Hence, a Type −II compensator, as shown in Fig. 3, is used for outer voltage loop.

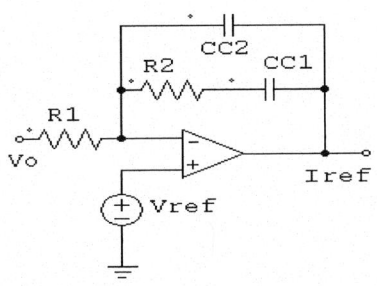

Fig. 3. Type-II Compensator.

### C. Compensator design steps

Step-1: Place a pole at the origin in order to realize high dc gain.
Step-2: Place a zero as high as possible.
Step-3: Place a pole at the corner frequency of capacitor $C_2$ ESR.
Step-4: Adjust gain of the compensator, $K_v$, such that the closed-loop system meets the desired stability margins (gain and phase margin's).

## IV. SIMULATION AND EXPERIMENTAL RESULTS

To verify the developed modeling and controller design, a 20 W HSCC system was designed to supply a constant load voltage of 28 V± 5% from a source voltage of 42 V. The power stage parameters are given in Table I. Controller is designed using pole-zero placement technique, its design steps are illustrated in preceding Section, and the design program is developed in MATLAB. For the above given converter parameters the open-loop control-to-inductor current, $G_{id} = \dfrac{\hat{i}_l(s)}{\hat{d}(s)}$, and control -to- output voltage, $G_{vd} = \dfrac{\hat{v}_0(s)}{\hat{d}(s)}$, transfer functions are plotted in Fig. 4 and 5.

The $G_{vd}(s)$ transfer function has one complex conjugate pair of poles, one real pole, two real zero's and all these are in the left half of jω-plane. The compensator, Type-II, is designed by adopting the design procedure. In the design process the current and voltage loop gains, $T_i(s)$ and $T_v(s)$, are utilized and their frequency response characteristics are shown in Fig. 6. From these characteristics it can be seen that the overall-loop gain, $T_l(s)$, is matching with the outer voltage loop-gain ($T_v$) until the crossover frequency and after that it is following the current loop-gain ($T_i$).

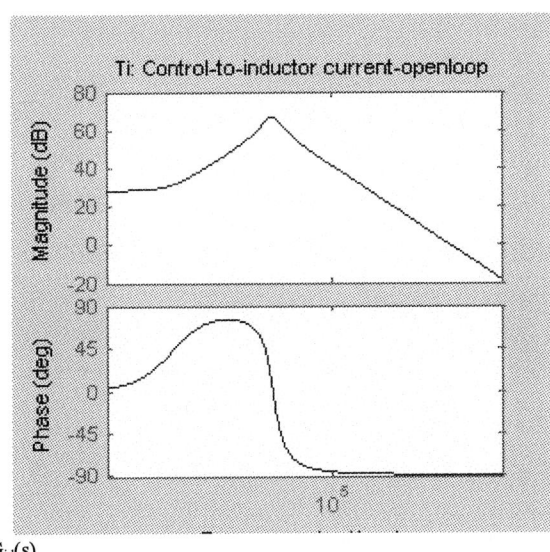

(a) $G_{id}(s)$

481

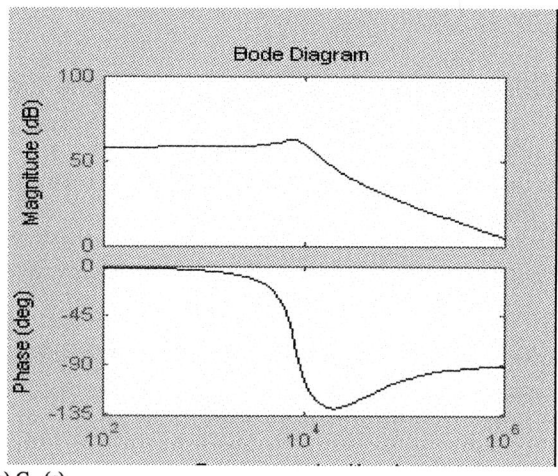

(b) $G_{vd}(s)$

Fig. 4. Bode plots of $G_{id}(s)$ and $G_{vd}(s)$.

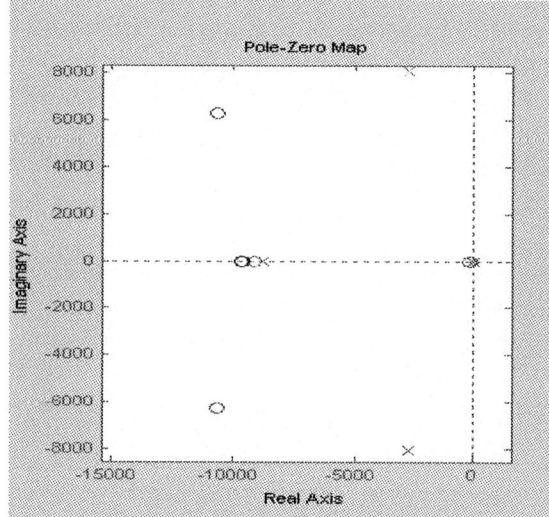

$T_l(s)$: red, $G_{vd}(s)$: blue $G_{id}(s)$: green

Fig. 5. Ploe-zero map of various small-signal transfer functions.

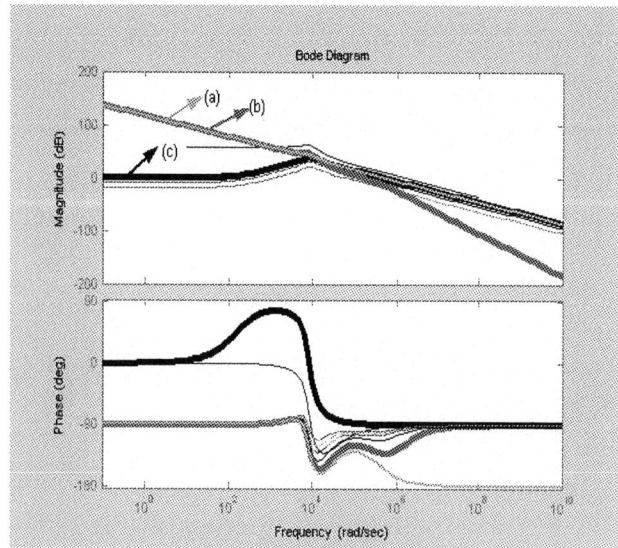

a: $T_l(s)$; b: $T_v(s)$; c: $T_i(s)$

Fig. 6. Frequency response characteristics of loop-gains.

The final design parameters of the controller for this operating condition are given in Table I. With these parameters the closed-loop converter system regulation capability is tested for (i) supply voltage change 42 → 35 V, (ii) load disturbance of 40 → 20 Ω. For one operating conditions the steady-state inductor current and load voltage waveform are shown in Fig. 7. From this figure it can be noted that the in the source current ripple is low as compared to simple buck topology. To find the large-signal response of the closed-loop converter system PSIM power electronics simulator [13] is used. For demonstration load voltage regulation against supply voltage and load disturbances are obtained for the above-mentioned cases and they are given in Fig. 8. Note that for all these cases the controller is regulating the load voltage equal to the reference voltage. But at the instant of disturbance the load voltage is undergoing the dynamics, drop in load voltage and response time is negligible, and finally settling to reference voltage of 28 V.

To verify the theoretical analysis and simulation results an experimental prototype HSCC circuit, with parameter values given above, has been built. MOSFET: IRF540, Diode: MUR820, Driver: IR2110, Opto-coupler: 6N136 were used in the prototype model. UC3825 IC is used to realize the peak-current-mode control. In the experiment it was observed that this converter for duty ratio's more than 50% showing unstable behavior. Slope compensation is provided to solve this problem. Experimental closed-loop converter system regulation capability is tested for: (i) supply voltage change 42 → 35 V, (ii) load disturbance of 40 → 20 Ω and these results are shown in Fig. 9. These results are closely matching with the simulation results shown in Fig. 8. Slight discrepancy between the simulation and experimental values are attributed to the following factors: (i) voltage drops in the switching devices, (ii) simulation modeling methods, (iii) error's in measuring instruments, etc.

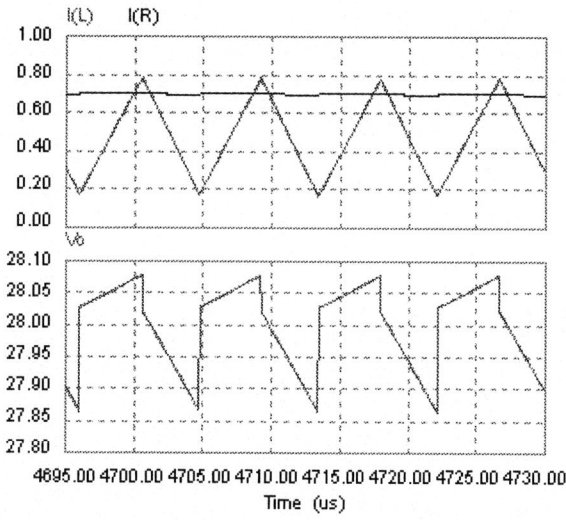

Fig. 7. Steady-state $i_L$, $i_0$, $v_0$ waveforms.

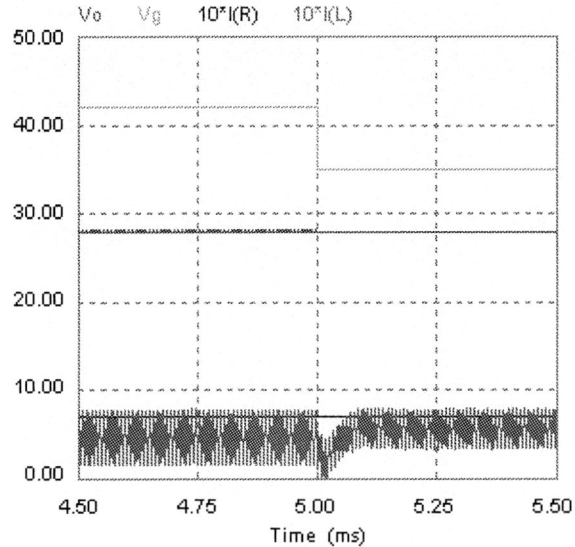

(a) supply voltage disturbance (42 → 35 V).

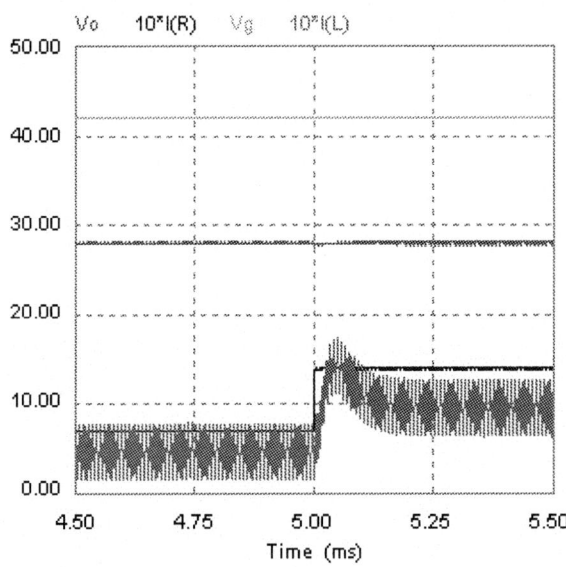

(b) load disturbance (40 → 20 Ω).
Fig. 8. Simulated dynamic response of the converter.

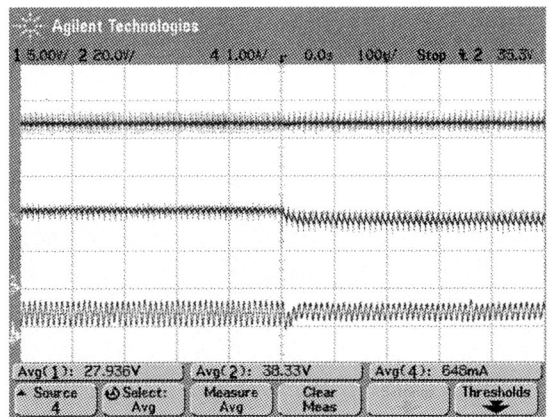

(a) supply voltage disturbance (42 → 35 V).

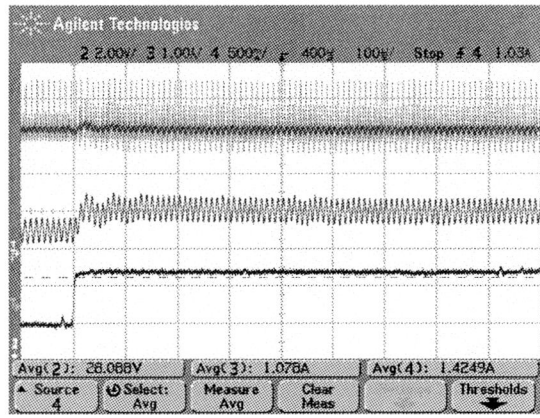

(b) load disturbance (40 → 20 Ω).
Fig. 9. Experimental dynamic response of the converter.

TABLE I
CONVERTER PARAMETERS

| Power stage parameters | Compensator parameters |
|---|---|
| L=100 μH | |
| r=0.034 Ω | |
| $C_1$ =100 μF | $R_1$=10  kΩ |
| $C_2$ =220 μF | $R_2$=5.3  kΩ |
| $r_{c1}$=0.3 Ω | |
| $r_{c2}$=0.25 Ω | $CC_1$ =100 pF |
| $R_L$=40 Ω | $CC_2$ =50  pF |

## V. CONCLUSIONS

A hybrid switched-capacitor step-down buck topology was proposed to realize low ripple content on the source side. Two-loop peak current-mode controller was used for regulation purpose. Step-by-step controller design procedure is established and a Type-II compensator was designed. Regulation capability of the proposed converter was demonstrated through simulation and experimental results. These results show that proposed converter capable of achieving the desired performance, both steady-state and dynamic, with simple controller.

## VI. REFERENCES

[1] Lamberechts. P.F., Bosgra, O.H.: 'The parameterization of all controllers that achieve output regulation and tracking', *Proceedings of the 30$^{th}$ IEEE Conference on Decision and control*, 1991, pp. 569-574.
[2] Banerjee, S., and Chakrabarthy, K.: 'Nonlinear modeling and bifurcations in the boost converter', *IEEE Trans.On Power Electronics*, 1998, Vol. 13, pp. 252-260.
[3] AL-Mothafar, M.R.D., and Hammad, K.A.: 'Small-signal modeling of peak current-mode controlled buck-derived circuits', *IEE Proc., Electr. Power Appl.*, 1999, Vol. 146, pp. 607-619.
[4] Lloyd Dixon, "Average Current Mode Control of Switching Power Supplies," *Unitrode application note*, U-140, 1990, pp. 1-14.
[5] Chunxiao. S., Lehman. B., and Sun. J.: 'Ripple effects on small signal models in average current mode control'. *Fifteenth Annual IEEE Applied power electronics Conference and Exposition*, 2000, pp. 818-823.
[6] Garcera. G., Pascual. M., and Figures. E.: 'Robust average current-mode control of multi-module parallel DC-DC PWM converter systems with improved dynamic response', *IEEE Trans.On Ind. Electronics*, 2001, Vol. 48, pp. 991-1005.

[7] J.-J. Shieh, "Closed-form oriented loop compensator design for peak current-mode controlled DC-DC regulators," IEE Proc.-Electr. Power Appl.. 2003, VoL 150(3), pp. 351-356.

[8] J.-J. Shieh, "Analysis and design of parallel-connected peak-current-mode-controlled switching DC/DC power supplies," *IEE Proc.-Electr. Power Appl.,* Vol. 151(4), 2004, pp. 434-442.

[9] J.-J. Shieh, "Peak-Current-Mode Based Single-Wire Current-Share Multi module Paralleling DC Power Supplies," *IEEE Trans. On Circuits and Systems*, Vol. 50, 2003, pp. 1564–1568.

[10] Wonseok Lim, Byungcho Choi, and Jian Sun, "Comparative Performance Evaluation of Current-Mode Controls Adapted to Asymmetrically-Driven Bridge-Type Pulse-Width Modulated DC-To-DC Converters," *Applied Power Electronics Conference and Exposition,* Vol. 2, 2005, pp. 1179 – 1185.

[11] R. B. Ridley, "A new, continuous-time model for current-mode control," *IEEE Trans. Power Electron.*, Vol. 6, 1991, pp. 271–280.

[12] T. Sunito, M. Rahkala, I. Gadoura, K. Zenger, "Dynamic Effects of Inductor Current Ripple in Peak-Current and Average-Current Mode Control," *IEEE Industrial Electronics Society Annual Conference (IECON)*, 2001, pp. 1072-1077.

[13] PSIM user manual, 2004.

## VI. BIOGRAPHIES

**M.Veerachary** was born in India in 1968. Mr. Chary completed his Dr. Eng from University of the Ryukyus, Okinawa, Japan in 2002. Since then he is with Dept. of Electrical Engineering, Indian Institute of Technology Delhi, India and currently he is an Associate Professor. His research interests are High frequency dc-dc conversion, design and analysis of satellite/spacecraft power conditioning systems, modeling and simulation of large power electronic systems, control theory application to power electronic systems, and intelligent controller applications to power supplies.

**Singamaneni Bala Sudhakar** was born in Prakasham district, A.P, India. He received the Bachelor degree from the R & N Engineering College, Ongole, JNT University, India, in 2005, and currently he is working towards Master of Technology degree in the Dept. of Electrical Engineering, IIT Delhi. His fields of interest are power electronic converters applications to small power supplies and controllers design for the switch-mode converters.

**2006 IEEE International Conference on Power Electronic, Drives and Energy Systems**

# Observer based current control of single-phase inverter in DQ rotating frame

B.Saritha, and P.A.Jankiraman, *Senior member, IEEE*

*Abstract- A modified current control scheme in D-Q rotating frame for single-phase inverters is presented in this paper to provide zero steady state error at fundamental frequency as well as waveform fidelity. The new procedure uses tuned observers to transform the physical quantities like "current" to real and imaginary values in stationary frame. The rotation matrix is used to transform from stationary frame to synchronous frame so that the transformed current values in steady state become DC variables in DQ frame. The controller design is similar to conventional DC/DC converters. The concept is validated by simulation in MATLAB and realized using FPGA EP1C3T100C7. The experimental results are provided for current control of single-phase inverters supplying R, RL and RC loads. The voltage control loop for UPS applications is also discussed in the paper.*

*Keywords- Single-phase inverter, synchronous frame, curve fitting observer, current control.*

## I. INTRODUCTION

Current control is an important issue for power electronic converters used for high performance motor drives.Hysteresis current controllers are widely used due to its robustness but they provide variable switching frequency with poor harmonic characteristics [1]. PI controllers are used due to the good harmonic performance but they must be tuned to suit particular loads to ensure stability. Analog-based controllers utilizing inductor current feedback is found in [2], [3], while capacitor current feedback topologies are found in [4], [5]. The controllers can be further subdivided as synchronous and stationary frame controllers. In a stationary frame, a new PI compensator had been considered to achieve zero steady state error but its realization limits the gain provided by the controller [6]. Voltage control scheme in the synchronous frame had been attempted using a grid simulator [7]. It uses a FIFO (stack) to get the imaginary (orthogonal) values. In [7], the stack size increases with increased demand in accuracy of the feedback real current. The transformation from stationary to synchronous frame amplifies errors due to inaccurate stationary frame values. In this paper, we use a curve-fitting observer to get real and imaginary orthogonal

current values. The observer acts as a filter for measurement noises in load current without introducing any pronounced delays. The two output signals of the observer after d-q transformation resemble dc signals and are used as feed back signals to control the real and orthogonal values of the current in a two-input current controller.

Regarding the voltage loop of the inverter, the voltage drop in the series inductor of the LC filter due to the harmonic currents can be considered to be equivalent to an external noise. Hence an inner loop, can be built around this noise signal to improve the signal to noise ratio.

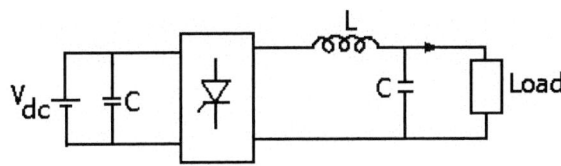

Fig. 1. Single phase inverter with LC filter.

## II. TRANSFORMATION TO SYNCHRONOUS FRAME

For a three-phase inverter, simple PI controllers in D-Q frame are sufficient to provide zero steady state error. However, this is not a case with single-phase converters (Fig.1), as we are required to an additional imaginary or orthogonal signal to move to DQ frame. The imaginary circuit is similar to the real circuit with the same circuit parameters with its current orthogonal to real current.

The real current is given by

$$I_R = I_m \sin(\omega t + \phi)$$

where $\phi$ is the phase delay and $I_m$ is the peak value of current. The imaginary orthogonal current is defined by

$$I_I = I_m \cos(\omega t + \phi)$$

where $\omega$ is the fundamental frequency.

The rotation matrix to transform from stationary to synchronous frame is given by

$$T = \begin{bmatrix} \sin \omega t & \cos \omega t \\ \cos \omega t & -\sin \omega t \end{bmatrix}$$

The dc quantities in {DQ} is given by

---

B.Saritha is doing her Ph.D. in Department of Electrical Engineering, Indian Institute of Technology, Madras .(e-mail: saritha_oct@yahoo.com).

Dr.P.A.Jankairaman is Professor in the Department of Electrical Engineering, Indian Institute of Technology, Madras 600036 India (e-mail: pajraman@ieee.org).

0-7803-9771-1/06/$25.00 ©2006 IEEE

$$\begin{bmatrix} I_d \\ I_q \end{bmatrix} = \begin{bmatrix} \sin \omega t & \cos \omega t \\ \cos \omega t & -\sin \omega t \end{bmatrix} \begin{bmatrix} I_m \sin (\omega t + \varphi) \\ I_m \cos (\omega t + \varphi) \end{bmatrix}$$

$$\begin{bmatrix} I_d \\ I_q \end{bmatrix} = \begin{bmatrix} I_m \cos \varphi \\ I_m \sin \varphi \end{bmatrix}$$

If the phase angle and magnitude of current coincide with the reference, then

$$\begin{bmatrix} I_d \\ I_q \end{bmatrix} = \begin{bmatrix} I_m \\ 0 \end{bmatrix}$$

The reference for $I_d$ is set at the desired peak value and the reference for $I_q$ is set at zero for the two inputs to the controller.

## III. CONSTRUCTION OF ORTHOGONAL VALUES

The load current is sensed using a Hall sensor and it is passed through a curve-fitting observer to get orthogonal values. The observer differential equation is given by

$$\begin{bmatrix} \dot{x}_1 \\ \dot{x}_2 \end{bmatrix} = \begin{bmatrix} 0 & \omega \\ -\omega & 0 \end{bmatrix} \begin{bmatrix} x_1 \\ x_2 \end{bmatrix} + \begin{bmatrix} d_1 \\ d_2 \end{bmatrix} e(t)$$

where $x_1$ = real current (i.e. $I_m \sin \omega t$)

$x_2$ = orthogonal current (i.e. $I_m \cos \omega t$)

Error $e(t)$= Measured load current $- x_1(t)$

$\omega$ = Fundamental frequency

The observer poles are placed at $5\omega$ and the values of $d_1$ and $d_2$ are given by

$d_1 = 10*\omega$ and $d_2 = 24*\omega$

The real and imaginary current values are obtained from the observer as $x_1$ and $x_2$. The curve-fitting observer is shown in Fig.2.

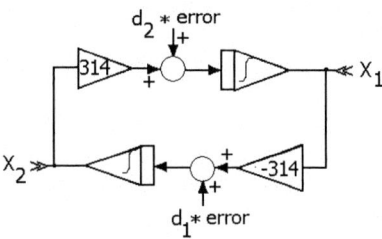

Fig. 2. Curve fitting observer with correction for $x_1$ and $x_2$.

## IV. CONTROLLER DESIGN

The PI controllers for $I_d$ and $I_q$ are tuned using Ziegler's Nichols procedure. The PI controller output represents duty ratio in DQ frame, is transformed to synchronous frame using inverse rotation matrix

$$\begin{bmatrix} E_R \\ E_Y \end{bmatrix} = \begin{bmatrix} \sin \omega t & \cos \omega t \\ \cos \omega t & -\sin \omega t \end{bmatrix} \begin{bmatrix} E_d \\ E_q \end{bmatrix}$$

The error value $E_Y$ is ignored as it corresponds to duty ratio applied to imaginary orthogonal circuit. A high frequency dither of frequency 5 kHz is added to $E_R$ and it is passed

through a signum function to eventually generate firing signals. The controller is shown in Fig.3.

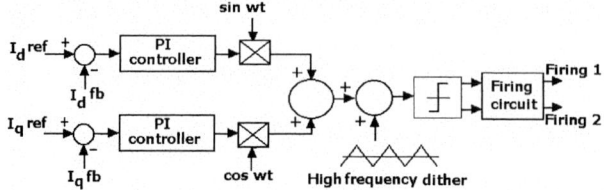

Fig. 3. Controller for generating firing signals to inverter.

## V. CURRENT CONTROL MODEL IN DQ FRAME

The error in $I_d$ and $I_q$ are processed through PI controller outputs to generate control signals $m_1$ and $m_2$. The gain involved in transferring control signals as voltage to the load is taken as unity.

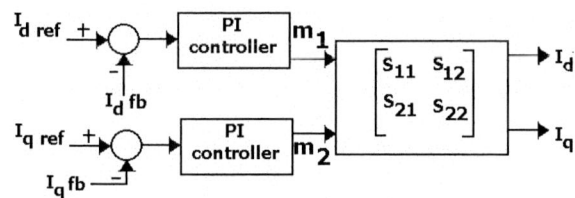

Fig. 4. Current control structure in synchronous (DQ) frame.

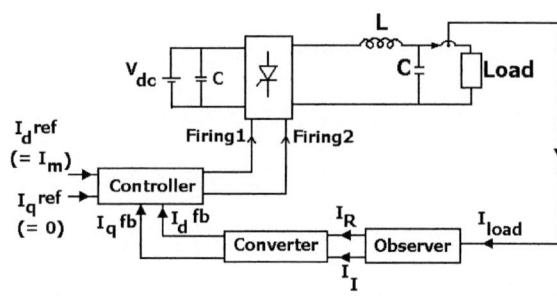

Fig. 5a. Overall current control schematic with inverter.

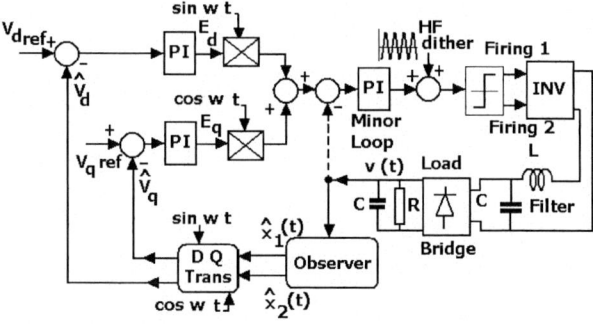

Fig. 5b. Overall voltage control schematic with minor loop.

The transfer function relating $I_d$ and $I_q$ with $m_1$ and $m_2$ is written in the Laplace domain as

$$\begin{bmatrix} I_d \\ I_q \end{bmatrix} = \begin{bmatrix} S_{11} & S_{12} \\ S_{21} & S_{22} \end{bmatrix} \begin{bmatrix} m_1 \\ m_2 \end{bmatrix}$$

Where $S_{11} = \dfrac{\left(S^2 LCR + S L + R\right)}{D}$

$S_{12} = \dfrac{\omega L + 2\omega LCRS}{D}$

$S_{21} = -S_{12}$

$S_{22} = S_{11}$

Where $D = S^4 L^2 C^2 R^2 + S^3 2L^2 CR +$
$S^2 \left(2LCR^2 + L^2 + 2\omega^2 L^2 C^2 R^2\right) +$
$S\left(2\omega^2 L^2 CR + 2LR\right) + \left(\omega^2 L^2 + R^2\right)$

The current control model in DQ frame is shown in Fig.4. The model can be used for tuning the PI controllers to get optimum performance.

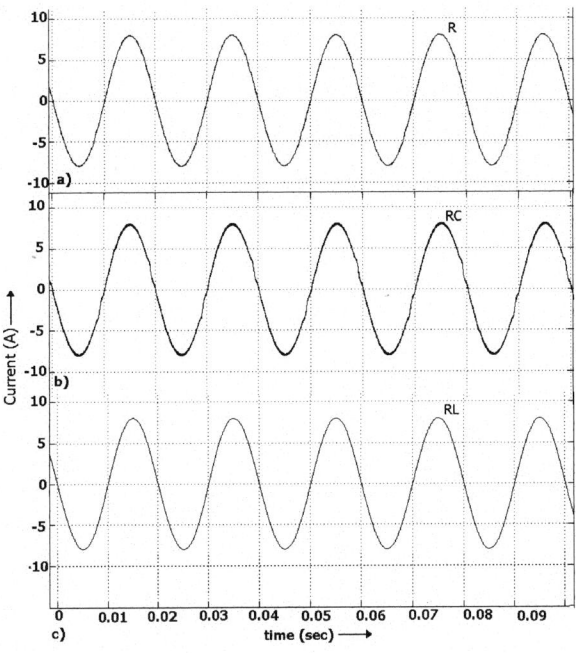

Fig. 6. Load current with a)R load b)RC load c)RL load.

## VI. SIMULATION

The overall current control schematic diagram is shown in Fig.5a.. In a similar manner the voltage control scheme based on similar ideas can be developed. The schematic for voltage control is shown in Fig 5b. The simulation was carried out in MATLAB. The load current waveform for various types of loads are shown in Fig.6.The load current coincides exactly with the reference current as shown in Fig.7. The settling of $I_d$ and $I_q$ is shown in Fig.8. The real and orthogonal current values from observer are shown in Fig.9.

Then, the voltage control loop of an inverter supplying a non-linear load was simulated. The non-linear load comprised of a diode bridge rectifier, supplying a resistive load (18$\Omega$) with a parallel capacitor (1500$\mu$F). The output voltage of the inverter appears to be distorted due to the voltage drop in the

series inductor pf the LC filter as shown in fig 10a. A minor loop can be built as shown in Fig.5b to minimise the effect of "noise" namely the harmonic drop. With this inner loop, the improvement in voltage waveform is shown in Fig.10b.

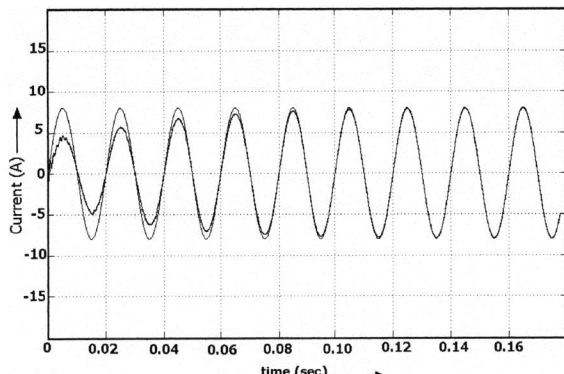

Fig. 7. Load current coincides with the reference current.

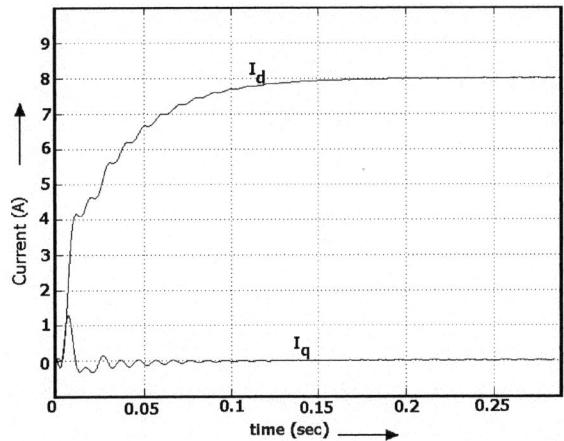

Fig. 8. Settling of current $I_d$ and $I_q$ of the inverter energising R load.

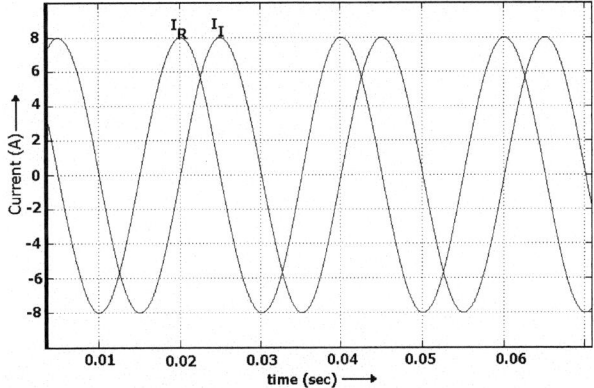

Fig. 9. Real and imaginary (orthogonal) current from observer.

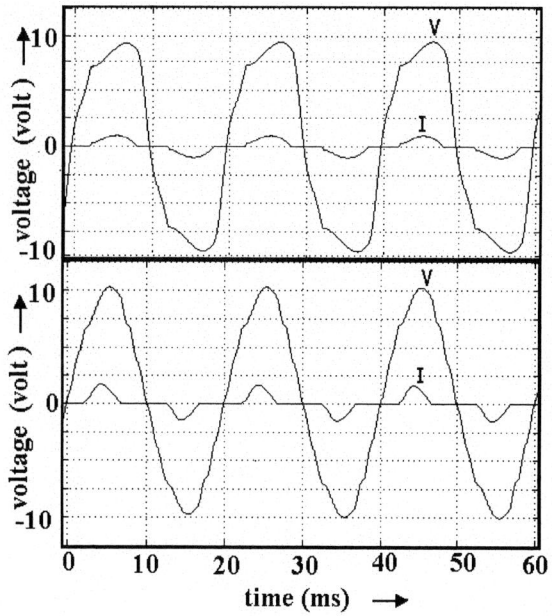

Fig. 10. Voltage across non-linear load (diode rectifier) a) without minor loop b)with minor loop in the controller.

## VII.    EXPERIMENTAL RESULTS

The experiment was carried out with the following parameters of single-phase inverter:

Sampling frequency: 20 kHz
Switching frequency: 5 kHz
Source voltage frequency: 50 Hz
DC link voltage: 12 V
DC link capacitor: 1000 μF
Filter Inductance: 0.3 mH
Filter capacitor: 96 μF
MOSFET: IRFP450

The control algorithm is realized by Verilog program in FPGA EP1C3T100C7. The sine and cosine components of the transformation matrix was realized as lookup tables in the FPGA driven at 50 Hz frequency-using a counter and clock arrangement. To begin with, a current-lopp was set up. The $I_d$ reference was selected as 2A and the control circuit was evaluated with R, RL and RC type of loads and load current waveforms are shown in Fig.11. The distortion factor in load current is given in Table 1. The settling of $I_d$ and $I_q$ is shown in Fig.12. The load current coincides exactly with the reference current as shown in Fig.13. The real and orthogonal current values from observer are shown in Fig.14.

Next, a voltage loop built for the inverter for driving a symmetrical non-linear load. The distorted voltage waveform without minor loop and the improvement in voltage waveform due to the presence of minor loop in the experimental set up are shown in fig15a and 15b. The Total Harmonic distortion (THD) in the waveform in the experimental inverter without inner loop was 13%; whereas with the inner loop the THD was found to be 4%.

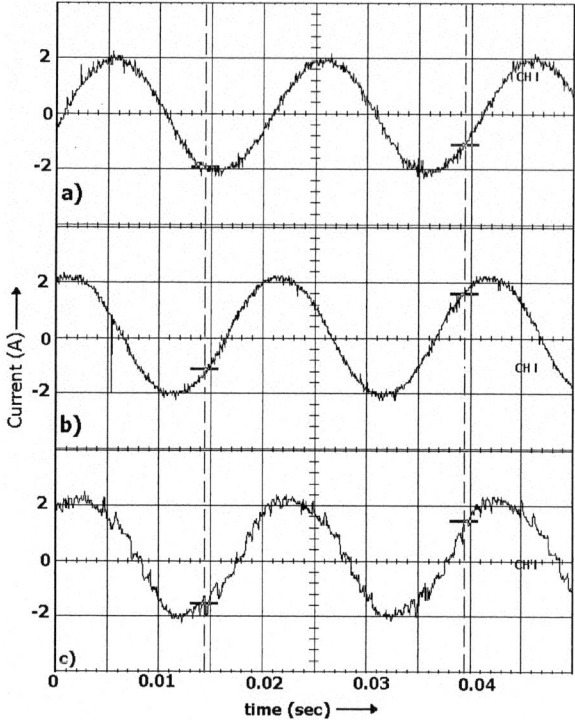

Fig. 11.  Sinusoidal load current in a)R b)RL c)RC load (experiment).

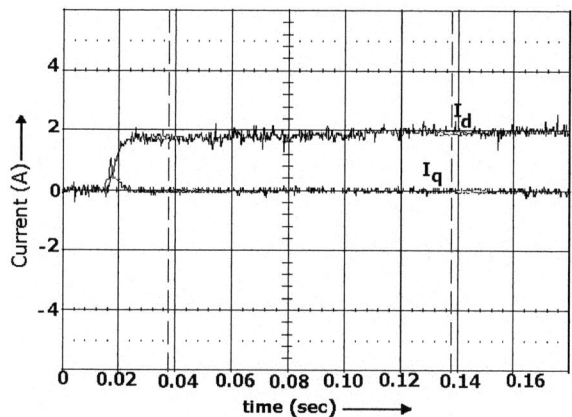

Fig. 12.  Settling of $I_d$ and $I_q$ of the inverter energising  RL load.

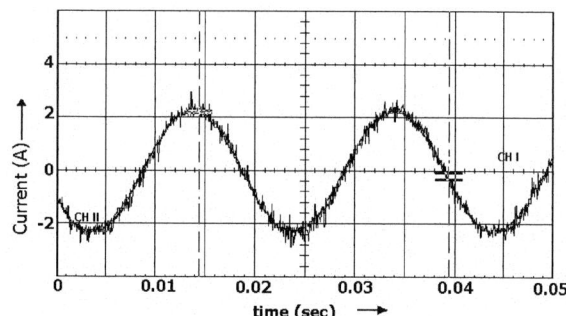

Fig. 13.  Load current coincides with the reference (experiment).

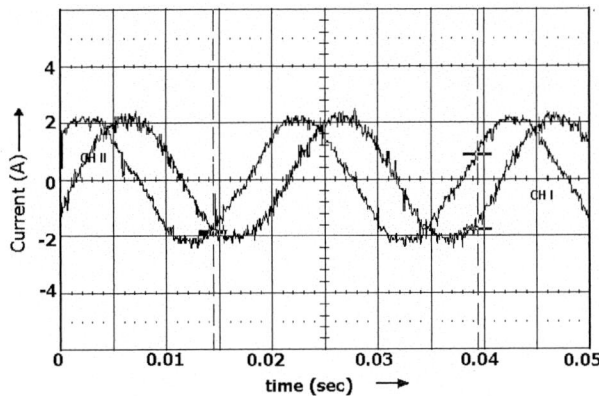

Fig. 14. Real and orthogonal currents from observer (experiment).

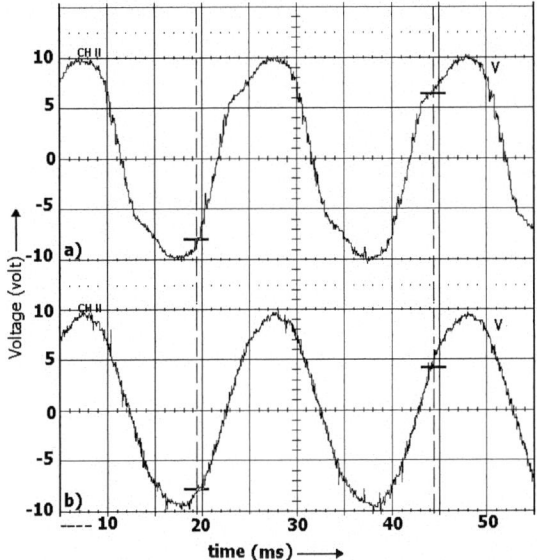

Fig. 15. Voltage across non-linear load (diode rectifier) a)without minor loop b)with minor loop in the controller (experiment).

TABLE I
DISTORTION FACTOR IN LOAD CURRENT

| Type of Load | Distortion factor |
|---|---|
| R load | 0.978 |
| RL load | 0.98 |
| RC load | 0.96 |

## VIII. CONCLUSION

Synchronous frame current regulators allow conventional DC compensation strategy to be used to achieve zero steady state error. The curve-fitting observer is used to generate real and orthogonal current values. The observer acts as a filter for measurement noises without introducing any pronounced delays. The current control algorithm with observer has been implemented in FPGA and the concept has been validated by both simulation and experiments.
While the synchronous frame voltage regulator in the outer loop achieves zero steady state error with fast transient response, an additional conventional minor voltage loop

incorporated in the control structure improves the output voltage waveform of the inverter. In stand-alone inverters, supplying rectifier loads, the above modifications are particularly useful. The asymptotic observer has two main functions namely: a) it tracks the real voltage signals b) generates an orthogonal signal, replacing a FIFO stack for generating an orthogonal signal. A tuned observer with poles placed suitably would also act as a filter for measurement noise without introducing any significant delay. The control scheme with inner loop and observer had been implemented in FPGA. The simulation and experimental results have shown significant waveform improvement under non-linear loads when the DQ frame method is combined with the conventional feed back procedure.

## IX. REFERENCES

[1] M.P.Kazmierkowski and L.Malesani, "Current control Techniques for 3-phase voltage source PWM converters: A survey", *IEEE trans. on Industrial Electronics*, vol.45, No.5, oct.1998, pp. 691-703.

[2] G.Venkataramanan, D.M.Divan and T.M.Jahns, "Discrete pulse modulation strategies for high-frequency inverter systems", in *Rec. IEEE-PESC Conf.*, Milwaukee, WI, 1989, pp.1013-1020.

[3] S.D.Finn, "A high performance inverter technology, architecture and applications", in *Rec. IEEE-APEC Conf.*, San Diego, CA, 1993, pp. 556-560.

[4] N.Abdel-Rahim and J.E. Quaicoe, "A single-phase voltage-source utility interface system for weak AC network applications", in *Rec. IEEE-APEC Conf.*, Orlando, FL, 1994, pp. 93-99.

[5] N.R.Zargari, P.D.Ziogas and G.Joos, "A two switch high performance current regulated DC/AC converter module", in *Proc. IEEE-IAS Annu.Meeting*, Seattle, WA, 1990, pp. 929-934.

[6] D.N.Zmood and D.G.Holmes, "Stationary frame current regulation of PWM inverters with Zero steady state error", *IEEE 30th Annual Power Electronics specialists conference* 1999, vol.2, 27 June-July 1, pp.1185-1190.

[7] Richard Zhang, Mark Cardinal, Paul Szczesmy and Mark dame, "A grid simulator with control of single-phase power converters in D-Q rotating frame", *IEEE 33rd Annual Power Electronics specialists conference* 2002, vol.3, 23-27 June, pp.1431-1436.

**2006 IEEE International Conference on Power Electronic, Drives and Energy Systems**

# MATLAB Simulation of current control of PMSM using single sensor technology

B. Saritha, and P. A. Jankiraman, *Senior Member, IEEE*

*Abstract*--Three phase winding current of Permanent Magnet Synchronous Motor (PMSM) is reconstructed from DC line current using a single sensor placed at the low potential D.C. return line of the inverter. Without using the load or motor parameters, the reconstruction is carried out by a curve-fitting observer, which is robust under steady and dynamic conditions. The reconstructed current is used for feedback in the current control loop. The control scheme was found to operate satisfactorily. Simulation results are presented in detail.

*Index Terms*-- Current Control, Curve fitting Observer, PMSM, Simulation, Single Sensor.

## I. INTRODUCTION

$\mathbf{V}$OLTAGE source inverters are widely used in many Industrial applications. Hall sensors are commonly used in two of the 3-Phase inverter output lines for current feedback. They are expensive and bulky to integrate inside the modules. To reduce the number of sensors, a single current sensor can be used in DC link. Moreover, single sensor technique eliminates imbalance errors due to multiple sensors. Reconstruction of phase currents from DC link current using active voltage vector timing had been reported [1-3]. This method encounters limitations when the duration of the active state becomes very short. When the active state vector timing is less than minimum sampling time, PWM scheme has to be modified periodically to update the phase current value [4]. Instead of using complicated strategies, predictive state observers have also been employed which use the R, L values of load [4]. In AC drives, three observers were used to reconstruct the phase current, with R, L values of machine updated periodically to take into account the non-linearity [5]. The R,L values of load were found necessary to reconstruct the phase current for feedback, using integrated current sensors on the lower side switches of the Inverter [6]. We propose a new method for reconstruction, which does not require any information about the load. A high-speed curve-fitting observer is used to estimate the phase currents from partial information obtained from DC link current. The

---

B.Saritha is doing her Ph.D. in Department of Electrical Engineering, Indian Institute of Technology, Madras .(e-mail: saritha_oct@yahoo.com).

Dr.P.A.Jankairaman is Professor in the Department of Electrical Engineering, Indian Institute of Technology, Madras 600036 India (e-mail: pajraman@ieee.org).

sampling instants for DC link current are determined using a simple logic circuit. The reconstructed phase current is fed back to the current control circuit to maintain a 3-phase sinusoidal load current.

## II. CURRENT CONTROL ALGORITHM

The complete single sensor control structure is divided into several functions as shown in Fig. 1. The PMSM is fed from a PWM voltage source inverter (Fig. 2). A single current sensor in the dc line and a position encoder mounted on the rotor shaft provide the feedback signals to the controller. All control functions are discussed in detail in the following sections.

### A. Position determination

The PMSM has a 2-$\phi$ position encoder with a synchronizing pulse Z to determine $\theta$. The phase A or B pulses drive an Up/Down counter to generate the address for the $\sin\theta$ and $\cos\theta$ lookup tables (LUT's).

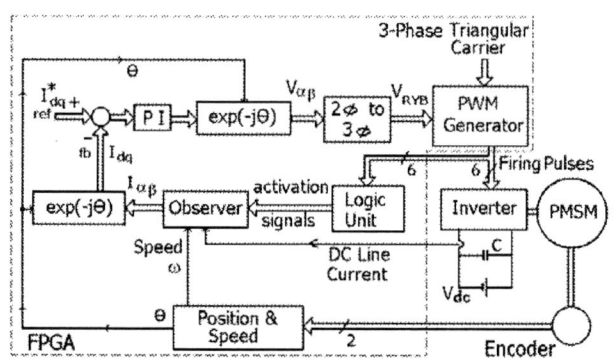

Fig. 1. Overall control structure of PMSM.

### B. Vector control

The q-axis reference current ( $I_q$ ) is set as a constant 8-bit number in the software. In the absence of field weakening, direct axis current demand ( $I_d$ ) is set to zero. The observer continuously provides the stator current in a stationary reference frame attached to the stator ( $I_{\alpha\beta}$ ). The stator to rotor frame vector rotation is used to compute the feedback current vector, $[I_d, I_q]t$. The values of $\sin\theta$ and $\cos\theta$, required for transformation are read off from the LUT's. The feedback signals are compared with the reference and the two error signals are passed through PI controllers. The PI controller output are operated by Rotor to Stator vector frame rotation

---

0-7803-9771-1/06/$25.00 ©2006 IEEE          490

matrix and a further $2\phi$ to $3\phi$ transformation to generate the voltage vector $V_{RYB}(t)$.

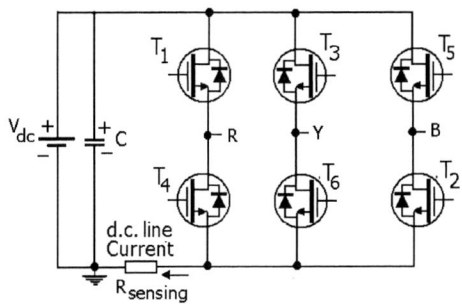

Fig. 2. Voltage source inverter with dc line current sensor.

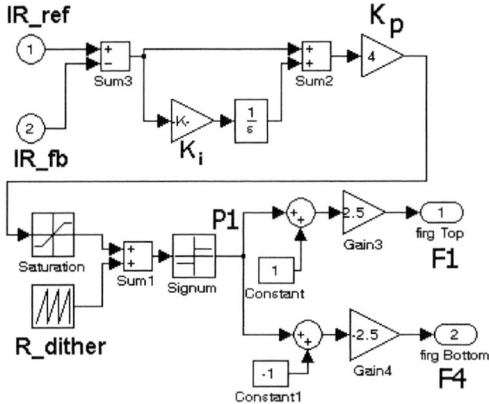

Fig. 3. Firing signal generation for R-phase of inverter.

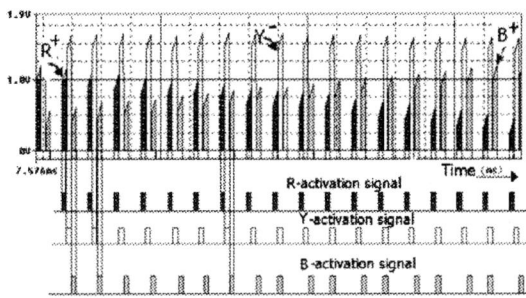

Fig. 4. Interleaved current samples and activation signals.

## C. Firing signal generation

Pulse width modulation is carried out using a three-phase dither arrangement, in which each of the three phase error signals $V_{RYB}$ is modulated by individual 8 kHz triangular dither signals, shifted in phase by 120° with each other.. The voltage vector $V_{RYB}$ after addition of three-phase dither signals is passed through three signum functions to generate control signals. Due to the $3\phi$ nature of dither signal, the control signals P1, P2 and P3 also exhibit 120° phase displacement. The control signal P1 is the firing signal F1 of MOSFET-T1 in the R-leg of the inverter and inverse of P1 is applied as firing signal F4 to MOSFET T4 (Fig.3). Similarly, P2 controls the firing of Y-leg and P3 controls the firing of MOSFETs in B-leg of the inverter.

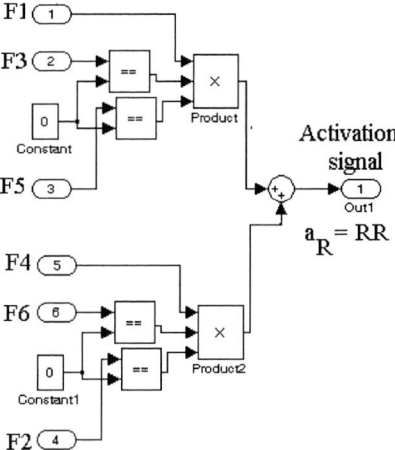

Fig. 5. Activation signal generation for observer ($a_R$ = RR = 1 or 0).

## D. Generation of Activation signals

A single current sensor had been placed on the grounded side of the DC link of the inverter. Since the DC link current contains the multiplexed phase current information, the link must be sampled at the correct instants to extract the phase current information. Let F1 indicate the firing signal applied to MOSFET T1. Similarly F2, F3, F4, F5 and F6 indicate the firing signals applied to corresponding MOSFETs. Under steady state, when the switch T1 is ON, T3 and T5 are OFF, the DC link current contains positive R-phase current information ($R^+$). When the switch T4 is ON, T6 and T2 are OFF, the DC link current contains negative R-phase current information ($R^-$). Consequently, to get R-phase current slices, whenever they exist, from the DC link current, the sampling signal is given by the following logic expression. R-Phase activation signal:

$$a_R = RR = F1.\overline{F3}.\overline{F5} + F4.\overline{F6}.\overline{F2}$$

TABLE I
SAMPLING SIGNALS FROM FIRING SIGNALS

| Firing signal condition | Information in DC Link Current | Activation (Enabling) Output Signal $a_i$ |
|---|---|---|
| $F1.\overline{F3}.\overline{F5}$ | $R^+$ | RR = $a_R$ =1 |
| $F4.\overline{F6}.\overline{F2}$ | $R^-$ | |
| $F3.\overline{F5}.\overline{F1}$ | $Y^+$ | YY = $a_Y$ =1 |
| $F6.\overline{F2}.\overline{F4}$ | $Y^-$ | |
| $F5.\overline{F3}.\overline{F1}$ | $B^+$ | BB = $a_B$ =1 |
| $F2.\overline{F6}.\overline{F4}$ | $B^-$ | |

Similarly we can write the logic expression for Y and B-phase activation signals. The sampling signals are generated from firing signals as shown in Table I. Only one phase current information is available at any time from DC link

491

current sensor. Also, the current information is not continuous. The interleaved nature of current samples and activation signals, which is mainly due to the 3-phase dither, are shown in Fig.4. The logic circuit for generating R-Phase activation signal is shown in Fig.5. From the non-uniformly sampled data, 3-$\phi$ signals are to be reconstructed. A curve-fitting observer to reconstruct the 3-$\phi$ current for feedback was simulated using MATLAB.

### III. CURVE-FITTING OBSERVER

The differential equation of a sinusoidal oscillator is given by

$$x'' = -\omega^2 x \qquad\qquad\text{-(1)}$$

Let the state variables be: $\begin{bmatrix} x_1 \\ x_2 \end{bmatrix}$ The desired steady-state

solution is given by $\quad x_1(t) = \sin \omega t; \; x_2(t) = \cos \omega t$

Where $\omega$ = speed of PMSM in rad/sec.

The oscillator is described in the state variable form as

$$\begin{bmatrix} x'_1 \\ x'_2 \end{bmatrix} = \begin{bmatrix} 0 & \omega \\ -\omega & 0 \end{bmatrix}\begin{bmatrix} x_1 \\ x_2 \end{bmatrix} \qquad\qquad\text{-(2)}$$

An observer for 3$\phi$ currents has been built around this oscillator, using the inverse Clarke transformation. Let:

$$\begin{bmatrix} \hat{I}_R \\ \hat{I}_Y \\ \hat{I}_B \end{bmatrix} = \begin{bmatrix} 1 & 0 \\ -\dfrac{1}{2} & \dfrac{\sqrt{3}}{2} \\ -\dfrac{1}{2} & -\dfrac{\sqrt{3}}{2} \end{bmatrix}\begin{bmatrix} \hat{x}_1 \\ \hat{x}_2 \end{bmatrix} \qquad\qquad\text{-(3)}$$

The oscillator can be modified to take into account the observer errors in R, Y and B currents. Only one error is considered at a time and this is achieved by using individual activation signals. The observer is described by (4):

$$\begin{bmatrix} \hat{x}'_1 \\ \hat{x}'_2 \end{bmatrix} = \begin{bmatrix} 0 & \omega \\ -\omega & 0 \end{bmatrix}\begin{bmatrix} \hat{x}_1 \\ \hat{x}_2 \end{bmatrix} + \begin{bmatrix} d_1 \\ d_2 \end{bmatrix}e^*(t) \qquad\text{-(4)}$$

where $\begin{bmatrix} \hat{x}_1 \\ \hat{x}_2 \end{bmatrix}$ are the estimates of $x_1$ and $x_2$.

$e^*(t) = e_i(t) \cdot a_{i,}$
where $e_i(t) = (e_R, e_Y, e_B)$ and
$a_i$ is the corresponding activation signal ($a_R, a_Y, a_B$)
$e_R(t) = I_{R\_m}(t) - I_{R\_estimate}(t)$
$e_Y(t) = I_{Y\_m}(t) - I_{Y\_estimate}(t)$
$e_B(t) = I_{B\_m}(t) - I_{B\_estimate}(t)$

The interleaved three-phase measured signal-slices from dc line current ($I_{R\_m}$, $I_{S\_m}$, $I_{T\_m}$) and the activation signals for each phase are shown in Fig.4. The determination of the observer coefficients when the error in R-phase remains activated (i.e., for the duration $a_R = 1$), is given below:

$$\begin{bmatrix} \hat{x}'_1 \\ \hat{x}'_2 \end{bmatrix} = \begin{bmatrix} 0 & \omega \\ -\omega & 0 \end{bmatrix}\begin{bmatrix} \hat{x}_1 \\ \hat{x}_2 \end{bmatrix} + \begin{bmatrix} d_1 \\ d_2 \end{bmatrix}_R [1 \quad 0]\begin{bmatrix} e_1 \\ e_2 \end{bmatrix} \quad\text{-(5)}$$

where, $E = \begin{bmatrix} e_1 \\ e_2 \end{bmatrix} = \begin{bmatrix} x_1(t) - \hat{x}_1(t) \\ x_2(t) - \hat{x}_2(t) \end{bmatrix}$ and $[1 \quad 0]$ is the first

row of the matrix (3) and the observation error vector obeys the following differential equation.

$$E' = \begin{bmatrix} 0 & \omega \\ -\omega & 0 \end{bmatrix}E - \begin{bmatrix} d_1 \\ d_2 \end{bmatrix}_R [1 \quad 0]E \qquad\text{-(6)}$$

$$E' = \begin{bmatrix} -d_1 & \omega \\ -\omega-d_2 & 0 \end{bmatrix}E$$

$$E' = A E \qquad\qquad\text{-(7)}$$

The characteristic equation is given by
$\det[\,sI - A\,] = 0$
i.e. $s^2 + d_1 s + d_2 \omega + \omega^2 = 0 \qquad\text{-(8)}$

The observer poles are selected to be
$s_1 = s_2 = -3\omega$ where $\omega$ = speed of PMSM, resulting in observer feed back constants:

$$\begin{bmatrix} d_1 \\ d_2 \end{bmatrix}_R = \begin{bmatrix} 6 \\ 8 \end{bmatrix} \qquad\qquad\text{-(9)}$$

Considering the error in Y-phase, the observer equation can be derived as:

$$E' = \begin{bmatrix} 0 & \omega \\ -\omega & 0 \end{bmatrix}E + \begin{bmatrix} d_1 \\ d_2 \end{bmatrix}_Y [-0.5 \quad 0.866\,]E \quad\text{-(10)}$$

Assuming that the observer poles are located at $s_1 = s_2 = -3\omega$, the observer feed back coefficients can be obtained as:

$$\begin{bmatrix} d1 \\ d2 \end{bmatrix}_Y = \begin{bmatrix} 3.8 \\ -9.2 \end{bmatrix} \qquad\qquad\text{-(11)}$$

Similarly, when the B-phase activation signal is present, the observer coefficients are:

$$\begin{bmatrix} d1 \\ d2 \end{bmatrix}_B = \begin{bmatrix} -9.9 \\ 1.2 \end{bmatrix} \qquad\qquad\text{-(12)}$$

Since the signals $a_R$, $a_Y$, $a_B$ are mutually exclusive as shown in Fig.4, the resulting three different observer modes can be combined without interaction to yield a composite $e_{x1}$ and $e_{x2}$ as shown in Fig.6. The tuned oscillator, which is a part of the observer, can also be seen in Fig.6. It is obvious that the signals $x_1(t)$ and $x_2(t)$ from the observer are the estimates of $(\alpha,\beta)$ components of the 3$\phi$ currents and they are recombined to form 3$\phi$ currents for feedback. While determining the Observer feed back coefficients, a fixed speed $\omega$ was assumed. However, the speed of the motor is a variable. This means that at any arbitrary $\omega$, to maintain the poles of the observer at $3\omega$, without changing the numerical values of $d_{ij}$ the observer block must be frequency-scaled. Alternatively, if the discrete version of the observer is employed, the sampling frequency should be made proportional to the speed $\omega$. It can be easily achieved for the PMSM by making use of the encoder pulses as the sampling pulses.

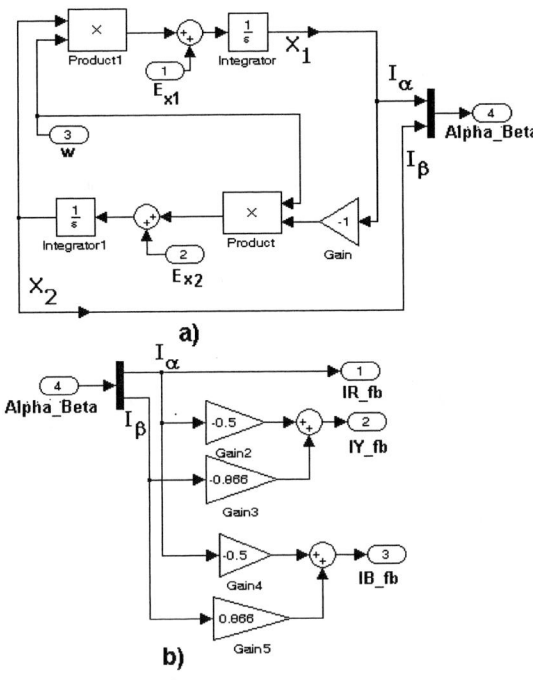

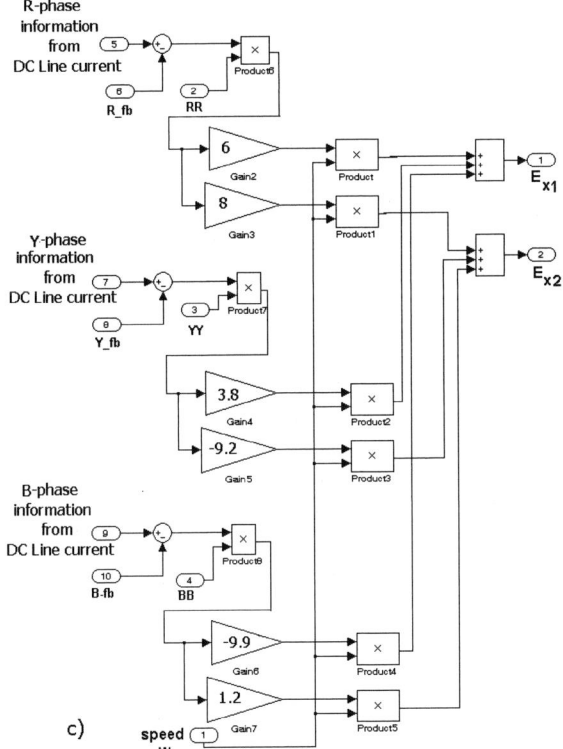

Fig. 6. a)Oscillator b)Inverse Clarke transformation c)Error generation.

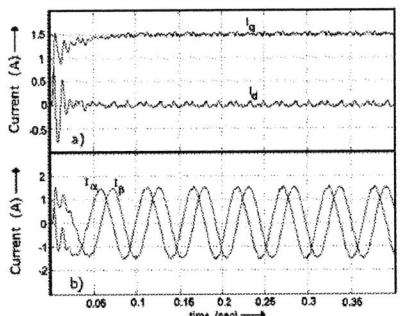

Fig. 7. a)$I_d$, $I_q$ componenets b)Alpha, Beta componenets from observer under steady load for Iqref = 1.5 A with electrical speed = 119 rpm.

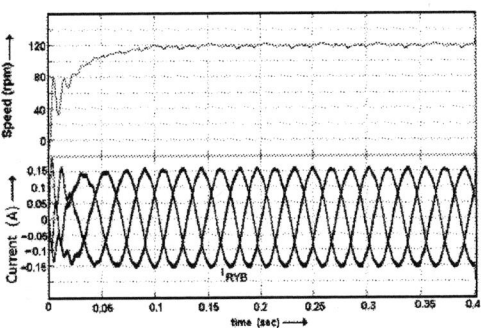

Fig. 8. a)Electrical speed b)Three-phase stator current of PMSM.

## IV.    RESULTS

The simulation was carried out with the following data:
Voltage Source Inverter data:
Battery Voltage     : 30V
Capacitor            : 1000 µf
Switches            : MOSFET IRFP450
PMSM data:
Resistance (Phase): 1.18 Ω
Inductance        : 4.4 mH
Moment of Inertia : 87 g-cm$^2$
Mechanical Time constant: 4.1 ms
Back-Emf constant : 5.2 V/kRPM
No. of poles        : 4

The response of observer for the reference current of 1.5A under steady load torque of 0.1Nm is shown in Fig.7. The 3-φ stator current of PMSM and its electrical speed is shown in Fig. 8.The observer was also tested for its robustness by applying sudden change in reference current at 0.4 sec and the response of the observer is shown in Fig.9. The load torque was suddenly changed from 0.1 Nm to 0.05 Nm at 0.4 sec and the observer was found to be stable under this condition also. As the system operates under current control mode, estimated Iq from observer returns back to the reference value quickly after a small disturbance as shown in Fig.10. The polarity reversal of Iq at 0.45 sec and the consequent speed reversal of PMSM is shown in Fig.11a. The robustness of the observer can be seen in Fig.11b and c in which the observer quickly reached the new equilibrium.

493

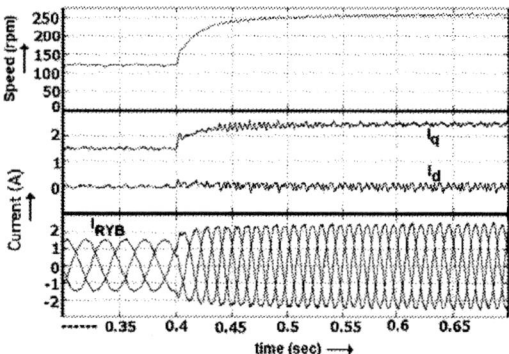

Fig. 9. Dynamic change in reference current a)Electrical speed of PMSM b)$I_q$, $I_d$ feedback from observer c) Three-phase reconstructed current from observer.

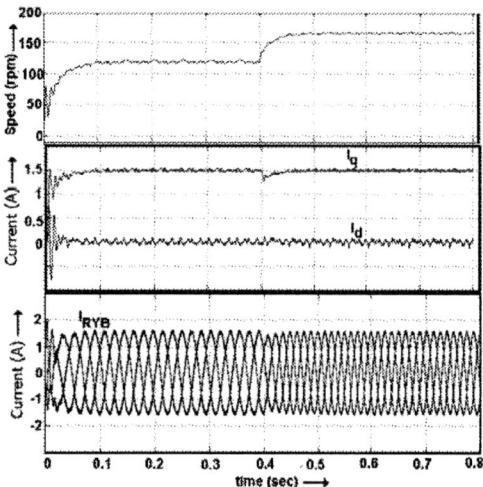

Fig. 10. Dynamic change in Load torque a)Electrical speed of PMSM b)$I_q$, $I_d$ feedback from observer c) Three-phase reconstructed current from observer.

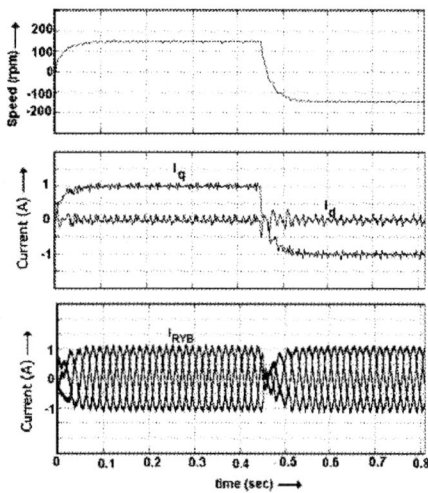

Fig. 11. Dynamic change in polarity of reference current a) a)Electrical speed b)$I_q$, $I_d$ feedback c) Three-phase stator current of PMSM.

The experimental graph depicting three-phase stator current of PMSM under unloaded condition is shown in Fig.12. At the start, the exact rotor position of PMSM becomes known, only

after a synchronising pulse Z occurs. Hence the PMSM starts with arbitary torque by this procedure. The PMSM reverses when the change in polarity of reference current $I_q$ occurs. The control loop was also found to be stable for sudden changes of load-torque on the PMSM.

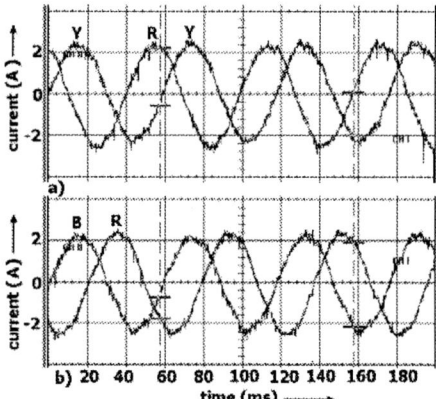

Fig. 12. Winding current under unloaded condition with current reference as 2.4 A a)RY b)BR current (experiment).

## V. CONCLUSION

An on-line method to estimate the stator current of PMSM by observing the DC line current has been presented in this paper. The desired sinusoidal 3-Phase current can be passed through motor, using the reconstructed current as the feed back signal. Even though only the current loop has been described, an outer speed control loop may also be installed around the current loop.

## VI. REFERENCES

[1] F.Blaabjerg, J.K.Pederson, U.Jaeger, P.Theogerson, "Single current sensor techniques in the DC link of three phase PWM-VS inverters-A review and the ultimate solution", *Proc. of IEEE Industry Applications Society Annual meeting*, Oct.1996, pp-512-522.

[2] F.Parasiliti, R.Petrella, M.Tursini, "Low cost phase current sensing in DSP based AC drives", *Proc. of IEEE International symposium on Industrial electronics*, Vol.3, 1999, pp.1284-1289.

[3] T.G.Habetler, D.M.Divan, "Control strategies for direct torque control using discrete pulse modulation", *Proc. of IEEE Industry Application Society Annual Meeting*, 1989, pp-512-522.

[4] Woo-Cheol Lee, Dong-Seok Hyun, Taeck-Kie Lee, "A Novel control Method for Three-Phase PWM Rectifiers usinbg a single current sensor", in *IEEE trans. on Power electronics*, Vol.15, No.5, September 2000.

[5] T.M.Wolbank, P.Macheiner, "Scheme to reconstruct phase current information of inverter fed AC drives", *IEE Electronics letters*, vol.38, 5,Feb 2002,pp:204-205

[6] Sibaprasad Chakrabati, Thomas M.Jahns, Robert D.Lorenz, "A current reconstruction algorithm for Three-phase Inverters using Integrated current sensors in the low side switches", *Proc. of IEEE Industry Application Society Annual Meeting*, 2003, vol.2, Oct 2003, pp:925-932.

**2006 IEEE International Conference on Power Electronic, Drives and Energy Systems**

# Novel Approach to Develop Behavioral Model Of 12-Pulse Converter

Amit Sanglikar, and Vinod John, *Member, IEEE*

*Abstract*–A novel approach to develop behavioral model of 12-pulse converter, which reduces the overall system simulation time, is presented. A detailed model of a 12-pulse converter with its control using circuit simulation tools such as PSPICE, MATLAB or SABER takes significant computer time for simulation. This is even more critical when the power converter is only a small part of a large system under study. A model based on mathematical equations derived under ideal conditions could be used, but the need to consider secondary effects such as continuous and discontinuous current or current commutation make this method much complicated due to the multiple modes of power converter operation. A regression based behavioral model was developed using a "Design of Experiment" (DOE) approach with very good statistical fit. This reduces simulation time drastically capturing also those aspects of the 12-pulse converter that are not normally included in models using other simplifications. This method could also be extended to the modeling of other complex power converter topologies and sub-systems allowing a much faster simulation of full installations.

*Index Terms*-- Simulation, power converter model, positive sequence solution, table lookup, design of experiments.

## I. INTRODUCTION

WITH increasing electrical power demands, and loading limitations of existing power distribution networks the trend is towards distributed energy generation. Increasing use of computerized information systems demand higher power quality from the grid. The need exists for detailed system level analysis using computer simulation of power generating units such as micro-turbine, fuel cell, and power quality improvement systems such as high power UPS, Dynamic voltage restorer (DVR) and distribution Statcoms interconnected with the grid or in islanded operation. This leads us to the development of models of various power system components that may be interconnected in complex manner. The power converter can be a relatively small component of the overall system. All node voltages and branch currents of the power converter may not be necessary for the study of the overall systems. One such power electronic

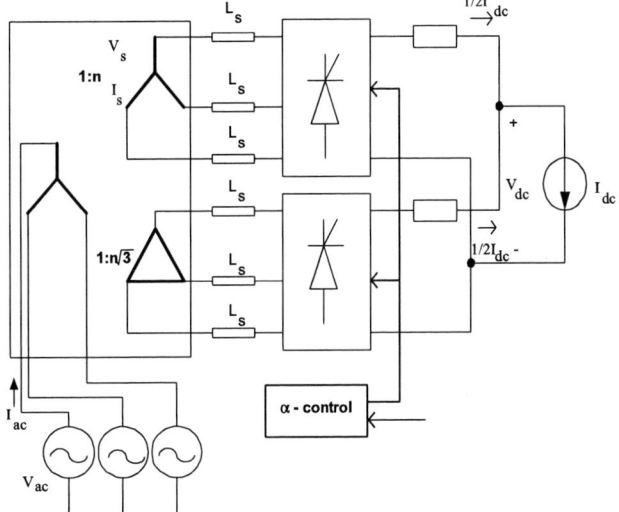

Fig. 1. Twelve-pulse converter used as rectifier front end for the UPS with the inverter side modeled as a current source load.

component is a large UPS, within which the 12-pulse converter is significant part of power electronics. For power-system level study a simplified representation of the 12-pulse rectifier is required [1]. A prime requirement is to make use of component models of the subsystems to perform analysis such as sizing, energy storage requirements overall power quality compatibility with loads.

System level analysis of distributed power generating units using circuit simulators involves system level modeling of various sub-systems, such as 12-pulse converter in a Megawatts UPS. An analytical model defined using mathematical equations in high level simulation languages takes less simulation time. Mathematical model developed with simplified assumption limits the scope of study because secondary effects such as continuous and discontinuous current, current overlap in the thyristors during commutation etc., can not be easily analyzed since mathematical model are derived under ideal conditions [2]-[4]. A detailed component level model of 12-pulse converter can be used to take into account these secondary effects [5]. However such component level model even by itself would take significant time for simulation. Hence, there is a need for developing a model that takes into account secondary effects and can be expressed in mathematical form in order to reduce simulation time. In this paper a regression based behavioral model of a 12-pulse

---

Amit Sanglikar is with Tektronix Engg. Development India Pvt. Ltd, Bangalore 560001, Karnataka, India, (e-mail: amit.sanglikar@tek.com).

Vinod John is with the Department of Electrical Engineering, Indian Institute of Science, Bangalore 560012, Karnataka, India. (e-mail: vjohn@ee.iisc.ernet.in).

0-7803-9771-1/06/$25.00 ©2006 IEEE

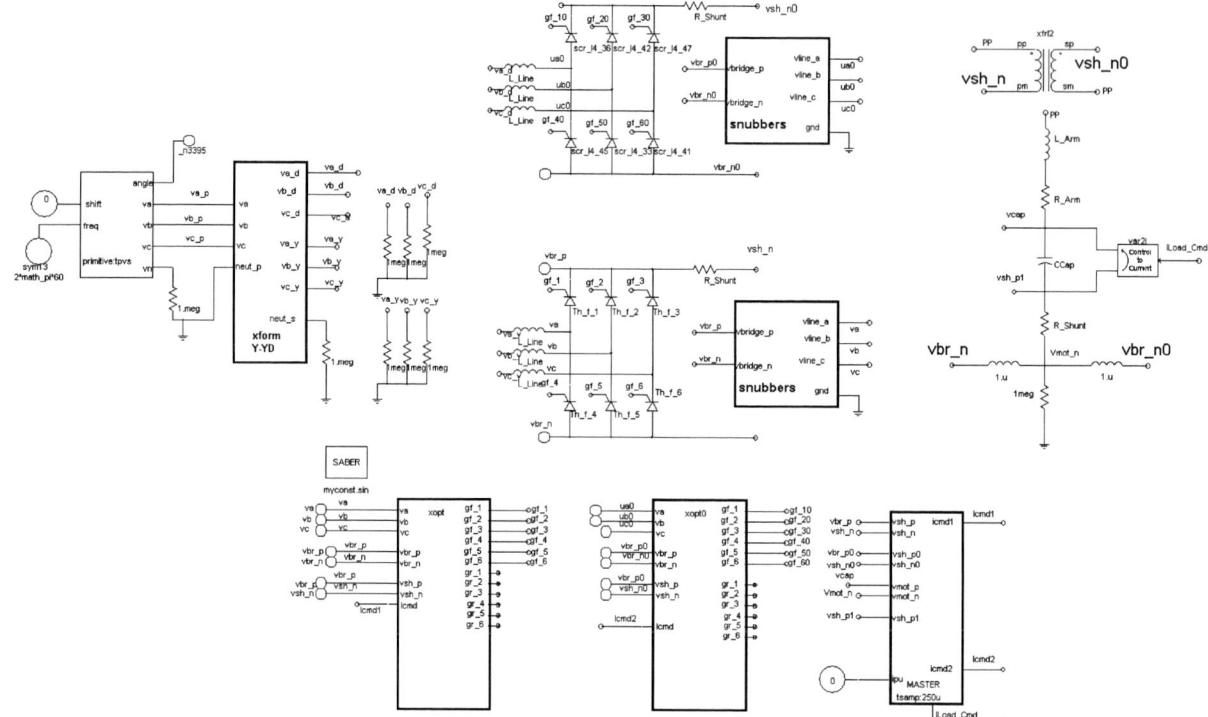

Fig. 2. Network component level circuit simulation model of a 12-pulse converter.

converter is described. The performance of the model is compared with simplified analytical models and detailed network models in terms of simulation speed and accuracy.

## II. MODELING APPROACH

Modeling of the power converter is based on a three step approach consisting of the following. First is a preliminary mathematical idealized model that is used for simplified understanding. The second step is to create a detailed component level model that captures all the required non-idealities. The third step is to obtain the regression model.

### A. Mathematical approach

A simplified analysis of the 12-pulse converter shown in Fig. 1 leads to the following equations for the input and output quantities.

$$V_{dc} = \frac{3\sqrt{2}}{\pi} V_s \cos\alpha - \frac{3\omega L_s}{\pi} \frac{1}{2} I_{dc} \quad (1)$$

$$\cos(\alpha + u) = \cos\alpha - \frac{\omega L_s}{\sqrt{2} V_{LL}} I_{dc} \quad (2)$$

$$DPF \approx \cos(\alpha + u) \quad (3)$$

$$V_s I_{s1} DPF = V_{dc} I_{dc} / 2 \quad (4)$$

$$I_{s1} = \frac{V_{dc} I_{dc}}{2V_s DPF} \quad (5)$$

Where, $\alpha$: firing angle
u: commutation interval
$I_{s1}$: fundamental-frequency current of $I_s$

n: effective transformer ratio (primary Wye and secondary Wye is 1:n)

These mathematical relations are based on following assumptions:

- Continuous conduction of DC bus current
- Sinusoidal, balanced and constant AC voltage
- Balanced AC side impedance
- Ideal circuit component.

A model based on mathematical equations derived under ideal conditions could be used for the system level study. However the need to consider secondary effects such as continuous and discontinuous current, and different modes of current overlap in the thyristors during commutation make this method of analysis complex [6]. Deriving the equations for quantities such as Total harmonic distortion (THD) requires complex expressions along with simplifying assumptions on the interactions between the input and output filters.

### B. Component level modeling

A detailed component level model of 12-pulse converter that takes into account secondary effects such as continuous and discontinuous current, current overlap in the thyristors during commutation etc. can be used. A detailed network component model of a 12-pulse converter implemented in SABER is shown in Fig. 2.

For the system level study that we are found that this model is not as useful for following reasons. One is that, this model alone already takes a long time to complete a simulation run in a computer simulation. It is not desirable for such a small component of a larger system to incur such a long model computation time. Also, very few variables of this model such as DC bus voltage ($V_{dc}$), input current, power factor, real power, reactive power, THD etc., are the main variables required from the overall system simulation and analysis.

Simulation using the component based model allows one to capture the operation of the power converter. The primary regression functions of the operation of the 12-pulse converter was derived in terms of the firing angle and the equivalent DC bus current.

### C. Regression based model

Regression methods can be used to develop simple empirical mathematical relationships between inputs and output variables. Regression provides an approximate means of obtaining the relationships between a response parameter and the factors that influence it when the true relationship is unknown, or is too difficult to derive. Regression involves mainly analyzing factors that affects response parameters and building equations that can be used to predict system behavior at interpolated operating points and conditions. All the input and output variables for the regression are evaluated at steady state operating conditions. Energy storage elements such as capacitors and inductors are not included in the regression equation. Hence after accounting for energy loss, the power flow between the input and the output in the circuit is balanced on an instantaneous basis in the regression model.

The first step to obtain the regression model is to identify the input (independent) variables and the output (dependent) variables. The complexity of the regression equation can rapidly increase as the number of input variables is increased. Hence, a minimum number of input variables should be used. The next step is to determine the operating range of each of the independent variables. Care should be taken to ensure that the results of the study make use of input variables within this range. Choosing a narrower operating range can help improve the accuracy of the model. The third step is to evaluate the granularity of the operating conditions under which the regression is evaluated. In case a large nonlinearity is expected in certain variables, then the number of evaluation of the system should be increased for those variables. Once these decisions are made, the simulations to obtain the regression equations is carried out using the detailed network model of 12-pulse converter a different points in the space and a non-linear multi-variable function that can be represented as a response surface was obtained. From these simulation results, the relations between system level parameters of the 12-pulse converter and the variables on which they depend are obtained as a polynomial function of the independent variables.

### III. REGRESSION BASED BEHAVIORAL MODEL OF 12-PULSE CONVERTER

Response parameters are system level variables for which empirical mathematical relation need to be obtained. In this case, DC bus voltage ($V_{dc}$), input AC current ($I_{ac}$), input power factor (pf), real and reactive power at AC input terminals, and input current THD are identified as response parameters

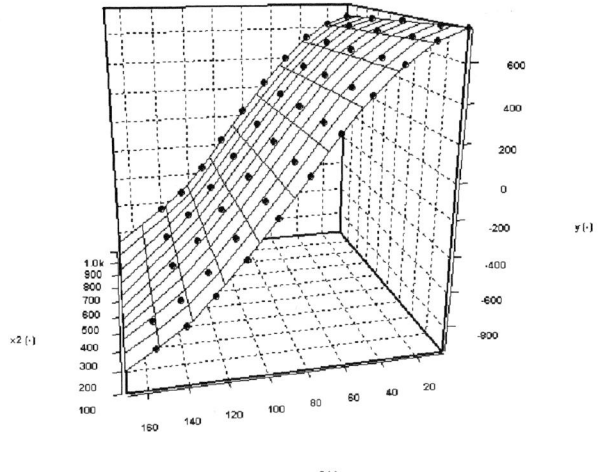

Fig. 3. Surface showing variation of DC bus voltage at different firing angle and DC bus current of 12-pulse converter for fixed input AC voltage2. Y-axis is DC bus voltage, X2 axis is DC bus current and X3 axis is firing angle.

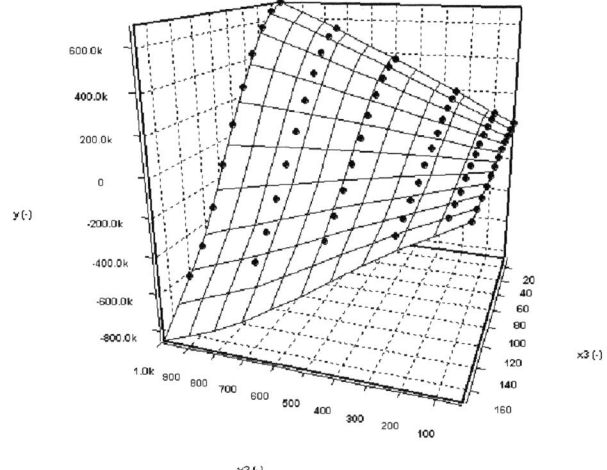

Fig. 4. Surface showing variation of real power at different firing angle and DC bus current of 12-pulse converter for fixed input AC voltage3. Y-axis is real power, X2 axis is DC bus current and X3 axis is firing angle.

for the system level study. Factors that influences response parameters are understood by analyzing simplified mathematical model, which are input AC voltage ($V_{ac}$), firing angle ($\alpha$), average DC bus current ($I_{dc}$). Since regression is essentially obtaining a curve fitting equation for set of data points, various data points representing relationship between response parameters and factors that influences response parameters are required for developing regression model of 12-pulse converter. Such data points were obtained by performing design of experiment (DOE) simulations on network component level model of 12-pulse converter at different points in the design space[1]. A full factorial DOE with

---

[1] Design space for influencing factors are as follows:
488.75 V $< V_{ac} <$ 661.25 V,
$5^0 < \alpha < 175^0$,
25 A $< I_{dc} <$ 1000 A.

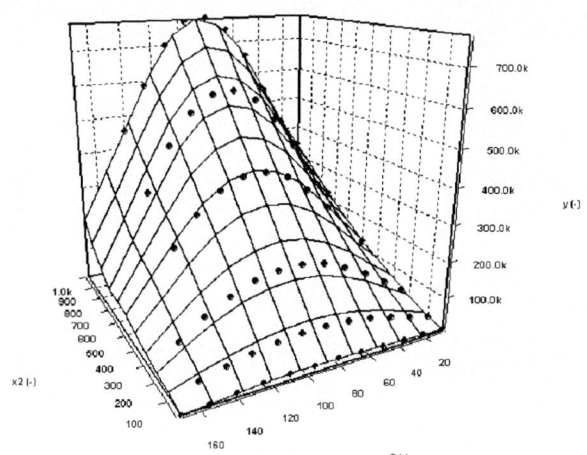

Fig. 5. Surface showing variation of reactive power at different firing angle and DC bus current of 12-pulse converter for fixed input AC voltage4. Y-axis is reactive power, X2 axis is DC bus current and X3 axis is firing angle.

6 levels of DC bus current, 12 levels of firing angle and 3 levels of Vac was performed to obtain data for the regression model. These data points obtained from simulation can be represented as non-linear multi-variable response surface. Response surface for some response parameters are shown in Fig. 3 through Fig. 5. Fig. 3 shows the variation of DC bus voltage with respect to firing angle indicating cosine relation, with a droop as a function of load current. Fig. 4 shows that for firing angle less that $90^0$ real power is positive and for firing angle greater that $90^0$ real power is reported as negative, as expected from the mathematical model. Fig. 5 shows reactive power is maximum when firing angle close to $90^0$ and reactive power approaching zero for firing angle of $0^0$ and $180^0$. Next step is to obtain curve fit equation; various statistical software tools such as MINITAB are available for such purpose. Curve fit equations are obtained for $V_{ac}$, $I_{ac}$, input power factor, real and reactive power at AC input terminals. A specimen third-order regression equation for DC bus voltage is shown below.

$$\begin{aligned}
V_{dc} = {} & (k0) + V_{ac}(k1) + I_{dc}(k2) + \alpha(k3) \\
& + V_{ac}I_{dc}(k4) + V_{ac}\alpha(k5) + I_{dc}\alpha(k6) + I_{dc}^2(k7) \\
& + \alpha^2(k8) + I_{dc}^2V_{ac}(k9) + I_{dc}^2\alpha(k10) + \alpha^2V_{ac}(k11) \\
& + \alpha^2I_{dc}(k12) + I_{dc}^3(k13) + \alpha^3(k14)
\end{aligned} \quad (6)$$

The regression coefficients for the above equation are in Appendix. Conducting a larger number of simulation runs allows the flexibility to incorporate higher order terms in the above equation. However, a larger number of simulation runs would require additional computation time. In the above equation some regression coefficients are not significant and hence are ignored. The statistical significance of each term in the equation provides the guidance as to whether to include the term in the equation. For example, the square term for Vac is ignored in (6). Such statistical significance can be evaluated by set confidence interval and the p-value obtained from hypothesis testing [7]. The equations are valid in the range of inputs where the detailed component simulations were carried out. Care should be taken not to extrapolate out of this range.

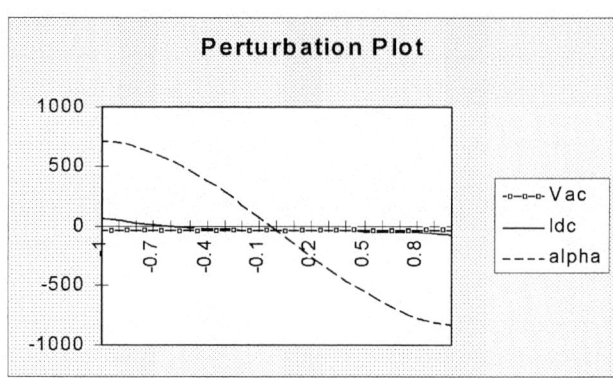

Fig. 6. Perturbation plot for DC bus voltage.

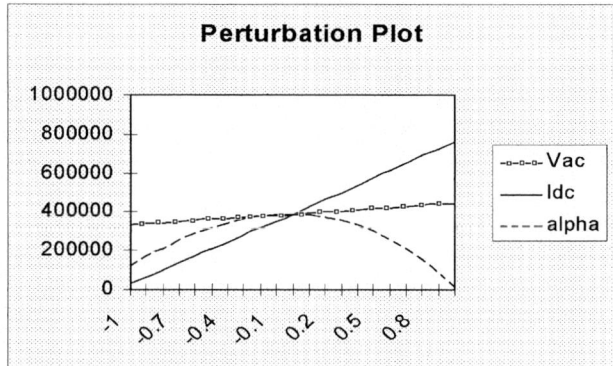

Fig. 7. Perturbation plot for reactive power.

The measure of goodness of curve fit is determined by the $R^2$ value of the regression equation [7]. A larger number of DOE runs and use of appropriate higher order terms for the regression equations gives a better $R^2$ value. The $R^2$ for most of the responses such as $V_{dc}$, $I_{ac}$, real and reactive power and power factor was more than 99% and THD was more than 80%. The mathematical expression captures the underlying physics of the system. This can be explained by analyzing perturbation plots. Fig. 6 shows perturbation plot for DC bus voltage. The variation of DC bus voltage with respect to firing angle shows cosine relationship. Similarly Fig. 7 shows perturbation plot for reactive power. The variation of reactive power with respect to firing angle is sine variation and with input AC voltage is linear. This variation observed is identical to what we observe in mathematical model derived under ideal conditions.

The regression based behavioral model of 12-pulse converter modeled in SABER is shown in Fig. 8, all regression equations are defined in simple high level simulation language. The output of model is a steady state numerical value of the response parameter and not time-domain waveform. The regression model represents on the quasi steady state operation of the system and the controller that is used to model the overall system as shown in Fig. 8 does not indicate the high speed response capability of the system. The control system includes an inner dc link current controller and outer dc bus voltage controller.

498

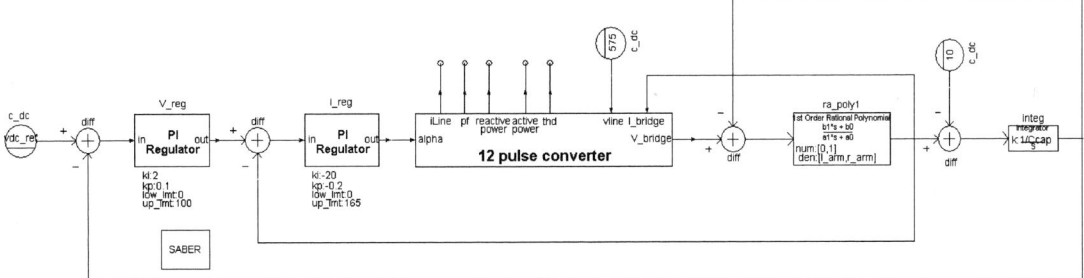

Fig. 8. Regression based behavioral model of 12-pulse converter.

Regression based model is significantly simplified and takes a small fraction of the original computational time. The computation time for the behavioral model is 0.02 seconds compared to 183 seconds for the detailed network model for 2 seconds of circuit simulation in time domain. This helps to reduce simulation time especially for system level analysis of multiple interconnected power converters. It is also possible to extend this type of regression model wherein number of distributed power units of various types is interconnected an a large number of system configurations. This model not only reduces simulation time significantly but also takes into account secondary effects such as continuous and discontinuous current, current overlap in the thyristors during commutation etc., which are would require multi-mode mathematical modeling approach.

Regression based behavioral model of 12-pulse converter developed here is not a generalized model. This model is valid only for component values (such as DC bus capacitor, filter inductor etc.) and other system parameters that are used in component level model and design space within which data points were collected. Furthermore, instantaneous time-domain waveform of response variables is not captured; the model calculates only steady-state numerical value of the response parameter, however slow control loop dynamics and slow time domain transients can be studied with the model.

A final system consisting of the grid power supply, standby genset, an UPS with a 12-pulse rectifier and inverter, and multiple flywheels for energy storage was studies with this approach. It was possible to study various operating scenarios, using this approach with significantly reduced simulation time. It was seen that the predicted operating scenarios matched well with tests conducted on a industrial power system under study.

## IV. CONCLUSION

A simplified regression based behavioral model of 12-pulse converter is presented. This model reduces simulation time significantly and also takes into account secondary effects that

are difficult to model using analytical mathematical modeling approach. However, this model is not a generalized model and detailed time-domain waveforms of response variables cannot be obtained. Preliminary study indicates the usefulness of the proposed novel modeling approach for system level analysis using computer simulation.

## V. APPENDIX

The UPS parameters considered are Vs =575V, frequency = 60Hz, nominal power = 500kW, Vdc nominal = 540V.

Regression coefficients for (6) are given by: k0 = -168.22; k1 = 1.64; k2 = 9e-2; k3 = 9.26; k4 = -7.89e-4; k5 = -1.29e-2; k6 = -2.92e-3; k7 = 4e-4; k8 = -1.41e-1; k9 = 5.17e-7; k10 = 1.15e-6; k11 = -3.6e-5; k12 = 1.2e-5; k13 = -4.21e-7; k14 = 5.8e-4.

## VI. REFERENCES

[1] Tsorng-Juu Liang, Jiann-Fuh Chen, Kuen-Jyh Chen, "Analysis of 12 pulse phase control AC-DC converter", IEEE 1999 International conference on Power electronics and Drives Systems, PEDS'99, Hong Kong, July 1999.

[2] Leland A. Schlabach, "Analysis of discontinuous current in a 12-pulse thyristor DC motor drive", *IEEE Trans. on Industry Applications*, Vol. 27, No. 6, Nov/Dec 1991.

[3] K. R. Padiyar, Sachidanand, A. G. Kothari, S. Bhattacharyya, A. Srivastava, "Study of HVDC controls through efficient dynamic digital simulation of converters", *IEEE Trans. on Power Delivery*, Vol. 4, No. 4, Oct. 1989.

[4] Keiju Matsui, Kazuo Tsuboi, Saburo Muto, Koji Iwata, "A dual thyristor converter reducing harmonics of power supply without input transformer," *Conference Record of the 1991 IEEE Industry Applications Society Annual Meeting*, Page(s): 925 -931 vol.1, 1991

[5] V. Rajagopalan, "Modeling and simulation of power electronic converters for power supplies", Industrial Electronics, Control, and Instrumentation, 1995, *Proceedings of the 1995 IEEE IECON 21st International Conference*, Page(s): 27 -32 vol. 1, 1995.

[6] T.H. Barton, *Rectifiers cycloconverters and AC Controllers*, Oxford Science Publications, Clarendon Press, 1994.

[7] T. Hill, P. Lewicki,. *Statistics Methods and Applications, Electronic Statistics Textbook*, StatSoft, Inc., Tulsa, OK, 2006. Available: http://www.statsoft.com/textbook/stathome.html.

**2006 IEEE International Conference on Power Electronic, Drives and Energy Systems**

# Simulation of PMSM VSI Drive for Determination of the Size Limits of the DC-Link Capacitor of Aircraft Control Surface Actuator Drives

M.Khatre, *Student Member, IEEE*, and Alan G. Jack, *Member, IEEE*

*Abstract --***This paper aims to contribute to the determination of the minimum acceptable value of the capacitor in the dc-link of drives suitable for aerospace actuation applications. As part of the analysis a complete simulation of a three-phase 3.6 kW permanent-magnet synchronous machine (PMSM) with a PWM rectifier-inverter drive for a full electric flap drive for a mid sized aircraft like an Airbus A-320 is presented. The effect of the size of the capacitor is examined and ground rules to establish minimum acceptable size are derived with the knowledge of exact switching states in the rectifier and inverter.**

*Index Terms--***DC-Link Capacitor, PMSM, PWM.**

## I. INTRODUCTION

ONE of the key drivers in the move towards the more electric aircraft is a reduction in the weight (and size) of the various actuation systems. At the same time safety requirements must be met and through-life costs reduced. The capacitor in the dc-link of the ac/ac converters used in prospective actuation systems is one of the most unreliable, bulky and heavy parts of the system. Forcing the required capacitance down to a minimum acceptable value can be seen to be contributing to all of the key objectives. The dc-link capacitor serves as a buffer between the rectifier and inverter stages and is essential to allow the two sides of the system to switch independently. On a simple level a design goal might be to maintain a "constant" dc link voltage irrespective of operating point or input conditions. In practice such an infinite goal can never be achieved and there is bound to be dc link voltage variation. They key question is just how much
is acceptable. The size of the capacitor is decided by the current it handles from the rectifier as well as from the inverter. Previous research to reduce the dc-link capacitor reported in [5] and [6] are based on voltage sensing on the dc-link and adding a compensating term to the controller.

---

This work is supported in part by the ORS Grant by Universities UK and the University of Newcastle Upon Tyne, UK.

Manas Khatre and Alan G. Jack are with the Department of Electrical Electronics and Computer Engineering at University of Newcastle upon tyne, NE2 2DL, UK. (e-mail: manas.khatre@ncl.ac.uk, alan.jack@ncl.ac.uk).

An iterative approach to determine lower limit of the dc-link capacitor with little concern over the rectifier side is reported in [7].Research involving elimination of dc-link capacitor is reported in [8]. The implementation of all the above methods either requires significant changes in circuit configuration which involves complex control and concern over system stability or they simply do not full-fill the reliability and safety specifications of aircraft actuator drives.

Specific requirements and data for the electric flap actuation application drive are taken from [1] which employs a front end rectifier and three independent single-phases, each supplied by a single phase bridge inverter, each of which has a 940μF in dc-link. That drive was configured to be fault tolerant in order to achieve the required flight dispatch specification. The flight safety of that flap system was actually provided using power-off brakes. The flight dispatch figures are related to economics rather the safety. It can be argued that the required reliability to meet flight dispatch can be reached by close control of the operating stress that the driven components are subjected to. One of the least reliable elements is the dc-link capacitor and if it can be drastically reduced in size more reliable technologies for the capacitor become applicable and reliability figures thereby improved and it is on this pretext that this study is predicated.

## II. DRIVE SIMULATION

(1) *Permanent-magnet motor:* A three-phase PMSM is chosen because of its high torque per unit volume. The application chosen is a mid sized aircraft and the specification calls for the flap to be fully extended or retracted within 15 seconds which corresponds to a peak power per flap of approximately 3.6kW and torque of 3.4kNm which is supplied using a PMSM operating at a rated speed of 10,000 r.p.m. via a step down gearbox [1]. Simulink (SimPower Systems toolbox) is used for the simulation of the complete drive. The motor is connected in star(Y) configuration (Fig. 5) and the motor's three phase back e.m.f. profile is created (as per the parameters given in the Table. I) as:-

$$Er = - \psi_m . \omega_e \, Sin \, (\theta)$$

$$Ey = - \psi_m . \omega_e \, Sin \, (\theta - 120°)$$

$$Eb = - \psi_m . \omega_e \, Sin \, (\theta - 240°)$$

0-7803-9771-1/06/$25.00 ©2006 IEEE

where rotor angle $\theta$ is determined as:-

$$\theta = \int \omega_e \, dt. \qquad (1)$$

For control purpose the above motor is modelled and simulated in the synchronous reference frame using two axis (d-q) theory for the machine where the d-axis and the q-axis transient equations used are [2][3]:-

$$V_d = i_d.R + \frac{d\psi_d}{d\theta} - \omega_e \, \psi_q \qquad (2)$$

$$V_q = i_q.R + \frac{d\psi_q}{d\theta} + \omega_e.\psi_d \qquad (3)$$

where $\qquad \psi_d = \psi_m + i_d.L_d \qquad (4)$

and $\qquad \psi_q = i_q.L_q \qquad (5)$

Here $V_d$ and $V_q$ are d, q axis voltages, $i_d$ and $i_q$ are d, q axis currents, $\omega_e$ is electrical speed of the motor in rad/s, $\psi_q$ and $\psi_d$ are the q-axis and d-axis flux linkages respectively in synchronous stationary reference frame. The motor's per-phase stator resistance is shown as R and $L_d$ and $L_q$ are the d, q axis stator inductances. Symbol $p$ is number of pole pair and $\psi_m$ is the peak flux linkage due to the magnets linking the stator. The electrical torque output from the motor is:-

$$T_e = \frac{3}{2} \, p. \, (\psi_d.i_q - \psi_q.i_d) \qquad (6)$$

To speed up the simulation process and to avoid unnecessary complication the motor dynamics are simulated as:-

$$T_e - T_L = J \frac{d\omega_r}{d\theta} \qquad (7)$$

where $T_L$ is load torque provided by flap loads, $\omega_r$ is the rotor speed in rev/min and J is the rotor inertia constant.

TABLE I

SYSTEM PARAMETERS [1]

| | |
|---|---|
| Motor power ($P_{mech}$) | 3.6 kW |
| Rated speed ($\omega_r$) | 10000 rpm |
| Motor rated torque ($T_e$) | 3.4 Nm |
| Pole pair ($p$) | 5 |
| Motor stator resistance (R) | 0.156 $\Omega$ |
| Motor Stator inductance (L) | 1.27 mH |
| r.m.s field flux linkage | 0.0258 Volt-seconds |
| DC-Link Voltage ($V_{dc}$) | 460 Volts |
| Aircraft available Supply | 200 Volts L-L @ 400 Hz |

To operate the motor under full vector control the d-axis component ($i_d$) of the motor current is controlled to zero and motor torque $T_e$ is controlled by $i_q$ in rotor synchronous reference frame.

Then using (4)-(6) the vector controlled motor torque and flux linkages will be given as:-

$$T_e = \frac{3}{2} \, p \, (\psi_d.i_q) \qquad (8)$$

and $\qquad \psi_d = \psi_m \qquad (9)$

and $\qquad \psi_q = i_q.L_q \qquad (10)$

In order to aid validation it is useful to determine parameters that can be expected in steady state. To supply the load of 3.4kNm via gear box the machine produces 3.4 Nm at rated speed. Now for a vector controlled drive the q-axis current $i_q$ requirement can be calculated using (8)-(9) and Table 1 parameters as :-

$$i_q = 12.42 \text{ Amp}$$

Using this demand current value and using (2)-(5) required input d, q axis voltages can be calculated for required vector control of the motor model as :-

$$V_d = -82.59 \text{ volts}$$

and $\qquad V_q = 193.1 \text{ volts}$

The three phase input voltages and currents to the motor model for vector control can be computed using Park transform matrices [3] as:-

$$i_r = -12.42 \sin(\theta)$$
$$i_y = -12.42 \sin(\theta - 120°)$$
$$i_b = -12.42 \sin(\theta - 240°)$$

and

$$V_r = -210 \sin(\theta + 23°)$$
$$V_y = -210 \sin(\theta + 23° - 120°)$$
$$V_b = -210 \sin(\theta + 23° - 240°)$$

The total input power for the vector controlled motor can be computed for this balanced system as:-

$$P_m = V_r \, i_r + V_y.i_y + V_b \, i_b = \frac{3}{2} \, (V_d.i_d + V_q.i_q)$$

as for vector control $i_d = 0$ then

$$P_m = \frac{3}{2} \, V_q.i_q = 3.6 \text{ kW}$$

(2) *PWM Inverter:* The three-phase inverter is designed with 6 IGBTs and diodes connected back to back. Switching of these IGBTs is done with three phase carrier pulse width modulation scheme as shown in Fig. 1, the PWM Switching frequency is set to 20 kHz for appropriate results.

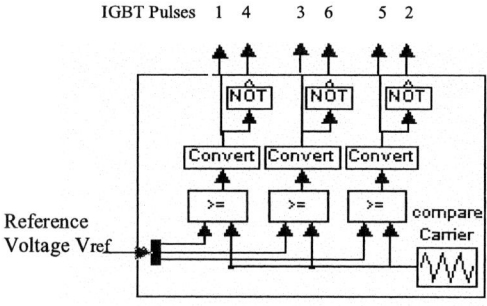

Fig. 1. IGBT Pulse Generation by Carrier PWM Scheme.

The vector controller with outer speed loop and inner current loop is simulated as in Fig. 2, which incorporates a voltage feed forward term.

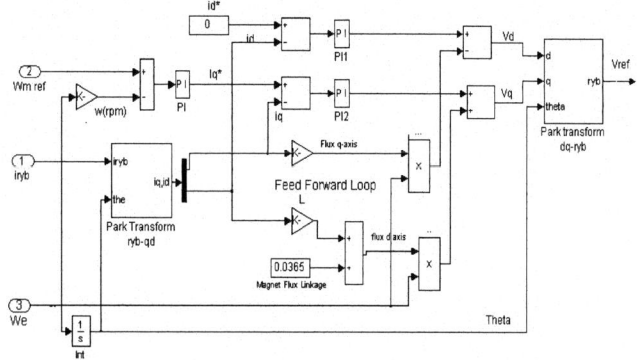

Fig. 2. Vector Controller with Feed-Forward Loop.

(3) *PWM Rectifier*: For drive operation with minimum dc-link capacitance a PWM IGBT "rectifier" supplying 460 Volts dc-link is chosen to get maximum control over the capacitor currents and better input power factor. The switching scheme for the rectifier IGBTs is the same as Fig. 1 and the PWM switching frequency is set to 12 kHz for appropriate results. The PWM rectifier is controlled for unity input power factor (UPF) operation by controlling the q-axis rectifier current ($i_q$) in supply frequency reference frame ($V_{qs}=0$) to zero as shown in Fig. 4 with the UPF controller in Fig. 3. This rectifier is supplied with a typical aircraft available power supply of 200 Volts line-line at 400Hz. Supply line inductances are chosen for 0.1p.u.voltage drop.

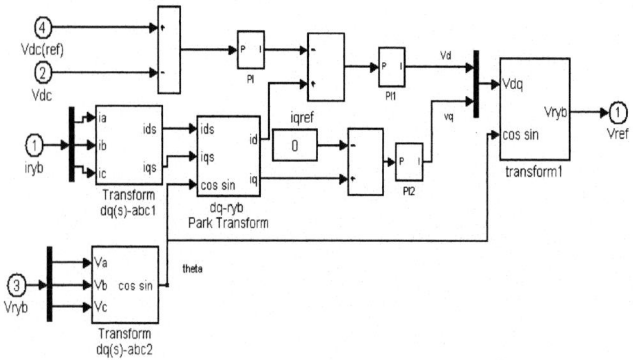

Fig. 3. UPF Controller for PWM Rectifier.

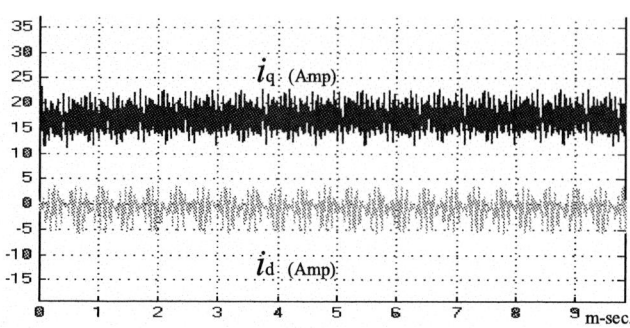

Fig. 4. UPF Rectifier d-q axis currents.

A complete dc-link rectifier-inverter drive schematic is shown in Fig. 5. The output waveforms of the above drive running at rated speed and supplying full rated torque with vector control currents is shown in Fig. 6.

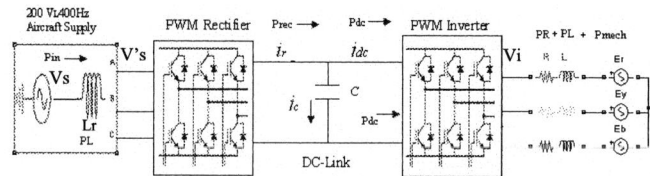

Fig. 5. PMW Rectifier-Inverter PMSM Drive Schematic.

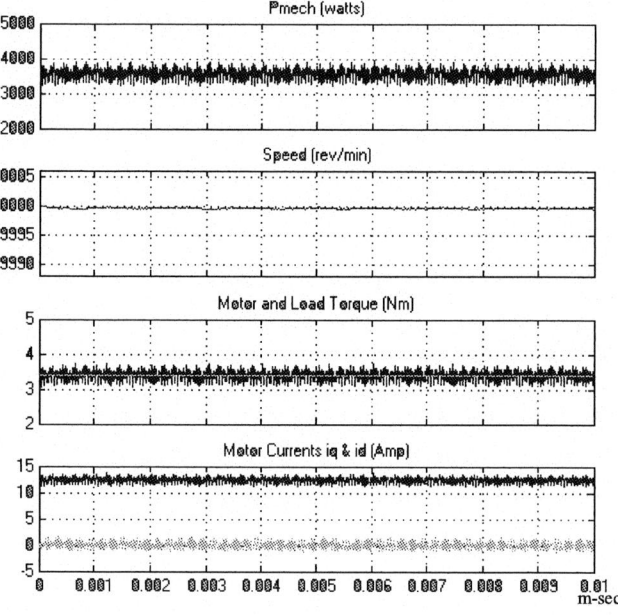

Fig. 6. Vector Controlled PMSM Output Waveforms.

When the drive is supplying a constant torque of 3.4Nm at rated speed, the motor output mechanical power $P_{mech}$ is constant along with three phase resistive loss $P_r$ in the motor. Due to the switching nature of the inverter bridge the inductive component of the power ($P_L$) that flows in and out of the motor is pulsed. If the inverter internal losses can be ignored then the addition of these three motor powers equals the power that is supplied by dc-link ($P_{dc}$) to the inverter as:-

$$P_{dc} = P_r + P_L + P_{mech} = V_{dc}.i_{dc} \qquad (11)$$

502

Similarly from the opposite end the power input to the drive can be equated to:-

$$P_{in} = P_{Lr} + P_{rec} \qquad (12)$$

where

$$P_{rec} = P_c + P_{dc} \qquad (13)$$

Power $P_{in}$ is supplied by the aircraft power supply, $P_{Lr}$ is the pulsed rectifier inductive power and $P_c$ is the power in dc-link capacitor. The maximum power handling capacity of the dc-link decides the required minimum capacitor value in dc-link. For calculation of these instantaneous powers flowing in and out of the dc-link capacitor an understanding of the switching patterns in rectifier and inverter is required. By matching supply and demand of power on both sides of the dc-link the capacitor value in the dc-link can be reduced.

### III. SWITCHING STATE DETERMINATION LOGIC

An online switching state logic is designed to know the instantaneous switching states of devices in the rectifier and in the inverter as shown in Fig. 7.

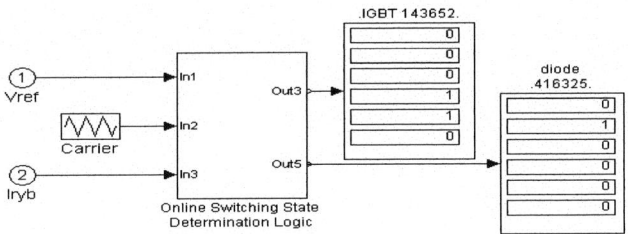

Fig. 7. Online Switching State Determination Logic.

The switching state determination logic takes the reference $V_{ref}$ from the inverter vector controller, measured motor currents and carrier signal information to produce instantaneous switching states of the IGBT and rectifier as in Fig. 8 which shows the current in IGBT1 in the inverter and the predicted online switching.

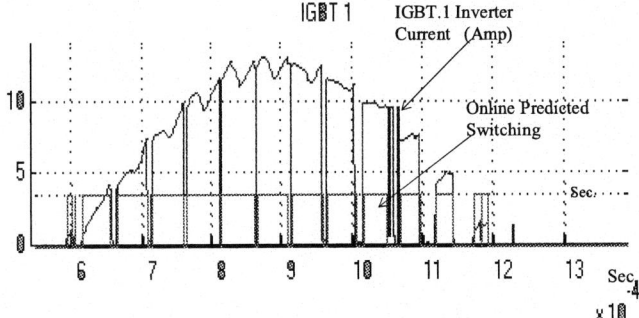

Fig. 8. Measured IGBT currents and switch prediction.

Now the various instantaneous voltage, currents and powers can be predicted from the switching information. From Fig. 5. :-

$$V_s = L_r \frac{di_s}{dt} + V'_s \qquad (14)$$

and

$$V_i = L \frac{di}{dt} + E \qquad (15)$$

where the three phase rectifier input terminal voltage $V'_s$ and inverter output terminal voltage $V_i$ can be computed from the knowledge of the switching states as [4] :-

$$V'_{sr} \& V_{ir} = \{2S_r-(S_y+S_b)\}.V_{dc}/3$$
$$V'_{sy} \& V_{iy} = \{2S_y-(S_r+S_b)\}.V_{dc}/3$$
$$V'_{sb} \& V_{ib} = \{2S_b-(S_r+S_y)\}.V_{dc}/3$$

where the switch logic states Sr, Sy and Sb are separately obtained from the rectifier and inverter phase current directions as in Fig. 9. For example if in phase ''R'' the current measured is negative that means the first leg of the bridge switch is connected to logic ''1'' and so on.

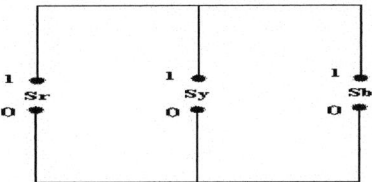

Fig. 9. Terminal Voltage Switch Logic.

With known profiles of supply voltages and motor back e.m.f now it is easy to calculate $\frac{di_s}{dt}$ using (14).Similarly the rectifier output current $i_r$ which is a function of three phase supply current can be written as:-

$$i_r = [S_r. i_{sr} + S_y. i_{sy} + S_b. i_{sr}] \qquad (16)$$

Inverter input current $i_{dc}$ can be written as a function of motor currents as:-

$$i_{dc} = [S_r. i_r + S_y. i_y + S_b. i_b] \qquad (17)$$

Now the dc-link capacitor currents can be calculated as:-

$$i_c = i_r - i_{dc} = c. \frac{dV_{dc}}{dt} \qquad (18)$$

So clearly the instantaneous ripple current goes into the dc-link capacitor is a function of the input and output phase currents. To ensure maximum power matching on both sides of rectifier and inverter, and therefore determine the lower limit of dc link capacitor value, the exact instantaneous power flow and demands at various points of the drive and instantaneous $\frac{di}{dt}$ can be calculated using (11)-(13).

### IV. DC-LINK CONSIDERATION

The drive operation at rated speed and rated load with a fully pre-charged (460 volts) dc-link capacitor of very high value such as 5000µF is quite smooth as this large capacitor provides significant isolation between the rectifier and the inverter stages. The dc-link variation ($\frac{dV_{dc}}{dt}$) with C=5000µF is quite low as shown in Fig. 10 below. UPF Controller of the rectifier in Fig. 3 can easily handle this low $\frac{dV_{dc}}{dt}$ and the reference voltage (M<1) thus the supply currents has low ripple.

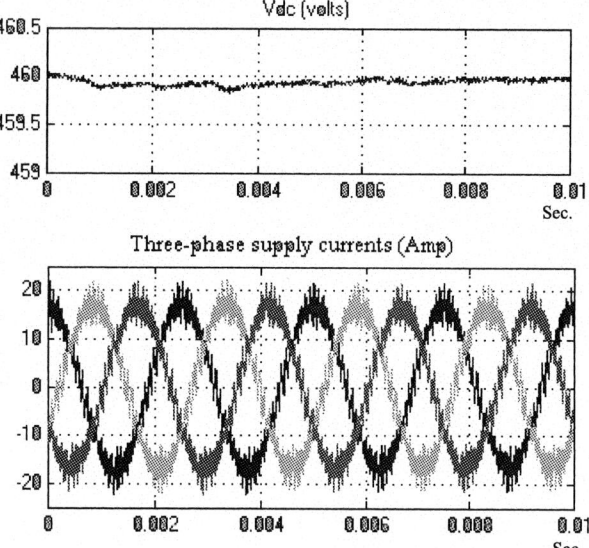

Fig. 10. $V$dc with 5000μF dc-link capacitor and supply currents.

As the capacitor's value is reduced with the same rectifier controller, the voltage ripple $\frac{dV_{dc}}{dt}$ increases and the harmonic content of the generated $V$ref increases causing the rectifier switching to go into over-modulation (M>1) region. With the same supply inductances the $\frac{di_s}{dt}$ remains the same irrespective of the value of the dc-link capacitor. Due to over modulation the switching "on" times increase and as do the rectifier volt-amperes. With a marginally sufficient value of the dc-link capacitor the drive still manages to remain stable with unity power factor at the rectifier side and vector control at the inverter side. As shown in Fig. 11, the dc-link voltage and supply current ripple with 235μF in dc-link are increased by about 50% in comparison to the results obtained with 5000μF. Note that there is no change in the power supplied by dc-link to the inverter ($P$dc is unchanged) for the same load at rated speed.

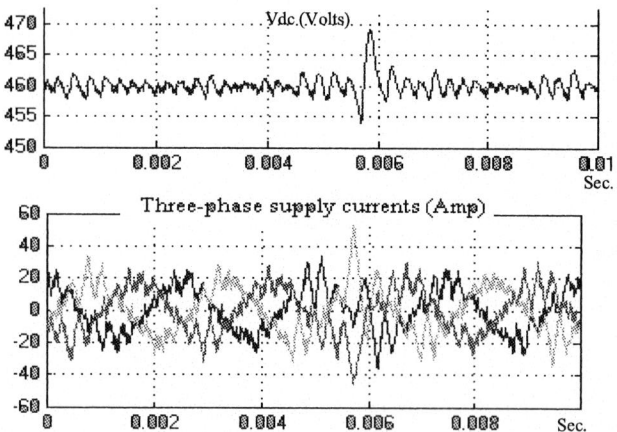

Fig. 11. $V$dc with 235μF dc-link capacitor and supply currents.

At 230μF the $\frac{dV_{dc}}{dt}$ becomes so high that with the same rectifier controller and drive parameters the dc-link becomes unstable and with it the whole drive as shown in Fig. 12 with C=230μF.

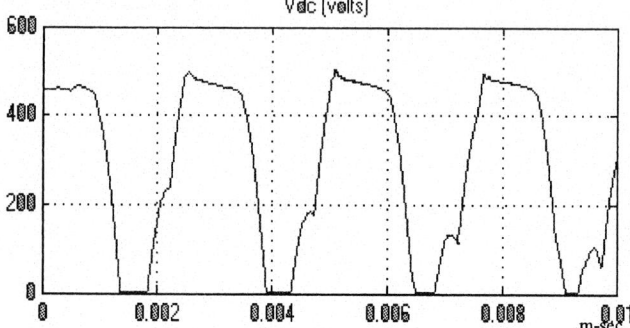

Fig. 12. Unstable Drive condition with $V$dc for 230μF dc-link capacitor.

So the simulation results shows that for a PMSM drive running at rated speed and supplying rated load torque the acceptable dc-link capacitor value heavily depends on the rectifier control supply current harmonics (and thus supply inductors) and rectifier volt-ampere rating. For the safety reasons practice is to keep these inductor voltage drops near to 0.1p.u value.

When the supply filter inductor values are reduced supply current harmonic content (TDH) is increased and it again forces the PWM to go in to over-modulation region (M>1) thus increasing the switching "on" time. The reduction in supply inductance causes the inductive power component $P$Lr to reduce. The volt-amperes of rectifiers are again increased but this time with less dc-link voltage ripple (Fig. 13) compares to Fig. 11 for the same 235μF capacitor. This might allow less capacitance in dc-link.

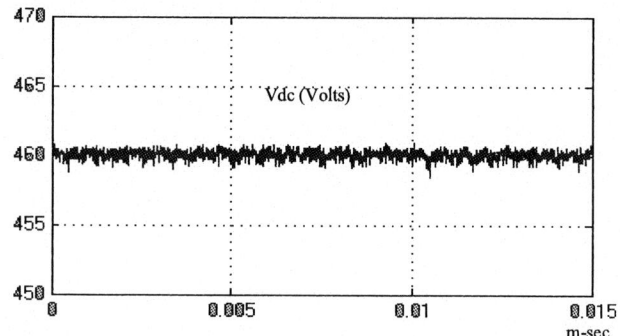

Fig. 13. $V$dc with supply inductance Lr =1e-4 H & 235μF dc-link capacitor.

If the supply inductor could be reduced (down to 0.03p.u inductor voltage drop) then the dc-link capacitor value can be dramatically reduced to just 85μF without affecting the drive output quality as the simulation results shown in Fig. 14. The impact of reducing the supply side inductance is that far higher currents follow a short circuit in the input converter. As stated earlier the feasibility of moving in this direction is a system issue and cannot be settled in the context of this paper. Suffice for the time being to note that dependence of capacitor size on input inductance.

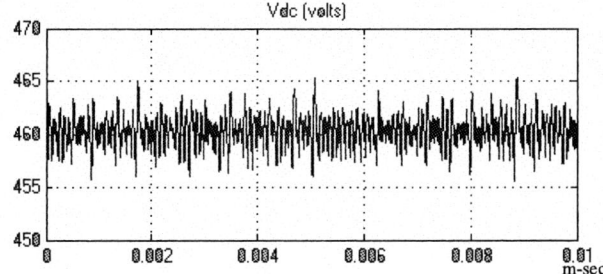

Fig. 14. *V*dc with supply inductance Lr =1e-4 H & 85µF dc-link capacitor.

## V. CONCLUSION

A complete three-phase PMSM drive simulation of a flap surface actuator drive has been described with fully switched front end rectifiers. Various cases of the drive with different dc-link capacitor values have been shown and with given specifications minimum acceptable dc-link capacitor value is established up to a point where control is lost. Factors affecting size of the dc-link capacitor were described and the results have been analysed using the simulation. The dependency of TDH, rectifier volt-ampere rating and supply inductor on the dc-link capacitor value is shown by simulation results. It has been shown by the result that the supply inductance drastically affect the size of the capacitor in dc-link if the system specifications remains the same. And this type of three-phase drive for control surface actuators with fully switched front end rectifiers can have huge weight and cost saving advantages in contrast to the three independent single-phase drive design in [1] which requires 940µF capacitor for each phase bridge inverter.

## V. REFERENCES

[1] J. W. Bennett, B. C. Mecrow, A. G. Jack, D. J. Atkinson, S. Sheldon, B. Cooper, G. Mason, C. Sewell, "Choice of Drive Topologies for Electrical Actuation of Aircraft Flaps and Slats," *Second International Conference on, IEEE PEMD Conf.* Publ. No. 498, pp. 332- 337, March 2004.

[2] P. Enjeti, J. F. Lindsay, and M. H. Rashid, "Stability and dynamic performance of variable speed permanent magnet synchronous motors," in *Proc. IECON*, 1985, pp. 749-754.

[3] P. Pillay, R. Krishnan, "Modelling, simulation and analysis of permanent magnet motor drives-Part I: The permanent magnet synchronous motor drive," *Industry Applications, IEEE Trans.* Volume 25, Issue 2, pp. 265 - 273 March-April 1989.

[4] A. Carlsson, "The back to back converter control and design," Ph.D. dissertation, Dept. Industrial EE & Automation., Lund Inst. of Tech., Sweden, 1998.

[5] B. G. Gu, K. Nam, "A Theoretical minimum DC-link capacitance in PWM converter-inverter systems," *Electric Power Applications, IEE Proceedings.* Volume 152, Issue 1, pp. 81 - 88 Jan. 2005.

[6] F. D. Kieferndorf, M. Forster, T. A. Lipo, "Reduction of DC-bus capacitor ripple current with PAM/PWM converter," *Industry Applications, IEEE Transactions*, Volume 40, Issue 2, pp.607 – 614,March-April 2004.

[7] M. N. Anwar, M. Teimor, "An analytical method for selecting DC-link-

capacitor of a voltage stiff inverter," *Industry Applications Conference, 37th IAS Annual Meeting.* Volume 2, pp. 803 - 810Oct. 2002.

[8] J. S. Kim, S. K. Sul, " New control scheme for AC-DC-AC converter without DC link electrolytic capacitor," *24th Annual IEEE Power Electronics Specialists Conference, 1993. PESC,* pp.300 – 306, June 1993.

**Alan G. Jack** received the Ph.D. degree for work on numerical analysis of electromagnetic (EM) fields in turbogenerators from Southampton University, Southampton, U.K., in 1975.
He is past head of the department and leader of the Newcastle Electric Drives and Machines Group at the University of Newcastle upon Tyne, Newcastle upon Tyne, U.K. He has been with Southampton University for more than 20 years, joining them from NEI Parsons, Newcastle upon Tyne, whom he was with for 13 years with roles from Craft Apprentice to Principal Design Engineer. He is the author of more than 80 papers in the area of electrical machines and drives. He holds the department's Chair in electrical engineering.

**Manas Khatre** received the M.Sc.degree in electrical power engineering at Newcastle University, Newcastle upon Tyne, U.K., in 2004, where he is currently pursuing the Ph.D.degree in engineering.
Currently, he is researching the machines and drives for electric actuators for use in aerospace applications.

**2006 IEEE International Conference on Power Electronic, Drives and Energy Systems**

# A Novel Soft Switched Improved Power Quality Converter Fed D.C. Motor Drive

M. B. Daigavane, Z. J. Khan, and H. M. Suryawanshi, *Member, IEEE*

*Abstract*--**In this paper, an attempt is made to improve the performance of DC drive by reducing motor ripples and discontinuous armature current operation and simultaneously maintain the high power factor and better input power quality in the entire load torque and current range of the motor. The improved power quality converter makes the input power factor unity over a wide operating shaft speed range, also reduces the total harmonic distortion (THD) of input supply current, and nearly ripple free the armature current and voltage waveform. Increase in switching frequency has the advantageous effects on both the amplitudes and the orders of load ripples and line harmonics besides having the advantage of lower armature series inductance requirement in case of controlled dc drives.**

*Index Terms*-- **DC motor drive, DSP, Power quality, Quality factor, Resonant converter, ZVS.**

## I. INTRODUCTION

POWER electronics engineers started to develop converter topologies that shape either the sinusoidal current or the sinusoidal voltage waveform, significantly reducing the switching losses. The key idea is to use a resonant circuit with a sufficiently high quality factor. The large number of single-phase resonant converter configuration have been proposed and analyzed in the literature [1]-[3]. However, as output increases, stress on the components becomes a limiting factor for these topologies. A very limited study on resonant converter fed dc drive system has been reported [4]. However studies on performance evaluation of a three phase SPRC fed dc drive system have not been reported in the recent literature. Balanced three phase resonant topologies offers substantial advantages, including low input/output ripple current due to the six-pulse operation of the inverter. The series-parallel resonant converter offers the optimum performance in terms of a reduction in series tank current at reduced load together with good voltage regulation under no load condition [5]-[7]. The proposed drive system is appropriate and most economical solution to preserve the present separately excited dc motors in industry compared with use of variable frequency ac drive technology. The block schematic of proposed drive is shown in Fig. 1.

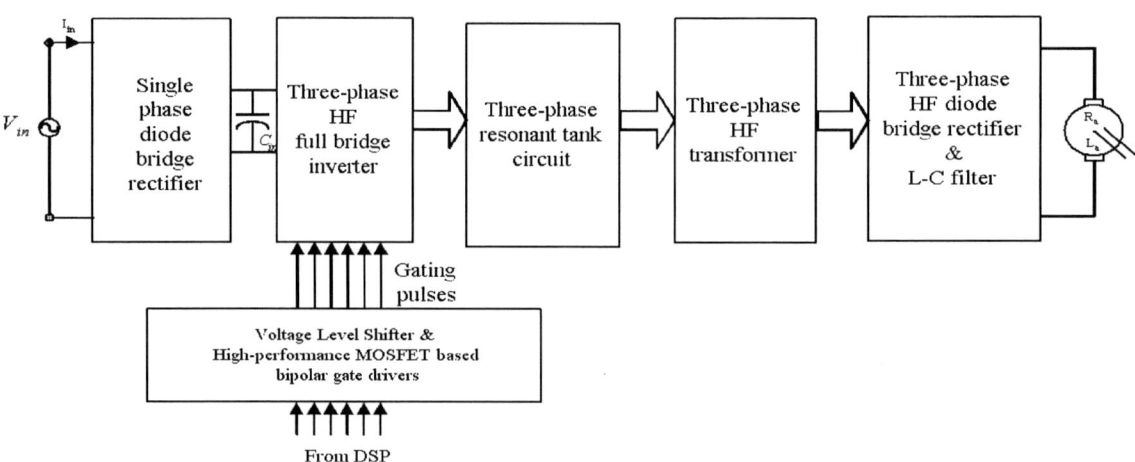

Fig. 1. Block schematic of proposed AC-DC resonant converter fed drive.

M.B. Daigavane is with the Department of Electronics and Power Engineering, B.D. College of Engineering, Sewagram 442102, India (e-mail: mdai@rediffmail.com).
Z.J.Khan is with Department of Electrical Engineering Rajeev Gandhi College of Engineering Research & technology, Chandrapur, India (e-mail: khanzj1@rediffmail.com).
H.M.Suryawanshi is with Department of Electrical Engineering, Visvesvaraya National Institute of Technology, Nagpur, India (e-mail: hms_1963@rediffmail.com).

## II. OPERATING PRINCIPLE

The proposed converter fed dc motor drive as shown in Fig. 2, the input supply to the converter is single-phase ac, which is rectified using an uncontrolled diode bridge rectifier, followed by a small DC link capacitor $C_{in}$ (i.e. high frequency by pass) connected to three phase series parallel resonant converter (SPRC). The dc output from SPRC is applied to the armature of separately excited dc motor. Three-phase converters are used for high power level. Also for same power level, three-phase resonant converter offers many advantages [4]-[5]. Hence in this paper, a three phase SPRC fed dc drive system is proposed.

## III. DESIGN OF CONVERTER

A resonant converter is designed for 2.5KW, 230V, and 1500-rpm dc motor. The converter is operated at 150 KHz switching frequency on full load condition to increase the power packing density. The proposed converter fed dc drive system operates in lagging power factor (pf) mode for the entire load range ensuring ZVS operation for all of the inverter switches.

The converter specifications are as follows:

Input voltage: $V_{in} = 230$ V (RMS)
Supply frequency: $f_{in} = 50$ Hz
Output Voltage: $V_O = 230$ V
Output Power: $P_O = 2500$ W
Resonant frequency: $f_r = 140$ KHz
Minimum switching frequency : $f_s = 150$ KHz

Design of converter is based on sine wave approximation for the waveforms, in which only fundamental components are considered. This approximation is valid since the resonant tank circuit acts as a tuned low pass filter allowing only the fundamental frequency component of current to pass through the tank circuit. Per phase equivalent circuit parameters of three phase series parallel resonant converter (SPRC) feeding motor load obtained by referring all the components on the secondary side of the HF transformer to its primary. The ac equivalent resistance $R_{ac}$ replaces the HF transformer, HF diode bridge rectifier, LC filter and motor Load ($R$-$L$-$E$).

The design procedure for converter is as follows:

Average dc input to the three phase inverter for the supply voltage of 230V, $V_{DC} = 207$.

For $180^0$ wide gating pulse control scheme, the phase voltage of the three phase HF inverter output is given by,

Inverter output voltage (line to line) = $V_{RY}$ is given by

$$V_{RY} = \sum_{n=1,3,5}^{\infty} \frac{4V_{DC}}{n\pi} \cos \frac{n\pi}{6} \sin n(\omega t + \pi/6) \quad (1)$$

$$Vph = \sum_{n=6k\pm1}^{\infty} \frac{2 \cdot V_{DC}}{n \cdot \pi} \cdot \sin n\omega t \quad (2)$$

For the star connected load, line to neutral voltage = $V_{RN}$ is given by

$V_{RN} = V_{RY} / \sqrt{3}$ with the delay of $30^0$

So the rms value of fundamental component of phase voltage ($V_{RN1}$) of the primary of HF transformer is given by

$$V_{RN1} = \left( \frac{\sqrt{2}}{\pi} \right) V_{DC} \quad (3)$$

RMS value of the fundamental frequency component of this voltage is 93.18V.

### Calculation of turns ratio

At resonance frequency, tank gain ($M$) is almost unity, the rms voltage of secondary of HF transformer = $V_{RN2}$ is given by

$$V_{RN2} = \frac{\pi V_{DC}}{3\sqrt{6}} \quad (4)$$

So,

turn ratio (primary to secondary) = $n_t = V_{RN1}/V_{RN2}$ (5)

To obtain 230V dc output voltage, the input phase voltage to the three-phase diode bridge rectifier should be $\left( 230 \cdot \pi / 3 \cdot \sqrt{6} \right)$ = 110V. Hence turns ratio of three-phase transformer is 93.18: 98.32

### Design of Tank

To find the values of tank circuit parameters, following expressions are used.

Quality factor ($Q$) of the resonant tank circuit is

$$Q = \frac{\omega_r L_1}{R_L'} \quad (6)$$

$$\omega_r = \frac{1}{\sqrt{L_1 \cdot C_1}} \quad (7)$$

$R_L' = $ Equivalent Load resistance referred to primary of the transformer.

$\omega_s$ = switching frequency and

$\omega_r$ = tank resonant frequency

$k = \pi^2/18$ for inductive filter.

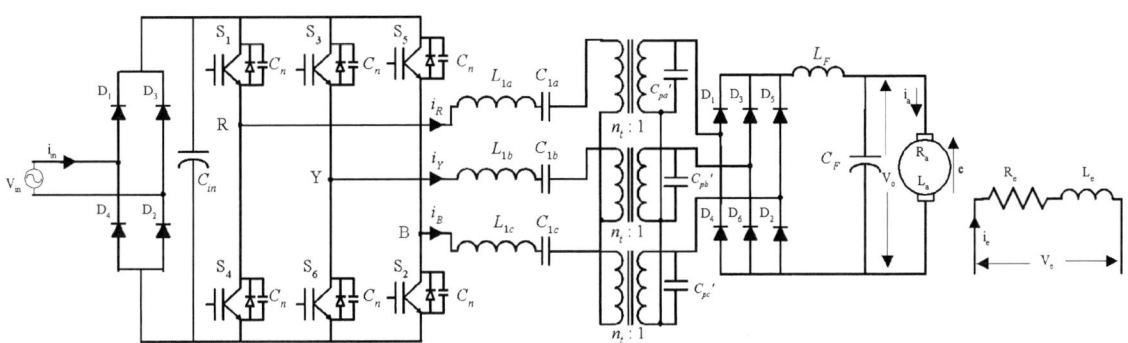

Fig. 2. Power circuit of proposed resonant converter fed drive.

The major design issues are choice of $C_1/C_p$ ratio and selection of quality factor-$Q$ of the resonant tank circuit. For $C_1/C_p$ ratio greater than 1, the frequency variation required for output voltage regulation is large [5]. The peak current through the inverter switching devices does not decrease with the load current for smaller ratios of $C_1/C_p$. It is therefore necessary to choose compromised value of $C_1/C_p$ to be equal to 1 [5]-[9] The KVA rating of the resonant tank circuit decreases as the quality factor-$Q$ decreases for a given $f_{sw}/f_{reso}$ ratio. This dictates the lower value choice for quality factor-$Q$. But it is observed that as the value of full load $Q$ is increased, there is a decrease in the peak inverter output current with a larger decrease in load current. However, this decrease is small for the values of quality factor - $Q$ greater than 4.

Considering the above constraints, following compromised design values are chosen for capacitor ratio $C_1/C_p$ and full load quality factor $Q$.

$$C_1/C_p = 1 \quad \text{and} \quad Q = 4$$

At full load condition the converter operated with switching frequency $f_{sw}$ of 150 KHz. For this condition component values of resonant tank circuit are calculated as below:
$L_1 = 106.66\,\mu\text{H}$, $C_1 = 0.0121\mu\text{f}$, $C_p = 0.0121\mu\text{f}$, $C_{p'} = 0.0133\mu\text{f}$.

## IV. CONVERTER CONTROL STRATEGY

DSP-TMS320LF2402 is used for generating six gating pulses for three-phase resonant inverter. A variable frequency, symmetrical, 180° [5]-[6] gating control scheme is achieved using this DSP. Fig. 3 shows Typical operating waveforms of the practical three-phase series-parallel resonant converter operating with gating pulse width lesser than180$^0$ (Typical value of the gating pulse width is 150$^0$). Performance of a three-phase SPRC fed dc drive system is evaluated with variable frequency control for controlling the output voltage of the resonant converter for maintaining the motor speed constant at different load torques.

Master DSP is used to realize the output voltage regulation and overload protection function under closed loop operation, while the slave DSP does the critical function of generating gating pulse pattern for HF inverter. Changing the content of the T1PR register can vary switching frequency.

## V. SIMULATION AND EXPERIMENTAL RESULTS

According to design in section III, Simulation studies are done using PSIM for variable load torque condition of dc motor,to investigate the control to output characteristic of three phase SPRC. The Fig. 4, shows the simulation results at full load. An experimental prototype model for three- phase SPRC is fabricated for verifying its performance. Power circuit designed with three-phase IGBT bridge module

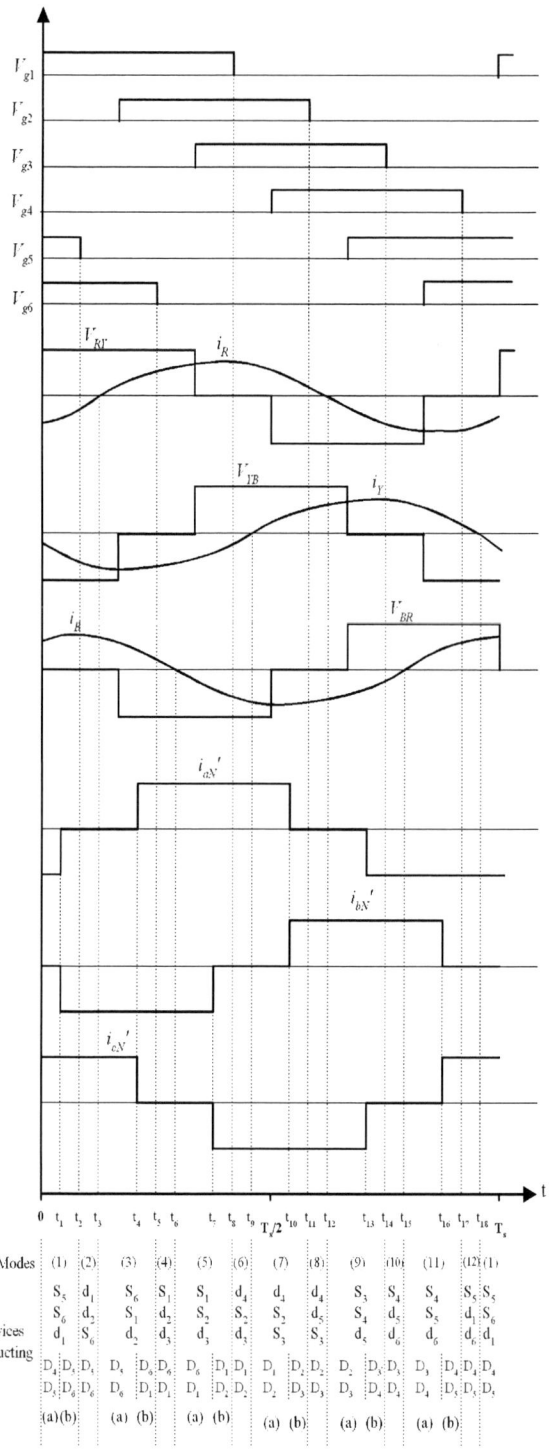

Fig. 3. Typical operating waveforms of the practical three-phase series-parallel resonant converter operating with gating pulse width lesser than180$^0$ (Typical value of the gating pulse width is 150$^0$).

BSM25GD120DN2 (EUPEC make) and HF diodes DESI60-12A.The practical values of reactive components are: $L=107.23\mu\text{H}$, $C_1 = 0.01\mu f$, $C_p = 0.01\mu f$, $L_F = 4.07\mu\text{H}$.
Fig. 5, show the experimental performance of the converter at full load condition. It is seen from Fig. 5 (a), that the current

maintains very high power factor throught the entire loading conditions with very low THD.

Fig. 5 (b) revels that the converter operates in ZVS mode at all loads, eliminating the turn on losses. Fig. 5(c) show waveforms of motor armature voltage and armature current.

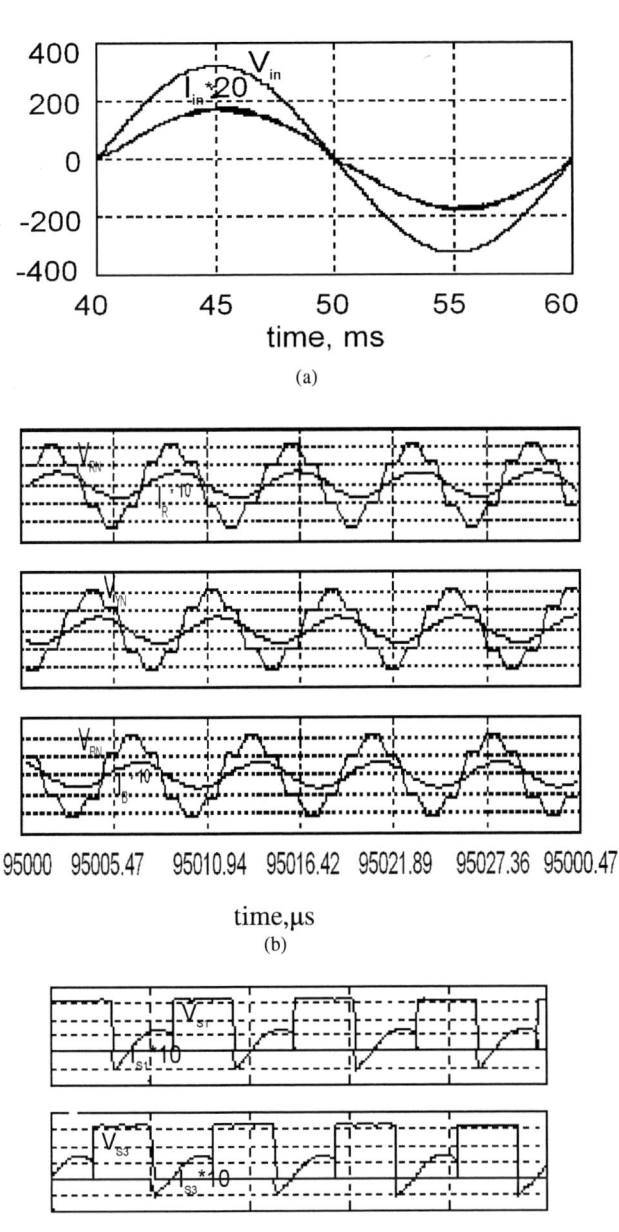

(a)

(b)

(c)

Fig. 4. Simulation results for the converter at full load condition:
a. Input voltage (V) and input line current (A).
b. Inverter output phase voltages and currents.
c. Inverter switch voltages and currents.

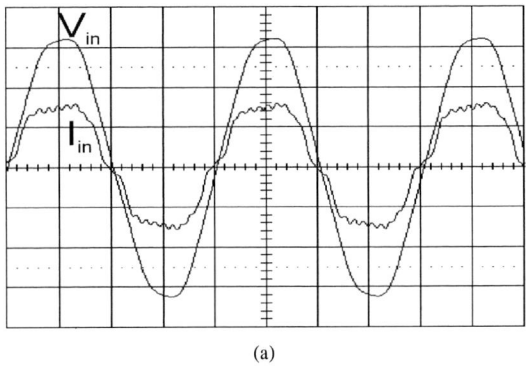

(a)

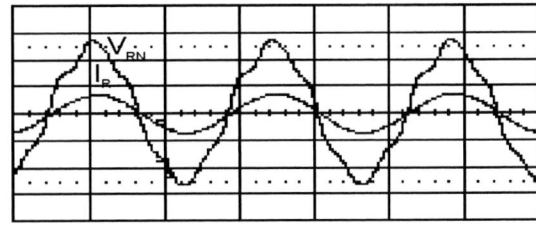

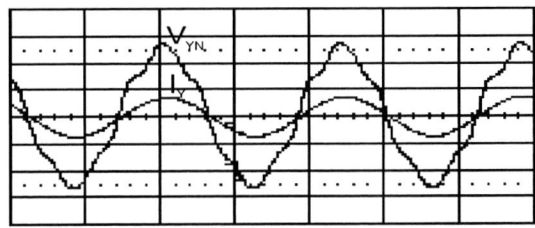

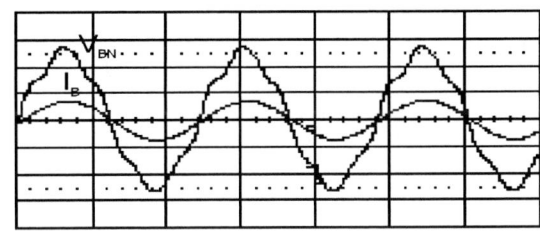

(b)

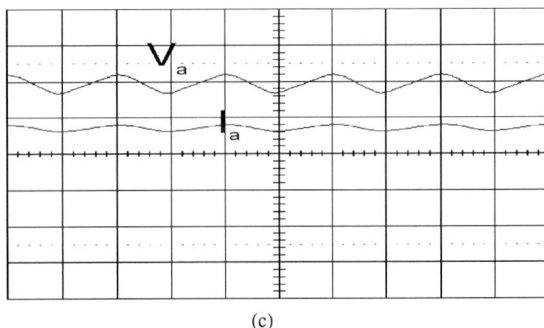

(c)

Fig. 5. Experimental waveforms for the converter at full load:
a. Input voltage and input line current –
   Scales: current 20A/division, , voltage 100V/division, time 5ms/division.
b. Phase voltage and phase current for three-phase H.F. inverter
   Scales: time 1μS/division, voltage 100V/division.
c. Motor armature voltage and armature current
   Scales: voltage 100V/division, current 20A/division.

## VI. CONCLUSION

A three-phase series parallel (SPRC) resonant converter is a popular choice for medium to high power levels variable speed dc motor drives, due to the improved power quality converter, the input power factor unity over a wide operating shaft speed range, the power density. This drive system is appropriate and most economical solution to preserve the present separately excited dc motors in industry compared with use of variable frequency ac drive technology. Application of this converter in dc drive has been investigated in this paper. Design of three phase SPRC that inherently operates at very high power factor, the performance on ac line side has also been improved in terms of quality of power drawn by the drive system.

## VIII. REFERENCES

[1] R. L. Steigerwald, "A Comparison of Half-Bridge Resonant Converter Topologies," *IEEE transactions on Power Electronics*, vol. 3, no. 2, pp. 174-182, Apr. 1988.

[2] K. H. Liu and F. C. Lee, "Zero-voltage switching technique in DC/DC converters," *IEEE Trans. Power Electronics*, vol. 5, no.3, pp. 293-304, 1990.

[3] Suryawanshi, H. M., Tarnekar, S.G., "Resonant Converter In High Power factor ,High Voltage DC Applications," *IEE Proc.- Electric Power Applications*, vol.145, no.4, pp. 307-314, Jul. 1988.

[4] C. C. Chong, C. T. Chan, and C. F. Foo, "Quasi resonant converter fed DC drive System," *1993 The European Power Electronics Association*, pp. 372-376.

[5] Bhat, A. K. S., and Zheng, R. L., "A three-phase series-parallel resonant converter- Analysis, Design, Simulation and Experimental results," *IEEE transactions on Industry applications*, vol.32, no.4, pp. 951-960 Jul.-Aug 1996.

[6] Bhat, A. K. S., and Zheng, R. L., "*Analysis and Design of a three-phase LCC-type resonant converter*," IEEE transactions on Aerospace and Electronic systems, vol.34, no.2, pp. 508-519, Apr. 1998.

[7] M. J. Schutten, R. L. Steigerwald, and M. H. Kheraluwala. " Characteristics of load resonant converter operated in a high-power factor mode," IEEE Transactions on power Electronics, vol. 7, no. 2, pp 304- 314, Apr.1992.

[8] Tanavade, S. S., Suryawanshi, H. M., Thakre, K. L., and Chaudhari, M. A., "Application of three phase resonant converter in high power DC supplies," *IEE Proc.-Electr. Power Applications*, vol. 152, no. 6, pp. 1401-1409, Nov. 2005.

[9] V. Belaguli, A. K. S. Bhat, "High power factor operation of DCM series- parallel resonant converter," *IEEE aerospace and electronic systems*, vol.. 35, no. 2, pp 602-613, Apr. 1999.

[10] Kazimierczuk, M. K., and Czarkowski D., *Resonant Power Converters*, John Wiley & Sons, Inc, New York, 1995.

[11] N. Mohan, T. Undeland, and W. Robins, *Power Electronics:Converters, Applications, and Design*. 2nd Ed. New York: Wiley, 1995.

[12] M. H. Rashid , *Power Electronics:, Applications, and Design*, 2nd Ed. PHI, 1995 .

[13] Siemens Matsushita components data book, *Ferrites and Accessories*, 1999.

[14] L. Umanand, S. R. Bhat, *Design of Magnetic components for switched mode power converters*, Wiley Eastern Limited, 1992 edition.

## VII. BIOGRAPHIES

**Manoj B. Daigavane** obtained the B.E.Degree in Power Electronics Engineering from Nagpur University, India in 1988. He received the M.S.Degree in Electronics and Control Engineering from Birla Institute of Technology and Science, Pilani (Raj) India in 1994. He also obtained the M.E. Degree in Power Electronics Engineering from Rajeev Gandhi University of Technology, Bhopal (M.P), India in 2001.

Since 1988,he has been with the Department of Electronics and Power Electronics Engineering, B. D. College of Engineering, Sewagram (Wardha), affiliated to the Nagpur University, India.He is currently Assistant Professor of Electronics and Power Engineering, where he is engaged in teaching and working towards the Ph.D.Degree. He has been responsible for the development of Electrical Machines and Power Electronics Laboratories He is a Member of the Institution of Engineers (India) and a Life Member of the Indian Society for technical Education.

**Zafar Jawed Khan** received the B.E.Degree in Electrical Engineering from Nagpur University, India in 1983.He received the M.Tech.Degree in Electrical Engineering from V.R.C.E Nagpur, India in 1986.He also obtained the Ph.D from Regional Engineering College, Warangal (A.P),India in 1996.

Since 1986,he has been with the Department of Electrical Engineering, Rajeev Gandhi College of Engineering Research &Technology, Chandrapur, affiliated to the Nagpur University, India, where he is Professor & Head of Electrical Engineering, engaged in teaching, maintenance and research in the area of power electronics and power system modeling and control. He has been Dean Faculty of Engineering & Technology, Nagpur University for past five years. He is a Fellow of the Institution of Engineers (India) and a Life Member of the Indian Society for Technical Education.

**H. M. Suryawanshi** graduated from Shivaji University in 1988 with a BE degree. He obtained M.Tech Degree in Electrical Engineering from Indian Institute of Science, Bangalore ,(India) in 1994. He also obtained Ph.D. Degree in Electrical Engineering from Nagpur University in 1998.

Presently he is working at Visvesvaraya National Institue of Technology, Nagpur (India) as Assistant Professor in Electrical Engineering and is engaged in teaching and conducting research in the area of power electronics. He was responsible for the completion of a number of research and development projects. His main areas of interest are Resonant converters, Active power filters, Power quality issues and Application of Power Electronics to Power systems, Power electronics for motor drives. He is a life member of ISTE as well as MIE (I).

**2006 IEEE International Conference on Power Electronic, Drives and Energy Systems**

# Generalized Discontinuous PWM Based Direct Torque Controlled Induction Motor Drive with a Sliding Mode Speed Controller

T. Brahmananda Reddy, J. Amarnath, D. Subbarayudu, and Md. Haseeb Khan

*Abstract*--This paper presents a Generalized Discontinuous pulsewidth modulation (GDPWM) algorithm for sensorless direct torque controlled induction motor drive, which ensures the high-performance operation, both in steady state and transient conditions. In the proposed method, the merits of basic direct torque control transient behavior are maintained, while the steady state operation is improved. As the bus-clamped sequences are being used, the switching losses of the inverter can be reduced. Also this method gives good performance over the conventional space vector strategy in the higher modulation indices. Then a sliding mode speed controller with an integral switching surface is proposed, which is robust under uncertainties caused by parameter variation and load torque disturbances. To validate the proposed method, the simulation results are presented.

*Index Terms*--direct torque control, GDPWM, sliding mode speed control.

## I. NOMENCLATURE

| | |
|---|---|
| $R_s$ , $R_r$ | stator and rotor resistances |
| $L_s$, $L_r$, $L_m$ | self and mutual inductances |
| $P$ | number of poles |
| $\overline{v}_s$ | stator voltage phasor |
| $\overline{i}_s$ , $\overline{i}_r$ | stator and rotor currents phasors |
| $\overline{\psi}_s$ , $\overline{\psi}_r$ | stator and rotor flux linkage phasors |
| $\omega_r$ | rotor electrical speed in radians |
| $T_e$ | electromagnetic torque |
| $B$ | friction coefficient |
| $J$ | inertia constant of the induction motor |
| $T_L$ | load torque. |
| $\omega_m$ | rotor mechanical speed in radians |

---

T. Brahmananda Reddy and Dr. D. Subba Rayudu are with Department of Electrical and Electronics Engineering, G. Pulla Reddy Engineering College, Kurnool, AP 518002, INDIA (e-mail: tbnr@rediffmail.com).

Dr. J. Amarnath is with Jawaharlal Nehru Technological University, Hyderabad, AP, INDIA (e-mail: amarnathjinka@yahoo.com).

Md. Haseeb Khan is with Department of Electrical and Electronics Engineering, Royal Institute of Technology & Science, Hyderabad, AP, INDIA.

## II. INTRODUCTION

RESEARCH interest in high-performance control strategies for induction motor drives has grown significantly over the last three decades due to some of their advantages, such as less maintenance, simple construction and mechanical robustness. This progress is remarkable especially after the invention of field oriented control (FOC) by F. Blaschke [1]. But this control scheme is quite complex due to reference frame transformation and also it is highly dependent upon the motor parameter and mechanical shaft speed. To mitigate these problems, a new control strategy for the torque control of induction motors was developed by I.Takahashi as direct torque control (DTC) [2] and by M. Depenbrock as direct self control (DSC) [3]. The DTC scheme is gaining popularity in industry due to its simplicity, good dynamic response and reduced machine parameter dependence [4]. The basic DTC scheme uses torque hysteresis controller, flux hysteresis controller and a switching table. This control strategy is based on selecting the proper voltage vector in order to maintain the torque and flux errors with in the prefixed hysteresis bands. Though the conventional DTC (CDTC) has good dynamic performance, it has few drawbacks such as steady state ripple in torque and flux and higher acoustical noise. In recent years several techniques have been developed to reduce the ripple and to get constant switching frequency. One of such technique is space vector pulsewidth modulation (SVPWM) [5]-[6]. In this method, for each sampling period, reference voltage vector has been generated and is used for the reduction of flux and torque ripples. In [7] DTC is developed based on the 7-zone hybrid SVPWM to reduce the ripples in torque, flux and current. In recent years, several discontinuous PWM (DPWM) methods are reported in [8]-[12]. In the DPWM methods, the modulation wave of a phase has at least one segment which is clamped to the positive or negative dc bus of the voltage source inverter. No modulation implies that there are no switching losses. An attempt toward unifying all discovered DPWM methods has lead to the development of generalized DPWM algorithm. Carrier based generalized DPWM (GDPWM) algorithms are given in [11-12]. Industrial applications exhibit significant uncertainties, so that performance may deteriorate, if conventional controller such as PI controller is used. For this

---

0-7803-9771-1/06/$25.00 ©2006 IEEE

reasons it is worth to develop controllers that have capabilities of to handle uncertainties caused by parameter variations. The sliding mode control can offer good performance against insensitivities to parameter variation and load disturbance [13]. Hence, to improve the speed performance under uncertainties, a sliding mode speed controller is used for FOC in [14].

The main objective of this paper is to develop space vector based GDPWM method for direct torque controlled induction motor drive to reduce the ripples and switching losses. Also to improve the speed performance of DTC, an integral switching surface sliding mode speed controller is proposed, which is robust under uncertainties caused by load torque disturbances.

### III. SVPWM ALGORITHM

With a three-phase voltage source inverter (VSI) there are eight possible operating states. The two states, where no power gets transferred from source to load are termed as 'null vectors' or 'zero states'. The other six states called active states. The active states can be represented by space vectors

$$V_k = \frac{2}{3}V_{dc} * e^{j(k-1)\frac{\pi}{3}}, \quad k = 1,2,...6 \qquad (1)$$

and dividing the space vector plane into six equal sectors as in Fig. 1.

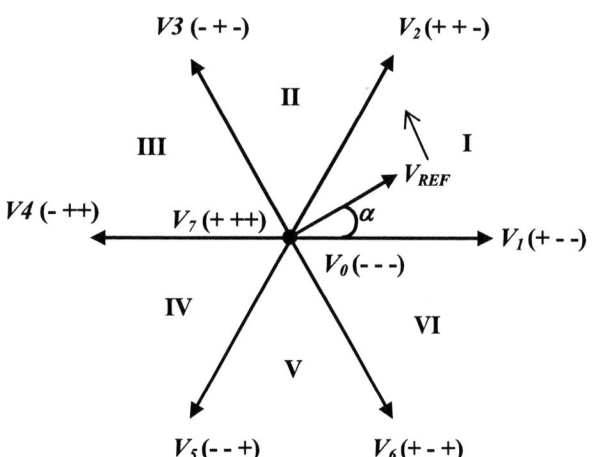

Fig. 1. Voltage space vectors of the three-phase inverter.

In the SVPWM strategy, the desired reference voltage vector is generated by time averaging the suitable discrete voltage space vectors within sampling period or subcycle $T_s$. For a given reference voltage $V_{REF}$ and angle $\alpha$ in first sector, the volt-time balance is maintained by applying the active state 1, active state 2 and two zero states together for durations $T_1$, $T_2$ and $T_Z$ respectively, as given in (2) [7].

$$T_1 = M * \frac{Sin(60^o - \alpha)}{Sin60^o} * T_s \qquad (2.1)$$

$$T_2 = M * \frac{Sin\alpha}{Sin60^o} * T_s \qquad (2.2)$$

$$T_Z = T_s - T_1 - T_2 \qquad (2.3)$$

Where 'M' is the modulation index, given by $M = \frac{3V_{REF}}{2V_{dc}}$

Also in SVPWM strategy, the total zero voltage vector time is equally distributed between $V_0$ and $V_7$. Also in this method, the zero voltage vector time is distributed symmetrically at the start and end of the sampling period in a symmetrical manner. Thus, SVPWM uses 0127-7210 in sector-I, 0327-7230 in sector-II and so on.

### IV. GDPWM ALGORITHM

Discontinuous PWM methods use the discontinuous type of zero sequence signals. In DPWM methods, during each sampling period, one phase ceases the modulation and the associated phase is clamped to the positive dc bus or negative dc bus. Hence, the switching losses of the associated inverter leg are eliminated. The performance of the PWM methods depends upon the modulation index. In the lower modulation range, the continuous PWM (CPWM) methods are superior to DPWM methods, while in the higher modulation range, the DPWM methods are superior to CPWM methods [11]. However at all the operating modulation indices, CPWM method has higher switching losses than DPWM methods. Utilizing the generalized phase shift in the DPWM methods, a carrier based PWM scheme, known as generalized discontinuous PWM (GDPWM) is proposed in [11]-[12].

The SVPWM algorithm employs equal division of zero voltage vector times within a sampling period. This paper sets forth a space vector based GDPWM algorithm, which uses the utilization of the freedom of zero state time division that results in different DPWM schemes. In the proposed method the zero state time will be shared between two zero states as $T_Z \mu$ for $V_0$ and $T_Z(1-\mu)$ for $V_7$ respectively, where $\mu$ lies between 0 and 1 [12]. The $\mu$ can be defined as $\mu = 1 - 0.5(1 + sgn(\cos 3(\omega t + \delta)))$. Where, $\omega$ is the angular frequency of the reference voltage, $sgn(y)$ is the sign function, $sgn(y)$ is 1, 0.0 and -1 when $y$ is positive, zero and negative respectively. The modulation phase angle is represented by $\delta$. When $\mu = 0$, any one of the phases is clamped to positive dc bus for 120 degrees and when $\mu = 1$, any one of the phases is clamped to negative dc bus for 120 degrees. When $\mu = 0$ and $\mu = 1$, DPWMMAX and DPWMMIN are obtained respectively and $\mu = 0.5$ gives the SVPWM algorithm. Variation of modulation phase angle $\delta$ yields to infinite number of DPWM methods. When $\delta = 0$, -$\pi/6$, -$\pi/3$, DPWM1, DPWM2 and DPWM3 can be obtained respectively [11]-[12]. The modulation waveforms of the different PWM methods are as shown in Fig. 2.

Thus, by varying $\mu$ and $\delta$, the switching time periods for zero voltage vectors can be changed, so that different DPWM methods can be obtained. After obtaining the switching times for active states and zero states, the reference voltage vector is calculated in an average sense within a subcycle.

512

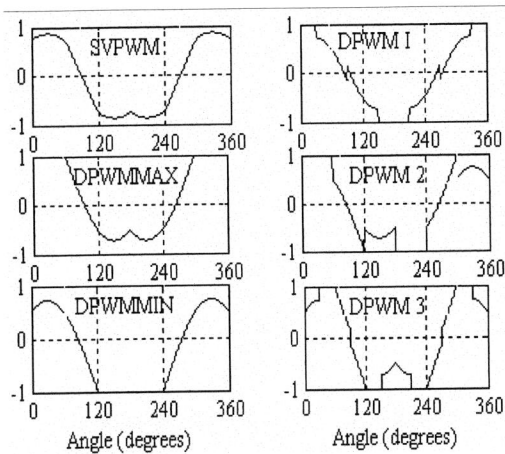

Fig. 2. Modulation waveforms of the different PWM methods.

## V. SLIDING MODE SPEED CONTROL ALGORITHM

To improve the speed performance under uncertainties, an integral switching surface sliding mode speed controller is proposed. The electromechanical equation of an induction motor is described as

$$J\frac{d\omega_m}{dt} + B\omega_m + T_L = T_e \qquad (3)$$

The electromechanical equation can be modified as

$$\dot{\omega}_m + a\omega_m + d = bT_e \qquad (4)$$

Where $a = \dfrac{B}{J}$, $b = \dfrac{1}{J}$ and $d = \dfrac{T_L}{J}$.

By considering (4) with uncertainties as

$$\dot{\omega}_m = -(a + \Delta a)\omega_m - (d + \Delta d) + (b + \Delta b)T_e \qquad (5)$$

$\Delta a$, $\Delta b$ and $\Delta d$ represents the uncertainties of the terms $a$, $b$ and $d$ introduced by system parameters $J$ and $B$. Now let us define the tracking speed error further as

$$e(t) = \omega_m(t) - \omega_m^*(t) \qquad (6)$$

where $\omega_m^*$ is the rotor reference speed command in angular frequency.

Taking derivative of (6) with respect to time yields

$$\dot{e}(t) = \dot{\omega}_m(t) - \dot{\omega}_m^*(t) = -ae(t) + f(t) + x(t) \qquad (7)$$

where the following terms have been collected in the signal $f(t)$,

$$f(t) = bT_e(t) - a\omega_m^* - d(t) - \dot{\omega}_m^*(t) \qquad (8)$$

and the $x(t)$, lumped uncertainty, defined as

$$x(t) = -\Delta a\omega_m(t) - \Delta d(t) + \Delta bT_e(t) \qquad (9)$$

Now, the sliding variable with integral component, is defined as

$$S(t) = e(t) - \int_0^t (h - a)e(\tau)d\tau \qquad (10)$$

where $h$ is a constant gain. Also in order to obtain the speed trajectory tracking, the following assumptions are made.

**Assumption-1:** The $h$ must be chosen so that the term $(h-a)$ is strictly negative and hence $h<0$.

Then the sliding surface is defined as follows:

$$S(t) = e(t) - \int_0^t (h - a)e(\tau)d\tau = 0 \qquad (11)$$

Based on the developed switching surface, a switching control that guarantees the existence of sliding mode, a speed controller is defined as

$$f(t) = h\, e(t) - \beta sgn(S(t)) \qquad (12)$$

Where $\beta$ is the switching gain, $S(t)$ is the sliding variable defined by (10) and $sgn(\cdot)$ is the sign function defined as

$$sgn(S(t)) = \begin{cases} +1, & \text{if } S(t) > 0 \\ -1, & \text{if } S(t) < 0 \end{cases} \qquad (13)$$

**Assumtion-2:** The gain $\beta$ must be chosen so that $\beta \geq |x(t)|$ for all time.

When the sliding mode occurs on the sliding surface (11), then, $S(t) = \dot{S}(t) = 0$ and the tracking error $e(t)$ converges to zero exponentially. Finally, the reference torque command $T_e^*$ can be obtained by substituting (12) in (8) as follows.

$$T_e^*(t) = \frac{1}{b}\left[ (h.e) - \beta\, sgn(S) + a\omega_m^* + \dot{\omega}_m^* + d \right] \qquad (14)$$

## VI. PROPOSED DTC

The block diagram of the proposed method is as shown in Fig. 3. In the proposed method, speed of the reference stator flux vector $\overline{\psi}_s^*$ is derived by the addition of slip speed and actual rotor speed. The actual synchronous speed of the stator flux vector $\overline{\psi}_s$ is calculated from the adaptive motor model.

After each sampling interval, actual stator flux vector $\overline{\psi}_s$ is corrected by the error and it tries to attain the reference flux space vector $\overline{\psi}_s^*$. Thus the flux error is minimized in each sampling interval. In this paper, the d-axis and q-axis components of the reference voltage vector are constructed by corresponding d-axis and q-axes stator flux error components respectively. The procedure is as follows:

Reference value of the d-axis and q- axis stator fluxes and actual value of the d-axis and q-axis stator fluxes are compared in the reference voltage vector calculator block and hence the error in the d and q-axes stator flux vectors is obtained as in (15).

$$\Delta\psi_{ds} = \psi_{ds}^* - \psi_{ds} \qquad (15.1)$$

$$\Delta\psi_{qs} = \psi_{qs}^* - \psi_{qs} \qquad (15.2)$$

The knowledge of flux error and stator ohmic drop allows the determination of appropriate reference voltage space vectors along d-axis and q-axis, which are given in (16)

$$V_{ds}^* = R_s i_{ds} + \frac{\Delta\psi_{ds}}{T_s} \qquad (16.1)$$

$$V_{qs}^* = R_s i_{qs} + \frac{\Delta\psi_{qs}}{T_s} \qquad (16.2)$$

These d-q components of the reference voltage vector are then fed to the GDPWM block, from where the gating pulses for the inverter are generated. Finally, to improve the speed performance against uncertainties caused by load variations, an integral switching surface sliding mode speed controller is used and this is shown by SMC block in the block diagram.

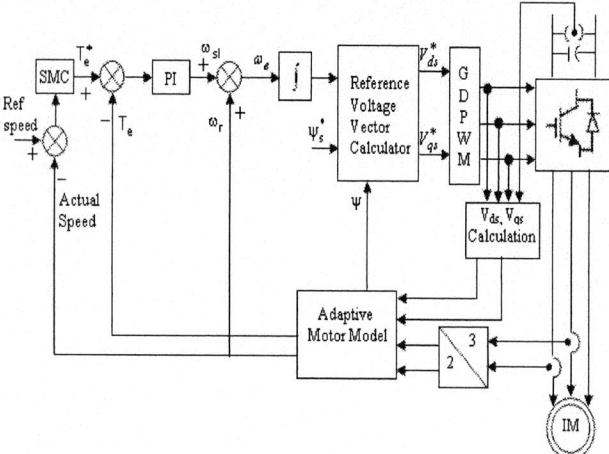

Fig. 3. Block diagram of proposed method.

## VII. SIMULATION RESULTS

To verify the proposed scheme, a numerical simulation has been carried out by using Matlab/Simulink. Sampling time of 125μs and ode 4 (Runge-Kutta) methods are used for a fixed step size of 10μs. For the simulation, the reference flux is taken as 1wb and starting torque is limited to 15 N-m. The induction motor used in this case study is a 1.5 KW, 1440 rpm, 4-pole, 3-phase induction motor having the following parameters:

$R_s = 7.83\Omega$, $R_r = 7.55\Omega$, $L_s = 0.4751H$, $L_r = 0.4751H$, $L_m = 0.4535$ H, $J = 0.06$ Kg.m$^2$ and $B = 0.01$ N-m.sec/rad

The values, which have chosen for sliding mode speed controller, are $h = -200$ and $\beta = 10$.

Various conditions such as starting, steady state, step change in load and speed reversal are simulated. The results for CDTC are shown in Fig. 4 - Fig. 6. From the simulation results of CDTC, it can be observed that, the ripple in torque, flux and current is more in steady state. To reduce these ripples and to get constant switching frequency, space vector based GDPWM algorithm is proposed in this paper. Among the various DPWM methods, DPWM 2 is known as minimum switching loss PWM method. It is suitable for induction motor drives operating near 30° lagging power factor angle at which the switching losses are minimum [11]. The simulation results for DPWM 2 based DTC are given in Fig. 7 –Fig. 11. From Fig. 8, it can be observed that, the ripple in torque, flux and current can be reduced with the proposed method compared to that of CDTC and hence acoustical noise can be reduced. To improve the speed performance of DTC under uncertainties caused by load changes, an integral switching surface sliding mode speed controller is proposed. Fig. 9 shows the transients during the load torque disturbances with proposed sliding

mode speed controller. From, which it can be observed that the speed tracking is not affected by the load torque disturbances. Fig. 10 shows the transients during speed reversal operation.

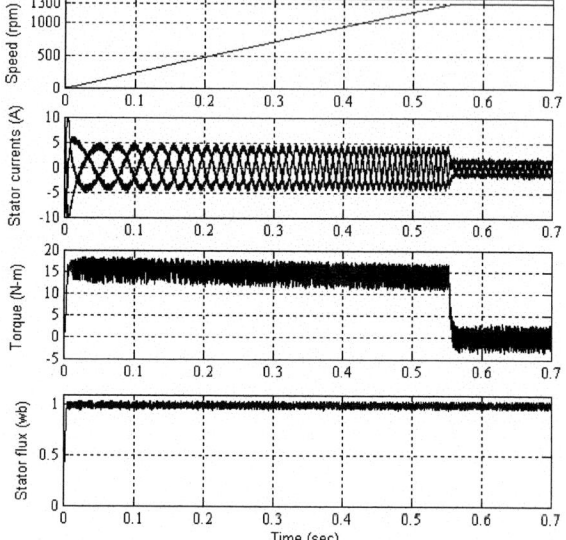

Fig. 4. Simulation results of CDTC: No-load Starting transients.

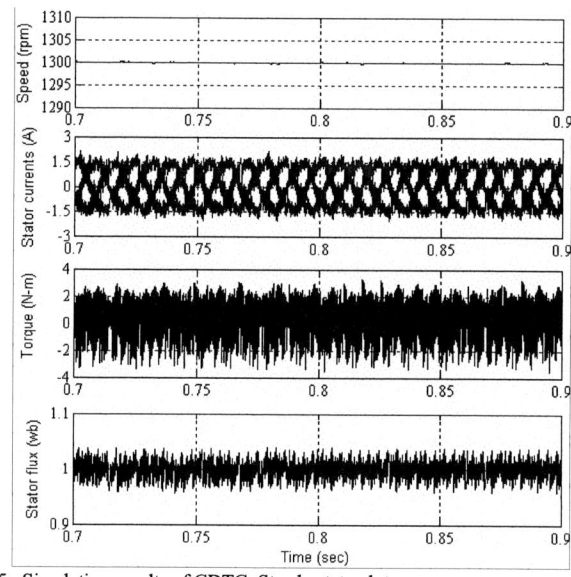

Fig. 5. Simulation results of CDTC: Steady state plots.

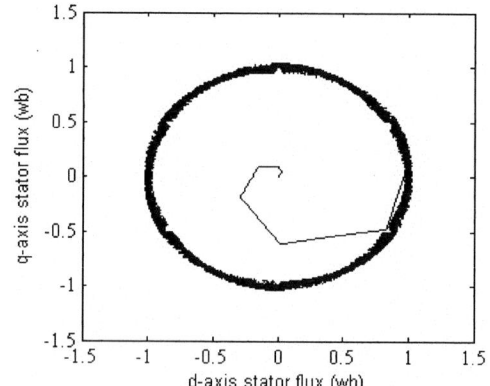

Fig. 6. Stator flux locus in CDTC.

514

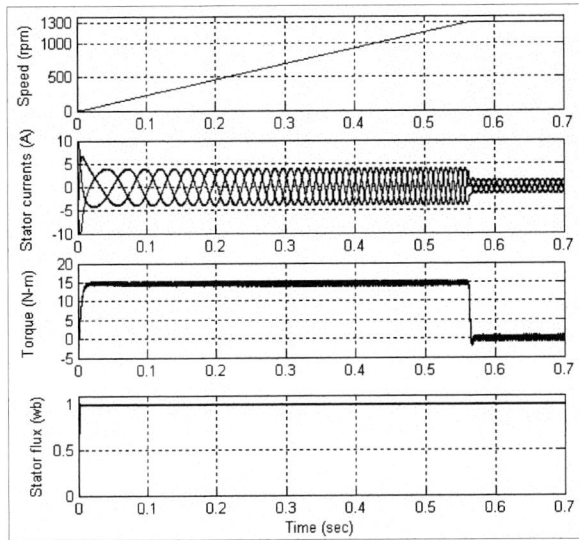

Fig. 7. Simulation results of DPWM 2 based DTC: Starting transients.

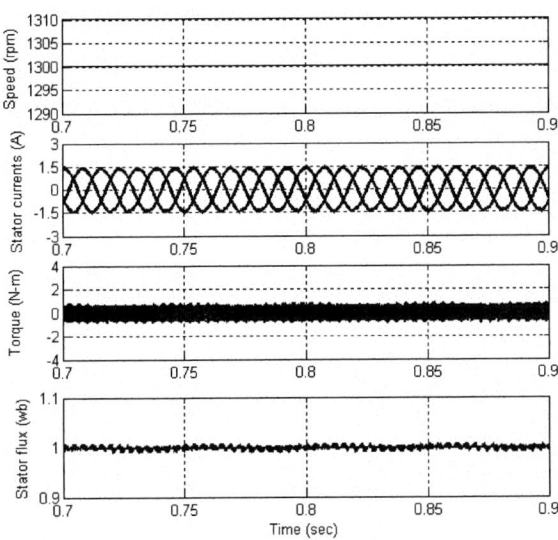

Fig. 8. Simulation results of DPWM 2 based DTC: Steady state plots.

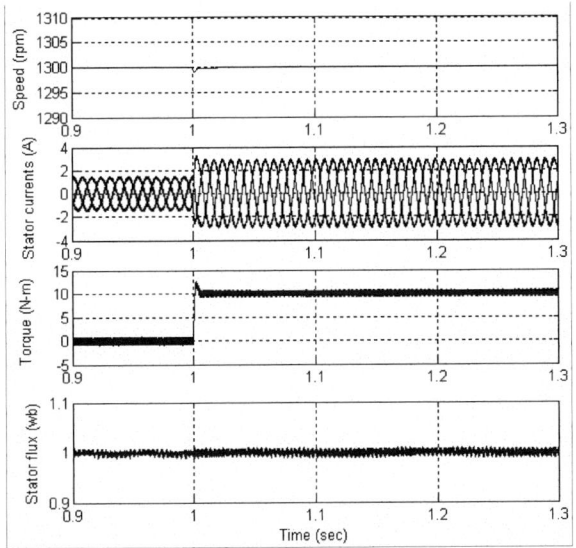

Fig. 9. Simulation results of DPWM 2 based DTC with SMC: A 10 N-m load is applied at 1sec.

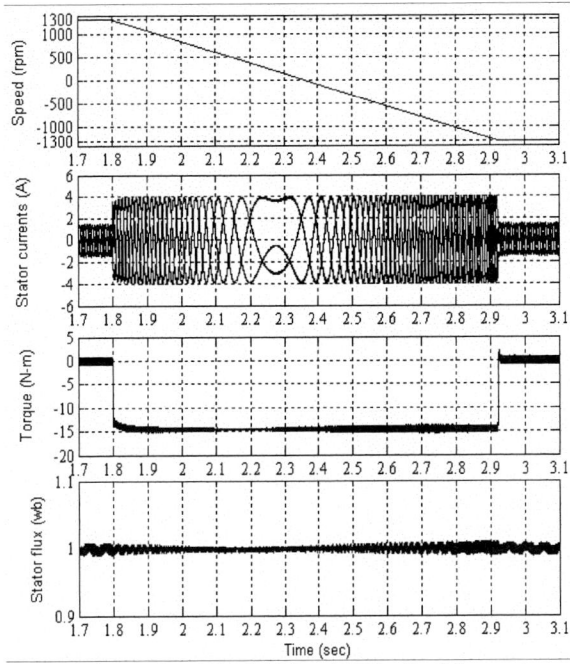

Fig. 10. Simulation results of DPWM 2 based DTC with sliding mode speed controller: speed is changed from +1300 rpm to -1300 rpm.

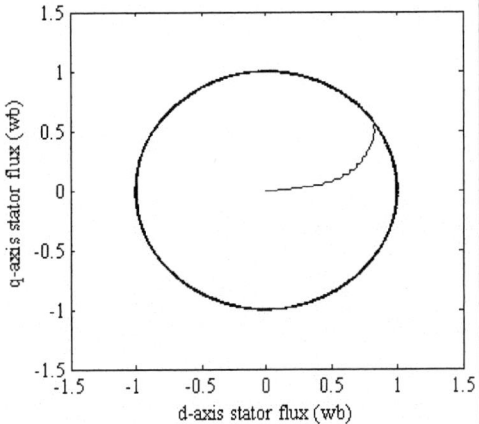

Fig. 11. locus of stator flux in DPWM 2 based DTC.

## VIII. CONCLUSIONS

The basic DTC is simple and gives quick torque response. But, it gives a considerable ripple in torque, flux and current in the steady state. Hence, the acoustical noise is more. In this paper, to reduce the ripple in steady state, a space vector based GDPWM algorithm has been proposed. In this algorithm, by varying a parameter, an infinite number of DPWM methods can be generated. From the simulation results, it can be observed that, with the proposed method, the ripple in torque, flux and current in the steady state can be reduced with compared to that of CDTC. Also DPWM methods give less switching losses. Finally, to improve the speed performance under uncertainties against load disturbances, an integral switching surface sliding mode speed controller is proposed. The simulation results show that, the proposed speed controller is robust against the variations in load torque. From the simulation results, it can be observed that, the proposed method gives good performance over the wide range of speeds. Also, the performance of the system is improved under load disturbances and in terms of ripple.

## IX. REFERENCES

[1] F. Blaschke, "The principle of field orientation as applied to the new TRANSVECTOR closed loop control system for rotating field machines" Siemens Review XXXIX, Vol No.5, pp. 217-220. 1972.

[2] Isao Takahashi, Toshihiko Noguchi, "A new quick-response and high-efficiency control strategy of an induction motor", IEEE Trans Ind Appl, Vol.IA-22, No.5, pp. 820-827, Sep/Oct, 1986.

[3] M. Depenbrock, "Direct-self control (DSC) of inverter-fed induction machine," *IEEE Trans. Power Electron.*, vol. 3, no. 4, pp. 420-429, Oct.1988.

[4] P.Tiitinen, M.Surandra, "the next generation motor control method, DTC Direct torque control", Proc on Power Electronics, Drives and Energy Systems for industrial growth, Vol. 1, pp. 37-43, 1996.

[5] Thomas G. Habetler, Francesco Profumo Michele Pastorelli and Leon M. Tolbert, "Direct Torque Control of Induction Machines Using Space Vector Modulation", *IEEE Trans. Ind. Appl.*, Vol. 28, No.5, pp. 1045-1053, September/October 1992.

[6] L.Tang, L.Zhong, M.F.Rahman, Y.Hu, "An investigation of a modified direct torque control strategy for flux and torque ripple reduction for induction machine drive system with fixed switching frequency" IEEE-IAS, pp. 837-844, 2002.

[7] T. Brahmananda Reddy, B. Kalyan Reddy, J. Amarnath, D. Subbarayudu and Md. Haseeb Khan, "Sensorless Direct Torque Control of Induction Motor based on Hybrid Space Vector Pulsewidth Modulation to Reduce Ripples and Switching Losses – A Variable Structure Controller Approach" *IEEE Power India Conference*, paper no. 208, April, 2006.

[8] A.M. Trzynadlowski and S. Legowski, "Minimum-loss vector PWM strategy for three-phase inverters", *IEEE Trans. Power Electron.*, vol.6, pp. 26-34, Jan, 1994.

[9] Ahmet M. Hava, Russel J. Kerkman, and Thomas A. Lipo, "Simple Analytical and Graphical Methods for Carrier-Based PWM-VSI Drives", IEEE *Trans. Power Electron.*, vol.4, no.1, pp. 49-61, Jan, 1999.

[10] K. Zhou and D. Wang, "Relationship between Space-Vector Modulation and Three-Phase Carrier-Based PWM: A Comprehensive Analysis", *IEEE Trans. Ind Electron.* Vol.49, no.1, pp. 186-196, Feb, 2002.

[11] Ahmet M. Hava, et al "A High-Performance Generalized Discontinuous PWM Algorithm" IEEE Trans Ind Appl, Vol. 34, No. 5, pp.1059-1071 September/October 1998.

[12] Olorunfemi Ojo, "The generalized discontinuous PWM scheme for three-phase voltage source inverters" IEEE Trans Ind Electron, Vol. 51, No. 6, pp.1280-1289, December, 2004.

[13] Vadim I. Utkin, "Sliding Mode Control Design Principles and Applications to Electric Drives" *IEEE Trans. on Ind. Electronics*, vol. 40, no. 1, Feb, 1993.

[14] Oscar Barambones, "A Robust Field Oriented Control of Induction Motor with Flux Observer and Speed Adaptation" Proc. IEEE-ETFA, pp. 245-252, 2003.ison methods" *in IEEE Proc. .ETFA'2003.*, pp. 245-252, 2003.

## X. BIOGRAPHIES

**T. Brahmananda Reddy** graduated from Sri Krishna Devaraya University, Anantapur in the year 2001, M.E from Osmania University, Hyderabad in the year 2003. He is currently pursuing Ph.D at Jawaharlal Nehru Technological University, Hyderabad, India. He is presently working as Assistant Professor in the Electrical and Electronics Engineering Department at G. Pulla Reddy Engineering College, Kurnool, Andhra Pradesh, India.

His research areas include PWM techniques, and control of electrical drives.

**Dr. J. Amarnath** graduated from Osmania University in the year 1982, M.E from Andhra University in the year 1984 and Ph.D from J.N.T.University, Hyderabad in the year 2001.

He is presently Professor and Head of the Electrical Engineering Department, JNTU College of Engineering, Hyderabad. He presented more than 60 research papers in various national and international conferences and journals. His research areas include Gas Insulated Substations, High Voltage Engineering, Power Systems and Electrical Drives.

**Dr. D. Subba Rayudu** received B.E degree in Electrical Engineering from S.V. University, Tirupati, India in 1960, M.Sc (Engg) degree from Madras University in 1962 and Ph.D degree from Indian Institute of Technology, Madras, India in 1977.

He is at present working as professor in Department of Electrical Engineering at G. Pulla Reddy Engineering College, Kurnool, India. His research interests include Electrical machines, Power Electronics and electrical drives.

**Md. Haseeb Khan** was born in Hyderabad, India in 1976. He received the B.Tech. degree in Electrical & Electronics Engineering from Kakatiya University, Warangal, India in 1999 and M.Tech. degree in Electrical Engineering from JNT University, Hyderabad in 2003, where he is currently pursuing the Ph.D. degree.

In 1999 he joined Noor college of engineering & technology as Assistant Professor. In 2003 he joined Royal Institute of Technology & Science, Chvella, India as Sr. Assistant Professor and since 2004 he is Associate Professor in the same Institute. His research interest includes power electronics, control of electric drives.

**2006 IEEE International Conference on Power Electronic, Drives and Energy Systems**

# Hardware-in-Loop Simulation of Direct Torque Controlled Induction Motor

P. K. Gujarathi, and M. V. Aware

*Abstract--* **In this paper, a method for design and implementation of direct torque control (DTC) strategy of induction motor based on hardware-in- loop simulation (HIL) is proposed. The hardware in loop simulation offers the rapid prototyping by reducing time in development and cost. The DTC controller for induction motor is designed and simulation is carried out to test the performance in Simulink/MATLAB.**

**A simple code generation tool interfaced to the MATLAB/SIMULINK by way of real time workshop (RTW) is used to convert the control model for a suitable hardware target. The space vector PWM based DTC control is realized by converting the control blocks in to C-code and implemented in dSPACE environment. The DTC controlled induction motor performance is tested using control desk. The results of offline simulation & real time simulation with laboratory prototype are presented.**

*Index Terms--* **Digital Signal Processor (DSP), Direct Torque Control (DTC), Hardware in loop (HIL), and Induction Motor.**

## I. INTRODUCTION

THE motor drive testing is expensive, power and time consuming. It is also having constraints to adjust designated parameters while keeping others constant for evaluating objective due to non-linearity of the motor drive under testing. On the other hand, pure software simulation can not explore the real situation of the motor drive due to its idealization and simplification. Over the past few years, a hybrid simulation approach combining software with hardware in loop has turned out to be an effective tool for real time simulation and evaluation of motor drive [1]-[4]. There are many ready to use software and hardware platforms available for this use.

In the present work, model of induction motor with Direct Torque Control (DTC) is simulated and performance is evaluated in Simulink/MATLAB. This simulation control blocks are used for integrating Hardware-In–Loop (HIL) environment for testing the induction motor. This is realized by using Real Time Workshop (RTW) to convert the Simulink blocks in to 'C' code and executed on DSP (TMS320C31) with dSPACE. Real time simulation is feasible

by interfacing actual motor and Voltage Source Inverter (VSI) to a DSP with feedback signals of motor terminal currents and voltages. While simulation is running, many variables can be observed in dSPACE control desk. This makes us to realize the actual working of the machine for a set control parameters. This offers unique advantage of rapid prototyping. This saves time and cost on account of developing the control schemes of the induction motor control. The scheme of HIL has following other advantages.

1. The system developed becomes flexible so that it can be used in multiple applications with minimal changes.
2. System debugging tools available to aid the end user as well as the system developers.
3. The induction motor model implemented in an easy to use environment that permits reuse and exchange of component models between architectures.
4. The computational platform can be expandable to permit future growth without adversely affecting the real time through put of the system.
5. Hardware and software interface is "user- friendly" so that the end user is comfortable using the system.

## II. DTC CONTROL INDUCTION MOTOR

In three phase induction machine, the instantaneous electromagnetic torque is proportional to the cross vectorial product of the stator flux linkage and stator current space vector.

$$T_e = \frac{3}{2} P \overline{\psi}_s \times \overline{i}_s \qquad (1)$$

$$\psi_s = |\overline{\psi}_s| e^{j\rho_s} \ , \quad i_s = |\overline{i}_s| e^{j\alpha_s}$$

$$T_e = \frac{3}{2} P |\overline{\psi}_s| \|\overline{i}_s| \sin(\alpha_s - \rho_s), \qquad (2)$$

$$T_e = \frac{3}{2} P |\overline{\psi}_s| \|\overline{i}_s| \sin(\alpha) \qquad (3)$$

Where,

$\Psi_S$ – Stator Flux Linkage

$\alpha_s$ - Sector Angle between $i_s$ and $\Psi_S$

$T_e$ –Electromagnetic Torque in Nm

$\rho_S$ – Angle between d-axis and stator flux linkages

$i_s$ – Fundamental Component of Source Current in Ampere

---

M. V. Aware is with the Department of Electrical Engineering, Visvesvaraya National Institute of Technology, Nagpur, INDIA. (email:mva_win@yahoo.com)

P. K. Gujarathi, is with the Department of Electrical Engineering, Visvesvaraya National Institute of Technology, Nagpur, INDIA. (email: pritamgujarathi@rediffmail.com).

**0-7803-9771-1/06/$25.00 ©2006 IEEE**      517

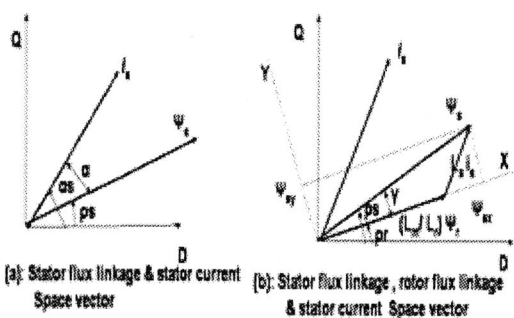

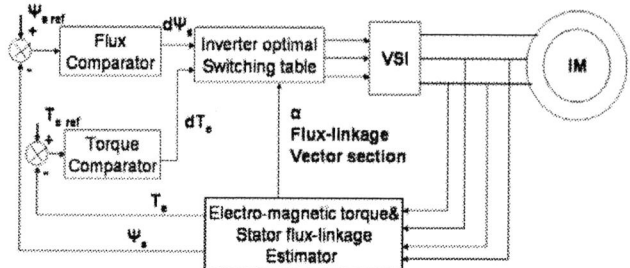

Fig.2. Direst Torque Control of Induction Motor.

Fig.1. Stator and rotor flux linkages relation in space vector.

The relation between the stator and rotor flux linkages is shown in the Fig.1.From above equations, the electromagnetic torque can be rapidly changed by changing angle between stator and rotor flux in the required direction. During short transient, the rotor flux is almost unchanged, thus rapid changes of electromagnetic torque can be produced by rotating the stator flux in the forward direction(phase advancing) or by rotating it in the negative direction (retarding), or by stopping it, according to demand of torque. To summarize that, torque can be quickly control by the stator flux-linkage space vector, which can be changed by using appropriate stator voltages. It can be seen that there is direct stator flux and electromagnetic torque control achieved by using the appropriate stator voltages [3].

The DTC system operation is based on space vector concept which is defined in a stationary reference frame fixed to the stator. The basic concept of direct torque control (DTC) of Induction-motor drives is to control directly the stator flux linkage and the electromagnetic torque by the selection of optimum inverter switching pattern using a switching vector look-up table. The selection is made to restrict the flux and torque errors within respective limits, to obtain fast torque response, low switching frequency, and low harmonic losses. The direct torque control allows very fast torque response and flexible control of an induction motor.

## III. UNIVERSAL CONTOLLER DEVELOPENENT

### A. Induction motor control

It is established fact that the variable speed drives are more efficient and widely adopted for the numerous applications. The dynamic response of the ac drives are improved at par with that of dc motor due to their new control methodology. The scalar and vector controls are used for the induction motor drives. These controls are now feasible by using high speed processors. The choice of the control scheme is tread off between the performance and cost.

These control methods are implemented through the power modulators. These consist of front end rectifier and inverters. These power semiconductor devices are configured according to the required power capacity and operating feasibility. These devices are interfaced to the controller by way of signal amplifiers and drivers with appropriate feedback control loop.

In DTC implementation, flux and torque estimators are used to obtain the feedback signals. The error in the torque and flux loop is generated by setting the hysteresis controllers. These signals are used for the voltage vector selection to obtain the proper inverter switching signal to inject the voltage space vector at the stator of the induction motor. This control scheme is shown in Fig. 2.

### B. Basic controller development environment

The control scheme for the given application is chosen by its suitability for a given application. It is essential to design the suitable control parameters involved in the scheme. The effectiveness of the control methodology is to be tested by way of simulation. The suitable software platform is required to make modeling and simulation to the accurate details. The finalization of the design parameters are obtained by fine tuning of the control parameters. The convenient control design tools may be MATLAB or C language.

In MATLAB, various control toolboxes are available to design control schemes. The SIMULINK expands the MATLAB with a block diagram interface for fast modeling and offline simulation. The use of C programming for the design of control algorithms are also way out for offline simulation.

The fast and smooth implementation of the control system on real time hardware is essential for rapid design iterations. This is achieved by way of generating the C source code. This task of C code generation is easily performed in SIMULINK by using Real Time Interface (RTI) tool. While implementing the C code, all the functions are necessary for real time implementation.

The real time simulation shall implant the C code to targeted hardware. The most useful processor, which has capability to execute and interface the environment need to be selected. These fast processors, such as digital signal processors (DSP) and a wide range of high-resolution I/O boards should handle the toughest control task. While control system is running on real time hardware, various easy to use PC software tools should help to monitor the process. The control parameters are tuned on-line without disturbing the closed-loop operation. The control scheme development stages are shown in the Fig. 3.

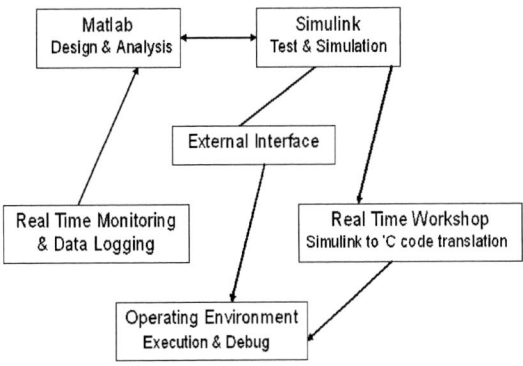

Fig.3. Control Development with dSPACE

## IV. PERFORMANCE OF DTC CONTROLLED INDUCTION MOTOR

The direct controlled induction motor performance is tested to investigate the dynamic performance under given constraints of the flux and torque estimator. This needs to be realized in real time so as to build the induction motor controller. There are two major steps involved in this process. The first one is to realize the performance in simulation to ascertain the perfection in the model of the induction motor and control scheme. Another step involves the separation of control part and builds the C-code for it along with the I/O interface. This involves the ADC and DAC signals from the DSP to be connected to actual hardware. This hardware involves the gating signal conditioning to the VSI and interfacing the current and voltage signals of the motor terminal. The dynamic performance of the DTC controlled is shown in the Fig. 4. The flux trace current waveforms are indicated in the Fig. 4 (a) and (d). The step change in the torque response is shown in Fig. 4 (c).

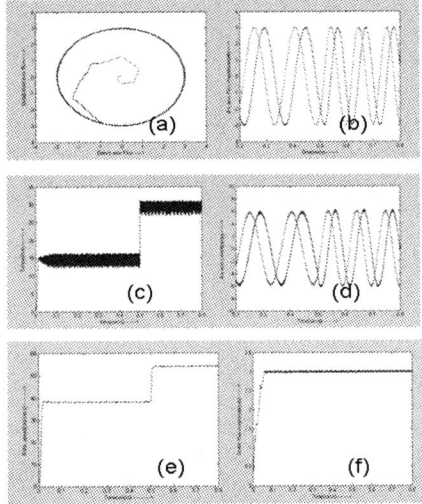

Fig.4. (a) Flux, (b) d-q axis flux, (c) Torque, (d) d-q axis Currents, (e) Speed and (f) air gap flux.

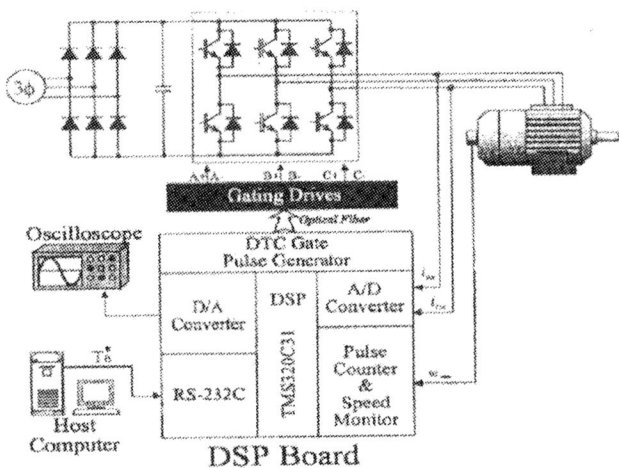

Fig.5.Control Set up of Hardware in Loop Simulation.

## V. HARDWARE-IN-LOOP IMPLEMENTATION

The 2 kW induction motor is connected through the VSI. This is built by using IPM module (FUJI-6MBP15JB060).The interfacing of DSP board to VSI is realized by isolation and gate driver circuit. The DSP is loaded with control algorithm by generating the code using RTW. This builds the complete control of DTC control induction motor with all the machine variables can be visualized in control desk of the dSPACE.

The DS1102 DSP Controller Board is ideally suited to perform all computation and I/O tasks. The TMS320C31 DSP executes the control algorithm. The inputs to the controller are the phase currents and the position sensor signal, the output from it is the phase voltage commands. With these commands, the digital-I/O subsystem generates PWM signals for the motor inverter. The voltage signal produced by the position sensor is captured and processed by the incremental encoder interface. The two 16-bit A/d Converters are responsible for the phase currents. The experimental set up is shown in Fig. 5.The processor board is located inside the PC and connected to hardware through the interfacing card with conditioned I/O signals. The board consists of various blocks as shown in Fig. 6.

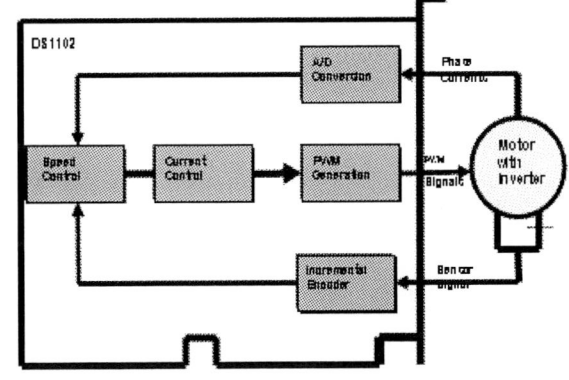

Fig.6. DSP Board DS-1102.

The real time execution is carried out by executing the build command from the SIMULINK menu. The solver used for this is EULAR and step size used for this is 0.0001.This generates the C code to down load it on the DSP. The variables and control signals are easily monitored by using the softwares from the dSPACE. The control and display of these variables are obtained in COCKPIT and TRACE software. The test results for the DTC controlled drive & the SPWM controller are presented by capturing these signals using TRACE are presented in Fig. 7 and 8. These results are in line with the simulated results.

The on line capture of torque and flux can be observed from Fig. 7. The set references are also shown in the Fig. 7.The components of the flux and flux trajectory is indicated in the Fig. 8. The motor terminal variables like motor current and speed is also captured in TRACE. These values are indicated in the Fig. 9. The terminal voltages of the motor after the d-q conversion are shown in the Fig. 10.These on line traces of the machine variables can also be observed by changing the set values. This is performed by changing the variables from the COCKPIT. The speed or the load reference can be set to any desirable value while simulation is running. This facility is used to check the dynamics of the system under design constraints.

The switching signals which are available from the DSP board are interfaced to IGBT based voltage source inverter. The output of the inverter is connected to the induction motor. The signals, currents and voltages from the terminals are captured by using LEM sensors. These are fed back to DSP board. This makes the induction motor to run under direct torque control. The generated signals are shown in the Fig. 11. The process of DSP download is also indicated in the background. A typical pattern of SPWM is shown in this Fig.11.

The experimental test results for the DTC controlled drive are presented by capturing these signals on CRO. The Fig. 12 indicates the motor terminal voltage and current of the induction motor. These results are similar to that of simulation results.

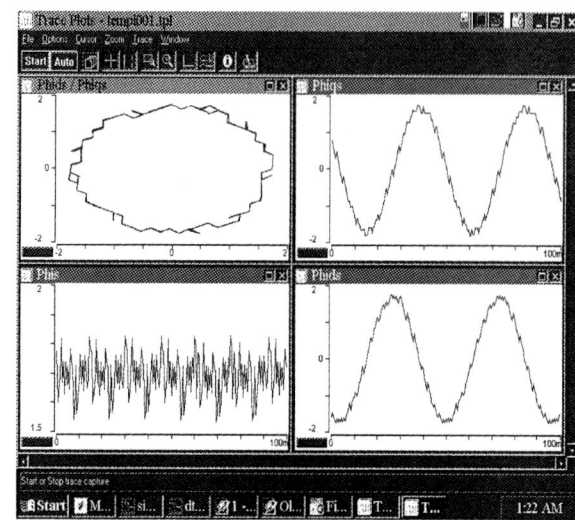

Fig.8.b) Flux trajectory, Flux develop, d-axis flux and q-axis flux.

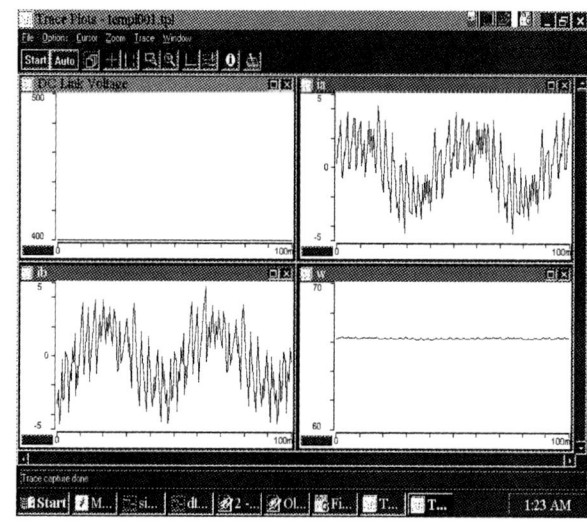

Fig.9.c) DC link voltages, motor current of phase a, motor current of phase b and Speed in rad/sec.

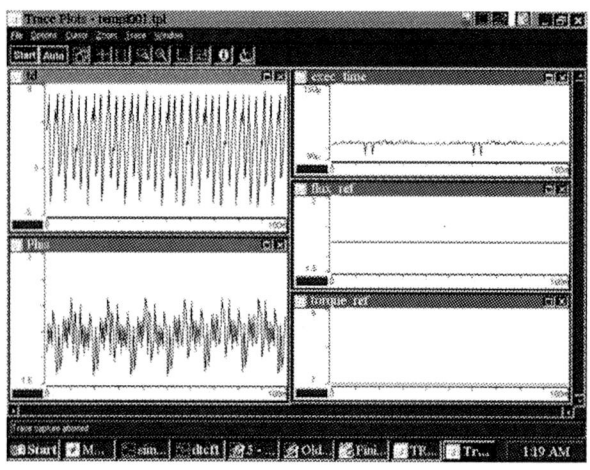

Fig.7. a) Torque developed, d-axis flux, execution time, flux reference and Torque reference.

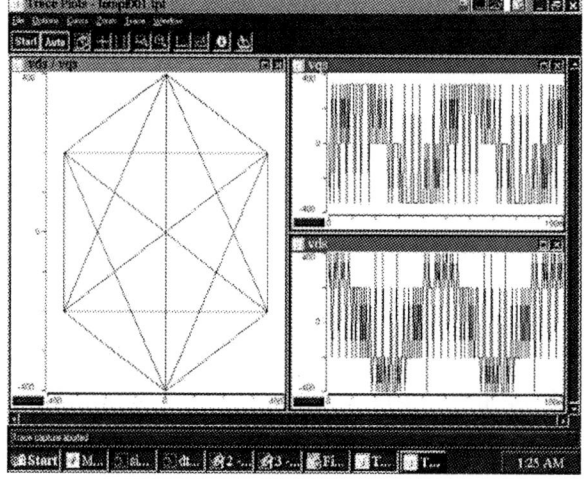

Fig.10.d) d-q axis voltage trajectory,q axis voltage and d axis voltage.

520

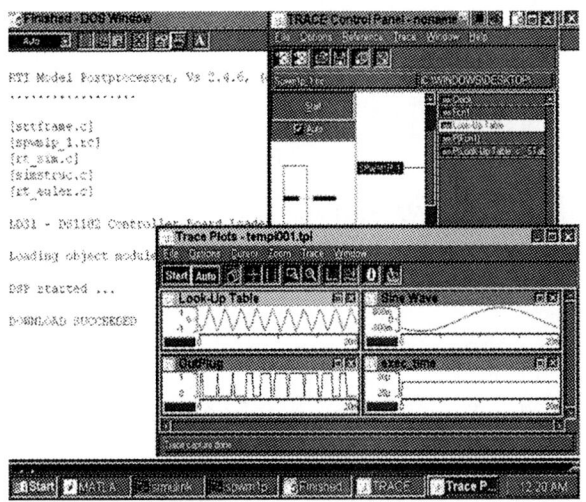

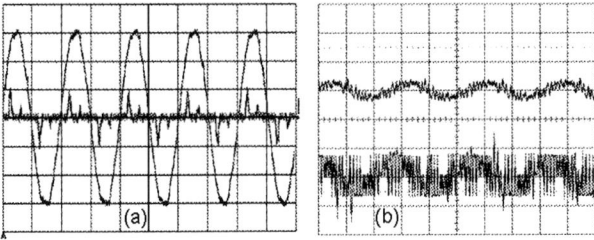

Fig.11.SPWM Generation

Fig.12. (a) Supply side Voltage and Current, (b) Motor Terminal Voltage and Curent.

## VI. CONCLUSION

In this paper, a reliable method for designing, developing and testing of direct torque controlled induction motor is presented. The new approach of Hardware-In-Loop (HIL) simulation of direct torque controlled induction motor using space vector PWM is carried out. A simple code generation tool interfaced to the MATLAB/SIMULINK by way of real time workshop (RTW) gives more flexibility to convert the control model suitable for DSP hardware. This tool is useful for faster testing and development of digital controllers for high performance electric drives. This is not only saves the time but also the cost. The required prototyping efforts are reduced as the control algorithm is working with the actual hardware. The risk of failure is also reduced as the all variables can be observed and tuned before hardware interface. Real time simulation also reduces the risk of discovering any error in the last stage of implementation. With the DSP-controller it is possible to improve the reliability and dynamic performance of the drive system. The generated code can be use on target hardware. The results obtained from real time simulation are experimentally validated on a 2-kW induction motor drive.

## VII. REFERENCES

[1] A. G. Jack, D.J. Atkinson, and H. J. Slate, "Real time emulation for power equipment development, part I: real time simulation" Proc. of Electric Power Applications, Vol. 145-2, pp.92-97, March 1998.

[2] H. J. Slate D.J. Atkinson, and A. G. Jack,, " Real time emulation for power equipment development, part II: Virtual machine" Proc. of Electric Power Applications, Vol. 145-3, pp.153-158,March 1998.

[3] R. Champagne, L. A. Dessaint, G. Sybille and B. Khodabakhchian, "An approach for real-time simulation of electric drives," in Proc, of IEEE Canadian Conference. 2000, pp.340-344.

[4] N. Sureshbabu, S. Seshagiri, A. Mwni and B. K. Powell, "On real time Simulation of induction motors." in Proc. of IEEE American Control Conference, 1999, pp.719-723.

[5] I. Takahashi, T. Noguchi, "A new quick response and high efficiency control strategy of an induction motor", IEEE Trans. Ind. Appl. Vol. IA-22, No.5, Sept/Oct 1986, pp.

[6] Control solutions, dSPACE, GmbH, Paderborn, Germany

[7] MATLAB, Users Guide, Mathworks, Inc. Natick, Mass, USA,1995

[8] SIMULINK, Users Guide,. Mathworks, Inc. Natick, Mass. USA,1995

[9] Real Time Workshop, Users Guide, Math works, Inc. Natick, Mass, 1995. USA

[10] B. K. Bose, "Power Electronics and ac drives", Printice Hall, 1986.

[11] P. Vas, "Vector control of AC machines", Oxford University Press, NY, 1990.

[12] Ned Mohan, "Power Electronics", John Wiley, Singapore, 1995.

**2006 IEEE International Conference on Power Electronic, Drives and Energy Systems**

# Near-Field Modeling and Prediction of Switched Mode Power Supply

Bai Feng, Niu Zhong-Xia, Shi Yu-Jie, and Zhou Dong-Fang

*Abstract* -- Two models are proposed to characterize and predict the near electromagnetic field of switched mode power supply (SMPS). One is abstracted from BUCK converter, which has only one high frequency current loop. The other is from BOOST converter, which has two high frequency current loops. Analytic representations of the electromagnetic near field for each model are presented. Finally, stabilized simulation results of the near field are given. The models and the simulation methods are of great values to the high-density SMPS's assembly.

*Index Terms* – Near-Field, Modeling, Prediction, SMPS.

## I. INTRODUCTION

When calculating radiation fields, the values of the fields over the radiation surface are used to compute the fields in the space surrounding the device. This space is typically split into two regions: the near field region and the far field region. The near field region is the region closest to the source, which is always much more complicated than far field. This is because in the far field region, when the dimension of the circuit is much smaller than the measure distance, the equations may be very simple in a rational way. While in the near field region many practical conditions could not be neglected.

The product is therefore in the near field of the antenna for certain of the lower frequency ranges of the regulatory limits and in the far field for the higher frequency ranges. The simple models used for predictions of radiated emissions assume that the measurement antenna is in the far field. However, the field structure of the emissions in the near field of an emitter is considerably more complicated than that in the far field. And there is no uniform algorithm to the near field calculation of SMPS.

In the low frequency range, the modeling of radiated emissions from the converter section of SMPS can be found in reference [1, 2]. But the model is fairly simple, and the solutions are transient in time domain.

---

Bai Feng is with the department of Communication Engineering, PLA Information Engineering University, Zhengzhou City, Henan 450002, China. (e-mail: thirdplanet@sohu.com)

Niu Zhong-Xia, Shi Yu-Jie and Zhou Dong-Fang are all with the department of Communication Engineering, PLA Information Engineering University, Zhengzhou City, Henan 450002, China.

This paper presents two models to characterize and predict the near field of SMPS, which are abstracted respectively from BUCK converter and BOOST converter. And gives the stabilized simulation results in frequency domain.

## II. SINGLE-LOOP MODEL OF SMPS

DC-DC converters are used to convert unregulated DC voltage to regulated or variable DC voltage at the output. They are widely used in SMPS and in DC motor drive applications. These converters are generally either hard-switched PWM types or soft-switched resonant-link types. There are several DC-DC converter topologies, the most common ones being buck converter, boost converter.

A BUCK converter is illustrated by referring to Fig.1(a), it is also called a step-down converter. The power switch $M$ acts as a high-frequency switch, it is repetitively closed for a time $t_{on}$ and opened for a time $t_{off}$. During $t_{on}$, the supply terminals are connected to the load, and power flows from supply to the load. During $t_{off}$, load current flows through the freewheeling diode $D$, and the load voltage is ideally zero. This converter is used to produce lower voltage at the load than the supply voltage.

As is illustrated in Fig. 1, the current flowing in the printed tracks of the converter section of an SMPS is considered to be the source of the device's radiated electromagnetic field. In most power electronics applications the maximal harmonic frequency doesn't exceed 100 MHz; therefore $\lambda min \approx 3m$, and the dimensions of the circuit are thus negligible compared to the wavelength. Then we can consider that the current is the same throughout the circuit.

Let $V$=28V, $L$=1mH, $C$=2$\mu$F, $R_{load}$=100$\Omega$. There are two high frequency harmonic currents loops, $i_1$ and $i_2$. Table I gives some representative harmonic currents of the two loops calculated by PSpice, from which we can see that $i_2$ is much less than $i_1$. Of course the electromagnetic energies radiated by $i_2$ are very faint, and can be ignored rationally. So we've got a single-loop model, as is illustrated in Fig. 1 (b).

The magnetic field and the electric field generated can be determined by:

$$\begin{cases} \vec{H} = \dfrac{1}{\mu_0} rot\vec{A} \\ \\ \vec{E} = -\dfrac{\partial \vec{A}}{\partial t} - gradV \end{cases} \tag{1}$$

0-7803-9771-1/06/$25.00 ©2006 IEEE

Where, $\vec{A}$ is magnetic vector potential, $V$ is electric scalar potential.

The two potentials are related by the Coulomb gauge:

$$div\vec{A} + \mu_0\varepsilon\frac{\partial V}{\partial t} = 0 \qquad (2)$$

Then, we can use $\vec{A}$ to calculate $\vec{E}$ and $\vec{H}$.

At the observation point $P$ (x, y, z), the magnetic vector potential is given by:

$$\vec{A} = \frac{\mu_0}{4\pi}\int_C \frac{i}{r}e^{-j\beta r}\,d\vec{l} \qquad (3)$$

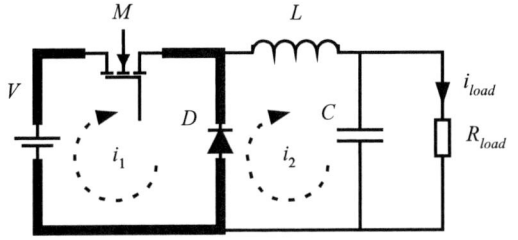

(a) Schematic of BUCK converter

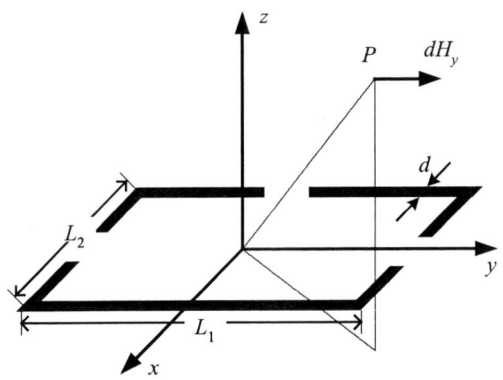

(b) Near-field radiated emission model of BUCK converter

Fig. 1. Model of BUCK converter

TABLE I
HARMONIC CURRENTS OF THE BUCK LOOPS

| $f$ (MHz) | $i_1$(mA) | $i_2$(mA) |
|---|---|---|
| 0.25 | 95.777 | 9.493 |
| 0.75 | 31.327 | 1.054 |
| 1.25 | 16.011 | 0.430 |
| 2.25 | 8.003 | 0.087 |
| 5.25 | 1.254 | 0.012 |
| 8.25 | 2.402 | 0.001 |
| 11.25 | 0.598 | 0.002 |
| 13.00 | 1.240 | 0.002 |

In most of the power electronics applications the maximal harmonic frequency doesn't exceed 100 MHz, then $\lambda_{min} \approx 3m$, so the dimensions of the circuit are negligible comparing to the wavelength. Then we can consider that the current $i$ is the same along the circuit, moreover we can perform the near-field approximation: $e^{-j\beta r} \approx 1$. Then (3) can be reduced to:

$$\vec{A} = \frac{\mu_0 i}{4\pi}\int_C \frac{d\vec{l}}{r} \qquad (4)$$

In the Fig. 1 (b), since the current is only in the direction of $X$ and $Y$, $A_z = 0$. $A_X$ and $A_Y$ can be calculated by:

$$A_1 = \frac{\mu_0 i}{4\pi}\int_{-L_1/2}^{L_1/2} \frac{dy_1}{\sqrt{(L_2/2-x)^2 + (y_1-y)^2 + z^2}} \qquad (5)$$

$$A_2 = \frac{\mu_0 i}{4\pi}\int_{-L_2/2}^{L_2/2} \frac{dx_1}{\sqrt{(x_1-x)^2 + (L_1/2-y)^2 + z^2}} \qquad (6)$$

$$A_3 = \frac{\mu_0 i}{4\pi}\int_{-L_1/2}^{L_1/2} \frac{dy_1}{\sqrt{(L_2/2+x)^2 + (y_1-y)^2 + z^2}} \qquad (7)$$

$$A_4 = \frac{\mu_0 i}{4\pi}\int_{-L_2/2}^{L_2/2} \frac{dx_1}{\sqrt{(x_1-x)^2 + (L_1/2+y)^2 + z^2}} \qquad (8)$$

And,

$$A_X = A_2 + A_4 = \frac{\mu_0 i}{4\pi}\ln\left[\frac{(a+F_1)(-b+F_4)}{(-b+F_2)(a+F_3)}\right] \qquad (9)$$

$$A_Y = A_1 + A_3 = \frac{\mu_0 i}{4\pi}\ln\left[\frac{(c+F_2)(-d+F_3)}{(-d+F_4)(c+F_1)}\right] \qquad (10)$$

Where,

$$a = \frac{L_1}{2}-x, b = \frac{L_1}{2}+x, c = \frac{L_2}{2}-y, d = \frac{L_2}{2}+y \qquad (11)$$

$L_1$ is the length of the loop area, while $L_2$ is the width. And,

$$\begin{cases} F_1 = \sqrt{a^2+c^2+z^2}, F_2 = \sqrt{b^2+c^2+z^2} \\ F_3 = \sqrt{a^2+d^2+z^2}, F_4 = \sqrt{b^2+d^2+z^2} \end{cases} \qquad (12)$$

Using (1) and (2), the magnetic and the electric fields are given by:

$$H_X = -\frac{iz}{4\pi}\left[\frac{1}{(c+F_2)F_2} + \frac{1}{(-d+F_3)F_3} - \frac{1}{(-d+F_4)F_4} - \frac{1}{(c+F_1)F_1}\right] \qquad (13)$$

$$H_Y = \frac{iz}{4\pi}\left[\frac{1}{(a+F_1)F_1} + \frac{1}{(-b+F_4)F_4} - \frac{1}{(-b+F_2)F_2} - \frac{1}{(a+F_3)F_3}\right] \qquad (14)$$

$$H_Z = \frac{iz}{4\pi}\left[\frac{b}{(c+F_2)F_2} - \frac{a}{(-d+F_3)F_3} - \frac{b}{(-d+F_4)F_4} + \frac{a}{(c+F_1)F_1} + \frac{c}{(a+F_1)F_1} - \frac{d}{(-d+F_4)F_4} - \frac{c}{(b+F_2)F_2} + \frac{d}{(-a+F_3)F_3}\right] \qquad (15)$$

$$E_X = -\frac{\mu_0}{4\pi}\frac{di}{dt}\ln\left[\frac{(a+F_1)(-b+F_4)}{(-b+F_2)(a+F_3)}\right] \qquad (16)$$

$$E_Y = -\frac{\mu_0}{4\pi}\frac{di}{dt}\ln\left[\frac{(c+F_2)(-d+F_3)}{(-d+F_4)(c+F_1)}\right] \qquad (17)$$

523

## III. DUAL-LOOP MODEL OF SMPS

A BOOST converter is illustrated by referring to Fig. 2 (a), it is also called a step-up converter. When the power switch $M$ is on, the inductor $L$ is connected to the DC source and the energy from the supply is stored in it. When the device is off, the inductor current is forced to flow through the diode $D$ and the load. The induced voltage across the inductor is negative. The inductor adds to the source voltage to force the inductor current into the load. This converter is used to produce higher voltage at the load than the supply voltage.

Here the two high frequency harmonic currents $i_1$ and $i_2$ are comparable in quantities. Let $V$=28V, $L$=0.1mH, $C$=2μF, $R_{load}$=100Ω. Table II shows the calculated results. So the electromagnetic energies radiated by the currents are both in count. A dual-loop model is illustrated in Fig. 2 (b). When calculating the near electromagnetic field, the dual-loop model can be seen as two single-loop models' superposition.

In the coordinate system of Fig. 2 (b), loop $i_1$ can be seen as a loop whose center is at the origin has translated $L_1/2$ in the reverse direction of axis-$y$. And equation (11) changes to:

$$a_1 = \frac{L_1}{2} - x, \ b_1 = \frac{L_1}{2} + x, \ c_1 = \frac{L_2}{2} - \left( y + \frac{L_1}{2} \right),$$

$$d_1 = \frac{L_2}{2} + \left( y + \frac{L_1}{2} \right)$$

And,

$$F_1' = \sqrt{a_1^2 + c_1^2 + z^2}, \ F_2' = \sqrt{b_1^2 + c_1^2 + z^2}$$

$$F_3' = \sqrt{a_1^2 + d_1^2 + z^2}, \ F_4' = \sqrt{b_1^2 + d_1^2 + z^2}.$$

According to equations (13) ~(17), $H_{X1}$, $H_{Y1}$, $H_{Z1}$, $E_{X1}$, $E_{Y1}$ are in the same form except for using $F_1' \sim F_4'$ to substitute $F_1 \sim F_4$.

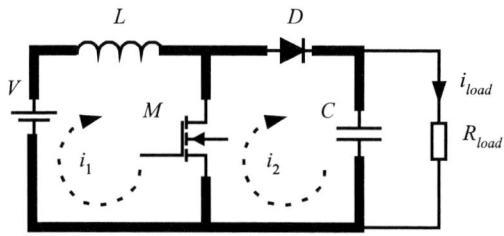

(a) Schematic of BOOST converter

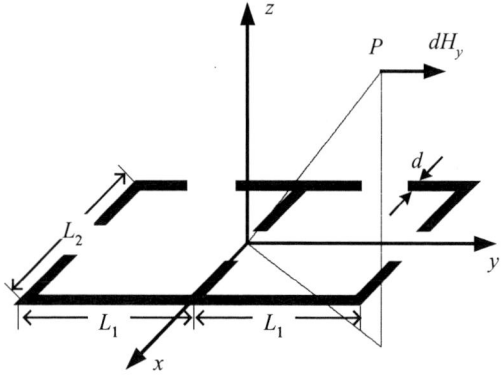

(b) Near-field radiated emission model of BOOST converter

Fig. 2.  Model of BOOST converter

### TABLE II
#### HARMONIC CURRENTS OF THE BOOST LOOPS

| $f$ (MHz) | $i_1$(mA) | $i_2$(mA) |
|---|---|---|
| 0.25 | 224.50 | 746.20 |
| 0.75 | 24.70 | 246.00 |
| 1.25 | 8.90 | 145.60 |
| 2.25 | 2.70 | 81.20 |
| 5.25 | 0.50 | 33.10 |
| 8.25 | 0.20 | 19.10 |
| 11.25 | 0.10 | 12.20 |
| 13.00 | 0.038 | 10.90 |

Homoplastically, in the coordinate system of Fig.2(b), loop $i_2$ can be seen as a loop whose center is at the origin has translated $L_1/2$ in the direction of axis-$y$. And equation (11) changes to:

$$a_2 = \frac{L_1}{2} - x, \ b_2 = \frac{L_1}{2} + x, \ c_2 = \frac{L_2}{2} - \left( y - \frac{L_1}{2} \right),$$

$$d_2 = \frac{L_2}{2} + \left( y - \frac{L_1}{2} \right). \ \text{And,}$$

$$F_1'' = \sqrt{a_2^2 + c_2^2 + z^2}, \ F_2'' = \sqrt{b_2^2 + c_2^2 + z^2}$$

$$F_3'' = \sqrt{a_2^2 + d_2^2 + z^2}, \ F_4'' = \sqrt{b_2^2 + d_2^2 + z^2}.$$

According to equations (13) ~ (17), we can derive $H_{X2}$, $H_{Y2}$, $H_{Z2}$, $E_{X2}$ and $E_{Y2}$ by analogy method using $F_1'' \sim F_4''$ to substitute $F_1 \sim F_4$.

Eventually, we've got the analytic representations of the near electromagnetic field as follows:

$$\begin{cases} H_X = H_{X1} + H_{X2} \\ H_Y = H_{Y1} + H_{Y2} \\ H_Z = H_{Z1} + H_{Z2} \\ E_X = E_{X1} + E_{X2} \\ E_Y = E_{Y1} + E_{Y2} \end{cases} \tag{18}$$

## IV. SIMULATION RESULTS

Although we have deduced the analytic representations of the near electromagnetic field, it's hard to calculate the exact radiated emissions of practical circuits. Here we use the Ansoft finite element analysis tools to simulate near field.

According to Fig. 1(b) and Fig. 2(b), let $L_1$=5cm, $L_2$=5cm, $d$=4mm. Fig. 3 and Fig. 4 give the stabilized solution's simulation results. Some representative points are selected from frequencies domain to simulate the near field 5 mm above the PCB plane, which is a very close plane to the circuit. The reason why we select such a close plane to study is that in solid assembly the different PCBs may be very closed to each other. The near field generated by one may influence the components on the other one. This is because a rather stronger electromagnetic field could affect the current carriers of the semiconductor devices and even lead to failure to the whole system.

The regions in the white line of dots are the main PCB tracks of the converter, which include one or two loop areas. The results show that in some particular areas, the electromagnetic fields are comparative strong. We should pay more attention on their influences.

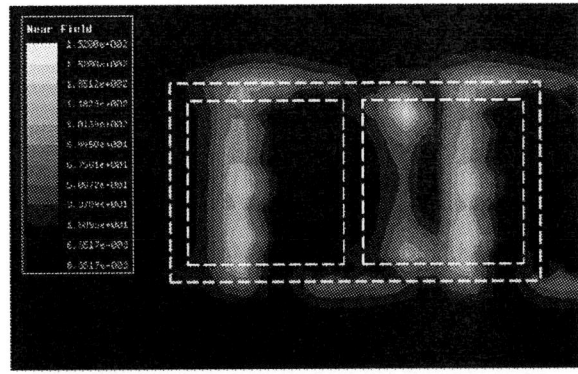

(a) Near electric field, $f$=1.25 MHz, $i_1$=8.9 mA, $i_2$=145.60 mA

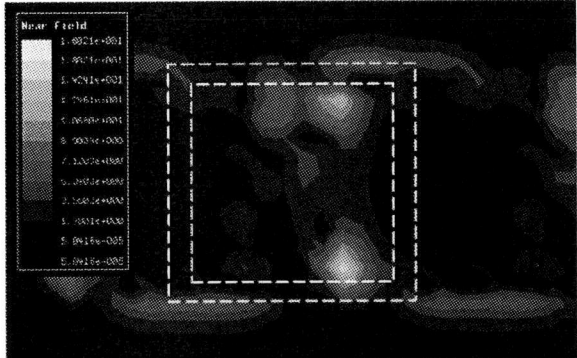

(a) Near electric field, $f$=1.25 MHz, $i_1$=16.011 mA

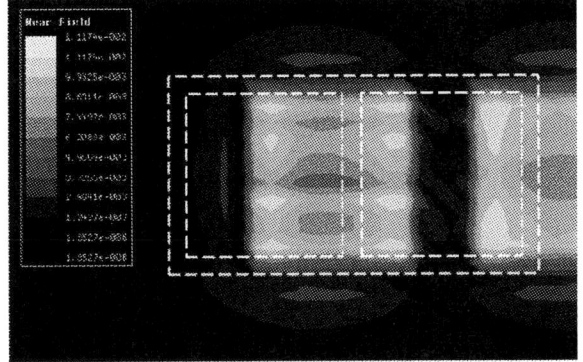

(b)Near magnetic field, $f$=1.25 MHz, $i_1$=8.9 mA, $i_2$=145.60 mA

Fig. 4. Simulation results of BOOST converter

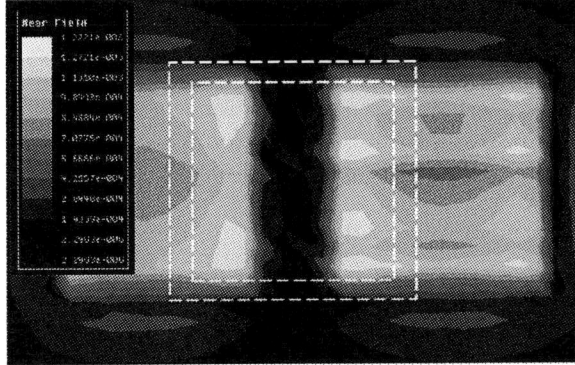

(b) Near magnetic field, $f$=1.25 MHz, $i_1$=16.011 mA

Fig. 3. Simulation results of BUCK converter

## V. REFERENCES

[1] Antonini, G.; Cristina, S.; Orlandi, A., "EMC characterization of SNIPS devices: circuit and radiated emissions model", IEEE Transactions on Electromagnetic Compatibility, 1996, Vo. 38, No.3, pp. 300-309

[2] Youssef, M;Roudet, J.; Marechal, Y., "Near-field characterisation of power electronics circuits for radiation prediction", Power Electronics Specialists Conference, PESC '97 Record., 28th Annual IEEE, 1997, Vol.2, pp. 1529-1534

**2006 IEEE International Conference on Power Electronic, Drives and Energy Systems**

# Power Electronic Circuit-oriented Model for the Fuel Cell System

Veerachary M, *Senior Member, IEEE*, and Arun Shailendra Kumar

*Abstract*— In this paper a fuel cell power electronics circuit-oriented simulator model development is discussed. This developed model can be employed in any power electronics simulators and thus useful for the power converter/ controller design. From the fundamental chemical properties, the relationship that describes the electrical behavior of cell is obtained first and then parameter affect is included. Circuit synthesis procedure is applied for this relationship and a model is developed using active sources, passive bilateral components and some function blocks that describes the non-linear variations. For validation various *v-i* characteristics are generated using PSIM power electronics simulator and then compared with the experimentally obtained measurements. The simulated and experimental characteristics are matching each other and thus validate the suitability and modeling approach. The model compatibility with power processing units is also verified and few simulation results are given for demonstration.

*Index Terms*—Circuit-oriented model, Polymer Electrolyte Membrane Fuel Cell.

## I. INTRODUCTION

FUEL Cells (FC's) convert chemical energy into electrical energy. There are several different types of FC's are available in the market [1]-[5] and the main difference essentially due to the type of chemical agent used, temperatures at which the electrochemical process takes place, amount of fuel and air requirements etc. Among all these FC's proton exchange membrane (PEM) are popular because of the advantages like (i) higher power density, (ii) smaller size for a given power rating, (iii) lower operating temperatures, (v) lower start-up time etc. These FC's are finding wide application in the power industry, mainly in the hybrid power systems.

Due to increasing demand for conventional energy sources like coal, natural gas and crude oil, the power industry is forced to develop new methods of generation schemes employing alternate energy sources [6]-[9]. Many such energy sources like wind energy and photovoltaic are now well developed, cost effective and are being widely used, while some others like fuel cells are in their advanced developmental stage. A hybrid power system employing all the above sources provides cost effective alternative solutions. In this hybrid system, the photovoltaic (PV) and wind (WD) are the prime power source, while the FC's acts as a secondary source. That is the FC's will come into operation only during peak demand period or no sun period.

However, design and control of such hybrid system, having multiple energy sources with different characteristics, is a complex task. Particularly, the power conditioning device, dc-dc converter or inverter, design and control plays a major role in the overall efficiency/ reliability of the system. The complexity in the design methodology can be eliminated partly by using the power electronic simulators. However, among the available power electronics simulators, about 95 % of them, are not having a ready-made power sources such PV, WD and FC's. Hence, there is a need to develop such sources electrical equivalent models. To fill this gap the author's have developed a user-friendly FC model in this paper, which can easily be realized in any power electronic simulator.

## II. CIRCUIT SIMULATOR MODEL FOR THE PEM FUEL CELL

Several types of fuel cells are available in the market and they are broadly classified into: (i) proton exchange membrane fuel cells (PEM), (ii) direct methanol fuel cells (DM), (iii) alkaline fuel cells, (iv) phosphoric acid fuel cells (PAFC), (v) molten carbonate fuel cell (MCFC), (vi) solid oxide fuel cell (SOFC). Although all above FC's capable of providing electrical power, but some of the above FC's takes a long time to begin producing electricity. However, PEM fuel cells capable of delivering power to load almost instantaneously, making them the ideal type of FC's for back-up power applications such as telecom infrastructure, data centers, and other mission critical load applications. However, to design power conditioner for backup applications, that uses PEMFC, preliminary simulation studies are essential to determine the feasibility of the available FC. As the PEMFC source is not the standard electrical power source, the designer has to model such system either in system level or circuit-level simulators. In any case, the designer should know the chemical behavior and how this chemical action is transformed into electrical energy production. Once this phenomenon is transparent then model development will not be problematic for the power supply designer. The fundamental relationship between FC

---

This work was supported in part by the IITD, under the New faculty Grant.

Veerachary. M is with Dept. of Electrical Engineering, Indian Institute of Technology Delhi, New Delhi, India (e-mail: mvchary@ee.iitdac.in).

Arun Shailendra Kumar is with the Department of Electrical Engineering, Indian Institute of Technology Delhi, New Delhi, India (e-mail: mvchary@ee.iitdac.in).

0-7803-9771-1/06/$25.00 ©2006 IEEE

terminal voltage and current was discussed in [7] and it is used here as basic equation for the power electronic equivalent circuit model (PECM) development. The single fuel cell voltage and current density capacity is small, and hence several fuel cells are connected in series parallel to meet the load voltage/current and power demand. Such combination is called FC stack. Its mathematical representation is based on single FC representation and it is simply multiplied by quantity equal to number cells in the stack.

$$V_{stack} = E_o^{'} - R^{'}i - b^{'} * \log(i) - m^{'} * e^{(n^{'}i)} \quad (1)$$

where $E_0^{'} = N \cdot E_0$; $E_0$-is the open circuit voltage of the single fuel cell, N- number of fuel cells. The above *v-i* relationship is non-linear and it includes the effect of (i) activation voltage drop: which mainly depends on equivalent activation resistance, (ii) ohmic voltage drop: resistance between electrodes and membrane essentially contributes to this value. Further, this also depends on cell temperature and current passing through it, (iii) concentration voltage drop: this voltage drop is due to water film covering the catalyst surfaces at the electrodes and its dependence is function of cell current and temperature, (iv) double layer charging effect: within the cell, in the process of chemical reaction two charged layers of opposite polarity are formed across the boundary between the cathode and membrane, called double layer and gives the property of super capacitor. Electrical equivalent circuit of the FC, including all the above affects, is proposed in [6]. This electrical equivalent circuit (EEC) is almost identical to the empirical equation defined by (1) but only the difference is that the physical properties of the FC are represented in terms of electrical circuit parameters. However, both the representations are accurate and useful for FC system studies, but the EEC form is more convenient for the power supply designer. The reason is very simple that performance prediction now becomes easier. In Fig. 1 $E^{'}$ is the open circuit voltage of the fuel cell, $R^{'}_h$ models the immediate ohmic voltage drops. $R^{'}_{cl}$ and $C^{'}_{cl}$ represent the double layer charge effect. From this EEC we can divide the circuit essentially into three parts, which are: (i) open circuit voltage block, (ii) FC thermal resistance variation block, (iii) double layer charge effect block. A simple schematic block diagram is shown in Fig. 2. Detailed discussion on the variation of thermal resistance and double layer charge effect was documented in [3]. The thermal dynamics cannot be neglected when a load change occurs over a long duration and its variation can be written as

$$R = \xi_1 - \xi_2 T + \xi_3 i \quad (2)$$

$$T = \theta_{ss}\left(1 - e^{\frac{-t}{\tau_t}}\right) + \theta_0 e^{\frac{-t}{\tau_t}} \quad (3)$$

Substituting above equation in Eqn. 2 we get a equation whose block diagram representation is given in Fig. 3.

The double layer charge effect gives the FC dynamic behavior and its transfer function can easily be determined by transforming the circuit corresponding to this portion and then applying laplace transform, which results into Eqn. 4.

$$\frac{v_{cl}(s)}{i_{cell}(s)} = \frac{R_{cl}}{(sR_{cl}C_{cl}+1)} \quad (4)$$

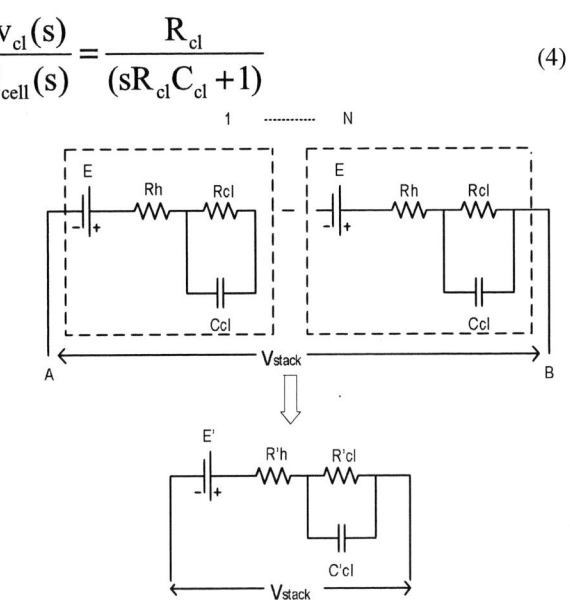

Fig. 1. FC stack EEC model.

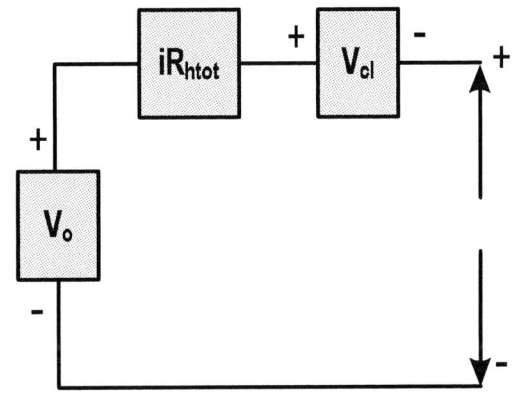

Fig. 2. FC stack diagram.

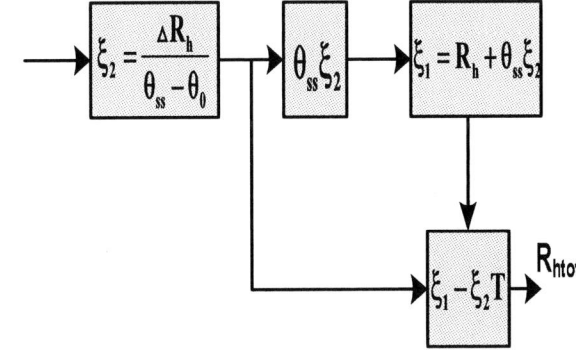

Fig. 3. Block diagram of FC thermal resistance variation.

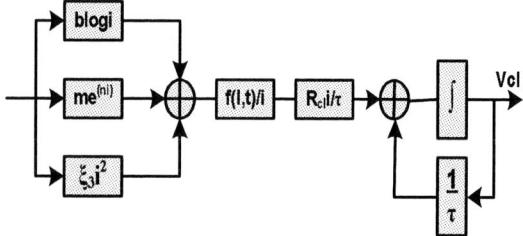

Fig. 4. Block diagram of FC double layer charge effect.

On simplification it gives the following equation.

$$sv_{cl}(s) = \frac{R_{cl}i_{cell}(s)}{\tau} - \frac{v_{cl}(s)}{\tau} \qquad (5)$$

Above equation block diagram representation is shown in Fig. 4. All the above blocks are non-linear blocks and its representation in the circuit-level simulator is not straightforward. Moreover, combining all these blocks together and modeling will loose the modularity and finally cause and effect study becomes complex task. To avoid some of these difficulties and to formulate simplified circuit-level FC model, individual model based development is employed in this paper.

The model development employs a modular approach, in which cause and effect parameter sensitivity incorporation becomes straightforward. The stages in the circuit-oriented model development are: (i) empirical fuel cell model formulation, (ii) synthesis of electrical equivalent circuit that emulates FC behavior, (iii) development of sub blocks to account for parameter sensitivity and then their integration. Adopting above-mentioned modeling stages a single fuel cell model is developed and its generalized circuit layout is shown in Fig. 5. In this figure the FC temperature, "Rhtot" block, was considered as constant value. However, in reality this temperature depends on the load and its inclusion is necessary to study the FC behavior. The "Rhtot" dependence is shown in Fig. 6, where-in "T_var" is the temperature variation block and it can be defined by taking linear voltage source or a look-up table. In most of the cases single cell is unable to meet the load demand, in terms of voltage and current capability, and few such cells are connected in series-parallel fashion to fulfill the load requirements. Since the FC model developed here is the most generalized one, series – parallel connection will not present any convergence problems while performing the simulations. The unique feature of the developed model is that, it uses a simple circuit-oriented blocks to realize the complex non-linearities. As a result the model becomes simple without loosing the accuracy of the FC v-i characteristic.

### III. SIMULATION RESULTS AND DISCUSSIONS

To verify the developed FC model a bench-mark 1.2 kW standard PEM cell parameters [3] are used in the simulations and they are given in Table I. Several simulations have been carried out and tested various operating conditions. In all the cases the FC circuit model is running without any problems.

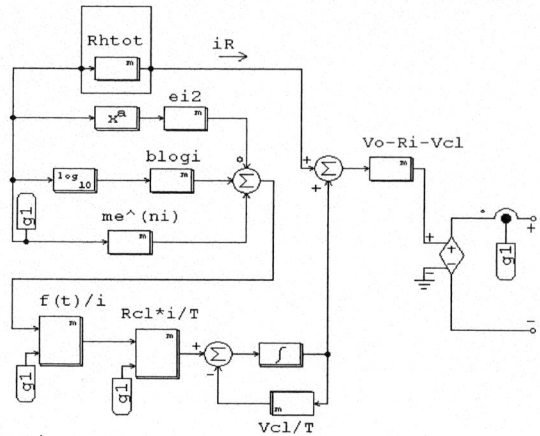

Fig. 5. EEC simulator model of FC (T: Constant).

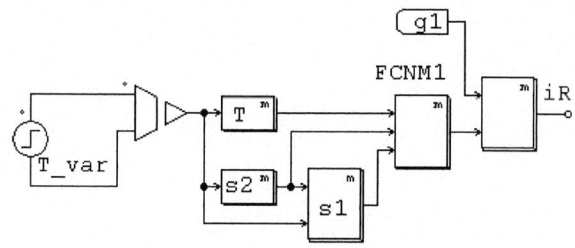

Fig. 6. EEC simulator model of the temperature effect.

Further, the simulated v-i characteristics are identical to the measured one. For demonstration the v-i characteristics of the developed FC model is plotted for rated conditions in Fig. 7 (blue trace). For comparison purpose the actual test data of the FC is also superimposed in the Fig. 7 itself (red trace). These observations show that the model is accurate and is representing the actual FC behavior. The discrepancy in the simulated v-i characteristics due to the use of constant thermal resistance is compared with the actual characteristic in Fig. 8. This result suggests that the FC model should be included with temperature variations.

To test the suitability of this developed model in the power conversion process, an example of FC supplied boost and buck converter regulated systems are considered here. The FC system is running in closed-loop, while the power electronic converter regulating the load voltage forms a constant power load. For illustration the converter load voltage is controlled by means of a proportional plus integral controller. Whenever any change in load on the converter then the closed-loop controller will come into action and changes the duty ratio in order to meet the load voltage reference requirement. Although the controller of the converter is fast enough to respond the load changes but the chemical process of the FC will not allow immediate reference tracking. Hence, during this transient time the FC takes some time to settle to new operating point. For demonstration, key sample load and FC source voltage characteristics are shown in Fig. 9 for boost, while Fig. 10 for buck converter.

TABLE I

| Parameters | Values |
|---|---|
| $E_0$ | 42 V |
| b | 58 mV/dec |
| m | 0.2 mV |
| n | 0.01 A$^{-1}$ |
| $\tau, \tau_t$ | 0.2451, 100 s |
| $\xi_3$ | 0.0015 m$\Omega$/A |

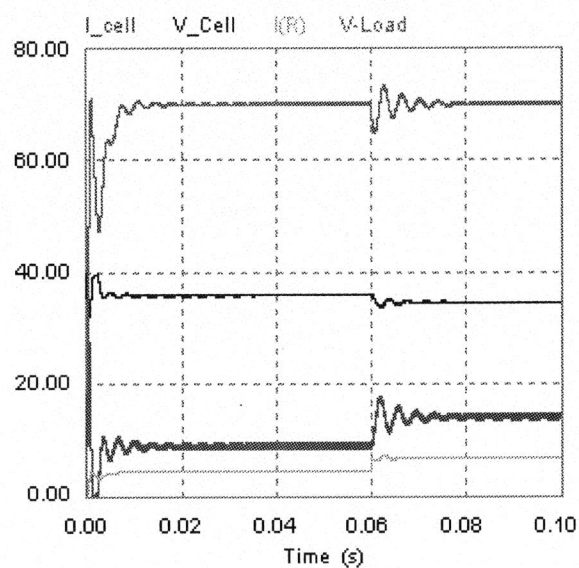

Fig. 9. Dynamic response of FC with boost regulator.

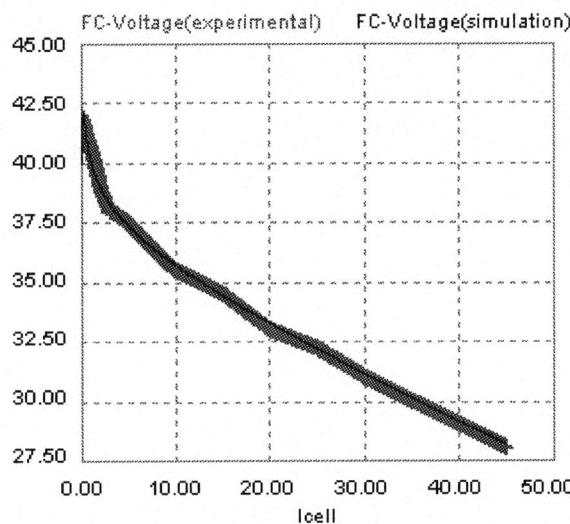

Fig. 7. Comparison of Fuel Cell v-i characteristics.

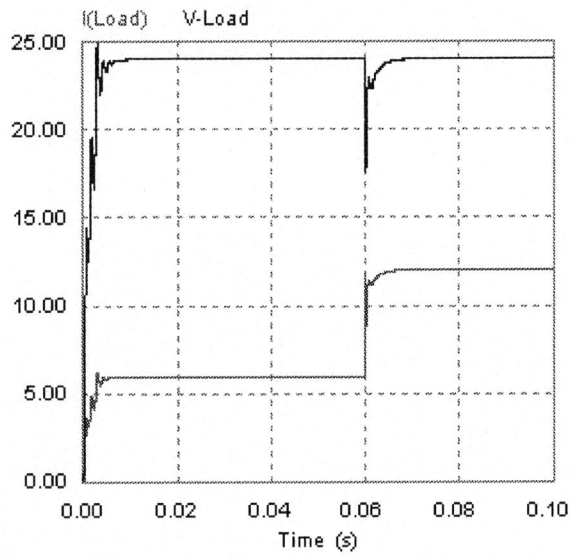

Fig. 10. Dynamic response of FC with buck regulator.

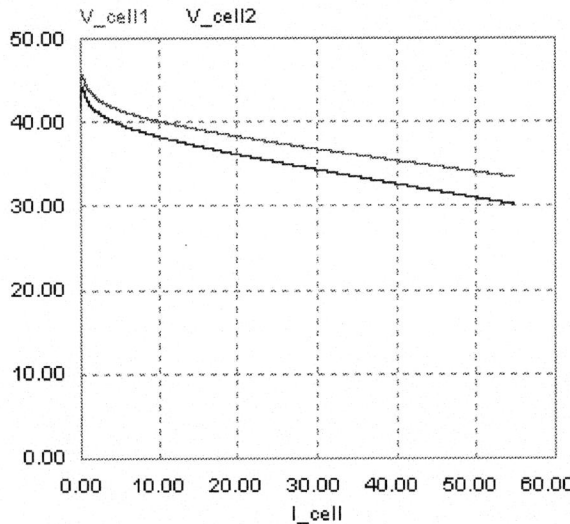

Fig. 8. Simulated FC v-i characteristics with and without T variation.

## IV. CONCLUSIONS

A simple circuit-oriented power electronic circuit-oriented model for the PEM fuel cell was developed. Combining the empirical model with electrical equivalent circuit model evolved this model. Since the model is modularity based one cause and effect implementation becomes easier. The suitability of the developed FC model was also verified for power conversion applications by considering boost and buck voltage regulators. In all these cases the developed FC model was working without any convergence problems.

## V. REFERENCES

[1] M Ceraolo, C Miulli, and A Pozio, "Modeling static and dynamic behavior of proton exchange membrane fuel cells on the basis of electrochemical description," *Journal of Power Sources*, 2003, Vol. 113, pp. 131–144.

[2] C Wang, M H Nehrir, "Dynamic models and model validation for Pem Fuel Cells using electrical circuit," IEEE Trans. on Energy Conversion, Vol. 20, 2005, pp. 442 – 451.

[3] X Kong, A M Khambadkone, S K Thum, "A hybrid model with combined steady-state and dynamic characteristics of PEMFC fuel cell stack." *Industry Applications Conference, Fourteenth IAS Annual Meeting. Conference Record*, Vol. 3, 2005, pp. 1618 – 1625.

[4] Jonathan J. Dogterom, Mark Kammerer, "Powering change with fuel cells in the telecommunication industry", *Proc. Of IEEE International Telecommunication Energy Conference*, 2005, pp. 401-405.

[5] Ellart de Wit, "Fuel cells for alternative critical backup power", *Proc. Of IEEE International Telecommunication Energy Conference*, 2005, pp. 311-313.

[6] Randall Gemmen, Parviz Famouri, "PEM fuel cell electric circuit model", *Power Electronics for Fuel Cells Workshop*, National Fuel Cell Research Center, University of California, 2002, pp. 8-9.

[7] J C Amphlett, R M Baumert, R F Mann, B A Peppley, P R Roberge, and J P Salvador, "A model predicting transient responses of proton exchange membrane fuel cells", *Journal of Power Sources*, 1995, Vol. 142, pp. 9–15.

[8] Kourosh Sedghisigarchi and Ali Feliachi, "Dynamic and transient analysis of power distribution systems with fuel cells-part i: Fuel cell dynamic model", *IEEE Transactions on energy conversion*, 2004, Vol. 19, pp. 423–428.

[9] P R Pathapati, X Xue, and J Tang, "a new dynamic model for predicting transient phenomena in a pem fuel cell system," *Renewable Energy*, 2005, Vol. 30, pp. 1–22.

[10] PSIM user manual, POWERSIM Technologies, 2004.

## VI. BIOGRAPHIES

**M.Veerachary** was born in India in 1968. Dr. Chary completed his Dr. Eng from University of the Ryukyus, Okinawa, Japan in 2002. Since then he is with Dept. of Electrical Engineering, Indian Institute of Technology Delhi, India and currently he is an Associate Professor. His research interests are High frequency dc-dc conversion, design and analysis of satellite/spacecraft power conditioning systems, modeling and simulation of large power electronic systems, control theory application to power electronic systems and intelligent controller applications to power supplies.

**Arun Shailendra Kumar was** born in Uttar Pradesh. He received the Bachelor degree from the Madhan Mohan Malviya College of Engineering, Gorakhpur, UP, India, in 2004, and currently he is working towards Master of Technology degree in the Dept. of Electrical Engineering, IIT Delhi. His fields of interest are design and analysis of Fuel cell systems and power electronic converters applications to hybrid energy storage systems.

**2006 IEEE International Conference on Power Electronic, Drives and Energy Systems**

# A Simplified Space-Vector Modulated Control Scheme for CSI fed IM drive

P.Parthiban, *Student Member, IEEE*, Pramod Agarwal, *Member, IEEE,* and S.P.Srivastava

*Abstract—* **Space vector modulation is the preferred PWM method for three-phase current source inverters. This paper presents an SVM implementation that is based on calculation of duty ratio required to synthesize the reference current. The steady state and dynamic results of the proposed strategy are presented. The proposed scheme has the advantage of space-vector modulation with fast dynamic response.**

*Index Terms*—Current-source inverter, space-vector pulse-width modulation.

## I. INTRODUCTION

THE rapid development of the switching frequency capabilities of power semiconductor devices requires faster, more accurate, and simpler modulation techniques. Among the different PWM techniques available, the Space Vector Modulation (SVM) has become the preferred method for three-phase converters [1].

This paper presents a simple space vector modulation algorithm for six switch current source inverter in which the duty ratios for each switching state is to be applied is obtained by solving two linear algebraic equations [5].

In this paper, first of all, the main circuit configuration and the bloack diagram of the control circuit are described. Next the SVM waveform generation method is explained. Finally the simulation results of the SVM CSI fed IM are presented.

## II. MAIN CIRCUIT AND CONTROL CIRCUIT CONFIGURATIONS

### A. Main Circuit Configuration and its Basic Operation

Fig. 1 shows the main circuit configuration of the CSI inverter. This inverter comprises a constant dc power supply,

---

P.Parthiban is with the Department of Electrical Engineering, Indian Institute of Technology, Roorkee, Uttranchal 247667, INDIA (e-mail: parthdee@gmail.com).

Pramod Agarwal is with the Department of Electrical Engineering, Indian Institute of Technology, Roorkee, Uttranchal 247667, INDIA (e-mail: pramgfee@iitr.ernet.in).

S.P. Srivastava is with Kumaon College of Engineering, Nainital, India, on leave from Indian Institute of Technology Roorkee, INDIA (e-mail: satyafee@iitr.ernet.in).

an inverter section to convert the dc power to the variable-voltage variable frequency ac power and an induction motor at the load side. In the inverter section six switches are used as to form a three-phase bridge circuit. Moreover, three capacitors are connected to the ac output terminals to absorb the overvoltages which occur when the current is cutoff.

The sinusoidally modulated current $I_i$ is produced in the three-phase bridge circuit composed of switches. The overvolatage absorption capacitors which are connected to the ac output terminals function as a filter, so the output current $I_l$ has a sinusoidal waveform.

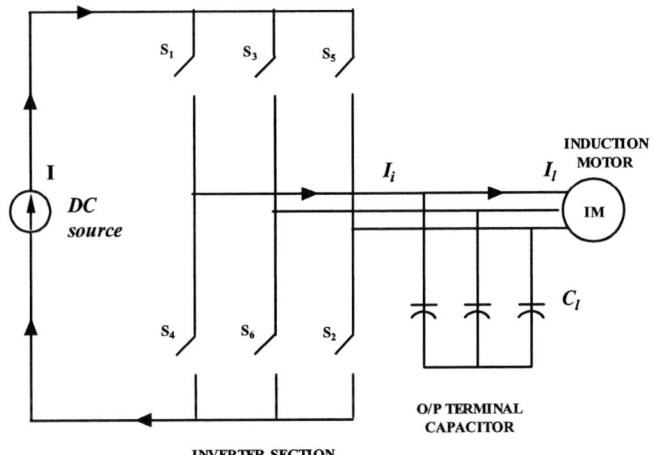

Fig. 1. Generalized six-switch three-phase CSI.

### B. Block diagram of control circuit

The block diagram given in fig. 2 illustrates a variable-speed induction motor drive using field-oriented control. The induction motor is fed by a current-controlled space vector modulated inverter. The motor speed $\omega$ is compared to the reference $\omega^*$ and the error is processed by the speed controller to produce a torque command $T_e^*$.

The stator quadrature-axis current reference $i_{qs}^*$ is calculated from torque reference $T_e^*$ as given in (1)

$$i_{qs}^* = \frac{2}{3}\frac{2}{p}\frac{L_r}{L_m}\frac{T_e^*}{\left|\psi_r\right|_{est}} \qquad (1)$$

where $L_r$ is the rotor resistance, $L_m$ is the mutual inductance and $\left|\psi_r\right|_{est}$ is the estimated rotor flux linkage given by (2)

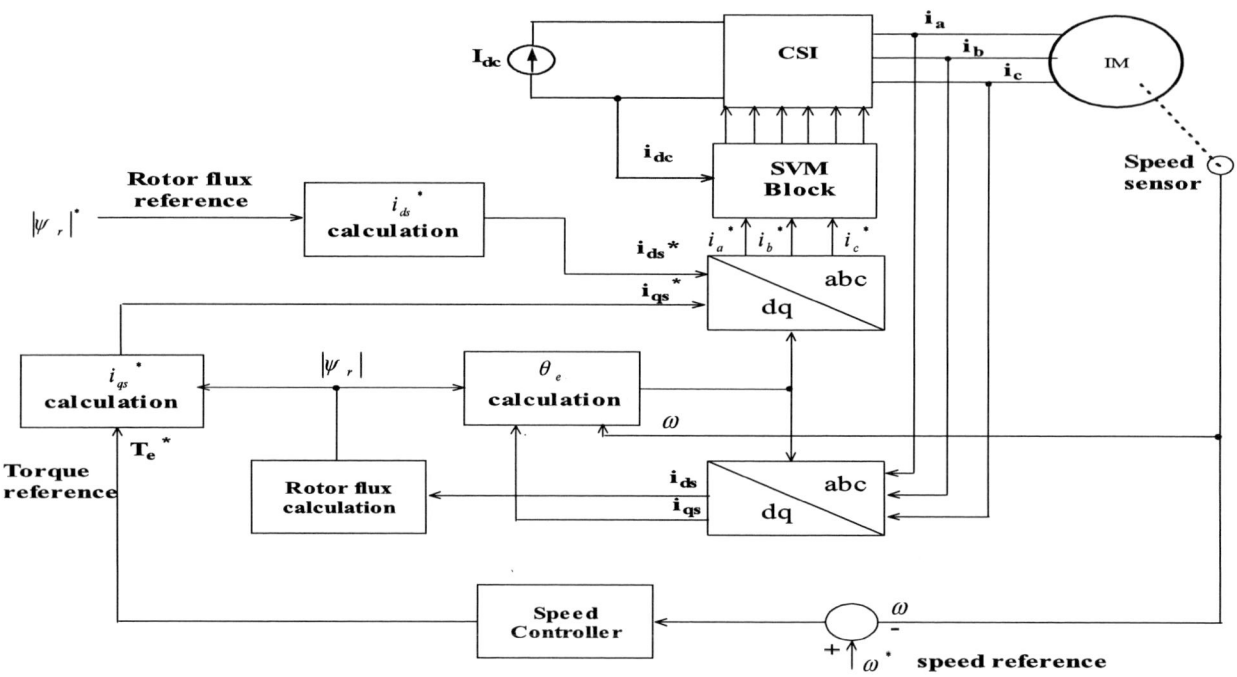

Fig. 2. Block diagram of Field oriented control SVM CSI fed IM drive.

$$\left|\psi_r\right|_{est} = \frac{L_m i_{ds}}{1 + \tau_r s} \tag{2}$$

where $\tau_r = \dfrac{L_r}{R_r}$ is the rotor time constant.

The stator direct-axis current reference $i_{ds}^*$ is obtained from rotor flux reference input $\left|\psi_r\right|^*$ as in (3)

$$i_{ds}^* = \frac{\left|\psi_r\right|^*}{L_m} \tag{3}$$

The rotor flux position $\theta_e$ required for coordinates transformation is generated from the rotor speed $\omega_m$ and $\omega_{sl}$ as in (4)

$$\theta_e = \int(\omega_m + \omega_{sl})dt \tag{4}$$

The slip frequency is calculated from the stator reference current $i_{qs}^*$ and the motor parameters in (5)

$$\omega_{sl} = \frac{L_m}{\left|\psi_r\right|_{est}} \frac{R_r}{L_r} i_{qs}^* \tag{5}$$

The $i_{qs}^*$ and $i_{ds}^*$ current references are converted into phase current references $i_a^*, i_b^*, i_c^*$ and then given as input to the SVM block.

The SVM block process the measured dc link current reference currents to produce the inverter gating signal.

The role of the speed controller is to keep the motor speed equal to the speed reference input in steady state and to provide a good dynamic during transients [2]-[4],[7].

### C. The Space Vector Modulation

A space current vector in complex notation is given by (6)

$$I_k = \frac{2}{3}(i_a + \vec{a}i_b + \vec{a}^2 i_c) \tag{6}$$

where $\vec{a} = e^{j\frac{2}{3}\pi}$ .

The PWM CSI uses six unidirectional switches to connect the dc current source to the load. Two switches in two different legs(one in the upper switch bridge($S_1$, $S_3$, $S_5$) and the other in the lower switch bridge($S_2$, $S_4$, $S_6$) in Fig.1) are allowed to conduct at any instant. It is noted that there should always exist a current path in the CSI. If not, a high voltage at open terminal is induced so that the switch might be destroyed.

The CSI can generate only six effective space vectors and three zero vectors. The zero vectors mean that both the upper and lower switches in the same arm are turn-on state simultaneously[8]. The space current vectors are illustrated in Fig. 3 [6].

The main function of SVM block is to generate six switching pulses to six switches of the CSI inverter so that the desired phase currents are obtained.

The input to the SVM block is the three reference currents

532

$i_a^*$, $i_b^*$, $i_c^*$ and the dc link current $I_{dc}$.

The dq component is calculated from (7)

$$\begin{bmatrix} i_d \\ i_q \end{bmatrix} = \frac{2}{3} \begin{bmatrix} 1 & -1/2 & -1/2 \\ 0 & \sqrt{3}/2 & -\sqrt{3}/2 \end{bmatrix} \begin{bmatrix} i_a \\ i_b \\ i_c \end{bmatrix} \qquad (7)$$

This transformation can be used with three phase currents associated with three-phase six-switch current source inverters. From which we can calculate $i_{ref}$, $\alpha$ as in (8),(9)

$$\left| i_{ref} \right| = \sqrt{i_d^2 + i_q^2} \qquad (8)$$

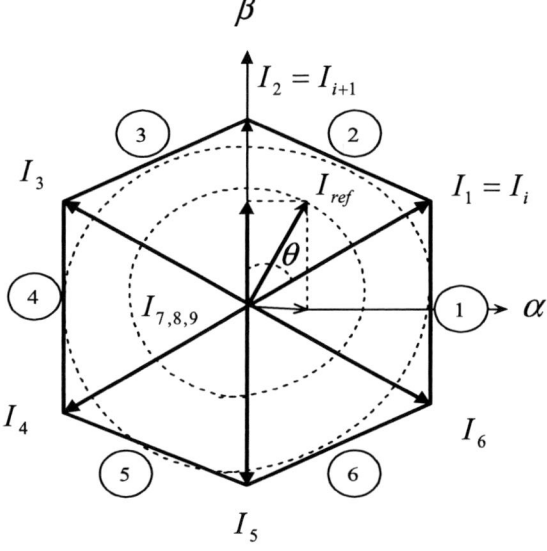

Fig. 3. Sectors and Space vectors in the complex plane.

$$\alpha = \tan^{-1}\left( \frac{i_q}{i_d} \right) \qquad (9)$$

Suppose $i_{ref}$ is in second sector, then $i_{ref}$ is synthesized by imposing current vectors $I_1$ and $I_2$ for duty ratios $d_1, d_2$, over a sampling period of $\Delta T$

$$\begin{aligned} i_{an}^* &= A_1 d_1 + A_2 d_2 \\ i_{bn}^* &= A_3 d_1 + A_4 d_2 \end{aligned} \qquad (10)$$

where $A_1$, $A_2$, $A_3$, $A_4$ are the dc link current values and it will be I or –I depending on the switching state (Table I). The above two equations are solved to get duty ratios $d_1, d_2$.

To summarize, the SVM block is realized based on the following steps:

TABLE I
CURRENT SOURCE INVERTER SWITCH COMBINATIONS AND ASSOCIATED SPACE VECTORS

| State(k) | On Switches | $i_{ia}$ | $i_{ib}$ | $i_{ic}$ | $I_k$ |
|---|---|---|---|---|---|
| 1 | 1,2 | I | 0 | -I | $\sqrt{3}(\sqrt{3}/2 + j1/2)I$ |
| 2 | 2,3 | 0 | I | -I | $j\sqrt{3}I$ |
| 3 | 3,4 | -I | I | 0 | $\sqrt{3}(-\sqrt{3}/2 + j1/2)I$ |
| 4 | 4,5 | -I | 0 | I | $-\sqrt{3}(\sqrt{3}/2 + j1/2)I$ |
| 5 | 5,6 | 0 | -I | I | $-j\sqrt{3}I$ |
| 6 | 6,1 | I | -I | 0 | $\sqrt{3}(\sqrt{3}/2 - j1/2)I$ |
| 7 | 1,4 | 0 | 0 | 0 | 0 |
| 8 | 3,6 | 0 | 0 | 0 | 0 |
| 9 | 5,2 | 0 | 0 | 0 | 0 |

Step 1: Determine $i_d$, $i_q$, angle ($\alpha$).

Step 2: Determine the sector in which angle $\alpha$ lies.

Step 3: Determine the duty ratios using Table I and (10)

Step 4: Determine the switching time of each transistor($S_1$ to $S_6$).

### D. Simulation Results

Fig. 4 shows the simulink model used for simulation using MATLAB/SIMULINK

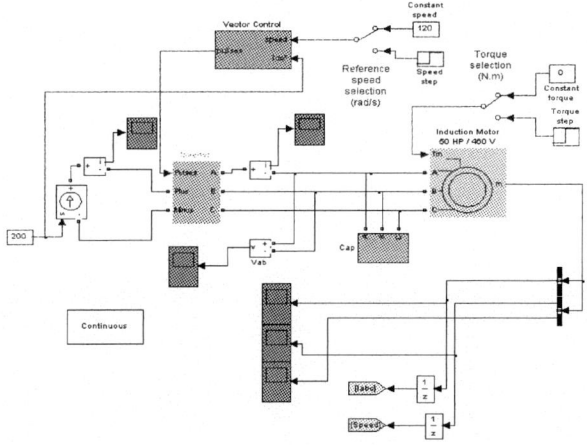

Fig. 4. MATLAB/SIMULINK control block diagram.

533

The system parameters used for simulation are as follows:
inverter filter capacitor : 200 $\mu$F,
induction motor : 3-phase, 4 poles, 460V(L-L), 50Hp,
$R_s = 0.087\Omega$, $R_r = 0.228\Omega$,
$L_s = L_r = 0.8mH$, $L_m = 34.7mH$,
sampling frequency : 20kHz.

The motor speed, electromechanical torque and currents observed during the starting of the induction motor drive are shown in Fig. 5

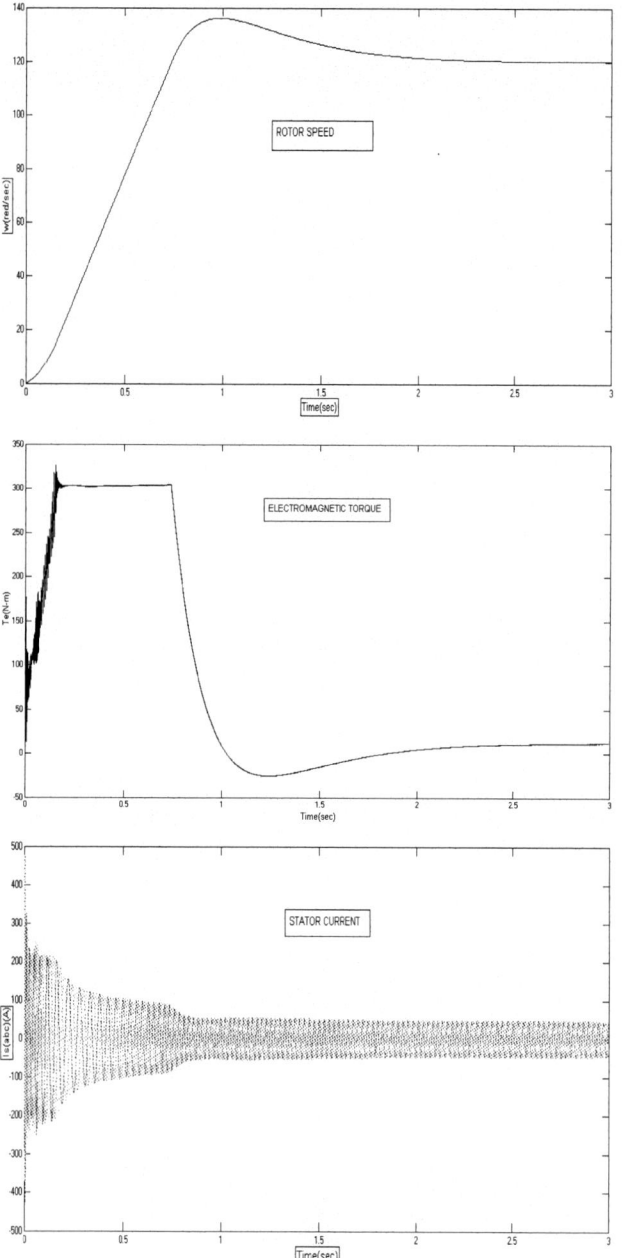

Fig. 5. Starting performance of the IM drive.

Fig. 6 shows the motor voltage, current and torque waveforms obtained when the motor is running at no load (torque = 0 N-m) at a speed of 120 rad/sec.

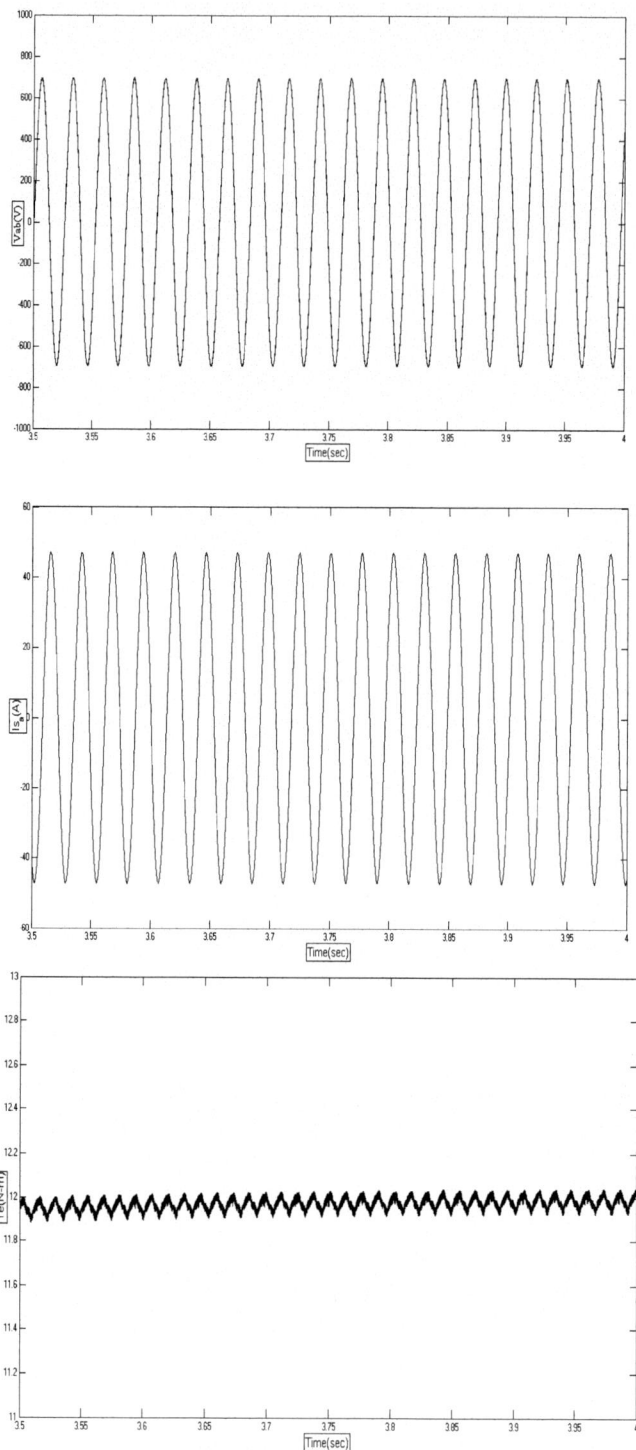

Fig. 6. Steady-state Motor Voltage, Current, and Torque performance waveforms.

The dynamic performance of the drive is studied by applying two changing operating conditions to the drive: a step change in speed reference and a step change in load torque.

The response of the induction motor drive to successive changes in speed reference and load torque is shown in Fig. 7

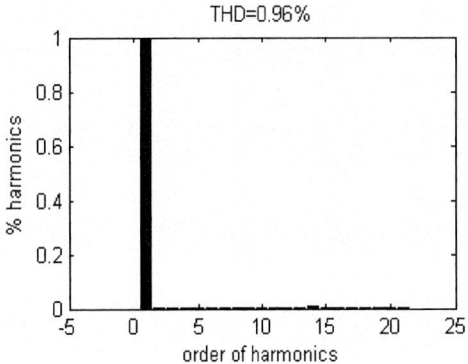

Fig. 8. Harmonic spectrum of phase 'a' line current.

The simulated results in Fig. 6 shows that the output voltage and load current are sinusoidal with low total harmonic distortion (THD = 0.96%)as shown in Fig. 8. The output filter capacitor absorbs most of the inverter output current harmonics, leaving only small amount of harmonic content in the load current.

## III. CONCLUSION

This paper presents an SVM implementation that is based on calculation of duty ratio required to synthesize the reference current. The advantage of the above scheme is that 1)It utilizes the SVM technique, which has fast dynamic response and stability 2)the measured THD values for load current are less than 2% that can satisfies the demand of the harmonic standard in industrial applications 3)the proposed control scheme simplifies the calculation process of the space vector modulation, which makes it possible to be implemented in low cost processors.

## IV. REFERENCES

[1] J.Holtz, "Pulse width modulation- A survey," *IEEE Trans. Ind. Electron,* vol. 39, pp. 410-420, Oct. 1992.

[2] Bimal K.Bose, *Modern Power Electronics and AC Drives.* (4th Indian Reprint): Pearson Education, 2002, p. 356-384.

[3] R.Krishnan, *Electric Motor Drives.*(2nd Indian Reprint): Printice-Hall of India, 2003, p.411-512.

[4] H.Le-Huy, "Case Study: Variable frequency induction motor drive," [online].Available:http://www.mathworks.com/access/helpdesk_r13/help/toolbox/physmod/powersys/stud_c13.html.

[5] Chern-Lin Chen, Che-Ming Lee, Rong-jie Tu and Guo-kiang Horng, "A novel simplified space-vector modulated control scheme for three-phase switch-mode rectifier," in *Proc. 1998 Power Electronics Specialists Conf.,* pp. 1358-1361.

[6] A.Bakhshai, J.Espinoza, G.Joos, H.Jin, "A combined artificial neural network and DSP approach to the implementation of space vector modulation techniques," in *Proc. 1996 Industry Applications Conference.,* pp. 934-940.

[7] M.Salo, H.Tuusa, "Experimental results of the current-source PWM inverter fed induction motor drive with an open-loop stator current control," in *Proc. 2003 Applied Power Electronics Conference and Exposition,* pp. 839-845.

[8] Dong-Choon Lee, Dong-Hee Kim, Dae-Woong Chung, "Control of PWM current source converter and inverter system for high performance induction motor drives," in *Proc. 1996 Industrial Electronics, Control, and Instrumentation conference.,* pp. 1100-1105.

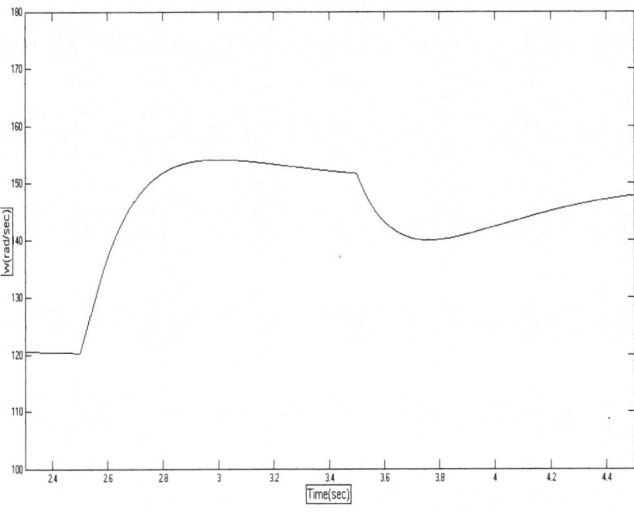

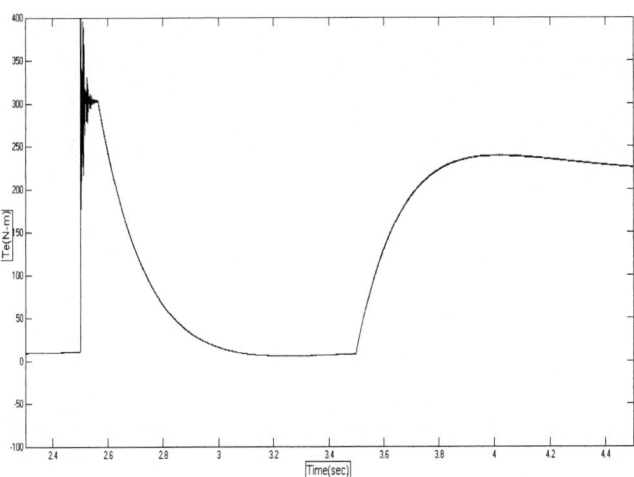

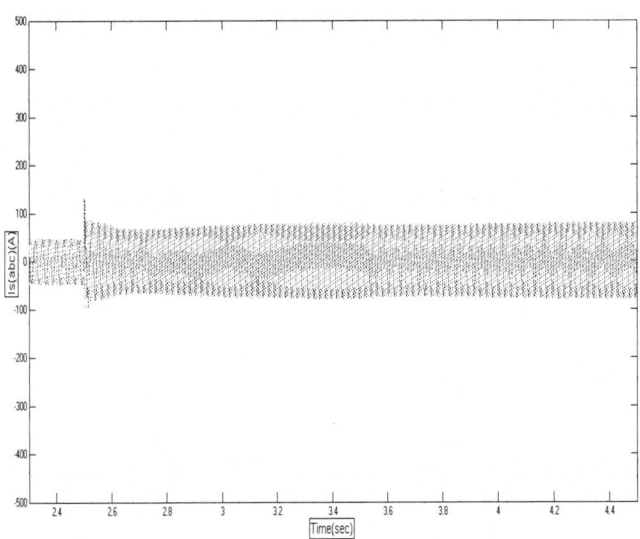

Fig. 7. Dynamic performance of the induction motor drive.

## V. BIOGRAPHIES

**P.Parthiban** was born in Salem, India, on May 24, 1980. He received the Bachelor's degree in Electrical and Electronics Engineering from Madras University, Chennai, India in 2001 and the Master's degree in Power Electronics and Drives from SASTRA Deemed University, Thanjavur, India in 2002. He is currently working towards the Ph.D degree at Indian Institute of Technology Roorkee, India.

From 2002 to 2003, he was with Sona College of Technology, Salem, as a Lecturer. His research interest includes power electronics and motor control.

**Pramod Agarwal** received the bachelor's, master's and Ph. D degrees in Electrical Engineering from the University of Roorkee (now, Indian Institute of Technology Roorkee), India in 1983, 1985, and 1995 respectively.

Currently he is with Indian Institute of Technology Roorkee, India, where he is a Professor in the Department of Electrical Engineering. His special fields of interests include electrical machines, power electronics, power quality, microprocessors and microprocessor-controlled drives, active power filters, high power factor converters, multilevel inverters, and dSPACE-controlled converters.

**S. P. Srivastava** received the bachelor's and master's degrees in Electrical Technology from I.T. Banarus Hindu University, Varanasi, India in 1976, 1979 respectively and the Ph. D degree in Electrical Engineering from the University of Roorkee, India in 1983.

Currently he is with Indian Institute of Technology (IIT) Roorkee, India, where he is an Associate Professor in the Department of Electrical Engineering. He is also a principal at the Kumaon Engineering College, Nainital, India, on taking leave from the IIT Roorkee. His research interests include power apparatus and electric drives.

**2006 IEEE International Conference on Power Electronic, Drives and Energy Systems**

# A Study on Design and Dynamics of Voltage Source Inverter in Current Control Mode to Compensate Unbalanced and Non-linear Loads

Mahesh K. Mishra, *Member, IEEE,* and K. Karthikeyan

*Abstract--*In this paper, a design of the voltage source inverter (VSI) in a hysteresis band current control mode is presented. A proper selection of VSI parameters is necessary to track the desired reference currents at the point of the common coupling to compensate unbalanced and non-linear loads. In this paper, the authors carry out a detailed study on the design parameters and dynamics of the VSI while tracking a reference compensator current. A simulation of three-phase, four-wire compensated system is carried out using PSCAD 4.2.0 to verify the proposed design methods.

*Index Terms--*Active power filter, design, dynamics, hysteresis band, interface inductance, power quality, switching frequency, voltage source inverter.

## I. INTRODUCTION

THE parameters and the dynamic performance of the VSI in current tracking mode play an important role in compensating unbalanced and non-linear load currents. The important parameters of VSIs are the level of dc link voltage, value of dc storage capacitor, value of interface inductor and hysteresis band. A proper selection of VSI parameters plays a crucial role in the operation of active power filter. The VSI is generally operated in hysteresis band current control mode to track desired reference current at the point of common coupling (PCC).

Before we go into the design details of the VSIs, it is necessary to have a look at some commonly used topologies [1]-[9]. These are illustrated in Fig. 1(a)-(e). The inductance $L_f$ in these figures represents the net inductance of the isolation transformer and/or additional external interface inductance. The switching losses of the inverter and the copper losses of the isolation transformer and/or external interface inductance are represented by a resistance $R_f$. The iron losses of the transformer are neglected. Each of the switches shown in Fig 1(a)-(e) has an IGBT and an anti- parallel freewheeling diode. Fig. 1(a) illustrates the H-bridge inverter topology. It

consists of three H-bridge VSIs that are connected to a common dc storage capacitor. Each VSI is connected to the power network at the PCC through an isolation transformer. The six output terminals of the transformer are connected in star. Also these six terminals can be connected in delta to compensate a delta connected load. In this case, each transformer is connected in parallel with the corresponding load [4]. The purpose of these transformers is to provide isolation between the inverter legs and to prevent the dc storage capacitor from being shorted by switches in different inverter legs. This topology however is not suitable for compensation of load current containing dc components in addition to ac components due to the presence of isolation transformers.

A three-phase, three-leg topology is shown in Fig. 1(b). It has six switches and a single dc storage capacitor. If we use this topology, the zero sequence currents in the load cannot be compensated and hence the zero sequence currents flow in the neutral wire (*N-n*) between the system and load. The zero sequence current thus returns to the ac distribution system. In addition, if the load is non-linear and contains harmonics, then these harmonics also enter the ac system, thus degrading the power quality. In this topology, the generation of the three compensator currents is not independent because $i_{fa}+i_{fb}+i_{fc}=0$. Hence this scheme is not suitable for three-phase, four-wire distribution system with loads containing zero sequence currents.

Three-phase, four-leg and single capacitor inverter topology shown in Fig. 1(c) is suitable for the elimination of dc as well zero sequence component from the source current, if the load current contains dc components [5]. Three of its legs are connected to three phases and the fourth leg is connected to the load and supply neutral through an interface reactance. The reference current for the fourth leg is the negative sum of three-phase load currents. This nullifies the effect of dc component in load current. To maintain adequate charge in the dc storage capacitor, a proportional integral (PI) controller is used to regulate the flow of real power from the source to compensator. When the compensator is working, zero sequence current containing switching frequency and harmonics is routed to path *n-n'*. Using fourth leg of inverter, the negative of zero sequence current ($-i_0$) is tracked. Certainly it needs a higher bandwidth VSI to track the reference neutral current as it contains harmonics due to non-

---

Mahesh K. Mishra is with the Department of Electrical Engineering, Indian Institute of Technology Madras, Chennai, India. (e-mail: mahesh@ee.iitm.ac.in).

K. Karthikeyan is a Ph.D scholar in Department of Electrical Engineering, Indian Institute of Technology Madras, Chennai, India. (e-mail: karthikeyan@ee.iitm.ac.in).

0-7803-9771-1/06/$25.00 ©2006 IEEE

linear loads. This increases the switching losses. If this current is not tracked properly, it will leave high switching frequency current components in the *N-n* path, which is not desirable. Also this control algorithm is quite complicated as it requires coordination between the switches in the fourth (neutral) leg and those in other three legs.

Fig. 1(d) shows a neutral-clamped inverter VSI topology. It consists of six IGBT switches and two identical dc storage capacitors. This topology is well equipped to compensate dc components of the load, but due to the presence of dc components in VSIs, the two dc capacitors are charged to different voltages. The total voltage of dc capacitors however is maintained constant by using a separate PI control loop. It is not only the dc current in load which can make drift in the dc capacitor voltages from the reference value, but also the unequal capacitance leakage currents, unequal delays in the semiconductor devices, asymmetrical charging of the capacitors during transients and due to measurement and signal conditioning circuits [6]-[7], [10].

To overcome the above problem, a neutral clamped inverter-chopper VSI topology is proposed in [8]-[9] as depicted in Fig. 1(e). The function of the chopper is to transfer energy between the two dc capacitors so as to make their voltages close to the reference value. The chopper circuit requires an additional control hence increases the control complexity.

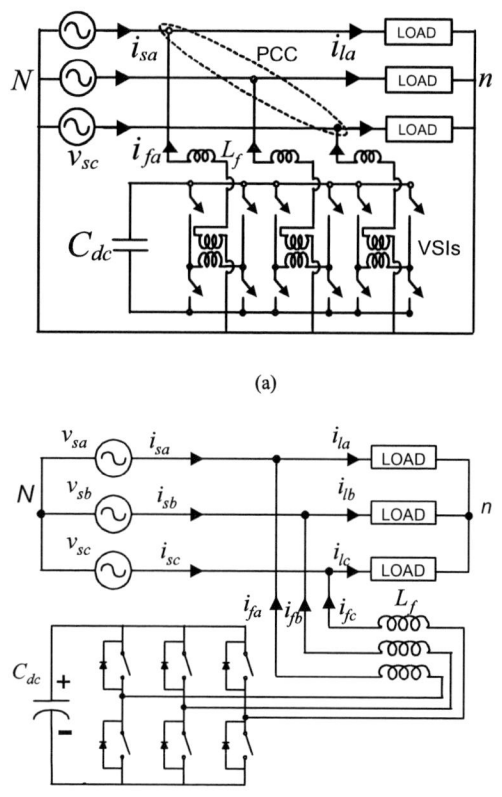

(a)

(b)

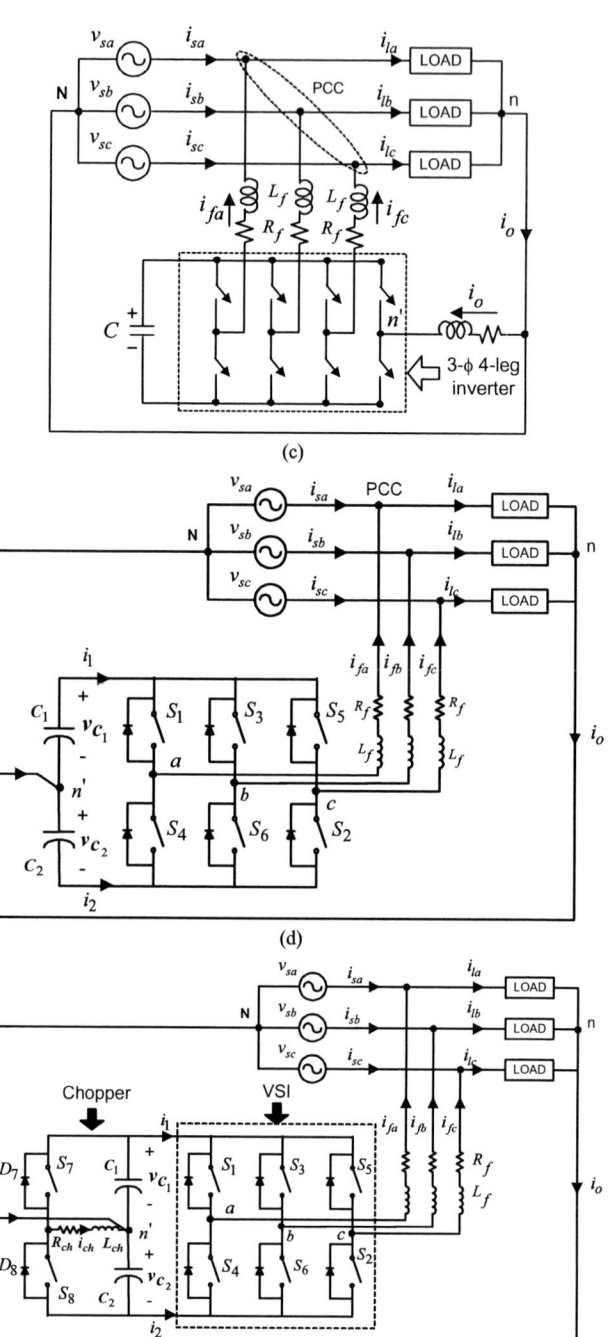

(c)

(d)

(e)

Fig. 1. VSI topologies (a) H-bridge VSI topology (b) Three-phase, three-leg VSI (c) Three-phase, four-leg VSI (d) Neutral clamped VSI (e) Chopper-inverter neutral clamped VSI.

In all above VSI topologies, there are some common parameters which must be carefully selected to provide satisfactory performance while tracking reference currents. These parameters are the dc link voltage ($V_{dc}$), dc capacitor ($C_{dc}$), interface inductance ($L_f$), hysteresis band ($h$) and switching frequency ($f_{sw}$). Although it is possible to select these parameters of the VSI circuit by trial and error, there is a need for a systematic study and investigation for the selection of these parameters considering the various constraints in the

538

system, to ensure the faithful tracking of the given reference compensator currents. This paper focuses on the above described issues.

## II. SWITCHING DYNAMICS OF VSI

Before we discuss the design of the VSI, it is necessary to have an understanding of the switching dynamics of the VSI while it tracks the given reference currents. Here, to make the analysis simple, a three-phase H-bridge VSI topology is chosen. Since for a three-phase four-wire system, each phase is independent of neutral, only the single-phase H-bridge inverter supported by a dc storage capacitor is considered for analysis. The circuit for the single phase VSI is shown in Fig. 2 (a). The switching dynamics for other topologies given in Section-I can be understood in a similar way. In Fig. 2(a), an arbitrary reference filter current is shown by dashed curve. The upper and lower limits of the reference currents are created by adding and subtracting hysteresis band $h$ to the filter reference current. To track the positive $i_{fref}$, at any time $t_1$, switches $S_{1a}$ and $S_{2a}$ are closed and $S_{3a}$ and $S_{4a}$ are opened. This connects $+V_{dc}$ to the inverter and the actual filter current rises from $(i_{fref} - h)$ to $(i_{fref} + h)$ through $i_{fref}$. Once it crosses the limit $(i_{fref} + h)$, the actual current has to be brought within the pre-defined band. To do this, the switches $S_{3a}$ and $S_{4a}$ are closed and the switches $S_{1a}$ and $S_{2a}$ are opened. However, if the actual current remains positive, the switches $S_{3a}$ and $S_{4a}$ will not conduct. Hence the actual filter current flows through diodes $D_{4a}$ and $D_{3a}$. It is important to note that during the negative slope of tracked current (from instant $t_1'$ to $t_2''$), control input $-V_{dc}$ is connected to the VSI through diodes $D_{4a}$ and $D_{3a}$ and not through switches $S_{3a}$ and $S_{4a}$.

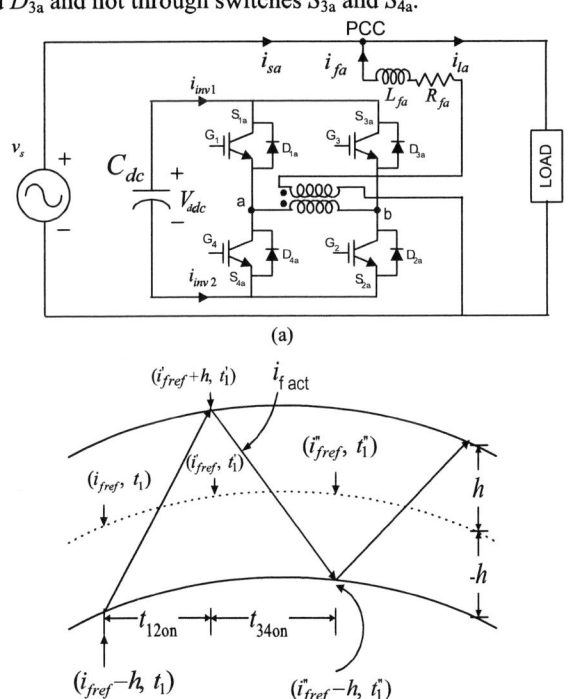

(a)

(b)

Fig. 2. (a) Single phase H-bridge topology to study dynamics and performance of VSI (b) Switching dynamics of VSI.

## III. DESIGN OF VSI PARAMETERS

As it has been discussed, in any one of the above mentioned VSI topologies, the important parameters are dc link voltage, the value of dc storage capacitor(s), the interface inductor and the hysteresis band within which the actual compensator current should follow the reference currents. A systematic study on design of these parameters of a three-phase H-bridge, VSI topology is presented in this section.

### A. Selection of DC Link Voltage

Detailed simulation and experimental studies are carried out to converge to an empirical relationship between the dc link voltage and switching/tracking performance of the compensator. In the simulation study, the dc link voltage is varied from 1.3 to 2 times the peak value of the ac system voltage. The total harmonic distortion (THD) of the compensated source current is taken as figure of merit. The simulation and the experimental results are tabulated in Table I and II respectively. It is observed from the simulation and the experimental study that when dc link voltage is approximately equal to 1.6 times the peak ac system voltage, with some variations, the THD is minimum. The study is carried out for various values of interface inductance varying from 10 mH to 40 mH. The value of dc capacitor in Fig. 2(a) is taken as 2200 μF. The system peak voltage ($V_m$) is considered as 172 volts. The ratio $m$ (defined as $V_{dc}/V_m$) is varied by changing the value of $V_{dc}$. The PI controller is used to regulate the dc link voltage at reference value $V_{dcref}$ [9].

TABLE I
SIMULATION STUDY: THD VARIATION WITH DC LINK VOLTAGE AND INTERFACE INDUCTANCE

| $L_f$(mH) \ $m$ | 1.30 | 1.45 | 1.60 | 1.75 | 2.00 |
|---|---|---|---|---|---|
| 10 | 3.00 | 2.67 | 1.88 | 2.21 | 2.67 |
| 20 | 2.27 | 1.61 | 1.10 | 1.38 | 1.25 |
| 30 | 3.36 | 2.66 | 2.31 | 2.47 | 2.4 |
| 40 | 4.89 | 4.22 | 3.2 | 2.98 | 2.12 |

TABLE II
EXPERIMENTAL STUDY: THD VARIATION WITH DC LINK VOLTAGE AND INTERFACE INDUCTANCE

| $L_f$(mH) \ $m$ | 1.30 | 1.45 | 1.60 | 1.75 | 2.00 |
|---|---|---|---|---|---|
| 10 | 5.35 | 4.87 | 4.43 | 4.41 | 4.17 |
| 20 | 5.68 | 5.24 | 4.79 | 5.35 | 4.89 |
| 30 | 6.53 | 6.74 | 6.06 | 5.87 | 5.36 |
| 40 | 6.85 | 7.15 | 6.2 | 6.43 | 6.19 |

The practical constraints imply that the dc link voltage should satisfy the following condition.

539

$$V_m \leq V_{dc} \leq V_{CE\,rated} \qquad (1)$$

where $V_{CE\,rated}$ is the rated value of collector to emitter voltage of the power switch . Empirically, it can be concluded that the dc link voltage can be expressed as,

$$V_{dc} = m\,V_{dc} = 1.6\,V_m \qquad (2)$$

### B. Selection of DC Storage Capacitor

Once the dc link voltage is arrived as (2), the selection of dc capacitor value is important. The dc capacitor value can be selected based on the transients or voltage sag/swell occurring in the system and its ability to regulate under this condition. Let us assume that the compensator in Fig. 1(a) is connected to an $X$ kVA system. The energy of the system in Joules per second is therefore given by $X \times 1000$ J/s. Let us further assume that, the VSI compensator deals with half (i.e. $X/2$) and twice (i.e. $2X$) kVA handling capacity under the transient conditions for $n$ cycles. Then the change in energy to be dealt by the dc capacitor is given as

$$\Delta E = (2X - X/2)\,n\,T \qquad (3)$$

Now this change in energy (3) should be supported by the energy of dc capacitor. Let us allow dc capacitor to change its voltage from $1.4V_{dc}$ to $1.8V_{dc}$. Hence we can write,

$$\frac{1}{2}C_{dc}\left[(1.8V_m)^2 - (1.4V_m)^2\right] = (2X - X/2)\,n\,T \qquad (4)$$

which implies that,

$$C_{dc} = \frac{2(2X - X/2)\,n\,T}{(1.8\,V_m)^2 - (1.4\,V_m)^2} \qquad (5)$$

For example, consider a 10 kVA system, i.e. $X = 10$ and system peak voltage $V_m = 325$ V, $n = 1/2$, $T = 0.02$ s. The value of $C_{dc}$ computed using (5) is 2218 $\mu$F. It is also possible to extend this relation for compensator in dynamic voltage restorer (DVR) mode to mitigate voltage sag/swell [11]. For voltage sag, $1.8V_m$ represents the maximum allowable dc link voltage. The term $nT$ is replaced by $T_{sag}$ and $1.4\,V_m$ is replaced by reference dc capacitor voltage which is $1.6\,V_m$. The value of the dc storage capacitor is given by,

$$C_{dc} = \frac{2(2X - X/2)\,T_{sag}}{(1.8\,V_m)^2 - (1.6\,V_m)^2} \qquad (6)$$

Similarly to withstand voltage swell, the value of dc storage capacitor can be computed as

$$C_{dc} = \frac{2(2X - X/2)\,T_{swell}}{(1.6\,V_m)^2 - (1.4\,V_m)^2} \qquad (7)$$

### C. Selection of Interfacing Inductor

Selection of the interface inductor is the next important step in VSI design. The proper selection of interface inductor plays a crucial role to give a satisfactory performance of the VSI in terms of the bandwidth (switching frequency) and hysteresis band enveloped over the reference quantity. In the Fig. 2(a) the supply system is assumed to be stiff and the resistance of the interface inductance is neglected. The control input voltage to track the reference current is given as

$$V_{ab} = S_{switch} V_{dc} - V_m \sin \omega t \qquad (8)$$

When $S_{switch} = 1$, $(V_{dc} - V_m \sin \omega t)$ is applied to $ab$ by conducting $S_1 - S_2$ or $D_1 - D_2$ and non-conducting $S_3 - S_4$ or $D_3 - D_4$. When $S_{switch} = -1$, implies $(-V_{dc} - V_m \sin \omega t)$ is applied to $ab$ by conducting $S_3 - S_4$ or $D_3 - D_4$ and non-conducting $S_1 - S_2$ or $D_1 - D_2$.

From Fig 2(b), it is found that during the positive slope of positive actual current, the input voltage to interface inductor is $(mV_m - V_m \sin \omega t)$, where $m = 1.6$ and $V_m$ is the peak of the system voltage. Using the Fig. 2(b), the relation for $t_{12\,on}$ or $t_{34\,off}$ is obtained as following [12].

$$t_{12\,on} = t_{34\,off} = L_f \frac{(i'_{fref} + h) - (i_{fref} - h)}{mV_m - V_m \sin \omega t} \qquad (9)$$

which is simplified to

$$t_{12\,on} = t_{34\,off} = \frac{\{(i'_{fref} - i_{fref}) + 2h\}L_f}{mV_m\left(1 - \dfrac{\sin \omega t}{m}\right)} \qquad (10)$$

Similarly during the negative slope of the positive actual current, the following relation is given for $t_{12\,off}$ or $t_{34\,on}$.

$$t_{12\,off} = t_{34\,on} = \frac{\{-(i''_{fref} - i'_{fref}) + 2h\}L_f}{mV_m\left(1 + \dfrac{\sin \omega t}{m}\right)} \qquad (11)$$

Table III shows the conduction details of various switching devices during the tracking of the reference current by the H-bridge VSI as shown in Fig. 2(a).

TABLE III
CONDUCTION DETAILS OF POWER SWITCHES

| Positive actual current ($i_{fact}$) | | | | Negative actual current ($i_{fact}$) | | | |
|---|---|---|---|---|---|---|---|
| Positive slope $t_{12\,on}$ (or) $t_{34\,off}$ | | Negative slope $t_{34\,on}$ (or) $t_{12\,off}$ | | Positive slope $t_{12\,on}$ (or) $t_{34\,off}$ | | Negative slope $t_{34\,on}$ (or) $t_{12\,off}$ | |
| ON Devices | OFF Devices | ON Devices | OFF Devices | ON Devices | OFF Devices | ON Devices | OFF Devices |
| $S_1 - S_2$ | $D_1 - D_2$ $S_3 - S_4$ $D_3 - D_4$ | $D_3 - D_4$ | $S_3 - S_4$ $S_1 - S_2$ $D_1 - D_2$ | $D_1 - D_2$ | $S_1 - S_2$ $S_3 - S_4$ $D_3 - D_4$ | $S_3 - S_4$ | $D_3 - D_4$ $S_1 - S_2$ $D_1 - D_2$ |

Thus when the actual filter current is positive with the positive slope, switches $S_1 - S_2$ are ON and the other devices are OFF during $t_{12\,on}$ or $t_{34\,off}$. Similarly the ON and OFF status of the switching devices can be explained with the help of the above table using the function parameters $t_{12\,on}$ or $t_{34\,off}$ and $t_{12\,off}$ or $t_{34\,on}$.

Consider slopes $s_1$ and $s_2$ such that

$$\begin{aligned} i'_{f\,ref} - i_{f\,ref} &= s_1 t_{12\,on} \\ i''_{f\,ref} - i'_{f\,ref} &= s_2 t_{12\,off} \end{aligned} \qquad (12)$$

Then using (10), (11) and (12), ON/OFF time for the switches $S_1/S_2$ and $S_3/S_4$ can be expressed as following.

$$t_{12\,on} = t_{34\,off} = \frac{2hL_f}{mV_m\left(1 - \dfrac{\sin \omega t}{m} - \dfrac{s_1 L_f}{mV_m}\right)} \qquad (13)$$

$$t_{12\,off} = t_{34\,on} = \frac{2hL_f}{mV_m\left(1 + \dfrac{\sin\omega t}{m} + \dfrac{s_2 L_f}{mV_m}\right)} \tag{14}$$

It is noted that at $\theta = \omega t$, the switching frequency of the inverter $f_{sw}$ can be found as following.

$$f_{sw} = \frac{mV_m}{2hL_f} \frac{\left(1 - \dfrac{\sin\theta}{m} - \dfrac{s_1 L_f}{mV_m}\right)\left(1 + \dfrac{\sin\theta}{m} + \dfrac{s_2 L_f}{mV_m}\right)}{\left(2 + \dfrac{(s_2 - s_1)L_f}{mV_m}\right)} \tag{15}$$

where $f_{sw} = 1/T$ and $T = t_{12\,on} + t_{12\,off}$ or $T = t_{34\,on} + t_{34\,off}$. To find out the maximum and minimum values of switching frequency, the following equality is solved.

$$\frac{df_{sw}}{d\theta} = -\frac{V_m}{2hL_f} \frac{\cos\theta\left(\dfrac{2\sin\theta}{m} + \dfrac{(s_2 + s_1)L_f}{mV_m}\right)}{\left(2 + \dfrac{(s_2 - s_1)L_f}{mV_m}\right)} = 0 \tag{16}$$

Generally the slopes of the tracking segments i.e. from $(i_{fref} - h, t_1)$ to $(i'_{fref} + h, t_1')$ and $(i'_{fref} + h, t_1')$ to $(i''_{fref} - h, t_1'')$ shown in Fig. 2(b) are very large as compared to $s_1$ and $s_2$. Therefore, $s_1$ and $s_2$ can be approximately taken as zero except at discontinuities. Then (16) can be simplified to

$$\frac{df_{sw}}{d\theta} = -\frac{V_m \sin 2\theta}{4hL_f} \tag{17}$$

Thus the solution of (17) can be written as following

$$\theta = \begin{cases} \pm n\pi \\ \pm(2n+1)\pi/2 \end{cases} \quad \text{for } n = 0, 1, 2, \ldots \tag{18}$$

It can be further shown that,

$$\frac{\partial^2 f_{sw}}{\partial\theta^2} = \begin{cases} -\dfrac{V_m}{2hL_f} \Rightarrow f_{sw} = f_{sw\max} \text{ at } \theta = \pm n\pi \\[2mm] \dfrac{V_m}{2hL_f} \Rightarrow f_{sw} = f_{sw\min} \text{ at } \theta = \pm(2n+1)\pi/2 \end{cases} \tag{19}$$

The maximum and minimum values of switching frequency can be computed for the above values of $\theta$ from (15). These values are given as follows

$$f_{sw\max} = \frac{mV_m}{4hL_f} \tag{20}$$

$$f_{sw\min} = \frac{mV_m}{4hL_f}\left(1 - \frac{1}{m^2}\right) \tag{21}$$

Since the maximum value of switching frequency is important for designing the interface inductor, hence the value of $L_f$ (in Henry) is given as,

$$L_f = \frac{mV_m}{4h f_{sw\max}} \tag{22}$$

### D. Design Example on Switching Characteristics

Consider a three-phase system with 220 V rms phase voltage. A typical value of $m$ is taken as 1.6 as discussed earlier (2), however it is varied from 1.4 to 1.8. The typical value of current hysteresis band is ±5% of the rated current and it is varied from ±4% to ±8%. Assuming 10 A of rated current, $2h = 1$ A for $h = \pm5\%$. The nominal switching frequency of IGBT is taken as 20 kHz. With these typical value of $m$, $h$, and $f_{sw}$, the interface inductor $L_f$ is computed using (22) and is found to be 13 mH. If the IGBT is operated at 10 kHz, the $L_f$ becomes 26 mH. The interface inductance $L_f$ is now varied from 10 mH to 30 mH and its effect on the switching performance is studied.

Fig. 3(a) shows the variation of switching frequency of the IGBT switch when $m$ varies from 1.4 to 1.8 with variation in system voltage from $-V_m$ to $V_m$. It is seen that for m=1.6, the minimum and maximum values of switching frequencies are 12.1 kHz and 20 kHz respectively. For $m$=1.8, there is an increase in minimum and maximum switching frequencies from 15.5 to 22.4 kHz. For $m$=1.4, there is decrease in the minimum and maximum switching frequencies from 8.5 to 17.5 kHz.

The variation of switching frequency for different values of $m$ as a function of phase angle ($\theta$) of voltage is illustrated in Fig. 3(b). Since over a full range of phase angle (i.e. $0° \le \theta \le 360°$), the system voltage attains zero value three times and positive peak and negative peak each one time, the switching frequency modulates from 8.5 to 22.5 kHz for values of m varying from 1.4 to 1.8. The intersection of plate with the switching frequency surface shows the operating switching frequencies for m=1.6.

Fig. 3(c) and (d) illustrate the variation of switching frequency of the IGBT switch as a function of hysteresis band ($h$) and interface inductance ($L_f$) with $\theta$. In Fig. 3(c), the hysteresis band is varied from 4% to 8% for constant value of m=1.6. As a result of this, $f_{sw}$ changes from 7.59 to 24.92 kHz. The intersection of vertical plate (for h=0.5 A) with switching surface shows the switching frequency varying from 12.15 to 19.94 kHz. In Fig. 3(d), $L_f$ is varied from 10 to 30 mH, $f_{sw}$ is plotted by changing $\theta$ and keeping m=1.6, h=0.5 A. It is evident that as $L_f$ increases from 10 to 30 mH, the $f_{sw}$ changes from 26 to 5.28 kHz. The intersection of vertical plate with the switching surface shows range of switching frequency 12.19 to 20 kHz. In all these illustrations, it is observed that at zero crossings of system voltages (i.e. at $\theta = \pm n\pi$), the switching frequency is maximum and at positive or negative peak of system voltages (i.e. at $\theta = \pm (2n+1)\pi/2$), the switching frequencies are minimum for any given set of parameters.

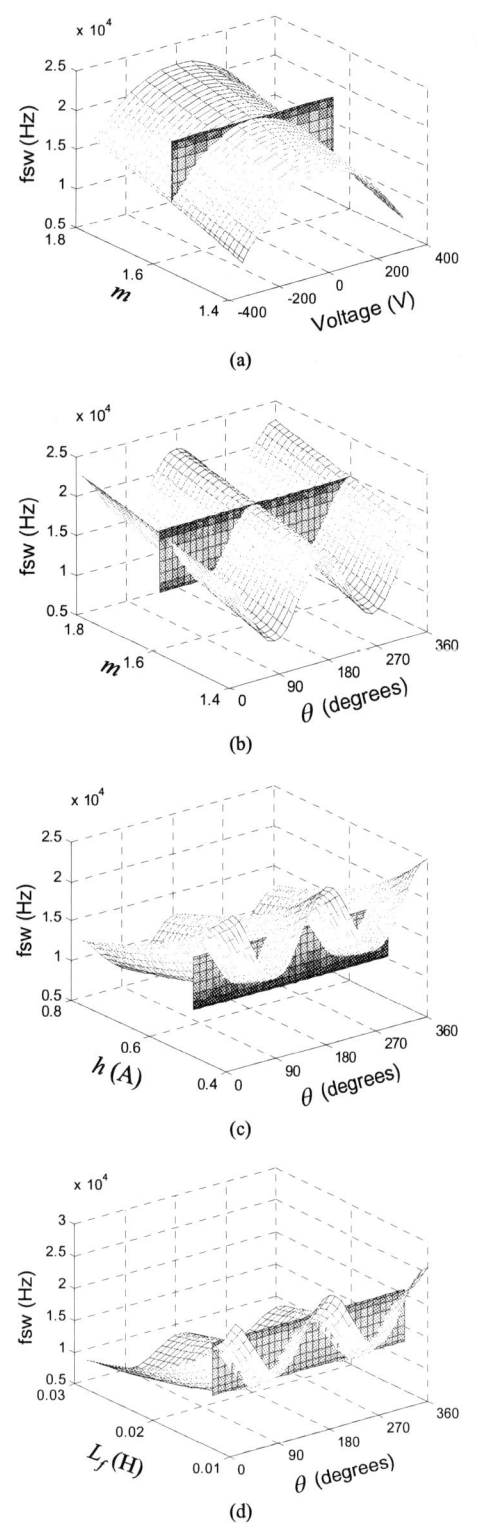

Fig. 3. Variation of switching frequencies (a) with $m$ and system voltage (b) with $m$ and $\theta$ (c) with $h$ and $\theta$ (d) with $L_f$ and $\theta$.

## IV. SIMULATION STUDIES

The load compensator is realized by using a H-bridge VSI topology as shown in Fig.1 (a). The load and the compensator are connected at the PCC. The load consists of a three-phase

unbalanced R-L and three-phase diode bridge rectifier feeding an R-L load. The compensator comprises of twelve IGBT switches, each with anti-parallel diode, three isolation transformers and a dc storage capacitor ($C_{dc}$). The secondaries of the isolation transformers are joined in star and the star point is connected to the neutral of the load ($n$) and source ($N$). The H-bridge VSIs are connected to PCC through interface inductors ($L_f$). The resistance of this inductor is denoted by $R_f$ and is modeled to account total losses in the inverter. The switching losses and ohmic losses in the actual compensator are denoted by $P_{loss}$. The compensator reference currents are extracted using theory of instantaneous symmetrical components [4]. Based on this theory, the reference currents are given by,

$$
\left.
\begin{aligned}
i_{fa}^* &= i_{la} - i_{sa} = i_{la} - \frac{v_{sa} + \gamma\left(v_{sb} - v_{sc}\right)}{\Delta}\left(P_{lavg} + P_{loss}\right) \\
i_{fb}^* &= i_{lb} - i_{sb} = i_{lb} - \frac{v_{sb1} + \gamma\left(v_{sc} - v_{sa}\right)}{\Delta}\left(P_{lavg} + P_{loss}\right) \\
i_{fc}^* &= i_{lc} - i_{sc} = i_{lc} - \frac{v_{sc1} + \gamma\left(v_{sa} - v_{sb}\right)}{\Delta}\left(P_{lavg} + P_{loss}\right)
\end{aligned}
\right\}
\tag{23}
$$

where $\gamma = \tan\phi/\sqrt{3}$, $\phi$ is the desired phase angle between supply voltages ($v_{sa}$, $v_{sb}$, $v_{sc}$) and compensated source currents ($i_{sa}$, $i_{sb}$, $i_{sc}$). The term $P_{lavg}$ in (23) is the dc or mean value of the load power. It is computed using a moving average filter that has an averaging time of half cycle or one cycle of supply voltage waveform depending upon whether load current contains odd or both odd and even harmonics. The term $P_{loss}$ in (23) accounts for the losses in the VSI while realizing the actual compensator and is computed using PI voltage control loop. It is a slow control loop which generates $P_{loss}$ at every positive zero crossing of phase-$a$ voltage based on the difference between the actual and reference voltage of the dc capacitor.

TABLE IV
SIMULATION PARAMETERS

| System Parameters | Values |
|---|---|
| Frequency | 50 Hz |
| Voltage (L-L) | 400 V, rms |
| Unbalanced and non-linear load | ▪ $Z_a$=40+$j$ 18.85 Ω, $Z_b$=25+$j$ 12.57 Ω, $Z_c$=90+$j$ 25.13 Ω<br>▪ Three-phase full bridge diode rectifier drawing dc load current of 5 A |
| DC storage capacitor | $C_{dc}$=2200 μF |
| Interface inductor | $R_f$=0.5 Ω, $L_f$=13 mH |
| PI controller gains | $K_P$=10, $K_I$=1 |
| Reference voltage | $V_{dc\,ref}$= 520 V |
| Hysteresis band | ± 0.5 A |
| Simulation time step | 2 μs |

The three-phase load consists of unbalanced impedances and a rectifier load drawing a constant current of 5 A as given in Table IV. These are shown in Fig. 4(a). The THD content in these load currents are 12.21 %, 9.17 % and 17.42 % in phase-$a$, $b$ and $c$ respectively.

The source currents after compensation are illustrated in Fig. 4(b). As seen from this figure, the source currents are balanced and sinusoidal. However the switching frequency components are present due to VSI operation. The currents have unity power factor relationship with respect to their phase voltage. The THDs of compensated source currents as shown in figure are less than 1%.

Fig 4 (c) shows the variation in switching frequencies of the VSI for the given parameters in Table IV. The phase-$a$ voltage ($v_{sa}$) and compensator current ($i_{fa}$) in phase-$a$ are also displayed to understand the variation of switching frequency with them. From the figure, the switching frequency is found to be varying between 11.3 kHz and 19.23 kHz. For same VSI parameters, minimum and maximum switching frequencies computed using (20)-(21) are $f_{min}$=12.1 kHz and $f_{max}$=20 kHz. It is further observed that maximum and minimum switching frequencies occur at zero crossings and peaks (positive and negative) of the system voltage. This is as demonstrated through Fig. 3(a)-(d) based on (15). The slight difference between actual and computed switching frequencies is due to the filter currents violating the hysteresis band limits within the simulation time step. It is seen that there are drastic change in switching frequency of the VSI due to its inability to track the sudden changes in the reference current. Also the formulations (15), (20)-(21) hold true for tracking reference current with smooth variation.

Fig. 4 (d) shows the switching pulses and switching frequency together. A close observation reveals that the variation in switching frequency is delayed by one switching period. This is because the switching frequency can be calculated only at the end of one switching period.

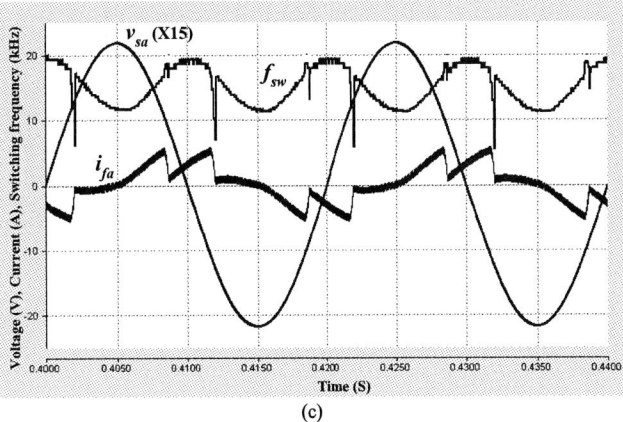

(c)

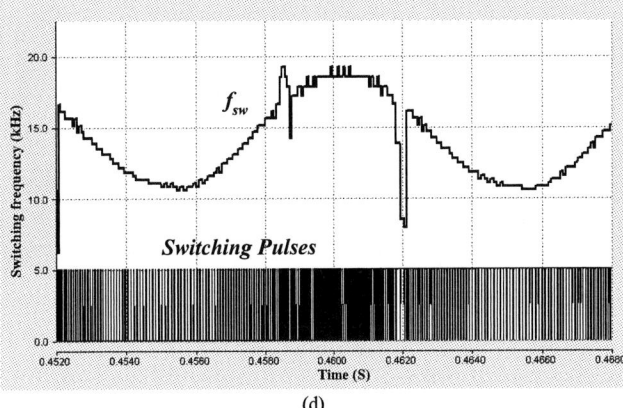

(d)

Fig. 4 PSCAD simulation results (a) Load currents (b) Source currents (c) Switching frequency variation with source voltage (d) switching frequency and switching pulses

## V. Conclusion

Various topologies of VSIs from application point of view are discussed in detail. In general for these configurations, the proper design of VSI parameters is essential for satisfactory performance of the compensator. The design of dc link voltage is confirmed with simulation as well experimental study. The analytical equations are derived to compute dc capacitor, interface inductance, hysteresis band and switching frequency. The simulation study for a three-phase, four-wire compensated system is carried out with the designed parameters using PSCAD. The switching frequency variation of the VSI is found similar to the theoretical values. The simulation results are given to evaluate the compensator performance.

## VI. References

[1] S. Iyer, A. Ghosh, and A. Joshi, "Inverter topologies for DSTATCOM applications – a simulation study", *Electric Power System Research*, Vol. 75, August 2005, pp. 161-170.

[2] B. Singh, K. Al-Hadded, and A. Chandra, "A review of active filters for power quality improvements," *IEEE Trans. Industrial Electronics*, Vol. 46, No. 5, Oct. 1998, pp. 960-971.

[3] M. El-Habrouk, M. K. Darwish, and P. Mehta, "Active power filters: a review," *IEE Proc. Electr. Power Appl.*, Vol. 147, No. 5, Sept. 2000, pp. 403-413.

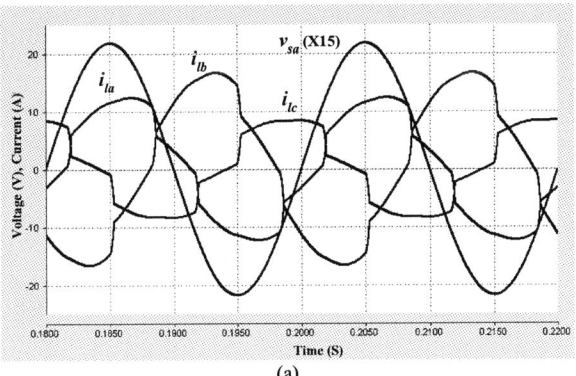

(a)

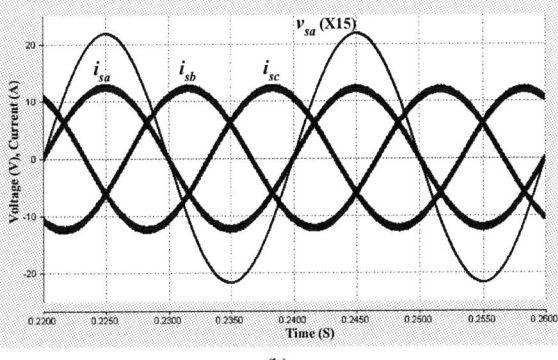

(b)

543

[4] A. Ghosh and A. Joshi, "A new approach to load balancing and power factor correction in power distribution system," *IEEE Trans. Power Delivery*, Vol. 15, No. 1, Jan. 2000, pp. 417-422.

[5] C. A. Quinn, N. Mohan, and H. Mehta, "Active filtering of harmonic currents in three-phase, four-wire systems with three-phase and single phase non-linear loads," in *Proc. 1992 Applied Power Elec. Conf.*, pp. 829-836.

[6] A. Nabae, I. Takahashi, and H. Akagi, "A new neutral-point-clamped PWM inverter," *IEEE Trans. Industrial Application*, Vol.17, No.5, Sept./Oct. 1981, pp. 518-523.

[7] M. Aredes, J. Häfner, and K. Heumann, "Three-phase four-wire shunt active filter control strategies," *IEEE Trans. Power Electronics*, Vol. 12, No. 2, March 1997, pp. 311-318.

[8] Mahesh K. Mishra, A. Ghosh, and A. Joshi, "A new STATCOM topology to compensate loads containing ac and dc components," *IEEE Power Engineering Society, Winter Meeting 2000*, Jan. 23-27, Singapore.

[9] Mahesh K. M., A. Joshi, and A. Ghosh, "Control schemes for equalization of capacitor voltages in neutral clamped shunt compensator," *IEEE Trans. Power Delivery*, Vol. 18, April 2003, pp. 538-544.

[10] Y. Chen, B. Mwinyiwiwa, Z. Wolanski, and Boon-Teck Ooi, "Regulating and equalizing dc capacitance voltages in multilevel STATCOM" *IEEE Trans. Power Delivery*, Vol. 12, No. 2, April 1997, pp. 901-907.

[11] Vilathgamuwa, D.M. Wijekoon, H.M., and Choi, S.S., "Interline dynamic voltage restorer: a novel and economical approach for multiline power quality compensation,", *IEEE Trans. Industry Applications*, Vol. 40, Issue 6, Nov.-Dec. 2004 pp. 1678 - 1685

[12] R. Srinivasan and R. Oruganti, "A unity power factor converter using half-bridge boost topology", *IEEE Trans. Power Electronics*, Vol. 13, No. 3, May 1998, pp. 487-500.

## VII. BIOGRAPHIES

**Mahesh K. Mishra** (S'2000-M'02) received his Bachelor of Technology from College of Technology, Patnagar, India and M.E. from University of Roorkee, India in 1991 and 1993 respectively. In Feb. 2002, he received the Ph.D. in Electrical Engineering from Indian Institute of Technology, Kanpur, India. He has teaching and research experience of about 12 years. For about 10 year he was a faculty in Electrical Engineering Department, Visvesvaraya National Institute of Technology, Nagpur, India. Currently he is an Assistant Professor in Electrical Engineering Department, Indian Institute of Technology Madras, India. His interests are in the areas of Power Distribution Systems, Power Electronics and Control systems.

**K. Karthikeyan** received his Bachelor of Engineering in Electrical and Electronics Engineering from Syed Ammal Engineering College-Ramanathapuram, Madurai Kamaraj University, India, in 2002 and Master of Engineering in power systems from College of Engineering Guindy-Chennai, Anna University, India in 2004. He is currently pursuing the Ph.D. degree in the Power Systems Hardware Laboratory, Department of Electrical Engineering, Indian Institute of Technology Madras, India. His fields of interest include power quality and power electronics control in power system.

**2006 IEEE International Conference on Power Electronic, Drives and Energy Systems**

# Optimal Voltage and Reactive Power Control Based on Multi-Objective Genetic Algorithm

Behzad Mirzaeian Dehkordi

*Abstract*--In this paper, a novel multi-objective optimization method based on Genetic Algorithm (MGA) is proposed. MGA is applied to optimize the three objective functions. The objective functions are designed for reactive power optimization control, such as, loss and/or invest cost minimization. The simulation results of the new method for IEEE-30 bus system are compared with the results obtained by the single weighted objective method, which shows better performance.

*Index Terms*—Genetic Algorithm, Optimization, Reactive Power Control.

## I. INTRODUCTION

REACTIVE power and voltage control [1] has been an attractive topic for research. There have been many different methods provided by researchers for solving the such as linear programming [2], expert systems [3] and fuzzy linear programming [4] as well as neural network [5]. In power systems, variations in loads and generators always exist. If the new system state has any over-voltages, under voltages or generator VAR limit violation, the system operator must take an appropriate action to alleviate the problem. These actions include the adjustment of the generator voltage, reactive power injection transformer taps, for this adjustments, it is better for operator to consider the minimization of power loss and minimization of the adjustment of the control variables. Obviously, the problem is a constrained optimization problem; Constraints are the upper and lower limit of voltage for the load bus and upper and lower limit of reactive power for the generator bus. A reactive power/voltage control problem is a non-linear multi-objective optimization problem and as such can be optimize by MGA.

The optimal power flow problem has been a traditional one in power system control and planning. Many methods have been successfully used in existing power system [6]-[11]. However, many power stations have been established in remote districts and ultra-high power transmission will be carried out. In addition, many distributed generation systems such as photovoltaic and fuel cell will be interconnected to electric power distribution systems in the future. Therefore,

system monitoring and control technology are required from a viewpoint of the voltage maintenance. Genetic algorithms (GA's) have been successfully applied to various optimization problems [12],[13].The application of GA's to multi-objective optimization has been reported in several research works,for example see Schaffer,[14] in 1985, Kursae, [15] in 1991, Horn et. al. [16] in 1994, Fonseca and Fleming, [17]-[19] in 1993, Murata,[20] in 1995, and Hisao and Tadahiko,[21] in 1998.

In this paper, MGA is applied to optimize the three objective functions. The objective functions are designed for reactive power optimization control in power system. The simulation results of the new method for IEEE-30 bus system are compared with the results obtained by the single weighted objective method [12], which shows better performance.

## II. MULTI-OBJECTIVE OPTIMIZATION BASED ON GENETIC ALGORITHM

There are three optimization methods:
Deterministic methods,
Stochastic methods,
Hybrid methods,
The deterministic methods are designed based on arithmetic rules and most of them are sensitive to initial points and need derivative of goal functions and also does not guarantee reaching the optimized point. Stochastic methods are based on statistic rules and can find the optimum point, but converge to the optimum point slowly. In hybrid method the initial points for deterministic methods are determined by stochastic methods. Genetic algorithms are powerful and broadly applicable stochastic search and optimization techniques. In the past few years the genetic algorithm community has turned much of its attention to optimization problems in industrial engineering, resulting in fresh body of research and applications. Genetic population of solutions, and are a class of general- algorithms can be used as a way to create an initial purpose search methods combining elements of directed and stochastic search space. In genetic algorithms, accumulated information is exploited by selection mechanism, while new regions of the search space are explored by means of genetic operators.

Optimization deals with the problem of seeking solutions over a set of possible choices to optimize criteria. If there is only one criterion to consider, it becomes a single-objective optimization problem, a type studied extensively for the past 50 years. If there is more than one criterion and they must be

---

Behzad Mirzaeian Dehkordi is with the Department of Electronic Engineering, Faculty of Engineering, Isfahan University, Isfahan, Iran (e-mail: mirzaeian@eng.ui.ac.ir).

0-7803-9771-1/06/$25.00 ©2006 IEEE

treated simultaneously, we have a multi-objective optimization problem. Multi-objective problems arise in the design, modeling and planning of many complex real systems in the many areas, such as: industrial production, urban transportation, capital budgeting, forest management, reservoir management, layout and landscaping of new cities, and energy distribution. Multi-objective optimization problems have been of increasing interest to researchers of various backgrounds since the early 1960s. Genetic algorithms have received problems, resulting in fresh body of research and applications known as genetic multi-objective optimizations. The inherent characteristics of genetic algorithms demonstrate why genetic search may be well-suited multi-objective optimization problems. The basic feature of genetic algorithms is multiple directional and global searches through maintaining a population of potential solutions from generation to generation.

The population-to-population approach is useful when exploring Pareto solutions. Genetic algorithms do not have many mathematical requirements and can handle all types of objective functions and constraints. Because of their evolutionary nature, this method can used to search for solutions without regard to specific inner workings of problem. Therefore, it is hoped that many more complex problems can be solved using genetic algorithms than using conventional methods.

The general multi-objective optimization problem can be stated as to find an n dimensional vector, X, such that:

$$Maximize \quad [f_1(x), f_2(x),....., \quad f_M(x)] \tag{1}$$
$$S.T. \quad x_{io} \leq x_i \leq x_{if} \quad i=1,..., n$$

Suppose that the number of population, $N$, is fixed and each solution of the multi-objective problem defined as a chromosome, $S_L$, with the length $L$.

In MGA, a probability function has been defined to select the best-fitted chromosomes for existing population. Mutation and recombination operators applied to create the new chromosomes for the new population. In the probability function, the objective functions are combined by constant weights ($K_m$), so that the chromosomes with best performances for all objective functions have more chances to be chosen for participation in the next generation. Meanwhile, chances for participation of other chromosomes are not null. The probability function is defined as

$$P_{L,i} = \frac{\prod\limits_{m=1}^{M}\left[C_m(S_{L,i})\right]^{K_m}}{\sum\limits_{L=1}^{N}\left[\prod\limits_{m=1}^{M}\left[C_m(S_{L,i})\right]^{K_m}\right]} \tag{2}$$

In this equation, $C_m(S_{L,i})$ is the fitness number of the $m$'th objective function for the $L$'th chromosome, $M$ is number of the objective functions, $N$ is the number of the chromosomes in population and, $K_m$, is the objective function weight which shows the goodness of the $m$'th objective function. In this

paper all the objective function weights are equal to one. The fitness number of the $m$'th objective function for $L$'th chromosome defined as

$$C_m(S_{L,i}) = \begin{cases} \dfrac{C_m^o(S_{L,i})}{\gamma_m} & if \quad C_m^o(S_{L,i}) \succ 0 \\ 0 & if \quad C_m^o(S_{L,i}) \leq 0 \end{cases}$$
$$L = 1,2,...,N \tag{3}$$
$$m = 1,2,...,M$$

In this equation $C_m^\circ(S_{L,i})$ is the fitness value for the $m$'th objective function and $\gamma_m$ is the summation of the positive fitness values. Fitness value for each objective function is defined as

$$C_m^o(S_{L,i}) = f_{objm}(S_{L,i}) - \mu_m$$
$$m = 1,2,..., M \tag{4}$$
$$L = 1,2,...., N$$

Where, $\mu_m$ is the mean value of the $m$'th objective function in population sequences.

## III. NEWTON RAPHSON METHOD

There are different methods of modeling physical processes and also different methods of parameter estimation. Except in linear problems, root-finding proceeds by iteration, and this is equally true in one or many dimensions. Starting from some approximate trial solution, a useful algorithm will improve the solution until some predetermined convergence criterion is satisfied. However, how crucially success depends on having a good first-guess for the solution, especially for multidimensional problems. Models vary according to the input-output relation and especially according to the statistical prosperities of the model error; most of them are explained in [6]. In our case the input-output relation is well defined by Newton-Raphson:

$f_i(x_1,x_2...x_n)=0$ are the functions used to be solved. We have Jakobian matrix as follows

$$J = \begin{bmatrix} \dfrac{\partial f_1}{\partial x_1} & \dfrac{\partial f_1}{\partial x_2} & \cdots & \dfrac{\partial f_1}{\partial x_n} \\ \dfrac{\partial f_2}{\partial x_1} & \dfrac{\partial f_2}{\partial x_2} & \cdots & \dfrac{\partial f_2}{\partial x_n} \\ . & . & . & . \\ . & . & . & . \\ \dfrac{\partial f_n}{\partial x_1} & \dfrac{\partial f_n}{\partial x_2} & \cdots & \dfrac{\partial f_n}{\partial x_n} \end{bmatrix} \tag{5}$$

The irritation begins with initial values. The new values for $X$ vector can be defined as:

$$X_{new} = -inv(J) * F$$
$$F = [f_1 \quad f_2 \quad ..... \quad f_n]^{-1} \tag{6}$$

While the difference between $X_{new}$ and $X_{old}$ is larger than a specified $\varepsilon$ this iteration goes on.

## IV. MATHEMATICS MODEL OF REACTIVE POWER OPTIMIZATION

Reactive power optimization control conventionally employs a suitable capacitance compensation and switching transformer tap settings to improved voltage quality, This must use optimization method to determine reactive power compensation capacity, compensation site and transformer tap setting as well as coordinating site and transformer tap setting as well as coordinating with each other. Mathematics models contain power flow constraint equations, variable constraint conditions and objective function.

### A. Power Flow Constraint Equations

Power flow constraint equations employ Newton-Raphson equations in polar coordinate form

$$P_i = V_i \sum_{j=1}^{N_B} V_j (G_{ij} \cos \delta_{ij} + B_{ij} \sin \delta_{ij})$$

$$Q_i = V_i \sum_{j=1}^{N_B} V_j (G_{ij} \sin \delta_{ij} - B_{ij} \cos \delta_{ij}) \quad (7)$$

Where $P_i$ and $Q_i$ represent real and reactive power, respectively, injected into network at bus $i$ (pu); $V_i$ represent mutual conductance and susceptance, respectively, between bus $i$ (pu); $G_{ij}$ and $B_{ij}$ represent mutual conductance and susceptance, respectively, between bus $i$ and bus $j$ (pu); $\delta_{ij}$ represents voltage angle difference between bus $i$ and bus $j$ (rad); $N_B$ represents set of number of total buses.

Real power loss in the system can be started as follow:

$$P_L = \sum_{i=1}^{N_B} V_i \sum_{j \in h} V_j (G_{ij} \cos \delta_{ij} + B_{ij} \sin \delta_{ij}) \quad (8)$$

Where $h$ represents set of number of busses adjacent to bus $i$, including bus $i$.

### B. Variable Constraint Conditions

In this paper, variables are divided into control variables and state variables. The transformer tap-setting $T$, generator bus voltages $V_E$ and reactive power source installations $Q_c$ are control variables so they are self-restricted. The load bus voltages $V_{load}$ and reactive power generations $Q_g$ are state variables, which are restricted by adding them as two objective functions.

Inequality constrains of control variables are written as,

$$T_{i\,min} \leq T_i \leq T_{i\,max}$$

$$Q_{cj\,min} \leq Q_{cj} \leq Q_{cj\,max} \quad (9)$$

$$V_{gk\,min} \leq V_{gk} \leq V_{gk\,max}$$

Inequality constrains of state variables are written as,

$$V_{i\,min} \leq V_i \leq V_{i\,max}$$

$$Q_{i\,min} \leq Q_i \leq Q_{i\,max} \quad (10)$$

Where $X_{i\,min}$, $X_{i\,max}$ represent the upper and lower limits of corresponding variables

### C. Objective Function

There are many aims in reactive power optimization control, such as, loss and/or invest cost minimization. Some constraints might be treated as penalty functions. This paper constructs objective functions as follow:

$$Min \quad f_1 = \sum_{i=1}^{N_B} V_i \sum_{j \in h} V_j (G_{ij} \cos \delta_{ij} + B_{ij} \sin \delta_{ij})$$

$$Min \quad f_2 = \sum \left( \frac{V_i - V_{i\,lim}}{V_{i\,max} - V_{i\,min}} \right)^2 \quad (11)$$

$$Min \quad f_3 = \sum \left( \frac{Q_i - Q_{i\,lim}}{Q_{i\,max} - Q_{i\,min}} \right)^2$$

On the equation, the first objective function is network loss; the second and third objective functions are penalty functions of voltage and reactive power violate limits.

$$if \ X_i > X_{i\,max}, \qquad then \quad X_{i\,lim} = X_{i\,max};$$

$$if \ X_i < X_{i\,min}, \qquad then \quad X_{i\,lim} = X_{i\,min};$$

$$if \ X_{i\,min} \leq X_i \leq X_{i\,max}, \quad then \quad X_{i\,lim} = X_i;$$

Where $X_{i\,lim}$ ($X_{i\,min}$ and $X_{i\,max}$) denote the upper and lower limits of integer value of variables.

## V. REACTIVE POWER OPTIMIZATION CONTROL BASED ON MGA

Discrete control variables are represented as a chromosome or string $X$

$$X = [T \,|\, Q_c \,|\, V_g]$$

$$= [T_1, T_2, \ldots \,|\, Q_{c1}, Q_{c2}, \ldots \,|\, V_{g1}, V_{g2}, \ldots] \quad (12)$$

Where $T_l$: tap position at $i$-th transformer.

$Q_{cj}$: No. of shunt compensator banks at $i$-th bus.

$V_{gk}$: position of generator voltage at $i$-th

Generator (capital letter implies integer values).

The first generation of chromosomes is created using uniform random variables

$$X_i = INT(rnd \times (X_{i\max} - X_{i\min} + 1)) + X_{i\min}$$

Where *rnd*: Generate uniform random value
   *0<rnd<1.*
   *INT* (*): Return the next lowest whole number of *.

## VI. ALGORITHM PROGRAM FLOWCHART

Fig. 1 and Fig. 2 show the proposed algorithm and its flowchart.

## VII . SIMULATION RESULTS

In this section, the IEEE-30 bus system is used to show the effectiveness of the algorithm. Partial parameters of the system are given in Table I. The variable limits of system are given in Table II. The rest results are compared with the results obtain in [22] named FSGA, the comparison results of these algorithms are given in Table III. In these Tables, all parameters are represented in pu.

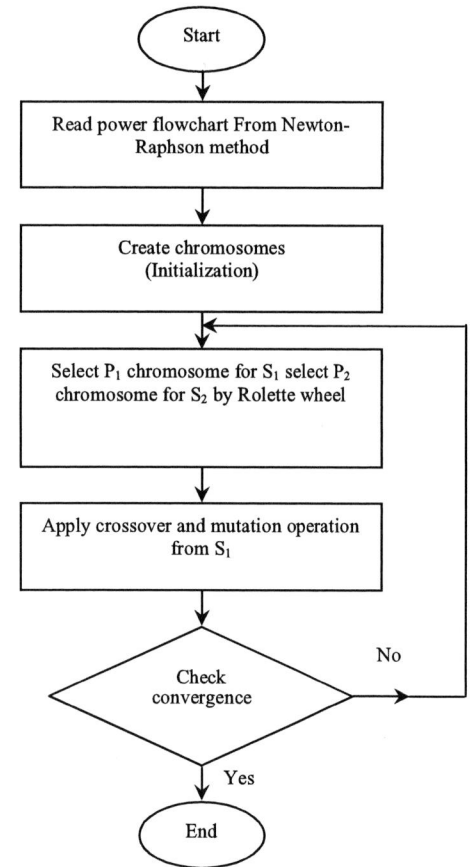

Fig. 1. Algorithm program flowchart.

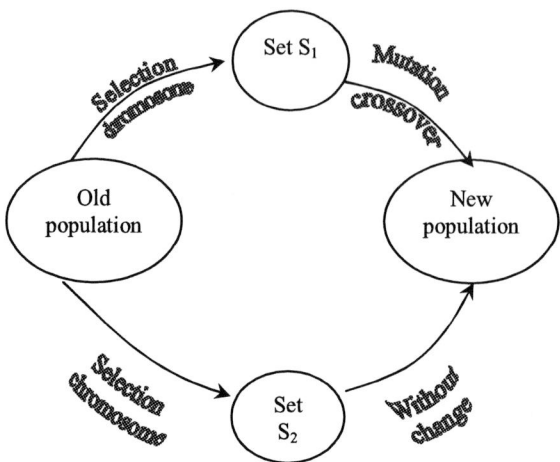

Fig. 2.   S$_1$ and S$_2$ set in GA.

TABLE I
PARTIAL PARAMETERS OF POWER SYSTEM

| items \ systems | IEEE-30 bus system |
|---|---|
| the number of generators | 6 |
| the number of buses | 30 |
| the number of branches | 41 |
| the number of under load tap setting transformers | 4 |
| the number of shunt compensator | 4 |
| the number of total real load (pu) | 5.668 |
| the number of total reactive load (pu) | 2.524 |

TABLE II
VARIABLES LIMIT

| systems \ items | bus | Ogmax | Ogmin | Vmax | Vmin |
|---|---|---|---|---|---|
| IEEE-30 bus system | 1 | 0.596 | -0.298 | | |
| | 2 | 0.480 | -0.24 | | |
| | 5 | 0.6 | -0.3 | 1.05 | 0.95 |
| | 8 | 0.53 | -0.265 | | |
| | 11 | 0.15 | -0.075 | | |
| | 13 | 0.155 | -0.078 | | |

TABLE III
COMPARISON OF TEST RESULTS

| items \ systems | IEEE-30 bus system | | |
|---|---|---|---|
| algorithm | PF | MGA | FSGA |
| network loss | 0.239 | 0.221 | 0.231 |
| reactive decrease | 0.436 | 0.420 | 0.430 |
| voltage violate limits node numbers | 2 | 0 | 0 |
| reactive violate limits node numbers | 0 | 0 | 0 |
| system average voltage | 0.961 | 1.002 | 1.035 |
| load node average power coefficient | 0.845 | 0.921 | 0.891 |
| computation time (min) | 2.54 | 1.34 | 1.56 |

548

## VIII. CONCLUSION

In this paper, a new method based on genetic algorithms for multi-objective optimization process is proposed. The method is successfully applied for optimization of three objective functions for optimal power reactive control in power systyem.The simulation results of the new method applied for IEEE-30 bus system are compared with the results obtained by the single weighted objective method [22], which shows better performance.

The new MGA method has shown promising success in the multi-objective optimization problems with constraints. Specially, it is suitable for optimization problems which consist of many constrained parameters and several contradictory objective function.

## IX. REFERENCES

[1]    Abdul Rahman,K.H. and S.M. Shahidehpour," A Fuzzy based optimal reactive power control," *IEEE Trans. on Power Systems*, vol.8,no.2, 1993.

[2]    Mamandur,K.R.C.,"Emergency adjustment to var control variables to alleviate over-voltages, under-voltage and generator var limit violations," *IEEE Trans. on Power Apparatus and Systems*, vol. PAS-101,no.5,1982.

[3]    Ananthapadmanabha T.,A.D. Kulkarni,A.S.G.Rao, R.Rao and K.parthasarathy, "Knowledge based expert system for optimal reactive power control in distribution system," *Electrical Power & Energy Systems* ,vol. 18, no.1,pp..27-31, 1996.

[4]    Tomsovic,K., "A Fuzzy linear programming approach to the reactive power/voltage control problem," *IEEE Trans. on Power System*,vol.7.,1992.

[5]    Su.C.T. and C.T. Lin, "Application of neural network and heuristic model for voltage-reactive power control," *Electric Power Systems Research*,vol.34,pp.143-148,1995.

[6]    H.W.Dommel & W.F.Tinney, "Optimal power flow solution," *IEEE Trans. Power Apparatus. Sys.* PAS-87, 1988.

[7]    A.M.Sasson , "Nonlinear programming Solution for load-flow, minimum-loss and econimic dispatching problems," *ibid,. PAS*-88, pp.399,1969.

[8]    R.A.Fernamdes,et.al, "Large scale reactive planning," *IEEE Trans. on PAS,pp.*1083,1983.

[9]    C.C.Liu, et.al, "An expert system assisting decision making of VQ control," *IEEE TRWRS-1* no.3, 1986.

[10]   K.Nara,et.al, "Implementation of genetic algorithm for distribution system loss minimum re-configuration," presented at the IEEE PES Summer Meeting, 9ISM 467-1 PWSR ,1991.

[11]   K.lba, "Reactive power optimization by genetic algorithm," *IEEE Trans. on PS,*pp. 685, 1994.

[12]   D.E.Goldberg, *Genetic Algorithms in Search, Optimization, and Machine Learning Reading,* Addison-Wesley,1989.

[13]   L.Davis,Ed., *Handbook of Genetic Algorithms,* New York: Van Nostrand Reinhold,1991.

[14]   J.D.Schaffer, "Multi-objective optimization with vector evaluated genetic algorithms," *in Proc. 1985 Ist. Conf. Genetic Algorithms,* pp.93-100.

[15]   F. Kursawe, "A variant of evolution strategies for vector optimization in parallel problem solving from nature," in *Proc. 1991 H.P. Schaeffer and R. Manner, Eds. Berlin, Germany, Springer-verlag,*pp.193-197.

[16]   J. Horn, N. Nafpliotis and D.E. Goldberg, "A niched pareto genetic algorithm for multi-objective optimization," *in Proc. 1994 Ist IEEE Int. Conf. Evolutionary Computat,pp.*82-86.

[17]   C.M. Fonseoa and P.J. Fleming , *An overview of evolutionary algorithms in multi-objective optimization,* Dept. of Automatic Control and Systems Eng.University of Sheffiel.UM U.K. Res. Rep. 527,1994.

[18]   C.M. Fonseca and P.J. Fleming, "Genetic algorithms for multi-objective optimization, discussion, and generalization, Genetic Algorithms," proc. fifth. IM, conf. 5. forrest. Ed. scan. Mareo, C*A/ Morgan Kaufmann:* 416-423.,1993.

[19]   C.M. Fonseca and P.J. Fleming, "Multi-objective genetic algorithms made easy, selection, sharing and making restriction,"*in Proc. 1995 Ist IEE/IEEE Int. conf. GA'S in Engineering, Systems. Innovations and Applications, Sheffield. U.K,*pp. 42-52.

[20]   T. Murata and H. Ishibuchi, "MOGAI multi-objective genetic algorithms," *in Proc. 1995 IEEE Int, Conf, Evolutionary Computed,*pp.289-294.

[21]   Hisao lshibuchi, T. Murata, "A multi-objective genetic local search algorithm and application to flow-shop scheduling," *IEEE Trans. on sys. Man and cyb.* Vol.28, NO.3: 392-403,1998.

[22]   M.Xiangping,L.Zhishan,Z,Huaguang, "Fast synthetic genetic algorithm and its application to optimal control of reactive power flow," *in Proc. 1998PowerSystemchnology,Proceeding,PowerCon'98,1998,Internation al Conference ,Volume 2,18-21,*pp.1454-1458.

**Behzad Mirzaeian Dehkordi** was born in Shahr-e-kord, in the year 1966. He received the B.Sc. of Electronics Engineering from Shiraaz University, Iran in 1985 and M.Sc. and Ph.D. of Electrical Engineering form Isfahan University of Technology in the years 1994 and 2000 respectively.
He is currently an assistant professor at the department of electrical and electronics engineering, the university of Isfahan.

His field of interests include power electronics and drives and power quality problems.

**2006 IEEE International Conference on Power Electronic, Drives and Energy Systems**

# Model Validation Studies in Obtaining Q-V Characteristics of P-Q Loads in Respect of Reactive Power Management and Voltage Stability.

G. Govinda Rao, and K. V. S. Ramachandra Murthy

*Abstract* – The paper largely addresses to investigate the validity of the Thevenin Model in obtaining Q-V characteristics of P-Q loads, the main consideration for analysis of the problem of voltage instability and reactive power management.

*Index Terms* - Reactive Power, Voltage Stability, Q-V characteristics.

## I. INTRODUCTION

THE problem of voltage instability is gaining more and more importance in developed countries, because of the unusual growth of power systems and insufficient or inefficient reactive power management. A little too frequent voltage collapse conditions in both western and eastern parts of U.S. in the recent past could be perhaps due to this. The subject voltage stability in contrast with the rotor angle stability is rather of recent times, only one or two decades old.

The problem of voltage instability has two facets. One, the dynamic case and the other the static case. As a static viability problem, the voltage stability is well analysed with the help of Q-V characteristics and the associated reactive power limit and voltage collapse. A theoretical study to obtain the reactive power limit for the voltage collapse usually is obtained through a model containing a voltage source feeding the load through a series reactance. The variation of the voltage magnitude at the load bus versus reactive power varying from leading through lagging is plotted to get the complete Q-V characteristic. The source in series with the impedance could be by itself a system or is expected to be the Thevenin equivalent of a power network feeding a P-Q load. How exact is the Thevenin equivalent of the entire network with respect to the load under investigation and how to construct the equivalent is the main concern of this paper.

## II. THE PROBLEM OF VOLTAGE STABILITY AND REACTIVE POWER MANAGEMENT

References [2] and [3] make a detailed presentation on this subject and [1] discusses in particular the voltage stability problem and analyses the Q-V characteristics of a P-Q load. While the major literature on this presents the analysis of Q-V

Dr. G. Govinda Rao, M.E., Ph.D.
(Former Professor of Andhra University).
Professor in E.E.E, G.V.P.College of Engg., Visakhapatnam
K. V. S. Ramachandra Murthy, Associate Professor in E.E.E,
G.V.P.College of Engg., Visakhapatnam, India.

characteristics in terms of an approximate formula derived for the voltage magnitude at the load bus, [1] derives an exact formula for the voltage, which has been the basis for computation of the Q-V characteristics of the Thevenin Model.

## III. ABOUT THE THEVENIN MODEL

A power network feeding a P-Q load, modeled as Thevenin Equivalent containing voltage source in series with an impedance is considered for obtaining Q-V characteristics. In fact using such Thevenin Model to investigate the performance of a test-subsystem, with the rest of the system as a Thevenin source is much more than a common practice in the state of the art.

Can a Power Grid, with respect to a load bus be represented as a Thevenin Equivalent? Although the power network contains linear elements, the load flow equations are non-linear because of the fact that powers are inputs and voltages are outputs. Thevenin model being exclusively true for linear networks, can not apparently be used in a non-linear environment. But it seems that several research investigations are being conducted to generalize certain facts, on the basis of a Thevenin Model. Should the Thevenin model be rejected? This question is answered in the affirmative showing its limitations.

Firstly the Thevenin Model is very largely used in Fault analysis. A base case load flow study is conducted on the Power Network and the voltage at the faulted bus is considered to be the $V_{th}=V_{oc}$. Then $Z_{bus}$ is formulated by replacing the generator buses with transient / sub transient impedances. The load currents, being very small with respect to fault currents are neglected and set to zero. That means, the diagonal elements of $Y_{BUS}$ corresponding to Load are not altered. But the elements of $Y_{BUS}$ corresponding to generator buses (P-V Buses) only are altered. The $Y_{BUS}$ so formed may be inverted to get $Z_{BUS}$ or $Z_{BUS}$ may be directly formed. The diagonal element of $Z_{BUS}$ corresponding to the faulted bus becomes the $Z_{th}$ to be used in fault calculations.

This way of getting $Z_{th}$ will not work in the present situation and is also shown that it will not work. However if the load buses are approximated to constant voltage sources and so much so, the generator buses (P-V buses) are also approximated to voltage sources and the thevenin impedance is computed at the load bus, the Thevenin Model works with

0-7803-9771-1/06/$25.00 ©2006 IEEE

a reasonable accuracy. The justification for this is that the voltage magnitudes at the various P-Q buses deviate, only in the small from the nominal voltages, for changes of load at any particular load bus. This is a presumption. Thus the $Z_{th}$ so obtained is indeed, the reciprocal of the diagonal element of the $Y_{BUS}$, corresponding to the load bus under investigation for its Q-V characteristic.

Case studies are conducted on standard sample networks of 3-Bus, 6-Bus 14-Bus and 39 bus systems, to establish the said fact.

## IV. CASE STUDIES CONDUCTED

A case study is considered in the following way. A power network is considered. A base case load flow is conducted, setting P=0 and Q=0 at the P-Q bus and Vth = Voc at the load bus obtained. Further, the bus is loaded for a given P, changing Q from high lagging to high leading. This load bus voltage magnitudes are computed and Q-V characteristics obtained, by conducting load flow studies The load flow program used in this work has been validated by testing it on systems with known results.

Now, in the same $Y_{BUS}$ used for Load Flow programme, the diagonal element is considered as the reciprocal of the Thevenin impedance. This implies that $Z_{th}$ is a short-circuit parameter but not the O.C. parameter. The Thevenin equivalent is used, to obtain Q-V characteristic of the P-Q load using the formulae derived in [1].

### A. Case Study 1

Case study 1 is conducted on a 3 bus system given in [6]. Single line diagram of 3 bus system is given in Fig. 1. Data is given in Table No. I. Bus No. 3 is P-Q load under investigation. For a given P of 1.4 pu. Q is varied between 0 and 4.0 and magnitude of voltage is obtained by Thevenin Model as well as by conducting the load flow study. Results are compared in Table II.

TABLE I
DATA FOR 3 BUS SYSTEM

**Line Data:**

| From Bus | To Bus | Resistance | Reactance | Susceptance |
|---|---|---|---|---|
| 1 | 2 | 0.02 | 0.08 | 0.02 |
| 1 | 3 | 0.02 | 0.08 | 0.02 |
| 2 | 3 | 0.02 | 0.08 | 0.02 |

**Bus Data**

| Bus No. | Type | Magnitude Of voltage | Angle | $P_G$ | $P_D$ | Q |
|---|---|---|---|---|---|---|
| 1 | Slack | 1.02 | 0.0 | - | - | - |
| 2 | Gen | 1.01 | -- | 0.5 | - | - |
| 3 | Load | -- | -- | -- | 1.4 | variable |

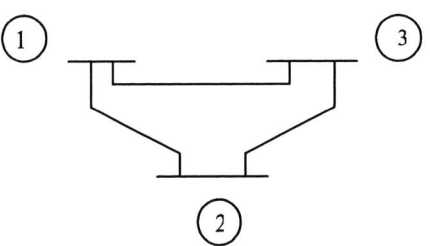

Fig. 1. Single Line Diagram for 3 Bus System.

TABLE II
Q-V ANALYSIS ON BUS NO. 3 OF 3-BUS SYSTEM

| $Q_3$ | Voltage obtained from Load Flow solution | Voltage obtained from Thevenin Model | Error |
|---|---|---|---|
| 0.0 | 1.0024 | 1.0001 | 0.0001 |
| 0.5 | 0.9797 | 0.9796 | 0.0001 |
| 1.0 | 0.9582 | 0.9581 | 0.0001 |
| 1.5 | 0.9356 | 0.9355 | 0.0001 |
| 2.0 | 0.9118 | 0.9117 | 0.0001 |
| 2.5 | 0.8864 | 0.8863 | 0.0001 |
| 3.0 | 0.8592 | 0.8591 | 0.0010 |
| 3.5 | 0.8297 | 0.8295 | 0.0020 |
| 4.0 | 0.7971 | 0.7969 | 0.0020 |

### B. Case Study 2

Case study 2 is conducted on a 6-bus system. Single line diagram of 6 bus system is given in Fig. 2. Bus No. 5 is P-Q load under investigation. For a given P of 0.7 pu Q is varied between –2 pu and 2.5 pu and magnitude of voltage is obtained by Thevenin Model as well as by conducting the load flow study. Results are compared in Table IV.

TABLE III
DATA FOR 6-BUS SYSTEM

**Line Data**

| From Bus | To Bus | Resistance | Reactance | Susceptance |
|---|---|---|---|---|
| 1 | 2 | 0.10 | 0.20 | 0.04 |
| 1 | 4 | 0.05 | 0.20 | 0.04 |
| 1 | 5 | 0.08 | 0.30 | 0.06 |
| 2 | 3 | 0.05 | 0.25 | 0.06 |
| 2 | 4 | 0.05 | 0.10 | 0.02 |
| 2 | 5 | 0.10 | 0.30 | 0.04 |
| 2 | 6 | 0.07 | 0.20 | 0.05 |
| 2 | 5 | 0.12 | 0.26 | 0.05 |
| 3 | 6 | 0.02 | 0.10 | 0.02 |
| 4 | 5 | 0.20 | 0.40 | 0.08 |
| 5 | 6 | 0.10 | 0.30 | 0.06 |

**Bus Data**

| Bus No. | Type | Magnitude Of voltage | Angle | $P_G$ | $P_D$ | $Q_D$ |
|---|---|---|---|---|---|---|
| 1 | Slack | 1.05 | 0.0 | - | - | |
| 2 | Gen | 1.05 | -- | 0.50 | -- | -- |
| 3 | Gen | 1.07 | -- | 0.60 | -- | -- |
| 4 | Load | -- | -- | | 0.70 | 0.70 |
| 5 | Load | -- | -- | | 0.70 | variable |
| 6 | Load | -- | -- | | 0.70 | 0.70 |

551

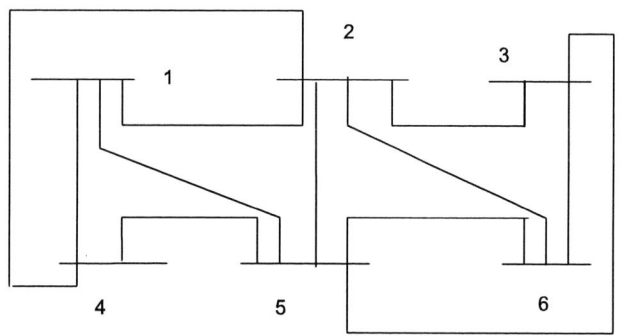

Fig. 2. Single Line Diagram for 6 Bus System.

### TABLE IV
### Q-V ANALYSIS ON BUS NO. 5 OF 6 BUS SYSTEMS

| $Q_5$ | Voltage obtained from Load Flow solution | Voltage obtained from Thevenin Model | Error |
|---|---|---|---|
| -2.0 | 1.1473 | 1.1414 | 0.0059 |
| -1.5 | 1.1209 | 1.1162 | 0.0047 |
| -1.0 | 1.0930 | 1.0897 | 0.0033 |
| -0.5 | 1.0637 | 1.0619 | 0.0018 |
| 0.0 | 1.0326 | 1.0325 | 0.0001 |
| 0.5 | 0.9994 | 1.0012 | 0.0018 |
| 1.0 | 0.9635 | 0.9675 | 0.0040 |
| 1.5 | 0.9242 | 0.9308 | 0.0066 |
| 2.0 | 0.8804 | 0.8904 | 0.0100 |
| 2.5 | 0.8321 | 0.8446 | 0.0125 |

### C. Case Study 3

Case study 3 is conducted on a 14 bus system. Single line diagram of 14 bus system is given in Fig. 3. Bus No. 9 is P-Q load under investigation. For a given P of 0.478 pu. Q is varied between –3.0 pu and 2.0 and magnitude of voltage is obtained by Thevenin Model as well as by conducting the load flow study. Results are compared in Table VII.

### TABLE V
### GENERATOR AND LOAD DATA FOR 14 BUS SYSTEM

**Generation and Load Data**

| Bus No. | Type | Magnitude Of voltage | Angle | $P_G$ | $P_D$ | $Q_D$ |
|---|---|---|---|---|---|---|
| 1 | Slack | 1.06 | 0.0 | - | -- | -- |
| 2 | Gen | 1.045 | -- | 0.40 | 0.217 | 0.127 |
| 3 | Gen | 1.070 | | 0.20 | 0.112 | 0.075 |
| 4 | Gen | 1.010 | -- | 0.00 | 0.942 | 0.190 |
| 5 | Gen | 1.090 | -- | 0.00 | -- | -- |
| 6 | Load | -- | -- | -- | -- | -- |
| 7 | Load | -- | -- | -- | 0.295 | 0.166 |
| 8 | Load | -- | -- | -- | 0.076 | 0.016 |
| 9 | Load | -- | -- | -- | 0.478 | variable |
| 10 | Load | -- | -- | -- | 0.090 | 0.058 |
| 11 | Load | -- | -- | -- | 0.035 | 0.018 |
| 12 | Load | -- | -- | -- | 0.061 | 0.016 |
| 13 | Load | -- | -- | -- | 0.135 | 0.058 |
| 14 | Load | -- | -- | -- | 0.149 | 0.050 |

### TABLE VI
### LINE DATA FOR 14 BUS SYSTEM

**Line Data**

| From Bus | To Bus | Resistance | Reactance | Susceptance |
|---|---|---|---|---|
| 1 | 2 | 0.01938 | 0.05917 | 0.0264 |
| 2 | 4 | 0.04699 | 0.19797 | 0.0219 |
| 2 | 9 | 0.05811 | 0.17632 | 0.0187 |
| 1 | 8 | 0.05403 | 0.22304 | 0.0246 |
| 2 | 8 | 0.05695 | 0.17388 | 0.0170 |
| 4 | 9 | 0.06701 | 0.17103 | 0.0173 |
| 9 | 8 | 0.01335 | 0.04211 | 0.0064 |
| 8 | 3 | 0.00000 | 0.25202 | 0.0000 |
| 9 | 6 | 0.00000 | 0.20912 | 0.0000 |
| 6 | 5 | 0.00000 | 0.17615 | 0.0000 |
| 7 | 9 | 0.00000 | 0.55618 | 0.0000 |
| 7 | 6 | 0.00000 | 0.11001 | 0.0000 |
| 7 | 10 | 0.03181 | 0.08450 | 0.0000 |
| 3 | 11 | 0.09498 | 0.19890 | 0.0000 |
| 3 | 12 | 0.12291 | 0.25581 | 0.0000 |
| 3 | 13 | 0.06615 | 0.13027 | 0.0000 |
| 7 | 14 | 0.12711 | 0.27038 | 0.0000 |
| 10 | 11 | 0.08205 | 0.19207 | 0.0000 |
| 12 | 13 | 0.22092 | 0.19988 | 0.0000 |
| 13 | 14 | 0.17093 | 0.34802 | 0.0000 |

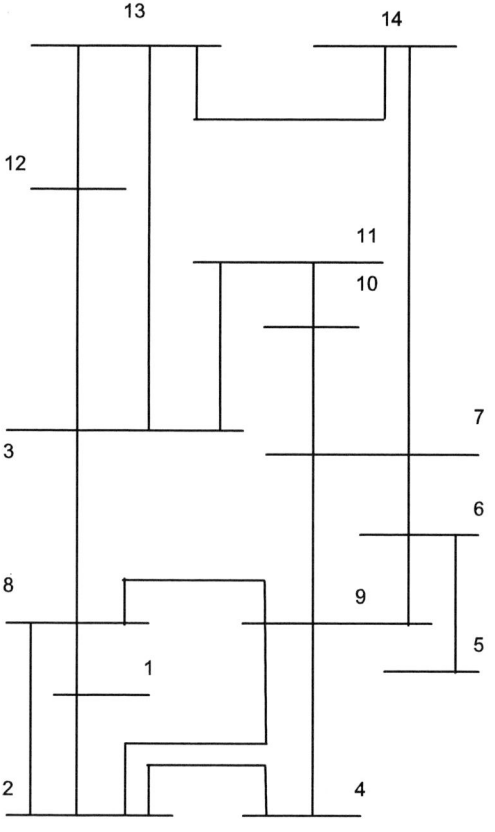

Fig. 3. Single Line Diagram for 14 Bus System.

### TABLE VII
### Q-V ANALYSIS ON BUS NO. 9 OF 14-BUS SYSTEM

| $Q_9$ | Voltage obtained from Load Flow solution | Voltage obtained from Thevenin Model | Error |
|---|---|---|---|
| -3.0 | 1.1373 | 1.0981 | 0.0392 |
| -2.5 | 1.1205 | 1.0877 | 0.0328 |
| -2.0 | 1.1032 | 1.0771 | 0.0261 |
| -1.5 | 1.0854 | 1.0663 | 0.0191 |
| -1.0 | 1.0669 | 1.055 | 0.0118 |
| -0.5 | 1.0478 | 1.0439 | 0.0039 |
| 0.0 | 1.0280 | 1.0324 | 0.0044 |
| 0.5 | 1.0072 | 1.0206 | 0.0134 |
| 1.0 | 0.9855 | 1.0085 | 0.0230 |
| 1.5 | 0.9627 | 0.9960 | 0.0333 |
| 2.0 | 0.9386 | 0.9832 | 0.0446 |

*C. Case Study 4*

Case study 4 is conducted on a 39 bus system taken from [7]. Bus No. 24 is P-Q load under investigation. For a given P of 3.086 pu., Q is varied between –3.0 pu and 2.5 and magnitude of voltage is obtained by Thevenin Model as well as by conducting the load flow study. Results are compared in Table VIII.

### TABLE VIII
### Q-V ANALYSIS ON BUS NO. 24 OF 39-BUS SYSTEM

| $Q_{24}$ | Voltage obtained from Load Flow solution | Voltage obtained from Thevenin Model | Error |
|---|---|---|---|
| -3.0 | 1.0222 | 0.9998 | 0.0225 |
| -2.5 | 1.0163 | 0.9973 | 0.0190 |
| -2.0 | 1.0102 | 0.9948 | 0.0154 |
| -1.5 | 1.0041 | 0.9923 | 0.0118 |
| -1.0 | 0.9979 | 0.9897 | 0.0082 |
| -0.5 | 0.9916 | 0.9872 | 0.0044 |
| 0.0 | 0.9852 | 0.9846 | 0.0006 |
| 0.5 | 0.9788 | 0.9820 | 0.0032 |
| 1.0 | 0.9722 | 0.9794 | 0.0071 |
| 1.5 | 0.9656 | 0.9768 | 0.0112 |
| 2.0 | 0.9589 | 0.9742 | 0.0153 |
| 2.5 | 0.9520 | 0.9716 | 0.0196 |

## III. DISCUSSION OF RESULTS

Normally, in several investigations, as the size of the system increases, the error gets lower. Here, it seems to be the other way. Because, the comparison of the results of a non-linear system and those of an approximated linear system cannot be equal on theoretical grounds. As already pointed out neither the sources specified by P and V nor the loads can be treated as ideal voltage sources. But load flow studies reveal that the changes in bus voltage magnitudes when Q at a specific load bus is varied will not change in the large. Thus, the Thevenin Model can be used to study the Q-V characteristics. However, large variations in Q and large systems should caution the investigator.

That the $Z_{th}$ should not be computed as for the fault analysis is established by the results presented in the table IX corresponding to the 14 bus system, in which only one result is shown. The error can be seen to be 22.9percent and similar

is the case when Q is varied and also for other systems. Thus, the results are not at all in agreement with those obtained from Load flow studies. Infact, this conclusive inference may be considered as the major contribution of the work presented in this paper. The error appears to increase with the system size and also with wide variations in Q.

### TABLE IX
### COMPARISION OF VOLTAAGE OF 9TH BUS

| Load on 9th Bus for Q-V analysis | Voltage obtained from Load flow solution | Voltage obtained with $Z_{th}$ taken from $Z_{bus}$ |
|---|---|---|
| 0.478+j0 | 1.0280 | 0.8009 |

## IV. CONCLUSIONS

Investigations have been carried out to validate the Thevenin Model to model a non-linear system with some presumptions. Particularly, how to compute Thevenin impedance has been presented and discussed. That this Thevenin impedance cannot be the same as the one that is determined for fault analysis is an interesting observation that has been discovered. The Thevenin Model presented works with a very reasonable accuracy for normal range of Q-variations. The error appears to increase with the system size and with also with wide variations in Q.

## V. ACKNOWLEDGEMENTS

The authors gratefully acknowledge B. R. M. Gandhi and C.V.K. Bhanu for helpful discussions and providing some data.

## VI. REFERENCES

[1] G. Govinda Rao, D. Thukaram, and H.P.Khincha, "Some Reflections on Q-V Characteristics of the Loads, " in *Proc, 1998 IEEE ,on Energy Management and Power Delivery Conf.,* vol. 1, pp79-84.

[2] C. Taylor, *Power Systems Voltage Stability,* Vol. I. New York: McGraw-Hill , 1994, pp.27-34.

[3] T. J. E. Miller, *Reactive power control in Electric systems,* New York: John Wiley, 1982, pp. 6-20.

[4] William D. Stevenson, Jr. *Elements of Power System Analysis,* 4th ed.., McGraw-Hill, 1982. pp 193-226.

[5] Peter W. Sauer, M. A. Pai *Power System Dynamics and Stability,* Low Price ed., Pearson Educational Inc. , 1998 pp278-279.

[6] I. J. Nagrath, D. P. Kothari, *Power System Engineering,* 1st ed., Tata McGraw-Hill Publishing Company Ltd., New Delhi, 1994. pp226

[7] K. R. Padiyar, *Power System Dynamics, Stability and Control,* 2nd ed., B. S. Publications, Hyderabad, 2002. pp 547-550.

## VII. BIOGRAPHIES

G.Govinda Rao was born in Srikakulam, Andhra Pradesh, India on January 19th, 1939. He graduated in Electrical Engineering from the College of Engineering, Kakinada, affiliated to Andhra University in 1959. He obtained his M.E. Degree from Andhra University in 1966 and Ph.D. frm I.I.Sc, Bangalore in 1981. He had his teaching career at Andhra University college of Engineering for four decades and is currently employed in Gayatri Vidya Parishad College of Engineering, Visakhapatnam, India. Govinda Rao received the Best Teacher Award from the A.P. State Govt. in 1996 and is a Felllow of Institute of Engineers, India and a Member of the Indian Society of Technical Education. His research interests are in the areas of Fundamentals of Circuit Analysis and Power System Analysis and Stability.

K.V.S.Ramachandra Murthy was born in Kakinada, Andhra Pradesh, India on May 17th, 1972. He did his graduation in Electrical Engineering and M.Tech in Power Systems from Regional Institute of Technology, Jamshedpur affiliated to Ranchi University in 1994 and 2002 respectively. He is pursuing his Ph.D. from Jawaharlal Nehru Technological University, Hyderabad, Andhra Pradesh, India. He had his experience as Maintenance Engineer in Easy Call Communications, Hyderbad for about four years and Teaching Experience of 6 years. He is currently employed as Associate Professor in Electrical Engineering Department in Gayatri Vidya Parishad College of Engineering, Visakhapatnam, India. His research interests are in the Management of Electrical Energy.

**2006 IEEE International Conference on Power Electronic, Drives and Energy Systems**

# Simulation Study of a Shunt Active Power Filter Using Nonlinear Least Squares Harmonic Extraction Technique

R. Chudamani□ K. Vasudevan□*Member, IEEE,* and C.S. Ramalingam□*Member, IEEE*

*Abstract--* **In the control strategies for an Active Power Filter, determination of harmonic components of the load current is the most important stage. In this paper the behaviour of the shunt active power filter using the 'Nonlinear Least Squares' estimation technique for extracting the harmonic components of the load current is studied. The performance of active power filter under steady state and dynamic load conditions are analyzed. Simulation results depict a satisfactory performance even under step load changes.**

*Index Terms--* **Active Power Filter, Nonlinear least squares**

## I. INTRODUCTION

ACTIVE Power Filters (APF) for improving the power quality has become a matured technology in recent years.
They are also being used in several countries. Shunt APF is used to eliminate the current harmonics which are generated by loads like adjustable speed drives and electric arc furnace etc. that employ power switching devices. The shunt APF is a configuration that is used to prevent load current harmonics from appearing on the utility side. The basic principle behind its operation is the injection of current harmonics required by the load so that the utility side is relieved of those. This requires determination of the harmonic components in the load current. Therefore□determination of harmonic components of the load current is the most important part of the active power filter.

The APF performance is mainly dependent on how quickly and accurately the harmonic components are extracted from the load current. Many harmonic extraction algorithms are available and their responses under steady state conditions have been explored by researchers profoundly but the performance under dynamic load conditions needs to be studied. When sudden load change occurs the algorithm introduces a delay in tracking the reference currents and this delay causes active power flow through the APF. Therefore□ the delay in tracking the reference currents should be as small as possible.

The techniques proposed in [□-3] employ a low pass filter. However□design of a low pass filter is an inherent trade-off

between filter lag□attenuation of harmonics and the speed of response. To solve this problem□techniques using artificial neural networks (ANN) [4□5] and adaptive filters [6] are introduced to the area of active filtering□but they also suffer from disadvantages. The ANN requires large number of training data and is prone to errors if the current waveform deviates much from the samples used for training. In adaptive filtering approach accurate system frequency information is required.

A novel method to extract the load current harmonics with minimum delay even under dynamic load conditions is proposed in [7]. This method first estimates the power line frequency using the principle of Nonlinear Least Squares (NLS) estimate and uses this estimated frequency in the Fourier model of the current signal to extract the harmonic components. The scope of the work reported in [7] is limited to the extraction of harmonic components alone and the actual performance of the APF using this approach is not shown.

In this paper the performance of the shunt active power filter using the NLS algorithm is studied through simulation carried out in SABER. The nonlinear load shown in Fig. □ is considered for simulation study. Fig. 2(ii) shows the waveform of the current drawn by the nonlinear load.

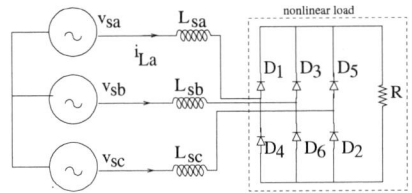

Fig. □ Nonlinear load.

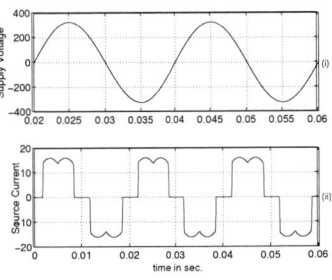

Fig. 2. (i) Supply Voltage (ii) Waveform of the current drawn from the source.

Correponding author: R. Chudamani□email: ee03d0□9@ee.iitm.ac.in
All authors are with the Department of Electrical Engg.□Indian Institute of Technology Madras□India

0-7803-9771-1/06/$25.00 ©2006 IEEE

In section II a brief description of the NLS algorithm is given. In section III the performance of the active power filter for various conditions of supply voltage and load are simulated and the results are analyzed. In section IV the merits and demerits of this approach and its practical implementation are discussed.

## II. NONLINEAR LEAST SQUARES APPROACH

Any periodic signal can be expressed as a Fourier series, i.e., as a linear combination of harmonically related sinusoids. The current drawn by a nonlinear load can therefore be written as

$$i(t) = a_0 + \sum_{n=1}^{\infty} a_n \cos n\omega_0 t + b_n \sin n\omega_0 t \qquad (1)$$

where $\sqrt{a_n^2 + b_n^2}$ is the amplitude of the $n$th harmonic component and $\omega_0$ is the fundamental frequency in radians/sec. Since $i(t)$ does not usually contain a DC component, $a_0 = 0$. In practice, we limit the number of harmonics to be estimated to a finite number $N$. This converts (1) to an approximation. This is particularly true if measurement errors and/or additive noise are present. In order to determine the harmonic contents of the load current, we have to estimate $a_n$ and $b_n$ for $n=1,2,...,N$. In this expression, we have $2N$ unknown parameters; if $\omega_0$ were also not known precisely, we have one additional parameter. The procedure to obtain these unknowns may be explained as follows. Let us say that $\omega_0$ is known. To solve for $a_n$ and $b_n$, we assume that $i(t)$ is known at $M$ uniformly sampled points, i.e., at $i(t_k)$ for $k=0,1,...,M-1$. This leads to the following set of $M$ equations:

$$i(t_k) \simeq a_0 + \sum_{n=1}^{N} a_n \cos n\omega_0 t + b_n \sin n\omega_0 t \qquad (2)$$

where $k=0,1,...,M-1$.
In vector notation (2) can be written as

$$\underline{H}\underline{a} \simeq \underline{b} \qquad (3)$$

where $\underline{b}$ is the vector of samples $i(t_k)$ and $\underline{a}$ is the vector of unknowns $a_n$ and $b_n$. The number of samples $M$ required to solve these simultaneous equations is equal to $2N$. But in the presence of measurement and/or additive noise, an overdetermined system of equations i.e., $M \geq 2N$ will give a better solution. Since the system is overdetermined, the solution to the equations converges in a least square sense when $\underline{a}$ is given by the following equation.

$$\underline{a} \simeq (\underline{H}^T \underline{H})^{-1} \underline{H}^T \underline{b} \qquad (4)$$

If the system frequency $\omega_0$ is known, then the problem is a linear least squares problem. If $\omega_0$ is not known precisely, the linear least squares problem is transformed into a nonlinear one [8]. Even though $2N+1$ unknowns are present, the linearly entering $2N$ amplitude variables can be eliminated, resulting in a one-dimensional nonlinear least squares problem. The standard trick is to eliminate the linear variables by substituting (4) into (3), resulting in (5).

$$H(\underline{H}^T \underline{H})^{-1} \underline{H}^T) \simeq \underline{b} \qquad (5)$$

The error vector $e$ is given by

$$\underline{e} \simeq (I - H(\underline{H}^T \underline{H})^{-1} \underline{H}^T)\underline{b} \qquad (6)$$

The system frequency $\omega_0$ is considered to be the one that minimizes $\| \underline{e} \|_2^2$. Since we have only one parameter, a simple grid search is enough to locate the minimum. Once $\omega_0$ is estimated, (4) can be used to estimate $\underline{a}$ and hence $a_n$ and $b_n$. With these estimated values of $a_n$ and $b_n$ the fundamental is constructed and subtracted from the load current to get the harmonic components. The behaviour of the error norm, for a particular set of sampled waveform values, as the frequency search is done, is shown in Fig. 3.

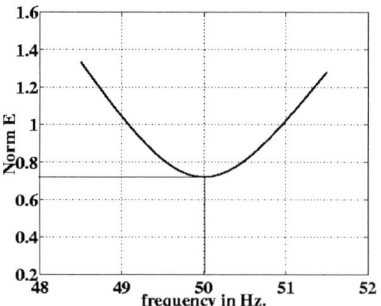

Fig. 3. Variation of $\| \underline{e} \|_2^2$ with frequency.

It is seen that the norm reaches a minimum at a particular frequency, which is taken as the estimated frequency. It is concluded in [7] that a window length of one fundamental period for frequency estimation and a quarter for harmonic extraction give satisfactory results.

## III. PERFORMANCE OF SHUNT ACTIVE POWER FILTER USING NLS ALGORITHM

In this section the behaviour of the APF for various conditions of supply voltage viz., (i) Sinusoidal balanced voltage with constant but unknown frequency (ii) Nonsinusoidal supply voltage and (iii) Sinusoidal voltage with varying frequency is presented. The effect of load variations is also considered and the results are given. Fig. 4 shows the active power filter based harmonic elimination scheme, used for the simulation study.

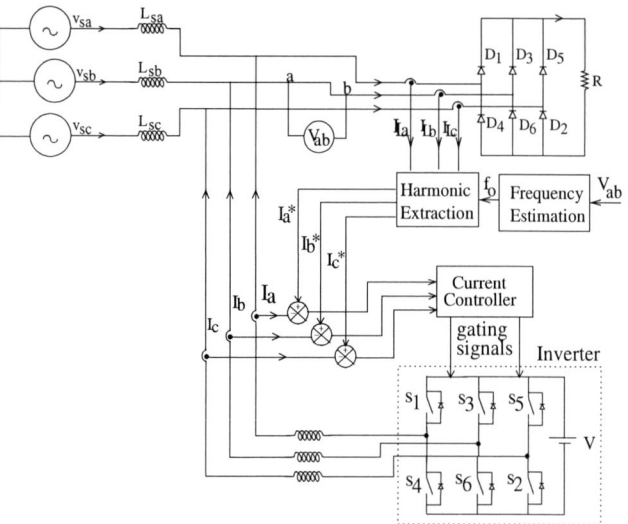

Fig. 4. Active Power Filter Based Harmonic Compensation Scheme.

## A. Performance with Sinusoidal Voltage with Constant but Unknown Frequency

The simulation results obtained when the supply voltage is balanced and sinusoidal are shown in Fig. 5. It is observed from Fig. 5(ii) that the source current has almost become sinusoidal and the Total Harmonic Distortion (THD) is reduced to 2.37% which is well within the permissible limit of 5%.

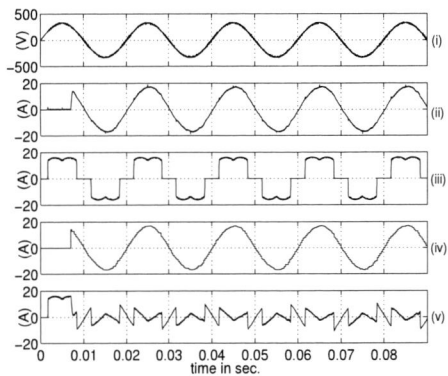

Fig. 5. Simulation results with NLS method (i) Voltage at PCC (ii) Source current after compensation (iii) Source current before compensation (iv) Fundamental component (v) Harmonic components.

## B. Performance with Nonsinusoidal Supply Voltage

The simulation results obtained when the nonlinear load is connected to the voltage supply with THD of □0% is shown in Fig. 6. When there are harmonic distortions in the supply voltage□the frequency estimation is affected□but this problem could be eliminated by increasing the window length or the number of harmonics in the voltage signal model. This aspect has been explored in detail in [7]. Therefore the extraction of harmonic components of the load current is not much affected. This is evident from the value of THD of current waveform which is found to be 2.48%.

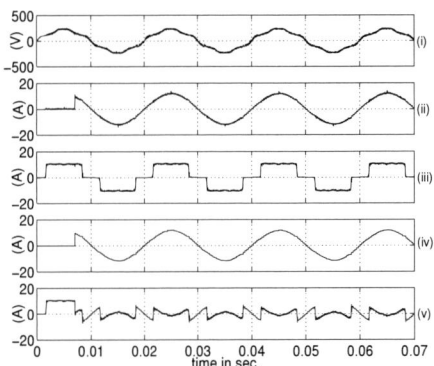

Fig. 6. Simulation results with nonsinusoidal supply voltage (i) Voltage at PCC (ii) Source current after compensation (iii) Source current before compensation (iv) Fundamental component (v) Harmonic components.

## C. Performance with Sinusoidal Voltage with Varying Frequency

Sinusoidal supply with varying frequency is considered□to study its effects on the APF performance. The variation in the frequency is taken to be from 48.5 Hz to 5□5 Hz over a period of 0.3 s in a manner as shown in Fig. 7□which also shows the performance of the frequency estimator. It can be seen from Fig. 7 that as the grid frequency varies□the algorithm is able to track the variation. The performance of the APF with this frequency variation is shown in Fig. 8. The results show that good compensation is achieved by the APF in spite of variation in the supply voltage.

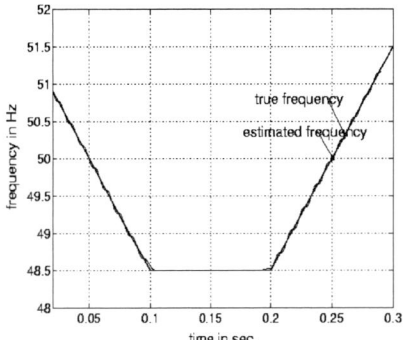

Fig. 7. Variation of actual and the estimated frequency with time.

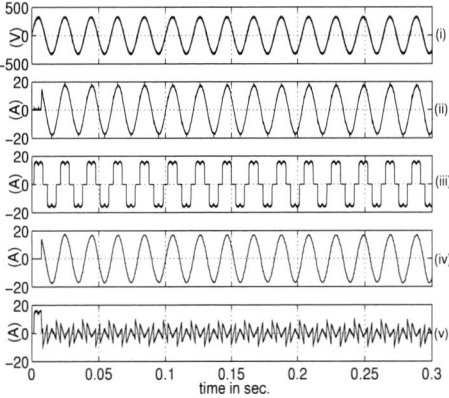

Fig. 8. Simulation results with varying frequency utility supply (i) Voltage at PCC (ii) Source current after compensation (iii) Source current before compensation (iv) Fundamental component (v) Harmonic components.

## D. Response to Sudden Load Changes and Continuously Varying load

The NLS method is capable of providing harmonic information with about a quarter cycle delay as explained in section II. In order to demonstrate this□response of the APF to a step load change is shown in Fig. 9. Sudden load changes occur at 6 ms and □0 ms. It is observed from the simulation results that the algorithm tracks the reference currents within a quarter of a fundamental period. Further□the response of the APF to continuously varying load is also simulated and the results are shown in Fig. □0. It is observed that the algorithm tracks the fundamental current quickly. Since the delay in tracking the reference currents is minimum□the active power flow through the active filter is reduced. This can be seen in comparison with the active power flow through the APF when a low pass filter is used to extract the fundamental component. Fig. □□shows the average power flow from the DC bus of the

557

filter. It can be seen that the power flow is much reduced when NLS method is used.

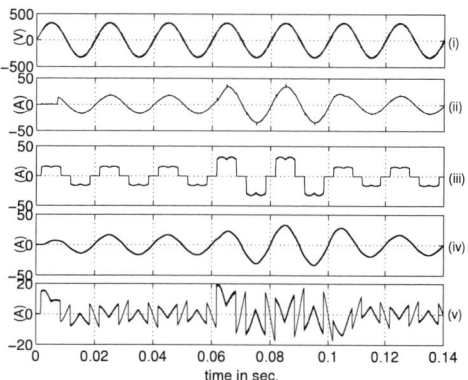

Fig. 9. Simulation results with step load change (i) Voltage at PCC (ii) Source current after compensation (iii) Source current before compensation (iv) Fundamental component (v) Harmonic components.

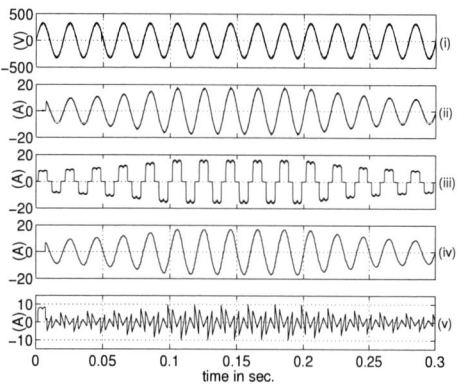

Fig. 10. Simulation results with continuously varying load (i) Voltage at PCC (ii) Source current after compensation (iii) Source current before compensation (iv) Fundamental component (v) Harmonic components.

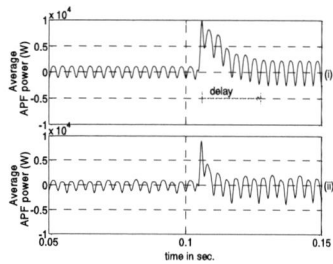

Fig. 11. Simulation results with varying load (i) Voltage at PCC (ii) Source current after compensation (iii) Source current before compensation (iv) Fundamental component (v) Harmonic components.

## IV. DISCUSSION

The shunt active power filter using the NLS algorithm is studied under various conditions of supply voltage and load. The simulation results are analyzed. In order to reduce the active power flow through the APF during sudden load changes, quick estimation of the reference currents is essential. It is observed from the results that the harmonic currents are extracted with a minimum delay of a quarter of a fundamental cycle.

For the practical implementation of the NLS algorithm, a single dimensional grid search can be made to locate the fundamental frequency. It can be shown that $I - H(H^{T}H)^{-\circ}H^{T}$ is a constant matrix, the values of which can be stored in memory. Multiplication of this matrix with the voltage/current vector directly yields the error vector, which makes the practical implementation of this algorithm simple. For fundamental extraction $(H^{T}H)^{-\circ}H^{T}$ should be multiplied with the sampled current vector. Since the harmonic components can be extracted by subtracting the fundamental component from the load current, it is sufficient if $a_\circ$ and $b_\circ$ are known. Therefore multiplication of the rows of $(H^{T}H)^{-\circ}H^{T}$ that correspond to $a_\circ$ and $b_\circ$ alone need to be multiplied with the current vector and not the entire matrix. This drastically reduces the number of multiplications. Since the waveform exhibits half wave symmetry, only odd harmonics can be included in the current model which will further reduce the computations.

## V. CONCLLUSION

In this paper, performance of an APF using the NLS algorithm has been studied. Briefly reviewed in section II, the NLS algorithm consists of two parts-- frequency estimation and subsequent harmonic extraction. The major advantage of this technique is that it tracks the harmonic currents quickly even under sudden load changes. This has been demonstrated by simulation results presented in this paper. Further, it can also extract harmonic components explicitly. The NLS algorithm is computation intensive. Various features of the algorithm that could simplify calculation, have been discussed in the paper. Also, with the availability of high speed digital signal processors and the programmable logic devices like FPGAs, one could think of implementing a complex algorithm such as NLS algorithm. Further, the grid frequency varies very slowly within a narrow range enabling us to go for a simple grid search for frequency estimation. Practical implementation of this algorithm is currently in progress.

## VI. REFERENCES

[1]  Luis A. Moran, Juan W. Dixon and Rogel R. Wallace," A Three-phase Active Power Filter Operating with Fixed Switching Frequency for Reactive and Current Harmonic Compensation", *IEEE Transactions on Industrial Electronics*, Vol. 42, No. 4, August 1995.

[2]  Hirofumi Akagi, Yoshihira Kanazawa and Akira Nabae," Instantaneous Reactive Power Compensators Comprising Switching Devices without Energy Storage Components", *IEEE Trans. on Industry Applications*, Vol. IA-20, No.3, May/June 1984, pp.625-63.

[3]  Vasco Soares, Pedro Verdelho and Gil Marques,"Active Power Filter Control Circuit based on the Instantaneous Active and Reactive Current $i_q$-$i_d$ method", *Power Electronics Specialists Conference*, 1997, PESC'97 Record, 28th annual *IEEE* Vol. 2, 22-27 June 1997, pp.1096-1101.

[4]  M.Rukonuzzaman, Katsumi Nishida and Mutsuo Nakaoda, "DSP Control Shunt APF with Harmonic Extraction with Adaptive Neural Network", *Industry Applications Conference*,2003,38th *IAS Annual meeting, Conference record of the* Vol. 2, 12-16 Oct. 2003, pp.215-22.

[5]  L.L. Lai, C.T. Tse, W.L. Chan, A.T.P. So,"Real-time Frequency and Harmonic Evaluation using Artificial Neural Networks", *IEEE Transactions on Power Delivery*, Vol.14, No.1, January 1999, pp.52-59.

[6] Sami Valiviita, Seppo J. Ovaska, "Delay less Method to Generate Current Reference for Active Filters", *IEEE Transactions on Industrial Electronics*, Vol. 45, No. 4, August 1998, pp.559-569.

[7] R. Chudamani, Krishna Vasudevan and C.S. Ramalingam," Nonlinear Least Squares Current Estimator for Three Phase Loads", IEEE Conference on Industrial Technology, Dec.15-17, Mumbai, India pp.581-585.

[8] S.M.Kay, *Fundamentals of Statistical Signal Processing: Estimation Theory*, Prentice-Hall, Englewood Cliff, NJ, 1993.

## VII. BIOGRAPHIES

**R. Chudamani** completed her B.E in Electrical Engineering from National Institute of Technology Surat, India in 1990. She obtained her M.Tech (Power Electronics) in Electrical Engineering from IIT, Delhi in 1997. She is working as a lecturer in the Department of Electrical Engineering, N.I.T., Surat, Gujarat, India. Presently she is doing Ph.D under the guidance of Dr. Krishna Vasudevan, in the Department of Electrical Engg., IIT Madras.

**K. Vasudevan** completed his B.Tech in Electrical Engineering (Power) from IIT, Madras in 1989. He obtained his M.E in Electrical Engineering from IISc in 1991, where he was awarded a gold medal for being the best ME student. He worked from March 1991 to June 1992 at Kirlosaker Electric Company at Mysore, where he was with the R&D (Static Power Sources Group) and was in involved in the design and development of UPS systems. After a Ph.D degree in Electrical Engineering in 1996, he was employed in the R&D division of Lucas TVS from July 1996 to December 1998 where he was involved in the design and performance improvement approaches for automotive alternators. He is currently Associate Professor at the Department of Electrical Engineering at IIT Madras. In this position he has been active in the areas of teaching, research and industrial consultancy. His research interests are in the areas of power electronics and drives.

**C.S. Ramalingam** obtained his BE from Madras University (1985), M.Tech from IIT Kharagpur (1987), and the Ph.D degree in Electrical Engineering from the University of Rhode Island (1995). In 1988 he was an Associate Lecturer at VLB Janaki Ammal College of Engineering and later in Kumaraguru College of Technoloy, both in Coimbatore. From 1995 to 2001 he was with the DSPS R&D Centre at Texas Instruments, Dallas, working in the areas of Speech Recognition and Speech Coding. He is currently a faculty in the Electrical Engineering Department at IITM. His areas of interest are Signal Processing, Speech Recognition, and Speech Coding.

**2006 IEEE International Conference on Power Electronic, Drives and Energy Systems**

# Comparison of Synchronous Detection and I.Cosφ Shunt Active Filtering Algorithms

G. Bhuvaneswari, *Senior Member, IEEE*, Manjula G. Nair, and Sathish Kumar Reddy

*Abstract—***In this paper, two shunt active filtering algorithms, namely, synchronous detection technique and newly proposed I.cosφ algorithm have been simulated and implemented in hardware and a comparison has been brought out. The proposed algorithm is found to work satisfactorily under various system operating conditions. The hardware implementation of both the schemes have been done and the proposed scheme is found to be simple and working satisfactorily.**

*Index Terms--*Power quality, Shunt active power filters.

## I. INTRODUCTION

POWER quality has received a great deal of attention lately, with the increased use of power electronic converters in adjustable speed drives (ASDs), uninterruptible power supplies (UPS) etc. There are power quality standards that define the maximum allowable limit of distortions in voltage and current waveforms in the power supply [1]. Active power filters are one of the most important remedial measures to solve power quality problems. Various well-established control schemes [2-5] exist for the generation of compensation signals for active power filters. In this paper, a voltage source inverter (VSI) based three-phase shunt active filter is implemented in simulation as well as in hardware using two different controllers, namely, the synchronous detection (SD) controller and the newly proposed I.cosφ controller. The results are compared under different operating conditions to bring out the merits of one scheme over the other.

## II. ACTIVE POWER FILTERING ALGORITHMS

The compensation signals generated by active power filters can be based on control schemes developed in either the time-domain or the frequency-domain. The SD algorithm [4-5] is one of the time-domain techniques. It generates reference compensation currents for the shunt active filter based on the real power consumed by individual phases. The authors had earlier proposed [6] I.cosφ control algorithm, which is simple and easy-to-implement. In this algorithm, the product of the amplitude of the fundamental load current |I| and the

---

G.Bhuvaneswari, Manjula G.Nair and Sathish Kumar Reddy are with the Department of Electrical Engineering, I.ndian Institute of Technology, New Delhi-110016. INDIA. Their e-mail addresses are bhuvan@ee.iitd.ac.in, cp.manju@gmail.com and sat_341@yahoo.co.in.

displacement power factor |cosφ| is set as the amplitude of the reference mains current which will be in phase with the corresponding phase voltage. The reference compensation currents to be delivered by the active power filter are computed as the difference between the actual load current and the reference source current. A provision for self-supporting DC bus for the active filter is also incorporated in the algorithm.

### A. Synchronous Detection Algorithm

In the synchronous detection (SD) algorithm [4], the average real power consumed by the load with respect to the three phases gives the desired mains currents, assuming them to be balanced and in-phase with the supply voltages after compensation. The reference compensation signals are then derived as the difference between the load currents and the desired mains currents.

The loading on individual phases of the system, from the three-phase mains voltages and load currents are computed as:

$$p = \begin{bmatrix} e_a & e_b & e_c \end{bmatrix} \begin{bmatrix} i_{La} \\ i_{Lb} \\ i_{Lc} \end{bmatrix}$$

The average value of 'p' is then computed as $P_{dc}$, by passing 'p' through a low pass filter. $P_{dc}$ is divided among the three phases of the system as:

$$P_a = \frac{P_{dc} \cdot E_a}{E_{tot}} \; ; \; P_b = \frac{P_{dc} \cdot E_b}{E_{tot}} \; ; \; P_c = \frac{P_{dc} \cdot E_c}{E_{tot}} \quad \text{where}$$

$E_a$, $E_b$ and $E_c$ are the amplitudes of the three-phase mains voltages and $E_{tot}$ is the sum of $E_a, E_b$ and $E_c$.

To achieve unity power factor at the source end, the desired mains currents in the three phases are computed as:

$$i_{ma} = \frac{2 \cdot e_a \cdot P_a}{E_a^2} \; ; \; i_{mb} = \frac{2 \cdot e_b \cdot P_b}{E_b^2} \; ; \; i_{mc} = \frac{2 \cdot e_c \cdot P_c}{E_c^2}$$

The reference compensation currents to be supplied by the active filter are, therefore, the differences between the desired mains currents and the actual load currents.

i.e.; $i^*_{Ca} = i_{ma} \sim i_{La}$ ; $i^*_{Cb} = i_{mb} \sim i_{Lb}$ ; $i^*_{Cc} = i_{mc} \sim i_{Lc}$

### B. Proposed I.cosφ Algorithm

In the I.cosφ algorithm, the desired mains current is assumed to be the product of the magnitude I.cosφ and a unit amplitude sinusoidal wave in phase with the mains voltage. The mains is required to supply only the active portion of the

---

0-7803-9771-1/06/$25.00 ©2006 IEEE      560

load current as the shunt active power filter is expected to provide compensation for the harmonic and reactive portion of the three-phase load current, and also for any imbalance in the three-phase load currents. Hence, only balanced currents will be drawn from the mains which will be purely sinusoidal and in phase with the mains voltages.

The reference compensation currents for the shunt active filter are thereby deduced as the difference between the actual load current and the desired source current in each phase.

$$i_{a(comp)} = i_{La} - i_{sa(ref)}; \qquad i_{b(comp)} = i_{Lb} - i_{sb(ref)};$$

and $\qquad i_{c(comp)} = i_{Lc} - i_{sc(ref)}.$

where, the desired (reference) source currents in the three phases are given as,

$$i_{sa(ref)} = \left| I_{s(ref)} \right| \times U_a = \left| I_{s(ref)} \right|.\sin \omega t \, ;$$

$$i_{sb(ref)} = \left| I_{s(ref)} \right| \times U_b = \left| I_{s(ref)} \right|.\sin (\omega t - 120^\circ) \, ;$$

$$i_{sc(ref)} = \left| I_{s(ref)} \right| \times U_c = \left| I_{s(ref)} \right|.\sin (\omega t + 120^\circ)$$

$U_a$, $U_b$ and $U_c$ are the unit amplitude templates of the phase to ground source voltages in the three phases respectively.
$U_a = 1.\sin \omega t$ ; $U_b = 1.\sin(\omega t - 120^\circ)$ ; $U_c = 1.\sin(\omega t + 120^\circ)$.

The magnitude of the desired source current $|$ Is(ref) $|$ can be expressed as the average of the magnitudes of the real components of the fundamental load currents in the three phases.

$$\left| I_{s(ref)} \right| = \frac{\left| \mathrm{Re}(I_{La}) \right| + \left| \mathrm{Re}(I_{Lb}) \right| + \left| \mathrm{Re}(I_{Lc}) \right|}{3}$$

$$= \frac{\left| I_{La} \right|.\cos \phi_a + \left| I_{Lb} \right|.\cos \phi_b + \left| I_{Lc} \right|.\cos \phi_c}{3}$$

## III. SIMULATION RESULTS

The system considered for simulation, is a three-phase balanced source of 400 V connected to a thyristorized bridge rectifier feeding a resistive load of 150 Ω, operating at a firing angle of 60º. This load draws a highly non-linear current rich in harmonics with a substantial reactive power requirement. A three-phase, VSI based shunt active power filter is connected to the system for reactive power compensation and harmonic elimination. The simulation is done in the SIMULINK/MATLAB environment.

Fig.1 and Fig.2 show the simulation results under the ideal condition of balanced source and balanced load. Both the algorithms are found to give satisfactory results under this condition.

Fig.3 and Fig.4 show the simulation results under the unbalanced-source balanced-load condition. Here again, both the controllers are able to provide satisfactory compensation making the three phase currents absolutely balanced..

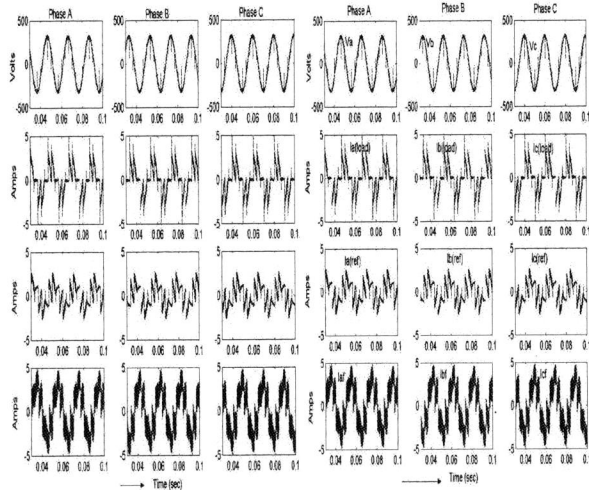

Fig. 1. Simulation results for balanced-source and balanced-load condition with (i) SD algorithm and (ii) I.cosφ algorithm: (a) Phase voltage (b) Load current (c) Ref. compensation current (d) Actual filter current.

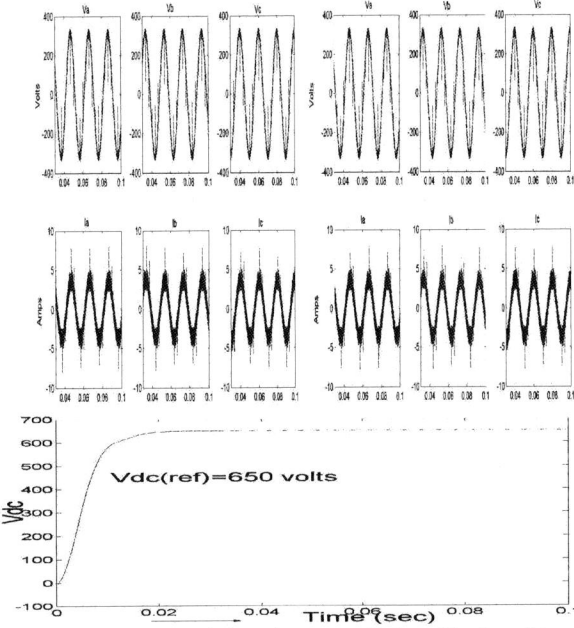

Fig. 2. Simulation results for balanced-source and balanced-load condition with (i) SD algorithm and (ii) I.cosφ algorithm: (a) Phase voltage (b) Source current after compensation (c) DC Link capacitor voltage, V_dc.

Fig.5 and Fig.6 show the simulation results under the unbalanced-distorted-source condition with a balanced RL load. The SD algorithm fails to give a satisfactory compensation as seen from the source currents which are still non-sinusoidal / distorted after compensation. The inability of SD algorithm for efficient compensation under distorted source voltages has been reported earlier [5]. The source currents are obtained with the help of I.cosφ controller are perfectly sinusoidal and at unity power factor, as seen from Fig.6. This accounts for the differences in the THD values depicted in Table1 below.

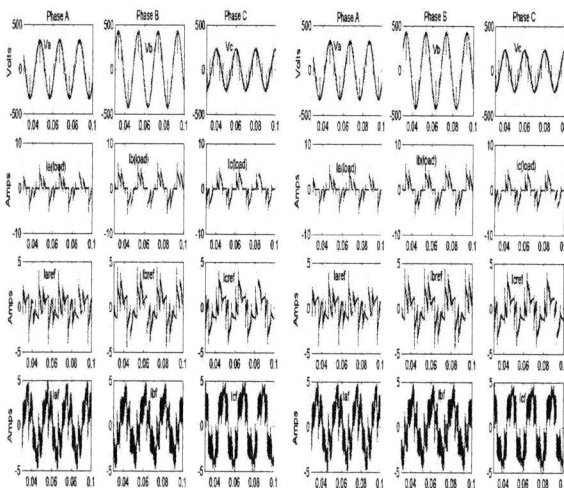

Fig. 3. Simulation results for unbalanced source voltage condition for (i) SD algorithm and (ii) I.cosφ algorithm: -(a) Phase voltage (b) Load current (c)Reference compensation current (d) Actual filter output current.

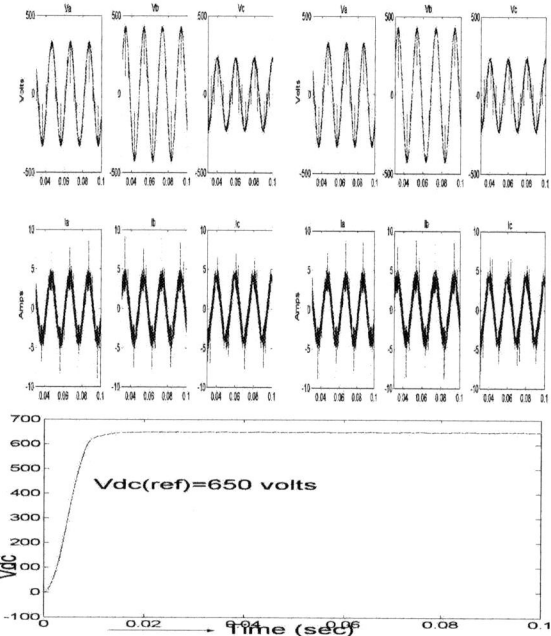

Fig.4. Simulation results for unbalanced source voltage condition for (i) SD algorithm and (ii) I.cosφ algorithm : (a) Phase voltage (b) Source current after compensation (c) DC Link capacitor voltage, Vdc.

Fig.7 and Fig.8 show the simulation results under the balanced-source unbalanced-load condition. The I.cosφ controller is again found to work better than the SD controller as seen from the THD values in Table.1.

TABLE I

TOTAL HARMONIC DISTORTION IN SORCE CURRENT BEFORE AND AFTER COMPENSATION USING SD AND ICOSΦ ALGORITHMS WITH THYRISTOR CONVERTER LOAD

| Algorithms/ operating conditions | Before any compensation | SD | \|ICosΦ\| |
|---|---|---|---|
| Balanced source | 25.21% | 3.51% | 3.57% |
| Unbalanced source | 26.44% | 2.54% | 2.37% |
| Unbalanced , distorted source with R-L load | 5.33% | 11.43% | 0.42% |
| Balanced source with unbalanced load | 66.46% | 7.62% | 3.05% |

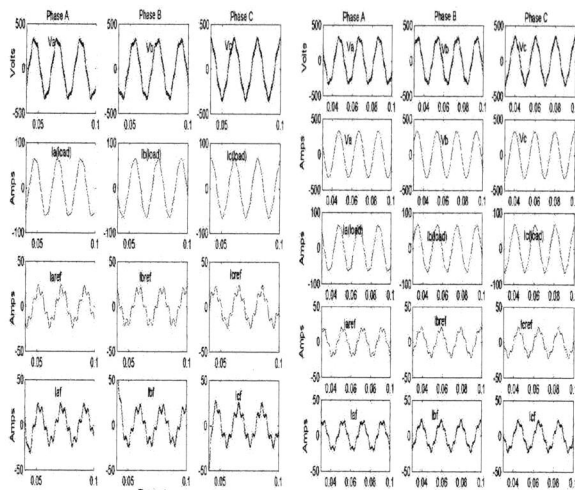

Fig. 5. Simulation results for unbalanced-distorted source voltage condition with RL load for (i) SD algorithm and (ii) I.cosφ algorithm: -(a) Phase voltage (b) Load current (c)Reference compensation current (d) Actual filter output current.

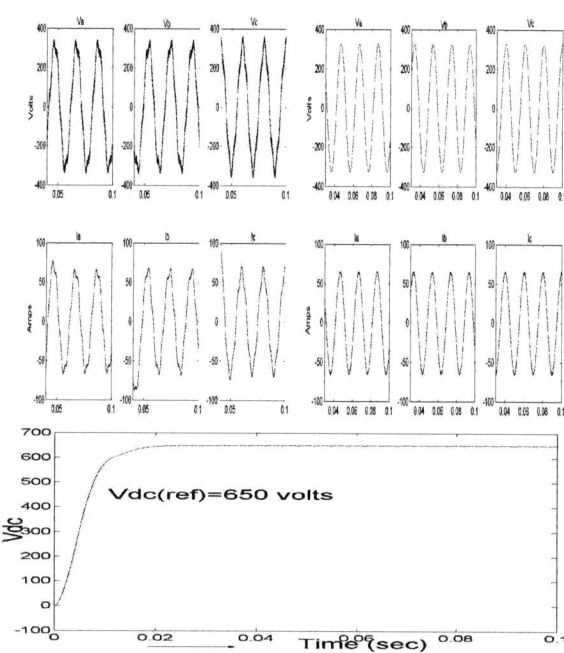

Fig.6. Simulation results for unbalanced-istorted source voltage condition with RL load for (i) SD algorithm and (ii) I.cosφ algorithm : (a) Phase voltage (b) Source current after compensation (c) DC Link capacitor voltage, Vdc

Table 1 shows the comparison of results obtained from SD and I.cosφ algorithms under various system operating conditions. It can be seen that I.cosφ algorithm performs better especially under distorted source condition and unbalanced load condition.

## IV. HARDWARE IMPLEMENTATION RESULTS

Both the schemes have been implemented in hardware as well. The synchronous detection algorithm has been implemented in ADMC 401 environment whereas the IcosΦ algorithm has been implemented using both analog circuits and ADMC 401 digital signal processor. The experimental results are shown here for the diode rectifier load under the balanced-source, balanced-load condition. Fig.9 shows the three-phase load currents along with the source voltage for the diode rectifier load. Figs. 10 and 11 show the compensation currents and source currents obtained using SD algorithm. Figs. 12 and 13 show the waveforms obtained using I.cosφ algorithm. Both the SD and I.cosφ controllers give satisfactory compensation as seen from the results. After compensation, the THD of the source currents reduce to about 5% from the original value of about 25%. In terms of implementation aspects, the IcosΦ algorithm is simpler as compared to SD algorithm.

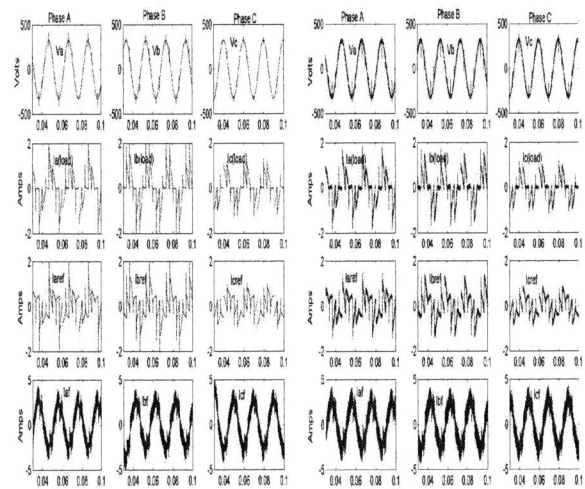

Fig. 7. Simulation results for load unbalance condition for (i) SD algorithm and (ii) I.cosφ algorithm : (a) Phase voltage (b) Load current (c) Desired source current (d)Reference compensation current (e) Actual filter output current.

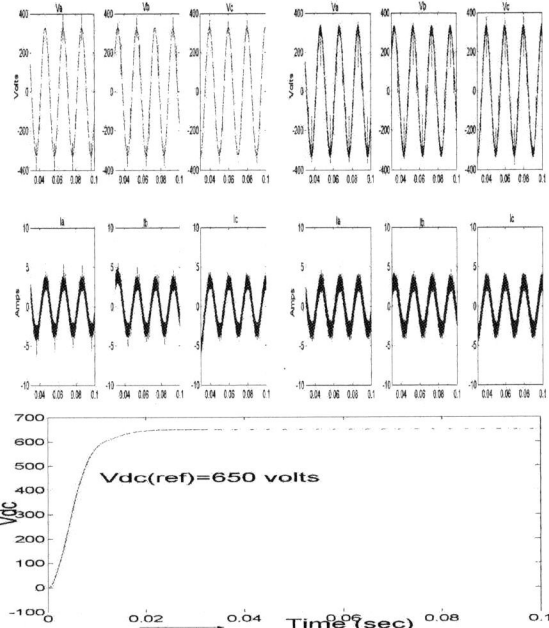

Fig. 8. Simulation results for load unbalance condition: (a) Phase voltage (b) Source current after compensation (c) DC link capacitor voltage, Vdc.

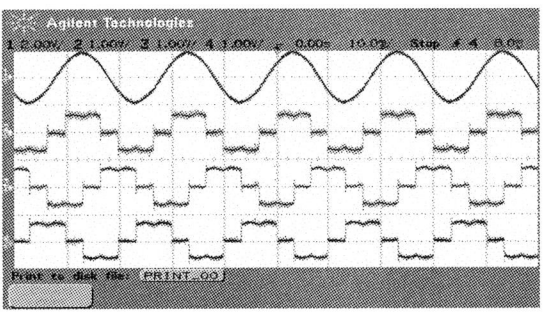

Fig. 9. Three-phase load currents with diode rectifier load.

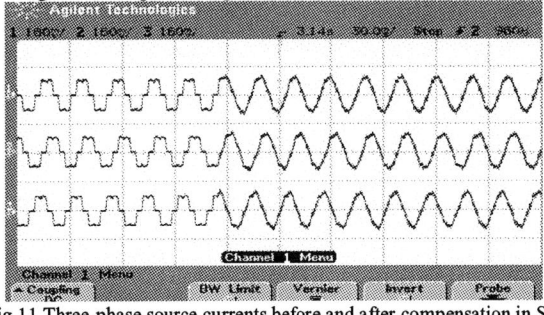

Fig. 10. Three-phase compensation currents with SD algorithm.

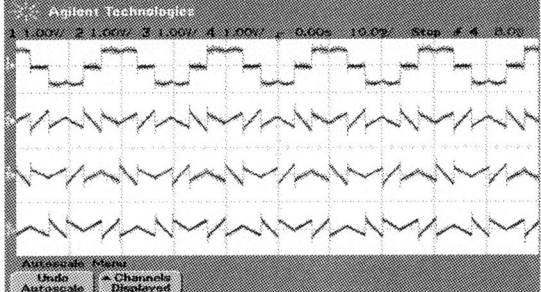

Fig.11 Three-phase source currents before and after compensation in SD algorithm

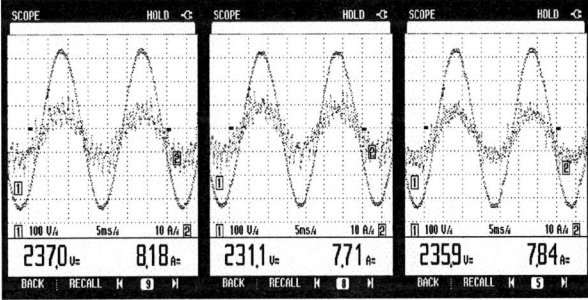

Fig. 12. Three-phase compensation currents with I.cosφ algorithm.

Fig. 13. Three-phase source currents after compensation with I.cosφ algorithm.

563

## V. CONCLUSIONS

The three-phase shunt active filter is simulated and implemented in hardware using synchronous detection (SD) controller and the newly proposed I.cosϕ controller. Both are found to work satisfactorily. Despite the fact that I.cosϕ controller is easier to implement, it is found to work better under distorted source voltage and unbalanced load conditions.

## REFERENCES

[1] IEEE Guide for harmonic control and reactive compensation of Static Power Converters, *IEEE Standard* 519-1992.

[2] Hirofumi Akagi, Yoshihira kanazawa and Akira Nabae, "instantaneous reactive power compensator comprising switching power devices without energy storage components," *IEEE Transactions on Industry Applications,* Vol. 1A-20, May/June 1984, pp. 625-630.

[3] S.Bhattacharya, D.M. Divan and B.Banerjee, "synchronous frame harmonics isolator using active filter," *European Power Electronics Conference, 1991,* pp.3-30, 3.35, Firenzi, Italy.

[4] C.E. Lin, C.L.Chen,C.L.Huang, "Calculating approach and implementation for Active filters in unbalanced three phase system using synchronous detection method", IEEE IECON'92, San Diego, Nov 19-21,1992, pp. 374-380.

[5] H.L. Jou, "Performance comparison of the three-phase active-power-filter algorithms", *Proc. IEE Conf. on Gener.,Transm.,Distrib.*, pp. 646-652, 1995.

[6] Manjula G. Nair and G. Bhuvaneswari, "A novel shunt active filter algorithm – simulation and analog circuit based implementation", *Special issue on Power Quality, International journal of Energy Technology and Policy (IJETP)*, 2006, vol 4, 1/2, pp. 118-125.

## VI. BIOGRAPHIES

**G.Bhuvaneswari** graduated from College of Engg., Anna University, Chennai in 1985. She obtained her Masters' and Doctoral degree from Indian Institute of Technology, Madras. She is currently working as an Associate professor in the Department of Electrical Engineering, I.I.T., Delhi. She is a senior member of IEEE and a Life Fellow of IETE. Her areas of interest are Power electronics, Drives and Power Quality.

**Manjula G.Nair** is a Ph.D. student in the Department of EE, I.I.T., Delhi. She is working a faculty member in Amrita Institute of Technology, Coimbatore, India. Her areas of interest are Power System, Power Electronics, ANN and Fuzzy Control.

**Sathish Kumar Reddy** obtained his M.Tech. degree from the Department of EE, I.I.T, Delhi in May 2006. His areas of interest are Power electronics, Drives and Power Quality.

**2006 IEEE International Conference on Power Electronic, Drives and Energy Systems**

# A Nonlinear Control Method for SSSC to Improve Power System Stability

Majid Poshtan, *Member, IEEE*, Brij N. Singh, *Member, IEEE*, and Parviz Rastgoufard, *Member, IEEE*

*Abstract*—This paper presents an investigation on transient stability of power systems equipped with a Flexible Alternating Current Transmission System (FACTS) device. A Static Synchronous Series Compensator (SSSC) is considered as a FACTS device. In a power system transmission, the SSSC has two functions; first, it compensates reactive power, second, it improves transient stability. The second functionality of the SSSC system is due to its capability to raise the maximum transferable electric power ($P_{max}$) from generator to infinite bus. To raise the maximum transferable electric power ($P_{max}$), the SSSC actively and appropriately changes the line reactance. Since the control of the power system is a nonlinear problem, therefore, a nonlinear controller for the SSSC is designed and presented in this paper. The proposed nonlinear controller of the SSSC increases critical clearing time of the power systems' faults and damps out rotor oscillations of a single machine connected to an infinite bus. Unlike linear controllers, performance and operating region of the proposed nonlinear controller of the SSSC system is not restricted to a close vicinity of the generator operating point, but includes the entire region of operation. The Matlab simulations for computational analysis and the Lyapunov criteria of stability for analytical investigations are used to examine the response time and robustness of the proposed nonlinear controller.

*Index Terms*--Nonlinear control, FACTS, SSSC, Stability, Critical clearing time, Transient Energy Function

## I. INTRODUCTION

VARIOUS kinds of faults in a power systems transmission line change the network topology and may cause an imbalance between generated and transmitted electric energy. In general, this imbalance condition causes undesired rotor angle oscillations [1]. A possible but expensive solution of this problem is to improve the transmission system reliability by installing redundant transmission lines. In addition to the cost, the right of way, environmental concerns, and economical proposition are some of the important issues with the installation of redundant transmission lines. Among several solid state control techniques, a cost effective and reliable method to improve the transient stability can be

achieved by using a Solid State Series Compensator (SSSC), which is one of the FACTS devices [2]. Besides reactive power compensation [2-4], the SSSC can also raise the critical clearing time of fault and damps out the post-fault rotor angle oscillations. In a nonlinear power system, the rotor angle oscillations suppressing capability of the SSSC can be fully utilized with a nonlinear controller.

This paper presents a nonlinear controller for the SSSC to raise the maximum transferable electric power from generator to infinite bus by actively and appropriately changing the post-fault line reactance [2]. The nonlinear controller tracks the generator's angle trajectory and brings it to a new stable operating point. This is achieved by a suitable change in the line reactance ($X_L$) via fast acting SSSC system. Contrary to linear controllers [5], the performance of the nonlinear controllers is not restricted to a close vicinity of the operating point [6] of transmission and generator systems, but includes the entire region of operation [6-7]. This operating region may be spread over $0 < \delta < 180°$. The quantity $\delta$ is power angle between two points in a power system. In this paper $\delta$ is power angle between a generator and infinite bus. The generator could be a single machine or a combination of machines representing a generating area of the power system. The performance evaluation of the proposed nonlinear controller of the SSSC is carried out by both analytical method and computational simulations in MATLAB. Numerical simulation as well as the Lyapunov criteria of stability analysis is applied to examine the response time and robustness of the proposed nonlinear controller. The simulation results are given to show that the SSSC system considerably increases the critical clearing time of the fault, retains system's stability, and keeps the generator free from rotor angle oscillations.

## II. TEST POWER SYSTEM

Fig.1 shows a single-line diagram of the test power system equipped with the SSSC. The system voltages at the generator side and the infinite bus have an equal magnitude of 1 pu ($V_3 = V_2 = 1$). For analysis of the proposed test power system, it is considered that the SSSC system has three operating modes; pre-fault, faulted, and post-fault. In pre-fault mode, the SSSC compensates the reactive power of the transmission system. As fault occurs, the SSSC is switched off to be appropriately activated by the nonlinear controller immediately after the fault is cleared. This is necessary to accomplish an appropriate

---

Majid Poshtan is with Dept of Elect Eng, The Petroleum Institute, P.O. Box 2533, Abu Dhabi, UAE (e-mail: mposhtan@pi.ac.ae).

Brij N. Singh and Parviz Rastgoufard are with the Department of Elect Eng and Computer Science, Tulane University, New Orleans, LA 70118, USA (e-mails: singh@eecs.tulane.edu and parvizr@tulane.edu).

---

0-7803-9771-1/06/$25.00 ©2006 IEEE

line reactance in the post-fault power transmission system. From Fig.1, the equivalent line reactance before the fault ($X_L$) is expressed as: $X_L = X_{L1} \parallel X_{L2}$. Fig.2 shows the phasor diagram of the test power system equipped with the SSSC.

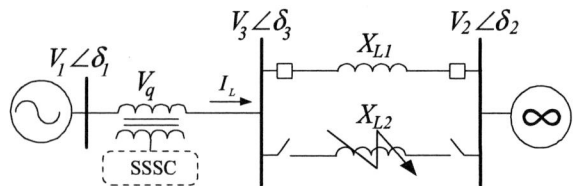

Fig. 1. Single line diagram of SSSC equipped test power system.

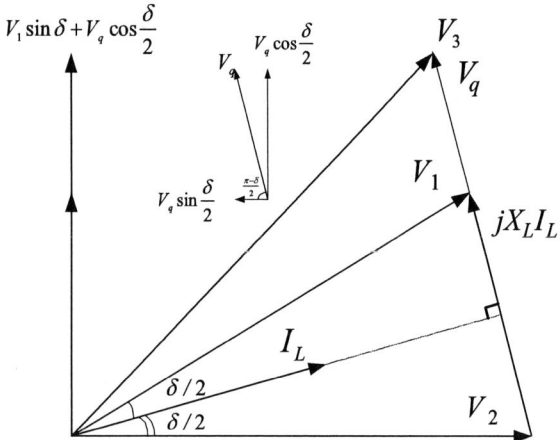

Fig. 2. Phasor diagram of SSSC equipped test power system.

From Fig.2, the expressions in Eq.1 describe voltage components of each bus.

$$\begin{cases} \overline{V}_1 = V_1 \cos\delta + jV_1 \sin\delta \\ \overline{V}_q = -V_q \sin\dfrac{\delta}{2} + jV_q \cos\dfrac{\delta}{2} \\ \overline{V}_3 = \overline{V}_1 + \overline{V}_q \\ \overline{V}_2 = V_2 + j0 \end{cases} \quad (1)$$

In Eq.1, the bar superscript is used to indicate phasor quantities. The real power ($P_e$) and reactive power ($Q_e$) flowing to infinite bus are expressed in Eq.2.

$$\begin{cases} P_e = \left(V_2 V_q \cos(\delta/2) + V_2 V_1 \sin(\delta)\right)/X_L = P + P_{inj} \\ Q_e = \left(V_2 V_1 \cos(\delta) - V_2 V_q \sin(\delta/2) - V_2^2\right)/X_L = Q + Q_{inj} \end{cases} \quad (2)$$

In Eq.2, subscript inj is used to denote injected power from the SSSC to transmission line. Therefore, depending upon power flow direction between SSSC and transmission line, quantities $P_{inj}$ and $Q_{inj}$ acquire positive or negative sign. The quantities ($P_e$) and ($Q_e$) consist of two components as follows:

Power exchange between generator and infinite bus :

$$\begin{cases} P = V_2 V_1 \sin(\delta)/X_L \\ Q = [V_2 V_1 \cos(\delta) - V_2^2]/X_L \end{cases}$$
$$(3)$$

Power exchange between SSSC and transmission line :

$$\begin{cases} P_{inj} = V_2 V_q \cos(\delta/2)/X_L \\ Q_{inj} = -V_2 V_q \sin(\delta/2)/X_L \end{cases}$$

Here, quantity $\delta$ is phase difference between voltages $V_1$ (generator voltage) and $V_2$ (infinite bus voltage), which is also called as power angle. The above equations and the equal area criteria of Fig.3 are used to perform the numerical and analytical analysis of the post-fault test power system. In a single machine system, the pre-fault, the faulted, and the post-fault real power output of the generator is determined by expression: $P = V_1 V_2 \sin\delta/X_L$. In the presence of a fault in either of the lines in Fig.1, the generator gets accelerated due to insufficient power transmission capacity. The net acceleration energy of generator defined in Eq.4 [1], increases rapidly during the fault period ($0 < t < t_c$) and decreases immediately after the fault is cleared at the instant, $t = t_c$.

$$E_i(t, t_c) = \begin{cases} KE = P_m * (\delta_c - \delta_o) & 0 \le t < t_c \\ PE = \displaystyle\int_{\delta_o}^{\delta_x} P_{\max} \sin(\delta) d\delta & t > t_c \end{cases} \quad (4)$$

In Eq.4, $t_c$ is fault clearing time, and $\delta_o$ is the pre-fault power angle. If the fault is cleared before the critical clearing time ($t_{cc}$) and the net kinetic energy (KE) is equal to the net potential energy (PE), the post-fault test power system remains stable. Contrarily, if the fault is cleared after critical clearing time ($t_c > t_{cc}$), the post-fault test power system becomes unstable due unwanted energy imbalance.

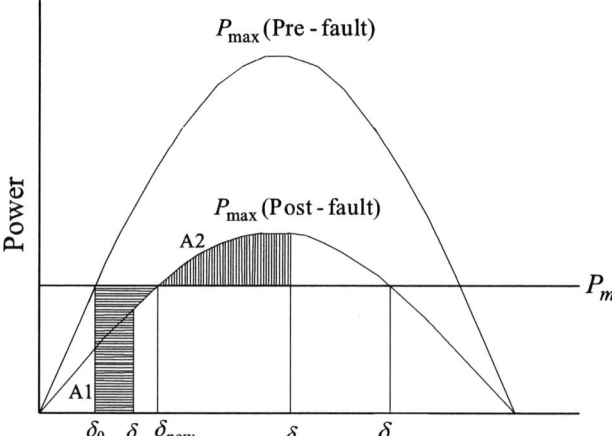

Fig. 3. Power-angle characteristics of test power system.

A suitable controller is needed to reduce the post-fault angular velocity ($\omega$) of generator and return the power angle ($\delta$) to the stable region ($0° < \delta < 90°$). This can be realized by maximizing the transferred electric power from generator to infinite bus, thereby, relieving the generator from undesired kinetic energy accumulated during the fault. Based on the fast response and flexibility of advanced power electronic devices, the SSSC is potentially a practical device to control the transferable electric power ($P_e$) from the generator to infinite bus. The control function (law) of the SSSC system is derived using nonlinear dynamic equations of the test power system shown in Fig.1.

## III. DYNAMIC EQUATIONS

Eq.5 is the dynamic equation of the test power system

(Fig.1).

$$m\ddot{\delta} + D\dot{\delta} = P_m - P_e \qquad (5)$$

where, $P_e = V_1 V_2 \sin\delta / X_L$ and $\ddot{\delta} = \omega$ = angular deviation of generator speed from synchronous value.

The effect of damper winding is considered in Eq.5 as the rate of change of angular velocity from the synchronous value. The damper winding produces a breaking power proportional to the relative movement between the air-gap field (produced by stator) and the damper winding field. This phenomenon helps to damp out rotor angle ($\delta$) oscillations, however an excessive heat may be generated due to damper winding copper loss ($I^2 R$). Therefore, for completely damping out rotor angle oscillations, excitation level (amount of current $I$) of damper winding may be one of limiting factors for safe and efficient operation of generators connected in a power system.

The conventional approach to control the state variables ($\delta$ and $\dot{\delta}$) of a generator is to design a feedback controller for the input mechanical power through its governor [5]. In Eq.5, substitution of $P_e = |V_1||V_2| \sin\delta / X_L$ and rearranging the terms lead Eq.6. Here, Eq.6 represents state space equation for a system with input mechanical power ($P_m$) as a controlled input.

$$\begin{bmatrix} \dot{\delta} \\ \ddot{\delta} \end{bmatrix} = \begin{bmatrix} \dot{\delta} \\ -(1/M)\left(|V_1||V_2|\sin(\delta)/(X_L) + D\dot{\delta}\right) + (1/M)(P_m) \end{bmatrix} \qquad (6)$$

or

$$\begin{bmatrix} \dot{\delta} \\ \ddot{\delta} \end{bmatrix} = \begin{bmatrix} \dot{\delta} \\ -(1/M)\left(|V_1||V_2|\sin(\delta)/(X_L) + D\dot{\delta}\right) \end{bmatrix} + \begin{bmatrix} 0 \\ (1/M) \end{bmatrix}(P_m) \qquad (7)$$

An alternative to Eq.7 is proposed in Eq.8, where the input control signal is an electrical signal ($1/X_L$) instead of a mechanical signal ($P_m$).

$$\begin{bmatrix} \dot{\delta} \\ \ddot{\delta} \end{bmatrix} = \begin{bmatrix} \dot{\delta} \\ (1/M)\left(P_m - D\dot{\delta}\right) \end{bmatrix} + \begin{bmatrix} 0 \\ -(1/M)\, V_1 V_2 \sin(\delta) \end{bmatrix}\left(\frac{1}{X_L}\right) \qquad (8)$$

In Eq.8, the mechanical input $P_m$ is assumed to remain constant, because as compared to electrical dynamics of the system, the mechanical dynamics of a power system is significantly slow. Therefore, quantity $P_m$ may be considered as a constant over one sampling interval of electrical quantities ($1/X_L$, in this case) of the power system.

A desired value of ($1/X_L$) can be realized by the SSSC as described in Section 4. This desired value is found by the nonlinear controller and is expressed in Eq.9 (see Appendix).

$$u(\delta, \dot{\delta}) = \frac{1}{X_L} = \frac{M\dot{\delta}^2 \sin(\delta) - P_m \cos(\delta) + D\dot{\delta}\cos(\delta) + w}{-V_1 V_2 \sin(\delta)\cos(\delta)} \qquad (9)$$

In section 5 it is demonstrated that the proposed nonlinear controller for the SSSC brings the test power system (represented by Eq.8) to a new equilibrium point after it has been subjected to a large disturbance.

## IV. STATIC SYNCHRONOUS SERIES COMPENSATOR

The proposed control scheme of the SSSC is shown in Fig.4 and necessary mathematical equations of the inverter (SSSC power circuit) control scheme have been formulated in one of our papers [2].

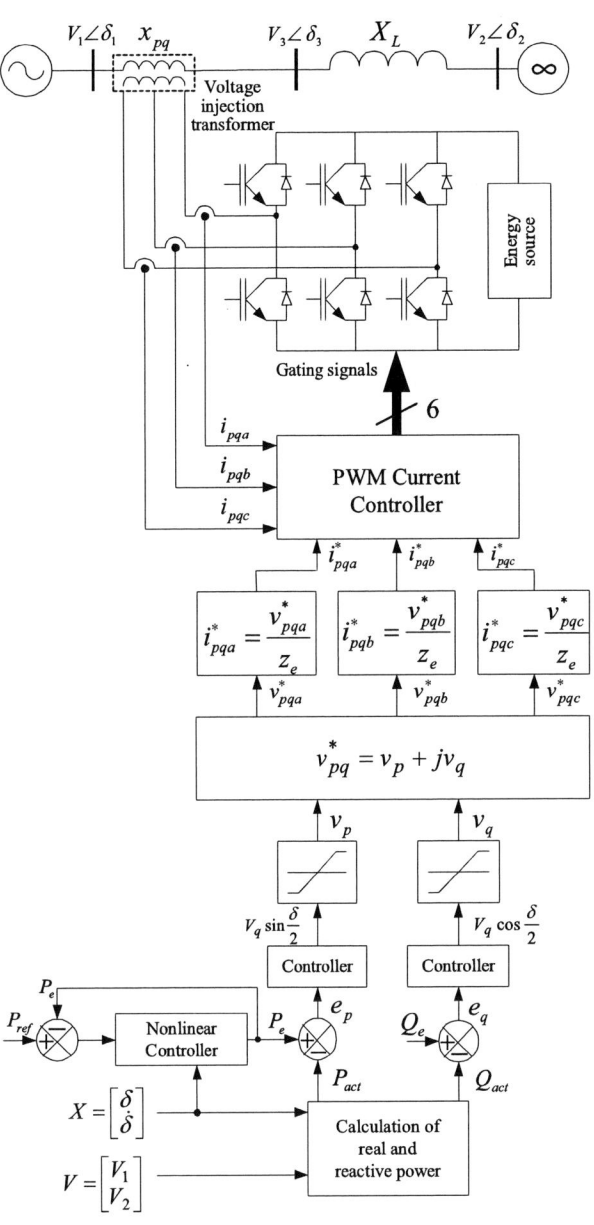

Fig. 4. Control scheme of static synchronous series compensator.

It is noted that besides Static Compensator (STATCOM) and Unified Power Flow controller (UPFC) [8-9], the SSSC is one of the advanced FACTS devices. The SSSC system equipped with an energy source in its DC bus can supply or absorb the reactive and real power to or from the power system transmission line. Contrary to conventional series compensation methods, the SSSC can control power flow of the transmission line without originating classical network resonance and oscillations [10-13]. The SSSC equipped power transmission system will have a greater potential energy (transmission capability) margin resulting in a quick return of the synchronous generators to a new steady state condition without rotor angle oscillations. In this investigation, it is shown that the SSSC makes the power system more flexible and capable to handle large disturbances, which could otherwise experience instability in absence of the SSSC system.

As seen from Fig.4, the success of the SSSC in controlling the power flow and the admittance of the transmission network is contingent on design of an appropriate controller. This controller changes the line reactance by implementing a variable reactance ($x_q$) in series with the transmission line [14]. The variable reactance realized by the SSSC is expressed in Eq.10.

$$x_q = \frac{v_q}{i_L} \tag{10}$$

In Equation (10), $v_q$ is the voltage SSSC injects in series with the transmission line and $i_L$ is the line current. The magnitude ($V_q$) of the instantaneous SSSC voltage $v_q$ varies according to limitations of the devices used in its power circuit. In general, we could have $-0.135V_1 < V_q < +0.135V_1$, as suggested by Hingorani [11]. In this investigation, $V_1$ is the generator bus voltage but in general, for a large scale power system $V_1$ could be magnitude of the transmission line voltage. Since $X_L$ changes with the changing values of $x_q$, therefore, maximum power transmission capability (see Fig.3), $P_{max} (= \frac{V_1 V_2}{X_L})$ can be changed by using a SSSC in series with the transmission line.

## V. DESIGN OF NONLINEAR CONTROLLER FOR SSSC

The developed block diagram of the system including nonlinear controller is shown in Fig.5. Eq.8 is a nonlinear differential equation and it expresses system's dynamics. It is difficult to design a state feedback controller using Eq.8, thus we propose and develop (see Appendix) a state transformation as given in Eq.11.

$$\begin{bmatrix} z_1 \\ z_2 \end{bmatrix} = \begin{bmatrix} \sin(\delta) \\ \dot{\delta}\cos(\delta) \end{bmatrix} \tag{11}$$

By applying the proposed transformation, the nonlinear Eq.8 can be rewritten as the linear Eq.A12 (see Appendix) and given here under in Eq.12.

$$\begin{bmatrix} \dot{z}_1 \\ \dot{z}_2 \end{bmatrix} = \begin{bmatrix} 0 & 1 \\ 0 & 0 \end{bmatrix} \begin{bmatrix} z_1 \\ z_2 \end{bmatrix} + \begin{bmatrix} 0 \\ 1 \end{bmatrix} \left( \frac{w}{M} \right) \tag{12}$$

In Eq.12, quantity $w$ is state feedback as expressed in Eq.13.

$$w = -Mkz_2 \tag{13}$$

Having developed the SSSC control law stated in Eq.13, it is pertinent to carry out stability analysis of the SSSC equipped test power system to demonstrate and subsequently utilize advantages of FACTS device. Using Lyapunov's stability criteria, the stability of the test power system subjected to a large disturbance is investigated in Section 6.

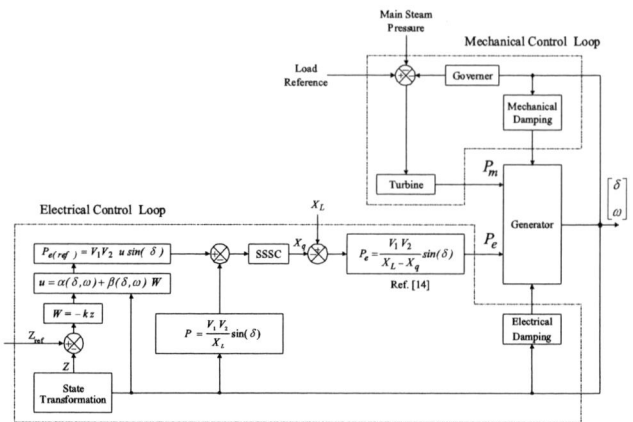

Fig. 5. Nonlinear control scheme of test power system with SSSC.

## VI. STABILITY EVALUATION OF TEST POWER SYSTEM

This section deals with the stability evaluation of the test power system. However, before we begin mathematical formulations for stability analysis, it is deemed necessary to highlight the inherent features and advantages of the SSSC system.

### A. Stability Improvement via SSSC System

The developed nonlinear controller of the SSSC is integrated with the mathematical model of the test power system shown in Fig.1. The obtained integrated model is used to find out system's Transient Energy Function (TEF). It is well known from literature [5] that the loss of a transmission line will instantaneously decrease the transferable electric power capacity of the survived transmission system between generating units to load centers or infinite bus, whereas, for next few seconds the mechanical input power to generating units remains constant at its pre-fault value. This causes an imbalance between mechanical input power and transmitted electric power with latter being smaller than the former. The power imbalance during the fault causes the generator's kinetic energy to increase, hence driving the generator towards an unstable region of operation ($\delta \geq 90°$). In Fig.1, the post-fault generator remains stable if extra energy stored in the generator during the fault is completely absorbed by transmission system. It is quite possible that post-fault power system may settle down to a new stable operating point provided the fault is cleared quickly and the post-fault power transmission capacity is restored to a desired level. On the other hand, a slow switching scheme employed for fault clearance may not prevent the generator to become unstable due to unmanageable and significant energy imbalance. Therefore, it can be stated that the stability problem could become more acute if the transmission capacity is not restored to a required level in a post-fault power system. It can be seen from Fig.3, the post-fault maximum transmission capability ($P_{max}$) is closer to mechanical input power $P_m$ than $P_{max}$ of the pre-fault transmission system. Therefore, the fault has to be cleared in time to ensure that Area A1 stays equal to area A2. Contrary to this, the test power system equipped with the SSSC doesn't exhibit this constraint as we can control maximum transmission capability ($P_{max}$) by lowering the line reactance to a desired value. However, this change in the line reactance has to be in compliance with the operating voltage limit of the SSSC.

Active control of the line reactance to raise or lower the transmission capacity makes the SSSC a practical device for contingency management leading to a proper energy balance in a post-fault test power system. The SSSC has not only a potential to increase the stability margin but also prevents any sub-synchronous resonance and classical oscillations in a post-fault power system. Apart from above stated features, in a power system equipped with the SSSC, fault clearing time is significantly increased while retaining its stability in post-fault operation. To validate this claim about the SSSC, we next focus on Lyapunov's Transient Energy Function (TEF) of the proposed system shown in Fig.1.

### B. Mathematical Formulation for Stability Assessment of Test Power System

In a SSSC equipped power system, the TEF could become a good pointer for stability provided an appropriate mathematical formulation is carried out to reveal inherent features of the SSSC system. The TEF is the time integral of the swing equation of the test power system stated in Eq.5. The lower limit of this integration is the instant of the fault ($t = 0$, $\delta = \delta_0$), and the upper limit is the time instant when $\dot\delta$ becomes zero. The upper limit can not exceed $t_{max}$ that correspond to $\delta_{max}$ ($= \pi - \delta_0$). The transient energy contained in the test power system is a time integral of accelerating power (Eq.5). Therefore, the time integral of Eq.5 provides TEF of the test power system as expressed below;

$$\text{TEF} = \int_0^{t_x} (M\ddot\delta + D\dot\delta + \frac{V_1 V_2 \sin(\delta)}{X_L} - P_m)dt \quad (14)$$

Substitution of the nonlinear controller of Eq.9 in Eq.14 results in the TEF stated in Eq.15 and Eq.15 includes properties of the SSSC.

$$\text{TEF} = \int_0^{t_x} \{M\ddot\delta + D\dot\delta + \frac{M\dot\delta^2 \sin(\delta) - P_m \cos(\delta) + D\dot\delta \cos(\delta) - k\dot\delta \cos(\delta)}{-\sin(\delta)\cos(\delta)}\sin(\delta) - P_m\}dt \quad (15)$$

On following the steps given in Appendix (Eqs.A8 to A11) and substituting $w = -Mk\dot\delta \cos(\delta)$, we get following equation for the TEF.

$$\text{TEF} = \int_0^{t_x} \left\{ M\frac{d}{dt}[\dot\delta \cos(\delta)] - Mk\dot\delta \cos(\delta) \right\}dt$$
$$= M\dot\delta \cos(\delta(t_c)) - M\dot\delta \cos(\delta(t_0)) - Mk \sin(\delta(t_c)) + Mk \sin(\delta(t_0)) \quad (16)$$
$$= M\dot\delta_c \cos(\delta_c) - M\dot\delta_0 \cos(\delta_0) - Mk \sin(\delta_c) + Mk \sin(\delta_0)$$

In worst case scenario $\delta_c = \delta_{max}$ with $\delta_{max} > \delta_{cc}$. At $\delta = \delta_{max}$, $\omega$ ($= \dot\delta = d\delta/dt$) becomes zero. Therefore, Eq.16 can be simplified and this simplification results in Eq.17 as follows:

$$\text{TEF} = -Mk \sin(\delta_{max}) + Mk \sin(\delta_0)$$
$$= -Mk \sin(\pi - \delta_0) + Mk \sin(\delta_0) \quad (17)$$
$$= 0$$

It is found from Eq.17 that even for the worst case ($\delta_c = \delta_{max}$), the SSSC equipped test power system remains stable as TEF is zero. Therefore, our claim is validated and it is confirmed that the SSSC system is capable of stabilizing the post-fault test power system. Having developed control law for the SSSC system along with stability analysis, it is imperative to use the developed control model for performance simulation using MATLAB.

## VII. PERFORMANCE EVALUATION AND MATLAB SIMULATION

The MATLAB simulation program provides the performance parameters of the test power system including the voltage, current, TEF, injected power, and trajectory of voltage generated by the SSSC. Through a series of simulations we determined that fault critical clearing time ($t_{cc}$) for the test power system is 0.75 seconds. The obtained results are portrayed to reveal the inherent capabilities of the SSSC system. We simulated the test power system (Fig.1) under following operating conditions:

(a) Without SSSC and fault is cleared after the critical clearing time ($t_c > t_{cc}$),
(b) Without SSSC and fault is cleared within the critical clearing time ($t_c < t_{cc}$),
(c) With SSSC and fault is cleared much after the critical clearing time ($t_c \gg t_{cc}$).

### A. Case A - Without SSSC and fault is cleared after the critical clearing time ($t_c > t_{cc}$)

In this case, the test power system (Fig.1) operates in open loop, the fault in line 2 occurs at $t = 0$ second and is cleared at $t_c = 0.75$ seconds. The states' trajectory is shown in Fig.6 (a); it is a curve of state variable $\dot\delta$ ($= \omega$) with respect to state variable $\delta$. As seen from Fig.6(a) that system states do not converge to a stable equilibrium point, indicating that the test power system without SSSC is not stable if fault clearing time is elapsed. The post-fault operating parameters given in Fig.7 (a) confirm the information contained in Fig.6 (a).

### B. Case B - Without SSSC and fault is cleared within the critical clearing time ($t_c < t_{cc}$)

In this case, the test power system operates in open loop, the fault in line 2 occurs at $t = 0$ second to be cleared at $t_c < 0.75$ seconds. From the states' trajectory (Fig.6b) and post-fault quantities (Fig.7b), it is evident that the test power system remains stable and state variables converge to a new steady state condition after passing through a series of rotor angle oscillations. The new steady state value ($\delta_{new}$) of the power angle is at the center of the eye of the state's trajectory.

### C. Case C - With SSSC and fault is cleared much after critical clearing time ($t_c \gg t_{cc}$)

In this case, the test power system operates in closed loop in conjunction with the SSSC, the fault in line 2 occurs at $t = 0$ seconds. Through a series of simulations we have determined that system remains stable even if the fault is cleared at $t = 1.36$ seconds. Therefore, the SSSC equipped test power system remains stable even for the condition $t_c \gg t_{cc} = 0.75$ seconds. This indicates that the SSSC system yields a significant (81%) improvement (an increase) in the critical clearing time. Also, from the states' trajectory (Fig.6c) it is evident that the test power system with the SSSC converges to a new stable condition ($\delta_{new}$) without encountering the unwanted phenomena of rotor angle oscillations.

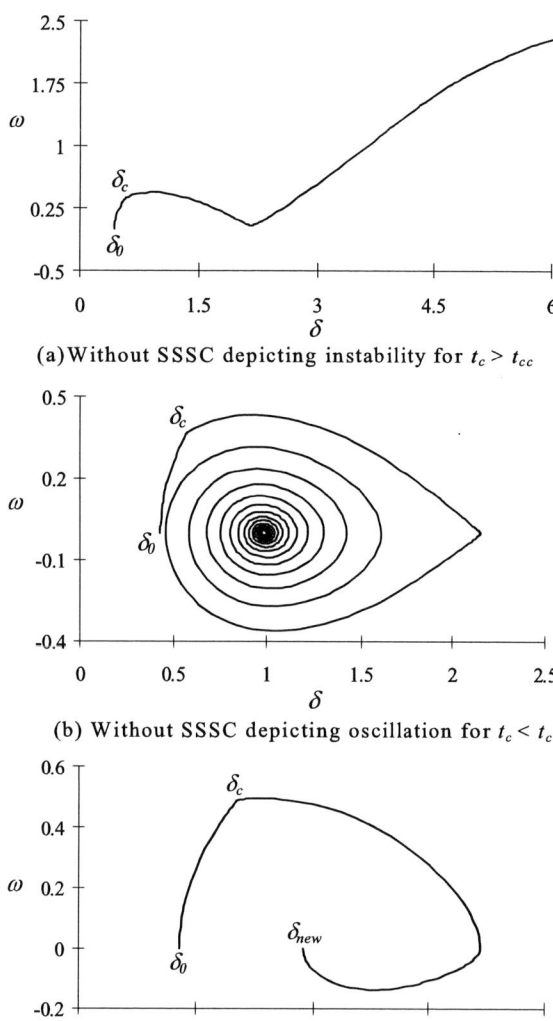

(a) Without SSSC depicting instability for $t_c > t_{cc}$

(b) Without SSSC depicting oscillation for $t_c < t_{cc}$

(c) With SSSC depicting mitigation of SSR and improved stability for $t_c \gg t_{cc}$

Fig. 6. Variations of $\delta$ and $\omega (= \dot{\delta})$ for test power system.

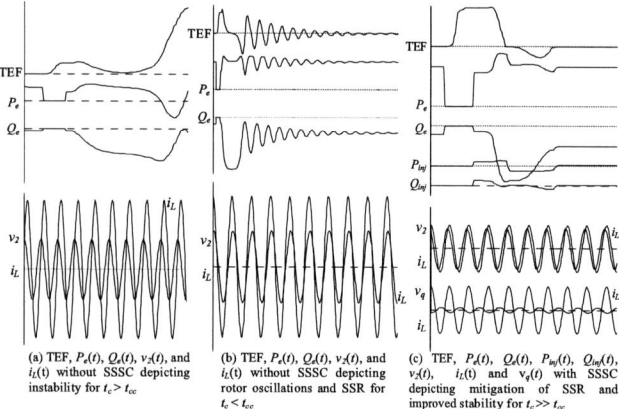

(a) TEF, $P_e(t)$, $Q_e(t)$, $v_2(t)$, and $i_L(t)$ without SSSC depicting instability for $t_c > t_{cc}$

(b) TEF, $P_e(t)$, $Q_e(t)$, $v_2(t)$, and $i_L(t)$ without SSSC depicting rotor oscillations and SSR for $t_c < t_{cc}$

(c) TEF, $P_e(t)$, $Q_e(t)$, $P_{inj}(t)$, $Q_{inj}(t)$, $v_2(t)$, $i_L(t)$ and $v_q(t)$ with SSSC depicting mitigation of SSR and improved stability for $t_c \gg t_{cc}$

Fig. 7. Post-fault response of test power system with and without SSSC.

The post-fault quantities shown in Fig.7c confirm that the SSSC equipped test power system is free from rotor angle oscillations. Fig.8 shows the trajectory of the two components of injected voltage ($V_q$) in series with the transmission line.

As mentioned in section 4 that depending upon power flow requirement, the injected voltage ($v_q$) could vary in the range of between $\pm - 0.135 V_l$. Here, $V_l$ is magnitude of the transmission line voltage. Therefore, Fig.8 confirms this flexibility in injected voltage when a SSSC is used as a FACTS device to mitigate post-fault rotor oscillations and Sub-synchronous Resonance (SSR) problem.

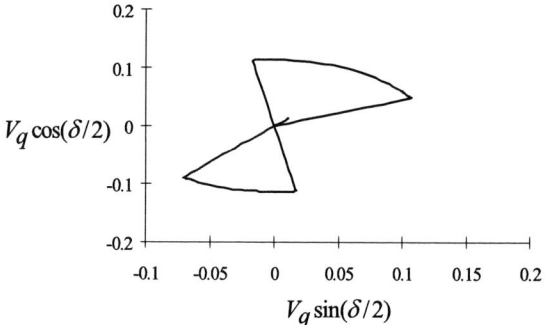

Fig. 8. Post-fault trajectory of injected voltage $V_q$ produced by SSSC.

## VIII. CONCLUSIONS

The dynamic equations of the test power system have been developed in the form of state space model. The proposed nonlinear controller for the Static Synchronous Series Compensator (SSSC) system has been developed. It has been found that the test power system in conjunction with the SSSC converges to a new steady state post-fault position without encountering unwanted classical oscillations and Sub-synchronous Resonance. MATLAB simulation results have confirmed viability of the SSSC system for effective control of the test power system. The Lyapunov's criterion of stability has been used to verify enhanced stability of the SSSC equipped test power system. In the proposed nonlinear controller, the fast acting SSSC system has used the line reactance ($X_L$) as a control variable. This is contrary to conventional controllers, wherein a slow varying mechanical input power ($P_m$) to generator is considered as control variable. It has been found that due the fast possible changes in line reactance, there is a significant (81%) improvement in critical clearing time of the fault. Therefore, it is confirmed that the test power system in conjunction with the SSSC exhibits desired stability as opposed to a system without a FACTS device. It has been observed that the SSSC system offers the inherent advantages in a post-fault power system. Nevertheless, under normal operating conditions of a power system, the SSSC system can be used for reactive power reserve management. Therefore, it can be stated that the FACTS devices including the SSSC system may have excellent economic proposition in the future power systems operating under deregulated conditions.

## IX. APPENDIX

In this section state transformation matrix expressed in Eq.11 is derived. The first new state $z_1$ is an arbitrary function of the nonlinear state $\delta$, but independent from $\dot{\delta}$. It can be seen that our candidate $z_1 = \sin(\delta)$ satisfies the condition of

570

nonlinear control law of $\partial z_1 / \partial z_2 = 0$, because it is a function of $\delta$ only. The second new state $z_2$ is expressed in Eq.A1.

$$z_2 = \nabla z_1 \times f = \dot{\delta} \cos(\delta) \qquad (A1)$$

To prove the controllability [15] of Eq.8, there must exist a region $\Omega$ such that the vector fields $\{g, [f, g]\}$ are linearly independent, and $\{g\}$ is involutive. This condition can be verified by finding the rank of controllability matrix of Eq.A2.

$$\begin{bmatrix} \dot{\delta} \\ \ddot{\delta} \end{bmatrix} = \begin{bmatrix} \dot{\delta} \\ (1/M)\left( P_m - D\dot{\delta} \right) \end{bmatrix} + \begin{bmatrix} 0 \\ -(1/M) \ V_1 V_2 \sin(\delta) \end{bmatrix} \left( \frac{1}{X_L} \right) \qquad (A2)$$

Eq.A2 can be rewritten as follows:

$$\begin{bmatrix} \dot{\delta} \\ \ddot{\delta} \end{bmatrix} = f(\delta, \ \dot{\delta}) + g(\delta, \ \dot{\delta}) u \qquad (A3)$$

The controllability matrix is expressed in Eq.A4.

$$[g \quad [f,g]] = \begin{bmatrix} 0 & (1/M) \ V_1 V_2 \sin(\delta) \\ -(1/M) \ V_1 V_2 \sin(\delta) & - V_1 V_2 \ \dot{\delta} \ \cos(\delta) + (D/M^2) \ V_1 V_2 \sin(\delta) \end{bmatrix} \qquad (A4)$$

Eq.A4 shows that the controllability matrix and has a full rank for $0 < \delta < 180°$. Therefore, the test power system is controllable within the range of $0 < \delta < 180°$. This observation is very valid because synchronous machine operates as a generator within the range of $0 < \delta < 180°$.

Input control signal ($u = 1/X_L$) to the SSSC is expressed as:

$$u = \alpha(\delta, \ \dot{\delta}) + \beta(\delta, \ \dot{\delta}) w \qquad (A5)$$

where,

$$\begin{cases} \alpha(\delta, \ \dot{\delta}) = -\dfrac{\nabla(\nabla z_1 \times f) \times f}{\nabla(\nabla z_1 \times f) \times g} \\ \beta(\delta, \ \dot{\delta}) = -\dfrac{1}{\nabla(\nabla z_1 \times f) \times g} \end{cases} \Longrightarrow \begin{cases} \alpha(\delta, \ \dot{\delta}) = \dfrac{M\dot{\delta}^2 \sin(\delta) - P_m \cos(\delta) + D\dot{\delta}\cos(\delta)}{-V_1 V_2 \sin(\delta)\cos(\delta)} \\ \beta(\delta, \ \dot{\delta}) = \dfrac{1}{-V_1 V_2 \sin(\delta)\cos(\delta)} \end{cases} \qquad (A6)$$

On substituting Eq.A6 into Eq.A5, we get control signal for the SSSC, which is expressed below:

$$u(\delta, \ \dot{\delta}) = \frac{M\dot{\delta}^2 \sin(\delta) - P_m \cos(\delta) + D \ \dot{\delta} \ \cos(\delta) + w}{-V_1 V_2 \sin(\delta) \cos(\delta)} \qquad (A7)$$

On substituting function ($u(x) = 1/X_L$) in the system dynamic equation (Eq.5, section 3), we get closed loop equation (Eq.A8) for the test power system.

$$M\ddot{\delta} + D\dot{\delta} = P_m - \left[ \frac{M\dot{\delta}^2 \sin(\delta) - P_m \cos(\delta) + D\dot{\delta}\cos(\delta) + w}{-V_1 V_2 \sin(\delta)\cos(\delta)} \right] V_1 V_2 \sin(\delta) \qquad (A8)$$

On simplification, Eq.A8 turns to Eq.A9.

$$M\ddot{\delta} + D\dot{\delta} = P_m - P_m + D\dot{\delta} + \frac{M\dot{\delta}^2 \sin(\delta) + w}{\cos(\delta)} \Rightarrow M\ddot{\delta} = \frac{M\dot{\delta}^2 \sin(\delta) + w}{\cos(\delta)} \qquad (A9)$$

On multiplying both sides with $\cos(\delta)$, we get;

$$M\ddot{\delta}\cos(\delta) = M\dot{\delta}^2 \sin(\delta) + w \Rightarrow M\ddot{\delta}\cos(\delta) - M\omega\dot{\delta}\sin(\delta) = w \qquad (A10)$$

As we see the left side of Eq.A10 is a complete derivative of the term $M\dot{\delta}\cos(\delta)$. Therefore, Eq.A10 can be rewritten as follows;

$$M\frac{d}{dt}(\dot{\delta}\cos(\delta)) = w \text{ and } \dot{\delta}\cos(\delta) = z_2 \Rightarrow M\frac{d}{dt}z_2 = M\dot{z}_2 = w \qquad (A11)$$

On expressing state equations in terms of new state variables ($z_1$, and $z_2$), we get Eq.A12.

$$\begin{bmatrix} \dot{z}_1 \\ \dot{z}_2 \end{bmatrix} = \begin{bmatrix} 0 & 1 \\ 0 & 0 \end{bmatrix} \begin{bmatrix} z_1 \\ z_2 \end{bmatrix} + \begin{bmatrix} 0 \\ 1 \end{bmatrix} \left( \frac{w}{M} \right) \qquad (A12)$$

Eq.A12 is used in section 5 to develop control law for the SSSC system.

## X. REFERENCES

[1] P. Rastgoufard, A. Yazdankhah, and R. A. Schlueter, "Multi-machine equal area based power system transient stability measure," *IEEE Trans. on Power Systems*, vol. 3, no. 1, pp. 188-196, 1988.

[2] B. N. Singh, A. Chandra, K. Al-Haddad, and B. Singh "Performance of sliding mode and fuzzy controllers for static synchronous series compensator," *The Journal of IEE Proceedings on Generation, Transmission and Distribution*, vol. 146, no. 2, pp. 200-206, 1999.

[3] B. N. Singh, A. Chandra, and K. Al-Haddad, "A DSP based indirect current controlled STATCOM-Part I: Evaluation of current control techniques," *The Journal of IEE Proceedings on Electric Power Applications*, vol. 147, no. 2, pp. 107-112, 2000.

[4] Y. Xiao, Y. H. Song, C. C. Liu, and Y. Z. Sun, "Available transfer capability enhancement using FACTS devices," *IEEE Trans. on Power Systems*, vol. 18, no. 1, pp. 305-312, 2003.

[5] P. Kundur, *Power Systems Stability and Control*, McGraw-Hill, Inc., 1993.

[6] M. Poshtan, P. Rastgoufard, and M. Wu, "Nonlinear control application to enhance fault clearing time in power systems," in *Proc. 1999 IEEE SSST Conference*, pp. 138-142.

[7] M. Wu, P. Rastgoufard, and M. Poshtan, "Application of feedback linearization control to small scale power systems stability analysis," in *Proc. 1999 IEEE SSST Conference*, pp. 213-217.

[8] L. Gyugyi, C. D. Schauder, and K. K. Sen, "Static synchronous series compensator: A solid-state approach to the series compensation of transmission lines," *IEEE Trans. on Power Delivery*, vol. 12, no. 1, pp. 406-417, 1997.

[9] B. N. Singh, A. Chandra, and K. Al-Haddad, "A DSP based indirect current controlled STATCOM-Part II: Multi-functional capabilities," *The Journal of IEE Proceedings on Electric Power Applications*, vol. 147, no. 2, pp.113-118, 2000.

[10] B. N. Singh, B. Singh, A. Chandra, and K. Al-Haddad, "A new control scheme of hybrid series active filter," in *Proc. 1999 IEEE PESC*, pp. 249-254.

[11] N. G. Hingorani, "Flexible AC transmission systems," *IEEE Spectrum*, no. 4, pp 40-45, 1993.

[12] K. K. Sen, "SSSC-static synchronous series compensator: theory, modeling, and applications," *IEEE Trans. on Power Delivery*, vol. 13, no. 1, pp. 241-246, 2001.

[13] L. Gyugyi, "Unified power flow control concept for flexible AC transmission systems," *The Journal of IEE Proceedings on Generation, Transmission and Distribution*, vol. 139, no. 4, 1992.

[14] "Overview about FACTS: Applying flexibility to your electric power system," http://www.abb.com/electricutilities.

[15] J. J. E. Slotine and W. Li, *Applied Nonlinear Control*, New Jersey: Prentice Hall, 1991.

**2006 IEEE International Conference on Power Electronic, Drives and Energy Systems**

# An Improved Power Flow Analysis Technique with STATCOM

Annapurna Bhargava, Vinay Pant, and Biswarup Das

*Abstract* - In this paper, a simple and efficient algorithm for incorporating STATCOM losses in the load flow calculation is proposed. In this technique, both the STATCOM switching loss and the step-down transformer ohmic loss are accounted for and are calculated iteratively. To reduce the computational burden, the system load flow equations and the STATCOM equations are solved separately. The proposed algorithm has been tested on IEEE 30 bus and IEEE 57 bus power systems. Based on the results obtained on different operating conditions, the proposed algorithm has been found to possess good convergence characteristics. The influence of different factors on the convergence characteristics of the proposed algorithm has also been investigated in detail.

*Index Terms* - FACTS, Load Flow, STATCOM.

## I. NOMENCLATURE

$I_{si}$ : STATCOM current in p.u.
$\theta_{si}$ : Phase angle of STATCOM current in degree
$P_{si}$ : Active power of STATCOM in p.u. (for replenishing STATCOM losses)
$Q_{si}$ : Reactive power of STATCOM in p.u.
$P_i$ : Injected active power at ith bus in p.u.
$Q_i$ : Injected reactive power at ith bus in p.u.
$R_{si}$ : Leakage resistance of stepdown transformer in p.u.
$X_{si}$ : Leakage reactance of stepdown transformer in p.u.
$V_i$ : System bus voltage in p.u.
$\delta_i$ : System bus voltage angle in degree
$V_{dc}$ : STATCOM capacitor dc voltage in p.u.
$V_{si}$ : STATCOM output voltage in p.u.
$\alpha_{si}$ : Phase angle of STATCOM output voltage in degree
$\alpha_{stat}$ : Phase difference between $V_i$ & $V_{si}$ in degree
$\theta_{stat}$ : Phase difference between $V_i$ & $I_{si}$ in degree
$G_c$ : Admittance representing the switching losses of the converter
$m_0$ : Pulse width modulation index

## II. INTRODUCTION

Annapurna Bhargava is a Research Scholar in Electrical Engineering Department, IIT, Roorkee, India (e-mail: ab_eck@sify.com).
B. Das is in Electrical Engineering Department, Indian Institute of Technology, Roorkee, India (e-mail: biswafee@iitr.ernet.in ).
Vinay Pant is in Electrical Engineering Department, Indian Institute of Technology, Roorkee, India (e-mail: vpantfee@iitr.ernet.in ).

THE static compensator (STATCOM) [1] is one of the most prominent members in the family of flexible AC transmission system (FACTS) [2] devices, which is connected in shunt to the transmission grid. Its basic purpose is to provide voltage support to the transmission/distribution grid through controlled reactive power injection. However, apart from the voltage support, it is also used to provide damping to the transmission grid for enhancing the stability of the system.

For assessing the impact of the STATCOMs in controlling the grid voltage, power flow study is necessary. Moreover, in the planning stage, to determine the ratings of the STATCOMs, among others, repeated load flow studies are carried out. Also, in a stability study, load flow solution is required to establish the initial operating point. Thus, power flow studies are indeed one of the most fundamental studies necessary to be carried out before implementing any STATCOM in a power system.

Traditionally, a STATCOM has been represented in a power flow study as a PV bus [3]. In this approach, generally the internal losses in the STATCOM (both switching losses as well as the ohmic losses of the step down transformer) are neglected and as a result, the specified real powers at these PV buses are set to zero. As the internal losses are neglected, this method, although very popular, is not very accurate. In [4] an improved model of the STATCOM has been proposed to consider the ohmic losses of the step down transformer in the power flow solution method. This method also suffers from inaccuracy as it neglects the switching losses in the STATCOM. In [5] another procedure for including STATCOM losses in power flow calculation is presented. However, in this method the Jacobian matrix of the system is changed to incorporate STATCOM and its losses. Due to changes in the main load flow Jacobian the above method may not possess excellent convergence characteristics for large power systems. In [6] detailed equations for steady state and transient analysis for the STATCOM have been presented but no numerical results have been given to illustrate the application of these equations in load flow solution.

To address the above limitations, in this paper, an improved load flow technique is presented, in which, both the switching losses and the transformer ohmic losses are taken into consideration. Moreover, the proposed

0-7803-9771-1/06/$25.00 ©2006 IEEE

technique does not need to make any change in the main load flow Jacobian, which, in turn, ensures good convergence characteristics of this method. Also, to reduce the computational burden, the main load flow equations and the STATCOM equations are solved separately. The effectiveness of this proposed technique has been tested on IEEE 30 bus and IEEE 57 bus power systems. The paper is organized as follows. In Section II, the detail STATCOM equations are described. In Section III, the proposed algorithm is presented in detail. Section IV describes the results and relevant discussions of this work and lastly, Section V concludes the paper.

## III. STATCOM MODEL

A schematic diagram and the equivalent circuit of the STATCOM are shown in Fig. 1 and Fig. 2 respectively. From Fig. 2, the equations for the $i^{th}$ bus with STATCOM can be written as follows:

$$S_i = -P_{Li} - P_{si} + j(-Q_{Li} - Q_{si}) \qquad (1)$$

$$P_i = -P_{Li} - P_{si} \qquad (2)$$

$$Q_i = -Q_{Li} - Q_{si} \qquad (3)$$

$$I_{si} = (P_{si} - jQ_{si})/V_i^* \qquad (4)$$

$V_{zsi}$ is the drop across $Z_{si}$ and is given as

$$\overline{V}_{Zsi} = \overline{Z}_{si}.\overline{I}_{si} \qquad (5)$$

$$Z_{si} = R_{si} + jX_{si}$$

$$V_{si} = V_i - V_{Zsi} \qquad (6)$$

$$V_{si} = K.V_{dc} \qquad (7)$$

$$P_{si} = G_c.V_{dc}^2 + I_{si}^2.R_{si} \qquad (8)$$

Where, K is given by [6], $K = \left(\sqrt{\frac{3}{8}}\right)m_0$. It is to be noted that, the value of the parameter $G_c$ is important in this model because it defines dc voltage dynamics and directly affects the capacitors charging and discharging time constant [6].

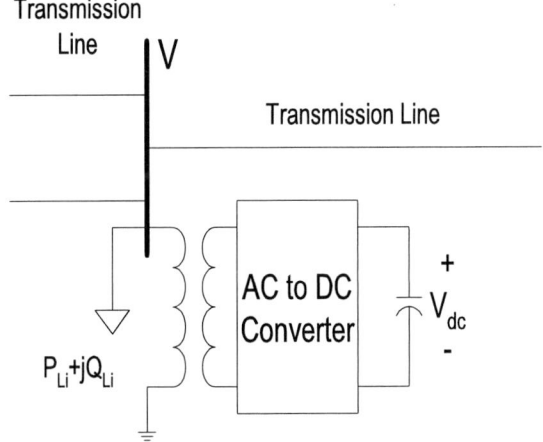

Fig. 1. Schematic diagram of STATCOM.

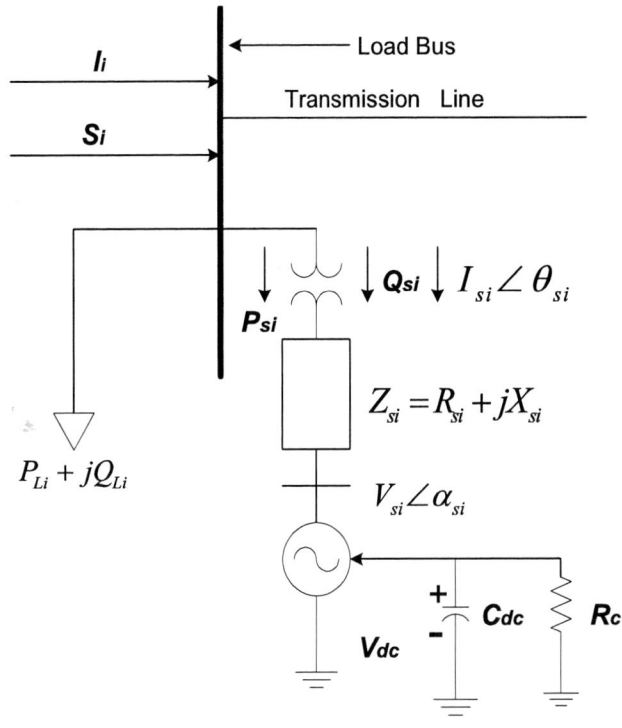

Fig. 2. Equivalent circuit of STATCOM.

## IV. PROPOSED ALGORITHM

*Steps*

i. Assume an initial value of $P_{si}$ (active power drawn by STATCOM to supply its losses). In the present study it has been assumed to be equal to 0.001 p.u.

ii. The system buses, at which the STATCOMs are assumed to be placed, are made *PV* buses.

iii. Set $P_{si\,old} = P_{si}$ at all the STATCOM buses.

iv. On the STATCOM buses, the specified voltage is set according to the desired bus voltage and the upper and lower limits of reactive power are set following the STATCOM ratings.

v. The specified real powers at these above *PV* buses are calculated using equation (2).

vi. Run normal load flow to obtain system bus voltages, phase angles and injected reactive powers.

vii. Calculate new value of ($P_{si\,new}$) at each of the STATCOM buses using the following procedure:

a. Calculate $Q_{si}$ from equation (3)

b. Calculate $I_{si}$ from equation (4)

c. Calculate drop in $Z_{si}$ from equation (5)

d. Calculate $V_{si}$ and its angle $\alpha_{si}$ from equation (6)

e. Calculate $\alpha_{stat}$ (angle responsible for supply the losses)

$$\alpha_{stat} = \delta_i - \alpha_{si}$$

573

f. Calculate $V_{dc}$ from equation (7)

g. Calculate $P_{si\,new}$ from equation (8)

viii. Calculate error in $P_{si}$ at all the STATCOM buses.

$$\Delta P_{si} = P_{si\,new} - P_{si\,old}$$

ix. If $\Delta P_{si} \leq \varepsilon$ (where $\varepsilon = 0.0001$) then the load flow converges.

If convergence criteria is not satisfied then set $P_{si} = P_{si\,new}$ and go to step (vi).

## V. RESULTS AND DISCUSSION

The effectiveness of the proposed algorithm has been tested by carrying out numerous case studies in IEEE 30 bus and IEEE 57 bus power systems. For deciding the appropriate locations of the STATCOMS in these three systems, initially load flow studies (without any STATCOM) have been carried out at the base loading conditions in these two systems. From the results of load flow solutions, buses with relatively low voltage magnitude have been picked up and subsequently STATCOMs have been assumed to be placed at these particular buses. With this approach, 3 and 4 STATCOMs have been assumed to be installed in IEEE 30 bus and IEEE 57 bus system respectively and the detail of these STATCOMs are given in the appendix. The proposed algorithm outlined in section III has been applied to both these systems at different loading levels and some illustrative numerical results for IEEE 30 bus system at the base loading condition are given below in Table I and Table II.

Table I and II show that at base loading condition, all the three STATCOMs in the 30 bus system operate in capacitive mode to maintain the respective bus voltages at 1.0 p.u. The capacitive mode of operation is indicated by negative value of $Q_{si}$ (consistent with the direction of $Q_{si}$ as shown in Fig. 2) as well as positive value of $\theta_{stat}$ for each STATCOM. Moreover, for each STATCOM, $\alpha_{stat}$ is small which indicates that in steady state, the output voltage of the STATCOM $V_{si}$ slightly lags the system bus voltage $V_i$ so that the STATCOM absorbs a small amount of real power from the system bus to replenish its internal losses and to keep the capacitor voltage constant. Also $\theta_{stat}$ is nearly equal to □0 degrees, which indicates that the injected current is nearly in quadrature with the system bus voltage $V_i$ and hence principally reactive in nature. The amount of real power absorbed by each STATCOM is also shown in Table I and for comparison, these quantities have also been calculated from the equations given in [6] (as shown by $P_{si}^{*}$ in Table I). It is observed that the losses calculated by these two approaches almost match with each other. Similar results have also been obtained for IEEE 57 bus system and hence these are not repeated here. The improvements in the voltage profiles achieved in these two systems (at base loading condition) obtained with the proposed algorithm are shown in Figs. 3

and 4 respectively.

### TABLE I

SOLVED STATCOM PARAMETERS IN IEEE 30 BUS SYSTEM

| Bus | $Q_{si}$ | $I_{si}$ | $V_{si}$ | $V_{dc}$ | $P_{si}$ | $P_{si}^{*}$ |
|-----|----------|----------|----------|----------|----------|--------------|
| 7 | -0.78274 | 0.7827 | 1.0□00 | 1.2112 | 0.0020 | 0.001□7 |
| 26 | -0.0□856 | 0.0□86 | 1.0113 | 1.1237 | 0.00015 | 0.00014□ |
| 30 | -0.0□438 | 0.0□44 | 1.010□ | 1.1232 | 0.00015 | 0.00015 |

### TABLE II

SOLVED STATCOM PARAMETERS IN IEEE 30 BUS SYSTEM (CONTD...)

| Bus | $V_i$ | | $\delta_i$ | $\theta_{si}$ | $\alpha_{si}$ | $\theta_{stat}$ | $\alpha_{stat}$ |
|-----|-------|------|------------|---------------|---------------|-----------------|-----------------|
| | without STATCOM | With STATCOM | | | | | |
| 7 | 0.864□□ | 1.00 | -13.833 | 76.030 | -13.□6□ | 8□854 | 0.136 |
| 26 | 0.81666 | 1.00 | -1□110 | 70.802 | -1□128 | 8□□13 | 0.018 |
| 30 | 0.81883 | 1.00 | -1□812 | 70.0□7 | -1□82□ | 8□□0□ | 0.017 |

The proposed algorithm has also been found to be equally effective for different loading conditions in both these test systems and the results obtained are in complete agreement with the expected behavior of the STATCOMs. Some of the representative results obtained in IEEE 30 bus system are shown in Table III. During light load condition case, i.e. when the load on the system is reduced to less than 50% of the normal loading condition, the system experiences an over voltage condition and as a result, some or all of the STATCOMs operate in the 'inductive mode', thereby absorbing reactive power. When the loading on the system is very low (about 10%), one of the STATCOMs (7th) reaches its inductive limit and hence, the voltage of the corresponding system bus can not be maintained at the desired level (1.0 p.u.) as this bus acts as a $PQ$ bus and not as a $PV$ bus. The inductive mode of operation is indicated by the positive values of $Q_{si}$ (consistent with the direction of $Q_{si}$ as shown in Fig. 2). Similarly, under the overloaded condition, the STATCOMs operate in the 'capacitive mode', thereby supplying reactive power to the system and in this process, some or all of the STATCOMs hit their corresponding reactive power generation limits resulting in uncontrolled voltages at their corresponding system buses. Here it is found that for the chosen limits (rating of STATCOM) the STATCOMs are able to maintain the system bus voltages at the desired level in the loading range of 30% to 100%. Similar behavior has also been observed in IEEE 57 system.

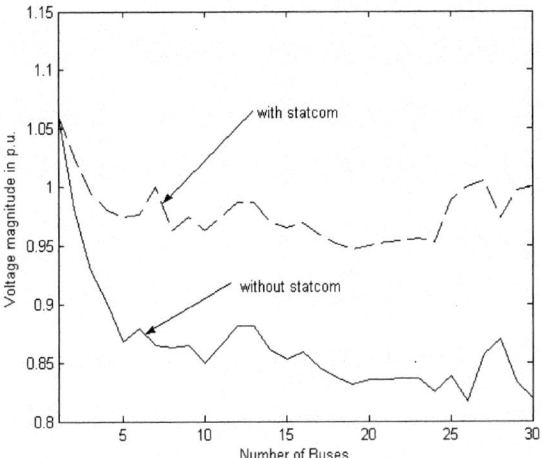

Fig. 3. Comparison of voltage profile in IEEE 30 bus system.

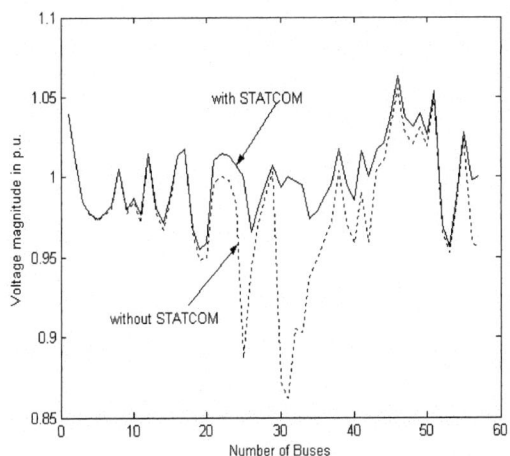

Fig. 4. Comparison of voltage profile in IEEE 57 bus system.

### TABLE III

RESULTS OF THE IEEE 30 BUS SYSTEM AT DIFFERENT OPERATING
CONDITIONS

| Bus no | 10% loading | | 110% loading | | 130% loading | |
|---|---|---|---|---|---|---|
| | $V_i$ | $Q_{si}$ | $V_i$ | $Q_{si}$ | $V_i$ | $Q_{si}$ |
| 7 | 1.015 | 0.300 | 0.□78 | -0.700 | 0.□11 | -0.700 |
| 26 | 1.000 | 0.071 | 1.000 | -0.112 | 0.□86 | -0.200 |
| 30 | 1.000 | 0.076 | 1.000 | -0.114 | 0.□84 | -0.200 |

The convergence characteristics of the proposed algorithm are shown in Fig. 5. In this figure, the number of iterations needed by the proposed algorithm at different loading levels is shown for both these test systems. It is observed that for low and normal loading conditions in the system, the proposed algorithm has excellent convergence characteristics. However, under over-loaded conditions,

the proposed method tends to need higher number of iterations to converge.

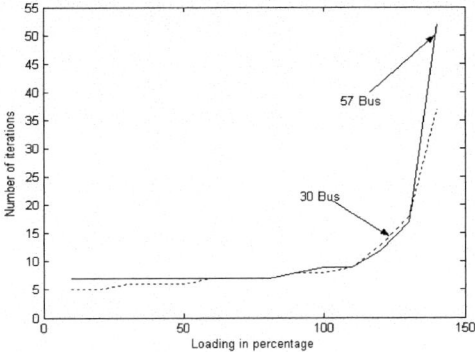

Fig. 5. Convergence characteristics of the proposed algorithm.

It has been observed that the value of $G_c$ affects the convergence characteristic of proposed algorithm quite significantly. The number of iterations needed by the proposed algorithm (in IEEE 30 bus system) for different values of $G_c$ are shown in Fig. 7. It has been found that for the value of $R_c$ (reciprocal of $G_c$) above 124 p.u, the algorithm converges, but below this value it diverges as shown in Fig. 6. Moreover, below this critical value (124), the number of iterations required by the proposed algorithm is more or less independent of the value of $R_c$. Now it is to be noted that a value of $R_c$ equal to 124 or less essentially represents a case of very high switching loss in the STATCOM, which is quite impractical. Hence, it can be said that for all practical levels of switching loss in the STATCOM, the proposed algorithm has an excellent convergence characteristics.

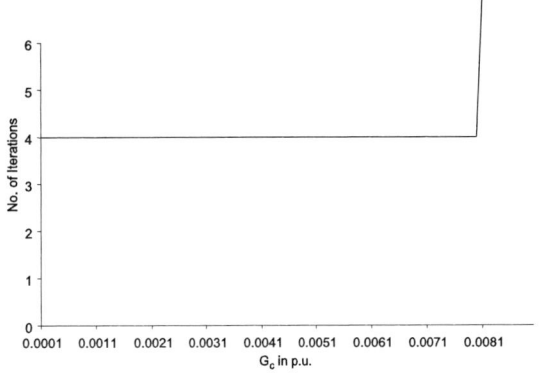

Fig. 6. Variation in no. of iterations with $G_c$.

With the variation of the value of $G_c$, the angle $\theta_{stat}$ also varies considerably. For higher value of $G_c$ (higher value of switching loss), the STATCOM absorbs more real power to replenish its losses and as a result, this angle deviates more and more from $\square 0°$. A typical variation of $\theta_{stat}$ with variation of $G_c$ is shown in Fig. 7 for the STATCOM at bus no. 30 in the IEEE 30 bus system.

Similar to $G_c$, the convergence characteristics of the proposed algorithm also depends on the value of $R_{si}$. Fig.

575

8 shows the number of iterations needed by the proposed algorithm for different values of $R_{si}$ for both these test systems. It has been observed that with increasing $R_{si}$, the number of iterations needed by the algorithm for achieving convergence also increases. Moreover, with increasing $R_{si}$ (increasing ohmic loss in the transformer), the angle $\theta_{stat}$ also deviates increasingly from $0^o$ as the STATCOM absorbs more real power to replenish its increasing transformer ohmic loss. A typical variation of $\theta_{stat}$ vis-à-vis $R_{si}$ is shown in Fig. □ for the STATCOM at bus no. 7 in the IEEE 30 bus system. It is observed that $\theta_{stat}$ varies almost linearly with $R_{si}$. Similar trend has also been observed at all the other STATCOM buses in both the test systems.

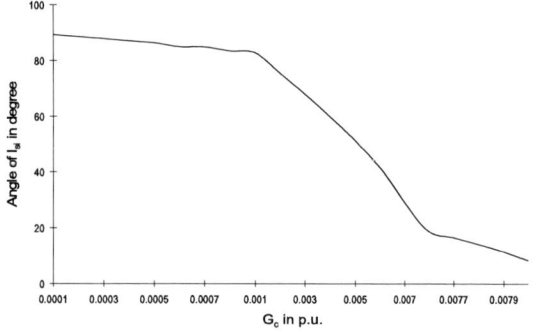

Fig. 7. Variation of $\theta_{stat}$ with $G_c$.

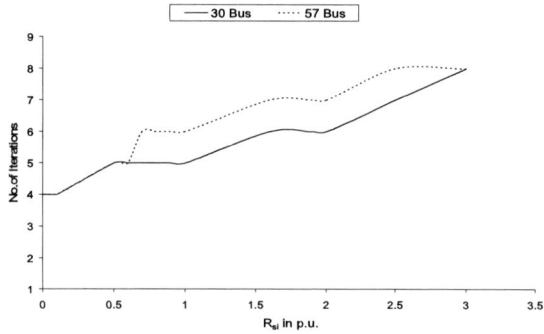

Fig. 8. Variation in no. of iterations with $R_{si}$.

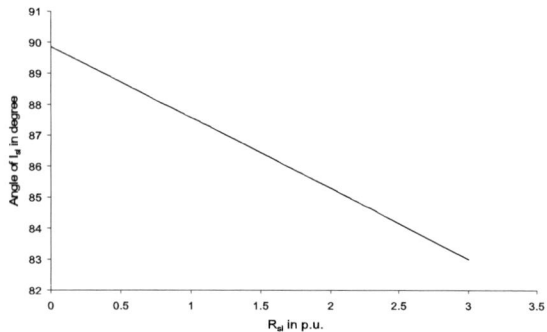

Fig. □ Variation of $\theta_{stat}$ with $R_{si}$.

## VI. CONCLUSION

In this paper, a simple, two step algorithm for incorporating STATCOM losses in the load flow calculation is developed. Based on the detail case studies carried out on IEEE 30 bus and IEEE 57 bus power systems, the main conclusions of this work can be summarized as follows:

a. The proposed technique can incorporate both the switching losses and the transformer ohmic losses into the power flow solution quite efficiently.

b. The proposed algorithm has a satisfactory convergence property over a wide range of loading conditions.

c. For low and normal loading conditions, the computational effort needed by the proposed algorithm is quite moderate.

For practical levels of switching losses (SL) in the STATCOM circuit, the convergence characteristic of the proposed technique is very good and independent of SL.

## VII. REFERENCES

[1]. Kalyan K. Sen, "STATCOM-STATic synchronous COMpensator: Theory, Modeling and Applications", Power Engineering Society Winter Meeting, IEEE, Vol. 2, 31 Jan-4 Feb 1□□, pp.1177-1183.

[2]. L. Gyugyi and N. G. Hingorani, "Understanding FACTS: Concepts and technology of flexible AC transmission systems", Willey John & Sons, Incorporated, New York, 1□□□

[3]. D. J. Gotham and G. T. Heydt, "Power flow control and power flow studies for systems with FACTS devices", IEEE Trans. on power systems, Vol. 13, No.1, February 1□8, pp. 60-66.

[4]. Z. Yang, C. Shen, M. L. Crow and L. Zhang, "An improved STATCOM model for power flow analysis", Power Engineering Society Summer Meeting, IEEE, Vol. 2, 16-20 July 2000, pp. 1121-1126.

[5]. Edvina Uzunovic, Claudio A. Canizares and John Reeve, "Fundamental frequency model of static synchronous compensator", North American Power Symposium (NAPS), Laramie, Wyoming, October 1□□7, pp. 4□54.

[6]. Claudio A. Canizares, Massimo Pozzi, Sandro Corsi and Edvina Uzunovic, "STATCOM modeling for voltage and angle stability studies", International Journal of Electrical Power & Energy Systems, Vol. 44, Issue 6, November 2003, pp. 421-422.

## VIII. APPENDIX

STATCOM parameters:
$G_c = 0.0001$ p.u, $R_{si} = 0.003$ p.u, $X = 0.145$ p.u, $K = 0.$□

STATCOM Ratings and Locations
(i) IEEE 30 bus system:
   Rated  MVAR = -30 MVAR   70 MVAR (7th bus)
                   -20 MVAR   20 MVAR
                     (26th bus, 30th bus)

(ii) IEEE 57 bus system:
   Rated  MVAR = -10 MVAR   20 MVAR (25th bus)
                   -10 MVAR   15 MVAR (31st bus)
                   -10 MVAR   10 MVAR
                     (42nd bus, 57th bus)

# Author Index

## A

| | |
|---|---|
| Abiri, E. | 298 |
| Abrishamifar, A. | 298 |
| Acarnley, P. P | 1020 |
| Achari, V. T. Sadasivan | 122 |
| Adachi, Shun-ichi | 310 |
| Adhinarayanan, T. | 942 |
| Adya, A | 892 |
| Afjei, E | 151 |
| Agarwal, P. | 274 |
| Agarwal, Pramod | 86, 531, 673 |
| Agarwal, Vivek | 443, 460 |
| Agrawal, Pramod | 467, 473 |
| Ahmad, Mukhtar | 179 |
| Ahmadian, J. | 678 |
| Ahsan, Faisal M. | 602 |
| Aktarujjaman, M. | 753 |
| Alam, Shahabur | 729 |
| Ali, Iqbal | 1041 |
| Amaresh, K. | 662 |
| Amarnath , J. | 424 |
| Amarnath, J. | 511 |
| Ambusaidi, K. A. | 389 |
| Anand, Keerthi | 401 |
| Anuradha, K. | 718 |
| Arvindan, A.N. | 315 |
| Asghar, Ali | 46 |
| Asghar, M. S. Jamil | 1143 |
| Aware, M.V. | 517, 782, 395 |

## B

| | |
|---|---|
| B, Isha T | 824 |
| Babu, K. Hari | 267 |
| Bajaj, Harbans L. | 1074 |
| Balasubramanian, R. | 771 |
| Banerjee, S. | 20 |
| Barati, Hassan | 959 |
| Basavaraja, B. | 219 |
| Basu, K. P. | 46 |
| Bates, Ian | 871 |
| Bathaee, S. M. T. | 1096 |
| Behera, Ranjan K. | 430 |
| Beig, A. R. | 7 |
| Benedict, Eric | 740 |
| Bhalodi, Kalpesh H. | 467 |
| Bhargava, Annapurna | 572 |
| Bhat, A. H. | 274 |
| Bhudamani, RM | 555 |
| Bhuvaneswari, G. | 39, 560, 647, 652 |
| Bodkhe, S. B. | 395 |

## C

| | |
|---|---|
| C, Sreekumar | 443 |
| C., Rohith Kumar H. | 1069 |
| Chanana, Saurabh | 1013 |
| Chandra, A. | 608 |
| Chandra, Ambrish | 583 |
| Chandra, D. | 370 |
| Chatterjee, Dheeman | 818 |
| Chatterjee, J.K. | 602, 812 |
| Chatterjee, Kishore | 460 |

| | |
|---|---|
| Chaturvedi, Ganesh Dutt | 876 |
| Chaturvedi, P. K. | 473 |
| Chaudhari, M. A. | 348 |
| Chaudhuri, N. Ray | 1090 |
| Chen, Po-Jia | 199 |
| Chiang, Sheng-Feng | 199 |
| Choudhuri, S. Ghatak | 204 |
| Chowdhury, S. | 90 |
| Chowdhury, S. P. | 90 |
| Chunduru, V. | 1154 |

## D

| | |
|---|---|
| Dahiya, Surender | 620 |
| Daigavane, M. B. | 506 |
| Dalvand, H. | 577 |
| Dananjayan, P. | 331 |
| Das, Anandarup | 602 |
| Das, Biswarup | 572 |
| Das, G.Tulasi Ram | 449 |
| Das, Shyama P. | 360, 430 |
| Dash, P. K. | 1064, 1127 |
| Dash, Subhransu Sekhar | 966 |
| Dehkordi, B. Mirzaeian | 54 |
| Dehkordi, Behzad Mirzaeian | 545, 101 |
| Devadoss, Surendran | 871 |
| Devasahayam, Robert | 122, 131 |
| Dixit, T. V. | 430 |
| Dobariya, C. V. | 792 |
| Dong-Fang, Zhou | 11, 522 |
| Donyavi, F. | 657 |
| Dwivedi, Sanjeet | 342 |

## E

| | |
|---|---|
| Easwarlal, C. | 15 |
| Ehsan, M. | 1096 |
| Ehsan, Mehdi | 959 |
| Ekram, S. | 171 |
| Elangovan, S. | 683 |
| Ethny, S. A. | 1020 |
| Etingov, P. V. | 909 |

## F

| | |
|---|---|
| Farhadi, A. | 1137 |
| Fazil, M. | 50 |
| Feng, Bai | 11, 522 |
| Fernandez, E. | 898 |
| Fotuhi-Firuzabad , Mahmud | 959 |
| Fotuhi-Firuzabad, M. | 1096 |

## G

| | |
|---|---|
| G, Subhash Joshi T | 636 |
| Gairola, Sanjay | 336, 701 |
| Garg, Vipin | 647, 652 |
| Gaur, Prerna | 96 |
| Gautam, Ashutosh | 1036 |
| Gayathri, M. S. L. | 842 |
| Geethalakshmi, B. | 331 |
| Ghawghawe, N. D. | 985 |
| Gholami, A. | 1053 |
| Ghose, T | 888 |
| Ghosh, Arindam | 181 |

# Author Index

Giaouris, D. .................... 20, 1020
Gilreath, Phil ........................ 1024
Goedtel, A. .................... 918, 926
Goel, Ankur ......................... 735
Goel, Manish ........................ 836
Gopakumar, K. ...................... 256
Gopila, M. ............................ 15
Gounden, N. Ammasai ............. 854
Goyal, Devendra .................... 135
Goyal, Himani ...................... 902
Gujarathi, P. K. .................... 517
Gupta , Sushma .................... 1084
Gupta, Ajai ......................... 801
Gupta, H.O. .................. 673, 1069
Gupta, I. ........................... 1069
Gupta, J.R.P ....................... 892
Gupta, Karunesh K ................ 1150
Gupta, R. A. ........................ 210
Gupta, Ranjan K. ................... 707
Gupta, S P .................. 1031, 1059
Gupta, Swapnil ..................... 888
Gusev, B.A. ......................... 383

## H

Hangal, A. ........................... 50
Hanmandlu, M. ..................... 902
Haque, Ahteshamul ............... 1143
Hasani, S. .......................... 657
Hidayat, Nabil M .............. 310, 307
Hojabri, H. ......................... 631
Huang, Rongjun .................... 860

## I

Iqbal, Atif .......................... 179

## J

Jack, Alan G. ....................... 500
Jadid, Shahram ..................... 970
Jain, D.K. .......................... 620
Jain, Shailendra K ................. 473
Jaiswal, V. .......................... 50
Jalalifar, M. .................... 59, 193
Jalilian, A. ............... 25, 678, 1137
Jamali, S. .......................... 1114
Janakiraman, P. A. ...... 641, 490, 485
Jaswant, L .......................... 712
Jaswanti, ........................... 992
Jeyabharath, R. ..................... 117
Jha, A.N. ........................... 735
Jha, Aman Kumar .................. 267
Jhang, Jyun-Jhong ................. 199
Jie, Shi Yu- .................... 11, 522
John, Vinod .................. 495, 740
Joseph, Aby ........................ 636
Joseph, Achari, .................... 131
Joseph, C. C. ...................... 122
Joshi, R R .......................... 303
Joshi, R. R. ........................ 210

## K

K, Aroul. ........................... 378

K, Kaushik ........................ 1007
K, Unnikrishnan A. ................ 636
K, Vadirajacharya .................. 673
Kadwane, S.G. ..................... 888
Kanabar, M. G. .................... 792
Kangsanant, Theo .................. 871
Kank, Amogh ....................... 975
Kapoor, Rajiv ...................... 620
Karan, B. M. .................. 267, 888
Karthikeyan, A. ................... 1120
Karthikeyan, K. .................... 537
Kasal, Gaurav Kumar ......... 64, 847
Kashem, M. A. ............... 830, 753
Kastha, D. ......................... 824
KATO, Yoshito ................ 307, 310
Kazemi ,A. ......... 970, 1053, 1107, 1114
Khadkikar, V. ...................... 608
Khan, M. Ahfaz ................... 1047
Khan, M.Rizwan .................... 179
Khan, Md. Haseeb .................. 511
Khan, Z. J. ........................ 506
Khaparde, S. A. ................... 792
Khatre, M. ......................... 500
Khincha, H. P. ................ 936, 996
Kiranmai, K. S. Phani ............. 354
Kothari, D. P. ................. 902, 882
KotharI, M. L. .................... 1090
Kottayil, Sasi K .................. 747
Krishnaswami, Hariharan .......... 707
Kulkarni, S. V. .................. 1007
Kulkarni, S. V. .................... 975
Kumar, A. D. Raj .................. 718
Kumar, Amit ....................... 888
kumar, Anil ........................ 370
Kumar, Arun Shailendra ........... 526
Kumar, Ashok ...................... 620
Kumar, Ashwani ................... 1013
Kumar, CH.Siva .................... 759
Kumar, M. Vijaya .................. 262
Kumar, Manish ..................... 620
Kumar, Mukesh ............... 107, 126
Kumar, Parveen ..................... 70
Kumar, Rajneesh .................. 1150
Kumar, Vinod ......... 303, 1031, 1059, 1069
Kumbhar, G. B. ..................... 975

## L

Lakshminarasamma, N. ............. 455
Lakshminarayanan, Sanjay ......... 256
Lavanya, K. ........................ 463
Lavanya, V. ........................ 854
Le, An D.T. ........................ 830
Ledwich , G. ....................... 753
Ledwich, G. ........................ 830
Lee, Fu-Shin ....................... 199
Lei, Yung-Tsung .................... 199
Lone, Shameem Ahmad ............. 865

## M

M, Veerachary ......... 354, 413, 479, 526
Ma, D.D. ............................ 20
Mabel, M. Carolin .................. 898
Madhusudan, ....................... 836

# Author Index

Mahajan, D. .................................................. 171
Mahato, S. N. .................................................. 80
Mahdian , B. .................................................. 1107
Mallesham, Gaddam .................................................. 401
Manjunath, H. V. .................................................. 1150
Manna, Manpreet Singh .................................................. 948
Marwaha, Anupma .................................................. 30, 948
Marwaha, Sanjay .................................................. 30, 948
Marzband, M. .................................................. 1, 625
Masoudi, M. .................................................. 657
Masoum, M.A.S. .................................................. 678, 970, 1053
Maswood, Ali. I. .................................................. 366
Mazumder, Sudip K. .................................................. 860
Meleshin, V.I. .................................................. 383
Mirzaeian, B. .................................................. 193
Mishra, Mahesh K. .................................................. 537
Mishra, S. .................................................. 735, 979
Mittal, A. P. .................................................. 892, 96
Moallem, Peyman .................................................. 101
Modi, P. K. .................................................. 473
Mohan, D. Madhan .................................................. 1131
Mohan, Ned .................................................. 250, 707, 765
Mohapatra, K.K. .................................................. 765
Mohapatra, Krushna K .................................................. 250
Moharana , A. K. .................................................. 1127
Mohod , S. W. .................................................. 395, 782
Mokhtari, H. .................................................. 596, 631, 657
Mondal, Gopal .................................................. 256
Moradi, H. .................................................. 151
Morris, Stella .................................................. 46
Mufti, Mairaj-ud-Din .................................................. 865
Muni, B.P. .................................................. 718
Muni, Bishnu P. .................................................. 111, 667
Murthy, K. V. S. Ramachandra .................................................. 550
Murthy, S. S. .................................................. 7, 39, 842, 1084, 836
Muthuselvan, N.B. .................................................. 966

## N

Nabav, S.M.H .................................................. 970, 1053
Nagamani, C. .................................................. 1120
Naidu, Kiran .................................................. 39, 842
Nair, Manjula G. .................................................. 560
NAKAMURA, Masaaki .................................................. 307
Nakamura, Masaaki .................................................. 310
Narayanan, G. .................................................. 231
Natarajan, S.P. .................................................. 373
Negnevitsky, M. .................................................. 753, 830
nezhad, S. M. Saghaeian .................................................. 193
Nirody, J. S. .................................................. 75

## O

Ovchinnikov, D.A. .................................................. 383

## P

P, Vinodh Kumar .................................................. 747
Pai, M. A. .................................................. 818
Palanisamy, V. .................................................. 15
Panda, A. K. .................................................. 378
Panda, G. .................................................. 1064
Panda, R. C. .................................................. 463
Pandi, V. Ravikumar .................................................. 695
Panigrahi, B. K. .................................................. 695, 724, 1064, 1127, 1131

Panigrahi, Ms. K. .................................................. 1127
Pant, Vinay .................................................. 572
Parmar, K.P.Singh .................................................. 882
Parthiban, P. .................................................. 531
Patel , Rajesh .................................................. 1031
Payam, A. Farrokh .................................................. 54, 59, 193, 215
Perumal, B.Venkatesa .................................................. 812
Peterson, Maryclaire .................................................. 280, 1024
Phattanasak, M. .................................................. 407
Pickert, V. .................................................. 20, 389
Pinto, A. J. P. .................................................. 7
Poshtan, Majid .................................................. 565
Pradhan, A K .................................................. 70
Prakash, Anupama .................................................. 1036
Prasad, P.V.N. .................................................. 175, 225, 759

## R

Rahimi, M. .................................................. 596
Rahmati, A. .................................................. 298
Raj, C. Thanga .................................................. 86
Rajagopal, K. R. .................................................. 146, 157, 163, 168, 182, 187
Rajaram, M. .................................................. 117
Ramakrishnan, K. .................................................. 244
Ramalingam, BMSM .................................................. 555
Ramanarayanan, V. .................................................. 455
Ramesh, L. .................................................. 90
Ramrathnam, .................................................. 842
Ranjbar, A. M. .................................................. 931, 1079
Ranjbar, M. .................................................. 931
Rao, E.V. Chandra Sekhara .................................................. 175
Rao, G. Govinda .................................................. 550
Rao, M.V. Ramana .................................................. 225
Rao, Polimera Malleswara .................................................. 854
Rao, S. Eswar .................................................. 667
Rao, T.K. Nagaraja .................................................. 787
Rastgoufard, Parviz .................................................. 565
Ravi, Jally .................................................. 237
Ravi, N. .................................................. 50, 171
Ravichandran, M. H. .................................................. 122, 131
Ravikumar, B. .................................................. 936
Ravindranath, G. .................................................. 175
Reddy, K. V. V. .................................................. 1013
Reddy, Sathish Kumar .................................................. 560
Reddy, T. Brahmananda .................................................. 262, 424, 511
Reddy, Y.V. Siva .................................................. 262
Rlavi, Jally .................................................. 141

## S

S, Meera K .................................................. 747
Sadati, N. .................................................. 931
Sadhukhan, Gautam .................................................. 70
Sagar, Prem .................................................. 1002
Saha, A. K. .................................................. 90
Saha, R. .................................................. 1102
Saini, R P .................................................. 801, 776
Samantaray, S. R. .................................................. 1064
Sanavullah, M.Y. .................................................. 15
Sanglikar, Amit .................................................. 495
Sanjeevikumar, P. .................................................. 331
Sankar, V. .................................................. 662
Saritha, B. .................................................. 485, 490
Sarkar, Arghya .................................................. 689

# Author Index

Sarma, A.V.R.S. ............................759
Sarma, D.V.S.S.Siva ....................219
Sathaiah, Chippa .........................157
Satish, T. .....................................765
Saxena, Rakesh ..........................1047
Sekhar, K.Chandra ......................449
Selvajyothi, K. ............................641
Sengupta, S. ................................689
Seni, ..........................................830
Serni, P. J. A. ...................918, 926
Seshachalam, D. ..........................370
Shaikholeslami, A. ..............1, 625
Shakarami, M. R. ..........................25
Sharma , Deepak ........................1074
Sharma ,V.K. ...............................315
Sharma, A.K. ..............................1047
Sharma, Deepen ..........................413
Sharma, K. Manjunatha ..............787
Sharma, M P .......................801, 80
Sharma, V. K. ..............................436
Sharma, Vishnu K .......................460
Shateri, H. ................................1114
Shazreen, ....................................740
Shenoy, T. P. .................................75
Shet, Vinayak N. ..............286, 292
Sheth, Nimit K. ........182, 146, 187
Shirani, A. R. ............................1079
Shyamala, P. ...............................996
Silva, I. N. da ...................918, 926
Singh , Ravindra Kumar .....321, 417
Singh, B. P. ...............107, 126, 342
Singh, Bhim ........39, 64, 96, 107, 126, 204, 237, 336, 342
Singh, Brij N. ...........280, 565, 1024
Singh, Chanan ............................806
Singh, G. K. ................................776
Singh, S. P. ...................................80
SinghT , Fhim .............................141
Singla, Bhoj Raj ...........................30
Sinha, S. K. ................................724
Siva, U. ......................................842
Siva, Uddanti ...............................39
Sivakumaran, T.S. .......................373
Sivanagaraju, S. ..........................662
Sivanandakumar, D. .....................244
Skandari, R. ................................298
Slabharwal, Slatish Chander ........35
Solanki, Jitendra .........................614
Soleymani, S. ...................931, 1079
Somasundaram, P. ........................966
Song, Y. H. ...................................90
Soni, K. M. .................................436
Sood, Vijay K. ............................729
Srinivasu, B. ...............................225
Srivastava, S. P. ...................86, 531
Srividhya, S. .............................1120
Subbarayudu, D. ................424, 511
Subramanian, N. .......................1154
Sudhakar, Singamaneni Bala .......479
Suryawanshi, H. M. .............348, 506
Sydulu, M. ..................................942

## T

TAKAHASHI, Nobuo ....................307

Takahashi, Nobuo .......................310
Tandon, A. K. ..............................836
Tekwani, P.N ...............................256
Tewolde, Meharegzi .....................360
Thakre, K. L. ..............................985
Thakur, T. .........................712, 992
Thakur, Tripta .............................797
Thomas, Mini S. ..............1036, 1041
Thukaram, D. ............936, 953, 996
Tofighi, A. ................................1107
Toliyat, H. .................................151
Tripathi, R. K. .............................370
Tripathy, M. ................................979
Tseng, Shao-Chun ........................199

## U

Umamaheswari, B. ........................463
Upadhyay, Parag ..................163, 168

## V

Vaitheeswaran, N. ........................771
Vasudevan, JM .............................555
Veena, P. ....................................117
Venugopal, S. ..............................231
Verma, Vishal ..............................589
Virulkar , V. B. ............................395
Vithal, JVR .................................667
Vittal, K.P. .................................787
Voropai , N. I. .............................909
Vyjayanthi, C. ..............................953

## W

Wadhwani , Sulochana ................1059
Wadhwani, A. K. ................210, 1059
Wang, Lingfeng ...........................806
Wani, M. G. ................................436

## Y

Yadav, K. B. ................................776
Yesuratnam, G. ............................953
Yokozeki , Ichiro .........................310

## Z

Zahawi, B. .................20, 389, 1020
Zhong-Xia, Niu ...................11, 522
Zué, Aslain Ovono ........................583

9780780397712

# 2006 IEEE International Conference on Power Electronics, Drives and Energy Systems for Industrial Growth

New Delhi, India
12-15 December 2006

IEEE Catalog Number:    CFP06PED-POD
ISBN:                 978-0-78039-771-2

# 2006 IEEE International Conference on Power Electronics, Drives and Energy Systems for Industrial Growth

New Delhi, India
12 – 15 December 2006

Volume 2 of 2

| IEEE Catalog Number: | 06TH8899 |
|---|---|
| ISBN: | 0-7803-9771-1 |

**Copyright © 2006 by The Institute of Electrical and Electronics Engineers, Inc.**
**All Rights Reserved**

*Copyright and Reprint Permissions:* Abstracting is permitted with credit to the source. Libraries are permitted to photocopy beyond the limit of U.S. copyright law for private use of patrons those articles in this volume that carry a code at the bottom of the first page, provided the per-copy fee indicated in the code is paid through Copyright Clearance Center, 222 Rosewood Drive, Danvers, MA 01923.

For other copying, reprint or republications permission, write to IEEE Copyrights Manager, IEEE Operations Center, 445 Hoes Lane, Piscataway, New Jersey USA 08854. All rights reserved.

| | |
|---|---|
| IEEE Catalog Number: | 06TH8899 |
| ISBN: | 0-7803-9771-1 |
| LOC: | 2006928206 |

**Additional Copies of This Publication Are Available from:**

IEEE Service Center
445 Hoes Lane
Piscataway, NJ 08854
IEEE Service Center
445 Hoes Lane
Piscataway, NJ 08854
Phone:          (800) 678-IEEE
                    (732) 981-1393
Fax:          (732) 981-9667
E-mail:          customer-service@ieee.org

# Table of Contents

**A program for harmonic modeling of distribution network transformers and determination of loss in the transformers and the amount of decrease of their life**..............1
*M. Marzband, A. Shaikholeslami*

**Novel Integral Cycle Voltage Controller for Self Excited Induction Generators**..............7
*S. S. Murthy, A. J. P. Pinto, A. R. Beig*

**EMI Modeling and Simulation of High Voltage Planar Transformer**..............11
*Bai Feng, Niu Zhong-Xia, Shi Yu-Jie, Zhou Dong-Fang*

**Graphical Estimation of Optimum Weights of Iron and Copper of a Transformer**..............15
*C. Easwarlal, V. Palanisamy, M.Y. Sanavullah, M.Gopila*

**Nonlinear Behavior of Self-excited Induction Generator Feeding an Inductive Load**..............20
*D.D.Ma, B. Zahawi, D. Giaouris, S. Banerjee, V. Pickert*

**Effects of Different Voltage Sags on Three-Phase Transformers**..............25
*M. R. Shakarami, A. Jalilian*

**Design and Transient Analysis of Cage Induction Motor Using Finite Element Methods**..............30
*Bhoj Raj Singla, Sanjay Marwaha, Anupma Marwaha*

**Methodology for Estimating Performance Characteristics of Three Phase Induction MotorOperating Direct-on-Line or with Six Pulse Inverter**..............35
*Slatish Chander Slabharwal*

**Design of Squirrel Cage Induction Motors for Traction Applications**..............39
*S. S. Murthy, Bhim Singh, G. Bhuvaneswari, Kiran Naidu, Uddanti Siva*

**Effect of Sequential Phase Energization on the Inrush Current of a Delta Connected Transformer**..............46
*K. P. Basu, Ali Asghar, Stella Morris*

**Accurate Performance Prediction of Three-Phase Induction Motor by FEM Using Separate Saturation Curves for Teeth and Yoke**..............50
*V. Jaiswal, M. Fazil, A. Hangal, N. Ravi*

**Nonlinear Sliding-Mode Controller for Sensorless Speed control of DC servo Motor Using Adaptive Backstepping Observer**..............54
*A. Farrokh Payam, B. Mirzaeian Dehkordi*

**Robust Speed Sensorless Control of Doubly-Fed Induction Machine Based on Input-Output Feedback Linearization Control Using a Sliding-Mode Observer**..............59
*A. Farrokh Payam, M. Jalalifar*

**Adaline Based Control of Solid State Voltage Regulator for Isolated Asynchronous Generators**..............64
*Bhim Singh, Gaurav Kumar Kasal*

**Development of a Prototype Controller for PMDC Motor Based Portable Telemetry Tracking System for Defense Application**..............70
*Parveen Kumar, A K Pradhan, Gautam Sadhukhan*

**Design & Development of a High Performance Electronic Starter for Single- Phase Induction Motors**..............75
*T. P. Shenoy, J. S. Nirody*

**Transient Analysis of a Single-Phase Self- Excited Induction Generator using a Three- Phase Machine feeding Dynamic Load**..............80
*S. N. Mahato, M. P. Sharma, S. P. Singh*

**Performance Analysis of a Three-Phase Squirrel-Cage Induction Motor under Unbalanced Sinusoidal and Balanced Non-Sinusoidal Supply Voltages**..............86
*C. Thanga Raj, Pramod Agarwal, S. P. Srivastava*

**Efficiency Optimization of Induction Motor Using a Fuzzy Logic Based Optimum Flux Search Controller**..............90
*L. Ramesh, S. P.Chowdhury, S.Chowdhury, A. K. Saha, Y. H. Song*

**Observer Based Position and Speed Estimation of Interior Permanent Magnet Motor**..............96
*Bhim Singh, Prerna Gaur, A.P.Mittal*

**Genetic Algorithm Based Optimal Design of Switching Circuit Parameters for a Switched Reluctance Motor Drive**..............101
*Behzad Mirzaeian-Dehkordi, Peyman Moallem*

**Reduction of Cogging Torque in PMBLD Motor with Reduced Stator Tooth Width and Bifurcated Surface Area Using Finite $lement Analysis**..............107
*Zx Somanathamv flxVxKxflrasadv and 3x/xZajkumarx*

# Table of Contents

**A Novel Phasor Diagram of Interior Permanent Magnet Synchronous Motors based on Spiral Vector Theory**................................ 111
*Bishnu P. Muni*

**A Novel DTC Strategy of Torque and Flux Control for Switched Reluctance Motor Drive**.................................... 117
*R. Jeyabharath, P.Veena, M.Rajaram*

**Remedial Strategies for the Minimization of Cogging torque in PMBDC Motor possessing Material Saturation** ........................... 122
*M. H. Ravichandran, V. T. Sadasivan Achari, C. C. Joseph, Robert Devasahayam*

**Fuzzy Pre-compensated PI Controller for PMBLDC Motor Drive**........................................................... 126
*Mukesh Kumar, Bhim Singh, B. P. Singh*

**A Slimplified Design M'ethodology for Slwitched Reluctance M'otor using analytical and Finite Llement M'ethod** ........... 131
*'0Y0fiavichandranq V0T0fladasivan 4chariq F0F0=osephq fiobert jevasahayam*

**Computer Aided Design of Permanent Magnet Brushless DC Motor for Hybrid Electric Vehicle Application** ................................. 135
*Bhim Singh, Devendra Goyal*

**Design and Analysis of a 3 kVA,28 Permanent Magnet Brushless Alternator for Light Combat Aircraft** ........................................ 141
*Fhim SinghT , Jally Rlavi*

**Estimation of Core Loss in a Switched Reluctance Motor Based on Actual Flux Variations** ........................................ 146
*Nimit. K. Sheth, K. R. Rajagopal*

**A Novel Hybrid Brushless dc motor/Generator for Hybrid Vehicles Applications** .......................................... 151
*E. Afjei, H. Toliyat, H. Moradi*

**Computer Aided Design and FE Analysis of a PM BLDC Hub Motor** ................................................ 157
*K. R. Rajagopal, Chippa Sathaiah*

**Effect of Armature Reaction and Skewing on the Performance of Radial-flux Permanent Magnet Brushless DC Motor** .............. 163
*Parag Upadhyay, K. R. Rajagopal*

**Torque Ripple Minimization of Interior Permanent Magnet Brushless DC Motor Using Rotor Pole Shaping** ......................... 168
*Parag Upadhyay, K. R. Rajagopal*

**Design and Development of a In-Wheel Brushless D.C. Motor Drive for an Electric Scooter** ................................. 171
*N. Ravi, S. Ekram, D. Mahajan*

**Comparative Study of Laminated Core Permanent Magnet Hybrid Stepping Motor with Soft Magnetic Composite Core Claw Pole Motor**............................................................................................ 175
*E.V. Chandra Sekhara Rao, P.V.N. Prasad, G. Ravindranath*

**A Doubly Fed Induction Motor as High Torque Low Speed Drive** ........................................................ 179
*Mukhtar Ahmad, M.Rizwan Khan, Atif Iqbal*

**Performance of Doubly Salient Permanent Magnet Motors for Parallel and Tapered Rotor Poles**........................................ 182
*Nimit K. Sheth, K. R. Rajagopal*

**Improved Torque Profile of a Doubly Salient Permanent Magnet Motor using Skewed Rotor Teeth and Sinusoidal Excitation**................................................................................................................ 187
*Nimit. K. Sheth, K. R. Rajagopal*

**Dynamic Modeling and Simulation of an Induction Motor with Adaptive Backstepping Design of an Input-Output Feedback Linearization Controller in Series Hybrid lectric Vehicle** ................................................ 193
*M. Jalalifar, A. Farrokh Payam, B. Mirzaeian, S. M. Saghaeian nezhad*

**Prototyping of a Precision Mechanism Using a Hybrid-Driven Piezoelectric Actuator** ........................................ 199
*Fu-Shin Lee, Yung-Tsung Lei, Sheng-Feng Chiang, Jyun-Jhong Jhang, Shao-Chun Tseng, Po-Jia Chen*

**DSP Based Implementation of Vector Controlled Induction Motor Drive using Fuzzy Pre-compensated Proportional Integral Speed Controllers**................................................................................................ 204
*Bhim Singh, S. Ghatak Choudhuri*

**Optimal Controller for High Frequency AC-Link Converter Induction Motor Drive System** ................................. 210
*R. A. Gupta, A. K. Wadhwani, R. R. Joshi*

**An Adaptive Backstepping Controller for Doubly-Fed Induction Machine Drives**................................................ 215
*A. Farrokh Payam*

**Application problem of PWM AC drives due to long cable length and high dv/dt** ................................... 219
*B.Basavaraja, D.V.S.S.Siva Sarma*

**Adaptive Controller Design for Permanent Magnet Linear Synchronous Motor Control System**................................ 225
*B. Srinivasu, P.V.N. Prasad, M.V. Ramana Rao*

*iv*

# Table of Contents

**An Overmodulation Scheme for Vector Controlled Induction Motor Drives** .................................................. 231
*S. Venugopal, G. Narayanan*

**Modified Direct Torque Control of Matrix Converter Fed Induction Motor Drive** .................................. 237
*Bhim Singh, Jally Ravi*

**LMI Based Digital State Feedback Controller for a Wound Rotor Induction Drive with Guaranteed Closed Loop Stability** ........................................................................................................................................................... 244
*D. Sivanandakumar, K. Ramakrishnan*

**Open-End Winding Induction Motor Driven With Matrix Converter For Common-Mode Elimination** ........... 250
*Krushna K Mohapatra, Ned Mohan*

**Elimination of Common Mode Voltage and Fifth and Seventh Harmonics in a Multilevel Inverter fed IM Drive using 12-Sided Polygonal Voltage Space Phasor** ............................................................................. 256
*Sanjay Lakshminarayanan, Gopal Mondal, P.N Tekwani, K. Gopakumar*

**A New Space Vector Pulsewidth Modulation for Reduction of Common Mode Voltage in Direct Torque Controlled Induction Motor Drive** ....................................................................................................................... 262
*Y.V. Siva Reddy, T. Brahmananda Reddy, M. Vijaya Kumar*

**Parallel Power Flow AC/DC Converter with High Input Power Factor and Tight Output Voltage Regulation for Universal Voltage Application** ........................................................................................................................... 267
*Aman Kumar Jha, K. Hari Babu, B. M. Karan*

**A Generalized Space Vector Modulation with Simple Control technique for Balancing DC-Bus Capacitor Voltages of a Three-Phase, Neutral-Point Clamped Converter** ............................................................. 274
*A. H. Bhat, P. Agarwal*

**A Novel Load Compensator for a 12-pulse Diode Converter** ..................................................................... 280
*Maryclaire Peterson, Brij N. Singh*

**Resonant Operated Buck Converter with Reduced Device Switching Stress with Power Factor Improvement** ................................................................................................................................................................ 286
*Vinayak N. Shet*

**A High Power Factor Forward Flyback Converter with Input Current Waveshaping** ............................... 292
*Vinayak N. Shet*

**A Fuzzy Logic Controller for Direct Power Control of PWM Rectifiers with SVM** ................................... 298
*R. Skandari, A. Rahmati, A. Abrishamifar, E. Abiri*

**DSP-Based Matrix Converter Operation Under Various Abnormal Conditions with Practicality** ............. 303
*Vinod Kumar, R R Joshi*

**Improvement of an input waveform of a Neutral Point Type Step-down Converter** ................................. 307
*Yoshito KATO, Masaaki NAKAMURA, Nabil M. Hidayat, Nobuo TAKAHASHI*

**Development of Neutral-Point Type Converter and Application to Electronic Ballast** .............................. 310
*Nabil M Hidayat, Masaaki Nakamura, Yoshito Kato, Nobuo Takahashi, Shun-ichi Adachi, Ichiro Yokozeki*

**Hysteresis-Band Current Control of a Four Quadrant AC -DC Converter giving IEEE 519 compliant performance at any Power Factor** ........................................................................................................................ 315
*A.N.Arvindan, V.K.Sharma*

**Multiphase Inverter Topology and its Modulation Technique for Optimal Harmonic Output** .................. 321
*Ravindra Kumar Singh*

**A PWM Current Source Rectifier with Leading Power Factor** ................................................................. 331
*B. Geethalakshmi, P. Sanjeevikumar, P. Dananjayan*

**A Novel Harmonic Mitigation Converter for Variable Frequency Drives** ................................................ 336
*Bhim Singh, Sanjay Gairola*

**Performance Comparison of High Frequency Isolated AC-DC Converters for Power Quality Improvement at Input AC Mains** ........................................................................................................................................ 342
*Bhim Singh, B.P. Singh, Sanjeet Dwivedi*

**Single-Phase Resonant Converter with Active Power Filter** .................................................................... 348
*M. A. Chaudhari, H. M. Suryawanshi*

**PV Power Tracking Through Utility Connected Single-Stage Inverter** .................................................... 354
*K. S. Phani Kiranmai, Veerachary. M*

**A Novel Control of Bi-Directional Switches in Matrix Converter** ........................................................... 360
*Meharegzi Tewolde, Shyama P. Das*

# Table of Contents

**PWM SHE Switching Algorithm for Voltage Source Vnverter** .................................................... 366
*Ali. I. Maswood*

**New Fuzzy logic Controller for a Buck Converter** .................................................... 370
*D. Seshachalam, R. K. Tripathi, D. Chandra, Anil kumar*

**Development of Conventional Control of Parallel Loaded Resonant Converter -Simulation and Experimental Evaluation** .................................................... 373
*T.S.Sivakumaran, S.P.Natarajan*

**A Novel Technique to Reduce the Switching Losses in a Synchronous Buck Converter** .................................................... 378
*A. K. Panda, Aroul. K*

**Transformer Core Unbalancing Issue in a Full-Bridge DC-DC Converter with Current Doubler Rectifier** .................................................... 383
*B.A. Gusev, V.I. Meleshin, D.A. Ovchinnikov*

**Computer Aided Analysis of Fault Tolerant Multilevel DC/DC Converters** .................................................... 389
*K. A. Ambusaidi, V. Pickert, B. Zahawi*

**Auto Voltage Balancing in High Power DC-DC Converter** .................................................... 395
*S. B. Bodkhe, V. B. Virulkar , S. W. Mohod , M.V. Aware*

**Inrush Current Control of a DC/DC Converter Using MOSFET** .................................................... 401
*Gaddam Mallesham, Keerthi Anand*

**A ZVT Boost Converter using an Auxiliary Resonant Circuit** .................................................... 407
*M. Phattanasak*

**Adaptive Hysteretic Control of 3rd Order Buck Converter** .................................................... 413
*Veerachary M, Deepen Sharma*

**A Novel Topology for Multiple Output DC-DC Converters for One Cycle Control** .................................................... 417
*Ravindra Kumar Singh*

**New Hybrid SVPWM Methods for Direct Torque Controlled Induction Motor Drive for Reduced Current Ripple** .................................................... 424
*T. Brahmananda Reddy, J. Amarnath , D. Subbarayudu*

**Analysis of Experimental Investigation of Various Carrier-based Modulation Schemes for Three Level Neutral Point Clamped Inverter-fed Induction Motor Drive** .................................................... 430
*Ranjan K. Behera, T. V. Dixit, Shyama P. Das*

**High Frequency SMPS Based Inverter With Improved Power Factor** .................................................... 436
*M. G. Wani, V. K. Sharma, K. M. Soni*

**Comparison of Mode Switched Controllers for a Pseudo Continuous Current Mode Boost Converter** .................................................... 443
*Sreekumar C, Vivek Agarwal*

**Multi-level inverter for Induction Motor Drive** .................................................... 449
*K.Chandra Sekhar, G.Tulasi Ram Das*

**A Unified Model For Auxiliary Switch Commutated DC-DC Converters** .................................................... 455
*N. Lakshminarasamma, V. Ramanarayanan*

**Novel Pulse Power Supply Operating at High Input Power Factor** .................................................... 460
*Vishnu K Sharma, Kishore Chatterjee, Vivek Agarwal*

**System Identification and controller tuning rule for DC-DC converter using ripple voltage waveform** .................................................... 463
*K. Lavanya, B. Umamaheswari, R. C. Panda*

**Space Vector Modulation with DC-Link Voltage Balancing Control for Three-Level Inverters** .................................................... 467
*Kalpesh H. Bhalodi, Pramod Agrawal*

**Investigations on Different Multilevel Inverter Control Techniques by Simulation** .................................................... 473
*P. K. Chaturvedi, Shailendra K Jain, Pramod Agrawal, P. K. Modi*

**Peak-Current Mode control of Hybrid Switched Capacitor Converter** .................................................... 479
*Veerachary M, Singamaneni Bala Sudhakar*

**Observer based current control of single-phase inverter in DQ rotating frame** .................................................... 485
*B.Saritha, and P.A.Jankiraman*

**MATLAB Simulation of current control of PMSM using single sensor technology** .................................................... 490
*B. Saritha, P. A. Jankiraman*

**Novel Approach to Develop Behavioral Model Of 12-Pulse Converter** .................................................... 495
*Amit Sanglikar, Vinod John*

# Table of Contents

**Simulation of PMSM VSI Drive for Determination of the Size Limits of the DC-Link Capacitor of Aircraft Control Surface Actuator Drives** ..................................................................................................... 500
*M.Khatre, Alan G. Jack*

**A Novel Soft Switched Improved Power Quality Converter Fed D.C. Motor Drive** .......................... 506
*M. B. Daigavane, Z. J. Khan, H. M. Suryawanshi*

**Generalized Discontinuous PWM Based Direct Torque Controlled Induction Motor Drive with a Sliding Mode Speed Controller** ...................................................................................................................................... 511
*T. Brahmananda Reddy, J. Amarnath, D. Subbarayudu, Md. Haseeb Khan*

**Hardware-in-Loop Simulation of Direct Torque Controlled Induction Motor** ................................ 517
*P. K. Gujarathi, M. V. Aware*

**Near-Field Modeling and Prediction of Switched Mode Power Supply** ........................................ 522
*Bai Feng, Niu Zhong-Xia, Shi Yu-Jie, Zhou Dong-Fang*

**Power Electronic Circuit-oriented Model for the Fuel Cell System** ............................................ 526
*Veerachary M, Arun Shailendra Kumar*

**A Simplified Space-Vector Modulated Control Scheme for CSI fed IM drive** ............................... 531
*P.Parthiban, Pramod Agarwal, S.P.Srivastava*

**A Study on Design and Dynamics of Voltage Source Inverter in Current Control Mode to Compensate Unbalanced and Non-linear Loads** ........................................................................................................ 537
*Mahesh K. Mishra, K. Karthikeyan*

**Optimal Voltage and Reactive Power Control Based on Multi-Objective Genetic Algorithm** .......... 545
*Behzad Mirzaeian Dehkordi*

**Model Validation Studies in Obtaining Q-V Characteristics of P-Q Loads in Respect of Reactive Power Management and Voltage Stability.** ........................................................................................... 550
*G. Govinda Rao, K. V. S. Ramachandra Murthy*

**Simulation Study of a Shunt 5ctive Power Filter Using Nonlinear Least Squares Harmonic Extraction Technique** ...................................................................................................................................... 555
*RM Bhudamani, JM Vasudevan, BMSM Ramalingam*

**Comparison of Synchronous Detection and I.Cosf Shunt Active Filtering Algorithms** ................... 560
*G. Bhuvaneswari, Manjula G. Nair, Sathish Kumar Reddy*

**A Nonlinear Control Method for SSSC to Improve Power System Stability** .................................. 565
*Majid Poshtan, Brij N. Singh, Parviz Rastgoufard*

**An Improved Power Flow Analysis Technique with STATCOM** .................................................. 572
*Annapurna Bhargava, Vinay Pant, Biswarup Das*

**Design of a Current Hybrid Filter Including Active and Variable Passive Filters** ........................... 577
*H. Dalvand*

**Grid Connected Photovoltaic Interface with VAR Compensation and Active Filtering Functions** .... 583
*Aslain Ovono Zué, Ambrish Chandra*

**Design and Implementation of a Current Controlled Parallel Hybrid Power Filter** ........................ 589
*Bhim Singh, Vishal Verma*

**Active Power Filter Control in Three-Phase four-wire Systems using Space Vector Modulation** ...... 596
*H. Mokhtari, M. Rahimi*

**Operation of a 12-pulse converter in closed loop for controlled P-Q operation** ............................. 602
*Faisal M. Ahsan, J.K. Chatterjee, Anandarup Das*

**A Novel Structure for Three-Phase Four-Wire Distribution System Utilizing Unified Power Quality Conditioner (UPQC)** ........................................................................................................................ 608
*V. Khadkikar, A. Chandra*

**Load Compensation for Diesel Generator Based Isolated Generation System Employing DSTATCOM** ... 614
*Bhim Singh, Jitendra Solanki*

**Automatic Classification of Power Quality Events Using Multiwavelets** ....................................... 620
*Surender Dahiya, D.K. Jain, Manish Kumar, Ashok Kumar, Rajiv Kapoor*

**Power quality monitoring at the industrial, commercial and educational centers of Mazandaran province and presenting the related solution** ................................................................................... 625
*M. Marzband, A. Shaikholeslami*

# Table of Contents

A New Power Quality Enhancement Method for Two-Phase Loads .......................................................................... 631
*H. Hojabri, H. Mokhtari*

Three level STATCOM Based Power Quality Solution for a 4 MW Induction Furnace ........................................ 636
*Unnikrishnan A.K, Aby Joseph, Subhash Joshi T G*

Analysis and Simulation of Single Phase Composite Observer for Harmonics Extraction ................................ 641
*K. Selvajyothi, P. A. Janakiraman*

Third Harmonic Current Injection for Power Quality Improvement in Rectifier Loads ................................... 647
*Bhim Singh, Vipin Garg, G.Bhuvaneswari*

Polygon Connected 15-Phase AC-DC Converter for Power Quality Improvement ........................................... 652
*Bhim Singh, Vipin Garg, G.Bhuvaneswari*

Power Quality Standards and Their Application to a Granite Factory ................................................................ 657
*S. Hasani, F. Donyavi, M. Masoudi, H. Mokhtari*

Minimization of Losses in Radial Distribution System by using HVDS ............................................................. 662
*K. Amaresh, S. Sivanagaraju, V. Sankar*

SVPWM Switched DSTATCOM for Power Factor and Voltage Sag Compensation ......................................... 667
*Bishnu P. Muni, S. Eswar Rao, JVR Vithal*

Unified Constant frequency Integration Control of Universal Power Quality Conditioner ............................ 673
*Vadirajacharya K, Pramod Agarwal, H.O.Gupta*

Application of a Boundary Model to Assess Power Quality Cost Function ........................................................ 678
*J. Ahmadian, A. Jalilian, M.A.S. Masoum*

Active Power Filter Solution without PLL for Fluctuating Industrial Load ..................................................... 683
*S. Elangovan*

A Novel Digital Signal Processing Algorithm for On-line Assessment of Power System Frequency ............... 689
*Arghya Sarkar, S. Sengupta*

An Evolutionary Algorithm Approach to Estimate the Parameters of Power Quality Signals ......................... 695
*V. Ravikumar Pandi, B. K. Panigrahi*

A 36-Pulse AC-DC Converter for Line Current Harmonic Reduction ............................................................. 701
*Bhim Singh, Sanjay Gairola*

A Unified Analysis of CCM Boost PFC for Various Current Control Strategies ............................................. 707
*Ranjan K. Gupta, Hariharan Krishnaswami, Ned Mohan*

Minimum Loss Configuration of Power Distribution System ........................................................................... 712
*L Jaswant, T. Thakur*

Control of Cascaded H-Bridge Converter based DSTATCOM for High Power Applications ......................... 718
*K. Anuradha, B.P.Muni, A. D. Raj Kumar*

Detection and Classification of Non-stationary Power disturbances in Noisy Conditions ............................... 724
*B. K. Panigrahi, S. K. Sinha*

3-Phase Fault Current Limiter for distribution systems ................................................................................... 729
*Vijay K. Sood, Shahabur Alam*

Power Flow Control of a Solid Oxide Fuel-Cell for Grid Connected Operation ............................................. 735
*Ankur Goel, S. Mishra, A.N. Jha*

An Universal Interconnection System to Connect Distributed Generation to the Grid .................................... 740
*Vinod John, Eric Benedict, Shazreen*

Transient Fault Response of Grid Connected Wind Electric Generators ......................................................... 747
*Vinodh Kumar P, Meera K S, Sasi K Kottayil*

Black Start with DFIG Based Distributed Generation after Major Emergencies ............................................ 753
*M. Aktarujjaman, M.A. Kashem, M. Negnevitsky, G. Ledwich*

Fuzzy Logic Based Control of Wind Turbine Driven Squirrel Cage Induction Generator Connected to Grid ............... 759
*CH.Siva Kumar, A.V.R.S.Sarma, P.V.N. Prasad*

Speed Sensor-less Direct Power Control of a Matrix Converter Fed Induction Generator for Variable Speed Wind
Turbines .......................................................................................................................................................... 765
*T. Satish, K.K. Mohapatra, Ned Mohan*

# Table of Contents

**Stochastic Model for Optimal Selection of DDG by Monte Carlo Simulation**.......................................771
*N. Vaitheeswaran, R. Balasubramanian*

**Capacitive Self-Excitation in a Six-Phase Induction Generator for Small Hydro Power  An Experimental Investigation**...............776
*G. K. Singh, K. B. Yadav, R. P. Saini*

**Grid Power Quality with Variable Speed Wind Energy Conversion**........................................782
*S.W. Mohod, M. V. Aware*

**Investigations on Combined Operation of Industrial Distribution System and utility in Distributed Generation Environment**.......................................787
*K. Manjunatha Sharma, K.P. Vittal, T.K. Nagaraja Rao*

**Rotor Speed Stability Analysis of Constant Speed Wind Turbine Generators**.........................792
*M. G. Kanabar, C. V. Dobariya, S. A. Khaparde*

**Performance Evaluation of Indian Electric Power Utilities Based on Data Envelopment Analysis**.........................797
*Tripta Thakur*

**Modelling of Hybrid Energy System for Off Grid Electrification of Clusters of Villages**.........................801
*Ajai Gupta, R P Saini, M P Sharma*

**PSO-Based Multidisciplinary Design of A Hybrid Power Generation System With Statistical Models of Wind Speed and Solar Insolation**.........................806
*Lingfeng Wang, Chanan Singh*

**SVPWM Implementation in dSPACE for Generalized Impedance Controller Used for Self Excited Induction Generation System**.........................812
*B.Venkatesa Perumal, J.K. Chatterjee*

**Trajectory Sensitivity Analysis in Distributed Generation Systems**.........................818
*Dheeman Chatterjee, Arindam Ghosh, M. A. Pai*

**Steady State Performance Of A Stand-Alone Variable Speed Constant Frequency Generation System Using A New Build Up Algorithm**.........................824
*Isha T B, D. Kastha*

**Control Strategy of Distributed Generation for Voltage Support in Distribution Systems**.........................830
*M. Negnevitsky, G. Ledwich, An D.T. Le, M. A. Kashem, Seni*

**A Steady State Analysis on Voltage and Frequency Control of Self-Excited Induction Generator in Micro-Hydro System**.........................836
*Bhim Singh, S.S. Murthy, Madhusudan, Manish Goel, A. K. Tandon*

**A Novel Digital Control Technique of Electronic Load Controller for SEIG Based Micro Hydel Power Generation**.........................842
*S. S. Murthy, Ramrathnam, M. S. L.Gayathri, Kiran Naidu, U. Siva*

**Analysis and Design of Voltage and Frequency Controllers for Isolated Asynchronous Generators in Constant Power Applications**.........................847
*Bhim Singh, Gaurav Kumar Kasal*

**A Simple Controller using Line Commutated Inverter with Maximum Power Tracking for Wind-Driven Grid-Connected Permanent Magnet Synchronous Generators**.........................854
*V. Lavanya, N. Ammasai Gounden, Polimera Malleswara Rao*

**A High-power High-frequency and Scalable Multi-megawatt Fuel-cell Inverter for Power Quality and Distributed Generation**.........................860
*Sudip K. Mazumder, Rongjun Huang*

**Integrating a Redox Flow Battery System with a Wind-Diesel Power System**.........................865
*Shameem Ahmad Lone, Mairaj-ud-Din Mufti*

**Hydrocarbon Fuel Based Micro Battery Power System**.........................871
*Surendran Devadoss, Theo Kangsanant, Ian Bates*

**Analysis, Design and Development of Single Switch Forward Buck AC-DC Converter for Low Power Battery Charging Application**.........................876
*Bhim Singh, Ganesh Dutt Chaturvedi*

**A Novel Approach for Eco-Friendly and Economic Power Dispatch using MATLAB**.........................882
*D.P.Kothari, K.P.Singh Parmar*

**Real Time Based PI-like Fuzzy Controller for DC Servomotor**.........................888
*S.G. Kadwane, Swapnil Gupta, B.M. Karan, T Ghose, Amit Kumar*

# Table of Contents

**Neural Network Based DSTATCOM Controller for Three-phase, Three-wire System** ...... 892
*Bhim Singh, A. Adya, A. P. Mittal, J.R.P Gupta*

**Analysis of the Influence of Control Parameters on Wind Farm Output: a Sensitivity Analysis using ANN Modelling** ...... 898
*E. Fernandez, M. Carolin Mabel*

**An Advanced Control Scheme for Micro Hydro Power Plants** ...... 902
*M. Hanmandlu, Himani Goyal, D. P. Kothari*

**Application of Fuzzy Logic PSS to Enhance Transient Stability in Large Power Systems** ...... 909
*P. V. Etingov, N. I. Voropai*

**Neural Approach for Automatic Identification of Induction Motor Load Torque in Real-Time Industrial Applications** ...... 918
*A. Goedtel, I. N. da Silva, P. J. A. Serni*

**Speed Estimation for Sensorless Technology Using Recurrent Neural Networks and Single Current Sensor** ...... 926
*A. Goedtel, I. N. da Silva, P. J. A. Serni*

**Electricity Price Forecasting Using Artificial Neural Network** ...... 931
*M. Ranjbar, S. Soleymani, N. Sadati, A. M. Ranjbar*

**A New Approach for Fault Location Identification in Transmission system using Stability Analysis and SVMs** ...... 936
*D. Thukaram, H. P. Khincha, B. Ravikumar*

**Fast and Effective Algorithm for Economic Dispatch with Prohibited operating zones** ...... 942
*T. Adhinarayanan, M. Sydulu*

**Computation & Analysis of End Region EM Force for Electrical Rotating Machines using FEM** ...... 948
*Manpreet Singh Manna, Sanjay Marwaha, Anupma Marwaha*

**Optimal Reactive Power Dispatch based on Voltage Stability Criteria in a Large Power System with AC/DC and FACTs Devices** ...... 953
*D.Thukaram, G. Yesuratnam, C. Vyjayanthi*

**Location of Unified Power Flow Controller and its Parameters settin for congestion Management in Pool M arket Model** ...... 959
*Hassan Barati, Mehdi Ehsan, Mahmud Fotuhi-Firuzabad*

**Security Enhancement of Optimal Power Flow using Genetic Algorithm** ...... 966
*N.B. Muthuselvan, P. Somasundaram, and Subhransu Sekhar Dash*

**Congestion Management in Nodal Pricing With Genetic Algorithm** ...... 970
*S.M.H Nabav, Shahram Jadid, M.A.S. Masoum, A. Kazemi*

**Coupled Magneto-Mechanical Field Computations** ...... 975
*Amogh Kank, G. B. Kumbhar, S. V. Kulkarni*

**Optimizing Voltage Stability Limit and Real Power Loss in a Large Power System using Bacteria Foraging** ...... 979
*M. Tripathy, S. Mishra*

**Application of Power Flow Sensitivity Analysis and PTDF for Determination of ATC** ...... 985
*N. D. Ghawghawe, K. L. Thakre*

**Application of Tabu-Search Algorithm for Network Reconfiguration in Radial Distribution System** ...... 992
*T. Thakur, Jaswanti*

**Comparative Studies of Transient and Steady State Analysis for a Typical 765kV/400kV EHV Transmission System in Indian Power System** ...... 996
*D. Thukaram, H. P. Khincha, P. Shyamala*

**A Finite Element Modeling and Simulation Method for Time-Varying Field-Circuit Problems** ...... 1002
*Prem Sagar*

**A Wavelet Based Numerical Technique for Electromagnetic Field Analysis** ...... 1007
*Kaushik K, S. V. Kulkarni*

**Frequency Linked Pricing as an Instrument for Frequency Regulation Market and ABT Mechanism** ...... 1013
*K. V. V. Reddy, Ashwani Kumar, Saurabh Chanana*

**Induction Machine Fault Identification using Particle Swarm Algorithms** ...... 1020
*S. A. Ethny, P. P Acarnley, B. Zahawi, D. Giaouris*

**A Novel Technique for Identification and Condition Monitoring of Nonlinear Loads in Power Systems** ...... 1024
*Phil Gilreath, Maryclaire Peterson, Brij N. Singh*

# Table of Contents

**Real-Time Identification of Distributed Bearing Faults in Induction Motor** .................... 1031
*Rajesh Patel , S P Gupta, Vinod Kumar*

**Integration of IEDs Using Legacy and IEC61850 Protocol** ............................................ 1036
*Anupama Prakash, Mini S. Thomas, Ashutosh Gautam*

**Ethernet Enabled Fast and Reliable Monitoring, Protection and Control of Electric Power Substation** ......................... 1041
*Iqbal Ali, Mini S. Thomas*

**Expert System for Power Transformer Condition Monitoring and Diagnosis** ..................... 1047
*M. Ahfaz Khan, A.K. Sharma, Rakesh Saxena*

**Evaluation of Leakage Current Measurement for Site Pollution Severity Assessment** .......... 1053
*S.M.H Nabavi, A. Gholami, A. Kazemi, M.A.S. Masoum*

**Detection of Bearing Failure in Rotating Machine Using Adaptive Neuro-Fuzzy Inference System** .................... 1059
*Sulochana Wadhwani , A.K. Wadhwani, S P Gupta, Vinod Kumar*

**Discrimination between Inrush current and Internal Faults using Pattern Recognition Approach** .................... 1064
*B. K. Panigrahi, S. R. Samantaray, P. K. Dash, G. Panda*

**Stepwise Restoration of Power Distribution Network under Cold Load Pickup** .................. 1069
*Vishal Kumar, Rohith Kumar H.C., I. Gupta, H.O. Gupta*

**Power Sector Reforms in India** ...................................................................................... 1074
*Harbans L. Bajaj, Deepak Sharma*

**A New Structure for Electricity Market Scheduling** ...................................................... 1079
*S. Soleymani, A. M. Ranjbar, A. R. Shirani*

**Modelling of STATCOM Based Voltage Regulator for Self-Excited Induction Generator with Dynamic Loads** ........................ 1084
*Bhim Singh, S. S. Murthy, Sushma Gupta*

**Optimum Design of UPFC Controllers Using GEA: Decoupled Real & Reactive Power Flow and Damping Controllers** ............ 1090
*N. Ray Chaudhuri, M. L. KotharI*

**Application of Static Synchronous Series Compensator to Dam Sub-Synchronous Resonance** ...................... 1096
*M. Ehsan, M. Fotuhi-Firuzabad, S. M. T. Bathaee*

**A New 24-Pulse STATCOM for Voltage Regulation** ...................................................... 1102
*Bhim Singh, R. Saha*

**A Nonlinear Fuzzy PID Controller for CSI-STATCOM** ................................................. 1107
*A. Kazemi, A. Tofighi, B.Mahdian*

**Distance Relay Tripping Characteristic in Presence of UPFC** ......................................... 1114
*S. Jamali, A. Kazemi, H. Shateri*

**Investigations on Boundaries of Controllable Power Flow with Unified Power Flow Controller** .................... 1120
*S. Srividhya, C. Nagamani, A. Karthikeyan*

**VSC Based HVDC System for Passive Network with Fuzzy Controller** ........................... 1127
*A. K. Moharana , Ms. K. Panigrahi, B. K. Panigrahi, P. K. Dash*

**Voltage Regulation and Power Flow Control of VSC Based HVDC System** ....................... 1131
*Bhim Singh, B. K. Panigrahi, D. Madhan Mohan*

**Modeling and Simulation of Electromagnetic Conducted Emission Due to Power Electronics Converters** ...................... 1137
*A. Farhadi, A. Jalilian*

**Evaluation of Operational Characteristics Of Electronic Ballasts For Metal-Halide HID Lamps** ...................... 1143
*Ahteshamul Haque, M. S. Jamil Asghar*

**Active Power Filter Control Algorithm using Wavelets** .................................................. 1150
*Karunesh K Gupta, Rajneesh Kumar, H. V. Manjunath*

**Effects of Power Lines on Performance of Home Control System** .................................... 1154
*V. Chunduru, N. Subramanian*

*xii*

**2006 IEEE International Conference on Power Electronic, Drives and Energy Systems**

# Design of a Current Hybrid Filter Including Active and Variable Passive Filters

H. Dalvand, *Student Member, IEEE*

*Abstract*--This paper presents a new scheme for parallel hybrid power filter with variable passive part. The proposed model uses a current variable inductor in passive part designed by the software of finite elements method (FEM). The aim of this scheme is reducing power rating of active part of hybrid power filter. To evaluate its performance, this method is simulated by using MATLAB Simulink Power System Toolbox. Simulation results show the performance of the proposed method with quite satisfaction to eliminate harmonics and reactive power components from utility current. Then to evaluate the cost of proposed model, an economical comparison between this model and conventional one is done. This comparison shows that the cost of proposed model is less than the conventional one. Therefore this method can eliminate harmonics of load current with less cost when load current varies.

*Index Terms*— 1-Power electronics, 2- Power filters, 3- Power quality, 4- Power system harmonics, 5- Power system simulation.

## I. INTRODUCTION

POWER filters play an important role in reducing harmonic contamination in power lines. In the past twenty years, the proliferation of nonlinear loads such as static power converters, arc furnaces and others, have resulted in a variety of undesirable phenomena in the operation of power systems, which in many cases cannot be solved with passive LC filters. The basic difference between LC filters and active filters is that the active filters have the capability to compensate randomly varying currents. However, the active filters have problem derived from their practical implementation. Among them, is to build a large-capacity PWM-inverter with fast current response and low losses. Another one is that they need to be designed for high power, because they inject the required harmonic current under nominal voltage.

As an alternative to mitigate the problems of passive and pure active filters in parallel, hybrid filters have been

---

H. Dalvand is currently pursuing the M.S. degree of electrical engineering in Amirkabir University of Technology, Tehran, Iran. (h_dalvand@ieee.org).

proposed. They are composed of passive LC filters connected in series to an active power filter [1-5]. Hybrid topology significantly improves the compensation characteristics of simple passive filters; make the active power filter available for high-power applications, at a relatively lower cost.

There are various modulation methods proposed for such hybrid power filter configurations, but in terms of quick current controllability and easy implementation hysteresis band current control method has the highest rate among other current control methods such as sinusoidal PWM. In principle increasing inverter operation frequency helps to get a better compensating waveform. However there are device limitations and increasing the switching frequency causes increasing switching losses, audible noise and EMF related problems. The range of frequencies used is based on a compromise between these two different factors [6-11]. In this paper, the control of switching frequency is realized by introducing an adaptive hysteresis band current control algorithm.

Also in this paper, a method is proposed to reduce the cost of such filters. By this method, it is possible to decrease the ratings of active filter by utilizing variable passive filters. To consider the passive filters as variable ones, there is a gap which is changed by load varying. These filters are designed by the finite element method software (OPERA). This software gives the appropriate air gap to a certain inductance which is required for the filter to be tuned for a definite load current.

## II. PASSIVE PART

This part includes two single tuned filters per phase. These filters are tuned for fifth and seventh harmonics. The inductance of each filter has air gap that varies with current. When the load current varies, these gaps change and therefore inductance varies. Because conventional passive filters are tuned only for a certain load current, if the load current varies, these filters can not completely eliminate fundamental current harmonics and therefore active filter should compensate the extra fundamental current harmonics. But by this variable structure, passive filters can eliminate the fundamental harmonics by variable load. As a result it is not necessary to compensate the extra fundamental current harmonics by active filter when load current is varying. For this purpose, three simple topologies are considered as shown in fig. 1.

The structure in fig. 1a has a very simple layout but it can be used only for balanced three phase loads and it is not possible to be applied for unbalanced loads. The topology shown in fig.1b can be utilized in unbalanced condition but it has a very complicated control method. Therefore, the structure in fig. 1c is chosen because of its simple control and use in single phase for compensating unbalanced current. The moveable part is varied by an electrical motor and fixed with mechanical bolts in certain positions [12].

For two considered currents, two gaps are reached with the FEM software (OPERA). These gaps are shown in table 1 and 2 respectively for 5th and 7th harmonics.

The magnetic flux density for the gaps and currents of 5th and 7th harmonics obtained by this software are illustrated in fig. 2 and 3 respectively.

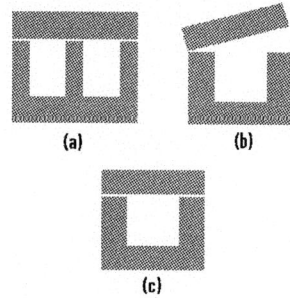

Fig. 1. The proposed topologies for passive filters

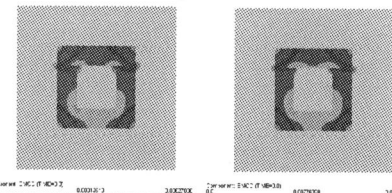

Fig. 2. The magnetic flux density obtained by OPERA relevant to 5th harmonic for 4.13 and 3.03mm air gaps respectively.

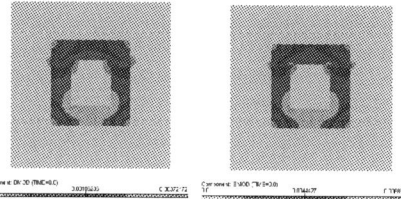

Fig. 3. The magnetic flux density obtained by OPERA relevant to 7th harmonic for 8.63 and 6.33mm air gaps respectively.

TABLE I
THE AIR GAPS FOR 5TH HARMONIC

| Inductance (mH) | Load current | Air gap (mm) |
|---|---|---|
| 1.9 | 22.6 | 4.13 |
| 2.59 | 43.8 | 3.03 |

TABLE II
THE AIR GAPS FOR 7TH HARMONIC

| Inductance (mH) | Load current | Air gap (mm) |
|---|---|---|
| 0.91 | 22.6 | 8.63 |
| 1.24 | 43.8 | 6.33 |

## III. ACTIVE PART

### A. Synchronous d-q-0 reference frame based compensation

Three phase load currents have already been transformed to the synchronous reference frame (a-b-c to d-q-0 transformation). A high pass filter is used to extract the dc component representing the fundamental frequency of the currents. The coordinate transformation from three-phase load currents ($i_{La}$, $i_{Lb}$, $i_{Lc}$) to the synchronous reference frame based load currents ($i_{Ld}$, $i_{Lq}$, $i_{L0}$) is obtained as follows:

$$\begin{bmatrix} i_{Ld} \\ i_{Lq} \\ i_{L0} \end{bmatrix} = \begin{bmatrix} \cos(\omega t) & \cos(\omega t - (2\pi/3)) & \cos(\omega t + (2\pi/3)) \\ -\sin(\omega t) & -\sin(\omega t - (2\pi/3)) & -\sin(\omega t + (2\pi/3)) \\ 1/\sqrt{2} & 1/\sqrt{2} & 1/\sqrt{2} \end{bmatrix} \times \begin{bmatrix} i_{La} \\ i_{Lb} \\ i_{Lc} \end{bmatrix}$$

(1)

The high pass filter to remove the dc component of load current should only be applied to the $i_{Ld}$ current. Q axis current ($i_{Lq}$) is applied to inverse transformation to compensate reactive power. Zero axis current ($i_{L0}$) must be used when the voltages are distorted or unbalanced and sinusoidal current are desired. In this study, it is not investigated.

The dc side voltage of active part of filter should be controlled and kept at a constant value to maintain the normal operation of the inverter. Because there is energy loss due to conduction and switching power losses associated with the diodes and IGBTs of the inverter in active part, which tend to reduce the value of $V_{dc}$ across capacitor $C_{dc}$. A feedback voltage control circuit needs to be incorporated into the inverter for this reason. The difference between the reference value, $V_{ref}$ and the feedback value ($V$dc), an error function first passes a PI regulator and the output of the PI regulator is subtracted from the d axis value of the harmonic current components. Synchronous d-q-0 reference frame based compensation algorithm, described above, is depicted in Fig. 3,.

Reference filter currents ($i_{abc}^{*}$) are determined negatives of the outputs of the inverse transformation matrix (d-q-0 to a-b-c) [9-11].

### B. The adaptive hysteresis band current controller

The hysteresis band current control technique has proven to be most suitable for all the applications of current controlled voltage source inverters in hybrid power filters. The hysteresis band current control is characterized by unconditioned stability, very fast response, and good accuracy [13]. On the other hand, the basic hysteresis technique exhibits also several undesirable features; such as uneven switching frequency that causes acoustic noise and difficulty in designing input filters [14].

The conventional hysteresis band current control scheme used for the control of hybrid power filter line current is composed of a hysteresis around the reference line current. The reference line current of the hybrid power filter is referred to as $I_c^*$ and actual line current of the hybrid power filter is referred to as $I_c$.

The hysteresis band current controller decides the switching pattern of hybrid power filter [1]. The switching logic is formulated as follows:

If $i_{ca} < (i_{ca}^* - HB)$ upper switch is OFF and lower switch is ON for leg "a" (SA=1).

If $i_{ca} > (i_{ca}^* + HB)$ upper switch is ON and lower switch is OFF for leg "a" (SA = 0).

The switching functions SB and SC for phases B and C are determined similarly, using corresponding reference and measured currents and hysteresis bandwidth (HB).

The switching frequency of the hysteresis band current control method described above depends on how fast the current changes from the upper limit of the hysteresis band to the lower limit of the hysteresis band, or vice versa. The rate of change of the actual active power filter line currents vary the switching frequency, therefore the switching frequency does not remain constant throughout the switching operation, but varies along with the current waveform.

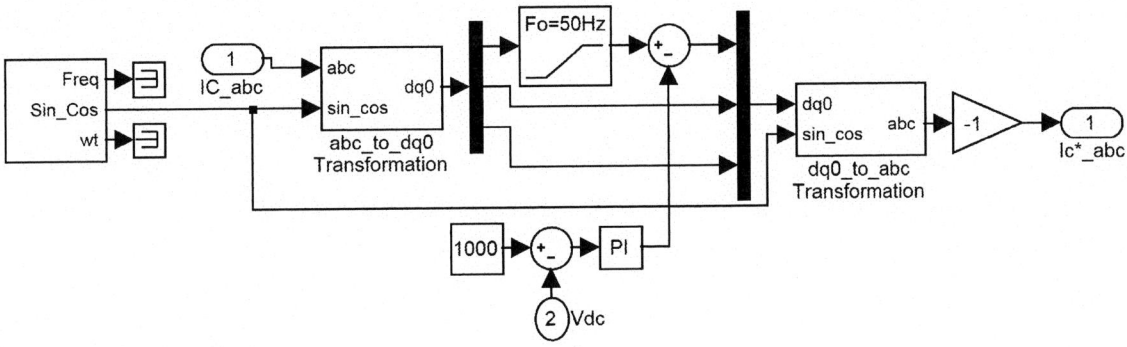

Fig. 4, Synchronous d-q-0 reference frame based compensation algorithm.

Furthermore, the line inductance value of the hybrid power filter and the dc link capacitor voltage are the main parameters determining the rate of change of hybrid power filter line currents.

The switching frequency of the hybrid power filter system also depends on the capacitor voltage and the line inductances of the hybrid power filter configuration.

The bandwidth of the hysteresis current controller determines the allowable current shaping error. By changing the bandwidth the user can control the average switching frequency of the hybrid power filter and evaluate the performance for different values of hysteresis bandwidth. In principle, increasing the inverter operating frequency helps to get a better compensating current waveform. However, there are device limitations and increasing the switching frequency causes increased switching losses, and EMI related problems. The range of switching frequencies used is based on a compromise between these factors.

The hysteresis-band current control method is popularly used because of its simplicity of implementation, among the various PWM techniques. Besides fast-response current loop and inherent-peak current limiting capability, the technique does not need any information about system parameters. However, the current control with a fixed hysteresis band has the disadvantage that the switching frequency varies within a band because peak-to-peak current ripple is required to be controlled at all points of the fundamental frequency wave [14]. But interesting improved versions of this technique are presented in literature [14].

Fig. 4, shows the PWM current and voltage waves for phase a. The currents $i_a$ tends to cross the lower hysteresis band at point 1, where upper side IGBT of leg "a" is switched on.

The linearly rising current ($i_{ca}^+$) then touches the upper band at point 2, where the lower side IGBT of leg

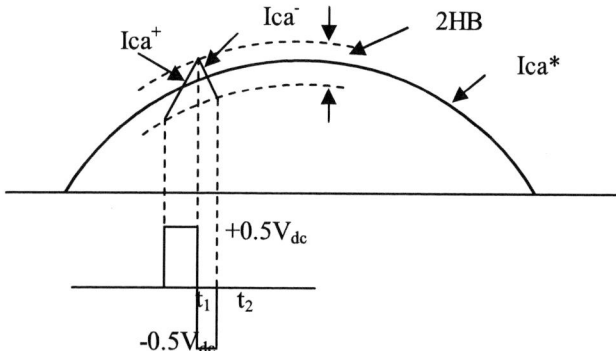

Fig. 5, Current and voltage waves with hysteresis band current control

"a" is switched on. The following equations can be written in the respective switching intervals $t_1$ and $t_2$ from Fig. 4.

$$\frac{di_{ca}^+}{dt} = \frac{1}{L}(0.5V_{dc} - V_s) \qquad (2)$$

$$\frac{di_{ca}^-}{dt} = -\frac{1}{L}(0.5V_{dc} + V_s) \qquad (3)$$

From the geometry of Fig. 4, can be written,

$$\frac{di_{ca}^+}{dt}t_1 - \frac{di_{ca}^*}{dt}t_1 = 2HB \qquad (4)$$

$$\frac{di_{ca}^-}{dt}t_2 - \frac{di_{ca}^*}{dt}t_2 = -2HB \qquad (5)$$

$$t_1 + t_2 = T_c = \frac{1}{f_c} \qquad (6)$$

Where $t_1$ and $t_2$ are the respective switching intervals, and $f_c$ is the switching frequency.

Adding (4) and (5) and substituting (6), it can be written

$$t_1\frac{di_a^+}{dt} + t_2\frac{di_a^-}{dt} - \frac{1}{f_c}\frac{di_{ca}^*}{dt} = 0 \qquad (7)$$

Subtracting (5) from (4), we get

$$4HB = t_1\frac{di_{ca}^+}{dt} - t_2\frac{di_{ca}^-}{dt} - (t_1 - t_2)\frac{di_{ca}^*}{dt} \qquad (8)$$

Substituting (3) in (8), gives

$$4HB = (t_1 + t_2)\frac{di_{ca}^+}{dt} - (t_1 - t_2)\frac{di_{ca}^*}{dt} \qquad (9)$$

Substituting (3) in (7), simplifying

$$(t_1 - t_2) = \frac{di_{ca}^*/dt}{f_c(di_{ca}^+/dt)} \qquad (10)$$

Substituting (10) in (9), gives

$$HB = \left\{ \frac{0.125V_{dc}}{f_c L}\left[1 - \frac{4L^2}{V_{dc}^2}\left(\frac{v_s}{L} + m\right)^2\right]\right\} \qquad (11)$$

Where $f_c$ is modulation frequency, $m = di_{ca}^*/dt$ is the slope of command current wave. Hysteresis band (HB) can be modulated at different points of fundamental frequency cycle

to control the switching pattern of the inverter. For symmetrical operation of all three phases, it is expected that the hysteresis bandwidth (HB) profiles $HB_a$, $HB_b$ and $HB_c$ will be same, but have phase difference.

The adaptive hysteresis band current controller changes the hysteresis bandwidth according to instantaneous compensation current variation ($di_c/dt$) and $V_{dc}$ voltage to minimize the influence of current distortion on modulated waveform. In this paper, the adaptive hysteresis band current controller, proposed by Bose [14] for electrical machine drives given by Eq. (11), is adapted to hybrid power filter.

Eq. (11) shows the hysteresis bandwidth (HB) as a function of modulation frequency, supply voltage, dc capacitor voltage and slope of the $i_c^*$ reference compensator current wave. Hysteresis band can be modulated as a function of $V_{dc}$ and m so that the modulation frequency $f_c$ remains nearly constant. This will improve the PWM performances and hybrid power filter substantially. Block diagram of variable hysteresis band current controller created by s-functions in Matlab is shown in Fig. 5,. The produced pulses are sent to IGBT inverter.

## IV. RESULTS AND DISCUSSION

Block diagram of hybrid power filter simulated by the MATLAB Simulink Power System Toolbox is shown in Fig. 6. Also the design specifications and the circuit parameters used in the simulation are indicated in Table III. For consideration a variable load current, the firing angle of a thyristor rectifier is changed from $60^0$ to $30^0$. This change is done to investigate the ability of variable structure of passive filters.

For comparison of conventional and proposed variable passive filters, first a conventional one is implemented for compensation the variable load current. This filter includes two fixed LC filters which are tuned for 5th and 7th harmonic. The supply current waveforms before and after compensation by this filter are shown in figures 6 and 7 respectively. From Fig. 7, it is clear that the filter can not completely compensate these harmonics by changing the load current. But by applying the variable passive filter, the predominant harmonics can be removed from the supply current. Fig. 8 shows this current after compensating by the proposed passive filter.

Because the passive filters can not compensating all of the undesirable harmonics, the hybrid power filters are applied for this purpose.

First the conventional filter is considered. This structure includes fixed tuned passive filters and an active filter. The passive part of this filter can not eliminate the whole 5th and 7th harmonics currents while the load current changed as mentioned above. Therefore, the active part should be imposed to tolerate the extra 5th and 7th harmonics currents which are not observed by the passive ones. As a result the ratings of active part increase dependent on these extra currents.

In the next step, the hybrid filter includes variable passive filters which vary by load current changes. Consequently, the

580

active part does not carry the extra currents and any added rating does not need to its operation correctly.

The load current waveform is shown in Fig. 8. The utility power source current after the harmonic compensation by conventional and the proposed hybrid filters are illustrated in Fig. 9 and 10 respectively.

The results of simulation study of different power filters presented in this paper are shown Table IV. From this table, it is clear that the proposed structure is found quite satisfactory to eliminate harmonics and reactive power components from utility current.

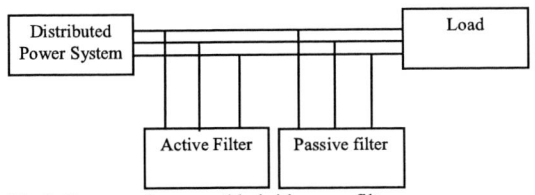

Fig. 6. Block diagram of proposed hybrid power filter.

TABLE III
DESIGN SPECIFICATIONS AND CIRCUIT PARAMETERS

| ac supply voltage | 380V |
|---|---|
| fundamental frequency | 50Hz |
| supply resistance | 0.001Ω |
| supply inductance | 0.6 mH |
| inverter dc voltage (Vdc) | 1000V |
| Thyristor load resistance | 8Ω |
| Thyristor load inductance | 0.1mH |
| Switching frequency | 12kHz |
| Coupling inductance | 1mH |

TABLE IV
HARMONIC LEVELS OF SUPPLY CURRENT BEFORE AND AFTER
COMPENSATION BY DIFFERENT POWER FILTERS

| filter type | THD of supply current | |
|---|---|---|
| | (Load current 22.6 A) (%) | (Load current 43.8 A) (%) |
| without filter | 27.95 | 25.05 |
| conventional passive | 9.83 | 19.03 |
| proposed passive | 11.14 | 10.88 |
| conventional hybrid | 4.05 | 3.97 |
| proposed hybrid | 4.37 | 4.43 |

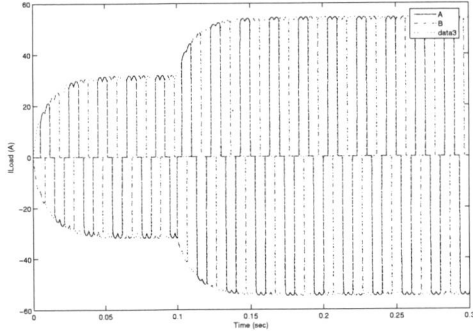

Fig. 6. Supply current without any filter (load current).

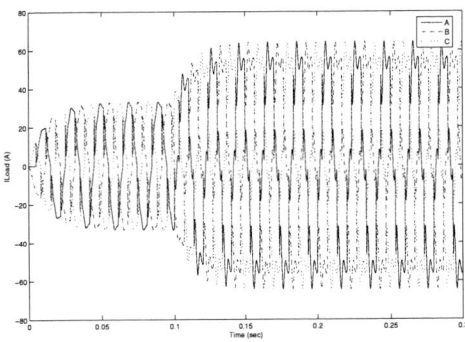

Fig. 7. Supply current after compensating by conventional passive power filter.

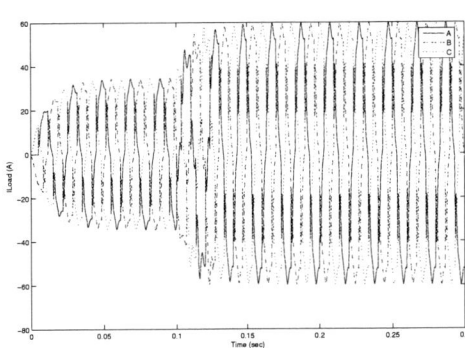

Fig. 8. Supply current after compensating by the proposed passive power filter.

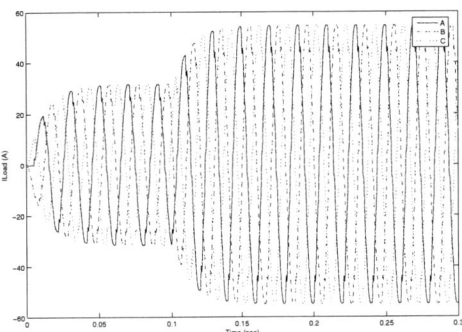

Fig. 9. Supply current after compensating by conventional hybrid power filter.

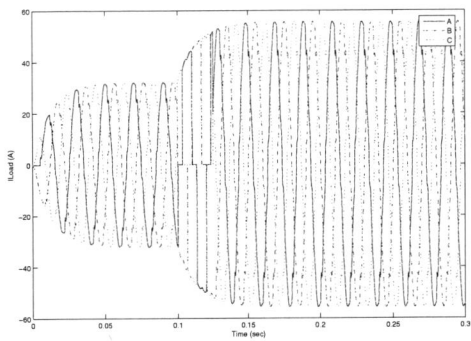

Fig. 10. Supply current after compensating by the proposed hybrid power filter.

## V. ECONOMICAL ASSESSMENT

Because of the variable structure of the passive part, a motor to change the air gap and mechanical bolt and clamp to maintain it in fixed positions are needed. The cost difference between this filter and the conventional one is about 200US$ for the preassigned ratings.

The most expensive part of an active filter is its inverter. Therefore, by consideration of this part current and voltage, the required rating of the inverter of the conventional type is about 5.5kW. The cost of Telemecanique Iran one for this rating is about 1000US$ whereas for the proposed filter, it is about 550US$ for 1.1kW. This difference is caused by the remainder of $5^{th}$ and $7^{th}$ harmonics currents which should be carried by them. As a result, the expense difference of the active part is approximately: 1000-550=450US$.

At the end, it is obvious that the difference of the cost of these filters is obtained by subtracting the cost difference of the passive part from the active one. Therefore, the proposed filter for preassigned ratings is 450-200=250US$ cheaper than conventional one.

## VI. DISCUSSION

In this paper the operation of a variable hybrid power filter is assessed. The proposed filter includes variable and active filters. The passive part is designed by the finite element method software (OPERA) which varies by load current changes. Therefore, it can compensate the dominant harmonics ($5^{th}$ and $7^{th}$) completely and the active part is not obliged to carry the extra harmonic currents caused by load current. This filter is compared with a conventional hybrid filter to examine its ability in unbalanced condition. As a result the simulation results show that the proposed method can eliminate harmonics and balance the supply current. Consequently, this filter meets IEEE 519 standard recommendations on harmonics levels. At the end, an economical comparison is done between a conventional hybrid power filter and the proposed one with same ratings. The results of this assessment show that by the proposed design, it is possible to save about 250$ in the use of variable hybrid power filter with respect to conventional one.

## VII. ACKNOWLEDGMENT

The author gratefully acknowledges the contributions of Nahid Chalyavi for her work on the original version of this document.

## VIII. REFERENCES

[1] B. Singh, K. Haddad, and A. Chandra, "A new control approach to three-phase active filter for harmonics and reactive power compensation," IEEE Trans. Power Syst., 1998, 13, (1), pp. 133–138.

[2] F. Z. Peng, H. Akagi, and A. Nabae, "A new approach to harmonic compensation in power systems- A combined system of shunt passive and series active filters," IEEE Trans. Ind. Apple., 1990, 26, (6), pp. 983-990.

[3] A. V. Zyl, J. H. R. Enslin, W. H. Steyn, and R. Spee, "A new unified approach to power quality management," Proceedings of PESC, 1995, CA, USA, pp. 183-188.

[4] L. Moran, J. Dixon, and R. Wallace, "A three-phase active power filter operating with fixed switching frequency for reactive power and current harmonic compensation," IEEE Trans. Ind. Electronics., 1996, 42, (4), pp. 402-408.

[5] S. Bhattacharya, D. M. Divan, and P. T. Chang, "Hybrid solutions for improving passive filter performance in high power applications," IEEE Trans. Ind. Apple., 1997, 33, (3), pp. 732-747.

[6] F. Libano, J. Cobos, and J. Uceda, "Simplified control strategy for hybrid active filters," Proceedings of PESC, 1997, Saint-Louis, USA, pp. 1102-1108.

[7] L. F. C. Monteiro, and M. Aredes, "A Comparative Analysis among Different Control Strategies for Shunt Active Filters," V INDUSCON - Conference de Appliances Industrials, Salvador BA, (3-5) July 2002, Brazil, pp. 345-350.

[8] F. B. Libano, D. S. L. Simonetti, and J. Uceda, "An overview on hybrid active filters," COBEP, 1995, vol. 1, pp. 333-337.

[9] H. Akagi, Y. Kanazawa, and A. Nabae, "Generalized theory of the instantaneous reactive power in three-phase circuits," Proc. JIEE-IPEC-Tokyo, 1983, pp. 1375.

[10] H. Akagi, Y. Kanazawa, and A. Nabae, "Instantaneous reactive power compensators comprising switching devices without energy storage components," IEEE Trans. Ind. Apple., May/June 1984, 1A-20, (3), pp. 625-630.

[11] S. Bhattacharya, D. M. Divan, and B. B. Banerjee, "Synchronous frame harmonic isolator using active series filter," EPE, 1991, vol. 3, pp. 30-35.

[12] M. Mirsalim, A. Doroudi, and M. Halati, *Analysis of Electrical Machines with Finite Element Method (in Persian)*, Amirkabir University of Technology Publisher, 2002, pp. 90-95.

[13] J. Holtz, "Pulsewidth modulation for electronic power conversion," Proc. IEEE, 1994, 82, (8), pp. 1194–1214.

[14] S. Buso, S. Fasolo, L. Malesani, and P. Mattavelli, "A dead beat adaptive hysteresis current control," IEEE Trans. Ind. Apple., 2000, 36, (4), pp. 1174–1180.

[15] B. K. Bose, "An adaptive hysteresis band current control technique of a voltage feed PWM inverter for machine drive system," IEEE Trans. Ind. Electron, 1990, 37, (5), pp. 402–406.

## IX. BIOGRAPHIES

**Hedayatollah Dalvand** was born on 1981 in Iran. He received the B.S. degree in electrical engineering from K. N. Toosi University of Technology, Tehran, Iran in 2002. Dalvand is currently pursuing the M.S. degree of electrical engineering in Amirkabir University of Technology, Tehran, Iran. His research interests are power filters, reactive power control, FACTS devices, finite elements method in power systems and optimization of electrical motors.

# Grid Connected Photovoltaic Interface with VAR Compensation and Active Filtering Functions

Aslain Ovono Zué, *Student Member, IEEE*, Ambrish Chandra, *Senior Member, IEEE*

*Abstract--* **A 1kWp (peak kilowatt) grid connected photovoltaic (PV) interface is presented. The main functions of that interface are shunt active filtering and VAR compensation of nonlinear load. The controller algorithms are based on maximum power point tracking (MPPT). The indirect current control technique is used. The Matlab simulation shows the validity of the control strategy under environmental and load changes and compliance with power quality requirement of IEEE 519-1992.**

*Index Terms--* **indirect current control technique, maximum power point tracking, photovoltaic, shunt active filter, VAR compensation.**

## I. INTRODUCTION

WITH the deregulation of energy market, independent producers with competitive rates can now access the transmission and distribution grid with small scale power stations closed to the end users. The word commonly used to describe that small power asset is distributed generation. Distributed generation has the advantage of saving money since there is no need of building new and long transmission lines from generation to distribution. Therefore losses associated with transmission are avoided. The excess of energy generated by those small assets is sent to the grid and sold to the utility's owner. Instead of building new large scale fossil based power plants which are the mainly responsible for greenhouse gases emission in the atmosphere and global warming, small power assets based on renewable energy sources or less pollutant fuel such as natural gas used with microturbines can be good alternatives for a sustainable development.

The increase of energy demand and the decline of natural resource reserves worldwide leading to high volatility of oil prices should encourage the emergence of new technologies or the improvement of existing ones. Solar energy is a clean and available source of energy for distributed power generation. Photovoltaic is the term used to describe a device capable of converting directly the energy contained in photons of light into an electrical voltage. Photovoltaic systems have the advantage of being maintenance free and the fuel (sun) is inexhaustible. When connected to the grid, PV systems require a power conditioning unit that ensures the perfect compliance of PV systems operation with IEEE standard 519-1992 [1]. The kernel of the power conditioning unit is the inverter which converts direct current to alternative current. Inverter operation can inject high frequency harmonic into the grid and cause electromagnetic interference with sensitive loads connected to the grid.

With appropriate inverter control and passive output filter, interference can be reduced. Nowadays with a lot of nonlinear loads being introduced with digitalization of process, the grid current becomes distorted or pollute. These nonlinear loads consume reactive power too. Reactive power is necessary since most of loads are inductive. But it is also responsible for voltage drop and it limits the active power transmitted. Therefore, reactive power should be compensated locally.

The idea of a solar power conditioning device capable of both reducing greenhouse gases emission and harmonic current drawn by nonlinear load while compensating the reactive power is very interesting.

Many utility interfaces have been proposed. The major problem with these interfaces is the dependency upon irradiance. The interface should be disconnected from the grid when irradiance is low or during night time. The proposed interface is operative 24 hours a day whatever the environmental conditions are. Increasing the use factor of grid connected PV systems has been proposed by some researchers. In [2] a single phase three wire photovoltaic energy conversion system is proposed. The proposed system in [2] acts as a solar generator on sunny days and as active filter on rainy or cloudy days. The control is just valid for a single phase system and there are two different control loops depending on the system acting as active filter or solar generator. In [3] a multi function grid connected PV system with VAR compensation is proposed. The control system in [3] uses synchronous frame PI controllers and a two stage configuration with a dc boost converter for maximum power point tracking and the inverter for synchronization with grid.

---

The authors would like to thank Natural Sciences and Engineering Research Council of Canada (N.S.E.R.C.) for providing financial support for this research work.

A. Ovono Zué and A. Chandra are with the Department of Electrical Engineering, Ecole de Technologie Supérieure, Université du Québec, Montréal QC H3C-1K3, Canada (email : aslain.ovono-zue.1@ens.etsmtl.ca; chandra@ele.etsmtl.ca ) ).

---

0-7803-9771-1/06/$25.00 ©2006 IEEE

The two stage topology is more expensive than the one stage topology because it requires more devices and the control requires a supplementary loop. The transformation of coordinates from a stationary frame to a synchronous frame is computation time consuming. In addition, the control algorithm proposed in [3] is only usable with linear loads. In [4] a PV power generation system with shunt active filtering function is presented. The control strategy is based on *d-q* transformation to compensate the negative and positive component of current harmonics. This control algorithm is complex. In this paper the conditioning unit serves as both shunt active filter and solar generator. The current control used in this paper is based on indirect current control proposed by Singh et al (1998). A brief review of photovoltaic cell mathematical model is presented in section IIA. Then maximum power point of PV array is defined. The computing algorithm of maximum power point is presented in section IIB. Finally the control strategy and simulation results are presented and discussed in sections IIE and III respectively.

## II. TOPOLOGY DESCRIPTION AND CONTROL

The topology of the proposed grid connected PV interface uses a full bridge three phase sine PWM, bidirectional voltage source inverter for MPPT, VAR and harmonics compensation without voltage boost dc converter. An L filter for removing the inverter output current ripple and the control are shown in Fig.6. The PV array rated power is $1kW_p$.

### A. PV array model

PV module consists of PV cells connected in a given way in parallel or series depending on the PV module ratings. A single module rating is limited to few hundreds watts. When higher power is required PV modules are connected in series and in parallel to obtain PV array. The theoretical model of a PV array is deducted from that of a single PV cell. This model is useful to simulate the PV array behaviour with Matlab/Simulink. The well known model [5] of a PV cell is as follow:

$$I = I_{LG} - I_{os}\left\{\exp\left[\alpha(V + IR_s)\right] - 1\right\} - (V + IR_s)/R_{sh} \quad (1)$$

Where,

$$I_{OS} = I_{OR}\left[T/T_r\right]^3 \exp\left[\beta(1/T_r - 1/T)\right] \quad (2)$$

$$I_{LG} = \left[I_{SCR} + K_I(T - 25)\right](\lambda/1000) \quad (3)$$

$$\alpha = (q/(AkT)) \quad (4)$$

$$\beta = (qE_{G0})/(Ak) \quad (5)$$

Where,

| | |
|---|---|
| $I_{OS}$ | cell reverse saturation current |
| T | cell temperature in K |
| k | Boltzmann's constant |
| q | electronic charge |
| $K_I$=0.0017 A/°C | short circuit current temperature coefficient at $I_{SCR}$ |
| λ | solar irradiance in W/m$^2$ |

| | |
|---|---|
| $I_{SCR}$ | short circuit current at 25˚C and 1kW/m$^2$ |
| $I_{LG}$ | light generated current |
| $E_{G0}$=1.1 eV | band gap of silicon |
| A=1.92 | ideality factor |
| $T_r$=301.18˚ | reference temperature |
| K | |
| $I_{OR}$ | cell saturation current at $T_r$ |
| $R_{sh}$ | cell shunt resistance |
| $R_s$ | cell series resistance |
| I-V | cell output current and voltage |

From the above equations the solar cell equivalent circuit is made of current source with a shunt resistance in parallel with a diode and a series resistance (Fig.2). The series resistance represents the contact resistance associated with the bond between the cell and its wire leads [6] and the resistance of the semiconductor itself. The shunt resistance takes into account the output voltage drop caused by the shading effect on some cells in a PV string. The I-V characteristic is implemented in Simulink as three inputs (voltage, irradiance and cell temperature), single output dc current programmable source as follow:

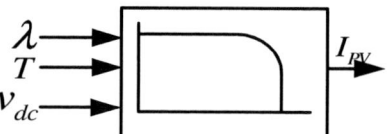

Fig.1. PV Array Simulink model.

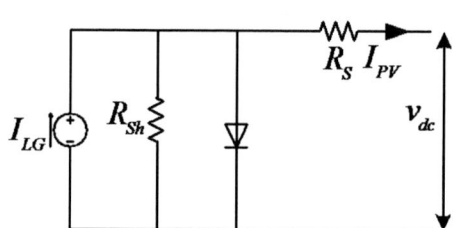

Fig.2. PV Cell equivalent circuit.

The P-V characteristic of a solar cell is not as rectangular (Fig.3) as one might think but shows a maximum power point (Fig.4). The fill factor is used to characterize module performance. The fill factor shows how well module P-V characteristic suits in a rectangle of $V_{OC}$ length and $I_{SCR}$ width. Since most of the incident energy absorbed is converted to heat, PV cells are hotter than ambient environment. Manufacturers do provide the difference between ambient temperature and cell temperature as NOCT (normal operating cell temperature). Both NOCT and fill form are useful for PV systems design.

### B. Maximum power point tracking

The maximum of a given continuous function is the point at which its derivative is null. Equation (1) of current is very nonlinear meaning that P-V function is also nonlinear. Taking the derivative of PV cell output power the following equation

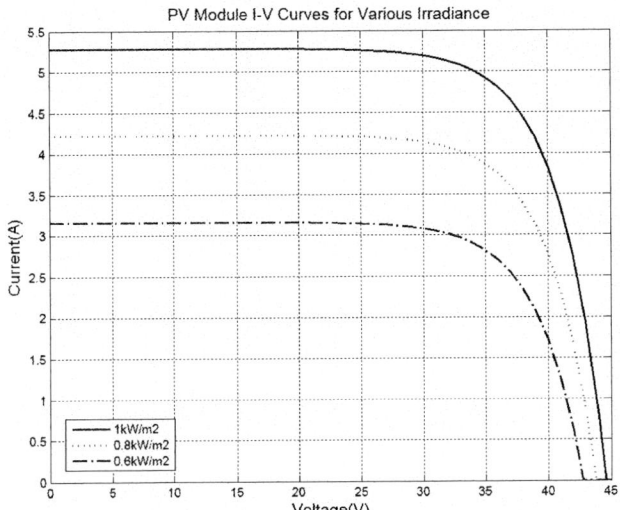

Fig.3. PV module I-V Characteristic.

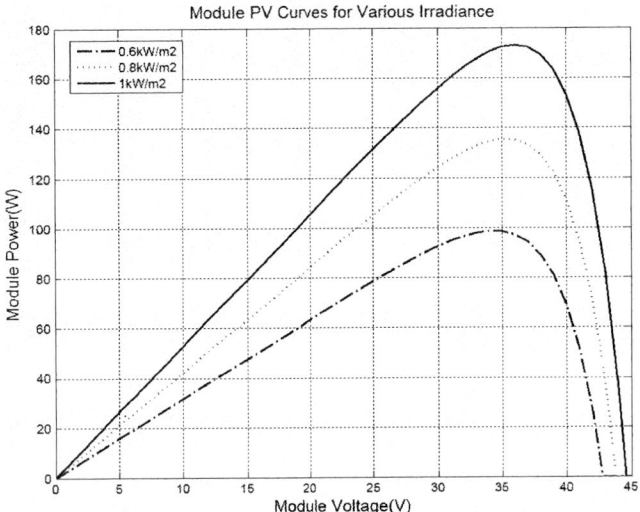

Fig.4. PV module P-V Characteristic.

TABLE I
SIMULATION PARAMETERS

| Utility voltage | $V_{ga}$ | 60Vrms |
|---|---|---|
| Utility frequency | $f_n$ | 60Hz |
| Inverter power | $P_n$ | 1kVA |
| Switching frequency | $f_s$ | 4.5kHz |
| Output inductance | $L_C$ | 0.8mH |
| Dc bus capacitor | $C_{dc}$ | 1.5mF |
| PV array | | |
| - PV module power | | 170W$_P$ |
| - No of modules in series | $N_S$ | 6 |
| - No of modules in parallel | $N_p$ | 1 |

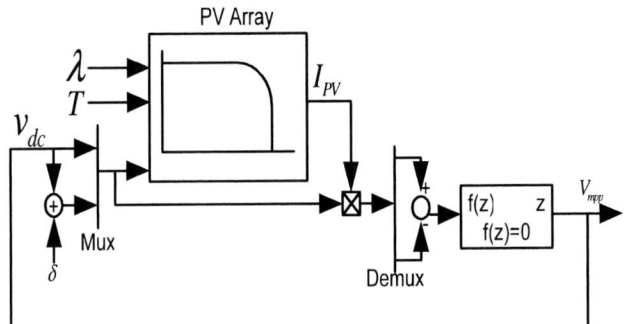

Fig.5. MPP location with Simulink.

is obtained:

$$\frac{dP}{dV} = \frac{I}{V} + \frac{dI}{dV} = 0 \qquad (6)$$

Equation (6) can be solved easily with numerical algorithm such as the perturbation and observation, the incremental conductance, hill climbing, Newton algorithm etc. Some of these methods do not necessitate knowledge of PV model. The MPP is found only by comparing the sensed PV output power of one sample with the power of the previous sample and then locate the MPP. The simulation performed in this paper uses a Simulink block named solve [7] based on Newton algorithm to locate the maximum power point for any weather condition. The solve block adjusts its output z until its input f(z)=0. In this simulation z is the maximum power point voltage. The maximum power point voltage ($V_{mpp}$) is found after perturbation of the PV output power with a small amount $\delta$ (Fig.5). The problems faced with most of algorithms failing to locate the MPP when a sudden change of irradiance occurs are avoided. No need of additional programming but just connecting few Simulink blocks to locate the MPP.

### C. The power circuit

The power circuit depicted in Fig.6 consists of a photovoltaic array with a dc link capacitor on the dc side. The dc capacitor is used to steady the dc link voltage. On the ac side are a three leg two level inverter sine wave PWM controlled, a passive filter $L_C$ with its internal resistance $R_C$ at each phase and the grid with its internal series inductance $L_s$. The $L_C$ filter attenuates the inverter output current high frequency component due to the switching of the inverter. The grid is assumed balanced and perfectly sinusoidal. The nonlinear load consists of a three phase rectifier bridge with an RL load ($R_L$, $L_L$). The load is drawing harmonics current and reactive power from the grid. The PV grid connected interface should compensate the current harmonic drawn by the load such a way that the grid current remains a pure sinusoid and in phase with the grid voltage. Reactive power compensation and harmonic compensation are ensured through the interface inverter with the control algorithm. The parameters of the power circuit simulated are summarized in table I.

### D. Maximum Power Point Tracking Control

The proposed interface has two control loops. An outer loop for dc link voltage control and maximum power point tracking, an inner loop for active filter output current control (Fig.6). The dc loop time constant is considered greater than that of the current loop so as both loops can be studied separately.

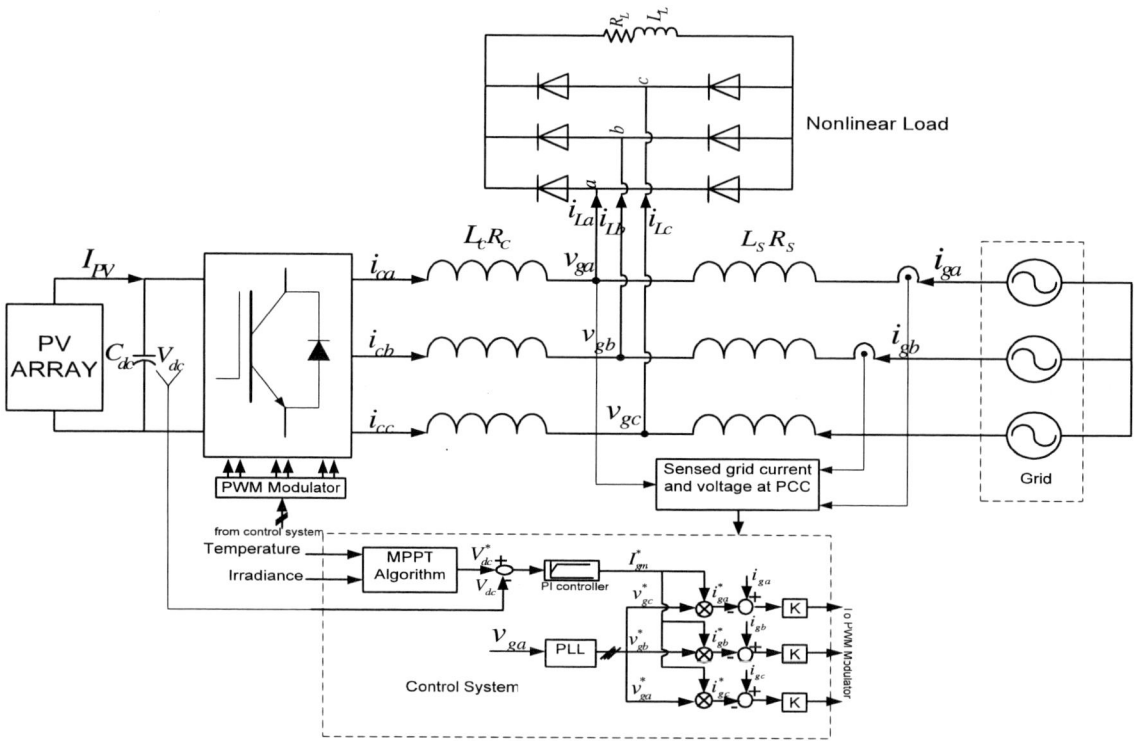

Fig.6. PV Utility Interface used as a Shunt Active Filter.

Neglecting the inverter internal losses and assuming a unit power factor, the equilibrium between inverter input and output power leads to the following equations in dq coordinates [8]:

$$C_{dc}\frac{dv_{dc}}{dt} = I_{PV} - \frac{3}{2}[d_d i_{cd}] \qquad (7)$$

Where, $d_d$ & and $d_q$ ($d_q$=0 since the power factor is assumed equal to unity) represent the switching state functions in synchronous coordinates and $i_{cd}$ represents the direct component of inverter output current in $d$-$q$ coordinates. The auxiliary input current is defined as:

$$i^*_{dc} = C_{dc}\frac{dv_{dc}}{dt} \qquad (8)$$

$$i^*_{dc} = C_{dc}(dv_{dc}/dt)^* + K_P(v^*_{dc} - v_{dc}) + K_I \int(v^*_{dc} - v_{dc})dt \qquad (9)$$

Let's define the dc link voltage error as:

$$e_{dc} = (v^*_{dc} - v_{dc}) \qquad (10)$$

From (8)-(10) the following error dynamics is derived:

$$\ddot{e}_{dc} + (K_P/C_{dc})\dot{e}_{dc} + (K_I/C_{dc})e_{dc} = 0 \qquad (11)$$

The proportional and integral gains $K_P$ and $K_I$ respectively are found by assigning desired poles to the left half plane i.e. taking the Laplace's transform of (11) and equalizing it with the desired characteristic polynomial below:

$$s^2 + 2\xi\omega_n s + \omega_n^2 \qquad (12)$$

Where $\xi$ is the damping factor and $\omega_{ni}$ is the natural frequency. Then the gains are found as:

$$\begin{cases} K_P = 2\xi\omega_n C_{dc} \\ K_I = \omega_n^2 C_{dc} \end{cases} \qquad (13)$$

Using the technical optimum [9], the damping factor is set to (0.707) so as to lower the overshot to the step change of dc link voltage reference. The dc voltage reference is the voltage at MPP computed with the chosen MPPT algorithm.

E. Indirect Current Control Technique

The indirect current control technique has been chosen because of its superior performances in terms of low power dissipation, lower THD and fewer sensors required compared to that of the direct current control technique [10].

The output of the PI controller of maximum power point tracking loop is considered as the amplitude of the grid current $I_{gm}$. The grid current magnitude $I_{gm}^*$ is then multiplied with three phase unit sine waves generated with phase lock loop of the measured three phase grid voltage $v_{ga}^*$, $v_{gb}^*$, $v_{gc}^*$, resulting in three phase reference grid currents denoted $i_{ga}^*$, $i_{gb}^*$ and $i_{gc}^*$. Since the system is balanced, the measure of two phase's currents is more than enough. In the indirect current control technique the three phase reference grid currents are considered as the three phase active filter reference currents and the three sensed grid currents $i_{ga}$, $i_{gb}$, $i_{gc}$ as active filter currents. The measured three phase grid currents are considered as the measured active filter output currents.

586

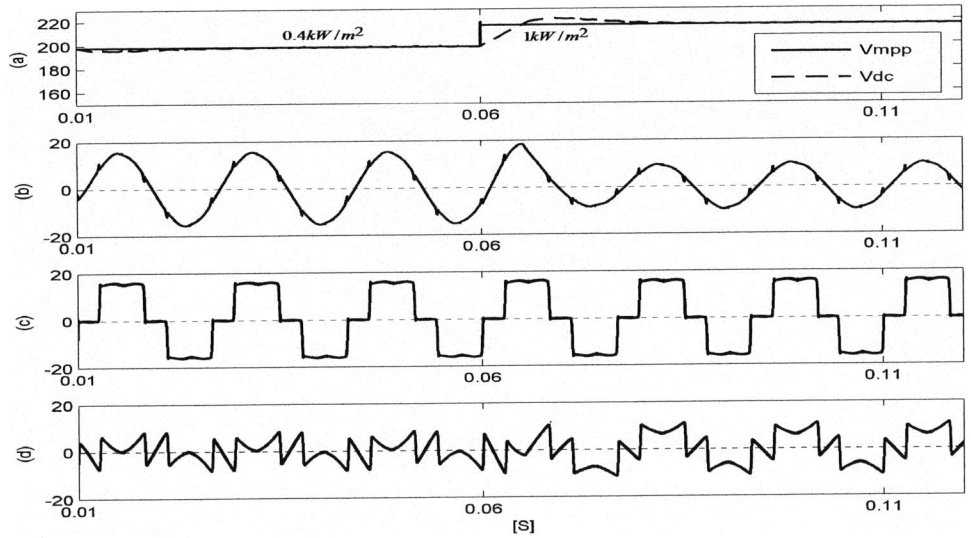

Fig.7. Simulation Results of the PV Utility Interface Topology (a) MPPT (b) $i_{gc}$ (c) $i_{Lc}$ (d) $i_{cc}$.

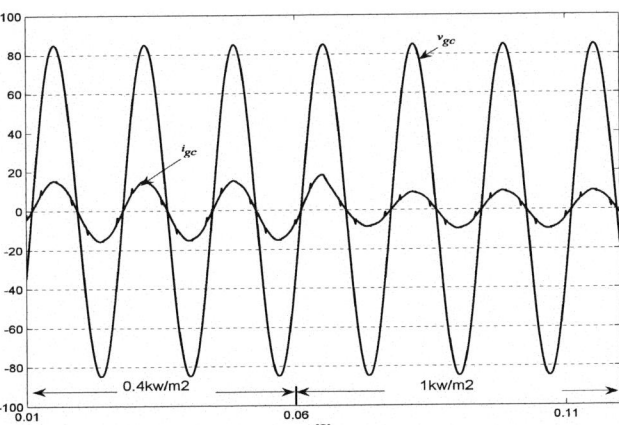

Fig.8. Grid Current and Voltage at PCC.

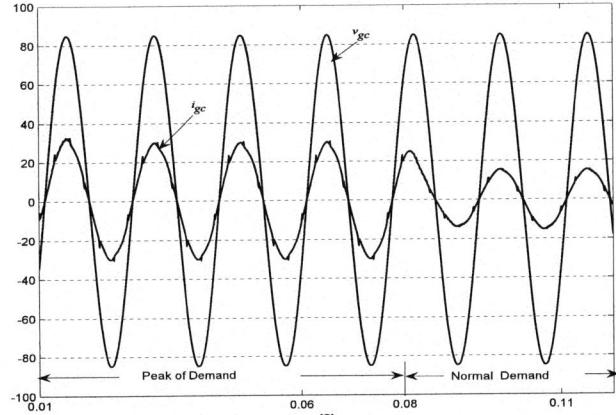

Fig.9. Grid Current and Voltage after load increase.

While in the direct current control technique the two phase load currents are also measured and then subtracted from the reference grid currents resulting to three phase reference output currents of the shunt active filter. The main difference in the algorithm used in this paper with [10] is that the three phase reference output currents of shunt active filter are generated by subtracting the three phase reference grid currents $i_{ga}^*$, $i_{gb}^*$ and $i_{gc}^*$ from the measured grid currents $i_{ga}$, $i_{gb}$ and $i_{gc}$ . The three phase shunt active filter currents are compared with a triangular waveform carrier signal to generate the PWM switching signals for the shunt active filter devices.

Both grid and photovoltaic supply the load. The nonlinear load current for each phase is therefore:

$$i_{Labc} = i_{cabc} + i_{gabc} \qquad (14)$$

## III. SIMULATION RESULTS

The simulation has been performed under a step change of irradiance and a constant cell temperature of 25°C. At 0.06 second irradiance is increased from 0.4kW/m² to 1kW/m² as shown in Fig.7. The dc bus voltage reference is the voltage at

the MPP computed with MPPT algorithm. The Fig.7 (a) shows that the maximum power point voltage is tracked for this change of irradiance. The grid current THD is lower than 5% meaning compliance with IEEE-519-1992 (see table II). Grid current and voltage are in phase (Fig.8) i.e. the VAR compensation is ensured by the inverter. Active filter output current has a fundamental component. Therefore the shunt active filter does not compensate only nonlinear load harmonic current but also injects active power from the PV array to the nonlinear load. The shunt active filter and the grid supply active power to the load. The increase of irradiance leads to the increase of active current supplied by shunt active filter and the decrease of grid current. The grid connected interface is still operative at night time but the active power exchanged is just the amount needed to load the dc link capacitor and to supply internal losses of the power converter. The grid currents THDs obtained for various irradiance are listed in table II below. For all the range of irradiance simulated the grid current THD are well within the IEEE standard tolerance. This is at odd with [11] where the increase of current THD with the decrease of solar irradiance

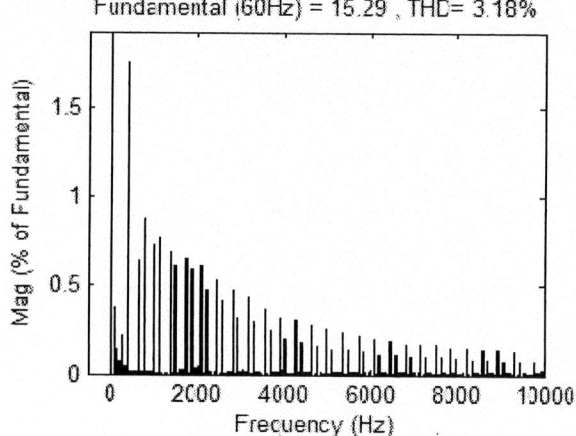

Fundamental (60Hz) = 15.29 , THD= 3.18%

Fig.10. Grid Current Spectrum.

TABLE II
THD VALUES

| Irradiance (kW/m$^2$) | THD (%) |
|---|---|
| 1 | 4.6 |
| 0.8 | 3.9 |
| 0.6 | 3.5 |
| 0.4 | 3.2 |

is shown experimentally for different single phase grid connected PV inverters.

To see the transient behavior of the solar shunt active filter, the load had first been increased from its previous simulated value, step decreasing load resistance $R_L$, under a constant irradiance and temperature. At 0.08 second the load had been set back to its previous value. It can be noticed in Fig.9 that grid current harmonic and reactive power compensations are still operative. Grid current THD is still in compliance with [1]

## IV. CONCLUSION

The proposed PV grid connected interface has two control loop, an inner or faster loop for active filter output current control and an outer loop for dc link voltage control or maximum power point tracking. Dc voltage reference is the voltage at maximum power point which is computed with Newton algorithm. The dc link voltage is controlled with a PI while the active filter current is controlled with a proportional regulator using the indirect current control technique. Knowledge of PV array mathematical model enables its simulation with Matlab/Simulink. The MPPT and unit displacement factor are still ensured even if irradiance varies. The topology acts as a shunt active filter with MPPT and VAR compensation while injecting active power to the nonlinear load at the same time.

The utilization of the interface is maximized (24h). The increase of THD output current with the decrease of irradiance commonly encountered with grid connected PV interface is avoided. When irradiance increases, the grid current decreases meaning PV system and grid are supplying

current to the load. The current spectrum for an irradiance of 0.4kw/m2 is in compliance with IEEE-519-1992. The load has been increased from its rated value current harmonic and VAR compensation are still performed while tracking the maximum power point of PV array.

## V. REFERENCES

[1] *IEEE Recommended Practices and Requirements for Harmonic Control in Electrical Power Systems*, IEEE Standard 519-1992, NY, 1993.

[2] Kuo Yeong-Chau, Liang Tsorng-Juu; Chen Jiann-Fuh "A high-efficiency single-phase three-wire photovoltaic energy conversion system" *IEEE Trans on Industrial Electronics*, vol. 50, n 1, pp 116-122, Feb. 2003,.

[3] Huajun. Yu, Junmin. Pan, and An. Xiang, "A multi-function grid-connected PV system with reactive power compensation for the grid," *Solar Energy*, vol. 79, no. 1, pp. 101-106, July. 2005

[4] Nak-gueon Sung Jae-deuk Lee; Bong-tae Kim; Minwon Park; In-keun Yu "Novel concept of a PV power generation system adding the function of shunt active filter" *in proc. 2002 Power Engineering Society Transmission and Distribution Conf.*, pp. 1658-63

[5] E. Koutroulis, K. Kalaitzakis, N.C. Voulgaris, " Development of a microcontroller-based, photovoltaic maximum power point tracking control system" *IEEE Trans on Power Electronics*, vol. 16, no. 1, pp. 46-54, January. 2001

[6] Gilbert. M. Masters, *Renewable and Efficient Electric Power System*, New Jersey: Wiley & Sons, 2004, p. 464-477.

[7] A. Murray Thomson "Reverse osmosis desalination of seawater powered by photovoltaic without batteries," PhD thesis, Loughborough Univ., pp.98-101. , 2003

[8] M. Raoufi, M. T. Lamchich, " Average current mode control of a voltage source inverter connected to the grid: application to different filter cells " *Journal of Electrical Engineering*, vol. 55, no. 3-4, pp. 77-82, 2004

[9] Mendalek, N. Al-Haddad K.; Dessaint L.A.; Fnaiech F. "Nonlinear control strategy applied to a shunt active power filter" *in proc. 2001 Power Engineering Society Conf.*, pp 1877-1882.

[10] B.N. Singh, A. Chandra and K. Al-Haddad, "Performance comparison of two current control techniques applied to an active filter" in *Proc. 1998 International Conference on Harmonics and Quality of Power*, pp 133-138

[11] Sidrach De Cardona M. and Carretero J. "Analysis of the current total harmonic distortion for different single-phase inverters for grid-connected pv-systems" *Solar Energy Materials and Solar Cells*, vol. 87, n 1-4, pp 529-540, May. 2005

[12] Richard. C. Dorf, *Technology, Humans, and Society: Toward a sustainable world*, San Diego, California: Academic Press, 2001

## VI. BIOGRAPHIES

**Aslain Ovono Zué** graduated as an electromechanical engineer from Ecole Polytechnique de Masuku, Gabon in 2001 and is presently pursuing his M.ing degree in the department of electrical engineering, Ecole de Technologie Supérieure, Université du Québec, Montréal Québec. His employment experience includes oilfield mechanical engineering and oilfield operation engineering from 2002 to 2004. His research interest includes grid connected photovoltaic interfaces, power quality and distributed generation.

**Ambrish Chandra** (SM'99) received the B.E. degree from the University of Roorkee, Roorkee, India, the M.Tech. degree from the Indian Institute of Technology (I.I.T.), New Delhi, India, and the Ph.D. degree from University of Calgary, Calgary, AB, Canada, in 1977, 1980, and 1987, respectively.He worked as a Lecturer and later as a Reader at University of Roorkee, presently I.I.T. Roorkee. Since 1994, he has been a professor with the Electrical Engineering Department, Ecole de Technologie Supérieure, Université du Québec, Montréal, QC, Canada. His main research interests are power quality, active filters, static reactive power compensation and flexible AC transmission systems (FACTS). Dr Chandra is member of Ordre des Ingénieurs du Québec.

**2006 IEEE International Conference on Power Electronic, Drives and Energy Systems**

# Design and Implementation of a Current Controlled Parallel Hybrid Power Filter

Bhim Singh, *Senior Member, IEEE*, and Vishal Verma, *Member, IEEE*

*Abstract--*This paper deals with the implementation of an indirect current control of Parallel Hybrid Filter System (PHF) with rectifier load to eliminate generated harmonics. The compensation principle, design and PHF implementation are discussed in detail. The proposed control enhances the performance of the passive filters under unbalance situations existing in the passive filters under fault conditions. Moreover, due to limited capacity of VSI a upper limit for such compensation may be imposed by proposed indirect current control based on synchronous reference frame employing decomposition of current in synchronously rotating reference frame (SRF) is used to control PHF. The PHF is controlled using decomposition of load and source currents. The implementation of the control scheme is carried out on dSpace with digital signal processor (DSP) for developed prototype of PHF. Passive filters are developed keeping in view of its operation with PHF, so that, the total harmonic distortion (THD) in the source current gets minimized along with minimized fundamental frequency current sinking through the filter section. The design has been carried out by genetic algorithm (GA), which optimizes the solution for multiple constraints. Operation of PHF under dynamic change of load is also investigated through simulations under MATLAB and through experimentation.

*Index Terms--* Harmonics, Compensators, Harmonic Filters.

## I. INTRODUCTION

INCREASED use of nonlinear loads produces current and voltage harmonics in the power system. Harmonic currents generated by nonlinear loads not only increase rms current, but also create voltage drop at harmonic frequencies distorting the voltage waveform at the point of common coupling (PCC). These distortions are magnified due to the interaction between capacitors and transformers causing harmonic resonance.

Passive filters are traditionally used to absorb harmonic currents because of low cost and simple robust structure. But they provide fixed compensation and create system resonance [1-4]. The filtering characteristics of passive filters are determined by the impedance ratio of the supply and the passive filter and are often difficult to design [5,6].

Hybrid filters effectively mitigate the problems of both passive filters and pure active filter solutions and provide cost effective and practical harmonic compensation approach, particularly for high power nonlinear loads. The combination of low cost passive filters and control capability of small

rating active filter effectively improve the compensation characteristics of passive filters and hence reduce the rating of the active filters (<5%), compared to pure shunt or series active filter solutions [1-3].

The parallel hybrid power filter system consists of small rating active filter in series with tuned passive filter set. The set of passive filters consisting of tuned filters at $5^{th}$, $7^{th}$ harmonics of the fundamental frequency and high pass filter element. The series connected active filter(AF) is controlled to act as a harmonic compensator for the load by constraining all the harmonic currents to sink into passive filters [1-4,7-8]. By actively improving the compensation characteristics of the tuned passive filters, the need for precise tuning of the passive filters is greatly reduced and the design of the passive filter becomes insensitive to supply impedance [3]. The compensation of harmonics in the source current through enhancement of the compensation characteristics of the passive filters also eliminates the chances of resonance [4]. The topology is suited for the harmonic compensation of the load connected to stiff supply. For large and diverse type of nonlinear loads, a single parallel hybrid power filter can be installed.

This configuration effectively provides compensation of current harmonics and limited supply voltage distortions, since it acts as a harmonic voltage source, compensating voltage drop in passive filters at harmonic frequencies at PCC [1-4,9]. However, the distortions in the utility voltage will add to the required voltage injected, and hence the required rating of the active filter may get increased.

The performance of PHF is very much dependent on how the reference compensating signals are estimated. PHF has been reported mostly with Instantaneous Reactive Power (IRP) theory [2,10] and Synchronous Reference Frame (SRF) theory [3,4,8,10,11,14]. These schemes have computed the reference components through subtraction of positive sequence fundamental current component from the load/source current. These control schemes look very attractive for their simplicity and ease of implementation, but lack in providing adequate solution under extreme or severe condition of harmonics, and unbalance or fault conditions in passive filter branch or their combinations with limited power rating of VSI, employed as AF in PHF. In such cases to safeguard the PHF hardware the protection scheme bypasses the series AF that leaves the system to the mercy of unwanted disturbances for passive filters alone causing further damage. The earlier reported control schemes do not suggest solution for proper operations of passive filters under their unbalance conditions (e.g. one of the branch in a phase of the passive filter goes out) in PHF. Moreover PWM control of PHF needs feed-forward compensation for phase lag created by LC ripple filter and transformer impedance.

---

Bhim Singh is with the Department of Electrical Engineering, Indian Institute of Technology, Delhi, INDIA(e-mail: bsingh@ee.iitd.ac.in).

Vishal Verma is with the Department of Electrical Engineering, Delhi College of Engineering, Delhi, INDIA (e-mail: vishalverma1@hotmail.com).

---

0-7803-9771-1/06/$25.00 ©2006 IEEE

This paper deals with design aspects of PHF and its implementation with indirect current control. The passive filters are designed for an optimal selection of components so that, the total harmonic distortion (THD) in the source current gets minimized along with minimized fundamental frequency current sinking through the filter section. The design has been carried out by GA, which optimizes the solution for multiple constraints [15,16]. An SRF based control scheme is used to decompose current into four parts; positive sequence active current (at fundamental frequency), positive sequence reactive current (at fundamental frequency) current at harmonic frequencies, and negative sequence current (at fundamental frequency). The control determines extreme or severe condition of harmonics, and unbalance or fault conditions in passive filter branch or their combinations and provide limited compensation respecting the limited power rating of the VSI, employed as AF in PHF. Both load and source currents are decomposed and the reference current is constructed with positive sequence active current derived from load, reactive component derived from source current and negative sequence component form load current, thus, avoiding any disturbances to compensation characteristics of passive filter and compen-sation of unbalance in load current. Moreover, indirect current control of PHF eliminates the need of any feedforward compensation and compensates the voltage drop caused due to impedance offered by passive filters at harmonic frequencies. Proposed scheme can be easily implemented on digital processors and employs least calculations. The in-phase unit templates of the synchronized signals are generated by filtering of voltage signals using a second order digital filter implemented on the DSP along with dSpace. The distortion in the voltage is filtered and reference signal is obtained without delay by filtering the sequence component ahead of the present phase of the signal. The scheme is simulated under the MATLAB environment using Simulink and power system blockset (PSB) toolboxes and the results are verified by implementation of the control scheme in real time using DSP. The performance of the proposed scheme has been demonstrated through simulated and experimental results.

## II. System Configuration

Fig 1 shows the block diagram of the PHF system, which consists of small rated series active filter(SAF) with coupling transformer connected in series with tuned passive filters together connected in shunt. The set of passive filter(PF) include tuned branches of 5th and 7th harmonic frequencies, and a high pass filter at corner frequency of 11th harmonic frequency. Passive filters are designed to result in minimum supply rms current at full load [15,16]. The combined loads include voltage fed type load, like variable frequency ac motor drives. The active filter in this topology is controlled in such a way that the ac mains current is shaped to have only fundamental frequency component, by sucking harmonics through passive filters. The active filter acts as a harmonic voltage source, compensating voltage drop in passive filter at harmonic frequencies at PCC, thereby producing almost a

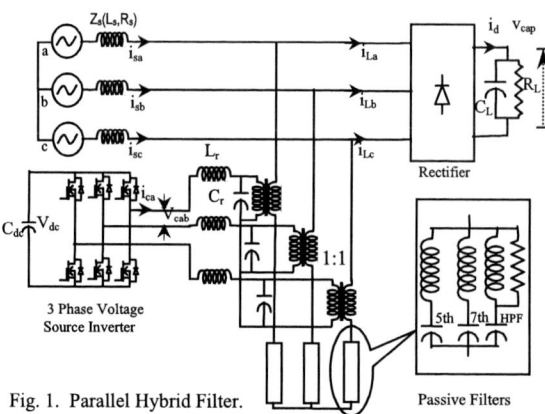

Fig. 1. Parallel Hybrid Filter.

short circuit across passive filter at harmonic frequencies. The DC bus of the system has been supported through a separate three-phase low power rectifier connected to the ac system. The externally supported DC bus voltage holds advantage in terms of compensation during fault conditions in any leg of the passive filter. Under such unbalanced conditions the self-support feature may deteriorate the performance of PHF.

## III. Control Scheme

The main objective of control scheme is to control the SAF in such a way, that, zero impedance is presented for the harmonic currents flowing in the passive filters and a high resistance at fundamental frequency. This forces the entire harmonics of load current to flow into passive filter.

The proposed control of SAF depends on decomposition scheme applied to net load and source current, selectivity of positive sequence active, reactive and negative sequence components, weightages for extreme conditions of harmonics and fault conditions in passive filters. Following section deals with basic scheme of decomposition of source and load current and overall control scheme for indirect current control of PHF. A hysteresis current controller is used to switch the bottom and top devices of the voltage source inverter of the active filter.

### A. Basic Theory

It is proposed to use SRF theory to decompose the source currents instantaneously into real and reactive components of positive and negative sequence of currents both at fundamental frequency and harmonic frequencies. The SRF isolator extracts the fundamental component of the source ($i_{sa}$, $i_{sb}$ and $i_{sc}$) / load currents ($i_{La}$, $i_{Lb}$ and $i_{Lc}$) by transformation from a-b-c to d-q reference frame. Through following transform $i_{sa}$, $i_{sb}$ and $i_{sc}$ are decomposed to different components; similarly $i_{La}$, $i_{Lb}$ and $i_{Lc}$ can also be decomposed. In the synchronously rotating reference frame, the positive sequence components at fundamental frequency ($\omega_1$), are transformed to DC quantities and all harmonic and negative frequency components undergo a frequency shift of 50Hz.

$$\begin{bmatrix} i_{s\alpha} \\ i_{s\beta} \end{bmatrix} = \sqrt{\frac{2}{3}} \begin{bmatrix} 1 & -\frac{1}{2} & -\frac{1}{2} \\ 0 & \frac{\sqrt{3}}{2} & -\frac{\sqrt{3}}{2} \end{bmatrix} \begin{bmatrix} i_{sa} \\ i_{sb} \\ i_{sc} \end{bmatrix} \qquad (1)$$

$$\begin{bmatrix} i_{sd}^+ \\ i_{sq}^+ \end{bmatrix} = \begin{bmatrix} \cos(\omega_1 t) & \sin(\omega_1 t) \\ -\sin(\omega_1 t) & \cos(\omega_1 t) \end{bmatrix} \begin{bmatrix} i_{s\alpha} \\ i_{s\beta} \end{bmatrix} \qquad (2)$$

SRF isolator extracts the DC quantities by low pass filters (LPF) for each $i_{sd}$ and $i_{sq}$, realized by moving averager at 100Hz. The extracted DC components $i_{sdcD}^+$ and $i_{sdcQ}^+$ are transformed back into first $\alpha$-$\beta$ frame and then into a-b-c coordinates to obtain net positive sequence fundamental components as shown below:

$$\begin{bmatrix} i_{s1\alpha}^+ \\ i_{s1\beta}^+ \end{bmatrix} = \begin{bmatrix} \cos(\omega_1 t) & -\sin(\omega_1 t) \\ \sin(\omega_1 t) & \cos(\omega_1 t) \end{bmatrix} \begin{bmatrix} i_{sdcD}^+ \\ i_{sdcQ}^+ \end{bmatrix} \qquad (3)$$

Whereas, the real and the reactive decomposition of the positive sequence fundamental frequency current($i_{s1R\alpha}^+$, $i_{s1X\beta}^+$) can be easily made from d-q frame, thus the a-b-c coordinates of real and reactive component at fundamental frequency can be evaluated as detailed below:

$$\begin{bmatrix} i_{s1R\alpha}^+ \\ i_{s1R\beta}^+ \end{bmatrix} = \begin{bmatrix} \cos(\omega_1 t) & -\sin(\omega_1 t) \\ \sin(\omega_1 t) & \cos(\omega_1 t) \end{bmatrix} \begin{bmatrix} 0 \\ i_{sdcQ}^+ \end{bmatrix} \qquad (4)$$

$$\begin{bmatrix} i_{s1X\alpha}^+ \\ i_{s1X\beta}^+ \end{bmatrix} = \begin{bmatrix} \cos(\omega_1 t) & -\sin(\omega_1 t) \\ \sin(\omega_1 t) & \cos(\omega_1 t) \end{bmatrix} \begin{bmatrix} i_{sdcD}^+ \\ 0 \end{bmatrix} \qquad (5)$$

$$\begin{bmatrix} i_{s1Ra}^+ \\ i_{s1Rb}^+ \\ i_{s1Rc}^+ \end{bmatrix} = \sqrt{\frac{2}{3}} \begin{bmatrix} 1 & 0 \\ -\frac{1}{2} & \frac{\sqrt{3}}{2} \\ -\frac{1}{2} & -\frac{\sqrt{3}}{2} \end{bmatrix} \begin{bmatrix} i_{s1R\alpha}^+ \\ i_{s1R\beta}^+ \end{bmatrix} \qquad (6)$$

$$\begin{bmatrix} i_{s1xa}^+ \\ i_{s1xb}^+ \\ i_{s1xc}^+ \end{bmatrix} = \sqrt{\frac{2}{3}} \begin{bmatrix} 1 & 0 \\ -\frac{1}{2} & \frac{\sqrt{3}}{2} \\ -\frac{1}{2} & -\frac{\sqrt{3}}{2} \end{bmatrix} \begin{bmatrix} i_{s1X\alpha}^+ \\ i_{s1X\beta}^+ \end{bmatrix} \qquad (7)$$

Similarly for negative sequence fundamental component can be extracted by rotating the frame in the opposite direction, i.e. executing the following transformation.

$$\begin{bmatrix} i_{sd}^- \\ i_{sq}^- \end{bmatrix} = \begin{bmatrix} \cos(\omega_1 t) & -\sin(\omega_1 t) \\ \sin(\omega_1 t) & \cos(\omega_1 t) \end{bmatrix} \begin{bmatrix} i_{s\alpha} \\ i_{s\beta} \end{bmatrix} \qquad (8)$$

And then, DC quantities are extracted by a LPF in the similar fashion. The DC quantity would amount to the negative sequence component under unbalanced conditions. The DC components so extracted ($i_{sdcD}^-$, $i_{sdcQ}^-$) are transformed back into $\alpha$-$\beta$ and then into a-b-c coordinates to obtain the negative sequence fundamental components as shown below:

$$\begin{bmatrix} i_{s1\alpha}^- \\ i_{s1\beta}^- \end{bmatrix} = \begin{bmatrix} \cos(\omega_1 t) & \sin(\omega_1 t) \\ -\sin(\omega_1 t) & \cos(\omega_1 t) \end{bmatrix} \begin{bmatrix} i_{sdcD}^- \\ i_{sdcQ}^- \end{bmatrix} \qquad (9)$$

$$\begin{bmatrix} i_{s1a}^- \\ i_{s1b}^- \\ i_{s1c}^- \end{bmatrix} = \sqrt{\frac{2}{3}} \begin{bmatrix} 1 & 0 \\ -\frac{1}{2} & \frac{\sqrt{3}}{2} \\ -\frac{1}{2} & -\frac{\sqrt{3}}{2} \end{bmatrix} \begin{bmatrix} i_{s1\alpha}^- \\ i_{s1\beta}^- \end{bmatrix} \qquad (10)$$

*B. Control Scheme*

Fig.2 shows the flow of various control signals and control scheme based on the decomposed components. The DC bus of the VSI is fed from a separate source and is held constant.

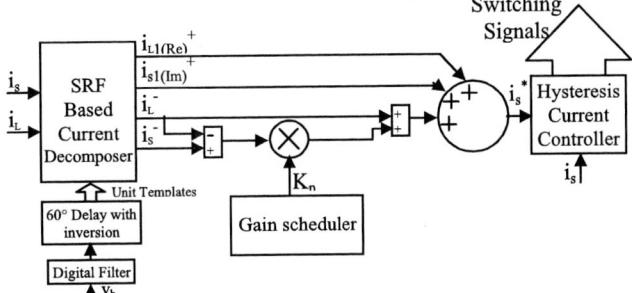

Fig. 2. Proposed Control Scheme for operation of PHF.

The reference current for desired compensation ($i_{sa}^*$, $i_{sb}^*$, $i_{sc}^*$) is derived from different components of source current and load current. The gain scheduler block is programmed to provide the flexibility of limited depth unbalance compensation (LDUC) in passive filter currents, which can be afforded by the PHF under extreme conditions. For full compensation the gain is assigned a value of '0', and when compensation is not desired it would be assigned the value '1'. The gain decides the depth of compensation desired by the system to circumvent the overloading of SAF by adjustment of the gain. in reverse order. The gains is always being assigned value less than '1'. This scheme facilitates the control of PHF for LDUC by indirect current control through hysteresis PWM current controller.

The PWM gating pulses for the IGBTs in VSI of SAF are generated by indirect current control using hysteresis current controller over reference supply currents ($i_{sa}^*$, $i_{sb}^*$, $i_{sc}^*$) and sensed supply currents ($i_{sa}$, $i_{sb}$, $i_{sc}$). The switching frequency is limited to 12.8kHz to avoid the controller to enter limit cycles while simulation and kept fixed through constant sampling and processing rate of DSP during implementation. The controlled compensation current is injected such that the supply current follows the reference current. Hence source current becomes close to reference currents estimated by the proposed scheme.

## IV. DESIGN OF PHF

The following section deals with selection and design of various components of the PHF system, viz, ripple filter, shunt passive filters and the VSI. The ripple filter has been designed keeping in mind that the active filter component is to be inserted in the shunt branch after energization of the passive filters. Whereas, design of the shunt passive filters has been carried out for harmonic reduction, reactive power reserve (10% for the diode rectifier load) in shunt passive filters with special emphasis to net rms source current reduction when filters are inserted in the circuit for their specific use in hybrid filter configuration has been done using GA based optimization [16].

*A. Design of Ripple Filter*

It is important to design an effective passive ripple filter($L_r$,$C_r$) to suppress the switching ripples of voltage or current generated by VSI used in series active filters. The purpose of $L_r$ and $C_r$ connected in the secondary of the CTs is to suppress the switching voltage ripples of the IGBTs based VSI. with switching frequency $f_s$(12.8kHz). The SAF of PHF is inserted to the shunt branch through the CTs by switching first the lower IGBT switches of the VSI on followed by opening of the by pass switch across CT, strictly after energization of passive filters. The value of $L_r$ thus would determine the voltage drop it creates for short time when the SAF of PHF is inserted in the circuit. Typically the voltage drop should be minimum and must not exceed 2% at fundamental frequency. Choosing the bases to be 398.37V, 10kVA, 50Hz. For an allowable drop of voltage 0.98% of the nominal line-to-neutral voltage of 230V, and keeping the rating of passive filters to be 25% of the rated kVA, the $L_r \approx 4.1$mH. Keeping the corner frequency of around 5.3kHz the value of $C_r \approx 0.22\mu$F.

Verifying the design form the fact that the switching voltage ripples ($V_r$) generated by VSI see impedance equivalent to impedance due to inductance ($L_r$) in series with parallel combination of impedance caused by of $C_r$ and $Z_{eq}$. Where, $Z_{eq}$ is the sum of the source impedance and the passive filter impedance seen from the secondary of the CTs, and, can be represented as

$Z_{eq} = (n_2/n_1)^2 (Z_s + Z_F)$

In order not to allow the current ripple to enter the ac system impedance seen by the leg containing the $C_r$ to be lesser than that of $Z_{eq}$. For the $L_s$=2.52mH (5%) and $Z_F$ =11.2 Ω (13.8%)and $n_2/n_1$=1:1, the impedance $1/2\pi f_s C_r$ comes out to be 56.52Ω, and $Z_{eq}$ is evaluated to be 213.87Ω. Thus the design is satisfactory for PHF.

### B. Design of Shunt Passive Filters

The component of harmonic current flowing towards the source is obtained through division of current based on the ratio of admittance of source to the sum of admittance of source and net admittance offered by passive filter set. The main goal is to minimize the actual harmonic current flowing towards the source, which is subjected to system constraints.

By minimizing of net harmonic current flow toward the source side (determined by minimizing the passive filter impedance with respect to source impedance) and maximizing the filter impedance at fundamental frequency, the insertion of passive filter would result in the net rms current reduction, due to more reduction in harmonic currents in source currents then increase of fundamental frequency current sinking in the passive filters.

#### 1) Minimizing harmonic source current:

The harmonic current minimization is the prime objective of the algorithm. The minimization of harmonic current has been ensured by minimizing the net harmonic current contents in the source current. This can be represented by summation of all harmonic currents at different harmonic frequencies as:

$$I'_h = \sqrt{\sum_{h=5,7,..} (I'_{sh})^2} \qquad (11)$$

#### 2) Minimizing fundamental current in passive filters:

The constraint of maximizing the passive filter impedance at fundamental frequency which is given by:

$$\text{Max.} \left[ \frac{|Z_1|}{\gamma} \right] \quad \text{or} \quad \text{Min.} \left[ \gamma \cdot \frac{1}{|Z_1|} \right] \qquad (12)$$

where, $\gamma$ is a multiplication factor ($\geq 1$) that represents the sensitivity of the filter to the influence of the impedance at fundamental frequency.

#### 3) Reactive Power:

The other operational constraint is to accumulate a reserve of reactive power for stable operation of power system. The summation of the filter's reactive power has been kept equal to the total reactive power compensation required for the system.

$$\sum_{m=5,7,11....} Q_{Fm} = Q_c \qquad (13)$$

where m denotes the filter legs tuned at resonant points.

#### 4) Objective Function:

Incorporating the main objective, i.e., minimization of harmonic currents in mains along with the system constraints of maximizing the filter impedance at fundamental frequency and net reactive power requirement, constitutes the objective function as:

$$ObjF = \sqrt{\sum_{h=5,7,11} (I'_{sh} + \alpha P1 + \beta P2)^2} \qquad (14)$$

where

$$P1 = \left[ \gamma \cdot \frac{1}{|Z1|} \right] \qquad (15)$$

$$P2 = \left( Q - \sum_{m=5,7,11....} Q_{Fm} \right)^2 \qquad (16)$$

The penalty functions $P_1$ and $P_2$ whose value depend upon the degree to which constraints are violated and α, β the multiplication constants control the influence of the penalty functions over objective function, optimize the design against major constraints for smoother operation of passive filters

### C. Design of VSI used as Series Active Filter

Design of VSI is primarily done with an assumption that no harmonic current escapes towards the source, i.e. the entire voltage drop across the passive filters at harmonic frequencies is compensated by the SAF of PHF. For worst case the entire voltage may be assumed to be dropped across the VSI through the CT. The output voltage of the $S_eAF$ should be discussed in peak value instead of rms values. Thus for worst-case $\sqrt{2}*2$(factor of safety)*p.u. voltage and current rating may be considered for selecting the voltage and current rating of the IGBT switch of the VSI. Thus for present case IGBT switches must be rated for 1200V, 15A.

## V. MATLAB BASED SIMULATION

The proposed control scheme and compensation by PHF is simulated under MATLAB-PSB environment to estimate its' performance under dynamic change of load and under unbalance conditions in the passive filter(opening of branch/branches of passive filter set). The load consists of diode rectifier with smoothing filter capacitor on DC side to study the effectiveness of the proposed scheme. The rectifier fed load has been modeled as diode rectifier with smoothing capacitance of $500\mu$ F feeding

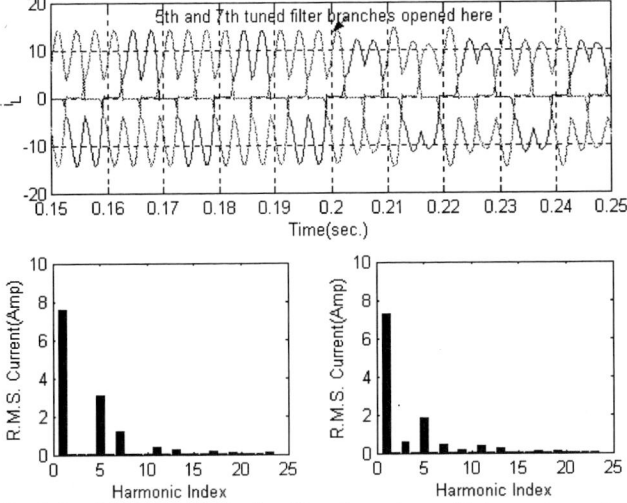

Fig. 3. Load current with Passive filter alone along with harmonic spectrum before and after the unbalance in passive filters respectively.

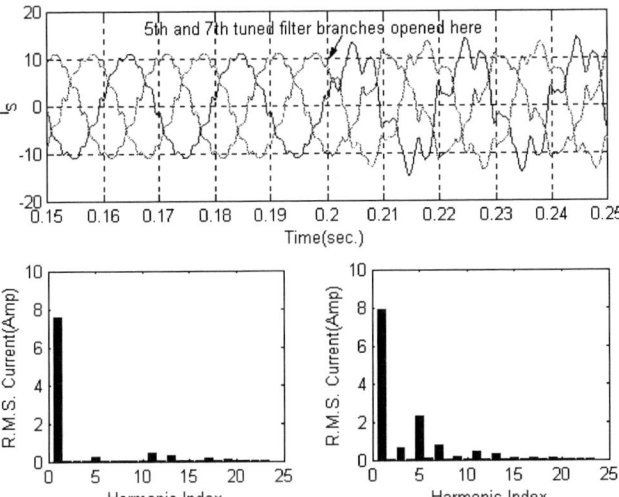

Fig. 4. Source current with Passive filter alone along with harmonic spectrum before and after the unbalance in passive filters respectively.

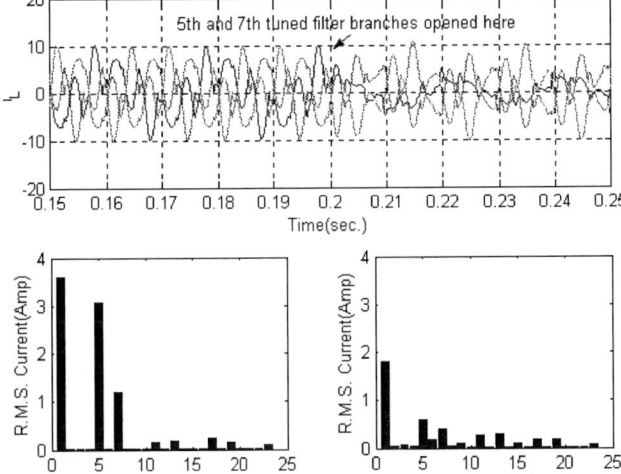

Fig. 5. Passive filter current when used alone along with harmonic spectrum before and after the unbalance in passive filters respectively.

resistive load, which represents the real power consumed by the load. This equivalent resistive load corresponds to 2.62kW load

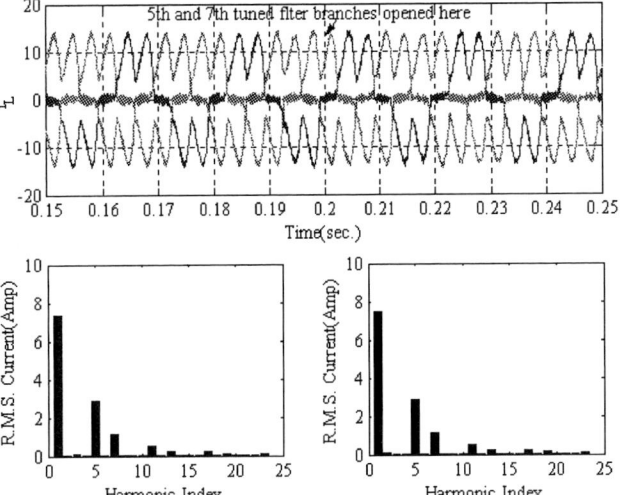

Fig. 6. Load current with PHF along with harmonic spectrum before and after the unbalance in passive filters respectively.

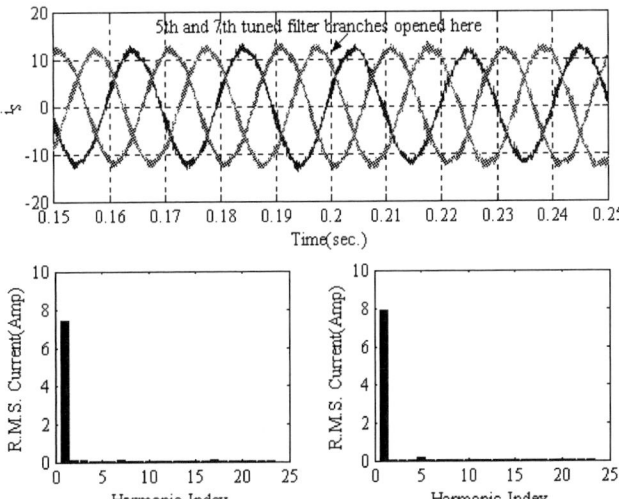

Fig.7. Source current with PHF along with harmonic spectrum before and after the unbalance in passive filters respectively.

at a line voltage of 192.52V. The performance of the PHF with proposed scheme has been computed and analyzed under dynamic change of load and under unbalance conditions in the passive filter through THD of source and load current.

The compensation of reactive power is intentionally avoided for proper operation of passive filters in PHF. The performance of the proposed scheme is evaluated for its application to unbalanced operation of passive filters. In the present study the $5^{th}$ and $7^{th}$ tuned branches of passive filter are taken out/opened due to blowing of fuses. Figs.3-5 show load current, source current and passive filter current alongwith harmonic spectrum (before /after unbalanced operation of passive filters) respectively. The rms values of currents and %THDs have been given in the Table I. It may be observed that under such conditions not only the passive filters, but also the load currents become unbalanced. The performance of PHF in the similar situations has been depicted in Figs.6-7. It may observed that by insertion of SAF of PHF the (%THD of the source current drastically improves to a maximum of 3.3% which conform to IEEE-519 standard.

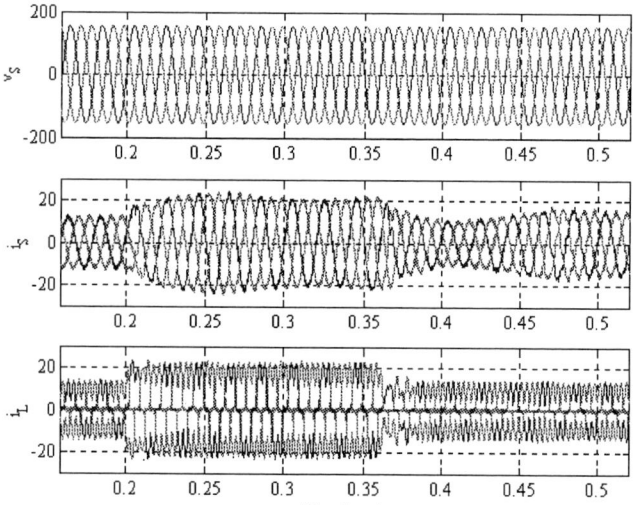

Fig. 8. Performance of PHF under load dynamics (Initially load is 2.62kW, at t=0.2s the load is increased to 5.23kW and at t=0.36s is again reverted to 2.62kW.

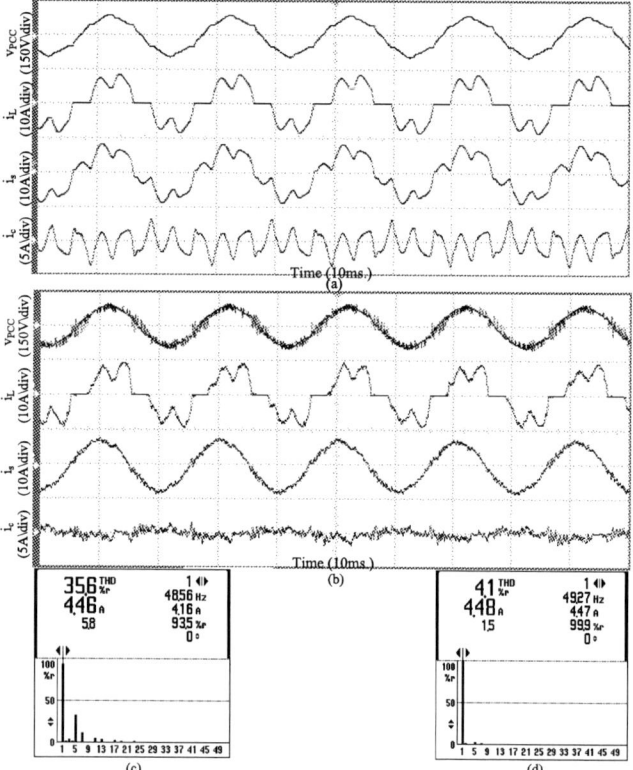

Fig. 9 (a-d), Steady state performance of passive filters alone and of PHF with diode rectifier for harmonic compensation along with harmonic spectrum of load and source currents (coupling transformer 1:1).

It may also be observed from Figs.5-6 and Table I that even under unbalanced condition of passive filters the shunt branch draws balanced currents from PCC, as a result, the current becomes balanced on source side and characteristics of load is also prevented from disturbance. The details of the rms currents and %THD of load current and source current under balanced and unbalanced conditions of passive filter have been presented for respective phases in Table-I. Fig.7 shows the simulated performance of PHF for dynamic change of load. The load is initially kept at 2.62kW and is changed

to5.23kW at t=0.2s. It may be observed that the source current remains sinusoidal through the operation. The load is again changed from 5.23kW to 2.62kW at t=0.36s to study the reverse dynamics also. %THD of source current remains below 4% through out its' operation even under load change. Thus simulation results validate the utility and effectiveness of the proposed scheme for dynamic change of load.

## VI. HARDWARE REALIZATION

A three phase experimental prototype of the shunt passive filters and SAF for PHF has been fabricated in the laboratory to verify the performance of proposed scheme. VSI has been built with IGBT switches with interface inductor($L_r$) and filter capacitor($C_f$). The switching signals for the VSI have been generated by hysteresis current controller. The set of passive filter with tuned branches of 5th and 7th harmonic frequencies, and a high pass filter at corner frequency of 11th harmonic have been fabricated as per design specifications. The parameters are given in Table II. Nonlinear load is realized by 3 phase diode rectifier. The DC bus of SAF has been maintained by separate source at 250V, and the source voltage is kept at 110V(rms).

The control algorithm has been implemented in real time on dSPACE with TMS320F240 DSP. The voltage signals are sensed by CV-3-1500 LEM voltage sensors and current signals by CT 100-s hall effect current sensors. The signals are sampled in simultaneous sampling mode by inbuilt ADCs. The loop for the real time implementation has been realized through interrupt routine synchronized with sampling signal to ADC. The sampling rate, therefore loop time has been fixed at 78.125μs to realize a constant frequency hysteresis controller at 12.8kHz. The out of phase unit template for phase 'a' has been derived from 'b' phase voltage input signals filtered by 2nd order Chebyshev filter with appropriate delay. Experimental results of the proposed scheme have been obtained for steady state to validate the simulation exercise. Fig.9a shows the compensation performance of the passive filter alone and Fig.9b shows the effectiveness of PHF for compensation of harmonics in the load current for phase 'a' for diode rectifier load. The injection transformer ratio corresponds to 1:1 with a load of 4.5A. The representation of

TABLE I
%THD AND RMS VALUE OF LOAD CURRENT, SOURCE CURRENT WITH PF AND PHF FOR OPERATION OF PHF (SIMULATION STUDY)

| Current | | RMS Current (A) | | | % THD | | |
|---|---|---|---|---|---|---|---|
| | | Ph. a | Ph. b | Ph. c | Ph. a | Ph. b | Ph. c |
| Load Current (with PF alone) | B | 8.25 | 8.25 | 8.25 | 44.79 | 44.85 | 44.86 |
| | U | 7.53 | 8.06 | 8.23 | 28.28 | 41.47 | 42.07 |
| Source Current with PF | B | 7.56 | 7.56 | 7.56 | 8.36 | 8.36 | 8.37 |
| | U | 8.34 | 6.94 | 8.83 | 32.88 | 21.5 | 21.4 |
| Passive Filter current | B | 4.89 | 4.90 | 4.90 | 92.3 | 93 | 93.06 |
| | U | 2.02 | 3.87 | 4.74 | 49.51 | 96 | 83.6 |
| Load Current with PHF | B | 8.14 | 8.15 | 8.14 | 42.37 | 43.24 | 42.27 |
| | U | 8.06 | 8.15 | 8.09 | 42.82 | 41.72 | 42.82 |
| Source Current with PHF | B | 7.90 | 7.95 | 7.97 | 2.97 | 3.34 | 3.09 |
| | U | 7.41 | 7.6 | 7.8 | 3.13 | 3.7 | 4.7 |

channels in the scope depicting experimental result is as follows: 1st channel -Phase voltage of the source, 2nd channel- load current, 3rd channel -source current and 4th channel - compensating current of the SAF of PHF. It can be seen that source current has become near sinusoid with THD of 4.0%. The harmonic spectrum of load and source currents has been given in Figs.9c-d respectively. The harmonic spectra are recorded on FLUKE 434 three-phase power quality analyzer. The results are summarized in Table III. The %THD in source current validates the effectiveness of proposed scheme conforming to IEEE-519 standard.

TABLE II

SET OF PARAMETERS FOR THE PROTOTYPE SHUNT PASSIVE FILTERS

|  |  |  |  | $R_h$ | 2.5Ω |
|---|---|---|---|---|---|
| $C_5$ | 40μF | $C_7$ | 25μF | $C_{11}$ | 40μF |
| $L_5$ | 10 mH | $L_7$ | 8.2 mH | $L_{11}$ | 2 mH |
| $R_5$ | 0.24Ω | $R_7$ | 0.2Ω | $R_{11}$ | 0.17Ω |

TABLE III

%THD AND RMS VALUE OF LOAD CURRENT , SOURCE CURRENT WITH PF AND PHF FOR OPERATION OF PHF (EXPERIMENTAL STUDY)

| Current | % THD | RMS Current (A) |
|---|---|---|
|  | Other Load | Other Load |
| Load Current | 35.1 | 4.46 |
| Source Current with PF | 19.2 | 4.58 |
| Source Current with SHF | 4.0 | 4.48 |

## VII. CONCLUSIONS

A new current control technique based based current decomposition by SPF theory with indirect current control has been proposed for PHF. The observed performance of the PHF has demonstrated the ability of the proposed control technique to compensate the current harmonics and unbalance in the passive filter branch within the power capacity of the VSI. It has been observed that the system has a fast dynamic response and is able to keep the %THD of the source current well below the limit specified by the IEEE 519 standards. The proposed configuration is effective to keep the rms value of the supply current lower than or equal to the load current. The scheme has an advantage of simplicity and flexibility of selection of components to be computed, and appropriate compensation by PHF. Under unbalanced operation of passive filters it is recommended that harmonic compensation can be effectively achieved for limited power capacity of the VSI of PHF using indirect current control, as per proposed scheme. The scheme offers befitting solution amidst uncertainties in power system due to current quality problems.

## VIII. APPENDIX

THE PARAMETERS OF THE CONSIDERED SYSTEM FOR SIMULATIONS ARE:

| Line Impedance | $L_s$=0.5 mH, $R_s$=0.1Ω |
|---|---|
| Ripple Filter | $C_r$=0.1μF, $L_r$= 5 mH |
| DC Bus Voltage | 250V |
| Max Switching Frequency | 12.8 kHz |
| Mains Voltage / phase, line frequency | 110V, 50 Hz |

## IX. REFERENCES

[1] M. Rastogi, N. Mohan, A.A. Edris, "Hybrid-Active Filtering of Harmonic Currents in Power Systems," *IEEE Trans. on Power Delivery*, vol.10, no.4, pp.1994-2000, Oct. 1995.

[2] H. Fujita, H. Akagi, "A Practical Approach to Harmonic Compensation in Power Systems-Series Connection of Passive, Active Filters," *IEEE Trans. on Ind.. Appl.*, vol.27, no.6, pp.1020-1025, Nov./Dec. 1991.

[3] S. Bhattacharya, P. T. Cheng and D. M. Divan, "Hybrid Solutions for Improving Passive Filter Performance in High Power Applications," *IEEE Trans. on Ind. Appl.*, vol. 33, no. 3, pp. 732-747, May/June 1997.

[4] H. Fujita, T. Yamasaki and H. Akagi, "A hybrid active filter for damping of harmonic resonance in industrial power systems," *IEEE Trans. on Power Electronics*, vol.15, no. 2, pp. 215-222, March 2000.

[5] D.A. Gonzalez and J.C. Mccall, "Design of Filters to Reduce Harmonic Distortion in Industrial Power Systems", *IEEE Trans. Industry Applications*, Vol. IA-23, No. 3, pp 504-511, May-Jun. 1987.

[6] J.C. Das, "Passive filters - potentialities and limitations," *IEEE Transactions on Ind. Appl.*, vol. 40, no.1, pp. 232–241, Jan.-Feb. 2004.

[7] N. Balbo, D. Sella, R. Penzo, G. Bisiach, D. Cappellieri, L. Malesani, A. Zuccato, "Hybrid Active Filter for Parallel Harmonic Compensation," in *Proc. EPE'93*, 1993, pp.133-138.

[8] P.T. Cheng, S. Bhattacharya and D. Divan, "Experimental Verification of Dominant Harmonic Active Filter for High Power Applications", *IEEE Trans. on Ind. Appl.*, vol. 36, pp.567-577, March/April, 2000.

[9] L. Gyugyi, E.C. Strycula, "Active AC Power Filters", IEEE IAS Meeting 1976, pp 529-535.

[10] F.B. Libano, J.A.Cobos, J. Uceda, "Simplified control strategy for Hybrid Active Filters", *IEEE PESC Record*, 1997, pp 1102-1108.

[11] D. Rivas, L. Moran, J. Dixon and J. Espinoza, "A simple control scheme for hybrid active power filter," *IEE Proc. GTD*, vol. 149, no. 4, pp. 485-490, July 2002.

[12] B.R. Lin, B. R. Yang and H. R. Tsai, "Analysis and operation of hybrid active filter for harmonic elimination," *Electric Power Systems Research*, vol. 62, pp. 191-200, 2002.

[13] D. Rivas, L. Moran, J. Dixon and J. Espinoza, "Improving passive filter compensation performance with active techniques," *IEEE Trans. on Industrial Electronics*, vol. 50, no. 1, pp. 161-170, Feb. 2003.

[14] B. Singh, and V. Verma, "An Indirect Current Control of Hybrid Power Filter for Varying Loads," *IEEE Trans. on Power Delivery*, vol. 21, No. 1, pp. 178-184, Jan. 2006.

[15] IEEE Guide for Application and Specification of Harmonic Filters IEEE Std 1531-2003.

[16] V, Verma, S. Shanker, B. Singh, A. Chandra, and Kamal Al- Haddad, "Genetic Algorithm Based Design of Passive Filters for Offshore Applications," *Proc. of IEEE PCIC'04*, 2004, pp 55-62.

## X. BIOGRAPHIES

**Bhim Singh** (SM'99) was born in Rahamapur, U. P., India in 1956. He received B. E. (Electrical) degree from the University of Roorkee, India in 1977 and M. Tech. and Ph. D. degrees from the Indian Institute of technology (IIT), New Delhi, in 1979 and 1983, respectively. In 1983, he joined as a Lecturer and in 1988 became a Reader in the Dept. of Elect. Engg, University of Roorkee. In December 1990, he joined as an Assistant Professor, became an Associate Professor in 1994 and full Professor in 1997 at the Dept. of Elect. Engg., IIT Delhi. His field of interest includes power electronics, electrical machines and drives, active filters, static VAR compensator. Prof. Singh is a Fellow of INAE, IE (I) and IETE, a Life Member of ISTE, SSI and NIQR and Senior Member IEEE..

**Vishal Verma** (M'04) was born in Bareilly, U.P., India in 1966. He received B. E. (Electrical) degree from the G.B. Pant University of Ag. & Tech, Pantnagar, India in 1989 and M.Tech. and Ph.D degrees from the Indian Institute of technology (IIT), New Delhi, in 1998 and 2006 respectively. His field of interest includes power electronics, drives, active filters, and power quality issues. In 1991, he joined as a Assistant Professor in the Dept of Elect. Engg., the G.B. Pant University of Ag. & Tech, Pantnagar. and in 2004 he joined Delhi College of Engineering, Delhi. Mr. Verma is a member of ISTE and Life Member CES(I).

**2006 IEEE International Conference on Power Electronic, Drives and Energy Systems**

# Active Power Filter Control in Three-Phase four-wire Systems using Space Vector Modulation

H. Mokhtari, *Member, IEEE*, and M. Rahimi

*Abstract*--In this paper, by extending Space Vector Modulation (SVM) technique to three-phase four-wire systems, a new strategy is developed for the control of Active Power Filters (APFs). It is shown that the conventional SVM method cannot compensate for the current in the neutral wire in a three-phase four wire system. Simulations have been performed using PSCAD/EMTDC software. Simulation results are provided to prove the ability of the proposed technique in compensating the zero-sequence current.

*Index Terms*--Active power Filter, Space vector modulation, Three-phase four-wire systems

## I. NOMENCLATURE

$s$ : switching vector

$s_i^+$ : switching function

$v_{in}^f$ : filter phase-i output voltage

$v_{in_{ref}}^f$ : phase-i reference output voltage

$\vec{u}^f(t)$ : filter output vector

$\vec{u}_{ref}^f(t)$ : filter reference output vector

$T_1, T_2$ : time intervals corresponding to the times for having output vector $\vec{u}_i$ and $\vec{u}_{i+1}$

$T_0, T_{00}$ : time intervals zero vectors

$u_{0_{ref}}^f$ : zero reference variable

$u_0^f$ : zero variable

$V_d$ : dc link voltage

$T_S$ : sampling period

$\vec{u}_7, \vec{u}_8$ : zero vectors in complex plane

$v_{if}^0, i_{if}^0$ : phase-i filter zero-sequence voltage and current

$v_{if}^{1,2}, i_{if}^{12}$ : phase-i filter non-zero sequence voltage and current

$p_{af}^0$ : instantaneous zero-sequence power

$i_d^0$ : dc bus current due to ac zero-sequence current

$i_{fn}$ : zero-sequence filter current

$i_{sn}$ : zero-sequence source current

$i_{d\_a}^{1,2}$ : dc bus current due to phase-a non-zero sequence current

$v_{c1}^{1,2}$ : dc-link voltage viewed from the non-zero sequence currents on the ac side of the filter

$v_{c1}, v_{c2}$ : dc voltage of spilt capacitors

$\hat{V}_s$ : peak voltage source

$L_s$ : source inductance

$L_f$ : filter inductance

## II. INTRODUCTION

Pulse Width Modulation (PWM) techniques have been employed in APFs for cancellation of harmonics generated by nonlinear loads. Most of the previous works have been focused on carrier based PWM techniques mainly triangular ones [1]-[3]. With the development of high speed microprocessors, Space Vector Modulation (SVM) has become one of the most important PWM methods for three-phase converters [4]-[11]. Studies have shown that SVM-based PWM methods have several advantages over carrier-based ones such as lower Total Harmonic Distortion (THD), higher efficiency, easier digital implementation and wider linear modulation range [12], [13]. It is shown that in the linear range of modulation, the higher the modulation index, the lower the inverter dc link voltage in the APF.

Most of the previous works deal with the application of APFs in three-phase three-wire systems [12], [14-15]. In many cases, electricity is supplied through three-phase four wire systems in distribution networks. Therefore, application of APFs in such systems is attracting interest among researchers. However, there is not much in the literature discussing the problem of neutral wire current compensation in three-phase four-wire SVM-based PWM modulator systems. This paper, while presents a new control strategy for an SVM-based APF, thoroughly discusses the issue of neutral current compensation. After a brief review of SVM principles, the paper extends the same idea to a four-wire system. The main problem in doing so is to accurately

---

H. Mokhtari is Associate Professor of Sharif University of Technology Tehran, Iran. (e-mail: mokhtari@sharif.ir).

Mohsen Rahimi is a Ph.D. Student at Sharif University of Technology, Tehran, Iran.

---

0-7803-9771-1/06/$25.00 ©2006 IEEE

compensate for the neutral wire current as well as for the phase current harmonics. To do so, it first addresses the issue of modeling technique in three-phase four-wire systems. New equations are derived for the calculation of the duty cycle of semiconductor switches to compensate for the neutral wire. To verify the proposed concept, several simulations are carried out, and the results are provided. Also, it is shown that the conventional SVM methods cannot compensate for the neutral wire in three-phase four-wire systems but the suggested system can do it.

## III. SVM-PWM Technique

Fig.1 depicts a three-phase Voltage Source Inverter (VSI) which can be used in an APF system. Knowing that (1) the two switches in each leg of the inverter cannot be gated at the same time, and (2) at least one switch must be "on" due to the system inductances, the switching functions can be defined as follows:

$$s_i^+ = \begin{cases} 1 & T_i^+ \ on \\ 0 & T_i^+ \ off \end{cases} \qquad i = a,b,c \qquad (1)$$

Therefore, filter output voltages are:

$$v_{in}^f = (s_i^+ - \frac{1}{2})V_d \qquad i = a,b,c \qquad (2)$$

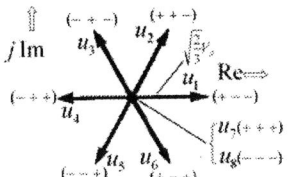

Fig. 1. A three-phase VSI.

If the switching vector $s$ is defined as $s = (s_a^+ \ s_b^+ \ s_c^+)$, there will be eight switching combinations starting from (0,0,0) to (1,1,1). The space vectors corresponding to instantaneous phase voltages can be calculated as [7]:

$$\vec{u}(t) = \sqrt{\frac{2}{3}}(v_{an} + \lambda v_{bn} + \lambda^2 v_{cn}) \quad , \quad \lambda = e^{j\frac{2\pi}{3}} \qquad (3)$$

Substituting from (2) into (3), yields the filter output voltage vector as:

$$\vec{u}^f(t) = \sqrt{\frac{2}{3}}(s_a^+ + \lambda s_b^+ + \lambda^2 s_c^+)V_d \qquad (4)$$

Each state $s$ of the switching vector corresponds to the vector $\vec{u}^f$. Fig. 2 depicts vectors u1 to u8 corresponding to

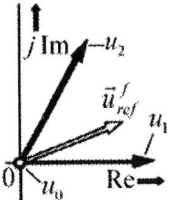

Fig. 2. SVM switching vectors.

states (0,0,0) to (1,1,1). Vectors u7 and u8 correspond to states (1,1,1) and (0,0,0) and are called zero vectors, i.e. $\vec{u}_0$.

Similarly, filter reference voltages can also be determined using a vector with an amplitude and an angular frequency of:

$$\vec{u}_{ref}^f(t) = \sqrt{\frac{2}{3}}(v_{an_{ref}}^f(t) + \lambda v_{bn_{ref}}^f(t) + \lambda^2 v_{cn_{ref}}^f(t)) \qquad (5)$$

where $v_{an_{ref}}^f$, $v_{bn_{ref}}^f$ and $v_{cn_{ref}}^f$ are the reference phase voltages.

The reference voltages are obtained from active filter current controller block. In fact, the reference voltages are the output current regulators that are fed to the PWM block for switch gating generation. The control scheme of the active filter is explained in [4]. SVM-PWM is a method with fixed switching frequency. With a suitable linear combination of the vectors corresponding to filter output voltages, the desired output can be obtained. Fig. 3 shows the case in which the reference voltage vector $\vec{u}_{ref}^f$ is in the first 60 degrees of the complex plane.

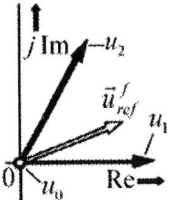

Fig. 3. Example of switching vectors – first 60 degree region.

Suppose that $\vec{u}_{ref}^f$ is between $\vec{u}_i$ and $\vec{u}_{i+1}$. Time durations $T_1$ and $T_2$ to output $\vec{u}_i$ and $\vec{u}_{i+1}$ within one sampling period are:

$$T_1 = \frac{\sqrt{2}|u_{ref}^f|}{V_d}T_s Sin(\frac{\pi}{3} - \theta)$$

(6)

$$T_2 = \frac{\sqrt{2}|u_{ref}^f|}{V_d}T_s Sin(\theta)$$

(7)

Where $\theta$ is the angle between $\vec{u}_{ref}^f$ and $\vec{u}_i$, $T_0$ is the time for applying a zero vector (u7 or u8) and $T_0 = T_S - T_1 - T_2$, and $T_S$ is the sampling period.

This method cannot be applied to a three-phase four-wire system. This is due to the fact that the zero sequence of reference voltages is not included in the reference vector $\vec{u}_{ref}^f$.

## IV. Extension of SVM to a Three-Phase Four-Wire System

As mentioned earlier, one of the important problems in a three-phase four-wire system is the compensation of neutral wire current. Since, the current in the neutral wire results in zero sequence voltages, therefore suitable zero sequence references must be defined. To do so, new zero vectors $\vec{u}_7$

and $\vec{u}_8$ are considered in defining the reference voltage based on a suitable strategy. In order to compensate for the current in the fourth wire, zero reference variable $u_{0_{ref}}^f$ is defined as:

$$u_{0ref}^f(t) = v_{an_{ref}}^f(t) + v_{bn_{ref}}^f(t) + v_{cn_{ref}}^f(t) \qquad (8)$$

As explained in previous section, $\vec{u}_i$ and $\vec{u}_{i+1}$ (the vectors close to $\vec{u}_{ref}^f$) are applied for durations of $T_1$ and $T_2$ such that the average voltage over a sampling period can follow $\vec{u}_{ref}^f$. To compensate for $u_{0_{ref}}^f$, $\vec{u}_7$ or $\vec{u}_8$ is applied for a duration of $T_0$ depending on the polarity of $u_{0_{ref}}^f$. Therefore, a zero variable $u_0^f$ is defined as follows:

$$u_0^f(t) = v_{an}^f(t) + v_{bn}^f(t) + v_{cn}^f(t) \qquad (9)$$

Substituting (2) into (9) yields:

$$u_0^f = (s_a^+ + s_b^+ + s_c^+ - \frac{3}{2})V_d \qquad (10)$$

or:

$$u_0^f = \begin{cases} +\frac{1}{2}V_d & if \quad \vec{u}^f = \vec{u}_2, u_4 \ or \ \vec{u}_6 \\ -\frac{1}{2}V_d & if \quad \vec{u}^f = \vec{u}_1, u_3 \ or \ \vec{u}_5 \\ \pm\frac{3}{2}V_d \begin{cases} + \ if \ \vec{u}_0 = \vec{u}_7 \\ - \ if \ \vec{u}_0 = \vec{u}_8 \end{cases} \end{cases} \qquad (11)$$

To compensate for the zero sequence of the load current, the average of $u_0^f$ over one sampling period must be equal to $u_{0_{ref}}^f$, therefore;

$$T_1(\pm\frac{1}{2}V_d) + T_2(\mp\frac{1}{2}V_d) + T_0(\pm\frac{3}{2}V_d) = T_s u_{0_{ref}}^f \qquad (12)$$

For example, if $\vec{u}_{ref}^f$ is close to $\vec{u}_1$ and $\vec{u}_2$ and $u_{0_{ref}}^f > 0$, (12) changes to:

$$T_1(-\frac{1}{2}V_d) + T_2(+\frac{1}{2}V_d) + T_0(+\frac{3}{2}V_d) = T_s u_{0_{ref}}^f \qquad (13)$$

Parameter $T_{00}$ is also defined as $T_{00} = T_s - (T_1 + T_2 + T_0)$. During $T_{00}$ vector $\vec{u}_7$ or $\vec{u}_8$ is generated in such a way that the number of switching is minimized. $T_1$ and $T_2$ in a three-phase four-wire system is different from those obtained for a three-phase three-wire system. Consider the equivalent circuit of one phase of a three-phase four-wire system shown in Fig.4. If the average values over one switching cycle for the zero sequence of voltage and current are $v_{af}^0$ and $i_{af}^0$, and the average values for non-zero sequence components of phase-a are $v_{af}^{1,2}$ and $i_{af}^{1,2}$, the instantaneous power delivered by the zero sequence components of current in phase-a of the ac system is:

$$p_{af}^0 = v_{af}(t)i_{af}^0(t) \qquad (14)$$

Since $i_d^0 = i_{fn} = 3i_{af}^0$, the effect of zero sequence current on dc link voltage can be found from the following power balance equation.

$$v_{c1}^0 i_d^0 = p_{af}^0 + p_{bf}^0 + p_{cf}^0 \qquad (15)$$

Considering:

$$v_{if} = v_{if}^{1,2} + v_{if}^0 \qquad i = a,b,c \qquad (16)$$

$$v_{af}^{1,2} + v_{bf}^{1,2} + v_{cf}^{1,2} = 0 \qquad (17)$$

then; $\quad v_{c1}^0 = v_{af}^0,$

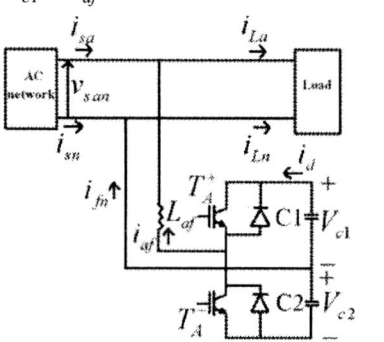

Fig. 4. Equivalent circuit for phase-a.

This means that on the ac side of the filter and from the zero sequence point of view, the average dc-link voltage is $v_{af}^0$. Since there are two separate algorithms for compensating the zero sequence and non-zero sequence of load currents, one can conclude that all sequences are compensated on the ac side. Fig. 5 depicts filter equivalent circuits on the ac and dc sides for the zero sequence and non-zero sequence components of phase-a current.

It can be seen that there is a coupling between the two circuits. $i_{d\_a}^{1,2}$ is the contribution of phase-a non-zero sequence current on the dc side, and $v_{c1}^{1,2}$ is the dc-link voltage viewed from the non-zero sequence currents on the ac side of the filter. By writing the power balance equation between the dc and ac sides as:

$$v_{c1}i_d^{1,2} + v_{c2}i_d^{1,2} + 3v_{c1}i_{af}^0 = \sum_{k=a,b,c} v_{kf} i_{kf} \qquad (18)$$

and knowing that:

$$\sum_{j=a,b,c} v_{jf} i_{jf} = \sum_{k=a,b,c} (v_{kf}^{1,2} i_{kf}^{1,2} + v_{kf}^0 i_{kf}^0) \qquad (19)$$

from Fig. 5 the power balance equation between the dc and ac sides can be written as:

$$v_{af}^0(2i_d^{1,2} + 3i_{af}^0) + v_{c1}^{1,2}(3i_{af}^0 + 2i_d^{1,2}) = \sum_{k=a,b,c}(v_{kf}^{1,2} \cdot i_{kf}^{1,2} + v_{kf}^0 \cdot i_{kf}^0) \qquad (20)$$

From Equations (18) and (20) and knowing that $v_{c1} = v_{c1}^0 + v_{c1}^{1,2}$, $v_{c1}^0 = v_{af}^0$ and $v_{c1} = v_{c2}$, (21) is concluded.

$$v_{c1}^{1,2} = v_{c1} - v_{af}^0 \qquad (21)$$

Equation (21) means that the average dc-link voltage viewed from the non-zero sequence of the ac side of the filter is $v_{c1} - v_{af}^0$ over one switching cycle. In fact, this is equivalent to replacing $v_{c1}$ with $v_{c1} - v_{af}^0$ or $v_d$ with $v_d - 2v_{af}^0$ in the three-phase three-wire system of Fig 1. Therefore, Equation (4) changes to (22) in a four-wire system.

$$\vec{u}^f(t) = \sqrt{\frac{2}{3}}(s_a^+ + \lambda s_b^+ + \lambda^2 s_c^+)(V_d - 2v_{af}^0) \qquad (22)$$

Similarly, time intervals $T_1$ and $T_2$ are:

$$T_1 = \frac{\sqrt{2}|u_{ref}^f|}{V_d - 2v_{af0}} T_s Sin(\frac{\pi}{3} - \theta) \qquad (23)$$

$$T_2 = \frac{\sqrt{2}|u_{ref}^f|}{V_d - 2v_{af0}} T_s Sin(\theta) \qquad (24)$$

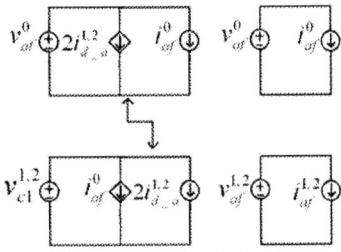

Fig. 5. Filter ac and dc side equivalent circuits for phase-a.

## V. SVM MODELING IN AN APF SYSTEM IN A FOUR-WIRE SYSTEM

Fig. 6 depicts the power circuit of an APF in a three-phase four-wire system. AC link inductances shown in Fig. 6 prevent high frequency switching harmonics from entering the ac system and thus reduces current ripple.

SVM modeling in a three-phase four-wire system can be done in two steps. In the first step, harmonic components and the reactive component of load current are compensated. This step consists of:

    1. determining reference space vector $\vec{u}_{ref}^f$

    2. finding space position of $\vec{u}_{ref}^f$ in the complex plane

    3. calculating $T_1$ and $T_2$

    4. selecting switching states.

In the second step, the zero-sequence component is compensated. This step consists of:

    1. determining zero reference variable $u_{ref}^0$

    2. determining the sign of $u_{ref}^0$

    3. finding $T_0$

    4. selecting switching states.

To select switching states for each upper switch (Ta+,Tb+,Tc+) a state table is formed. With the instantaneous values found during stages 1 to 3, "on" or "off" state of the switch is selected. States of the lower switches are opposite to those of the upper ones. Table 1 shows the states of Ta+.

Index "n" determines the 60-degrees region in which $\vec{u}_{ref}^f$ is located.

Index "i" corresponds to time periods of $T_1$ (i=1), $T_2$ (i=2), $T_0$ (i=3,4), and $T_{00}$ (i=5) where i=3 corresponds to the

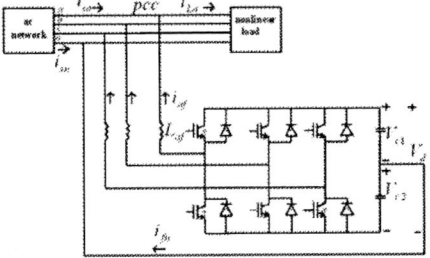

Fig. 6. Power circuit of a three-phase four-wire APF.

switching vector S=(0,0,0) and i=4 corresponds to the switching vector S=(1,1,1).

TABLE I
STATES FOR SWITCH TA+

| n \ i | 1 | 2 | 3 | 4 | 5 |
|---|---|---|---|---|---|
| 1 | 1 | 1 | 0 | 1 | 1 |
| 2 | 1 | 0 | 0 | 1 | 0 |
| 3 | 0 | 0 | 0 | 1 | 0 |
| 4 | 0 | 0 | 0 | 1 | 0 |
| 5 | 0 | 1 | 0 | 1 | 1 |
| 6 | 1 | 1 | 0 | 1 | 1 |

## VI. SIMULATION RESULTS

This section presents the simulation results to verify the new control strategy introduced in this paper. System parameters (source impedance, load and active filters parameters) are taken from [17] in which the detail design procedure is given. The load of each phase is composed of parallel combination of a single-phase full-wave rectifier and an inductive load. The load current in phase-a is:

$$i_{la} = 39.8 Sin(\omega t - 24°) + \sum_{n=3,5,...} \frac{36.4}{n} Sin(n\omega t) \qquad (25)$$

System parameters ($\hat{V}_s$ peak voltage source, $L_s$ source inductance and $L_f$ filter inductance) are:

$$\hat{V}_s = 162.5 \, V, \ L_s = 0.1 \, mH, \ L_f = 1.5 \, mH$$

Fig. 7 shows the results for phase-a when the SVM method described in Section 2 is used. As it can be seen, the filter is not able to compensate for the zero-sequence components.

Fig. 8 shows the variations of dc link capacitor voltages ($\Delta Vc1$, $\Delta Vc2$, $\Delta Vd$) with respect to the reference dc bus voltages ($V_{C1}^{ref} = 170V$, $V_{C2}^{ref} = 170V$, $V_d^{ref} = 340V$) when the conventional SVM method is used. As it can be seen, large voltage variations have resulted in dc-bus over voltages. The results for phase-a when the new SVM method developed for a three-phase four-wire system is employed are shown in Fig. 9. This figure shows that the new strategy is able to fully compensate for the harmonics, reactive power as well as the zero-sequence components. Fig. 10 shows the variations of dc-link capacitor voltages with respect to the reference dc-bus voltages when the new SVM method developed for a three-phase four-wire system is employed. This figure indicates that the resulted over/under voltages on the dc-bus are not large.

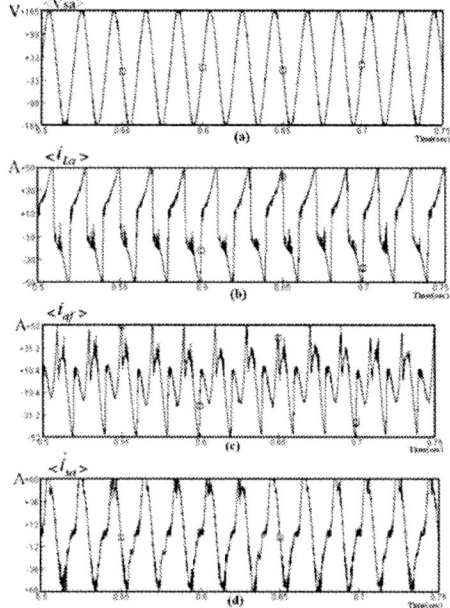

Fig. 7. Conventional SVM method (a) source voltage (b) load current (c)filter current (d)source current.

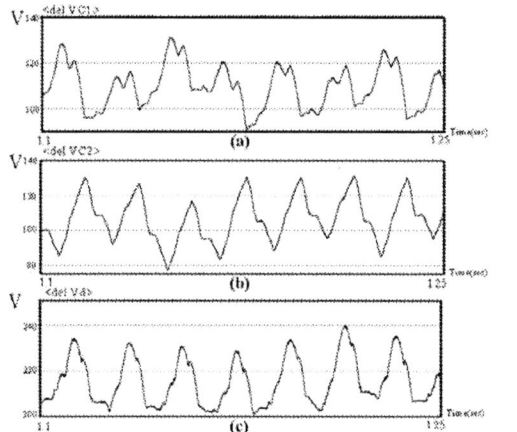

Fig. 8. Conventional SVM method: variations of dc link capacitor voltages with respect to the reference dc-bus voltages (a) $\Delta$ Vc1 (b) $\Delta$ Vc2 (c) $\Delta$ Vd.

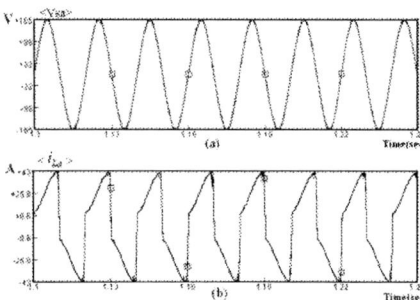

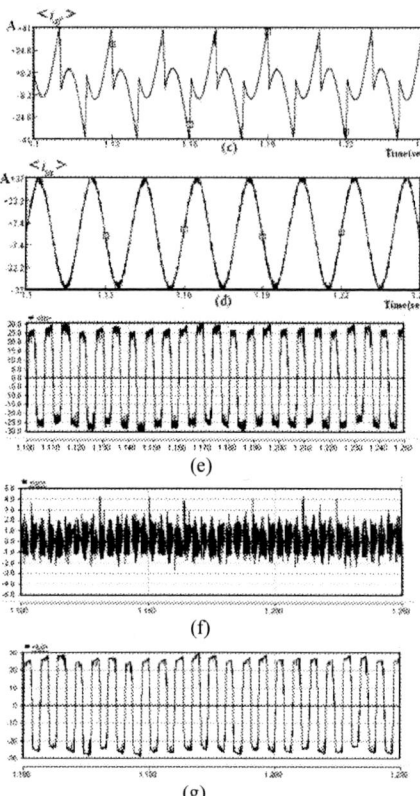

Fig. 9. New SVM method (a) source voltage (b)load current (c) filter current (d) source current (e) zero sequence filter current (f) zero sequence source current (g) load neutral wire current.

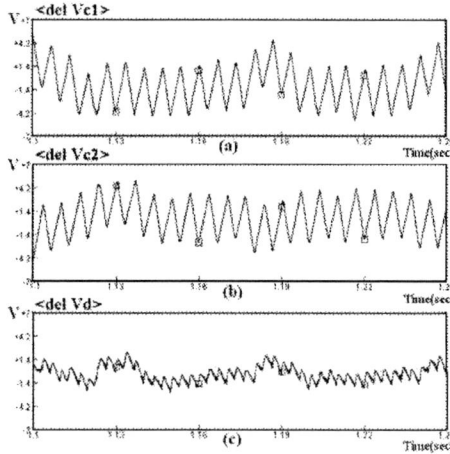

Fig. 10. New SVM method: variations of the dc-link capacitor voltages with respect to the reference dc-bus voltages (a) $\Delta$ Vc1 (b) $\Delta$ Vc2 (c) $\Delta$ Vd.

## VII. CONCLUSIONS

In this paper, conventional SVM technique is modified to be used in a three-phase four-wire system. The application of the method is investigated in an APF. It is shown that the conventional SVM method cannot compensate for the zero-sequence components of load current. This is due to the fact that the effects of zero sequence signals are not considered in defining the reference signals. However, the new strategy can compensate not only for line harmonics but also for the neutral wire. In the proposed method, a new reference variable is defined for the zero-sequence voltages, and selection of zero-voltage vectors $\vec{u}_7/\vec{u}_8$ is based on the polarity of the new zero reference variable. Simulation results for different operating conditions verify the method effectiveness.

## VIII. References

[1] Singh, B. and et. al. ; "Design and implementation of a new control algorithm for a three-phase active filter," *Applied Power Electronics Conference and Exposition, 1997. APEC '97 Conference Proceedings 1997, Twelfth Annual, Vol. 1, 23-27 Feb. 1997.*

[2] Singh, Aredes, M.; Hafner, J.; Heumann, K.; "Three-phase four wire shunt active filter control strategies," *IEEE Trans. on Power Electronic, Vol. 12, No. 2, March 1997.*

[3] Kuo, H.-H.; Yeh, S.-N.; Hwang, J.-C.; "Novel analytical model for design and implementation of three-phase active power filter controller," *Electric Power Applications, IEE Proceedings- , Vol. 148 No. 4, July 2001.*

[4] H. W. V. Brocker, H. C. Skudenly,.and G. Stanke, "Analysis and realization of a pulse width modulation based on the voltage space vectors," *in Conf. Rec. IEEE-IAS Annu. Meeting, Denver, CO, 1986, pp. 244-251.*

[5] O. Ogasawara, H. Akagi, and A. Nabel, "A novel PWM scheme of voltage source inverters based on space vector theory," *in Proc. EPE European Conf. Power Electronics and Applications, 1989, pp. 1197-1202.*

[6] M. Depenbrok, "Pulse width control of a 3-phase inverter with no sinusoidal phase voltages," *in Proc. IEEE-IAS Int. Semiconductor Power Conversion Conf., Orlando, FL, 1975, pp. 389-398.*

[7] J. Holtz, "Pulse width modulation – A survey," *in Proc. IEEE PESC'92, 1992, pp. 11-18.*

[8] J. Holtz, W. Lotzkat, and A. Khambadkone, "On continuous control of PWM inverters in the over modulation range including the six-step mode," *in Proc. IEEE IECON'92, 1992, pp. 307-312.*

[9] B. K. Bose and H. A. Sutherland, "A high performance pulse width modulator for an inverter-fed drive system using a microcomputer, " *IEEE Trans. Ind. Applicat., Vol. 19, pp. 235-243, Mar./Apr. 1983.*

[10] A. M. Hava, R. Kerkman, and T. A. Lipo, "Carrier-based PWM-VSI over modulation strategies: Analysis, comparison, and design, " *IEEE Trans. Power Electron. Vol. 13, pp. 674-689, July 1998.*

[11] D. W. Chung, J. S. Kim, and S.K. Sul, "Unified voltage modulation technique for real-time three-phase power conversion," *IEEE Trans. Ind. Applicat., Vol. 34, pp. 374-380, Mar./Apr. 1998.*

[12] Espinoza, J.R.; joos, G.; Jin, H. ; "Modeling and implementation of space vector PWM techniques in active filter applications," *Computers in Power Electronics, 1996., IEEE Workshop on , 11-14 Aug. 1996.*

[13] K. Zhou and D. Wang, "Relationship between Space-Vector modulation and Three-Phase carrier-based PWM: a comprehensive analysis," *IEEE Trans. Ind. Electron. Vol. 49, No. 1, Feb. 2002.*

[14] Nava-Segura, A.; Linares-Flores, J.; "Transient analysis of a vector controlled active filte," *Industry Applications Conference, 2000. Conference Record of the 2000 IEEE, Vol. 4, 8-12 Oct. 2000.*

[15] Moon, G.-W.; "Predictive current control of distribution static compensator for reactive power compensation," *Transmission and Distribution, IEE Proceedings- , Vol. 146. No. 5, Sept. 1999.*

[16] Holtz, J.; "Pulse width modulation for electronic power conversion," *Proceedings of the IEEE, Vol. 82, No. 8, Aug. 1994.*

[17] M. Rahimi ; "Study the Parameters Affecting the Design of a Three Phase Active Filter and Proposing an Optimized Design in order to Reduce Filter Size," M.Sc. Thesis, Sharif University of Technology, Iran, December 2003.

## IX. Biographies

**Hossein Mokhtari** (M'98) was born in Tehran, Iran. He received the B.Sc. degree in electrical engineering from Tehran University, Tehran, Iran, in 1989, the M.A.Sc. degree from the University of New Brunswick, Fredericton, NB, Canada, in 1994, and the Ph.D. degree in electrical engineering from the University of Toronto, Toronto, ON, Canada, in 1998. Currently, he is an Associate Professor at Sharif University of Technology, Tehran, Iran. He was a Consultant Engineer in dispatching projects with Electric Power Research Center (EPRC), Tehran.

**Mohsen Rahimi** was born in Isfahan, Iran. He received his B.Sc. degree in electrical engineering from Isfahan University of Technology, Isfahan, Iran, in 2001, and his M.Sc. degree from Sharif University of Technology, Tehran, Iran, in 2003. Currently, he is a Ph.D. Student at Sharif University of Technology, Tehran, Iran.

**2006 IEEE International Conference on Power Electronic, Drives and Energy Systems**

# Operation of a 12-pulse converter in closed loop for controlled P-Q operation

Faisal M. Ahsan, J.K.Chatterjee, *Member IEEE*, and Anandarup Das[1]

*Abstract--* **In this paper, a new four quadrant active-reactive power controller is presented. The proposed system consists of a 12-pulse converter operating under asymmetrical firing angle operation, integrated to a combination of tuned passive filters. The study shows that by suitable choice of firing angles of the constituent 6-pulse converters, the active and reactive power exchanged by the 12-pulse converter can be independently controlled. The 12-pulse converter always absorbs reactive power from the supply while it can either absorb from or supply active power to the supply. On the other hand, under fundamental frequency of operation, the passive filters act as a source of reactive power. The combined system of 12-pulse converter and passive filter thus acts as a four quadrant P-Q controller connected at the point of common coupling between the supply and the load. Simulated results followed by experimental verifications show that such a system can always make the supply operate at unity power factor besides maintaining system voltage and frequency constant.**

*Index Terms--* **active and reactive power control, asymmetrical firing, four quadrant P-Q controller, 12-pulse converter.**

## I. INTRODUCTION

IN recent years, with the increased use of power semiconductor switches, power quality related problems like poor power factor, excessive harmonics injected into the supply, distorted supply voltage (sag and swell) and current profile etc. have assumed serious proportion. Many sensitive equipments and devices, adjustable speed drives are prone to these disturbances in the supply system and may suffer from malfunction and failure.

Among several other methods for power quality improvement is asymmetrical firing control of a 12-pulse converter system. Although the information regarding basic operation of a 12-pulse converter in asymmetrical mode is available in the literature [1]-[3], however, detailed investigation of such a system in terms of controlled P-Q operation has not been carried out so far. In the present work,

the 12-pulse converter is made to operate as a four quadrant P-Q controller designed to maintain 1) the supply power factor unity, thereby making voltage at point of common coupling (PCC) constant in terms of magnitude and phase; 2) active power input constant, thereby making the system immune to frequency perturbation; and 3) input current harmonic distortion at the source within the acceptable limit as prescribed in IEEE 519 standard.

In this arrangement, the setup consists of a 3-phase balanced supply with finite source impedance feeding a linear load, shown in Fig. 1. The 12-pulse converter feeding a constant current load and integrated with a combination of passive filters is connected at the point of common coupling (PCC) between the source and the load. The 12-pulse converter can draw reactive power while it has the capability to absorb or supply active power, both of which can be independently controlled. The passive filters can supply only reactive power while drawing negligible amount of active power. When such a system is connected at the PCC between the source and load, then it can supply the entire reactive power demanded by the load; thus making the source to operate at unity power factor. On the other hand, whenever the active power demand at the load side changes suddenly causing a possible change in frequency in the system, the 12-pulse converter can supply or absorb the excess active power thereby making the system immune to frequency perturbation. In addition to this, the passive filters maintain the harmonics injected into the source within acceptable limits. Thus, the entire system can act as a four quadrant p-q controller.

---

Faisal M. Ahsan is with the Department of Electrical Engineering, Indian Institute of Technology Delhi, New Delhi, 110016 India (e-mail: faisalmahsan@gmail.com).

J.K. Chatterjee is a Professor in the Department of Electrical Engineering, Indian Institute of Technology Delhi, New Delhi, 110016 India (e-mail: jkc@iitd.ac.in. ).

Anandarup Das is with the Department of Electrical Engineering, Indian Institute of Technology Delhi, New Delhi, 110016 India (e-mail: d_anandarup@rediffmail.com ).

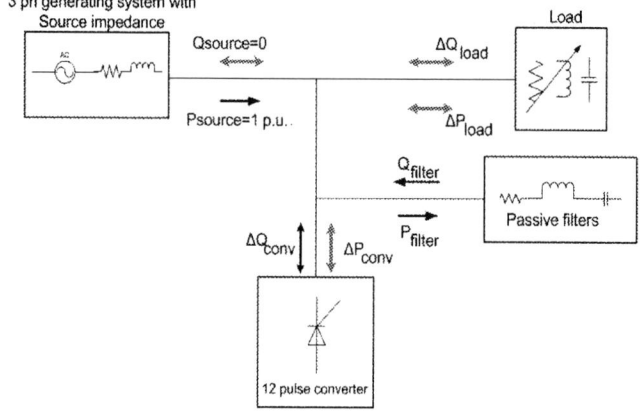

Fig. 1. 12-pulse converter and passive filter combination connected at PCC.

0-7803-9771-1/06/$25.00 ©2006 IEEE

## II. THEORY

A 12-pulse converter consists of two 6-pulse converters connected in cascade as shown in Fig. 2. The asymmetrical firing pulse operation is achieved by firing two converters independently, with one converter synchronized with the star connected and the other with the delta connected supply.

A theory for controlled active and reactive power operation of an asymmetrically triggered 12-pulse converter is given below. The essential aspect of such analysis is to achieve constant current condition on the converter DC bus, which is shown in Fig. 2.

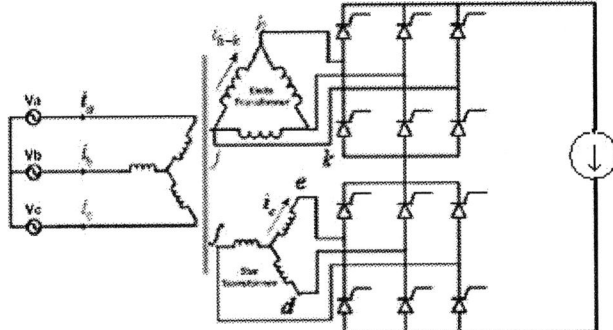

Fig. 2. Cascade connection of a 12-pulse converter.

From the theory of 12-pulse converters, in absence of source inductance, the DC bus voltage contribution due to each constituent 6-pulse converter is given by,

$$V_{di} = \frac{3V_{max}}{\pi} \cos(\alpha_i) \qquad (1)$$

where, "$V_{max}$" is the peak line voltage of the input supply and "$\alpha_i$" is the firing angle of the constituent converter (i =1, 2).

Multiplying DC load current "$I_d$" on both sides of (1) leads to

$$P_{di} = P_{max} \cos(\alpha_i) \qquad (2)$$

$$P_{max} = \frac{3V_{max}I_d}{\pi} \qquad (3)$$

where, "$P_{max}$" is the active power transfer through each converter. In a similar way, the reactive power drawn by the converters is given by,

$$Q_{di} = P_{max} \sin(\alpha_i) \qquad (4)$$

The active power "$P_d$" and reactive power "$Q_d$", drawn by the resultant 12-pulse converter is the sum of the contribution made by each converter and can be expressed as

$$P_d = P_{max} (\cos \alpha_1 + \cos \alpha_2) \qquad (5)$$

$$Q_d = P_{max} (\sin \alpha_1 + \sin \alpha_2) \qquad (6)$$

The power locus thus obtained in the above manner can be plotted in a P-Q diagram as shown in Fig. 3. The inner semicircle with centre $O$ and radius $P_{max}$ is the P-Q locus of one six-pulse converter. The outer semicircle with centre $O$ and radius $2P_{max}$ is the P-Q locus of the 12-pulse converter, obtained by varying the firing angles "$\alpha_1$" and "$\alpha_2$" from $0^0$ to $180^0$ under the condition $\alpha_1 = \alpha_2$.

For constant var operation of the 12-pulse converter, a straight line parallel to the x-axis can be drawn passing through points BCDE in Fig. 3. This operation can be achieved by a number of combinations of firing angle of the constituent converters which are shown in Fig. 3. Similarly, different combination of firing angles can produce constant active power operation of the converter. The area of operation is limited between the outer semicircle and the inner semicircles centered at ($+P_{max}$, 0) and ($-P_{max}$, 0).

It is to be noted that corresponding to a particular point in the P-Q plane, there exists a unique combination of firing angles "$\alpha_1$" and "$\alpha_2$". This is illustrated in Fig. 4. Point S in Fig. 4 corresponds to the VA at which the 12-pulse converter is to be operated as decided by the system requirement. The phasor OS makes an angle "$\theta$" with P-axis. From the isosceles triangle OAS and assuming the 12-pulse converter rating to be 1 p.u., it can be concluded from simple geometry that,

$$\alpha_1 = \theta - \cos^{-1}(\overline{OS}) \qquad (7)$$

$$\alpha_2 = 2\theta - \alpha_1 \qquad (8)$$

where, $\overline{OS}$ is the length of phasor OS.

If the point S is in the second quadrant, then (7) gets modified to

$$\alpha_1 = \theta + \cos^{-1}(\overline{OS}) \qquad (9)$$

while (8) remains the same. Thus, by knowing the point S in the P-Q diagram, the firing angle of the two converters can be readily determined. For example, corresponding to point S in the Fig. 4, P=0.1p.u. and Q=0.6p.u. From (7) and (8), the values of "$\alpha_1$" and "$\alpha_2$" are found to be $27.97^0$ and $133.07^0$ respectively. In this way, it is possible to operate this converter under constant P or constant Q mode.

## III. SIMULATION RESULTS

In this section, computer simulated results of the proposed system with MATLAB/SIMULINK is presented. The 12-pulse converter, passive filter and the load are connected to a 3-phase 120V, 50Hz balanced utility grid as shown in Fig. 5. The aim of this simulation is to show that under changing power requirement of the load, the active power drawn from the grid is always held at 810W (1p.u.), while the reactive power drawn is zero. Active power and lagging reactive power absorbed by a component is taken to be negative, while power supplied by the component is taken as positive.

The active and reactive power exchanged by the load and the passive filters are calculated from the instantaneous p-q theory proposed in [4] and is converted to per unit. Since the aim is to reduce the reactive power input from the source to zero, hence the reference reactive power "$q_{ref}$" for the 12-pulse converter is calculated as,

$$q_{ref} = q_{filter} + q_{load} \qquad (10)$$

where, "$q_{filter}$" and "$q_{load}$" are the reactive powers consumed by the filter and load respectively. In a similar way, the reference active power "$p_{ref}$" for the 12-pulse converter is calculated as,

$$p_{ref} = 1 - p_{filter} - p_{load} \qquad (11)$$

where, "$p_{filter}$" and "$p_{load}$" are the active power absorbed by filter and load respectively.

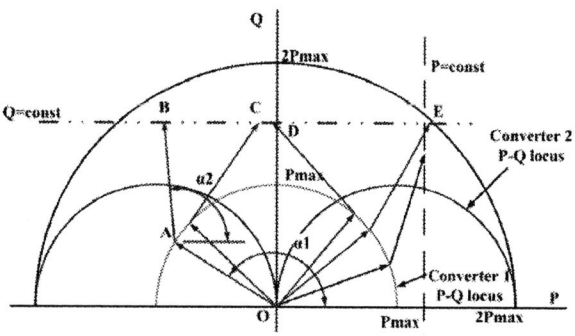

Fig. 3. Power locus of a 12-pulse converter for controlled P-Q operation.

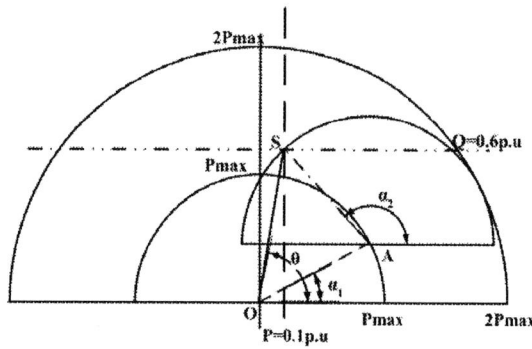

Fig. 4. Calculation of firing angles for a particular P and Q.

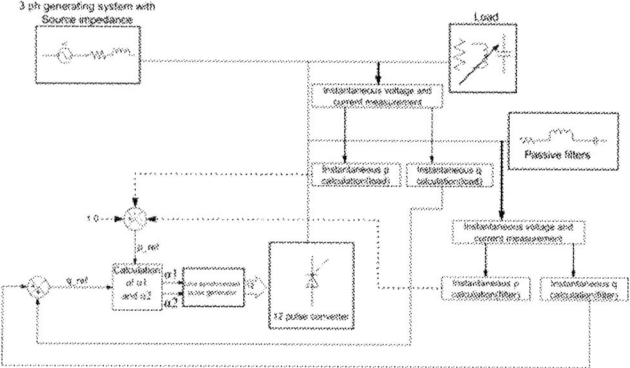

Fig. 5. Closed loop controlled P-Q operation of the proposed system.

The active power delivered from the source is 1 p.u. The reference values of active and reactive power for the 12-pulse converter are filtered and used to calculate "$\alpha_1$" and "$\alpha_2$" as per (7), (8), and (9).

TABLE I
PASSIVE FILTER COMPONENTS USED IN SIMULATION

| Passive filter | Inductance | Capacitance | Resistance |
|---|---|---|---|
| 5th | 10.13mH | 40μF | 0.39ohm |
| 7th | 10.33mH | 20μF | 0.56ohm |
| 11th | 2.093mH | 40μF | 0.18ohm |
| 13th | 2.950mH | 20μF | 0.01ohm |

The design of passive filter is well documented and an established procedure [5], [6]. The passive filter has to absorb the (6n±1) order current harmonics generated due to asymmetrical firing of the 12 pulse converter. Moreover, the passive filters supply reactive power to PCC at fundamental frequency. Single tuned series RLC passive filters for 5th, 7th, 11th, and 13th harmonics are designed and their values tabulated in Table I. Parallel resonance with the source inductance, which is kept at 5% of the base impedance, is eliminated by selecting these values of filter components. The presence of filter capacitors and the source inductance of the star-delta transformer prevent notches from appearing in the voltage profile at PCC. The total reactive power supplied by the filter capacitors at rated voltage is 0.68p.u.

In order to show the effectiveness of the proposed converter as a four quadrant P-Q controller load perturbations comprising of both active and reactive powers were applied.

Table II shows the different loads connected to the system at different instants. Table III shows the converter "$q_{ref}$" and "$p_{ref}$" calculated from (10) and (11) respectively; the corresponding values of "$\alpha_1$" and "$\alpha_2$" are calculated from (7) and (8). It also shows the quadrant of operation of the combined system of 12- pulse converter and passive filters.

TABLE II
LOAD CONNECTED AT DIFFERENT INSTANTS

| t (secs) | Load | |
|---|---|---|
| | P (p.u) | Q (p.u.) |
| 0 | 0.74 | 0.12 lead) |
| 0.25 | 1.17 | 0.12 (lead) |
| 0.50 | 1.23 | 0.12 (lag) |
| 0.75 | 0.92 | 0.37 (lag) |

At t=0s, the 12-pulse converter absorbs the reactive power supplied by the passive filters and the load and also active power to maintain the active power supplied by the source equal to 1 p.u. and reactive power supplied by the source equal to zero suggesting unity power factor operation. Since the combined system of the 12-pulse converter and the passive filters absorbs both active and reactive power therefore the

604

operation is in the 1ˢᵗ quadrant of the P-Q plane.

At t=0.25s, the 12-pulse converter supplies the extra active power required by the load but continues to absorb reactive power. The source supplies rated active power at unity power factor. Since the combined system of the 12-pulse converter and the passive filters supplies active power but absorbs the reactive power therefore the operation is in the 2ⁿᵈ quadrant of the P-Q plane.

TABLE III
CONVERTER PARAMETERS

| T (secs) | 12 pulse Converter | | Firing Angles | | Quadrant |
| | Pref (p.u.) | Qref (p.u.) | $\alpha 1$ (deg) | $\alpha 2$ (deg) | |
|---|---|---|---|---|---|
| 0 | 0.27 | 0.79 | 40.0 | 104.6 | I |
| 0.25 | -0.14 | 0.81 | 136.8 | 67.4 | II |
| 0.50 | -0.21 | 0.56 | 165.4 | 60.0 | III |
| 0.75 | 0.09 | 0.33 | 5.0 | 147.5 | IV |

"-"value of "$p_{ref}$" here indicates that active power is being supplied by the 12- pulse converter.

At t=0.5s, the nature of load has been changed to lagging therefore it requires reactive power from the source. This reactive power is supplied by the passive filters instead and hence the source continues to operate at unity power factor. The 12-pulse converter continues to supply the additional active power demanded by the load maintaining the source active power at 1 p.u. Since at this point the combined P-Q controller supplies both active and reactive power the region of operation is the 3ʳᵈ quadrant in the P-Q plane.

At t =0.75s the total load connected to the system is such that the 12-pulse converter absorbs active power and the passive filters supply the desired reactive power by the load. Therefore the source supplies rated active power at unity power factor. Since at this point the combined P-Q controller absorbs active power but supplies reactive power the quadrant operation is the 4ᵗʰ quadrant in the P-Q plane.

Fig. 6 and 7 show the variation of active and reactive powers of the filter, source, converter and load. It can be seen that active power supplied by the source remains at 1 p.u and the reactive power supplied by the source is zero suggesting unity power factor operation. Fig. 8 shows the variation in firing angles of the twelve pulse converter. Fig. 9 shows the voltage at PCC and current profiles of source, 12-pulse converter, load and passive filters from t=0s to t=1.0s. After the load perturbations die out, the peak of the voltage and the source current becomes constant while the phase angle between them remains zero. This indicates that the active power drawn from the source is always constant while the reactive power drawn is zero.

The total harmonic distortion (THD) of the 12-pulse converter input current varies with the variation in firing angles "$\alpha_1$" and "$\alpha_2$" and is minimum when $\alpha_1=\alpha_2$. The THD of the 12-pulse converter input current immediately before t=0.25s, t=0.5s, t=0.75s and t = 1.0s are 19.30%, 20.28%, 37.74% and 47.73% respectively. However, with the passive

filter components listed in Table I, the THD of the source current remains between 1.37% and 1.95% at these instants of time which are within the IEEE 519 standard.

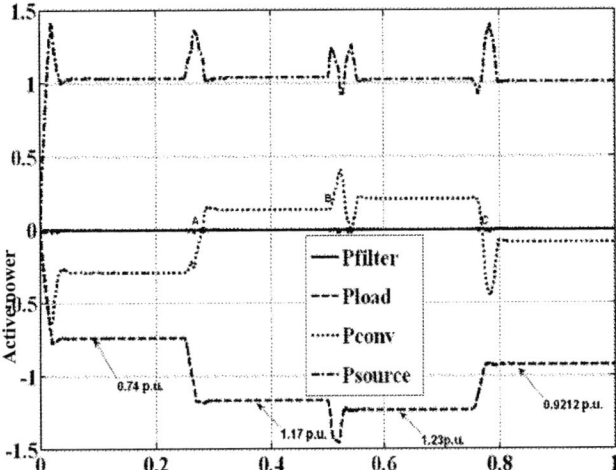

Fig. 6. Variation of source, 12-pulse converter, load and filter active powers.

**Time (sec)**

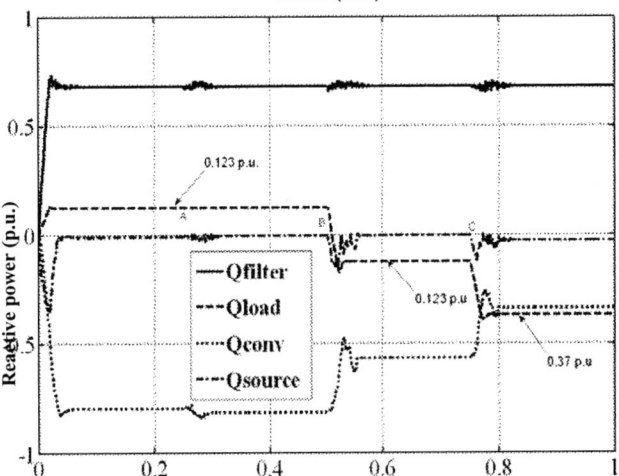

Fig. 7. Variation of source, 12-pulse converter, load and filter reactive powers.

**Time(sec)**

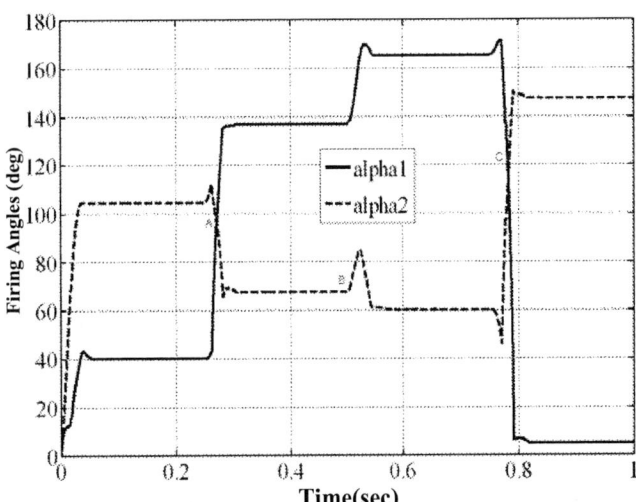

Fig. 8. Variation of firing angles "α1" and "α2" with changing loads.

605

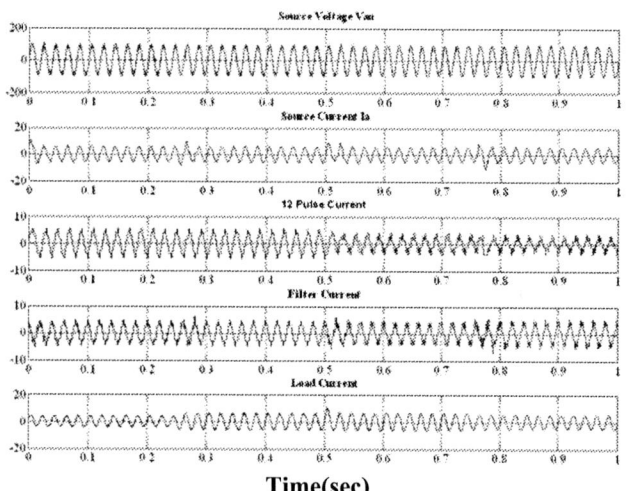

Fig. 9. Profiles of voltage at PCC (A), source current (B), 12-pulse converter input current(C), filter current (D) and load current(E) from t=0 to t=1.0s.

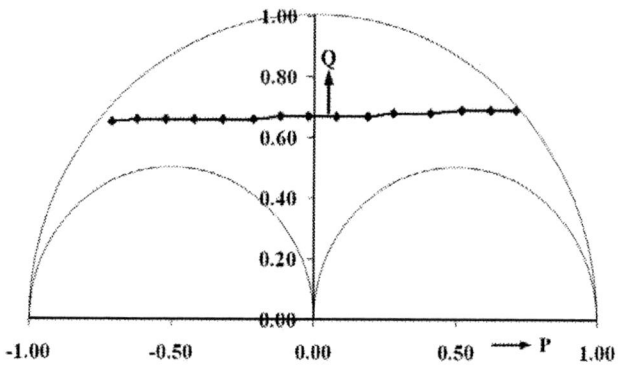

Fig. 10. Experimentally obtained constant var operation of a 12-pulse converter.

## IV. EXPERIMENTAL RESULTS

To demonstrate the validity of the proposed approach experimentally, a prototype of the system is developed in the laboratory. The system rating is kept at 675VA with voltage at PCC at $100V_{L-L}$ and source impedance at 6% of the base impedance. The dc side load current of the 12-pulse converter is maintained constant at 5A. The passive filter component values used in the experimental setup are listed in Table IV.

TABLE IV
PASSIVE FILTER COMPONENTS USED IN EXPERIMENT

| Passive filter | Inductance | Capacitance | Resistance |
|---|---|---|---|
| $5^{th}$ | 10.3mH | 40uF | 0.45 ohm |
| $7^{th}$ | 10.4mH | 20uF | 0.46 ohm |
| $11^{th}$ | 2.0mH | 42uF | 0.11 ohm |
| $13^{th}$ | 1.5mH | 40uF | 0.1 ohm |

Constant var operation of the 12-pulse converter is achieved by setting "$q_{ref}$" of the 12-pulse converter at +0.68 p.u. and "$p_{ref}$" varied from +0.7 p.u. to -0.7 p.u. The corresponding values of "$α_1$" and "$α_2$" are calculated. "$P_{exp}$" and "$Q_{exp}$" are the active and reactive power exchanged by the 12-pulse converter. These are plotted in Fig. 10 which shows the operation of the 12-pulse converter absorbing constant var from PCC.

In the next set of experiments, the variation of active and reactive power profile of the 12-pulse converter is shown under changing loads, which ensures that source supplies rated active power at unity power factor. Fig. 11(a) and (b) show the experimental waveforms obtained before and after an active load perturbation of 0.12 p.u. Initially the 12-pulse converter absorbs 0.53 p.u. active and 0.67 p.u. reactive power corresponding to "$α_1$" and "$α_2$" equal to $18^0$ and $86^0$.

Note that, the power drawn by the 12-pulse converter is the

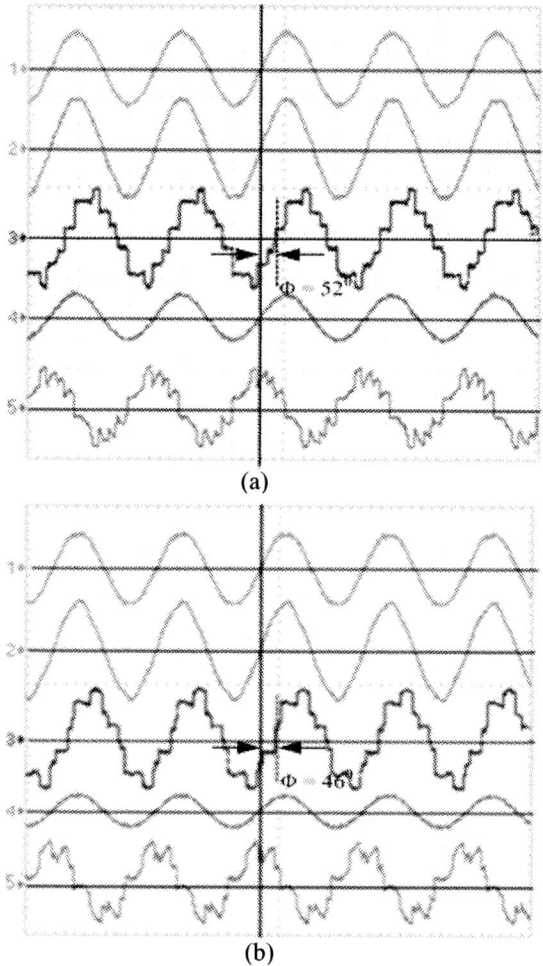

Fig. 11(a). Steady state operation before an R-load perturbation (b) Steady state operation after the load perturbation. (1) Voltage at PCC (100V/div); (2) Source current (5A/div); (3) 12-pulse converter input current (5A/div); (4) Load current (5A/div); (5) Filter current (5A/div).

sum of the power drawn by the 12-pulse converter and the losses incurred in star-delta transformer and the devices. This compensates the reactive power supplied by the filter capacitors, while drawing rated active power from the supply. The phase angle between source voltage and source current is zero indicating unity power factor operation. After the active load perturbation of 0.12 p.u., the firing angle of the 12-pulse converter changes to $21^0$ and $71^0$; thus the active power drawn by the 12-pulse converter increases by an amount equal to the active load rejection. The peak of the source voltage and current as well as the angle between them remain same before and after the load perturbation. With a change in firing angle of the 12-pulse converter, the 12-pulse converter current magnitude and its harmonic content changes which gets reflected in the filter current profile. The THD of the input current of the 12-pulse converter before and after the load perturbations are 19.2% and 22.5% respectively, however, because of the passive filters, the source current THD is below 5% under both conditions.

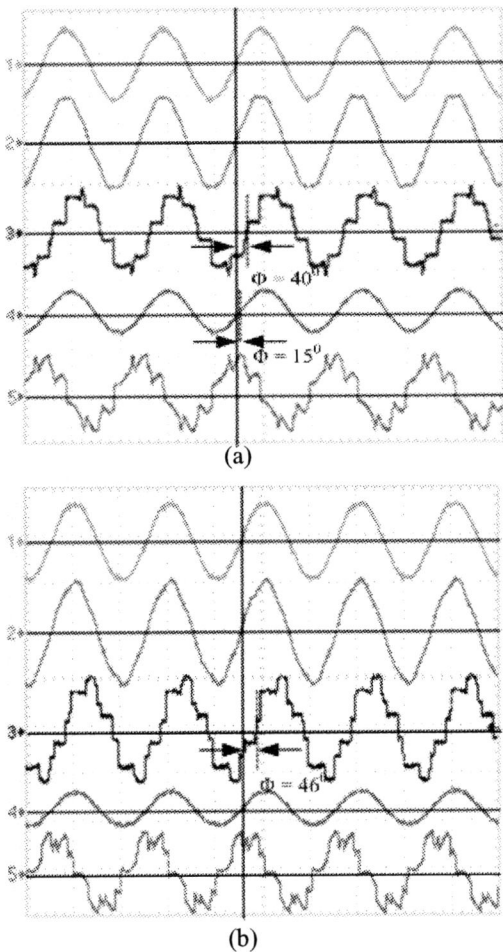

(a)

(b)

Fig. 12(a). Steady state operation before an R-L load perturbation (b) Steady state operation after the load perturbation. (1) Voltage at PCC (100V/div); (2) Source current (5A/div); (3) 12-pulse converter input current (5A/div); (4) Load current (5A/div); (5) Filter current (5A/div).

The experimental waveforms for an R-L load are shown in Fig. 12(a) and 12(b). Initially the R-L load absorbs 0.46 p.u. active and 0.12 p.u. reactive power. After the perturbation the load absorbs 0.35 p.u. active power only, which is same as the post perturbation condition in R load. In spite of var rejection by the load, the source always operates at unity power factor

condition. The 12-pulse converter adjusts its firing angles from $(6^0, 76^0)$ to $(21^0, 71^0)$ to accommodate the changing var condition. The close resemblance between Fig. 11(b) and Fig. 12(b) show the similarity of the post perturbation condition in R load and R-L load.

## V. CONCLUSIONS

A new four quadrant P-Q controller is proposed in this paper, which is based on the asymmetrical firing angle operation of a 12-pulse converter integrated to a combination of passive filters. When such a system is connected at the PCC between source and load, it maintains source power factor unity besides keeping voltage and frequency constant. This inexpensive system can thus have applications in power quality improvement in FACTS operation, adjustable speed drives, and isolated generation systems.

## VI. REFERENCES

[1] W. Mc Murray, A Study of Asymmetrical Gating Phase Controlled Converters, IEEE Transactions on Industry Applications, 8(3), 1972, 289-295.

[2] D.A.Deib and H.W.Hill, Optimal Firing-Angle Control of Cascaded HVDC Converters for Minimum Reactive Power Demand, Proceedings of IECON, 1993, 15-19.

[3] A.K. Gaja, J.K. Chatterjee, and P.S. Modi, "Constant Var Operation of Asymmetrical Firing Angle Controlled 12-Pulse Converter" Proceedings of the IASTED International Conference, Energy and Power Systems, April18-20, 2005, Krabi, Thailand.

[4] H.Akagi, Y. Kanazawa and A.Nabae, " Instantaneous Reactive Power Compensators Comprising Switching Devices without Energy Storage Components," IEEE Transactions on Industry Applications, Vol. 20, No. 3, May, 1984, pp. 625-630.

[5] J.C. Das, "Passive Filters-Potentialities and Limitations," IEEE Transactions on Industry Applications, vol. 40, no. 1, January, 2004. pp. 232-241.

[6] Power System Harmonics by J Arrillaga, D. Bradley, P Bodger, John Wiley & Sons, 1985.

## VII. BIOGRAPHIES

**Faisal M. Ahsan** received his B.Tech Degree in Electrical Engineering from Aligarh Muslim University, Aligarh and is currently doing M. Tech in Power Electronics Electrical Machines and Drives at IIT Delhi, New Delhi India. His research interests include embedded systems, power electronics and power quality.

**Jayanta Kr. Chatterjee (M'77)** received the B.E. and M.E. degrees in electrical engineering from the University of Roorkee, Roorkee, India, in 1967 and 1969, respectively and Ph.D. degree from the University of Bristol, Bristol, UK, in 1975.Currently; he is a Professor in the Department of Electrical Engineering, Indian Institute of Technology Delhi., New Delhi, India. His research interests include embedded systems, power electronics, power quality, brushless generation, and motor control.

**Anandarup Das** received his B.E Degree in Electrical Engineering from Bengal Engineering College Kolkata and and has recently completed his M.Tech in Power Electronics and Machine Drives at IIT Delhi New Delhi India. His research interests are Power Electronics and Power quality.

607

**2006 IEEE International Conference on Power Electronic, Drives and Energy Systems**

# A Novel Structure for Three-Phase Four-Wire Distribution System Utilizing Unified Power Quality Conditioner (UPQC)

V. Khadkikar, *Student Member, IEEE*, and A. Chandra, *Senior Member, IEEE*

*Abstract*--This paper presents a novel structure for three-phase four-wire (3P4W) distribution system utilizing Unified Power Quality Conditioner (UPQC). The 3P4W system is realized from three-phase three-wire (3P3W) system where the neutral of series transformer used in series part UPQC is considered as the fourth wire for 3P4W system. A new control strategy to balance the imbalanced load currents is also presented in this paper. The neutral current that may flow towards transformer neutral point is compensated by using a four-leg voltage source inverter (VSI) topology for shunt part. Thus the series transformer neutral will be at virtual zero potential during all operating conditions. The simulation results based on MATLAB / Simulink are presented to show the effectiveness of proposed UPQC based 3P4W distribution system.

*Index Terms*--Active Power Filter (APF), Four-Leg Voltage Source Inverter (VSI) Structure, Three Phase Four Wire System, Unified Power Quality Conditioner (UPQC).

## I. INTRODUCTION

THE use of sophisticated equipments/loads at transmission and distribution level has been increased considerable in recent years due to the developments in the semiconductor device technologies. These devices need clean power in order to function properly. At the same time these device involving switching operation generates current harmonics resulting a polluted distribution system. The power electronics based devices have been used to overcome the major power quality problems [1]. To provide the balance, distortion free and constant magnitude power to sensitive load and at the same time to restrict the harmonics, imbalance and reactive power demanded by the load and hence to make the overall power distribution system more healthy the Unified Power Quality Conditioner is one of the best solutions [6]-[11].

The three-phase four-wire (3P4W) distribution system can be realized by providing the neutral conductor along with the three power lines from generation station or by utilizing a delta-star transformer at distribution level. The UPQC installed for 3P4W application generally considers 3P4W supply [9]-[11]. This paper proposes a new topology/structure

---

V. Khadkikar and A. Chandra are with Département de génie électrique, École de technologie supérieure, Montréal, Québec, H3C 1K3, Canada. (e-mails: v_khadkikar@yahoo.com, chandra@ele.etsmtl.ca).

that can be realized in UPQC based applications, in which the series transformer neutral, used for series inverter, can be used to realize 3P4W system even if the power supplied by utility is 3P3W. This new functionality using UPQC could be useful in future UPQC based distribution system.

The imbalanced load currents are very common and yet an important problem in 3P4W distribution system. This paper deals with the imbalanced load current problems with a new control approach, in which, the fundamental active power demanded by each phases are computed first and these active powers are then redistributed on each phase. Thus the proposed control strategy makes the imbalanced load currents as perfectly balanced source currents using UPQC.

The proposed 3P4W distribution system realized form existing 3P3W UPQC based system is discussed in section II. The proposed control strategy for balancing the imbalanced load currents is explained in the section III. The major simulation results are given in the section IV and finally section V concludes the paper.

## II. PROPOSED 3P4W DISTRIBUTION SYSTEM UTILIZING UPQC

Generally a three-phase four-wire (3P4W) distribution system is realized by providing the neutral conductor along with the three power conductors from generation station or by utilizing a three-phase delta-star ($\Delta$-Y) transformer at distribution level. Fig. 1 shows a 3P4W network in which the neutral conductor is provided from the generating station itself, whereas, Fig.2 shows a 3P4W distribution network considering $\Delta$-Y transformer. Assuming a plant site where three-phase, three-wire UPQC is already installed to protect a sensitive load and to restrict any entry of distortion from load side towards utility, as shown in Fig. 3. Now, if, we want to upgrade the system from 3P3W to 3P4W system due to installation of some single phase loads, in such cases, if the distribution transformer is close to the plant under consideration then utility would provide the neutral conductor from this transformer without major cost involvement. In certain ceases this may appear as a costly solution, since, the distribution transformer may be not situated in close vicinity. Recently, the utility service providers are putting more and more restrictions on current THD limits, drawn by non-linear loads, to control the power distribution system harmonic pollution. At the same time the use of sophisticated equipments/loads has been increased significantly, needs clean

---

0-7803-9771-1/06/$25.00 ©2006 IEEE

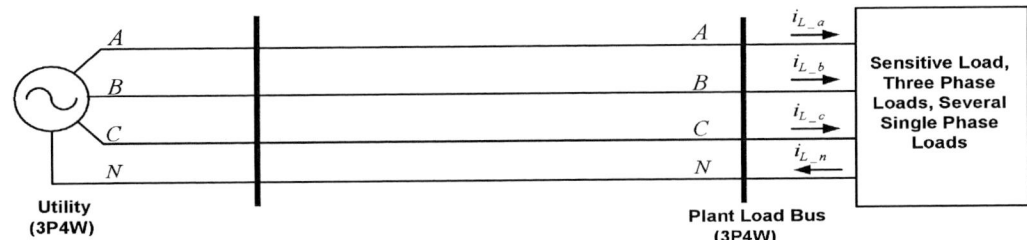

Fig. 1. Three-Phase Four-Wire Distribution System: Neutral Provided from Generation Station.

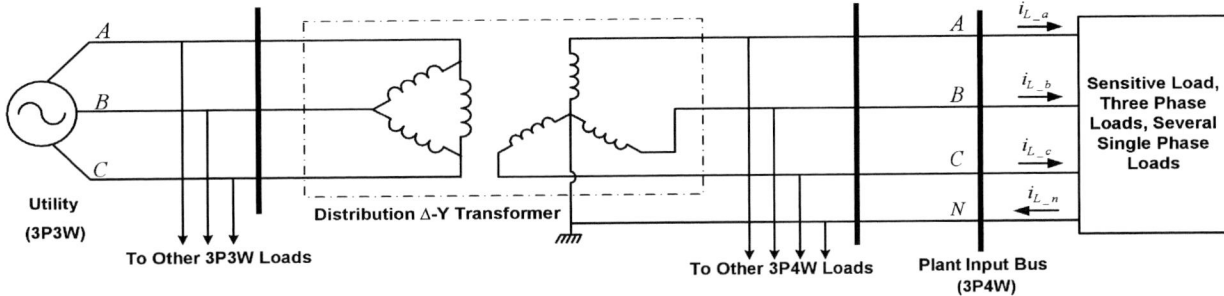

Fig. 2. Three-Phase Four-Wire Distribution System: Neutral Provided from Delta-Star Transformer.

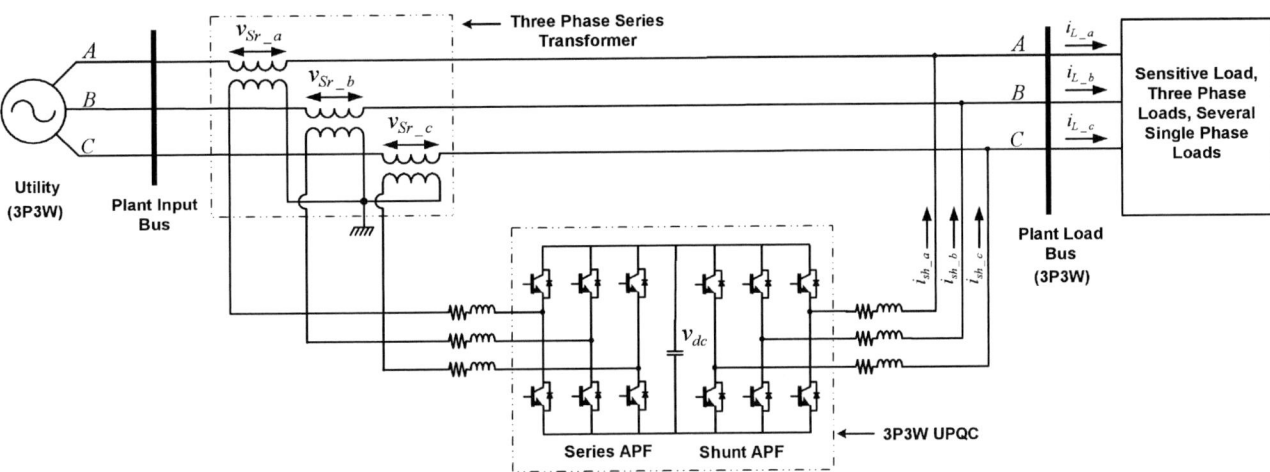

Fig. 3. Three-Phase Three-Wire UPQC Structure.

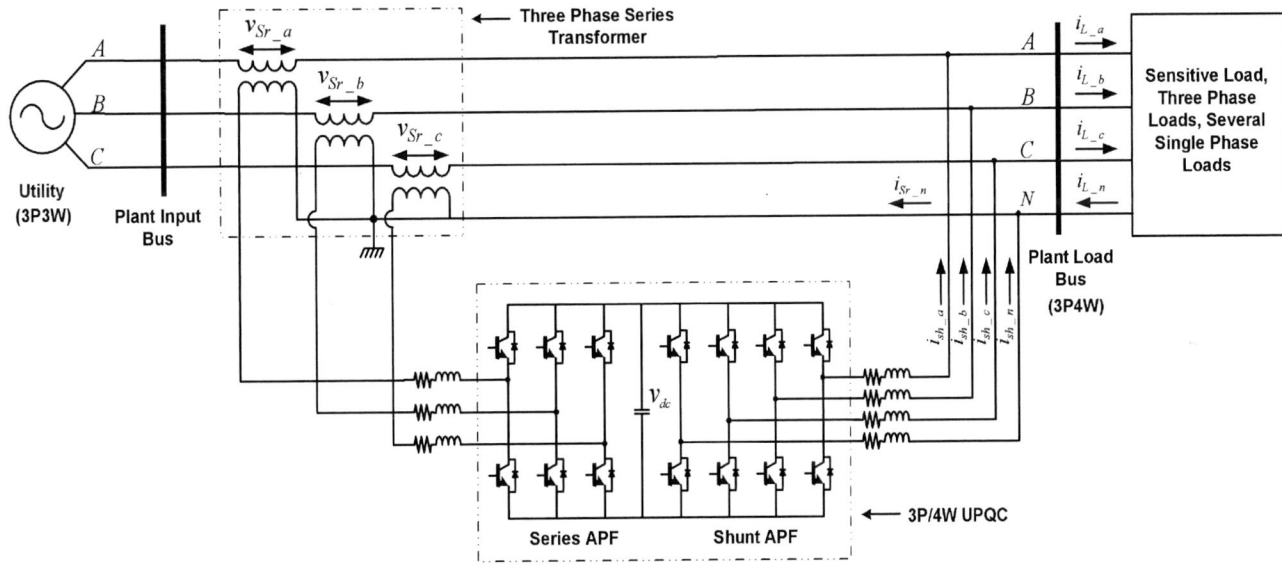

Fig. 4. Proposed There-Phase Four-Wire System Realized from Three-Phase Three-Wire System Utilizing UPQC.

609

power for their proper operation. Therefore, in the future distribution system and the plant/load centers consisting of UPQC would be common. Fig.4 shows the proposed novel 3P4W topology that can be realized from 3P3W system. This proposed system has all the advantages of general UPQC in addition to easy 3P3W system to 3P4W system expansion. Thus the proposed topology may play an important role in the future 3P4W distribution system for more advanced, UPQC based plant/load center installation, where, utility would be having an additional option to realized 3P4W system just by providing three-phase three-wire supply.

As noticed from Fig. 3, the UPQC should necessarily consist of three-phase series transformer in order to connect one of the inverters in the series with the line to function as controlled voltage source. If we could use the neutral of three-phase series transformer as the neutral wire to realize the three-phase four-wire system, then, 3P4W system can easily be achieved from a three-phase three-wire system (Fig.4). The neutral current, present if any, would flow through this fourth wire towards transformer neutral point. This neutral current can be compensated by using split capacitor topology [2], [9]-[10] or four-leg VSI topology for shunt inverter [2], [11]. The four-leg VSI topology requires one addition leg as compared split topology. The neutral current compensation in four-leg VSI structure is much easier than the split capacitor, since the split capacitor topology essentially needs two capacitors and an extra control loop to maintain zero voltage error difference between both the capacitor voltages, resulting more complex control loop to maintain the dc bus voltage at constant level.

In this paper a four-leg VSI topology is considered to compensate the neutral current flowing towards transformer neutral point. A fourth leg is added on the existing 3P3W UPQC, such that, the transformer neutral point will be at virtual zero potential. Thus the proposed structure would help to realize 3P4W system from 3P3W system at distribution load end. This would eventually result in easy expansion from 3P3W system to 3P4W system. A new control strategy to generate balanced reference source currents under imbalanced load condition is also proposed in this paper and is explained in next section.

## III. UPQC CONTROLLER

The control algorithm for series APF is based on Unit Vector Template Generation scheme [7], whereas, the control strategy for shunt APF is discussed in this section. Based on the load on 3P4W system the current drawn from the utility can be imbalanced. In this paper a new control strategy is proposed to compensate the current imbalance present in the load currents by expanding the concept of single-phase p-q theory [5]-[6]. According to this theory a signal-phase system can be defined as a pseudo two-phase system by giving $\pi/2$ lead or $\pi/2$ lag, i.e. each phase voltages and currents of original three-phase system can be considered as three independent two-phase systems. These resultant two-phase systems can be represented in $\alpha$-$\beta$ coordinates and thus the p-q theory applied for balanced three-phase system [3], can also be used for each phase of imbalanced system independently. The actual load voltages and load currents are considered as $\alpha$-axis

quantities, whereas, the $\pi/2$ lead load or $\pi/2$ lag voltages and $\pi/2$ lead or $\pi/2$ lag load currents are considered as $\beta$-axis quantities. In this paper $\pi/2$ lead is considered to achieve two-phase system for each phase. The major disadvantage of p-q theory is that it gives poor results under distorted and/or imbalanced input/utility voltages [4]-[5]. In order to eliminate these limitations the reference load voltage signals extracted for series APF are used instead of actual load voltages.

For phase-a, the load voltage and current in $\alpha$-$\beta$ coordinates can be represented by $\pi/2$ lead as –

$$\begin{bmatrix} v_{La\_\alpha} \\ v_{La\_\beta} \end{bmatrix} = \begin{bmatrix} v^*_{La}(\omega t) \\ v^*_{La}(\omega t + \pi/2) \end{bmatrix} = \begin{bmatrix} V_{Lm}\sin(\omega t) \\ V_{Lm}\cos(\omega t) \end{bmatrix} \quad (1)$$

$$\begin{bmatrix} i_{La\_\alpha} \\ i_{La\_\beta} \end{bmatrix} = \begin{bmatrix} i_{La}(\omega t + \varphi_L) \\ i_{La}[(\omega t + \varphi_L) + \pi/2] \end{bmatrix} \quad (2)$$

Where, $v^*_{La}(\omega t)$ represents the reference load voltage and $V_{Lm}$ represents the desired load voltage magnitude.

Similarly, for phase-b, the load voltage and current in $\alpha$-$\beta$ coordinates can be represented by $\pi/2$ lead as –

$$\begin{bmatrix} v_{Lb\_\alpha} \\ v_{Lb\_\beta} \end{bmatrix} = \begin{bmatrix} v^*_{Lb}(\omega t) \\ v^*_{Lb}(\omega t + \pi/2) \end{bmatrix} = \begin{bmatrix} V_{Lm}\sin(\omega t - 120^o) \\ V_{Lm}\cos(\omega t - 120^o) \end{bmatrix} \quad (3)$$

$$\begin{bmatrix} i_{Lb\_\alpha} \\ i_{Lb\_\beta} \end{bmatrix} = \begin{bmatrix} i_{Lb}(\omega t + \varphi_L) \\ i_{Lb}[(\omega t + \varphi_L) + \pi/2] \end{bmatrix} \quad (4)$$

And for phase-c, the load voltage and current in $\alpha$-$\beta$ coordinates can be represented by $\pi/2$ lead as –

$$\begin{bmatrix} v_{Lc\_\alpha} \\ v_{Lc\_\beta} \end{bmatrix} = \begin{bmatrix} v^*_{Lc}(\omega t) \\ v^*_{Lc}(\omega t + \pi/2) \end{bmatrix} = \begin{bmatrix} V_{Lm}\sin(\omega t + 120^o) \\ V_{Lm}\cos(\omega t + 120^o) \end{bmatrix} \quad (5)$$

$$\begin{bmatrix} i_{Lc\_\alpha} \\ i_{Lc\_\beta} \end{bmatrix} = \begin{bmatrix} i_{Lc}(\omega t + \varphi_L) \\ i_{Lc}[(\omega t + \varphi_L) + \pi/2] \end{bmatrix} \quad (6)$$

By using the definition of 3-phase p-q theory for balanced three-phase system [3], the instantaneous power components can be represented as –

Instantaneous active power,

$$p_{L,abc} = v_{L,abc\_\alpha} \cdot i_{L,abc\_\alpha} + v_{L,abc\_\beta} \cdot i_{L,abc\_\beta} \quad (7)$$

Instantaneous reactive power,

$$q_{L,abc} = v_{L,abc\_\alpha} \cdot i_{L,abc\_\beta} - v_{L,abc\_\beta} \cdot i_{L,abc\_\alpha} \quad (8)$$

Considering phase-a, the phase-a instantaneous load active and instantaneous load reactive power can be represented by,

$$\begin{bmatrix} p_{La} \\ q_{La} \end{bmatrix} = \begin{bmatrix} v_{La\_\alpha} & v_{La\_\beta} \\ -v_{La\_\beta} & v_{La\_\alpha} \end{bmatrix} \cdot \begin{bmatrix} i_{La\_\alpha} \\ i_{La\_\beta} \end{bmatrix} \quad (9)$$

Where, $p_{La} = p_{La}^- + p_{La}^\sim \quad (10)$

$$q_{La} = q_{La}^- + q_{La}^\sim \quad (11)$$

In (10) & (11), $p_{La}^-$ and $q_{La}^-$ represent the DC components that are responsible for fundamental load active and reactive power, whereas, $p_{La}^\sim$ and $q_{La}^\sim$ represent the AC components that are responsible for harmonic powers. The phase-a fundamental instantaneous load active and reactive power components can be extracted from $p_{La}$ and $q_{La}$ by using a low pass filter (LPF), respectively.

$\therefore$ Instantaneous fundamental load active power for phase -a,

$$p_{La,1} = p_{La}^- \quad (12)$$

Instantaneous fundamental load reactive power for phase-a,

610

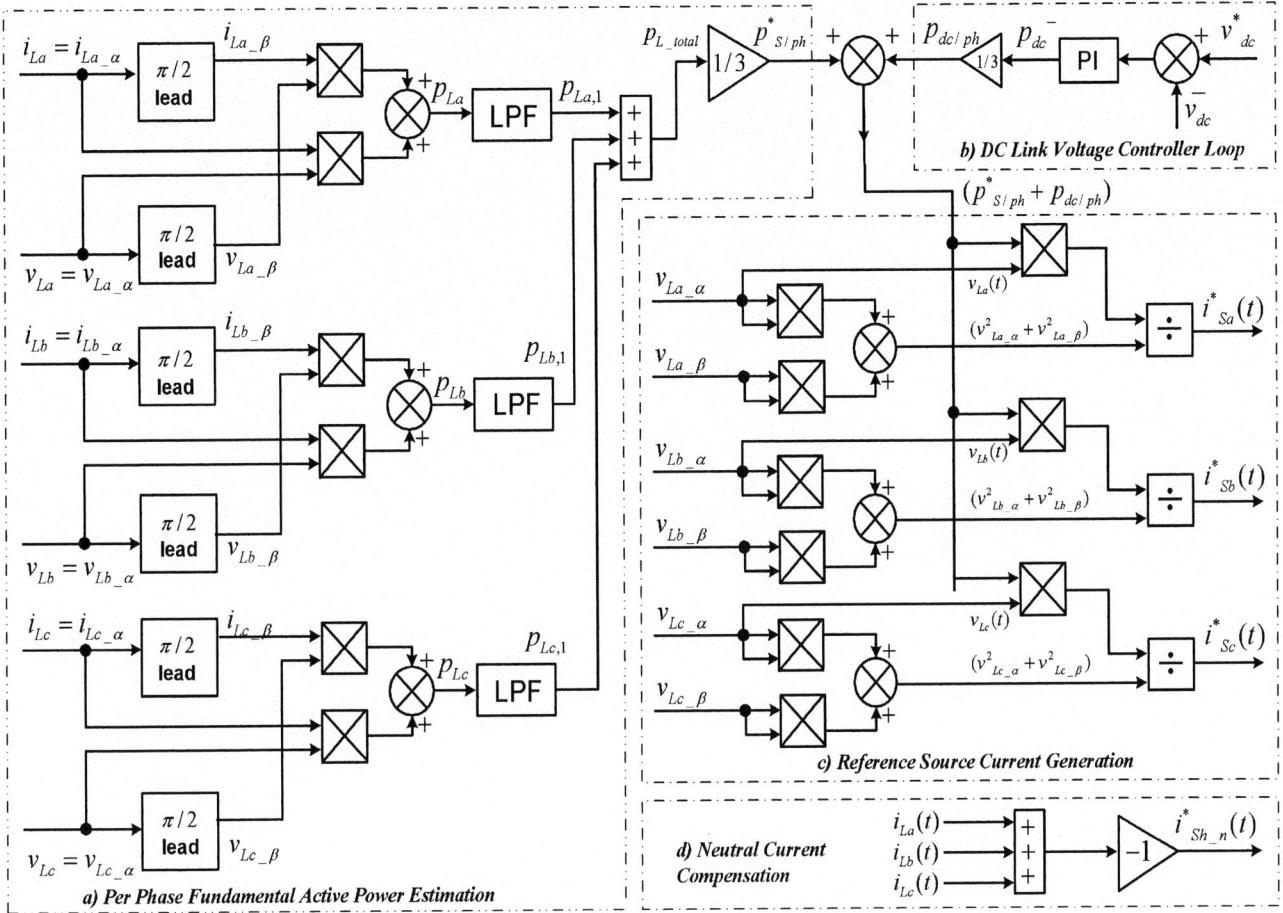

Fig. 5. Shunt Active Filter Control Block Diagram: a) Proposed Balanced Per Phase Fundamental Active Power Estimation, b) DC Link Voltage Control Loop, c) Reference Source Current Generation, and d) Neutral Current Compensation

$$q_{La,1} = q_{La}^- \qquad (13)$$

Similarly, the fundamental instantaneous load active and the fundamental instantaneous load reactive powers for phase–b and phase–c can be calculated as –

Instantaneous fundamental load active power for phase –b,

$$p_{Lb,1} = p_{Lb}^- \qquad (14)$$

Instantaneous fundamental load reactive power for phase–b,

$$q_{Lb,1} = q_{Lb}^- \qquad (15)$$

Instantaneous fundamental load active power for phase –c,

$$p_{Lc,1} = p_{Lc}^- \qquad (16)$$

Instantaneous fundamental load reactive power for phase–c,

$$q_{Lc,1} = q_{Lc}^- \qquad (17)$$

Since the load current drawn by each phase may be different due to different loads that may present inside plant, therefore, the instantaneous fundamental load active power and instantaneous fundamental load reactive power demand for each phase may not be the same. In order to make this load imbalanced power demand, seen from the utility side, as a perfectly balanced, fundamental three-phase active power, the imbalanced load power should be properly redistributed between Utility – UPQC – Load; such that, the total load seen by the utility would be linear and balanced load. The

imbalanced or balanced reactive power demanded by the load should be handled by shunt APF. The aforementioned task can be achieved by summing instantaneous fundamental load active power demands of all the three phases and redistributing it again on each utility phase, i.e. from (12), (14) and (16),

$$p_{L,total} = p_{La,1} + p_{Lb,1} + p_{Lc,1} \qquad (18)$$

$$p_{S/ph}^* = \frac{p_{L,total}}{3} \qquad (19)$$

Equation (19) gives the redistributed per phase fundamental active power demand that each phases of utility should supply in order to achieve perfect balanced source currents. From (19) it is evident that under all the conditions, the total fundamental active power demanded by the loads would be equal to the total power drawn from the utility but with perfectly balanced way even the load currents are imbalanced. Thus the reference compensating currents representing a perfectly balance three-phase system can be extracted by taking inverse of equation (9),

$$\begin{bmatrix} i_{Sa\_\alpha}^* \\ i_{Sa\_\beta}^* \end{bmatrix} = \begin{bmatrix} v_{La\_\alpha} & v_{La\_\beta} \\ -v_{La\_\beta} & v_{La\_\alpha} \end{bmatrix}^{-1} \cdot \begin{bmatrix} p_{S/ph}^* + p_{dc/ph} \\ 0 \end{bmatrix} \qquad (20)$$

In (20), $p_{dc/ph}$ is the precise amount of per phase active power that should be taken from the source in order to

611

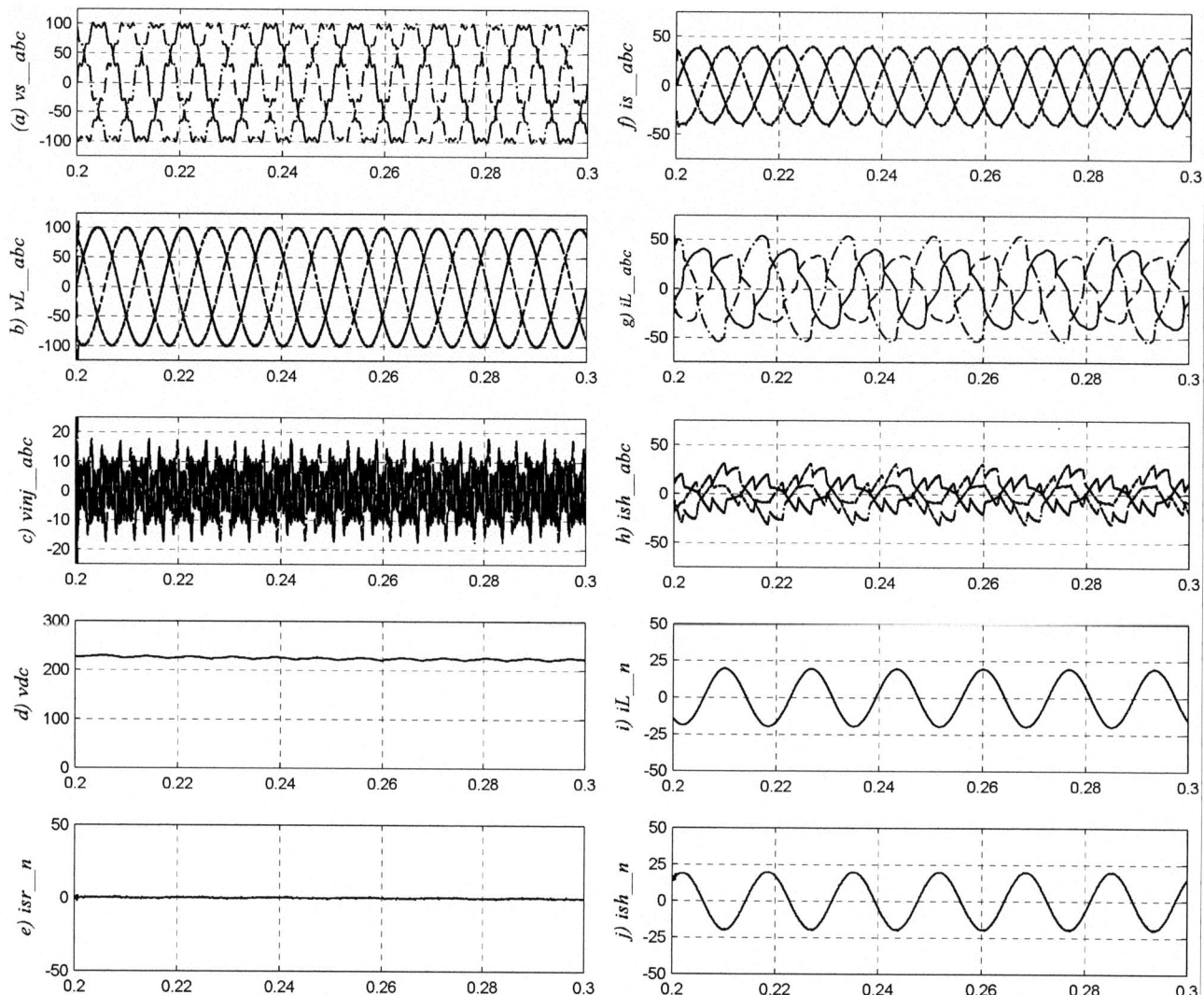

Fig. 6. Simulation Result – Proposed 3P4W UPQC Structure: a) Utility Voltage ($v_{S\_abc}$), b) Load Voltage ($v_{L\_abc}$), c) Injected Voltage ($v_{inj\_abc}$), d) DC Link Voltage ($v_{dc}$), e) Neutral Current Flowing Towards Series Transformer ($i_{Sr\_n}$), f) Source Current ($i_{S\_abc}$), g) Load Current ($i_{L\_abc}$), h) Shunt Compensating Current ($i_{Sh\_abc}$), i) Current Flowing Through Load Neutral Wire ($i_{L\_n}$), and j) Shunt Neutral Compensating Current ($i_{Sh\_n}$).

maintain the dc link voltage at constant level and to overcome the losses associated with UPQC. The oscillating instantaneous active power $p_{La}\tilde{}$ should be exchanged between the load and shunt APF. The reactive power term ($q_{La}$) in (20) is considered as zero, since, the utility should not supply load reactive power demand. In the above matrix the $\alpha$–axis reference compensating current represents the instantaneous fundamental source current, since, $\alpha$–axis quantities belong to the original system under consideration and the $\beta$–axis reference compensating current represents the current that is at $\pi/2$ lead with respect to the original system.

$$\therefore \quad i^{*}_{Sa}(t) = \frac{v_{La\_\alpha}(t)}{v^{2}_{La\_\alpha} + v^{2}_{La\_\beta}} \cdot \left[ p^{*}_{S/ph}(t) + p_{dc/ph}(t) \right] \quad (21)$$

Similarly, the reference source current for phase–$b$ and phase–$c$ can be estimated as,

$$i^{*}_{Sb}(t) = \frac{v_{Lb\_\alpha}(t)}{v^{2}_{Lb\_\alpha} + v^{2}_{Lb\_\beta}} \cdot \left[ p^{*}_{L/ph}(t) + p_{dc/ph}(t) \right] \quad (22)$$

$$i^{*}_{Sc}(t) = \frac{v_{Lc\_\alpha}(t)}{v^{2}_{Lc\_\alpha} + v^{2}_{Lc\_\beta}} \cdot \left[ p^{*}_{L/ph}(t) + p_{dc/ph}(t) \right] \quad (23),$$

The reference neutral current signal can be extracted simply by adding all the sensed load currents, without actual neutral current sensing, as-

$$i_{L\_n}(t) = i_{La}(t) + i_{Lb}(t) + i_{Lc}(t) \quad (24)$$

$$i^{*}_{sh\_n}(t) = -i_{L\_n}(t) \quad (25)$$

The proposed balanced per phase fundamental active power estimation, dc link voltage control loop based on PI regulator, the reference source current generation as given by (21)-(23), and the reference neutral current generation are shown in Fig.5. (a), (b), (c), and (d), respectively.

IV. SIMULATION RESULTS

The simulation results for proposed UPQC based 3P4W topology are shown in Fig. 6 (a)–(j). The MATLAB /

Simulink is used as a simulation tool. The utility voltages are assumed to be distorted with voltage THD of 9.5%. These distorted voltages profile is shown in Fig. 6 (a). The UPQC should maintain the voltage at load bus at desired value and free from distortion. The plant load is assumed to be the combination of a balanced three-phase Diode Bridge Rectifier followed by a R-L load, which acts as a harmonic generating load, and three different single-phase loads on each phase, with different load active and reactive power demands. The resulting load current profile is shown in Fig. 6 (g), has THD of 12.15%.

The shunt APF is turned ON first at time t=0.1 sec such that it maintains the dc link voltage at set reference value, here 220 V, not shown in the Fig. 6. At time t = 0.2 sec, the series APF is put into the operation. The series APF injects the required compensating voltages through series transformer, making the load voltage free from distortion (THD= 1.5%) and at desired level as seen from Fig. 6 (b). The series APF injected voltage profile is shown in Fig. 6 (c). Simultaneously, the shunt APF injects the compensating currents to achieve the balanced source current, free from distortion, as discussed in pervious section. The compensated source currents are shown in Fig. 6 (f), are perfectly balanced with the THD of 2.3%. The current injected by shunt APF are shown in Fig. 6 (g). Since the load on the network is imbalanced in nature, the neutral current may flow through neutral conductor towards series transformer neutral point. The load neutral current profile is shown in Fig. 6 (i). As noticed from Fig. 6 (e), shunt APF effectively compensates the current flowing towards transformer neutral point. Thus the series transformer neutral point is maintained at virtual zero potential. The compensating current injected through fourth leg of shunt APF is shown in Fig. 6 (j).

## V. CONCLUSION

A new three-phase four-wire topology for distribution system utilizing UPQC has been proposed in this paper. This proposed topology would be very useful to expand the existing 3P3W system to 3P4W system where UPQC is installed to compensate the different power quality problems, which may play an important role in future UPQC based distribution system. A new control strategy to generate the balanced reference source current under imbalanced load condition is also presented in this paper. The MATLAB / Simulink based simulation results show that the distorted and imbalanced load currents seen form utility side, act as perfectly balanced source currents and free from distortion. The neutral current that may flow towards transformer neutral point is effectively compensated such that the transformer neutral point is always at virtual zero potential.

## VI. APPENDIX

The system parameters are given here:
$V_S$ = 100 V (peak, fundamental), $f$ = 60 Hz, $L_S$ = 0.1 $mH$, $L_{sh}$ = 3 $mH$, $R_{Sh}$= 0.1 Ω, $L_{Sr}$ = 3 mH, $R_{Sr}$ = 0.1 Ω, $C_{dc}$ = 5000 μF.
Plant Loads:
i) Three-Phase Diode Bridge Rectifier followed by R-L load with R = 10 Ω and L = 5 $mH$ . ii) Three single-phase loads

with 1000 $w$ & 600 $Var$, 750 $w$ & 400 $Var$ and 1400 $w$ & 1200 $Var$ demand on phase $a$, $b$, and $c$, respectively.

## VII. REFERENCES

[1] B. Singh, K. Al-Haddad and A. Chandra, "A Review of Active Power Filters for Power Quality Improvement", IEEE Trans on Industrial Electronics, Vol. 45, No.5, Oct 1999, pp. 960-971.

[2] Quinn C.A., and Mohan N., "Active filtering of harmonic currents in three-phase, four-wire systems with three-phase and single-phase nonlinear loads", APEC Conference Proceeding, 23-27 Feb, 1992, pp. 829-836.

[3] H. Akagi, Y. Kanazawa, and A. Nabae, "Instantaneous Reactive Power Compensators Comprising Switching Devices Without Energy Storage Components", IEEE Trans. on Industry Applications, Vol.IA-20, No.3, May/June 1984, pp.625-630.

[4] Komatsu Y. and Kawabata T., "A control method of active power filter in unsymmetrical and distorted voltage system", Proc. of IEEE Power Conversion Conference, Volume 1, 3-6 Aug. 1997 Page(s):161 – 168.

[5] Haque M.T., "Single-phase PQ theory", IEEE 33rd Power Electronics Specialists Conference, 2002. Vol. 4, June 2002, pp. 1815 – 1820.

[6] Correa J.M., Chakraborty S., Simoes M.G., and Farret, F.A., "A single phase high frequency AC microgrid with an unified power quality conditioner", Industry Applications Conference, 2003, 38th Annual Meeting. Vol. 2, 12-16 Oct 2003, pp. 956-962.

[7] Khadkikar V., Chandra A., Barry A. O. and Nguyen, T.D, "Application of UPQC to Protect a Sensitive Load on a Polluted Distribution Network", IEEE PES General Meeting 2006, 18-22 June 2006, pages-6.

[8] Khadkikar V., Chandra A., Barry A. O. and Nguyen T.D, "Conceptual Analysis of unified power quality conditioner (UPQC)", IEEE International Symposium on Industrial Electronics Conference on 9-13 July, 2006.

[9] Aredes M., Heumann K., and Watanabe E.H., "An universal active power line conditioner." Power Delivery, IEEE Transactions on, Volume: 13 Issue: 2, April 1998 Page(s): 545 -551.

[10] Faranda, R., and Valade I., "UPQC compensation strategy and design aimed at reducing losses." Proceedings of the 2002 IEEE-ISIE International Symposium on, Volume: 4, 8-11 July 2002, pp. 1264-1270.

[11] Chen G., Chen Y., and Smedley K.M., "Three-phase four-leg active power quality conditioner without references calculation", IEEE APEC Conference Proceedings, Vol. 1, 2004, pp. 587-593.

## VIII. BIOGRAPHIES

**Vinod Khadkikar** was born at Aurangabad (MH, INDIA) in 1978. He received his Bachelor of Engineering and Master of Technology degrees in Electrical Engineering from Dr. B.A.M University, Aurangabad and IIT Delhi, India in 2000 and 2002, respectively. Presently he is pursuing his Ph. D. in the Department of Electrical Engineering, École de Technologie Supérieure, Montréal, Canada. His research interests are Application of Power Electronics in Distribution Systems, Power Quality Analysis, Active Power Filters, etc.

**Ambrish Chandra** (SM'99) was born in India in 1955. He received B.E. degree from the University of Roorkee, India, M. Tech. degree form I.I.T., New Delhi, India, and Ph.D. degree from University of Calgary, Canada, in 1977, 1980, and 1987, respectively.

He worked as a Lecturer and later as a Reader at University of Roorkee. Since 1994 he is working as a Professor in Electrical Engineering Department at École de technologie supérieure, Universié du Québec, Montreal, Canada. His main research interests are power quality, active filters, static reactive power compensation, and flexible AC transmission systems (FACTS). Dr. Chandra is a senior member of IEEE and member of the Ordre des Ingénieurs du Québec, Canada.

**2006 IEEE International Conference on Power Electronic, Drives and Energy Systems**

# Load Compensation for Diesel Generator Based Isolated Generation System Employing DSTATCOM

Bhim Singh, *Senior Member, IEEE*, and Jitendra Solanki

*Abstract*— **This paper presents the control of distribution static compensator (DSTATCOM) for reactive power, harmonics and unbalanced load currents compensation on diesel generator (DG) set for isolated system. The control of DSTATCOM is achieved using least mean square (LMS) based Adaline. The Adaline is used to extract the balanced positive sequence real fundamental frequency component of load current and PI (Proportional Integral) controller is used to maintain the constant voltage at the dc bus of voltage source converter (VSC) working as DSTATCOM. Switching of VSC is achieved by controlling source currents to follow reference currents using hysteresis based PWM control. The scheme is simulated under MATLAB environment using Simulink and PSB block set toolboxes for linear and non linear loads. The modeling is performed for 3 phase, 3 wire star connected synchronous generator coupled to diesel engine, along with the 3 leg VSC working as DSTATCOM. The results verify the effectiveness of the control of DSTATCOM for load compensation and optimal operation of DG set.**

*Index Terms*—**Diesel generator Set, Distribution Static Compensator (DSTATCOM) , load Compensation.**

## I. INTRODUCTION

INSTALLATION of diesel engine based electricity generation unit (DG set) is widely used practice to feed power to crucial equipments in remote areas [1,2]. DG sets used for these purposes are loaded with unbalanced and nonlinear loads such as power supplies in telecommunication equipments and medical equipments. The source impedance of DG set is high and the unbalance and distorted currents leads to the unbalanced and distorted three-phase voltages at PCC. All these factors lead to the increased fuel consumption and reduced life of the DG sets. This forces to operate these DG sets with derating, which results into increased cost of the system. Instead of this a DSTATCOM [2] can be used with three-phase DG set to feed unbalanced loads without dearting the DG set and within the same cost involved. Moreover, DSATCOM can provide compensation for harmonics and reactive power that facilitates to the load the DG set upto its full kVA rating.

The performance of DSTATCOM is very much dependent on the method of deriving reference compensating signals. Instantaneous reactive power theory, modified p-q theory,

synchronous reference frame(SRF) theory, instantaneous $i_d$-$i_q$ theory, and method for estimation of reference currents by maintaining the voltage of dc link are generally reported in the literature for estimation of reference currents for DSTATCOM through subtraction of positive sequence fundamental current component from the load current [3-6]. These techniques are based on complex calculations and generally incorporates low pass filter which results in delay in computation of reference currents and therefore leads to slow dynamic response of the DSTATCOM. In this paper a fast and simple neural network based control scheme is used to estimate reference source currents for control of DSTATCOM.

This paper presents a DSTATCOM for load compensation of a diesel generator set to enhance its performance. The control of DSTATCOM with capabilities of reactive power, harmonics and unbalanced load compensation is achieved by LMS algorithm [7,8] based Adaline. The Adaline is used to extract the positive sequence fundamental frequency real component of the load current. The dc bus voltage of the VSC is supported by a PI controller, which computes the loss component of the current in DSTATCOM. The extraction of reference currents using Adaline involves the calculation of weights, these weights are a measure of peak of fundamental frequency current component of the load current. The life of DG set is enhanced in absence of unbalance and harmonic currents. The modeling of the DG set is performed using synchronous generator, speed governor and excitation control system. The system is simulated under MATLAB environment using Simulink and PSB Block-set toolboxes. The results for 30kVA DG set with linear load at 0.8 lagging pf and nonlinear load with different load dynamics and unbalanced load conditions are presented to show the effectiveness of the DSTATCOM-DG set system.

## II. SYSTEM CONFIGURATION

Fig.1 shows the configuration of the system for 3-phase 3-wire DG set feeding to variety of loads. A 30kVA system is chosen to demonstrate the working of the system with the DSTATCOM. The DSTATCOM consists of the IGBT (Insulated gate bipolar junction transistor) based 3-phase 3-leg VSC system. The load current is tracked using Adaline based reference generator, which in conjunction with hysteresis based PWM current controller provides switching signals for the VSC of DSTATCOM. It controls source currents to follow the reference currents. The parameters of the synchronous generator are salient pole, 415V, 30kVA, 4 pole, 1500rpm, 50 Hz, $X_d$=1.56pu, $X_d$'=0.15pu $X_d$''=0.11pu, $X_q$=0.78, $X_q$'=0.17,

---

Bhim Singh and Jitendra Solanki are with Department of Electrical Engineering, Indian Institute of technology Delhi, New Delhi, 110016 India. (e-mail: bsingh@ee.iitd.ac.in, ejitendra@yahoo.com).

0-7803-9771-1/06/$25.00 ©2006 IEEE

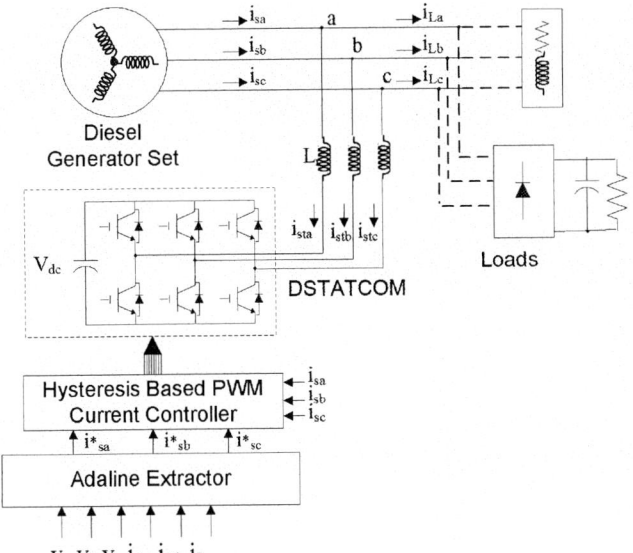

Fig. 1. Basic configuration of the DG set with DSTATCOM.

TABLE I
SYSTEM SPECIFICATIONS

| Load | Linear | Delta Connected R-L load of 37.5kVA at 0.8pf |
|------|--------|-----------------------------------------------|
|      | Non-linear | 30kW Diode bridge converter with LC filter at output with L=2mH and C=500μF |
| Voltage Source Converter | | DC link capacitor C$_{dc}$=10000μF, AC inductor=3mH, Ripple Filter: C=10μF and R=8Ω, f$_s$=20kHz. |

X$_q$''=0.6, H$_s$=0.08. the other critical parameters are given in Table I.

## III. CONTROL ALGORITHM

The operation of the system requires DG set to supply real power needed to the load and some losses (switching losses of devices, losses in reactor and dielectric losses of dc capacitor) in the DSTATCOM. Therefore, the reference source current used to decide the switching of the DSTATCOM, has two parts one is real fundamental frequency component of the load current, which is being extracted using Adaline and another component, which corresponds to the losses in the DSTATCOM, are estimated using a PI controller over dc voltage of DSTATCOM. Fig. 2a shows the scheme for the implementation of reactive, unbalanced and harmonic currents compensation. The output of the PI controller is added to the weight calculated by the Adaline to maintain the dc bus voltage of the DSTATCOM.

### A. Extraction of Real Positive Sequence Fundamental Frequency Current from Load Current

The basic theory of the proposed decomposer is based on Least Mean Square (LMS) algorithm [8] and its training through Adaline, which tracks the unit vector templates to maintain minimum error. The basic concept of theory used here can be understood by considering the analysis in single-phase system which is given as under. For an AC system the

supply voltage may be expressed as:

$$v_s = V \sin \omega t \tag{1}$$

Load current consists of active current (i$_p$), reactive current (i$_q$) for positive sequence, negative sequence current ( i$^-$ ) and harmonic frequency current (i$_h$) can be decomposed in parts as:

$$i_L = i_p^+ + i_q^+ + i^- + i_h \tag{2}$$

The control algorithm is based on the extraction of current component in phase with the unit voltage template. To estimate fundamental frequency positive sequence real component of current, the unit voltage template should be in phase with the system voltage and should have unit amplitude. The unit voltage template derived from the system phase voltage can be represented as:

$$u_p = v_s / V \tag{3}$$

For proper estimation of current components of load current, unit voltages templates must be undistorted. In case of voltage being distorted, the zero crossing of phase voltage is detected to generate sinusoid (sinωt) vector template, synchronized with ac mains. The signal is generated from look-up table by adjustment of delay to track the change in frequency of the ac mains.

The initial estimate of active part of current for single-phase can be chosen as:

$$i_p = W_p u_p \tag{4}$$

where weight ( W$_p$ ) is estimated using Adaline. The weight is variable and changes as per the load current and magnitude of phase voltage. Scheme for estimating weights corresponding to fundamental frequency real component of current (for three-phase system), based on LMS algorithm tuned Adaline tracks the unit vector templates to maintain minimum error. The estimation of weight is given as per the following iterations:

$$W_{p(k+1)} = W_{p(k)} + \eta\{i_{L(k)} - W_{p(k)}u_{p(k)}\}u_{p(k)} \tag{5}$$

The value of η(convergence coefficient) decides the rate of convergence and accuracy of estimation. The practical range of convergence coefficient lies in between 0.1 to 1.0. Three-phase reference currents corresponding to positive sequence real component of load current may be computed as:

$$i_{pa}^+ = W_p^+ u_{pa}; \; i_{pb}^+ = W_p^+ u_{pb}; \; i_{pc}^+ = W_p^+ u_{pc} \tag{6}$$

$$W_p^+ = (W_{pa}^+ + W_{pa}^+ + W_{pc}^+)/3 \tag{7}$$

For proper estimation of reference current signals, the weights are averaged to compute the equivalent weight for positive sequence current component in the decomposed form. The averaging of weights helps in removing the unbalance in the current components.

### B. PI Controller for maintaining constant DC bus voltage of DSTATCOM

To compute the second component of reference active current, a reference dc bus voltage is compared with sensed dc bus voltage of DSTATCOM. The comparison of sensed dc bus voltage to the reference dc bus voltage of VSC, results in a

615

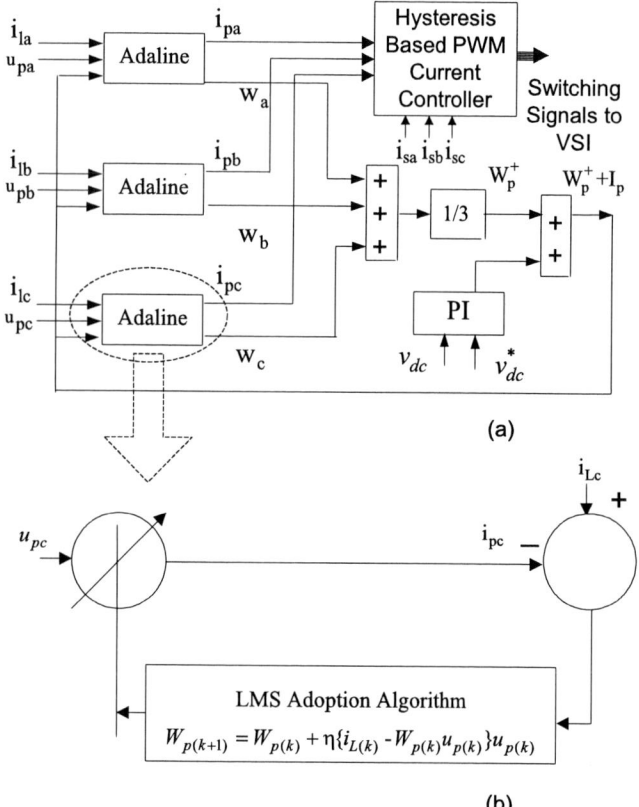

(a)

(b)

Fig. 2(a)-(b). Control block diagram of the reference current extraction scheme.

voltage error, which in the nth sampling instant is expressed as:

$$v_{dcl(n)} = v_{dc(n)}^* - v_{dc(n)} \qquad (8)$$

The error signal, $v_{dcl(n)}$, is processed in a PI controller and output $\{ I_{p(n)} \}$ at the nth sampling instant is expressed as:

$$I_{p(n)} = I_{p(n-1)} + K_{pdc} \{ v_{dcl(n)} - v_{dcl(n-1)} \} + K_{idc} v_{dcl(n)} \qquad (9)$$

where $K_{pdc}$ and $K_{idc}$ are the proportional and integral gains of the PI controller.

The output of PI controller accounts for the losses in DSTATCOM and it is considered as loss component of the current, which is added with the weight estimated by the Adaline corresponding to fundamental frequency positive sequence reference active current component. Therefore, the total real reference current has component corresponding to the load and component corresponding to feed the losses of DSTATCOM, is expressed as:

$$i_{sa}^* = (W_p^+ + I_p)u_{pa}; \; i_{sb}^* = (W_p^+ + I_p)u_{pb}; \\ i_{sc}^* = (W_p^+ + I_p)u_{pc} \qquad (10)$$

These 3-phase currents are considered reference supply currents $i_{ref}$ ($i_{sa}^*, i_{sb}^*$ and $i_{sc}^*$) and along with sensed source currents $i_{act}$ ($i_{sa}, i_{sb}$ and $i_{sc}$) these are fed to the hysteresis based PWM current controller to control the supply currents to follow these reference currents. The switching signals generated by PWM current controller force the actual supply currents to acquire shape close to the reference supply current.

This indirect current control results in control of the slow varying source current (as compared to DSTATCOM currents) and therefore requires less computational efforts. Moreover this scheme automatically compensates the computational delay caused by the processor. The switching signals are generated on the following logic:

if ($i_{act}$)>($i_{ref}$+hb) upper switch of the leg is ON and lower switch is OFF

if ($i_{act}$)<($i_{ref}$+hb) upper switch of the leg is OFF and lower switch is ON

where hb is hysteresis band around the reference current $i_{ref}$.

The weights are computed online by LMS algorithm. The update equation of weights based on LMS algorithm is described in eqn. (5) for each phase. The structure of such Adaline is depicted in Fig. 2b. Weights are averaged not only for averaging at fundamental frequency but this cancel out the sinusoidally oscillating components in weights present due to harmonics in the load current. The averaging of weights in different phases is shown in Fig. 2a. Thus Adaline is trained at fundamental frequency of a particular sequence in-phase with voltage. Figs.2a-b show the detailed scheme implemented for control of DSTATCOM.

Due to the unbalance in the load currents the second harmonic ripple is produced in the dc bus voltage. This ripple has to be filtered out before feeding the signal to the PI controller; otherwise this may cause the harmonic components in the supply currents. For this purpose the dc bus voltage is filtered using a low pass filter (LPF). Since the major amount of the reference current (load current component) is computed using Adaline based extractor, the effect of the delay caused by the LPF is negligible in practical cases.

## IV. MATLAB SIMULATION

Fig. 3 shows the MATLAB model of the DSTATCOM-DG set isolated system. The modeling of the DG set is carried out using star connected synchronous generator of 30kVA, controlled by the speed governor and excitation system. The linear load applied to the generator is at 0.8 lagging pf which is modeled as the delta connection of the series combination of the resistance and inductance (R-L) models. The nonlinear load is modeled using discrete diodes connected in bridge with

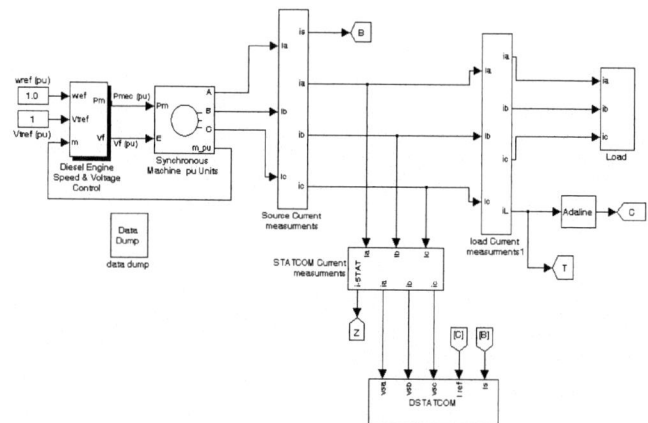

Fig. 3. MATLAB based simulation model.

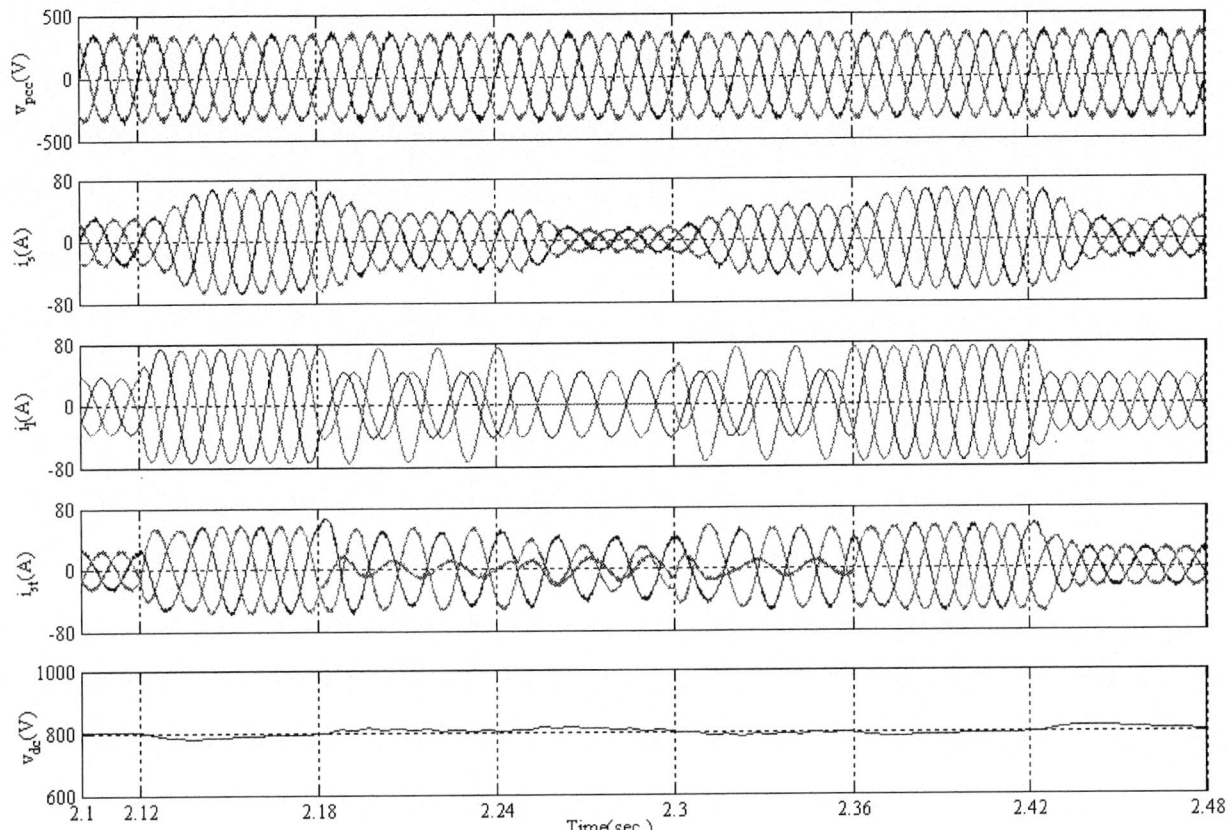

Fig. 4. Dynamic performance of the DSTATCOM-DG isolated system with linear load.

capacitor filter and resistive load on the dc side. The unbalance is realized by disconnecting phase-a from the diode bridge. The simulation is carried out in continuous mode at 1*10-6 step size with ode15s(stiff/NDF).

## V. RESULTS AND DISCUSSION

The simulation of the DSTATCOM-DG set isolated system is carried out with different types of loads i.e. linear R-L load, nonlinear load i.e. diode bridge converter load. The load compensation is demonstrated for these types of loads using DSTATCOM system for isolated DG set. The following observations are made on the basis of obtained simulation results under different system conditions.

### A. Operation with Linear Load

Fig. 4 shows the dynamic performance of the DSTATCOM-DG system. From t=2.10sec to 2.12sec a 3-phase 18.75kVA load at 0.8pf is being connected. At t=2.12sec the load is increased upto 37.5kVA at 0.8pf. The real power supplied by the DG set is 30kW and the reactive power is supplied by the DSTATCOM. At t=2.18 sec unbalanced is introduced by taking off the load from phase a, it can be easily observed that even if the load currents ($i_L$) are unbalanced the source currents ($i_s$) are still balanced. At t=2.24sec load is taken out from b phase also and even in this condition the DSTATCOM system is able to balance the DG set currents. For time t=2.3sec to t=2.48sec these dynamics are shown in the reverse sequence of events. The dc bus voltage is well maintained at 800V during the complete range of operation and the small sag

and swell in the voltage at the load change are compensated by PI controller action.

### B. Operation with Non-Linear Load

Fig. 5 shows the performance of the DG set with DSTATCOM - DG set under nonlinear loading conditions. The load on the system is kept 15.0kW initially for time t=2.1sec to2.12sec. The load compensation in terms of harmonic mitigation is also being provided by the DSTATCOM during this condition. Load is increased to 30kW at t=2.12sec. At t=2.18 sec the unbalance is introduced and therefore the load is reduced to 16.4kW. At t=2.30sec, phase-a is reconnected again to the diode bridge and the load is reduced to its initial value (15.6kW) at t=2.42sec to demonstrate the dynamics in reverse sequence of events. The harmonic spectrum of the phase-a load and supply currents are shown in Fig. 6a-c for peak load condition. %THD values of voltage at PCC, load and source currents are given in Table II for light load and peak load conditions. The high value of the %THD of the voltage at PCC is due to the high source impedance of the generator. The improvement in the voltage waveform is achieved using a ripple filter employed at the DG set terminal comprising of a capacitance-resistive high pass filter. The DG set currents and voltages are observed to be almost sinusoidal and balanced and operating at unity power factor.

## VI. CONCLUSION

The proposed control of DSTATCOM has been able to

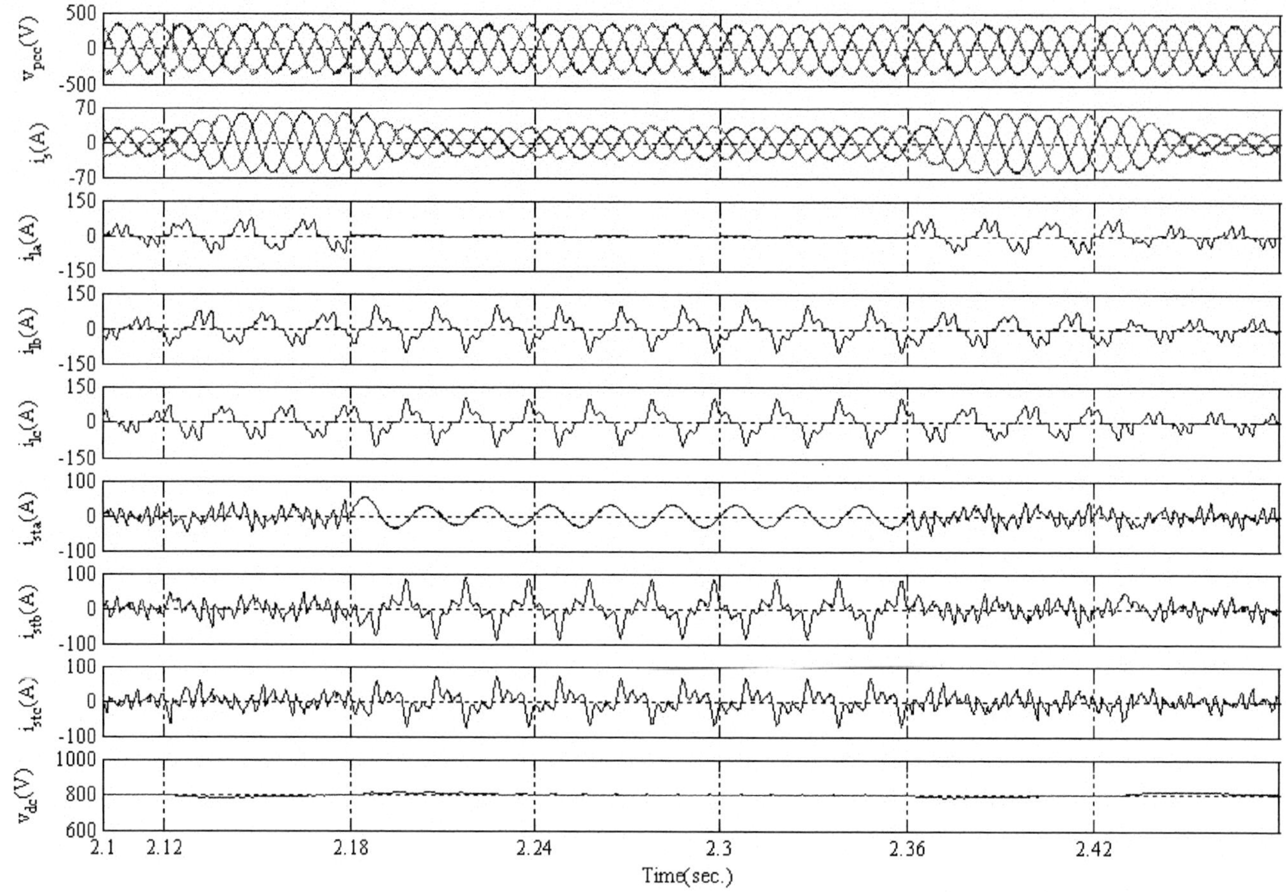

$v_{pcc}$(V)
$i_s$(A)
$i_{La}$(A)
$i_{Lb}$(A)
$i_{Lc}$(A)
$i_{sta}$(A)
$i_{stb}$(A)
$i_{stc}$(A)
$v_{dc}$(V)
Time(sec.)


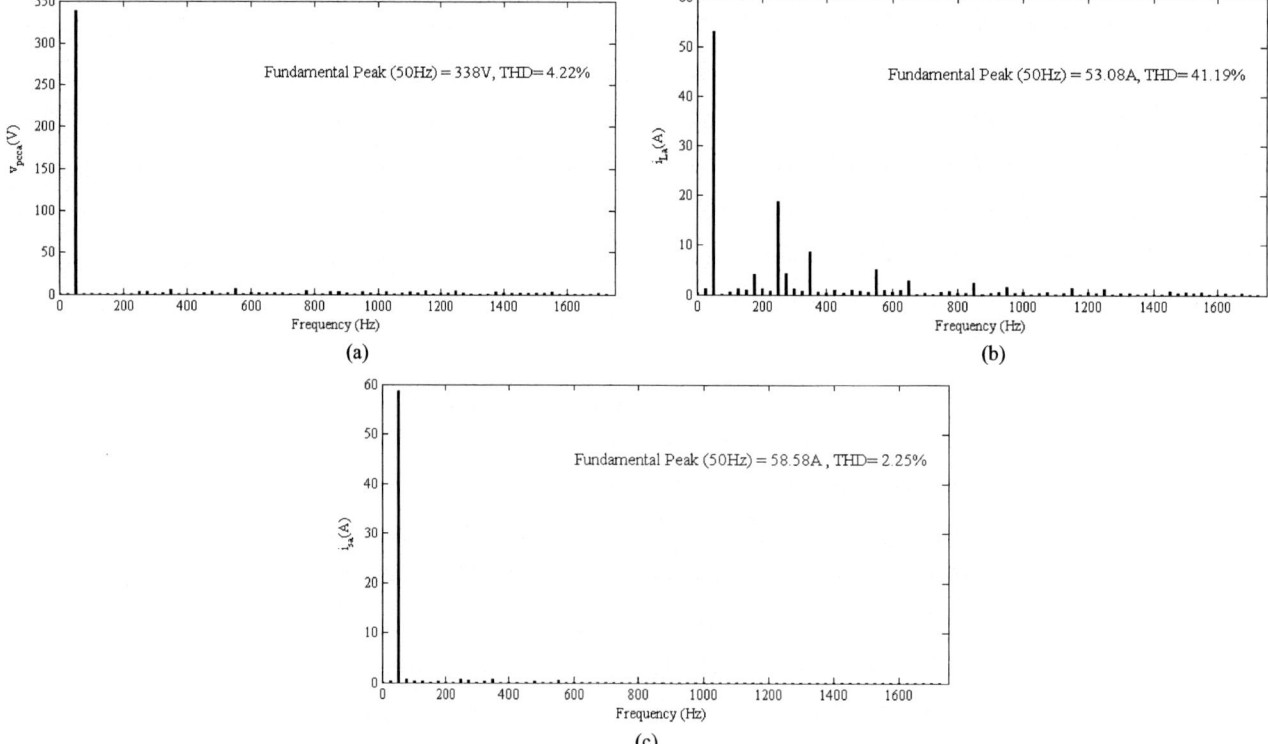

Fig. 5. Dynamic performance of the DSTATCOM-DG isolated system with non-linear load.

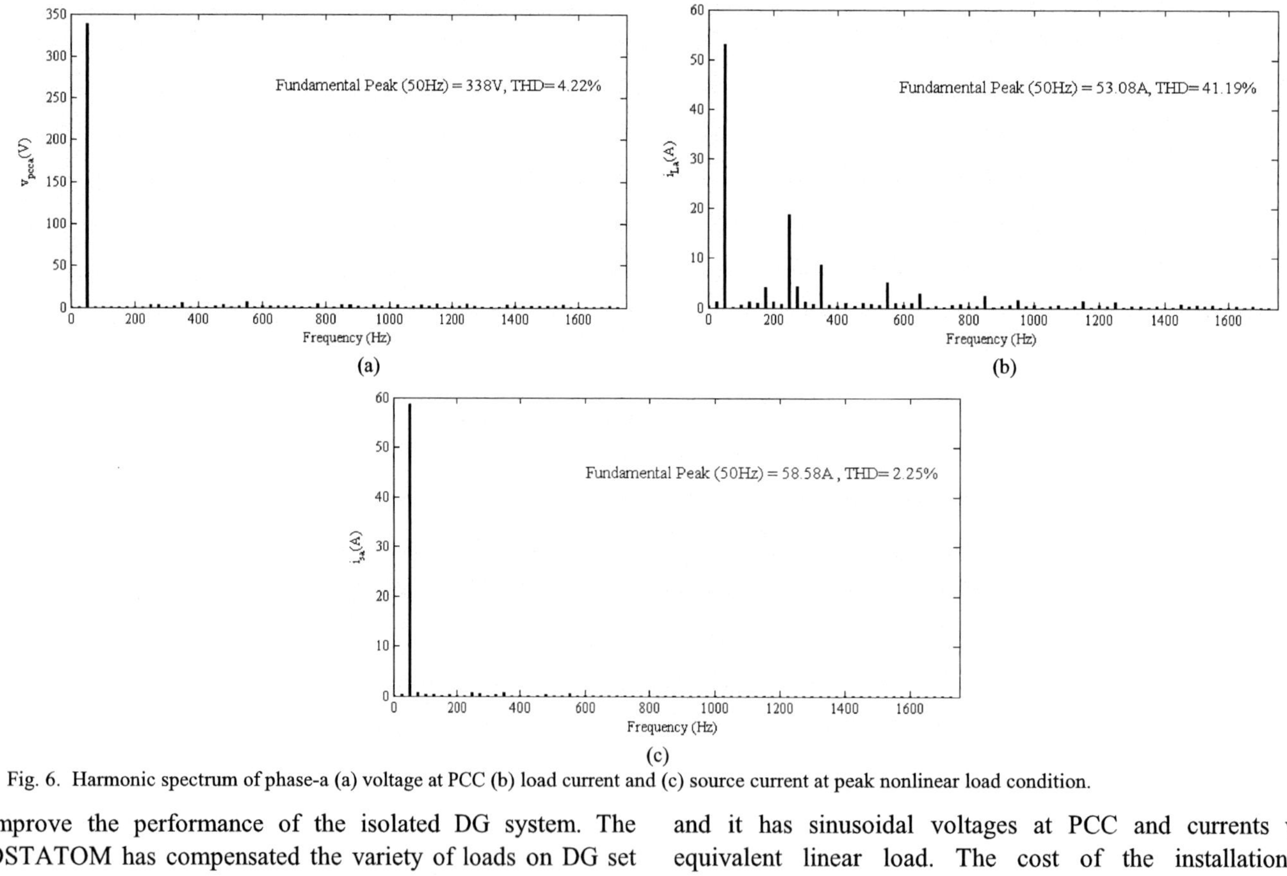

Fundamental Peak (50Hz) = 338V, THD= 4.22%
$v_{pcca}$(V)
Frequency (Hz)
(a)
Fundamental Peak (50Hz) = 53.08A, THD= 41.19%
$i_{La}$(A)
Frequency (Hz)
(b)
Fundamental Peak (50Hz) = 58.58A , THD= 2.25%
$i_{sa}$(A)
Frequency (Hz)
(c)


Fig. 6. Harmonic spectrum of phase-a (a) voltage at PCC (b) load current and (c) source current at peak nonlinear load condition.

improve the performance of the isolated DG system. The DSTATOM has compensated the variety of loads on DG set and it has sinusoidal voltages at PCC and currents with equivalent linear load. The cost of the installation of

TABLE II
%THD OF THREE –PHASE VOLTAGE SAT PCC, LOAD CURRENTS AND
SOURCE CURRENTS WITH NON-LINEAR LOADS

| Condition | | % THD | | | RMS Voltage (V) and Currents (A) | | |
|---|---|---|---|---|---|---|---|
| | | Ph. a | Ph. b | Ph. c | Ph. a | Ph. b | Ph. c |
| Light Load Condition | $V_{pcc}$ | 3.49 | 3.72 | 3.43 | 239.5 | 239.5 | 239.6 |
| | $i_L$ | 69.99 | 68.85 | 70.76 | 23.2 | 23.2 | 23.3 |
| | $i_S$ | 3.73 | 3.72 | 4.10 | 20.3 | 20.3 | 20.4 |
| Peak Load Condition | $V_{pcc}$ | 4.22 | 4.19 | 4.35 | 239.4 | 239.5 | 239.5 |
| | $i_L$ | 41.19 | 38.81 | 38.89 | 42.5 | 42.4 | 42.6 |
| | $i_S$ | 2.25 | 2.00 | 2.14 | 41.4 | 41.3 | 41.4 |

DSTATCOM system with DG set can be compensated as it leads to less initial and running cost of DG set.

## REFERENCES

[1] IEEE Standard Criteria for Diesel-Generator Units Applied as Standby Power Supplies for Nuclear Power Generating Stations, IEEE Std 387-1995, pp. 1 – 85.

[2] B. Singh, A. Adya, A.P. Mittal and J.R.P. Gupta, "Performance of DSTATCOM for Isolated Small alternator feeding non-linear loads," in *Proc. of International Conf. on Computer Application in Electrical Engineering Recent Advances*, 2005, pp. 211-216.

[3] E. Acha, V.G. Agelidis, O. Anaya-Lara and T.J.E. Miller, Power electronic control in electrical systems, England, Newnes, 2002.

[4] Akagi H., Kanazawa Y. and Nabae A., "Generalized theory of the instantaneous reactive power in three-phase circuits," in *Proc. of IEEE Proceedings IPEC*. Tokyo, 1983, pp. 821-827.

[5] A. Chandra, B. Singh, B.N. Singh and K. Al-Haddad, "An improved control algorithm of shunt active filter for voltage regulation, harmonic elimination, power-factor correction, and balancing of nonlinear loads," *IEEE Transactions on Power Electronics*, Vol. 15, No. 3, May 2000, pp. 495 – 507.

[6] G.D. Marques, "A comparison of active power filter control methods in unbalanced and non-sinusoidal conditions," in *Proc. of IEEE Annual Conference of the Industrial Electronics Society*, Vol. 1, 1998, pp. 444 –449.

[7] B. Widrow and A. Mchal Lehr, "30 years of adaptive neural networks perception, madaline and backpropagation," *IEEE Proceedings*, Vol. 78 pp. 1415-1442, September 1990.

[8] B. Widrow, J.M. McCool, and M. Ball, "The complex LMS algorithm," *IEEE Proceedings*, Vol. 63, no 4, 1975, pp-719-720.

**Bhim Singh** (SM'99) was born in Rahamapur, U. P., India in 1956. He received B. E. (Electrical) degree from the University of Roorkee, India in 1977 and M. Tech. and Ph. D. degrees from the Indian Institute of Technology (IIT), New Delhi, in 1979 and 1983, respectively. In 1983, he joined as a Lecturer and in 1988 became a Reader in the Department of Electrical Engineering, University of Roorkee. In December 1990, he joined as an Assistant Professor, became an Associate Professor in 1994 and Professor in 1997 at the Department of Electrical Engineering, IIT Delhi. His field of interest includes power electronics, electrical machines and drives, active filters, static VAR compensator, and analysis and digital control of electrical machines. Dr. Singh is a Fellow of Indian National Academy of Engineering (INAE), the Institution of Engineers (India) [IE (I)], and the Institution of Electronics and Telecommunication Engineers (IETE), a Life Member of the Indian Society for Technical Education (ISTE), the System Society of India (SSI), and the National Institution of Quality and Reliability (NIQR) Senior Member of Institute of Electrical and Electronics Engineers (IEEE).

**Jitendra Solanki** was born in Agra, Uttar Pradesh, India in 1981. He has received B-Tech degree in 'Electrical Engineering' from G. B. Pant University of Ag. & Tech., Pantnagar, India in 2004 and M-Tech degree in 'Power Electronics, Electrical Machines and Drives' from IIT Delhi, India in 2006. He is presently with GE Global Research, GE India Technology Center, Bangalore, India. His research interest includes application of power electronics in power system and electric drives.

# Automatic Classification of Power Quality Events Using Multiwavelets

Surender Dahiya, D.K. Jain, Manish Kumar, Ashok Kumar, and Rajiv Kapoor

*Abstract--***Multiwavelets technique is here proposed to classify the Power Quality (PQ) events. This leads to easy extraction of feature set. In the proposed classification scheme initially the events are detected from the test data in accordance with the IEEE standards. Then, two sub-classifiers, namely, chi-square distribution and heuristic classifier with different confidence levels have been used along with the Dempster-Shafer (DS) class for final decision making.**

*Index Terms--***Event Classification, Multiwavelets, Power Quality.**

## I. INTRODUCTION

POWER quality refers to the measure, analysis and improvement of bus voltage to be a sinusoid at rated voltage and frequency[1-2]. There are different types of transient disturbances that leads to malfunction of sensitive equipments, failure of protection and relaying and malfunction of electronic control. Accurate and robust on line monitoring of power quality is essential to take measure to mitigate the power quality problem. Techniques using the conventional signal processing tools have been developed to detect and classify the events. These are fast fourier transforms (FFT), Kalman filters. FFT performs well for estimation of periodic signals in stationary state; however, it is not effective for detection of sudden or fast changes in waveform e. g. transients or voltage dips.

The wavelets based analysis [3-7] has been proposed to study power system non-stationary harmonics distortion. Wavelets have been used to decompose the signal in different frequency sub-bands and study separately its characteristics. As described, Wavelets performs better with non-periodic signals that contain short duration impulse components as is typical in power systems transients. A number of 1D wavelets have been applied to power system events, like, Daubechies, Dyadic, Coiflets, Moriet and Symlets and the results were

---

Surender Dahiya and Dr. D. K Jain are with Department of Electrical Engineering,C.R. State College of Engineering, Murthal, Haryana, India. dahiyasurender@hotmail.com, jaindk66@yahoo.co.in .
Dr. Ashok Kumar and Dr. Rajiv kapoor are with Y.M.C.A. Institute of Engineering & Technology, Faridabad, Haryana, India.
ashokarora123@yahoo.com, raj_himani12@rediffmail.com .

found more suitable for power systems studies.

Wavelets are a useful tool for signal processing applications such as image compression. Until recently, only scalar wavelets were known i.e. wavelet generated by one scaling function. But with more than one scaling function it leads to multiwavelets. Multiwavelets have several advantages in comparison to scalar wavelets. Such features as short support, orthogonality, symmetry, and vanishing moments are known to be important in signal processing. A scalar wavelet cannot possess all these properties at the same time. On the other hand, a multiwavelets system can simultaneously provide perfect reconstruction while preserving length (orthogonality), good performance at the boundaries and a high order of approximation (vanishing moment). Thus multiwavelets offer the possibility of superior performance for pattern recognition applications, compared with scalar wavelets. In this paper, multiwavelets technique is proposed to analysis the power quality events and Dempster- Shafer has been used to take an intelligent and faster anniversaries. The section-2 contains the classification system. Section-3 explains the Multiwavelet Analysis, Section-4 gives classification and decision making. Section-5 gives Dempster-Shafer Technique, Section-6 explains results and the Section-7 gives conclusion.

## II. PROPOSED POWER QUALITY CLASSIFICATION SYSTEM

The proposed scheme for automatic classification system is shown in Fig. 1.

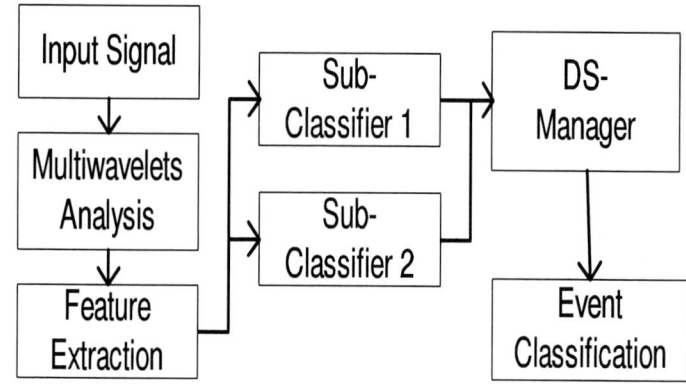

Fig. 1. Scheme of automatic classification.

Input signal is initially applied to pre-processing stage. In this block estimation of the signal components is preformed. Then, an algorithm for signal segmentation is applied e.g. separating the pre-event, or transition and post-event stages.

Feature extraction can be done through wavelets or Kalman filter. Wavelets are mainly used to quantify features for different types of power system events. The proposed power quality events classification system uses multiwavelets analysis on the input signal. The classification stage is based on defined rules, e.g. knowledge based expert systems or any logic to discriminate different types of events. The features, based on the IEEE standards, are extracted after decomposition. These features are then compared with the IEEE standard and side-by-side their chi-square distribution factors are calculated.

Both the classifiers forward output to the DS classifier which acts as the intelligent decision maker which specifies the class of the power quality events.

### III. MULTIWAVELETS ANALYSIS

Multiwavelets are a new addition to the body of wavelet theory. Realizable as matrix-valued filter banks leading to wavelet bases, Multiwavelets offer simultaneously orthogonality, symmetry, and short support, which is not possible with scalar 2-channel wavelet systems. After reviewing this recently developed theory, the use of Multiwavelets in a filter bank setting for discrete time signal and pattern recognition has increased. Multiwavelets differ from scalar wavelets systems as they require two or more input streams to the Multiwavelets filter bank.

As in the scalar wavelet case, the theory of multiwavelets is based on the idea of Multi-resolution Analysis (MRA). The difference is those multiwavelets have several scaling functions. The standard multi-resolution has one scaling function $\varphi(t)$ and corresponding wavelet:

1. $\{0\} .... \subset V_{-1} \subset V_0 \subset V_1 \subset ...... \subset L^2(R)$
2. $\cap_j V_j = \{0\}$
3. $f(t) \in V_j \Leftrightarrow f(2t) \in V_{j+1}$
4. $f(t) \in V_0 \Rightarrow f(t-k) \in V_0$

There exists a function $\phi(t)$, called the scaling function, such that $\{\phi(t-k)\}$ is an orthogonal basis of $V_0$. The translates $\phi(t-k)$ are linearly are independent and produce a basis of the subspace $V_0$.

There is one wavelet w (t). It's translates w (t-k) produce a basis of the "detail" subspace $W_0$ to give $V_1$:

$V_1 = V_0 \oplus W_0$

For multiwavelets, the notion of MRA is the same expect that now a basis for $V_0$ is generated by translates of N scaling functions $[\phi_1(t-k), \phi_2(t-k), .... \phi_N(t-k)]$. The vector $\phi(t) = [\phi_1(t), ........, \phi_N(t)]^T$, will satisfy a matrix dilation equation:

$$\Phi(t) = \sum_k C[k]\phi(2t - k)$$

The coefficient C[k] is N by N matrices instead of scalars. Associated with these scaling functions are N wavelets $w_1(t), ..., w_N(t)$, satisfying the matrix wavelet equation:

$$W(t) = \sum_k D[k]\Phi(2t - k).$$

Again, $W(t) = [w_1(t), ..... w_N(t)]^T$ is a vector and the D[k] are N by N matrices. A very important Multiwavelets system was constructed by Geronimo, Hardin, and Massopust (GHM). Their system contains the two scaling functions $\phi_1(t)$, $\phi_2(t)$ and the two wavelets $w_1(t)$, $w_2(t)$. The dilation and wavelet equation for this system have four coefficients:

$$\Phi(t) = \begin{bmatrix} \phi_1(t) \\ \phi_2(t) \end{bmatrix} = C[0]\Phi(2t) + C[1]\Phi(2t-1) + C[2]\Phi(2t-2) + C[3]\Phi(2t-3)$$

$$W(t) = \begin{bmatrix} w_1(t) \\ w_2(t) \end{bmatrix} = D[0]\Phi(2t) + D[1]\Phi(2t-1) + D[2]\Phi(2t-2) + D[3]\Phi(2t-3)$$

There are four remarkable properties of the GHM scaling functions:

- They each have short support (the intervals [0, 1] and [0, 2]).
- Both scaling functions are symmetric, and the wavelets form a symmetric /anti-symmetric pair.
- All integer translates of the scaling functions are orthogonal.
- The system has second order of approximation (locally constant and locally linear functions are in $V_0$).

Both Fourier and Wavelet analysis was applied for extracting distinct features for various types of events as well as for characterizing the events. Based on FFT and wavelet analysis eight different features were extracted: the Fundamental Component, Phase Angle Shift, Total Harmonic Distortion, Number of Peaks of the Wavelet Co-efficient, Oscillation Number of the Missing Voltage, Lower Harmonics Distortion and Oscillation Number of the RMS Variation.

### IV. CLASSIFICATION AND DECISION MAKING

Features are the keys of a classifier on the basis of which the classifier is working; the light features set effects the performance of the classifier. The strong features set increases the efficiency of the classifier that leads to precise classification. So the features set are the key to get that best classification. The feature set is chosen here after rigorous analysis of the power quality events and are according to the IEEE. The Sub Classifier I and then Sub Classifier II are designed using the statistical distribution tool i.e. known as Chi-Square Distribution because of its fastness to find the distribution among the data.

Feature Set:

*1) Peak*

This denotes the maxima when the signal is undergoing transition i.e. from low to high and then high to low.

*2) Trough*

This denotes the minima when the signal is undergoing

transition i. e. from high to low and then low to high.

### 3) Rise-Time (Upper)

Rise time is the time taken by the signal to rise from the trough/zero-crossing (above or at the zero level) to reach the peak.

### 4) Fall Time (Upper)

This is the amount of time taken by the signal from the peak to either reach the trough (if above the zero level) or the zero crossing.

### 5) Duration (Upper)

Duration (Upper) = Rise-Time (Upper) + Fall-Time (Upper).

### 6) Rise-Time (Lower)

Rise time is the time taken by the signal to rise from the trough (below the zero level) to reach the zero-crossing.

### 7) Fall Time (Lower)

This is the amount of time taken by the signal from the zero-crossing to reach the trough.

### 8) Duration (Lower)

Duration (Lower) = Rise-Time(Lower) + Fall-Time (Lower).

### 9) Amplitude

It is Highest Value of the signal from the zero-level.

### 10) Frequency

Frequency = 1/ (Duration (Upper) + Duration (Lower))

For the classification of the impulsive-transient we have to consider the rise time and the decay time of the maximum magnitude and the duration of the event. Various events will have different combinations of the features

The Sub-Classifier I acts on the bases of IEEE standards. In these standards, following are the information which we have to extract from the signal. The chi-square coefficients of the each sample are given as:

The square of the difference of the word and mean of all the data is summed and divided by variance of that particular script. This gives unique coefficient regarding the distribution of the event related data. This is the factor which distinguishes the various patterns. This factor gives the unique factor of concern of each even; hence, this factor is used as a classifier.

## V. DEMPSTER-SHAFER TECHNIQUE

After getting the classification from both the classifier, a manager is required to distinguish between them. The Dempster-Shafer Technique is used for this purpose. Dempster-Shafer Theory (DST) is a mathematical theory of evidence. In a finite discrete space, Dempster-Shafer theory can be interpreted as a generalization of probability theory where probabilities are assigned to sets as opposed to mutually exclusive singletons. In traditional probability theory, evidence is associated with only one possible event. In DST, evidence can be associated with multiple possible events, e.g., sets of events. As a result, evidence in DST can be meaningful at a higher level of abstraction without having to resort to assumptions about the events within the evidential set. Where the evidence is sufficient enough to permit the assignment of probabilities to single events, the Dempster-Shafer model collapses to the traditional probabilistic formulation.

## VI. RESULTS AND DISCUSSIONS

### A. Model of Power Quality Issues

The model of power quality events are simulated in MATLAB for different events as shown in Fig. 2.

### B. Decomposition of the Power Quality Events

GHM Scaling Functions and Wavelets using Multiwavelets with the different level of decomposition for sample event namely impulsive transients are shown in Fig. 3.

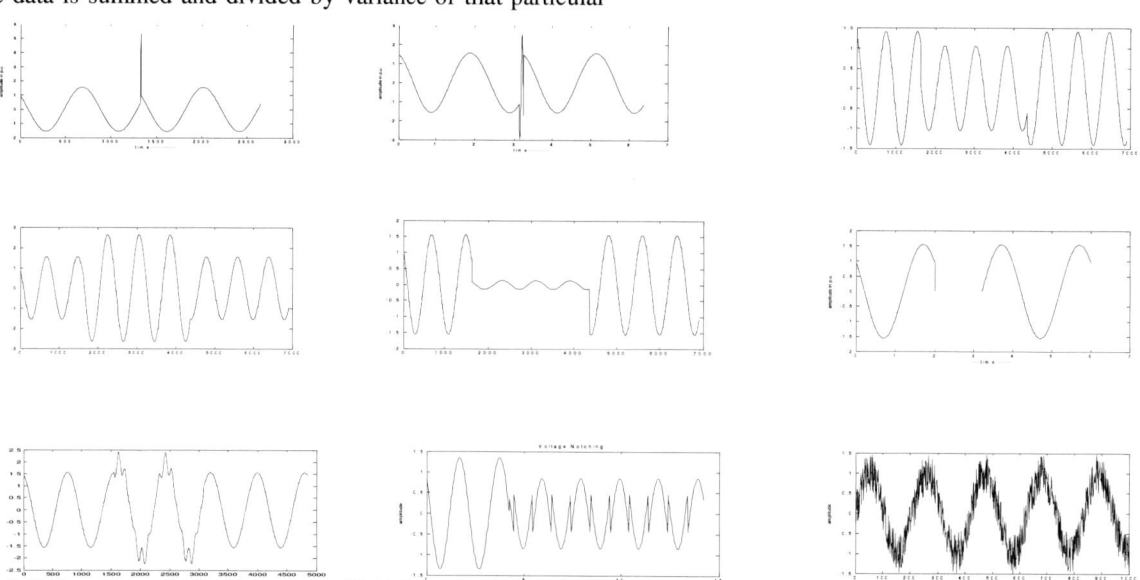

Fig. 2. Different Power Quality Events as specified below respectively: (a) Impulsive – Transient (b) Impulsive – Oscillatory (c) Voltage Sag (d) Voltage Swell (e) Interruption (f) Outage (g) Harmonics (h) Voltage Notch (i) Noise.

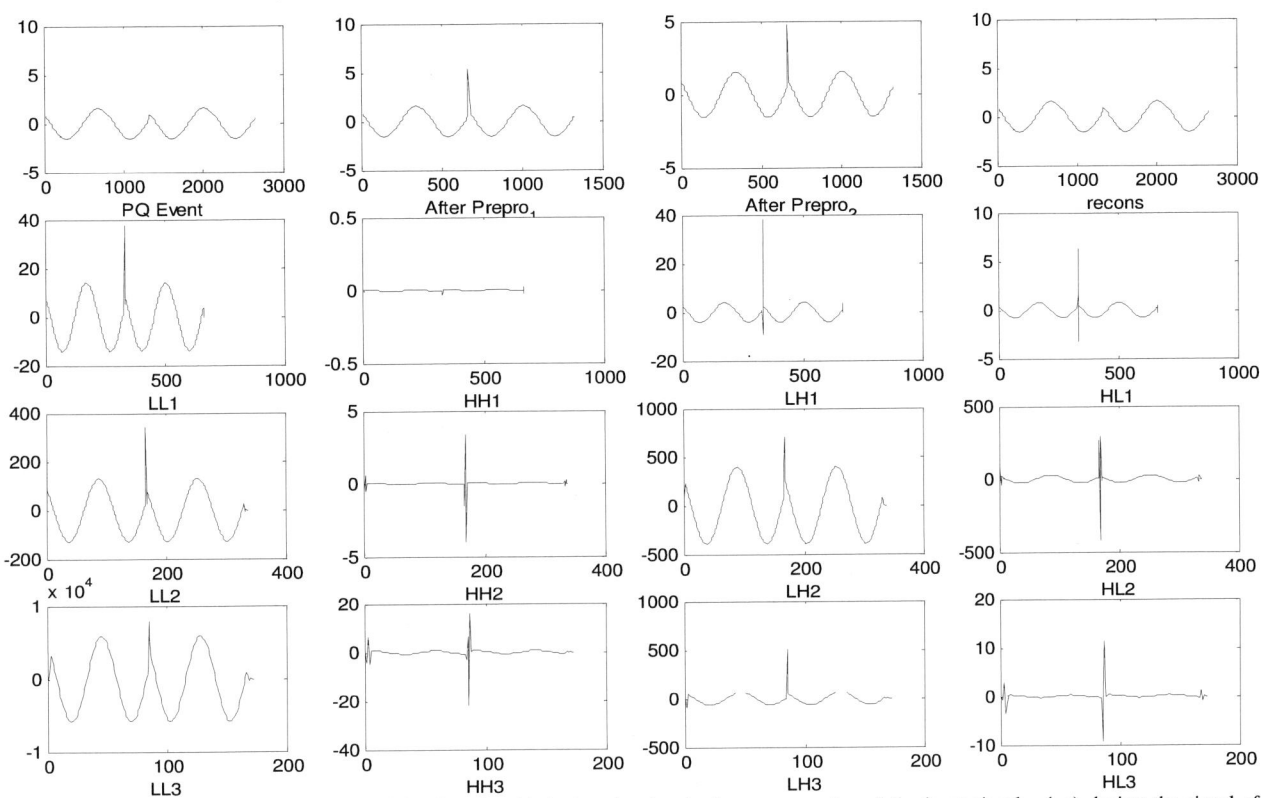

Fig. 3. a) Sample Power Quality Event is Impulsive-Transient b) depicts the signal after preprocessing of the input signal - 1 c) depicts the signal after preprocessing of the input signal – 2 d) depicts the reconstruction of the signal after full decomposition e), f), g) h) depicts the decomposition of the signal at level -1. LL1- depicts Low-Low Filter, HH1-depicts High-High Filter, LH1-depicts Low-High Filter, HL1-depicts High-Low Filter. i), j), k), l) depicts the decomposition of the signal at level 2. LL2- depicts Low-Low Filter, HH2-depicts High-High Filter, LH2-depicts Low-High Filter, and HL2-depicts High-Low Filter. m), n), o), p) depicts the decomposition of the signal at level -3. LL3- depicts Low-Low Filter, HH3-depicts High-High Filter, LH3-depicts Low-High Filter, HL3-depicts High-Low Filter.

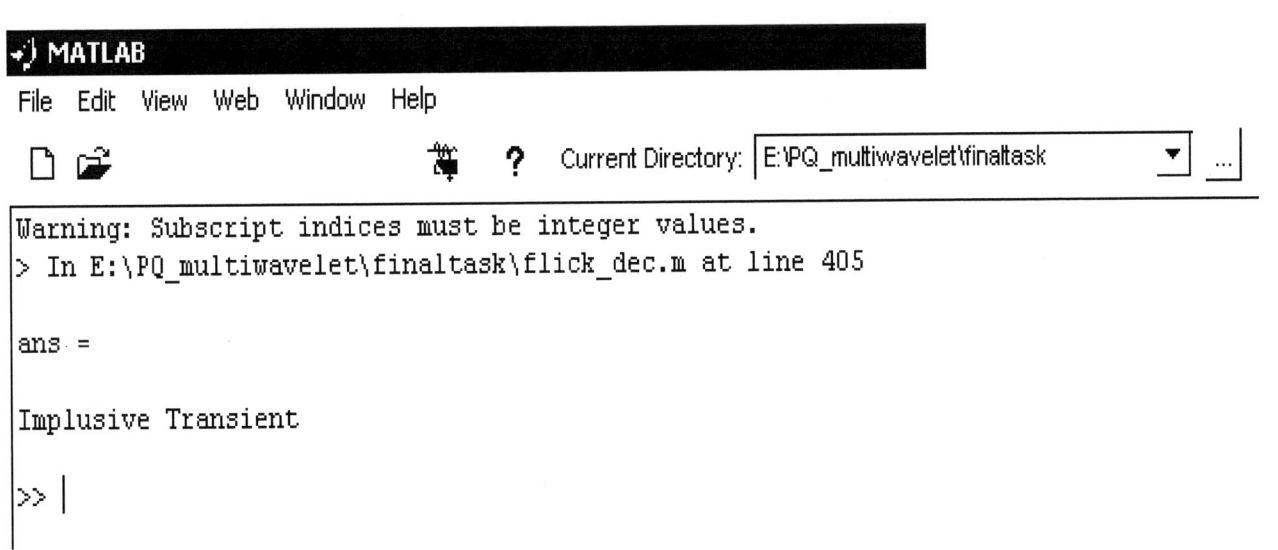

Fig. 4. Depict the output i.e. Impulsive Transient.

## C. Recognition Results

After the decomposition of the input signal, the features are extracted and event is recognized. The results are shown in Fig. 4. Like wise other power quality events have been precisely classified with the help of proposed method.

## VII. CONCLUSION

In this paper, investigation on the application of two-dimensional multiwavelets for the power quality event recognition has been carried out. The required recognition time will be very small. The performance of the multiwavelets is based on the selection of the wavelets, which can be seen from the results of the GHM Multiwavelets. The standard table for the comparison is used to ensure the correctness of the recognition process. The proposed system (which is based on GHM scaling functions and wavelets) is 93 % in case of the Power Quality Events.

## VIII. REFERENCES

[1] Bollen Math H.J. "Understanding Power Quality Problems-Voltage Sags and Interruptions", IEEE Press Series On Power Engineering, 2000.

[2] Heydt G.T. " Electric Power Quality" Star in a Circle Pub. 1994.

[3] Santos S., Powers E.J., Gray W.M., and Parsons A.C., "Power Quality Disturbance Waveform-based Neural Classifier ---Part 1: Theoretical Foundation" IEEE Trans. On Power Delivery, vol. 15, no. 1 pp.222-228, 2000.

[4] Santos S., Powers E.J., Gray W.M., and Parsons A.C., "Power Quality Disturbance Waveform-based Neural Classifier ---Part 2: Application", IEEE Trans. On Power Delivery, vol. 15, no. 1 pp.229-235, 2000.

[5] Gaouda A.M., Kanoum S.H., Salama M.M.A., and Chikhani A.Y., "Pattern Recognition Applications for Power System Disturbance Classification", IEEE Trans. On Power Delivery, vol. 17, no.3, pp. 677-683, 2002.

[6] Rieder P., Gotze J. and Nossek J.A., "Multiwavelet Transforms Based on Several Scaling Functions", IEEE , pp.393-396,1994.

[7] Dash P.K., Chilukuri M.V. and Panigrahi B.K., "Power Quality Analysis and Classification Using a Generalized Phase Corrected Wavelet Transform", IEE Conference Pub. No. 487 on Power Electronics and Drives, pp. 610-616, 2002.

**2006 IEEE International Conference on Power Electronic, Drives and Energy Systems**

# Power quality monitoring at the industrial, commercial and educational centers of Mazandaran province and presenting the related solution

M. Marzband, *Student, IEEE,* and A. Shaikholeslami, *Student, IEEE*

*Abstract*--The magnetic current of the transformers harmonics, the alternating switches, the increase of usage of power electronic devices in the computers, control process, adjustable speed drives and nonlinear loads has severely monitoring of the power quality the amount of present distortions in the network has been specified for the user, then by using it can present solutions such as filter design the change of the main harmonic producing sources in the system and etc. in this paper, several samples of measurements done at different existing centers at Mazandaran network which has been measured by the advanced Agilent device, is considered. We divide the centers into industrial, commercial and educational centers, and we consider the main harmonic generating sources at each section. From the obtained harmonic spectrum by using the SIMULINK/MATLAB software, we analyze the present figure.

*Index Terms*--Power quality monitoring, Harmonic generation sources, Power networks, Monitoring device, Harmonics, FFT

## I. INTRODUCTION

FIVE types of distortion in the waveform can be distinguished as follows: [1]-[2]
1- Offset 2- inter-harmonics 3- harmonics 4- notching 5- noise
We can divide the events that take place in the system into two categories. 1- Events that is temporary. 2- Events that exist in the system's stable state. From the first category we can name 1- the transients that take place in less than 0.5 cycles. 2- Excess and luck of voltage 0.5 cycles- 1 minute. 3- Long variations of the voltage more than one minute. 4- The Flicker. [3]-[4]-[5]

## II. DISCUSSION AND INVESTIGATING THE RESULTS OF THE MEASUREMENTS

### A. Harmonics Generated at Commercial Centers

Electronic Feed Sources: Personal computers, accessories such as printers and copying devices, work stations, all have

---

Part of the financial support of this work was supported by the Department of electrical engineering of the Mazandaran University and other part was supported by National Petrochemical Company.

M. Marzband is with National Petrochemical Company, PARS SPECIAL ECONOMIC/ENERGY ZONE, ASSALOUYEH, BUSHEHR, IRAN, PO.BOX NO.75391-115, TEL: +98 7727323250-4, FAX: +98 7727323255 (e-mail: m_marzband2005@yahoo.com).

A. sheikhol-Eslami is with the Department of Electrical Engineering, Mazandaran University, Babol, Iran (e-mail: abdolahinegar@yahoo.com).

electrical circuits for changing AC voltage to DC voltage; that are fed by microelectronic devices. Line current measurement from the branch of a circuit that only has computer load has been shown in Fig. 1. In Figures 2 & 3 by using Agilent device the voltage and single-phase current are measured from the electricity board of the computer site of the Technical-Engineering School of Mazandaran province and Yahya-Nejad Hospital. Fast switching and the high entry impedance in the IGBT transistors causes the inverters to generate unwanted circulative currents, these currents cause electromagnetic interference. [6]-[7]

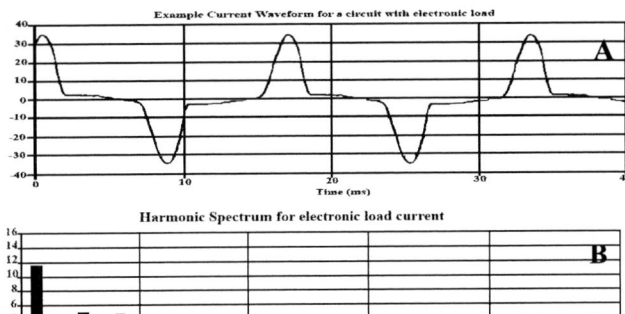

Fig. 1. A. Phase current waveform B. The related spectrum for electronic feed sources.

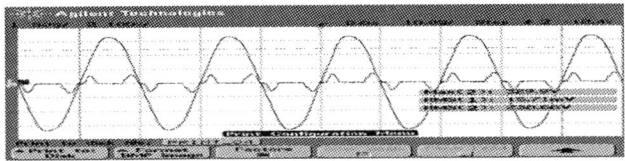

Fig. 2. Data taken from the computer site of Mazandaran University's Technical-Engineering School.

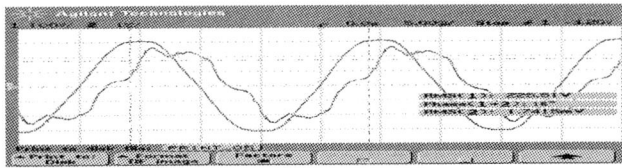

Fig. 3. Phase S with 240A current.

In the Figures 4 to 6 you see the voltage and current waveforms related to main panel of the school of Electrical Engineering. In Fig. 4 first the capacitance bank was separated from the network. As is seen from the waveform 4, we have a low harmonic distortion. But when a capacitance bank of

---

0-7803-9771-1/06/$25.00 ©2006 IEEE

55KVar was added to the network, we saw the distortion in the waveform. You can observe this fact in Fig. 5 When the total capacitive bank with the capacity 105KVar was added to the network, this distortion has increased. The capacitors are not harmonic generators but using capacitors in a network that has harmonic voltage or current causes the resonance effect to take place in the electrical network. The capacitive reactance decreases with frequency increase and the inductive reactance increase with frequency increase. As a result at a specific frequency by the name of resonance and the resonance effect takes place. When resonance take place depending on whether the resonance of the LC series circuit or the LC parallel circuit takes place the inductance's current amplitude or the voltage of the two sides of the capacitor may become a lot. Using the capacitor in the power network is generally for power factor correction and both kinds of series or parallel capacitors or a combination of them may be present. About the series circuit, at the resonance frequency the network's total impedance reduces to the present resistance's impedance and in the case of smallness of this resistive component lot of currents passes the circuit. In the parallel circuit, at the resonance frequency the total network's impedance is very large, as a result due to a very small excitation. a large circulative current passes between the inductor and the capacitor and the voltage on the two sides of the network becomes a lot. If a network's resonance frequency that has inductive-capacitive property be close to one at the harmonics generated in the circuit, very large currents or voltages in network having harmonics. Such a phenomenon can cause the distraction of the capacitive bank, incorrect and repeated performance of the capacitors plugs and the breaking of the electric insulation in the cables. As soon as we add capacitive bank to the network we have the voltage swell effect. This effect is a result of low term voltage increase for a period of several cycles. The amount of voltage increase is between 10 to 80 percent. Sudden break of a large load or the charge of a capacitive bank can cause this effect to take place. Also voltage swelling can take place because of single-phase error between the earths that causes the voltage of the two other phases to increase. [8]-[9]-[10]

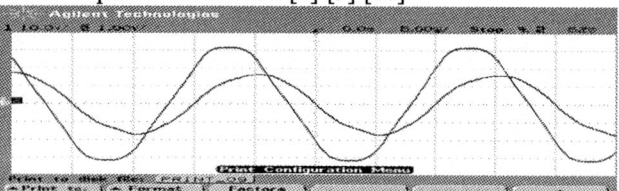

Fig. 4. Current and Voltage of the phase A Without the Presence of Capacitance Bank of 105Kvar.

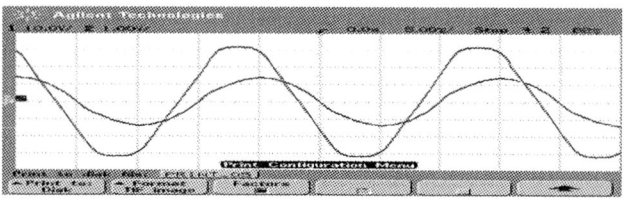

Fig. 5. Current and Voltage of the phase A With the Presence of Capacitance Bank of 55Kvar.

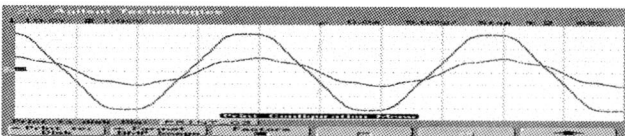

Fig. 6. Current and Voltage of the phase A With the Presence of Capacitance Bank of 105Kvar.

In Fig. 7 you observed the 20 kilovolt current and voltage of the Technical School. As you observe from the waveforms phase A of this network has the fifth harmonic with low amplitude; but also phase of B in addition to the fifth harmonic has the seventh harmonic and higher harmonics. By noting that the measuring time has been 10 o'clock in the morning and it is the opening time of mechanical laboratories and the start-up of CNC and milling machines. Having these harmonics is not far from expectation.

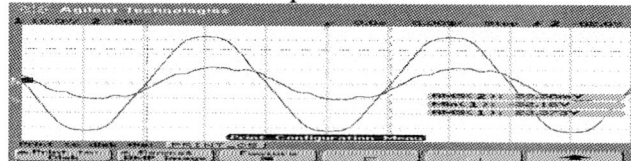

Fig. 7. current and voltage of the phase B of the 20KV network.

In Fig. 8 you observe the current and voltage of Babol's (city) Science and Technology University. Here you observe the effect of the presence of the capacitive bank. In Fig. 9 you observe the current waveform's harmonic spectrum. From the harmonic spectrum you can find out which harmonics and inter-harmonics have the most amplitude. As you observe from the graphed harmonic spectrum by the MATLAB/SIMULINK software, the amplitude of the third and fifth harmonics is very significant.

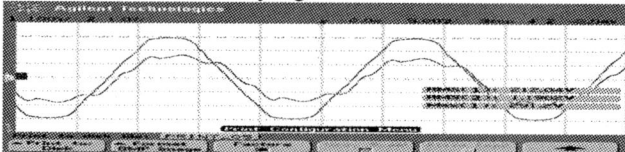

Fig. 8. Data taken with the presence of the capacitive bank.

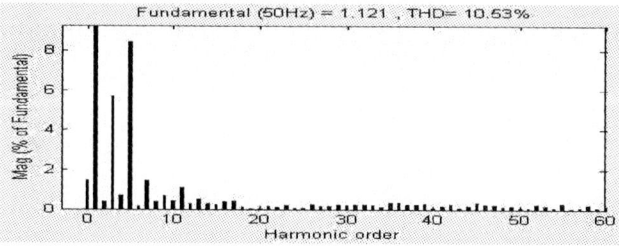

Fig. 9. Current Waveform and the related Harmonic Spectrum in the Low Range

Also an experiment has been done in the machine laboratory. In this experiment a motor that consumed 5A current was connected in series low resistance to a high power resistance low resistance. So by reading the voltage on its two sides we can observe the current's waveform. In figures 10 and 11 you observe the related waveform and the harmonic spectrum. As you observe the harmonic spectrum in addition to odd harmonics we have add harmonics also. Because of the dent's ripple or ripple in the rotating machine's voltage waveform the non-sinusoidal flux distortion in the air gap and the magnetizing currents that pass the magnetizing coils can also cause the generation of these harmonics.

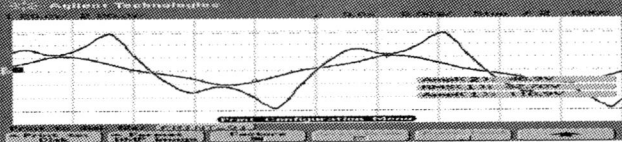

Fig. 10. Voltage and current on the two sides of the high power load in the machine laboratory.

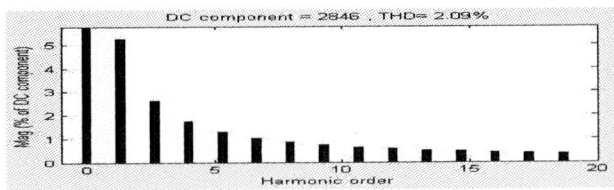

Fig. 11. Harmonic spectrum related to current at low frequencies.

Fluorescent Lamp: The relationship between the fluorescent lamp's voltage and current is nonlinear. Because its electrical characteristic is non-linear. Fluorescent lamps cause significant increase in the odd order harmonic currents. The generated harmonic current by the fluorescent lighting systems are severely effected by the type of lamp blasts that are selected. Because fluorescent lamps with magnetic blasts draw non-sinusoidal currents, recently more attention has been paid to electronic blasts. The harmonic generation characteristic in the electronic blasts is much higher than 5 to 35 percent. The generated harmonic current in the electronic blasts because of the performance of the single-phase dyadic bridge rectifier is only observed in the electronic feed sources. The correction circuit of passive's power coefficient is usually used for reducing distortion surfaces in the entrance current. The light circuit's line current uses a common type of electronic blast as shown in Fig. 12 Also in Fig. 13 you observe the voltage and current waveforms related to Babol's (city) health center. In this center our major load is fluorescent lamps. [11]-[12]

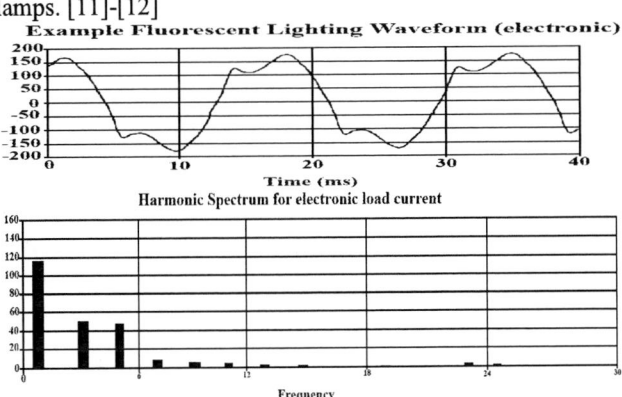

Fig. 12. Phase current and its related spectrum for fluorescent lamps with electronic blast.

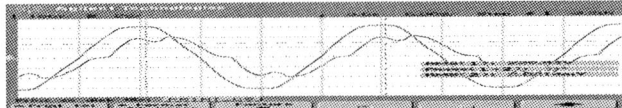

Fig. 13. Phase S with 65A current.

In Fig. 14 you observe the current and voltage of the different phases of Babol's (city) School Department. In this center in addition to computer, fluorescent lamps also exist. The harmonic spectrum of these waveforms has been mentioned in the following paragraphs. Also the harmonic spectrum of the taken waveforms has been mentioned in Fig. 15 Also in these waveforms we can observe the inter-harmonics effect. When distortion in the waveform causes the generation of sinusoidal components with frequencies that are not an integer coefficient of the main harmonic's coefficient such an effect is called inter-harmonics. This effect can be because of the remote control system's performance. As it is observed from the waveforms, in the harmonic spectrum, the amplitude of the fifth and nineteenth harmonics is significant. Of course mentioning this point is also essential that the school department's main panel was common with a private hospital. Also in the hospital in addition to the mentioned devices special monitors and medical devices exist that are nonlinear. In the S phase with 40A current, the fifth, seventeenth and nineteenth harmonics are significant. The current's THD value is averagely 13. The amount of each of the phases is observed in the figures. [13]

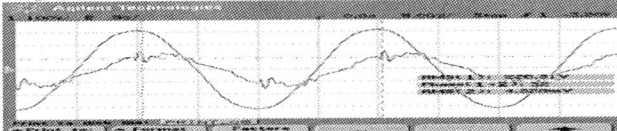

Fig. 14. Phase R with 30A current.

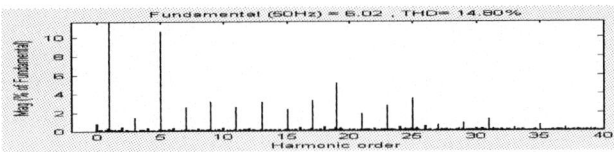

Fig. 15. Current waveform and its related harmonic spectrum.

The waveform of Fig. 16 is related to Science and Technology University. This waveform has been taken at the university's work peak hour the time of opening of the laboratory and holding of the classes. As is observed from the waveforms as a result of the presence of fluorescent lamp's major loads and the start-up of motor in the laboratory we observe the voltage in the Flicker waveform. When a capacitance bank was added to the network, we saw the distortion in the waveform as has been observed in fig. 17.

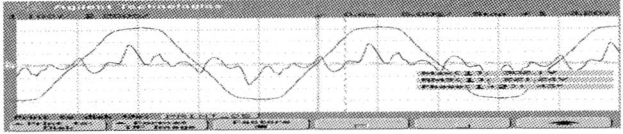

Fig. 16. Measurement with the presence of capacitive bank.

627

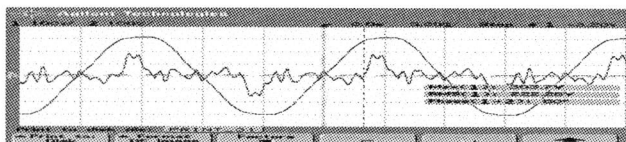

Fig. 17. Measurement without capacitive bank.

Harmonics Generated at Industrial Centers: Inductive motors are by the ASDs. A sample ASD for Heat Ventilation Air Conditioner systems is connected to on AC power system by a three phase diode bridge rectifier circuit [6]. The entrance current's waveform for a small ASD at one HVAC has been shown in Fig. 20 Also UPS systems with three-phase entrance rectifiers have similar characteristics.

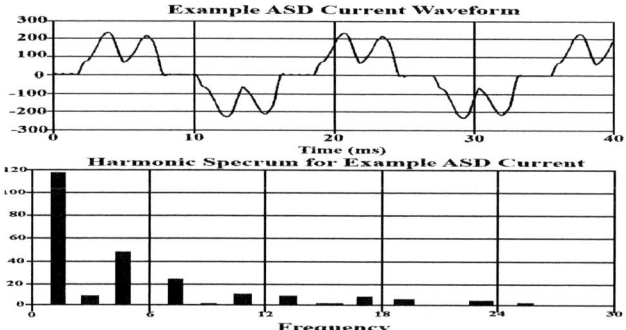

Fig. 18. Entrance Current and ASD Harmonic Spectrum for HVAC Applications.

Also in Figures 21 and 22 you observe the current and voltage waveforms related to Sari's (city) Akand Industrial Town's Cast Iron Factory. In this center the major load is the inductive furnace. Because of the existence of inductive furnace in this factory's system and using power electronic devices with high switching we have the effects of voltage distortion, inter-harmonics and noise in the waveform. Irregular sparks in the furnace causes the current conduction in the positive and negative half cycle not to be the same and this fact causes the power network connected to electric furnace to have harmonic pollution also. Harmonic currents that are measured for a scrap furnace or heater is generated when they are voltage is applied to the arc's electrodes and furnace's transformers. These harmonic currents are injected into the system and if the system doesn't have resonance in any of the dominant harmonics frequencies no problem would be generated. But if resonance exists at one of the above mentioned harmonics the related harmonic current can cause the excitation of the resonance circuit and can generate very large voltages. This fact damages the installed equipments in the system and will cause their destruction. [14]

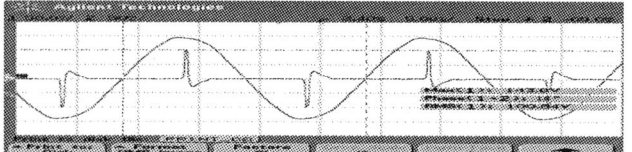

Fig. 19. Measurement of the Data with the Presence of Arc and the Capacitance Bank.

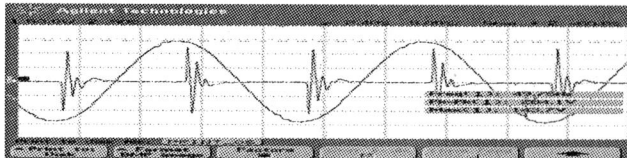

Fig. 20. Measurement of the Data with the Presence of Arc and without the Capacitance Bank.

The unbalanced performance of the furnaces in the unit has caused some problems sometimes. in this factory. That is as follows: 1- Numerous defects in the transformers. 2- The defectiveness of the capacitors and the repeated burning of the capacitor's plugs. One of the reasons for the highness of the harmonics amplitude can be the existence of resonance in the network when the feeding transformer is on a specific pulse. This suggestion can be given that when the unit's process conditions are changed in such a manner that when the network works with this transformer specific pulse the capacitors are withdrawn from the network. Also in Fig. 23 you observe the current and voltage waveforms related to Babol's (city) ice making factory. In here the HVAC[4] system was installed.

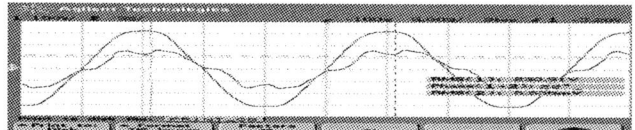

Fig. 21. Phase T with 26A Current.

In Figures 24 to 26 the current and voltage waveforms related to Asia Automobile Development Factory. In this factory the motor loads of welding, press and other devices exist. As a result they will have a significant effect on increasing harmonics to the network. Because of the very high voltage unbalance in this factory a significant current passes its null wire. High currents in the null wire in the four wire three phase systems causes the temperature increase in the null's conductor. Some loads like the switching feed sources, produce many third harmonics and in the three phase system, the third harmonic as opposed to the main harmonic that neutralize each other. are added together in the null wire and so the null wire's current can because up to 1.7 times the phases current. We can present several solutions regarding the ground currents excess problem: 1- The null wire conductor rate be increases. 2- Double null wire conductor is applied. 3- Having null conductor for each phase's conductor. 4- A zig zag transformer on the load side is chosen for the reduction of the null conductor's effect. 5- A series filter is installed for eliminating third harmonic current in the null wire. The improper distribution of the current in the conductor that includes the skin effect and the side by side effect; The skin effect increases with the frequency increase and the conductor's diameter. causes their temperature increase and the cable's insulation life reduction. In the waveforms the effect of transient impulsive voltage is observed. This effect is

---

[4] Heat Ventilation Air Conditioner System

known commonly as switching surge or the voltage spike. This effect takes place as a result of current's switching off not being adjusted, capacitors switching, lightening or the system's errors. In addition to this you observe the transient oscillatory voltage in the waveform. Transient oscillatory voltage is a sudden effect with a frequency outside the electricity's frequency that is known as ringing. The factors generating this effect are similar to impulse's transient voltage. As a result of starting-up large motors we have the voltage sag effect in the factory's electricity network. This effect has a short term loss in the voltage for a period of several cycles. The amount of voltage loss is between 10 to 20 percent and its duration is from half a cycle to one minute. Mostly voltage sag takes place as a result of staring-up large motors or the occurrence of an error in the adjacent power lines.

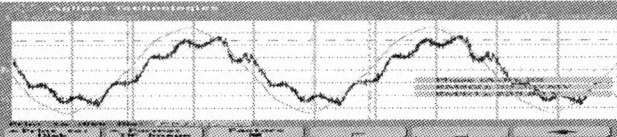

Fig. 22. Phase with 180A Current.

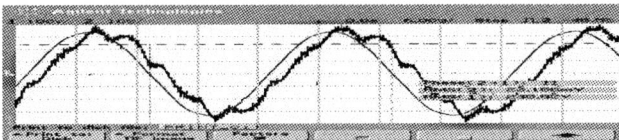

Fig. 23. Phase with 350A Current.

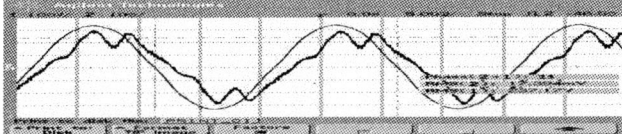

Fig. 24. Phase with 400A Current.

In figures 27 to 29 voltage and current waveforms related to Hasanpour Flower Factory are observed. In this factory, motor loads with high power existed. Also the controllers of these motors are electronic. Also the capacitance bank installed in this factory had an important effect on increasing harmonics. Also in these waveforms like the previous waveforms we have the impulse transient voltage and transient oscillatory voltage. Because of using different loads with different nominal loads, reducing transformers were used in this factory. The transformers have both linear and nonlinear characteristic. Impedance connector that is placed in the circuit in series is linear and magnetizing impedance that is placed in parallel in the circuit is nonlinear.

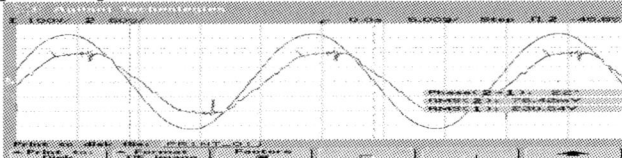

Fig. 25. Phase with 480A Current.

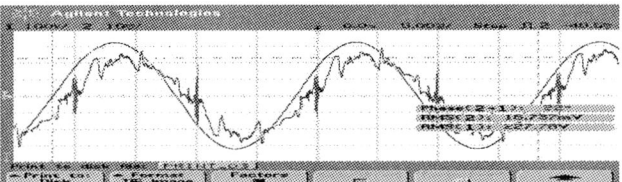

Fig. 26. Phase with 480A Current.

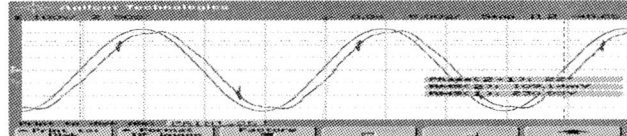

Fig. 27. Phase with 490A Current.

Also in Fig. 30 to 32 voltage and current waveforms related to Edalat-Qoreishi Paddy Beating Factory is observed. The waveform shows a very high harmonic content. According to the maintenance engineers saying several problems were found when the dejunctors were disconnected. The load current that has distortion at low level error, effects the error current and it is possible that it may have a high percentage of current distortion. But high level error currents are not effected by the load current distortion. When harmonic current exists, this current cans a higher $\frac{di}{dt}$ when passing null zero and as a result the ability for the key to switch off becomes more difficult. Current with 50 percent harmonic limits the dejunctor's switching off coil's ability in turning off the spark. In case the key is disconnected longer than the designated time, the error current is present in a longer period that can cause the switch to reconnect.

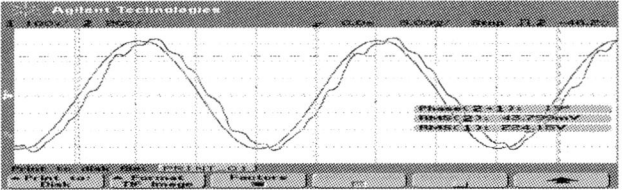

Fig. 28. Phase with 115A Current.

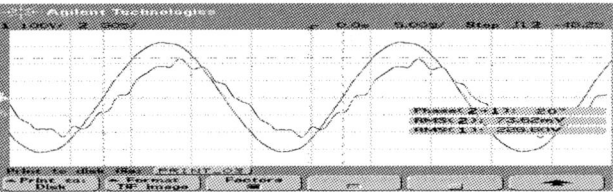

Fig. 29. Phase with 115A Current.

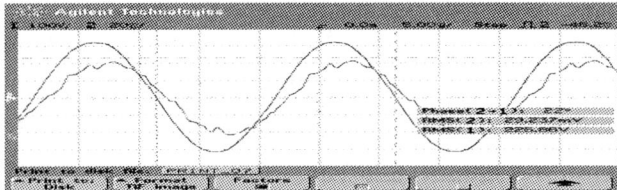

Fig. 30. Phase with 125A Current.

## III. CONCLUSION

In this paper it has been tried to analyze samples that have been taken from the main panel of each center and also to present some solutions. Mean while in this paper the effect of capacitance bank over the increase of harmonic generation has been investigated. The investigation done gives us the following results:

1- Harmonic pollution more than the allowable limit can cause problems and protective devices and damage to some of the devices. That by monitoring on time and by on time modifications such events can be prevented.

2- It is suggested that before taking a decision regarding installation of capacitors in the factory's electricity network, the consequences resulting from generating resonance resulting from the harmonics be investigated.

3- Power quality monitoring shall be done at different times and or by using registering devices that have memory.

4- The major distortion generating centers in the network are identified and solutions for decreasing these distortions be presented.

5- What is obtained in the laboratories in house hold and industrial consumptions are respectively the fifth and then the seventh harmonics and also for house hold and commercial loads the maximum component is the fifth harmonic and after it is the third harmonic.

6- The application of power coefficient correction capacitors is increasing and these capacitors will cause the change of system's frequency characteristic and the change of resonance at some of the frequencies and also the increase of harmonic voltage and as a result it is needed that at some of the network's points harmonic filters be installed. In other words the capacitors don't generate harmonics under any circumstances but their installation as modifier of power coefficient increases potential problems of the power coefficient and their presence in the inductive circuit essentially prepares the network for local resonance, general resonance and or magnifying the harmonics. These harmonics in the capacitor cause thermal and extra losses, explosion as a result of very high current, undesired phase relation between the applied harmonic and voltage and the increase of 10 percent of the phase voltage and the capacitor's life decreases as a result of Corona's Effect.

7- By noting the amplitude of the fifth and seventh harmonics of some of the measured centers we suggest placing filter for eliminating them.

8- Simultaneous using of several power quality measurement devices for investigating the mutual effect of the harmonics of different centers over the network is suggested.

## IV. REFRENCES

[1] T.Gomez, San Roman, J.Roman, "Power Quality Regulation in Argentina: Flicker and Harmonics", IEEE Trans, On Power Delivery, Vol.13, No.3, July 1998, PP.895-901.

[2] Paul Doig, Utility Marketing Manager, power measurement, creator of ION Technology

[3] McLellan, D.W.et al, "Telephone Interference from power Line Loads Regulated by Semiconductor Control Devices, "IEEE/IAS Annual Meeting, 1975.

[4] IEEE Working Group, "Power Line Harmonics Effects on Communication Line Interference," IEEE Trans, Vol. PAS-104, No. 9, Sept 1985.

[5] A. Teshome, W. Jewell, R Egbert, "Effect of harmonics on the Performance of AC Power Systems", 22nd annual edition of power conference, Still Water, Oklahoma, October 30- 31, 1989.

[6] Mahesh M. Swamy, Steven L. Rossiter, Michael C. Spencer, Michael Richardson, "Case Studies on Mitigating Harmonics in ASD Systems to Meet IEEE519-1992 Standards" IEEE I&CPS Conference 1994, pp. 685-692.

[7] B.C.Smith, N.R.Watson, A.R.Wood. J.Arrillaga, "Steady State Model of the AC/DC Converter in the Harmonic Domain", IEE Proc Gener, Trans Dist. Vol.142, No.2, March 1995.

[8] A.Mansoor, W.M.Grady, P.T.Staats, R.S.Thallam, M.T.Doyle, M.JSamotyj, "Predicting the Net Harmonic Currents Produced by Large Numbers of Distributed Single-Phase Computer Loads", IEEE Trans on Power Delivery, vol.10, No. 4 October 1995.

[9]M. Wright, P. Dembele, "A study of harmonic series-resonance in industrial plants with linear loads", Proceedings of the 7th international Conference on Harmonics and Quality of Power, Las Vegas, 1996, PP.307-313.

[10] A. M. Dan and A. Mohacsi, "Computer simulation of a three-phase AC electric arc furnace and its reactive power compensation", in Proceedings of IEEE, 1994., PP. 415-421.

[11] Collantes-Bellido, T. Gómez, "Identification and modeling of a three phase arc furnace for voltage disturbance simulation", IEEE Trans, Power Delivery, vol. 12, no. 4, Oct 1997.

[12] Javad Rohi, Seyed Ali Nabavi Niyaki, Abdolreze Shaikholeslami, hosian mohammadian, "Power quality in electrical systems", 1999, 964-6433-04-9, Babolsar, Mazandaran University.

[13] "IEEE Command Practice and Requirement for Harmonic Control in Electrical Power System", IEEE Standard 519-1992, 12 April 1993.

[14] "General guide on harmonics and interharmonics measurements and instrumentation for power supply systems and equipment connected there to", IEC 1000-4-7, 1991.

## V. BIOGRAPHIES

**Mosa Marzband** received his M.S. the electrical engineering faculty of the school of engineering of Mazandaran University, Iran in 2005. Currently he is working as a senior engineer in the electrical projects of Borzoye Petrochemical Company in the Pars Special Economic Energy Zone. His current interests are harmonics and power quality. He can be connected at m_marzband2005@yahoo.com.

**A.sheikhol-Eslami** was born in Iran on 1956. He received the B.S from Mazandaran University Iran, in 1979 and M.S ad PhD degree in electrical engineering from Strathclyde University U.K. in 1989. Since 1989, he has been Assistant Professor of Mazandaran University Iran. His research interests are power quality, power electronics and reacting power control.

**2006 IEEE International Conference on Power Electronic, Drives and Energy Systems**

# A New Power Quality Enhancement Method for Two-Phase Loads

H. Hojabri, and H. Mokhtari, *Member, IEEE*

*Abstract* — **This paper describes a new compensating system for the flicker and harmonics generated by a highly-varying load, e.g. a welding machine, connected between two phases of a three-phase system. The proposed system is an active filter with two-phase structure employed in a three-phase system. The system also employs simple and effective control strategies for dc-bus regulation and current control. The proposed system is a package with less switches compared to the conventional three-phase systems. It results in no increase in switch ratings, thus giving a more economic compensator.**

*Index Terms* — **active power filter, current harmonic compensation, flicker mitigation, two-phase loads.**

## I. INTRODUCTION

TWO-phase loads such as resistance spot welders and induction furnaces are in extensive use in industry. Spot welders are extensively used in automotive industry for welding two sheets of metal. These loads are unbalanced and cause severe flicker and harmonics. In some cases, 20% voltage drop may even be observed [1]. For Compensating this type of loads, two classic methods are common. The first one is using capacitor banks to decrease system impedance. The second method is using a TCR (Thyristor Control Reactor) for compensating the reactive part of current and voltage flicker [2]. These two methods are not only expensive but also cannot compensate for current harmonics. For complete compensation, active power filters can be used. But, active filters are so expensive that small industries cannot afford.

Fig. 1 shows possible topologies for a three-phase shunt active power filters structure [3]. Figs. 1a and b are used in three-phase three-wire systems, and Fig. 1c and d are used in three-phase four-wire systems.

As it can be seen from Fig. 1, a filter with (m-1) leg is sufficient for compensating an m wire system. The reason is that there are m-1 independent currents, and the sum of the currents in an m-wire system adds up to zero. Although the structure shown in Fig. 1b employs two switches less than the filter shown in Fig. 1a, in this configuration, each capacitor voltage should be larger than the line voltage, and the dc-bus voltage should be larger than twice the line voltage. Therefore, switches voltage rating is twice larger than those employed in the filter of Fig. 1a. The dc-bus voltage in Fig. 1a should be larger than the line voltage. Therefore, the filter shown in Fig. 1b is not economical.

In this paper, by using the filter shown in Fig.1 b in a new state and without increasing switches voltage rating, a two-phase load, which generates flicker and harmonics, can be compensated. This configuration is suitable for a two-phase load which is connected to a three-phase four wire systems. This filter can also be used for complete compensation of three-phase loads in three-phase three-wire systems. However, the switches voltage rating and the DC-bus voltage becomes two times larger than those in conventional three-phase filters making it less attractive in large industries with high power three-phase loads.

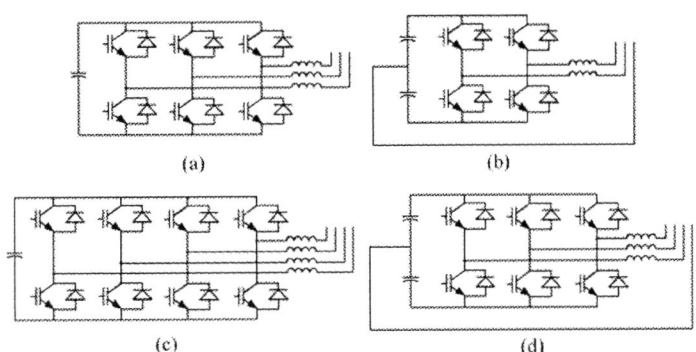

Fig. 1.  Three phase active power filter structures.

---

H. Hojabri is Phd. Student of Sharif University of Technology
Tehran, Iran, (e-mail: hhojabri@yahoo.com).
H. Mokhtari is Associate Professor of Sharif University of Technology
Tehran, Iran. (e-mail: mokhtari@sharif.ir).

In this paper, a single 20KVA controlled spot welder machine is used as the load. This load generates several power quality problems such as current imbalance, current harmonic, reactive current, voltage flicker and power change making it a good candidate for studying compensating methods and active filters control. This load is modeled as shown in Fig. 2.

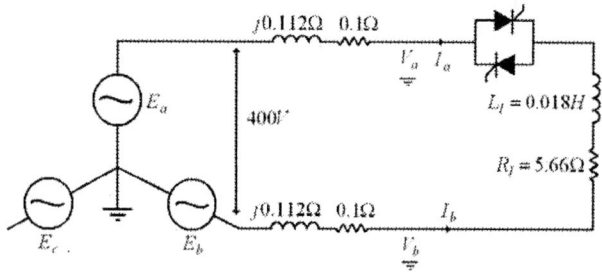

Fig. 2. Spot welder model.

## II. TWO PHASE COMPENSATION

Fig. 3a depicts the phasor diagram of a two-phase load such as a spot welder. The reactive part of the load current is the main cause of voltage flicker. As it can be seen from Fig. 3a, if phase (a) current has no reactive part, phase (b) current is $30^\circ$ lead relative to its voltage which may cause flicker. The relation between the load reactive power and load terminal voltage is given in Eq. 1 [4].

$$\frac{\Delta V}{V} \approx -\frac{Q_L}{S_{sc}} = I_q X_s \qquad (1)$$

If the reactive part of these two phase currents is compensated, the voltage flicker is compensated at the price of non-zero neutral current (Fig. 3b). By injecting harmonic using the filter, load current harmonic can also be compensated.

By using the two-phase filter of Fig. 1b with its C terminal grounded, the two-phase load can be compensated (Fig. 4). This type of compensation will not balance the load, and the neutral current flows causing load to become unbalanced.

Design of this filter is similar to that of a conventional three phase one. The filter inductance and dc-bus voltage should be selected using Eqs. 2 and 3 in away that the current harmonic level and the compensation speed are acceptable [5].

$$\frac{I_{af,sw}}{I_1} = \frac{V_{af,sw}}{2\pi f_{sw} L_f I_1} < 5\% \qquad (2)$$

$$\sqrt{2}V_\phi + L_f . \{\max\left|\frac{di}{dt}\right|_{filter}\} < 0.5 V_{dc} \qquad (3)$$

Filter dc-bus capacitors should be designed such that the dc-bus voltage ripple is at a desired level. The DC-bus capacitors should be designed based on Eq. 5 such that the dc-bus voltage variation is within the acceptable range, i.e.:

$$W_{Ceq}(V_{dc0} + \Delta V_{dc}) - W_{Ceq}(V_{dc0}) = \max\left\{\int (P_{source} - P_{load})\right\} \qquad (4)$$

$$C_{eq} > \frac{\max\left\{\int (P_{source} - P_{load})\right\}}{2V_{dc0}\Delta V_{dc,max}} \qquad (5)$$

Where $C_{eq} = C_1 \| C_2 = 0.5C_1 = 0.5C_2$ and $\Delta V_{dc,max}$ is the maximum dc-bus voltage ripple.

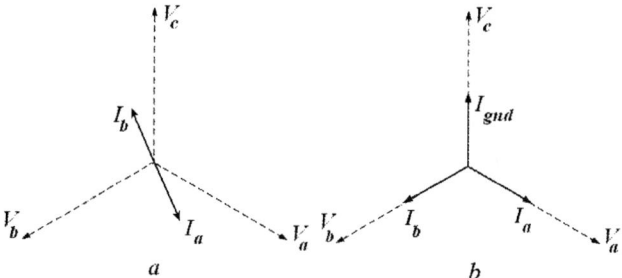

Fig. 3. Load phasor diagram a) before and b) after compensation.

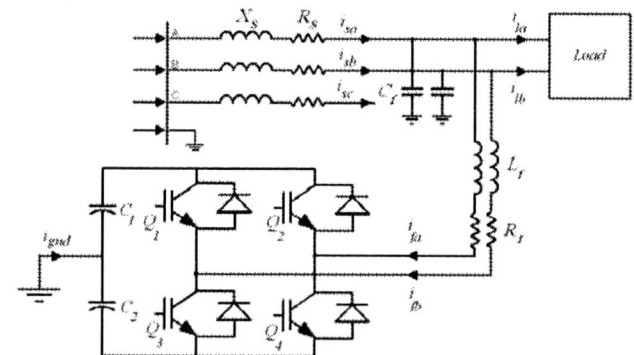

Fig. 4. Two phase active power filter structure.

For the current detection, the capacitor voltage control method is used. After compensation, currents of two phases are in-phase with their corresponding voltages. Therefore, shapes of the source current are fixed and only their amplitude must be specified. If the current amplitude is less than its desired value, source average power will be less than load average power, and this difference should come from the filter capacitor, therefore the dc voltage will drop. If current amplitude is more than its desired value, source dc power must be more than load dc power, and this difference will charge the filter capacitor increasing the dc-bus voltage. Therefore, controlling capacitor voltage would determine source currents amplitude.

After determining source current, by subtracting load current, filter reference current will be obtained if the load current is known. The block diagram in Fig. 5 is used for the capacitor voltage control and filter reference current detection. With this compensation technique, load ac power cannot be completely compensated. Therefore, the dc-bus voltage has smaller ripples compared to the one with three phase complete load compensation. Other detection methods such as instantaneous power theory proposed by Akagi can also be used for filter reference detection [6,7].

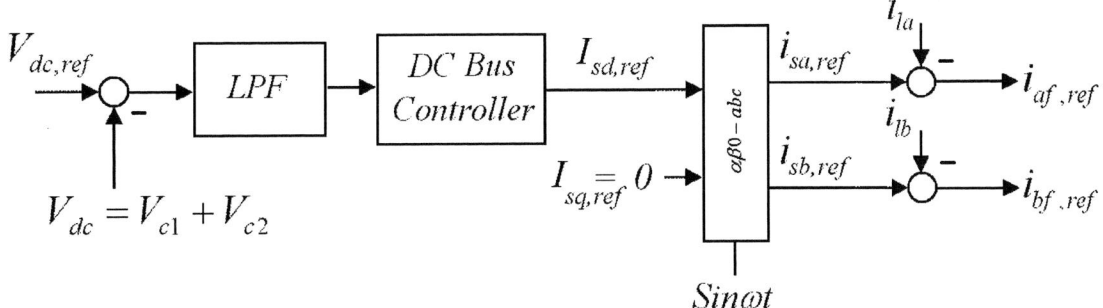

Fig. 5. Compensation system block diagram.

### III. CAPACITORS VOLTAGE REGULATION

The block diagram in Fig. 5 only controls the dc-bus voltage. It cannot control the voltage of each capacitor. If the losses for dc bus capacitors are not equal, then one of them will be discharged and the other one will be over charged until its voltage becomes $V_{dc,ref}$ resulting in inverter malfunction. To regulate capacitors voltages, a dc term current can be added to the filter reference current. Table 1 shows the variation of the filter dc-bus capacitors versus the filter current status.

TABLE I
CAP. VOLTAGE VS. FILTER CURRENT

| | $i_{af}\uparrow$ | $i_{af}\downarrow$ |
|---|---|---|
| $i_{af}>0$ | $V_{c2}\downarrow$ | $V_{c1}\uparrow$ |
| $i_{af}<0$ | $V_{c2}\uparrow$ | $V_{c1}\downarrow$ |

As Table 1 shows, when the filter current is positive and $Q_3$ is on, to increase the filter current, $V_{C2}$ must be decreased. To decrease the filter current when $Q_1$ is on, $V_{C1}$ must be increased.

When the filter current is negative and $Q_3$ is on, to increase the filter current, $V_{C2}$ must be increased. If $Q_1$ is on to decrease the filter current, $V_{C1}$ must be decreased. Therefore, by adding a dc term to the filter current, the time during which $i_{af}$ is positive will be lengthened resulting in increase in $V_{C1}$ and a decrease in $V_{C2}$. By subtracting a dc term from the filter current, the time during which $i_{af}$ is negative will be lengthened resulting in an increase in $V_{C2}$ and a decrease in $V_{C1}$. Fig. 6 shows this method.

To improve this method, the relation between capacitors voltage and filter current must be obtained. In one switching period, during which phase currents can be assumed constant, capacitors small voltage changes are equal to:

$$\delta V_{C1} = \int_{DT_s} \frac{I_{fa}}{C_1} dt = \frac{I_{fa}DT_s}{C_1} \tag{6}$$

$$\delta V_{C2} = -\int_{D'T_s} \frac{I_{fa}}{C_2} dt = -\frac{I_{fa}D'T_s}{C_2} \tag{7}$$

$$\delta V_C = \delta V_{C1} - \delta V_{C2} = \frac{I_{fa}DT_s}{C_1} + \frac{I_{fa}D'T_s}{C_2} \tag{8}$$

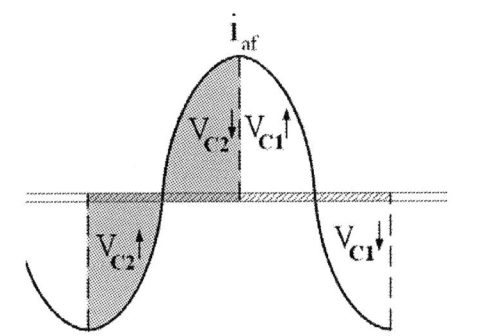

Fig. 6. Adding DC term to filter current.

Assuming ($C_1=C_2=C$), Eq. 8 can be written as follows:

$$\Delta V_C = \int \delta V_C = \int \frac{I_{fa}T_s}{C} \tag{9}$$

Therefore, the transfer function between $\Delta\bar{V}_C$ and $\bar{I}_{fa}$ can be obtained taking Laplas transform from Eq. 9, i.e.:

$$\frac{\Delta\bar{V}_C}{\bar{I}_{fa}} = \frac{T_s}{Cs} \tag{10}$$

Hence, by using a time constant as a low pass filter with a small gain and by adding a dc current to the active filter reference current, the difference between capacitors voltages can be controlled. This control method can be used for controlling the voltage difference between the two capacitors of the dc bus of all filters with split capacitors. Fig. 7 depicts a block diagram of the proposed compensating system.

### IV. SIMULATION RESULTS

The filter is used for compensating a spot welder machine load. System parameters are given in Table 2.

Simulation results are given through Figs. 8 to 12. Figs. 8 and 9 depict the load, source and filter currents for phase (a) and (b) respectively. Fig. 10 shows the neutral current which is in-phase with phase-C voltage which its amplitude is equal to that of other phase currents.

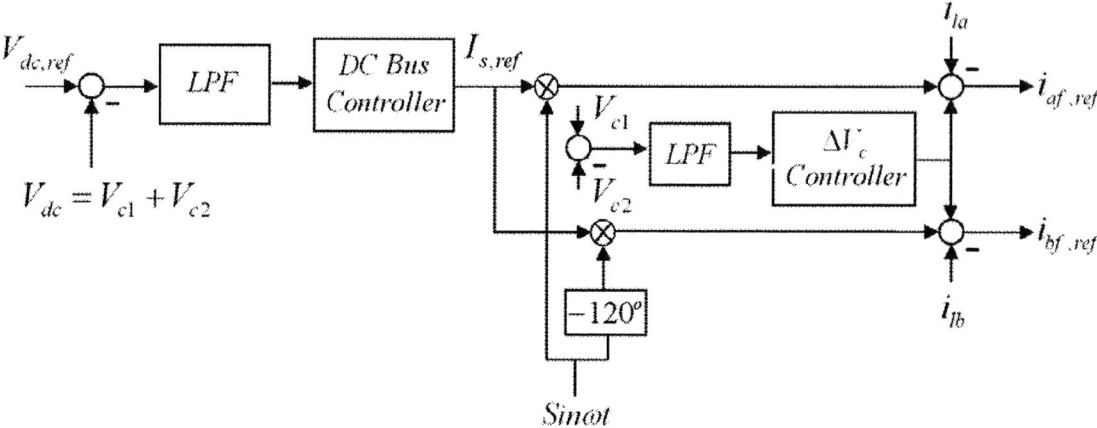

Fig. 7. Compensation system block diagram.

TABLE II
COMPENSATION SYSTEM PARAMETER

| | |
|---|---|
| $V_s$ | 400V |
| $X_s$ | $j0.112\Omega$ |
| $R_s$ | $0.1\Omega$ |
| $L_f$ | 6mH |
| $R_f$ | $0.2\Omega$ |
| $C_1, C_2$ (DC bus Cap.) | 800µF |
| $C_f$ (Output LPF Cap.) | 15µF |
| $C_1$ Loss Resistance | 2000Ω |
| $C_2$ Loss Resistance | 4000Ω |
| $f_{sw}$ | 10KHz |

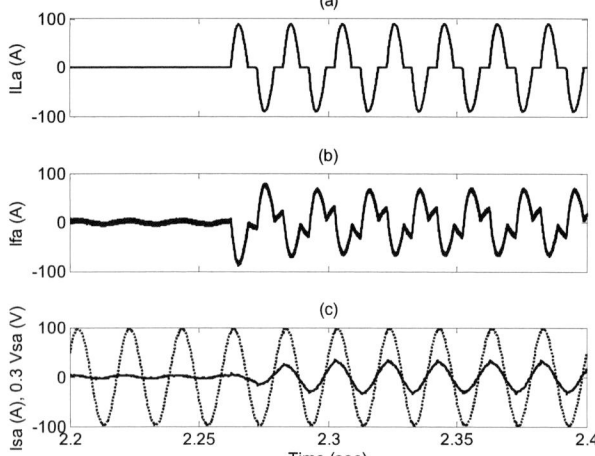

Fig. 8. Phase (a) Load, filter and source current.

Fig. 11a and b show the load and source active power. As it can be seen from this figure, the compensation method can also compensate for the oscillatory part of the load power reducing the source power oscillations.

Figs. 11c and d show the reactive power of the load and source. As these figures indicate, the source reactive power is reduced. Fig. 12 shows the capacitors voltages and their difference. Capacitors losses are modeled with different resistors, therefore, their voltages must be different. However this difference is compensated by the proposed controller. The dc current injected by the filter is of low amplitude and has negligible effect on system and transformer saturation.

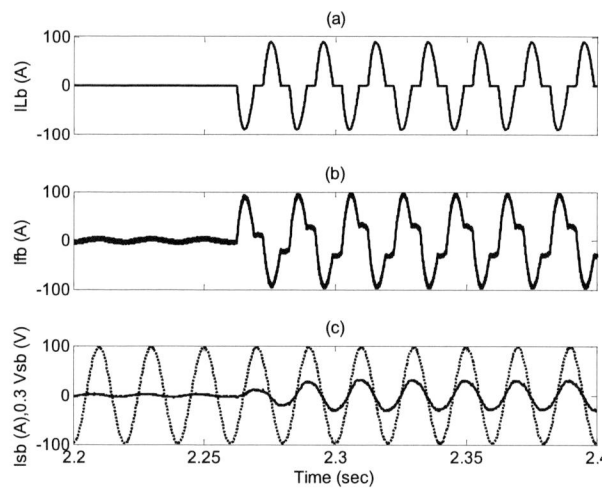

Fig. 9. Phase (b) Load, filter and source current.

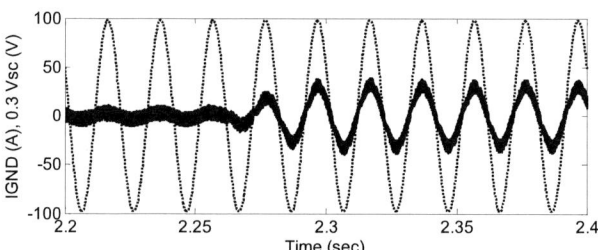

Fig. 10. Neutral point current.

634

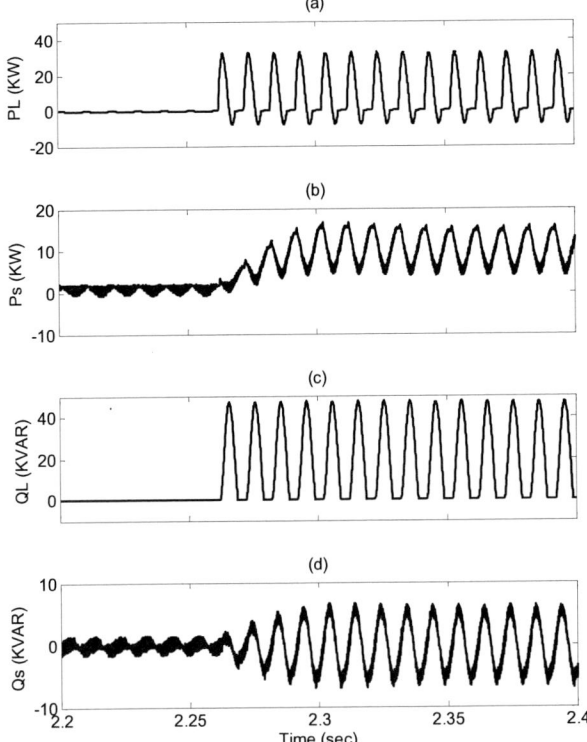

Fig. 11. Load and source power.

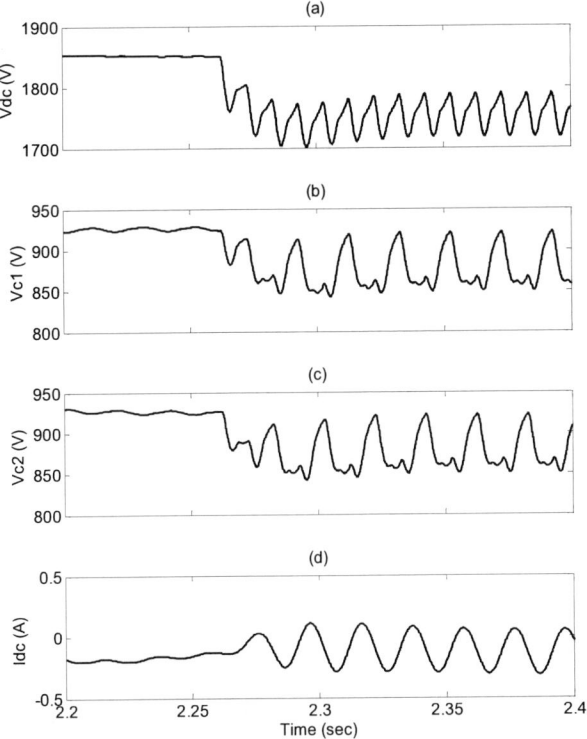

Fig. 12. Vdc, $V_{C1}$, $V_{C2}$ and $\Delta V_C$ during load change.

## V. CONCLUSION

This paper describes a new and economic method for power quality improvement in small industries with two-phase

loads. The system is a two-phase active filter with effective control strategies which results in flicker and harmonic reduction. A new control method for balancing dc bus capacitors voltage for active filter with split capacitor in their dc bus is also proposed.

## VI. REFERENCES

[1]. T. L. Baldwin, T. Hogans, S. D. Henry, F. Renovich and P. T. Latkovic, "Reactive Power Compensation for voltage Control at resistance welder", *IEEE Trans. On Ind. Appl.*, Vol. 41, Nov./Dec. 2005, pp. 1485 – 1492

[2]. M. M. Morcos and J. C. Gomez, "Flicker Source and Mitigation", *IEEE Power Eng. Review*, Vol. 22, Nov. 2002, pp. 5 – 10

[3]. B. Singh, K. Al-Haddad and A. Chandra, "A review of active filters for power quality improvement", *IEEE Trans. On Ind. Electron.*, Vol. 46, No. 5, Oct. 1999, pp. 960 – 971

[4]. T. J. E. Miller, "Reactive power control in electric systems", John Wiley, 1982.

[5]. T. Thomas, K. Haddad, G. Joos and A. Jaafari, "Design and Performance of Active Power Filters", *IEEE Ind. Appl. Magazine,* Vol 4, Sep./Oct. 1998, pp. 38 – 46

[6]. H. Akagi and et. al., "Control strategy for active power filters using multiple voltage source PWM converters", *IEEE Trans. On Ind. Appl.*, 1986, Vol. 22, No. 3, p. 460

[7]. T. C. Green and J. H. Marks, "Control techniques for active power filters", *IEE Electric Power Appl. Proc.*, Vol. 152, No. 2, Mar. 2005, pp. 369 – 381.

**2006 IEEE International Conference on Power Electronic, Drives and Energy Systems**

# Three level STATCOM Based Power Quality Solution for a 4 MW Induction Furnace

Unnikrishnan A.K, *Senior Member, IEEE*, Aby Joseph, and Subhash Joshi T.G., *Member, IEEE*

*Abstract–* **The paper explains the power quality study conducted in a 4MW mini steel plant, having fault level of 250MVA, to investigate the critical electrical problems caused by the harmonics and the development of a STATCOM as a solution for harmonic mitigation. A 500Kvar STATCOM is realized with a current controlled three level voltage source converter (VSC). The hardware configuration and the control strategy have been dealt in detail. Site installation and experimental results verifying STATCOM's performance are also presented.**

*Index Terms–* **Digital Signal Processor (DSP), IEEE 519-1992 Std., induction furnace, Power Quality (PQ), Pulse Width Modulation (PWM), STATCOM, Total Demand Distortion (TDD), Three level converter, Unit vector, Voltage Source Converter (VSC).**

## I. INTRODUCTION

AS technology advances, the electrical loads are multiplying in numbers and complexity. This throws tremendous challenge to the quality of power supply system. Power electronic converters are examples of such types of loads. Steel mills, which employ induction furnaces for melting scrap iron is one of the industrial areas where the use of such power converters is inevitable. The tuned filters or passive compensators are the traditional solution for harmonics issues. Since they are tuned for a fixed frequency they are not affective for varying harmonics spectrum. The induction furnace is a typical example of load, which generates harmonics in different spectrum based on the configuration of the controlled rectifier i.e., 12-pulse for heating mode or 6-pulse for sintering mode. Passive filters are also susceptible to sinking the harmonics injected by other loads in the grid. To overcome these shortcomings of passive filters various active power filter configurations have been reported [1]. Among these configurations, shunt active filters with two-level or multilevel converter have been recognized as one of the viable solutions for harmonic compensation. The

---

This project is funded by Department of Information Technology, Govt. of India, under administrative approval No. 28(5) 099 –IAD, dated 24-08-01.

Unnikrishnan A.K. is with Centre for Development of Advanced Computing (C-DAC), Trivandrum, Kerala, India. Phone: + 91 471 2723333, Fax: + 91 471 2723456, (e-mail: unnikrishnan@cdactvm.in)

Aby Joseph is with Centre for Development of Advanced Computing (C-DAC), Trivandrum, Kerala, India (e-mail: abypj@cdactvm.in)

Subhash Joshi T.G. is with Centre for Development of Advanced Computing (C-DAC), Trivandrum, India (e-mail: subhash@cdactvm.in)

use of two-level converter is limited to medium voltage levels due to higher voltage stress on switching devices and the effective switching frequency. But multilevel converters [2] can be used in higher voltage and power applications to obtain better performance for the given switching frequency and lower voltage stress on switching devices. This paper explains the power quality study conducted in a mini steel plant to investigate the critical problems caused by the harmonics current drawn by induction furnace and development of 500kVAR three-level STATCOM as a solution for harmonic mitigation.

## II. THE TEST SITE

The plant selected for study is located in northern Kerala, India. The plant consists of a rolling mill and a casting unit. There is an 11kV under ground cables from the utility sub-station and connected to High Voltage Distribution Board (HVDB). Two feeders are taken out from HVDB. One feeder is connected to induction furnace installation via 11kV/850-850V,Δ/Δ-Y, 5MVA transformer with a fault level of 250MVA.The induction furnace installation consists of a 12 pulse controlled rectifier, dc reactor followed by a single-phase medium frequency inverter feeding the furnace coil. Second feeder is connected to the auxiliary loads, which includes cooling pump motors, cranes, welding machines, lightings etc. The furnace load has power factor correction capacitors also. The single-line diagram showing main power distribution of the plant is shown in Fig.1.

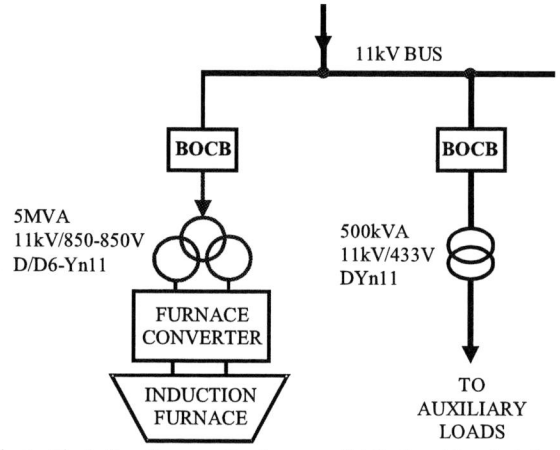

Fig. 1. Single-line diagram of main power distribution at the steel plant.

0-7803-9771-1/06/$25.00 ©2006 IEEE

The plant has three induction furnaces, one of 9-ton and other two of 4-ton capacity each. Reports regarding the failure of the protection and relaying equipments and over heating of the distribution transformers supplying for the steel plant in substation premises was the inspiration for the field study. Continuous monitoring and data logging of different signals were carried out and a harmonics database was prepared. The measuring point used for data logging is shown in Fig. 2.

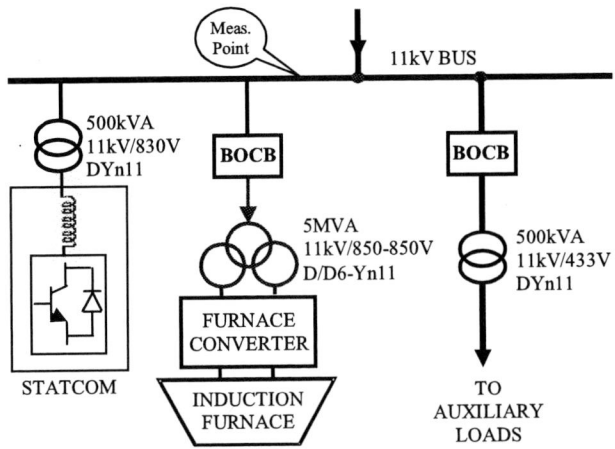

Fig. 2. Single-line diagram of STATCOM connection to the main power distribution at the steel plant.

From the harmonic measurements, it was found that 4MW 12-pulse converter which feeds 9-ton furnace in casting plant draws a current from grid with Total Demand Distortion (TDD) of 8.6%, which exceeds the IEEE 519-1992 recommendations. The harmonic spectrum shows that 11th and 13th harmonics, having TDD of 6.4% and 4.5% respectively, are predominant in the line current while the furnace is under operation. To bring down the TDD within recommended level, STATCOM is identified as the appropriate solution [3].

### III. THE HARDWARE CONFIGURATION

STATCOM detailed in this paper is realized with a voltage source converter (VSC) connected to the 3- Phase grid of 830V, 50Hz, through a series inductor, as shown in Fig. 2. IGBT based three-level neutral-point clamped converter [2] topology, is used for realizing VSC.

The STATCOM is rated for 500kVAR and having DC bus voltage of 1700V.

For constructional simplicity, each phase leg of the converter module is integrated as a stack, which contains IGBT modules and their gate drivers mounted on a heat sink, fans for forced air cooling, thermistors mounted on the heatsink near the device for temperature monitoring and protection. In order to minimize the stray inductance of the converter circuit a sandwiched multi-layer bus bar architecture is used. The system consists of 300A, 1400V IGBT modules as switching devices, three-phase series inductor of 0.44mH and a DC capacitor bank of effective capacitance 7050 µF. The inductance of series inductor take care the leakage of transformer also. The power stack assembly of one of the phase is shown in Fig. 3.

Fig. 3. Power stack assembly of one phase.

### IV. CONTROL STRATEGY

Synchronous d-q reference frame based control strategy [4] is adopted to control the STATCOM, which requires sensing of grid voltage and generation of unit vectors for the orientation. But the presence of harmonics in the grid voltage and noise in voltage sensor restricts its direct use. To reduce the effect of harmonics and noise in the unit vector, simple and efficient method is adopted [5].

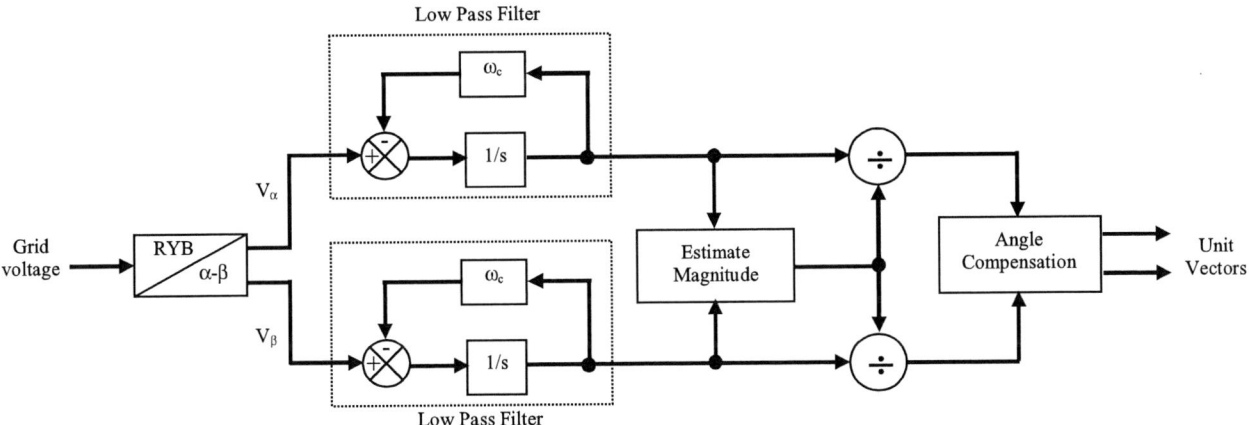

Fig. 4. Block diagram of unit vector generation.

The method used for the construction of unit vector is shown in Fig. 4. The sensed grid voltages are filtered using a first order digital low pass filter whose corner frequency is $\omega_c$. In the present work, $\omega_c$ is chosen to be equal to the nominal grid frequency (50Hz). So, after filtering, the percentage of $h^{th}$ order harmonics of the sensed grid voltage is reduced by a factor of $\sqrt{\dfrac{2}{h^2+1}}$ (assuming fundamental voltage is always 100%). It is clear that $5^{th}$ and higher order harmonics, which can normally be present in the line-to-line grid voltages, are reduced considerably using this low pass filter. High frequency noise gets eliminated almost completely. Finally, dividing the filter output signals with their magnitude generates the required unit vectors. However, this low pass filter introduces a phase lag to the unit vector generation. For nominal grid frequency (50Hz), this phase lag is $45^0$ and can be easily compensated.

The block diagram of the control strategy is shown in Fig. 5. Converter and load currents are oriented along the grid voltage using the unit vectors. After orientation, the fundamental components of d-axis and q-axis load currents are extracted using low pass filters. Knowing the fundamental d and q components of the load currents, the harmonics component is estimated by subtracting the fundamental component from the total d and q load currents. The estimated harmonic components of the load current are used to generate the references for the d and q axis current controllers. A PI controller is used to regulate the dc bus voltage at a constant level. The output of this controller generates the reference d-axis current required to keep a steady dc bus. This reference current added with the d-axis component of the harmonics in the load current to generate the total reference for the d-axis current controller. The reference to the q-axis current controller is the q-axis harmonic component of the load current.

After extracting the harmonics from the load current it is passed through a lead lag compensator. The lead lag compensator will compensate the phase lag incurred in the sensor, ADCs etc.

The sensed converter current is transformed into synchronous d-q reference frame. The d and q axis converter current is then compared with the d and q references to get the error, which is fed to the PI-controller. The remaining terms of the control equations "(1) and (2)," are added at the output of the PI-controller to get the voltage references of d and q axis. These references are then transferred back into 3φ stationary frames and fed to a PWM modulator to control the operation of the VSC. Space vector modulation technique is used for switching the IGBT [6][7][8]. The control algorithm is implemented on a digital signal processor, TMS 320F240, based controller.

The control law along the d-q axis is given below.

$$V_d = i_d R + L\frac{di_d}{dt} - \omega L i_q + V_g \tag{1}$$

$$V_q = i_q R + L\frac{di_q}{dt} + \omega L i_d \tag{2}$$

where, $V_d$ and $V_q$ are d and q axis voltage command. L and R are the inductance and resistance of the coupling inductor. $i_d$ and $i_q$ are the d and q axis current of the converter.

## V. RESULTS

Fig. 6 shows the fully assembled STATCOM installed at the mini steel plant. Field trial had been carried out to validate the performance of both the hardware and the control strategy. It was observed that the source current TDD is brought down to 4.1% from 8.6% at full load of the furnace after the installation. The $11^{th}$ and $13^{th}$ harmonics are brought down from 6.4% to 2.8% and 4.5% to 2% respectively.

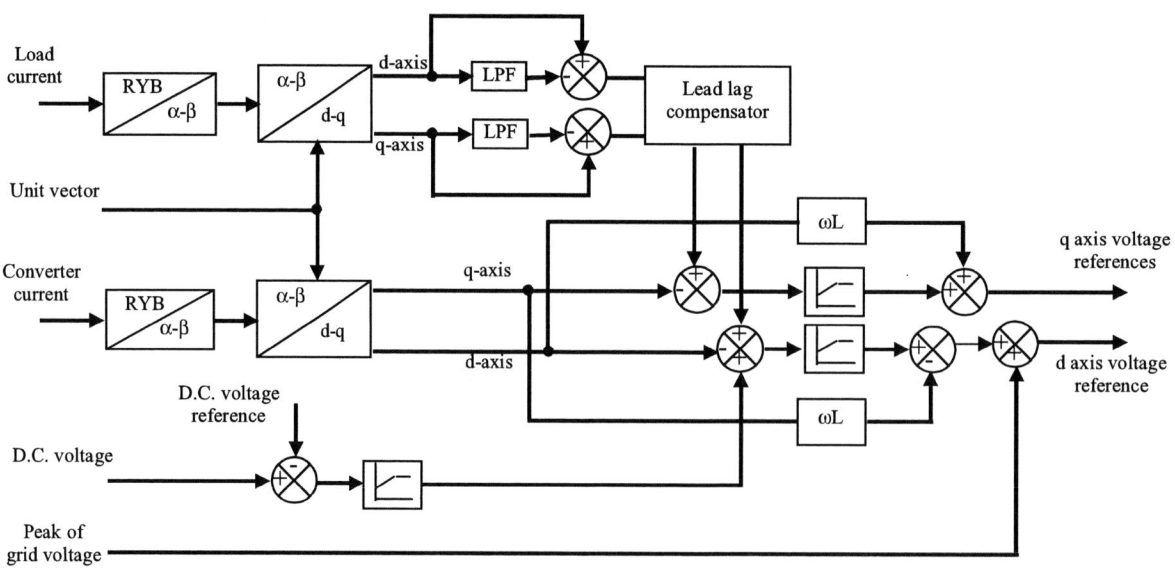

Fig. 5. Synchronous D-Q reference frame based control strategy.

It is observed that the STATCOM under operation brought down TDD and individual harmonic levels of the source current well within the limit stipulated by IEEE-519-1992 standards.-The summary data given in Table 1 clearly show the performance of the STATCOM. Source current and the source voltage waveforms are shown in Fig. 7 and Fig. 8 respectively.

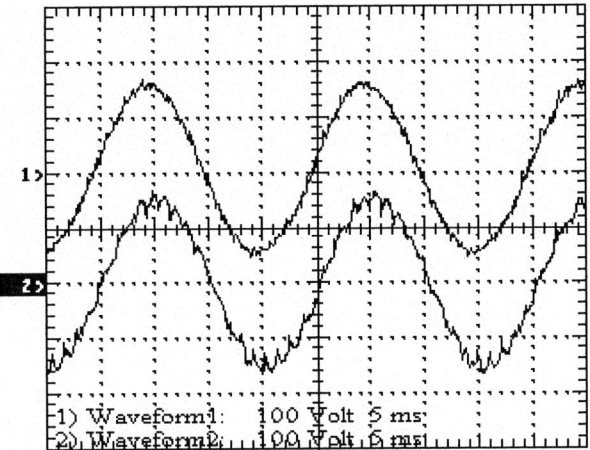

Fig. 8. 11kV incomer voltage measured at secondary (110V side) of PT, with compensation (1) and without compensation (2).

## VI. REFERENCES

[1] W.M. Grady, M.J. Samotyj and A.H. Noyola, "Survey of active power line conditioning methodologies", in *IEEE Trans. Power Delivery*, vol. 5, pp. 1536-1542, July 1990.

[2] A. Nabae, I. Takahashi, and H. Akagi, "A new neutral-point-clamped PWM inverter," *IEEE Trans. on Industrial Applicatiosn*, vol. IA-17, no.5, pp. 518-523, Sep./Oct. 1981.

[3] H. Akagi et al., "Instantaneous Reactive Power Compensators Comprising Switching Devices Without Energy Storage Components," *IEEE Trans. on Industry Applications*, Vol.IA-20, No.3, pp. 625-630, May/June, 1984.

[4] K.R.Padiyar , V.Ramanarayanan , P.S Sensarma , "A Statcom for composite power line conditioning ," *IEEE International Conference on Industrial Technology, ICIT, held at Goa, India in January 2000.*

[5] T.G. Subhash Joshi, Aby Joseph,Gautam Poddar, "Active Power Factor Correction for Highly Fluctuating Industrial Load", *in Proc. of NPEC-2003, held at IIT, Bombay.*

[6] J. H. Seo, C. H. Choi and D. S. Hyun, "A new simplified space-vector PWM method for three-level inverters," *IEEE Trans. Power Electronics*, Vol. 16(4), pp. 545-550, July 2001.

[7] F. Wang, "Sine-triangle versus space-vector modulation for three-level PWM voltage-source inverters," *IEEE Trans. Industry Application*, Vol. 38(2), pp 500-506, March/April 2002.

[8] T.G Subhash Joshi, A.S Haneesh, G.Narayanan, V.T Ranganathan, "A Computationally Efficient PWM Algorithm for Multilevel Inverters", *in Proc. Of NPEC- 2003, held at IIT, Bombay.*

Fig. 6. STATCOM assembly.

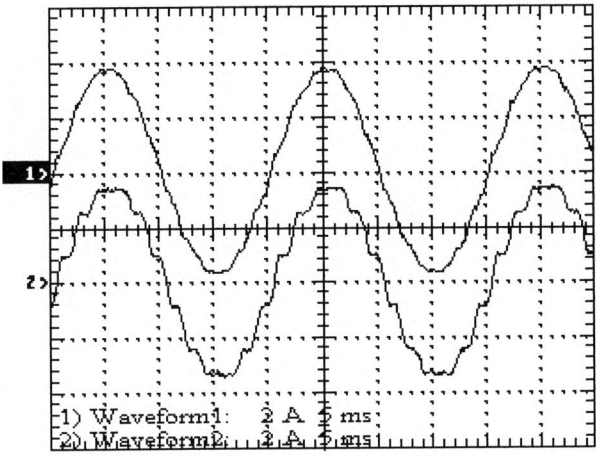

Fig. 7. Compensated (1) and Un-compensated (2) grid current.

TABLE I

CURRENT HARMONICS AS % OF F.L FUNDAMENTAL (TDD)

|  | STATCOM OFF | STATCOM ON |
|---|---|---|
| Fundamental | 100.00 | 100.00 |
| Harmonic 5 | 0.29 | 0.19 |
| Harmonic 7 | 0.35 | 0.21 |
| Harmonic 11 | 6.40 | 2.80 |
| Harmonic 13 | 4.50 | 2.00 |
| Harmonic 17 | 0.75 | 0.12 |
| Total (I TDD) | 8.60 | 4.10 |

## VII. BIOGRAPHIES

**Unnikrishnan A. K.** (M'90–SM'99) received the B.Sc. degree in electrical engineering from Regional Engineering College, Calicut, India, in 1978, and

the M.Tech. degree in electrical engineering with specialization in control guidance and instrumentation from the Indian Institute of Technology, Madras, India, in 1991. During 1979–1986, he was with Kerala State Electronics Development Corporation, Trivandrum, India, in various capacities in the Industrial Electronics Projects Division. Since 1986, he has been with the Power Electronics Group, Electronics Research and Development Centre of India (ER&DCI), Trivandrum, India, which is currently C-DAC, working on various development projects. Currently, he is Additional Director of the Power Electronics Group, C-DAC, Trivandrum. Mr. Unnikrishnan served the IEEE Kerala Section as Secretary during 1999–2001, vice Chair during 2005 and Chair during 2006.

**Aby Joseph** received the B.Tech degree from Mahatma Gandhi University, Kerala, India, in 1994, and the M.Tech degree from the University of Kerala, Kerala, India, in 1997. He is currently working as Deputy Director in the Power Electronics Group, C-DAC, Trivandrum, India.

**Subhash Joshi T.G.** (M'06) received the B.Tech. degree in electrical engineering from Government College of Engineering, Cannanore, India, in 1998, and the M.Tech. degree in electrical engineering with specialization in Power Electronics from Regional Engineering College, Calicut, India, in 2002. He is currently working as Scientist-B in the Power Electronics Group, C-DAC, Trivandrum, India.

**2006 IEEE International Conference on Power Electronic, Drives and Energy Systems**

# Analysis and Simulation of Single Phase Composite Observer for Harmonics Extraction

K. Selvajyothi, and P. A. Janakiraman, *Member, IEEE*

*Abstract*– The method of extracting individual harmonics from repetitive signals using multiple asymptotic observers is analyzed in this paper. For proper operation of such a composite observer, it is sufficient if the fundamental frequency is approximately made known because, the implementation exhibits sufficient pass-bandwidth centered on the assumed fundamental as well as the harmonics. However, the settling period and the bandwidth of the observer depend on how far the observer poles have been placed from the origin of the s-plane or the z-plane. The effect of small deviations from the central fundamental frequency of the observer, on the magnitudes and phase angles of the extracted signals is discussed. These parameters assume significance when the composite observer is used in feed back loops. Simulation and experimental results are incorporated.

*Index Terms*– Bandwidth, Composite Observers, Harmonics extraction, Phase shift .

## I. INTRODUCTION

THERE is an ever-increasing demand for measurement and elimination of supply harmonics [1] introduced by un-interruptible power supplies, adjustable speed drives, non-linear loads like the arc furnaces. Harmonic currents flowing into the power systems, cause interference with communication lines, malfunctioning of meters, relays, equipment heating, and data loss and over voltages. The total harmonic distortion (THD) of voltage in industry should not exceed 5% as per the guidelines given in the IEEE Standard.519-1992.

Though fixed passive filters improve power factor, they suffer from disadvantages like the shifting of the resonant point with load and the inability to filter out harmonics of varying strength. To monitor and maintain the power quality, there has been an increasing interest in devising harmonic detection and extraction algorithms and devices over the last decade [1]–[14]. The voltage and current harmonic content can be somewhat time varying due to continuous changes in the system configuration and the load conditions [2].Conventionally, extraction of individual harmonics is based on the Fast Fourier Transform (FFT) [3], though

---

K. Selvajyothi is Ph.D. Scholar in Indian Institute of Technology, Madras, Chennai - 600 036, India (phone: 91-044-22575410; e-mail: ee05d002@ee.iitm.ac.in).

Dr. P. A. Janakiraman is Professor in the Department of Electrical Engineering, Indian Institute of Technology, Madras, Chennai - 600 036, India. (Phone: 91-044-22575409; e-mail: pajraman@ee.iitm.ac.in).

efficient in stationary conditions loses accuracy under time-varying conditions [2]. A number of algorithms, e.g., least-square techniques [4], [5], Kalman filtering [7], Parseval's relation and the energy concept [8], artificial neural networks [9], adaptive infinite impulse response line enhancer [10] and advanced signal processing methods [11] have been proposed to extract and measure harmonics under time-varying conditions. Active filters or harmonic current compensation circuitry [12] have also been employed. Selective harmonic elimination using Luenberger observer has also been reported. In [13] an enhanced phase-locked loop (EPLL) was employed as the main building block. It is slower compared to the DFT procedures, while showing accurate performance in frequency-varying environments and immunity to noise. Notch filters without PLL has also been used [14], which however employs many multipliers.

This paper presents the analysis of the composite harmonic extraction procedure, based on multiple Luenberger observers, designed for a central frequency, 50Hz. This composite observer is reasonably accurate and robust regarding time varying harmonic content and frequency. However large frequency deviations from the nominal value of 50Hz can degrade the performance to some extent. Analysis in the frequency domain is presented which aims at the determination of the sensitivity of the amplitude and the phase errors of the extracted harmonics.

## II. SINGLE FREQUENCY OBSERVER

A linear second order continuous time oscillating system at the fundamental frequency $\omega_1$, can be described by:

$$\dot{X}_1 = A_1 X_1$$
$$y_1 = C_1^T X_1 \tag{1}$$

where $A_1 = \begin{bmatrix} 0 & \omega_1 \\ -\omega_1 & 0 \end{bmatrix}$ and $[X_1] = \begin{bmatrix} x_{11} \\ x_{12} \end{bmatrix}$

$$C_1 = [1 \ 0]$$

The system output y is a sinusoidal signal, which can be followed by the well-known Luenberger Observer:

$$\dot{\hat{X}}_1 = A_1 \hat{X}_1 + D_1(y_1 - C_1^T \hat{X}_1) \tag{2}$$

The error between the system and the observer state vectors is given by:

$$E_1 = X_1 - \hat{X}_1 \tag{3}$$

The error dynamic equation is

$$\dot{E}_1 = (A_1 - D_1 C_1^T) E_1 = A'_1 E_1 \tag{4}$$

0-7803-9771-1/06/$25.00 ©2006 IEEE

has all its eigen-values in the left-half plane. The observer design refers to the selection of the constant gain vector D1, using, for example, the pole placement method. Ideally the 50 Hz observer follows a 50 Hz input sine wave without any magnitude or phase error.

## III. DESIGN OF CONTINUOUS COMPOSITE OBSERVER

A periodic signal of frequency $\omega_1$, rich in harmonics can be modeled as the sum of sine waves of frequency $m.\omega_1$, emanating from subsystems as shown in fig.1 is defined by:

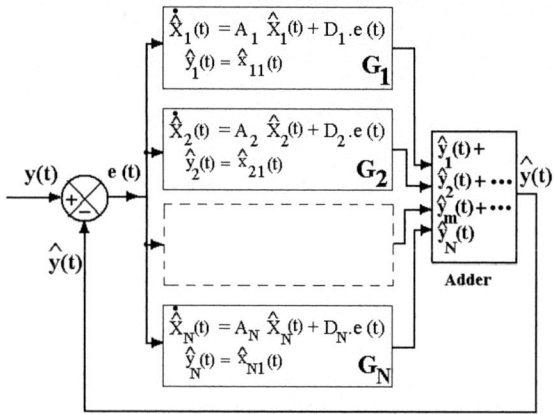

Fig. 1. Structural Schematic of the Composite Observer.

$$\dot{X}_m = A_m X_m$$
$$y_m = C_m^T X_m \quad ; \quad m = 1,2,3......N. \quad (5)$$

$$A_m = \begin{bmatrix} 0 & m\omega \\ -m\omega & 0 \end{bmatrix} \quad (6)$$

The state vector of the $m^{th}$ block is:

$$X_m(t) = \begin{bmatrix} x_{m1}(t) \\ x_{m2}(t) \end{bmatrix} \quad (7)$$

and

$$C_m = \begin{bmatrix} 1 & 0 \end{bmatrix} \text{ i.e., } y_m(t) = x_{m1}(t) \quad (8)$$

The structure of a single unit of the composite observer is shown in Fig.2.

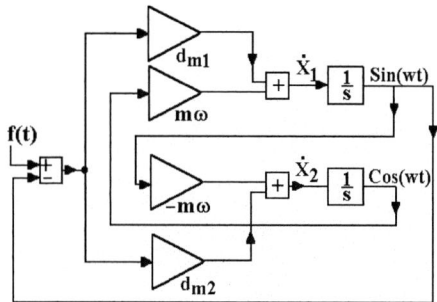

Fig. 2. Structure of a Single unit of the Composite Observer.

The composite 2N dimensional system describing the entire arrangement is:

$$\dot{X} = A X$$
$$y = C^T X \quad (9)$$

where

$$A = \begin{bmatrix} A1 & 0 & 0 & 0 & 0 & - & - & 0 \\ 0 & A2 & 0 & 0 & 0 & - & - & 0 \\ 0 & 0 & A3 & 0 & 0 & - & - & 0 \\ 0 & 0 & 0 & A4 & 0 & - & - & 0 \\ 0 & 0 & 0 & 0 & A5 & - & - & 0 \\ - & - & - & - & - & - & - & - \\ - & - & - & - & - & - & - & - \\ 0 & 0 & 0 & 0 & 0 & - & - & A_N \end{bmatrix} \quad (10)$$

and $[C]^T = \begin{bmatrix} 1 & 0 & 1 & 0 & - & - & 1 & 0 \end{bmatrix}$ (11)

The composite [2N x 1] constant gain vector D, to be determined using the specified pole locations is:

D = [(d11,d12) (d21,d22) (dm1,dm2) (dN1,dN2)]T  (12)

The closed loop poles are assumed to be equi-dominant and are located at

$$s = (-a\omega1 + j \, m.\omega1). \quad (13)$$

The observation speed and the bandwidth of the observer at the various notches m. $\omega1$, increase with the factor "a", the real part of the observer poles.

## IV. CLOSED LOOP FREQUENCY RESPONSE OF COMPOSITE OBSERVER

For the sake of simplicity, we consider a composite observer of the continuous type, shown in Fig.1.in which poles were placed as per eq (13). Each unit in the observer acts like a notch filter accepting the corresponding tuned frequency and rejecting all other harmonic frequencies. The frequency response for the closed loop system, measured at various channels show interesting features. The transfer function of the $m^{th}$ channel is:

$$\frac{Y_m(s)}{Y(s)} = \frac{G_m(s)}{1 + \sum\limits_{k=1}^{N} G_k(s)} = \frac{G_m(s)}{\Delta(s)} = H_m(s)$$

For the sake of simplicity, the Bode plots of H1(s) and H7(s) for a = 0.2 shown in fig.3a, and fig. 3b respectively.

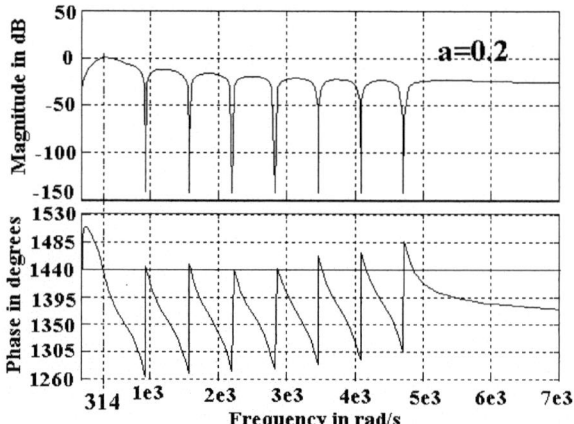

Fig. 3a. Notch Filter like Behaviour of H1 (jω).

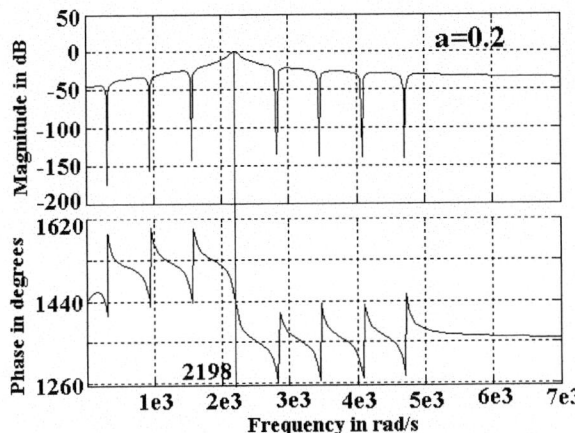

Fig. 3b. Notch Filter like Behaviour of H7 (jω).

The plot for H1(jω) shows that at the fundamental frequency of 314 rad/sec, the response is a maximum at 0dB. The effective phase shift is 0° and strong attenuation is seen at other harmonics.

The plot for H7(jω) shows that at the 7th harmonic namely 2198 rad/sec, the response is a maximum at 0 dB with an effective phase shift of 0□and strong attenuation is seen at the fundamental, 3rd,5th,9th,11th,13th and 15th harmonics.

The Bode plots of H1(s) and H7(s) for another value of 'a = 0.5' are shown in fig.3c, and fig. 3d respectively. This corresponds to "fast observer poles". The gains at the respective frequencies are 0dB and the phase shifts are 0□. However, the tuning is broader at these frequencies, suggesting increased bandwidth.

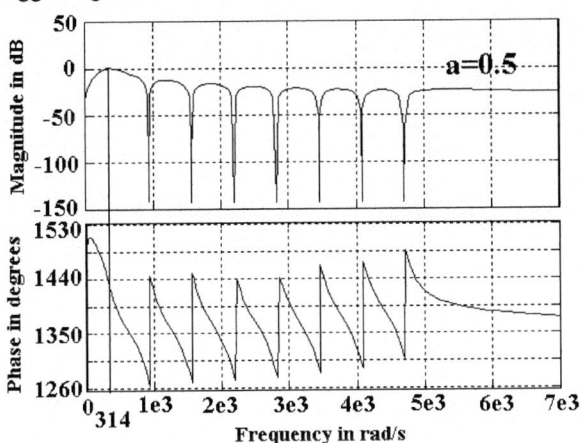

Fig. 3c. Notch Filter like Behaviour of H1 (jω) for a=0.5.

## V. VARIATION OF BAND WIDTH & PHASE SHIFT FOR HARMONICS

The bandwidth of the various closed loop blocks Hm(s), around the respective central harmonic frequencies is a function of the "a" that controls the location of the observer poles.

In Fig.4a the magnified view of frequency response of H1(jω) around 50Hz is shown for a = 0.5 and a = 0.2. The magnified

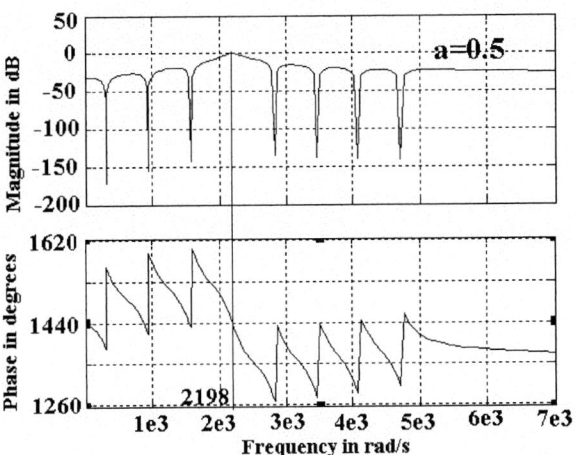

Fig. 3d. Notch Filter like Behaviour of H7 (jω) for a=0.5.

view of the frequency response of H7(jω) around 350 Hz is shown in Fig.4b for "a" = 0.5 and "a"= 0.2.

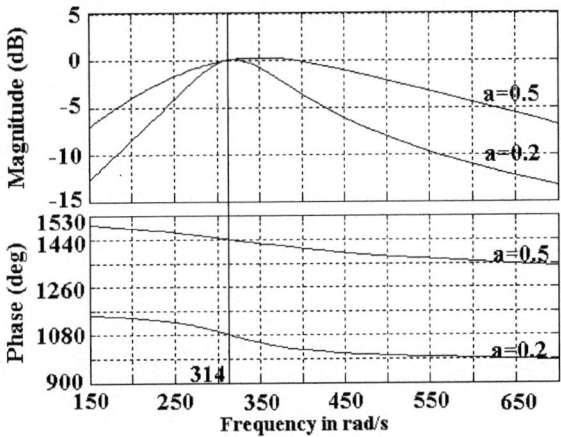

Fig. 4a. Frequency response of H1 (jω) for a=0.5 and a=0.2.

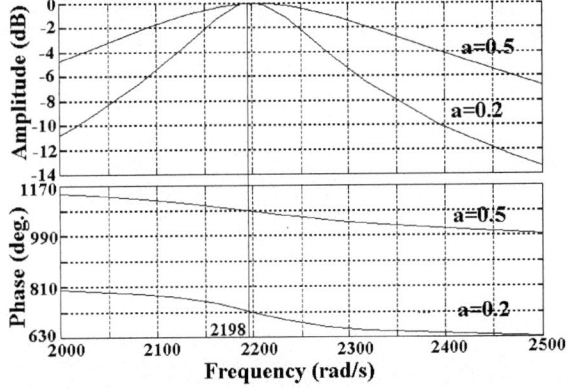

Fig. 4b. Frequency response of H7 (jω) for a=0.5 and a=0.2.

The BW of H1(jω) and H7(jω) and phase shift for ±1.5% deviation from the central frequency are shown in the Table I. For the blocks, descending values of "a" can be assigned so that the fundamental is estimated faster than the harmonics (for example, a1= 0.5; a3=0.45; a5=0.4; a7=0.35; a9=0.3;

643

a11=0.25; a13=0.2; a15=0.15). Alternatively, the factor "a" could be ascending (a1 = 0.50; a3= 0.55; a5= 0.60; a7= 0.65; a9=0.70; a11= 0.75; a13=0.80; a15=0.85). The descending "a" decreases the total harmonic distortion. The observer was tested using a square wave. The THD of voltage in each of the extracted harmonic signals is shown in the Table II.

TABLE I
PERFORMANCE OF THE OBSERVER FOR ±1.5% FROM THE CENTRAL FREQUENCY

| Extracted Signal | a | BW rad/s | For ±1.5% deviation from centr. Freq. | |
|---|---|---|---|---|
| | | | \|.\| dev dB | φ dev |
| Fund. | 0.2 | 127 | -0.1023 | 4.71° |
| | 0.5 | 315 | -0.0755 | 2.1° |
| | 1.0 | 460 | -0.0713 | 1.45° |
| 7H | 0.2 | 120 | -0.9304 | 31° |
| | 0.5 | 310 | -0.2707 | 13.3° |
| | 1.0 | 490 | -0.1087 | 9.2° |

TABLE II
PERCENTAGE THD IN EACH OF THE HARMONICS EXTRACTED

| Extracted Signal | Equal a* | Descending a** | Ascending a*** |
|---|---|---|---|
| Fund. | 0.6645 | 0.6705 | 0.6537 |
| 3H | 1.657 | 1.556 | 1.708 |
| 5H | 7.175 | 5.882 | 8.178 |
| 7H | 6.074 | 4.315 | 7.657 |
| 9H | 9.537 | 5.739 | 13.3 |
| 11H | 14.06 | 7.038 | 20.95 |
| 13H | 18.03 | 7.419 | 27.68 |
| 15H | 20.25 | 6.466 | 30.22 |

In Table III, the sensitivity of the BW and the phase shift are shown for both descending "a" and ascending "a"(entries in bold).

TABLE III
SENSITIVITY OF THE BW AND THE PHASE SHIFT

| Extracted Signal | BW rad/s | For ±1.5% deviation from centr. Freq. | |
|---|---|---|---|
| | | \|.\| dev dB | φ dev |
| Fund. | 317 | -0.0758 | 2° |
| | **311** | **-0.0745** | **2°** |
| 3H | 278 | -0.1116 | 7° |
| | **330** | **-0.0907** | **6°** |
| 5H | 240 | -0.2496 | 12° |

The phase error for the nth harmonic is 'n' times the phase error for the fundamental. The plots are shown in Figs 4c, 4d, and 4e.

## VI. SIMULATION RESULTS

A square wave at 50Hz, whose Fourier series is given by:
$$f(t) = \frac{4}{\pi} \sum_{m=1,3,5,\ldots} \frac{1}{m} \sin(m\omega t)$$ is taken for illustration. The

continuous and discrete Composite Observer and the Fourier transform gave identical results as shown in Table IV.

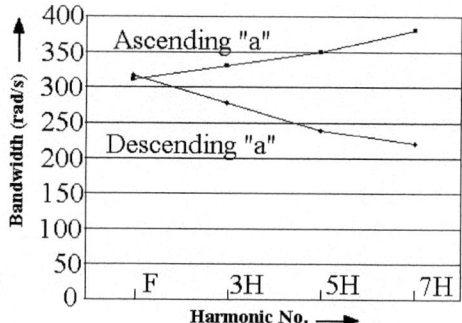

Fig. 4c. Phase angle deviation for ± 1.5% freq.Change from 50Hz.

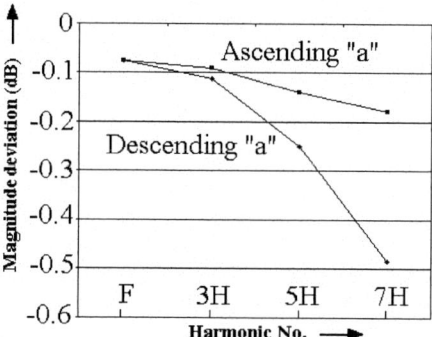

Fig. 4d. Magnitude deviation for ±1.5% freq. Change from 50Hz.

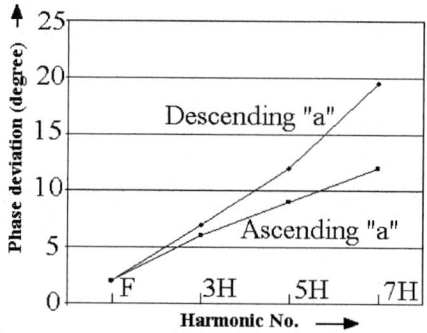

Fig. 4e. Variation of Bandwidth for ascending and descending 'a'.

TABLE IV
AMPLITUDE OF HARMONICS - ACTUAL AND SIMULATION

| Extracted Signal | Simulation | | Actual |
|---|---|---|---|
| | MATLAB | DSP builder | |
| Fund. | 1.273 | 1.270 | 1.273 |
| 3H | 0.424 | 0.400 | 0.424 |
| 5H | 0.255 | 0.254 | 0.255 |
| 7H | 0.182 | 0.173 | 0.182 |
| 9H | 0.141 | 0.143 | 0.141 |
| 11H | 0.116 | 0.115 | 0.116 |
| 13H | 0.098 | 0.098 | 0.098 |
| 15H | 0.085 | 0.088 | 0.085 |

The observer extracts about 94.37% of the harmonics up to the 15th. However the small errors seen in the simulation are due to the left over harmonics in the signal.

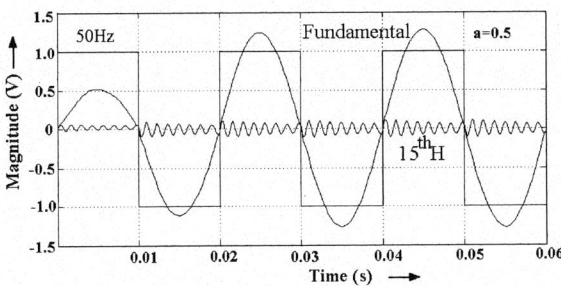

Fig. 5a. Fund.&15thHarmonic Extracted from 50Hz Sq. wave.

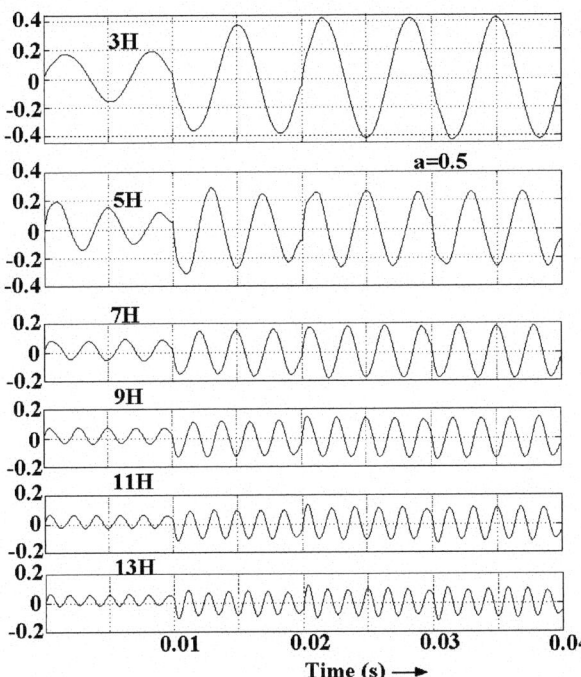

Fig. 5b. Different Harmonics Extracted from 50Hz Sq. wave.

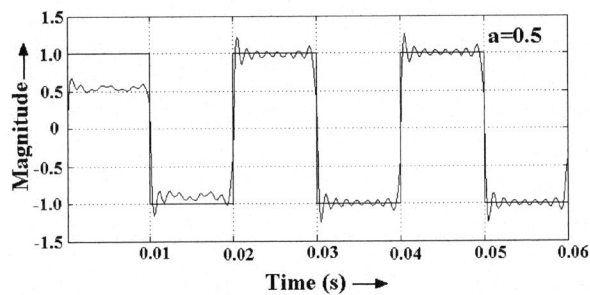

Fig. 6. Reconstructed Sq. using Extracted Harmonics from the 50Hz Sq.

## VII. EXPERIMENTAL RESULTS

Using DSP builder the signal to be analyzed and the discrete observer are designed. Analyzing 8 odd harmonics up to 15th the steady state response of the system for the extracted fundamental, 3rd and 5th harmonic for a 50 Hz Square wave with a=0.5 are shown in fig.7.

Similarly, the splits of Square Wave signals of frequencies 48.54Hz and 51.81Hz signals are shown in fig 8 and fig 9 respectively.

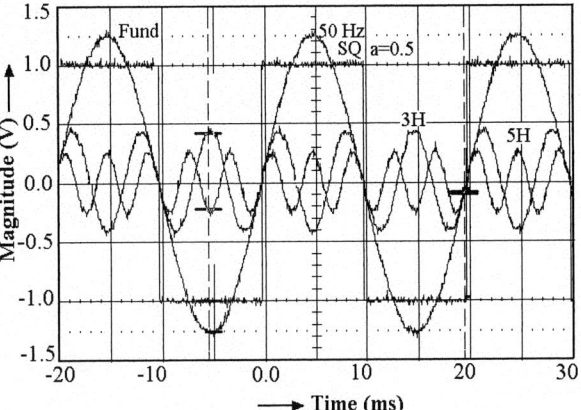

Fig. 7. 50Hz Square wave, Fund., 3H & 5H - Experimental

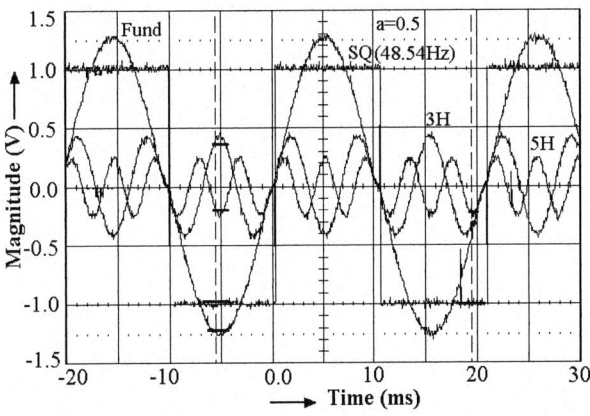

Fig. 8. 48.54Hz Square wave, Fund., 3H & 5H - Experimental

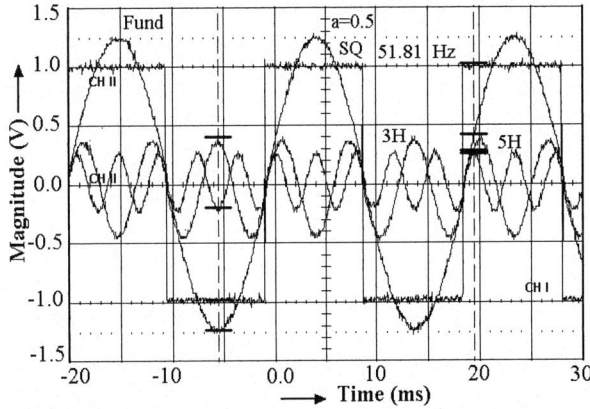

Fig. 9. 51.81Hz Square wave, Fund.,3H & 5H - Experimental

## VIII. CONCLUSION

The performance of a continuous composite observer for extracting harmonics from periodic waveforms by simulation

645

and experiments has been presented. The steady state phase error depends on frequency deviations from the central value. Even though the design frequency was 50Hz, for values of 'a ≅ 0.5' the composite observer gave acceptable performance regarding BW and phase errors for input frequencies in the range 47.5Hz to 52.5 Hz. Including more harmonics would certainly reduce the distortions at the cost of increased computation burden. It is seen that as the observer poles are shifted away from the origin, the BW increases and the phase shift decreases. Smaller values of "a", give sharper tuning, lower distortion in the extracted waveforms and smaller leakage of harmonics into the neighbouring blocks, at the cost of higher phase shift and lower stability margin.

## IX. REFERENCES

[1] S. L. Clark, P. Famouri, and W. L. Cooley, "Elimination of supply harmonics," *IEEE Ind. Applic Mag.*, vol. 3, no. 2, pp. 62–67, Mar./Apr.1997.

[2] Baghzouz, Y.; Burch, R.F.; etal, "Time-varying harmonics: Part I—Characterizing measured data," *IEEE Trans. Power Del.*, vol. 13, no. 3, pp. 938–944, Jul. 1998.

[3] T. A. George and D. Bones, "Harmonic power flow determination using the fast Fourier transform," *IEEE Trans. Power Del.*,vol.6, no.2, pp.530–535, Apr. 1991.

[4] I. Kamwa and R. Grondin, "Fast adaptive schemes for tracking voltage phasor and local frequency in power transmission and distribution systems," *IEEE Trans. Power Del.*,vol. 7,no. 2,pp. 789–795, Apr. 1992.

[5] M. Bettayeb and U. Qidwai, "Recursive estimation of power system harmonics," *Elect. Power Syst. Res.*, vol. 47, pp. 143–152, 1998.

[6] S.-L. Lu, C. E. Lin, and C.-L. Huang, "Power frequency harmonic measurement using integer periodic extension method,"*Elect.PowerSyst.Res.*,vol.44,pp.107–115, 1998.

[7] A. Girgis,W. B. Chang, and E. B. Makram,"A digital recursive measurement scheme for on-line tracking of power system harmonics," *IEEE Trans. Power Del.*, vol. 6, no. 3, pp. 1153–1160, Jul. 1991.

[8] C. S. Moo, Y. N. Chang, and P. P. Mok, "A digital measurement scheme for time-varying transient harmonics," in Proc. *IEEE/Power Eng.* Soc. Summer Meeting, 1994, Paper no. 94 SM 490-3-PWRD.

[9] P. K. Dash, S. K. Patnaik, A. C. Liew, and S. Rahman, "An adaptive linear combiner for on-line tracking of power system harmonics," in Proc. *IEEE/Power Eng.* Soc. Winter Meeting, Baltimore, MD, Jan. 21–25, 1996, Paper no. 96WM 181-8-PWRD.

[10] J. F. Chicharo and H. Wang, "Power system harmonic signal estimation and retrieval for active power filter applications," *IEEE Trans. Power Electron.*, vol. 9, no. 6, pp.580–586, Nov. 1994.

[11] T. Lobos, Leonowicz, J. Rezmer "Harmonics and Interharmonics Estimation Using Advanced Signal Processing Methods", *IEEE Trans. on power del.*, Vol. 18, No. 2, Apr. 2003.

[12] Andrea Tilli, Fabio Ronchi, Alberto Tonielli " Shunt active filters: selective compensation of current harmonics via state observer", IECON 02, 5-8 Nov. 2002 Vol. 2, pp.874- 879, Inspec Acc No.7731895

[13] Masoud Karimi-Ghartemani, and M. Reza Iravani "Measurement of Harmonics/Inter-harmonics of Time-Varying Frequencies", *IEEE Trans. Power Del.*, vol. 20, no. 1, Jan. 2005.

[14] Karimi- Ghartemani, M.; Mojiri, M.; Bakhshai, A.R "A technique for extracting time-varying harmonics based on an adaptive notch filter", *Control Applications*, Proceedings of 2005 *IEEE Conference* on 28-31 Aug.2005 pp.624 – 628, Dig. Obj. Ident. 10.1109/CCA.2005.1507196.

**2006 IEEE International Conference on Power Electronic, Drives and Energy Systems**

# Third Harmonic Current Injection for Power Quality Improvement in Rectifier Loads

Bhim Singh, *Senior Member, IEEE,* Vipin Garg, *Member, IEEE* and G.Bhuvaneswari, *Senior Member, IEEE*

*Abstract*— **This paper presents a novel method of reducing harmonic currents at the point of common coupling (PCC) in a three-phase diode bridge rectifier type utility interface. The proposed configuration consists of a current injection device, a newly designed current injection network and a properly tuned passive shunt filter. It also presents the design details of the current injection network and the passive filter. The proposed configuration performs well under varying loads on the rectifier. The effect of load variation on rectifier is also studied to demonstrate the effectiveness of the proposed ac-dc converter. A set of power quality indices on input ac mains and on dc bus is also presented and discussed in detail.**

*Index Terms*— **Harmonic current injection, rectifier load, power quality improvement.**

## I. INTRODUCTION

**W**ITH the proliferation of solid state power converters, the use of three-phase diode bridge rectifiers for interfacing to the electric utility has increased in many applications such as variable frequency ac drives, high power induction heating equipment, power supplies etc. The non-linear characteristics of diodes cause injection of harmonic currents in the supply system resulting in voltage distortion, increased losses. malfunction of sensitive electronic equipments etc [1]. Such problems have led to national and international standards and recommendations that limit the harmonic currents. An IEEE Standard 519 [2] has been established in 1981 as the "Recommended Practices and Requirements for Harmonic Control in Electrical Power System" giving limits on voltage distortion

Different techniques for harmonic current reduction have been proposed in the literature [3-6]. The active waveshaping techniques making use of fast digital processors along with complicated control circuitry result in high switching losses due to its operation at high switching frequency [3]. On the other hand, passive waveshaping techniques are robust, rugged and do not need complex control arrangement [4]. The current injection technique is one of these passive waveshapng techniques and is very effective in harmonic mitigation. The concept of current injection used for harmonic reduction has been given by Bird and Marsh[5] and later it has been used and modified by many researchers resulting in its different configurations [6-8]. Pejovic and Janda have reported one topology for third harmonic current injection for rectifier loads using optimal current programming [9]. Later it has been

Bhim Singh and G. Bhuvaneswari are with Department of Electrical Engg, I.I.T.Delhi, New Delhi, India -110016.
(E-mail: bhim_singh@yahoo.com,bhuvan225@gmail.com).
Vipin Garg is with Indian Railways, Baroda House, New Delhi.
(E-mail: vipin123123@gmail.com).

shown that it is possible to obtain sinusoidal input current using only passive elements [10]. But these cannot provide regulated dc output voltage. Pejovic and Janda [11] have proposed one configuration with passive elements for harmonic reduction. This topology is suitable only for fixed loads on the rectifier. But, in practice, the load on the rectifier varies similar to that of a variable frequency ac motor drive that it is feeding through an inverter. Under this condition, THD of ac mains current increases beyond IEEE-519 Standard [2].

This paper presents a new configuration for third harmonic current injection based network suitable for varying rectifier loads. The proposed configuration consists of a passive shunt filter (of small rating) at the input, current injection device and a current injection network. The current injection device has been realized with a zig-zag transformer to result in reduced magnetics rating.

The effect of load variation on the rectifier is also studied. The proposed ac-dc converter is able to provide almost unity power factor during wide variation of the rectifier load. A set of tabulated results is also presented giving the comparison of different power quality indices such as total harmonic distortion (THD) and crest factor of ac mains current (CF), power factor (PF), displacement factor (DPF) and distortion factor (DF), THD of supply voltage at PCC. The evaluation of the system efficiency is also presented under load varying condition.

## II. DESCRIPTION OF THE PROPOSED SCHEME

Fig.1 shows the schematic diagram of the proposed configuration for third harmonic current injection. It mainly consists of a current injection device, a passive filter and a current injection network.

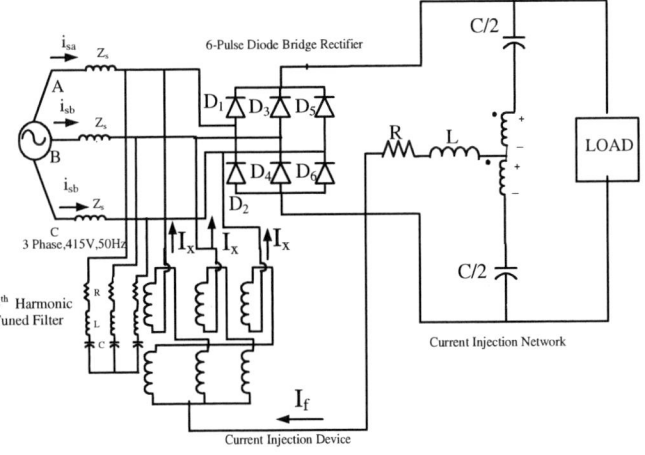

Fig.1. Proposed configuration for third harmonic current injection for rectifier loads.

## A. Current Injection Device

To reduce the rating of magnetics, zig-zag transformer is used to act as current injection device. The zig-zag transformer does not need any secondary windings, thus resulting in reduction in its rating. The zig-zag transformer has a very high (magnetizing) impedance for the positive and negative sequence voltages, therefore, it appears as an open circuit for these voltage components. The total third harmonic current generated by the current injection network $I_f$ is equally divided among the three phases. It enters into the ac side through the zig-zag transformer. The total third harmonic current $I_x$ splits equally in the three legs of the transformer. The current injected on ac side is given by

$$I_x = I_f / 3 \qquad (1)$$

For transformer rating calculations, each winding of the zig-zag transformer gets the line to neutral voltage ($V_{l-l}/\sqrt{3}$) and the winding current is the third harmonic injected current $I_f/3$. Therefore, transformer VA rating is given by

Transformer VA rating = $3/2 \ (V_{l-l}/\sqrt{3})(I_f/3)$    (2)

The factor of ½ is taken as there is no secondary winding.
The output voltage $V_0 = 1.35 \ V_{l-l}$      (3)
Also from Ref [5] $I_o = I_f$, giving

$$P_0 = I_o \ V_0 = 1.35 \ I_f \ V_{l-l} \qquad (4)$$

From eqns.(2) and (4), one gets the transformer rating as:
Transformer VA rating = $0.2138 \ P_0$      (5)

## B. Design of Passive Filter

The ac mains current in a three-phase rectifier type load mainly consists of 5th harmonic current. Hence, a passive shunt filter tuned at 5th harmonic frequency is connected at the input. The passive shunt filter has been designed in accordance with IEEE Standard 1531-2003 [12]. Various issues involved with the design of the passive filter are elucidated here.

### 1) Design Equations:

The passive shunt filter is governed by the following design equations [13-14].
The impedance of the filter branch is given as:

$$Z = R + j \ (\omega L - 1/\omega C) \qquad (6)$$

At resonance the imaginary part becomes zero. Thus resonance frequency becomes as:

$$f_n = 1/ \{2\pi \ (LC)^{1/2}\} \qquad (7)$$

For $n$th harmonic, the inductor and capacitor impedances are $X_{ln} = n\omega L$ and $X_{cn} = 1/ (n\omega C)$. At resonance $X_{ln} = X_{cn}$.

Quality Factor: The quality of the filter is a measure of the sharpness of tuning. It is defined as:

$$Q = X_{ln} / R = X_{cn} / R \qquad (8)$$

The capacitance is decided mainly based on the approximate reactive power requirement, which is assumed as 10% of the drive rating. With this value of capacitance, the inductance is calculated from eqn.(7). The quality factor is considered as 30.

The 50 Hz kVA of the filter capacitor at power frequency (50Hz) is given as [14]:

$$kVA = 0.314159 \ C \ V_{l-l}^2 \qquad (9)$$

where $V_{l-l}$ is the nominal line to line voltage of the system at the point of connection of the filter. C is the capacitance of the filter capacitor, $X_{ln}$ is the reactance of inductor and $X_{cn}$ is the reactance of capacitor at the resonant frequency.

## C. Current Injection Network

Fig.1 shows the current injection network consisting of only passive components. Consider that the rectifier in Fig.1 is being supplied from a balanced three-phase supply system as:

$$v_1 = V_m \ Cos \ \omega t \qquad (10)$$
$$v_2 = V_m \ Cos \ (\omega t - 120^0) \qquad (11)$$
$$v_3 = V_m \ Cos \ (\omega t - 240^0) \qquad (12)$$

Each diode in the rectifier bridge conducts sequentially as per the cathode potential and all the diodes in the rectifier bridge conduct for $120^0$. Under these conditions, the voltages and currents on the rectifier dc side are of triplen frequency The proposed configuration omits the resistance in current injection network. Here, the current injection network consists of two capacitors, a transformer with unity turns ratio, an inductor and a resistor. To inject the third harmonic current on ac side, the capacitors and the inductor should satisfy the resonance condition:

$$LC = \frac{1}{72 \pi^2 f^2} \qquad (13)$$

The quality factor ($Q_{CI}$) of current injection network is defined as $Q_{CI} = (1/R) \ [\sqrt{\{L/(2C)\}}]$    (14)

It is observed that the input current THD decreases with decrease in $Q_{CI}$ factor. Practically, $Q_{CI}$ may be chosen in the range of 1 to 2. To provide the optimal current injection, the resistance of the resistor is given by

$$R = (\sqrt{3}/\pi) \ (V_m / I_{out}) \qquad (15)$$

where, $V_m$ is the peak magnitude of the phase voltage of input ac mains and $I_{out}$ is the output current. Thus the proposed configuration can be designed with these values of the components in the current injection network. Capacitors with value C/2 = 130μF were used. The inductor value chosen is 4.17mH with parasitic resistance of 0.6Ω. Thus the proposed configuration can be designed with these values of components in the current injection network.

## III. MATLAB BASED SIMULATION

The proposed third harmonic current injection based ac-dc converter alongwith the passive shunt filter has been simulated in MATLAB environment alongwith Simulink and Power

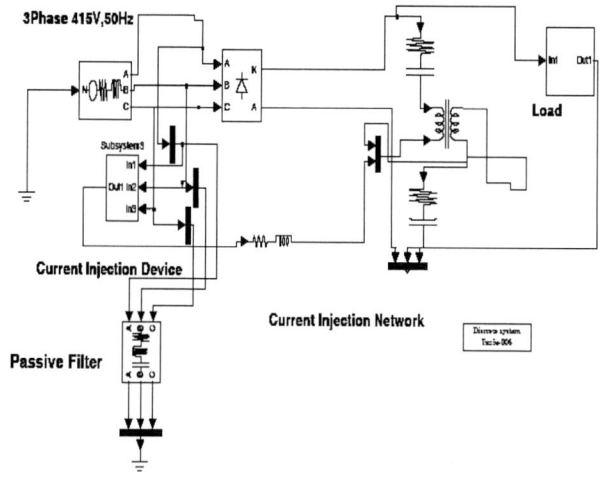

Fig.2. MATLAB model of the proposed configuration for third harmonic current injection for rectifier loads.

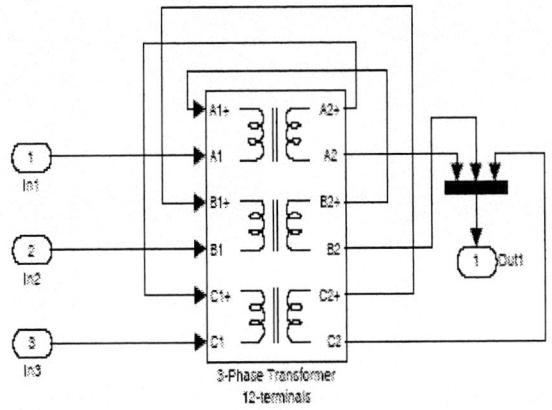

Fig.3. MATLAB based model of current injection device using a zig-zag transformer.

MATLAB model of the proposed ac-dc converter to improve various power quality indices. Fig.3 shows the MATLAB model of the zig-zag transformer used to act as the current injection device.

## IV. RESULTS AND DISCUSSION

To compare the performance of the proposed ac-dc converter with the existing 6-pulse fed system, a 6-pulse fed system, referred as Topology 'A' is also simulated under the same loading conditions. The supply current waveform at full load in a 6-pulse fed system is shown in Fig.4. Its harmonic spectrum is also shown, indicating the THD of ac mains current as 26.8%. The power factor is 0.956 as shown in Table-I.

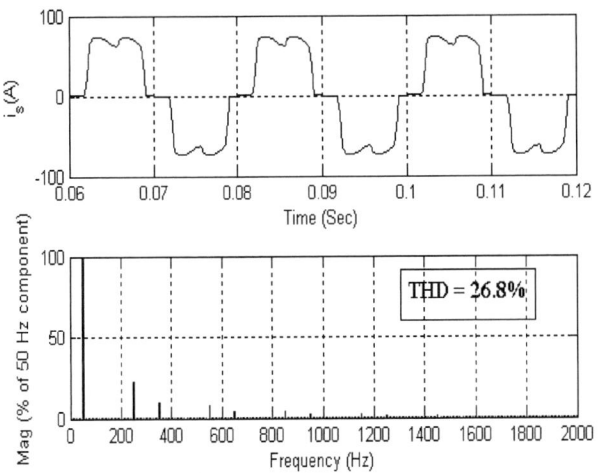

Fig. 4. AC mains current waveform alongwith its harmonic spectrum at full load for Topology 'A'.

The performance of the proposed system, referred as Topology 'B' is shown in Fig.5. The set of curves consists of supply voltage $v_s$(V), supply current $i_s$(A), dc link current $i_{dc}$(A) and the dc link voltage $v_{dc}$(V). These curves clearly show the similarity in the waveforms of supply voltage and supply current. It shows that the proposed ac-dc converter

behaves as a resistance emulator for the supply current. The supply current waveform and its harmonic spectrum at full load is shown in Fig.6, showing the THD as 3.99%. Moreover, all the harmonic components are less that 4% of the fundamental, thus easily qualifying for the IEEE standard 519 [2]. Under light load condition, the supply current waveform alongwith its harmonic spectrum is shown in Fig.7. Here again the THD of ac mains current is 3.82%.

To show the effect of the load variation, load is varied on the rectifier and different power quality indices are estimated. These parameters are shown in Table–II. It can be observed from the Table-II, that the proposed configuration results in near unity power factor in the wide operating range of the load. In this range the THD of ac mains current is always less than 5% . This is within the IEEE Standard 519 [2] limits for SCR >20. The variation of THD of ac mains current is shown in Fig.8. It again shows that the THD of ac mains current is as low as 2.33% at a supply current of 20.3A.

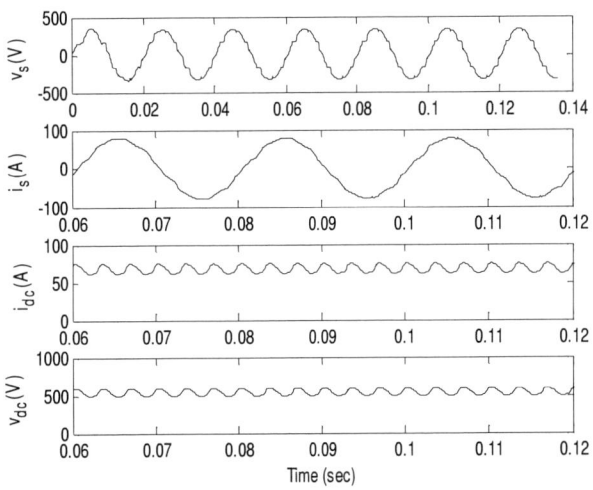

Fig.5. Waveforms of supply voltage $v_s$, supply current $i_s$ and DC link voltage $v_{dc}$.

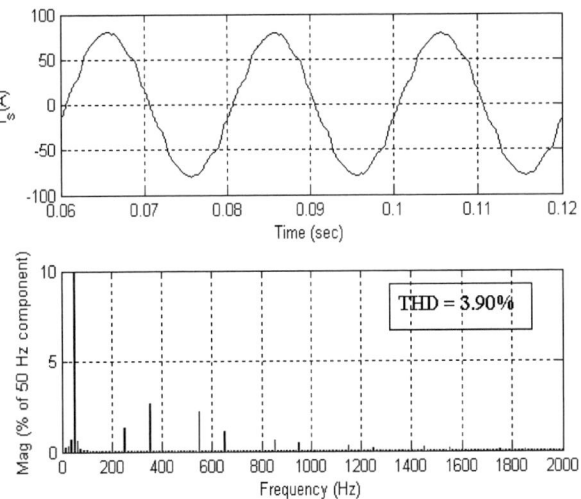

Fig. 6. AC mains current waveform alongwith its harmonic spectrum at full load for Topology 'B'.

TABLE I

COMPARISON OF POWER QUALITY PARAMETERS OF A RECTIFIER LOAD FED FROM DIFFERENT CONVERTERS

| Sr. No. | Topology | THD $V_s$ (%) FL | $I_s$ (A) | | THD of $I_s$ (%) | | DF | | DPF | | PF | | DC Link Voltage (V) Average | | DC Output Power (kW) | |
|---|---|---|---|---|---|---|---|---|---|---|---|---|---|---|---|---|
| | | | FL | LL | FL | LL | FL | LL | FL | LL | FL | LL | FL | LL | FL | LL |
| 1. | A(6-pulse) | 6.0 | 55.8 | 8.6 | 26.8 | 29.2 | .966 | .959 | .989 | .998 | .956 | .958 | 553 | 560 | 38.28 | 5.91 |
| 2. | B (Proposed) | 1.48 | 55.0 | 9.2 | 3.99 | 3.82 | .999 | .999 | .997 | .998 | .996 | .998 | 542 | 595 | 36.66 | 5.90 |

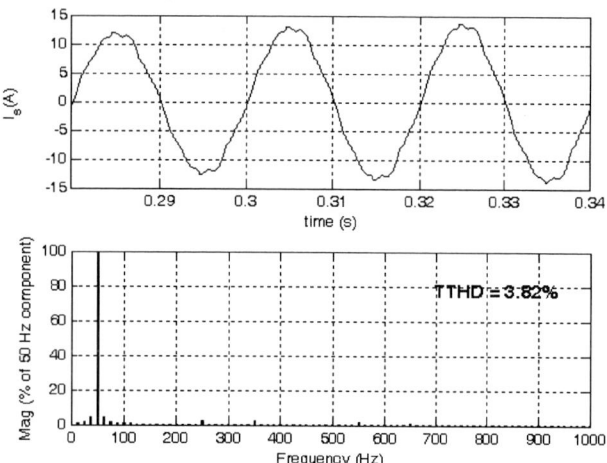

Fig. 7. AC mains current waveform alongwith its harmonic spectrum at light load (25% of full load) for Topology 'B'.

TABLE II

COMPARISON OF POWER QUALITY INDICES IN THE PROPOSED CONFIGURATION UNDER VARYING LOADS

| $I_s$(A) | $I_c$(A) | THD (%) | | CF of $I_s$ | DF | DPF | PF | $V_{dc}$ (V) |
|---|---|---|---|---|---|---|---|---|
| | | $I_s$ | $V_t$ | | | | | |
| 9.2 | 9.5 | 3.82 | 1.31 | 1.41 | .999 | .998 | .998 | 595 |
| 13.0 | 13.7 | 3.60 | 1.23 | 1.41 | .999 | 1.00 | .999 | 579 |
| 20.3 | 21.3 | 2.33 | 1.11 | 1.41 | .999 | 1.00 | .999 | 561 |
| 29.3 | 29.9 | 2.52 | 0.92 | 1.41 | .999 | 1.00 | .999 | 555 |
| 39.5 | 39.8 | 3.65 | 1.35 | 1.42 | .999 | .998 | .997 | 549 |
| 55.0 | 55.3 | 3.99 | 1.48 | 1.41 | .999 | .997 | .996 | 542 |

The power factor under this loading condition is 0.999. Hence, this converter gives its optimum performance under this loading condition. Even if the load is varied on both sides, again the THD of ac mains current is within IEEE Standard limits. This proves the reasonable load independence of THD of ac mains current in the proposed configuration.

To ensure that the installation of the passive filter does not cause the system loading, the current drawn from supply and current at the converter input are compared as shown in table-II. It can be observed from this table, that the supply current is always less than the converter input current.

The losses in the system have been calculated and the efficiency of the complete system is calculated under different loading conditions. The system efficiency varies between 89% and 93%. The efficiency under full load condition is observed to be 93%. The rating of passive components used in the passive filter is 5.5kVA, which comes out to be 14% of the output power of load.

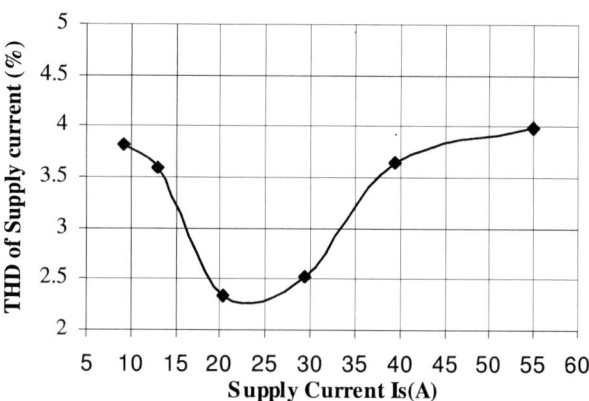

Fig. 8. Variation of THD of ac mains current with load in the proposed configuration.

## V. CONCLUSIONS

A third harmonic current injection based configuration has been proposed for varying rectifier loads. The proposed converter makes use of only passive components. The observed performance of the proposed configuration has demonstrated its ability to improve the THD of ac mains current and also the power factor under varying rectifier loads. The scheme has the advantage of simplicity and reliability in terms of components and the overall cost. The use of zig-zag transformer has resulted in reduction in rating of passive components. The proposed ac-dc converter has resulted in almost unity power factor in a wide variation of the load. The overall efficiency of the complete system is observed to be more than 90% under variable load conditions. There has been satisfactory improvement in the THD and crest factor of ac mains current.

## VI. APPENDIX

System Parameters:
Rated Power =10kW, Supply voltage =415V, 3-Phase, 50Hz,
Base Impedance = 0.18ohm
Passive Filter Details:
Components Rating
Inductor L =8.1mH, Resistance R = 0.4244ohm, Capacitance C = 50 µF
Filter kVAR rating = 5.5kVAR (14% of Load).

## VII. REFERENCES

[1] B.K.Bose, "Recent advances in power electronics," *IEEE Trans. On Power Electronics*, Vol.7, No.1, Jan.1992, pp. 2-16.

[2] *IEEE Guide for harmonic control and reactive compensation of Static Power Converters*, IEEE Standard 519-1992.

[3] B. Singh, B.N.Singh, A. Chandra, K.Al-Haddad, A. Pandey and D.P.Kothari, "A review of three-phase improved power quality ac-dc converters," *IEEE Trans. Ind. Electronics*, vol.51, pp.641-660, June 2004.

[4] Bhim Singh, G.Bhuvaneswari and Vipin Garg, "Analysis and Design of A Passive Filter for Power Quality Improvement in Three Phase Variable Frequency Induction Motor Drives" Proc. *IEEE Petroleum and Chemical Industry Conference*, Nov. 2004, New Delhi, pp 134-143.

[5] B.M.Bird and J.F.Marsh., "Harminic reduction in multiplex converters by triple frequency current injection," *Proc. IEE*, vol.116, no.10,Oct.1969.

[6] J.F.Baird and J.Arrillaga, "Harmonic reduction in dc ripple reinjection," *Proc. IEE*, vol. 127, no.5, Sept.1980.

[7] S.Kim, P.N. Enjeti, P.Packebush and I.Pitel "A new approach to improve power factor and reduce harmonics in a three-phase diode rectifier utility interface," *IEEE Trans. on Industry Applications*, Vol.30, No. 6, pp. 1557-1564, Nov./Dec.1994.

[8] P.Pejovic and Z.Janda, "Multipulse high power factor rectifier applying a novel current injection network," in *Proc., IEEE ICECS'01, 2001*, pp. 651-654.

[9] P.Pejovic and Z.Janda, "Optimal current programming in three-phase high power factor based on two boost converters," *IEEE Trans. on Power. Electron*. Vol.3, pp. 1152-1163, Dec.1998.

[10] W.B.Lawrence and W.Mielczarski, "Harmonic current reduction in a three-phase diode bridge rectifier," *IEEE Trans. on Ind. Electron*. Vol.39, pp. 571-576, Dec.1992.

[11] P.Pejovic and Z.Janda, "Low-harmonic three-phase rectifier applying current injection," *IEE Proc. Electr. Power Appl*. Vol.46, No.5, pp.545-551, Sept.1999.

[12] IEEE *Guide for application and specification of harmonic filters*, IEEE Standard 1573, 2003.

[13] J.Arrilaga, D.A.Bradley and P.S. Bodger, *Power System Harmonics*, New York: Wiley Interscience, 1985, ch.10, pp. 296-324.

[14] D. A. Gongalez, John C. Mccall, "Design of filters to reduce harmonic distortion in industrial power systems", *IEEE Trans. on Industry Applications*, Vol.IA-23, no.3, pp 504-511, May / June 1987.

## VIII. BIOGRAPHIES

**Bhim Singh** (SM'99) was born in Rahamapur, U. P., India in 1956. He received B. E. (Electrical) degree from University of Roorkee, India in 1977 and M. Tech. and Ph. D. degrees from Indian Institute of technology (IIT), New Delhi, in 1979 and 1983, respectively. In 1983, he joined as a Lecturer and in 1988 became a Reader in the Department of Electrical Engineering, University of Roorkee. In December 1990, he joined as an Assistant Professor, became an Associate Professor in 1994 and full Professor in 1997 at the Department of Electrical Engineering, IIT Delhi. His field of interest includes power electronics, electrical machines and drives, active filters, static VAR compensator, analysis and digital control of electrical machines. Prof. Singh is a Fellow of Indian National Academy of Engineering (INAE), Institution of Engineers (India) (IE (I)) and Institution of Electronics and Telecommunication Engineers (IETE), a Life Member of Indian Society for Technical Education (ISTE), System Society of India (SSI) and National Institution of Quality and Reliability (NIQR) and Senior Member IEEE (Institution of Electrical and Electronics Engineers).

**G. Bhuvaneswari** (SM'99) received M. Tech. and Ph. D. degrees from the Indian Institute of technology (IIT), Madras in 1988 and 1992, respectively. In 1997, she joined as an Assistant Professor at the Department of Electrical Engineering, IIT Delhi. Her field of interest includes power electronics, electrical machines and drives, active filters, and power conditioning. She is a fellow of Institution of Electronics and Telecommunication Engineers (IETE) and Senior Member IEEE (Institution of Electrical and Electronics Engineers).

**Vipin Garg** (M'05) was born in Kurushhetra, Haryana., India in 1972. He received B. Tech. (Electrical) and M.Tech. degrees from National Institute of Technology, Kurukshetra, India in 1994 and 1996 respectively and Ph. D. degree from Indian Institute of technology (IIT), New Delhi, in 2006 . In 1995, he joined as a Lecturer in the Department of Electrical Engineering, Regional Engg. College, Kurukshetra. In January 1998, he joined IRSEE (Indian Railways Service of Electrical Engineers) as an Assistant Electrical Engineer, became Divisional Electrical Engineer in 2002 and Deputy Chief Electrical Engineer in 2006. His field of interest includes power electronics, power conditioning, electrical machines and drives.

**2006 IEEE International Conference on Power Electronic, Drives and Energy Systems**

# Polygon Connected 15-Phase AC-DC Converter for Power Quality Improvement

Bhim Singh, *Senior Member, IEEE,* Vipin Garg, *Member, IEEE* and G.Bhuvaneswari, *Senior Member, IEEE*

*Abstract—* This paper presents a polygon connected autotransformer based 15-phase ac-dc converter for power quality improvement in vector controlled induction motor drives (VCIMD's). The proposed multi-phase ac-dc converter is based on a polygon connected autotransformer, designed for producing 15-phase voltages for effective harmonic reduction. The effect of load variation on drive is also studied to demonstrate the effectiveness of the proposed ac-dc converter. A set of power quality indices on input ac mains and on dc bus is also presented.

*Index Terms—*Autotransformer, multiphase AC-DC converter, power quality improvement.

## I. INTRODUCTION

THERE has been enormous increase in the use of variable frequency induction motor drives (VFIMD's) in various applications such as air conditioning, rolling mills, waste water treatment plants etc. These VFIMD's are generally used in vector control mode due to its advantages such as energy conservation, ease of control, less inrush current etc [1]. These VFIMD's are powered by some kind of electronic power processor, which generally consists of six-pulse diode bridge rectifier for rectification. These power processors have the side effect of injection of current harmonics into the ac mains, thereby, resulting in power pollution at the point of common coupling (PCC) [2]. The injection of harmonics leads to poor power factor operation, malfunction of sensitive electronic equipments connected at PCC. To have a control on the injection of these harmonics in utility side, An IEEE Standard 519 [3] has been established to impose strict restrictions on both utility and consumers.

To improve the power quality indices, different active as well as passive waveshaping techniques have been reported in the literature. The active waveshaping techniques such as active filters, hybrid filters, generally make use of fast commutating devices being operated at very high switching frequencies. This needs fast computing digital processors along with complicated control circuit arrangement. Moreover, these result in high switching losses due to high switching frequency operation. The passive wavehshaping techniques make use of passive filters or multipulse ac-dc converters for effective harmonic reduction.

The multipulse converters are simple, robust, rugged and more efficient [4]. For applications, where isolation is not required, autotransformer based configurations are found to be

more economical due to reduced magnetics, as the magnetics transfer only a small portion of the load power. Different configurations of 12-pulse and 18-pulse based ac-dc converters have been reported [5-9]. These converters fail to comply with the above standard. It is seen that as the number of pulses increases, the power quality indices improve. An ac-dc converter has been reported [10], which results in 28 steps in input current waveform. This configuration results in THD of supply current as 6.54%, which deteriorates at the reduced load.

This paper presents an autotransformer based 15-phase ac-dc converter for harmonic reduction. The design procedure of the autotransformer is given for producing 15-phase shifted voltages of equal magnitude. The proposed design technique provides flexibility in varying the transformer output voltage for different applications. This ac-dc converter results in elimination upto $29^{th}$ harmonic in the supply current.

A set of tabulated results giving the comparison of different power quality indices such as total harmonic distortion (THD) and crest factor of ac mains current (CF), power factor (PF), displacement factor (DPF) and distortion factor (DF), THD of supply voltage at PCC is presented for a load fed from an existing 6-pulse ac-dc converter, shown in Fig.1 and referred as Topology 'A' and proposed 15-phase ac-dc converter, shown in Fig.2 and referred as Topology 'B'.

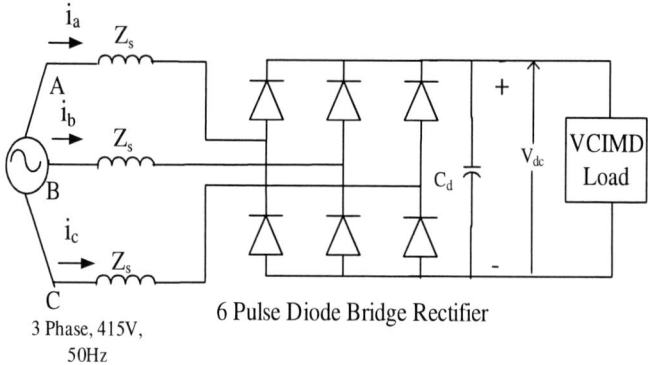

Fig. 1. Six-pulse diode bridge rectifier fed load (Topology 'A').

## II. DESIGN OF PROPOSED 15-PHASE AC-DC CONVERTER

For harmonic elimination, the required minimum phase shift is given by

Phase shift = $360^0$/ Number of phases

For achieving 15-phase ac-dc conversion, the phase shift between any two sets of nearby voltages should be of $24^0$ with respect to each other. For achieving the 15-phase operation, five sets of 3-phase voltages (phase shifted through an angle of $+24^0$) are produced.

The number of turns required for achieving these phase

---

Bhim Singh and G. Bhuvaneswari are with Department of Electrical Engg, I.I.T.Delhi, New Delhi, India -110016.
(E-mail: bhim_singh@yahoo.com,bhuvan225@gmail.com).
Vipin Garg is with Indian Railways, Baroda House, New Delhi.
(E-mail: vipin123123@gmail.com).

0-7803-9771-1/06/$25.00 ©2006 IEEE

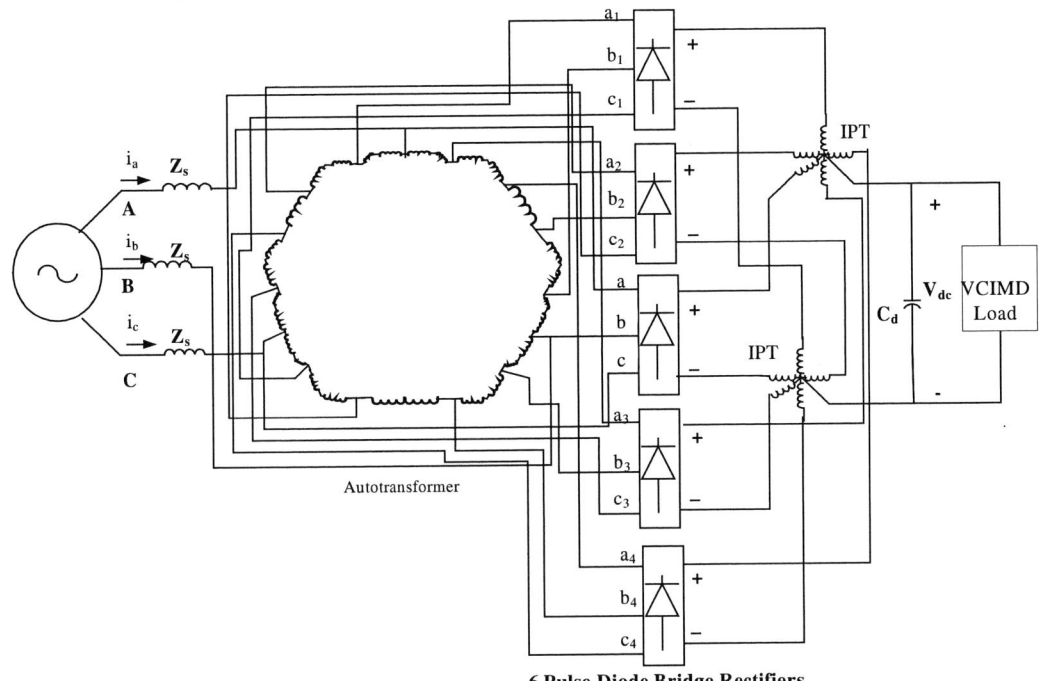

**6 Pulse Diode Bridge Rectifiers**

Fig. 2. Proposed 15-phase ac-dc converter feeding a VCIMD load.

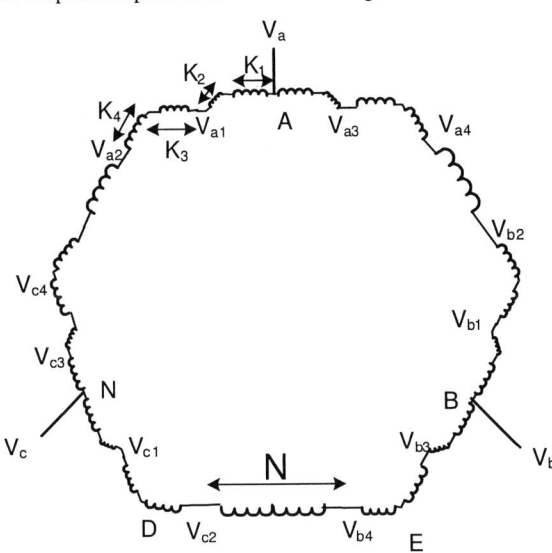

Fig. 3a. Proposed autotransformer winding connection diagram.

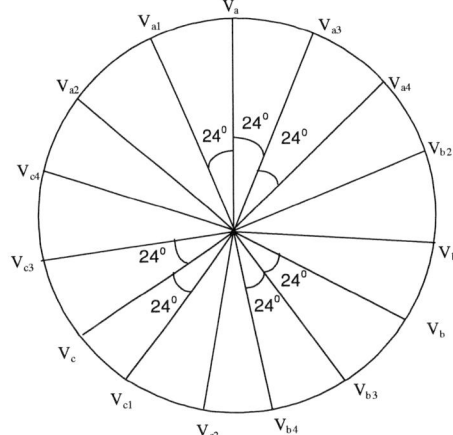

Fig. 3b. Phasor diagram of different phase voltages in the proposed autotransformer based 15-phase ac-dc converter.

$$V_{a3} = V \angle{-7.5^0}, \quad V_{b3} = V\angle{-127.5^0}, \quad V_{c3} = V\angle{112.5^0} \quad (8)$$
$$V_{a4} = V \angle{-22.5^0}, \quad V_{b4} = V\angle{-142.5^0}, \quad V_{c4} = V\angle{97.5^0} \quad (9)$$

Here, $V_a$, $V_b$ and $V_c$ are the phase voltages and V is the rms value of phase voltage

The angle between CB and CD in Fig.3(a) is $60^0$, similarly angle between BC and BE is $60^0$. Thus, from the parallelogram CBED in Fig.3(a), it can be written as:

$$K_1 \cos 60^0 + K_2 + K_3 \cos 60^0 + K_4 + N + K_1 \cos 60^0 + K_2 + K_3 \cos 60^0 = 1 \quad (10)$$
$$2K_2 + 2K_4 + K_1 + K_3 + N = 1 \quad (11)$$
$$\text{Thus N= 0.2404} \quad (12)$$

Using above equations $K_1$, $K_2$, $K_3$ and $K_4$ can be calculated. These equations result in $K_1 = 0.2059$, $K_2 = 0.0576$, $K_3 = 0.1127$ and $K_4 = 0.1629$ for the desired phase shift in autotransformer.

The five sets of three-phase voltages are given to five sets of three-phase diode rectifier bridges, which rectify these ac

shifts among different phase voltages are calculated as follows. Fig. 3a shows the winding connection diagram of the proposed autotransformer. Fig. 3b shows the corresponding phasor diagram of different phase voltages.

Consider phase 'a' voltages in Fig.3a as:

$$V_{a1} = V_a - K_1 V_{bc} + K_2 V_{ca} \quad (1)$$
$$V_{a2} = V_{a1} - K_3 V_{bc} + K_4 V_{ca} \quad (2)$$
$$V_{a3} = V_a + K_1 V_{bc} - K_2 V_{ab} \quad (3)$$
$$V_{a4} = V_{a3} + K_3 V_{bc} - K_4 V_{ab} \quad (4)$$

Assume the following set of voltages:

$$V_a = V\angle{0^0}, V_b = V\angle{-120^0}, V_c = V\angle{120^0} \quad (5)$$
$$V_{a1} = V \angle{7.5^0}, \quad V_{b1} = V\angle{-112.5^0}, \quad V_{c1} = V\angle{127.5^0} \quad (6)$$
$$V_{a2} = V \angle{22.5^0}, \quad V_{b2} = V\angle{-97.5^0}, \quad V_{c2} = V\angle{122.5^0} \quad (7)$$

653

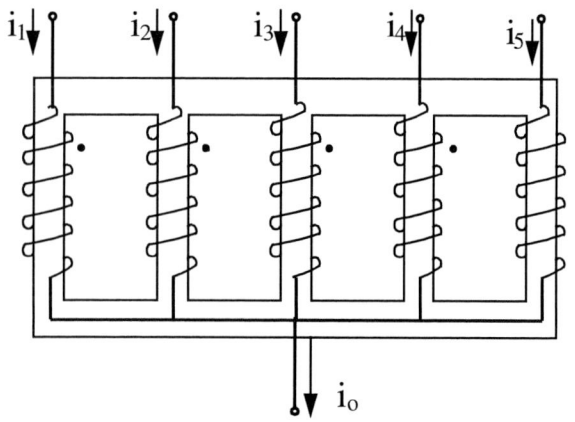

Fig. 4. Winding configuration of interphase transformer.

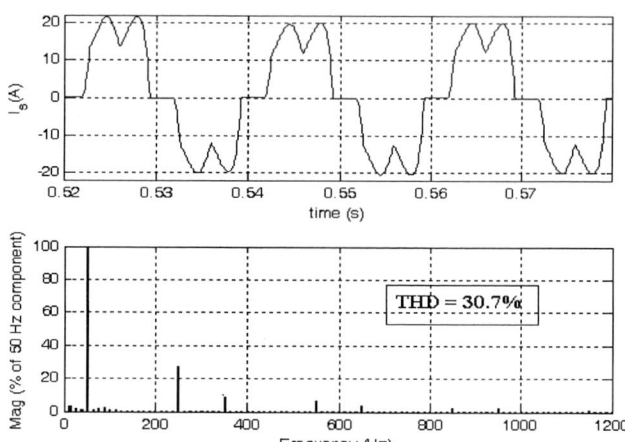

Fig. 6. AC mains current waveform of VCIMD fed by 6-pulse diode rectifier along with its harmonic spectrum at full load (Topology 'A').

voltages. These rectified voltages are then fed to the interphase transformers as shown in Fig.4. The function of an interphase transformer is to ensure symmetrical conduction of each diode, thus ensuring the proper harmonic elimination.

## III. MATLAB BASED SIMULATION

The complete system consisting of the proposed autotransformer, interphase transformer, rectifiers and the load has been designed, analysed, modeled and simulated in MATLAB environment using Simulink and Power System Blockset (PSB) toolboxes. Fig.5 shows the MATLAB model of the proposed autotransformer based ac-dc converter. The simulated results are shown in Figs. 6-11 and Tables I-II. The simulations on the proposed 15-phase ac-dc converter are conducted at three-phase line voltage of 415V, 50Hz, AC input and with a VCIMD load. The source impedance has been considered as 3% in all simulations.

## IV. RESULTS AND DISCUSSION

To compare the performance of the proposed multiphase ac-dc converter, first the performance of a 6-pulse ac-dc converter has been computed as shown in Table-I. It is shown that the THD of ac mains current at full load is 30.7%, which deteriorates to 71.6% under light load (20% of full load)

condition for Topology 'A'.

The performance of the proposed multi-phase ac-dc converter is shown in Fig.6. It consists of supply voltage $V_s$ (V), supply current $I_s$ (A) and dc link voltage $v_{dc}$(V). It can be observed that the THD of ac mains current at full load is 1.96%, as shown in Fig 7 and 8. Under light load condition the THD of ac mains current is 2.82 %, as shown in Fig 9. It is observed that all the harmonic components are less that 2% of the fundamental component. The power factor at full load is 0.991 and under light load is 0.986. Further, the load is varied and its effect on different power quality indices is studied and shown in Table-II. It can be observed from Table-II, that the proposed ac-dc converter results in THD of ac mains current less than 5%, which is within the IEEE Standard 519 [3] limits for SCR. Moreover it results in an almost unity power factor in a wide variation of load on the rectifier. It shows the efficacy of the proposed multi-phase ac-dc converter. It is also observed from Table-I, that there is remarkable improvement in the power quality indices on dc bus in terms of low ripple factor (RF). The ripple factor of dc link voltage has improved reasonably, which means that for the same allowable dc link voltage ripple, the size of dc link capacitor can be reduced, leading to saving in capital cost. The variation of % THD of ac mains current and power factor (PF) with load on VCIMD fed

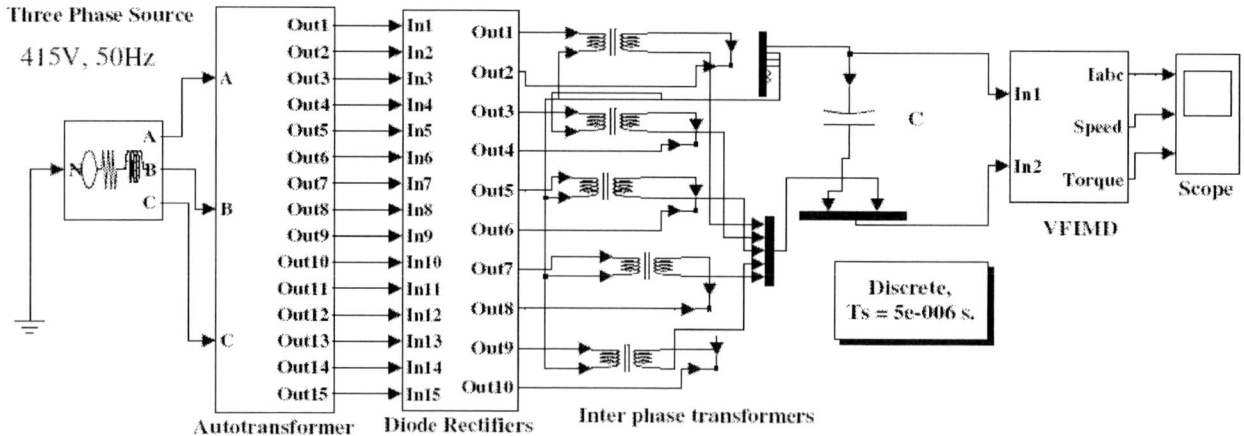

Fig. 5. MATLAB block diagram of proposed 15-phase ac-dc converter fed VCIMD (Topology 'B').

TABLE I
COMPARISON OF POWER QUALITY PARAMETERS OF A LOAD FED FROM DIFFERENT CONVERTERS

| Sr. No. | Topo logy | THD $V_s$ (%) FL | THD of $I_s$ (%) | | DF | | DPF | | PF | | DC Link Voltage | | | |
|---|---|---|---|---|---|---|---|---|---|---|---|---|---|---|
| | | | | | | | | | | | Average (V) | | RF (%) | |
| | | | FL | LL (20%) | FL | LL (20%) | FL | LL (20%) | FL | LL (20%) | FL | LL (20%) | FL | LL (20%) |
| 1. | A | 5.84 | 32.7 | 71.6 | .950 | 0.813 | .982 | 0.972 | .934 | 0.791 | 548 | 556 | 0.38 | 0.26 |
| 2. | B | 2.84 | 1.96 | 2.82 | .999 | 0.999 | .992 | 0.987 | .991 | 0.986 | 574 | 581 | 0.01 | 0.004 |

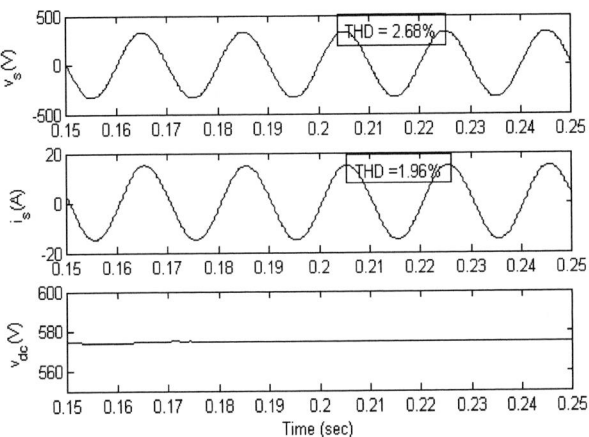

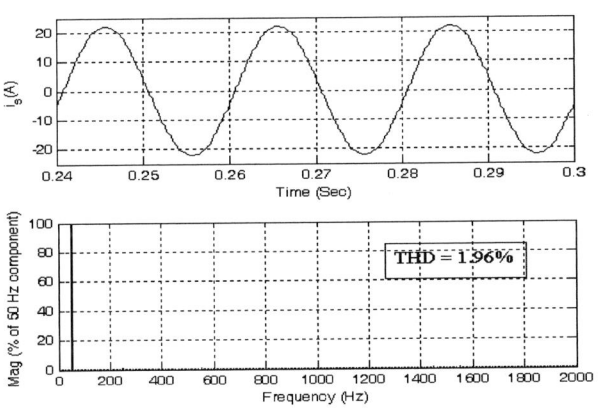

Fig. 7. Waveforms of supply voltage, supply current and dc link voltage.

Fig. 8. Supply current waveforms along with its harmonic spectrum at full load in proposed converter.

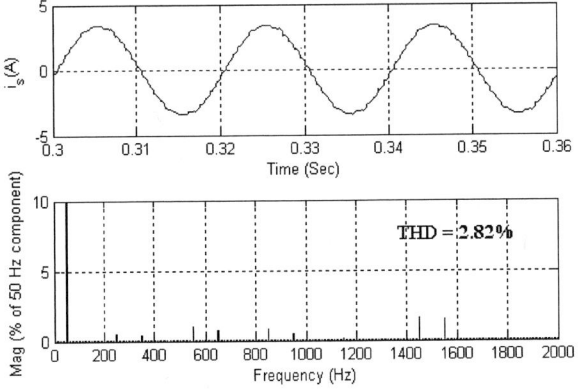

Fig. 9. Supply current waveforms along with its harmonic spectrum at light load in proposed converter.

TABLE II
COMPARISON OF POWER QUALITY INDICES OF PROPOSED 15-PHASE
(TOPOLOGY 'B')

| $I_s$(A) | THD (%) | | CF of $I_s$ | DF | DPF | PF | RF (%) | $V_{dc}$ (V) |
|---|---|---|---|---|---|---|---|---|
| | $I_s$ | $V_t$ | | | | | | |
| 2.40 | 2.82 | 0.90 | 1.42 | .999 | .987 | .986 | .004 | 581 |
| 4.50 | 2.40 | 1.46 | 1.42 | .999 | .991 | .990 | .005 | 579 |
| 6.12 | 2.30 | 1.84 | 1.41 | .999 | .992 | .991 | .006 | 578 |
| 8.05 | 2.20 | 2.25 | 1.41 | .999 | .982 | .991 | .018 | 576 |
| 11.8 | 1.96 | 2.84 | 1.41 | .999 | .992 | .991 | 0.01 | 574 |

from a six-pulse ac-dc converter and the proposed 15-phase ac-dc converter is shown in Figs.10 and 11 respectively, showing a remarkable improvement in these power quality indices for the proposed 15-phase ac-dc converter. On the magnetics front also, the proposed converter needs an autotransformer of 3.05kVA and interphase transformer of 0.75kVA, totaling the magnetics rating only to 37% of the drive rating.

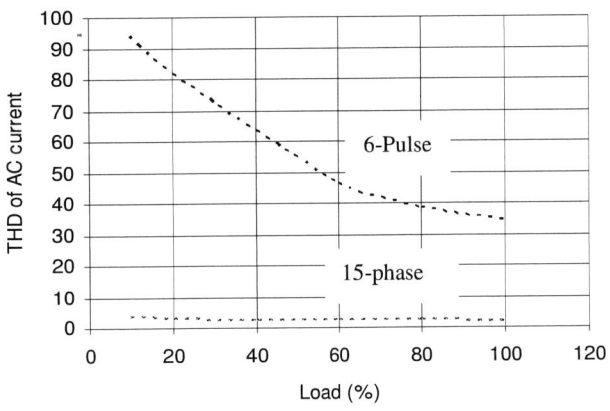

Fig. 10. Variation of THD of ac mains current with load on VCIMD in 6-pulse (Topology 'A'), and proposed 15-phase ac-dc converter (Topology 'B').

## V. CONCLUSIONS

The design, analysis, modeling and simulation of a novel multi-phase ac-dc converter have been carried out for rectifier load. It is seen that the proposed ac-dc converter results in THD of supply current less than 5% in a wide variation of load with a nearly unity power factor operation. The proposed 15-phase ac-dc converter may be used for applications, where harmonic reduction is more stringent such as variable frequency induction motors for waste water plant and air conditioning.

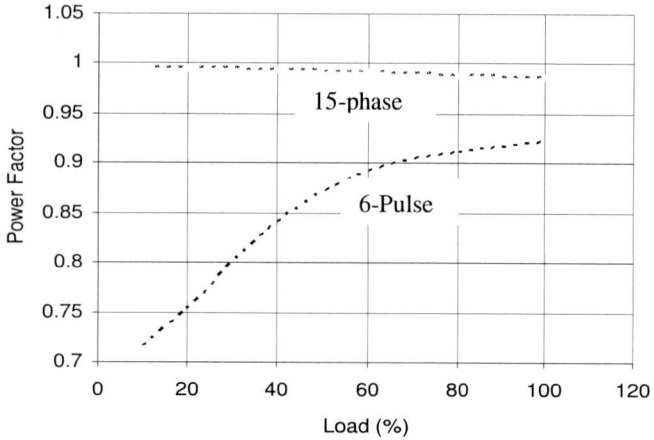

Fig. 11. Variation of power factor with load on VCIMD in 6-pulse (Topology 'A') and proposed 15-phase ac-dc converter (Topology 'B').

## VI. APPENDIX

System Parameters:

Rated Power =10kW, Supply voltage =415V, 3-Phase, 50Hz, Base Impedance = 0.18ohm

Magnetics Rating:

Autotransformer rating = 3.05kVA, Interphase transformer rating = 0.75kVA

## VII. REFERENCES

[1] P.Vas, *Sensorless vector and direct torque control*, Oxford University Press, 1998.

[2] B.K.Bose, "Recent advances in power electronics," *IEEE Trans. on Power Electronics*, Vol.7, No.1, Jan.1992, pp. 2-16.

[3] *IEEE Guide for harmonic control and reactive compensation of Static Power Converters*, IEEE Standard 519-1992.

[4] D. A. Paice, Power Electronic Converter Harmonics: Multipulse Methods for Clean Power, New York, IEEE Press, 1996.

[5] Bhim Singh, G.Bhuvaneswari and Vipin Garg, "A novel harmonic mitigator based 12-pulse rectification for vector controlled induction motor drives," *Int. J. Energy Technology and Policy*, Vol.4, Nos.1/2, 2006, pp.205-228.

[6] D.A. Paice, "Simplified wye connected 3-phase to 9-phase autotransformer," U.S. Patent No. 6,525,951 B1, Feb. 25, 2003.

[7] J.Ferens, H.D.Hajdinjak and S.Rhodes, "18-pulse rectification system using a wye connected autotransformer," U.S. Patent No. 6,650,557 B2, Nov.18, 2003.

[8]  G. R. Kamath, B. Runyan and Richard Wood, "A compact autotransformer based 12-pulse rectifier circuit," Proc. 2001, *IEEE*

*IECON, Conf.*, pp. 1344-1349.

[9]  K.Oguchi and T.Yamada, "Novel 18-pulse diode rectifier circuit with non-isolated phase shifting transformers," IEE Proc. Electr. Power Appl. Vol.14, No.1, pp.1-5, Jan.1997.

[10] C.L.Chen and G.K.Horng, "A new passive 28-step current shaper for three-phase rectification," *IEEE Trans. on Industrial Electronics*, Vol.47, No.6, pp1212-1219, December 2000.

## VIII. BIOGRAPHIES

**Bhim Singh** (SM'99) was born in Rahamapur, U. P., India in 1956. He received B. E. (Electrical) degree from University of Roorkee, India in 1977 and M. Tech. and Ph. D. degrees from Indian Institute of technology (IIT), New Delhi, in 1979 and 1983, respectively. In 1983, he joined as a Lecturer and in 1988 became a Reader in the Department of Electrical Engineering, University of Roorkee. In December 1990, he joined as an Assistant Professor, became an Associate Professor in 1994 and full Professor in 1997 at the Department of Electrical Engineering, IIT Delhi. His field of interest includes power electronics, electrical machines and drives, active filters, static VAR compensator, analysis and digital control of electrical machines. Prof. Singh is a Fellow of Indian National Academy of Engineering (INAE), Institution of Engineers (India) (IE (I)) and Institution of Electronics and Telecommunication Engineers (IETE), a Life Member of Indian Society for Technical Education (ISTE), System Society of India (SSI) and National Institution of Quality and Reliability (NIQR) and Senior Member IEEE (Institution of Electrical and Electronics Engineers).

**G. Bhuvaneswari** (SM'99) received M. Tech. and Ph. D. degrees from the Indian Institute of technology (IIT), Madras in 1988 and 1992, respectively. In 1997, she joined as an Assistant Professor at the Department of Electrical Engineering, IIT Delhi. Her field of interest includes power electronics, electrical machines and drives, active filters, and power conditioning. She is a fellow of Institution of Electronics and Telecommunication Engineers (IETE) and Senior Member IEEE (Institution of Electrical and Electronics Engineers).

**Vipin Garg** (M'05) was born in Kurushhetra, Haryana., India in 1972. He received B. Tech. (Electrical) and M.Tech. degrees from National Institute of Technology, Kurukshetra, India in 1994 and 1996 respectively and Ph. D. degree from Indian Institute of technology (IIT), New Delhi, in 2006 . In 1995, he joined as a Lecturer in the Department of Electrical Engineering, Regional Engg. College, Kurukshetra. In January 1998, he joined IRSEE (Indian Railways Service of Electrical Engineers) as an Assistant Electrical Engineer, became Divisional Electrical Engineer in 2002 and Deputy Chief Electrical Engineer in 2006. His field of interest includes power electronics, power conditioning, electrical machines and drives.

**2006 IEEE International Conference on Power Electronic, Drives and Energy Systems**

# Power Quality Standards and Their Application to a Granite Factory

S. Hasani, F. Donyavi, M. Masoudi, and H. Mokhtari, *Member, IEEE*

*Abstract*--This paper presents the results of a thorough analysis of applying power quality standards to a granite factory. It explains the procedure of determining power quality indices in a power system. The paper presents the field data and provides the results of statistical analysis in order to extract power quality indices. A comparison against standard limits is also given. The case study gives an overview of power quality in such customers and brings up some practical limitations when applying power quality standards.

*Index Terms*--flicker, granite load, harmonics, IEEE519, power quality, standard.

## I. INTRODUCTION

USE of Power-Electronic (PE) loads in industries has been increasing in the past few years. This is due to the fact that PE devices provide more flexibility in production lines, decrease energy consumption, and result in lower cost of products. However, these advantages are achieved at the price of causing Power Quality (PQ) issues for both utilities and customers [1-4]. Several works have been carried out to determine the cost of low power quality on different industry sectors from one hand. From the other hand, engineers try to mitigate the consequences of low PQ by using PQ enhancement techniques [1-4]. Therefore, it has become a necessity for both utilities and customers to perform PQ analysis in their systems in order to 1) know system current/voltage profiles 2) perform PQ enhancement methods and 3) prevent possible high cost malfunctions/interruptions in system.

The objectives of this paper are to explain standards or recommended practices for PQ analysis and apply them to a granite factory as a case study. A picture of power quality at such loads is given by providing both time-domain transient waveforms as well as time-series PQ data. Some proposals are also made to improve PQ. The paper also brings up concerns and issues attributed to the existing standard or recommended practices in applying to power systems.

S. Hasani, F. Donyavi,, and M. Masoudi are with West Azarbayjan Utility, Oroumieh, Iran (e-mail: sh592b@yahoo.com).

H. Mokhtari is with the School of Electrical Engineering, Sharif University of Technology, Azadi Ave., Tehran, Iran. (e-mail: mokhtari@sharif.edu).

## II. STANDARD PROCEDURES

Unfortunately, there are a few reports on procedures of determining PQ indices in industries. IEEE519 standard [5] recommends harmonic limits in power systems for voltage and current signals. Table I shows the limits for voltages at a Point of Common Coupling (PCC) as a function of the bus voltage. The limits for current harmonics are determined based on the PCC voltage and bus short circuit capability as Table II indicates. However, this standard is silent about how to determine the harmonic level at a PCC. A procedure is recommended by IEEE harmonic working group [6]. According to this guide, measurement is taken over a period of one week. Then, the Cumulative Probability of 95% (CP95) of the data is calculated and considered as the pollution level at the test node.

TABLE I
IEEE519 VOLTAGE HARMONIC LIMITS

| Bus Voltage at PCC | Individual Voltage Distortion (%) | Total Voltage Distortion THD (%) |
|---|---|---|
| 69 kV and below | 3.0 | 5.0 |
| 69.001 kV through 161 kV | 1.5 | 2.5 |
| 161 kV and above | 1.0 | 1.5 |

TABLE II
IEEE519 CURRENT HARMONIC LIMITS

| $I_{sc} / I_l$ | Individual Harmonic Order (Odd Harmonics) | | | | | |
|---|---|---|---|---|---|---|
|  | <11 | 11<h<17 | 17<h<23 | 23<h<35 | 35<h | THD |
| <20* | 4.0 | 2.0 | 1.5 | 0.6 | 0.3 | 5.0 |
| 20<50 | 7.0 | 3.5 | 2.5 | 1.0 | 0.5 | 8.0 |
| 50<100 | 10.0 | 4.5 | 4.0 | 1.5 | 0.7 | 12.0 |
| 100<1000 | 12.0 | 5.5 | 5.0 | 2.0 | 1.0 | 15.0 |
| >1000 | 15.0 | 7.0 | 6.0 | 2.5 | 1.4 | 20.0 |

Even harmonics are limited to 25% of the odd harmonics above.

$I_{sc}$ = maximum short-circuit current at PCC.
$I_l$ = maximum demand load current (fundamental frequency component) at PCC

IEC61000-4-15 [7] standard is consulted for flicker determination. The flicker level is calculated for all three phases, and the CP95% and the maximum value of the phase with the worst condition is determined.

## III. FIELD DATA

This section presents some of the filed data taken by the PQ analyzer. Specifications of the analyzer are given in the appendix. Fig. 1 depicts the voltage THD and current TDD data calculated every 10 minutes over a week period.

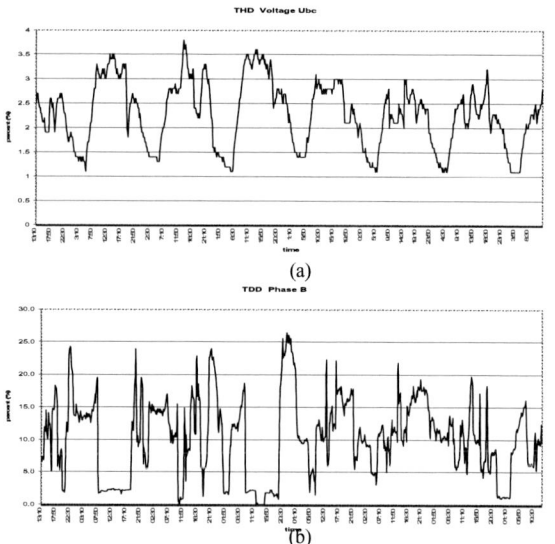

(a)

(b)

Fig. 1. Field data a) voltage THD data b) Current TDD data.

The data corresponding to the imbalance index is plotted in Fig. 2. Fig. 2-a shows the voltage imbalance and Fig. 2-b shows the current imbalance.

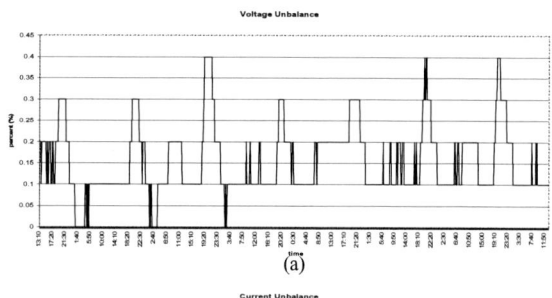

(a)

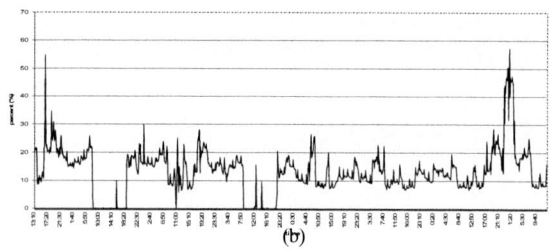

(b)

Fig. 2. Field data a) voltage THD data b) Current TDD data.

## IV. STATISTICAL ANALYSIS

This section provides statistical analysis in order to determine PQ indices according to [6,7]. Table III summarizes the results for the voltage THD. As it can be seen, the voltage THD is IEEE519 compliant.

TABLE III
VOLTAGE THD RESULTS SUMMARY

| Index | Maximum | CP 95 | IEEE519 Limit |
|---|---|---|---|
| THD Uab | 3.1 % | 2.9 % | 5 % |
| THD Ubc | 3.8 % | 3.4 % | 5 % |
| THD Uca | 3.3 % | 2.8 % | 5 % |
| THD Va | 2.9 % | 2.7 % | 5 % |
| THD Vb | 3.4 % | 3.1 % | 5 % |
| THD Vc | 3.8 % | 3.3 % | 5 % |

The procedure is extended to individual harmonics as well. Fig. 3 depicts the results for only one of the line voltages, i.e. Uab. Table IV shows the results for all three line voltages. The table shows the CP95 index as well the maximum value for all three phases.

Fig. 3. Individual harmonics for Uab line voltage.

TABLE IV
VOLTAGE INDIVIDUAL HARMONICS

| H# | Uab | | Ubc | | Uca | | |
|---|---|---|---|---|---|---|---|
| | Max | Cp95 | Max | cp95 | Max | Cp95 | Limit |
| 2 | 0 | 0 | 0 | 0 | 0 | 0 | 1.5 |
| 3 | 0.4 | 0.3 | 0.4 | 0.3 | 0 | 0 | 3 |
| 4 | 0 | 0 | 0 | 0 | 0 | 0 | 1.5 |
| 5 | 3.1 | 2.9 | 3.8 | 3.4 | 3.3 | 2.8 | 3 |
| 6 | 0 | 0 | 0 | 0 | 0 | 0 | 1.5 |
| 7 | 0 | 0 | 0 | 0 | 0 | 0 | 3 |
| 8 | 0 | 0 | 0 | 0 | 0 | 0 | 1.5 |
| 9 | 0 | 0 | 0 | 0 | 0 | 0 | 3 |
| 10 | 0 | 0 | 0 | 0 | 0 | 0 | 1.5 |
| 11 | 0 | 0 | 0 | 0 | 0 | 0 | 3 |
| 12 | 0 | 0 | 0 | 0 | 0 | 0 | 1.5 |
| 13 | 0 | 0 | 0 | 0 | 0 | 0 | 3 |

From Table IV, it can be seen that the 5th voltage harmonic is marginally higher than IEEE519 standard limits.

Table V shows the analysis results for the voltage imbalance. Based on IEEE1159 Standard limit, the ratio of the negative sequence to the positive sequence is the imbalance ratio. To determine the imbalance index, the CP95 of the imbalance value calculated for each day, and the maximum CP95 is selected.

TABLE V
VOLTAGE IMBALANCE INDEX

| Standard Limit | Imbalance Index |
|---|---|
| 2 % | 0.4 % |

From Table V, one can conclude that the voltage imbalance

is of no concern.

For the flicker level, short term and long term flicker, i.e. Pst and Plt, indices are considered. The power analyzer calculates only the Pst. Plt index is then calculated using the following equation:

$$Plt_i = \sqrt[3]{(Pst_i^3 + Pst_{i-1}^3 + ... + Pst_{i-11}^3)/12}$$

Fig. 4 depicts the results for flicker determination.

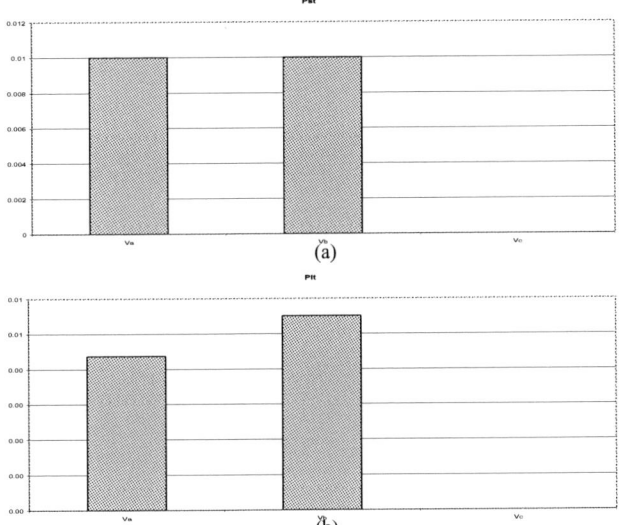

Fig. 4. Voltage flicker a) Pst b) Plt.

Based on IEC61000-4-15, the limits for Pst and Plt are 1 and 0.8 respectively. Therefore, it can be seen that the flicker level is not violating the standard limit.

Study is also performed to determine the quality of load current at the test location. This study includes current TDD and individual current harmonic level, i.e. HDi. The procedure to determine current harmonic indices is the same as that used for voltage harmonics, i.e. the CP95 value is determined. The results are summarized in Table VI and Fig. 5.

The ratio of the short circuit ratio to the fundamental load current is greater than 1000. Therefore, the maximum permitted limit for the TDD is 20% and for the individual harmonics is 15%.

TABLE VI
CURRENT TDD RESULTS SUMMARY

| Index | Maximum | CP 95 | IEEE519 Limit |
|---|---|---|---|
| TDD Ia | 18.1 % | 14.3 % | 20 % |
| TDDI b | **26.5** % | **21.1** % | 20 % |
| TDD Ic | **24.1** % | 17.3 % | 20 % |

Investigating the results given in Table VI and Fig. 5 reveals that the level of the 5th harmonic is higher than the limits specified by IEEE519 standard. The high level of the 5th harmonic has caused the TDD also to go above the standard limit. The maximum 5th harmonic is 25.6% and the highest TDD is 26.5%.

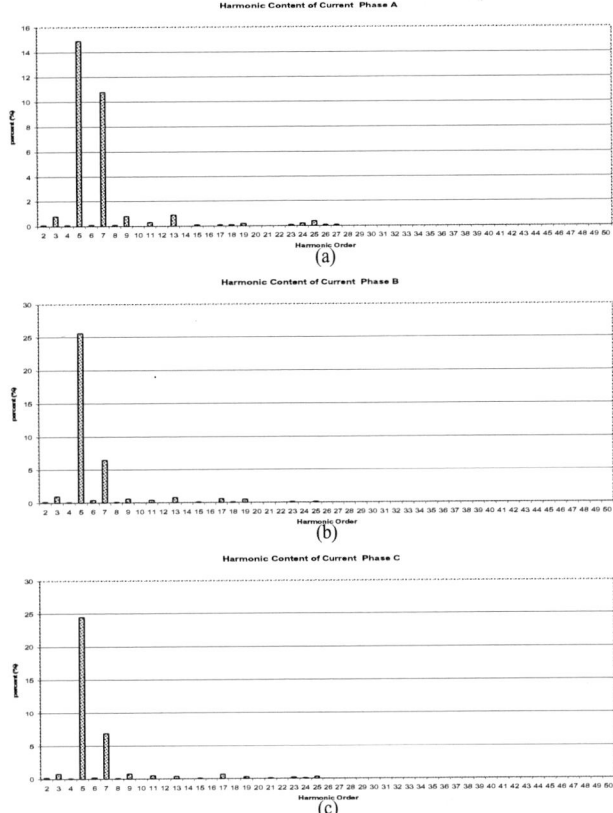

Fig. 5. Current harmonic results summary.

## V. LOAD FLUCTATION STUDY

In this section, the load profile is studied. Fig. 6 is the load active power, reactive power, and power factor in one of the phases, i.e. phase a. From this figure, it can be seen that the average load power factor is low. The load change is so high that the existing reactive power compensator cannot effectively increase the load power factor. This is due to the fact that contactor-based capacitor banks are not fast enough for reactive power compensation of highly varying loads.

## VI. TRANSIENT ANALYSIS

The power analyzer was also set in order to capture transients as well as steady state PQ parameters. The settings for transient capturing are selected based on IEEE1159 standard. The analyzer is capable of capturing up to fifty transients. Table VII summarizes transient analysis results. Out of 21 transients, 14 transients are due to the switching of capacitor banks.

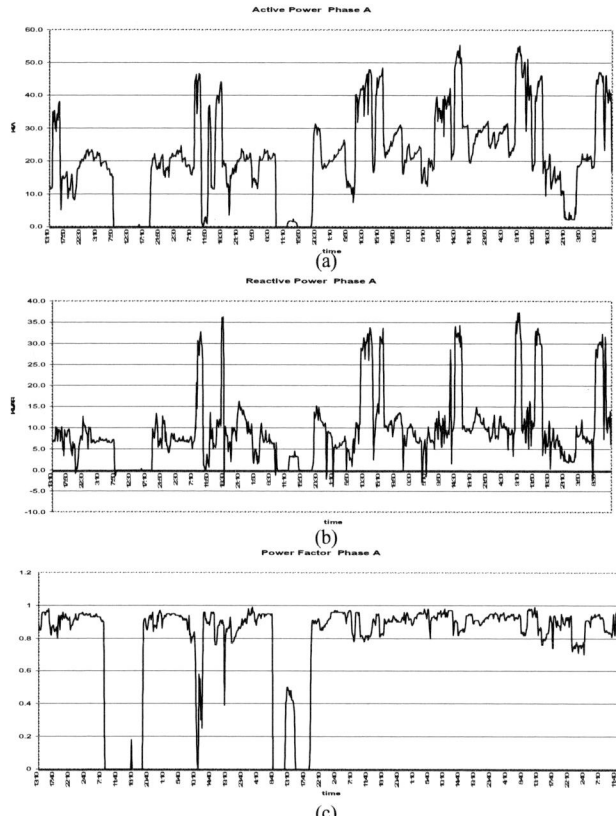

(a)

(b)

(c)

Fig. 6. Phase-a load (a) active power (b) reactive power (c) power factor.

TABLE VII
TRANSIENT ANALYSIS RESULTS SUMMARY

| capacitor bank switching | transformer switching | large load switching | other transients |
|---|---|---|---|
| 14 | 1 | 3 | 3 |

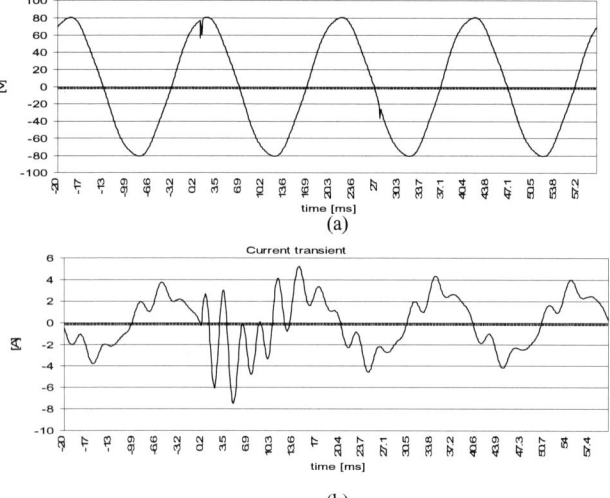

(a)

(b)

Fig. 7. Capacitor switching transient.

## VII. STANDARD LIMITAION

This section brings up some concerns about applying IEEE standards in determining PQ indices in power systems. The issues can be summarized as follows.

I. The CP95 factor does not seem to be a solid index in reflecting the harmonic pollution level. Fig. 8 depicts two different cases. In one case, 94% of the captured data are much lower than the standard and only 6% of the data exceeds the limit. In other case, 96% of the data is close to the limit and only 4% of the data violates the limit. Based on CP95 index, the first case is not IEEE519 compliant but the second one is. However, it is clear that the second case has more disturbing effects on system components from harmonic pollution point of view

.

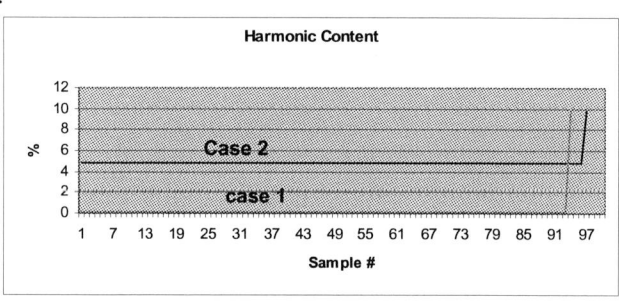

Fig. 8. CP95 index for two different cases.

II. The CP95 index is silent about the operating condition. In some cases, e.g. at light load conditions, the THD and TDD values may become larger than expected.

III. Harmonic emission limits in IEEE519 standard are the same for all the nodes with a voltage rating below 69 kV. However, it seems that theses limits must be different for low-voltage systems, e.g. 400V, and distribution systems, e.g. 20 kV.

IV. The CP95 of current harmonic cannot be easily related to the CP95 of the voltage harmonic.

## VIII. CONCLUSIONS

In this paper, power quality at a distribution node supplying a granite factory is investigated. The process is based on IEEE and IEC standard procedures. The results indicate that:

1. This type of load basically generates current harmonic pollution which results in voltage harmonic as well. Other steady state PQ parameters are within standard limits.

2. The rate of load change seems to be higher than the speed of reactive power compensating devices. Therefore, the average power factor is low. Switching of capacitor banks results in many voltage/current transients.

3. The standard procedure for determining PQ indices at a PCC has some limitations. The harmonic content index determination and the limits proposed by

660

IEEE519 have some practical shortcomings which are to be revised in the future.

Based on this survey, the quality of power at loads with the same characteristics can be improved by taking the following mitigating techniques.

1. Design and installation of a $5^{th}$ harmonic filter will decrease the current and voltage distortion limits. Simulations show that, with this filter, other harmonics will be also adequately suppressed.

2. For loads with high rate of change, the application of thyristor-switched capacitor banks can effectively improve the average power factor. Contactor-based capacitor banks are slow in operation and can cause other PQ problems such as voltage and current transients.

## IX. APPENDIX

The power quality analyzer used for this project is a CA8334 device with the following specifications:

256 samples/cycle

Accuracy: ±0.5% for rms calculation

±1.5% for power calculation

±1.5% for power factor calculation

±1% for harmonic and imbalance calculation

## X. REFERENCES:

[1] G. W. Massey, "Estimation method for power system harmonic effect on power distribution transformer," *IEEE Transaction on Industry Applications*, vol. 30, no. 2, pp. 485-489, 1994.

[2] Mark McGranaghan and Bill Roettger, "Economic Evaluation of Power Quality", IEEE Power Engineering Review, Feb. 2002, pp. 8-12.

[3] Geun-Joon Lee et. Al., "A Power Quality Index Based on Equipment Sensitivity, Cost, and Network Vulnerability", IEEE Trans. On Power Delivery, Vol. 13, No. 3, July 2004, pp. 1504-1510.

[4] G.T. Heydt, R. Ayyanar, R. Thallam, "Power Acceptability", IEEE Power Engineering Review, 2001.

[5] *Recommended Practices and Requirements for Harmonic Control in Electrical Power Systems,* IEEE Standard 519-1992.

[6] *Guide for Applying Harmonic Limits on Power systems*, Harmonic Working Group (IEEE PES T & D Committee) and SCC22-Power Quality, 1994.

[7] Flicker meter-functional and design specification, IEC Standard 61000-4-15, 1997.

## XI. BIOGRAPHIES

**Hossein Mokhtari** was born in 1969 in Tehran, Iran. He received his B.Sc. degree in electrical engineering from Tehran University, Tehran, Iran in 1989. He worked as a consultant engineer for Electric Power Research Center (EPRC) in Tehran in dispatching projects. In 1994, he received his M.A.Sc. degree from University of New Brunswick, Fredericton, N.B., Canada. He obtained his Ph.D. degree in electrical engineering from the University of Toronto in 1998. He is currently an associate professor at Sharif University of Technology, Tehran, Iran. His research interests include power quality and power electronics.

**Sasan Hasani** was born in July 1976 in Orumieh. He received his B.Sc. in electrical engineering from Shaihd Abbaspour University, Tehran, Iran. He is currently a project engineer working in transmission and distribution network division of West Azarbayjan Regional Electric Company, Orumieh, Iran.

**Farideh Donyavi** was born in 1966 in Orumieh. She received his B.Sc. in electronic engineering from Tabriz University, Tabriz, Iran. She is currently a project manager working in transmission and distribution network division of West Azarbayjan Regional Electric Company, Orumieh, Iran.

**Masoud Masoudi** was born in April 1951 in Orumieh. He received his B.Sc. in electrical engineering from Iran University of Science and Technology, Tehran, Iran. He obtained his masters degree in management from Orumieh University, Orumieh, Iran. He is currently the head of Engineering Department of West Azarbayjan Regional Electric Company.

**2006 IEEE International Conference on Power Electronic, Drives and Energy Systems**

# Minimization of Losses in Radial Distribution System by using HVDS

K. Amaresh, S. Sivanagaraju, and V. Sankar, *Member, IEEE*

*Abstract*—The loads in rural area are predominantly pump sets used for lift irrigation. The loads have low power factor and low load factor. Further, load density is low due to dispersal of loads. The existing distribution system consists of three-phase 11KV/433Volts distribution transformer with lengthy L.T Lines. In this system, the losses are high, voltage profile and reliability are unsatisfactory. In this paper, HVDS has been introduced with small capacity distribution transformers. A simple load flow technique has been used for solving radial distribution networks before and after implementation of HVDS. An advantage of implementing HVDS over LVDS system is discussed.

*Keywords*—Distribution Transformers, HVDS, LVDS, Losses, Load Flow Technique, Radial Distribution System, Real Power Flow, Reactive Power Flow.

## I. INTRODUCTION

THE Low Voltage Distribution System (LVDS), accounts for about 1/3rd of the total transmission and distribution losses in India. The main contributing factor of this loss is the present practice of using 3φ Distribution Transformer of considerable capacities, which leads to the use of very long LV lines. This is based on European practice. The LV line is extended to cater a group loads and useful particularly when catering loads of high load density. Presently, the LT lines are extended irrespective of voltage drops up to full capacity of distribution transformers and sometimes even above the transformer capacity ignoring the load of lines. This leads to severe voltage drops, high line losses accompanied with low power factor, chances of unauthorized connections etc.The practical and feasible solution is to eliminate or minimize the LV lines by changing over to HT distribution system, where the HT line is drawn as near the load as possible and small capacity 3φ/1φ distribution transformers are installed. This is best suited to meet the scattered loads. Several studies [1], [6] on this subject revealed that, distribution losses can be brought down considerably by suitable HT distribution System. In India, many State Electricity Boards have conducted studies

---

K. Amaresh, Assistant Professor in EEE, KSRMCE, Kadapa (e-mail – mad975@yahoo.com)
S. Sivanagaraju, Associate Professor in EEE, JNTUCE (Autonomous), Kakinada (e-mail – sirigiri70@yahoo.co.in)
V. Sankar, Professor in EEE, JNTUCE (Autonomous), Anantapur (e-mail – vs_jntucea@yahoo.com)

and proved that distribution losses can be brought down to levels prevailing in advanced countries by adopting this system.

In the existing LT distribution system 100KVA and 63KVA distribution transformers are provided and lengthy LT lines are laid till the consumer end as the consumer equipment is of 400V. In the proposed HVDS the existing LT lines are converted as 11KV lines and a small capacity transformer suitable for consumer load is provided at the nearest support of the line and LT supply is extended through AB cable.

In this paper, a method is proposed to find power losses, voltage drop for radial distribution system. This method has been applied for both LVDS and HVDS system. A simple load flow technique for solving radial distribution networks has been presented. The salient features of the method are presented below.

## II. LOAD FLOW METHOD FOR RADIAL NETWORK

The computational procedure required to determine the steady state operating conditions of a power system network is termed as load flow. Load flow calculation is an important tool in the area of transmission and distribution system. It is required for automated distribution systems for operation and control, planning and optimization etc,.This load flow technique is computationally efficient and is fast in convergence. The load flow techniques [2]-[5] are used for solving radial distribution system. A simple and efficient load flow technique [4] has been used for solving the both Low Voltage Distribution System (LVDS) and High Voltage Distribution System (HVDS). The effectiveness of the proposed method is tested with practical radial distribution system.

### A. Assumptions

It is assumed that the 3-phase radial distribution network are balanced and represented by their single line representation.

Consider 18-node practical radial rural distribution system in A.P, India whose single line diagram is shown in fig. 1.

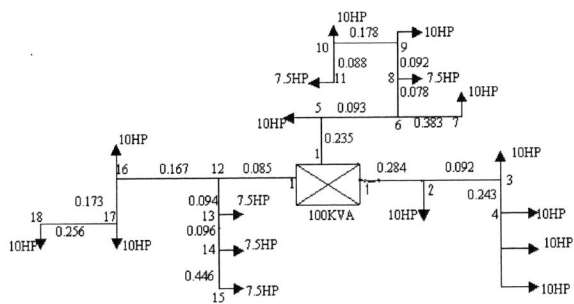

Fig. 1. Single line diagram of 18 bus LVDS system.

*(᭐) Mathematical Formulation;*

Considering, Fig. 1, i.e, a branch that is connected between nodes 1 and 2, is having a resistance $R_1$ and inductive reactance $X_1$

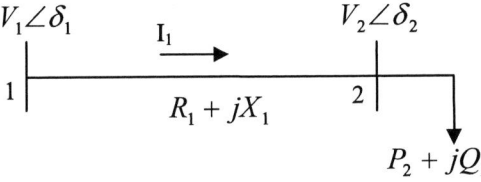

Fig. 2. Electrical equivalent of a typical branch connected between two nodes.

From Fig. 2, current following through the branch-1 is given by

$$I_1 = \frac{V_1\angle\delta_1 - V_2\angle\delta_2}{R_1 + jX_1} \qquad (1)$$

$$= \frac{P_2 - jQ_2}{V_2\angle\delta_2} \qquad (2)$$

Where, $V_1\angle\delta_1$, $V_2\angle\delta_2$ are the voltage magnitudes and corresponding phase angles at sending end node 1 and receiving end node 2 respectively.

$P_2$= Sum of the real power loads of all the nodes beyond node 2 plus the real load at node 2 itself plus the sum of real power losses of all branches beyond node 2.

$Q_2$ = Sum of the reactive power loads of all the nodes beyond node 2 plus the reactive load at node 2 itself plus the sum of reactive power losses of all branches beyond node 2.

From equations 1 and 2

$$\frac{P_2 - jQ_2}{V_2\angle\delta_2} = \frac{V_1\angle\delta_1 - V_2\angle\delta_2}{R_1 + jX_1}$$

$$|V_1||V_2|\left[\cos(\delta_1 - \delta_2) + j\sin(\delta_1 - \delta_2)\right] - |V_2|^2$$
$$= \left(P_2 - jQ_2\right) \times \left(R_1 + jX_1\right)$$

Separating, the real and imaginary parts

The real part is

$$|V_1||V_2|\cos(\delta_1 - \delta_2) = |V_2|^2 + P_2 R_1 + Q_2 X_1 \qquad (3)$$

The imaginary part is

$$|V_1||V_2|\sin(\delta_1 - \delta_2) = P_2 X_1 - Q_2 R_1 \qquad (4)$$

Squaring and adding equations 3 and 4

$$|V_1|^2 |V_2|^2 = |V_2|^4 + \left(P_2 R_1 + Q_2 X_1\right)^2$$
$$+ \left(P_2 X_1 - Q_2 R_1\right)^2 + 2|V_2|^2 \left(P_2 R_1 + Q_2 X_1\right) \quad (5)$$

$$|V_2|^4 + R_1^2 P_2^2 + X_1^2 Q_2^2 + 2P_2 Q_2 R_1 X_1$$
$$+ 2|V_2|^2 \left(P_2 R_1 + Q_2 X_1 - 0.5|V_1|^2\right) +$$
$$X_1^2 P_2^2 + R_1^2 Q_2^2 - 2P_2 Q_2 R_1 X_1 = 0$$

$$|V_2|^4 + 2|V_2|^2 \left(P_2 R_1 + Q_2 X_1 - 0.5|V_1|^2\right) +$$
$$\left(R_1^2 + X_1^2\right)\left(P_2^2 + Q_2^2\right) = 0 \qquad (6)$$

Equation 6 has a straightforward solution and do not depend on the phase angle, which simplifies the problem formulation. In a distribution system, the voltage angle is not important because the variation of voltage angle from the substation to the tail end of the distribution feeder is only a few degrees.

Note that from the two solutions of |V₂|, only the one considering the sign of the square root of the solution of the quadratic equation gives a realistic value. The same is applicable when solving for |V₂|.

Therefore from equation, 6 the solution of |V₂| can be written as:

$$|V_2| = \left\{ \left[ \left( P_2 R_1 + Q_2 X_1 - 0.5|V_1|^2 \right)^2 - \right. \right.$$

$$\left. \left( \left( R_1^2 + X_1^2 \right)\left( P_2^2 + Q_2^2 \right) \right) \right]^{1/2} -$$

$$\left. \left( P_2 R_1 + Q_2 X_1 - 0.5|V_1|^2 \right)^2 \right\}^{1/2} \qquad (7)$$

In general,

$$|V_{i+1}| = \left\{ \left[ \left( P_{i+1} R_j + Q_{i+1} X_j - 0.5|V_i|^2 \right)^2 - \right. \right.$$

$$\left. \left( \left( R_j^2 + X_j^2 \right)\left( P_{i+1}^2 + Q_{i+1}^2 \right) \right) \right]^{1/2} -$$

$$\left. \left( P_{i+1} R_j + Q_{i+1} X_j - 0.5|V_i|^2 \right)^2 \right\}^{1/2} \qquad (8)$$

Where,
Node no., i = 1,2…. n
Branch no., j = 1,2…n-1
n = total number of nodes

The active and reactive power losses in branch 'j' are given by

$$P_{loss_j} = R_j \times \frac{\left( P_{i+1}^2 + Q_{i+1}^2 \right)}{|V_{i+1}|^2} \qquad (9)$$

$$Q_{loss_j} = X_j \times \frac{\left( P_{i+1}^2 + Q_{i+1}^2 \right)}{|V_{i+1}|^2} \qquad (10)$$

The total active and reactive power of the system is

$$TPL = \sum_{j=1}^{n-1} P_{loss_j} \qquad (11)$$

$$TQL = \sum_{j=1}^{n-1} Q_{loss_j} \qquad (12)$$

(c) *Load flow calculation;*

Usually the substation voltage $V_1$ is taken as $V_1$=1.0 p.u. Initially, $P_{lossj}$ and $Q_{lossj}$ are set to 0 for all j. Then the initial estimate of $P_2$ and $Q_2$ will be the sum of the loads of all nodes beyond node 2 plus the local load of node 2.

For all the branches j=1,2…n-1 compute $P_{i+1}$ and $Q_{i+1}$, $V_{i+1}$, $P_{lossj}$ and $Q_{lossj}$ using equations 8, 9 and 10. This will complete one iteration. Update the real and reactive powers $P_{i+1}$ and $Q_{i+1}$ (which includes losses) and repeat the same procedure until all the voltage magnitudes are computed to a tolerance level of 0.0001 p.u in successive iterations. Based on the above ideas, the algorithm for load flow calculation is presented below.

## III. ALGORITHM FOR LOAD FLOW CALCULATION

1. Read line and load data of radial distribution system. Initialize TPL, TQL to zero. Assume node voltage as 1 p.u. Set convergence criterion $|V_i - V_{i+1}| \leq \in$ where € is the tolerance.
2. Set iteration count, iter = 1.
3. Calculate effective load at each node starting from the last node.
4. Initialize real power loss and reactive power loss vectors to zero.
5. Find effective losses at each node.
6. Calculate load at each node including losses.
7. Calculate node voltage, real power loss and reactive power loss of each branch using equations 8, 9 and 10 respectively.
8. Calculate the largest absolute value of change in voltage i.e.$|\Delta Vmax| \leq \varepsilon$ in successive iterations. If $|\Delta Vmax| \leq \varepsilon$ go to step 11 otherwise go to step 10.
9. Increment iteration number and go to step 6.
10. Calculate TPL and TQL using equations. 11 and 12.
11. Print voltages at each node, TPL, TQL and number of iterations.
12. Stop.

## IV. ILLUSTRATIVE EXAMPLE

To illustrate the effectiveness of the proposed method, consider an 18-node, practical radial distribution system whose single line diagram is shown in figure 1. The line and load data had been presented in Appendex-1.

## V. RESULTS

The load flow results of the proposed test system are given in tables I and II.

TABLE I
RESULTS FOR LOW VOLTAGE DISTRIBUTION SYSTEM

| Node | Voltage (P.u) | Actual Voltage |
|------|---------------|----------------|
| 1 | 1.0000 | 440.00 |
| 2 | 0.8973 | 394.83 |
| 3 | 0.8710 | 383.22 |
| 4 | 0.8190 | 360.35 |
| 5 | 0.9003 | 396.11 |
| 6 | 0.8677 | 381.78 |
| 7 | 0.8387 | 369.02 |
| 8 | 0.8465 | 372.46 |
| 9 | 0.8267 | 363.74 |
| 10 | 0.8019 | 352.85 |
| 11 | 0.7967 | 350.54 |
| 12 | 0.9691 | 426.40 |
| 13 | 0.9548 | 420.11 |
| 14 | 0.9450 | 415.79 |
| 15 | 0.9219 | 405.65 |
| 16 | 0.9340 | 410.98 |
| 17 | 0.9096 | 400.22 |
| 18 | 0.8914 | 392.20 |

664

The power losses are

TPL = 22.4400 KW

TQL = 6.3502 KVAR

TABLE II
RESULTS FOR HIGH VOLTAGE DISTRIBUTION SYSTEM

| Node | Voltage (p.u) | Actual Voltage (V) | LV side voltage (V) |
|---|---|---|---|
| 1 | 1.0000 | 11000.00 | 440.00 |
| 2 | 0.9998 | 10998.16 | 439.92 |
| 3 | 0.9998 | 10997.78 | 439.91 |
| 4 | 0.9997 | 10997.04 | 439.88 |
| 5 | 0.9999 | 10998.57 | 439.94 |
| 6 | 0.9998 | 10998.11 | 439.92 |
| 7 | 0.9998 | 10997.68 | 439.90 |
| 8 | 0.9998 | 10997.80 | 439.91 |
| 9 | 0.9998 | 10997.52 | 439.90 |
| 10 | 0.9997 | 10997.18 | 439.88 |
| 11 | 0.9997 | 10997.11 | 439.88 |
| 12 | 1.0000 | 10999.51 | 439.98 |
| 13 | 0.9999 | 10999.27 | 439.97 |
| 14 | 0.9999 | 10999.27 | 439.97 |
| 15 | 0.9999 | 10998.75 | 439.95 |
| 16 | 0.9999 | 10998.95 | 439.94 |
| 17 | 0.9999 | 10998.57 | 439.94 |
| 18 | 0.9998 | 10998.28 | 439.93 |

The power losses are

TPL = 0.0255 KW

TQL = 0.0095 KVAR

A. *Analysis of results*

The line losses of distribution lines (L.T.lines) in the existing LVDS and the 11KV line losses of the same lines after converting them as 11KV (H.T.lines) are tabulated below

| Loss Criteria | In LVDS | In HVDS | HVDS losses as % LVDS losses |
|---|---|---|---|
| TPL (KW) | 22.40 | 0.0255 | 0.1 |
| TQL (KVAR) | 6.3502 | 0.0095 | - |

The end user, either in the HVDS or in the LVDS operates his equipment at LT voltage only. Hence by comparison of Tables 1 and 2 it can be observed that an improvement of LV voltage up to a value of 20% or 87V. It is an accepted and established phenomenon that the electrical machines operate with higher efficiencies while operating at rated voltages than at the reduced voltages. The voltage improvement varies from 50V to 87V i.e. by 11 to 22%. The corresponding efficiency improvement is by 20 to 30%.

## VI. CONCLUSIONS

By comparing the above results, the HVDS system is better than LVDS system. As the voltages at the nodes are improved and total power losses are reduced, the efficiency of the motors on agricultural side is benefited. It is estimated that about 6% of total 20.47% T&D losses are lost in LT 3-phase distribution system. This is mainly due to the high capacity (100KVA & 63KVA) transformers serving the disbursed agricultural loads through long 440V Lines.

Reduction of these losses is due to conversion of the existing LVDS to HVDS. This is done through restructuring of the existing LVDS network to HVDS network and installation of 3-phase 11KV/440V, 25KVAand 15/16KVA transformer to serve the loads in 11KV agricultural feeders.

## VII. APPENDIX

The conductor used is squirrel, its
R = 1.376/km, X = 0.3896/km

the line data for Fig. 1.

| S.No | Line between the nodes | Distance (km) | Resistance (Ohms) | Reactance (Ohms) |
|---|---|---|---|---|
| 1 | 1-2 | 0.284 | 0.3907 | 0.1106 |
| 2 | 2-3 | 0.092 | 0.1265 | 0.0035 |
| 3 | 3-4 | 0.243 | 0.3343 | 0.0946 |
| 4 | 1-5 | 0.235 | 0.3233 | 0.0915 |
| 5 | 5-6 | 0.093 | 0.1279 | 0.0362 |
| 6 | 6-7 | 0.383 | 0.5270 | 0.1492 |
| 7 | 6-8 | 0.078 | 0.1073 | 0.0303 |
| 8 | 8-9 | 0.092 | 0.1265 | 0.0035 |
| 9 | 9-10 | 0.178 | 0.2449 | 0.0693 |
| 10 | 10-11 | 0.088 | 0.1210 | 0.0342 |
| 11 | 1-12 | 0.085 | 0.1169 | 0.0331 |
| 12 | 12-13 | 0.094 | 0.1293 | 0.0366 |
| 13 | 13-14 | 0.096 | 0.1320 | 0.0374 |
| 14 | 14-15 | 0.446 | 0.6136 | 0.1737 |
| 15 | 12-16 | 0.167 | 0.2297 | 0.0650 |
| 16 | 16-17 | 0.173 | 0.2380 | 0.0674 |
| 17 | 17-18 | 0.256 | 0.3522 | 0.0997 |

the load data for Fig. 1.

| Node | P(KW) | Q(KVAR) |
|---|---|---|
| 1 | 0.00 | 0.00 |
| 2 | 7.36 | 5.52 |
| 3 | 7.36 | 5.52 |
| 4 | 20.24 | 15.18 |
| 5 | 7.36 | 5.52 |
| 6 | 0.00 | 0.00 |
| 7 | 7.36 | 5.52 |
| 8 | 5.52 | 4.14 |
| 9 | 7.36 | 5.52 |
| 10 | 7.36 | 5.52 |
| 11 | 5.52 | 4.14 |
| 12 | 0.00 | 0.00 |
| 13 | 5.52 | 4.14 |
| 14 | 5.52 | 4.14 |
| 15 | 5.52 | 4.14 |
| 16 | 7.36 | 5.52 |
| 17 | 7.36 | 5.52 |
| 18 | 7.36 | 5.52 |

## VIII. REFERENCES

[1] D. L. Nickel, "Higher Voltage Distribution Systems", Transmission and Distribution magazine, Aug 1974, pp.52-56.

[2] M. Chen, et. al, " Distribution System Power Flow Analysis - a rigid Approach", IEEE PWRD Vol.6, NO3, pp. 1146-1152, July 1991.

[3] M.Srinivas, " Distribution load flows; a brief review", in Procedings IEEE PES Winter Meeting, Vol.2, Jan.2000, pp.942-945.

[4] S. Sivanagaraju, N. Sreenivasulu, M. Vijay kumar, New method of load flow solution of radial distribution networks, Proceedings All India Seminar, Power Systems; Recent Advances and Prospects in 21$^{st}$ Century, 2001, pp.226-234.

[5] W. H. Kersting, Distribution System Modeling and Analysis, Boca Raten, FL: CRC, 2002.

[6] S. S. Venkata, A.Pahwa, R.E. Brown, and R.D.Christie, "What Future Distribution Engineers need to learn", IEEE Trans. on Power Systems, Vol 19, 2004, pp.17-23.

## IX. BIOGRAPHIES

**K. Amaresh** was born in Kurnool District.AP, India on May 13, 1972. He graduated from the Visveswarayya Institute of Technology, Bangalore and Post graduation at JNTU Anantapur, India. His employment experience included the auma (India) Ltd., His field of interest included distribution system. Presently he is working as Assistant Professor in Department of EEE, KSRMCE, Kadapa, AP, India.

**S. Sivanagaraju,** was born in Kadapa, AP, India in 1970. Graduated in 1998, obtained M. Tech Degree in 2000 from IIT, Kharagpur and did his Ph.D from J.N.T. University in 2004. Working as Associate Professor in the Department of Electrical Engineering, J.N.T.U. College of Engineering. (Autonomous), Kakinada, Andhra Pradesh. He has received two National awards (Pandit Madan Mohan Malaviya memorial prize award and Best paper prize award) from the Institute of Engineers (India) for the year 2003-04.About 40 Publications in National and International Journals and Conferences to his credit. His area of interest is in Electrical Distribution Systems.

**V. Sankar** was born in 1958 at Machilipatanam, AP, India, is graduated in 1978, Masters in 1980 and did his Ph.D from IIT Delhi in 1994. he is referee of IEEE Transactions on Reliability, guided two Ph.D students. He is presently a member of Board of studies of JNTU and served as Head of EEE and vice –Principal in JNTU college of Engineering (Autonomous), Anantapur. His area of interest includes Reliability Engineering and its applications to Power Systems. Presently he is working as Professor in Department of EEE, JNTUCE, Anantapur, AP, India.

**2006 IEEE International Conference on Power Electronic, Drives and Energy Systems**

# SVPWM Switched DSTATCOM for Power Factor and Voltage Sag Compensation

Bishnu P. Muni, S. Eswar Rao, and JVR Vithal

*Abstract*—In a power distribution network of a power utility or industry, the reactive power management plays a major role in reducing distribution loss and maintaining constant distribution voltage. Voltage source PWM inverter based reactive power generator has fast response to reactive power demand and hence can be used for power factor improvement and voltage support. With the availability of new generation DSPs with inbuilt event manager module, the space vector PWM (SVPWM) switching patterns can be generated easily. This paper proposes SVPWM switched DSTATCOM for power factor and voltage sag compensation. The paper presents the simulation studies on SVPWM switched DSTATCOM in power factor control mode. It also gives the mathematical model of DSTATCOM in voltage sag compensation mode. The simulation results show that the space vector PWM strategy can be used for power factor and voltage compensation. Further, experimental studies on SVPWM switched DSTATCOM have been presented to validate the control philosophy.

*Index Terms*-- DSTATCOM, DSP control, Reactive power compensation, SVPWM, Voltage Sag

## I. INTRODUCTION

THE industrial loads with low power factor cause excessive loading of the distribution transformer and low voltage at the point of common coupling (PCC). The distribution static synchronous compensators (DSTATCOM) can help in generating leading reactive volt-ampere (VAR) at the PCC and avoid the problems associated with the low power factor [1]. The availability of high power self-commutating semiconductor devices, new control philosophies and high performance computing elements, like, micro-controller and DSP, are revolutionizing the electric power transmission and distribution systems. Shunt connected, PWM converter based DSTATCOMs are gaining popularity for improving power factor and power quality in industrial premises. In addition to the reactive power compensation, the DSTATCOM can also act as active filter or load conditioner to improve the power quality. DSTATCOMs are generally used for reactive power management in

distribution network with highly fluctuating loads, like arc furnace, rolling mills or high power induction motors, induction machine based wind electric generator, voltage regulation in distribution network etc.

Momentary power interruptions and voltage sags are most important power quality problems affecting industrial and large commercial customers. Voltage sags cause the malfunction of the modern process control, programmable logic control, and variable speed drives. Voltage sags can initiate tripping off for voltage-sensitive loads. Voltage sags are usually associated with a fault somewhere on power system network. Voltage sags are much more common since they can be associated with faults remote from the customer. Motor starting also results in voltage sags but the magnitudes are usually not severe enough to cause equipment malfunction. VSC based series compensators are the ideal choice for voltage sag compensator. DSTATCOMs are the most preferred device for voltage flicker compensation. Reference [6] presents the experimental study of advanced static VAR compensator for voltage sag mitigation. Kalman filter based unified approach for voltage sag and voltage flicker mitigation with DSTATCOM has been reported [7].

In a DSTATCOM, the voltage source is normally controlled with hysteresis or ramp comparison type current regulated PWM inverter [1-4]. The hysteresis and ramp comparison type current regulated PWM inverter can be implemented with conventional analog and digital circuit. The PWM switching pattern for DSTACOM can also be generated by comparing phase shifted reference voltage signal generated PLL and associated circuitry with triangular carrier wave. The application of SVPWM for DSTACOM is not explored fully. SVPWM technique offers several advantages like higher output voltage (15 % more than conventional sine – triangle modulation technique and hence, better utilization of DC link), lower harmonics contents and suitability for complete digital implementation with new generations DSPs.

This paper presents a novel DSTATCOM with SVPWM technique for generation of PWM pattern. The suitability of SVPWM technique for reactive power management in DSTATCOM has been verified by simulation by using SimPower System of MATLAB and Simulink. Further, the control logics for DC link voltage control, reactive power generation and PWM paper generation with SVPWM

---

Bishnu P. Muni (e-mail:bpmuni@bhelrnd.co.in), S. Eswar Rao (e-mail: eswar@bhelrnd.co.in) and JVR Vithal (e-mail: jvrv@bhelrnd.co.in) are with BHEL, Corporate R&D, Hyderabad – 500093, India (e-mail:bpmuni@bhelrnd.co.in).

0-7803-9771-1/06/$25.00 ©2006 IEEE

technique have been implemented on TMS320F240 based control card. Further, this paper also presents the mathematical analysis of DSTATCOM for voltage sag mitigation.

The paper has been organised in five sections. The second section gives the control methodology for power factor compensation and brief introduction of SVPWM technique; the third section gives modelling of DSTATCOM and simulation results of pf compensation, the fourth section gives the control methodology for voltage sag compensation, the fifth section gives the experimental results and the sixth section gives the conclusions.

## II. CONTROL METHODOLOGY PF COMPENSATION

The power circuit of the DSTATCOM and the instantaneous power flow in the DSTATCOM are shown in Figs.1 and 2, respectively. The source power has two components, viz. instantaneous active power and instantaneous reactive power. Similarly, the load power and inverter power also comprise of active and reactive components. For full reactive power compensation of the load, the inverter has to supply reactive power of the same magnitude, but of opposite sign. Thus, in such a case, the reactive power drawn from the source is zero. Hence, if reactive power supplied by the source is monitored and is controlled in closed-loop to maintain it at zero, then the desired objective of full VAR compensation or unity power factor can be achieved.

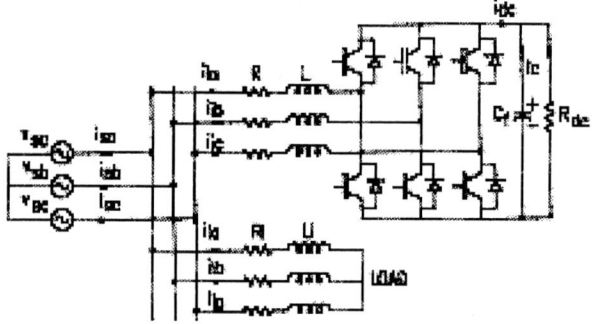

Fig. 1. Power Circuit of Distribution STATCOM.

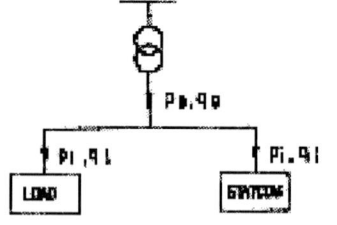

Fig. 2. Power Flow in a Distribution STATCOM.

The DSTATCOM can operate in variable VAR generation and power factor mode. The control block diagram of DSTATCOM in reactive power generation mode is shown in Fig. 3. The three-phase PCC voltages ($v_{sa}$, $v_{sb}$, $v_{sc}$) and source currents ($i_{sa}$, $i_{sb}$, $i_{sc}$) are measured and are converted to a

control level signals by using PTs, CTs and signal conditioning cards. These three-phase voltage and current signals at control level are converted into equivalent α-β axis (two axis stationary reference frame) components ($i_{s\alpha}$, $i_{s\beta}$, $v_{s\alpha}$, $v_{s\beta}$). The current or voltage signals in α-β (two-axes stationary) frame are converted to d-q (two-axes synchronously rotating reference) frame. In synchronously rotating reference frame, the direct axis current component is proportional to active power and quadrature axis current component is proportional to reactive power [2]. The equations involved for phase transformations are well known [1].

$$\begin{bmatrix} v_{s\alpha} \\ v_{s\beta} \end{bmatrix} = \begin{bmatrix} C \end{bmatrix} \begin{bmatrix} v_{sa} \\ v_{sb} \\ v_{sc} \end{bmatrix} \tag{1}$$

$$\begin{bmatrix} C \end{bmatrix} = \sqrt{\frac{2}{3}} \begin{bmatrix} 1 & -1/2 & -1/2 \\ 0 & \sqrt{3}/2 & -\sqrt{3}/2 \end{bmatrix} \tag{2}$$

$$\begin{bmatrix} v_d \\ v_q \end{bmatrix} = \begin{bmatrix} \cos(\theta) & \sin(\theta) \\ -\sin(\theta) & \sin(\theta) \end{bmatrix} \begin{bmatrix} v_\alpha \\ v_\beta \end{bmatrix} \tag{3}$$

$$\begin{bmatrix} v_\alpha \\ v_\beta \end{bmatrix} = \begin{bmatrix} \cos(\theta) & -\sin(\theta) \\ \sin(\theta) & \sin(\theta) \end{bmatrix} \begin{bmatrix} v_d \\ v_q \end{bmatrix} \tag{4}$$

Where,

$v_{sa}$, $v_{sb}$, $v_{sc}$: Source voltges

$v_\alpha$, $v_\beta$: Voltages in two-axes stationery frame

$v_d$, $v_q$: Voltges in two-axes synchronously rotating frame

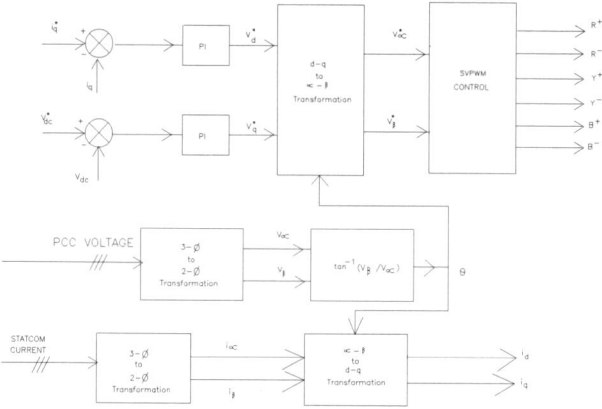

Fig. 3. Control Block Diagram of DSTATCOM in VAR Generator Mode.

In reactive power control loop, the reference, reactive power is compared with the actual reactive power generated by the STATCOM and the error in reactive power is

668

processed by the PI controller to give the direct axis component of converter reference voltage. The DC link voltage of the STATCOM should be selected such that the converter can deliver the rated reactive power in the desired range of grid voltage. The reference DC link voltage is compared with actual DC link voltage and the error in DC link voltage is processed by the PI controller to give quadrature axis component of converter reference voltage. The direct axis and quadrature axis reference voltages are transformed to two-axes stationary reference frame and fed to SVPWM block.

The SVPWM block receives the reference voltages in α-β frame and generates PWM switching signals. Depending on the magnitude and phase angle of the reference voltage space vector, the duration of the adjacent switching state vectors are calculated [8] and accordingly the PWM switching signals are generated by internal compare registers. The PWM switching signals generated by the DSP are given to IGBTs of VSC through respective gate drives.

SVPWM switching strategy has many advantages such as well-defined harmonic spectrum, optimum switching patterns, and higher ac output voltage compared sine triangle comparison method. Further, the present generation DSPs have special hardware feature for implementation of SVPWM switching strategy. For a three-phase voltage source converter there are totally eight possible switching patterns, out of which six are active and two are non-active or zero switching states. As shown in Fig. 4, six voltage space vectors divide the whole space into six sectors, 1 to 6. Except two zero vectors, $V_0$ and $V_7$, all other active space vectors have the same magnitude of $(2/3)V_{DC}$.

From Fig. 3, it is seen that the input to the SVPWM block is the inverter reference voltage vector in the α − β frame. According to the phase angle of the reference voltage vector in the α − β coordinates the sector in which the reference voltage vector is located can be easily found out. In SVPWM, the reference voltage vector is synthesized by the adjacent vectors of the located sector in order to minimize the switching times and to minimize the current harmonics.

The reference space vector $V^*$ can be decomposed as normalized $V_\alpha$ and $V_\beta$ in the α − β coordinates as given in eq. 5. The sector number of reference voltage space vector can be obtained from eq. 7.

Where,
V: Magnitude of reference voltage vector.
γ: Phase angle of $V^*$ in α − β coordinate
S: Sector number 1 to 6

If the reference space vector lies in sector I, the reference space vector can be decomposed as shown in Fig. 4. In order to synthesize $V^*$, the required durations of two adjacent vectors, $V_4$ and $V_6$, and of zero vectors in one switching period T are $T_I$, $T_2$ and $T_0$, respectively. $T_I$, $T_2$ and $T_0$ can be calculated by eq. 8. Relations for other cases when $V^*$ is

located in Sector 2 - 6 can be obtained by rotating the α − β coordinate shown in Fig. 5 anticlockwise by an angle $(S-1)\pi/3$. There are several methods for SVPWM to generate the desired PWM pulses. The PWM patterns are generated by the event manager module of DSP TMS320F240 [8]. In event manager module, a timer with a period of T/2 is set as PWM carrier timer and operates at continuous up/down counting mode. Using eq. 8, three counter values are calculated and loaded in timer. These three counter values are compared with the timer counting value in real to generate the PWM pattern. Fig. 5 shows the method of PWM generation when the reference space vector lies on 1st sector.

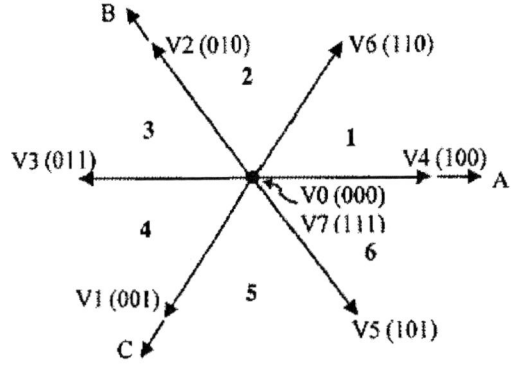

Fig. 4. Voltage Space vectors for VSC.

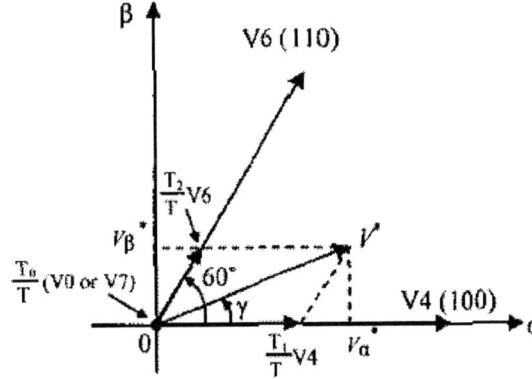

Fig. 5. Decomposition of Reference Space Vector in Sector I.

$$\begin{cases} V_\alpha^* = \dfrac{3\,|V^*|.\cos\gamma}{2V_{DC}} \\[2mm] V_\beta^* = \dfrac{3\,|V^*|.\sin\gamma}{2V_{DC}} \end{cases} \tag{5}$$

$$|V^*| = \sqrt{(v^*_d)^2 + (v^*_q)^2} \tag{6}$$

$$\gamma = \theta + \tan^{-1}\left(\frac{v^*_q}{v^*_d}\right) - (S-1).\frac{\pi}{3} \tag{7}$$

669

$$\begin{cases} T_1 = \left( v^*_\alpha - \dfrac{v^*_\beta}{\sqrt{3}} \right).\, T \\[2mm] T_2 = \dfrac{2}{\sqrt{3}}.\, v^*_\beta .\, T \\[2mm] T_0 = T - T_1 - T_2 \end{cases} \qquad (8)$$

$$\begin{cases} M_1 = \dfrac{T_0}{4} \\[2mm] M_2 = \dfrac{T_0}{4} + \dfrac{T_1}{2} \\[2mm] M_3 = \dfrac{T_0}{4} + \dfrac{T_1}{2} + \dfrac{T_2}{2} \end{cases} \qquad (9)$$

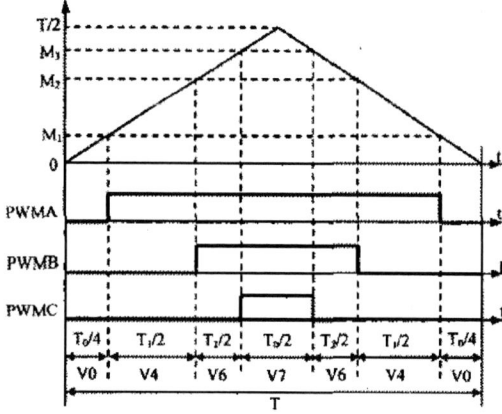

Fig. 6. Method of PWM Generation with Reference Space Vector in First Sector.

## III. MODELLING AND SIMULATION

Simulation studies have been carried out to validate the SVPWM switching pattern generation technique for DSTATCOM operating in unity power factor control mode. The entire power circuit and control circuit of DSTATCOM have been modeled and simulated using SimPower System of MATLAB & Simulink. In the present paper, the power system network has been modelled as an AC source feeding equivalent inductive or capacitive load. The DSTATCOM is controlled such that the source current is drawn in phase with the source voltage. The schematic of DSTATCOM is shown in Fig. 7.

Fig. 8 shows the source phase voltage and source current with the DSTATCOM operating in closed loop power factor control mode. The DSTATCOM is switched on to the network with an active and reactive load combination of 500 KW and 400 kVAR lagging. At 0.3 second, another active and reactive load of 400 kW and 300 kVAR is switched on to the network and at 0.4 second, it is switched off. It is observed that the source current is drawn in phase with the source

voltage both during steady state and transients due to sudden load change. The response of the DSTACOM is of the order of quarter cycle during step load change. Fig. 8 shows the source phase voltage, DSTATCOM phase current and DC link voltage. It is observed that the DSTATCOM draws leading reactive current to compensate lagging reactive power drawn by the load. The bottom trace in Fig. 9 shows the DC link voltage which is maintained at reference voltage of 1000 V DC.

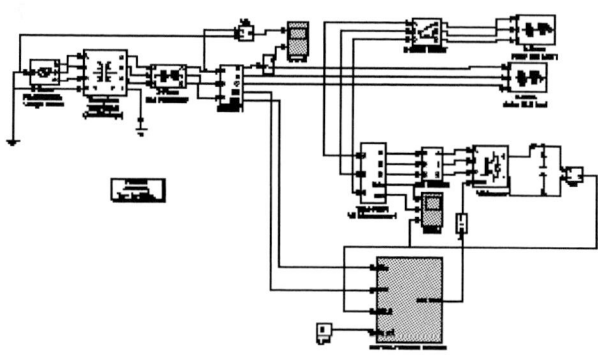

Fig. 7. Schematic of DSTATCOM in Power Factor Control Mode.

From the results of these simulation studies, it is observed that the source current is in phase with the source voltage at steady-state and transients, except for a brief period of nearly a quarter cycle. Thus, the response of DSTATCOM to reactive power demand is of the order of nearly one quarter cycle. The DC link voltage changes slightly at the time of step application or removal of load. The DSTATCOM draws current at nearly $90^0$ leading with respect to the load voltage. The simulation results show that the DSTACOM with SVPWM switching strategy is capable of meeting reactive power demand of the load both during steady state and transients.

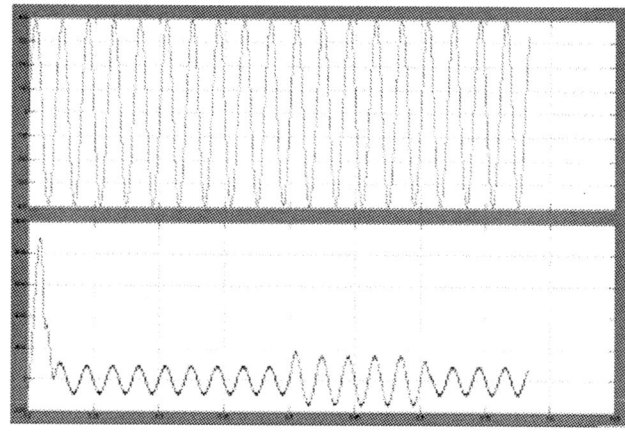

Fig. 8. Source phase voltage (top trace) and source phase current (Bottom trace) with the DSTATCOM in closed loop power factor control mode.

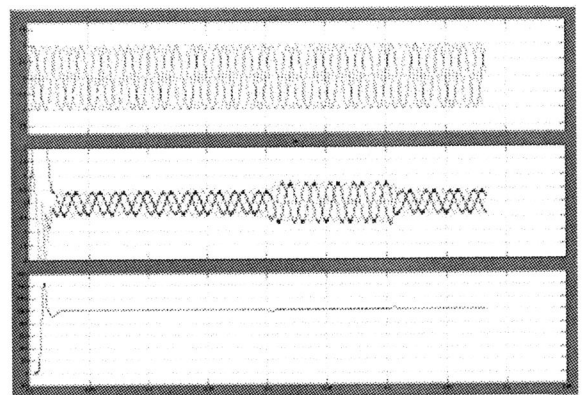

Fig. 9. Source phase voltage (top trace) and DSTACOM current (middle trace)and DC link voltage (bottom trace) with the DSTATCOM in closed loop pf control mode.

## IV. CONTROL METHODOLOGY FOR VOLTAGE SAG COMPENSATION

A voltage sag is an the reduction in rms value of the AC voltage, at the power frequency, for durations from a half-cycle to a few seconds [5]. Magnitude, duration and phase jump, if any are the three most important parameters for characterizing voltage sag. Motor starting and short-circuit faults in power systems are the two main causes of voltage sags. Motor starting produces shallow sags, but of longer duration. Short-circuit faults cause severe voltage sags, and has a major impact on power quality. DSTATCOM can adjust its reactive power output independently of its supply voltage. Thus, DSTACOM can increase its capacitive reactive power output during source voltage reduction by operating at its maximum capacity and thereby improving the voltage at PCC. The equivalent circuit of DSTATCOM in voltage compensation mode is shown in Fig. 10.

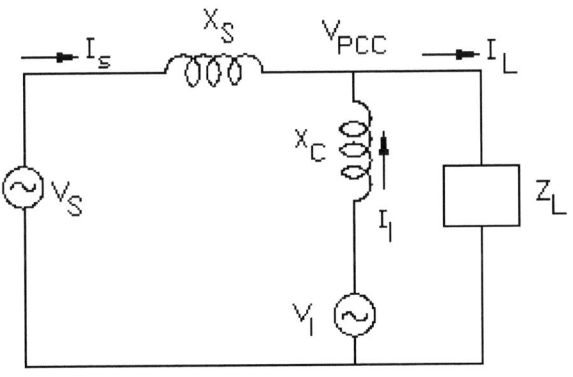

Fig. 10. Equivalent Circuit of DSTATCOM in Voltage Sag Compensation Mode.

Equations (10) to (12) for voltage sag mode of operation of DSTATCOM can be obtained from equivalent circuit by applying KVL and KCL.

$$I_L = I_S + I_I$$

$$= \frac{V_{PCC} - V_I}{X_C} + \frac{V_{PCC} - V_S}{X_S} \qquad (10)$$

$$V_{PCC} = I_L . Z_L \qquad (11)$$

$$V_S = I_S . X_S + V_{PCC} \qquad (12)$$

It can be observed that the amount of reactive current needed to maintain the PCC voltage at a given value during a voltage sag is inversely proportional to the system reactance $X_s$. The error in PCC voltage and set value can be processed by a PI controller to generate the quadrature component of the current. This signal is summed with quadrature component of load current to generate the quadrature component of the reference current. The DC capacitor voltage regulating loop gives the direct axis component of the reference current. The error in direct axis reference current and DSTATCOM generated direct axis current is processed by a PI controller to generate direct axis voltage reference. Similarly, the error in quadrature axis reference current and DSTATCOM generated quadrature axis current is processed by a PI controller to generate quadrature axis voltage reference. Direct and qaudrature axes reference voltages are transformed to α – β reference frame. The SVPWM block receives the reference voltages in α-β frame and generates PWM switching signals.

## V. EXPERIMENTAL RESULTS

A ±150 kVAR DSTATCOM has been developed using the control methodology mentioned in Section 2. The closed loop control of reactive power, DC voltage control loops and PWM switching signal generation based on space vector PWM approach, have been implemented on DSP TMS320F240. The control software for closed loop reactive power generation has been developed. Extensive experiments have been carried out on the DSTATCOM. The DSTATCOM has been tested in variable VAR generator mode. In variable VAR generator mode, the DSTATCOM has been tested for both lagging and leading power generation modes.

Fig. 11 shows the PCC voltage and DSTATCOM current when the DSTATCOM is operating lagging VAR generator mode. In leading VAR generator mode, the PCC voltage and DSTATCOM current is shown in Fig. 12. Extensive experiments have been carried out on the DSTATCOM for various steady state and transient operating conditions and the performance has been found to be satisfactory. The photograph of the DSTATCOM is shown in Fig. 13.

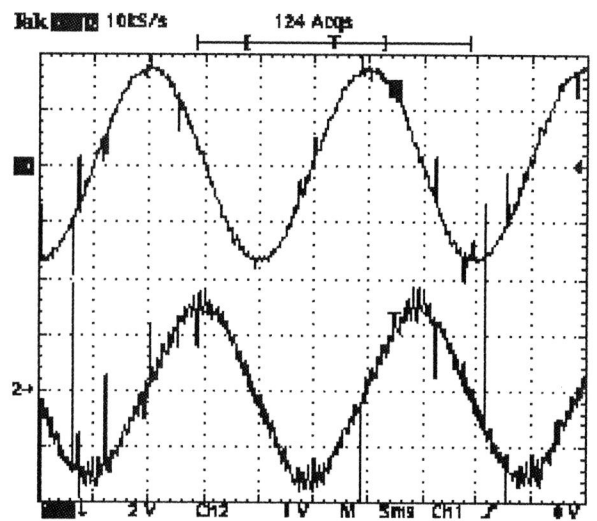

Fig. 11. DSTATCOM in Lagging VAR generation Mode (top trace: Source Fig. Voltage, Bottom trace: DSTATCOM Current.
Scale: 1 V = 100 V (voltage), 1 V = 100 A (Current) )

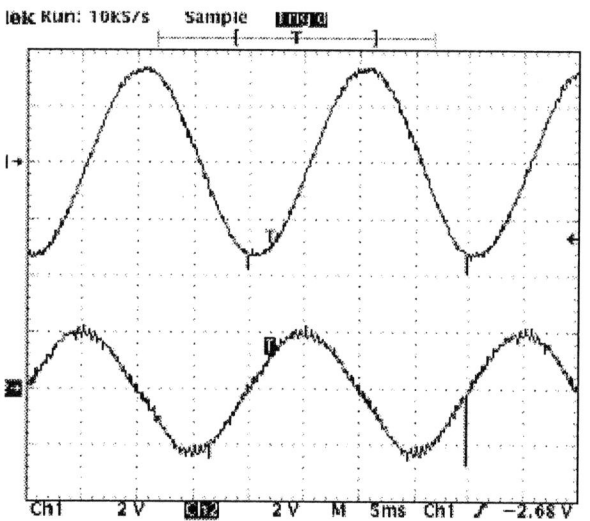

Fig. 12. DSTATCOM in Lagging VAR generation Mode (top trace: Source Voltage, Bottom trace: DSTATCOM Current).
Scale: 1 V = 100 V (voltage), 1 V = 100 A (Current))

Fig. 13. Photograph of the Developed DSTATCOM.

## V. CONCLUSIONS

With the proposed SVPWM based control logic, the PWM converter output voltages are controlled such that the reactive power generated by the DSTATCOM closely follows the reference reactive power and the DC link voltage is maintained at desired value. The simulation and experimental results show that, the proposed DSTATCOM controller with SVPWM switching pattern generation technique can be used for closed loop reactive power generation mode of operation. The entire control logics of DC link voltage control, reactive power control loop and SVPWM based PWM generation and protection logics have been implanted digitally on a TMS320F240 based DSP card. DSTATCOM has limited voltage sag compensation capability unlike VSC converter based series compensators.

## VI. ACKNOWLEDGMENT

The authors gratefully acknowledge the support of BHEL, Corporate R&D for providing facilities to carry out the present work and permission to present the paper in PEDES 2006.

## VII. REFERENCES

[1] Bishnu P. Muni, S. Eswar Rao et. al., "Development of ± 500 kVAR DSTATCOM for Distribution Utility and Industrial Applications", Conference Proceedings, IEEE, Region Ten Annual Conference, TENCON-03, 2003, pp278-282.

[2] C. D. Schauder and H. Mehta, "Vector Analysis and Control of Advanced Static VAR Compensator", IEE Proceedings – C, Vol 140, No. 4, July 1993.

[3] B. N. Singh, A. Chandra and K. AI-Haddad, "DSP-based Indirect-Current-Controlled STATCOM Part I: Evaluation of Current Control Techniques", IEE Proceedings, Electrical Power Applications, Vol. 147, No.2 March 2000, pp 107-112.

[4] H. Akagi, "New Trends in Active Filters s for improving power quality", IEEE-PEDES Conf. Record, 1996, pp 417-425.

[5] Transmission and Distribution Committee, "IEEE Guide for Service to Equipment Sensitive to Momentary Voltage Disturbances", IEEE Std 1250-1995

[6] A. Elnady, M. Magdy and A. Salama, "Unified Approach for Mitigating Voltage Sag and Voltage Flicker Using The DSTATCOM", IEEE Trans on Power Delivery, Vol. 20, No. 2, April2005, pp 992-1000.

[7] P. Wang, N. Jenkins, and M. H. J. Bollen, "Experimental Investigation of Voltage Sag Mitigation by an Advanced Static VAR Compensator," *IEEE Trans. Power Del.*, vol. 13, no. 4, pp. 1461–1467, Oct. 1998.

[8] "AC Induction Motor Control Using Constant V/Hz Principle and Space Vector PWM Technique with TMS320C240", Texas Instruments, Application Report (SPRA 284A).

**2006 IEEE International Conference on Power Electronic, Drives and Energy Systems**

# Unified Constant frequency Integration Control of Universal Power Quality Conditioner

Vadirajacharya K, *Student Member, IEEE*, Pramod Agarwal , *Member IEEE,* and H.O.Gupta, *Member IEEE*

*Abstract–* **Unified Power Quality Conditioner (UPQC) has the function of voltage as well as current harmonic suppression, reactive power compensation. A simple, low cost and high performance controller based on one cycle control is proposed to UPQC. The control method features constant switching frequency operation, minimum reactive and harmonic current generation, and simple analog circuitry. The simulation results verify the performance evaluation of proposed controller.**

*Index Terms--* **Power Quality, UCI, UPQC.**

## I. INTRODUCTION

THE recent development in power electronics led to developments of solid-state controllers, which are in extensive use with industry, commercial and domestic sectors. These solid-state controllers use nonlinear components such as diode rectifiers, thyristerised converters, choppers and cycloconverters. These nonlinear loads inject non-sinusoidal currents in to the power system, thus contributing to the degradation of power quality in the power system network [1]. There are varieties of disturbances reported ranging from sub cycle duration to long term steady state problems .These are supply line interruption, over voltage , under voltage, momentary interruption, voltage sag, voltage swell, current harmonics, EMI & EMC effects, flicker and voltage imbalance and is well documented in article [2,3]. Many publications proposed innovative techniques to alleviate the current harmonics produced by non-linear loads [4]. Some modifications are introduced to improve the performance of shunt active filter [5-6].Most of these publications tackled only the current or voltage harmonics in the distribution systems. Recently a unique device is being used to compensate both voltage as well as current distortion by single unit called Unified Power Quality Conditioner (UPQC) [7]. The UPQC is regarded as an integrated operation of active filters in series and parallel combinations. Its control technique is extended from that of

Vadirajacharya is a research scholar with the department of Electrical Engineering, Indian Institute of Technology Roorkee, India. (email- vadi_k@indiatimes.com)
Pramod Agarwal is with the department of Electrical Engineering, Indian Institute of technology Roorkee, India. (email- pramgfee.ernet.in)
H.O.Gupt is with the department of Electrical Engineering, Indian Institute of technology Roorkee, India. (email- hogfee.ernet.in)

series active and shunts active filters. Lot of research publications is available on control technique of UPQC [8-11]. The earlier control techniques were based on time domine or frequency domine with hystersis-based current control, PWM current or voltage control, deadbeat control, sliding mode of current control, fuzzy-based current control, etc .They are implemented, either through hardware or software (in DSP-based designs) to obtain the control signals for the switching devices [12]. All above control methods uses a controller to generate gating signals for inverters with a closed loop system. In 1991 K.M. Smedley developed a concept of One Cycle Control of Switching Converters which can be extended to control of active filters [13]. This method employs an integrator with reset as its core component to control the pulse width of APF converter so that its current / voltage is precisely opposite to the reactive and harmonic current / voltage of the nonlinear loads. In contrast to previously proposed methods, there is no need to generate a current reference for the control of the converter current, thus no need for a multiplier and no need to sense the ac line voltage, the APF current, or the nonlinear load current in case of shunt controller. Only one current sensor and one voltage sensor is used to sense the ac main current and the dc capacitor voltage. The control method features carrier free, constant switching frequency operation, minimum reactive and harmonic current/voltage generation, and simple analog circuitry [14-15]. In this paper analysis of single phase UPQC controlled through One Cycle Controller is presented. A brief description on configuration of single phase UPQC is given in Section-II. The control concept of Unified Constant Frequency Integration Control (UCI) is given in section –III. The control of UPQC through UCI is presented in section–IV .Simulation results are presented in section V.

## II. CONFIGURATION

A conventional UPQC topology comprises of integration of two active power filters connected to a common DC link bus. A simple block diagram of a typical UPQC is as shown in Fig.1.

0-7803-9771-1/06/$25.00 ©2006 IEEE

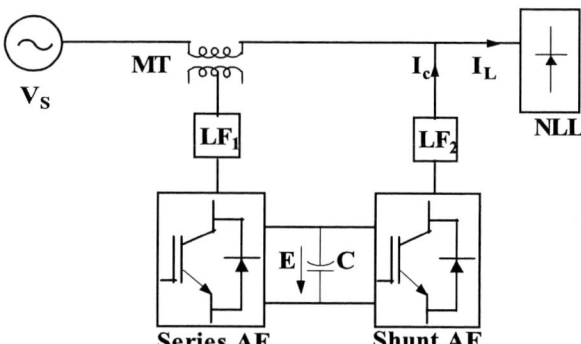

MT= matching Transformer, NLL= Non Linear Load

Fig. 1. Block diagram of a typical UPQC.

The first active filter connected in series through a matching transformer (MT) is called as series filters (SF). It is controlled as voltage generator. It provides harmonic isolation between a sub transmission system and a distribution system. In addition, it has capability of voltage imbalance compensation, voltage regulation and harmonic compensation at the utility-consumer point of common coupling (PCC). The second unit, which is connected in parallel with load, is called Shunt Filter or Active Rectifier (AR). It is controlled as a current generator. The main purpose of the Active Rectifier is to absorb current harmonics, compensate for reactive power and negative sequence current injected by the load. In addition, the voltage of the DC link capacitor is controlled to a desired value by this active filter. Third element of this power line conditioner is the energy storage device. A small amount of DC power supply is required to operate active power filter for harmonic compensation. The energy storage element like capacitor in VSI configuration and inductor in CSI configuration will function as DC power supply sources and hence does not demand any external power source. However in order to maintain constant DC voltage or current in the energy storage element a small fundamental current is drawn to compensate active filter losses.

### A. Single-phase UPQC configurisation

Configurisation of UPQC is mainly depending upon type of converters, nature of supply and load, level of compensation desired. The configurisation considered for analysis is as given below in Fig. 2 [16].

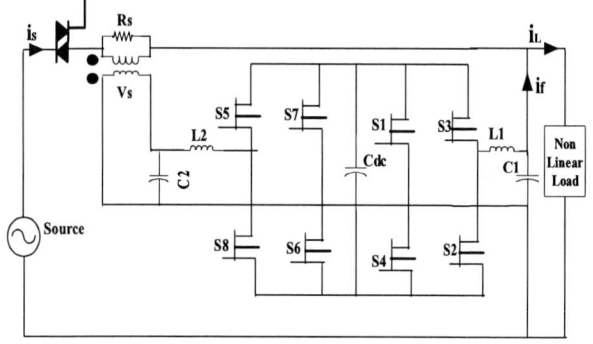

Fig. 2. Single phase Full Bridge UPQC.

It consists of two full-bridge bidirectional converters connected to a common DC link bus. The series bi-directional converter consists of four switches connected via a matching transformer in series with the AC line. The parallel bi-directional converter also consists of four switches. Two passive filters formed by $L_1$, $C_1$, $L_2$, and $C_2$ remove switching frequency harmonics from the output current of the parallel converter and the output voltage of the series converter, respectively. $L_1$ also acts as a link between the filter and the system. Parallel converter delivers its current to the system through this inductor. Load is a sensitive non-linear electronic device. The series converter compensates the voltage difference of the input and reference voltage and regulates the voltage of the load terminal. This converter has the ability to cancel voltage disturbances such as harmonics, voltage sags and swells, and spikes. It gives or absorbs active power in the case of voltage sags and swells, respectively. For compensating the voltage harmonic of the source side, it only delivers reactive power. Parallel inverter mitigates load current harmonics, compensates the reactive current, and draws a small component of the fundamental current to recharge the DC link capacitor.

### III. UCI PRINCIPLE

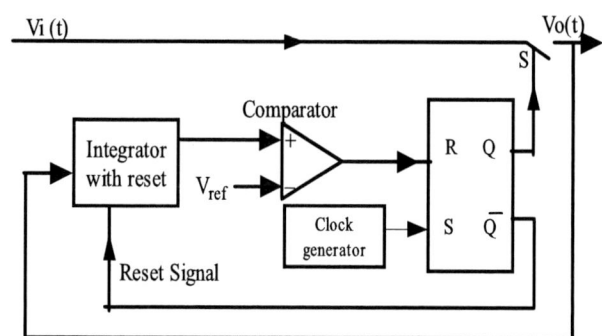

Fig. 3. Principle block diagram of One Cycle Control.

Principally Unified Constant Frequency Integration controller is based on OCC. It is a nonlinear pulse width modulation method. In contrast to the traditional PWM method, OCC control realizes PWM and nonlinear control in one go by modulating the slope of the saw tooth waveform. The implementation circuit is very simple. In Fig. 3, a basic control core is shown. Clock generates a periodic pulse train that sets the flop/flop at the beginning of the each switching cycle. Signal $V_o$, at the input of the integrator, is integrated and the output value is compared to signal $V_i$. When both signals at the two inputs of the comparator approach one another, the comparator changes its state, which in turn resets the flip/flop and the integrator to zero. This operation can be expressed mathematically as

$$1/RC \int_0^{dt} V_0 \, dt \doteq V_i \qquad (1)$$

Where T is the switching period and d is the duty ratio (the on time of the switch verses the switching period), R and C are the value of the resistor and capacitors of the integrator respectively. This process repeats each switching cycle. With this control circuit, the duty ratio of the switch is controlled

674

such that the chopped signal of $V_o$ has an average in each switching cycle that is equal to or proportional to signal $V_i$. Without loss of generality, if the integration constant is chosen to be the same as the switching period, the average of the chopped signal of $V_o$ in each switching cycle is equal to signal $V_i$. In other words, the duty ratio is modulated as

$$V_o d = Vi \qquad (2)$$

This establishes a solver for the first order polynomial function of d.

### A. System configuration of proposed UCI -UPQC

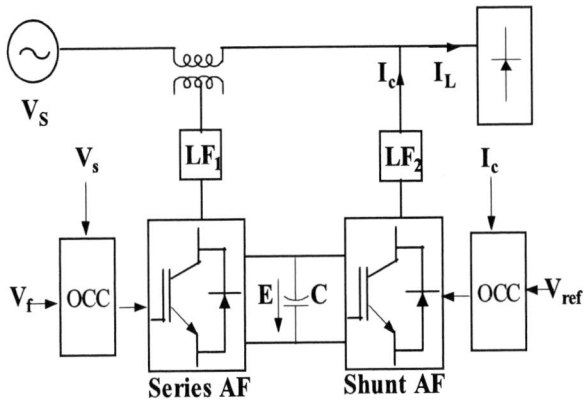

Fig. 4. System configuration of proposed UCI-UP'QC.

Fig. 4 shows the system configuration of proposed OCC-UP'QC. The OCC-UPQC contains a series connected voltage regulator (SF) and a shunt connected current compensator (AR) with switching filters $LF_I$ and $LF_2$ respectively. Controlled by the OCC controller, both SF and AR share a DC bus and have lower power rating compared with that of load. The DC bus capacitor is used as an energy buffer. The matching transformer MT connects SF in series in between the source and the load. The matching transformer MT is designed with low impedances to minimize the voltage drops and energy losses. The turn's ratio is determined by the maximal amplitude of the compensated voltage, the DC bus voltage level, and the power rating of the inverter. $Vs$ and $is$ indicate the voltage and current at the line side respectively. $Vc$ and $Ic$ are the compensation voltage and current generated by the UPQC respectively. Voltage $E$ is the DC bus voltage. Both SF and AR are realized with OCC to simplify the control and to ensure the robust performance.

## IV. CONTROL PRINCIPLES

For the flexibility of operation independently or jointly, the control circuit of the UCI-UPQC is designed as independent parts, i.e. the series voltage regulator (SF), the shunt current regulator (AR) are controlled individually.

### A. Principles of voltage quality improvement

The series filter controller contains an integrator with reset INT, a comparator   and an R-S flip-flop as shown in Fig. 5[15].

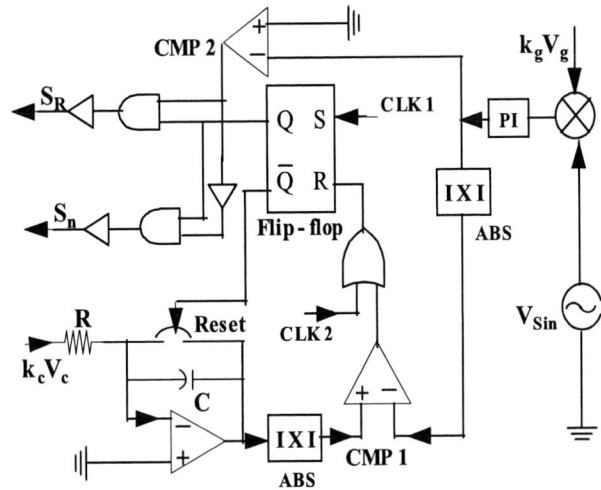

Fig. 5. Series Filter controller.

$V_g$,and $V_c$,are the controlled phase voltage and the output of the. UPQC (including the transformer turn ratio), Clock signal Clk sets the flip-flop at high frequency. The control circuit itself gives a desired voltage reference Vref. If Any voltage quality problem will result in a difference $V_C$ between the sensed voltage $V_g$, and the reference with open loop control, i.e.

$$V_e = Vsin - k_g V_g \qquad (3)$$

where $k_g$ is the voltage sensing ratio. If the compensation voltage $V_c$, is generated by the voltage source inverter (VSI) with OCC as    $Vc = (1/k1) Ve$, then the load side voltage $V_L$ will be

$$V_L = V_S + (1/k1) V_e = (1/k_1) V_{sin} \qquad (4)$$

Equation (4) indicates that although the supply voltage $V_g$, does not meet power quality standard, the voltage appeared at the load side will be sinusoidal and proportional to the reference $V_{sin}$. Voltage $V_c$ following the error signal $V_e$ is performed by the OCC controller, the R-S register is set when the clock arrives ($Q \rightarrow 1$), which is a command to turn on the switch S, in the case of PI output positive. In this switching state, the output of the converter $V_C$ equals to the negative DC bus voltage, i.e. $V_c = -E/2$. The output of the inverted integrator $V_{int}$ increases from its initial value zero. When $V_{ref}$, reaches the reference, an overturn occurs in the comparator, which resets the R-S flip flop ($Q \rightarrow 0$). Switch $S$ is then turned off at the same time the integrator is reset to zero. The Integrator restarts integration from zero after the reset. The duty ratio of the switches are determined by the equation

$$1/Ts \int_0^{dt} k_e V_c \, dt = V_e \qquad (5)$$

The left side of the equation is the average voltage in every switching cycle. Therefore, the output voltage of the UPQC $V_C$ follows the voltage error between the distribution line voltage $V_g$ and the reference $V_{sin}$. Since the switching period

675

Ts, is usually several thousand times smaller than the line cycle, the voltage compensation with OCC has fast transient response and high control precision. To achieve no-error steady-state load-side voltage, a PI controller is used and $V_g$ is sensed at the load side instead of the line side.

## B  Principles of current quality improvement

The shunt controller consists of a PI controller, an integrator with reset and a flip-flop. The control object of the Shunt APF is to provide the reactive and harmonic current required by the nonlinear load, so that the net current draws from the ac main is the fundamental active power used at the nonlinear load. From the viewpoint of the ac main, the nonlinear load with an active power filter in parallel imposes a linear resistive load to the ac power system in steady state. Therefore, an equivalent resistor is used to emulate the nonlinear load with an active power filter in parallel for ac main. The control goal of Shunt APF is

$$V_s = R_e * i_s \qquad (6)$$

The UPQC is operated at switching frequency of $f_S$. Assuming the value of the energy storage capacitor C is large enough so that voltage E is nearly constant in one switching cycle. Switching frequency $f_S$ is much higher than both the line frequency and the frequency of nonlinear load current. The value of energy storage capacitor voltage is given by

$$E = [1/ (1-2D)]*v_s \qquad (7)$$

Where D is duty ratio. $D = T_{on}/$ Ts. Combining equation 6 & 7 with a current sensing resistor $R_s$

$$R_S/R_E (1-2D)*E = i_S * R_S \qquad (8)$$

$$\text{If} \quad Vm = (R_S/ R_E) *E \qquad (9)$$

$$Vm (1-2D) = i_s * R_S \qquad (10)$$

Then control goal of active filter becomes

$$2DVm = Vm - R_S * is \qquad (11)$$

In each switching cycle, if the duty ratio is controlled to satisfy the equation (11) equation (6) is satisfied. In each line cycle, if the capacitor voltage is controlled to be constant from cycle to cycle, only the reactive power is processed in the shunt APF. The net current drawn from the ac main is equal to the fundamental active current required by the nonlinear load that has the same waveform as and in phase with line voltage. The reactive and harmonic current of nonlinear load is canceled from ac line current. The UCI controller as shown in fig 6 below is used to realize equation (10).

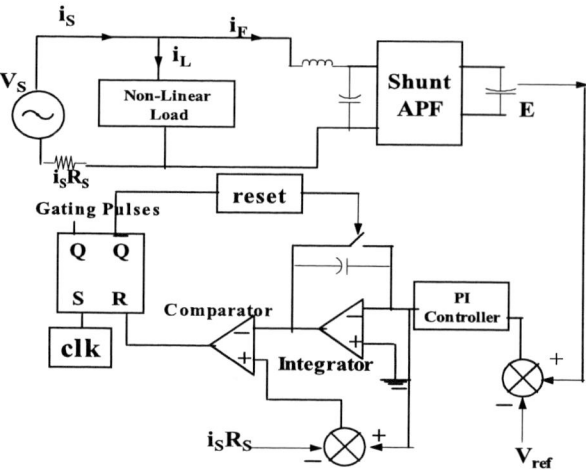

Fig. 6.  Shunt APF with UCI controller.

## V.  SIMULATION RESULT

The simulation results are shown in the Fig. 7 and 8. The shunt compensation is activated after 0.01 sec, while series compensation is activated after 0.1 sec. Fig 7 shows the performance of Shunt filter and fig 8 gives the performance of Series filter. The THD of non-linear load current was 31.25%. After compensated by proposed UPQC, the THD of AC mains line current is 2.02%. THD of load voltage after compensation is 10.53%.

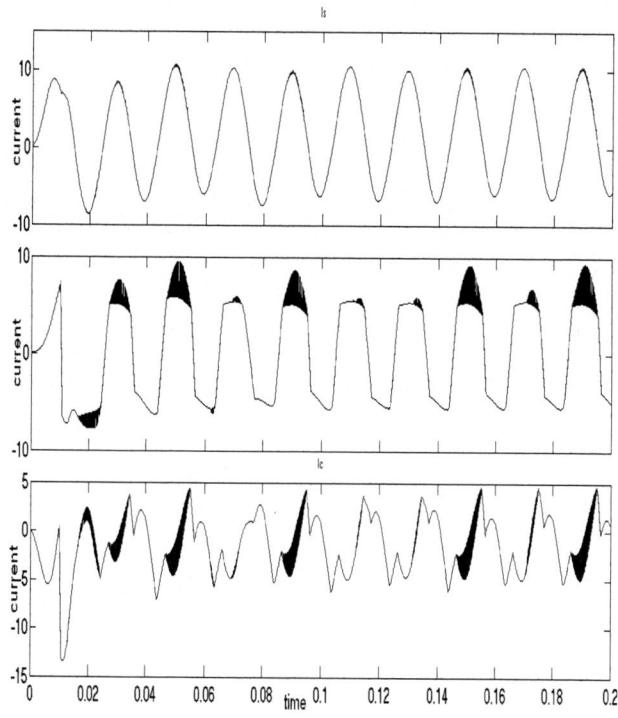

Fig. 7. Line, load and Shunt compensation current of UPQC.

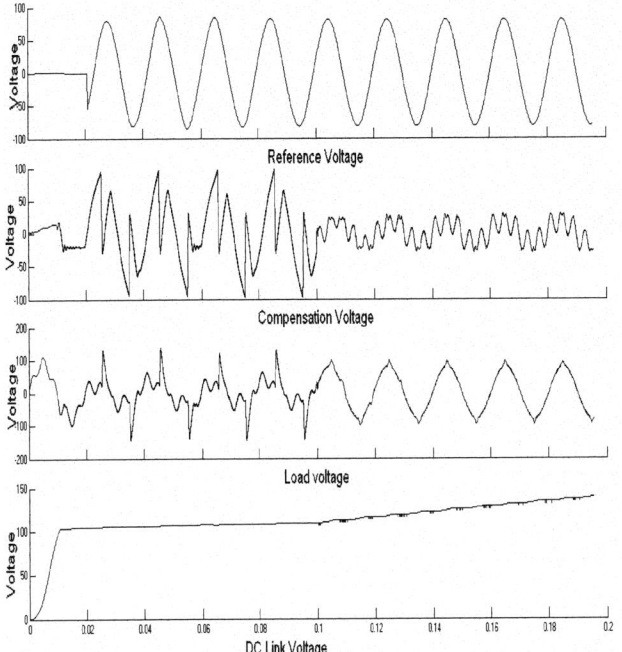

Fig. 8. Reference, Series compensation, Load and DC Link Voltage of UPQC.

## VI. CONCLUSION

The proposed UCI -UPQC has the ability to improve both voltage and load current quality, including reactive component. It provides a multifunctional, high performance, and reliable solution for total power quality control. It is suitable for commercial, industrial, space, and military applications, as well as distributed power generation.

## VII. REFERENCES

[1]  J. Arrilaga, D. A. Bardely and P. S. Bodger, "Power system harmonics", New York: John Wiley and Sons Ltd, 1985. pp. 199–214.

[2]  E. W. Gunther and H. Mehta, " A survey of distribution system power quality", IEEE transactions on Power Delivery, Volume 10, No. 1, January1995, pp 322-329.

[3]  V. H. Tahiliqni and H. Mehta, " Custom power utility's response to power quality issues", proceedings of the American Power Conference, April 1992,Vol- 2, pp 1561-1565.

[4]  L. Gyugi and E. C. stygula, " Active AC Power filter", IEEE/IAS Annual Meet 1976, pp 529-535.

[5]  F. Z. Peng and H. Akagi, "New Approach to Harmonic Compensation in Power System", IEEE PESC 1988, pp874-880.

[6]  F. Z. Peng, "A power Line Conditioner Using Cascade Multilevel Inverter for Distribution System", IEEE Trans. On Industry Applications, Vol. 34, No. 6, Nov/Dec 1998, pp1293-1299.

[7]  H. Fujita, H. Akagi, "The unified power quality conditioner: the integration of series active and shunt active filters" IEEE PESC, 96 record, 1996, pp494-501.

[8]  M. Hu and H. Chen, "Modeling and controlling of unified Power Quality Compensators" Proceedings of the 5th International Conference on Advances in Power System Control, Operation and Management, APSCOM 2000, Hong brig , 2000 pp 431-435.

[9]  Elnady, A. Gauda, M. M. A. Salama, "Unified Power Quality Conditioner with a novel control algorithm based on wavelet transform", Proceedings on Electrical and Computer Engineering 2001 Canadian Conference, Vol- 2,May 2001, pp 1041- 1045.

[10]  G. U. Jianjum, X. U. Dianguo, L. Hankui and G. Maozhang, "Unified Power Quality Conditioner (UPQC): the Principle, Control and applications", IEEE proceedings on Power Conversion Conference, Osaka, 2002, Volume.1, April 2002, pp80-85.

[11]  L. H. Tey, P. H. So and V. C. Chu, "Neural network controlled Unified Power Quality Conditioner for system harmonic compensation", Transmission and distribution conference and exhibition 2002, Asia pacific IEEE/PES, Volume 2, 6-10,Octobor 2002, pp 1038-1043.

[12]  Bhim Singh, A. H. Kamal and Ambrish Chandra, " A Review of Active Filters for Power Quality improvement", IEEE Transactions on Industrial Electronics,Vol-46,No-5, October 1999,pp 960-971.

[13]  K. M. Smeldy and S Cuk, "One–Cycle Control of Switching Converters", IEEE Power Electronics Specialist conference, Boston, June 1991, pp 1173-1180.

[14]  K. M. Smedley, L. Zhou and C. Qiao "Unified Constant Frequency Integration control of Active Power Filters-steady state and dynamics" IEEE Transactions on Power electronics, Vol- 16, No-3, May 2001, pp 428-431.

[15]  G. Chen, Y. Chen, L. F. Sanchez and K. M. Smedley, "Unified power quality conditioner for distribution system without reference calculations", in Proc. IPEMC-2004, Vol-3, pp 1201-1206.

[16]  A. Nasiri and A. Emadi, "Different topologies for Unified power Quality Conditioner", IEEE transactions on power Electronics, vol-3, pp 976-981.

## VIII. BIOGRAPHIES

**Vadirajacharya** graduated in Electrical Engineering from PDA college of Engineering Gulabarga, India in 1984. He completed his PG in Electrical Engineering from SGSITS, Indore, India in 1998. Currently he is pursuing research in the field of power quality converters at Electrical Engineering Dept Indian Institute of Technology Roorkee, India. His fields of interest include Power quality, and Energy auditing.

**Pramod Agarwal** obtained his Bachelor's degree in Electrical Engineering from University of Roorkee, now Indian Institute of Technology Roorkee (IITR), India. He received his PG and completed his PhD in Electrical Engineering from IITR in 1985 & 1995 respectively. Currently he is Professor in the department of Electrical Engineering, IITR. His fields of interest include Electrical Machines, Power Electronics, Power quality, Microprocessors and microprocessor-controlled Drives, Active power filters, High power factor Converters, Multilevel Converters, and application of dSPACE for the control of Power Converters.

**H.O.Guta obtained** his Bachelor's degree in Electrical Engineering from University of Jabalpur, India. He received his PG and completed his PhD in Electrical Engineering from University of Roorkee, India in 1975 and 1980 respectively. Currently he is Professor in the department of Electrical Engineering, Indian Institute of Technology Roorkee, India. His fields of interest include Database Management, Transformer, Power system analysis and optimization, System Engineering.

**2006 IEEE International Conference on Power Electronic, Drives and Energy Systems**

# Application of a Boundary Model to Assess Power Quality Cost Function

### J. Ahmadian, A. Jalilian, and M.A.S. Masoum

*Abstract*—In this paper a boundary model is defined to assess the power quality and loading cost function of power systems. Each bus is assigned a position in the space state model that is calculated using a simple equation. The state space representation of buses in the boundary model facilitates analysis of power system including the effects disturbances, linear and nonlinear loading, as well as reactive power compensation (e.g., using filters or FACTS devices) without the requirements for detailed power quality measurements. The proposed boundary model is used to investigate the impact of loading and compensation on a three bus system with linear and nonlinear loads.

*Index Terms*-- Boundary model, cost function, harmonic distortion, power quality, and nonlinear loads.

## I. INTRODUCTION

$D$EVELOPING of power electronic devices has increased their applications in different fields such as AC/DC converters, electric motor drives and other equipments. Most power electronic and switching devices produce harmonics and detriment the quality of electric power. This has raised many issues regarding the safe and optimal operation of power systems [1-13].

One of the most important problems in power systems is to obtain optimal operating points for different loading conditions with acceptable distortion levels as recommended by the corresponding standards such as IEEE-519. Conventional approaches to determine the power quality of power systems require detailed and tedious non-sinusoidal measurements at selected buses.

This article proposes a simple cost function to estimate the power quality of the system prior to (linear and nonlinear) loading and/or reactive power compensation. The approach uses a biaxial state space plane that is calculated from investment and commercial method as suggested by the Research Group of American Boston University [3]. After introducing the concept of boundary model, the proposed boundary model algorithm is implemented for a three-bus system with linear and nonlinear loads.

## II. POWER QUALITY PARAMETERS

Many power quality parameters and indices have been proposed and considered in the literature and standards [1]. This paper uses the total demand distortion of voltage and current as defined by Equations 1-4 [2]:

$$TDD_i = \frac{\sqrt{\sum_{h=2}^{\infty} I_h^2}}{I_n} \qquad (1)$$

where TDD$_i$ is Total Demand Distortion (TDD) of current,

$$TDD_v = \frac{\sqrt{\sum_{h=2}^{\infty} V_h^2}}{V_n} \qquad (2)$$

where TDD$_V$ is TDD of voltage,

$$TDD_{5i} = \frac{\sqrt{\sum_{h=1}^{\infty} I_{5h}^2}}{I_n} \qquad (3)$$

where TDD$_{5i}$ is TDD of multiple 5$^{th}$ harmonics of current,

$$TDD_{7i} = \frac{\sqrt{\sum_{h=1}^{\infty} I_{7h}^2}}{I_n} \qquad (4)$$

where TDD$_{7i}$ is TDD of multiple 7$^{th}$ harmonics of current.

## III. BOUNDARY MODEL

Boundary Model (BM) results from the approach proposed by the Research Group of Boston University in the field of trade and investment [3]. Horizontal and vertical axes are specified through two variables that are independently defined. The resultant coordinate system divides the sate space plane into four operating blocks (regions) as specified in figure 1.

Block I indicates that in this state space region, there is the possibility of loading with no limitations. Unites to be placed in this region, depending upon their distances (far or near) from blocks II and IV, create the chance of maneuver for designers. These unites don't need more investments, in other words, less investments is needed for them in comparison with other unites places in other blocks.

Block II shows the state space therein is the possibility of loading through certain limitations. Unites to be placed in this region have no chance of more loading in the direction of horizontal axis, so the chance of maneuver for designers is created only in the direction of vertical axis. These unites will require more investment, possibly high, to try to maneuver in the horizontal axis.

Block III shows the state space therein is no possibility of further loading without facing serious problems. Unites to be located here have no chance of loading in either the horizontal or the vertical axis. Therefore, the designer is unable to have a wide maneuver. This unites need high level of investment to maneuver.

Block IV displays state space therein is the possibility of loading through certain limitations. Unites to be placed in

---

Javad Ahmadian is a PhD student, A. Jalilian and M.A.S. Masoum are academic members of the Department of Electrical Engineering, Iran University of Science and Technology (IUST), Tehran, Iran.
(emails:Javad_Ahmadian@yahoo.com, jalilian@iust.ac.ir, m_masoum@iust.ac.ir).

0-7803-9771-1/06/$25.00 ©2006 IEEE

this limit have more chance for loading in the horizontal direction, so the designer is able to maneuver only in horizontal axis. These unites need to invest more to maneuver in vertical direction which would probably be of high amount.

This model depending on the number of variables could be multi axial, but the more axes we have, the more difficult is their analysis. In this case it is suggested that it would be better to transform the n-axial mode to biaxial mode, through essential strategies and definition of suitable functions. In this article horizontal axis is a variable which define the risk of feeder concerning quality of power, and the vertical axis specifies the variable of loading level (e.g., line current) and the associated limitations. Limits of these blocks are specifies using the power quality limits as specified by IEEE519 standard [14].

## IV. PURPOSE COST FUNCTIONS

Variables related to horizontal and vertical axes are values which must be determined as purpose cost functions. There are limits in BM related to the fourfold blocks shown in Fig. 1. These limits are also determined on the basis of strategies and policies of electric companies. This model must be defined for each bus of the distribution feeder, by which the entrance and exit of each load changes the position of each bus in the model of network.

Variable of the horizontal axle indicate the power quality limits while variables in the vertical variable axle show cost of loading. These are suggested as follows:

$$X = \max(D_i * PQ_i / PQ_s) \qquad (5)$$

$$Y = F_s(B * I_s, P_{loss}, PQ_i) \qquad (6)$$

where:
-X is the horizontal variable representing power quality,
-$D_i$ is the coefficient distance between bus and load,
-$PQ_i$ are power quality parameters arising from every load that introduced in Eqs. 1-4,
-$PQ_s$ are the standard values of power quality parameters based on IEEE519 [14],
-Y is the vertical variable representing $F_s$; the cost function for adding new loads to the network,
-$P_{loss}$ represents increase in losses (Eq. 8),
-$I_s$ is the effective current of power system,
-B is the per unit coefficient that can be calculated by

$$B = I_{base-l} / I_{base-sys} \qquad (7)$$

where $I_{base-1}$ and $I_{base-sys}$ are base load current and base system current, respectively.

$F_s$, is a nonlinear function, indicating system losses:

$$P_{loss} = \sum_i \sum_h R_{ih} I_{ih}^2 \qquad (8)$$

In Eq. 8, $I_{ih}$ is $h^{th}$ harmonic current of line i, and $R_{ih}$ is resistance of line i at the $h^{th}$ harmonic frequency which can be approximated by

$$R_h = R_{dc}(1 + 0/05\sqrt{h}) \qquad (9)$$

$R_{dc}$ is dc resistance of line.
Therefore, we have

$$F_s = [F_h + F_{loss} + \sum FPQ_i)]/[C_s * (S_n)] \qquad (10)$$

$$F_h = C_s * V_{rms} * I_{rms} \qquad (11)$$

$$F_{loss} = C_p * P_{loss} \qquad (12)$$

where

- $F_h$ is the cost corresponding to the reduced system capacity due to the $h^{th}$ harmonic current,
- $V_{rms}$, $I_{rms}$, $V_{1rms}$, $I_{1rms}$ are rms voltage and current of system and fundamental rms voltage and current of system respectively,
- $C_s$ cost of network installation per \$/MVA,
- $I_n$, $V_n$ are rated current and voltage,
- $S_n$ is rated complex power,
- $F_{loss}$ cost of loss,
- $C_p$ is coefficient of average cost of system,
- $\sum FPQ_i$ is sum of costs resulting form of power quality perversity (which is not considered in this paper).

## V. PROGRAM FLOWCHART

The proposed flowchart for the evaluation of power quality state space in shown in Fig.2. Based on this algorithm, variable of power quality of load will be obtained in term of Eqs. 1-4. Then variables X and Y will be calculated according to Eqs. 5-6. Based on the boundary model (Fig. 1), local change condition of each bus is distinguished in the model. The final step is to determine the new position of the bus (within the four blocks of Fig. 1) based on the changed condition.

If the position of bus doesn't change in exist block, rendering any service to load would be of no mistake. In the case that this change causes the bus to enter the 4th block, it shows that feeder is approaching to its rated capacity and as it enters to the second block, it determines that power quality of system has been weakened. Entrance to the third block has the result of both. However, such a feeder is operated in economic constraints, and must be added in the coming program list of the company to amplify electrification, or compensate the power quality with FACTS devices or filters.

If the entrance of load results in the exit of bus from the second block in horizontal direction, it shows that utility is bounded to compensate the power quality immediately, and by entrance of load, power quality in question will be faded. If loading result to exit from the fourth block in vertical direction, it reports excessive loading system that causes voltage drop and much loss. Consequently it is obligatory to establish a new feeder. The exit of third block in horizontal and vertical direction means that we need both establishing of the new feeder and compensation.

## VI. TEST TYPE NETWORK

Single line diagram of a real network is displayed in Fig.3. There are three buses with linear and nonlinear loads. While nonlinear loads of buses 1 and 2 are supplied by buses, different types of compensation are considerable by standing on these three buses.

Parameters of system are as displayed in Table I. Loads of buses are described in Table II. The following equations are used for the simulations:

$$v_k(t) = \sum_h V_{kh} \cos(h\omega t + h\varphi_k) \qquad (13)$$

$$i_k(t) = \sum_h I_{kh} \cos(h\omega t + h\theta_k) \qquad (14)$$

where $v_k$ and $i_k$ are voltage and current of bus k. Loads are connected to the network as specified in Table III.

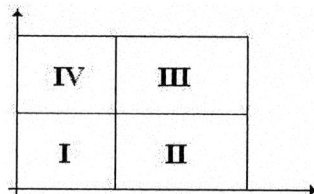

Fig. 1. Four blocks of the proposed Boundary Model (BM).

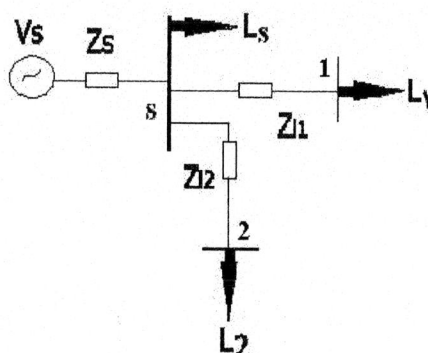

Fig. 3. Single line diagram of the three-bus Network.

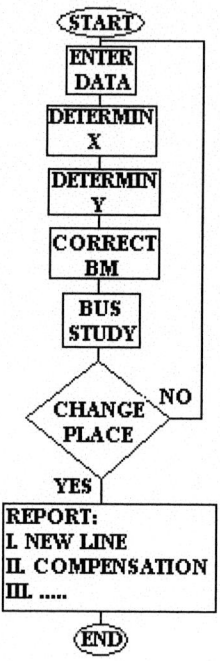

Fig. 2. The proposed boundary model flowchart.

TABLE I
NETWORK PARAMETERS (AT THE FUNDAMENTAL FREQUENCY)

| $V_s$ (KV) | S (MVA) | $Z_s$ (pu) | $Z_{l1}$ (pu) | $Z_{l2}$ (pu) |
|---|---|---|---|---|
| 63 | 100 | 0.0073+j0.0866 | 0.0001+j0.0074 | 0.0001+j0.0079 |

TABLE II
LOAD CURRENTS LOAD OF THE NETWORK (IN PER UNIT *0.007)

| Harmonic Order | Ls [pu] | L1[pu] | L2[pu] |
|---|---|---|---|
| Power Factor | 0.65 | 0.74 | 0.8 |
| 1 | 8 | 5 | 6 |
| 3 | - | 1.5 | 3 |
| 5 | - | 1.2 | 3 |
| 7 | - | 0.8 | 2 |
| 9 | - | 0.6 | 2.5 |
| 11 | - | 0.5 | 2 |
| 13 | - | 0.4 | 1.5 |
| 15 | - | 0.3 | 1.5 |
| 17 | - | 0.2 | 0.7 |
| 19 | - | 0.2 | 0.7 |
| 21 | - | 0.15 | 0.5 |
| 23 | - | 0.14 | 0.3 |
| 25 | - | 0.1 | 0.2 |

TABLE III
PERCENT OF LOAD IN THE NETWORK

| No | Bus S | Bus 1 | Bus 2 |
|---|---|---|---|
| 1 | 50 | 0 | 0 |
| 2 | 50 | 50 | 0 |
| 3 | 50 | 50 | 50 |
| 4 | 100 | 50 | 50 |
| 5 | 100 | 100 | 50 |
| 6 | 100 | 100 | 100 |

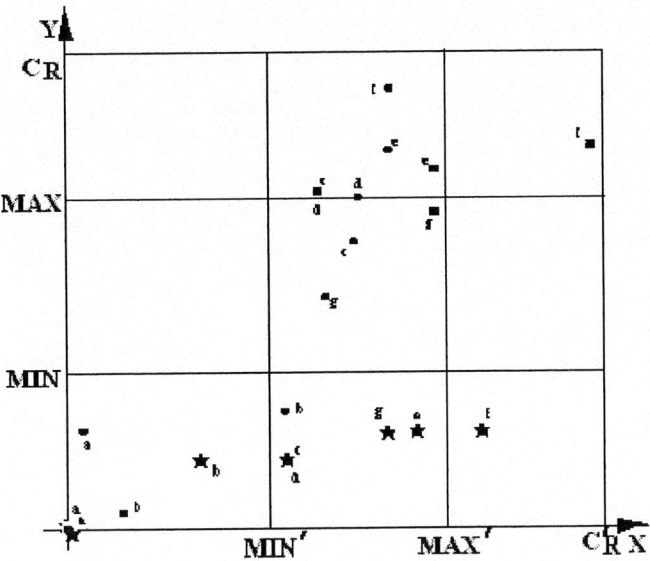

Fig. 4. State space of BM for different buses (* shows bus 1 and ■ shows bus 2; letters "a", "a", "b", "c", "d", "e", and "f" correspond to loading according to rows 1 to 6 of Table III; while "g" represents loading according to row 6 of Table III with compensation at bus 2.

## VII. RESULTS OF BOUDARY MODEL

The proposed boundary model algorithm (Fiq. 2 and Eqs. 5-6) is implemented for the three-bus network of Fig. 3. Loads entrance to buses of the network is described in Table III. Figure 4 shows the output results consisting of bus locations in state space before and after the compensation. In this Figure MIN', MAX' and CR' represent minimum, maximum, and critical limits for power quality (based on IEEE519 and the policy of electric utility), respectively, while MIN, MAX, CR correspond to minimum, maximum and outrange of investment for the network, that will be determined based on policy of electric utility. As specified in Fig. 4:

a) Entrance of the linear load to bus s only changes the position of this bus in power quality state space vertically. In other word, cost function has a value and power quality doesn't change.

b) Addition of nonlinear loads to bus 1 not only changes the position of this bus but also change the positions of other buses in the network. These changes are vertical and horizontal.

c) Addition of linear loads to bus 1 not only changes the position of this bus but also change the position of bus s. These changes are vertical.

d) Introducing linear and nonlinear loads to bus 2 will have the same results as parts (a) and (b).

e) According to the power quality state space, we can conclude that compensator is required for the network of Fig.3. In this case, one STATCOM is recommended. With finding optimum place for it, we come to the out come in bus 2. In the case of utilizing that, position of made space in bus s and bus 2 will change.

It should be mentioned that the optimal solution and the sizing of the STATCOM depends on system loading conditions. In this paper the load growing are of no attention and are relinquished.

## VIII. CONCLUSIONS

A boundary model is defined to assess the power quality and loading cost function of power systems. Using this model, users defines two functions representing "power quality" and "cost of network" for the orthogonal axels of state space.

The proposed model is applied to a real network that consists of three buses with linear and nonlinear loads. Different loading conditions are examined and the state space model is displayed for each case. In addition, the effects of compensator are also investigated. Main conclusions are:

1. Using the proposed BM it is possible to obtain the relation among various buses in the network relevant to different conditions of loading.

2. Entrance of linear load to a bus only changes the position of this bus in BM vertically. In other word, cost function has value and power quality doesn't change. The above bus has also these changes.

3. If nonlinear load is added to a bus, not only the position of all buses change in BM horizontally but also the position of the bus itself will change vertically.

4. Introducing additional loads on one bus with no problems does not guarantee safe operation of others bus of the network. Therefore, one must generate and examine the BM model for all buses of the system to achieve safe operation.

5. With the BM of Fig. 4, it is possible to determine which buses will be in trouble through loading. Consequently, through using the technical and economical computing, alternatives such as using FACTS devices or adding new network branches may be put under investigation.

## IX. REFRENCES

[1] F.Z. Peng, H. Akagi, and A. Nabae, "Compensation Characteristics of the Combined system of Shut Passive and Series Active Filters", IEEE, Trans. Industry Applications, vol. 29, No. 1, pp. 144-152, 1993.

[2] J. Hafner, M. Aredos and K. Heumann, "A shunt Active Power Filter Applied to High Voltage Distribution Lines", IEEE, Trans. Power Delivery, vol. 12, No. 1, pp. 266-272, 1997.

[3] Management Group of Beheshti University, "Strategy of COCACOLA Co. 1980-1990", Seminar of Management Faculty, Beheshti University, Iran, 2004.

[4] H. Akagi, "Control Strategy and Site Selection of a Shunt Active Filter for Damping of Harmonic Propagation in Power Distribution Systems", IEEE, Trans. Power Delivery, vol. 12, No.1, pp. 354-363, 1997.

[5] T. Larsson, C. Poumarede, "STATCOM, an Efficient Means for Flicker Mitigation", IEEE, Power System Conference 1998.

[6] T. Nakajima, "Operating Experiences of STATCOMs and a Three-Terminal HVDC System Using Voltage Sourced Converters in Japan", IEEE Electric Conference 2002.

[7] G. F. Reed..., "Application of a 5 MVA, 4.16 KV D-STATCOM System for Voltage Flicker Compensation at Seattle Iron & Metals", IEEE Power Electric Conference 2000.

[8] B. T. Ooi, G. Joos, X. Huang, "Operating Principles of Shunt STATCOM Based on 3-level Diode-Clamped Converters", IEEE Trans. On Power Delivery, Vol.14, No.4, Oct. 1999.

[9] C. Hochgarf, and R. H. Lasseter, "StatCom Controls for Operation with Unbalanced Voltages", IEEE Electric Conference 1997.

[10] Q. Yu, P. Li, " Overview of STATCOM Technologies", IEEE, International Conference on Electric Utility, DRPT, Hong Kong, April 2004.

[11] C. Schauder, "STATCOM for Compensation of Large Electric Arc Furnace Installations", IEEE Electric Conference 1999.

[12] G. Reed, J. Paserba, T. Croasdaile...,"The VELCO STATCOM-Based Transmission System Project", IEEE, Trans. On Power Delivery, Vol. 2, 2001.

[13] Stewall M. Ramcsay Patrick E. Cronin, "Using Distribution Static Compensator (D-STATCOMs) To Extend The Capability of Voltage-Limited Distributions Feeders" UMS Group 2001.

[14] IEEE Standard 519, IEEE recommended practices and requirements for harmonic control in electric power systems, IEEE-519, 1992.

**Javad Ahmadian** received the B.S. and M.S. degrees in Electrical Engineering from Isfahan University of Technology, Esfahan, Iran, in 1991 and 1996 respectively. He is currently a PhD student at Department of Electrical Engineering, Iran University of Science and Technology, Tehran, Iran. His areas of research interest are electric drives, power electronics and power quality.

**Alireza Jalilian** was horn in Yazd, Iran in 1961. He received his BSc degree in Electrical Engineering from Mazandran University, Iran in 1989 and his ME (Hons) and PhD degree in Electrical Engineering from University of Wollongong, Australia in 1992 and 1997 respectively. Dr Jalilian joined the power engineering group of the Department of Electrical Engineering of Iran University of Science and Technology (IUST) in 1998 as an academic member. Dr Jalilian's research interests are Power Quality causes, effects and mitigations.

**Mohammad A.S. Masoum** received his B.S., M.S. and Ph.D. degrees in Electrical and Computer Engineering in 1983, 1985, and 1991, respectively, from the University of Colorado at Boulder, USA. Currently, he is an Associate Professor at the Department of Electrical Engineering, Iran University of Science & Technology, Tehran, Iran.

**2006 IEEE International Conference on Power Electronic, Drives and Energy Systems**

# Active Power Filter Solution without PLL for Fluctuating Industrial Load

S. Elangovan, Grad *IEEE*

*Abstract* — **To improve the quality of electrical power to the consumers, many power electronic based solutions are available. In the case of active power factor compensation, grid tied voltage source inverter based STATCOM is one of the versatile solution. But in this system most of the control strategies demand generation of unit vector, based on the grid voltage for vector orientation. But presence of harmonics in grid voltages and the level of noise in the grid voltage sensing circuits will make estimation of unit vector difficult and PLL circuits will become a necessity. This paper presents a simple and efficient method to evaluate the unit vector without the use of PLL or any hardware filter. This method uses d-q reference frame control strategy for shunt active filter and has been tested with a 100 kVAR system. The experimental results are presented to validate the effectiveness of this control strategy.**

*Index Terms* - **Unit vector generation, Shunt active filter, digital controller using DSP**

## I. INTRODUCTION

Process industries use wide range of variable speed motor drives, ac plants, UPS systems and various power electronic converters to improve system efficiency and hence the productivity. These loads draw reactive power from the grid. Excess reactive power will result in poor voltage regulation and poor utilization of ac network, mainly the distribution transformer for the plant. These constraints make it necessary to compensate reactive power.

Traditionally, passive elements like ac capacitor banks are used for reactive power compensation. But the effectiveness of these capacitors is limited. Moreover they have some disadvantages like drawing fixed amount of leading reactive power irrespective of the load requirements. For continuously varying load, which is the case in most of the process industries, these fixed capacitor banks fail to control the power factor effectively.

Thyristor switched or mechanically switched capacitor banks are also used to improve this situation. But they too cannot maintain the power factor near unity when the load varies rapidly.

Also, the life span of these passive elements is less due to load harmonics, current sinking and these results in regular maintenance. Due to these limitations of passive filters, various active filters have been reported for reactive power compensation.

The basic block diagram of a three phase shunt active filter is shown in Fig. 1.This STATCOM is primarily a voltage source inverter connected to Point of control ,

Several control strategies, such as Instantaneous reactive power based hysteresis current controller [1], sliding mode controller [2]-[3], synchronous reference frame controller [4] etc. have been proposed and developed for a three phase voltage source inverter based shunt compensator. Most of these control strategies require computations using the grid voltage sensed at PCC. But the source harmonics and the noise in the feedback circuit for grid voltage restrict direct use of the voltage signals in the control algorithm. The harmonics present in the grid voltage can be filtered using hardware band-pass filter [4]. But, it is very difficult to design such a filter with precise choice of cut-off frequency that would separate out the fundamental without appreciable phase errors. Some work has been reported on the use of PLL to remove the effect of harmonics and noise in the grid voltage. But the design of a high performance PLL is not so easy when various non-idealities like multiple zero crossing in the grid voltage are occurring.

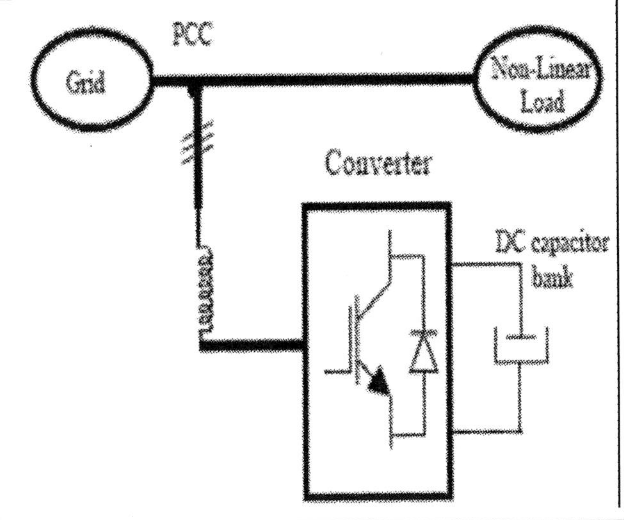

S. Elangovan is with Coimbatore Institute of Engineering and Information Technology, Narasipuram, Coimbatore, Tamilnadu, India (email : elangoeee@rediffmail.com , Mobile: (+91) 9994234061)

Fig. 1. Block diagram of shunt active power filter.

In this paper, synchronous d-q reference frame based control strategy is presented for a 100 kVAR two-level shunt active filter. The control law is derived for the compensation of reactive power drawn from the grid.

This paper also presents a simple and efficient method of unit vector generation, which will minimize the effect of harmonics and noise present in the grid voltage feedback.

## II. PROPOSED SCHEME

As mentioned earlier, the three phase shunt active filter is a fully controlled three phase boost converter, which is connected to the grid through a three-phase series choke. This converter supplies the required reactive power to the load to maintain the grid side power factor unity. The series choke separates the two voltage sources namely grid and inverter. The capacitor voltage is maintained at a constant value by closed loop control, which regulates the active power drawn from the grid and caters to the internal losses of the system.

Synchronous d-q reference frame based control strategy is proposed to control this converter. This control strategy requires sensing of grid voltage and generation of unit vectors for the orientation along the grid voltages. Fig. 2. explains the method adopted to reduce the effect of harmonics and noise in the grid voltage feedback.

The sensed grid voltages are filtered using a first order digital low pass filter whose corner frequency is $\omega_c$. In the present work, $\omega_c$ is chosen to be equal to the nominal grid frequency (50 Hz). So, after filtering, the percentage of $h^{th}$ order harmonics of the sensed grid voltage is reduced by a factor of $\sqrt{\frac{2}{(h^2+1)}}$ (assuming fundamental voltage is always 100%. It is clear that $5^{th}$ and higher order harmonics, which can normally be present in the line-to-line grid voltage, are reduced considerably using this low pass filter. High frequency noise gets eliminated almost completely. Finally, dividing the filter output signals with their magnitude generates the required unit vectors.

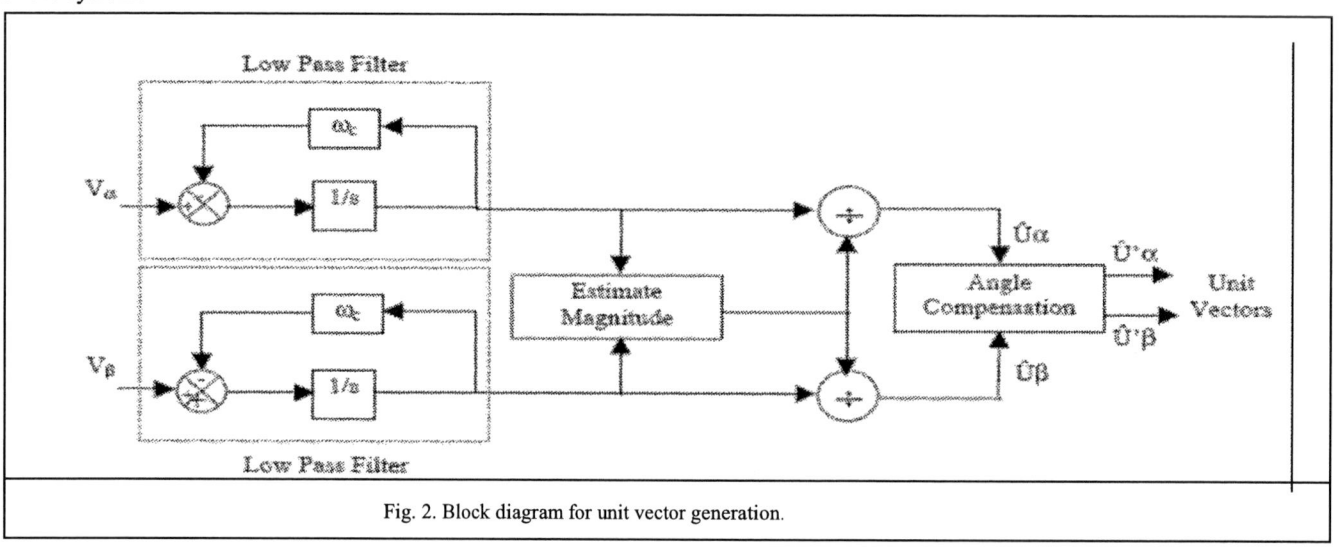

Fig. 2. Block diagram for unit vector generation.

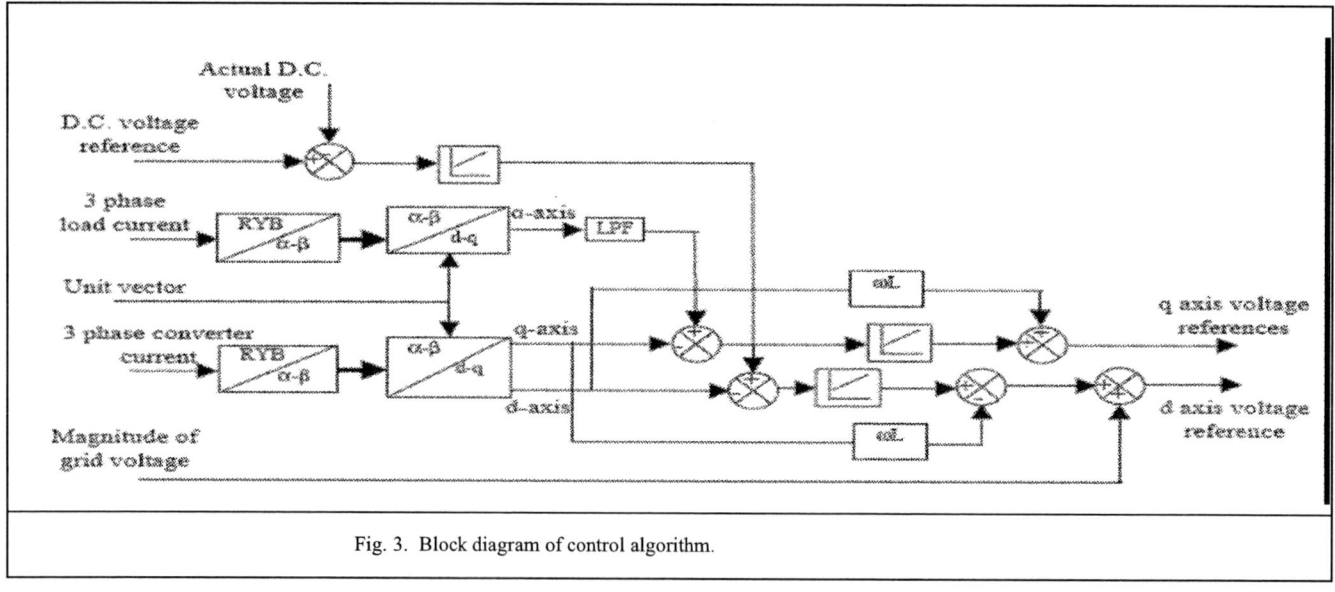

Fig. 3. Block diagram of control algorithm.

684

However, this low pass filter introduces a phase lag to the unit vector generation. For nominal grid frequency (50Hz), this phase lag is 45° and can be easily compensated (Fig. 2). However, the grid frequency varies with a small range (±2.5 Hz). So, there is a small phase error in unit vector generation if constant phase compensation is done. In the next section, synchronous d-q frame based control strategy is explained using this unit vector.

## III. CONTROL STRATEGY

The control scheme is presented in Fig. 3. The three phase currents of the load and the converter are transformed into the synchronously rotating reference frame using (1) and (2).

The Unit vector required for the transformation is generated with grid voltages as described earlier. In other words the first step in the control scheme is orienting the

$$\begin{bmatrix} i\alpha \\ i\beta \end{bmatrix} = \begin{bmatrix} 1 & -\dfrac{1}{2} & -\dfrac{1}{2} \\ 0 & \dfrac{\sqrt{3}}{2} & -\dfrac{\sqrt{3}}{2} \end{bmatrix} \begin{bmatrix} i_a \\ i_b \\ i_c \end{bmatrix} \quad (1)$$

$$\begin{bmatrix} i_d \\ i_q \end{bmatrix} = \begin{bmatrix} \cos\omega t & -\sin\omega t \\ \sin\omega t & \cos\omega t \end{bmatrix} \begin{bmatrix} \mathbf{i}_\alpha \\ \mathbf{i}_\beta \end{bmatrix} \quad (2)$$

converter and grid current along the grid voltage. The load current will be a composite current containing the fundamental and harmonics. After orientation, the fundamental components of d-axis and q-axis currents are the active and reactive parts respectively of the fundamental load currents. For grid reactive power compensation, this fundamental q-axis load current is used as the reference of q-axis current controller of the converter. There is a closed loop PI controller to maintain the dc bus constant. The output of this controller generates the reference of d-axis current controller of the converter. The control law [6]-[7] along the d and q axis is given below:

$$V_d = i_d R + L\frac{di_d}{dt} - \omega L i_q + V_g \quad (3)$$

$$V_q = i_q R + L\frac{di_q}{dt} - \omega L i_d \quad (4)$$

Here R and L are the resistance and inductance of the series choke. $V_d$ and $V_q$ are d and q axis voltage commands respectively. There are two PI current controllers to control the d-axis and q-axis current of the converter. The outputs of these current controllers are added with feed forward terms based on (3) and (4) to generate the d-axis and q-axis voltage references for the

converter. Finally, d-q voltage references are transformed back to 3-phase stationary voltage references using the unit vectors. These reference signals are fed to the PWM modulator to generate the gate pulses for the converter.

## IV. RESULTS AND DISCUSSIONS

### A. Simulation Results

The above said control algorithm is simulated using Matlab / Simulink simulation software and the following results are obtained.

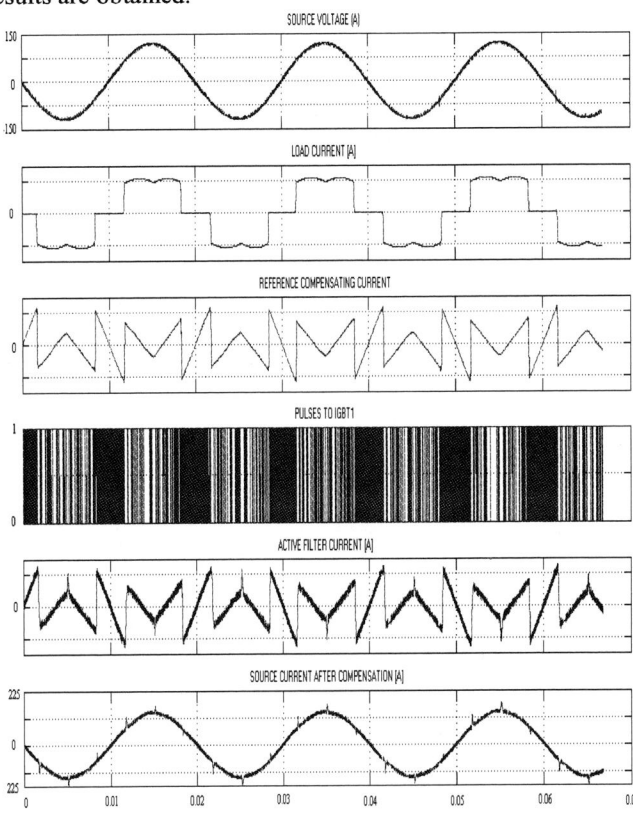

Fig. 4. Phase A scope showing the compensation of harmonics and improved power factor.

The simulation result clearly shows that the control algorithm improves the power factor and at the same time all the higher order harmonics are almost eliminated as in Fig. 4. The simulation results at other two phases namely phase B and phase C have also shown similar results.

### B. Hardware Implementation

The control algorithm explained above has been tested in a IPM (Intellectual Power Module) power electronics module consists of IGBT based inverter circuits, with a series inductor of 0.51mH and a DC bus capacitance of 28200µF. The DC bus voltage is 700V. The power module is air cooled. The filter is programmed to compensate up to 100kVAR.

The digital controller for this system is built around Texas DSP TMS320LF2407A. The digital controller has simultaneous sampling, 12 bit ADC inbuilt with a maximum conversion time of 6.6µsec. The carrier frequency is set to be 10 kHz. The controller incorporates a dead band of 5µsec in the gate pulses for complementary switches of each arm of the IGBT converter. In order to improve noise immunity and isolation, fiber optic links are used to transmit gate signals from the controller to the gate driver of the IGBT converter.

Load current and grid current waveforms are shown in Fig. 5 and Fig. 9. The load condition showing the requirement for compensation is shown in Fig. 6. The measurements shown in Fig. 6 and Fig. 10 show the power factor improvement attained by connecting the STATCOM. The harmonic analysis of grid voltage shown in Fig. 7 shows a THD of 2.4% with fifth harmonics coming up to 1.4%. The harmonic analysis of the load and grid current are shown in Fig. 8 and Fig. 11.

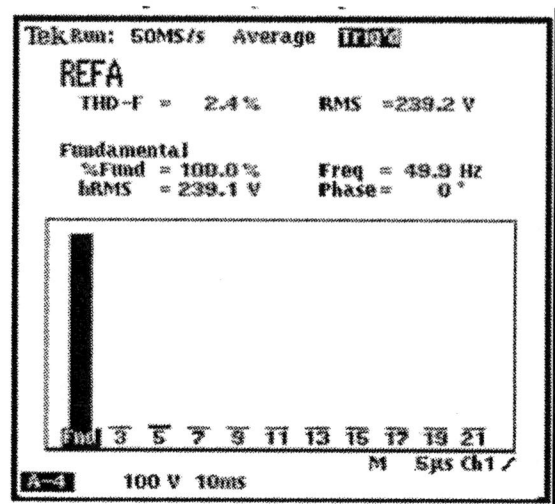

Fig. 7. Harmonic analysis of source voltage.

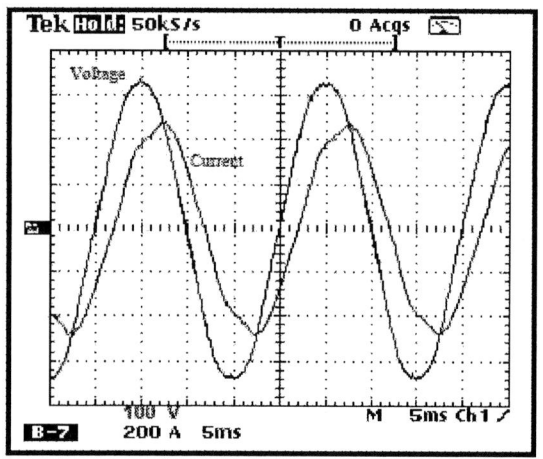

Fig. 5. Load current and grid voltages.

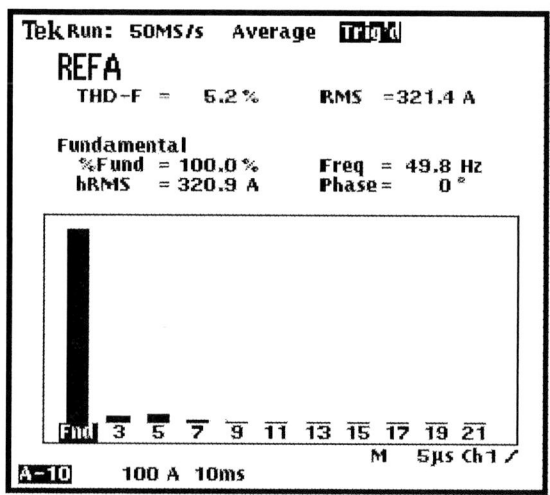

Fig. 8. Harmonic analysis of load current.

Fig. 6. Load condition showing requirement for compensation.

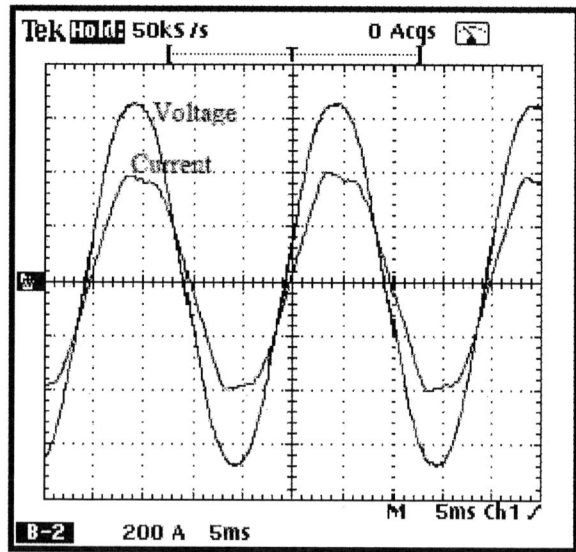

Fig. 9. Compensated grid current with grid voltage.

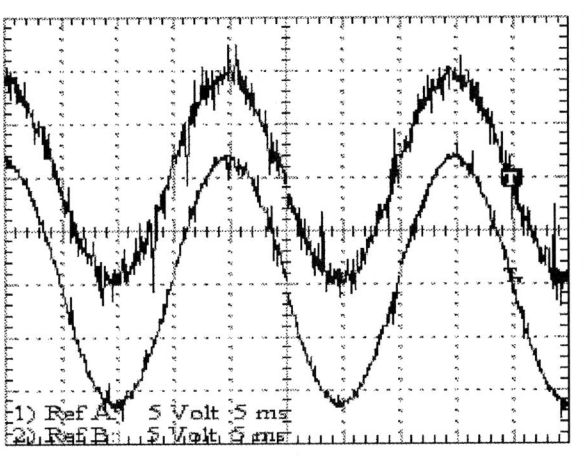

Fig. 13. Sensed grid voltages and extracted unit vector.

## V. CONCLUSION

This paper proposed an improved algorithm to generate unit vector from the sensed grid voltage with a reduced effect of harmonics and noise present in the feedback signal. This simplified algorithm was used in synchronous d-q reference frame method for the control of shunt active power filter. The experimental results established the effectiveness of the proposed algorithm

## VI. REFERENCES

[1] H. Akagi, Y. Kanazawa, A. Nabae, "Instantaneous reactive compensators comprising switching devices without Energy Storage components", IEEE Trans. On Industry Applications.. vol. IA-20, No.3, pp.625-630, May / June, 1984

[2] S. Saetieo R. Devaraj and D.A. Torrey, "TI design and implementation of a three phase active power filter based on sliding mode controoler", in Proc. Inst. Elect. Engg., Generation, Transm., Distrib., Vol.144, Nov. 1997, pp. 564-568

[3] Marthinus G. F. Gous, Hendrik J. Beukes, "Sliding mode control for a three phase shunt active power filter utilizing a four leg voltage source inverter", 35th Annual IEEE power elect. Sp. Conf., Aachen, Germany, 2004

[4] P. S. Sensarma, K. R. Padiyar and V. Ramanarayanan, "A STATCOM for composite power line conditioning" Proc. IEEE Int. Conf. on Industrial Technology, ICIT, Goa, Jan. 2000.

[5] A. Chandra, B. Singh, B.N. Singh and kamal Al- Haddad, " An Improved control algorithm of shunt active power filter for voltage regulation, harmonic elimination, power-factor correction and balancing of non-linear loads", IEEE Trans. Power Electron., Vol.15, No.3, May 2000.

[6] M. Labben-Ben Braiek, F. Fnaiech, Kamal Al-Haddad, and L. Yacobi, "Comparison of Direct current control techniques for a three phase shunt active power filter", Proc. IEEE conf. , 2002.

[7] Gary W. Chang and Tai-Chang Shee,, "A Novel Reference compensation current strategy for shunt active power filter control", IEEE Trans. On Power Delivery, Vol.19, No.4, October,2004.

Fig. 10. Grid condition showing reactive compensation.

Fig. 12. displays the inverter condition showing the reactive power injection. Fig. 13 shows the improvement achieved with the proposed algorithm for the estimation of unit vector.

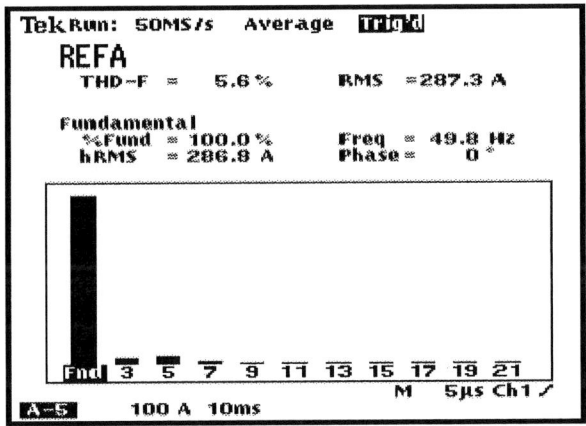

Fig. 11. Harmonic analysis of grid current

Fig. 12. Inverter condition showing reactive power injection

## VII. BIOGRAPHY

 **S. Elangovan** born in 1978 in India. He received bachelor degree in Electrical and Electronics Engineering from University of Madras, Tamilnadu, India, in 1999 and the M.E. degree in Power Electronics and drives from Government college of Technology, Coimbatore, Tamilnadu, India in 2006. Currently, he is faculty in the Department of Electrical Engineering, Coimbatore Institute of Engineering and Information Technology, Coimbatore, India. He is going to register for the Ph.D. Degree in power electronic from Government college of Institute of Technology, Coimbatore, India. His research interests are in the area of power quality, shunt active power filters, and direct torque control of induction motors.

**2006 IEEE International Conference on Power Electronic, Drives and Energy Systems**

# A Novel Digital Signal Processing Algorithm for On-line Assessment of Power System Frequency

Arghya Sarkar, *Member, IEEE*, and S. Sengupta, *Member, IEEE*

*Abstract*--An innovative, computationally efficient digital signal processing algorithm has been proposed to evaluate the fundamental frequency of a non-sinusoidal power system signal at every sample instant. This approach adopts backward difference approximation as second derivative FIR filter to estimate the instantaneous frequency. Besides this, an efficient square root algorithm, based on iterative Newton-Raphson Inverse (NRI) method is also presented to get high accuracy in fixed point processor. The method is illustrated and evaluated in real time by means of a Texas Instruments (TI) TMS320VC5416 digital signal processor (DSP). The experimental results confirm the validity and accurate performance of the proposed approach even under slow magnitude and frequency variations.

*Index Terms*-- Backward Difference Approximation, Butterworth Filter, Frequency, Measurement, Newton-Raphson Inverse Method, Non-sinusoidal Condition.

## I. INTRODUCTION

FREQUENCY information is one of the most important parameter in power system operation, control and protection. Load shedding, load restoration, generator protection from over speeding, detection of the generation-load out-of- step conditions and EHV line synchronism-checking scheme may in general be based on the small frequency deviation measurements. Moreover, from the instrumentation point of view, frequency measurement takes a great importance for sampling control in analog to digital conversion.

During recent years electric power systems are growing more and more complex. The use of distributed generation, the connection of non-linear loads and the presence of unexpected system faults generate lot of harmonics and noise. It is therefore essential to develop a reliable method that can measure fundamental power system frequency accurately in non-sinusoidal environment.

A variety of techniques and algorithms have been proposed in different literature for real time estimation of power system

frequency. Among them Zero Crossing Detection method or its modification by using the curve fitting of voltage sample [1] is the simplest approach but its accuracy deteriorated at the presence of high harmonic distortion or at transient condition. The widely used frequency assessment techniques are based on Discrete Fourier Transform (DFT) algorithm [2], [3]. However, in applying DFT the phenomena of aliasing, leakage and picketfence effects may lead to inaccurate estimation which are analyzed and investigate in [4], [5]. A revised digital algorithm called Smart Discrete Fourier Transform (SDFT) has been presented in [6] which provides exact solution of power frequency recursively. Although this approach is suitable for measurement of frequency over a wide range, the on-line application requires a trade off between accuracy and computational complexity. *Extended Kalman Filtering [7], Least Mean Square (LMS) algorithm [8], [9] and Wavelet Transform based algorithm [10] are examples of some well known alternative approach.* However accuracy of these methods have been influenced by one or more of the following factors: superimposed noise, non-linear static characteristic, computational complexity and slow response. The search for more accurate, computationally simple and robust algorithms still continues.

In this paper a simple and computationally efficient digital signal processing algorithm has been proposed that can measure the fundamental frequency of a distorted power signal accurately at every sample instant. The algorithm is implemented in real-time with a Texas Instruments (TI) TMS320VC5416 digital signal processor (DSP) along with the TI THS1206 12-bit 6 MSPS analog to digital converter. Accurate measurement of instantaneous fundamental frequency, both static and dynamic condition, has been observed utilizing the developed scheme. The results are presented in the text.

## II. FREQUENCY MEASUREMENT ALGORITHM

The development of frequency measurement algorithm involves design of a suitable method in the analog domain and transforming the design into the digital domain. Discretizations of continuous-time signals have been performed in such a way that a stable realizable structure is obtained.

---

Arghya Sarkar is with the MCKV Institute of Engineering, 243 G. T. Road (N), Liluah, Howrah-711204, India (e-mail- sarkararghya@yahoo.co.in)

S. Sengupta is with the Department of Applied Physics, University of Calcutta,92,A. P. C. Road,Kolkata-700009,India(e-mail- samarsgp@vsnl.net).

0-7803-9771-1/06/$25.00 ©2006 IEEE

## A. Frequency Measurement Algorithm in Analog Domain

A pure sinusoidal signal $z(t)$, with fundamental frequency $f_1$ and amplitude $Z_{Max1}$ can be expressed as

$$z(t) = Z_{Max1}\sin(2\pi f_1 t + \alpha_1) \tag{1}$$

The second derivative of $z(t)$ with respect to $t$ is given by

$$z''(t) = -4\pi^2 f_1^2 Z_{Max1}\sin(2\pi f_1 t + \alpha_1) \tag{2}$$

From (1) and (2), the fundamental frequency $f_1$ in analog domain can be obtained as

$$f_1 = \frac{1}{2\pi}\sqrt{-\frac{z''(t)}{z(t)}} \tag{3}$$

## B. Frequency Measurement Algorithm in Digital Domain

In order to get discrete counter part of (3) from non-sinusoidal power signal $x(n)$, the sampled data is first filtered by a bandpass filter to eliminate all harmonics, sub-harmonics and inter-harmonics distortions, keeping the pass band maximally flat. The output of this filter is a pure sinusoidal signal $y(n)$ with fundamental frequency $f_1(n)$ which is fed to a second derivative filter. The square root of ratio of second derivative filter output $y_D(n)$ to band pass filter output $y(n)$ give the fundamental angular frequency $\omega_1(n)$ of non-sinusoidal signal which is transformed to $f_1(n)$ by multiplying a constant $(1/2\pi)$. The block diagram of these discretization efforts has been shown in Fig. 1.

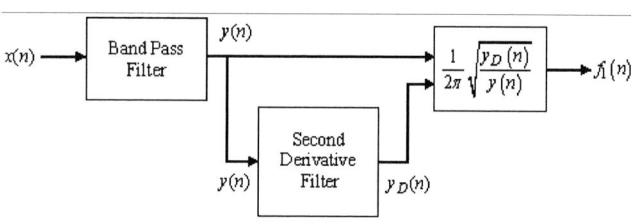

Fig. 1. Block diagram of frequency measurement algorithm.

The design of individual blocks has been given in the subsequent subsections.

### 1) Design of Band Pass Filter

In frequency measurement algorithm, bandpass filter has been utilized to extract fundamental frequency component from non-sinusoidal power signal. Among the various types of IIR and FIR filter the Butterworth characteristics provides the most efficient filters in terms of amplitude response. It yields the smaller filter order for a given set of specifications and should be the method of first choice in this filter design algorithm as it provides a flat response in pass band frequency range with reduced computational cost [11]. Butterworth IIR bandpass filter has been designed [12] with the following specifications:

F$_{stop1}$ = 10 Hz;          F$_{pass1}$ = 40 Hz;
F$_{pass2}$ = 70 Hz;          F$_{stop2}$ = 120 Hz;
Attenuation$_{stop1}$ = 60 dB;   Attenuation$_{stop2}$ = 80 dB.
Fig. 2, shows the amplitude versus frequency response of the

Butterworth bandpass filter for 6400 Hz sampling frequency.

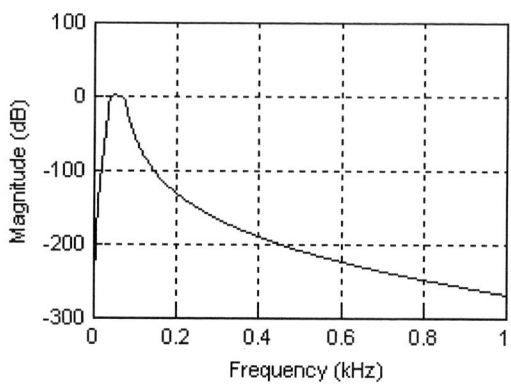

Fig. 2. Magnitude response of bandpass Butterworth filter.

### 2) Design of Second Derivative Filter

The simplest and traditional method of computing an estimate of the second derivative of a discrete time signal is to use the backward difference approximation from the theory of numerical analysis [13].

The continuous time signal $y_c(t)$ can be expanded using Taylor's series in derivative form as

$$y_c(t+\Delta t) = y_c(t) + \Delta t y_c'(t) + \frac{\Delta t^2}{2!} y_c''(t) + \frac{\Delta t^3}{3!} y_c'''(t)$$
$$+ \frac{\Delta t^4}{4!} y_c'''(\theta) \tag{4}$$

where $\Delta t$ is a small increment from a fixed point $t$ and $t \leq \theta \leq t + \Delta t$.

Using step size $2\Delta t$ and $3\Delta t$ in (4), the second derivative of $y_c(t)$ has been derived as

$$y_c''(t) \approx \frac{2y_c(t) - 5y_c(t-\Delta t) + 4y_c(t-2\Delta t) - y_c(t-3\Delta t)}{\Delta t^2} \tag{5}$$

In discrete time system the sample interval $\Delta t$ is small and finite, and the second derivative, at the right hand side of (5) can be replaced by a third order difference equation. Thus evaluating at $t = n\Delta t$ gives

$$y_c''(t)\Big|_{t=n\Delta t} \approx \frac{1}{\Delta t^2}\big[2y_c(n\Delta t) - 5y_c([n-1]\Delta t) + 4y_c([n-2]\Delta t)$$
$$- y_c([n-3]\Delta t)\big] \tag{6}$$

Considering $y(n) = y_c(n\Delta t)$, (5) can be written in following form in discrete time system

$$y''(n) \approx \frac{2y(n) - 5y(n-1) + 4y(n-2) - y(n-3)}{\Delta t^2} \tag{7}$$

$$y_D = -y''(n) \approx \frac{-2y(n) + 5y(n-1) - 4y(n-2) + y(n-3)}{\Delta t^2} \tag{8}$$

The backward difference approximation (8) is actually a stable, causal FIR filter of order three and can be expressed in terms of FIR filter coefficient as:

$$y_D(n) \approx \sum_{r=0}^{3} h(r)y(n-r) \tag{9}$$

where the filter coefficient

$$h(r) = \left[ \frac{-2}{\Delta t^2} \quad \frac{5}{\Delta t^2} \quad \frac{-4}{\Delta t^2} \quad \frac{1}{\Delta t^2} \right] \quad (10)$$

The designed second derivative filter can easily be realized using conventional cascade structure.

The error introduced in this second derivative approximation is due to truncation error $E_{trun}$ in Taylor's series expansion and round off noise $E_{round}$ in fixed point implementation. The total error $E$ of this approximation is given by

$$E = E_{trun} + E_{round}$$
$$= \frac{11\Delta t^2}{12} f'''(\theta) + \frac{12e}{\Delta t^2} \quad (11)$$

From (11) it has been observed that with the increase of $\Delta t$ the truncation error increases while the round off error decreases. This is illustrated in Fig. 3. For small value of $\Delta t$, roundoff error has an overriding influence on the total error.

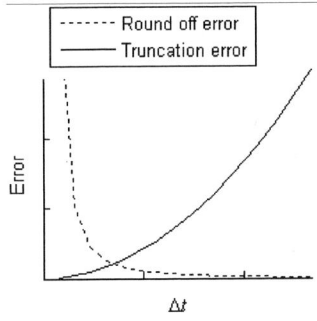

Fig. 3. Error in second derivative as a function of sampling frequency.

From (3) and (8), the fundamental frequency $f_1(n)$ of non-sinusoidal signal $x(n)$ at sampling instant $n$ can be obtained as:

$$f_1(n) = \begin{cases} A(n) & \text{if } y(n) \neq 0 \\ f_1(n-1) & \text{if } y(n) = 0 \end{cases} \quad (12)$$

where $A(n) = \dfrac{1}{2\pi} \sqrt{\dfrac{\sum\limits_{r=0}^{3} h(r)y(n-r)}{y(n)}} = \dfrac{1}{2\pi} \sqrt{\dfrac{y_D(n)}{y(n)}} \quad (13)$

From (11) and (13), it has been observed that obtained frequency is instantaneous, independent of zero-crossing of the measured signal and insensitive to white noise and harmonic pollution.

*3) Square Root Algorithm*

A venerable algorithm for computing the square root of a function is "Newton-Raphson Inverse" (NRI) iterative technique. This approach provides a fast and accurate convergence than other available algorithms if the initial guess is close to the required root [14]. To apply NRI method effectively in frequency measurement algorithm $A(n)$ is rearranged as follows

$$A(n) = \frac{y_D(n)}{2\pi} \frac{1}{\sqrt{y(n)y_D(n)}} \quad (14)$$

In NRI method the output of $1/\sqrt{y(n)y_D(n)}$ looks more linear and accurate when the input value $y(n)y_D(n)$ is restricted to range $0.25 < y(n)y_D(n) < 1$ [15]. In fixed-point format with sign and fractional bits, normalization of input value greater than one, using conventional arithmetic left shift method, increases the error by a factor of two per bit. In this paper a modified normalization method has been proposed that can be implemented efficiently in fixed point arithmetic. Since the maximum measurable fundamental frequency by the proposed scheme is considered to be 64 Hz, $y(n)y_D(n)$ is first multiplied by $6.1842 \times 10^{-6}$ so that resultant value is less than one. The output value is again multiplied by $3^m$ ($m$ is a constant term) is until $0.25 < 6.1842 \times 10^{-6} \times 3^m \times y(n)y_D(n) < 1$. After calculation of square root of this normalized input value and multiplying this by $y_D(n)$, the output result is denormalised by a factor, obtained from a small look up table (LUT), to acquire the frequency information. In this situation $A(n)$ takes the following form

$$A(n) = q(m)y_D(n)\frac{1}{\sqrt{G(n)}} \quad (15)$$

where $G(n) = 0.0089 \times 3^m \times y(n)y_D(n) \quad (16)$

and $q(m) = 3.9579 \times 3^{m/2} \quad (17)$

Using NRI method the output of $1/\sqrt{G(n)}$ can be approximated as

$$p(k+1) = 0.5p(k)\left[3 - G(n)p^2(k)\right] \quad (18)$$

where the variable k is the iteration index.

The initial guessed value $p(0)$, as defined by [15], is used in the first iteration

$$p(0) = \frac{1}{\left(\dfrac{2G(n)}{3} + 0.354167\right)} \quad (19)$$

A perfect realization of this algorithm has been shown in Fig. 4.

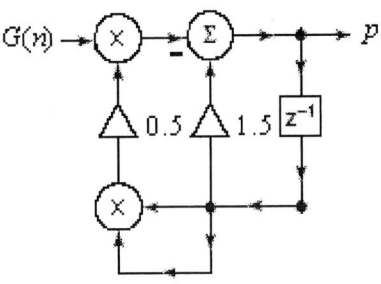

Fig. 4. Realization of the NRI method.

The error curve for two-iteration NRI square root method is shown in Fig. 5, which depicts the maximum error is roughly 0.0008% within this specified range.

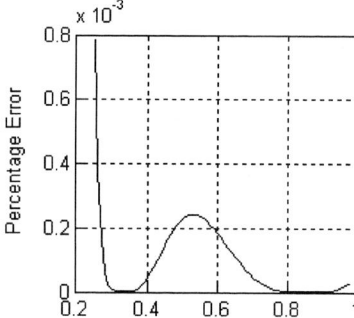

Fig. 5. Percentage error in square root calculation using NRI method.

## III. DSP FEATURES

The TMS320VC5402 fixed-point, digital signal processor (DSP) is based on an advanced modified Harvard architecture that has one program memory bus and three data memory buses. This processor provides an arithmetic logic unit (ALU), application-specific hardware logic, on-chip memory, and additional on-chip peripherals. The basis of the operational flexibility and speed of this DSP is a highly specialized instruction set. Separate program and data spaces allow simultaneous access to program instructions and data, providing the high degree of parallelism. Two read operations and one write operation can be performed in a single cycle. Instructions with parallel store and application-specific instructions can fully utilize this architecture. While floating point multipliers on other processors may allow direct multiplication of floating point values, this DSP processor executes single clock cycle integer multiplications. Optimization to use all integers is therefore necessary. However, if a loss of precision is allowable, this processor will actually execute an integer multiplication faster than a floating-point processor of a similar clock speed due to the parallel multiplier and accumulation units in the place of a pipelined multiplier [16].

## IV. PERFORMANCE ANALYSIS

The validity and performance of the proposed frequency monitoring system as shown in Fig. 1 has been observed by applying this method to measure the instantaneous frequency under static, dynamic, harmonic and noisy conditions. A programmable ac power source/power analyzer Agilent 6812B [17] has been utilized to generate required waveforms in different experimentation. The measuring signals are scaled down in such a fashion that the maximum values are less than one volt. These attenuated signals are fed to a Texas Instruments (TI) TMS320VC5416 digital signal processor (DSP) through the TI THS1206 12-bit 6 MSPS analog to digital converter at sampling frequency 6.4 kHz.

### A. Performance under static sinusoidal condition

The performance of proposed approach at stationary condition has been evaluated using static sinusoids ranging from 40 Hz to 64 Hz. The variation of percentage error with respect to time and frequency has been shown in Fig. 6, which reveals that the error exists within the maximum allowable range. The sources of errors in the above experiments have been identified as (a) Current transformer characteristics (b) Quantization error in analog to digital conversion and (c) Quantization error, truncation error and round off noise in the calculation part of the system.

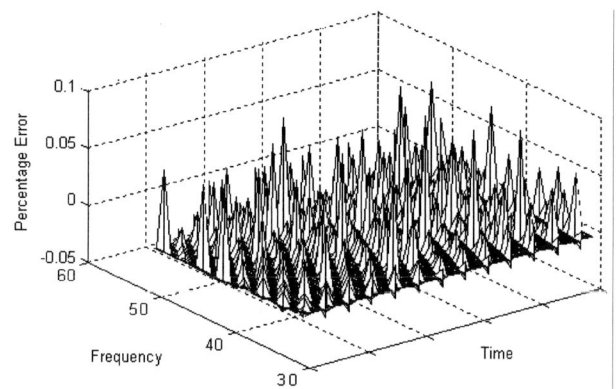

Fig. 6. Variation of percentage error with respect to time and frequency in case of sinusoidal signal.

### B. Performance under static non-sinusoidal condition

The developed algorithm has been tested using a stationary non-sinusoidal signal with 50 Hz fundamental frequency and harmonics of orders three (50 %) and five (30 %).

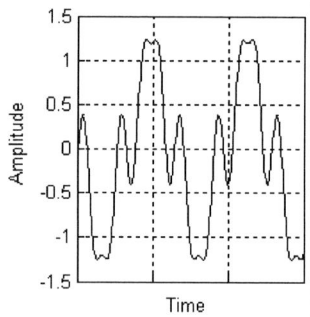

Fig. 7. Waveform of non-sinusoidal signal.

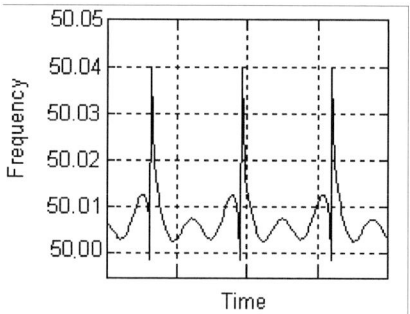

Fig. 8. Estimated frequency in case of non-sinusoidal signal.

The signal waveform and measured values have been presented in Fig. 7, and Fig. 8, respectively. From the evaluation result, it is explicit that a high accuracy and high rejection to harmonics can be achieved utilizing the proposed scheme.

### C. Performance under dynamic condition

#### 1) Performance during magnitude variance

The purpose of this test is to examine the effects of gradually varied signal magnitude to the frequency evaluation. A sinusoid with fixed 50 Hz frequency and gradually increasing amplitude has been used to perform this test. The measured frequency has been shown in Fig. 9, which demonstrates an acceptable transient response of the proposed approach.

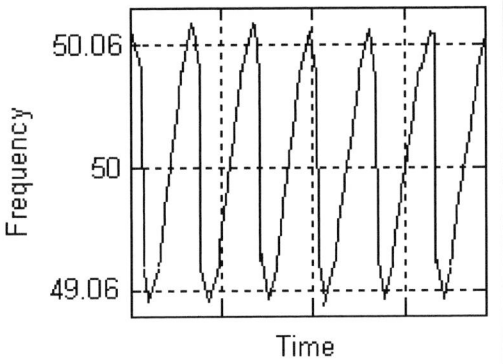

Fig. 9. Estimated frequency in case of slow magnitude variation.

#### 2) Performance during frequency variance

The proposed technique has been used to estimate the frequency of a fixed amplitude sinusoidal transient signal, of which the fundamental frequency gradually increases from 48 Hz to 52 Hz at 1 sec. Frequency estimation plot as shown in Fig. 10, indicates fast and accurate convergence of the results.

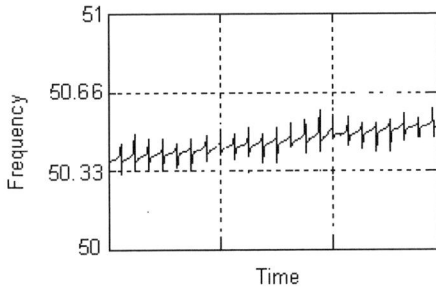

Fig. 10. Estimated frequency in case of slow frequency variation.

### D. Performance in presence of noise signal

As the response of original signal which consist 50 Hz fundamental component and zero mean, variance one Gaussian noise, the estimated frequency is depicted in Fig. 12, which exhibit that the proposed approach possesses a fair rejection to the noise signal. The signal waveform has been depicted in Fig. 11.

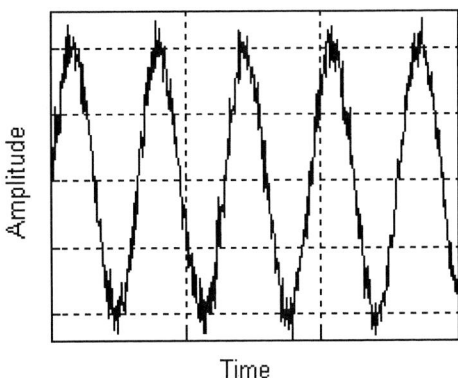

Fig. 11. Waveform of noisy sinusoidal signal.

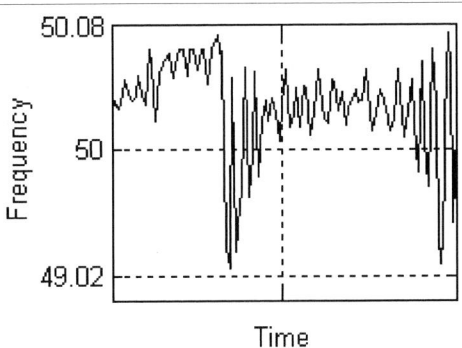

Fig. 12. Estimated frequency in case of noisy sinusoidal signal.

Although the developed system provides significant accuracy in slow variation of magnitude and frequency of the measuring signal, the application of the said technique has been restricted during abrupt changes in magnitude or frequency.

## V. CONCLUSIONS

The paper presents a new digital signal processing algorithm for estimation of fundamental frequency of distorted power signal at each sampling instant. The proposed scheme has been designed for a measurable fundamental frequency ranging from 40 Hz to 64 Hz. The backward difference approximation provides a very pragmatic and promising approach for fast estimation of the second derivative. For a given number of data points, though the central difference formula is more accurate than their forward or backward counter parts, from the causality stand point, backward difference formula is much more acceptable. Proposed NRI based square root algorithm provides a well tradeoff between accuracy and computational complexity in fixed point Digital Signal Processor.

The real time laboratory test results reveal that the proposed method can successfully be employed for on line monitoring of system fundamental frequency even at the presence of harmonics or low frequency and amplitude variations. The developed scheme is accurate, fast, computationally efficient, highly reliable and possesses enough flexibility to suit the requirement of different power system operation, control and protection systems.

## VI. REFERENCES

[1] M. M Begovic, and P. M. Djuric, "Frequency tracking in power networks in the presence of harmonics," *IEEE Trans. Power Delivery*, vol. 8, no. 2, pp. 480-486, 1993.

[2] M. Kezunovic, and P. Spasojevic, "New digital signal processing algorithms for Frequency Deviation Measurement," *IEEE Trans. Power Delivery*, vo1.7, pp. 1563-1573 , July 1992.

[3] D. Agrez, "Frequency estimation of the non-stationary signals using interpolated DFT," in *Proc. IEEE Instrumentation and Measurement Technology Conference*, pp. 925-930, 2002.

[4] T. S. Sidhu, and M. S. Sachdev, "An iterative technique for fast and accurate measurement of power system frequency," *IEEE Trans. Power Delivery*, vol. 13, no. 1, pp. 109-115, 1998.

[5] J. A. Iao, H. J. Altuve and I. Diaz, "A new digital filter for phasor computation—part1: theory," *IEEE Trans. Power Delivery*, vol. 13, no. 3, pp. 1026-1031, 1998.

[6] J. Z. Yang, and C. W. Liu, "A precise calculation of power system frequency and phasor," *IEEE Trans. Power Delivery*, vol. 15, no. 2, pp. 494-499, 2000.

[7] A. A. Girgis, and W. L. Peterson, "Adaptive estimation of power system frequency deviation and its rate of change for calculating sudden power system overloads," *IEEE Trans Power Delivery*, vol. 5, no. 2, pp. 585–594, Apr. 1990.

[8] I. Kamwa, and R. Grondin, "Fast adaptive schemes for tracking voltage phasor and local frequency in power transmission and distribution systems," *IEEE Trans. on Power Delivery*, vol. 7, no. 2, pp. 789–795, Apr. 1992.

[9] M. S. Sachdev, and M. M. Giray, "A least error squares technique for determining power system frequency, " *IEEE Trans. on Power Apparatus and Systems*, vol. PAS-104, no. 2, pp. 437–443, Feb. 1985.

[10] Lin, T.; Tsuji, M.; Yamada, E.; "A Wavelet approach to real time estimation of power system frequency," in *Proc. 40th SICE Annual Conference*, pp 58-65, Nagoya, 2001.

[11] A. Oppenheim, and R. Schafer, *Discrete-Time Signal Processing*, Englewood Cliffs, NJ: Prentice-Hall, 1989.

[12] T. W. Parks, C. S. Burrus, *Digital Filter Design*, New York: Wiley, 1987.

[13] T. J. Cavicchi, *Digital Signal Processing*, John Wiley and Sons, 2000.

[14] Mark Allie, and Richard Lyons, "A root of less evil," IEEE Signal Processing Magazine, pp. 94-96, March, 2006.

[15] E. Balagurusamy, *Numerical Methods*, Tata McGraw-Hill Publishing Company Limited, New Delhi, 2000.

[16] Texas Instruments, TMS320C54x DSP Reference Set (Vol-I, II, III, IV), June 1999.

[17] Agilent Technologies, User's Guide AC Power Solutions Agilent Models 6811B, 6812B, and 6813B, September 2004.

## VII. BIOGRAPHIES

**Arghya Sarkar** was born in West Bengal in India, on December 25, 1974. He obtained B. Sc. (Physics Honours), B. Tech. and M.Tech in Electrical Engineering from University of Calcutta, Kolkata, India.

His employment experience included the MCKV Institute of Engineering, Liluah, Howrah, West Bengal, India. His special fields of interest include Power Quality, DSP and Control System.

**Samarjit Sengupta (***MIEEE***)** obtained B. Sc. (Physics Honours), B. Tech., M.Tech and Ph.D. in Electrical Engineering from University of Calcutta, Kolkata, India.

His employment experience includes about nine years in different industries and the about sixteen years of teaching and research in the Department of Applied Physics, University of Calcutta, Kolkata, West Bengal, India. His special fields of interest include Power Quality and power system protection.

**2006 IEEE International Conference on Power Electronic, Drives and Energy Systems**

# An Evolutionary Algorithm Approach to Estimate the Parameters of Power Quality Signals

V. Ravikumar Pandi, and B. K. Panigrahi, *Member, IEEE*

*Abstract--*This paper presents an evolutionary algorithm approach to determine the amplitude, phase and frequency of a power quality signal. Genetic Algorithm (GA) and Particle Swarm Optimization (PSO) are the two evolutionary algorithms adopted to extract the above parameters of the power quality signal.

*IndexTerms--*Genetic Algorithm, Particle Swarm Optimization, Power Quality, Voltage flicker

## I. INTRODUCTION

THE estimation of amplitude and phase of fundamental, as well as harmonic signals has been one of the important tasks in measurement, control, relaying protection, distribution automation, and intelligent instrumentation of power system, such as power metering. Accurate power fundamental frequency is a necessity to check the state of health of the power index, and a guarantee for accurate quantitative measurement of power parameters, such as voltages, currents, active power, and energy, reactive power, and energy, and so on, in multifunction power meters under steady states. Lots of harmonics and noises in power systems are generated, owing to the increasing use of power-electronic devices and controllers in power transmission, industrial processes, and drives. It is more difficult to precisely estimate the fundamental frequency of power systems in presence of harmonics and noises than under sinusoidal condition. It is essential to seek and develop some effective algorithms for accurate estimation of the instantaneous fundamental frequency of power systems under non-sinusoidal conditions. Several methods like Fast Fourier Transform (FFT), Least Mean Square (LMS) estimation [1], Least Absolute Value (LAV) estimation are more common to estimate the above parameters. Besides that soft computing technique like Artificial Neural Network (ANN), Expert Systems (ES), and Genetic Algorithm (GA) [2-4] etc. have been applied for the purpose. In [5], the authors proposed an adaptive linear combiner for tracking the harmonics.

In this paper we have tried to extract the amplitude, phase of fundamental as well as that of the harmonics present in a power quality signal using two of the Evolutionary Algorithm (EA) techniques like, Genetic Algorithm (GA) and Particle Swarm Optimization (PSO).

---

V. Ravikumar Pandi is with Department of Electrical Engineering, IIT, Delhi, India (e-mail: ravikumarpandi@gmail.com).

B.K.Panigrahi is with Department of Electrical Engineering, IIT, Delhi, India (e-mail: bkpanigrahi@ee.iitd.ac.in ).

## II. MATHEMATICAL FORMULATION

The aim of the present paper is to find the amplitude, phase and frequency components present in a power quality signal. As a first step to look into the problem the power quality data is collected. Let $f_s$ is the sampling frequency for the digital data to be collected and N is the number of data points in one cycle of the signal. The objective function for the problem can be stated as

$$Min\ F = \sqrt{\frac{\sum_{i=1}^{k}(v(i) - \bar{v}(i))^2}{k}} \qquad (1)$$

Where k is the no of samples, v(i) is the estimated sample value at $i^{th}$ time interval and $\bar{v}$(i) is the actual sample value given in to the algorithm.

## III. OVERVIEW OF GA AND PSO

Genetic algorithms (GAs) are search algorithm based on the concept of natural selection [6]. GA is used for many non-linear optimization problems. J.Kennedy and R.C.Eberhart introduce a concept for the optimization of nonlinear functions using particle swarm methodology. The performance of particle swarm optimization using an inertia weight is compared with performance using a constriction factor is also explained. Developments and resources in the particle swarm algorithm are reviewed in [7]. Some improvements in the PSO algorithm is proposed in [8].

Genetic algorithms are exploratory search methods based on mechanics of selection and survival of fittest. They operate on string structures called chromosomes, typically a concatenated list of binary digits representing a encoding of the control parameters of a given problem. It works with a population of individuals and decisions taken are based on probabilistic rules. Generally, GAs consists of three basic operations, namely reproduction, crossover and mutation.

Reproduction comprises forming a new population, usually with the same total number of chromosomes, by selecting from members of the current population following a particular scheme. The higher the fitness, the more likely it is that the chromosome will be selected for the next generation. There are several strategies for selecting the individuals, example roulette-wheel selection, ranking methods and tournament selection. Here we use tournament selection. In tournament selection, 'n' individuals are selected at random manner from the population and the best among the 'n' is inserted into the new population for further genetic processing. This procedure is repeated until the mating pool is filled. The crossover

0-7803-9771-1/06/$25.00 ©2006 IEEE

operator is mainly responsible for the global search property of the GA. Two child strings are then generated from the parent strings in the process of crossover by complementing the child strings at selected bit positions in order to exchange the already existing information. Mutation is then applied on some of strings to introduce new information in the mating pool, with small probability. In order to maintain the best chromosome found so far, elitism is performed. It always maintains the best chromosome called elites in the mating pool.

Similar to other evolutionary algorithms, the Particle Swarm Optimization method conducts searches using a population of particles, corresponding to individuals. Each particle in the swarm represents a candidate solution to the problem. It starts with a random initialization of a population of individuals in the search space and works on the social behavior of the particles in the swarm like bird flocking, fish schooling and the swarm theory. Therefore, it finds the global optimum by simply adjusting the trajectory of each individual towards its own best location and towards the best particle of the swarm at each generation of evolution. However, the trajectory of each individual in the search space is adjusted by dynamically altering the velocity of each particle, according to the flying experience of its own and the other particles in the search space. This population based robust algorithm always ensures the convergence to the global optimum solution when compared to GA.

A. *Advantages of PSO*

1) PSO is easy to implement and there are few parameters to adjust.

2) Unlike GA, PSO has no evolution operators such as crossover and mutation.

3) In GAs chromosomes share information, so that the whole population moves like a one group, but in PSO only Gbest gives out information to others. It is more robust than that of GA.

4) PSO can be more efficient than GAs; that is, PSO often finds the solutions with fewer objective function evaluations than are required by GAs.

5) PSO uses payoff (performance index or objective function) information to guide the search in the problem space.

6) Unlike GA and other heuristic algorithms, PSO has the flexibility to control the balance between the global and local exploration of the search space. This unique feature of PSO overcomes the premature convergence problem and enhances the search capability.

B. *PSO Algorithm*

The position and the velocity of the $i^{th}$ particle in the d-dimensional search space can be represented as $Xi = [x_{i1}, x_{i2}, . . . , x_{id}]^T$ and $Vi = [v_{i1}, v_{i2}, . . . , v_{id}]^T$, respectively. Each particle has its own best position (pbest) $Pi(t) = [p_{i1}(t), p_{i2}(t), . . . , p_{id}(t)]^T$ corresponding to the personal best objective value obtained so far at generation 't'. The global best particle (Gbest) is denoted by $Pg(t) = [p_{g1}(t), p_{g2}(t), . . . , p_{gd}(t)]^T$, which represents the best particle found so far at generation 't' in the entire swarm. The new velocity of each particle is calculated as follows:

$$v_{ij}(t+1) = \omega v_{ij}(t) + c_1 r_1 (p_{ij}(t) - x_{ij}(t)) + c_2 r_2 (p_{gj}(t) - x_{ij}(t))$$
$$j = 1, 2, ..., d; \quad i = 1, 2, ..., n \tag{2}$$

Where $c_1$ and $c_2$ are constants named acceleration coefficients corresponding to the cognitive and social behavior, $\omega$ is called the inertia factor, n is the population size, $r_1$ and $r_2$ are two independent random numbers uniformly distributed in the range of [0, 1]. Thus, the position of each particle at each generation is updated according to the following equation:

$$x_{ij}(t+1) = x_{ij}(t) + v_{ij}(t+1)$$
$$i = 1, 2, ........n \text{ and } j = 1, 2, ........d \tag{3}$$

Equation (2) shows that the new velocity is updated according to its previous velocity and to the distance of its current position from both its best historical position and the global best position of the swarm. Generally, the value of each component in Vi can be clamped to the range $[V_{imin}, V_{imax}]$ to control excessive roaming of particles outside the search space $[X_{imin}, X_{imax}]$. Then the particle flies towards a new position according to (3). The process is repeated until a user-defined stopping criterion is reached.

In simple PSO method the inertia weight is made constant for all the particles in a single generation. But the most important parameter that moves the current position towards the optimum position is inertia weight ($\omega$). In order to increase the search ability, the algorithm should be redefined in the manner that the movement of swarm should be controlled by the objective function. In our Adaptive PSO, the fine adjustment of particle position is done for best particle in order to move the highly fitted particle (in our case minimization of objective function) slowly when compared to low fitted particle, for that select the different $\omega$ values for each particles according to their fitness function values between $\omega_{max}$ and $\omega_{min}$ as in the following form.

$$\omega = (\omega_{max}, ...., \omega_{min}) \tag{4}$$

Now the velocity of each particle is updated using (5), and if any updated velocity goes beyond Vmax, then limit it to Vmax using (6).

$$v_{ij}(t+1) = \omega_i v_{ij}(t) + c_1 r_1 (p_{ij}(t) - x_{ij}(t)) + c_2 r_2 (p_{gj}(t) - x_{ij}(t)) \tag{5}$$

$$v_{ij}(t+1) = sign(v_{ij}(t+1)) * min(|v_{ij}(t+1)|, V_{jmax}) \tag{6}$$

$$j = 1, 2, ..., d; \quad i = 1, 2, ..., n$$

The new particle position is obtained by using the (7) and if any particle position goes beyond the range specified, then it is adjusted to its boundary using (8).

$$x_{ij}(t+1) = x_{ij}(t) + v_{ij}(t+1), \quad j = 1, 2, ..., d; \quad i = 1, 2, ..., n \tag{7}$$

$$x_{ij}(t+1) = min(x_{ij}(t+1), range_{jmax}),$$
$$x_{ij}(t+1) = max(x_{ij}(t+1), range_{jmin}) \tag{8}$$

## C. The procedure of Adaptive PSO

Step 1: Get the input parameters like range [min max] for each variables, $c_1$, $c_2$, samples of the signal, Iteration counter=0, $V_{max}$, $\omega_{min}$ and $\omega_{max}$.

Step 2: Initialize n number of population of particles of dimension d with random positions and velocities.

Step 3: Increment Iteration counter by one.

Step 4: Evaluate the fitness function of all particles in the population, find particles best position pbest of each particle and update its objective value. Similarly find the global best position among all particles and update its objective value.

Step 5: If stopping criterion is met go to step (10). Otherwise continue.

Step 6: Evaluate the inertia factor according to (4), so that each particles movement is directly controlled by its fitness value.

Step 7: Update the velocity using (5) and correct it if $V_{new} > V_{max}$ using (6).

Step 8: Update the position of each particle according to (7) and if new position goes out of range set it to the boundary value using (8).

Step 9: The Elites are inserted in the first position of the new population in order to maintain the best particle found so far. Go to step (3).

Step 10: Output the Gbest particle and its objective value.

## IV. SIMULATION RESULTS

### A. Simulation of the Voltage Flicker Signal

The proposed APSO is simulated for estimating the flicker frequency and its magnitude and the result was compared with Genetic Algorithm approach.

$$V(t) = 1.0\cos(2\pi 50t) + 0.1\cos(2\pi 2t)\cos(2\pi 50t) \quad (9)$$

Equation (9) is taken to produce the samples where $A_1$=1.0, $A_{fl}$=0.1, $f_1$=50 and $f_{fl}$=2. In order to find the effectiveness of each algorithm we varied the number of samples as 16, 32, 64, 128 and 256. The estimation of these parameters for different samples is shown in Fig. 1-3, and the values are compared in Table I for 32 sample case. The original and reconstructed waveform using APSO is shown in Fig. 4. The percentage error is also given in Table I for these methods. The actual, 10dB white Gaussian Noise added waveform and reconstructed waveform is shown in Fig. 5. From the results shown, the proposed Adaptive particle swarm optimization works better than GA at all conditions.

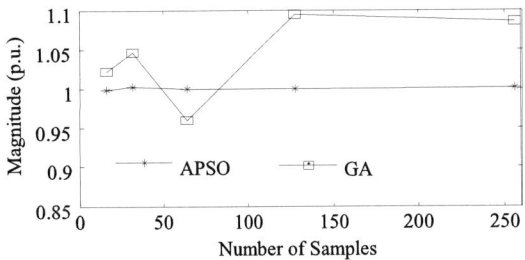

Fig. 1. Estimated value of Fundamental Magnitude.

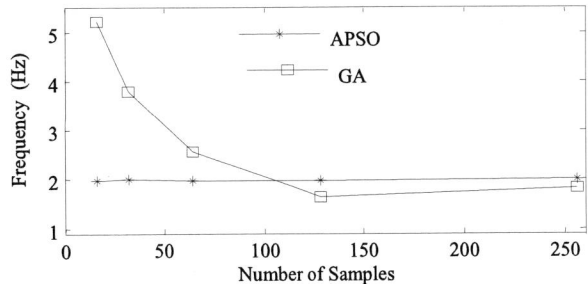

Fig. 2. Estimated value of Flicker Frequency.

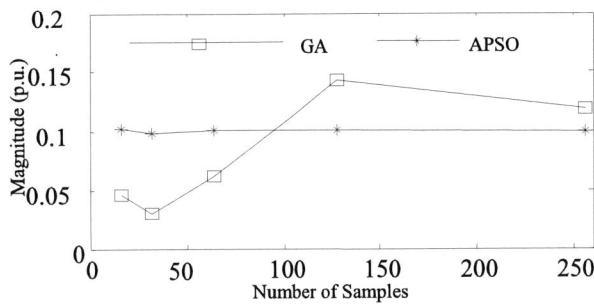

Fig. 3. Estimated value of Flicker Magnitude.

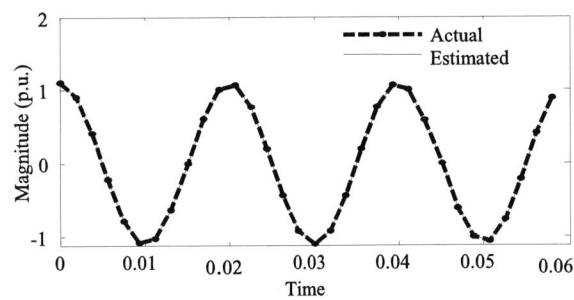

Fig. 4. Estimated Voltage Flicker Signal.

TABLE I
COMPARISON OF ESTIMATED VALUES FOR FLICKER SIGNAL [FOR 32 SAMPLES]

| Parameter | Actual Value | GA | | APSO | |
|---|---|---|---|---|---|
| | | Est. | Err % | Est. | Err % |
| $A_1$ | 1 | 1.0457 | 4.57 | 1.0013 | 0.13 |
| $A_{fl}$ | 0.1 | 0.0547 | 45.3 | 0.0987 | 1.3 |
| $f_1$ | 50 | 51.98 | 3.96 | 50 | 0.00 |
| $f_{fl}$ | 2 | 2.7839 | 39.2 | 2.0137 | 0.685 |

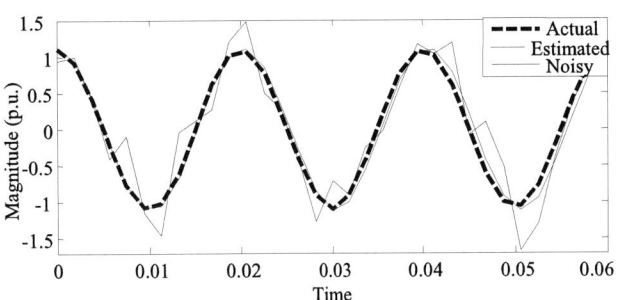

Fig. 5. Estimated Voltage Flicker Signal for 10dB Noise.

## B. Simulation of Harmonic Voltage Signal

Harmonic voltage signal model is considered as

$$v(t) = a\,e^{-bt} + \sum_{i=1}^{m} A_i \cos(\omega_i t + \phi_i) \qquad (10)$$

where $i = 1$ for the fundamental component and equals 2,3,..,m for harmonic components, m is the number of harmonics considered. The actual values taken for producing the test signal are as follows. a=0.2, b=0.02, $A_1$=1.414, $A_3$=0.707, $A_5$=0.3535, $\varphi_1$=60, $\varphi_3$=30, $\varphi_5$=25 and $\omega_1$=50. Here also for analyzing the efficiency of the algorithm different sampling rates are selected like 16, 32, 64, 128 and 256.

The results are shown in Fig. 6-17. Fig. 6 shows the Error comparison, it is observed that the proposed method converges more quickly for the error of the order of $1e^{-5}$. In order to ensure that the algorithm works for as many runs as possible the results of 50 different runs was shown in Fig. 7 and Fig. 8.

TABLE II
ESTIMATED VALUES OF AMPLITUDE, PHASE AND FREQUENCY OF HARMONIC SIGNAL [32 SAMPLES]

| Para meter | Act. Val | GA | | APSO | |
|---|---|---|---|---|---|
| | | Est. | Err % | Est. | Err % |
| $A_1$ | 1.414 | 1.4364 | 1.58 | 1.414 | 0.00 |
| $A_3$ | 0.707 | 0.6904 | 2.35 | 0.7071 | 0.01 |
| $A_5$ | 0.3535 | 0.2983 | 15.62 | 0.3536 | 0.03 |
| $\Phi_1$ | 60 | 69.8824 | 16.47 | 60 | 0.00 |
| $\Phi_3$ | 30 | 43.0592 | 43.53 | 30 | 0.00 |
| $\Phi_5$ | 25 | 45.8823 | 83.53 | 25 | 0.00 |
| $f_1$ | 50 | 48.725 | 2.55 | 50 | 0.00 |
| a | 0.2 | 0.2875 | 43.75 | 0.2 | 0.00 |
| b | 0.02 | 0.04 | 100.0 | 0.0204 | 2.00 |

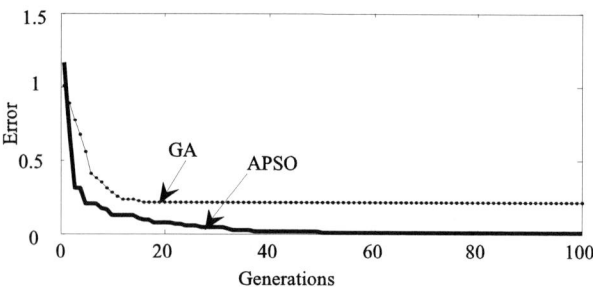

Fig. 6. Comparison of Error Vs Generation [for 40 dB White Gaussian noise level with values as in Table III].

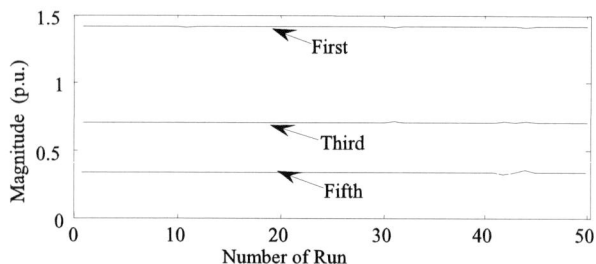

Fig.7. Estimate Magnitudes of 1st, 3rd & 5th Harmonic.

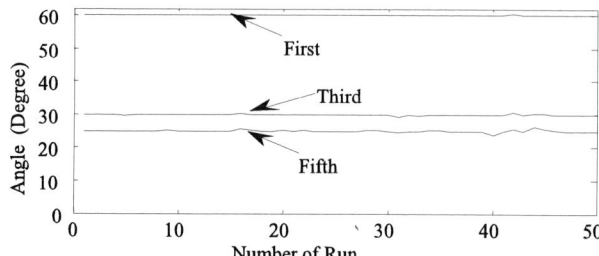

Fig. 8. Estimate Angles of 1st, 3rd & 5th Harmonic.

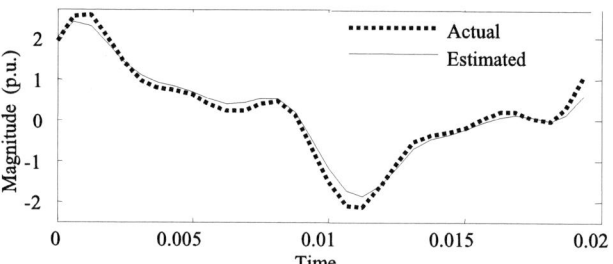

Fig. 9. Estimated Signal for 10 dB noise.

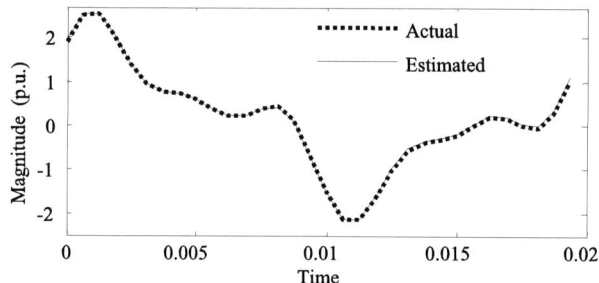

Fig. 10. Estimated Signal for 20 dB noise.

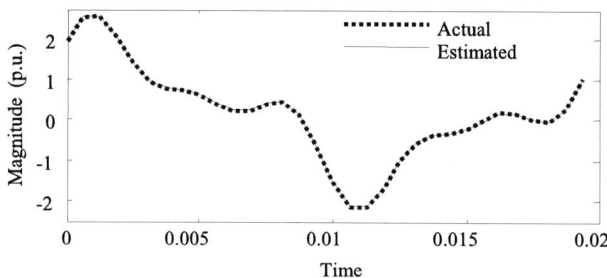

Fig. 11. Estimated Signal for 30 dB noise.

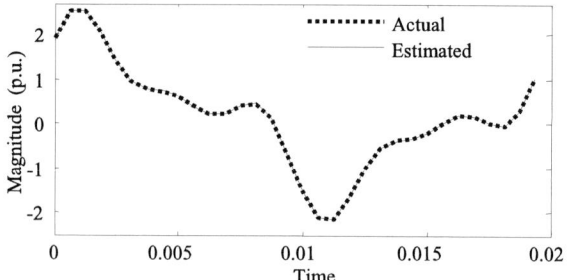

Fig. 12. Estimated Signal for 40 dB noise.

TABLE III
COMPARISON OF ACTUAL AND ESTIMATED PARAMETERS USING APSO FOR VARIOUS WHITE GAUSSIAN NOISE LEVELS

| Noise Level | $A_1$ | $A_3$ | $A_5$ | $\Phi_1$ | $\Phi_3$ | $\Phi_5$ | $f_1$ | a | b |
|---|---|---|---|---|---|---|---|---|---|
| Actual Value | 1.414 | 0.707 | 0.353 | 60.0 | 30.0 | 25.0 | 50.0 | 0.2 | 0.02 |
| 40 | 1.412 | 0.708 | 0.352 | 59.75 | 29.672 | 25.177 | 50.01 | 0.201 | 0.021 |
| 35 | 1.408 | 0.709 | 0.353 | 59.504 | 29.438 | 24.028 | 50.05 | 0.204 | 0.0242 |
| 30 | 1.418 | 0.705 | 0.343 | 59.88 | 29.48 | 22.4 | 50.09 | 0.1873 | 0.025 |
| 25 | 1.411 | 0.713 | 0.373 | 59.51 | 29.27 | 25.17 | 50.15 | 0.186 | 0.0261 |
| 20 | 1.423 | 0.712 | 0.341 | 61.27 | 33.5 | 28.98 | 49.76 | 0.174 | 0.0236 |
| 15 | 1.424 | 0.704 | 0.356 | 62.312 | 31.824 | 24.87 | 50.10 | 0.233 | 0.0297 |
| 10 | 1.38 | 0.821 | 0.254 | 58.476 | 23.55 | 33.9 | 50.25 | 0.214 | 0.020 |

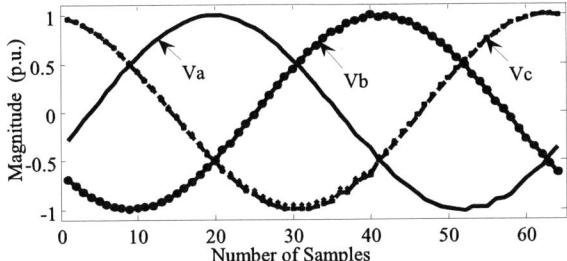

Fig. 13. Practical Voltage Signal.

TABLE IV
ESTIMATED VALUES OF AMPLITUDE, PHASE AND FREQUENCY FOR PRACTICAL DATA

| Para-meter | GA | | | APSO | | |
|---|---|---|---|---|---|---|
| | Va | Vb | Vc | Va | Vb | Vc |
| $A_1$ | 0.9717 | 0.9857 | 0.9723 | 1.00 | 0.998 | 0.9993 |
| $\Phi_1$ | -10.58 | -136.23 | 110.82 | -15.75 | -135.8 | 104.33 |
| $f_1$ | 48.68 | 49.02 | 48.14 | 49.95 | 49.99 | 49.97 |

For testing the algorithm for higher harmonics the following equation is consider for producing the data samples

$$v(t) = 1.0\sin(\omega t + 10^\circ) + 0.1\sin(3\omega t + 20^\circ) + 0.08\sin(5\omega t + 30^\circ)$$
$$+0.08\sin(9\omega t + 40^\circ) + 0.06\sin(11\omega t + 50^\circ) + 0.05\sin(13\omega t + 60^\circ)$$
$$+0.03\sin(19\omega t + 70^\circ) \qquad (11)$$

The Estimated waveform is shown in Fig. 14 and Estimated values are given in Table V.

Fig14   Estimated Voltage Signal

Fig. 14. Estimated Harmonic Voltage Signal.

TABLE V
ESTIMATION OF HARMONIC MAGNITUDE AND PHASE ANGLES

| Order of harmonic | Actual Magnitude | Estimated Magnitude | Actual phase | Estimated phase |
|---|---|---|---|---|
| 1 | 1 | 0.9985 | 10 | 9.9547 |
| 3 | 0.1 | .099 | 20 | 20.0012 |
| 5 | 0.08 | 0.08001 | 30 | 29.9967 |
| 9 | 0.08 | 0.079954 | 40 | 40.001 |
| 11 | 0.06 | 0.06997 | 50 | 49.9982 |
| 13 | 0.05 | 0.050019 | 60 | 60.0027 |
| 19 | 0.03 | 0.029914 | 70 | 69.99127 |

The practical noisy data was collected in the laboratory and the proposed algorithm is applied to the data for the extraction of frequency, phase and amplitude. The results are demonstrated in Fig. 15 to 17 and also reported in Table VI.

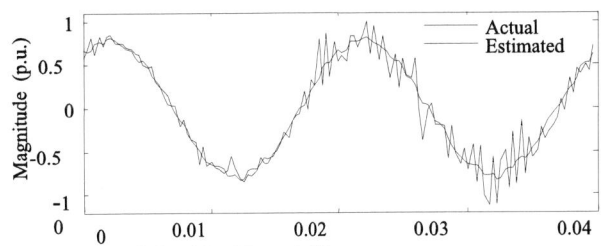

Fig. 15. Estimated Signal For Phase-A   Time.

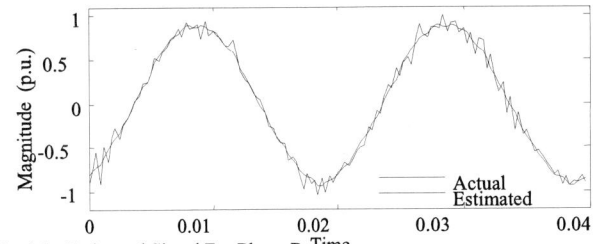

Fig. 16. Estimated Signal For Phase-B. Time

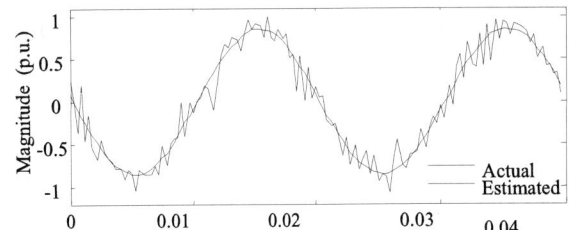

Fig. 17. Estimated Signal For Phase-C. Time

TABLE VI
ESTIMATED VALUES FOR REAL NOISY DATA

| Parameter | Phase A | Phase B | Phase C |
|-----------|---------|---------|---------|
| $A_1$ | 0.7816 | 0.9022 | 0.8531 |
| $f_1$ | 50.065 | 50.023 | 49.969 |
| $\Phi_1$ | 54.814 | -65.757 | 174.717 |

## V. CONCLUSION

This paper presents an evolutionary algorithm to estimate the flicker and harmonics present in a power signal. The effect of number of samples and noise present in the measured signal are studied with the proposed algorithm. The proposed APSO algorithm works better in above said situations.

## VI. REFERENCES

[1] T.Lobos, T.Kozina and H.J.Koglin, "Power System Harmonics Estimation using Linear least squares method and SVD." IEE proceedings on Generation, Transmission & Distribution, Vol.:148, No 6, Nov 2001, pp 567-572

[2] M.Bettayeb and U.Qidwai, "A Hybrid Least Squares-GA based Algorithm for Harmonic Estimation", IEEE Trans. On Power Delvcry, Vol.18,No.2,April2003,pp377-382.

[3] R.Aghazadeh, H.Lesani, M.Sanaye-Pasand and B.Ganji, "New technique for frequency and amplitude Estimation of Power Signals", in IEE Proc. of Gene. Transm. Distrb., Vol.152, no.3,May 2005.

[4] Khaled M.El-Naggar, Wael M. AL-Hasawi, " A Genetic based algorithm for measurement of Power System Disturbances", Electric Power System Research76 (2006) , pp 808-814.

[5] P.K.Dash, D.P.Swain, A.C.Liew and Saifur Rahman, "An Adaptive Linear Combiner for On-Line Tracking of Power System Harmonics", IEEE Trans. on Power Systems, Vol. 11, No. 4, November 1996,pp 1730-1735.

[6] D. E. Goldberg, *Genetic Algorithms in Search Optimization and Machine Learning.* Reading, MA: Addison-Wesley, 1989.

[7] Eberhart R C, Shi Y. "Particle swarm optimization: developments, applications and resources", Proc. Congress on Evolutionary Computation. Piscataway: IEEE, Soul, 2001.pp 81-86

[8] Yuhui Shi and R.C.Eberhart, "Fuzzy Adaptive Particle Swarm Optimization", Proceedings of Evolutionary Computation 2001, 27-30 May , vol. 1, pp 101-106.

## VII. BIOGRAPHIES

**B. K. Panigrahi** is working as an Assistant Professor in the Department of Electrical Engineering, IIT, New Delhi, India. Prior to joining IIT Delhi, he was working as Lecturer at University College of Engineering, Burla, Sambalpur, Orissa for about 13 years. The research interests of Dr. Panigrahi are in the areas of Intelligent control of FACTS devices, Application of advanced DSP techniques for Power Quality assessment.

**V. Ravikumar Pandi** received his B.E. degree from Madurai Kamaraj University, Tamilnadu in 2003 and M.E. degree from Annamalai University, Tamilnadu in 2005. He was worked as a Lecturer in Sri Venkateswara College of Engineering, Chennai. He is now working towards his Ph.D. degree at IIT Delhi, India. His research interest includes Numerical Optimization, Power system analysis with HVDC and FACTS and Soft Computing.

**2006 IEEE International Conference on Power Electronic, Drives and Energy Systems**

# A 36-Pulse AC-DC Converter for Line Current Harmonic Reduction

Bhim Singh, *Senior Member, IEEE,* and Sanjay Gairola

*Abstract*--In this paper, a novel transformer for 36-pulse AC-DC conversion is designed, modeled and developed to feed isolated varying DC loads. The proposed transformer is normally used for large current rating rectifiers such as electric-aircraft power supply, electrowinning, electrochemical processes, induction heating, drives, plasma torches, etc., where isolation is required mainly for stepping down the supply voltage. It consists of two paralleled 18-pulse AC-DC converters involving nine-phase shifted diode bridges. It improves power quality at AC mains and it meets IEEE-519 Standard requirements at varying loads.

*Index Terms*--36-pulse, fork connection, AC-DC converter, isolation transformer, power quality.

## I. INTRODUCTION

TWELVE-PULSE and eighteen-pulse AC-DC converter configurations are commonly used in several important processes like electrolysis, electro-winning, electro-chemical, welding etc [1]. Thyristor rectifiers have been used for large rated, controlled operation and the technology is well established. A combination of rectifier-chopper is also becoming popular for large controlled DC currents [1] where diode rectifiers are used at front end. The insulated gate bipolar transistors (IGBTs) or integrated gate commutated thyristors (IGCTs) based chopper systems are used for high power DC arc furnaces [2] and have capability to reduce flicker and increase productivity. These rectifier-chopper systems have many operational benefits over conventional rectifiers [3] such as fast dynamic response, low output ripple, improved power factor and efficiency. The issues involved in specifying rectifier-chopper- systems are unique [□] as IGBTs are used and currents are large. It is common practice to use multiple 12-pulse or 18-pulse units [5] fed from phase–staggered transformers to meet IEEE-519 standard [6] requirements as the total harmonic distortion of input line current (THD$_i$) of single unit is still high and may not qualify as clean power at high loads. The isolation transformers used for high currents can have different winding arrangements such as star, delta, fork, zig-zag, polygon etc. [7]. A six-pulse diode bridge converter that is very commonly used is shown in

Bhim Singh and Sanjay Gairola are with Department of Electrical Engineering, Indian Institute of Technology, Delhi, Hauz-Khas, New Delhi-110016, India (e-mail: bhimsinghr@gmail.com, sanjaygairola@gmail.com).

Fig. 1 and needs no explanation. Analysis of some winding arrangements for 18-pulse autotransformer rectifiers have been described by Burgos et. al. [8]. However, it is observed that the total harmonic distortion (THD) of input current is more than 5% when operating at light load or the source impedance is small.

DC power supplies for aerospace applications are typically rated 28V for commercial planes and 28V or 270V on military airplane. The aircraft engine is coupled to a generator (115 V/□00Hz AC) that feeds isolated transformer rectifier unit (TRU) to produce the desired DC voltage. Moreover, to meet stringent power quality specifications (US Navy limits THD$_i$ <3% for its special application needs) it is inevitable to go beyond 2□-pulse AC-DC converter system configuration [9]. The high pulse number multipulse AC-DC converters for motor drives have been described by Wu [10].

Therefore, it is suggested that higher pulse configuration must be used so that AC-DC converters meets IEEE-519 Standard requirements at varying loads. With this in view, a 36-pulse AC-DC converter is designed, modeled and developed in this work.

The developed 36-pulse rectifier is fed from delta/ fork transformer. As the power to the load is transferred at low voltage levels, the use of parallel bridge configuration is justified. The input transformer secondary is symmetric fork extending two sets of 9-phases for the two diode bridge converters.

In this paper, the proposed transformer is used to feed two diode bridges (each having nine legs) that are connected in parallel. This parallel connection produces 36-pulse configuration for AC-DC conversion. Detailed design of the transformer and resulting 36-pulse diode rectifier system is carried out to study the behavior of the AC-DC converter. The designed system is modeled and simulated in MATLAB to demonstrate its power quality improvement at AC mains. A laboratory prototype of proposed 36-pulse AC-DC converter is developed to validate the design and its simulation model.

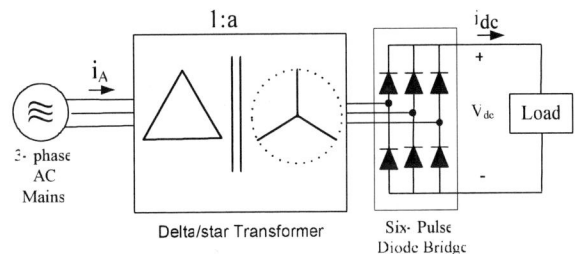

Fig. 1. A 6-pulse AC-DC converter using delta/star transformer.

## II. PROPOSED 36-PULSE AC-DC CONVERTER

The commonly used input transformers for 12/18-pulse converters have non-identical secondary and this leads to unbalance of output voltages as the turn ratio doesn't exactly match with the required voltage ratios. In this proposed configuration a single fork secondary is used which generates two sets of 9-phase supply for each bridge. The proposed 36-pulse AC-DC converter is shown in Fig. 2.

### A. Design of delta/fork transformer for 36-pulse AC-DC converter

Fig. 3 shows a schematic diagram of the proposed delta/fork transformer winding arrangement and its connection to two nine-leg diode bridges, DB1 and DB2. Two nine-leg diode bridge converters DB1 and DB2 are connected to two sets of nine-phase secondary at ($a_1$, $a_3$, $a_5$, $b_1$, $b_3$, $b_5$, $c_1$, $c_3$, $c_5$) and ($a_2$, $a_\square$, $a_6$, $b_2$, $b_\square$, $b_6$, $c_2$, $c_\square$, $c_6$) respectively. Fig. $\square$ depicts the graphical representation of the transformer secondary and angular position of various voltage phasors. Two sets of nine phase voltages are ($V_{B11}$, $V_{B12}$, $V_{B13}$, $V_{B1\square}$ $V_{B15}$, $V_{B16}$, $V_{B17}$, $V_{B18}$, $V_{B19}$) and ($V_{B21}$, $V_{B22}$, $V_{B23}$, $V_{B2\square}$ $V_{B25}$, $V_{B26}$, $V_{B27}$, $V_{B28}$, $V_{B29}$). These sets are displaced by 10° from each other and at -5° and +5° respectively from voltage $V_{sA}$. The number of turns for every winding is determined as a function of the required secondary voltage, $V_{sA}(=V_R)$. These secondary voltages, as marked in Fig. $\square$, are expressed by following relationships.

Let the transformation ratio of the transformer be as:

$$a = V_A / V_{sA} \tag{1}$$

Assume the set of three phase secondary voltages as:

$$V_{sA} = V_R \angle 0°, \ V_{sB} = V_R \angle -120°, \ V_{sC} = V_R \angle 120° \tag{2}$$

The required voltages for the converter I (DB1) are:

$$V_{B11} = V_R \angle \square 5, \ V_{B12} = V_R \angle 5°, \ V_{B13} = V_R \angle -35°,$$
$$V_{B1\square} = V_R \angle -75°, \ V_{B15} = V_R \angle -115°, \ V_{B16} = V_R \angle -155°, \tag{3}$$
$$V_{B17} = V_R \angle -195°, \ V_{B18} = V_R \angle -235°, \ V_{B19} = V_R \angle -275°$$

The required voltages for the converter II (DB2) are:

$$V_{B21} = V_R \angle 35°, \ V_{B22} = V_R \angle -5°, \ V_{B23} = V_R \angle -\square 5,$$
$$V_{B2\square} = V_R \angle -85°, \ V_{B25} = V_R \angle -125°, \ V_{B26} = V_R \angle -165°, \tag{$\square$}$$
$$V_{B27} = V_R \angle -205°, \ V_{B28} = V_R \angle -2\square 5°, \ V_{B29} = V_R \angle -285°$$

The values of constants $K_1$ to $K_6$ marked in Fig. 3 determine the secondary winding turns as a fraction of primary windings turns. These values can be determined by solving the following equations:

$$V_{B11} = K_1 V_{sA} - (K_\square + K_5)V_{sB} \tag{5}$$

$$V_{B21} = (K_1 + K_3)V_{sA} - K_\square V_{sB} \tag{6}$$

$$V_{B12} = (K_1 + K_2)V_{sA} - K_6 V_{sB} \tag{7}$$

$$V_{B22} = (K_1 + K_2)V_{sA} - K_6 V_{sC} \tag{8}$$

$$V_{B13} = (K_1 + K_3)V_{sA} - K_\square V_{sC} \tag{9}$$

$$V_{B23} = K_1 V_{sA} - (K_\square + K_5)V_{sC} \tag{10}$$

Eqns. (5-10) give the values of constants $K_1$ to $K_6$ for desired phase shift as

$$K_1 = 0.2988, K_2 = 0.6\square 71, K_3 = 0.1891, \ K_\square = 0.6623$$
$$K_5 = 0.15\square 2, K_6 = 0.1006 \tag{11}$$

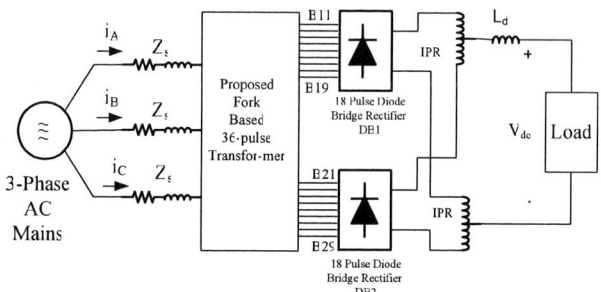

Fig. 2. Proposed delta/fork transformer based 36-pulse AC-DC converter.

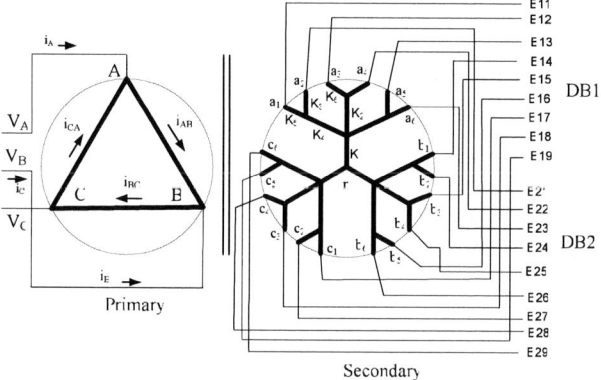

Fig. 3. Delta/fork transformer winding arrangement for 36-pulse AC-DC converter.

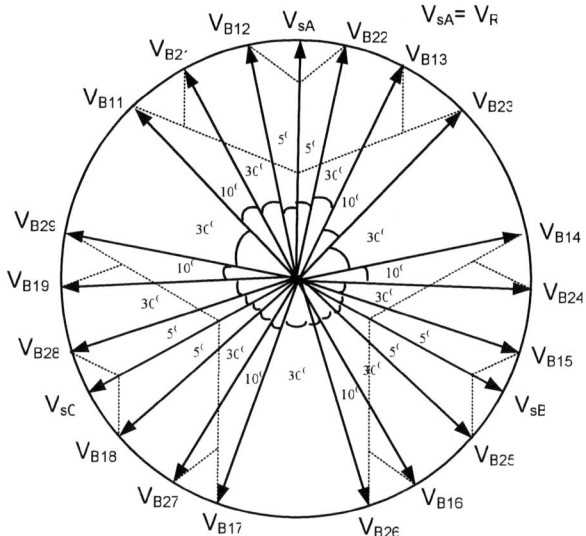

Fig. $\square$ Graphical representation of delta/polygon transformer secondary for 36-pulse AC-DC converter and phasor diagram.

Based on these constants the proposed transformer for 36-pulse AC-DC converter is designed and developed in the laboratory.

The rating of a transformer is dependent on the voltage across each winding and current through it. The winding voltages determine the core size while the currents determine

702

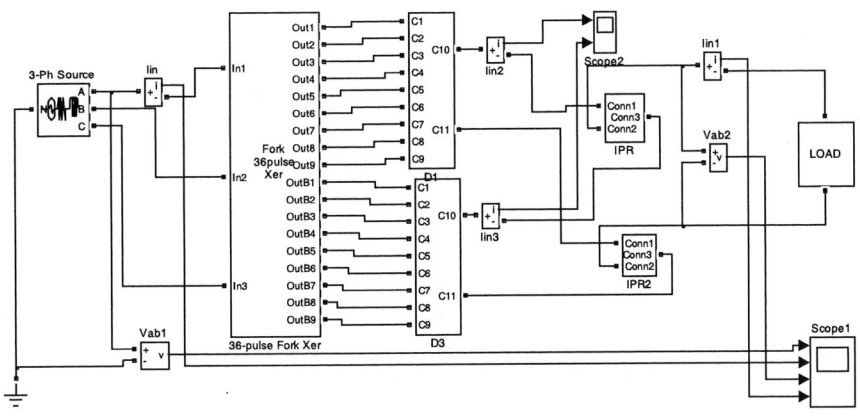

Fig.5(a). MATLAB model of the 36-pulse AC-DC converter.

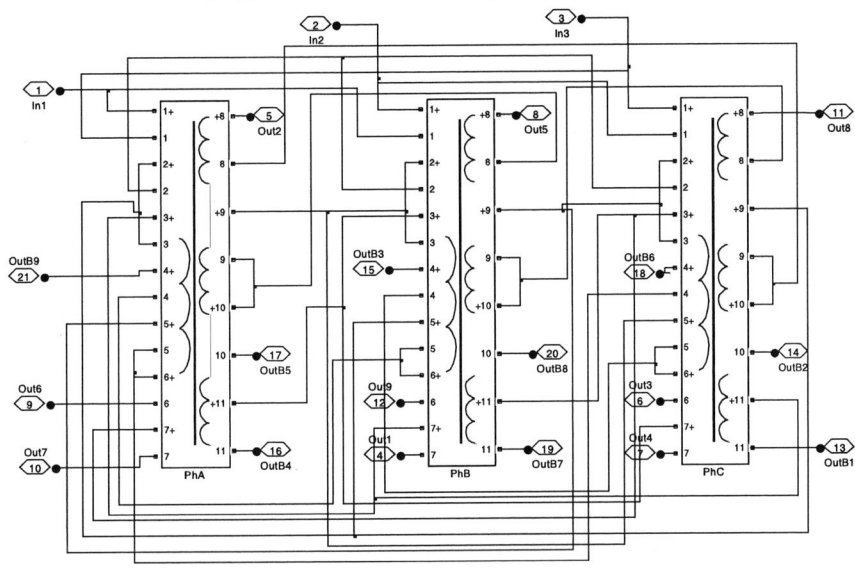

Fig. 5(b). MATLAB model of the proposed transformer for 36-pulse AC-DC converter.

conductor size and hence the two determine it's VA rating. The VA rating of these transformers is calculated as:

$$\text{VA rating} = 0.5 \sum (V_k I_k) \qquad (12)$$

Where, $V_k$ and $I_k$ are root-mean-square (r.m.s.) value of voltage across and current through $k^{th}$ winding. The same relation is used for estimating the transformer rating of these AC-DC converters.

## III. MATLAB BASED SIMULATION

The proposed 36–pulse AC-DC converter is simulated in MATLAB environment alongwith SIMULINK and Power System Blockset (PSB) toolboxes. The 36-pulse AC-DC converter system is fed from □5V, 50Hz AC supply. The DC load connected to the converter is considered to be a 300V, 20kW R-L load. The source impedance of 3% is considered in these designs. Fig. 5(a) shows the MATLAB model of the 36-pulse AC-DC converter to improve various power-quality indices and Fig. 5(b) shows model of the used transformer winding arrangement. The simulation results are shown in Figs. 6-12 and Table I-II. The ratings of input transformer and

IPT are estimated and these are found 12□2% and 0.85% respectively of the load rating, as given in Table III.

## IV. HARDWARE IMPLEMENTATION

A prototype of the proposed 36-pulse converter is developed in the laboratory for 6kW load fed from three-phase supply voltage of 230V, 50Hz. The transformation ratio is chosen as 1: 0.52. The detailed design data are given in Appendix. Extensive tests have been carried out to validate the model and test results have been recorded using 'Fluke □3B' power quality analyzer. The main transformer rating for the prototype is 6kVA while the total interphase reactor rating is quite small (62VA).

The recorded waveforms for the 36-pulse converter configuration are shown in Fig. 13 and data are given inTable III. The power measurements, AC mains voltage and current waveforms can be seen alongwith harmonic spectrum of AC mains current. The input current spectrum at light-load is shown in Fig. 13(a) and full-load in Fig. 13(b).

703

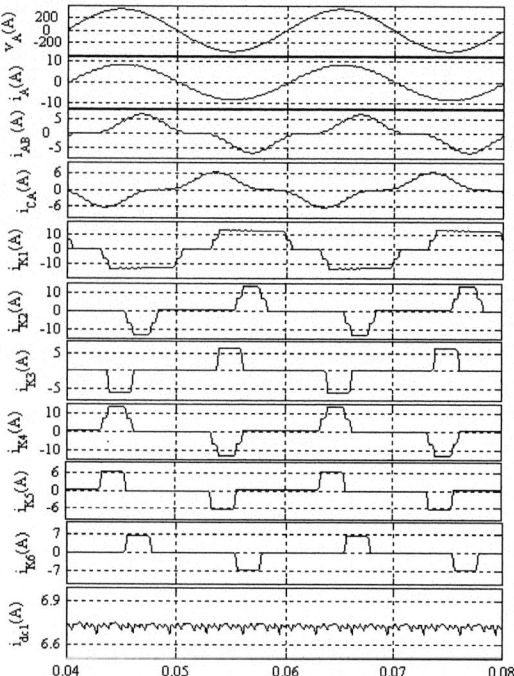

Fig. 6. The instantaneous values of input phase voltage ($v_A$), input AC mains current ($i_A$), primary winding currents ($i_{AB}$ and $i_{CA}$), winding-currents of secondary ($i_{K1}$ to $i_{K6}$), a bridge output current ($i_{dc1}$) of 36-pulse AC-DC converter.

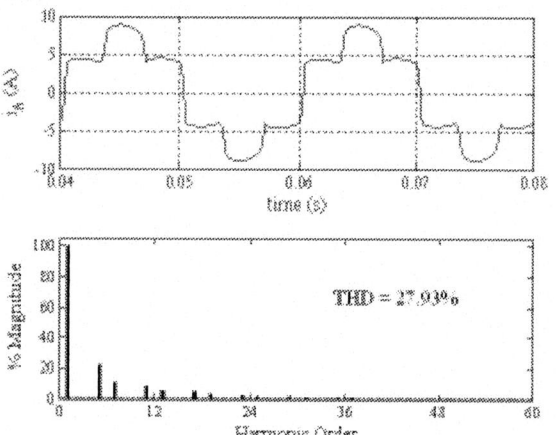

Fig. 7. Input current waveform of 6-pulse AC-DC converter at light-load and its harmonic spectrum.

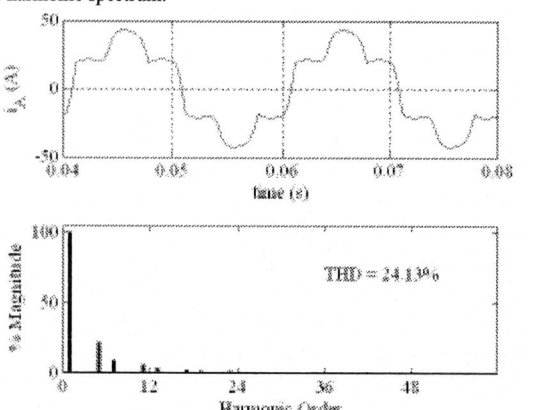

Fig. 8. Input current waveform of 6-pulse AC-DC converter at full-load and its harmonic spectrum.

## V. RESULTS AND DISCUSSION

The power quality indices obtained from simulations of proposed 36-pulse AC-DC converter is given in Table I. The various waveforms of the 36-pulse transformer are shown in Fig. 6 at light load to make the steps visible. It can be seen that the input current ($i_A$) has 36 steps in one cycle of AC supply. Two primary winding currents $i_{AB}$ and $i_{CA}$ are shown, which add algebraically to the supply current ($i_A$). Moreover there are the secondary winding currents ($i_{K1}$ to $i_{K6}$) which add to produce $i_{dc1}$. The current $i_{dc}$ is sum of two bridge output currents which can be seen in Fig. 9 along with output DC voltage, $V_{dc}$.

Figs. 7 and 8 show the input AC current waveform and its harmonic spectrum of a 6-pulse AC-DC converter at light-load and full-load respectively. The THD$_i$ of six-pulse AC-DC converter is of order of 2□13% which is not at all acceptable by IEEE-519 Standard requirements.

Fig. 9 shows the input AC phase voltage $v_A$, input AC current $i_A$, output DC voltage $v_{dc}$, and load current $i_{dc}$ waveform of the proposed 36-pulse AC-DC converter at light load (20% of full-load). Fig. 10 shows the AC mains current waveform $i_A$ and its harmonic spectrum at light load. These waveforms clearly show that 35th and 37th are the dominant

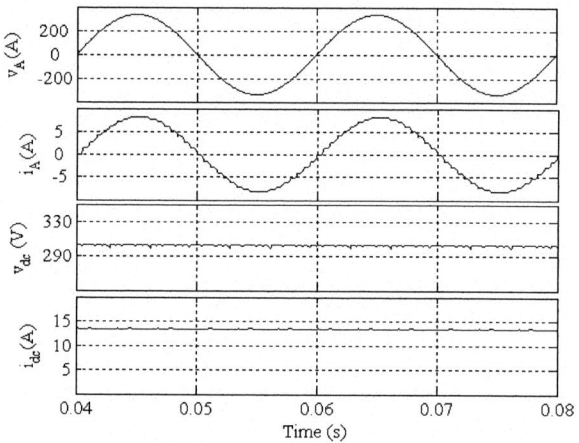

Fig. 9. Input and output voltage and current waveforms of 36-pulse AC-DC converter at light load.

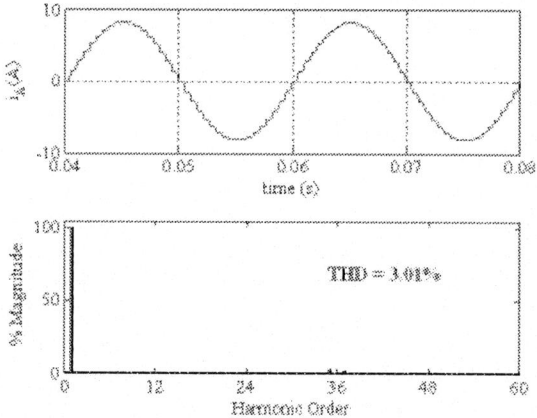

Fig. 10. Input current waveform of 36-pulse AC-DC converters at light load and its harmonic spectrum.

TABLE I

COMPARISON OF POWER QUALITY PARAMETERS OF PROPOSED 36-PULSE AC-DC CONVERTERS WITH VARYING LOAD

| Load | THD $V_{ac}$ (%) | AC Mains Current $I_{ac}$ (A) | THD of $I_{ac}$(%) | Distortion Factor, DF | Displace-ment Factor, DPF | Power Factor, PF | DC Voltage (V) | Load Current $I_{dc}$ (A) | Ripple Factor (%) |
|---|---|---|---|---|---|---|---|---|---|
| 20% | 0.7571 | 5.787 | 3.01 | 0.9995 | 0.998□ | 0.9979 | 302.9 | 13.□6 | 0.2377 |
| □0% | 1.07□ | 11.□6 | 2.169 | 0.9997 | 0.9972 | 0.9969 | 301.8 | 26.83 | 0.3279 |
| 50% | 1.155 | 1□28 | 1.855 | 0.9998 | 0.9967 | 0.9965 | 301.3 | 33.□8 | 0.3753 |
| 60% | 1.193 | 17.09 | 1.605 | 0.9998 | 0.9962 | 0.9960 | 300.8 | □0.10 | 0.□223 |
| 80% | 1.210 | 22.68 | 1.31□ | 0.9999 | 0.9951 | 0.9950 | 299.6 | 53.27 | 0.5217 |
| 100% | 1.21□ | 28.21 | 1.197 | 0.9998 | 0.9937 | 0.9935 | 298.□ | 66.30 | 0.62□ |

TABLE II

COMPARISON OF POWER QUALITY PARAMETERS OF THE 36-PULSE AC-DC CONVERTERS WITH 6-PULSE AC-DC CONVERTER AT FULL-LOAD AND LIGHT LOAD.

| Sr. No. | Topology | %THD $V_{ac}$ | % THD of $I_{ac}$, at | | Distortion Factor | | Displacement Factor | | Power Factor | | DC Voltage (V) | |
|---|---|---|---|---|---|---|---|---|---|---|---|---|
| | | | Full Load | Light Load | Full Load | Light Load | Full Load | Light Load | Full Load | Light Load | Full Load | Light Load |
| 1 | 6-pulse | □056 | 2□81 | 27.93 | .9815 | .99□□ | .9698 | .9630 | .9519 | .9575 | 279.2 | 29□□ |
| 2 | 36-pulse | 1.21□ | 1.197 | 3.01 | .9998 | .9995 | .9737 | .998□ | .9935 | .9979 | 298.□ | 302.9 |

harmonics and THD of AC mains current (THD$_i$) is 2.99%. Figs. 11 and 12 show the performance of 36-pulse AC-DC converter at full-load. The THD of AC current at full-load is observed to be 1.2% only, in a converter using transformer having single secondary. The value of AC mains current THD is found to vary in range 1.2% to 3% with the variation of load.

The improvement in power quality indices such as total harmonic distortion of supply current (THD$_i$), total harmonic distortion of supply voltage (THD$_v$), distortion factor (DF),

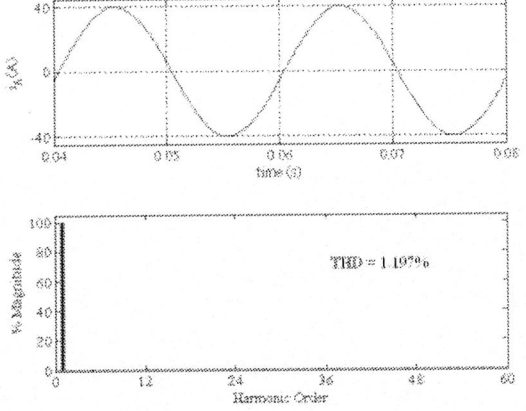

Fig. 11. Input and output voltage and current waveforms of 36-pulse AC-DC converter at full load.

Fig. 12. Input current waveforms of 36 pulse AC-DC converter at full load and its harmonic spectrum.

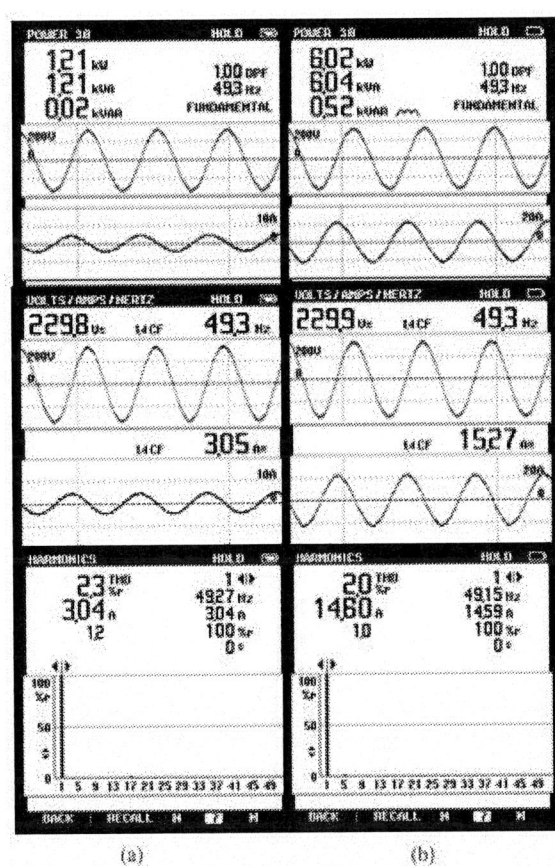

Fig. 13. Test result showing input power, voltage alongwith current waveforms and input current harmonic spectrum at (a) light load and (b) full-load.

TABLE III
COMPARISON OF MAGNETIC RATINGS IN DIFFERENT AC-DC CONVERTERS.

| Sr. No. | Topology | Main Transformer rating (% of load) | IPT rating (% of load) | Total rating of magnetics, (% of load) |
|---|---|---|---|---|
| 1 | 6-pulse | 108.0 | - | 108.0 |
| 2 | 36-pulse | 12□2 | 0.85 | 125.0 |

TABLE IV
TEST RESULTS SHOWING POWER QUALITY PARAMETERS OF 36-PULSE AC-DC CONVERTER.

| Load (kW) | THD $V_{ac}$ (%) | AC Mains Current $I_{ac}$ (A) | THD of $I_{ac}$ (%) | DPF | PF | $V_{dc}$ (V) | $I_{dc}$ (A) |
|---|---|---|---|---|---|---|---|
| 1.21 | 1.30 | 3.05 | 2.3 | 1.0 | 0.999 | 168.6 | 7.256 |
| 2.16 | 1.□0 | 5.38 | 2.5 | 1.0 | 0.999 | 160.9 | 12.86 |
| 3.08 | 1.30 | 7.79 | 2.2 | 1.0 | 0.999 | 151.□ | 18.□0 |
| □09 | 1.30 | 10.28 | 2.1 | 1.0 | 0.998 | 1□3.7 | 2□21 |
| 5.29 | 1.30 | 13.29 | 2.1 | 1.0 | 0.996 | 133.7 | 31.09 |

and power factor (PF) at different loads can be seen in Table I. It can be seen that the THD of input AC current of proposed 36-pulse AC-DC converter system is order of 1% to 3% and current waveform is almost sinusoidal and in phase with the supply voltage. The power factor is observed to be order of 0.993 at varying loads in proposed 36-pulse AC-DC converter as given in Table I. Moreover, Table II reveals that the voltage distortion at point of common coupling (PCC) is negligible as $THD_v$ remains less than 1.22%.

The comparison of power quality indices of proposed converter is made with 6-pulse AC-DC converter and shown in Table II. It can be observed that the 36-pulse AC-DC converter have much improved performance. The size of active magnetics involved in different converters is also given in Table III. It can be seen that a total of 125% of magnetics is needed to achieve THDi <3% in the 36-pulse AC-DC converter.

Test results shown in Fig. 13 validate the effectiveness of fork connected transformer based 36-pulse AC-DC converter. The THD of input current is 2.3% at light load while it is 2.0% at full-load. Table IV shows the test results demonstrating the improvement in power-quality parameters at varying load. The value of DPF and PF are observed to be close to unity. The stringent power quality requirements are met without using any filter at the input.

## V. CONCLUSIONS

The delta/fork connection based transformer used for a 36-pulse AC-DC converter system has single secondary and it provides balanced outputs voltages. Simulated and test results have demonstrated that only two interphase reactors are needed for the proposed 36-pulse AC-DC converter operation unlike the 18-pulse and 2□-pulse AC-DC converters where three and four IPR are needed, respectively. The resulting AC-DC converter system has exhibited a high level performance with clean power characteristics to be used in front-end rectifiers. The total harmonic distortion of input current is

observed to be much less than 5% at varying loads. The output voltage ripple is reduced to order less than 0.6% and input power factor is improved to the order of 0.99 at varying loads.

## VI. APPENDIX

The propose 36-pulse AC-DC converter is realized by three 2.2kVA, single-phase transformers and the design details are as follows.

Flux Density: 0.8Tesla, Current Density: 2.3A/mm$^2$,
Turns per volt: 0.88
E-Laminations: Length=23.5cm, Width=16cm
I-Laminations: Length=23.5cm, Width= □cm
Effective Area of cross-section of core=58cm$^2$ (7.6 cm X 8.6cm)
Autotransformer winding details-
Winding Number of turns Gauge of wire (SWG)

| Winding voltage | No. of Turns | Gauge of Wire (SWG) |
|---|---|---|
| $V_{AC}$ | 365 | 17 |
| $K_1 * V_R$ | 31 | 15 |
| $K_2 * V_R$ | 68 | 17 |
| $K_3 * V_R$ | 20 | 17 |
| $K_□ * V_R$ | 69.5 | 17 |
| $K_5 * V_R$ | 16 | 17 |
| $K_6 * V_R$ | 10.5 | 17 |

## VI. REFERENCES

[1] J. R. Rodriguez, J. Pontt, C. Silva, E.P. Wiechmann, P.W. Hammond, F.W. Santucci, R.Alvarez, R.Musalem, S.Kouro and P.Lezana, "Large current rectifiers: state of the art and future trends", *IEEE Trans. on Ind. Appl.*, vol. 52, no. 3, pp. 738-7□6, June 2005.

[2] P. Ladoux, G. Postiglione, H. Foch and J. Nuns, "A comparative study of AC/DC converters for high-power DC arc furnace", *IEEE Trans. on Ind. Appl.*, vol. 52, no. 3, pp. 7□7-757, June 2005.

[3] V. Scaini and T. Ma, "High-Current DC Choppers in the Metal Industry", *IEEE Industrial Applications Magazine*, pp. 26-33, March/April, 2002.

[□] P. S. Maniscalco, V. Scaini and W. E. Veerkamp, "Specifying DC Choppers Systems for Electrochemical Applications", *IEEE Trans. on Ind. Appl.*, vol. 37, no. 3, pp. 9□1-9□8, May/June, 2001.

[5] E. P. Wiechmann and P. E. Aqueveque, "Filterless high current rectifier for electrolytic applications", *in Proc. of IEEE conf. IAS 2005*, vol.1, pp.198-203, 2-6 Oct. 2005.

[6] IEEE Standard 519-1992, IEEE Recommended Practices and Requirements for Harmonic Control in Electrical Power Systems, *IEEE Inc.*, New York, 1992.

[7] Derek A. Paice, Power Electronic Converter Harmonics: Multipulse Methods For Clean Power, *IEEE Press*, New York, 1996.

[8] R. P. Burgos, A. Uan-zo-li, F. Lacaux, A Roshan, F.Wang and D. Boroyevich, "Analysis of New Step-Up and Step-Down 18-pulse Direct Asymmetric Autotransformer-Rectifiers", *in Proc. of IEEE conf. IAS-2005*, vol. 1, pp. 1□5-152, 2-6 Oct., 2005.

[9] R. Hammond, L. Johnson, A. Shimp and D. Harder, "Magnetic solutions to line current Harmonic reduction", *in Proc. of Conf. Power Con.-1994*, pp.35□-36□ Sept., 199□

[10] Bin Wu, High-Power Converters and AC Drives, IEEE Press, Wiley-Interscience, 2006.

**2006 IEEE International Conference on Power Electronic, Drives and Energy Systems**

# A Unified Analysis of CCM Boost PFC for Various Current Control Strategies

Ranjan K. Gupta, *Student Member, IEEE*, Hariharan Krishnaswami, *Student Member, IEEE*, and Ned Mohan, *Fellow, IEEE*

*Abstract*-- **A single phase Continuous Conduction Mode (CCM) Boost PFC can be classified by the method in which input current shaping is done: average-current, peak-current and valley-current controlled PFC. The output of voltage compensator is different in above current controllers for the same loading condition. In this paper, a mathematical derivation using Fourier analysis of input current is presented to determine the value of the compensator output. The analysis is applied to determine the boundary condition for CCM operation. Theoretical comparison of the above derivations is supported by simulation results. A small-signal model of the system is required to design the voltage compensator parameters. A unified method of deriving small-signal model and the transfer function of the outer voltage loop is proposed. A design example of a compensator for average-current controlled PFC is given and differences at various points in conjunction with other current controllers are presented.**

## I. INTRODUCTION

THE commonly used topology of single phase power factor correction (PFC) is a single-phase diode bridge followed by a dc-dc boost converter. There are two feedback loops in every single phase CCM boost PFC; the outer voltage loop determines the load requirement and input current reference; the inner current loop maintains the input current to be in phase with input voltage in addition to shaping it sinusoidal. Various control strategies have been reported in literature for single phase CCM boost PFC and these can be classified by the method in which they shape the input current: Average current controlled PFC, Peak current controlled PFC and Valley current controlled PFC. In average current controlled PFC (Average current mode controller [1], Non linear carrier control (NLC) PFC [2]) the average of input current is made sinusoidal. In peak current controlled PFC (Linear peak current mode controller [3], One Cycle controlled PFC [4]) the peak of the input current is controlled while in valley current controlled PFC (Predictive switching modulator (PSM) [5]) the valley of the input current is controlled and made sinusoidal in phase with input voltage. The output of

Ranjan K. Gupta and Hariharan Krishnaswami are graduate students in the Department of Electrical and Computer Engineering, University of Minnesota, Minneapolis, MN 55455 (e-mail: gupt0108@umn.edu, kris0136@umn.edu)

Ned Mohan is with the Department of Electrical and Computer Engineering, University of Minnesota, Minneapolis, MN 55455 USA (e-mail: mohan@umn.edu)

voltage control loop compensator gives the peak of reference current (average/valley/peak) in corresponding control strategies. A unified derivation of this compensator output for all types of current controllers is presented in this paper.

All these controllers can also be classified under two types of approaches to realize the current controller: Multiplier based approach and Integration reset based approach. A block diagram illustrating these two approaches is shown in Fig. 1. Average current mode control using a multiplier and current compensator is well known in literature [1]. The integration reset technique in relevance to control of DC to DC converters was introduced in one cycle control of switching converters [6]. Integration reset technique is also used in [2], [4], [5] and [7] to control input current. In this approach the output of voltage compensator is integrated and reset every switching cycle. The integration and reset procedure is applied in a manner to realize different duty functions $f_1(d)$ and $f_2(d)$ (shown in Fig.1) for various current control strategies. To design the voltage compensator it is necessary to know the small signal model of the system with respective current controllers. Different approaches have been adopted in [2], [5] and [7] to derive the small signal equivalent circuit. In this paper a unified method is proposed wherein the derivation is common to all types of current controllers. To design a CCM boost PFC for high performance at low loading condition the boundary of CCM and DCM operation has to be known to avoid the DCM operation. The boundary condition is derived for all the abovementioned control techniques.

Section II starts with the analysis of input current, derivation of the voltage compensator outputs and the boundary of CCM operation. Section III presents a unified approach to arrive at the small signal model for various current control techniques. The final section takes an example design of a CCM boost PFC and presents simulation results to confirm the analysis.

## II. ANALYSIS OF INPUT CURRENT

The ripple in the inductor current varies over a line cycle as given by (1). Variables are shown in Fig. 1.

$$\Delta i(\theta) = \frac{\hat{V}_{in} T_s}{L} \left( 1 - \frac{\hat{V}_{in}}{V_{dc}} \sin \theta \right) \sin \theta \tag{1}$$

where $\theta = \omega t$

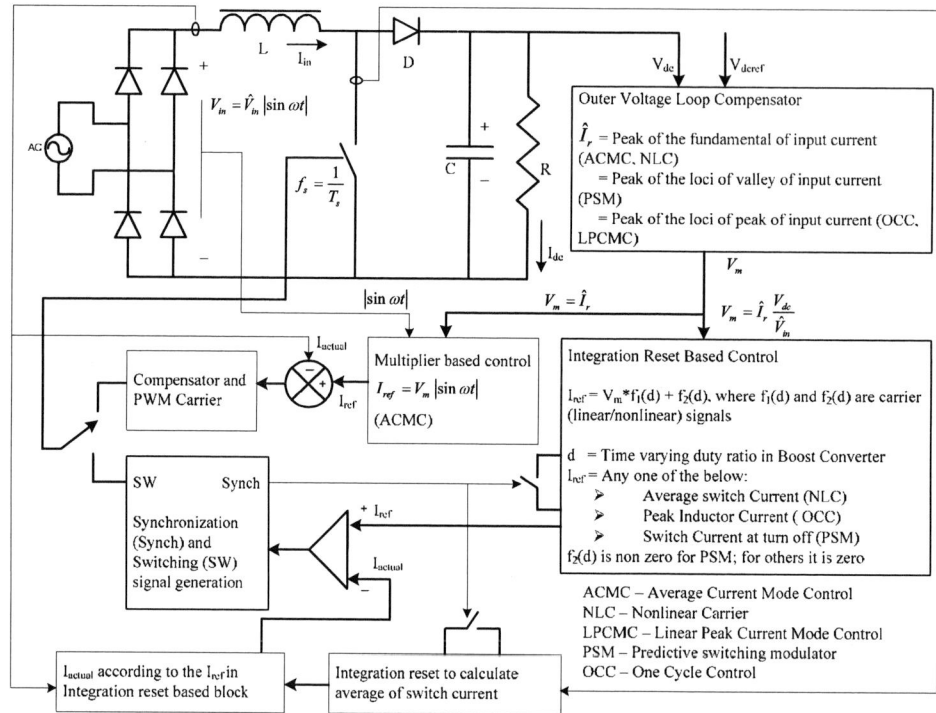

Fig. 1. Multiplier based and Integration reset based control

If the loading condition and the boost circuit parameters are same, then in all the control techniques introduced in section I, the input current will have same instantaneous ripple over a different peak of sinusoidal reference ($\hat{i}_r$, $\hat{i}_{rp}$) used in (2) and (5), where $\hat{i}_r$ and $\hat{i}_{rp}$ are the loci of valley and peak of input current respectively. The expression for the peak of sinusoidal reference is derived below:

### A. Valley Current Control

Using (1) the input current can be written as

$$I_{in}(\theta) = \begin{cases} \hat{i}_r \sin\theta + \dfrac{\hat{V}_{in}T_s}{2L}\left(1 - \dfrac{\hat{V}_{in}}{V_{dc}}\sin\theta\right)\sin\theta & 0 \le \theta \le \pi \\[3mm] \hat{i}_r \sin\theta + \dfrac{\hat{V}_{in}T_s}{2L}\left(1 + \dfrac{\hat{V}_{in}}{V_{dc}}\sin\theta\right)\sin\theta & \pi \le \theta \le 2\pi \end{cases} \quad (2)$$

Fourier analysis of above equation results into the magnitude of fundamental as:

$$I_{in1} = \left(\frac{1}{2}\left(\hat{i}_r + \frac{\hat{V}_{in}T_s}{2L}\right) - \frac{17}{3\pi}\frac{\hat{V}_{in}^2 T_s}{4LV_{dc}}\right) \quad (3)$$

Equating input and output power assuming lossless system

$$\hat{i}_r = 2\left(\frac{V_{dc}I_{dc}}{\hat{V}_{in}} + \frac{17}{6\pi}\frac{\hat{V}_{in}^2 T_s}{4LV_{dc}} - \frac{\hat{V}_{in}T_s}{4L}\right) \quad (4)$$

### B. Peak Current Control

Following the same procedure as in valley current control, the expression for input current and the output of voltage compensator loop is given by (5) and (6).

$$I_{in}(\theta) = \begin{cases} \hat{i}_{rp}\sin\theta - \dfrac{\hat{V}_{in}T_s}{2L}\left(1 - \dfrac{\hat{V}_{in}}{V_{dc}}\sin\theta\right)\sin\theta & 0 \le \theta \le \pi \\[3mm] \hat{i}_{rp}\sin\theta - \dfrac{\hat{V}_{in}T_s}{2L}\left(1 + \dfrac{\hat{V}_{in}}{V_{dc}}\sin\theta\right)\sin\theta & \pi \le \theta \le 2\pi \end{cases} \quad (5)$$

$$\hat{i}_{rp} = 2\left(\frac{V_{dc}I_{dc}}{\hat{V}_{in}} - \frac{17}{6\pi}\frac{\hat{V}_{in}^2 T_s}{4LV_{dc}} + \frac{\hat{V}_{in}T_s}{4L}\right) \quad (6)$$

The output of voltage loop compensator in peak/valley current controlled PFC can be written as (explained in Section III)

$$V_m = \frac{IV_{dc}}{\hat{V}_{in}} \qquad \text{where } I = \hat{i}_r \text{ or } \hat{i}_{rp} \quad (7)$$

The peak of sinusoidal reference will be highest in peak current control and lowest in valley current control which can be verified with (4) and (6). This results in different boundary conditions for CCM.

### C. CCM Boundary

For CCM operation in valley current control

$$\hat{i}_r \ge 0 \quad (8)$$

Using (4) and (8) the critical load is given by

$$I_{dc} \ge \frac{\hat{V}_{in}^2 T_s}{4LV_{dc}}\left(1 - \frac{17}{6\pi}\frac{\hat{V}_{in}}{V_{dc}}\right) \quad (9)$$

Boundary condition for peak current control can be written as

$$\hat{i}_{rp}\sin\theta - \Delta i(\theta) \ge 0 \quad (10)$$

Using (6) and (10) the minimum load for CCM operation is

708

$$I_{dc} \geq \frac{\hat{V}_{in}^2 T_s}{4LV_{dc}}\left(1+\frac{17}{6\pi}\frac{\hat{V}_{in}}{V_{dc}}\right) \qquad (11)$$

The boundary condition results along with compensator outputs are summarized in Table I. Results of average current mode control are also included [1]. Boundary condition for valley current control is same as given in [5].

TABLE I
$V_m$ AND CCM BOUNDARY

| Control | $V_m$ | Boundary for CCM |
|---|---|---|
| Valley Current | $2\dfrac{V_{dc}}{\hat{V}_{in}}\left(\dfrac{V_{dc}I_{dc}}{\hat{V}_{in}}+\dfrac{17}{6\pi}\dfrac{\hat{V}_{in}^2 T_s}{4LV_{dc}}-\dfrac{\hat{V}_{in}T_s}{4L}\right)$ | $I_{dc}\geq\dfrac{\hat{V}_{in}^2 T_s}{4LV_{dc}}\left(1-\dfrac{17}{6\pi}\dfrac{\hat{V}_{in}}{V_{dc}}\right)$ |
| Peak Current | $2\dfrac{V_{dc}}{\hat{V}_{in}}\left(\dfrac{V_{dc}I_{dc}}{\hat{V}_{in}}-\dfrac{17}{6\pi}\dfrac{\hat{V}_{in}^2 T_s}{4LV_{dc}}+\dfrac{\hat{V}_{in}T_s}{4L}\right)$ | $I_{dc}\geq\dfrac{\hat{V}_{in}^2 T_s}{4LV_{dc}}\left(1+\dfrac{17}{6\pi}\dfrac{\hat{V}_{in}}{V_{dc}}\right)$ |
| Average Current | $\dfrac{2V_{dc}I_{dc}}{\hat{V}_{in}}$ | $I_{dc}\geq\dfrac{\hat{V}_{in}^2 T_s}{4LV_{dc}}$ |

Comparing the load at boundary condition for the three different current control schemes in Table I it can be concluded that valley current control PFC has the largest range for CCM operation (also presented in [5]) and peak current control PFC has the smallest.

## III. SMALL SIGNAL MODEL

In CCM boost PFC the duty ratio (d) is expressed by the relation,

$$(1-d)=\frac{\hat{V}_{in}}{V_{dc}}|\sin\omega t| \qquad (12)$$

It is noted from (12) that $(1-d)$ term inherently contains the sinusoidal template. This term is multiplied (with the help of integration reset technique) to the output of voltage loop compensator to arrive at reference current for various current control schemes.

$$i_{inp}=(1-d)v_{mpcc} \qquad \rightarrow \text{peak currrent control} \qquad (13)$$

$$i_{inv}=(1-d)v_{mvcc} \qquad \rightarrow \text{valley currrent control} \qquad (14)$$

Since in Boost converter

$$\bar{i}_s=d\bar{i}_{in}$$

$$\bar{i}_s=d(1-d)v_{mnlc} \rightarrow \text{NLC (average current control)} \qquad (15)$$

$i_{inp}$ and $i_{inv}$ are the loci of peak and valley of input current in each switching cycle, $\bar{i}_{in}$ and $\bar{i}_s$ is the average of input and switch current over a switching cycle and $v_{mpcc}, v_{mvcc}, \mathbf{v}_{mnlc}$ are the outputs of voltage compensator in peak, valley and non linear carrier control PFC respectively. The right hand side of (13)-(15) represents a carrier signal which is generated as explained in [2], [4] and [5].

The output of voltage compensator can be expressed as $\bar{I}_{in}+\tilde{i}_{in}(t)$ in multiplier based control or $\bar{V}_m+\tilde{v}_m(t)$ in integration reset based control. Here $\tilde{i}_{in}(t)$ and $\tilde{v}_m(t)$ are the

small disturbances (less than 15Hz) over the steady state output $\bar{I}_{in}$ and $\bar{V}_m$ given by,

$$\tilde{i}_{in}(t)=\hat{i}_{in-}\sin(\omega_- t+\phi) \qquad (16)$$

$$\tilde{v}_m(t)=\hat{v}_m\sin(\omega_- t+\phi) \qquad (17)$$

where, $\omega_-$ is the small perturbation frequency below 15Hz. The expression of reference current for various control techniques is derived as follows:

### A. Peak /Valley Current Control

Input currents can be written as:

$$I_{in}^*(t)=i_{inp}-\frac{\Delta i(t)}{2} \qquad \rightarrow \text{Peak Current Control} \qquad (18)$$

$$I_{in}^*(t)=i_{inv}+\frac{\Delta i(t)}{2} \qquad \rightarrow \text{Valley Current Control} \qquad (19)$$

Using (1), (12)-(14) the expression for reference current for peak and valley current control is given by (20) and (21) respectively.

$$I_{in}^*(t)=V_{mpcc}\frac{\hat{V}_{in}}{V_{dc}}|\sin\omega t|-\frac{\hat{V}_{in}T_s}{2L}|\sin\omega t|(1-\frac{\hat{V}_{in}}{V_{dc}}|\sin\omega t|) \qquad (20)$$

$$I_{in}^*(t)=V_{mvcc}\frac{\hat{V}_{in}}{V_{dc}}|\sin\omega t|+\frac{\hat{V}_{in}T_s}{2L}|\sin\omega t|(1-\frac{\hat{V}_{in}}{V_{dc}}|\sin\omega t|) \qquad (21)$$

### B. Average Current Control

$$I_{in}^*(t)=\left[\bar{I}_{in}+\hat{i}_{in-}\sin(\omega_- t+\phi)\right]|\sin\omega t| \qquad (22)$$

$$I_{in}^*(t)=\left[\bar{V}_m+\hat{v}_m\sin(\omega_- t+\phi)\right]\frac{\hat{V}_{in}}{V_{dc}}|\sin\omega t| \qquad (23)$$

Equation (22) is for multiplier based control and (23) for integration reset based control (NLC).

The reference current expression derived above for various control technique is used in (24).

$$|V_{in}(t)|I_{in}^*(t)=V_{dc}\bar{I}_{dc} \qquad (24)$$

Perturbing (24) using $V_m=\bar{V}_m+\tilde{v}_m(t)$, $V_{dc}=\bar{V}_{dc}+\tilde{v}_{dc}$ and $I_{dc}=\bar{I}_{dc}+\tilde{i}_{dc}$ and equating the lower frequency term (i.e. below 15 Hz) results into the small signal model equations listed in Table II.

TABLE II
SMALL SIGNAL MODEL

| Control | $\mathfrak{R}$ | $\tilde{i}_{dc}$ |
|---|---|---|
| Average Current Control PFC* | $R$ | $\dfrac{\hat{V}_{in}\tilde{i}_{in}}{2V_{dc}}-\dfrac{\tilde{v}_{dc}I_{dc}}{V_{dc}}$ |
| Nonlinear Carrier Control PFC | $R/2$ | $\dfrac{\hat{V}_{in}^2\tilde{v}_m}{2V_{dc}^2}-\dfrac{2\tilde{v}_{dc}I_{dc}}{V_{dc}}$ |
| Valley Current Control PFC | $\dfrac{R}{2}\left\|\left(-1\middle/\sqrt{\dfrac{\hat{V}_{in}^2 T_s}{4V_{dc}^2 L}}\right)\right.$ | $\dfrac{\hat{V}_{in}^2\tilde{v}_m}{2V_{dc}^2}-\left(\dfrac{2I_{dc}}{V_{dc}}-\dfrac{\hat{V}_{in}^2 T_s}{4V_{dc}^2 L}\right)\tilde{v}_{dc}$ |
| Peak Current Control PFC | $\dfrac{R}{2}\left\|\left(1\middle/\sqrt{\dfrac{\hat{V}_{in}^2 T_s}{4V_{dc}^2 L}}\right)\right.$ | $\dfrac{\hat{V}_{in}^2\tilde{v}_m}{2V_{dc}^2}-\left(\dfrac{2I_{dc}}{V_{dc}}+\dfrac{\hat{V}_{in}^2 T_s}{4V_{dc}^2 L}\right)\tilde{v}_{dc}$ |

* Multiplier based

Using the results of $\tilde{i}_{dc}$ given in Table II, a generalized small signal model is constructed and shown in Fig. 2. The equivalent resistance $\mathfrak{R}$ that appears in parallel with the load resistance is different for the four current control techniques as given in Table II. The value of current source I is the first term of $\tilde{i}_{dc}$. The expression of $\tilde{i}_{dc}$ and the small signal model for non linear carrier control PFC is same as reported in [2]. A detailed derivation of small signal model for predictive switching modulator (valley current control) is also given in [5].

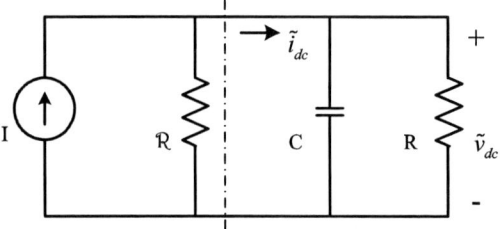

Fig. 2. Small signal equivalent circuit

## IV. SIMULATION RESULTS

A 600W PFC is designed and simulated in Saber© for peak and average current control. The PFC specifications and boost circuit parameters are listed below:

Input Voltage = 120 V AC RMS
Output Voltage = 300V
Power output (full load) = 600W
Line frequency = 60Hz
Switching frequency = 100 kHz
L = 2.39mH; C = 220μF

Fig. 3 shows the compensator output for peak current control PFC. Using (6), (7) and the specifications given above, the voltage compensator output is calculated as 12.77V. The magnitude of $V_m$ in Fig. 3 matches with the theoretical calculation. Fig. 3 also shows the input current and the input voltage waveforms to indicate unity power factor.

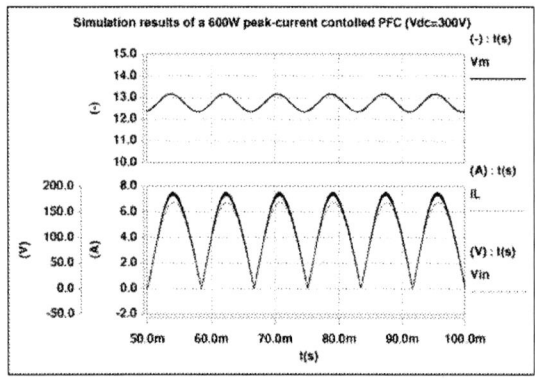

Fig. 3. $V_m$, $I_{in}$ and $V_o$ for peak current control PFC

Fig. 4 shows the response of the output voltage for a step input in $V_m$. From the small signal equivalent circuit, the time constant for the first order system for peak current control is given by:

$$\tau = \frac{C}{\dfrac{3}{R} + \dfrac{\hat{V}_{in}^2 T_s}{4V_{dc}^2 L}} \quad (25)$$

Substituting the values we get $\tau$ as 11ms. From Fig. 4, it is clear that the time constant matches the simulation result.

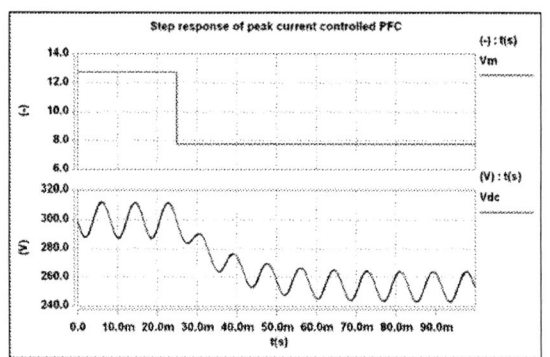

Fig. 4. Response of $V_{dc}$ for a step change in $V_m$ for peak current control PFC

Using the equations given in Table1 and 2 for average current control (multiplier based), the compensator output is calculated as 7.06A and the time constant for the small signal equivalent circuit is given by $RC/2 = 16.5ms$. Fig. 5 and 6 show the simulation results for average current mode control (multiplier based). It should be noted that for the same loading conditions, the output voltage will have a faster response for peak current mode control than average current mode control for a change in compensator output.

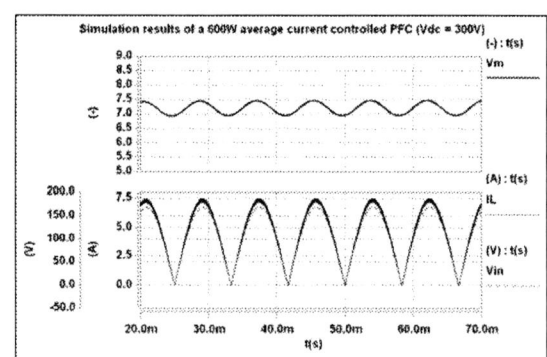

Fig. 5. $V_m$, $I_{in}$ and $V_o$ for average current control PFC

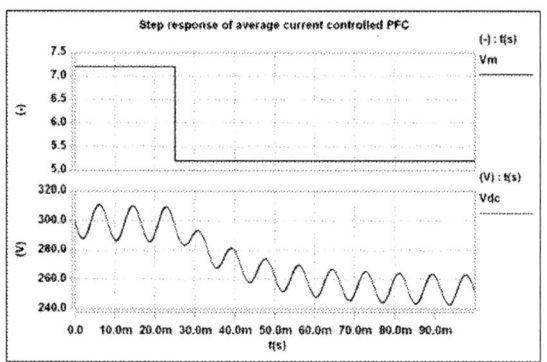

Fig. 6. Response of $V_{dc}$ for step change in $V_m$ for average current control PFC

## V. CONCLUSION

A unified method which uses Fourier analysis of input current to derive the expression for steady state output of voltage compensator and CCM boundary condition is proposed. The theoretical values calculated from the expressions derived in this paper and the values from the simulation results match. The small signal model derivation shows the similarity between multiplier and integration reset based control. It also gives a comparison of transient response among the three current controllers.

## VI. REFERENCES

[1]  Ned Mohan, *First Course in Power Electronics,* MNPERE, 2005.

[2]  D. Maksimovic, Yungtaek Jang and R. Erickson, "Nonlinear-carrier control for high power factor boost rectifiers," *Proc 1995 IEEE Applied Power Electronics Conf.,* vol. 2, pp. 635-641.

[3]  J. P. Gegner and C. Q. Lee, "Linear peak current mode control: A simple active power factor correction control technique for continuous conduction mode," *Proc 1996 IEEE Power Electronics Specialists Conf.,* vol. 1, pp. 196-202.

[4]  R. Brown and M. Soldano, "One cycle control IC simplifies PFC designs," *Proc 2005 IEEE Applied Power Electronics Conf.,* vol. 2, pp. 825-829.

[5]  S. Chattopadhyay, V. Ramanarayanan and V. Jayashankar, "A predictive switching modulator for current mode control of high power factor boost rectifier," *IEEE Trans. Power Electronics,* vol. 18, pp. 114-123, Jan. 2003.

[6]  K. M. Smedley and S. Cuk, "One-cycle control of switching converters," *Proc 1991 IEEE Power Electronics Specialists Conf.,* pp. 888-896.

[7]  Zheren Lai, K. M. Smedley and Yunhong Ma, "Time quantity one-cycle control for power-factor correctors," *IEEE Trans. Power Electronics,* vol. 12, pp. 369-375, Mar. 1997.

## VII. BIOGRAPHIES

**Ranjan K. Gupta** (S'06) is a graduate student, currently working on his Masters thesis in the department of Electrical and Computer Engineering at University of Minnesota, Minneapolis. He completed his Bachelor degree in Electrical Engineering from Indian Institute of Technology Roorkee, India in 2004. His research interests include Permanent magnet brushless DC motor design, Switched reluctance motors and Single phase power factor correction.

**Hariharan Krishnaswami** (S'06) is currently pursuing his Ph.D. degree in the Department of Electrical and Computer Engineering, University of Minnesota, Minneapolis. Between 2002-2006 he was with GE Healthcare, Bangalore, India as Senior Design Engineer. He obtained his M.S. degree in electrical engineering from Indian Institute of Science, Bangalore, India in 2002 and his Bachelor degree in electrical engineering from College of Engineering, Guindy, Anna University, India in 1999. His interests are in power converter topologies, power factor correction and soft-switching.

**Ned Mohan** (S'72–M'73–SM'91–F'96) is the Oscar A. Schott Professor of Power Electronics at the University of Minnesota, Minneapolis, where he has been since 1976. He has many patents and publications in the field of power electronics. He has written four books: *Power Electronics: Converters, Applications and Design,* (New York: Wiley 2002) coauthored with T. M. Undeland and W. P. Robbins; *Electric Drives: An Integrative Approach,* (Minneapolis, MN: MNPERE, 2003); *Advanced Electric Drives: Analysis, Control and Modeling Using Simulink,* (Minneapolis, MN: MNPERE, 2001); and *First Course of Power Electronics and Drives,* (Minneapolis, MN: MNPERE, 2005). Dr. Mohan received the Distinguished Teaching Award from the Institute of Technology, University of Minnesota.

**2006 IEEE International Conference on Power Electronic, Drives and Energy Systems**

# Minimum Loss Configuration of Power Distribution System

Jaswanti, *Member, IEEE,* and T. Thakur, *Member, IEEE*

*Abstract*---This paper presents a new method for minimum loss reconfiguration for radial power distribution system, in which the choice of the switches to be opened / closed is based on the calculation of voltage at the buses, real and reactive power flowing through lines, real power losses and voltage deviation, using distribution load flow (DLF) program. In the process of load flow calculation, two developed matrices bus injection to bus current (BIBC) and branch current to bus voltage (BCBV), and a simple matrix multiplication are used to obtain load flow solutions. A 33-bus radial distribution test system is taken as a study system for performing the test. The results reveal the speed and the effectiveness of the proposed method for solving the problem.

*Index Terms*---Combinatorial Optimization, Distribution network reconfiguration, Resistive line losses minimization.

## I. INTRODUCTION

LOSS minimization in power distribution system is one of the biggest challenges before power engineers. Transmission and distribution losses are as high as 20 to 30 percent of total power generation. Therefore, the challenge is more pronounced in case of distribution systems. Basic reason behind these huge power losses is resistive loss, as distribution systems are operated at much lower voltages as compared to transmission systems. So, operating current in distribution system is much more than that in transmission systems, and hence, larger power loss (resistive) in distribution systems as compared to transmission systems. So, in totality, we can say that, optimal operation of a distribution power networks has become an engineering challenge.

The concept of reconfiguring the topology of the distribution network to minimize losses can immediately be recognized as being cost efficient and consequently of interest to efficiency conscious electric utilities. Electric distribution networks are mostly an figured as radial for proper protection coordination, distribution feeders may be frequently reconfigure by opening and closing switches to while meeting all load requirements and maintaining a radial network. This requirement results in a very complicated non-linear integer optimization problem. The exact optimal solution of such a problem may be obtained only by adumbratively examining all possible switch options requiring prohibitively long computation time because the number of switch options is usually very large in a practical distribution network. Therefore many heuristic approximation methods have been proposed for efficiently solving the problem.

Jaswanti and T.Thakur are with the Department of Electrical Engineering, Punjab Engineering College (Deemed University), Chandigarh, India -160012 (e-mail: jaswanti98@yahoo.co.in; tilak20042005@yahoo.co.in)

A large number of papers has been published so for on various load flow approaches and reconfiguration of distribution networks [1-17].

This paper presents a new method for minimum loss configuration in radial power distribution networks. In the process of calculation, two developed matrices Bus Injection to Bus Current (BIBC) and Branch Current to Bus Voltage (BCBV), and a simple matrix multiplication are used to obtain load flow solutions. The solution get converged very early on; therefore execution time is very small. The results reveal the speed and the effectiveness of the proposed method for solving the problem.

The paper is organized as follows: Section 2 Mathematical Problem Formulation. Section 3 Algorithm Development. Section 4 Computation results from the proposed method are compared with other methods. Lastly, conclusion is given in Section 5.

## II. MATHEMATICAL PROBLEM FORMULATION

### A. Computation of Voltages at the Buses

The algorithm used to calculate the load flow is a novel one [15]. The only input data of this algorithm is the conventional bus-branch oriented data used by most utilities. Here the objective is to develop a formulation, which takes advantages of the topological characteristics of distribution systems, and solve the distribution load flow directly. It means that the time consuming L-U decomposition and forward backward substitution of the Jacobian matrix or the Y admittance matrix, required in the traditional Newton-Raphson and Gauss implicit Z matrix algorithms, are not necessary in the new development.

In order to obtain load flow solutions, first objective is to obtain voltages at the buses.

If $V^k$ is the voltages of the buses after $k^{th}$ iteration, then voltages at the buses after $(k+1)^{th}$ iteration is given by

$$V^{k+1} = V^k - \Delta V^k \qquad (1)$$

Here $\Delta V^k$ is change in bus voltages after two successive iterations.

### B. Real Power Flow

$$P_{ij} = \mathrm{Re}\, al[V_i\{(V_i - V_j)y_{ij}\}^*] \qquad (2)$$

Here $P_{ij}$ is the real power flowing through the line connecting $i^{th}$ and $j^{th}$ buses, $V_i$ and $V_j$ are the voltages of $i^{th}$ and $j^{th}$ bus respectively and $y_{ij}$ is the admittance of the line between $i^{th}$ and $j^{th}$ buses.

0-7803-9771-1/06/$25.00 ©2006 IEEE

## C. Reactive Power Flow

$$Q_{ij} = \text{Im}\,ag[V_i\{(V_i - V_j)y_{ij}\}^*] \qquad (3)$$

Here $Q_{ij}$ is the real power flowing through the line connecting $i^{th}$ and $j^{th}$ buses.

## D. Real Power Loss

$$\text{Loss} = \text{Real}\left\{ V_{ss} \sum_{j \in ss}\left[(V_{ss} - V_j)y_{ss,j}\right]^* - \sum_{j=1}^{N} PD_j \right\} \qquad (4)$$

Where $V_{ss}$ and $V_j$ in Eq. (2.4) refers to the voltages at main substation and bus j, respectively, $y_{ss,j}$ refers to the line admittance between the main substation bus and bus j, $PD_j$ refers to the real power load at bus j and N the number of buses in the RDS.

## E. Voltage Deviation Index (VDI)

In order to quantify the extent of violation of limits imposed on voltages at buses in a RDS, the following Voltage Deviation Index (VDI) has been defined.

$$VDI = \sqrt{\frac{\sum_{i=1}^{NVB}(V_{Li} - V_{LiLIM})^2}{N}} \qquad (5)$$

Subject to $\qquad V_{jMIN} \leq V_j \leq V_{jMAX} \qquad j \in 1 \text{ to } N$

Where NVB is the number of buses that violates the prescribed voltage limits and $V_{LiLIM}$ is the upper limit of the $I^{th}$ load bus voltage if there is upper limit violation or lower limit if there is a lower limit violation.

During reconfiguration, if the state of the system has voltage limit violations; the given solution must try and minimize the index VDI.

### III    ALGORITHM DEVELOPMENT

A sample distribution system drawn bellow is taken here to illustrate the methodology [6].

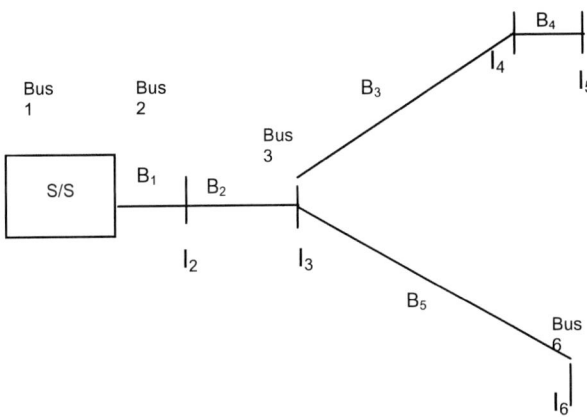

Fig. 1. Equivalent current injection based model of distribution network.

The distribution Networks, the equivalent-current-injection-based model is more practical as shown in figure 1.

For bus i, the complex load Si is expressed by

$$Si = (Pi + jQi) \quad i = 1 \ldots\ldots N \qquad (6)$$

Corresponding equivalent current injection at the $k^{th}$ iteration of solution is

$$Ii^k = Ii^k(Vi^k) + j\,Ii^i(Vi^k) = \left(\frac{Pi + jQi}{Vi^k}\right)^* \qquad (7)$$

Where $Vi^k$ and $Ii^k$ are the bus voltage and equivalent current injection of bus I at the $k^{th}$ iteration, respectively. $Ii^r$ and $Ji^i$ are the real and imaginary parts of the equivalent current injection of bus i at the $k^{th}$ iteration respectively.

## A. Relationship Matrix Development

A sample distribution system shown in fig. 1 is used as an example here. The power injection can be connected to the equivalent current injections by using equation 7 and relationship between the bus current injections and branch current can be obtained by applying Kirchoff's current law (KCL) to the distribution network. The branch currents can then be formulated as functions of equivalent current injections. For example the branch currents $B_1$, $B_2$ and $B_5$ can be expressed by equivalent current injections as

$$B_1 = I_2 + I_3 + I_4\,I_5 + I_6$$
$$B_3 = I_4 + I_5$$
$$B_5 = I_6 \qquad (8)$$

Therefore the relationship between the bus current injections and branch currents can be expressed as

$$\begin{bmatrix} B_1 \\ B_2 \\ B_3 \\ B_4 \\ B_5 \end{bmatrix} = \begin{bmatrix} 1 & 1 & 1 & 1 & 1 \\ 0 & 1 & 1 & 1 & 1 \\ 0 & 0 & 1 & 1 & 0 \\ 0 & 0 & 0 & 1 & 0 \\ 0 & 0 & 0 & 0 & 1 \end{bmatrix} \begin{bmatrix} I_2 \\ I_3 \\ I_4 \\ I_5 \\ I_6 \end{bmatrix} \qquad (9.1)$$

Above can be expressed in general form as

$$[B] = [BIBC][I] \qquad (9.2)$$

The relationship between branch currents and bus voltages can be obtained as follows:

$$V_2 = V_1 - B_1 Z_{12} \qquad (10.1)$$
$$V_3 = V_2 - B_2 Z_{23} \qquad (10.2)$$
$$V_4 = V_3 - B_3 Z_{34} \qquad (10.3)$$

Substituting (10.1) and (10.2) into (10.3), the equation (10.3) can be written as

$$V_4 = V_1 - B_1 Z_{12} - B_2 Z_{23} - B_3 Z_{34} \qquad (11)$$

From (11), it can be seen that the bus voltage can be expressed as a function of branch currents, line parameters and the substation voltage. Similar procedures can be performed another buses; therefore the relationship between branch currents and bus voltages can be expressed as

$$\begin{bmatrix} V_1 \\ V_1 \\ V_1 \\ V_1 \\ V_1 \end{bmatrix} - \begin{bmatrix} V_2 \\ V_3 \\ V_4 \\ V_5 \\ V_6 \end{bmatrix} = \begin{bmatrix} Z_{12} & 0 & 0 & 0 & 0 \\ Z_{12} & Z_{23} & 0 & 0 & 0 \\ Z_{12} & Z_{23} & Z_{34} & 0 & 0 \\ Z_{12} & Z_{23} & Z_{34} & Z_{45} & 0 \\ Z_{12} & Z_{23} & 0 & 0 & Z_{36} \end{bmatrix} \begin{bmatrix} B_1 \\ B_2 \\ B_3 \\ B_4 \\ B_5 \end{bmatrix} \qquad (12.1)$$

Above equation (12.1) can be expressed in general form as

$$\text{DeltaV} = [BCBV][B] \qquad (12.2)$$

## B. Building Formulation Development

Observing equation. (9), a building algorithm for BIBC matrix can be developed as follows:

Step 1) For a distribution system with m branch sections and n buses, the dimension of the BIBC matrix is $m^x(n-1)$.

Step 2) If a line section $B_k$ is located between bus I and bus j, copy the column of the $I^{th}$ bus of the BIBC matrix to the column of the $j^{th}$ bus and fill a +1 to the position of the k-th row and the $j^{th}$ bus column.

Step 3) Repeat procedure (2) until all line sections are included in the BIBC matrix. From equation (12), a building algorithm for BCBV matrix can be developed as follows.

Step 4) For distribution network with in branch section and n bus, the dimension of BC BV matrix is (n-1) x m.

Step 5) If a line section $(B_k)$ is located between bus i and bus j copy the row of the $i^{th}$ bus of BCBV matrix to the row of the $j^{th}$ bus and fill the line impedances (Zij) to the positions of the $j^{th}$ bus row and $k^{th}$ column.

Step 6) Repeat procedure (5) until all line sections one included in the BCBV matrix.

The algorithm can easily be expanded to a multiphase line section or bus. For example, if the line section between bus i and bus j is a three phase line section, the corresponding branch current Bi will be a 3x 1 vector and the +1 in the BI BC matrix will be a 3 x 3 identity matrix. Similarly if the line section between bus i and bus j is a three phase line section, the Zij in the BCBV matrix is a 3 x 3 impedance matrix.

It can also be seen that the building algorithm of the BIBC and BCBV matrices are similar. In fact these two matrices will be made by using same subroutine of our test program. Therefore the competition resources needed, can be saved. In addition the building algorithms are developed based on the traditional bus-branch oriented data base; thus the data preparation time can be reduced and the proposed method can be easily integrated.

## C. Solution Technique Developments

The BIBC and BCBV matrices are developed based on the topological structure of distribution systems. The BIBC matrix represents the relation shop between bus current injections and branch currents. The corresponding variations at branch currents, generated by the variations at bus current injection can be calculated directly by the BIBC matrix. The BCBV matrix represents the relationship between branch current and bus voltages. The corresponding variations at bus voltage, generated by the variations at branch currents can be calculated directly by the BCBV matrix. Combining equation (9.2) and (12.2), the relationship between bus current injections and bus voltages can be expressed as

$$[\Delta V] = [BCBV][BIBC][I]$$
$$[\Delta V] = [DLF][I] \qquad (13)$$

DLF is a multiplication matrix of BCBV and BIBC matrices and the solution for distribution load flow can be obtained by solving (8) iteratively as

$$Ii^k = Ii^k (Vi^k) + j\, Ii^i (Vi^k) = \left( \frac{Pi + jQ_i}{V_i k} \right)^*$$

$$[\Delta V^{K+1}] = [DLF][I^k]$$
$$[V^{k+1}] = [V^0] + [\Delta V^{k+1}] \qquad (14)$$

According to the research, the arithmetic operation, number for LU factorization is approximately proportional to $N^3$. For a large value of N, the LU factorization will occupy a large portion of the computational time. Therefore if the LU factorization can be avoided, the load flow method can save tremendous computational resource. From the solution technique described in this chapter, the LU decomposition and forward backward substitution of the Jacobian matrix are the Y admittance matrices are no longer necessary. Only the DLF matrix is necessary in solving load flow problem. Therefore above discussed method can save considerable computation resources and this feature make the proposed method suitable for online operation.

## IV. RESULTS AND ANALYSIS

Distribution Load Flow (DLF) program has been tested on 33-bus RDS given in Fig. 2. The load data, line details and the tie lines available for switching are given in [16]. Substation voltage is 12.66 KV and base MVA has been taken as 10 MVA.

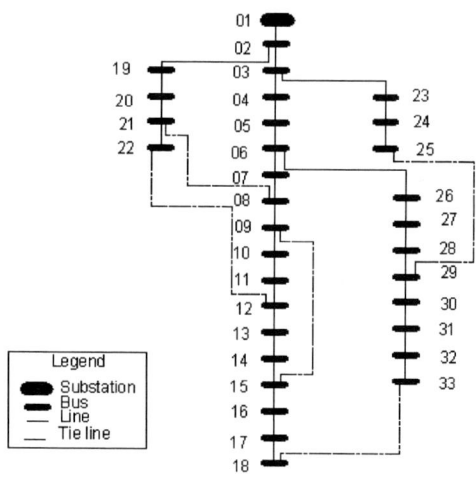

Fig. 2. A 33-bus radial distribution system.

System has five tie lines. The two configurations are termed as Base Configuration and Optimal Configuration respectively. Using DLF program voltages at the buses, real and reactive powers flowing through lines, real power loss and voltage deviation index (VDI) were calculated for the two configurations. ETAP simulation was performed and results obtained from the simulation proved the authenticity of the program developed.

Reduction in the real power loss, improved voltage deviation and increased bus voltages are the merits shown by the method used. This can be under stood by having a look on the following tables I and II:

714

TABLE I

| Case | Loss(DLF/given in [21]) in KW | VDI(DLF/given in [21]) | Worst Voltage(DLF/given in [21]) in p.u. |
|------|-------------------------------|------------------------|-------------------------------------------|
| Base | 201.42/211 | 0.0174/ 0.02489 | 0.9143/ 0.9038 |
| Optimal | 158.24/178 | 0.0039 0.0041 | 0.9388/ 0.9378 |

TABLE II

| Case | %Loss Reduction | % VDI Improvement | % Increment in Worst Voltage |
|------|-----------------|-------------------|------------------------------|
| Base | 4.5 | 30 | 1.16 |
| Optimal | 11.1 | 4.8 | 0.106 |

### A. Voltage comparison:

V1: Bus Voltages in per unit obtained from the DLF Program for Base Case.

V2: Bus Voltages in per unit obtained by using ETAP software.

TABLE III
COMPARISON OF DLF AND ETAP SOLUTIONS

| Bus No. | V1(p.u) | V2(p.u) |
|---------|---------|---------|
| 2 | 0.997 | 0.996 |
| 3 | 0.992 | 0.984 |
| 4 | 0.985 | 0.975 |
| 5 | 0.977 | 0.968 |
| 6 | 0.959 | 0.951 |
| 7 | 0.956 | 0.947 |
| 8 | 0.942 | 0.934 |
| 9 | 0.936 | 0.929 |
| 10 | 0.93 | 0.923 |
| 11 | 0.929 | 0.922 |
| 12 | 0.928 | 0.921 |
| 13 | 0.922 | 0.915 |
| 14 | 0.919 | 0.913 |
| 15 | 0.918 | 0.912 |
| 16 | 0.917 | 0.911 |
| 17 | 0.915 | 0.909 |
| 18 | 0.914 | 0.908 |
| 19 | 0.996 | 0.996 |
| 20 | 0.993 | 0.992 |
| 21 | 0.992 | 0.991 |
| 22 | 0.991 | 0.99 |
| 23 | 0.989 | 0.979 |
| 24 | 0.982 | 0.972 |
| 25 | 0.979 | 0.968 |
| 26 | 0.957 | 0.949 |
| 27 | 0.955 | 0.946 |
| 28 | 0.943 | 0.936 |
| 29 | 0.935 | 0.928 |
| 30 | 0.932 | 0.924 |
| 31 | 0.928 | 0.921 |
| 32 | 0.927 | 0.92 |
| 33 | 0.926 | 0.919 |

### A. Comparison Between the Bus Voltages

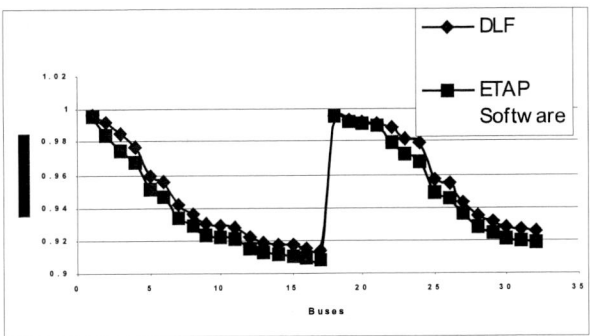

Fig. 3. Voltage comparison.

Solutions to the voltages at the buses obtained show that at each bus, voltage in case DLF program is better than those in case of ETAP simulation results. Worst bus voltage in case of ETAP is 0.908 and that in case of DLF method it is 0.914.Also the best voltage is higher in case of DLF solutions. Once the voltages become higher, the losses are bound to be reduced.. For the same load, power drawn in case of ETAP solutions is higher as compared to that obtained by DLF method. This only signifies the fact that losses in latter case have been reduced.

### B. Graphical Comparison of Base and Optimal Real and Reactive Power

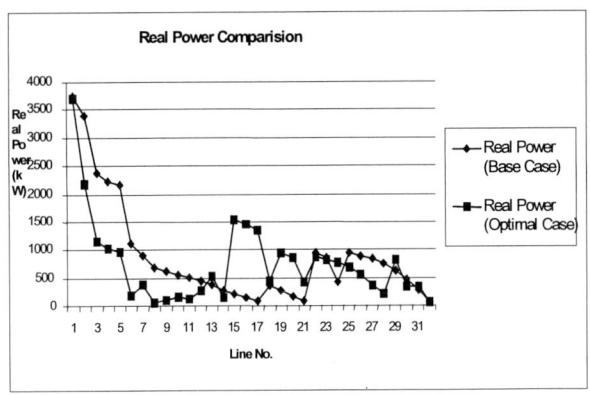

Fig. 4. Real power comparison.

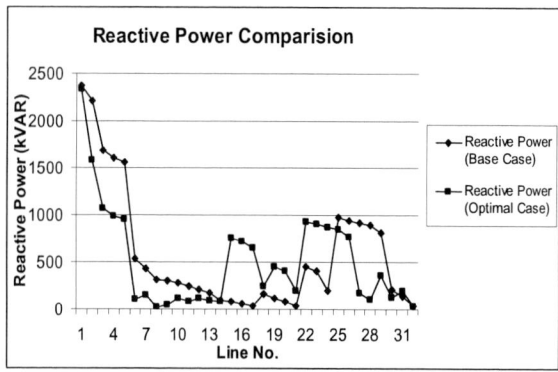

Fig. 5. Reactive power comparison.

Figure 4 and 5 are showing the comparisons between real and reactive power respectively flowing through the lines in the two cases. In optimal case lesser real power is required because the loss has been decreased. This was the objective to be achieved through reconfiguration. This discussion is equally applicable to reactive power comparison.

### C. Comparison of Base and Optimal Voltages

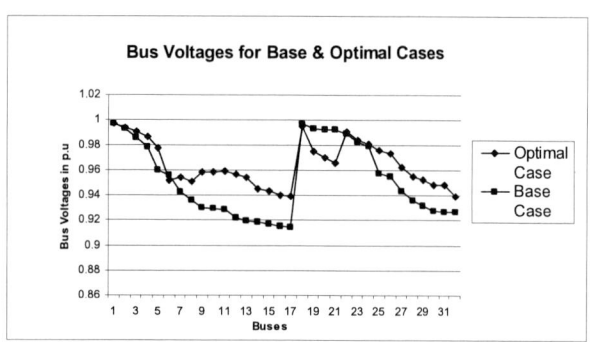

Fig. 6. Base and optimal voltage comparison.

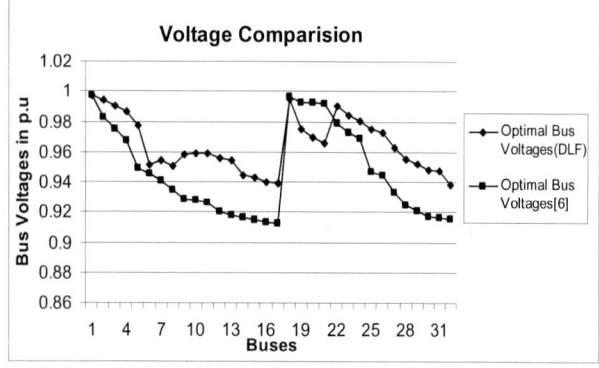

Fig. 7. Voltage comparison for optimal cases.

Fig 6 and 7compares the results obtained for the two cases considered. It is concluded from the figure that voltages at the

buses in case of optimal case is much better than that in the base case for majority of buses. Few buses have lower voltages (in case of optimal case) than that in base case. This is because of the fact that in the former case the structure of the network has been drastically changed as compared to that of later case.

Results obtained indicate that the approach to load flow solutions is much superior to the previous approaches such as used in [6] and ETAP software. For example even in base configuration the worst voltage is better than the worst voltage obtained through ETAP simulation. Also, voltages at the majority of buses are greater than those obtained by the other methods such as in [6] and ETAP simulation. Also, real and reactive powers drawn are lower for the same demand. This aspect leads the system to have lower losses and better voltage deviations index (VDI) as shown by the results.

Results shown and compared in table I and II. The voltage was improved by 4.5% and 11.1% in base case and optimal case respectively. Voltage deviation index was improved by 30% and 4.8% in base case and optimal case respectively. Similarly worst voltage was improved by 1.16% and 0.106% in base case and optimal case respectively.

## V. CONCLUSION

In this paper, a new method for minimum loss reconfiguration for radial power distribution system using DLF program is presented. Two matrices that are developed from the topological characteristics of distribution systems are used to solve distribution network load flow problems. The BIBC matrix represents the relationship between bus current injections and branch currents, and BCBV matrix represents the relationship between branch current and bus voltages. These two matrices are combined to form a direct approach for solving reconfiguration problems. The execution time is extremely smaller as compared to other recent methods reported in literature for radial distribution systems, such as fast decoupled and Gauss Implicit Z-matrix method. Here, we do not require to compute Z-matrix or jacobian.

Merit of used method is that it is very effective and solutions get conversed even in second equation there for the execution time of DLF program is quite small. This is the big advantage for distribution system where the load varies indiscriminately. Limitation of the program is that it can be used only for the radial distribution system and, not for meshed distribution systems and transmission systems.

## VI. REFERENCES

[1]. C.S.Cheng and D.Shirmohammadi, "A Three Phase Power Flow Method to Real-Time Distribution System Analysis," IEEE Transactions on Power Systems, Vol.10, Nov.1995.

[2]. R.D.Zimmermann and H.P.Chiang, "Fast Decoupled Power Flow foe Radial Distribution System," IEEE Transactions on Power Systems, Vol.10, Nov.1995.

[3]. Antonio Gomez Exposito Esther Romero Ramos, "Reliable Load Flow Technique for Radial Distribution Networks,"IEEE Trans on Power Syst, Vol.14, No.3, 1999.

[4]. Mesut E.Baran & Felix F. Wu, "Network Reconfiguration in Distribution Systems for Loss Reduction & Load Balancing," IEEE Transactions on Power Delivery, Vol.4, No.2, April 1989.

[5]. S.K.Goswami & S.K.Basu, "A New Algorithm for the Reconfiguration of Distribution Feeders for Loss Minimization," IEEE Transactions on Power Delivery, Vol.7, No.3, July1992.

[6]. Hoyong Kim, Yunseok Ko & Kung-Hee Jung, "Artificial Neural-Network Based Feeder Reconfiguration for Loss Reduction in Distribution Systems," IEEE Trans on Power Del, Vol.8, No.3, July1993.

[7]. Ji-Yuan Fan, L.Zhang & John D. McDonald, "Distribution Network Reconfiguration: Single Loop Optimization," IEEE Trans on Power Syst.Vol.11, No. 3, Aug.1996.

[8]. R.J.Sarfi, M.M.A.Salama, and A.Y. Chikhani, "Distribution system reconfiguration for system loss reduction: an algorithm based on network partitioning theory.", IEEE Trans on Power Syst,vol.11,Feb.1996.

[9]. Jin-C Wang, Hsiao-Dong Chiang & G.R.Darling, "An Efficient Algorithm for Real Time Network Reconfiguration in Large Scale Unbalanced Distribution Systems," IEEE Trans on Power Syst.Vol.11, No. 1, Feb.1996.

[10]. G.J.Peponis, M.P.Papadopoulas and N.D.Hatziragyrioue,"Optimal operation of distribution networks," IEEE Trans on Power Syst, vol.11, February 1996.

[11]. Tsai-Siang Chen & Jen-T. Cherng, "Optimal Phase Arrangement of Distribution Transformers Connected to a Primary Feeder for System Unbalance Improvement & Loss Reduction Using a Genetic Algorithm," IEEE Trans on Power Syst.Vol.15, No. 3, Aug. 2000.

[12]. K.N.Miu, H.D.Chiang, and R.J.Mcnulty, "Multitier service restoration through network reconfiguration and capacitor control for large scale radial distribution networks,"IEEE Trans on Power Syst, vol.15, Aug.2000.

[13]. Antonio Augugliaro, L.Dusonchet, M.G.Ippolito & E.R Sanseverino, "Minimum Loss Reconfiguration of MV Distribution Networks Through Local Control of Tie Switches," IEEE Transactions on Power Delivery,Vol.8,No.3,July2003.

[14]. A.Moussa, M.El-Gammal, E.N.Abdallah, and A.I.Attia, "A Genetic Based Algorithm For Loss Reduction in Distribution Systems", Resarch and Energy Conservation Sector in Alexandria Electricity Distribution Company, Alexandria, Egypt.

[15]. Jen-Hao Teng, "A Direct Approach for Distribution System Load Flow Solutions," IEEE Trans on Power Del, Vol.8, No.3, July2003.

[16].B.Venkatesh, Rakesh Ranjan & H.B.Gooi, "Optimal Reconfiguration of Radial Distribution Systems to Maximize Loadability," IEEE Transactions on Power Systems.Vol.19, No1, Feb. 2004.

[17].A. Merlin & H.Back, "Search for a Minimum Loss Operating Spanning Tree Configuration in an Urban Power Distribution System," Proc. 5th Power Systems Computer Conference (P.S.C.C) Cambridge, 1975.

## VII. BIOGRAPHIES

Mrs. Jaswanti graduated in Electrical Engineering from Punjab Engineering College, Chandigarh, India in 1993. She got her master of Engineering from same institute in 1997. She is currently doing her Ph.D in power system. She is member of IEEE. She is also a associate member of IEI and ISTE. Her main research interests are power distribution system operation, analysis and control.

Dr. Tilak Thakur is born in 1963. He graduated from B.I.T. Sindri, in Electrical engineering in 1987. He completed his Post graduation in Power System from the same institute and achieved his Ph.D in Electronic Instrumentation from Indian School of Mines, Dhanbad in the area of SCADA in 1999.

He served as a lecturer in B.I.T., Sindri and NERIST Arunachal Pardesh. Presently, he is Assistant Professor in the Department of Electrical Engineering, Punjab Engineering College (PEC), Chandigarh, India.. He has a teaching experience of more than 15 years. He is member of IEEE. He is involved in active Research in Power System Automation and Control

**2006 IEEE International Conference on Power Electronic, Drives and Energy Systems**

# Control of Cascaded H-Bridge Converter based DSTATCOM
# for High Power Applications

K. Anuradha, B.P.Muni, and A. D. Raj Kumar

*Abstract*—This paper presents the simulation studies on a Cascaded H-Bridge converter based Distribution Static Synchronous Compensator (DSTATCOM) for improving the power quality of a distribution system. Voltage source converter based DSTATCOM has been established as the most preferred solution for management of reactive power in distribution utilities and for improving voltage regulation, power factor and power quality in industries. For high power applications, cascaded H-Bridge converter is the most ideal choice compared to two-level inverter with series connected power devices. In the presnt work DSTATCOM controller is designed using DQ0 modelling for reactive power management and thereby improving the power factor in distribution systems. The dc link voltage and the three phase load currents are ued as feed back signals for the controller and it is designed in such a way that DSTATCOM is able to supply the reactive current demanded by the load both during steady state and transient conditions using sinusoidal pulse width modulation control.

*Index Terms*—**Cascaded H-Bridge Converter, DSTATCOM, Reactive power compensation, Sinusoidal PWM.**

## I. INTRODUCTION

RECENTLY, with the growth of industry manufacturers and population, electric power quality becomes more and more important The availability of high power self-commutating switching devices, high speed digital signal processors and discovery of advanced control philosophies have paved the way for application of DSTATCOM in distribution utilities and improving power quality in industries. The main function of a DSTATCOM is to inject or absorb reactive power to the grid for improving power factor and voltage regulation. Further, with change in control approach, the DSTATCOM can be used as an active filter and a dynamic uninterruptible power source. As the power rating

---

K.Anuraha is with the Department of Electrical and Electronics Engineering,VNRVJIET, JNTU, Hyderabad-500072,India.(e-mail:anuyalavarti@yahoo.co.in)

Bishnu P. Muni (e-mail: bpmuni@bhelrnd.co.in), is with BHEL, Corporate R&D, Hyderabad – 500093, India

A.D.Rajkumar is with the Department of Electrical Engineering, Osmania University, Hyderabad-- 500007 ,India (e-mail:adjkumar@yahoo.com).

increases, high voltage switching devices have to be used to reduce the current rating of the converter. High voltage switching devices are comparatively costlier and cannot be switched at high switching frequency. To overcome these problems several new inverter topologies have been used in high voltage FACTS, custom power equipment and industrial drives. Multilevel converters based on neutral point clamped philosophy and cascaded H-Bridge converter are widely used for high power conditioning applications.

A cascaded H-bridge converter has less number of devices compared to same voltage level neutral point converter. Cascaded converters are modular in nature. The power and voltage rating as well waveform quality can be enhanced by cascading several H-bridges.[1]

Most of the published literature on DSTATCOM is based on two-level [2], [3] or neutral point clamped multi level inverters. The adoption of H-bridge Cascaded converters for DSTATCOM applications leads to reduced injection of harmonics and low cost. Low voltage power switching devices are available at less cost from several manufactures. The major advantages of the H-bridge cascaded converters are scalable power rating, modularity, and cost effectiveness. The output voltage of the cascaded H-bridge converter is the summation of the output voltage of the individual H-bridges. By increasing the number of series H-bridge converters, the output voltage of the converter can be increased, while the switching frequency of the individual H-bridge converter can be decreased to achieve the same output waveform quality.

## II. CASCADED INVERTERS WITH SEPARATE DC SOURCES

A cascaded H-bridge converter consists of a series of H-bridge (single-phase full-bridge) inverters per phase. The general function of the cascaded H-bridge converter is to synthesize a desired voltage from several Isolated DC Sources (IDCS). Fig. 1 shows single-phase structure of the cascaded inverter with IDCS [4]. Each IDCS is connected to a single-phase-full-bridge inverter. For DSTATCOM applications, the DC link capacitor of each H-bridge can be maintained at desired level by closed loop control by drawing power from the grid. Each inverter level can generate three different voltage outputs, $+V_{dc}$, $0$ and $-V_{dc}$ by connecting the dc source to the ac output side by different combinations of the four switches $S_1$, $S_2$, $S_3$ and $S_4$. To obtain $+V_{dc}$ switches $S_1$ and $S_4$ are turned on. Turning on switches $S_2$ and $S_3$ yields $-V_{dc}$. By turning on $S_1$ and $S_2$ or $S_3$ and $S_4$ the output voltage is

0-7803-9771-1/06/$25.00 ©2006 IEEE

0. The ac output of each of the different level full-bridge inverters is connected in series such that the synthesized voltage waveform is the sum of the individual inverter outputs. The number of output phase voltage levels in a Cascaded Inverter is defined by m=2s+1; where s is the number of dc sources. An example phase voltage waveform for a 11-level Cascade-Inverter with five SDCSs and five full bridges is shown in Fig. 2. The phase voltage $V_{an}=v_1+v_2+v_3+v_4+v_5$. A three phase cascaded H-bridge converter consists of three numbers of series connected H-bridge inverters, connected in star or delta. The star connected cascaded H-bridge converters are popular because of higher output voltage and better waveform quality compared to delta connected scheme with same numbers of converter units per phase. The power circuit of a 3-level cascaded H-bridge converter is shown in Fig. 3.

### A. Generation of Sinusoidal PWM pulses

In most of the published literature stepped wave modulation is used as a control strategy for Cascaded Multilevel Converters in very high power applications like STATCOM. In the present paper the most popular method of generating the PWM pattern that is Sinusoidal PWM is applied for cascaded H-bridge converter as the controller is designed for power quality improvement. The generation of PWM signals for the switches $s_1$, $s_3$, $s_2$, $s_4$ of a single H-Bridge inverter of Fig. 1 are shown in Fig. 4.

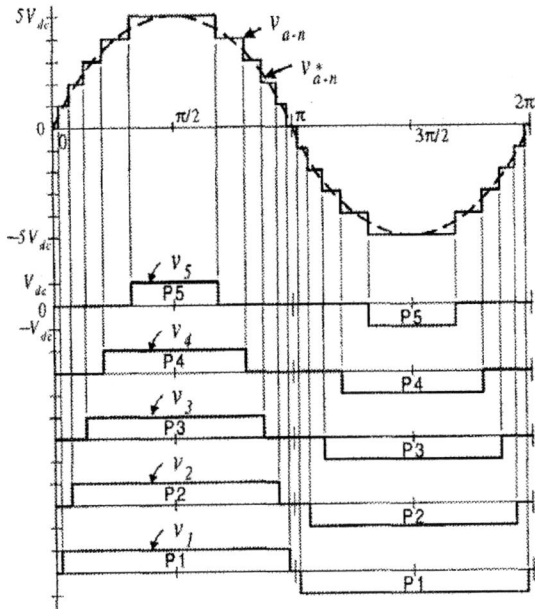

Fig. 2. Output voltage waveform of an 11-level cascade inverter.

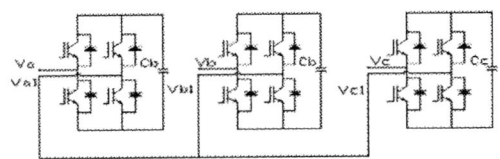

Fig. 3. Three-level cascaded H-bridge converter.

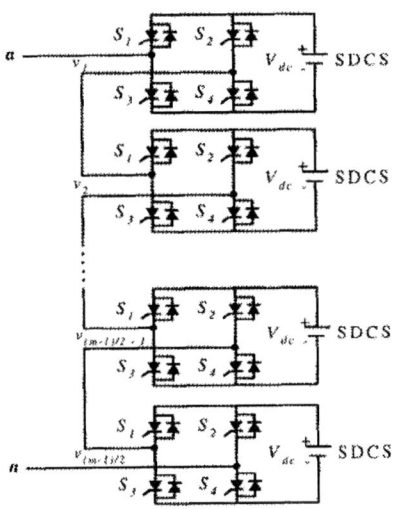

Fig. 1. Single phase structure of a multilevel cascaded inverter.

Here gating signals for a pair of devices ($s_1$, $s_3$) of a H-Bridge are obtained by comparing the sinusoidal modulating signal with triangular carrier wave. The frequency of modulating signal is kept at 50 Hz. And that of carrier signal is kept at 1080 Hz. Using a comparator the modulating signal and carrier signals are compared and gating pulses are provided to switch $s_1$ when the amplitude of modulating wave is more than the triangular wave. The complements of gate pulses of switch s1 are applied to switch $s_3$ of the same leg. For another pair of devices ($s_2$, $s_4$) gating signals are obtained by comparing the inverse of the sinusoidal modulating signal with triangular carrier wave.

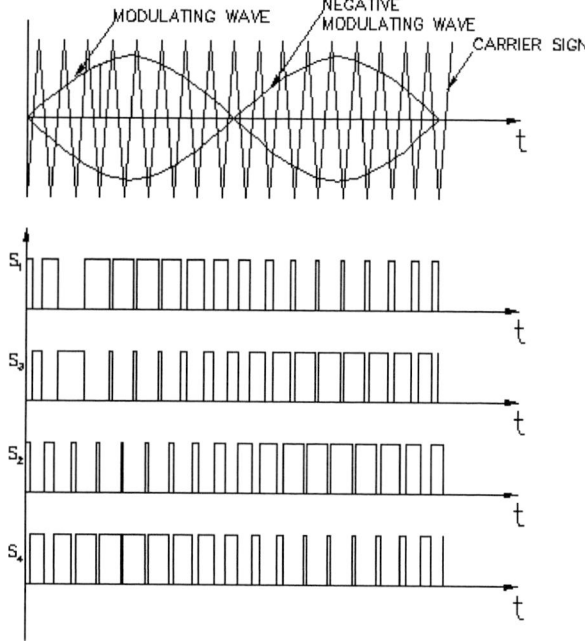

Fig. 4. Generation of Sinusoidal PWM signals for a single H-Bridge Inverter.

## III. DSTATCOM

The DSTATCOM is a static compensator consisting of IGBT or GTO based voltage source converter to provide voltage stabilization, power factor correction, flicker compensation, harmonic compensation and a host of other power quality solutions for both utility and industrial applications. The basic principle of reactive power generation by a voltage converter is the same as that of a conventional rotating synchronous machine. The basic voltage source converter scheme for reactive power generation is shown schematically in the form of a single line diagram as shown in Fig. 5. From a DC input voltage source, provided by the charged capacitor C, the converter produces a set of controllable three-phase output voltages of the same frequency as that of the AC power system. Each output voltage can be controlled both in magnitude and phase angle, which is coupled to the corresponding AC system voltage through a relatively small (0.15 – 0.2 p.u.) tie reactance (which in practice may be provided by per phase leakage inductance of the coupling transformer).

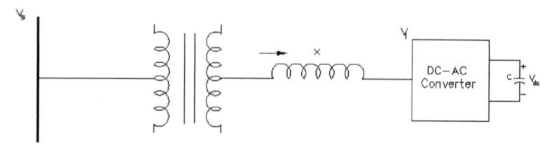

Fig. 5. Power circuit of DSTATCOM.

The active and reactive powers (P,Q) exchanged between the system and the converter are given by the following equations.

$$P = \frac{V_s V_i}{X} Sin\delta \qquad (1)$$

$$Q = \frac{V_s}{X}(V_s - V_i Cos\delta) \qquad (2)$$

where $\delta$ =Phase angle between $V_S$ and $V_i$ and X is the total reactance (of the reactor and transformer if any).

## IV. MODELING AND CONTROL

The cascaded H-bridge converter based DSTATCOM for power factor improvement is simulated in MATLAB & Simulink.[5]. The schematic of an entire system is shown in Fig. 6. The DSTATCOM basically consists of three main parts: a cascaded H-bridge converter with separate DC capacitors, the coupling inductors, and a closed loop controller. The coupling inductor in each phase serves both as a converter output-current filter and an inductive coupler between the mains phase and the respective converter phase voltage.

### A. Control strategy

The control strategy for DSTATCOM involves the measurement of three phase voltages at the source and currents at the load , inverter and DC link voltage only.

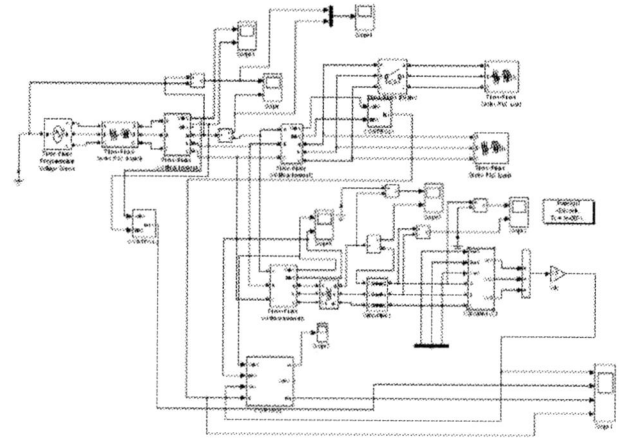

Fig. 6. The cascade H-bridge converter based DSTATCOM.

The instantaneous power flow in the STATCOM is shown in Fig. 7. The source power has two components, viz. instantaneous active power proportional to $i_d$ and instantaneous reactive power proportional to $i_q$. Similarly, the load power and inverter power also comprise of active and reactive components. For full reactive power compensation of the load, the inverter has to supply reactive power of the same magnitude as the load but have opposite sign. Thus, in such a case, the reactive power drawn from the source is zero.

$$q_s = q_l + q_i \qquad (3)$$

Where  $q_s$ - reactive power supplied by the Source,

$q_l$ – reactive power absorbed by the Load and

$q_i$ - reactive power absorbed by the Inverter

Under complete compensation,

$$q_i = -q_l \qquad (4)$$

Therefore, here the reactive power absorbed by the load which is proportional to $i_q$ is monitored and set as reference ($i_{qref}$) for the controller of Cascaded three level converter in closed loop so that DSTATCOM supplies the reactive power demanded by the load. Thus the desired objective of full VAR compensation can be achieved.

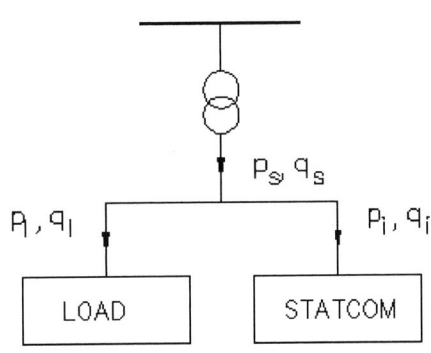

Fig. 7. Power flow in a Distribution STATCOM.

The inverter has to draw active power from the grid to maintain the DC link voltage at a desired level. This can be achieved by DC link voltage control loop. The control block diagram of the cascaded three-level converter-based DSTATCOM is shown in Fig. 8.

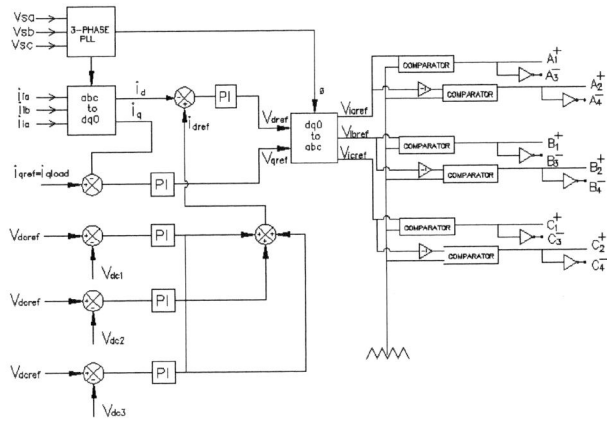

Fig. 8. The control block diagram of the cascaded three-level converter-based DSTATCOM.

The three- phase source voltages ($V_{sa}$, $V_{sb}$, $V_{sc}$) are applied to three-phase Phase Locked Loop (PLL) to synchronize the three-phase voltages at the converter output with the zero crossings of the fundamental component of the supply phase voltages. The PLL provides the synchronous reference angle $\phi$ required by the abc-dq0 (and dq0-abc) transformation. The three phase cascaded inverter currents ($i_{ia}$, $i_{ib}$, $i_{ic}$) are converted into equivalent direct axis and quadrature axis component currents ($i_d$, $i_q$) by using the following equations [6].

$$\begin{bmatrix} i_d \\ i_q \\ i_o \end{bmatrix} = \frac{2}{3} \begin{bmatrix} \sin(\omega t) & \sin(\omega t - \frac{2\pi}{3}) & \sin(\omega t + \frac{2\pi}{3}) \\ \cos(\omega t) & \cos(\omega t - \frac{2\pi}{3}) & \cos(\omega t + \frac{2\pi}{3}) \\ \frac{1}{2} & \frac{1}{2} & \frac{1}{2} \end{bmatrix} \begin{bmatrix} i_a \\ i_b \\ i_c \end{bmatrix} \qquad (5)$$

In order to maintain the reactive power drawn from the source as zero, the output currents of the cascaded H-bridge inverter are controlled in such a way that the inverter supplies the required load reactive power. Thus for a power factor of unity at source end , the load reactive power sets the reference for inverter control which sets $i_q$ reference ($i_{qref}$) as $I_{qload}$. The reactive current supplied by the inverter ($i_q$) is subtracted from the reference value ($i_{qref} = I_{qload}$) to obtain the error in reactive current for full compensation. This error signal is passed through a PI controller block to obtain the reference voltage signal ($V_{qref}$), which is fed to the dq0-abc transformation block.

The reference for $i_d$ ($i_{dref}$) comes from the DC link voltage PI controller, which maintains the DC link voltage ($V_{dc}$) at reference value ($V_{dcref} = 4000$ V). The active current supplied by the source ($i_d$) is subtracted from the reference value ($i_{dref}$) and this error signal is processed through a PI controller block to obtain the reference voltage signal ($V_{dref}$), which goes as another input for dq0-abc transformation. PI compensators for current and voltage loops are tuned to give the optimum performance.

The output voltage signals of transformation block (dq0-abc) act as reference voltages ($V_{ia}$, $V_{ib}$, $V_{ic}$) for PWM signal generators of cascaded H-bridge converter. These signals are compared with a triangular carrier wave to obtain PWM signals for Cascaded H-bridge inverter phases, respectively.

## V. SIMULATION STUDIES

To verify the control performance, the proposed controller with the three-level cascaded inverter based DSTATCOM is simulated. For the simulation study a three-phase source is treated as the primary distribution substation and the distribution line is considered as the lumped inductance in series with the resistance. The entire system was simulated with a steady load application along with a sudden change of load by connecting or disconnecting an additional load through a three-phase circuit breaker. The phase voltage and line-to-line voltage of H-bridge cascaded inverter are shown in Fig. 9.

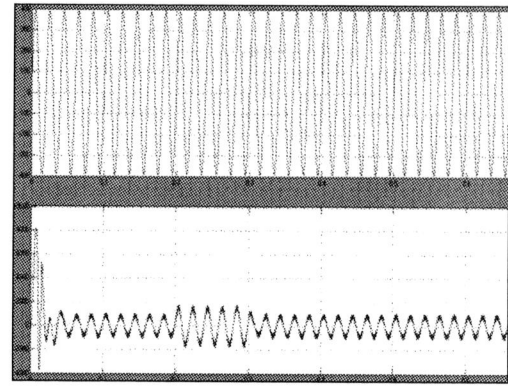

Fig. 10. Source phase voltage (top trace) and source phase Current (bottom trace) with DSTATCOM in closed loop power factor control mode.

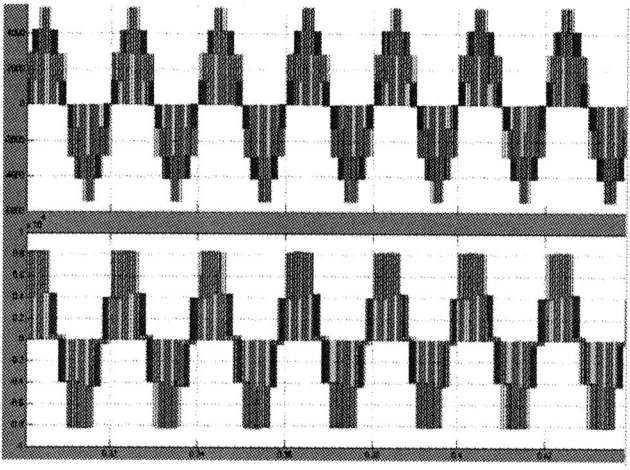

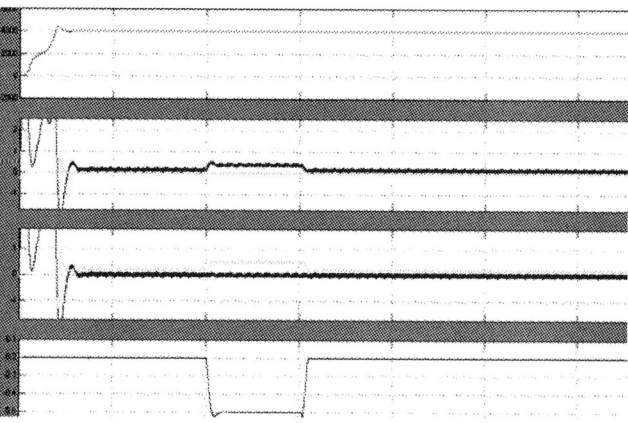

Fig. 11. DC link voltage ($V_{dc}$) (Top or First Trace), direct and quadrature axis source currents (Second Trace) ,inverter currents $I_d$ and $I_q$ (Third Trace) and load reactive current (Bottom Trace).

Fig. 9. The phase voltage (top trace ) and line-to-line voltages of H-bridge cascaded inverter.

Fig. 10 shows the source phase voltage and current. The DSTATCOM is switched with a steady load of 1.25 MVA with a power factor of 0.8 lag. At 0.2 seconds an additional load 2MVA with power factor of 0.8 lag is switched on. At 0.3 seconds the additional load is switched off .It is observed that the source phase current is in phase with source phase voltage both during steady state and transient conditions. The change in amplitude of source current is because of active power drawn by the additional load. Fig.11. shows the DC link voltage, direct and quadrature axis component of source current, inverter current and load reactive current during the steady state and transient operating state of the DSTATCOM. It is observed that the net reactive power drawn from the source is zero in steady state and transient conditions and load reactive current is supplied by the DSTATCOM. Fig.12. shows three individual capacitor voltages of three phase Cascaded H-Bridge Converter. It is seen from the Figure that Individual Capacitor Voltages are maintained constant.

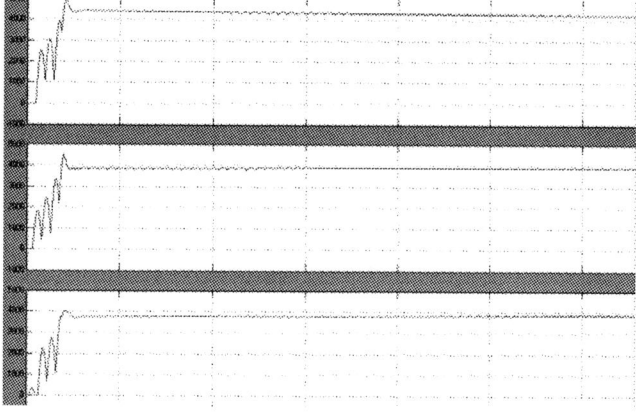

Fig . 12. Individual Capacitor voltages of three level Cascaded H-Bridge Inverter.

## VI. CONCLUSIONS

The paper presents the principle of operation of cascaded H-bridge converter and simulation studies on cascaded converter based DSTATCOM using Sinusoidal PWM control. It is observed that the DSTATCOM is capable of supplying the reactive power demanded by the load both during steady state and transient operating conditions. The harmonics in cascaded H-bridge three-level inverter current are less compared to two-level inverter operating at same switching frequency.

## VII. REFERENCES

[1] Jih-Sheng Lal, Fang Zheng Peng," Multilevel Converters – A New Breed of Power Converters", IEEE Transactions on Industry Applications, Vol.32, no.3, pp.509,1996.

[2] Muni B.P., Rao S.E., Vithal J.V.R., Saxena S.N., Lakshminarayana S., Das R.L., Lal G., Arunachalam M., "DSTATCOM for Distribution Utility and Industrial Applications", Conference Proceedings, IEEE, Region Tenth Annual Conference, TENCON-03. Page(s): 278- 282 Vol.1

[3] Bishnu P. Muni, S.Eswar Rao, JVR Vithal and SN Saxena, "Development of Distribution STATCOM for power Distribution Network" Conference Records, International conference on "Present and Future Trends in Transmission and Convergence", New Delhi, Dec.2002,pp. VII_26-33.

[4] F.Z. Peng, J. S. Lai, J.W. Mckeever, J. Van Coevering, "A Multilevel Voltage – Source inverter with Separate dc sources for Static Var Generation" IEEE Transactions on Industry Applications, Vol. 32, No. 5, Sep 1996, pp1130-1138.

[5] K.Anuradha, B.P.Muni, A.D.Rajkumar," Simulation of Cascaded H-Bridge Converter Based DSTATCOM" First IEEE Conference on Industrial Electronics and Applications, May 2006, pp 501-505.

[6] R. Krishnan, Electric Motor Drives, Modeling, Analysis, and control, Pearson Education, 2001.

[7] Narain G.Hingorani, Laszlo Gyugyi, Under standing FACTS, IEEE Press, 2001.

## VIII. BIOGRAPHIES

K.Anuradha was born in India on 11th July, 1970. She obtained B.E degree from Andhra University, Visakhapatnam, India in 1992 and M.E degree from Osmania University, Hyderabad, India in 1997. From 1997 she is in the faculty of Electrical Engineering at VNR Vignana Jyothi Institute of Engg. & Technology, Hyderabad, India. Currently she is working towards her Ph.D. at Osmania University, Hyderabad, India. Her research interests are in the areas of Power Electronics, Reactive Power Control and Electrical Machines.

Bishnu P. Muni was born in India on November 16, 1961. He obtained B. Sc.(Engg.) in Electrical Engineering from Regional Engineering College, Rourkela in the year 1983. Subsequently, he obtained Ph. D from IIT, Bombay in 1997. Since 1983, he is working with BHEL, Corporate R&D in the area of power electronics for industrial, traction, power system and distributed generation applications.

Rajkumar A.D. was born on 30th December 1947, Hyderabad, India. He obtained the B.Sc., B.E., and Ph.D. degrees from Osmania University, India in 1965, 1968 and 1983 respectively. He was awarded the M.Tech. Degree from Indian Institute of Technology, Madras, Chennai, India in 1971. He joined the Osmania University in 1972 and rose to become Professor of Electrical Engineering in 1987. He has published several papers in Conferences and Journals in India and abroad. He is presently the Dean of the Faculty of Engineering at Osmania University, Hyderabad, India. His present research interests are Cascaded Inverters for DSTATCOM applications, Frequency Response Analysis of Transformers for Condition Monitoring. He is a Fellow of the Institution of Engineers (India).

**2006 IEEE International Conference on Power Electronic, Drives and Energy Systems**

# Detection and Classification of Non-stationary Power disturbances in Noisy Conditions

B. K. Panigrahi, *Member, IEEE*, and S. K. Sinha

*Abstract--* This paper presents a new approach to detect and classify the non-stationary power quality disturbances using advanced digital signal processing approaches like Wavelet Transform (WT) and S-Transform (ST) in noisy conditions. WT is a powerful tool to detect the power quality disturbances, but the application of WT is limited in noisy conditions due to its degradable performance. Hence in this paper an on-line detection of power quality disturbances is proposed using ST, which shows potential effect in noisy conditions.

*Index Terms--* Power quality, S-Transform, Wavelet Transform

## I. INTRODUCTION

IN recent years, power quality has become a significant issue for both utilities and customers. Power quality issues and the resulting problems are the consequences of the increasing use of solid state switching devices, non-linear and power electronically switched loads, unbalanced power systems, lighting controls, computer and data processing equipment as well as industrial plant rectifiers and inverters. These electronic type loads cause quasi-static harmonic dynamic voltage distortions, inrush, pulse type current phenomenon with excessive harmonics and high distortion. In order to improve electric power quality, the sources and causes of such disturbances must be known before appropriate mitigating action can be taken. However, in order to determine the causes and sources of disturbances, one must have the ability to detect and localize these disturbances. Wavelet transform (WT) [1] is widely used now for the power quality assessment. [2],[3] are the few initial papers which explores the power of WT in analyzing non-stationary signals in power systems. Its adequacy for this particular application was further confirmed by other authors. Wavelet based online disturbance detection for power quality applications are discussed in some research work. Several types of Wavelet have been applied to detect the power quality events and the results are nicely reported, Although WT has been extensively used for detection of power quality disturbances, but in many of the cases the effect of electrical noise is not adequately considered. A de-noising scheme is proposed for enhancing wavelet based power quality monitoring system. Detection and classification of power quality disturbance is also presented where WT is combined with short-time correlation transform (STCT).

---

B.K.Panigrahi is with Department of Electrical Engineering, IIT, Delhi, India (e-mail: bkpanigrahi@ee.iitd.ac.in ).

S. K. Sinha is with National Power Training Institute, Badarpur, Delhi, India (e-mail : sunilkumarsinha2005@yahoo.co.in)

Although a lot of research achievements have been reported for the detection and classification of power quality disturbances, the objective of detecting the disturbance in noisy environment and correctly classifying the nature of disturbance is still a challenging one. This paper aims to propose one of the powerful digital signal processing technique, S-Transform [4],[5] for the detection of the power quality disturbance both in normal and noisy environment. As the ST is able to extract the three important features of a power signal (i) fundamental component of the signal ( i.e. either 50 or 60 HZ) (ii) frequency content of the signal and (iii) the stationary phase of the signal, we are inspired to use these three features of the disturbance signal for accurate classification of the PQ disturbance.

Like WT, S-Transform (ST) [6]-[8] is a powerful signal processing technique, which explores the possibility of better time-frequency representation of a signal. It is an invertible time-frequency spectral localization technique that combines elements of wavelet transform and Short Time Fourier Transform (STFT). The S-transform uses an analysis window whose width scales inversely with frequency thereby providing a frequency dependent resolution. It has been successfully applied in many fields of research.

## II. S- TRANSFORM

Stockwell *et al.* [5] proposed a new windowed Fourier transform called the S transform, as an extension to the ideas of the Gabor transform and the wavelet transform. The S - transform of a signal $x(t)$ is defined as

$$S(\tau,f) = \int_{-\infty}^{\infty} x(t)g(\tau-t)\exp(-j2\pi f\tau)d\tau \qquad (1)$$

where

$$g(t) = \frac{1}{\sigma\sqrt{2\pi}}\exp\left(-\frac{t^2}{2\sigma^2}\right) \qquad (2)$$

and

$$\sigma(f) = \frac{1}{|f|} \qquad (3)$$

Combining (1) to (3) gives

$$S(\tau,f) = \int_{-\infty}^{\infty} x(\tau)\frac{|f|}{\sqrt{2\pi}}\exp\left(-\frac{(\tau-t)^2 f^2}{2}\right)\exp(-j2\pi ft)dt \qquad (4)$$

The normalization factor of $\dfrac{|f|}{\sqrt{2\pi}}$ ensures that when integrated over all $\tau$, $S(\tau,f)$ converges to $X(f)$, the Fourier transform of $x$.

---

0-7803-9771-1/06/$25.00 ©2006 IEEE

$$\int_{-\infty}^{\infty} S(\tau, f)d\tau = \int_{-\infty}^{\infty} x(t) \exp(-j2\pi ft)\, dt = X(f) \qquad (5)$$

It is clear that $x(t)$ can be obtained from $S(f, t)$. Thus the S-transform is invertible. In (2), $S$ denotes the $S$-transform of $x$, which is a continuous function of time $t$ and the frequency is denoted by $f$; and the quantity $\tau$ is a parameter which controls the position of the Gaussian window on the $t$-axis. The scaling property of the Gaussian window is similar to that of the scaling property of continuous wavelets, because one wavelength of the Fourier frequency is always equal to one standard deviation of the window. The $S$-transform, however, is not a wavelet transform, because the oscillatory parts of the $S$-transform is provided by the complex Fourier sinusoid, which does not translate with the Gaussian window when $\tau$ is changed. As a result, the shapes of the real and imaginary parts of the $S$-transform change as the Gaussian window translates in time. True wavelets do not have this property because their entire waveform translates in time with no change in shape. Thus, the $S$-transform is conceptually a hybrid of short-time Fourier analysis and wavelet analysis, containing elements of both but falling entirely into neither category.

The $S$-transform has an advantage in that it provides multi resolution analysis while retaining the absolute phase of each frequency. This has led to its application for detection and interpretation of events in time series in a variety of disciplines. In this expression of $S$ – transform the scalable Gaussian window (the product of $\dfrac{|f|}{\sqrt{2\pi}}$ and the real exponential) localizes the complex Fourier sinusoid, giving the S-transform analyzing function (the term in braces in (4)). The CWT provides time resolution by translating its whole analyzing function (the wavelet) along the time axis. The S-transform is different because only the amplitude envelope of the analyzing function (the window) translates; the oscillations are given by the fixed Fourier sinusoid, which does not depend on $\tau$. Since the local oscillatory properties of the analyzing function determine the phase of the local spectrum, the S-transform can be considered as having "fixed" phase reference. The fixed-phase reference gives the S-transform some advantages over wavelet transforms.

The S-transform is a linear operation on the signal $y(t)$. If additive noise is added to the signal $y(t)$, it can be modeled as $y_{noise}(t)=y(t)+\eta(t)$. The operation of the S-transform leads to

$$S\{y_{noisy}(t)\} = S\{y(t)\} + S\{\eta(t)\} \qquad (6)$$

The S-spectrum of Gaussian white noise varies as $\sqrt{|f|}$, thereby noise peaks having larger amplitude at higher frequencies on the S-spectrum. Hence to have a better performance at noisy conditions, the Gaussian window width varies as $\sqrt{\dfrac{1}{|f|}}$ instead of $\dfrac{1}{|f|}$. Thus the time average of the resulting S-spectrum tends to vary as $\sqrt[4]{\dfrac{1}{|f|}}$, which leads to smaller noise peaks at high frequencies.

## III. DETECTION AND FEATURE EXTRACTION OF POWER QUALITY DISTURBANCES USING S - TRANSFORM

The output of the ST is an N by M matrix called the S-matrix whose rows pertain to frequency and whose columns pertain to time. Each element of the S-matrix is complex valued. The S-matrix can be represented in a time-frequency plane similar to that of the wavelet transform.

The S-transform performs multi resolution on a time varying power network signal, because its window width varies inversely with frequency. This gives high time resolution at high frequency and high frequency resolution at low frequency.

The signals are simulated using MATLAB7.0. The sampling frequency is (64x50) i.e. 3.2 kHz. Eleven types of power quality disturbances are simulated and the features of all the types of disturbances are extracted from the S – matrix. To demonstrate the features extracted and the capability of the S-transform, two types of disturbances i.e. voltage sag and voltage swell along with the features are presented in Figures 1 and 2. Fig. 1 (a) shows the voltage sag signal, the dotted line on the figure is the feature extracted from the S-matrix, which represents the maximum value of the S-matrix at a particular time. Fig. 1(b) represents the contour of the S-matrix, which gives the complete visualization of the voltage sag. Fig 1(c) is the stationary phase of the signal, derived from the S-matrix. Fig. 1(d) represents the magnitude of the frequency components present in the signal. Similarly Fig 2(a) to 2 (d) represents the above described features for a voltage swell disturbance.

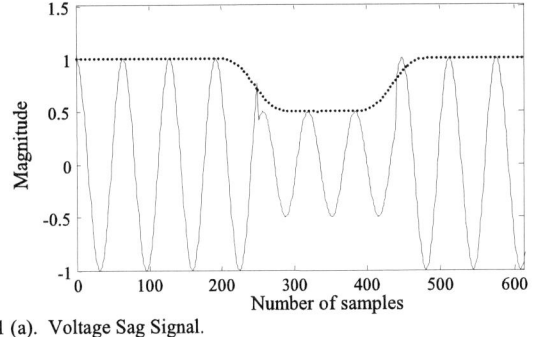

Fig. 1 (a). Voltage Sag Signal.

Fig. 1 (b). Contour Of The S-Matrix.

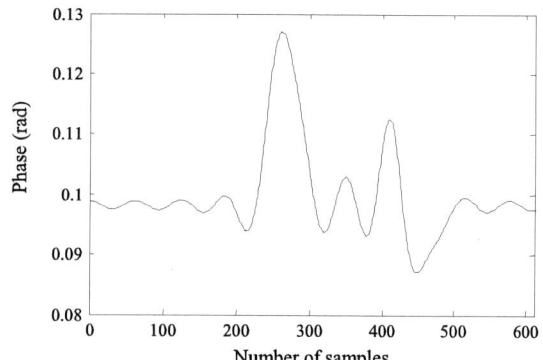

Fig. 1 (c). Stationary Phase Of The Signal.

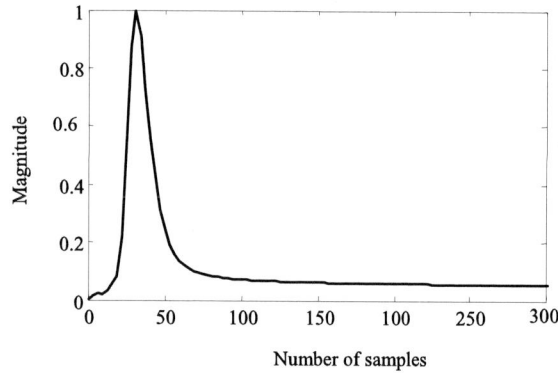

Fig. 1(d). Magnitude Of The Frequency Components.

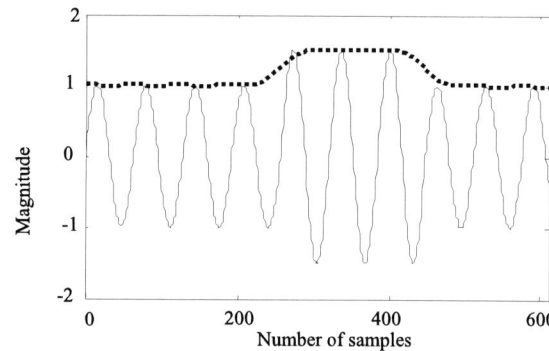

Fig. 2(a). Voltage Swell Signal.

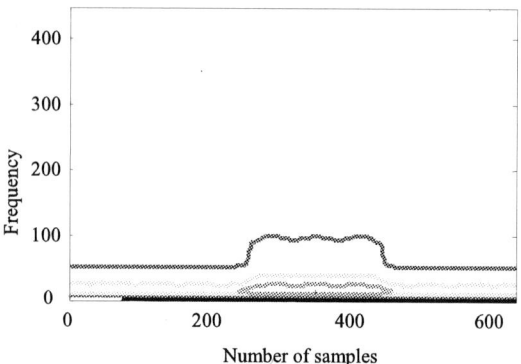

Fig. 2(b). Contour Of The S-Matrix.

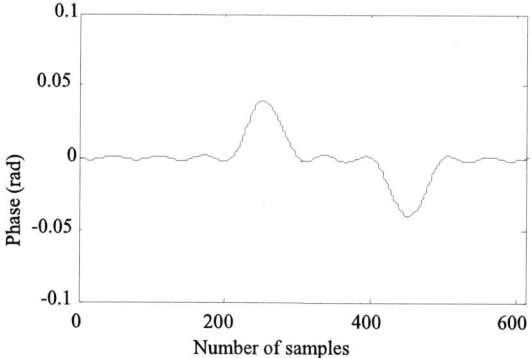

Fig. 2(c). Stationary Phase of the Signal.

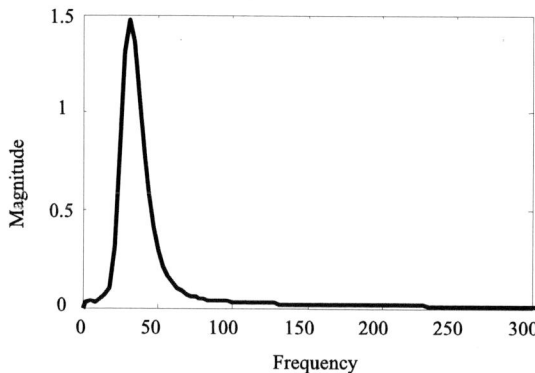

Fig. 2(d). Magnitude Of The Frequency Components.

Fig. 3(a) to Fig. 3(d) shows the detection capability of Wavelet Transform. It is observed that although WT is able to detect the voltage sag in normal conditions its performance for the detection of the event degrades in the presence of noise.

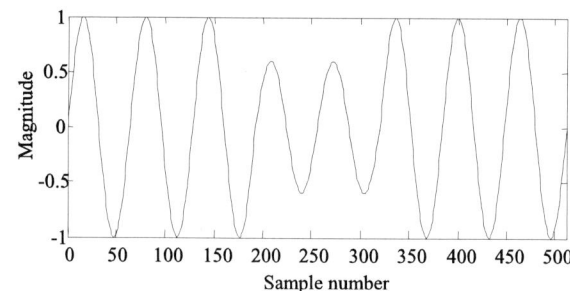

Fig. 3(a). Voltage Sag signal.

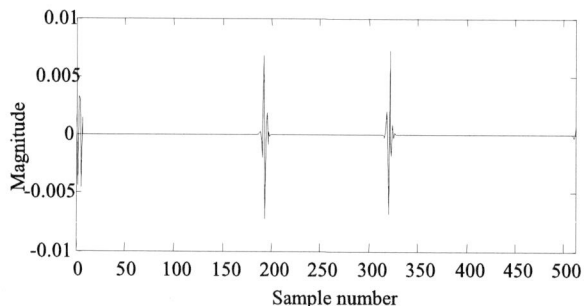

Fig. 3(b). D1 of Wavelet decomposition of the signal.

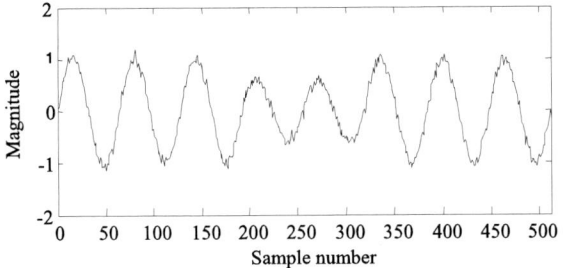

Fig. 3(c). Voltage sag signal with SNR 20 dB.

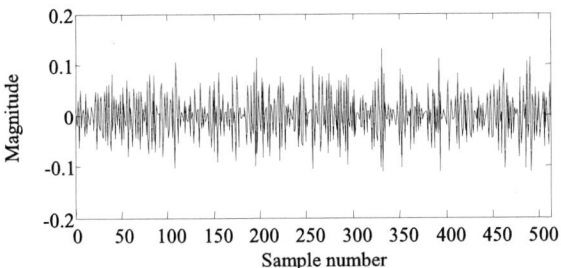

Fig. 3(d). D1: 1st level of Wavelet decomposition of the signal.

Feature extraction is done by applying standard statistical techniques to the contours of the S-matrix as well as directly on the S-matrix. These features have been found to be useful for detection, classification or quantification of relevant parameters of the PQ disturbance signals. The power network signal is normalized with respect to a base value, which is the normal value without any disturbance.

1. Standard deviation of the feature which corresponds to the maximum value of the S-matrix at each sample
2. Energy of the feature which corresponds to the maximum value of the S-matrix at each sample
3. Standard deviation of the frequency contour.
4. Standard deviation of the Phase contour

## IV. Classification Of PQ Disturbances Using PNN

The PNN model is one of the supervised learning networks, and has the following features distinct from those of other networks in the learning processes.

- It is implemented using the probabilistic model, such as Bayesian classifiers.
- A PNN is guaranteed to converge to a Bayesian classifier provided that it is given enough training data.
- No learning processes are required.
- No need to set the initial weights of the network.
- No relationship between learning processes and recalling processes.
- The difference between the inference vector and the target vector are not used to modify the weights of the network.

The learning speed of the PNN model is very fast, making it suitable for fault diagnosis and signal classification problems in real time

## V. Simulation Results

### A. Pure Signals

Eleven classes (C1-C11) of different PQ disturbances are taken for classification and they are named as

C1→ Normal
C2→ Pure Sag
C3→ Pure Swell
C4→ Momentary Interruption (MI)
C5→ Harmonics
C6→ Sag with Harmonic
C7→ Swell with Harmonic
C8→ Flicker
C9→ Notch
C10→ Spike
C11→ Transient

Based on the feature extraction by the S-Transform method, 4 dimensional feature sets for training and testing are constructed. The dimensions here describe the different features resulting from S-Transform. All the data sets of features for various classes are applied to PNN for automatic classification of PQ events. 100 data sets are taken for each class and the result is given in Table I.

TABLE I

CLASSIFICATION RESULTS OF PNN

|     | C1  | C2 | C3 | C4  | C5  | C6 | C7 | C8 | C9 | C10 | C11 |
|-----|-----|----|----|-----|-----|----|----|----|----|-----|-----|
| C1  | 100 | 0  | 0  | 0   | 0   | 0  | 0  | 0  | 0  | 0   | 0   |
| C2  | 0   | 94 | 0  | 3   | 0   | 3  | 0  | 0  | 0  | 0   | 0   |
| C3  | 0   | 0  | 97 | 0   | 0   | 0  | 3  | 0  | 0  | 0   | 0   |
| C4  | 0   | 0  | 0  | 100 | 0   | 0  | 0  | 0  | 0  | 0   | 0   |
| C5  | 0   | 0  | 0  | 0   | 100 | 0  | 0  | 0  | 0  | 0   | 0   |
| C6  | 0   | 5  | 0  | 0   | 0   | 95 | 0  | 0  | 0  | 0   | 0   |
| C7  | 0   | 0  | 4  | 0   | 0   | 0  | 96 | 0  | 0  | 0   | 0   |
| C8  | 4   | 0  | 0  | 0   | 0   | 0  | 0  | 96 | 0  | 0   | 0   |
| C9  | 0   | 0  | 0  | 0   | 0   | 0  | 0  | 0  | 91 | 0   | 9   |
| C10 | 0   | 0  | 0  | 0   | 0   | 0  | 0  | 0  | 0  | 100 | 0   |
| C11 | 0   | 0  | 0  | 0   | 0   | 0  | 0  | 0  | 0  | 0   | 100 |

**Overall Accuracy: 97.18**

### B. Performance of ST and PNN under Noise Environment

In an electrical power distribution network, the practical data consists of noise; therefore, the proposed approach has to be analyzed under noise environment. Gaussian white noise is widely considered in the research of power quality issues. In the pure data, the noise data with different levels of noises with signal to noise ratio (SNR) ranging from 10 to 40 dB are added and applied to S-Transform for the feature extraction. Then these features (mean a, mean b, standard deviation a, standard deviation b and energy) are applied to PNN for automatic classification. The classification results are presented in Table II for different noise levels and the same is shown in Fig. 4 also. Since 10dB is a severe noise, the classification accuracy rate degrades, but with other noise levels (20-40 dB) PNN can effectively classify different kinds of PQ disturbances.

TABLE II

CLASSIFICATION RESULTS WITH NOISE DATA MIXED WITH PURE DATA

| Training data | | 50% | 60% | 70% | 80% |
|---|---|---|---|---|---|
| Testing data | | 50% | 40% | 30% | 20% |
| Average accuracy rate | 10 dB | 81.52% | 82.7% | 83.02% | 84.05% |
| | 20 dB | 91.49% | 92.2% | 93.45% | 94.05% |
| | 30 dB | 94.77% | 95.47% | 96.2% | 96.86% |
| | 40 dB | 96.2% | 96.9% | 97.45% | 97.74% |

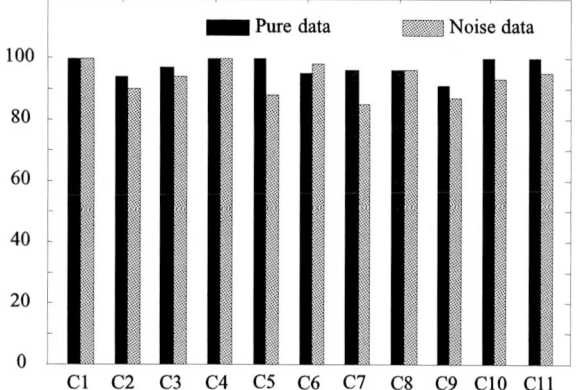

Fig. 4. Classification Accuracy.

## VI. CONCLUSIONS

This paper presents S- Transform approach for the detection of power quality disturbances. The features extracted by applying ST to the non-stationary power signals are used to classify the power quality event. It is observed that the ST is able to detect the disturbance in the presence of noise. Further the classification accuracy in the presence of noise is improved.

## VII. REFERENCES

[1] S. Santoso, E.J.Powers, W.M.Grady and P.Hofmann," Power Quality Assesment Via Wavelet Transform Analysis", IEEE Trans. On Power Delivery, Vol. 11, No.2, pp. 924-930, April 1996.

[2] M. Karimi, H. Mokhtari and M.R. Iravani," Wavelet based online disturbance detection for power quality applications", IEEE trans. On Power Delivery, Vol. 5, Oct 2000, pp 1212-1220.

[3] H. Mokhtari, M.K.Ghartemani and M.R. Iravani," Experimental performance evaluation of a wavelet-based on-line voltage detection method for power quality application, IEEE trans. On Power Delivery, Vol. 17, Jan 2002, pp 161-172.

[4] H. Zang, P. Liu and O.P.Malik," Detection and Classification of power quality disturbances in noisy conditions", IEE Proc. Gener. Transm. Distrib, vol 150, No.5, Sept. 2003, pp. 567572.

[5] Stockwell, R.G., Mansinha ,L., and Lowe, R.P., "Localization of the complex spectrum :The S-transform" , *IEEE Trans. On Signal Process*, vol.44, No.4pp.998-1001,April-1996.

[6] Pinnegar, C.R and Mansinha, Lalu., "The S-transform with windows of arbitrary and varying window", *Geophysics*,vol-68,No-1,pp.381-385, 2003.

[7] Mansinha, L., Stockwell., R..G. and Lowe, R.P., "Pattern analysis with two dimensional spectral localization: Application of two dimensional S-transforms*", Physica* A,239, pp.286-295, 1997.

[8] P.K.Dash, B.K.Panigrahi, G Panda, "Power quality analysis using S-transform", *IEEE Trans. On Power Delivery*,2002.

**2006 IEEE International Conference on Power Electronic, Drives and Energy Systems**

# 3-Phase Fault Current Limiter for distribution systems

Vijay K. Sood, *Fellow, IEEE,* and Shahabur Alam

*Abstract --***An EMTP RV based study to limit fault currents in a distribution system using a 3-phase version of a Solid-State Fault Current Limiter (SSFCL) is presented. For demonstration purposes, two types of distribution systems have been considered: a single-source radial system and a multiple-source distribution system with a bus-tie. Simulations have been done for different types of faults to show that the SSFCL is effective for limiting fault currents. Comparisons are also made with a single-phase version of the SSFCL which had been previously presented.**

*Index Terms --***Solid state fault current limiter**

## I. INTRODUCTION

IN a power distribution system, feeder overloading can occur inadvertently due to disturbances. To meet growing customer demand, system expansion is often required; this implies that existing equipment (i.e. breakers, transformers) be replaced by higher capacity versions. A replacement requires additional system renovation work and often it is both difficult and expensive to get a higher-current interrupting breaker with high-speed interrupting capability within 2-3 cycles.

A Solid State Fault Current Limiter (SSFCL) [1,2,7] is investigated to control the short circuit capacity of a substation from exceeding the interrupting capacity of downstream devices (e.g. circuit breaker, transformer etc.) as system short circuit capacity is increased. This paper investigates the model of a 3-phase version of a SSFCL for two typical distribution system configurations using the EMTP RV simulation package. An alternative 1-phase version of a SSFCL for these two typical systems was presented earlier [3].

A distribution system can be fed either from a single-source radial system or from a multiple-source bus-tie system [4,5]. In a single-source radial system, all feeder lines are connected to a bus and power is supplied to this bus from a single transformer (Fig. 1). In the case of a multiple-source bus-tie system, feeder lines are connected to several buses which are then connected through a tie [6]. Power may be supplied to these buses from several transformers (Fig. 2). The FCL system is implemented by inserting a SSFCL downstream to the transformer of the

single-source radial system; in the case of the multiple-source distribution system, the FCL is inserted in the bus-tie itself.

The paper is structured as follows: In section II, the mode of operation of the SSFCL for 1-phase and 3-phase versions is explained, and a comparison between them is made. In section III, simulation results for system faults on two common distribution structures are presented. This is followed by some tests on parametric sensitivity analysis for critical components of the FCL. And finally, conclusions are presented in section IV.

## II. METHODOLOGY

### A. The operating principle of a FCL

The operating principle of a FCL is to insert a current-limiting ac reactor in series with the main feeder line during the fault period (Fig. 3). The FCL is composed of a current-limiting reactor $L_{ac}$ and a parallel transformer with a normally-closed circuit breaker, CB in its secondary winding. Under normal operation, the transformer primary winding acts as a vir-

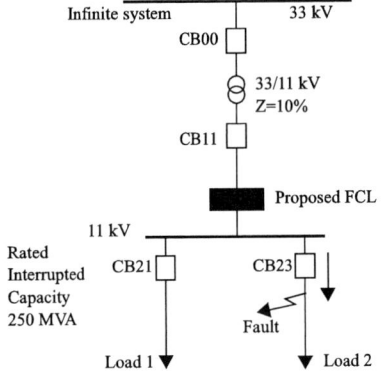

Fig. 1. Distribution system fed from single source.

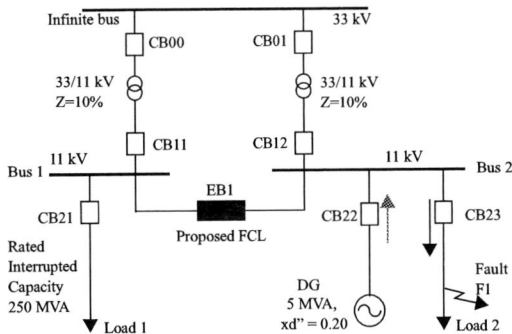

Fig. 2. Distribution system fed from multiple sources and bus-tie.

---

This work was supported by Natural Sciences and Engineering Research Council of Canada under Grant 4518.

Vijay K. Sood and Shahabur Alam are both with the Department of Electrical and Computer Engineering, Concordia University, Montreal, Qc, H3G 1M8, (e-mail: vijay@ece.concordia.ca).

0-7803-9771-1/06/$25.00 ©2006 IEEE

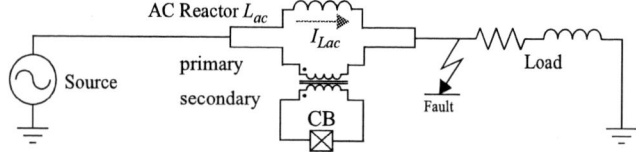

Fig. 3. Operating principle of FCL

tual short circuit due to the closed CB, and the resultant ampere-turn balance requirement for the transformer windings. When a fault downstream from the FCL is detected, the breaker CB is opened. This causes the transformer primary winding to be open-circuited as the secondary winding now carries no current. Due to the resultant ampere-turn balance requirement for the transformer windings, this implies that the primary winding current is also greatly reduced - to the magnetizing current only. Therefore, fault current is forced through the parallel ac reactor $L_{ac}$ and the fault current will be limited.

As the breaker CB is relatively expensive, it is replaced with an equivalent electronic "switch" composed of a GTO thyristor bridge and dc reactor load (Fig. 4). The transformer can then be made to operate either as a short- or open-circuit with the assistance of the thyristor bridge and dc reactor load.

In the GTO thyristor bridge type of FCL, an ac line reactor $L_{ac}$ is connected in parallel with the bridge. But connecting the bridge circuit directly in series with a feeder line means that it is necessary to use high-voltage GTO thyristors. Since the maximum voltage rating of a high-power GTO thyristor currently available is about 9 kV and, operating multiple devices in series to build a high-voltage, high-current valve is both complex and expensive, it is desirable to avoid this by using a step-down transformer and connect the bridge on the low-voltage side of the transformer. This can greatly reduce the cost of the FCL, even though now a transformer is required.

Thus, the primary winding of the transformer is in series with the main feeder line. And the current limiting reactance $L_{ac}$ (and its inherent resistance $R_{ac}$) is in parallel with the transformer primary winding. In this arrangement, the bridge circuit, composed of four GTO valves T1–T4, is connected to the secondary winding of the transformer. A dc reactor $L_{dc}$ is connected as a load to this bridge circuit. During normal operation of the FCL, a steady state current flows through the dc reactor. The dc reactor $L_{dc}$ (and its inherent resistance $R_{dc}$) presents a small load impedance; hence, the secondary of the transformer is virtually short-circuited. The transformer itself is designed to have a low leakage impedance.

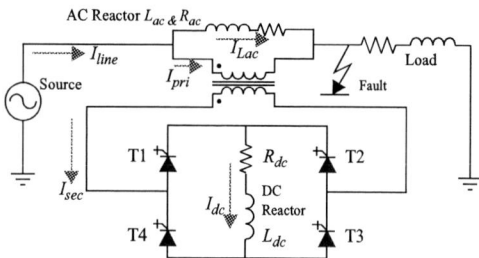

Fig. 4. Bridge type SSFCL with transformer.

During normal operation, firing pulses are sent to all four valves T1–T4 to turn them ON. Consequently, they act as diodes and natural commutation takes place in the bridge circuit. The bridge performs full wave rectification on the induced secondary voltage, and a dc current, mean value $I_{dc}$, is established in reactor $L_{dc}$. Unfortunately, due to the resistance $R_{dc}$, a steady state power loss, equal to $I_{dc}^2*R_{dc}$, ensues in the load. As the primary winding of the series connected transformer is in parallel with reactor $L_{ac}$, the fault current initially flows through mainly the transformer primary winding and only a tiny proportion flows through the parallel reactor $L_{ac}$. At the same time, a large inrush current $I_{dc}$ will attempt to flow through the secondary winding of the transformer and subsequently through the dc reactor $L_{dc}$. This large sudden change in dc current is highly impeded by the reactor $L_{dc}$, due to a large back emf caused by $V_{dc} = -L_{dc}.dI_{dc}/dt$ and hence the impedance $Z_{dc} = V_{dc}/I_{dc}$ is very high. This impedance is reflected onto the transformer primary by a factor $N^2*Z_{dc}$, where $N$ is the turns ratio of the transformer. Consequently, the transformer primary current is then forced through the parallel limiting reactor $L_{ac}$. Notice that during the pre-fault, steady-state period, the value of $V_{dc} = -L_{dc}.dI_{dc}/dt$ is equal to zero as $I_{dc}$ is a constant.

Resistances $R_{dc}$ and $R_{ac}$ are necessary components as they are needed to dissipate fault energy. As such, the power loss needs to be paid for, and therefore, these are conflicting requirements. However, the resistance $R_{ac}$ incurs a power loss only during the fault period, whereas the resistance $R_{dc}$ incurs the loss during steady state operation as well as during the fault period. Consequently, it is preferable that the fault energy be dissipated primarily in $R_{ac}$ instead of $R_{dc}$. The quality factor of the inductors $L_{ac}$ is, therefore, chosen to be low (i.e. 15-30) and for $L_{dc}$ it is chosen to be high (i.e. 150-200).

TABLE I
COMPARISON OF TWO TYPES OF SSFCL

| 1-phase bridge rectifier | 3-phase bridge rectifier |
|---|---|
| SSFCL uses three, 1-phase rectifiers occupies more area/space. | SSFCL uses one, 3-phase bridge rectifier occupies less area/space. |
| A failure in one module of the SSFCL does not affect the other phases. | A failure in the SSFCL affects all three phases. |
| More expensive approach | Less expensive approach. |
| 12 valves are used, with their associated snubber circuits, etc. | 8 valves are used, with their associated snubber circuits, etc. |
| 3 dc reactors (30 mH each) needed | Only 1 dc reactor (30 mH) needed |
| In dc reactors, three energy dissipating resistors are used; so total energy loss may be higher. | In dc reactor, one energy dissipating resistor is used, so total energy loss may be lower. |
| More than 23 kA short circuit current. System is unbalanced during short circuit period. | Up to 23 kA short circuit current. System is balanced during short circuit period. |
| A fault in one phase has limited effect on the unfaulted phase(s). | A fault in one phase has impact on the unfaulted phase(s). |
| Controller design is easier. | Controller design is more complex. |
| AC reactor loss is higher. | AC reactor loss is less. |

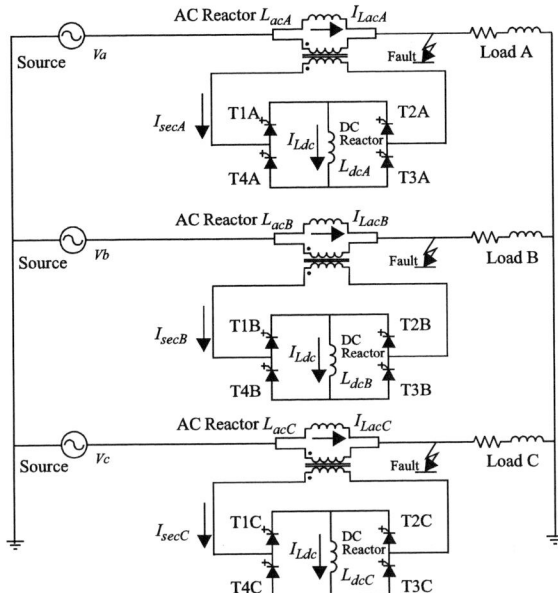

Fig. 5a. 1-phase bridge type SSFCL

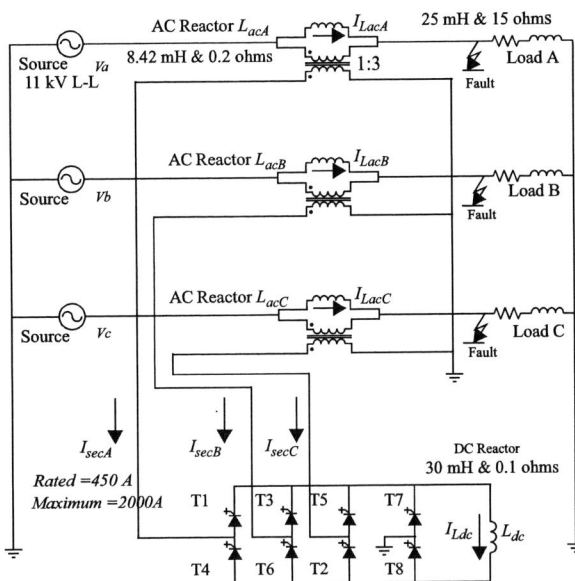

Fig. 5b. 3-phase bridge type SSFCL

There are two variants possible for the utilization of the SSFCL: (a) a 1-phase bridge type SSFCL (Fig. 5a), and (b) a 3-phase bridge type SSFCL (Fig. 5b). A comparison of the main features of these two variants is made in Table I.

### B. Mode of operation of the 1-phase version of SSFCL

The mode of operation of a 1-phase version of the SSFCL for a 3-phase system (Fig. 5a) is similar to the above description made with the aid of Fig. 4.

### C. Mode of operation for the 3-phase version of SSFCL

The operational behavior of the 3-phase version of SSFCL is seen in Fig. 5b. The period 0-0.3 s represents the PRE-FAULT normal period, the period 0.3-0.5 s represents the FAULT period, and the period 0.5-0.7 s represents the POST-FAULT period. During PRE-FAULT period of operation, all six valves T1-T6 are gated ON (Fig. 6a) and, as a result, they operate as diodes. Also, the free-wheeling valves T7-T8 are gated OFF (Fig. 6c). In Figs. 6a and 6c, a "1" represents the gates are ON and a "0" represents that the gates are OFF. The source current $I_{pri}$ (Fig. 6d) mostly flows through the primary transformer winding and a very small current $I_{Lac}$ (Fig. 6b) flows through the parallel inductors ($L_{ac\_A}$, $L_{ac\_B}$, $L_{ac\_C}$). Since there is current in the transformer secondary winding, $I_{sec}$ (Fig. 6f), this will eventually pass through the rectifier circuit, and a dc current $I_{dc}$ (Fig. 6h) will flow through the dc inductor $L_{dc}$ and its internal resistance $R_{dc}$. Therefore, there will be a power loss (typically, 1.26*100/505=0.25% of nominal power i.e. (2.1*3)=6.3 kW) during normal operation.

In the event of a fault occurring (as at 0.3 s), the fault current will flow through both the transformer primary winding (peak current more than 3500 A) and the parallel fault current limiting reactance (peak value more than 2500 A). As a result,

a large current will flow through the secondary winding (with a value of current more than 1200 A pk) and subsequently through the dc reactor. This sudden change in current $di/dt$ will result in a very high back emf $V_{Ldc}=L_{dc}di/dt$. Hence, the apparent impedance $Z_{Ldc}=V_{Ldc}/I_{Ldc}$, where $I_{Ldc}$ is the inductance current, will be reflected back into the primary winding of the transformer by a factor $N^2*Z_{Ldc}$ (where $N$ is the turns ratio of the transformer windings). Since this is very high, it will force the fault current (within one-quarter cycle) to flow through the bypass parallel limiting inductance ($L_{ac}$) instead of the primary winding of the transformer, thus limiting the current.

During the FAULT period (from 0.3-0.5 s) in any phase(s), firing pulses to the main valves T1-T6 of the faulted phase(s) are turned OFF. This causes the transformer secondary to be effectively open-circuited. Since the ampere-turn balance of the transformer windings must be respected, the fault current is forced to flow through the bypass inductors $L_{ac\_A}$, $L_{ac\_B}$ and/

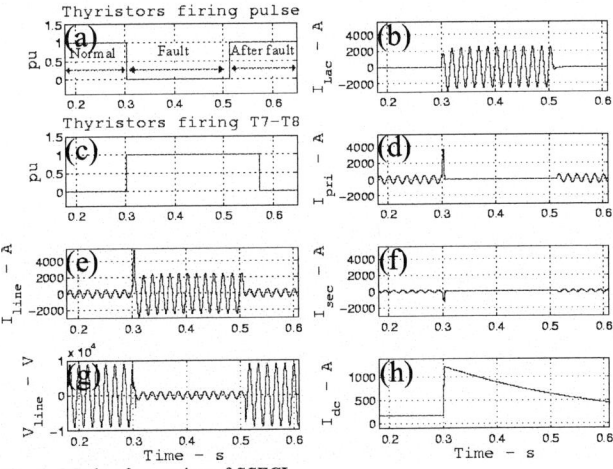

Fig. 6. Mode of operation of SSFCL

731

or $L_{ac\_C}$ until the fault is removed. Due to the higher impedance of inductances $L_{ac\_A}$, $L_{ac\_B}$ and/or $L_{ac\_C}$ as compared to the primary winding of the transformer, the fault current is limited. Meanwhile, valves T7-T8 are turned ON to provide a free-wheeling path for the dc reactor current $I_{dc}$. In this example, the dissipation time constant $T_{dc}=L_{dc}/R_{dc}$ is selected to be about 0.3s. In the POST-FAULT period, system conditions rapidly return to pre-fault normal levels.

### D. Design of Rectifier Controller

To dissipate the pre-fault energy stored in the dc reactor, a free-wheeling circuit is used. When a fault occurs in any phase(s) the corresponding pair(s) of GTO valves T1-T6 are turned OFF, and the valve pair T7-T8 are turned ON for free-wheeling purposes. Consequently, the stored energy in the dc reactor $L_{dc}$ is dissipated in its resistance $R_{dc}$, and the dissipation time constant $T_{dc} = L_{dc}/R_{dc}$. The rectifier controller design is shown in Fig. 7. The sensed phase current ($I_a$, $I_b$ or $I_c$) is first converted to a pu value, and then its magnitude is compared to a reference value $I_{ac\_ref} = 1600$ A. If the phase current is greater than the reference current, then the comparator outputs a logic signal to turn OFF the relevant phase thyristor pair, as per Table II. Similarly, the dc reactor current $I_{dc}$ is sensed, and compared to a reference value of $I_{dc\_ref} = 500$ A. If the dc reactor current is greater than its reference current, the comparator emits a signal to turn ON the thyristor pair T7 & T8 (Table II).

### III. SIMULATION RESULTS AND DISCUSSIONS

#### A. Comparison of fault currents without/with SSFCL

Table III depicts the case of the radial system when subjected to either a 1-phase, 2-phase or 3-phase fault for 12 cycles without any limiter. The fault current levels are observed to be: 1-ph fault $I_{dc}$ =1226 A pk; 2-ph fault $I_{dc}$ =3093 A pk, and 3-ph fault $I_{dc}$ =3950 A pk after the initial first cycle transient. In a radial system, the most common faults are the asymmetric faults (1-phase and 2-phase to ground). A comparison of the three fault cases with a fault limiter of the 3-phase type are shown in the bottom signals in Fig. 8. Within the first quarter-cycle after fault detection, the limiter starts limiting and the transient current. After the first cycle, the limiter is able to reduce the currents to a stable value of about 2495 A pk. Similar results with the bus-tie system were observed.

#### B. SSFCL located in a radial line feeder

For the radial system fed from a 33/11 kV, 10 MVA transformer, simulations are carried out for typical faults with a fault duration of 0.2 s. The SSFCL limited the fault current to the desired value (2495 A pk) within one-quarter cycle of fault occurrence. The fault impedance used for the circuit is 0.3 ohm and 1 mH with a load of 25 mH and 15 ohms. The simulation results for (a) 1-phase, (b) 2-phase and (c) 3-phase to ground fault are shown in Fig. 9. From a summary of results (Table III), extracted from Fig. 9, it can be seen that the dc reactor

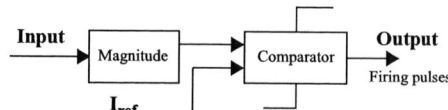

Fig. 7. Rectifier control circuit (in conjunction with Table II)

TABLE II
RECTIFIER CONTROL CIRCUIT DATA (SEE FIG. 7)

| Input | $I_{ref}$ | Output |
|---|---|---|
| $I_a$ | $I_{ac\text{-}ref}$ = 1600 A | T1 and T4 |
| $I_b$ | $I_{ac\text{-}ref}$ = 1600 A | T3 and T6 |
| $I_c$ | $I_{ac\text{-}ref}$ = 1600 A | T5 and T2 |
| $I_{dc}$ | $I_{dc\text{-}ref}$ = 500 A | T7 and T8 |

current (which indicates fault energy to be dissipated) is lowest for a 1-phase fault, and highest for a 3-phase fault.

#### C. SSFCL located in a bus-tie feeder

For the bus-tie system fed from a 33/11 kV, 10 MVA transformer, simulations are carried out for typical faults with a fault duration of 0.2 s. The SSFCL limited the fault current to the desired value (3302 A pk) within one-quarter cycle of fault occurrence. The fault impedance used for the circuit is 0.3 ohm, 1 mH with a load of 25 mH and 15 ohms. The simulation results for (a) 1-phase, (b) 2-phase and (c) 3-phase to ground faults are shown in Fig. 10. From Table IV results, which are extracted from Fig. 10, it can be seen that the dc reactor current (which is an indication of the fault energy to be dissipated) is lowest for a 1-phase fault, and highest for a 3-phase fault

TABLE III
SSFCL OPERATION IN RADIAL SYSTEM

| Parameter steady state values | 1-ph fault | 2-ph fault | 3-ph fault |
|---|---|---|---|
| $I_{dc}$ - (A) pk | 1226 | 3093 | 3950 |
| $I_{ac}$ - (A) pk | 2900 | 2881 | 2988 |
| $I_{line}$ - (A) pk | 5255 | 11180 | 13120 |
| $P_{dc}$ - (kW) pk | 150.3 | 956.7 | 1560.3 |

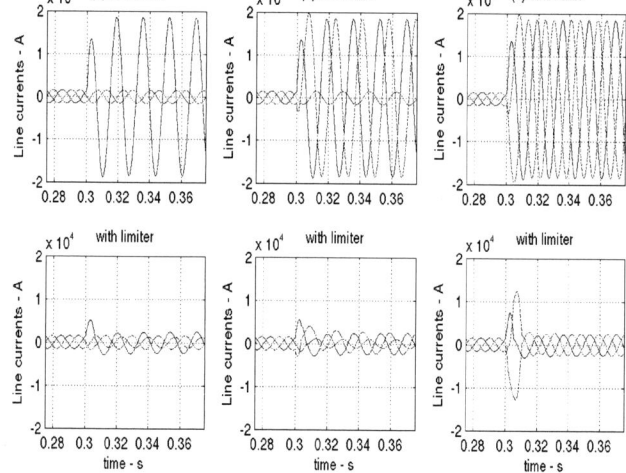

Fig. 8. Comparison of line current fault levels without and with limiter. Upper signals are without limiter and lower signals are with limiter in action.

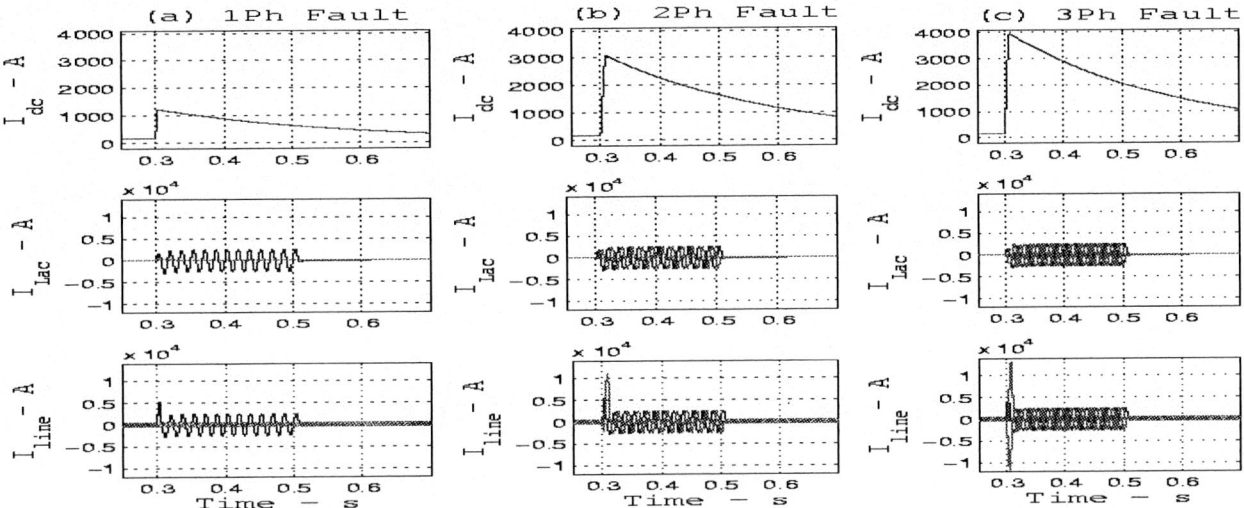

Fig. 9. SSFCL operation in 11 kV radial system

TABLE IV
SSFCL OPERATION IN BUSTIE SYSTEM

| Parameter steady state values | 1-ph fault | 2-ph fault | 3-ph fault |
|---|---|---|---|
| $I_{dc}$ - (A) pk | 1279 | 2990 | 3849 |
| $I_{ac}$ - (A) pk | 2740 | 2697 | 2798 |
| $I_{line}$ - (A) pk | 5829 | 11420 | 13330 |
| $P_{dc}$ - (kW) pk | 163.6 | 894.0 | 1481.5 |

### D. Optimization of dc reactor $L_{dc}$ parameters

The dc reactor plays an important role in the dissipation of the fault energy. The two parameters of the dc reactor i.e. $L_{dc}$ and $R_{dc}$ combine to give the decay time constant $T_{dc} = L_{dc}/R_{dc}$. Unfortunately, a finite value of $R_{dc}$ implies a steady-state power loss $P_{dc} = I_{dc}^2.R_{dc}$ in the dc reactor. Using the design values of $L_{dc} = 30$ mH at a typical quality factor $Q = wL_{dc}/R_{dc} = 200$ gives a $R_{dc}= 0.1$ ohms. Hence, the steady-state power loss in the dc reactor is $P_{dc} = 1.4$ kW, for the radial system with

$I_{dc}$ = 500A. In addition, the steady-state conduction loss in two valves is about at 0.7 kW. A total power loss of 2.1 kW is high (in terms of operating energy loss and heat dissipation requirements). The use of new high-temperature, super-conducting materials to aid in the fabrication of the dc reactor is one possible solution. However, the time constant $T_{dc} = 300$ ms.

Optimization of the dc reactor parameters is needed to trade-off between conflicting requirements i.e. time constant, steady-state and peak values of $I_{dc}$, and the power loss. This is shown in Fig. 11a for a variation in four values of $L_{dc}$, and in Fig. 11b for four variations in $R_{dc}$. A compromise value of $L_{dc}$=30 mH and $R_{dc}$=0.1 ohms is used to give a time constant $T_{dc} = 300$ ms, average $I_{dc}$=167 A and $I_{dc} = 1226$ A pk. Note that reducing $R_{dc}$ increases the average $I_{dc}$ value.

### E. Optimization of ac reactor $L_{ac}$ parameters

The ac reactor plays an important role in dissipation of the fault energy. The parameters of the ac reactor i.e. $L_{ac}$ and $R_{ac}$,

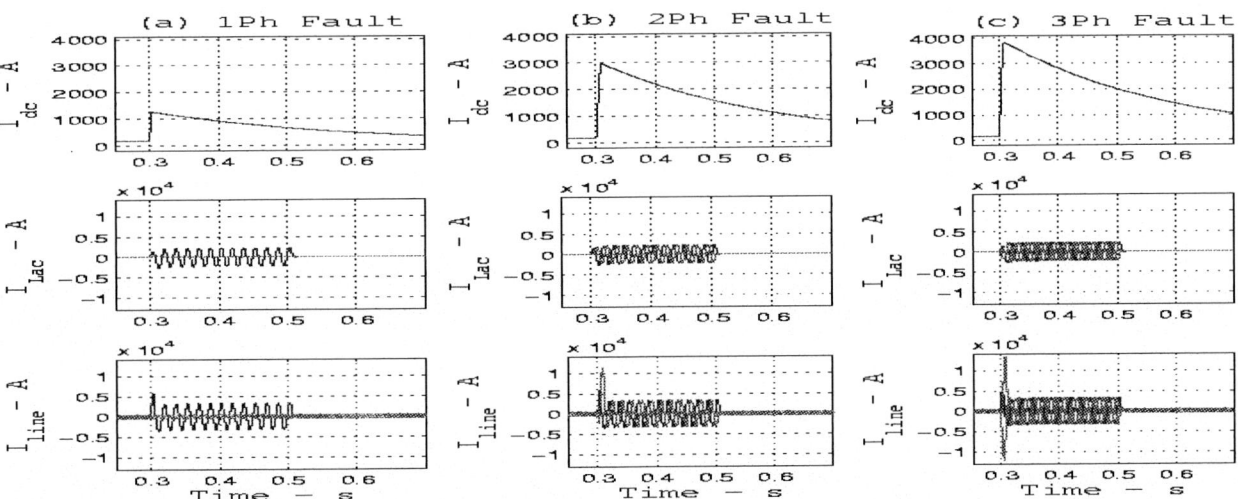

Fig. 10. SSFCL operation in 11 kV bus-tie system

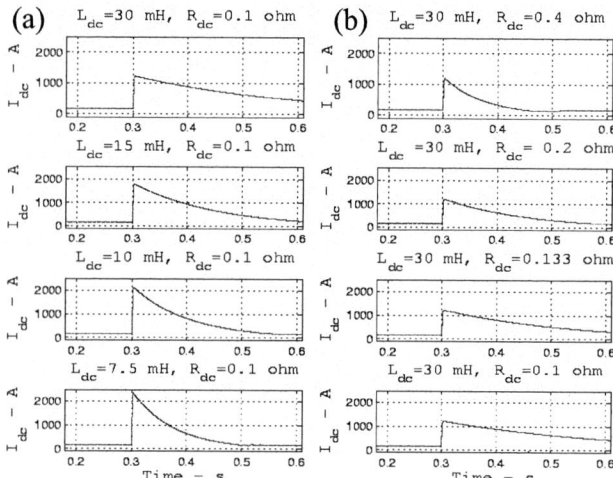

Fig. 11. Optimization of parameters of the dc reactor

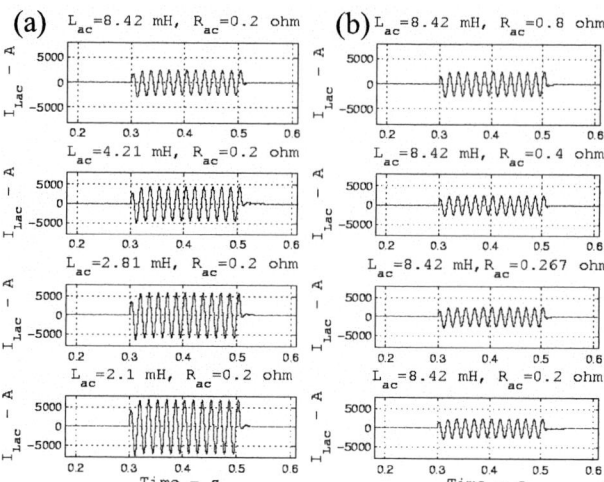

Fig. 12. Optimization of parameters of the ac reactor

give the decay time constant $T_{ac}=L_{ac}/R_{ac}$. A finite value of $R_{ac}$ implies a power loss $P_{ac}=I_{ac}^2.R_{ac}$ in the ac reactor. Using the design values of $L_{ac}$=8.42 mH at a typical quality factor $Q=wL_{ac}/R_{ac}$=16 gives a resistance $R_{ac}$=0.2 ohms. Hence, the steady-state power loss in the dc reactor is $P_{dc}$=1.4 kW, for the radial system studied here with $I_{ac}$=1600A. In addition, the steady-state conduction loss in two valves is estimated at 0.7 kW. This total power loss of 2.1 kW is expensive (in terms of operating energy loss and heat dissipation requirements). The use of new high-temperature, super-conducting materials to aid in the fabrication of the dc reactor is one possible solution. However, the time constant $T_{ac}$= 42 ms.

Optimization of the parameters of the dc reactor is needed to trade off between conflicting requirements such as the time constant, peak values of $I_{ac}$, and the power loss. This is shown in Fig. 12a for a variation in four values of $L_{ac}$, and in Fig. 12b for four variations in $R_{ac}$. A suitable compromise value of $L_{ac}$=8.42 mH and $R_{ac}$=0.2 ohms is selected to give a time constant $T_{ac}$=42 ms, average $I_{ac}$=505 A and $I_{ac}$ = 1588 A pk (as shown in Fig. 12a, row 1). A comparison of the variation of $R_{ac}$, keeping $L_{ac}$ constant, is shown in Fig. 12b. It is noted that reducing $R_{ac}$ increases the average $I_{ac}$ value.

## IV. CONCLUSION

The paper presents results from an EMTP RV based study of a SSFCL with a 3-phase bridge rectifier using GTO valves. The advantage of the FCL is its rapid response time (within one-quarter cycle) to limit the fault current. Its principle disadvantage is the steady-state power loss. Two versions of the SSFCL (1-phase and 3-phase) have been discussed and their relative merits have been compared. Two typical distribution systems have been considered for SSFCL use: (1) a single-source radial system, and (2) a multiple-source distribution system with a bus-tie. The system evaluation has been carried out with typical line faults such as 1-phase, 2-phase and 3-phase to ground faults. When any one of these faults occur, the

SSFCL operates, within one-quarter cycle, to insert the bypass fault limiting reactor(s) which limit the short circuit current flowing to main feeder(s). The unique feature of these two SSFCLs is that they can control fault current of any phase, or any combination of phases by keeping unfaulted line(s) uninterrupted. Following are the main observations:

- Without SSFCL, a very high fault current of about 20 kA pk flows through the feeder line(s). With the SSFCL, the fault current is limited at below 2500 A pk value.
- Under normal operation, the load current of 505 A pk flows through the line and only 1.26 A pk (which is 0.25% of load current) flows through the bypass reactor.
- The SSFCL limits the fault current to the design value within one-quarter cycle from fault occurrence.
- A fault in one or more phases has limited or no effect on the unfaulted phase(s).

## V. REFERENCES

[1]. G. Chen, D. Jiang et al, "A New Proposal for Solid State Fault Current Limiter and Its Control Strategies", IEEE Power Engineering Society General Meeting, 2004, Vol 2, pp. 1468-1473, 6-10 June 2004.

[2]. G. Chen., D. Jiang, Z. Wu, Z. Lu, "Simulation study of bridge type solid state fault current limiter for High-Voltage Power Network", Power Engineering Society General Meeting, 2003, IEEE, Vol.4, no., pp.- 2526 Vol. 4, 13-17 July 2003

[3]. V.K. Sood, R. Amin and M. Salam, "EMTP RV-based study of Solid-State Fault Current Limiter for distribution system", 2006 IEEE Power India conference with IEEE PES as technical sponsor, April 10-12, 2006, Inter-Continental The Grand, N. Delhi.

[4]. A.J. Power, "An overview of transmission fault current limiters", IEE Colloquium on Fault Current Limiters - A Look at Tomorrow, pp.1/1-1/5, 8 Jun 1995

[5]. P.G. Slade, J.-L. Wu, E.J.Stacey, W.F. Stubler, R.E. Voshall, J.J. Bonk, J.W. Porter, L. Hong, "The utility requirements for a distribution fault current limiter", IEEE Trans. on Power Delivery, Vol.7, No.2, pp.507-515, Apr 1992

[6]. J.C. Das, "Limitations of fault-current limiters for expansion of electrical distribution systems", IEEE Trans. on Industry Applications, Vol.33, No.4, pp.1073-1082, Jul/Aug 1997

[7]. T. Ueda, M. Morita, H. Arita, Y. Kida, Y. Kurosawa, T. Yamagiwa, "Solid-state current limiter for power distribution system", IEEE Trans. on Power Delivery, Vol.8, No.4 pp.1796-1801, Oct 1993

**2006 IEEE International Conference on Power Electronic, Drives and Energy Systems**

# Power Flow Control of a Solid Oxide Fuel-Cell for Grid Connected Operation

Ankur Goel, S. Mishra, *Senior Member, IEEE*, and A.N. Jha, *Member, IEEE*

*Abstract*-- **The objective of this paper is to control the real and reactive power feed to a grid from a Solid-Oxide fuel cell plant independently. The modification in the thermal dynamic model of the Solid-Oxide fuel cell is carried out to include the simulation time step. The controller performance analysis is done on a fuel-cell model developed in MATLAB/SIMULINK platform. It is found from simulation that the control scheme proposed is quite capable of controlling the real and reactive power flow from the fuel cell to the grid independently.**

*Index Terms*-- **Distributed Generation, Dynamic Modeling, Fuel Cell, PID Controller, Solid-Oxide.**

## I. INTRODUCTION

IN a deregulated power system scenario, the power produced from distributed generators will play a major role in satisfying the load demand on a grid. Most likely fuel cells and micro-turbines will be the dominant grid-connected distributed generators. The efficiency of power conversion can be increased by integrating the fuel cell with a gas turbine. The fuel cell plants are good choice for distributed generation because they are modular, efficient and environmental friendly.

Two types of fuel-cells are likely to be used as power plants namely solid-oxide fuel cells (SOFCs) and molten carbonate fuel cells (MCFCs) [1]. These fuel cells operate at high temperature and generate electricity at or near load site. One important aspect concerning the application of the DG technology is to ensure that a suitable dynamic model should consider the electro-chemical-thermodynamic process and electrical performance. Different authors have proposed a variety of models for fuel cells [1-4]. The inputs to the fuel cells are hydrogen and air (for oxygen) [1]. A comprehensive dynamic model of a SOFC, comprising of thermal dynamics, has been reported recently [1]. However, there are certain issues in the model which need to be addressed as follows:

i. The thermal dynamic block in [1] has been developed based on a value of output temperature that has been predicted after a time interval of 'relaxation time' which is of the order of 150-200 sec [5]. However, the SIMULINK model employs very small time steps. Therefore, there is a need to reformulate the model.

ii. The fuel processor dynamics have been neglected, which should also be included.

Ankur Goel is with Galgotia's College of Engg. and Technology Greater Noida and working for Phd from Department of Electrical Engg., Indian Institute of Technology, Delhi, India (e-mail: iitd.ankurgoel@gmail.com)

S.Mishra, and A.N.Jha are with the Department of Electrical Engineering, Indian Institute of Technology, Delhi, India (email: sukumar@ee.iitd.ac.in, and anjha@ee.iitd.ac.in,)

El-sharkh *et. al* [6-7] have proposed a control scheme for real and reactive power regulation of PEM fuel cell. However, the scheme does not decouple the real and reactive power output from the fuel cell. The decoupling of real and reactive power is essential for proper functioning of the grid. The real and reactive power mismatch between supply and demand will initiate frequency and voltage variation in the grid, respectively. Therefore, independent regulation of real and reactive power is essential. In view of this, apart from modifying the SIMULINK model the paper also proposes a novel decoupled control strategy that independently regulates the real and reactive power feeds to the grid.

Fuel cells are dc voltage sources connected to electric power networks through dc/ac inverters. The current flowing in the circuit is due to the flow of electrons produced by chemical reactions at the anode and cathode, respectively. On the other hand the dc voltage is a function of temperature, partial pressure of reactants and Gibbs free energy, related by Nernst's equation. In case of a dc system the real power is the product of voltage and current. Hence, regulating current will influence the real power generated from the fuel cell. As the current is produced by the chemical reaction of hydrogen and oxygen, regulating hydrogen flow can control the real power. Similarly, by proper choice of modulation index of the inverter it will be possible to regulate the reactive power. Therefore, the objectives of this paper are as follows:

i. To present a modified thermal dynamic block to overcome the existing limitations.

ii. To incorporate the fuel processor dynamics into a SIMULINK model of a SOFC.

iii. To design a decoupled real and reactive power control strategy for SOFC.

## II. FUEL CELL SYSTEM MODEL

Padulles *et. al* [8] created a simulation model of a solid oxide fuel cell (SOFC) power plant intended for a power system analysis package. This SOFC model has been modified in [1] by incorporating the thermal dynamics, concentration and activation loss block into it. Sedghisigarchi *et. al* [1] have developed the thermal dynamic block with reference to [5], where the output temperature is derived based on input temperature and other parameters of the fuel cell, after a laps of relaxation time. As already discussed in the introduction this needs some modification. Moreover, in a fuel cell the hydrogen gas is produced by processing some hydrocarbon fuel such as propane, natural gas and methanol in a reformer/fuel processor [9]. The consideration of its dynamics is very vital to make the simulation model realistic. Therefore, this section deals with the modeling details of the two new blocks and the other blocks are same as that of [8, 1].

## A. Thermal Dynamic Block

Depending on the instantaneous energy loss in the fuel cell the temperature will either increase or decrease from the present state. The relationship between input and output temperature, at a particular loss, following a relaxation time of around 200 sec is already presented in [5]. Assuming a linear relationship between input and output quantity the amount of increase in temperature from the present state can be calculated as follows:

$$T_o = T + \left( \frac{T_{in} + \Delta T - T}{t} \right) dt \qquad (1)$$

Where,

$T_o$    is the output temperature.

$T_{in}$   is the initial temperature at starting/no load.

$\Delta T$   is the rise in temperature from $T_{in}$, at a particular loss, that will occur after a laps of the relaxation time.

T    is the present temperature of the fuel cell under load.

t    is the relaxation time.

dt   is the SIMULINK time step.

The SIMULINK diagram of the thermal dynamic block is presented in Fig.1

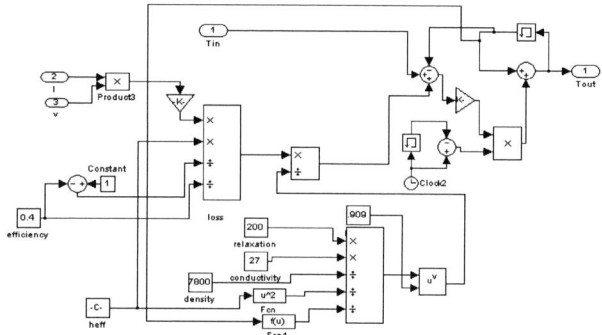

Fig. 1. Temperature dynamic block of SOFC.

Here care should be taken so that the memory block used from $T_o$ and clock should be initialized to $T_{in}$ and '0' respectively.

## B. Reformer Block

In [6, 7], the authors have proposed the model of a reformer based on a second order transfer function to convert methane to hydrogen. Rajkaruna *et al.* in [10], shows that fuel process approximated by first-order transfer functions is suitable for simulation point of view. So, our model can be written in mathematical form as follows-

$$\frac{q_{H_2}}{q_{methane}} = \frac{C_v}{(\tau_1)s + 1} \qquad (2)$$

Where,

$q_{methane}$   Methane flow rate [kmol/s]

$C_v$   Conversion factor [kmol of $H_2$/kmol of methane]

$\tau_1$   Reformer time constants [s].

A proportional integral (PI) controller is used to control the flow rate of methane in the reformer [6, 7]. Oxygen flow is determined using the hydrogen-oxygen flow ratio $r_{H-O}$. The SIMULINK diagram of the reformer along with controller block is depicted in Fig.2.

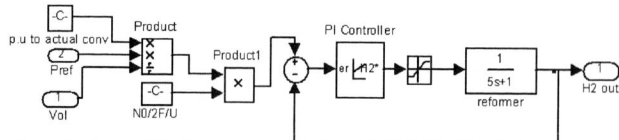

Fig. 2. Reformer Block.

## C. Power Conditioning Block

This is the interface between fuel cell output dc voltage and the ac grid. The dc voltage is converted to ac by means of PWM inverter, which has also the capability of varying the magnitude and phase angle of the ac output voltage. The ac voltage produced is stepped up by means of a transformer and is feed to the grid. The model of this system is presented in [6, 7]. The output voltage and the output power as a function of the modulation index and the phase angle can be written as:

$$V_{ac} = mV_{cell} \angle \delta \qquad (3)$$

$$P_{ac} = \frac{mV_{cell}V_s}{X} \sin(\delta) \qquad (4)$$

$$Q = \frac{(mV_{cell})^2 - mV_{cell}V_s \cos(\delta)}{X} \qquad (5)$$

Where,

$V_{ac}$   ac output voltage of the inverter

$V_{cell}$   Fuel cell dc output voltage

m   Inverter modulation index

$\delta$   Phase angle of the ac voltage $mV_{cell}$

$P_{ac}$   ac output power from the inverter

Q   Reactive output power from the inverter

$V_s$   Grid voltage

X   Reactance of the line and transformer connecting the inverter to the grid

In this formulation assumption is made that system has a large X/R ratio, and the grid voltage is taken as the reference with its magnitude being constant at 1 p.u. The 1 p.u. voltage at the input of the transformer is considered as 400 V. Based on the above equations the block diagram of the interfacing block is shown in Fig.3.

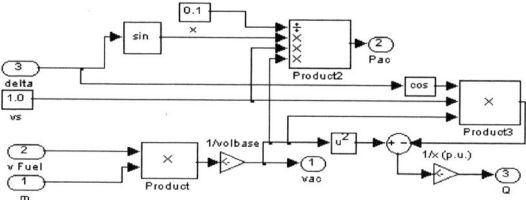

Fig. 3. Grid interfacing block.

## III. DECOUPLED CONTROL SCHEME

Fuel cell is a device which converts chemical energy to electrical energy. The reactions between $H_2$, $O_2$ and CO (produced during reforming) are involved to have circulation of electron in the load circuit, which constitute the current. Therefore, the current in case of fuel cell is influenced by the hydrogen flow rate. As a result, if the current feed back is derived from the load demand, as proposed in [6, 7] it may not yield to a proper controller. Moreover, in case of fuel cell power plant the aim may not be to regulate the ac voltage output and the real power flow. As it is already connected to a grid and grid is supplying the power, a better choice will be to control independently both real and reactive power output from the fuel cell. Therefore, in this paper a decoupled control scheme for both real and reactive power is proposed.

### A. Real Power Controller

Dividing the reference power with the dc voltage generates a reference current which when multiplied by a constant (derived from the chemical property and no. of stacks) depicts the hydrogen flow reference. The mathematical representation is as follows,

$$\frac{P_{ref}}{V_{cell}} = I_{dcref} \qquad (6)$$

$$q_{H2ref} = \frac{I_{dcref} \times N_0}{2 \times F \times U} \qquad (7)$$

Where,

$P_{ref}$ Reference real power to be tracked

$V_{dc}$ Fuel cell dc output voltage

$N_0$ No of stacks in series

$F$ Faraday's constant (Coulombs per kilo-mole)

$U$ Hydrogen utilization factor of fuel cell.

The derived '$q_{H2ref}$' is used by the PI controller to regulate the methane input flow rate so that the reformer output should track to '$q_{H2ref}$'. The hydrogen produced is further multiplied by the inverse constant of (7) (i.e $2 \times F \times U/N_0$) to decide the actual current that the fuel cell will circulate. Assuming a loss less inverter, we get

$$P_{ac} = P_{dc} = V_{cell} I_{dc} \qquad (8)$$

Therefore, combining (4), (7) and (8) we get

$$\sin(\delta) = \frac{2 \times F \times U \times X}{m \times V_s \times N_0} q_{H2} \qquad (9)$$

From (9) the voltage phase angle of the inverter with respect to the grid can be found out. As this has been derived only from $P_{ref}$ the controller will track to the reference command. The control scheme along with the reformer dynamics is already presented in Fig.2.

### B. Reactive Power Controller

The modulation index of the PWM scheme has the capability to vary the magnitude of inverter output voltage. As the reactive power influences the voltage at large, it will be a good choice to regulate the modulation index of the inverter to have a control over the reactive power. Therefore, a PI controller is used with ($Q_{ref} - Q_{actual}$) as error and the output being change in modulation index ($\Delta m$). The control system schematic diagram is presented in Fig.4.

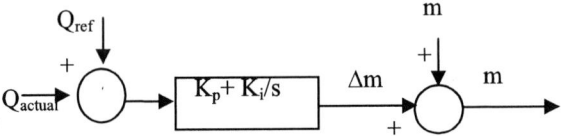

Fig. 4. Reactive Power Controller.

## IV. SIMULATION RESULTS

The SIMULINK model of the fuel cell incorporating all its control structure and blocks are presented in Fig.5. The parameters used in the model are presented in Table-I. This section establishes the correctness of the newly designed thermal dynamic block. Moreover, the robustness of the decoupled control scheme is proved by considering different types of disturbances.

### Case I:

The initial temperature of the fuel cell at no load is kept at 1273 °k. At 100 sec the real power reference is made 0.3 p.u. It is further increased to 0.6 p.u. at 1600 sec and again brought back to 0.3 p.u. at 3100 sec. The fuel cell's temperature is kept under control to avoid damage to their delicate ceramics. The simulation result for this case is presented in Fig. 5. As the real power is varied and brought back to its initial condition following a disturbance, the temperature should also follow the same path provided large time is allowed to stabilize. From Fig.5 it is clear that the temperature is also coming to the original situation following the disturbance. This establishes that the temperature model is correct.

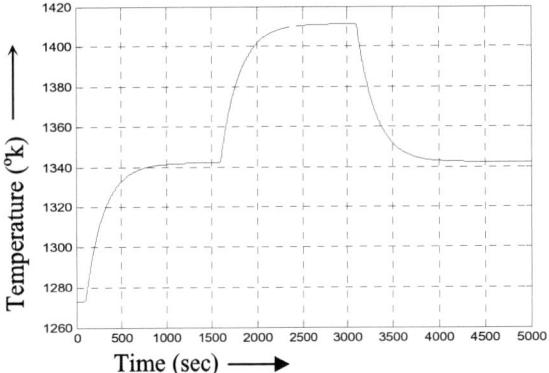

Fig. 5. Temperature variation curve.

TABLE I
PARAMETERS USED IN SIMULATION

| | |
|---|---|
| Faraday's constant (F) | 96484600 C/kmol |
| No load voltage $(E_0)$ | 1V |
| Number of cells $(N_0)$ | 384 |
| Utilization factor $(U)$ | 0.8 |
| Hydrogen valve constant $(K_{H_2})$ | 0.843 mol/(s.atm) |
| Water valve constant $(K_{H_2O})$ | 0.281 mol/(s.atm) |
| Oxygen valve constant $(K_{O_2})$ | 2.52 mol/(s.atm) |
| Hydrogen time constant $(\tau_{H_2})$ | 26.1s |
| Water time constant $(\tau_{H_2O})$ | 78.3 s |
| Oxygen time constant $(\tau_{O_2})$ | 2.91s |
| Reformer time constant $(\tau_1)$ | 5s |
| Conversion factor $(C_v)$ | 1 |
| Line & Transformer reactance referred to LV side $(X_{actual})$ | 0.16Ω |
| Operating temperature | 1273°k |
| Base Voltage L.V side $(V_r)$ | 400 v |
| $H_2$-$O_2$ flow ratio $(r_{H-O})$ | 1.145 |
| Base KVA | 100 |

*Case II:-*

The real power reference is varied from no load to 0.4 p.u. at 100 sec and the response of the model is obtained. At this condition the reactive power reference is kept intentionally at 0 p.u to show the decoupling strength of the scheme. The performance of the scheme is presented in Fig.6

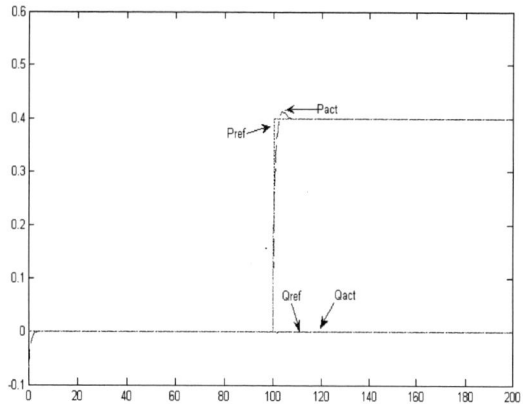

Fig. 6. Real Power Tracking Performance.

From Fig.6 it is well established that the decoupled controller tracks the real power change command without affecting the reactive power. The transient that are occurring in reactive power curve is due to the sudden change in fuel cell voltage and phase angle $\delta$ following a real power change. The variation of fuel cell voltage in this case is shown in Fig.7. The nature of this curve is different from that shown in [1]. This is mainly due to the assumption made in [1] that the input air and fuel flow is constant, which in this paper are varied through a controller.

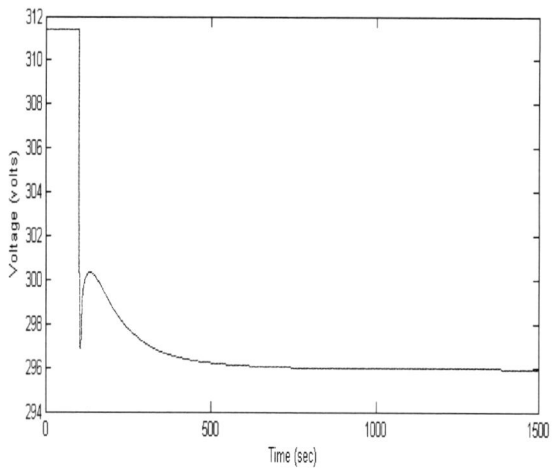

Fig. 7. Fuel Cell Voltage Variation.

*Case III:*

The reactive power reference is changed from 0 p.u. to 0.5 p.u at 100 sec keeping the real power reference at 0 p.u. The performance of the scheme is presented in Fig.8. In this case it is found that there is absolutely no transient in real power this is because of the fact that the dc voltage will not change during variation in reactive power. Moreover, the tracking

performance of the proposed reactive power controller is quite acceptable.

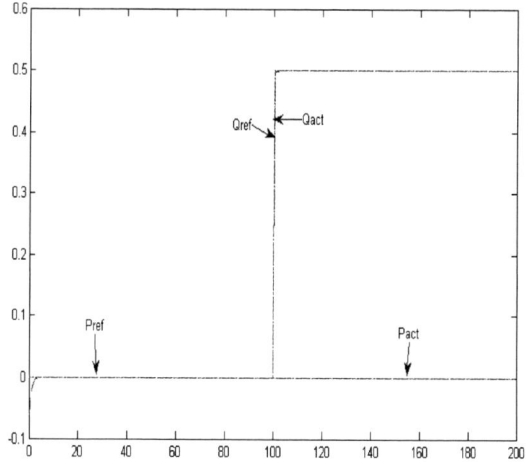

Fig. 8. Reactive Power Tracking Performance.

*Case IV:*

From the previous cases it is established that the controller performance is quite satisfactory when one of the power is changed. However, situations will arise when there will be requirement of a change in both the power. Therefore, in this case both the real and reactive power reference is changed from 0 p.u. to 0.6 and 0.3 p.u. respectively. The performance of the controller for this simultaneous change is depicted in Fig.9.This shows that when both the powers are changed simultaneously, the controller satisfactorily tracks the individual reference powers.

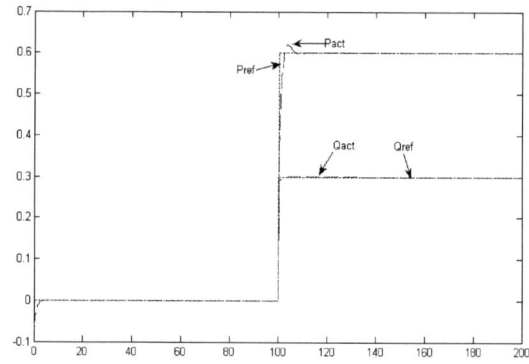

Fig. 9. Both Real & Reactive Power Tracking Performance.

*Case V:*

In a specific time span of a day there could be many variation in real and reactive power demand. Therefore, multiple excursion of both real and reactive power reference is adopted to simulate a more analogues situation as that of the real practice. The real and reactive power tracking performances are depicted in Fig.10 and Fig.11 respectively. Further, the control action i.e. rate of flow of hydrogen input in case of real power regulation is presented in Fig.12 which clearly indicates a smooth variation. A smooth control action is required for proper functioning of the controller. Similarly,

738

the variation of modulation index is depicted in Fig.13, which also has a smooth characteristic.

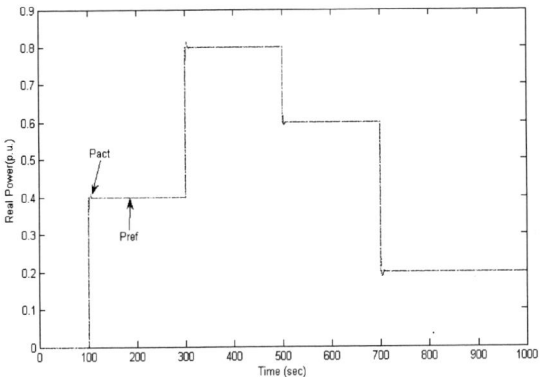

Fig. 10. Multiple Variation: Real Power Tracking Performance.

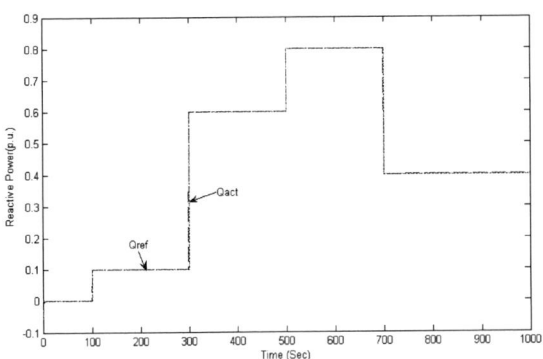

Fig. 11. Multiple Variation: Reactive Power Tracking Performance.

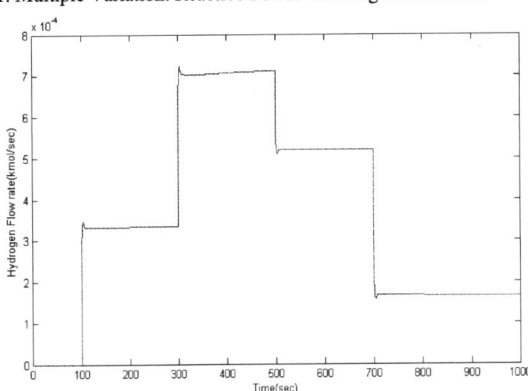

Fig. 12. Hydrogen output flow rate regulation.

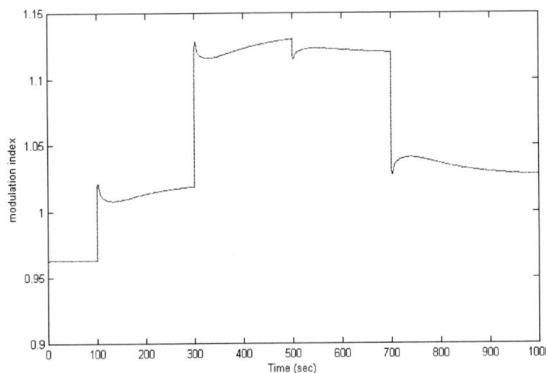

Fig. 13. Modulation Index Variation Curve.

## V. CONCLUSIONS

This paper presents a decoupled control strategy to regulate both real and reactive power feeds from a grid connected Solid-Oxide Fuel Cell power plant. The main contribution of this paper can be summarized as follows:

- The thermal dynamic model is modified to be more accurate and realistic.
- A decoupled control strategy for real and reactive power regulation is obtained by varying
  i. Hydrogen flow rate for real power and
  ii. Modulation index for reactive power control

The model was implemented in SIMULINK and the performance of the decoupled scheme is presented. From analysis of the performance curves it is found that the control strategy is suitably robust and also decoupled in nature.

## VI. REFERENCES

[1] K. Sedghisigarchi, Ali Feliachi, "Dynamic and Transient Analysis of power distribution systems with fuel cells - Part I: Fuel-cell Dynamic Model," *IEEE Trans. on Energy Conversion*, Vol, 19, No.2, June 2004, pp 423-428E. H. Miller, "A note on reflector arrays," *IEEE Trans. Antennas Propagat.*, to be published.

[2] D. J. Hal and R. G. Colclaser, "Transient Modeling and Simulation of Tabular Solid Oxide Fuel Cell," *IEEE Trans. on Energy Conversion*, Vol.14, No.3, Sept-1999, pp.749-753.

[3] Y.H. Kim, and S.S. Kim, "An Electrical Modeling and Fuzzy Logic Control of a Fuel Cell generation System," *IEEE Trans. on Energy Conversion*, Vol.14, No.2, June-1999, pp 239-244.

[4] C.J. Hatziadoniu, A.A. Lobo, F. Pourboghrat, and M. Daneshdoost, "A Simplified Dynamic Model of Grid-Connected Fuel-Cell Generators," *IEEE Trans.. on Power Delivery*, Vol.17, No. 2, April-2002, pp 467-473.

[5] E. Achenbach, "Response of a solid fuel cell to load change," *J.Power Sources*, Vol.57, 1995, pp.105-109.

[6] M.Y. El-Sharkh, A. Rahman, M.S. Alam, A.A. Sakla, "Analysis of Active and Reactive Power Control of a Stand-Alone PEM Fuel Cell Power Plant," *IEEE Trans. on Power Systems*, Vol.19, No. 4, Nov-2004, pp 2022-2028.

[7] M.Y. El-Sharkh, A. Rah man, M. S. Alam, " Neural networks-based control of active and reactive power of a stand-alone PEM fuel cell power plant," *J.Power Sources* Vol. 135, 2004, pp 88-94.

[8] J. Padulles, G. W. Ault, and J. R. McDonald, "An integrated SOFC plant dynamic model for power system simulation," *J. Power Sources*, Vol. 86, 2000, pp.399-408.

[9] M.Y. El-Sharkh, A. Rahman, M.S. Alam, A.A. Sakla, T. Thomas "A dynamic model for a stand-alone PEM fuel cell power plant for residential applications," *J. Power Sources*, Vol.138, 2004, pp.199-204.

[10] Y. H. Li, S. S. Choi, S. Rajkaruna ,"An Analysis of the control and operation of a solid oxide fuel cell power plant in an isolated system" *IEEE Trans. on Energy Conversion*, Vol, 20, No.2, June 2005, pp 381-387.

**2006 IEEE International Conference on Power Electronic, Drives and Energy Systems**

# An Universal Interconnection System to Connect Distributed Generation to the Grid

Vinod John, *Member IEEE,* Eric Benedict, *Member IEEE,* and Shazreen Meor Danial

*Abstract*—**Interconnection equipment between distributed energy resources (DER) and the grid is typically custom designed by the distributed generation (DG) equipment manufacturer or integrated by engineering firms using subcomponents such as relays, sensors and switchgear. The DER Switch described in the paper has integrated all of the required equipment for the DER interconnection into a single package that is designed to be compliant with IEEE 1547 and UL 1741 standards. A 480V, 200A, circuit breaker based DER Switch prototype with a digital signal processing (DSP) board was designed, built and tested. The objective was to create a standard, flexible universal interface switch for distributed energy resources so that single or multiple DER systems like wind turbines or solar arrays can be connected to a utility. The resulting interconnection switch design is DER technology neutral and can be used for inverter and machine DG applications.**

*Index Terms*— **Distributed generation, universal interconnection, synchronization, islanding, protective relaying.**

## I. INTRODUCTION

INTEGRATION of distributed energy resources (DER) with the electric power system is increasingly seen as a technology that can change the traditional method of electrical power delivery and can provide multiple advantages to energy customers, energy suppliers, and society overall [1]. Numerous promising generation, storage, and load management technologies are under development or are entering early commercialization stages. It is becoming increasingly apparent that new systems level technology and functionality are necessary to unlock the full potential of the emerging DER technology and to ensure a broad acceptance of DER systems as a key component in the overall energy delivery system. This paper describes an universal grid interconnection prototype that integrates multiple functions of switching, sensing, control,

---

This work was supported by the National Renewable Energy Laboratory (NREL) and California Energy Commission (CEC) under NREL Subcontract ZAT-4-32616-05.

Vinod John is with the Department of Electrical Engineering, Indian Institute of Science, Bangalore 560032, Karnataka, India, (e-mail: vjohn@ee.iisc.ernet.in).

Eric Benedict and Shazreen Meor Danial are with Northern Power Systems, Waitsfield Vermont 05602, USA, (e-mail: ebenedict@northernpower.com and smeor@northernpower.com).

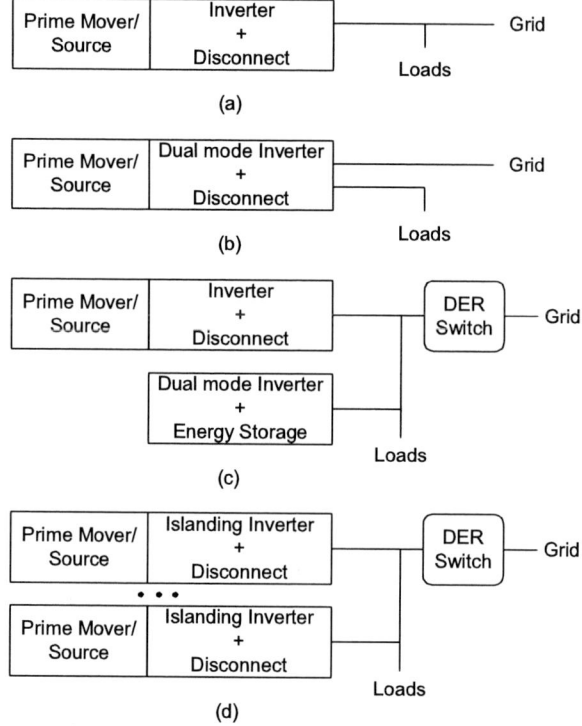

Fig. 1. DG architecture showing (a) Grid parallel configuration. (b) Dual mode inverter with grid parallel and stand alone capability and do not require additional DER Switch. (c) Upgrade of grid parallel configuration to provide standalone backup capability. (d) Multiple DG configuration that requires additional DER switch.

protection and communications into a single package.

## II. TECHNICAL APPROACH

The approach taken to create the universal interconnection system was to define the system architecture, and to create a DER switch specification and perform design and component selection for prototype fabrication. It was seen that existing switch configurations offer basic solutions for DER interconnections. The DER Switch design addresses some of the issues and concerns for the interconnection especially related to intentional islanding. In addition, a DER Switch can be used to improve the services provided by the DER in terms of load power quality [2]-[4]. These improvements are provided through the combination of the DER Switch's advanced controller and higher possible switching speeds along with DGs with appropriate control capabilities. There are many DER applications where the interconnection switch is integrated with the DER itself. This approach can be effective in applica-

0-7803-9771-1/06/$25.00 ©2006 IEEE

tions where a cost effective DER package can be built and shipped out to a customer for installation. This is especially true for low power single-phase applications. However different customers may need different combinations of DER equipment to meet their needs. To broaden the range of DER options there can be advantages to decouple the DG from the utility interconnection switch. Two related questions need to be considered in this context. The first is, when does a separate DER Switch make sense? And the second is, what applications require the DER Switch to be a separate entity from the DG? This requires one to look at specific DG system architectures where such a separate switch is desirable.

The DG can be operated either as a grid connected or a stand-alone configuration. A basic grid connected DG system as shown in Fig. 1(a) does not require an additional interconnection device under normal operation. If the operation of the DG system as an intentional island is required, then there are multiple architectures that can be utilized. Some inverter based DG systems are available from DG vendors with dual mode capability as shown in Fig. 1(b). These systems have some internal energy storage capacity, which is utilized to obtain an acceptable response to step load changes. These dual mode DG systems may have explicit external or internal switchgear that is used to transfer between standalone and grid parallel operation. When the external switchgear is present, the inverter typically needs to know the open/closed status of the switch on an instantaneous basis and this requires a high-speed control interface between the switch and inverter controller.

Fig. 1(c) shows a grid connected DG system that has been upgraded to operate in standalone mode by adding an inverter with intentional islanding capability and a DER Switch. This configuration can be used to upgrade a grid parallel DG configuration to one that has intentional islanding capability. This configuration can have a higher cost because it makes use of two inverters. However, a more flexible system configuration is achieved because the two inverters and the DER Switch can operate with some physical separation between them and without the necessity for high-speed control interconnections. Replacing the dual mode DG inverter controls with controls that can seamlessly operate in grid connected and islanded modes of operation will further optimize the overall system [3]. Such a configuration can be extended to the case of multiple DG systems that operate together in a Microgrid power network. This architecture that is shown in Fig. 1(d), leads to a simpler overall system design. In case of the configurations that utilized the DER Switch, the interconnection protection functions, such as specified in IEEE 1547, should reside at the DER Switch to prevent the power converter from unwanted trips in situations where intentional islanding is required. The islanding or dual mode inverters should have adequate provision to facilitate this coordination.

In general, the DER Switch lends itself to more advanced power network architectures. A representative system indicating DER assets, loads, the DER Switch and the area electric

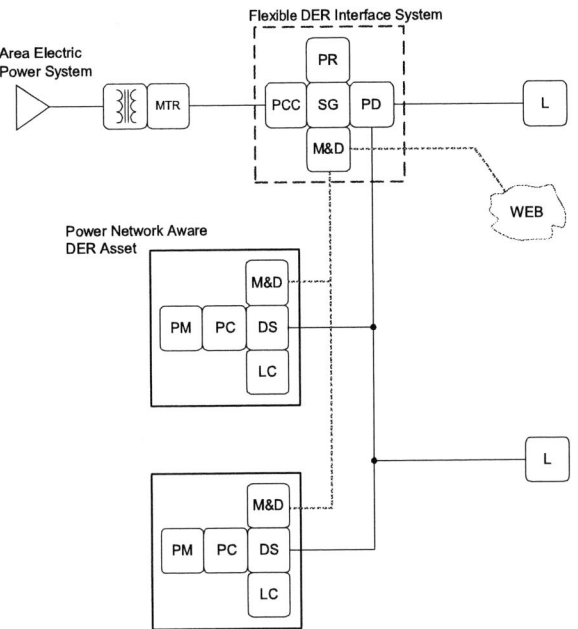

Fig. 2. DER power network block diagram indicating the target role of the DER Switch (dashed blue line envelope). MTR – meter, PCC – point of common coupling, PR – protective relay, SG – switchgear, PD – power distribution, M&D – monitoring and diagnostics, PM prime mover, PC – power conversion, DS – DG switchgear, LC – local control, L – loads.

power system is shown in Fig. 2. The DER Switch consists of point of common coupling, switchgear, protective relay, monitoring and diagnostics and power distribution components. The communication interface can be low bandwidth between the multiple DG and the DER Switch (shown in the red dotted line) and would be used for SCADA and energy management system functions. High-speed control interconnections and communications would not be required to operate such a power network.

### A. Hardware Requirements

A DER interconnection with the grid that meets the application requirements described in the previous section requires a flexible hardware concept. A traditional implementation of such a concept will involve switch hardware, voltage and current sensing devices, protective relays, a controller with diagnostic and monitoring functions, a communications processor, power supplies and other components.

The DER Switch aggregates the control functions in a digital signal processor. The hardware flexibility is retained in the DER Switch with the additional capability of replacing a circuit breaker with solid-state switches. In the case of a solid-state switch, additional breakers are used to obtain high fault interrupt rating under any internal faults and to meet Basic Impulse Insulation Level (BIL) targets when disconnected. A solid-state version of the DER Switch would be able to switch at higher speeds and provide higher power quality. The ability of the DER Switch platform to provide a range of interconnection speeds offers the flexibility to match the application requirements. The circuit breaker based DER Switch design was selected for the prototype because it was the simplest and lowest cost technology that could test all of the relay, IEEE 1547

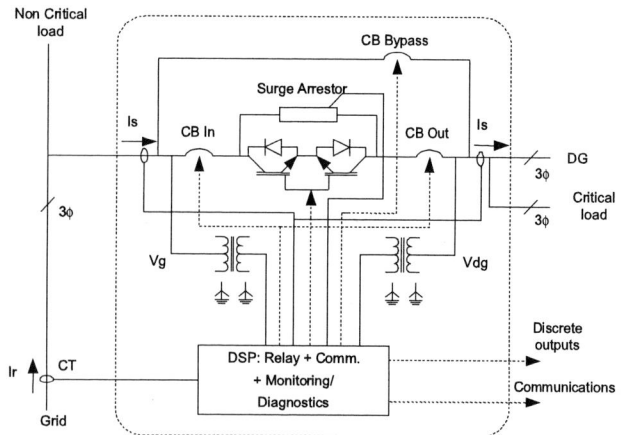

Fig. 3. One-line schematic of a semiconductor based DER Switch showing switch hardware, sensor locations and grid and load connections.

TABLE I
ANALOG INPUT AND OUTPUT CHANNELS USED IN THE DER SWITCH DSP CONTROLLER.

| Analog Input | Number of channels |
|---|---|
| Current sensor – (Is) phase A, B, C, N | (4x) |
| External CTs – (Ir) phase A, B, C | (3x) |
| Grid side sensing – (Vg) phase A, B, C | (3x) |
| DG side sensing – (Vdg) phase A, B, C | (3x) |
| Snubber circuit common and differential voltage | (2x) |
| Control voltage sensing (V24) | (1x) |
| Spare multiplexed inputs for application-related requirements | (2x) |
| Analog Output | |
| Spare multiplexed outputs for application-related requirements | (2x) |

requirements [5] and power quality functions [2]. In the semiconductor DER Switch designs, input and output circuit breakers are used as backup protection in case the semiconductor switch fails in order to allow the DER Switch to still be able to disconnect DG from the grid. The control of the DER Switch is designed to be switch technology neutral. In addition, the same control system can be used for a DER Switch which implements the switch using either power semiconductor based components or using a circuit breaker.

The high-speed capability of the semiconductor switch and DSP allows clearing times in the range of fractions of a millisecond. This allows the capability of zero fault current contribution to the grid and the possibility of operation with network protectors that need zero reverse power flow [4]. When the DER Switch is used in combination with DER assets that can seamlessly pickup loads in cases of grid disconnection, then high power quality is available to the load. For less demanding and cost critical applications, the same design with circuit breaker switch hardware can be used.

The internal current sensors within the DER Switch are capable of measuring both ac and dc current with a high bandwidth. The external analog inputs for current measurement assume the use of 5Arms secondary CT. Analog inputs for grid and DG ac voltage measurements are made using 120Vrms nominal secondary VT. The one-line schematic indicating analog signal sensor locations is shown in Fig. 3. Sixteen analog channels that are directly available in the DSP board for high speed sampling are utilized in the controller. Table I list the analog input and outputs used in the DER Switch controller.

Digital inputs to the DSP are optoisolated and debounced. The primary inputs from the switch controller are start, stop and reset. The application related digital inputs are interfaced with 24V DC relay coils that can be externally energized. The two inputs are 1) Trip signal where a high signal indicates an external command to the DER Switch to transition to the Disconnect state, and 2) Two additional spare channels are available and can be configured either as active high or active low. Additional internal digital input signals are used to monitor the

semiconductors', the circuit breakers', and the contactors' fault and status indicators.

The digital outputs of the DSP are opto-isolated. Additional interposing relays with Normal Close/Normal Open dry contacts with surge protection are used to provide isolation and voltage surge ratings. The two application-related output contacts available externally are the 'Connected or Disconnect status' of the DER Switch and an 'auxiliary switch' output. The auxiliary switch output can be used to open any circuit breaker in series with the DER Switch or trigger another DER Switch or circuit breaker to obtain a transfer switch or other more complex power switching configurations.

Optional analog zero to 5V output voltage signals are available on the DSP controller board. The signals are centered at 2.5V and 5V represents the maximum rated output capacity. The analog output signals implemented in the DER Switch are three phase real and reactive power. Two additional spare channels can be used for the application-related signals.

### B. Control Requirements

The control functions are computed based on the raw analog and digital inputs to the DSP. A small amount of filtering is provided for EMI and noise rejection. Additional control inputs are possible through a Human Machine Interface (HMI) on a remote computer through serial communications interface. Control functions evaluate these inputs to achieve interconnection protection, to meet the IEEE 1547 standard requirements, and to evaluate ambient power quality. DG protection is left to the DG controls and is not included in the DER Switch. The DSP controller provides on-off commands for semiconductor switches and control of the circuit breakers within the DER Switch. Additional spare analog and digital outputs can be used for overall power system integration. The evaluated values of the control algorithms are available through the HMI for energy management functions.

### 1) DER Switch State Machine

The state machine controls the operation of the DER Switch. The primary operating states of the DER Switch are to connect the grid and DER power network or to stay disconnected. A number of the other control states are used for the startup sequence, faults and bypass operations. Fig. 4 shows the state diagram for the DER Switch controller, which is used for all the DER Switch options. It is also possible to select

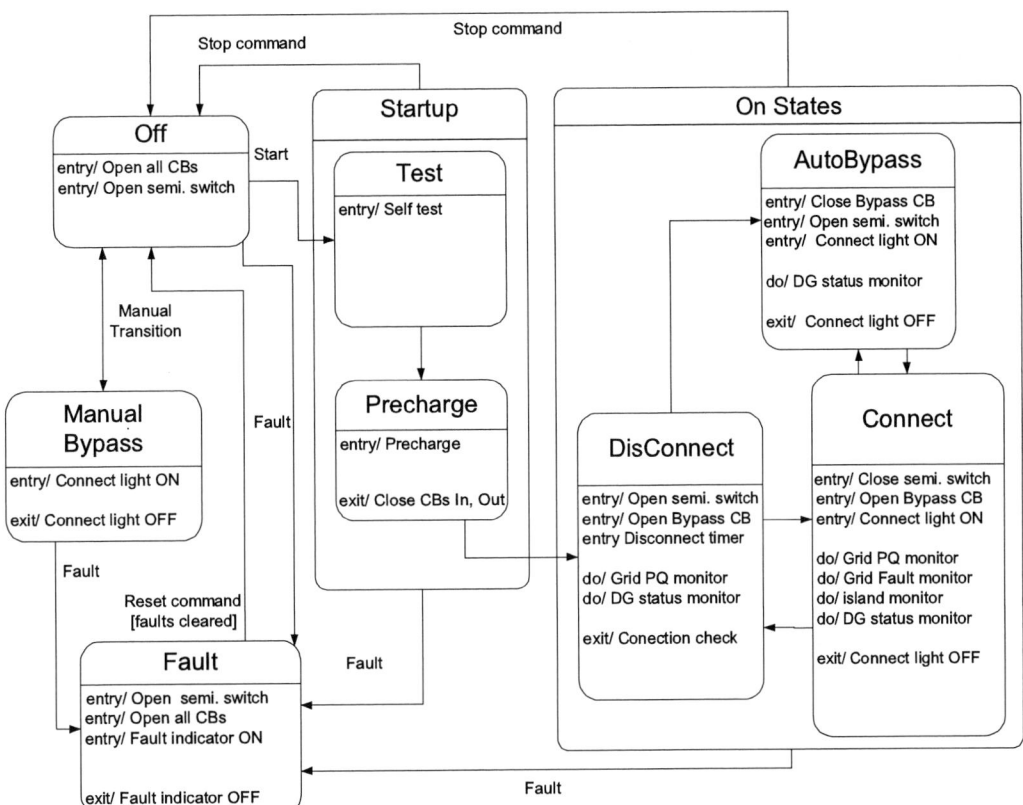

Fig. 4. Control state diagram for the DER Switch.

modes of operation of the DER Switch to obtain different behavioral characteristics within the operating states. The behavior of the DER Switch when operating in the controller states is described below.

The DER Switch following power-up and reset of the controls is considered to be the Off State. The DSP controller conditions for the Off state are: no faults are latched, all alarms are enabled, waveform capture snaplog is enabled, all semiconductors are disabled, and all discrete outputs are off. Any change in the mode settings can be done only in this state. Any CB that stays closed in this state triggers a fault. A startup sequence can be initiated from this state. The unit can transition to the Manual Bypass state if the Manual mode is set.

The condition for Manual Bypass State is similar to the Off state with the exception that the Bypass CB can be closed. A fault is triggered if the Input and Output CBs are closed in this state. There is no event-related change of state in this mode.

A start command in the Off state initiates the transition to the Test state. The motorized input and output circuit breakers should be ready to close. The grid and DER side voltages are monitored to see if they are in the nominal range and the voltage phase lock occurs on at least one side. If voltage is absent from both sides, then the unit goes to Fault, else the unit goes to the Precharge state. The Precharge state is present for semiconductor based DER Switch. The precharge contactor is closed and the voltage snubber clamp capacitor is charged up to the peak ac line to line voltage. The input and output circuit breakers are closed if the clamp voltage is above a minimum

level. The precharge contactor is opened after this. A fault is generated if the clamp voltage is too low or if the input and output circuit breakers fail to close, subsequent to precharge.

In the Disconnect state the DER Switch checks to see if DER is present in the system. The DER status is obtained with a bit that is set high when DER is present or low when DER is absent. If it is present and the synchronization functions evaluate to True then the DER Switch transitions to Connect State. In the case that no DER is present, the controller checks to see if the DER side represents dead bus and the DER Switch transitions to the AutoBypass state. If there is a request to shut down the DER without de-energizing the loads through the supervisory control system, then the synchronization functions are evaluated. When the synchronization functions are true, the DER Switch transitions to the AutoBypass state. This feature is used only if there is an external supervisory controller for the DG and DER Switch system, which will ensure that the DG status is off within a short duration after reaching the AutoBypass state.

In the Connect State the DER Switch checks for power quality, anti-islanding other IEEE1547 functions and fault events. If any of these events are true, the DER Switch transitions to the Disconnect state. In case the DG is shut off (DG status is absent), the DER switch transitions to the AutoBypass state.

In the AutoBypass state the DER Switch checks to see if the DER is reconnected to the system. The DER Switch operates like a regular circuit breaker in this state and does not open in case of any power quality problems. In case DER status indicates that it is going to be reconnected, the DER Switch transitions to the Connect state. It is expected that the DER status will be updated before the reconnection by the Supervisory control system. In case the supervisory control is not in place, the AutoBypass state is not utilized. If the DER Switch entered the AutoBypass state because of a Request for DER shutdown, the control system should ensure that this request does not persist for a long time. The controller should time out this request and indicate a warning in case the request continues to persist.

When a fault occurs, the DER Switch can shut down rapidly into a safe state. Within this state, all CBs are open and the semiconductor switches are off. The snubber clamp circuit that is used to protect the semiconductor switch is discharged. This is the state that the DER switch enters on initial power up. The semiconductor based designs have extra input and output circuit breakers which provides backup protection in case the semiconductor-based switch fails.

In addition to the control states additional modes are provided in the controller such as: Auto/Manual mode, Local/Remote mode, Test mode and additional modes based on hardware attributes of the system.

*2) Control Functions*

The relay functions implemented in the DER Switch are based on typical requirements from DG projects and IEEE 1547. Additional functions are implemented for monitoring and diagnostics. Warning message and activation thresholds are provided for the control functions. Most of the relay functions trigger a transition to the Disconnect state. Exceptions are the synchronization and deadbus reclose functions that are used as enable signals for the DER Switch to reconnect. The relay functions are provided with options to enable each algorithm independently as individual functions or as a combination of functions.

The evaluation of the algorithm to check for synchronization between the grid side voltage and DG side voltage is performed in the Disconnect state. This function checks to verify that the voltage amplitude (for all three phases), frequency and phase angle are within an acceptable window to enable the closing of the DER Switch. Other DER Switch control functions are also evaluated to be true to enable the closing. Closing of the switch is only done in DER system when it is in Auto mode and if the synchronization enable is valid and reconnection enable is evaluated to be true. There are two modes for the synchronization function: 1) The absolute mode allows for synchronization when voltage, frequency and phase error magnitudes are small. 2) The Positive mode allows synchronization when frequency and phase error are small and has positive values. This is to prevent any reverse power surge during synchronization and connection from causing any trips based on fast reverse power relay calculations. The voltage magnitude and phase comparison is performed at the highest computation rate of the DSP. This ensures that any sudden jump in voltage or magnitude on either side of the DER Switch does not cause any false synchronization.

Dead bus reclosing relay function is provided in the DER Switch so that it can black start the loads connected to the DER side of the switch, when there is no DG connected to the system. A voltage threshold and time delay is provided for coordination. This relay checks to verify that closing occurs only under a dead bus condition where the voltage is below threshold on all three phases and when DG status is OFF.

The voltages on the grid and DER side are monitored to be within acceptable ranges. A threshold and time delay is provided for coordination of protection functions. Separate voltage thresholds and time delays for the grid side and DER side are provided. An event in the Connect state will make the DSP controller transition to the Disconnected state. Any event measured on the DER side when in the Disconnect state results in an alarm to the supervisory control system.

Frequency is measured from the three phase voltage measurement on the grid and DER side. Separate frequency thresholds and time delays are provided for coordination on the grid and DER side. An event in the Connect state will make the DER Switch to transition to the Disconnected state. Any event measured on the DER side when in the Disconnect state triggers an alarm to the supervisory control system.

Additional protection functions checks for phase rotation direction, missing phase information or lack of signal on the phase voltage. In addition, this function checks if the controller's internal data is synchronized with the grid and DER operating frequencies. An event in the Connect state will make the DER Switch to transition to Disconnect state. An event in the Disconnected state will prevent the operation of synchronization function.

The DER Switch controller provides both instantaneous and time over current relay functions. Additional neutral and ground time overcurrent relay functions are also implemented. The neutral current relay function can be used for 4 wire applications. The trip threshold levels and time delay before tripping are independently adjustable. The calculations for this algorithm are performed when in Connect state. Any triggered event makes the controller transition to the Disconnect state.

The DER Switch controller evaluates single and three phase power flow at the switch and at a remote location. Remote measurement is possible if additional CTs are wired into the DER Switch controller. Independent thresholds are available for switch and remote reverse power. The power measurement is compared with an adjustable threshold and time delay for coordination. If the power flow crosses the threshold then an event is set. The user can select if the crossing occurs in the positive or negative direction for the event to occur. An event in the Connect state will cause the DER Switch to transition to the Disconnect state.

The IEEE 1547 standard contains control requirements that should be satisfied for the interconnection of DER to the grid [5]. Other standards such as UL1741 and State standards such as California Rule 21 reflect many of the underlying concerns that are addressed by the IEEE 1547 series standards [6]. The main control functions required for the DER Switch to meet the IEEE 1547 standard relate to: under and over voltage, under and over frequency, harmonics, DC injection,

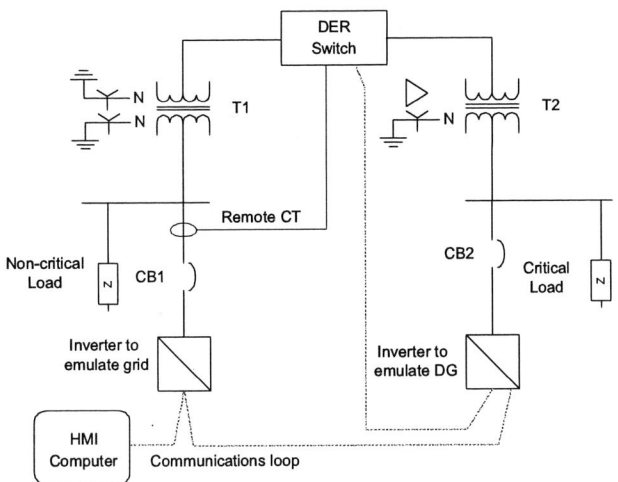

Fig. 5. Setup for testing the DER Switch using inverter to emulate grid and DG along with critical and non-critical loads.

anti-islanding, synchronization, and reconnection timing. The reverse power flow monitored at the point of aggregate loading is used to detect unintentional islanding. The DER Switch is programmed with relay functions to be fully compliant with the IEEE 1547 standard and has the flexibility to meet additional requirements. In addition to the above relay functions it is possible to manually set a digital bit to simulate the occurrence of an event. This was used to verify the proper operation of the DER Switch controller during testing.

## III. RESULTS

The tests performed included general commissioning followed by the various operational tests. The general commissioning tests were done to ensure the switch's manufacturing integrity. The operational tests concentrated on confirming the operation of the various control algorithms including the prototype's performance, relay functions, IEEE 1547 functions and power quality functions. A representative test power system indicating DER assets, loads, the DER Switch and the area electric power system is shown in Fig. 5. The test facility consists of two power converters, one which is used to emulate the grid and the second which is used to emulate the DG. Transformers and loads in the connection can be switched in to study various grounding options, power flow, and grid conditions.

Fig. 6 shows the connection transition of the circuit breaker. When the switch is not connected, the waveforms at the grid and DG side are distinctive since the DG inverter was not generating a perfect sine wave due to the test setup arrangement. However, the voltage magnitudes in RMS basis and frequency were almost identical and are within the synchronization window. The step signal indicates the command given from the switch controller to the circuit breaker. Scope channels 1 and 2 are the grid and DG voltages, respectively. It was observed that the both voltages would drift and eventually be in phase, which allowed the switch to close. When the switch is connected the voltage seen at the grid and DG outputs are equal and the switch would allow current to flow. The

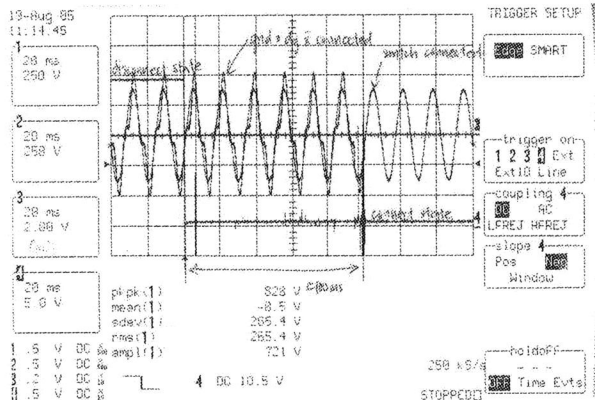

Fig. 6. Grid and DG side voltage and command to circuit breaker waveforms that indicate the DER Switch connection transition.

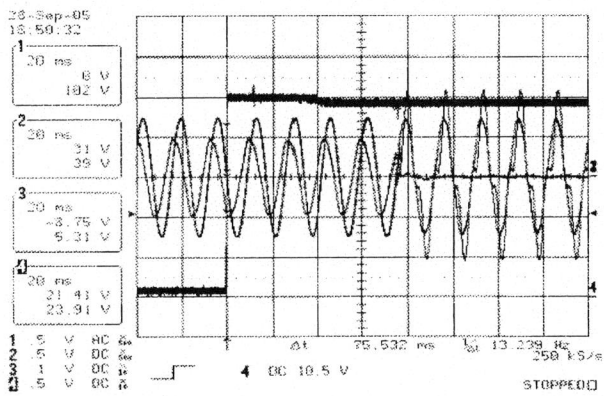

Fig. 7. Grid and DG side voltage, current, and command to circuit breaker waveforms showing switch disconnection transition.

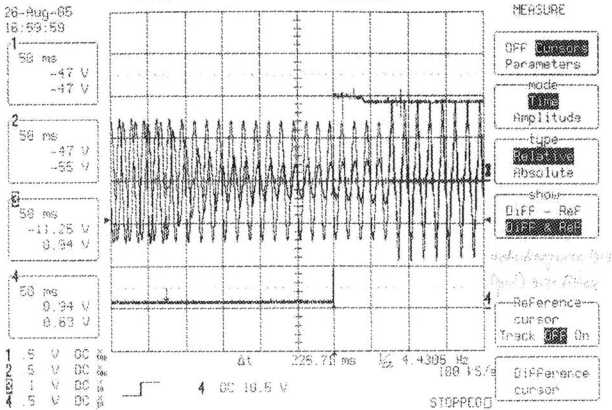

Fig. 8. Response of the system to a frequency event resulting in the DER Switch opening.

switch disconnection transition, which can be seen in Fig. 7, has the similar idea but opposite to the description provided above. The waveforms in Fig. 8 are the grid and DG voltages and current through the switch and the digital command to the circuit breaker. It is observed in the plot that the current waveform ramped down when there was a frequency change of either the grid or the DG inverters. The tripping of the switch occurred after a delay corresponding to the frequency relay coordination settings.

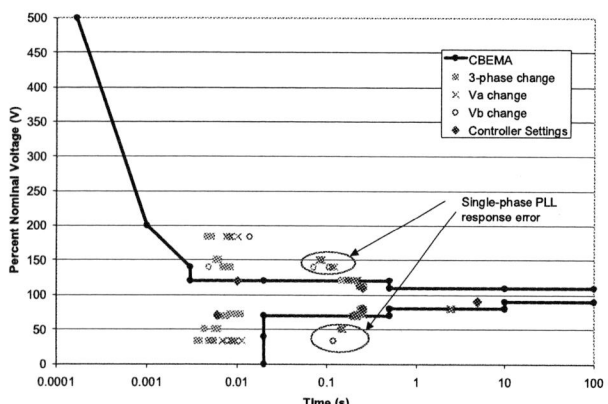

Fig. 9. CBEMA curve requirements, the DSP controller settings and test results performed on three phase and single phase basis.

The DER Switch had the capability of rapidly disconnecting the grid and DG terminals from each other if the switch senses any power quality disturbance according to the ITIC/CBEMA curve requirements. A few tests were performed to verify that the switch responds as expected, which is to disconnect the grid and DG terminals, when a CBEMA power quality event occurs. The magnitude and time response were also confirmed to meet the specified values. The power quality function in the switch controller is represented by 5 magnitude and time response settings that can be adjusted by the user via the HMI. Actual values that approximate the CBEMA curve based on the DSP controller voltage and timer settings were tested. From the initial observation, the switch managed to respond to the specified interruption voltage within the specified time setting. The time response, however, was limited by the circuit breaker disconnection time. A graph of the CBEMA curve with the test data overlaid is shown below in Fig. 9. The results obtained include the 3-phase and single-phase responses. The blue line represents the CBEMA curve, purple dots represents the actual magnitude and time settings specified in the Engineering HMI and the three other marker types indicate the response of the switch to the magnitude change. The majority of the test points responded within the vicinity of the specified magnitude and time settings. However, there are a few responses, as noted by the ovals on the plot, which occurred with a longer transition time than expected. Additional simulations were done to investigate this behavior. Response to very fast sub-cycle, single-phase voltage fluctuation is a very demanding requirement. Synchronous reference frame transformation used to generate phase information from the phase lock loop (PLL) algorithm caused the reaction delay especially when dealing with unbalanced 3-phase voltages. It was observed that the PLL responded cleanly to the three-phase voltage test, permitting the fastest reactions designed into the algorithm. The breaker opening delay, a mechanical limitation, prevented the unit from optimally complying with the CBEMA curve. On average, the breaker would fully disconnect no faster than 90ms after the most severe transients. It was seen from the test result that a three phase response setting can be programmed to meet the CBEMA voltage curve characteristics. However, the same

setting can be less sensitive for the single phase CBEMA event. Detailed results of all the tests are published in [7].

## IV. CONCLUSIONS

The test results resulted in successful meeting of the IEEE 1547 requirements. The complete control of the DER Switch is implemented in a single DSP package. The DSP activates switch operation, performs protective relaying functions and is compatible with the HMI software for enterprise energy management. The integration of all these functions into a single DSP helps achieve the overall equipment reduction target. The use of standard commercially available components for the design and use of components well within the ratings margin helps achieve the reliability goals for the switch. More detailed tests and analysis would be required to obtain the exact reliability characteristics of the DER Switch. The decoupling of the interconnection switch from the DG equipment allows for fast repair with the only limitation being that of spares availability. The ability to identify unintentional islanding situations in a fast and reliable way and to resynchronize was demonstrated during the test program. The present HMI implemented in the DER Switch is an engineering interface that provided full flexibility in making changes to settings. This also allows easy testing and data collection from the prototype DER Switch.

## V. ACKNOWLEDGMENT

The authors gratefully acknowledge the contributions of J. Lynch, I. Vihinen, B. Freeman, K. Kumar, and G. Kreis for their help at various stages of the DER Switch project.

## VI. REFERENCES

[1] F.Z. Peng, "Editorial – Special issues on Distributed generation," *IEEE Transactions on Power Electronics*, Vol. 19, No 5, pp. 1157-1158, Sept. 2004.

[2] Information Technology Industry Council, CBEMA/ITIC curve. Available: http://www.itic.org/archives/iticurv.pdf

[3] Robert Lasseter, Abbas Akhil, Chris Marnay, John Stephens, Jeff Dagle, Ross Guttromson, A. Sakis Meliopoulous, Robert Yinger, and Joe Eto, "Integration of distributed energy resources. The CERTS Microgrid Concept" (April 1, 2002). Lawrence Berkeley National Laboratory. Paper LBNL-50829. Available: http://repositories.cdlib.org/lbnl/LBNL-50829

[4] Consolidated Edison company of New York, "Specification EO-2115: Handbook of General Requirements for Electrical Service to Dispersed Generation Customers," 2005.

[5] *IEEE Standard for Interconnecting Distributed Resources with Electric Power Systems,* IEEE Std. 1547 - 2003, June 2003.

[6] US Department of Energy, Distributed Energy Program, Interconnection standards development. Available: http://www.eere.energy.gov/de/interconnection_stan_dev.html

[7] J. Lynch, V. John, S.M. Danial, E. Benedict, I. Vihinen, B. Kroposki and C. Pink, "Flexible DER Utility Interface System, Final report," NREL/TP-560-39876, Aug. 2006. Available: http://www.nrel.gov/docs/fy06osti/39876.pdf

**2006 IEEE International Conference on Power Electronic, Drives and Energy Systems**

# Transient Fault Response of Grid Connected Wind Electric Generators

Vinodh Kumar.P, Meera K S, and Sasi K Kottayil

*Abstract--* The paper deals with simulation studies on grid connected wind electric generators (WEG) employing Squirrel Cage Induction Generator (SCIG) and Doubly Fed Induction Generator (DFIG). Their dynamic responses to wind speed variations and transient faults on transmission line are studied.

*Index Terms--* -WT-DFIG, WT-SCIG, power system stability, MATLAB/Simulink, WEG, performance coefficient, grid.

## I. INTRODUCTION

GRID-connected wind electricity generation is showing the highest rate of growth of any form of electricity generation, achieving global annual growth rates in the order of 20 - 25%. It is doubtful whether any other energy technology is growing, or has grown, at such a rate. Global installed capacity was 47.6 GW in the year 2004 and 58.9 GW in 2005 [1], [2]. Wind power is increasingly being viewed as a mainstream electricity supply technology. Its attraction as an electricity supply source has fostered ambitious targets for wind power in many countries around the world.

Wind power penetration levels have increased in electricity supply systems in a few countries in recent years; so have concerns about how to incorporate this significant amount of intermittent, uncontrolled and non-dispatchable generation without disrupting the finely-tuned balance that network systems demand [3]. Grid integration issues are a challenge to the expansion of wind power in some countries. Measures such as aggregation of wind turbines, load and wind forecasting and simulation studies are expected to facilitate larger grid penetration of wind power.

In this paper simulation studies on grid connected wind electric generators (WEG) employing (i) Squirrel Cage Induction Generator (SCIG) and (ii) Doubly Fed Induction Generator (DFIG) have been carried separately. Their dynamic response to disturbances such as variations in wind speed, occurrence of fault etc. have been studied, separately for each type of WEG.

---

Vinodh Kumar Peruri is M.Tech student in the Department of Electrical and Electronics Engineering, Amrita Vishwa VidyaPeetham, Coimbatore-105 (e-mail: vk_peruri@yahoo.co.in)

Sasi K Kottayil is Professor in he Department ofElectrical and Electronics Engineering, Amrita Vishwa VidyaPeetham, Coimbatore-105 (e-mail : kk_sasi@ettimadai.amrita.edu)

K.S.Meera is with Central Power Research Institute, Bangalore-80. (e-mail: meera@powersearch.cpri.res.in)

## II. POWER SYSTEM STABILITY AND GRID INTERGRATION CODES

Power system stability may be broadly defined as that property of a power system that enables it to remain in a state of equilibrium under normal operating conditions and to regain an acceptable state of equilibrium after being subjected to disturbance [3,4].

### A. Transient stability

Transient stability can be defined as the ability of a power system to maintain synchronism when subjected to severe transient disturbances. The loss of synchronism (first swing stability) is a primary concern during transient simulations. It may occur immediately after the fault due to the initial acceleration or deceleration of the rotors of the machines. Time domain simulations have been traditionally used to assess the first swing stability.

### B. Dynamic Stability

The loss of synchronism can also occur after a few swings due to poor damping of system oscillations. This phenomenon is also known as "oscillatory stability" or "small-signal stability" or "dynamic stability. Many utilities use time-domain simulations to determine if the oscillations are present and how well damped they are. It is convenient to monitor voltages or active and reactive power at the generator terminals and along the major transmission paths.

### C. Grid Integration

The *grid integration code* for windfarms proposed by American Wind Energy Association (AWEA) includes the following four technical areas of concern [5]:

1. Low voltage ride-through (LVRT) capability of WEGs: The AWEA standard requires that the machine stay connected for voltages at the terminals as low as 15% of nominal per unit for approximately 0.6 s.
2. Supervisory control and data acquisition (SCADA) equipment for remote control to accommodate reliable scheduling and forecasting information exchange.
3. Reactive power capability: windfarms should be capable of operating over a power factor range of ± 0.95.
4. *WEG simulation models:* AWEA recommended that utility operators and WEG manufacturers participate in a formal process for developing, updating, and improving engineering models and turbine specifications used for modeling grid-connected windfarms.

---

0-7803-9771-1/06/$25.00 ©2006 IEEE

The work presented in this paper includes monitoring of critical values of voltage and power as well as stability assessment at several points in the simulated model of a wind farm–grid interface. MATLAB/Simulink software is used for the study.

## III. TRANSIENT PERFORMANCE OF GRID CONNECTED WT-SCIG

For evaluation of transient performance of grid connected SCIG that has wind turbine as prime mover, named as WT-SCIG, the following disturbances are applied to the system: (a) variation in the wind speed and, (b) faults in the system.

### A. Varying wind speed:

In this moderl, three WEGs each of capacity 250 kW are connected to a common grid but the three units are subjected to three different wind profiles. The behavior of each WEG is studied in case of a change in wind speed.

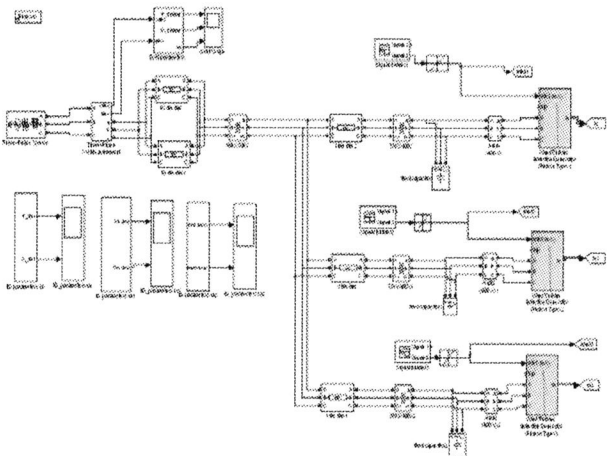

Fig. 1. Multiple WT-SCIGs connected to common grid.

The windfarm model is shown in Fig 1. Each 400 V rated WT-SCIG is connected to 66 kV grid through two transformers with ratings of 400V/33 kV and 33 kV/66 kV. There are two transmission lines included in the system - one is of 1 km length and is in between the two transformers, and the other is of 10 km in length and is in between 33kV/66kV transformer and the grid.

The schematic block diagram of WT-SCIG is shown in Fig 2. When the wind speed exceeds the rated speed, the blade pitch angle control will be activated (with the help of PI controller in Fig 2) so that blade pitch angle will increase thereby regulating the output of the WEG to the rated power. The parameters of the blocks used in this section are given in APPENDIX .

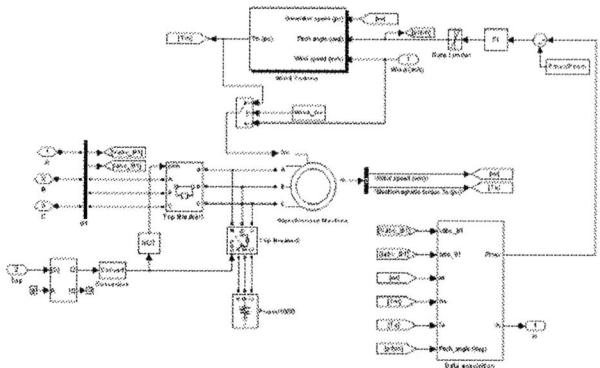

Fig. 2. Schematic block diagram of WT-SCIG.

### B. Wind Turbine Model:

The model is based on the steady-state power characteristics of the turbine. The output power of the turbine is given by the following equation [6]:

$$P_m = C_P(\lambda,\beta)\frac{\rho A}{2}v^3 \qquad (1)$$

where

$P_m$ -Mechanical output power of the turbine (W)

$C_p$ - Performance coefficient of wind turbine

$\rho$ - Air density (kg/m$^3$)

v -Wind speed (m/s)

A - Turbine swept area (m$^2$)

$\lambda$ - Tip speed ratio of the turbine $=\dfrac{\omega R}{v}$

$\beta$ - Blade pitch angle (deg)

$\omega$ - Angular speed of the turbine blade in rad/sec.

R - Radius of the turbine's blade (m).

The generic equation of a wind turbine is as follows [7]:

$$C_P(\lambda,\beta) = c_1\left(\frac{c_2}{\lambda} - c_3\beta - c_4\right)e^{\frac{-c_5}{\lambda}} \qquad (1)$$

The coefficients are $c_1 = 0.5176$, $c_2 = 116$, $c_3 = 0.4$, $c_4 = 5$, $c_5 = 21$ and $c_6 = 0.0068$.

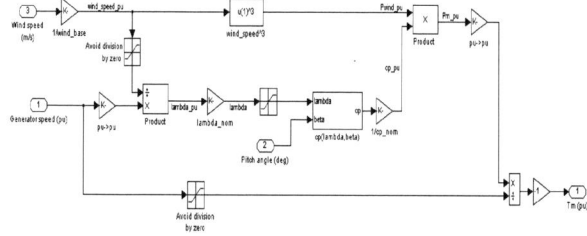

Fig. 3. Schematic block diagram of wind turbine.

The schematic diagram of wind turbine is shown in Fig 3 and Cp-$\lambda$ characteristics of the turbine for different values of blade pitch angle are shown in Fig 4.

748

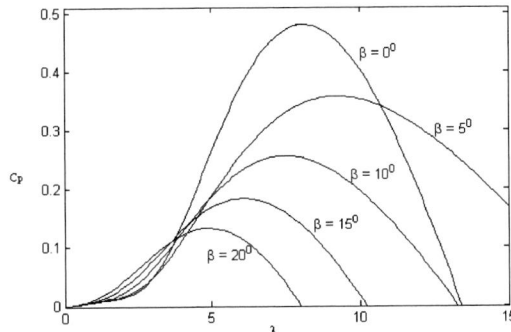

Fig. 4. Variation of Cp with $\lambda$ for different values of $\beta$ .

## C. Simulation results:

In practice, all the WEGs in a windfarm may not face the same wind speed at the same point of time. So the wind profile chosen in this simulation for each machine is different from those of the other two. Fig 6 shows the wind profiles for the WEGs.

The cut-in, rated and cut-off wind speeds of the simulated WT-SCIG have been found to be 5.6 m/s, 9 m/s and 23 m/s respectively. The pitch angle control is active above 9 m/s in order to regulate the power output at 250 kW. Fig 5 shows the transient response in the active power generation P and the reactive power consumption Q of the three units.

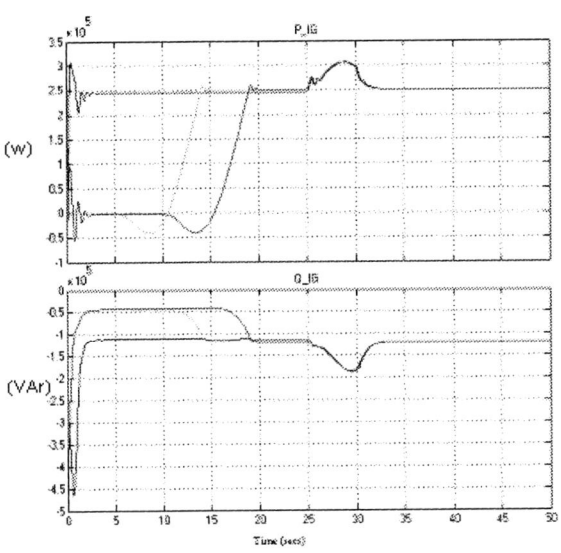

Fig. 5. Active and Reactive power at the terminals of SCIGs.

Fig 6 shows, along with the wind speed profiles, the pitch angle variation in each of the WT-SCIG. When the wind speed rises above 9 m/s at the 25th second the generator output also starts rising. However, the pitch angle controller comes into action and within 5 seconds the output of the generator is brought back to the rated level.

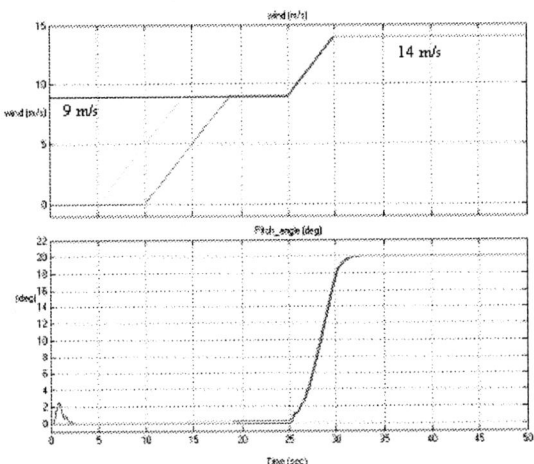

Fig. 6. Wind speed and the pitch angle for each SCIG.

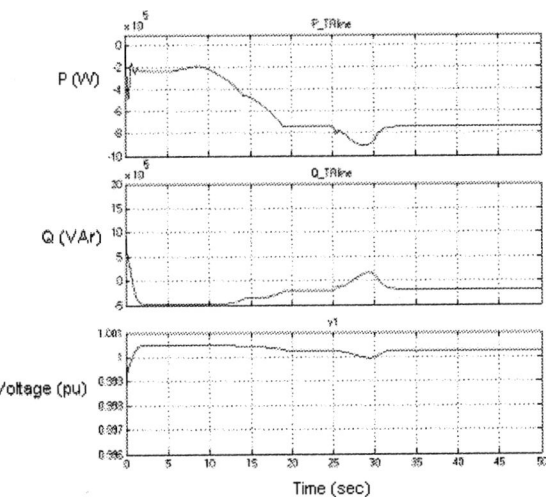

Fig. 7. Active Power, Reactive power and Voltage at Grid terminals.

Fig 7 shows the total power injected to the grid by the windfarm, which is about 725 kW. Each SCIG is provided with a fixed capacitance of 125 kVAr for reactive power compensation.

## D. Results with faults simulated in the System:

Different types of faults (L-G, L-L-G and L-L-L-G) are simulated on the transmission line that connects the windfarm with the grid and the performance is studied while keeping the wind speed constant.

A double line to ground transient fault (L-L-G) is applied at the middle of the 10 km transmission line. The fault is applied at the 35th second and its duration is 0.7 seconds. Fig 8 shows the plots of wind speed and pitch angle response. The rotor speed $\omega$ and the mechanical torque $Tm$ of the wind turbine are shown in Fig 9. A negative value of $Tm$ indicates torque produced by the turbine.

749

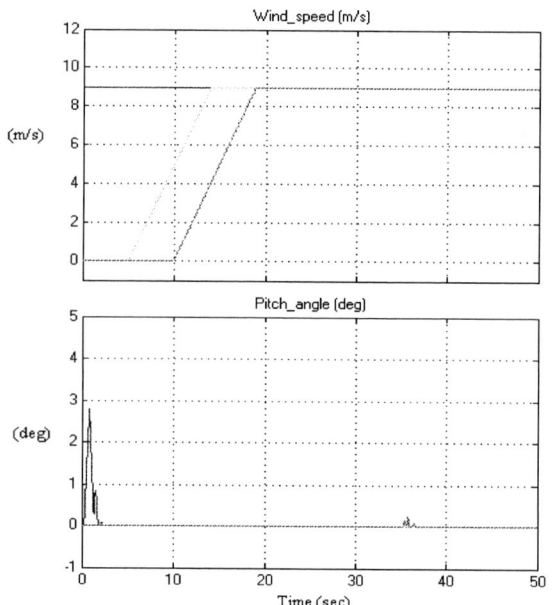

Fig. 8. Wind speed profiles and pitch angle variation of WT-SCIGs subjected to line fault.

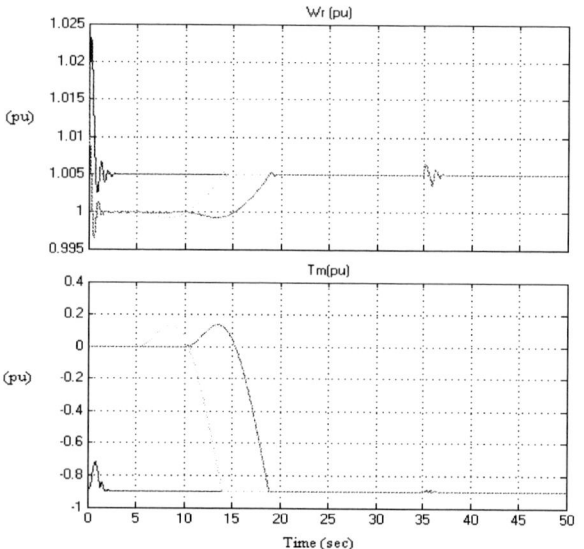

Fig. 9. Rotor speed and mechanical torque of WT-SCIGs subjected to line fault.

Fig 10 shows the corresponding P and Q oscillations of the generators. The power and voltage observed at the grid terminals of the windfarm are shown in Fig 11. It can be observed that there is a voltage dip of 38% during the fault.

## IV. TRANSIENT PERFORMANCE OF GRID CONNECTED WT-DFIG

The doubly fed induction generator (DFIG) is a wound rotor induction machine with a four-quadrant AC-DC-AC converter setup connected between the rotor terminals and the grid [8,9]. With the machine-side converter it is possible to control the stator power factor besides controlling the torque and the speed at the machine shaft, while the grid-side converter keeps the DC-link voltage constant. Fig 12 shows the schematic block diagram of WT-DFIG, which is studied.

The parameters are given in APPENDIX

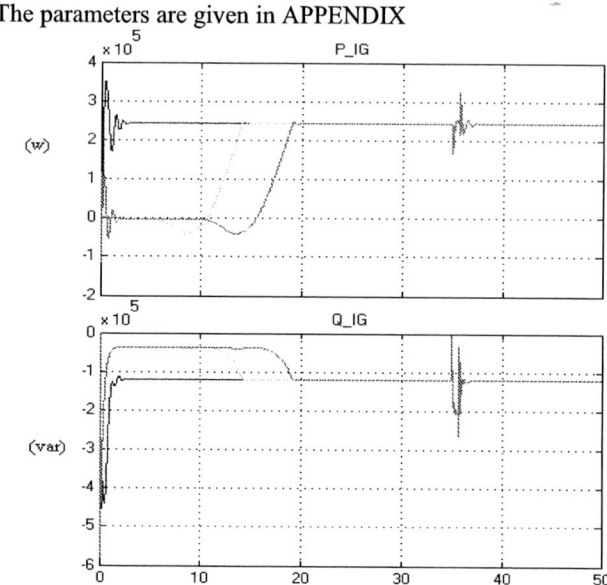

Fig. 10. Real, and reactive power of WT-SCIGs when subjected to line fault.

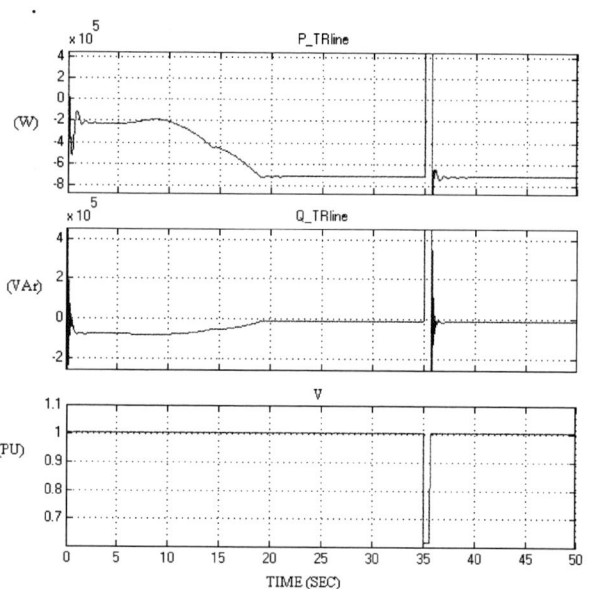

Fig. 11. Transient fault response in real power, reactive power and voltage at the grid terminals.

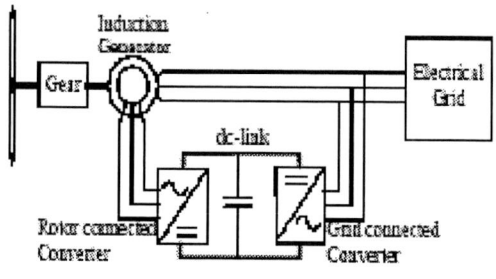

Fig. 12. Block diagram of DFIG.

750

## A. Tracking Characteristics

In DFIG the power output is controlled in order to follow a pre-defined power-speed characteristic, named tracking characteristic. Tracking Characteristic is represented by the ABCD curve superimposed on the mechanical power characteristics of the turbine obtained at different wind speeds as shown in Fig 13. The actual speed of the turbine ($\omega_r$) is measured and the corresponding mechanical power of the tracking characteristic is used as the reference power for the power control loop. The four points define the tracking characteristic: A, B, C and D. From zero speed to A the reference power is zero. Between A and B the tracking characteristic is a straight line, the speed at B must be greater than that at A. Between B and C the tracking characteristic is the locus of the maximum power of the turbine (maxima of the power-speed curves). The tracking characteristic is a straight line from C to D. The power at D is 1 p.u and the speed at D must be greater than that at C. Beyond point D the reference power is a constant at 1 p.u.

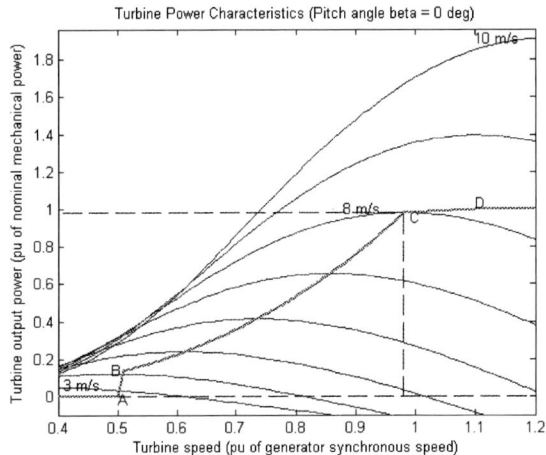

Fig. 13. Tracking characteristics of the Turbine.

## B. Simulation results of WT-DFIG connected to grid

The rated wind speed and the rated power of each of the WT-DFIG are 9 m/s and 240 kW respectively. As shown in Fig 14, the three units of WT-DFIG are included in the simulated windfarm. These three machines start generation at different instants, as the wind profile applied to each machine is different. The performance curves are shown in Fig. 15.

A single Line to Ground (L-G) fault is simulated at the 35th second, with fault duration of 5 seconds. As seen from the plots in Fig 16, sudden variations are there in both P and Q. Compared tyo WT-SCIG, it can be seen that the settling time after the fault clearance is very less.

It is also verified that the WT-DFIG can be made to consume less or no reactive power from the grid by properly tuning the parameters of the rotor side converter.

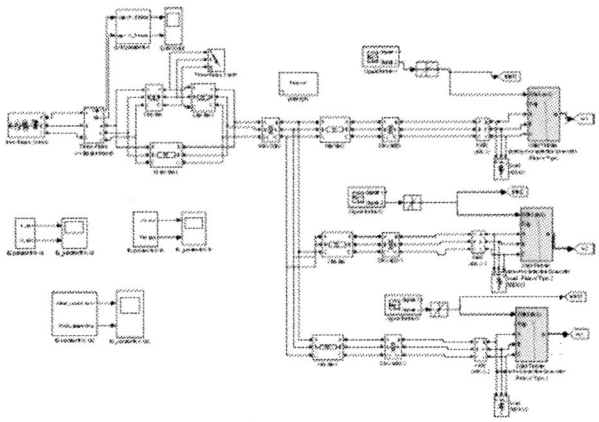

Fig. 14. Multiple DFIGs connected to Grid.

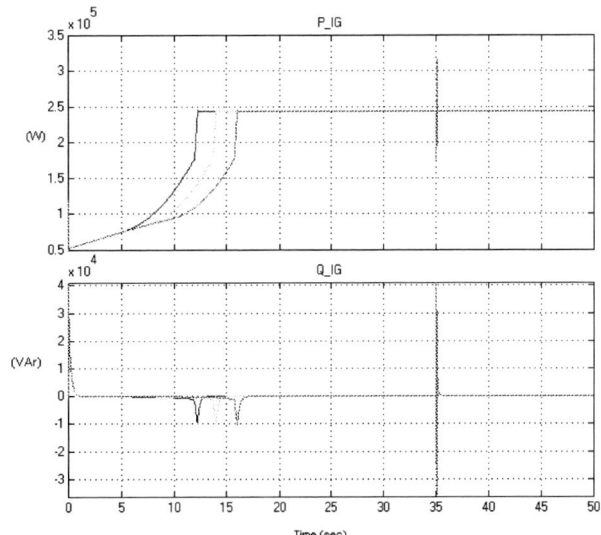

Fig 15. Real and reactive power of WT-DFIGs.

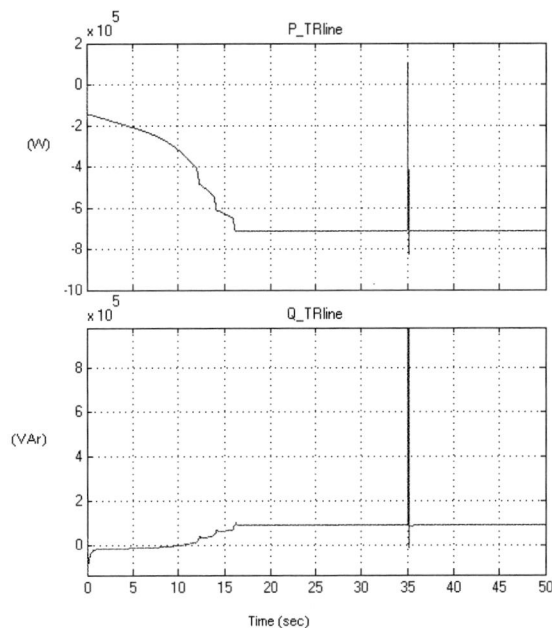

Fig. 16. Real and reactive power variations at terminals of the grid connected WT-DFIG under fault condition.

## V. CONCLUSION

The paper has presented a study on grid-connected wind electric generators, WT-SCIG and WT-DFIG, subjected to transient faults on transmission lines connected to the windfarm. All simulations are performed in MATLAB/Simulink. Results obtained from the study provide useful information regarding the behavior of these generators when connected to grid and helps a comparison between the two types.

## VI. REFERENCES

[1] Energy Information Administration / Annual Energy Outlook 2005 Reference case forecast pp-1-3
[2] World Wind Energy Association WWEA - http://www.wwindea.org/
[3] D P Kothari, I J Nagrath, "Modern Power System Analysisysis", 3$^{rd}$ Edition, Tata Mcgraw-Hill, New Delhi, 2003.
[4] P.Kundur, "Power Systems Stability and Control", EPRI, McGraw-Hill, ISBN 0-7803-3463-9, edition 1993.
[5] The American Wind Energy Association  - http://www.awea.org/
[6] Gary L. Johnson, *Wind Energy Systems*,1985, Prentice-Hall, Inc.,Englewood Cliffs, New Jersey 07632
[7] MATLAB/Simulink version 7.1
[8] Ian Norheim, kjetil uhlen, john olav tande, trond toftvaag, "Doubly Fed Induction Generator model for power system simulation tools", SINTEF Energy Research, Nordic Wind Power Conference, 2004
[9] S. Seman, J. Nirianen, S.Kanerva and A. Arkkio, "Analysis of a 1.7MVA Doubly Fed Wind- Power Induction Generator during Power Systems Disturbances", IEEE Transactions 2002.

## VII. APPENDIX

TABLE I
WT- SCIG PARAMETERS

| Rated Power | 250 kW |
|---|---|
| L-L Voltage | 400 V |
| Frequency of operation | 50Hz |
| Stator Resistance | 0.004843 p.u |
| Stator Inductance | 0.1248 p.u |
| Rotor Resistance | 0.00437p.u (ref.to primary) |
| Rotor Inductance | 0.1791 p.u (ref. to primary) |
| Magnetizing inductance | 6.77 p.u |
| Inertia constant H(s) | 5.04 |
| Friction factor | 0.01 p.u |
| No of poles | 6 |
| Synchronous speed | 1000 rpm |

TABLE II
GRID PARAMETERS

| Ph-ph RMS Voltage | 66 kV |
|---|---|
| Frequency | 50 Hz |
| 3-ph short-circuit level at base Voltage | 200 MVA |
| X/R ratio | 2 |

TABLE III
WIND TURBINE PARAMETERS

| Air density | 1.08 kg/m$^3$ |
|---|---|
| Radius | 0.25 m |
| Rated wind speed | 9 m/s |
| Maximum pitch angle | 45 deg |

| Maximum rate of change of pitch angle | 0.5 deg/s |
|---|---|
| Pitch angle controller gain [Kp, Ki] | [5, 25] |

TABLE IV
WT-DFIG PARAMETERS

| Rated Power | 250 kW |
|---|---|
| L-L Voltage | 400 V |
| Frequency of operation | 50Hz |
| Stator Resistance | 0.00706 p.u |
| Stator Inductance | 0.171 p.u |
| Rotor Resistance | 0.005 p.u (ref. to primary) |
| Rotor Inductance | 0.156 p.u (ref. to primary) |
| Magnetizing inductance | 2.9 p.u |
| Inertia constant H(s) | 5.04 |
| Friction factor | 0.01 p.u |
| No of poles | 6 |
| Synchronous speed | 1000 rpm |

# 2006 IEEE International Conference on Power Electronic, Drives and Energy Systems

# Black Start with DFIG Based Distributed Generation after Major Emergencies

M. Aktarujjaman, *Student Member, IEEE,* M.A. Kashem, *Senior Member, IEEE,* M. Negnevitsky,
*Member, IEEE, a*nd G. Ledwich, *Senior Member, IEEE*

*Abstract*—Grid connected distributed generation (DG) increases reliability and additional benefits for consumers as well as utilities. The stable and reliable operation of a power system is necessary after major emergencies (or blackouts) following a major system event. Distributed generation may be capable of black start and contribute to fast restoration process at medium to low voltage level. A large scale voltage and frequency excursions may occur during the process of black start with distributed generation due to low inertia and intermittency in power generation. Energy storage integrated with DG can absorb initial impact of central generation and ensure smooth load pick-up during the restoration of a system. In this paper, the process of black start with a doubly fed induction generator (DFIG) based wind turbine is addressed and energy storage in DC link of DFIG is used for fast restoration after blackout. A control system has been developed for the process of black-start with DFIG. A sequence of actions for black start procedure is presented and tested.

*Index Terms*—Black Start, Distributed Generation, Islanding Operation, Doubly-fed Wind Turbine, Storage System.

## I. INTRODUCTION

THE contribution of renewable based distributed generation has been increasing dramatically into the power system for last two decades. There are many types of renewable resources available, such as wind, solar, biomass etc. Among them, wind power has been growing faster than expected due to technology advantages, available resources, and environmental benefits. The presence of wind generation system into the low voltage level deteriorates the power quality and reliability due to its fluctuating nature of power generation [1]. Therefore, extra power source is necessary for participating in black start procedure after major emergency. Doubly-fed induction generator (DFIG) is an attractive option in wind power generation because of the variable speed operation, control of the real and reactive power, and reduced rating of converters

---

This research has been financially supported by the Australian Research Council under ARC Linkage Grant K0014223 "Integration of Distributed and Renewable Power Generation into Electricity Gird Systems", collaboration with Aurora Energy, Tasmania.

Md. Aktarujjaman, Dr. Mohammad A. Kashem, and Assoc. Prof. Michael Negnevitsky are with the School of Engineering, University of Tasmania, Hobart, Australia; and Prof. Gerard Ledwich is with the School of Engineering Systems, Queensland University of Technology, Brisbane, Australia, (emails: mda0@utas.edu.au; M.Kashem@utas.edu.au; Michael.Negnevitsky@utas.edu.au; g.ledwich@qut.edu.au).

0-7803-9772-X/06/$20.00©2006 IEEE

[2-3]. A battery storage system can be connected in the DC link of the two back-to-back PWM converters of DFIG to improve supply quality [4]. System reliability and supply continuity can be ensured by DFIG based distributed generation [5].

In the past, wind turbine was not considered for operation with grid systems for the reasons such as lack of fault-ride through capability and frequency variation due to uncontrollable input resource and power generation [6]. Now a days, advance technologies and better control systems allow wind turbine to operate with grid systems. Stabilization of system frequency is necessary to ensure the participation of wind generator for islanding operation as well as black start. Blackout may occur due to unavoidable large system disturbances. For the case of large disturbances, system may face steady state and dynamic instability, which may violate the system's contingency, resulting to complete blackout [7].

This paper proposes a technique for operation of DFIG with a storage system in its DC link to participate restoration process after major emergencies. The proposed technique can operate DFIG in emergency situation such as islanding operation and can participate black start process.

## II. SEQUENCE OF ACTIONS FOR BLACK START WITH DFIG

Utility grids are frequently interrupted by transmission system events such as lightning strikes, equipment failures, and downed power lines. These system events sometimes lead to a total blackout. To achieve a successful black start, the storage system in DC link of a DFIG can be used. Line side converter of DFIG is used to provide real and reactive power to the system. Storage can be used to minimize the initial impact of black start and speed up the re-energizing the transmission network [8]. In addition, the impact of load pick-up during the restoration process can be absorbed by the DC link storage system.

In a sequence of black start, a set of rules and actions needs to be followed, which should be identified in advance [9]. The control of DFIG with energy storage system will deal with building isolated network, forming and operating island, controlling voltage and frequency, and connecting island and re-synchronization with upstream network or grid system [10]. These tasks must be completed speedily to avoid large voltage and frequency excursions. Black start process can be established through the following three major stages.

0-7803-9771-1/06/$25.00 ©2006 IEEE

## A. Building low voltage network

A set of tasks need to be carefully organized for building low voltage network. This can be considered as an initial requirement for the black start process. The tasks are often considered a series of processes such as energizing electrical paths and distribution transformers, and starting of DFIG, etc.

### 1) Disconnect all loads:

To prevent large voltage and frequency excursions, it is necessary to disconnect both controllable and uncontrollable loads from the terminal of the DFIG.

### 2) Energizing dead transmission line:

Energizing electrical paths has to be done with extreme care. This process includes protection issues, steady state voltage, transient voltage, and dynamic voltage [7].

#### a) Steady state voltage issue

Capacitive charging current will dominate when energizing unloaded line. It is necessary to have the capability of the generator to absorb the reactive power for reducing excessive voltage rise at the point of generator terminal.

#### b) Transient voltage issue

In a weak system, transient over-voltages can be experienced depending on the generation capacity, length of the transmission line, transformer characteristics, etc [7].

#### c) Dynamic voltage issue

Harmonic resonance and steady state voltage rise are the main reasons for dynamic voltage rise, which may prolong for several seconds [7].

#### d) DFIG startup

DFIG can be started with the help of DC link storage system. The line side converter provides the amount of reactive support for starting-up the DFIG.

## B. Forming Island

At the beginning local load will be supplied by the DFIG-storage system. If multiple DGs are present in a system, then more than one island may be formed. They need to be synchronized with each other by matching phase sequence, frequency and voltage of both sides of the switch and form a larger island. The synchronization prevents large transient currents and power exchanges. In island, system control provides voltage and frequency stabilization.

### 1) Connecting load:

Two type loads are considered which are controllable and uncontrollable. Controllable loads are automatically disconnected from the system during the event of frequency deviation to keep the system's frequency within the limit. When the control system observes frequency more than the nominal frequency, the enough power is available for connecting additional load to the system. Depending on the generation capacity of DFIG, the amount of additional load to be connected can be determined.

## C. Re-synchronization

After forming island, system needs to be resynchronized with the upper-stream network by following the conditions of synchronization. It is necessary to change the islanded system's control strategy to avoid conflict between grid system and DFIG. The control of DFIG-storage system needs to be changed from voltage control mode to PQ control mode at the end of re-synchronization process [10].

## III. CONTROL METHODOLOGY OF DFIG FOR BLACKSTART AND ISLANDING OPERATION

Fig.1 shows a typical arrangement of a doubly fed induction generator (DFIG) equipped wind turbine with back to back PWM converter and a battery storage system connected to the DC link. The voltage controller of line side converter is used to utilize the DC link storage system in the process of black start and islanding operation.

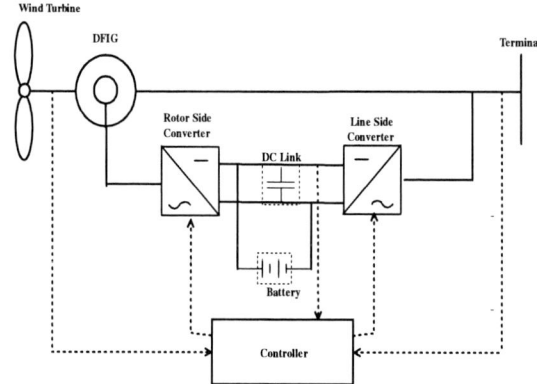

Fig. 1. Diagram of DFIG wind turbine with storage system.

Rotor Side Converter (RSC) controls the magnitude and phase angle of the voltage applied to the rotor circuit of the DFIG to create enough electromagnetic torque. The control structure of the rotor side converter is described in [2-4, 6]. The line side converter is used for controlling the DC link voltage, regardless of the magnitude and direction of the rotor power. A filter is added to the line side converter and shown in Fig.2.

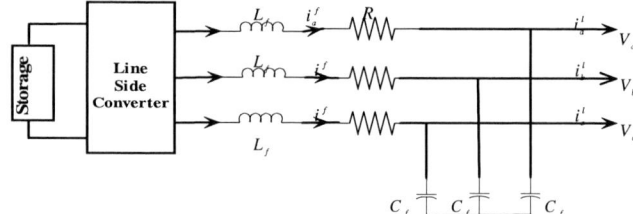

Fig. 2. RLC filter with line side Converter equivalent circuit.

Fig.3 shows the block diagram of the voltage control technique for the line side converter. The voltage control is necessary to reduce the initial impact of black-start.

At the beginning of the black start, line side converter is used to absorb initial impact of the process of energizing electrical paths. The storage system in the DC link of DFIG provides active power to maintain the constant load voltage through the line side converter. This converter performs the voltage and current control using PI regulators.

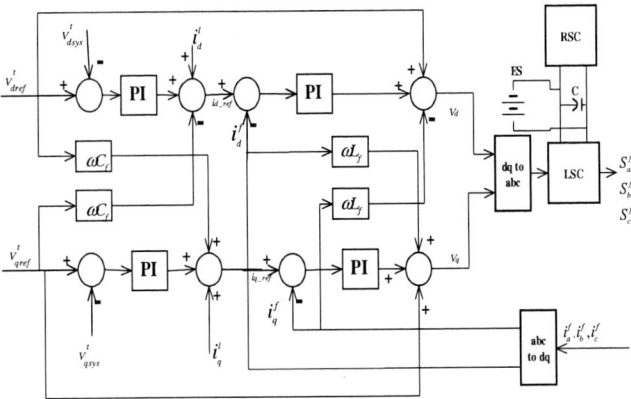

Fig. 3. Voltage control of line side converter.

The current control can be expressed mathematically as [11]:

$$i_{d\_ref} = k_p \Delta V_d^t + k_i \int \Delta V_d^t dt. + i_d^l - \omega C_f V_{qref}^t \quad (1)$$

$$i_{q\_ref} = k_p \Delta V_q^t + k_i \int \Delta V_q^t dt. + i_q^l + \omega C_f V_{dref}^t \quad (2)$$

Where, $\Delta V_d^t = V_{dref}^t - V_{dsys}^t$ and $\Delta V_q^t = V_{qref}^t - V_{qsys}^t$. $\Delta V_d^t$ and $\Delta V_q^t$ are the d-axis and q-axis voltage errors, respectively. $i_d^l$ and $i_q^l$ are the d-axis and q-axis line currents, respectively. $C_f$ is the capacitance of the filter.

The voltage control can be expressed mathematically as [11]:

$$V_d = k_p \Delta I_d^t + k_i \int \Delta I_d^t dt. + V_{dref}^t - \omega L_f i_q^f \quad (3)$$

$$V_d = k_p \Delta I_q^t + k_i \int \Delta I_q^t dt. + V_{qref}^t + \omega L_f i_d^f \quad (4)$$

Where, $\Delta I_d^t = i_{dref} - i_d^f$ and $\Delta I_q^t = i_{qref} - i_q^f$. $\Delta I_d^t$ and $\Delta I_q^t$ are the d-axis and q-axis current errors, respectively. $V_{dref}^t$ and $V_{qref}^t$ are the d-axis and q-axis references of the terminal voltage, respectively. $L_f$ is the inductance of the filter.

The proposed control system provides power to the load and keeps voltage and frequency close to nominal values. The control logic is implemented by employing active power-frequency droop characteristic and reactive power-voltage droop characteristic, shown in Fig.4 (a) and 4 (b) respectively, to keep the system frequency and voltage within the limits. The frequency and voltage droop characteristic can be mathematically represented as [12]:

$$f_L = f_o - D_p P \quad (5)$$

$$V_L = V_o - D_q Q \quad (6)$$

Where, $P$ and $Q$ are the real and reactive power outputs of DFIG, $D_p$ and $D_q$ are the frequency and voltage droop slopes respectively, and $f$ and $v$ are the nominal values of the frequency and voltage, respectively.

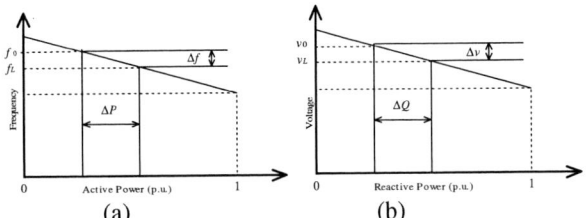

Fig. 4. Typical frequency and voltage droop characteristics.

When the islanded system is resynchronized with the grid system, the grid frequency and grid voltage are imposed to the system. The control system of line side converter implies PQ control method. The PQ control will allow to inject available generating power to the system [6].

Doubly fed induction generator uses field oriented control for both converters, which allows independent control of the electromagnetic torque and stator reactive power. Load connection in small islanded system often experiences frequency deviation in the system. Fig.5 shows the closed loop control model for DFIG. The effect of changes in system frequency and subsequent inertial responses of the DFIG are modelled and incorporated in the control system [13].

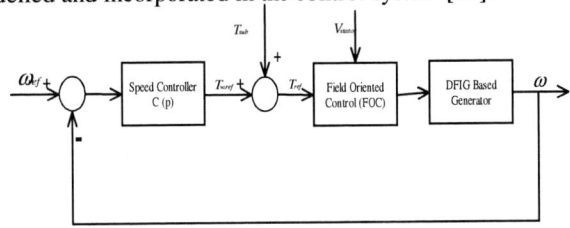

Fig. 5. Closed loop frequency control model for DFIG based DG in variable speed operation.

It can be seen in Fig.5 that field oriented controller (FOC) measures the stator voltage and maintains the constant electromagnetic torque during the restoration process. One of the inputs of the field oriented controller is the torque reference which is the combination of output of speed controller and primary frequency support system. The primary frequency system provides additional signal for the field oriented control system to compensate frequency variation. It is noted that frequency is the key issue in the process of restoration.

Fig.6 shows the detailed control actions of speed controller. Speed controller controls the turbine rotational speed. To produce reference turbine torque, control system uses power-speed curve of the wind turbine for producing reference mechanical speed according to available wind power. Actual turbine speed is measured and compared with the reference mechanical speed [14]. A PI regulator is used to compensate the error and produce the reference electromagnetic torque for converter controller of the DFIG.

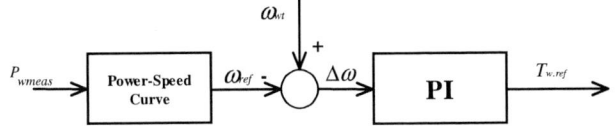

Fig. 6. Control action of speed controller.

As seen in Fig.6, the difference between the actual and reference turbine speed is fed to the speed controller and the controller produces the reference torque $T_{w.ref}$, which is used by the field oriented controller of the converter. This can be mathematically expressed as

$$Tw.ref = k_p \Delta\omega + k_i \int \Delta\omega dt \qquad (7)$$

Where, $\Delta\omega$ is the speed variation, $k_p$ and $k_i$ are the proportional and integral gain constants of the PI controller, respectively.

During the frequency control support, the dynamics of the turbine can be expressed by the torque difference and rotor inertia, which can be expressed as:

$$J \frac{d\omega_{wt}}{dt} = T_a - T_e \qquad (8)$$

Where, J is the inertia of the rotor and $\omega_{wt}$ is the rotational speed, and $T_a$ is the aerodynamic torque which depends on wind speed, pitch angle, and tip speed ration.

Kinetic energy is stored in the rotating mass of the wind turbine which can be expressed as:

$$E = \frac{1}{2} J\omega_{wt}^2 \qquad (9)$$

The inertia constant, H is often used to quantify the stored kinetic energy of the wind generator and can be defined as:

$$H = \frac{E}{S_n} = \frac{J\omega_{wt}^2}{2S_n} \qquad (10)$$

Where, $S_n$ is the rated apparent power (MVA). The inertia constant, H describes the number of seconds that the generator can supply rated power from its stored kinetic energy [15]. Typical inertia constant for wind turbines is roughly in the range of 2-6 seconds. Inclusion an energy storage system in DFIG increases the value of inertia, which depends on the storage size.

Additional torque signal, $T_{sub}$ given in Fig.5 is derived based on inertial effect of turbine as shown in Fig.7. The control system of Fig.7 supports primary frequency control of the DFIG [15]. In this control system, it is seen that when the system frequency starts to fall, the rotor speed of DFIG starts to decelerate. The control system provides additional torque signal as a function of the deviation of the system frequency, $\Delta f$ and of the rate of the change of the system frequency, $df_{sys}/dt$. $k_{inertia}$ and $k_{wt}$ are the inertia constant and primary frequency controller constant, respectively. The control signal, $T_{sub}$ is generated only when the grid frequency violates the frequency limit during the process of system restoration.

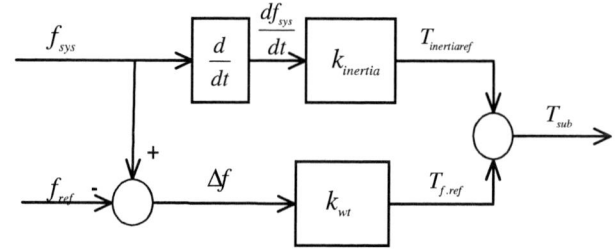

Fig.7. Additional torque signal for supporting primary frequency control.

## IV. CASE STUDY

### A. System description

The test system shown in Fig.8 is used to test and validate the proposed control technique for black start after major emergencies. The parameters for the test system are given in Appendix. As shown in Fig.8, low voltage network consists of load group -1, -2, -3 and -4. A circuit breaker associated with distribution transformer is responsible for separating or connecting LV network (0.415 kV) with upstream (11 kV) network.

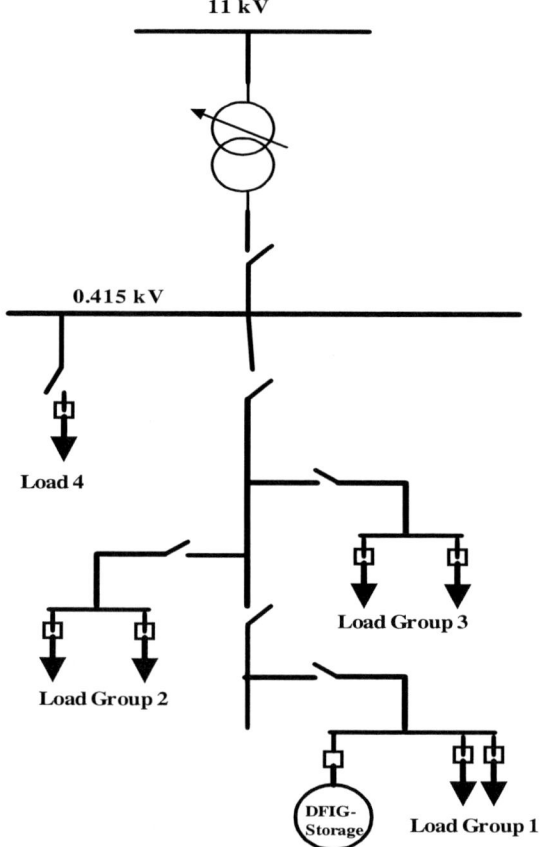

Fig. 8. Test system.

### B. Simulation results and discussions

Simulations are conducted to evaluate the sequence of actions of the black start. Simulation results show the system responses and controller behaviors during the process of system restoration. Fig.9 shows the responses of DFIG controller for real power, terminal voltage, frequency and current, and DC link voltage during load pick-up and synchronization. The re-synchronization has been performed after checking the conditions for synchronization.

The process of smooth load pick-up is necessary to build-up low voltage network. When the frequency goes above nominal frequency, the load pick-up will start and continue until the DFIG-storage system is capable to supply power to the load. The load inclusion in the islanded system is shown in Fig.9(a). Load group-2 and load group-3 are connected to the islanded network at t = 2 sec and t = 4 sec, respectively after simulation start. Voltage and frequency variations are observed during load inclusion and shown in Figs.9(b) and 9(c), respectively. Current of the islanded system increases as a response of load inclusion and is shown in Fig.9(b). The voltage controller of the line side converter plays a major role for controlling the voltage and frequency of the islanded system. The controller responses can be observed through the system's voltage and frequency responses during islanding operation. Islanded system is operated during the period of t = 0 to t = 6 sec. Fig.9(b) and 9(c) show that load pick-up and islanding operation can be successfully performed by using the proposed control technique.

Fig.9(e) shows the responses of the DC link voltage during black start process. DC link voltage experiences large variation during islanding operation. This is because during load pick-up and islanding operation, storage system in the DC link injects or absorbs extra power to minimize power imbalance. After synchronization with the grid (at t = 6 sec), the grid system controls the frequency and matches the power imbalance.

When main returns, islanded system is synchronized with the grid. Synchronization occurs at t = 6 sec. After synchronization, DFIG-storage system changes its control mode from voltage control to PQ control. At the same time system of the changing of the control mode, voltage and frequency are imposed to the low voltage network by the upstream system and DFIG-storage system provides available generating power into the system.

## V. CONCLUSIONS

In this paper, a control methodology is presented for black start participation of DFIG based wind generation system. A sequence of black-start process is developed and also presented. Restoration process is performed in three stages. In the first stage, DFIG-Storage absorbs initial impact of black start during energizing electrical paths. Second stage includes controlling islanded system's voltage and frequency by using voltage control of line side converter. Third stage includes load restoration and resynchronization. Load restoration depends on the available generation of DFIG. The proposed control methodology has been tested on a low to medium voltage network. Simulation results show that the developed control strategy for DFIG with DC link storage system can provide strong capability for fast restoration of a network from black start. This control approach can also be implemented in

the case where uncontrollable primary resources are being used for supplying power to the load.

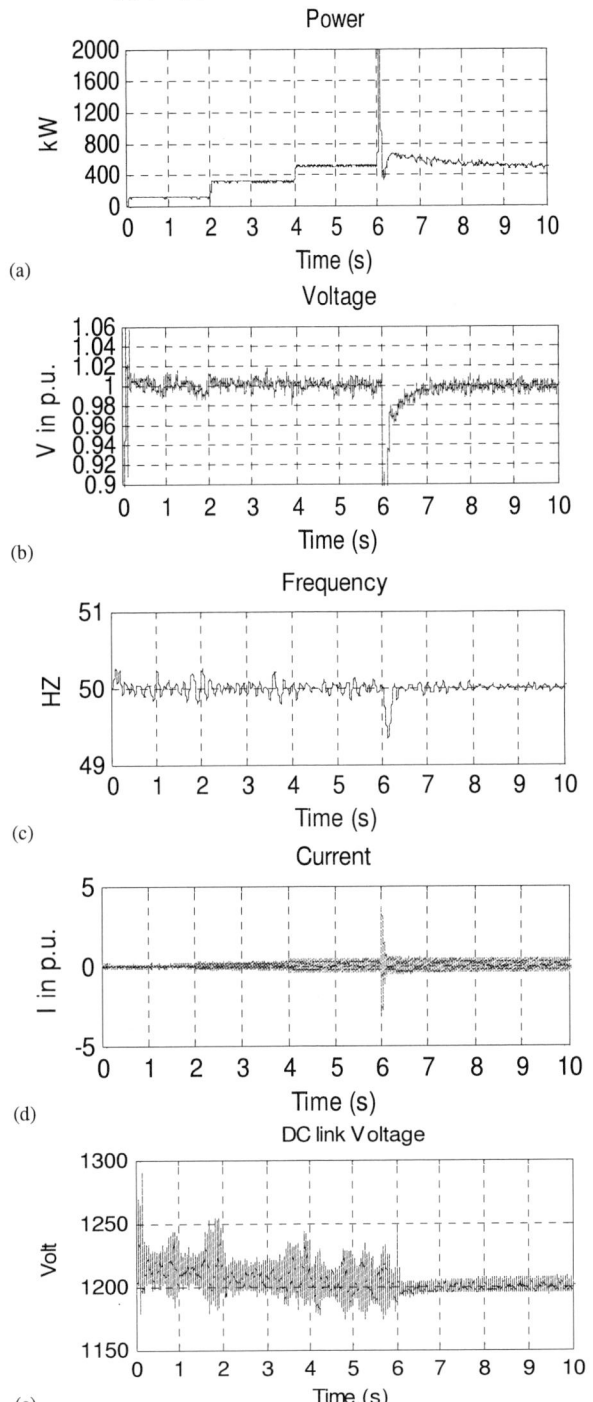

Fig. 9. Responses of restoration process: (a) Power (b) Voltage (c) Frequency (d) Current and (e) DC link Voltage of the system.

## VI. APPENDIX

The specifications of test system shown in Fig.7 are given below:

**Wind turbine:**
Rating:    1500 kW

## Generator: Doubly Fed Induction Generator

Stator resistance    $R_s = 0.00706$ p.u

Stator inductance    $L_s = 0.171$ p.u

Magnetizing inductance $L_m = 2.9$ p.u

Inertia constant   H=3.0

Poles     P=6

## Storage:

Equivalent capacitance   $C_b = 1800$ F

## Inverter: Each

Each rated:        450 kW

## Line parameters:

Positive- sequence resistance $R_1 = 0.1153$ $\Omega/km$

Zero-sequence resistance    $R_0 = 0.413$ $\Omega/km$

Positive-sequence inductance $L_1 = 1.05e\text{-}03$ $H/km$

Zero-sequence inductance    $L_0 = 3.32e\text{-}03$ $H/km$

Positive-sequence capacitance $C_1 = 11.33e\text{-}09$ $F/km$

Zero-sequence capacitance    $C_0 = 5.01e\text{-}09$ $F/km$

Load group 1 = 100 kW

Load group 2 = 200 kW

Load group 3 = 200 kW

Load group 4 = 200 kW

## VII. References

[1] Z. Chen and E. Spooner, "Grid Power Quality with Variable Speed Wind Turbines", IEEE Trans. On Energy Conversion, Vol. 16, No. 2, June 2001, pp. 148-154.

[2] A. Miller, E. Muljadi, and D. Zinger, "A Variable Speed Wind Turbine Power Control", IEEE Trans. On Energy Conversion, Vol. 12, No. 2, June 1997, pp. 181-186.

[3] R. Peña, J. C. Clare, and G. M. Asher, "Doubly fed induction generator using back-to-back PWM converters and its application to variable speed wind-energy generation," Proc. Inst. Elect. Eng., Elect. Power Appl., vol.143, no. 3, pp. 231–241, May 1996.

[4] C. Abbey and G. Joss, "Integration of Energy Storage with a Doubly-Fed Induction Machine for Wind Power Generation", IEEE PESC 2004, Aachen, Germany, July 2004.

[5] M. Banejad, G. Ledwich and M.A. Kashem, "Operation of Power System Islands", Australasian Universities Power Engineering Conference (AUPEC 2005), 25-28 Sept. 2005, Hobart, Australia.

[6] F. Hughes, O. Anaya-Lara, N. Jenkins, and G. Strbac, "Control of DFIG-Based Wind Generation for Power Network Support", IEEE Transactions on Power Systems, Vol.20, No.4, 2005, pp. 1958-1966.

[7] T.S. Sidhu et. al., "Protection Issues During System Restoration", IEEE Transactions on Power Delivery, Vol. 20, No. 1, January 2005, pp. 47-56.

[8] W.R. Lachs, and H. Tabatabaci-Yazdi, "Energy Storage in Power Systems", IEEE Int. Conference on Power Electronics and Drive Systems, PEDS'99, July 1999, pp.843-848.

[9] J.J. Ancona," A Framework for Power System Restoration Following a Major Power Failure", IEEE Transactions on Power Systems, Vol. 10, 1995, pp. 1480-1485.

[10] J.A. Lopes, C.L. Moreira and F.O. Resemde," Microgrid Black Start and Islanded Operation",15th Power Systems Computation Conference, August 22-26, 2005.

[11] B. Han, B. Bae, H. Kim and S. Baek, "Combined Operation of Unified Power-Quality Conditioner with Distributed Generation", IEEE Transactions on Power Delivery, Vol. 21, No. 1, January 2006, pp. 330-338.

[12] J.A. Lopes, C.L. Moreira and A.G. Madureira," Defining Control Strategies for Microgrids Islanded Operation", IEEE Transactions on Power Systems, Vol. 21, No.2, May 2006, pp. 916-924.

[13] A. Mullane and M. O'Malley, "The Inertial Responses of Induction Machine Based Wind Turbines", IEEE Transactions on Power Systems, Vol. 20, No.3, August 2005, pp. 1496-1505.

[14] A. Mullane, G. Bryans, and M. Malley, "Kinetic Energy and Frequency Response Comparison for Renewable Generation Systems", International Conference on Future Power Systems, 16-18 November 2005. Amsterdam, Netherland.

[15] J. Morren, S.W.H. de Hann, W.L. Kling, and J.A. Ferreira," Wind Turbine Emulating Inertia and Supporting Primary Frequency Control", IEEE Transactions on Power Systems, Vol. 21, No.1, February 2006, pp. 433-434.

## VIII. Biographies

**Md. Aktarujjaman** (M'00) received the B.Sc.TE. from the Islamic University of Technology, Bangladesh, in 2000. He obtained Masters in Information Technology from Charles Sturt University, Australia in 2002. He is currently pursuing the Ph.D degree at the University of Tasmania. His special fields of interests are power system analysis, renewable energy, distributed generation, power system control and protection.

**M. A. Kashem** (A'00–SM'05) received the Ph.D. degree from Multimedia University, Malaysia, in 2001. Currently, he is a Senior Lecturer at the School of Engineering, University of Tasmania, Australia. He was associated with the Queensland University of Technology, Australia as a Postdoctoral Research Fellow from 2000 to 2002. Previously, he also worked for Multimedia University as a Lecturer for three years. His special fields of interests include distributed generation, renewable energy, distribution system automation, power system planning, and artificial intelligence. He has published more than 50 technical papers in these areas.

**Michael Negnevitsky** (M'95) received the B.S.E.E. (Hons.) and Ph.D. degrees from Byelorussian University of Technology, Minsk, Belarus, in 1978 and 1983, respectively. Currently, he is an Associate Professor in the School of Engineering at the University of Tasmania, Hobart, Australia. From 1984 to 1991, he was a Senior Research Fellow and Senior Lecturer in the Department of Electrical Engineering, Byelorussian University of Technology. After arriving in Australia, he was with Monash University, Melbourne, Australia. His interests are power system analysis, power quality, and intelligent systems applications in power systems. Dr. Negnevitsky is a Chartered Professional Engineer, a Senior Member of the Institution of Engineers Australia, and a Member of CIGRE AP36 (Electromagnetic Compatibility), Australian Technical Committee.

**Gerard Ledwich** (M'73–SM'92) received the Ph.D. in electrical engineering from the University of Newcastle, Australia, in 1976. He has been Chair Professor in Electrical Asset Management at Queensland University of Technology, Australia, since 1998. He was Head of electrical engineering at the University of Newcastle from 1997 to 1998. Previously, he was associated with the University of Queensland from 1976 to 1994. His interests are in the areas of power systems, power electronics, and controls. Prof. Ledwich is a Fellow of the Institution of Engineers Australia.

**2006 IEEE International Conference on Power Electronic, Drives and Energy Systems**

# Fuzzy Logic Based Control of Wind Turbine Driven Squirrel Cage Induction Generator Connected to Grid

CH.Siva Kumar, A.V.R.S.Sarma, and P.V.N. Prasad

*Abstract* – **The paper deals with the fuzzy logic based control of wind turbine driven squirrel cage induction generator connected to grid. The controller is proposed to maintain the speed of the squirrel cage induction generator constant for different wind speeds. The proposed fuzzy logic controller is used with several rules to implement the control strategy for the induction generator. For verification of the control, the estimators like speed, torque and flux are not required. These state variables are the feedback signals from the model of the induction machine. The computer simulation results show the controller is satisfactory in operation of induction generator with the significant improvement in the output power.**

*Index Terms* – **Fuzzy Logic Controller, Induction generator, Squirrel Cage Induction Generator, Wind Turbine.**

## I. INTRODUCTION

GENERATION of electrical energy mainly so far has been from thermal, nuclear, and hydro plants. They have continuously degraded the environmental conditions. Increasing the rate of depletion of conventional energy sources and degradation of environmental conditions has given rise to increased emphasis on renewable energy sources. Due to clean and economical energy generation, a huge number of wind farms are going to be connected with the existing network in the near future. The high penetration of wind turbines in the power system has been closely related to the advancement of the wind turbine technology and control. The control of the system, that generates power from an unsteady input as the wind, presents a formidable problem. The wind speed varies from time to time due to gusts.

Wind turbines often do not take part of in voltage and frequency control and if a disturbance occurs, the wind turbines are disconnected and reconnected when normal operation has been resumed. Thus not with standing the presence of wind turbines, frequency and voltage are maintained by controlling the large power plants as would have been the case without any wind turbines present. This is possible as long as wind power penetration is still low. However tendency to increase the amount of electricity generated from wind can be observed.

---

CH.Siva Kumar (e-mail:ch_siva_kumar@rediffmail.com), A.V.R.S.Sarma (e-mail:avrs2000@yahoo.com), and P.V.N. Prasad (e-mail::polaki@rediffmail.com) are with the Department of Electrical Engineering, University College of Engineering., Osmania University, Hyderabad, India.

Therefore the penetration of wind turbines in electrical power systems will increase, they may begin to influence overall power system behavior and it will no longer be possible to run a power system by only controlling large scale power plants. Use of an Induction machine as a generator is becoming more popular for the renewable sources [1]. It is therefore important to study the behavior of wind turbine driven Induction Generator in an electrical power system

The squirrel cage induction generator (SCIG) is quite useful in wind turbine system due to its simpler, robust & brushless construction, lower initial & run time costs and lower maintenance cost. These generators are suitable for both grid connected and stand-alone applications. In grid connected mode, the excitation current to the induction generator is supplied through grid.

In this paper d-q model of induction machine is considered and a fuzzy logic controller is used to control speed of SCIG connected to grid. Following a change in wind speed the turbine output ($T_m$) changes. This change in turbine output causes change in power output, change in rotor speed and change in rotor losses.

The fuzzy logic controller affects electrical torque $T_e$ through d-axis stator current $I_{ds}$ in accordance with the changes in wind turbine output $T_m$. This controller controls the speed of SCIG by converting excess kinetic energy in rotor to electrical energy. Simulink is used to simulate the system with and without fuzzy logic controller and used to control the speed of the SCIG at different wind speeds. The results indicate effectiveness of the fuzzy logic controller in improving performance of SCIG connected to grid.

## II. INDUCTION MACHINE MODEL

The induction machine d-q or dynamic equivalent circuit [2] is shown in Fig.1.

According to this model, the modeling equations in flux linkage forms are as follows.

$$\frac{dF_{qs}}{dt} = \omega_b \left[ v_{qs} - \frac{\omega_e}{\omega_b} F_{ds} - \frac{R_s}{X_{ls}} (F_{qs} - F_{qm}) \right] \quad (1)$$

$$\frac{dF_{ds}}{dt} = \omega_b \left[ v_{ds} + \frac{\omega_e}{\omega_b} F_{qs} - \frac{R_s}{X_{ls}} (F_{ds} - F_{dm}) \right] \quad (2)$$

0-7803-9771-1/06/$25.00 ©2006 IEEE

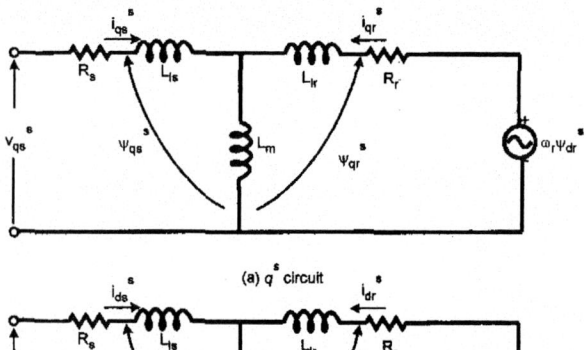

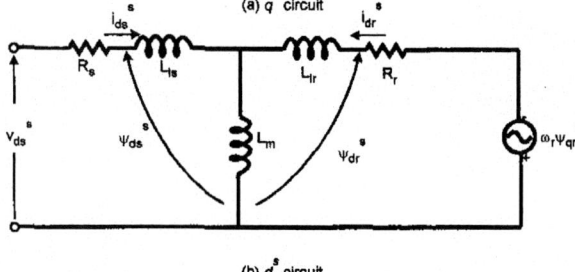

Fig. 1. Dynamic or d-q Equivalent Circuit of an Induction Machine.

$$\frac{dF_{qr}}{dt} = -\omega_b \left[ \frac{(\omega_e - \omega_r)}{\omega_b} F_{dr} + \frac{R_r}{X_{lr}} (F_{qr} - F_{qm}) \right] \quad (3)$$

$$\frac{dF_{dr}}{dt} = -\omega_b \left[ -\frac{(\omega_e - \omega_r)}{\omega_b} F_{qr} + \frac{R_r}{X_{lr}} (F_{dr} - F_{dm}) \right] \quad (4)$$

$$F_{qm} = X_m \left[ \frac{(F_{qs} - F_{qm})}{X_{ls}} + \frac{(F_{qr} - F_{qm})}{X_{lr}} \right] \quad (5)$$

If,
$$X_{ml} = \frac{1}{\frac{1}{X_m} + \frac{1}{X_{ls}} + \frac{1}{X_{lr}}}$$

$$F_{qm} = \frac{X_{ml}}{X_{ls}} F_{qs} + \frac{X_{ml}}{X_{lr}} F_{qr} \quad (6)$$

$$F_{dm} = \frac{X_{ml}}{X_{ls}} F_{ds} + \frac{X_{ml}}{X_{lr}} F_{dr} \quad (7)$$

$$i_{qs} = \frac{F_{qs} - F_{qm}}{X_{ls}} \quad (8)$$

$$i_{qr} = \frac{F_{qr} - F_{qm}}{X_{lr}} \quad (9)$$

$$i_{ds} = \frac{F_{ds} - F_{dm}}{X_{ls}} \quad (10)$$

$$i_{dr} = \frac{F_{dr} - F_{dm}}{X_{lr}} \quad (11)$$

$$T_e = \frac{3}{2} \left( \frac{P}{2} \right) \frac{1}{\omega_b} (F_{ds} i_{qs} - F_{qs} i_{ds}) \quad (12)$$

$$T_e = T_m + J \frac{d\omega_m}{dt} = T_m + \frac{2}{p} J \frac{d\omega_r}{dt} \quad (13)$$

where, the suffixes d & q represent d-axis & q-axis quantities respectively , suffixes s & r represent stator & rotor quantities respectively, m represents the suffix for mutual induction.. The other notations of the system are as follows.

R     resistance  ($\Omega$),
$\omega_b$     base electrical frequency (radians /s),
$\omega_e$     stator supply frequency (radians /s),
$\omega_r$     rotor electrical speed (radians /s),
$\omega_m$     rotor mechanical speed, (radians /s) (= (2/p) ($\omega_r$)),
$X_{ls}$     stator leakage reactance,
$X_{lr}$     rotor leakage reactance,
$T_m$     mechanical torque,
$T_c$     electromagnetic torque,
$J$     moment of inertia (kg.m$^2$),
$p$     number of poles,
$V_{ds}$     d axis stator voltage,
$V_{qs}$     q axis stator voltage,
$V_{qr}$     q axis rotor voltage,
$V_{dr}$     d axis rotor voltage,
$I_{ds}$     d axis stator current,
$I_{qs}$     q axis stator current,
$I_{qr}$     q axis rotor current,
$I_{dr}$     d axis f rotor current,
F     flux linkages

It is assumed that $v_{qr} = v_{dr} = 0$. To solve these equations, they are arranged in state-space form with $F_{qs}$, $F_{ds}$ $F_{qr}$. and $F_{dr}$ as state variables.

### III. WIND TURBINE MODEL

The wind turbine converts the energy contained by the wind into mechanical energy. The rotor output power $P_t$ [3] in watts is modeled using the following equation

$$P_t = 0.5 c_p (\lambda, \theta) \rho A v^3 \quad (14)$$

where,
$c_p (\lambda, \theta)$ is power coefficient,
$\lambda$ is tip speed ratio,
$\theta$ is blade pitch angle in radians,
$\rho$ is air density in kg/m$^3$,
$A$ is area swept by the rotating blades of wind turbine in m$^2$
$v$ is velocity of the wind in m/sec and

Fig. 2, shows the relationship between the output power and rotor speed for various wind velocities of the simulated wind turbine.

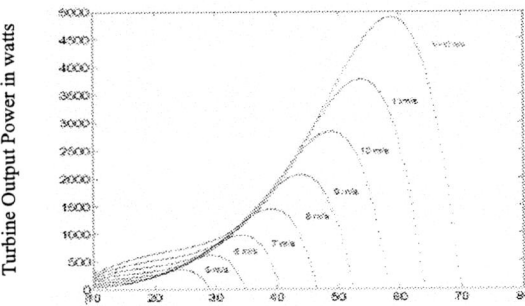

Wind Turbine Speed in Revolutions per Second

Fig. 2. Wind Turbine Output Power versus Speed characteristics.

## IV. FUZZY LOGIC CONTROLLER

The control algorithm of a process that is based on fuzzy logic or fuzzy interference system [6] is defined as a fuzzy control. Fuzzy control is basically an adaptive and nonlinear control, which gives robust performance for a linear or non-linear plant with parameter variation.

To implement the fuzzy control [2], [4] strategy rules were framed. The inputs are the rotor speed and its incremental change in error. The input and the output membership functions are shown in Fig. 3, Fig. 4 and Fig. 5 respectively.

The output is the current amplitude and it is denoted by $I_{ds}(n)$ with seven fuzzy sets for each. Fuzzification is done using continuous universe of discourse and the defuzzification is done using "Centroid" method.

Membership Functions are defined as follows:

Input    1)    Error: The range of error is -1 to 1

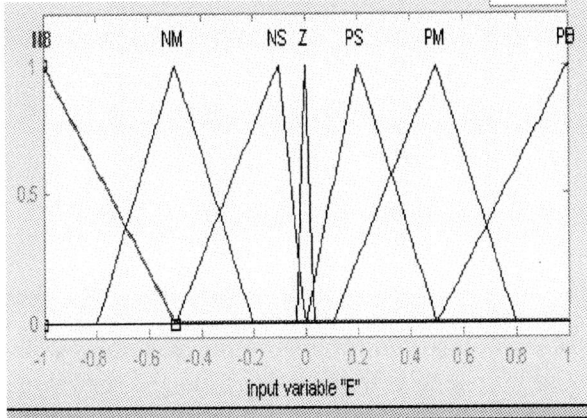

Fig. 3. Membership Function of Input 1 (Rotor Speed).

2)    Change in Error: The range of change in error is -1 to 1.

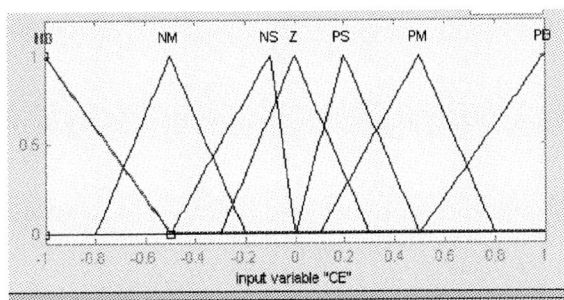

Fig. 4. Membership Function of Input 2. (Incremental Change in Error)

Output:   The output range is -1 to 1

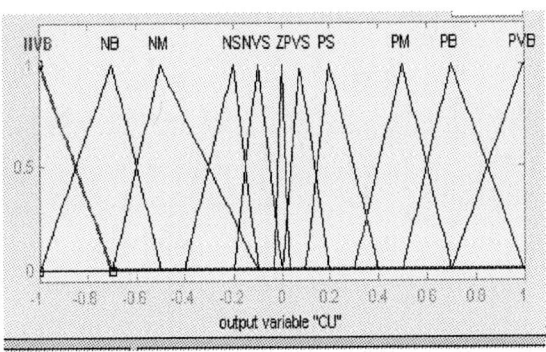

Fig. 5. Membership Function of Output. ( Current Signal)

The fuzzy sets are defined as follows.

NB:  Negative Big
NM:  Negative Medium
NS:  Negative Small
Zero: Zero
PS:  Positive Small
PM:  Positive Medium
PB:  Positive Big
NVB: Negative Very Big
NVS: Negative Very Small
PVB: Positive Very Big
PVS: Positive Very Small

## V. SIMULATION AND RESULTS

The single line diagram of the system and the complete Simulink block diagram model of the systems are shown in Fig. 6 and Fig. 7 respectively. Simulink model is developed with a fuzzy logic controller for the simulation of SCIG connected to grid. The simulation model of the system is shown in Fig. 8.

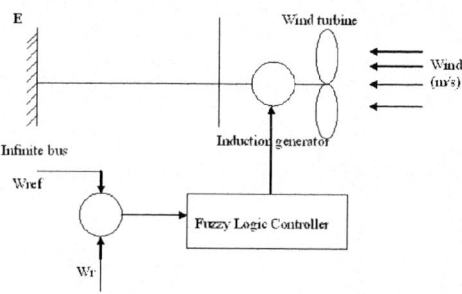

Fig. 6. Single line diagram of FLC based Wind Turbine Driven SCIG connected to the grid.

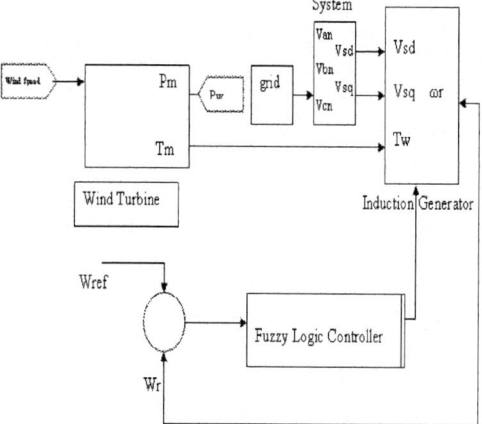

Fig. 7. Simulink block diagram of the model.

The single line diagram of the system and the complete Simulink block diagram model of the systems are shown in Fig. 6 and Fig. 7 respectively. A simulink model is developed with a fuzzy logic controller for the simulation of SCIG connected to grid. The simulation model of the system is shown in Fig. 8.

The initial wind speed is taken as 6 m/s and step changes in wind speed of 8m/s and 11 m/s are assumed at 6 sec. and 8 sec. respectively as shown in Fig. 9. Fig. 10 shows the characteristics of the electrical torque of squirrel cage induction generator.

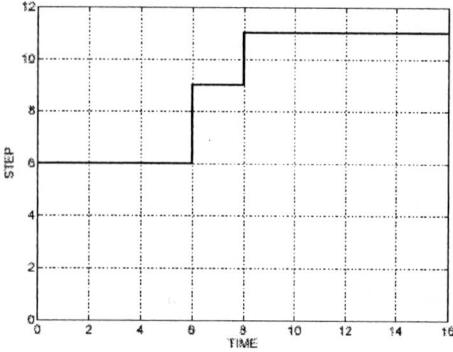

Fig. 9. Wind speed

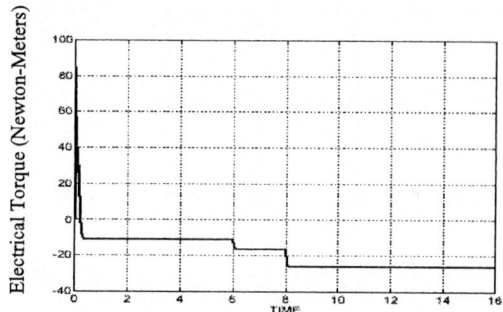

Fig. 10. Characteristic of Electrical Torque.

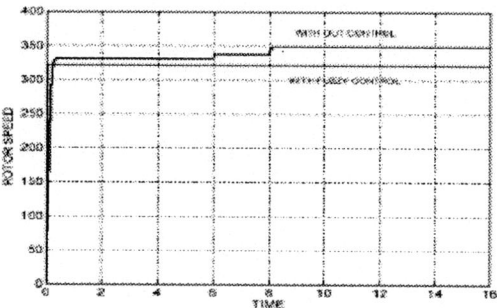

Fig. 11. Characteristic of Rotor Speed with and without control.

In Fig. 11, the variation of rotor speed characteristic versus time with and without the fuzzy logic controller is compared. In Fig. 12, the output power variations with and without the fuzzy logic controller is compared.

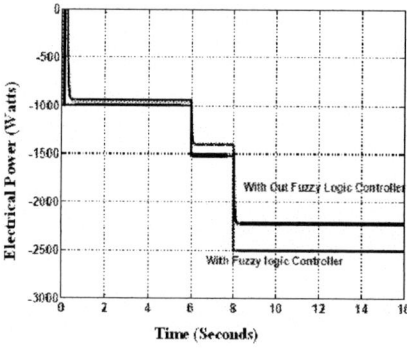

Fig. 12. Characteristic of Power versus time with and without Controller.

Following a change in wind speed, the turbine output changes. This change in turbine output causes (i) change in output power as it is evident from fig.12. (ii) Change in rotor speed which is evident from fig. 11 and (iii) Change in rotor losses. A fuzzy logic controller affects electrical torque $T_e$ through $I_{ds}$ in accordance with the changes in wind turbine output. This fuzzy logic controller controls the speed of squirrel cage induction generator by converting excess kinetic energy in rotor to electrical energy.

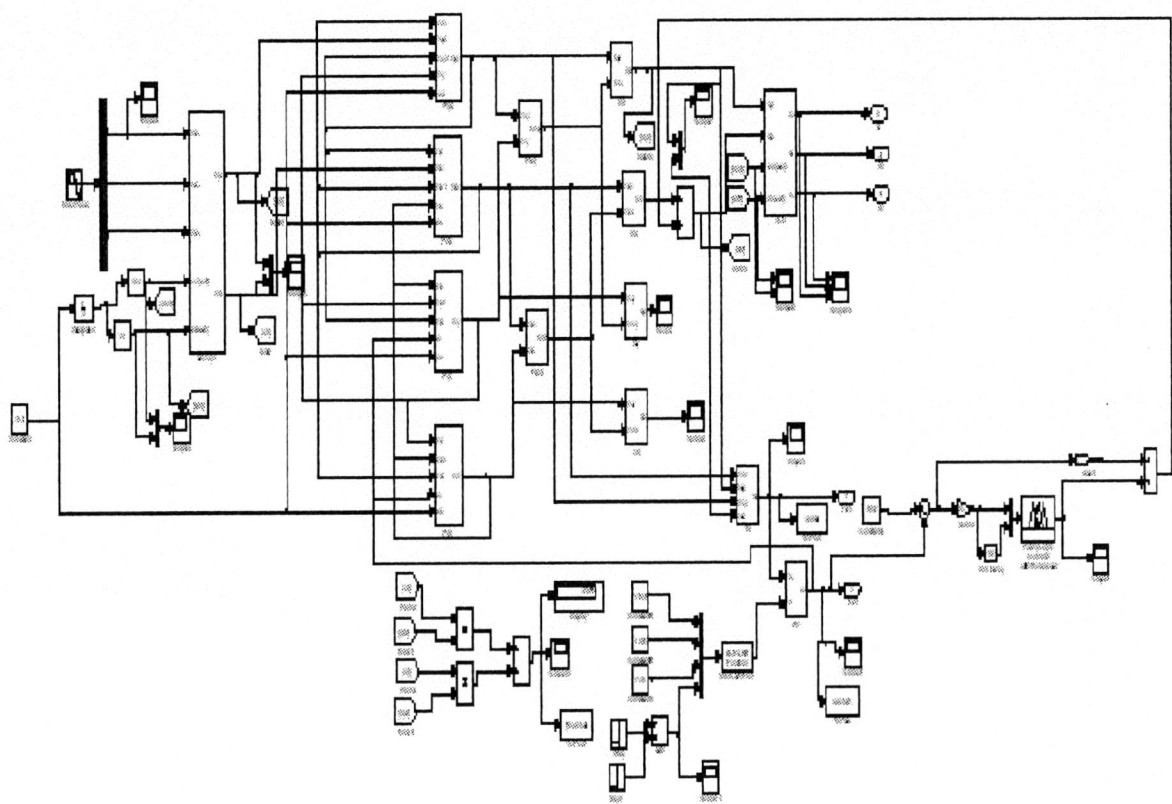

Fig. 8. Model of Wind Turbine Driven Induction Generator with Fuzzy Logic Controller.

## VI. CONCLUSIONS

Fuzzy logic controller is effectively used to maintain a constant speed of squirrel cage induction generator at three different wind speeds. The performance of fuzzy logic controller based induction generator is compared with that of generator without controller. The losses in the rotor are reduced, which increase the efficiency of the machine. From the results it is also observed that there is an increase in the power output by more than 10% at a wind speed of 11 m/s.

## VII. APPENDIX

$R_s$ = 7.55 (Ω),
$R_r$ = 7.85 (Ω),
$X_{ls}$ = 3.14 (Ω),
$X_{lr}$ = 3.14 (Ω),
$X_m$ = 86.35 (Ω),
$p$ = 4, $J$ = 0.06 kg.m$^2$,
$\rho$ =1.08 kg/m$^3$ ,and
wind turbine radius R=1.525 m.

## VIII. REFERENCES

[1] R. C. Bansal, T. S. Bhatti, and D. P. Kothari, " Bibliography on the Application of Induction Generators in Nonconventional Energy Systems," *IEEE Transactions on Energy Conversation*, Vol. 18, no. 3, pp. 433 – 439, Sep. 2003.

[2] Bose B.K., *"Modern Power Electronics and AC Drives"*, PH, New Jersy, 1986.

[3] J.G.Slootweg, S.W.H. de Haan, H.Palimder, W.G Klingl, "Aggregated Modeling of Wind Parks with Variable Speed Wind Turbines in Power System Dynamics Simulations", *14<sup>th</sup> PSCC, Sevilla*, 24-28 June 2002.

[4] Marcelo Godoy Sim˜oes, Bimal K. Bose, and Ronald J. Spiegel, "Design and Performance Evaluation of a Fuzzy-Logic-Based Variable-Speed Wind Generation System, *IEEE Transactions on Industry Applications*, Vol. 33, No. 4, July/August 1997.

[5] Ned Gulley and J.S Roger Jang, "Fuzzy Logic TOOLBOX for Use with MATLAB*", The MATHWORKS* Inc, 1999.

[6] A.Kaufmann, Introduction to the theory of Fuzzy Sbusets.- Volume I New York, Academic Press, 1975.

[7] R.C.Bansal ,"Three phase Self-Excited Induction Generators: An Overview*" IEEE transactions on Energy Conversion*, 2005.

[8] Paul . C. Krause, Oleg Wasynczuk, Scott D. Sudhoff, "Analysis of Electric Machinery and Drive Systems" IEEE Press Power Engineering Series, A John Wiley & Sons ,Inc. Publication.

## IX. Biographies

**CH.Siva Kumar** graduated in Electrical & Electronics Engineering in 1999 and received M.E degree in Power Systems from Osmania University, Hyderabad in 2003. He is a life member of Indian Society for Technical Education. At present, he is serving as Academic Consultant in the Department of Electrical Engineering, Osmania University, Hyderabad, India. His special fields of interest include Power System Operation and Control, Renewable Energy Sources, Artificial Intelligence and Distribution SCADA systems.

**A.V.R.S.Sarma** graduated in Electrical & Electronics Engineering in 1976 and received his M.Tech degree in Power Systems in 1978 from Regional Engineering College, Warangal., India. He received his Ph.D degree in Electrical Engineering in 1995 from Osmania University. His areas of interest are Power System Dynamics, Distribution Systems and HVDC Transmission. He is a member of Institution of Engineers (India). At present, he is serving as Professor in the Department of Electrical Engineering, Osmania University, Hyderabad, India.

**P.V.N.Prasad** graduated in Electrical & Electronics Engineering from Jawaharlal Nehru Technological University, Hyderabad in 1983 and received M.E in Industrial Drives & Control from Osmania University, Hyderabad in 1986. He served as faculty member in Kothagudem School of Mines during 1987 - 95 and presently serving as Associate Professor in the Department of Electrical Engineering, Osmania University, Hyderabad. He received his Ph.D in Electrical Engineering in 2002 from Osmania University. His areas of interest are Reliability Engineering and Computer Simulation of Electrical Machines & Power Electronic Drives. He is a member of Institution of Engineers (India) and Indian Society for Technical Education. He is recipient of Dr.Rajendra Prasad Memorial Prize, Institution of Engineers (India), 1993 - 94 for best paper. He has got over 30 publications in National Journals and International Conferences & Symposia and presented technical papers in Thailand and Italy.

**2006 IEEE International Conference on Power Electronic, Drives and Energy Systems**

# Speed Sensor-less Direct Power Control of a Matrix Converter Fed Induction Generator for Variable Speed Wind Turbines

T. Satish, *Student Member, IEEE*, K.K. Mohapatra, *Member, IEEE*, and Ned Mohan, *Fellow, IEEE*

*Abstract*--In this paper, the conventional direct torque control (DTC) scheme for induction motor drives is extended to directly control the active power (DPC) delivered to the grid by a wind turbine driven squirrel cage induction generator (SCIG). The SCIG is interfaced to the grid through an AC-AC matrix converter. A constant switching frequency based direct power control scheme with flux and power controllers is proposed. A closed loop Luenberger observer and an adaptive speed estimator are used to estimate the torque, stator flux magnitude and phase, and rotor speed of the induction generator. The simplified carrier based modulation method is used to synthesize the pulse width modulated output voltages from the matrix converter and to control the power factor of the currents on the grid side. Simulation results are presented to support the theoretical discussion.

*Index Terms*--DTC, Induction machines, Matrix Converter, Wind power generation.

## I. INTRODUCTION

THE escalating capacities of wind turbine installations and increasing power ratings of the wind turbines, demands intensive research on the power electronic converters used to interface the wind generators to the grid and their control. A vast majority of the turbines being installed are the variable speed type because of their superiority over the fixed speed turbines. Various topologies of the variable speed wind turbine systems are proposed in the literature [1]. This paper focuses on the so called full power converter topology, shown in Fig. 1, in which the power electronic converter is connected between the grid and the stator terminals of the induction generator. This implies that the rating of the converter is the comparable to that of the wind turbine and hence this topology is suitable for wind turbines in the small power range. The power electronic interface used is an AC-AC matrix converter which is a direct link converter capable of generating output 3-phase sinusoidal AC voltages of any frequency from an

input 3-phase AC system without any intermediate energy storage elements. Having eliminated the large dc link energy storage element the matrix converter, also known as an all silicon converter, can be designed in a compact size. Thus the conventional voltage link converter in the full power converter topology of wind turbines can be replaced by the matrix converter. The output voltages in the matrix converter have three levels corresponding to the three input phase voltages and hence the problems associated with bearing currents and bearing failures are reduced when compared with the two-level inverters. The simplified carrier based modulation method [3], [4] is used to generate the pulse width modulated output voltages from the matrix converter.

Direct Power Control (DPC) is the control strategy in which the error in the reference power and the actual power is utilized to generate the voltage command directly as in conventional DTC drives. This method reduces the number of PI controllers used when compared to the vector control based variable speed wind turbine generator systems [5]. The DPC like DTC is a stator flux based control technique, having the advantages of robustness and fast controls [6]. With the carrier based modulation of the matrix converter, a constant switching frequency based control is implemented different from the variable switching frequency based hysteresis power controllers [7], [8].

## II. MATRIX CONVERTER EXCITED SCIG SYSTEM

Fig. 1 shows the topology of the full power converter based wind turbine system excited by an AC-AC matrix converter. As can be seen, the matrix converter is connected between the stator terminals and the terminals of the power grid. One of the advantages of using the matrix converter is being able to control the power factor of the input currents irrespective of the power factor of the load. Thus in this case, the reactive power requirement of the squirrel cage induction generator can be met, while injecting active power at unity power factor into the grid. Reactive power can be injected into the grid at the expense of decreased stator terminal voltage.

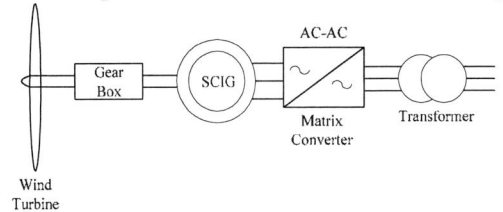

Fig. 1. Full power converter variable speed wind generation system

---

The work reported in this paper is supported by grants from the Office of Naval Research grant No. N00014-05-1-0291 and Xcel Energy.

Satish Thuta is a graduate student at the University of Minnesota, Minneapolis, USA. (email: thuta002@umn.edu).

K.K.Mohapatra is a Post Doctorate at the University of Minnesota, Minneapolis, USA (email: mohap002@umn.edu).

Ned Mohan is with the Department of Electrical Engineering, University of Minnesota, Minneapolis, USA. (email: mohan@umn.edu).

0-7803-9771-1/06/$25.00 ©2006 IEEE

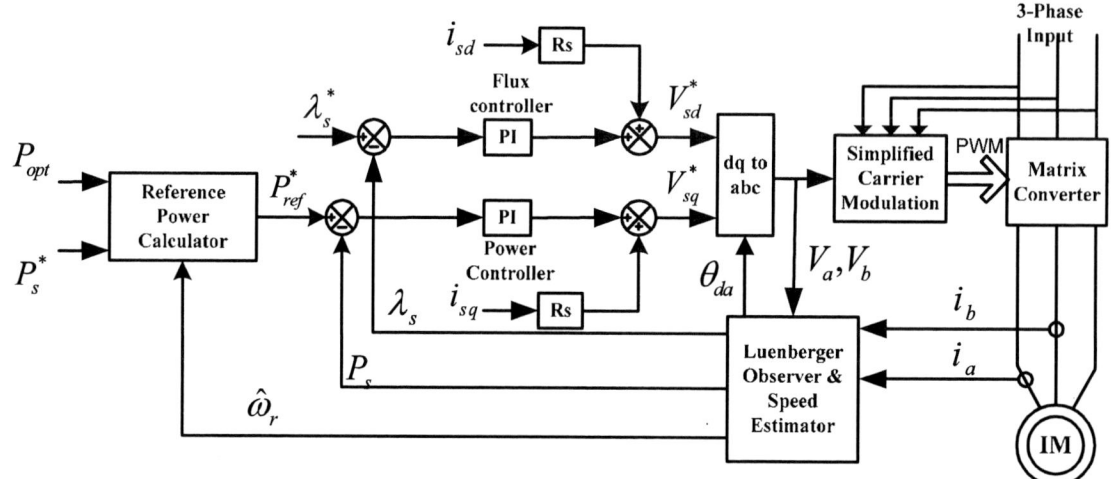

Fig. 2. Block diagram of Direct Power Control (DPC) system

## III.  DIRECT POWER CONTROL (DPC) SYSTEM

The block diagram of the proposed DPC system is shown in Fig.2. The flux and the power controllers give the reference dq-axis voltages depending on the errors in the actual and the reference values. These voltages are transformed into the three phase system and are input as the reference voltages to the matrix converter. The commanded voltages are synthesized from the matrix converter using the simplified carrier based modulation algorithm. A closed loop Luenberger observer calculates the stator flux magnitude and phase, and the active power of the generator from the sensed currents and the reference voltages. The estimated flux and power are feedback to the PI controllers. A speed estimator estimates the rotor speed based on the errors in the actual and the estimated currents. Feed-forward compensation is used to compensate for the stator resistance drop.

## IV.  OBSERVER AND SPEED ESTIMATOR

A closed loop Luenberger observer is used to estimate the stator flux, the electromagnetic torque and the power. This provides better estimation over a wide speed range with good dynamic response.

### A. Motor Model

To overcome the problem of integrator drift in flux estimation with the voltage based model, where the stator flux is estimated as an integration of the voltage after the stator resistance drop, a current based model is used in this paper to estimate the rotor fluxes and the stator fluxes are then calculated based on the estimated currents and the motor parameters. Hence the state vector of the state-space model of the induction motor is chosen as $\left[ i_{sd}, i_{sq}, \lambda_{rd}, \lambda_{rq} \right]^T$. The model of the induction motor in the stationary reference frame with the above choice of state vector is given in (1).

$$\dot{x} = \underbrace{\begin{bmatrix} -\left(\dfrac{R_s}{\sigma L_s}+\dfrac{(1-\sigma)R_r}{L_r}\right)I_2 & \dfrac{L_m}{\sigma L_s L_r}\left(\dfrac{R_r}{L_r}I_2 - \omega_r J\right) \\ \dfrac{L_m R_r}{L_r}I_2 & -\dfrac{R_r}{L_r}I_2 + \omega_r J \end{bmatrix}}_{A} x + \underbrace{\begin{bmatrix} \dfrac{1}{\sigma L_s}I_2 \\ O_2 \end{bmatrix}}_{B} u \qquad (1)$$

$$y = \underbrace{\begin{bmatrix} I_2 & O_2 \end{bmatrix}}_{c} x$$

Where $x = \left[ i_{sd}, i_{sq}, \lambda_{rd}, \lambda_{rq} \right]^T$ is the state-vector,

$u = \left[ V_{sd}, V_{sq} \right]^T$ is the input vector of stator dq-axis voltages,

$y = \left[ i_{sd}, i_{sq} \right]$ is the output vector of stator dq-axis currents,

$\sigma = 1 - \dfrac{L_m^2}{L_s L_r}$ is defined as the equivalent leakage.

$I_2 = \begin{bmatrix} 1 & 0 \\ 0 & 1 \end{bmatrix}$ and $J = \begin{bmatrix} 0 & -1 \\ 1 & 0 \end{bmatrix}$.

### B. Observer Model

The pole locations of the observer are chosen as in [9] using the Pole Placement Method (PPM) to have a faster error dynamics and at the same time improving the robustness to parameter variations and disturbances in the measured signals. The modeling equations and the closed loop gain matrix of the observer are shown below.

$$\dot{\hat{x}} = \hat{A}\hat{x} + Bu + L(y - \hat{y})$$
$$\hat{y} = c\hat{x} \qquad (2)$$

Where $\hat{x} = \left[ \hat{i}_{sd}, \hat{i}_{sq}, \hat{\lambda}_{rd}, \hat{\lambda}_{rq} \right]^T$ is the vector of estimated states.

$\hat{y} = \left[ \hat{i}_{sd}, \hat{i}_{sq} \right]$ is the estimated output vector and

$L$ is the feedback gain matrix of the closed-loop observer.

The choice of $L$ using the pole placement design method is given as

$$L = \begin{bmatrix} l_1 I_2 + l_2 J \\ l_3 I_2 + l_4 J \end{bmatrix} = \begin{bmatrix} l_1 & -l_2 & l_3 & -l_4 \\ l_2 & l_1 & l_4 & l_3 \end{bmatrix}^T \qquad (3)$$

766

Where

$$l_1 = -(k-1)\left(\frac{R_s}{\sigma L_s} + \frac{(1-\sigma)R_r}{\sigma L_r} + \frac{R_r}{L_r}\right) \tag{4}$$

$$l_2 = (k-1)\omega_r$$

$$l_3 = (k^2-1)\left\{-\left(\frac{R_s}{\sigma L_s} + \frac{(1-\sigma)R_r}{\sigma L_r}\right)\frac{\sigma L_s L_m}{L_r} + \frac{L_m R_r}{L_r}\right\} + \frac{\sigma L_s L_m}{L_r}(k-1)\left(\frac{R_s}{\sigma L_s} + \frac{(1-\sigma)R_r}{\sigma L_r} - \frac{R_r}{L_r}\right)$$

$$l_4 = -(k-1)\omega_r \frac{\sigma L_s L_m}{L_r}$$

Where $k$ is a proportionality constant that determines the closed loop bandwidth of the observer and $k \geq 1$.

### C. Speed Estimator

The model of the observer in (2) requires information of the rotor speed to estimate the stator currents and flux and to calculate the observer gains in (4). The rotor speed is estimated from the error in the actual and the estimated output states [9] as shown in (5).

$$\frac{d\hat{\omega}_r}{dt} = K_p\left(e_{sd}\hat{\lambda}_{rq} - e_{sq}\hat{\lambda}_{rd}\right) + K_I \int\left(e_{sd}\hat{\lambda}_{rq} - e_{sq}\hat{\lambda}_{rd}\right)dt \tag{5}$$

Where $e_{sd} = i_{sd} - \hat{i}_{sd}$ is the error in the stator d-axis current,

$e_{sq} = i_{sq} - \hat{i}_{sq}$ is the error in the stator q-axis current,

$\hat{\omega}_r$ is the estimated rotor speed.

Since the observer requires information of the rotor speed, the bandwidth of the PI controller in (5) used in the speed estimator should be higher than the closed loop observer since both the observer and the speed estimator act on the same errors, $e_{sd}$ and $e_{sq}$. This imposes a constraint on the choice of $k$ in (4) which is chosen to be a compromise between accuracy of state estimation and speed estimation. Nevertheless, for wind farm applications, since the shaft speed does not vary over a wide range, the Luenberger observer and the speed estimator described above give a good dynamic response over the whole range of operation.

## V. CONTROLLER DESIGN

The proposed DPC scheme uses two controllers to regulate the commanded flux and the power delivered. This section presents an approximate procedure for the design of the proportional and integral gains of the controllers.

### A. Design of Flux Controller

The flux model of the machine can be derived from its d-axis equivalent circuit in the stator flux oriented reference frame shown in Fig. 3. The back-emf along the direct axis in the stator flux oriented reference frame is zero, since the stator voltage is along the q-axis. Hence the equation governing the stator flux model is represented in (6).

$$V_{sd} = \frac{d\lambda_{sd}}{dt} + R_s * i_{sd} = \frac{d\lambda_s}{dt} + R_s * i_{sd} \tag{6}$$

With feed-forward compensation, the output d-axis reference voltage $V_{sd}^*$ from the flux controller can be obtained as

$$V_{sd}^* = V_{sd} - i_{sd}R_s = \frac{d\lambda_s}{dt} \tag{7}$$

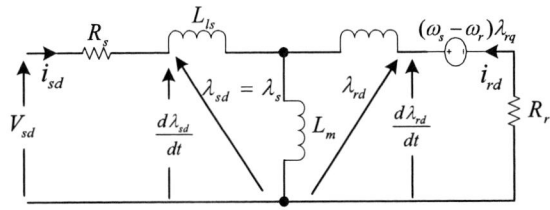

Fig.3. d-axis equivalent circuit in stator flux oriented reference frame

Thus the system is reduced to an integrator as shown in Fig.4 implying that the flux can be controlled with a simple proportional controller.

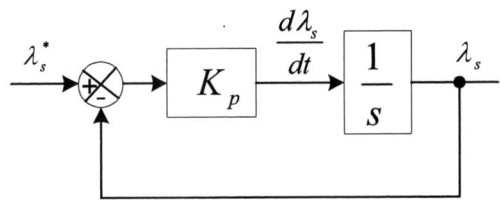

Fig.4. Closed loop control system for stator flux.

### B. Design of Power Controller

The power model of the machine can be obtained from the torque model, since the mechanical power developed in the machine is given by (8). Thus the power model can be derived from the q-axis equivalent circuit in the stator flux oriented reference frame shown in Fig. 5.

$$P = T_{em} * \hat{\omega}_{r_{mech}} = T_{em} * (2/p) * \hat{\omega}_r \tag{8}$$

where $p$ is the number of poles of the machine.

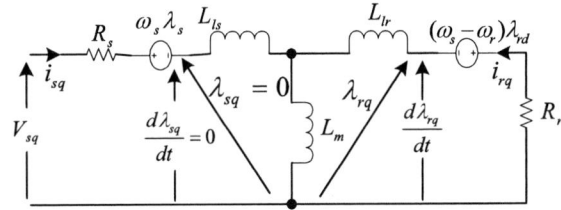

Fig.5. q-axis equivalent circuit in stator flux oriented reference frame

With stator flux orientated along the d-axis, $\lambda_{sq}$ is zero, thus the q-axis model equations can be obtained as shown in (9) and (10).

$$L_{eq}\frac{di_{sq}}{dt} + R_{eq}i_{sq} = \omega_s . \lambda_{rd} \tag{9}$$

$$\omega_s = \frac{V_{sq} - R_s * i_{sq}}{\lambda_s} \tag{10}$$

Where $L_{eq} = \frac{L_{lr}L_s}{L_m} + L_{ls}$ and $R_{eq} = R_r\frac{L_s}{L_m}$

Equation (9) can be represented in transfer function form as

$$i_{sq} = \frac{\omega_s \times \lambda_{rd}}{R_{r\_eq}\left(\tau_{eq}s + 1\right)} \tag{11}$$

Where $\tau_{eq} = \frac{L_{ls\_eq}}{Rr\_eq}$ is the time constant.

767

Thus the developed torque is given by (12)

$$T_{em} = \frac{p}{2} i_{sq} \lambda_s = \frac{p}{2} \frac{(V_{sq} - i_{sq} R_s) \times \lambda_{rd}}{R_{r\_eq}(\tau_{eq}s + 1)} \qquad (12)$$

The generated mechanical power can be obtained by substituting (12) in (8) and is given in (13)

$$P_m = \frac{(V_{sq} - i_{sq} R_s) \times \lambda_{rd}}{R_{r\_eq}(\tau_{eq}s + 1)} * \hat{\omega}_r \qquad (13)$$

With feed-forward compensation, the generated mechanical power is given as (14)

$$P_m = \frac{V_{sq}^* \times \lambda_{rd}}{R_{r\_eq}(\tau_{eq}s + 1)} * \hat{\omega}_r \qquad (14)$$

Expressing the rotor speed in terms of the synchronous speed and the slip $s$, we have

$$\hat{\omega}_r = \omega_s(1 - s) = \frac{V_{sq} - i_{sq} R_s}{\lambda_s}(1 - s) \qquad (15)$$

Substituting (15) in (14), the active power transferred to the stator after feed forward compensation can be obtained as

$$P_s = \frac{V_{sq}^{*2} \times \lambda_{rd}}{\lambda_s R_{r\_eq}(\tau_{eq}s + 1)} = V_{sq}^* * i_{sq} \qquad (16)$$

It can be inferred from (16) that the power delivered to the stator of the induction machine is a non-linear function of voltage along the q-axis with stator flux orientation. Since the system being controlled is a first order system, a PI controller can be used to control the power directly as shown in Fig. 6.

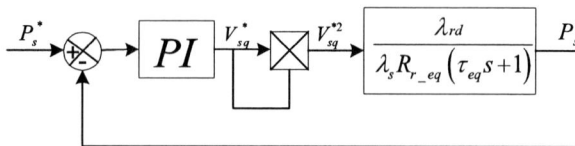

Fig.6. Closed loop feed forward compensated control system for torque.

## VI. CARRIER BASED MODULATION OF MATRIX CONVERTER

The reference three phase PWM voltages obtained from the controllers are generated from the matrix converter using the simplified carrier based modulation. The modulation functions are derived by representing the matrix converter as three-level inverter. The duty ratios of output phase-A are shown below.

$$d_{aA} = \Delta + D_a(t) + k_A \cos(\omega_i t - \rho)$$
$$\quad - [\{\max(k_A, k_B, k_C) + \min(k_A, k_B, k_C)\}/2]\cos(\omega_i t - \rho)$$
$$d_{bA} = \Delta + D_b(t) + k_A \cos(\omega_i t - 2\pi/3 - \rho)$$
$$\quad - [\{\max(k_A, k_B, k_C) + \min(k_A, k_B, k_C)\}/2]\cos(\omega_i t - 2\pi/3 - \rho)$$
$$d_{cA} = \Delta + D_c(t) + k_A \cos(\omega_i t - 4\pi/3 - \rho)$$
$$\quad - [\{\max(k_A, k_B, k_C) + \min(k_A, k_B, k_C)\}/2]\cos(\omega_i t - 4\pi/3 - \rho)$$
$$(17)$$

Where $D_a, D_b, D_c$ are the offset-duty ratios which appear as common mode voltages in the output voltages and $\Delta = (1 - \{D_a(t) + D_b(t) + D_c(t)\})/3$ is the offset duty ratio added to make the sum of the duty ratios in a switching cycle

equal to 1. The switching signals can be obtained from the duty ratios by carrier comparison as shown in Fig. 7.

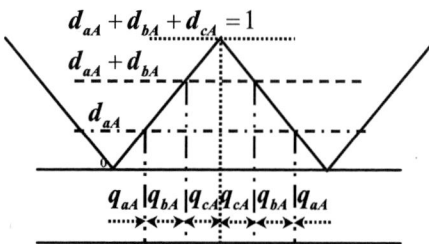

Fig. 7. Generation of Switching signals for switches connected to phase-A using carrier-PWM

## VII. RESULTS & DISCUSSION

The proposed direct power control scheme is simulated by modeling the individual components of the system in Fig.2 using the Matlab Simulink toolbox. Simulation results are presented for a 3Hp induction motor. The parameters of the flux and the power controllers are designed using the design procedure described in Section-V. The switching frequency of the matrix converter is chosen to be 10 KHz. The simulation results showing the state estimation, speed estimation as well as the performance of the closed loop power control system are shown in Figs. [8-13].

Fig. 8 shows the actual and the estimated dq-axis stator currents with the closed loop gains of the Luenberger observer chosen such that the observer is twice as fast as the actual system. The motor initial conditions are set such that it supplies rated torque at t = 0. It can be observed from the plots, that the estimated currents track the actual currents in less than 0.3sec. Fig. 9 shows the plot of the actual and the estimated values of speed and dq-axis stator fluxes with the same initial conditions. It can be observed that the estimated speed matches the actual speed in less than 0.2sec. To illustrate the dynamic performance of the observer and the speed estimator, Fig. 10 shows the plots of actual and estimated speed and stator fluxes when the load on the motor is changed from no-load to full load.

Fig. 11 shows the plots of the stator active power and reactive power, for the case in which the stator flux is established initially and power commands of 500W and 1000W are given to the power controller at 0.5 and 0.9 sec respectively. The initial speed of the generator is zero and is maintained constant at 150rad/s from 0.2sec. As can be seen the stator power instantaneously changes to the commanded value. While the active power is delivered to the grid, the induction machine draws reactive power from the output end of the matrix converter. Fig. 12 shows the 3-phase stator currents and the grid voltage and current in phase-a during the same dynamic conditions as in Fig. 8. During all the transient conditions the grid side voltage and current differ in phase by $180^0$ implying that current is flowing from the induction generator to the grid and the power factor is unity. This is one of the advantages of using the matrix converter that the input power factor can be controlled independent of load power factor and the simplified carrier based control used in this

768

paper makes this task easier to implement. Fig. 13 shows the reference and the actual values of the stator flux and the active power. As mentioned previously, the flux in the machine is established initially and it can be observed that the flux remains constant at the commanded value during all the transients. The second plot in Fig. 13 shows the reference and actual power generated by the induction machine. Thus the stator power is equal to the generated power after subtracting the stator copper losses and can be observed from Fig. 11to be slightly less than that in Fig. 13.

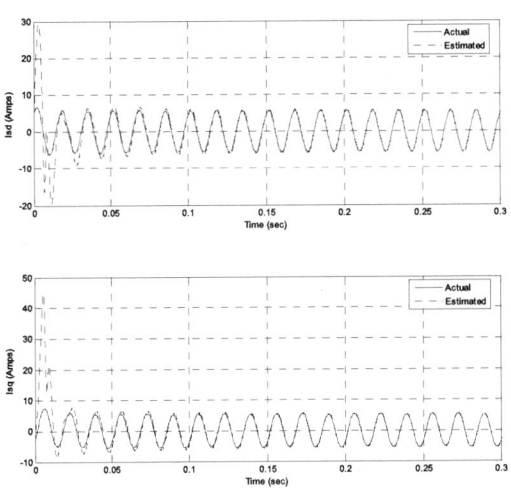

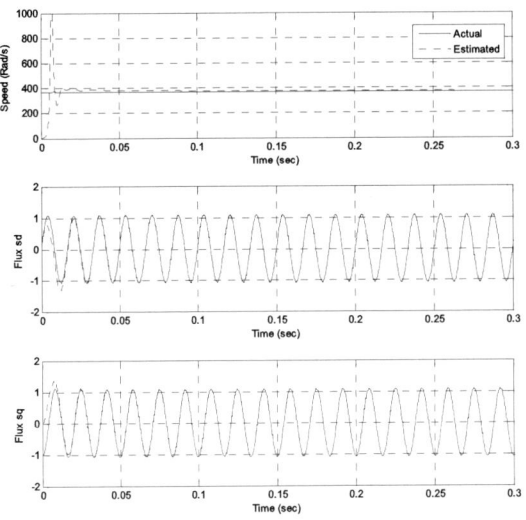

Fig. 8. Actual and Estimated dq-axis stator currents

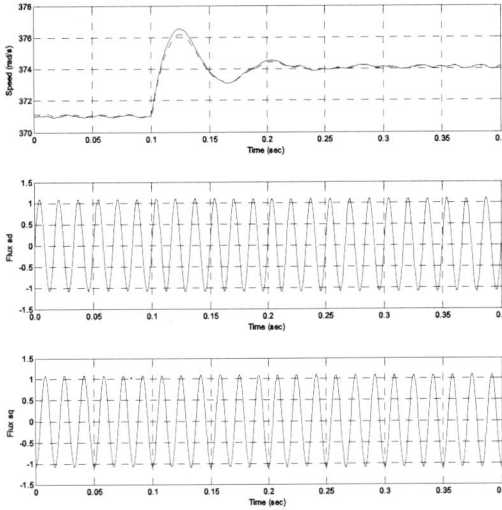

Fig. 10. Actual and estimated speed, dq-axis stator fluxes during Load change

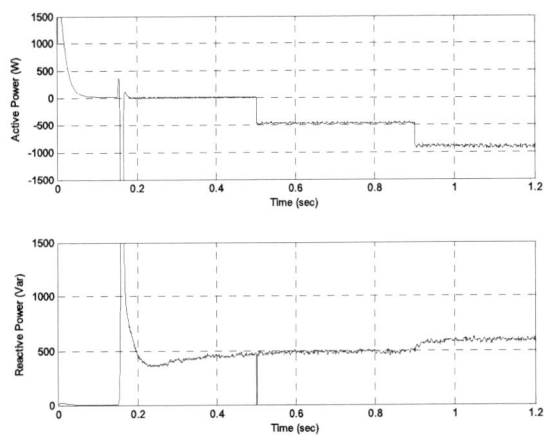

Fig. 11. Plots of stator active and reactive power

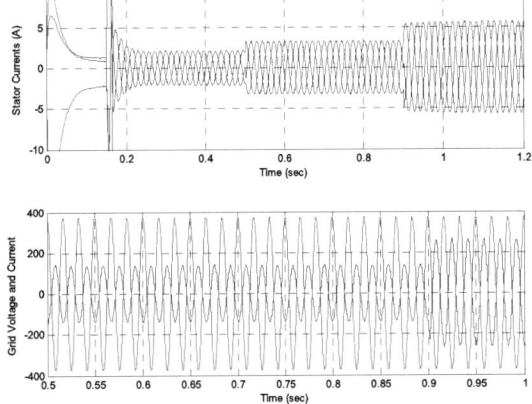

Fig. 9. Actual and estimated speed, dq-axis stator fluxes

Fig. 12. Plots of Stator 3-phase currents, Grid voltage and current

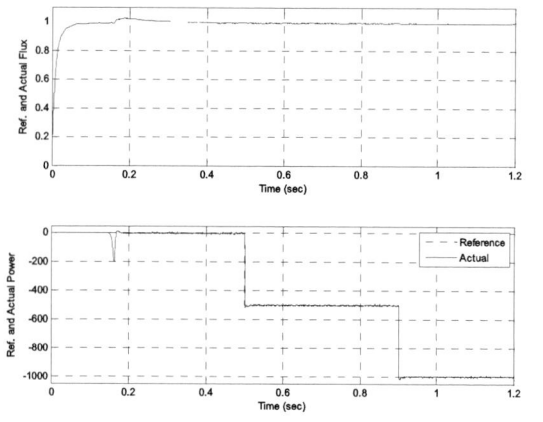

Fig. 13. Plots of Commanded and actual stator flux and active power

## VIII. CONCLUSIONS

A Direct Power Control (DPC) scheme is presented that directly controls the active power delivered by an induction generator connected to the grid. An AC-AC matrix converter is used as the power electronic interface between the grid and the induction generator. The closed loop Luenberger observer and the speed observer used to estimate the motor states and the speed is found to have much better dynamic response than the open-loop observers. The power delivered by the generator is found to be a non-linear function of the q-axis voltage in the stator flux oriented reference frame. The modulation functions of the matrix converter are chosen such that the power factor on the grid side is maintained unity. Simulation results of the proposed scheme are presented for a 3Hp squirrel cage induction machine (3HP, 440-V, 4-pole, 60Hz, $R_s = 1.77\Omega$, $R_r = 1.34\Omega$, $L_s = 0.3826H$, $L_r = 0.3808H$, $L_m = 0.3687H$, $J_{eq} = 0.025kg\text{-}m^2$) for different dynamic conditions.

## IX. REFERENCES

[1] T. Ackermann, "Wind power in power systems," England, John Wiley & sons Ltd, 2005, pp. 56.

[2] A. Alesina, M. Venturini, "Solid-state power conversion: A Fourier analysis approach to generalized transformer synthesis," *IEEE Trans. on Circuits and Systems,* Vol: 28, Issue: 4, pp.319 – 330, Apr 1981.

[3] N. Mohan, K.K. Mohapatra, P. Jose, A. Drolia, G. Aggarwal, T. Satish, "A Novel carrier based PWM scheme for matrix converters that is easy to implement," Conf. record of PESC'05, pp. 2410-2414.

[4] T. Satish, K.K. Mohapatra, N. Mohan, "Modulation Methods Based on a Novel Carrier-Based PWM Scheme for Matrix Converter

operation under Unbalanced Input Voltages," Conf. record of APEC'06, pp. 127-132.

[5] L. Zhang, C. Watthanasarn, W. Shepherd, "Application of a matrix converter for the power control of a variable-speed wind-turbine driving a doubly-fed induction generator," IECON'97, vol. 2, no. 2, pp. 906-911, Nov. 1997.

[6] M. P. Kazmierkowski, R. Krishnan, F. Blaabjerg. Control in Power Electronics – Selected Problems, Academic press, ISBN 0-12-402772-5,ch. 3, 2002.

[7] R. Datta, V.T. Ranganathan, "Direct power control of grid connected wound rotor induction machine without rotor position sensors," *IEEE Trans. On Pwr. Elect.,* vol. 16, pp. 390-399, May 2001.

[8] Lie Xu and P. Cartwright, "Direct active and reactive power control of DFIG for wind energy generation," *IEEE Trans. Energy Conversion,* vol. 21, no. 3, pp. 750-758, Sep. 2006.

[9] H. Kubota, K. Matsuse, "Speed sensor-less field oriented control of induction machines," Proc. IEEE IECON'94, pp. 1611-1615, Nov.1994.

## X. BIOGRAPHIES

**Satish Thuta** (SM'04) received the M.Tech degree from the Indian Institute of Technology Delhi, New Delhi, India in 2004, and is pursuing Ph.D. in the Department of Electrical Engineering at the University of Minnesota, Minneapolis, USA. His research interests include, Power Electronics, Motor Drives, PWM techniques, Switch Mode Power Supplies and Control Systems.

**Krushna K. Mohapatra** (SM'00-M'04) received the M.Tech degree in electrical engineering from the Indian Institute of Technology, Kharaghpur, India, in 1996, Ph.D. degree from the Indian Institute of Science, Bangalore, India, in 2004, and is currently a Post Doctorate in the Electrical Engineering Department at the University of Minnesota, Minneapolis, USA. He was a Design and Development Engineer in National Radio and Electronics Company Ltd., from 1995 to 2000. His research interests are in the area of power converters, PWM strategies, and motor drives.

**Ned Mohan** (M'73-SM'91-F'96) received the Ph.D. degree in electrical engineering from the university of Wisconsin, Madison, in 1973. He is the Oscar A. Schott professor of power electronics at the University of Minnesota, Minneapolis, where he has been teaching since 1976. He has numerous patents and publications in the field of power electronics, electric drive and power systems. He has written five textbooks. Prof. Mohan is a recipient of the Distinguished Teaching Award presented by the Institute of Technology at the University of Minnesota, Minneapolis.

**2006 IEEE International Conference on Power Electronic, Drives and Energy Systems**

# Stochastic Model for Optimal Selection of DDG by Monte Carlo Simulation

N.Vaitheeswaran, and R.Balasubramanian, *Senior Member, IEEE*

*Abstract*—Decentralized Distributed Generation (DDG) is considered as a viable option for electrification of rural areas where there is no access to electric network. An approach has been presented in this paper for optimal selection of micro gas turbine based Distributed Generator meant for standalone application. The decision support model incorporates the stochastic nature of opportunity cost of capital and uncertain gas fuel price scenario. Expected Net Present Value (NPV) of all future costs is estimated for a set of DDGs by Monte Carlo simulation and optimal DDG unit having minimum life cycle cost has been arrived at. A numerical case example has been illustrated in this paper

*Index terms*--DDG, Monte Carlo Simulation, Net Present Value, Stochastic process,

## I. NOMENCLATURE

$DG$    Set of different DG units under evaluation ($G_1,G_2$ $G_3$.... $G_I$)

$i$    Index of DG units

$t$    Index of planning horizon in years for DGs

$k$    Index of fuel price scenarios for each year

$I$    Number of DGs under evaluation

$T$    Life of each DG unit in number of years

$K$    No of intervals in a year

$C_i$    Capital cost of $i^{th}$ distributed generation unit in Rupees (Indian Currency)

$HR_o^i$    Heat rate of $i^{th}$ unit in Kcal/KWH(design value)

$HR_k^i$    Heat rate of $i^{th}$ distributed generation unit in Kcal/KWH at $k^{th}$ interval

$d_k$    Linear Heat Rate deterioration index for $k^{th}$ interval

$P_k^t$    Gas price in Rs per Million Kcal for kth interval in $t^{th}$ year, (a random variable)

$g^t$    Gas price escalation index in $t^{th}$ year (first year is taken as base year, hence $g^1=1$ )

$NPV(Gi)$ Expected Net Present Value of $i^{th}$ DG

$Mi^t$    Annual maintenance cost of $i^{th}$ DG in $t^{th}$ year

$Oi^t$    Annual fuel cost of electricity produced by $i^{th}$ DG in Rupees in $t^{th}$ year.

$E_k^t$    Energy requirement for $k^{th}$ interval of $t^{th}$ year in KWH

$e_t$    Linear energy demand escalation for $t^{th}$ year

$r_t$    Discount rate on capital cost (a random variable)

---

N.Vaitheeswaran is with Power Management Institute of NTPC Ltd., New Delhi, India. (e-mail: vaity94@rediffmail.com)

R. Balasubramanian is with Indian Institute of technology, New Delhi, India.(e-mail : balu@ieee.org)

## II. INTRODUCTION

THE Distributed Generators are small size, localized generation sources that delivers electricity directly to the consumers. The rating of these generators range from 10KW to 10 MW. DGs are dispersed over geographical locations widely. The practice of installation of small generators, both renewable and conventional, networked with distribution system as well as on standalone basis, is increasingly becoming a trend these days. Over the years, different technology options for DGs have been made commercially available in the energy markets. These include diesel units, gas turbine, wind turbine, fuel cell, micro/mini hydel e.t.c.

The advancements made in technology and the potential benefits associated with DG generation like cost effectiveness and environmental friendly nature are making small-scale, on-site power generation a widely acceptable proposition. In deregulated power markets in developed countries, the competition induced at retail level has acted as a key driver for DGs. The DG penetration in distribution networks has indeed created complex operational problems. Hence the research efforts in developed economies have been directed at integration of DGs with the existing distribution networks, improving power quality through embedded generations, minimizing the environmental impact through renewable DGs. Thus, many research works focusing on optimizing the choice of different DGs to decide the best alternative between network improvement, cost of energy not served and quality considerations, have been published [1],[2],[3]. The methodology for integrated distribution optimization model has been discussed in [4]. The work of Rodrigo Palma-Bhenke *et al i*n [5] addresses the day ahead planning of a distribution company considering DGs in energy acquisition market. Stochastic natures of DGs connected to distribution networks, by Monte Carlo simulation have been applied for adequacy assessment in [6] and for analysis of steady state performance in [7]. In all the above cases the interest of DGs is driven by the need for increasing power quality, requirements for rescheduling investments on T&D and environmental reasons.

In developing countries like India, primary need for DG is driven by the priority for accelerated rural electrification. Many interior villages have no access to electricity and it is not feasible to extend the T&D network to such areas. Stand-alone fossil fuel fired DGs are increasingly considered as a feasible alternative for electrification these remote villages in the country. National policy of federal government in India has accorded top priority towards large scale electrification of such areas [8], [9].

0-7803-9771-1/06/$25.00 ©2006 IEEE

These days, micro gas turbine based DG technology is increasingly emerging as an option for stand alone application. This paper is an effort to model the cost economics of micro gas turbine based DGs, wherein electricity is generated at the consumer end, thereby avoiding transmission and distribution costs. Optimal selection of DG involves an assessment of different cost elements. In deregulated markets application of DGs, the costs related to reliability and power quality aspects are also to be denominated as components. For stand alone DGs under consideration for rural electrification, the capital cost and operating cost are the deciding factors. Therefore, the real-life problem in the Indian scenario is to have a computational model for optimal selection of DG. Deterministic approaches to cost evaluation and sensitivity analysis of different types of DG units are discussed in [10]. Many deterministic models have been developed for cost evaluation and selection of DG. This work develops a methodology for optimal selection of micro gas turbine (GT) based DG, for standalone application, incorporating in fuel price uncertainty and randomness in opportunity cost of capital. Both random parameters are simulated by Monte Carlo simulation [11].

The organisation of the paper involves description of the problem in section III. Mathematical formulation of the problem involving stochastic and deterministic variables, followed by Monte Carlo simulation process in section IV. Step by step procedure for solving the problem is covered in V. Demonstration of a practical situation is illustrated with a numerical case example in VI.

## III. PROBLEM OVERVIEW

The practical problem for a planner, in the electrification of remote and interior places, is to decide the selection of cost effective DG unit from amongst the available ones in the market. The type of DG technology under consideration is fossil fuel based DG of micro gas turbine technology. The decision maker has to choose a DG, which would incur minimum life cycle cost for generation in the life span of DG. The unit should also meet future load requirement. This involves analysis of engineering economics of different DGs available and comparing them. Today, micro GTs of varied techno-commercial specifications, even for same unit sizes, is available from many vendors. In a comparison between two competitive DGs, one may have higher efficiency (lower fuel cost), but higher capital cost and maintenance cost than the other. These parameters have to be balanced against each other in evaluating the economic merits. This assessment gets complicated with the stochastic nature of fuel price and randomness in the discount rate on capital.

The objective of this paper is to find the most cost effective DG unit, by estimating the expected life cycle cost of all DG units. The approach considers the capital cost, discount rate on capital considering its randomness, DG efficiency, stochastic fuel price and maintenance cost, while considering the site specific load constraint and other deterministic elements.

## IV. MATHEMATICAL FORMULATION

The set '$DG$'={$G_1$, G2............$G_I$} is the possible class of micro Gas Turbine units available for evaluation. The objective of this model to find the optimal DG shown by (1)

$$\underset{DG}{Min} \left\{ NPV(G_1), NPV(G_2),.........NPV(G_i)....NPV(G_I) \right\} \quad (1)$$

$$NPV(G_i) = C_i + \sum_{t=1}^{T} \left( O_i^{t} + M_i^{t} \right)/(1 + r_t)^t \quad (2)$$

$$\forall \ i = 1,2,......I$$

$$O_i^t = \left\{ \sum_{k=1}^{K} E_k^t * g^t * P_k^t * HR_k^i \right\} \quad (3)$$

$$HR_k^i = d_k * HR_o^i \quad (4)$$

$$E_k^t = e_t * E_k^1 \quad (5)$$

The expression (2) determines the Net Present Value of the life cycle cost of each DG, which combines the effect of stochastic parameters of capital cost and operational cost. The operational cost for different intervals, accounting for deterministic heat rate degradation and stochastic gas price is found from (3). Equation (4) incorporates the heat rate degradation function, as the heat rate of Combustion Turbine increases with the operating hours, signifying drop in efficiency with progressing of intervals. Annual maintenance at the end of each year is assumed to restore the heat rate to original design value. The linear escalation of energy demand is shown by (5). $r_t$ and $P_k^t$ are random variables generated by Monte Carlo simulation, the distribution of which is obtained from their past trend.

## V. OPTIMIZATION PROCEDURE

Flow chart of the computational model is shown in Fig-1. The computer implementation of simulation process is summarized below:

1.  Determine the probability distributions of random variables - fuel price and discount rate on capital- from the past trend of gas price index and market interest rate respectively.
2.  Generate the stochastic fuel price for all scenarios $k$ by Monte Carlo sampling..
3.  Generate the stochastic value of discount rate on capital cost for each year by Monte Carlo sampling process.
4.  Estimate the expected NPV of the life cycle cost of all DGs for each sampling.
5.  Repeat the steps 2,3,4 until the simulation converges, which is indicated by coefficient of variation.
6.  When the coefficient of variation of expected Net Present Value reaches below the preset mark, stop simulation and choose the optimal DG unit

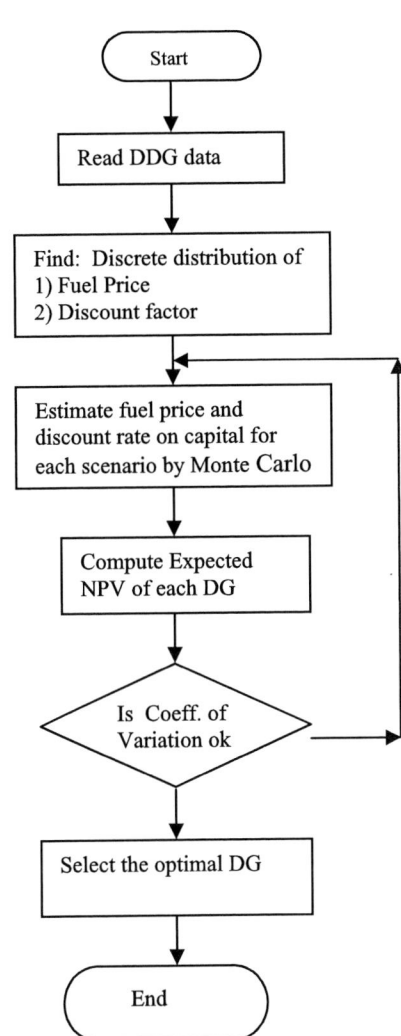

**TABLE I**
**DG UNIT DATA (MICRO GAS TURBINE)**

| DG Unit number | Capacity (KW) | Annual Maintenance Cost in Rs. | Capital cost Rs.'000 | Design Heat Rate Kcal per KWH |
|---|---|---|---|---|
| 1 | 100 | 60000 | 2000 | 2800 |
| 2 | 100 | 75000 | 1800 | 2950 |
| 3 | 110 | 60000 | 2200 | 2750 |
| 4 | 110 | 70000 | 2100 | 2770 |
| 5 | 125 | 80000 | 2400 | 2730 |
| 6 | 125 | 70000 | 2300 | 2700 |

**TABLE II**
**PROBABILTY DISTRIBUTION OF YEARLY DISCOUNT VALUE**

| Scenario No. | Prob-ability | Discount Value (%) | Scenario No. | Discount Value (%) | Prob-ability |
|---|---|---|---|---|---|
| 1 | .06 | 7 | 6 | 10.0 | .09 |
| 2 | .09 | 7.5 | 7 | 10.5 | .12 |
| 3 | .10 | 8.0 | 8 | 11.0 | .07 |
| 4 | .12 | 8.5 | 9 | 12.0 | .10 |
| 5 | .11 | 9.0 | 10 | 13.0 | .14 |

**TABLE III**
**PROBILITY DISTRIBUTION OF BASE GAS PRICE**

| Scenario No. | Prob-ability | Fuel price (Rs) | Scenario No. | Fuel price (Rs) | Prob-ability |
|---|---|---|---|---|---|
| 1 | .06 | 500 | 6 | 615 | .13 |
| 2 | .13 | 520 | 7 | 640 | .13 |
| 3 | .10 | 555 | 8 | 670 | .08 |
| 4 | .10 | 585 | 9 | 690 | .09 |
| 5 | .09 | 600 | 10 | 700 | .09 |

Fig.1. Flow Chart of Monte Carlo procedure for Life Cycle Cost Evaluation of DGs.

## VI. CASE EXAMPLE

Application of the technique discussed above has been illustrated with a real life example. Table I gives the data pertaining to 6 numbers of DGs under evaluation. The DG, which minimizes the Net Present Value of life cycle cost considering the capital cost and recurring running cost elements needs to be selected. The life span of all DGs is taken as 20 years and the prior distribution of present worth discount factor based on market interest is shown in table II. Similarly the prior distribution representing the stochastic variation base fuel price of different scenario is provided in table III. Average energy equivalent to a load of 55 KW for different intervals 'k' in the first year, linear rural load growth is @3 % per annum, Gas price linear escalation component is @ 3% per year, based on gas price index analysis. Heat rate degradation of 5% between two successive maintenance intervals is considered, for all DGs. Annual maintenance is assumed to restore the heat rate to initial design value.

Monte Carlo run was repeated for coefficient of variation less than 1% for all DGs. The sample value of discount rate on capital and gas price are depicted in Fig.2 and Fig.3 respectively. In the test case is the optimal unit is DG-3, as shown in Fig.4. The convergence of Monte Carlo process for the test case is given in Fig.5. The sensitivity of optimality to increased load growth to 4% is also tested (Fig.6) and the optimality is same for marginal change in the load growth in the test case. Computer implementation of the model has been done using Excel- Solver.

The 95% confidence intervals of mean, variance and standard deviation of the maximum expected revenue, corresponding to optimum station availability are :

$E[R]$ : [3,36,50000- 3,39,50000]

$Var[R]$ : [2.6228564x10$^{10}$ – 3.0912855x 10$^{10}$]

$Std[R]$ : [227390 – 244610]

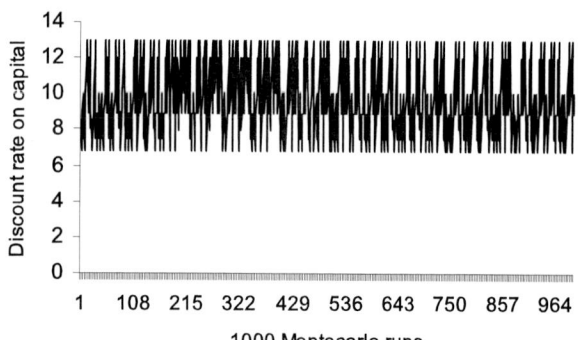

Fig. 2. Sample value of Discount rate on Capital for different Monte Carlo Runs.

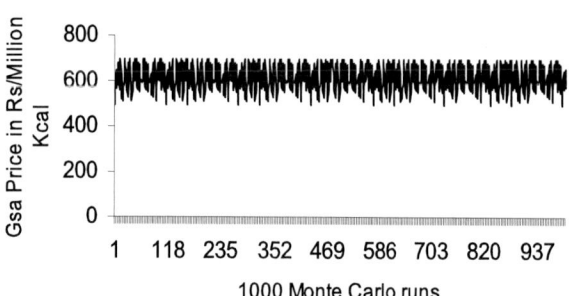

Fig. 3. Sample value of base Gas price for different Monte Carlo Runs.

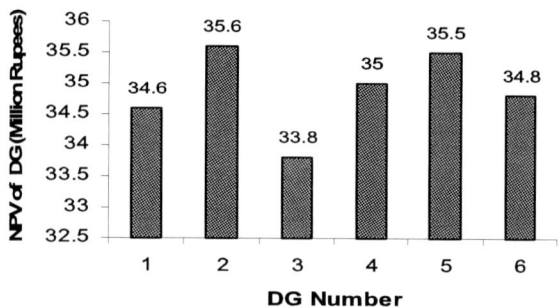

Fig. 4. Expected NPV of different DDG.

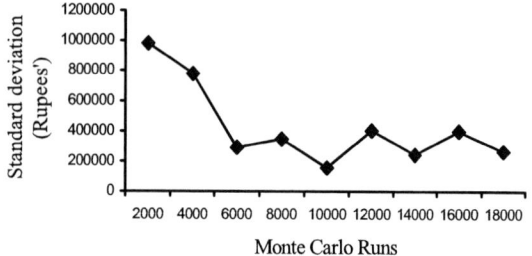

Fig. 5. Convergence Process of Monte Carlo Runs.

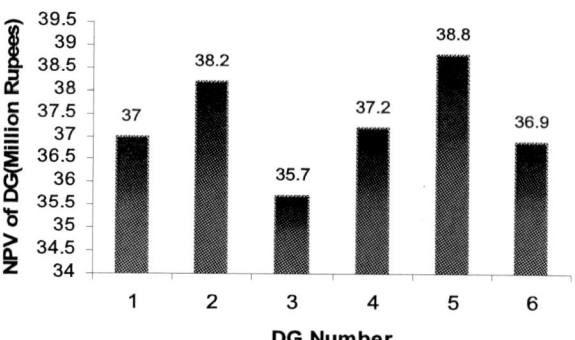

Fig. 6. Expected NPV of DG with 4% annual demand growth.

## VII. CONCLUSION

The general computational method described in this paper addresses least cost planning from a practical view point to deliver electricity to remote rural areas. The approach for selection of least cost DG was based on Expected Net Present Value minimization of life cycle cost. The formulation considered the randomness of opportunity cost arising from variation in discount rate on capital. The stochastic behavior of fuel cost due to gas price uncertainty is also incorporated in the model. The combined application of the two random parameters would make this model a practical decision support tool in the hands of planners.

## VIII. REFERENCES

[1] Gianni Celli, Emilio Ghiani, Susanna Mocci, Fabrizio Pilo, "A Multi objective Evolutionary Algorithm for Sizing and Siting of Distributed Generation", *IEEE Trans. Power Syst.*,vol. 20, pp. 750-757, May 2005.

[2] Paul M. Sotkiewicz and J. Mario Vignolo, "651Allocation of Fixed Costs in Distribution Networks With Distributed Generation", *IEEE Trans. Power Syst.*,vol. 21, pp. 639-652, May 2006.

[3] Víctor H. Méndez Quezada, Juan Rivier Abbad, and Tomás Gómez San Román, "Assessment of Energy Distribution Losses for Increasing Penetration of Distributed Generation", *IEEE Trans. Power Syst.*,vol. 21, pp. 533-540, May 2006.

[4] Walid El-Khattam, Y.G.Hegazy and M.M.A. Salama, " An Integreted Distributed Generation Optimisation Model for Distribution Planning", *IEEE Trans. Power Syst.*,vol. 20, pp. 1158-1165, May 2005.

[5] Rodrigo Palma- Bhenke, Jose Luis Cerda A, Luis, and Vargs, " A Destributio Company Energy Acqasition Market Model with Integration og Distributed Generation and Load Curtailment Options", *IEEE Trans. Power Syst.*,vol. 20, pp. 1719-1727, November 2005.

[6] Y.G.Hegazy,M.M.A.Salama,andA.Y.Chikhani, "Adequacy assessment of distributed generation systems using Monte Carlo simulatin,". *IEEE Trans. Power Syst.*,vol. 18, pp. 48-52, Feb.2003

[7] Walid El-Khattam, Y. G. Hegazy, and M. M. A. Salama, "Investigating Distributed Generation Systems Performance Using Monte Carlo Simulation", *IEEE Trans. Power Syst.*,vol. 21, pp. 524-532, May 2006.

[8] Acts and Notifications.[Online]. Available : http://powermin.nic.in/

[9] Distributed Generation. [Online]. Available: http://powermin.gov.in/JSP_SERVLETS/internal.jsp

[10] H.L.Willis and W.G.Scott, *Distributed Power Generation. Planning and Evaluation*, 1st ed. New York: Marcel Dekker, 2000.

[11] Roy Billington and Wenyuan Li, *Reliability Assesment Electric Power System using Monte Carlo Method*, Newyork: Plenum Press, 1994

## IX. BIOGRAPHIES

**N.Vaitheeswaran** obtained his B.Tech from Calicut University, India in 1983 After receiving M.Tech from Indian Institute of Technology, New Delhi in

2002, he is pursuing PhD in the same institution in the area of restructured power markets. Currently he is working as a faculty member at Power Management Institute of NTPC, in India. He has been with NTPC for over 20 years and has diverse experience in various fields of power generation like commissioning, operation, and plant commercial management.

**R.Balasubramanian** obtained his PhD from Indian Institute of Technology, Kanpur, India in 1976. He is currently the NTPC Chair Professor at Indian Institute of Technology, Delhi. He has many national and international publications to his credit. His areas of interest include power system planning, operation, control and optimization and development of computational models in restructured electricity market

**2006 IEEE International Conference on Power Electronic, Drives and Energy Systems**

# Capacitive Self-Excitation in a Six-Phase Induction Generator for Small Hydro Power – An Experimental Investigation

## G. K. Singh, K. B. Yadav, and R. P. Saini

*Abstract*—Results of an experimental investigation dealing with the behavior of a split wound 6-phase self-excited induction generator (SEIG) with capacitor excitation in conjunction to a stand-alone small hydro power scheme is presented. As predicted, the scheme is found offering high reliability, improved capability of voltage build-up and load withstanding along with acceptable efficiency. Results obtained show that the system has sufficient capability for practical use in stand-alone power generation with small hydropower.

*Index Terms*—Experimental investigation, Hydraulic turbine, Multi-phase machine, Split-wound, Six-phase SEIG, Self-excitation, Small hydropower.

## I. INTRODUCTION

IN recent years, owing to the increased emphasis on renewable energy resources, development of suitable isolated power generators driven by energy sources such as wind, small hydroelectric, biogas, etc. has assumed greater significance. Due to its reduced unit cost, brushless rotor construction, absence of separate source of excitation, ruggedness, operational and maintenance simplicity, and self-protection against severe overloads and short circuits, a capacitor self-excited induction generator has emerged as a suitable candidate of isolated power source. The induction generator's ability to generate power at varying speed facilitates its application in various modes such as self-excited stand-alone (isolated) mode; in parallel with synchronous generator to supplement the local load, and in grid connected mode [1].

Electric power systems have largely developed as three-phase systems, although high phase order (in excess of three) machine construction and power transmission have been considered for last several years. With the growth of

---

G. K. Singh is with the Department of Electrical Engineering, Indian Institute of Technology, Roorkee-247 667, India.
(e mail: gksngfee@iitr.ernet.in)

K. B. Yadav is with Alternate Hydro Energy Centre, Indian Institute of Technology, Roorkee-247 667, India
(e mail: yadavbkrishna@rediffmail.com)

R. P. Saini is with Alternate Hydro Energy Centre, Indian Institute of Technology, Roorkee-247 667, India (e mail: rajsafah@iitr.ernet.in)

increasingly sophisticated design methods and increased importance of economic, environmental and several other factors, the multi-phase systems are being considered as one of the potential alternatives to conventional three-phase systems.

With the use of higher phase in transmission, the inter-phase insulation requirement, spacing, conductor surface gradient and noise levels are reduced considerably. As a result, a multi-phase line with smaller dimensions can be used to transmit a larger amount of power covering entire range of transmission voltages [2]. Because of the potential benefits resulting from the use of a phase order higher than three in transmission, some interest has also grown in the area of multi-phase machine [3]. For machine drive applications, multi-phase system could potentially meet the demand for high power electric drive systems, which are both rugged and energy-efficient. High phase number drives possess several advantages over conventional three-phase drives such as: reducing the amplitude and increasing the frequency of torque pulsation, reducing the rotor harmonic currents, reducing the current per phase without increasing the voltage per phase, lowering the dc link current harmonics, higher reliability and increased power in the same frame. The high phase order drive is likely to remain limited to specialized applications where high reliability is demanded such as electric /hybrid vehicles, aerospace applications, ship propulsion, and high power application where a combination of several solid state devices form one leg of the drive.

The investigations spread over the last two decades indicate the technical and economic viability of using the number of phases higher than three in induction machine. The research in this area is still in infancy, yet some extremely important findings have been reported in the literature indicating general feasibility of multi-phase systems [4-6]. A generator scheme based on the dual stator winding induction machine with displaced power and control three-phase winding is presented in [7] in which, power and control winding have the same number of poles. References [8, 9] deal with the double stator machine with extended rotor common to both stators. In all the three cases, output is three-phase. Recently, two papers have reported on modeling and analysis of six-phase self-excited induction generator [10, 11]. However, so far as the authors have been able to ascertain, practical applications of multi-phase (comprising of more than the conventional three

---

0-7803-9771-1/06/$25.00 ©2006 IEEE

phases) induction generator in hydropower scheme are still unreported. This paper, therefore, presents detailed experimental investigations of self-excitation process at no-load condition, and behavior of six-phase SEIG on switching-in of pure resistive load without and with series compensation.

## II. POWER GENERATION SCHEME

Fig.1 shows the schematic diagram of proposed power generation scheme employing a 1.1 kW prototype 6-pole, 6-phase SEIG. The prototype is coupled to a hydro turbine as a stand-alone generator with excitation provided at each 3-phase set along with a 6 /3-phase transformer connected to output terminals of the 6-phase SEIG. Y-connected capacitor banks of per phase capacitance 108 μF, determined by synchronous speed test were used for purpose [3].

Since the conventional supply and uses are three phase and single phase, it seems necessary to mention here that the combination with two three-phase windings displaced 30 degree in phase is the configuration of greatest practical interest for very large generators as it permits re-combination of two three-phase power in the step-up transformer bank without the need for increased transformer KVA rating for phase shifting. This indicates the need for a coordinated design of the 3-winding step-up transformer or of two separate transformers. In the study, it has been found that transformer cost increments (relative to a three-phase application of the same KVA and voltage rating) could be limited to 5 % or less in instances when single-phase transformer banks were necessary anyhow due to the rating or high-side voltage level. Even with the further cost increments associated with the need for two sets of isolated three-phase bus (of half-current), these costs did not seem excessive relative to the improved reliability that was expected due to decreased vibratory forces on the windings and other advantages [12, 13].

## III. PROTOTYPE DESCRIPTION AND TEST RIG

In order to investigate the performance of proof-of-concept system for stand-alone power generation scheme, a detailed experimentation was performed on a hydraulic turbine driven induction machine. For this purpose, three-phase, 1.1 kW, 415 V, 50 Hz, 960 rpm, 36-slot squirrel cage induction motor was used.

All the 72 armature coil terminals were taken out on a connection table mounted on the top of the machine body to yield alternative winding scheme for different number of pole and phases. The tests were performed on the test machine with winding reconfigured for six-pole, six-phase (semi 12-phase), and for different loading conditions under different modes of operation.

Fig. 2 shows the laboratory set up implemented for experimental evaluation of developed 6-phase SEIG in a small hydro power test rig, It consists of two identical service pumps, each of 150 litre /sec discharge capacity at the head of 10 metre connected to a 5 kW cross flow turbine (η = 56 %) through pipe line. Depending on the requirement, these may be connected in series as well in parallel. Control valves provided at the pipe line and turbine are used to vary flow and head. The cross-flow turbine coupled mechanically to MPSEIG facilitates its operation at adjustable speed, which is sensed and displayed by a micron (model: 2176) speed transducer and digital encoder (range: 60-9999 rpm) unit.

The power generation was achieved by driving the induction generator at different speeds by aforesaid drive, as the voltage is induced due to capacitor self-excitation. A 3-phase, Y-connected purely resistive load banks were used to load the SEIG adequately. The set-up was instrumented to monitor the voltage, current, power and frequency at the desired locations of network. A Power Quality Analyzer (Fluke 43B) was utilized throughout to capture various transient and steady state waveforms during experimentation.

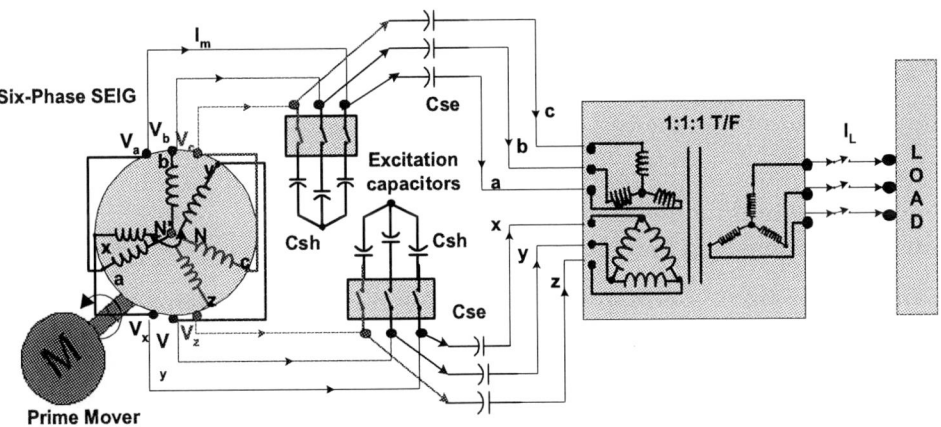

Fig. 1. Schematic diagram of generator system employing a split wound six-phase SEIG.

Fig. 2.    Laboratory setup for generator system evaluation.

The electric power generated is determined by the available hydropower, which is related to the head and discharge of water as follow

$$P_{mech} = \rho\, g\, H\, Q \qquad (1)$$

where,

$\rho$ (=1000 kg/m$^3$) is the density of water,

$g$ (=9.81 m/sec$^2$) is the acceleration due to gravity,

$H$ (in m) is the head of water,

$Q$ (in m$^3$/ sec) is the discharge of water through the rectangular weir.

This available power will be converted by the hydro turbine in mechanical power. As a turbine has efficiency lower than 1, the generated power (output of MPSEIG) will be a fraction of the available gross power.

## IV.  EXPERIMENTAL SYSTEM EVALUATION

### A. Parameter Identification through Zero Output Tests

No load and blocked rotor test were conducted to determine the parameters of the test machine reconfigured for 6-pole, 6-phase operation. Parameters so determined are given in    Table 1.

TABLE 1
PARAMETERS OF 6-POLE, 6-PHASE SEIG

| Parameters | Value (per Phase) |
|---|---|
| Stator resistances: Rs1 = Rs2 | 4.125 Ω |
| Stator leakage inductances: Ls1=Ls2 | 21.6 mH |
| Rotor resistance (R$_r$') | 8.79 Ω |
| Rotor leakage inductance (L$_r$') | 43.3 mH |
| Mutual inductance (L$_m$) | 234.6 mH |

'Synchronous Speed Test' was also conducted to determine magnetization characteristics of machine by driving the rotor at synchronous speed corresponding to base frequency, and monitoring input current and power at different input voltage across the stator windings.

### B. Experimentation and Discussion of Results

(i)  The no-load terminal voltage of 6-phase SEIG was determined by plotting the generator magnetization characteristic and the capacitor load line on a single set of axes. The intersection of the two curves is the point at which the capacitor bank exactly supplies the reactive power demanded by the generator. Fig. 3 shows the no-load terminal voltage on vertical axis corresponding to cross sectional points for these two set of capacitors respectively.

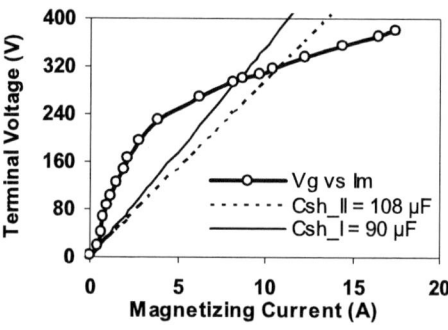

Fig. 3.  Magnetization characteristics of 6-phase SEIG with Y-connected two capacitor banks at 1000 rpm.

(ii) Figs. 4 (a) and 4 (b) present the no-load characteristics of 6-phase SEIG regarding its terminal voltage build-up and collapse with respect to prime mover speed. It was investigated for two values of shunt capacitor banks with $C_{sh\_I}$ = 90 µF and $C_{sh\_II}$ = 108 µF respectively. The transient voltage and current profile at the outset of self-excitation and collapse are shown in Fig. 5, when excitation capacitor banks are switched on /off across both three-phase winding sets of six-phase SEIG.

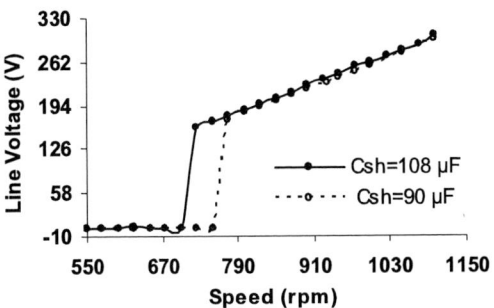

(a)   Terminal voltage build-up

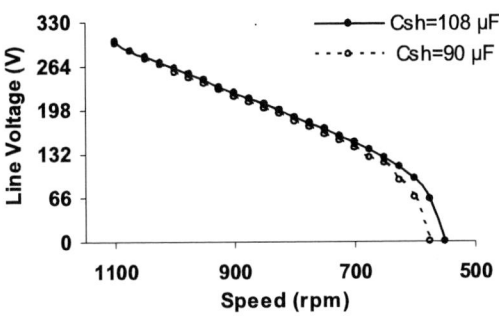

(b)   Terminal voltage collapse

Fig. 4. No-load characteristics of 6-phase SEIG with Y-connected two capacitor banks at 1000 rpm.

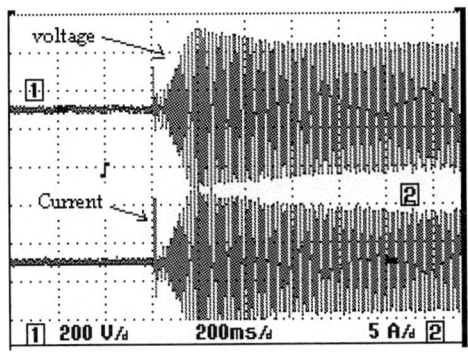

(a)   Terminal voltage build-up

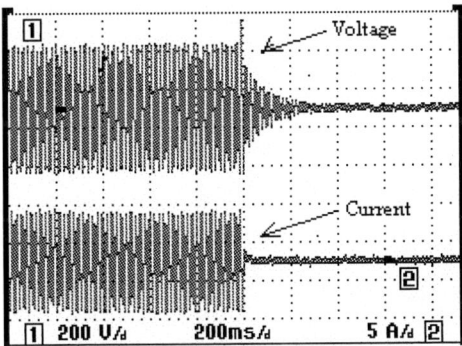

(b)   Terminal voltage collapse

Fig. 5. Transient no-load characteristics of 6-phase SEIG with Y-connected two capacitor banks across set abc at 1000 rpm.

(iii) Fig. 6 (a) to (e) illustrates the load characteristics of 6-

phase SEIG in simple shunt mode showing the variation of terminal voltage and load current with output power under the resistive load by maintaining its speeds at 1100, 1000, 900 rpm for 6-phase configuration. With the proper value of shunt capacitor bank, it is observed the 6-phase SEIG successfully delivered 1.15, 0.85, and 0.57 kW at 1100, 1000, 900 rpm without the voltage collapse with comparatively improved regulation at all speeds. It also withstood more than two times load current safely with better power conversion efficiency.

(iv) Load characteristics of 6-phase SEIG showing the effect of series capacitances are shown in Fig. 7 (a) for $C_{se} = 0, 108, 216$ µF with $C_{sh\_II} = 108$ µF and in Fig. 7 (b) for $C_{se} = 0, 108, 216$ µF with $C_{sh\_I} = 90$ µF against the resistive load at prime mover speed 1000 rpm. It was interesting to observe that generator delivered 0.9 kW power in all three cases without voltage collapse, rather voltage regulation was found better with $C_{se} = 0$, i.e. simple shunt mode compared to the short shunt mode operation.

Fig. 7(c) illustrates the effect of shunt capacitances on resistively loaded 6-phase SEIG for $C_{sh} = 90$ µF, $C_{sh} = 108$ µF by restricting its operation at 1000 rpm. Performance of prototype was examined for two values of shunt capacitor banks. The power delivery capability of six-phase configuration was found to be far better with less variation in load voltage throughout the loading range as compared to its three-phase counterpart.

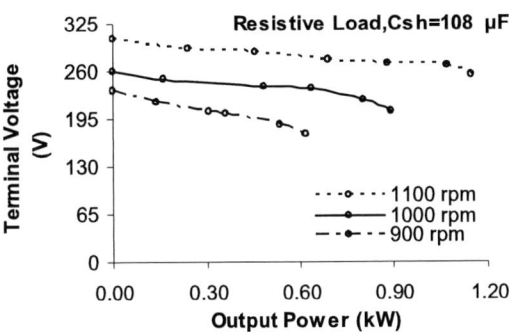

(a)   Variation of terminal voltage across winding set abc

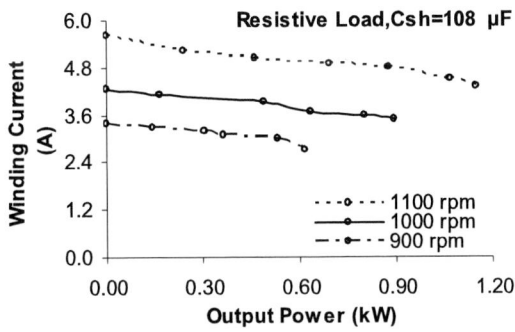

(b)   Variation of current in winding set abc

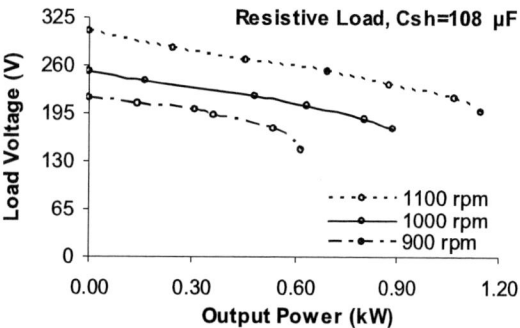

(c)    Variation of load voltage

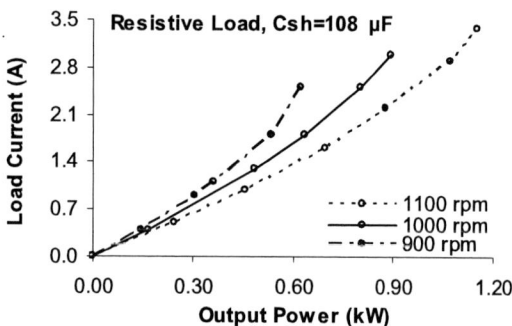

(d)    Variation of load current

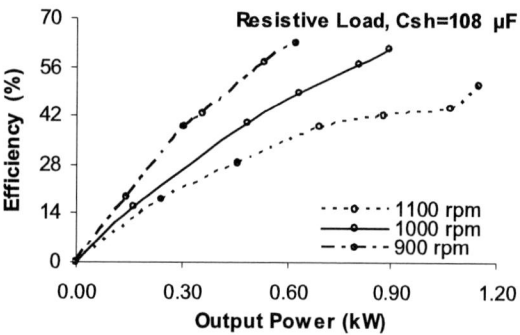

(e)    Variation of Power Conversion Efficiency

Fig. 6. Load characteristics of 6-phase SEIG at 1100, 1000, 900 rpm with Csh= 90 μF.

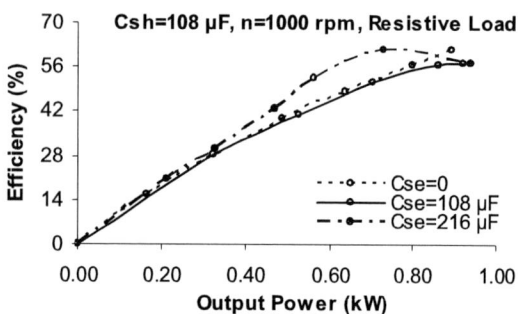

(a)    Effect of Series Capacitances

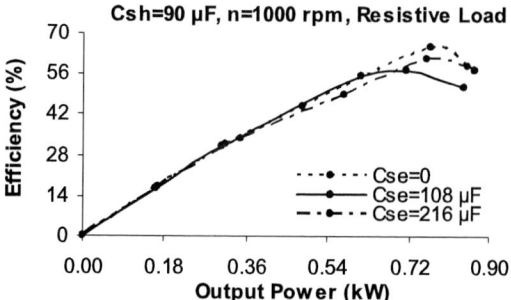

(b)    Effect of series capacitances

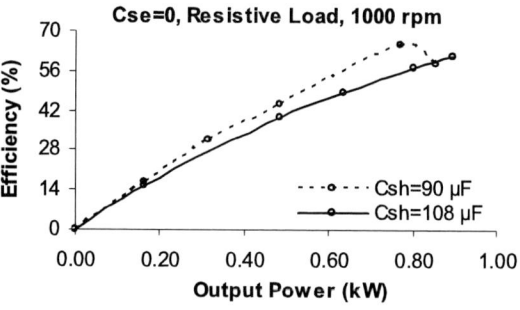

(c)    Effect of shunt capacitances

Fig. 7. Load characteristics showing effect of (a), (b) series    capacitance ($C_{se}$= 0, 108, 216 μF) and (c)   shunt  capacitances ($C_{sh}$= 90 μF, $C_{sh}$=108 μF) respectively.

## V.  Conclusion

The detailed experimental investigation on hydro turbine driven 6-phase SEIG is reported in this paper. Self-excitation under no-load condition and loading performance under typical resistive load is elaborated. It is found that some of the major drawbacks such as voltage collapse and situations that results total demagnetization regarding the operation of a three-phase SEIG are improved in this case. Though the results of experimental investigations conducted on three-phase SEIG are not included in this paper, a comparative study conducted on a three-phase and six-phase SEIG reveals that at reasonable amount of load:

> Variation in 6-phase SEIG terminal voltage and current is less;
> Variation in load voltage in case of 6-phase SEIG is less;
> Sensitivity of 6-phase SEIG with respect to speed during self-excitation is less;
> Power output of 6-phase SEIG is more (approximately double);
> Power conversion efficiency of 6-phase SEIG is more;
> Overall efficiency of the hydropower scheme employing 6-phase SEIG is more.
> Load withstanding capability was almost double with a considerable enhancement in power conversion efficiency compared to the 3-phase counterpart in the same frame for different mode of operation.

With  the  known  advantage  of  SEIG  i.e.  of-the-self

availability, operational simplicity, excellent power quality and feasibility of optimum efficiency operation with better voltage and frequency regulation, the proposed generation system should find wide acceptability for small hydropower schemes.

## VI. REFERENCES

[1] G. K. Singh, "Self-Excited Induction Generator Research – A Survey," Electric Power Systems Research, 2004, vol. 69, pp. 107-114.

[2] S. N. Tiwari, G. K. Singh and A. S. B. Saroor, "Multiphase Power Transmission Research – A Survey," Electric Power Systems Research, 1992, vol. 24, pp. 207-215.

[3] M. Jones and E. Levi, "A Literature Survey of State-of-the-Art in Multi-Phase AC Drives," in Proc. 36th Univ. Power Eng. Conf. UPEC, 2002, Stafford, U.K., pp. 505-510.

[4] G. K. Singh, "Multi-Phase Induction Machine Drive Research – A Survey," Electric Power Systems Research, 2002, vol. 61, pp. 139-147.

[5] T. M. Jahns, "Improved Reliability in Solid-State AC Drives by Means of Multiple Independent Phase-Drive Units," IEEE Trans. Ind. Applicat., May /June 1980, vol. IA-16, pp. 321-331.

[6] E. A. Klingshirn, "High Phase Order Induction Motors—Part I: Experimental Results," IEEE Trans. Power App. Syst., Jan. 1983, vol. PAS- 102, pp. 54-59.

[7] O. Ojo and I. E. Davidson, "PWM-VSI Inverter-Assisted Stand-Alone Dual Stator Winding Induction Generator," IEEE Trans. Energy Conversion, 2000, vol. 36, pp. 1604-1611.

[8] D. Basic, J. G. Zhu and G. Boardman, "Transient Performance Study of a Brushless Doubly Fed Twin Stator Induction Generator," IEEE Trans. Energy Conversion, 2003, vol. 18, pp. 400-408.

[9] [9] D. Levy, "Analysis of a Double-Stator Induction Machine used for a Variable-Speed /Constant-Frequency Small-Scale Hydro /Wind Electric Power Generator," Electric Power Systems Research, 1986, vol. 11, pp. 205-223.

[10] G. K. Singh, K. B. Yadav and R. P. Saini, "Modeling and Analysis of Multi-Phase (Six Phase) Self-Excited Induction Generator," in Proc. IEEE Conf. ICEMS'2005, the Eighth International Conference on Electrical Machines and Systems, 2005, China, pp. 1992-1927.

[11] G. K. Singh, K. B. Yadav and R. P. Saini, "Analysis of a Saturated Multi-Phase (Six-Phase) SEIG," International Journal of Emerging Electric Power System, Vol. 7 [2006], No. 2, Article 5, 2006.

[12] E. F. Fuchs and L. T. Rosenberg, "Analysis of an Alternator with Two Displaced Stator Windings," IEEE Trans. Power Apparatus and Systems, 1974, vol. 93, pp. 1776-1786.

[13] C. H. Holley and D. M. Willyoung, "Stator Winding Systems with Reduced Vibrating Force for Large Turbine Generators," IEEE Trans. Power Apparatus and Systems, 1970, vol. 89, pp. 1922-1934

## VII. BIOGRAPHIES

**G. K. Singh** received the B.Tech. degree from G. B. Pant University of Agriculture and Technology, Pantnagar, India, in 1981, and the Ph.D. degree from Banaras Hindu University, Varanasi, India, in 1991, both in electrical engineering. He worked in industry for nearly five and a half years. In 1991, he joined M. N. R. Engineering College, Allahabad, India, as a Lecturer.

In 1996, he moved to the University of Roorkee, Roorkee, India. Currently, he is a Professor in the Electrical Engineering Department, Indian Institute of Technology, Roorkee, India. He has been involved in design and analysis of electrical machines in general and high-phase-order ac machines in particular. He has coordinated a number of research projects sponsored by the CSIR and UGC, Government of India. He has served as a Visiting Associate Professor in the Department of Electrical Engineering, Pohang University of Science and Technology (POSTECH), Pohang, Korea. Prof. Singh received the Pt. Madan Mohan Malaviya Memorial Medal and the Certificate of Merit Award 2001–2002 of The Institution of Engineers (India).

**K. B. Yadav** received the B. Sc. (Eggs.) and M. Sc (Engg.) degree in Institute of Technology, Jamshedpur, India, in 1989 and in 2002 respectively. Since 1996, he is serving as faculty in Department of Electrical Engineering, National Institute of Technology, Jamshedpur, Jharkhand, India.

Currently, he is pursuing his Ph.D. Degree under QIP scheme of MHRD, Govt. of India at Alternate Hydro Energy Centre, Indian Institute of Technology Roorkee, India. His main research interests are multi-phase machines, and renewable energy source generation.

**R. P. Saini** received his B.E., M.E. and Ph.D. degree in Mechanical Engineering from University of Roorkee, India in the year 1982, 1989 and 1996 respectively. Currently, he is a Senior Scientific Officer in the Alternate Hydro Energy Centre, Indian Institute of Technology, Roorkee.

His fields of interest include Renewable Energy Systems, Environment Auditing / Energy Conservation.

**2006 IEEE International Conference on Power Electronic, Drives and Energy Systems**

# Grid Power Quality with Variable Speed Wind Energy Conversion

S.W. Mohod, Member, *IEEE,* and M.V.Aware, *Member, IEEE*

*Abstract-* **In the grid new renewable resources are added to extract more power. This adds more of problems to grid connection. They are voltage fluctuations and harmonic distortion. In this paper various interfacing topologies are analyzed to get wind turbine power within the norms specified in IEC 61400-21. In this paper, a simplified control strategy is used. The front-end voltage source inverter is operated in hysteresis current control mode. The reference signals are derived from one of the phase voltage. The main objective of the proposed control is the three phase supply currents both in its waveform, magnitude and phase to follow three phase reference signals. When this is achieved, ideally the supply current will then be always sinusoidal, with robust control over its magnitude and phase, irrespective of the harmonics and unbalance of the load demand or the supply voltage system. This confirms nearly unity power factor on supply side with active and reactive power support from the wind turbine side.**

*Index Terms-* **Distortion, Harmonics, Hysteresis Control, Power Quality, Reactive Power, Wind Power.**

## I. INTRODUCTION

ELECTRICAL utility grid system cannot readily accept connection of new generation plant without strict conditions placed on voltage regulation due to real power fluctuation and reactive power generation or absorption, and on voltage waveform distortion resulting from harmonic currents injected by non-linear elements of the plant. The standard IEC 61400-21 (*International Electro technical Commission*) guidelines for power quality issues related to wind turbine [1]-[3]. Major issues related to the power quality are

a) Voltage fluctuation on grid

b) Switching operation of wind turbine on grid

c) Voltage dips on grid.

d) Reactive Power

e) Harmonics

In conventional variable speed wind turbine connection to the

---

S.W Mohod is with Electronics Engineering Department, Ram Meghe Institute of Technology & Research, Badnera, Amravati Maharashtra India (e-mail: shardmohod@rediffmail.com).

M.V. Aware is with the Department of Electrical Engineering, Visveshvaraya National Institute of Technology, Nagpur, Maharashtra, INDIA (e-mail: mva_win @ yhaoo.com).

grid is as shown in Fig. 1.

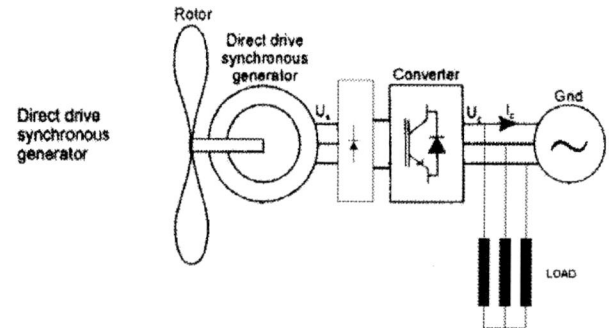

Fig. 1. Wind Energy Conversion System.

## II. WIND POWER CONVERTER MODELING

The wind power conversion system is shown in Fig. 1. The building blocks for this include the wind turbine, converter inverter system and grid having non-linear load. The study of steady state and dynamic performance is carried out by modeling these components in MATLAB / SIMULINK with appropriate control scheme.

### A. Turbine-Generator and rectifier Modeling

The variable speed turbine with generator is typically having characteristics as indicated in Fig. 2.

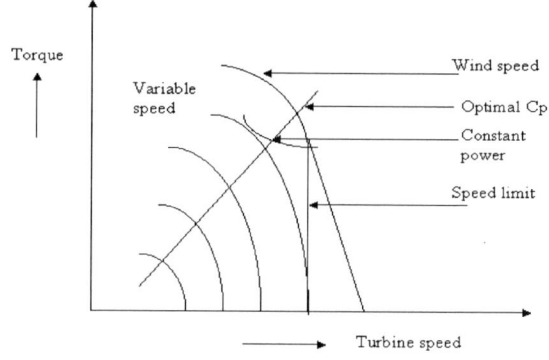

Fig. 2. Variable Speed Operating Curve.

This is interfaced with the rectifier unit to get dc bus voltage. The variable dc source having characteristics with in the normal operating region is shown in Fig. 3.

---

0-7803-9771-1/06/$25.00 ©2006 IEEE

The dc link voltage $V_d$ and current $I_d$ are related and presented as

$$E_d = V_d - I_d.R \qquad (1)$$

where $E_d$ equivalent regenerated voltage and $V_d$ is the dc bus voltage. The power flow is represented with dc bus current $I_d$ for the constant dc bus voltage. The equivalent resistance R represents for the generator and rectifier.

The dc equivalent can be expressed as a function of frequency and power from the turbine. The power equation for the wind turbine is presented as under.

$$P_{windturbine} = 0.5 C_p \rho A V^3 \qquad (2)$$

where $\rho$ (kg/m$^3$) is the air density and A (m$^2$) is the area swept out by turbine blade, Cp is the power coefficient, depends on type and operating condition of wind turbine.

This is modeled in SIMULINK as a variable dc source connected to the dc bus having capacitor across it and represented in Fig. 3.

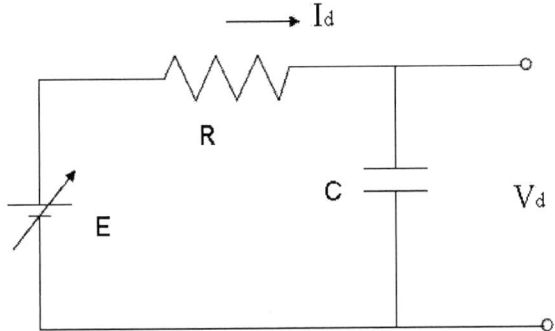

Fig. 3. Equivalent Model for Generator-Rectifier.

### B. Inverter Modeling

The generator and rectifier system is uncontrolled and requires the very efficient power electronics system to transfer the power to the grid. Many types of the power electronics interfaces topologies are available [4]-[6]. The main control objective is to track wind speed to capture and transfer the maximum power to the grid. The major parameter that is sensitive to the power flow across the dc link is the dc bus voltage. This requires being constant to operate the voltage source inverter satisfactorily. The dc bus voltage can be maintained constant by regulating the turbine control. The demand of load as when it is increased or decrease can be felt by change in the dc bus voltage. This is incorporated in the control to regulate the wind turbine. This enables the system to support the varying real power of the load demand. The voltage source inverter is operated in current control mode.

The three-phase inverter models with IGBT switches are built using Power System Block Set of SIMULINK. The dc bus capacitor rating is chosen to withstand the circulating currents while operating with more of reactive power support. The dc bus voltage is more than the peak of the inverter output voltage. The interface of the inverter with the grid is through the transformer. This is represented as equivalent inductor in the simulation.

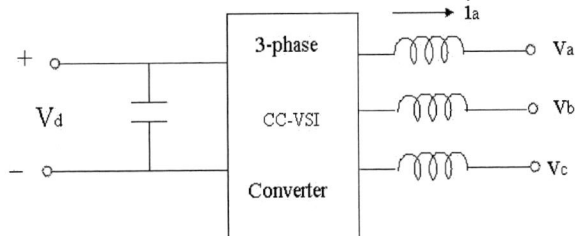

Fig. 4. Voltage Source Inverter Model.

### C. Grid with Non linear Load Modeling

A simple radial power system grid having non-linear load is used for the study. The three phase three wire 50 Hz supply to the dc drive is modeled in the SIMULINK using PSB. The power flow during the steady state and under dynamic condition is incorporated to test the efficiency of the designed controller. The system configuration with the load and wind generator is shown in the Fig. 5.

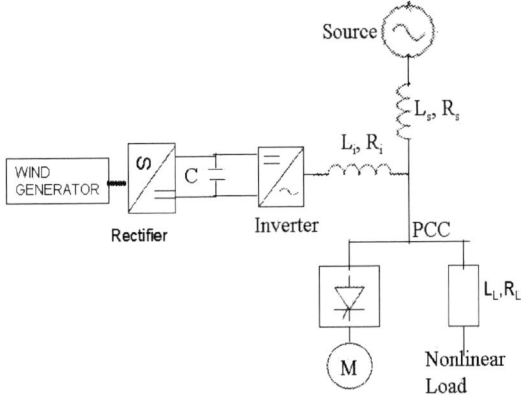

Fig. 5. System with nonlinear load ( Thyristor Controlled d.c. Drive).

### III. PROPOSED CONTROL SCHEME

The voltage applied at the point of common coupling is $v'_{sa}$ $v'_{sb}$ $v'_{sc}$ with coupling transformer resistance ($R_i$) and inductance ($L_i$). The DC voltage ($V_d$) is placed across the capacitance (C) of the inverter. The basic circuit is shown in Fig. 2. The current controlled mode of inverter operation is presented as

$$\frac{di_{ia}}{dt} = -(R_i)i_{ia}/L_i + (v'_{sa} - v_{ia})/L_i \qquad (3)$$

$$\frac{di_{ib}}{dt} = -(R_i)i_{ib}/L_i + (v'_{sb} - v_{ib})/L_i \qquad (4)$$

$$\frac{di_{ic}}{dt} = -(R_i)i_{ic}/L_i + (v'_{sc} - v_{ic})/L_i \qquad (5)$$

$$\frac{dv_{dc}}{dt} = (i_{ia}S_A + i_{ib}S_B + i_{ic}S_c)/C \qquad (6)$$

where $v_{ia,b,c}$ and $v'_{sa,b,c}$ are the inverter and PCC voltages respectively. Switching signals are obtained by comparing reference currents $i^*_{sa}$, $i^*_{sb}$, $i^*_{sc}$ with actual currents $i_{sa}$, $i_{sb}$, $i_{sc}$.

783

Current errors $\Delta i_a$, $\Delta i_b$, $\Delta i_c$ are applied to the hysteresis controllers that produces the correct signal to switch the power electronics switches ON and OFF. The characteristics of the switching function $S_A= f(\Delta i_a)$ of a hysteresis controller for phase A of the inverter is shown in Fig. 6. The characteristic constitutes a hysteresis loop block that can be described as

$$S_A = \begin{cases} 0 ....if ...\Delta i_a < -\dfrac{h}{2} \\ 1 ....if ...\Delta i_a > \dfrac{h}{2} \end{cases} \qquad (7)$$

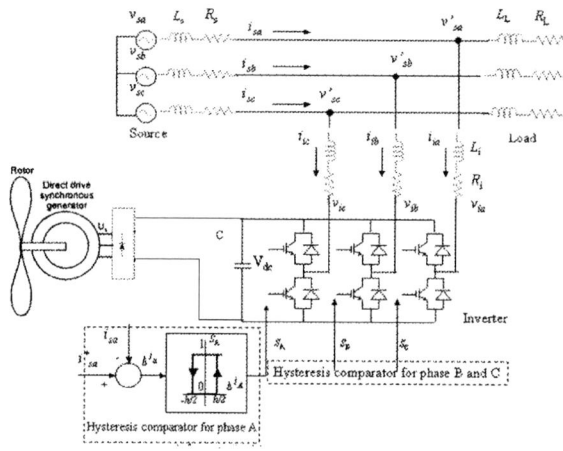

Fig. 6. Basic Control Scheme for Inverter in Wind Conversion System.

where $h$ denotes the width of the loop and $S_A$=0, 1 indicates the state of switch.

The required source currents are to be controlled to be sinusoidal and in phase with the mains voltage in spite of the load characteristics. Therefore the reference current for the comparison must be derived from the source voltage. These currents can be expressed as

$i_{sa} = I\sin(\omega t)$

$i_{sb} = I\sin(\omega t - 120^0)$ $\qquad$ (8)

$i_{sc} = I\sin(\omega t - 240^0)$

where, $I$ is proportional to the magnitude of the filtered source voltage of phase 'a'. This ensures that the source current is controlled to be sinusoidal irrespective of whether the source voltage is unbalanced. The magnitude of the reference current is controlled to ensure that the DC capacitor voltage remains constant [7].

In the proposed controller, the active power of the load in the steady state should be supplied from the source only and therefore the inverter will only provide the necessary reactive power. Among the various current control techniques, the hysteresis current control is the simplest and most commonly used method. The hysteresis control scheme provides excellent dynamic performance and inherent peak current limiting capability. The main drawback of hysteresis control schemes is variable switching frequency [8]-[9].

## IV. SYSTEM PERFORMANCE

The proposed scheme with its controller action is simulated in MATLAB / SIMULINK. The basic scheme involves the variable active and reactive load at PCC. The performances under various test conditions are presented below. Only one-phase voltages and current is indicated to get more clarity in the results.

### A. Steady state performance

The wind generation system is connected with the grid having inductive load. The simulation is performed with the system parameters given in the Table I.

TABLE I
SIMULATION PARAMETERS FOR THE WIND-GERERATION SYSTEM

| Source Voltage | 200Volts | Load (KW) | 3000KW |
|---|---|---|---|
| Wind-Generating Capacity | 10KW | Load KVAR | 1500Var |
| D.C.Motor Rating | 3KW | Operating frequency | 50 HZ |
| Inverter DC link Voltage | 400Volt | Inverter Rating | 10 KW |

The results of source current and source voltage are shown in the Fig. 7 (a). The load and inverter current is also shown in Fig. 7 (b). The power flow is shown in Fig. 8. This is representing the real and reactive power flow at point of common coupling.

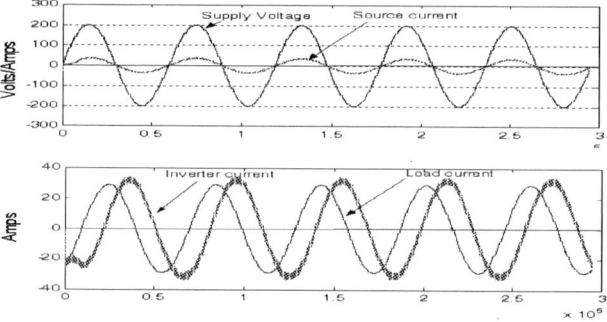

Fig. 7.(a) Supply Voltage and Source current,(b)load and inverter.

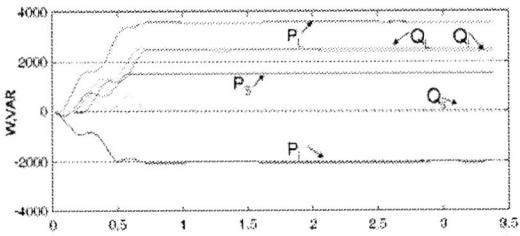

Fig. 8. Power Flow in Steady State ($P_s$, $P_l$, $P_i$) Real Power and ($Q_s$, $Q_l$, $Q_i$) Reactive Power form source, load and inverter.

The source current is maintained in phase with the source voltage indicating the unity power factor. The power flow indicates the wind power contribution to support the load. The reactive power is totally taken care by the inverter current as shown in Fig. 7 (b).

### B. Performance with step change in load

The dynamic performance is carried out by step change in load. The load connected to the system is inductive. The control loop with PI controller is designed to hold the set value of the dc voltage across the dc bus constant. The performance of the designed controller is shown in the Fig. 9 (a) and (b) with power exchange during this process is shown in the Fig. 10.The control action under this can regulate the source side current. This ensures the utilization of the power distribution network to its optimum rating.

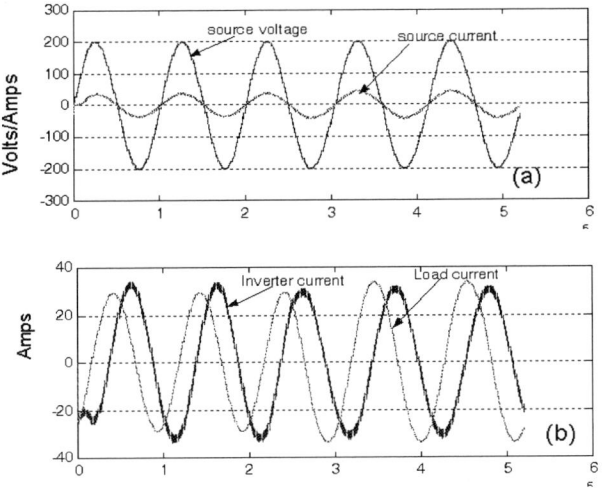

Fig. 9. Load Step Change (a -source voltage and current, b-load current and inverter current).

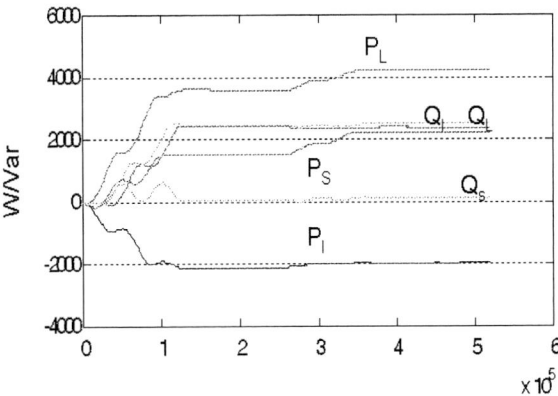

Fig.10. Real and Reactive Power under step load change.

### C. Performance under non-linear load

The system is supplying the dc drive load. This dc drive is thyristor controlled and causing the distortion in the supply current. The inverter compensates the distorted load current.

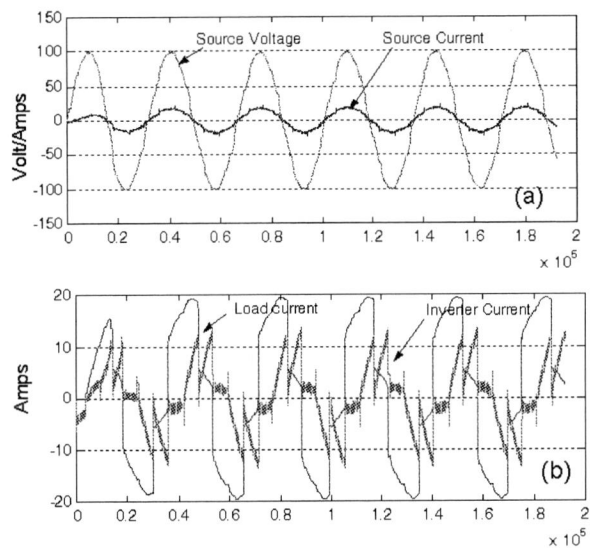

Fig. 11. Performance under nonlinear load (a- source voltage and current b- load and inverter current).

The control with this scheme makes the incoming source side current in phase with the source voltage representing the unity power factor performance. The steady state performance is indicated in Fig. 11(a) and (b). The non-linear load is supported for its reactive power demand by the inverter. The unity power factor and real power support make this scheme more promising to improve the grid power quality. The experimental study is conducted to investigate the effectiveness of proposed scheme.

## V. EXPERIMENTAL RESULTS

The operation of the combined is tested with the inverter connected with the distribution network at the PCC. The control algorithm is implemented through the digital signal processor (DSP TMS320C31). The inverter built with the IGBT switches. The performance of the connected system is shown in Fig. 12. The non-linear load with out wind power connection is shown in Fig. 12 (a). The action of the connected wind turbine is shown in Fig. 12 (b). This indicates the improvement in source current. It is sinusoidal in nature and in phase with the voltage. This maintains the unity power factor on the source side. The real power from the wind generator is expected when the load demand is increasing. This exchange is closely control by allowing full exploiting the thermal limit of the source side power network. The real power support is indicated as it can be observed from the inverter current. This not only tries to compensate the harmonic current but also provides the reactive power component demands from the load.

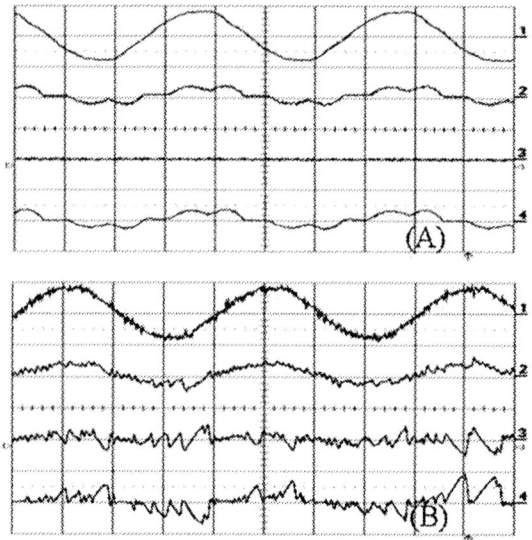

Fig. 12. Source voltage, source current, inverter current and load current (a-without active filter action. b-with active filter action).

## VI. CONCLUSION

This wind conversion system is performing satisfactorily for a given load condition while maintaining the power quality norms. The front end control of VSI can be implemented on DSP with DC link voltage and source voltage as a feedback signals. Application of such control gives reactive power support so that the grid voltage is maintained constant. Also the real power is pushed from wind side into the interconnected system.

This relives the burden of the existing power system so that it can have more transient stability margin. The experimental results are in full agreement with the simulation results. This also satisfies the IEC 61400 21 norms for the power quality.

## VII. ACKNOWLEDGMENT

Author's thanks for support provided by the Electrical Drives Laboratory, Electrical Engineering Department of Visvesvaraya National Institute of Technology, Nagpur (INDIA). This work is carried out as a part of research project.

## VIII. REFERENCES

[1]  W.E Reid-' Power Quality Issues –Standards and Guidelines', *IEEE Trans. on Industry Application*, pp 625-632, 1996
[2]  Thiringer, T. Petrup 'Power Quality Impact of sea located Hybrid Wind Park', *IEEE Tran. on Energy Conversion* , pp 23-25 , 2001
[3]  J. O .Q. Tande 'Applying power quality characteristics of wind turbine for Assessing impact on voltage quality', *Wind Energy*, pp 52, 2002
[4]  A. B .Raju , B.G. Fernades , K. Chatterjee .' A UPF Power Conditioner with Maximum Power Point Tracker for Grid Connected Variable Speed Wind Energy Conversion System' conference, pp 107-112, 2002
[5]  Z. Chen, E. Spooner,'Grid Power Quality with Variable Speed Wind Turbines', *IEEE Trans* 2001on *Energy Conversion*, vol.16, No .2, pp 148-154, June 2001
[6]  Juan Manel Carrasco, 'Power Electronic System for Grid Integration of Renewable Energy Source: A Survey', *IEEE Trans on Industrial Electronics*, vol. 53, No. 4, pp 1002-1014, August 2006
[7]  Hossein Madadi Kojabadi, Liuchen Chang, 'Development of a Novel Wind Turbine Simulator for Energy Conversion System Using an Inverter-Controlled Induction Motor ', *IEEE Trans .on Energy Conversion*, vol. 19 No. 5, Sept. 2004.:
[8]  Z. Lubosny, 'Wind Turbine Operation in Electric Power System' Advance Modeling, New York: Springer, 2003
[9]  Ned Mohan, T. M. Undelend, William. Robins-'Power Electronics' Converters, Application & Design, pp 464-502 ,1995
[10]  Mannual, MATLAB / SIMULINK user guide, MathWorks inc. USA, 1995

**2006 IEEE International Conference on Power Electronic, Drives and Energy Systems**

# Investigations on Combined Operation of Industrial Distribution System and utility in Distributed Generation Environment

K. Manjunatha Sharma, K.P. Vittal, *Member, IEEE*, and T.K. Nagaraja Rao

*Abstract*— The deregulation regime has given lot of scope for independent power producers to feed power to the utility nearest to the load center. This feature provides opportunities for lot of improvements in distribution system operation. In this paper investigations have been done for integrated operation of the practical MRPL industrial distribution plant having its own captive generation to act as distributed generation source serving neighboring utility loads. The performance of the system is simulated in both cases of industry acting as source and sink depending on the load cycle. The simulation results will serve as useful tool for decision making in power trading during ABT regime.

*Index Terms*— ABT regime, Distributed Generation, Independent Power Producers, Power Flow Studies, Power Trading, Voltage Profile.

## I. INTRODUCTION

THE power distribution networks have recently acquired enormous importance in order to achieve efficient operation of electrical system. Distribution Automation (DA) is a promising tool to address the problems associated with the operation and control of the networks.. Because of the complexity of the network, their simulation studies is always a challenging field of research. Recently developments in distribution automation have brought the opportunity to offer higher quality service by means of the implementation of new operation functions. Many applications, such as network optimisation, reactive-power planning, feeder reconfiguration, state estimation, short- circuit analysis are incorporated in the

---

This work was supported by Ministry of Human Resource and Development under Research and Development Scheme.

K, Manjunatha Sharma is with the Department of Electrical and Electronics Engineering, National Institute of Technology Karnataka, Surathkal – 575025, India (e-mail: manjusuma@yahoo.com)

Dr. K.P. Vittal is Assistant Professor in the Department of Electrical and Electronics Engineering, National Institute of Technology Karnataka, Surathkal – 575025, India (e-mail: vital_nitk@yahoo.com)

T.K. Nagaraja Rao is working as plant engineer at MRPL, Mangalore and PG scholar in the Department of Electrical and Electronics Engineering, National Institute of Technology Karnataka, Surathkal – 575025, India (e-mail: tknagaraj@rediffmail.com)

design of distribution automation scheme effectively. While carrying out analysis based on simulation of distribution networks, building the appropriate model is very much essential to obtain results matching with the field values. The industrial distribution system having its own captive power generation need to focus on the power generated by the generators and the power consumed by the load during varying load cycle and to achieve optimum system losses. This aspect necessitates the modeling of the entire distribution system and performing power flow analysis at various typical instants in order to arrive at network operation and control strategies.

The load flow calculation for electrical systems is one of the most vital aspect and forms the knowledge base for the energy management in the system.. Its accuracy is crucial to power system security and stable operation. The possible errors during the power flow may be due to the following factors :

- The mathematical model of each component (such as: transformers, distribution feeders) is not consistent with the actual situation.
- The parameters of each component is not consistent with the actual situation.

Mathematically the load flow requires a solution of a system of simultaneous nonlinear equation.

The radial distribution system power flow has also received large amount of work by researchers. The aim is at arriving at faster convergence, making the technique suitable for on-line applications. The system data poses severe constraints on the computations. The forward backward substitution method has been accepted and widely used because of simplicity and less computation time [1].

The distribution system need to be monitored carefully for efficient and safe operation. This aspect is much essential in industrial plant where continuity and quality of supply need to be ensured. The control strategies to be used in industrial distribution system are governed by the load forecasting, power flow analysis, network reconfiguration and protection methodologies. The simulation of the industrial distribution network under various operating conditions help the decision making process and appropriate control action can be commanded for system improvement [4].

---

0-7803-9771-1/06/$25.00 ©2006 IEEE

The reforms in the distribution sector has been launched globally by deregulation policy. In India, Electricity Act 2003 [7] which has come into force from 2003 aims at restructuring of the distribution system. The act provides for establishment of competitive distribution companies, independent power producers, strict guidelines for power quality. Since the establishment of this act regular amendments are being done, and the nation is moving towards implementation of the act. The Government of India is taking all possible measures to improve the performance of the distribution system to stay tuned with global trends. Since past two years the act has resulted in better service, good governance from electricity regulatory commissions and more benefits are expected in forthcoming years.

The deregulation policy emphasizes on Distributed Generation or Co- generation where in generators are hooked on to the distribution network to improve the system performance. As accepted at international level since 2003, distributed generation has brought dawn of a new era [2] [3]. The technical benefits of distributed generation are reduced line losses, voltage profile improvement, increased overall energy efficiency, enhancement of system reliability and security, improved power quality, relived T & D congestion. The economic benefits are deferred investments for upgrades or facilities, reduced O & M costs of DG technologies, increased security for critical loads, peak load shaving.

The power trading depending on the load cycle can be done efficiently with this simulation tool [5][6], which is a quite important aspect in ABT regime. This paper presents the simulation studies on Industrial Distribution System which comprises of its own captive generation and industrial load and this installation is serving as distribution generation source feeding the neighboring utility. The simulation results validate the feasibility of the integrated operation and the effective operation of both the industry and utility owing to diversified load cycles.

## II. SKM POWER TOOLS PACKAGE

The Load Flow Study conducted on SKM Power Tools Package [8], predicts the overall apparent real and reactive power distribution throughout a power system, including associated losses. Additionally, the study calculates the voltage drop through each branch impedance component, and the associated voltages at each bus or node in the electrical system.

The solution depends on the system topology, combined with knowledge of associated branch impedances and load data. The formation of the appropriate matrices and, through optimal ordering and standard matrix algebra techniques, solves for the dependent variables. The power flow solution technique used is the double current injection method. In this method, the first estimate assumes no losses and calculates the current flows in each branch, given the load values and system nominal voltages. Subsequently the system losses are calculated, the voltage drop is determined for branch and bus.

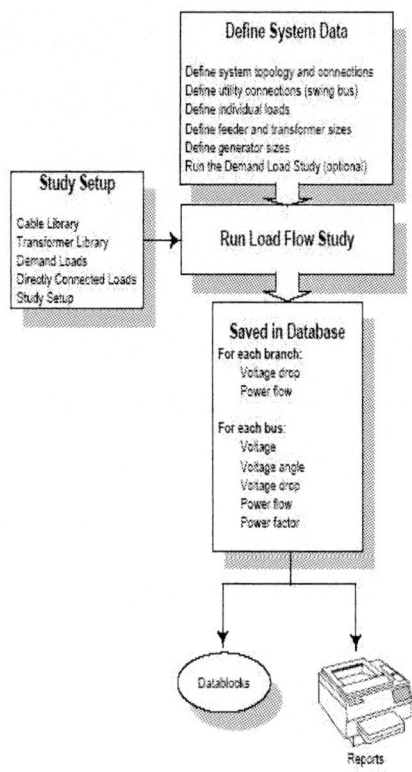

Fig. 1. Flow chart with features pertaining to load flow studies.

Given this new voltage at each bus, the load currents are re-calculated, and the iterative process begins. The new currents develop new losses in the branches and thus new voltage drops in each branch and bus. The iterative process continues until there is little change in the voltage at each bus between estimates, and convergence is achieved. Transformer primary and secondary tap settings and transformer off-nominal voltages are considered in the steady-state load flow solution. These transformer settings are to be dynamically controlled through automation scheme to obtain optimum operation of the system.

The load flow solution takes into account load characteristics to calculate the apparent load flow conditions in the distribution system. The load flow conditions are solved in harmony with solution of the voltage conditions at each load bus. The type of loads are specified and the system losses significantly influence the results of the load flow and voltage drop calculations. The modeling of the load points are crucial in deciding the condition of the network. Hence it is necessary to account for different characteristics of loads present in the plant. The load cycle variations also poses a challenge in the network operation owing to different pattern of loads coming over. Hence the strategies of network operation need to be decided on the load demand level as well as characteristics of the load at any typical instants.

The total voltage drop in any one branch or the total bus voltage drop is calculated as per NEC standards, USA. Thus, it is critical to know the voltage drop in each branch of the power system, and the total voltage drop from the source of supply to the bus in the branch circuit. The voltage drop calculations are incorporated directly into the calculation of the steady-state load flows. Before carrying out the load flow study it is required to fulfill the following requirements :

· Topology, and connections are to be defined

· Utility connection (swing bus) is to be defined

· Individual loads to be defined

· Feeder and transformer sizes are to be defined

· Generator sizes are to be defined

### III. SHORT CIRCUIT STUDIES

The short circuit study models the current that flows in the power system under abnormal conditions and determines the prospective fault currents in an electrical power system. These currents must be calculated in order to adequately specify the ratings of the protection apparatus. The study results are also used to selectively coordinate time current characteristics of electrical protective devices.

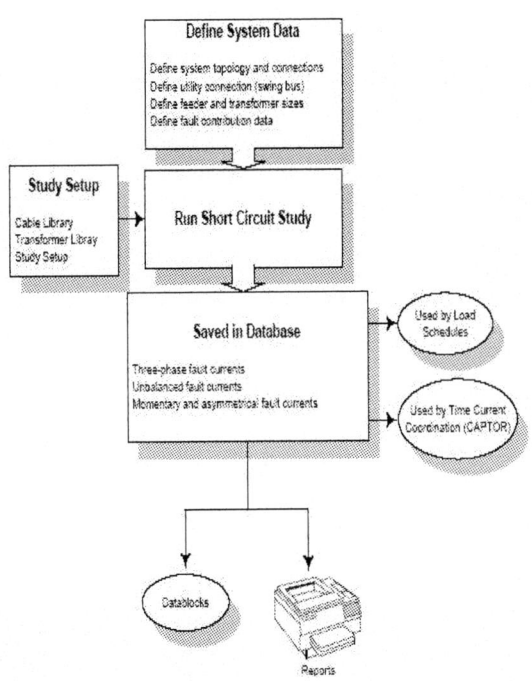

Fig. 2. Flow chart with features pertaining to short circuit studies.

The short circuit study requires all the data pertaining to the power flow studies, in addition the fault contribution data need to be specified. The model developed will be useful for

various types of fault analysis in the distribution system. During short circuit study run, check is done for appropriate feeder sizes and lengths, and transformer sizes in the library. If the data is inappropriate or missing, error and warning messages are shown in the study Run dialog box which are included in the report. The positive-sequence impedance of the cable and one-way circuit length is specified. The modeling is done treating negative-sequence impedance as equal to the positive-sequence impedance. If a cable's zero-sequence impedance is zero, the short circuit study uses the positive-sequence value. Cable positive and zero sequence impedances may be selected from the cable library, or can be defined them in the component editor.

If the cable user defined, then specific cable impedance in ohms per 1000 feet or ohms per 1000 meters need to be specified. Cable lengths must be entered in the same units as the cable impedance data (feet or meters). Cable impedances are unaffected by the wire circuit description characteristics.

### IV. MOTOR TRANSIENT ANALYSIS

The flow chart provides a quick overview of the necessary steps in motor transient analysis.

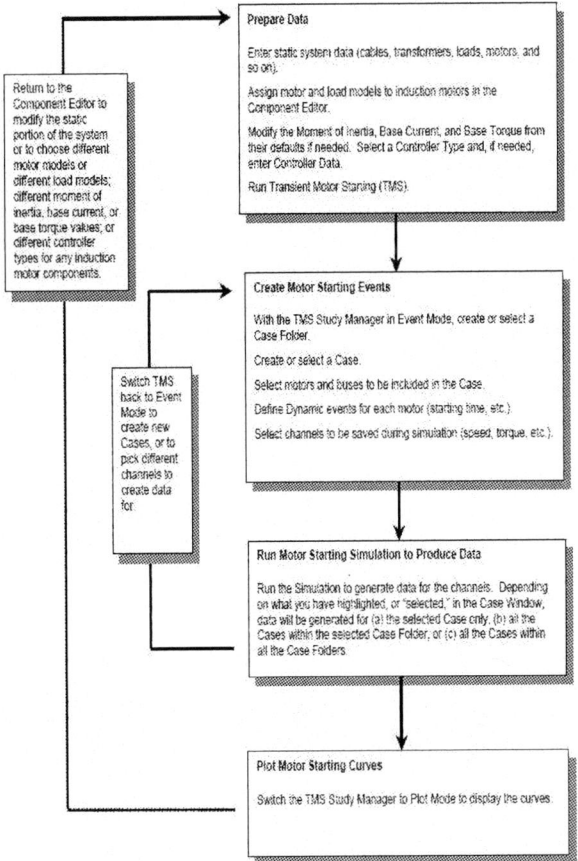

Fig. 3. Flow chart with features pertaining to motor starting studies.

- Data Preparation: system data - cables, transformers, all types of loads and devices.
- assigning dynamic motor and load models of induction motors in the component editor; and running Transient Motor Starting engine
- Creating cases: This include, with TMS in Event Mode, creating or selecting a Case; selecting motors and buses for including in the dynamic Case events. Defining Dynamic events for each motor (starting time, etc.). Selecting motor data for saving during simulation (speed, torque, etc.).
- Running TMS study engine to produce data : This includes running the Study engine to generate data for the channels. Depending on requirement "selected," data will be generated for (a) the selected case only, (b) all the cases within the selected case folder, or (c) all the cases within all the case folders.
- Plotting motor starting curves : This was included for switching TMS to plot mode to display the plot curves. (For creating new cases, or to pick different channels to create data for, return to step of creating cases. To choose different motor models, different load models, different moment of inertia values, or different controller types for any induction motor components, return to data preparation step.)

## V. CASE STUDIES AND RESULTS

The electrical power distribution system of Mangalore Refineries and Petrochemicals Ltd. (MRPL) is simulated using SKM power tools. The system comprises of the generators, transformers, switchgears, cables, motor loads. The plant layout has been categorised as Phase – I , Phase – II. These Phase – I and II can operate independently and also in integrated manner. The industry can either import power from the utility or export power to it. Considering these probabilities the case studies have been formulated as below.

- Case Study 1 : Phase – I Generators and Load Exists
- Case Study 2 : Phase – II Generators and Load Exists
- Case Study 3 : Both Phase – I and Phase – II Generators and Load Exists, Integrated Operation of Phase – I and Phase – II.
- Case Study 4 : Integrated Operation of Phase – I and Phase – II with Power Import from the Utility Grid
- Case Study 5 : Integrated Operation of Phase – I and Phase – II with Power Export to the Utility Grid

In each of the above cases, the parameters of the plant are tuned to meet the power flow constraints. The simulation results obtained are matching with the field results for the normal configuration considered as case study 3. The voltage profile is found conforming to the industrial standards.

The lay out of the MRPL industrial system and power flow results obtained are presented in the following section.

Case Study 1 :
Generation - STG 1 : 7.69 MW, STG 2 :13.00 MW
   Plant load demand : 20.57 MW,
   System losses : 0.12 MW.

Case Study 2 :
Generation - STG 3: 6.73 MW, STG 4 :5.20 MW,
   STG 5 : 5.20 MW,
   Plant load demand : 16.90 MW,
   System losses : 0.23 MW.

Case Study 3 :
Generation – STG 1 : 14.00 MW, STG 2 : 12.68 MW,
   STG 3 : 9.01 MW, STG 4 : 0 MW,
   STG 5 : 14.0 MW,
   Plant load demand : 49.27 MW,
   System losses : 0.42 MW.

Case Study 4 :
Generation – STG 1 : 15.57 MW, STG 2 : 0 MW,
   STG 3 : 9.01 MW, STG 4 : 0 MW,
   STG 5 : 14.00 MW,
   Import from Utility : 11.30 MW,
   Plant load demand : 49.27 MW,
   System losses : 0..60 MW.

Case Study 5 :
Generation – STG 1 : 24.16 MW, STG 2 : 0 MW,
   STG 3 : 13.00  MW, STG 4 : 0 MW,
   STG 5 : 15.00 MW,
   Export to Utility :2.28 MW,
   Plant load demand : 49.27 MW,
   System losses : 0..60 MW.

In all the configurations, the network control in done so as to maintain good voltage profile and ensuring scope to handle contingent conditions arising in the industrial plant.

Short circuit analysis indicated that the protection system is capable of handling the fault conditions both under independent operation of the plant and combined operation of the plant and utility.

The schematics of the Phase I and II of MRPL electrical power distribution system is shown in Fig. 4 and Fig. 5.

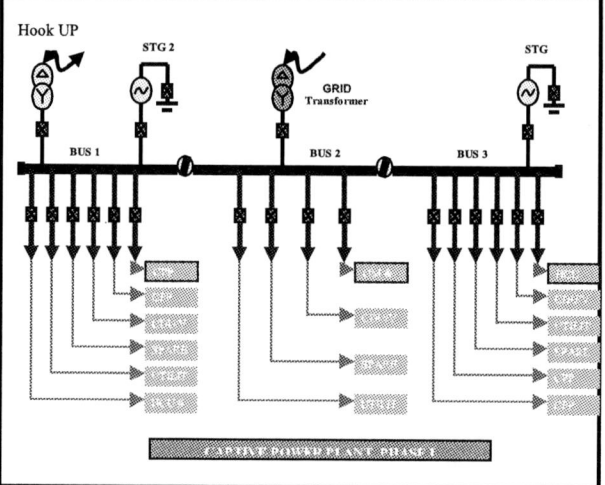

Fig. 4. MRPL industrial power distribution system – lay out of phase – I.

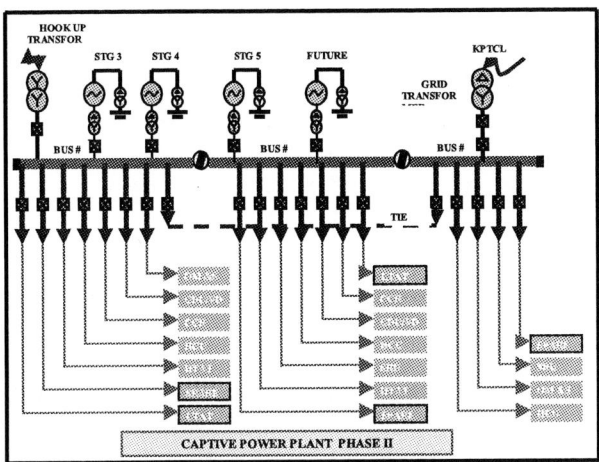

Fig. 5. MRPL industrial power distribution system – lay out of phase – II.

Motor starting analysis indicated that the parameters of the network are well within limits with the combined operation of the plant and utility and varying load cycle can be handled effectively with network control strategies. It is planned to extend the analysis for techno-economic feasibility studies.

## VI. ACKNOWLEDGEMENT

The authors acknowledge the support extended by MRPL and NITK, Surathkal authorities for carrying out this work.

## VII. CONCLUSIONS

In this paper, simulation studies have been carried out for performance analysis of Industrial Plant serving the neighboring utility, acting like distributed generation source. In ABT regime, power trading is a vital factor for determination of the plant operation and efficiency. The simulation results have shown the feasibility of such operation. The detailed analysis for exploring the economics of the combined operation of the plant and the utility is proposed in future.

## VIII. REFERENCES

[1] D. Thukaram, H.M. Wijekoon Banda, Jovitha Jerome, "A Robust power flow algorithm for distribution systems," *Electrical power system research*, vol 50, pp. 227-236, 1999.

[2] Hans B Puttgen, Paul R Macgregor, Frank C Lambert, "Distributed Generation : Semantic Hype or the Dawn of a new era? ", *IEEE Power and Energy Magazine,* pp 22-29, Feb. 2003

[3] Roger Dettmer, "Empowering Distributed Generation", *IEEE Review*, pp51-53, Jan. 2003

[4] Terence hazel Isabelle Condamine, Fabrice Audemard "Facilitating Plant operation and maintenance using an electrical network monitoring and control system simuation tool", in *Proc. 2004 IEEE Petroleum and Chemical Industry Technical Conf.,* pp 361 – 369.

[5] Hugo Morais, Marilo Cardoso, "A Decision Support Simulation Tool for Virtual Power Producers," presented at IEEE Int. Conf. on future Power Systems, Netherlands, 2005

[6] G Strbac, J Mutale, D Pudjianto, "Pricing of Distribution Networks with Distributed Generation," presented at IEEE Int. Conf. on future Power Systems, Netherlands, 2005

[7] Ministry of Power, Government of India, "Electricity Act 2003.",

[8] SKM System Analysis Inc., USA, "SKM Power Tool for Windows".

**2006 IEEE International Conference on Power Electronic, Drives and Energy Systems**

# Rotor Speed Stability Analysis of Constant Speed Wind Turbine Generators

M. G. Kanabar, *Student Member, IEEE*, C. V. Dobariya, *Student Member, IEEE*, and S. A. Khaparde, *Senior Member, IEEE*

*Abstract*—This paper presents an analysis of rotor speed stability of a constant speed wind turbine generator with active stall control. To analyze the rotor speed stability, a 3-phase short circuit fault on a sample system with a constant speed WTG has been simulated using DIgSILENT software package. From the simulation results, it has been shown that the operating point of wind turbine generator has influence on rotor speed stability. Using this phenomenon, how a constant speed wind turbine generator with active stall controls the operating point has been reported. Negative pitch control strategy of active stall controlled wind turbine generator to enhance rotor speed stability has been described with the help of simulation results.

*Index Terms*—active stall control, constant speed (squirrel cage) induction generators, power system stability, wind turbine generator (WTG).

## I. INTRODUCTION

AMONG all renewable resources, wind power is the most booming renewable technology all over the world. India ranks fourth in the world with total installed capacity of more than $5,000$ MW [1]. Most of the wind turbine generators (WTGs) installed in India are constant speed (squirrel cage) induction generators. This is because of their robustness, mechanical simplicity and low price. However, a constant speed WTG always demands reactive power, hence reactive power compensation is needed.

During grid disturbance near to a WTG, severe voltage sag in the connecting network may cause a significant reduction in active power generation and rise in rotor speed. After voltage recovery, the rotor speed of the induction generator may be so high that it does not return to the pre-fault value. This may lead to rotor speed stability problem [2].

As the penetration level of constant speed WTGs increases, it is very much important to maintain the rotor speed stability during low voltage at point of common coupling (PCC). This is referred as low voltage ride through (LVRT) capability of a WTG. Normally, LVRT requirements are stringent in regions with high penetration of the wind power. The specific requirements like voltage level and duration of fault differ from country to country [3], [4].

A controlled constant speed WTG consumes reactive power, and this consumption ramps up drastically during faults. Therefore, such a WTG does not possess LVRT capability.

M. G. Kanabar, C. V. Dobariya, and S. A. Khaparde are with the Department of Electrical Engineering, Indian Institute of Technology Bombay, Mumbai, India, (emails: mital@ee.iitb.ac.in, chandv@ee.iitb.ac.in, sak@ee.iitb.ac.in,).

Consequently, it has to be disconnected from the grid due to rotor speed instability. The rotor speed stability of a constant speed WTG can be improved by active stall control. In transient condition, active stall controller controls the pitch angle ($\beta$) in negative direction to reduce turbine torque, this action helps to reduce acceleration in the rotor speed, and improves the rotor speed stability.

In [5]–[8] efforts have been put to model the WTGs, and to understand the behavior of constant speed WTGs. Pitch controller of wind turbine for optimum generation control has been explained in [9]. Application of the pitch control mechanism of a WTG for voltage recovery, subsequent to a short-circuit fault has been explained in [10]. Comparative study of active stall control with pitch control strategies for constant speed WTGs are discussed in [5], [11]. Transient stability of a fixed speed wind turbines has been evaluated in [12], [13].

This paper evaluates the rotor speed stability of a constant speed WTG with active stall connected to a sample system. The system has been simulated using DIgSILENT Power-Factory, which is a computer aided engineering tool for the analysis of industrial, utility, and commercial electrical power systems [14]. Using negative pitch control strategy of active stall controlled WTG, the enhancement of the rotor speed stability has been demonstrated for the sample system.

The organization of the paper is as follows. In the section II, the phenomenon of rotor speed stability has been described. Negative pitch control strategy of active stall control based constant speed WTG is described in section III. Validation of 1.3 MW squirrel cage induction generator model of DIgSILENT is detailed in section IV. Simulation of a squirrel cage induction generator with capacitor bank connected to the sample system, and effect of negative pitch control strategy on rotor speed stability are discussed in section V. Section VI concludes the paper.

## II. ROTOR SPEED STABILITY OF A SQUIRREL CAGE INDUCTION GENERATOR

When a grid connected induction generator is subjected to a nearby fault, due to severe voltage sag, its rotor may accelerate to very high speed far from the system frequency. This phenomenon is related to the power system stability, but it is not covered in conventional stability concepts, such as, rotor angle stability, voltage stability and frequency stability [2]. Because of induction generator, stability of a WTG cannot be classified under rotor angle stability phenomenon. After fault clearance, the system voltage may be recovered to a new allowable value, but the speed of induction generator may rise to unaccepted value. Hence, this cannot be classified under the voltage stability phenomenon. The frequency of the system after fault clearance may be acceptable, so it cannot be classified under frequency stability phenomenon. Thus, this kind of stability phenomenon is referred as rotor speed stability.

According to [2], *Rotor speed stability refers to the ability of an induction (asynchronous) machine to remain connected to the electric power system and running at a mechanical speed close to the speed corresponding to the actual system frequency after being subjected to a disturbance.*

The rotor speed stability of a constant speed WTG depends on the several factors such as, machine operating point, short-circuit power at PCC, distance from the fault location, rotor inertia etc [12]. Out of these factors, active power output (operating point) can be controlled by pitching wind turbine blades. By reducing output power of a constant speed WTG, reactive power drawn by the WTG reduces (due to reduction in slip). And, hence it may improve rotor speed stability margin of the WTG. Using this phenomenon, enhancement of the rotor speed stability of a constant speed WTG with active stall is explained in the next section.

## III. ACTIVE STALL CONTROL OF A CONSTANT SPEED WTG

The active stall control based constant speed WTG (referred as Type-A2 wind turbine technology) has control of pitch in the negative direction (i.e. between $-90^o$ to $0^o$) with respect to pitch control based WTG (Type-A1). The rate of negative pitch control is normally less than $5^o$ per second. Although the pitch rate may exceed $10^o$ per second during emergencies [5].

Fig. 1 shows the difference in the direction of blade rotation between the pitch controlled and the active stall controlled constant speed WTGs. In the figure, the chord line is the straight line connecting the leading and trailing edges of an airfoil. The plane of rotation is the plane in which the blade tips lie as they rotate. The pitch angle ($\beta$) is the angle between the chord line of the blade and the plane of rotation. And, the angle of attack ($\alpha$), is the angle between the chord line of the blade and the relative wind or the effective direction of air flow.

In active stall control, at low wind speeds, the machine is usually controlled to pitch their blades similar to a pitch controlled machine. When the machine reaches its rated power value, the blades are pitched in the direction opposite from what a pitch controlled machine does, in order to control output power. This needs pitch angle $\beta$ to be decreased typically by a small amount only. Hence, the rating of pitch drives is less for active stall control, as compare to pitch control [11]. Therefore, the cost and complexity are less for active stall control, comparatively.

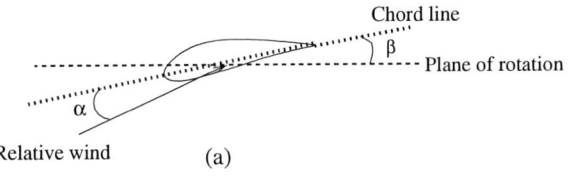

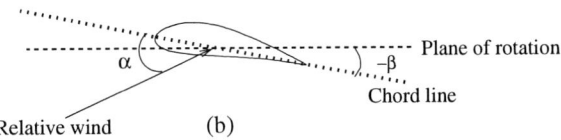

Fig. 1. Power generation control methods: (a) pitch control, (b) active stall control.

### A. Negative pitch control strategy for constant speed WTGs

If a fault occurs close to a constant speed WTG, the voltage at the generator terminals of the wind turbine drops, which results in the reduction of active power. If a wind turbine controller does not attempt to reduce the mechanical power input, the turbine accelerates during the fault. If a wind turbine has no means of controlling its power, then critical clearing time will be very short [12]. Hence, an active stall controlled WTG has to change the pitch angle as quickly as possible. This will eventually change $\alpha$. By increasing $\alpha$, the active power generation can be reduced. In a constant speed WTG, the reactive power demand depends on the active power generation also. Consequently, the rotor speed and reactive power reduce, which enhances the rotor speed stability.

In this paper, negative pitch control strategy has been simulated using DIgSILENT based built-in torque controller, to analyze the improvement in the rotor speed stability of a constant speed WTG. According to this strategy, during the fault, the turbine torque reduces as the $\beta$ increases in negative direction. This will reduce the rate of rise of the rotor speed. Hence, the rotor speed stability of a constant speed WTG can be maintained, and WTG remains connected to the grid. After fault clearance, if rotor speed is not recovered to its nominal value, then negative pitch angle does not return to its pre-fault (initial) value.

### IV. VALIDATION OF AN INDUCTION MACHINE MODEL OF DIgSILENT SOFTWARE

DIgSILENT PowerFactory is optimized to handle large amount of data, e.g., very large systems with thousands of buses/machines, appropriate initialization of models, etc. DIgSILENT PowerFactory is using A-stable integration methods with optionally adaptive, error controlled step-length. For RMS simulations, steps length may vary between ms and minutes with precise event handling. For EMT simulations, variable step length my vary between a couple of $\mu$s and some ms [14].

## A. Modeling of a squirrel cage induction generator in MAT-LAB

The equations used for squirrel cage induction generator modeling are described as follows [5]:

**Stator Equations:**

$$V_{qs} = -i_{qs}R_s + \omega_s\Psi_{ds} + \frac{d}{dt}\Psi_{qs} \qquad (1)$$

$$V_{ds} = -i_{ds}R_s - \omega_s\Psi_{qs} + \frac{d}{dt}\Psi_{ds} \qquad (2)$$

**Rotor Equations:**

$$V_{qr} = 0 = -i_{qr}R_r + S\omega_s\Psi_{dr} + \frac{d}{dt}\Psi_{qr} \qquad (3)$$

$$V_{dr} = 0 = -i_{dr}R_r - S\omega_s\Psi_{qr} + \frac{d}{dt}\Psi_{dr} \qquad (4)$$

**Flux Linkage are:**

$$\Psi_{qs} = -(L_{s\sigma} + L_m)i_{qs} - L_mi_{qr} \qquad (5)$$

$$\Psi_{ds} = -(L_{s\sigma} + L_m)i_{ds} - L_mi_{dr} \qquad (6)$$

$$\Psi_{qr} = -(L_{r\sigma} + L_m)i_{qr} - L_mi_{qs} \qquad (7)$$

$$\Psi_{dr} = -(L_{r\sigma} + L_m)i_{dr} - L_mi_{ds} \qquad (8)$$

Electrical Torque equation is:

$$T_e = \Psi_{qr}i_{dr} - \Psi_{dr}i_{qr} \qquad (9)$$

The equation of motion of the generator is:

$$\frac{dw_r}{dt} = \frac{T_{mech} - T_{elect}}{2H_m} \qquad (10)$$

where, $V$ is the voltage, $i$ is the current, $R$ is the resistance, $S$ is the slip, $L$ is the inductance, and $\Psi$ is the flux linkage. The subscripts $d$ and q stand for direct and quadrature component, respectively. And, the subscripts $r$ and $s$ stand for rotor and stator, respectively. The indices m and $\sigma$ are mutual and leakage, respectively.

These equations have been simulated using MATLAB software package. In both the softwares induction generator model has been executed for motor start-up condition. However, induction generator model available in the DIgSILENT does not have exact similar rating of the model simulated in MATLAB. Hence, the validation of the DIgSILENT result has been done qualitatively, as shown in Fig. 2 and Fig. 3.

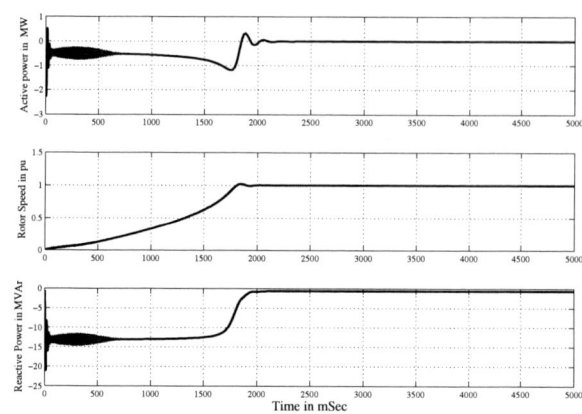

Fig. 2.   Waveforms of active power, speed and reactive power of induction generator from the MATLAB.

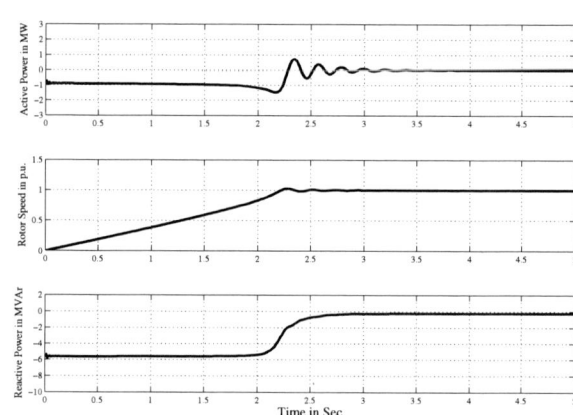

Fig. 3.   Waveforms of active power, speed and reactive power of induction generator from the DIgSILENT.

## V. SIMULATION OF A GRID CONNECTED WTG WITH ACTIVE STALL CONTROL

As shown in Fig. 4, a 1.3 MW constant speed WTG has been connected to a medium voltage (MV) distribution network. Modeling of WTG with capacitor bank connected to a sample system has been simulated using DIgSILENT PowerFactory. For rotor speed stability analysis, a 3-phase severe fault has been created on the line-1 at 3 seconds. The fault has been cleared by removing that line from the system at 3.12 seconds. To analyze the behavior of the WTG during this grid disturbance, the quantities of induction generator, such as active power generation (in MW), rotor speed (in p.u.), reactive power generation (in Mvar), and generator terminal voltage (in p.u.) are plotted. The effect of operating point and negative pitch control on rotor speed stability is illustrated in next sub-section.

794

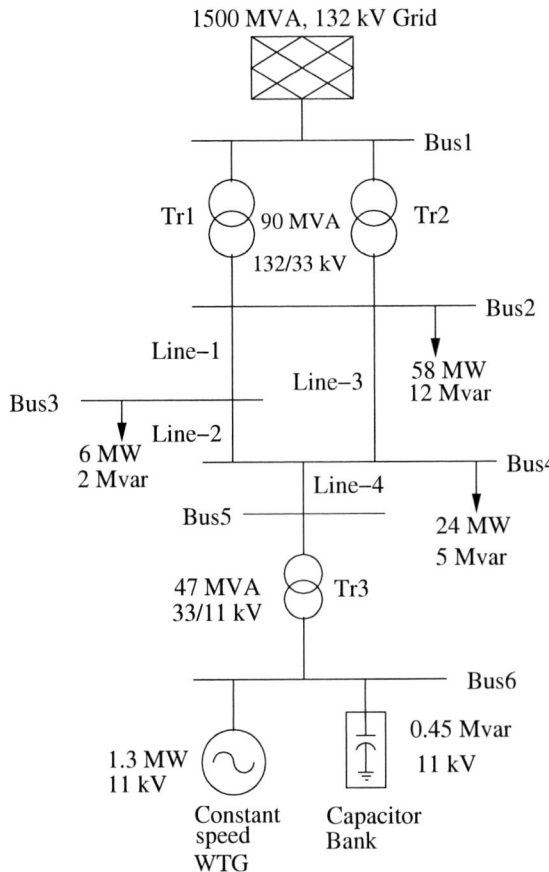

Fig. 4.   A sample system with a 1.3 MW constant speed WTG.

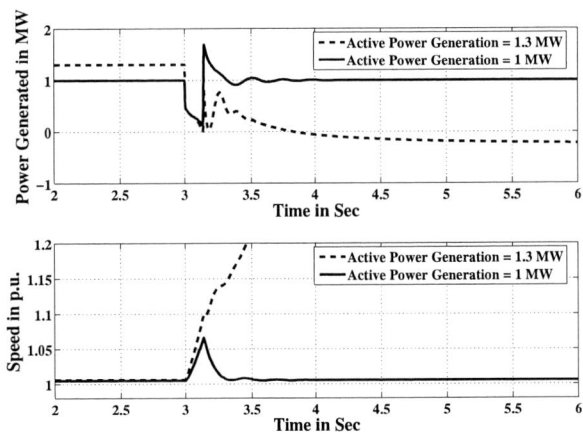

Fig. 5.   Effect of operating point on rotor speed stability.

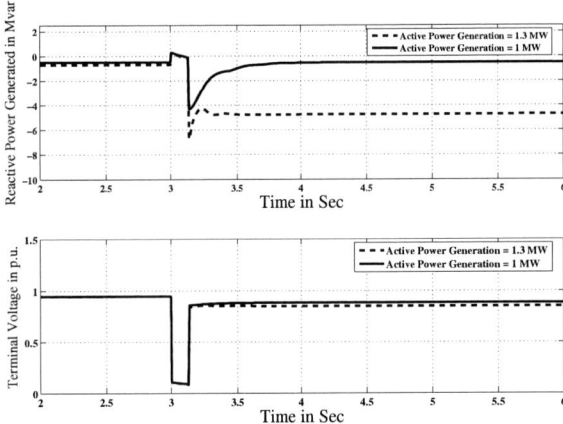

Fig. 6.   Effect of operating point on rotor speed stability.

## A. Effect of operating point on the rotor speed stability

As mentioned earlier, as the power output of a constant speed WTG decreases, its rotor speed stability increases. It can be deduced from Fig. 5 and Fig. 6, if power output of the WTG reduces from 1.3 MW to 1 MW (due to low wind velocity), the rotor speed stability of a constant speed WTG increases. That is because, the reactive power drawn by the induction generator is proportional to the active power generation, rotor speed and terminal voltage. Fig. 6 show that 1.3 MW power output of WTG draws more reactive power, as compared to 1 MW power output. Hence, when the fault is created at 3 sec, the instantaneous reactive power drawn at 1.3 MW power generation is 6.5 Mvar, as compare to 4.5 Mvar at 1.0 MW. When the fault is cleared at 3.13 seconds, and at 1.3 MW power generation continue to draw large amount of reactive power due to very high rotor speed. This will reduce active power generation to zero, as shown in Fig. 5 and Fig. 6. Consequently, the generator protection system isolates the WTG from the system.

## B. Effect of negative pitch control on the rotor speed stability

With the help of negative pitch control, the mechanical torque of a constant speed WTG can be reduced. During the grid disturbance, system voltage sags, and hence active power supplied by WTG decreases. Because of that, rotor speed increases and WTG draws very high reactive power, as shown in Fig. 7 and Fig. 8. In this case, the WTG power output is 1.3 MW (at constant rated speed). Critical clearing time 3.12 sec for this WTG is obtained by executing the simulation several times. Using negative pitch control strategy, the WTG will remain stable, eventhough fault is cleared after 3.12 sec. That is because, the active power generated is reduced from 1.3 MW to 1.2 MW till 3.2 sec, by controlling pitch angle $\beta$ in negative direction (as explained in section III). Consequently, the speed, and hence the reactive power drawn reduces, this improves the rotor speed stability of a constant speed WTG as shown in Fig. 7 and Fig. 8.

## VI. CONCLUSIONS

The rotor speed stability margin of constant speed WTG is higher at lower active power operating point. With negative pitch control strategy, the operating point can be controlled

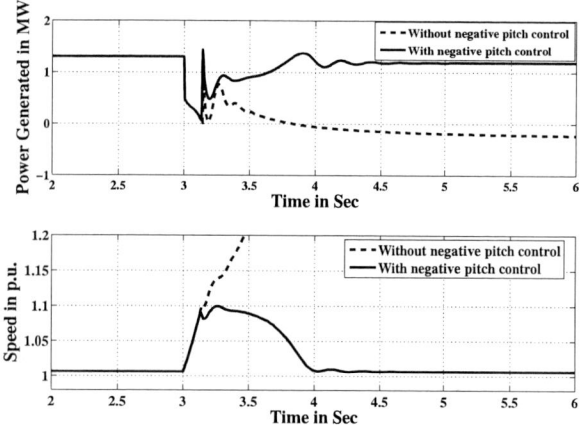

Fig. 7. Effect of negative pitch control on rotor speed stability.

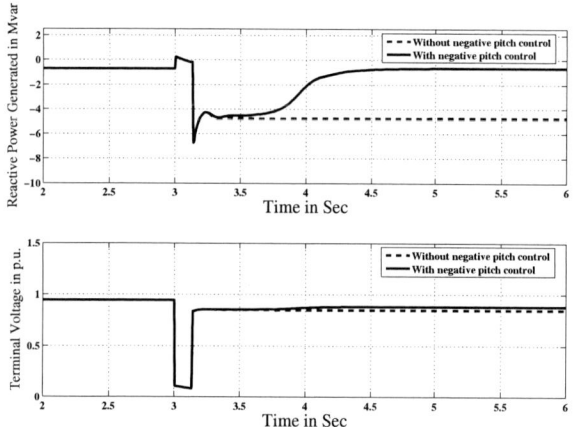

Fig. 8. Effect of negative pitch control on rotor speed stability.

under faulty conditions, thus, helping to maintain stability as shown in the result section. However, such control will be limited by the severity of the fault and ramping rate of pitch control. The work can be extended to coordinate the control with FACTS devices.

### REFERENCES

[1] (2006). [Online]. Available: http://www.windpowerindia.com
[2] O. Samuelsson and S. Lindahl, "On speed stability," *IEEE Trans. Power Syst.*, vol. 20, no. 2, pp. 1179–1180, May 2005.
[3] R. Zavadil, N. Miller, A. Ellis, and E. Muljadi, "Making connections," *IEEE Power Energy Mag.*, pp. 26–37, Nov. 2005.
[4] "Large scale integration of wind energy in the european power supply: analysis, issues and recommendations," European Wind Energy Association, Tech. Rep., Dec. 2005.

[5] T. Ackermann, *Wind Power in Power Systems.* England: John Wiley & Sons, Ltd., 2005.
[6] J. M. Rodrguez, "Incidence on power system dynamics of high penetration of fixed speed and doubly fed wind energy systems: Study of the spanish case," *IEEE Trans. Power Syst.*, vol. 17, no. 4, pp. 1089–1095, Nov. 2002.
[7] J. G. Slootweg, "Wind power: Modelling and impact on power system dynamics," Ph.D. dissertation, Delft University of Technology, the Netherlands, 2003.
[8] J. G. Slootweg, H. Polinder, and W. L. Kling, "Initialization of wind turbine models in power system dynamics simulations," in *Proc. International Power Tech. Conference*, Porto, 2001.
[9] R. G. de Almeida, E. D. Castronuovo, and J. A. P. Lopes, "Optimum generation control in wind parks when carrying out system operator requests," *IEEE Trans. Power Syst.*, vol. 21, no. 2, pp. 718–725, May 2006.
[10] T. Sun, Z. Chen, and F. Blaabjerg, "Voltage recovery of grid-connected wind turbines after a short-circuit fault," in *Proc. IEEE Industrial Society, IECON*, vol. 3, Nov. 2003, pp. 2723–2728.
[11] P. Mutschler and R. Hoffmann, "Comparison of wind turbines regarding their energy generation," in *Proc. IEEE Power Electronics Specialist Conference*, June 2002.
[12] P. Ledesma, J. Usaola, and J. L. Rodriguez, "Transient stability of a fixed speed wind farm," *Renewable Energy*, vol. 28, pp. 1341–1355, May 2003.
[13] C. Jauch, P. Sorensen, I. Norheim, and C. Rasmussen, "Simulation of the impact of wind power on the transient fault behavior of the nordic power system," *Electric Power System Research*, vol. 77, pp. 135–144, 2006.
[14] (2006). [Online]. Available: http://www.digsilent.com

**M. G. Kanabar** (S'05) is currently pursuing his M. Tech. degree at Electrical Engineering Department, Indian Institute of Technology Bombay, India. He received his B.E. degree from Birla Vishwakarma Mahaviddhyalaya, Vallabh Vidyanagar, Gujarat, India in the year 2003. His research area includes Grid integration issues for Distributed Generation, Renewable Energy Technologies, FACTS etc.

**C. V. Dobariya** (S'05) received B. E. degree from Nirma Institute of Technology, Ahmedabad, India in the year 2002. Currently, he is working towards his M. Tech. degree at Electrical Engineering Department, Indian Institute of Technology Bombay, India. His research area includes analysis and control of Distributed Generation and MicroGrid.

**S. A. Khaparde** (M'87, SM'91) is Professor, Department of Electrical Engineering, Indian Institute of Technology Bombay, India. He is member of Advisory Committee of Maharashtra Electricity Regulatory Commission (MERC). He is on editorial board of *International Journal of Emerging Electric Power Systems (IJEEPS)*. He has co-authored books on Computational Methods for Large Sparse Power System Analysis: An Object Oriented Approach, as well as, Transformer Engineering, published by Kluwer Academic Publishers and Marcel Dekker, respectively. His research area includes Distributed Generation and power system restructuring.

**2006 IEEE International Conference on Power Electronic, Drives and Energy Systems**

# Performance Evaluation of Indian Electric Power Utilities Based on Data Envelopment Analysis

Tripta Thakur

*Abstract-.* **This paper presents a framework for carrying out the productivity analysis of State Electricity boards (SEBs), which have been mainly responsible for the generation, distribution and transmission of electricity in India. Productivity Performances of twenty-six utilities was evaluated using Malmquist index for the time period 1996-2002. The results indicate positive productivity growths for SEBs, while also exhibiting the fact that these utilities are inefficient in operation and can be significantly improved.**

*Index Terms--***Data Envelopment Analysis (DEA), Productivity, State Electricity Boards**

## I. INTRODUCTION

**P**OWER Sector reforms are transforming the structures and operating environments of electricity industries across developed and developing countries in order to increase competition, and to enable the introduction of private capital in the sector. Performance evaluation plays a crucial role in structural reforms in facilitating an understanding of the behavior of electric utilities, and also in defining regulatory policies for both transmission and distribution. Benchmarking models for electricity distribution have been introduced in UK and US, and have become common throughout Latin America and Europe.

Power reform in developing countries have been necessitated by the pressing need for improvement in the existing power services as several of these countries are plagued by sub-optimal sector performances. Huge demand-supply gap is often a universal problem in developing countries and the distribution sectors are frequently

financially crippled [1]. Serious cash flow constraints result in palpable curtailment of the otherwise much needed investment in expansion and maintenance of services, and ultimately manifest in poor sector performances. High distribution losses, poor management, low market densities, poor metering and billing practices and weak institutions are some of the common problems besieging the electricity sectors in the developing nations.

Performance evaluation plays a crucial role in structural reforms in facilitating an understanding of the behaviour of electric utilities, and also in defining regulatory policies for both transmission and distribution. Benchmarking models for electricity distribution have been introduced in Europe and US [2], [3], and have become common throughout the Latin America [4]. However, for developing countries few studies have so far been reported. No detailed performance analysis has so far been reported for the Indian electric sector, despite the fact that the sector has undergone reforms since 1991, and has further accelerated the process of change with the recently enacted Electricity Act, 2003 [5]. The current paper is amongst the first productivity analyses carried out for the SEBs, which are the vertically integrated utilities mainly responsible for generation, transmission and distribution of power in India1.

This paper is structured as follows: A brief introduction is provided in section I which is followed by Section II presents methodology adopted in this paper, while section III analyses the data used. Section IV then presents the results of analysis Section V provides a brief conclusion to this paper.

## II. METHODOLOGY

In the recent times, DEA is receiving increasing importance as a tool for evaluating and improving the performance of manufacturing and service operations. DEA is a multi-factor productivity analysis for measuring the relative efficiencies of a homogenous set of decision-making units (DMUs) that perform similar tasks by consuming multiple inputs to produce multiple outputs.

---

The author is grateful to the commonwealth Association (UK) for providing support for carrying the above research. Author is grateful to the Center for Energy Petroleum Mineral Law and Policy (CEPMLP), University of Dundee,UK, for providing financial support for attending this conference. The support of IIT-Delhi and MANIT, Bhopal, India fare also being acknowledged.

Tripta Thakur is currently with the Center for Energy Petroleum Mineral Law and Policy (CEPMLP), University of Dundee, UK and is a teaching faculty at MANIT, Bhopal, India. (e-mail: tripta_thakur@yahoo.co.in, ttkulshresta@dundee.ac.uk

---

1 Electric distribution is still a monopoly of SEBs in India with a few private distribution utilities and two state Delhi and Orissa as an exception [1].

0-7803-9771-1/06/$25.00 ©2006 IEEE

*A.. Data Envelop Analysis (DEA)*

In DEA, the data are enveloped by a piecewise linear frontier in such a way that radial distances to the frontier are minimized. The basic model of DEA, the CCR model, was proposed by [6]. The CCR model was formulated as a linear programming (LP) problem concerned with, say, n decision making units (DMUs), electric utilities in the present analysis, which use varying quantities and combination of inputs Xi (i=1,…s) to produce varying quantities and combinations of outputs Yj (j=1,…m).The most common form of measurement of efficiency in case of a single output and single input framework is the ratio output/input. In case of multiple outputs and inputs, it is a weighted combination of outputs to weighted combination of inputs, known as virtual outputs and virtual inputs, where the weights are derived from data instead of being fixed in advance. Efficiency of each DMU is measured and hence n optimization exercises are carried out. The following problem is solved to obtain the values of input weights ($v_i$) and output weights ($u_r$) as variables:

max
$$\vartheta_o = \frac{\sum_r \{u_r y_{ro}\}}{\sum_i \{v_i x_{io}\}}$$

s.t.
$$\frac{u_1 y_{1j} + \dots + u_s y_{sj}}{v_1 x_{1j} + \dots + v_m x_{mj}} \leq 1$$

where j=1,…,n

$v_1, v_2,\dots v_m \geq 0$,

$u_1, u_2,\dots u_s \geq 0$,

The constraints imply that the ratio of "virtual output" to "virtual input" should not exceed 1 for every DMU. The objective is to obtain weights $v_i$ and $u_r$ that maximize the ratio for DMU$_o$. The optimal objective value $\theta^*$ is at most 1. However, multiple solutions might exist for the above problem. Hence it is transformed into a linear programming problem using transformation. The problem can be considered as an input minimization problem by normalizing the linear combination of inputs consumed by the concerned DMU as:

$\max_{u\,v}$
$$\sum u_r y_{ro}$$

s.t.
$$\sum v_i x_{io} = 1$$

$$\sum_r y_{rj} - \sum v_i x_{ij} \leq 0 \qquad \forall j = 1,\dots,n$$

The maximum value of the objective function is 1, when the DMU is efficient. This is also called the multiplier problem as the aim is to derive the optimal multipliers $v_i$'s and $u_r^*$s. The dual to the above problem is called the envelopment problem, which is easier to solve with lesser number of constraints. The envelopment problem is

min. $\vartheta_o$

s.t.
$$\sum_j \lambda_j x_{ij} \leq \vartheta_o x_{io}$$

$$\sum \lambda_j y_{ij} \geq y_{io}$$

$$\lambda_i \geq 0$$

Here $\theta_o$ signifies the extent to which the inputs need to be reduced to bring them on the best practice frontier. The $\lambda_j$s are the intensity variables to indicate the intensity with which the DMU being scored is related to the DMUs in the efficient facet. The output-oriented problem can be formulated similarly by normalizing the linear combination of outputs.

To allow for variable returns to scale [7] added the convexity constraint to the optimization problem in (3), by restricting the summation of the intensity variables to 1. The LP problem now becomes:

min. $\vartheta_o$

s.t.
$$\sum_j \lambda_j x_{ij} \leq \vartheta_o x_{io}$$

$$\sum_j \lambda_i = 1,$$

$$\sum \lambda_j y_{ij} \geq y_{io} \quad \lambda_i \geq 0 \qquad \forall j = 1,\dots,n$$

In solving the CCR and the BCC models, optimal weights in the multiplier problem can attain the value zero. In the envelopment problem also, slacks may exist. Hence the non-archimedean $\varepsilon$ is introduced in the problem, by restricting the multipliers to take values greater than $\varepsilon$ in the multiplier problem and in the envelopment problem, it is included in the objective function as:

min. $\vartheta_o - \varepsilon(\sum_i S_i^- + \sum_{ir} S_r^+)$

s.t.
$$\vartheta_o x_{io} - \sum_j \lambda_j x_{ij} - S_i^- = 0$$

$$\sum_r \lambda_j y_{ij} - S_r^+ = y_{ro}$$

$$\sum_j \lambda_j = 1, \quad \lambda_j, S_i^-, S_r^+ \geq 0$$

*B. Malmquist Productivity Index*

The DEA techniques can be used to calculate Malmquist Index of productivity change over time, assuming the underlying technology is constant returns to scale (CRS). The Malmquist total factor productivity (TFP) index measures the TFP change between two data points by calculating the ratio of the distances of each point relative to a common technology. The distance function in terms of the above analysis can be defined as $\{D_t(x_t,y_t)\}^{-1} = \theta_t$

Malmquist input oriented TFP change index between period s and period t is given by: tfpch = effch × techch

$$effch = \frac{d_i^t(y_t, x_t)}{d_i^s(y_s, x_s)}$$

$$techch = \left[\frac{d_i^s(y_t, x_t)}{d_i^t(y_t, x_t)} \times \frac{d_i^s(y_s, x_s)}{d_i^t(y_s, x_s)}\right]^{1/2}$$

where, tfpch signifies change in total productivity, which is caused by the joint influence of effch, i.e. the change in efficiency from period s to t and techch, the geometric mean of the shift in technology between the two periods, evaluated at $x_t$ and also at $x_s$. The value of the indices greater than one signifies increase in productivity.

## III. DATA ANALYSIS

The Planning commission of the government of India has till 2002 published annual performance reports for the SEBs on yearly basis [8]. These have been the source for the cost data. The physical data for various states were obtained from the General Reviews published by the Central Electricity Authority, Government of India on yearly basis [9].

### A.. Input/output selection

Selecting the appropriate inputs and outputs constitute the most important task of evaluating the performances. No universally applicable rational template is available for selection of variables. However, in general, the inputs must reflect the resources used and the outputs must reflect the service levels of the utility and the degree to which the utility is meeting its objective of supplying electricity to consumers. A study of standard literature reveals significant insights into the choice of variables. The most widely used variables based on international experience have been outlined in the literature [10].

The input/output selection for the present study was made in view of those parameters that directly affect the consumers in terms of cost of electricity supply. Total cost (Totex), which represents the cost incurred by the utility to supply electricity to the ultimate consumers. In this study three outputs, namely, the energy sold, consumer numbers and length of distribution network, are employed in line with their common usage in the available literature and because these are able to represent the activity levels of the utilities and have a bearing on the cost.

Table 1 shows the change in the input during 1996-2002, It is clear from the Table 1 that change in the input total cost from 1996 to 2002 is 143.46%, while the changes in output parameters are relatively less and amount to 28.62%, 21.89% and 24.89% for Number of Customers, Distribution line length and Energy sold respectively.

TABLE I
AVERAGE OUTPUT AND INPUT VALUES FOR 1996-97 AND 2001-02

|  | Year 1996 | Year 2002 | Change |
|---|---|---|---|
| *Outputs* |  |  |  |
| Number of Customers (Million) | 3.37 | 4.33 | 28.62% |
| Distribution line length (circuit Kms) | 192710.04 | 234895.23 | 21.89% |
| Energy sold (Mkwh) | 10121.15 | 12640.81 | 24.89% |
| *Input* |  |  |  |
| Total Cost (Rs, Million) | 21622.37 | 52642.47 | 143.46% |

## IV. RESUTS AND DISCUSSIONS

The results of Malmquist indices for the electricity utilities in each state are reported in Table 2. Table 2 shows the multiplicative decomposition of the Malmquist productivity index into technical efficiency change that is movement towards the production frontier termed as the 'catching up' effect and the pure technological change for the period 1996-97 to 2001-02. The result in Table 2 clearly shows the positive change in the overall electricity sector. Numbers greater than one indicate productivity growth, while the numbers smaller than one show regress and all SEBs shows progress as all of them has productivity index more than one. This shows that the overall the electric utilities in India have improved productivities since 1996-97, and investments in this industry took place with technical progress.

Fig. 1 describes the mean annual TFP growths for each SEB between 1996/97 and 2001/2002. The SEBs attained average TFP growth of 192%. However there are large variations within the SEBs. The results indicate a rapid increase in TFP productivity for all SEBs; however the growth is not persistent at the same rate for the all the SEB. The large maximal growth of Himachal Pradesh is striking- indicating an improvement of 338% in total productivity, with catching up improvement of 184% and frontier change of 183%. Other 9 best performing SEBs having growth index of more than 2 are all big states. The less performing states are the smaller ones, mainly the union territories with the exception of Bihar.

## V. CONCLUSION

This study attempted to carry out a performance analysis of the government owned and operated State Electricity Boards, the main electric power utilities, in India. A non-parametric approach to frontier analysis was adopted and efficiency scores for the different SEBs were determined using DEA. Malmquist productivity index was employed to analyse the productivity changes over the periods from 1996-97 to 2001-02. Total factor productivity was decomposed into the effects of technical change and pure efficiency change. It is clear from the analysis that SEBs have attained productivity progress from 1996-97 to 2001-02 mainly due to investments in the electricity sector. The technical progress also occurred, but was rather subdued mainly due to the organization structures and lack of competition in the sector. The SEBs have reported productivity growths, but have been unable to sustain their growth and this need urgent attention of the policy-makers. Individually, their operations are relatively inefficient and there is a lot of scope for efficiency improvement.

TABLE II
PRODUCTIVITY DEVELOPMENT DURING 1997-2002

| DMUs | Malmquist Index | Efficiency Change | Frontier Shift |
|------|-----------------|-------------------|----------------|
| Andhra Pradesh | 2.407 | 1.321 | 1.822 |
| Assam | 1.792 | 0.971 | 1.845 |
| Bihar | 1.334 | 0.795 | 1.679 |
| Delhi | 1.298 | 0.775 | 1.674 |
| Gujarat | 1.763 | 1.144 | 1.541 |
| Haryana | 1.773 | 1.034 | 1.715 |
| Himachal Pradesh | 3.389 | 1.847 | 1.835 |
| Jammu & Kashmir | 1.547 | 0.930 | 1.663 |
| Karnataka | 2.562 | 1.367 | 1.875 |
| Kerala | 2.333 | 1.210 | 1.927 |
| Madhya Pradesh | 1.524 | 0.901 | 1.691 |
| Maharashtra | 1.722 | 1.075 | 1.601 |
| Meghalaya | 1.721 | 1.011 | 1.702 |
| Punjab | 1.542 | 0.983 | 1.569 |
| Rajasthan | 2.616 | 1.533 | 1.706 |
| Tamil Nadu | 1.697 | 0.943 | 1.800 |
| Uttar Pradesh | 2.540 | 1.542 | 1.648 |
| West Bengal | 2.692 | 1.634 | 1.647 |
| Arunachal Pradesh | 2.462 | 1.177 | 2.092 |
| Goa | 2.226 | 1.302 | 1.710 |
| Manipur | 1.657 | 0.915 | 1.810 |
| Mizoram | 1.701 | 0.879 | 1.935 |
| Nagaland | 1.193 | 0.650 | 1.834 |
| Pondicherry | 1.441 | 0.982 | 1.468 |
| Sikkim | 1.972 | 1.000 | 1.972 |
| Tripura | 1.089 | 0.582 | 1.873 |
| *Average* | *192%* | *110%* | *176%* |

# VI. REFERENCES

[1] Thakur Tripta. A Distribution Sector Reforms In India: The Tasks Ahead", International Journal of Global Energy Issues (IJGEI), 23(2/3), pp.196 –217, 2005

[2] Burns, P. and Weyman-Jones, T."Cost Functions and Cost Efficiency in Electricity Distribution: A Stochastic Frontier Approach", Bulletin of Economic Res., 48(1), pp. 41-64,1996

[3] Pahwa A., Feng X., and Lubkeman D. "Performance Evaluation of Electric Distribution Utilities based on Data Envelopment Analysis", IEEE Trans. on Power Systems 17(3),2002

[4] Estache, A, M. A. Rossi, and C. A. Ruzzier. "The Case for International Coordination of Electricity Regulation: Evidence from the Measurement of Efficiency in South America Washington, D.C.", Journal of Regulatory Economics, 25(3), pp. 271-295.,2004

[5] Thakur Tripta, Deshmukh S.G., Kaushik S.C, and Kulshrestha Mukul.. "Impact assessment of the Electricity Act 2003 on the Indian power sector", Energy Policy 33(9), pp.1187-1198,2005

[6] Charnes, A., Cooper, W. W., & Rhodes, E. "Measuring the efficiency of decision-making units", European Journal of Operational Research, 2, pp. 429-444,1978

[7] Banker R.D., Charnes A, and Cooper WW. "Some models for estimating Technical and Scale Inefficiencies in Data Envelopment Analysis", Management Science, 30, pp. 1078-1092,1984

[8] Annual Report on the working of state electricity boards and electricity departments (2006, March) [Online]. Available: http://planningcommission.nic.in/reports/genrep/reportsf.htm

[9] General Review, All India Electricity Statistics. Central Electricity Authority, Ministry of power, India, 2005

[10] Jamasb T., Pollitt M."Benchmarking and Regulation: International electricity experience", Utilities Policy, 9,pp. 107-130, 2003

# VII. BIOGRAPHY

**Tripta Thakur** has graduated in Electrical Engineering from MITS, Gwalior, India. She has a Masters degree in Power Electronics from IIT Kanpur, and is a doctoral researcher at IIT Delhi. She is a recipient of the commonwealth fellowship award at the Center for Energy Petroleum Mineral Law and Policy (CEPMLP), University of Dundee, UK. She is also teaching as an Assistant Professor in the Electrical Engineering Department at the National Institute of Technology, MANIT, Bhopal.

She has published widely about power sector reforms and performance evaluation of electric utilities in International Journals and Conferences.

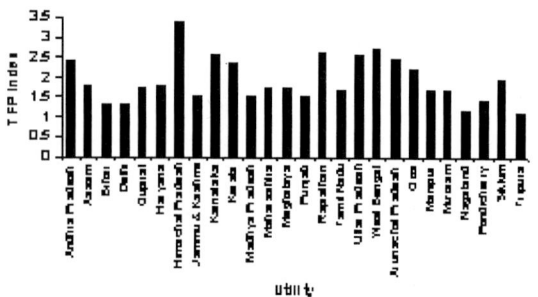

Fig. 1.   Annual TFP growth between 1996-97 and 2001-2002.

**2006 IEEE International Conference on Power Electronic, Drives and Energy Systems**

# Modelling of Hybrid Energy System for Off Grid Electrification of Clusters of Villages

Ajai Gupta, R P Saini, and M P Sharma

*Abstract--* Hybrid energy systems are increasingly being applied in areas where grid extension is considered uneconomical. Their costs can be minimized through proper equipment sizing and load matching. This paper reports the results of optimization of hybrid energy system model for remote area in India. For this purpose, the Jaunpur block of Uttaranchal state of India has been selected as remote area. The model is developed with the objective of minimizing cost function based on demand and potential constraints. The model has been optimized using LINDO software 6.10 version. From the economic analysis, the capital cost, cost of energy for different types of resources, optimized cost of hybrid energy system are determined. In order to consider the fluctuation in the discharge and power generation from SHP, the EPDF has been varied from 1.0 to 0.0. The EPDF is electric power delivery factor and also called optimizing power factor and is maximum equal to1.

*Index Terms*—Renewable Energy, Hybrid Energy System (HES), Off-grid Electrification, Stand-Alone Energy System Design, Small Hydro Power, Solar Photovoltaic System, Wind Energy Conversion System, Biogas Usage, Remote Area Power Generation, Power-generation Economics, Optimization Methods, EPDF

## I. INTRODUCTION

THERE are only about 7.36% of over 15761 villages in Uttaranchal that are not yet electrified. These villages are either on hill or in remote areas. Jaunpur block, consisting of clusters of villages, a hill-tribe remote area in district Tehri Garhwal, is one of them. The village is accessible by road from the city. There are 8291 households with population of 50,636. The houses are close to each other and most of the households use candles and kerosene lamps for lighting in early morning and in the evening. Peoples are facing the main needs such as good type of food facility & drinking water and electricity. Only primary school is available in some villages. Peoples have not higher education. Only one zone of this block is selected as a case study in this work [1].

A hybrid energy system consists of two or more energy systems, an energy storage system, power conditioning equipment and a controller [2]. A hybrid energy system may or may not be connected to the grid. Examples of energy systems commonly used in hybrid configurations are small wind turbines, photovoltaic systems, micro hydro, biomass, diesel generator, fuel cells and stirling engines. The Hybrid Energy System (HES) has received much attention over the past decade. It is a viable alternative solution as compared to systems, which rely entirely on hydrocarbon fuel. Apart from the mobility of the system, it also has longer life cycle. In particular, the integrated approach [3]-[5]. makes a hybrid system to be the most appropriate for isolated communities such as remote area in Uttaranchal.

For off-grid power supplies, a diesel generator (DG) system is most attractive because of its low capital, although operation and maintenance cost is high and engine emissions can pollute an environment [6]-[7]. In some cases, diesel generator is not cost effective due to high fuel transportation cost to remote area.

For systems employing totally clean renewable energy, high capital cost is an important barrier. However, we can produce green power by adding different renewable energy sources to diesel generator and battery, which is called a hybrid system. This kind of system can compromise investment cost, diesel fuel usage cost and also operation and maintenance costs. Many hybrid systems sizing have been studied and optimized by economic analysis based on system life cycle cost and cost of energy [8]-[10].

Optimization of a hybrid energy system is site specific and it depends upon the resources available and the load demand. The aim of this study is to identify the most economic and appropriate power supply system for a selected remote rural area.

## II. LITERATURE REVIEW CONCLUSIONS

Literature review reveals that the modeling of hybrid energy system and their application in decentralized mode are quite limited. The models applied, are based on one of the available resources. While the literature has focused on one or two available resources. Further, attempts for developing optimum energy mix of different resources for meeting the energy needs of the rural people are also limited.

Application of models for matching the projected energy demand with mix of sources at decentralized level is limited. The models developed so far mainly focus on rural areas and not individual villages, clusters of villages, blocks, or district. Very limited efforts are reported for block level planning and are not based on any optimization approach [11]-[13].

---

Ajai.Gupta, R.P.Saini, and M.P.Sharma are with Alternate Hydro Energy Centre, Indian Institute of Technology, Roorkee, U.A-247667 India (phone: 9411153723; e-mail: ajai_ms2002@ yahoo.co.in)

0-7803-9771-1/06/$25.00 ©2006 IEEE

## III. PROBLEM FORMULATION

From the literature, it is observed that lot of work has been carried out for modelling of hybrid energy sources. The renewable energy sources considered under these studies are mainly solar, biomass, and wind but a very little work has been reported for the modelling of hybrid renewable energy sources involving small hydro power (SHP)/micro hydropower in combination with conventional system.

The Uttaranchal State is one of the richest states in hydropower having potential of about 254 MW at 47 sites. In addition, thousands of sites in the range of micro hydro also exist.

The main objective of the study is to develop a model of Hybrid Energy System with more emphasis on small hydro power (SHP)/micro hydro for a remote rural area in cost effective manner.

## IV. METHODOLOGY

### A. Study Area

The remote rural area for the study was Jaunpur block of district Tehri Garhwal of Uttaranchal state, India. The area comprises of major hilly and the fertile area under forest with scattered households. The area has been considered by Uttaranchal Renewable Energy Development Agency (UREDA) to be remote and not economically viable for electrification by grid extension. The Jaunpur block (study area) divided into four zones of clusters of villages and only 9% of the total villages are unelectrified, which have been considered for the present study as the best candidate for electrification by decentralized hybrid energy systems consisting of biomass, micro hydro, solar, wind, diesel generator. There is a primary school in every gram sabha, consisting of 6 to 7 villages. The literacy level in Jaunpur block is 16% for women and 62% for men. The total literacy rate of the Jaunpur block is 38.9%.

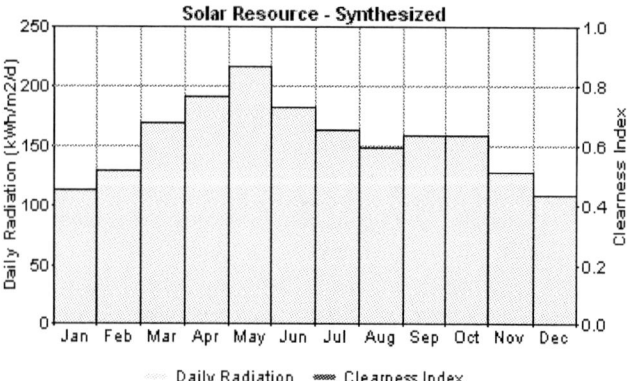

Fig. 1. Monthly average solar radiations in study area.

The study area has adequate sunshine, low to moderate wind speeds, falling water is available 8-10 months in a year frequently, and biomass resource is available in sufficient amount. Monthly average horizontal solar radiations

(insolation in kW/m$^2$) in study area is shown in Fig. 1. Monthly average wind speed distribution at 10m height recorded at study area (30$^0$ 04'-N, 78$^0$ 37'-E) [Details of Bachlikhal Metrological Station] is shown in Fig. 2. During a day wind speed varies from 2.99 m/s to 6.37 m/s.

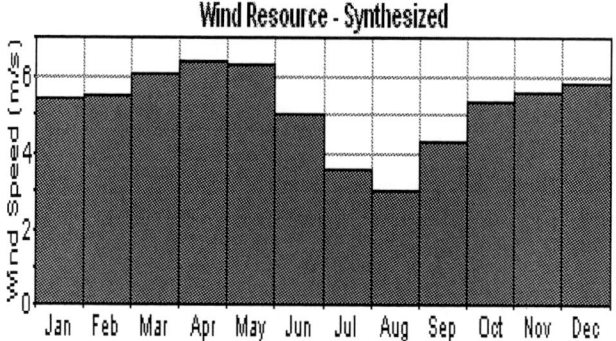

Fig. 2. Monthly average wind speed in study area.

### B. Assessment of Energy Potential and Energy Demand

The energy resources data shows that the biomass constitutes maximum potential (641,385 kWh/yr) followed by micro hydro (128,166 kWh/yr), solar (22,363 kWh/yr), and wind energy (15,251 kWh/yr). Though the actual exploitation of potential will depend upon the system configuration. The total potential including all the renewable energy resources considered is about 807,165 kWh/yr. The total estimated demand is 810,000 kWh/yr. This means that the energy demand of the area cannot be fully exploiting the available resources. Therefore diesel-generator option of potential 350,400 kWh/yr is also included.

Therefore, the entire electricity generation will be 1157565 kWh/yr and accordingly the model consisting of micro hydro power (MHP), biomass energy system (BES), solar photovoltaic system (SPV), wind energy system (WES), diesel-generator set (DG) has been considered. The unit costs have been calculated using standard procedures described by [14]-[15] and are based on capital cost of installed capacity, operational & maintenance costs, life of plants etc. used for calculation of each resource.

The energy needs of the area have been identified as domestic, agricultural, transportation, and motive power for small-scale industries [16]-[17]. The energy demands in different sectors calculated on the basis of data collected from survey.

Fig. 3. Indicates the individual load in different sectors of 12-unelectrified villages of the study area with total load 810,000 kWh/yr, out of which lighting load 291,978, cooking load 396,563, other domestic load such as T.V. & fans are 51,786 & 42,964 kWh/yr. Agriculture and motive/industries loads are 135 and 26,576 kWh/yr. Figure 3. Also shows the demand of 12 un-electrified villages, out of which the lighting load is required maximum in the village no. V9 but cooking load needed maximum in all the villages. This graph also shows that the trend of consumption in the study area.

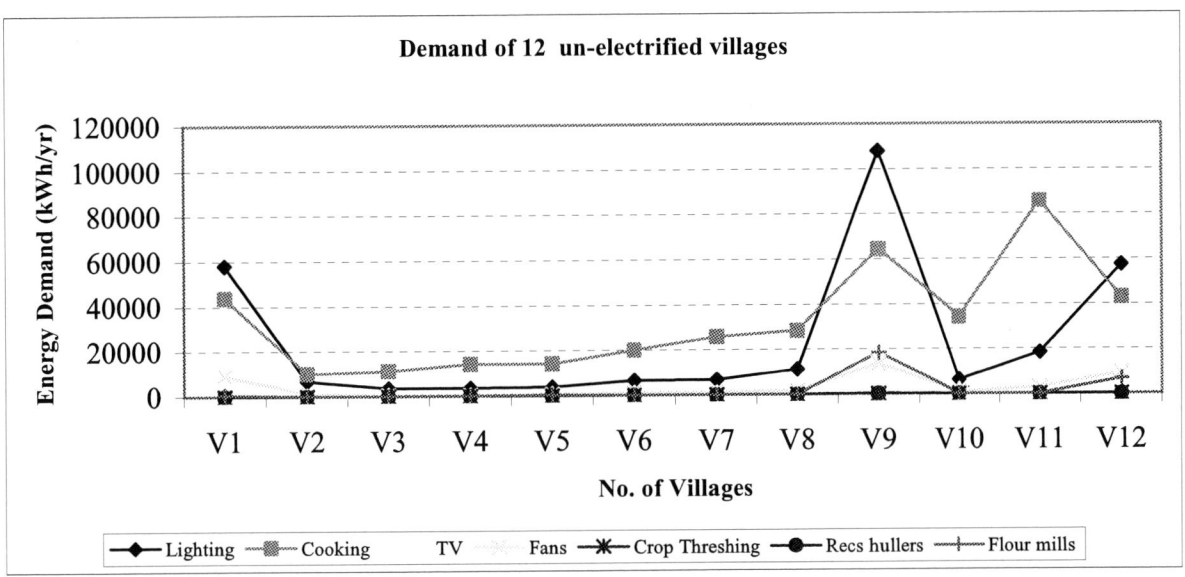

Fig. 3. Monthly load profile of unelectrified villages.

## V. MODEL FORMULATION

A hybrid energy system model has been constructed for the major end uses lighting mainly and other uses also. The general model can be formulated on the basis of linear programming as [18]-[20]:

Minimize: $\quad TC = \sum C_{ij} \times X_{ij}$

Subject to: $\quad \sum X_{ij} = D_j$

$\qquad\qquad \sum X_{ij}/\eta_{ij} \leq Si$

$\qquad\qquad X_{ij} \geq 0$

Where, TC is the total cost of providing energy for all end uses for operation of the system; $C_{ij}$, cost/unit of the $i_{th}$ resource option for $j_{th}$ end use (Rs/kWh); $X_{ij}$, optimal amount of the $i_{th}$ resource option for $j_{th}$ end use (kWh); $D_j$, total energy for $j_{th}$ end use (kWh); $S_i$, availability of the $i_{th}$ resource option for $j_{th}$ end use (kWh); $\eta_{ij}$, conversion efficiency for the $i_{th}$ resource option for $j_{th}$ end-use.

The effective cost per unit of energy ($C_{ij}$) for each of the proposed resource-device combinations (variables) is an important factor in the optimization model. The main objective is to optimize an objective function subject to a set of constraints. The cost function governs the optimal mix in such a manner that resources with lesser-cost function share the greater of the total energy demand in an attempt to optimize (minimize) the objective function.

To account for the fluctuation in the total energy delivered by the SHP, a term known as effective power delivery factor (EPDF) has been introduced, which may be defined as the ratio of the power obtained per year to the maximum power available per year.

## VI. OPTIMIZATION RESULTS AND DISCUSSION

The unit cost of energy of different resources and optimized cost for hybrid energy systems (HES) using LINDO software 6.1 version [21] are shown in Table I. and Table II.

TABLE I
UNIT COST OF ENERGY FOR DIFFERENT RESOURCES

| Sl. No. | Type of Energy Resource | | Cost of Energy (Rs/kWh) |
|---------|-------------------------|---|-------------------------|
| 1. | Micro Hydro Power | (MHP) | 1.50 |
| 2. | Solar Photovoltaic | (SPV) | 15.27 |
| 3. | Wind Energy System | (WES) | 3.50 |
| 4. | Biomass Energy System | (BES) | 3.10 |
| 5. | Diesel Generator | (DG) | 15.48 |

TABLE II
UNIT COST OF HYBRID ENERGY SYSTEM

| System Type | Installed capacity (kW) | Resource Fraction (kWh) | Optimal cost of hybrid system (Rs) | COE of Hybrid System (Rs/kWh) |
|-------------|-------------------------|-------------------------|------------------------------------|-------------------------------|
| MHP | 19.26 | 115464.87 | 4045098 | 4.99 |
| SPV | 2 | 20146.84 | | |
| WES | 3 | 12200.79 | | |
| BES | 70 | 543546.62 | | |
| DG | 60 | 118640.87 | | |

The results shown are for specified parameters, which can vary for individual customers, as well as, from area to area. It can be seen that the least economical system is the stand-alone micro hydro generation system (1.50/kWh) as it has to be run all the time in order to meet the load demand constantly. On the other hand, the most expensive system is the stand-alone diesel generator (15.48/kWh) because it has the highest replacement costs. It needs to replace the generator once every

803

TABLE III
VARIATION OF UNIT COST WITH EPDF

| SI. No. | EPDF | Unit Cost (Rs/kWh) | Load Distribution | | | | | Energy Availability |
|---|---|---|---|---|---|---|---|---|
| | | | MHP (kWh) | SPV (kWh) | WES (kWh) | BES (kWh) | DG (kWh) | |
| 1. | 1.0 | 4.99 | 115464.87 | 20146.84 | 12200.79 | 543546.62 | 118640.87 | MHP: 19.26 kW |
| 2. | 0.9 | 5.19 | 103918.37 | 20146.84 | 12200.79 | 543546.62 | 130187.36 | SPV: 2 kW |
| 3. | 0.8 | 5.39 | 92371.89 | 20146.84 | 12200.79 | 543546.62 | 141733.84 | WES: 3 kW |
| 4. | 0.7 | 5.59 | 80825.40 | 20146.84 | 12200.79 | 543546.62 | 153280.34 | BES: 70 kW |
| 5. | 0.6 | 5.79 | 69278.921 | 20146.84 | 12200.79 | 543546.62 | 164826.82 | DG: 60 kW |
| 6. | 0.5 | 5.99 | 57732.40 | 20146.84 | 12200.79 | 543546.62 | 176373.31 | |
| 7. | 0.4 | 6.189 | 46185.945 | 20146.84 | 12200.79 | 543546.62 | 187919.79 | |
| 8. | 0.3 | 6.38 | 34639 | 20146.84 | 12200.79 | 543546.62 | 199466.78 | |
| 9. | 0.2 | 6.58 | 23092.97 | 20146.84 | 12200.79 | 543546.62 | 211012.76 | |
| 10. | 0.1 | 6.78 | 11546.48 | 20146.84 | 12200.79 | 543546.62 | 222559 | |
| 11. | 0.0 | 6.98 | 0.0 | 20146.84 | 12200.79 | 543546.62 | 234105 | |

three years while the hybrid energy systems needs to replace the generator only once every six years. So the stand-alone system will cost more money than it is necessary.

Regarding the biomass energy it is clear that potential of biomass is sufficient with second lowest cost of energy. In order to fully utilize the biomass resource, one is required to explore the possibility of generating electricity using biomass gasifier engine system in decentralized mode because the cost of generation from the individual resource is Rs 3.10/kWh followed by wind energy system (3.50/kWh), solar photovoltaic system (15.27/kWh).

Table III. Shows the optimization results of hybrid energy model for different values of EPDF from 1.0 to 0.0 and graphically represented in Fig.4. It gives the share of different sources with different EPDF value. It indicates that an EPDF of 1.0, the SHP plant deliver its maximum energy to the load. Similarly, an EPDF of 0.9 gives 10% reduction in the energy delivery capability of the SHP plant. In the present model, it is felt also that the breakdown/nonfunctioning/fluctuation in the discharge of SHP may drastically affect the unit cost of hybrid energy system model. With decrease in EPDF from 1.0 to 0.0, the unit cost of generation increases from 4.99 to 6.98.

## VII. CONCLUSIONS

Given the fact that a hybrid energy system consisting two or more energy system has the advantage of stability and supplying the power on sustainable basis. The objective of the electrification at the study area can be achieved by making use of wind, solar, micro hydro and biomass with diesel generator. The information about local wind, solar, micro hydro and biomass indicates that a feasible hybrid energy system can be planned, modeled and designed for the above purpose. The collected data of the various energy sources were analyzed in order to plan for the structure of the system.

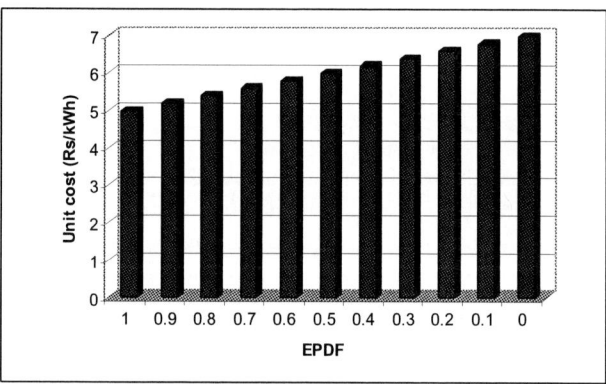

Fig. 4. Variation of Unit cost with EPDF.

## VIII. REFERENCES

[1] www.upcl.org
[2] R. Ramakumar, N. G. Butler, and A. P. Podriguez, "Economic aspects of advanced energy technologies," in *Proc. 1993 of IEEE* vol. 81, No. 3, pp. 318-332, March 1993.
[3] R. Ramakumar, I. Abouzahr and K. Asenyayi, "A Knowledge-Based approach to the Design of Integrated Renewable Energy Systems," *IEEE Trans. on Energy Conversion,* vol. 7, No. 4, pp. 648-657, 1992.

[4] R. Ramakumar, I. Abouzahr, K. Krishnan and K. Ashenayi, "Design Scenario for integrated Renewable Energy Systems", *IEEE Trans. on Energy Conversion*, Vol. 10, No. 4, pp. 736-746, December 1995.

[5] E. S. Gavanidou, A. G. Bakirtzis, "Design of a stand alone system with renewable energy sources using trade off methods," *IEEE Trans. on Energy Conversion*, Vol. 7, No. 1, pp. 42-48, March 1992.

[6] C. V. Nayar, W. B. Lawrance, S. J. Phillips, "Solar/wind/diesel hybrid energy systems for remote areas," in *Proc. of IEEE* pp. 2029-2034, 1989.

[7] A. G. Bakirtzis, E. S. Gavanidou, "Optimum operation of a small autonomous system with unconventional energy sources," *Electric Power Systems Research*, vol. 23, pp. 93-102, 1992.

[8] A. Rosenthal, S. Durand, M. Thomas and H. Post, "Economic analysis of PV hybrid power system: Pinnacles National Monument," in *Proc. of IEEE Photovoltaic Specialists Conf.*, pp. 1269-1272.

[9] W.D. Kellog, M.H. Nehrir, G. Venkataramanan, and V. Gerez, "Generation unit sizing and cost analysis for stand-alone wind, photovoltaic, and hybrid wind/pv systems", *IEEE Transactions on Energy Conversion*, vol. 13, No. 1, pp. 70-75, March 1998.

[10] W. Durisch and P. Aebli, "Economics of a photovoltaic electricity supply system in a small remote village in Southern Jordan," in *Proc. of World Renewable Energy Congress* VI, Part II, July, pp. 795-800.

[11] S. Jebaraj and S. Iniyan, "A review of energy models," *Renewable & Sustainable Energy Reviews*, vol. xx, pp. 1-31, 2004.

[12] R. B. Hiremath, S. Shikha and N. H. Ravindranath, "Decentralized energy planning; modeling and application," *Renewable & Sustainable Energy Reviews*, [Online]. Available: http://www.sciencedirect.com

[13] T. Nakata, "Energy-economic models and the environment," *Progress in Energy and Combustion Science*, vol. 30, pp. 417-475, 2004.

[14] R. Ramakumar and W.L. Hughes, "Renewable energy sources and rural development in developing countries," *IEEE Transactions on Education*, vol. E-24, No. 3, pp. 242-251, August 1981.

[15] M. Ashari, C.V. Nayar and W.W.L. Keerthipala, "Optimum operation strategy and economic analysis of a photovoltaic-diesel-battery-mains hybrid uninterruptible power supply," *Renewable Energy*, vol. 22, pp. 247-254, 2001.

[16] H. C. de. Coninck, K. J. Dinesh, A. Kets, S. Maithel, P. Mohanty, and H. J. de Vries, "Providing electricity to remote villages-Implementation models for sustainable of India's rural power, " Energy research centers of Netherlands, ECN Rep. ECN-C-05-037, July 2005.

[17] M. G. Green, "How to generate electricity in remote areas: A simple guide to choosing the right technology," M.S. thesis, Dept. Mech. Eng., Texas Univ., Austin, 2002.

[18] T. Markvast, "Sizing of hybrid photovoltaic-wind energy systems," *Solar Energy*, Vol. 57, No. 4, pp. 277-281, 1996.

[19] S. Iniyan, K. Sumathy, "An optimal renewable energy model for various end-uses," *Energy*, Vol. 25, pp. 563-575, 2000.

[20] S. Iniyan, K. Sumathy, "The application of Delphi technique in the linear programming optimisation of future renewable energy options for India," *Biomass and BioEnergy*, Vol. 24, pp. 39-50, 2003.

[21] www.lindo.com

## IX. BIOGRAPHIES

**Ajai Gupta** was born in Bareilly, India on May 2, 1974. Ajai Gupta received B. Tech in Electrical Engineering from I.E.T. Rohilkhand University, Bareilly in 2000 and M. Tech in Instrumentation and Control from Aligarh Muslim University, Aligarh in 2004 respectively. Currently he is a Research Scholar at Alternate Hydro Energy Centre, Indian Institute of Technology, Roorkee, India. Before joining in a PhD course, he worked as a lecturer at Vishveshwarya Institute of Engineering & Technology, India. He has interest in field of Modelling of Hybrid Energy System, IRES, Artificial Intelligence applications in power system.

**Dr. R. P. Saini** obtained B.E. in Mechanical Engineering from University of Mysore, India in 1982, ME and PhD from University of Roorkee in 1989 and 1996 respectively. Presently he is serving as a Senior Scientific Officer at Alternate Hydro Energy Centre, Indian Institute of Technology, Roorkee. He is a life member of International Association of small hydro, Indian society of

continuing Engineering Education and Solar Energy Society of India. His interest includes Small hydropower development, Renewable energy technologies and Solar PV system design and applications.

**Dr. M. P. Sharma** has been working as Senior Scientific Officer at Alternate Hydro Energy Centre, Indian Institute of Technology, Roorkee since the last 20 years. His area of research are renewable energy sources with special reference to Modeling of IRES, Hybrid Energy Systems, modeling of induction generators, Energy Conversation and Environment Impact Assessment of renewable and other projects.

**2006 IEEE International Conference on Power Electronic, Drives and Energy Systems**

# PSO-Based Multidisciplinary Design of A Hybrid Power Generation System With Statistical Models of Wind Speed and Solar Insolation

Lingfeng Wang, *Student Member, IEEE,* and Chanan Singh, *Fellow, IEEE*

*Abstract*— With the increasing concerns on air pollution and global warming, the clean green renewable sources of energy are expected to be playing more significant role in the global energy future. Multi-source hybrid power generation systems are representative applications of the renewables' technology. In this investigation, wind turbine generators, photovoltaic panels, and storage batteries are used to build a grid-linked generation system which is optimal in terms of multiple criteria including cost, reliability, and emissions. Multidisciplinary design facilitates the decision maker to make more rational evaluations. A set of tradeoff solutions can be obtained using the multidisciplinary approach, which offers many design alternatives to the decision-maker. A customized particle swarm optimization algorithm is developed to derive these non-dominated solutions. A grid-linked hybrid power system is designed based on the proposed approach. Furthermore, due to the unpredictability of wind speed and solar insolation, autoregressive moving average (ARMA) models are adopted to reflect the stochastic characteristics of wind speed and solar insolation. Sensitivity studies are also carried out to examine the impacts of different weather conditions and economic rates.

## I. INTRODUCTION

RENEWABLE sources of energy are playing an increasing role in the current global energy map. The public attention is focused on these renewable technologies as environmentally sustainable and convenient alternatives. Wind power and solar power are the two most widely used renewable sources of energy since they feature certain merits as compared with the conventional fossil-fuel-fired generation. Unfortunately, these renewable sources of energy are essentially intermittent and quite variable. As a result, it is possible that power fluctuations will be incurred since both power sources are highly dependent on the weather conditions. To reflect the stochastic characteristics of wind and solar power, an autoregressive moving average (ARMA) time series is used to model the wind speed and solar insolation at different time instants. To mitigate or even cancel out the fluctuations, energy

The work reported in this paper was partially supported by The National Science Foundation (Grant No. ECS0406794: Exploring the Future of Distributed Generation and Micro-Grid Networks.)

L. F. Wang is a Ph.D student in the Electrical and Computer Engineering Department, Texas A&M University, College Station, TX, 77843. He is an IEEE student member and his major research interest includes optimization of power system operations and planning. (e-mail: l.f.wang@ieee.org).

C. Singh is a Professor in the Electrical and Computer Engineering Department, Texas A&M University, College Station, TX, 77843. He is an IEEE Fellow and his major research interests include power system reliability and electric power quality. (e-mail: singh@ece.tamu.edu).

storage technologies such as storage batteries (SBs) are very often employed.

In this paper, we employ a multidisciplinary approach to handle the hybrid system design problem by taking into account multiple design objectives including economics, reliability, and pollutant emissions. Multidisciplinary design helps the decision-maker to determine reasonable tradeoffs. In such hybrid system designs, not all the criteria can be monetized. Thus, multidisciplinary design provides a viable way to reach tradeoffs among these design objectives with different preferences. For this purpose, a metaheuristics called particle swarm optimization is extended accordingly in order to derive a set of non-dominated solutions with sufficient diversity for decision-making support.

## II. HYBRID POWER GENERATION SYSTEMS

As shown in Fig. 1, a typical hybrid generation system comprises different power sources including wind turbine generators (WTGs), PV panels (PVs), and storage batteries (SBs). These power sources have different impacts on cost, environment, and reliability. In a hybrid generation system, they are integrated together and complement one another in order to serve the load while satisfying certain economic, environmental, and reliability criteria. The hybrid system can be operated autonomously or connected to the utility grid whose power is assumed from the conventional fossil-fuel-fired generators (FFGs).

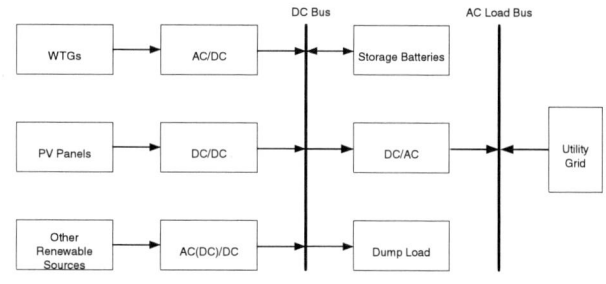

Fig. 1. Configuration of a typical hybrid generation system.

### A. Wind Turbine Generators (WTGs)

Wind speed at a specific hour is related to the wind speed in the previous hour. Here an autoregressive moving average

(ARMA) time series is used to model the wind speed.

$$
\begin{aligned}
y_{w_t} =\ & \phi_{w_1} * y_{w_{t-1}} + \phi_{w_2} * y_{w_{t-2}} + \ldots + \phi_{w_n} * y_{w_{t-n}} \\
& + \epsilon_{w_t} - \theta_{w_1} * \epsilon_{w_{t-1}} - \theta_{w_2} * \epsilon_{w_{t-2}} \\
& - \ldots - \theta_{w_m} * \epsilon_{w_{t-m}}
\end{aligned}
\qquad \text{(II.1)}
$$

where $y_{w_t}$ is the time-series value at time $t$, $\phi_{w_i}(i = 1, 2, \ldots, n)$ and $\theta_{w_j}(j = 1, 2, \ldots, m)$ are the autoregressive and moving average parameters of the model, respectively. $\{\epsilon_t\}$ is the normal white noise with zero mean and a variance of $\sigma_{w_\epsilon}^2$. The hourly wind speed $V_t$ at hour $t$ can be derived from the mean wind speed $\mu_{w_t}$, its standard deviation $\sigma_{w_t}$, and the time-series value $y_{w_t}$ as follows:

$$
V_t = \mu_{w_t} + \sigma_{w_t} * y_{w_t}. \qquad \text{(II.2)}
$$

### B. Photovoltaic generation (PV)

Insolation embodies the available solar energy that can be converted to electricity. The factors that have influences on insolation are the intensity of the sunshine and the operating temperature of the PV panels. In a similar fashion, solar insolation at a specific hour is correlated with the insolation during its preceding hour. Here an autoregressive moving average (ARMA) time series is used to model the solar insolation:

$$
\begin{aligned}
y_{s_t} =\ & \phi_{s_1} * y_{s_{t-1}} + \phi_{s_2} * y_{s_{t-2}} + \ldots + \phi_{s_n} * y_{s_{t-n}} + \epsilon_{s_t} \\
& - \theta_{s_1} * \epsilon_{s_{t-1}} - \theta_{s_2} * \epsilon_{s_{t-2}} - \ldots - \theta_{s_m} * \epsilon_{s_{t-m}}
\end{aligned}
\qquad \text{(II.3)}
$$

where $y_{s_t}$ is the time-series value at time $t$, $\phi_{s_i}(i = 1, 2, \ldots, n)$ and $\theta_{s_j}(j = 1, 2, \ldots, m)$ are the autoregressive and moving average parameters of the ARMA model, respectively. $\{\epsilon_t\}$ is a normal white noise with zero mean and a variance of $\sigma_{s_\epsilon}^2$. The hourly insolation $H_t$ at hour $t$ can be derived from the mean insolation $\mu_{s_t}$, its standard deviation $\sigma_{s_t}$, and the time series value $y_{s_t}$ as follows:

$$
H_t = \mu_{s_t} + \sigma_{s_t} * y_{s_t}. \qquad \text{(II.4)}
$$

### C. Storage Batteries (SBs)

Energy storage such as batteries can be used to stabilize the DG (Distributed Generation) output, provide ride-through capacity, and render a non-dispatchable DG unit dispatchable. The wind and solar power are both sporadic in a certain sense since PV can not provide solar power without sunshine and WTG has no power output when there is no sufficiently strong wind blowing. Therefore, it is appealing to think about the energy storage to compensate for the power fluctuations from these renewables.

## III. PROBLEM FORMULATION

The objective of this study is to achieve a grid-linked hybrid generation system, which should be appropriately designed in terms of economic, reliability, and environmental measures subject to physical and operational constraints/strategies [1]-[4].

### A. Design objectives

- Objective 1: Costs

Cost estimation is crucial in decision-making for constructing a hybrid power plant. The total cost $COST(\$/year)$ includes initial cost, operational and maintenance (OM) cost for each type of power source, and the salvage value of each equipment should be deducted [1]-[4]:

$$
COST = \frac{\sum_{i=w,s,b}(I_i - S_{P_i} + OM_{P_i})}{N_p} + C_g \qquad \text{(III.5)}
$$

where $w, s, b$ indicates the wind power, solar power, and battery storage, respectively; $I_i$, $S_{P_i}$, $OM_{P_i}$ are the initial cost, present worth of salvage value, and present worth of operation and maintenance cost for equipment $i$, respectively; $N_p(year)$ is the life span of the project; and $C_g$ is the annual costs for purchasing power from the utility grid. Here we assume that the life time of the project does not exceed those of both WTGs and PV arrays.

1) For the WTGs,

$$
I_w = \alpha_w A_w \qquad \text{(III.6)}
$$

where $\alpha_w(\$/m^2)$ is the initial cost of WTGs; the present worth of the total salvage value is

$$
S_{P_w} = S_w A_w \left(\frac{1+\beta}{1+\gamma}\right)^{N_p} \qquad \text{(III.7)}
$$

where $S_w(\$/m^2)$ is the salvage value of WTGs per square meter, $\beta$ and $\gamma$ are the inflation rate and interest rate, respectively; the present worth of the total operation and maintenance cost (OM) in the project life time is

$$
OM_{P_w} = \alpha_{OM_w} * A_w * \sum_{i=1}^{N_p}\left(\frac{1+\nu}{1+\gamma}\right)^i \qquad \text{(III.8)}
$$

where $\alpha_{OM_w}(\$/m^2/year)$ is the yearly OM cost per unit area and $\nu$ is the escalation rate.

2) For the PV panels, the initial cost is

$$
I_s = \alpha_s A_s \qquad \text{(III.9)}
$$

where $\alpha_s(\$/m^2)$ is the initial cost; the present worth of the total salvage value is

$$
S_{P_s} = S_s A_s \left(\frac{1+\beta}{1+\gamma}\right)^{N_p} \qquad \text{(III.10)}
$$

where $S_s(\$/m^2)$ is the salvage value of PVs per square meters of PV panels; the present worth of the total operation and maintenance cost (OM) in the project life time is

$$
OM_{P_s} = \alpha_{OM_s} * A_s * \sum_{i=1}^{N_p}\left(\frac{1+\nu}{1+\gamma}\right)^i \qquad \text{(III.11)}
$$

where $\alpha_{OM_s}(\$/m^2/year)$ is the yearly OM cost per unit area and $\nu$ is the escalation rate.

3) For the storage batteries, since their life span is usually shorter than that of the project, the total present worth of capital investments can be calculated as follows:

$$I_b = \alpha_b * P_{b_{cap}} * \sum_{i=1}^{X_b} (\frac{1+\nu}{1+\beta})^{(i-1)N_b} \quad \text{(III.12)}$$

where $N_b$ is the life span of SBs; $X_b$ is the number of times to purchase the batteries during the project life span $N_p$; the salvage value of SBs is ignored in this study; and the present worth of the total OM cost in the project life time is calculated as follows:

$$OM_{p_b} = \alpha_{OM_b} * P_{b_{cap}} * \sum_{i=1}^{N_p} (\frac{1+\nu}{1+\gamma})^i \quad \text{(III.13)}$$

where $\alpha_{OM_b}(\$/kWh/year)$ is the yearly OM cost per kilowatthour.

4) For the grid-linked system design, the annual cost for purchasing power from the utility grid can be calculated as follows:

$$C_g = \sum_{t=1}^{T} P_{g,t} * \varphi \quad \text{(III.14)}$$

where $P_{g,t}(\$/year)$ is the power purchased from the utility at hour $t$; $\varphi(\$/kWh)$ is the grid power price; and $T$ (8760 hours) is the operational duration under consideration.

- Objective 2: Reliability
Here Energy Index of Reliability (EIR) is used to measure the reliability of each candidate hybrid system design. EIR can be calculated from Expected Energy Not Served (EENS) and the total energy demanded (E) as follows:

$$EIR = 1 - \frac{EENS}{E} \quad \text{(III.15)}$$

The $EENS(kWh/year)$ for the duration under consideration $T$ (8760 hours) can be calculated as follows [1]-[4]:

$$EENS = \sum_{t=1}^{T} (P_{b_{min}} - P_{b_{soc}}(t) - P_{sup}(t)) * U(t) \quad \text{(III.16)}$$

where $U(t)$ is a step function, which is zero when the supply exceeds or equals to the demand, and equals to one if there is insufficient power in period $t$; $P_d(t)$ is the load demand during hour $t$, $P_{sup}(t) = P_{total}(t) - P_d(t)$ is the surplus power in hour $t$, $P_{total}(t)$ is the total power from WTGs, PVs, and FFGs during hour $t$:

$$P_{total}(t) = P_w(t) + P_s(t) + P_g(t) \quad \text{(III.17)}$$

$P_{b_{soc}}(t)$ is the battery charge level during hour $t$, and $P_{b_{min}}$ is the minimum permitted storage level, the term $P_{b_{soc}}(t) - P_{b_{min}}$ indicates the available power supply from batteries during hour $t$; and provided that there is insufficient power in hour $t$,

$$P_g(t) = \kappa * (P_d(t) - P_w(t) - P_s(t) - P_b(t)) \quad \text{(III.18)}$$

where $\kappa \in [0,1]$ indicates the portion of purchased power with respect to the hourly insufficient power; or else, $P_g(t) = 0$. Note that no generator failures and unexpected load deviations are considered in calculating the EENS, which in this study is all contributed by the fluctuations of renewable power generation.

- Objective 3: Pollutant emissions
Emissions can be modeled through functions that associate emissions with power production for generating units:

$$PE = \alpha + \beta * \sum_{t=1}^{T} P_{g,t} + \gamma * (\sum_{t=1}^{T} P_{g,t})^2 \quad \text{(III.19)}$$

where $\alpha$, $\beta$, and $\gamma$ are the coefficients approximating the generator emission characteristics.

### B. Design constraints

There is a set of constraints that should be satisfied throughout system operations for any feasible solution [1]-[4].

- Constraint 1: Power balance constraint
For any period $t$, the total power supply from the hybrid generation system must supply the total demand $P_d$ with a certain reliability criterion. This relation can be represented by

$$P_w(t) + P_s(t) + P_b(t) + P_g(t) \geq (1-R)P_d(t) \quad \text{(III.20)}$$
$$P_w(t) + P_s(t) + P_b(t) + P_g(t) - P_{dump}(t) \leq P_d(t) \quad \text{(III.21)}$$

where $P_w$, $P_s$, $P_b$, $P_g$, $P_{dump}(t)$, and $P_d$ are the wind power, solar power, charged/discharged battery power, power bought from grid, dumped power, and total load demand, respectively; $R$ is the ratio of the maximum permissible unmet power with respect to the total load demand in each time instant. The transmission loss is not considered in this investigation.

The output $P_{WTG}$ $(kW/m^2)$ from WTGs for wind speed $V_t$ can be calculated as

$$P_{WTG} = \begin{cases} 0, & V_t < V_{ci} \\ a * V_t^3 - b * P_r, & V_{ci} \leq V_t < V_r \\ P_r, & V_r \leq V_t \leq V_{co} \\ 0 & V_t > V_{co} \end{cases} \quad \text{(III.22)}$$

where $a = \frac{P_r}{V_r^3 - V_{ci}^3}$, $b = \frac{V_{ci}^3}{V_r^3 - V_{ci}^3}$, $P_r$ is the rated power, $V_{ci}$, $V_r$, and $V_{co}$ are the cut-in, rated, and cut-out wind speed, respectively. The real electric power from WTGs can be calculated as follows:

$$P_w = P_{WTG} * A_w * \eta_w \quad \text{(III.23)}$$

where $A_w$ is the total swept area of WTGs and $\eta_w$ is the efficiency of WTGs.

The output power $P_s(kW)$ from PV panels can be calculated as follows:

$$P_s = H * A_s * \eta_s \quad \text{(III.24)}$$

where $H(kW/m^2)$ is the horizontal irradiance, $A_s$ is the PV area, and $\eta_s$ is the efficiency of PV panels.

808

- Constraint 2: Bounds of design variables
The swept area of WGTs should be within a certain range:

$$A_{w_{min}} \leq A_w \leq A_{w_{max}} \qquad \text{(III.25)}$$

Similarly, the area of PV arrays should also be within a certain range:

$$A_{s_{min}} \leq A_s \leq A_{s_{max}} \qquad \text{(III.26)}$$

The state of charge (SOC) of storage batteries $P_{b_{soc}}$ should not exceed the capacity of storage batteries $P_{b_{cap}}$ and should be larger than the minimum permissible storage level $P_{b_{min}}$; the total SB capacity should not exceed the allowed storage capacity $P_{b_{cap}max}$; and the hourly charge or discharge power $P_b$ should not exceed the hourly inverter capacity $P_{b_{max}}$. As a result,

$$P_{b_{min}} \leq P_{b_{soc}} \leq P_{b_{cap}} \qquad \text{(III.27)}$$

$$0 \leq P_{b_{cap}} \leq P_{b_{cap}max} \qquad \text{(III.28)}$$

$$P_b \leq P_{b_{max}} \qquad \text{(III.29)}$$

The amount of power bought from utility grid should be within a certain range:

$$P_{g_{min}} \leq \sum_{t=1}^{T} P_{g,t} \leq P_{g_{max}}, \qquad \text{(III.30)}$$

where $P_{g_{min}}$ and $P_{g_{max}}$ are the minimum and maximum power allowed to be bought from the utility grid, respectively.

The coefficient $\kappa$ indicates the portion of purchased power from utility grid with respect to the insufficient power:

$$0 \leq \kappa \leq 1 \qquad \text{(III.31)}$$

### C. Problem statement

In summary, for the grid-link system design, the objective of optimum design for renewable hybrid generation system is to simultaneously minimize $COST(A_w, S_w, P_{b_{cap}}, \kappa)$ and $PE(A_w, S_w, P_{b_{cap}}, \kappa)$, as well as maximize $EIR(A_w, S_w, P_{b_{cap}}, \kappa)$, subject to the constraints (III.20)–(III.31). The design parameters that should be derived include WTG swept area $A_w(m^2)$, PV area $A_s(m^2)$, total battery capacity $P_{b_{cap}}(kWh)$, and the ratio of power purchased from grid $\kappa$.

## IV. OPERATION STRATEGIES

The power outputs from WTGs and PVs have the highest priorities to feed the load. Only if the total power from wind and solar systems is insufficient to satisfy the load demand, the storage batteries can be discharged a certain amount of energy to supply the load. If there is still not enough power to supply the load, a certain amount of power will be purchased from the utility grid. Furthermore, if there is any excess power from WTGs and PVs, the batteries will be charged to store a certain permissible amount of energy for future use. If there is surplus power from WTGs and PVs even after feeding the load and charging the SBs, the dump load will consume the spilled power.

## V. MECHANISM OF PARTICLE SWARM OPTIMIZATION

Particle swarm optimization (PSO) is a population-based stochastic optimization procedure inspired by certain social behaviors in bird groups and fish schools [5]. Assume $x$ and $v$ denote a particle position and its speed in the search space. Therefore, the $i$-th particle can be represented as $x_i = [x_{i_1}, x_{i_2}, \ldots, x_{i_d}, \ldots, x_{i_M}]$ in the $M$-dimensional space. Each particle continuously records the best solution it has achieved thus far during its flight. This fitness value of the solution is called $pbest$. The best previous position of the $i$-th particle is memorized and represented as $pbest_i = [pbest_{i_1}, pbest_{i_2}, \ldots, pbest_{i_d}, \ldots, pbest_{i_M}]$. The global best $gbest$ is also tracked by the optimizer, which is the best value achieved so far by any particle in the swarm. The best particle of all the particles in the swarm is denoted by $gbest_d$. The velocity for particle $i$ is represented as $v_i = (v_{i_1}, v_{i_2}, \ldots, v_{i_d}, \ldots, v_{i_M})$. The velocity and position of each particle can be continuously adjusted based on the current velocity and the distance from $pbest_{i_d}$ to $gbest_d$:

$$\begin{aligned}
v_{i_d}^{(t+1)} &= w * v_{i_d}^{(t)} + c_1 * \text{rand}() * (pbest_{i_d} - x_{i_d}^{(t)}) \\
&\quad + c_2 * \text{Rand}() * (gbest_d - x_{i_d}^{(t)}), \qquad \text{(V.32)}
\end{aligned}$$

$$x_{i_d}^{(t+1)} = x_{i_d}^{(t)} + \chi * v_{i_d}^{(t+1)}, i = 1, 2, \ldots, N, d = 1, 2, \ldots, M. \qquad \text{(V.33)}$$

where $N$ is the number of particles in a swarm, $M$ is the number of members in a particle, $t$ is the counter of generations, $\chi \in [0, 1]$ is the constriction factor which controls the velocity magnitude, $w$ is the inertia weight factor, $c_1$ and $c_2$ are acceleration constants, rand() and Rand() are uniform random values in a range $[0, 1]$, $v_i^{(t)}$ is the velocity of particle $i$ in generation $t$, and $x_i^{(t)}$ is the current position of particle $i$ in generation $t$.

## VI. THE PROPOSED APPROACH

A Constrained Mixed-Integer Multi-Objective PSO (CMI-MOPSO) is developed to derive a set of non-dominated solutions.

### A. CMIMOPSO

- Mixed-integer PSO: Since the target problem involves the optimization of system configuration, integer numbers are used to indicate the unit sizing. The standard PSO is in fact a real-coded algorithm, thus some revisions are needed to enable it to deal with the binary-coded optimization problem. In the discrete binary PSO [6], the relevant variables are interpreted in terms of changes of probabilities. A particle flies in a search space restricted to zero and one in each direction and each $v_{i_d}$ represents the probability of member $x_{i_d}$ taking value 1. The update rule governing the particle flight speed can be modified accordingly by introducing a logistic sigmoid transformation function:

$$S(v_{i_d}) = \frac{1}{1 - e^{-v_{i_d}}} \qquad \text{(VI.34)}$$

The velocity can be updated according to this rule: If $rand() < S(v_{i_d})$, then $x_{i_d} = 1$; or else $x_{i_d} = 0$. The maximum allowable velocity $V_{max}$ is desired to limit the probability that member $x_{i_d}$ will take a one or zero value. The smaller the $V_{max}$ is, the higher the chance of mutation is for the new individual.

- Multi-objective PSO: In this study, since a multi-objective optimization problem is concerned, the standard PSO algorithm is also modified accordingly to facilitate a multi-objective optimization approach, i.e., multi-objective particle swarm optimization (MOPSO). The Pareto-dominance concept is used to appraise the fitness of each particle and thus determine which particles should be chosen as the non-dominated solutions. For this purpose, the archiving mechanism is used to store the non-dominated solutions throughout the optimization process. The best historical solutions found by the optimizer are absorbed continuously into the archive as the non-dominated solutions generated in the past. Furthermore, to enhance the solution diversity, some diversity preserving measures such as fuzzified global best selection, and niching and fitness sharing are taken.

- Constrained PSO: In the proposed method, a natural constraint checking procedure called rejection strategy is adopted to deal with the imposed constraints. When an individual is evaluated, the constraints are first checked to determine if it is a feasible candidate solution. If it satisfies all of the constraints, it is then compared with the non-dominated solutions in the archive. The concept of Pareto dominance is applied to determine if it is eligible to be chosen to store in the archive of non-dominated solutions. As long as any constraint is violated, the candidate solution is deemed to be infeasible. This procedure is simple to implement but it turns out to be quite effective in ensuring the solution feasibility while not significantly slowing down the search.

### B. Representation of candidate solutions

The design variables including WTG swept area, PV area, amount of power purchased from the grid, and total SB capacity are encoded as the position value in each dimension of a particle. Several member positions indicate the coordinate of the particle in a multi-dimensional search space. Each particle is considered as a potential solution to the optimal design problem, since each of them represents a specific configuration of the hybrid generation system. Excluding $P_{b_{cap}}$, all the remaining positions are real-coded. The $i$-th particle (i.e., candidate design) $D_i$ can be represented as follows:

$$D_i = [P_{w,i}, P_{s,i}, P_{b_{cap},i}, \kappa_i], \quad i = 1, 2, \ldots, N \quad \text{(VI.35)}$$

where $\kappa$ is ratio of power bought from the grid with respect to the deficit power, and the total SB capacity $P_{b_{cap}}$ is encoded using three binary bits.

## VII. SIMULATION AND EVALUATION OF THE PROPOSED APPROACH

In this section, the tradeoff solutions are derived by the developed optimization procedure and some sensitivity studies

TABLE I

THE DATA USED IN THE SIMULATION PROGRAM.

| System parameters | Values |
|---|---|
| Efficiency of WTG ($\eta_w$) | 50% |
| Efficiency of PV ($\eta_s$) | 16% |
| Efficiency of SB ($\eta_b$) | 82% |
| Inflation rate ($\beta$) | 9% |
| Interest rate ($\gamma$) | 12% |
| Escalation rate ($\nu$) | 12% |
| Life span of project ($N_p$) | 20 years |
| Life span of WTG ($N_w$) | 20 years |
| Life span of PV ($N_s$) | 22 years |
| Life span of SB ($N_b$) | 10 years |
| PV panel price ($\alpha_s$) | 450\$/$m^2$ |
| WTG price ($\alpha_w$) | 100\$/$m^2$ |
| SB price ($\alpha_b$) | 100\$/$KWh$ |
| PV panel salvage value ($S_s$) | 45\$/$m^2$ |
| WTG salvage value ($S_w$) | 10\$/$m^2$ |
| OM costs of WTG ($\alpha_{OM_w}$) | 2.5\$/$m^2$/$year$ |
| OM costs of PV panel ($\alpha_{OM_s}$) | 4.3\$/$m^2$/$year$ |
| OM costs of SB ($\alpha_{OM_b}$) | 10\$/$KWh$ |
| Cut-in wind speed ($V_{ci}$) | 2.5 m/s |
| Rated wind speed ($V_r$) | 12.5 m/s |
| Cut-out wind speed ($V_{co}$) | 20.0 m/s |
| Rated WTG power ($P_r$) | 4.0 kW |
| Period under observation ($T$) | 8760 hours |
| Maximum swept area of WTGs ($A_{w_{max}}$) | $10,000 m^2$ |
| Minimum swept area of WTGs ($A_{w_{min}}$) | $400 m^2$ |
| Maximum area of PV panels ($A_{s_{max}}$) | $200 m^2$ |
| Minimum area of PV panels ($A_{s_{min}}$) | $8,000 m^2$ |
| Maximum conversion capacity ($P_{b_{max}}$) | 3 kWh |
| Minimum storage level ($P_{b_{min}}$) | 3 kWh |
| Rated battery capacity ($P_{b_r}$) | 8 kWh |
| Maximum total SB capacity ($P_{b_{max}}$) | 40 kWh |
| Price of utility grid power ($\varphi$) | 0.12\$/kWh |

are carried out.

### A. System parameters

The data used in the simulation program are listed in Table I [2]. The hourly wind speed patterns, the hourly insolation conditions, and the hourly load profile are shown in Figure 2.

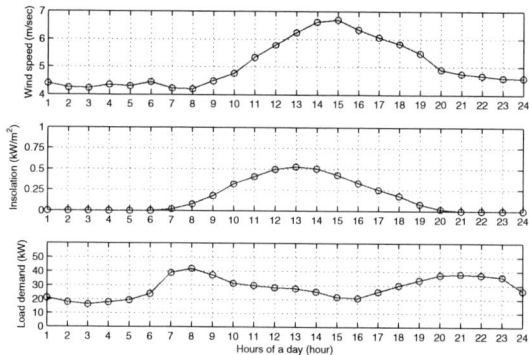

Fig. 2. Hourly mean wind speed, insolation, and load profiles.

### B. PSO parameters

In the simulations, both the population size and archive size are set to 100, and the maximum number of iterations is set

TABLE II

TWO ILLUSTRATIVE NON-DOMINATED SOLUTIONS FOR TRI-OBJECTIVE
OPTIMIZATION.

| Variables/objectives | Design 1 | Design 2 |
|---|---|---|
| $A_w(m^2)$ | 810 | 680 |
| $A_s(m^2)$ | 40 | 40 |
| $P_{b_{cap}}(kWh)$ | 16 | 16 |
| $\kappa$ | 0.34 | 0.58 |
| Cost (\$/year) | 7,919.55 | 7,012.10 |
| EIR | 0.9579 | 0.9521 |
| Emissions (ton/year) | 17.0836 | 59.8295 |

to 500. The acceleration constants $c_1$ and $c_2$ are chosen as 1. Both turbulence factor and niche radius are set to 0.02. The inertia weight factor $w$ decreases when the number of generations increases:

$$w = w_{max} - \frac{w_{max} - w_{min}}{iter_{max}} \times iter \qquad (\text{VII.36})$$

where $iter_{max}$ is the maximum number of iterations and $iter$ is the current number of iterations. The simulation program is coded using C++ and executed in a 2.20 GHz Pentium-4 processor.

C. Simulation results

The Pareto-optimal fronts evolved using the proposed approach for bi- and tri-objective optimization problems are shown in Figure 3, and two illustrative non-dominated solutions are listed in Table II.

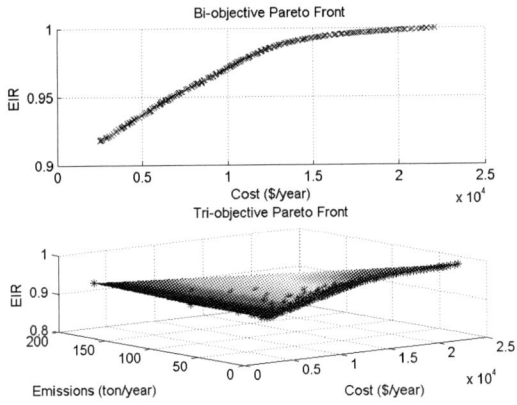

Fig. 3. Pareto fronts for bi- and tri-objective optimization scenarios.

D. Sensitivity to system parameters

The results of sensitivity analysis are illustrated in Figure 4 and Figure 5, respectively. From Figure 4, we can appreciate the importance of site locations for a wind power plant. The mean wind speeds are changed by different multiplication factors (MFs). The simulation results shown in Figure 5 fit with the cost estimation equations described in Section III.

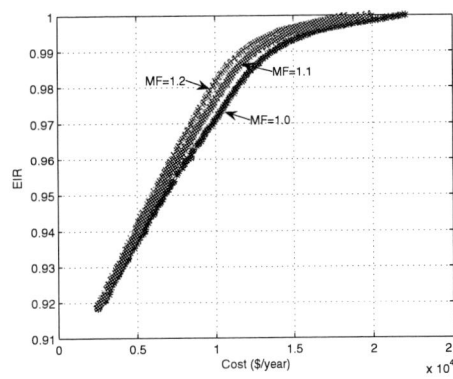

Fig. 4. Pareto fronts obtained from different mean wind speeds.

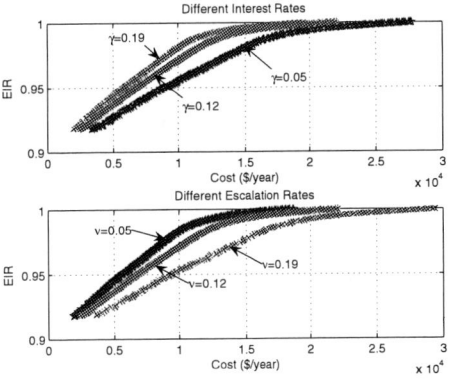

Fig. 5. Pareto fronts obtained from different economic rates.

VIII. CONCLUDING REMARKS

In this paper, a hybrid power generation system including wind power and solar power is designed on the basis of cost, reliability, and emission criteria. ARMA model is used to forecast the wind speed and solar insolation based on a set of time-series data. In the future work, more system uncertainties such as generator failures will also be considered.

REFERENCES

[1] Chedid, R., Akiki, H., and Rahman, S. (1998). A decision support technique for the design of hybrid solar-wind power systems, *IEEE Transactions on Energy Conversion*, Vol. 13, No. 1, March, pp. 76–83.

[2] Chedid, R. and Rahman, S. (1997). Unit sizing and control of hybrid wind-solar power systems, *IEEE Transactions on Energy Conversion*, Vol. 12, No. 1, March, pp. 79–85.

[3] Kellogg, W., Nehrir, M. H., Venkataramanan, G., and Gerez, V. (1996). Optimal unit sizing for a hybrid wind/photovoltaic generating system, *Electric Power Systems Research*, Vol. 39, pp. 35–38.

[4] Kellogg, W., Nehrir, M. H., Venkataramanan, and G., Gerez, V. (1998). Generation unit sizing and cost analysis for stand-alone wind, photovoltaic, and hybrid wind/PV systems, *IEEE Transactions on Energy Conversion*, Vol. 13, No. 1, March, pp. 70–75.

[5] Kennedy, J. and Eberhart, R. (1995). Particle swarm optimization, *IEEE Proceedings of the International Conference on Neural Networks*, Perth, Australia, pp. 1942–1948.

[6] Kennedy, J. and Eberhart, R. (1997). A discrete binary version of the particle swarm optimization, *IEEE Proceedings of the International Conference on Neural Networks*, Perth, Australia, pp. 4104–4108.

**2006 IEEE International Conference on Power Electronic, Drives and Energy Systems**

# SVPWM Implementation in dSPACE for Generalized Impedance Controller Used for Self Excited Induction Generation System

B.Venkatesa Perumal, and J.K.Chatterjee, *Member, IEEE*

*Abstract*— **In this paper space vector pulse width modulation implementation using dSPACE 1104 to control the operation of a Generalized Impedance Controller (GIC) has been explained with the hardware results. The GIC is used for regulating terminal voltage and frequency of a Self Excited Induction Generator (SEIG). It consists of an inverter having DC link battery with AC side coupling inductance. A space vector control algorithm has been developed; by writing a code in MATLAB programming environment and it has been implemented in dSPACE 1104 kit. This paper shows the implementation of SVPWM algorithm in dSPACE 1104 to investigate the open loop performance of the GIC-SEIG system and the results are presented.**

*Index Terms*-- **frequency controller, generalized impedance controller (GIC) and self-excited induction generator (SEIG).**

## I. INTRODUCTION

THREE-phase squirrel-cage induction machine operated as SEIG has gained popularity due to its ruggedness, ease of maintenance, reduced unit cost. The fundamental problem with the SEIG is poor voltage and frequency regulation [2].In order to regulate its terminal voltage and frequency, the active and reactive power level at terminals of the SEIG has to be maintained constant. Recent break-through in new types of VAR compensators has led to enormous development in self-excited induction generator control technology. All attempts have been made to regulate these parameters in a desired preprogrammed manner under varying source power and load volt-ampere conditions. Integrated control of prime mover, generating machine and power controller and the interplay between the three, while in operation under steady state and dynamic conditions have attracted greater attention.

In this paper, attempt has been made to study the integrated SEIG-GIC system performance behavior under open loop condition. The GIC switches are controlled independently by the dSPACE. The space vector pulse width modulation, program has been implemented in dSPACE 1104 kit. First

---

B.Venkatesa Perumal is with the Department of Electrical Engineering, , Indian Institute of Technology Delhi, New Delhi, India (e-mail: bvperumal@gmail.com).

J. K. Chatterjee is with the Department of Electrical Engineering, Indian Institute of Technology Delhi, New Delhi, India (e-mail: jkc@ee.iitd.ac.in).

section of the paper is consisting of system description, followed by SVPWM generation algorithm implementation in dSPACE 1104 kit. Subsequently the integrated SEIG-GIC system has been analyzed with source and load perturbation. Novel observation of the present work is reported in this paper.

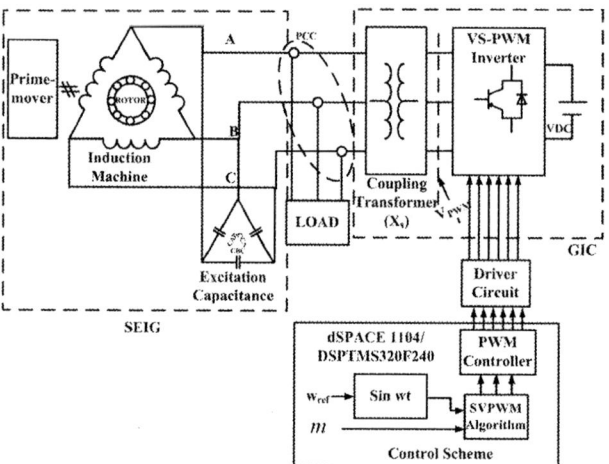

Fig. 1. Schematic diagram of SEIG with GIC.

## II. SYSTEM DESCRIPTION

Fig. 1 shows the schematic diagram of an integrated SEIG–GIC system. In the Fig. 1 a three-phase SEIG with a bank of delta-connected fixed excitation capacitors, the GIC and a balanced resistive load are connected at the point of common coupling (PCC). The GIC is a pulse width modulated bi-directional voltage source inverter with a bank of battery connected at its DC bus having voltage magnitude "$V_{DC}$". It is connected to the PCC through an interconnecting transformer, having leakage reactance "$X_s$". As stated in [3] the GIC offers variable controlled impedance across the SEIG terminals according to the values of the modulation index "$m$" of the GIC and the phase angle "$\delta$", between the fundamental component of the GIC voltage "$V_{PWM}$" and the SEIG terminal voltage "$V_{ac}$". In the diagram the prime mover represents a controllable variable speed mechanical power source, which can simulate the source perturbation conditions. A dSPACE 1104 kit, has been used to provide the electrical commands at desirable switching instants for the GIC power modules, which are identified according the values of the GIC modulation index *"m"* and phase angle *"δ"*. The resident software in dSPACE 1104 kit generates control pulses, which are amplified by the driver IR2130 and given to the gate of the IGBT switches. Fig. 3 shows the flowchart of the complete program implemented in dSPACE.

### III. SVPMW PROGRAM IN DSPACE

Fig. 2 shows control algorithm implemented in dSPEACE for the integrated SEIG-GIC system. The inputs to the space vector algorithm block are the three phase voltage signals. These three phase voltage signals are converted into two-phase quantities by applying the clark's transformation. By using these two signals, sector is identified. The duty ratios are calculated by using angle *"δ"* and modulation index *"m"* (see Fig. 2). Using both duty ratio and sector number, switching signals has been constructed. Fig. 4 shows flowchart the space vector pulse width modulation program implemented in dSPACE. The results for different values of modulation index *"m"* are given below to show the effectiveness of the space vector modulation program implemented in dSPACE. Fig. 5(a)-(b) shows the space vector program output waveform for different values of modulation index *"m"* (0.5 and 1.0). These waveforms are captured using the data acquisition system in dSPACE 1104kit.

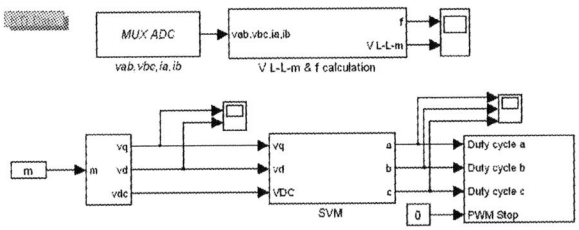

Fig. 2. SVPWM algorithm implemented in dSPEACE for the integrated SEIG-GIC system open loop analysis in MATLAB/Simulink environment.

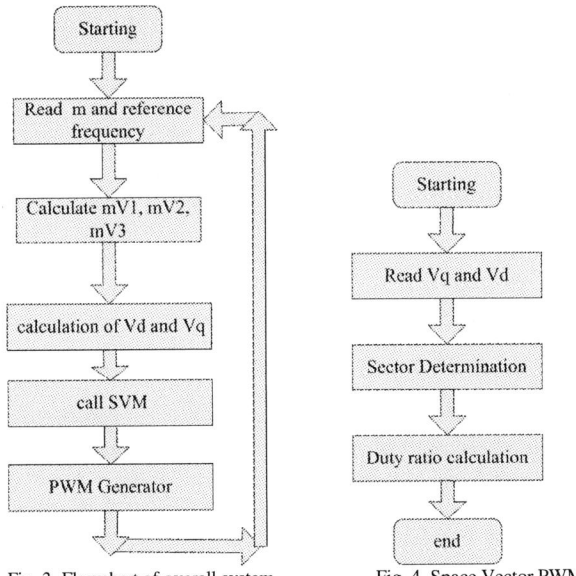

Fig, 3. Flowchart of overall system operation.

Fig. 4. Space Vector PWM flowchart.

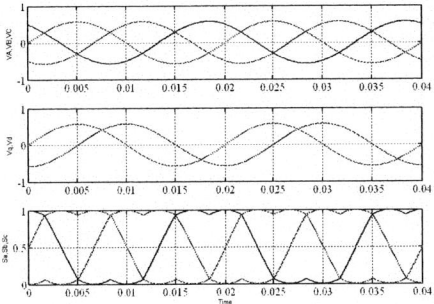

Fig. 5(a). SVPWM block output for modulation index *"m"* =0.50 in dSPACE.

Fig. 5(b). SVPWM block output for modulation index *"m"* =1 in dSPACE.

### IV. RESULTS AND DISCUSSION

Initially the prime mover speed $\omega_r$ is set to 158.65rad/sec, which is slightly above the synchronous speed of the 50 Hz, four-pole induction machine. While operating with the three-phase capacitor bank the SEIG voltage builds up to a steady state peak value ($V_{ABIG}$) to 405volts(l-l) and frequency of its terminal voltage is held at 50 Hz, which is measured by using PLL. The GIC is connected to the SEIG terminals, with fundamental frequency of its output voltage at 50Hz and its modulation index *"m"* is set at *"$m_{crit}$"*=0.75 so that the voltage amplitude ratio *"r"*= *"$V_{PWM}$ / $V_{ac}$"* = 1 and the relative phase angle between the SEIG and the fundamental component of the GIC output voltage waveforms *"δ"*=0°. The fundamental output frequency of the GIC is set equal to 50 Hz, which forms a constant frequency bus at the SEIG terminals. The SEIG draws leading reactive power supplied by the excitation capacitors. This sets the base operating condition for the SEIG-GIC system around which the investigations of the system behavior following application of perturbations of the kind stated below. In the following investigation, the transient behavior of integrated SEIG-GIC system has been studied when the speed, resistive load and the GIC parameter *"m"* perturbations are independently applied. Among the GIC parameters, perturbation is given only to modulation index *"m"*; no perturbation is given to *"δ"*.

In the experiment following perturbations are considered:
i) Insertion of three-phase balanced resistive load of 230-ohm/phase at the PCC,
ii) Increase in the prime mover speed from 158.65-rad/sec to 160 rad/sec,

iii) Increase in the prime mover speed from 160 rad/sec to 162 rad/sec and

iv) Insertion of balanced resistive load of 120-ohm/phase at the PCC.

Following each perturbation, in dSPACE 1104 kit, the peak of the SEIG terminal voltage is instantaneously detected and they are shown as an envelop in Fig. 6(a)-10(a). The frequency of the SEIG terminal voltage waveform is calculated by using PLL and they are shown in Fig. 6(b)-10(b).

Fig. 6- Fig. 10 show various plots during the transients following the abovementioned five case of application of perturbations in consideration. The description of various waveforms for each case in consideration, are given below. For the transients following the resistive load perturbation Fig. 6(a)–(c) are plotted against time. They respectively show (a) variation in the SEIG peak line voltage during the transient (b) variation in frequency of the SEIG terminal voltage during the transient, (c) scaled line voltage waveforms of the SEIG and the GIC.

In the following subsections, the explanation for the effects of each type of perturbation on all the parameters of the SEIG-GIC system, as listed above, are given in detail.

### A. Three Phase Balanced Resistive Load (230 ohm/phase) Perturbation

The SEIG peak line voltage and frequency, around 405 volts and 50 Hz are shown in Fig. 6(a) and (b) respectively, before apply the three phase balanced resistive load perturbation. The GIC parameter modulation index "m" is set at 0.75 and its D.C bus having a 120 volts independent source.

At time t=4.5 sec, a three-phase balanced resistive load of 230-ohm/phase is connected across the SEIG output terminals (PCC). Due to this additional active power loading, momentarily the difference between machine shaft power supplied by the prime mover and electrical power generated by the machine causes the frequency to fall to about 49.2Hz (see Fig. 6(b)). With the GIC terminal frequency remaining the same at 50 Hz, due to the increase in the frequency differential between the SEIG and the GIC output voltages, which leads to growth in "$\delta$" in the positive direction. The SEIG starts injecting active power (370 watts) along with the GIC into the load. This leads to growth in the magnitudes of SEIG stator and rotor current. As a consequence, under open loop condition, the peak SEIG terminal voltage falls to about 330 volts (see Fig. 6(a)), due to voltage drop in its stator and rotor impedances "$I_sZ_s$" and "$I_rZ_r$". Due to fall in the SEIG terminal voltage the reactive power supplied by the excitation capacitor goes down, so the resultant lagging reactive power drawn into the SEIG terminal also goes down. With "m" of the GIC remaining the same the amplitude ratio "r"=1.19 between the SEIG and the GIC output voltages increases thus letting the GIC act as an equivalent capacitor and it draws leading reactive power from the SEIG, consequently, the terminal voltage of the SEIG is restored to a higher value. This process continues until the active and reactive power balance at the terminals of SEIG. At the end of transient the SEIG peak line voltage settles to 340 volts (see Fig. 6(a)) and its frequency becomes equal to the set GIC fundamental frequency 50 Hz (see Fig. 6(b)). The active and reactive power supplied by the SEIG settles to about 120 watts and 1450 lagging var. Fig. 6(c) shows the SEIG and GIC line voltage waveforms. In the Fig. 6(c) GIC voltage waveform is leading the SEIG voltage waveform by "$\delta$" =18°, this indicates the GIC supplies the active power to load.

### B. Prime Mover Speed Rise Perturbation (158.65 rad/sec to 160 rad/sec)

The peak SEIG line voltage and frequency are about 340 volts and 50 Hz (see Fig. 7(a) and (b)). The GIC modulation index "m" and the D.C bus are still set at 0.75 and 120 volts respectively. The SEIG delivers 120 watts of the active power and it draws about 1450-var lagging, when the GIC and balanced there phase resistive load of 230 ohm/phase connected across its terminal prior to the application of the speed perturbation from 158.65 rad/sec to 160 rad/sec.

At time t=7.5 sec, the prime mover speed of the SEIG is raised from 158.65 rad/sec to 160 rad/sec. Due to rise in speed, the difference between shaft power supplied to the machine by the prime mover and the electrical power generated by the machine causes the frequency to rise to 50.25Hz (see Fig. 7(b)). With the GIC terminal frequency remaining the same at 50 Hz, due to sudden rise in the frequency differential between the SEIG and the GIC output voltages and consequent growth in the relative phase angle "$\delta$" between the SEIG and the GIC terminal voltages, in the negative direction, this causes an increase in the active power supplied by the SEIG and also a fall in the active power supplied by the GIC. Under open loop condition, the peak SEIG terminal voltage increases to 380 volts (see Fig. 7(a)). With "m" of the GIC remaining the same the amplitude ratio "r" between the SEIG and the GIC output voltages decreases to 1.09, thus the GIC draws smaller leading reactive power. However, with reduced "$\delta$", the fall in leading reactive power drawn by the GIC becomes smaller. Due to this reactive power supplied by the excitation capacitor will go high, so the resultant leading reactive power supplied by the SEIG goes high.

This process continues until the active and reactive power balance across the terminals of SEIG. At the end of transient the peak SEIG line voltage settles to around 380 volts (see Fig. 7(a) ) and its frequency is coming back to GIC set frequency 50 Hz (see Fig. 7(b)). The active and reactive power supplied by the SEIG is around 250 watts and 1770 var.

In the Fig. 7(c) GIC voltage waveform is leading the SEIG voltage waveform. As compared to the previous case the lead angle is reduced to "$\delta$"= 4°, which indicates the GIC, supplies smaller active power to the load.

### C. Prime Mover Speed Rise Perturbation (160 rad/sec to 162rad/sec)

The peak SEIG line voltage and frequency are about 380 volts and 50 Hz (see Fig. 8(a) and (b)). The GIC modulation index "m" and the D.C bus are still set at 0.75 and 120 volts respectively. The SEIG delivers 250 watts of the active power and it draws about 1770-var lagging. When balanced there phase resistive load of 230 ohm/phase connected across its terminal prior to the application of the speed perturbation from 160 rad/sec to 162 rad/sec

814

At time t= 6.5 sec., the prime mover speed of the SEIG is raised from 160 rad/sec to 162 rad/sec. This is done to make final "δ"=0°. Due to further rise in speed, the difference between shaft power supplied to the machine by the prime mover and the electrical power generated by the machine causes the frequency to rise to 50.25Hz (see Fig. 8(b)). With the GIC terminal frequency remaining the same at 50 Hz, due to rise in the frequency differential between the SEIG and the GIC output voltages and consequent growth in the relative phase angle "δ" between the SEIG and the GIC terminal voltages, in the negative direction, this causes an increase in the active power supplied by the SEIG and also a further fall in the active power supplied by the GIC.

Under open loop condition, the peak SEIG terminal voltage further increases to 405 volts (see Fig. 8(a)). With "m" of the GIC remaining the same the amplitude ratio "r" between the SEIG and the GIC output voltages decreases to 1.0 and the "δ" also reduces to 0°. Thus, the GIC neither draws active nor reactive power. Due to this reactive power supplied by the excitation capacitor will go high, so the resultant leading reactive power supplied by the SEIG goes high. This process continues until the active and reactive power balance across the terminals of SEIG. At the end of transient the peak SEIG line voltage settles to around 405 volts (see Fig. 8(a)) and its frequency is comes back to GIC set frequency 50 Hz (see Fig. 8(b)). The active and reactive power supplied by the SEIG is around 440 watts and 2275 var.

In the Fig. 8(c) GIC voltage waveform is leading the SEIG voltage waveform. As compared to the previous case the lead angle is reduced to "δ"=0°, which indicates the GIC, supplies negligible active power to the load.

### D. Three Phase Balanced Resistive Load (120 ohm/phase) Perturbation

The peak SEIG line voltage and frequency are about 405 volts and 50 Hz (see Fig. 9(a) and (b)). The GIC modulation index "m" and the D.C bus are still set at 0.75 and 120 volts respectively. The SEIG delivers 450 watts of the active power and it draws about 2275-var lagging. When the GIC and balanced there phase resistive load of 230 ohm/phase connected across its terminal prior to the application of additional load perturbation of 120 Ω/phase.

At time t= 2 sec, additional three phase balanced resistive load of 120 ohm/ phase is connected across the SEIG output terminals. This is done to increase the "δ" to larger positive value, so that, the flow of active power from the GIC in the negative direction be experimentally observed. Due to additional loading, momentarily the difference between machine shaft power supplied by the prime mover and electrical power generated by the machine causes the frequency to fall to about 48.7Hz (see Fig. 9(b)). With the GIC terminal frequency remaining the same at 50 Hz, due to the increase in the frequency differential between the SEIG and the GIC output voltages, which leads to growth in "δ" in the positive direction. The SEIG starts injecting greater active power (850 watts) along with the GIC into the load. Under open loop condition, the peak SEIG terminal voltage falls to

about 250 volts (see Fig. 9(b)), Due to fall in the SEIG terminal voltage the reactive power supplied by the excitation capacitor goes down, so the resultant lagging reactive power drawn into the SEIG terminal also goes down. With "m" of the GIC remaining the same the amplitude ratio "r"=1.33 between the SEIG and the GIC output voltages increases thus letting the GIC draw greater leading reactive power from the SEIG, consequently, the terminal voltage of the SEIG is restored to a higher value. This process continues until the active and reactive power balance at the terminals of SEIG. At the end of transient the SEIG peak line voltage settles to 300 volts (see Fig. 9(a)) and its frequency becomes equal to the set GIC fundamental frequency 50 Hz (see Fig. 9(b)). The active and reactive power supplied by the SEIG settles to about 520 watts and 1200 lagging var.

In the Fig. 9(c) GIC voltage waveform is leading the SEIG voltage waveform by "δ" =18°, this indicates, the GIC supplies the active power.

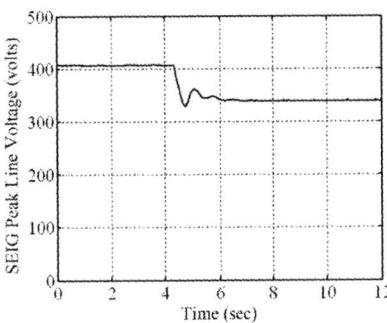

Fig. 6(a). SEIG line voltage peak during 230 ohm/phase connected at PCC.

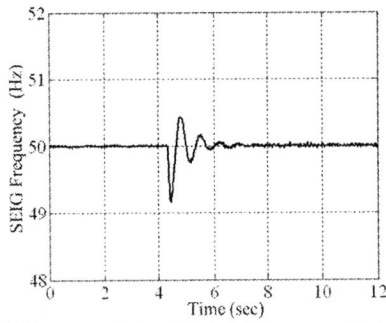

Fig. 6(b). SEIG frequency during 230 ohm/phase connected at PCC.

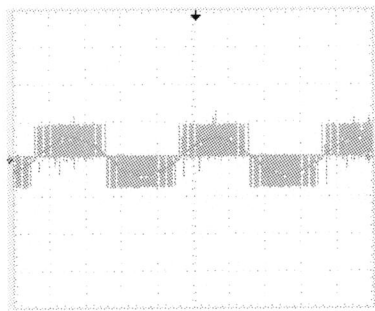

Fig. 6(c). SEIG and GIC line voltage waveform with 250 Ω/phase resistive load.

815

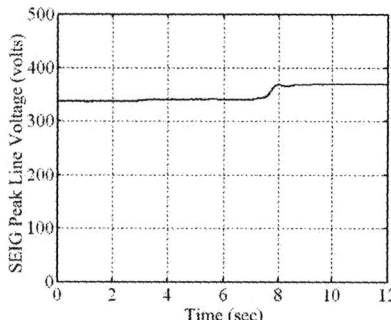

Fig. 7(a). SEIG line voltage peak during speed rise 158.65 rad/sec to160 rad/sec.

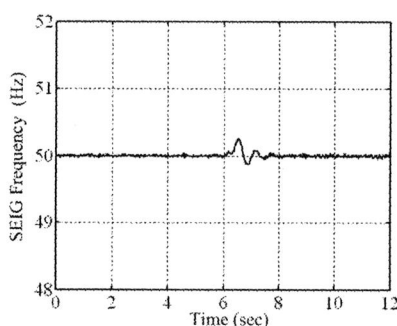

Fig. 8(b). SEIG frequency during speed rise 160 rad/sec to 162rad/sec.

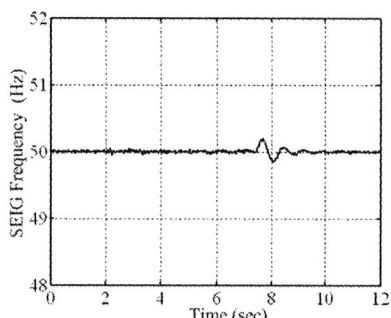

Fig. 7(b). SEIG frequency during speed rise 158.65 rad/sec to160 rad/sec.

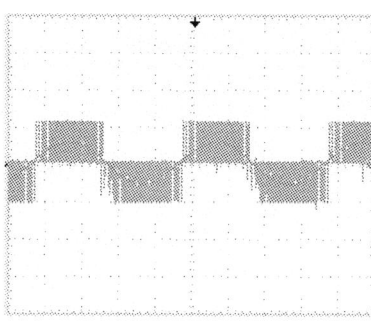

Fig. 8(c). SEIG and GIC line voltage waveform speed raised from 160 rad/sec to 163 rad/sec.

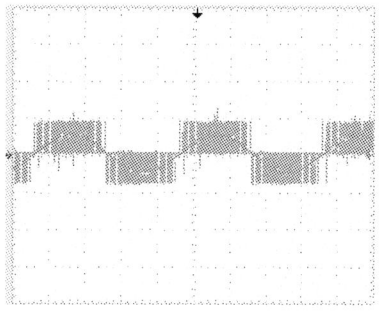

Fig. 7(c). SEIG and GIC line voltage waveform speed rise from 157 rad/sec to 160 rad/sec.

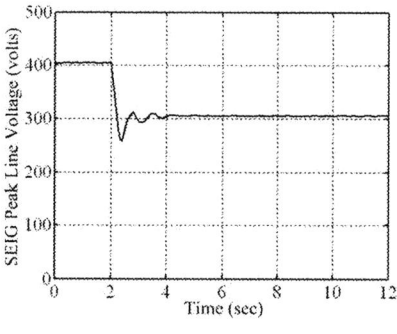

Fig. 9(a). SEIG line voltage peak during 120 ohm/ phase connected.

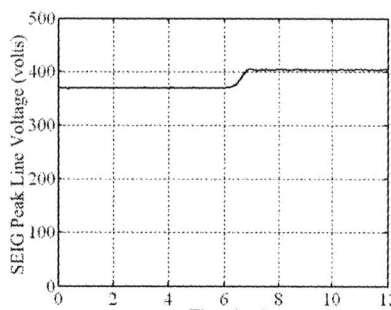

Fig. 8(a). SEIG line voltage peak during speed rise 160 rad/sec to 162 rad/sec.

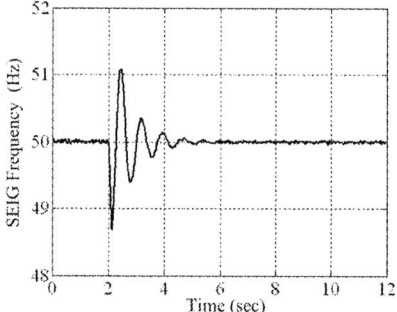

Fig. 9(b). SEIG frequency during 120 ohm/ phase connected.

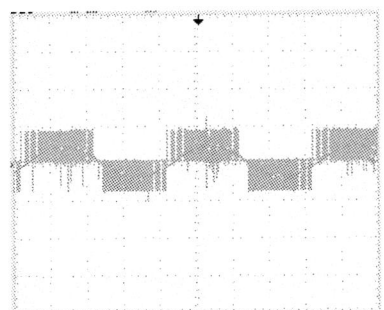

Fig. 9(c). SEIG and GIC line voltage waveform after 120 ohm/phase connected across the SEIG.

## V. CONCLUSION

The space vector pulse width modulation code has been developed in MATLAB/Simulink programming environment and it has been implemented by suing dSPACE 1104 kit for SEIG-GIC system. The performance analysis of a SEIG-GIC system, from the point of view of the SEIG terminal voltage and frequency has been presented in this paper. The results, of the open loop operation of the SEIG-GIC system, show the effects of load and speed perturbations on SEIG terminal voltage and frequency. From the results of source/load perturbations, it is apparent that under steady state condition, following each perturbation, there is no deviation in output frequency of the SEIG. The final outcome of the observations has been, for such operation of SEIG- GIC system, the frequency of SEIG output voltage is controlled by the output frequency "setting" of the GIC, irrespective of the nature and magnitude of the perturbation applied and even under open loop operation of the system.

## VI. APPENDIX-I

### A. . Induction Machine Parameters:

Rating of the induction machine 5 HP, 475V, 7A, 4 pole and its parameters are as follows: $r_s$ =6.5 ohm; $r_r$=3.4 ohm; $L_{ls}=L_{lr}$= 5.10mH;

Magnetizing inductance variation with magnetizing current

$$M = 0.1956 - 0.122i_m + 0.039i_m^2 - 0.0025i_m^3$$

### B. Coupling Transformer:

$R_C$ = 2 ohm; $L_C$ = 0.1H;

## VII. REFERENCES

[1] Yen-Shin Lai and Jian-Ho Chen, "A New Approach to Direct Torque Control of Induction Motor Drives for Constant Inverter Switching Frequency and Torque Ripple Reduction", IEEE Trans. on Energy Convertion, vol.16, no.3, pp 220-227,Sept. 2001.

[2] Bhim Singh and L. B. Shilpakar, "Analysis of a novel solid state voltage regulator for a self-excited induction generator," IEE Proc. Transm.Distrib , vol. 145, no. 6. pp. 647-655, Nov. 1998.

[3] J. K. Chatterjee, B.V.Perumal and N.R.Gopu,"Analysis of operation of Self excited induction generator with Generalized Impedance Controller", to appear IEEE Trans. on Energy Conversion (Available in IEEE xplorer site ).

[4] R. Bonetrt and S.Rajakaruna, "Self-excited induction generator with excellent voltage and frequency control," IEE Proc. Transm. Distrib , vol. 145, no. 1. pp. 33-39, Jan. 1998.

[5] J. K. Chatterjee, P. K. S. Khan, A. Anand and A. Jindal, , "Performance evaluation of an electronic lead-lag VAr compensator and its application in brushless generation," IEEE Power Electronics and Drive Systems Conference, vol. 1 , pp. 59 -64, May 1997.

[6] T. Ahmed, E. Hiraki, M. Nakaoka, O. Noro, " Three-phase self-excited induction generator driven by variable-speed prime mover for clean renewable energy utilizations and its terminal voltage regulation characteristics by static VAr compensator," IEEE Industry Applications conference, vol. 2 , pp. 693 -700, Oct. 2003.

[7] Shashank Wekhande and Vivek Agarwal, "Simple control for a wind driven induction generator," IEEE Industry Application Magazine, pp. 44 -53, March/April.

[8] S. C. Kuo and L. Wang, "Analysis of voltage control for a self-excited induction generator using a current-controlled voltage course inverter (CCVSI)," IEE Proc. Transm. Distrib , vol. 148, no. 5. pp. 431-438, Sep.2001.

[9] D. Seyoum, M. F. Rahman, C. Grantham, , " Terminal voltage control of awind turbine driven isolated induction generator using stator oriented field control," IEEE Applied Power Electronics Conference and Exposition, vol. 2 , pp. 846 -854, Feb. 2003.

[10] Bhim Singh , S.S.Murthy and Sushma Gupta, "Modeling and Analysis of STATCOM Based Voltage Regulation for Self-Excitation Induction Generator with Unbalanced Loads" , TENCON 2003. Conference on Convergent Technologies for Asia-Pacific Region, vol. 3, pp. 1109 – 1114, Oct. 2003.

[11] E. G. Marra and J. A. Pomilio, "Self-excited induction generator controlled by a VS-PWM bidirectional converter for rural applications," IEEE Trans. on Industry Application, vol. 35, no. 4,pp. 877-883,July/Aug. 1999.

[12] G.E.Valdarannma,P.Mattavalli and A.M.Stankonic, "Reactive power and Unbalance compensation using STATCOM with Dissipativity Based Control", IEEE Trans. on Control System Technology, vol. 19, no. 5,pp. 598-608,sep. 2001.

[13] Paul C. Krause, "Analysis of Electrical Machinery," McGraw-Hill Book Company.

## VIII. BIOGRAPHIES

**B.Venkatesa Perumal** received his B.E Degree in Electrical and Electronics Engineering from Madras University, Chennai and M.E Degree from Bharathidasan University, Tiruchirapalli, India. Currently he is a research scholar in IIT Delhi New Delhi India. His research interests are Power Electronics, Embedded systems, Power quality, and Brushless generation.

**Jayanta Kr. Chatterjee** received his B.E and M.E degrees in Electrical Engineering from University of Roorkee and Ph.D. degree from University of Bristol, UK. Currently he is Professor in the Department of Electrical Engineering, IIT Delhi New Delhi India. His research interests are Embedded systems, Power electronics, Power quality, Brushless generation and Motor control.

**2006 IEEE International Conference on Power Electronic, Drives and Energy Systems**

# Trajectory Sensitivity Analysis in Distributed Generation Systems

Dheeman Chatterjee, *Student Member, IEEE,* Arindam Ghosh, *Fellow, IEEE,* and M. A. Pai, *Fellow, IEEE*

*Abstract* – This paper discusses the use of trajectory sensitivity analysis (TSA) in distributed generation (DG) systems. It is shown that the method can be helpful in the determination of influence of parameters such as line reactance, clearing times, exciter gain and mechanical power input on the transient stability of the system. This may be used to find the critical parameter values and also to locate DGs in the system. The variation of system stability with single and multiple DG locations is studied in the IEEE 16-machine 68-bus system. Suitably placed DGs are found to help bus voltage profiles also.

*Index Terms* – Distributed generation, Trajectory sensitivity analysis, transient stability margin.

## I. INTRODUCTION

ECONOMIC and environmental constraints are often putting restrictions on the building of large generating stations and transmission lines. Connecting small and medium sized generating units to the existing system is emerging as the alternative. These units are usually established near the load centers and are known as distributed generators (DG). DGs can be of various types, like squirrel cage induction generators driven by wind turbine, fuel cells connected to the system through power electronic converters or synchronous generators driven by a combustion turbine [1]. With the increasing share of DG generation, its impact on the stability of power systems becomes more important [2,3].

The standard method for dynamic security assessment of power systems is the transient energy function (TEF) based methods. However, these methods become increasingly complex when detailed models are considered. Sensitivity theory can be introduced to overcome this problem by carrying out trajectory sensitivity analysis (TSA) [4]. The technique to extend the method for systems with both continuous and discrete equations (hybrid systems) is discussed in [5]. A method for reducing the number of trajectory sensitivity calculations to get the most effective control is described in [6]. The use of TSA in effective application of a FACTS device in a multimachine power system is discussed in [7].

In this paper, combustion turbine driven synchronous generators are used as DG. Flux-decay model of generators with fast exciter is considered. At first TSA is carried out in an SMIB system to show its effectiveness in assessment of transient stability margin. Critical parameter values are identified with the help of TS. Next the method is applied in a large

D. Chatterjee is with the Department of Electrical Engineering, Indian Institute of Technology, Kanpur, UP, 208016,India. Email: dheeman@iitk.ac.in

A. Ghosh is with School of Engineering Systems, Queensland University of Technology, Brisbane, Qld, 4001, Australia. Email: a.ghosh@qut.edu.au

M.A. Pai is with Dept of Electrical & Computer Engineering, University of Illinois, 1406 W Green St, Urbana, IL 61801, USA. Email: mapai@uiuc.edu

multi-machine system to study impacts of DG on system stability. DGs are connected to load buses through a transformer and supply the additional load at that bus. The power level of the combustion turbine is 50 to 100 MW. Therefore in this paper the generations of the DGs are kept within this range. The effect of DGs on the voltage profile of the load buses is also studied. The system considered for study is the IEEE 16-machine 68-bus system.

## II. TRAJECTORY SENSITIVITY ANALYSIS AND MODELLING OF POWER SYSTEMS

### A. Trajectory Sensitivity Analysis

Let us consider a system described by a set of differential and algebraic equations (DAEs)

$$\begin{aligned}
\dot{x} &= f(x, y, \lambda), \quad x(t_0) = x_0 \\
0 &= g(x, y, \lambda), \quad y(t_0) = y_0
\end{aligned} \tag{1}$$

where $x$ is the state vector, $y$ is a vector of algebraic variables and $\lambda$ is a vector of system parameters. The sensitivities of state trajectories with respect to system parameters can be found by perturbing $\lambda$ from its nominal value $\lambda_0$. The equations for trajectory sensitivity can be found as [4]

$$\begin{aligned}
\dot{w}_1 &= \left[\frac{\partial f}{\partial x}\right]w_1 + \left[\frac{\partial f}{\partial y}\right]w_2 + \left[\frac{\partial f}{\partial \lambda}\right], \quad w_1(t_0) = 0 \\
0 &= \left[\frac{\partial g}{\partial x}\right]w_1 + \left[\frac{\partial g}{\partial y}\right]w_2 + \left[\frac{\partial g}{\partial \lambda}\right], \quad w_2(t_0) = 0
\end{aligned} \tag{2}$$

where $w_1 = \partial x/\partial \lambda$ and $w_2 = \partial y/\partial \lambda$. Solving (1) and (2) simultaneously, we get $x$, $y$ and the sensitivities $w_1$ and $w_2$.

However, the sensitivities can be found in a simpler way by using the numerical method. Let us choose only one parameter, i.e., $\lambda$ becomes a scalar and the sensitivities with respect to it are studied. Two values of $\lambda$ are chosen (say $\lambda_1$ and $\lambda_2$). The corresponding state vectors $x_1$ and $x_2$ respectively are then computed. Now the sensitivity is defined as

$$Sens = \frac{x_2 - x_1}{\lambda_2 - \lambda_1} = \frac{\Delta x}{\Delta \lambda} \tag{3}$$

If $\Delta\lambda$ is small, the numerical sensitivity is expected to be very near to the analytically calculated trajectory sensitivity value. The trajectory sensitivity is therefore calculated numerically.

### B. Power System Model

In this paper, the synchronous machines are represented by the flux-decay model as shown in Fig. 1 (a). A simplified static exciter model with one gain and one time constant is considered, as in Fig. 1 (b) [8,9].

The internal voltage of the generator and its angle, $E_i \angle \phi_i$ are given by

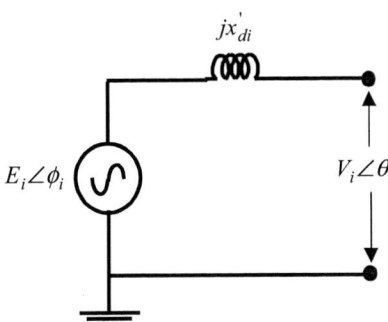

Fig. 1(a). Flux-decay model of generator.

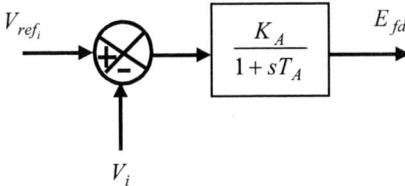

Fig. 1(b). Static exciter model.

$$E_i \angle \phi_i = \left[ \left( 1 - \frac{x'_{di}}{x_{qi}} \right) V_i \sin(\delta_i - \theta_i) + j E'_{qi} \right] e^{j(\delta_i - \pi/2)} \quad (4)$$

The generator and exciter dynamics for an $m$-machine system are described by the following equations [8,9]

$$\frac{d\delta_i}{dt} = \omega_s \Delta \omega_{r_i}, \quad i = 1, \ldots, m \quad (5)$$

$$2H_i \frac{d\Delta \omega_{r_i}}{dt} = P_{m_i} - P_{e_i} - K_{D_i} \Delta \omega_{r_i}, \quad i = 1, \ldots, m \quad (6)$$

$$T'_{do_i} \frac{dE'_{q_i}}{dt} = -\frac{x_{d_i}}{x'_{d_i}} E'_{q_i} + \left( \frac{x_{d_i}}{x'_{d_i}} - 1 \right) V_i \cos(\delta_i - \theta_i) + E_{fd_i}, \quad (7)$$
$$i = 1, \ldots, m$$

$$T_{A_i} \frac{dE_{fd_i}}{dt} = -E_{fd_i} + \left( V_{ref_i} - V_i \right) K_{A_i}, \quad i = 1, \ldots, m \quad (8)$$

$$P_{e_i} = E'_{q_i} V_i \sin(\delta_i - \theta_i) / x'_{di} + 0.5 \left( 1/x_q - 1/x'_d \right) V_i^2 \sin 2(\delta_i - \theta_i), \quad i = 1, \ldots, m \quad (9)$$

where, $\delta$ is the angular position of the rotor, $\Delta \omega_i$ per unit speed deviation of rotor, $\omega_s$ is the synchronous speed, $H$ is the inertia constant, $K_D$ is the damping coefficient, $P_m$ is the mechanical power input, $P_e$ is the electrical power output, $x_d$ and $x_q$ are the d-axis and q-axis synchronous reactance, $x'_d$ is the d-axis transient reactance, $E'_q$ is the q-axis component of the voltage behind the transient reactance $x'_d$, $T'_{do}$ is the d-axis open circuit time constant, $E_{fd}$ is the exciter voltage, $K_A$ and $T_A$ are the gain and time constant of the exciter, $V$ is the terminal voltage of the machine in per unit and $\theta$ is its phase angle.

The dynamics of the network and the stator windings are neglected and the network (containing $n$ buses) is represented by a set of algebraic equations for $i=1, \ldots, n$.

$$P_{L_i} = \sum_{j=1}^{n+m} |V_i||V_j| [G_{ij} \cos(\theta_i - \theta_j) + B_{ij} \sin(\theta_i - \theta_j)] \quad (10)$$

$$Q_{L_i} = \sum_{j=1}^{n+m} |V_i||V_j| [G_{ij} \sin(\theta_i - \theta_j) - B_{ij} \cos(\theta_i - \theta_j)] \quad (11)$$

$G_{ij}$ and $B_{ij}$ are the network transfer conductance and admittance respectively of the augmented $Y_{BUS}$ matrix where the admittance corresponding to the transient reactance of the machines are included along with the normal $Y_{BUS}$ [8]. $P_{Li}$ and $Q_{Li}$ are the real and reactive powers loads respectively.

### C. Quantification of Trajectory Sensitivity of power systems

The sensitivity of state variables of power systems, e.g., the generator rotor angle ($\delta$) and per unit speed deviation ($\Delta \omega_i$) can be computed as in (3) with respect to some parameter $\lambda$. These sensitivities give us information about the effect of change of parameter on individual state variables and hence on the generators of the system to which the particular state variable correspond. However, to know the overall system condition, we need to sum up all these information and develop a suitable metric. To achieve this goal, the norm of the sensitivities is first calculated. The sensitivity norm, $S_N$ for the $m$ machine system of section II(B) is given as

$$S_N = \sqrt{\sum_{i=1}^{m} \left[ \left( \frac{\partial \delta_i}{\partial \alpha} - \frac{\partial \delta_j}{\partial \alpha} \right)^2 + \left( \frac{\partial \Delta \omega_{r_i}}{\partial \alpha} \right)^2 \right]} \quad (12)$$

where the $j^{th}$ machine is taken as the reference. Then a new term $\eta$ (ETA) is introduced [10], which is defined as

$$\eta = \frac{1}{\max(S_N)} \quad (13)$$

As the system moves towards instability, the oscillation in trajectory sensitivity (TS) will be more resulting in larger values of $S_N$. This will result in smaller values of $\eta$. Ideally $\eta$ should be zero at the point of instability. Therefore, the value of $\eta$ gives us an indication of the distance from instability.

### D. Connection of Distributed Generator in an existing system

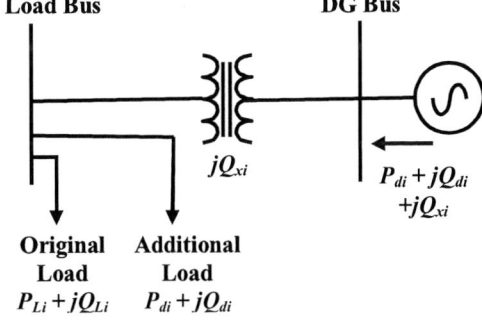

Fig. 2. Connection of the DG with the system load bus.

819

Let us consider that some additional load $(P_{di} + jQ_{di})$ is to be supplied at one of the buses of the power system. The DG is connected to that particular bus as shown in Fig. 2 through a transformer [2]. The DG supplies the additional load as well as the loss in the transformer (given by $jQ_{xi}$), so that the steady state power flow in the original network and the generation of the existing generators remain the same.

## III. TSA OF AN SMIB SYSTEM

Trajectory sensitivity analysis is carried out in a SMIB system (Fig. 3). The generator is connected to the infinite bus through a single line represented by the reactance $x_l$. A three-phase fault is considered at the generator terminal (point A) and is cleared after $t_{cl}$ seconds. The initial values of different parameters of the system are given in Table I.

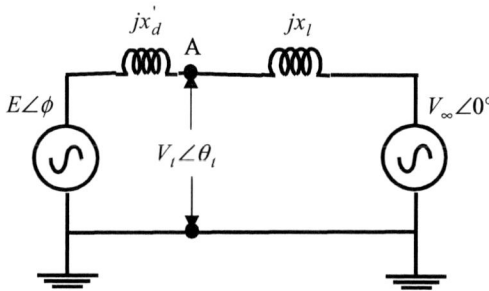

Fig. 3. Schematic diagram of the SMIB system.

TABLE I
SMIB SYSTEM PARAMETERS

| Quantities | Values |
|---|---|
| System frequency | 60 Hz |
| Generator terminal voltage magnitude | 1.05 per unit |
| Infinite bus voltage magnitude | 1.00 per unit |
| Inertia constant ($H$) | 5.0 MJ/MVA |
| Damping coefficient ($D$) | 2.0 s/rad |
| $d$-axis synchronous reactance ($x_d$) | 1.000 per unit |
| $q$-axis synchronous reactance ($x_q$) | 0.960 per unit |
| $d$-axis transient reactance ($x'_d$) | 0.123 per unit |
| $d$-axis open circuit time constant ($T'_{do}$) | 5.0 |
| Exciter gain ($K_A$) | 10.0 |
| Exciter time constant ($T_A$) | 0.4 |

### A. Trajectory sensitivity with respect to fault clearing time

At first, the sensitivities of state variables with respect to the fault clearing time ($t_{cl}$) are computed. Sensitivity norm and $\eta$ are calculated using (12-13). The mechanical power ($P_m$) is chosen to be 0.9 per unit. Initially the value of $t_{cl}$ is taken as 0.10s. Then it is increased in steps till the critical clearing time ($t_{cr}$) is reached. The TS and $\eta$ are computed at each step. Next the value of line reactance $x_l$ is changed and the whole procedure is repeated. Values of $\eta$ for different $t_{cl}$ and $x_l$ are shown in Table II. The corresponding values of $t_{cr}$ are also given.

We know that increase in $t_{cl}$ causes deterioration in the stability condition of the post-fault system. It can be observed from the Table that the value of $\eta$ decreases in all the cases with increase in $t_{cl}$. Further, the value of $\eta$ also decreases with increase in $x_l$. It can be observed that the corresponding values of $t_{cr}$ also decrease, which indicates a deterioration of stability condition. Therefore we can conclude that the value of $\eta$ suc-

cessfully reflects the stability condition of the system and can be used as a measure of transient stability margin. When the system becomes unstable, the $\eta$ goes to zero. Critical values of parameter can be identified utilizing this fact. Fig. 4 shows the variation of $\eta$ with $x_l$ for two values of clearing time, $t_{cl} = 0.10$s and $t_{cl} = 0.15$s. It can be found from the plots that the critical values of line reactance for these two $t_{cl}$ are 0.68 and 0.79 p.u., i.e. the system becomes unstable when the line reactance is increased beyond these values.

TABLE II
VALUE OF $\eta$ FOR DIFFERENT LINE REACTANCE

| Line reactance ($x_l$) | $\eta$ | | Critical clearing time ($t_{cr}$) |
|---|---|---|---|
| | $t_{cl} = 0.10$s | $t_{cl} = 0.15$s | |
| 0.40 | 0.1601 | 0.1236 | 0.230s |
| 0.45 | 0.1565 | 0.1134 | 0.220s |
| 0.50 | 0.1519 | 0.0943 | 0.205s |
| 0.55 | 0.1448 | 0.0665 | 0.190s |

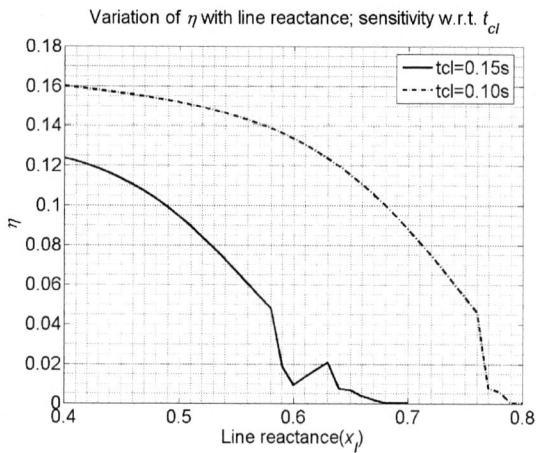

Fig. 4. Variation of $\eta$ with line reactance.

### B. Trajectory sensitivity with respect to Exciter Gain

Next, let us investigate the impact of variation of mechanical power input ($P_m$) on stability. For this, TS is computed with respect to the gain of the exciter ($K_A$). The values of $\eta$ are computed for increasing $P_m$. The variation of $\eta$ with $P_m$ for two values of clearing time, $t_{cl} = 0.10$s and $t_{cl} = 0.15$s are shown in Fig. 5(a).

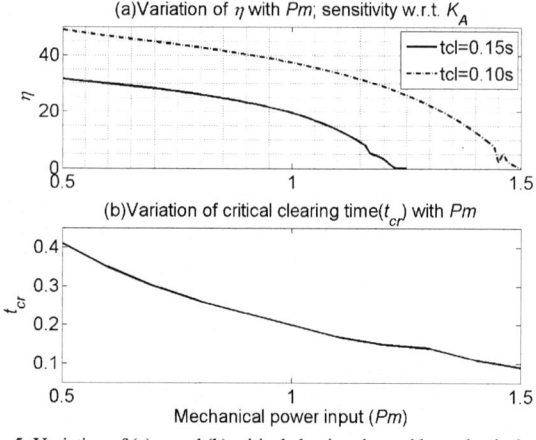

Fig. 5. Variation of (a) $\eta$ and (b) critical clearing time with mechanical power.

820

The value of $\eta$ decreases with increase of $P_m$, indicating deterioration in stability. To verify this, the critical clearing time for different values of $P_m$ are also computed and plotted in Fig. 5(b). A decreasing $t_{cr}$ with increase in $P_m$ supports the trend shown by $\eta$. The critical values of $P_m$ are found (from Fig. 5(a)) to be 1.5 and 1.23 p.u. for clearing time 0.10s and 0.15s respectively.

## IV. USE OF TSA IN THE CHOICE OF DG LOCATION IN A MULTI-MACHINE SYSTEM

### A. The system under study

The system studied in this paper is the IEEE 16-machine 68-bus system [11] shown in Fig 6. A 3-phase fault is simulated in one of the lines of the system. The fault is cleared by isolating the line from both ends after $t_{cl}$ seconds. Then the post-fault system is simulated for a longer duration (say 5-10 seconds) to observe the nature of the transients.

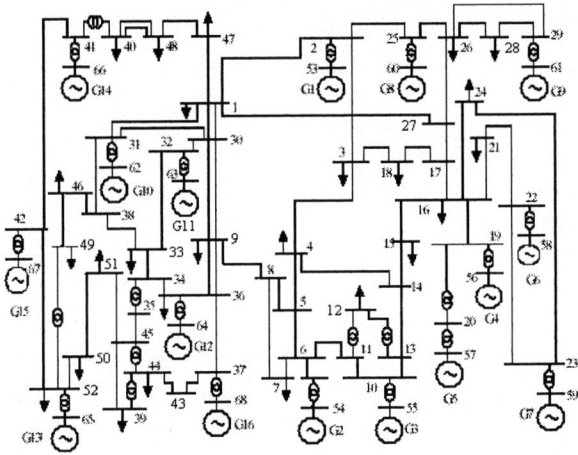

Fig. 6. Single line diagram of the IEEE 16 machine 68-bus system.

### B. Variation of stability condition with DG location

Since the location of the extra load is not under control of the utility, the location of the DG can be decided within a small zone as per the requirement of the utility. To clarify further, suppose a load demand arises near bus 5 of the 68-bus system. Supplying this load from a bus as far as say 33 or 28 is not a feasible option. However, it can be supplied from some nearby buses like 8, 4 or 12, if that is deemed to be favourable for the overall system condition instead of connecting the load at bus 5. Choice of location of the DG within a small group of buses located close to each other can be made using TS on the basis of transient stability condition of the system as the criteria. Four small groups of buses of the 68-bus system are considered here as zones. The constituent buses of the zones are given below:
Zone 1: 15,16,17,18,19,20,21,22,23,24,27,56,57,58,59
Zone 2: 4,5,6,7,8,9,10,11,12,13,14,54,55
Zone 3: 1,2,3,17,18,25,26,27,28,29,53,60,61
Zone 4: 1,9,30,31,32,33,34,35,36,38,62,63,64

Let us investigate the variation of stability condition of the system for different possible locations of the DG in a particular zone. At first, a fault of duration 0.1s is considered at bus

16 and it is cleared by isolating line 16-21. This line is in zone 1. The value of the additional load is taken as $(0.5 + j0.2)$ p.u.. This load and the DG are simulated at different load buses of zone 1 and the trajectory sensitivity and $\eta$ are computed. The $\eta$ values are normalized with respect to the $\eta$ for the system without any DG (termed as base case). Hence a normalized $\eta$ value of more than 1 indicates an improved stability condition and vice versa. The values of normalized $\eta$ for different DG locations are shown in Table III. Usually DGs are connected at places in the system far from the existing generators. Therefore, the load buses connected to existing generator transformers are not considered as DG locations. The fault bus (here bus 16) is also not considered as DG location. Next the additional load is increased to $(1.0 + j0.4)$ p.u. and the TS and $\eta$ are computed for different DG locations as before. The values of normalized $\eta$ are shown in Table III.

It can be observed from the Table that the value of normalized $\eta$ is more than 1 for most of the DG locations. For example the value of $\eta$ is 1.0511when the DG is at bus 24. This indicates that the transient stability condition of the system is improved from the base case. The maximum value of normalized $\eta$ (1.0739) is obtained when the load and the DG are connected to bus 21. It can also be seen that the value of $\eta$ decreases in most of the cases with increase in the additional load to $(1.0+j0.4)$ p.u. For example, the maximum $\eta$ (for DG in bus 21) reduces to 1.0436, indicating deterioration in stability with connection of larger additional load. Further, the value of $\eta$ decreases to 0.9981(<1.0) when the DG is at bus 27 thus indicating deterioration in stability even below the base case condition.

TABLE III
VALUE OF $\eta$ FOR DIFFERENT DG LOCATIONS

| DG location | Load =(0.5+j*0.2) p.u. | Load =(1.0+j*0.4) p.u. |
|---|---|---|
| Bus 15 | 1.0114 | 1.0095 |
| Bus 16 | 1.0227 | 1.0114 |
| Bus 17 | 1.0114 | 1.0038 |
| Bus 18 | 1.0114 | 1.0038 |
| Bus 21 | **1.0739** | **1.0436** |
| Bus 24 | 1.0511 | 1.0360 |
| Bus 27 | 1.0038 | 0.9981 |

To check whether $\eta$ truly reflects the stability condition or not, the fault duration is increased up to the critical clearing time ($t_{cr}$), i.e. the maximum value of fault clearing time for which the system retains stability. The $t_{cr}$ is found to be 0.215s for the system without any DG. When a DG is connected to bus 21 and additional load is $(0.5+j0.2)$ p.u., $t_{cr}$ increases to 0.240s. To verify, the plot of relative rotor angle $\delta_{4-16}$ is shown in Fig. 7 for these two cases for a clearing time of 0.22s. It can be seen that the relative rotor angle diverges for the system without DG whereas it remains bounded when DG is placed in bus 21. The value of $t_{cr}$ for increased additional load (1.0+j0.4 p.u.) is also computed. It is found that $t_{cr}$ decreases to 0.235s with increase in load, which is in accordance with the corresponding decrease in $\eta$.

Next, the fault is considered in line 7-6 of zone 2, line 26-25 of zone 3 and line 33-34 of zone 4 (one at a time). The stability is assessed for different DG locations in the respective zones and for different additional load amounts by computing TS and $\eta$ as before. The values of normalized $\eta$ for the three cases (with additional load of 0.5+j0.2 p.u.) are shown in Table IV. It can be observed from the Table that for a fault in line 7-6, the maximum normalized $\eta$ (1.0259) is obtained for two DG locations, bus 4 and bus 14. Therefore, these two are the best possible locations in this case. For fault in lines 26-25 and 33-34, the introduction of DG causes deterioration of system stability for some of the locations, as indicated by the values of normalized $\eta$ which are less than 1. However, the best possible locations in these two zones are bus 28 (normalized $\eta$=1.0893) and bus 30 (normalized $\eta$=1.0179) respectively.

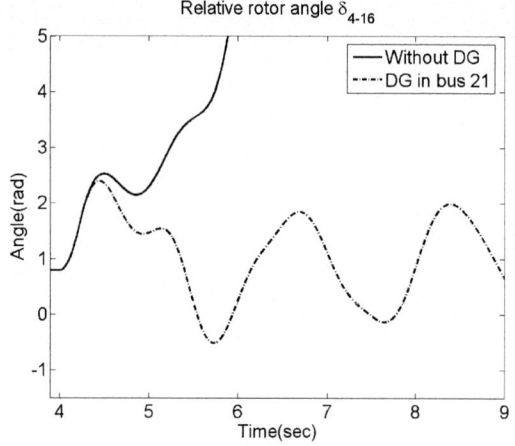

Fig. 7. Response of relative rotor angle $\delta_{4-16}$ for a fault in line 16-21 with clearing time 0.22s.

TABLE IV
VALUE OF $\eta$ FOR DIFFERENT DG AND FAULT LOCATIONS

| DG loc. | Fault in 7-6 | DG loc. | Fault in 26-25 | DG loc. | Fault in 33-34 |
|---------|--------------|---------|----------------|---------|----------------|
| Bus 4   | **1.0259**   | Bus 1   | 0.9833         | Bus 1   | 1.0149         |
| Bus 5   | 1.0173       | Bus 3   | 0.9893         | Bus 9   | 1.0067         |
| Bus 8   | 0.9849       | Bus 17  | 1.0143         | Bus 30  | **1.0179**     |
| Bus 9   | 0.9968       | Bus 18  | 1.0131         | Bus 34  | 0.9948         |
| Bus 11  | 1.0183       | Bus 26  | 1.0131         | Bus 35  | 0.9933         |
| Bus 12  | 1.0205       | Bus 27  | 1.0143         | Bus 38  | 1.0119         |
| Bus 13  | 1.0216       | Bus 28  | **1.0893**     |         |                |
| Bus 14  | **1.0259**   |         |                |         |                |

### C. Stability condition with multiple DG

Till now the stability condition of the 68-bus system has been studied with a single DG placed in different locations. However, in a large system there may be multiple DGs supplying additional loads at different places of the system. Let us now extend our study to such situations. Suppose 4 DGs are placed simultaneously at the four best possible locations of the four zones, i.e. buses 21, 4, 28 and 30. The additional load is taken as 0.5+j0.2 p.u. in all four locations. Now fault is simulated in the four locations mentioned in the previous subsection, one at a time. The values of normalized $\eta$ found in these conditions are given in Table V along with the values of $\eta$ for single DG for comparison.

It can be observed that with four DGs placed simultaneously, the values of normalized $\eta$ are more than 1 for all four fault locations. This indicates that the system stability conditions with the 4 DGs are better than that of the system without any DG. It can also be seen that for fault in line 33-34, the

stability condition of the system becomes better with simultaneous application of four DGs in comparison with the application of single DG, which is apparent from the increase in normalized $\eta$ value from 1.0179 to 1.0418. Similarly, for a fault in line 26-25, the normalized $\eta$ increases from 1.0893 (single DG) to 1.0976 (multiple DG). However, for a fault in line 16-21, the stability condition deteriorates on simultaneous application of DGs in comparison to the single DG condition, though the normalized $\eta$ is more than 1 (1.017) indicating a system more stable than the base case.

To verify the results, the clearing time of the fault in line 26-25 is increased up to the $t_{cr}$. The value of $t_{cr}$ is 0.40s when only a single DG is connected to bus 28. When all four DGs are connected, the value of $t_{cr}$ for fault at the same location increases to 0.42s. This increase in $t_{cr}$ is in accordance with the higher value of $\eta$ in the corresponding case. The plot of relative rotor angle $\delta_{9-16}$ is shown in Fig. 8 to support this result. The fault is in line 26-25 and the clearing time is 0.41s. It can be seen that $\delta_{9-16}$ diverges in case of single DG whereas it remains bounded and hence retains stability when multiple DGs are used.

TABLE V
COMPARISON OF $\eta$ VALUES FOR SINGLE AND MULTIPLE DG

| Fault in line | Single DG | | Multiple DG | |
|---------------|-----------|------|-------------|------|
|               | DG location | $\eta$ | DG location | $\eta$ |
| 33-34 | 30 | 1.0179 | 21,4, | 1.0418 |
| 26-25 | 28 | 1.0893 | 28,30 | 1.0976 |
| 7-6   | 4  | 1.0259 | (simultane- | 1.0259 |
| 16-21 | 21 | 1.0739 | ously) | 1.0170 |

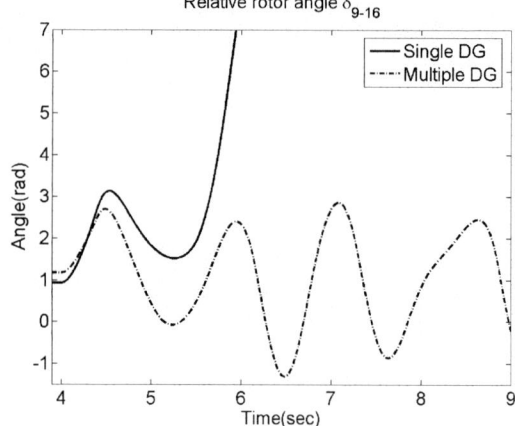

Fig. 8. Response of relative rotor angle $\delta_{4-16}$ for a fault in line 26-25 with clearing time 0.41s.

### D. Effect of DG on load bus voltages

Another point of interest is the effect of adding DG on the voltage profile of the system. Fault and subsequent line isolation often results in voltage sag in some of the buses. If there is large amount of induction motor load, then such under voltage condition results in higher reactive power consumption and hence further voltage reduction. This may even lead to voltage instability. Let us investigate whether the installation of DGs can provide any help in restricting such probabilities. Fig. 9(a) shows the variation of the voltages of load bus 21 for a fault in line 7-6. Plot of voltage for the system without any DG and for system with DG at all four places mentioned in the previous sub-section are shown. Similarly, Fig. 9(b) shows the voltage of bus 4 for a fault in line 16-21. It can be clearly observed from the figures that the introduction of the DGs has two effects

i) The voltage deep during the fault period gets reduced. This is helpful to improve voltage stability of the system.

ii) Oscillation of the voltage in the post-fault system also decreases, thus giving a comparatively smoother voltage profile.

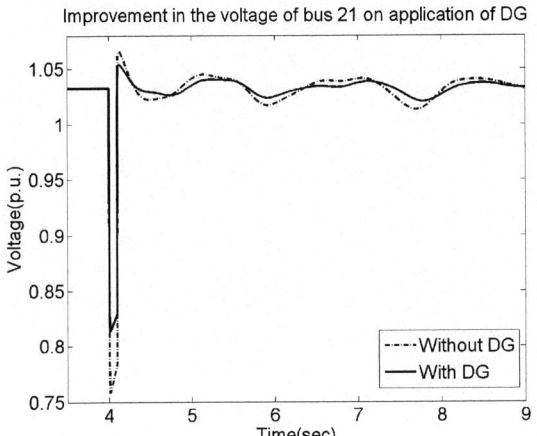

Fig. 9(a). Variation of voltage of bus 21 for a fault in line 7-6.

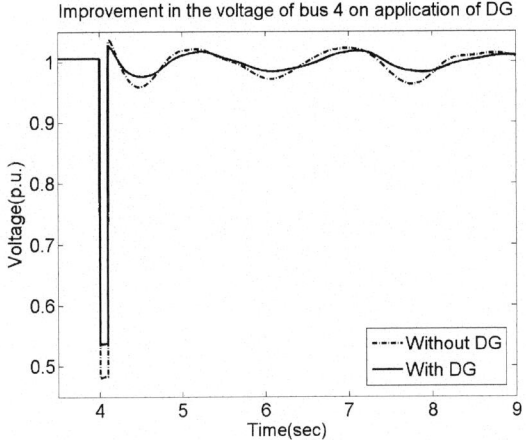

Fig. 9(b). Variation of voltage of bus 4 for a fault in line 16-21.

## VI. CONCLUSIONS

This paper presents the trajectory sensitivity analysis of a power system containing distributed generators. The DGs used here are synchronous generators with combustion turbine. The inverse of the maximum value of the norm of the sensitivities ($\eta$) of state variables like generator rotor angle and rotor speed deviation is used as a measure of transient stability margin. The dependence of system stability on the location of DGs is shown using TS and $\eta$.

The impacts of various parameters on system stability are assessed in an SMIB system. This may help in identifying the suitable location of DGs. TS is used to find the critical values of different parameters. TSA is carried out in a large multimachine power system to observe the effect of different locations of DG. It is found that this method can be used to choose the location of the DG within a small group of buses located close to each other. DGs may also help in improving voltage stability and post-fault voltage profile of load buses.

## ACKNOWLEDGEMENT

The last author acknowledges the support of the National Science Foundation under grant number CNS 05-40237.

## REFERENCES

[1] A.M. Azmy and I. Erlich, "Impact of distributed generation on the stability of electrical power system," *Proc. IEEE PES General Meeting,* 12-16 June 2005, Vol. 2, pp 1056-1063.

[2] J.G. Slootweg and W.L. Kling, "Impacts of distributed generation on power system transient stability," *Proc. IEEE PES General Meeting,* 6-10 June 2004, Vol. 2, pp 2150-2155.

[3] M.K. Donnelly, J.E. Dagle, D.J. Trudnowski, and G.J. Rogers, "Impacts of the distributed utility on transmission system stability", *IEEE Transactions on Power Systems,* Vol. 11, No. 2, pp 741-46, 1996.

[4] M. J. Laufenberg and M. A. Pai, "A new approach to dynamic security assessment using trajectory sensitivities," *IEEE Trans. Power Systems,* Vol. 13, No. 3, pp. 953-958, 1998.

[5] I. A. Hiskens and M. A. Pai, "Trajectory sensitivity analysis of hybrid systems," *IEEE Trans. Circuits & Systems - Part 1: Fundamental Theory & Applications,* Vol. 47, No. 2, pp. 204-220, 2000.

[6] K. N. Shubhanga and A. M. Kulkarni, "Determination of effectiveness of transient stability controls using reduced number of trajectory sensitivity computations," *IEEE Trans. Power Systems,* Vol. 19, No. 1, pp. 473-482, 2004.

[7] D. Chatterjee and A Ghosh, "TCSC Control Design for Transient Stability Improvement of a Multi-Machine Power System using Trajectory Sensitivity", accepted for publication in *Electric Power Systems Research.*

[8] P. W. Sauer and M. A. Pai, *Power System Dynamics and Stability,* Pearson Education (Singapore) Pte. Ltd., Indian branch, Delhi, India.

[9] K. R. Padiyar, *Power System Dynamics, Stability and Control,* BS Publications, Hyderabad, 2002.

[10] M. A. Pai and T. B. Nguyen, "Trajectory sensitivity theory in nonlinear dynamical systems: some power system application," *Stability and Control of Dynamical Systems with Applications: A Tribute to Anthony N. Michel, D. Liu and P. J. Antsaklis,* Eds. Birkhauser Boston, 2003.

[11] Power System Toolbox, version 2.0, Cherry Tree Scientific Software.

**Dheeman Chatterjee** (S'04) obtained his B.E. and M.E. degree in Electrical Engineering from Bengal Engineering College, Howrah, W.B., India in 1996 and 2002 respectively. He is currently a Ph.D. student at the Department of Electrical Engineering, IIT Kanpur. His area of interest is Power System studies.

**Arindam Ghosh** (S'80-M'83-SM'93-F'06) received his Ph.D. in Electrical Engineering from University of Calgary, Alberta, Canada in 1983. Currently he is a Professor of Power Engineering at Queensland University of Technology, Brisbane, Australia. Prior to this, he has been with IIT Kanpur for 21 years. His interests are in Control of Power Systems and Power Electronic devices.

**M. A. Pai** (F'86) received the Bachelor's degree from the University of Madras, India, in 1953, and the Master's and Ph.D. degrees from the University of California, Berkeley, in 1958 and 1961, respectively. Currently, he is Professor Emeritus of Electrical and Computer Engineering at the University of Illinois at Urbana-Champaign, where he has been since 1981. He was on the faculty of the Indian Institute of Technology, Kanpur from 1963 to 1981.

**2006 IEEE International Conference on Power Electronic, Drives and Energy Systems**

# Steady State Performance Of A Stand-Alone Variable Speed Constant Frequency Generation System Using A New Build Up Algorithm

Isha T.B, and D. Kastha, *Member, IEEE*

*Abstract*-- The steady state performance of a stand-alone variable speed constant frequency generation system using slip ring induction motor and two back-to-back connected IGBT based PWM converters is investigated in this paper. A new algorithm is developed which builds the dc link voltage up from a low initial value to the desired level without undue stress either on the machine or on the power converters. In steady state, the dc link voltage, machine flux, output voltage and frequency are tightly regulated as the load and the machine speed are varied over wide (sub synchronous to almost twice the synchronous speed) range. The control algorithm was implemented on a TMS320C31 based DS1102 DSP board which generated the control voltages for the PWM modulators.

*Index Terms*--Bi-directional converters, digital signal processor, double output induction generator, stand-alone system, steady state performance, sub-synchronous speed, VSCF generation.

## I. NOMENCLATURE

$v_{ds}$, $v_{qs}$, $v_{dr}$, $v_{qr}$, $i_{ds}$, $i_{qs}$, $i_{dr}$, $i_{qr}$, $\lambda_{ds}$, $\lambda_{qs}$, $\lambda_{dr}$, $\lambda_{qr}$ : d-axis and q-axis components of voltages and currents. The second subscript '*s*' refers to the stator and '*r*' refers to rotor. A superscript '*s*' denotes a stator reference frame and '*e*' refers to a synchronously rotating reference frame. A '*\**' refers to a reference value and a primed variable is one with respect to the rotor.

$i_{di}$, $i_{qi}$: d-axis and q-axis current components of stator side inverter

$i_{dl}$, $i_{ql}$: d-axis and q-axis components of load current

$\omega_r$: rotor speed

$\omega_e$: stator frequency

$C$: capacitance value

$v_d$: dc link voltage at any time

$V_d$: dc link voltage at steady state

$V_{d0}$: initial dc link voltage.

$L_m$: magnetizing inductance

$L_s$: stator inductance

$L_r$: rotor inductance

## II. INTRODUCTION

IN most of the isolated power generators using squirrel cage induction machine, a variable terminal capacitance is required to maintain constant terminal voltage, which is a distinct disadvantage. It is known that the energy capture capability of constant speed constant frequency (CSCF) systems is smaller compared to their VSCF counterpart. Current and proposed VSCF generators are mainly based on synchronous and induction machine technology. Both wound rotor and cage rotor induction machines can be connected to a constant voltage constant frequency grid via an ac-dc-ac link. Schemes based on wound rotor machine have the advantage that the power converters need to be rated only at a fraction of the machine power. In addition, more than rated power can be extracted from the machine without overheating [1]. In cage rotor machines where a terminal capacitor bank is used to supply the reactive power, the voltage magnitude and frequency at the machine terminal varies widely with prime mover speed and load power factor due to self-excitation. Therefore, to accommodate a reasonable variation in load and prime mover speed, the converter rating will be considerably higher. Several isolated power generation schemes based on the phenomenon of self-excitation have been reported. A novel self-excitation strategy using a single d.c side capacitor and a voltage source inverter is described in [3] and [4]. Here the load is connected to the d.c. side of the inverter. However, the scheme can be easily adapted to a.c. loads by connecting another inverter to the d.c. link. In [5] and [6], the analysis and experimental results of a stand alone VSCF generation system using slip ring induction machines have been described. A similar system in the grid-connected mode has been presented in [5]. Though vector control scheme is used in this, dynamic performance is seen to be poor inherent with vector controlled schemes. Also a large dc link voltage (550 volts for a 220 volts load voltage output) is used, which increases the voltage rating of the converters. And the isolated system described in [6] assumes a pre-charged dc link capacitor at 550 volts. The present paper proposes a novel control strategy incorporating these aspects. At steady state, the dc link voltage is maintained at a much lower level

---

D. Kastha is with Department of Electrical Engineering, Indian Institute of Technology, Kharagpur, West Bengal, India (e-mail: kastha@ee.iitkgp.ernet.in).

Isha T.B. is with the Department of Electrical and Electronics Engineering, Amrita Viswavidyapeetham, Ettimadai, Coimbatore, Tamilnadu (e-mail: isha_t_b@yahoo.co.in).

0-7803-9771-1/06/$25.00 ©2006 IEEE     824

compared to [6] thereby reducing the converter rating by almost 45%. The steady state simulation and exeperimental results are provided for a very wide range of speed and load currents at an internal power factor angle of $45^0$.

## III. STEADY STATE ANALYSIS

Fig. 1 (a) and (b) show the schematic diagram and block diagram of the proposed system respectively. Positive polarity convention of important system variables is shown in Fig. 1(b). It consists of a slip ring induction machine, two voltage source PWM converters and a d.c. link capacitor. The speed variation of the motor is brought about by an open loop speed controlled d.c. motor. Initially, the induction machine is run at no load with a small d.c. voltage applied across the capacitor. A small current from the charged capacitor is injected to the rotor winding through the rotor side converter. The stator side converter is voltage controlled. As a result of stator and rotor winding currents, machine flux and torque are produced. As the machine runs, mechanical energy from the d.c. machine is converted into electrical energy and stored in the d.c. link capacitor through both stator and rotor side converters. When the capacitor voltage reaches a predetermined reference value, both converters are controlled in such a way that the net power flowing to the capacitor is zero. At this point, the load switch along with the filter is turned ON. At steady state, the system operates exactly as a conventional slip power recovery drive except that the combined reactive power demand of the machine and load is supplied by both the converters instead of the grid. A d-q axes model of the machine in the synchronously rotating reference frame is used for dynamic analysis. The system equations in a synchronously rotating reference frame are: -

$$v_{qs}^e = r_s i_{qs}^e + p\lambda_{qs}^e + \omega_e \lambda_{ds}^e \quad (1)$$

$$v_{ds}^e = r_s i_{ds}^e + p\lambda_{ds}^e - \omega_e \lambda_{qs}^e \quad (2)$$

$$v_{qr}^{'e} = r_r' i_{qr}^{'e} + p\lambda_{qr}^{'e} + (\omega_e - \omega_r)\lambda_{dr}^{'e} \quad (3)$$

$$v_{dr}^{'e} = r_r' i_{dr}^{'e} + p\lambda_{dr}^{'e} - (\omega_e - \omega_r)\lambda_{qr}^{'e} \quad (4)$$

Considering the d-axis of the synchronously rotating reference frame to be oriented along the stator flux,

i.e., $\quad \lambda_{ds}^e = \lambda_s = \lambda, \quad \lambda_{qs}^e = p\lambda_{qs}^e = 0 \quad (5),$

then the machine equations reduce to

$$i_{ds}^e = \frac{\lambda}{L_s} - \frac{L_m}{L_s} i_{dr}^{'e} ; \ i_{qs}^e = \frac{L_m}{L_s} i_{qr}^{'e} \quad (6)$$

$$v_{qs}^e = \omega_e \lambda_s - \frac{L_m}{\tau_s} i_{qr}^{'e} ; \ v_{ds}^e = \frac{\lambda_s}{\tau_s} - \frac{L_m}{\tau_s} i_{dr}^{'e} + p\lambda_s \quad (7)$$

$$v_{qr}^{'e} = r_r' i_{qr}^{'e} + \frac{L_r' L_m - L_m^2}{L_s}\left[pi_{qr}^{'e} + (\omega_e - \omega_r)i_{dr}^{'e}\right] + \frac{L_m}{L_s}(\omega_e - \omega_r)\lambda_s \quad (8)$$

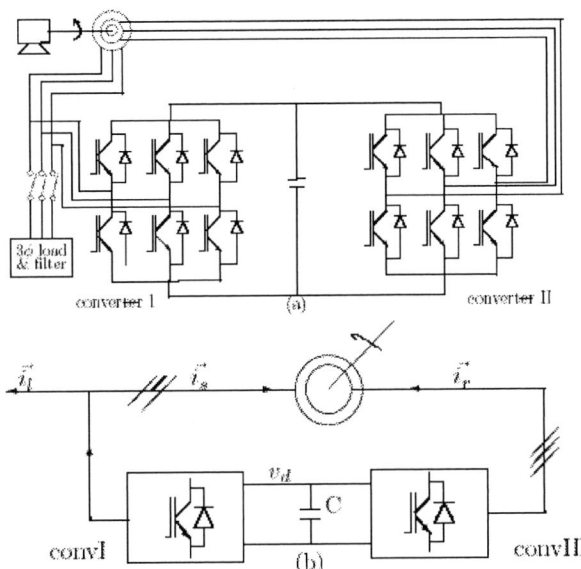

Fig. 1. Schematic and Block diagrams of the proposed system; (a) Schematic (b) Block diagram.

$$v_{dr}^{'e} = r_r' i_{dr}^{'e} + \frac{L_r' L_s - L_m^2}{L_s}\left[ri_{dr}^{'e} - (\omega_e - \omega_r)i_{qr}^{'e}\right] + \frac{L_m}{L_s} p\lambda_s$$

where $\tau_s = \dfrac{L_s}{r}$ and $r = r_r' + \dfrac{L_m}{L_s \tau_s}$

At steady state, $p = 0$, $\lambda_s = \lambda$. Hence

$$v_{ds}^e = \frac{\lambda}{\tau_s} - \frac{L_m}{\tau_s} i_{dr}^{'e} ; \ v_{qs}^e = \omega_e \lambda - \frac{L_m}{\tau_s} i_{qr}^{'e} \quad (9)$$

$$v_{qs}^{e2} + v_{ds}^{e2} = v_s^2 \quad (10)$$

Eqn. (10) can be used to determine the stator voltage at any prime mover speed and load condition. Parameters of the machine used for simulation study and experimental validation are listed in Table 1. To maintain constant d.c. link voltage at steady state, the net power flowing out of the converters must be zero. i.e.,

$$v_{qs}^e i_{qi}^e + v_{ds}^e i_{di}^e + v_{qr}^{'e} i_{qr}^{'e} + v_{dr}^{'e} i_{dr}^{'e} = 0 \quad (11)$$

From Fig. 1(b), $i_{qi}^e = i_{qs}^e + i_{ql}^e$; $\ i_{di}^e = i_{ds}^e + i_{dl}^e \quad (12)$

Substituting (12) in (11), an equation of the form

$$(i_{qr}^{'e} - a)^2 + (i_{dr}^{'e} - b)^2 = c^2 \quad (13)$$

is obtained, where $\quad a = \dfrac{\omega_r L_m \tau_s \lambda + L_m L_s i_{ql}^e}{2(r_r' L_s \tau_s + L_m^2)},$

825

$$b = \frac{2L_m\lambda + L_m L_s i_{dl}^e}{2(r_r' L_s \tau_s + L_m^2)} ; \quad c = \frac{1}{2(r_r' L_s \tau_s + L_m^2)}\sqrt{c_1 - c_2} ;$$

$$c_1 = c_{1a} + c_{1b}; c_{1a} = L_m^2\left[(\omega_r \lambda \tau_s + L_s i_{ql}^e)^2\right];$$

$$c_{1b} = L_m^2\left[(2\lambda + L_s i_{dl}^e)^2\right]; \qquad c_2 = c_{2a} + c_{2b};$$

$$c_{2a} = 2(r_r L_s \tau_s + L_m^2); \qquad \qquad \text{and}$$

$$c_{2b} = \left[2\lambda^2 + 2L_s\lambda(\tau_s \omega_e i_{ql}^e + i_{dl}^e)\right]$$

Under steady state operation, $i_{qr}^{'e}$ and $i_{dr}^{'e}$ must satisfy equation (13). However, it will be of interest to find out the steady state rotor current vector $i_r'$ that minimizes total kVA rating of the converters. A computer program was written to find out the optimum rotor current vector for different load and prime mover speed by analytical simulation. The optimum steady state performance characteristics of the system using a 2.5 hp machine are presented in Fig. 2. However, the experimental validation of the optimum converter rating is remaining.

TABLE 1
PARAMETERS OF THE 2.5 HP INDUCTION MACHINE

| Motor rating: 220Volts 2.5HP 50Hz 1370Rpm 3Phase | |
| --- | --- |
| stator resistance ($r_s$) | 1.036 Ohms |
| rotor resistance($r_r$) | 1.734 Ohms |
| rotor leakage reactance($X_{lr}$) | 1.779 Ohms |
| magnetizing reactance($X_m$) | 33.38 Ohms |
| Core loss resistance($R_m$) | 115.92 Ohms |

## IV. TRANSIENT ANALYSIS

DC link voltage build up may be analyzed as follows:
Initial system conditions are:
$$\omega_r = \omega_{r1} > 0; ; \cdots \omega_e = 0; \cdots \lambda_s = 0; \cdots v_d = V_{d0} > 0.$$

Target system conditions will be:
$$\omega_r = \omega_{r1}; \cdots v_d = V_d; \cdots \lambda_s = \lambda; \cdots \omega_e = \omega_e^*$$

From machine equations (2) and (3), if
$$v_{ds}^e = \frac{\lambda}{\tau_s} - \frac{L_m}{\tau_s}i_{dr}^{'e},$$

then $\lambda_s = \lambda(1 - e^{-\frac{t}{\tau_s}})$. If $v_{qs}^e = \omega_e^* \lambda_s - \frac{L_m}{\tau_s}i_{qr}^{'e}$,

then, $\omega_e = \omega_e^*$. Power flowing out of the capacitors is:
$$P_c = \frac{3}{2}\left(v_{qs}^e i_{qi}^e + v_{ds}^e i_{di}^e + v_{qr}^e i_{qr}^e + v_{dr}^e i_{dr}^e\right) = -\frac{1}{2}\frac{d}{dt}\left(Cv_d^2\right)$$

where, $i_{qi}^e = i_{qs}^e + i_{ql}^e; \quad i_{di}^e = i_{ds}^e + i_{dl}^e$

Substituting for $v_{qs}^e, v_{ds}^e, i_{qi}^e, i_{di}^e$ in terms of $i_{qr}^{'e}$ we reach

at an equation of the form $-\frac{1}{2}\frac{d}{dt}\left(Cv_d^2\right)^2 = I(t)$. Thus we get,

$$v_d^2 = V_{d0}^2 - \frac{2}{C}\int_0^t I(\Psi)d\Psi$$

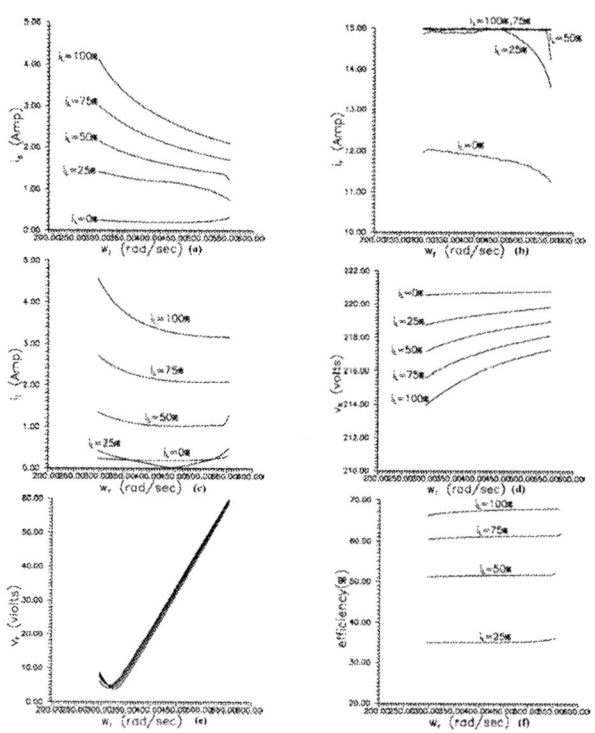

Fig. 2. Steady state performance of the system for different rotor speeds, load currents and the internal power factor angle = $45^0$
(a) stator current; (b)rotor current; (c) stator side inverter current;
(d) stator voltage; (e) rotor voltage; (f) machine efficiency.

For a sinusoidal PWM inverter,

$$v_s^2 \leq \frac{v_d^2}{4}; \quad v_r^2 \leq \frac{v_d^2}{4} \qquad (14)$$

Any control algorithm for d.c. link voltage build up must satisfy the inequalities of (14). The control algorithm for voltage build up is as follows: stator side inverter is voltage controlled with

$$v_{qs}^{e*} = \omega_e^* \lambda_s - \frac{L_m}{\tau_s}i_{qr}^{'e} \text{ and } v_{ds}^{*e} = \frac{\lambda}{\tau_s} - \frac{L_m}{\tau_s}i_{qr}^{'e} \quad (15)$$

rotor side inverter is current controlled with

$$i_{dr}^{'*} = I_{dr}^{'*}\left(1 - e^{-\frac{t}{\tau_d}}\right); \quad i_{qr}^{'*} = I_{qr}^{'*}\left(1 - e^{-\frac{t}{\tau_q}}\right) \quad (16)$$

The following parameters of the controller were chosen by trial and error.

826

$$I_{dr}^{'*} = \frac{\lambda}{4L_m}; I_{qr}^{'*} = \frac{I_{rated}}{4}; \tau_d = \frac{\tau_s}{4}; \tau_q = \frac{\tau_s}{4}; \omega_e^* = \frac{\omega_r}{4}$$

The rotor side current controller is considered to be ideal. i.e., $i_{qr}^{'e} = i_{qr}^{'*}$ and $i_{dr}^{'e} = i_{dr}^{'*}$. It was seen that the system was building up with an initial voltage $V_{d0}=11$ volts across the capacitor initially. Fig. 3 shows the computed build up transient of the system.

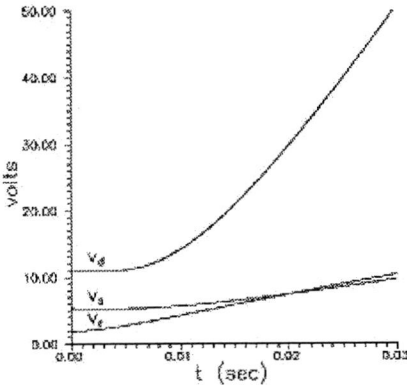

Fig. 3. Computed waveforms of $v_d$, $v_s$ and $v_r$ during d.c. link voltage buildup.

It is observed that with above-mentioned choice of $v_{qs}^*, v_{ds}^*, i_{qr}^{'*}, i_{dr}^{'*}, V_{d0}$ and $\omega_e^*$, the inequalities of (14) are satisfied. However, d.c. link voltage control consists of two different algorithms for the generation of $i_{dr}^{'e}$ and $i_{qr}^{'e}$ and a switching scheme between the two. During build up, $i_{qr}^{'e}$ and $i_{dr}^{'e}$ are generated according to equation (16). When d.c. link voltage crosses its reference value, the second algorithm becomes effective. In this algorithm, $i_{dr}^{'*}$ is set to its optimum value such that the load on the converters are minimized for a given rotor speed and load current. The optimum value of $i_{dr}^{'*}$ is obtained from offline calculation and stored in a table. On the other hand, $i_{qr}^{'*}$ is the output of the closed loop d.c. link voltage controller. The d.c. link voltage dynamics is speed dependant. Fig. 4 shows the block diagram of the complete controller, which is sufficiently self-explanatory. However, this has negligible effect on system performance. The unit vectors for field orientation are obtained from the following equations:

$$\lambda_{ds}^s = L_s i_{ds}^s + L_m i_{dr}^{'s} \text{ and } \lambda_{qs}^s = L_s i_{qs}^s + L_m i_{qr}^{'s}.$$

Therefore, $\lambda_s = \sqrt{\lambda_{ds}^{s2} + \lambda_{qs}^{s2}}$ and $\theta_e = \cos^{-1}\frac{\lambda s_{ds}^s}{\lambda_s}$

from which the unit vectors $\sin\theta_e$ and $\cos\theta_e$ can be calculated. It can be shown [7] that the controller performance is robust against machine parameter variation. The d.c. link voltage dynamics is speed dependant. However, this has negligible effect on system performance.

## A. Simulation Verification

The complete system along with the controller and load was simulated using PC SIMNON. To start the simulation experiment, the d.c. link capacitor voltage was initialized to 11 volts. The d.c. link voltage builds up to the reference value of 370 volts in 0.3 seconds as seen in Fig. 5. The build up experiment was simulated for both super and sub-synchronous speeds.

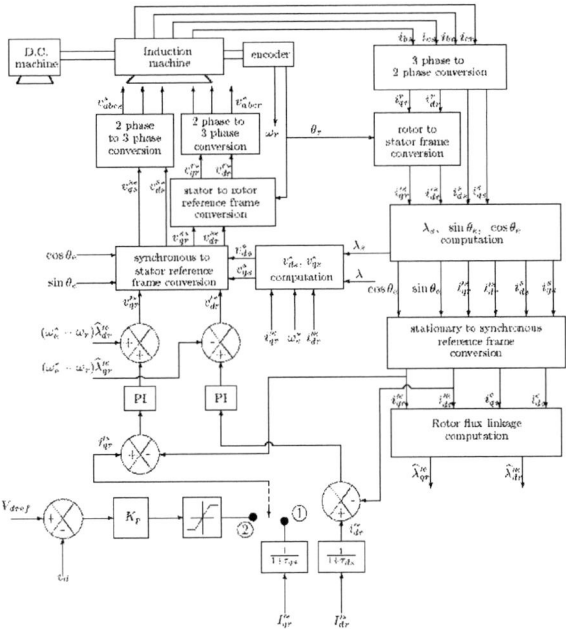

Fig. 4. Block diagram of the control system.

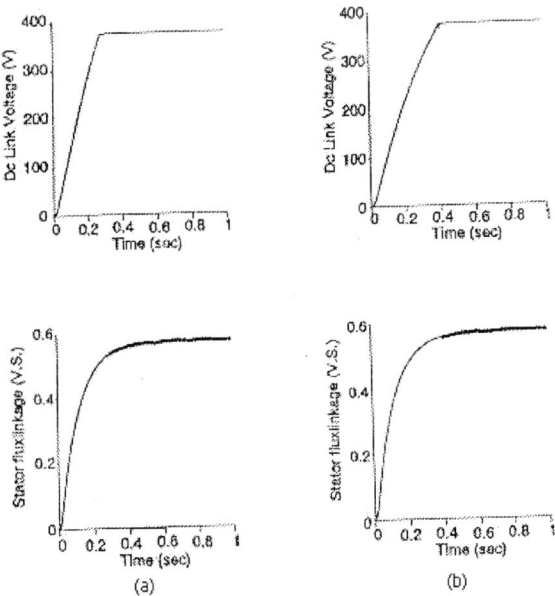

Fig. 5. DC link voltage and stator flux linkage during d.c. link voltage build up (a) for super synchronous speed; (b) for sub synchronous speed.

After the d.c. link voltage reaches the reference value, the d.c. link voltage control algorithm is switched to closed loop

827

control. No appreciable switching transient is seen in Fig. 5. Also Fig. 5(a) and (b) shows the voltage build up transient at super synchronous and sub synchronous speed respectively. It is evident that variation in rotor speed has no qualitative effect on d.c. link votage dynamics. Fig. 6(a) and (b) shows load voltage and current waveforms for super and sub synchronous speeds respectively. As before, variation in rotor speed has negligible effect on these waveforms. Steady state performance of the system was also verified by simulation. For this experiment, the load parameters (resistance and inductance) were adjusted such that the load current magnitude and internal power factor angle remains constant. Rotor side d-axis reference current $i_{dr}^{'*}$ was set to the optimum value for each operating condition so that the converters operate at mininmum kVA.

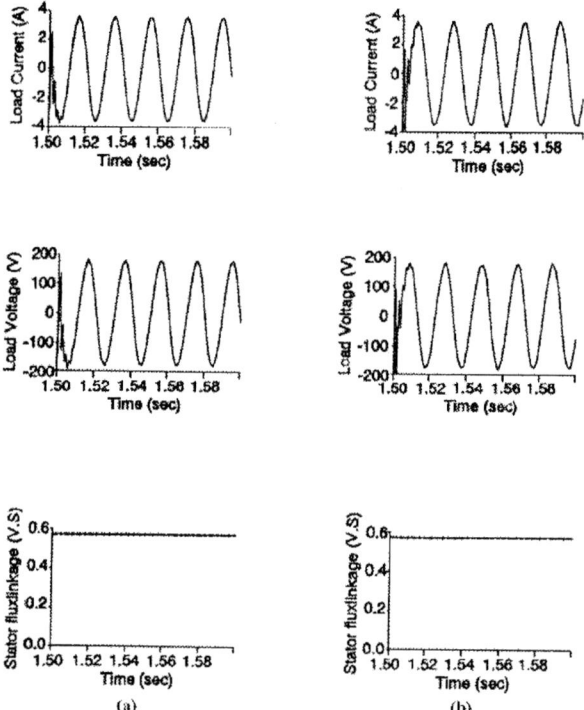

Fig. 6. Load current, voltage and machine stator flux linkage for super and sub synchronous speeds. (a) for super synchronous speed (b) for sub synchronous speed.

### B. Experimental Results

The performance of the VSCF generation system has been extensively tested on a laboratory set up. A block diagram of the experimental set up is shown in Fig. 7. The control algorithm was implemented on a TMS320C31 based DS1102 DSP board which generated the control voltages for the PWM modulators. For experimental verification of the system, the following procedure is adopted. The hardware circuit includes a manually operated switch and a software switch provided through the DSP. With the manual switch kept in the 'OFF' position, the d.c. machine is started. The d.c.link capacitor is charged to a small voltage (between 10 to 20 volts) through an auxiliary single-phase diode bridge rectifier. The reference values of the d.c.link voltage $v_d$, output frequency $\omega_e$, initial values of $i_{dr}^{'*}$, $i_{qr}^{'*}$ and the controller gains are also communicated to the real time control program.

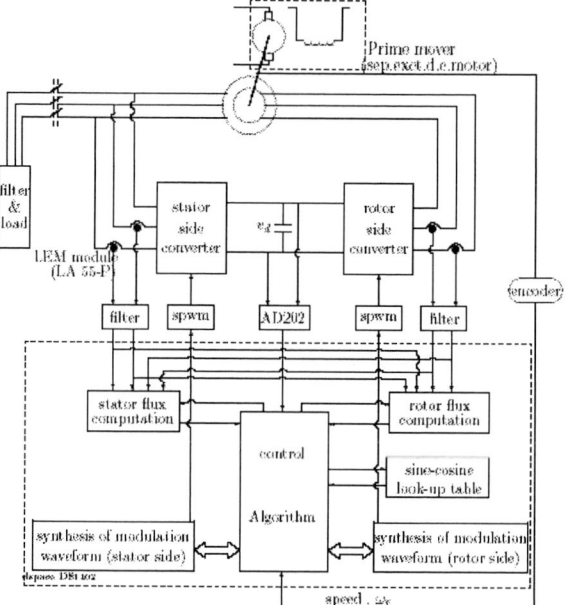

Fig. 7. Schematic diagram of the experimental setup.

The software 'START' switch is made 'ON' along with the manual start switch to start the process of d.c. link voltage build up. The DSP starts sending the modulated control waveforms to the hardware as given in the software description and the whole system becomes operational. Once the system attains steady state, the filter switch (LC filter) is turned 'ON' first, followed by an RL load. A small RC snubber is connected across both the switches to protect the converters from any switching surge. For the purpose of experimental verification of the system, the reference output frequency was set at 20 Hz so as to limit the d.c. link voltage to 160 volts only which was required to maintain the rated machine flux without saturating the stator side PWM inverter. To obtain an output voltage at 50 Hz, the d.c. link voltage has to be increased to 360 volts which could have been detrimental to the IGBT modules rated at 600 volts. If space vector modulation technique is employed to the stator side inverter instead of SPWM technique, the d.c. link voltage requirement for the same 50 Hz output frequency drops to 310 volts which is perfectly safe for inverters made with 600 volts devices. In any case, this practical problem has no bearing on the ability of the system to operate at 50 Hz (or any other frequency for that matter) output frequency. The steady state characteristics of the VSCF generating system obtained from experimental verification is shown in Fig. 8. In Fig. 8, it is seen that the regulation of the voltage generated ($v_s$) is good. The d.c. link voltage is found to be constant and so also, the frequency of generation. The variation of the rotor voltage and

828

current with speed seems to be the same as that obtained by analytical simulation. Fig. 9 shows load voltage, machine stator and rotor current waveforms. The line voltage waveform had an RMS value of 90 volts and a frequency of 20 Hz for which it was set. The results presented in this paper are for output frequency of 20 Hz, which is seen to remain constant (thus verifying the basic VSCF concept) for wide variation in rotor speed and electrical load.

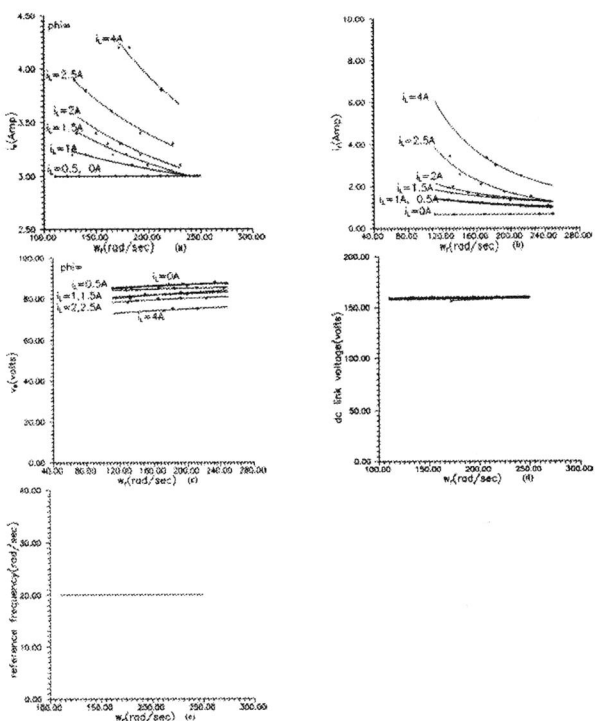

Fig. 8. Steady state characteristics obtained through experimental validation
(a)stator current; (b) rotor current; (c) stator voltage; (d) rotor voltage; (e) d.c. link voltage; (f) frequency of generation.

Thus, the experimental results seem to be in good agreement with the results obtained analytically as shown in Fig. 2.

## V. CONCLUSION

The steady state performance of the system is analyzed. The design procedure developed here ensures minimum converter rating without exceeding the thermal limit of the machine. Since the stator flux linkage is constant, good terminal voltage regulation and machine efficiency are achieved. In this paper, analysis and performance prediction of the proposed system has been presented. However, machine wind turbine interaction has not been studied. Certain aspects while operating with a wind turbine, particularly, the issue of maximum power tracking has received substantial attention in current literature. In this respect the authors would like to state that the power to be generated in an isolated generating system is determined by the connected load unless there exists an energy reservoir like a battery bank. Using an

auxiliary load, serve any useful purpose, because any load (auxiliary or not) that serves any useful purpose will have its own demand curve which is not necessarily controllable from the generator. On the other hand, if storage units (like battery banks) are used, the design of the system will change substantially to warrant fresh treatment. This was deemed to be outside the scope of this paper.

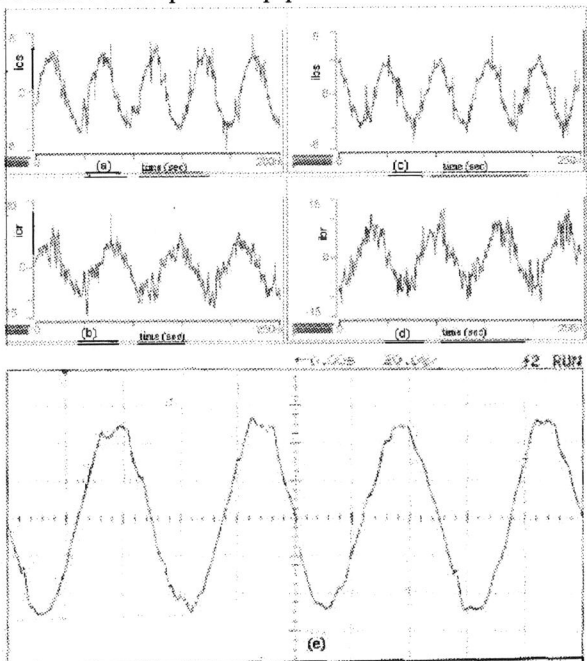

X-axis: time; scale: 20msec/div.
Y-axis: voltage(L-L); scale: 50V/div.
Fig. 9. Experimental waveforms of machine currents and load voltage
(a) stator c phase current; (b) rotor c phase current; (c) stator b phase current;
(d) rotor b phase current; (e) line to line voltage.

## VI. REFERENCES

[1] Z.M. Salameh , L.F. Kazda. "Analysis of the steady state performance of the Double Output Induction Generator" IEEE transactions on Energy Conversion, vol.EC-1, No.1 March 1986, pp. 26-32.

[2] Ziyad Salameh, Sunway Wang "Microprocessor control of Double Output Induction Generator" IEEE transactions on Energy Conversion, vol.4, No.2 June 1988, pp. 172-176.

[3] S.N. Bhadra, K. Venkataratnam, A. Manjunath "Study of Voltage Build Up in a Self-excited Variable Speed Induction Generator/Static Inverter System with DC side capacitor" IEEE proceedings of PEDES, January 1996, pp. 964-970.

[4] D.W. Novotny, D.J. Gritter, G.H. Studtman "Self-excitation in Inverter Driven Induction Machines" IEEE transactions on Power Apparatus and Systems, vol.PAS-96, No.4, July/August 1977, pp. 1117-1125.

[5] R. Pena, J.C. Clare, G.M. Asher "Doubly fed induction generator using back-to-back PWM converters and its application to variable speed windenergy generation" IEE Proc.-Electr.Power Appl., Vol. 143, No. 3, May 1996, pp.231-242.

[6] R. Pena, J.C. Clare, G.M. Asher "A doubly fed induction generator using back-to-back PWM converters supplying an isolated load from a variable speed wind turbine" IEE Proc.-Electr.Power Appl., Vol. 143, No. 5, September 1996, pp.380-387.

[7] Isha T.B. "Analysis, Design and DSP-based Implementation of a Stand-alone Variable Speed Constant FrequencyWound Rotor Induction Generator" PhD thesis, I.I.T. Kharagpur, June 2001.

**2006 IEEE International Conference on Power Electronic, Drives and Energy Systems**

# Control Strategy of Distributed Generation for Voltage Support in Distribution Systems

An D.T. Le, M.A. Kashem, *Senior Member, IEEE*, M. Negnevitsky, *Member, IEEE*, and G. Ledwich, *Senior Member, IEEE*

*Abstract*—Voltage problem is always a critical issue in operating a distribution system. The uncertainties of load distribution and variation have introduced a great complexity to the task of maintaining system voltage within the permitted range. In this paper, small-scale generator, known as distributed generation (DG), is employed in the system and acting as a voltage regulator. The output of DG is controlled in such a way that acceptable level of electrical supply quality is achieved with a reasonable operating cost. The DG controller is tested with a non-uniformly varying load on the time domain basis. Simulations have been conducted with both short term (few hundred seconds) and long term (weekly) load data to validate the proposed control technique. Results have proved that the system is well functioning not only technically but also economically.

*Index Terms*—Distributed Generation, Distribution Systems, Voltage Control, and Voltage Regulation.

## I. INTRODUCTION

DELIVERY of an acceptable voltage level to customers has always been a major obstacle of power system design and operation [1]. Inadequate voltage control may lead to out of the acceptable voltage limits, voltage violation, and eventually cause malfunctions or damages to the connected equipment. Moreover, it also limits the amount of load that can be supplied in the system. Currently, a number of voltage regulation devices, such as On-load tap changing transformers, shunt capacitors, compensators, etc., are in charged of controlling voltage of the distribution system. However, it is expected that in the near future, the distributed generation (DG) system will also gradually take part in the regulation process.

A partial solution for global climate change is believed to come from the use of DGs to supply electricity to residential, commercial and industrial sectors [2]. A study by the Electric Power Research Institute (EPRI) shows that by 2010, 25% of

the new generation will be from DG [3]. Various technologies are being used in DG applications with a variable degree of success, such as [4-5]:

- Photovoltaic, wind turbines, fuel cells with the advantages of pollution free and no green house effects.
- Reciprocating engines with the ability of dispatch.
- Internal combustion engine generators are widely used for increased reliability and peak shaving applications.

Further advances in power electronics and digital control technology have allowed more possibility and better flexibility of using these types of generators, not only as an addition to the electric energy production, but also as the enhancement to the voltage quality and power system stability [6].

The control issues of DG systems for voltage regulation have been addressed by many researchers. Authors in [7] have proposed an approach to limit the excess steady-state voltage-rise using consumer load control, specific to wind generation. Authors in [8] have assessed the operation and control of DG, as well as the dynamics of regulator-DG, DG-DG for single wire earth return (SWER) system. Villacci *et al.* [9] have proposed a computational architecture for the voltage regulation of distribution networks with DG, using adaptive local learning-based methodology. In [10], network voltage regulation is obtained by controlling the target voltage of automatic voltage control relays. Authors in [11] have described a novel excitation control method for DG, which can provide voltage support to the network by altering reactive power generation. Authors in [12] have proposed a voltage regulation coordination method of DG in distribution system using load-tap changing transformer (LTC) and line drop compensator (LDC). In [13], authors have developed a voltage control algorithm for grid-connected DGs based on active and reactive power control.

In this paper, the DG will be integrated in the distribution feeder as a voltage support device. A voltage control technique for DG systems has been developed to ensure the best performance of radial distribution systems. The technical and economic feasibility of the proposed technique has been verified through several case studies of practical distribution networks.

## II. VOLTAGE CORRECTION BY DG

Modern customers are more aware of the necessity of having reasonable supply quality at their busbars. Thus, the goal of maintaining voltage within the specification in any circumstances of load and generation variations has turned out

---

This research has been financially supported by the Australian Research Council under ARC Linkage Grant K0014223 "Integration of Distributed and Renewable Power Generation into Electricity Gird Systems", collaboration with Aurora Energy, Tasmania.

Dr. Mohammad A. Kashem, Ms. An D.T. Le, and Assoc. Prof. Michael Negnevitsky are with the School of Engineering, University of Tasmania, Hobart, Australia, and Prof. Gerard Ledwich is with the School of Engineering Systems, Queensland University of Technology, Brisbane, Australia, (emails: dtale@utas.edu.au; M.Kashem@utas.edu.au; g.ledwich@qut.edu.au; Michael.Negnevitsky@utas.edu.au;).

0-7803-9771-1/06/$25.00 ©2006 IEEE

to be very important. The integration of DG into distribution networks alters the power flow and as a result, changes voltage profile of the system. Majority of DGs installed by power utilities produce real power only, while capacitors, which are often used for voltage support, inject reactive power only. However, previous studies have revealed that DG operating with both real ($P_G$) and reactive power ($Q_G$) will result in more benefit for the utilities as well as for customers.

Any power system can be expressed by a Thevenin equivalent two-bus system, as shown in Fig.1(a). This two-bus system has a Thevenin voltage source, E, and a Thevenin equivalent impedance (R + jX). A DG of ($P_G$ +j$Q_G$) and a load of ($P_L$ +j$Q_L$) are connected at the receiving end bus. Fig.1(b) represents the phasor diagram of the network.

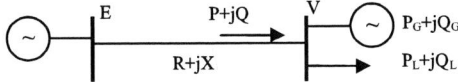

Fig. 1(a). One-line diagram of two-bus system.

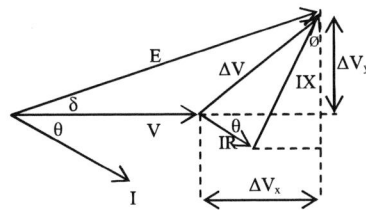

Fig. 1(b). Phasor diagram of two-bus system.

According to Fig.1(b), we obtain the real and imaginary components of voltage drop in the presence of DG, respectively, as:

$$\Delta V_x = \frac{RP + XQ}{V} \tag{1}$$

$$\Delta V_y = \frac{XP - RQ}{V} \tag{2}$$

By substituting $P = P_L - P_G$ and $Q = Q_L - Q_G$, we get:

$$\Delta V_x = \frac{R(P_L - P_G) + X(Q_L - Q_G)}{V} \tag{3}$$

$$\Delta V_y = \frac{X(P_L - P_G) - R(Q_L - Q_G)}{V} \tag{4}$$

On the other hand, when DG is not present in the system ($P_G = 0$, $Q_G = 0$), the real and imaginary components of voltage drop are:

$$\Delta V_x' = \frac{RP_L + XQ_L}{V} \tag{5}$$

$$\Delta V_y' = \frac{XP_L - RQ_L}{V} \tag{6}$$

Thus, the reduction of voltage drop, i.e. voltage improvement, due to DG connection, can be obtained by substracting (3) from (5) and (4) from (6):

$$V_x^I = \frac{RP_G + XQ_G}{V} \tag{7}$$

$$V_y^I = \frac{XP_G - RQ_G}{V} \tag{8}$$

Therefore, the magnitude and phase angle of voltage improvement in term of DG power, respectively, are:

$$\left| V^I \right| = \frac{1}{V} \sqrt{\left( R^2 + X^2 \right)\left( P_G^2 + Q_G^2 \right)} \tag{9}$$

$$\theta^I = \tan^{-1} \left( \frac{XP_G - RQ_G}{RP_G + XQ_R} \right) \tag{10}$$

Equations (9) and (10) reveal that the integration of DG into distribution system results in a change in voltage, which depends on the amount of real and reactive power injection from a DG.

For any particular system, there is an optimal ratio between $P_G$ and $Q_G$ injection, which can improve the system voltage profile at the best. By considering the impact of real and reactive power injections of the DG on system voltage, the most effective operating point of the DG can be determined. This ratio can be determined by the voltage sensitivity of lines as in (11):

$$\frac{P_G}{Q_G} = \frac{\partial V / \partial P_G}{\partial V / \partial Q_G} \tag{11}$$

We know that the voltage at the receiving end with DG is:

$$V = E - \Delta V \tag{12}$$

where, $\Delta V = \Delta V_x + j\Delta V_y$.
By substituting (12) into (11), with $\Delta V$ given in (3) and (4), we obtain [14]:

$$\frac{\frac{\partial V}{\partial P_G}}{\frac{\partial V}{\partial Q_G}} = \frac{R\sqrt{\frac{E^4}{4} + E^2[R(P_G - P_L) + X(Q_G - Q_L)] - [X(P_G - P_L) - R(Q_G - Q_L)]^2} + \frac{1}{2}E^2 R - X^2(P_G - P_L) + RX(Q_G - Q_L)}{X\sqrt{\frac{E^4}{4} + E^2[R(P_G - P_L) + X(Q_G - Q_L)] - [X(P_G - P_L) - R(Q_G - Q_L)]^2} + \frac{1}{2}E^2 X - R^2(Q_G - Q_L) + RX(P_G - P_L)} \tag{13}$$

We also note that the real and reactive power from a DG can be altered by changing the fuel injection and field excitation, respectively [8].

### III. Voltage Regulation Method with DG

In this paper, the voltage control of a system is achieved by injecting active power together with reactive power. The ratio of real and reactive power generation of DG is always kept at a constant level such that maximum voltage support to the system is obtained, as discussed in previous section.

The control system uses the measurement of current at the connection point of DG and the voltage at the most voltage-unsafe bus to derive the control parameter. The control parameter is actually the power output from DG. A question arises here is how to define the most unsafe bus in term of voltage. In the distribution networks, which are mostly radial, the voltage decreases along the feeder and the lowest voltage normally occurs at the end. The remote end voltage, therefore, is used in long radial system as the most voltage-unsafe bus and can be utilised to generate the feedback signal for voltage controller.

The DG controller has been designed based on the concept of the proportional and integral (P-I) controller. This controller has the output signal $M$ [15]:

$$M = K_p \Delta V + K_I \int \Delta V \qquad (14)$$

where, $K_P$ and $K_I$ are the proportional and integral constant, respectively, and $\Delta V$ is the voltage error. Voltage error ($\Delta V$) is defined as the difference in magnitude between the remote voltage in the system ($V_m$) and the reference voltage ($V_r$), plus a tolerance factor $\varepsilon$ as given in (15). The reference voltage is normally set at the lower voltage limit.

$$\Delta V = |V_m| - |V_r| + \varepsilon \qquad (15)$$

The values of $K_P$ and $K_I$ should be chosen so that sufficient voltage correction is obtained, and at the same time, they do not cause instability for the controller.

As the controller will control DG switch on/off and also real and reactive power generation, the control action of DG controller consists of the followings: switch-on, switch-off, increasing output, decreasing output, and doing nothing. The block diagram of the voltage regulation method is shown in Fig.2:

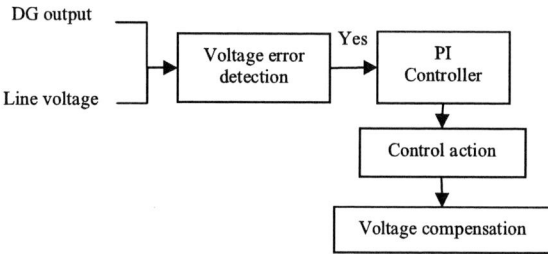

Fig. 2. Block diagram of voltage control algorithm.

The proposed voltage regulation algorithm is described in details below:

**Step 1:** Determine current output level of DG and calculate the voltage at the remote end, $V_m$. This can be done either by online measurement or load flow solution in conjunction with some basic system knowledge (i.e. feeder currents, line parameters, load distribution).

**Step 2:** Compare $V_m$ with the upper and lower voltage limits. If $V_m$ is within the voltage limits, controller action will be 'doing nothing'. Otherwise, go to step 3.

**Step 3:** Calculate the voltage error, $\Delta V$, by using (15).

**Step 4:** $\Delta V$ is then fed into the controller as the feedback signal and controller output is determined by using (14). Controller action can be one of the followings:
- Switch on: DG is currently off and power output of DG is required.

- Increasing output: DG is currently on, output power of DG is required to be increased, and DG is not going to exceed the maximum output power.
- Decreasing output: DG is currently on, output power of DG is required to be decreased, and DG is not going to operate under the minimum output power.
- Switch off: DG is currently on and DG is going to operate under the minimum output power.
- Doing nothing: otherwise.

Periodically, the remote end voltage is checked to detect any violation of the voltage limit and Step 1 to Step 4 are performed to determine the controller action and DG response. Also, to avoid unnecessary switching on/off of the generator, which is very costly to some kinds of DG technologies, a time delay is introduced in the DG controller.

## IV. DESCRIPTION OF TEST SYSTEM

The test system is derived from a practical distribution system of Aurora Energy, Tasmania. The system has a main feeder of 37.5 km long with several laterals originated from Richmond Substation, as shown in Fig.3. The voltage and Thevenin impedance of the substation are 11 kV and (1.6868 + j2.22) $\Omega$, respectively. The distribution line has the impedance of (0.1857 + j0.3345) $\Omega$ per km.

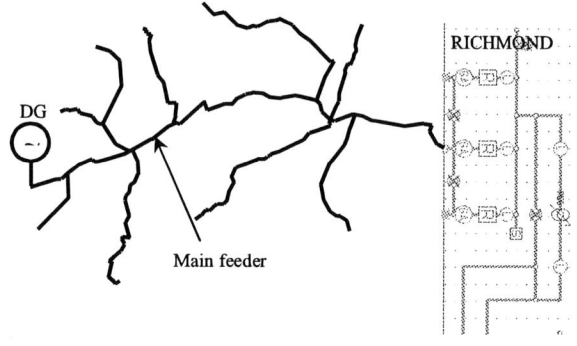

Fig. 3. Test system.

Simulations are conducted using MATLAB 7.0. Base voltage and base MVA used in calculations are 11 kV and 1 MVA, respectively.

## V. SIMULATION RESULTS

The control algorithm has been applied to the test system with different loading conditions to validate its feasibility. A synchronous DG of 200 kVA is placed at the remote end of the feeder and keeps the voltage within ± 5%, as the obligation of Tasmanian utilities. For this particular case of DG placement, the voltage is determined by local measurement at the point of DG connection. Two case studies have been examined:

a) Short term load data: The total system load is increasing from 1.6 MVA to 1.86 MVA in 200 seconds, which is considered as worst load scenario.

b) Long term load data: The load will be changing at different days of the week and also it is different during weekdays and weekends.

Since different loads have different load characteristics,

peak demands occur at different time of the day. Also, other factors such as types of load (i.e. residential, commercial or industrial), days in a week, temperature, etc. greatly impact on the load profile. For those reasons, in both case studies, the load is generated with the characteristics of stochastic in time and magnitude.

*A. Short term load data:*

In this section, the worst case scenario of a system with short-term load has been studied. Total load of the feeder increases more than 16% within 200 seconds and reaches the level of 1.865 MVA at the end, as shown in Fig.4. Since the system without DG can support the total load of 1.295 MVA with no voltage violation, the system examining here is considered as working in high loading condition. During this time, the load pattern is varying with a random variation of electricity demand for individual customers.

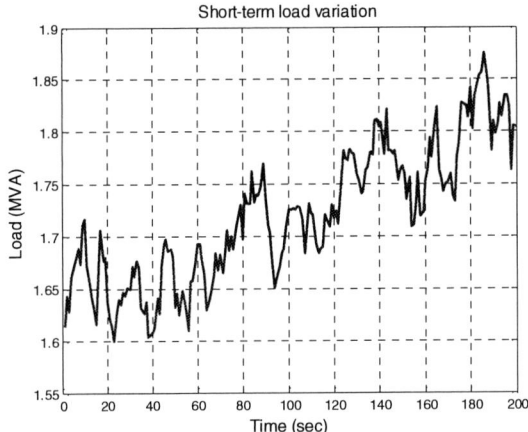

Fig. 4. Worst load scenario.

Fig.5 shows minimum voltage of the system without DG. Due to the load changes, the voltage varies and falls below the voltage threshold of 0.95 p.u. Minimum voltage is 0.9315 p.u., which is approximately 2% below the lower limit.

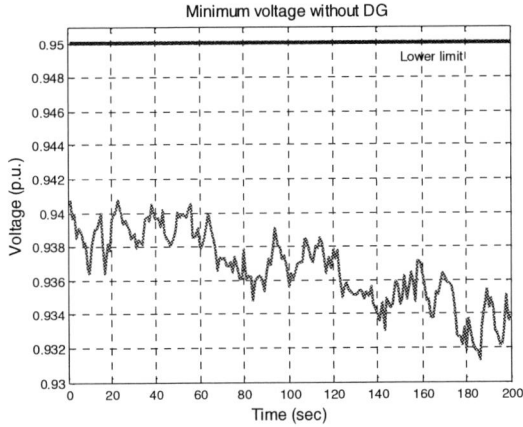

Fig. 5. Minimum system voltage without DG.

Fig.6 shows the DG response corresponding to the load pattern. As we can see from the figure, the DG is off for the first few seconds due to the time delay. After that, DG is on for the rest of the simulation time to improve the voltage

profile. We can also see that the ratio of real and reactive power of DG is maintained at the optimal ratio for the best voltage support.

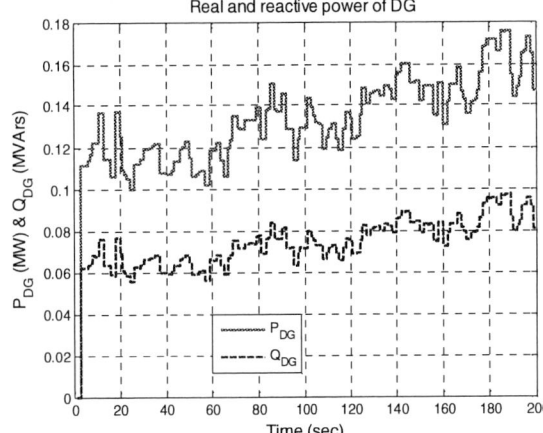

Fig. 6. DG response.

In Fig.7, the minimum voltage of the system with DG is shown. It can be seen that there is only 1.5% of time where the voltage is below 0.95 p.u., and voltage is significantly improved as compared to the case of without DG.

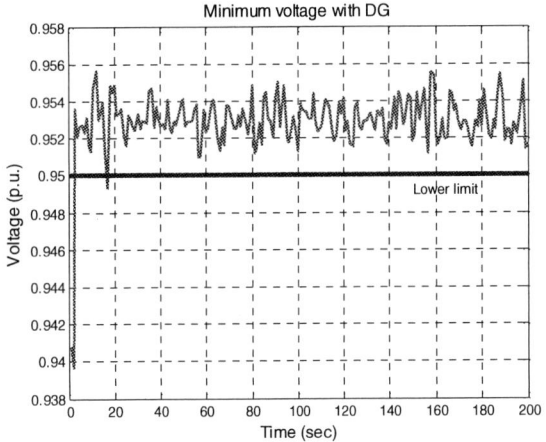

Fig. 7. Minimum system voltage with DG.

*B. Long term load data*

Fig.8 shows a weekly load curve of the system, where peak load of each day occurs at around 8:00 pm to 8:30 pm. The load curve is generated by scaling load data of a standard day by a random coefficient. Although the coefficient for each day is random, it should follow the general rule, when daily peak load is higher during weekdays compared to that of weekends. In this case, the load has maximum peak on Thursday and minimum peak on Saturday.

833

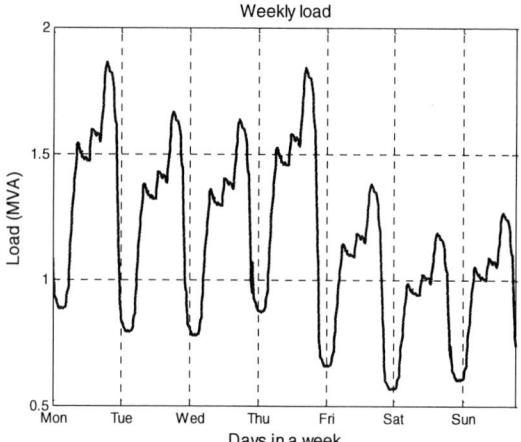

Fig. 8. Weekly load curve.

Fig.9 presents the minimum voltage of the system with and without DG presence. It can be observed that when DG is not available, the system suffers with low voltage for 16 hours on Monday, 14 hours and 45 minutes on Tuesday, 12 hours 30 minutes on Wednesday, 15 hours and 30 minutes on Thursday, and 2 hours and 30 minutes on Friday. During Saturday and Sunday, no voltage problem occurs. Therefore, in total, the voltage is below lower limit for more than 36% of time is DG is not present. However, Fig.9 also reveals that the voltage is always above 0.95 p.u. if DG is integrated.

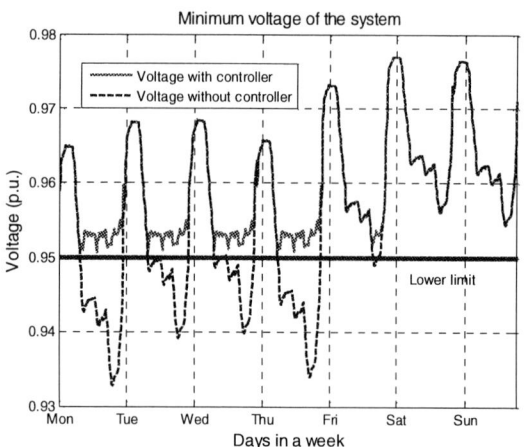

Fig. 9. Minimum voltage in a week.

In Fig.10, the load level and DG output level are shown to validate the economic feasibility of the control system. We can see that the DG is operated during peak hours of weekdays and off during weekends. Total energy contributed by DG in a week is 4.75 MWh, which is 2.5% of the total energy required by the load. Assume that the electricity sold to customers is at 25 dollars/MWh and the DG operating cost is 30 dollars/MWh. The customers only have to pay for approximately 3% more for a much better supply quality.

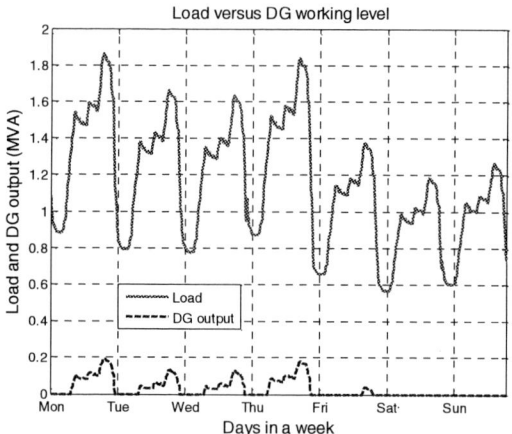

Fig. 10. Load versus DG output.

Another significant cost introduced by DG is the capital investment. This cost could be allocated wholly to the utilities, or partly to customers and partly to the utilities, by agreement. Although this payment is relatively high, it can be justified in long term operation of the DG system.

## VI. CONCLUSIONS

This paper has proposed a technique of voltage regulation using DG system. A voltage control methodology has been presented for optimal power injection from DG. Simulations are performed on a distribution feeder with short term and long term load data. The results show that the system without DG is less capable of meeting voltage requirement, especially during peak hours in a day. The proposed control method provides an effective solution for poor voltage problem, even in the case of heavy loading and high variation in load demand. In the economic aspect, the cost of DG investment and operation can be justified not only by the voltage support capability, but also by other benefits, such as reliability improvement, line loss reduction, etc.

## VII. ACKNOWLEDGEMENT

The authors gratefully acknowledge the support and cooperation of Aurora Energy personnel in providing data and advice on the operation of distribution systems.

## VIII. REFERENCES

[1] R. O'Gorman, M.A. Redfern, and H. Al-Nasseri, "Voltage Control for Distribution Systems", *2004 IEEE Power Engineering Society General Meeting*, 6-10 June 2004, Vol.1, pp. 662 – 667.

[2] J.H. Choi, and J.C. Kim, "Advanced Voltage Regulation Method of Power Distribution Systems Interconnected with Dispersed Storage and Generation Systems", *IEEE Transactions on Power Delivery*, April 2001, Vol. 16, Issue 2, pp. 329 – 334.

[3] T. Niknam, A.M. Ranjbar, A.R. Shirani, "Impact of Distributed Generation on Volt/Var Control in Distribution Networks", *2003 Power Tech Conference Proceedings, Bologna*, 23-26 June 2003, Vol.3.

[4] Y. Baghzouz, "Voltage Regulation and Overcurent Protection Issue in Distribution Feeders with Distributed Generation – A case study", *2005 Proceedings of the 38th Annual Hawaii International Conference on System Sciences*, 03-06 Jan, 2005.

[5] N. Jenkins, R. Allan, P. Crossley, D. Kirschen, and G. Strbac, "Embedded Generation", ISBN: 0-85-2967748, INSPEC, Inc., 2000.

[6] M. Dai, M.N. Marwali, J.W. Jung, and A. Keyhani, "Power Flow Control of a Single Distributed Generation Unit with Nonlinear Local Load", *2004 Power Systems Conference and Exposition*, 10-13 Oct., 2004, Vol. 1, pp. 398 – 403.

[7] N.C. Scott, D.J. Atkinson, and J.E. Morrell, "Use of Load Control to Regulate Voltage on Distribution Networks with Embedded Generation", *IEEE Transactions on Power Systems*, May 2002, Vol. 17, Issue 2, pp. 510 – 515.

[8] M.A. Kashem; G. Ledwich; "Distributed generation as Voltage support for single wire Earth return systems*", IEEE Transactions on Power Delivery*, July 2004, Vol. 19, Issue 3, pp. 1002 – 1011.

[9] D. Villacci, G. Bontempi, and A. Vaccaro, "An Adaptive Local Learning-Based Methodology for Voltage Regulation in Distribution Networks with Dispersed Generation", *IEEE Transations on Power Systems*, Aug. 2006, Vol. 21, Issue 3, pp. 1131 – 1140.

[10] C.M. Hird, H. Leite, N. Jenkins, and H. Li, "Network voltage controller for distributed generation", *IEE Proceedings on Generation, Transmission and Distribution*, March 2004, Vol. 151, Issue 2, pp. 150 – 156.

[11] K. Pandiaraj and B. Fox, "Novel Voltage Control for Embedded Generators in Rural Distribution Networks", *2000 Proceedings of the International Conference on Power System Technology PowerCon*, 4-7 Dec., 2000, Vol. 1, pp. 457 – 462.

[12] T.E. Kim, and J.E. Kim, "Voltage Regulation Coordination of Distributed Generation System in Distribution System", *IEEE Power Engineering Society Summer Meeting*, 15-19 July 2001, Vol. 1, pp. 480 – 484.

[13] E.F. Mogos, and X. Guillaud, "A Voltage Regulation System for Distributed Generation", *IEEE PES Power Systems Conference and Exposition, 2004*, 10-13 Oct. 2004, Vol.2, pp. 787 – 794.

[14] An D.T. Le, M.A. Kashem, M. Negnevitsky and G. Ledwich, "Maximising Voltage Support in Distribution Systems by Distributed Generation", *2005 IEEE TENCON Conference*, Melbourne, Australia, 21-25 November, 2005.

[15] K. Ogata, "Modern Control Engineering", ISBN: 0-13-0609072, Prentice Hall, 2001.

## IX. BIOGRAPHIES

**Ms An D.T. Le** was born in Vietnam in 1982. She received the B.E. (Hons.) from the University of Tasmania, Australia, in 2004. She is currently pursuing the PhD degree at the University of Tasmania. Her special fields of interests are power system analysis, renewable energy, distributed generation, power system control and protection.

**Dr. M. A. Kashem** received the Ph.D. degree from Multimedia University, Malaysia, in 2001. Currently, he is a Senior Lecturer at the School of Engineering, University of Tasmania, Australia. He was associated with the Queensland University of Technology, Australia as a Postdoctoral Research Fellow from 2000 to 2002. Previously, he also worked for Multimedia University as a Lecturer for three years. His special fields of interests include distributed generation, renewable energy, distribution system automation, power system planning, and artificial intelligence. He has published more than 50 technical papers in these areas.

**Assoc/Prof. M. Negnevitsky** received the B.S.E.E. (Hons.) and Ph.D. degrees from Byelorussian University of Technology, Minsk, Belarus, in 1978 and 1983, respectively. Currently, he is an Associate Professor in the School of Engineering at the University of Tasmania, Hobart, Australia. From 1984 to 1991, he was a Senior Research Fellow and Senior Lecturer in the Department of Electrical Engineering, Byelorussian University of Technology. After arriving in Australia, he was with Monash University, Melbourne, Australia. His interests are power system analysis, power quality, and intelligent systems applications in power systems. Dr. Negnevitsky is a Chartered Professional Engineer, a Senior Member of the Institution of Engineers Australia, and a Member of CIGRE AP36 (Electromagnetic Compatibility), Australian Technical Committee.

**Prof. Gerard Ledwich** received the Ph.D. in electrical engineering from the University of Newcastle, Australia, in 1976. He has been Chair Professor in Electrical Asset Management at Queensland University of Technology, Australia, since 1998. He was Head of electrical engineering at the University of Newcastle from 1997 to 1998. Previously, he was associated with the University of Queensland from 1976 to 1994. His interests are in the areas of power systems, power electronics, and controls. Prof. Ledwich is a Fellow of the Institution of Engineers Australia.

# A Steady State Analysis on Voltage and Frequency Control of Self-Excited Induction Generator in Micro-Hydro System

Bhim Singh, *Senior Member, IEEE*, S.S. Murthy, *Life Senior Member, IEEE*, Madhusudan, *Member, IEEE,* Manish Goel, and A. K.Tandon

*Abstract*--This paper presents a steady state analysis of Self-Excited Induction Generator (SEIG) operating with an Electronic Load Controller (ELC) for regulating its voltage and frequency under varying load condition. The ELC consists of a rectifier and a chopper circuit whose operation generates harmonics on AC side of the SEIG system. To achieve an adequate performance characteristics of the SEIG with ELC information of harmonic contents and real power is necessary. In this paper a complete description of the AC current harmonics generated in the ELC operation and their effects on the performance of SEIG is presented in detail.

*Index Terms*--Electronic Load Controller, Micro Hydro Systems, Self-Excited Induction Generator.

## I. INTRODUCTION

DEVELOPMENT of micro-hydro systems in favourable locations for generating electricity can play an important role in the improvement of socio-economic conditions in rural areas. They can meet the energy requirement of far-flung remote rural areas, where it is not economically prudent to extend the grid. Micro-hydro units require less investment and can be set-up by an individual user or village Panchyats. Use of SEIG driven by uncontrolled hydro turbine with ELC and its advantages over synchronous generator coupled to controlled turbine is already reported in the literature [1-9]. In the absence of the turbine governor when the electrical load on the generator decreases, the turbine begins to accelerate and increases generated frequency. Similarly, an increase in electrical load on the generator causes deceleration in the turbine speed and the frequency decreases. ELC is used in micro-hydro system working as a stand-alone generator unit. ELC regulates the voltage and frequency of the generator through monitoring of consumer load and adding an additional controllable load called dump load, so that, the total output power of SEIG remains equal to its rated power. Several methods have been developed for design and analysis of SEIG operating with different types of load controllers. A binary weighted load controller (BWLC) consists of dump load

resistors divided in the ratio of 1:2:4:8 of total per phase dump load [2]. The values of resistors in binary weighted manner are determined on the basis of the rated capacity of the SEIG. However, BWLC in three-phase system require large number of resistive elements for binary switching and their associated complex switching circuits.

A thyristor controlled three-phase bridge converter based terminal impedance controllers are also reported in the literature [3-5]. This controller offers a wide range of control in dump load to balance the power generated by SEIG and power absorbed by consumer load. The firing angles of thyristors in the bridge are adjusted for controlling the power dissipated in the dump load. However, the phase angle control (PAC) of thyristor gives an increased burden of reactive power on the SEIG and also distorts the SEIG terminal voltage waveform due to harmonic currents drawn by the converter. The modeling, analysis, design and implementation of rectifier/chopper based ELC for SEIG in micro-hydro applications are also described [6-7,9]. In the present analysis based on MATLAB the harmonic effects of the load controller on the power generated by the SEIG is studied in detail under varying power demand of consumers.

## II. SYSTEM DESCRIPTION

The SEIG along with ELC system is shown in Fig.1 [9]. It consists of a delta connected SEIG, which is feeding power to a delta connected consumer load, a capacitor bank which

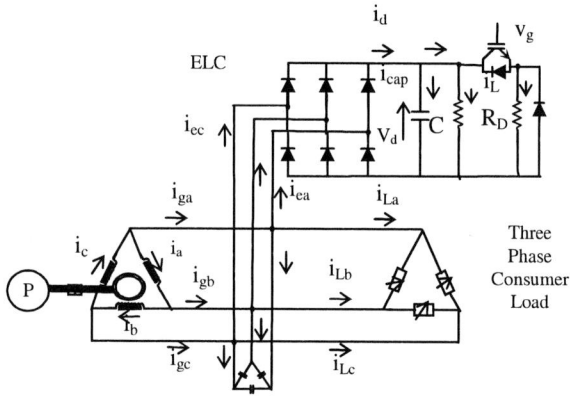

Fig. 1. A schematic diagram of SEIG and ELC system.

supplies the maximum amount of reactive power required by the SEIG, ELC and load. The ELC consists of a three-phase uncontrolled diode bridge and a step down chopper. The

---

Bhim Singh and S. S. Murthy are with Department of Electrical Engineering, Indian Institute of Technology, Delhi, India-110016 (e-mail: bsingh@ee.iitd.ac.in), (e-mail: ssmurthy@ee.iitd.ac.in).

Madhusudan, Manish Goel and A. K. Tandon, are with the Department of Electrical Engineering, Delhi College of Engineering, Delhi-110016, India (e-mail: madhusudan_s2@rediffmail.com).

diode bridge converts the generated SEIG voltage into unidirectional voltage, which is filtered through an electrolytic capacitor. The chopper switch converts the filtered DC voltage into a variable DC to control the power in dump load and hence to keep the total output power of SEIG constant. The load controller control circuit senses the SEIG terminal voltage and accordingly controls the power in dump load.

## III. STEADY STATE MATHEMATICAL MODEL

The steady-state performance analysis of SEIG with ELC system is based on the selection of an equivalent circuit of the system at the fundamental frequency. The equivalent circuit which is chosen for the analysis is shown in Fig. 2. It consists of the standard steady state per phase model of SEIG with an additional variable impedance $(Z_1)$ in parallel with the main consumer load $(R_L)$. The impedance $(Z_1)$ represents the real and reactive power absorbed by the bridge rectifier/chopper based ELC at fundamental frequency which is calculated using the Fourier analysis of the waveform of the line current of the

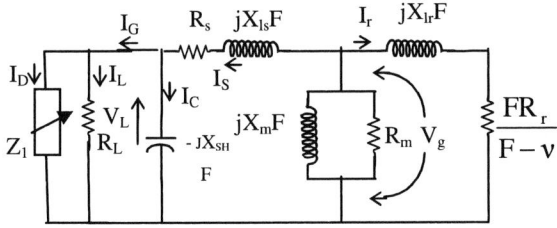

Fig. 2. Steady-state equivalent circuit of SEIG-ELC system.

three-phase bridge rectifier.

The equivalent circuit has been transformed to the base frequency by introducing the parameters F and ν, where F is the per unit frequency and ν is the per unit speed. Since, the equivalent circuit consists of only passive circuit elements, the total voltage drop around the loop must be zero.

$$I_s Z = 0 \qquad (1)$$

Therefore, for successful voltage build-up since Is ≠ 0, hence

$$Z = 0 \qquad (2)$$

where, $Z = Z_{rm} + Z_s + Z_{LC}$ $\qquad (3)$

and, $Z_r = \{ R_r F /(F - v) + jFX_{lr} \} \}$ $\qquad (4)$

$$Z_m = jFX_m R_m /(R_m + jFX_m) \qquad (5)$$

$$Z_L = \frac{R_L Z_1}{R_L + Z_1} \qquad (6)$$

$$Z_{sh} = -jX_{sh} / F \qquad (7)$$

$$Z_{rm} = Z_r Z_m /(Z_r + Z_m) \qquad (8)$$

$$Z_s = R_s + jFX_{ls} \qquad (9)$$

$$Z_{LC} = Z_L Z_{sh} /(Z_L + Z_{sh}) \qquad (10)$$

The subscripts s, r, l, m, sh, L denote stator, rotor, leakage, magnetizing, shunt, and load quantities. Since, Z is a complex quantity, hence the net impedance of the system may be written as :

$$Z = Re(F, X_m) + Im(F, X_m) \qquad (11)$$

where, Re is the real part of Z and Im is its imaginary part, both expressed as functions of F and $X_m$.

For performance evaluation of SEIG the complex variable in eqn. (11) must be solved in order to obtain the per unit frequency F and magnetising reactance $X_m$. Several methods [8] are reported to determine $X_m$ and F. In the present analysis a method based on function minimization technique has been used to reduce computational effort involved in the conventional technique. After F and $X_m$ have been determined, the air gap voltage is found from the magnetization characteristic of the machine obtained experimentally through synchronous speed test[8]. The magnetization characteristic and rating of machine used in the analysis are given in Appendix. With known values of $V_g$, $X_m$, F, ν, $X_{sh}$, $R_L$ and $Z_1$, the SEIG performance with ELC are determined by making use of eqns. (12-18).

$$I_s = V_g / Z_{rm} \qquad (12)$$

$$I_r = V_g / Z_r \qquad (13)$$

$$V_L = V_g - I_s Z_s \qquad (14)$$

$$I_L = V_L / R_L \qquad (15)$$

$$I_D = (I_S - I_C - I_L) \qquad (16)$$

$$P_C = 3|I_L|^2 R_L \qquad (17)$$

$$P_D = P_G - P_C \qquad (18)$$

where, $V_g$, $V_L$, $P_G$, $P_C$ and $P_D$ are respectively are air gap voltage, terminal voltage, output power of SEIG, consumer load power and power in dump load.

## IV. DESIGN AND CONTROL OF ELC

There are a number of electronic load controller circuits based on various methods [1-6] used to dissipate the power in dump load, so that balancing between the electrical output power of the SEIG and input power to the hydraulic turbine is obtained. With variation of consumer load the load controller has to change the effective dump load resistance, so that

$$P_G = P_C + P_D \qquad (19)$$

The power in dump load depends on the duty cycle of the chopper and is given as :

$$P_D = \frac{(\delta V_d)^2}{R_D} \qquad (20)$$

where δ is duty cycle, $V_d$ is DC output voltage of diode bridge rectifier and $R_D$ is dump load resistance, corresponding to rated capacity of the generator. The rating of dump load resistance is given by:

$$R_D = \frac{V_d^2}{P_G} \qquad (21)$$

A design procedure for selection of the rating of devices used in power electronic circuit is described for a SEIG of 3.7 kW, 415 V, which has been considered for the purpose of simulation as well as experimental studies. It is assumed that SEIG is driven by an unregulated turbine and it is excited through shunt excitation capacitance bank which supplies the

837

maximum amount of reactive power require by the SEIG and load.

### A. Design of Rectifier and Chopper Circuits

The rating of diode bridge rectifier and chopper switch depends on the rated voltage and power of the SEIG. The DC output voltage of rectifier corresponding to rated voltage of SEIG is given by:

$$V_d = \frac{3\sqrt{2}\,V_{LL}}{\pi} \qquad (22)$$

$$= 1.35\,V_{LL} = 1.35 \times 415 = 560.25\ \text{V}$$

where, $V_{LL}$ is line to line rms voltage of SEIG.
For 10% variation in SEIG terminal voltage is as:
$V_{LL} = 415 \pm 41.5 = 456.5\ \text{V (rms) to } 373.5$ .
Hence the peak line voltage $V_p = \sqrt{2} \times 456.5\ \text{V} = 645.58\ \text{V}$.

The current rating of diodes and IGBT in chopper circuit is decided by ac line current at rated output power of rectifier.

$$I_s = \frac{P_G}{\sqrt{3}V_{LL}} = \frac{3.7 \times 1000}{\sqrt{3} \times 415} \cong 5\text{A} \qquad (23)$$

Considering, distortion factor of three-phase diode bridge rectifier as 0.955.

The AC side input current of ELC, $I_s \cong \dfrac{5}{0.955} = 5.23\text{A}$ (24)

Moreover, crest factor for three-phase rectifier is from 1.4 to 2.0, therefore the peak input current of ELC

$$= 2 \times I_s = 2 \times 5.2 = 10.47\text{A} \qquad (25)$$

From the above design calculations, it is found that a rectifier circuit and IGBT with the voltage ratings of at least 650 V and current rating of 10 A are necessary for ELC power circuit. Considering a safety factor of two for the power devices, the voltage and current ratings of diode rectifier and IGBT are chosen to the order of 1200 V and 20 A.

### B. Design of Filter Capacitor at DC Bus

Using the 10% voltage fluctuation, the dc-link capacitance can be calculated using the power and energy stored in the capacitor as follows:

$$C \geq \frac{2.17 \times 10^{-3}\,P_G}{\left(2 \times 1.35 V_{LL} \times \Delta V - \Delta V^2\right)} \qquad (26)$$

where $\Delta V$ is the variation in the DC link average voltage.

Therefore, the calculated value of filter capacitor for 10% voltage variation in DC link voltage is 300 μF. The nearest commercially available specification of capacitor is 500 μF and 650V.

### C. Control Scheme of ELC

A block diagram representation of control scheme of ELC is shown in Fig. 3. The control circuit of ELC senses the SEIG terminal voltage ($V_L$) and compares with a reference signal ($V_{ref}$), which is proportional to rated terminal voltage of SEIG. The difference voltage is process in a proportional integral (PI) controller.The output of PI controller is compared with a sawtooth carrier waveform of 3 kHz frequency to generate the PWM switching signal for IGBT switch. The output of PI controller, when compared to sawtooth waveform produces the PWM switching signals for IGBT.

## V. HARMONIC ANALYSIS OF ELC

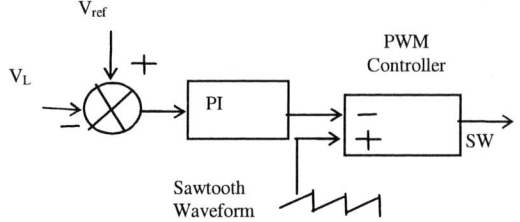

Fig. 3.  Control Scheme for ELC.

The steady-state performance analysis of SEIG-ELC system is performed by determining the amount of fundamental and harmonic real and reactive power absorbed by the diode bridge rectifier and chopper circuits. The Fourier analysis of the line currents of the three-phase bridge rectifier is performed to determine the harmonic currents through the ELC. These harmonic currents are important to determine the rating of the excitation capacitor banks for the SEIG, and also the rating of solid state devices in the rectifier as well as the chopper circuit.

The Fourier analysis for the line currents of ELC is described as:

$$a_n = \frac{1}{\pi}\begin{bmatrix} \int_{\pi/3}^{2\pi/3} ICos(\theta - \pi/3)Cos\,n\theta\,d\theta \\[4pt] + \int_{2\pi/3}^{\pi} ICos\left(\theta - \frac{5\pi}{6}\right)Cos\,n\theta\,d\theta \\[4pt] - \int_{4\pi/3}^{5\pi/3} ICos\left(\theta - \frac{3\pi}{2}\right)Cos\,n\theta\,d\theta \\[4pt] - \int_{5\pi/3}^{2\pi} ICos\left(\theta - \frac{11\pi}{6}\right)Cos\,n\theta\,d\theta \end{bmatrix} \qquad (27)$$

$$b_n = \frac{1}{\pi}\begin{bmatrix} \int_{\pi/3}^{2\pi/3} ICos(\theta - \pi/3)Sin\,n\theta\,d\theta \\[4pt] + \int_{2\pi/3}^{\pi} ICos\left(\theta - \frac{5\pi}{6}\right)Sin\,n\theta\,d\theta \\[4pt] - \int_{4\pi/3}^{5\pi/3} ICos\left(\theta - \frac{3\pi}{2}\right)Sin\,n\theta\,d\theta \\[4pt] - \int_{5\pi/3}^{2\pi} ICos\left(\theta - \frac{11\pi}{6}\right)Sin\,n\theta\,d\theta \end{bmatrix} \qquad (28)$$

where,

$$I = \frac{V_L}{R_D} \ and\ R_D' = \frac{R_D}{\delta} \qquad (29)$$

The rms value of $n^{th}$ harmonic current is defined as:

$$I_n = \sqrt{a_n^2 + b_n^2} \qquad (30)$$

The fundamental frequency impedance of ELC is obtained as:

$$Z_1 = \frac{V_L}{I_1} \qquad (31)$$

The real power absorbed by the dump load at fundamental frequency depends on the real part of the fundamental frequency impedance of the ELC. The real power consumption in the ELC at fundamental frequency is given as:

$$P_{D1} = 3I_1^2 \, \text{Re}[Z_1] \qquad (32)$$

where, $\text{Re}[Z_1]$ represents the real part of the fundamental frequency impedance of the ELC.

Also, harmonic power in the dump load is given by

$$P_{Dn} = P_D - P_{D1} \qquad (33)$$

The total harmonic distortion (THD) of the ELC line current is defined as:

$$THD = \frac{\sum I_n}{I_1} \qquad (34)$$

The equations (27-34) are used to determine the harmonic content and the associated harmonic power drawn by the ELC.

## VI. RESULTS AND DISSCUSSION

The steady-state performance analysis of SEIG-ELC system under varying load conditions is presented. Fig.4 shows the SEIG terminal voltage (V_L), current (I_G), frequency (F). It is

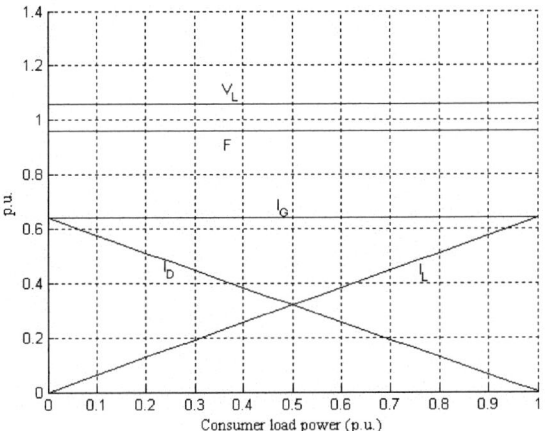

Fig. 4. Steady-state performance of SEIG with electronic load controller.

observed that as the net output power of SEIG is constant, the terminal voltage (V_L), current (I_G) and the frequency (F) are constant. The increase in the consumer load current (I_L) causes a corresponding decrease in the current (I_D) of the dump load to ensure a constant output power from the SEIG.

Fig.5 shows the variation of fundamental real power in dump load (P_D1), harmonic component of ELC power (P_Dn), duty cycle (δ) with varying output power of consumer load. It is observed that as the consumer load power increases, the corresponding the dump load power P_D1 decreases, in order to keep the net output power from the SEIG constant.

The component of the harmonic power (P_Dn) absorbed by the ELC is maximum at the zero consumer load and it decrease linearly as the consumer load power increases. It is also observed that as the consumer output power increases the duty

cycle of the chopper decreases and at rated power in consumer load, the duty ratio reduces to very small value.

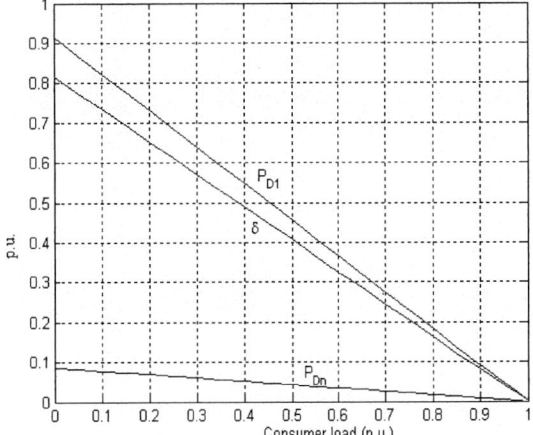

Fig. 5. Steady-state performance of SEIG with electronic load controller.

Fig.6 shows the harmonic spectrum of the ELC line current, when consumer load absorb only 80% of the rated output power of the SEIG. The harmonic spectrum represents the magnitude of the harmonic orders, which are multiple of

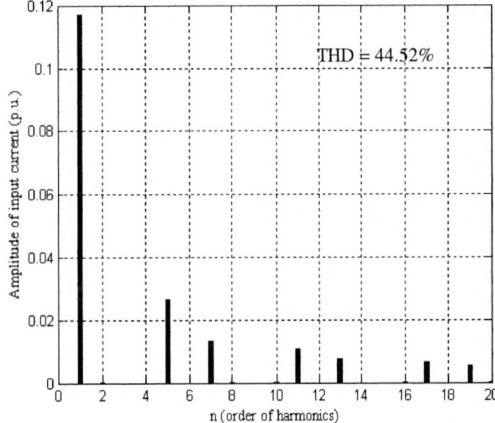

Fig. 6. Harmonic spectrum of ELC line current at 80% consumer load.

the fundamental frequency of the $5^{th}$, $7^{th}$, $11^{th}$, $13^{th}$, $17^{th}$, $19^{th}$, etc. The magnitude of the harmonics in per unit of the fundamental is simply the reciprocal of the harmonic order. Moreover, the THD in the line current of ELC is 44.52%, while the consumer load is 80% of the rated output power of the SEIG.

Fig. 7 and Fig. 8 show the harmonic spectrum of the ELC line current, when consumer load absorb 60% and 20% of the rated output power of the SEIG respectively. It is clear from the spectrum of the ELC line currents that as the consumer load decreases more power are absorbed in the dump load and magnitude of the fundamental as well as harmonic currents are enhanced accordingly.

Fig.9 shows the variation of the fundamental frequency impedance of the ELC with varying duty cycle according to

839

consumer load. At large value of duty cycle, since more power is absorbed by the ELC the net impedance offered per phase by ELC reduces.

Fig.10 shows the effective variation of fundamental

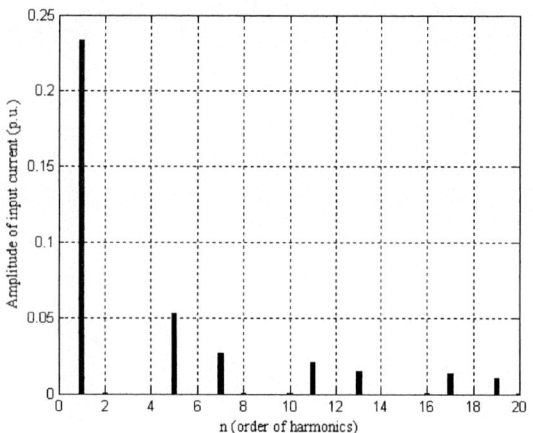

Fig. 7. Harmonic spectrum of ELC line current at 60% consumer load.

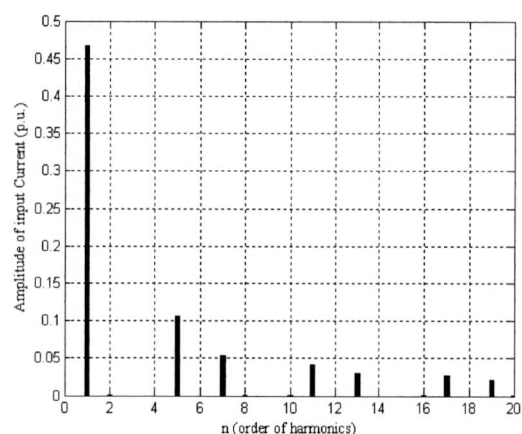

Fig. 8. Harmonic spectrum of ELC line current at 20% consumer load.

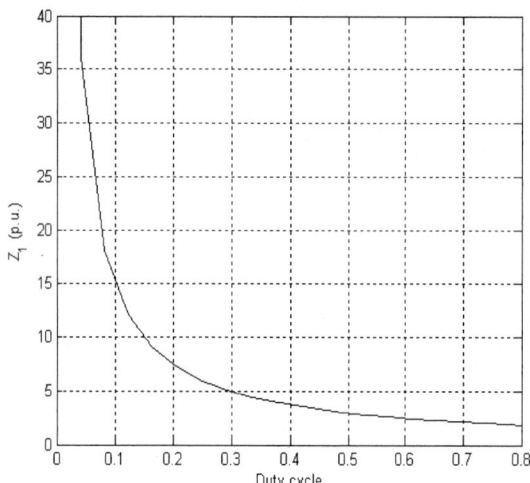

Fig. 9. Variation of fundamental frequency impedance with duty cycle.

frequency impedance of ELC with varying consumer load power. It is observed that as the consumer load power increases the effective fundamental frequency impedance at

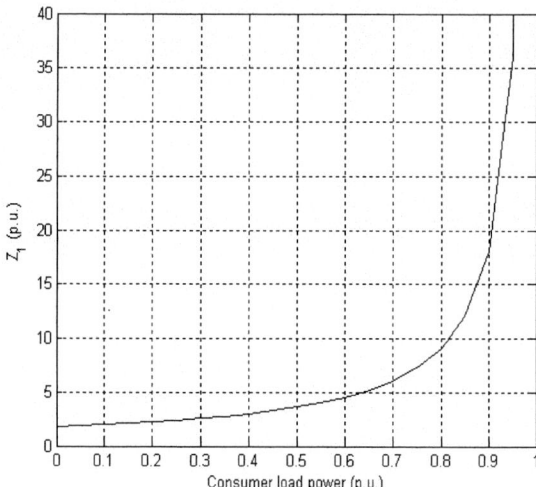

Fig. 10. Variation of fundamental frequency impedance with consumer load power.

the input terminals of the ELC increases to reduce the power absorbed by the ELC, so that the net output power of SEIG remains constant resulting-in the constant voltage and constant frequency operation of SEIG under varying consumer load.

## VII. CONCLUSION

A load controller using a three-phase uncontrolled bridge rectifier and chopper converter has been specially modeled in MATLAB environment, for controlling the voltage and frequency of micro hydro turbine driven SEIG under varying load conditions. The harmonic analysis of the SEIG-ELC system is performed under steady state operating condition and the harmonic effect of ELC on the performance on SEIG is presented and analyzed. The controller is modeled as a variable impedance. The harmonic analysis of ELC find use in design of suitable rating excitation capacitors bank and also solid-state voltage and frequency regulating schemes for the SEIG.

## VIII. REFERENCES

[1] D. B. Watson and R. M. Watson, "Microprocessor control of a self-excited induction generator," *Int. J. Elect. Engg. Edu.* Vol. 22, pp. 69-82, 1985.

[2] J. M. Elder, J. T. Boys and J. L. Woodward, "Integral cycle control of stand-alone generators," *IEE Proc.* Vol. 132, Part C, No. 2, pp. 57-66, 1985.

[3] R. Bonert and G. Hoops, "Stand alone induction generator with terminal impedance controller and no-turbine control," *IEEE Trans. on Energy Conversion,* Vol. 5, No. 2, pp. 28-31, 1990.

[4] D. Henderson, "An advanced electronic load governor for control of micro-hydroelectric generation," *IEEE Trans. on Energy Conversion,* Vol. 13, No. 3, pp. 300-304, 1998.

[5] R. Bonert and S. Rajakaruna, "Self-excited induction generator with excellent voltage and frequency control", *IEE Proc. Gener Transm. Distrib.,* Vol. 145, No. 1, pp. 33-39, 1998.

[6] B. Singh, S. S. Murthy and S. Gupta, "Analysis and implementation of an electronic load controller for a self-excited induction

generator", *IEE Proc. Gener. Transm. Distrib.* Vol. 151, No. 7, pp. 51-60, 2004.

[7] B. Singh, S. S. Murthy and S. Gupta, "Analysis and design of electronic load controller for a self-excited induction generator", *IEEE Trans. On Energy Conversion,* Vol.21, No.1, pp. 285-93, *2006.*

[8] S. S. Murthy, O. P. Malik and A. K. Tandon, "Analysis of self-excited induction generator", *IEE Proc.* Vol. 129, Part C No. 6, pp. 260-65, 1982.

[9] S. S. Murthy, Bhim Singh, A Kulkarni, Shivarajan and Sushma Gupta, "Field experience on a novel pico-hydel system using self-excited induction generator and electronic load controller", *IEEE, IAS Magazine* 2006.

## IX. APPENDIX

**Machine Rating:** Three-phase squirrel cage induction machine 3.7 kW (5 HP), 415 V, 7.6 A, 1500 rpm, 4-Poles, 50 Hz, delta connected, $R_s = 0.0533$ p.u, $R_r = 0.061$ p.u, $X_{ls} = X_{lr} = 0.087$ p.u.

The coefficients of polynomial in magnetization characteristics of induction machine are given as:

$$V_g / F = K_1 - K_2 X_m$$

where, $K_1 = 1.628$ and $K_2 = 0.342$.

## X. BIOGRAPHIES

**Bhim Singh (SM'99)** was born in Rahamanpur (U.P.), India, in 1956. He received B.E.(Electrical) degree from the University of Roorkee, Roorkee, India, in 1977 and M.Tech. and Ph.D. degrees from Indian Institute of Technology(IIT), New Delhi, India, in 1979 and 1983, respectively.

In 1983, he joined Department of Electrical Engineering, University of Roorkee, as a Lecturer. In 1988, he became a Reader. In December 1990, he joined Department of Electrical Engineering, IIT, New Delhi, India, as an Assistant Professor. He became an Associate Professor in 1994 and Professor in 1997. His fields of interest include power electronics, electrical machines and drives, active filters, static compensators, and analysis and digital control of electrical machines.

Prof. Singh is a Fellow of the Indian National Academy of Engineering, Institution of Engineers (India), Institution of Electronics and Telecommunication Engineers, a Life Member of the Indian Society for Technical Education, System Society of India, and National Institution of Quality and Reliability.

**SS Murthy (SM'87,LS'2003)** received his B E (1967), M Tech (1969) and PhD (1974) degrees from Bangalore University, IIT Bombay and IITDelhi respectively. He joined as a Faculty Menber at IIT Delhi in 1970 and steadily rose to become Professor in 1983. He undertook visiting assignments at Univ. of New Castle,UK, (1975-76), Univ. of Calgary, Canada, (1980-82), Kirloskar Electric. Co. Bangalore (1985-86).

He was Director of Electrical Research and Development Association, Baroda (1990-92) and National Institute of Technology Karnataka, Surathkal (2003-05).

His fields of interest includes Electrical machines, Drives, Power Electronics applications, Renewable Energy, Energy Conservation and Engineering education.

He is a Fellow of Indian National Academy of Engineering, Institution of Engineers (India), Institution of Engineering and Technology (UK), a Life Member of the Indian Society for Technical Education.

**Madhusudan** was born in Ghazipur (U.P.) India, in 1968. He received B.Sc. (Electrical Engineering) degree from the Faculty of Technology, Dayalbagh Educational Institute, Dayalbagh Agra, India, in 1990 , M.E. degree from the University of Allahabad, India, in 1992 and Ph.D. degree from University of Delhi, Delhi , India in 2006. In 1992, he joined the Department of Electrical Engineering, NERIST, Nirjuli, Arunachal Pradesh, as a Lecturer.

In June 1996, he joined Electrical Engineering Department, IET Lucknow, India as a Lecturer. In March 1999, he joined Department of Electrical Engineering, Delhi College of Engineering, Delhi, India as an Assistant Professor. He is currently an Assistant Professor in Department of Electrical Engineering, University of Delhi, India.

His research interests are in area of modeling and analysis of electrical machines, voltage control aspects of Self–Excited Induction Generator, power electronics and drives.

He is a member of the Institution of Engineers (IE), India, member of Institution of Electronics and Telecommunication Engineers, New Delhi, India. He is also a member of IEEE, USA and a Life Member of the Indian Society for Technical Education, New Delhi, India.

**Manish Goel** was born in Sonepat, (Harayana), India, in 1984. He received his B.E (Electrical Engineering), from Delhi College of Engineering, Delhi, of Faculty of Technology, University of Delhi, India in 2006. His area of interest are modelling and simulation of electrical machines, MATLAB, C/C++, AUTOCAD etc.

**A.K.Tandon** was born in Kota, (Rajasthan), India, in 1945. He received B.Sc.(Electrical Engineering) degree from, Faculty of Technology, Dayalbagh Educational Institute, Dayalbagh Agra, India, in 1967, M.Tech. from Indian Institute of Technology, Bombay, India in 1969 and Ph.D. degree from Indian Institute of Technology (IIT), New Delhi, India, in 1984. After having gained industrial experience of 2 years, he joined Department of Electrical Engineering, Delhi College of Engineering, as a lecturer in 1971, where he became an Assistant Professor in 1982, and Professor in 1994. His fields of interest are analysis and control of electrical machines and drives.

Prof. Tandon is a Fellow of the Institution of Electrical Engineers (IEE, U.K.), Fellow of the Institution of Engineers (IE), India, and a Fellow of Institution of Electronics and Telecommunication Engineers, New Delhi, India. He is also a senior member of IEEE (USA), Life Member of the Indian Society for Technical Education, New Delhi, India, and member of Computer Society of India. Prof. Tandon is a chancellor medalist and holds a patent in his name alongwith Prof. C.S Jha and Prof. S.S. Murthy of IIT Delhi, India.

**2006 IEEE International Conference on Power Electronic, Drives and Energy Systems**

# A Novel Digital Control Technique of Electronic Load Controller for SEIG Based Micro Hydel Power Generation

S. S. Murthy, *Life Senior Member, IEEE,* Ramrathnam, *Member, IEEE,* M. S. L.Gayathri, Kiran Naidu, and U. Siva

*Abstract*--This paper presents the dynamic and steady state performance of a stand alone Self Excited Induction Generator (SEIG) with digitally controlled Electronic Load Controller feeding single phase and three phase loads. The values of the capacitances are chosen to ensure self-excitation of the machine and to minimize the unbalance between the stator voltages. The excitation capacitors, ELC and load combined with the d-q model of the machine, together with the saturation are used to predict the dynamic behavior of the SEIG. This paper mainly deals with the viability of using digitally controlled ELC, which is more compact, reliable and cost effective for providing effective voltage regulation

*Index Terms*--Electronic Load Controller (ELC), Self-Excited Induction Generator (SEIG).

## I. INTRODUCTION

IN developing countries, small (pico) hydro projects producing power outputs in the range 1-10kW are found attractive to electrify remote locations where utility power is well out of reach. Consumer loads connected to these small hydro schemes are normally single-phase lighting loads. 3-phase SEIG is prescribed [1] for small hydro schemes provided appropriate phase balancing and excitation capacitors are used [3].

Apart from many advantages already reported [2] SEIG has improved performance under line short circuit compared to synchronous generator since the drop in voltage under short circuit automatically reduces the excitation and limits the short circuit current [5].

The behavior of SEIG feeding dynamic loads differs considerably from that under static loads as the former experiences voltage dips and inrush currents. Further, the performance deteriorates under unbalanced conditions due to excessive heating, insulation stress, winding stress, and shaft vibrations caused by unequal phase currents and voltages. Therefore, wider acceptance of the SEIG is dependent on the methodology to be adopted to overcome the poor voltage and frequency regulation, its capability to handle dynamic loading,

The authors Prof. S.S.Murthy, (Email: ssmurthy@ee.iitd.ac.in), M.S.L.Gayathri (gayathriiitd@gmail.com), Kiran Naidu and Uddanti Siva are with the Department of Electrical Engineering, Indian Institute of Technology, IIT Delhi, New Delhi, India-110016.

and its performance under unbalanced The voltage and frequency are maintained within acceptable values by connecting resistive ballast, which maintains the sum of the consumer load and the ballast load at a constant value.

This paper presents a novel digital control technique for the ELC and analyses the steady state and transient behavior of an uncontrolled micro-hydro-turbine-driven SEIG-ELC system feeding both dynamic and static loads.

## II. 3-PHASE SEIG FEEDING 1- PHASE LOADS

A schematic diagram of the developed SEIG-ELC system is shown in Fig. 1. It consists of a three-phase SEIG driven by a constant power prime mover (typically, an uncontrolled micro-hydro turbine). The excitation capacitors are connected at the terminals of the SEIG, which have a fixed value to result in rated terminal voltage at full load.

Since the input power is nearly constant, output power of the SEIG must be held constant at all loads. Any decrease in load may accelerate the machine and raise the voltage and frequency levels to prohibitively high values to affect other connected loads. The power in surplus of the consumer load is dumped in a resistance through an ELC connected at the terminals of the SEIG [1].

A variable mark-space ratio chopping approach has been adopted for the ELC because it produces a variable unity power factor load with just single ballast. Waveform distortion resulting from the chopping action is reduced by the action of the excitation capacitors [1].

Integral control is used which produces no stability problems. The time constant is variable to enable the controller response to be tuned to the inertia of the turbine generator system and the characteristics of the generator. It is also used to minimize cyclic voltage fluctuations due to uneven turbine power output. In this circuit, the electronic switch is operated at a high frequency, thus chopping the rectified AC voltage. Varying the duty ratio of the switch can change the effective resistance of the ballast load. Due to the inductance of the generator, the current drawn from the generator is nearly sinusoidal with a superimposed high-frequency ripple component.

The electronic switch senses and directly controls voltage rather than frequency. When a load is connected to the generator the voltage will decrease and the ELC will respond

0-7803-9771-1/06/$25.00 ©2006 IEEE

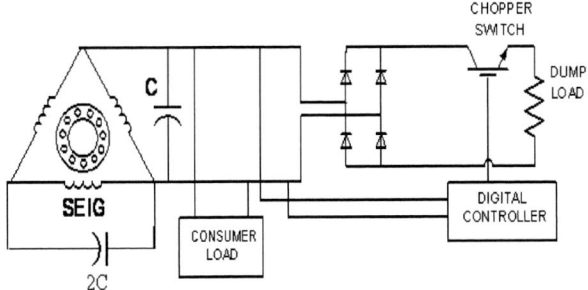

Fig. 1. 3-phase SEIG with digitally controlled ELC.

by reducing the ballast load. The reduced load will cause an almost instantaneous frequency increase due to reduced slip and an additional gradual increase in frequency due to rising turbine speed. The rising frequency results in an increasing voltage and a corresponding increasing load. A stable operating point will be reached when the load on the turbine has increased sufficiently to match the power output of the turbine.

### A. Dynamic Modeling of SEIG Feeding 1-Phase Loads

For the purpose of digital simulation, the model equations defining dynamic model of the induction generator and load are represented by a set of differential equations [4]. Subjected to the given constraints, these equations are solved to obtain instantaneous values of desired quantities. An isolated SEIG excited by a capacitor bank and supplying electric power to a three phase static load is represented by d-q axis.

The voltage current relationship of the system can be expressed as in (1).

$$[v] = [R][i] + [L]p[i] + [G][i]\omega_r \tag{1}$$

Which can be expressed as in current derivative form as (2):

$$p[i] = [L]^{-1}\{[v] - [R][i] - \omega_r[G][i]\} \tag{2}$$

Where

$$[v] = [v_{ds}\, v_{qs}\, v_{dr}\, v_{qr}]^T \tag{3}$$

$$[i] = [i_{ds}\, i_{qs}\, i_{dr}\, i_{qr}]^T \tag{4}$$

$$[R] = diag[R_s\, R_s\, R_r\, R_r]^T \tag{5}$$

$$[L] = \begin{bmatrix} L_{ls} + L_{md} & L_{dq} & L_{md} & L_{dq} \\ L_{dq} & L_{ls} + L_{md} & L_{dq} & L_{mq} \\ L_{md} & L_{dq} & L_{lr} + L_{md} & L_{dq} \\ L_{dq} & L_{mq} & L_{dq} & L_{lr} + L_{md} \end{bmatrix} \tag{6}$$

$$[G] = \begin{bmatrix} 0 & 0 & 0 & 0 \\ 0 & 0 & 0 & 0 \\ 0 & -L_{mi} & 0 & -(L_{mi} + L_{lr}) \\ L_{mi} & 0 & L_{mi} + L_{lr} & 0 \end{bmatrix} \tag{7}$$

The effect of cross saturation is considered with the presence of $L_{dq}$, $L_{md}$ and $L_{mq}$ terms in the elements of matrix [L] The electromagnetic torque balance equation is:

$$T_{shaft} = T_e + J(2/P))p\omega_r \tag{8}$$

The relationship between shaft torque and speed is represented by linear curve given in (9).

$$T_{shaft} = a - b*\omega_r \tag{9}$$

Where a=3370 and b=10 for the machine considered

### B. Excitation Capacitor Equations

The matrix $[v_s]$ comprising the direct and quadrature axes components of the stator voltage is the set of differential variables derivable from the terminal capacitance C and its charge $[q_c]$ as $[v_s]=[q_c]/C$, which on differentiation with respect to time gives

$$p[v_s] = [i]/C \tag{10}$$

$$[i] = [i_s] - [i_l] \tag{11}$$

### C. Load Side Equations

The model equations of the R-L load are represented by

$$[v_s] = L_l * p[i_l] + R_L[i_l] \tag{12}$$

Hence the SEIG is represented by nine first order differential equations. Since the magnetization characteristic of SEIG is nonlinear due to saturation, the magnetizing current must be calculated at each step of integration. Given the updated condition from the previous step, the magnitude of magnetizing current, $i_m$ is calculated as

$$i_m = \sqrt{(i_{ds} + i_{dr})^2 + (i_{qs} + i_{qr})^2} \tag{13}$$

The magnetizing inductance $L_m$ can then be evaluated from magnetizing characteristics plotted between $L_m$ and $i_m$.

### D. Modelling of Digitally Controlled ELC

The rectifier circuit with chopper can be represented by the mathematical equations given in (14), (15).

$$pi_d = (v_m - v_d - 3 * R_s * i_d)/(3 * L_s) \tag{14}$$

$$i_{ld} = (v_d * S)/R_d \tag{15}$$

Where $v_d$ is DC output voltage of rectifier.

$i_{ld}$ is dump load current.

$i_d$ is rectifier output dc current.

$v_x$ is max AC line voltage.

$R_s$ and $R_d$ are source and dump load resistances respectively.

$L_s$ is the source inductance.

S=1 if the switch is on.

S=0 if the switch is off.

The set of above mentioned differential equations are solved by Runge-Kutta method. Simulations have been carried out for a 7.5kW SEIG with single-phase UPF and 0.8pf loads. Fig.2, Fig. 3, Fig. 4 shows the transient waveforms of three-phase terminal voltages Va, Vb, Vc, three phase generator current, capacitance currents under voltage buildup without employing ELC.

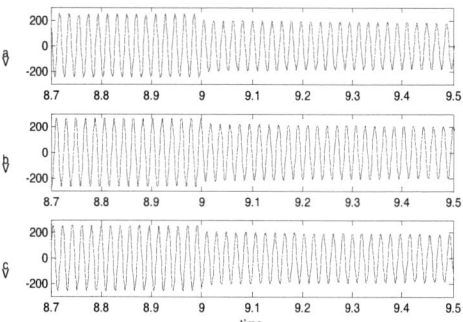

Fig. 2. Transient Three phase voltage waveforms of SEIG without ELC and application of 3kW resistive load.

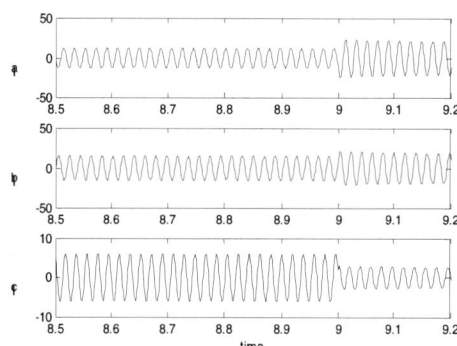

Fig. 3. Transient Three phase current waveforms of SEIG without ELC and application of 3kW resistive load.

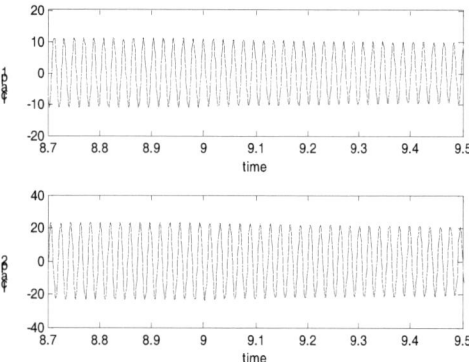

Fig. 4. Transient capacitor current waveforms of SEIG without ELC and application of 3kW resistive load.

## III. IMPLEMENTATION OF DIGITAL CONTROL

Fig.5 shows the block diagram of the digital control circuit of ELC. The ELCs reported so far used analog control, which exhibited certain problems like lack of reliability, limited duty cycle variation and large in size. This paper presents a new digital controller, which is found to be more compact, reliable and which provides better voltage regulation. This controller is being explored for 3kW load for both laboratory and field trials.

The main functions of this control circuit include,

1) To extract the duty cycle of the chopper switch if the supply voltage is more than the rated value.

2) To provide triggering pulses to the TRIACs in order to switch on additional capacitances if the supply voltage is less than the rated value.

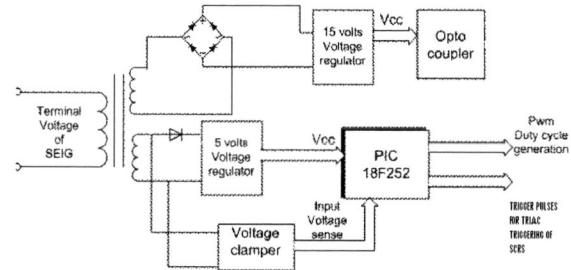

Fig. 5. Block diagram of digital control circuit of ELC.

In order to provide these functions, the instantaneous value of the SEIG terminal voltage is sensed for every two cycles and the error signal is generated by comparing the voltage with reference value. Depending upon the value of the error signal the PIC microcontroller is programmed to provide the PWM pulses to the switch.

The algorithm to provide voltage regulation using this digital control circuit is given below:

1. Sense the terminal voltage of SEIG for each millisecond.
2. Compare the voltage with reference voltage i.e., 230 V.
3. If the error voltage is positive, PWM pulse should be given to the chopper switch.
4. If the error voltage is negative, extra capacitance should be switched by providing trigger pulses to TRIACs of two capacitors.
5. Update the display for every 100msec.

## IV. DESIGN OF CONTROL CIRCUIT

The voltage rating of the uncontrolled rectifier and chopper switch will be the same and dependent on the RMS AC input voltage and average value of the output DC voltage.

For a 3 kW load, the RMS value of the input voltage is 230 V and the average value of the DC voltage is

$$Vdc = 0.9 * V_l = 0.9 * (230 + 10\%) = 207 V \qquad (16)$$

An over voltage of 10% of the rated voltage is considered for the transient condition and RMS AC input voltage will be 230+23=253 volts.

The current rating of the single-phase uncontrolled rectifier and chopper switch (IGBT) is decided by the active AC input current and average value of the DC current. The active component of the AC input current for a single-phase rectifier is calculated as

$$I_{ac} = \frac{P}{V_l} = \frac{3000}{230} = 13.04 \, A \tag{17}$$

In a single-phase uncontrolled rectifier, the distortion factor is 0.9 and the crest factor varies from 1.8 to 3, peak current is given by

$$I_{peak} = \frac{13.04 * 2}{0.9} = 14.5 \, A \tag{18}$$

The rating of dump load resistance is calculated by

$$R_d = \frac{(207)^2}{3000} = 14.283 \, \Omega \tag{19}$$

In the developed load controller, the DC link capacitor is not required since in the presence of capacitor the diodes draw non-linear current from the input circuit. This introduces harmonics in the SEIG output voltage and results in poor power factor. In order to improve the voltage quality, DC link capacitor is avoided. The microcontroller provides the flexibility of changing the PI controller parameters for providing duty cycle over a wide range.

## V. RESULTS AND DISCUSSION

Simulation was carried out on the developed prototype of the SEIG-ELC system to verify the validity of the derived mathematical models. The induction generator is coupled to a closed loop speed-controlled converter-fed dc motor drive. The transient waveforms of the three phase terminal voltages, load current are shown in the Fig.6 and Fig.7 to Fig.11. From the load characteristics of the machine it can be observed the unbalance in the voltages and current can be decreased by the appropriate choice of capacitances and dump load. Samples of the correct formats for various types of references are given below.

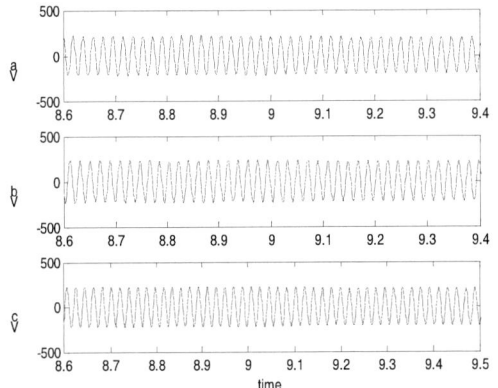

Fig. 6. Transient Three phase voltage waveforms of SEIG with ELC at the sudden application of 3kW resistive load.

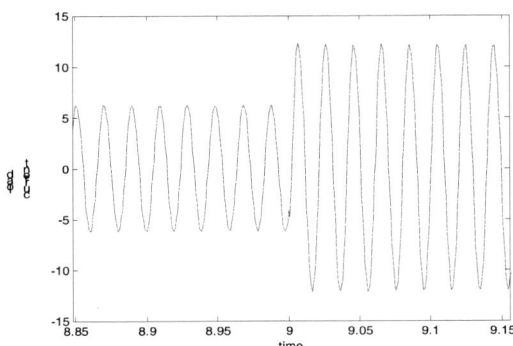

Fig. 7. Transient Three phase Load current waveforms of SEIG with ELC at the sudden application of 3kW resistive load.

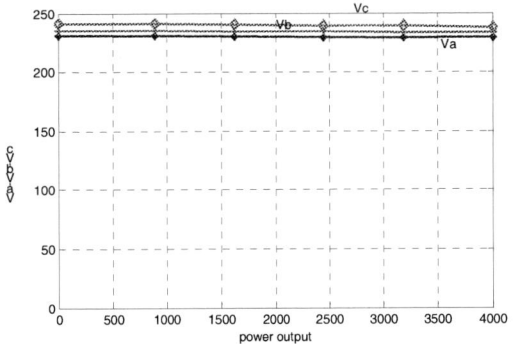

Fig. 8. Three phase voltage Vs Load characteristics of SEIG feeding 1-phase resistive load.

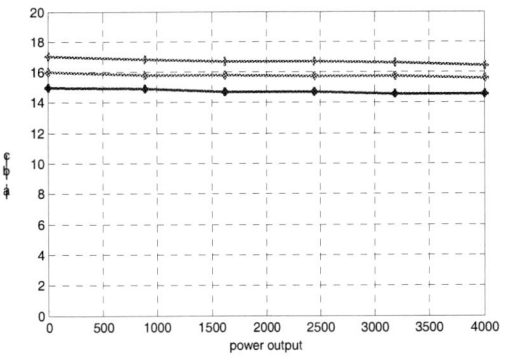

Fig. 9. Three phase current Vs Load characteristics of SEIG feeding 1-phase resistive load.

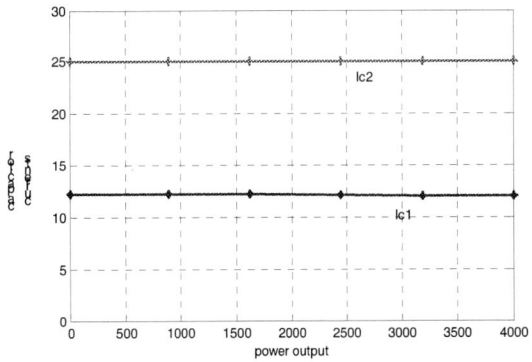

Fig. 10. Capacitor current Vs Load characteristics of SEIG feeding 1-phase resistive load.

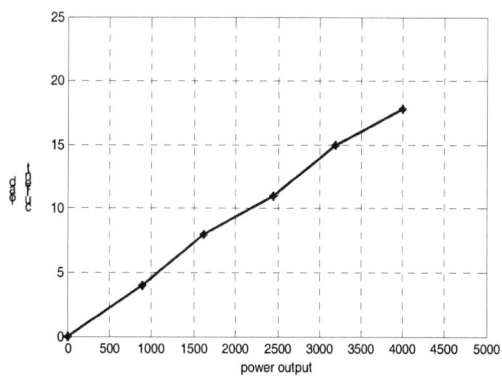

Fig. 11. Load current Vs output characteristics of SEIG feeding 1-phase resistive load.

## VI. CONCLUSIONS

A new SEIG voltage regulation scheme with digitally controlled ELC has been proposed which is found to be more compact, cost effective, and reliable and provides quick response. It is proven by a detailed modeling and analysis that proposed scheme can regulate the output voltage from no load to full load range with high performance. This controller is being explored for both laboratory and field trials. It can sense the change in voltage and adjusts the duty cycle of the chopper switch within few milliseconds. It also provides TRIAC triggering of additional capacitances in order to provide better power factor. The system has been successfully fabricated and preliminary tests carried out.

## VII. APPENDIX

TABLE I
GENERATOR RATING AND PARAMETERS

| PARMAETER | VALUE |
|---|---|
| Rated Power | 7.5 KW |
| Rated Line to Line Voltage | 400/230 V |
| Rated Frequency | 50 Hz |
| Number of poles (P) | 4 poles |
| Stator Resistance (Rs) | 1 Ω |
| Stator Leakage Reactance (Xls) | 1 Ω |
| Rotor Resistance (Rr) | 0.77 Ω |
| Rotor leakage Reactance (Xlr) | 1 Ω |
| Excitation capacitance (C1) | 70μF |
| Excitation capacitance (C2) | 140 μF |

Fig. 12. Pictorial view of developed digital control circuit.

TABLE II
COMPONENTS USED IN HARDWARE IMPLEMENTATION

| COMPONENT | RATING |
|---|---|
| IGBT | 600V, $I_{C}$ = 20A; $V_{ceon}$=1.72V |
| Current sensing resistor | 01Ω/3W |
| Gate driver | Low side IR-2121 |
| Microcontroller | PIC18F252 |
| Display Module | ODM (Oriole Display Module) |
| Crystal oscillator | 10MHz |

## VIII. REFERENCES

[1] Murthy.S.S, Bhim Singh, Sushma Gupta, Ashish Kulkarni, and R.Siva Rajan: "Field Experience on a Novel pico hydel system to supply power to remote locations", *Industrial Applications Magazine*, July 2006, pp-65-76.

[2] Murthy.S.S, Malik, O.P., and Tandon, A.K.: "Analysis of self-excited induction generator", *In Proc. IEE 1982 Gener. Trans. Distrib,* Vol. 129, (6), pp. 260–265.

[3] S.S. Murthy, B. Singh, S. Gupta and B.M. Gulati, "General steady-state analysis of three-phase self-excited induction generator feeding three-phase unbalanced load/single-phase load for stand-alone applications", *In Proc. of 2003 IEE Gener.Transm. Distrib,* Vol 150, No. I, Jan 2003.

[4] Bhim Singh, S. S. Murthy, and Sushma Gupta, "Transient analysis of self-excited induction generator with electronic load controller (ELC) supplying static and dynamic loads", *IEEE Trans. on Industry Applications,* vol. 41, no. 5, September 2005.

[5] Murthy, S.S., Malik, O.P., and Walsh, P.: "Capacitive VAR requirements of self-excited induction generators to achieve desired voltage regulation". Presented at IEEE Industrial and Commercial Power System Conf., Milwaukee, USA, June 1983

[6] Grantham, C., Sutanto, D., and Mismail, B.: "Steady- state and transient analysis of self-excited induction generators", *In Proc. of IEE Electr. Power Appl.,* 1989, 136, (2), pp. 61–68

[7] Elder, J.M., Boys, J.T., and Woodward, J.L.: "Integral cycle control of stand-alone generators", *In Proc. of IEE Gener. Transm. Distrib,* 1985, 132, (2), pp. 57–66.

**2006 IEEE International Conference on Power Electronic, Drives and Energy Systems**

# Analysis and Design of Voltage and Frequency Controllers for Isolated Asynchronous Generators in Constant Power Applications

Bhim Singh, *Senior Member, IEEE*, and Gaurav Kumar Kasal

*Abstract--* **This paper deals with the analysis and design of voltage and frequency controllers for asynchronous generators to be used in isolated constant power applications such as pico and micro hydro sites. These controllers are basically load controllers which maintain the load power constant at generator terminals which in turns maintain the system frequency constant. A set of load controllers are designed, modeled and simulated in MATLAB using Simulink and PSB (Power System Block-set) toolboxes to demonstrate their performance.**

*Index Terms--* **Isolated Asynchronous Generators, Electronic Load Controller, Battery Energy Storage System, Decoupled Load Controller, Pico Hydro System.**

## I. INTRODUCTION

ASYNCHRONOUS machines (AMs) are robust, inexpensive compared with DC and wound rotor synchronous machines, require little maintenance, and have high power weight ratio [1-3]. Despite these favorable features, commercialization of an asynchronous machine as an isolated asynchronous generator is still a bottleneck because of its unsatisfactory voltage and frequency regulation, even when driven under constant speed and feeding varying loads. In view of this, a number of attempts have been made in the area of development of voltage and frequency regulator of isolated asynchronous generators (IAGs). Attempts are also made to improve voltage and frequency regulation through machine design modification [3, 4], using passive components [5, 6] and using active components [7, 8]. Classification of controllers is also given in the literature on the basis of constant [9-17] as well as variable power applications [7, 8]. Here analysis and design of new types of voltage and frequency controllers for isolated asynchronous generator is dealt in constant power applications driven by uncontrolled pico hydro turbines, in which the generator is operated in single point operation where voltage, frequency and excitation of the machine is kept constant under all operating conditions.

The concept behind the operation of these new types of voltage and frequency controllers is that for regulating the voltage and frequency, controller takes up the difference between the active power provided by the uncontrolled turbine and the power consumed by the varying loads. These controllers are broadly classified as an electronic load

Bhim Singh and Gaurav Kumar Kasal are with the Dept. of Electrical Engineering, Indian Institute of Technology Delhi, 110016 New Delhi India (e-mail: bhimsinghr@gmail.com , gauravkasal@gmail.com)

controller (ELC), such an arrangement is applicable only to smaller hydraulic power station, where water use is not a primary concern, as the real power absorbed by the electronic load controller is dissipated as heat in the resistor or can be used to charge a battery.

In view of this a set of electronic load controllers (ELCs) are designed, modeled and simulated to study their behavior. These controllers are classified as binary weighted load controller, controlled rectifier based controller and uncontrolled rectifier based controller, novel electronic load controller, battery energy storage system based load controller and decoupled controller. A systematic design approach is presented and these ELCs are modeled and simulated results are given to provide suitable guidelines for the design of such systems.

## II. CLASSIFICATION OF CONTROLLERS

Fig 1 shows the classification of electronic load controllers used as voltage and frequency controller for isolated asynchronous generators driven by constant power uncontrolled pico hydro turbines. They are classified here according to their chronological development and improved performance. These are classified here into mainly four categories.

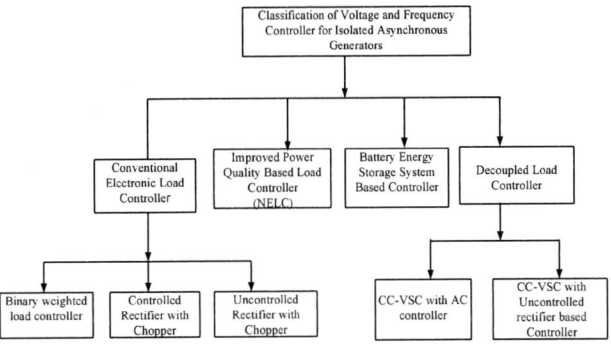

Fig. 1. Classification of voltage and frequency controllers for isolated asynchronous generator in constant power pico hydro system.

### A. Conventional Electronic Load controllers (ELC)

As shown in Fig. 1 conventional ELC are classified as binary weighted load controller (BWLC), thyristor based impedance controller and uncontrolled rectifier based electronic load controller. BWLC is four step binary weighted load controller using integral cycle control of switched resistors with zero voltage switching of fixed steps to make low cost generating system, but it generates instability in the system as well as control in steps not continuous. Controlled

0-7803-9771-1/06/$25.00 ©2006 IEEE 847

rectifier based electronic impedance controller shown in Fig 2(a), consists of controlled thyristor bridge and chopper with dump load. In this proposed scheme an impedance controller consumes the difference in active power provided by the uncontrolled turbine and the power is used by the consumer loads. But due to consumption of reactive power and need of thyristor firing circuit, uncontrolled rectifier (diode bridge based) with chopper and dump load based ELC is proposed as depicted in Fig 2(b). In the above mentioned controllers it is observed that they are having problems of high total harmonic distortion (THD). Therefore a need of an improved electronic load controller with power quality point of view is emerged recently. In view of this, following advanced load controllers are proposed which are given in following sections.

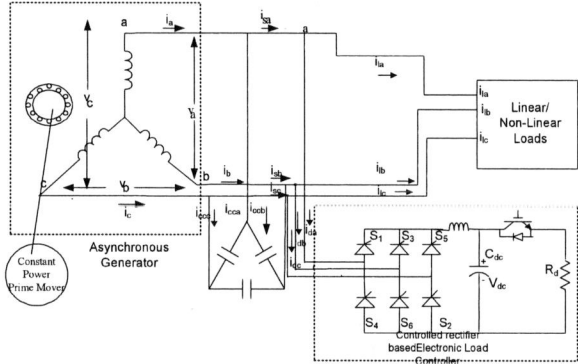

Fig. 2 (a). Controlled rectifier based electronic load controller asynchronous generator in constant power pico hydro system.

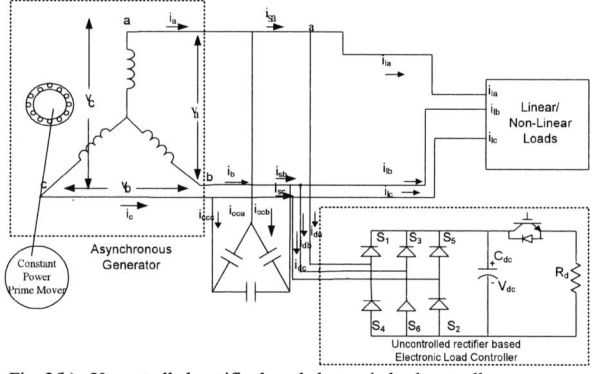

Fig. 2(b). Uncontrolled rectifier based electronic load controller.

### B. Novel Electronic Load Controller (NELC)

Novel electronic load controller shown in Fig. 3, consists of CC-VSC (current controlled voltage source converter) with DC chopper and dump load at its DC link. The output of the VSC is connected through the AC filtering inductor to the IAG terminals. The DC bus capacitor is used to filter voltage ripples and provides self supporting DC bus. DC chopper is used to control dump power in the controller due to varying in consumer loads. Excitation capacitor for the generator is selected to generate rated voltage of isolated asynchronous generator at no load. The additional reactive power requirement of the generator and consumer load is fulfilled by the controller. This controller acts as voltage and frequency regulator along with harmonic mitigator and load balancer.

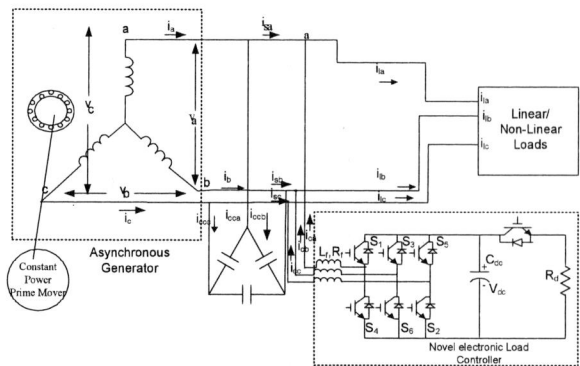

Fig. 3. Novel electronic load controller.

### C. Battery Energy Storage System (BESS) Based Controller

Battery energy storage system (BESS) based voltage and frequency (VF) controller consists of CC-VSC and battery at its DC link as shown in Fig.4. This can be a replacement of NELC because wastage of power in dump load resistance now it can be used for charging the battery while during peak consumer loads additional requirement of active power may be fulfilled by battery such that controller transfers active as well as reactive powers. BESS type VF controller is used for load leveling, load balancing, reactive power compensation and harmonic elimination.

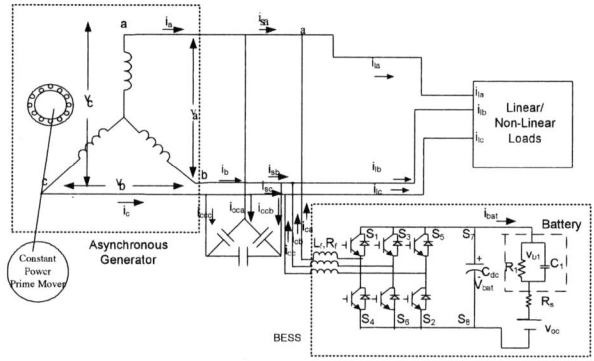

Fig. 4. Battery energy storage system based load controller.

### D. Decoupled Load Controllers

Decoupled voltage and frequency controller (DVFC) or decoupled load controller is a combination of STATCOM and conventional electronic load controller (ELC), in which STATCOM functions as a voltage regulator, load balancer and harmonic eliminator while ELC regulates the active power which in turn maintains the system frequency constant.

Advantages of such type of controller are that it avoids the flow of active power current through CC-VSC. Two type of decoupled load controllers are shown here one consists of IGBTs based CC-VSC and AC controller shown in Fig. 5(a) while other one consists of IGBTs based CC-VSC and uncontrolled rectifier based ELC as shown in Fig 5(b). Advantage of later one is that it does not require complicated firing circuit needed for AC voltage controller as well as it does not need additional reactive power, and it is improved than conventional ELC because harmonic distortion is compensated by STATCOM.

848

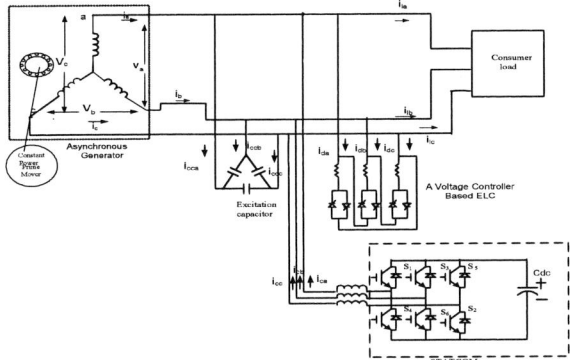

Fig. 5(a). Decoupled load controller using STATCOM with AC controller.

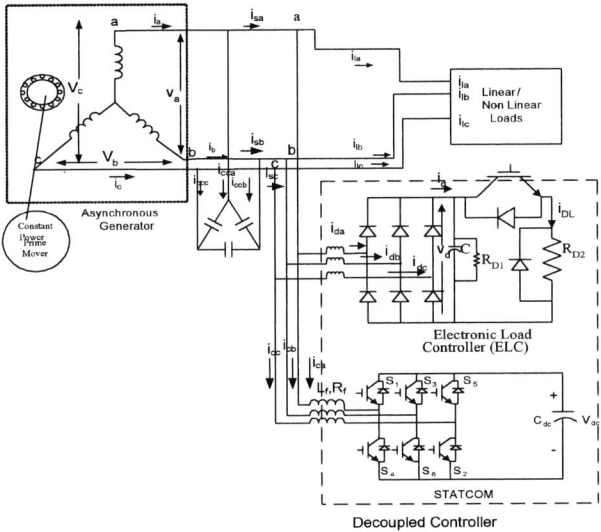

Fig. 5(b). Decoupled load controller using CC-VSC with uncontrolled rectifier based electronic load controller.

### III. DESIGN OF ELECTRONICS LOAD CONTROLLERS

The design of a set of electronic load controllers is given for the selection of their components.

#### A. Conventional Electronic Load Controllers

In this controller (shown in Fig 2(b)) the value of dump load resistance ($R_d$) is estimated using dump power and voltage across it. If the controller is designed for rated power of the generator then dump load resistance can be calculated as:

$$R_d = (V_{dc})^2/P_{gen} \qquad (1)$$

Where $P_{gen}$ is rated power of asynchronous generator and $V_{dc}$ is defined as average voltage at DC bus.

$$V_{dc} = 3\sqrt{2}V_{LL}/\pi \qquad (2)$$

Where $V_{LL}$ is input line to line r.m.s voltage at ELC terminals. An over voltage of x% of rated voltage is considered for the transient conditions then the peak AC voltage as:

$$V_{peak} = (\sqrt{2}) V_{LL} (1 + x/100) \qquad (3)$$

The current rating of uncontrolled rectifier and chopper switch is decided by the active component of input AC current and calculated as [20, 21]:

$$I_{AC} = P_{gen}/ (\sqrt{3} V_{LL}) \qquad (4)$$

Because of drawing approximately quasi-square current, hence the distortion factor is ($3/\pi = 0.955$) then input AC current of ELC may be obtained as:

$$I_{DAC} = I_{AC}/0.955 \qquad (5)$$

The crest factor (CF) of the AC current drawn by uncontrolled rectifier with capacitive filter varies from 1.4 to 2.0 hence the AC input peak current may be calculated as.

$$I_{peak} = 2 I_{DAC} \qquad (6)$$

The value of DC link capacitance of ELC is selected on the basis of ripple factor. The relation between the value of DC link capacitance and ripple factor (RF) for 3-phase uncontrolled rectifier is as [20, 21]:

$$C_{dc} = \{1 / (12 f R_d)\} \{1+1/ (\sqrt{2} RF)\} \qquad (7)$$

From these above equations voltage and current rating of uncontrolled rectifier ($v_{db}$, $i_{db}$), chopper switch (IGBT) ($v_{ch}$, $i_{ch}$)and filter capacitor ($c_{dc}$), have been calculated for a 7.5 kW, 415V, 50 Hz, asynchronous generator, by considering an over voltage of 10%, ripple factor (RF) of 5%. Calculated as well selected values of controller component rating are given in Table–I. Based on these calculated values, the selected values of components are decided, which depend upon the availability and consideration of safety factor etc.

TABLE I
COMPONENT RATING OF UNCONTROLLED RECTIFIER BASED CONVENTIONAL ELC

|  | ($v_{db}$) (V) | $I_{db}$ (A) | $V_{ch}$ (V) | $I_{ch}$ (A) | ($R_d$) ($\Omega$) | $C_{dc}$ ($\mu$F) |
|---|---|---|---|---|---|---|
| Calculated | 645.5 | 21.85 | 645.5 | 21.85 | 41.88 | 615 |
| Selected | 900 | 25 | 900 | 25 | 30 | 650 |

#### B. Novel Electronic Load Controller

Component rating of NELC (shown in Fig.3) includes rating of VSC and chopper switch at DC link of VSC. VSC rating is calculated by determining the additional ($Q_{AR}$) needed by generator at full load [8].

$$Q_{AR} = (Q_R - Q_0) \qquad (8)$$

Where $Q_R$ is required reactive power at full load and $Q_0$ is required reactive power at no load to maintain rated voltage of AG.

Current rating of the NELC is decided by the apparent power ($S_A$) which is given as:

$$S_A = \sqrt{( P_{gen}^2 + Q_{AR}^2)} \qquad (9)$$

Current rating of the NELC corresponding to apparent power is as

$$S_A = \sqrt{3} V_{LL} I_{NELC} \qquad (10)$$

Where $I_{NELC}$ is the NELC r.m.s current. DC bus voltage must be more than the peak of line voltage for satisfactory PWM control as [20, 21]:

$$V_{dc} = (2\sqrt{2}) (V_{LL}/\sqrt{3})/m_a \qquad (11)$$

Where $m_a$ is the modulation index with maximum value of 1. Filtering inductance ($L_f$) of the VSC can be calculated on the basis of allowable current ripple ($i_{cr(p-p)}$) as:

$$L_f = (\sqrt{3}/2)m_a v_{dc}/(6 a f_s i_{cr(p-p)}) \qquad (12)$$

Where $f_s$ is the switching frequency. During transients current rating is likely to vary from 120% to 180% of the steady state value [8]. In inductance calculation the current rating of 120% (a=1.2) of steady state current is considered during dynamics. Voltage drop across AC inductor ($V_d$) is calculated as:

$$V_d = 2\pi f L_f I_{NELC} \qquad (13)$$

849

Rating of DC bus capacitor is determined on the basis of energy transfer through capacitor during transient and permissible dip in DC bus voltage.

$$0.5C_{dc}\{(V_{dc}^2 - V_{dcl}^2)\} = \sqrt{3}\,V_{LL}\,I_{NELC}\,t \qquad (14)$$

Where $V_{dc}$ and $V_{dcl}$ are the nominal DC link voltage and DC link voltage under sudden dip respectively on application of load, while '$V_{LL}$' and '$I_{NELC}$' are rated rms line voltage and current and 't' is sampling time.

Voltage and current rating of the solid state devices (IGBTs) are decided based on maximum voltage across the devices and current through the devices. Under dynamic condition, an overshoot in terminal voltage of the IAG is considered to be x%. Maximum AC voltage is calculated as:

$$= \sqrt{2}\,\{(1+x/100)\,V_{LL} + V_d\} \qquad (15)$$

Current rating is calculated by considering safety factor of 1.25 as :

$$= 1.25\,\{I_{cr(p-p)} + I_{s(peak)}\} \qquad (16)$$

The voltage and current rating of the chopper switch is decided by the DC bus voltage and active power dump into dump load resistor. The current rating of chopper is calculated as:

$$I_{cs} = P_{gen}/V_{dc} \qquad (17)$$

The value of $R_d$ is selected based on rated power of isolated asynchronous generators (IAG) and DC bus voltage. The value of $R_d$ is as:

$$R_d = (V_{dc})^2/P_{gen} \qquad (18)$$

From these equations, voltage and current rating of inverter ($v_{in}$, $i_{in}$) chopper ($v_{ch}$, $i_{ch}$) inductor ($L_f$) dump load ($R_d$) and capacitor ($C_{dc}$) are calculated and based on practical constraints the value of components are selected and these are given in Table-II. Here an overshoot in terminal voltage and ripple current is considered to be order of 10% while dip in DC bus voltage is taken around 8%.

TABLE II
COMPONENT RATING OF NELC

|  | $V_{dc}$ (V) | $i_{in}$ (A) | $v_{in}$ (V) | $i_{ch}$ (A) | $v_{ch}$ (V) | $(L_f)$ mH | $(R_d)$ ($\Omega$) | $C_{dc}$ ($\mu$F) |
|---|---|---|---|---|---|---|---|---|
| Calculate | 678 | 23 | 678 | 12.3 | 678 | 4.65 | 65 | 2500 |
| Selected | 700 | 25 | 900 | 15 | 900 | 5 | 50 | 3000 |

*C. Battery Energy Storage Based Controller*

BESS based controller consists of CC-VSC with battery at its DC link. In Fig 4 Thevenin equivalent circuit of battery based model [18, 19] is shown at DC link of controller. The terminal voltage of the equivalent battery is obtained as:

$$V_{bat} = (2\sqrt{2}/3)V_{LL} \qquad (19)$$

Where $V_{LL}$ is the line rms voltage.

Since the battery is an energy storage unit, its energy is represented in kWh when a battery is used to model the battery unit, the capacitance can be determined from

$$C_1 = (kWh\ 3600 * 10^3)/0.5(v_{ocmax}^2 - v_{ocmin}^2) \qquad (20)$$

In given Thevenins equivalent model $R_s$ is the equivalent resistance (external + internal) of parallel/series combination of battery which is usually a small value. The parallel circuit of $R_1$ and $C_1$ is used to describe the energy and voltage during charging or discharging. $R_1$ in parallel with $C_1$, represents self discharging of battery, since the self discharging current of a battery is small, the resistance $R_1$ is large. Rating of voltage

source converter (VSC) and its component can be similarly determined as computed for novel electronic load controller.

TABLE III
COMPONENT RATING OF BESS

|  | $V_{dc}$ (V) | $i_{in}$ (A) | $v_{in}$ (V) | $(L_f)$ mH | $C_1$ (F) | $R_1$ (k$\Omega$) | $C_{dc}$ ($\mu$F) |
|---|---|---|---|---|---|---|---|
| Calculated | 678 | 23 | 678 | 4.65 | 2400 | 2.5 | 2500 |
| Selected | 700 | 25 | 900 | 5 | 3000 | 3 | 3000 |

From these equations, different parameters of battery ($R_1$, $C_1$) voltage and current rating of CC-VSC are selected and given in Table-III, while rating of VSC is calculated in similar manner case of NELC. Here considering that battery is having 7 kW for 3 hrs peaking capacity, and variation in voltage of 672V-720V.

*D. Decoupled Load Controller*

Decoupled load controller or decoupled voltage and frequency controller (DVFC) is a combination of the STATCOM and ELC. Design of STATCOM consists of voltage and current rating of IGBTs of voltage source converter, value of DC link capacitor ($C_{dc}$) and value of filtering inductors ($L_f$) which are similarly calculated as in case of NELC. While design of ELC is described in section '*A*'. Decoupled controller is having advantages of less component rating of VSC because of avoiding flow of active power through it. Table-I and Table-IV show the component rating of

TABLE IV
COMPONENT RATING OF STATCOM FOR DVFC

|  | $V_{dc}$ (V) | $i_{in}$ (A) | $v_{in}$ (V) | $(L_f)$ mH | $C_{dc}$ ($\mu$F) |
|---|---|---|---|---|---|
| Calculated | 678 | 6.67 | 678 | 9.5 | 2611 |
| Selected | 700 | 15 | 900 | 10 | 4000 |

uncontrolled rectifier based conventional ELC as well as STATCOM respectively. Calculation is made for same machine with considering an overshoot in terminal voltage and dip in DC bus voltage of 10% and 8% respectively, and allowable ripple factor RF of 5%, with peak ripple current of VSC around 10%.

IV. MATLAB BASED MODELING

After selection of components of these voltage and frequency controllers (VFCs), the MATLAB models of these VFCs are developed along with their control scheme to study their steady state and dynamic behavior. Figs. 6-9 show the MATLAB based simulation model of these controllers. Fig 6 shows the model of conventional electronic load controller using diode bridge rectifier and chopper switch at its DC link. MATLAB based simulation model of NELC is depicted in Fig. 7 using IGBTs based VSC and chopper switch at its DC link. Battery energy storage system based controller is modeled using voltage source converter and Thevenin equivalent battery model at DC link and it is shown in Fig. 8. Similarly Fig 9 represents the developed MATLAB based simulation model of decoupled load controller. A 7.5kW, 415 V, 50Hz Y-connected asynchronous generator is used for

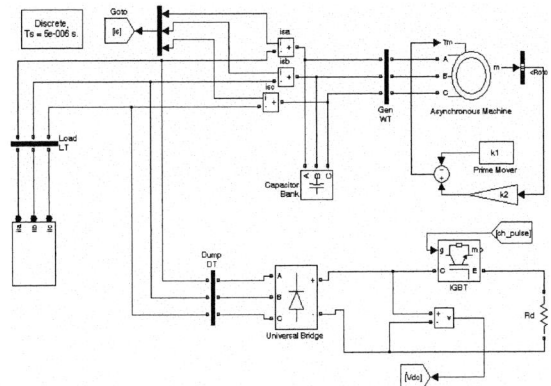

Fig. 6. MATLAB based simulation model for Conventional ELC – IAG system.

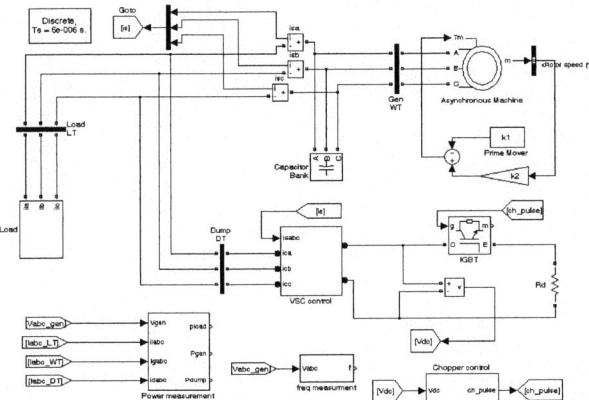

Fig. 7. MATLAB based simulation model for novel ELC – IAG system.

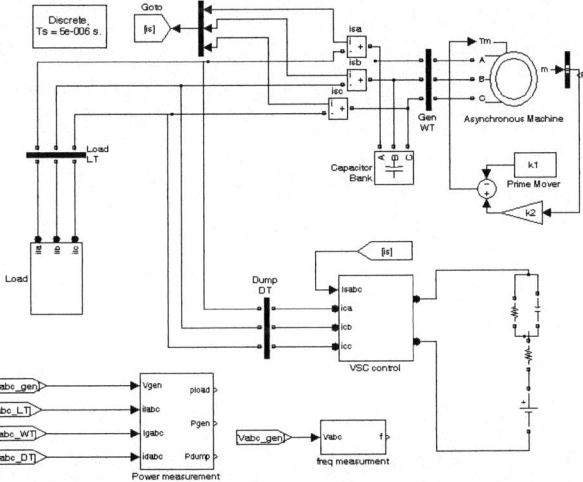

Fig. 8. MATLAB based simulation model for BESS – IAG system.

simulation and parameters of the machine are given in Appendix. The saturation characteristic, which is obtained from synchronous speed test is considered in the model of AG system. Simulation is carried out in discrete mode at 5e-6 step size with ode23tb (stiff/TR-BDF-2) solver.

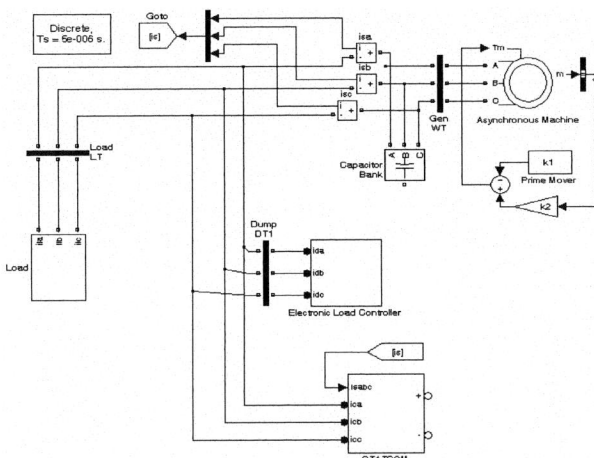

Fig. 9. MATLAB based simulation model for DVFC – IAG system.

## V. RESULTS AND DISCUSSION

After careful design and selection of component rating and parameters of these VF controllers, these controllers for IAGs are modeled and simulated in MATLAB along with Simulink and PSB toolboxes and the salient points from the obtained results are observed here.

Fig. 10 shows the simulated results of uncontrolled rectifier based electronic load controller (ELC). Here excitation capacitor is chosen corresponding to maintain the rated voltage at rated load (7.5kW) and ELC dumps the difference in generated power and consumer load. At 3 sec a resistive load of 5kW is applied then ELC currents ($i_{da}$, $i_{db}$, $i_{dc}$) are reduced which shows the power balancing aspect of the controller which in turns regulates the system frequency (f). However it is also observed that due to non-linear nature of ELC, generator currents ($i_a$,$i_b$,$i_c$) and voltages ($v_a$,$v_b$,$v_c$) are distorted under reduced consumer loads. When consumer load is removed then generated power ($P_{gen}$) is fully dumped into ELC ($P_{dump}$) such that active power at generator terminals remains constant.

Fig. 11 shows the simulated results of NELC-IAG system. Here it is shown that at application of full load ($i_{labc}$) at 2.5sec the controller current ($i_{cabc}$) is reduced to maintain the power constant. At 2.6 sec one phase and at 2.7 sec another phase of load are opened so the load becomes unbalanced and ripples are observed in DC bus ($v_{dc}$), which shows the load balancing aspects of the controller. The voltage ($v_{abc}$) at generator terminals is maintained constant and sinusoidal. Therefore the controller functions as load balancer, voltage and frequency controller and harmonic eliminator.

Fig. 12 shows the simulated results for battery energy storage based voltage and frequency controller for isolated asynchronous generator. Before applying the consumer load it is observed that the battery consumes total generated active power ($P_{gen}$). At 2sec when the consumer load ($P_{load}$) of 9kW is applied, battery starts discharging and provides additional active power to the load. Similarly the waveforms of source voltage ($v_{abc}$), generator currents ($i_{abc}$), capacitor current ($i_{cca}$),

851

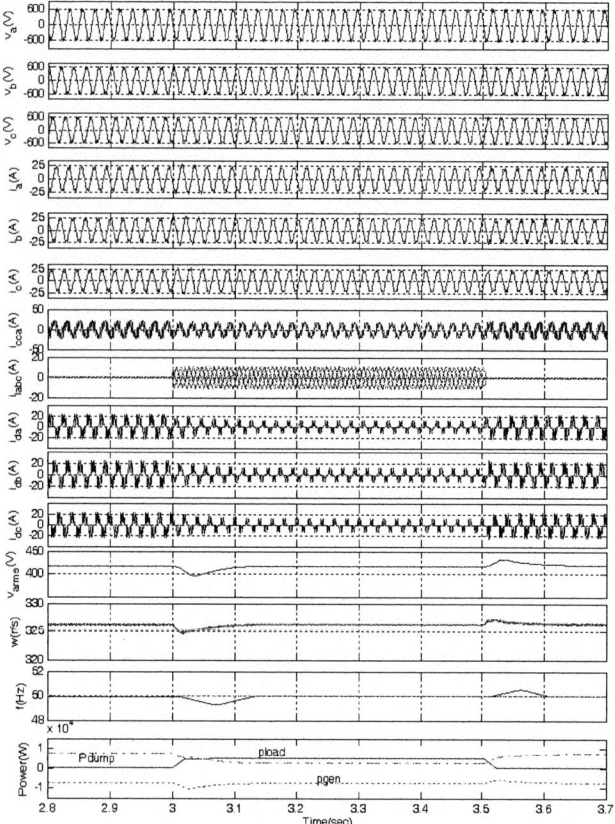

Fig. 10. Dynamic performance of a 7.5kW IAG with uncontrolled rectifier based electronic load controller.

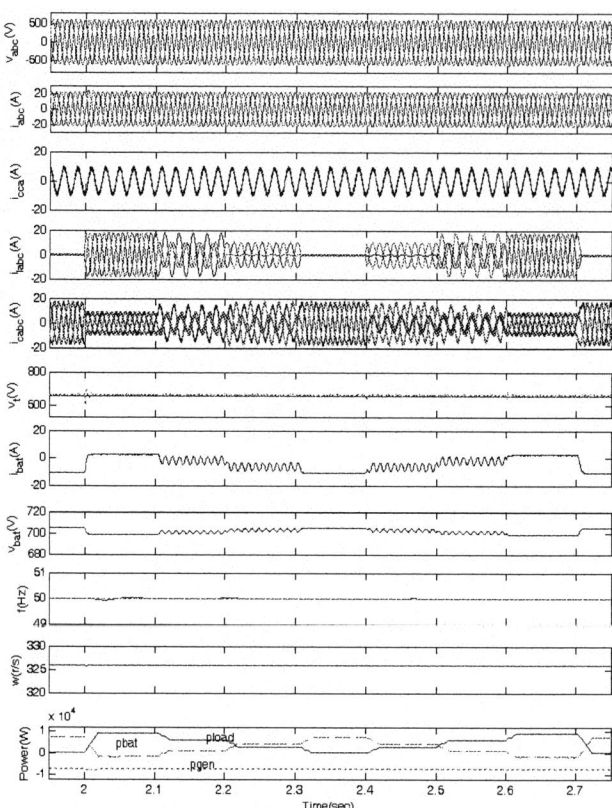

Fig. 12. Dynamic performance of a 7.5kW IAG with battery energy storage system based voltage and frequency controller.

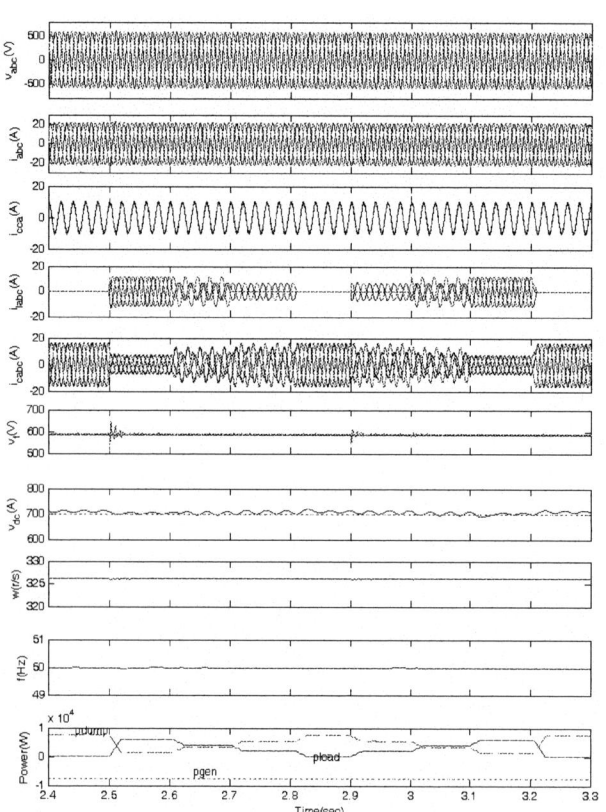

Fig. 11. Dynamic performance of a 7.5kW IAG with novel electronic load controller.

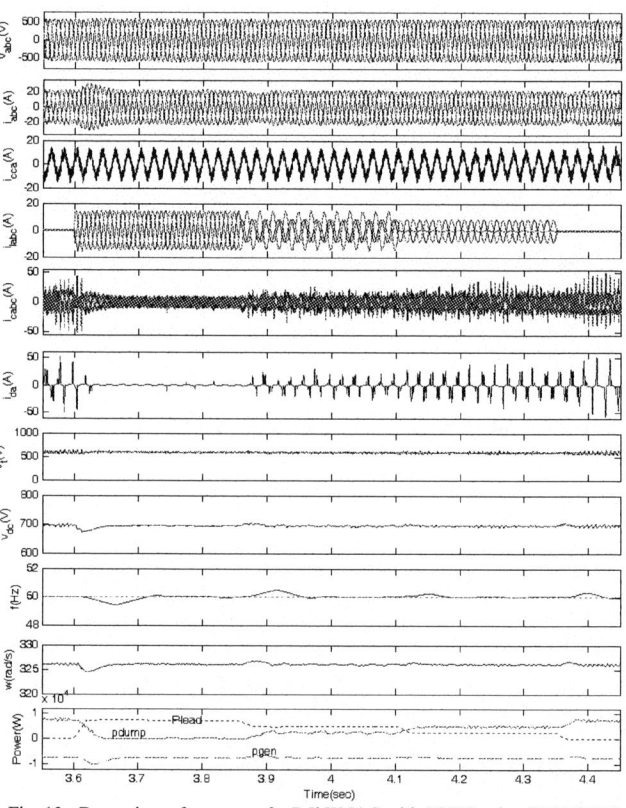

Fig. 13. Dynamic performance of a 7.5kW IAG with DVFC using STATCOM and uncontrolled rectifier based ELC.

load current ($i_{labc}$), controller currents ($i_{cabc}$), amplitude of source peak voltage ($v_t$), battry current ($i_{bat}$) and voltage ($V_{bat}$), frequency and speed (f,w) are shown for various dynamic conditions.

Fig. 13 shows the simulated results for decoupled load controller which is combination of STATCOM and uncontrolled rectifier based electronic load controller (ELC). Transient waveforms of generator voltage ($v_{abc}$), currents ($i_{abc}$), capacitor currents ($i_{cca}$), load current ($i_{labc}$), STATCOM currents ($i_{cabc}$), load controller current ($i_{da}$), amplitude of transient ($v_t$), DC link voltage of STATCOM ($v_{dc}$), frequency and speed (f,w) and variation in power ($P_{dump}$, $P_{gen}$, $P_{load}$) are shown for various operating conditions. As the load is applied at 3.6 sec, the controller current is reduced to maintain the system frequency constant. During load unbalancing at 3.85 sec and 4.1 sec, DC link charging and discharging are observed which shows load balancing features of STATCOM. In comparisons to other controllers it has comparatively reduced rating because active current does not flow through the current controlled voltage source converter (CC-VSC).

## VI. CONCLUSIONS

A detailed design procedure for a set of voltage and frequency controllers has been given for uncontrolled rectifier based ELC, improved power quality based novel ELC, battery energy storage based controllers and decoupled load controller along with their MATLAB simulations. Using the design criteria given here the values of the AC inductor, DC link capacitor, DC link voltage, converter rating and dump load resistance, battery parameters have been calculated and selected on the basis of considering their performance, safety factors and availability of the component rating. After the selection of different parameters like filtering inductor, DC bus capacitor, DC bus voltage and battery parameters the simulation results of different voltage and frequency controllers with IAGs system have been obtained and comparative study of their performance has also been made in detail. In conventional ELC excitation capacitor is selected corresponding to full rating of the generator to generate rated voltage, in other three cases it has been selected corresponding to no load condition to generate the rated voltage and additional reactive power requirement is met by the voltage source converter. Here it is also concluded that because of wasting the power in resistive dump load conventional ELC, NELC and DVFC can be used only for low power applications while BESS based VF controller may be used for high power applications.

## VII. APPENDIX

**A.** The parameters of 7.5kW, 415V, 50Hz, Y-connected, 4-pole asynchronous machine are given below.

$R_s = 1\Omega$, $R_r = 0.77\Omega$, $X_{lr} = X_{ls} = 1.5\Omega$, J = 0.1384 kg-m$^2$

$L_m = 0.14$ ($I_m < 3.26$)

$L_m = 9e-5 I_m^2 - 0.0077 I_m + 0.17$ ($3.16 < I_m < 12.5$)

$L_m = 0.064$ ($I_m > 12.5$)

**B.** Prime Mover characteristic for 7.5 kW machine

$T_{sh} = K_1 - K_2\,\omega_r$, $K_1 = 3300$, $K_2 = 10$

## VIII. REFERENCES

[1] B. Singh, "Induction generator –a prospective," *Electric Machines and Power Systems*, vol.23,pp 163-177,1995.

[2] R.C. Bansal, T.S. Bhatti and D.P. Kothari, "Bibliography on the application of induction generator in non conventional energy systems," *IEEE Trans. on Energy Conversion*, vol. EC-18, no.3, pp.433-439, September 2003.

[3] S. S. Murthy, H. S. Nagaraj and A. Kuriyan, "Design-based computational procedure for performance prediction and analysis of self-excited induction generators using motor design packages," *IEE Proc.*, Vol. 135, Pt. B, No. 1, pp. 8-16, January 1988.

[4] J. Faiz, A. A. Dadgaro, S. Horning and A. Keyhani, "Design of a three-phase self-excited induction generator," *IEEE Trans. on Energy Conversion*, Vol. 10, No. 3, pp. 516-523, September 1995.

[5] L. Wang and J. Y. Su, "Effects of long-shunt and short-shunt connections on voltage variations of a self-excited induction generator," *IEEE Trans. on Energy Conversion*, Vol. 12, No. 4, pp. 368-374, December 1997.

[6] B. Singh, L. Shridhar and C. S. Jha, "Improvements in the performance of self-excited induction generator through series compensation," *IEE Proc.-Gener. Transm. and Distrib*, Vol. 146, No. 6, pp.602-608, November 1999.

[7] S. Wekhande and V. Agrawal, "A new variable speed constant voltage controller for self-excited induction generator," *Electric Power Systems Research*, Vol. 59, pp. 157-164, 2001.

[8] Bhim Singh, S.S. Murthy and Sushma Gupta, "Analysis and design of STATCOM based based regulator for self excited induction generator," *IEEE Trans. on Energy Conversion*, Vol. 19, No. 4, pp. 783-790, December 2004.

[9] J. M. Elder, J. T. Boys and J. L. Woodward, "Integral cycle control of stand-alone generators," *IEE Proc*, vol. 132, Pt. C, no. 2, pp. 57-66, March 1985.

[10] R. Bonert and S. Rajakaruna, "Self-excited induction generator with excellent voltage and frequency control," *IEE Proc.-Gener. Transm. Distrib.*, vol. 145, no. 1, pp. 33-39, January 1998.

[11] B. Singh, S.S. Murthy and Sushma Gupta, "Analysis and implementation of an electronic load controller for a self excited induction generator" *IEE Proc.-Gener. Transm. Distrib.*, Vol. 151, No. 1, pp. 51-60, January 2004.

[12] B. Singh, S.S. Murthy and Sushma Gupta, "Transient analysis of self excited induction generator with electronic load controller supplying static and dynamic loads" *IEEE Trans. on Industry Applications*, vol. 41, no. 5, pp.1194-1204, Sept 2005

[13] B.Singh, S.S. Murthy and Sushma Gupta, "An improved electronic load controller for self excited induction generator in micro-hydel applications", in *Proc.of IEEE Annual Conference of the Industrial Electronic Society*, vol. 3, Nov. 2003, pp. 2741-2746.

[14] B.Singh, S.S. Murthy and Sushma Gupta, "A voltage and frequency controller for self-excited induction generators" *Electric Power Components and Systems*, vol., 34, pp 141-157, 2006.

[15] R.S Bhatia., D.K Jain., B. Singh and S.P. Jain, "Battery energy storage system for power conditioning", in *Proc. of National Power Sys. Conf. NPSC-2004*, vol-1, 27-30, pp. 86-91.

[16] Luiz A.C. Lopes and Rogerio G. Almeida, "Wind-driven induction generator with voltage and frequency regulated by a reduced rating voltage source inverter", *IEEE Trans. on Energy Conversion*, vol. 21, no. 2, pp. 297-304, June 2006.

[17] E. G. Marra and J. A. Pomilio, "Self excited induction generator controlled by a VS-PWM bi-directional converter for rural application," *IEEE Trans. on Industry Applications*, vol. 35, no. 4, pp. 877-883, July/August 1999.

[18] Z. M. Salameh, M. A. Casacca and W.A. Lynch, "A mathematical model for lead-acid batteries" *IEEE Trans. Energy Conversion*, vol.7, no.1, pp.93-97, March 1992.

[19] Z. M. Salameh, M. A. Casacca and W.A. Lynch, "Determination of lead-acid battery capacity via mathematical modeling technique" *IEEE Trans. Energy Conversion*, vol.7, no.3, pp.442-446, Sept 1992.

[20] M.H. Rashid, "*Power Electronics, Circuits, Devices and Applications,*" Pearson Prentice Hall Private Limited, Singapore, Third Edition, 2004.

[21] N. Mohan, T.M. Undeland and W.P. Robbins, "*Power Electronics: Converters, Applications and Design,*" John Willey and Sons, Singapore, Third Edition, 2004.

**2006 IEEE International Conference on Power Electronic, Drives and Energy Systems**

# A Simple Controller using Line Commutated Inverter with Maximum Power Tracking for Wind-Driven Grid-Connected Permanent Magnet Synchronous Generators

V. Lavanya, N. Ammasai Gounden, and Polimera Malleswara Rao

*Abstract* - **A simple wind energy conversion scheme for Permanent Magnet Synchronous Generators (PMSG) has been proposed using three-phase line commutated inverter with maximum power tracking for the first time. The controller extracts maximum power from the wind and feeds it to the three-phase utility grid. A closed loop scheme employing a PI controller and supplementary controller has been modeled in the power system blockset platform and the complete system has been simulated for different wind velocities. A prototype of the proposed system is built in the laboratory and the simulated results are experimentally verified.**

*Index terms* – **line commutated inverter, Maximum Power Point Tracking, wind driven PMSG.**

## I. NOMENCLATURE

| | |
|---|---|
| E | DC link voltage (V) |
| $I_{dc}$ | DC link current (A) |
| $I_{rms}$ | RMS value of phase current (A) |
| $K_P$ | Proportional gain |
| $K_I$ | Integral gain |
| $L_{DC}$ | DC link inductance (H) |
| $R_{DC}$ | DC link resistance (Ω) |
| $T_m$ | Mechanical torque input to PMSG (Nm) |
| V | Wind speed (m/s) |
| $V_I$ | Input voltage to the inverter (V) |
| $V_{rms}$ | RMS terminal phase voltage of PMSG (V) |
| $\alpha_I$ | Firing angle for line commutated inverter (deg) |
| $\varphi$ | Electromagnetic flux (wb) |
| $L_q, L_d$ | Quadrature and direct axis inductance of PMSG respectively (H) |

## II. INTRODUCTION

IT is well known that the development of renewable energy sources is strongly encouraged now-a-days due to fast depletion of traditional energy sources and the environmental pollution caused by them. The wind power generating system is one of the most useful generating systems, which harnesses the natural energy. Directly interfacing the wind energy systems to the utility gives rise to problems such as voltage fluctuations and harmonics associated with the pulsating torque. This problem is overcome by using a power electronic interface between wind energy system (WES) and the utility [1]. In the above paper, a self excited induction generator has been used. The reactive power burden on the self excitation capacitor bank is reduced by using an uncontrolled diode bridge rectifier followed by a line commutated inverter. This necessitates the use of a supplementary controller to maintain the dc link voltage within predetermined values.

With the advent of high power density permanent magnets, the synchronous generators can be built with a very high power density compared to wound rotor type resulting in the elimination of an external dc supply for excitation, the slip rings and brushes. Compared with induction generator (IG), permanent magnet synchronous generator (PMSG) does not require any excitation capacitors for building up of voltage. Though the existing rating of PMSG is meant for its use in self excited (isolated) applications, recent trend is to go in for grid connected operation of the PMSG and corresponding higher rating machines are being designed [2]–[5]. Further, recent researches have focused on how to get maximum power from wind using permanent magnet machines [6]–[9].In these schemes, diode bridge rectifier followed by dc-dc converter or inverter has been used. All the schemes invariably employ forced commutation for the dc-dc converter / inverter. In the present paper, a closed loop controller employing line commutated SCR inverter for extracting maximum power from wind–driven grid-connected PMSG has been proposed. The inherent advantage of self latching property of SCRs has been exploited in the proposed scheme.

## III. PROPOSED SCHEME

The block diagram schematic of the proposed wind energy conversion scheme (WECS) is shown in Fig. 1. It consists of a wind turbine driving a PMSG interfaced to the utility grid through a power electronic interface. The variable frequency, variable magnitude ac voltage available at the PMSG terminals is first rectified using diode bridge rectifier and the

---

The authors are with the Department of Electrical and Electronics Engineering , National Institute of Technology, Tiruchirappalli – 620 015, India (email : ammas@nitt.edu)

0-7803-9771-1/06/$25.00 ©2006 IEEE

dc power is then transferred to the utility using the line commutated inverter.

The dc link voltage and current are sensed and are applied to a multiplier. The output of the multiplier gives the actual dc link power $P_o$ which is compared with the reference power $P_{ref}$ and the difference between these two powers is fed as input to the PI controller. The output of the PI controller modifies the firing angle such that the error gets minimized. In the supplementary controller, the voltage at the dc link is sensed and compared with the reference voltage (32.4V). The error is used for adjusting the firing angle such that the output voltage of PMSG is limited to rated voltage (24V).

where $V_{ML}$ = maximum line voltage of transformer primary.

The transformer at the output of the inverter adjusts the level of the voltage and current for grid interface.

### B. Analysis of Line Commutated Inverter

A three-phase fully controlled bridge converter shown in Fig.2 can be operated in two modes namely rectifier and inverter modes. When the firing angle $\alpha$ is between $0^o$ and $90^o$, the converter is said to be in rectification mode and when $\alpha$ is between $90^o$ and $180^o$, it is said to be in inversion mode. In the proposed scheme, the converter operates as an inverter.

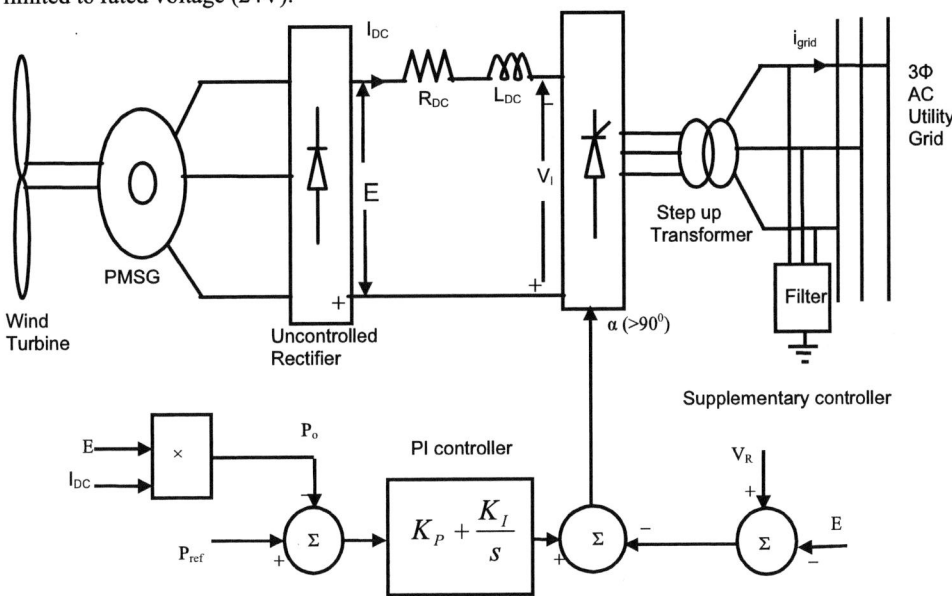

Fig. 1. The block diagram schematic of the proposed wind energy conversion scheme.

### A. Power Electronic Interface

It consists of a diode bridge rectifier, an inductor and the line commutated inverter. The diode rectifier converts the variable magnitude, variable frequency voltage at the PMSG terminals to dc supply and the dc link voltage can be calculated by

$$E = \frac{3V_a}{\pi} = \frac{3\sqrt{6}V_{rms}}{\pi} \qquad (1)$$

The inductor reduces the ripple content in the dc link current $I_{DC}$, which is governed by the differential equation

$$\frac{dI_{dc}}{dt} = \left(\frac{1}{L_{DC}}\right)\left[E - V_I - R_{DC}I_{DC}\right] \qquad (2)$$

where $V_I$ is the input dc voltage of the inverter given by

$$V_I = \frac{3V_{ML}}{\pi}\cos\alpha \quad . \qquad (3)$$

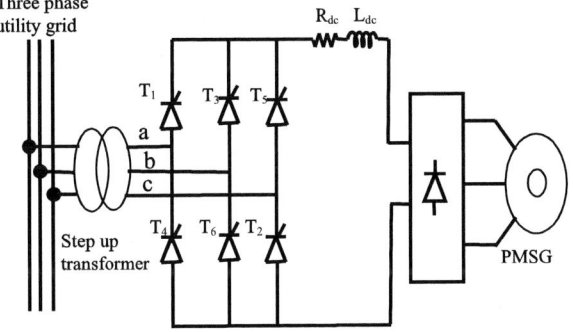

Fig. 2. Circuit of three-phase fully controlled bridge converter.

If the thyristors are numbered as shown in Fig. 2, the firing sequence is $T_6T_1$, $T_1T_2$, $T_2T_3$, $T_3T_4$, $T_4T_5$, and $T_5T_6$. So, if the load is capable of supplying power, then the direction of power flow can be reversed by reversal of the dc voltage, the current direction being unchanged. The delay angle $\alpha$ must be greater than $90^o$. In the present case, no extra effort is required to synchronize the inverter output frequency with that of the grid supply. This of course is possible only with SCR

855

converters. The average output voltage $V_{dc}$ is hence given by,

$$V_{dc} = \frac{3\sqrt{3}}{\pi} V_m \cos \alpha \tag{4}$$

(i) *Harmonics* :

The rms value of $n^{th}$ harmonic current in phase-a is given by,

$$I_{an} = \frac{1}{\sqrt{2}}\left(a_n^2 + b_n^2\right)^{1/2} = \frac{2\sqrt{2}}{n\pi} I_{dc} \sin\frac{n\pi}{3} \tag{5}$$

The rms value of the fundamental current is

$$I_{a1} = \frac{\sqrt{6}}{\pi} I_{dc} = 0.7797 I_{dc} \tag{6}$$

The total rms current in phase 'a' is given by,

$$I_a = \left[\frac{2}{2\pi}\int_{\pi/6+\alpha}^{5\pi/6+\alpha} I_{dc}^2 \ d(\omega t)\right]^{1/2} = I_{dc}\sqrt{\frac{2}{3}} = 0.8165 I_{dc} \tag{7}$$

Now, harmonic factor (HF) or total harmonic distortion (THD) is given by,

$$HF = \left[\left(\frac{I_a}{I_{a1}}\right)^2 - 1\right]^{1/2} = 0.3108$$

Hence,

$$THD = 31.08\% \tag{8}$$

## IV. CONTROLLER FOR MAXIMUM POWER TRACKING

For extracting maximum power from the WES, the firing angle of the inverter is adjusted in closed loop. The maximum power available at the dc link of WES is given by [1]

$$P_{Max} = f(V) = -3.0 + 1.08V - 0.125 \ V^2 + 0.842 \ V^3 \ W \tag{9}$$

and a reference power at the rectifier output is

$$P_{ref} = \eta_g \eta_r \ P_{Max} \tag{10}$$

where $\eta_g$ and $\eta_r$ are conversion efficiencies of the generator and rectifier respectively. The actual output power, $P_o$ is compared with the reference power and any mismatch is used to change the firing angle $\alpha_I$ of the inverter as follows:

$$\alpha_I = (P_{ref} - P_o)[K_P + K_I / s] \tag{11}$$

where $K_P$ and $K_I$ are the proportional and integral stage gains respectively.

The optimum values for $K_P$ and $K_I$ have been arrived at by trial and error method. The values have been chosen taking in

to account the range of mechanical torque of the wind turbine. This range will represent the variation in wind velocity (prime mover speed) with which the system has to operate. In the proposed scheme, the P and I controller gains ($K_P = 0.3$ and $K_I = 7$) have been chosen for operating the system with mechanical torque varying from 10Nm to 20Nm. However, if the range of prime mover speed is different, re-tuning of PI controller has to be done.

## V. SIMULATION RESULTS

The complete model of the proposed WECS is modeled using MATLAB simulink blocks in PSB platform. It consists of blocks for PMSG, uncontrolled diode bridge rectifier, line commutated inverter, power grid and closed loop controllers. The dc link resistance and inductance together with the output of PMSG and rectifier block act as load on the line commutated inverter. The parameters of PMSG were obtained as R= 1.1Ω, $L_d$= 4.4e-03H, Lq = 4.4e-03 H, φ= 0.2 wb; for the dc link, $R_{dc}$=1Ω, $L_{dc}$ =11e-03 H; for the filter capacitor, C= 60e-06 F, R = 0.2Ω. The d-q model of PMSG is available as a built-in block in MATLAB power system block set.

The PMSG is driven by wind turbine and as wind velocity increases, the mechanical torque output of the turbine increases. The closed loop model of the proposed scheme is simulated for different wind velocities and the results are given along with experimental readings in the next section.

## VI. EXPERIMENTAL INVESTIGATION

The experimental setup of the closed loop scheme consists of a 3-phase, 12 pole PMSG, an uncontrolled rectifier, a line commutated inverter, a step up transformer and a controller to generate firing pulses to the thyristors. The PMSG is driven by a separately excited dc machine to simulate the variable speed wind turbine. The PMSG is rated for 24V, 30A, and 500 rpm. An auto transformer of 400V, 15A has been connected between the line commutated inverter and the grid. A three-phase SCR converter has been fabricated and a microcontroller firing scheme has been developed to trigger the SCRs. The firing angle of the inverter is automatically adjusted in the closed loop to feed maximum power to the grid. Care has been taken to see that the firing angle is kept above 90° in order to facilitate inverter operation. The sequence of firing the SCRs depends on the band of α; viz., $T_5$ $T_6$, $T_6$ $T_1$, $T_1$ $T_2$, $T_2$ $T_3$, $T_3$ $T_4$, $T_4$ $T_5$ if α is between 90° and 120° and for α > 120°, the sequence is $T_4$ $T_5$, $T_5$ $T_6$, $T_6$ $T_1$, T1 $T_2$, $T_2$ $T_3$, $T_3$ $T_4$.

The closed loop controller has been fabricated using a micro controller PIC16F876 together with an analog multiplier IC (MPY634KP).The voltage and current are sensed from the grid side using potential transformer (PT) and current transformer (CT) respectively and are fed as inputs to the multiplier IC. The inputs are multiplied and the output consists of two components, a dc component and an ac component. The dc component corresponds to the active power fed to the grid and the ac component varies with time, but at twice the frequency of input. So the output is filtered and only the dc component is obtained. The output of the filter after passing through the buffer is given to the microcontroller circuit where it gets compared with the power at the previous instant. After the comparison the firing angle gets changed accordingly to track the maximum power. The power fed to the grid is maximum when the power at the present instant and the power at the previous instant are same. The connection of multiplier IC along with the signal conditioning circuit is shown in Fig. 3.

with filter as the THD reduces to 7.8%.The validity of the controller can be ascertained by the close agreement between experimental and simulated waveforms shown in Figs. 4-6. The ZCD pulse and the firing pulses to first pair of SCRs corresponding to a maximum power point (182 W) are shown in Fig.8.

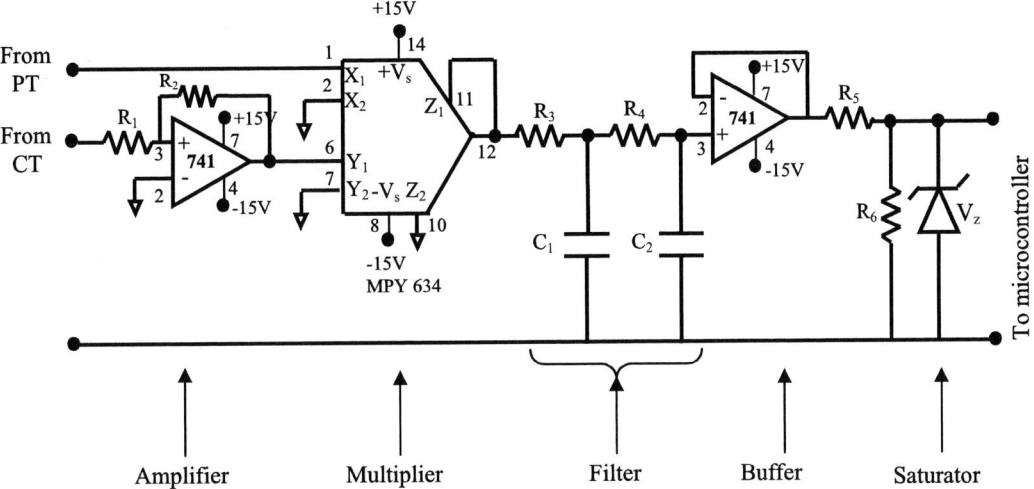

Fig. 3. Multiplier with the signal conditioning circuit.

$R_1=R_2=12\Omega$; $R_3=1\ \Omega$; $R_4=15\ \Omega$; $C_1=C_2=47\ \mu F$.

The results obtained from the experimental investigation and simulation study of the proposed scheme are furnished in Table I for comparison. It is seen that there is very close agreement between the two, which ensures the validity of the proposed scheme.

It is seen that the time delay in the firing of SCRs after the zero crossing detection is 6.1 ms. Hence the firing angle delay at which the maximum power is fed to the grid is 6.1*(360/20) = 109.8°.

TABLE I
COMPARISON OF SIMULATION AND EXPERIMENTAL RESULTS
MECHANICAL TORQUE = 20 NM

| Parameters | Simulation Results | Experimental Results |
|---|---|---|
| Firing angle at which max power occurs, $\alpha$ | 111° | 109.8° |
| DC voltage, $V_{dc}$(in V) | - 21.8 | - 22.0 |
| DC link current, $I_{dc}$(in A) | 9.7 | 9.5 |
| Grid current, $I_{rms}$(in A) | 0.75 | 0.75 |
| Three-phase active power fed to the grid, $P_{grid}$ (in W) | - 180 | - 182 |

The oscillographic waveforms of dc link voltage and the dc link current for a mechanical torque of $T_m = 20$ Nm are shown in Fig. 4. The corresponding waveforms obtained using simulation study are also given along with experimental ones for comparison. Similarly the voltage and current waveforms at the output of the inverter obtained experimentally are shown in Fig.5 along with the corresponding simulated waveforms. The observed and simulated waveforms of voltage and current at the grid with a capacitor bank (60 μF) connected are given in Fig.6. The harmonic spectrum of the grid current obtained with and without filter is shown in Fig.7. It can be observed that THD without filter is 32.8% which is nearly same as given by (8). The grid current waveform is almost sinusoidal

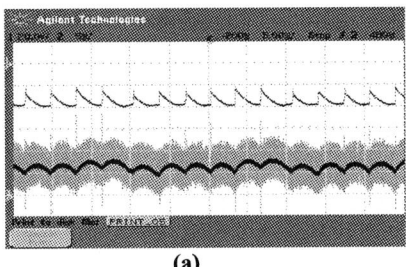

(a)

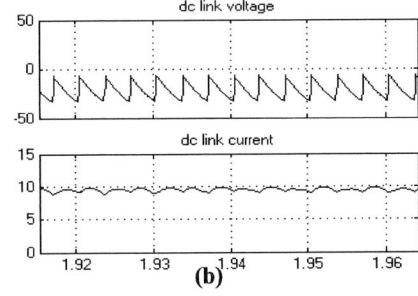

(b)

Fig. 4.   Dc link voltage and dc link current. (a) Experimental (b) Simulated.

857

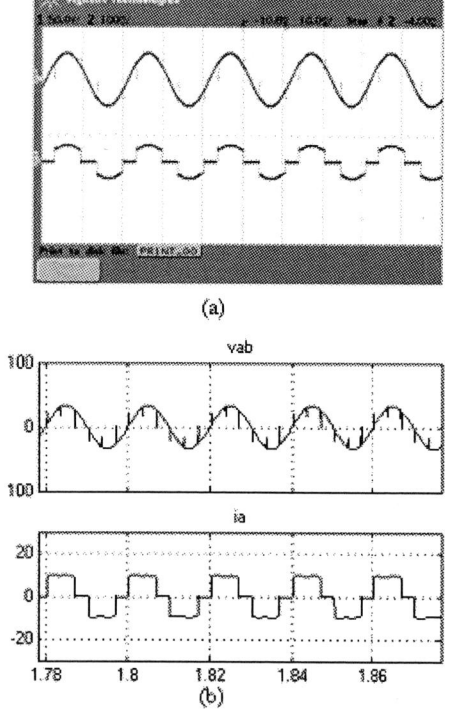

(a)

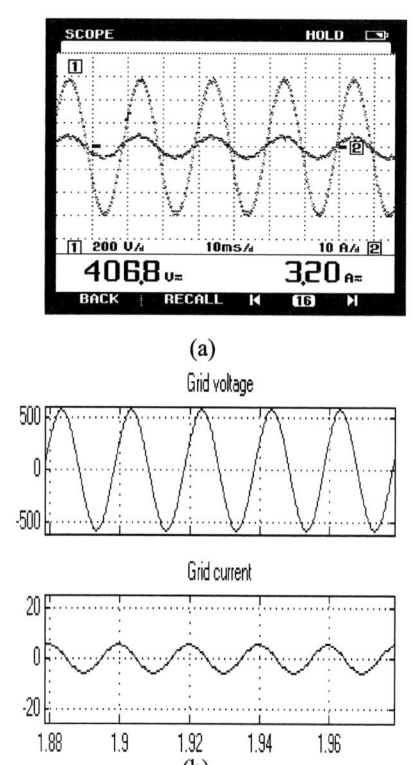

(b)

Fig. 5. Voltage and current waveform at the output of the inverter. (a)Experimental. (b) Simulated.

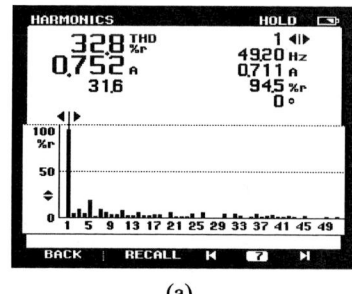

(a)

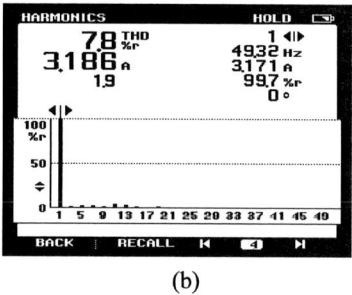

(b)

Fig. 7. Harmonic spectrum of the grid current. (a) without filter. (b) with filter.

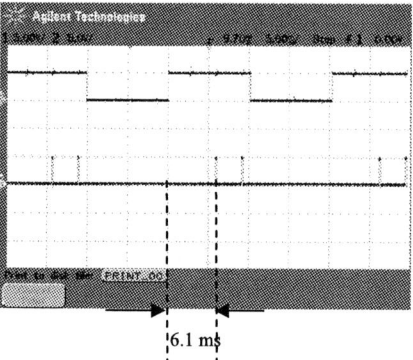

Fig. 8. ZCD pulse and firing pulse to $T_5$ , $T_6$.

## VII. CONCLUSION

A simple closed loop scheme employing a three phase line commutated inverter has been developed for interfacing wind driven PMSG with the utility grid. Simulation studies have been carried out to get the various parameters of the scheme such as active and reactive powers, dc link voltage and current and the firing angle corresponding to the maximum power for a given wind velocity. It has been observed from the results that greater the wind velocity, larger is the firing angle for maximum power. Experimental set up has been built using a laboratory size PMSG of 750 W and a PIC microcontroller has been programmed to generate the trigger pulses for the SCRs and the firing angle is automatically adjusted to feed maximum active power to the grid. The THD obtained for the grid current waveform is also closely matching with the analytical value, which further enhances the correctness of the

(a)

Fig. 6. Voltage and current waveform at the grid with filter.(a) Experimental. (b) Simulated.

858

controller. However, due to losses in the dc inductor and the autotransformer interposed between the converter and utility grid, the output power fed to the grid is somewhat less. This can be increased by selecting an inductor with low losses and operating the transformer at rated load condition, which is possible with higher capacity PMSG.

## VIII. ACKNOWLEDGMENT

The analog multiplier IC (MPY 634KP) used in the closed loop controller of this work was supplied as free samples by Texas Instruments, India. The authors gratefully acknowledge the same. The authors also thank the centre for Energy and Environmental Science and Technology (CEESAT) of NIT, Tiruchirapalli for providing the SCRs used in the fabrication of the inverter. The authors also thank V.Manimaran and T.Suresh for their assistance in conducting the experiment.

## IX. REFERENCES

[1] R.M.Hillowala and A.M.Sharaf, "A Utility Interactive Wind Energy Conversion Scheme with an Asynchronous DC link using a Supplementary Control Loop," IEEE Transactions on Energy Conversion, vol. 9, No.3, September 1994, pp. 558 -563.

[2] Pragasen Pillay and Ramu Krishnan, "Modeling, Simulation and Analysis of Permanent Magnet Motor Drives, part I: The Permanent Magnet Synchronous Motor Drive," IEEE Transactions on Industry Applications, vol. 25,No.2,March 1989,pp. 265 -273.

[3] R.Krishnan and Geun-Hie Rim, "Performance and Design of a Variable Speed Constant Frequency Power Conversion Scheme with a Permanent Magnet Synchronous Generator," IEEE Conference Proceedings, 1989, pp.45 – 50.

[4] Z.Chen and E.Spooner, "Grid Interface Options for Variable-speed, Permanent-Magnet Generators," IEE Proc.-Electr. Power Applications, vol. 145, No. 4, July 1998, pp.273-283.

[5] Jianyi Chen, Chenmagot V.Nayar and Longya Xu, "Design and Finite-Element Analysis of an Outer-Rotor Permanent-Magnet Generator for Directly Coupled Wind Turbines," IEEE Transactions on Magnetics, vol. 36, No. 5, September 2000, pp.3802-3809.

[6] Jia Yaokin, Yang Zhongqing and Cao Binggang, " New Maximum Power Point Tracking Control Scheme for Wind Generation," IEEE Transactions on Industry Applications, vol.2,2002, pp. 144-148.

[7] Kenji Amei, Yukichi Takayasu, Takahisa Ohji and Masaaki Sakui, " A Maximum Power Control of Wind Generator System Using a Permanent Magnet Synchronous Generator and a Boost Chopper Circuit," IEEE PCC-Osaka 2002, pp. 1447-1452.

[8] R.Esmaili, L.Xu and D.K.Nichols, "A New Control Method of Permanent Magnet Generator for Maximum power Tracking in Wind Turbine Application," IEEE PES General Meeting 2005,, pp. 1-6.

[9] A.B.Raju, K.Chatterjee and B.G.Fernandes, "A simple Maximum power point tracker for grid connected variable speed wind energy conversion system with reduced switch count power converters," IEEE PESC-2003,vol. 2, pp. 748-753.

## X. BIOGRAPHIES

N.AmmasaiGounden was born in Coimbatore, TamilNadu, India, on October 5, 1955. He received the B.E. degree from the College of Engineering, Guindy, India (Madras University) in 1978, and the M.E. degree in control systems from P.S.G.College of Technology, Coimbatore, India (Madras University) in 1980. He received the Ph.D. degree from the Bharathidasan University, Tiruchirappalli, India, in 1990.

Currently, he is a professor with the Department of Electrical and Electronics Engineering, National Institute of Technology, Tiruchirappalli where he has been since 1982. His areas of interest are power electronic applications in renewable energy systems, energy conversion and use of programmable digital controllers in hybrid renewable systems.

V.Lavanya was born in Tiruchirappalli, TamilNadu, India, on March 10, 1980. She received the B.E. degree from Sona College of Technology, Salem, India (Madras University) in 2001, and M.Tech. degree in power systems from National Institute of Technology, Tiruchirappalli, India in 2006. Currently, she is a Lecturer with the Department of Electrical and Electronics Engineering, K.S.R.College of Engineering, Tiruchengode, India. This work was carried out when she was working towards her M.Tech. degree.

Polimera Malleswara Rao was born in Palakollu, Andhra Pradesh, India on May 21, 1980. He received the B.E. degree from GITAM college of Engineering (Andhra University), Vishakhapatnam, India in 2002 and the M.Tech. degree in power systems from National Institute of Technology, Tiruchirappalli, India in 2005. Currently, he is a Developer in ERP with Cognizant Technology Solutions, Chennai.

**2006 IEEE International Conference on Power Electronic, Drives and Energy Systems**

# A High-power High-frequency and Scalable Multi-megawatt Fuel-cell Inverter for Power Quality and Distributed Generation

Sudip K. Mazumder, *Senior Member, IEEE*, and, Rongjun Huang, *Student Member, IEEE*

*Abstract--* **Recent R&D in high-voltage and high-power SiC power semiconductor devices have enabled the possibility of a radical shift in the design of high-power inverters from one that is based primarily on line-frequency switching to one that operates between 10 to 20 kHz [1]. It is based on this ongoing advancement, we propose a 1-10 MW and 20 kHz inverter that is fed from a fuel-cell stack and delivers active and reactive power to a utility grid at 12.47 kV and 60 Hz. The inverter eliminates the need for bulky line-frequency transformer and reactive components and has less volume and lower cost. Further, owing to the modular structure of the inverter, it is scalable for higher power and higher voltage applications.**

*Index Terms*—Fuel-cell Inverter, High power, High Frequency, Distributed Generation.

## I. INTRODUCTION

IN addition to residential usages with relatively low power rating, fuel-cells are increasingly considered for 1-10 megawatt applications to support utility distribution circuit or supply stand alone loads [2]. Similar as the low-power case, fuel-cell stacks in this power range most likely will have low voltage (less than 1 kV dc) and high current output characteristics. On the other hand, the desired output voltage is much higher (commonly 12.47 kV ac for the distribution systems in US). The huge gap between these two values make the design of power electronics inverters, which should have high input current and high output voltage handling capabilities, a really challenge.

The existing inverters with aforementioned power ratings usually operate at low switching frequency (around 1 kHz or less) due to the limited turn on/off performances of the high voltage power devices, resulting in bulky and costly magetics and filters. The recent progress on semiconductor devices especially on SiC based high voltage devices makes inverter design with high-frequency switching at 10 kHz or higher a

---

S. K. Mazumder is an Associate Professor in the Department of Electrical and Computer Engineering, University of Illinois at Chicago, Chicago, IL 60607 USA (e-mail: mazumder@ece.uic.edu). He is also the Director of the Laboratory of Energy and Switching-electronics Systems (LESES).

R. Huang is with the Department of Electrical and Computer Engineering, University of Illinois at Chicago, Chicago, IL 60607 USA (e-mail: rhuang2@uic.edu).

possible and potentially more favorable option than the low-frequency based design [1].

Connecting multiple power electronics modules with their inputs in parallel and outputs in series is a more viable option rather than using single module inverter due to the lack of power semiconductor devices which have such high V-I ratings. The former option can also have relatively higher reliability and redundancy. Isolation among different modules is required to avoid short-circuit and usually achieved by using transformers rather than using multiple input fuel-cell sources since the latter option may not be available and usually not cost effective for the power rating of 1-10 MW range.

The conventional high-power dc/ac inverter topologies usually require multiple isolated dc/dc converters with their outputs in series to obtain a high voltage dc bus to feed a following two-level or multi-level dc/ac inverter. Alternatively, a multi-cell cascaded inverter can be used since isolated dc sources are available. In any case, inductors and capacitors are required for these so-called "rectifier type" inverters to stabilize the dc bus and sink the current if diode rectifier are used at the front. A typical "rectifier type" inverter module is shown in Figs. 1(a), which contains a two-leg full bridge isolated dc/dc converter and a three-phase voltage source inverter. The existing of relatively low-life time components L and C on the dc bus will not only increase the cost but also reduce the system reliability.

Another set of topologies [3, 4], referred to as "cycloconverter type converter", reduce the system complexity by removing the dc stage. The circuit shown in Figs.1(b) composes of a two-leg full bridge inverter followed by a single-phase high-frequency transformer and a three-phase PWM cycloconverter. Extension for the higher power can be achieved by two ways: (i) connecting the all modules' transformer secondary sides in series and feeding the obtained ac bus into a three-phase cycloconverter, or (ii) cascading the single-phase cycloconverter outputs to form one phase output. Although it has lesser power conversion stage, the number of active components required especially for the high voltage secondary side is relatively high due to the bidirectional switches for cycloconverters.

This paper proposes a three-phase single-source high-frequency link isolated rectifier type inverter without using any inductor or capacitor in its main circuit for the high-power

---

0-7803-9771-1/06/$25.00 ©2006 IEEE

fuel-cell based distributed generation applications. As compared to the "cycloconverter type" inverter (Figs. 1(b)), the proposed topology has less number of active switchers to achieve three phase output. The circuit schematic of one module of the proposed inverter is shown in Figs. 1(c). It comprises three stages of power conversion: 1) a high-frequency (10 kHz) sinusoidal phase-shift- modulated zero-voltage turn-on full-bridge inverter, which interfaces to a low-voltage and high-current fuel-cell stack; 2) a three-leg diode rectifier that transforms the bipolar ac voltage at the secondary of the high-frequency transformer to a unipolar 20 kHz pulsating waveform (which has a 6-pulse envelope); and 3) an ac/ac converter that converts the pulsating output of the rectifier to an line-frequency ac output using pulse-width modulation. The high-frequency transformers in Figs. 1(c) provide galvanic isolation, boost the input-side voltage, and enable series connection of multiple modules without short circuit.

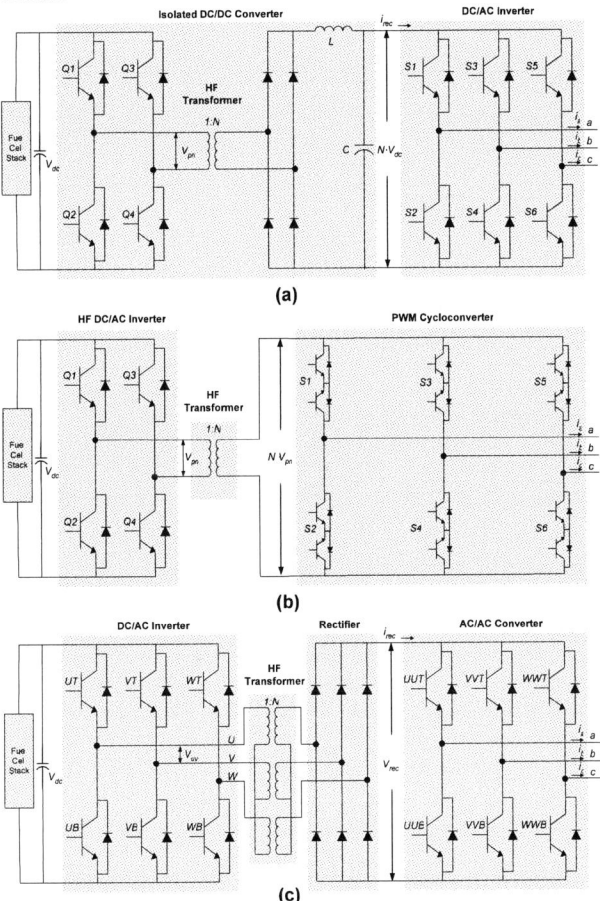

Figs. 1. Schematics of a typical "rectifier type" inverter module with constant dc bus (a), the "cycloconverter type" inverter module (b), and the **proposed** "rectifier type" inverter module without constant dc bus (c). For each type of converter, only one module is shown.

The final stage ac/ac inverter is switched in a hybrid manner. Only at one-third of one period time the devices are operating with high-frequency (20 kHz) while they stay at either ON or OFF for the rest of time. This fundamental difference in switching strategies of the inverter leads to lower switching losses. The inverter can support a load power factor angle up to ±30 degrees. Adding anti-parallel switchers to the rectifier diodes provides four-quadrant operation. The inverter in Figs. 1(c) can be easily extended to higher voltage or higher as will be described in section IV.

## II. CIRCUIT CONFIGURATION AND OPERATION

### A. Three-phase dc/ac Inverter

Fig. 2 illustrates the generation of switch gate signals for the proposed converter. The bottom switches are controlled complimentarily to the upper ones hence they are not described further. Three gate-drive signals UT, VT and WT for primary side devices are obtained by phase shifting a square wave with respect to a 10 kHz square wave signal Q (shown in Figs. 3(b)). Q is synchronous with a 20 kHz saw-tooth carrier signal, shown in Figs. 3(a). The phase differences are modulated sinusoidally using three 60 Hz references a, b and c respectively. Two gate signals for phase U and V are plotted in Figs. 3 (c) and (d). Since carrier frequency is much higher than the reference frequency, UT, VT and WT will be square wave with the frequency of 10 kHz and their phases are modulated. The obtained output line-line voltages at the primary side of the transformers are bipolar waveforms. $V_{uv}$ is plotted in Figs. 3 (e) as an example. After passing through high-frequency transformers, they are rectified by a three-leg diode bridge at the secondary side to obtain a unipolar PWM waveform, which has six-pulse as envelop, Its waveform is shown in Figs. 3 (g) and the mathematic expression are:

$$V_{rec} = N \cdot V_{dc} \cdot \mathbf{MAX}\left(|UT - VT|, |VT - WT|, |WT - UT|\right) \quad (1)$$

$$UT = \overline{Q \otimes PWM_a} \quad (2)$$

$$VT = \overline{Q \otimes PWM_b} \quad (3)$$

$$WT = \overline{Q \otimes PWM_c} \quad (4)$$

Where, $PWM_x$ ($x$ = a, b or c) denotes the binary comparator output between reference and carrier for phase $x$. Symbol "$\otimes$" stands for XNOR operation. N is the transformer turns ratio.

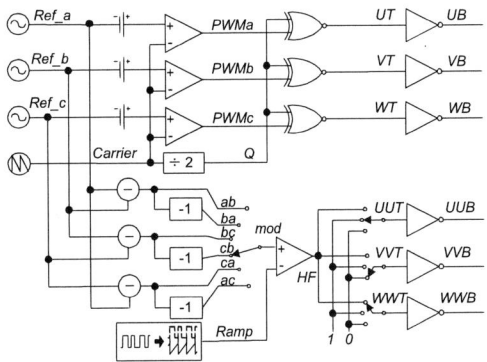

Fig.2. Diagram of gate drive signal generation for the proposed inverter.

861

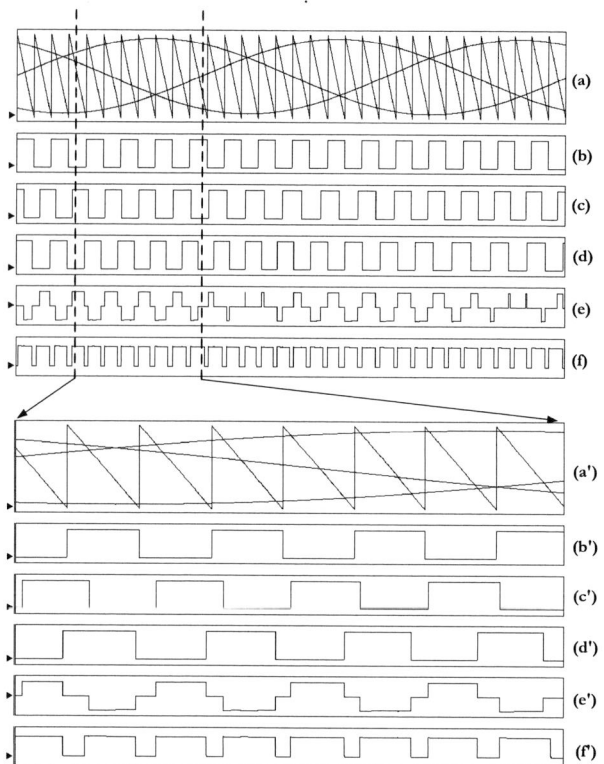

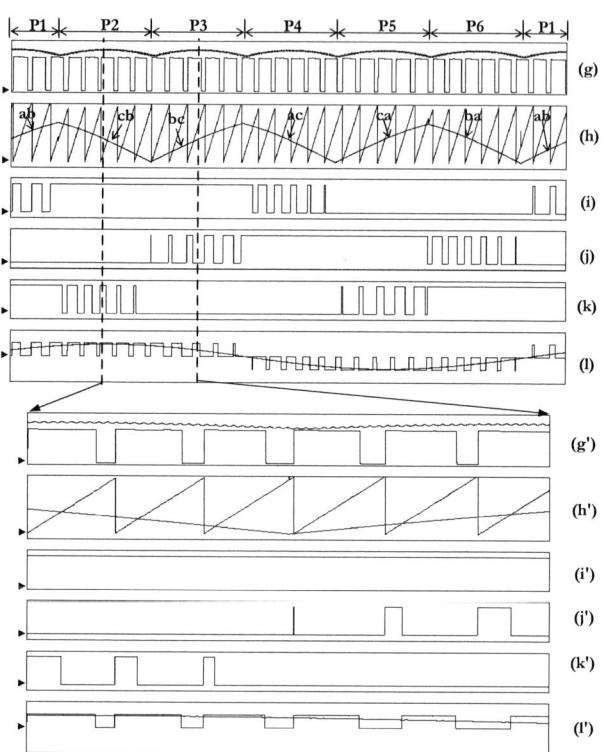

Figs. 3. Key waveforms of the **primary** dc/ac inverter in one cycle and enlarged view of the interval between two dot lines; (a) three-phase sine wave references and carrier signal; (b) Q: square ware with half frequency of the carrier; (c) UT: gate signal for the upper switch of phase U; (d) VT: gate signal for the upper switch of phase V; (e) $V_{uv}$: output of phase U and V; and (f) $V_{rec}$: output waveform of the rectifier.

Figs. 4. Key waveforms of the **secondary** side ac/ac inverter in one cycle and enlarged view of the interval between two dot lines. (g) $V_{rec}$: output PWM waveform of rectifier with six-pulse envelop; (h) *mod*: modulated signal and *ramp*: the carrier which is synchronous with (g); (i) UUT: gate signal for the top switch of phase a; (j) VVT: gate signal for the top switch of phase b; (k) WWT: gate signal for the top switch of phase c; and (l) PWM output of the line-line voltage $V_{ab}$ and its envelop.

Divide the six-pulse rectified waveform into six segments named P1~P6 as shown in Figs 4. (g).The rising and falling edges of $V_{rec}$ are different for different segment. Figs. 3 (a)'-(f)' show a particular time interval within segment 2, where the rising and falling edges of $V_{rec}$ (marked as $\uparrow V_{rec}$ and $\downarrow V_{rec}$) are determined by the edges of UT and VT respectively. Other cases are summarized in Table I.

TABLE I
THE EDGE DEPENDENCE OF THE RECTIFIER OUTPUT ON GATE SIGNALS

|  | P1 | P2 | P3 | P4 | P5 | P6 |
|---|---|---|---|---|---|---|
| $\uparrow V_{rec}$ | wt | ut | ut | vt | vt | wt |
| $\downarrow V_{rec}$ | vt | vt | wt | wt | ut | ut |

### B. Switching Strategy for Three-phase ac/ac Inverter

Similarly to the case of three-phase ac/dc rectifier, the rectified PWM output is contributed respectively by $V_{wv}$, $V_{uv}$, $V_{uw}$, $V_{vw}$, $V_{vu}$, and $V_{wu}$ at each segment from P1 to P6. The bottom part of the Fig. 2 shows the diagram of generating switching signals for three upper switchers of secondary side ac/ac inverter. During each segment, every switch will be either: permanently ON ("1"), permanently OFF ("0") or toggling with 20 kHz ("HF"). The switching pattern for the upper three switches in each segment for one cycle period is summarized in Table II.

The switch positions illustrated at Fig. 2 are for the case of segment P2. Since the rectifier output has the same shape as $V_{uv}$ within this interval, the line-line voltage $V_{ab}$ at the output side of the ac/ac inverter can be directly obtained by keeping switchers UUT and VVT at ON and OFF status respectively. Another line-line voltage $V_{cb}$, however, needs to be achieved by operating switches on the third leg WWT and WWB in a high-frequency way where modulated signal ("*mod*") is the difference between reference c and b and the carrier signal ("*ramp*") is a 20 kHz saw-tooth waveform synchronized with the PWM output of the rectifier. The key waveforms are shown in Figs. 4. The mathematic express for three line-line voltages are given as follows:

$$V_{ab} = V_{rec} \cdot (UUT - VVT) \tag{5}$$

$$V_{cb} = V_{rec} \cdot (WWT - VVT) \tag{6}$$

$$V_{ca} = V_{cb} - V_{ab} \tag{7}$$

The proposed switching strategy is the best option for resistive load, because the peaks of the currents follow the peaks of the fundamental voltages. Therefore each phase leg does not switch just when the current is at its maximum value thereby minimizing switching losses.

### TABLE II
#### SWITCHING PATTERN FOR UPPER SWITCHES OF THE AC/AC INVERTER

|  | P1 | P2 | P3 | P4 | P5 | P6 |
|---|---|---|---|---|---|---|
| $V_{rec}$ (g) | $V_{wv}$ | $V_{uv}$ | $V_{uw}$ | $V_{vw}$ | $V_{vu}$ | $V_{wu}$ |
| UUT (i) | HF | ON | ON | HF | OFF | OFF |
| VVT (j) | OFF | OFF | HF | ON | ON | HF |
| WWT (k) | ON | HF | OFF | OFF | HF | ON |
| mod (h) | ab | cb | bc | ac | ca | ba |
| $i_{rec}$ | $(i_a+i_c, i_c)$ or $(-i_b, i_c)$ | $(i_a+i_c, i_a)$ or $(-i_b, i_a)$ | $(i_a+i_b, i_a)$ or $(-i_c, i_a)$ | $(i_b+i_a, i_b)$ or $(-i_c, i_b)$ | $(i_b+i_c, i_b)$ or $(-i_a, i_b)$ | $(i_c+i_b, i_c)$ or $(-i_a, i_c)$ |
| $i_{rec} > 0$ | $i_b<0\rightarrow$ $<30°$ lagging | $i_a>0\rightarrow$ $<30°$ lagging | $i_c<0\rightarrow$ $<30°$ lagging | $i_b>0\rightarrow$ $<30°$ lagging | $i_a<0\rightarrow$ $<30°$ lagging | $i_c>0\rightarrow$ $<30°$ lagging |
|  | $i_c>0\rightarrow$ $<30°$ leading | $i_b<0\rightarrow$ $<30°$ leading | $i_a>0\rightarrow$ $<30°$ leading | $i_c<0\rightarrow$ $<30°$ leading | $i_b>0\rightarrow$ $<30°$ leading | $i_a<0\rightarrow$ $<30°$ leading |

## III. LOAD POWER FACTOR

The proposed converter along with the specific modulation scheme allows a certain phase displacement between the output current and voltage without using any dc link capacitor or additional active switch. It is up to 30 degree for a balanced three-phase system with the assumption that the output phase currents are line frequency sine-waves.

At any time, either one or two of all three upper switches of the ac/ac converter will conduct. The current $i_{rec}$ will swing between one phase current and the sum of this current and another phase current depending on which segment the converter is locating at. For example, $i_{rec}$ pulsates between $i_a$ and $i_a+i_c$ at segment P2 since UUT is always on, VVT is always off, while WWT is switching. Similarly, $i_{rec}$ varies between $i_a$ and $i_a+i_b$ for segment 3. Row 7 of table II summarizes cases for all segments.

The fact that $i_{rec}$, the output current of the diode rectifier, should not less than zero results in limitation on load power factor. The maximal phase discrepancies for all segments are list in Table II row 8. In any case, a maximal 30 degree leading or lagging load can be supported. By adding anti-parallel active switches for all six diodes, the converter can achieve four-quadrant operations without any reactive components in the main circuit.

## IV. EXTENSION FOR HIGHER POWER APPLICATIONS

N modules can be easily connected together to achieve N times higher power rating by paralleling their inputs and connected their outputs of rectifiers in series. All modules will share currents from the common fuel-cell stack. The final stage ac/ac inverter can remain the same two-level structure, as shown in Figs. 5 (a), in which the switch can be implemented using either single high voltage rating device or several low voltage rating devices in series. It can also be configured as a multi-level structure. Figs. 5 (b) show an alternative configuration for the final ac/ac stage using a three-level diode-clamped (NPC) inverter. The voltage stresses of devices will be reduced by half as compared to the case of two-level.

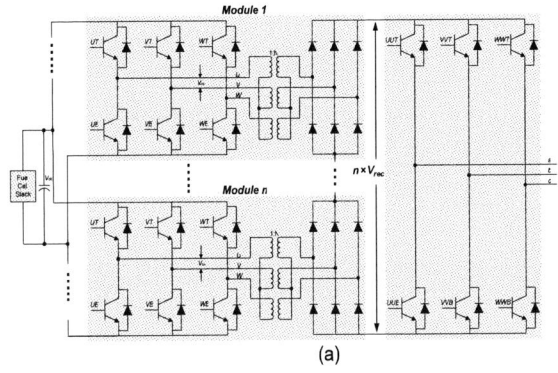

(a)

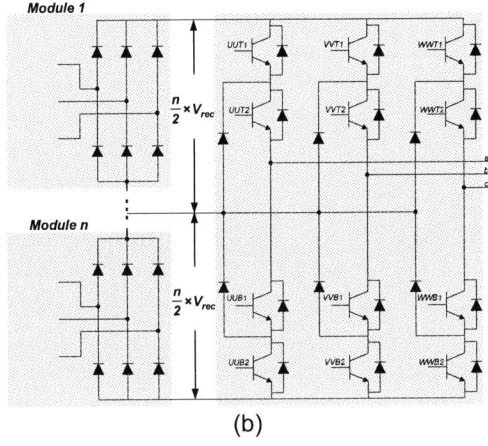

(b)

Figs. 5. Two possible structures to scale up for higher power applications: N modules of dc/ac inverters and rectifiers (a) followed by a two-level ac/ac inverter, and (b) followed by a three-level NPC inverter.

## V. SIMULATION AND PARAMETRIC RESULTS

Figs. 6 show SABER simulation results of a 1 MVA inverter for 12.47 kV distribution circuit. The nominal fuel-cell stack voltage is 600V. Four dc/ac inverters and rectifier modules are used in the simulation, followed by a two-level inverter (Figs. 5 (a) with n equals 4 and N equals 10). As shown in Figs. 6, the converter can handle 30-degree inductive and capacitive loads. The variation of THD value with respect to the switching frequency (frequency of the output PWM voltage) for (i) the proposed inverter and (ii) the topology shown in Fig.1(a) with naturally sampled sine-ramp modulation are plotted in Fig. 7. Results show that the THD

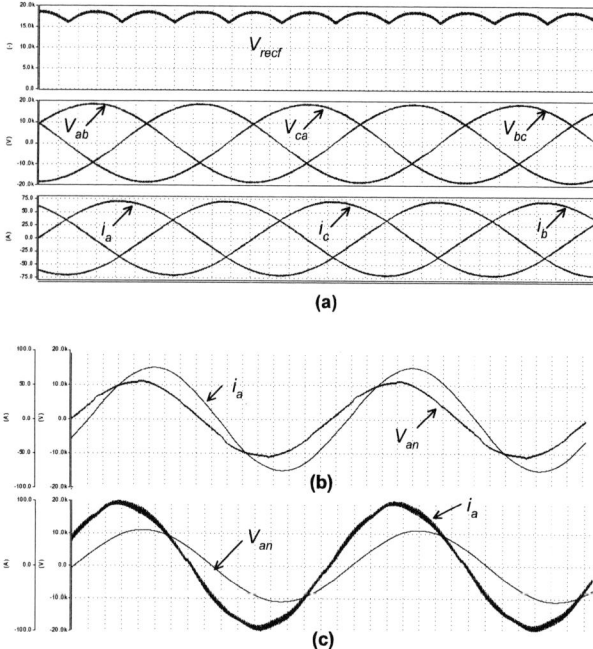

**(a)**

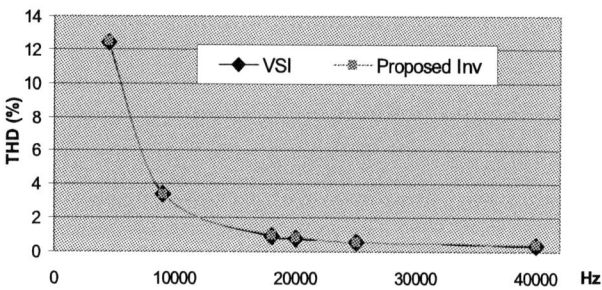

**(b)**

**(c)**

Figs. 6. Simulation results of the proposed inverter for 12.47 kV power systems with 1 MVA rating. (a) With resistive load, (b)inductive load , and (c) capacitive load. Waveforms included in (a) from top to bottom are envelop of the rectifier output waveform, three-phase line-line voltages and phase currents. In (b) and (c) phase voltages and currents are plotted for non unity power factor load.

Fig. 7. THD values of the conventional dc/ac inverter with dc bus (shown in Fig. 1(a)) and proposed inverter under the same switching frequency variation. A three-phase filter with 1.5mH inductance and 5uF capacitance is used for both cases.

performance of the proposed inverter is comparable to the conventional dc/ac inverter with constant dc bus.

## VI. CONCLUSION

A high-frequency link three-phase dc/ac inverter is proposed in this paper for mega-watt range applications, as a potential substitute for the traditional low-frequency power electronics inverter. The premise of high frequency operation at high power relies on the availability of high voltage SiC power devices, currently under developmental progress. The proposed inverter structure and the novel "hybrid" switching strategy achieve low-voltage dc to high-voltage ac conversion without the need of constant dc bus. It reduces the switching

losses of the ac/ac inverter and enables the scalability for higher power higher voltage applications. The proposed inverter has less volume, low cost, and relatively high efficiency. The operating principle of the inverter and its scalability options for high power (> 300 MW) FutureGen application [5] are outlined in this paper and basic performances of the inverter are demonstrated.

## VII. REFERENCES

[1] J.W. Palmour et al., (2005). Large Area Silicon Carbide Power Devices on 3 inch wafers and Beyond. Cree Inc., Durham, NC. [Online]. Available: www.gaasmantech.org/Digests/2005/2005papers/13.1.pdf

[2] Department of Energy, (2005). SBIR solicitation: A 5MW Fuel-cell dc-ac Inverter with Reactive Power Management Functionality [Online]. Available: http://www.science.doe.gov/sbir/solicitations/FY%202006/15.FE3.htm

[3] M. Matsui, et al, "High-Frequency Link dc/ac Converter with Suppressed Voltage Clamp Circuits- naturally Commutated Phase Angle Control with Self Turn-off Devices", IEEE Trans. on Industry Applications. Vol. 32 Page(s):293 – 300, 1996

[4] S.K. Mazumder et al., "Fuel cell power conditioner for stationary power system: towards optimal design from reliability, efficiency, and cost standpoint", Keynote Paper, ASME Third International Conference on Fuel Cell Science, Engineering and Technology, Yipsilanti, Michigan, FUELCELL2005-74178, May 23-25, 2005.

[5] http://www.futuregenalliance.org/

**2006 IEEE International Conference on Power Electronic, Drives and Energy Systems**

# Integrating a Redox Flow Battery System with a Wind-Diesel Power System

Shameem Ahmad Lone and Mairaj-ud-Din Mufti

*Abstract*--Energy storage devices are required for power quality maintenance in stand alone power systems like wind-diesel ones. A redox flow battery system has many virtues which make its integration with a wind-diesel power system attractive. This paper proposes the integration of a redox flow battery system with a typical multi-machine wind-diesel power system for simultaneous voltage and frequency regulation. The redox flow battery is connected to wind park bus through a current controlled voltage source converter based on hysterisis current control. Keeping in view the non-linear and time varying nature of the hybrid wind-diesel-redox flow battery system, neuro-adaptive control is proposed for active/reactive power modulation of the redox flow battery.

*Index Terms*—Wind-Diesel, Redox Flow Battery, Four Quadrant Control, Neural Control

## I. INTRODUCTION

WIND power has achieved a significant level of penetration in the power generation market in recent years. According to the figures released very recently by the Global Wind Energy Council (GWEC) Brussels Belgium, the year 2005, saw the installation of 11,769 megawatts (MW), which represents a 44.4 % increase in annual additions to the global market, up from 8,207 MW in the previous year.

Most of the wind power installations are tied to strong grids, where the inherent stiffness of the network minimizes the power quality concerns. However, the introduction of the wind power into weak electricity grids like diesel grids is challenging. These weak grids are typically electrical islands which are not linked to mainland grid. In such systems the introduction of wind generation can create power quality problems even before 20% penetration is reached. Stand alone wind-diesel power systems are also prone to load disturbances due to overall low system inertia. However, energy storage devices [1] can allow wind derived power to be a good percentage of the overall generation in weak electricity grids. Compared with other energy storage devices the redox flow battery (RFB) has a number of advantages which make its integration with a wind-diesel power system very attractive.

Shameem Ahmad Lone is with the department of electrical Engineering Department of National Institute of Technology Srinagar, India 190006 and is working towards his Ph.D degree.(email: sadial_14@yahoo.com)

Mairaj-ud-Din Mufti is with the department of Electrical Engineering Department of National Institute of Technology Srinagar, India 190006. email : muftimd@yahoo.com Fax no. +91 194 2420475.

An RFB system has a high speed response and is not aged by frequent charging and discharging. The battery efficiency increases when charging/discharging period becomes shorter unlike the classical batteries. An RFB system is environmental friendly, it operates at ambient temperature, has a short duration overload capacity and a long service life. An introduction to operation of RFB and its salient features are reported in [2], where a preliminary study of the wind-diesel-RFB system has been presented. In this paper we study the interfacing of a redox flow battery system with the wind-park bus of a typical wind –diesel system through a current controlled voltage-source converter. For the modulation of active/reactive power of the RFB system two neuro-adaptive control loops are proposed, keeping in view the non-linear and time varying nature of the hybrid power system. The effectiveness of the proposed scheme is then presented using the simulation studies.

## II. SYSTEM ARCHITECTURE

The system under study is shown in Fig.1 [3]. It comprises of ten wind turbines coupled to induction generators, two diesel engines driving synchronous generators (equipped with automatic voltage regulators (AVRS) and speed governors), transmission network, capacitor banks for reactive power compensation of induction generators and the village distribution network. In addition, we have the proposed RFB connected to the wind park bus, through a CC-VSI.

## III. NEURO-ADAPTIVE CONTROL OF REDOX FLOW BATTERY

Neural network (NN) based adaptive control systems with on-line learning have the capability to cope up with the challenges of non-linearity and time varying nature of the controlled system. This type of control can maintain the consistent performance of the system in presence of unknown and time varying parameters of the controlled system. The neuro-adaptive control is therefore best suited for hybrid wind-diesel-RFB plant. In view of this we have proposed in this paper the use of neuro-adaptive control [4] for active / reactive power modulation of the RFB system. The control scheme is depicted in Fig. 2. There are two control loops i.e. frequency control loop and voltage control loop. The control objective of loop1 is to bring $y_1$ (the frequency deviation at wind-park bus) to zero. Similarly loop2 aims at bringing the controlled variable $y_2$ (which is the voltage deviation at wind park bus )

0-7803-9771-1/06/$25.00 ©2006 IEEE

to zero. Each controller computes the control signal based on an optimal control law.

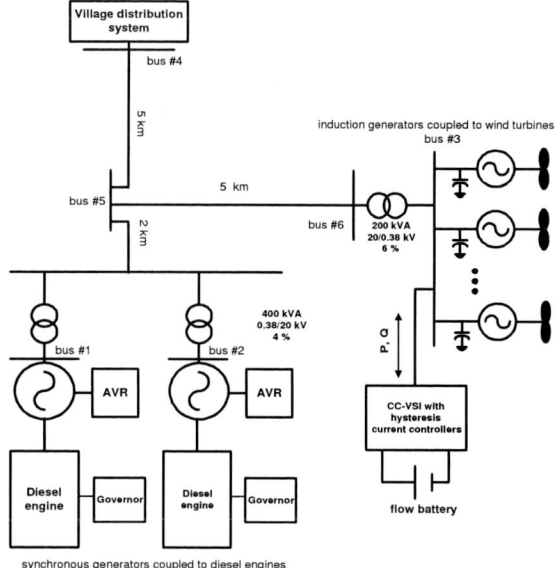

Fig. 1. System Architecture.

## A. Neural Estimators

The relationship governing the dynamics of loop 1 is modelled by a linear two-layered neural network. The neural network learns the dynamics using the Widrow-Hoff delta rule. The input layer consists of (n+m) elements [4]. The (n+m) inputs are the input (injected active power) and output (frequency deviation) signals at the previous sampling instants. The output layer has only one element whose output is the predicted output. The neural estimator for estimating the injected active power/frequency deviation relationship is shown in Fig. 3.

The weight vector at the sampling instant k defined by

$$W(k) = [w_1(k) \ w_2(k) .... w_n(k) \quad w_{n+1}(k) \ .. \ w_{n+m}(k)]^T \quad (1)$$

is updated using Widrow-Hoff rule as per the following :

$$W(k+1) = W(k) + \frac{\alpha e(k+1) X(k)}{\varepsilon + x^T(k) X(k)} \quad (2)$$

Where X(k) is the neural network input vector which is given by:

$$X(k) = [x_1(k) \ x_2(k)........ x_n(k) \quad x_{n+1}(k) .........x_{n+m}(k)]^T$$

$$= [-y(k)-y(k-1).....y(k-n+1) \ u(k).......u(k-m+1)]^T \quad (3)$$

and $\alpha$ ( 0,2) is the reduction factor. The control $\varepsilon$ is chosen to be close to zero.

In (3) y represents the output (frequency deviation) and u is the control signal (active power injected by the storage device). It is important to note that the estimation is performed online in the presence of disturbances acting on the system so

that no separate disturbance model [4] is required in the model.

The output predicted by the neural network can be calculated as:

$$\hat{y}(k) = \sum_{i=1}^{n+m} w_i(k-1) x_i(k-1) \quad (4)$$

The adaptation algorithm of (2) minimizes the error between the measured output, y(k) and the predicted output, $\hat{y}(k)$ .

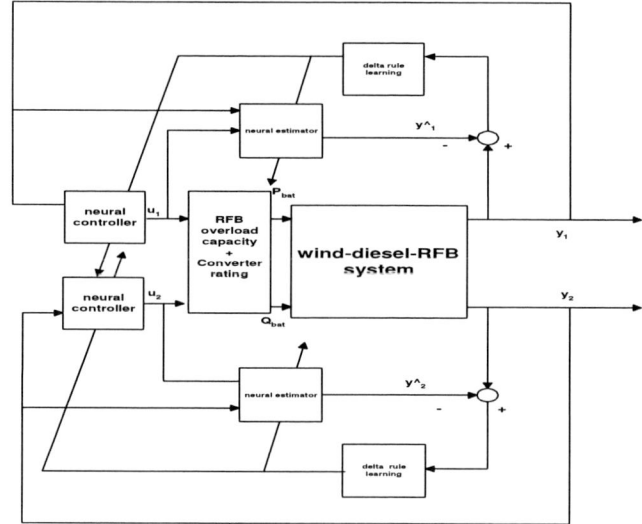

Fig. 2. Neuro-adaptive control scheme of the RFB integrated with a wind-diesel power system.

The dynamic relationship between the voltage deviation and the reactive power injected at the wind-park bus is modelled by a similar neural-network.

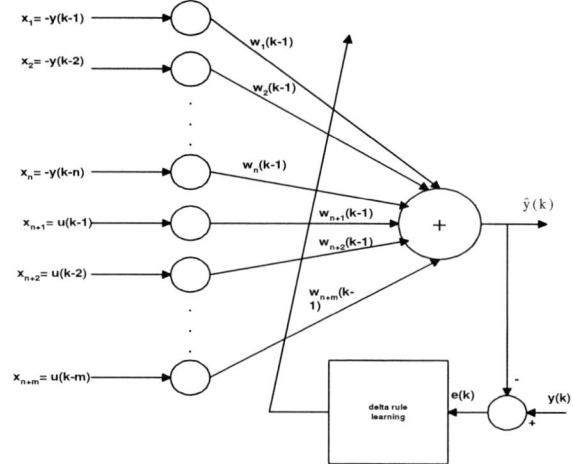

Fig. 3. Neural Estimator.

## B. Neural Controller

The dynamics learnt by the neural estimators is used to adjust the connection weights of the neural controllers. Each neural controller consists of two-layered neural network with n+m-1 elements in the input layer and one element in the output layer. The connection weight vector $W'(k)$ associated with the neural controller is related to the connection weight vector $W(k)$ of the corresponding neural estimator and the relationship depends on the type of control strategy. In this paper we have proposed an optimal control strategy which minimizes the performance index of (5).

$$J(k) = \hat{y}^2(k+1) + \lambda[u(k) - qu(k-1)]^2 \qquad (5)$$

where $\lambda$ is the weighing constant and q can assume a value of 1/0.

The control action which minimizes $J(k)$ can be derived by undergoing the following steps.

I.  Obtain the expression for $\hat{y}(k+1)$ from the neural estimator model of (4).

II. Substitute the resulting expression of $\hat{y}(k+1)$ in (5)

III. Set $\dfrac{\partial J(k)}{\partial u(k)}$ to zero.

Once the above procedure is followed a neural controller of Fig. 4 is obtained. The weight vector $W'(k)$ of the neural controller is related to the elements of weight vector $W(k)$ of the corresponding neural estimator through the following relation:

$$W'(k) = c_1[w_1(k) \ w_2(k) .. w_n(k) c_2 + w_{n+2}(k) \cdots w_{n+m}(k)]^T$$

..............(6)

Where $\quad c_1 = \dfrac{w_{n+1}(k)}{w_{n+1}^2(k) + \lambda}$ and $c_2 = -\dfrac{\lambda q}{w_{n+1}(k)}$

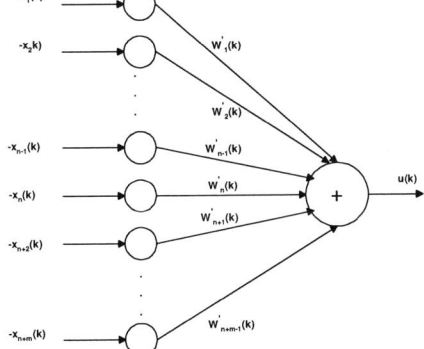

Fig. 4. Neural Controller.

## IV. SIMULATION MODEL

In order to carry out the simulation studies to investigate the effectiveness of the proposed scheme a unified model was developed in a manner detailed in [1] and the RFB battery was represented by the model shown in Fig. 5.

In this figure $C_o$ simulates the state of charge (SOC), $i_o$ simulates the self discharge characteristics, $R_1$ is the contact resistance and two parallel RC combinations simulate the domain characteristics of the battery [5]-[6]. The converter interfacing the RFB with the wind park bus uses six bidirectional switches that are capable of conducting the current in both the directions. For the sake of analysis, wind park bus is represented by its equivalent circuit as shown in Fig. 5.

## V. CONTROL SCHEME

The sequence of actions for frequency control loop at instant k is as follows:

1. Measure $y_1(k)$ i.e. the frequency deviation at wind park bus.
2. Use the neural network of Fig. 3 to compute the predicted value $\hat{y}_1(k)$.
3. Compute error $e_1(k) = \hat{y}_1(k) - y_1(k)$ and use delta rule to calculate new weights w(k).
4. Compute the per unit control signal $u_1(k)$ as per Fig. 4 which represents the desired active power $P_{bat}$ to be exchanged between the wind park bus and the redox flow battery system.

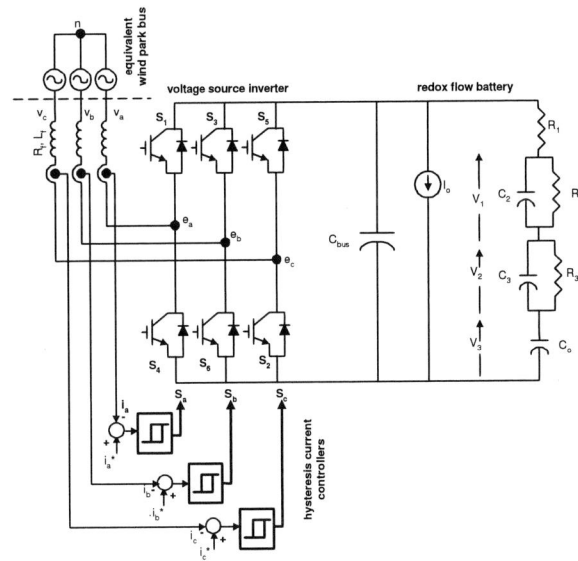

Fig. 5. Redox flow battery model with Hysteresis current controller.

The similar sequence of actions takes place for the voltage control loop. However keeping in view the different range of values for frequency and voltage deviations a scaling factor [7] is introduced for voltage control .Thus in voltage control loop $y_2 = k_v *$ voltage deviation at wind park bus. The control signal $u_2(k)$ represents the desired reactive power $Q_{bat}$ to be exchanged between the wind-park bus and the RFB system.

The exchange of desired active and reactive powers is exercised through the current controlled voltage source converter (CC- VSC) based on hysteresis current control. The scheme works in the following manner.

The reference currents $i_a^*$, $i_b^*$ and $i_c^*$ which should flow from the wind park bus to the RFB system to accomplish the desired active/reactive power exchange are calculated as per (7).

$$\begin{bmatrix} i_a^* \\ i_b^* \\ i_c^* \end{bmatrix} = \frac{\sqrt{2}}{3} \begin{bmatrix} 1 & 0 \\ -\frac{1}{2} & \frac{\sqrt{3}}{2} \\ -\frac{1}{2} & \frac{\sqrt{3}}{2} \end{bmatrix} \begin{bmatrix} i_\alpha \\ i_\beta \end{bmatrix} \qquad (7)$$

where

$$\begin{bmatrix} i_\alpha \\ i_\beta \end{bmatrix} = \begin{bmatrix} V_\alpha & V_\beta \\ -V_\beta & V_\alpha \end{bmatrix}^{-1} \begin{bmatrix} P_{bat} \\ Q_{bat} \end{bmatrix}$$

$V_\alpha$ and $V_\beta$ are the α-β components of the wind park bus voltage.

Depending on the actual currents $i_a$, $i_b$ and $i_c$ the ON/OFF switching pattern of the gate drive signals to the IGBTs, generated from the hysteresis comparator [8], is represented mathematically as:

if $i_a < (i_a^* - hb)$ switch $S_1$ off and switch $S_4$ on , $S_A=0$
if $i_a > (i_a^* + hb)$ switch $S_1$ on and switch $S_4$ off , $S_A=1$
if $i_b < (i_b^* - hb)$ switch $S_3$ off and switch $S_6$ on , $S_B=0$
if $i_b > (i_b^* + hb)$ switch $S_3$ on and switch $S_6$ off , $S_B=1$
if $i_c < (i_c^* - hb)$ switch $S_5$ off and switch $S_2$ on , $S_C=0$
if $i_c > (i_c^* + hb)$ switch $S_5$ on and switch $S_2$ off , $S_C=1$

Where hb is the current band of the hysteresis current controller and $S_i$,i=A,B,C are called as the switching functions.

## VI. SIMULATION STUDIES

The various disturbances to which the wind diesel system under study is subjected to, include wind power disconnection, active and reactive power load changes and the turbulent wind. In all the case studies the system is assumed to be in steady state prior to the occurrence of disturbance i.e. (t < 0). This condition is established by carrying out load flow studies and running some initial condition programs. Time domain simulations are performed using MATLAB programs.

### A. Wind Power Disconnection

The objective of this case study is to examine the behaviour of the power system with abrupt change in penetration level. Initially the wind turbines are generating 100 kW and the power system load is 312 kW/ 200 kVAR, while the system frequency is 50 Hz. At t=to the wind park along with associated capacitor banks is disconnected. According to Fig. 6 & Fig.7, the power quality of the wind-diesel system is not acceptable [3].However the same figures reveal that with the introduction of RFB unit at wind park bus, the quality of the power supplied by the hybrid system is quite satisfactory.

### B. Load Disturbance

Change in load is a noteworthy source of perturbation in stand-alone power systems. This is because of the fact that a stand-alone power system has low inertia and is not able to contain the variations in system frequency and voltage unlike the utility grid. A change in static load is modelled as a change in network topology. In the studies performed here the load profile is assumed to be the one shown in Table I. The wind power is assumed to remain constant at 50 Kw in this case study. The simulation results pertaining to this disturbance are depicted in Fig.8 & Fig.9. It is observed from these results that the association of RFB unit with the wind-diesel system helps in maintaining the power quality in this case also.

TABLE I

LOAD PROFILE

| Time (sec) | Active Power Kw | Reactive Power Kvar |
|---|---|---|
| t < 0 | 260 | 182 |
| 0 < t ≤ 10 | 360 | 200 |

### C. Turbulent Wind Disturbance

The objective of this case study is to examine the behaviour of the hybrid power system when the wind turbines operate in turbulent wind. This mode of operation is very common for wind power based power systems. Initially the wind turbines are supplying 50 kW of power and the load on the power system is taken as 260 kW/182 kVAR. For t > to, the wind turbines operate in turbulent wind mode and the power supplied by the wind turbines varies as shown in Fig.10. Fig.11 & Fig. 12 clearly show the positive impact of RFB on the power quality under turbulent wind.

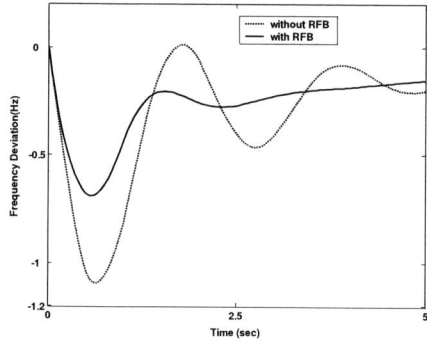

Fig. 6. Frequency deviation (Hz) due to wind power disconnection.

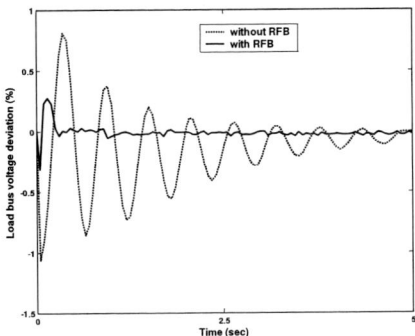

Fig. 7. Load Bus Voltage Deviation (%) due to wind power disconnection.

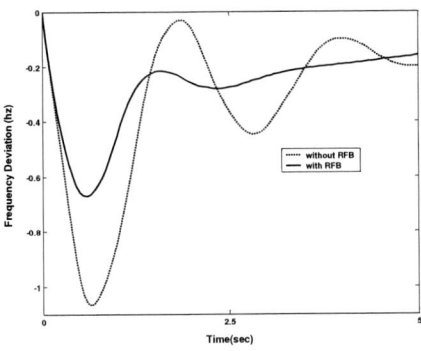

Fig. 8. Frequency deviation (Hz) due to load disturbance.

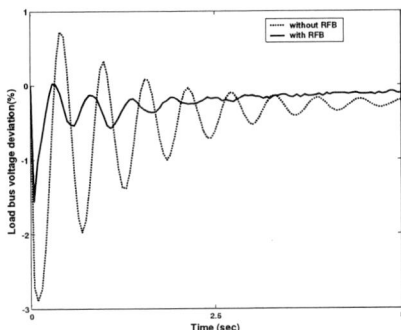

Fig. 9. Load Bus voltage deviation (%) due to load disturbance.

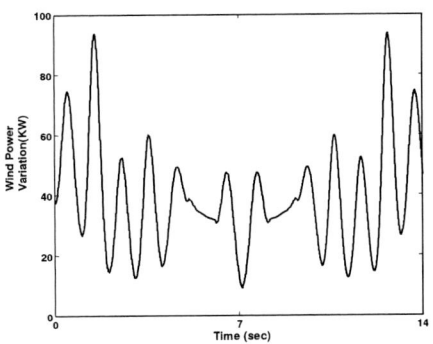

Fig. 10. Turbulent wind to which the system is subjected.

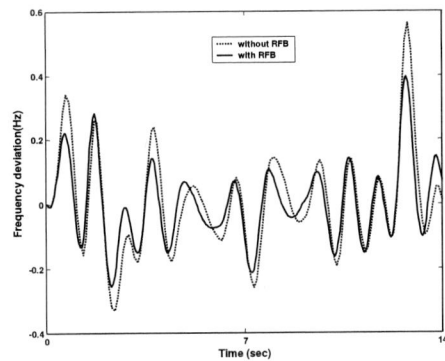

Fig. 11. Frequency deviation (Hz) due to turbulent wind.

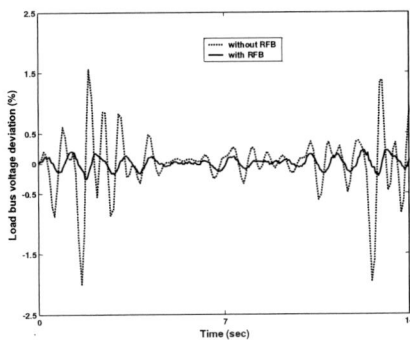

Fig. 12. Load bus voltage deviation (%) due to turbulent wind.

## VII. CONCLUSION

In this paper impact of an intelligently controlled redox flow battery on the power quality of wind-diesel system subjected to various disturbances, is studied. The intelligent control scheme for RFB makes use of two decoupled neural adaptive controllers. The necessary models for studying the impact of the proposed scheme are developed and programmed using MATLAB. The simulation results show that the proposed scheme helps to maintain the power quality of the system under various perturbations.

## VIII. APPENDIX

System Data: Base Power 400 kVA for the whole system
Synchronous generators:
Nominal power (kW) = 240 Nominal voltage= 380 V
Nominal apparent power (kVA)= 240    $r_s$=0.067  $X_d$=260%  $X_q$=156  $X_d'$=22%  $X_d''$=15%  $X_q''$=22.5 %  $T_{do}'$=1.182sec  $T_{do}''$= 0.015sec  $T_{qo}''$=1.18 sec

Induction generators:
Nominal power= 12(kW) Nominal voltage= 380V Nominal speed = 1545 rpm  $r_s$=0.022 p.u.  $x_s$= 0.181 p.u.  $x_m$ =1.605 p.u.  $r_r$=0.022 p.u  $x_r$=0.181 p.u  $H_A$= 0.259 sec

Redox flow battery:
RFB capacity = 15kW,
RFB equivalent circuit parameters:
$i_o$ = 0.536 A, $R_1$=1.14 Ω, $R_2$=0.15 Ω, $R_3$=0.42 Ω, $C_o$=1333.33 F, $C_1$=20 F $C_2$=142.86 F

Converter parameters:
$R_f$= 0.025Ω, $L_f$ = 1.2mH, hb = 0.03, $C_{bus}$ = 5000μF converter rating = 50 kVA.

Control Parameters
Frequency control Loop: $\lambda$ = 5,  $\varepsilon$ = 0.001,  $\alpha$ = 1, q = 0
Voltage Control Loop: $\lambda$ = 6.5, $\varepsilon$ = 0.001, $\alpha$ = 1,  q = 1 $K_v$=12

## IX. REFERENCES

[1] Mufti,M.D., Balasubramanian , R., Tripathy S.C. and Lone S.A (2003), Modelling and control of weak power systems supplied from diesel and wind. International Journal of Power and Energy systems, vol 23 no 1 pp (24-36).

[2] Lone Shameem Ahmad and Mufti Mairaj ud-din, Redox Flow Batteries: Modelled for Power Quality improvements in autonomous Wind-diesl Power systems. Wind Engineering vol 28, No. 5, 2004, pp.577-586.

[3] Kariniotakis, G.N, Stavrakakis , G.S., and. "A General Simulation Algorithm for Accurate Assessment of Isolated Diesel-Wind Turbine System Integration Part I and II , IEEE Transactions on Energy Conversion , Vol. 10 No.3,September, 1995, pp. 577-590.

[4] V. Etxebarria " Adaptive Control of Discrete systems Using Neural Networks" IEE Proc. Control Theory Appl Vol 141 No.4 July 1994. pp. 209-215.

[5] Ecomoto, K., Sasaki, T., Shigematsu, T. and Deguchi, H. Evaluation study about redox flow battery response and modeling, IEEJ Trans., Power Eng., 122-B, 2002, pp. 554-560.

[6] Sasaki, T., Kadoya , T. and Enomoto, K.(2004) Study on load frequency control using redox flow batteries. IEEE Transactions on Power Systems, 19, 2003 pp.660-667.

[7] Lim, C. M. " Neural Network control for Damping of Multi-Mode Oscillations in a Power System" Proc. Of the IEEE Region 10 Int Conference Vol. 2, 19-22 aug, 2001,pp. 658-663.

[8] Lone Shameem Ahmad and Mufti Mairaj ud-din " Power quality Improvement in a wind based stand-alone power systems" Proc. of the Int. Conf. on Wind Energy : Trends and Issues, NITTTR Bhopal, India,.5-7 Jan, 2007 , pp. TP-VII/12-TP-VII/24.

## X. BIOGRAPHIES

**Shameem Ahmad Lone** has obtained his M.E degree in Measurements and Instrumentation from Electrical Engineering Department of IIT Roorkee in Jan 1998. Presently he is with the Electrical Engineering Department of NIT Srinagar India and is working towards his Ph.D. in the area of Modelling and control of stand –alone power systems using fast acting storage devices.

**Mairaj-ud-Din Mufti** obtained his M.Tech and Ph.D degrees from Indian Institute of Technology, New Delhi in 1991 and 1998 respectively. Currently he is with the Electrical Engineering Department of NIT Srinagar India. His main fields of interest are power system dynamics and control, intelligent control and energy conversion.

**2006 IEEE International Conference on Power Electronic, Drives and Energy Systems**

# Hydrocarbon Fuel Based Micro Battery Power System

Surendran Devadoss[1], *Student member, IEEE*, Theo Kangsanant[2], and Ian Bates[3]

*Abstract* — This paper describes the possibilities of a liquid fuel battery system that can have an energy density/life more than or comparable to conventional battery for electronic communication application. Micro liquid fuel engine driven micro generator is described and discussed in the paper. An analysis on State-of-art technologies is provided and a way to improve energy and power density of the micro-generator is identified and explained. The generator parameters and the factors to be considered for efficient power generation are investigated and the limiting conditions highlighted. Requirements of the system to perform similar to a dry battery and supply a variable load are defined.

*Index Terms-- Index Terms* — MEMS power, Microgen, Battery, MEMS generator, Microgen, Power MEMS.

## I. NOMENCLATURE

| | |
|---|---|
| $N_m, \varpi_m$ | Rotor Shaft speed, Angular speed |
| $P_m, \eta_m$ | Mechanical power and efficiency |
| $ED_S, \eta_e$ | System energy density & Electrical efficiency |
| $P_{op}, P_{losses}$ | Generator output power, Losses |
| $\tau_m, \tau_g$ | Mechanical and Electrical torque |
| $B_g, \lambda_w$ | Flux density at air gap, Flux linkage |
| $F_m, R_{path}$ | MMF, Net Reluctance in flux path |
| $N_t, A_w$ | No of turns, Winding area |
| $e(t), v(t)$ | Induced EMF and terminal voltage |
| $i_a, \forall_{EnGen}$ | Current, Engine-Generator volume |
| $\forall_S, \forall_{tank}$ | Net System volume, Tank volume |

## II. INTRODUCTION

WITH the advent of increased electronic devices for communication and GPS systems and to allow quick and reliable position location in a building or city, the energy

---

Surendran Devadoss is with the School of Electrical and Computer Engineering, RMIT University, Melbourne 3001, Australia (e-mail: surendran_80@ieee.org).

Theo Kangsanant is with the School of Electrical and Computer Engineering, RMIT University, Melbourne 3001, Australia.

Ian Bates is the Director of Microtechnology Center, RMIT University, Melbourne 3001, Australia

available in a conventional batteries is used to supply electronic devices. The typical energies available in AA and AAA batteries are shown in shown in Table.1 and the best battery (Li/Mno2) can supply a mobile phone typically for 4 hours when it is fully operational with nominal voltage of 3.2V and 1400mAh and supply a GPS system for 0.5 hrs. Rechargeable AA Ni-Cd batteries can supply a similar mobile phone for <2 hrs and <0.3 hrs for a GPS systems. New ways of energy sources which can achieve high energy supply than existing batteries is desired if the full potential of modern equipment is to be achieved.

TABLE I
ENERGY AVAILABLE IN AA AND AAA BATTERIES [1] [2]

| Typical battery types | Energy densities (W-h/kg) | Typical Energy densities (W-h/m3) | Volume (cm3) | Weight (g) | Typical Energy (W-h) |
|---|---|---|---|---|---|
| Lithium Manganese (Li/Mno2) (AA) | 230 | 430 | 10.5 | 19.5 | 4.2 |
| Lithium Manganese (Li/Mno2) (AAA) | 230 | 430 | 5.2 | 9.3 | 2.2 |
| Alkaline (AA) | 170 | 470 | 8.4 | 24 | 2.6 |
| Rechargeable Ni-Cd (AA) | 40 | 115 | 10.5 | 27.2 | 1.2 |

## III. NEXT GENERATION ENERGY SYSTEMS

Koeneman *et al* [3] have compared different sources of energy with radioactive fuels and hydrocarbon fuels are found to have the highest energy densities known to humans of 4.17e8 kW-h/kg(Uranium) and 9.72e6 kW-h/m3 respectively. Radioactive systems require suffer from very low system efficiencies and thus have very low energy densities[4]. Hydrocarbon fuels have high specific energy of 9.72e3 kW-h/m3 and 13.39 kW-h/kg. A hydrocarbon fuel (Butane) in a volume same as an AA battery (AA battery means Rechargeable Ni-Cd AA battery considered for illustration and applies to rest of the paper) the paper for all further comparisons more than 86 times the energy as in the considered AA battery as shown in fig.1. So,

0-7803-9771-1/06/$25.00 ©2006 IEEE

to achieve a higher energy density system by 1-8 times a battery a method for converting chemical energy available in hydrocarbon fuels to electrical energy has to be used and also they should achieve a system efficiency of 1-10%. Rotary engine-generator integrated one such system used to convert the hydrocarbon fuel chemical energy to electrical energy. For the engine-generator system to provide energies better than the AA batteries they need to be able to fit in a volume of similar less than that of a battery and to allow more space for fuel and the system should be able to achieve an over all efficiency of 1-3%.

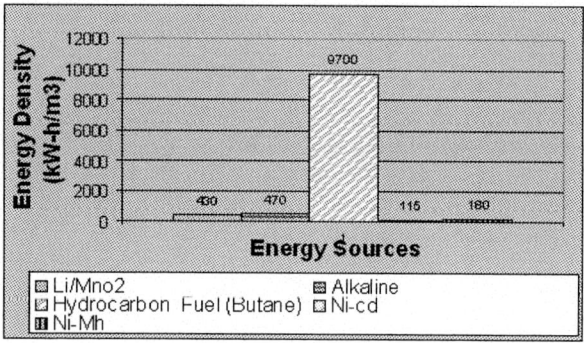

Fig. 1. Specific energy density in battery and hydrocarbon fuels.

## IV. CURRENT RESEARCH

The size of the engine–generator system has to fit into the volume of an AA battery typically 5*1.45*1.45 cm shown in Table. II to achieve high energy density and they need to be in millimetre and micrometer scales. Recent advancement in micromachining technology had made the manufacturing of micro structures with high aspect ratios (length to width ratio) and dimensions in the range of 300 microns – 1mm possible [3][4]. The technology advancement has also provided ways to fabricate 3-D structures using micro electronics fabrication technologies as low as 1-10 microns in thickness with non-assembly of parts for bulk manufacturing [2]-[4]. There are perceptible numbers of millimeter and micrometer scale combustion driven power generation systems developed by various research laboratories and the hydrocarbon energy is used to power drive micro meter scale fuel cells, IC engines, thermoelectric etc.

Micro rotary IC engine have better efficiency than its counterparts and have efficiency as high as 0.5-1% and micro generator efficiencies of 30-40% [4], [5]. Another notable advantage of Combustion driven systems is that the $CO_2$ and $H_2O$ emission and heat release from micro meter engine devices is very small and 4 times lesser than the $CO_2$ released from a normal breathing from humans. Recent advancement in fabrication techniques had shown that these micrometer scale engine components can be developed with high yield using conventional EDM (Electron Discharge machining) and DRIE (Deep Reactive Ion Etching) processes [5] for non–assembly bulk manufacturing. The papers available in literature for micro engine integrated generators gives much importance and

focus on engine and its development micro generators as a solution prospective to be added to the engine [3]-[5]. Recent research results from publications has shown engine-generator integrated devices can be fabricated with rotor dimensions of in the scale of 900μm*2.4mm*2.4mm with 1% efficiency [3]-[5]. However, a micro-engine with better efficiency is desirable to improve the overall efficiency of an integrated system.

## V. LIQUID BATTERY SYSTEM- THE REQUIRED ELEMENTS

Fig.2.shows an envisaged model of a liquid battery system developed under research. The system consists of an engine-generator block included with energy management and control logics and a fuel tank block with selected fuel of which its size determines the net energy density of the system. In the current technology the micro generator are integrated into engines to compromise engine efficiency which affects the net electrical energy produced by the system and the efficiency of the generator which is a key to determine the energy density available in the system

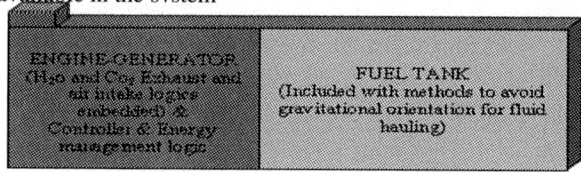

Fig. 2. Envisaged Model of a Liquid Fuel battery system.

It can be seen from Table II that the existing micro engine generator technology can produce an energy density of ~58.2 kW-h/m3 and ~37 W-h/kg which is 0.5 times volumetric and 0.9 times gravimetric than that produced in the considered AA battery with a volume of 5*1.45*1.45 cm with an generator efficiency of 30-40% and a net efficiency of 0.5% as shown in existing system schematics in fig.3

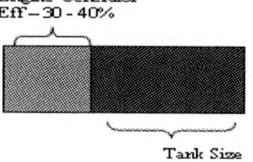

Fig. 3. Generator Technology Schematics –Illustrative Existing Engine with specific energy of the system of 58.2 kW-h/m3 (figure not in scale).

If, the desired energy density to be achieved is 2.X where 'X' represent an unknown better energy density value than the existing system which is comparable or more to conventional AA batteries. The value of 'X' can be obtained depending on the objective function. As an illustration to achieve 2 times the energy density of the battery then 'X' can be calculated as

*Energy density (Existing) * (1/2) = 2 * X*
*Rearranging to obtain X, Then, X = 4.*

To obtain an energy density twice of AA battery, the energy density of the system has to be improved by a factor of 5. The system density can be defined as in (1) as

872

TABLE II
COMPARISON BETWEEN CONVENTIONAL BATTERIES, TYPICAL HYDROCARBON FUEL AND EXISTING ENGINE-GENRATOR DEVICES

| S.NO | Typical Energy Density Volumetric (kW-h/m3) Gravimetric (kW-h/kg) | Typical Density (Kg/m3) | Energy density ratio compared to AA battery | Typical System Efficiency |
|---|---|---|---|---|
| Hydrocarbons fuels 1. Hydrogen – SOA 2. Ethanol | 1. 9.7e3 kW-h/m3 109e3 kW-h/kg 2. 5.3e3 kW-h/m3 6.7e3 kW-h/kg | Hydrogen – 0.089 Ethanol – 790 | 1.74 ->kW-h/m3 1100 -> kW-h/kg 2.40.8 -> kW-h/m3 111.7 -> kW- h/kg | Depending on the system using the energy source |
| Duracell-[2] 1. AA LiMno2 battery 5.2*1.45*1.45 cm (AA) & 3.2V and 1400 mAh 2. Rechargeable Ni-Cd 5*1.4*1.4 & 1.5 volts & 20 Ah | 1. 430 kW-h/m3 230 W-h/kg 2. 170 kW-h/m3 40 W-h/kg | 1. 1850 2. 2600 | 1. 1 ->kW-h/m3 1 ->kW-h/kg 2. 1 ->kW-h/m3 1 ->kW-h/kg | 98 - 99% with 1-2% internal resistance or leakage losses |
| Existing micro engine – generator devices using hydrogen fuels [6] Volume: Gen - 1.1* 8*2.5cm System – 2.64*1.92*6cm (1.4 times linear scale and 36 times Engine-Generator volume includes fuel ) | 58.2 kW-h/m3 36.4 W-h/kg | Engine-Generator = 4000 Fuel =0.09 (Hydrogen) System density Hydrogen =0.3*4000+0.7*0.09= 1600 1 Ethanol = 1200 | (Hydrogen) (1/4) ->kW-h/m3 (1/2.5) ->kW-h/kg | Engine – 0.9 - 1.5% Generator – 30-40% System Efficiency = 5% |

$$ED_s = \frac{\eta_s * P_{op}}{\forall_s} \text{ kW-h/m3} \qquad (1)$$

Where,

$\eta_s = \eta_e * \eta_{gen}$, $\forall_s = \forall_{Engen} + \forall_{fuel}$, $\forall_{Engen}$ also includes the controller and energy management logics.

This paper focuses on the extension to existing technologies in generator development as seen in fig..4 to improve the net system energy density by the following ways

- Reducing volume of the generator by a factor of 3 and reducing system density. This approach also increases the space for the tank per given system volume and improves net mechanical output in the shaft.
- Improving efficiency in the range of 70-80% the integrated engine-generator device efficiency can be improved by 1-2%.
- Increasing power output by a factor of 2-3 per volume of generator by optimal selection of generator parameters.

Then volumetric and gravimetric energy density then becomes 194 kW-h/m3 and 300 W-h/kg respectively, which are 1.7 times and 7.5 times greater than the considered AA battery and 0.5 times volumetric than that of a conventional Li/Mno2 battery and illustrative schematics of a envisaged improved engine-generator system is shown in fig.4.

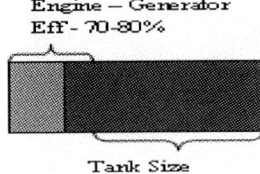

Fig. 4. Envisaged device illustrative Schematics achieving an energy density of 194 kW-hr/m3 (figure not in scale).

## VI. MICRO ENGINES

Development of rotary micro engines has been reported previously by various research groups producing a mechanical power output of 3 to 10 Watts [7] with an engine rotor radius of 10-13mm respectively. The efficiency of the engine in reported architecture is in the range of 1% with most of losses are due to the thermal heat losses produced in the engine. An engine with better efficiencies would be desirable to improve the net efficiency of the system. The scope of the paper refrains from explaining in depth on engine development but suggest the reader to look at the references [2]-[4]. The rotary engines reported in recent research can exhibits almost a constant torque characteristics[2], [7] under various speeds. Thus by increasing the speed of the engine provides a proportional increase in mechanical power output available in the shaft and the relation ship also illustrated in the (2).

$$P_{mech} = \tau \frac{d\Theta}{dt} = \tau \omega_m = 0.12 * \tau * n_m \text{ Watts} \qquad (2)$$

The engine speed–torque characteristic is important for the generator development to transfer maxim power from engine shaft to generator shaft .The micro generators developed are either integrated into the engine or integrated to the shaft which comes with compromise on energy density due to reduction in net energy due to heat in the earlier and increase in volume in the later both affecting the net energy density of the integrated system.

## VII. MICRO GENERATORS

The micro generators that are developed at micro scale can be classified using by the way EMF is generated using faraday's law of electromagnetic induction as in Eq.3 their principle are classified on the basis of the techniques used to achieve the change in flux as per (3) &(4) as Radial, Axial, Reluctance and Transverse flux machines

$$e = N_t \cdot \frac{d\phi_w(x,t)}{dt} = \frac{d\lambda_w(x,t)}{dt} \quad (3)$$

Where, $x$ is the direction of flux responsible for producing emf $e$ and $e$ is a vector quantity along the defined direction. The flux linking the winding is then given by (3) as

$$\phi_w = \frac{F}{R_{path}} = B_g(\theta,t).A_w \quad (4)$$

Where $Bg$ is the air gap flux density linking the winding and variable over a time $t$ to create change in flux to induce emf E. The generated power available at the supply end can be written in (5) as

$$P_{op} = \frac{1}{t}\int_0^t v(t)i_a(t).dt = \eta_{Generator}P_{ip} = P_{ip} - P_{losses} \quad (5)$$

The power density or energy delivering capacity of the generator is given by

$$\rho_g = \frac{1}{\forall_g t}\int_0^t v(t)i_a(t).dt = \frac{1}{\forall_g t}\int_0^t v(t)i_a(t).dt \quad (6)$$

With $\phi_w, \lambda_w, R_{path}$ is flux in winding, Flux linkage, and reluctance of the flux path over the air gap. The change in magnetomotive force (MMF) as in (4) on the windings can be achieved by either varying the flux $\phi_w$ along the rotational path with respect to time to (or) changing the reluctance of the flux path thus achieving the change in flux required(Reluctance machine) to induce voltage in the winding as in Eq.4. Power generated in a generating machine is proportional to square of the flux density and proportional to speed to achieve. To achieve a higher power the net flux linking the winding or the speed of the generator needs to be improved. However, the maximum speed that could be achieved is set by the stress formed due to torsion forces and the rating of bearings. The generators developed needs to keep the forces down and achieve planar development to suit microelectronic manufacturing technologies with reduced assembly. The ratings of the generator also should posses the capability to take up as many turns as possible to improve power and keep the maximum short circuit current under the permissible value to reduce losses to a minimum value.

The micro generators reported in published papers either are reluctance machines [6] or axial flux machines [8]. The micro-generator configuration is chosen to suit the engine requirements which compromise for engine efficiency. Suggestion for improvement for the generators to improve efficiency and energy density are provided by the authors of the reported micro-generators[6], [8]. However the suggestions outline to effective utilization and optimization of electric and magnetic circuit to achieve better energy density and efficiency.

## VIII. FACTORS TO BE CONSIDERED FOR DEVICE INTEGRATION

Micro-generator can be directly coupled to the shaft of the engine or integrated to the engine rotor. The choice of integration is a compromise between efficiencies and energy density achievable. Integration of micro generator directly coupled to engine shaft an in traditional methods can improve the efficiency of the generator but increases the volume of the system thus reducing the total energy density and integrating the micro generator into the engine rotor reduces the volume and also reduces the efficiency of the system due to the generated efficiency affected by heat losses due to the engine.

At micro levels integration of micro generator with engine causes complications in alignment of the engine and generator shaft and by integrating the generator inside engine [5] makes integration easier and allows non-assembly of structures possible for bulk manufacturing using microelectronics fabrication technologies. Existing integrated systems can produce an max output of 300µWatt [6] and energy density of 58.2 kW-h/m3. An analysis of a system using improved generator parameters in a typical volume of 0.7 times AA battery (7*4*.25cm) and a modest efficiency of 0.7% are done. Results show that the system can generate an output of 2.5 mWatts and an energy density of 80-90 kW-h/m3 at 30000 rpm by choosing suitable Engine-Generation proportion in the net system volume as shown in fig.4 and 5. The energy density when multiplied with the constant (1/0.7) becomes 126 kW-h/m3 which is 1.2 times than the AA battery considered. This increase in energy density can supply more power and improve the operating time of the GPS system by 10-15% and in similar scale with related electronic applications.

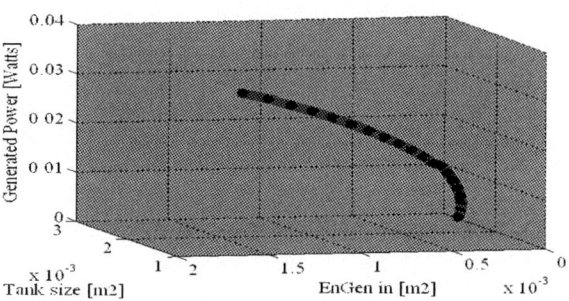

Fig. 5. Generated power for the integrated device with 0.7% efficiency at 30,000 rpm.

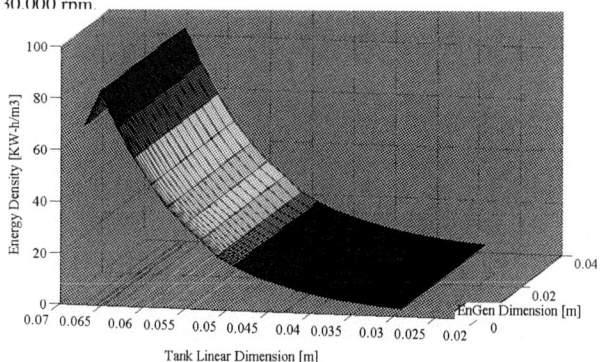

Fig. 6. Energy Density available n the system with 0.7% efficiency.

The generated power can be improved further by increasing the speed but there is a limit that it can be achieved which is set by different factors including stress developed at the rotor at high speeds. Similarly, the energy density of the system can further be increased by reduction of net system volume but has to be compromised over maximum power generated at lower volumes.

## IX. FUNCTIONAL REQUIREMENTS TO SUPPLY VARIABLE LOAD

One of the functional characteristics of the battery is such that it supplies power when the load is on and switches off when the load is disconnected. The engine-generator systems generate a constant power and to attain similar characteristics to a battery it should be able to supply variable load. Suitable control strategies need to be included in the engine-generator system to meet load variations and other functional requirements to work like a battery. The power of the generator can be controlled by controlling speed in real time. But, effective control of speed to supply variable load is a complex solution for which its practical realization at micro scale is difficult. In such conditions an improved power management circuit with storage to supply variable load needs to be considered. Super capacitors and secondary batteries are potential entities for storage; however detailed analysis of the requirements to suit the variable load would guide to selection of storage and efficient method for supplying variable load.

## X. CONCLUSION AND FUTURE WORK

The following had been presented in this paper which covers
1. The need for new alternative sources to achieve high energy density than conventional batteries to power modern electronic equipments is discussed.
2. The energy density available in hydrocarbon fuels and conventional batteries are compared and the possibilities of a liquid fuel battery system with an energy density more than or comparable to conventional AA batteries as power source for electronic communication application are explained.
3. The engine and the generator available in the state-of-art technology are conversed proposing possibilities of improvement in generator efficiency and output power to improve the net energy density of the system with energy densities comparable to battery and net energy of 1-2 times than that of the compared AA batteries .
4. The requirements for the device to supply variable load is discussed.
5. Future improvement includes research in the most efficient development of micro generator with efficiencies close to 70-80% as mentioned in the paper to improve the energy density of existing system by 40% from its original value and attain maximum energy levels better than the battery expected in Section.VIII. FEA magnetic and electric circuit analysis, thermal and stress analysis of the system would be continuing step to substantiate the expected results currently obtained. Experimental validation for a

manufactured prototype would be done and method to feed variable load be developed in future.

## XI. REFERENCES

[1] H.-A. Kiehne, *Battery technology handbook*. New York: Dekker, 2003.
[2] Duracell, "Duracell Data Sheet," in *Duracell Product Data Sheets at http://www.duracell.com/oem/productdata/default.asp*.
[3] I. J. B.-V. Paul B. Koeneman, and Kristin L. Wood, "Feasibility of Micro Power Supplies for MEMS," *Journal of microelectromechanical systems*, vol. Vol .6, pp. pp 355-362, 1997.
[4] J. G. Vican, B.F., Dryer, F.L., Milius, D.L., Aksay, I.A. and Yetter, R.A, "Development of a Microreactor as a Thermal Source for MEMS Power Generation," presented at Proceedings of the Twenty-Ninth International Symposium on Combustion, Sapporo,Japan, July 21-26,2002.
[5] C. H. Lee, K. C. Jiang, P. Jin, and P. D. Prewett, "Design and fabrication of a micro Wankel engine using MEMS technology," *Microelectronic Engineering*, vol. 73-74, pp. 529-534, 2004.
[6] M. K. Senesky, "Electromagnetic Generators for Portable power Applications," in *Electrical and Computer Engineering*. Berkeley: Universtity of California, Spring 2005, pp. 175.
[7] A. P. P. A. Carlos Fernandez-Pello, Kevin Fu, David C. Walthers, Aaron Knobloch, Fabian martinez, Conrad Stoldt, Roya Maboudian. Seth Sanders, Dorian Liepmann, "MEMS Rotary Engine Power System," presented at Proc. Int. Workshop on Power MEMS, Tsukuba, Japan, 2002.
[8] A. S. Holmes, G. Hong, and K. R. Pullen, "Axial-flux permanent magnet machines for micropower generation," *Microelectromechanical Systems, Journal of*, vol. 14, pp. 54-62, 2005.

## XII. BIOGRAPHIES

Surendran Devadoss received his B.E from Amrita Vishwa Vidyapeetham Deemed University, India and M.E from RMIT University, Australia.
He is currently working for his PhD in Electrical Engineering as a student in Micro technology centre, RMIT University. He is a member of IEEE and Magnetic society and a reviewer for transactions and journal papers for Magnetics society. His research interest is in areas of MEMS Electrical machine design, Micro Energy scavenging, FEA structural, Electromagnetic and permanent magnet machines, FPGA's and microprocessors.

Theo Kangsanant received his B.E (Hons) from University of Western Australia, Australia and M.Sc and PhD from University of Manchester, U.K
He is currently the Associate professor for Automation in the school of electrical and computer engineering, RMIT university. His research interest are in Robotics, Motion control, CNC machines, Condition monitoring and Diagnostics Systems

Ian Bates received his Dip EE from Swinburne University, Australia and B.E (Hons) and M.E (Hons) from University of Melbourne
He is currently the Director of Microtechnology centre, RMIT University. His research interests are in Power Systems and Control Engineering, Microtechnology Systems, Systems Engineering

**2006 IEEE International Conference on Power Electronic, Drives and Energy Systems**

# Analysis, Design and Development of Single Switch Forward Buck AC-DC Converter for Low Power Battery Charging Application

Bhim Singh, *Senior Member, IEEE,* and Ganesh Dutt Chaturvedi

*Abstract*— In this paper a low power single switch high power factor AC-DC converter is designed and developed for battery charging. Desired features for battery charger are low cost, fast charging, high power factor, high efficiency, minimum ripple and high reliability are fully achieved using proposed circuit. The converter using single switch forward buck converter topology and operating in discontinuous conduction mode (DCM) is designed and developed to achieve inherent power-factor correction in voltage follower mode. The design equations are derived and calculations are performed to achieve circuit parameters. The test results on developed converter are compared with simulated results to validate proposed design. A prototype for 20.25W battery charger is built and tested to verify the design and analytical model. Satisfactory performance is obtained from experimental results.

*Index Terms*-- Single switch forward buck converter, DCM operation, power factor, efficiency.

## I INTRODUCTION

SINGLE ended converters such as the forward, flyback, Cuk, SEPIC, ZETA and others are often chosen for implementing simple low cost and low power converters. The use of only one switch and relatively simple control circuit required are strong reasons for their choice [1-3]. The conventional forward buck converter available generally for more than 50W rating with efficiency up to 80%. The paper describes the operation characteristics of this AC-DC converter in detail and the experimental result conforms that this AC-DC converter can achieve efficiency more than 80% with sufficient suppression of output voltage ripple for less than 50W output power.

A detailed analysis and design is carried out for single switch forward buck AC-DC converter with high frequency isolation in discontinuous conduction mode (DCM) of operation [4]. This converter is designed for the power rating of 20.25W with 13.5V output voltage. To verify and investigate the design and performance at preliminary stage, simulation study of forward converter is accomplished in DCM operation for input AC voltage 220V at 50Hz using PSIM6.0 platform. Simulation results show high quality steady state performance from 20% to 100% loading conditions with power factor of order of 0.986. It is observed that DCM

Bhim Singh and Ganesh Dutt Chaturvedi are with Department of Electrical Engg, I.I.T.Delhi, New Delhi, India- 110016 (e-mail: bhimsinghr@gmail.com, gd7feb@yahoo.co.in).

operation of forward topology is most suitable for low power applications, where these converters present excellent features of power factor correction with very simple control scheme with one voltage feedback loop only [5]-[8].

Prototype of proposed converter is built and tested to verify the analytical predictions. The components are selected with closest specifications available in market, satisfactory performance is obtained from experimental results.

## II CIRCUIT AND OPERATION

Fig. 1 shows the proposed single switch forward buck converter in DCM operation. This converter consists of the AC power source $V_{in}$, input EMI filter of inductor $L_f$ and capacitor $C_f$, full-wave rectifier FWR, high frequency transformer with two primary windings $N_1$ and $N_3$ one secondary winding $N_2$, high frequency diodes $D_1$, $D_2$ and $D_r$, Output filter of inductor $L_o$ and capacitor $C_o$.

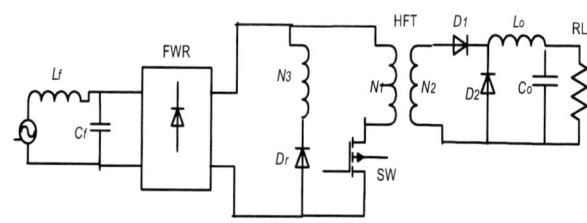

Fig. 1. Practical single switch forward buck converter.

The input filter is required to reduce the ripple in the input current and power factor correction. A large value of input capacitor distorts input current waveform as input current becomes discontinuous due to the fact that reactive energy of capacitor $C_f$ can not be fed back to input supply in presence of one-directional diode bridge. Thus a small value of input filter capacitor should be selected. In a practical forward converter, the transformer magnetizing current must be selected for the proper converter operation, otherwise the stored energy in the transformer core would result in converter failure. To allow the transformer magnetic energy to be recovered and feedback to input supply, it requires a third demagnetizing winding $N_3$ and diode $D_r$.

In discontinuous conduction mode, voltage follower approach is applied for the PWM control of the converter, which needs output voltage sensing only. The output voltage regulation is provided by the feedback loop as shown in Fig. 2, where the output sensing voltage $V_o$ is compared with a

0-7803-9771-1/06/$25.00 ©2006 IEEE

reference V$_{oref}$ value and the error is amplified in a proportional integral (PI) controller which is compared with a saw-tooth ramp V$_s$, thus providing the pulses to power switch.

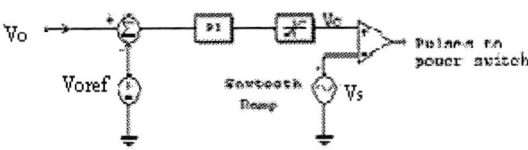

Fig. 2.  Practical voltage follower approach circuit for PWM control.

Therefore, this circuit is controlled by the change of on- time interval and constant switching frequency f$_s$.

## III. DESIGN CONSIDERATIONS

The different parts of the forward converter system are modeled using basic equations. The complete model of the forward buck converter can be divided into following sections:

### A. Voltage Gain

Initially, assuming a transformer to be ideal, when the switch is ON, D$_1$ becomes forward biased and D$_2$ is reverse biased. Therefore output inductor voltage is as:

$$V_L = \frac{N_2}{N_1}V_d - V_o \qquad\qquad 0 < t < t_{on} \qquad (1)$$

Which is positive. Therefore, inductor current i$_L$ increases. When the switch is turned OFF, the i$_L$ circulates through the diode D$_2$, and

$$V_L = -V_o \qquad\qquad t_{on} < t < T_s \qquad (2)$$

Which is negative and therefore causes i$_L$ to decrease linearly. Equating the integral of the inductor voltage over one time period to zero using (1) and (2) yields

$$\frac{V_o}{V_{in}} = (N_1/N_2)D \qquad (3)$$

Eqn.(3) shows that the voltage ratio in the forward converter is proportional to the switch duty ratio D and inversely proportional to the transformer primary to secondary turn ratio $N_1/N_2$.

### B. Input EMI Filter

For calculating the value of capacitor and inductor for EMI filter design first determine DC link capacitor C$_f$ :

$$C_f = P_{in}(1 - D_{max}\sqrt{2}V_{in}.(2fl).(\Delta V_{dc})_{(max)}) \qquad (4)$$

Where $P_{in}$ is input power and $fl$ is line frequency.

Minimum input filter inductance L$_{in}$ (min):

$$L_{in}(min) = \frac{V_{in} (1-D)}{(\Delta I f_s)} \qquad (5)$$

D being the switching duty cycle, f$_s$ is switching frequency and ΔI the change in output current.

### C. Transformer Primary and Secondary Turns

Transformer primary to secondary turn ratio $N_1/N_2$ can be determined as:

$$V_o = (N_1/N_2)D V_{in} \qquad (6)$$

The minimum primary turns of high frequency transformer can be calculated as

$$N_{1\,(min)} = \frac{(V_o D_{max})}{(A_e f_s \Delta B)} \qquad (7)$$

Where V$_o$ is voltage of DC link capacitor, A$_e$ is area of core and ΔB is magnetic flux density.
Secondary turns N$_2$ can be calculated as:

$$\frac{N_1}{N_2}V_o = \frac{V_o D}{(V_o + V_f)} \qquad (8)$$

Where V$_f$, being forward voltage drop of rectifier diode.

### D. Critical Inductance and Ripple Factor

Critical inductance calculated to operate the converter in DCM or CCM (L$_{cr}$):

$$L_{cr} = \frac{R (1 - D_{max})}{2 f_s} \qquad (9)$$

Where L$_{cr}$= Critical inductance, R is load resistance, D$_{max}$ is maximum duty ratio of converter and f$_s$ is switching frequency. Here L$_o$ is selected smaller than the L$_{cr}$ to operate the converter in DCM operation.
Ripple factor of output inductor current (K$_{rf}$) can be defined as:

$$K_{rf} = \frac{\Delta I}{2I_o} \qquad (10)$$

Where ΔI is change in output current I$_o$.

### E. Output Filter Capacitor and Ripple Current

The minimum output capacitor value C$_o$ can be determined from the following equation, which is:

$$\frac{\Delta V_c}{V_o} = \frac{(1-D)}{8 L_o C_o (f_s^2)} \qquad (11)$$

Ripple current of output capacitor Icmax can be calculated as:

$$Icmax = \frac{K_{rf} (Io)}{\sqrt{3}} \qquad (12)$$

These equations are used to get design data, which are used in the model of the proposed forward buck converter in PSIM6.0 to analyze its steady state and dynamic behavior of converter. Based on the above design equations the designed data is summarized in Table I.

## IV. MODELING AND SIMULATION

On the basis of obtained design, the simulation of converter carried out in DCM operation. Fig.3 shows PSIM model of forward buck converter in DCM operation. As discussed already, it uses voltage mode control. It mainly consists EMI input filter and transformer with one extra reset winding. The converter consists of PWM control using voltage follower approach. Simulated results are shown in Figs. 4-8 and Table

II summarized the design parameters of forward buck converter.

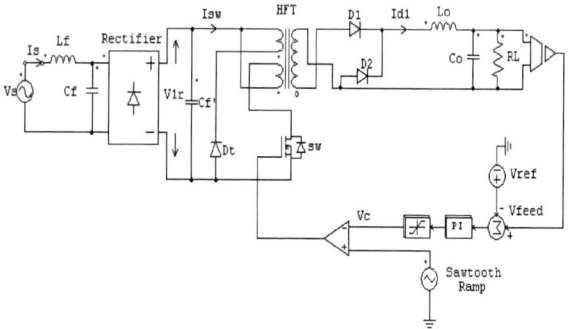

Fig. 3. PSIM model of single switch forward buck AC–DC converter in DCM.

TABLE I

CIRCUIT SPECIFICATIONS FOR SINGLE SWITCH FORWARD BUCK CONVERTER

| | |
|---|---|
| Input Voltage, $V_{in}$ (RMS)@ 50Hz | 220V |
| Output Power, $P_o$ | 20.25W |
| Output voltage, $V_o$ | 13.5V |
| Switching frequency, $f_s$ | 50kHz |
| Maximum Duty Ratio, $D_{max}$ | 0.5 |
| Critical Inductance to operate, $L_{cr}$ | 45 µH |
| DC link capacitor, $C_f$ | 11.6µF |
| Ripple factor of output inductor current, $K_{rf}$ | 0.005 |
| Transformer primary to secondary turn ratio $N_1 / N_2$ | 20 |
| Minimum primary turns $N_1$ (min) | 60 |
| Secondary turns $N_2$ | 3 |
| Minimum output capacitor $C_o$ | 14.5µF |
| Minimum turns in output inductor $N_{L1}$ (min) | 3 |
| Max. diode voltage on secondary side $V_d$ (max) | 25.9V |
| Minimum input filter inductance $L_{in}$ (min) | 2.073mH |

TABLE II

DESIGN PARAMETERS OF FORWARD BUCK CONVERTER

| Components | DCM Operation |
|---|---|
| Transformer Turn ratio | 20: 1 |
| Transformer Magnetizing Inductance | 3.5mH |
| Output Capacitor | 20.5mF |
| EMI Filter | 5.5mH, 180nF, 20nF |

## V. HARDWARE IMPLEMENTATION

A 20.25W isolated single switch forward buck converter prototype is developed with 13.5V output voltage having transformer isolation with 50kHz switching frequency and PFC in DCM operation. It includes the control supply and protections such as overload shutdown, input under voltage and output over voltage. Voltage follower approach is applied for the control using PWM controller chip UC3843. Hardware implementation is carried out using the parameters designed and verified through simulation results. The photograph of experimental setup of the forward converter is shown in Fig. 9.

Extensive tests are conducted on developed prototype of the converter and test results are chosen in Figs. 10-16. High input power quality is achieved with improved power factor and reduced source current distortion at full load. Controller shows

fast dynamic response with line and load disturbance with output voltage displacement of 1% for 4W to 20.25 W load changes. These results show the suitability of DCM operation in lower power applications. Table III summarized the power quality observation via comparison between experimental and simulation results of forward buck converter.

The components used in hardware implementation of forward buck converter are summarized in Table IV with detail description.

TABLE III

POWER QUALITY OBSERVATION FOR FORWARD BUCK CONVERTER

| Quantity | Experimental Results | | Simulation Results | |
|---|---|---|---|---|
| | Output Power | | Output Power | |
| | 4W | 20.25W | 4W | 20.25W |
| Power Factor | 0.974 | 0.986 | 0.965 | 0.982 |
| Voltage Ripple | 0.3% | 1% | 0.5% | 1.9% |
| Efficiency | 78.7% | 82.6 % | 78.5% | 82.4% |
| THD | 14.2% | 12.1% | 14.8% | 12.2% |

TABLE IV

SPECIFICATIONS OF COMPONENTS USED IN HARDWARE IMPLEMENTATION OF FORWARD BUCK CONVERTER

| Component | Company | Description |
|---|---|---|
| Rectifier Diode | General Semiconductor | 1N5408 (Vrrm=1000V, If(max)=1A) |
| MOSFET | International Rectifier | 2SK962 (N Channel, VDSS=900V, Id=3A) |
| High Frequency Diode | On Semiconductor | MBR2045CT (Two Diode, Vrrm=45V, If (max)=20A) |
| Transformer | EPCOS | EE25, (Ferrite Core), Material-N67 |
| Output Inductor | EPCOS | EE20, (Ferrite Core), Material-N67 |
| PWM Controller Chip | Unitrode | UC3843 |
| Optocoupler Chip | Fairchild semiconductor | 4N35 |

## VI. RESULT AND DISCUSSION

Fig.4 and Fig.8 show the PSIM6.0 simulated input voltage and current waveforms at 100% and 20% load respectively. From these curves, it is clear that the input current follows the input voltage and the circuit behaves as resistor emulator. The simulated output voltage waveforms with less than 2% peak-to-peak voltage ripple at full load shown in Fig. 5. In order to analyze the control loop functionality and dynamic behavior of the converter, in terms of source voltage and source current are shown in Fig. 6 for the sudden application of 100% load and then removal of load (20%-100%-20% load change) in DCM operation. Output voltage and current are shown in Fig. 7. Where regulated output voltage observed even at load change, so excellent steady state performance is achieved for forward buck converter.

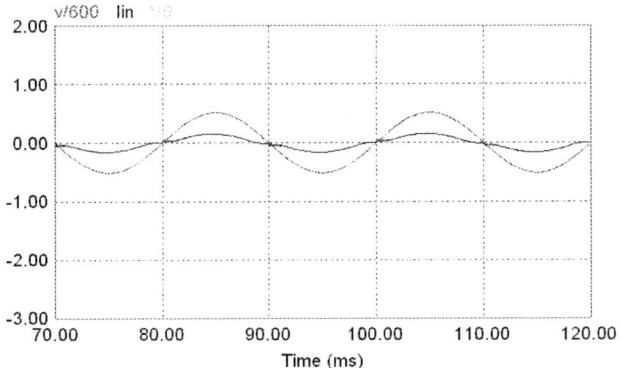

Fig. 4. Input voltage and Current waveform at 100% load.

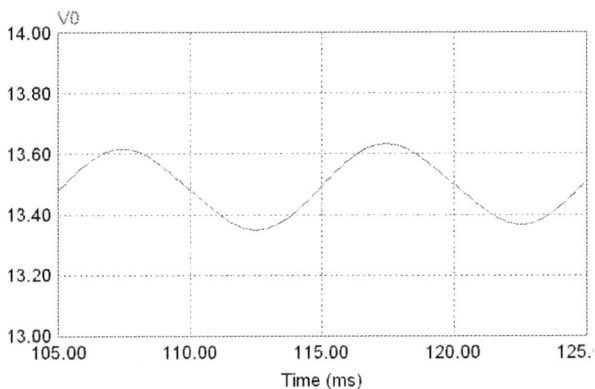

Fig. 5. Output voltage waveform at 100% load.

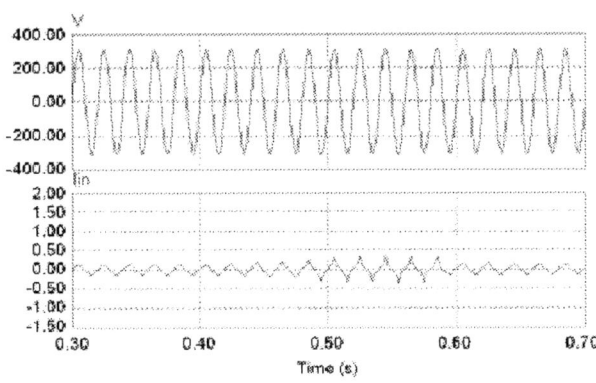

Fig. 6. Source voltage and Current for load application and removal at 100% load.

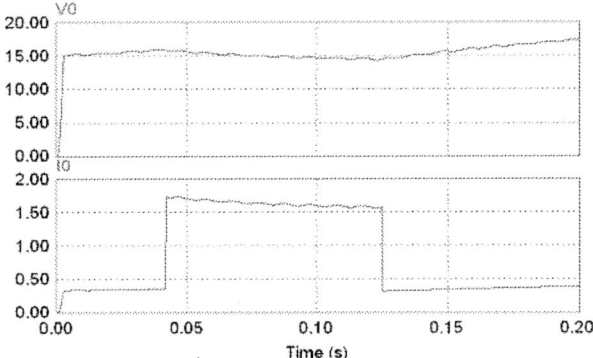

Fig. 7. Output voltage and Current for load application and removal at 100% load.

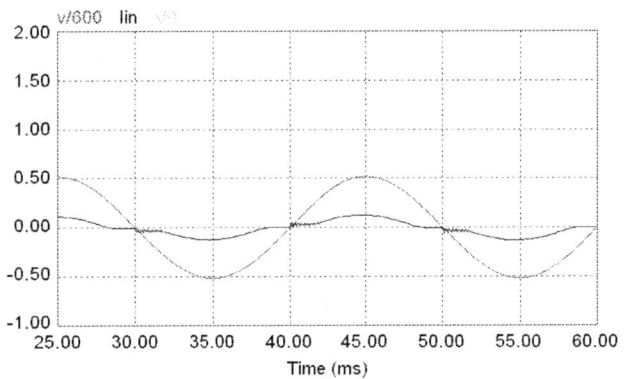

Fig. 8. Input voltage and Current waveform at 20% load.

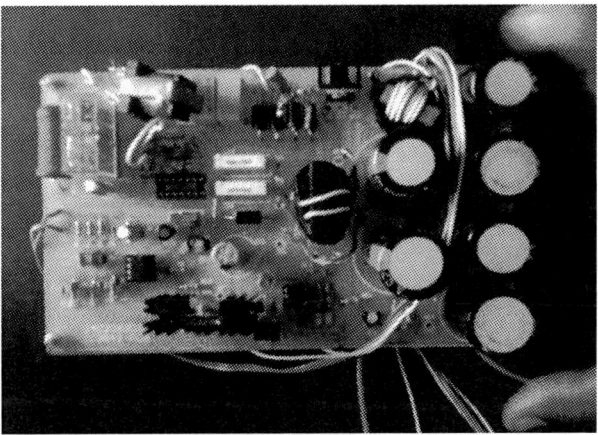

Fig. 9. Experimental setup of forward buck converter.

Fig. 9 shows the photograph of experimental setup of the forward converter. It includes the control supply and protections such as overload shutdown, input under voltage and output over voltage. To verify with simulation input voltage and current waveforms at 100% and 20% load, the experimental waveforms are shown in Fig.10 and Fig.14. Practically high power factor 0.986 achieved which is better than simulated value at full load.

The output DC voltage and current at full load shown in Fig.11 and enlarged view of output DC voltage shown in Fig.12 with nearly 135mV or 1% peak-to-peak ripple in practical implementation. Consequently, it is clear that the voltage follower control scheme is quit effective to suppress the output voltage ripple in proposed single switch forward buck converter in DCM operation.

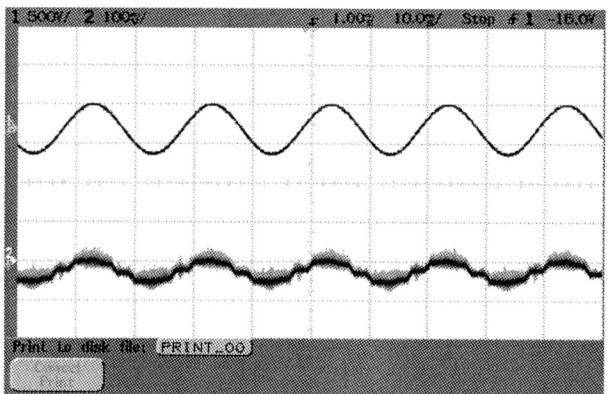

Fig. 10. Input voltage and Current waveform at 100% load Scales: 400V/div, 0.5A/div and 5ms/div.

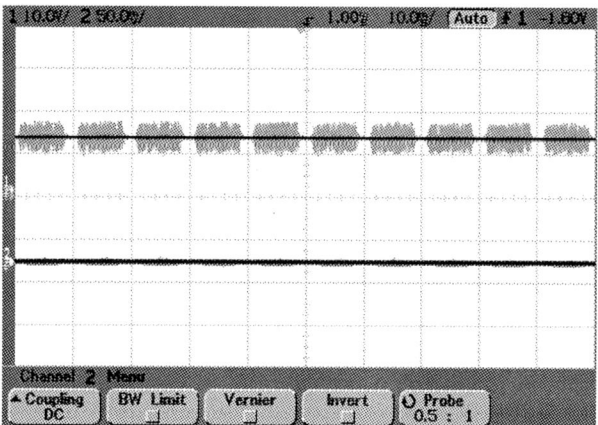

Fig. 11. Output voltage and Current waveform at 100% load Scales: 5V/div, 0.5A/div and 1ms/div.

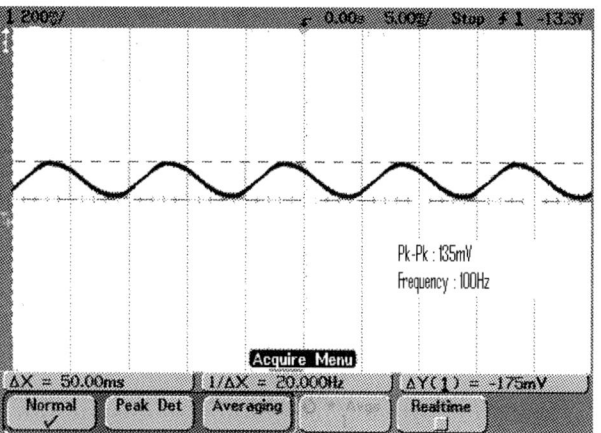

Fig. 12. Enlarged Output voltage waveform at 100% load Scale: 5V/div and 1ms/div.

On the other hand the switch voltage and current waveform can be seen in Fig.13. Where peak to peak value of switch voltage and current observed 600V and 340mA respectively, which satisfied the design value with simulated waveform.

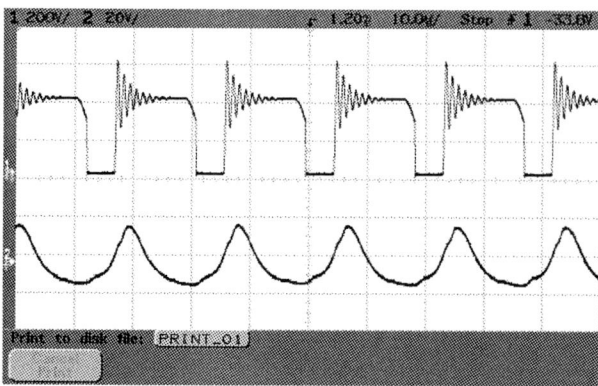

Fig. 13. Switch voltage and Current waveform at 100% load Scales: 200V/div, 0.2A/div and 20µs/div.

The forward buck converter with sudden change of load from 4W to 20.25W and then removal of load from 20.25W to 4W verified. Fig.15 shows the output voltage and current waveforms and Fig.16 shows the source voltage and current waveforms under dynamic condition. These results show the fast response of PI controller, where output voltage is regulated to 13.5V after sudden application and removal of load at constant source voltage 220V RMS.

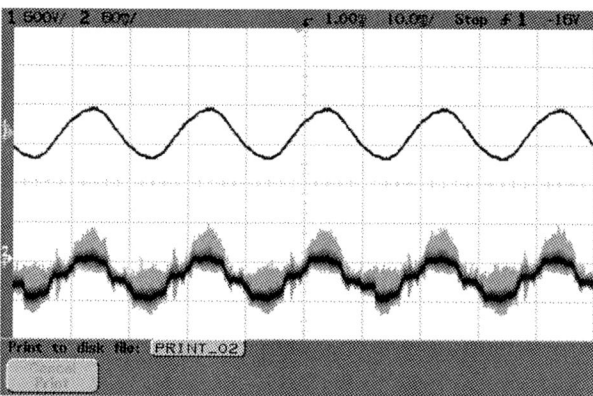

Fig. 14. Input voltage and Current waveform at 20% load Scales: 400V/div, 0.2A/div and 5ms/div.

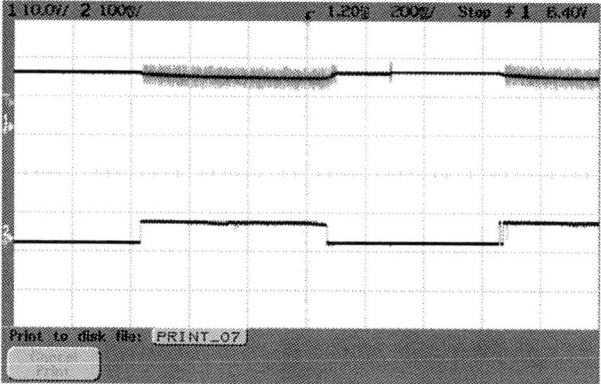

Fig. 15. Output voltage and Current for load application and removal at 100% load. Scales: 5V/div, 0.5A/div and 1ms/div.

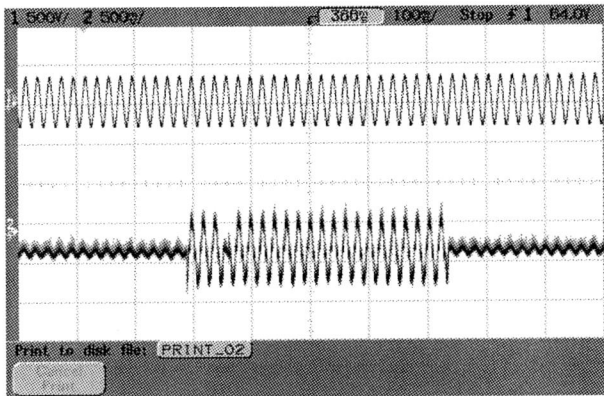

Fig. 16. Source voltage and Current for load application and removal at 100% load. Scales: 600V/div, 0.5A/div and 0.5ms/div.

The converter source current THD, power factor, efficiency and output voltage at 10% to 100% load are summarized in Table V, which demonstrates the improved power quality and high efficiency of the forward buck converter.

TABLE V
THD, PF, EFFICIENCY AND OUTPUT VOLTAGE RIPPLE AT DIFFERENT OUTPUT POWER

| Output power (W) | Source Current THD (%) | Power Factor | Efficiency (%) | Output Voltage Ripple (%) |
|---|---|---|---|---|
| 2 | 14.3 | 0.974 | 78.1 | 0.3 |
| 4 | 14.2 | 0.974 | 78.7 | 0.3 |
| 6 | 14.1 | 0.977 | 78.8 | 0.4 |
| 8 | 13.8 | 0.979 | 78.9 | 0.5 |
| 10 | 13.3 | 0.982 | 80.1 | 0.6 |
| 12 | 13.1 | 0.983 | 80.3 | 0.7 |
| 14 | 12.8 | 0.983 | 80.7 | 0.9 |
| 16 | 12.5 | 0.984 | 81.0 | 0.9 |
| 18 | 12.3 | 0.985 | 81.6 | 1.0 |
| 20.25 | 12.1 | 0.986 | 82.6 | 1.0 |

## VII. CONCLUSIONS

The simulated and experimental results have revealed the improved performance of proposed converter in low power battery charging applications. It has been observed from the results that the power factor at AC input mains observed to be 0.986 and full load efficiency order of 82.6%. The most suitable circuit parameters are obtained in both steady state and dynamic characteristics. Experimental results for line and load disturbances show fast dynamic response of control circuit with voltage dip of only 1.1% for sudden load change from 20% to 100%. Thus design and simulation studies carried out are verified experimentally. Hardware implementation shows the efficiency, power factor and output voltage ripple improvement as compared to simulation results.

## VIII. REFERENCES

[1] R. Carbone and A. Scappatura, "AHigh Efficiency Passive Power Factor Corrector for Single-Phase Bridge Diode Rectifiers," in *IEEE Proc. of Power Eelectronics Specialists Conference*, 2004, pp.1627-1630.

[2] Ningliang Mi, Boris Sasic, Jon Marshall and Steve Tomasiewicz, "A Novel Economical Single Stage Battery Charger with Power Factor Correction," in *Proc. of IEEE Conf.*, 2004, pp.760-763.

[3] Tsai-Fu Wu, Chien-Chih Chen, Chih- Lung shen and Cheng-Nan Wu, "Analysis, Design and Practical Considerations for 500 W Power Factor Correctors', in *IEEE –Transactions on Aerospace and Electronic Systems, Vol. 39, No.3*, pp.961-975, July 2003.

[4] Kin- siu Fung, Wing- Hung Ki and Philip K. T. Mok , "Analysis and Measurement of DCM Power Factor Correctors," in *Proc. IEEE Power Eelectronics Specialists Conference*, 1999, pp.709-714.

[5] Yan Liu and Gerry Moschopoulos, "A Single Stage AC-DC Forward Converter with Input Power Factor Correction and Reduced DC Bus Voltage," in *Proc. of IEICE / IEEE INTELEC'03*, Oct. 19-23, 2003, pp.132-139.

[6] J. Sebastian, A. Fernandez, P. Villegas, J. A. Martinez and E. de la Cruz, "A New Low-Cost Control Technique for Power Factor Correctors Operating in Continuous Conduction Mode," in *Proc. IEEE Power Eelectronics Specialists Conference*, 2002, pp.1132-1136.

[7] Chongming Qiao and Keyue Ma Smedley, "A Topology Survey of Single-Stage Power Factor Corrector with a Boost Type Input-Current-Shaper," *IEEE- Transactions on Power Electronics ,Vol. 16, No.3* , pp.360-368, May 2001.

[8] E. Burkin Yu and Nigof B. M., "Power Factor Corrector," in *Proc. IEEE* Modern Techniques and Technology, 2000. MTT 2000. Proceedings of the VI International Scientific and Practical Conference of Students, Post graduates and Young Scientists, pp.76-78.

## VIII. BIOGRAPHIES

**Bhim Singh** (SM'99) was born in Rahamanpur, U.P. India in 1956. He received B.E. (Electrical) degree from University of Roorkee, India in 1977 and M.Tech. and Ph.D. degrees from Indian Institute of technology (IIT) , New Delhi, in 1979 and 1983, respectively. In 1983, he joined as a Lecturer and in 1988 became a reader in department of Electrical Engineering, University of Roorkee. In December 1990, he joined as an assistant Professor, became an Associate Professor in 1994 and Professor in1997 at Department of Electrical Engineering, IIT Delhi. His field of interest includes power electronics, electrical machines and drives, active filters, static VAR compensator, analysis and digital control of electrical machines. Prof Singh is a Fellow of Indian National Academy of Engineering (INAE), Institution of Engineers (India) (IE) (I) and Institution of Electronics and Telecommunication Engineers (IETE), a life Member of Indian Society for Technical Education (ISTE), System Society of India (SSI) and national Institution of Quality and Reliability (NIQR) and Senior Member of IEEE (Institute of Electrical and Electrical and Electronics Engineers).

**Ganesh Dutt Chaturvedi** was born in Chanderi, M.P. India in 1976. He received B.E. (Electronics and Communication) degree from R.K.D.F Institute of Science and Technology, Bhopal, India in 1999. In 2000, he joined as a Research Trainee and in 2001 became a Design Engineer in Associated Electronics Research Foundation, Noida. Presently he is a research scholar in the Department of Electrical Engineering, IIT Delhi, pursuing for his MS(Research) degree. His field of interest includes power quality, low power converter design and digital control.

**2006 IEEE International Conference on Power Electronic, Drives and Energy Systems**

# A Novel Approach for Eco-Friendly and Economic Power Dispatch using MATLAB

D.P.Kothari, *Senior Member, IEEE*, and K.P.Singh Parmar

*Abstract*--The paper proposes an optimization approach to determine the most appropriate generation schedule of a thermal power plant. The problem is formulated considering two objectives: Operating cost and $NO_x$ emission. The weighting method is used to simulate the trade-off relation between the conflicting objectives in the non-inferior domain. Decision maker uses Fuzzy set theory to choose the best alternative among non-inferior solutions. MATLAB platform is used to provide the ON-line solution of the problem. The effectiveness of the methodology is demonstrated through an example. The thermal generation schedule obtained is economic and environment friendly.

*Index Terms*-- Decision making, Economic power dispatch, Fuel cost, Fuzzy set, Multi-objective optimization, Weighting method.

## I. INTRODUCTION

THIS paper provides a multi-objective thermal generation schedule for a thermal power plant. The economic load dispatch is very important optimized problem solution in power system operations for allocating generation among the committed units such that the system constraints imposed are satisfied and energy requirements in terms of Rupees per hour (Rs/h) are minimized [3]-[4]. Multi-objective optimization problem is the methodology which simultaneously satisfies multiple contradictory criterion. The most important objectives which are to be satisfied simultaneously in a large complex power system are economic operation, minimal impact on environment, reliability and security.

The classical economic dispatch problem is to operate electric power systems so as to minimize the total fuel cost. This single objective can no longer be considered alone due to the environmental concerns that arise from the emissions produced by fossil fueled electric power plants. In fact, the Clean Air Act Amendments have been applied to reduce SOx and $NO_x$ emissions from such power plants. Accordingly, emissions can be reduced by dispatch of power generation to minimize emissions instead of or as a supplement to the usual cost objective of economic dispatch [5].

Environmental/economic dispatch is a multi-objective problem with conflicting objectives because pollution is conflicting with minimum cost of generation. Various techniques have been proposed to solve this multi-objective problem whereby most researchers have concentrated on the deterministic problem.

Economic dispatch calculates the cost of generation based on data relating fuel cost and power output. This cost function is approximated by a quadratic equation with cost coefficients. In conventional economic dispatch the coefficients are assumed to be deterministic, but in real-world situations, these data are subjected to inaccuracies and uncertainties.

There has been much research using the deterministic approach to solve the environmental/economic dispatch problem. Gent and Lamont introduced the minimum-emission dispatch concept where they developed a program for on-line steam unit dispatch that results in the minimizing of $NO_x$ emission [5]. These authors introduced the mathematical representation of $NO_x$ emission of steam generating units and used a Newton-Raphson convergence technique to obtain base points and participation factors. Presently, there is a growing trend towards the formulation of decision-making problems in terms of Multi-objective. The classical method of economic power dispatch is still being used in the power industry, since today very rigorous and sophisticated method dispatching have not led to the significant difference in the financial benefits. Therefore in this paper, classical economic dispatch problem is formulated as Multi-objective problem. Fundamental to Multi-objective analysis is Pareto optimum concept, also known as non-inferior solution. Qualitatively, a non-inferior solution of a Multi-objective problem is one where any improvement of one objective function can be achieved only at the expense of another. Weighting technique is used to generate non-inferior solutions. An attempt is also made to apply fuzzy set theory for decision making to get the comprised optimal generation dispatch [1]. The practical viability of the multi-objective thermal scheduling problem has been demonstrated on a sample system with six-thermal generating units for various load levels.

## II. FORMULATION OF MULTI-OBJECTIVE THERMAL POWER DISPATCH PROBLEM-WEIGHTING METHOD

The multi-objective thermal dispatch problem is defined to minimize the operating cost and the $NO_x$ emission level of the thermal plant while meeting total real power load plus real

---

D.P.Kothari is Professor, Centre for Energy Studies, Former Director In-charge, Indian Institute of Technology Delhi, Hauz Khas, New Delhi,110016,India.(E-mail: dkothari@ces.iitd.ernet.in )

K.P.Singh Parmar is Senior Faculty, Electrical Engineering Department, JSS Academy of Technical Education, NOIDA (UP), 201301, India.(E-mail: kpsingh_jss@rediffmail.com, phone (91)9811220732)

---

0-7803-9771-1/06/$25.00 ©2006 IEEE

power transmission losses [1]. Mathematical problem is defined as:

Minimize operating cost

$$F_1 = \sum_{i=1}^{NU} \left( a_{1i} P_{gi}^2 + b_{1i} P_{gi} + c_{1i} \right) \quad Rs/h \tag{1}$$

Minimize $NO_x$ emission level

$$F_2 = \sum_{i=1}^{NU} \left( d_{2i} P_{gi}^2 + e_{2i} P_{gi} + f_{2i} \right) \quad Kg/h \tag{2}$$

Subject to:

(i) To ensure real power balance

$$\sum_{i=1}^{NU} P_{gi} = \left( P_D + P_L \right) \tag{3}$$

(ii) To satisfy real power generation limits of generating units

$$P_{gi}^{\min} \leq P_{gi} \leq P_{gi}^{\max} \quad ; i=1, 2, \ldots, NU \tag{4}$$

Where

NU is the number of thermal generating units.

$a_{1i}$, $b_{1i}$ and $c_{1i}$ are cost coefficients.

$d_{2i}$, $e_{2i}$ and $f_{2i}$ are $NO_X$ emission coefficients.

$P_{gi}$ is the real power generation by ith unit (at ith bus)

$P_D$ is the power demand

$P_{gi}^{\min}$ is the lower limit of ith generator output

$P_{gi}^{\max}$ is the upper limit of ith generator output.

$P_L$ is the transmission losses.

$P_L$ is approximated in terms of B-coefficients as:

$$P_L = B_{00} + \sum_{i=1}^{NU} \left( B_{i0} P_{gi} \right) + \sum_{i=1}^{NU} \sum_{j=1}^{NU} \left( P_{gi} B_{ij} P_{gj} \right) \quad MW \tag{5}$$

To generate the non-inferior solution to the multi-objective problem, the weighting method is applied [1]-[3]. This method converts multi-objective problem into scalar optimization problem as:

Minimize $$\sum_{k=1}^{M} \left( w_k F_k \right) \tag{6}$$

Subject to:

Equations (3)-(4) and

$$\sum_{k=1}^{M} w_k = 1; (w_k \geq 0) \tag{7}$$

Where, M is the number of objectives and $w_k$ are levels of normalized weights. The augmented objective function is

$$L = \sum_{k=1}^{M} \left( w_k F_k \right) + \lambda \left( P_D + P_L - \sum_{i=1}^{NU} P_{gi} \right) \tag{8}$$

Where, $\lambda$ is Lagrange multiplier.

The optimality conditions are described by taking partial derivatives of augmented objective function with respect to decision variables.

$$\frac{\partial L}{\partial P_{gi}} = \sum_{k=1}^{M} \left( w_k \frac{\partial F_k}{\partial P_{gi}} \right) + \lambda \left( \frac{\partial P_L}{\partial P_{gi}} - 1 \right) = 0 \tag{9}$$

$$\frac{\partial L}{\partial \lambda} = P_D + P_L - \sum_{i=1}^{NU} P_{gi} = 0 \tag{10}$$

These equations are inherently non linear. The Newton-Raphson method is used to find the solution.

## III. DECISION MAKING

The imprecise nature of the decision maker's judgment is considered therefore it is obvious to assume that the decision maker may have fuzzy or imprecise goals for each objective function. Membership functions are used to define the fuzzy sets [1]. These functions represent the degree of membership in some fuzzy set values ranging from 0 to 1. The membership value '1' means full compatibility with the sets, while '0' indicates incompatibility. The decision maker must detect membership function $\mu(F_i)$; $i=1,2,\ldots,M$ in a subjective manner after considering the minimum and maximum values of each objective function together with the rate of increase of membership satisfaction. We assume that $\mu(F_i)$; $i=1,2,\ldots,M$ is a strictly monotonic decreasing and continuous function defined as:

$$\mu(F_i) = \begin{cases} 1 & if, \quad F_i \leq F_i^{\min} \\ \dfrac{F_i^{\max} - F_i}{F_i^{\max} - F_i^{\min}} & if, \quad F_i^{\min} < F_i < F_i^{\max} \\ 0 & if, \quad F_i \geq F_i^{\max} \end{cases} \tag{11}$$

Where, $F_i^{\max}$ and $F_i^{\min}$ are the maximum and minimum values of ith objective function. The value of membership function indicates how close (in the range from 0 to 1) a non-inferior solution is to satisfy the $F_i$ objective. The accomplishment of each non-inferior solution can be rated with respect to all the K non-inferior solutions by normalizing its accomplishment over the sum of the accomplishments of K non-inferior solutions as follows:

$$\mu_D^k = \frac{\sum\limits_{i=1}^{M} \mu(F_i^k)}{\sum\limits_{k=1}^{K} \sum\limits_{i=1}^{M} \mu(F_i^k)} \tag{12}$$

The membership function $\mu_D^k$ for non-inferior solutions in a fuzzy set represents the fuzzy cardinal priority ranking of the non-inferior solutions. The solution corresponding to the maximum value of $\mu_D^k$ in the fuzzy set so obtained can be chosen as best solution.

## IV. ALGORITHM

1. Choose cost coefficients, emission coefficients, B-coefficients, load demand, ε (convergence), maximum allowed iterations, M (number of objectives), NU (number of thermal generating units) and K (number of inferior solutions), etc.

2. Set iteration for non-inferior solutions, k=1.

3. Increment count, k=k+1.

4. If (k ≥ K) GOTO step 17.

5. Generate weights, $w_i$ (i=1, 2,......, M).

6. Compute initial values of $P_{gi}$ ( i=1,2,......,NU ) and λ by presuming that $P_L$= 0.
7. Assume that no thermal generating unit has been fixed either at lower limit or at upper limit.
8. Set iteration counter IT=1.
9. Compute Hessian and Jacobian matrix elements.
10. Check the limits of generators and fix up as following
    If $P_{gi}$ is less than $P_{gi}^{min}$; then set $P_{gi}=P_{gi}^{min}$
    If $P_{gi}$ is more than $P_{gi}^{max}$; then set $P_{gi}=P_{gi}^{max}$.
11. Deactivate row and column of Hessian matrix and row of Jacobian matrix representing the thermal generating unit whose generation is fixed either at lower limit or at upper limit. This is done so that fixed thermal generating units cannot participate in allocation.
12. Apply Gauss elimination/matrix inverse method to find $\Delta P_{gi}$ (i =1, 2, …., R) and Δλ.Where, R is the number of thermal generating units which can participate in allocation.
13. Check either

$$\sqrt{\sum_{i=1}^{R}\left(\Delta P_{gi}\right)^2+\left(\Delta\lambda\right)^2}\leq\varepsilon$$

or

$$\sqrt{\sum_{i=1}^{R}\left(\frac{\partial L}{\partial P_{gi}}\right)^2+\left(\frac{\partial L}{\partial\lambda}\right)^2}\leq\varepsilon$$

If convergence condition is met then GOTO step 16,
Check IT is greater than ITMAX, if yes, GOTO step 16 (it means the procedure proceeds without obtaining required convergence).
14. Modify control variables, $P_{gi}^{new}=P_{gi}+\Delta P_{gi}$ (i=1, 2, …,R) and $\lambda^{new}=\lambda+\Delta\lambda$.
15. IT=IT+1, $P_{gi}=P_{gi}^{new}$, $\lambda=\lambda^{new}$ and GOTO step 9 and repeat.
16. Record as non inferior solution and compute $F_k$ (k=1,2,….,M) and transmission loss and GOTO step 3 for another non inferior solution.
17. STOP.

## V. TEST SYSTEM AND RESULTS

The validity of the proposed method is illustrated on a sample system comprising six thermal generating units. A short range thermal load scheduling problem of 24 h duration has been undertaken. Fig. 1 shows that entire optimization period has 24 time intervals and each interval is of one hour. The operating cost and NO$_x$ emission characteristics of thermal generating units are depicted in Table V.The B-coefficients for the calculation of transmission losses are given in Table VI. A MATLAB code is developed to perform entire calculations and to generate the Tables I-IV. We have demonstrated the entire calculations for the first time interval with 505 MW power demand. The non-inferior solutions are obtained for the various combinations of weights; see Table I and corresponding generation schedules obtained are given in Table II.To decide the best solution, minimum and maximum values of objective functions are required. Minimum values

of objectives are obtained by giving full weightage to one of the objectives and neglecting others. When the given weightage value is '1' it means that full weightage is given to the objective and when the weightage is '0' objective is neglected. Owing to the conflicting nature of the objectives NO$_x$ will have the maximum value when operating cost has minimum value and vice versa. The minimum and maximum values obtained are given below:

$F_1^{min}$ =28592.89 Rs/h, and $F_1^{max}$=29133.75 Rs/h
$F_2^{min}$=291.25 Kg/h, and $F_2^{max}$=330.21Kg/h

Using equation. (11), the membership functions of operating cost and NO$_x$ emission objectives corresponding to each non-inferior solution are obtained and are given in Table III.Using Equation (12), the normalized function $\mu_D^k$ of each non-inferior solution is generated, see Table III. A non-inferior solution that attains the maximum value of normalized membership function $\mu_D^k$ is the best solution. From Table III, solution number 10, having weights $w_1$=0.1 and $w_2$=0.9 shows the maximum value of $\mu_D^k$, i.e. 0.1177.The operating fuel cost and NO$_x$ emission corresponding to the solution number 10 is considered as the best solution. The most appropriate thermal generation schedule will be corresponding to solution number 10, see Table II.Similar method is repeated for 24 time intervals and a most appropriate thermal generation schedule is prepared as shown in Table IV.

TABLE I
NON-INFERIOR SOLUTIONS FOR TWO OBJECTIVES

| S.No. | $w_1$ | $w_2$ | IC (Rs/MWh) | Fuel cost (Rs/h) | No$_x$ emission (Kg/h) | Trans. losses (MW) |
|-------|-------|-------|-------------|------------------|------------------------|--------------------|
| 1 | 1.0 | 0.0 | 50.5173 | 28592.89 | 330.21 | 26.20 |
| 2 | 0.9 | 0.1 | 45.5670 | 28592.93 | 329.55 | 26.21 |
| 3 | 0.8 | 0.2 | 40.6165 | 28593.09 | 328.74 | 26.23 |
| 4 | 0.7 | 0.3 | 35.6656 | 28593.43 | 327.74 | 26.25 |
| 5 | 0.6 | 0.4 | 30.7142 | 28594.13 | 326.48 | 26.28 |
| 6 | 0.5 | 0.5 | 25.7620 | 28595.53 | 324.82 | 26.33 |
| 7 | 0.4 | 0.6 | 20.8087 | 28598.36 | 322.55 | 26.40 |
| 8 | 0.3 | 0.7 | 15.8531 | 28604.61 | 319.28 | 26.55 |
| 9 | 0.2 | 0.8 | 10.8931 | 28620.61 | 314.16 | 26.85 |
| 10 | 0.1 | 0.9 | 5.9210 | 28675.90 | 305.20 | 27.75 |
| 11 | 0.0 | 1.0 | 0.8825 | 29133.75 | 291.25 | 33.50 |

## TABLE II
### THERMAL GENERATION SCHEDULES (MW) CORRESPONDING TO NON - INFERIOR SOLUTIONS

| S.No. | Unit-1 | Unit-2 | Unit-3 | Unit-4 | Unit-5 | Unit-6 |
|-------|--------|--------|--------|--------|--------|--------|
| 1 | 52.35 | 10 | 37.17 | 77.94 | 206.51 | 147.21 |
| 2 | 52.42 | 10 | 37.53 | 78.06 | 205.92 | 147.26 |
| 3 | 52.51 | 10 | 37.96 | 78.21 | 205.21 | 147.32 |
| 4 | 52.62 | 10 | 38.51 | 78.39 | 204.32 | 147.39 |
| 5 | 52.76 | 10 | 39.22 | 78.63 | 203.18 | 147.48 |
| 6 | 52.95 | 10 | 40.17 | 78.94 | 201.65 | 147.59 |
| 7 | 53.23 | 10 | 41.52 | 79.38 | 199.53 | 147.72 |
| 8 | 53.68 | 10 | 43.59 | 80.03 | 196.35 | 147.87 |
| 9 | 54.50 | 10 | 47.14 | 81.14 | 191.06 | 148.00 |
| 10 | 56.47 | 10 | 54.68 | 83.40 | 180.49 | 147.70 |
| 11 | 68.62 | 10 | 81.86 | 91.06 | 146.77 | 140.17 |

## TABLE III
### DECISION MAKING USING FUZZY SET THEORY

| S.No. | Fuel cost (Rs/h) | $No_x$ emission (Kg/h) | $\mu$-cost | $\mu$-$No_x$ | $\mu_D^{\ k}$ |
|-------|------------------|------------------------|-----------|-------------|---------------|
| 1 | 28592.89 | 330.21 | 1.0000 | 0.0000 | 0.0791 |
| 2 | 28592.93 | 329.55 | 0.9999 | 0.0170 | 0.0804 |
| 3 | 28593.09 | 328.74 | 0.9996 | 0.0377 | 0.0820 |
| 4 | 28593.43 | 327.74 | 0.9989 | 0.0634 | 0.0840 |
| 5 | 28594.13 | 326.48 | 0.9976 | 0.0958 | 0.0865 |
| 6 | 28595.53 | 324.82 | 0.9951 | 0.1384 | 0.0896 |
| 7 | 28598.36 | 322.55 | 0.9898 | 0.1965 | 0.0938 |
| 8 | 28604.61 | 319.28 | 0.9783 | 0.2805 | 0.0996 |
| 9 | 28620.61 | 314.16 | 0.9487 | 0.4119 | 0.1076 |
| 10 | 28675.90 | 305.20 | 0.8465 | 0.6418 | 0.1177 |
| 11 | 29133.75 | 291.25 | 0.0000 | 1.0000 | 0.0791 |

## VI. CONCLUSION

The conventional economic thermal power dispatch method allocates generation schedules to the individual thermal generating units based upon deterministic operating cost function and $NO_x$ emission function. The load demand is considered constant, but in practice it is random. Such generation schedules result in the lowest cost or lowest emission. The solution set of the formulated problem is non-inferior due to contradictions between the objectives taken. In order to generate non-inferior solutions of the multi-objective optimization problem, weights required to be varied in a systematic manner.

The weighting method is used to simulate the trade-off relation between the conflicting objectives in the non-inferior domain. Decision maker uses Fuzzy set theory to choose the best alternative among non-inferior solutions. The generation of non-inferior solution requires enormous amount of computation time when the objectives are more. A program is developed to solve the problem on the MATLAB platform which provides faster on-line thermal generation schedule. Its effectiveness is demonstrated through an example. The proposed method provides the following technical supports:

(a). On-line thermal generation schedule with the help of the software developed using MATLAB .

(b). Allows explicit trade-off among objective functions.

(c). Provides the decision maker (power system operator) to get the efficient optimal solution from the non inferior set.

(d). Provides economic and environment friendly thermal generation scheduling.

A more comprehensive analysis may be carried out by incorporating risk aversion in the optimal power flow problems. The operating cost, $NO_x$, $SO_x$ emission and security index objectives may be considered for the optimal power flow.

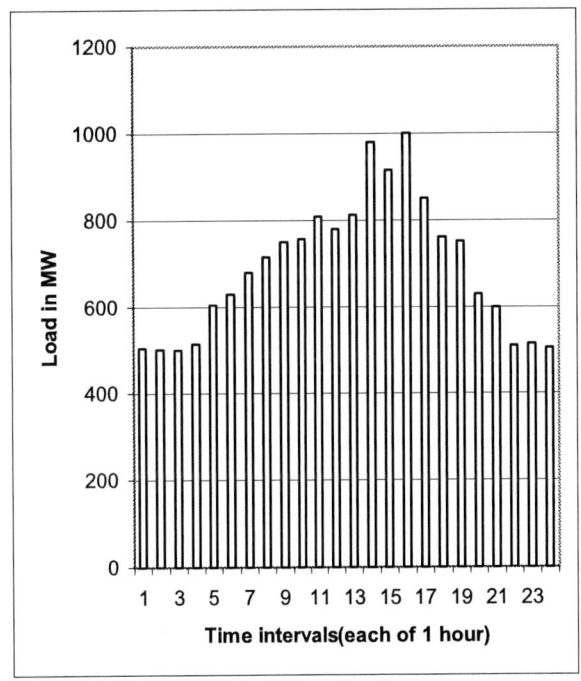

Fig. 1. Load curve for a day.

## TABLE IV
### OPTIMAL THERMAL GENERATION SCHEDULES (MW) FOR 24 HOURS

| Time interval | Demand (MW) | Unit-1 | Unit-2 | Unit-3 | Unit-4 | Unit-5 | Unit-6 | Fuel cost (Rs/h) | $NO_X$ emission (Kg/h) |
|---|---|---|---|---|---|---|---|---|---|
| 1 | 505.0 | 56.47 | 10.00 | 54.68 | 83.40 | 180.49 | 147.70 | 28675.90 | 305.20 |
| 2 | 502.0 | 55.98 | 10.00 | 54.34 | 82.89 | 179.37 | 146.81 | 28523.57 | 302.41 |
| 3 | 501.0 | 55.82 | 10.00 | 54.23 | 82.72 | 178.99 | 146.51 | 28472.86 | 301.48 |
| 4 | 515.0 | 58.08 | 10.00 | 55.82 | 85.10 | 184.24 | 150.67 | 29185.78 | 314.73 |
| 5 | 605.0 | 72.95 | 10.00 | 65.85 | 100.60 | 218.69 | 177.57 | 33923.95 | 415.17 |
| 6 | 630.0 | 77.20 | 10.00 | 68.58 | 104.95 | 228.49 | 185.08 | 35289.41 | 447.91 |
| 7 | 680.0 | 85.86 | 10.00 | 73.99 | 113.72 | 248.41 | 200.15 | 38087.55 | 519.97 |
| 8 | 715.0 | 92.04 | 10.00 | 77.72 | 119.90 | 262.61 | 210.74 | 40101.21 | 575.77 |
| 9 | 750.0 | 98.33 | 10.00 | 81.42 | 126.13 | 277.01 | 221.37 | 42161.65 | 636.13 |
| 10 | 757.0 | 99.60 | 10.00 | 82.15 | 127.38 | 279.92 | 223.50 | 42579.47 | 648.76 |
| 11 | 808.0 | 108.99 | 10.00 | 87.46 | 136.55 | 301.36 | 239.06 | 45682.85 | 746.53 |
| 12 | 780.0 | 103.81 | 10.00 | 84.55 | 131.51 | 289.53 | 230.51 | 43966.03 | 691.59 |
| 13 | 812.0 | 109.73 | 10.00 | 87.87 | 137.28 | 303.06 | 240.28 | 45930.74 | 754.64 |
| 14 | 980.0 | 115.48 | 87.60 | 81.07 | 161.29 | 325.00 | 281.25 | 54639.95 | 980.69 |
| 15 | 915.0 | 107.59 | 81.00 | 80.62 | 136.93 | 325.00 | 242.90 | 50604.97 | 854.27 |
| 16 | 1000.0 | 116.96 | 89.06 | 79.61 | 170.65 | 325.00 | 295.39 | 55935.27 | 1029.86 |
| 17 | 850.0 | 115.95 | 10.00 | 90.93 | 142.06 | 325.00 | 249.61 | 48280.14 | 839.72 |
| 18 | 760.0 | 100.15 | 10.00 | 82.47 | 127.92 | 281.17 | 224.41 | 42759.13 | 654.23 |
| 19 | 751.0 | 98.51 | 10.00 | 81.52 | 126.31 | 277.43 | 221.68 | 42221.22 | 637.93 |
| 20 | 630.0 | 77.20 | 10.00 | 68.58 | 104.95 | 228.49 | 185.08 | 35289.41 | 447.91 |
| 21 | 600.0 | 72.11 | 10.00 | 65.30 | 99.73 | 216.74 | 176.07 | 33653.49 | 408.88 |
| 22 | 510.0 | 57.27 | 10.00 | 55.25 | 84.25 | 182.36 | 149.19 | 28930.44 | 309.93 |
| 23 | 515.0 | 58.08 | 10.00 | 55.82 | 85.10 | 184.24 | 150.67 | 29185.78 | 314.73 |
| 24 | 505.0 | 56.47 | 10.00 | 54.68 | 83.40 | 180.49 | 147.70 | 28675.90 | 305.20 |

## TABLE V
### PROFILE OF SIX THERMAL GENERATING UNITS

| Unit No. | Cost Coefficients | | | $NO_X$ emission coefficients | | | Generation limits (MW) | |
|---|---|---|---|---|---|---|---|---|
| i | $a_{1i}$ | $b_{1i}$ | $c_{1i}$ | $d_{2i}$ | $e_{2i}$ | $f_{2i}$ | $P_{gi}^{min}$ | $P_{gi}^{max}$ |
| 1. | 0.15247 | 38.53973 | 756.79886 | 0.00419 | 0.32767 | 13.85932 | 10 | 125 |
| 2. | 0.10587 | 46.15196 | 451.32513 | 0.00419 | 0.32767 | 13.85932 | 10 | 150 |
| 3. | 0.02803 | 40.39655 | 1049.99770 | 0.00683 | -0.54551 | 40.26690 | 35 | 225 |
| 4. | 0.03546 | 38.30553 | 1243.53110 | 0.00683 | -0.54551 | 40.26690 | 35 | 210 |
| 5. | 0.02111 | 36.32782 | 1658.56960 | 0.00461 | -0.51116 | 42.89553 | 130 | 325 |
| 6. | 0.01799 | 38.27041 | 1356.65920 | 0.00461 | -0.51116 | 42.89553 | 125 | 315 |

## TABLE VI
### TRANSMISSION LOSS COEFFICIENTS ($MW^{-1}$)

| | | | | | |
|---|---|---|---|---|---|
| 0.002022 | -0.000286 | -0.000534 | -0.000565 | -0.000454 | 0.000103 |
| -0.000286 | 0.003243 | 0.000016 | -0.000307 | -0.000422 | -0.000147 |
| -0.000533 | 0.000016 | 0.002085 | 0.000831 | 0.000023 | -0.000270 |
| -0.000565 | -0.000307 | 0.000831 | 0.001129 | 0.000113 | -0.000295 |
| -0.000454 | -0.000422 | 0.000023 | 0.000113 | 0.000460 | -0.000153 |
| 0.000103 | -0.000417 | -0.000270 | -0.000295 | -0.000153 | 0.000898 |

## VII. REFERENCES

[1]. D. P. Kothari, and J.S.Dhillon, "Power System Optimization," India: Prentice Hall, 2004, pp.321-386.

[2]. J.S.Dhillon, and D.P.Kothari, "The Surrogate worth trade-off approach for multi-objective thermal power dispatch problem," Electric Power Systems Research, Vol. 56, pp.103-110, 2000.

[3]. Y.S. Brar, J.S.Dhillon, and D.P.Kothari, "Multi-objective load dispatch by fuzzy logic based searching weightage pattern," Electric Power Systems Research, Vol. 63, pp.149-160, 2002.

[4]. Y.S. Brar, J.S.Dhillon, and D.P.Kothari, "Genetic-fuzzy logic based weightage pattern search for multi-objective load dispatch problem," Asian Journal of Information Technology, Vol. 2, No. 4, pp. 365-373, 2003.

[5]. Gent M.R., and Lamont J.W., "Minimum emission dispatch," IEEE Trans. Power Apparatus and Systems, Vol.6, pp. 2650-2660, 1971.

## VIII. BIOGRAPHIES

**D.P.Kothari** is Professor of Electrical Engineering, Former Head, Centre for Energy Studies, and Former Director In-charge, Indian Institute of Technology Delhi, New Delhi, India. He was Principal, VRCE, Nagpur, India during 1997-98.He was a visiting Fellow in 1982-83 and in 1989 at Royal Melbourne Institute of Technology, Melbourne, Australia. He obtained his B.E., M.E., and Ph.D. degrees from BITS, Pilani, India and has been involved in teaching and research since 1977 at IIT, Delhi. He has published and presented more than 600 papers in prestigious National and international journals and conferences. Dr. Kothari is also a Fellow of the Institution of Engineers (India) and senior member, IEEE. He has co-authored more than 17 books including Power system Engineering, Modern Power system Analysis, Basic Electrical Engineering and Electric Machines. His research interests include power system control, optimization, unit commitment, reliability and energy conservation.

**K.P.Singh Parmar** is a senior faculty member, Electrical Engineering Department, JSS Academy of Technical Education, NOIDA (UP), India. He obtained his B.E. (Hons) from Govt. Engineering College Rewa (MP), India, and M.Tech. from Indian Institute of Technology Delhi, India, and has been involved in teaching and research since 2001.He has taught various subjects of Electrical Engineering including Electric machines, Power system Analysis, Electric drives, Control systems, Electric networks and systems, and Basic system Engineering. He has published and presented several papers in prestigious National and international journals and conferences. Mr.Parmar is a life member, ISTE, India, and member, IETE, India. His research interests include power quality, power system control, energy conservation, unit commitment, generation scheduling, and power management.

**2006 IEEE International Conference on Power Electronic, Drives and Energy Systems**

# Real Time Based PI-like Fuzzy Controller for DC Servomotor

S.G. Kadwane, Swapnil Gupta, B.M. Karan, T Ghose, and Amit Kumar

*Abstract*--Buck converter is highly nonlinear system because of its inherent switching. When a buck converter is cascaded with DC servomotor for controlling the speed, conventional design approach (like PID, Deadbeat...) for speed control can not be applied. This paper proposes design of fuzzy PI like controller for speed control of small dc servo motor cascaded with buck converter. The proposed fuzzy logic controller is first programmed in C language and results are compared with Fuzzy Inference System (FIS) editor in Matlab. This C program is then executed in Code Composer Studio (CCS) for real time implementation on DSP processor TMS320LF2407A. Cascaded buck converter has an added advantage over digital to analog converters, that, it can be principally extended for motors of higher ratings only by changing the component values of buck converter.

*Index Terms*-- DC servomotor, Fuzzy logic control, Speed control.

## I. INTRODUCTION

DUE to its excellent speed control characteristics DC servo motor has been widely used in industry even though its maintenance costs are higher than the induction motor. As a result, speed control of DC motor has attracted considerable research and several methods have evolved. By controlling DC motors accurately, they can overlap many applications of stepper motors. The cost of the control system depends on the accuracy of the encoder and the speed of the processor.

This paper addresses real time DC motor speed control by adaptive PI like fuzzy logic controller using the TMS320LF2407A digital signal processor [1]. The rapid and revolutionary progress in power electronics and micro electronics in recent years has made it possible to apply modern control technology to the area of motor and motion control. The speed of DC motor can be easily and efficiently controlled if we cascade the dc motor with Switched-Mode Power Supply (SMPS). Conventionally, a SMPS is stabilized by monitoring variables such as the output voltage or inductor

---

This prototype is prepared in control system (research) laboratory of Birla Institute of Technology, Mesra, Ranchi (India)

S. G. Kadwane is with EEE Department BIT Mesra, Ranchi (India) (e-mail: sgkadwane@gmail.com).

Dr. B. M. Karan is HOD EEE and Advisor (Projects) BIT Mesra

Dr. T. Ghose is with EEE Department BIT Mesra, Ranchi

Mr. Amit Kumar and Swapnil Gupta are the ME Students of BIT Mesra.

current, and using these measurements to govern the duty ratio of the switching process.

A compensation network is used to process the feedback signal generated by the measurements, shaping the overall frequency response of the converter and providing good transient response characteristics. To design this compensation network, a transfer function, or mathematical model, of the power stage of the converter is required. If this model can be manipulated into a form that is linear and that does not change over time (a linear time-invariant (LTI) model) then many simple and effective classical methods can be used to design the controller. This is conventionally the approach adopted by converter designers.

Unfortunately, the technique of pulse width modulation (PWM) used in switched-mode devices to regulate the output power is neither time-invariant nor linear[8]; the model of the device changes depending on the state of the switching transistor and can also depend on the (time-varying) input voltage and load impedance[8]. Because the switching frequency of the PWM device is typically much greater than that of the output filter, its inter-cycle behavior can be well approximated by a time average and the small output ripple ignored in any analogue model. Likewise, in a digital control scheme, a sampled data model can be derived that samples the measured variables at the end of each switching cycle.

The issue of time-variance is therefore dealt with equally in both cases, neither scheme being any better or worse than the other, and the averaged models work to a good degree of accuracy. The non-linear nature of the PWM scheme can be handled with fuzzy logic, because fuzzy rule based systems has emerged as powerful computing tool for handling nonlinearity and imprecision. A Fuzzy Knowledge Based Systems (FKBCS) for closed loop system enhances the performance, reliability, and robustness of control by incorporating knowledge which cannot be accommodated in the analytic model upon which the design of control algorithm is based, and that is usually taken care of by manual modes of operation, or by other safety and ancillary logic mechanisms [7].

Most of the fuzzy logic controllers are implemented in real time by conventional lookup tables [1]. In this paper Digital Signal Processor TMS320LF2407A was used for loading and running codes of program for real time control. The desired speed of the motor can be adjusted on line by changing the set point speed. This avoids shutting down the program recompiling and executing the same codes each time the parameters are adjusted. The results show considerable improvement in startup response and response to set speed changes.

0-7803-9771-1/06/$25.00 ©2006 IEEE

## II. SPEED CONTROL SCHEME

The speed control scheme is shown in Fig 1. The TMS320LF2407A DSP calculates the present speed by counting the pulses form the incremental encoder attached to motor shaft. The calculated speed is then compared with set point speed. The error and change of error are the two inputs to which are Fuzzified for Fuzzy controller. The set are rules are defined in fuzzy inference engine which gives Fuzzified output depending on the no of rules fired. According to the defuzzified output it varies the PWM duty cycle by changing its Compare register 1 (CMPR1) contents.

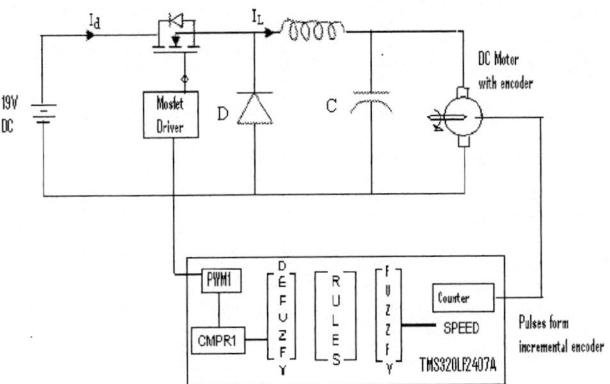

Fig. 1. Speed control scheme.

The TMS320LF2407A DSP has many special features for the control applications [2] & [3]. It has Event Manager that is specially designed for the motor control and motion control applications
. The general-purpose timer1 in EVA (Event Manager A) is used in continuous up/down count mode for the symmetric PWM generation. The general-purpose timers 3 and 4 of EVB are used for the speed measurement purpose.          The PWM output coming from the DSP will be of 3.3V. This voltage is fed to the Mosfet driver for giving PWM input to the gate of Mosfet.

The controlled output voltage from the buck converter is given as the input to the DC motor. According to the given input voltage the motor starts rotating. The incremental encoder connected to the motor gives pulses which are fed to counter in DSP processor.

The Event Manager B (EVB) is used for the speed measurement purpose. The timer 3 is configured for some periodic interrupt. The timer 4 is configured to have the optical encoder pluses as the external clock source and its counter is set to continuous UP count mode. Timer 4 is synchronized with timer 3. The value of the timer 4 counter is taken and is converted to speed. This gives the present speed in rpm. According to the difference between the required speed and the present speed the CMPR1 register content is modified by the Fuzzy controller algorithm

## III. STRUCTURE OF FUZZY LOGIC CONTROLLER (FLC)

The structure of proposed PI like FKBC mainly consists of normalization, fuzzification, membership function

definition, rule base, defuzzification and denormalization. Structure of FLC is designed to control the speed of DC Servomotor [4] using Mamdani-style fuzzy inference system. We defined input and output variables of fuzzy inference system as following.

$$e(k) = y_{sp} - y(k)$$
$$de(k) = e(k) - e(k-1)$$
$$du(k) = u(k) - u(k-1)$$

where $e(k)$ is error at any instant k, $y_{sp}$ is the set point speed, $y(k)$ is the actual speed, $de(k)$ is the change in error in the $k^{th}$ instant, $du(k)$ is the control output that corresponds to duty cycle of PWM. The membership functions for input variables (e, de) are shown in are taken as triangular. Five fuzzy sets namely (NM, NS, Z, PS, and PM) are selected for each input. Membership function of output variable (e) with linguistic terms are defined in Table I. All these Fuzzy Sets are selected as overlapping isosceles triangles. Two input variables, error (e) and change of error (de) are used in this fuzzy logic system. The single output variable (du) is change in duty cycle of PWM output.

The fuzzy output du changes the pulse duty which determines motor speed so that stable speed and torque may be maintained in case of starting or load changing of motor The fuzzy rule base to control the speed of DC Servo Motor is composed of following 25 rules as shown in Table II

### TABLE I
### LINGUISTIC TERMS DEFINITIONS

| Term | Definition |
|------|------------|
| NM | Negative Medium |
| NS | Negative Small |
| ZE | Zero |
| PS | Positive Small |
| PM | Positive Large |

### TABLE II
### RULE BASE DEFINITION

| De \ Error | NM | NS | ZE | PS | PM |
|------------|----|----|----|----|----|
| NM | NM | NM | NM | NS | ZE |
| NS | NM | NM | NS | ZE | PS |
| ZE | NM | NS | ZE | PS | PM |
| PS | NS | ZE | PS | PM | PM |
| PM | ZE | PS | PM | PM | PM |

PI like FKBC consist of following rules [7] of the form
If e is <property symbol > and de is <property symbol > then du is <property symbol >.
The 25 rules can be plotted in a surface shown in Fig 2 which is a plot of output surface for change is output verses error and change in error.

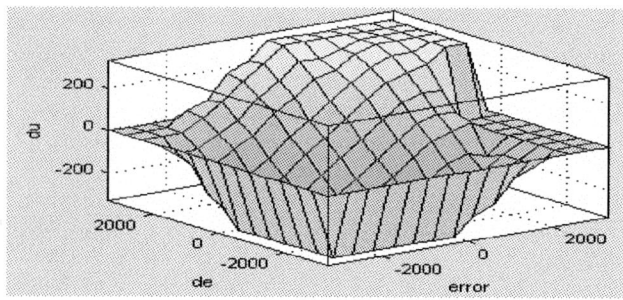

Fig. 2. Surface plot of change in output (du) verses inputs (e and de).

## IV. EXPERIMENTAL RESULTS

The proposed scheme is implemented experimentally with following component specifications:

*DC motor* : 24V, 2.1A, servomotor 8225D

*Buck converter:*

Switching frequency : 20 KHz
Inductor : 3.33 mH, 0.1 ohms
Mosfet : IRF 840
Mosfet driver : TLP 250
Diode : UF5407
Capacitor : 1000μF, 35V
DSP : TMS320LF2407A

The design parameters for buck converter are such that input voltage is 19V dc and maximum output of 10V. Accordingly the results are noted for motor speed up to 4000 RPM. The X-axis represents time and Y axis represents terminal voltage.

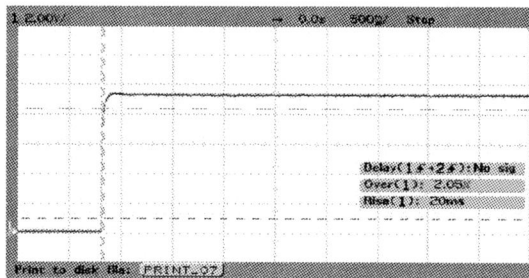

Fig. 3. Startup response with set speed of 4000 RPM.

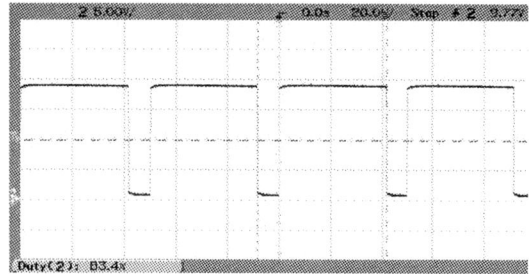

Fig. 4. Duty ratio for the motor speed of 1000 RPM.

Fig 3 shows the duty cycle for operating speed set to 4000 RPM. While Fig 4 shows the startup response with very small overshoot. This result is quite improved with that shown in [9].

## V. CONCLUSION AND FUTURE WORK

This paper proposes the real time controller design for speed control of dc servo motor with cascaded buck converter. The codes are written in CCS and executed on DSP processor. The fuzzy PI like control shows considerable improvement in startup response and response to set point changes in speed. The future work is to further improve the response through dynamic programming by generating the membership functions through genetic algorithm. This approach can be used in wide range of applications in industries

## VI. REFERENCES

[1] Dusan Gleich, Mira Milanovic, Suzana Uran, Franic Mihalic." Digitally controlled buck converter" IEEE power electronics society ISCAS-2004,V944-V947
[2] TMS320LF/LC240XA DSP CONTROLLRS SYSTEM AND PERIPHERALS USER'S GUIDE. Literature number SPRU357B. TEXAS INSTRUMENTS.
[3] TMS320C2x/C2xx/C5x OPTIMIZING C COMPILER USER'S GUIDE. Literature number SPRU018. TEXAS INSTRUMNETS.
[4] Mathew George, Jr. "Fuzzy Logic: An Overview of the Latest Control Methodology" SPRA028. TEXAS INSTRUMENTS.
[5] Y.F. Li and C.C. Lau. "Development of Fuzzy Algorithms for Servo Systems," IEEE Control Systems Magazine. April, 1989, pp. 65-71.
[6] K. Self. "Designing with Fuzzy Logic," IEEE Spectrum. November, 1990, pp. 42-44, 105.
[7] D. Driankov, H Hellendoorn, M. Reinfrank" An Introduction to Fuzzy control"second edition, 1996, Springer. pp145-241.
[8] Collin Holland "DSP improves control of SMPS ". http://www.embedded.com/story/OEG20021014S0002
[9] Sumant.G.Kadwane, Someswara Phani Vepa, B.M.Karan, T Ghose, "Converter Based DC Motor Speed Control Using TMS320LF2407A DSK" IEEE conference on Industrial Electronics and Applications, ICIEA, Singapore, 25-27 May 2006, pp: 117-121.

## VII. BIOGRAPHIES

**S. G. Kadwane** has done his B.E. from Nagpur University and M.E. degree from Pune University. Presently he is Lecturer in Department of Electrical and Electronics Engineering Birla Institute of Technology, Mesra, India and pursuing his PhD form the same and in charge of control system laboratory . His areas of interest are control system design, Neuro-fuzzy applications in power electronic converters through digital signal processor.

**Amit Kumar** has done his Bachelor of Technology (B.Tech) from JNTU A.P, India. and Master of Engineering from B.I.T, Mesra, India. His area of research is adaptive control, neural networks, DSP. He is presently working in LG electronics Noida India.

**Dr. B. M. Karan** is the Head of the Department in Department of Electrical & Electronics Engineering of Birla Institute of Technology, Mesra, India. Presently he is also working as Advisor in the same institute. His areas of interest include soft-computing, digital signal processing and system biology

**Dr. T. Ghose** obtained B.E., M.Tech., Ph.D. degrees from B.I.T., Rajshahi, Bangladesh, University of Calcutta, India, Jadavpur University, India respectively. Presently he is reader in the Department of Electrical & Electronics Engineering of Birla Institute of Technology, Mesra, India. He has published about 30 research papers. His areas of interest include Application of soft computing Techniques, Power system analysis and control

**Swapnil Gupta** has done his Bachelor of Technology (B.E.) from Rewa Engg. College, Rewa, M.P.,India and Master of Engineering from B.I.T, Mesra, India. His area of research is adaptive control, Fuzzy, Genetic Algorithm, DSP. He is presently working in CTS, Kolkota, India

**2006 IEEE International Conference on Power Electronic, Drives and Energy Systems**

# Neural Network Based DSTATCOM Controller for Three-phase, Three-wire System

Bhim Singh, *Senior Member, IEEE,* A. Adya, *Student Member,* A. P. Mittal, *Member, IEEE,* and J.R.P Gupta, *Member, IEEE*

*Abstract--*This paper deals with the design, analysis and simulation of neural network based DSTATCOM controller. Conventional controllers being fixed structure provide optimum performance only over a limited range of operating conditions for which they are designed. Alternate controllers based on fuzzy logic and neural networks are more robust and can be designed to operate well under a wide range of operating conditions. This paper highlights the performance of artificial neural network (ANN) based DSTATCOM controller with respect to PI controllers.

*Index Terms--* ANN controller, DSTATCOM, PI controller, power quality.

## I. INTRODUCTION

INCREASED emphasis on power quality both on industrial and domestic front have led to the development of custom power devices specifically DSTATCOM, DVR and UPQC [1-7]. DSTATCOM is a shunt compensating device having ability to provide voltage regulation as well as load compensation [5-7]. Conventional PI based control is the simplest and the most common type of control used in industrial systems till date. Its structure is fixed; so it may not perform in an optimum manner under varying load conditions. Hence, a need arises for the development of new controllers based on fuzzy logic, artificial neural networks, genetic algorithms[4, 8]. This paper presents a procedure for design of ANN controllers for a three-phase, three-wire system with DSTATCOM compensator. The performance of ANN controller with respect to its PI based counterpart is reported for few power quality problems. Simulation of DSTATCOM with conventional (PI) and ANN based controllers are carried out using standard simulation software such as MATLAB along with its toolboxes.

---

Bhim Singh is with the Department of Electrical Engineering, Indian Institute of Technology, Delhi, Hauz Khas, New Delhi, INDIA (e-mail:bhimsinghr@gmail.com).

A.Adya is with the Department of Electrical and Electronics Engineering, MAIT, IP University, New Delhi, India (e-mail: alkaadya@gmail.com).

A. P. Mittal is with the Department of Instrumentation and Control Engineering, Netaji Subhas Institute of Technology, Dwarka,, New Delhi, India (e-mail: alok@nsit.ac.in).

J.R.P Gupta is with the Department of Instrumentation and Control Engineering, Netaji Subhas Institute of Technology, Dwarka, New Delhi, India (e-mail: jrpg83@yahoo.com).

In the past, attempts have been made on electric power quality problems and many solutions have been suggested to improve power quality in electric distribution systems [5-7]. A number of compensators have been reported for power factor correction, voltage regulation and load balancing using lossless passive elements (L and C) and active elements solid state CSI (current source inverter) and VSI (voltage source inverter). Development of DSTATCOM for three-wire system is reported in the literature in recent years [1, 5, 7] with PI based controller. Many other control techniques such as instantaneous reactive power theory, power balance theory, indirect current control technique [2, 3] etc. have been used. Recently the use of ANN and fuzzy control techniques has increased multifold. ANN based controllers provide extremely fast processing ability as they are based on parallel processing. They have the capability to realize complicated nonlinear mapping from input-output space to output space. Hence, they are being extensively applied to power system problems [4]. This paper presents the modeling and simulation of DSTATCOM used for compensation of three-phase, three-wire supply feeding linear (R-L) lagging power-factor loads. The simulated results are shown for power factor correction, load balancing and voltage regulation.

## II. SYSTEM CONFIGURATION

The system considered consists of a three-phase ac voltage source feeding varying lagging power factor load (R-L load). DSTATCOM compensator can be effectively used for improving the power factor of the system or voltage regulation and load balancing. Fig.1a shows the block diagram representation of the system.

DSTATCOM system consists of a standard three-phase IGBT based VSI bridge with the input ac inductors and a dc bus capacitor to obtain a self-supporting dc bus (Fig.1b). A three-phase ac supply representing the grid feeds power to loads. The load on the system may be linear or non-linear. In the present case, three-phase linear load of lagging power factor is considered for power factor correction and voltage regulation.

## III. CONTROL SCHEME USING PI CONTROLLERS

Indirect current control scheme [2, 3] is used and three-phase reference source currents are computed using dc bus voltage ($V_{dc}$) and three-phase ac voltages ($v_{ta}$, $v_{tb}$, $v_{tc}$). The complete control scheme is represented in Fig.2. The grid supplies active power of the load currents and other active power components of current to maintain the average voltage

---

0-7803-9771-1/06/$25.00 ©2006 IEEE

of the dc bus capacitor of DSTATCOM to a constant value. This second component of the supply current is required to meet switching losses of VSI bridge and the loss of DC bus

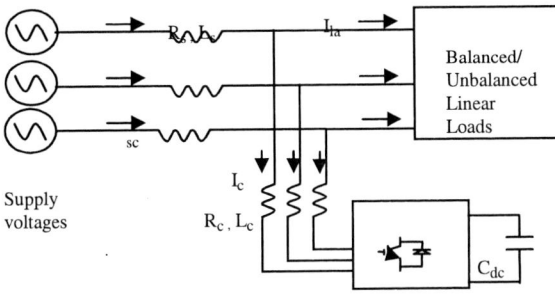

Fig. 1a. Block diagram representation of the system.

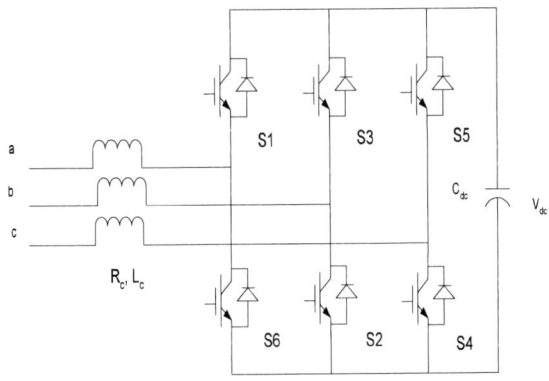

Fig. 1b. Block diagram representation of three leg DSTATCOM.

capacitor under steady state conditions. The amplitude of in-phase reference supply current ($I_{spdr}$) is computed using one PI controller over the average value of DC link voltage ($v_{dc}$). In-phase components of the reference supply currents ($i_{sadr}$, $i_{sbdr}$, $i_{scdr}$) are computed by multiplying peak value of supply currents ($I_{spdr}$) with the in-phase unit current templates ($u_a$, $u_b$, $u_c$) derived from the ac terminal voltages ($v_{tan}$, $v_{tbn}$, $v_{tcn}$). Second PI controller is realized over the ac terminal voltage ($v_{tm}$).

The quadrature component of reference supply current ($I_{spqr}$) is computed using second PI controller over the average value of ac terminal voltage ($v_{tm}$). Quadrature components of the reference supply currents ($i_{saqr}$, $i_{sbqr}$, $i_{scqr}$) are computed by multiplying peak value of supply currents ($I_{spqr}$) with the quadrature unit current templates ($w_a$, $w_b$, $w_c$) derived from the in-phase unit current templates ($u_a$, $u_b$, $u_c$). The total reference supply currents ($i_{sar}$, $i_{sbr}$, $i_{scr}$) are obtained by adding the respective in-phase ($i_{sadr}$, $i_{sbdr}$, $i_{scdr}$) and quadrature components ($i_{saqr}$, $i_{sbqr}$, $i_{scqr}$).

The control scheme with ANN based controller is modified as shown in Fig.4. The ANN based controller is developed with four inputs viz. dc link voltage, load currents in three-phases and one output ($I_{spdr}$) for power-factor correction. Additional input i.e. voltage at PCC is also required for voltage regulation

to obtain both the in-phase and quadrature reference current components ($I_{spdr}$ and $I_{spqr}$). The PI controller over the dc link voltage and ac terminal voltage are replaced by the ANN controller. The structure of ANN based controller is decided by off-line training of over 5000 sets of input-output points to extract the weights and biases of the trained network.

## IV. MODELING OF DSTATCOM SYSTEM

The various steps involved in the control scheme are modeled as follows.

### A. Computation of In-Phase Components of Reference Supply Current

The amplitude of in-phase component of reference supply currents ($I_{spdr}$) is computed using PI controller over the average value of dc bus voltage of the DSTATCOM ($v_{dc}$) and reference dc voltage ($v_{dcr}$) as:

$$I_{spdr(n)} = I_{spdr(n-1)} + K_{pd}\{v_{de(n)} - v_{de(n-1)}\} + K_{id}\,v_{de(n)} \qquad (1)$$

where $v_{de(n)} = v_{dcr} - v_{dca(n)}$ denotes the error in $v_{dc}$ calculated over reference $v_{dcr}$ and average value of $v_{dc}$ at the $n^{th}$ sampling instant. $K_{pd}$ and $K_{id}$ are proportional and integral gains of the dc bus voltage PI controller.

The in-phase components of the reference supply currents ($i_{sadr}$, $i_{sbdr}$, $i_{scdr}$) are computed using the in-phase unit current templates ($u_a$, $u_b$, $u_c$) derived from the ac terminal voltages ($v_{tan}$, $v_{tbn}$, $v_{tcn}$) as:

$$u_a = v_{ta}/\,V_{tm}$$
$$u_b = v_{tb}/\,V_{tm}$$
$$u_c = v_{tc}/\,V_{tm} \qquad (2)$$

where $V_{tm}$ is amplitude of the supply voltage is and it is computed as:

$$V_{tm} = [2/3(v_{tan}^2 + v_{tbn}^2 + v_{tcn}^2)]^{1/2} \qquad (3)$$

The instantaneous values of in-phase component of reference source currents ($i_{sadr}$, $i_{sbdr}$, $i_{scdr}$) are computed as:

$$i_{sadr} = I_{spdr}\,u_a$$
$$i_{sbdr} = I_{spdr}\,u_b$$
$$i_{scdr} = I_{spdr}\,u_c \qquad (4)$$

### B. Computation of Quadrature Components of Reference Supply Currents

The amplitude of quadrature component of reference supply currents is computed using another PI controller over the amplitude of supply voltage ($v_{tm}$) and its reference ($v_{tmr}$). The following equation can be written for computing $I_{spqr}$.

$$I_{spqr(n)} = I_{spqr(n-1)} + K_{pq}\{v_{ae(n)} - v_{ae(n-1)}\} + K_{iq}\,v_{ae(n)} \qquad (5)$$

where $v_{ae(n)} = v_{tmr} - v_{tm(n)}$ denotes the error in $v_{tmn}$ calculated over reference $v_{tmn}$ and average value of $v_{tmnr}$ and $K_{pq}$ and $K_{iq}$ are the proportional and integral gains of the second PI controller.

The two PI controllers are required for evaluating the amplitude of in-phase and quadrature components of the reference supply current.

The quadrature unit current templates ($w_a$, $w_b$, $w_c$) are derived from in-phase unit current templates ($u_a$, $u_b$, $u_c$) [2, 3]. Three-phase quadrature components of the reference supply currents ($i_{saqr}$, $i_{sbqr}$, $i_{scqr}$) are computed using the output of second PI controller ($I_{spqr}$) and quadrature unit current vectors ($w_a$, $w_b$, $w_c$) as:

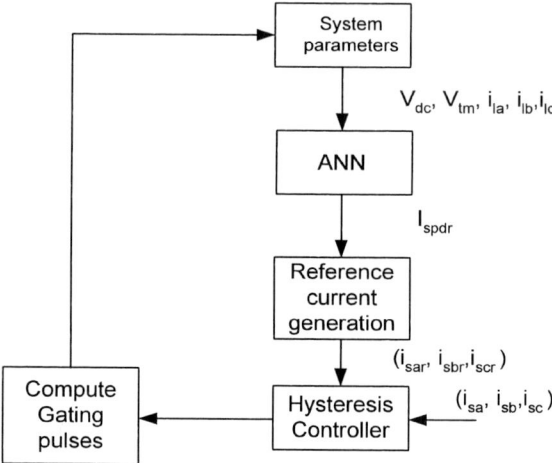

Fig. 4. Block diagram for ANN based control system.

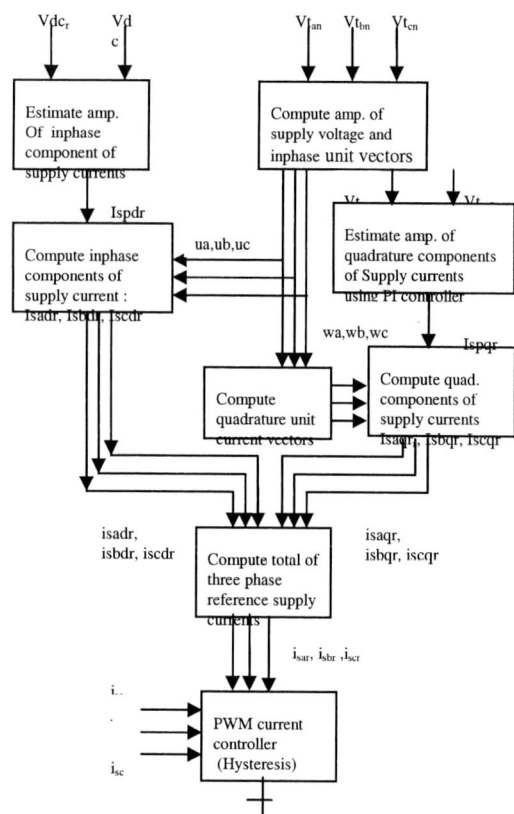

6 Gating signals for IGBTs of DSTATCOM

Fig. 3. Control scheme for DSTATCOM using PI controllers.

$i_{saqr} = I_{spqr} w_a$ ,
$i_{sbqr} = I_{spqr} w_b$ ,
$i_{scqr} = I_{spqr} w_c$          (6)

### C. Computation of Reference Supply Currents

Three-phase reference supply currents ($i_{sar}$, $i_{sbr}$, $i_{scr}$) are computed by adding in-phase ($i_{sadr}$, $i_{sbdr}$, $i_{scdr}$) and quadrature components of supply currents ($i_{saqr}$, $i_{sbqr}$, $i_{scqr}$) as:

$i_{sar} = i_{sadr} + i_{saqr}$
$i_{sbr} = i_{sbdr} + i_{sbqr}$
$i_{scr} = i_{scdr} + i_{scqr}$          (7)

A hysteresis PWM current controller is employed over the reference supply ($i_{sar}$, $i_{sbr}$, $i_{scr}$) and sensed supply currents ($i_{sa}$, $i_{sb}$, $i_{sc}$) to generate gating pulses of IGBTs of the DSTATCOM. These gating pulses control switching on and off of all the devices of voltage source inverter (VSI) bridge used as DSTATCOM.

### V. DEVELOPMENT OF ANN BASED CONTROLLER

A number of ANN architectures have been proposed in literature [4] but the feed-forward ANN is the most commonly

used. In this type of ANN, there a number of neuron elements organized in the form of layers. Scaled data is fed into the network at the input layer. The number of hidden layers and the neurons in each layer depends on the complexity of the problem. In this paper, the steady state error (SSE) is obtained each time by varying the number of neurons in the hidden layer. It has been observed that n=18 in the hidden layer provides the lowest SSE and good convergence accuracy and high convergence rate. It is important to suitably decide the architecture of ANN controller because the errors in outputs from each sample are fed back to the ANN as inputs for next sample's training. Accumulation and propagation of errors is minimized with efficient training and ANN design. Moreover, the spectrum of training is made over a wide range of operating conditions and a set of over 5000 input-output conditions are generated for the system with the PI controller action.

The training algorithm used is Lavenberg- Marquedart (LM) technique which is simple yet more efficient than the most popular back error propagation (BP) technique. Good convergence is the main advantage of LM over the BP and once the ANN was designed and trained, it is tested over a set of operating points not included in the training data.

### VI. SIMULATION AND RESULTS

Both conventional and ANN based controllers have been developed and tested using MATLAB under varying loads for the same system conditions. Simulation results for PI and ANN (Fig.5- Fig.10) controllers are taken for power factor correction (with load change and with load balancing) and voltage regulation along with load balancing.

### A. Power Factor Correction and Load balancing

Fig.5 shows the performance of the three-phase, three-wire system for load R=5ohms, L=7.25mH having 0.8 lagging pf. It is observed that the supply currents are in phase with the supply voltages even though the load currents have lagging pf. The DSTATCOM is able to improve the power-factor of supply currents to unity. Phase 'a' of load is thrown off during

894

t=0.35sec to t=0.45sec. With PI controller, the peak overshoot in dc link voltage is of the order of 70V at time t=0.35sec and the undershoot in dc link voltage is order of 60V observed at t=0.45sec.

The same load conditions when simulated with ANN controller as shown in Fig.6. The ANN controller resulted in peak overshoot in DC link voltage of the order of 40V at time t=0.35sec and the undershoot in DC link voltage is order of 30V observed at t=0.45sec.

It is observed that there is a considerable reduction in peak overshoot and undershoot voltage. Thus, there is tremendous improvement in the transient response with ANN controller. The comparison is illustrated in Table I.

TABLE I
COMPARISON OF CONTROLLER RESPONSES

|  | Overshoot(V) | Undershoot(V) |
| --- | --- | --- |
| Overshoot | 70 | 60 |
| Undershoot | 40 | 30 |

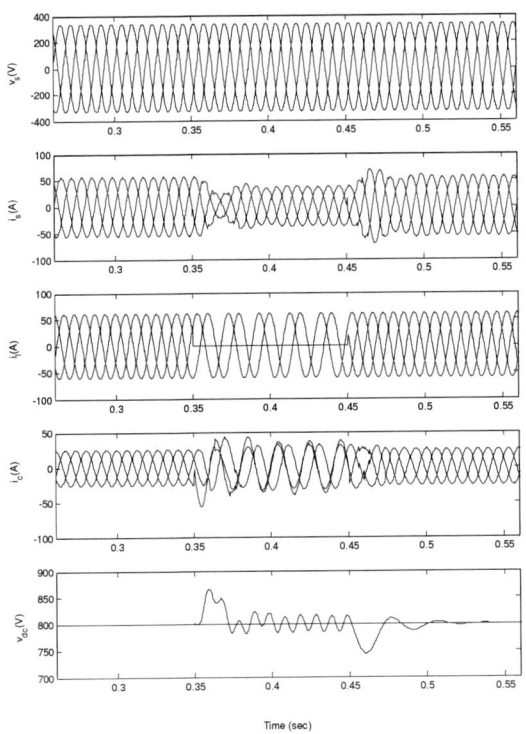

Fig. 5. Performance with PI control for power factor correction and load balancing.

### B. Power factor correction and Load Change

Fig.7 shows the performance of the three-phase, three-wire system for load R=5ohms, L=7.25mH having 0.8 lagging pf. It is observed that the supply currents are in phase with the supply voltages even though the load currents have 0.8 lagging pf.

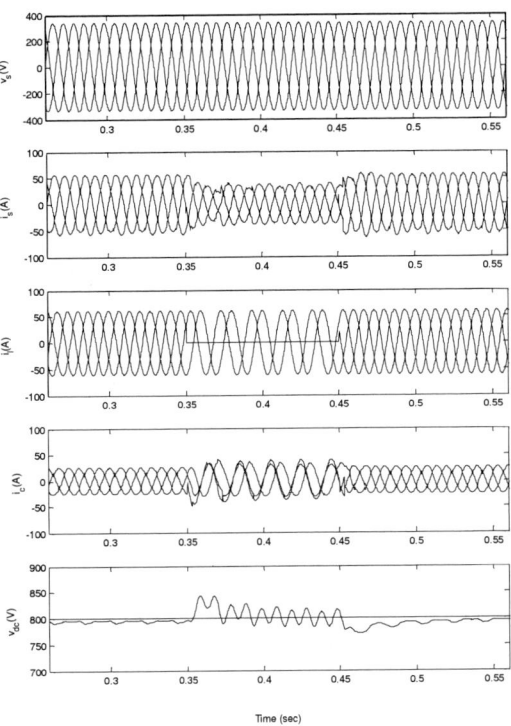

Fig. 6. Performance with ANN controller for power factor correction and load balancing .

The DSTATCOM is able to improve the power-factor of supply currents to unity. The load is now changed to R=6.25ohms, L=9.1mH and then changed back to original load at t= 0.45sec. With PI controller, the peak overshoot in DC link voltage is of the order of 35V at time t=0.3sec and the under-shoot in DC link voltage is order of 30V observed at t=0.45sec.

The same load conditions have been simulated with ANN controller as shown in Fig.8. The ANN controller has resulted in peak overshoot in DC link voltage of the order of 10V at time t=0.3sec and the under-shoot in DC link voltage is order of 20V observed at t=0.45sec. It is observed that there is reduction in peak over-shoot and undershoot voltage and improvement in the dynamic response with ANN controller.

### C. Voltage Regulation and Load Balancing

Fig.9 shows the performance of the three-phase, three-wire system for load R=5ohms, L=7.25mH having 0.8 lagging pf for voltage regulation. It is observed that the supply currents are slightly leading in phase with respect to the supply voltages even though the load currents have lagging pf. The DSTATCOM is able to improve the voltage at the point of common coupling to a reference voltage value of 338V. At t=0.34sec, one phase of the load is thrown off and this is reinserted at t=0.44sec. As a result of this load change, the dc link voltage as well as PCC voltage undergo a change. Fig.9 shows the system response with two PI controllers tuned to

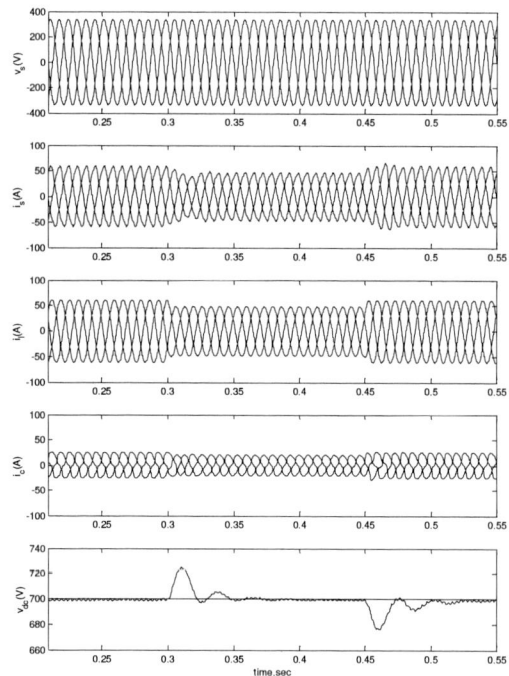

Fig. 7. Performance with PI controller for power factor correction and load balancing.

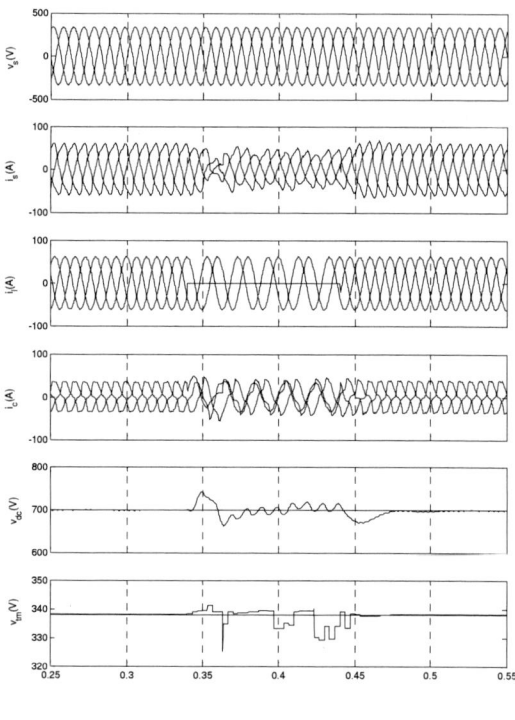

Fig. 9. Performance with PI controller for voltage regulation and load balancing.

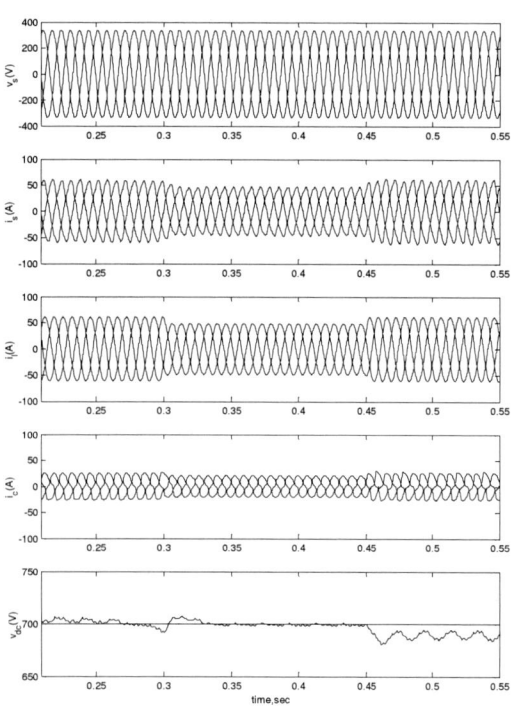

Fig. 8. Performance with ANN controller for power factor correction.

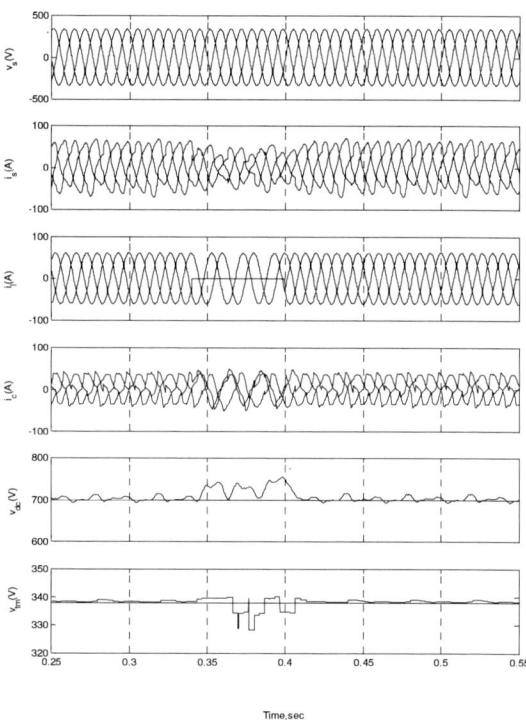

Fig. 10. Performance with ANN controller for voltage regulation and load balancing.

896

Their optimum values. One controller is used for the regulation of dc link voltage and the other for PCC voltage. Fig.10 shows the similar response for voltage regulation as well as load balancing with ANN controller. The ANN based controller has been trained, tested and finally applied for the same system configuration. The load disturbance is applied from t=0.34sec to t=0.4sec. It is observed that the magnitude of overshoot in dc link voltage is slightly reduced as compared to the PI system response; however the under-shoot magnitude is reduced to a large extent. The PCC voltage regulates faster with ANN controller and the extent of voltage dip is reduced. Thus, the system response improves considerably with ANN controller.

## VII. CONCLUSIONS

Simulation results for comparison of performance of two types of controllers viz. ANN and PI controllers for DSTATCOM control used for load compensation have been carried out. Two cases viz. power factor correction with load balancing and power factor correction with load change have been studied in detail. Simulated responses for voltage regulation and load balancing have been presented. The results have shown that ANN controller is found to provide better, more robust control of DSTATCOM as compared to PI controller which performs well only in a limited range of operation. The transient response of the system has improved with ANN based controller resulting in lower amplitude of over-shoot and undershoot.

## VIII. APPENDIX

The three-phase, three-wire system parameters are:
Supply parameters: 230V(rms/ph), 50Hz
Load parameters : $r_l=5\Omega$, $l_l=7.25mH$.
DSTATCOM parameters: $r_c=0.1\Omega$, $l_c=2.5mH$,
$C_{dc}=4700\mu F$, $h_b=0.5A$
PI controller parameters: $K_{pd}=0.33$, $K_{id}=0.08$
$K_{pa}=0.025$, $K_{ia}=0.05$

## IX. REFERENCES

[1] O. Lara and E. Acha, "Modeling and analysis of custom power systems by PSCAD/ EMTDC," *IEEE Transactions on Power Delivery*, vol. 17, No.1, Jan. 2002, pp. 266-272.

[2] B.N. Singh, A. Chandra and K. Al-Haddad, "DSP based indirect-current controlled STATCOM -I Evaluation of current controlled techniques," *IEE Proc. on Electric Power Applications, V*ol. 147-2, March 2000, pp.107-112.

[3] B.N.Singh, A. Chandra and K. Al-Haddad, "DSP based indirect-current controlled STATCOM -I Evaluation of current controlled techniques,"

*IEE Proc. on Electric Power Applications, V*ol. 147-2, March 2000, pp.113-118.

[4] Y.Zhang, O.P. Malik, G.P.Chen, "Artificial neural Network Power System Stabilizers in multi-machine power system environment," *IEEE Transactions on Energy Conversion*, vol. 10, No.1, March 1995, pp. 147-154.

[5] T. J. E. Miller, Reactive Power Control in Electric Systems, Toronto, Wiley, 1982.

[6] N.G.Hingorani and L.Gyugyi, *Understanding FACTS*, IEEE Press, Delhi, 2001.

[7] A.Ghosh and G.Ledwich, *Power Quality Enhancement Using Custom Power Devices*, Kluwer Academic Publishers, London, 2002.

[8] B. Singh, V. Verma, J. Solanki, A. Chandra and K.Al Haddad, "Neural network controlled power conditioner with battery energy storage feature for isolated offshore power system" *Proc. 2004 IEEE PCIC'04 Conf.*, pp. 127-133.

[9] G. Reed, M. Takeda, F. Ojima, A. Sidell, R. Chevus and C. Nebecker, "Application of 5Mva, 4.16kV DSTATCOM for voltage flicker at Seattle Iron and Metals," in *Proc.2000 IEEE Power Engineering Society Summer Meeting*, Vol.3, pp. 1605-1611.

## X. BIOGRAPHIES

Bhim Singh was born in Rahamapur, India, in 1956. He received the B.E. (Electrical) degree from the University of Roorkee, Roorkee, India, in 1977, and the M.Tech. and Ph.D. degrees from the Indian Institute of Technology (IIT), New Delhi, India, in 1979 and 1983, respectively. In 1983, he joined the Department of Electrical Engineering, University of Roorkee, as a Lecturer. He became a Reader there in 1988. In December 1990, he joined the Department of Electrical Engineering, IIT, New Delhi, India, as an Assistant Professor. He became an Associate Professor in 1994 and a Professor in 1997. He is recipient of JC Bose and BK Bose awards of IETE. His fields of interest include power electronics, electrical machines and drives, active filters, static VAR compensators, analysis and digital control of electrical machines. Prof. Singh is a Fellow of the Indian National Academy of Engineering (INAE), Institution of Engineers (India) (IE(I)), and Institution of Electronics and Telecommunication Engineers (IETE) and a Life Member of the Indian Society for Technical Education (ISTE), System Society of India (SSI), and National Institution of Quality And Reliability (NIQR) and Senior Member of Institute of Electrical and Electronics Engineers (IEEE).

Alka Adya graduated from Delhi College of Engineering in 1996 with BE degree, MTech in Power Sytsems from IIT Delhi in 2001 and submitted her Ph.D.in University of Delhi in 2006. He is currently working as a Sr. Lecturer in MAIT Delhi. Her research interests include control of power systems, custom power devices, power quality.

A. P. Mittal graduated from M.M.M Engg. College, Gorakhpur, ME from University of Roorkee in 1980 and Ph.D. from IIT Delhi in 1991. He is currently working as a Professor in Netaji Subhas Institute of Technology, Delhi. He is a Fellow of the Institution of Engineers (India). His research interests include active filters, FACTS, electric drives.

J.R.P. graduated from Muzaffarpur Institute of Technology and received his BSc.(Engg).Degree in 1972 and Ph.D. from University of Bihar in 1983. He is currently working as a Professor in Netaji Subhas Institute of Technology, Delhi. He is a Fellow of the Institution of Engineers (India). His research interests include power electronics, power quality, electric drives.

# Analysis of the Influence of Control Parameters on Wind Farm Output: a Sensitivity Analysis using ANN Modelling

E. Fernandez, and M. Carolin Mabel

*Abstract*--Wind energy planners are interested in studies that highlight the impact of control input parameters on the output of wind farms. Yet, there are few studies highlighting such investigations. It has been observed that wind energy programs are being actively pursued in most developing countries. In India, one of the states that is actively involved in wind energy power generation programs is Tamil Nadu. Within this state, Muppandal area is one of the identified regions where wind farms concentration is being encouraged.

*Index Term*-- ANN models, Impact Assessment, Sensitivity Analysis, Wind farms, Wind Power Generation.

## I. INTRODUCTION

THE state of Tamil Nadu (India) presently has an installed wind turbine capacity of 2507.3 MW in the private sector and 19.4 MW as government demonstration projects. The total installed wind energy capacity in Tamil Nadu as on 31st December 2005 is thus 2526.7 [1]. The wind farms are primarily located in the Muppandal and the Coimbatore areas. More than 50% of the total capacity is installed in the Muppandal area, the rest being located in Coimbatore district. The growth of wind farm capacity is expected to be of the order of 100 MW/year the coming years, showing the importance being given to such energy projects in Tamil Nadu. Annual average wind speeds of 18-20 km/h are found, and the wind season starts in March and continues until November. Annual capacity factors of 20-22% have been observed. The present capacity of conventional power supply is 9591.6 MW from thermal, nuclear and hydro power stations. The power system is connected to the other three states (Karnataka, Andhra Pradesh and Kerala) in the Southern region through 400 kV lines. Since the penetration of wind power generation is about 26% in respect to the total power generation of the state, it is clear that wind power is and will continue to be an important renewable energy resource for the state. Wind energy assessment is therefore an important task for design of additional wind farm capacity that is likely to be required in the near future.

The authors are with Indian Institute of Technology Roorkee, Roorkee, Uttranchal – 247667, INDIA (e-mails: eugfefee@iitr.ernet.in, carolin_mabel@yahoo.co.in).

## II. THE PROBLEM INVESTIGATED

A number of controlling variables are associated with wind power generation. Most studies on wind farm energy generation focus on the output of wind farms on a statistical basis without any attempt to extract meaningful information for improving wind farm productivity. It has been felt by the authors the statistical database normally available at wind farms can be applied to study the impact of certain identified control variables on the system performance, and thereby gain valuable insight into the mechanics of wind power generation. As research information on this issue is somewhat scarce, it was felt appropriate to carry out an exploratory investigation. Hence, the present study addresses itself to this issue.

While various methodologies can be employed for this purpose, it is felt that the Artificial Neural Networks (ANN) methodology can be usefully adapted for the purpose. ANN's have been tried for wind energy related studies and several reports are available in the literature [2-13] However, there are virtually no reported studies that attempt to apply ANN's for sensitivity analysis of a few control input parameters on the wind energy output of wind farms. The present paper is therefore an attempt in this direction.

Although the study takes up wind farms located in Muppandal district of the state of Tamil Nadu, it is believed that the findings are general and are applicable to all wind farms regardless of location.

## III. CHOICE OF CONTROL VARIABLES

Various control variables can be identified for investigating the performance of wind farms. Some of the many possible variables include:

1) Wind speed
2) Latitude of wind site
3) Height above mean sea level
4) Hub height of wind turbines
5) Generation hours
6) Temperature and Humidity
7) Installed capacity
8) Reactive energy consumption
9) Installed capacitor kVAR

A practical study needs to identify a few selected variables capable of influencing the wind farm output. This is necessary in order to reduce the amount of input data needed to model the wind farms with acceptable accuracy. After some preliminary examinations relating to the availability of field data of input variables, it was eventually decided to select the following variables for the study:

1) Monthly Generation Hours
2) Installed kVAR in wind farm
3) Reactive energy (kVARh) consumed per month
4) Installed kW Capacity of the wind farm

Historical data pertaining to 5 wind farms in Muppandal district was collected in relation to the four variables so chosen. The data spans period of three years (from 2003 to 2005, both inclusive). The collected data was then processed and applied for training and subsequent testing of an ANN Model. Table I gives the details of the wind farms included in the study while Table II provides the details of the ANN model used for simulations using MATLAB toolbox.

TABLE I
DETAILS OF WIND FARMS IN MUPPANDAL DISTRICT (TAMIL NADU) USED FOR THE STUDY

| Sl. No | Capacity of Wind Farm | No. of Wind Turbines Installed |
|---|---|---|
| 1 | 1.0 | 4 |
| 2 | 1.75 | 7 |
| 3 | 2.25 | 9 |
| 4 | 8.08 | 31 |
| 5 | 11.50 | 40 |

TABLE II
DETAILS OF THE ANN MODEL USED FOR THE PRESENT STUDY

| Sl. No. | Feature | Details |
|---|---|---|
| 1. | Architecture of the ANN Model | Three layer network * 4 input nodes * single hidden layer of 3 nodes * a single node output |
| 2. | Training data set patterns used | 130 |
| 3. | Testing patterns used | 44 |
| 4. | Training error | 6.5E-03 |
| 5. | Testing error | 1.34E-02 |

IV. VALIDATION OF THE ANN MODEL

Fig. 1 shows the differences in the actual and predicted values for the testing patterns used. A good agreement has been observed, which is reflected in the low model testing error of $1.3 \times 10^{-2}$.

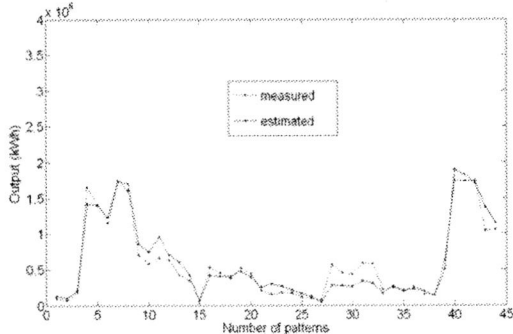

Fig. 1. Accuracy of the ANN model.

The developed and tested ANN model is then applied for investigating the sensitivity analysis of the four control variables. For the base case, the mean values of the control parameter have been taken as reference values. The sensitivities are expressed as percentage variations of the output (kWh/month) as a function of the percentage variations of the control variables. The results are then discussed for obtaining relevant conclusions.

V. RESULTS AND DISCUSSIONS

The developed ANN model was applied to carry out a sensitivity analysis in regard to the impact of the control variables on the wind farm output. The mean values of each of the control variable, when applied to the model yield an output taken as the base or reference value for the sensitivity analysis. Using this approach, the sensitivities of the output to each of the control input variables are obtained ranging over a ± 20 % variation. Fig. 2 shows the sensitivities graphically.

The sensitivities of the individual variables chosen for the analysis are discussed as follows:

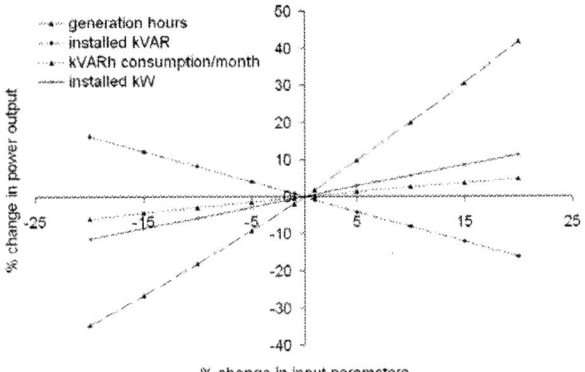

Fig. 2. Sensitivity analysis results for the control variables.

899

## A. Impact of monthly generation hours on wind farm output

It is seen from the variations indicated in Fig. 2 that the generation hours have the most significant impact with an approximate gradient of +2.0. The change in the percentage output is almost linear with respect to that for the variable .A 20 % decrement in the monthly generation hours results in a fall of 34.55% in the monthly energy output. Likewise, an increase of 20% in the monthly generation hours shows an increase of about 41.89% in the output. Thus on an average, the approximate impact of the monthly generation hours is quite significant and needs to be well appreciated in the estimation of the wind energy output.

## B. Impact of installed kVAR on wind farm output

A decrement of 20% in this variable result in a rise of 16.278% in the monthly wind energy output, while an increase of 20% in the installed kVAR shows a fall of about 16.194% It may be observed that the ANN model shows an inverse relation between the installed kVAR capacity and the energy output of the wind farm. The possible reason for this observation is that the wind farms in Tamil Nadu are required by grid regulations to maintain a kVAR capacity of at least 50 %. Presumably, the wind farm is over compensated in regard to lagging reactive power correction and operates at high power factor (lead). In this situation, any decrease in the kVAR capacity implies an increase in the power factor magnitude leading to a rise in generated power. A further increase in the kVAR capacity will, however, lead to possibly a worsening of the already leading power factor power magnitude which would result in a lower power output. (Since the maximum power is generated at unity power factor). An approximate gradient of −1.0 appears to define the observed sensitivity of the variable.

## C. Impact of Reactive kVARh consumed on wind farm output

The reactive power consumed by the wind farms will be dependent on the design of the machines and the associated power converters. Presently in Muppandal district, most of the wind farms make use of grid excited induction generators, which draw a large reactive power from the grid. This leads to a weakening of the grid voltage level and forced shutdowns during critical load demand periods. It is noticed that this variable does not have a strong impact on the output-generated energy. The fall in output for a 20% decrement in the variable is 5.88% while an equivalent increase in the consumed kVARh leads to a rise of 4.81% in the output. The association of the variable with the output is a positively related one having an approximate gradient of + 0.25.

## D. Impact of kW installed capacity of wind farm on wind farm output

This variable is positively correlated with the wind farm output, as would perhaps be evident. The greater the size of the wind farm, the greater will be the expected energy output. However, the impact of this variable on the output is seen to be lesser that that due to the generation hours. For a 20 % decrement in the variable, the expected output fall by about 11.45% while an equivalent increase in the variable results in an increase of almost the same magnitude (11.25%). The variation is almost linear with a gradient of approximately +0.5.

## VI. CONCLUSIONS

A study was carried out to determine the impact of a set of selected control input variables on the output of wind farms. The study involved the collection of field data of 5 wind farms in the Muppandal district of Tamil Nadu, a concentrated area of wind power generation in the state. Data collected over a period of three years was analyzed for sensitivity of the control variables on the output. To do so an ANN model was developed with high prediction accuracy. The analysis shows that the number of generation hours in the wind farms is the most significant control variable in affecting the wind farm output, and, must therefore be maximized for enhancing wind energy output The installed capacity of the wind farm is also an important parameter. The remaining parameters do not affect the output significantly.

## VII. REFERENCES

[1] Central Power Authority (CEA) report statistics, New Delhi, Government of India, 2006.

[2] M.C. Alexiadis, P.S. Dokopolous, H.S. Sahsamauoglou, and I.M. Manaousaridis, "Short Term Forecasting of Wind Speed and Related Electric Power", *Solar Energy*, vol. 63, no. 1, pp 61-68, 1998.

[3] D.A. Becharakis, and P.D. Sparis, "Simulation of Wind Speeds at Different Heights using Artificial Neural Networks," *Wind Engineering*, vol. 24, no. 2, pp. 127-136, 2000.

[4] D.A. Becharakis, and P.D. Sparis, "Correlation of Wind Speeds between Neighbouring Measuring Stations", IEEE Trans. Energy Conversion, vol. 19, n(2), pp. 400-406, June 2004 .

[5] Ertugul, C, Arcahhoglu, E, Cavusogla, A.,and Akbiyi B(2005).," A Classification Mechanism for Determining Average Wind Speed and Power in Several Regions of Turkey Using Artificial Neural Network", Renewable Energy, Vol. 30, pp 227-239.

[6] Flores P., Tapia A., and Tapia G (2005), "Application of a Control Algorithm for Wind Speed Prediction and Active Power Generation." Renewable Energy, Vol. 30. pp.523-536.

[7] Giraud F and Salameh, Z.M. (1999) "Neural Network Modelling of the Gust Effects on a Grid Interactive Wind Energy Conversion System with Battery Storage", Electrical Power System Research, Vol. 50(3), pp 155-161.

[8] Kariniotakis G.N., Stavrakakis G.S. and Nogaret, E.F (1996), "Wind Power Forecasting Using Advanced Neural Network Models", IEEE Transaction on Energy Conversion, EC Vol. 11(4), Dec 1996.

[9] Li S., Wunsch, D.C., O'Hair, E. and Giesselmann, M.G. (2001a), "Using Neural Networks to Estimate Wind Turbine Power Generation", IEEE Transaction on Energy Conversion EC Vol. 16(3), pp. 276-282, Sep 2001.

[10] Li S., Wunsch, D.C., O'Hair E. and Giesselmann M.G (2001b), "Comparative Analysis of Regression and Artificial Neural Network Models for Wind Turbine Power Curve Estimation", ASME Journal of Solar Energy Engineering, Vol. 123, Nov 2001.

[11] Mohandes, M.A., Rehman, S. and Halawani, T.O. (1998), "A Neural Network Approach for Wind Speed Prediction", Renewable Energy, Vol. 13(3), pp. 345-354.

[12] Sorensen, P. et al (2000) "Power Quality and Integration of Wind Farms in Weak Gridsin India", Report of Riso National Laboratory Roskilde, Denmark.

[13] Stefses A., (2000) "A Comparison of Various Forecasting Technologies Applied to Mean Hourly Wind Speed Time Series", Renewable Energy, Vol. 21, pp. 23-35.

## VIII. BIOGRAPHIES

E. Fernandez did his post graduation in Power Systems Engineering from Motilal Nehru Regional Institute of Technology, Allahabad (India) in 1998 and his PhD in Energy Planning from the Indian Institute of Technology, Roorkee (India) in 2004. He served for a brief period from 1986-1990 as Assistant Professor at the Electrical Engineering Department, Faculty of Engineering, University of Jodhpur (Rajasthan, India). Presently he is an Assistant Professor in the Electrical Engineering Department at the Indian Institute of Technology, Roorkee. Dr. Fernandez's research interests include: Renewable Energy System, Energy Planning, Small Scale Electric Power Generation Systems, and Grid connected Wind Energy Conversion Systems. He has a number of research publications in these areas in Conferences.

**M. Carolin Mabel** received the B.E. degree in Electrical and Electronics Engineering from Manonmaniam Sundaranar University, Tamilnadu, India in 1994 and M.E. degree in Power Systems from Annamalai University, Tamilnadu, India in 1998. She worked as Lecturer in C.S.I. Institute of Technology, Tamilnadu, India till June 2004.

Currently she is a Research Scholar in the Department of Electrical Engineering, Indian Institute of Technology Roorkee, Uttranchal, India. Her fields of interest include neural network applications, wind energy studies, and power system economic dispatch.

# An Advanced Control Scheme for Micro Hydro Power Plants

M.Hanmandlu, *Member, IEEE*, Himani Goyal, and D.P.Kothari, *Senior Member, IEEE*

*Abstract*--Micro hydropower plants are emerging as a major renewable energy resource today as they do not encounter the problems of population displacement and environmental problems associated with the large hydro power plants. However, they require control systems to limit the huge variation in input flows expected in rivulets over which these are established to produce a constant power supply. This paper proposes an electric servomotor as a governor for a micro hydro power plant especially those plants that are operated in isolated mode. An advanced controller is developed combining four control schemes for the control of the governor following the concept that the control action can be split up into linear and non linear parts. The linear part of this controller contains an adaptive Fast Transversal Filter (FTF) algorithm and normalized LMS (nLMS) algorithm. The non-linear part of the controller incorporates Fuzzy PI and a neural network. The new controller has a superior performance over other control schemes.

*Index Terms*-- Frequency Control, Fuzzy Control, Least Mean Square Methods, Load Modeling, Neural Networks, Nonlinear Systems, Power Generation, Proportional Controller, Servomotors, Stochastic Processes.

## I. INTRODUCTION

IN an electric power system, consumers require uninterrupted power at rated frequency and voltage. To maintain these parameters within the prescribed limits, controls are required on the system. Voltage is maintained by the control of excitation of the generator and frequency is maintained by eliminating the mismatch between generation and load demand. Since frequency is an indicator of the energy balance in the system, the problem of maintenance of constant frequency is analyzed in this paper. A novel scheme is proposed for the speed control of hydro turbines. This scheme regulates the flow of water being fed to the turbine in accordance with the load perturbations thereby maintaining the frequency of the system at the desired level.

### A. Conventional Governors

Conventional Governor Systems can be classified as mechanical-hydraulic governors, electro hydraulic governors or mechanical types. Mechanical hydraulic governors are

---

[1]

[1] Authors acknowledge the financial support received from DRDO, New Delhi.
M.Hanmandlu is with Department of Electrical Engg. IIT Delhi, India, email: mhmandlu@ee.iitd.ac.in,
Himani Goyal and D.P.Kothari are with Centre for Energy Studies, IIT Delhi,India,email:goyalhimani@yahoo.com,himani.g@rediffmail.com

sophisticated devices which are generally used in large hydro power systems. They require heavy maintenance and are expensive to install, making their usage in micro hydro power plants uneconomical. Electro hydraulic governors are complex devices needing precision design and are expensive. Mechanical governors incorporate a massive fly ball arrangement and usually do not provide flow control. They require an elaborate set of complex guide vanes, inlet valves and jet deflectors. Hence conventional governing systems because of their cost and complexity are not ideally suited for installing at the isolated areas that are not grid connected. The current trend is therefore to use load side regulation.

### B. Electronic Load Controller for Water Turbines

Electronic load controllers govern the turbine speed by adjusting the electrical load on the alternator. As lights and electrical appliances are turned on and off, the electronic controller varies the amount of power fed into a 'ballast' load. The load controller therefore maintains a constant electrical load on a generator in spite of changing user loads. This permits the use of a turbine with no flow regulating devices and the governor control system. Load controllers however waste precious energy that can be used gainfully. Also they do not carry out flow control implying that the mineral rich water is made to spill away which could have been diverted at high head for irrigation purposes. Henderson [18], [38], [40] describes development of a microprocessor based electronic load governor for micro hydroelectric power plants. The governor maintains the speed of the set by adjusting an electrical ballast load connected to the generator terminals, thus balancing the total electrical load torque with the hydraulic input torque from the turbine.

### C. Servomotor as a Governor

In the proposed control system an electric servomotor is used as a governor [19]. An electric servo motor is a precision electric motor whose function is to cause motion in the form of rotation or linear motion in proportion to a supplied electrical command signal. Type Zero servomechanism is used in the proposed system. A feedback control system of Type Zero is generally referred to as a regulator system. Such systems are designed primarily to maintain the controlled variable as constant at a certain desired value despite disturbances. Here the controlled variables are the frequency and the turbine power. The electric servo motors are preferable for the control of micro hydro power systems as they have a simple design, require less maintenance and are less expensive than conventional governors.

### D. Literature Survey

Glattfelder [12] has advocated use of a compensating element in addition to a speed regulator as a speed governor for low head hydro units in an isolated grid. Hagihara et al. [13] consider the effect of derivative gain and other governor parameters on the stability of a hydraulic turbine supplying an isolated load. Pereira et al. [14] propose the addition of eddy current brake to help governor achieve greater degree of control. Schniter et al. [15] make use of the adjustable blade angles to achieve maximum operating efficiency at a given load. Tano and sannomiya [31] have developed an integrated digital control panel for control, protection and supervisory functions. Malik et al. [16] recommend the use of microprocessor based governor for frequency measurement and control and Thappar et al. [32] present the distributed microprocessor based control for stable frequency and voltage output. Djukanovic et al. [17] propose the non linear multivariable control using adaptive network based fuzzy inference system (ANFIS)

### E. Need for a New Governing System

The new Governing System should be relatively inexpensive, simple to operate and easy to maintain by incorporating the flow control. Therefore, we will explore a set of advanced control schemes for better load frequency control.

In this work, we will discuss how the control action can be split up into linear and non linear components as this will allow the selection of appropriate control schemes for the micro hydro power plant. In the light of this concept we will explore different control schemes consisting of combinations of linear and nonlinear parts to come out with a scheme that meets the desired performance in terms of peak overshoot and steady state error. We will also show that the chosen scheme is stable. Thus this work will pave the way for the design of a suitable controller for any type of plant by an appropriate choice of control components.

The organization of this paper is as follows: Section II gives formulation of the state space model for the hydro plant. Some of the advanced control schemes are discussed and a new controller is proposed based on the concept of splitting up of control action in Section III. Performance of this new controller is compared with other control schemes to illustrate its effectiveness in Section IV. Conclusions are drawn in Section V. Data assumed for the model is given in Appendix.

## II. FORMULATION OF PLANT MODELS FOR MHP PLANT

The approximate transfer function for the servo motor based governor is considered for the analysis and is given by

$$G(s) = \frac{1}{(1+sT_1)(1+sT_2)} \qquad (1)$$

where, $T_1$ = mechanical time constant, $T_2$ = electrical time constant. In addition, unity gain is applied as a feedback. A PI controller with the following transfer function is superimposed on the servomotor based governor:

$$G(s) = K_{pl} + \frac{K_i}{s}$$

Where $K_{pl}$ = Proportional constant, $K_i$ = Integral constant

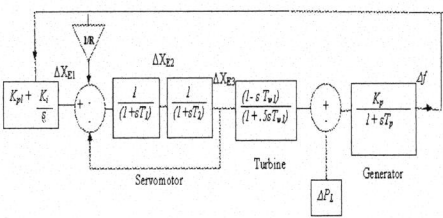

Fig. 1. Model of a Micro hydro power plant using Servomotor as a Governor.

The block diagram of micro hydro power (MHP) plant is shown in Fig.1. This plant can be reduced to a simpler representation as in Fig. 2 by employing partial fractions.

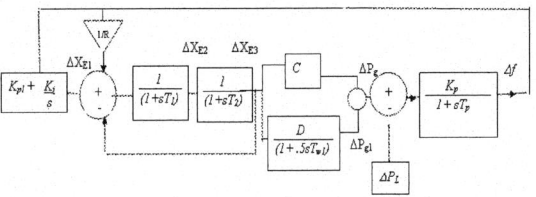

Fig. 2. Simplified representation using Partial Fractions.

The differential equations for the governor can be written as:

$$\frac{d}{dt}\Delta X_{E1} = K_i \Delta f + K_{pl}\frac{d}{dt}\Delta f \qquad (2)$$

$$\frac{d}{dt}\Delta X_{E2} = \frac{-1}{T_2}\Delta X_{E2}\frac{-1}{T_2}[\Delta X_{E3} + \frac{1}{R}\Delta f - \Delta X_{E1}] \qquad (3)$$

$$\frac{d}{dt}\Delta X_{E3} = \frac{-1}{T_3}\Delta X_{E3} + \frac{1}{T_3}\Delta X_{E2} \qquad (4)$$

The change in power generated is given by:

$$\Delta P_g = C[\Delta X_{E3}] + \Delta P_{g1} \qquad (5)$$

The differential equations for the hydro turbine are as follows:

$$\frac{d}{dt}\Delta P_{g1} = \frac{-1}{0.5T_{w1}}\Delta P_{g1} + \frac{D}{0.5T_{w1}}[\Delta X_{E3}] \qquad (6)$$

$$\frac{d}{dt}\Delta f = \frac{-1}{T_p}\Delta f + \frac{K_p}{T_p}[\Delta P_{g1} + C.\Delta X_{E3} - \Delta P_L] \qquad (7)$$

In view of the above, Eqn. (2) becomes

$$\frac{d}{dt}\Delta X_{E1} = K_i\Delta f + K_{pl}[\frac{-1}{T_p}\Delta f + \frac{K_p}{T_p}[\Delta P_{g1} + C.\Delta X_{E3} - \Delta P_L]] \qquad (8)$$

The system dynamics described by the above set of differential equation (2)-(8) appears in the state space form as:

$$\underline{\dot{X}} = [A]\underline{X} + [B]\underline{\mu} + [\Gamma]\underline{p} \qquad (9)$$

where X, $\mu$ and p are the state, control and disturbance vectors respectively and [A], [B] and [Γ] are constant matrices of appropriate dimension. The Eqn. (9) is given by:

903

$$
\begin{bmatrix}
\dot{\Delta f} \\[4pt]
\dot{\Delta P_{g1}} \\[4pt]
\dot{\Delta X_{E3}} \\[4pt]
\dot{\Delta X_{E2}} \\[4pt]
\dot{\Delta X_{E1}}
\end{bmatrix}
=
\begin{bmatrix}
\dfrac{-1}{T_p} & \dfrac{K_p}{T_p} & \dfrac{C\,K_p}{T_p} & 0 & 0 \\[8pt]
0 & \dfrac{-1}{.5T_{W1}} & \dfrac{D}{.5T_{W1}} & 0 & 0 \\[8pt]
0 & 0 & \dfrac{-1}{T_2} & \dfrac{1}{T_2} & 0 \\[8pt]
\dfrac{-1}{RT_1} & 0 & \dfrac{-1}{T_1} & \dfrac{-1}{T_1} & \dfrac{1}{T_1} \\[8pt]
\dfrac{K_I K_{p1}}{T_p} & \dfrac{K_p K_{p1}}{T_p} & \dfrac{K_p K_{p1} C}{T_p} & 0 & 0
\end{bmatrix}
\begin{bmatrix}
\Delta f \\[4pt]
\Delta P_{g1} \\[4pt]
\Delta X_{E3} \\[4pt]
\Delta X_{E2} \\[4pt]
\Delta X_{E1}
\end{bmatrix}
+
\begin{bmatrix}
\dfrac{-K_p}{T_p} \\[6pt]
0 \\[6pt]
0 \\[6pt]
0 \\[6pt]
\dfrac{-K_p K_{p1}}{T_p}
\end{bmatrix}
\Delta P_L
$$

(10)

In the solution of Eqn. (10), we take C=-2 and D=3.

## III. ADVANCED CONTROL

PI controllers provide a good control action for micro hydro power plants. However the control action can be further improved using advanced control methods. In these methods, the representation and adaptation of information are the key issues to reduce complexities and to eliminate the heuristic procedures in process control. Moreover, good transient and steady state responses for different operating points of the processes can be achieved. These advanced techniques include the Fuzzy PI, the nLMS algorithm, the FTF algorithm and a Neural Network (NN). Here NN incorporates an adaptive algorithm, which is a combination of nLMS and gradient descent algorithms. Fuzzy techniques drastically reduce the development time and cost for the synthesis of nonlinear controller for dynamical systems. Triangular membership functions are generally preferred. The nLMS algorithm is used in many applications where the input signals are subject to widely fluctuating power levels causing gradient noise amplification, which in turn affects the stability, convergence and steady-state properties of the LMS algorithm. The advantage of nLMS in micro hydro power plant is that it adapts the gain to its optimal value, resulting in fast, stable convergence. The FTF algorithm, a well-known tool in the field of signal processing, is also applied for control. In FTF algorithm, projection techniques and vector space methods are used to derive a fixed order transversal least squares filter. A new controller is now proposed using a combination of the above mentioned techniques with a view to minimize the peak overshoot and achieve the early settling time by tapping the advantages of individual components. The schematic of the proposed controller is shown in Fig. 3.

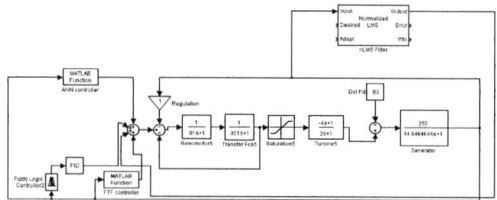

Fig. 3. The model of the proposed scheme.

### A. Different control combinations

The new controller incorporates the advanced control techniques to cater to the plant linearities and nonlinearities. The underlying concept is that the non-linear part of the

controller operates on non-linear error signals to make them adhere to the set point and the linear part of the controller tries to maintain the linearity between the two points. This concept advocated by Srivastava in her thesis [34] is adapted here to justify the composition of the new controller. From this work it is possible to choose appropriate linear and nonlinear components of controller such that the desired performance is achieved. Moreover, it has been noticed that each technique retains its advantage and contributes to the overall performance of the controller. Keeping this in mind, our control scheme shown in Fig. 4 is devised. In this, nLMS and FTF constitute the linear part and Fuzzy PI and NN constitute the non linear part of the controller.

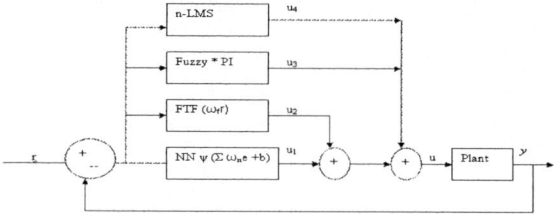

Fig . 4. The block diagram of the proposed Scheme of control.

The combined output of the controller is

$$u(k) = u_1(k) + u_2(k) + u_3(k) + u_4(k) \qquad (11)$$

where $u_1(k)$ is the output of NN, $u_2(k)$ is the output of FTF, $u_3(k)$ is the output of Fuzzy *PID and $u_4(k)$ is the output of n-LMS.

(a) *Scheme of control employing FTF and Neural Networks*

A combination of FTF algorithm and a neural network from [19]-[21] is shown in Fig. 5. The weights of the neural network are adjusted by n-LMS and gradient descent algorithms.

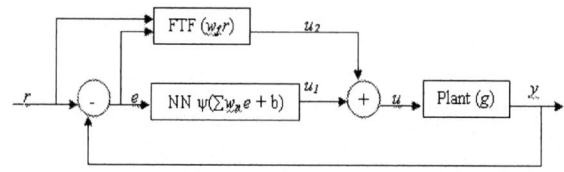

Fig. 5. A Scheme of Control employing FTF and Neural Networks.

The output of the NN, denoted by $u_1(k)$ is

$$u_1(k) = \psi\left[\sum_{i=1}^{M} w_n(i)e(i) + b\right] \qquad (12)$$

where $M$ is the number of samples taken at a time, $e = r - y$ is the error, $y$ is the actual output, $r$ is the set point. The somatic gain denoted by $\psi$ is the advanced feature that improves the performance of the neural network. The NN used here is composed of only input and output layers. The weights of the NN are adjusted by the following equation [4]:

$$w_n(k+1) = w_n(k) + \Delta w_n(k) + \gamma \frac{\partial V}{\partial w_n} \qquad (13)$$

$$\Delta w_n(k) = \alpha' e(k) * e(k) \qquad (14)$$

where $V = \frac{1}{2}e^2$ and $\alpha' = \dfrac{\alpha}{\|e\|^2}$, with $\alpha$ and $\gamma$ being the step

sizes used in $\alpha$ - LMS and gradient learning respectively.

*Off line parameter optimization using ANN*

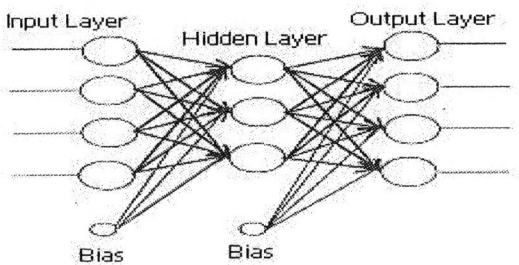

Fig. 6. Block diagram of MLP feed forward ANN.

The training of parameters using Multi Layer Perceptron (MLP) feed forward ANN (See Fig. 6) is now discussed. Inputs to the ANN are different values of regulation parameter R, water starting time $T_w$ and nominal loading. Output of ANN is the desired proportional gain $K_p$ and integral gain, $K_i$. Prior to conducting training, a set of input–output patterns is first prepared. The network is trained until a good agreement between predicted gain settings and the actual gains is reached. Once the network is adequately trained, the network is again tested to ensure that it can adequately predict the correct gain settings for the inputs that are not included in the training set. The results show that predictions by ANN are in good agreement with the training data. It is observed that parameter optimization of $K_p$ and $K_i$ is independent of the magnitude of the step disturbance.

*The FTF controller part*

The output of the FTF, denoted by $u_2(k)$ is

$$u_2(k) = w_{f1} \times r(k) + w_{f2} \times r(k-1) \qquad (15)$$

where $w_{f1}$ and $w_{f2}$ are the FTF weights. The computation of these weights can be found from [7].

(b)    *Fuzzy PI Control*

A schematic diagram of the Fuzzy PI controller is given in Fig. 7.

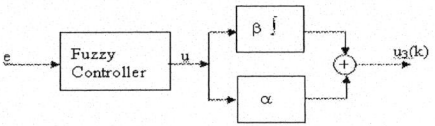

Fig. 7. Scheme employing Fuzzy PI control.

The output of a product-sum fuzzy controller is of the form [9].[10],[35],[36],[37]:

$$u = A + PE \qquad (16)$$

where $E = k_e\, e = e_i$, $A$ is the input membership function, $e$ is the error , $u$ is the output of fuzzy controller, $k_e$ is the scaling factor and $P$ is equivalent proportional term [9]. Hence, the control input to the plant can be approximated by

$$
\begin{aligned}
u_3(k) &= \alpha.\,[A + PE] + \beta\!\int (A+PE).dt \qquad (17)\\
&= \alpha.A + \alpha PE + \beta A\, t + \beta P\!\int k_e.e.dt\\
&= \alpha.A + \alpha PE + \beta A\, t + \beta P\; k_e\!\int.e.dt\\
&= \alpha.A + \alpha P\, k_e e + \beta A\, t + \beta P\; k_e\!\int edt
\end{aligned}
$$

where $\alpha$ is the weight on PD type fuzzy controller and $\beta$ is the weight on PI type fuzzy controller. Here, the fuzzy controller becomes a parameter cum time-varying PI controller with its equivalent proportional control and integral control components being $\alpha Pk_e$ and $\beta Pk_e$ respectively.    Hence this Fuzzy PI controller behaves as the PI type fuzzy controller. In (15) the derivative of $e$ for the fear of accentuating the stochastic disturbance is not used.

(c)    *Normalised LMS Control*

The weight update in $\alpha$ -LMS rule is given by

$$W_{k+1} = W_k + \alpha.\, \mathcal{E}_k.X_k/|X_k|^2 \qquad (18)$$

The time index or adaptation cycle number is $k$, $W_{k+1}$ is the next value of the weight vector, $W_k$ is the present value of the weight vector and $X_k$ is the present input pattern vector. The present linear error $\mathcal{E}_k$ is the difference between the desired response $s_k$ and the linear output $r_k = W_k^T.X_k$

Hence

$$\Delta \mathcal{E}_k = \Delta(s_k - W_k^T.X_k) = -X_k^T.\Delta W_k \qquad (19)$$

Rewriting the $\alpha$ -LMS rule from Eqn. (19), we have

$$\Delta Wk = W_{k+1} - W_k = \alpha.\, \mathcal{E}_k.X_k/|X_k|^2 \qquad (20)$$

In view of Eqn. (20), Eqn. (19) is written as

$$\Delta\, \mathcal{E}_k = -\, \alpha.\, \mathcal{E}_k.X_k^T.X_k / |X_k|^2 = -\, \alpha.\, \mathcal{E}_k \qquad (21)$$

But in our case $\Delta f = \Delta e_k$; so the output $u_4(k) = W_k^T.\, e_k$ .

Stability and speed of convergence depend on the parameter $\alpha$. For input pattern vectors independent in time, stability is ensured for most practical purposes if $0 < \alpha < 2$. Choosing $\alpha$ greater than 1 results in over corrections hence practical range of $\alpha$ is $0.1 < \alpha < 1.0$.

IV. COMPARISON OF PERFORMANCE

Exhaustive and comprehensive simulations are carried out on micro hydro power plants using the proposed controller designated as *"g"* as well as different combinations of the control schemes *(a- f)* with a view to come up with an

905

efficient combination that can reduce the peak overshoot and ensure quick stabilization of the supply frequency. The flow chart of the controller scheme useful for implementation purpose is shown in Fig. 8. The data model employed is given in Appendix. The performance of ANN using off-line optimization is depicted in Fig. 9.The plots of $\Delta f$ vs. $t$ for different control schemes are shown in Fig. 10 to ascertain their relative performance. The composition of various schemes is as follows:

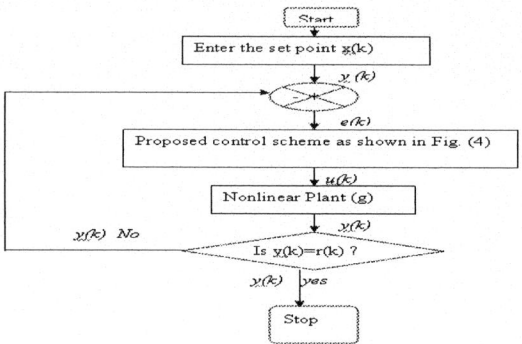

Fig. 8. Flowchart.

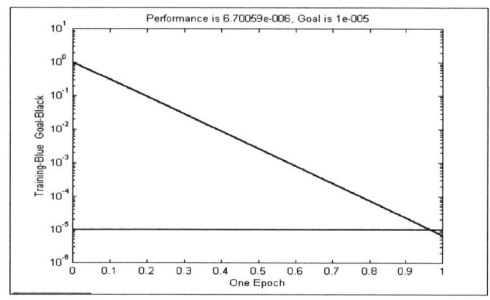

Fig. 9. Performance of ANN.

$a$: PI ; $b$: nLMS + PI; $c$: FTF +PI; $d$ : nLMS + NN+PI;
$e$: *PI ; $f$: nLMS+NN+ FTF+PI ; $g$ : nLMS + NN+ FTF + *PI.

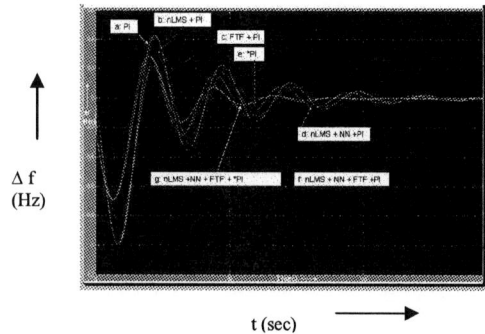

Fig. 10. Comparison of different control schemes.

The results of simulations demonstrate that in general the application of the advanced control techniques using the linear and non linear parts improves the performance of the micro hydro power plants as compared to a simple PI controller. The proposed controller "g" outperforms over all other controllers by achieving the least peak overshoot and the early settling time as shown in Table 1 where it may be observed that all controllers (i.e., d, f, & g) having both linear and non linear parts have less Squared Error (SE). In Table 1, we denote the under shoot and over shoot by $\xi_u$ and $\xi_0$ respectively and the settling time by $t_s$.

TABLE I
COMPARISON OF CONTROL COMBINATIONS (HERE 1: BEST; 7: WORST)

| Type | max $\{\xi_o, \xi_u\}$ | $t_s$ | $1^{st}$ $\xi_u$ | $2^{nd}$ $\xi_o$ | $3^{rd}$ $\xi_u$ | J=Σ (SE) |
|---|---|---|---|---|---|---|
| a | 7 | 7 | 0.99 | 0.427 | 0.353 | 1.38178 |
| b | 4 | 5 | 0.834 | 0.291 | 0.267 | 0.90322 |
| c | 6 | 6 | 1 | 0.423 | 0.317 | 1.342765 |
| d | 3 | 4 | 0.6929 | 0.2869 | 0.2545 | 0.68026 |
| e | 5 | 2 | 0.993 | 0.22 | 0.2449 | 1.106379 |
| f | 2 | 3 | 0.691 | 0.28055 | 0.22439 | 0.64025 |
| g | 1 | 1 | 0.69 | 0.2798 | 0.16655 | 0.589780 |

Table 1 indicates guidelines for the choice of control components. We will first try a linear component one after another and see the performance. Which ever is the best we will select that one. Next we repeat the experiment with one non-linear component at a time. We will analyze the performance. This will be continued with next set of linear components followed by the non-linear components until we get the desired combination of control components.

## V. CONCLUSIONS

The maintenance of desired power generation and frequency of micro hydro power plants using flow control is the main theme of this paper. An electric servomotor is suggested as a governor for these plants. A state space model of the plant is derived. A new scheme is proposed for the control of micro hydro power plants. This scheme consists of dividing the controller action into two parts: Linear and Non-linear. The non-linear part of the controller incorporates Fuzzy PI and NN. The linear part is based on the FTF and nLMS algorithms. The concept behind splitting the controller is that any nonlinear error signal is better represented by linear and nonlinear parts. The performance of the new controller and several others with different linear and linear components is compared. The new controller is found to be superior to that of other control techniques in terms of reducing the sum of the squared error. Thus this work makes an important contribution in terms of splitting of control action and providing guidelines for the choice of new control schemes. We have also implemented the proposed controller on small hydro power plants and the results are found to be equally good as those of MHP.

There are reservations in some circles on the use of sophisticated control schemes for the control of micro hydro power plants. However, we justify this study on the ground that analysis of different schemes and their combinations will lead to better understanding of the control action and will pave the way to a practically viable scheme. For example an alternative to the proposed controller in which all the individual schemes are simultaneously functioning could be to switch them in a particular sequence of linear and nonlinear parts so as to achieve the similar performance but at the reduced power consumption and utility.

## VI. APPENDIX

### Data for the Model
The following data is considered for constructing the model.
1. Total rated capacity  : 50 kW
2. Normal operating Load: 25kW
3. Inertia Constant $H$: 7.75 seconds        (2<H<8)
4. Regulation $R$:  10 Hz / pu kW        ( 2<R<10)

*Assumption:*  Load - frequency dependency is linear. Nominal Load = 48 % = 0.48; $\Delta$ Pd = 3 % = 0.03.

The damping parameter [4]-[7], $D = \partial \text{Pd}/ \partial \text{ f}$

$$= \frac{0.48 \times 25}{60 \times 50} = 0.004 \text{ pu kW / Hz}$$

Generator parameters are:

$$\text{Kp} = \frac{1}{D} = 250 \text{ Hz / pu  kW}$$

$$\text{Tp} = \frac{2 \times H}{f^0 \times D} = 64.64 \text{ seconds}$$

The open loop transfer function of a servomotor is given by [27]-[29]
$$G(s)H(s) = \frac{K_n \ K_a K_g / K_c}{(1+T_f S)(1+T_m s)}$$

where,
$K_a$ = net control field amperes per volt actuating error signal,
$K_g$ = No-load amplidyne terminal voltage per net
Control field current,
$T_f / R_f = L_f$ = Time constant of quadrature field
of amplidyne, seconds,

$T_m = \dfrac{JR_a}{K_T K_e}$ = Time constant of motor and load, seconds,

$K_c$ = Motor volts per radian per second of motor,
$K_n$ = Voltage from tachometer per radian per second of motor.
For our model, we choose the following values:-
$K_n$ =1; $K_a K_g / K_e$ =1 ; $T_f$ =0.001 seconds and
$T_m$ =0.01 seconds
PI Controller parameters: $K_{pl}$ = 0.056, $K_i$ = -0.002

## VII. ACKNOWLEDGMENT

The authors gratefully acknowledge the financial support received from the Defence Research Development Organisation (DRDO), Matcalfe House, New Delhi, under the project titled, "Adaptive Control Algorithm design for Servo Control System of Beam Directing Assembly" .

## VIII. REFERENCES

[1] IEEE Committee Report, Dynamic models for steam and hydro turbines in power system studies, *IEEE Trans. on Power Apparatus and Systems*, 1973, pp 1904-1915.

[2] Working Group on Prime Movers,  Hydraulic turbine and turbine control models for system dynamic studies, *IEEE Tran, on Power Systems*, 7, 1992, pp. 167-179.

[3] S. Kumpati , Narendra and Kannan Parathasarathy,  "Identification and control of dynamic system using neural network", *IEEE Trans. on Neural Networks*, Vol. 1, no.1, , March 1990, . pp.4-27.

[4] Zi-Qin Wang, T Michael, Manry, and Jeffery L. Schiano,  "LMS Learning Algorithms: Misconceptions and New Results on Convergence," *IEEE Trans. on Neural Network*, vol. 11, No. 4, Jan. 2000.

[5] L.Ljung , M.Morf and D.D.Falconer, "Fast Calculation of Gain Matrices for Recursive Estimation Schemes,"*International Journal of Control*,Vol.27,pp. 1-17,January 1978.

[6] D.D.Falconer and  L.Ljung, " Application of Fast Kalman Estimation to Adaptive Equalization," *IEEE  Trans. On Communications*, Vol. COM-26, pp-1439-1445, October 1978.

[7] J.M.Cioffi and T.Kailath, "Fast Recursive Least –Squares Transversal Filters for Adaptive Filtering, *IEEE Trans. On Acous. Speech and Signal Processing*", Vol. ASSP-32, pp-304-338, April 1984.

[8] O. Adetona, and L.H. Keel, "A New Method for the Control of Discrete Nonlinear dynamic Systems Using Neural Networks", *IEEE Trans. on Neural Network*s, Vol. 11, no.1, pp.102-112, January 2000.

[9] Wu Zhi Qioa, and Masaharu Mizumoto, "PID type fuzzy controller and parameters adaptive method", *Fuzzy Sets and Systems*, 78 (1996) 23-35.[26]

[10] Onur Karasakal, Yesil, Guzelkaya, Eksin, "Implementation of a New Self-Tuning Fuzzy PID Controller on PLC", *Turk J. Elec Engin*, Vol. 13, No. 2, 2005, pp. 277-286.

[11] Woodward,J.L. "Hydraulic-Turbine Transfer Function for use in Governing Studies", proc. IEE, Vol. 115,pp. 424-426,1968.

[12] A.H. Glattfelder, J. Rettich, "Frequency control for low-head hydro units in isolated networks", Water Power & Dam Construction, pp.42-46, March 1988.

[13] S. Hagihara, H. Yokota ,& K. Goda, K. Isobe,( Nov./Dec, 1979 ), "Stability of a hydraulic turbine Generating Unit Controlled By P.I.D. Governor", *IEEE Transactions on Power Apparatus and Systems*, Vol. PAS-98, No.6, pp. 2294-2297.

[14] L. Pereira, (November 1981), 'Induction generators for small hydro plants', Water Power & Dam Construction, pp.30-34.

[15] Philip Schniter, L. Wozniak, (June 1995), 'Efficiency based optimal control of Kaplan hydro generators', *IEEE  Transactions on Energy Conversion*, Vol.10, No.2,   pp.348-353.

[16] O.P. Malik,  G.S. Hope, G.C. Hancock, Li. Zhaohui, Ye Luqing, Shouping,WEI.,( September 1991), 'Frequency Measurement for use with a Microprocessor-Based Water Turbine Governor', *IEEE Transactions on Energy Conversion*, Vol. 6, No.3, pp. 361-366.

[17] M. Djukanovic, M. Novicevic, D.J. Dobrijevic, B. Babic, Dejan J.Sobajic, Yoh-Han Pao, (December 1995),'Neural-Net Based Coordinated Stabilizing Control For the Exciter and Governor Loops of Low head Hydropower Plants', *IEEE transactions on Energy Conversion*, Vol.10, No.4, pp.760-767.

[18] D.S. Henderson, 1992, "An Advanced Electronic Load Governor for Control of Micro Hydroelectric Power Generation", IEEE Transactions on Energy Conversion, Vol. 13, No.3, September 1998.

[19] P.M. Andersson , and A.A. Fouad, , *Power System Control & Stability*, 2nd ed., John Wiley & Sons, Inc., 111, River Street,  Hoboken, NJ., IEEE Series on Power  Engineering ,2003.

[20] T.S. Bhatti, R.C. Bansal, and D.P. Kothari, *Small hydro power plants*, Dhanpat Rai & Co., New Delhi ,2004.

[21] J.R. Carstens, *Automatic Control Systems & Components*, Prentice –Hall, Inc., New Jersey.U.S.A, 1990.

[22] O.I. Elgerd, *Electric energy systems theory: An Introduction* , Tata McGraw – Hill, New Delhi, 1982.

[23] A. Harvey, A. Brown, P. Hettiarachi and A. Inversin, *Micro Hydel Design Manual, A Guide to Small Scale Water Power  Schemes*, Intermediate Technology Publications, 1993.

[24] P. Kundur, *Power system stability and Control*, Tata-McGraw Hill Co. 1221, Avenue of the Americas, New York, NY, 1994.

[25] M. Madan Gupta and Dandina H. Rao, *Neuro-Control Systems, Theory, Principles and Applications*, IEEE Press, May 1994.

[26] S. Thomas Alexander, *Adaptive Signal Processing: Theory and Applications*, Springer-Verlag New York, 1986.

[27] J.M.Zurada, , St. Paul, MN (Editor): *Introduction to Artificial Neural Systems* , West Publishing, 1992.

[28] G.S. Brown, and D.P. Campbell, " Principles of Servomechanism-Dynamics & Synthesis of Closed-Loop Control Systems", John Wiley & Sons, New York, Chapmen & Hall, Limited,London,1958.

[29] Chestnut,H and Mayer, R.W., "Servomechanism & Regulating System Design", Volume I, John Wiley & Sons,Inc.New York,London, 1959.

[30] M.L.Honig, "Recursive Fixed Order Covariance Least Squares Algorithms", *Bell Sys. Tech. J.*, Vol.62,pp-2961-2992, December 1983.

[31] H. Tano, Y. Sannomiya,( February 13-22, 2001),'Integrated Digital Control panel for Hydroelectric Power Stations', *International Course on Planning, technology Selection and Implementation of Small Hydro Power Project*, A.H.E.C, IIT, Roorkee, India, pp. 461-496.

[32] Rakesh Thapar, Michael Ruane, David A. Perreault,( February 13-22, 2001 ), 'Microprocessor Controller for a Small Unattended Hydroelectric System', *International Course on Planning, technology Selection and Implementation of Small Hydro Power Project*, A.H.E.C, IIT, Roorkee, India.

[33] S.Srivastava, M.S. Singh , M.Hanmandlu and A.N. Jha,"Development of a New Controller for NonLinear Systems ," Communicated to Journal of Soft Computing in October 2003.

[34] S.Srivastava, M.S. Singh and A.N. Jha, "Identification of Non Linear Systems using FTF and α-LMS algorithms", *Proceedings of NSC*, IIT Kharagpur, December 2003,INDIA.

[35] O. Ghanayem, and L. Reznik " Excitation Control of a Synchronous generator Using an On-Linear Adaptive Fuzzy Logic Controller Structure ", Proc. of the *6th IEEE International Conference on Fuzzy Systems*, July 1-5, 1997, Barcelona, Spain, IEEE neural Networks Council, Vol. 3,pp. 1493-1498.

[36] Bernard Widrow, "30 Years of Adaptive Neural Networks perceptron, Madaline, and Backpropagation", *Proceedings of the IEEE*, Vol. 78, no.9, September 1990, pp. 1415-1442.

[37] Huang,T Hung-Yuan Chung and Jin-Jye Lin, " A Fuzzy PID Controller Being Like Parameter varying PID", *1999 IEEE International Fuzzy Systems Conference proceedings*, August 22-25, 1999, Seoul, Korea, pp.I-269- 276.

[38] D.S. Henderson, "Recent Developments of an Electronic Load Governor for Micro Hydroelectric Generation", Proceedings HIDROENERGIA 93, Munich, Germany, 1993.

[39] S.Srivastava, *Modeling and Control of Non-Linear Systems Using Intelligent Tools*, PhD Thesis, IIT Delhi, 2005.

[40] D.S. Henderson, 1992, *A three phase electronic load governor for micro hydro generation*, Thesis for the degree of Doctor of Philosophy, The University of Edinburgh.

[41] IEEE Guide for control of small hydroelectric power plants, Std 1020-1988.

## IX. BIOGRAPHIES

**M.Hanmandlu** is with Deptt. Of Electrical Engineeriing, IIT Delhi. He received B.E.from Osmania University in 1973, M.Tech from JNTU, REC Warangal in 1976, and Ph.D. from IIT Delhi in 1981. He held different positions in IIT Delhi before becoming a Professor in 1997. His research interests include fuzzy modeling, Document processing, Image processing and Computer Vision.

**Himani Goyal** is pursuing her Ph.D. degree from Indian Institute of Technology, Delhi. She did her BE in Electronics Engineering from R.A.I.T., University Of Mumbai, Mumbai, India in 1995 and M.Tech. in Alternate Hydro Energy Systems from, IIT Roorkee in 2000-2001. She is working in the area of Advanced Control of Small Hydro Power plants and has published and presented a number of papers in international journals and conferences

**D.P.Kothari** is with Center for Energy Studies, Indian Institute of Technology, Delhi. A Senior Fellow of IEEE, Prof. Kothari has published/presented nearly 500 papers in national and international journals/conferences. He has authored/co-authored more than 18 books. His research interests include power system control, optimization, and reliability and energy conservation.

**2006 IEEE International Conference on Power Electronic, Drives and Energy Systems**

# Application of Fuzzy Logic PSS to Enhance Transient Stability in Large Power Systems

P. V. Etingov, *Member, IEEE,* and N. I. Voropai, *Senior Member, IEEE*

*Abstract*– The paper presents an application of fuzzy logic power system stabilizers (FLPSS) to enhance transient stability in large electric power systems. A two-stage technology of FLPSS adaptation is considered taking into account the real conditions of a power system. Self-organizing artificial neural network (ANN) is used for clusterization of the test disturbances. A genetic algorithm (GA) is applied to tuning parameters of FLPSS. ANN is used on-line to adapt FLPSS to changes in operating conditions. The studies have been conducted using the models of a 14-bus multi-machine power system and regional Siberian power system as the study cases. The results obtained are presented.

*Index Terms*– Power System Stabilizer, Fuzzy Logic, Genetic Algorithms, Neural Networks, Power System Stability.

## I. INTRODUCTION

THE problem of power system stability improvement is very important in terms of a market environment. It is well known that power system stabilizers (PSSs) on generation units are effective tools to damp electromechanical oscillation and enhance transient stability in electric power systems.

However, the long-term operating experience has shown that conventional PSSs do not always completely use the capabilities of excitation system. The reason is a complicated selection of proper settings of controllers. Also conventional PSSs cannot provide high performance in all operating conditions, because parameters of controllers are fixed and tuned only for a specific operation point.

Recently, new approaches to designing PSSs have been proposed as a result of developing computer technologies including Artificial Intelligence (AI) technique such as Expert Systems [1], ANN [2, 3], Fuzzy Logic [4-13] and GA [7, 8]. The most progress in the field of AI application has been made in the development of FLPSSs. To date FLPSSs have evolved from mathematical modeling on simple two-machine power systems to operation in real power system [13].

The FLPSS has several advantages compared to a conventional PSS, for instance, better robust properties, simple structure and capability to knowledge accumulation. However, the problems of FLPSS optimal tuning and adaptation as well as FLPSSs application in large power systems have not been fully investigated.

A main problem of constructing a FLPSS is finding optimum parameters, such as values of scaling factors, shape of membership functions and values of rule base table. In the conventional design method, the FLPSS settings are chosen by trial-and-error methods and human expert knowledge. However, these approaches do not allow one to obtain an optimal control and moreover, they are time consuming.

Another problem is adaptation of parameters to changing operating conditions. FLPSSs with fixed parameters have advanced robustness. However, they have not demonstrated satisfactory results under a variety of operating conditions, as is shown in many papers [9, 11, etc].

Many authors proposed different ways of solving the problem of FLPSS tuning, including ANN [5, 6], GA for optimization problems [7, 8], adaptive fuzzy linearization [9], etc.

This paper suggests a somewhat new view on the problem of FLPSS constructing. A GA is considered for optimal tuning of the FLPSS settings and the ANN is applied to on-line adaptation of FLPSS parameters. Besides, classification and grouping techniques are used to find operating conditions and disturbances critical for power system stability.

The efficiency of the proposed methodology of the FLPSS adaptation is studied using the models of a 14-bus multi-machine power system and the regional Siberian power system.

## II. PROBLEM FORMULATION

### A. Main Idea

The commonly known structure of FLPSS is illustrated in

---

The study was supported by the Grant of Russian Leading Scientific School #2234.2003.8 and by the Grant of the President of the Russian Federation for young scientists MK-3105.2005.8

N. I. Voropai is with Energy Systems Institute, 130, Lermontov str, 664033 Irkutsk, Russia (e-mail: voropai@isem.sei.irk.ru).

P. V. Etingov is with Energy Systems Institute, 130, Lermontov str, 664033 Irkutsk, Russia (e-mail: etingov@ieee.org).

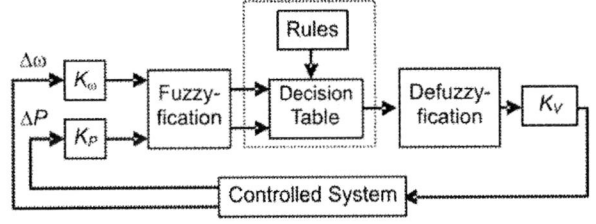

Fig. 1. Fuzzy Logic PSS Structure.

Fig. 1. Controller generally uses a scheme with two inputs (speed deviation Δω and acceleration power ΔP), blocks of fuzzyfication / defuzzyfication and a decision table. The operation principle of such devices is well-known and described in many publications [4, 12, etc].

An uncontrolled GA is used in many papers to form a rule base table and membership functions of FLPSS, as it is noticed in [14]. It leads that each cell of the rule base table can take any value that is unnecessary and frequently results in a scrambled rule base. Also, such an approach leads to messy overlapping membership functions. The terms "scrambled rule base" and "messy overlapping membership function" were used in [14]. These kinds of FLPSSs have certain disadvantages, for instance, they are not clear for understanding and sometimes operate incorrectly.

In this paper the GA with imposed restrictions is used that allow the well-formed rule base tables and membership functions to be received. The purpose of the proposed approach is getting near to optimal, simple and comprehensible controller.

The problem of FLPSS adaptation in a large EPS can be decomposed into two subproblems (two stages):

• preliminary (off-line) tuning under various test operating conditions (including various network topologies) and disturbances;

• on-line adaptation parameters of the FLPSS to changes in the operating conditions of power system.

The first subproblem (stage) is not strongly sensitive to time limit. Effective optimization techniques can be applied to solve this subproblem. Analytical and heuristic techniques are used. However, the number of probable operating conditions and disturbances is very big in the bulk power systems. Therefore, optimization of FLPSS parameters under all operating conditions can be an unsolvable problem. To solve this problem the set of disturbances that can be dangerous to transient stability is classified to identify the groups of disturbances with similar nature of transients. Thus, classification of the disturbances allows to perform the FLPSS tuning only for one selected representative from each group. A GA is used for tuning of the FLPSS settings. Optimization is carried out for several major disturbances selected on the basis of previous classification. As a result of FLPSSs tuning, the optimal values of membership functions, rule base table and scaling factors are obtained. Simultaneously a feed-forward backpropagation ANN is trained by optimal parameters of the FLPSS obtained under different operating conditions of power system.

The second subproblem (stage) has stringent requirements to the time limit. Therefore this subproblem is solved using the ANN preliminarily trained at the first stage. The ANN automatically identifies power system operating conditions and adapts parameters of FLPSS in the case of operating conditions changing.

Decomposition of the problem offers possibilities to solve the fuzzy PSS adaptation problem more efficiently.

### B. Stage 1(Off-line)

#### 1) Classification Technique

At the beginning operating conditions and disturbances are classified. Then, a representative selection of the studied conditions is carried out, which calls for PSSs tuning.

Consider initial conditions under which FLPSSs are preliminarily tuned, supposing that the problem of their effective location in the bulk power system is solved by one method or another.

Let $S = \{s_1,...,s_M\}$ be a set of power system schemes (normal, post–emergency, maintenance), $Z = \{z_1,...,z_L\}$ be a set of conditions for the studied power system which are determined by the typical points of the load curves of consumers and power plant units and $V = \{v_1,...,v_K\}$ be a set of disturbances (short circuits, emergency tripping of generations, transmission lines and consumers etc.) which can superpose on each combination from $s_m \in S$ and $z_l \in Z$. At some combinations $\{s_m, z_l, v_k\}$, where $m = \overline{1,M}$; $l = \overline{1,L}$; $k = \overline{1,K}$, the power system stability is maintained and for some other combinations it can be violated.

Determine the conditions to be studied that represent some subset of the combinations $\{s_m, z_l, v_k\}$, for which the power system instability is possible. Here it is supposed that FLPSS adaptation to the states dangerous in terms of power system stability disturbance will also be efficient for the states which are not dangerous in terms of power system stability disturbance. Besides, it is obvious that PSS is not the only means providing power system stability, in some studied test conditions the emergency control scheme should be taken into account.

Since there are very many possible combinations $\{s_m, z_l, v_k\}$ for complex power system, the classical model of the power system dynamics with pairwise analysis of equations of mutual motion generators operate under the assumption that motion of the remaining part of the system is applied for their estimation [15]. The elements of the square matrix $[w_{ij}]$ $i, j = \overline{1,n}$, where $n$ is the number of generators in the power sytem scheme, reflect dynamic interaction of generators for the considered combinations.

One of the most objective parameters, that do not require a numerical integration, is initial mutual accelerations of generators.

The value of initial mutual accelerations of generators $i$ and $j$ can be calculated as:

$$\frac{d^2\delta_{ij}^{(0)}}{dt^2} = \frac{d^2\delta_i^{(0)}}{dt^2} - \frac{d^2\delta_j^{(0)}}{dt^2} , \qquad (1)$$

where $\frac{d^2\delta^{(0)}}{dt^2}$ is an absolute value of initial acceleration of generator;

Analysis of the matrix $[w_{ij}]$ allows:

1) identification of the combinations $\{s_m, z_l, v_k\}$ that are not dangerous in terms of stability and hence will not be studied further;

2) classification of the combinations $\{s_m, z_l, v_k\}$ into clusters with respect to the similarity of system responses, such that for the cluster $\{s_f, z_s, v_p\}...,\{s_h, z_r, v_q\}$, where $f, h \in M$; $s, r \in L$; $p, q \in K$, we have $F(w_{ijfsp}) \approx F(w_{ijhrq}), \forall w_{ij}$, where $F$ is some operator. Let us denote the considered cluster by $G_d$, $d \in D$, $D$ is the number of clusters;

The self-organizing ANN was considered to solve this problem (Fig. 2). The matrix $[w_{ij}]$ is fed to the input layer of the ANN consisting of $n(n-1)/2$ neurons. The number of neurons in an output layer corresponds to the number of groups (subsets) for which the set of disturbances will be divided. Therefore, varying the number of output neurons we can vary the number of groups. After training the self-organizing ANN is able to recognize and classify groups of similar input vectors.

3) selection of a representative combination $\{s_{md}, z_{ld}, v_{kd}\}$ for each cluster, the set of which represents the studied test conditions, under which tuning of FLPSS is required.

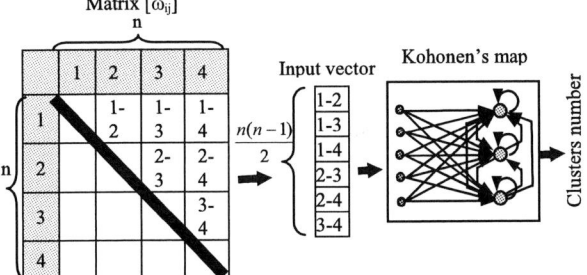

Fig. 2. Classification of emergences by self-organizing ANN.

*2) Tuning of Rule Base Table and Scaling Factors*

Restrictions were applied to the GA to produce a well-formed rule base like in [14]. The value of each rule can change in a certain range relative to a value given in the default table (Fig. 3), where 1 corresponds to NB, 2–NM, 3–NS, 4–ZR, 5–PS, 6–PM, 7–PB. The value of the marked cells should be fixed (Fig. 3). Such measures allow us to reduce optimization time and exclude obviously incorrect rules.

Each cell (except fixed cells) may take values in the range $\pm\Delta$ of the values taken by default. The variable $\Delta$ can accept integer values in a specified range, which depends on the size of the table and amount of the output linguistic variables. The value of $\Delta$ equal to 1 is enough in the case of 25 rules. Binary values were used for chromosomes coding. Instead of coding the rule base table values, its deviations $\Delta$ from default are coded. Two bits are sufficient for each rule. In total it is necessary to adjust 22 rules, because 3 rules are fixed. Also, it is very important to correctly choose input scaling factors $K_\omega$ and $K_P$ (Fig. 1).

*3) Coding of Membership Functions*

Fig. 4 shows the principle of membership functions coding

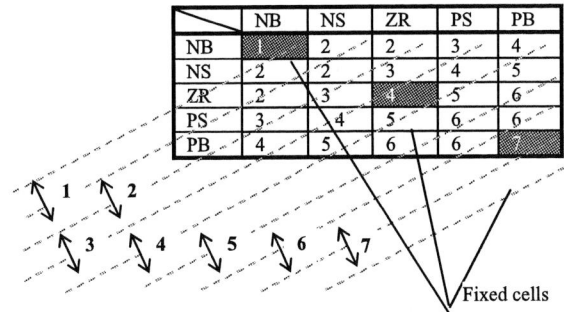

Fig. 3. Rule base table restriction.

on the example of the FLPSS with seven fuzzy subsets and triangle membership functions. In this case, it is necessary to adjust only four parameters to produce well-formed membership functions. This kind of membership functions is characterized by the following feature: left and right corners are fixed at the points described by the apices of other triangles. The apices of extreme subsets and the apex of subset "Zero" should be fixed at the points "-1", "1" and "0". As is shown in Fig. 4, five bits are allocated on each adjustable parameter $X_1$, $X_2$, $X_3$, $X_4$.

Besides, the following constraints were used:

$$X_1 < X_2 < X_3 < X_4 \qquad (2)$$

$$X_1, X_2 \in d_1,\ X_3, X_4 \in d_2 \qquad (3)$$

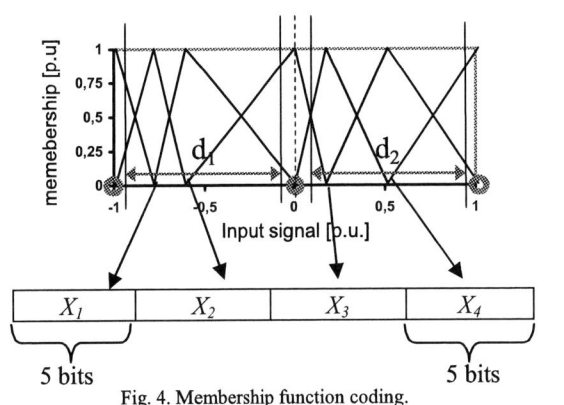

Fig. 4. Membership function coding.

*4) Fitness Function*

A fitness function for disturbance $m$ was chosen based on the following criterion:

$$J_k^{(m)} = \sqrt{\frac{1}{T} \sum_{i=1}^{N} \int_{t=0}^{T} \left(\omega_0 - \omega_{i_k}^{(m)}(t)\right)^2 dt}, \qquad (4)$$

where $\omega_{i_k}^{(m)}(t)$ is a rotor speed in time interval $t$; $\omega_0$ is a synchronous speed; $T$ is time of integration; $N$ is the number of generators; $k$ is the index of an individual in population.

As the purpose of optimization is evaluating the influence of FLPSS on a transient process under several disturbances the following criterion was proposed:

$$J^j_{k_\Sigma} = \sqrt{\frac{1}{M} \sum_{m=1}^{M} \left( J^{(m)}_{k_j} - J^j_0 \right)^2} , \qquad (5)$$

where $M$ is the quantity of disturbances ; $j$ is the index of generation; $J_0$ is the minimum average value in population:

$$J_0 = \min_{k=1..K} \left( \sum_{m=1}^{M} J^{(m)}_k \bigg/ M \right).$$

$K$ is the number of individuals in population.

Application of (5) instead of simple summation allows a more exact contribution of each disturbance to the evaluation (fitness) function to be considered.

Also scaling of the fitness function value was applied:

$$J^j_{\text{scal}_k} = J^j_{k_\Sigma} - \min_{k=1..K} \left( J^1_{k_\Sigma} \right) \cdot \alpha , \qquad (6)$$

where $\alpha$ is the coefficient of correction.

*5) Characteristics of GA*

Hybrid initialization was used to generate the initial population of individuals. Part of population is randomly created, taking into account the above restrictions. The other part can be created based on the previous experience, expert knowledge, etc. The correct choice of the population size is very important. With a small population, there is a fast degeneration, but if the size is too big, the speed of optimization is strongly reduced. Therefore it is necessary to find a compromise decision satisfying both criteria. Obtained experimentally, the optimum size of population is about 50-100 individuals. Furthermore, the GA was updated to check "twins" in populations, in case a pair of similar individuals is detected, one of them leaves.

The one-point crossover and roulette wheel technique were used. The probability of crossover is 0.85 and the probability of mutation is 0.1.

*C. Stage 2 (On-line)*

The FLPSSs with fixed parameters show a good robustness to changes in the operating conditions of a power system. However, they do not always ensure the desired performance in all operating conditions. Adaptive FLPSSs do not have such a disadvantage.

We used a feed-forward backpropagation network to FLPSS adaptation (Fig. 5). Parameters of operating conditions are fed for input of the net, such as values of active and reactive power of the generator units and loads. The values of FLPSS parameters are received at the network output. Thus, we have a regulator with adaptation ability and a neural network is used to store all possible kinds of FLPSS settings.

The network can be trained by any known method, for example via function approximation, pattern association or with the help of GA. The training process requires a set of examples of proper network behavior- network inputs and target outputs. Properly-trained backpropagation networks tend to give reasonable answers for the inputs they have never seen.

This generalization property makes it possible to train a network on a representative set of input/ target pairs and get

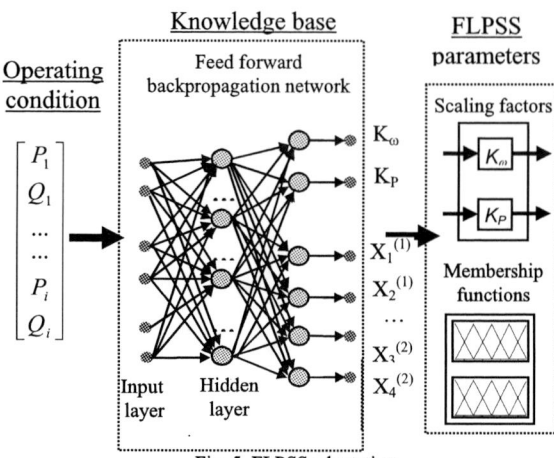

Fig. 5. FLPSS adaptation.

good results without training the network on all possible input/ output pairs [16]. Training is off-line for the set of operating conditions. For each of them a genetic optimization is done and then the obtained optimum settings of the controller are used for training the network.

Thus, ANN application allows the new near-optimal parameters of the controller to be instantly received on-line in case of a change in operating conditions.

## III. SIMULATION RESULTS

*A. Studies on the 14-bus power system*

The special software "PAU" was used for calculations [12]. This program allows modeling of transient process, and also it includes special blocks for modeling fuzzy PSS and tuning by GA.

The multi-machine test power system (Fig. 6) consists of five generators. The most important generators 101,201,203 are equipped with FLPSSs. Based on irregularity of load and generation location, two areas can be separated in the network: the first one (with power shortage) is in the neighborhood of generators 1, 3 and 101, and the second one (with power surplus) is in the neighborhood of generators 201

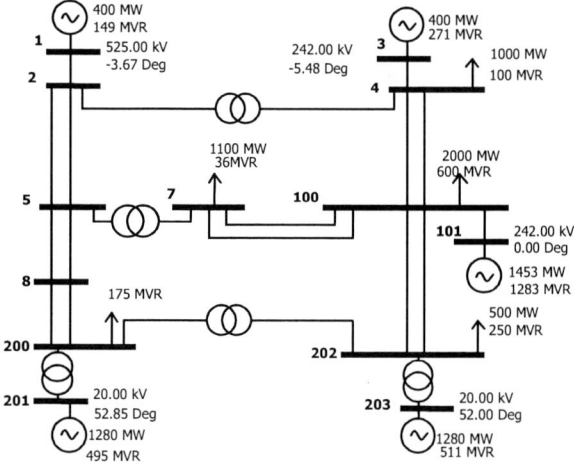

Fig. 6. The 14-bus multi-machine test power system.

and 203.

Two-circuit transmission lines (5–8, 8–200 and 100–202), which connect these areas, are heavily loaded even in pre-emergency conditions. Therefore disconnection of even one circuit in any of these lines is a strong disturbance which can provoke a loss of the power system transient stability.

At the first stage, the test power system was analyzed. A list of disturbances dangerous to transient stability was formed (Table I). Only strong disturbances were selected. Then they were classified and grouped to determine a set of the disturbances with similar nature of transients. For that purpose, initial mutual accelerations of generators were calculated for each disturbance. This information is used to train the self-organizing ANN.

The result of disturbances classification is given in Fig. 7. The visual analysis has proved the accuracy of classification results.

*1) Off-line Simulation*

Simultaneous tuning was carried out for FLPSSs installed on generators 101, 201 and 203. The controllers with five fuzzy subsets were considered. Therefore, it is necessary to adjust only two parameters of membership functions ($X_1$ and $X_2$) for each input signal. In the beginning we studied the test power system in the base (nominal) operations conditions shown in Fig. 6.

Training has been done for disturbances 2, 11, 12, 14 and 15 (Table I). As a result the optimal rule base tables, scaling factors and membership functions (Table II-III) were obtained.

Then different kinds of disturbances were simulated to test the efficiency of the proposed optimization technique. The performance indices $J$ were calculated by (4) for each test disturbance. Three cases were considered:

TABLE I.
LIST OF TEST DISTURBANCES

| Outage of double-circuit line | | | | | |
|---|---|---|---|---|---|
| Number | Branch (Node) | Number | Branch (Node) | Number | Branch (Node) |
| 1 | 100–202 | 2 | 5–8 | 3 | 8–200 |
| 4 | 4–100 | 5 | 7–100 | | |
| Outage of transformer | | | | | |
| 6 | 200–202 | 7 | 2–4 | 8 | 5–7 |
| Outage of one circuit of double-circuit line | | | | | |
| 9 | 100–202 | 10 | 5–8 | 11 | 8–200 |
| 12 | 4–100 | | | | |
| 0,1s 3-phase short circuit in node | | | | | |
| 13 | 100 | 14 | 8 | 15 | 7 |
| Load step change | | | | | |
| 16 | 100 (+25%) | 17 | 100 (-25%) | | |
| Generator step change (-25%) | | | | | |
| 18 | 101 | 19 | 203 | 20 | 3 |
| Outage of double-circuit line + autoreclosing | | | | | |
| 21 | 100–202 | 22 | 5–8 | 23 | 8–200 |
| Outage of one circuit of double-circuit line + autoreclosing | | | | | |
| 24 | 100–202 | 25 | 5–8 | | |

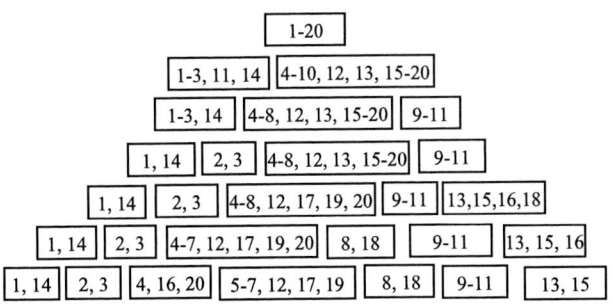

Fig. 7. Disturbances classification result.

TABLE II.
THE RULE BASE TABLE FOR G-101, 201, 203 RECEIVED AFTER GA OPTIMIZATION

| Speed deviation $\Delta\omega$ | Acceleration power $\Delta P$ | | | | |
|---|---|---|---|---|---|
| | NB | NS | ZR | PS | PB |
| Generator 101 ($k_\omega$=1,1 $k_p$=5,7) | | | | | |
| NB | NB | NB | NB | NS | ZR |
| NS | NB | NB | NM | PS | PM |
| ZR | NB | NM | ZR | PS | PB |
| PS | NS | PS | PM | PM | PB |
| PB | ZR | PS | PM | PB | PB |
| Generator 201(203) ($k_\omega$=1 $k_p$=4,1) | | | | | |
| NB | NB | NB | NB | NM | NS |
| NS | NS | NB | NS | NS | ZR |
| ZR | NM | NS | ZR | PM | PM |
| PS | NS | ZR | PM | PB | PB |
| PB | PS | PM | PB | PB | PB |

TABLE III.
MEMBERSHIP FUNCTIONS PARAMETERS

| Generator | Input signal $\Delta P$ | | Input signal $\Delta\omega$ | |
|---|---|---|---|---|
| | $X_1$ | $X_2$ | $X_1$ | $X_2$ |
| 101 | -0.4 | 0.25 | -0.48 | 0.74 |
| 201 (203) | -0.229 | 0.1 | -0.41 | 0.57 |

– generators equipped with conventional PSSs;
– generators equipped with FLPSS with expert settings [9];
– generators equipped with FLPSS tuned by GA.

These simulations show higher efficiency of an optimized FLPSS in comparison with conventional PSS. In Figs. 8–9 some examples of system response are shown for the cases with conventional PSS, expert FLPSS and FLPSS tuned by GA.

However, these calculations were performed for certain initial operating conditions. Therefore, the robustness of controllers was estimated.

For this test initial operating conditions have been changed according to consumer loads and power plant units twenty-four hour curves. Estimations were made for conventional PSS and GA-tuned FLPSS.

The tests have shown that FLPSS has more robust properties than the conventional PSS. The performance index of FLPSS changes a little at small and medium variations of operating conditions, but in some cases an essential

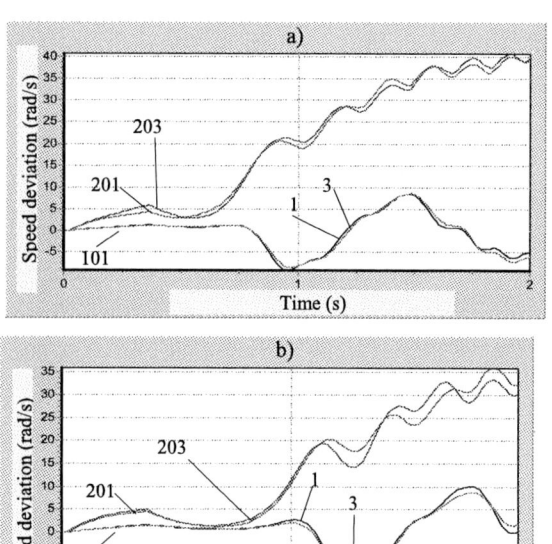

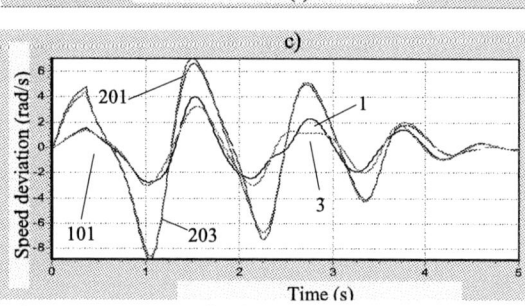

Fig. 8.System responses for disturbance 21: a) for conventional PSS; b) for FLPSS with expert settings; c) for FLPSS tuned by GA.

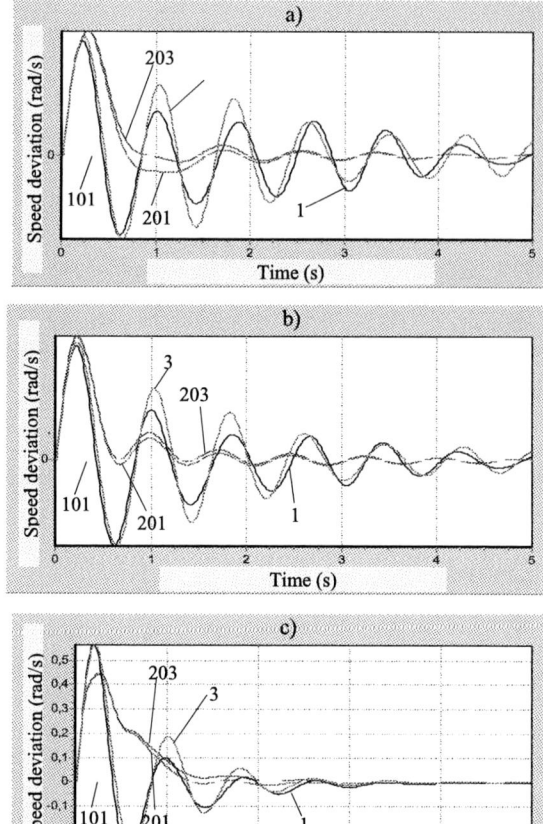

Fig. 9.System responses for disturbance 18: a) for conventional PSS; b) for FLPSS with expert settings; c) for FLPSS tuned by GA.

deterioration of the performance has been observed. This proves the necessity to retune (adapt) the FLPSS parameters.

### 2) On-line Simulation

The adaptation of scaling factors and membership functions and the fixing of rule base table are suggested for reducing the number of output neurons. The number of hidden neurons strongly depends on amount of teaching information. Thus, we used ANN consisting of 18 input neurons (values active/reactive power of 5 generator units and 4 loads), 6 output neurons (values of $k_\omega$, $k_p$, $X_{1P}$, $X_{2P}$, $X_{1\omega}$, $X_{2\omega}$) and 5 hidden neurons. The number of input nodes depends on available measurements and information.

The input information should characterize current operating conditions of the power system.

Training has been carried out for various operating conditions according to winter and summer 24-hour load curves, as well as for some emergency conditions. After ANN training, the adaptive FLPSS was estimated for the same operating conditions as described in the previous section. The summary performance indices were calculated according to (5) for adaptive FLPSS, FLPSS with fixed parameters and for conventional PSS (Fig. 10). The adaptive FLPSSs have shown the best performance in all power system operating conditions.

### B. Studies on the Siberian power system

The simplified model of regional Siberian power system "IrkutskEnergo" is shown in Fig. 11. It is characterized by a large amount of powerful hydro- and turbogenerators equipped with fast-acting excitation systems with analogous PSS and long high-voltage transmission lines. The test model consists of more than 200 nodes (including 40 generators) and more than 250 transmission lines and transformers. Simplified connections are shown by dashed lines (See Fig. 11). Long transmission lines and large power flows may lead to violation of the power system transient stability.

The list of test disturbances dangerous to transient stability

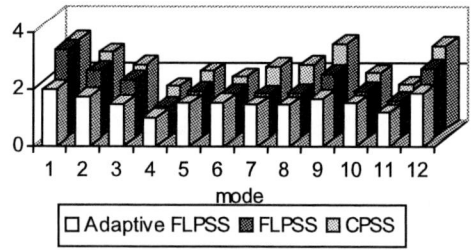

Fig. 10. Comparison of performance indices for adaptive FLPSS, FLPSS and CPSS (conventional PSS).

is presented in Table IV. Only strong disturbances were selected, namely, disturbances in electric network with the rated voltage of 500 kV. The result of disturbances classification is given in Fig. 12.

Then a GA was applied for FLPSS tuning installed on generators of Bratsk hydro power plant (node 1500). This is the most powerful power plant and it regulates the frequency in the power system. As a result of the GA optimization the optimal rule base tables, scaling factors and membership functions were obtained (Table V, Fig. 13).

The simulations have demonstrated higher efficiency of an optimized FLPSS in comparison with other types of PSS (see Table VI). The calculated values of summary index $J_\Sigma$ have also confirmed this conclusion.

In Fig.14 the system responses are shown for disturbance 16 for conventional PSS, expert FLPSS and FLPSS tuned by GA. In this case, the conventional PSS does not maintain the transient stability of the power system. Therefore, special emergency control automation has been used. It disconnects some electrical customers in deficient subsystem and part of power generating units in surplus subsystem.

Application of FLPSS prevents loss of transient stability without involving any automatic system and consequently, provides power supply to electrical customers.

### TABLE IV
### LIST OF TEST DISTURBANCES

| Line disconnection | | | | | |
|---|---|---|---|---|---|
| Number | line | Number | line | Number | line |
| 1 | 501 | 2 | 569 | 3 | 560 |
| 4 | 572 | 5 | 571 | 6 | 561 |
| 7 | 564 | 8 | 563 | 9 | 568 |
| 10 | 565 | 11 | 569+570 | 12 | 561+562 |
| 13 | 501+502 | | | | |

| Two-phase short circuit + emergency line disconnection | | | |
|---|---|---|---|
| number | node | duration | line |
| 14 | 1500 | 0.12 | 561 |
| 15 | 2500 | 0.12 | 572 |
| 16 | 1505 | 0.12 | 569, 570, 560 |
| 17 | 2540 | 0.12 | 501 |
| 18 | 6503 | 0.12 | 563 |
| 19 | 6500 | 0.12 | 565 |
| 20 | 6510 | 0.12 | 564 |

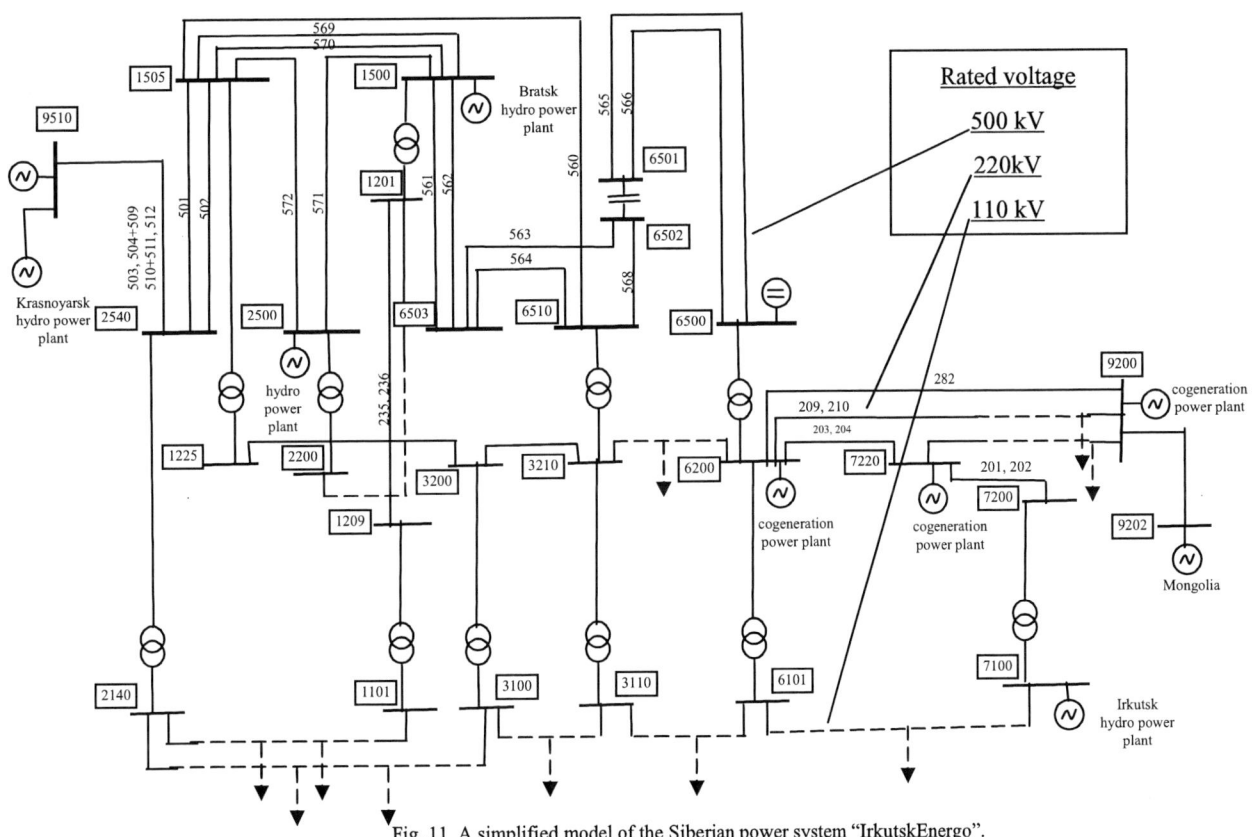

Fig. 11 A simplified model of the Siberian power system "IrkutskEnergo".

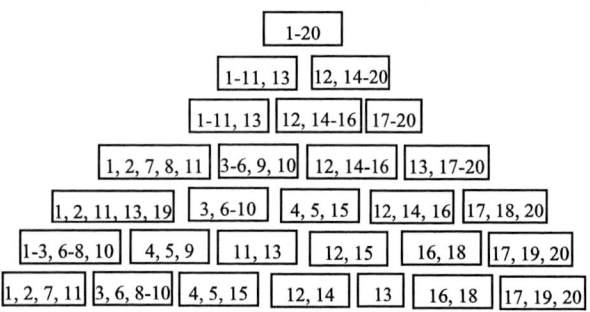

Fig. 12. Disturbances classification result.

TABLE V
THE RULE BASE TABLE OF FLPSS RECEIVED AFTER
OPTIMIZATION BY GA

| Speed deviation $\Delta\omega$ | $k_w= 1,1;\quad k_p= 4,2$ | | | | |
|---|---|---|---|---|---|
| | Acceleration power $\Delta P$ | | | | |
| | NB | NS | ZR | PS | PB |
| NB | NB | NS | NS | NS | ZR |
| NS | NM | NM | NM | NS | PS |
| ZR | NB | ZR | ZR | ZR | PB |
| PS | ZR | PS | PM | PM | PB |
| PB | NS | PS | PM | PB | PB |

a)

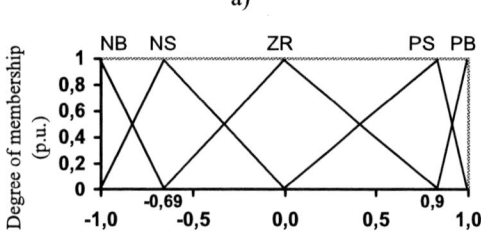

Speed deviation, $\Delta\omega$ (p.u.)

b)

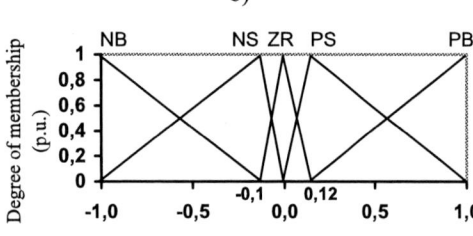

Acceleration power, $\Delta P$ (p.u.)

Fig. 13. Membership functions of input variables obtained after optimization by GA: a) for input signal $\Delta\omega$; b) for input signal $\Delta P$.

a)

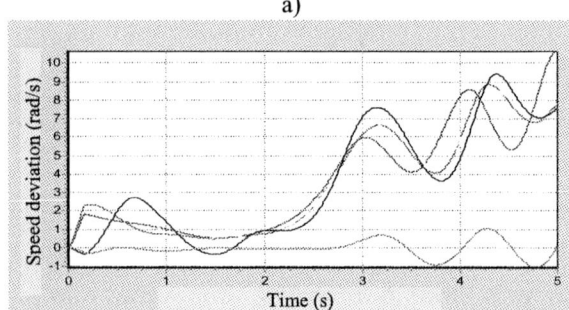

b)

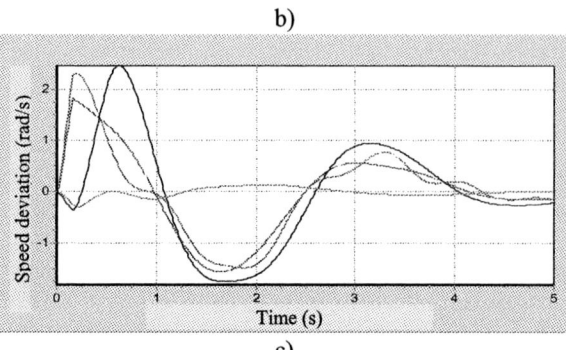

c)

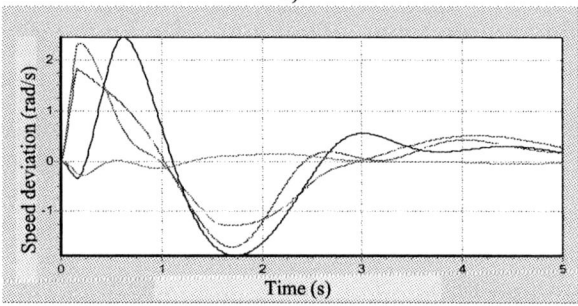

Fig. 14. System responses for disturbance 16: a) for conventional PSS; b) for FLPSS with expert settings; c) for FLPSS tuned by GA.

TABLE VI
PERFORMANCE INDEXES OF PSSs

| Number of disturbance | Conven-tional PSS | FLPSS | |
|---|---|---|---|
| | | Before optimization | After optimization |
| 1 | 13.0 | 11.4 | 11.4 |
| 2 | 5.5 | 4.6 | 4.8 |
| 3 | 10.6 | 10.1 | 8.4 |
| 4 | 20.2 | 20.7 | 16.8 |
| 5 | 18.0 | 18.8 | 14.7 |
| 6 | 9.6 | 8.9 | 7.4 |
| 7 | 6.9 | 6.2 | 5.0 |
| 8 | 9.3 | 8.4 | 7.0 |
| 9 | 10.2 | 9.5 | 8.1 |
| 10 | 10.8 | 9.9 | 8.2 |
| 11 | 15.7 | 13.1 | 15.0 |
| 12 | 25.5 | 24.0 | 23.2 |
| 13 | - | - | - |
| 13+autoreclosing | 31.9 | 29.9 | 27.2 |
| 14 | 34.3 | 31.9 | 29.9 |
| 15 | 37.9 | 36.6 | 36.8 |
| 16 | - | 31.0 | 29.0 |
| 17 | 20.1 | 16.9 | 15.6 |
| 18 | 24.6 | 22.3 | 20.9 |
| 19 | 23.4 | 21.5 | 19.6 |
| 20 | 23.6 | 21.5 | 20.0 |
| $J_\Sigma$ | 13.76 | 9.27 | 8.93 |

## IV. CONCLUSION

This paper presents tuning and adaptation procedures for FLPSS. The studies on multi-machine test power systems show that the adaptive FLPSS has better performance in comparison with conventional PSS. The conclusions of the paper can be summarized as follows:

1. The adaptive FLPSS is the effective means to enhance transient stability in large power systems. A large number of simulations for different types of disturbances have shown that the adaptive FLPSS prevents stability violations and damps oscillations in power system more efficiently in comparison with conventional PSS.

2. Proposed GA optimization technique is highly effective for tuning FLPSS. The studies have shown that the performance of FLPSS tuned by GA is better as compared to FLPSS with expert settings.

3. The preliminarily trained ANN applied for FLPSS adaptation allows one to maintain a high efficiency of controller in all variety of power system operating conditions and topologies.

4. Self-organized ANN allows various expected situations to be classified into the groups thus making possible to correctly select the training information for FLPSS.

## V. REFERENCES

[1] T. Hiyama, "Rule-based stabilizer for multi-machine power system", *IEEE Transactions on Power Systems*, vol. 5, No. 2, 1990, pp.403–409.

[2] Y. Zhang, O.P. Malik, G.P. Chen, "Artificial neural network power system stabilizers in multi-machine power system environment", *IEEE Transactions on Energy Conversion.*, vol. 10, No. 1. 1995, pp. 147-153.

[3] D.K. Chaturvedi, O.P. Malik, "Generalized neuron-based adaptive PSS for multimachine environment", *IEEE Transactions on Power Systems*, vol. 20, No. 1, 2005, pp. 358-366

[4] E. Handschin, W. Hoffmann, F. Reyer, e.a., "A new method of excitation control based on fuzzy set theory", *IEEE Transactions on Power Systems*, vol. 9, No. 1, 1994, pp.533-539

[5] H.-C. Chang and M.-H. Wang, "Neural network based self organizing fuzzy controller for transient stability of multi-machine power system", *IEEE Transactions on. Energy Conversion*, vol. 10, No. 2, 1995, pp. 339-347.

[6] You Ruhua; H.J. Eghbali, M.H. Nehrir, "An online adaptive neuro-fuzzy power system stabilizer for multimachine systems", *IEEE Transactions on Power Systems*, vol. 18, No. 1, 2003, pp. 128-135

[7] J. Wen, S. Cheng, O.P. Malik, "A synchronous generator fuzzy excitation controller optimally designed with a genetic algorithm", *IEEE Transaction on Power Systems.*, vol. 13, No. 3, 1998, pp. 884-889.

[8] P. Lakshmi, M. Abdullah Khan, "Stability enhancement of a multimachine power system using fuzzy logic based power system stabilizer tuned through genetic algorithm", *International Journal of Electrical Power and Energy Systems*, vol. 22, No. 2, 2000, pp.137-145.

[9] Liang Zhishan, Pan Kaigan and Zhang Huaguang, "Robust adaptive fuzzy excitation control of multi-machine electric power system",

*Proceedings of POWERCON'98 Conf.*, Beijing, China, Aug. 18-22, vol.2, 1998, pp. 804-808.

[10] T. Abdelazim, O.P. Malik, "An adaptive power system stabilizer using on-line self-learning fuzzy systems" *Proceedings of PES General Meeting'03*, Toronto, Canada, July 13-17, vol. 3, 2003, pp. 1715-1720.

[11] N.I. Voropai, P.V. Etingov, "Two-stage adaptive fuzzy PSS application to power systems". *Proceedings of International Conference on Electrical Engineering ICEE'2001*, Xi'an, China, July 22-26, vol. 1, 2001, pp. 314-318.

[12] N.I. Voropai, D.N. Efimov, D.B. Popov, P.V. Etingov "Fuzzy logic stabilizer modeling in the transients simulation software", *Proceedings of International Conference IEEE ISAP'2001*, Budapest, Hungary, June 18-21, 2001, pp. 315-320.

[13] T. Hiyama, "Development of fuzzy logic power system stabilizer and further studies". *Proceedings of International Conference on Systems, Man and Cybernetics IEEE SMC'99*, Tokyo, Japan, October 12–15, 1999, pp. 545–550.

[14] F. Cheong, R. Lai, "Constraining the optimization of a fuzzy logic controller using an enhanced genetic algorithm", *IEEE Transactions on Systems, Man and Cybernetics*, Part B, vol. 30, Feb 2000, pp. 31–46.

[15] Agarkov O.A., Voropai N.I., Abramenkova N.A. and Zaslavskaya T.B., "Structural analysis in power system stability studies", *Proceedings of 10th PSCC.*, Graz, Austria, Aug.30-Sept. 3, 1990, pp. 152-159.

[16] Neural Network Toolbox User's Guide, MathWorks, Inc, 1998.

## VI. BIOGRAPHIES

**Nikolai I. Voropai** (M'1996, SM'1998) is Director of the Energy Systems Institute (Siberian Energy Institute until 1997) of the Russian Academy of Sciences, Irkutsk, Russia. He is also Head of Department at Irkutsk Technical University. He was born in Belarus in 1943. He graduated from Leningrad (St. Petersburg) Polytechnic Institute in 1966 and has been with the Siberian Energy Institute since. N.I.Voropai received his degree of Candidate of Technical Sciences at Leningrad Polytechnic Institute in 1974, and Doctor of Technical Sciences at the Siberian Energy Institute in 1990. His research interests include: modeling of power systems; operation and dynamic performance of large interconnections; reliability, security and restoration of power systems; development of national, international and intercontinental electric power grids. N.I.Voropai is a member of CIGRE, a Senior Member of IEEE, and a member of PES. He is the IEEE PES Region 8 Zone East Representative.

**Pavel V. Etingov** (M'05) is a senior researcher at the Energy Systems Institute of the Russian Academy of Sciences, Irkutsk, Russia. He was born in 1976 in Irkutsk. He graduated with honors from Irkutsk State Technical University specializing in electrical engineering in 1997. He was a fellow at the Swiss Federal Institute of Technology in 2000-2001. P.V. Etingov received his PhD degree in 2003 from the Energy Systems Institute of the Russian Academy of Sciences. His research interests include stability analysis of electric power systems, emergency control, FACTS devices and application of artificial intelligence to power systems. He has been the secretary of the IEEE PES Russian Chapter since October 2005.

**2006 IEEE International Conference on Power Electronic, Drives and Energy Systems**

# Neural Approach for Automatic Identification of Induction Motor Load Torque in Real-Time Industrial Applications

A. Goedtel, I. N. da Silva, *Member, IEEE*, and P. J. A. Serni

*Abstract*—**Induction motors are widely used in several industrial sectors. However, the dimensioning of induction motors is often inaccurate because, in most cases, the load behavior in the shaft is completely unknown. The proposal of this paper is to use artificial neural networks as a tool for dimensioning induction motors rather than conventional methods, which use classical identification techniques and mechanical load modeling. Since the proposed approach uses current, voltage and speed values as the only input parameters, one of its potentialities is related to the facility of hardware implementation for industrial environments and field applications. Simulation results are also presented to validate the proposed approach.**

*Index Terms*—**Induction motors, load modeling, neural networks, parameter estimation, system identification.**

## I. INTRODUCTION

THREE-phase Induction Motors(TIM) are indispensable in many industrial applications involving the conversion of electrical energy to mechanical energy. The extensive use of this motor is frequently associated with its simple rugged structure, easy maintenance, adaptation to several load situations, and economical operation when well dimensioned. However, when the load behavior is unknown, the selection of the proper induction motor for a determined application becomes a difficult task since the usual procedure is an experimental trial in the specific application. If a particular motor presents current measures with a value over the nominal

---

This work was supported in part by the FAPESP and CNPQ under Grant (06/56093-3) and (14236/2005-4).

A Goedtel is with the Electrical Engineering Department (EESC) at University of São Paulo (USP) Av. Trabalhador São-carlense, 400 CEP 13566-590, São Carlos, SP, Brazil, Phone: +55 (16) 33739363; Fax: +55 (16) 33739363; e-mail: agoedtel@ sel.eesc.usp.br.

I. N. da Silva, is with is with the Electrical Engineering Department (EESC) at University of São Paulo (USP) Av. Trabalhador São-carlense, 400 CEP 13566-590, São Carlos, SP, Brazil, Phone: +55 (16) 33739363; Fax: +55 (16) 33739363; e-mail: insilva@ sel.eesc.usp.br.

P. J. A. Serni is with the Electrical Engineering Department (DEE) at State University of São Paulo (UNESP), CP 473, CEP 17033-360, Bauru, SP, Brazil, Phone: +55 (14) 31036115; Fax: +55 (14) 31036116; e-mail: paulojas@feb.unesp.br.

and speed under the admissible nominal value, the choice of this motor is considered to be inadequate. The next step, would then be to substitute this motor with another, more powerful than the first one.

Practical and mathematical analysis demonstrates that a three-phase induction motor, which is working in an over-dimensioned way, presents a reduction in its power factor and a decrease in its efficiency [8,9]. On the other hand, three-phase induction motors working in an under-dimensioned way present overheating and a drastic reduction in their useful life.

A study carried out at CEMIG (Electrical Energy Company of Minas Gerais State – Brazil), with 3425 three-phase induction motors, in several industrial sectors, showed that 28.7% of them were over dimensioned and 5.9% of them were under dimensioned. Another study, at COPEL (Electrical Energy Company of Parana State – Brazil) with 6108 three-phase induction motors showed that 37.75% were working over dimensioned [1].

Thus, in this paper the purpose of load torque estimation of induction motors has three main objectives. The first and most important of these is to provide information contributing to the correct dimensioning of the motor. The second objective is to provide data relating load behavior of the motor shaft in order to determine its efficiency and performance [18]. Thirdly, load torque estimation assists in the design of efficient schemes for the control of induction motors in their transient and steady states [16,17].

The conventional methods used for determining load torque, from transient to steady state, are based on direct and indirect techniques. The direct techniques consist of using winding torque meters between the motor and the load. According to [19], the use of these meters requires physical longitudinal displacing between the motor and the load. The disadvantage associated with winding torque meters is that the high start torque demanded by some loads requires an over dimensioning of the sensor element, thus reducing their sensitivity. Moreover, the meters must be carefully aligned to the motor shaft in order to avoid flexions that may reduce the useful life of these instruments, making their installation procedures slow and expensive.

On the other hand, indirect techniques for determining load torque are based on mathematical models that represent the electromechanical dynamics of the motor in relation to the load [4,7]. The use of these techniques has allowed load

---

0-7803-9771-1/06/$25.00 ©2006 IEEE

torque, from transient to steady state, to be obtained by solving differential equations that depend on knowledge of several electrical and mechanical parameters of the motor, such as rotor resistance, magnetizing inductance, rotor inertia moment, damping constant, etc. However, obtaining of these parameters in an industrial environment can be an arduous and complicated task. Moreover, the algorithms for implementing such models require extensive computations for real-time applications [6].

In order to overcome these difficulties, some load torque estimation methods, using linearized dynamical models, have been developed [3,8,9,18]. However, these models estimate load torque only for the steady state. Moreover, the computational complexity involved in the implementation of these models in terms of hardware can also make them difficult to use in real-time applications. Other models for load torque estimation, such as those based on state observer and sliding modes, have been particularly developed for control purposes, which make them difficult to apply in dimensioning of induction motors.

Recently, several methods using neural networks in problems involving induction motors have provided efficient results, particularly in the field of control [4-7]. This paper proposes an efficient neural approach to load torque estimation in induction motors. The proposed neural approach, in contrast with the methods based on linearized dynamic models, can be used to estimate load torque from transient to steady state.

The main advantage of an artificial neural network (ANN) is its ability to approximate nonlinear functional relationships [15]. When ANNs are implemented to perform load torque estimation, complex nonlinear equations, rigorous mathematical modeling, and modeling errors can be avoided altogether. The high nonlinearity involved with the load torque estimation process by mathematical models is implicitly embedded in the network itself, i.e., in the links (or interconnections) between the network layers. Indeed, neural networks can be viewed as a method for nonlinear adaptive system identification, relying on pattern recognition for the load torque identification process. Moreover, the resulting system can be successfully implemented for real-time applications.

Thus, the above considerations indicate that methodologies based on artificial neural networks can be effectively applied in two basic situations [15]. The first situation is when the relationship between the input and output variables of the process is difficult to model by conventional identification techniques. The second situation is related to implementation facilities in hardware for real-time applications. In fact, the massively parallel nature of neural networks makes them potentially fast for the computation of tasks. This same characteristic makes a neural network well suited for implementation using very-large-scale-integrated (VLSI) technologies, such as those explored in DSP.

Thus, the potential contributions of the proposed approach are the following: i) facility of implementation in hardware,

which makes it appropriate for real-time applications; ii) the system inputs can be obtained using simpler sensor devices since the input variables are current, voltage and speed signals; iii) for industrial applications, the proposed methodology is able to estimate load torque, even when the electrical and mechanical parameters of the motor, such as resistance, magnetizing inductance, rotor inertia moment, and damping constant, are unknown or inaccessible; iv) reduction of computational effort since the computations involved are reduced to the elementary matrix operations; v) the method helps to check if a motor is appropriate for a particular application.

This paper proposes, therefore, the use of artificial neural networks as an efficient method for load torque estimation with the purpose of better dimensioning the induction motor, and also for optimizing control systems, whose variables of interest are torque and failure prediction in mechanical systems. The proposed approach has excellent potential for applications in industrial environments, where load torque estimation may be required to verify if an induction motor and its start strategy are appropriate for a particular application.

The organization of the present paper is as follows. In Section II the modeling aspects involved with the induction motor are presented. Section III outlines the main principles related to the identification process using artificial neural networks. In Section IV, the methodology and simulation technique proposed in this study are described. Simulation results are presented in Section V to demonstrate the validity and performance of the developed approach. Finally, in Section VI, the key issues raised in the paper are summarized and conclusions are drawn.

## II. MODELING ASPECTS OF THE INDUCTION MOTOR

The first step involved in the design of an artificial neural network is to compile a set of input-output patterns in order to adjust its internal parameters. This procedure is also known as the training process and it must be ensured that the network is exposed to sequences of patterns representing the desired behavior in the analyzed system.

For the purpose of generating training patterns for load torque estimation, various simulations were carried out using Matlab/Simulink. The induction motor model used in the simulations was developed by [3] and it is accepted and used by many researchers as a model that is very close to physical reality [10, 17, 20]. This model takes into account various aspects involved with the electromechanical dynamics of the motor, allowing its behavior from the transient to the steady state to be simulated for several operating configurations.

The main nonlinearities considered in the simulations were the skin effect and core saturation. The skin effect is maximized during the transient time and the core saturation is directly related on the current [19]. When the current has values over its nominal value the magnetizing curve passes from a linear region to a saturated region [3,19]. Due to the short time of the simulation, the effect of temperature was not considered in the simulations.

The schematic diagram representing the input-output configuration used in induction motor simulations is presented in Fig. 1. The machine parameters, such as voltage, electric parameters of rotor and stator, load and rotor inertia moments, and load torque, are the inputs of the model. The electric current, electromagnetic torque and rotor speed are outputs of the induction motor model. These variables will be used in the training process of the neural network.

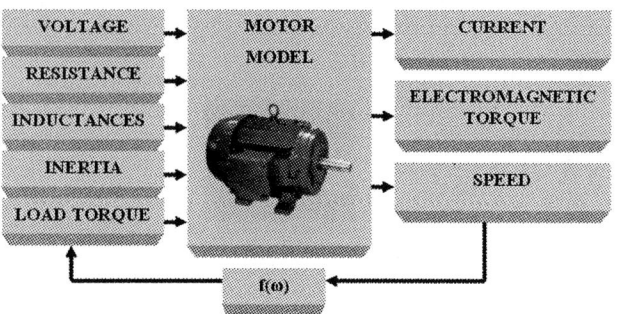

Fig. 1. Schematic diagram representing the input-output configuration used to simulate the induction motor.

The schematic diagram presented in Fig. 1 to obtain input-output data is applicable to all three-phase asynchronous induction motors. As mentioned before, the induction motor model used for generation of the input-output vectors was that proposed by [3], which is able to reproduce electromechanical behavior from the transient to the steady state in a precise way.

Table I shows the induction motor parameters used in the simulations involved with the machine model. From the diagram presented in Fig. 1, it was possible to produce enough data for the training process of the artificial neural network.

TABLE I
INDUCTION MOTOR SPECIFICATION AND LOAD PARAMETERS.

| Standard Line – 4 Poles – 60Hz – 220/380V | |
| --- | --- |
| Power | 1 hp |
| Stator Start Resistance | 10.17 ($\Omega$) |
| Stator Steady State Resistance | 12.40 ($\Omega$) |
| Rotor Start Resistance | 5.80 ($\Omega$) |
| Rotor Steady State Resistance | 6.95 ($\Omega$) |
| Stator Start Inductance | $1.77 \times 10^{-2}$ (H) |
| Stator Steady State Inductance | $2.05 \times 10^{-2}$ (H) |
| Rotor Start Inductance | $1.10 \times 10^{-2}$ (H) |
| Rotor Steady Start Inductance | $4.84 \times 10^{-2}$ (H) |
| Magnetizing Start Inductance | 0.606 (H) |
| Magnetizing Steady State Inductance | 0.546 (H) |
| Rotor Inertia Moment | $2.71 \times 10^{-3}$ (kg.m$^2$) |
| Synchronous Speed | 188.49 rad/s |
| Nominal Split | 3.8 % |
| Nominal Torque | 4.1 Nm |
| Maximum Torque | 11.89 Nm |
| Start Torque | 10.25 Nm |

## III. IDENTIFICATION USING NEURAL NETWORKS

Identification using artificial neural networks has shown promise for the solution of a series of problems involving power systems. More specifically, the use of ANN has provided alternative schemes to handling problems related to electrical machines [11]-[14]. In this study, ANNs were applied to estimate torque demanded by the load coupled onto the induction motor shaft.

The appropriate specification of an induction motor requires a knowledge of the load to be coupled onto its shaft. The lack of this information is compensated by the following procedure: current and speed are measured and, if they have values outside the range specified for that induction motor, then this motor will be substituted by another one, which is, sometimes, 100% more powerful. It is also known that induction motors working in an over-estimated manner increase losses and present a low power factor. As a consequence of all these factors, a significant amount of electrical energy can be lost.

Therefore, the main objective here is in using artificial neural networks to estimate load behavior on the motor shaft. In this study, a multilayer perceptron network was used, which was trained with a backpropagation algorithm [15]. This training algorithm has two basic steps: the first one, called propagation, applies values to the ANN inputs and verifies the response signal in its output layer. This value is then compared with the desired signal for that output. The second step occurs in the reverse way, i.e., from the output to the input layer. The error produced by the network is used in the adjustment process of its internal parameters (weights and bias) [15].

The basic element of a neural network is the artificial neuron (Fig. 2), which is also known as the node or processing element.

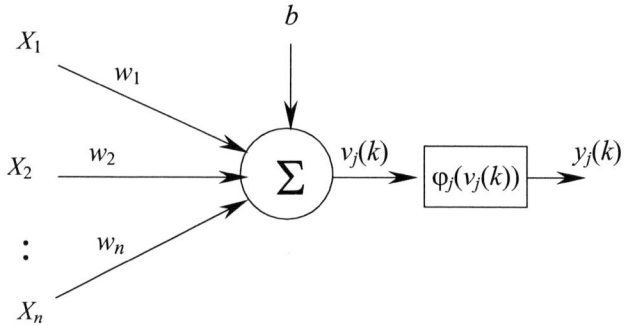

Fig. 2. Representation of the artificial neuron.

The artificial neuron illustrated in Fig. 2 can be modeled mathematically as follows:

$$v_j(k) = \sum_{i=1}^{n} X_i . w_i + b \qquad (1)$$

$$y_j(k) = \varphi_j(v_j(k)) \qquad (2)$$

where:

$n$ is the number of input signals of the neuron;

$X_i$ is the $i$-th input signal of the neuron;

$w_i$ is the weight associated with the $i$-th input signal;

$b$ is the threshold associated with the neuron;

$v_j(k)$ is the weighted response (summing junction) of the $j$-th neuron with respect to the instant $k$;

$\varphi_j(.)$ is the activation function of the $j$-th neuron;

$y_j(k)$ is the output signal of the $j$-th neuron with respect to the instant $k$.

Each artificial neuron is able to compute the respective output signal from input signals. The activation functions used to calculate the output signal are typically nonlinear. Neural networks that process analog data, which are also involved in this application, have often used sigmoid or hyperbolic tangent activation functions.

Adjusting the network weights ($w_j$) associated with the $j$-th output neuron is done by computing of the error signal linked to the $k$-th iteration or $k$-th input vector (training example). This error signal is provided by:

$$e_j(k) = d_j(k) - y_j(k) \qquad (3)$$

where $d_j(k)$ is the desired response to the $j$-th output neuron.

Adding all squared errors produced by the output neurons of the network with respect to $k$-th iteration, we have:

$$E(k) = \frac{1}{2} \sum_{j=1}^{p} e_j^2(k) \qquad (4)$$

where $p$ is the number of output neurons.

For an optimum weight configuration, $E(k)$ is minimized with respect to the synaptic weight $w_{ji}$. The weights associated with the output layer of the network are therefore updated using the following relationship:

$$w_{ji}(k+1) \leftarrow w_{ji}(k) - \eta \frac{\partial E(k)}{\partial w_{ji}(k)} \qquad (5)$$

where $w_{ji}$ is the weight connecting the $j$-th neuron of the output layer to $i$-th neuron of the previous layer, and $\eta$ is a constant that determines the learning rate of the backpropagation algorithm.

The adjustment of weights belonging to the hidden layers of the network is carried out in an analogous way. The necessary steps for adjusting the weights associated with the hidden neurons can be found in [15].

## IV. METHODOLOGY AND SIMULATION TECHNIQUE

The methodology using the neural approach proposed in this study is presented in Fig. 3, which illustrates the block diagram involving the identification process necessary to estimate load torque using the neural approach.

### A. Industrial Load Modeling

The main industrial loads are pre-sorted into four groups, i.e., quadratic, linear, inverse and constant. According to [2], these groups incorporate most of the industrial loads coupled onto the shafts of induction motors.

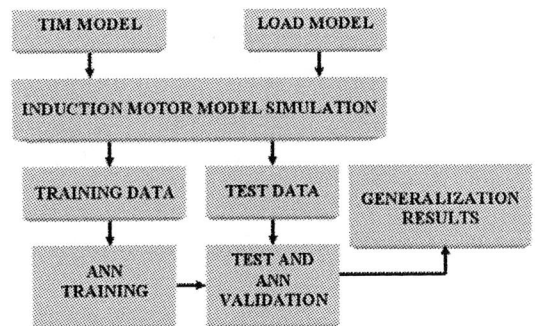

Fig. 3. Block diagram involving the identification process.

The loads are represented by mathematical functions dependent on speed and are introduced into the model presented in Fig. 1 in order to simulate the TIM behavior from transient to steady state. Table II describes these mathematical functions.

TABLE II
MATHEMATICAL FUNCTIONS FOR INDUSTRIAL LOADS

| | |
|---|---|
| Quadratic | $f(\omega) = T(\omega) = K + a.\omega^2$ |
| Linear | $f(\omega) = T(\omega) = K + a.\omega$ |
| Constant | $f(\omega) = T(\omega) = K$ |
| Inverse | $f(\omega) = T(\omega) = a\varepsilon^{-b\omega} + K$ |

For linear and quadratic loads, the constant $K$ is related to the initial torque in $\omega_{t=0+}$. In relation to constant loads, the $K$ value remains the same throughout the simulation period. For inverse loads, the $K$ value represents the torque to steady state, which means $\omega_{t=\infty}$.

The inertia moment is also an important variable in the induction motor dimensioning procedure since its value is rarely given to the electrical designer. In this study the inertia moment has been considered in the TIM mathematical model simulation. The minimum value is the inertia moment on the motor shaft running without a load (manufacturer's data) and the maximum value, according to Brazilian norms, is described by:

$$J = 0.04 P_n^{0.9} \cdot p^{2.5} \qquad (6)$$

where:

$J$ is the inertia moment in kg.m$^2$;

$P_n$ is the nominal power of the motor in kW;

$p$ is the number of pole pair.

An association was made between the minimum inertia moment for the lowest load (5% of nominal torque) and the maximum value for the highest load applied to the shaft during simulations.

The simulation of the TIM model proposed in Section II produces the data required for the ANN training process. The input data of the multilayer perceptron network are motor speed, RMS current value and voltage. The output to be computed by the artificial neural network is the load torque.

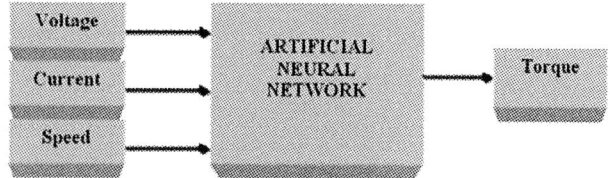

Fig. 4. Training structure for multilayer perceptron network.

Fig. 5 shows an example of linear load simulation. The quadratic load is identical to the linear load. The only difference is the alteration of the straight lines, which define the linear load, to parabolic curves. In this region, limited by curve 1 and curve 26, the ANN is trained to make the load torque estimation.

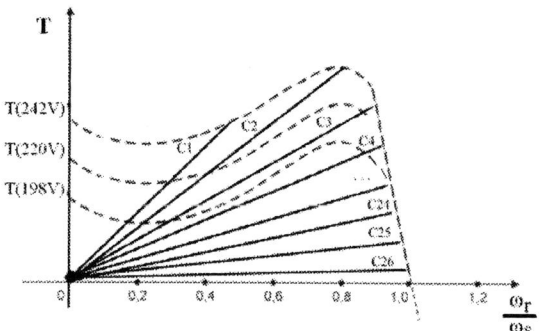

Fig. 5. Curves of linear load torque.

Fig. 6 shows the simulation of a constant load limited by curves $C_1$ to $C_{26}$. The inverse load has the same region mapped by this simulation, but the curves in this case are exponentials.

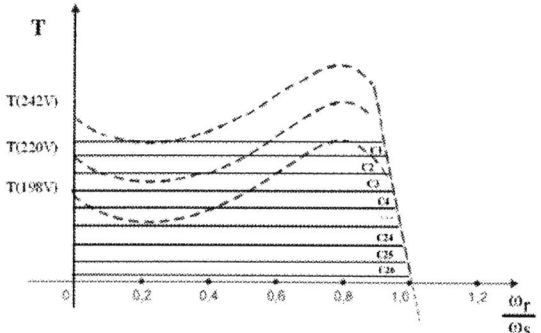

Fig. 6. Curves of constant load torque.

It is very important to highlight the difference between the torques produced by the induction motor along the source voltage variation. The value of torque for transient and steady states depends on the source voltage [9]. To reproduce this torque variation and its effects in the TIM behavior, this study simulates $C_1$ to $C_{26}$ in the range of -10% to +10% of source voltage variation in steps of 2%. There are 11 discrete zones to map the whole voltage variation. The total number of simulations for each kind of load is the product of the number of zones by the number of curves, which means 286 simulations.

## B. Training Process of the Neural Network

During the training process, the input-output pairs representing the process behavior are sequentially presented to the network in order to minimize the error produced by its output. For each load type and voltage range considered in this paper, several curves were obtained for the training process by using the model presented in Fig. 1. Each curve set, simulating a specific load from 5% to 250% of nominal torque, is composed of a vector constituted of 100 input-output pairs, which represent the transient and steady behavior of the motor for one specific load and a voltage range.

This study divides the TIM source voltage into discrete rates from -10% to +10% (198V,..., 242V), which generate a total of 11 operating regions. The main objective of this division is to improve the training efficiency and to allow the simulation of real industrial environment. A representative diagram of this procedure is presented in Fig. 7. The division into load classes (linear, quadratic, inverse and constant) increases the system modularity and reduces the computational effort involved in ANN training.

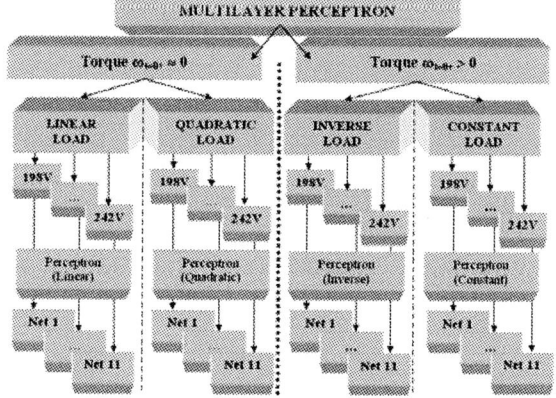

Fig. 7. Block diagram of the training procedure.

Thus, as an illustrative example, for the case of a quadratic load and a specific voltage range (198V,...,242V) 26 simulations (curves) were generated, which simulate the motor's behavior from 5% to 250% of nominal torque, where 13 were used for the training process and 13 for the testing process. Each training set, composed of 13 curves, is constituted by 100 input-output pairs, which are sequentially grouped to produce the training matrix. Each voltage range and each load type has a specific neural network described in Fig. 7 as Net 1 to Net 11.

After the training process, the network is able to estimate load torque curve from sequential values of speed, current and voltage. In this case, the testing process used to validate the proposed approach consists of using other operating configurations that were absent during the training process.

For linear, quadratic and inverse loads, the ANN generalization results have reached satisfactory values with 5 neurons in the first hidden layer, 25 neurons in the second hidden layer and 1 neuron in the output. The neural structure for the constant load is smaller than the structure used for the other loads. This load has shown satisfactory generalization results with 5 neurons in the first hidden layer, 15 neurons in the second hidden layer and 1 neuron in the output.

## V. SIMULATION RESULTS

Simulation results are presented in this section. It must be highlighted that the input data set (voltage, current and speed) presented to the network did not participate in the training process. The output of the network is the torque estimation.

### A. Quadratic Load

For quadratic loads 26 simulations were carried out, from transient to steady state of the TIM, using each source voltage scale. In this data set 13 simulated curves were used in the training process and the other 13 in the network validation process. In these simulations, the induction motor specified in Table I was used as a reference.

Fig. 8 shows the generalization results obtained for a quadratic load demanding 0.5 Nm at the start of the machine and 150% of nominal torque in its steady state. This motor (Table I) works in an under-dimensioned way and its useful life will probably be reduced.

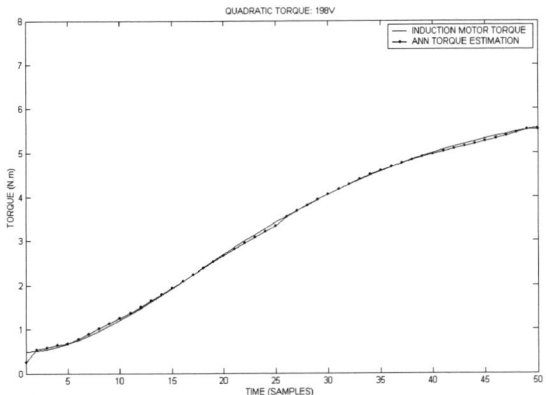

Fig. 8. TIM submitted to quadratic load and 198V source voltage.

Fig. 9 shows the generalization results for an induction motor (Table I) submitted to a quadratic load and 215 V source voltage. In this simulation, despite the 5% error between the desired value and the ANN output, it is possible to infer the satisfactory dimensioning of the motor in this case.

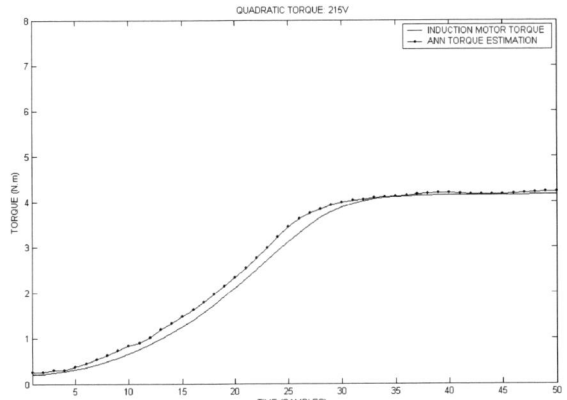

Fig. 9. TIM submitted to quadratic load and 215V source voltage.

### B. Linear Load

The training procedure for linear loads was similar to that for quadratic loads. A total of 26 simulated curves were used, of which 13 participated in the training process and 13 in the validation process.

Fig. 10 shows the simulation results for a linear load with a 220 V source voltage. In steady state this load demands 180% of the nominal torque (Table I). In this case, the TIM should be exchanged for another of 2 hp, which presents nominal torque of 8.3 Nm and is the best option for an alternative, considering the motors produced by Brazilian suppliers.

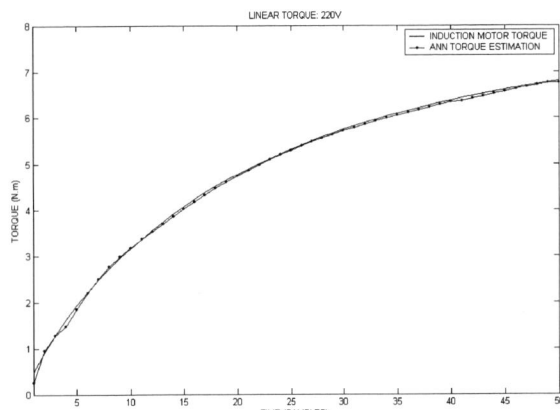

Fig. 10. TIM submitted to a linear load and 220V source voltage.

Fig. 11 shows the simulation results of a TIM (Table I) submitted to a linear load and 215V source voltage. The load demands about 90% of the nominal torque in steady state. The analysis of this motor, considering the torque from transient to steady state, indicates that it is well dimensioned.

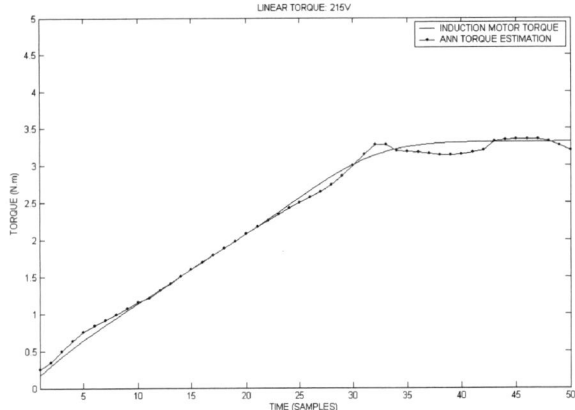

Fig. 11. TIM submitted to a linear load and 215V source voltage.

### C. Constant Load

For constant loads, a total of 20 simulations from transient to steady state were carried out for each source voltage range. The simulations were divided into two groups, i.e., 10 were used in the ANN training process and the other 10 in the ANN validation process.

923

The simulations demanding higher torque and supplied with a low source voltage presented problems in the transient state. To solve this question the range of this type of kind was reduced from 5% to 150% of nominal torque. Fig. 12 shows the simulation results of a 1 hp TIM (Table I) with a source voltage of 198V. The curve obtained from the simulation shows an under-dimensioned motor that should be changed for another of 1.5 hp.

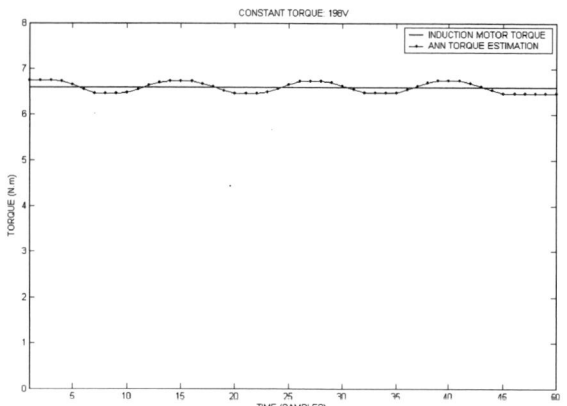

Fig. 12. TIM submitted to a constant load and 198V source voltage.

Fig. 13 shows the simulation of a TIM with a source voltage of 211 V. The constant torque imposed on the motor was 95% of the nominal torque.

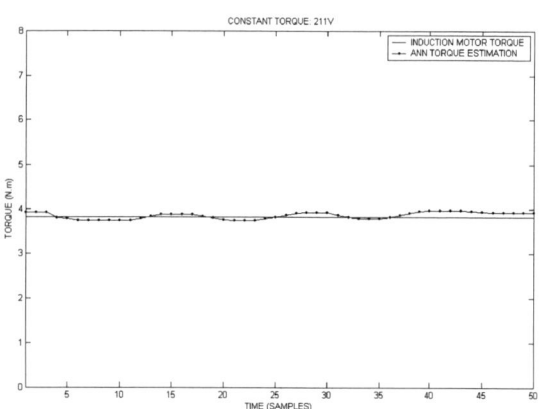

Fig. 13. TIM submitted to a constant load and 211V source voltage.

*D. Inverse Load*

In the inverse load training process 14 simulations were used for each voltage range. A total of 7 curves were used for ANN training and the other 7 curves in the validation process.

Fig. 14 shows the generalization results of an inverse load applied to the TIM shaft (Table I) with a 198V source voltage. This load starts with 140% of the nominal torque and reduces the torque as the speed increases. The load reduces asymptotically to 2.0 Nm in the steady state. This motor may be considered well dimensioned in this application.

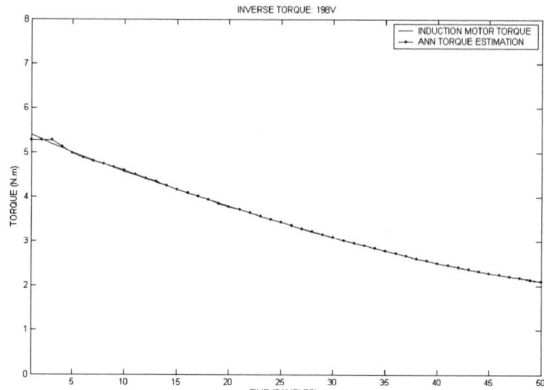

Fig. 14. TIM submitted to an inverse load and 198V source voltage.

Fig. 15 shows another simulation, also with an inverse load, but where the induction motor has a 220V source voltage. The initial torque is 170% of the nominal value and reduces to a value close to the nominal torque in steady state (4.1 Nm). This TIM may also be considered well dimensioned. In this case there is one situation that must be analyzed, i.e., if the source voltage is reduced then the motor may be difficult to start. In order to solve this problem, it is common for this motor to be erroneously substituted by another with higher power.

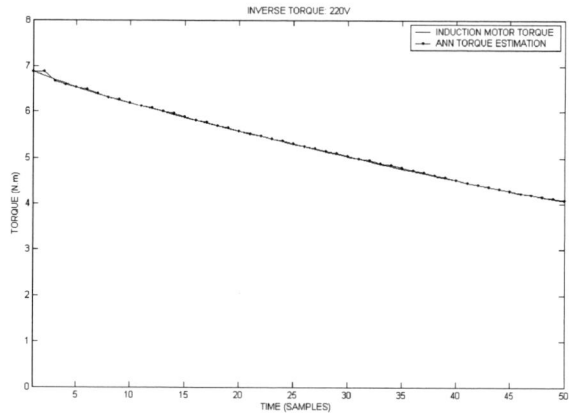

Fig. 15. TIM submitted to an inverse load and 220V source voltage.

The simulation results presented in this section confirm the usefulness of artificial neural networks for estimating the torque applied to the three-phase induction motor shaft. The simulations have shown good generalization results over a wide torque range.

An analysis of all results and curves provided by the proposed approach reveals that the load torque estimation error was smaller than 5% for most of the simulations. This performance may be considered quite appropriate because induction motors are usually designed to support torque tolerances over 5% depending on the motor work cycle [2].

## VI. CONCLUSIONS

This paper describes the application of artificial neural networks to load torque estimation used mainly for industrial purposes. The simulation results are considered very satisfactory. The torque is estimated from 5% to 250% of the TIM nominal torque for quadratic and linear load situations. For inverse and constant loads the estimation goes from 5% to 150% of the TIM nominal torque. The methodology used in this study can be applied to other loads and other types of motors.

The method developed in this study can be used as an alternative tool to identify the behavior of loads applied to an induction motor shaft and to help the designer to choose the best motor for each application. Characteristics of this technique like portability and easy practical implementation are worthy of note. Another advantage is that this method requires no knowledge of electrical and mechanical parameters of the motor after the neural network training process. The parameters are implicit in the network weights.

The main justifications for the use of this technique are the following: i) torque estimation is made directly by an artificial neural network, ii) the method may be used as an auxiliary tool in the control and dimensioning of a TIM, iii) the method helps to check if a motor is over estimated or under dimensioned, iv) the proposed approach only requires simple measurements of current, voltage and speed, v) the practical implementation of this approach can be carried out either with hardware or software. Therefore, the method here developed contributes significantly towards the reduction of electric energy losses and the increase of the power factor, which are derived from bad dimensioning of electrical motors.

Finally, for industrial applications involving real-time processes, the proposed methodology has excellent potential for hardware implementation using DSP, which can be successfully integrated with the design of efficient schemes for control of motors in their transient and steady states.

## VII. REFERENCES

[1] L. C. Marach, "A Methodology for Overestimated Motors Substitution," *Modern Electricity Magazine* (in Portuguese), pp. 220-228, Aug. 2001.

[2] L. P. C. Dias and O. S. Lobosco, *Electrical Motors: Selection and Application* (in Portuguese), McGraw-Hill, 1988.

[3] C. M. Ong, *Dynamic Simulation of Electric Machinery Using Matlab/Simulink*, Prentice Hall, 1997.

[4] T. C. Chen and T. T. Sheu, "Model reference neural network controller for induction motor speed control," *IEEE Transactions on Energy Conversion*, vol. 17, no. 2, pp. 157-163, 2002.

[5] F. J. Lin and R. J. Way, "Adaptive fuzzy-neural-network control for induction spindle motor drive," *IEEE Transactions on Energy Conversion*, vol. 17, no. 2, pp. 507-513, 2002.

[6] C. Y. Huang, T. C. Chen and C. L. Huang, "Robust control of induction motor with a neural-network load torque estimator and a neural-network identification," *IEEE Transactions on Industrial Electronics*, vol. 46, no. 5, pp. 990-997, 1999.

[7] T. T. Sheu and T. C. Chen, "Self-tuning control of induction motor drive using neural network identifier," *IEEE Transactions on Energy Conversion*, vol. 14, no. 4, pp. 881-886, 1999.

[8] E. Fitzgerald, C. Kingsley and A. Kusko, *Electric Machinery*, McGraw-Hill, 1975.

[9] L. Kosov, *Electrical Machinery and Transformers*, Prentice-Hall, 1972.

[10] J. B. Song and K. S. Byun, "Design of time delay controller using variable reference model," *Control Engineering Practice*, vol. 8, pp. 581-588, 2000.

[11] J. Campbell and M. Summer, "Practical sensorless induction motor drive employing an artificial neural network for online parameter adaptation," *IEE Proc. Electr. Power Appl.*, vol. 149, no 4, July 2002.

[12] S. Kolla and L. Varatharasa, "Identifying three-phase induction motor faults using artificial neural networks," *ISA Transactions*, vol. 39, pp. 433-439, 2000.

[13] D. Kukolj, F. Kulic and E. Levi, "Design of speed controller for sensorless electric drives based on AI techniques: a comparative Study," *Artificial Intelligence in Engineering*, vol. 14, pp. 165-174, 2000.

[14] A. L. Orille, G. M. Sowilam, "Application of Neural Networks for Direct Torque Control," *Computers & Industrial Engineering*, no. 37, pp. 391-394, 1999.

[15] S. Haykin, *Neural Networks*, 2nd Edition, Prentice-Hall, 1999.

[16] P. Vas, *Artificial Intelligence Based Electrical Machines and Drives*, Oxford University Press, 1999

[17] P. Vas, *Sensorless Vector and Direct Torque Control*, Oxford University Press, 1998.

[18] Y. El-Ibiary, "An accurate low-cost method for determining electric motors' efficiency for the purpose of plant energy management," *IEEE Transactions on Industry Applications*, vol. 39, pp.1205-1210, 2003.

[19] A. Goedtel, *Load Torque Estimation in Induction Motor Shaft Using Artificial Neural Networks*, M.Sc. Thesis (in Portuguese), State University of São Paulo (UNESP\PPGEI), Brazil, 2003.

[20] P. C. Krause, O. Wasynczuk and S. D. Sudhoff, *Analysis of Electric Machinery and Drive Systems*, 2nd Edition, John Wiley, 2002.

## VIII. BIOGRAPHIES

**Ivan Nunes da Silva** was born in São José do Rio Preto, Brazil, in 1967. He received both M.Sc. and Ph.D. degree in electrical engineering from the State University of Campinas, Brazil, in 1995 and 1997, respectively. Currently he is an Associate Professor at the University of São Paulo (USP), Brazil. His research interests are within the fields of artificial neural networks, power systems, nonlinear optimization, identification and control.

**Alessandro Goedtel** was born in Arroio do Meio, Brazil, in 1972. He graduated in electrical engineering from the Federal University of Rio Grande do Sul and received M.Sc. degree in industrial engineering from the São Paulo State University, in 2003. Currently he is a Ph.D. student in electrical engineering at the University of São Paulo (USP). His research interests are within the field of intelligent systems and electrical machinery.

**Paulo José Amaral Serni** was born in Botucatu, Brazil in 1957. He received both his M.Sc. and Ph.D. degree in electrical engineering from the State University of Campinas, Brazil, in 1992 and 1999, respectively. Currently he is an Assistant Professor at the State University of São Paulo, Brazil. His research interests are within the fields of artificial neural networks, power electronics, electrical machines, data acquisition systems and instrumentation.

**2006 IEEE International Conference on Power Electronic, Drives and Energy Systems**

# Speed Estimation for Sensorless Technology Using Recurrent Neural Networks and Single Current Sensor

A. Goedtel, I. N. da Silva, *Member, IEEE,* P. J. A. Serni

*Abstract*—The use of sensorless technologies is an increasing tendency on industrial drivers for electrical machines. The estimation of electrical and mechanical parameters involved with the electrical machine control is used very frequently in order to avoid measurement of all variables involved in this process. The cost reduction may also be considered in industrial drivers, besides the increasing robustness of the system, as an advantage of the use of sensorless technologies. This work proposes the use of artificial neural networks to estimate one of the most important variables in the induction motor control schemes: the speed. Simulation results are presented to validate the proposed approach.

## I. INTRODUCTION

THE Three-Phase Induction Motors (TIM) are used in many industrial sectors as leading element to convert electrical in mechanical energy. The control strategies of these machines make use of electronic drivers based on sensorless technologies.

The sensorless technology is an increasing tendency in the TIM control. The conventional control methods, which are based on direct measurement of the machine variables like torque and speed, have some disadvantages besides the cost involved.

The speed can be measured with optical encoders, electromagnetic resolvers or brushless d.c. tachogenerators. However, the use of these electromechanical devices presents some limitations in their application, like the increasing cost of the driver, reduced mechanical robustness, low noise immunity, they affect the machine inertia and require a special attention in hostile environments [1].

---

This work was supported in part by the FAPESP and CNPQ under Grant (06/56093-3) and (14236/2005-4).

A Goedtel is with the Electrical Engineering Department (EESC) at University of São Paulo (USP) Av. Trabalhador São-carlense, 400 CEP 13566-590, São Carlos, SP, Brazil, Phone: +55 (16) 33739363; Fax: +55 (16) 33739363; e-mail: agoedtel@ sel.eesc.usp.br.

I. N. da Silva, is with is with the Electrical Engineering Department (EESC) at University of São Paulo (USP) Av. Trabalhador São-carlense, 400 CEP 13566-590, São Carlos, SP, Brazil, Phone: +55 (16) 33739363; Fax: +55 (16) 33739363; e-mail: insilva@ sel.eesc.usp.br.

P. J. A. Serni is with the Electrical Engineering Department (DEE) at State University of São Paulo (UNESP), CP 473, CEP 17033-360, Bauru, SP, Brazil, Phone: +55 (14) 31036115; Fax: +55 (14) 31036116; e-mail: paulojas@feb.unesp.br.

The use of sensorless techniques is found mainly in high performance applications like vector-controlled drives and direct torque controlled drivers [1-6]. The main approaches are open-loop estimators using monitored stator and voltage currents, state observers, model reference adaptive systems and estimators using artificial intelligence [1]. Various approaches were proposed where the speed estimation is based on applied voltage, line current and frequency [14].

It was mentioned in the work of Joachim Holtz, in 2002, [15] that the research was concentrated on the elimination of the speed sensor at the machine shaft without deteriorating the dynamic response of the control system. Many sensorless techniques and control strategies are discussed in this work, such as: i) the model reference of adaptive system based on the rotor flux, ii) the model reference based on the induced voltage, iii) feedforward control of stator voltages, iv) rotor field orientation with improved stator model, v) accurate speed estimation based on rotor slot harmonics, vi) low speed estimation by weakening field and vii) sensorless using signal injection [15].

The proposal of this work is to present an alternative approach of the speed estimation of the induction motor using artificial neural networks in a recurrent configuration. The inputs of the network are RMS current, speed estimated in the output with feedback and delay by *n*-samples, where *n* is an integer number.

The efficient estimation of speed of the induction motor is also considered using artificial neural networks in a recurrent configuration with one single sensor: the current. The method proposed is based on off-line training considering different types of load and wide range of voltage applied to the induction motor. The current is quite simple to measure for the driver data acquisition system and the control processing effort is reduced to simple matrix solving after the neural network is trained. The use of speed estimator is the basis of some feedback-controlled drivers [1].

The use of artificial neural network has the following advantages: i) the ANN, after trained, is reduced to a weight matrix; the matrix solving requires lower computation effort compared to equation solving; ii) the machine parameters are implied in the network weights; iii) the portability of the solution in lower cost hardware.

Recently, several methods using neural networks in problems involving induction motor have provided efficient results, particularly in the field of control [8-11]. An artificial

0-7803-9771-1/06/$25.00 ©2006 IEEE

neural network is a massively parallel-distributed processor that has a natural propensity for storing experiential knowledge and making it available for use. The main advantage of an ANN is its ability to approximate nonlinear functional relationships.

This work is organized in five sections. The modeling aspects of the TIM are presented in the second section. The third section presents the basis of artificial neural networks. The fourth section presents the methodology and simulation results and, in the fifth section, the conclusions of this work are finally presented.

## II. MODELING ASPECTS OF THE INDUCTION MOTOR

The first step involved in the design of an artificial neural network is to compile a set of input-output patterns in order to adjust its internal parameters. This procedure is also known as training process and it must be ensured that the network is exposed to sequences of patterns representing the desired behavior of the analyzed system.

For the purpose of generating training patterns for load torque estimation, various simulations were carried out using Matlab/Simulink. The induction motor model used in the simulations was developed by [12] and it is accepted and used by many researchers as a model that is very close to physical reality [1,16,17]. This model takes into account various aspects involved in the electromechanical dynamics of the motor, allowing its behavior from transient to steady state to be simulated in several operating configurations.

The main nonlinearities considered in the simulations were the skin effect and core saturation. Since the skin effect depends on slip, it is then maximized during the transient time and also when higher loads are connected to the TIM shaft. Core saturation is directly related to the current value [7]; when the current has values over its nominal value the magnetizing curve goes from a linear region to a saturated region [7,12].

More specifically, this saturation behavior was reproduced by a Matlab/Simulink block using two straight lines. The first line (lower bound) has considered a direct relationship between flux and current of the motor ($d\psi/di=1$), where $\psi$ is the mutual flux and $i$ is the stator RMS current. The second line has taken into account a lower increase in the flux with respect to the current increase, which defines a second region between current and flux that has a lower inclination when compared with the first one ($d\psi/di < 1$). The transition between these two regions (non-saturated and saturated) is implemented by an exponential function that smoothes the shift from one region to another and also avoids abrupt discontinuities in $\psi(i)$. When the current increases its value over the nominal value, the output flux of the block has its value decreased by a constant value to simulate the saturation effect [12].

Thus, the implemented saturation block imposes upper and lower bounds on a signal. When the input signal is within the range specified by the lower limit and upper limit parameters, the input signal passes through unchanged. When the input signal is outside these bounds, the signal is clipped to the upper or lower bound. Due to the short time of the simulation, the effect of temperature was not considered in the simulations.

The schematic diagram representing the input-output configuration used in induction motor simulations is presented in Fig. 1. The machine parameters, such as voltage, electric parameters of rotor and stator, load and rotor inertia moments, and load torque, are the inputs of the model. The electric current, electromagnetic torque and rotor speed are outputs of the induction motor model. These variables will be used in the training process of the neural network.

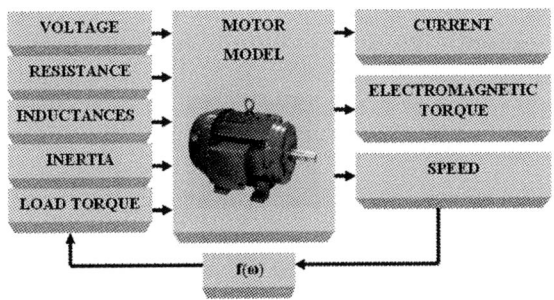

Fig. 1. Schematic diagram representing the input-output configuration used to simulate the induction motor.

The schematic diagram presented in Fig. 1 can be applied to all three phase induction to obtain input-output data. As mentioned before, the induction motor model used for generation of the input-output vectors was that proposed by [7], which is able to reproduce electromechanical behavior from the transient to the steady state in a precise way. Table I presents the electrical and mechanical parameters used in TIM simulations.

TABLE I
INDUCTION MOTOR SPECIFICATION AND LOAD PARAMETERS.

| Standard Line – 4 Poles – 60Hz – 220/380V | |
|---|---|
| Power | 1 hp |
| Stator Start Resistance | 10.17 ($\Omega$) |
| Stator Steady State Resistance | 12.40 ($\Omega$) |
| Rotor Start Resistance | 5.80 ($\Omega$) |
| Rotor Steady State Resistance | 6.95 ($\Omega$) |
| Stator Start Inductance | $1.77 \times 10^{-2}$ (H) |
| Stator Steady State Inductance | $2.05 \times 10^{-2}$ (H) |
| Rotor Start Inductance | $1.10 \times 10^{-2}$ (H) |
| Rotor Steady Start Inductance | $4.84 \times 10^{-2}$ (H) |
| Magnetizing Start Inductance | 0.606 (H) |
| Magnetizing Steady State Inductance | 0.546 (H) |
| Rotor Inertial Moment | $2.71 \times 10^{-3}$ (kg.m$^2$) |
| Synchronous Speed | 188.49 rad/s |
| Nominal Split | 3.8 % |
| Nominal Torque | 4.1 Nm |
| Maximum Torque | 11.89 Nm |
| Start Torque | 10.25 Nm |

## III. IDENTIFICATION USING NEURAL NETWORKS

Identification using artificial neural networks has shown promise for the solution of a series of problems involving power systems. More specifically, the use of ANN has provided alternative schemes to handle problems related to electrical machines [18,19]. In this study, ANNs were applied to estimate speed in the induction motor shaft.

Therefore, the main objective here is the use of artificial neural networks to estimate load behavior on the motor shaft. In this study, a multilayer perceptron network was used, which was trained with a backpropagation algorithm [13]. This training algorithm has two basic steps: the first one, called propagation, applies values to the ANN inputs and verifies the response signal in its output layer. This value is then compared with the desired signal for that output. The second step occurs in the reverse way, i.e., from the output to the input layer. The error produced by the network is used in the adjustment process of its internal parameters (weights and bias) [13]. The basic element of a neural network is the artificial neuron (Fig. 2), which is also known as the node or processing element.

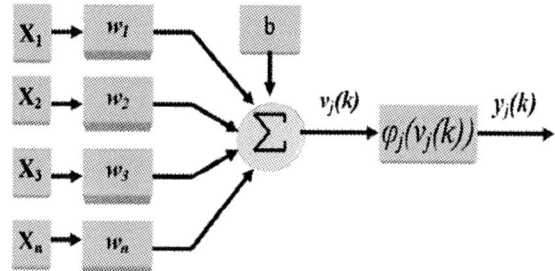

Fig. 2. Representation of the artificial neuron.

The artificial neuron illustrated in Fig. 2 can be modeled mathematically as follows:

$$v_j(k) = \sum_{i=1}^{n} X_i . w_i + b \qquad (1)$$

$$y_j(k) = \varphi_j(v_j(k)) \qquad (2)$$

where:
$n$ is the number of input signals of the neuron;
$X_i$ is the $i$-th input signal of the neuron;
$w_i$ is the weight associated with the $i$-th input signal;
$b$ is the threshold associated with the neuron;
$v_j(k)$ is the weighted response (summing junction) of the $j$-th neuron with respect to the instant $k$;
$\varphi_j(.)$ is the activation function of the $j$-th neuron;
$y_j(k)$ is the output signal of the $j$-th neuron with respect to the instant $k$.

Each artificial neuron is able to compute the respective output signal from input signals. The activation functions used to calculate the output signal are typically nonlinear. Neural networks that process analog data, which are also involved in this application, have often used sigmoid or hyperbolic tangent activation functions.

Adjustment of the network weights ($w_j$) associated with the $j$-th output neuron is done by computing the error signal linked to the $k$-th iteration or $k$-th input vector (training example). This error signal is provided by:

$$e_j(k) = d_j(k) - y_j(k) \qquad (3)$$

where $d_j(k)$ is the desired response to the $j$-th output neuron.

Adding all squared errors produced by the output neurons of the network with respect to $k$-th iteration, we have:

$$E(k) = \frac{1}{2} \sum_{j=1}^{p} e_j^2(k) \qquad (4)$$

where $p$ is the number of output neurons.

For an optimized weight configuration, $E(k)$ is minimized regarding the synaptic weight $w_{ji}$. The weights associated with the output layer of the network are, therefore, updated using the following relationship:

$$w_{ji}(k+1) \leftarrow w_{ji}(k) - \eta \frac{\partial E(k)}{\partial w_{ji}(k)} \qquad (5)$$

where $w_{ji}$ is the weight connecting the $j$-th neuron of the output layer to $i$-th neuron of the previous layer, and $\eta$ is a constant that determines the learning rate of the backpropagation algorithm.

The adjustment of weights belonging to the hidden layers of the network is carried out in an analogous way. The necessary steps for adjusting the weights associated with hidden neurons can be found in [15].

## IV. METHODOLOGY AND SIMULATION RESULTS

The methodology using the neural approach proposed in this study is presented in Fig. 3, which illustrates the block diagram involving the identification process necessary to estimate load torque using the neural approach.

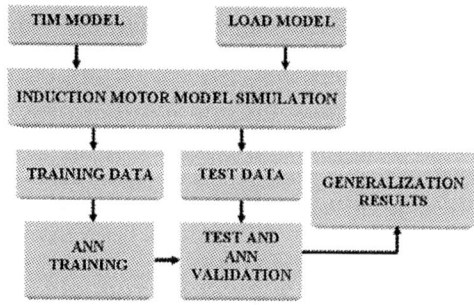

Fig. 3. Block diagram involving the identification process.

Fig. 4 shows the block diagram proposed for ANN recurrent operation. During the training process, the input-output pairs representing the system behavior are sequentially presented to the network in order to minimize the error produced by its output. For each load value and fixed voltage (220V), several curves were obtained for the training process by using the model presented in Fig. 1. Each curve set, simulating a specific load from 5% to 150% of nominal torque, is composed of a vector made of 60 input-output pairs, which represent the transient and steady behavior of the motor for one specific load and a voltage range.

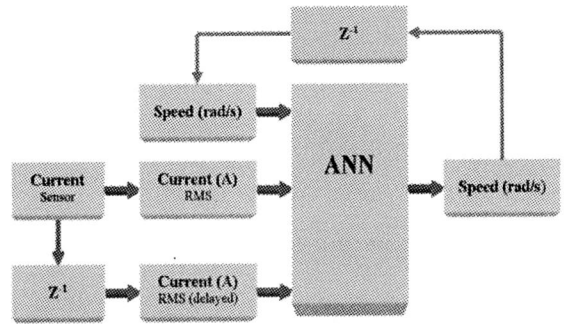

Fig. 4. Block diagram involving the identification process.

The speed used as input during the ANN training process was delayed in 10 samples. The first 10 samples were assumed as zero. The following samples were the desired value of speed, generated by the TIM simulation, presented to the ANN as input.

Fig. 5 presents the simulation results of five curves where each one is a constant torque applied to the induction motor shaft varying from 0.75 to 5.5 Nm (18.29% to 134% of nominal torque). The five test curves, which did not take part in the training set, were presented to an ANN with five neurons in the first hidden layer, 5 neurons in the second hidden layer and one neuron in the output layer.

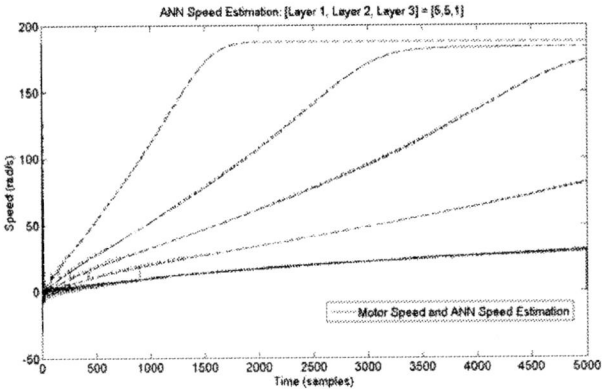

Fig. 5. Generalization results using 5 neurons in the first hidden, 5 neurons in the second hidden layer and 1 neuron in the output.

The same curves used in the simulation presented in Fig. 5 are used to generate the Fig. 6. However, the data was presented to a different ANN structure. The ANN structure of Fig. 6 has 5 neurons in the first hidden layer, 10 neurons in the second hidden layer and 1 neuron in the output layer. The results of Fig. 6 are not as precise as presented in Fig. 5. This structure, in this case, reaches better performance with fewer neurons in the second hidden layer, which demand lower computation effort to estimate de TIM speed.

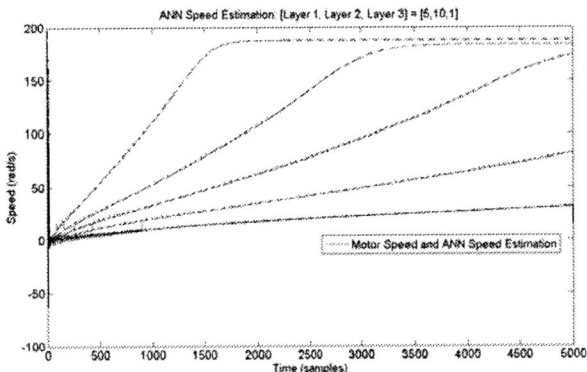

Fig. 6. Generalization results using 5 neurons in the first hidden layer, 5 neurons in the second hidden layer and 1 neuron in the output.

Fig. 7 presents a long-term simulation. The induction motor goes from transient time to steady state with a constant load applied to its shaft of 0.5 Nm. In 3 seconds (3000 samples at 1k samples rate) the torque is increased to 2 Nm, which is half of the nominal torque of the motor. In five seconds the torque was increased to 5 Nm. This simulation was presented to the structure described in Fig. 4 with 5 neurons in the first hidden layer, 5 neurons in the second hidden layer and 1 neuron in the output layer.

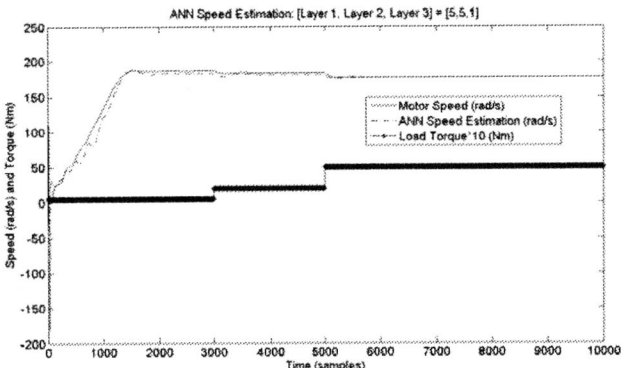

Fig. 7. Long term simulation using 5 neurons in the first hidden layer, 5 neurons in the second hidden layer and 1 neuron in the output.

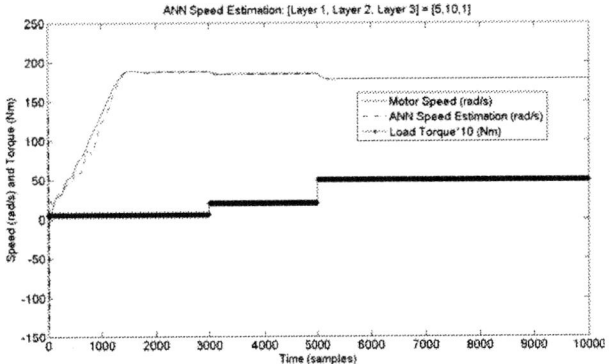

Fig. 8. Long term simulation using 5 neurons in the first hidden layer, 10 neurons in the second hidden layer and 1 neuron in the output.

The same long-term simulation data was presented to the second structure, which has good generalization results in the short-term simulation illustrated in Fig. 6. Figure 8 presents the long-term simulation presented to the structure with 5 neurons in the first hidden layer, 10 neurons in the second hidden layer and 1 neuron in the output layer.

## V. CONCLUSION

The simulation results confirm the use of ANN networks for speed estimation produced by the induction motor. The proposed method requires just computational effort for the ANN weight adjustment during off-line training. After the training process, based on PC platform, the synaptic weights may be transferred to the electronic driver. The proposed method reduces the setup time of the driver rather than on-line parameter adaptation, which makes the proposed approach valuable for industrial application.

The proposed approach only requires simple measurements of current and the practical implementation of this approach can be carried out either by hardware or by software. The speed estimation is also made by a recurrent artificial neural network.

The main justifications for the use of this technique are the following: i) speed estimation is made directly by an artificial neural network, ii) the method may be used as an auxiliary tool in the speed estimation of a TIM, iii) the proposed approach only requires simple measurements of current. Therefore, the method developed here contributes significantly towards the sensorless technologies and its application in the induction motor control.

## VI. REFERENCES

[1] P. Vas. *Sensorless Vector and Direct Torque Control*, Oxford University Press, USA, 1998.

[2] L. B. Brahin, "Motor Speed Identification via Neural Networks". *IEEE Industry Applications Magazine*, vol. 1, pp. 28-32, 1995.

[3] W. Dazhi, G. Shusheng and W. Kenan, "A Neural Network Base Adaptive Estimator for Speed-Sensorless Control of Induction Motor". Proc. of the 4th World Congress on Intelligent Control and Automation, vol. 4, pp. 2812-2816, 2002.

[4] M. Mohamadian, E. P. Nowicki, E. Evanik, F. Ashrafzadeh, R. Sachdeva and A. Chu, "A Novel Neural Network Controller and its Efficient DSP Implementation for Vector Controlled Induction Motor Drives". IEEE 37th IAS Annual M., vol. 2, pp. 1455-1462, 2002.

[5] P. P. Cruz and J. J. R. Rivas, "A Small Neural Network Structure Application in Speed Estimation of an Induction Motor Using Direct Torque Control". IEEE 32nd Annual Power Electronics Specialists Conference, vol. 2, pp. 823-827, 2001.

[6] S. H. Kim, T. S. Park, J. Y. Yoo and G. T. Park, "Speed-Sensorless Vector Control of an Induction Motor Using Neural Network Speed Estimation". *IEEE Transactions on Industrial Electronics*, vol. 48, pp. 609-614, 2001.

[7] A. Goedtel, "Load Torque Estimation in Induction Motor Shaft Using Artificial Neural Networks", Master Thesis (in Portuguese), State University of São Paulo (UNESP/PPGEI), Brazil, 2003.

[8] T. C. Chen and T. T. Sheu, "Model reference neural network controller for induction motor speed control," *IEEE Transactions on Energy Conversion*, vol. 17, no. 2, pp. 157-163, 2002.

[9] F. J. Lin and R. J. Way, "Adaptive fuzzy-neural-network control for induction spindle motor drive," *IEEE Transactions on Energy Conversion*, vol. 17, no. 2, pp. 507-513, 2002.

[10] C. Y. Huang, T. C. Chen and C. L. Huang, "Robust control of induction motor with a neural-network load torque estimator and a neural-network identification," *IEEE Transactions on Industrial Electronics*, vol. 46, no. 5, pp. 990-997, 1999.

[11] T. T. Sheu and T. C. Chen, "Self-tuning control of induction motor drive using neural network identifier," *IEEE Transactions on Energy Conversion*, vol. 14, no. 4, pp. 881-886, 1999.

[12] C. M. Ong, *Dynamic Simulation of Electric Machinery Using Matlab/Simulink*, Prentice Hall, 1997.

[13] S. Haykin, *Neural Networks*, 2nd Edition, Prentice-Hall, 1999.

[14] B. K. Bose, *Power Electroncis and Variable Frequency Drives*, IEEE Press, 1997.

[15] J. Holtz, "Sensorless Control of Induction Motor Drives", *Proceedings of the IEEE*, vol. 90, no. 8 pp. 1359-1394, 2002.

[16] J. B. Song, K.S. Byun, Design of time delay controller using variable reference model, *Control Engineering Practice*, no. 8, pp. 581-588, 2000.

[17] P. C. Krause, O. Wasynczuk, S.D. Sudhoff, *Analysis of Electric Machinery and Drive Systems*, 2nd Edition, John Wiley, 2002.

[18] J. Campbell and M. Summer, "Practical sensorless induction motor drive employing an artificial neural network for online parameter adaptation," IEE Proc. Electr. Power Appl., vol. 149, no. 4, July 2002.

[19] A. L. Orille, G. M. Sowilam, "Application of Neural Networks for Direct Torque Control," *Computers & Industrial Engineering*, no. 37, pp. 391-394, 1999.

## VII. BIOGRAPHIES

**Ivan Nunes da Silva** was born in São José do Rio Preto, Brazil, in 1967. He received both M.Sc. and Ph.D. degree in electrical engineering from the State University of Campinas, Brazil, in 1995 and 1997, respectively. Currently he is an Associate Professor at the University of São Paulo (USP), Brazil. His research interests are within the fields of artificial neural networks, power systems, nonlinear optimization, identification and control.

**Alessandro Goedtel** was born in Arroio do Meio, Brazil, in 1972. He graduated in electrical engineering from the Federal University of Rio Grande do Sul and received M.Sc. degree in industrial engineering from the São Paulo State University, in 2003. Currently he is a Ph.D. student in electrical engineering at the University of São Paulo (USP). His research interests are within the field of intelligent systems and electrical machinery.

**Paulo José Amaral Serni** was born in Botucatu, Brazil in 1957. He received both his M.Sc. and Ph.D. degree in electrical engineering from the State University of Campinas, Brazil, in 1992 and 1999, respectively. Currently he is an Assistant Professor at the State University of São Paulo, Brazil. His research interests are within the fields of artificial neural networks, power electronics, electrical machines, data acquisition systems and instrumentation;

**2006 IEEE International Conference on Power Electronic, Drives and Energy Systems**

# Electricity Price Forecasting Using Artificial Neural Network

M. Ranjbar, S. Soleymani, N. Sadati, and A. M. Ranjbar

*Abstract--* In the restructured power markets, price of electricity has been the key of all activities in the power market. Accurately and efficiently forecasting electricity price becomes more and more important. Therefore in this paper, an Artificial Neural Network (ANN) model is designed for short term price forecasting of electricity in the environment of restructured power market. The proposed ANN model is a four-layered perceptron neural network, which consists of, input layer, two hidden layers and output layer. Instead of conventional back propagation (BP) method, Levenberg- Marquardt BP (LMBP) method has been used for the ANN training to increase the speed of convergence. Matlab is used for training the proposed ANN model, also it is performed on the Ontario electricity market to illustrate its high capability and performance.

*Index Terms--* Artificial Neural Network, Electricity Market, Price Forecasting.

## I. INTRODUCTION

RECENT changes in the electricity industry in several countries have led to a less regulated and more competitive energy market. In this new structure, price of electricity has been the key of all activities in the power market. Price forecasting, with dependable accuracy helps power suppliers in selling up rational offers in the short term.

Generally there are two groups of forecasting models, traditional models and modern techniques (such as ANN, Fuzzy logic) [1]. Traditional price forecasting models are time series and regression analysis. In recent years, artificial intelligent methods are more commonly used for price forecasting. Among these methods, ANN method is a simple and powerful tool for forecasting. The reason should be the ANN capability to learn the complex input-output relationship through a supervised training process with historical data.

We know that many factors are impacting electricity price, in which some factors are more important than others are, and practically, we can only consider those more important factors. So, it will be very useful to study which factors really impact on price and to what extent. The factors, which affect

M. Ranjbar, S. Soleymani, N. Sadati, and A. M. Ranjbar are with the Electrical Engineering Department, Sharif University of Technology, Tehran, Iran. Also A. M. Ranjbar is with the Niroo Research Institute, Tehran, Iran (e-mail: ranjbar@gmail.com, ssoleymani@mehr.sharif.edu, nsadati@sharif.edu, amranjbar@sharif.edu, ranjbar@nri.ac.ir )

on the electricity price, are line limit, load pattern, bidding pattern and generator outage. In the power market, the load pattern is an effective parameter on the bidding behavior of Generating Companies (Gencos). Therefore we can consider the historical price and system load as the factors which impact on price.

The performance of the ANN approach is greatly affected by the selection of its inputs. Day type, historical price data and the amount of load have been identified for electricity price forecasting.

In this paper we mainly describe short-term price forecasting of electricity in the environment of restructured power market. This paper is organized as follows:

## II. PRICE FORECASTING BY ANN

This section describes the modules, which should be considered to design a good neural network model for short term price forecasting.

### A. Input Selection

The aim of the input selection in the case of ANN is finding optimal input parameters. Using optimal inputs would result in smaller ANN with more accuracy and convergence speed. Parameters, which affect on the electricity price can be categorized into day type (the day of a week), historical price data and the amount of demand (system load). The most effective lags (price of the previous hours) are selected by correlation analysis.

### B. Training

The ANN training process requires a set of examples with proper network behavior (network inputs and target outputs). During training, the weights and biases of the ANN are iteratively adjusted to minimize the network performance function. the selected training method for the new ANN models is Levenberg-Marquart back propagation (LMBP), which is a network training function that updates weight and bias values according to Levenberg-Marquardt optimization. This method is an improve Gauss-Newton method that has an extra regularization term to deal with the additive noise in the training samples. In comparison to LMBP, conventional back propagation methods are often too slow for practical problems.

Neurons in the hidden and output layers have nonlinear transfer function known as the " tangent sigmoid" (tansig):

0-7803-9771-1/06/$25.00 ©2006 IEEE

$$f(x) = \frac{2}{1 + \exp(-2x)} - 1 \qquad (1)$$

the weighted inputs received by a tansig node are summed and passed through this function to produce an output. The tansig function generates outputs between –1 and +1 and its inputs should be in the same range. So, it is necessary to limit the ANN inputs and target outputs. Mean-standard deviation and minimum (min)-maximum (max) normalization methods have been tested and min-max method has been selected:

$$X_{normalized} = \frac{X_{actual} - X_{min}}{X_{max} - X_{min}} \times 2 - 1 \qquad (2)$$

This normalization method has also the advantage of mapping the target output to the non-saturated sector of tansig function. This process helps to improve the accuracy of both the training and forecasting modes.

### C. Output and Hidden layers

The ANN models have the output layer. In the model of price forecasting the output is the hourly price, so the output layer has only one neuron.

The number of hidden layers and the number of neurons in each layer are selected whereas the best results are obtained.

### D. Performance Evaluation

In order to evaluate performance of the ANN models, we would compare its forecasts with those of alternative methods.

There are different alternative methods for this purpose. Among these methods, Mean Absolute Percentage Error (MAPE) is widely used to evaluate the performance of price forecasting. The MAPE is defined as follows:

The Percentage Error (PE) is defined as

$$PE = (X_{forecasted} - X_{actual}) / X_{actual} \quad 100\% \qquad (3)$$

and the APE is:

$$APE = |PE| \qquad (4)$$

then, the MAPE is given as:

$$MAPE = \frac{1}{N} \sum_{i=1}^{N} APE_i \qquad (5)$$

where

$X_{forecasted}$ : the forecasted value of price

$X_{actual}$ : the actual value of price

N : number of trained data

### III. PROPOSED ANN MODEL

In this section, we propose an ANN model to price forecasting and use Matlab for training the ANN. ANN provides a very powerful tool for analyzing factors that could impact electricity prices. For using ANN in price forecasting, we identify parameters that would fit a predefined mathematical formula based on historical data and use the resulting models to predict future electricity price based on actual inputs. The proposed ANN model is a four-layered perceptron neural network, which consists of, input layer, two

hidden layers and output layer. The input layer includes the parameters, which affect on hourly electricity price, where are historical price data and the system load. The ANN output is the hourly price, so the output layer has only one neuron. Instead of conventional back propagation (BP) method, Levenberg- Marquardt BP (LMBP) method has been used for the ANN training to increase the speed of convergence.

### IV. SIMULATION RESULTS

The proposed ANN model is performed on Ontario electricity market. The data for Ontario power market including system loads and unconstrained market clearing price (MCP) from 1/1/2003 to 12/31/2003 is adopted [6]. The load and price curves are given in Figs.1 and 2, respectively.

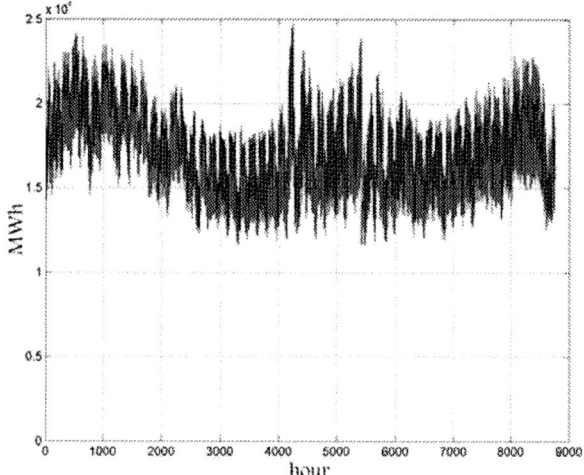

Fig. 1. Load curve of Ontario power market from 1/1/2003 to 12/31/2003

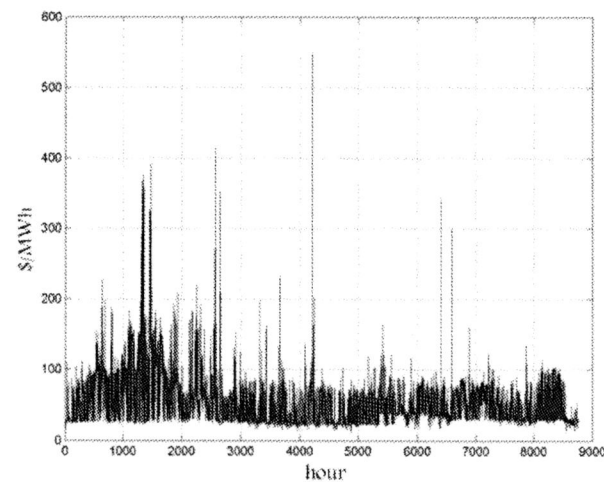

Fig. 2. Price curve of Ontario power market from 1/1/2003 to 12/31/2003

As previously mentioned, the inputs of the ANN model is the historical price data and the system load. The most effective lags (price of the previous hours) are selected by correlation analysis [2]. The lags, $l \in L = \{1,2,3,24,25,48,49,72,73,96,97,120,121,144,145,168\}$ were the selected lags based on this analysis. At the first one hidden layer and then two hidden layers is selected and the results show that two hidden layers have better performance than one hidden layers. The same experience was repeated about choosing the best number of neurons for each hidden layer. Table I, shows the result of different ANN models based on their different number of neurons in each layer. The results show that two hidden layers with 11 and 5 neurons respectively, has the better performance for short term electricity price forecasting. Fig. 3 shows the designed ANN model for this case study.

TABLE I
THE PERFORMANCE OF DIFFERENT MODELS

| Model | Neurons of hidden layer 1 | Neurons of hidden layer 2 | MAPE |
|-------|---------------------------|---------------------------|--------|
| 1 | 8 | 3 | 18.968 |
| 2 | 8 | 4 | 20.188 |
| 3 | 8 | 5 | 18.684 |
| 4 | 9 | 4 | 18.881 |
| 5 | 9 | 5 | 18.691 |
| 6 | 10 | 5 | 18.632 |
| 7 | 11 | 4 | 18.71 |
| 8 | 11 | 5 | 18.51 |
| 9 | 11 | 6 | 19.19 |

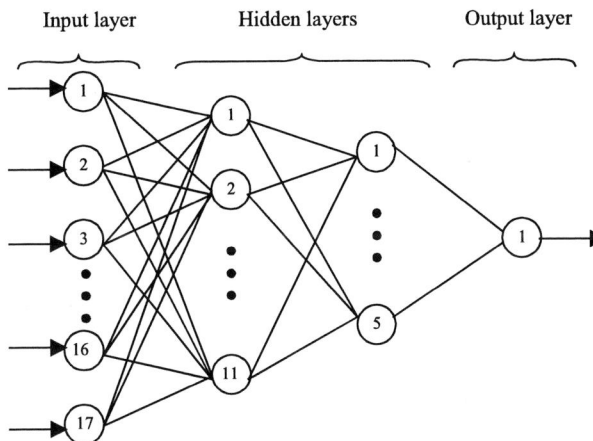

Fig. 3. The ANN model of Ontario electricity market

The proposed ANN model is trained with the historical data of year 2003. In order to evaluate performance of ANN model, we would compare its forecasts with those of the actual hourly electricity price. Mean Absolute Percentage Error (MAPE) is used to evaluate its performance. This model is used to forecast the hourly electricity price of year 2004.

Figs 4, 5, and 6 show the different examples of up to a week forecasting performance of proposed model.

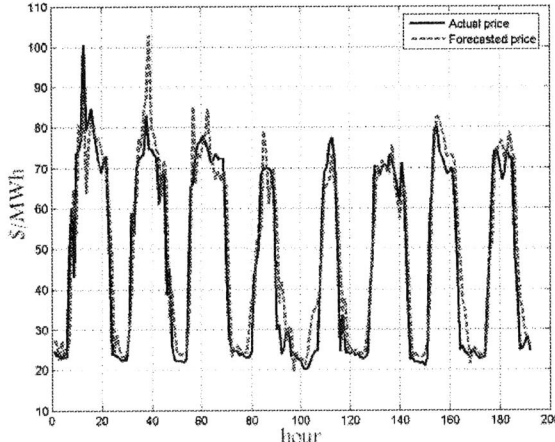

Fig. 4. Actual and forecasted electricity price from 2 July to 9 June 2004 (MAPE=12.68)

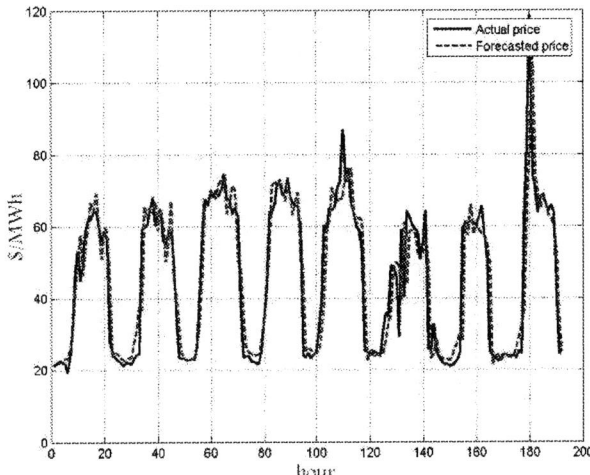

Fig. 5. Actual and forecasted electricity price from 28 July to 4 Augest 2004 (MAPE=11.36)

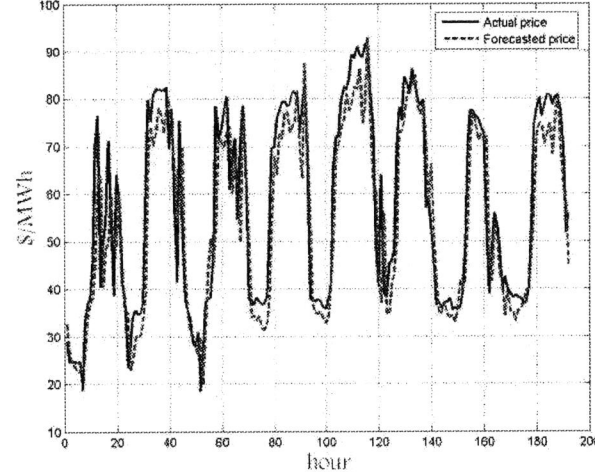

Fig. 6. Actual and forecasted electricity price from 7 September to 14 September 2004 (MAPE=11.22)

As shown in these Figs, except the hours of price spikes, the forecasted electricity price in the other hours follows its actual value with a high accuracy. To improve the accuracy of forecasting in the spike prices, we develop a data pre-processing for eliminating price spiked and then a data post processing is used to recover the original price.

Generally, there are two pre-processing methods for eliminating price spikes: limiting price spikes and excluding price spikes. In this paper, the limiting price spikes is used for eliminating price spikes. For limiting price spikes, we consider the following two options: First, we set an upper limit, $\overline{X}$, on price. In other words, in pre-processing, if the price is larger than $\overline{X}$, it will be set to $\overline{X}$. For example, if the price is larger than 200 \$/MWh, we set it to 200 \$/MWh, Accordingly, the training and testing performances are improved. Second, we set an upper limit of price, $\overline{X}$.

However, instead of setting prices higher than $\overline{X}$ to $\overline{X}$, we offer the following pre-processing scheme [1]:

$$
X_{pre} = \begin{cases} X & if\ X \le \overline{X} \\ \overline{X} + \overline{X}\ln(\dfrac{X}{\overline{X}}) & if\ X > \overline{X} \end{cases} \tag{6}
$$

We also introduce a post-processing scheme for recovering the original price after curtailing its spikes. The post-processing scheme for a forecasted price $X_{pre}$ is as follows:

$$
X_{pre} = \begin{cases} X_{pre} & if\ X_{pre} \le \overline{X} \\ \overline{X}\exp(\dfrac{X_{pre} - \overline{X}}{\overline{X}}) & if\ X_{pre} > \overline{X} \end{cases} \tag{7}
$$

where $X_{pre}$ is the forecasted price and $X_{post}$ is the modified forecasted price (i.e. after post-processing). Accordingly both the training and testing performances will be improved as shown in Figs 7, 8, and 9 which are corresponding to Figs 4, 5, and 6.

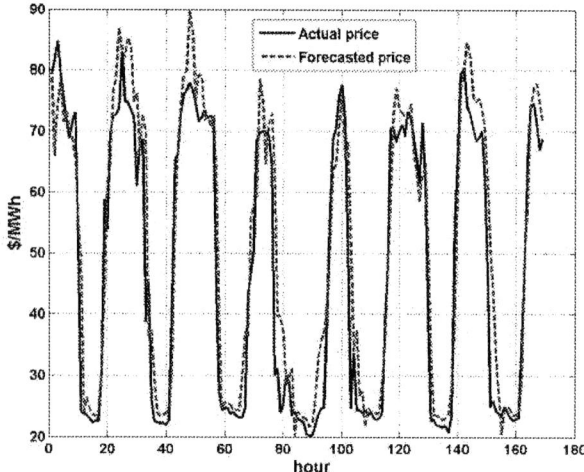

Fig. 7. Actual and forecasted electricity price from 2 July to 9 June 2004 after data pre- processing (MAPE=9.51)

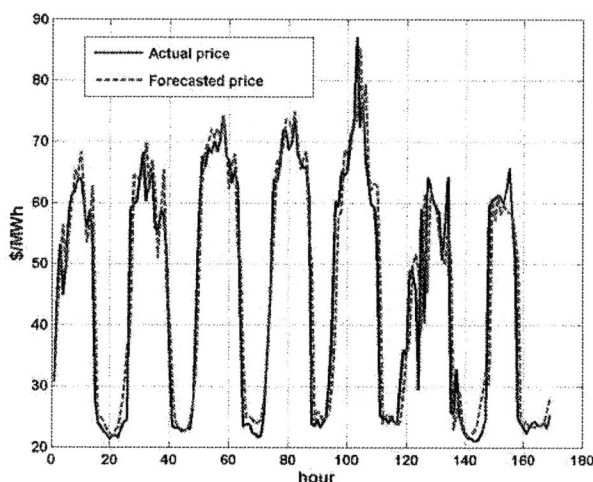

Fig. 8. Actual and forecasted electricity price from 28 July to 4 Augest 2004 after data pre- processing (MAPE=8.1)

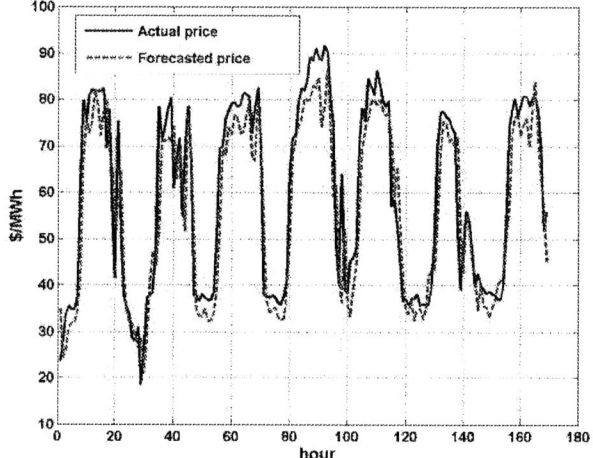

Fig. 9. Actual and forecasted electricity price from 7 September to 14 September 2004 after data pre- processing (MAPE=7.52)

## V. CONCLUSIONS

Recent changes in the electricity industry have led to a less regulated and more competitive energy market. In these conditions the forecasting electricity price is the key of all activities in the power market, such as developing the bidding strategies for generating companies. Therefore this paper presents a model for the short term electricity price forecasting using ANN. To improve the accuracy of the model, a data pre-processing and post-processing scheme are introduced for price spikes. Accordingly, both the training and testing performances was improved. The results show that, the ANN model is a good tool for price forecasting compared to other simple methods in terms of accuracy as well as convenience.

## VI. REFERENCES

[1]  H.Y.Yamin, S.M.Shahidehpour, Z.Li " Adaptive short-term electricity price forecasting using artificial neural networks in the restructured power markets", Electrical power and energy systems, PP 571-581

[2]  J. Bastin, J. Zhu, V. Banunarayanan, R. Mukerji " Forecasting energy prices in a competitive market", IEEE Computer Applications in Power, July 1999.

[3]  Bastian J, Zhu J, Banunarayanan V,Mukerji R, " Forecasting energy prices.in a competitive market", IEEE Comput Appl.Power 1999;July:40-5.

[4]  B. R. Szkuta, L. A. Sanabria and T. S. Dillon, "Electricity Price Short-Term Forecasting using Artificial Neural Networks", IEEE Transactions on Power System, Vol. 14, no. 3, pp. 851-857, 1999.

[5]  C. Gao, E. Bompard, R. Napoli, H. Cheng, "Price forecast in the competitive electricity market by support vector machine", http://www2.polito.it/eventi/apfa5/Abstract/Abstract_Bompard.pdf.

[6]  www.ieso.ca.

## VII. Biographies

**Mona Ranjbar** received the B.S. degree in the Electrical Engineering Department of Sharif University of Technology, Tehran, Iran. Her areas of research include market power monitoring, and electricity market issues.

**Soodabeh Soleymani** received her B.S and M.S degrees in electrical engineering from the Sharif University of Technology. Since 2004, She is pursuing her PhD program at the same university. Her area of research includes power market simulation, and market power monitoring in deregulated power systems.

**Nasser Sadati** received his PhD degree in Electrical Engineering from Cleveland State University. Since 1990, he has been with the Sharif University of technology, Tehran, Iran, where he is currently a professor in the department Electrical Engineering and the Director of Intelligent Systems Laboratory.

**Ali Mohammad Ranjbar** received the M.S. and Ph.D. degrees in Electrical Engineering from Tehran University in 1967 and Imperial College of London University in 1975, respectively. Since then, he has been at Sharif University of Technology, Department of Electrical Engineering, where he is currently a full professor. Dr. Ranjbar is the Editor-in-Chief of Journal of Electrical Science and Technology since 1989 and the Director of Niroo Research Institute since 1996 too. His main research interests are in the areas of electric power systems protection and operation, and electrical machine.

**2006 IEEE International Conference on Power Electronic, Drives and Energy Systems**

# A New Approach for Fault Location Identification in Transmission system using Stability Analysis and SVMs

D. Thukaram, *Senior Member, IEEE*, H.P. Khincha, *Senior Member, IEEE*, and B. Ravikumar

*Abstract*--This paper presents a new approach to the location of fault in the high voltage power transmission system using Support Vector Machines (SVMs). A knowledge base is developed using transient stability studies for apparent impedance swing trajectory in the R-X plane. SVM technique is applied to identify the fault location in the system. Results are presented on sample 3-power station, a 9-bus system illustrate the implementation of the proposed method.

*Index Terms*--Power system transient stability, SVMs, Transmission lines fault location.

## I. INTRODUCTION

IDENTIFICATION of fault location is a desirable feature in any protection scheme. The increased complexities of modern power transmission system have raised the great importance of fault location research studies in recent years. The restoration can be expedited if the location of a fault is either known or can be estimated with reasonable accuracy. Locating the faults on the transmission line accelerates line restoration, reduce the outage time, operating cost, customer complaints and maintain system stability.

Fault location techniques can be classified into those that use data from just one end of the transmission line and those that use data from both ends of the line. Protective relays are basically single-ended fault locators. The accuracy of single ended fault locators is affected by the assumptions that are made about the fault impedance, the source impedance and the in feed into the fault from the remote end source. Two- or multi-ended fault location techniques are therefore more accurate than single-ended methods. The unknown fault resistance can be eliminated from the line model equations to estimate the location of the fault. Although two ends algorithms may present a better performance, single end algorithms have advantages from the commercial viewpoint

[1]–[2]. This is mainly due to the extra-complexity associated with two ends algorithms including communication and synchronization between both ends as well as the increase in the cost. Thus, fault location techniques using single-ended data could be more attractive for researchers.

The conventional single-ended fault location methods can be broadly classified into two types: (a) impedance method; (b) Traveling wave method. The impedance method determines the fault position by measuring the impedance from the relay end to the fault point, taking assumption that the line impedance is linear to distance. Obviously, it is based on power frequency measurement and affected by many factors of power frequency phenomena, such as fault path resistance, line loading, and source parameters, etc. Therefore, the accuracy of the impedance based fault location methods is limited [3]–[4]. The traveling wave method is independent of fault resistance, pre-fault loading, and source parameters; however, present traveling wave-based fault location methods cannot accurately separate the traveling wave which reveals the fault position from other waves of different frequencies, although wavelet transform is used for signal processing. Moreover, traveling wave method could not reliably locate the fault occurring at the point closely near the bus [5]–[7]. High frequency transients based methods are same to traveling wave based essentially, and suffer same limitation. In the neural network based fault location method currently under development, it was found, during feature extraction, that using the power frequency signals is far more accurate in fault location than using the traveling wave frequency signals.

In this paper, a new method based on the application of support vector machine (SVM) network and swing characteristics in R-X plane. The SVM, the solution of universal feed forward networks, pioneered by Vapnik [8, 9], is known as the excellent tool for classification and regression problems of good generalization performance. A knowledge base is developed using transient stability studies for apparent impedance swing trajectory in the R-X plane. SVM technique is applied to identify the fault location in the system from the relay location. Results are presented on sample 3-power station, a 9-bus system illustrate the implementation of the proposed method.

---

D. Thukaram is with the Department of Electrical Engineering, Indian Institute of Science, Bangalore-560012, INDIA (dtram@ee.iisc.ernet.in).

H.P. Khincha is with the Department of Electrical Engineering, Indian Institute of Science, Bangalore-560012, INDIA (hpk@ee.iisc.ernet.in). and

B. Ravikumar is with the Department of Electrical Engineering, Indian Institute of Science, Bangalore-560012, INDIA (ravi@ee.iisc.ernet.in).

0-7803-9771-1/06/$25.00 ©2006 IEEE

## II. PROPOSED APPROACH

Power system protection at the transmission level is based on distance relaying. The apparent impedance seen by the distance relay at substation on a transmission line connecting the nodes i and j, and having flow $P_{ij}+jQ_{ij}$ is given as

$$Z_R = \left[ \frac{P_{ij}}{P_{ij}^2 + Q_{ij}^2} + j \frac{Q_{ij}}{P_{ij}^2 + Q_{ij}^2} \right] |V_i|^2$$

Where, all quantities refer to positive sequence values. Thus, the quadrant of $Z_R$ depends only on the direction of $P$ and $Q$ flows.

Following a disturbance, the distance relays at different locations will observe different swing characteristics in R-X plane, and if a swing trajectory enters zone-1, then it is considered as severe enough to cause system instability, the relay will also trip if the swing stays depending on the Time Dial Setting (TDS). The characteristics trajectory will also be different for different points on a line.

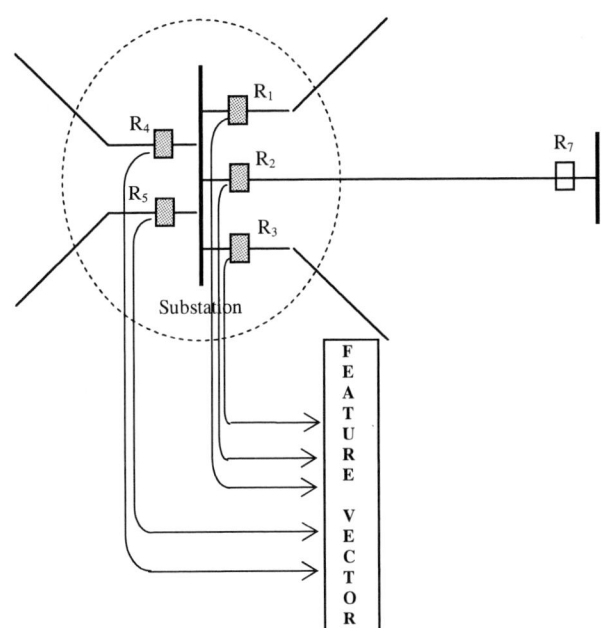

Fig. 1(a). Collection of feature vectors at a substations.

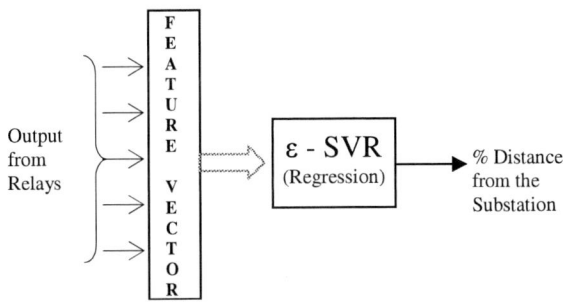

Fig. 1(b). Feeding the input vector to the SVM regression block.

In this paper, this information is intelligently utilized for identifying the distance of the disturbance from the relay position. Support Vector Machines are used to capture the underlying concept/model between the fault location and the swing trajectory characteristics. The feature vector consists of the apparent impedances seen by the different relays located on lines connected to the substation at different instances of time. Suppose a stable system at time $t_0$ is subjected to 3-ph fault on a line at time $t_1$ and cleared at time $t_2$. During this period, the possible events can be opening of the faulted line, and/or some generators may fall in out-of synchronism and/or some load rejection can happen. The apparent impedances observed by the relays mounted on the substation lines at different time instances are considered together as the feature vector. Fig. 1(a), shows the collection of apparent impedances values observed by different relays and Fig. 1(b), shows the training/testing of the SVM. Fig. 1, collectively shows the implementation of the proposed method at a substation.

### III. BRIEF REVIEW ON SUPPORT VECTOR MACHINES

Support Vector Machines are a new learning-by-example paradigm spanning a broad range of classification, regression, and density estimation problems. They were first introduced by Vapnik *et. al* [8]-[10] and are described in more detail by in B. Schölkopf *et. al.*[11] The roots of this approach, the so-called support vector (SV) methods of constructing the optimal separating hyperplane for pattern recognition. The SV technique was generalized for nonlinear separating surfaces in [11], and it was further extended for constructing decision rules in the non separable case. The training task involves optimization of a convex cost function conveying to a technique without local minima.

#### A. Support Vector Classification

The problem of classification consists of estimating a function $f : R^N \to \{\pm 1\}$ using $l$ *i.i.d.* input–output training data $(X_1, y_1), \ldots, (X_l, y_l) \in R^N \times \{\pm 1\}$ from a data set $D$ such that $f$ classifies correctly unobserved data $(x, y)$ (i.e., $f(x) = y$ for examples $(x, y)$ generated from the some underlying probability distribution $P(x, y)$). In other words, the loss function $L$ can be defined by (1)

$$L\left(y_i, f\left(x_i\right)\right) = \left|1 - y_i f\left(x_i\right)\right|_+ \tag{1}$$

Where $|val|_+ = \max\{0, val\} \quad val \in R$.

A brief review of support vector classification (SVC) [12-15] is presented in this section; when data is linearly separable there exists a vector $w \in R^N$ and a scalar $b \in R$ such that $y_i\left(w \cdot x_i + b\right) \geq 1$ for all patterns in the training set $(i = 1, \ldots l)$. The optimal hyperplane separates points lying on opposites classes yielding to the maximum margin separation. A separating hyperplane which generalizes well can be found by solving the following quadratic programming (QP) problem (for $i = 1, \ldots l$):

Minimize $\dfrac{1}{2}\|w\|^2$ (2)

subject to $y_i(w \cdot x_i + b) \geq 1, \quad \forall i.$

This constrained optimization problem is solved by constructing a Lagrangian

$$\lambda_P(w, b, \alpha) = \frac{1}{2}\|w\|^2 - \sum_{i=1}^{l} \alpha_i(y_i(w \cdot x_i + b) - 1) \quad (3)$$

The Lagrangian has to be minimized with respect to the primal variables $w$ and $b$ and maximized with respect to the dual variables $\alpha_i$. The Karush–Kuhn–Tucker (KKT) conditions lead to find the solution vector in terms of the training patterns, $w = \sum_{i=1}^{l} \alpha_i y_i x_i$ for some $\alpha_i \geq 0$. Notice that $\alpha_i \neq 0$ only for a subset of the training patterns, precisely those few vectors that lie on the margin, called the support vectors (SVs). Under certain conditions, a kernel function $K(.,.)$ can be found such that $K(x_i, x_j) = x_i \cdot x_j$. An SVM uses then the convolution of the scalar product to build, in input space, the nonlinear decision function

$$f(x) = \text{sgn}\left(\sum_{i=1}^{l} \alpha_i y_i K(x, x_i) + b\right) \quad (4)$$

Where $b$ is found from the primal constraints and is computed by $\alpha_i(y_i(w \cdot x_i + b) - 1) = 0, i = 1,\dots l$, such that $\alpha_i$ is not zero and sgn is the signal function.

When the training data is not linearly separable, a separating hyperplane does not exist. Besides, when real data sets are used, SVMs can fit noise and outliers leading to poor generalization. Thus, a hard margin classifier is no longer adequate. Introducing a soft margin, the learning task is essentially the same as indicated in (2) except for the introduction of the penalty term $C$ and the slack variable $\xi$. The classifier tries then to separate the data by minimizing the objective function

Minimize $\dfrac{1}{2}\|w\|^2 + \dfrac{C}{l}\sum_{i=1}^{l}\xi_i$

subject to $y_i(w \cdot x_i + b) \geq 1 - \xi_i,$ (5)

$0 \leq \alpha_i \leq C/l, \quad \xi_i \geq 0$

for $i = 1,\dots,l$. In this sense, it acts by controlling the classifier capacity and the number of training errors. In other words, the task is now to minimize the sum of errors $\sum_{i=1}^{l}\xi_i$ in addition to $\|w\|^2$. Again this optimization problem can be transformed into a QP problem. The value of $C$ can be found by experimentation in a validation set and cannot be determined from either the model or the data set.

### B. $\varepsilon$ -Support Vector Regression

The problem of Regression consists of estimating a function $f : R^N \rightarrow R$ using $l$ i.i.d. input–output training data set of data points $\{(X_1, y_1),\dots,(X_l, y_l)\} \in R^N \times R$. The

standard form of support vector regression [10], [16] is:

Minimize $\dfrac{1}{2}w^T w + C\sum_{i=1}^{l}\xi_i + C\sum_{i=1}^{l}\xi_i^*$

subjected to $\left[w^T\phi(x_i) + b - y_i\right] \leq \varepsilon + \xi_i,$ (6)

$\left[y_i - w^T\phi(x_i) - b\right] \leq \varepsilon + \xi_i^*,$

$\xi_i, \xi_i^* \geq 0, i = 1,\dots,l.$

The duel is:

$$\underset{\alpha,\alpha^*}{\text{Min}} \quad \frac{1}{2}(\alpha - \alpha^*)^T Q(\alpha - \alpha^*) + \varepsilon\sum_{i=1}^{l}(\alpha_i + \alpha_i^*)$$

$$+ \sum_{i=1}^{l} y_i(\alpha_i - \alpha_i^*) \quad (7)$$

Subjected to $\sum_{i=1}^{l}(\alpha_i - \alpha_i^*) = 0.0,$

$0 \leq \alpha_i, \alpha_i^* \leq C, i = 1,\dots,l;$

where $Q_{ij} = K(x_i, x_j) \equiv \phi(x_i)^T \phi(x_i)$

The decision function is $\sum_{i=1}^{l}(-\alpha_i + \alpha_i^*)K(x_i, x_j) + b$ (8)

### C. Kernel Choice:

The use of kernel methods [17] provides a powerful way of obtaining nonlinear algorithms capable of handling non-separable data sets in the original input space. The basic idea is to construct a mapping into a higher dimensional feature space by the use of reproducing kernels. The kernel function is a positive definite function $R^n \times R^n$ to $R^n$ that defines an embedding of input patterns into feature vectors.

Training is carried out with different types of kernel function like Linear Kernel $K(x, y) = x^1 \cdot y$, Polynomial Kernel $K(x, y) = (\gamma(x \cdot y) + r)^{Degree}$ with $\gamma > 0$, Radial Basis Kernel $K(x, y) = \exp(-\gamma\|x - y\|^2)$ where $\gamma > 0$ related with the kernel width, Sigmoid Kernel $K(x, y) = \tanh(\gamma(x \cdot y) + r)$ where $\gamma$ and $r$ are kernel parameters.

### D. SVM Model Selection:

In any predictive learning task, such as classification, an appropriate representation of examples as well as the model and parameter estimation method should be selected to obtain a high level of performance of the learning machine. Under the SVM's approach, the usually parameters to be chosen are the following:

1) the penalty term $C$ which determines the tradeoff between the complexity of the decision function and the number of training examples misclassified;

2) the mapping function $\phi$ ; and

3) the kernel function such that $K(x_i, x_j) = \phi(x_i) \cdot \phi(x_j)$.

## IV. System Studies

In our approach to fault location, we have considered a sample 9-bus system as shown in Fig.2. The system has three generator busses 1, 2 and 3, three load busses 5, 6 and 8. Each generating bus is considered to have two machines. The double circuit line connected between busses 5 and 7 is considered for studies. The relays $R_1$, $R_2$ and $R_3$ at the substation 5 are observed during the fault for obtaining the training and test patterns. Three phase fault disturbances on one of the line between busses 5 and 7 are simulated using the stability program developed by Prof. D. Thukaram at IISc, Bangalore. Training and test patterns are generated for SVM by creating and clearing faults at various locations on the line at different instances of time.

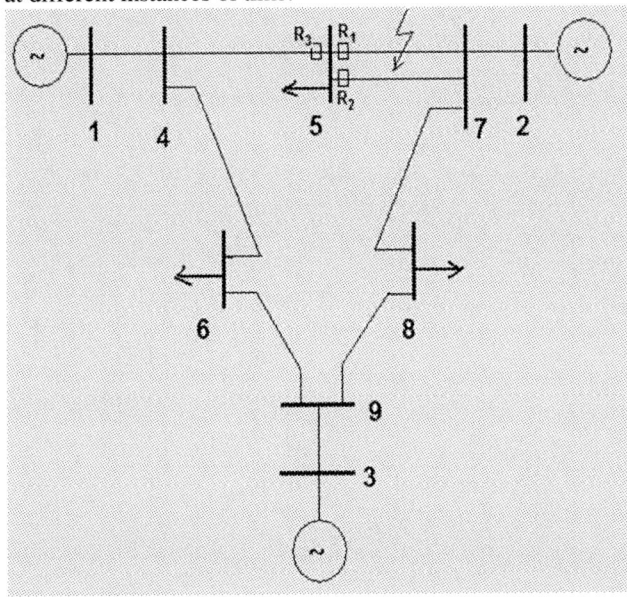

Fig. 2. Single line diagram of the 9-bus system considered for studies.

Training patterns are generated for three types of cases. In the first, normal case, the base system is considered and the faults are created at 5%, 15 %, 25%...85%, 95% distances from the bus 5. At first instance of time, 3-ph fault is created at time 0.025 seconds and cleared three different instances of time, at 0.105sec, 0.125 sec and 0.145 sec. in this case, we assumed that during fault no other simultaneous disturbances have happened. In the second case, the above faults are created at the above distances and cleared at the above same clearance time but during clearance times, one of the generator at the bus 2 is out-of synchronism. During the third case, the fault distances and fault clearance times are same as the base case, but the load at the bus 5 is rejected from the system. The swing trajectories of the relay $R_2$ on line 5-7 for two different fault locations and clearance times are shown in Fig.3. For a fault at 35% distance from the bus 5, Fig. 3a, and Fig. 3c, shows the trajectories with fault clearance times 0.125 sec and 0.145 sec respectively. Fig. 3b, and Fig. 3d, are for a fault at 75% distance with the above two clearance times. Test patterns are generated for the above three cases by creating the faults at 20%, 40%, 60% and 80% distance from the bus 5.

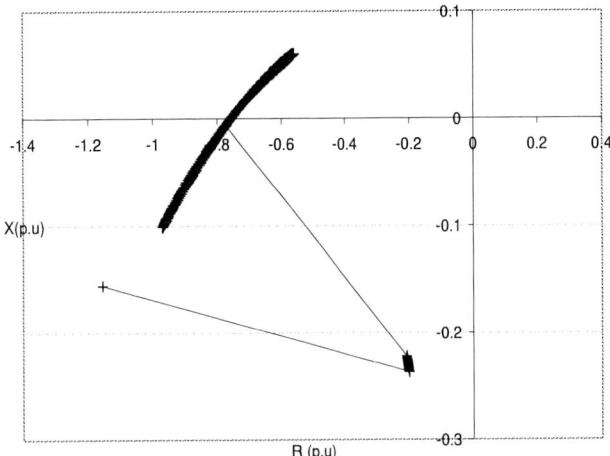

Fig. 3(a). Swing trajectory in R-X plane for the 3-phase fault on line 5-7 at 35% distance from 5, created at 0.025sec and cleared at 0.1250sec.

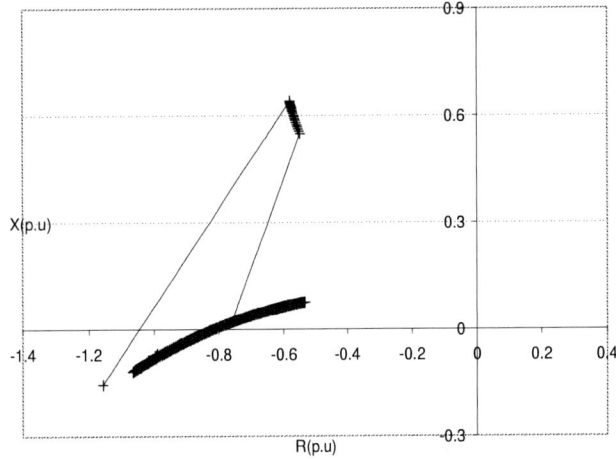

Fig. 3(b). Swing trajectory in R-X plane for the 3-phase fault on line 5-7 at 75% distance from 5, created at 0.025sec and cleared at 0.1250sec.

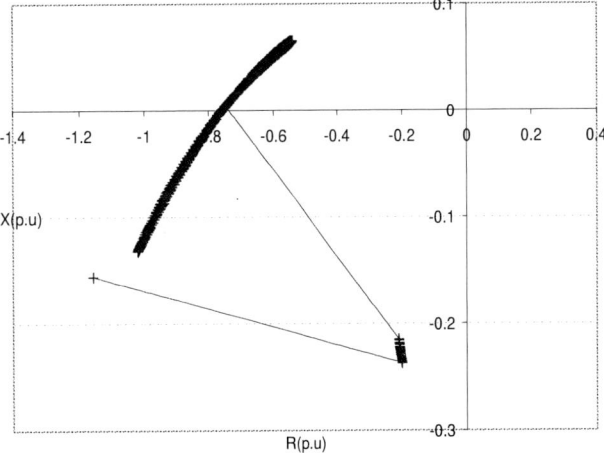

Fig. 3(c). Swing trajectory in R-X plane for the 3-phase fault on line 5-7 at 35% distance from 5, created at 0.025sec and cleared at 0.1450sec.

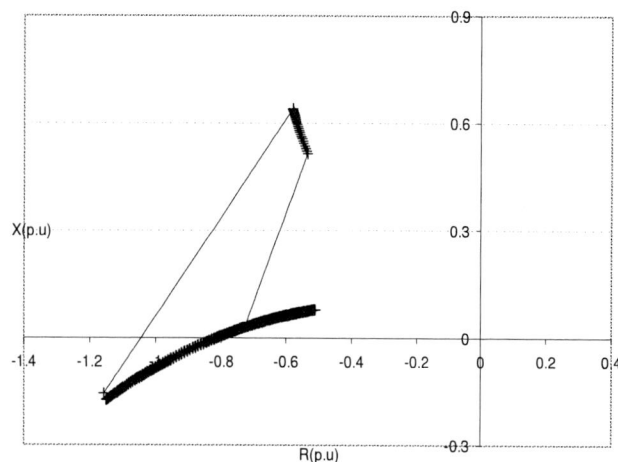

Fig. 3(d). Swing trajectory in R-X plane for the 3-phase fault on line 5-7 at 75% distance from 5, created at 0.025sec and cleared at 0.1450sec.

## A. Model Design:

In this paper LIBSVM [18]-[19] is used for training the support vector machine. Training the SVM requires the selection of parameters which influence the model performance. Therefore to achieve a good model, the parameters $C$ and kernel function have to be chosen correctly. To model the $\varepsilon$-SVR, two types of kernels, polynomial and RBF are chosen. The value of the penalty parameter range is trailed between 1 and 1000. Aforementioned, $\gamma$ is an important parameter for both the kernel functions. The range of $\gamma$ chosen for training is [0.01, 5]. The stopping tolerance for solving the optimization problem is set to 0.0001.

Table I shows the Mean Squared Error (MSE) for different combinations of the parameters. From the Table I, RBF kernel is performing with good accuracy in identifying fault location. The estimated fault location obtained form the $\varepsilon$-SVR are shown in Fig 4, for the target distances 20%, 40%, 60% and 80%. Fig. 4a, shows the results when RBF kernel with C=100, $\varepsilon$=0.0001 and $\gamma$=0.05 is used. Fig. 4b, shows the output results obtained form the $\varepsilon$-SVR when polynomial kernel is used with degree=2, C=10, $\varepsilon$=0.0001 and $\gamma$=0.01. Fig. 4c, shows the output results with polynomial kernel with degree=3, C=1000, $\varepsilon$=0.0001 and $\gamma$=0.5 is used.

TABLE I
OUTPUT RESULTS FROM THE ε-SVR WITH DIFFERENT PARAMETERS

| Kernel 1. Poly. (Degree) 2. RBF | C | ε | γ | Mean Squared Error (MSE) | Squared Correlation Coeff. (SCC) |
|---|---|---|---|---|---|
| 2 | 100 | 0.0001 | 0.05 | 0.000594 | 0.994655 |
| 2 | 10 | 0.001 | 0.05 | 0.000605 | 0.994575 |
| 2 | 1000 | 0.01 | 0.01 | 0.00065 | 0.992085 |
| 2 | 10 | 0.001 | 0.1 | 0.000851 | 0.995208 |
| 1(2) | 10 | 0.0001 | 0.01 | 0.001679 | 0.991819 |
| 1(3) | 1000 | 0.001 | 0.5 | 0.001952 | 0.99643 |
| 1(2) | 100 | 0.0001 | 0.01 | 0.002912 | 0.974131 |
| 1(2) | 10 | 0.01 | 0.05 | 0.003066 | 0.973071 |
| 1(2) | 10 | 0.0001 | 0.5 | 0.00415 | 0.964221 |
| 2 | 10 | 0.0001 | 0.5 | 0.010859 | 0.87998 |

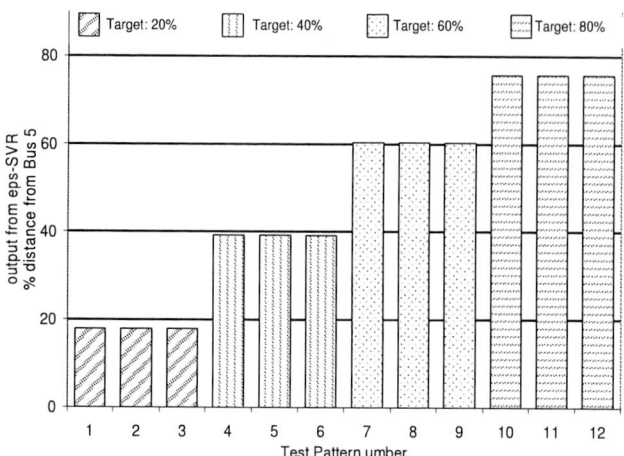

Fig. 4(a). RBF kernel with [C, $\varepsilon$, $\gamma$] = [100, 0.0001, 0.05] is used.

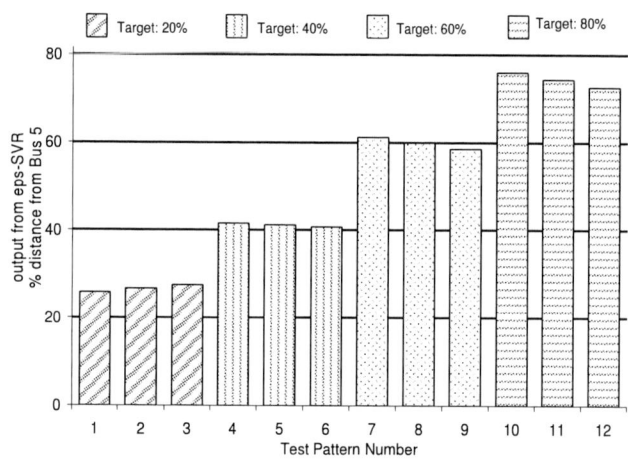

Fig. 4(b). Polynomial kernel (d=2) with [C, $\varepsilon$, $\gamma$] = [10, 0.0001, 0.01] is used.

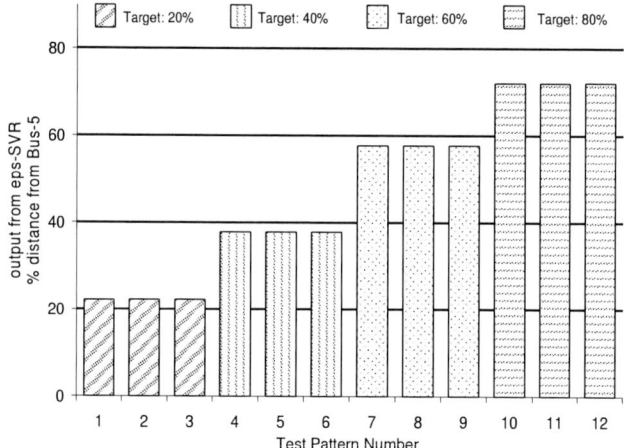

Fig. 4(c). Polynomial kernel (d=3) with [C, $\varepsilon$, $\gamma$]=[1000, 0.0001, 0.5] is used.

Fig. 4. Estimated target values from ε-SVR for the test patterns with target distances 20%, 40%, 60% and 80%.

## V. CONCLUSIONS

A new approach based on the swing characteristics along with SVM can be better used for fast identification fault location at a substation. The use of Support Vector Machines as a powerful tool for fault location identification is presented. Support Vector Machines with RBF kernel is able to learn the underlying concept between swing trajectory and fault location. From the results, we conclude that, the knowledge base developed from the R-X plots by using the transient stability studies during fault contain significant information about the location of fault.

## VI. REFERENCES

[1] T. Adu, "A new transmission line fault locating system," IEEE Trans. Power Delivery, vol. 16, pp. 498–503, Oct. 2001.

[2] Distance Protection Relay for Transmission Lines, 7SA522: User Catalog, Catalog SIP 4.2, Siemens AG, 1999.

[3] V. Cook, "Fundamental aspects of fault location algorithms used in distance protection," Proc. Inst. Elect. Eng. C, vol. 133, no. 6, pp. 354–368, 1986.

[4] L. Eriksson, M. Saha, and G. D. Rockefeller, "An accurate fault locator with compensation for apparent reactance in the fault resistance resulting from remote end infeed," IEEE Trans. Power App. Syst., vol. PAS-104, pp. 424–436, Feb. 1985.

[5] A. O. Ibe and B. J. Cory, "A traveling wave based fault locator for two and three-terminal networks," IEEE Trans. Power Delivery, vol. 1, pp. 283–288, Apr. 1986.

[6] G. B. Ancell and N. C. Pahalowathitha, "Effects of frequency dependence and line parameters on single-ended traveling wave based fault location schemes," Proc. Inst. Elect. Eng. C, vol. 139, no. 4, pp. 332–342, 1992.

[7] F. H. Magnago and A. Abur, "Accurate fault location with wavelets," IEEE Trans. Power Delivery, vol. 13, pp. 1475–1480, Oct. 1998.

[8] Vapnik, V., Statistical Learning Theory. Wiley, New York, NY, 1998.

[9] A. Smola and B. Scholkopf. (1998) A Tutorial on Support Vector Regression, Neurocolt Tech. Rep. NV2-TR-1998-030. [Online]. Available: http://www.neurocolt.com

[10] V. N. Vapnik, The Nature of Statistical Learning Theory. New York: Springer-Verlag, 1995.

[11] B. Schölkopf, C. Burges, and A. Smola, "Advances in Kernel Methods—Support Vector Learning" Cambridge, MA: MIT Press, 1999.

[12] Sastry, P.S., "An introduction to Support Vector Machine, Chapter from computing and information sciences: Recent Trends", Narosa Publishing House, New Delhi 2003.

[13] Burges, C. J. C, "A tutorial on support vector machines for pattern recognition," Data Mining and Knowledge Discovery, vol. 2, no. 2, pp. 955-974, 1998.

[14] Joachims, T, Making large-scale SVM learning practical. Advances in Kernel Methods - Support Vector Learning, Cambridge, MA, 1998. MIT Press.

[15] Platt, J., "Fast Training of Support Vector Machines using Sequential Minimal Optimization, Advance in Kernel Methods: Support Vector Learning," pp. 185 – 208, MIT Press, Cambridge, MA, 1999. Available at: http://www.research.microsoft.com/users/jplatt/smo-book.pdf.

[16] Vladimir N. Vapnik, Steven E. Golowich, and Alex Smola. Support vector method for function approximation, regression estimation and signal processing. http://citeseer.ist.psu.edu/vapnik96support.html

[17] C. Campbell, "Kernel methods: A survey of current techniques," Neurocomput., vol. 48, pp. 63–84, 2002.

[18] Chih-Chung Chang and Chih-Jen Lin, LIBSVM: a library for support vector machines, 2001. Software available at http://www.csie.ntu.edu.tw/~cjlin/libsvm

[19] http://www.csie.ntu.edu.tw/~cjlin/papers/libsvm.pdf

## VII. BIOGRAPHIES

**D. Thukaram** (SM'90) received the B.E. degree in Electrical Engineering from Osmania University, Hyderabad in 1974, M.Tech degree in Integrated Power Systems from Nagpur University in 1976 and Ph.D. degree from Indian Institute of Science, Bangalore in 1986. Since 1976 he has been with Indian Institute of Science as a research fellow and faculty in various positions and currently he is Professor. His research interests include computer aided power system Analysis, reactive power optimization, voltage stability, distribution automation and AI applications in power systems.

**H. P. Khincha** (SM'82) received the B.E. degree in Electrical Engineering from Bangalore University in 1966. He received M.E. degree in 1968 and Ph.D. degree in 1973 both in Electrical Engineering from the Indian Institute of Science, Bangalore. Since 1973 he has been with Indian Institute of Science, Bangalore as faculty where currently he is Professor. His research interests include computer aided power system analysis, power system protection, distribution automation and AI applications in power systems.

**B. Ravikumar** received the B.Tech degree in Electrical & Electronics Engineering from Nagarjuna University (A.P) in 2002 and M.Sc (Engg.) degree in Electrical Engineering at the Indian Institute of Science, Bangalore in 2004. Currently he is pursuing his PhD from Indian Institute of Science, Bangalore, India. His research interests include Computer aided Power System analysis, Distribution Automation, AI techniques applications to Power Systems.

**2006 IEEE International Conference on Power Electronic, Drives and Energy Systems**

# Fast and Effective Algorithm for Economic Dispatch with Prohibited operating zones

T. Adhinarayanan[1], and M. Sydulu[2]

*Abstract* -- **This paper presents a very fast and effective "λ-logic" based algorithm for Economic dispatch (ED) problem when some of the on-line generating units have prohibited operating zones. The operating region of units having prohibited zones is broken into isolated feasible sub-regions, which results in multiple decision spaces for the economic dispatch problem. The optimal solution will lie in one of the feasible decision spaces and can be found using the conventional λ-δ iterative method based on the equal-incremental-cost criterion in each of the feasible decision spaces. However, for a system with a large number of decision spaces, such exhaustive search method would not be acceptable in real time operation due to high computational time requirement. In this paper, a two executions of "λ-logic based" algorithm is used. It uses an efficient approach to determine the most advantageous space using the average power of prohibited operating region. The proposed method is very efficient for both small and large size of generating units with prohibited operating zones in the problem and it remarkably reduces the computation time. The proposed algorithm has been tested successfully on 4 units, 5 units and 15 units with some of the on-line units having prohibited operating zones and test results are reported in this paper.**

*Index Terms* – **Economic dispatch, Decision spaces, Prohibited operating zones.**

## I INTRODUCTION

THE purpose of Economic dispatch is to allocate generation levels to the various generating units in the system so as to meet the load demand in the most economic way without violating any system or individual unit constraints. The importance of economic dispatch is to get maximum usable power using minimum resources. Conventionally, the economic dispatch problem of a power system is solved in the environment of unit commitment and real time operation plants by assuming that each of the dispatchable on-line units can be regulated continuously between its minimum generation limit, $P_{min}$, and its maximum generation limit, $P_{max}$. In practical systems, however some of the on-line units may have prohibited operating zones due to the physical limitations of power plant components such as vibrations in a shaft bearing in a certain operating region etc. For a unit with prohibited zones, its operating region, [$P_{min}$,

$P_{max}$], will be broken into several isolated sub-regions. It can only be dispatched to one of the isolated sub-regions in practical operation.

The isolated sub-regions will form multiple decision spaces and result in a very challenging task for determining the optimal economic dispatch. Several methods have been proposed in the literature [1]-[4], [6]-[9], to solve ED Problem with units having prohibited operating zones. This paper presents a new algorithm based on "λ-logic" for determining the optimal loading of generators having prohibited zones. The fuel cost curve of units with prohibited zones are broken into several isolated feasible sub-regions. These isolated feasible or infeasible with respect to system demand and the optimal solution of the dispatch problem will reside in one of the feasible spaces. Lee and Breipohl [1] presented an approach that selects a small set of classified feasible decision spaces and determines the feasible optimal dispatch by performing the conventional λ-δ iterative search on each of the classified spaces. This approach may be subject to high computational time due to its requirement of performing the conventional λ-δ iterative dispatch on each of the classified spaces that could be a notably big number when several on-line units possess prohibited zones. Fan and McDonald [2] presented a method that determines a small advantageous decision spaces and selects the most advantageous decision space among them. The optimal solution in the most advantageous decision space is obtained by performing the conventional λ-δ iterative method. In [6], ED problem with prohibited zones is solved using standard evolutionary programming algorithm (SEPA) taking the generator outputs as decision variables and fuel cost as fitness function. If the optimal generating schedule lies in the prohibited zone, then they are re-dispatched to nearest boundary of the prohibited zone. Major drawbacks of this algorithm are very slow and inconsistent convergence, large number of iterations, more number of decision variables and indeterministic stopping criteria, etc. Recently, ED problem with prohibited zones is solved using fast computation evolutionary programming algorithm (FCEPA) [7] taking the system lambda as decision variable and power mismatch as fitness function without considering prohibited zone along with Participation factor method to estimate a penalty cost for each selected feasible advantageous space.

In this paper, a two executions of "λ-logic" based approach is used to determine the optimal loading of generators having prohibited zones. It determines the optimum dispatch without considering prohibited zone in the first execution. If all the units are within limits and out of the prohibited operating zones, the optimal dispatch is obtained. If any of the units/all fallen within prohibited zones check the power output with prohibited zone limits. If the power output of prohibited zone unit is greater or lesser than the average of

---

[1]. T. Adhinarayanan is with the Department of Electrical Engineering, National Institute of Technology, Warangal, A.P-506004 INDIA (e-mail: adhi_71@yahoo.co.in)

[2] Prof.M.Sydulu is with the Department of Electrical Engineering, National Institute of Technology, Warangal, A.P-506004 INDIA (e-mail: msydulu@nitw.ac.in)

0-7803-9771-1/06/$25.00 ©2006 IEEE

power of prohibited zone, choose the upper or lower limits respectively. once the outputs of such generators are set to upper or lower limit of prohibited zones, add the output power of these generators and estimate remaining load yet to be supplied, the remaining load would be economically dispatched by second execution of "λ-logic" based approach. This method is faster and efficient and requires only two executions of "λ-logic" based approach for small as well as for large size problems.

## II PROBLEM FORMULATION

The objective of ED problem is to determine the generation levels for all on-line units which minimize the total fuel cost, while satisfying a set of constraints can be stated as follows:

*Objective Function*

$$\text{Minimize } F_T = \sum_{i \, \varepsilon \, \Omega}^{n} F_i(P_i) \qquad (1)$$

$$F_i(P_i) = a_i + b_i P_i + c_i P_i^2 \qquad \text{'or'} \qquad (2)$$

$$F_i(P_i) = a_i + b_i P_i + c_i P_i^2 + d_i P_i^3 \qquad (3)$$

Where

| | |
|---|---|
| $F_T$ | Total fuel cost ($/hr); |
| $F_i(P_i)$ | Fuel cost of unit i |
| | Where i=1, 2,…k. |
| $a_i, b_i, c_i, d_i$ | fuel cost coefficients of unit i |
| $P_i$ | power generation of unit i (MW) |

*Constraints*

( i ) Power-balance constraint

$$\sum_{i \, \varepsilon \, \Omega} P_i = P_D \qquad (4)$$

Where

| | |
|---|---|
| $\Omega$ | set of all dispatchable on-line units |
| $P_D$ | system demand |

(ii) Generation limit constraint

$$P_{imin} \le P_i \ge P_{imax} \quad i \, \varepsilon \, (\Omega - \omega) \qquad (5)$$

The additional constraints for units with prohibited operating zones can be described as follows

$$P_{imin} \le P_i \ge P_{i,1}^{\ l} \quad \text{or}$$
$$P_{i,k-1}^{\ u} \le P_i \ge P_{i,k}^{\ l} \quad , k=2,…,n_i \text{ or}$$
$$P_{i,n_i}^{\ u} \le P_i \ge P_{i,max} \quad i \, \varepsilon \, \omega \qquad (6)$$

Where

| | |
|---|---|
| $P_{imin}$ | Unit i's minimum generation limit in MW |
| $P_{imax}$ | Unit i's maximum generation limit in MW |
| $P_{i,k}^{\ l}$ | Lower bound of kth prohibited zone of unit i |
| $P_{i,k}^{\ u}$ | Upper bound of kth prohibited zone of unit i |
| $n_i$ | No. of prohibited zones in unit i. |
| $\omega$ | Set of units having prohibited zones |

The constraints in (6) implies that if an unit has $n_i$ prohibited zones its operating region will be broken into $n_i + 1$ isolated

feasible sub-regions, resulting in multiple decision spaces for the ED problem. The number of total disjoint decision spaces is given by

$$N = \prod_{i \, \varepsilon \, \omega} ( n_i + 1 ) \qquad (7)$$

The optimal solution will reside in one of the feasible decision spaces.

## III PROPOSED APPROACH

The proposed method for ED consists of two executions of "λ-logic" and the various steps involved in the proposed approach are outlined below.

**Step I** Solve the economic dispatch (ED) problem without considering the prohibited zones using "λ-logic" based algorithm and obtain the optimum dispatch for all the dispatchable units and as well as dispatchable units without prohibited zones. Calculate the average power of unit prohibited zones for all the $n_i$ zones.

**Step II** If the power output of all the units are within limits and fall out of the prohibited operating zones, the optimal dispatch is obtained for given $P_D$ . Then got to step V. If any one of the units/all fallen within prohibited zones check the 'power output' with prohibited zone limits. If the 'power output' of that unit is greater or lesser than the average power of prohibited zone, choose the upper or lower limits respectively.

**Step III** Once the outputs of generators with prohibited zones are known, output power of these generators and estimate remaining load yet to be dispatched among other generators (without prohibited zones)

**Step IV** Now economically dispatch the remaining load by the second execution of "λ-logic" based algorithm for the units without prohibited zones.

**Step V** Stop the calculation process.

**"λ-logic" Algorithm [5]:**

It consists of two stages i.e., (i) pre-prepared power demand data (PPD) using λ-logic and (ii) calculation of solution $P_1, P_2, …, Pk$ for specified Power demand ($P_D$ ). The first stage involves a systematic approach with fixed number of steps and offers unique PPD (Pre-prepared power demand data) for K units. This PPD acts as a big source in reducing the computational burden of ED of K-units. This PPD remains unaltered for all values of $P_D$ variations (ranging from $P_{Dmin}$ and $P_{Dmax}$). The second stage consists of calculation of $P_1$, $P_2,… P_k$ for given $P_D$ by calculating Incremental cost for the corresponding load from PPD. This PPD consists of all λ values along with corresponding demand. By knowing the intervals in which $P_D$ is located, the corresponding $\lambda_{final}$ can be evaluated. This $\lambda_{final}$ can be used for evaluating the optimal generation.

For Economic Dispatch of K thermal units of a plant, the ED condition is

$$dF_1/ dP_1 = dF_2/ dP_2 = \ldots\ldots = dF_k/ dP_k = \lambda \qquad (8)$$

For Quadratic fuel cost function:
$$dF_i/ dP_i = \lambda_i = \alpha_i + \beta_i P_i \qquad (9)$$

For Cubic fuel cost function:
$$dF_i/ dP_i = \lambda_i = \alpha_i + \beta_i P_i + \gamma_i P_i^2 \qquad (10)$$

where

$\lambda_i$ — Incremental cost of unit i \$/MWhr

$\alpha_i = b_i$, $\beta_i = 2*c_i$, $\gamma_i = 3*d_i$ – Coefficients of Incremental cost function of unit i.

The following (11) and (12) are useful for calculating $P_i$ for a specified $\lambda$ value and $P_i$ is given by

For Quadratic fuel cost function:
$$P_i = (\lambda - \alpha_i)/ (2* \beta_i) \qquad (11)$$

For Cubic Fuel cost function:
$$P_i = (-\beta_i \pm \sqrt{(\beta_i^2 - 4* \gamma_i *( \alpha_i - \lambda))}/ (2* \gamma_i)) \qquad (12)$$

## IV. RESULTS

The above algorithm is tested on 4 units [1], 5 units [2] and 15 units [1] test system data. The effectiveness of the proposed approach is demonstrated by solving the test cases as shown below:

### 4.1 Test case 1

There are four on-line units with the following input-output characteristics:

$$F_i (P_i) = 500 + 10P_i + 0.001P_i^2 \ \$/hr$$

$$100 \text{ MW} \leq P_i \geq 500 \text{ MW}, \ i = 1,2,3,4 \qquad (13)$$

$P_D = 1390$ MW

Each of unit 1 and unit 2 have two prohibited zones. Units 3 and 4 have none. The prohibited zones are described in Table I:

TABLE I
PROHIBITED ZONES

| Unit | Zone 1 (MW) | Zone 2 (MW) |
|------|-------------|-------------|
| 1 | [200,250] | [300,350] |
| 2 | [210,260] | [310,360] |

TABLE II
AVERAGE POWER OF PROHIBITED ZONES

| Unit | Zone 1 (MW) | Zone 2 (MW) |
|------|-------------|-------------|
| 1 | 225 | 325 |
| 2 | 235 | 335 |

Using the "$\lambda$-logic" based algorithm, the desired optimal generation without considering prohibited operating zones is found to be:

$P_1 = P_2 = P_3 = P_4 = 347.50$ MW.

From Table III, it can be seen that unit 1 and 2 falls in prohibited zone 2. Both Unit 1 and 2 falls in second zone of each unit and the optimal generation for each unit is above average power (325 MW and 335 MW respectively), so set the power output of each unit to upper limits of prohibited zones

$P_1 = 350$ MW, $P_2 = 360$ MW

TABLE III
OPTIMAL GENERATION

| Unit | Generation levels Without Prohibited zones (MW) | Final Generation levels Considering Prohibited zones (MW) |
|------|------------------------------------------------|------------------------------------------------------------|
| 1 | 347.5 | 350 |
| 2 | 347.5 | 360 |
| 3 | 347.5 | 340 |
| 4 | 347.5 | 340 |

The remaining load (1390 – 350 - 360= 680 MW) should be met by other units 3 and 4 by second execution of "$\lambda$-logic" based algorithm[5]. The power output of unit 3 and 4 is found to be 340 MW each

Total fuel cost = 16383.3 \$/hr.

### 4.2 Test case 2

There are five on-line units with the following input-output characteristics:

$$F_i (P_i) = 350 + 8P_i + 0.001P_i^2 + 1.0*10^{-6} P_i^3 \ \$/hr$$

$$120 \text{ MW} \leq P_i \geq 450 \text{ MW}, \ i = 1,2, 3,4,5 \qquad (14)$$

The input-output cost function of each unit is represented by a cubic function $a_i + b_i P_i + c_i P_i^2 + d_i P_i^3$. Unit 1, 2 and 3 have prohibited zones and units 4 and 5 have none. The prohibited zones are defined in Table IV when $P_D = 1175$ MW.

944

### TABLE IV
### PROHIBITED ZONES

| Unit | Zone 1 (MW) | Zone 2 (MW) |
|------|-------------|-------------|
| 1 | [240,275] | [315,375] |
| 2 | [210,270] | [300,390] |
| 3 | [200,250] | [290,370] |

The average power of unit prohibited zones for all the $n_i$ zones is calculated and shown below in Table V.

### TABLE V
### AVERAGE POWER OF PROHIBITED ZONES

| Unit | Zone 1 (MW) | Zone 2 (MW) |
|------|-------------|-------------|
| 1 | 257.5 | 345 |
| 2 | 240 | 345 |
| 3 | 225 | 330 |

### TABLE VI
### OPTIMAL GENERATION

| Unit | Generation levels Without Prohibited zones (MW) | Final Generation levels Considering Prohibited zones (MW) |
|------|------------------------------------------------|----------------------------------------------------------|
| 1 | 235 | 235 |
| 2 | 235 | 210 |
| 3 | 235 | 250 |
| 4 | 235 | 240 |
| 5 | 235 | 240 |

From Table VI, it can be seen that unit 1 is in the feasible region, while each of unit 2 and 3 falls in a prohibited zone. For Unit 2, $P_2$ is below "average power" of Prohibited zone and $P_2 = 210$ and for unit 3, $P_3$ is above the "average Power" of prohibited zone and $P_3 = 250$ MW.

The remaining load (1175–235–210–250 = 480 MW) should be met by other units 4 and 5 by second execution of "λ-logic" based algorithm[5]. The power output of unit 4 and 5 is found to be 240 MW each.

Total fuel cost = 11492.51 $/hr.

Table VII presents the results obtained using the λ-δ iterative method [2], SEPA [6], FCEPA [7] and the proposed method. From Table VII, it can be seen that the optimal fuel cost obtained with the proposed method is exactly same as that of results obtained with [2,6,7]. It can be observed that the optimal generation obtained with the proposed method is different from that of results obtained with [2,6,7]. With the

proposed method there is a reduction in computation burden compared to the methods [2,6,7], since there is no need of calculating penalty costs for each feasible solution as in the methods shown in [2,7].

### TABLE VII
### COMPARISON OF GENERATION ALLOCATION OF ON-LINE UNITS

| Unit | λ-δ iterative method [2] | SEPA [6] | FCEPA [7] | Proposed method |
|------|--------------------------|----------|-----------|-----------------|
| $P_1$(MW) | 238.33 | 240.0 | 238.33 | 235.0 |
| $P_2$(MW) | 210.0 | 210.0 | 210.0 | 210.0 |
| $P_3$(MW) | 250.0 | 250.0 | 250.0 | 250.0 |
| $P_4$(MW) | 238.33 | 223.07 | 238.33 | 240.0 |
| $P_5$(MW) | 238.33 | 251.93 | 238.33 | 240.0 |
| Fuel cost ($/hr) | 11492.51 | 11493.23 | 11492.5 | 11492.51 |

SEPA  - Standard Evolutionary Programming Algorithm.
FCEPA- Fast Computation Evolutionary Programming
        Algorithm

The test cases 1 and 2 consist of 4 units and 5 units respectively. The test case 1 is quadratic fuel cost function and test case 2 is cubic fuel cost function. The proposed method is working effectively for both type of functions, offering optimal generation and fuel cost within two executions of "λ-logic" based algorithm with no need of big search space.

### 4.3 Test case 3

There are 15 on-line units to supply a system bus-bar demand of 2,650 MW.

The unit characteristics are described in Table VIII .

### TABLE VIII
### CHARACTERISTICS OF ON-LINE UNITS

| Unit | $a_i$ ($/hr) | $b_i$ ($/MW hr) | $c_i$ ($/MW$^2$ hr) | $P_{min}$ (MW) | $P_{max}$ (MW) |
|------|------|------|------|------|------|
| 1 | 671.03 | 10.07 | 0.000299 | 150 | 455 |
| 2 | 574.54 | 10.22 | 0.000183 | 150 | 455 |
| 3 | 374.59 | 8.80 | 0.001126 | 20 | 130 |
| 4 | 374.59 | 8.80 | 0.001126 | 20 | 130 |
| 5 | 461.37 | 10.40 | 0.000205 | 150 | 470 |
| 6 | 630.14 | 10.10 | 0.000301 | 135 | 460 |
| 7 | 548.20 | 9.87 | 0.000364 | 135 | 465 |
| 8 | 227.09 | 11.50 | 0.000338 | 60 | 300 |
| 9 | 173.72 | 11.21 | 0.000807 | 25 | 162 |
| 10 | 175.95 | 10.72 | 0.001203 | 20 | 160 |
| 11 | 186.86 | 11.21 | 0.003586 | 20 | 80 |
| 12 | 230.27 | 9.90 | 0.005513 | 20 | 80 |
| 13 | 225.28 | 13.12 | 0.000371 | 25 | 85 |
| 14 | 309.03 | 12.12 | 0.001929 | 15 | 55 |
| 15 | 323.79 | 12.41 | 0.004447 | 15 | 55 |

In Table VIII, the input-output cost function of each unit is represented by a quadratic function $a_i+b_iP_i+c_iP_i^2$. Among all the on-line units, units 2, 5, 6, and 12 have prohibited operating zone as described in Table IX:

TABLE IX
PROHIBITED ZONES

| Unit | No. of zones | Zone1(MW) | Zone2(MW) | Zone3(MW) |
|------|------|-----------|-----------|-----------|
| 2 | 3 | [185,225] | [305,335] | [420,450] |
| 5 | 3 | [180,200] | [260,335] | [390,420] |
| 6 | 3 | [230,255] | [365,395] | [430,455] |
| 12 | 2 | [30,55] | [65,75] | |

The average power of each unit prohibited zones for all the $n_i$ zones is calculated and shown below in Table X.

TABLE X
AVERAGE POWER OF PROHIBITED ZONES

| Unit | Zone1(MW) | Zone2(MW) | Zone3(MW) |
|------|-----------|-----------|-----------|
| 2 | 205.0 | 320.0 | 435.0 |
| 5 | 190.0 | 297.5 | 405.0 |
| 6 | 242.5 | 380.0 | 442.5 |
| 12 | 42.5 | 70.0 | |

TABLE XI
OPTIMAL GENERATION

| Unit | Generation levels Without Prohibited zones (MW) | Final Generation levels Considering Prohibited zones (MW) |
|------|------|------|
| 1 | 455.0 | 437.8 |
| 2 | 455.0 | 455.0 |
| 3 | 130.0 | 130.0 |
| 4 | 130.0 | 130.0 |
| 5 | 317.835 | 335.0 |
| 6 | 460.0 | 460.0 |
| 7 | 465.0 | 465.0 |
| 8 | 60.0 | 60.0 |
| 9 | 25.0 | 25.0 |
| 10 | 20.0 | 20.0 |
| 11 | 20.0 | 20.0 |
| 12 | 57.166 | 57.1 |
| 13 | 25.0 | 25.0 |
| 14 | 15.0 | 15.0 |
| 15 | 15.0 | 15.0 |

From Table XI, it can be seen that units 2, 6 and 12 are in the feasible region, while unit 5 falls in a prohibited zone (zone 2). Unit 5 generation is above average power of prohibited zone, so set the power level to upper limit of zone 335 MW. The remaining load (2650–455–335–460–

57.16=1342.83 MW) should be met by other units by second execution of "$\lambda$-logic" based algorithm[5]. The power output of remaining units is shown in Table XI.
Total fuel cost = 32,545 $/hr.

Table XII presents the results obtained using the method [1, 2], SEPA [6], FCEPA [7] and the proposed method. From Table XII, it can be seen that the optimal fuel cost with the proposed method is same as that of results obtained with [1, 2,6,7] but with reduced computational burden.

The analysis shows that the proposed method is reliable and uses a simple novel approach to select the operating regions. It requires only two executions of the "$\lambda$-logic" approach, one for determining the optimal solution without considering the prohibited operating regions and to select the feasible solutions for the units having prohibited zones as well as to determine the remaining load and the second step is to determine optimal generation of remaining units.

TABLE XII
COMPARISON OF GENERATION ALLOCATION OF ON-LINE UNITS

| Unit | $\lambda$-$\delta$ iterative method [1,2] (MW) | SEPA [6] (MW) | FCEPA [7] (MW) | Proposed method (MW) |
|------|------|------|------|------|
| $P_1$(MW) | 450.0 | 446.98 | 450.0 | 437.8 |
| $P_2$(MW) | 450.0 | 451.50 | 450.0 | 455.0 |
| $P_3$(MW) | 130.0 | 130.0 | 130.0 | 130.0 |
| $P_4$(MW) | 130.0 | 130.0 | 130.0 | 130.0 |
| $P_5$(MW) | 335.0 | 335.02 | 335.0 | 335.0 |
| $P_6$(MW) | 455.0 | 456.11 | 455.0 | 460.0 |
| $P_7$(MW) | 465.0 | 464.91 | 465.0 | 465.0 |
| $P_8$(MW) | 60.0 | 60.0 | 60.0 | 60.0 |
| $P_9$(MW) | 25.0 | 25.0 | 25.0 | 25.0 |
| $P_{10}$(MW) | 20.0 | 20.0 | 20.0 | 20.0 |
| $P_{11}$(MW) | 20.0 | 20.01 | 20.0 | 20.0 |
| $P_{12}$(MW) | 55.0 | 55.46 | 55.0 | 57.1 |
| $P_{13}$(MW) | 25.0 | 25.0 | 25.0 | 25.0 |
| $P_{14}$(MW) | 15.0 | 15.01 | 15.0 | 15.0 |
| $P_{15}$(MW) | 15.0 | 15.0 | 15.0 | 15.0 |
| Fuel cost ($/hr) | 32544.9 | 32545.20 | 32544.9 | 32,545 |

V. CONCLUSION

This paper presents an efficient, fast and effective approach for solving the economic dispatch problem when some of the on-line generating units have prohibited operating zones. When some of the on-line units have prohibited zones, multiple decision spaces will be formed for the economic dispatch problem. It requires high computational time for solving the problem due to the need of determination of advantageous set of decision spaces. The proposed approach uses two execution of 'λ-logic method'. This method can go as a best contribution in the area of economic dispatch. In contrast to other conventional methods, this approach gives a promising value of power for providing improved economic dispatch. The proposed method reduces the computational burden and is superior to many of available techniques of economic dispatch. The method yields a true optimal solution for large -scale systems without any convergence problems, which has been illustrated through several test cases for both quadratic as well as cubic fuel cost function.

## VI. REFERENCES

[1] F. N. Lee and A. M. Breipohl, "Reserve constrained economic dispatch with prohibited zones," *IEEE Trans. Power Syst.*, vol.8, pp.246-254, Feb. 1993.

[2] J.Y.Fan and J.D.McDonald, "A practical approach to real time economic dispatch considering unit's prohibited zones," *IEEE Trans. Power Syst.*, vol. 9, pp. 1737-1743, Nov. 1994.

[3] S.O.Orero and M.R.Irving, "Economic dispatch of generators with prohibited zones: a genetic algorithm approach," *IEE Proc.-Gener.,Trans.,Distrib.*, vol. 143, pp. 529-534, Nov. 1996.

[4] A.J Wood and B.F Wollenberg, "Power generation, operation, and control," 2nd ed., John Wiley and sons, 1996.

[5] Maheswarapu Sydulu, "A very fast and effective non-iterative "λ- logic based" algorithm for economic dispatch of thermal units," *Proc. IEEE Tencon Int conf.*, vol. 2, pp. 1434 – 1437, Sept. 1999.

[6] T.Jayabarathi, G.Sadasivam and V.Ramachandran, "Evolutionary programming based economic dispatch of generators with prohibited operating zones," Electric *power syst. Res.* , vol. 52, pp. 261-266, Feb 1999.

[7] P.Somasundaram, K.Kuppusamy and R.P. Kumudini devi, "Economic dispatch with prohibited operating zones using fast evolutionary programming algorithm," *Electric power syst. Res.*, vol. 70, pp. 245-252, Dec. 2003.

[8] R.Naresh, J..Dubey and J.Sharma, "Two-phase neural network based modeling framework of constrained economic load dispatch," *IEE Proc.-Gener.,Trans.,Distrib.*, vol. 151, pp. 373-378, May. 2004.

[9] Zwc-Lee Gang, " Particle swarm optimization to solving the Economic dispatch considering the generator constraints," *IEEE Trans. Power Syst.*, vol. 18, pp. 1187-1195, Nov. 1994.

## VII. BIOGRAPHIES

**T.Adhinarayanan** received the B.E. degree in Electrical and Electronics Engineering at Arunai Engineering College, Tiruvannamalai, India, in 2002 and M.E degree in Power systems engineering at Crescent Engineering College, Chennai, India, in 2004. He is currently pursuing the Ph.D. degree in the Department of Electrical Engineering, National Institute of Technology, Warangal, India. His Research interests include Power system operation and control, Metaheuristic Applications to Power systems.

**Prof.M. Sydulu** obtained his B.Tech (Electrical Engineering,1978), M.Tech (PowerSystems,1980), Ph.D (Electrical Engineering –Power Systems,1993), all degrees from Regional Engineering College, Warangal, Andhra Pradesh, INDIA. His areas of interest include Real Time power system operation and control, ANN, fuzzy logic and Genetic Algorithm applications in Power Systems, Distribution system studies, Economic operation, Reactive power planning and management, Metaheuristic Applications to Power systems. Presently he is working as Professor and Head of Electrical Engineering Department, National Institute of Technology, Warangal (formerly RECW).

**2006 IEEE International Conference on Power Electronic, Drives and Energy Systems**

# Computation & Analysis of End Region EM Force for Electrical Rotating Machines using FEM

Manpreet Singh Manna, Sanjay Marwaha, and Anupma Marwaha

*Abstract*-**In the recent designs of electrical rotating machines, the emphasis are to have high capacity and electrical loading without comparable increase in size. Therefore the end-winding support structure becomes more significant to withstand with large stress developed due to electromagnetic forces.**
**The end winding support structure of large rotating machine is designed to ensure that the stress in copper or insulation of the end windings do not exceed acceptable limits under steady state and transient conditions. The reliability of end-region structure is proportional to the EM forces therefore predetermination of the parameters of these forces may quit useful for the selection of end ring material and deciding the insulation level for the proposed structures. In this paper, 3-D Finite Element Method has been used to compute and analyze the electromagnetic forces across the end region structure of rotating machines. The computed results are in good agreement with the reported results.**

*Index Terms*-**End winding, Electromagnetic force distribution, Finite element method,**

## I. INTRODUCTION

$\text{T}$HE electromagnetic forces experienced by the end windings and its supporting structure of induction machines during the transient current of a line starting circuit can be large enough to cause serious trouble if the proper selection of insulation and materials is not done. Usually during starting transient operation a high voltage induction motor generates a current with high magnitude and long interval. The end windings are subjected to immense electromagnetic force due to this starting current. This undesired force causes vibration and adversely affects insulation stability of end windings. The electromagnetic forces can be several ten or hundred times of the rating values, and as a consequence of same many severe problems may arise related to insulation fault of winding which is due to sudden voltage slopes, trembling and

---

This work was supported in part by the Computational Laboratory, Department of Electrical & Instrumentation Engineering, SLIET, Longowal.

Manpreet Singh Manna is with Electrical & Instrumentation Engineering Department of Sant Longowal Institute of Engineering & Technology, Longowal (Punjab), India. (e-mail: manpreetsinghmanna@yahoo.com )

Sanjay Marwaha is with Electrical & Instrumentation Engineering Department of Sant Longowal Institute of Engineering & Technology, Longowal (Punjab), India. (e-mail: marwaha_sanjay@yahoo.co.in.)

Anupma Marwaha is with Electronics & Communication Engineering Department of Baba Hira Singh Bhattal Institute of Engineering & Technology, Lehragaga, India. (email: marwaha_anupama@yahoo.co.in)

electromagnetic disturbance. Broadly two methods are used for calculation of end winding force one is the analytical method by using biot-savart law and the other the other is numerical analysis by finite element method (FEM). Although reference papers of the former method are widely published, this method cannot derive an exact solution as well as force distribution on the end winding because of difficulty in representing the geometry of end windings. Because of this problem, Khan, Buckley, and Brooks introduced a quasi-three-dimensional (3-D) (quasi-3D) FEM for end windings of a generator model [1]. Wen, Yao, and Tegopoulos calculated the electromagnetic force acting on the stator end windings of a turbo generator [2], which is based on the magnetic field also computed by the quasi-3D FEM method. However, in quasi-3D FEM, it is very difficult to consider variation and passage of the magnetic flux in the peripheral direction and it is impossible to consider effect of a teeth structure of an electrical machine.

Many tools that can calculate the force of the end windings more correctly have been developed as follows. Calvert [3] calculated the force of the coil in the turbo generator using an approximated model that is based on Biot-Savart's law in 1930. Harrigngton [4] calculated the end winding force by an approximating method similar to Calvert.
The applied machine was 2 poles turbo generator in 1952. Ashworth and Hammond [5] modeled the end winding by cylindrical current sheets and calculated the magnetic field of the end winding, primarily in order to calculate the end turn leakage reactance, which is related with the stray load loss in 1961. They used the Bessel function to have analytical solutions. Lambrecht 1983, calculated end winding forces under 3-phase short-circuit conditions by using mix of analytical and numerical techniques. [6], [7].

## II. MODELLING AND COMPUTATION OF EM FORCE AT END WINDING STRUCTURE

As the electromagnetic force experienced by the end winding of the induction machine is high. It is due to the larger transient current flow for the long period at the starting operation of induction machine. The main cause for development of noise and vibration in the induction machine is the electromagnetic force, which acts on the end winding. These effects are harmful for the insulation of machine. Therefore it is essential to design strong structure for the end winding. There are mainly three factors for designing the structure for the end winding support for IM. [8]
   a)   Steady state withstands capability for large duration.
   b)   Transient force withstanding capability.

0-7803-9771-1/06/$25.00 ©2006 IEEE                948

c) Fretting (homogeneous response across the structure).

TABLE 1
SPECIFICATIONS OF ANALYSIS MODEL

| | |
|---|---|
| Power | 950 KW |
| Rated voltage | 6.6 KV |
| Speed | 30 r.p.s |
| No. of pole | 4 |
| Connection | Y (Star) |
| Coil pitch | 10 |
| Rated current/starting current | 101/661 Ampere |
| No. of slots/pole/phase | 4 |
| Diameter of the ring | 10 mm |

Building on this model, the work reported here takes the state of art of end region calculations a stage further, its main features are as fallows:

a) The same approach is used to calculate end-region flux densities and stator end winding forces for the starting current flow in the one phase whereas other two phases carries half of that current in opposite direction.

b) The electromagnetic and geometric modeling is demanding strict attention to rules and procedures than reported till now.

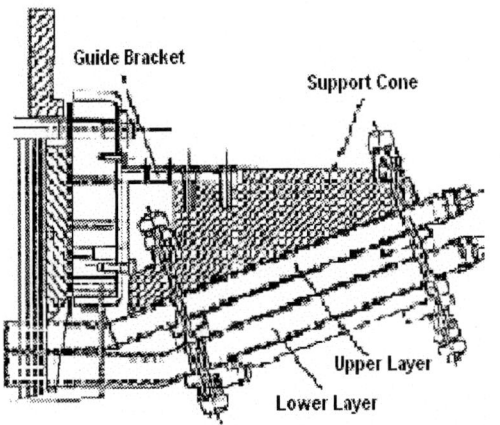

Fig. 1. Stator end winding support structure.

The starting current is 6-7 times higher than rated current where as the transient time to the normal rated speed is approximately 8.5 sec. Fig. 1 shows schematic layout of the starter end winding and the support structure whereas the Fig. 2 shows the complete view of the end winding in which is easily shown the two layers of the stator coils one leg which is nearest bore (upper layer) and leg furthest from bore (lower layer). The finite element method is used here, which thus permits an accurate and detailed comprising the electromagnetic forces effect on the end winding structure including supporting ring even in the nearby of end core region.

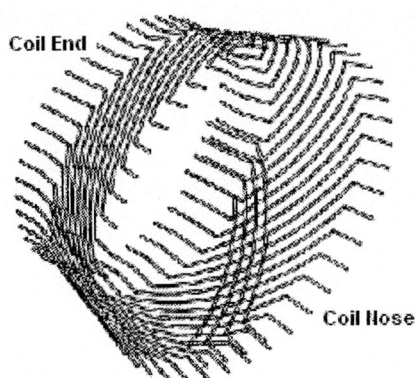

Fig. 2. Complete view of the end-winding hoop.

The finite-element discretisation of end-winding support structure consisting nearly 5200 nodes and 8100 linear triangular elements are shown in the Fig. 3.

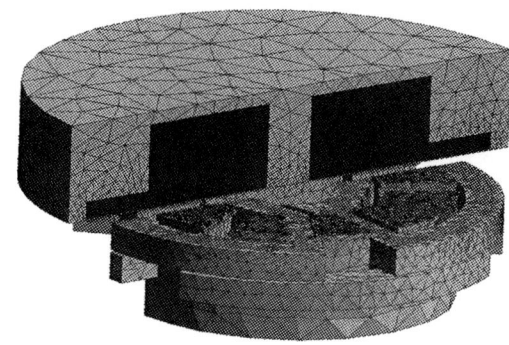

Fig. 3. Stator end winding support structure.

The studies were carried out using a separate finite element modeling at the end region in which the rotor body and holding ring surfaces were treated as eddy current boundaries. For the simplicity of analysis and to save the calculation time, 1 pole model with half periodic boundary conditions (magnetic field intensity at master boundary plane and at slave boundary plane) is used. The calculation of electromagnetic forces is calculated at the instant when largest starting current flows in R Phase while the other two phases B & Y were at the half and opposite current of phase R as shown in Fig. 3.

For the simplicity of the analysis the one segment of the end winding turn distributed in 47 parts as shown in Fig. 4. The 4th coil is selected for the calculation because the magnetic force is highest in this coil. After completing the one end winding we fixed the input conditions same as here and rotating the phase current to 30 degree in clockwise direction for get the force distribution of another end winding. Finally, the whole procedure for the calculation of force distribution could achieve after completing the 12 iterations. There are number of tetrahedral elements taken for the analysis for this model.

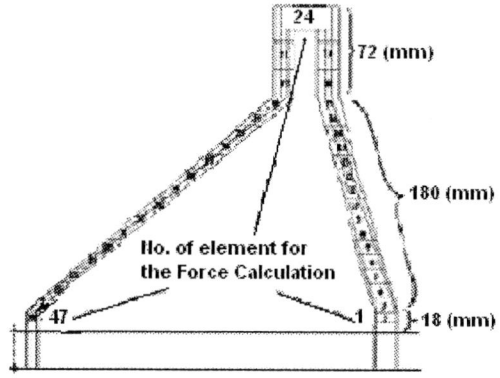

Fig. 4. Reference work of end winding for force calculation.

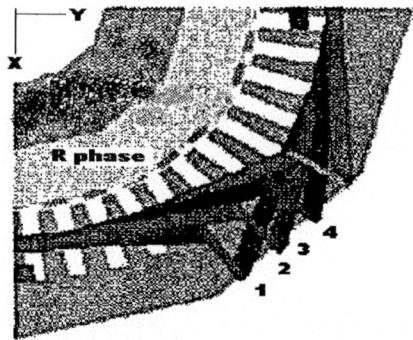

Fig. 5. Reference work of end-region structure for EM force calculations.

## III. ANALYSIS OF END WINDING STRUCTURE FOR EM FORCES

The results by 3-D Finite Element Method shows that 1st and 2nd coil of the upper coil and 46th and 47th coil of the lower coil, where the core is near has larger force then the element of the other regions due to flux concentration. This analysis could be difficult rather impossible for study the accurate effect of the flux concentration in the vicinity of the core by other method viz. analytical approximation.

## IV. STRESS ANALYSIS OF END WINDING

The stress analysis of an end winding part with a supporting ring is performed by applying electromagnetic force distribution and compared with yield stress of the support ring material in order to determine the reliability of the support structure at the end winding of the induction motor. Fig. 5 shows the maximum stress part. The insulation stability of the end winding is also investigated by comparing the simulation results. The insulation fault stress value by a bending moment test. The mechanical characteristics analyzed and found that radial displacement of 2nd and 4th coil of Phase R is same as 24% as shown in Fig. 6. But the stress analysis results as shown in Fig. 7. It is found that the maximum stress of end winding is 34%, which is highest then other coils moreover; the stress of a support ring is 6%. As the maximum stress ratio is 0.323 (here the conductor material of copper with yield stress 72-340 MPa have taken) the design of end winding is within the very reliable level compared with the yield stress as shown in bar diagram in Fig. 8. The magnetic force distribution R phase group is shown in Fig. 9.

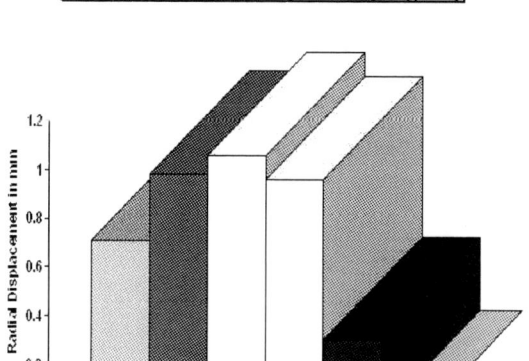

Fig. 6. The results of magnetic field distribution across end winding.

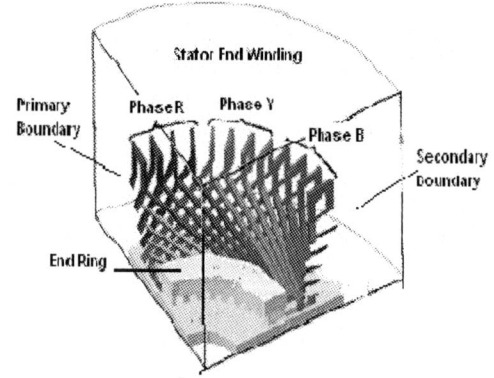

Fig. 7. 3D FEM Model with single pole for EM force calculation.

950

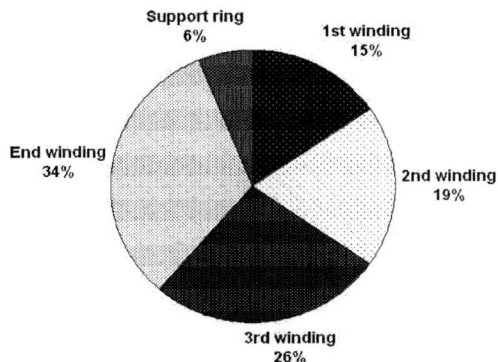

**Stress Ratio**

Fig. 8. Stress Ratio from the Stress Analysis.

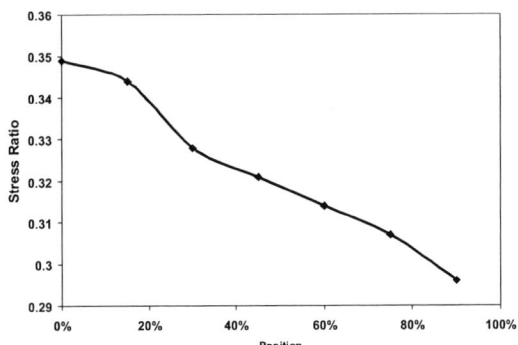

Fig. 9. Magnetic force distribution of R phase group (for end winding).

TABLE 1
SUPPORT RING STRESS ANALYSIS

| Position | Displacement in mm | Stress (Mpa) | Stress Ratio |
|---|---|---|---|
| 0% | 0.146 | 24.1 | 0.349 |
| 15% | 0.135 | 23.76 | 0.344 |
| 30% | 0.121 | 23.31 | 0.328 |
| 45% | 0.103 | 22.78 | 0.321 |
| 60% | 0.089 | 22.1 | 0.314 |
| 75% | 0.072 | 21.22 | 0.307 |
| 90% | 0.059 | 19.89 | 0.296 |

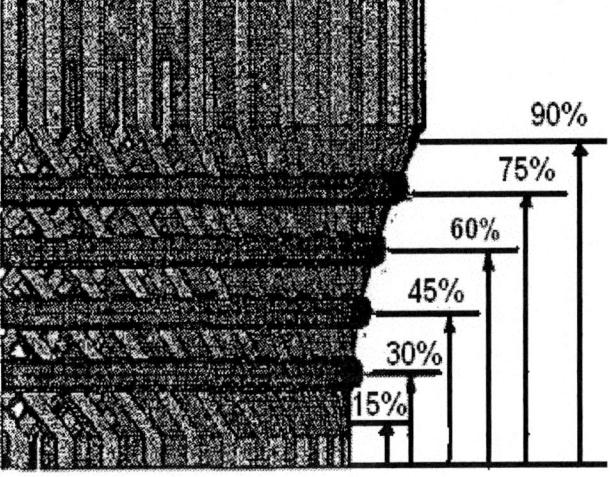

Fig. 10. Support ring stress analysis.

## V. STRESS ANALYSIS AND COMPUTATION OF SUPPORT RING

The support ring stress analysis is done at the various positions and its results are shown in the Table 1. The Fig. 10 shows the various positions of the supporting ring, which has been divided in the six different positions. When the supporting ring is installed the equivalent stress is reduced by 11% and the radial displacement is also reduced 28%. Therefore it is observed the closer the position of the supporting ring to the coil is more beneficial. It is also significant to select proper material for the ring to bear mechanical stress or to share the stress appears at the end winding. Furthermore, the reported analytical results by S. Williamson [10] has been compared with proposed 3D FEM and found the significant differences between both with same frequency as shown in the Fig. 11. The three dimensional calculations are 36% higher than analytical calculations. The other advantage of three-dimensional model that the part of the rotor bars inside the iron core is also modeled. This allows the analysis of the current redistribution when the bar is leaving the iron core. The Fig. 12 shows the current distribution in the bar at 50 Hz. Even this method is useful for determination of overfluxing of power transformers. [11].

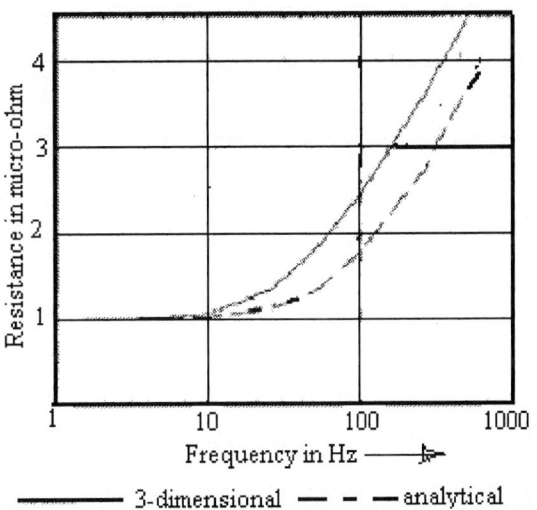

Figure. 11. Variation of the ring resistance with same frequency using analytical and three-dimensional computations.

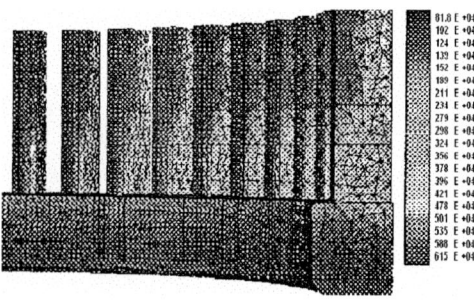

Fig. 12. Current distribution in the bar ends and the end-region.

## VI. CONCLUSIONS

The durability of the end ring structure is mainly dependent upon the magnitude of electromagnetic stresses and condition become more adverse during transient conditions. In this paper, the electromagnetic field distribution of the end winding and its supporting part of rotating machine have been analyzed. The computed results using 3-D Finite Element Method have been compared with the reported results and these are in good agreement. It has been observed that the analytical approximation is sensible only at region where the core is not saturated, but the 3D FEM analysis is applicable even adjacent to the core and any kind of complex domain may be analyzed with ease. It has been analyzed that the closer position of the supporting ring to the coil is more beneficial. It is also significant to select proper material for the ring to bear mechanical stress or to share the stress that appears at the end winding. The dielectric strength of the insulation must be significant to tolerate the abrupt voltage gradients during the transient conditions.

## VII. ACKNOWLEDGMENT

Support for this work is provided by Computational Laboratory part of Electrical & Instrumentation Engineering Department of Sant Longowal Institute of Engineering & Technology, Longowal, India.

## VIII. REFERENCES

[1]   X. Wen, R. Yao and J. A. Tegopoulos, "Transient Quasi- 3D method in the transient electromagnetic field calculation of end region of turbo-generator" , *IEEE Transactions on Magnetics*, Vol. 30, No. 5, Pt, II.1994, 3709-3712.

[2]   X. Wen, G. Gu and R. Yao, "Transient quasi-3D model for electromagnetic field calculation in the end region of turbo-generator", *ICEM'94*, September, 1994.

[3]   I. F. Calvert, "Forces on Turbine Generator Stator Winding," *AIEE* March, 1931, 178-196.

[4]   D. Harrington, "Forces in machine end windings," *AIEE* October, 1952, 849-859.

[5]   D. S. Ashworth P. Hammond, "Calculation of the magnetic field of rotating machines, Part 2 - the field of Turbo- generator End windings," *IEE* No. 3489 S, March, 1961, 527-538.

[6]   H. Berger, W. Fritz, and P.Juen, "Wickelopfbeanspruchung bei Fehlsynchronisation von Turbogenratoren", *ETZ-A*, vol 96, No. 4, pp 188-194, 1975.

[7]   D. Lambrecht and H. Berger, "Integrated End-Winding Ring Support for water-cooled Stator Winding", *IEEE Trans PAS*, vol PAS – 102, April, 1983, 998-1006

[8]   J T Park, K C Kim, "Calculation of the End Winding Force for Electrical Rotating Machines", *IEEE*, February, 2003, 1640 – 1645.

[9]   P C Krause, *Analysis of Electric Machinery* New York: Mc Graw Hill, 1987.

[10]  S. Williamson and M.A. Mueller: "Calculation of the impedance of rotor cage end rings," in *Proc.*1993 IEE Vol. 140, Pt. B, No. 1, 1993, pp. 51-60.

[11]  M S Manna, S Marwaha and Gurmeet Singh, *"3D FEM Analysis For Over Fluxing in Power Transformer"* in *Proc.*2006 CIEEPS Computational Intelligence to Emerging Electric Power system Conf., pp.255-258.

## IX. BIOGRAPHIES

**M. S. Manna** born on March 1971 at Patiala. Er. Manna is Senior Lecturer in the Department of Electrical and Instrumentation Engineering, SLIET, Longowal. Completed his BE (Electrical & Electronics) from Mysore University in 1993, ME (Power & Machines) from Thapar Inst. of Engg. & Tech. in 2000. He is Life member of ISTE and Institution of Engineers (India). Having forty six publications in National and International standards. His area of interest is Power Systems and Energy Generation and Management.

**Sanjay Marwaha** born on April 1966 at Nahan. Dr. Marwaha is Professor and Head in the Department of Electrical and Instrumentation Engineering at SLIET, Longowal. He did his BE (Electrical Engg.) from Gorakhpur University, Gorakhpur in 1988, ME (Power Systems) from Panjab University, Chandigarh in 1990 and Ph.D. from GNDU, Amritsar in 2000. He is life member of ISTE and Member, Institution of Engineers (India) and has published around 60 research papers in National and International journals/conferences of repute. His area of interest includes Design and Analysis of Electromagnetic Devices, Power Systems and HV Engg. Electrical and Electronic Measurements and Instrumentations, Industrial Electronics and Microwave Engineering.

**Anupma Marwaha** born on April 1969 at Chandigarh. Presently she is working as Asstt. Prof. and Head in the Department of Electronics and Communication Engg. at BHSBIET, Lehragaga. She did her BE (Electronics Engg.) from Panjab University, Chandigarh in 1990 and M. Tech. (Electronics and Communication Engg.) from Kurukshetra University in 1992 and Ph.D. from GNDU, Amritsar in 2003. She is life member of ISTE and Member, Institution of Engineers (India) and has published around 50 research papers in National and International journals/conferences of repute. Her area of interest includes Communication Systems, Microwave and Antennas and application of Finite Difference Time Domain and Finite Element tools in design of various electromagnetic structures.

**2006 IEEE International Conference on Power Electronic, Drives and Energy Systems**

# Optimal Reactive Power Dispatch based on Voltage Stability Criteria in a Large Power System with AC/DC and FACTs Devices

D.Thukaram, *Senior Member IEEE*, G.Yesuratnam, and C.Vyjayanthi

*Abstract--* **An algorithm for optimal allocation of reactive power in AC/DC system using FACTs devices, with an objective of improving the voltage profile and also voltage stability of the system has been presented. The technique attempts to utilize fully the reactive power sources in the system to improve the voltage stability and profile as well as meeting the reactive power requirements at the AC-DC terminals to facilitate the smooth operation of DC links. The method involves successive solution of steady-state power flows and optimization of reactive power control variables with Unified Power Flow Controller (UPFC) using linear programming technique. The proposed method has been tested on a real life equivalent 96-bus AC and a two terminal DC system under normal and contingency conditions.**

*Index Terms--* **AC/DC system, L-index, UPFC, Voltage stability.**

## I. INTRODUCTION

IN big developing countries like India, HVDC transmission is becoming an acceptable alternative to AC and is providing an economic solution for bulk power transfer over long distances. Considerable work has been reported in the literature in regard to integrated AC-DC system performance evaluation procedures, notably for load flow and stability studies [1,2]. There is very limited work in the area of reactive power control in AC/DC systems [3]. Even though DC transmission lines carry no reactive power, real power flow into the converters is accompanied by some reactive power flow because of the phase control. The considerations in the operation of a DC transmission system are to satisfy the need for reactive power at the terminals, maintain good voltage profile and improve voltage stability. In a day-to-day operation it may be beyond the operator's scope to take any control decision during emergencies. However, the operator can use various control devices like on load tap changers, generator excitations, Switchable Var Compensators and also FACTs devices like Static Var Compensators, UPFCs to restore the system to normal conditions. These control variables are optimized for the purpose of improving voltage stability of the system. In an AC/DC power system these control variables have to be optimized in a coordinated manner taking account of reactive power requirements at the DC terminals.

The most comprehensive device emanated from the FACTs initiative is the UPFC. The UPFC regulates the active and reactive power control as well as adaptive to voltage magnitude control simultaneously or any combination of them. Controlling the power flows in the network, under normal and network contingencies, help to reduce flows in heavily loaded lines, reduce system power loss, improve stability and performance of the system [4].

Reference [3] has given a method for co-ordinated optimum allocation of reactive power in AC/DC power system with an objective of enhancement of steady state voltage stability based on the L-index [5]. In this an algorithm is proposed for optimization of reactive power control variables using linear programming technique. The objective selected for reactive power optimization is to minimize the sum of the squares of the L indices of all the load buses of the system. The amount of complexity and computational effort involved is very much high with this objective. Hence to overcome this another objective of minimization of the sum squared voltage deviations of the load buses has been given in the reference [6]. The algorithm given in reference [3] did not consider any FACTs devices. This algorithm gives satisfactory results under peak load conditions, but under contingencies like line outages it may not give satisfactory results. In this connection the proposed algorithm is very much reliable.

This paper is mainly concerned with development of a method for co-ordinated optimum allocation of reactive power in AC/DC power systems using FACTs device UPFC, with an objective of minimization of the sum of the squares of the voltage deviations of all the load buses [6]. An algorithm is proposed for optimization of reactive power control variables using linear programming.

## II. VOLTAGE STABILITY ANALYSIS USING L-INDEX AND MINIMUM SINGULAR VALUE

### A. L-Index

Consider a system where, n=total number of busses, with 1,

---

D. Thukaram is with Department of Electrical Engineering, Indian Institute of Science, Bangalore-560012, INDIA (dtram@ee.iisc.ernet.in)

G. Yesuratnam is with Department of Electrical Engineering, Indian Institute of Science, Bangalore-560012, INDIA (ratnam@ee.iisc.ernet.in)

C.Vijayanthi is with Department of Electrical Engineering, Indian Institute of Science, Bangalore-560012, INDIA (jayanthi@ee.iisc.ernet.in)

---

0-7803-9771-1/06/$25.00 ©2006 IEEE

2... g generator busses (g), g+1, g+2... g+s SVC busses (s), g+s+1... n the remaining busses (r=n-g-s) and t =number of OLTC transformers.

A load flow result is obtained for a given system operating condition, which is otherwise available from the output of an on-line state estimator. Using the load flow results, the L-index [5] is computed as

$$L_j = \left| 1 - \sum_{i=1}^{g} F_{ji} \frac{V_i}{V_j} \right| \qquad (1)$$

Where $j=g+1...$ $n$ and all the terms within the sigma on the RHS of (1) are complex quantities. The values $F_{ji}$ are obtained from the $Y$ bus matrix as follows

$$\begin{bmatrix} I_G \\ I_L \end{bmatrix} = \begin{bmatrix} Y_{GG} & Y_{GL} \\ Y_{LG} & Y_{LL} \end{bmatrix} \begin{bmatrix} V_G \\ V_L \end{bmatrix} \qquad (2)$$

Where $I_G, I_L$ and $V_G, V_L$ represent currents and voltages at the generator nodes and load nodes. Rearranging (2) we get

$$\begin{bmatrix} V_L \\ I_G \end{bmatrix} = \begin{bmatrix} Z_{LL} & F_{LG} \\ K_{GL} & Y_{GG} \end{bmatrix} \begin{bmatrix} I_L \\ V_G \end{bmatrix} \qquad (3)$$

Where $F_{LG} = -[Y_{LL}]^{-1}[Y_{LG}]$ are the required values. The L-indices for a given load condition are computed for all load busses.

For stability, the bound on the index $L_j$ must not be violated (maximum limit=1) for any of the nodes j. For a given network, as the load/generation increases, the voltage magnitude and angles change, and for near maximum power transfer condition, the voltage stability index $L_j$ values for load buses tend to close to 1, indicating that the system is close to voltage collapse. The stability margin is obtained as the distance of L from a unit value i.e. (1-L).

### B. Minimum Singular Value

Some researchers have proposed Minimum Singular Value (MSV) of the load flow jacobian [7] as a measure of voltage stability. At the point of voltage collapse, no physically meaningful load flow solution is possible as the load flow jacobian becomes singular. At this point, the minimum singular value becomes zero. Hence the distance of the minimum singular value from zero at an operating point is a measure of proximity to voltage collapse.

### III. APPROACH

At the beginning of the reactive power optimization in AC/DC power systems, a satisfactory initial operating condition for the DC system is selected based on the control strategies, viz., constant power control, constant current control, and constant voltage control applicable at the DC terminals. As shown in Fig. 1, a solution for DC system is first obtained in block 2 and then the voltage, active and reactive power requirements at the DC terminals are computed. Defining these requirements at the AC side of the converter/inverter transformers an AC power flow solution with UPFC is obtained in block 3. Now the terminal transformer taps are computed and their range is checked for the satisfactory solution of the AC/DC system in block 4. If the transformers tap change is not satisfactory or the voltage

stability has to be further improved, reactive power optimization for the AC system is carried out in block 5 with suitable terminal conditions. At this stage, a check for the AC/DC system satisfactory condition is performed in block 6. If the solution is still not satisfactory, modifications in the initial conditions of the DC system are made with suitable changes in the firing angles as shown in block 7 and the process in blocks 2-7 is repeated. Finally, the nearest practical possible tap settings are selected for the transformers at the AC/DC terminal and the final AC/DC power-flow solution is obtained in block 8.

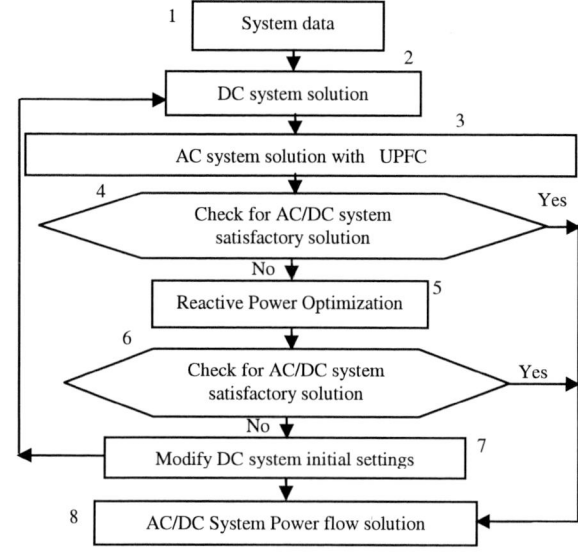

Fig. 1. Major computational blocks of the proposed approach.

### IV. DESCRIPTION OF MODEL

#### A. Converter representation

A general AC/DC terminal and its equivalent circuit is shown in Fig.2.The basic equations describing the converter with its firing angle, tap controls and the DC network are summarized based on the per-unit system selected as follows:

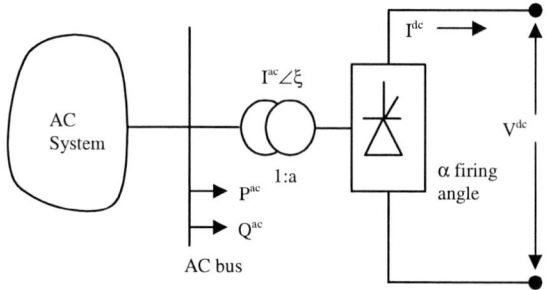

Fig. 2 . Equivalent circuit of DC terminal.

*AC system base quantities*

$P_{base}^{ac}$ = 3-phase power

$V_{base}^{ac}$ =Line-to-line RMS value

$$I_{base}^{ac} = P_{base}^{ac} / \sqrt{3} V_{base}^{ac}$$

*DC system base quantities*

$$P_{base}^{dc} = P_{base}^{ac} ; V_{base}^{dc} = K_b V_{base}^{ac} ; I_{base}^{dc} (\sqrt{3} / K_b) I_{base}^{ac}$$

Where $K_b = (3\sqrt{2}/\pi) nb$

$n_b$ is the number of series - connected bridges in a terminal. The direct voltage and power at the converter are given by

$$V^{dc} = aV^{ac}\cos(\alpha) - R_c I^{dc} \qquad (4)$$

$$P^{dc} = V^{dc} I^{dc} \qquad (5)$$

Where $R_c$ is commutation resistance, a is the transformer tap setting and $\alpha$ the firing angle. Neglecting the losses in the converter and its transformer and equating the expression for powers on the AC and DC sides, the equation for power factor angle ($\psi$ -$\xi$) is given by

$$V^{dc} = aV^{ac}\cos(\psi - \xi) \qquad (6)$$

and for the reactive power flowing from the AC bus into the converter terminal is

$$Q^{dc} = P^{ac}\tan(\psi - \xi) \qquad (7)$$

Where $\psi$ is the alternating voltage angle and $\xi$ is the alternating current angle

A practical operating scheme for a DC system using local terminal controls is to have the DC- system voltage determined at one terminal and the other terminals are provided with scheduled power or current settings. To keep the reactive power consumption of the converter and the losses low the firing angles should be small. But to maintain phase control and reliable commutation, a minimum control angle should be maintained.

### B. UPFC equivalent circuit

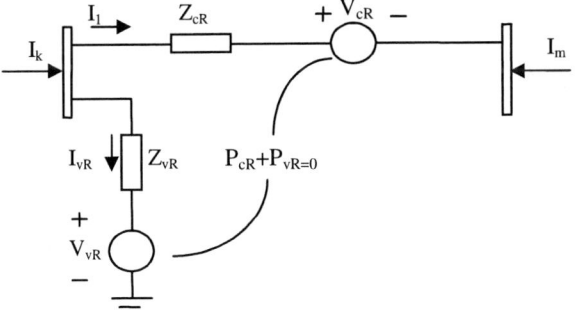

Fig. 3. UPFC Equivalent circuit.

The UPFC equivalent circuit for a steady-state model as shown in Fig.3, has been used in the evaluation of system performance.

The equivalent circuit consists of two ideal voltage sources,

$$V_{cR} = V_{cR}(\cos\theta_{cR} + j\sin\theta_{cR})$$

$$V_{vR} = V_{vR}(\cos\theta_{vR} + j\sin\theta_{vR}) \qquad (8)$$

Where $V_{vR}$ and $\theta_{vR}$ are the controllable magnitude ($V_{vR\ min} \le V_{vR} \le V_{vR\ max}$) and angle ($0 \le \theta_{vR} \le 2\pi$) of the parallel voltage source. The magnitude of $V_{cR}$ and angle $\theta_{cR}$ of the series voltage source are controlled between limits ($V_{cR\ min} \le V_{cR} \le$

$V_{cR\ max}$) and angle ($0 \le \theta_{vR} \le 2\pi$) respectively [8].

### C. Load model

A composite load model, a combination of the ZIP model and exponential model, is considered. Active and reactive power loads are modeled as a function of voltage at the bus. The functions considered are

$$P_{L_i}^{actual} = P_{Lo_i}^{no\ minal}(A_0 + A_1 V + A_2 V^2 + A_3 V^{ep}) \qquad (9)$$

$$Q_{L_i}^{actual} = Q_{Lo_i}^{no\ minal}(R_0 + R_1 V + R_2 V^2 + R_3 V^{eq}) \qquad (10)$$

Where $A_0, R_0, A_1, R_1, A_2, R_2, A_3, R_3$ denote the portion of total load proportional to constant power, constant impedance, constant current and exponential of voltages with ep ,eq given values.

### D. AC/DC power-flow solution method

Considering a DC system

Where m represents the total number of DC terminals

p represents the number of terminals with constant power control

c represents the number of terminals with constant current control and

m = (p+c+1) the terminal with voltage control.

It is assumed that 1, 2,..., p are the constant power control terminals, p+1,p+2,...., p+c are the constant current control terminals, and m the voltage controlled terminal. The algebraic sum of the direct currents flowing into the DC network must be zero and therefore

$$\sum_{k=1}^{m} I_k^{dc} = 0 \qquad (11)$$

The direct voltages at terminals other than the voltage controlled terminal are given by,

$$[V_{bus}] = [R_{bus}][I_{bus}] + [V_m] \qquad (12)$$

$$where [V_{bus}]^t = [V_1^{dc}, V_2^{dc}, ....., V_p^{dc}, V_{p+1}^{dc}, ..., V_{p+c}^{dc}]$$

$$[I_{bus}]^t = [I_1^{dc}, I_2^{dc}, ....., I_p^{dc}, I_{p+1}^{dc}, ...., I_{p+c}^{dc}]$$

$$[V_m]^t = [V_{m1}^{dc}, V_{m2}^{dc}, ....., V_{mp}^{dc}, V_{mp+1}^{dc}, ...., V_{mp+c}^{dc}]$$

[$R_{bus}$] is the bus resistance matrix of the DC network with voltage controlled terminal as reference

$V_m^{dc}$ is the scheduled voltage at the voltage controlled terminal

$I_{p+1}^{dc}, ....., I_{p+c}^{dc}$ are the scheduled currents at the controlled terminals

$I_1^{dc}, ....., I_p^{dc}$ are computed currents at the power controlled terminals ($I^{dc} = P^{dc}/V^{dc}$)

Using an iterative technique the solution of these equations (12) is obtained for the values of direct currents, voltages and powers at all the DC terminals. For the terminals with power control and current control it is common practice to co-ordinate the tap control with phase control so that the terminal will operate at some direct voltage below its own minimum firing (ignition or extinction) angle characteristic to avoid frequent mode shifts from occurring with normal alternative voltage fluctuations. Thus the direct voltage equation for the

955

terminals with power control and current control is modified as

$$V^{dc} = M \left[ aV^{ac} \cos \alpha - R_c I_{dc} \right] \qquad (13)$$

where M is a coefficient typical of 0.97 for 3% voltage margin. Substituting the values of $\alpha$, $R_c$, $V^{dc}$, $I^{dc}$ and M, the values of $aV^{ac}$ for all the terminals are obtained from (4) and (12). Substituting the values of $aV^{ac}$ into (6) the power factor angles ($\psi$-$\xi$) at all the terminals are obtained. The active and reactive powers flowing from the AC bus to the converter terminals are computed from (5) and (7), respectively. Now the AC power-flow solution is obtained with the defined values of P, Q at the AC/DC terminals. This solution provides the voltage conditions at all the AC buses. Knowing the values of $aV^{ac}$, i.e. the product of converter station transformer tap and AC-bus voltage from the DC-system solution and values of $V^{ac}$ from the AC system solution, the tap settings of the converter transformers are determined. If the tap settings violate the limits, modifications such as a change in scheduled voltage $V^{dc}$ at the voltage-controlled DC terminal, a change in control angle $\alpha$ and optimisation of the reactive power schedule in the AC system to obtain improved values of $V^{ac}$ at the AC/DC terminals are effected and the procedure to obtain AC/DC system solution repeated.

## V. DESCRIPTION OF THE REACTIVE POWER OPTIMIZATION PROBLEM

Minimisation of voltage deviations of all the load buses in a system forms the basis for the reactive power optimisation problem. The model uses linearised sensitivity relationships to define the problem. The constraints are: the linearised network performance equations relating to control and dependent variables and the limits on the control variables. Then the model selected for the reactive power optimisation uses linearised sensitivity relationships to define the optimisation problem. The objective is to minimize the sum of the squares of the voltage deviations of all the load buses for the system is given by

$$\upsilon_e = \sum_{j=g+1}^{n} (V_j^{desired} - V_j^{actual})^2 \qquad (14)$$

Where $V_j^{desired}$ is the desired value of the voltage magnitude at the $j^{th}$ load bus. $V_j^{desired}$ is usually set to be 1.0 pu.
The control variables are:

- The transformer tap settings ($\Delta T$)
- The generator excitation settings ($\Delta V$)
- The Switchable VAR Compensator (SVC) settings ($\Delta Q$)

These variables have their upper and lower limits. Changes in these variables affect the distribution of the reactive power and therefore change the reactive power at generators, the voltage profile and thus the voltage stability of the system.
The dependent variables are:

- The reactive power outputs of the generators ($\Delta Q$)
- The voltage magnitude of the buses other than the generator buses ($\Delta V$)

These variables also have their upper and lower limits. In mathematical form, the problem is expressed as:

$$\text{Minimize } \nu_e = Cx \qquad (15)$$

$$\text{Subject to} \qquad b^{min} \leq b = Sx \leq b^{max}$$

$$\text{and} \qquad x^{min} \leq x \leq x^{max}$$

Where C is the row matrix of the linearized objective function sensitivity coefficients, S the linearized sensitivity matrix relating the dependent and control variables, b the column matrix of linearized dependent variables, x the column matrix of the linearized control variables, $b^{max}$ and $b^{min}$ are the column matrices of the linearized upper and lower limits on the dependent variables and $x^{max}$ and $x^{min}$ are the column matrices of linearized upper and lower limits on the control variables.

The linear programming technique is now applied to the above problems to determine the optimal settings of the control variables.

### A. Computation of Sensitivity Matrix (S)

The sensitivity matrix S relating the dependent and control variable is evaluated [6] in the following manner. Considering the fact that the reactive power injections at a bus does not change for a small change in the phase angle of the bus voltage, the relation between the net reactive power change at any node due to change in the transformer tap settings and the voltage magnitudes can be written as

$$\begin{bmatrix} \Delta Q_g \\ \Delta Q_s \\ \Delta Q_r \end{bmatrix} = \begin{bmatrix} A_1 & A_2 & A_3 & A_4 \\ A_5 & A_6 & A_7 & A_8 \\ A_9 & A_{10} & A_{11} & A_{12} \end{bmatrix} \begin{bmatrix} \Delta T_t \\ \Delta V_g \\ \Delta V_s \\ \Delta V_r \end{bmatrix} \qquad (16)$$

Then, transferring all the control variables to the right hand side and the dependent variables to the left hand side and rearranging:

$$\begin{bmatrix} \Delta Q_g \\ \Delta V_s \\ \Delta V_r \end{bmatrix} = [S] \begin{bmatrix} \Delta T_t \\ \Delta V_g \\ \Delta Q_s \end{bmatrix} \qquad (17)$$

### B. Computation of Objective Function $V_{desired}$

$(\nu_e = \sum_{j=g+1}^{n} (V_j^{desired} - V_j^{actual})^2)$ Sensitivities (C) with respect to

### Control Variables

Consider a system where, k = total number of control variables with 1, 2...t number of OLTC transformers, t+1, t+2...t+g generator excitations and t+g+1...k SVCs, (k=t+g+s)

$$\frac{\partial \nu_e}{\partial T_m} = \sum_{j=g+1}^{n} 2(V_j^{desired} - V_j^{actual})(-S_{jm}) \qquad (18)$$

Where m=1,2...t for calculating the objective function sensitivities with respect to transformer taps, and $S_{jm}$ is corresponding elements in equation (17)

$$\frac{\partial \nu_e}{\partial V_m} = \sum_{j=g+1}^{n} 2(V_j^{desired} - V_j^{actual})(-S_{jm}) \qquad (19)$$

Where m= t+1,t+2...t+g for calculating the objective function sensitivities with respect to generator excitations

$$\frac{\partial \nu_e}{\partial Q_m} = \sum_{j=g+1}^{n} 2(V_j^{desired} - V_j^{actual})(-S_{jm}) \qquad (20)$$

Where m= t+g+1...k for calculating the objective function sensitivities with respect to SVCs.

## VI. SYSTEM STUDIES

An AC/DC system of two-terminal DC and 96 AC buses, typical of Indian grid equivalent system [3] including the voltage levels of 220 and 400 kV as shown in Fig. 4, has been considered for studies. There are 20 generators in the system connected at buses 1–13, 15–19, 95 and 96. The AC/DC converter stations are connected at buses 29 and 32. The DC system data is given in Table I. There are 20 generators, 18 tap regulating transformers and 95 transmission lines in the system. About 30 buses are considered as switchable VAr compensator buses. The system has about 12345.8 MW, 6410.0 MVAr peak load. For most of the contingencies it has been observed that the placement of UPFC on the line connected between the buses 39 and 86 gives better results as compared to other lines. Hence this line is selected for UPFC placement.

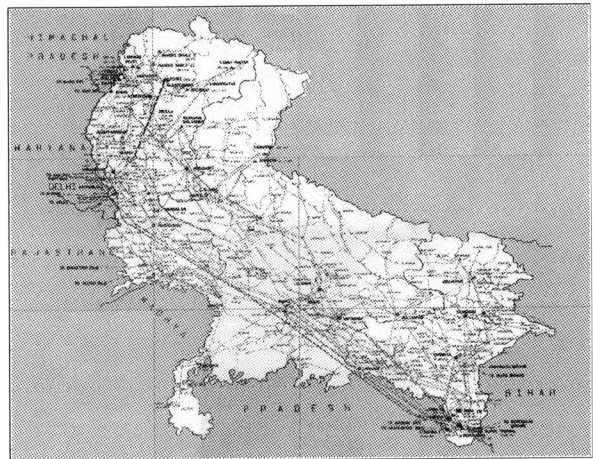

Fig. 4. AC/DC System of two-terminal DC and 96 AC buses.

### A. Case 1: line outage between buses 27 and 28

With this contingency, for a peak load condition, the power flow results for this case show a low voltage profile in the system with the voltages of about 48 buses not being within acceptable limits (0.95–1.05 p.u.). There are 13 generators exceeding the maximum Q limits and no generator Q is exceeding the minimum limit. As indicated in Table II, the minimum singular value before optimization is 0.20491. The

TABLE I
DC SYSTEM DATA

|  | Sending end | Receiving end |
|---|---|---|
| Transformer secondary (kV) | 219.0 | 216.0 |
| MVA rating | 465.0 | 460.0 |
| $X_c$ in pu | 0.1900 | 0.1900 |
| Tap max (pu) | 1.10 | 1.10. |
| Tap min (pu) | 0.90 | 0.90 |
| Tap step (pu) | 0.0125 | 0.0125 |
| P specified in MW | 1540.0 | 1500.0 |
| Commutating resistance (pu) | 0.00535 | 0.00541 |
| $R_{dc}$ line pu | 0.00137 |  |

minimum voltage is 0.732 pu at bus number 55. The sum of squares of voltage stability L-indices $\sum L^2$ is 7.2002. The sum squared voltage deviations of all the load buses is 0.64807. Initially the proposed algorithm for reactive power optimization has been applied to improve the situation without using any UPFC. Then the results are obtained by placing the UPFC on the line connected between the buses 39 and 86. The step-size taken for both the regulating transformers and generators excitations is 0.0125 p.u. The VAr compensation at the selected places is initially assumed to be zero. After four iterations of the VAr optimization the voltages at all the buses have been brought within the satisfactory operable limits (0.95–1.05 p.u.). As indicated in Table II, after the optimization the minimum voltage has been improved to 0.978 with UPFC and to 0.959 without UPFC and the sum of square of voltage stability L-indices $\sum L^2$ is reduced to 4.4566 without UPFC where as this value is only 2.5376 with UPFC. The minimum singular value after optimization is 0.26788 and the sum squared voltage deviations of all the load buses is improved to 0.03488. The corresponding values are 0.28193 and 0.03408 with UPFC. All these results indicate that there is an improvement in voltage stability margin with the given optimization algorithm. The improvement is much better with the presence of UPFC. The transmission loss of the system also reduced from 491.96 MW (initial value) to 382.44 MW without UPFC and to 363.53 MW with UPFC as indicated in Table II. After optimization all the generators reactive power outputs Q are not brought within the limits, still some of the generators Q were exceeding maximum limits (generator 6, 12 and 17 as highlighted in Table III), if there is no FACTs device. Where as with the presence of UPFC the reactive power outputs of all the generators have been brought within their limits during optimization as indicated in Table III.

TABLE II
SUMMARY OF RESULTS WITH THE OPTIMAL CONTROLLER SETTINGS

|  | Initial values | Without UPFC | With UPFC |
|---|---|---|---|
| Ploss in MW | 491.96 | 382.44 | 363.56 |
| MSV | 0.20491 | 0.26788 | 0.28193 |
| $\sum L^2$ | 7.2002 | 4.4566 | 2.5376 |
| $V_{min}$ | 0.732 (at bus 55) | 0.959 | 0.978 |
| $\upsilon_e$ | 0.64807 | 0.03488 | 0.03408 |

TABLE III
REACTIVE POWER OUTPUT OF SOME CRITICAL GENERATORS

| Gen bus no. | Reactive power output | | | QG max limit |
|---|---|---|---|---|
|  | Initial values | Without UPFC | With UPFC | in mvar |
| 6 | 450.92 | **219.6** | 177.5 | 206.0 |
| 9 | 491.54 | 311.0 | 223.4 | 330.0 |
| 10 | 317.72 | 149.0 | 132.0 | 248.0 |
| 11 | 64.72 | 27.4 | 19.0 | 30.0 |
| 12 | 239.04 | **145.5** | 82.5 | 135.0 |
| 13 | 179.90 | 80.4 | 16.8 | 96.0 |
| 15 | 133.68 | 72.3 | 29.3 | 99.0 |
| 16 | 247.58 | 102.3 | 96.0 | 160.0 |
| 17 | 187.11 | **89.7** | 45.5 | 80.0 |
| 18 | 739.88 | 361.4 | 435.2 | 586.0 |
| 19 | 410.74 | 282.8 | 204.2 | 297.0 |
| 95 | 186.92 | 119.6 | 56.9 | 120.0 |
| 96 | 533.89 | 395.6 | 388.6 | 469.0 |

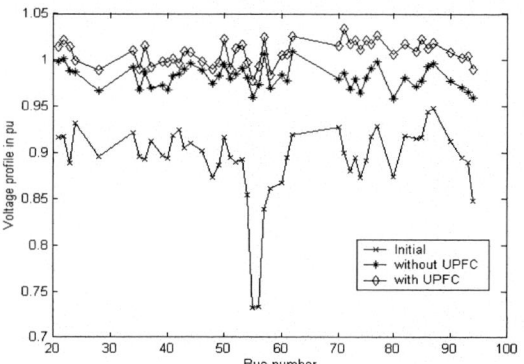

Fig. 5. Bus voltage profile before and after optimization with and without FACTs devices

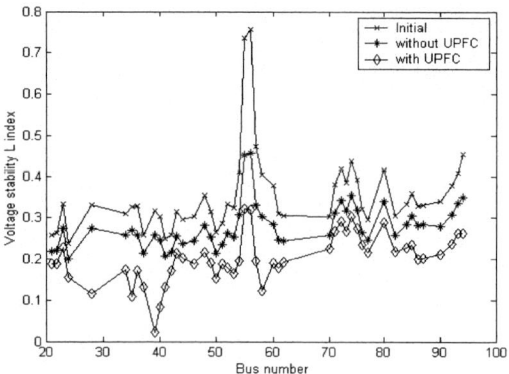

Fig. 6. Voltage stability L index profile before and after optimization with and without FACTs devices

The summarized results, initial and after optimization (with and without UPFC), for the given system are indicated in Table II and Table III. The load bus voltage profiles and voltage stability indices before and after optimization with and without UPFCs are shown in Fig. 5, and Fig. 6, respectively. In this case the DC terminal at bus 32 is considered as receiving end power control.

## VII. CONCLUSIONS

An algorithm for optimum allocation of reactive power in AC/DC system using FACTs devices, with an objective of improving the voltage profile and also voltage stability of the system has been presented. The developed algorithm has been tested on typical sample systems and on a practical real-life equivalent system of a 96-bus AC and a two-terminal DC system with UPFC. The proposed algorithm is giving encouraging results for improving the operational conditions of the system under normal and network contingencies.

## VIII. REFERENCES

[1] Fudeh, H., and Ong, C.M., "A simple and efficient AC-DC load-flow method for multiterminal DC systems", *IEEE Trans. Power Appar. Syst.*, 1982, 101, pp. 4381–4396

[2] Arrillaga, J., and Smith, B. "AC-DC power system analysis" (IEEE, London, 1998)

[3] Thukaram D., Jenkins L. and Visakha K., "Optimum allocation of reactive power for voltage stability improvement in AC-DC power systems", *IEE proceedings on Gener. Transm. Distrib.* vol.153, no.2, March 2006, pp. 237-246.

[4] Thukaram D., Jenkins L. and Visakha K., "Improvement of system security with Unified Power Flow Controller at suitable locations under network contingencies of interconnected systems", *IEE proceedings on Gener. Transm.Distrib.*, vo.152, no.5, September 2005, pp. 682-690.

[5] Kessel, P., and Glavitsch, H., "Estimating the voltage stability and loadability of power systems", *IEEE Trans. Power Deliv.*, 1986, 1, (3), pp. 586–1599

[6] Thukaram Dhadbanjan and G.Yesuratnam (2006) "Comparison of Optimum Reactive Power Schedule with Different Objectives Using LP Technique," *International Journal of Emerging Electric Power Systems*: Vol.7:No.3, Article 2.

[7] P.A.Lof. G.Anderson. D.J. Hill., "Voltage stability indices for stressed power systems," *IEEE Trans. Power Syst.* 8 (1) (February 1993).

[8] Fuerte-Eaquivel, C.R.,and Acha. E, "Unified Power Flow Controller: a critical comparison of Newton-Raphson UPFC algorithms in power flow studies", *IEE Proc.Gen.Trans.Distrib.*, 1997, 144, (5), pp.437-444.

## IX. BIOGRAPHIES

**D Thukaram** (SM, 90) received the B.E. degree in Electrical Engineering from Osmania University, Hyderabad in 1974, M.Tech degree in Integrated Power Systems from Nagpur University in 1976 and Ph.D. degree from Indian Institute of Science, Bangalore in 1986. Since 1976 he has been with Indian Institute of Science as a research fellow and faculty in various positions and currently he is Professor. His research interests include computer aided power system Analysis, reactive power optimization, voltage stability, distribution automation and AI applications in power systems.

**G. Yesuratnam** received the B.Tech degree in Electrical and Electronics Engineering from Jawaharlal Nehru Technological University, Hyderabad, India in 1995. He received M.Tech degree in 1998 form National Institute of Technology, Warangal in the field of Power Systems. Currently he is working towards his Ph.D in the Department of Electrical Engineering, Indian Institute of Science, Bangalore. His research interests include computer aided power system analysis, reactive power optimization, voltage stability and AI applications in power systems.

**C Vyjayanthi** received the B.Tech degree in Electrical and Electronics Engineering from Jawaharlal Nehru Technological University, Hyderabad, India in 2001. She received M.E degree in 2005 form Anna University, Chennai in the field of Power Systems. Currently she is working towards her PhD in the Department of Electrical Engineering, Indian Institute of Science, Bangalore. Her research interests include computer aided power system analysis, deregulation, reactive power optimization, voltage stability.

**2006 IEEE International Conference on Power Electronic, Drives and Energy Systems**

# Location of Unified Power Flow Controller and its Parameters setting for Congestion Management in Pool Market Model Using Genetic Algorithm

Hassan Barati, Mehdi Ehsan, and Mahmud Fotuhi-Firuzabad, *Senior Member, IEEE*

*Abstract*--In this paper, AC Optimal Power Flow combined with UPFC has been used to manage the congestion of transmission lines in a restructured power system with pool market model. The modeling of Unified Power Flow Controller (UPFC) has been adopted based on bipolar model and power injection method. To determine an appropriate location for UPFC as well as to set its parameters, an approach based on genetic algorithm has been suggested. The modified IEEE 14-bus system is used to determine the effectiveness and applicability of the proposed method and results are discussed.

*Index Terms*--Restructured Power Systems, Pool Market, Congestion Management, UPFC, AC-OPF, Genetic Algorithm.

## I. INTRODUCTION

IN recent years, major changes have been introduced into the structure of electric power utilities all over the world. The reason for this was to improve efficiency in the operation of the power system by means of deregulating and restructuring the industry and opening it up to private competition. Unlike many other commodities, electricity can not be stored easily, and the transportation of electricity is constrained by physical laws which have to be satisfied at all times in order to maintain the reliability and security of the power system. The presence of constraints generally leads to higher marginal costs and reduced revenues, unwanted voltage profile and congestion in transmission lines.

Congestion management deals with the relief of congested transmission networks in modern restructured electricity markets. Various congestion management schemes suitable for different electricity market structure have been reported [1]. Environmental, right-of-way and cost problems are major hurdles for power transmission network expansion in power

This work is sponsored by Islamic Azad University - Science & Research Branch, Tehran, Iran.

H. Barati is with the Department of Engineering, Islamic Azad University - Science & Research Branch, Tehran, Iran. (e-mail: barati216@gmail.com).

M. Ehsan is with the Dept. of Electrical Engineering, Sharif University of Technology, Tehran, Iran. (e-mail: ehsan@sharif.edu).

M. Fotuhi-Firuzabad is with the Dept. of Electrical Engineering, Sharif University of Technology, Tehran, Iran. (e-mail: fotuhi@sharif.edu).

systems. Patterns of generation that results in heavy flows tend to incur greater losses, and to threaten stability and security, ultimately making certain generation patterns economically undesirable. Hence, there is an interest in better utilization of available capacities by installing Flexible AC Transmission System (FACTS) devices such as UPFC. By using these devices, one can control the power flow in transmission lines without rescheduling the generation trend or topological changes in network in away not to violate the thermal limits, but to increase the loadability of the system, reduce the system losses, improve the stability of the network, reduce the cost of generation and fulfill contractual requirements.

Different approaches to using these devices have been suggested to remove or reduce those critical congestions. Each of these approaches have been used in a given congestion management. In general, to use FACTS devices, the following have to be considered: Suitable place for installing FACTS and How to set FACTS parameters.

The sensitivity-based approach for finding a suitable placement of FACTS devices has been used in pool market model for congestion management [2]-[7]. Cai et al. [8],[9] have proposed a genetic algorithm based approach to determine the suitable type of FACTS devices and its optimal location in deregulated electricity market (pool market model). In [10], the researchers have described the application of FACTS devices to deal with combined active and reactive congestion management under a deregulated environment. Lima et al. [11] have presented a method for the optimal allocation of Thyristor-Controlled Phase Shifting Transformers (TCPST's) within a network using a mixed integer linear programming technique. Carsten [12] has described the optimization method based on Sequential Quadratic Programming (SQP) including models of FACTS devices suitable for a gradient based security constrained optimization - the reduction of power losses and the maximization of power system loadability - in liberalized energy markets.

Congestion of transmission lines have prevented the deregulated power market to achieve its objectives, cheaper energy for consumers because congestion cost has been added to consumers Locational Marginal Price (LMP) besides

0-7803-9771-1/06/$25.00 ©2006 IEEE

generation cost. Occurrence of congestion has made LMP at certain locations for beyond their original LMP. The installation of FACTS devices have been proposed as an alternative to relieve congestion and reduce energy price for consumers (or reduced LMP).

Srivastava and Verma [13] have presented the impact of (TCSC, SVC) on transmission pricing in a deregulated electricity spot markets. Huang et al. [14] have used the Lagrange multipliers method to minimize the total biding cost for the determining problems of congestion in which an LMP based scheme is proposed for FACTS devices utilization charge and its pricing in congestion management.

In the pool market model, the market operator collects the electric power bids from the suppliers as well as from the consumers. These bids are within a certain time interval. When the bids are submitted, the market operator runs the OPF program taking the network constraints into consideration. The objective of this OPF program is to minimize the total costs, which means the maximization of the social welfare, and covering the desired load to required value [9].

In this article, AC-OPF combined with UPFC has been used to manage the congestion of transmission lines. To determine an appropriate location for UPFC as well as to set the relevant parameters in a restructured power system with pool market, an approach based on genetic algorithm has been offered. The modeling of UPFC has been adopted in terms of the power injection method. To show the effect of UPFC presence in congestion management the following three states have been compared and surveyed:

- Running OPF with no presence of UPFC without taking lines capacity constraints into account.
- Running OPF with no presence of UPFC with taking lines capacity constraints into account.
- Running OPF with presence of UPFC with taking lines capacity constraints into account.

The organization of the article is as follows: modeling the UPFC in terms of bipolar model and power injection, load flow equation of injection pi-model, Optimal Power Flow Incorporating UPFC, placement and setting parameters of UPFC using genetic algorithm, numerical example and results discussion.

## II. UPFC MODELING IN TERMS OF BIPOLAR MODEL AND POWER INJECTION METHOD

Power injection method is one of the common ways in modeling steady state of FACTS devices in load flow studies; in this method modeling of each FACTS devices is possible. On the one hand, these devices are converted into active and reactive power injection in the bus; these injection powers can be changed into controlling parameters according to each type of FACTS devices.

### A. Bipolar Model of Line

This model has been shown Fig. 1 in terms of pi-model as well as the relevant equations are:

$$\begin{bmatrix} V_{1L} \\ I_{1L} \end{bmatrix} = \begin{bmatrix} A_L & B_L \\ C_L & D_L \end{bmatrix} \begin{bmatrix} V_{2L} \\ I_{2L} \end{bmatrix}$$

$$A_L = (y_l + y)/y_l$$
$$B_l = -1/y_l$$
$$C_l = y.(2 + y/y_l)$$
$$D_l = -(y_l + y)/y_l$$

$$(1)$$

In which, $y_l = R_l + jX_l$ and $y = jB_c/2$.

### B. Bipolar model of UPFC

A UPFC scheme using voltage sourced converters is shown in Fig. 2 The UPFC is able to control, simultaneously or selectively, all parameters affecting power flow in the transmission line (i.e. voltage, impedance and phase angle). Alternatively, it can independently control both the real and reactive power flow in the line.

Fig. 3 shows a bipolar UPFC scheme as well as the relevant equations are:

$$\begin{bmatrix} V_{1F} \\ I_{1F} \end{bmatrix} = \begin{bmatrix} A_F & B_F \\ C_F & D_F \end{bmatrix} \begin{bmatrix} V_{2F} \\ I_{2F} \end{bmatrix} + \begin{bmatrix} S_{11} & S_{12} \\ S_{21} & S_{22} \end{bmatrix} \begin{bmatrix} V_{se} \\ V_{sh} \end{bmatrix}$$

$$(2)$$

$$\begin{bmatrix} A_F & B_F \\ C_F & D_F \end{bmatrix} = \begin{bmatrix} 1 & -z_{se} \\ y_{sh} & -(1 + z_{se}.y_{sh}) \end{bmatrix}, \begin{bmatrix} S_{11} & S_{12} \\ S_{21} & S_{22} \end{bmatrix} = \begin{bmatrix} -1 & 0 \\ -y_{se} & -y_{sh} \end{bmatrix}$$

in which, $A_F.D_F - B_F.C_F = 1$ and $y_{sh}, y_{se}$ the related impedance of series and shunt transformers of UPFC as well as its internal power loss.

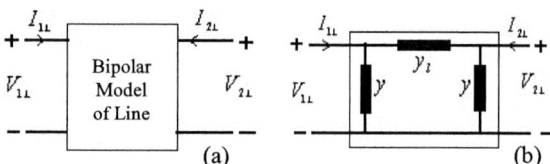

Fig. 1. Bipolar model of line: a- General scheme, b- Pi-model of line.

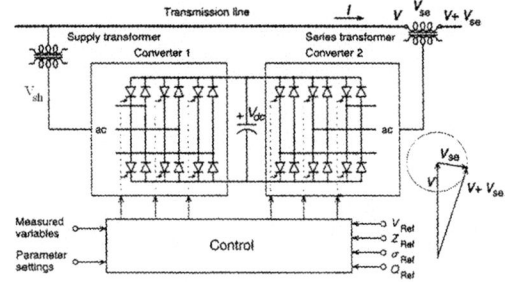

Fig. 2. A UPFC scheme using voltage sourced converters.

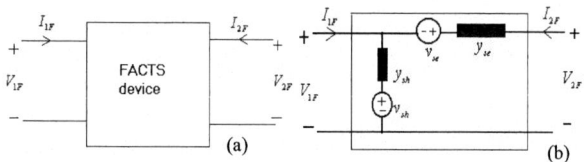

Fig. 3. Bipolar model of UPFC.

## C. Bipolar model of UPFC/Line

Fig. 4 shows the UPFC bipolar model installed in line as well as the relevant equations are:

$$\begin{bmatrix} V_{1LF} \\ I_{1LF} \end{bmatrix} = \begin{bmatrix} LF_{11} & LF_{12} \\ LF_{21} & LF_{22} \end{bmatrix} \begin{bmatrix} V_{2LF} \\ I_{2LF} \end{bmatrix} + \begin{bmatrix} S_{11} & S_{12} \\ S_{21} & S_{22} \end{bmatrix} \begin{bmatrix} V_{se} \\ V_{sh} \end{bmatrix} \tag{3}$$

In which,

$$LF_{11} = A_F.A_L - B_F.C_L, \quad LF_{12} = A_F.B_L - B_F.D_L$$

$$LF_{21} = C_F.A_L - D_F.C_L, \quad LF_{22} = C_F.B_L - D_F.D_L$$

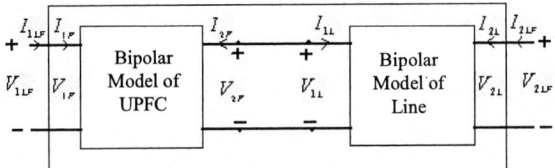

Fig. 4. Bipolar Model of UPFC/Line.

From (3), $I_{1LF}$ and $I_{2LF}$ can be express in terms of $V_{1LF}$, $V_{2LF}$, $V_{se}$ and $V_{sh}$.

$$\begin{bmatrix} I_{iLF} \\ I_{jLF} \end{bmatrix} = \begin{bmatrix} LF_{22}/LF_{12} & (LF_{12} - LF_{11}.LF_{22}/LF_{12}) \\ 1/LF_{12} & -LF_{11}/LF_{12} \end{bmatrix}.$$
$$\begin{bmatrix} V_{iLF} \\ V_{jLF} \end{bmatrix} + \begin{bmatrix} LF_{22}/LF_{11} - y_{sh} & -y_{sh} \\ 1/LF_{12} & 0 \end{bmatrix} \begin{bmatrix} V_{se} \\ V_{sh} \end{bmatrix}$$

$$\begin{bmatrix} I_{iLF} \\ I_{jLF} \end{bmatrix} = \begin{bmatrix} A_{ii} & A_{ij} \\ A_{ji} & A_{jj} \end{bmatrix} \begin{bmatrix} V_{iLF} \\ V_{jLF} \end{bmatrix} + \begin{bmatrix} B_{ii} & B_{ij} \\ B_{ji} & B_{jj} \end{bmatrix} \begin{bmatrix} V_{se} \\ V_{sh} \end{bmatrix} \tag{4}$$

Fig. 5 shows the UPFC model installed in line using current sources and corrected admittances obtained in terms of (4).

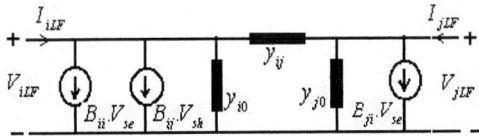

Fig. 5. Modeling of UPFC- inserted line using an equalent Pi-model and current source.

$$y_{i0} = (LF_{22} + 1)/LF_{12} = (y_{se}.y + y_{sh}.y_T)/y_T$$
$$\quad = G_{yi0} + jB_{yi0}$$
$$y_{ij} = -1/LF_{12} = y_{se}.y_l/y_T = G_{yij} + jB_{yij}$$
$$y_{j0} = (-LF_{11} + 1)/LF_{12} = y.(y_{se} + 2y_l + y)/y_T \tag{5}$$
$$\quad = G_{yj0} + jB_{yj0}$$
$$B_{ii} = G_{Bii} + jB_{Bii} = y_{se}(y_l + y)/y_T$$
$$B_{ij} = G_{Bij} + jB_{Bij} = -y_{sh}$$
$$B_{ji} = G_{Bji} + jB_{Bji} = -y_{se}y_l/y_T$$
$$B_{jj} = G_{Bii} + jB_{Bjj} = 0$$

### III. Load Flow Equations of Injection Pi-Model

Considering Fig. 5, Pi-model with the apparent power injection relevant to UPFC-inserted line has been shown in Fig. 6.

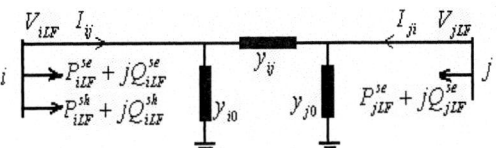

Fig. 6. The injection pi-model of UPFC-Embedded line.

$$S_{iLF}^{sh} = V_{iLF}.(B_{ij}.V_{sh})^* = P_{iLF}^{sh} + jQ_{iLF}^{sh}$$
$$S_{iLF}^{se} = V_{iLF}.(B_{ii}.V_{se})^* = P_{iLF}^{se} + jQ_{iLF}^{se}$$
$$S_{jLF}^{se} = V_{jLF}.(B_{ji}.V_{se})^* = P_{jLF}^{se} + jQ_{jLF}^{se} \tag{6}$$

The constraints imposed by device limits, which should be enforced in power flow calculations, include magnitude and phase angle of the series injected voltage $|V_{se}|\angle\delta_{se}$, magnitude and phase angle of the shunt injected voltage $|V_{sh}|\angle\delta_{sh}$, and real power between the shunt and series converters ($P_{se}$, $P_{sh}$). Due to the fact that the active power needed by the series converter is provided from the AC power system by the shunt converter through the dc link, the active power delivered to the shunt converter ($P_{sh}$) must satisfy the active power needed by series converter ($P_{se}$). Consequently,

$$P_{sh} - P_{se} = 0, \quad P_{sh} = \text{Re}\{V_{sh}.I_{sh}^*\} \quad P_{se} = \text{Re}\{V_{se}.I_{se}^*\} \tag{7}$$

### IV. Optimal Power Flow Incorporating UPFC

Optimal Power Flow (OPF) is one of the most important operational functions of the modern day energy management system. The purpose of the optimal power flow is to find the optimum generation among the existing units, such that the total biding cost is minimized while simultaneously satisfying the power balance equations and various other constraints in the system. Various algorithms have been reported to solve power flow and optimal flow for power systems equipment with various FACTS devices, that new control variables and control objective equations are usually added in conventional power flow equation [14]-[16].

In this article, the OPF in presence of UPFC has been formulated as follows in which generation cost has got minimized and the congestion relief was achieved, and the maximum demand load was met.

$$Min \quad \sum_{i=1}^{N_g} C_{gi}.P_{gi} \tag{8}$$

Subject To: Equality constraints,
(If UPFC is installed between buses i-j)

$$P_i^G - P_i^D = P_i^{se} + P_i^{sh} + P_i^0$$
$$Q_i^G - Q_i^D = Q_i^{se} + Q_i^{sh} + Q_i^0$$
$$P_j^G - P_j^D = P_j^{se} + P_j^0$$
$$Q_j^G - Q_j^D = Q_j^{se} + Q_j^0 \tag{9}$$

For the other buses

$$P_i^G - P_i^D = P_i^0, \quad Q_i^G - Q_i^D = Q_i^0 \qquad (10)$$

For power flow in UPFC converters

$$P_{sh} - P_{se} = 0 \qquad (11)$$

Inequality constraints:
Bus voltage limits and active power generators limits.

$$V_i^{min} \leq V_i \leq V_i^{max}, \quad P_{gi}^{min} \leq P_{gi} \leq P_{gh}^{max} \qquad (12)$$

Apparent power flow limit of lines (From end and To end)

$$\left| S_{ij}^f \right| \leq S_{ij}^{max}, \quad \left| S_{ij}^t \right| \leq S_{ij}^{max} \qquad (13)$$

UPFC control parameters limits

$$\begin{aligned} V_{se}^{min} \leq V_{se} \leq V_{se}^{max}, \quad \delta_{se}^{min} \leq \delta_{se} \leq \delta_{se}^{max} \\ V_{sh}^{min} \leq V_{sh} \leq V_{sh}^{max}, \quad \delta_{sh}^{min} \leq \delta_{sh} \leq \delta_{sh}^{max} \end{aligned} \qquad (14)$$

If the installation place and UPFC parameter are well-defined, the OPF with UPFC will be similar to a conventional AC power flow. Furthermore, the placement and the setting of UPFC parameters have been used based on genetic algorithm.

## V. PLACEMENT AND SETTING PARAMETERS OF UPFC USING GENETIC ALGORITHM

This article has used genetic algorithm to place and set the UPFC parameters to minimize the generation cost as well as the congestion relief in network transmission. UPFC includes four parameters: $(\delta_{se}, V_{se})$ and $(\delta_{sh}, V_{sh})$, Concerning the relation $P_{se} = P_{sh}$, and the parameters $(\delta_{sh}, \delta_{se}, V_{se})$, one can calculate $V_{sh}$.

The genetic algorithm works in terms of chromosomes, so they have to be precisely designed to optimize the capability of genetic algorithm. Each chromosome, here, consists of 4 genes (Fig. 7). The first gene carries the number of the line with the function of UPFC installation place. This gene contains a number between 1 to $N_{Line}$ (the No. of network lines). The second to the fourth genes contain $(\delta_{sh}, \delta_{se}, V_{se})$, respectively. The range of these genes loads are as follows:

$$0 \leq V_{se} \leq 0.25^{rad}, \quad 0 \leq \delta_{se} \leq 2\pi^{rad}, \quad 0 \leq \delta_{sh} \leq 2\pi^{rad}$$

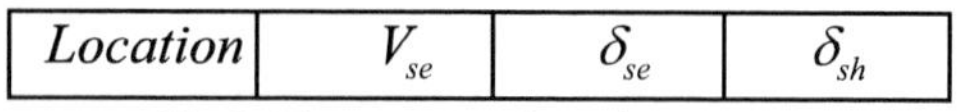

| Location | $V_{se}$ | $\delta_{se}$ | $\delta_{sh}$ |

Fig. 7. The structure of one chromosome

Depending on the number of the transmission network lines ($N_{Line}$), a primary population as many as $2N_{Line}$ of chromosomes is formed (Fig. 8). A set of numbers, are randomly assigned to each chromosome genes.

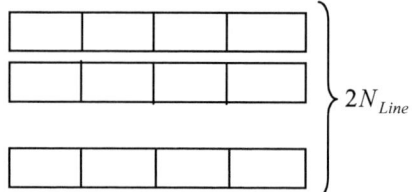

Fig. 8. The structure of the primary population

The amount of fitness function represents the value (fitness) of each chromosome and the selection standard is in terms of the valuable chromosomes. For each chromosome, the optimal power flow is carried out and the results of power flow studies are sorted in terms of fitness function. From the sorted population, $N_{Line}$ chromosomes of value are separated and form new publications. If the best chromosome as the "elite" one is separated and genetic operations (crossover & mutation) take place between the elite and other reminder chromosomes, it might be possible to have access to more desired chromosomes (with better fitness). Since the aim is to achieve the improved states, the best chromosome, by the choice of a high rate of mutation, we can provide the way for the development of more elite chromosome, and we can reduce the time required to meet the goals, or even to supply opportunities to survey more elite chromosomes.

## VI. NUMERICAL EXAMPLE AND THE ANALYSIS OF THE RESULTS

The standard IEEE 14-bus system with 5 generators, 12 loads, 20 lines and transformers has been assumed as follows:
- Pool market model
- Only energy generation market is available, Consumers demands are fixed and to be supplied.
- Some suggest price in energy market is used for congestion management.
- Energy quantity-price bids of generators are one-step based.
- A modified version of power simulation software: Matpower 1.0 is employed [19]. For the intended research, Matpower has been extended by incorporating the mathematical model of UPFC.

The energy quantity-price bids of generators and the consumers' demands are shown in Tables I and II, respectively.

TABLE I
ONE STEP ENERGY QUANTITY-PRICE BIDS OF GENERATORS

| Generation No. | Bus No. | Price ($/MWh) | Generation (MW) |
|---|---|---|---|
| 1 | 1 | 5 | 110 |
| 2 | 2 | 10 | 90 |
| 3 | 3 | 25 | 30 |
| 4 | 6 | 15 | 70 |
| 5 | 8 | 20 | 50 |

TABLE II
THE CONSUMERS' DEMANDS

| Load No. | Bus No. | Demand (MVA) |
|---|---|---|
| 1 | 2 | 21.7+j12.7 |
| 2 | 3 | 94.2+j19 |
| 3 | 4 | 47.8-j3.9 |
| 4 | 5 | 7.6+j1.6 |
| 5 | 6 | 11.2+j7.5 |
| 6 | 7 | 0.0+j0.0 |
| 7 | 8 | 29.5+j16.6 |
| 8 | 10 | 9+j5.8 |
| 9 | 11 | 3.5+j1.8 |
| 10 | 12 | 6.1+j1.6 |
| 11 | 13 | 13.5+j5.8 |
| 12 | 14 | 14.9+j5 |

In order to illustrate the impacts of UPFC for congestion management, the results of OPF have been studied in the following three cases on the IEEE 14-bus system:

- Case 1 - lines limits are ignored and UPFC has not been utilized.
- Case 2 - lines limits have been considered, whereas the UPFC has not been utilized.
- Case 3 - lines limits have been considered and the UPFC has been inserted in the network (line 1-2) and its parameters adjusted (with genetic algorithm).

In the continuation of the process, the results obtained from the computerize simulation in terms of the three cases stated above have been presented. The series-shunt impedance and parameters of UPFC are:

$Z_{se} = 0.01 + j0.1^{pu}$, $Z_{sh} = 1 + j10^{pu}$, $V_{sh} = 0.2576^{pu}$,

$\delta_{sh} = 2.3220^{rad}$, $V_{se} = 0.0642^{pu}$, $\delta_{se} = 0.3297^{rad}$.

The range of all buses voltage is between $0.95^{pu}$ and $1.05^{pu}$. The buses voltage and their relevant Lagrange multipliers are shown in Table III. The active powers generators and their relevant Lagrange multipliers are shown in Table IV. In buses with generators if $p_{gi-min} < P_{gi} < P_{gi-max}$, the suggested price of the generators is the determinant factor in $\lambda_i$ bus (or $LMP_i$). Under the condition that $P_{gi} \geq P_{gi-max}$ or $P_{gi} \leq P_{gi-min}$, $\lambda_i$ bus (or $LMP_i$) is determined by the system, so is the $PQ$ bus. The Locational Marginal Price (LMP) of buses is shown in Table V. As the total network load has been assumed to be fixed, the total generation in network is almost fixed and the same in the above-stated three cases. The Total Congestion Charge (TCC) is obtained from the difference between the revenue collection from network loads and generation cost paid to generators.

$$TCC = \sum_{i=1}^{Nb} P_i^D \lambda_i - \sum_{i=1}^{Nb} P_i^G \lambda_i \qquad (15)$$

The General Information Regarding the Network is shown in Table VI.

The maximum capacity for all the lines has been taken as $65^{KVA}$. Taking active and reactive power losses into account, the apparent power flow is not the same at the two ends of line. Apparent power flow in line and the relevant Lagrange multipliers for the three cases 1, 2 and 3 are shown in Tables VII, VIII and IX, respectively.

TABLE IV
THE ACTIVE POWERS GENERATORS AND THEIR RELEVANT LAGRANGE MULTIPLIERS

| | Bus No. | 1 | 2 | 3 | 6 | 8 |
|---|---|---|---|---|---|---|
| Case 1 | $\mu_{P_{g-max}}$ | 9.56 | 4.92 | - | - | - |
| | $P_g$ | 110 | 90 | 0 | 66ŀ | 0 |
| | $\mu_{P_{g-min}}$ | - | - | 8.98 | - | 4.32 |
| Case 2 | $\mu_{P_{g-max}}$ | - | 11.44 | - | 3.34 | - |
| | $P_g$ | 104.45 | 90 | 1 | 70 | 1.52 |
| | $\mu_{P_{g-min}}$ | - | - | - | - | - |
| Case 3 | $\mu_{P_{g-max}}$ | 9.54 | 5.02 | - | - | - |
| | $P_g$ | 110 | 90 | 0 | 66.5 | 0 |
| | $\mu_{P_{g-min}}$ | - | - | 9.18 | - | 4.34 |

TABLE VI
THE GENERAL INFORMATION REGARDING THE NETWORK

| | Case 1 | Case 2 | Case 3 |
|---|---|---|---|
| Total Active Losses (MW) | 7.567 | 7.966 | 11.448 |
| Total Reactive Losses (MVar) | 2.4 | 5.75 | 0.55 |
| Total Generation (MW) | 266.56 | 266.966 | 266.521 |
| Total Generation Cost ($/h) | 2448.39 | 2527.62 | 2447.81 |
| Total Congestion Charge ($/h) | 115.663 | 1834.02 | 92.019 |

TABLE III
THE BUSES VOLTAGE AND THEIR RELEVANT LAGRANGE MULTIPLIERS

| | Bus No. | 1 | 2 | 3 | 4 | 5 | 6 | 7 | 8 | 9 | 10 | 11 | 12 | 13 | 14 |
|---|---|---|---|---|---|---|---|---|---|---|---|---|---|---|---|
| Case 1 | $\mu_{V_{max}}$ | 121.77 | - | - | - | - | 84.50 | - | 19.9 | - | - | - | - | - | - |
| | $\|V_i\|^{pu}$ | 1.050 | 1.04 | 1.010 | 1.02 | 1.02 | 1.050 | 1.037 | 1.05 | 1.03 | 1.02 | 1.03 | 1.0ⁱ | 1.0ⁱ | 1.009 |
| | $\mu_{V_{min}}$ | - | - | - | - | - | - | - | - | - | - | - | - | - | - |
| Case 2 | $\mu_{V_{max}}$ | - | - | - | - | - | 278.2 | - | 80ⁱ | - | - | - | - | - | - |
| | $\|V_i\|^{pu}$ | 0.998 | 0.98 | 0.950 | 0.97 | 0.98 | 1.050 | 1.017 | 1.05 | 1.01 | 1.01 | 1.02 | 1.03 | 1.03 | 0.997 |
| | $\mu_{V_{min}}$ | - | - | 13.042 | - | - | - | - | - | - | - | - | - | - | - |
| Case 3 | $\mu_{V_{max}}$ | - | - | - | - | - | 22.465 | - | 7.07 | - | - | - | - | - | - |
| | $\|V_i\|^{pu}$ | 1.046 | 1.05 | 1.025 | 1.03 | 1.04 | 1.050 | 1.033 | 1.05 | 1.02 | 1.02 | 1.03 | 1.03 | 1.03 | 1.006 |
| | $\mu_{V_{min}}$ | - | - | - | - | - | - | - | - | - | - | - | - | - | - |

TABLE V
THE LOCATIONAL MARGINAL PRICE (LMP) OF BUSES

| Case | $LMP_1$ | $LMP_2$ | $LMP_3$ | $LMP_4$ | $LMP_5$ | $LMP_6$ | $LMP_7$ | $LMP_8$ | $LMP_9$ | $LMP_{10}$ | $LMP_{11}$ | $LMP_{12}$ | $LMP_{13}$ | $LMP_{14}$ |
|---|---|---|---|---|---|---|---|---|---|---|---|---|---|---|
| 1 | 14.56 | 14.92 | 16.02 | 15.53 | 15.25 | 15.00 | 15.68 | 15.68 | 15.75 | 15.73 | 15.44 | 15.27 | 15.42 | 15.96 |
| 2 | 5.000 | 21.44 | 25.00 | 20.05 | 18.23 | 18.34 | 20.00 | 20.00 | 19.97 | 19.82 | 19.17 | 18.72 | 18.96 | 19.98 |
| 3 | 14.54 | 15.02 | 15.82 | 15.52 | 15.26 | 15.00 | 15.66 | 15.66 | 15.74 | 15.72 | 15.43 | 15.27 | 15.42 | 15.95 |

### TABLE VII
#### APPARENT POWER FLOW IN LINE AND THE RELEVANT LAGRANGE MULTIPLIERS-CASE 1

| From Bus | "From" End $\mu_{sf}$ | $\lvert S_f \rvert$ | "To" End $\mu_s$ | $\lvert S_t \rvert$ | To Bus |
|---|---|---|---|---|---|
| 1 | | 70.25 | | 68.99 | 2 |
| 1 | - | 40.72 | - | 40.11 | 5 |
| 2 | | 66.90 | - | 64.91 | 3 |
| 2 | - | 42.34 | - | 41.41 | 4 |
| 2 | - | 27.98 | - | 27.98 | 5 |
| 3 | - | 30.63 | - | 31.15 | 4 |
| 4 | - | 61.54 | - | 61.79 | 5 |
| 4 | - | 15.42 | - | 15.41 | 7 |
| 4 | - | 9.68 | - | 9.50 | 9 |
| 5 | - | 18.84 | - | 18.10 | 6 |
| 6 | - | 20.05 | - | 19.70 | 11 |
| 6 | - | 9.54 | - | 9.4 | 12 |
| 6 | - | 24.94 | - | 24.41 | 13 |
| 7 | - | 7.93 | - | 8.03 | 8 |
| 7 | - | 17.49 | - | 17.34 | 9 |
| 9 | - | 10.93 | - | 10.88 | 10 |
| 9 | - | 6.65 | - | 6.52 | 14 |
| 10 | - | 16.19 | - | 16.33 | 11 |
| 12 | - | 3.16 | - | 3.14 | 13 |
| 13 | | 13.54 | - | 13.29 | 14 |

### TABLE VII
#### APPARENT POWER FLOW IN LINE AND THE RELEVANT LAGRANGE MULTIPLIERS-CASE 2

| From Bus | "From" End $\mu_{sf}$ | $\lvert S_f \rvert$ | "To" End $\mu_{st}$ | $\lvert S_t \rvert$ | To Bus |
|---|---|---|---|---|---|
| 1 | 19.33 | 65.00 | - | 64.23 | 2 |
| 1 | - | 39.47 | - | 38.60 | 5 |
| 2 | 6.19 | 65.00 | - | 62.84 | 3 |
| 2 | - | 41.41 | - | 40.38 | 4 |
| 2 | - | 27.49 | - | 26.68 | 5 |
| 3 | - | 30.37 | - | 31.07 | 4 |
| 4 | - | 60.04 | - | 60.37 | 5 |
| 4 | - | 16.65 | - | 17.03 | 7 |
| 4 | - | 7.69 | - | 7.73 | 9 |
| 5 | - | 2.76 | - | 2.76 | 6 |
| 6 | - | 21.78 | - | 21.23 | 11 |
| 6 | - | 9.91 | - | 9.76 | 12 |
| 6 | - | 26.18 | - | 25.56 | 13 |
| 7 | - | 19.02 | - | 19.63 | 8 |
| 7 | - | 16.37 | - | 16.24 | 9 |
| 9 | - | 9.60 | - | 9.59 | 10 |
| 9 | - | 4.27 | - | 4.22 | 14 |
| 10 | - | 17.26 | - | 17.54 | 11 |
| 12 | - | 3.46 | - | 3.44 | 13 |
| 13 | - | 14.60 | - | 14.20 | 14 |

### TABLE IX
#### APPARENT POWER FLOW IN LINE AND THE RELEVANT LAGRANGE MULTIPLIERS-CASE 3

| From Bus | "From" End $\mu_{sf}$ | $\lvert S_f \rvert$ | "To" End $\mu_{st}$ | $\lvert S_t \rvert$ | To Bus |
|---|---|---|---|---|---|
| 1 | - | 59.68 | - | 53.75 | 2 |
| 1 | - | 51.95 | - | 50.67 | 5 |
| 2 | - | 48.96 | - | 47.94 | 3 |
| 2 | - | 32.52 | - | 31.88 | 4 |
| 2 | - | 17.96 | - | 17.80 | 5 |
| 3 | - | 20.96 | - | 21.34 | 4 |
| 4 | - | 61.35 | - | 61.70 | 5 |
| 4 | - | 15.39 | - | 15.40 | 7 |
| 4 | - | 8.96 | - | 8.88 | 9 |
| 5 | - | 4.95 | - | 5.00 | 6 |
| 6 | - | 20.13 | - | 19.74 | 11 |
| 6 | - | 9.59 | - | 9.45 | 12 |
| 6 | - | 25.08 | - | 24.54 | 13 |
| 7 | - | 9.70 | - | 9.86 | 8 |
| 7 | - | 17.87 | - | 17.70 | 9 |
| 9 | - | 10.33 | - | 10.29 | 10 |
| 9 | - | 6.11 | - | 6.00 | 14 |
| 10 | - | 16.12 | - | 16.28 | 11 |
| 12 | - | 3.19 | - | 3.17 | 13 |
| 13 | - | 13.58 | - | 13.31 | 14 |

In case 1, as the lines capacity constraints has been ignored, the energy has been purchased from the generators at a lower suggested price $(G_1, G_2, G_6)$. Besides, the generators of higher bids $(G_3, G_8)$ lacked any sails. The LMP buses are almost close to each other, and their partial differences lies in the constrains including the bus voltage and active Power generators constraints. Since $0 < P_{g6} < 70^{MW}$, $LMP_6$ is equal to the bid suggest for $G_6$. But the generator powers in buses of 1, 2, 3 and 8 have been at the extreme values, and the relevant Lagrange multiplies are not zero; consequently, their relevant LMP like other buses are determined by the system. In this state, lines 1-2, 2-3 and 4-5, have the maximum Power flow. The total generation and congestion costs are $2448.39^{\$/h}$ and $115.663^{\$/h}$, respectively.

In case 2, line capacity constraint has been considered. Regarding the Table VII in lines 1-2, 2-3, the congestion has taken place, and the relevant Lagrange multipliers are opposite to zero. This congestion avoids receiving higher energy power form the cheaper generators. Moreover, the more expensive generators $(G_3, G_8)$ have been used in power flow. The total generation cost has been increased from $2448.39^{\$/h}$ to $2527.62^{\$/h}$ compare to case 1. LMPs in buses differ from each other. The active power constraints of generators have been met in buses 1,3 and 8, so the LMPs in those buses are the same suggested bids from generators, which are $5^{\$/MWh}$, $25^{\$/MWh}$ and $20^{\$/MWh}$, respectively. But the Lagrange multipliers $\mu_{Pg2-max}$ and $\mu_{Pg6-max}$ in buses 2 and 6 are not zero, and the generation power in those buses are $90^{MW}$ and $70^{MW}$. Therefore, $LMP_2$ and $LMP_6$ are different

from the suggested bids from generators. Compared with case 1, the total congestion cost increase from $115.663^{\$/h}$ to $1834.02^{\$/h}$.

In case 3, with presence of UPFC in an appropriate placement and their parameter setting, one can obtain an equal condition similar to case 1. Without rescheduling of generators in relation to case 1, only the cheaper generators have contributed to power flow. Furthermore, No purchase has taken place from more expensive generators. The total generation cost has been reduced form $2527.62^{\$/h}$ to $2447.81^{\$/h}$. The congestion of lines 1-2 and 2-3 has got released, and the power flow through these lines has gone down to the permitted limits. The total congestion cost has decreased from $1834.02^{\$/h}$ to $92.019^{\$/h}$. The voltage profile of buses has improved similar to case 1. The system buses LMPs are approximately close to each other and are the same as case 1. Therefore, with regard to the UPFC, the limitations resulted from the line capacity constraints such as lines congestion, generation cost, congestion cost, voltage profile, and buses LMP will be resolved

## VII. CONCLUSION

- FACTS devices provide the effective control over the parameters affecting the power flow in transmission systems. Consequently, they can be used in congestion management.
- The modeling of UPFC was used in terms of bipolar model and power injection method. This modeling accords with the conventional AC-OPF.

- To determine an appropriate placement and to set the UPFC parameters, a genetic algorithm can successfully be applied.
- Following merits were shown by application of UPFC in congested lines in the IEEE 14-bus system, reduction in generation cost and lines congestion cost, release of lines congestion and improvement in the network voltage profile.

## VIII. REFERENCES:

[1] Ashwani Kumar, S. C. Srivastava, S. N. Singh, "Congestion Management in Competitive Power Market: A Bibliographical Survey", Electric Power Systems Research 76 (2005) 153-164).

[2] S. N. Singh, K. S. Verma, H. O. Gupta, " Optimal Power Flow Control in Open Power Market using Unified Power Flow Controller", Power Engineering Society Summer Meeting, 2001. IEEE, Volume: 3, 15-19 July 2001, Pages:1698–1703.

[3] S. N. Singh, A. K. David, "A New Approach for Placement of FACTS Devices in Open Power Markets", Power Engineering Review, IEEE, Vol. 21, Issue: 9, Sep. 2001, Pages:58 – 60.

[4] S. N. Singh, "Location of FACTS Devices for Enhancing Power Systems' Security", Power Engineering, 2001, LESCOPE'01, 2001 Large Engineering System Conference on, 11-13 July 2001, Page: 162-166.

[5] S. N. Singh, A. K. David, "Congestion Management by Optimizing FACTS Device Location"; Electric Utility Deregulation and Restructuring and Power Technology ,Proceeding; DRPT 2000 ,pp..23-28, April 2000.

[6] S.N. Singh, A. K. David, "Placement of FACTS Devices in Open Power Market", Proceeding of the 5th International Conference on Advances in Power System Control, Operation and Management (APSCOM 2000), Hong Kong, 29 Oct.-1 Nov. 2000, pp173-177.

[7] K. S. Verma, S. N. Singh, H. O. Gupta, "Location of United Power Flow Controller for Congestion Management", Electric Power System Research Vol. 58, Issue 2, 21 June 2001, pp 89-96.

[8] L. Cai, I. Erlich, G. Stamtsis, Y. Luo, "Optimal Choice and Allocation of FACTS Devices in Deregulated Electricity Market using Genetic Algorithms", IREP 2004, Cortina, Italy, Aug. 2004.

[9] L. J. Cai, I. Erlich, G. Stamtsis, "Optimal Choice and Allocation of FACTS Devices in Deregulated Electricity Market Using Genetic Algorithms", IEEE PES Power Systems Conference of Exposition, New York City, USA, 10-13 Oct. 2004.

[10] S. Phichaisawat, S. H. Song, X. L. Wang, X. F. Wang, " Combined Active and Reactive Congestion Management with FACTS Devices", Electric Power Component and System, 30:1195-1205, 2002, Taylor & Francis Group.

[11] F. G. M. Lima, J. Munoz, I. Kokar, F. D. Galiana, "Optimal Location of Phase Shifter in a Competitive Market by Mixed Integer Linear Programming", 14th PSCC, Sevilla, 24-28 June 2002, session 43, paper 2, page 1.

[12] Carsten Lehmkoster, "Security Constrained Optimal Power Flow for an Economical Operation of FACTS-Devices in liberalized Energy Markets", IEEE Trans. on Power Delivery, Vol. 17, No. 2, April 2002, PP. 603-608.

[13] S. C. Srivastava, R. K. Verma, "Impact of FACTS Devices on Transmission Pricing in A De-Regulated Electricity Market" International Conference on Electric Utility Deregulation and Restructuring and Power Technologies 2000, April 2000, London.

[14] G. M. Huang, P. Yan, "Establishing Pricing Schemes for FACTS Devices in Congestion Management" Power Engineering Society General Meeting, IEEE, volume:2, 13-17 July 2003, Pages: 1030

[15] Abdel-Moamen M.A., Naragana Prasad Padhy, "Optimal POWER Flow Incorporating FACTS Devices _Bibliography and Survey", POWER Engineering society winter Meeting, 2001, IEEE, Vol.:2, 28 Jan.-1Feb.2001, Pages 23-28.

[16] A. Kazemi, H. Andami, "FACTS Devices in Deregulated Electric Power Systems: A Review", The 2nd International conference on Electric utility Deregulation, Restructuring and Power technologies, IEEE DRPT 2004,Hong Kong ,5-8 April 2004.

[17] M. I. Alomoush, "Exact Pi-Model of UPFC Inserted Transmission Lies in Power Flow Studies", Power Engineering Review, IEEE, Vplume: 22, Issue: 12, Des. 2002.

[18] M. I. Alomoush, "Derivation of UPFC DC Load Flow Model with Examples of its use in Restructured Power Systems", IEEE Trans. On Power Systems, Vol. 18, No. 3, Aug. 2003.

[19] R. D. Zimmermann, D. Gan, "Matpower a LAB power system simulation package", User's Manual, Version 1.0, 1997.

## IX. BIOGRAPHIES

**Hassan Barati** received the BSc degree in Electronic Engineering from Isfahan University of Technology, Iran, in 1992. He obtained his MSc degree in Electrical Power Engineering from Tabriz University, Iran, in 1996. He is now the Ph.D. student in Islamic Azad University – Science & Research Branch, Tehran, Iran. His research interest is application of FACTS devices for congestion management in restructured power systems and power quality.

**M. Ehsan** received BSc & MSc in Electrical Engineering from Technical College of Tehran University 1963 and DIC & Ph.D. from Imperial College, University of London in 1977. He is currently a professor in E. E. department of Sharif University of Technology. His research interests include power system dynamics, application of expert systems in operation, control and economic aspects.

**M. Fotuhi-Firuzabad** received BSc & MSc degrees in Electrical Engineering from Sharif University of Technology and Tehran University in 1986 and 1989 respectively and MSc and Ph.D. degrees in Electrical Engineering from the University of Saskatchewan in 1993 and 1997. He is currently Associate Professor and head of the Department of Electrical Engineering in Sharif University of Technology, Tehran, Iran. His research interests include Power System Reliability, Power System Analysis, Unit Commitment and Power System Operation, Computer Application in Power Systems, Renewable Energy.

**2006 IEEE International Conference on Power Electronic, Drives and Energy Systems**

# Security Enhancement of Optimal Power Flow using Genetic Algorithm

N.B. Muthuselvan, *Member, IEEE*, P. Somasundaram, and Subhransu Sekhar Dash

*Abstract*— The paper presents a genetic-algorithm (GA) based OPF algorithm for identifying the optimal values of generator active-power output and the angle of the phase-shifting transformer. The locations of phase shifters are selected based on sensitivity analysis. To overcome the shortcomings associated with the representation of real and integer variables using the binary string in the GA population, the control variables are represented in their natural form. Also crossover and mutation operators which can deal directly with integers and floating point numbers are used. Simulation results on IEEE 30-bus is presented and compared with the results of other approaches.

*Index Terms*— Genetic algorithms, Optimal Power Flow, Security Enhancement.

## I. INTRODUCTION

WITH the continued increase in the demand for the electrical energy with little addition to transmission capacity, security assessment and control have become important issues in power system-operation. Security assessment deals with determining whether or not the system operating in a normal state can withstand contingencies. If the present operating state is found to be insecure, action must be taken to prevent limit violation in the contingency state.

Transmission-line overload can be alleviated by rerouting power flows in the system. A change in line flow can be caused by an appropriate change in phase angles and magnitude of bus voltages, which are usually referred to as state variables. The state variables can, in turn, be modified by a variation in generated power. In [1], the linearized relationship between power flow in the overloaded lines and the generated power has been used to reschedule the power generation. A computationally simple algorithm has been developed in [2] for real-time security control. In [3], a fuzzy set theory based approach has been proposed for overload alleviation through real power-generation rescheduling. For secure operation of the system without any limit violation, complete modeling of the system through load flow equations and operational constraints is necessary [4]. This paper presents an optimal power flow with phase shifter for overload

N.B. Muthuselvan is with SSN College of Engineering, Chennai, TN 603110, India (e-mail: bmselva@yahoo.com)

P. Somasundaram is with College of Engineering, Chennai, TN 600025, India (e-mail: mpsomasundarm@yahoo.com )

Subhransu Sekhar Dash is with SSN College of Engineering, Chennai, TN 603110, India (e-mail: munu_dash_2k@yahoo.com)

0-7803-9772-X/06/$20.00 ©2006 IEEE.

alleviation. The locations of the phase shifters are identified based on sensitivity analysis.

The OPF solution gives the optimal settings of all controllable variables for a static power-system-loading condition. A number of mathematical-programming-based techniques have been proposed to solve the OPF problem. These include the gradient method [4], 6], Newton method and linear programming [7], [8]. In [9], [10] mixed-integer linear programming has been applied to identify the location of the phase shifter and FACTS devices in order to improve the loadability of the system. The gradient and Newton methods suffer from the difficulty in handling inequality constraints. To apply linear programming, the input–output function is to be expressed as a set of linear functions, which may lead to loss of accuracy. Also, difficulties are encountered in incorporating directly the discrete variables related to the phase-shifting transformers. In [11], a rule-based OPF with phase shifter has been proposed to alleviate the line overload. The principal shortcoming of a rule-base approach is that the construction of rules requires extensive help from skilled knowledge engineers. Also, it does not provide a continuous fabric over the solution space.

Recently, global-optimisation techniques such as the genetic algorithm have been proposed to solve the optimal power-flow problem [12], [14]. A genetic algorithm [15] is a stochastic search technique based on the mechanics of natural genetics and natural selection. It works by evolving a population of solutions towards the global optimum through the use of genetic operators: selection, crossover and mutation. The traditional binary-coded genetic algorithm has number of difficulties in dealing with continuous search spaces [13]. In this paper, the decision variables are represented in their natural form. Also, crossover and mutation operators which can operate directly with integer and floating-point numbers are presented. The proposed approach is illustrated through a corrective action plan for a few harmful contingencies in the IEEE 30-bus.

### A. Severity Index

The severity of a contingency to line overload may be expressed in terms of the following severity index, which express the stress on the power system in the post contingency period:

$$severity\ index\ I_{sl} = \sum_{l \in L_o}^{n} \left( \frac{S_l}{S_l^{max}} \right)^{2m} \tag{1}$$

where

0-7803-9771-1/06/$25.00 ©2006 IEEE

$I_{sl} = flow\,in\,line\,l\,(MVA)$

$S_l^{max} = rating\,of\,line\,l\,(MVA)$

$L_0 = set\,of\,overload\,lines$

$m = int\,eger\,exponent$

The line flows in (1) are obtained from Newton–Raphson load-flow calculations. While using the above severity index for security assessment, only the overloaded lines are considered to avoid masking effects. For IEEE 30-bus system considered in this work, we have fixed the value of m as 1. To determine the best location for installing a phase shifter, a sensitivity analysis is conducted.

### B. Mathematical Formulation of Optimal Power Flow Problem

The conventional formulation of the optimal-power-flow (OPF) problem determines the optimal settings of control variables such as real power generations, generator terminal voltages, transformer tap settings and phase-shifter angles while minimising an objective function such as fuel cost given in (3).

$$F_T = \sum \left( a_i P_{gi}^2 + b_i P_{gi} + C_i \right) \qquad (2)$$

During security control, the prime task of the power-system operator would be to remove the line overload. Hence, the minimum severity index is taken as the objective function in this paper. The minimisation problem is subjected to the constraints

(i)      Load-flow constraints:

$$P_i = V_i \sum_{j=1}^{N_b} V_j \left( G_{ij} \cos\theta_{ij} + B_{ij} \sin\theta_{ij} \right), i = 1,2,...N_b, i \neq s \quad (3)$$

$$Q_i = V_i \sum_{j=1}^{N_b} V_j \left( G_{ij} \cos\theta_{ij} - B_{ij} \sin\theta_{ij} \right), i = 1,2,...N_{pq} \quad (4)$$

(ii)      Voltage constraints:

$$\underline{v_i} \leq v_i \leq \overline{v_i}\, i \in N_b \qquad (5)$$

(iii)      Unit Constraints

$$\underline{P_{gi}} \leq P_{gi} \leq \overline{P_{gi}}\, i \in N_g \qquad (6)$$

$$\underline{Q_{gi}} \leq Q_{gi} \leq \overline{Q_{gi}}\, i \in N_g \qquad (7)$$

(iv)      Phase-shifting transformer constraint:

$$\underline{\phi_i} \leq \phi_i \leq \overline{\phi_i}\, i \in N_l \qquad (8)$$

The power-flow equations are used as equality constraints and the inequality constraints are the limit on active and reactive power generations, phase-shifting-transformer setting, busbar voltage magnitudes and apparent power flows in branches.

## II. GENETIC ALGORITHM IMPLEMENTATION

When applying Gas to solve a particular optimization problem, two main issues must be addressed:
(a)   representation of decision variable; and
(b)   formation of the fitness function
These issues are explained in this Section.

### A. Problem representation

In solving the OPF problem for security-control applications, two types of variable need to be determined by the optimization algorithm: generator active-power generation $P_{gi}$ and generator terminal voltages $V_{gi}$ which are continuous variables, and the phase angle of the phase-shifting transformer $\phi_i$, which is a discrete variable. Each individual in the population represents candidate OPF solutions. We consider the generator active power, generator terminal voltages and the phase angle of the phase shifting transformer as the control variables. These variables are represented in natural form. With this representation, a typical chromosome of the OPF problem looks like the following:

| 97.5 | ... | 250.70 | 0.975 | ... | 0.90 | -1 | ... | 1 |
|---|---|---|---|---|---|---|---|---|
| $P_{g2}$ | | $P_{gn}$ | $V_{gl}$ | | $V_{gm}$ | 1 | | n |

The use of floating-point numbers in GA representation has a number of advantages like lesser memory requirement and no loss in precision by discretisation to binary or other values.

### B. Fitness function

GA searches for the optimal solution by maximizing a given fitness function. In the OPF problem under consideration, the objective function is to minimize the severity in the post-contingency state satisfying the equality and inequality constraints. With the inclusion of the penalty function the new objective function becomes

$$Min f = I_s + P_s + \sum_{i=1}^{N_{pq}} P_{vi} + \sum_{i=1}^{N_g} P_{Qi} \qquad (9)$$

Here $P_s$, $P_{vi}$ and $P_{Qi}$ are the penalty terms for the reference bus bar generator active power limit violation, load bus bar voltage limit violation, and reactive power generation limit violation, respectively. These quantities are defined by the equations

$$P_s = \begin{cases} K_s \left( P_s - P_s^{max} \right)^2 & if\, P_s > P_s^{max} \\ K_s \left( P_s - P_s^{max} \right)^2 & if\, P_s < P_s^{max} \\ 0 & otherwise \end{cases} \qquad (10)$$

$$VP_j = \begin{cases} K_v \left( V_j - V_j^{max} \right)^2 & if\, V_j > V_j^{max} \\ K_v \left( V_j - V_j^{max} \right)^2 & if\, V_j < V_j^{max} \\ 0 & otherwise \end{cases} \qquad (11)$$

$$QP_j = \begin{cases} Kq \left( Q_j - Q_j^{max} \right)^2 & if\, Q_j > Q_j^{max} \\ K_q \left( Q_j - Q_j^{max} \right)^2 & if\, Q_j < Q_j^{max} \\ 0 & otherwise \end{cases} \qquad (12)$$

where $K_s$, $K_v$, and $K_q$ are the penalty factors. The success of the approach lies in the proper choice of these   penalty parameters. Using the above penalty-function approach, one has to experiment to find a correct combination of penalty parameters $K_s$, $K_v$, and $K_q$. However, to reduce the number of penalty parameters, the constraints are often normalized and only one penalty factor R is used.

During the GA run, GA searches for a solution with the maximum fitness function value. Hence, the minimization objective function is transformed to fitness function to be maximized as $Fitness = \dfrac{k}{1+f}$ where k is a constant. In the denominator a value of 1 is added to avoid division by zero in case of complete overload alleviation.

## III. SIMULATION RESULTS

Two different cases were considered for the study. In the first case proposed GA based algorithm was applied to obtain the optimal control variables in IEEE 30 bus system under normal conditions. In the second case, the proposed algorithm was applied to alleviate overloads underline outage through generator rescheduling and Phase Shifting transformers. The GA code was written in Matlab and executed on a PC with a Pentium IV processor.

### A. Case 1: Optimal scheduling for the base case

IEEE 30-bus system has 6 generators and 41 transmission lines. The generator and transmission-line data relevant to the system are taken from [5]. The upper and lower voltage limits at all busbars except slack were taken as 1.10 and 0.95 respectively. The slack busbar voltage was fixed to 1.06 p.u. Here the contingencies are considered and the GA-based algorithm was applied to find the optimal scheduling of the power system for the base case loading condition given in [5]. The objective function in this case is minimization of the total fuel cost. Generator active-power outputs and the generator-busbar terminal voltages were taken as the optimization variables. The optimization variables are represented as floating-point numbers in the GA population.

The initial population was randomly generated between the variable's lower and upper limits. Tournament selection was applied to select the members of the new population. Blend crossover and uniform mutation were applied on the selected individuals. The performance of GA generally depends on the GA parameters used, in particular the crossover and mutation probabilities Pc and Pm, respectively. The performance of a GA for various crossover and mutation probabilities in the ranges 0.6–1.0 and 0.001–0.1, respectively, was therefore evaluated.

The optimal values of control variables along with the real-power generation of the slack busbar generator and the reactive power generation of all the generating units are given in Table 1. The minimum cost obtained in the proposed algorithm is near to the minimum cost of \$ 802.43 /hr, reported in [5] using the gradient method. Also it is found that state variables satisfy the lower and upper limits.

TABLE I
BASE CASE SOLUTION OF IEEE 30 BUS SYSTEMS

| Busbar | Generated Power | | Busbar voltage magnitude (p.u.) |
|---|---|---|---|
| | Real (MW) | Reactive (MW) | |
| 1 | 179.39 | -11.33 | 1.0443 |
| 2 | 48.83 | 9.90 | 1.0317 |
| 5 | 21.84 | 23.73 | 1.0052 |
| 8 | 21.75 | 41.13 | 1.0125 |
| 11 | 12.05 | 31.14 | 1.0495 |
| 13 | 12.36 | 40.19 | 1.0426 |

Fuel Cost: 803.05 $/hr

### B. Case 2: Overload alleviation through generation rescheduling

In this case, the GA-based algorithm is used for corrective control under a contingency state. Contingency analysis was conducted under base-load conditions to identify the harmful contingencies. From the contingency analysis, it was found that line outages 1–2, 1–3, 3–4 and 2–5 have resulted in overload on other lines. The power flow on the overloaded lines and the calculated value of severity index for each contingency are given in Table 2.

From this Table it is found that line outage 1–2 is the most severe one, and results in overloading on three other lines. Sensitivity analysis was carried out to identify the suitable locations of the phase shifter to alleviate the line overload. The four locations identified for each contingency are given in Table 3.

The GA-based OPF algorithm was applied to alleviate the line overload in all four severe-contingency cases. To test the ability of the proposed algorithm to alleviate overload under severe conditions the real and reactive load in all the load busbars were increased to 1.2 times the baseload condition. The maximum reactive-power generation of all the generators was also increased correspondingly. Generator active power and the phase-shifting transformer are taken as the control variable. The limits of the phase shifters were taken as $\pm 10^0$. The minimum severity index was taken as the objective function of the GA. The algorithm was run for a maximum of 25 generations and was made to stop if the targeted value of $I_S = 0$ was reached.

TABLE II
CONTINGENCY ANALYSIS IEEE 30-BUS SYSTEM

| Outage Line | Overloaded lines | Line flow (MVA) | Line flow limit (MVA) | Severity Index |
|---|---|---|---|---|
| 1-2 | 1-3 | 191.24 | 130 | 5.6228 |
| | 3-4 | 182.13 | 130 | |
| | 4-6 | 110.07 | 90 | |
| 1-3 | 1-2 | 182.44 | 130 | 3.0478 |
| 3-4 | 1-2 | 179.64 | 130 | 2.9584 |

968

| | 2-6 | 66.56 | 65 | |
|---|---|---|---|---|
| 2-5 | 2-6 | 77.36 | 65 | 2.7252 |
| | 5-7 | 101.43 | 70 | 2.7252 |

TABLE III
PHASE SHIFTER LOCATION IN IEEE 30 BUS SYSTEM

| Line Outage | 1-2 | 1-3 | 3-4 | 2-5 |
|---|---|---|---|---|
| Phase- Shifter Location | 2-4 | 6-7 | 6-7 | 1-2 |
| | 2-5 | 2-5 | 2-5 | 2-6 |
| | 2-6 | 5-7 | 2-6 | 2-4 |
| | 4-6 | 4-12 | 5-7 | 4-6 |

## IV. CONCLUSIONS

This paper has proposed a flexible optimisation tool based on a genetic algorithm to aid power-system operators. The application of this tool for scheduling the power system during the normal operation and to schedule the power system controls during contingencies has been presented. The line overloads were relieved through rescheduling of generator outputs and adjustment of the phase angle of a phase-shifting transformer. To alleviate the line overloads effectively, the phase-shifter location is identified based on sensitivity analysis. In this paper, the problem of discretisation in the representation of the decision variables in the binary-coded GA has been alleviated by employing floating-point numbers to represent the generator loadings and integers for the phase angles. A modified form of crossover and mutation operations to deal with the real and integer variables has been presented. Compared with a conventional GA, the proposed algorithm occupies less computer space and takes less CPU time and is well suitable for real-time applications.

## V. REFERENCES

[1.] Medicherla, T.K.P., Billington, R., and Sachdev, M.S.: 'Generation rescheduling and load shedding to alleviate line overloads-analysis', IEEE Trans., 1979, PAS-98, (6), pp. 1876–1884
[2.] Lachs, W.R.: 'Transmission line overloads: real-time control', Proc. IEEE, 1987, 134, (5), pp. 342–347
[3.] Udupa, A.N., Purushothama, G.K., Parthasarathy, K., and Thukaram, D.: 'A fuzzy control for network overload alleviation', Electr. Power Energy Syst., 2001, 23, pp. 119–129
[4.] Monticelli, A., Pereira, M.V.F., and Granville, S.: 'Security-constrained optimal power flow with post-contingency corrective rescheduling', IEEE Trans., 1987, PWRS-2, (1), pp. 175–182
[5.] Alsac, O., and Scott, B.: 'Optimal load flow with steady state security', IEEE Trans., 1974, PAS-93, pp. 745–751
[6.] Lee, K.Y., Park, Y.M., and Ortiz, J.L.: 'Fuel-cost minimization for both real and reactive power dispatches', IEE Proc., 1984, 131C, (3), pp. 85–93
[7.] Mangoli, M.K., and Lee, K.Y.: 'Optimal real and reactive power control using linear programming', Electr. Power Syst. Res., 1993, 26, pp. 1–10
[8.] Scott, B., Alsac, O., Bright, J., and Paris, M.: 'Further developments in LP-based optimal power flow', IEEE Trans., 1990, PWRS-5, (3), pp. 697–711
[9.] Lima, G.M. et al.: 'Phase shifter placement in large-scale systems via mixed Integer programming', IEEE Trans., 2003, PWRS-18, (3), pp. 1029–1034
[10.] Kumar, A., and Parida, S.: 'Enhancement of power system loadability with location of FACTS controllers in competitive electricity markets using MILP'. Proc. Int. Conf. Energy, Information Technology and Power Sector, Kolkata, Jan. 2005, pp. 515–523
[11.] Momoh, J.A., Zhu, J.Z., Boswell, G.D., and Hoffman, S.: 'Power system security enhancement by OPF with phase shifter', IEEE Trans., 2001, PWRS-16, (2), pp. 287–293
[12.] Goldberg, D.: 'Genetic algorithms in search, optimization and machine learning' (Addison–Wesley, 1989)
[13.] Lai, L.L.,Ma, J.T., Yokayama, R., and Zhao,M.: ' Improved genetic algorithms for optimal power flow under both normal and contingent operation states', Int. J. Electr. Power Energy Syst., 1997, 9, (5), pp. 287–292
[14.] Paranjothi, S.R., and Anburaja, K.: 'Optimal power flow using refined genetic algorithms', Electr. Power Compon. Syst., 2002, 30, pp. 1055–1063
[15.] Devaraj, D., and Yegnanarayana, B.: 'A combined genetic algorithm approach for optimal power flow'. Proc. 11th National Power Systems Conf., Bangalore, India, 2000, Vol. 2, pp. 524–528
[16.] Eshelman, L.J., and Schaffer, J.D.: 'Real-coded genetic algorithms and interval schemata' (D. Whitley, 1993), pp. 187–202
[17.] IEEE 118-bus system (1996), (Online) Available at //www.ee. washington.edu.

## VI. BIOGRAPHIES

N.B. Muthuselvan obtained his BE degree in electrical and electronics engineering and MBA degree from Madurai Kamaraj University, India. He obtained his ME degree in power systems from Anna University. He is working as a Lecturer, in SSN College of Engineering, Chennai, India. His current research area is in Applications of Computational Intelligence in Power System problem.

P. Somasundaram obtained his B.E. degree in electrical and electronics engineering and his M.E. in power systems from Madras University and Annamalai University, India, respectively. He also obtained his Phd degree in electrical engineering from Anna University. Currently, he is a Lecturer in the Department of Electrical and Electronics Engineering, Anna University, India. His field of interest is artificial intelligence applications to power system optimization.

S.S.Dash obtained his Bachelor, Master and PhD degrees in Electrical Engineering field . His current research interests concerns Power system Stability and Modeling of FACTS devices. He holds more than ten years of research and teaching experience.

**2006 IEEE International Conference on Power Electronic, Drives and Energy Systems**

# Congestion Management in Nodal Pricing With Genetic Algorithm

S.M.H Nabav, Shahram Jadid, M.A.S. Masoum, *Senior Member, IEEE,* and A. Kazemi

*Abstract*-- **Congestion cost allocation is an important issue in congestion management. This paper presents a genetic algorithm (GA) to determine the optimal generation levels in a deregulated market. The main issue is congestion in lines, which limits transfer capability of a system with available generation capacity. Nodal pricing method is used to determine locational marginal price (LMP) of each generator at each bus. Simulation results based on the proposed GA and the Power World Simulator software are presented and compared for 3-bus and 5-bus test systems.**

*Index Terms*-- **Congestion management, nodal pricing, deregulated power systems, Genetic Algorithm and optimal bidding strategy.**

## I. INTRODUCTION

TRANSMISSION congestion will most likely occur when many transactions or scheduled/forced outages exist in the power system. Optimal allocation of congestion cost is usually performed using the branch [1] or the node [2] allocation techniques. The branch allocation method first assigns a system congestion cost (to the congested branch) and then allocates the branch congestion cost (for each transaction) by using sensitivity or tracing methods [3–6]. The node allocation method directly allocates the system congestion cost to the nodes.

Locational market power is a well-known and studied issue in power systems [7-8]. The pricing system (e.g., nodal price or locational marginal price (LMP)) plays an important role in the congestion management [9-12]. A "fixed transmission right" model for the congestion management [9]. [10] presents two approaches based on the pool model and the bilateral model. An operation decision support software system covering the functionality of transmission dispatch and congestion management system (TDCMS) [11].

In recent years the methods of managing transmission congestion have been under intense scrutiny [12-19]. Presently there are two distinct congestion management systems widely being employed: nodal pricing [18] and zonal pricing methods [13]. Many of these congestion management models are based on optimal power flow algorithms [13-14]. Genetic algorithms (GAs) are widely accepted as an effective optimization method to solve various types of large-scale engineering problems. Different congestion management methods based on genetic algorithms are presented in [15-19]. The problem of building optimal bidding strategies for generation companies has also been formulated as a stochastic optimization model and solved by the well-known Monte Carlo simulation method and Genetic Algorithms [15].

This paper proposes a new genetic algorithm to minimize the total system cost with congestion in a deregulated market. Simulation results of the genetic algorithm (for 3-bus and 5-bus test systems) are compared with those provided by the Power World Simulator [21] to demonstrate the applicability and efficiency of the proposed method for congestion management in a deregulated electricity market.

## II. NODAL PRICING

In the nodal pricing method, system operator sets the electricity price and determines the dispatch level of each generator based on the requested marginal supply bids which is a monotonically increasing curve indicating supplier's individual preference on the production amount at various prices. Consequently, the price and dispatch amount are used to determine an economically optimal operating point while respecting system constraints. Nodal pricing method of managing transmission congestion is based on the computation of location marginal price at each individual node of the power system. Nodal pricing applies spatial spot pricing theory on a real time basis to derive a bus-by-bus LMP [18].

### A. Locational Marginal Price (LMP)

LMP is the marginal cost of supplying the next increment of electric energy at a specific bus, considering the generation marginal cost and the physical aspects of the transmission system. Marginal pricing reflects the cost to serve the next increment of load in a system that is economically dispatched. Three factors influencing LMP are "marginal cost to operate

---

S. M. H. Nabavi is with the Department of Electrical Engineering, Iran University of Science & Technology, Tehran, Iran, 16846 I.R.I. (e-mail: H_nabavi@ee.iust.ac.ir).
S. Jadid is with the Department of Electrical Engineering, Iran University of Science & Technology, Tehran, Iran, 16846 I.R.I.(e-mail: Jadid@iust.iust.ac.ir).
M. A. S. Masoum is with the Department of Electrical Engineering, Iran University of Science & Technology, Tehran, Iran, 16846 I.R.I. (e-mail: M_Masoum@iust.iust.ac.ir).
A. Kazemi is with the Department of Electrical Engineering, Iran University of Science & Technology, Tehran, Iran, 16846 I.R.I. (e-mail: Kazemi@iust.iust.ac.ir).

---

0-7803-9771-1/06/$25.00 ©2006 IEEE

generation", "total load" and "cost of delivery". LMP is defined as [18]:

$$LMP = \text{generation marginal cost} + \text{congestion cost} + \text{cost of marginal losses} \quad (1)$$

LMP is the dual variable for the equality constraint at a node (e.g., sum of injections and withdrawals is equal to zero). Both loss and congestion components are always zero at the reference bus. Therefore, the price at the reference bus is always equal to the energy component. LMPs will not change if the reference bus is allocated. However, all three components of LMP dependent on the selection of the reference bus due to the dependency of sensitivities on the location of reference bus. In fact, LMP is the additional cost for providing additional MW at a certain bus [20].

Using LMP, buyers and sellers experience the actual price of delivering energy to locations on the transmission systems. If the line flow constraints are not included in the optimization problem, LMPs will be the same for all buses. This is the marginal cost of the most expensive dispatched generation unit (marginal unit). In this case, no congestion charges apply. However, if any line is constrained, LMPs will vary from bus to bus and may cause congestion charges [19].

### B. LMP Price Calculation Procedures

At any bus $i$, $LMP_i$ is composed of three components; marginal generation price at the reference bus ($LMP_i^{\text{ref}}$), loss component ($LMP_i^{\text{loss}}$), and congestion component ($LMP_i^{\text{cong}}$):

$$LMP_i = LMP_i^{\text{ref}} + LMP_i^{\text{loss}} + LMP_i^{\text{cong}} \quad (2)$$

where values of the three components are based on the selection of reference bus. The last two components are given as:

$$LMP_i^{\text{loss}} = (DF_i - 1)LMP_i^{\text{ref}} \quad (3A)$$

$$LMP_i^{\text{cong}} = -\sum_{k \in K} GSF_{ik}\beta_k \quad (3B)$$

*Where*

- $DF_i$ is the delivery factor of bus $i$ relative to reference bus (e.g., a measure of the portion of the next MW generation at buses $i$ that is delivered to the reference bus).
- $GSF_{ik}$ is the generation shift factor for bus $i$ on line $k$ (e.g., ratio of the change in flow of line $k$ to the change in generation of bus $i$). All generation shift factors at the reference bus are equal to zero [19].
- $k$ is the set of congested transmission lines.
- $\beta_k$ is the constraint cost of line $k$, defined as:

$$\beta = \frac{\text{Re}duction\ in\ total\ \cos t}{Change\ in\ constraint's\ flow}.$$

## III. PROBLEM FORMULATION

### A. Objective Function

LMP is determined by solving the following optimization problem:

$$\min_{Q_{G_i}} \sum_{G_i} C_{G_i}(P_{G_i}) \quad for\ i = 1,\dots,n \quad (4)$$

where $P_{G_i}$ is the amount of dispatched generation at node $G_i$ and $C_{G_i}$ is the total cost of generation at node $G_i$ (expressed in terms of $P_{G_i}$).

### B. Constraints

The optimization problem is subjected to a number of constrains that are discussed next.

*Load flow constraint* (total generation=system load):

$$\sum_{G_i} P_{G_i} = \sum_{D_i} P_{D_i} \quad for\ i = 1,\dots,n \quad (5)$$

*Transmission line flow constraints* (e.g., power flow on line $l$ is within the maximum line rating):

$$|F_l| = \left| H_{l_{G_i}}P_{G_i} + H_{l_{D_i}}P_{D_i} \right| \le F_l^{\max}\ for\ i = 1,\dots,n \quad (6)$$

*Generation limit constraints* (e.g., dispatch amount at node $G_i$ is within the maximum rating of the corresponding generator):

$$P_{G_i}^{\min} \le P_{G_i} \le P_{G_i}^{\max} \quad for\ i = 1,\dots,n \quad (7)$$

For simplicity, DC power flow is used for computing the flows on each line of the system. The DC power flow equations in matrix notation are written as:

$$B\delta = P_{G_i} - P_{D_i} \quad (8)$$

where $\delta$ is the voltage angle vector, $P_{G_i}$ is the real power generation vector for buses $G_i$ and $P_{D_i}$ is the real power load vector for buses $D_i$. Therefore, flow vectors for lines can be computed as:

$$F_1 = H\delta \quad (9)$$

where $H$ is the linearized flow matrix for the system.

If the generation cost of supplier $G_i$ (e.g., $C_{G_i}$) is a function of the output given by:

$$C_{G_i}(P_{G_i}) = a + bP_{G_i} + cP_{G_i}^2 + dP_{G_i}^3 \quad (10)$$

Then, assuming perfect competitive market conditions, the optimal supply bid by supplier $G_i$ (e.g., $MC_i$) is the marginal cost bid given by:

$$MC_i = b + cP_{G_i} + dP_{G_i}^2 \quad (11)$$

## IV. PROPOSED GENETIC ALGORITHM

Genetic algorithms use the principle of natural evolution and population genetics to search and arrive at a high quality near global solution. During each iterative procedure (referred to as generation), a new set of strings with improved performance is generated using three GA operators (namely reproduction, crossover and mutation).

### A. Structure of Chromosomes

In this paper, the chromosome structure for GA consists of MC substrings of binary numbers (Fig. 1), where MC denotes amount of generation at each generator bus.

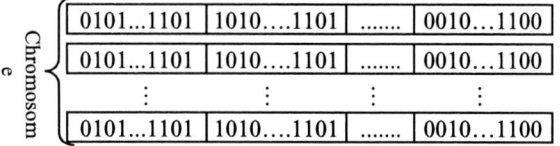

Fig. 1. Proposed chromosome structure for the genetic algorithm.

### B. Proposed Fitness Function

In this paper, exponential penalty functions are used as the fitness function to combine the objective and constraints:

$$F_{fitness} = \frac{1}{F_L \cdot F_G} \tag{12A}$$

$$F_L = \prod_{j=1}^{b} F_{line\ flow} \tag{12B}$$

$$F_G = \prod_{j=1}^{m} F_{generation\ limit} \tag{12C}$$

where b and m are the number of branches and generators in the power system, respectively. The proposed penalty functions $F_L$ and $F_G$ are shown in Fig. 2.

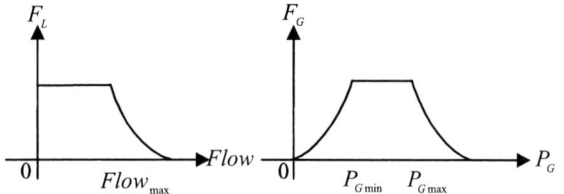

Fig. 2. Proposed penalty functions used to compute fitness (Eq. 12); (a) for $F_{line\ flow}$, (b) for $F_{generation\ Limit}$.

## V. SOLUTION METHODOLOGY

The cost minimization problem is solved using the proposed genetic algorithm, as follows:

*Step 1:* input system parameters (e.g., system topology, line and load specifications, maximum and minimum generation limits, and line flow limits). Input the initial population with $N_{chrom}$ chromosomes.

*Step 2:* Set initial counter and parameter values (e.g., $N_{ch} = N_{it} = N_R = 1$ and $F_{MIN}$ = a high number)

*Step 3 (Fitness Process):*

Step 3A: Run DC power flow for chromosome $N_{ch}$ and save outputs.

Step 3B: Compute proposed penalty functions (Fig. 2) using outputs of the DC power flow. Compute fitness functions (Eq. 12) for chromosome $N_{ch}$. Set $N_{ch} = N_{ch} + 1$

Step 3C: If $N_{ch} \leq N_{chrom}$ go to Step 3A.

*Step 4 (Reproduction Process):*

Step 4A: Define total fitness as the sum of all fitness values for all chromosomes.

Step 4B: Select a percentage of "roulette wheels" for each chromosome, which is equal to the ratio of its fitness value to the total fitness value.

Step 4C: Improve generation by rolling the "roulette wheel" $N_{chrom}$ times. Select a new combination of chromosomes.

*Step 5 (Crossover Process):*

Step 5A: Select a random number ($RND_1$) for mating two parent chromosomes.

Step 5B: If $RND_1$ is between 0.01 and 0.3 then combine the two parents, generate two offspring and go to Step 5D.

Step 5C: Else, transfer the chromosome with no crossover.

Step 5D: Repeat steps 5A to 5C for all chromosomes.

*Step 6 (Mutation Process):*

Step 6A: Select a random number ($RND_2$) for mutation of one chromosome.

Step 6B: If $RND_2$ is between 0.01 and 0.1 then apply the mutation process at a random position and go to Step 6D.

Step 6C: Else, transfer the chromosome with no mutation.

Step 6D: Repeat Steps 6A to 6C for all chromosomes.

*Step 7 (Updating Populations):* Replace the old population with the improved population generated by Steps 2 to 6. Check all chromosomes, if there is any chromosome with $F_L = 1$, $F_G = 1$ and $F_F < F_{MIN}$, set $F_{MIN} = F_F$ and save it. Set $N_{it} = N_{it} + 1$.

*Step 8 (Convergence):* If all chromosomes are the same or the maximum number of iterations is achieved ($N_{it} = N^{max}$), then print the solutions and stop, else go to Step 2.

## VI. SIMULATION RESULTS

### A. Simulation of the 3-bus System

The 3-bus system of Fig. 3 [18] is used to study the performance of the proposed GA. System data and line ratings are given in [18]. $P_{G_1}$ and $P_{G_2}$ are computed using the proposed GA algorithm. The selected GA parameters for generation number, crossover, and mutation are 30, 30% and 10%, respectively. Optimal solution is achieved after 20 iterations.

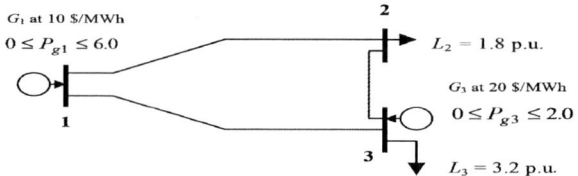

Fig. 3. 3-bus system diagram.

Table I shows simulation results (the system total cost) for the 3-bus system with and without the congestion mode. The Power World Simulator [20] is used to demonstrate that the proposed method is efficient for congestion management in a deregulated electricity market. Fig. 4 shows the generators incremental curve and Fig. 5 presents simulation of the optimal Generation with minimum cost in congestion mode.

TABLE I
GENERATION LEVELS OF THE 3-BUS SYSTEM BEFORE AND AFTER APPLYING THE PROPOSED GENETIC ALGORITHM

| Bus No. | LMP($/MW) and $P_G$ | | | | Total system cost | |
| | without congestion | | with congestion | | without congestion management | with congestion management |
|---|---|---|---|---|---|---|
| 1 | 10 | 5 | 10 | 3.9 | | |
| 2 | 10 | 0 | 15 | 0 | 1000$ | 610$ |
| 3 | 10 | 0 | 20 | 1.1 | | |

## B. Simulation of the 5-bus System

Performance of the proposed algorithm is also demonstrated using the 5-bus system of Fig. 6 with congestion. System data and line ratings are given in [18]. The GA algorithm is used to compute $P_{G_1}$ and $P_{G_2}$. The selected GA parameters for generation number, crossover, and mutation are 50, 30% and 10%, respectively. After 30 iterations, optimal values of generation are achieved. Table II shows system total cost for the 5-bus system in congestion mode and without any congestion. Application of the Power World Simulator shows the proposed method is efficient for congestion management in a deregulated electricity market. Fig. 7 shows the generators incremental curve and Fig. 8 illustrates simulation of the optimal generation with minimum cost in the congestion mode.

TABLE II
GENERATION LEVELS FOR THE 5-BUS SYSTEM BEFORE AND AFTER APPLYING THE PROPOSED GENETIC ALGORITHM

| Bus No. | LMP/ $P_G$ without congestion | | LMP/ $P_G$ with congestion | | Total system cost | |
|---|---|---|---|---|---|---|
| 1 | 30 | 110 | 15 | 110 | without congestion management | with congestion management |
| 2 | 30 | 100 | 30 | 35 | | |
| 3 | 30 | 90 | 30 | 65 | 27000$ | 17550$ |
| 4 | 30 | 0 | 30 | 90 | | |
| 5 | 30 | 600 | 17 | 600 | | |

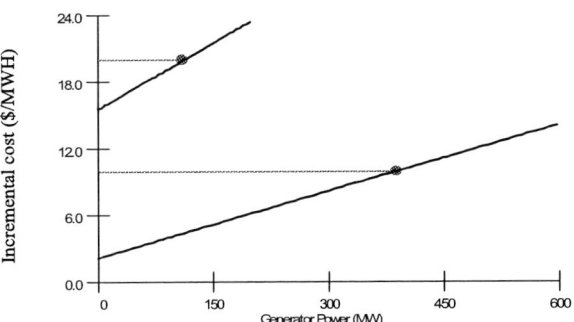

Fig. 4. Generators incremental curve for the 3-bus system.

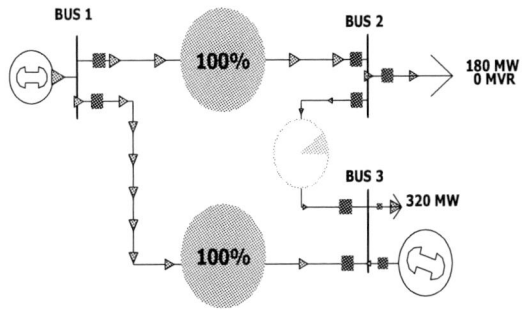

Fig. 5. Optimal generation of the 3-bus system with minimum cost in congestion mode.

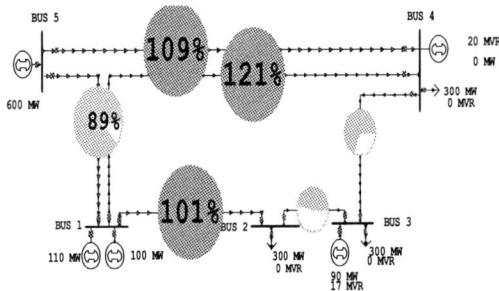

Fig. 6. 5-bus system diagram with congestion.

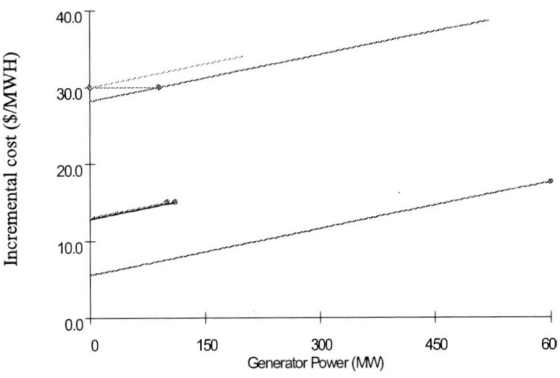

Fig. 7. Generators incremental curve of the 5-bus system.

973

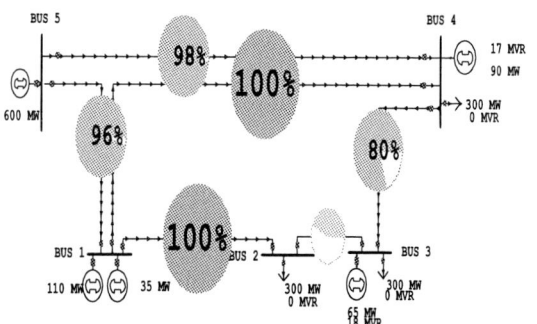

Fig. 8. Optimal generation of the 5-bus system with minimum cost in congestion mode.

## VII. CONCLUSIONS

In this paper a genetic algorithm is proposed and implemented to minimize the total cost of generation in power systems. Nodal pricing method is used to determine marginal cost at each bus. The locational marginal cost of each bus is determined by adding marginal cost of reference bus and the cost due to congestion. Simulation results of the proposed GA for sample systems with there and five buses indicate a significant reduction in system's total cost to be provided by the consumers (Tables I and II). Sample 3-bus and 5-bus systems are also simulated using the Power World Simulator software to illustrate the fine accuracy of the proposed approach for congestion management in deregulated environments.

## VIII. REFERENCES

[1] H. Singh, S. Hao, and A. Papalexopoulos, "Transmission congestion management in competitive electricity markets", *IEEE Trans. on Power Systems*, 1998, Vol. 13, No. 2, pp. 672–680.

[2] A.G. Bakirtzis, "Aumann-Shapley transmission congestion pricing", *IEEE Power Eng. Rev.*, 2001, Vol. 21, No.2, pp. 67–69.

[3] N.S. Rau, "Transmission loss and congestion cost allocation – an approach based on responsibility", *IEEE Trans. on Power Systems*, 2000, Vol. 15, No. 4, pp. 1401–1409.

[4] H. Rudnick, R. Palma, and J.E. Fernandez, "Marginal pricing and supplement cost allocation in transmission open accesses", *IEEE Trans. on Power Systems*, 1995, Vol. 10, No. 2, pp. 1125–1132.

[5] D. Shirmohammadi, P.R. Gribik, and E.T.K. Law, 'Evaluation of transmission network capacity use for wheeling transactions', *IEEE Trans. Power Systems*, 1989, Vol. 4, No. 4, pp. 1405–1413.

[6] M.E. Baran, V. Bannsunarayanan, and K.E. Garren, 'A transaction assessment method for allocation of transmission services', *IEEE Trans. Power Systems*, 1999, Vol. 24, No. 3, pp. 920–928.

[7] H. Singh, "Market power mitigation in electricity markets," Game Theory Applications in Electric Power Markets, *IEEE PES Tutorial*, TP-136-0, Feb. 1999, pp. 70–76.

[8] R. Ethier, R. Zimmerman, T. Mount, W. Schulze, and R. Thomas, "A uniform price auction with locational price adjustments for competitive electricity markets," *Int. J. Elect. Power Energy System*, Vol. 21, No. 2, pp. 103–110, 1999.

[9] M.J. Alomoush and S.M. Shahidehpour, "Fixed Transmission Right for Zonal Congestion Management," *IEE Proceedings-Generation. Transmission & Distribution*, Vol. 146, No. 5, Sep. 1999, pp. 471-476.

[10] H. Singh, S. Hao and A. Papalexopoulos, "Transmission Congestion Management in Competitive Electricity Markets," *IEEE Trans. on Power Systems*, Vol. 13, No. 2, May 1998, pp. 672- 680.

[11] D. Shirmohammadi, B. Wollenberg, A. vojdani, P. Sandrin, M. Pereira, F. Rahimi, T. Schneider and B. Stott, "Transmission Dispatch and Congestion Management in the Emerging Energy Market Structures," *IEEE Trans. on Power Systems*, Vol. 13, No. 4, Nov. 1998, pp. 1466-1474.

[12] T. Gedra, "On Transmission Congestion and Pricing", *IEEE Transactions on Power Systems*, Vol. 14, No 1, pp. 241-248, February 1999.

[13] X. Wang and Y.H. Song, "Apply lagrangian relaxation to multi – zone congestion management", *IEEE 2001*, Con. 2001, pp. 399-404.

[14] I.J.P. Arriaga, F.J. Rubio, J.F. Puerta and J.M. Arceluz, "Marginal Pricing of Transmission Services: An Analysis of Cost Recovery", *IEEE Transactions on Power Systems*, Vol. 10, No. 1, February 1995.

[15] M. Li, F. Wen, N. Yixin and F.F Wu, "Optimal bidding strategies for generation companies in electricity markets with transmission capacity constraints taken into account", *IEEE Power Engineering Society General Meeting*, 2003, Vol. 4, pp. 13-17, July 2003.

[16] J.M. Ramirez and G.A. Marin, "Alleviating congestion of an actual power system by genetic algorithms", *Power Engineering Society General Meeting, 2004. IEEE*, 6-10 June 2004, Vol. 2, pp. 2133-2141.

[17] M. Saguan, S. Plumel, P. Dessante, J.M. Glachant and P. Bastard, "Genetic algorithm associated to game theory in congestion management"; Probabilistic Methods Applied to Power Systems, 2004 International Conference on, 12-16 Sept. 2004, pp. 415-420.

[18] M. Shahidehpour, H. Yamin and Z. Li, "Market Operations in Electric Power Systems: Forecasting, Scheduling, and Risk Management", John Wiley & Sons, Inc., 2002, ISBNs: 0-471-44337-9 (Hardback); 0-471-22412-X (Electronic).

[19] A.J. Wood and B.F. Wollenberg, "Power Generation, Operation, and Control", New York: Wiley, 1996.

[20] R. Coutu, "Locational Marginal Pricing"; ISO New England's Wholesale Energy Market (Intermediate) - WEM 201, Nov 28-30, 2006.

[21] http://www.powerworld.com.

## IX. BIOGRAPHIES

**Seyyed Mohammad Hossein Nabavi** was born in Tabriz, Iran, in 1979. He received his M.Sc. degree in Electrical Engineering from university of Science & Technology, Tehran, Iran, in 2004. He is currently a PHD Student in electrical engineering department of Iran University of Science and Technology, Tehran, Iran. His research interests are reliability, optimization in power system, and control of FACTS devices.

**Shahram Jadid** was born in Tehran, Iran, in 1962. He received his PHD degree in electrical engineering from, Indian Institute of Technology, India in 1993. He is currently an associate professor in electrical engineering department of Iran University of Science and Technology, Tehran, Iran. His research interests are Power systems, Energy management and deregulated power system.

**Mohammad A. S. Masoum** received his B.S., M.S. and Ph.D. degrees in Electrical and Computer Engineering in 1983, 1985, and 1991, respectively, from the University of Colorado at Boulder, USA. Currently, he is an Associate Professor at Iran University of Science & Technology, Tehran, Iran. Dr. Masoum is a senior member of IEEE and his research interests include: Optimization, Power Quality and Stability of Power Systems, Drives and

Power Electronics

**Ahad Kazemi** was born in Tehran, Iran, in 1952. He received his M.Sc. degree in electrical engineering from Oklahoma State University, U.S.A in 1979. He is currently an associate professor in electrical engineering department of Iran University of Science and Technology, Tehran, Iran. His research interests are reactive power control, power system dynamics, stability and control and FACTS devices.

**2006 IEEE International Conference on Power Electronic, Drives and Energy Systems**

# Coupled Magneto-Mechanical Field Computations

Amogh Kank, G. B. Kumbhar, and S. V. Kulkarni, *Member, IEEE*

*Abstract*— Electromagnetic force calculation is essential in the analysis and design of any electrical machine. There are two kinds of forces, the magnetization force and the J×B force, which play an important role in the behavior of any electrical machine. This work involves development of finite element code for the calculation of electromagnetic force on a permeable material. Maxwell's Stress Tensor (MST) method has been implemented to calculate the force. This has been used to analyze the forces on an embedded conductor slot of a rotating machine.

*Index Terms*— Local virtual work, Lorentz force, Maxwell stress tensor, finite element method.

## I. INTRODUCTION

**E**LECTRICAL machine analysis using Finite Element Method (FEM) has gained importance with increasing complexity of analysis and need for optimized and efficient designs. One of the most important specifications in the analysis is the electromagnetic force or torque acting on various parts of the machine.

Magnetization forces and $\overline{J} \times \overline{B}$ forces are two forces which are important for the analysis of electrical machines. These forces can be computed by various methods:

1) Lorentz force method
2) Maxwell Stress Tensor (MST) method
3) Principle of virtual work
4) Method of equivalent sources

This paper gives an overview of these methods of force calculation. It also presents the results of force computation by the Maxwell stress tensor method using developed code and by using a commercial FEM software for a test geometry. The analysis of forces acting on iron and conductor in a rotor slot of a rotating electrical machine is, subsequently, reported.

## II. METHODS OF FORCE CALCULATION

### A. Lorentz Force Method

In this method, the total force acting on a body is obtained by integrating the force acting on each differential current carrying element due to a magnetic field over all differential current carrying elements,

$$\overline{F}_v = \int_V (\overline{J} \times \overline{B}) \mathrm{d}V \tag{1}$$

where $\overline{J} \times \overline{B}$ is the force density in the conductor.

Amogh Kank, G. B. Kumbhar, and S. V. Kulkarni are with the Department of Electrical Engineering, Indian Institute of Technology Bombay, Mumbai–400076, India (e-mails: amogh@ee.iitb.ac.in, ganeshk@ee.iitb.ac.in, svk@ee.iitb.ac.in).

### B. Maxwell Stress Tensor

Maxwell stress tensor is widely used for electromagnetic force computations. This method and its related computational aspects such as error performance are discussed in [1], [2] and [3]. A quantity called the stress tensor is defined in this method whose divergence is the force density in the volume of the body on which the force has to be determined. Applying the divergence theorem to the stress tensor, we can consider Maxwell stress as a surface force density which, when integrated over the surface enclosing the body, gives the total force acting on it. The choice of surface should be such as to satisfy certain performance criterion as well as to improve the accuracy of the results. The expression for the stress tensor is,

$$T_{ij} = \frac{1}{\mu}(B_i B_j - \frac{1}{2}B^2 \delta_{ij}) \tag{2}$$

where $(i, j)$ can take values $(x, y, z)$. $\delta_{ij}$ is 1 if $i = j$ and zero otherwise. The same can be written in terms of the force density vector as,

$$\overline{F} = \iint (\frac{1}{\mu}\overline{B}(\overline{B}.\overline{n}) - \frac{1}{2\mu}|\overline{B}|^2 \overline{n})\mathrm{d}s \tag{3}$$

where $\overline{n}$ is the normal unit vector to the surface under consideration.

### C. Virtual Work Method

The method of virtual work for electromagnetic force calculation is based on the generalized principle of virtual displacement [4], [5]. The movable part is assumed to be displaced and the rate of change in the stored magnetic energy with respect to the displacement gives the force acting on the body. The displacement is not an actual physical displacement of the body and hence, it is known as virtual displacement. One precaution which has to be taken while virtually displacing the body is that the flux linkage has to be kept constant throughout the motion. The implementation of this method can be both at the level of displacement of the whole body or displacement of elements or nodes. If the nodes are displaced, then the method is known as local virtual work method. The expression for the magnetic energy stored in the field is

$$W = \int_V \int_0^B \overline{H}.\mathrm{d}\overline{B}\mathrm{d}V \tag{4}$$

where V is the volume of the field region, $\overline{B}$ is the flux density, and $\overline{H}$ is the magnetic field intensity. The force acting on a node which is virtually displaced is then given by,

$$\overline{F}_k = -\frac{\partial W}{\partial q}\hat{q} \tag{5}$$

0-7803-9771-1/06/$25.00 ©2006 IEEE

where $q$ is the virtual displacement along $\hat{q}$.

### D. Equivalent Sources Method

This method uses equivalent magnetizing currents. The theory and implementation of the method is discussed in [6] and [7]. It uses the principle that there is physical existence of microscopic atomic current loops in any material, particularly ferromagnetic material, which experience the magnetization force in the presence of a magnetic field [8], which eventually gets transferred to the body. Conventionally, the field intensity produced by these atomic current loops is taken care of by introducing the concept of relative permeability for isotropic materials without hysteresis. The relative permeability accounts for the atomic current loops in the computation of flux inside the ferromagnetic material. In the equivalent sources method, the relative permeability inside the ferromagnetic material is taken to be the same as that of air, and the atomic current loops are accounted for by using equivalent current loops. Thus, instead of considering the presence of actual atomic current loops, we can find the total force acting on the body by calculating the forces acting on these fictitious equivalent sources and they turn out to be the same as the actual forces.

The magnetic behavior of a ferromagnetic material can be described by,

$$\overline{B} = \mu_0 \overline{H} + \overline{M} \tag{6}$$

where $\mu_0$ is the magnetic permeability of vacuum and $\overline{M}$ is the magnetization. For soft magnetic materials, $\overline{M}$ is induced due to external field and is a function of $\overline{H}$,

$$\overline{M} = (\mu_r - 1)\mu_0 \overline{H} \tag{7}$$

where $\mu_r$ is the relative permeability which may be constant or a function of $\overline{H}$. In materials involving permanent magnet, $\overline{M}$ corresponds to the remnant magnetic field. The governing equation is:

$$\nabla \times \frac{1}{\mu_0}\overline{B} = \overline{J} + \frac{1}{\mu_0}\nabla \times \overline{M} \tag{8}$$

where $\overline{J}$ is the conduction current density. The second term on the right-hand side has the same effect as the conduction current, hence, it is referred to as the equivalent magnetizing current $\overline{J_m}$. The force can be calculated by a formula similar to the one used in the Lorentz force method. It should be noted that $\overline{J_m}$ exists only at the boundaries.

In another approach, the magnetic material with permeability $\mu$ is replaced by a non-magnetic material having a superficial distribution of magnetic charges [9] and the force density is calculated as the product of the superficial surface charge density and the calculated surface magnetic field intensity.

### III. MAXWELL STRESS TENSOR METHOD

The expression of stress tensor in (2) results into a $3 \times 3$ tensor matrix,

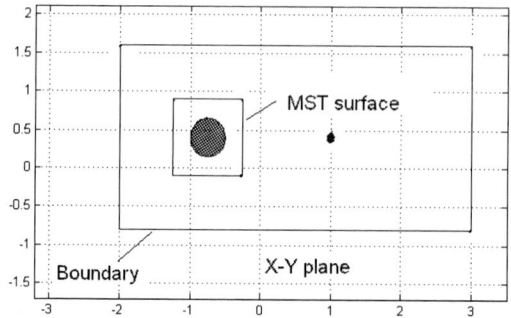

Fig. 1.   Geometry of problem for magnetization force calculation.

$$
\begin{bmatrix} T_{xx} & T_{xy} & T_{xz} \\ T_{yx} & T_{yy} & T_{yz} \\ T_{zx} & T_{zy} & T_{zz} \end{bmatrix} =
$$
$$
\begin{bmatrix} \frac{1}{\mu}(B_x B_x - \frac{1}{2}B^2) & \frac{1}{\mu}(B_x B_y) & \frac{1}{\mu}(B_x B_z) \\ \frac{1}{\mu}(B_y B_x) & \frac{1}{\mu}(B_y B_y - \frac{1}{2}B^2) & \frac{1}{\mu}(B_y B_z) \\ \frac{1}{\mu}(B_z B_x) & \frac{1}{\mu}(B_z B_y) & \frac{1}{\mu}(B_z B_z - \frac{1}{2}B^2) \end{bmatrix}
\tag{9}
$$

In this method the body on which the force has to be calculated is surrounded by an imaginary surface. After meshing the region, the edges of some elements (or faces of the elements in case of 3D) will lie on the chosen imaginary surface. Now these tiny elemental surfaces can be resolved into three components parallel to the three coordinate surfaces. The $x$, $y$ and $z$ components of the elemental surface are parallel to $y - z$, $x - z$ and $x - y$ planes respectively. $T_{xx}$ gives the surface force density on the x-component (as $i = x$) of the surface along the x-direction (as $j = x$), $T_{xy}$ gives the surface force density on x-component of the surface along the y-direction and so on. Similarly $T_{yx}$, $T_{yy}$ and $T_{yz}$ give surface force densities on the y-component of the surface along the $x$, $y$ and $z$ directions respectively, and $T_{zx}$, $T_{zy}$ and $T_{zz}$ give surface force densities on the z-component of the surface along the $x$, $y$ and $z$ directions respectively. It is to be noted that the above procedure gives the force density along the positive x- or y- or z-direction. However, we require the surface force density in the direction away from the enclosed region. Hence, all the values in the above MST matrix have to be negated, if the direction away from the enclosed region is along negative coordinate axis. The above procedure is equivalent to evaluating the integral given by (3) over the enclosing Maxwell surface.

The Maxwell stress tensor gives both types of forces, namely, $\overline{J} \times \overline{B}$ forces and magnetization forces. If the Maxwell surface does not enclose any current carrying part, then it gives the magnetization force acting on the enclosed body. If the surface encloses a current carrying part along with some permeable material, then it gives the addition of the $\overline{J} \times \overline{B}$ force

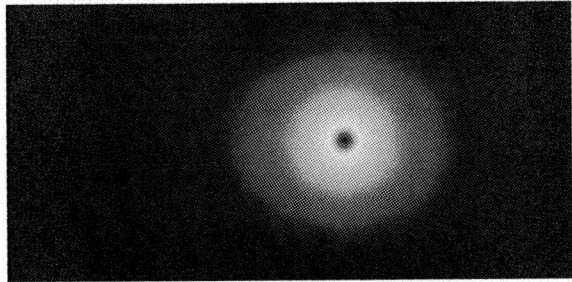

Fig. 2. Plot of $A_z$ in magnetization force calculation.

acting on the current carrying part and the magnetization force acting on the permeable material.

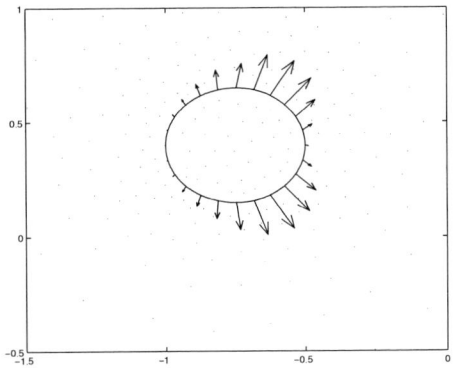

Fig. 3. Force distribution on the ferromagnetic cylinder.

### IV. SIMULATION RESULTS

Fig. 1 shows a typical test geometry. There are two entities involved. A ferromagnetic cylinder on the left-hand side has a relative magnetic permeability $\mu_r$ of 1000. The cylindrical conductor, which is on the right-hand side, carries a current only in the z-direction with a current density of $10^7 A/m^2$ (corresponding to a typical short-circuit condition) and it has a $\mu_r$ of 1. The ferromagnetic cylinder is enclosed in a rectangular surface known as MST surface (having one meter depth in the z-direction). Maxwell's stress, when integrated over this rectangular surface, using (3), gives the total force acting on the ferromagnetic cylinder. Since the Maxwell's surface does not enclose any current carrying part, the force obtained is the magnetization force acting on the permeable material (this force is, sometimes, referred to as the saliency force [3]). The boundary condition imposed is the flux parallel condition at the outermost boundary. Fig. 2 shows the surface plot of the magnetic vector potential. Since the ferromagnetic cylinder has a high $\mu_r$, it experiences the magnetization force. Fig. 3 shows the force acting on the ferromagnetic cylinder. As can be seen from Fig. 3, forces act mainly at the air-iron interface of the ferromagnetic material where the gradient of $\mu_r$ has a nonzero value. FEM codes were developed in MATLAB to calculate these forces. The value of the magnetization force obtained from the codes is 7.58 N/m (attractive in nature) as compared to 7.59 N/m obtained from a commercial FEM software, while the analytical value is 14.7 N/m which is obtained using,

$$F = \frac{\mu_r - 1}{\mu_r + 1} \frac{\mu_0 I^2}{2\pi d} \frac{a^2}{d^2 - a^2} \quad (10)$$

where $I$ is the current through the cylindrical conductor, $a$ is the radius of the cylindrical conductor and $d$ is the distance between the centers of the two cylindrical structures. For the considered case, these values are: $a = 0.25$ m, $d = 1.75$ m and radius of current carrying cylinder is 0.05 m. The relatively large difference between the analytical value and the values obtained using FEM analyses can be attributed to the simplifications involved while deriving the analytical expression.

The results of FEM analyses are summarized in table I.

TABLE I

COMPARISON OF SIMULATION RESULTS

|  | Maxwell Stress Tensor Method (using commertial FEM software) | Maxwell Stress Tensor Method (using code) |
|---|---|---|
| Magnetization forces | 7.59 N/m | 7.58 N/m |

### V. PRACTICAL APPLICATION - EMBEDDED CONDUCTOR IN ROTOR SLOT

Fig. 4 shows a representative geometry of a rotor slot in a rotating machine. The relative magnitudes and behaviors of the magnetization force acting on the high permeability rotor core and $\overline{J} \times \overline{B}$ force acting on the embedded rotor conductor are analysed. It is found, from the results of the simulation, that the magnetization force acting on the high permeability rotor is dominant as compared to the $\overline{J} \times \overline{B}$ force acting on the embedded rotor conductor. This is a welcome phenomenon since the rotor core can now withstand much higher forces than the rotor conductors. The ratio of the magnetization force to the $\overline{J} \times \overline{B}$ force on the rotor conductor is found to be about 23 for this geometry. Also, the resultant of the forces acting on the vertical edges of the teeth is in the same direction as that of the $\overline{J} \times \overline{B}$ force acting on the rotor conductor. The field distribution is shown in Fig. 5. The origin of the magnetization force for this configuration can be explained as follows. Let us consider that, initially, there is no current in the rotor conductor. The current flowing through the stator conductor will, thus, produce a symmetrical magnetic field around the rotor conductor, that is, the field on either side of the rotor conductor is similar. From Fig. 4, it can be seen that the three slot edges face the rotor conductor. Under the assumption of no rotor current, two opposite edges (the vertical edges) of the slot would have the same flux distribution along them. The magnetization force density is now given by the expression,

$$\overline{F}_{mag_v} = \int_V (-\frac{1}{2}|\overline{H}|^2 \nabla \mu) dV \quad (11)$$

977

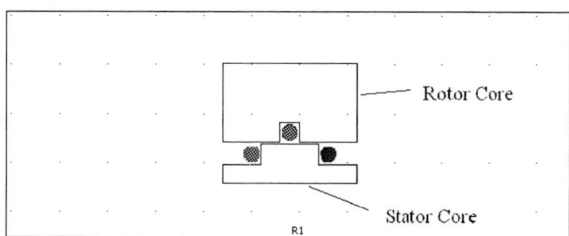

Fig. 4.   Geometry of embedded slot.

Fig. 5.   Field plot of embedded slot.

$\nabla\mu$ has a nonzero value only at the interface between air and the permeable material. Hence, $\nabla\mu$ is nonzero only at the three edges. Hence, the magnetization force acts only on these three edges. The force on the horizontal edge is radial and hence does not contribute to the torque. According to (11), the magnetization forces acting on the two vertical edges are equal and opposite and hence, cancel each other. When current flows through the rotor conductor, the flux densities along the vertical edges will no longer remain equal and hence, a resultant force will act on the rotor core which is dominant as compared to the $\overline{J}\times\overline{B}$ force acting on the rotor conductor.

## VI. CONCLUSION

We have reviewed various methods of force computation. The Maxwell stress tensor method has been studied in detail. Electromagnetic forces are calculated using this method. A FEM code based on **A**-formulation has been developed for this purpose. The magnitudes of the forces obtained from the MATLAB code and the commercial FEM software have been compared with that obtained from the analytical formula.

The analysis of forces on a rotor slot region has also been done. It is found that there are two components of torque which act along the same direction. One component is due to the $\overline{J}\times\overline{B}$ force acting on the rotor conductor while the other is due to the magnetization force acting on the edges of the rotor core, and the latter component is found to be much higher in magnitude.

## VII. REFERENCES

[1] A. N. Wignall, A. J. Gilbert, and S. J. Yang, "Calculation of forces on magnetised ferrous cores using the maxwell stress tensor method," *IEEE Trans. Magn.*, vol. 24, no. 1, pp. 459–462, 1988.

[2] T. Onuki, S. Wakao, and H. Saito, "Improvement in the calculation of electromagnetic force by the FEM," *IEEE Trans. Magn.*, vol. 30, pp. 1863–1866, July 1994.

[3] Z. W. Shi and C. B. Rajanathan, "A method of reducing errors in the calculation of electromagnetic forces by Maxwell stress summation method," *IEEE Trans. Magn.*, vol. 35, no. 3, pp. 1358–1369, May 1999.

[4] J. L. Coulomb and G. Meunier, "Finite element implementation of virtual work principle for magnetic or electric force and torque computation," *IEEE Trans. Magn.*, vol. 20, no. 5, pp. 1894–1896, Sept. 1984.

[5] A. Benhama, A. C. Williamson, and A. B. J. Reece, "Virtual work approach to the computation of magnetic force distribution from finite element field solutions," *IEE Proceedings-Electric Power Applications*, vol. 147, no. 6, pp. 437–442, Nov. 2000.

[6] T. Kabashima, A. Kawahara, and T. Goto, "Force calculation using magnetising current," *IEEE Trans. Magn.*, vol. 24, no. 1, pp. 451–454, Jan. 1988.

[7] G. Henneberger, P. K. Sattler, and D. Shen, "Nature of the equivalant magnetising current for the force calculation," *IEEE Trans. Magn.*, vol. 28, no. 2, pp. 1068–1071, Mar. 1992.

[8] H. H. Woodson and J. R. Melcher, *Electromechanical Dynamics Part II: Fields, Forces, and Motion.* New York: John Wiley & Sons, Inc., 1968.

[9] J. P. A. Bastos and N. Sadowski, *Electromagnetic Modeling by Finite Element Method.* New York: Marcel Dekker, 2003.

**Amogh V. Kank**   is presently pursuing his M.Tech in the Department of Electrical Engineering , Indian Institute of Technology, Bombay, India with specialization in Power Electronics and Power Systems. He obtained his B.E. from Sardar Patel College of Engineering, Andheri, Mumbai in Electrical Engineering. He is currently working in the area of computational electromagnetics, in particular, magneto-mechanical analysis.

**G. B. Kumbhar**   received his B.E. degree in Electrical Engineering from Government College of Engineering, Karad, Maharashtra, India, in 1999 and M.Tech degree from the Indian Institute of Technology-Madras, Chennai, in January 2002. He is presently a research scholar at the Indian Institute of Technology-Bombay, Mumbai, India, working in the area of coupled formulations in transformers.

**S. V. Kulkarni**   is Associate Professor, Department of Electrical Engineering, Indian Institute of Technology, Bombay, India. Previously, he worked at Crompton Greaves Limited and specialized in the design and development of transformers up to 400 kV class. He has authored a book titled *Transformer Engineering: Design and Practice* published by Marcel Dekker Inc. He is the author of more than 70 professional publications in reputed journals and conferences. He was the recipient of the Young Engineer Award (2000) from the Indian National Academy of Engineering. He has also received the Career Award for Young Teachers from All India Council for Technical Education in 2001. His research interests include transformer design and analysis, computational electromagnetics and distributed generation.

**2006 IEEE International Conference on Power Electronic, Drives and Energy Systems**

# Optimizing Voltage Stability Limit and Real Power Loss in a Large Power System using Bacteria Foraging

M. Tripathy, and S. Mishra, *Senior Member, IEEE*

*Abstract--* **An optimal allocation of transformer taps and reactive power injections at some selected locations is carried out with a view to maximize the Voltage Stability Limit (VSL) of a multi machine power system, simultaneously with the objective of real power loss minimization. This issue is formulated as a non-linear equality and inequality constrained multi-objective optimization problem. A new evolutionary algorithm known as Bacteria Foraging is applied for solving, first the optimum settings of the transformer taps only and then with selected reactive power injections at some buses, so that the power system operates more securely achieving both the objectives.**

*Index Terms--* **Bacteria Foraging, Continuation Power Flow, Optimal Power Flow**

## I. INTRODUCTION

SCHEDULING various controls of any power system is carried out with the help of Optimal Power Flow(OPF) in such a manner that certain objective function like real power loss can be optimized. When some operating equipment and security requirement, limit constraints are forced on the solution, the problem becomes a constrained optimization problem which has been solved from different perspectives, such as, studying the effects of load increase/decrease on voltage stability/power flow solvability, generation rescheduling to minimize the cost of power generation, controls such as taps, shunts and other modern VAR sources adjustments to minimize real power losses in the system etc.. The OPF is solved by varieties of methods, i.e., Successive Linear Programming (SLP) [1], Newton based non-linear programming method [2], and with varieties of recently proposed Interior Point Methods (IPM) [3],[4],[5], and in most of the cases the only objective is to reduce the transmission loss in the system. But, it is a well known fact that a secure operation of power system is not possible unless the optimization problem takes into account the system voltage security in its solution. The Continuation Power Flow (CPF) [6] gives information regarding, how much percentage overloading the system can withstand before a possible voltage collapse. In [7], the authors have successfully incorporated the CPF problem in to an OPF problem so that both the issues can be addressed simultaneously. In this paper, the maximum percentage overloading ($\lambda_{max}$) the system can withstand is defined as Voltage Stability Limit(VSL), and

incorporated along with the objective of real power loss minimization making the problem multi-objective. The main disadvantage with the conventional techniques of OPF solution lies in the fact that they are highly sensitive to starting points owing to a non-monotonic solution surface. This problem can be much worse particularly when it is a multi variable one, but, it can be properly dealt by applying evolutionary techniques to solve them [8], [9]. In [9] authors have applied Particle Swarm Optimization (PSO) to the problem of OPF. Such algorithms, based on food searching behavior of species (like birds etc.), compute both global and local best positions at each instant of time, to decide the best direction of search.

This paper employs a new algorithm from the family of evolutionary computation, known as Bacteria Foraging Algorithm (BFA), to solve a combined CPF-OPF problem of real power loss minimization and VSL maximization of the system. BFA has been recently proposed [10] and further applied for harmonic estimation problem in power systems [11]. The algorithm is based on the foraging behavior of E.coli bacteria present in human intestine. Several transformer tap positions along with numbers of reactive power injections at some selected buses in a power system are simultaneously optimized as control variables, so that the multiple objectives are fulfilled, keeping an eye to all specified constraints. The results so obtained show its strength in solving highly non-linear *epistatic* problems. The main objectives of this paper is to optimize the values of transformer taps so that

1. The Real power transmission loss is minimized.
2. Multiple objectives of minimum loss and maximum VSL are fulfilled.
3. Further improvement in both the objectives could be achieved by optimizing the transformer tap values simultaneously with the amount of reactive power injections at some selected buses.

## II. BACTERIA FORAGING: A BRIEF OVERVIEW

The idea of BFA is based on the fact that natural selection tends to eliminate animals with poor foraging strategies and favor those having successful foraging strategies. After many generations, poor foraging strategies are either eliminated or reshaped into good ones. The *E. coli* bacteria that are present in our intestines have a foraging strategy governed by four processes namely Chemotaxis, Swarming, Reproduction, Elimination and Dispersal [10].

### A. Chemotaxis

This process is achieved through swimming and tumbling. Depending upon the rotation of the flagella in each bacterium it decides whether it should move in a predefined direction

---

The authors are with the department of Electrical Engineering at Indian Institute of Technology, Delhi, India. The authors acknowledge the financial support provided by AICTE, India under its Young Teachers Award scheme-2004 to Dr S. Mishra and logistic support given to Mr. M.Triapthy by U.C.E. Burla, Orissa, India., in order to enable them to complete this research project.

---

0-7803-9771-1/06/$25.00 ©2006 IEEE

(swimming) or an altogether different direction (tumbling), in the entire lifetime of the bacterium. To represent a tumble, a unit length random direction, say $\phi(j)$, is generated; this will be used to define the direction of movement after a tumble. In particular

$$\theta^i(j+1,k,l)=\theta^i(j,k,l)+C(i)\phi(j) \qquad (1)$$

Where $\theta^i(j,k,l)$ represents the $i^{th}$ bacterium at $j^{th}$ chemotactic $k^{th}$ reproductive and $l^{th}$ elimination and dispersal step. $C(i)$ is the size of the step taken in the random direction specified by the tumble. 'C' is termed as the 'run length unit'

### B. Swarming

It is always desired that the bacterium which has searched optimum path of food should try to attract other bacteria so that they reach the desired place more rapidly. Swarming makes the bacteria congregate into groups.

### C. Reproduction

The least healthy bacteria die and the other healthiest bacteria each split into two bacteria, which are placed in the same location. This makes the population of bacteria constant.

### D. Elimination and Dispersal

It is possible that in the local environment, the life of a population of bacteria changes either gradually or suddenly due to some other influence. Events can kill or disperse all the bacteria in a region. This process can possibly destroy the chemotactic progress, but in contrast they also assist it, since dispersal may place bacteria near good food sources. Elimination and dispersal helps in reducing the behavior of *stagnation*,( i.e. being trapped in a premature solution point or local optima). [10]-[11].

### III. PROBLEM STATEMENT

*Problem*: To solve a voltage secure Real Power Loss Minimization of 10-machine New England power systems.

The one line diagram of the system is shown in Fig. 1. The system data in detail, including the 12 transformers' nominal tap values are given in [12].

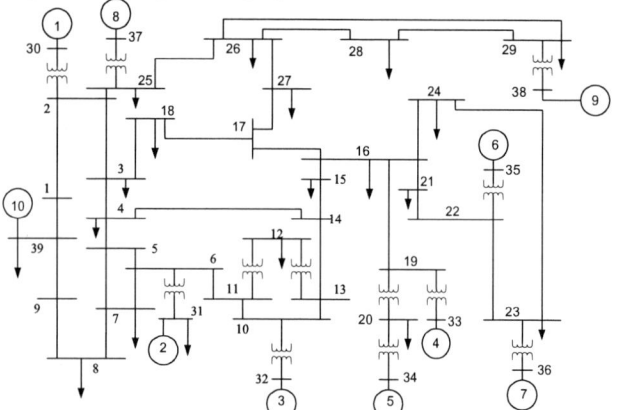

Fig. 1. New England Power System Layout.

### B. Optimal Power Flow (OPF): Problem Formulation

The OPF problem is a static constrained non-linear optimization problem, the solution of which determines the optimal settings of control variables in a power network respecting various constraints. Hence, the problem is to solve

a set of nonlinear equations describing the optimal solution of power system. It is expressed as:

$$\text{Minimize} \quad F(x,u) \qquad (2)$$
$$\text{Subject to} \quad \begin{aligned} g(x,u)=0 \\ h(x,u) \leq 0 \end{aligned}$$

The objective function F is real power loss of the mesh connected multi machine test system. $g(x,u)$ is a set of non-linear equality constraints to represent power flow and $h(x,u)$ is a set of non-linear inequality constraints ( i.e. bus voltages, transformer/line MVA limits, etc.). Vector 'x' consists of dependent variables and $u$ consists of control variables. For the above problem the control variables are the transformer tap values, and the amount of reactive power injections at some selected buses.

### C. OPF Formulation Considering Voltage Stability Limit

The same objective of real power loss minimization is augmented with maximization of VSL. The VSL can be calculated through CPF, which introduces a load parameter defined as the percentage increase of generation and load from its base value. The resulting load and generation equation in terms of the load parameter is given below.

$$\begin{aligned} P_{Li} &= P_{Li0}(1+\lambda) \\ Q_{Li} &= Q_{Li0}(1+\lambda) \\ P_{Gi} &= P_{Gi0}(1+\lambda) \end{aligned} \qquad (3)$$

The load parameter can be increased till the system just reaches the verge of instability which is also known as the 'Notch point' of PV-Curve. The maximum value of the load parameter ($\lambda_{max}$) is termed as VSL. The objective is to

$$\text{Optimize} \quad F(x,u,\lambda_{max}) \qquad (4)$$
$$\text{Subject to} \quad g(x,u)=0$$
$$h(x,u) \leq 0$$

The function to be optimized now can be represented as

$$F(x,u,\lambda_{max})= G(x,u) + V(\lambda_{max}) \qquad (5)$$

Where,
$$G(x,u) = \text{Real Power Loss}$$
$$V(\lambda_{max})=(1/\lambda_{max})$$

The solution of CPF is done with the help of a suitably chosen continuation parameter. With the increase of '$\lambda$', a new solution point is predicted first and then corrected in usual predictor and corrector steps [6]. Since the objective is to maximize the VSL, so its reciprocal is added to the original cost function of real power loss so that the overall cost function can be minimized.

### IV. IMPROVED BACTERIAL FORAGING: THE ALGORITHM

The BF algorithm suggested in [10,11] is modified as discussed below, so as to expedite the convergence.

980

1) In [11], the author has taken the average value of all the chemotactic cost functions, to decide the health of particular bacteria in that generation, before sorting is carried out for reproduction. In this paper, instead of the average value, the global minimum value of all the chemotactic cost functions till that point is retained for deciding the bacterium's health. This speeds up the convergence, because in the average scheme [11], it may not retain the fittest bacterium.

2) For swarming, the distances of all the bacteria in a new chemotactic stage is evaluated from the global optimum bacterium till that point, and not the distances of each bacterium from rest others as suggested in [10,11]

The algorithm is discussed here in brief. For details refer [10,11].

1. Several variables like Number of bacteria (S), Number of parameters (p), Swimming length $N_s$, Numbers of iterations in a chemotactic loop ($N_c$), Number of reproduction loop ($N_{re}$), Number of elimination and dispersal events ($N_{ed}$) and its probability( $P_{ed}$), etc., are initialized.

2. The iterative algorithm starts with following steps.

This section models the bacterial population chemotaxis, swarming, reproduction, elimination and dispersal (initially, $j=k=l=0$). For the algorithm updating $\theta^i$ automatically results in updating of 'P'.

a) Begin Elimination-dispersal loop: $l=l+1$
b) Begin Reproduction loop: $k=k+1$
c) Begin Chemotaxis loop: $j=j+1$

i) Compute the cost function for each bacterium and add on the cell-to-cell attractant effect for swarming behavior between bacteria.

$$J_{sw}(i, j, k, l) = J(i, j, k, l) + J_{cc}(\theta^i(j, k, l), P(j, k, l)) \quad (6)$$

Where, $P(j,k,l)$ is the location of bacterium corresponding to the global minimum cost function out of all the generations and chemotactic loops till that point and $J_{cc}(\theta, P(j, k, l))$ is given by (7).

$$J_{cc}(\theta, P(j, k, l)) = \sum_{i=1}^{S} J_{cc}^{i}\left(\theta, \theta^i(j, k, l)\right)$$
$$= \sum_{i=1}^{S}\left[-d_{attract} \exp\left(-\omega_{attract} \sum_{m=1}^{p}(\theta_m - \theta_m^i)^2\right)\right] \quad (7)$$
$$+ \sum_{i=1}^{S}\left[h_{repelent} \exp\left(-\omega_{repelent} \sum_{m=1}^{p}(\theta_m - \theta_m^i)^2\right)\right]$$

where, $d_{attract}, \omega_{attract}, h_{repelent}, \omega_{repelent}$ are different coefficients that are to be chosen judiciously.

ii) Take the decision to Tumble or Swim and also let the direction be decided also. This process generates

another direction for the bacteria to search food for its sustenance. If maximum chemotactic growth length ($N_c$) is not reached then go back to step (i) and calculate the cost for new set of bacteria, else, end the Chemotaxis loop.

d) Reproduce the best half of bacteria and replace them with the worst half without affecting its number (S). If maximum number of reproductions ($N_{re}$) is not reached then go back to step (b), else, end the Reproduction loop.

e) Eliminate the whole set of bacteria and generate entirely new ones by dispersing them in to an altogether new direction, in the case it is probable to do so. (i.e., if a random number is > $P_{ed}$). If maximum number of elimination ($N_{ed}$) is not reached then go to step (a), else, end the iterations.

The flow chart of the improved algorithm is shown in Fig.2.

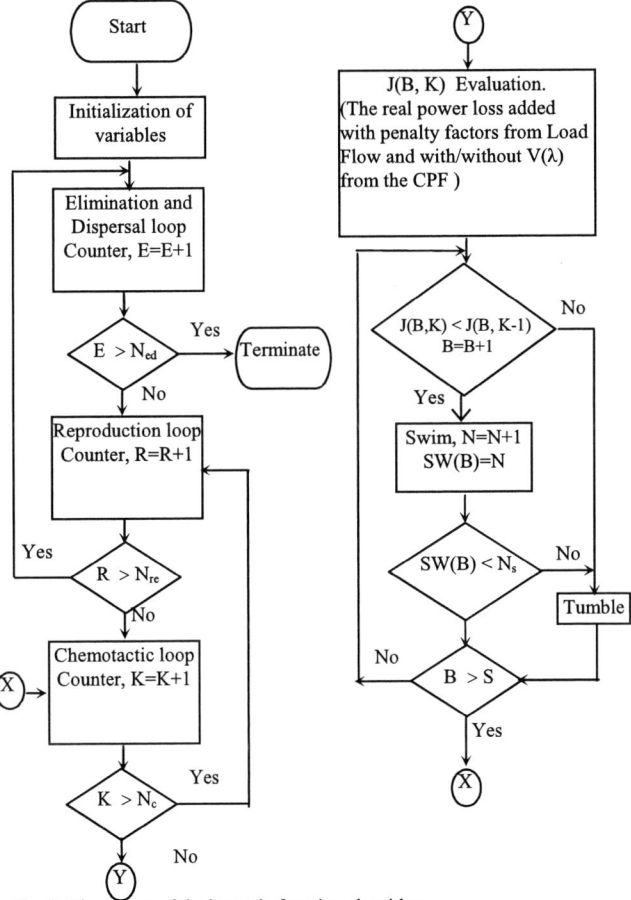

Fig. 2. Flow chart of the bacteria foraging algorithm.

## VI. SIMULATION AND RESULTS

The objective function to achieve minimization of real power loss without violating various operating limits is formulated by introducing penalty factors which accounts for specified limits of bus voltages, transformer MVA and transmission line MVA capacities. These penalty factors are added to the total real power loss in the system. This is depicted in (8).

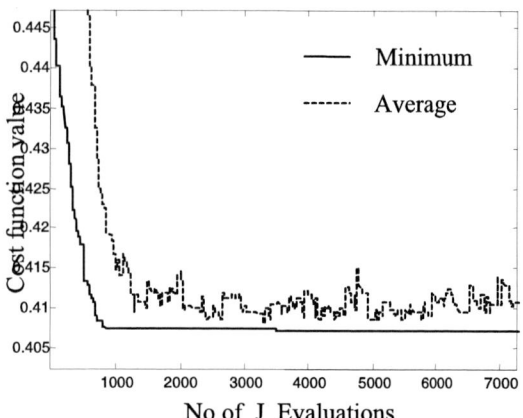

Fig. 3. Reproduction Schemes Proposed(Min) vs [13](Average).

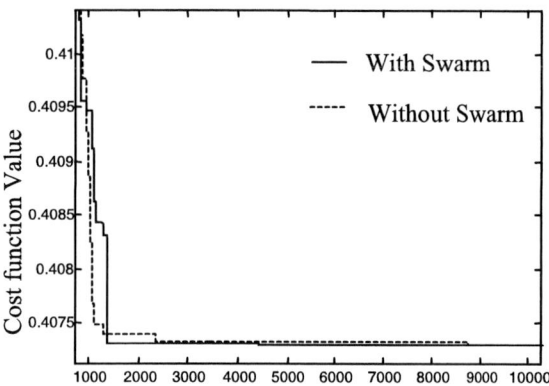

Fig. 4. Performance of Algorithms.

$$F = pf_1 + pf_2 + pf_3 + of$$

where,   $of$ = Real Power Loss

$$pf_1 = 10 * abs\,(sign\,(V_{\min} - 0.9) - 1)$$
$$\qquad + 10 * abs\,(sign\,(V_{\max} - 1.1) + 1)$$
$$pf_2 = 10 * abs(sign(trans_{\max} - 15) + 1)$$
$$pf_3 = 10 * abs(sign(line_{\max} - 20) + 1) \qquad (8)$$

$pf_1$, $pf_2$, and $pf_3$ are the penalty factors added with the real power loss ($of$), so that a constrained solution is achieved. $V_{\max}$ and $V_{\min}$ are respectively the maximum and minimum limits of bus voltages for all the buses [12]. Similarly $trans_{\max}$ and $line_{\max}$ are, respectively, the maximum MVA limits of the transformers and lines in the system. The values of $trans_{\max}$ and $line_{\max}$ are chosen at double the maximum nominal values of respective quantities. The formulation of penalty factors can be clearly understood with the help of an example. If all the bus voltage solution so obtained for a particular set of control variable, are within the $V_{\max}$ and $V_{\min}$ limits then the value of "$pf_1$" would be zero, otherwise it will be either 10 or 20 depending on whether one or both the upper and lower

limits have been violated. Thus, for a minimization problem the solution would always avoid the condition which violates the limits. The methodology adopted for optimization with BFA technique is discussed here in brief. For multi-objective case, the objective function can be formulated as

$$F = pf_1 + pf_2 + pf_3 + of + V(\lambda_{\max}) \qquad (9)$$

The optimization is carried out in three steps as given below.
1. Single objective of real power loss minimization, with only transformer taps are variables (Denoted as BFAS)
2. Multi objective of loss and VSL, with only transformer taps taken as variables. (Denoted as BFAM)
3. Multi objective of loss and VSL, with both transformer taps and reactive power support at some selected buses are taken as variables. (Denoted as BFAQ)

*A. Optimization with only Transformer taps as control variables (Single objective of Loss, BFAS):*

Bacteria Foraging (*Without Swarming*): Initially, to study the convergence behavior of the algorithm, the swarming effect is excluded from it. The algorithm is run for number of times with different combinations of number of bacteria (S) and the number of chemotactic loops (N_c). The convergence of the algorithm occurs when the minimum of cost function values among all bacteria becomes equal to their average for several successive reproduction loops. It was found that S=6 and N_c=4 gives the fastest convergence. A comparison of convergence by taking the average value of each bacterium [11] in the chemotactic stage to that of global minimum for reproduction is presented in Fig.3. It is found that with the proposed scheme the algorithm converges faster.

Moreover, it is also observed that with the 'average' scheme the algorithm convergence is very sensitive to the value of S, N_c and the run length unit coefficient C. For some combinations of these values the 'average' scheme has a tendency of oscillating around the solution point (as in Fig.3). This phenomenon is avoided when the global minimum bacterium is retained before reproduction.

Bacteria Foraging (*With Swarming*):
As established above swarming is included now considering the global minimum. To choose the parameters of swarming, the algorithm is run for different values of $d_{attract}$, $\omega_{attract}$, $h_{repelent}$ and $\omega_{repelent}$. It was found that these values when chosen as 2.0, 0.2, 2.0, and 10 respectively, give the fastest convergence. As shown in Fig.4, the convergence of the algorithm is faster when swarming effect is included as compared to that without swarming. Also it is observed that, though the minimization of real power loss is almost same with both the techniques, there is difference in the values of tap positions to which the algorithms have converged.

With nominal values of taps, the real power loss and VSL were found out to be 0.4483 p.u. and 0.8050 respectively. For BFAS, which allocates transformer taps for

982

single objective of real power loss only, the CPF solution is carried out at the end of optimization considering the optimized tap values. However, for BFAM and BFAQ, both of which consider the multiobjective of loss minimization and VSL maximization simultaneously, the CPF solution is done for each bacteria, so that $\lambda_{max}$ could be obtained, to be used to formulate the multiobjective cost function given in (9).

TABLE I
OPTIMIZATION OF TRANSFORMER TAPS WITH
VARIOUS SCHEMES

| Line No: | Taps (BFAS) | Loss (p.u.) | VSL | Taps (BFAM) | Loss (p.u.) | VSL | Taps (BFAQ) | Loss (p.u.) | VSL |
|---|---|---|---|---|---|---|---|---|---|
| 2 - 30 | 1.05 | | | 1.05 | | | 1.05 | | |
| 10 - 32 | 1.15 | | | 1.05 | | | 1.05 | | |
| 12 - 11 | 1.00 | | | 1.10 | | | 1.10 | | |
| 12 - 13 | 1.00 | | | 1.00 | | | 1.00 | | |
| 19 - 33 | 1.10 | 0.4071 | 0.9290 | 1.10 | 0.4303 | 0.9988 | 1.10 | 0.4258 | 1.0000 |
| 19 - 20 | 1.00 | | | 1.10 | | | 1.10 | | |
| 20 - 34 | 1.10 | | | 0.95 | | | 0.95 | | |
| 22 - 35 | 1.05 | | | 1.00 | | | 1.00 | | |
| 23 - 36 | 1.05 | | | 1.05 | | | 1.05 | | |
| 25 - 37 | 1.05 | | | 1.05 | | | 1.05 | | |
| 29 - 38 | 1.05 | | | 1.05 | | | 1.05 | | |
| 31 - 6 | 0.90 | | | 0.95 | | | 0.95 | | |
| Reactive Power Injections at Buses 7,8,12 and 20 (in p.u.) | | | | | | | | | |
| Bus 7 | Nil | | | Nil | | | 0.0535 | | |
| Bus 8 | Nil | | | Nil | | | 0.2526 | | |
| Bus 12 | Nil | | | Nil | | | 0.4990 | | |
| Bus 20 | Nil | | | Nil | | | 0.4308 | | |

*B. Optimization of both real power loss and VSL with only Transformer taps as control variables (BFAM):*

With an objective to optimize both real power loss and VSL the cost function is modified. The reciprocal of VSL is added to the real power loss. The optimization is carried out by BFA, with only transformer tap values being considered as variables. The optimized transformer tap values, along with the corresponding real power loss and VSL values are given in TABLE-I. It is seen that the VSL value has improved, although the real power loss has increased marginally. However, the sum of real power loss and the reciprocal of VSL together have reduced, when the multi objective function is considered.

*C. Optimization of both real power loss and VSL with both Transformer taps and reactive power injections taken as control variables (BFAQ):*

For locating the candidate buses where the reactive powers need to be injected, the bus voltage solutions with the optimized tap values obtained from both BFAS and BFAM are studied. It is seen that with BFAS, the bus voltage magnitudes of only bus number 7,8,12 and 20 are below 0.95

and with BFAM, the voltage magnitude of only bus 20 has reduced further to 0.92, while those of rest others have improved considerably. So a decision was taken to locate reactive power injection at all these buses, with the maximum value of these injections being limited to 0.5 p.u. for the 100 MVA base systems. From TABLE-I, it is seen that though the VSL value remains almost same as that of BFAM case, but the real power loss has reduced.

After solving the CPF, the PV-curve for the weakest bus ( The bus which undergoes maximum voltage deviation during a CPF solution) in the solution process is drawn w.r.t the load parameter ($\lambda$). Fig. 5, depicts these PV-curve for weakest bus, in the cases of Nominal, BFAS, BFAM and BFAQ optimized variable conditions.

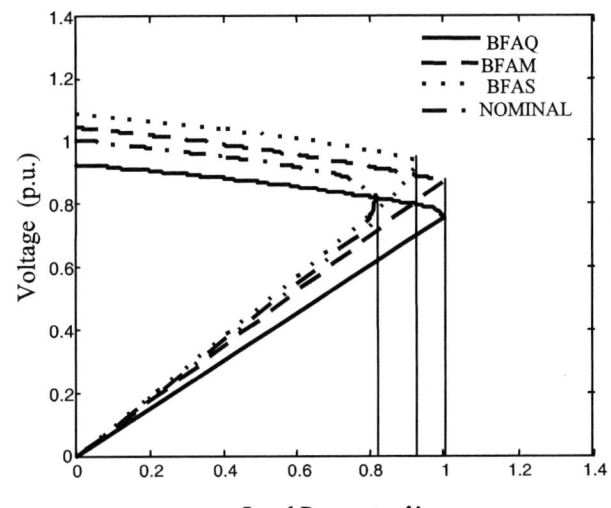

Fig. 5. PV Curve of weakest bus for different schemes.

VII. CONCLUSIONS

In any multiobjective optimization problem, it is seen that optimizing the cost function for any one objective generally gives a sub optimal solution for the other. However, in this work, bacteria foraging is used for allocating, transformer taps and reactive power injections at some buses with a view to minimize the real power loss and improve the Voltage Stability Limit (VSL) of the system simultaneously, so that the combined objective can be made optimal. As can be verified from the result, that for single objective of real power loss, the VSL has deteriorated, but for both the multiobjective cases though the loss has increased marginally but the combined cost function has reduced with considerable improvement in VSL. With BFAQ, though there is almost negligible improvement in VSL as compared to BFAM, but it has managed to reduce the cost function still further due to reduced real power loss. Therefore, it can be concluded that for secure operation of power system, the transformer taps should be scheduled in a manner to improve upon the voltage stability of the system, and further, with selective reactive power injections the cost paid for the deteriorated loss reduction can be recovered. This becomes quite a complex nonlinear problem, which has been solved effectively by the proposed Bacteria foraging algorithm.

## VIII. REFERENCES

[1] P.Ristanovic, "Successive Linear Programming Based OPF Solution", *Optimal Power Flow: Solution Techniques, Requirements and Challenges, IEEE Power Engineering Society,* 1996, pp. 1-9.

[2] D.Sun *et al.,* "Optimal Power Flow by Newton Approach," *IEEE Trans. Power App. Syst.,* vol. PAS-103, No.10, pp. 2864-2875, Oct. 1984.

[3] S.Granville, "Optimal Power Dispatch Through Interior Pont Methods," *IEEE Trans. Power Syst.,*Vol. 9, No.4, pp.1780-1787, Nov.1994.

[4] G.Torres and V.Quintana, "An Interior Point Method for Non-linear Optimal Power Flow using voltage rectangular coordinates," *IEEE Trans. Power Syst.,*Vol. 13, No.4, pp.1211-1218, Nov.1998.

[5] J.L.Martinez Ramos, A.G.Exposito and V.Quintana, "Transmission Loss Reduction by Interior Point Methods: implementation issues and practical experience," *IEE Proc. Gen. Trans. Distrib.,*Vol. 152, No.1, pp.90-98, Jan.2005.

[6] V.Ajjarapu, C.Christy, " The Continuation Power Flow: A Tool for Steady State Voltage Stability Analysis," *IEEE Trans. Power Syst.,* Vol. 7, No.1, pp.416-423, Feb. 1992.

[7] F.Milano, C.A.Canizares and A.J.Conejo, " Sensitivity-based Security-constrained OPF Market Clearing Model," *IEEE Trans. Power Syst.,* Vol. 20, No.4, pp. 2051-2060, Nov. 2005.

[8] J.Yuryevich and Kit.Po.Wong, "Evolutionary Programming Based Optimal Power Flow Algorithm," *IEEE Trans. Power Syst.,* Vol. 14, No.4, pp. 1245-1250, Nov. 1999.

[9] Ahmed A. A.Esmin, G.Torres and A.C.Z.de Souza, "A Hybrid Particle Swarm Optimization Applied to Loss Power Minimization," *IEEE Trans. Power Syst.,* Vol. 20, No.2, pp.859-866, May.2005.

[10] K.M.Passino, "Biomimicry of Bacterial Foraging for Distributed Optimization and Control", *IEEE Control System Magazine,* pp.52-67, June 2002.

[11] S.Mishra, " A hybrid least square-fuzzy bacteria foraging strategy for harmonic estimation", *IEEE Trans. Evolutionary Computation,* vol.9, No. 1, pp. 61-73, Feb. 2005.

[12] M.A. Pai, *Energy Function Analysis for Power System Stability.* Norwell, MA: Kluwer

## IX. BIOGRAPHIES

**M. Tripathy** is with University College of Engineering Burla, Orissa, India and presently pursuing his Ph.D degree (part-time) in the Department of Electrical Engineering, at Indian Institute of Technology, Delhi, India. His field of interest is intelligent control application to power system dynamics.

**S. Mishra** received the BE degree from University College of Engineering, Burla, Orissa, India and the M.E. and Ph. D. Degree from Regional Engineering College, Rourkela, Orissa, India, in 1990, 1992, and 2000, respectively. In 1992, he joined the department of Electrical Engineering, University College of Engineering, Burla, as a lecturer and subsequently became a reader in 2001. Presently he is an assistant professor at the department of Electrical Engineering, Indian Institute of Technology, Delhi, India.

He has been honored with many prestigious awards such as INSA young scientist medal-2002, INAE Young Engineer's award-2002, recognition as DST Young Scientist 2001-2002, Carrier Award for young teachers by AICTE, etc. His interests are in soft computing applications to power system control and power quality.

**2006 IEEE International Conference on Power Electronic, Drives and Energy Systems**

# Application of Power Flow Sensitivity Analysis and PTDF for Determination of ATC

## N.D.Ghawghawe, and K.L.Thakre

*Abstract-* **Power system restructuring throughout the world is undergoing various reforms in the power business. One of the objectives of restructuring process is to allow open access of transmission network to the market participants thereby creating a competitive power market. However it may cause the congestion and further threaten the security of the network. This situation may be avoided by providing the information about available transfer capability of the network to the market players (buyers and sellers). Available transfer capability (ATC) of transmission network is the additional amount of power transfer that can be effected between the buyer bus and seller bus without loss of security. For proper system operations under the various transactions, ATC calculations must be reasonably accurate and fast enough. This paper describes a novel approach of application of sensitivity analysis and power transfer distribution factor for the determination of ATC. Advantages of the proposed approach are also discussed.**

*Index Terms*—**Available transfer capability, Deregulation, Network congestion, Power transfer distribution factor, Sensitivity analysis.**

## I. INTRODUCTION

DEREGULATED framework has been replacing the traditional vertically integrated structure of power supply system. One of the objectives of deregulation is to create competitive markets to trade electricity [1]-[4]. So as to supply the loads, under optimal conditions, generation companies with cheaper power are preferred to transmit their power over the transmission network. In doing so, one or more components of transmission system may operate at or beyond their limits thereby causing network congestion and further threatening the security of the network. It has made researchers to work out on various issues related to planning and controlling of transmission infrastructure in the competitive markets of electricity trading. Lot of research work for determination of available transfer capability is carried out in last few years.

In order to have an open access in the restructured power market while satisfying the transmission constraints,

N. D. Ghawghawe is research scholar at VNIT Nagpur and is working as Asst. Professor with Department of Electrical Engg. Govt. College of Engg. Amravati (M.S.) India (e-mail: g_nit@rediffmail.com).

Dr. K. L. Thakre is Professor in Department of Electrical Engineering and is with Visvesvaraya National Institute of Technology Nagpur (M.S), India (e-mail: k_thakre@yahoo.com).

information about the generation capacity and the transmission capability of the system must be made available to the market participants. In the power market, the ISO can check the capability of the transmission paths. A key concept in the restructuring of the electric power industry is the ability to quantify accurately and rapidly the capabilities of the transmission system. Transmission network transfer capability is limited by a number of different mechanisms, including thermal, voltage, and stability constraints [5].

Thus utilities which provide transaction services for wholesale customers, must know about the information on ATC of their transmission networks. Such information will help power marketers, sellers and buyers in reserving transmission services. ATC must be rapidly updated for new capacity reservations, schedules or transactions.

Determination of ATC involves the determination of the existing power flows, which are sensitive to the power transactions between buyer and seller. For every change in power transaction, ATC must be evaluated. Power transfer distribution factor (PTDF) method is used by many utilities for determination of ATC. In these methods, AC load flow methods such as Newton Raphson are used to calculate the change in power flow for the respective change in a transaction. Due to iterative process of load flow methods, the computation time for ATC determination has been observed to be increased.

Objective of this paper is to propose the use of power flow sensitivity analysis for determination of changes in line flows and PTDFs to determine ATC. The method being proposed in this paper has two important advantages over the conventional method of ATC determination which uses AC load flow methods. These are-

i) Improved accuracy and ii) Higher speed of computation

The paper is divided in three parts. i) ATC definitions and its determination ii) Proposed method of ATC determination and iii) Test results

## II. AVAILABLE TRANSFER CAPABILITY (ATC)

### A. Overview

Available Transfer Capability (ATC) is an important term in restructured power system that affects the planning and controlling of transmission infrastructure. ATC of transmission network is the measure of unutilized capacity of the transmission network, which can be made available for further transactions to the market participants without loosing its

0-7803-9771-1/06/$25.00 ©2006 IEEE

security [6]. According to NERC report [7], Available Transfer Capability (ATC) is the amount of the transfer capability remaining in the physical transmission network for further commercial activity over and above already committed uses. As stated by NERC, ATC has to be calculated for the transaction between two buses or areas for specified time duration. Since the system conditions continuously change, ATC must be continuously updated.

Mathematically ATC can be given as,

ATC=TTC-TRM - (CBM+ETC)          (1)

where TTC- Total Transfer Capability

TRM-Transmission Reliability Margin

ETC-Existing Transmission Commitments

CBM-Capacity Benefit Margin

Graphically these values are represented in Fig. 1.


B.  *Methods of ATC Determination:*

Various mathematical models have been developed by the researchers to determine the ATC of the transmission system.

ETSO (European Transmission System Operators) has specified the procedure for TTC computation between a pair of neighboring control areas A and B. Starting from a base case exchange (BCE) and by means of a generation increase in area A and decrease in B, the power flow rises until insecurity is reached. The TTC from A to B is equal to sum of BCE and the maximum increase [8]-[9].

CPFLOW [10] is another tool available for the determination of TTC (or ATC). It utilizes a continuation power flow algorithm for the calculation of maximum loadability of electric power system. As reported in the paper, accuracy of results is obtained with negligible computational time.

Recently the methods of Power Transfer Distribution Factor (PTDF) using DC power flow and AC power flow are derived to calculate ATC. In DCPTDF method [11], DC load flow i.e. a linear model, is considered. This method is fast but does not provide the accuracy.

Ashvinikumar et. al. [12] have used ACPTDF for determination of ATC of a practical system case. It considers the determination of power transfer distribution factors, computed at a base case load flow using sensitivity properties of NRLF Jacobian. Similar method with CEED approach has been used by and Venkatesh et. al.[13]. In a method proposed by the researchers [14], linear approximation is considered for calculating PTDFs and ATCs of the transmission network. TTC evaluation by modified Jacobian matrix is proposed by researchers [15].

NERC has proposed the standard method for evaluation of ATC using PTDFs. An algorithm for this method is given below-

i) Compute initial system conditions (base case) such as, bus voltages, bus angles, power flows and current flows using a load flow method.
ii) Apply the real power transaction between source and sink bus
iii) Again run the load flow and obtain the change in real power flows for the applied transaction
iv) Determine the PTDFs for this transaction and calculate Total Transfer Capability (TTC) at base condition.
v) Finally evaluate ATC by subtracting ETC, TRM and CBM from TTC.

In this paper, the change in power flows are directly determined by sensitivity analysis from which PTDFs and ATC are evaluated.

### III.  ATC DETERMINATION USING SENSITIVITY ANALYSIS AND PTDF

A.  *Sensitivity Analysis-*

Earlier research on the application of sensitivity analysis in power system belongs to Peschon, Piercy, Tinney and Tveit [16]. They introduced two methods. First one can be applicable to normal power flow problems for small changes in the variables such as active generation and second method considers the minimization of objective function satisfying some constraints such as power flow equation. Similar research was carried out by Thanikchalam[17], Dillon [18], Gribik et al [19] and Talaq J. H. [20].

*1)  Mathematical Formulation-*

The method being proposed, uses PTDFs as determined in the standard method suggested by NERC except for the fact that the change in the line flows are determined by sensitivity analysis. Instead of repeated power flow to determine $\Delta P_{ij\,(mn)}$ as described in ACPTDF, it is determined directly by the sensitivity analysis in the proposed method. Also, sensitivities of voltages and currents are determined simultaneously, which are further used to determine the changes in power flows.

Considering the generalized equations of the form-

$$g(\mathbf{x},\mathbf{u},\mathbf{p}) = 0 \qquad (2)$$

where $\mathbf{g}$ is 2N dimensional vector

and N is number of buses

The variables mentioned in equation (2) can be categorized as-

i) Dependent Variables (x): These are the controlled variables and are unknown. $\mathbf{x}$ is a 2N dimensional vector.
ii) Independent Variables (u): These are the operating

variables or imposed variables of the system. **u** is an M dimensional vector.

iii) Parameter Variables (p): These are uncontrollable variables and are normally specified in the power flow problem.

Depending upon the variables to be determined, the variables in the power flow problem can be selected as x, u and p. One might be interested in controlling K variables out of the 2N variables.

If, $\mathbf{x_0}$, $\mathbf{u_0}$ and $\mathbf{p_0}$ are the initial state vectors, rewriting Equation (2) as-

$$g(\mathbf{x}_0,\mathbf{u}_0,\mathbf{p}_0)=0 \qquad (3)$$

The changes $\Delta\mathbf{x}$ corresponding to small changes $\Delta\mathbf{u}$ and $\Delta\mathbf{p}$, will satisfy the new equations,

$$g(\mathbf{x}_0+\Delta\mathbf{x},\mathbf{u}_0+\Delta\mathbf{u},\mathbf{p}_0+\Delta\mathbf{p})=0 \qquad (4)$$

Expanding (4) by Taylor's series and neglecting higher order terms,

$$g(\mathbf{x}_0+\Delta\mathbf{x},\mathbf{u}_0+\Delta\mathbf{u},\mathbf{p}_0+\Delta\mathbf{p})=g(\mathbf{x}_0,\mathbf{u}_0,\mathbf{p}_0)+\mathbf{g}_x\Delta\mathbf{x}+\mathbf{g}_u\Delta\mathbf{u}+\mathbf{g}_p\Delta\mathbf{p}$$
$$(5)$$

where, $\mathbf{g}_x$, $\mathbf{g}_u$ and $\mathbf{g}_p$ are the partial derivatives of g w.r.t. x, u and p respectively and are given by,

$$\mathbf{g_x} = \frac{\partial(g_1,g_2,g_3.....g_{2N})}{\partial(x_1,x_2,x_3.....x_{2N})}$$

$x_1,x_2,x_3.....x_{2N}$ are the elements of **x**

$$\mathbf{g_u} = \frac{\partial(g_1,g_2,g_3.....g_{2N})}{\partial(u_1,u_2,u_3.....u_M)}$$

$u_1,u_2,u_3.....u_M$ are the elements of **u**

$$\mathbf{g_p} = \frac{\partial(g_1,g_2,g_3.....g_{2N})}{\partial(p_1,p_2,p_3.....p_{2N})}$$

$p_1,p_2,p_3.....p_{2N}$ are the elements of **p**

When changes are small, solution for $\Delta\mathbf{x}$ will be,

$$\Delta\mathbf{x} = \mathbf{S}_u\Delta\mathbf{u} + \mathbf{S}_p\Delta\mathbf{p} \qquad (6)$$

where $\mathbf{S}_u$ and $\mathbf{S}_p$ are the sensitivities of x w.r.t. u and p respectively and are obtained as-

$$\mathbf{S}_u = -\mathbf{g}_x^{-1}\mathbf{g}_u \qquad (7)$$

$$\mathbf{S}_p = -\mathbf{g}_x^{-1}\mathbf{g}_p \qquad (8)$$

If p variables are not changed then (6) can be re-written as,

$$\Delta\mathbf{x}=\mathbf{S}_u\Delta\mathbf{u} \qquad (9)$$

The set of dependent and independent variables can be chosen as per the system requirement and problem formulation. Some of the parameters of a type may belong to the set of dependent whereas remaining parameters of same type may belong to the set of independent variables.

*2) Determination of Voltage Sensitivities at buses -*

Power flow equations are comprising of 6 variables namely P, Q, V, $\theta$, Y and $\phi$. All the variables can be assumed to be obtained or specified at the base condition. The variables Y and $\phi$ are normally specified and are constant. The other variables are not always constant and they are either specified or determined, depending upon the type of buses. The variables for which changes are specified are grouped as independent variables and the variables which are determined against these changes are grouped as dependent variables.

For the slack bus, V-$\theta$ are specified and P-Q are subjected to change. For generator bus, P-V are specified and Q-$\theta$ are subjected to change. For load buses, P-Q are specified and V-$\theta$ are changed.

Now consider the power system of N buses and B branches. Power flow equations can be described by-

$$P_i = V_i\sum_{j=1}^{N}V_jY_{ij}\cos(\theta_i-\theta_j-\phi_{ij}) \qquad (10)$$

$$Q_i = V_i\sum_{j=1}^{N}V_jY_{ij}\sin(\theta_i-\theta_j-\phi_{ij}) \qquad (11)$$

$i \in$ n

where,

n=$N_s$ ......for slack bus
n=$N_g$ ....for generator bus
n=$N_l$ .......for load bus
s,g and l-slack, generator and load bus number representation
G - number of generator buses
L - number of load buses
$N_s$ - Slack bus number ($N_s$=1)
$N_g$ - Generator bus numbers ($N_g$=2,3.....G+1)
$N_l$ - Load bus numbers ($N_l$=G+2,G+3.....N)
$P_i = P_{Gi}- P_{Li}$ and $Q_i = Q_{Gi}-Q_{Li}$
$P_{Gi}$ - active power generated at node i
$P_{Li}$ - active power consumed by the load at i
$Q_{Gi}$ - reactive power generated at node i
$Q_{Li}$ - reactive power consumed by the load at i
$V_i$ - voltage at node i
$V_j$ - voltage at node j
$\theta_i$ - phase angle at node i with $\theta$i arbitrarily set equal to zero.
$\theta_j$ - phase angle at node j with $\theta$j arbitrarily set equal to zero

$Y_{ij}$ - magnitude of elements of node admittance matrix

$\phi_{ij}$ - angles of elements of node admittance matrix

For this system, 2N equations will be formed and are arranged in the form as given in (2).

A set of 2N variables can be selected as dependent variables ($\mathbf{x}$) and remaining as independent variables ($\mathbf{u}$). Consider that only M independent variables are changed and for these changes, it is desired to obtain the changes in the real and reactive power at slack buses, reactive power and angles at generator buses and voltages and angles at load buses.

Equation (10) and (11) can be written in the form,

$$g(V_i, V_j, \theta_i, \theta_j, P_i, Q_i, Y_{ij}, \phi_{ij}) = 0 \qquad (12)$$

Let,

$P_s, P_g, P_l \in P_i$
$Q_s, Q_g, Q_l \in Q_i$
$V_s, V_g, V_l \in V_i$
$\theta_s, \theta_g, \theta_l \in \theta_i$

Grouping the variables of (12) as,

x=$P_s$, $Q_s$, $Q_g$, $\theta_g$, $V_l$, $\theta_l$
u=$V_s, \theta_s, P_g, V_g, P_l, Q_l$
p=$Y_{ij}$, $\phi_{ij}$

From (9), the changes in dependent variables can be obtained.

$$\left[ \Delta P_s, \Delta Q_s, \Delta Q_g, \Delta \theta_g, \Delta V_l, \Delta \theta_l \right] = \left[ \mathbf{S} \left[ \Delta V_s, \Delta \theta_s, \Delta P_g, \Delta V_g, \Delta P_l, \Delta Q_l \right] \right.$$
(13)

where $\mathbf{S}$ is the sensitivity matrix of order 2Nx2N and can be obtained as given by (7).

For slack bus and generator buses following substitution can be made in (13),

$$\Delta V_s = \Delta V_g = \Delta \theta_s = 0 \qquad (14)$$

After determining the changes in the load bus voltages, load bus angles and generator bus angles from (13) and with the substitutions from (14) all the bus voltages and angles can be arranged as,

$$\left[ \mathbf{\Delta V}, \mathbf{\Delta \theta} \right] = \left[ \Delta V_s, \Delta V_g, \Delta V_l, \Delta \theta_s, \Delta \theta_g, \Delta \theta_l \right] \qquad (15)$$

*3) Determination of Current Sensitivities in the Lines-*

It is well known that the changes in voltage angles and voltage magnitudes are related to branch currents $I_{ij}$. These currents in complex form can be expressed as,

$$I_{ij} = Y_{ij}(V_i(\cos \theta_i + i \sin \theta_i) - V_j(\cos \theta_j + i \sin \theta_j)) \qquad (16)$$

where,

$$\mathbf{Y_{ij}} = Y_{ij} \angle \phi_{ij}$$

and $I_{ij} \in B$

Equation (16) can be written in the form,

$$g_{ij}\left(I_{ij}, Y_{ij}, \phi_{ij}, V_i, V_j, \theta_i, \theta_j \right) = 0 \qquad (17)$$

Grouping the variables of (17) as,

x=$I_{ij}$
u= $V_i$, $V_j$, $\theta_i$, $\theta_j$ (i.e. V-θ variables at all buses)
p=$Y_{ij}$, $\phi_{ij}$

Sensitivities of $I_{ij}$ for the changes in $V_i$, $V_j$, $\theta_i$, $\theta_j$ can be obtained from (9) as,

$$\left[ \Delta I_{ij} \right] = \left[ \mathbf{R} \right]\left[ \mathbf{\Delta V}, \mathbf{\Delta \theta} \right] \qquad (18)$$

where R is sensitivity matrix obtained by (7) which is given as,

$$\mathbf{R} = -g_{ijx}^{-1} g_{iju}$$

with

$g_{ijx}$- Jacobian of $g_{ij}$ w.r.t. x (i.e. $I_{ij}$)
$g_{iju}$-Jacobian of $g_{ij}$ w.r.t. u (i.e. $V_i$, $V_j$, $\theta_i$, $\theta_j$)

Substituting from (15), (18) can be rewritten as,

$$\left[ \Delta I_{ij} \right]_{Bx1} = \left[ \mathbf{R} \right]_{Bx2N}\left[ \Delta V_s, \Delta V_g, \Delta V_l, \Delta \theta_s, \Delta \theta_g, \Delta \theta_l \right]_{2Nx1} \qquad (19)$$

where, $\Delta V_s = \Delta V_g = \Delta \theta_s = 0$

Thus for a change in power ($\Delta P_k$) at generator and load buses, corresponding changes in bus voltages and line currents can be obtained using (15) and (19) respectively. The corresponding change in power flow $\Delta P_{ij}$ over the line i-j can be determined thereafter.

*B. Power Transfer Distribution Factors (PTDFs)-*

PTDFs are the sensitivity factors of various components of power system network for specified changes in bus powers. Consider a real power transaction is carried out between the buses m and n after the base power flow is obtained. Let m be the source bus and n be the sink bus. For this change in bus powers at m and n, the changes in the line power flows can be obtained as explained above.

Power transfer distribution factors of i-j elements, for the transaction between m-n will be given as,

$$(PTDF)_{ij(mn)} = \frac{\Delta P_{ij}}{\Delta T_{mn}} \qquad (20)$$

where,

988

$\Delta T_{mn}$ - Change in transaction between m and n.

$\Delta P_{ij(mn)}$ -Change in real power flow of line i-j for transaction between m-n, obtained by sensitivity analysis.

For the power transfer of $\Delta T_{mn}$ from the bus m to bus n, the entries in $\Delta P_g$ and $\Delta P_l$ of (13) will be zero with the exception of the following,

$$\Delta P_m = \Delta T_{mn} \text{ and } \Delta P_n = -\Delta T_{mn} \quad (m, n \neq 1)$$

### C. Available Transfer Capability-

For a particular transaction, ETC, TRM and CBM are specified. Hence TTC represents Available Transfer Capability (ATC).

The mathematical formulation of Total Transfer Capability (TTC) or ATC at base case, between bus m and n using line flow limit criterion, is stated as-

$$TTC_{mn} = \min\left( \frac{P_{ij}^{\max} - P_{ij}^0}{PTDF_{ij(mn)}}, ij \in B \right) \quad (21)$$

where,

$P_{ij}^{\max}$ - MW power limit (thermal limit) of a line between i-j

$P_{ij}^0$ - Base case power flow in the line between bus i- j

$PTDF_{ij(mn)}$ - Power Transfer Distribution Factor for line i-j

for the real power transaction between the buses m and n.

$B$ - Total number of branches.

After considering the Existing Transfer Commitments (ETC), ATC will be calculated as,

$$ATC = TTC - ETC \quad (22)$$

Subtracting TRM and CBM, final value of ATC will be,

$$ATC = TTC - TRM - (ETC + CBM) \quad (23)$$

## IV. CASE STUDY

The proposed method is applied to a 6-bus system (Fig.5). For this system, ATCs are obtained for various transactions. The ATC values are also determined by conventional NRLF method and PWS 8.0. Bus 1 is the slack bus, 2 and 3 are the generator buses. Bus 1 is slack bus. 2 and 3 are generator buses. 4, 5 and 6 are the load buses. Real power transactions are carried out between 2-5, 2-4, 2-6, 3-5 and 3-6. For these transactions, corresponding change in load bus voltages, line currents and power flows are determined and new operating values are obtained by the sensitivity analysis. These values are used to determine the PTDFs and further ATC. Transactions with different power transfers are also applied between the same bus pair 2-5.

### A. Algorithm

Following algorithm can be carried out for the determination of ATC of the network.

i) Compute initial system conditions (base case) such as, bus voltages, bus angles, power flows and current flows using Newton-Raphson Load Flow method.

ii) Apply the real power transaction between m and n.

iii) Obtain the change in real power flows by carrying out sensitivity analysis as described in III

iv) Determine the PTDFs and calculate Available Transfer Capability (TTC) at base condition using (21).

v) Finally evaluate ATC by subtracting ETC from TTC. Updating the values of power flows after the transaction is applied and then substituting them in place of base power flow in (21) ATC can be evaluated directly.

## V. RESULT

We have developed the program SENSATC in MATLAB. SENSATC is a generalized program, which can compute the line flows, change in line flows, PTDFs and ATC using sensitivity analysis for various transactions. The program also features the determination of ATC quantities for any number of transactions applied across the bus pair. For this determination, program has to be run only once. To compare the results as obtained by SENSATC, another program named as NRATC is developed which can compute these quantities using conventional Newton Raphson AC load flow method. To calculate ATC values for various transactions using NRATC program, it has to be run every time. Thus computation speed of the proposed method of sensitivity analysis can be justified over the use of conventional load flow method of calculating ATC values. Similar results are obtained using Power World Simulator 8.0. Comparisons of results obtained by various methods for the test system are given in Table-II to IV. Variation of PTDFs for a line L2-4, in graphical form by various methods is shown in Fig.2. Fig.3 represents the transfer capabilities of lines whose TC represents ATC of the system for the respective transaction. Fig.4 represents the ATC of the system for transactions T1 to T5, using three different methods. The program is also tested on IEEE 30-bus system.

From tables II to IV it is observed that, the PTDFs, line TCs and ATC values as obtained by different methods are close to each other. The method (SAP) being proposed by us, involves the use of voltage and current sensitivities to determine the change in power flows. As reported by researchers [11-14], sensitivity analysis gives the reasonable accuracy and speed of computation in power system applications. This fact can be justified with the proposed method in which the improved results of ATC are obtained. It is also verified that for the transactions applied across the same bus pair, line PTDFs as obtained by sensitivity analysis are found to be almost equal. Which when are obtained by Power World Simulator or conventional NR load flow method, are seen to be little deviated. Due to additional programming feature, for such transactions applied in steps, the quantities such as PTDF and ATC can be evaluated only once thereby reducing the overall time of evaluation. The proposed method not only gives the improved results but also sorts the lines, most sensitive to the power transactions. The aim of this paper is to use sensitivity analysis to determine PTDF and further to evaluate ATC. Comparison of ATC evaluated using different methods is mentioned in Table-IV. The study is restricted, by considering only a thermal limit. However voltage and stability limits can be included by changing the algorithm and modifications in the program.

## VI. CONCLUSION

Major obstacle in the deregulated environment, is the congestion of transmission network. It can be avoided by providing the exact information of ATC to the market participants. In case of smaller ATC, the system operators can charge higher transmission prices for further transactions. These charges may be further utilized for improvement of transmission infrastructure thereby improving ATC. This may include the use of advanced techniques, such as the installation of FACTS devices. However it needs an accurate determination of ATC and the amount by which transmission capability to be improved. The most sensitive lines are also required to be sorted out. This is possible with the help of the proposed method and the program SENSATC

Another important feature of the proposed method is its application in planning stage and real time controlling of transmission network.

The program can be further used to determine the transmission capability to be improved in case of network congestion and to estimate the modifications required in system parameters.

## VII. ABBREVIATIONS

SAP-Proposed Sensitivity Analysis and PTDF
CNR-Conventional Newton Raphson AC Load Flow
PWS-Power World Simulator 8.0

## VIII. TABLES

TABLE I
BILATERAL TRANSACTIONS

| Transaction No. | Source Bus (Seller) | Sink Bus (Buyer) | Transaction Amount (MW) |
|---|---|---|---|
| T1 | 2 | 5 | 10 |
| T2 | 2 | 4 | 8 |
| T3 | 2 | 6 | 15 |
| T4 | 3 | 5 | 10 |
| T5 | 3 | 6 | 5 |

TABLE II
POWER TRANSFER DISTRIBUTION FACTORS

| Line | Method | Transaction No. | | | | |
|---|---|---|---|---|---|---|
| | | T1 | T2 | T3 | T4 | T5 |
| Line 1-2 | SAP | -0.1114 | -0.13 | -0.045 | -0.0568 | 0.0096 |
| | CNR | -0.1104 | -0.13 | -0.043 | -0.0557 | 0.0099 |
| | PWS | -0.108 | -0.129 | -0.042 | -0.054 | 0.006 |
| Line 1-4 | SAP | -0.0259 | 0.2161 | -0.011 | -0.0073 | 0.0077 |
| | CNR | -0.025 | 0.2172 | -0.01 | -0.0062 | 0.0079 |
| | PWS | -0.025 | 0.2063 | -0.009 | -0.006 | 0.006 |
| Line 1-5 | SAP | 0.2271 | -0.03 | 0.091 | 0.1411 | 0.0052 |
| | CNR | 0.2284 | -0.029 | 0.093 | 0.1428 | 0.0055 |
| | PWS | 0.22 | -0.03 | 0.089 | 0.139 | 0.004 |
| Line 2-3 | SAP | 0.174 | 0.0452 | 0.263 | -0.2454 | -0.1568 |
| | CNR | 0.1704 | 0.0404 | 0.262 | -0.247 | -0.1562 |
| | PWS | 0.174 | 0.0413 | 0.265 | -0.246 | -0.158 |
| Line 2-4 | SAP | 0.1985 | 0.684 | 0.079 | 0.1133 | -0.0066 |
| | CNR | 0.2036 | 0.6911 | 0.08 | 0.1156 | -0.0073 |
| | PWS | 0.201 | 0.6963 | 0.078 | 0.115 | -0.006 |
| Line 2-5 | SAP | 0.307 | 0.0744 | 0.126 | 0.1773 | -0.0038 |
| | CNR | 0.3068 | 0.0736 | 0.126 | 0.1776 | -0.0038 |
| | PWS | 0.31 | 0.0725 | 0.126 | 0.18 | -0.004 |
| Line 2-6 | SAP | 0.1965 | 0.0514 | 0.483 | -0.1084 | 0.1779 |
| | CNR | 0.192 | 0.0455 | 0.482 | -0.1103 | 0.1787 |
| | PWS | 0.195 | 0.045 | 0.485 | -0.109 | 0.178 |
| Line 3-5 | SAP | 0.1897 | 0.0412 | -0.077 | 0.3893 | 0.1222 |
| | CNR | 0.1882 | 0.0442 | -0.085 | 0.4001 | 0.1271 |
| | PWS | 0.191 | 0.0438 | -0.084 | 0.4 | 0.126 |
| Line 3-6 | SAP | -0.0202 | -0.002 | 0.34 | 0.361 | 0.7206 |
| | CNR | -0.019 | -0.004 | 0.345 | 0.3542 | 0.7176 |
| | PWS | -0.019 | -0.004 | 0.345 | 0.354 | 0.718 |
| Line 4-5 | SAP | 0.1709 | -0.115 | 0.067 | 0.1051 | 0.001 |
| | CNR | 0.1659 | -0.123 | 0.065 | 0.1027 | 0.0014 |
| | PWS | 0.165 | -0.124 | 0.065 | 0.103 | 0.0016 |
| Line 5-6 | SAP | -0.1695 | -0.042 | 0.196 | -0.2578 | 0.1118 |
| | CNR | -0.1593 | -0.038 | 0.193 | -0.2326 | 0.119 |
| | PWS | -0.161 | -0.037 | 0.191 | -0.233 | 0.118 |

TABLE III
TRANSFER CAPABILITIES

| Line | Method | Transaction No. | | | | |
|---|---|---|---|---|---|---|
| | | T1 | T2 | T3 | T4 | T5 |
| Line 1-2 | SAP | -2136.2 | -1824.4 | -5272.2 | -4179.0 | 24690.0 |
| | CNR | -2154.1 | -1834.4 | -5461.9 | -4261.4 | 23946.0 |
| | PWS | -2221.7 | -1863.2 | -5702.1 | -4433.3 | 39805.0 |
| Line 1-4 | SAP | -9226.5 | 1093.1 | -21825.0 | -32445.0 | 31033.0 |
| | CNR | -9535.1 | 1087.4 | -24889.0 | -38112.0 | 30027.0 |
| | PWS | -9493.2 | 1141.5 | -27370.4 | -39523.3 | 39508.3 |
| Line 1-5 | SAP | 531.5 | -4132.6 | 1334.6 | 862.5 | 23582.0 |
| | CNR | 529.0 | -4195.1 | 1314.2 | 852.4 | 22419.0 |
| | PWS | 549.7 | -4112.3 | 1363.3 | 875.8 | 30777.5 |
| Line 2-3 | SAP | 509.2 | 1991.6 | 329.4 | -378.6 | -582.1 |
| | CNR | 520.9 | 2229.9 | 330.2 | -376.2 | -584.0 |
| | PWS | 510.1 | 2185.9 | 326.9 | -377.9 | -577.8 |
| Line 2-4 | SAP | 464.9 | 130.0 | 1187.0 | 822.8 | -14201.0 |
| | CNR | 453.6 | 128.6 | 1169.4 | 806.2 | -12916.0 |
| | PWS | 465.1 | 129.1 | 1209.2 | 820.3 | -15920.0 |
| Line 2-5 | SAP | 181.4 | 781.2 | 451.3 | 321.1 | -15260.0 |
| | CNR | 181.3 | 789.0 | 451.8 | 320.5 | -15544.0 |
| | PWS | 180.7 | 807.4 | 454.2 | 318.4 | -14785.0 |
| Line 2-6 | SAP | 739.2 | 2861.5 | 290.3 | -1370.2 | 823.6 |
| | CNR | 757.8 | 3233.6 | 290.6 | -1345.9 | 819.8 |
| | PWS | 745.9 | 3267.8 | 289.1 | -1362.4 | 823.1 |
| Line 3-5 | SAP | 870.9 | 4054.7 | -2175.8 | 419.6 | 1364.0 |
| | CNR | 879.0 | 3777.0 | -1979.7 | 408.1 | 1310.8 |
| | PWS | 864.8 | 3811.0 | -2004.0 | 407.7 | 1321.0 |
| Line 3-6 | SAP | -4112.4 | -40147.0 | 230.5 | 220.9 | 110.6 |
| | CNR | -4400.5 | -20671.0 | 226.5 | 225.3 | 111.1 |
| | PWS | -4407.4 | -22288.0 | 226.9 | 226.0 | 111.4 |
| Line 4-5 | SAP | 485.1 | -743.1 | 1256.7 | 796.5 | 82078.0 |
| | CNR | 501.0 | -695.2 | 1287.3 | 815.3 | 60244.0 |
| | PWS | 503.0 | -692.0 | 1293.9 | 811.7 | 52895.0 |
| Line 5-6 | SAP | -1002.6 | -4030.0 | 838.8 | -660.2 | 1493.9 |
| | CNR | -1062.1 | -4396.5 | 851.7 | -730.7 | 1403.1 |
| | PWS | -1050.0 | -4473.1 | 860.1 | -728.6 | 1414.0 |

## TABLE IV
### AVAILABLE TRANSFER CAPABILITY

| Method | Transaction No. | | | | |
|--------|------|------|------|------|------|
|        | T1 | T2 | T3 | T4 | T5 |
| SAP | 181.4 | 129.99 | 230.5 | 220.849 | 110.64 |
| CNR | 181.327 | 128.58 | 226.5 | 225.283 | 111.13 |
| PWS | 180.71 | 129.15 | 226.9 | 226.017 | 111.36 |

## IX. FIGURES

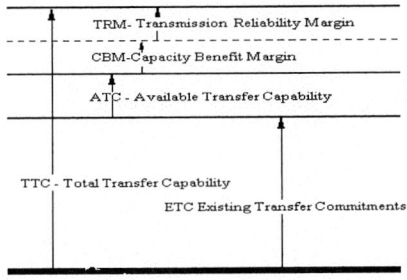

Fig. 1. Graphical Representation of ATC and other associated terms.

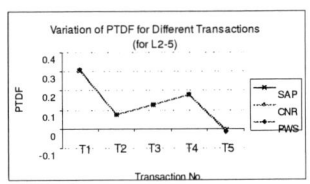

Fig. 2. Variation of PTDF for Different Transactions (for L2-5).

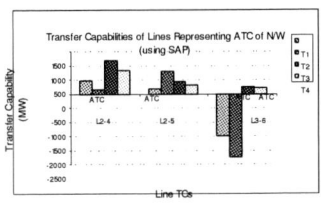

Fig. 3. Transfer capabilities of L2-4, L2-5,L3-6 (using SAP).

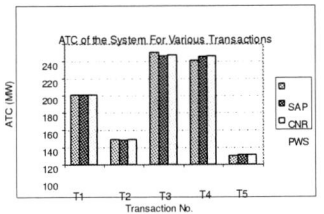

Fig. 4. Available Transfer Capability.

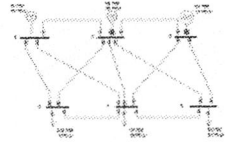

Fig. 5.   6-Bus Test System.

## X. REFERENCES

[1] Hugh Rudnick, "Planning in a deregulated environment in developing countries : Bolivia, Chile, Peru", *IEEE Power Engineering Review* 16 (7) (1996) 18-19.

[2] R.D.Tabors, "Lessons from the UK and Norway", *IEEE Spectrum* 33(8)(1996) 45-49.

[3] A.Srivastava, and M.Shahidehpour, "Restructuring choices for the Indian power sector", *IEEE Power Engineering Review* (2002) 25-29.

[4] K. Bhattacharya, M.H.J. Bollen, and Dalder J.E. , "*Operation of Restructured Power Systems*" Kluwer Academic Publishers, Boston, 2001.

[5] Peter W. Sauer , "Technical Challenges of Computing Available Transfer Capability (ATC) in Electric Power Systems" *Proceedings, 30th Annual Hawai International Conference on System Sciences*, Jan. 7-10,1997 pp.1-5

[6] S.C.Srivastava, "Transmission System Management in Restructured Electricity Markets" *Proceedings of International Tutorial and Seminar on Power Transmission Research Interests and Challenges organized by Central Power Research Institute, Banglore, India.* Dec. 2005

[7] North American Electric Reliability Council (NERC), "Available Transfer Capability Definitions and Determination", NERC Report, June 1996.

[8] ETSO-"Definitions of Transfer Capabilities in Liberalised Electricity Markets" final report April 2001.

[9] M. Shahidehpour, and M.Almoush "*Restructured Electric Power System,*" Marcel- Deccer Publishers pp.158-159.

[10] H.D.Chiang, A.J.Fluek, K.S.Shah, and N.Balu "CPFLOW: a practical tool for tracing power system steady state stationery behavior due to load and generation variations", *IEEE Transactions on Power Systems*, Vol.10 No.2, May 95.

[11] R.Christie, B.F.Woolenberg, and I.Wangestin, "Transmission Management in Deregulated Environment" *IEEE Proceedings* 88(2000) 170-195.

[12] Ashwani kumar, S.C.Srivastava, and S.N.Singh, "ATC determination in a competitive electricity market using AC distribution factors", *Electric Power Components and Systems* 32 (2004) 927-939

[13] Venkatesh P., Gnanadass R., and Padhy Narayan Prasad, "Available Transfer Capability Determination Using Power Distribution Factors" Journal of Emerging Electric Power Systems Vol. 1(2) 2004, article1009.

[14] C.Ejebe Gabriel, G.Waight James, Santos-Nieto, and F.Tinney William, "Fast calculation of linear available capability", *IEEE Transactions on Power Systems* 15 (3) (2000) 955-960.

[15] G.Sombuttwilailert, and B. Eua-Arporn, "A Novel Sensitivity Analysis for Total Transfer Capability Evaluation", *Proceedings 22nd IEEE Power Engineering Society International Conference on Power Industry Computer Applications (PICA –2001)* 20-24 May 2001 pp 342-347.

[16] Peschon J., Piercy D.S., Tinney W.F., and Tveit O.J. "Sensitivity in Power Systems" *IEEE Trans. on PAS* Vol. 87, No. 8 pp. 1687-1696, August 1968.

[17] Thanikachalam A., and Tudor J. R. "Optimal Rescheduling of Power for System Reliability" *IEEE Trans. on PAS*, Vol.90, pp. 2186-2192. 1971.

[18] Dillon T.S., and "Rescheduling, Constrained Participation Factors and Parameter Sensitivity in the Optimal Power Flow Problem", *IEEE Trans. on Power Apparatus and Systems*, Vol. PAS- 100 no. 5, pp. 2628-2634. 1981.

[19] Gribik, Shirmohammadi D., Hao S., Thomas C.L. "Optimal Power Flow Sensitivity Analysis" IEEE Winter meeting, Atlanta, Georgia, USA 1990, pp. 969-976.

[20] Talaq J.H., Ferial, and M.E. EI Hawary, "A Sensitivity Analysis Approach to Minimum Emissions Power Flow" *IEEE Transactions on Power Systems*, Vol. 9 No.1, pp. 436-442. Feb1994

**2006 IEEE International Conference on Power Electronic, Drives and Energy Systems**

# Application of Tabu-Search Algorithm for Network Reconfiguration in Radial Distribution System

T. Thakur, *Member, IEEE,* and Jaswanti, *Member, IEEE*

*Abstract* ---This paper presents the application of Tabu Search (TS) as Meta-heuristic Method for network reconfiguration problems in radial distribution system (RDS). This work has been tested on 33-bus RDS with five tie lines. The main advantage of TS with respect to conventional Genetic algorithm and Simulation annealing lies in the intelligent use of the past history of the search to influence its future search procedures. The effectiveness of the proposed method is demonstrated on typical network reconfiguration problems. The results reveal the speed and effectiveness of the proposed method for solving the problem.

*Keywords*---Combinatorial optimization, Loss minimization, Tabu search.

## I. INTRODUCTION

ABOUT 30 to 40% of total investment in the electrical sector goes to distribution systems, but nevertheless, they have not received the technological impact in the same manner as the generation and transmission systems. Many of the distribution networks work with minimum monitoring systems, mainly with local and manual control of the capacitors, sectionalizing switches and voltage regulators; and without adequate computational support for the system's operators [1]. Nevertheless, there is an increasing trend to automate distribution systems to improve their efficiency and service reliability [2]. Automation is possible due to the advance microprocessor control technology, to its increasing cost reduction and due to its joint use with telecommunications technologies. With the aid of these technologies, it is possible to monitor substations and feeders to reconfigure feeders and to control voltage, reactive power and loss [3]-[4].

Considerable researches have been carried out in power distribution network configuration since 1975 and are reported through research papers; their approaches are broadly classified under following three categories:
1. Classical methods combined with heuristic, where we find branch-and-bound and feeder-pair (loop) quadratic programming decomposition methods.

2. Heuristic based methods, where we find papers in branch-exchange; application of the compensation-based power flow technique to reduce Merlin and Back central idea; and more recently, used for other practical operational constraints like protection requirement and limited number of switching operations.
3. Modern heuristic methods using artificial intelligence, where we find papers in artificial neural network (ANN), expert systems, simulated annealing (SA), evolutionary programming, and genetic algorithm (GA).

Recently, it is found that modern heuristic methods such as SA [6], GA [7]-[9] and Tabu Search (TS) [10] can be used for large combinational optimization problem of distribution system. TS belong to a family of methods, which also include SA, and GA. TS explores the whole solution space definitely based on the local search in which controlled up-hill move is admitted.

The roots of the TS go back to the 1970's; it was first presented in its presentable form by Glover [11] later it was formalized by him; the basic idea was sketched by Hansen. Up to now, TS is a strategy with more functions for solving combinatorial optimization problem and is applied to various fields to obtain high quality solutions within reasonable computing time. The TS method is built upon a descent mechanism of a search process, which biases the search toward, points with lower objective function values [12]-[14]. However special features can also be added to avoid being trapped in the local minima. Basic component requirement to implement the TS are: Moves and Selection, Tabu List Aspiration Criterion Intensification and Diversification. A large number of papers has been published so for on tabu search algorithm for various combinatorial optimization solution [15]-[18].

This paper presents the application of TS as meta-heuristic method for network reconfiguration problems in radial distribution system. This work has been tested on 33-bus RDS. System has five tie lines. The main advantage of TS with respect to conventional Genetic algorithm and Simulation annealing lies in the intelligent use of the past history of the search to influence its future search procedures.

## II. PROBLEM DESCRIPTION

Rather than using a single configuration, as in more conventional algorithms, a family of concurrent configurations of the network is kept throughout the optimization process of RDS. This allows for a more comprehensive search on the

---

Dr. T.Thakur and Jaswanti are with the Department of Electrical Engineering, Punjab Engineering College (Deemed University), Chandigarh, India –160012.
(e-mail: tilak20042005@yahoo.co.inT; jaswanti98@yahoo.co.in)

0-7803-9771-1/06/$25.00 ©2006 IEEE

space of configurations. Although the process could be actually mapped on a parallel machine, this is not necessarily so, and serial machine can be used as well.

Initially, a family of configurations is obtained with a modified Garver's algorithm; the idea is to obtain a first configuration using the algorithm as is, and then turning tabu active (i.e., forbidden) some essential attribute (e.g., line additions or removal) of this configuration leading to other configuration belonging to different regions of the search space. Garver's algorithm uses a transportation model (Linear Programming), i.e. a model in which only Kirchhoff's current law is taken into account. When this is done the algorithm, although approximate, is able to pinpoint the main transmission trunks of the expanded network. These configurations normally contain the most interesting attributes which will eventually be found in optimal solutions which makes Garver's approach an extremely handy intializer for the optimal search process. Although the proposed approach, as well as GA and SA algorithms in general, can be randomly initialized, Garver's initialization has proved to be final solution quality.

### III. TABU SEARCH METHOD

Tabu Search is a strategy with more functions for solving combinatorial optimization problem [15] and is applied to various fields to obtain high quality solutions within reasonable computing time. The Tabu Search method is built upon a descent mechanism of a search process, which biases the search toward, points with lower objective function values. However special features can also be added to avoid being trapped in the local minima. Basic features needed to implement the Tabu Search are as follows:

#### A. Moves and Selection
In the network reconfiguration problem, the process of searching from a vector of trial solution to the other is called a 'Move'. A vector of trial solution denotes the candidate feeder's switches, their position, power losses, the voltage magnitude, and the objective function value corresponding to this Move, and the frequency counter. The frequency counter is used to indicate total times that the solution has been visited. The solution structure is represented as:

Structure Solution {feeders switches, their position, open / close condition, power losses, voltage magnitude, objective function value, frequency counter}.

#### B. Tabu List
In order to avoid returning to the local optimum just visited, the reverse move that is detrimental to achieving the optimum solution must be forbidden. This is done by storing this move in a data structure, such as a finite length first-in-first-out structure, called Tabu list. The elements of Tabu list are called 'Tabu Moves'. With the help of Tabu Moves, we can keep the search bias toward point with lower objective function values and escape from local optimum solution.

#### C. Aspiration Criterion
Since, the Tabu List may forbid certain worthy or interesting moves possibly leading to a better solution than the best one found so far. An aspiration criterion is used to allow Tabu Moves to be released if they are judged to be worth or interesting. In other words, the aspiration criterion is to allow "excellent" Tabu moves to be selected if the aspiration level is attained.

#### D. Intensification and Diversification
In order to obtain optimal solution, Tabu Search use Intensification and Diversification techniques. The former means intensification of the search in the neighborhood of the sub-optimal solution, the latter means diversification of the search to so far unexplored regions of the solution space. If intensification is missing, the search becomes an iterated random sampling, if diversification is missing, the process may be trapped in a sub-optimal region.

The frequency counter in the solution structure is used in the intensification and diversification techniques. The frequency counter denotes the times the solution (or the moves) having been visited throughout the solution process

### IV. SOLUTION ALGORITHM FOR NETWORK RECONFIGURATION

In this section, a step by step TS based solution algorithm for the network reconfiguration problem is presented.

1. Input system data, input system configuration, network data and parameter setting (e.g., lower and upper bounds on operating voltage, tabu list size, etc.)
2. Conduct sensitivity analysis. We select those nodes which have maximum impact on the system real power losses with respect to the nodal reactive power as the candidate locations to install feeder.
3. Generate an initial feasible solution state.
   a) Randomly select a solution state from the solution space.
   b) For each load level, execute the distribution power flow the check feasibility. If any constraint is violated, go to 1), otherwise proceed to next step.
4. Perform Tabu Search procedure.
   a) Select best move as next move direction from the move set based on the objective function evaluated.
   b) Check feasibility as in step 3-2. If not feasible, go to 1); otherwise proceed to next step.
5. Update the best solution state. If the move is not tabu based lead to another solution state better than so far visited, or if the move is tabu but aspiration criterion is attained, then update the best solution state.
6. Set/ release tabu move. Record the executed move as tabu move in the tabu list, or release the tabu move if the aspiration criterion is attained.
7. Check the stopping criterion. If the change of the objective function value of successive best so far solution is less than a given value, then stopping criterion is satisfied and go to next step; otherwise, return to step 4, and continue the tabu search procedure.

8. Output the optimal solution state. Output of the solution algorithm include minimizing power losses, ensuring voltage quality, service reliability assurance, lesser solution time, and maximum loadability.

## V. COMPARATIVE RESULTS OF VARIOUS METHODS USED

Network reconfiguration for optimization solution A sample example of 33-bus Radial Distribution System (RDS) is considered with the load data, transmission line details, and data of tie lines as presented in Figure 1. with a single line diagram [2]. The system suffers with following constraints in the formulation of the problem within the proposed solution procedure:

- The radial structure of the network must be maintained in each new structure.
- All loads must be served.
- Current magnitude of each branch (feeder, laterals, and switches) must lie with their permissible ranges to maintain power quality.
- Voltage magnitude at each section must lie with their permissible range.

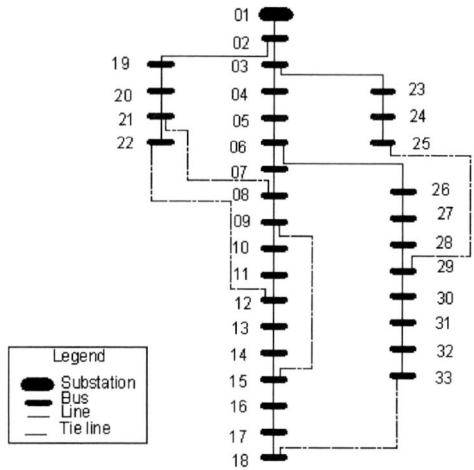

Fig. 1. A 33-bus radial distribution system.

Network reconfiguration for RDS can be formulated as follows:

Minimize fc = ∑ Loss ( i to n)

Where, n : number of branch,
i : Loss at branch i

Subject to all above constraints. These constraints can be checked using a search method.

Table I shows the comparison of results of the algorithms. The results indicate that the TS can generate the optimal solution.

TABLE I
COMPARISON OF ALGORITHMS

| Sr N. | Methods Used | Merits | Demerits |
|---|---|---|---|
| 1. | Simulated Annealing | It has more potential to find the global optimum, through this can't be proven. It is readily applied on decision tree or branch exchange methods, by adding an extra outer loop to the basic algorithm. | It take extra analysis time, so it is more applicable to planning rather than operations |
| 2. | Genetic Algorithm | These are most attractive for parallel processing environment, | These are not reliable for a more detailed model. This method is limited for few iterations of analysis. |
| 3. | Tabu Search | Comparing the results of TS with those of ANN methods reveals that the proposed solution methodology can offer the nearly optimal solution to the reconfiguration problem within less computing time which also directly affect the economic cost of system. | |

TABLE II
COMPARISON OF LOSS MINIMIZATION BY TS AND OTHER METHOD

| | Execution time per search (iteration sec.) | Average number of 3 phase load flow calculation per search | Minimum number search for obtaining optimal solution |
|---|---|---|---|
| Other Method | 20 | 86.6 | 12 |
| TS | 4 | 14.6 | 4 |

Table II shows the comparison results of loss minimization by TS and other method for unbalanced network of RDS [17]. According to the results, TS is 5 times faster than other method for one search and TS is 15 times faster than other method to obtain the optimal solution.

## VI. CONCLUSION

This paper proposes the application of TS as meta-heuristic method for network reconfiguration problems in radial distribution system. The proposed methods has compared modern heuristic algorithms genetic algorithm, simulation annealing and tabu search, for network reconfiguration. This work has been tested on 33-bus RDS with five tie lines.

Particularly nice feature of TS is that, like all approaches based on Local Search, it can quite easily handle the "dirty" complicating constraints that are typically found in real-life applications. It is thus, a really practical approach. It is not,

994

however, a panacea: every reviewer or editor of a scientific journal has seen more than his/her share of failed TS heuristics. These failures stem from two major causes: an insufficient understanding of fundamental concepts of the method but also, more often than not, a crippling lack of understanding of the problem at hand. One cannot develop a good TS heuristic for a problem that he/she does not know well! This is because significant problem knowledge is absolutely required to perform the most basic steps of the development of any TS procedure, namely the choice of a search space and of an effective neighborhood structure. If the search space and/or the neighborhood structure are inadequate, no amount of TS expertise will be sufficient to save the day. All meta-heuristics need to achieve both depth and breadth in their searching process; depth is usually not a problem for TS, which is quite aggressive in this respect (TS heuristics generally find pretty good solutions very early in the search), but breadth can be a critical issue. To handle this, it is extremely important to develop an effective diversification scheme. So, a properly designed distribution system alone can render efficient and fault-free service to the consumers and at the same time reduce distribution losses to the minimum economically optimum level.

## VII.  REFERENCES

[1]. Tilak Thakur & Jaswanti, "Study and Characterization of Power Distribution System Network Reconfiguration.," IEEE PES Transmission and Distribution conference & Exposition, Latin America 2006.

[2]. B.Venkatesh, Rakesh Ranjan, and H.B. Gooi, " Optimal Reconfiguration of Radial Distribution Systems to Maximize Loadability, " IEEE Trans. on PWRS-19, No. 1, February 2004, pp. 260-266.

[3]. Yasuhiro Hayashi & Junya Matsuki, "Loss Minimum Configuration of Distribution System Considering N-1 Security of Dispersed Generators," IEEE Transactions on Power Systems.Vol.19, No14, Nov 2004.

[4]. Miroslav M. Vigovic, Bransilav Radibratovic, Frank C. Lambert, "Multiobjective Volt-Var Optimization in Power Systems," Proc. 37th Hawaii International Conference on System Sciences, 2004.

[5]. Jen-Hao Teng, "A Direct Approach for Distribution System Load Flow Solutions," IEEE Transactions on Power Delivery, Vol.8, No.3, July2003.

[6]. Young-Jae Jeon, Jae-Chul Kim, Jin-O Kim, Joong-Rin Shin, and Kwang Y. Lee, "An Efficient Simulated Annealing Algorithm for Network Reconfiguration in Large-Scale Distribution Systems," IEEE Trans. on Power Delivery, Vol. 17, No. 4, October 2002, pp. 1870-1874.

[7]. A.Moussa, M.El-Gammal, E.N.Abdallah, and A.I.Attia, "A Genetic Based Algorithm For Loss Reduction in Distribution Systems", Resarch and Energy Conservation Sector in Alexandria Electricity Distribution Company, Alexandria, Egypt.

[8] A.Moussa, M.El-Gammal, E.N.Abdallah, and A.I.Attia, "A Genetic Based Algorithm For Loss Reduction in Distribution Systems", IEEE Trans. Power Del., vol. 4, no., 2, pp. 447-453, May 2000.

[9] Koichi Nara, Atsushi Shiose, Minoru Kitagawa & Toshihisa Ishihara, "Implementation of Genetic Algorithms for Distribution Systems Loss Minimum Reconfiguration," IEEE Trans. Power Syst., vol. 7, no. 3, pp. 1044-1051,Aug.1992.

[10]. F.Glover, "Future paths for integer programming and links to artificial intelligence," Comput. Oper. Res., vol. 13. no. 5, pp.533-549, 1986.

[11]. F.Glover, and Manuel Laguna, "Tabu Search".

[12]. Ramon A. Gallego, Alcir Jose Monticelli and Ruben Romero, "Optimal Capacitor Placement in Radial Distribution Networks," IEEE Trans. on PWRS-16, No.4, Nov. 2001, pp. 630-637.

[13]. Ramon A. Gallego, and Aleir J. Monticelli,"Tabu Search Algorithm for Network Synthesis," IEEE Trans PWRS-15, No. 2, May 2000, pp. 490-495.

[14]. Sakae Toune, hiroyuki Fudo, Takamu Genji, Yoshikazu Fukuyama, and Yosuke Nakanishi, " A Reactive Tabu Search for Service Restoration in Electric Power Distribution Systems," IEEE International Conference on Evolutionary Computation, Alaska, May 1998.

[15]. Yann-Chang Huang, Hong-Tzer Yang and Ching-Lien Huang, " Solving the Capacitor Placement Problem in a Radial Distribution System Using Tabu Search Approach ," IEEE Trans. on PWRS-11, No. 4, Nov. 1996, pp. 1868-1873.

[16]. Ramon A. Gallego, Ruben Romero, and Aleir J. Monticelli, "Tabu Search Algorithm for Network Synthesis," IEEE Trans. on PWRD-15, No. 2, May 2000, pp. 490-495.

[17]. Yoshikazu Fukuyama, " Reactive Tabu Search for Distribution Load Transfer Operation," IEEE PES Winter Meeting in Singapore, January 2000, pp. 1-6.

[18]. Takanobu Asakura, Toshiki Yura, Naoki Hayashi, and Yoshikazu Fukuyama, "Long-term Distribution Network Expansion Planning Considering Multiple Construction Plans," IEEE.

## VIII.  BIOGRAPHIES

Dr. Tilak Thakur is born in 1963. He graduated from B.I.T. Sindri, in Electrical engineering in 1987. He completed his Post graduation in Power System from the same institute and achieved his Ph.D in Electronic Instrumentation from Indian School of Mines, Dhanbad in the area of SCADA in 1999. He served as a lecturer in B.I.T., Sindri and NERIST Arunachal Pardesh. Presently, he is Assistant Professor in the Department of Electrical Engineering, Punjab Engineering College (PEC), Chandigarh, India. He has a teaching experience of more than 15 years. He is involved in active Research in Power System Automation and Control. He is member of IEEE.

Mrs. Jaswanti graduated in Electrical Engineering from Punjab Engineering College, Chandigarh, India in 1993. She got her master of Engineering from same institute in 1997. She is currently doing her Ph.D in power system. She is member of IEEE. She is also a associate member of IEI and ISTE. Her main research interests are power distribution system operation, analysis and control.

**2006 IEEE International Conference on Power Electronic, Drives and Energy Systems**

# Comparative Studies of Transient and Steady State Analysis for a Typical 765kV/400kV EHV Transmission System in Indian Power System

D. Thukaram, *Senior Member, IEEE*, H. P. Khincha, *Senior Member, IEEE*, and P. Shyamala

*Abstract*--This paper describes an approach for the analysis and design of 765kV/400kV EHV transmission system which is a typical expansion in Indian power grid system, based on the analysis of steady state and transient over voltages. The approach for transmission system design is iterative in nature. The first step involves exhaustive power flow analysis, based on constraints such as right of way, power to be transmitted, power transfer capabilities of lines, existing interconnecting transformer capabilities etc. Acceptable bus voltage profiles and satisfactory equipment loadings during all foreseeable operating conditions for normal and contingency operation are the guiding criteria. Critical operating strategies are also evolved in this initial design phase. With the steady state over voltages obtained, comprehensive dynamic and transient studies are to be carried out including switching over voltages studies. This paper presents steady state and switching transient studies for alternative two typical configurations of 765kV/400 kV systems and the results are compared. Transient studies are carried out to obtain the peak values of 765 kV transmission systems and are compared with the alternative configurations of existing 400 kV systems.

*Index Terms*--Line loadability, Power Transfer Capability, Switching over voltages, Voltage stability.

## I. INTRODUCTION

LINE loadability and overvoltages are two important aspects that must be addressed when upgradation of transmission system is contemplated. Methods generally acceptable in the industry and utility, convenient to estimate maximum line loading limits are to be adopted. Shunt compensation can cause considerable variations in loadability characteristics of lines fed by high impedance sources. Var sources such as Flexible AC Transmission systems (FACTs) increase line loadability. It is generally accepted that voltage drop of 5% between terminals of sending end and receiving end (not including source impedances) and angular separation

of 40° to 44° between sources (at sending and receiving ends) are good guidelines [1].

The reliable operation of any electrical power system is determined to a great extent by the amplitude, duration and frequency of the transient voltages appearing in different places in the network. Also the insulation level of EHV and UHV ac systems is largely determined by the magnitude of switching overvoltages. Switching overvoltages are therefore a focal point in carrying studies for these systems. Switching transients are fast transients that occur in the process of energizing transmission line, load capacitances immediately after a power source is connected to the network. Power transformers, surge arresters and circuit breakers are equipments which are first affected by overvoltages. Digital computer tool such as Electro Magnetic Transients Program (EMTP) [2] which is universally accepted as industry standard for computation of both switching and temporary over voltages is used in this paper. At the planning stage of the 765 kV systems, the insulation level of apparatus is to be decided on the basis of peak value of transient over voltages, so enormous numbers of cases are considered to arrive at the maximum magnitude. The values obtained are compared with the existing 400kV transmission system. Methods are proposed for reducing the overvoltages, so that apparatus used for the present system are compatible. Also the variation of magnitude and shape of voltages with increase in system voltage and others are discussed in this paper. Of the several methods to mitigate transient overvoltages, pre-insertion resistors are used.

## II. INDIAN SCENARIO

The scenario of power systems development is currently undergoing drastic changes. Viewed against the backdrop of multifold increase in power demand these changes are characterized by,
➢ Installed and available generation
➢ Addition of EHV transmission lines especially 400kV/765kV
➢ Advanced systems such as HVDC system and FACTs
➢ Induction and indigenization of associated technologies
➢ Development of analytical techniques for proper planning.

In India the available generation and installed capacity has been increased many folds in the last two decades. It is

---

D Thukaram is with Department of Electrical Engineering, Indian Institute of Science, Bangalore-560012, INDIA (dtram@ee.iisc.ernet.in)

H P Khincha is with Department of Electrical Engineering, Indian Institute of Science, Bangalore-560012, INDIA (hpk@ee.iisc.ernet.in)

P Shyamala is with Department of Electrical Engineering, Indian Institute of Science, Bangalore-560012, INDIA (shyam@ee.iisc.ernet.in)

0-7803-9771-1/06/$25.00 ©2006 IEEE

estimated that India would need a total installed capacity of 212GW by 2012 (Eleventh National Power Survey) [3]. This power demand calls for a coordinated evolution of strategies for transmission and generation planning among the electricity boards, regional and state, experts in academic institutions and consultants. Bulk power transmission has come a long way as far as kV level is concerned. Since late 1977, the 220kV and 132/110kV transmission systems are being overlaid by 400kV systems. The decision to boost kV level to 400kV and associated problems was successfully tackled by indigenous know-how.    Reckoning 765/800 kV as the horizon year and faced with a challenge of catering a power demand of 212 GW, power system planners are at a branch point viz., to further reinforce the 400kV network or to go for the next higher kV level. The decision in this regard will also consider HVDC option. The ac transmission options available are:

1. 400kV circuit reinforcement (multi-circuit lines)
2. 400kV lines with quad conductor bundles
3. FACTs compensated 400kV lines
4. Higher phase order
5. Higher level of voltage(765 kV/800 kV) for power transmission

The second and third options, (series compensation and quad conductor bundles) are short-term measures which raise power transmission capability of 400kV lines to some extent. In comparison with the required power delivery capabilities they are, therefore of limited advantage. Higher phase order transmission is still in its infancy even in the developed countries. Hence a choice has to be made between the first and last options. Among others, the choice would be influenced by techno-economics associated with right-of-way requirements, environmental and other problems.

A part of typical Indian grid system considered for analysis is shown in Fig. 1, and Fig. 2. Fig. 1, indicates the 765kV single circuit transmission line system and Fig. 2, represents the 400kV double circuit system. Both the systems are represented down to 220kV with appropriate interconnections.

### III. ASSESSMENT OF STEADY STATE PERFORMANCE

Assessment of steady-state performance of the system includes:

➢ Analysis of loadability characteristics
➢ Pre- and post-charging studies to determine steady-state overvoltage
➢ Compensation requirements

Detailed power flow analysis is carried out to address the first three aspects. Three major factors that can limit the power-transfer capability or loading of a transmission line are:

• Thermal constraints,
• Line voltage drop, and
• Steady-state stability margin

The thermal capability of typical bundled-conductor (two or more sub conductors in each phase) arrangement at EHV and UHV voltage levels generally exceeds, by a significant margin.

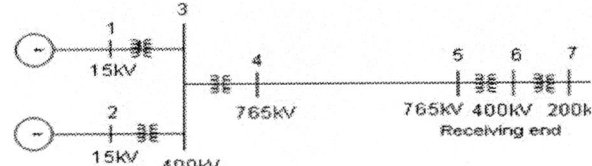

Fig. 1. Configuration 1.

Fig. 2. Configuration 2.

Thus for EHV and UHV transmission lines, practical limitations to line loadability are imposed by line voltage drop and steady-state stability margin considerations. Dunlop et al. [4] give the details of a conceptual study of UHV transmission line loadabilities. The steady state stability limitation is usually defined in terms of the desired margin between the maximum power transfer capability ($p_{max}$) of the system and the operating level ($p_{op}$) as

$$\% \text{ stability margin} = \frac{p_{\max} - p_{op}}{p_{\max}} \times 100$$

This margin is chosen so as to provide stable operation following a variety of credible contingencies which may cause steady-state and/or transient voltage increases in a given line loading. Such changes in loading may be caused by line-switching operations, generation dispatch, and by transient disturbances such as temporary faults or loss of generation. L – Index [5] is used in order to estimate voltage stability margin of a particular load bus in the system. The process involves an evaluation of voltage stability condition of a system by computing the L - index for load buses.

*A. L-Index*

Consider a system where, n=total number of busses, with 1, 2... g generator busses (g), g,g+1,g+2... n the remaining busses. A load flow result is obtained for an operating condition, from which L-index is computed as

$$L_j = \left| 1 - \sum_{i=1}^{i=g} F_{ji} \frac{V_i}{V_j} \right|$$

Where j=g+1... n and all the terms within the sigma on the RHS of above equation are complex quantities. The values $F_{ji}$ are obtained from the Y bus matrix as follows

$$\begin{bmatrix} I_G \\ I_L \end{bmatrix} = \begin{bmatrix} Y_{GG} & Y_{GL} \\ Y_{LG} & Y_{LL} \end{bmatrix} \begin{bmatrix} V_G \\ V_L \end{bmatrix}$$

Where $I_G$, $I_L$, $V_G$, $V_L$ represent currents and voltages at the generator nodes and load nodes. Rearranging, we get

$$\begin{bmatrix} V_L \\ I_G \end{bmatrix} = \begin{bmatrix} Z_{LL} & F_{LG} \\ K_{GL} & Y_{GG} \end{bmatrix} \begin{bmatrix} I_L \\ V_G \end{bmatrix}$$

where $F_{LG}=[Y_{LL}]^{-1}[Y_{LG}]$ are the required values. The L-indices for a given load condition are computed for all load buses. The point at which Lj closed to a predetermined maximum limit (=1) indicates the maximum possible connected load to a bus termed as maximum loadability and its value close to zero indicates near no load condition. The stability of the complete system is given by global indicator, L= maximum of Lj for all j (Load buses). An L-index value away from 1 and close to zero indicates an improved system security. The stability margin is obtained as the distance of L from a unit value i.e. (1-L).

*B. Steady State Analysis*

Two schemes with configuration 1 are considered for analysis, one from a typical Indian grid [6] and the other from the literature [7]. Scheme 1 is an Indian grid system considered with transmission line between Anpara and Unnao charged at 765 kV. Steady state analysis is carried out on scheme 1 with shunt reactors of 150 MVAR at bus 4 and 300 MVAR at bus 5. The variation of voltage at bus 7 as the load is increased is shown in Fig. 3. As the load is increasing, voltage is decreasing and L–index is increasing. In this case, as the load is increased above 700 MW voltage is very low, which is not acceptable. However stability limit is obtained at 800 MW, beyond which the voltage collapse is occurring. The scheme 2 of configuration 1, is considered with line parameters obtained from the American Electric Power (AEP). Power flow analyses are carried out for various values of shunt reactors and the appropriate value is selected such that the voltages at the load buses are within the acceptable limits. Variation of L–index at bus 7 as the load is varying for scheme 2 is shown in Fig. 4.

Analyses are carried out for configuration 2 i.e. 400kV double circuit with 80 MVAR reactors at both sending and receiving end. The loadability curve showing the variation of voltage at receiving end as the load is varied is given in Fig. 5. It was concluded that the maximum load that can be carried by the 400kV system is much less (around 600MW) as compared to the 765kV system (around 800MW).

## IV. TRANSIENT OVER VOLTAGE STUDIES

Transient overvoltages are usually a significant factor at transmission voltages above 400 kV [8]. In EHV and UHV systems there are a number of switching operations, such as overvoltages produced during the switching of reactors, capacitors and transformers. These can readily be limited by surge diverters [9] and are therefore not considered here. Of the other switching operations, line closing and reclosing generally produce the larger overvoltages and consequently we concentrate on line energization in this paper. At higher transmission voltages, overvoltages caused by switching may become significant, because arrester operating voltages are relatively close to normal system voltage and lines are usually long so that the energy stored on the lines may be large. Overvoltages will put the transformers into saturation, causing core heating and copious harmonic current generation. Circuit breaker called upon to operate during periods of high voltage will have reduced interrupting capability.

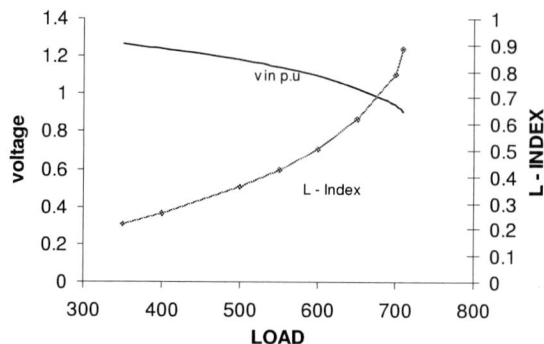

Fig. 3. Variation of voltage and L-Index at load bus 7 as the load is increasing for scheme 1 of configuration1

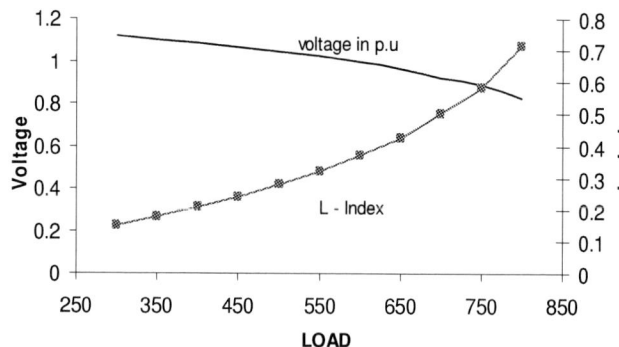

Fig. 4. Variation of voltage and L-Index at load bus 7 as the load is increasing for scheme 2 of configuration 1.

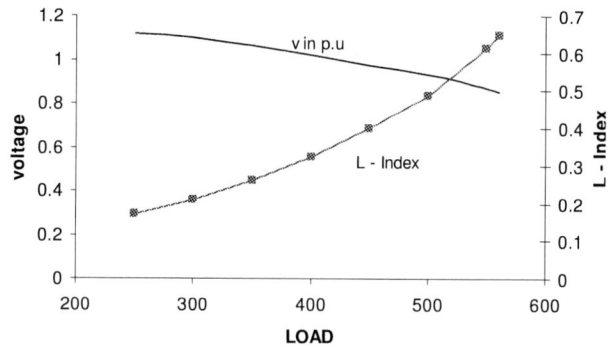

Fig. 5. Variation of voltage and L-Index at load bus 5 as the load is increasing for configuration 2.

At some voltage even the ability to interrupt line-charging current will be lost. Thus transient voltage magnitudes must be restricted to a safe value less than or equal to withstand capacity of the apparatus. In practical system a number of factors affect the overvoltages due to energization or reclosing. The influence of various factors can be grouped into three broad categories, such as strong, medium and weak as given below.

Strongly Influencing Parameters
- Line Length
- Degree of shunt compensation
- Line parameters
- Total short circuit level
- Line termination

- Trapped charges when PIR is not used
- Value of closing resistors
- Insertion times of closing resistors
- Pole closing instants
- Nature of source-inductive or complex

Medium Influencing Parameters
- Total pole closing span
- Trapped charges when PIR is used
- Frequency dependence of line parameter
- Saturation of reactor

Weak influencing Parameters
- Corona of lines

Of all these parameters, transients are most affected by line length, degree of shunt compensation, line parameters, and short circuit level. Switching surge magnitudes can be controlled by using any one or combination of the following methods:

o Circuit breaker with Pre – insertion resistor(PIR)
o Metal Oxide Arrester(MOA)
o Point on wave controlled switching
o Controlled switching taking into account residual flux.

Because of higher power consumption, space adequacy and along with the significant addition of transients due to opening and closing of PIR's and greater mechanical complexity, their usage is being reduced [10]. On the other hand PIR's are cheaper when compared to others like metal oxide arresters. Nevertheless, the resistor equipped circuit breakers are more expensive. These complex breakers show mechanical mal-functions as the most common cause of circuit breaker failures [11]. Controlled switching is being used which reduces magnitudes of switching overvoltages considerably.

*A. Transient studies results*

Transient analysis is carried out on three systems. They are

1. Scheme 1 of 765 kV system ( Fig. 1)
2. Scheme 2 of 765 kV system (Fig. 1)
3. 400 kV system (Fig. 2)

765 kV systems differ in line parameters as the line parameters are the factors which affect the overvoltages strongly. This difference is due to variation in the tower designs and other related parameters. Though the interconnecting systems are large networks, for transient analysis the systems with an equivalent source, transmission line and a load are considered. Thus, sample systems for both the schemes of UHV and 400kV are represented as shown in Fig. 6. Base values for 765kV system in p.u are

Base (MVA)        -        100 MVA
Line-to-line voltage -        765 kV
Phase voltage        -        (765*√2) / √3 kV  (1 p.u.)

Extensive EMTP simulation studies are to be carried out for planning of transmission systems. Studies are carried out by considering the range of various parameters such as,

Source strength: 1000–10000MVA in step of 1000MVA

- Line length: 300–1000 km in step of 50 km
- Switching angle: 0–90º in step of 30º
- Without PIR, with PIR upto 500 ohms in steps of 50 ohms

Initially analyses are carried out on scheme 1 of 765 kV systems. Fig. 7, illustrates the switching transient for scheme 1 of 765KV system at bus 3 (Receiving End) when line is energized without reactor and without PIR. The peak overvoltage observed is 2.99 p.u. Transients can heavily be damped by pre-insertion resistor (PIR). The optimum pre-insertion resistance is not only a function of the line shunt compensation but also the short circuit power of the feeding network and the line length. Shunt reactors along with the PIR reduce the overvoltage significantly; this is shown in Fig. 8. The peak overvoltage in this case reduces to 2.16 p.u.

Controlled switching of transmission lines can be used to mitigate the switching overvoltages during energization. If switching takes place at the voltage maximum i.e. at 90º, the voltage at first oscillate along the whole line length to almost twice the value of the system voltage. Over voltage can be limited by controlled switching of circuit breaker as shown in Fig. 9, in which line closing is done at 0º. Maximum transient over voltage will be more at 90º than at 0º for same line length and source strength. Overvoltages for the switching angles 90º and 0º of various sub cases are shown in Table I.

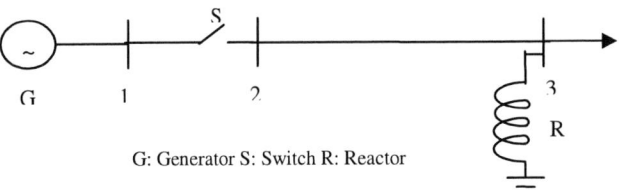

G: Generator S: Switch R: Reactor

Fig. 6. Equivalent system for switching transients

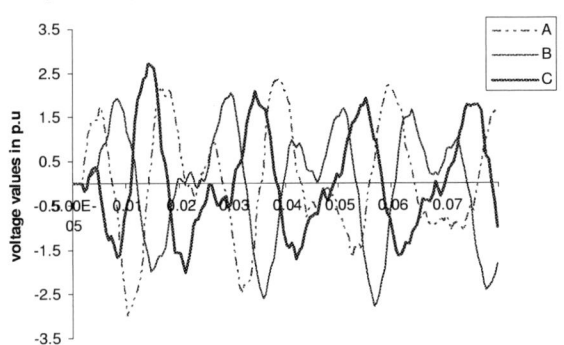

Fig. 7. Switching transient overvoltage for scheme 1 of 765KV system, without reactor at bus 3 and without PIR. Max Peak absolute value is 2.99, 2.77, and 2.71p.u. in phase A, B and C, respectively with switching angle90º.

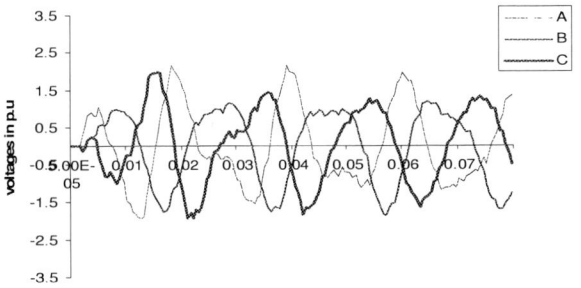

Fig. 8. Switching transient overvoltage for scheme 1 of 765KV system at bus 3, with reactor at bus 3, with PIR. Max Peak absolute value is 2.16, 1.87, and 1.97 p.u. in phases A, B and C, respectively with switching angle 90º.

Similarly, Transient over voltages for the scheme 2 of 765kV system (Fig. 5) is obtained for the following cases:

1. Without shunt reactor and without PIR
2. With shunt reactor and without PIR
3. Without shunt reactor and with PIR
4. With shunt reactor and with PIR.

The switching overvoltage curve for the case 4 is shown by Fig. 10. All the above cases are repeated for switching angle extreme values i.e. 90° and 0° and the results are summarized in Table II. From the steady state voltage stability studies it was observed that power carried by 400kV system is much less than that carried by UHV system. It can be overcome by usage of intermediate substation i.e. at a distance of 230km from the source end. Thus the system switching transients at receiving end with reactor and with PIR are shown in Fig. 11. Studies on 400kV system are carried out for different cases i.e. with or without 50 MVAR reactor at receiving end and with or without PIR. Table III gives the comparison results for the 400kv system. All the cases considered are described below. The case numbers in bracket correspond to scheme 2 of configuration 1.

**Case 1(5):** Transient analysis on scheme 1(2) of 765kV system considering without reactor at bus 3 and without PIR.

**Case 2(6):** Transient analysis on scheme 1(2) of 765kV system with 100 MVAR reactor at bus 3 and without PIR.

**Case 3(7):** Transient analysis on scheme 1(2) of 765kV system considering without reactor at bus 3 and with PIR of 300 ohms.

**Case 4(8):** Transient analysis on scheme 1(2) of 765kV system considering with 100 MVAR reactors at bus 3 and with PIR of 300 ohms.

**Case 9:** Transient analysis on 400 kV systems considering without reactor at bus 3 and without PIR.

**Case 10:** Transient analysis on 400 kV systems considering with 50 MVAR reactor at bus 3 and without PIR.

**Case 11:** Transient analysis on 400 kV systems considering without reactor at bus 3 and with PIR of 300 ohms.

**Case 12:** Transient analysis on 400 kV system considering with 50 MVAR reactor at bus 3 and with PIR of 300 ohms.

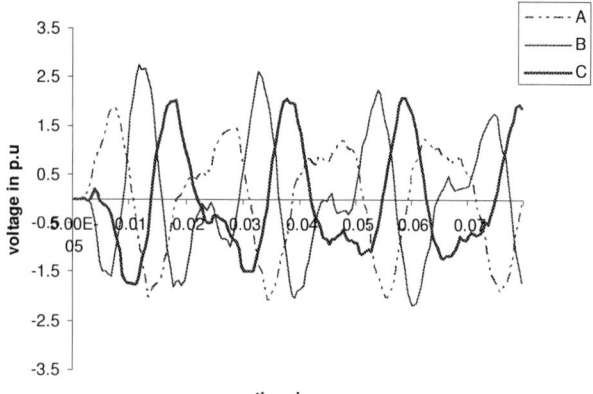

Fig. 9. Switching transient overvoltage for scheme 1 of 765KV system, without reactor at bus 3, without PIR. Max Peak absolute value is 2.09, 2.74, and 2.07 p.u. in phase A, B and C, respectively with switching angle 0°.

TABLE I

COMPARISON OF P.U PEAK VOLTAGES FOR SWITCHING ANGLES 90° AND 0° UNDER VARIOUS CASES FOR SCHEME 1 OF 765 kV SYSTEMS.

| Case No | Switching angle | | | | | |
|---|---|---|---|---|---|---|
| | 90° | | | 0° | | |
| | a | b | c | a | b | c |
| 1 | 2.99 | 2.77 | 2.71 | 2.09 | 2.74 | 2.07 |
| 2 | 2.94 | 2.09 | 2.63 | 1.85 | 2.62 | 1.95 |
| 3 | 2.36 | 2.19 | 2.14 | 2.06 | 2.24 | 1.98 |
| 4 | 2.16 | 1.87 | 1.97 | 1.80 | 1.94 | 1.81 |

TABLE II

COMPARISON OF P.U PEAK VOLTAGES FOR SWITCHING ANGLES 90° AND 0° UNDER VARIOUS CASES FOR SCHEME 2 OF 765 kV SYSTEMS.

| Case No | Switching angle | | | | | |
|---|---|---|---|---|---|---|
| | 90° | | | 0° | | |
| | a | b | c | a | b | c |
| 5 | 3.00 | 2.53 | 2.77 | 2.03 | 2.73 | 2.12 |
| 6 | 2.94 | 2.02 | 2.65 | 1.81 | 2.61 | 1.98 |
| 7 | 2.31 | 2.10 | 2.18 | 2.03 | 2.15 | 1.94 |
| 8 | 2.07 | 1.77 | 2.01 | 1.81 | 1.96 | 1.76 |

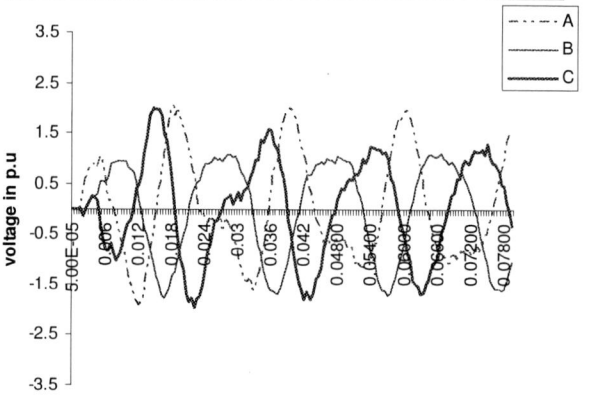

Fig. 10. Switching transient overvoltage for scheme 2 of 765KV system at bus 3, with reactor at bus 3, with PIR. Max Peak absolute value is 2.07, 1.77, and 2.01 p.u. in phase A, B and C, respectively with switching angle 90°.

TABLE III

COMPARISON OF P.U PEAK VOLTAGES FOR SWITCHING ANGLES 90° AND 0° UNDER VARIOUS CASES FOR 400 kV SYSTEM

| Case No | 90° | | | 0° | | |
|---|---|---|---|---|---|---|
| | a | b | c | a | b | c |
| 9 | 1.76 | 2.07 | 1.26 | 1.62 | 1.97 | 1.51 |
| 10 | 1.89 | 2.00 | 1.21 | 1.62 | 1.80 | 1.46 |
| 11 | 1.38 | 1.39 | 1.15 | 1.19 | 1.25 | 1.31 |
| 12 | 1.32 | 1.27 | 1.15 | 1.13 | 1.14 | 1.18 |

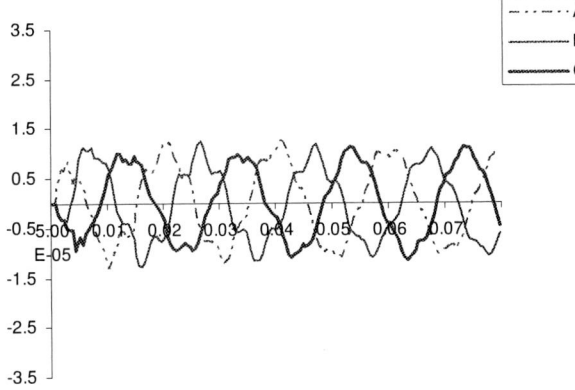

Fig. 13. Switching transient overvoltage for scheme 2 of 765KV system at bus 3, with reactor at bus 3, with PIR. Max Peak absolute value is 2.07, 1.77, and 2.01 p.u. in phase A, B and C, respectively with switching angle 90°.

### B. Comparison of Transient Analysis Results

In the case of 765kV, power transfer capability is more, however it results in greater transients as compared to 400kV, EHV systems. Peak values are considered for extreme cases as they are important for insulation coordination. In case of 765 kV systems the maximum peak overvoltages are higher compared with the case of 400kV.

## V. CONCLUSIONS

Transmission system expansion with a new 400kV double circuit and an alternative with 765kV single circuit are considered for studies. Preliminary basic studies for steady state and transient state are carried out. It was observed that the loadability level of 765kV is almost three times as compared to 400kV system based on voltage stability, L – Index. Transient analysis study for line energization case is done for 765 kV systems by varying the factors such as switching angle; source strength etc. and methods are proposed for limiting them. Analyses are carried for various system parameters and results obtained are compared with the 400 kV systems. Detailed transient studies are to be carried out for the proper design of EHV/UHV transmission systems. Thus , this paper is an indicative of comparison of 765kV and 400 kV systems considered.

## VI. REFERENCES

[1] Harold N. Scherer, G S Vassell, Transmission of Electric Power at Ultra-High Voltages: Current Status and Future Prospects, proc of IEEE(1985) 1252-78.

[2] Herman W. Dommel, Techniques for analyzing electromagnetic transients, IEEE comput. Appl. Power(1997) 18-21.

[3] www.expert-eyes.org/power/index.html - Power Survey

[4] R. D. Dunlop, R Gutman, and P. P. Marchenko, "Analytical development of loadability characteristics for EHV and UHV transmission lines," IEEE Trans. Power App. Syst. Vol. PAS-98, pp 606 – 617, Mar / Apr. 1979.

[5] D. Thukaram, K. Parthasarathy, H. P. Khincha, Narendranath Udupa and A. Bansilal. "Voltage Stability improvement: case studies of Indian power networks," Electric power systems research, Vol.44, 1998, pp. 35-44.

[6] D Thukaram, B S Sharma, UPSEB Lucknow, Overvoltage studies for UPSEB 765kV ANPARA-UNNAO line operated at 400kV, technical report, Second Workshop & conference on EHV Technology Bangalore, Aug 7 – 10, 1989.

[7] A J Samuelson, " American Electric Power 765 kV transmission line project," IEEE Trans. Power App. Syst. Vol.PAS-88, May 1969,pp.703-709.

[8] K. Ragaller, Surges in High voltage Networks, Plenum Press, 1980.

[9] Cigre working group, switching overvoltages in EHV and UHV systems with special reference to closing and reclosing of transmission lines, Electra 30(1973) 70-122.

[10] C.D. Tsirekis1 and N.D. Hatziargyriou2, Control of Shunt Capacitors and Shunt Reactors Energization transients, International Conference on Power Systems Transients – IPST 2003 in New Orleans, USA.

[11] K Joseph, et al., Controlling switching surges on 1100kV transmission systems, IEEE Trans. Power App. System PAS – 89 vol. 8, 1970, pp 1752 – 1762.

[12] Alessandro Clerici, G. Ruckstuhl, A. Vian, et at., Influence of shunt reactors on switching surges, IEEE Trans. Power App. Syst. PAS – 89, vol 8, 1970, pp 1727 – 1736.

## VII. BIOGRAPHIES

**D. Thukaram (*SMIEEE*)** received the B.E. degree in Electrical Engineering from Osmania University, Hyderabad in 1974, M.Tech degree in Integrated Power Systems from Nagpur University in 1976 and Ph.D. degree from Indian Institute of Science, Bangalore in 1986. Since 1976 he has been with Indian Institute of Science as a research fellow and faculty in various positions and currently he is Professor. His research interests include computer aided power system Analysis, reactive power optimization, voltage stability, distribution automation and AI applications in power systems.

**H. P. Khincha (*SMIEEE*)** received the B.E. degree in Electrical Engineering from Bangalore University in 1966. He received M.E. degree in 1968 and Ph.D. degree in 1973 both in Electrical Engineering from the Indian Institute of Science, Bangalore. Since 1973 he has been with Indian Institute of Science, Bangalore as faculty where currently he is Professor. His research interests include computer aided power system analysis, power system protection, distribution automation and AI applications in power systems.

**P. Shyamala** received the B.E. degree in Electrical & Electronics Engineering from Jawaharlal Nehru Technological University (JNTU) in 2005. Currently ursuing M.Sc (Engg.) degree in Electrical Engineering at the Indian Institute of Science, Bangalore. Her research interests include Computer aided Power System analysis, Designing of High Voltage Transmission Systems, AI techniques and its applications to Power Systems.

**2006 IEEE International Conference on Power Electronic, Drives and Energy Systems**

# A Finite Element Modeling and Simulation Method for Time-Varying Field-Circuit Problems

M. Nabi

*Abstract*—**A finite element (FE) computational modeling and solution scheme is presented for field-circuit systems involving general three-phase external circuits with possibly nonlinear circuit elements. The method is illustrated for a triggered regulator circuit and is an extension of an approach presented recently for similar problems but with linear circuits.**

## I. INTRODUCTION

A large number of electrical systems involve electromagnetic fields interacting with external circuits [1], [2]. In most practical cases, the geometries involved are complicated, ruling out analytical solution for the electromagnetic field and necessitating a numerical method like the finite element method (FEM). For transient field-circuit problems, the FE models are large systems of ordinary differential equations (ODEs). The combined systems of equations for the coupled problem are generally solved through time-stepping schemes [3]. For nonlinear problems, the equations to be solved at each time-step are nonlinear, requiring large linear systems to be solved many times iteratively. Thus any prior reduction of the overall size of the model a profitable alternative. In this paper, a computational approach suggested recently in [4] for linear time-harmonic eddy current problems and in [5] for transient circuit-field problems with linear passive circuit elements is further extended to a regulator circuit with triggered thyristors. The numerical properties of the reduced overall coupled model for such problems are also discussed briefly.

## II. THE FIELD-CIRCUIT COUPLED SYSTEM

The system considered is composed of three bars as shown in Fig.1 coupled to each other through their magnetic fields and having eddy current induced in them. Typically they represent massive conductors in the circuit with eddy currents and mutual coupling. The three bars are coupled to a threephase unbalanced triggered circuit with lumped parameters as shown in Fig.2. This results in a strongly coupled overall nonlinear field-circuit system. While the field

---

The author is with the Department of Electrical Engineering, Indian Institue of Technology Delhi, Hauz Khas, New Delhi-110016,India(e-mail: mnabi@ee.iitd.ac.in

is modeled by the FE method, the circuit is modeled through the standard KCL-KVL equations.

### A. Reduced Order FE Modeling of the Field

The electromagnetic field described by the vector magnetic potential A is governed by the time-varying eddy current equation given as

$$\nabla \partial \left( \frac{1}{\pi} \nabla \partial A \right) \cong 0 \, \omega \left( \frac{\partial A}{\partial t} \cdot g \right) \tag{1}$$

where $g$ denotes an externally applied voltage gradient and other symbols have standard interpretations.

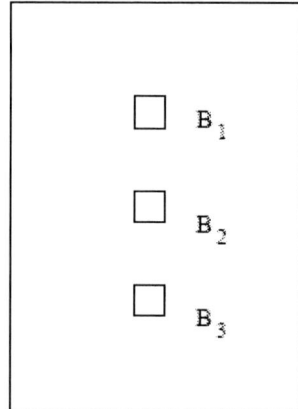

Fig. 1. Cross-section of three Bars.

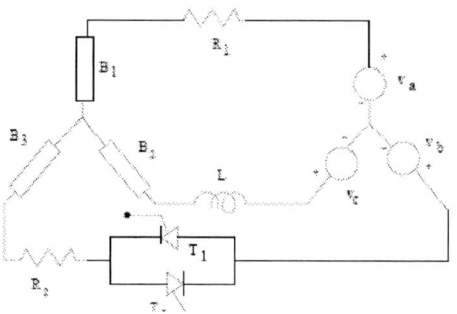

Fig.2. The Overall Circuit.

Obviously, in Fig.1, $\sigma$ and $g$ are both zero over the air region $\ddot{E}_1$, having nonzero values only inside the bar region $\ddot{E}_2$ or on the interface $\Gamma_{12}$. A reduced-order FE modeling for such electromagnetic field problems was suggested in [4] for time-harmonic problems and extended to coupled field-circuit problems with linear passive circuit elements only in [5]. Here, the field region is modeled along the same lines, as outlined below, and then coupled to the external nonlinear triggered circuit.

For the two-dimensional geometry considered, A is directed along the $z$-axis everywhere and hence can be treated as a scalar. The FE model for the electromagnetic field region is

$$[K . S\frac{d}{dt}]\overline{A} \cong 0\overline{fg} \qquad (2)$$

where $\overline{A}$ is the vector of nodal values of A, and $\overline{g}$ is a vector of length 3 containing voltages across the three bars. The coefficient matrices $K$ and $S$ are of the form

$$K \cong \begin{bmatrix} D_1 & D_2 & 0 \\ D_2^T & D_3 . F_1(\theta) & F_2(\theta) \\ 0 & F_2^T(\theta) & F_3(\theta) \end{bmatrix} \qquad (3)$$

$$S \cong \begin{bmatrix} 0 & 0 & 0 \\ 0 & E_1(\varpi) & E_2(\varpi) \\ 0 & E_2^T(\varpi) & E_3(\varpi) \end{bmatrix} \qquad (4)$$

Here the three blocks of rows and columns correspond to nodes inside $\ddot{E}_1$ (air), on interface $\Gamma_{12}$ and inside $\ddot{E}_2$, which are $n_1$, $n_{12}$ and $n_2$ in number. Combining the last two rows of the coefficient matrices as shown in (3) and (4), the global FE equations of (2) become of the form

$$\begin{Bmatrix} G_1 \\ G_{21} . G_{22}(\varpi)\frac{d}{dt} . G_{23}(\theta) \end{Bmatrix} x \cong \begin{Bmatrix} 0 \\ 0 f_2(\varpi)\overline{g}(t) \end{Bmatrix} \qquad (5)$$

where $G_1$ is a constant $n_1 \times (n_1 + n_{12} + n_2)$ matrix and $G_{21}, G_{22}, G_{23}$ are $(n_{12}+n_2) \times (n_1+n_{12}+n_2)$ real matrices, $G_{21}$ being constant. The corresponding static problem is

$$\begin{Bmatrix} G_1 \\ G_{21} . G_{23}(\theta) \end{Bmatrix} x \cong \begin{Bmatrix} 0 \\ 0 f_2(\varpi)\overline{g}* \end{Bmatrix} \qquad (6)$$

For the purpose of characterizing the solution of the above and reduction of the FE model, the following theorem can be proved [5]. The theorem allows in general for nonlinear material parameters in the bars $\ddot{E}_2$.

*Theorem 1:* If $x^*$ is the solution to the static problem (6) for some reluctivity distribution $\theta \cong \theta_1$ and conductivity distribution $\sigma = \sigma_1$ over $\ddot{E}_2$, then the solution $x(t)$ of (5) for

any instant and any arbitrary distribution of $\theta$, $\sigma$ over $\ddot{E}_2$ is given by

$$x(t) \cong x* . y(t) \qquad (7)$$

if $y(t)$ satisfies

$$\begin{Bmatrix} G_1 \\ G_{21} . G_{22}(\varpi)\frac{d}{dt} . G_{23}(\theta) \end{Bmatrix} y(t) \cong$$
$$\begin{Bmatrix} 0 \\ 0 f_2(\varpi)\overline{g}(t) . f_2(\varpi_1)\overline{g}* . G_{23}(\theta_1)x*0G_{23}(\theta)x* \end{Bmatrix} \qquad (8)$$

Now, it can be seen that for all instants of time and any distribution of reluctivity and conductivity over $\ddot{E}_2$, the condition $G_1 y = 0$ always hold. Hence $y$ can be expressed as $y \cong P\tilde{y}$, where $P \angle \mathbb{R}^{(n_1 . n_{12} . n_2)\partial(n_{12} . n_2)}$ is the kernel of $G_1$, and $\tilde{y}$ is a time-varying vector of length $(n_{12} + n_2)$. The kernel $P$ of $G_1 = [D_1 D_2 0]$ can be taken as

$$\ker(G_1) \cong P \cong \begin{bmatrix} P_1 & 0 \\ P_2 & 0 \\ 0 & I \end{bmatrix} \qquad (9)$$

such that the matrix $[P_1^T P_2^T]^T$ with full column rank is a basis of $ker[D_1 D_2]$. With this form of $P$, substituting (7) and $y \cong P\tilde{y}$ in the second set of equations, yields finally a reduced set of differential equations

$$\begin{Bmatrix} M_{21} . M_{22}(\varpi)\frac{d}{dt} . M_{23}(\theta) \end{Bmatrix} \tilde{y}(t) \cong$$
$$\begin{Bmatrix} 0 \\ 0 f_2'(\varpi)\overline{g}(t) . f_2'(\varpi_1)\overline{g}* . G_{23}'(\theta_1)x*0G_{23}'(\theta)x* \end{Bmatrix} \qquad (10)$$

where

$$M_{21} \cong \begin{Bmatrix} 0P_1^T D_1^T P_1 . P_2^T D_3 P_2 & 0 \\ 0 & 0 \end{Bmatrix}$$

$$M_{22}(\varpi) \cong \begin{Bmatrix} P_2^T E_1(\varpi)P_2 & P_2^T E_2(\varpi) \\ E_2^T(\varpi)P_2 & E_3(\varpi) \end{Bmatrix}$$

$$M_{23}(\theta) \cong \begin{Bmatrix} P_2^T F_1(\theta)P_2 & P_2^T F_2(\theta) \\ F_2^T P_2(\theta) & F_3(\theta) \end{Bmatrix} \qquad (11)$$

$$f_2' \cong \begin{bmatrix} P_2^T & 0 \\ 0 & I \end{bmatrix} f_2 \qquad (12)$$

**1003**

The above system is of size $(n_{12} + n_2)$, i.e number of nodes only inside on the bars. This system is taken as the FE model of the electromagnetic field, with $\tilde{y}$ being the relevant field variable.

### B. Dynamic Equations for the Bars

The current through an area element $\ddot{E}_m^{(e)}$ in the $m$-th bar can be computed as

$$\tilde{I}_m^{(e)} \cong 0 \int_{Z_m^{(e)}} \varpi\left(\frac{\div A}{\div t} \cdot g_m\right) dZ_m^{(e)}$$

$$\cong 0 \, \overline{f}^{(e)T} \frac{d}{dt} \overline{A}^{(e)} \, 0 \, a^{(e)} g_m \tag{13}$$

Where $a^{(e)} \cong \int_{Z_m^{(e)}} \varpi \, dZ_m^{(e)}$. The partial derivative has been changed to an ordinary derivative since the vector $\overline{A}^{(e)}$ containing values of $A$ at fixed nodes varies only with time. Combining all such equations for all the elements in the $m$-th bar, we have the total current through the bar as

$$\tilde{I}_m \cong 0 \, \overline{f}^{(gm)T} \frac{d}{dt} \overline{A} \, 0 \, a_m g_m \tag{14}$$

where $\overline{A}$ as defined in Eq. 2 is the global vector of nodal values of $A$, and the scalar $a_m$ is the integral of the product of conductivity and area over the $m$-th bar. Writing the expressions for all three conducting bars one below another, we have

$$I_A \cong 0 \, \overline{f}^T \frac{d}{dt} \overline{A} \, 0 \, a\overline{g} \tag{15}$$

where $I_A$ is the vector of the three bar currents. Now, noting that $\overline{f}$ has nonzero entries only for the nodes on the bars, and using (7) and $y \cong P\tilde{y}$, finally yields [5]

$$f_2^T(\varpi) \frac{d}{dt} \tilde{y} \cdot a\overline{g} \cdot I_A \cong 0 \tag{16}$$

which connects the filed variables in $\tilde{y}$ to the rest of the circuit through the vector $\overline{g}$.

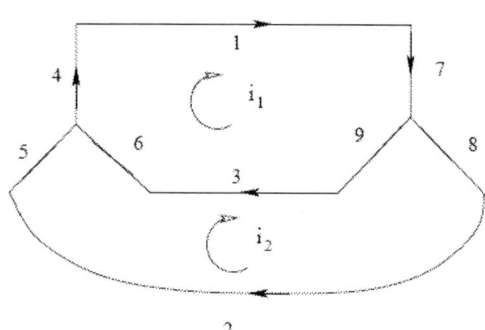

Fig.3. The circuit graph with thyristor conducting.

### B. The Triggered Circuit

The circuit as shown in Fig.2 is excited by the three-phase source voltages $V_a$, $V_b$ and $V_c$. The source $V_b$ is connected to the bar $B_3$ through the bi-directional thyristor-pair $T_1$ and $T_2$. Clearly, the current through $B_3$ is nonzero only when either one of the thyristors is conducting. Each thyristor gets switched off when the current through it falls to zero, and is triggered on after a delay of radians from the *current*-zero instant. Accordingly, for the time-stepping computation, two different circuit models are used. For the state when at least one of thyristors is conducting, the circuit consists of two loops with its graph as shown in Fig.3 and the loop incidence matrix being

$$B_1 \cong \begin{Bmatrix} 1 & 0 & 1 & 1 & 0 & 1 & 1 & 0 & 1 \\ 0 & 1 & 01 & 0 & 1 & 01 & 0 & 101 \end{Bmatrix} \tag{17}$$

For the state when both are thyristors are switched off, the branch 2 is opened and the circuit consists of only the upper loop. Hence the corresponding loop incidence matrix is

$$B_2 \cong \underline{\mathbb{1}} \; 0 \; 1 \; 1 \; 0 \; 1 \; 1 \; 0 \; 1 \overline{\phantom{xx}} \tag{18}$$

It may be noted that emf continues to be induced at the bar $B_3$ even for this state due to the changing magnetic field. Now, the set of branches are classified into four groups - those containing the resistance elements only, those with inductances only, those with the bars and those with voltage sources. Correspondingly the columns of loop- incidence matrices $B_k$, $k = 1, 2$ are also partitioned into four sets of columns as $B_i \cong [B_{Rk} \; B_{Lk} \; B_{Ak} \; B_{sk}]$ and similarly the branch currents and voltages. With this, the KCL equations for both the states become

$$[I_R^T \; I_L^T \; I_A^T]_k^T \cong i^T [B_R \; B_L \; B_A]_k \tag{19}$$

where $k = 1, 2$ and $i$ is the vector of loop currents. It should be noted that for the case $k = 1$, $i$ has two components while for $k = 2$, $i$ has only one component, i.e. the current of the upper loop. Moreover, the current in the upper loop varies continuously across instants when the thyristor branch change from conducting to open or vice versa. Similarly, the KVL equations are

$$[B_R \; B_L \; B_A]_k \begin{Bmatrix} v_R \\ v_L \\ v_A \end{Bmatrix}_k \cong B_{sk} v_s \tag{20}$$

Now, the branch voltages and currents are related by $v_R \cong Z_R I_R$ and $v_L \cong Z_L \frac{d}{dt} I_L$ for the two states (i.e. for $k = 1, 2$), where $Z_R, Z_L$ are the diagonal matrices of the resisitances and inductances involved. Further, the vector $v_A$ of voltage drops across the bars are related to the vector $\overline{g}$ of the gradient as simply $v_A = -l\bar{g}$ for length of the bars. Using these, the KVL equation (20) becomes

$$(\tilde{Z}_L \frac{d}{dt} . \tilde{Z}_R)_k i \, 0 \, B_{Ak}\bar{g} \cong B_{sk}V(t)/l \qquad (21)$$

where $\tilde{Z}_{Lk} \cong (B_L Z_L B_L^T)_k$ and $\tilde{Z}_{Rk} \cong (B_R Z_R B_R^T)_k$ for k= 1,2 are the effective circuit inductance and resistance matrices for the two states of the thyristors. It may be noted that for a general nonlinear circuit, these matrices are dependent on the currents in the circuit. In the present case, the nonlinearity is of a binary nature, with $\tilde{Z}_L$ and $\tilde{Z}_R$ taking two sets of values for the two states of the thyristor branch.

*D. The Coupled System*

Finally, combining the equations (10), (16) and (21), the combined system of differential equations for the overall coupled system is

$$\begin{Bmatrix} M_{21} . & M_{22}(\varpi)\frac{d}{dt} . & M_{23}(\theta) & f_2(\varpi) & 0 \\ & f_2^T(\varpi)\frac{d}{dt} & a(\varpi) & B_A^T \\ & 0 & B_A & 0(\bar{Z}_L\frac{d}{dt} . \bar{Z}_R) \end{Bmatrix}$$

$$\partial \begin{bmatrix} \tilde{y} \\ \bar{g} \\ i \end{bmatrix} \cong \begin{bmatrix} f_2(\varpi_1)\bar{g}* . G_{23}(\theta_1)x*0G_{23}(\theta)x* \\ 0 \\ 0 \, B_s V(t)/l \end{bmatrix} \quad (22)$$

In the above, the first block of equations describes the filed evolution, with matrices $M$ being sparse symmetric matrices much smaller than he original FE matrices for the field. The second block describes the coupling of the field and the circuit while the third block governs the circuit.

### III. TIME-STEPPING SIMULATION

The above system of differential equations (21) is solved through the backward-difference time- stepping scheme. As the time-stepping proceeds, the current in the lower loop is checked for reaching zero. When it reaches zero, the considered $\tilde{Z}_L$, $\tilde{Z}_R$ are changed from the thyristor conducting i.e. two-loop case to thyristor-off i.e. one loop case. This is again changed back after an angular displacement of  radians. The FE code written and used to model the field has been validated against standard commercial FE packages. Simulations for the coupled system are then carried out for a wide range of frequencies and values of the triggering delay angle . For one such simulation, Fig. 4 shows the computed currents in the three bars with the circuit parameters $L = 2H$, $r_1 = 10\,Z$ , $r_2 = 20\,Z$ , frequency 50 Hz and timestep size $h = 0.005$ sec. The three curves show the three bar-currents with

the smaller purely sinusoidal curve being the current through the middle bar, shown scaled (increased) due to its relatively much small magnitude compared to the other two. The curve with intermittent zero values corresponds to the current through the bar connected to the thyristors, as expected.

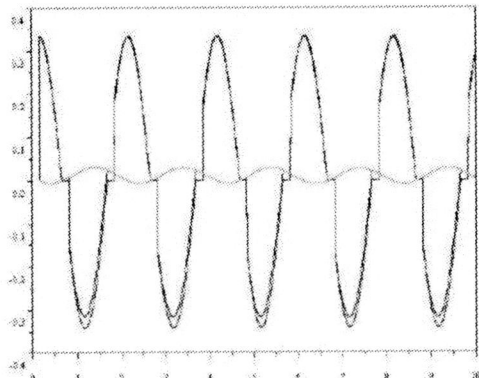

Fig. 4. Bar currents µ y-axis: Currents (A), x-axis: Angle (σ radians).

*A. Numerical properties of model*

It may be noted that in the presented methodology, the major saving is in terms of reduction of the size of the system of differential equation, and hence the size of the system of algebraic equations to be solved at each time-step. It may be recalled that the $M$ matrices in (22) of size same as number of nodes inside or on the surface of the bars, while in the conventional method it would be same as total number of nodes in the entire FE mesh over the entire region. Secondly, it can be seen from the equations (11) that for various choices of $P_1$ and correspondingly $P_2$, the eigenvalues of the $M$-matrices can be relocated, thus affecting the numerical properties of the overall differential equation system, and allowing for larger step sizes while preserving stability of the time-stepping routine. Further investigations on these lines, are however beyond the scope of the present paper.

### IV. CONCLUSION

The paper reports the extension of a FE modeling approach for coupled field- circuit problems reported recently, to threephase nonlinear triggered circuits involving thyristors. Such systems occur widely in various power system and machine drives applications. The presented method allows a significant decrease of the computational effort involved in FE coupled field-circuit modeling and simulation of such systems. The development of the model for such triggered circuits are shown, along with simulation results with the reduced model. The prospects of further improvement of the numerical properties are also mentioned. A simple

### V. REFERENCES

[1] I. A. Tsukerman, A.Konrad, G. Muenier and J. C. Vder qqdglhth/ÈF r xs dng# Field-circuit problems, trends and dff r p s dvkp hqw/$#IEEE Trans. Magn.*, vol. 29, pp. 1701µ1704, March 1993.

[2]  H. D. Gersem, R. Mertens, D. Lahaye, S. Vandewalle, and K. Hameyer, ÈVroxudrq# Vwdwhjlhv#irud#Wdqvlhqw#Ilhog-Flufxlw#Frxsdhg#V|wwlp v#Š# *IEEE Trans. Magn.*, vol. 36, no. 4, pp. 1531µ1534, Jul. 2000.

[3]  I. A. Tsukerman,        A. Konrad, G. Bedrosian, and M. V. NÌFkdul/#ÈD Survey of Numerical Methods for Transient Eddy Current Suredup v#Š# *IEEE Trans. Magn.*, vol. 29, no. 2, pp. 1711µ1716, Mar. 1993.

[4]  P l#XÌ#Qdel/#VÌ#Y l#Nxoduqdl/#dqg#Y l#UÌ#Vxdv/#ÈQr yho#P rghdqj #dqg# Solution Approach for Repeated Finite-Element Analysis of Eddy-Fxuhqw#V| wwlp v#Š#*IEEE Trans. Magn.*, vol. 40, no. 1, pp. 21µ28, Jan. 2004.

[5]  M. Nabi, S. V. Kulnduql/#DÌ#NÌ#J xs wd/#dqg#J l#EÌ#Nxp ekdu/#ÈDq# Improved FE Computational Scheme for Transient Circuit-Field Frxs dhg#V| wwlp v#Š#*Int. Jour. of Comp. Meth. in Engg. Sc & Mech.*, vol. 7, pp.313-322, 2006.

**2006 IEEE International Conference on Power Electronic, Drives and Energy Systems**

# A Wavelet Based Numerical Technique for Electromagnetic Field Analysis

Kaushik K, *Student Member, IEEE*, and S. V. Kulkarni, *Member, IEEE*

*Abstract*— Analysis of electromagnetic fields in electric machines has been traditionally carried out using the Finite Element Method (FEM). However, in the recent past, other meshless numerical techniques have evolved and been successfully applied in electromagnetics, to both high- and low-frequency problems. In this paper, we describe a solution scheme for a low-frequency electromagnetics problem using the wavelet–Galerkin technique. We also illustrate the application of this numerical technique to two example problems, the results of which highlight the feasibility of the technique.

*Index Terms*— Computational electromagnetics, Galerkin method, meshless methods, wavelet bases.

## I. INTRODUCTION

NUMERICAL analysis of electromagnetic fields in an electric machine is essentially carried out for evaluating the performance characteristics of the machine including any effect of external circuit elements. The analysis can also be used for validating a design or simply for a better understanding of the flux distribution in the machine. It can be carried out either on a simplified cross-sectional model of the machine involving some approximations or on an accurate three-dimensional model at the cost of increased complexity. Post-processing of the results can yield forces, torques, inductances or other physical quantities of interest. The Finite Element Method (FEM) has been the principal technique in such analysis for electromagnetic field computation, although other techniques such as the Boundary Element Method (BEM) or the Method of Moments (MoM) have also been employed for problems in both the high- and low-frequency domains, even if to a lesser extent.

Electromagnetic devices are, in general, difficult to model as they may have multiple complexities, such as, non-uniformity in material properties, material anisotropy, non-linear magnetisation characteristics, and complicated geometries and boundaries. In addition, the electromagnetic fields may be coupled with other physical quantities, such as, currents in external circuits, mechanical displacements, and thermal fields. The performance of FEM in the computation of magnetic fields with one or more of the above complications has been quite well established and there have been several successful applications. However, the inherent requirement of meshing the geometry in FEM based solutions implies that there are

Kaushik K and S. V. Kulkarni are with the Department of Electrical Engineering, Indian Institute of Technology Bombay, Mumbai–400076, India (e-mails: kaushikk@ee.iitb.ac.in, svk@ee.iitb.ac.in).

certain limitations in this method, which become significant under special conditions:

- Curved boundaries require the meshing to be very fine for a reasonable approximation of the geometry and solution.
- Adaptive computation techniques in FEM invariably require refinement of the mesh in regions of large gradients in the solution.
- Significant improvements in accuracy of the solution in FEM requires the mesh to be composed of higher order elements which increase the computational complexity.

These limitations apply not just to computational electromagnetics but also FEM-based field computations in other areas of engineering. Hence, there is a strong motivation for development of alternative techniques of field computation which will strive to overcome the above limitations of FEM. Meshless (also referred to as meshfree) methods, which do not involve any discretisation of the domain, are one of the potential candidates.

Wavelet based numerical methods are one among several possible meshless methods. They were introduced, amongst earliest, in [1]–[3] as numerical techniques for solving partial differential equations (PDEs). Among several available wavelets, the Daubechies wavelet, introduced in the landmark paper [4], was used in these works because of its several remarkable properties such as compact support, orthogonality and vanishing moments. The wavelet based numerical technique is, basically, a Galerkin method with wavelets as the basis and test functions. The integrals that arise in the Galerkin method have a closed form solution for the Daubechies wavelets. These integrals are known as the *connection coefficients* and their computation was outlined in [5] in which the integrals were computed over the entire unbounded domain. The computations for the bounded version of the integrals have been indicated in [6].

One of the earliest implementation of the wavelet-Galerkin method to a two-dimensional problem was shown in [7]. Subsequently, applications to problems in multiple disciplines were illustrated, for example, [8]–[12]. Wavelet based numerical solutions have been under consideration in the electromagnetics community too in recent times [13]. They have been shown to be a promising alternative and several applications have been shown in [14]–[16]. In these works, the integral formulation of the electromagnetics problem is solved using a wavelet-like basis function and the Green's function approach. In low-frequency electromagnetics, a wavelet based technique to solve the differential formulation of an electromagnetics problem was shown in [17].

0-7803-9771-1/06/$25.00 ©2006 IEEE

In this paper, we present a wavelet-Galerkin method for the computation of low-frequency electromagnetic fields in two-dimensional domains. Unlike earlier works reported in literature, we implement boundary conditions by adapting the strategy presented in [18] to two-dimensional problems. This method involves a fictitious expansion of the domain which, subsequently, leads to a simple FEM-like implementation of the boundary conditions. In order to verify feasibility of the technique, we solve a two-dimensional Laplace's problem on a square-shaped domain with Dirichlet boundary conditions. We compare the results obtained with the analytical solution, which exists for this case. We also illustrate the application of this method to two problems – study of skin effect phenomenon in the bar-plate configuration, which is a standard low-frequency test geometry and leakage field computation in a single-phase three-limb transformer.

The following is the organisation of the paper. Section II gives a brief overview on wavelets. In section III, we briefly discuss the computation of connection coefficients. This computation requires the moments of the scaling function, whose computation is also described. The wavelet-Galerkin method as applied to the two-dimensional **A**-formulation for low-frequency electromagnetic field computations is explained in section IV. We present the various results obtained from our work in section V. Finally, in section VI, we conclude our paper.

## II. WAVELET THEORY: AN OVERVIEW

Wavelets can be thought of as a class of basis functions which can be used to approximate any function in a given functional space. A functional space is a collection of functions which possess a well-defined mathematical structure. The Hilbert space, for example, is a complete functional space with an inner product defined on it.

The functional space of square-integrable functions is denoted as $L^2(\mathbb{R})$. Let $\psi$ be a function belonging to this space and defined as,

$$\psi_{m,n}(\cdot) \equiv 2^{m/2}\psi(2^m \cdot -n) \qquad m, n \in \mathbb{Z} \tag{1}$$

The variation of the indices $m$ and $n$ leads to the dilation and translation of this function. A function $f$ from $L^2(\mathbb{R})$ can be approximated in terms of $\psi$ as,

$$f = \sum_{m\in\mathbb{Z}} \sum_{n\in\mathbb{Z}} c_{m,n}\psi_{m,n} \tag{2}$$

where $c_{m,n}$ are the coefficients of expansion of $f$ in terms of $\psi$ and,

$$c_{m,n} = \langle f, \psi_{m,n} \rangle_{L^2} = \int_{-\infty}^{+\infty} f(x)\overline{\psi_{m,n}(x)}dx \tag{3}$$

The basis $\psi$ is a wavelet if it has the following two properties:

1) Zero average value,

$$\int_{-\infty}^{+\infty} \psi(x)dx = 0$$

2) Admissibility,

$$\int_{-\infty}^{+\infty} \frac{|\hat{\psi}(\omega)|^2}{|\omega|}d\omega < \infty$$

An elegant and natural framework for the construction of such wavelet bases is the *multiresolution analysis* (MRA).

An MRA in $L^2(\mathbb{R})$ is defined as a sequence $\{V_m\}$ of closed subspaces of $L^2(\mathbb{R})$ having the following properties:

1) $\cdots \subset V_2 \subset V_1 \subset V_0 \subset V_{-1} \subset V_{-2} \subset \cdots$
2) $f(\cdot) \in V_m \Leftrightarrow f(2^{-m}\cdot) \in V_0 \quad \forall m \in \mathbb{Z}$
3) $f(\cdot) \in V_0 \Rightarrow f(\cdot - n) \in V_0 \quad \forall n \in \mathbb{Z}$
4) $\bigcup_{j\in\mathbb{Z}} V_j = L^2(\mathbb{R})$
5) $\bigcap_{j\in\mathbb{Z}} V_j = \{0\}$
6) There exists an orthonormal basis for $V_0$ $\{\phi_{0,n}(\cdot) : \phi_{0,n}(\cdot - n) \in V_0 \; \forall n \in \mathbb{Z}\}$. This function $\phi$ is called the scaling function of the MRA.

The basic philosophy behind the MRA is that whenever a collection of closed subspaces satisfies the above six properties, then there exists an orthonormal wavelet basis $\{\psi_{m,n} : \psi_{m,n} \in L^2(\mathbb{R}) \; \forall m, n \in \mathbb{Z}\}$ such that every function $f$ in $L^2(\mathbb{R})$ can be represented as,

$$P_{m-1}f = P_m f + \sum_{n\in\mathbb{Z}} \langle f, \psi_{m,n} \rangle \psi_{m,n} \tag{4}$$

where $P_m$ is the orthogonal projection of $f$ onto $V_m$,

$$P_m f = \sum_{n\in\mathbb{Z}} \langle f, \phi_{m,n} \rangle \phi_{m,n} \tag{5}$$

Consequently, the scaling function has to satisfy a two-scale equation known as the dilation or the refinement equation,

$$\phi(\cdot) = \sum_n a_n \phi(2 \cdot -n) \tag{6}$$

In an MRA space, the basis $\psi$ can be constructed entirely from the scaling function $\phi$ alone [19]. Thus, the scaling function serves as the starting point for the construction of wavelet families. The scaling function is also not required to be orthonormal and a weaker condition of it being a Riesz basis is sufficient for the construction of the MRA.

The scaling function also has some additional characterisations, namely,

1) Normality of the scaling function,

$$\int_{-\infty}^{+\infty} \phi(x)dx = 1$$

2) Partition of unity,

$$\sum_{k\in\mathbb{Z}} \phi(x - k) = 1 \qquad \forall x \in \mathbb{R}$$

3) Moments of the scaling function,

$$\mathcal{M}_p = \int_{-\infty}^{+\infty} x^p\phi(x)dx \qquad p \in \mathbb{N}$$

The Daubechies family of wavelets is an example of compactly supported orthonormal wavelets. Daubechies showed that if $\psi$ is a wavelet with $p$ vanishing moments, and $\psi \in L^2(\mathbb{R})$, then $\psi$ has a support of size larger than or equal to $2p - 1$. The Daubechies wavelet has this minimum size and is, thus, optimal and has its support in $[-p + 1, p]$. The scaling function corresponding to the Daubechies wavelet has its support in $[0, 2p - 1]$. The Daubechies wavelet with support $N = 2p - 1$ is referred to as $DN$. Fig. 1 shows the Daubechies wavelet with three vanishing moments and the associated scaling function.

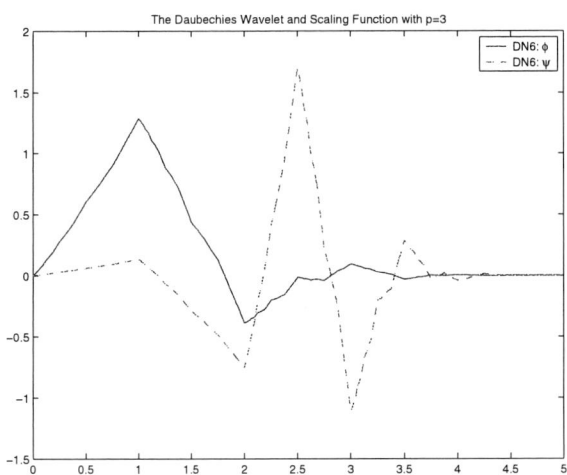

Fig. 1. The Daubechies wavelet D6 and its associated scaling function.

### III. COMPUTATION OF CONNECTION COEFFICIENTS AND MOMENTS OF SCALING FUNCTION

Connection coefficients are integrals that arise in the solution of PDEs using the wavelet-Galerkin method. Connection coefficients are classified into two- and three-term connection coefficients, depending on the number of terms in the integrand. The determination of these integrals by numerical quadrature may lead to large errors in the result as the wavelets involved are highly oscillatory. However, there exist well defined closed-form expressions for the evaluation of the connection coefficients with the Daubechies wavelet and its derivatives in the integrand. Reference [5] is comprehensive on the computation of these connection coefficients. We describe here only the important results for the sake of completeness.

Let us denote the derivative of the scaling function as,

$$\phi_n^{(d)} \equiv \frac{d^d}{dx^d}(\phi(x - n)) \tag{7}$$

The two-term connection coefficient, which is required in this work, in its most general form, is defined as,

$$\Lambda_{l,m}^{d_1 d_2} \equiv \int_{-\infty}^{+\infty} \phi_l^{d_1}(x)\phi_m^{d_2}(x)dx \qquad d_1, d_2 \in \mathbb{Z}^+ \tag{8}$$

For a given $d_1$ and $d_2$, $\Lambda^{d_1 d_2}$ is non-zero only when,

$$2 - N \leq l \leq N - 2, \quad 2 - N \leq m \leq N - 2 \quad \text{and} \quad |l - m| \leq N - 2$$

The Daubechies scaling function differentiated $d$ times also satisfies a dilation equation similar to (6),

$$\phi^d(x) = 2^d \sum_{n=0}^{N-1} a_n \phi^d(2x - n) \tag{9}$$

where $N$ is the support of the Daubechies wavelet $DN$ and $a_n$ are the wavelet filter coefficients.

The moments of the scaling function, $\mathcal{M}_p$, can be computed using a recurrence relation [6],

$$\mathcal{M}_p = \frac{1}{2^{p+1} - 2} \sum_{i=0}^{N-1} \sum_{j=1}^{p} \binom{p}{j} a_i i^j \mathcal{M}_{p-j} \tag{10}$$

with the condition that $\mathcal{M}_0 = 1$.

Thus, the two-term connection coefficients can be determined from the linear system,

$$A\Lambda^{d_1 d_2} = \frac{1}{2^{d-1}}\Lambda^{d_1 d_2} \tag{11}$$

where, $d = d_1 + d_2$ and,

$$A_{l:q} = \sum_p a_p a_{q-2l+p} \tag{12}$$

This linear set of equations are known as the *scaling equations*. However, these equations are homogeneous and hence, the following normalisation condition is required to solve for the connection coefficients,

$$\sum_i \mathcal{M}_i^d \Lambda_l^{0,d} = d! \tag{13}$$

### IV. WAVELET-GALERKIN METHOD FOR A-FORMULATION

Computation of electromagnetic fields in low-frequency electromagnetics requires the solution of Maxwell's equations over the domain of interest. However, fields are seldom obtained by directly solving the Maxwell's equations, instead, the electric and magnetic potentials are solved for using an appropriately formulated version of the Maxwell's equation.

The computation of magnetic fields at low-frequencies is carried out by solving the magnetic vector potential formulation or the **A**-formulation of the Maxwell's equations over the domain of interest,

$$\frac{1}{\mu}\nabla^2 \mathbf{A} - \sigma \frac{\partial \mathbf{A}}{\partial t} = -\mathbf{J} \tag{14}$$

where **J** is the current density of the source in some subdomain. An unique solution for **A** can be obtained by specifying either its magnitude over the entire domain boundary (Dirichlet boundary condition) or a combination of its derivative over some part of the boundary (Neumann boundary condition) and the Dirichlet boundary condition over the rest of the boundary.

In the case of two-dimensional domains, assuming that the source current is normal to the domain, the magnetic vector potential has only one component, namely, the component along the normal direction. Let us assume that this normal component is $A_z$, in which case, (14) reduces to,

$$\frac{1}{\mu}\left(\frac{\partial^2 A_z}{\partial x^2} + \frac{\partial^2 A_z}{\partial y^2}\right) - \sigma\frac{\partial A_z}{\partial t} = -J_z \qquad (15)$$

We solve (15) using the wavelet-Galerkin method wherein, we approximate $A_z$ as,

$$A_z = \sum_i \sum_j a_{ij}\phi_{ij}(x,y) \qquad (16)$$

where $\phi_{ij}(x,y)$ is the two-dimensional scaling function associated with a two-dimensional wavelet and is obtained as the tensor product of the scaling functions along x- and y-directions,

$$\phi_{i,j}(x,y) = \phi_i(x) \otimes \phi_j(y) \qquad (17)$$

where,

$$\phi_{m,n}(\cdot) = 2^{m/2}\phi(2^m \cdot -n) \qquad (18)$$

In the Galerkin's method, the test function for weighting the residual is the same as the basis function which approximates the unknown potential. For the low-frequency electromagnetics problems in this work, we solve for only the time-harmonic or static fields. In the first case, the time-derivative in (15) can be replaced by $j\omega$, where $j = \sqrt{-1}$ while in the latter, it is not present.

Using (16), (17) and (18) in (15), and by the Galerkin's technique, we obtain, after taking the appropriate inner products, the following equation.

$$\frac{1}{\mu}a_{ij}\left[\int \phi_i^{(2)}(x)\phi_p(x)\mathrm{d}x \otimes \int \phi_q(y)\phi_j(y)\mathrm{d}y \right.$$
$$\left. + \int \phi_i(x)\phi_p(x)\mathrm{d}x \otimes \int \phi_q^{(2)}(y)\phi_j(y)\mathrm{d}y\right]$$
$$-\sigma j\omega a_{ij}\left[\int \phi_i(x)\phi_p(x)\mathrm{d}x \otimes \int \phi_q(y)\phi_j(y)\mathrm{d}y\right]$$
$$= -J_z\int \phi_p(x)\mathrm{d}x \int \phi_q(y)\mathrm{d}y \qquad (19)$$

which leads to the linear system,

$$[\Lambda]\{a\} = \{b\} \qquad (20)$$

where $\Lambda$ and $b$ are respectively the *global coefficient matrix* and the *global force vector*.

In order to implement boundary conditions, one approach is to determine the coefficients $a_{ij}$ corresponding to those basis functions whose support lie on the boundary by seperately solving a system of linear equations, and then using these known values of $a_{ij}$ in (20) to solve for the other coefficients. This approach is followed in [17]. However, in this work, we adopt a different approach for implementation of the boundary conditions. This is an extension of the method used in [18] for one-dimensional problems to two-dimensional ones. In this method, we expand the domain fictitiously so as to ensure that the support of the wavelets lie entirely within this fictitious, larger domain. As a result, the integrals in (19) will have the limits of integration from $-\infty$ to $+\infty$. Although this leads to

an increase in the size of the linear system in (20) as compared to the case where the limits of integration would have been been the physical boundary limits of the domain, there are two distinct advantages. One is that the connection coefficients described in section III can be used in the wavelet-Galerkin method in the same form as described. More importantly, the Dirichlet boundary conditions can now be implemented in a very simple and intutive manner at the physical boundary of the domain, as in the case of FEM. Thus, at the nodes $i$ corresponding to the physical boundary of the domain, we have,

$$a_{ii} = 1, \qquad a_{ij} = 0 \ \forall \ i \neq j \quad \text{and} \quad b_i = 0$$

## V. Simulations and Results

In order to investigate the feasibility of the wavelet-Galerkin method, we tested the formulation on a two-dimensional Laplace's equation with a known analytical solution. The problem is defined as follows,

$$\nabla^2 u = 0 \qquad \forall x, y \in [0,1]$$

$$u|_{x=0} = u|_{y=0} = u|_{x=1} = 0 \ \& \ u|_{y=1} = 1$$

The analytical solution for this problem is,

$$u(x,y) = \sum_{n \in \mathbb{N}_{odd}} \frac{4}{n\pi} \frac{\sin(n\pi x)\sinh(n\pi y)}{\sinh(n\pi)}$$

The wavelet-Galerkin solution for this problem was obtained by a method similar to that outlined in section IV. MATLAB codes were written for the implementation of the wavelet-Galerkin formulation.

Fig. 2 shows the contour plot of the analytically obtained solution while Fig. 3 shows the plot as obtained from the wavelet-Galerkin method. The D6 wavelet was used in the numerical solution and the potential $u$ was determined on an uniformly chosen grid of $16 \times 16$ points in the domain. The two plots match each other closely except near the boundary having the non-zero boundary condition. The difference in the plots at this boundary can be explained on the basis of the fact that the analytical solution involves an infinite series. As a result, the solution does not exactly converge to the imposed boundary value even with a large number of terms in this series while obtaining the analytical solution.

We next illustrate the feasibility of applying the wavelet-Galerkin method to the computation of low-frequency electromagnetic fields by solving two problems. The first problem involves determining the skin effect phenomenon in a bar-plate geometry. This is a standard test geometry consisting of a high permeability conducting plate and a bar carrying a high current. The D6 wavelet was used in the numerical technique and the time-harmonic magnetic vector potential $A_z$ was determined on an uniformly spaced grid of $64 \times 64$ points. A higher resolution is required in order to effectively capture the low value of depth of penetration of the field in the plate. The contour plot of $A_z$ is shown in Fig. 4.

Fig. 5 shows the contour plot of the leakage field for a single-phase three-limb transformer. Simulation was carried

1010

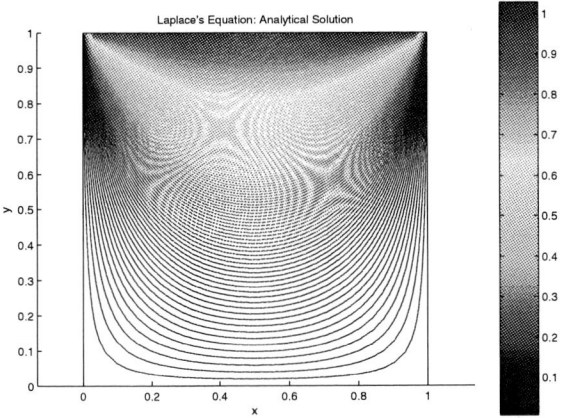

Fig. 2. Contour plot of the analytical solution to the Laplace's equation.

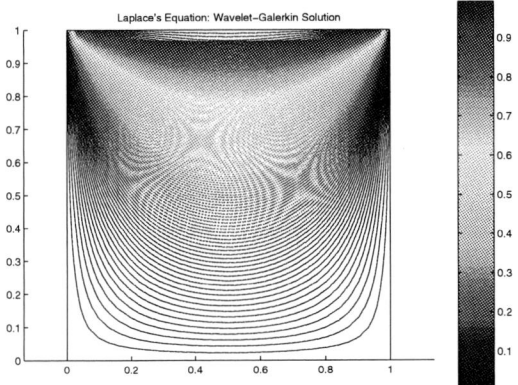

Fig. 3. Contour plot of the wavelet-Galerkin solution to the Laplace's equation.

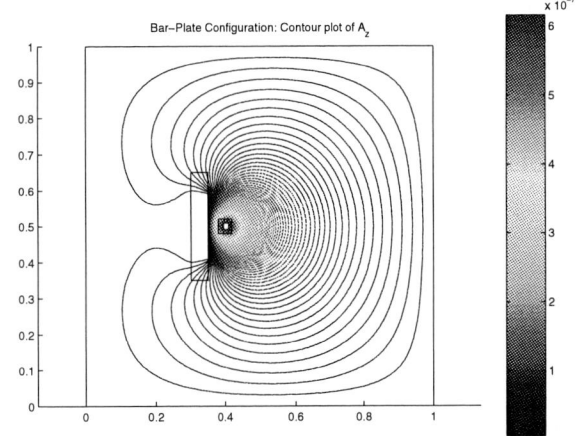

Fig. 4. Contour plot of $A_z$ showing skin effect phenomenon in a bar-plate configuration.

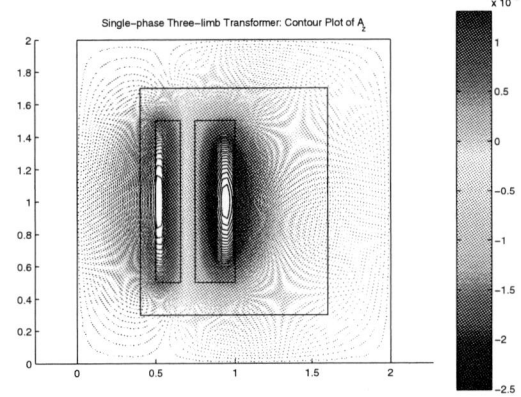

Fig. 5. Contour plot of $A_z$ showing the leakage of flux in a half-symmetric geometry of a single-phase three-limb transformer.

out for only a half-symmetric model of the transformer. Furthermore, the determination of the magnetic leakage flux requires only a magnetostatic solution. Thus, the time-derivative term in (15) is not included in the formulation. We used the D6 wavelet in this case too as it has the least number of vanishing moments among all Daubechies wavelets which are at least twice differentiable and, hence, represents the worst case choice of wavelet in terms of regularity. The results obtained in this case too are on expected lines.

## VI. CONCLUSION

The wavelet-Galerkin method has been outlined in this paper for the numerical computation of low-frequency electromagnetic fields. An alternative method for application of the boundary conditions has also been proposed. The application of this technique has been illustrated to two low-frequency electromagnetics problems. In both problems, the magnetic vector potential was obtained by solving the **A**-formulation of the magnetics problem. The results obtained are in good agreement with the expected solution for the problems

and thus, establishes the feasibility of the wavelet-Galerkin method. With this, future work will require evaluation of its performance as compared with other numerical methods.

### REFERENCES

[1] J.-C. Xu and W.-C. Shann, "Galerkin-wavelet methods for two-point boundary value problems," *Numerische Mathematik*, vol. 63, no. 1, pp. 123–144, 1992.

[2] S. Qian and J. Weiss, "Wavelets and the numerical solution of partial differential equations," *Journal of Computational Physics*, vol. 106, pp. 155–175, 1993.

[3] K. Amaratunga, J. R. Williams, S. Qian, and J. Weiss, "Wavelet-Galerkin solutions for one dimensional partial differential equations," *International Journal for Numerical Methods in Engineering*, vol. 37, no. 16, pp. 2703–2716, 1994.

[4] I. Daubechies, "Orthonormal bases of compactly supported wavelets," *Communications on Pure and Applied Mathematics*, vol. 41, pp. 909–996, 1988.

[5] A. Latto, H. L. Resnikoff, and E. Tenenbaum, "The evaluation of connection coefficients of compactly supported wavelets," in *Proceedings of the French-USA Workshop on Wavelets and Turbulence*, Princeton University. New York: Springer-Verlag, 1994.

[6] M.-Q. Chen, C. Hwang, and Y.-P. Shih, "The computation of wavelet-Galerkin approximation on a bounded interval," *International Journal for Numerical Methods in Engineering*, vol. 39, pp. 2921–2944, 1996.

[7] Raymond O. Wells Jr. and X. Zhou, "Wavelet solution for the Dirichlet problem," *Numerische Mathematik*, vol. 70, pp. 379–396, 1995.

[8] J. Ko, A. J. Kurdila, and M. S. Pilant, "A class of finite element methods based on orthonormal, compactly supported wavelets," *Computational Mechanics*, vol. 16, pp. 235–244, 1995.

[9] M.-Q. Chen, C. Hwang, and Y.-P. Shih, "A wavelet-Galerkin method for solving population balance equations," *Computers and Chemical Engineering*, vol. 20, no. 2, pp. 131–145, 1996.

[10] W. E. Hutchcraft, L. A. Harrison, R. K. Gordon, and J.-F. Lee, "The numerical solution of elliptic problems using compactly supported wavelets," in *Proceedings of the Thirtieth Southeastern Symposium on System Theory*, Mar. 1998, pp. 102–106.

[11] E. B. Lin and X. Zhou, "Connection coefficients on an interval and wavelet solutions of Burgers equation," *Journal of Computation and Applied Mathematics*, vol. 135, no. 1, pp. 63–78, Oct. 2001.

[12] X. Chen, S. Yang, J. Ma, and Z. He, "The construction of wavelet finite element and its application," *Finite Elements in Analysis and Design*, vol. 40, no. 5-6, pp. 541–554, 2004.

[13] M. N. O. Sadiku, C. M. Akujuobi, and R. C. Garcia, "An introduction to wavelets in electromagnetics," *IEEE Microwave*, vol. 6, no. 2, pp. 63–72, June 2005.

[14] R. L. Wagner and W. C. Chew, "A study of wavelets for the solution of electromagnetic integral equations," *IEEE Trans. Antennas Propagat.*, vol. 43, no. 8, pp. 802–810, Aug. 1995.

[15] G. Pan, "Orthogonal wavelets with applications in electromagnetics," *IEEE Trans. Magn.*, vol. 32, no. 3, pp. 975–983, May 1996.

[16] ——, "Wavelets: A promising approach to linear and nonlinear modeling of electromagnetic problems," in *Proceedings of the International Conference on Microwave and Millimeter Wave Technology*, Aug. 1998, pp. 23–25.

[17] Y. Shiyou, N. Guangzheng, Q. Jingen, and L. Ronglin, "Wavelet-Galerkin method for computation of electromagnetic fields," *IEEE Trans. Magn.*, vol. 34, no. 5, pp. 2493–2496, Sept. 1998.

[18] D. Lu, T. Ohyoshi, and L. Zhu, "Treatment of boundary conditions in the application of wavelet-Galerkin method to a SH wave problem," Mar. 1996. [Online]. Available: citeseer.ist.psu.edu/84953.html

[19] I. Daubechies, *Ten Lectures on Wavelets*. Philadelphia: SIAM, 1994.

**Kaushik K** (S'06) is currently pursuing his Master's Degree in Technology with specialisation in Power Electronics and Power Systems from Indian Institute of Technology Bombay, Mumbai, India. He received his Bachelor's degree in Engineering with specialisation in Electrical and Electronics Engineering from Anna University, Chennai, India in the year 2005. He was a Young Engineering Fellow at the Department of High Voltage Engineering, Indian Institute of Science, Bangalore in the summer of 2004. He was also a Summer Research Fellow at the Department of Electrical Engineering, Indian Institute of Technology Bombay, Mumbai, India in the summer of 2005.

**S. V. Kulkarni** (M'99) is Associate Professor, Department of Electrical Engineering, Indian Institute of Technology Bombay, India. Previously, he worked at Crompton Greaves Limited and specialised in the design and development of transformers up to 400 kV class. He has authored a book titled *Transformer Engineering: Design and Practice*, published by Marcel Dekker Inc. He has authored more than 70 professional publications in reputed journals and conferences, and was recipient of the Young Engineer Award (2000) from the Indian National Academy of Engineering. He has also received the Career Award for Young Teachers from All India Council for Technical Education in 2001. His research interests include transformer design and analysis, computational electromagnetics and distributed generation.

**2006 IEEE International Conference on Power Electronic, Drives and Energy Systems**

# Frequency Linked Pricing as an Instrument for Frequency Regulation Market and ABT Mechanism

K.V.V. Reddy, Ashwani Kumar, and Saurabh Chanana

*Abstract--*In this paper, a market framework for Automatic Balance Services (ABS), both for primary as well as secondary regulation is implemented. The market structure is based on the assumption that the balance providers automatically respond to frequency dependent price signals sent by the ISO, when there is a frequency excursion in the system. An optimization models are implemented to obtain the market clearing prices for the individual regulation services. These obtained optimum frequency regulation contracts are incorporated within a dynamic simulation model of a two control-area system, and frequency regulation has been utilized as the important signal to maintain grid discipline under the new tariff structure under ABT regime to maintain grid frequency. This paper signifies the role of Economic Load Dispatch as a useful tool for state beneficiaries in reducing their costs and promoting merit order dispatch under the new ABT Regime. The results have been demonstrated through studies made on a 14 Bus system

*Index Terms--* Automatic balance services, Frequency linked pricing, frequency regulation, dynamic model, Availability Based Tariff, Economic Dispatch, Unscheduled Interchange.

## I. INTRODUCTION

IN deregulated power systems the Independent System Operator (ISO) provides system frequency control services in three broad ways [1]:

- a primary regulation service from generating units that respond to frequency changes within a few seconds;
- a secondary regulation service from generating units that respond to signals from the ISO within 5 to 10 minutes; and
- a secondary regulation service from loads (customers) that respond to signals from the ISO within 5 to 10 minutes.

Real-time frequency-linked price signals have been used earlier and have been shown to achieve improved frequency regulation and control performances [2], [3]. Berger and Schweppe [2] demonstrated that real time pricing in the presence of system dynamics could aid in load-frequency

control. In [3] a pricing scheme was suggested that gives the importing area a signal in terms of increased price for any increment in power drawn, from scheduled value. The increase in price is viewed as a penalty that discourages deviation from the scheduled power flow and thereby ensures grid discipline.

A detailed discussion of the potential system control problems and issues, associated with frequency control in deregulation, has been provided in [4], [5]. Two possible structures for load frequency control have been proposed therein, the 'charged' and the 'bilateral'. In the first structure, the ISO would purchase real power from the providers (can be generators or customers) on a near-to real-time basis. This structure is already being practiced in some countries, for example the Scandinavian countries. The later structure could be where the ISO has no obligation to provide frequency control and customers purchase load matching contracts from their energy provider directly. In [6] and [7] a framework for price-based operation of AGC in deregulated electricity markets has been developed. Bilateral contracts, poolco based transactions and area regulation contracts have been considered for AGC. Most of the ISOs in the US and in the Union for the co-ordination of Transmission of Electricity (UCTE) in Europe have an AGC system in place, for their secondary regulation. On the other hand, the Nordel and the UK systems, which are not part of UCTE, have opted for manual secondary control. The manual secondary control service is known as the *balance service* (or by similar names) since it continuously seeks to balance the system generation and consumption. The ISO of these respective countries accepts bids by volume (in MW) and price ($/MWh) from generators willing to quickly (max 10 minutes) increase or decrease their generation, or even consumers willing to increase or decrease their consumption. The bids for regulation are arranged in price order to form a "staircase" for each operating hour. When regulation is needed, the ISO activates the most profitable bid for regulating up or down. At the end of each hour, the regulation price is determined in accordance with the most expensive measure taken during upward regulation (the purchase of balance energy), or the cheapest measure taken during downward regulation (the sell of balance energy), used during the hour. The final regulation price applies to all selected balance service providers [8].

Jhong and Kankar presented design of an automatic balance service market, which can act as an effective tool for system frequency control in deregulated electricity markets [10].

---

K.V.V. Reddy is M. Tech. student at NIT Kurukshetra, and completed his Ph.D. recently. He is with ABB presently. Email: vishnuvardhan.reddy@in.abb.com. Ashwani Kumar is A.P in the Department of Electrical Engineering. Email: ashwa_ks@yahoo.co.in. S. Chanana is with Electrical Engg. Department. Email: s_chanana@rediffmail.com.

0-7803-9771-1/06/$25.00 ©2006 IEEE

Introduction of Availability Based Tariff (ABT) along with Electricity Act 2003 was perhaps the most significant and definitive step taken in the Indian power sector so far to bring more efficiency and focus to this vital infrastructure sector. ABT concerns itself with the tariff structure for bulk power and is aimed at bringing about more responsibility and accountability in power generation and consumption through a scheme of incentives and disincentives. As per the notification [11], ABT is applicable to only central generating stations having more than one SEB/State/Union Territory as its beneficiary. The new tariff regime aims at inducing discipline at the generation and consumption end through adequate monetary incentives. These incentives will encourage generators to produce more during peak load hours and curtail generation adequately during off-peak hours on one hand and discourage consumers from overdrawing on the other hand.

The most significant aspect of ABT is the splitting of the existing monolithic energy charge structure into fixed and variable cost components. This will act as an incentive for power trading which shall (ideally) conclude in a self-regulating power market regime [12]. It is also expected to promote the concept of ELD (Economic Load Dispatch) among power generators [13]. In the current scenario, the generators tend to produce as much as they can irrespective of the demand side of the power equation. Under ABT, generators will need to ramp up and ramp down generation based on the declared generation schedule given by the RLDC (Regional Load Dispatch Center).

In this paper, a market framework for Automatic Balance Services (ABS), both for primary as well as secondary regulation has been utilized [10]. The frequency signal obtained from the dynamic simulation has been utilized as an important input for the grid discipline under the ABT regime in the Indian Power sector. In this paper an attempt has been made to analyze a 14 Bus system to illustrate that how states can minimize their cost by strategically re-dispatching / rescheduling their load under the given frequency conditions. For this purpose, the conventional formulation of Economic Load Dispatch [14] has to be modified to take into account the frequency dependent component of ABT. In this study, the analysis is carried under static conditions and we have considered frequency as an external input to the system assuming that redispatch of load in the area under consideration will have little impact on the grid frequency.

## II. METHODOLOGY

### II.1 PRIMARY REGULATION

The scheme considers that $\gamma_{PR, i}$ is the quantity per unit market price of primary frequency regulation offered by a genco '$i$'. Similarly, $\eta_{PR,i}$ is the price per unit of frequency deviation, offered by the genco '$i$' [10]. If unit-1 offers to increase its generation by 1.95% for a 1% increase in price, and assuming a market price of \$150/MWh, and can be represented as:

$$\gamma_i = \frac{1.95\% \, p.u.MW}{150\$/MWh.1\%} = 0.013 \frac{p.u.MW}{\$/MWh} \qquad (1)$$

Equation (1) means that a genco bidding $\gamma_i = 0.013$p.u MW/\$/MWh is offering 0.013p.u.MW per unit price change. The market is settled based on the price and quantity offers from all gencos, as discussed in the optimization model later. The ISO determines the primary regulation market price $P_{PR}$, specified in per unit of frequency excursion, from the generators' bid price parameters $\gamma_{PR,i}$ and $\eta_{PR,i}$ ($i=1,...,n$) as well as their location in the system. The contracted gencos receive the market price $P_{PR}$. Primary regulation services are activated by $P_{PR}$ that is sent when there is a frequency excursion in the system. The generating units contracting this service are prepared to increase or reduce their generation instantaneously as per the following relation:

$$\Delta P_{gi} = -P_{PR}.\gamma_{PR\,i}.\Delta f_j \qquad (2)$$

A dynamic model of gencos has been utilized that would participate and respond to the ISO's frequency-linked price signal, for primary frequency control within a control area [10]. Each contracted genco is available to provide $\gamma_{PR,i}$ amount of primary control service which is activated by the primary frequency regulation market price signal $P_{PR}$.

The optimization model to determine the primary regulation market price can be formulated as an objective function to minimize the payment for primary regulation service.

$$\text{Minimise PRobj} = \sum_i U_{PR,i}.\gamma_{PR,i}.P_{PR} \qquad (3)$$

Constraints:

$$\sum_i U_{PR,i}.\gamma_{PR,i}.LF_i \geq D_{PR} \qquad (4)$$

$$U_{PR,i}.\eta_{PR,i} \leq P_{PR} \qquad \forall i \in n \qquad (5)$$

$$U_{PR,,i} = \begin{cases} 1 \\ 0 \end{cases}$$

1 for Bid offer of generator i selected

0 for Bid offer of generator i not selected

$U_{PR,i}$ is a binary variable associated with the selection of a particular bid offer for primary regulation.

### II.2. SECONDARY REGULATION

These services are contracted in order to restore the system frequency to nominal (say 50 Hz) after a disturbance. This can be achieved either by increasing/reducing an area's total generation or by reducing / increasing an area's total demand. The secondary regulation market design includes bids from customers also, in addition to those from gencos. The market is settled in the same way as that for the primary regulation market. The proposed structure of bid price and quantity for the participants of the secondary-up and -down regulation market is presented in [10]. The secondary regulation bid $\eta_{SR}$ is the price per unit of energy while $\gamma_{SR}$ is the quantity offered per unit of price.

*SECONDARY REGULATION SERVICE FROM GENERATORS*: Consider the case of gencos contracting for secondary regulation services. The contracted gencos respond to the price-linked signal sent out by the ISO according to their response rates depending on their technical characteristic or on economical criteria. Frequency control in this case can be achieved by

1014

controlling generation in an area through a real-time secondary regulation. The price-linked control signal is of the form:

$$\Delta\rho\,(t) = -P_{SR,j} \int ACE_j\,(t)\,dt \qquad j = area\,1,\,2; \qquad (6)$$

$P_{SR}$ is the area's secondary regulation price that is obtained from the market settlement model, to be discussed next. The generators contracted to respond to the control signal increase or decrease their generation as per the following relation:

$$\Delta P_{gi} = \gamma_{SR,i}\,.\,\Delta\rho \qquad (7)$$

Secondary Regulation Service from Customers: In this case, the contracted customers respond to the price-linked control signal sent by the ISO by increasing or reducing their demand, according to their response rates that may depend on their load characteristic or economical criteria. The price-linked signal is the same, as described in (5). The loads contracted to respond to this signal, increase or decrease their load as per the following relation:

$$\Delta P_{di} = -\gamma_{SR,\,Di}\,.\,\Delta\rho \qquad (8)$$

Generators and customers can provide the regulation bid, quantity $\gamma_{SR,i}^{UP}$ at price $\eta_{SR,i}^{UP}$ for up-regulation, and $\gamma_{SR,i}^{DN}$ at price $\eta_{SR,i}^{DN}$ for down regulation. After receiving the bid information, the ISO determines the up regulation market price $P_{SR}^{UP}$ and the down-regulation market price $P_{SR}^{DN}$, such that the payment for secondary regulation service is minimal and the social welfare is maximized. In this case also, the loss factors associated with the location of each bid are taken into account within the optimization framework [10].

The problem is formulated as to minimize payment for secondary regulation service. The optimization problem has been solved using GAMS 21.3 LP solver.

$$\text{Minimize: } SRobj = \sum_i U_{SR,i}^{up}\,.\,\gamma_{SR,i}^{up}.\,P_{SR}^{up} + \sum_i U_{SR,i}^{DN}\,.$$

$$\gamma_{SR,i}^{DN}.\,P_{SR}^{DN} \qquad (9)$$

s.t. constraints:

$$\sum_i U_{SR,i}^{up}\,.\,\gamma_{PR,i}^{up}.\,LF_i \ge D_{PR}^{u} \qquad (10)$$

$$U_{SR,i}^{up}\,.\,\eta_{SR,i}^{UP} \le P_{SR}^{UP} \qquad \forall i \in n \qquad (11)$$

$$\left|\sum_i U_{SR,i}^{DN}\,.\,\gamma_{SR,i}^{DN}.\,LF_i\right| \ge D_{SR}^{DN} \qquad (12)$$

$$\left|U_{SR,i}^{DN}\,.\,\eta_{SR,i}^{DN}\right| \le P_{SR}^{DN} \qquad \forall i \in n \qquad (13)$$

$$U_{SR,i} = \begin{cases} 1 \\ 0 \end{cases}$$

1 for Bid offer of generator i selected
0 for Bid offer of generator i not selected
$U_{SR,i}^{UP}$ and $U_{SR,i}^{DN}$ are binary variables associated with the selection of a bid for secondary regulation services.

II.3. AVAILABILITY BASED TARIFF

Availability Based Tariff comprises of three components: (a) Capacity Charge (b) Energy Charge (c) Unscheduled Interchange (UI) Charge

*A. Capacity Charge*

This component represents the fixed cost and is linked to the availability of the plant, i.e., its capability to deliver MWs on a day-by-day basis. The total amount payable to the generating company over a year towards the fixed cost would depend on the average availability of the plant over the year. In case the average actually achieved over the year is higher than the specified norm for plant availability the generating company would get a higher payment. In case, the average availability achieved is lower the payment will be lower, hence the name Availability Based Tariff.

*B. Energy Charge*

This component of ABT comprises of the variable cost, i.e. the fuel cost of the power plant for generating energy as per given schedule for the day. Therefore, this energy charge is not according to the actual generation but only for scheduled generation.

*C. UI Charge*

In case there are deviations from schedule this third component of ABT comes into picture. Deviations from schedule are determined in 15-minute time blocks through special metering and priced according to the system condition prevailing at that time. If the frequency is above 50 Hz, UI rate will be small, and if its below 50 Hz it will be high. As long as the actual generation / drawl is according to the given schedule, the third component of ABT is zero. In case of over-drawl beneficiary has to pay UI charge according to the frequency dependent rate specified. According to CERC's ABT Order [11], this rate varies linearly between 50.5 Hz and 49.02 Hz as shown in Fig, 1. Earlier the maximum charge, during periods when frequency dropped below 49.02 Hz, was limited to 450 paise per kWh. But now it has been increased to 600 paise per kWh. UI Rate at different frequencies is listed in Table I.

TABLE I
EXAMPLE OF UI RATE AT DIFFEENT FREQUENCIES

| Frequency | UI Rate (paise/kWh) |
|---|---|
| 48.5 | 600 |
| 49.0 | 600 |
| 49.5 | 405 |
| 50.0 | 203 |
| 50.5 | 0 |
| 51.0 | 0 |

II.4 BENEFITS OF ABT

Besides promoting competition, efficiency and economy and leading to more economically viable power scenario, ABT has been able to pave way for high quality power with more reliability and availability through enhanced grid discipline.

➢ By giving incentives for enhancing the output capability of power plants, it enables more consumer load to be met during peak hours.

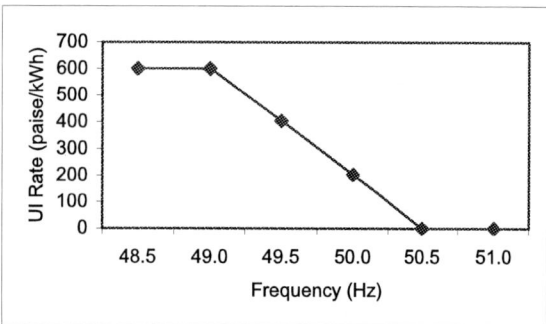

Fig. 1. UI Rate as a function of Grid Frequency.

➤ By separating fixed charges based on availability from variable charges backing down during off peak hours no longer results in financial loss of generating stations. Therefore earlier incentive for not backing down and raising system frequency during off-peak hours no longer exists.

➤ By charging separately for unscheduled interchanges the problem of over drawl during peak load condition, resulting in lowering of frequency has been controlled. UI rate is high during the low frequency condition which discourages the over drawl of power.

The ABT mechanism would provide vast scope of unscheduled interchange of as and when available energy on future. In case of embedded IPPs and licensees within a state the state participating in regional power pool as a control area can save / earn UI charge by strategically re-dispatching / scheduling generation.

### III. CASE STUDY

A two-area interconnected power system is now considered to examine the proposed market dynamics and the performance of frequency-linked prices on regulation services. The dynamic model for the two area system has been presented in [10]. It is assumed that the frequency regulation service providers provide their regulation service to their respective area's ISO only. A dynamic model of a two-area interconnected power system with the proposed primary and secondary control mechanisms have been considered for the simulation. System Performance with Primary and Secondary Regulation Services and the dynamic models are developed using SIMLINK and MATLAB. Market clearing price has been calculated using GAMS NLP solver [9]. Now the selected sets of primary and secondary regulation service providers in two control areas are incorporated for the combined frequency regulation simulation. Two perturbations are considered in Area-A during a 150-seconds simulation period. The first is a 1% increase in load at time t=0 and the second is a 5% increase in load at time t=12 second.

### III.1 SIMULATION RESULTS WITH PRIMARY REGULATION SERVICE

Primary Regulation: In Area-A, it is assumed that there are 10 generators willing to provide the primary regulation service. Their corresponding bid offers, $\eta_{PR,i}$ and $\gamma_{PR,i}$ and those selected by the ISO,

The primary regulation service providers are selected based on the optimization model discussed in section II. The primary regulation market price is calculated using GAMS solver [9] and price is 0.26 $/MWh-Hz and the total primary regulation energy cleared in the market is 0.8 p.u.MW. In Area-B, it is assumed that there are 8 generators willing to provide the primary regulation service. The primary regulation bid offers, $\eta_{PR,i}$ and $\gamma^{PR,i}$, and those selected by the ISO obtained from optimization problem [10].

Secondary Regulation: 11 providers, comprising 6 generators and 5 customers willing to provide the secondary regulation services in Area-A were considered. The secondary up and down bid prices $\eta_{SR,i}^{UP}$ and $\eta_{SR,i}^{DN}$ and the respective bid quantities $\gamma_{SR,i}^{UP}$ and $\gamma_{SR,i}^{DN}$, and those bids selected by the ISO are shown in Table III. In Area-B it is assumed that 9 providers, comprising 5 generators and 4 loads, are willing to provide secondary regulation services.

Figure 2 shows the plot of frequency as primary regulation services are activated. From figure 2 it is observed that the system frequency settles to a new steady-state value with the primary regulation service but does not comes to the initial value. The primary regulation helps arrest any further fall in frequency. The primary regulation performances of the selected primary regulation providers, generators on bus 1,5,6,9, are plotted in figure 3. From the figure 3, it is observed that the frequency dip results instantaneous increase in generation taking place with in a few seconds by the action of governor to match the load demand. The change in generation is proportional to their quantity offer selected from the primary regulation service.

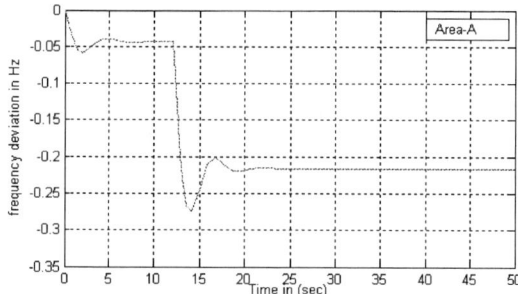

Fig. 2. Area-A Frequency Plot with Primary Regulation.

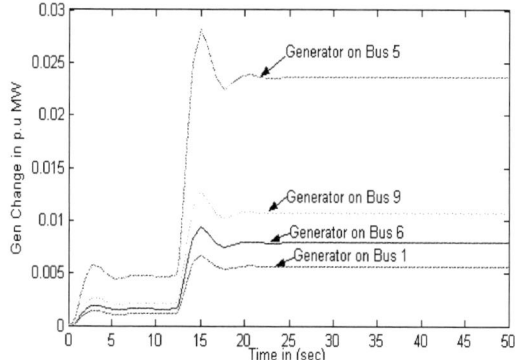

Fig.3. Area-A Primary Control through Generators.

### III. 2 SIMULATION RESULTS WITH PRIMARY AND SECONDARY REGULATION SERVICE

Figure 4 shows the plot of the frequency following the two perturbations with primary and secondary regulation services activated. It is seen that from the figure 3 that the frequency is brought back to the nominal (50 Hz) through secondary regulation in about 2 minutes. Due to primary regulation service it will goes to new steady state value, but after the activation of secondary regulation service I the presence of proportional plus integral controllers, it settles to initial value. Figure 4, 5, and 6 shows the plots for the generators' participation in the frequency control in Area-A.

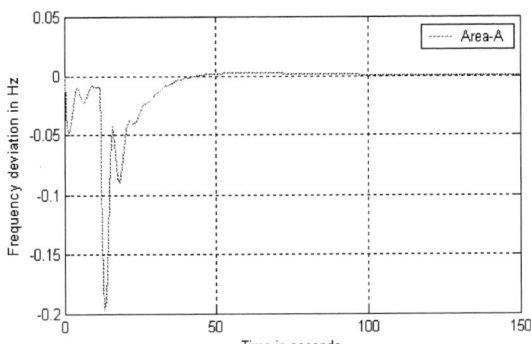

Fig. 4. Area-A Frequency Plot with Primary and Secondary Regulation.

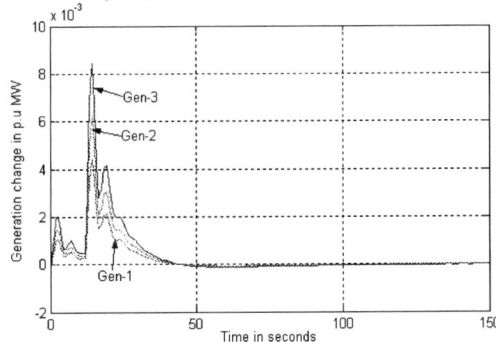

Fig. 5. Area-A Primary Control Action through Generators1-3.

Figure 5 shows primary control responses of three generators that only provide primary regulation. Due to primary regulation service the generation of generators increases instantaneously to match load demand by action of governor response after some time it goes to zero because secondary regulation activates. Figure 6 shows the plot of the generator output that provides both primary and secondary regulation. From the figure 6, it is observed that due to primary regulation the overshoot is occur but activation of secondary regulation it comes to new steady state-value. Figure 7 shows the plot of the generator output that provides only secondary regulation. From the above figures it is observed that due to only secondary regulation the generation increased to match the load

demand.

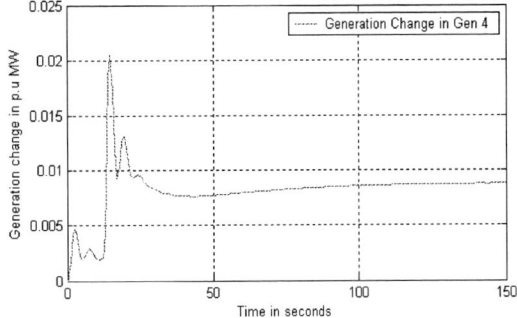

Fig. 6. Area-A Primary and Secondary Control Action from Generator-4.

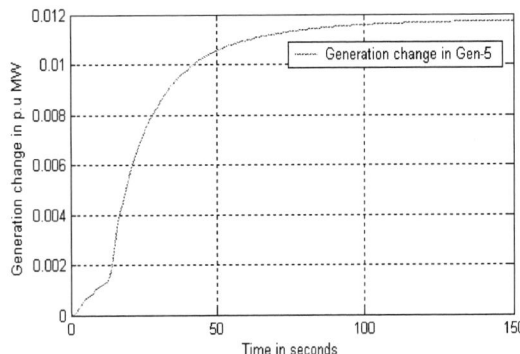

Fig. 7. Area-A Secondary Control Only from Generator-5.

### III.4. ECONOMIC DISPATCH OF POWER UNDER ABT REGIME

The frequency signal obtained form the frequency regulation services can be utilized for deciding the power share to be provided to different states and optimal output can be obtained. In this paper, an optimization approach based on frequency regulation service has been formulated and a sample 14 Bus-system is taken which represents a control area or a state in this case. It is illustrated through this paper that states can use Economic Load Dispatch (ELD) as a tool to reduce their overall costs and even earn UI Charges in some cases. The formulation of ELD problem in this case will include a frequency dependent component of cost in the objective function. This component will be a function of Unscheduled Interchange at grid as shown below.

$$\min C_{gr}(P_{gr}) + \sum_i C_i(Pg_i) \tag{14}$$

subject to

$$P_{gr} + \sum Pg_i - P_L - P_D = 0 \tag{15}$$

$$P_{gr} = SI + UI \tag{16}$$

$$P_{gi}^{\min} \le P_{gi} \le P_{gi}^{\max} \tag{17}$$

$$S_l \le S_l^{\max} \tag{18}$$

Here $C_i$ represents the cost of $i^{th}$ generator $C_{gr}$ represents the cost function for grid which can be further broken down in two components.
Therefore,

1017

$$C_{gr} = \text{UI Charge} + f(\text{SI}) \qquad (19)$$

where

$$\text{UI Charge} = \text{UI} * \text{UIRate} \qquad (20)$$

and UI Rate is dependent on frequency as shown in Fig1.

$P_{gr}$     the grid power output

$P_L$     transmission system losses

$P_D$     net demand

SI     Scheduled Interchange

UI     Unscheduled Interchange

$Pg_i$     power output of $i^{th}$ generator

$S_l$     flow of $l^{th}$ line

$Pg_i^{\min}, Pg_i^{\max}$     limits on output of $i^{th}$ generator

$S_l^{\max}$     limit on flow of $l^{th}$ line

While formulating the above equations for modified ELD problem the following assumptions have been made.

➢ Grid interconnection to the system is represented by a generator of very high rating and fixed terminal voltage.

➢ Re-dispatch of load in the area under consideration will have negligible influence on the frequency conditions in the grid.

➢ The dispatch is calculated for every 15-minute slot during which the system conditions are assumed to be static.

There are total four generators in the area, which compete with the grid to meet the total load of 399MW in a given 15 minute slot. 14 Bus Case diagram, Generator cost data, Grid cost and UI Rate is given in Appendix A. It is assumed that the system is connected to grid via bus no. 2 only, Therefore the grid interconnection is replaced by a generator of very high rating at bus no. 2. Power output of Generators and Grid at different frequencies as a result of economic dispatch is shown in Fig.8.

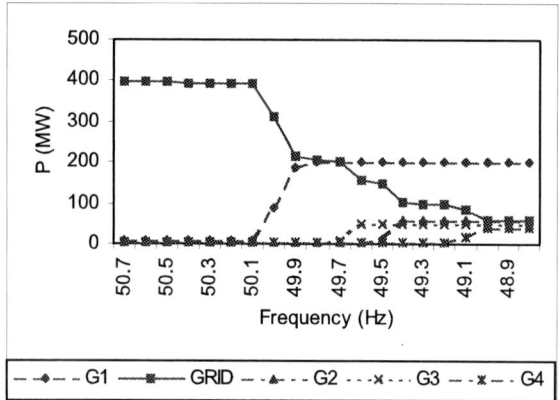

Fig. 8. Load Dispatch at different frequencies.

We can make some important observation from this study

At a frequency above 50 Hz it will be beneficial for state to draw more power from grid as UI Rate is less and Grid power is coming cheap as compared to any other source.

➢ At frequencies below 49.6 Hz Grid power is becoming costlier, therefore state will be benefited by under-drawing from grid. They can now look for other sources of power,

which are now cheaper as compared to grid power. As shown in Fig. 8, G3, G2 and then G4 also start participating in load dispatch as frequency drops below 49.6 Hz.

➢ Participation of generators in Dispatch is determined by there cost effectiveness. As the frequency is moving down first the cheapest generator (G1 in this case) will participate and then the next cheaper (G3) and so on.

➢ As the frequency is moving up or down, by strategically re-dispatching rather than sticking to the scheduled interchange, states can save cost. As illustrated in Fig 9., the cost of dispatch on sticking to SI remains at $5802.5 at all frequencies, as UI is zero. Whereas cost reduces on deviating from schedule on the basis of ED at high and low frequencies

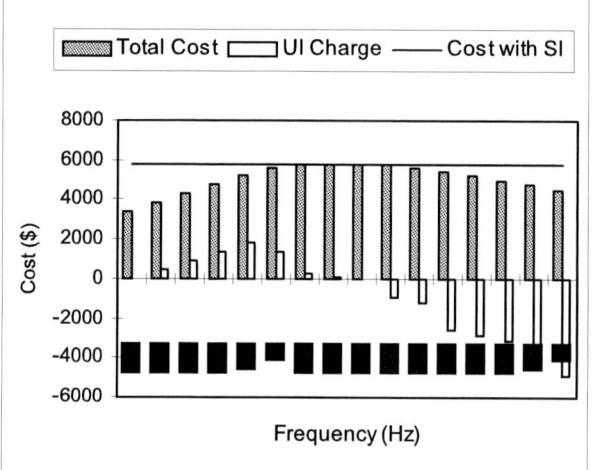

Fig. 9. Variation of System costs with frequency.

➢ At low frequency states can even earn UI by under drawing from grid. But to achieve this the must either be able to curtail their load or have access to some other generators who may be IPPs, licensees or state owned having higher cost as compared to grid at normal frequency.

## IV. CONCLUSIONS

In this paper, frequency regulation issues based on frequency linked pricing has been presented in the context of power sector deregulation. A frequency-linked bidding structure for the frequency regulation service market is implemented based on the methodology in [10]. The participating parties are requested to respond to price signals sent out by the ISO, based on their bid offers. The frequency signal obtained from dynamic study has been utilized in ABT mechanism to drastic improvements in the grid frequency condition. This study shows that beneficiary states can use economic load dispatch as a tool to minimize their payments and promote economic efficiency under this new ABT regime.

During the high frequency conditions they can draw more power from the grid as grid power is coming cheap. Under low frequency conditions they can under-draw from the grid and look for other sources of power, which are now cheaper.

In this way they can not only minimize their costs but also help in reducing frequency deviations in the grid.

## V. REFERENCES

[1] Automatic Generation Control, NERC Standard BAL-005-0-2005, April 2005. Available: http://www.nerc.com.

[2] W. Berger and F. C. Schweppe, "Real-Time Pricing To Assist in Load Frequency Control," IEEE Trans. Power Syst., vol. 4, No. 3, Aug. 1989, pp. 920–926.

[3] K.Battacharya ,D.Chattopadhyay and J.Parikh, "Real-time adaptive pricing for load frequency control in an interconnectd power system", International Journal of Power and Energy Systems,vol.18,1998,pp.102-109.

[4] R. D. Christie and A. Bose, "Load Frequency Control in Hybrid Electric Power Markets," in Proc. IEEE, Int. Conf. Contr. App., vol. 11, Dearborn, MI, Aug. 1996, pp. 432–436.

[5] E. Nobile, A. Bose, and K. Tomsovic, "Feasibility of A Bilateral Market for Load Following," IEEE Trans on Power System, vol. 16, Nov 2001, pp. 782–787.

[6] J. Kumar, K. Ng, G. Sheble, "AGC Simulator For Price-Based Operation Part 1: A Model," IEEE Transactions on Power Systems, vol. 12, No. 2, May 1997, pp.527-532.

[7] J. Kumar, K. Ng, G. Sheble, "AGC Simulator For Price-Based Operation Part II: Case Study Results," IEEE Transactions on Power Systems, vol. 12, No. 2, May 1997, pp.533-538.

[8] Svenska Kraftnät, The Swedish Electricity Market and the Role of Svenska Kraftnät, Second Edition, 1997. (available at www.svk.se).

[9] GAMS Release 2.50, a user's Guide, GAMS Development Corporation, 1999.

[10] J. Jhong and K. Bhattacharya, "Frequency linked pricing as an instrument for frequency regulation in deregulated electricity markets", IEEE PES General Meeting, Toronto, July 2004, pp. 566-571.

[11] ABT Order, CERC, Jan 2000. [Online]. Available: http://www.cercind.org

[12] Introduction to Availability Based Tariff, Kalki Communications Technology, Bangalore, India. [Online]. Available: http://www.kalkitech.com/downindex/IntroductionTo ABT.pdf.

[13] Ashok Banerjee, "Electricity Distribution: Issues in Multi-Year Tariff and Private Investment and Availability Based Tariff," presented at the Workshop on Electric Power Distribution: Reforms, Automation and Management, IIT Kanpur, May 10-14,2004.

[14] A.J. Wood and B.F. Wollenberg, *Power Generation Operation and Controls*, John Wiley and Sons, 1996.

## VI. BIOGRAPHIES

**K.V.V. Reddy** is M. Tech. student at NIT Kurukshetra and is pursuing his M. Tech. dissertation. His research interests include power system restructuring, Load frequency control, and optimization.

**Ashwani Kumar** received his M.Tech. in Power Systems from Punjab University, Chandigarh in 1994. He is with Department of Electrical Engeering at NIT-Kurukshetra, Haryana and presently pursuing his Ph.D under QIP in Electrical Engineering Deptt. at IIT-Kanpur, India. His research interest includes power system deregulation and power system optimization.

**Saurabh Chanana** received his B.Tech. in Electrical Engineering from NIT Kurukshetra in 1996 and M. Tech. in Power Systems from the same institute in 2002. He is currently serving as Lecturer in Department of Electrical Engineering at NIT Kurukshetra, Haryana. His research interest includes power system restructuring and economics. He is a life member of ISTE

**2006 IEEE International Conference on Power Electronic, Drives and Energy Systems**

# Induction Machine Fault Identification using Particle Swarm Algorithms

S. A. Ethny, P. P Acarnley, B. Zahawi, *Senior Member, IEEE*, and D. Giaouris, *Member, IEEE*

*Abstract*—**The principles of a new technique using particle swarm algorithms for condition monitoring of the stator and rotor circuits of an induction machine is described in this paper. Using terminal voltage and current data, the stochastic optimization technique is able to indicate the presence of a fault and provide information about the location and nature of the fault. The technique is demonstrated using experimental data from a laboratory machine with both stator and rotor winding faults.**

*Index Terms*—**Condition monitoring, induction machine, stochastic optimization, swarm algorithms.**

## I. INTRODUCTION

CONVENTIONAL induction machine condition monitoring techniques [1] usually involve the use of embedded sensors to measure, for example, temperature or vibration and help detect a developing fault. There has also been considerable interest in detecting winding and other machine faults by current signature analysis of stator current waveforms [2]. This involves frequency-domain analysis of data gathered under steady-state operating condition and may involve the calculation of quantities such as input power [3] or machine negative sequence components [4]. More recently, other fault detection methods using data acquired during speed transients [5] and estimation of machine parameters [6] have also been suggested.

This paper describes a new technique for machine condition monitoring and fault identification from terminal and rotor position data obtained during transient operation. In this method, a stochastic search is carried out using particle swarm algorithms to estimate values of winding resistance which give the best possible match between the performance of the faulty experimental machine and its mathematical model, thus identifying both the location and nature of the winding fault.

## II. SCHEMATIC DESCRIPTION OF THE NEW METHOD

Fig. 1 shows a schematic diagram of the new fault identification technique. Terminal voltage and rotor position data from a laboratory induction machine is used as the input

The authors are with the School of Electrical, Electronic & Computer Engineering, Newcastle University, Newcastle upon Tyne NE1 7RU, UK (email: bashar.zahawi@ncl.ac.uk).

to a transient ABCabc model to calculate the three stator currents. These calculated currents are then compared to the actual measured currents to produce a set of current errors that are integrated then summed to give an overall calculation error.

When the machine is in its healthy state, its effective parameters correspond to the model parameters and the calculation error is small. If a fault occurs in the machine's windings its electrical parameters are of course modified and when the measured stator currents are compared to the calculated currents there will be a large calculation error giving a fast indication that a fault of some type is present. Fault identification is carried out by adjusting the model parameters, using a stochastic search method, such as particle swarm algorithms, to minimize the error. The new set of model parameters then defines the nature and location of the fault, for example, an increased value of resistance for stator winding b, indicates a developing open-circuit condition in that circuit, and so on.

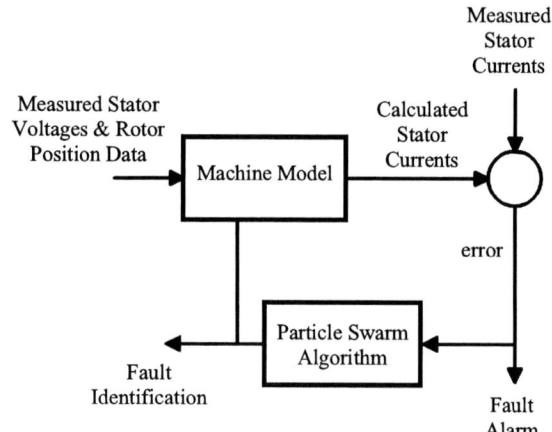

Fig. 1. Schematic representation of the fault identification technique using particle swarm algorithms.

### A. Particle Swarm Optimization

Particle Swarm Algorithms (PSA) is an evolutionary computation technique [7] inspired by social behavior of bird flocking. Like other evolutionary optimization techniques such as Genetic Algorithms, it is based on a population of randomly generated potential solutions that are dynamically adjusted in an iterative process in the search for an optimum solution. Unlike GA techniques however, the particle swarm

0-7803-9771-1/06/$25.00 ©2006 IEEE      1020

algorithm is not based on the idea of the survival of the fittest. Members of the population with lower fitness functions do survive during the optimization process and can potentially visit any point in the search space.

Each bird or member of the population in a PSA $\mathbf{X}_i$ is treated as a point in the N-dimensional space representing the optimization problem, so that:

$$\mathbf{X}_i = (x_{i1}, x_{i2}, \ldots, x_{iN}) \text{ for } i = 1, 2, \ldots, M \qquad (1)$$

where N is the number of variables and M is the number of particles that form the population.

The position of each particle within the search space is a potential result that can be evaluated in accordance with a given performance function to assess the fitness value of that member of the population. In addition to its position within the search space, each particle is free to fly with a velocity $\mathbf{V}i$ that is continuously adjusted in accordance with the flying history (i.e. position and speed) of the particle itself and of other members of the population.

$$\mathbf{V}_i = (v_{i1}, v_{i2}, \ldots, v_{iN}) \text{ for } i = 1, 2, \ldots, M \qquad (2)$$

The dynamic equations for the particle swarm algorithm are given by:

$$v_{in}^{k+1} = w v_{in}^k + a_1 b_1^k \left( p_{in}^k - x_{in}^k \right) + a_2 b_2^k \left( p_{gn}^k - x_{gn}^k \right) \qquad (3)$$

$$x_{in}^{k+1} = x_{in}^k + v_{in}^{k+1} \qquad (4)$$

where $n = (1, 2, \ldots, N)$, $i = (1, 2, \ldots, M)$, $a_1$ and $a_2$ are positive constants used to fine tune the operation and convergence of the algorithm, $b_1$ and $b_2$ are random numbers between 0 and 1, $p_{gn}$ is the position of the best particle in the flock, i.e. the bird with the best position and $w$ is a weighting function that determines the extent to which previous velocities can influence the current velocity of the particle. A large value of w assists the exploration of new areas by the flock, whereas a small value will restrict or narrow the search area for fine tuning purposes.

To apply this concept to condition monitoring of an induction machine or for machine parameter identification [8], each individual $\mathbf{X}_i$ in the bird population represents one set of values of the machine winding resistances ($R_{sa}$, $R_{sb}$, $R_{sc}$, $R_{ra}$, $R_{rb}$ and $R_{rc}$) where the resistance values must lie within a pre-defined search space. In this paper the resistance values are allowed to vary from 0.1x to 5x their nominal values.

### B. Mathematical Model of the Motor

Assuming that the machine has a smooth air-gap, the three-phase machine equations can be written in the natural ABCabc reference frame (Equation 5), where $v_A$, $v_B$, $v_C$, $i_A$, $i_B$, $i_C$ are stator winding voltages and currents, $v_a$, $v_b$, $v_c$, $i_a$, $i_b$, $i_c$ are rotor winding voltages and currents, $R_A$, $R_B$, $R_C$ are stator winding resistances, $L_s$ is stator winding self inductance, $R_a$, $R_b$, $R_c$ are rotor winding resistances, $L_r$ is rotor winding self inductance,

$M_s$ is the mutual inductance between pairs of stator windings, $M_r$ is the mutual inductance between pairs of rotor windings, $M$ is the peak value of rotor position dependent mutual inductance between stator/rotor winding pairs, $\theta_1$ is the rotor position angle measured in electrical radians, $\theta_2 = \theta_1 + 2\pi/3$ and $\theta_3 = \theta_1 + 4\pi/3$.

Only the stator and rotor winding resistances are separately defined, and subsequently adjusted during the search routine. Of course many faults also have an impact on machine inductance parameters and to obtain an exact match between measured and modeled armature currents (Fig. 1) under fault conditions it would be necessary to include the inductance parameters in the search. However, the aim of this work is not to completely identify the faulted machine parameters, but rather to demonstrate the new technique by using it to identify the location and type of rotor and stator series winding faults. Other machine faults could of course be included by extending the search to take in a wider range of machine parameters.

Because the six winding resistances may have different values, there is no advantage in seeking to transform the machine equations into an alternative reference frame. Instead the six winding voltage equations in (5) are simply subjected to the constraints imposed by winding connection (star or delta) and short-circuiting of the secondary, then solved by numerical integration.

### III. EXPERIMENTAL RESULTS

A three-phase, 50 Hz, 240 V, 2-pole wound-rotor induction motor rated at 1.5 kW was used to obtain experimental results for both healthy and faulted operating conditions. Both the stator and rotor windings of the machine are delta connected, though the rotor delta is short-circuited between all three terminals, giving effectively three independent short-circuited windings. Standard tests (dc resistance, no-load and locked rotor tests) were carried out to determine the nominal values of the machine parameters, giving the following results:

$R_A = R_B = R_C = 3.47 \ \Omega$, $R_a = R_b = R_c = 4.3 \ \Omega$, $L_s = 0.29$ H, $L_r = 0.47$ H, $M_s = 0.14$ H, $M_r = 0.23$ H and $M = 0.36$ H.

For each of the operating conditions discussed below, data was collected over a time window of 0.4 sec, with a sampling interval of 1 ms, as the machine accelerated from rest following the direct switch on of the 3-phase supply voltage. The acquired data was then processed off-line using the particle swarm algorithm to determine the effective resistances of the six windings, using values of 0.5, 1 and 1 for the constants w, $a_1$ and $a_2$, respectively.

### A. Initial Test using Healthy Machine

Initially, a test was carried out with the healthy, un-faulted machine to ensure that no spurious fault indications would

$$
\begin{bmatrix} v_A \\ v_B \\ v_C \\ v_a \\ v_b \\ v_c \end{bmatrix}
=
\begin{bmatrix}
R_A + L_s p & M_s p & M_s p & Mp\cos\theta_1 & Mp\cos\theta_2 & Mp\cos\theta_3 \\
M_s p & R_B & M_s p & Mp\cos\theta_3 & Mp\cos\theta_1 & Mp\cos\theta_2 \\
M_s p & M_s p & R_C & Mp\cos\theta_2 & Mp\cos\theta_3 & Mp\cos\theta_1 \\
Mp\cos\theta_1 & Mp\cos\theta_3 & Mp\cos\theta_2 & R_a + L_r p & M_r p & M_r p \\
Mp\cos\theta_2 & Mp\cos\theta_1 & Mp\cos\theta_3 & M_r p & R_b + L_r p & M_r p \\
Mp\cos\theta_3 & Mp\cos\theta_2 & Mp\cos\theta_1 & M_r p & M_r p & R_c + L_r p
\end{bmatrix}
\begin{bmatrix} i_A \\ i_B \\ i_C \\ i_a \\ i_b \\ i_c \end{bmatrix}
\qquad (5)
$$

arise and also to illustrate the behavior of the particle swarm algorithm. The three graphs in Figs. 2-4 show the two sets of estimated winding resistances and the error produced by the existing solution. About 25 investigations of potential solutions were required to obtain convergence of the two sets of estimated resistances to common values. The calculation error falls from a maximum value of 28 A.s, before gradually reducing to 5.5 A.s. These values of calculation error must be considered in the context of peak currents in the three stator windings reaching 60A throughout the 0.4s data window. The simplicity of the motor model means that it would be unreasonable to expect the calculation error to reduce to zero, even with a larger number of investigations.

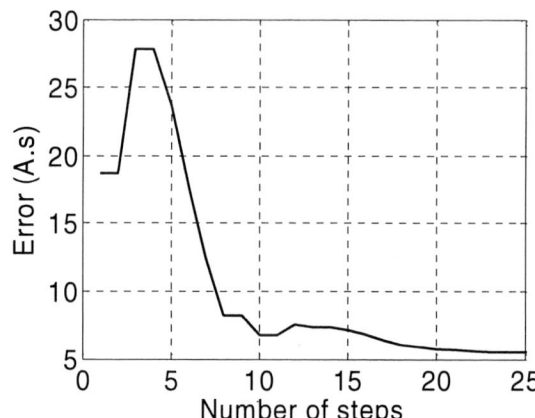

Fig. 4. Current estimation error for healthy operating conditions.

### B. Search for Rotor Winding Series Fault

A 5Ω resistance was then placed in series with rotor winding *a* to mimic a developing rotor winding open-circuit fault. The operation of the particle swarm algorithm as it estimates the winding resistances is illustrated in Figs. 5-7. A clear trend is established very quickly, with the estimated resistance of rotor winding *a* being noticeably higher than that of the other two rotor resistances. These trends become firmly established over the subsequent 25 investigations and highlight the presence of a developing open-circuit fault in rotor winding *a*. The calculation error in this case falls from a maximum value of 16 A.s, before gradually reducing to just less than 6.8 A.s.

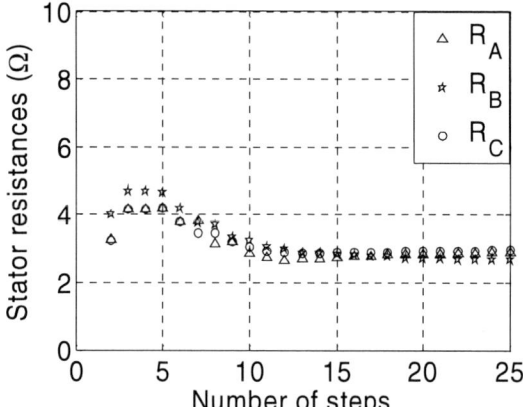

Fig. 2. Stator resistance estimation for healthy operating conditions.

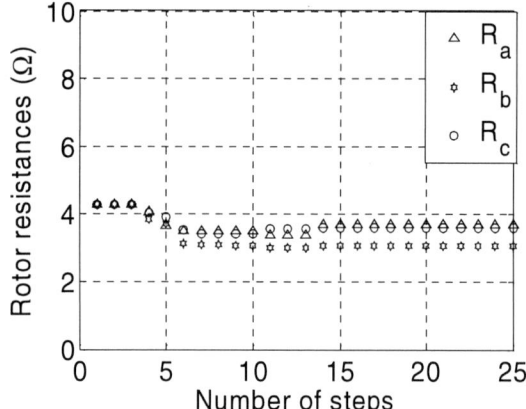

Fig. 3. Rotor resistance estimation for healthy operating conditions.

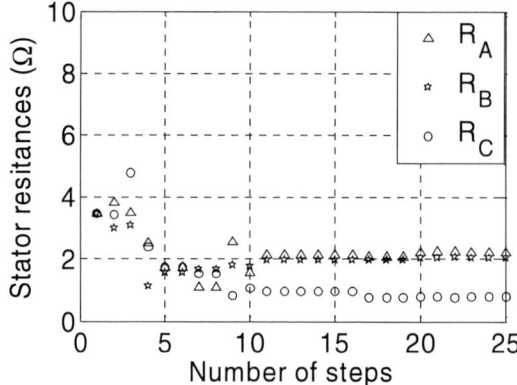

Fig. 5. Stator resistance estimation for operation with rotor winding fault.

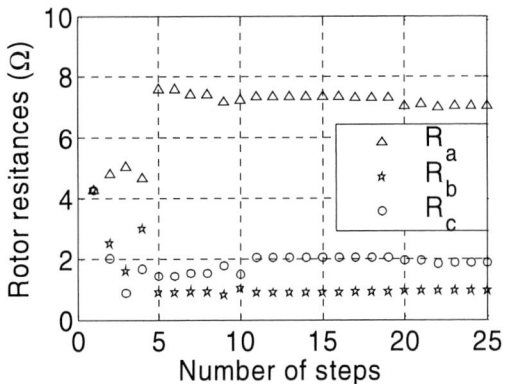

Fig. 6. Rotor resistance estimation for operation with rotor winding fault.

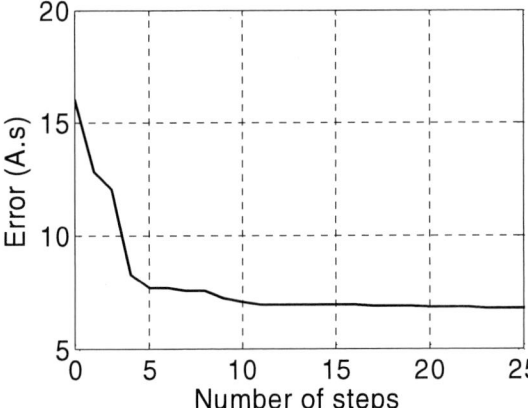

Fig. 7. Current estimation error for operation with rotor winding fault.

## IV. CONCLUSIONS

A new condition monitoring technique based on particle swarm optimization is shown to be able to identify the type and location of a motor winding series fault. Because the technique uses time-domain data, there is no requirement for the machine to be in a steady-state operating condition: in fact data acquired during a starting transient may be more helpful in discriminating between healthy and fault conditions.

The general scheme, described here for a wound rotor induction motor, is capable of being further developed by including in the machine model an appropriate set of equations to describe the secondary circuits of a cage induction machine. Other machine faults, such as inter-phase and inter-bar faults, could be included by extending the search to cover a wider range of machine parameters, including the inductances of the machine

## V. REFERENCES

[1] P. J. Tavner, and J. Penman, *Condition monitoring of electrical machines*, Research Studies Press, 1987, Letchworth, England, (1987).
[2] M. E. H. Benbouzid. "A review of induction motors signature analysis as a medium for faults detection", *IEEE Transactions on Industrial Electronics,* vol. 47, pp. 984-992, 2000.

[3] A. J. Trzynadlowski, and E. Ritchie. "Comparative investigation of diagnostic media for induction motors: a case of rotor cage faults", *IEEE Transactions on Industrial Electronics,* vol. 47, pp. 1092-1099, 2000.
[4] F. C. Trutt, J. Sottile, and J. L. Kohler. "Online condition monitoring of induction motors", *IEEE Transactions on Industry Applications,* vol. 38, pp. 1627-1632, 2002.
[5] H. Douglas, P. Pillay, A. K. Ziarani. "Broken rotor bar detection in induction machines with transient operating speeds", *IEEE Transactions on Energy Conversion,* vol. 20, pp. 135-141, 2005.
[6] M. S. N. Said, M. E. H. Benbouzid, and A. Benchaib. "Detection of broken bars in induction motors using an extended Kalman filter for rotor resistance sensorless estimation", *IEEE Transactions on Energy Conversion,* vol. 15, pp. 66-70, 2000.
[7] J. Kennedy, and R. C. Eberhart, "Particle Swarm Optimization", in Proc. 1995 IEEE International Conference on Neural Networks, pp. 1942-1948.
[8] C. Picardi, and N. Rogano, "Parameter Identification of Induction Motor Based on Particle Swarm Optimization", in Proc. 2006 International Symposium on Power Electronics, Electrical Drives, Automation and Motion, pp. 32-37.

## VI. BIOGRAPHIES

**Salah Ethni** was born in Tripoli in Libya on December 18, 1971. He received his BSc in Electrical and Electronic Engineering from the Bright Star University of Technology, Libya, in 1994 and his MSc from the Newcastle University, UK, in 2004. He is currently conducting his PhD studies at Newcastle University, UK.

**Paul Acarnley** received the BSc and PhD degrees in Electrical Engineering from Leeds University, UK, in 1974 and 1977, respectively, and an MA degree from Cambridge University, UK in 1978. After seven years in the Department of Engineering at Cambridge University, he joined the Power Electronics, Drives and Machines Group at Newcastle University, UK, in 1986. He authored 'Stepping Motors: a guide to theory and practice' which was first published by the IEE in 1982 with the 4th edition appearing in 2002. In 2003 Paul Acarnley founded RESEEDS (Research Engineering Education Services) based in Stonehaven, Scotland. He is also a Research Professor in the School of Engineering at Robert Gordon University, Aberdeen, UK and Emeritus Professor and Research Advisor at Newcastle University, UK. His principal research interest is in the control of electric drives, including work on state and parameter estimation applied to torque, current, temperature, speed and position estimation in motors, and temperature estimation in power electronic devices.

**Bashar Zahawi** received his BSc and PhD degrees in Electrical and Electronic Engineering from the University of Newcastle, England, in 1983 and 1988, respectively. From 1988 to 1993 he was a design engineer at Cortina Electric Company Ltd, a UK manufacturer of large ac variable speed drives and other power conversion equipment. In 1994, he was appointed as a Lecturer in Electrical Engineering at the University of Manchester and in 2003 he joined the School of Electrical, Electronic & Computer Engineering at Newcastle University, where he is currently the Director of Postgraduate Studies. His research interests include power conversion, variable speed drives and the application of nonlinear dynamical methods to transformer and power electronic circuits.

**Damian Giaouris** received the diploma of Automation Engineering from the Automation Department, Technological Educational Institute of Thessaloniki, Greece, in 2000, the MSc degree in Automation and Control with distinction from Newcastle University in 2001 and the PhD degree in the area of control and stability of induction machine drives in 2004. His research interests include advanced nonlinear control & estimation of electromagnetic devices, and nonlinear phenomena in power electronic converters. He is currently a lecturer in Control Systems at Newcastle University, UK.

**2006 IEEE International Conference on Power Electronic, Drives and Energy Systems**

# A Novel Technique for Identification and Condition Monitoring of Nonlinear Loads in Power Systems

Phil Gilreath, *Member, IEEE*, Maryclaire Peterson, *Student Member, IEEE*, and
Brij N. Singh, *Member, IEEE*

*Abstract*—This paper deals with Centroid-Concordia patterns for characterization of harmonic producing three-phase loads in a power system distribution network. The three-phase currents ($i_a$, $i_b$, and $i_c$) at the Point of Common Coupling (PCC) are sensed and processed through the Concordia mathematical formulation resulting in two-phase orthogonal currents ($i_\alpha$-$i_\beta$). The currents ($i_\alpha$-$i_\beta$) are used to obtain the Concordia patterns needed for centroid calculation. The centroid of the Concordia pattern reveals characteristics of the load connected at the PCC. In the majority of cases, the developed Concordia patterns do not differ much from each other as the centroid remains at the origin; this leads to a failure if conventional Concordia patterns are used to discern faults and condition monitoring of loads at power system distribution. Therefore, we proposed modified Concordia pattern methods. In the modified Concordia pattern, we exploit pattern symmetry around the $\alpha$-$\beta$ axes of the transform currents ($i_\alpha$-$i_\beta$). The computed value of the centroid of the modified Concordia pattern can be used to develop a drift pattern of the centroid location. Using the drift pattern over time, deterioration of system condition can be monitored and discerned. The drift pattern will allow us to develop the mathematical formulation for load modeling. This is an important aspect of the proposed investigation due to an increasing proliferation of nonlinear loads on aging power system network. The proposed method will be simple, non-invasive, and require reduced data sets and memory; therefore, it may prove to be easier for real-time load modeling, condition monitoring, and condition prediction for an ever changing nature of loads on power system networks.

*Index Terms*—Real-time condition monitoring, nonlinear loads, unbalanced power systems, centroid, Concordia pattern

## I. INTRODUCTION

THE scope of this work began with the focus on how power quality at the load satisfies IEEE 519 and IEC standards [1]. Historically, the majority of the loads connected at the Point of Common Coupling (PCC) have been linear in nature [2]. The majority of problems came from customers who had a high degree of cycling in their systems (such as the on/off cycling of direct on line induction motors) [3-5]. With

Phil Gilreath, Maryclaire Peterson, and Brij N. Singh are with Department of Electrical Engineering and Computer Science, Tulane University, New Orleans, LA 70118, USA (e-mails: elvisattulane@yahoo.com, mpeters2@tulane.edu, singh@eecs.tulane.edu).

the advent of power electronics, the problem of harmonics due to nonlinear loads has become increasingly apparent. With the rapid proliferation of automated equipment and electronic converter based power processing, it has become essential to develop real-time condition monitoring, condition prediction, and load modeling techniques. These techniques should be designed with a frame work that the measured signals and processed data could be used to predict the type of load connected at a power system downstream.

In automated controlled equipment and electrical systems, the power electronic converters are used to obtain the desired current and voltage waveform to drive the machines, processes etc. However, these power electronic converters introduce substantial power quality problems such as current and voltage harmonic distortion, poor power factor, and load unbalance (in case of a single-phase load connected on a three-phase four-wire system). As compared to electrical machines and linear loads, power electronic converter technology is a relatively new field and consequently, has fewer techniques for harmonic detection, identification, load modeling, and prediction of the load behavior. The ability to diagnose the type of harmonics and load unbalance introduced by linear (loads with frequent on/off operation such as a pilot light, etc.) and nonlinear loads that exist within the system becomes increasingly more complicated.

Having assessed the problems of harmonics and the requirement for condition monitoring, condition prediction, and load modeling techniques, this paper proposes a modified Concordia pattern for characterization of harmonic producing three-phase loads in a power system distribution network. This paper is divided as follows. Section 2 details a brief background on the studies performed on harmonic detection methods for the three-phase distribution systems. Section 3 presents the mathematical model of the modified Concordia pattern. Section 4 deals with centroid calculation. Section 5 and Section 6, respectively, cover simulated and experimental results. Section 7 provides a brief conclusion and important points pertaining to the future extension of this investigation.

## II. BACKGROUND

Harmonic distortion characteristics have been a concern in the power industry for decades [6-8]. The introduction of new loads with nonlinear characteristics and loads requiring

0-7803-9771-1/06/$25.00 ©2006 IEEE

frequent cycling (on/off operation) to the power grid has been the primary cause of induced harmonics, poor power-factor, and load unbalance. Ironically, the power electronic switching devices which are used for improving the power quality also contribute to the harmonic distortion when used in different applications. These problems (induced harmonics, poor power-factor, and load unbalance) result from the constant on/off switching patterns produced from the power electronics devices, such as thyristors in high power AC-DC converters.

The effects of harmonics have been widely investigated [9-13]. Early on, the power grid that is distributed to individual homes and companies was switched from DC to AC for reasons of lowering the current values and equipment costs while still supplying power with acceptable power quality as per the standards set forth by IEEE and IEC [1].

However, with the advent of numerous types of AC loads, the problems of current and voltage harmonics, load unbalance, and poor power-factor became a priority issue.

Total harmonic content is the summation of the root mean square (RMS) of various harmonic components excluding the fundamental (50 Hz or 60Hz) and is expressed by:

$$\sqrt{\sum_{h=2}^{\infty} I_h^2}.$$

The percentage of Total Harmonic Distortion (THD) is expressed as;

$$THD = \frac{\sqrt{\sum_{h=2}^{\infty} I_h^2}}{I_1} * 100 \qquad (1)$$

where, $I_1$ is the fundamental component.

The computer and telecommunication systems use semiconductor switch mode power supplies that convert the utility AC voltage to controlled and desired DC voltages with relatively lower values. These nonlinear power supplies behave as high current pulsating loads. These current pulses create significant non-sinusoidal voltage drop across the distribution line leading to a distorted voltage wave shape at the PCC. The current distortion travels back into the power system and can affect other equipment connected at the PCC. As seen in Fig. 1, the power system distribution network consists of single-phase/three-phase, linear/nonlinear, and balanced/unbalanced loads.

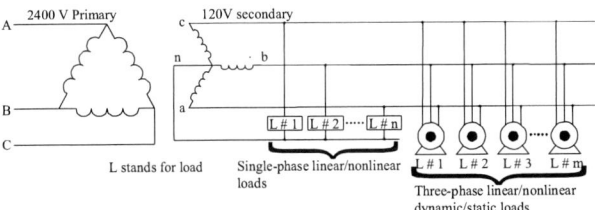

Fig. 1. Model of a distributed network.

Depending upon the operation at the consumer site, the load nature on the distribution network keeps changing. Therefore, load modeling and load characterization are required for better planning and operation of the power system.

In general, residential loads are single-phase and smaller in rating. Contrary to this, industrial loads are three-phase and larger in rating. A single-phase supply system is derived from a three-phase four-wire system by connecting line-to-line or line-to-neutral, and the single-phase current is derived from the three-phase single line source. For example, a 230 VAC three-phase power source would feed 230 VAC single-phase equipment from line-to-line, while feeding a 120 VAC single-phase equipment from line-to-neutral.

A characteristic of single-phase nonlinear loads is the predominant third harmonic. A group of single-phase nonlinear loads connected between different lines (a, b, and c) and neutral (n) draw harmonic current with dominant triplen harmonics (Fig. 2). Due to the co-phasal nature of triplen harmonics, the third multiple harmonic currents are added together with higher order harmonics (such as $5^{th}$, $7^{th}$, etc.) to return to the supply system via a neutral conductor. Large quantities of 120 VAC nonlinear load currents drawn from a three-phase four wire supply system will cause an excessive loading (sometimes 175% of phase wire capacity), which leads to heating and in some cases, destruction of the neutral conductor.

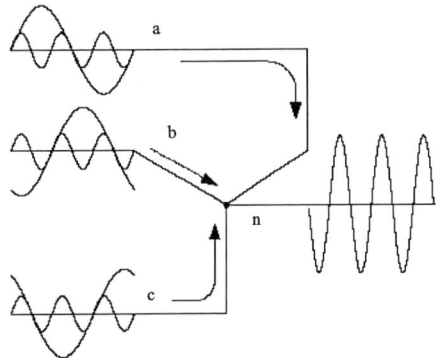

Fig. 2. Neutral current due to harmonics generated by single-phase nonlinear loads.

Single-phase nonlinear loads placed between lines (for example, a load placed between line a and b, line b and c, and line c and a) will form a close delta connection resulting in circulation of the triplen harmonics via these loads. This prevents them from flowing back to the supply system. However, non co-phasal harmonics still flow to the supply system via phase conductors. Contrary to single-phase loads, the three-phase loads do not include triplen (co-phasal) harmonics. The absence of triplen harmonics in a three-phase load is attributed to balanced load currents. This is one of the many reasons for preference of three-phase power distribution systems operating with balanced loading conditions.

The distortion in the current and voltage waveform produced by the nonlinear loads may not cause substantial problems for themselves but may affect the performance of other consumer loads. The current harmonics seek to flow through the minimum impedance path working as a current sink. For example, a capacitor bank used by a consumer to improve power-factor offers significantly reduced impedance for higher harmonics (such as $5^{th}$, $7^{th}$, and beyond). Therefore,

1025

nonlinear loads connected at the PCC can definitely produce harmful effects (such as overheating, inflated electricity bill, etc.) to consumers with linear loads and a capacitor bank.

Non-linear loads encompass load demands that are not of a constant magnitude. For example, turning on a desk lamp or a stove requires a constant value load supplied to the filament or the heating element. With the need for improved power quality and control capability, modification of the voltage/current waveforms is desirable. The voltage/current waveforms are controlled for the purpose of improving the power quality for sensitive equipment, such as computers, or modifying the input frequency for controlling the speed of an induction motor. Modification of the voltage/current waveforms had been made possible with the introduction of power electronics. Power electronic converters have the capability of using high frequency semiconductor switching devices to manipulate the magnitude and shape of the output waveform. The problem with the high frequency switching converters is that the voltage/current pulses that are required for each individual switch need to be computed in DSP for feedback and feed forward control system. Often, it is the discontinuous voltage/current pulse which is the cause of the harmonic propagation back to the PCC. Three-phase uncontrolled rectifiers are used for converting the AC voltage to a DC source with a relatively fixed value. However, numerous applications such as telecommunication systems, computers, lights, etc. need variable DC voltage requiring different types of DC-DC converters. An increase in the load at the output of an uncontrolled rectifier results in an increase in the harmonic content in the input current. Therefore, the input current to the rectifiers are load dependent, which change over time. The harmonics present in the voltages and currents at the PCC not only affect the electrical system but also produce Electromagnetic Interference (EMI) with communication systems, thereby, affecting those systems' performance. To avoid EMI problems and mitigate voltage and current harmonic distortion, IEEE 519 standard can be used as a guideline for solutions of these power quality problems.

A detailed literature survey and review identifies numerous possibilities of loads on a three-phase four-wire supply system [2-12]. This makes distributed load systems a complex network inherited with several operational problems, such as harmonics and ever-changing nature of the loads. This results in a nearly insurmountable task. However, due to the advancement in computing technology, the proposed method (Concordia patterns coupled with centroid calculations) could become a stepping stone for load modeling, load prediction, anomaly detection, and awareness of required repair and maintenance.

## III. Concordia Transforms and Patterns

The three-phase currents ($i_a$, $i_b$, and $i_c$) sensed at the PCC could be used to generate unique graphical patterns [14]. These patterns will reveal information about a load connected in a downstream power system network. These patterns will be used to disseminate voltage and current magnitude in a two

coordinate (two-phase) frame of reference (α-β frame). The three-phase currents ($i_a$, $i_b$, and $i_c$) are transformed into the α-β reference frame using the Concordia transform stated in Equations 2 and 3.

$$i_\alpha = \sqrt{\frac{2}{3}} i_a - \frac{1}{\sqrt{6}} i_b - \frac{1}{\sqrt{6}} i_c \qquad (2)$$

$$i_\beta = \frac{1}{\sqrt{2}} i_b - \frac{1}{\sqrt{2}} i_c \qquad (3)$$

It is clear from the above two equations that for three-phase time varying currents ($i_a$, $i_b$, and $i_c$) in a stationary frame, the Concordia transform results in two-phase time varying currents ($i_\alpha$-$i_\beta$) in a stationary frame. Therefore, as indicated in Fig. 3, the real-time dynamic behavior of the load system contained in the three-phase currents ($i_a$, $i_b$, and $i_c$) is replicated using the two-phase currents ($i_\alpha$-$i_\beta$) without any delay and filtering requirement, which occurs in the case of the Park Transformation. Therefore, it can be said that the proposed Concordia transform (Equations 1 and 2) and the resulting pattern has the potential to provide real-time condition monitoring without any delay.

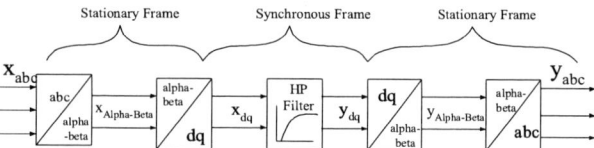

Fig. 3. Phase reference frame transformation.

Using the Concordia transform method, the α-β coordinate values of the Concordia pattern are obtained, which are circular in shape as shown in Fig. 4 for a three-phase balanced linear load. This figure also portrays the flow of information from three-phase input currents ($i_a$, $i_b$, and $i_c$) to the Concordia patterns depicted at the bottom of Fig. 4.

Although shown in Fig. 4, this investigation is not focused on modeling and prediction of three-phase systems without voltage and current harmonics. Contrarily, this interest of this investigation is to exploit the Concordia transform to develop Concordia patterns for a power system distribution network plagued with voltage and current harmonics, which are a characteristic of nonlinear loads. The currents for a three-phase balanced power distribution network without harmonics are expressed in Equation 4 as follows;

$$\begin{aligned} i_a &= I_m \sin \omega t \\ i_b &= I_m \sin(\omega t + 2\pi/3) \\ i_c &= I_m \sin(\omega t - 2\pi/3) \end{aligned} \qquad (4)$$

where, $i_a$, $i_b$, and $i_c$ are three-phase instantaneous currents. Quantity $I_m$ is the peak value of the currents and quantity $\omega t$ is equal to $2\pi f t$, where $f$ is the frequency of the current waveforms. The harmonic currents for a three-phase balanced power distribution network are expressed in Equation 5.

1026

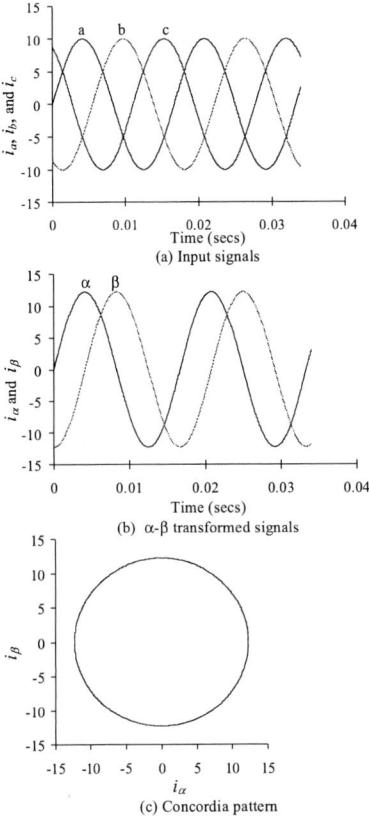

Fig. 4. Circular shaped Concordia pattern for balanced three-phase load.

$$i_a{}' = I_{mk} \sin k\omega t$$
$$i_b{}' = I_{mk} \sin(k\omega t + k \cdot 2\pi/3) \quad (5)$$
$$i_c{}' = I_{mk} \sin(k\omega t - k \cdot 2\pi/3)$$

where, $i_a{}'$, $i_a{}'$, and $i_a{}'$ are three-phase instantaneous currents with harmonics. Variable $k$ is used to represent the harmonic multiples (3rd, 5th, 7th, etc.). The total currents ($i_{aT}$, $i_{bT}$, and $i_{cT}$) in a three-phase balanced power distribution network are expressed in Equation 6 as follows;

$$i_{aT} = i_a + i_a{}'$$
$$i_{bT} = i_b + i_b{}' \quad (6)$$
$$i_{cT} = i_c + i_c{}'$$

In real-time implementation, Hall-effect current probes are used to sense the three-phase currents with harmonics ($i_{aT}$, $i_{bT}$, and $i_{cT}$). These currents are inputs to the Concordia transform as stated in Equations 2 and 3, wherein $i_a$ is replaced by $i_{aT}$, $i_b$ by $i_{bT}$, and $i_c$ by $i_{cT}$. Fig. 5 depicts the processing steps starting from the sensed signals ($i_{aT}$, $i_{bT}$, and $i_{cT}$) to the Concordia pattern.

The centroid of the Concordia patterns depicted in Figs. 4 and 5 does not reveal any relevant information. This is due to the centroid remaining at the origin for the Concordia patterns depicted in Figs. 4 and 5. Therefore, a clear difference between the balanced power systems with and without harmonics can not be discerned. This makes the traditional centroid analysis method incapable of revealing any meaningful information and therefore, a modification in the

traditional method is required to obtain useful Concordia patterns.

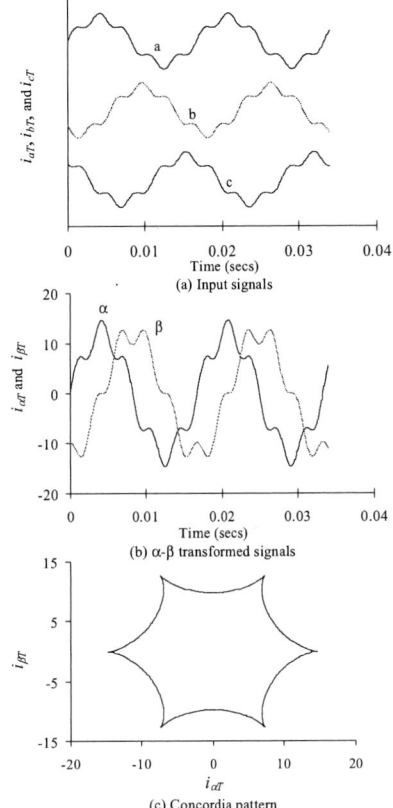

Fig. 5. Concordia pattern with 5th positive sequence harmonics in three-phase current.

## IV. METHODOLOGY FOR MODIFIED CONCORDIA PATTERNS AND CENTROID CALCULATIONS

The modified Concordia patterns will be used to calculate the centroid. The value of the centroid will be used to discern meaningful information to reveal the current condition of the system. The modified Concordia patterns are obtained using a two-step method. In this two-step method, the first step takes the time varying three-phase currents ($i_a$, $i_b$, and $i_c$) on an $a$-$b$-$c$ frame and transforms them into two-phase currents ($i_\alpha$ and $i_\beta$) on an $\alpha$-$\beta$ frame. Either of the following two combinations can be used in a no fault system to glean information about the harmonic content. These combinations are given in Equations 7 and 8; $i_\alpha$ with absolute value of $i_\beta$ or $i_\beta$ with absolute value of $i_\alpha$ to obtain the modified Concordia patterns.

The system under current investigation is considered to have only harmonics but no faults and/or unbalance. However, the proposed method can be extended to cover systems plagued with faults, unbalance, and harmonics.

$$i_\alpha = \sqrt{\frac{2}{3}}i_{aT} - \frac{1}{\sqrt{6}}i_{bT} - \frac{1}{\sqrt{6}}i_{cT} \text{ and } \left|i_\beta\right| = \left|\frac{1}{\sqrt{2}}i_{bT} - \frac{1}{\sqrt{2}}i_{cT}\right| \quad (7)$$

$$\left|i_\alpha\right| = \left|\sqrt{\frac{2}{3}}i_{aT} - \frac{1}{\sqrt{6}}i_{bT} - \frac{1}{\sqrt{6}}i_{cT}\right| \text{ and } i_\beta = \frac{1}{\sqrt{2}}i_{bT} - \frac{1}{\sqrt{2}}i_{cT} \quad (8)$$

1027

The second step transforms the two coordinate frame values of currents ($i_\alpha$ and $i_\beta$) into a one point centroid value using the formulation given in Equations 9 and 10.

$$I\alpha_{CM} = \frac{\sum_{i=1}^{N} m_i i_\alpha}{M} \qquad (9)$$

$$I\beta_{CM} = \frac{\sum_{i=1}^{N} m_i i_\beta}{M} \qquad (10)$$

where, $I\alpha_{CM}$ is the centroid of the modified Concordia Pattern for the $\alpha$-axis and $I\beta_{CM}$ is the centroid of the Concordia Pattern for the $\beta$-axis. Quantity M is the summation of the unit masses. For the sake of computational simplicity, in a digital sampling system each instantaneous current value is assigned a unit mass. The term $i_\alpha$ and $i_\beta$ represent the instantaneous value of current in $\alpha$-$\beta$ frame and $m_i$ is the unit mass for each data point. Therefore, the centroid calculation of Concordia patterns takes three-phase currents at input and computes a one point value of the centroid for further observation and analysis of the systems under real-time condition monitoring.

## V. SIMULATION RESULTS AND PERFORMANCE EVALUATION OF PROPOSED METHOD

This section deals with a detailed description of the simulation results for a three-phase system plagued with 5th and 7th harmonics and its comparison to a balanced power system without any harmonics. Fig. 6 shows the composite Concordia patterns for a system plagued with 5th and 7th positive and negative sequence harmonics. A quick comparison of all parts (a to d) in Fig. 6 reveals that the centroid of the Concordia patterns remains at the origin; therefore, meaningful information to predict the types (negative and positive sequence) and order (5th, 7th, and beyond) of the dominant harmonics in a power system can not be discerned. As described earlier, to avoid this ambiguity we take recourse to the modified Concordia patterns described in section 4. Before continuing the investigation, it is advantageous to develop a benchmark modified Concordia pattern for a three-phase balanced system that will be compared to the Concordia patterns for systems plagued with harmonics. This will demonstrate the value and potential of the proposed new method.

Fig. 7 shows the modified Concordia pattern of a three-phase balanced system without any harmonics. It is observed from this figure that the modified Concordia pattern is a semicircle with the centroid far away from the origin. This leads to meaningful information which reveals the state of the system being monitored. Although this paper investigates a three-phase system with harmonics, Fig. 7 will be used as the ideal system condition, which works as a benchmark for comparison to a system plagued with harmonics.

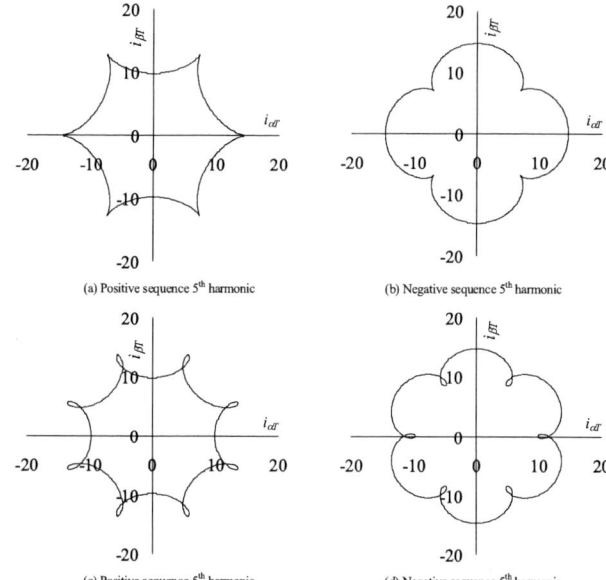

(a) Positive sequence 5th harmonic    (b) Negative sequence 5th harmonic

(c) Positive sequence 5th harmonic    (d) Negative sequence 5th harmonic

Fig. 6.  Concordia patterns of a 5th and 7th harmonics dominant nonlinear load.

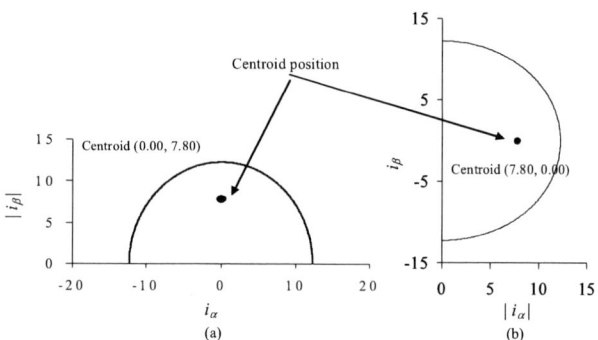

Fig. 7.  Modified Concordia pattern for a three-phase balanced system.

Figs. 8 and 9 show the modified Concordia patterns for a three-phase system plagued with positive and negative sequence 5th harmonics, respectively. Figs. 10 and 11 depict the modified Concordia patterns for a three-phase system plagued with positive and negative sequence 7th harmonics, respectively. These figures use Equations 4-6 for pre-processing of the sensed current signals ($i_{aT}$, $i_{bT}$, and $i_{cT}$), and the current signals $i_{\alpha T}$ and $i_{\beta T}$ are obtained using Equations 7-8 to plot the modified Concordia patterns as per the procedure described in section 4.

Before, discussing the modified Concordia pattern for the system plagued with 5th, 7th, and higher order harmonics, it is worthwhile to consider Fig. 6. As stated above, the centroid of the Concordia patterns shown in Fig. 6 remain at the origin irrespective of the order of the harmonics, thereby concealing information pertaining to the state and nature of the system being observed.

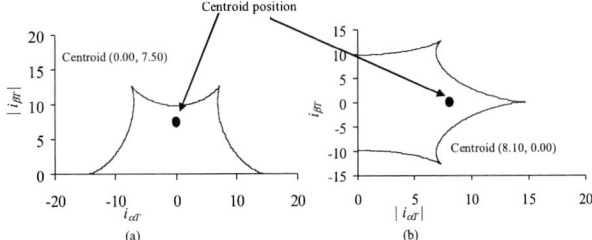

Fig. 8. Modified Concordia pattern for a three-phase system with 5th positive sequence harmonics.

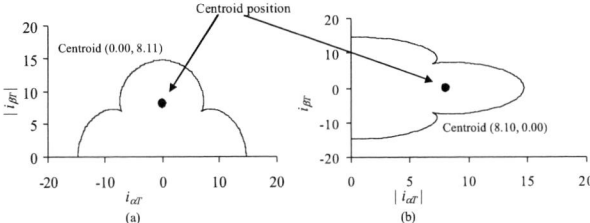

Fig. 9. Modified Concordia pattern for a three-phase system with 5th negative sequence harmonics.

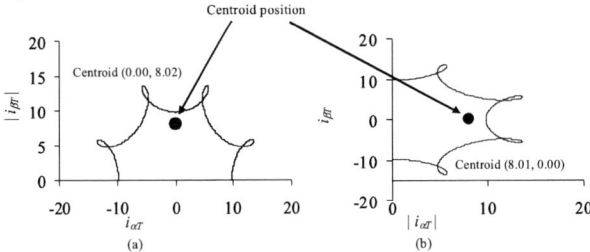

Fig. 10. Modified Concordia pattern for a three-phase system with 7th positive sequence harmonics.

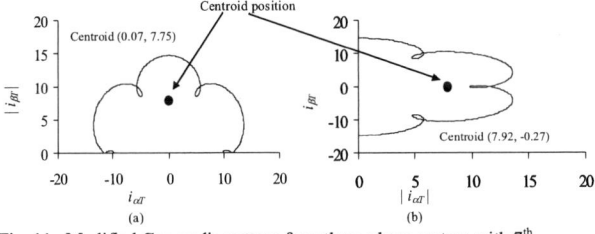

Fig. 11. Modified Concordia pattern for a three-phase system with 7th negative sequence harmonics.

Contrary to Fig. 6, the modified Concordia patterns shown in Figs. 8-11 exploit the symmetrical properties of the patterns shown in Fig. 6. As observed from Fig. 6, the Concordia patterns have a repeating nature with the positive and negative sequence $5^{th}$ and $7^{th}$ harmonics producing ring shaped Concordia patterns with $(n+1)$ valleys (positive sequence) and $(n-1)$ bumps (negative sequence). Here, $n$ is the order of the harmonics, with $n=5$ for $5^{th}$ harmonics and $n=7$ for $7^{th}$ harmonics. As observed from Figs. 4 and 7, the Concordia patterns for a balanced system without harmonics do not encounter any valley or bump type of distortion. The number of bumps and valleys in the ring shaped Concordia patterns will go up as the order of harmonics increases. These Concordia patterns will exactly follow the trend in the number of bumps and valleys as per the order of harmonics.

From Figs. 8-11, it is clear that the centroid of the modified Concordia patterns for harmonic producing loads is not at the origin and has different values as compared to the modified Concordia pattern depicted in Fig. 7, which is for a three-phase system without harmonics. It can be stated that the developed new method for condition monitoring is capable of differentiating cases with and without harmonics. However, further computational efforts are required to have a very clear differentiation between various sets of harmonics. For example, it is easy to differentiate between $5^{th}$ and $7^{th}$ harmonic systems and systems without harmonics; however, to differentiate between the Concordia patterns for $5^{th}$ and $7^{th}$ harmonics, further computational efforts are required. For quick comparison, the values of the centroid are given in Figs. 7-11.

To progress towards a real world application, a prototype laboratory experimental set-up has been developed consisting of three Hall Effect current sensors and a DSpace DSP system for real-time processing and computation of the sensed signals ($i_a$, $i_b$, and $i_c$). An experimental investigation on a DSP based condition monitoring system for a load without harmonics has been carried out. Although our experimental investigation is for sinusoidal balanced loads, the concept of the proposed modified Concordia pattern and real-time computation of its centroid value remains the same as described in previous sections. Therefore, it can be said that the proposed new method works well for a wide variety loads on a power system distribution network.

## VI. EXPERIMENTAL RESULTS

This section deals with a simple investigation on the proposed Concordia method for condition monitoring of a three-phase system without harmonics. Fig. 12 depicts the experimental results obtained using a DSpace control desk in conjunction with MATLAB based graphical user interface.

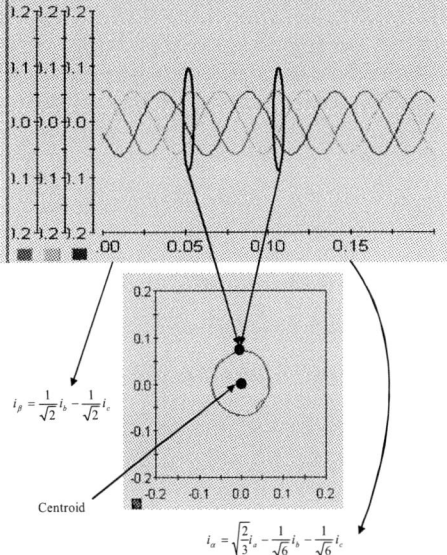

Fig. 12. Experimental Concordia Transform process for a balanced three-phase system.

1029

As clear from Fig. 12, the proposed Concordia pattern and centroid analysis is a cycle by cycle process (the dot on the circumference of the circular shaped Concordia pattern completes/represents one rotation in one cycle of the ac wave of the three-phase currents), which facilitates condition monitoring without any time delay. This is very important from a real world standpoint as there are some critical systems which need urgent attention from the operator before a catastrophic failure takes place.

## VII. CONCLUSION

In this paper we have demonstrated the need for an efficient method to predict and diagnose harmonic producing loads in a power system distribution network. It has been established that by converting the real-time currents to the modified Concordia pattern and calculating the centroid, information detailing the type and magnitude of the harmonic could be gleaned. It has been found that the centroid of the Concordia pattern reveals characteristics of the load connected at the PCC. As it has been shown in the previous sections, the developed Concordia patterns fail to characterize the loads due to lack of substantial difference among the centroids of the different types of loads. Therefore, a modified Concordia pattern method has been proposed. In the modified Concordia pattern, pattern symmetry is exploited around the $\alpha$-$\beta$ axes of the transform currents ($i_\alpha$-$i_\beta$). It is expected that the computed value of the centroid of the modified Concordia pattern can be used to develop a drift pattern of the centroid location. By storing the centroid data instead of the sensed current values, a significant reduction in the memory requirements can be achieved to develop a simple real-time condition monitoring system for complex loads connected in a power system distribution network.

In the future, this method will be combined with induced faults to determine not only the type of harmonics and their magnitude but also the types of faults in the system. It has been found that the proposed method is simple, non-invasive, and requires reduced data sets and memory and therefore proves to be an easier tool for real-time load modeling, condition monitoring, and condition prediction.

## VIII. REFERENCES

[1] IEEE 519-1992, "IEEE Recommended Practices and Requirements for Harmonic Control in Electrical Power Systems (ANSI)," IEEE, New York

[2] W. M. Grady and S. Santoso, "Understanding Power System Harmonics," *IEEE Power Engineering Review*, pp.8-11, November 2001.

[3] R. Birch, G. Chang, C. Hatziadoniu, M. Grady, Y. Liu, M. Marz, T. Ortmeyer, S. Ranade, P. Riberio, and W. Xu, "Impact of Aggregate Linear Load Modeling on Harmonic Analysis: A Comparison of Common Practice and Analytical Models," *IEEE Transactions on Power Delivery*, vol. 18, no. 2, , pp. 625-630, April 2003.

[4] W. Price, C. W. Taylor, G. J. Rogers, K. Srinivasan, C. Concordia, M. K. Pal, K. C. Bess, P. Kundur, B. L. Agrawal, J. F. Luini, E. Vaahedi, and B. K. Johnson, "Standard Load Models for Power Flow and Dynamic Performance Simulation," *IEEE Transactions on Power Systems*, vol. 1, no. 3, pp. 1302-1313, August 1995.

[5] IEEE Committee Report, "Load Representation for Dynamic Performance Studies," *IEEE Transactions on Power Systems*, vol. 8, no. 2, pp. 472-482, May 1993.

[6] E. L. Owen, "A History of Harmonics in Power Systems," *IEEE Industry Applications Magazine*, pp. 6-12, Jan./Feb. 1998.

[7] R. G. Ellis, "Harmonic Analysis of Industrial Power Systems," *IEEE Transactions on Industry Applications*, vol. 32, no. 2, pp. 417-421, March/April 1996.

[8] G. T. Heydt, "Electric Power Quality: A Tutorial Introduction," *IEEE Computer Applications in Power*, pp. 15-19, January 1998.

[9] A. Kwasinski, P. T. Krein, and P. L. Chapman, "Time Domain Comparison of Pulse-Width Modulation Schemes," *IEEE Power Electronics Letters*, vol. 1, no.3, pp. 64-68, September 2003.

[10] W. Xu, E. E. Ahmed, X. Zhang, and X. Liu, "Measurement of Network Harmonic Impedances: Practical Implementation Issues and Their Solutions," *IEEE Transactions on Power Delivery*, vol. 17, no. 1, pp.210-216, January 2002.

[11] Y. Jiang and A. Ekstrom, "General Analysis of Harmonic Transfer Through Converters," *IEEE Transactions on Power Electronics*, vol. 12, no. 2, pp. 287-293, March 1997.

[12] Task Force on Harmonics Modeling and Simulation, "Characteristics and Modeling of Harmonic Sources-Power Electronic Devices," *IEEE Transactions on Power Delivery*, vol. 16, no. 4, pp. 791-800, October 2001.

[13] J. Arrillaga, and N. R. Watson, *Power System Harmonics*, West Sussex, England: John Wiley & Sons, 2003.

[14] F. Zidani, M. E. H. Banbouzid, D. Diallo, and M. S. Nait-Said, "Induction Motor Stator Faults Diagnosis by a Current Concordia Pattern-Based Fuzzy Decision System," *IEEE Transactions on Energy Conversion*, vol. 18, no. 4, pp. 469-475, December 2003.

**2006 IEEE International Conference on Power Electronic, Drives and Energy Systems**

# Real-Time Identification of Distributed Bearing Faults in Induction Motor

Rajesh Patel , S P Gupta, and Vinod Kumar

*Abstract--* **This paper assesses the effectiveness and reliability of monitoring techniques for real-time fault detection in bearings of induction motor. The bearing failure modes and the characteristic bearing frequencies associated with the physical construction of the bearings are examined. Experimental results of vibration spectra with different faults are included, for a 7.5kW Cage Induction Motor, to demonstrate the effectiveness of the proposed scheme. Several ball bearings with known faults have been used in the experimental work.**

*Index Terms--* **Condition monitoring, fault detection, bearing faults, vibration measurement, signals processing.**

## I. INTRODUCTION

CONDITION monitoring and fault diagnostics is useful for avoiding breakdowns and ensuring long life of electrical rotating machines. Bearing problems account for over 40% of all faults in induction motors. Bearing faults have been traditionally detected at incipient stage through vibration and stator current monitoring. A successful bearing condition monitoring scheme must be able to detect location of faults and their severity level. This work provides an insight into real-time detection of bearing faults though identification of harmonic frequency bands, rather than characteristic frequencies, in the vibration spectrum when the fault is distributed over the bearing surface, as in the case of electrical fluting, corrosion, overloading, misalignments etc.

## II. REAL-TIME BEARING FAULTS

Most fault detection algorithm for rolling element bearings are designed to detect characteristic fault frequencies, corresponding to faults in balls, outer race, inner race and bearing cage. However, the faults which are distributed, such as roughness, do not correspond to above characteristic frequencies. This work initiates identification of distributed faults due to bearing surface roughness and uneven dents on balls. Experimental results reveal the nature of vibration harmonic frequency bands which do not have any relationship

to characteristic frequencies that have hitherto been used for bearing fault identification. Some sample faults examined in this work are shown in Fig. 1 and Fig. 2. They belong to bearings, which have been picked up from the rejected lot of damaged bearings by user industries.

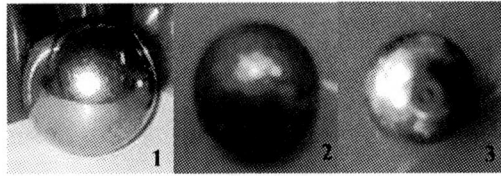

Fig. 1. Parts of Damaged Bearings (Bearing No. 6308) (1) minor dents on balls (2) one ball has severe roughness (3) large dent on a ball.

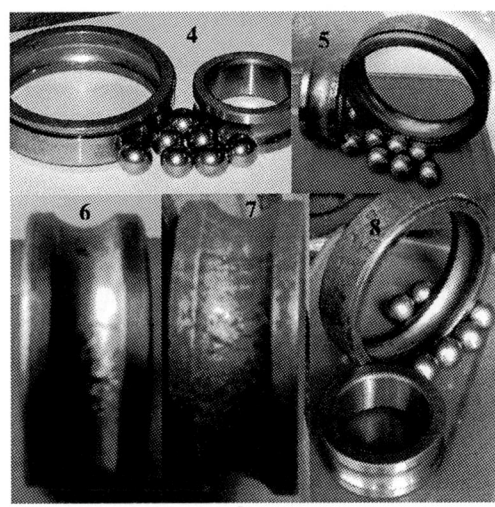

Fig. 2. Parts of Damaged Bearings (Bearing No. 6308) (4) minor dents on balls & races due to bearing current (5) dents on balls and races due to corrosion (6) inner race rough surface (7) inner race severely rough (8) outer race severely rough.

## III. MONITORING SET-UP

For the purpose of online monitoring and diagnosis of the machine condition, a PC based monitoring system is developed in this work. The monitoring system has high accuracy, fast response, high degree of reliability and rugged construction for industrial application. It acquires signals of vibration, supply voltages, currents, temperature and speed of the test motor. The transducers used in this investigation are as follows:

---

This work is financially supported by the Ministry of Human Resource Development, Government of India.

The authors are with the Department of Electrical Engineering, Indian Institute of Technology Roorkee, Roorkee, UA 247667, INDIA Phone: 91-1332-286233 (e-mail: rmp04dee@iitr.ernet.in).

0-7803-9771-1/06/$25.00 ©2006 IEEE

1. Piezo-electric accelerometer for vibration acceleration measurements.
2. Hall Sensor type current sensor for stator current monitoring.
3. Non-contact type speed sensor for speed monitoring.
4. Potential Transformer for supply voltage measurements.

The specifications of the transducers and interfacing system are given in Appendix-1.

## IV. MONITORING PROCEDURE

Fault study of eight different type of distributed faults related to bearings have been examined in the present study. For this purpose bearing No. 6308 suitable for a 7.5 kW Cage type Induction Motor is selected. The faults are introduced in the bearing, one at a time, by reassembling the bearing with the damaged part. The motor is then run with this bearing and vibration signals are recorded. This process is repeated for all the eight faults under study. In each run, the operating condition of the motor was kept same with regard to the supply voltage and the amount of load.

While the condition monitoring system acquires signals of voltage, current, temperature, speed and vibration, it is the vibration signal only which is included for discussion here. The vibration signal is obtained from the bearing cap. The raw vibration signal is suitably amplified. All the signals collected from the test machine are stored, after amplification, in the PC via DAQ card. The signals are sampled at high sampling frequencies, so as to get the desired frequency resolution. The vibration signal so picked up from the bearing cap for different faults are shown in Fig. 3. The acceleration scale is selected to be same in all the cases for comparison. Further, the vibration record for a healthy bearing is also included. It is noted that vibrations increase by different degree for different faults in comparison to healthy bearing. They are pronounced in cases 2 and 7 which correspond to severe roughness of ball and inner race surfaces, respectively.

## V. VIBRATION ANALYSIS TECHNIQUES

### A. Time domain analysis

From the vibration signal the following signal detectors, which define signal characteristics, are calculated for the purpose of fault identification:

### 1) RMS value

The RMS (Root Mean Square) value of the vibration acceleration can be used for primary health investigation of the machine [1]. In the present work the RMS value of the vibration signal is calculated from the acquired instantaneous vibration signal under different loading value on induction motor. The root-mean-square (RMS) of a variate $X$, is the square root of the mean squared value of $x$:

$$rms = \sqrt{\frac{1}{N}\sum_{i=1}^{N}(x_i - \mu)^2}$$

where, $N$ is number of samples, $x_i$ is the amplitude of individual sample and $\mu$ is the mean value of samples. It is found that RMS value of the vibration increases with increase in degree of fault level, but not on the amount of load. Table I gives the RMS value for different faults. The motor was run in each case at no load.

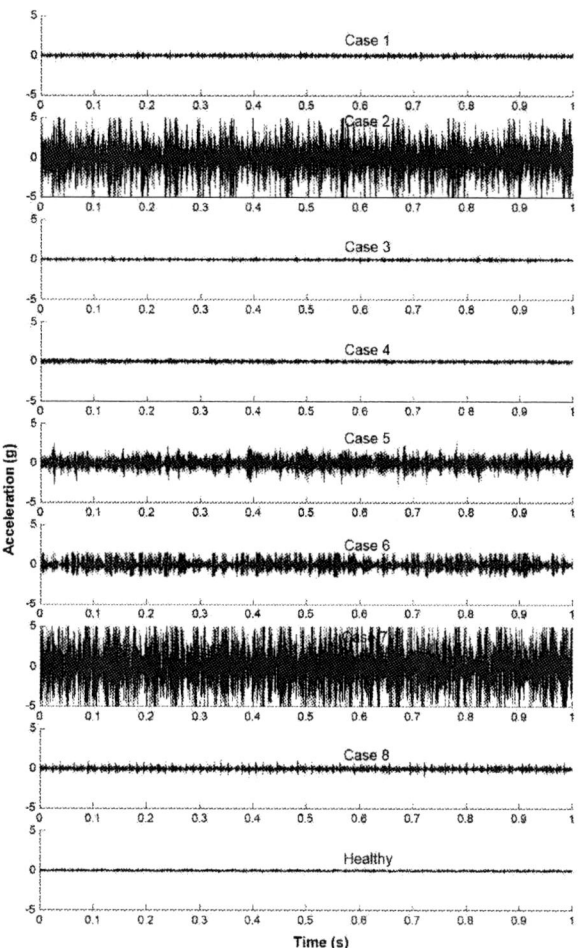

Fig. 3. Vibration Signal of various cases of defective bearings of Fig. 1 and healthy bearing.

### 2) Crest Factor

The crest factor, which is the ratio of peak value to the RMS value, yields a measure of spikiness of a signal. Crest factor of radial vibration signal is often used to indicate the rolling bearing faults. Table I gives the average of RMS value and crest factor of the vibration signal recorded from the each test run. It is observed that crest factor for healthy bearing is more as compared to that of damaged bearing, in many cases.

### 3) Skewness

Skewness is a measure of symmetry, or more precisely, the

lack of symmetry about its mean. A distribution, or data set, is symmetric if it looks the same to the left and right of the center point of Gaussian distribution. The skewness of normal distribution is zero, and any symmetric data should have skewness near zero (Table I (a), healthy bearing). Negative values for the skewness indicate data that are skewed left and positive values for the skewness indicate data that are skewed right. The skewed left; means that the left tail is heavier than the right tail. Similarly, skewed right means the right tail is heavier than the left tail.

$$c(skewness) = \frac{\frac{1}{N}\sum (x_i - \mu)^3}{\sigma^3}$$

### 4) Kurtosis

A more recent development in the state of art of bearing fault detection is statistically based parameter called Kurtosis. It is a measure of whether the data are peaked or flat relative to a normal distribution. That is, data sets with high kurtosis tend to have a distinct peak near the mean, decline rather rapidly, and have heavy tails. Data sets with low kurtosis tend to have a flat top near the mean rather than a sharp peak. A uniform distribution would be the extreme case. Positive kurtosis indicates a "peaked" distribution and negative kurtosis indicates a "flat" distribution

$$k(kurtosis) = \frac{\frac{1}{N}\sum_{i=1}^{N}(x_i - \mu)^4}{\sigma^4}$$

The Kurtosis technique has the major advantage that the calculated discriminate takes a value, which is independent of load or speed conditions. It has been found that the Kurtosis factor for undamaged bearing is 3. In general, the initial appearance of flaws is marked by an increase in the value of Kurtosis. As the damage become more severe, the values falling back towards 3.

### 5) Variance

For a single variant $X$ having a distribution $P(X)$ with *known* population mean $\mu$, the population variance $var(X)$, commonly also written $\sigma^2$, is defined [8] as

$$\sigma^2 \equiv \langle (X - \mu)^2 \rangle,$$

where $\mu$ is the population mean and $\langle X \rangle$ denotes the expectation value of $X$. For a discrete distribution with $N$ possible values of $x_i$, the population variance is therefore

$$\sigma^2 = \sum_{i=1}^{N} P(x_i)(x_i - \mu)^2,$$

The variance is therefore equal to the second central moment $\mu_2$.

### 6) Standard Deviation

Standard deviation is a statistical term that provides a good indication of volatility. The standard deviation $\sigma$ of a probability distribution is defined [8] as the square root of the variance $\sigma^2$,

$$\sigma = \sqrt{\langle x^2 \rangle - \langle x \rangle^2} = \sqrt{\mu_2' - \mu^2},$$

where $\mu = \bar{x} = \langle x \rangle$ is the mean, $\mu_2' = \langle x^2 \rangle$ is the second raw moment, and $\langle f \rangle$ denotes an expectation value.

TABLE I(A)
VIBRATION SIGNAL ANALYSIS: TIME-DOMAIN FEATURES WITH HEALTHY BEARING

| RMS | Crest factor | Skewness | Kurtosis | Variance | STD |
|---|---|---|---|---|---|
| 0.0778 | 3.6842 | -0.004 | 3.0657 | 0.0061 | 0.0778 |

TABLE I(B)
VIBRATION SIGNAL ANALYSIS: TIME-DOMAIN FEATURES WITH DEFECTIVE BEARINGS

| Cases | RMS | Crest factor | Skewness | Kurtosis | Variance | STD |
|---|---|---|---|---|---|---|
| 1 | 0.1184 | 5.0266 | -0.184 | 3.4951 | 0.014 | 0.1183 |
| 2 | 1.845 | 2.711 | -1.188 | 3.329 | 3.404 | 1.8449 |
| 3 | 0.095 | 4.381 | 0.0423 | 3.8765 | 0.009 | 0.0948 |
| 4 | 0.1483 | 2.7018 | 0.0022 | 2.8528 | 0.022 | 0.1483 |
| 5 | 0.6936 | 3.6212 | -0.100 | 3.1561 | 0.4811 | 0.6936 |
| 6 | 0.6535 | 2.2953 | 0.3556 | 2.8262 | 0.4269 | 0.6533 |
| 7 | 2.2368 | 2.2353 | 0.0763 | 2.7649 | 5.0032 | 2.2367 |
| 8 | 0.2059 | 4.5416 | 0.094 | 4.4109 | 0.0424 | 0.2059 |

### 7) Probability Distribution Function

To analyze the raw vibration signal, we plotted for motor monitor data a probability distribution of the vibration signals. A probability distribution is a histogram relating to the number of times vibration acceleration peak occurs during defined time period. A machine in good condition has a vibration probability distribution that depicts roughly Gaussian (bell shaped) curve. For a defect in machine, more varied peaks occur several times throughout the sample as seen in cases as compared to healthy one as shown in Fig. 4.

### B. Frequency domain analysis

Signal processing techniques such as filtering, averaging and Fourier Transform are used to extract more the useful information from the acquired vibration signal.

The vibrations due to mechanical forces are produced from the rotor of machine itself. The sources of this kind of force in any rotating machine are dynamic rotor unbalance, stator and rotor rubbing and rolling motion of the bearing. A considerable proportion of these rotor forces are transmitted to the stator via the bearings [3]. The characteristic frequencies, which are indicative of the defect of such a bearing, depend upon geometrical size of its parts. The test motor bearing characteristics are calculated [5] and given as Table II.

TABLE II
KEY FREQUENCIES OF THE BEARING USED IN TEST MOTORS

| Bearing Type (6308) D = 65 mm, d = 15.081 mm, n = 7, β = 0° | | | | |
|---|---|---|---|---|
| Vibration Frequency | fo | fi | fb | fc |
| Magnitude (Hz) | 65 | 105 | 32 | 5 |

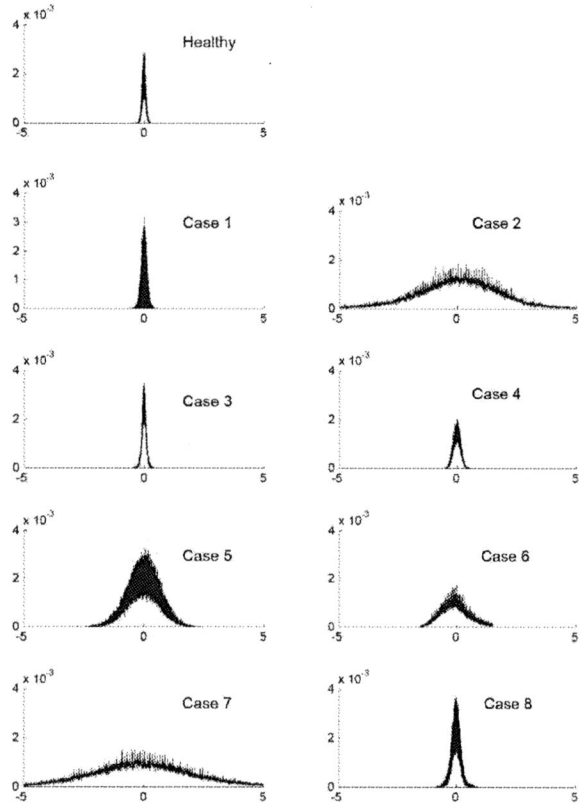

Fig. 4. Probability Distribution Function (PDF) of healthy bearing and various cases of defective bearings of Fig. 1.

*1) Calculation of vibration spectrum*

In order to eliminate the noise from the recorded signal, spectrum averaging is used. In spectrum averaging, the signal is divided into parts and FFT is calculated for each such part. The average of these FFT's is called the average spectrum, which is free from any noise. For this purpose 100K samples of the vibration signal are recorded for each run. A rectangular time window of width 2000 samples (which gives the frequency resolution of 1 Hz) is taken to obtain the time domain slices of the vibration signal. As the beginning and ending of the window produces side lobes in the frequency domain, which may give inaccurate results, therefore an overlap between two successive signal segments is used. It is found that an overlap of 500 samples gives adequate results.

The average spectrum is then obtained by averaging the FFT of individual vibration signal segments.

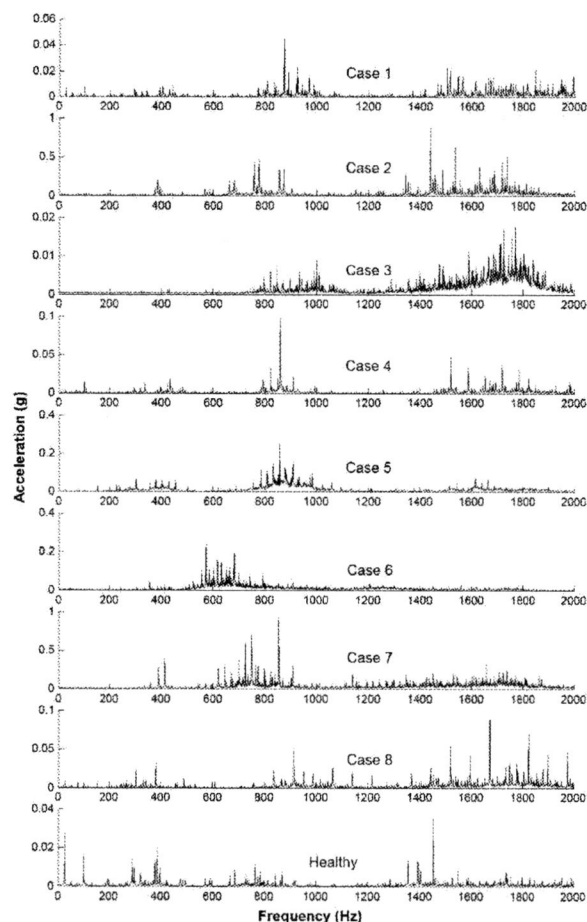

Fig. 5. Vibration Spectrum of various cases of defective bearings of Fig. 1 and healthy bearing.

*2) Extraction of important vibration harmonics*

Vibration spectrum obtained using above method contains large amount of spectral information as shown in Fig. 5. In Fig. 5, vibration signal and corresponding spectrum are plotted for different fault cases i.e. for healthy bearing, and faulty. We can observe that the dominant frequency components of the vibration signal for each case are different. Therefore machine condition can be known by monitoring not only characteristics frequency components, but also need to observe the frequency band widely. It changes differently for fault of different nature. To extract these useful frequencies which are coupled to the specific fault type, a feature extraction technique is needed. For offline fault analysis authors have developed programs in MATLAB.

VI. CONCLUSIONS

The results illustrate the nature of vibration harmonics due to (a) ball roughness (b) surface roughness and (c) severely damaged bearing components. It is established that they are unrelated to characteristic bearing frequencies and appear in the form of frequency bands in the vibration harmonic spectrum. Standard commercially available machinery fault analyzers are found unsuitable to categorize the nature of these kind of bearing faults.

## VII. Acknowledgment

The damaged bearings were provided by Vimal Steel Rolling Mill, Rishikesh, India, TATA Bearings Ltd, Kharagpur, India and NBC Bearings, Jaipur, India to carryout experimentation. Their support is thankfully acknowledged. The authors are also thankful to the Department of Electrical Engineering, IIT Roorkee for providing excellent laboratory facilities.

## VIII. Appendix – 1

The specification of the transducers and interfacing system used in this investigation are as follows:

1. National Instruments make multifunctional DAQ Card (Model: DAQ-AI-16E-4, 16 Channel, 250 kS/s, 12-bit), with BNC 2100 Series Terminal Block, which is compatible to PCMCIA port of Laptop Computer.
2. Vibration Sensor: PCB Make Piezo-electric accelerometer (Range ± 40g, Sensitivity 50 mV/g, frequency response 5 kHz) with signal conditioner and amplification module.
3. Current Sensor: Clamp-on Current Probe with inbuilt DC Power supply(Hall Sensory Technology, Sensitivity 10mv/V, frequency response DC-2kHz)
4. Non-Contact type Speed Sensor (Opto-Speed Sensor) with TTL Pulse output
5. Suitable Thermo couples at three different location on machine with signal conditioner
6. Potential Transformer (Voltage Ratio 100:1, Isolation type)

## IX. References

[1] P. J. Tavner, and J Penman, *Condition Monitoring of Electrical Machines*, UK: John Wiley & Sons Inc., 1987.
[2] M. Y. Chow, *Methodologies of Using Neural Network and Fuzzy Logic Technologies for Motor Incipient Fault Detection*, USA: World Scientific Publication Company, 1998.
[3] S.A. AL Kazzaz, "Intelligent Diagnostic and Monitoring of Electrical Drives," Ph.D. Thesis, Dept. Elect. Engg., IIT Roorkee, 2001.
[4] A.G. Parlos, K. Kyusung, and R. Bharadwaj, "Detection of induction motor faults-combining signal-based and model-based techniques," in *Proc. of the American Control Conference* May 2002, Vol. 6, pp. 4531 – 4536.
[5] J. R Stack, T. G. Habetler, and R G. Harley, "Fault classification and fault signature production for rolling element bearings in electric machines," IEEE Trans. Industry Applications, Vol. 40, No. 3, pp. 735-739, 2004.
[6] J. R Stack, T. G. Habetler, and R G. Harley, "Experimentally generating faults in rolling element bearings via shaft current," *IEEE Trans. Industry Applications,* Vol. 41, No. 1, pp 25- 29, 2005.
[7] Yu Azovtsev, A Barkov, and I Yudin, "Automatic diagnostics and condition prediction of rolling element bearings using enveloping methods," *published in proceedings of the 18th annual meeting of the Vibration Institute,* June, 1994, Available: http://www.vibrotek.com/article.php?article=articles/new94vi/index.htm.
[8] MathWorld -- A Wolfram most extensive Web Resource, Available: http://mathworld.wolfram.com/.
[9] J. Antoni, and R. B. Rajdall, "The spectral kurtosis: application to the vibratory surveillance and diagnostics of rotating machines," *Elsevier Journal of Mechanical Systems and Signal Processing*, Vol. 20, pp.08-331, 2006.
[10] G. Dalpiaz, and A. Rivola, "Condition monitoring and diagnostics in automatic machines; comparison of vibration analysis techniques," *Elsevier Journal of Mechanical Systems and Signal Processing*, Vol. 11, pp. 53-73, 1997.

## X. Biographies

**Rajesh Patel** was born in India in 1977. He did B.E. (Electrical) and M.E. (Electrical, Automation and Control) respectively in 1998 and 2000. He is Assistant Professor in the Department of Electrical Engineering in C.U. Shah College of Engineering and Technology, Wadhwan City, INDIA. He is presently pursuing Ph.D. in Electrical Engineering Department at IIT Roorkee in the field of Condition Monitoring and Fault Diagnosis of Electrical Machines.

**S P Gupta** was born in India in 1950. He did B.Sc. Engg (Electrical), M.E. (Advanced Electrical Machines) and Ph.D. respectively in 1971, 1973 and 1986. He has 32 years of teaching and research experience including two years of industrial experience. He is presently Head of Electrical Engineering Department at IIT Roorkee. His fields of interest includes Condition Monitoring of Electrical Machines, Power Electronics and Electrical Drives.

**Dr. Vinod Kumar** is Professor in Department of Electrical Engineering, Indian Institute of Technology, Roorkee (India). His area of interest is ECG signal processing and analysis, digital signal processing, transducer instrumentation, machine condition monitoring, medical instrumentation, telemedicine, medical informatics. He has published around 60 papers in national and international journals. He has also published 56 papers in national and international conferences. He is recipient of many academic awards. He has guided around 14 Ph.Ds and 70 M.Tech. Theses.

**2006 IEEE International Conference on Power Electronic, Drives and Energy Systems**

# Integration of IEDs Using Legacy and IEC61850 Protocol

Anupama Prakash[1], Mini S. Thomas[2], *Senior Member, IEEE,* and Ashutosh Gautam[3]

*Abstract*—The use of numerical relays and Intelligent Electronics Devices (IEDs) as source of information lays greater emphasis on communication protocols. In the past, manufacturers developed their own proprietary protocols that suited their products the best. This created difficulties for system suppliers, system integrators and utilities who wanted to use different products from multiple vendors and make them communicate with each other, which is difficult to achieve This paper describes the integration of IEDs from multiple vendors using Modbus and IEC60870-5-103 (T103) communication protocol to the bay controller. In order to achieve complete interoperability an IED is integrated on the station bus using IEC61850 protocol. On the basis of integration results an effort has been made to compare different features of these protocols.

*Index Terms*— Bay controller, De facto standards, Ethernet, IEC61850, IEC60870-5-103, IED, Interoperability, Modbus protocol, Proprietary, Time synchronization.

## I. INTRODUCTION

THE implementation of *Substation Automation* is increasing at a rapid pace all over the world. The mechanical relays are being replaced by the microprocessor-based digital intelligent electronic devices (IED) in electric utility systems for better and advanced communications.

The traditional control system architecture consists of a central remote terminal unit connected to secondary devices, typically protection relay, via point-to-point wiring, and possibly augmented by various serial communication links. This kind of architecture suffers from susceptibility to a single point of failure, proprietary equipment, as well as complex cabling and wiring. Further, protocols from multiple manufacturers, many of which are proprietary are likely to be used on the serial communication links [1]. However, a multitude of proprietary and de facto standard communication protocols used by the IEDs has prevented utilities from achieving true and total integration of protection, control and data acquisition functions. The main challenge therefore is achieving interoperability among a diverse group of IEDs and functions.

The integration of multifunctional intelligent electronic devices in complex substation or power plant automation systems requires the development of a standard protocol that will meet the requirements of protection, control, and monitoring, recording and metering functions. The emerging IEC61850 standard for communication networks and systems in substations allow the development of high-speed peer-to-peer communications based applications, as well as distributed metering, control and protection solutions based on sampled analog values [2-6].

An existing substation will have equipment running on multiple protocols, and IEC61850 devices are generally introduced when new bays are added. An IEC61850 system can also be built up through gradual upgrade/replacement of existing devices, without waiting for the substation to be extended. The devices would be upgraded bay by bay, and all would be communicating through the station unit.

Some new substations may require a mixture of IEC61850 equipment and non-IEC61850 equipment, and hence a migration path to enable the former equipment to be expanded and the latter equipment to be phased out is adopted. A system suitable for this migration generally comprises a station unit to which, both devices running on non IEC61850 protocols and on IEC61850 are connected. The protocol conversions are carried out conveniently. The phasing out of non-IEC61850 devices can be carried out in two ways replacement of the old devices with new ones or Upgrade of the old devices. The second method is normally adopted for recently manufactured devices [7].

The objective of the paper is to study and understand the various bay level and substation level protocols and integrate IEDs from multiple vendors using these protocols on the different ports of the bay controller and on the station bus. On the basis of these integration results an effort has been made to compare the distinct features of these protocols.

## II. SUBSTATION COMMUNICATION

An Electric Power Substation generally has three hierarchical levels namely station level, bay level and process level as shown in Fig. 1. For the proper functionality of substation automation system it is necessary to have appropriate communication between these levels so that data and information can be exchanged at every level.

---

[1] **Anupama Prakash,** student in the department of electrical Engineering, Jamia Millia Islamia, New Delhi, India.
(e-.mail: anupamaprakash@hotmail.com).

[2] **Mini S. Thomas,** Professor and Head of department of Electrical Engineering, Jamia Millia Islamia, New Delhi, India
(e-mail: mini@ieee.org).

[3] **Ashutosh Gautam,** senior manager in AREVA, New Delhi, India

0-7803-9771-1/06/$25.00 ©2006 IEEE

Communication between various levels is carried out through various buses. Telecontrol bus is used to transfer data between the substations. Station bus carries out communication within the substation using protocol UCA2.0 or IEC61850. The legacy bus transfers data between IEDs and the bay controller using various protocols such as IEC60870-5-103 (T103) and Modbus.

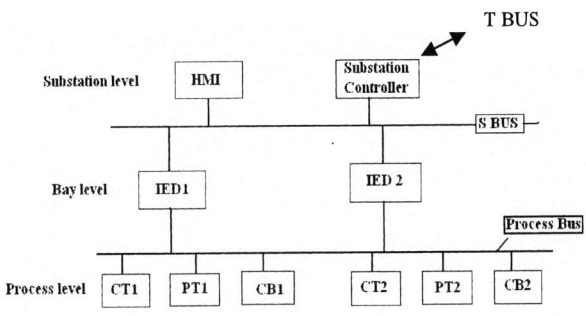

Fig. 1. Typical structure of automated substation.

However relay IEDs using single protocol from a single vendor are normally integrated to the bay controller. Here IEDs from different vendors using different protocols are integrated to the bay controller, which is the need of the hour. With the emergence of IEC61850 protocol instead of using bay controller and connecting IEDs to this bay controller, an IED is directly integrated on the station bus and can thus communicate directly with other devices placed on the station bus.

*A. IED Communication on Legacy Bus Network*

Legacy networks use "Master / Slave" protocols where the bay controller is the Master and the IEDs are the "slaves". Between "master" and "slave" either the Master transmits a request to a slave and waits for the response or the master transmits a command to all the slaves connected to the network and they perform it without sending any response. When IED is integrated to the bay controller, the bay controller first initializes the communication link and the communication protocol. It then carries out certain function such as general interrogation, time synchronization of IEDs, polling data from IEDs, handling Control sequences, disturbance file management, network supervision in order to extract complete and correct information from IEDs. Various international standard groups have developed many standard protocols to carry out communication on the legacy bus. The paper includes the use of these standard protocols (Modbus and T103) to integrate IEDs to the bay controller.

Modbus protocol is one of the oldest amongst communication protocol. It's an asynchronous, master-slave communication protocol. The Modbus protocol allows a master device to read and to write data bit-by-bit or word-by-word and to access the event recordings in the slave device.

The T103 protocol is based on the three-layer model physical layer, link layer, and application layer. The physical layer uses a fiber optic or a copper-wire based system that provides binary symmetric and memory less transmission. The link layer consists of a number of link transmission procedures, using explicit Link Protocol Control Information (LPCI) that is capable of carrying Application Service Data Units (ASDUs) as link user data. The link layer uses a selection of frame formats to provide the required integrity, efficiency, and convenience of transmission. The application layer contains a number of application functions that involve the transmission of ASDUs between source and destination.

T103 protocol is a master slave system and is unbalanced. In substation automation scenario the protective relays are slaves and the substation controller is a master. The communication speed is 9.6 kb/s or 19.2 kb/s; the physical interface may be RS-485, RS-232 or fiber optics. Status indication, measurement values, time-tagged events, control commands and clock synchronization of all protection IEDs through substation controller can be done via the T103.

*B. IED Communication on Station Bus*

In order to achieve complete interoperability; the ability for IEDs, to exchange information, irrespective of their vendors a communication protocol IEC61850 has been developed which is used at the station bus at present, allowing the IEDs to be directly integrated to the station bus. The drawback of most of the other existing standards is that the data of the functions, the services and the communication protocols are all mixed together. In other words, these standards define specific bit-sequences for the messages. Bit-sequences are protocol-dependent and protocol is technology-dependent. If communication technology changes, the bit-sequences would need to be adjusted accordingly. It would be undesirable to have a new communication standard or replace a substation automation system completely, whenever the communication technology changes. IEC 61850 addresses this issue by separating the data of the functions, services and communication protocols.

IEC 61850 is based on Ethernet, 100Mbit/s Ethernet is expected to be 'norm', but slower networks may be used in less demanding substations. It allows fast peer-to-peer data transfers by means of the multicast messages and also eliminates physical wiring between devices of most suppliers in substation. It possess properties such as collision avoidance, optimization of the message being transmitted, priority management and are readily available without the need of further detailed design but requires some extra precaution in the electronics hostile substation environment.

IEC61850 have client server systems instead of master-slave system. Client-server communication leads to better performance, as data is spontaneously sent to the client without the polling from a master device. Data transmission may be initiated by the change in the data value, and the change criteria may be adjusted from remote.

1037

## III. CONFIGURATION OF BAY CONTROLLER

Substation integration and automation can be broken down into five levels. The lowest level is the power system equipment, such as transformers and circuit breakers. The three levels in the middle are IED implementation, IED integration, and substation automation applications [8]. All electric utilities are implementing IEDs in their substations. The usage of multifunctional communicable IEDs communicating on a wide variety of open protocols has made bay controllers more important. Bay controllers communicate on Ethernet LAN at station level and also have the hardwired information. They act as a perfect slave device for a master SCADA, and also are effective masters for multivendor slave protection, control and monitoring devices.

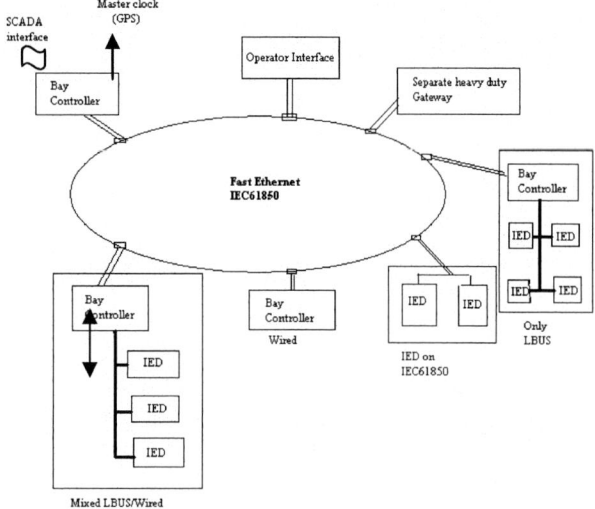

Fig. 2. Communication levels in substation.

All the bay controllers, Operator Interface (OI) sever and OI client communicates using protocol IEC61850 on the station bus. The IEDs are integrated to the Bay controller through the legacy bus using multiple legacy protocols as shown in Fig.2.

In order to implement substation automation AREVA PACiS system has been used in this paper where PACiS stands for Protection, automation and control integrated solution [9]. PACiS system is defined completely by three topologies. *System Topology* consists of device composition that manages the electrical process. The *electrical topology* consists of the electrical process definition in term of typed electrical devices (transformer, disconnector, circuit-breaker etc) that are connected through bus bars or lines. The *graphical topology* consists of the mimic and their graphical animation descriptions that appear at substation control points (operator interface) and bay control points (Bay controller local HMI).

'Site object' is composed of 'Substation' object as shown in Fig.3. A substation is constituted of 'voltage level' objects, each of them corresponding to an electrical partitioning of the

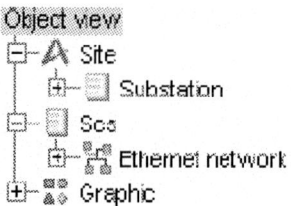

Fig. 3. General architecture of PACiS system.

substation by voltage level value (in kV). A voltage level is an aggregation of 'Bay' object, grouping of electrical devices, called module. There are different kinds of bays: feeder, transformer, busbar, bus coupler, bus section, capacitor bank and generic bay. Final electrical components are modules composing bays. There are different kinds of modules: circuit breaker, switchgear, transformer, motor, generator, battery, capacitor, inductor, converter and generic module. An extra module exists to describe substation external connection (external line). Every level of the electrical topology, except site level, can own data points

In PACiS system configuration, Bay controller is concerned with the three topologies as

System Topology (Scs): Bay controller is a direct subcomponent of the Ethernet network used for communication at station bus level.

Electrical Topology (Site): Bay controller manages bays and relevant modules or substation information.

Graphical Topology (Graphic): Bay controller can own an LCD display used for animated graphical bay panel representation.

When the bay controller is integrated to the Ethernet network, some of its attributes are set and verified as given in Fig. 4.

| Attributes of C264 | | | | |
|---|---|---|---|---|
| General | BI filtering | Measurements | Counter | Miscalleneous |
| short name | | | | C264 |
| long name | | | | C264 Computer |
| rack model | | | | 80 TE |
| date format | | | | DD/MM/YYYY |
| spare | | | | No |
| **External synchronisation** | | | | |
| synchronisation source | | | | None |
| **TCP/IP addressing** | | | | |
| TCP/IP address | | | | 10.106.170.221 |
| network name | | | | IED |

Fig. 4. General attributes of bay controller.

PACiS bay controller computers are composed of boards, responding to specific functions. Each board is configured as per requirement. A communication channel is a physical port available on CPU or the Basic Interface Unit (BIU) board and it is configured in accordance to the communication protocol.

Once the bay controller is configured IEDs are integrated on the legacy port of the bay controller. Depending upon the legacy protocol used legacy network is added and its attributes are set. The addressing of various data points is done on the legacy network.

## IV. INTEGRATION OF IEDs

IEDs are either integrated to the bay controller using legacy bus protocols or to achieve complete interoperability they are directly integrated to the station bus using IEC61850. The paper includes the integration of four IEDs. CM4000, a measuring IED, S42, a protection IED, from Schneider, both are integrated on the legacy port of the bay controller using the legacy protocol Modbus. The third IED, P638, is a protection IED of AREVA is integrated on the legacy port of the bay controller using T103 protocol. The fourth IED P145, a protection IED of AREVA is integrated on the station bus using IEC61850 as shown in Fig. 5.

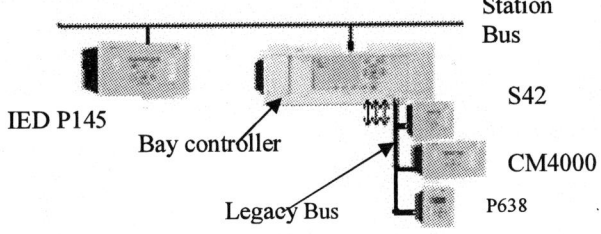

Fig. 5. Integration of IEDs on station and legacy bus.

### A. Networking IED on the Legacy Network

The IEDs are integrated on the legacy port of the bay controller and thus transfers data to the bay controller using multiple legacy protocols. The electrical data points (for example breaker failure) are created in the electrical topology of the system as shown in Fig.6. These data points are linked to their profiles (shows whether it is set or reset) created in the system topology.

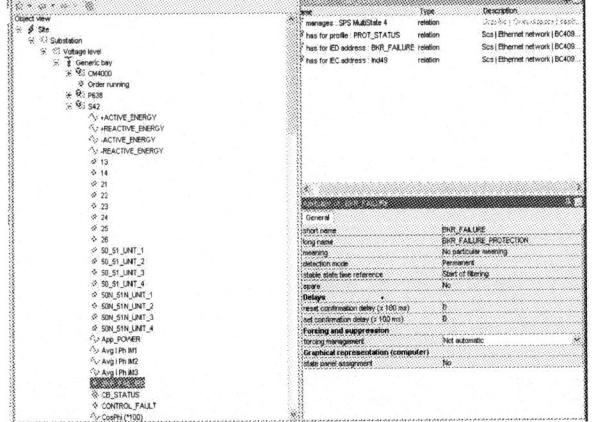

Fig. 6. Data point in system topology.

Since this data is available at the IED therefore these data points are to be linked to their particular IED address as shown in Fig. 7.

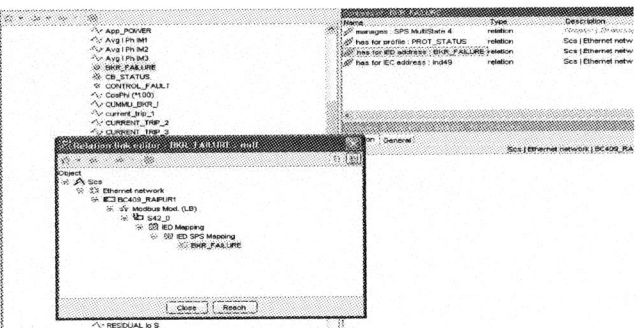

Fig. 7. Linking data point to its IED mapping.

These data points are also linked to their graphical interface (where one can see the value) in the graphical topology as shown in Fig. 8.

Fig. 8. Graphical representation of data point (CB healthy)

Once the database is ready appropriate voltage and current values are injected to the IEDs, which in turn process these values and the processed values are seen at the HMI.

### B. Networking IED on the Station Bus

To achieve complete interoperability IEDs are directly integrated on the station bus which supports protocol IEC61850, protocol has standardized the names and data of all monitoring, control and protection functions of substation. IEC61850 has data models, which have set of data belonging to the function. With the availability of standard data models, all the required data points are created in the electrical topology and are directly linked to their counterparts, which are defined in the protocol. Thus in case of IEC61850 there is no need of mapping the addresses of various data points.

## V. COMPARISON OF PROTOCOLS

IEDs with different communication protocols have been in use for quite sometime. However no effort has been made to compare the features of the protocols. On the basis of integration work done, an effort has been to compare the following protocols.

### A. Comparison between Modbus and IEC60870-5-103 (T103)

(1) *Time Synchronization*: T103 has standard format for time synchronization and supports this feature for all types of IEDs available from different vendors. Thus T103 is normally used to integrate protection IED as in protection system, time plays an important role and this protocol supports time synchronization. Modbus has no standard format for time synchronization and hence this feature is not available for IEDs from different vendors.

(2) *Disturbance File*: The disturbance recording function records the instantaneous values (samples) of the currents and voltages to visualize fast analogue changes around a failure

1039

for later analysis of network problems. T103 supports disturbance file feature and supports disturbance file upload from IEDs. Modbus supports only standard functions of automatic uploading. Since disturbance file does not have a standard format it is not supported by Modbus.

(3) *Number of Measurement Values Read*: Using T103 one can read set of 10 measurement values only which are fixed by software setting. Proper addressing is done in Modbus protocol and thus one-can read/write values for each register or coil.

(4) *Counter Feature*: Using IEC T103 one can not read energy values as it does not have counter feature. Modbus has counter feature and thus one can read energy values using this protocol.

(5) *Data Format*: In case of T103 protocol, decoding of data at the bay controller level is possible only if the documentation is available thus this protocol is used for IEDs either from the same vendor as that of the bay controller or for IEDs having proper documentation. Modbus can be used to integrate IEDs supplied by different vendors as using this protocol one can read any bit/byte/register.

(6) *Reset Command*: T103 supports one reset command for resetting LED and relay. Modbus supports different commands for LED reset and RELAY reset.

*B. Comparison between IEC60870-5-103 and IEC61850*

T103 defines the use of simple Information Objects, consisting of an ADDRESS+VALUE+TIME TAG (Optional) for individual single points in the Substation. IEC 61850 defines more elaborate Information Objects, each relating to a complete piece of equipment (multipoint entity) in the Process Plant (more elaborately Automated Substation).

The IEC 61850 enables the various Automation functions to be broken down into elementary functions, which may then be allocated to various pieces of electronic automation equipment distributed within the substation and interconnected by substation internal highways.

The transmission of messages, containing the more elaborate Information Objects, between the central Controlling station and the remote Substations requires broadband communication media, because each message contains more data than the simpler T103messages.Similarly broadband media are required for the Substation Internal highways.

All the above observations have been made during actual commissioning of the IEDs.

## VI. CONCLUSION

This paper is an attempt at substation automation and integration of intelligent devices with the substation. It aimed at configuring a bay controller and integrating protection and metering IEDs from multiple vendors using Modbus and T103 communication protocols. To achieve interoperability among a diverse group of IEDs and functions standard, an attempt has been made to integrate a protection IED on the station bus using IEC61850 communication protocol.

The CM4000, measuring IED and S42, a protection IED are integrated using Modbus. The third IED P638, a protection IED is integrated using T103 protocol. All these IEDs have been configured and successfully integrated to the bay. The fourth IED P145, a protection IED, has been successfully integrated on the station bus using IEC61850, the latest communication protocol. This IED can communicate to the bay controllers directly through station bus

Once the IEDs have been successfully integrated, various voltage and current values are injected to these IEDs to achieve desired results, which can be viewed through graphical interface. On the basis of observations made during actual commissioning of IEDs, an effort has been made to compare the protocols.

## VII. REFERENCES

[1] Zahra Moravej and D.N. Vishwakerma, "Integrated Digital Control and Protection for Modern Substation", presented at the 3rd Int. Conf Power System Protection and Automation, New Delhi, India, pp.59-66.

[2] H. Schubert, C. and D. Wong, "Streaming Operation And Maintenance in Substation by means of IEC61850", in *Proc. 2004 Power System Protection and Automation Conf.*, pp.10-20.

[3] Tushar M. Dhande , "Communication Infrastructure for Substation Automation", presented at the 3rd Int. Conf Power System Protection and Automation, New Delhi, India, pp.33-41.

[4] Lars Anderson, Klaus-Peter Brand, "The Benefits of the Coming Standard IEC 61850 for Communication in Substations," presented at the. Conf Power System Protection, Johannesburg, South Africa, Nov.2000, pp.65-71.

[5] Lars Anderson, Christoph Brunner, "Substation Automation Based on IEC 61850 with new process-close Technologies", in *Proc. 2003 IEEE Bologna Power Tech. Conf.*, pp.6-11.

[6] Phil G Beaumont, "New Trends in Protection Relays & Substation Automation Systems", in *Proc. 2002 IEEE Power Engineering Society Transmission and Distribution Conf.*, pp.609-612.

[7] C. Honga and G Wong, 'IEC 61850: Open Communication in Practice in Substation," in *Proc. 2004 IEEE Power Engineering Society Power System Conf.*, pp.618-623.

[8] John McDonald, "Substation automation – IED integration and availability of information", *IEEE Power and Energy magazine*, vol. 1, pp.22-31, Mar.-Apr. 2003.

[9] AREVA Manual on MiCOM hardware, MiCOM functional components, MiCOM application.

## VIII. BIOGRAPHIES

**Anupama Prakash** graduated in Electrical Engineering from Madan Mohan Malviya engineering college, Gorakhpur and just completed her M.Tech from Jamia Millia Islamia in Electrical Power System Management.

**Mini S. Thomas** (M-88, SM-99), graduated from University of Kerala in 1984, completed her M. Tech from IIT Madras in 1986 (both with gold medals) & PhD from IIT Delhi in 1991, all in Electrical Engineering. Her employment experiences include Regional College, Calicut, Kerala, Delhi College of Engineering, New Delhi and presently as professor in the Faculty of Engineering and Technology, Jamia Millia Islamia, New Delhi. Mini S. Thomas received the prestigious 'Career Award' for young teacher, instituted by AICTE, Govt. of India, for the year 1999. She has published over 30 papers in International/National Journals and Conferences. Her current research interests are in SCADA/EMS system and intelligent protection of power systems.

**Ashutosh Gautam,** graduated from the Delhi College of Engineering in 1987 with a degree in Electrical Engineering. He completed his M.S. (Controls & Robotics) from USA in the year 1989 and joined Areva in 1990. He is currently General Manager for Substation Automation.

**2006 IEEE International Conference on Power Electronic, Drives and Energy Systems**

# Ethernet Enabled Fast and Reliable Monitoring, Protection and Control of Electric Power Substation

Iqbal Ali[1], *Member, IEEE,* Mini S. Thomas[2], *Senior Member, IEEE*

*Abstract— Communication in the Electric Power Substation is very crucial for the entire power network reliability. This paper investigates the opportunities offered by the Ethernet technology and proposes a new Ethernet based Substation Communication Network (SCN) Architecture for fast and reliable communication as required by Monitoring, Protection and Control functions in an Electric Power Substation.*

*Index Terms-- Algorithm, Distributed Control, Ethernet, LAN, Monitoring, Protocol, Protection, Substation Automation System.*

## I. INTRODUCTION

THE reliability of the large, geographically distributed Power network critically depends on the reliable and fast operation of Generation, Transmission and Distribution Substations. Substation Automation System (SAS) is dedicated to the monitoring, protection and control of all the equipment, circuit breakers, transformers and buses of a substation and its associated feeders. The communication within SAS is crucial as the operations in a substation require time-critical data exchange.

A survey on communication technologies for electric system automation is presented in [1]. K.P. Brand, C. Brunner and W. Wimmer [2] has given the Design of IEC 61850 based Substation Automation System according to customer requirements. Application of peer-to-peer communication for monitoring, protection and control for a Distribution Substation is described in [3]. T. S. Sidhu [4] has given the implementation of some Control and Automation functions of Power System Substation according to IEC 61850 standard.

An Electric Power Substation (EPS) generally has three hierarchical levels, namely station level, bay level and process level. Till very recently, private serial communication was being used within EPS at the bay level and conventional parallel copper wiring at the process level to connect a process within the switchyard to the switchgears. The new substation

---

[1]**Iqbal Ali**, is Lecturer in the Department of Electrical Engineering, Jamia Millia Islamia, New Delhi, India
(e-mail: iqali_in@yahoo.com).

[2] **Mini S. Thomas**, is Professor and Head of the Department of Electrical Engineering, Jamia Millia Islamia, New Delhi, India
(e-mail: mini@ieee.org).

communication standard IEC 61850 now provides a comprehensive global standard for all communication needs in the substation at all levels [5-8].

This paper investigates the communication requirements, in terms of the data throughput, for the implementation of functions of an SAS and proposes an Ethernet based Substation Communication Network (SCN) Architecture. This SCN is shown to be accommodative for the services offered by the standard IEC 61850.

The following section estimates the worst case data traffic load on to the Ethernet highway within an EPS, and discusses the high level of redundancy, availability and reliability of the proposed SCN architecture.

## II. DATA TRAFFIC LOAD IN SUBSTATIONS

The speed of response of the time-critical functions of an SAS, related to monitoring, protection and control, directly depends on the data throughput of the communication network, laid down in that EPS. To decide upon the usability of the Ethernet network in an EPS, the worst case data throughput is calculated for computationally most demanding relaying algorithms i.e. Impedance / Distance Relaying. Since, the basic feature of any relaying algorithm is its 'Speed of Response'. The maximum speed, which has some practical significance in the existing technology scenario, is one fourth of the period of the fundamental frequency and this is also dependent on the fault transients and the entire fault clearing process [9]. Thus the fastest desirable response time of any relay algorithm is approximately 4-5 milliseconds for 50 Hz and 60 Hz power systems respectively.

To make secure decision for relaying, at least three data samples during the period of 4 ms or 5 ms are required [9]. Therefore, analog input sampling frequency required should be at least twelve times the fundamental frequency of the analog input signals, which gives sampling intervals of 1.4 ms to 1.6 ms for 60 Hz and 50 Hz systems respectively. Since the Relaying Algorithms, typically consisting of couple of thousands machine language instructions, must execute during the time between samples. Increasing the analog input sampling frequency to double or triple the above minimum rates, gives algorithm execution time of 0.5 ms to 0.7 ms. Hence, the average instruction execution time will be about 250 ns.

---

0-7803-9771-1/06/$25.00 ©2006 IEEE

Considering the margin of safety, the average instruction execution time desired for sampling rate of 2160 Hz can be taken as 100 nanoseconds. As a conclusion, Intelligent Electronic Devices (IEDs) or microprocessor based relays having average instruction execution time of 100 ns to 300 ns can only be considered for relaying algorithms.

Further, in any EPS, for the implementation of Metering, Monitoring, Control and Protection functions as well as for coordination and planning, it must be able to handle measurement data, administrative data, trip data and file transfers. Every measurement point in an EPS requires three sets of values, one for each phase. A typical setup generally has 8-12 measurement points, from which the data is required to be sent to 2-4 destinations for distributed applications. A gross estimate of the data traffic load, in any EPS, can be made by multiplying sampling rate, number of phases, number of data points and number of destinations. This way substation network will have about 140000 packets per second. A standard size of 32 bytes is taken for a measurement data packet and the total protocol overhead is of 60 bytes. Thus a standard packet size will be 736 bits long. Packet size if multiplied by packets per second on the substation network gives an estimate of total data volume of about 103 Mbps, which is slightly more than the gross capacity of fast Ethernet i.e. 100 Mbps.

There are various techniques, which help in reducing the data volume put on to Ethernet in an EPS e.g. Multiplexing, Switched Fast Ethernet, letting one measurement node transmit data for all the three phases instead of one etc.. Hence, it can be concluded here that the Ethernet possesses sufficient performance characteristics to meet the real-time communication demand within the EPS.

There exist various network simulator packages for simulating the whole SCN and enable the calculation of the performance parameters (e.g. round trip delay) of the designed network under various traffic conditions. OPNET is one such simulator package [10]. OPNET is object-oriented and therefore, the users can create new objects from the existing general purpose objects at will.

After verifying the usability of Ethernet, the proposed SCN architecture is discussed next in section-III.

## III. SUBSTATION COMMUNICATION NETWORK ARCHITECTURE

Generally, there are three levels in a Substation Automation System (SAS), namely; 1) Station level, having User Interface (UI) computer and the Network Control Center (NCC) gateways (GW). 2) Bay level, having bay units for protection and control and 3) Process level, having direct connection with the switchyard equipments.

The SCN architecture proposed in this paper considers the communication system between these levels as consisting of switched Ethernet as shown in Fig.1. This SCN Architecture is laid keeping in view of the time critical requirements of the Monitoring, Control and protection functions in an EPS. The notable features of this proposed SCN architecture,

are provision for alternative data paths in the event of a fault, protection of SAS from single-point-of-failure and provision for local data concentrators, which enhances the overall reliability of the SAS.

### A. Redundant Paths

Fig. 1. shows the Substation Communication Network (SCN) architecture of a Substation Automation System. It can be observed here that the formation of a single switched Ethernet ring provides alternative data path, at the bay level, in case of the link failure, where as formation of two independent rings, in which each bay level IED has a separate connection port to each ring through an Ethernet switch, provides even higher level of redundancy and zero recovery time in case of any link failure, as here two rings operate in parallel.

### B. No Single-point-of-failure

The Communication Network Architecture of a substation automation system may comprise of a single central switch to which all the IEDs, operator terminal (i.e. UI) and gateways may be connected, thereby forming a star type network topology. The central switch in the star type network becomes a single-point-of-failure for the complete communication network in SAS. Hence, the availability of the proposed SCN is much higher than the star type communication networks.

Moreover, it can support distributed functions like interlocking, autoreclosure, synchrocheck etc., where more than one IEDs need to cross communicate for data sharing, in addition to the communication between the IEDs and station level computer.

### C. Provision for Local Data Concentrators

Ethernet protocol is designed to operate on "Best-effort" or CSMA/CD (Collision Sense Multiple Access / Collision Detect) philosophy. Hence, it is non-deterministic in nature. Therefore, the response-time and throughput is not known in advance. For the SCN architecture presented here, local processing and the corresponding data traffic of any bay is isolated through the provision of a separate Ethernet switch for each bay. This provision of Ethernet switch keeps the number of collisions, on to the Ethernet highway, to its minimum possible. This, helps in maintaining high data speed, throughput and reliability during vertical communication i.e. the communication from the operator station to the process level and vice versa for the implementation of time-critical distributed control and protection functions.

### D. No Gateways for Inter IED Communication

The absence of gateways for inter-IED communications saves time elapsed during protocol conversion. This will result in significant increase in the speed with which different time-critical distributed functions related to monitoring, protection and control could be implemented and executed.

### E. Enhanced Reliability

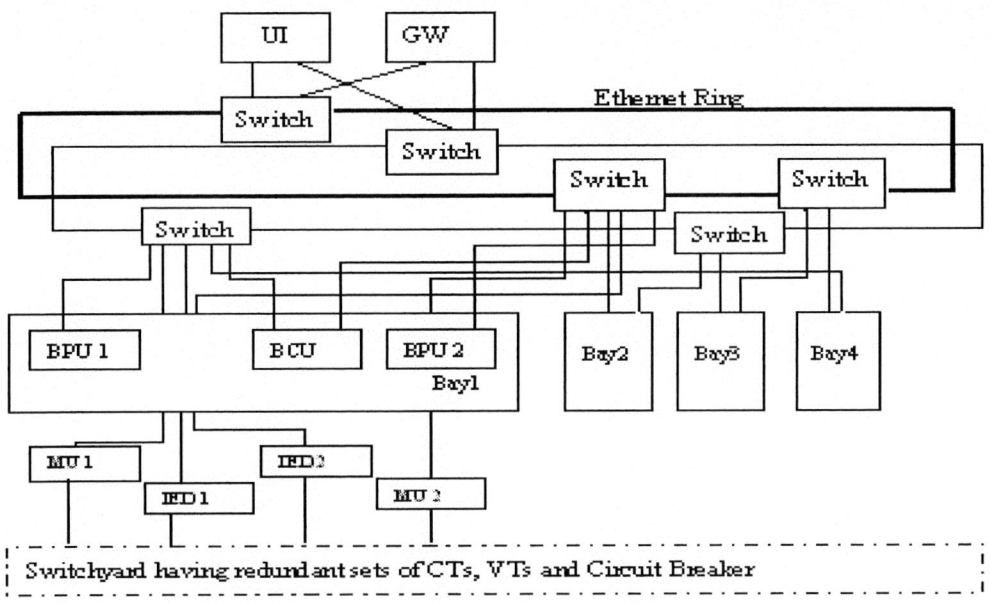

Fig. 1. Communication Network Architecture in a SA System; MU- Merging Unit, BPU-Bay Protection Unit, BCU-Bay Control Unit.

For the SCN shown in Fig. 1., the case of a breaker trip signal is considered and supposes it is initiated from IED 1 of the bay 1. In case of the failure of local circuit breaker, the trip signal will pass on directly to the breaker IED of the other bay, through process level switches, for appropriate protection action instead of traveling higher in the hierarchical level. Here, the IEDs are expected to have some inbuilt breaker fault protection logic. Thus, it may be observed that, the overall reliability and performance of the time-critical and distributed functions like interlocking, fast auto-reclosure and slow auto-reclosure, breaker failure, interrupting or reverse blocking and synchrocheck etc. is enhanced. This enhanced reliability and performance of local as well as distributed functions, related to Monitoring, Protection and Control in an EPS, may be credited to the communication path redundancy, ring network topology, switched Ethernet network at bay level as well as process level, gateway free inter IEDs communication.

This switched Ethernet based SCN architecture presented in this paper is encouraged by the advent of the global and future proof communication standard IEC 61850, which supports interoperability. Through the split of application and communication stack and through the standardized permitted extension of the supporting functions, the long-term validity of the standard is guaranteed. This approach leads to lower investments, less training efforts and lower life-cycle costs of substation automation systems. The implementation of some of the important control and protection applications using IEC 61850 have been discussed in [4].

The next section will describe the important features of the new substation communication standard IEC 61850 and its benefits to the users. Data models and abstract services definitions are shown to be accommodative by the SCN architecture of Fig. 1., which is presented in this paper.

## IV. THE COMMUNICATION STANDARD IEC 61850: FEATURES AND BENEFITS

Standard IEC 61850 [5] is the result of collaborative work of IEEE and IEC to develop a common standard for substation communication. This standardization process also involved some leading product/system manufacturers and some major utilities. The primary objective this group was to develop a communication protocol for substation communication, which will ensure Interoperability; the ability for IEDs, irrespective of their vendors, to exchange information, free configuration and long-term stability i.e. ability to follow the progress in communication technology as well as system requirements.

To meet the basic requirements of the standardization process, that is interoperability and to be future proof , the IEC 61850 standard is built over OSI seven layer model. The data model, services and applications related to the power system substation are built above the seventh layer (Application Layer) of the OSI model. This ensures that the substation communication can evolve with the evolution of communication technology. Fig. 2. shows this approach to the standardization process and Fig. 3. shows the interface between the IEC 61850 object model and OSI stack.

The users of IEC 61850 are Electric utilities, consultants and manufacturers. All the users get the benefits in various areas of substation automation e.g. Open energy-market and international activities, Specification, Design, Manufacturing, Installation and Commissioning, Operation and Maintenance, Asset Management etc.

### A. Open Energy Market and International Activities

In the present scenario of deregulated and restructured environment of the electric power industry, utilities find it harder to forecast power generation and demand, and often face an uncertain future. IEC 61850 has standardized the

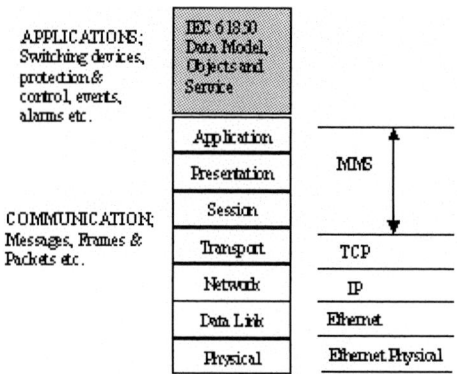

Fig. 2. IEC 61850 approach to standardization.

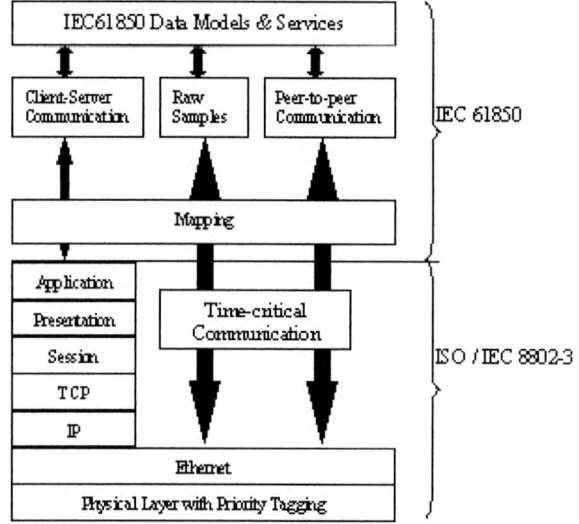

Fig. 3. Interface between IEC 61850 and OSI stack.

nomenclature for all the anticipated functions of a substation automation system and their data. The standard has also standardized the services and protocols, thereby eliminating the strategic unknowns in substation automation and brings certainty to the management of substations and the vital data necessary for the secure operation of the network.

Consultants and power utilities around the world have long been operating internationally. IEC 61850 was established with the requirements of substations worldwide in focus. Consultants and utilities adopting IEC 61850 can apply their expertise on the standard and equipment to most networks in the world and consequently broaden their client base.

### B. Specification

The standard IEC 61850 has standardized the names of all the monitoring, control and protection functions and their data of an SAS. For example, the name of Time-Overcurrent Protection function is PTOC. Circuit Breaker is also a function and its name is XCBR. XSWI is for Isolator or

earthing switch, TCTR is for Instrument Transformer, YLTC is for Power Transformer, CSWI is for Switch Control, CILO is for Interlocking, MMXU for Measuring Unit, ATCC for Automatic Tap Changer and IHMI is for Human Machine Interface or Operator Place etc. A data model is a set of data belonging to a function. For example, the data model of a circuit breaker includes 1) Mode, i.e. *enabled, blocked, disabled, under test* etc. 2) Name plate, i.e. the technical details of the switch controller. 3) Position, i.e. *open, closed, intermediate* etc. Most substations have more than one circuit breaker and the final data name comprises the standardized parts and user-specified parts. The latter parts allow the utility to identify their own circuit breakers. In IEC 61850 standard, the standardized function names such as PTOC and XCBR are called logical nodes. There are approximately 90 logical nodes, covering functions such as switchgear, protection, measurement and so forth. Objects, such as Pos, and attributes are added to them to form standardized data names.

The standard specifies a set of generic abstract services which cover all the data transfers required within a substation. The services include secure transfers of large block of data for reporting purposes and fast transfers of small data blocks such as trip. The standard maps these abstract services and the standardized data onto real protocols. These real protocols are Ethernet, Transmission Control Protocol/Internet Protocol (TCP/IP) and Manufacturing Message Specification (MMS).

When communication technology changes, it is necessary only to map the abstract services and the standardized data to the new protocols.

Thus the chances of ambiguity or misinterpretation between the consultants and the bidders are unlikely at the specification stage of a substation automation project.

### C. Design

Before IEC 61850, the specifications about the data models were not very specific rather were made sufficiently general, which encouraged a wide variety of manufacturers supporting different protocols to bid. The standardized data models of IEC 61850 are understood by all manufacturers and can be stated in the specification. These models can be carried forward directly into the design stage, saving time and reducing errors. For a substation automation based on Ethernet network, as one shown in Fig. 1., the existing infrastructure can be reused with little extra design work. In general, hardwire design of the substation automation system (SAS) is simplified because of no gateways are required among the IEDs, Fewer components in the SAS means less time spent on co-ordination and fewer review meetings.

The XML based Substation Configuration description Language (SCL) of IEC 61850 [7] defines the description of a Single Line Diagram (SLD) of a substation and creates a file called System Specification Description (SSD) file. Preparing the SLD in the form of SSD file reduces misunderstanding and enables automatic processing of this file for consistency checks of quotations against specifications. However, the SSD file does not define specific details of functions implementations and their interactions.

The Ethernet backbone is often light and easy-to-handle optical fibres. This new communication platform means a significant reduction in the number of multi-core cables and the accessories in most substations, and hence a reduction in design efforts.

### D. Manufacturing

SCL of IEC 61850 [7], has very much simplified the task of configuration of a substation automation system. It has partly become automated, as the manufacturer-independent information can be freely exchanged. Less time is needed for co-ordination and an SAS can be up-and-running quickly. Since the chance of errors in the information exchange is lower, less rework and retest are expected. Working to a common single standard, the test personnel need not learn and test unfamiliar protocols, and can expend their efforts more fruitfully on finding and removing faults in the automation functions. In the factory, an IEC 61850 SAS can be tested complete with Ethernet cabling exactly like that on site. Any errors found can be fixed quickly and easily. Tests on a complete SAS with conventional parallel wiring can be carried out only on site, where the rectification of errors may be more difficult and time-consuming.

### E. Installation and Commissioning

As Ethernet based communication networks requires fewer cables and accessories, installation time is reduced. Less cabling means that the chance of connection/termination errors is lower. Over 95% of the local area networks of offices are now Ethernet based. Standard tools for checking Ethernet systems are readily available and can also be applied in substations during commissioning.

Owing to the use of TCP/IP, the Human Machine Interface (HMI) may be moved around in the substation and plugged into any Ethernet switch during commissioning to display the data of the whole substation. In particular, many tests involves checking 'cause' and 'effect', which take place at different locations respectively in the substation. The data view at a single location can help speed up such tests. If wireless local area network is used, the mobility of the HMI would be further enhanced.

The technical experts mostly operate from the office, where they are close to their own databases, which help them in their decision making process, and can serve other personnel instantly. Thus the specialists can take part in commissioning without leaving the their desks and can read site data via the Internet, guiding and offering advice to site personnel. When a device is out of operation during commissioning, technical experts can take help of the simulators. This helps avoid delays on the critical path and keep the project on course.

### F. Operation and Maintenance

Internet based monitoring applications are now available to operate remotely from the office, therefore now one can adjust settings of devices, interrogate devices and retrieve data sitting right on the office desk. This drastically reduces number of site visits for data retrieval and inspection purpose and therefore, personnel in offices and on sites can exchange information more smoothly and efficiently throughout the enterprise.

The general performance of the SAS is considerably improved because the delay in data flow for peer-to-peer communication, due to gateways, for the implementation of distributed monitoring, control and protection function is no more there. Multicasting supported by Ethernet enables the messages to be sent to multiple destinations simultaneously, hence faster than the polling techniques. 100 Mbps is the state-of-the-art data rate of Ethernet in substation communication and can hardly be achieved by other protocols. Priority Tagging of Ethernet enables important messages to be sent faster.

The availability of the SAS is expected to be higher because IEC 61850 advocates distributed intelligence. Thus functions such as interlocking, can be implemented without a central coordinating device. Peer-to-peer communication ensures, that a function still runs normally even after one or more devices have failed.

The SCN, as shown in Fig. 1., consist of Ethernet switches, which help in isolating the individual bay level data traffic and drastically reduces the number of collisions, therefore throughput is higher.

IEC 61850 also advocates publish-subscribe approach, therefore substation events are reported as soon as they occur, hence faster. Whereas in master-slave approach, polling interval may be so large that the transmission of messages takes longer than desired.

Maintenance efforts in the SAS include the training of the staff to learn communication protocols and repair or replacement of the IEDs. As IEC 61850 is a single protocol stack for SAS, the maintenance task of SAS will be easier for the electric utilities personnel. Training effort is reduced because the staff need not learn many protocols. New IEDs can be added with the minimal operational impact on existing IEDs. This means shorter downtime of the SAS when the substation is extended or when old equipment is replaced with new equipment.

Security of the SAS in any Electric utility demands, that only authorized personnel should have access to the data related to operations and maintenance. Such security measures can be built by using existing methods in information technology. Furthermore, TCP/IP forms the information highway linking the offices and substations, which help in creating these security related applications.

### G. Asset Management

Asset management involves evaluating the present equipment condition and then establishing replacement and purchase plans for the future. IEC 61850 relieves some of the burdens in asset management, as it adopts a fixed set of mainstream communication protocols i.e Ethernet and TCP/IP. In particular, the data models defined in IEC 61850 are long lasting and Ethernet is highly forward and backward compatible, in addition to possessing a favourable cost-to-performance ratio. So one does not need to think twice for the communication technology to be used. Therefore, any

1045

upgrade or replacement of an SAS will clearly cost less than before.

IEC 61850 defines system life-cycle, which covers the management of product versions and the support after product discontinuation. Utilities can continue to run an equipment replacement schedule without speculating on the future of the products.

The interoperability of IEDs, advocated by IEC 61850, is expected to last for many product generations. Conformance testing forms a long-term reference basis for the confirmation of interoperability of IEDs. The documentation of the tests carried out and product version numbers help the utilities in keeping track of the interoperability characteristics of each IED, which are important for managing equipment in the long term. The extension of the data models of IEC 61850 to monitoring substation equipment is taking place, helping utilities in moving away from the time-based maintenance to the more efficient condition-based maintenance.

## V. SUBSTATION COMMUNICATION NETWORK (SCN) ARCHITECTURE AND IEC 61850

The proposed SCN in this paper is switched Ethernet based and provides Ethernet highway for communication at all levels in an EPS. Since, the primary objectives of the substation communication standard IEC 61850, is to develop a communication protocol for substation communication, which can ensure interoperability, free configuration and long-term stability. It is observed that, these noble objectives of the standard IEC 61850 can be easily mapped on to the proposed SCN of Fig. 1. Higher level of redundancy, availability and reliability provided by the SCN of this paper have already been discussed. The implementation of distributed automation functions related to monitoring, protection and control on this SCN architecture will result in minimal time delay, as there exist no gateway for inter IED communication.

## VI. CONCLUSIONS

The implementation of time-critical functions related to the monitoring, protection and control of SAS in an EPS of any electric utility requires very high-speed communication network. The justification for the performance characteristics of Ethernet is presented in terms of its handling of the worst case substation data traffic. A switched Ethernet based SCN architecture is presented and its higher availability and reliability are discussed. The features and benefits of the

global and future proof substation communication standard IEC 61850 are also discussed and it is observed that the services offered by the IEC 61850 may be mapped over the proposed SCN architecture presented in this paper.

## VII. REFERENCES

[1] V.C. Gungor and F.C. Lambert, "A Servey on Communication Networks for Electric System Automation", ELSEVIER Journal of Computer Network 50 (2006) 877-897.

[2] K.P. Brand, C. Brunner and W. Wimmer, "Design of IEC 61850 Based Substation Automation Systems According to Customer Requirements", B5-103. Session 2004 © CIGRE.

[3] Reckerd D., Vico J., "Application of Peer-to-Peer Communication for Protection and Control at Seward Distribution Substation", 58th Annual Conference for Protective Relay Engineers, 5-7 April 2005 Page(s): 40 - 45

[4] Sidhu, T. S., Gangadharan, P. K., "Control and Automation of Power System Substation using IEC 61850 Communication", Proc. IEEE Conference on Control Applications, August 28-31, 2005, Toronto, Canada.

[5] IEC 61850: "Communication Networks and Systems in Substations", 2002 – 2005 (www.iec.ch)

[6] IEC 61850 Communication Networks and Systems in Substations, Part 5: Communication Requirements for Functions and Device Models.

[7] IEC 61850-6 Communication Networks and Systems in Substations, Part 6: Substation Configuration Language for Communication in Electrical Substations related to IEDs.

[8] IEC 61850 Communication Networks and Systems in Substations, Part 7-2: Basic Communication Structure for Substations and Feeder Equipment, 1999.

[9] A. G. Phadke and J. S. Thorp, "Computer Relaying for Power System", Taunton, Somerset, England, Research Studies Press, and Wiley, New York, 1988.

[10] OPNET, Modeler, http://www.mil3.com.

## VIII. BIOGRAPHIES

**Mini S. Thomas** (M-88, SM-99), graduated from University of Kerala in 1984, completed her M.Tech from IIT Madras in 1986 (both with gold medals) & PhD from IIT Delhi in 1991, all in Electrical Engineering. Her employment experiences include Regional College, Calicut, Kerala, Delhi College of Engineering, New Delhi and presently as Professor in the Faculty of Engineering and Technology, Jamia Millia Islamia, New Delhi. Mini S. Thomas received the prestigious 'career Award' for young teacher, instituted by AICTE, Govt. of India, for the year 1999. She has published over 30 papers in International/National Journals & Conferences. Her current research interests are in SCADA/EMS system and intelligent protection of power system.

**Iqbal Ali** (M-04): graduated from Zakir Hussain College of Engineering & Technology, AMU, Aligarh. Received his M.Tech. degree from Indian Institute of Technology, Roorkee. He is lecturer in the Department of Electrical Engineering, Jamia Millia Islamia, New Delhi, India. His current research interests are in SCADA/EMS system, Substation Automation Systems and Power System Communication.

**2006 IEEE International Conference on Power Electronic, Drives and Energy Systems**

# Expert System for Power Transformer Condition Monitoring and Diagnosis

M. Ahfaz Khan, A.K. Sharma, and Rakesh Saxena

*Abstract*-- **In this paper a prototype expert system is proposed. This expert system utilizes dissolved gas in oil analysis techniques to monitor and diagnose for fault condition of transformer. The Roger's Four Ratio Method, NTT Flag point method, generation rate ratio method and Total dissolve combustible Gas method is used in this expert system. Both the Roger's four ratio and the NTT flag point method are implemented. These two methods are fuzzefied. Which is increasing the accuracy of both methods, as the generation rate ratio method is use to analyzed the trend of fault gases. All the four methods are used together into this expert system, to achieve greater accuracy in analyzing a fault condition monitoring and diagnose of transformers. The "MATLAB" based expert system is developed. This expert system tested with some faulty transformer cases. 96% accuracy is obtained.**

*Index Terms* – **Diagnose, Expert system, Monitoring, and Power Transformer.**

## I. INTRODUCTION

THE Oil Immerged Power Transformer are key-stone in transmission and distribution system the failure of one transformer may cause significant effect for electrical utilities. Failure, coming without warning causes large outage and economic losses.

What one needs is a diagnostic technique to assist any degradation of the insulation materials and aging criteria to detect whether or not maintenance on equipment is needed. Therefore the ability to evaluate the lifetime of Oil immerged power transformer is important. To detect the existence of incipient transformer faults, before they become catastrophic is highly essential uses a numerous techniques, both online and offline. The online includes winding vibration, acoustic measurement of corona, temperature monitoring, gas in oil monitoring using DGA and offline includes Partial discharge, transfer function, recovery voltage measurements, Degree of polymerization and furan analysis of cellulose insulation to assist the condition assessment of oil Immerged transformer [17]. Among these techniques the dissolved gas in oil analysis

Rakesh Saxena is with Shri Govindram Seksaria Institute of Technology and Science Indore (rakeshsaxena@hotmail.com).
A.K.Sharma is with Jabalpur Engineering College Jabalpur (arvindksharma_2000@rediffmail.com).
M. Ahfaz Khan is with Indore Institute of Science and Technology Indore (khan.ahfaz@gmail.com).

(DGA) technique is quite simple, non- intrusive and inexpensive method. During normal operation of oil-paper insulated power transformer the absorb moisture and their dielectric strength deteriorates with the level of moisture content. Typically 0.5% moisture in paper and 20 ppm in oil [12].

Electrical, mechanical and thermal stress results incipient faults, in the form of arcs or spark resulting imperfect insulation. Joints loose electrical connections, stranding of conductors or hot spot due to abnormally high current densities in conductors or Partial discharge due to gaseous bubble around high thermally stressed parts of solid insulation [12].

Due to these incipient electrical and thermal fault the transformer oil is oxidized and degraded. As a result gases produced, which are either partially or entirely dissolved in the transformer oil. The gases dissolve in the transformer oil are classify into two category, fault related gases and non-faulty gases. Generally, the fault related gases are hydrogen ($H_2$), methane ($CH_4$), ethane ($C_2H_6$), ethylene ($C_2H_4$), acetylene ($C_2H_2$), carbon monoxide ($CO$) and carbon dioxide ($CO_2$) and the non-faulty gases are Nitrogen and Oxygen. The fault related gases give an early warning and indicate the type of faults that are developing in the transformer and appropriate action can be taken [1]-[4].

There are numerous methods to interpret the transformer fault by these, Dissolve gas in oil. Some common methods are Roger IEC, Dorrenburg method, Key gas method, NTT flag point method and Total dissolves combustible gases. There is no one method, which is base on mathematical formulation. These methods have been developed on year of experience in fault diagnosis. Hence they vary from utility to utility [1]-[6].

## II. TRANSFORMER FAULT INTERPRETATION

The gases relies in transformer oil due to electrical and thermal fault in particular pattern. Various techniques have been used to interpret incipient fault in oil immerged transformer. The interpretation of incipient fault is out line by IEEE std. C37.104-1991 [2]. Degrade of mineral oil to different molecular weight gases is closely related to both thermal and electrical stress. The low molecular weight gases are produced at temperature below $500^{\circ}C$ and releasing large amount of gases. These gases include hydrogen and methane,

0-7803-9771-1/06/$25.00 ©2006 IEEE

and small amount of higher molecular weight gases include ethane and ethylene. The higher molecular weight gases concentration will increase, that of the lower molecular weight gases if temperature increases above 500°C.The concentration of acetylene will be pronounced if temperature increased 700°C to 1800°C. Transformer fault interpretations based on the gas in oil analysis are derived generally from the fact that, if key gas volume and few combination of gas ratio exceed a prescribed limit then a fault can be expected in the transformer. Some common techniques are Dorrenburg ratio method, Roger ratio method, Key gas method and NTT flag point method [1]-[6].

## III. DEVELOPMENT OF EXPERT SYSTEM

Analytical approaches have been used over the years for solving several power system planning, operation and control problems. However the mathematical formulations of real world problems are derived under few restrictive assumptions.

The solution of large-scale power system problem even suffers with these assumptions. In other words there are many uncertainties in various power system problems because power system are very complex and influenced by unexpected events.

This is main reason of the present interest in expert system technique in power system. Which is quite easy to handle these uncertainties in power system.

Expert system is able to handle mass data during alert and contingencies state. Which reduce the workload for human expertise operation. Expert system is computer programs that are drive from a branch of computer science research called artificial intelligence. Which is concerned with the concepts and methods of symbolic inference or reasoning where the human knowledge experience and expertise are coded into computer program for solution of problems.

The main advantages of the expert system are that it enables even a non-expert user to determine competently about the state of the insulated system. The insulation behavior in further operation even without the necessity of consulting the matter with top experts [17], [18].

The object of this paper is to developing a rule base expert system that can performs accurately diagnosis of transformer. Expert system of computer program that capture an experts decision-making knowledge. The non-expert users of the system interact with it in much the same way they would with a human experts to get answers of these questions.

Expert system is reserved for program that use engineer knowledge to solve problems. That normally requires expert person this expert knowledge is stared as data or rules within the computer and use when needed to solve problems. The expert system has several advantages over conventional programs. It performs task using decision-making logic, basic algorithms and boundary conditions [17], [18].

The building blocks of expert system consists of two principal part

- Knowledge (rule) Base.
- Reasoning or inference engine.

### A. Knowledge Base

The knowledge base use knowledge representation formalizes and organizes the knowledge. One widely used representation is the production rule or simple rule. A rule consist of an IF part and THEN part. The IF part lists a set of conditions in some logical condition. If the part of the rule is satisfied consequently. The THEN part can be concluded or its problem solving action taken. For diagnosis of transformer, the knowledge base incorporates DGA interpretation method. In order to make use of the expertise this is embodied in the knowledge base. The gases production in oil immerged transformer are imprecise or vogues which are not define in DGA method. The certain combination of gases is not defined in DGA method accurately therefore the fuzzy logic concept in oil immerge transformer associated with the uncertain of the gases combination.

Fuzzy logic can handle complex and imprecise or vogues problem, which DGA method cannot. The main building blocks of fuzzy logic system beginning with the fuzzy membership function, the fuzzy rules, the fuzzy inference and the defuzzification process [15]. The Roger ratio method is used, the ratios of gas pair are used in this method, Typically, four ratio ($CH_4/H_2$, $C_2H_6/CH_4$, $C_2H_4/C_2H_6$ and $C_2H_2/C_2H_4$) are used for sufficient accuracy however there is some limitation of this method.

The fuzzy logic uses crisp fuzzy value to quantify the ratio results into possibilities of codes combination for Roger ratio method. Table I is an example illustrating, the difference in coding process for both the traditional and fuzzify Roger's ratio method an example uses ratio $CH_4/H_2$ to demonstrate the difference between methods [7], [10]-[13].

TABLE I
RANGE AND CODE FOR TRADITIONAL AND FUZZY ROGER METHOD
FOR RATIO $CH_4/H_2$

| Roger method | | Fuzzify Roger method | |
|---|---|---|---|
| Range | Code | Range | Coding |
| < = 0.1 | 5 | < 0.2 | Very low |
| > 0.1 & < 1 | 0 | > 0.1 & < 1 | Low |
| > = 1 & < 3 | 1 | > 0.9 & < 3 | Medium |
| > = 3 | 2 | > 2.9 | High |

After the fuzzy number of $CH_4/H_2$, $C_2H_6/CH_4$, $C_2H_4/C_2H_6$ and $C_2H_2/C_2H_4$ has been obtained reference to the fault coding interpretation of Roger method are used to classify the transformer fault condition.

In the process of fuzzifying NTT flag point method allow a range of possibilities or gradient in transition between two different conditions. So that it is able to diagnose transformer with grater age variation. The range of flagpoint can be fuzzifying in the manner that can be understand by an example of Hydrogen gas. In NTT flagpoint method the maximum limit for Hydrogen is specify 1500 ppm. Which has been range of 1125 ppm to 1875 ppm. The Generation rate method applies the principal of fault indication with the concentration of individual fault gas. It indicates the rate at which the possibility of fault is progressing. In all the above-mentioned diagnosis method, the possibility of an incipient fault can determined, but the operational condition of the transformer unknown. The operational conditions of transformer are closely related to the maintenance cost, reliability and lifespan of the transformer. The total dissolve combustible gas method is used to analyze and classify the transformer into various operational conditions. The operation conditions are varying useful to many electrical utilities. As they can plan future development and load distribution, outage plan for a specific distribution area. Expert system whose knowledge is represented in rule form is called rule-based system.

### B. *Inference Engine*

The inference engine is program that uses rule-based system to arrive at conclusions regarding a given problem. Since an expert system is based on a body knowledge that may cover many aspects of a given problem. It may make a single recommendation or several recommendations arranged in order of likelihood and will explain the logical basis for each [17], [18].

The Inference engine incorporates all the above mention DGA method and unique diagnosis result. Fig. 1 shows a Bolding black of inference engine.

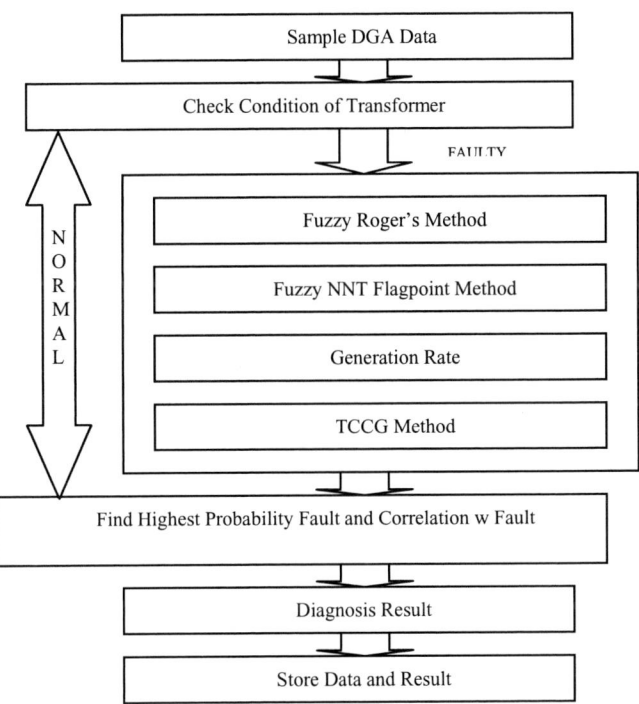

Fig. 1. Building black of Inference engine.

In the inference engine combination of their methods will allow for expert filed knowledge. The user needs to enter transformer DGA data in form of parts per million for all gases. Test the condition of transformer TDGC analyses classify the condition of transformer into two cases where the transformer is faulty or faultless. If the transformer is faultless the diagnosing procedure in not required and the result will be display ``Normal operation condition''. It allow user to save processing time, If the transformer faulty. Then Expert System allows to diagnosis the transformer with various method fault condition of transformer. In the diagnosis section of the Expert System the transformer which is classified under faulty condition, but the specific fault was unknown.There are five major faults for a large transformer [6]-[10].

- Partial discharge (low intensity electrical discharge)
- Sparking (Medium intensity electrical discharge)
- Arcing (high intensity electrical discharge)
- Local over heating (Thermal fault)
- Severe over heating (Thermal fault)

In this expert system fuzzy roger, generation rate and fuzzy NTT flag point method are uses to diagnosis the transformer. Find out the fault with the highest possibility.

## IV. RESULT

The Proposed expert system is demonstrated here by insulation diagnosis perform. As the expert system is correlate the various electrical faults, Thermal faults and operational condition of transformer. That allows the best understanding the status of the transformer.

Once invoked, the expert system will required the transformer background information from the user. These include the serial number other detail of the transformer, its identity, its date of manufacturing or maintenance, its MVA and KV rating. Its also required some information of the oil sample; its date of sampling, samples taken from top or bottom of the transformer. The DGA data in ppm volume the user interface for the input is shown in figure 2.

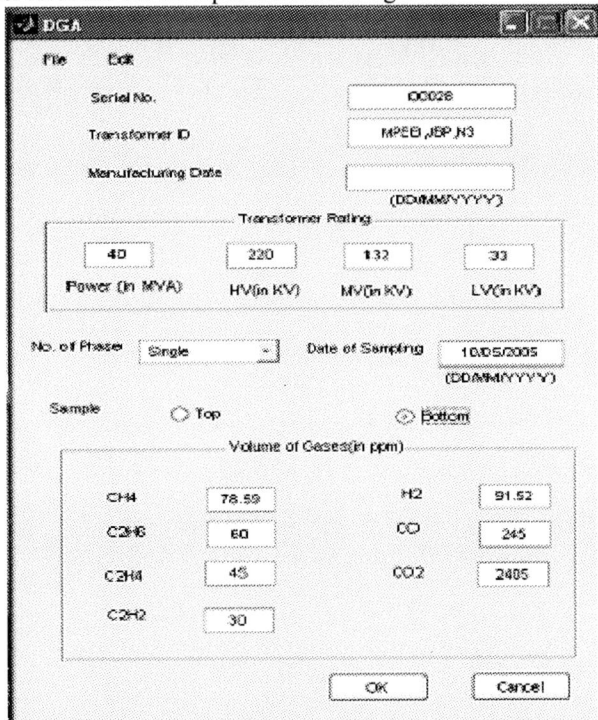

Fig. 2. Input window of Expert System.

Once the test detail of the transformer is fed to the database of the expert system. The inference engine is commanded to perform the diagnosis procedure based on the various fuzzified DGA method.

The expert system follows the basic flow chart as shown in fig.1. And arrive at it conclusion regarding the fault type and condition of the transformer.

The expert system displays the result of the estimated fault of the transformer. It also displays Recommended Actions, Causes of Fault as well as Supplementary Analysis.

This expert system is tested with some faulty cases of the transformer. The comparative result of Roger Ratio method, Flag Point method and expert system in terms of percentage are shown in Table- II

TABLE II

COMPARATIVES RESULT OF ROGER RATIO, FLAG POINT AND EXPERT SYSTEM METHOD

| Method / Result | Roger Ratio | Flag Point | Expert System |
|---|---|---|---|
| Correct | 66.67% | 46.67% | 96.67% |
| Incorrect | 13.33% | 0% | 3.33% |
| Unable to diagnosis | 20% | 53.33% | 0% |

The six comparative case studies of faulty transformer are included here.

*1) Case Study*
H2=91; CH4=78.59; C2H6=60.50; C2H4=45; C2H2=30; CO2=2405; CO=245; (in ppm)
*Traditional Roger Ratio Method Result:*
　　　# Flashover, no power flow through
*NTT Flagepoint Method Result:*
　　　#Local Overheating
*Expert Systems Results:*
Fault:
# Arcing without involvement of Cellulose
Causes of Fault:
# On-Load Tap Changers
# HV contacts
# Bushings and short circuit of winding
Recommended Actions:
# Moderate Gas concentration, exercise extreme caution and determine load dependence
# High Gas concentrations, exercise extreme caution.
# Extreme High Gas concentration, strongly advice removing unit from service Arcing (primary) causing Severe Overheating (secondary), not involving Cellulose
Supplementary Analysis:
# Monitor for hotspots and sample for traces of metal in oil
*2) Case Study*
H2=3784; CH4=422; C2H6=9; C2H4=69; C2H2=1; CO=757; (in ppm)
*Traditional Roger Ratio Method Result:*
# Undefined
NTT Flagepoint Method Result:
# Corona, Arcing, Sparking

Expert Systems Results:
Fault:
# Partial discharge without involvement of Cellulose
Causes of Fault:
# Maybe due to deterioration of insulation,
# Reducing of transformer's dielectric strength
Recommended Actions:
# Moderate Gas concentration, exercise caution and determine load dependence
# High Gas concentration, exercise caution, determines load dependence and plan　outrage
# Extreme High Gas concentration, prepare backup transformer Partial Discharge without involvement of Cellulose
Supplementary Analysis:
# Monitor of Acoustic Emission and Radio Interference of Partial Discharge
　　*3) Case Study*
H2=300;CH4=1066;C2H6=1886;C2H4=280; C2H2=65; CO=507; (in ppm)
Traditional Roger Ratio Method Result:
# Undefined
*NTT Flagepoint Method Result:*
# Sparking, Local Overheating, Severe Overheating
*Expert Systems Results:*
Fault:
# Local Overheating without involvement of Cellulose
Causes of Fault:
 # May due to the constant Overloading and failure in Cooling Devices
Recommended Actions:
# Moderate Gas concentration, exercise caution and determine load dependence
# High Gas concentration, exercise caution, determines load dependence and plan outrage
# Extreme High Gas concentration, and prepare backup transformer
Supplementary Analysis:
# Infra Red monitoring for possible hotspot
　　*4) Case Study*
H2=305, CH4=691, C2H6=0, C2H4=648, C2H2=192, CO=657, CO2=0 (in ppm)
*Traditional Roger Ratio Method Result:*
#Core and tank circulating currents, overheated joints
*NTT Flagepoint Method Result:*
Sparking, Severe Overheating, Arcing
*Expert Systems Results:*
Fault:
# Arcing without involvement of Cellulose
Causes of Fault:
 # On-Load Tap Changers, HV contacts, bushings and short circuit of winding
Recommended Actions:
　# Moderate Gas concentration, exercise extreme caution and determine load dependence
# High Gas concentration, exercise extreme caution.

1050

# Extreme High Gas concentration, strongly advice removing unit from service Arcing (primary) causing Severe Overheating (secondary), not involving Cellulose
Supplementary Analysis:
# Monitor for hotspots and sample for traces of metal in oil

*5) Case Study*
H2=1, CH4=22, C2H6=0, C2H4=7, C2H2=6, CO=5, CO2=0 (in ppm)
*Traditional Roger Ratio Method Result:*
# Undefined
*NTT Flagepoint Method Result:*
# Normal
*Expert Systems Results:*
Fault:
# Arcing without involvement of Cellulose
Causes of Fault:
 # On-Load Tap Changers, HV contacts, bushings and short circuit of winding
Recommended Actions:
# Moderate Gas concentration, exercise extreme caution and determine load dependence
# High Gas concentration, exercise extreme caution.
# Extreme High Gas concentration, strongly advice removing unit from service Arcing (primary) causing Severe Overheating (secondary), not involving Cellulose
Supplementary Analysis:
# Monitor for hotspots and sample for traces of metal in oil

*6) Case Study*
H2=80, CH4=619, C2H6=276, C2H4=2438, C2H2=8, CO=268, CO2=2952 (in ppm)
*Traditional Roger Ratio Method Result:*
# Higher than normal temperature in insulation
*NTT Flagepoint Method Result:*
# Sparking, Local overheating, Severe Overheating
*Expert Systems Results:*
Fault:
# Sparking without involvement of Cellulose
Causes of Fault:
# Consistence Sparking due to loose HV connections or undersize contacts
Recommended Actions:
# Moderate Gas concentration, exercise caution and determine load dependence
# High Gas concentration, exercise caution, determines load dependence.
# Extreme High Gas concentration, consider removing from service and advice manufacturer
 Sparking (primary) causing Severe Overheating (secondary), not involving cellulose
 Supplementary Analysis:
# Infra-red emission monitoring for determining of hotspots

## V. CONCLUSION

Expert system is demonstrated to perform insulation diagnosis in an effective and reliable transformer for condition monitoring and diagnosis. The Expert system over comes the difficulty of using the Roger's ratio method, which fails to interpret many cases; especially the borderline case and it also demonstrate the condition of transformer. This Expert system gives the information about recommend actions, causes of fault and supplementary diagnosis method that should be performs to further determine the fault condition and damage.

The Expert system is expected to be useful to even inexperienced maintenance engineers for quick and reliable insulation condition.

## VI. ACKNOWLEDGMENT

The authors gratefully acknowledge the late Shri R.D.Sharma for his motivation. Author also acknowledge the support and facilities provided by Department of Electrical Engineering Jabalpur Engineering College Jabalpur, High Voltage laboratory Department of Electrical Engineering Shri Govindram Seksaria Institute of Technology and Science Indore and Madhya Pradesh State Electricity Board Jabalpur.

## VII. REFERENCES

[1] M. Duval, Hydro-Qubec, It Can Save Your Transformer, *IEEE Electrical Insulation Magazine*, Vol.5, No.6, Nov/Dec 1989.

[2] IEEE Std C57.104-1991 - *revision of IEEE C57.104-1978* "IEEE Guide for the Interpretation of Gases Generated in Oil-Immersed Transformers"

[3] D 3612–96, Standard Test Method for Analysis of Gases Dissolved in Electrical Insulating oil by Gas Chromatography, *The American Society for Testing and Materials*, 1996, USA

[4] M. Duval, "Fault Gases Formed in Oil-Filled Breathing EHV Power Transformers. The Interpretation of Gas Analysis Data," *IEEE-PES Con$ Paper C 74-476-8*, 1974.

[5] R.R. Roger, "IEEE and IEC Codes to Interpret Incipient Faults in Transformers, Using Gas-In-Oil Analysis," *IEEE Tram. Elect. Insu.l*, Vol. EI-13, No. 5, pp. 348-354, 1978.

[6] E. Dornenburg, W. Strittmatter, "Monitoring Oil-Cooled Transformer by Gas Analysis", Brown Boveri Review, Vol. 61, No. 5, 1970 pp. 238.

[7] Yann- Chang Huang, Hong-Tza Yang, Ching-Lien Hnany, "Developing a New Transformer Fault Diagnosis System Through Evolutionary Fuzzy Logic", *IEEE Trans. on Power Delivery*, Vol.12, No2, April 1997.

[8] P. S. Pugh and H. H. Wagner, "Detection of Incipient Fault in Transformer by Gas Analysis", AIEE Transaction, Vol. 80,1961,pp. 189-195

[9] J.J. Kelly, "Transformer Fault Diagnosis by Dissolved Gas Analysis*"*, *IEEE Trans. On Industry Applications*, Vol. A-16 No.6, 1980 pp.777-782.

[10] O. Roizman and V. Davydov, "Application of Fuzzy Set Theory to Power System Analysis and Control," *FLAMOC'96, International discourse on &qv logic and management of complexify*, January 15-18, Sydney, 1996, pp. 145-150.

[11] O. Roizman and V. Davydov, " Neuro-Fuzzy Computing for Large Power Transformers Monitoring and Diagnostics" *EPRI SEDE VIII , New Orleanes* , February 22$^{nd}$, 2000

[12] S. M. Islam, Tony Wu and Gerard Ledwich "A Novel Fuzzy Logic Appoach to Traqnsformer Fault Diagnosis, *IEEE Transactions on Dielectrics and Electrical Insulation* , Vol. 7 No. 2 April 2000.

[13] K. Tomsvic, M. Tapper, T. Ingvasson, "A Fuzzy Information Approach to Integrating Different Transformer Diagnostic Methods", *IEEE Trans. On Power Delivery*, Vol. 8 No. 3, July 1993,pp1638-1646

[14] Y C Huang, C. M. Huang, "Evolving Wavelet Networks for Power Transformer Condition Monitoring", *IEEE Trans. On Power Delivery*, Vol.17 No. 2, April 2002, pp.412-416.

[15] G. J. Klir, T. A. Folger, " Fuzzy Sets and Uncertainty and Information ", Printic- Hall, New Jersey, 1988.

[16] Q. Su, C. Mi,L.L.Lai, P. Aution,"A Fuzzy Dissolved Gas Analysis Method for the Diagnosis of Multiple Incipient Faults in a Transformer", *IEEE Trans. On Power Delivery*, Vol.15 No. 2, may 2000, pp. 593-598.

[17] T.K. Saha, Prithwiraj Purkait, "Investigation of an Expert System for the Condition Assessment of Transformer Insulation Based on Dielectric Response Measurements*", IEEE Transactions on Power delivery,* Vol. 19 No. 3 July 2004.

[18] C.E. Lin, J.M. Ling and C.L. Huang, "Expert System for Transformer Diagnosis Using Dissolved Gas Analysis" *IEEE Transactions on Power Delivery*, Vol. 8, No 1, Jan 1993, pp. 231

[19] D. W. Patterson, "Introduction to Artificial Intelligence and Expert Systems," Printic- Hall, New Jersey, 1988

## VII. BIOGRAPHIES

**Mohammad Ahfaz Khan** was born in Narsinghpur in the India on October 26, 1976; He graduated and post graduated from the Rajiv Gandhi Technical University Bhopal in 2001 and 2005.

He is working as Lecturer in Indore Institute of Science and Technology Indore, His special fields of interest is High Voltage Engineering, AI Application

**Rakesh Saxena** was born in India on July 22, 1963; He graduated from the Jiwaji University Gwalior in 1984, post graduated and Ph. D. From Devi Ahilya University Indore in 1987 and 2002.

He is working as Professor in Shri Govindram Seksaria Institute of Technology and Science Indore

His special fields of interest is Power Electronics, Electric Drives, and High Voltage Engineering

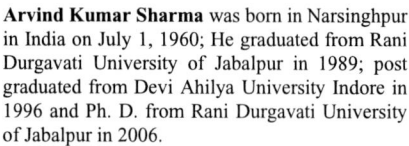

**Arvind Kumar Sharma** was born in Narsinghpur in India on July 1, 1960; He graduated from Rani Durgavati University of Jabalpur in 1989; post graduated from Devi Ahilya University Indore in 1996 and Ph. D. from Rani Durgavati University of Jabalpur in 2006.

He is working as Reader in Jabalpur Engineering College Jabalpur, and he is core coordinator of Technology Information Forecasting Assessment Council (TIFAC).

His special fields of interest is Power Electronics, Power System and High Voltage Engineering

**2006 IEEE International Conference on Power Electronic, Drives and Energy Systems**

# Evaluation of Leakage Current Measurement for Site Pollution Severity Assessment

S.M.H Nabavi, A. □holami, A. Kazemi, and M.A.S. Masoum, *Senior Member, IEEE*

*Abstract--* **Flashover of insulators in transmission and distribution systems may cause costly outages for power companies and their customers. Industrial and/or coastal pollution of external insulation is a major cause for such events at the normal power frequency voltage of the systems. The power companies are now facing increasing competition resulting in pressure to lower the cost and to increase the system reliability. Different methods have been applied in the past to overcome or reduce the problems with flashover on insulators. Methods which should provide reliable data under real physical conditions. In this paper several measuring methods to evaluate the pollution levels on outdoor insulators are described. According to this comparison, Leakage Current Measurement 'LCM' method is a reliable method for measurement leakage current in outdoor insulators and surge arresters.**

*Index-Terms--***Reliability, Leakage Current Measurement, Site Pollution Severity, Flashover.**

## I. INTRODUCTION

FLASHOVER of high voltage insulators results in the reduction of reliability of power systems and irretrievable losses to the power networks. The possibility of flashover in contaminated environments depends on the type of pollution and duration of the time that an insulator is placed in a polluted environment. The occurrence of pollution-born flashover on insulator surface generally consists of the following stages [1, 13]:

- Settling of pollutants on the insulator surface
- Compounding soluble pollutants with rainwater and formation of a conductive layer
- Formation of leakage current
- Insulator surface's getting hot and formation of a dry area
- Partial discharge and the occurrence of flashover.

Therefore, inappropriateness of insulation designing will be

---

S. M. H. Nabavi is with the Department of Electrical Engineering, Iran University of Science & Technology, Tehran, Iran, 16846 I.R.I. (e-mail: H_nabavi@ee.iust.ac.ir).

M. A. S. Masoum is with the Department of Electrical Engineering, Iran University of Science & Technology, Tehran, Iran, 16846 I.R.I. (e-mail: M_Masoum@iust.iust.ac.ir).

A. Kazemi is with the Department of Electrical Engineering, Iran University of Science & Technology, Tehran, Iran, 16846 I.R.I. (e-mail: Kazemi@iust.iust.ac.ir).

resulted, flashover and consequently in system outages. Hence, identifying different factors causing insulation surface pollution and dealing with their unfavorable effects have an important role in increasing the reliability of the network [1, 13, 15, 17, 21]. The major consequence of pollution is the reduction of insulation in high-voltage transmission lines and substations. During recent years, many studies have been carried out to improve methods of pollution measurement, including IEC815, IEC507, IEC383, IEC1106, ANSI C290, DE0441, VDE0218, and ASTMD2302. For this purpose, the relationship between Equivalent Salt Deposit Density ESDD/ Non-soluble Salt Deposit Density NSDD and LCM in high-voltage insulators have been studied [1, 6, 11]. The performance of spiral-shaped insulators and their electric resistance against pollutants is rather weak [20]. In a study carried out on pollution measurement, analysis of performance of ceramic insulators under polluted (estimation) circumstances is based on ESDD and LCM [12]. The relationship and the confirmation coefficient between flashover voltage and leakage current have been analyzed in [10]. Preparing the pollution map affecting insulation based on the values obtained from pollution measurement with ESDD and Directional Dust Deposit □auges (DD□) can also prove useful in designing insulator of high voltage substations and transmission lines [13]. Among other issues related to the methods of insulation pollution measurements are the relationship between ESDD and resistance of high voltage insulator's surfaces [2, 8, 9] and estimation of Salt Deposit Density (SDD) content in terms of leakage current peak [3]. In this paper, the LCM method to find site pollution severity affecting insulation has been explained and its capabilities have been compared with those of other common techniques. In the second part, the significance of pollution measurement is studied and then, in part three, common methods of pollution measurement are spelled out. Identifying pollution severity based on leakage current measurement and comparing it with other methods are presented in parts four and five, respectively.

## II. SI□NIFICANCE OF MEASURIN□ SITES POLLUTION AFFECTIN□ INSULATION

High voltage insulators are used to separate different voltage levels. □radually, sites pollution settles on the insulator surface and an electrolyte layer forms on it under the influence of environment and weather conditions such as moisture and

---

0-7803-9771-1/06/$25.00 ©2006 IEEE

rainfall [22, 23]. This layer extends over time and in some cases, such as in the case of inappropriate insulation designing, leads to flashover and system outage. Thus, identification of the factors causing insulation surface pollution, pollution severity measurement, and tackling its unfavorable effects has an important role in increasing the network reliability. The role of these is highly remarkable particularly in environments with high site pollution. According to the statistics already available, about 70% of high voltage transmission lines errors result from the inappropriate performance of insulation. The major consequence of pollution in these areas is the reduction of insulation in high voltage transmission lines and substations [1, 15, 17]. Identifying site pollution severity affecting insulation in these areas is crucial not only for sound designing of insulation of high voltage lines and substations but also for selecting insulation type and the suitable program for the maintenance of insulation (preventive practices such as washing and using special covers). Various methods to identify site pollution severity have already been proposed, each facing technical and procedural limitations.

### III. PROPOSED METHODS TO MEASURE POLLUTION

Weather changes and environmental pollution have direct influences on the performance of insulations. The type of pollutants and the duration of time the insulator is exposed to pollution are among other parameters affecting system insulation. Various methods are applied to identify the content and type of pollutants. These methods are described below:

#### A. ESDD Method

By definition, the density of equivalent salt deposit density (ESDD) equals an amount of sodium chloride which, solved in water, will change water's conductivity to the level equal to that resulting from the solution of polluted deposits gathered from insulator surface divided by the insulator's surface area (mg/cm$^2$) [1]. This method is generally used for calculating average pollution based on average density of soluble salt. Pollution measuring station of an insulator chain in this method consists of seven disk insulators as in figure 1 [13] (The number of insulators, of course, may vary between five and twelve). These insulators do not contain electricity and are used to gather pollution under site conditions. The dirt on the surface of insulators is washed up according to the schedule in figure 1. Pollution index and ESDD can be calculated according to the parameters of conductivity index of water, water temperature, and the volume. Equation 1 illustrates pollution in 20° C. Conductivity ($\sigma_t$) is measured by conductivity measurement probe in the temperature t.

$$\sigma_{20} = \sigma_t[1 - 0.02277(t-20)e^{0.01956(t-20)}] \quad (1)$$

where:

$\sigma_t$: conductivity index measured ($\mu S / cm$)

t: temperature of solution ($^{\circ}C$)

$\sigma_{20}$: revised conductivity index for $20^{\circ}C$

Equation 2 works out solution hardness ($S_a$) in $20^{\circ}C$

Equation 3 works out equivalent salt deposit density (ESDD) in terms of mg/cm$^2$

$$S_a = (5.7 \times \sigma_{20})^{1.03} \quad (2)$$

$$ESDD = S_a \times \frac{V}{A} \quad (3)$$

where:

V: volume of distilled water

A: area of washed surface of insulator

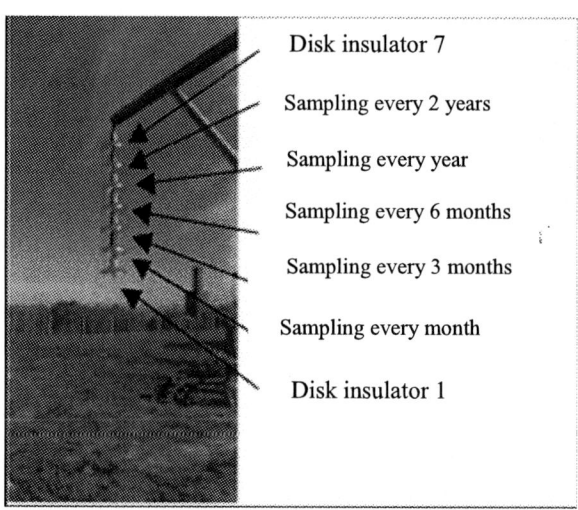

Fig. 1. Pollution gathering station in ESDD method.

The shortcomings with this method are as follows:
- It is costly
- It requires a lot of time to do the experiment and it is impossible to measure maximum pollution
- The measures pollution index is low compared with its real value after rainfall and natural washing of the insulator
- There is the possibility of human-made errors
- The effect of electricity-containing of the insulator in absorbing pollution is not considered
- Such special conditions as momentary pollution (conductive fog) and pollution stemming from metallic particles in industrial environments cannot be monitored.

In this method, insulators 1 and 7 are installed only to support other insulators, and pollution is not removed from them.

#### B. NSDD/ESDD Method

This method is the developed form of ESDD method, in which non-soluble pollution content in available samples is measured. This content finally is normalized according to the insulator area and its unit is mg/cm$^2$.

#### C. DDG Method

This method was first chosen by research institute of Electricity Supply Commission (Eskom) in 1974 to examine insulator pollution. As shown in figure 2, DD□ includes four vertical split pipes and a pot below each pipe to gather pollution. The pipes are placed along the four geographical directions, north, south, east, and west. To facilitate international comparison of the results, the size of DD□ cracks must be the same in all places to be tested [12, 13, 15]. DD□ pollution index, according to equation (4), is the average of the four conductivities ( $\mu S / cm$ ) obtained from the four directions for a 30-day month in 500 cc normalized washing water.

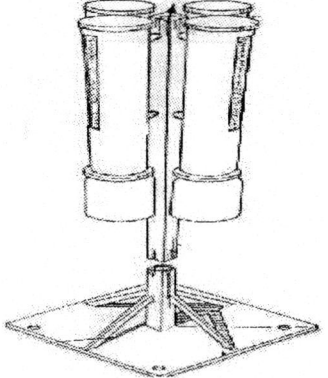

Fig. 2. Pollution gathering machine in DD□ method.

Normalized conductivity and average conductivity are worked out from the equations (4) and (5), respectively. [13]

$$\sigma_N = C(V/500)(30/N) \qquad (4)$$

$$\text{Average conductivity} = \frac{\sigma_N + \sigma_S + \sigma_E + \sigma_W}{4} \qquad (5)$$

where:

$\sigma_N , \sigma_E , \sigma_W , \sigma_S$ : normalized conductivity indexes in north, east, west, and south ( $\mu S / cm$ )

C: conductivity ( $\mu S / cm$ )

V: volume of distilled water (ml)

N: number of days when the insulator has been under investigation

DD□ method, in large scale, has the following advantages over other techniques:

• DD□ does not need any feeding resource or energy resource
• DD□ is cheap and allows for extensive measurements along a transmission line
• Calculations can be done easily and are repeatable
• There is no need for the existence of an insulator in doing DD□ tests
• it is equivalent to ESDD
• it identifies the direction that produces the most pollution
• it requires no maintenance other than cleaning after the tests
• it is not influenced by rainfall

The only shortcoming with DD□ is the inaccessibility of insulator's self-filtering features and the effect of insulator profile in dirt settling on the insulator surface. This method faces limitations in high-rain areas, where the primary conductivity index of the rain water in the container is not known. Therefore, a higher index may be acceptable. However, in areas with little rain but excessive fog, the actual pollution severity is larger than that of indicated by DD□. Another point is that the DD□ machine must be installed somewhere away from trees or any other barriers influencing natural airflow. Studies done already reveal that the results

from both ESDD and DD□ are equal if the site weather conditions are also considered. Thus, they can be considered equivalent in terms of the practical results. Table 1 illustrates site pollution severity obtained from DD□ and ESDD. [13] Figure 3 shows correlation coefficient and ESDD variations of correlation coefficient in terms of DD□. It shows a direct relationship between these two methods.

TABLE I

| Pollution severity | DD□ | | ESDD |
| --- | --- | --- | --- |
| | Monthly average | Monthly maximum | Monthly maximum |
| | ( $\mu S / cm$ ) | | $mg / cm^2$ |
| Light | 0-75 | 0-175 | 0.06< |
| Average | 76-200 | 176-500 | 0.06 - 0.12 |
| Heavy | 201-350 | 501-850 | -0.24 0.12> |
| Very heavy | 350> | 850> | 0.24 > |

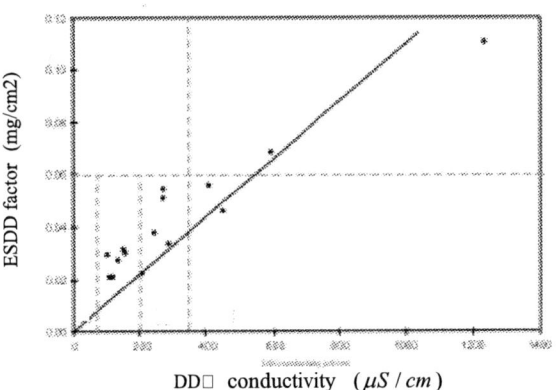

Fig. 3. Correlation and variation graph of ESDD in terms of DD□.

## IV. IDENTIFYIN□ SITES POLLUTION SEVERITY BASED ON MEASURIN□ LEAKA□E CURRENT

To identify exact site pollution index, the tests must be carried out in real conditions of insulation location. Since it is extremely difficult to provide a natural environment to measure the real value of pollution and since measurement must be done in absolutely real conditions, using an online method seems necessary. In this method, the leakage current of the insulator under nominal voltage is the basis for defining pollution index [6, 11, 16, 18, 19]. Defining pollution severity based on a leakage current measurement is the optimal solution for these problems. It is described below.

### A. Leakage Current Measurement System

The general systematic structure of leakage current measurement machine is illustrated in figure 4. A collar-shaped ring is placed in the end of the insulator near to the earth. The leakage current sensor is placed between the insulator and the ring to create a closed-loop current. The sensor, which has a high performance speed, work based on Hall's effect current transformer [15]. The leakage current made on the insulator surface passes through the sensor and flows toward the earth. Figure 5 illustrates the sensor performance based on Hall's effect and magnetic fields.

Sensor's input impedance is of a very small value. Sensor's output it directly connected to the central unit of the formation recording, as in figure 6. This unit consists of the analog - to - digital (A/D) converter and a microprocessor to gather information, which records leakage current indexes in all the insulators being tested. All the adopted and saved information can be transferred through RS232 port serial or modem [4, 5, 15]. For sufficient information, the sampling frequency related to the A/D converter is usually selected for 20 kHz. The central unit is also responsible for recording the phase voltage magnitude to the earth and line current for each of the phases [14].

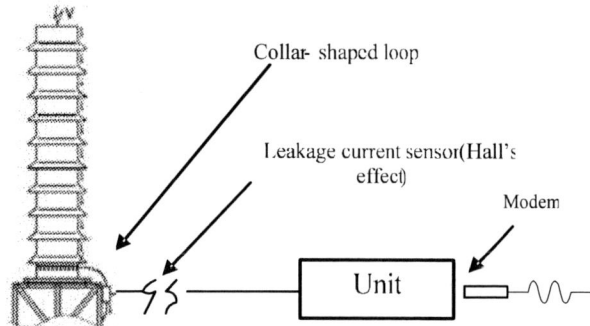

Fig. 4. General systematic structure and measuring equipment of leakage current.

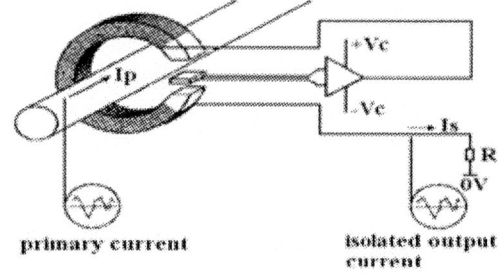

Fig. 5. Sensor's performance based on Hall's effect.

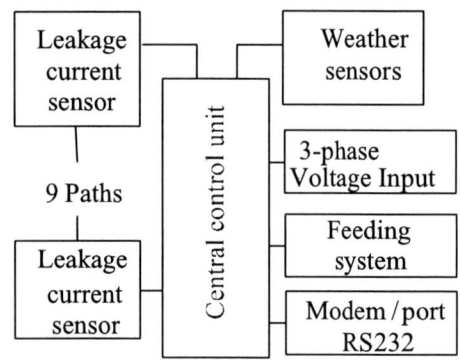

Fig. 6. Central unit of information recording.

Since the weather conditions like humidity and temperature have an essential role in creating leakage current, it is necessary for the measurement system to be equipped with

sensors of temperature and humidity and wind velocity gauges. The output of the sensors also is connected to the central unit of information recording and is saved.

### B. Analysis of Leakage Current Results

Studies show that it is possible to monitor the current and voltage online in LCM method [3, 8, 14, 15]. So the relation between them can be shown through figure 7 and 8 [15]. It is observed in this figure that the relation between the leakage current and the voltage of insulator's two ends is non-linear [15]. Maximum leakage current, however, occurs when voltage is also maximum. The current gradually increases during the first quarter of voltage curve period, and then the leakage current highly increases in a critical point. But during the second quarter, variations of leakage current curve remain sinusoidal and follow voltage variation with a less linear gradient. The current is zero in voltage zero point and this process recurs in the negative cycle too.

Since ESDD/ NSDD are common methods of measurement and since site pollution severity in related standards is defined on the basis of the parameter of this method, the relationship between these two methods and LCM method must be determined [1, 3, 6, 7, 8, 9]. The graph of flashover voltage variations in terms of ESDD leakage current index are shown in figures 9 and 10, respectively. Both are of the same exponential form and much similar [12].

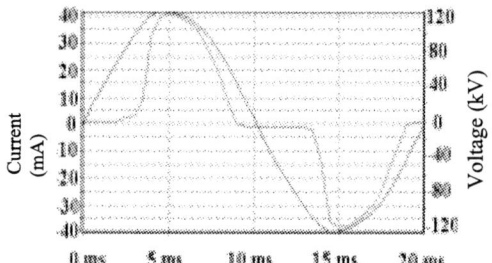

Fig. 7. Leakage current and voltage of insulators two ends.

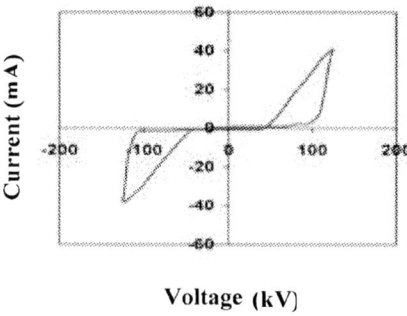

Fig. 8. Leakage current variations in terms of voltage of insulator's two ends .

### C. Characteristics of Leakage Current Measurement Method

Monitoring pollution using leakage current measurement is generally carried out for the following reasons:

- Defining site pollution coefficient to properly design insulation of high-voltage substations and lines

- Measuring pollutions severity to specify polluted areas and to prepare pollution maps
- On-line monitoring of pollution to specify insulators washing time and to create flashover
- Comparing performance and behavior of different insulators (in view of shape and length) and the quality of insulator materials under the same polluted environment.

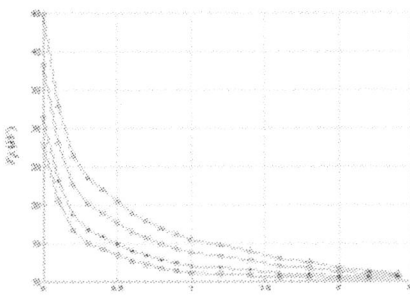

Fig. 9. Flashover voltage variations in terms of ESDD.

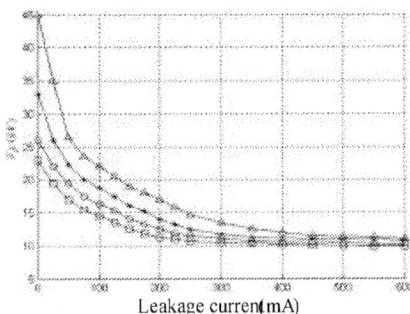

Fig. 10. Flashover voltage variations in terms of leakage current.

Regarding to the trend of flashover [13,15,18], it is observed that polluting soluble material is combined with air moisture or rain water and forms an electrolyte layer of the insulator surface. In common measuring methods like ESDD/ NSDD or DD□, the index obtained from the content of polluting soluble material is the only basis for defining site pollution severity and designing insulation. Factors like moisture or rainfall content are not considered although they have a major role in lowering insulation level in polluted areas. Therefore, these methods can not be highly efficient in polluted areas with high average moisture contents. Records of exploitation in coastal areas of southern Iran approve the fact that insulation level in polluted and very wet areas is lower than that of in equally polluted but dry areas. Insulation problems are more serious in such areas. Since flashover incidence is directly attributed to leakage current magnitude (or to the average content within a few seconds), common methods cannot submit precise information about insulator performance and leakage current created. Therefore, for the purpose of raising power system reliability, it is crucial to use a method free of the mentioned problems but capable of measurement in particular states. In leakage current method, the effects of all the factors involved in insulation, such as polluting soluble and non-soluble materials, moisture absorbed by the insulator, and other environmental conditions are taken into account [18]. This method is applicable in all environmental conditions like in conductive fog and for all pollutants, including industrial and metallic ones. In summary, advantages of LCM method include:

- On-line analysis of insulator behavior in the polluted environment
- Possibility of measuring maximum and minimum indices of pollution
- Lack of results being influenced by rainfall
- Studying the effect of insulator profile in the process of the settling of deposit on the insulator surface
- Precise measuring of the pollution content in conditions where common methods like ESDD/ NSDD or DD□ are not applicable because of the existence of conductive fog
- Testing under nominal voltage
- Possibility of testing without removing electricity from the power system
- Testing in real conditions of exploitation
- Decrease in the effect of physical and human errors during the test
- Lack of necessity to attend test site to gather information (possibility of remote controlling)
- Lack of need for complicated calculations and possibility of repetition
- Possibility of preparing exact pollution map and optimal designing of insulation of high-voltage transmission lines and substations
- Access to momentary profile of voltage and current
- Applicability for measuring leakage current of surge arrester
- Machine's being equipped with sensors of moisture, temperature, and wind velocity
- Easiness of comparing results internationally
- Applicability for environments with industrial pollutants like metallic particles

## V. CONCLUSION

Defining site pollution severity affecting insulation through common methods such as ESDD/ NSDD or DD□, though easy to apply, encounters limitations in some instances. The results of the studies reveal that in addition to factors measured in these methods, like ESDD index, other factors, including weather conditions, especially moisture, greatly decrease insulation level in polluted areas. These factors are not considered in common methods. Furthermore, these methods are not capable of measuring pollution in such conditions as conductive fog or non-soluble conductive pollutants like metallic particles. Specifying site pollution index using leakage current method allows identify pollution severity in natural conditions, i.e. conditions where insulators are actually exploited. This method raises the system reliability by indicating exact pollution content. According to the studies already done, this system is equipped with sensors of moisture, temperature, and wind velocity as well as sensors of current and voltage. LCM method allows to prepare the exact pollution maps and optimal designing of insulator of transmission lines and substations. In addition, the information required to access insulator self-refining features and estimate insulation duration can be obtain from LCM method.

The cost necessary for using LCM potentials is insignificant considering its advantages and preventing flashover and its consequent risks. Regarding to these advantages and limitations of commons methods like ESDD/NSDD and DD□ in coastal areas of southern Iran, this method can prove useful for designing insulation of high-voltage transmission lines and substations in these areas, where pollution and unfavorable environmental conditions like high moisture exist.

## VI. REFERENCES

[1] Montoya, □.; Ramirez, I.; Montoya, J.I.; "Correlation Among ESDD, NSDD and Leakage Current in Distribution insulators"; □eneration, Transmission and Distribution, IEE Proceedings- Volume 151, Issue 3, 15 May 2004 Page(S):334 - 340

[2] Fierro-Chavez, J.L.; Ramirez-Vazquez, Z.; Montoya-Tena, □.; "on-Line Leakage Current Monitoring of 400Kv insulator Strings in Polluted Areas"; □eneration, Transmission and Distribution, IEE Proceedings- Volume 143, Issue 6, Nov. 1996 Page(S):560 - 564

[3] Kanashiro, A.□.; Burani, □.F.; "Leakage Current Monitoring of insulators Exposed To Marine and industrial Pollution Electrical insulation"; Conference Record of the 1996 IEEE international Symposium on Volume 1, 16-19 June 1996, Page(S): 271 - 274 Vol.1

[4] Ziyu Zhao; Hengkun Xie; Zongren Peng; "A Newly Developed Leakage Current Monitoring System For the Assessment of Insulator Pollution Severity"; Electrical insulating Materials; international Symposium on 17-20 Sept. 1995 Page(S):319 - 322

[5] Sugawara, N.; Hokari, K.; "Leakage Resistance Data Acquisition System For Porcelain insulators Along the Coast"; Properties and Applications of Dielectric Materials, Proceedings of the 4th international Conference on Volume 2, 3-8 July 1994 Page(S):503 - 506 Vol.2

[6] □uan Zhicheng; Cui □uoshun; "A Study on the Leakage Current Along the Surface of Polluted insulator"; Properties and Applications of Dielectric Materials, Proceedings of the 4th international Conference on Volume 2, 3-8 July 1994 Page(S):495 - 498 Vol.2

[7] Schwardt, W.H.; Holtzhausen, J.P.; Vosloo, W.L.; "A Comparison Between Measured Leakage Current and Surface Conductivity During Salt Fog Tests [Power Line insulator Applications]"; 7th Africon Conference in Africa; Volume 1, 2004 Page(S):597 - 600 Vol.1

[8] Matsuo, H.; Fujishima, T.; Yamashita, T.; Hatase, K.; "Relation Between Leakage Impedance and Equivalent Salt Deposit Density on An insulator Under A Saltwater Spray"; IEEE Transactions on Electrical insulation; Volume 6, Issue 1, Feb. 1999 Page(S):117 - 121

[9] Salam, M.A.; Aamer, K.; Hamdan, A.; Hamdan, N.; "Study the Relationship between the Resistance and ESDD of A Contaminated insulator A Laboratory Approach"; Properties and Applications of Dielectric Materials, Proceedings of the 7th international Conference on Volume 3, 1-5 June 2003; Page(S): 1032 - 1034 Vol.3

[10] Kaidanov, F.; Munteanu, R.; "Investigations of Leakage Currents Along Polluted and Wetted insulators and their Correlation With Flashover Voltages"; Electrical and Electronics Engineers in Israel, Eighteenth Convention of 7-8 March 1995

[11] Tjokrodiponto, W.; Sebo, S.A.; Sakich, J.D.; Tiebin Zhao; "Leakage Current Magnitudes and Wave shapes Along Polymer insulators"; Electrical insulation and Dielectric Phenomena, IEEE 1997 Annual Report, Conference on Volume 2, 19-22 Oct. 1997; Page(S):390 393 Vol.2

[12] Ramirez-Vazquez, I.; Fierro-Chavez, J.L.; "Criteria For the Diagnostic of Polluted Ceramic insulators Based on the Leakage Current Monitoring Technique"; Electrical insulation and Dielectric Phenomena, Annual Report Conference on Volume 2, 17-20 Oct. 1999 Page(S):715 - 716 Vol.2

[13] Pietersen, D.; Holtzhausen, J.P.; Vosloo, W.L.; "An investigation into the Methodology To Develop An insulator Pollution Severity Application Map For South Africa"; 7th Africon Conference in Africa Volume 2, 2004 Page(S):697 - 703 Vol.2

[14] "insulator Pollution Monitor"; Diagnostic Monitoring Systems and Services; January 1999

[15] "Measurements and interpretations Concerning Leakage Currents on Polluted High Voltage insulators"; E. halassinakis, C.□. aragiannopoulos;

institute of Physics Publishing Measurement Science and Technology; Meas. Sci. Technol. 14;2003; Page(S):421–426

[16] "Surge Arrester Monitor Experience From Condition Monitoring of Metal Oxide Surge Arresters in Service"; Diagnostic Monitoring Systems and Services; January 1999

[17] Abdelaziz, E.O.; Javoronkov, M.; Abdelaziz, C.; Fethi, □.; Zohra, B.; "Prevention of the interruptions Due To the Phenomena of the Electric insulators Pollution"; Control, Communications and Signal Processing, First international Symposium on 2004; Page(S):493 - 497

[18] Lee, C.J.; "Field Experience and Pollution Monitoring of Composite Long Rod insulators"; Reliability of Transmission and Distribution Equipment, Second international Conference on the 29-31 Mar 1995 Page(S):204 - 209

[19] "Leakage Current Measuring Arrangement"; Software & Hardware Manual; Furukawa Electric Institute of Technology; Budapest, Hungary; 1997

[20] Boudissa, R.; Djafri, S.; Haddad, A.; Belaicha, R.; Bearsch, R.; "Effect of insulator Shape on Surface Discharges and Flashover Under Polluted Conditions"; Dielectrics and Electrical insulation, IEEE Transactions on Volume 12, Issue 3, June 2005 Page(S):429 - 437

[21] Vosloo, W.L.; Holtzhausen, J.P.; "the Electric Field of Polluted insulators"; Africon Conference in Africa, IEEE Africon. 6th Volume 2, 2-4 Oct. 2002; Page(S):599 - 602 Vol.2

[22] Vosloo, W.L.; Holtzhausen, J.P.; "the Effect of thermal Characteristics of Power Line insulators on Pollution Performance"; Africon Conference in Africa, IEEE Africon. 6th Volume 2, 2-4 Oct. 2002 Page(S):609 – 612 Vol.2

[23] J. Farzaneh-Dehkordi, J. Zhang, and M. Farzaneh; "Experimental Study and Mathematical Modeling of Flashover on EHV insulators Covered With Ice"; 61st Eastern Snow Conference Portland, Maine, USA 2004

## VII. BIO□RAPHIES

**Seyyed Mohammad Hossein Nabavi** was born in Tabriz, Iran, in 1979. He received his M.Sc. degree in Electrical Engineering from university of Science & Technology, Tehran, Iran, in 2004. He is currently a PHD Student in electrical engineering department of Iran University of Science and Technology, Tehran, Iran. His research interests are reliability, optimization in power system, and control FACTS devices.

**Ahmad Gholami** was born in Tehran, Iran, in 1955. He received his PHD degree in electrical engineering from UMIST, U.K in 1986. He is currently an associate professor in electrical engineering department of Iran University of Science and Technology, Tehran, Iran. His research interests are High Voltage systems and Isolators.

**Ahad Kazemi** was born in Tehran, Iran, in 1952. He received his M.Sc. degree in electrical engineering rom Oklahoma State University, U.S.A in 1979. He is currently an associate professor in electrical engineering department of Iran University of Science and Technology, Tehran, Iran. His research interests are reactive power control, power system dynamics, stability and control and FACTS devices.

**Mohammad A. S. Masoum** received his B.S., M.S. and Ph.D. degrees in Electrical and Computer Engineering in 1983, 1985, and 1991, respectively, from the University of Colorado at Boulder, USA. Currently, he is an Associate Professor at Iran University of Science & Technology, Tehran, Iran. Dr. Masoum is a senior member of IEEE and his research interests include: Optimization, Power Quality and Stability of Power Systems.

**2006 IEEE International Conference on Power Electronic, Drives and Energy Systems**

# Detection of Bearing Failure in Rotating Machine Using Adaptive Neuro-Fuzzy Inference System

Sulochana Wadhwani , A.K. Wadhwani, S P Gupta and Vinod Kumar

*Abstract-* **This paper proposes a novel approach for bearing health evaluation using Lempel-Ziv Complexity and time domain statistical parameters in conjunction with ANFIS. Compared to conventional techniques the presented approach works well for a non linear physical system and is thus suited for condition monitoring of machine system under varying operating and loading conditions. The performance of this technique is investigated through experimental study of realistic vibration signals. The results demonstrate that complexity analysis and time domain parameters in conjunction with ANFIS provide an effective measure forbearing health evaluation**

*Index Terms*—**Bearing fault, Fault detection, ANFIS.**

## I. INTRODUCTION

FAULT detection of motor system is inseparably related to the diagnosis of bearing assembly as bearings are most prone to fault occurrence. Since most of the phenomenon occurring in the bearing assembly of motors is in some way represented as motor vibration, therefore one of the most important signals to consider in most motor fault detection scheme is motor bearing vibration. Vibration-based signal analysis in the time and frequency domains has been a major technique for the detection and diagnosis of machine defects, such as bearing failures. In the time domain, statistical parameters such as root mean square, peak value, kurtosis, crest factor, and skewness have been calculated to detect the existence as well as growth of failures [1],[2]. Frequency domain techniques such as the cepstrum [1] and enveloping analysis [3], are useful for identifying the characteristic defect frequencies. Hence the frequency and time domain parameters can be successfully employed for predictive maintenance of machines [4]. But a common drawback of these parameters is that they are sensitive to interference from noise and undesired signals. In recent years, time- frequency methods, such as Wigner-Ville distribution [5], and wavelet analysis [6],[7] have been increasingly used for machine fault detection and shown to be effective complementary

techniques to the conventional time and frequency analysis. A common drawback of these techniques, however, lies in the assumption that the physical system being monitored possesses linear transfer functions, where the principle of superposition applies

A number of techniques have been developed to analyze non-linear system dynamics such as Lempel Ziv complexity measure and Correlation Dimension. Various researchers have used the complexity measure to examine the signals from a non-linear system such as EEG signals [8], [9]. Complexity analysis has also been engaged effectively in machine health evaluation [10],[12]. Hence a new system in which different parameters are used in combination and each parameter weighted differently can be a better health evaluation tool than the ones discussed above.

Recently artificial intelligent techniques, such as expert system, neural network, fuzzy logic and genetic algorithm, have been employed to assist the diagnostic task to correctly interpret the fault data. Motivated by the results in each of these areas and the potential for mutual progress in computational modeling, an integration of these concepts is very important [I].

ANFIS is a product combining the fuzzy inference system with neural networks. The fuzzy inference system is used widely in fuzzy control since it can deal with structural knowledge. Neural networks usually don't deal with structural knowledge, but it has the function of self-adapting and self-learning, by learning a lot of data, it can estimate the relations between the data of input and output. ANFIS fully makes use of the excellent characteristics of neural network and fuzzy inference system, and is widely applied in fuzzy control. This paper employs the capabilities of adaptive neuro-fuzzy system as a promising intelligent paradigm for characterizing the bearing failure in rotating electrical machines. The network architecture used for fault diagnosis consists of 7 inputs corresponding to the 3 time domain parameters and four frequency features from vibration spectrum and 4 outputs corresponding to 4 respective faults, i.e. defective balls, defective inner race, defective outer race and no defect.

## II. ANFIS: INTRODUCTION

ANFIS is a class of adaptive network, which are functionally equivalent to Fuzzy Inference Systems. It uses a hybrid-learning algorithm to identify the membership function

---

Sulochana Wadhwani and Dr. A.K. Wadhwani are with Madhav Institute of Technology and Science, Gwalior (M.P.)
(e-mail:Sulochana_wadhwani11@rediffmail.com ).
Dr S.P. Gupta and Dr. Vinod Kumar are with the Department of Electrical Engineering, IIT, Roorkee, Roorkee-247 667, India

0-7803-9771-1/06/$25.00 ©2006 IEEE

parameters of single-output, Sugeno type fuzzy inference systems (FIS). A combination of least-squares and back propagation gradient descent methods are used for training FISmembership function parameters to model a given set of input/ output data. The basic structure of a fuzzy inference system maps input characteristics to input membership functions (mf), input mf to rules, rules to a set of output characteristics, output characteristics to output mf, and the output mf to a single-valued output or a decision associated with the output. In a conventional fuzzy inference system, an expert who is familiar with the tar-get system to be modeled determines the number of rules. Incases where there are no experts available, the number of mf assigned to each input is chosen empirically. Also, the fuzzy inference system is applied to modeling systems whose rule structure is essentially predetermined by the user's interpretation of the characteristics of the variables in the model.

The steps for implementing ANFIS are shown are:

Layer 1: Each node in this layer generates membership grades of a linguistic label (low, medium, high etc.). For instance, the node function of the $k^{th}$ node may be generalized bell membership function: Here, the shape of the membership functions depends on parameters, and changing these parameters will change the shape of the membership function. Instead of just looking at the data to choose the mf parameters, it can be chosen automatically using ANFIS.

$$\mu_{Ai}(x) = \exp\left\{-\left(\frac{x-c_i}{a_i}\right)^2\right\} \qquad (1)$$

where $\{a_k, b_k, c_k\}$ is the parameter set that changes the shapes of the membership function. Parameter in this layer are referred as premise parameters.

Layer 2: Each node in this layer calculates the firing strength of a rule by multiplication

$$O_k^2 = w_i = \mu_{Ak}(x) \times \mu_{Bk}(y), i = 1,2\ldots\ldots \qquad (2)$$

Layer 3: Every node k in this layer calculates the ratio of the k-th rules firing strength to the total of all firing strengths.

$$O_k^3 = \overline{w_k} = \frac{w_i}{w_1 + w_2}, k = 1,2\ldots\ldots \qquad (3)$$

Layer 4: Node k in this layer compute the contribution of k- th rule toward the overall output with the following node function:

$$O_k^4 = \overline{w_k} f_k = \overline{w_k}(p_k x + q_k y + r_k) \qquad (4)$$

Where, $\overline{w_k}$ is the output of layer 3 and $\{p_k, q_k, r_k\}$ is the parameter set. Parameters in this layer are referred to as the consequent parameters.

Layer 5 : the single node in this layer computes the overall output as the summation of contribution from each rule:

$$O_k^5 = \sum_k \overline{w_k} f_k = \frac{\sum_k w_k f_k}{\sum_k w_k} \qquad (5)$$

Each node in layer 1 represents a fuzzy membership function; after combining them in layer 2, the firing strengths $w_k$ are normalized in layer 3. The generic node in layer 4 implements a linear approximator. The output of the net is obtained by merging the crisp conclusions of the rules.

## III. DATA ACQUISITION & FEATURE EXTRACTION

The vibration signals were recorded from the bearing housing of the test motor with the help of piezo-resistive accelerometer. The signal was sampled at 4 KHz using a 12-bit A/D converter. The sensitivity of the vibration accelerometer wais 100mV/g and the resonant frequency was 800 kHz. Therefore, an amplifier of gain 20 and low pass filter with cut-off frequency of 1 kHz was used before the sampling is done. The use of filter eliminates unwanted signal in the high frequency region. The signals were recorded for four bearing conditions: healthy, inner race defect, outer race defect and ball defect. These faults were introduced artificially in the bearing. For each bearing condition 200 data series were collected with each data series consisting of 20,000 samples.

Two reliable time domain parameters rms level, kurtosis and skewness along with normalized complexity number were selected for analyzing the bearing health condition. The major advantage of using time domain parameters is their relatively simple computational procedure and short computation time.

### A. RMS Level

Common approach to vibration monitoring is to measure the overall intensity of vibration signal by the estimate of Root Mean Square (R.M.S.) level of the time record. The RMS level of the discrete time signal is calculated as

$$rms\ value = \sqrt{\frac{1}{N}\sum_{i=1}^{N}(x_i - \mu)^2} \qquad (6)$$

where N=number of samples, $x_i$ is the amplitude of individual sample and $\mu$ is mean value of the time record samples.

### B. Skewness

Skewness is a major of symmetry or more precisely, the lack of symmetry about its mean. A distribution, or data set, is symmetric if it looks the same to the left and right of the centre point .The skewness of normal distribution is zero and any symmetric data should have a skewness near zero. Negative values for the skewness indicates data that are the skewed left and positive values for the skewness indicates data that are the skewed right. By skewed left ,mean that the left tail is heavier than the right tail. Similarly, skewed right mean that the right tail is heavier than the left tail.

1060

Mathematically, skewness of the data can be obtained as

$$skewness = \frac{\frac{1}{n}\sum(x_i - \mu)^3}{\sigma^3} \qquad (7)$$

### C. Kurtosis

It is a measure of whether the data are peaked or flat relative to normal distribution. That is data sets with high kurtosis tend to have a distinct peak near the mean, decline rather rapidly, and have heavy tails. Data sets with low kurtosis tend to have a flat top near the mean rather than sharp peak .A uniform distribution would be the extreme case. Positive kurtosis indicates a "peaked" distribution and negative kurtosis indicates a "flat" distribution.

$$kurtosis = \frac{\frac{1}{n}\sum(x_i - \mu)^4}{\sigma^4} \qquad (8)$$

### D. Complexity Number

Complexity refers to the number of discrete patterns that a system can be said to be composed of which when put together in a sequence result in the original system. It thus enables us to indicate the degree of randomness or orderliness of a time series. Several methods have been proposed to calculate it. Some of them are explained here:-

A.N. Kolmogorov was the first to suggest a general approach to estimating complexity of symbolic sequences (Kolmogorov, 1965). He proved that there exists an optimal algorithm or a program for the generation of the original sequence. He proposed to use the length of the shortest binary program which, when fed into a given algorithm, will cause it to produce a specified sequence, as a measure for the complexity of that sequence with respect to the given algorithm. If the length of the program is large we can say that the complexity of the sequence is large. Thus, the works of Kolmogorov and Chaitin marked the beginning of algorithmic complexity.

Pincus (1991) introduced Approximate Entropy as a complexity measure. Given groups of N points in a series, the approximate entropy is related to the probability that two sequences which are similar for N points remain similar at the next point. It is a probabilistic method of determining the complexity, developed to quantify the amount of regularity in the data without any *a priori* knowledge about the system generating them. It is a nonnegative number that will distinguish among data sets, with larger numbers indicating more irregularity, unpredictability, and randomness.

In 1976 Lempel and Ziv proposed and explored another approach to the problem of the complexity of a specific sequence. They linked the complexity of a specific sequence to the gradual buildup of new patterns along the given sequence. According to their theory, complexity measure is related to the number of distinct phrases and the rate of their occurrence along the sequence. It consists of a simple parsing algorithm whose task is to recognize newly encountered phrases during its scanning of a given sequence. In this algorithm a new phrase is established as the shortest substring which has not occurred previously, where the search for previous occurrences may be restricted or generalized in the modified algorithms in different ways. For example, by considering only a fixed number of preceding symbols, by considering only complete previously established phrases (Lempel-Ziv Incremental Parsing Algorithm), by allowing a number (not more than a fixed threshold) of previous occurrences of the phrase (Generalized Lempel-Ziv Algorithm ) etc.

There are many other methods available to calculate the complexity of a system like Sample Entropy, Wavelet Entropy, Fourier Entropy, permutation entropy etc. LZ Complexity Analysis has its appeal in that a single parsing is highly efficient and the algorithm is very easy to implement.

As the method for complexity evaluation, we have chosen the concept of complexity of a finite symbolic sequence, introduced by Lempel and Ziv (Lempel and Ziv, 1976). While studying complexity, we are interested in the detection of the regularities underlying the vibration signals. In this method, the coarse-graining of the signals obtained is accomplished by transforming it into a sequence of symbols, here 0s and 1s. This is done by choosing a suitable threshold value, say mean, $\mu$.. A sample is set to 1 if it is greater than the threshold, else to 0. The signal data is thus converted into a sequence of 0s and 1s. Another way to obtain a sequence of symbols for the data may be by choosing two thresholds. Say the maxima and minima of the data under observation are M and m, respectively. The two thresholds then will be $T_1 = \pi$-min and $T_2$=max-$\mu$. A sample is then set to 0 if it is less than $T_1$, to 1 if it is between $T_1$ and $T_2$ and to 2 if it is greater than $T_2$. The sequence hence obtained is then parsed from the first symbol to the last and the complexity number $c(n)$ is increased by 1 each time a new sequence is encountered. Normalized complexity measure $C(n)$ gives a complexity measure which is independent of the length of the sequence.

The value of $C(n)$ lies between 0 and 1 because it is a normalized parameter. For example- Let us consider the sequence obtained from course-graining is:
$S = 111110011111110011000000011101$. The distinct patterns obtained as per LZ Algorithm are 1, 11110, 01, 1111110, 0110, 0000001,110. The complexity number c(n) is therefore 7 for n=30, here. For different values of signal length n, the value of c(n) will differ. Complexity measure will become clear after the following example: Let us consider a new sequence 0 0 1 1 1 1 0 0. the procedure of computing its complexity is given in Table 1

The flowchart for computing complexity number of a given sequence is given in Fig.1.

The diagnostically important features are extracted from the vibration signal recorded for four bearing condition i.e. bearing with no fault, bearing with defective balls, bearing with defective inner race and bearing with defective outer race. A group of 50 vibration signals are recorded for each bearing condition. The features are extracted from the signals and median and standard deviation for extracted features are given in Table 2.

TABLE I

COMPUTATION OF COMPLEXITY NUMBER OF THE SEQUENCE

0 0111100

| s.n. | a | b | Ab | abp | Is b ∈ abp | C(n) |
|---|---|---|---|---|---|---|
| 1. | 0 | - | 0 | - | No | 1 |
| 2. | 0 | 0 | 00 | 0 | Yes | 1 |
| 3. | 0 | 01 | 001 | 00 | No | 2 |
| 4. | 001 | 1 | 0011 | 001 | Yes | 2 |
| 5. | 001 | 11 | 00111 | 0011 | Yes | 2 |
| 6. | 001 | 111 | 001111 | 00111 | Yes | 2 |
| 7. | 001 | 1110 | 0011110 | 001111 | No | 3 |
| 8. | 0011110 | 0 | 00111100 | 0011110 | Yes | 3 |

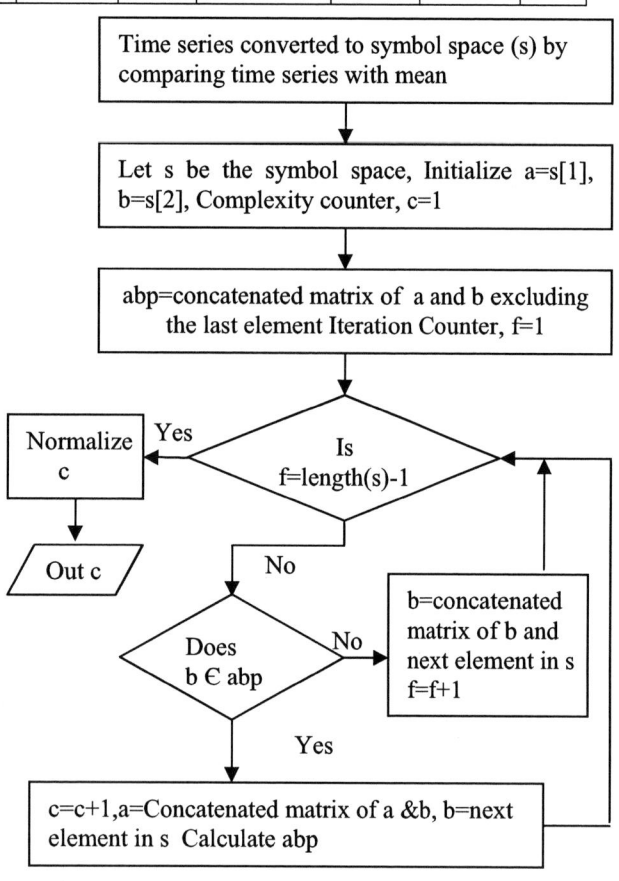

Fig 1. Flowchart for obtaining Normalized Complexity Number.

TABLE II

MEAN VALUES OF EXTRACTED FEATURES (UNDER FULL LOAD)

| s.n | Fault Type | RMS level (g) | | Skewness | | Kurtosis | | Normalized complexity number | |
|---|---|---|---|---|---|---|---|---|---|
| | | mean | std | mean | std | mean | std | mean | std |
| 1. | Normal | 0.04 | 0.01 | 0.04 | 0.02 | 3.00 | 0.12 | 0.55 | 0.02 |
| 2. | Ball Defect | 0.12 | 0.07 | -0.33 | 0.18 | 5.86 | 0.27 | 0.65 | 0.05 |
| 3. | Defective Inner race | 0.61 | 0.01 | -0.56 | 0.13 | 2.81 | 0.09 | 0.83 | 0.08 |
| 4. | Defective Outer race | 0.65 | 0.01 | -1.02 | 0.13 | 4.22 | 0.13 | 0.96 | 0.09 |

## IV. RESULT AND CONCLUSION

To build a derived fuzzy knowledge model based on ANFIS for estimating the health of bearing, two types of tuning i.e., model structure tuning and parameter tuning are required. Structure tuning concerns the structure of the rules: input and output variables selection, variable universe of discourse partition, linguistic labels determination, and type of logical operation to compose each rule. Parameter tuning is concerned with the fine adjustment of the position of all membership functions, their shape is controlled by *premise parameters* and the Takagi–Sugeno type [17], [18] if-then rules to be extracted controlled by the *consequent parameters*.

Four vibration derived parameters namely rms level, Kurtosis, Skewness and C(n) are the input variables of the model, and used to estimate the bearing health. Three fuzzy sets are defined on each of the input spaces, corresponding to linguistic *low, medium* and *high* for each variable.

Three membership functions are associated with each input, so that the input space is partitioned into eighty one fuzzy subspaces, each of which is governed by a fuzzy if-then rule. The *premise* part of a rule delineates a fuzzy subspace, while the *consequent* part specifies the output within this fuzzy subspace. The weighted average of the outputs of these eighty one fuzzy subspaces, i.e., the final output of the model, is a index between 0.0 and 1.0, which represents the degree of severity of fault and the type of fault. The output of the model is bearing health. Thus, the estimation process may be viewed as a mapping from the input space to the output space, which maps rms level, Kurtosis, Skewness and Complexity measure C(n) to bearing health. The membership function that is chosen with input variables is pi membership function. The performance of other membership functions such as triangular, gbell and trapezoidal is also observed. It is observed that most suitable performance is found with pi membership function. ANFIS employs an efficient hybrid learning procedure that combines back-propagation method and the least squares estimation to tune the parameters both of the membership functions and the Takagi–Sugeno type rules. As a conventional way of setting parameters in a fuzzy system, the *premise parameters* are set in a way that the membership function can cover the universe of discourse completely, with sufficient overlapping.

The proposed ANFIS is trained with 200 input vectors. The average error after 60 epochs was found very low of the range 0.000643. The developed model then tested with testing data set which was not used during training and an average error of 0.003 was observed. A total of 26 rules are formed to give the correct classification of bearing health with the accuracy of 99%.

The proposed model has been found to be an effective tool for bearing health analysis. Experimentation was done with various membership functions but the best results were obtained using pi membership function. Research is being continued to analyze vibration signals from different defect locations and on different types of bearings, to systematically validate the utility of this technique.

## V. REFERENCES

[1] N. Tandon, "A comparison of some vibration parameters for the condition monitoring of rolling element bearings," *J. Int. Meas. Confederation*, vol. 12, no. 3, pp. 285–289, 1994.

[2] R. Heng ,M. Nor, "Statistical analysis of sound and vibration signals for monitoring rolling element bearing condition," *Appl. Acoust.*, vol. 53, no. 1–3, pp. 211–226, 1998.

[3] R. Jones, "Enveloping for bearing analysis," *Sound Vibration*, vol. 30, pp. 10–15, 1996.

[4] Renwick T. Jhon , Babson E. Paul, " Vibration Analysis – A proven technique as a Predective Maintenance Tool", *IEEE Trans. On Industry Applications*, Vol. IA-21, no. 2, pp. 324-332, March/April 1985.

[5] W. J. Staszewski, K. Worden, G. R. Tomlinson, "Time-frequency analysis in gearbox fault detection using the wigner-ville distribution and pattern recognition," *Mech. Syst. Signal Processing*, vol. 11, no. 5, pp. 673–692, 1997.

[6] L. Eren , M. J. Devaney, "Bearing Damage detection via wavelet packet decomposition of stator current," in *Proc. 19th IEEE Instrumentation Measurement Technology Conf.*, vol. 1, Anchorage, AK, 2002, pp. 109–113

[7] Paya B. A., Esat. I., Badi M.N.M., "Artificial Neural Network Based Fault Diagnostics of Rotating Machinery using Wavelet Transforms as Preprocessor", Mechanical Systems and Signal Processing, Vol. 11, No. 5, pp. 751-765, September 1997.

[8] X. S. Zhang , R. J. Roy, "EEG complexity as a measure of depth of anesthesia for patients," *IEEE Trans. Biomed. Eng.*, vol. 48, pp. 1424–1433, Dec. 2001.

[9] J.W. Zhang, C. X. Zheng, "EEG complexity measurement of focal ischemic cerebral injury," in *Proc.20th Annu. Int. Conf. IEEE Engineering in Medicine and Biology Society*, vol. 4, Hong Kong, China, 1998, pp. 2027–2029.

[10] L. S. Qu , J. D. Jiang, "The complexity analysis of vibration signals of large rotating machinery," *J. Xi'an Jiaotong Univ.*, vol. 32, no. 6, pp. 31–35, 1998.

[11] Lempel, J. Ziv, "On the complexity of finite sequences," *IEEE Trans. Inform. Theory*, vol. IT-22, pp. 75–81, Jan. 1976.

[12] Yan Ruqiang, R. X. Gao, "Complexity as a measure for machine health evaluation", IEEE Transactions on Instrumentation and Measurement, vol 53, issue 4, pp. 1327 -1334, Aug. 2004

[13] C.-T. Lin, C. S. G. Lee, "Neural-network-based fuzzy logic control and decision system," *IEEE Trans. Comput.*, vol. 40, pp. 1320–1336, Dec. 1991.

[14] J.-S. R. Jang, "ANFIS: Adaptive-network-based fuzzy inference system," *IEEE Trans. On Systems, Man, and Cybernetics*, vol. 23, no. 3, pp. 665–684, 1993..

[15] S. Altug, M.-Y. Chow, H. J. Trussell, "Fuzzy inference systems implemented on neural architectures for motor fault detection and diagnosis,"*IEEE Trans. Ind. Electron.*, vol. 46, pp. 1069–1079, Dec. 1999.

[16] J.-S. R. Jang, C. T. Sun, "Neuro-fuzzy modeling and control," *Proc. IEEE*, vol. 83, pp. 378–406, Mar. 1995.

[17] T. Takagi, M. Sugeno, "Fuzzy identification of systems and its applications to modeling and control," *IEEE Trans. Syst., Man, Cybern.*, vol. SMC-15, pp. 116–132, 1985.

[18] M. Sugeno, G. T. Kang, "Structure identification of fuzzy model," *Fuzzy Sets Syst.*, vol. 28, pp.15–33, 1988.

## VI. BIOGRAPHIES

**Sulochana Wadhwani** (M'1888, F'17) was born in 1969 in India. She received bachelor degree in Electrical Engineering M. E. degree in control system from GEC Jabalpur in 1991and 1993 respectively. Currently, she is faculty in the Department of Electrical Engineering, Madhav Institute of Technology and Science, Gwalior, India. Her research interests are in the area of application of artificial intelligence for condition monitoring of industrial drives. She has received many awards including best paper award from Institution of Engineers.

**A. K. Wadhwani** born in 1966 in India. He received bachelor degree in Electrical Engineering from Bhopal University, India, in 1987, M. E. degree in Measurement & Instrumentation from University of Roorkee in 1993 and Ph.D from Indian Institute of Technology, Roorkee in 2003 respectively. Since 1988 he is employed in Electrical Engineering Department, MITS Gwalior. Dr Wadhwani is the life member of ISTE. His areas of interest are Biomedical Engineering, Measurement and Instrumentation and Energy Conservation.

**Dr. S.P. Gupta** obtained his ME and PhD degrees from the University of Roorkee in 1775 and 1986 respectively. He joined Jyoti Limited Baroda in 1975 and worked in the area of design and development of Multi-speed Inducction motors for two years. Dr Gupta joined the Electrical Engineering Department of University of Roorkee in 1976. He worke as post doctoral fellow at the Electrical and Electronic Engineering Department, University of Bristol (England), in 1989-90. Dr. Gupta worked at the University of Puerto Rico as an associate professor in 1990-91 and did teaching and research for two semesters. Presently, Dr. Gupta is Professor and Head in Electrical Engineering Department at IIT Roorkee. He has guided 9 doctoral and more than 20 Master's thesis and has more than 50 research publications in international reputed journals and conference proceedings. He has undertaken a number of consultancy and sponsored projects from industries and government departments. He was the member of expert committee of National Board of Accreditation (NBA) for evaluation of Electrical Engineering programme for number of institutuins.

**Vinod Kumar** Obtained his BSc (Electrical Engg) Hons from Punjab University in 1973. ME (Measurement & Instrumentation) Hons and PhD degree from University of Roorkee in 1975 and 1984 respectively. He joined The electrical Engineering department of University of Roorkee in 1975. He has published more than 125 research paper, guided 70 Masters thesis, supervised 12 PhD thesis He has undertaken large number of consultancy & sponsored projects from industries and government departments. He is a senior member of IEEE, USA He was elected as Fellow of the Institution of Engineers (I), Institution of Electronics and Telecommunication Engineers and a member of various professional bodies. He has received many awards including Khosla Medal and best paper awards of the Institution of Engineers. He has also conducted many courses, workshops for the benefit of faculty and field engineers. Dr Kumar served the institute as associate dean academic, director/coordinator AVRC and coordinator Information Super Highway Centre. He is presently head continuing education centre. His areas of interests are Measurement and Instrumentation, Medical Instrumentation, Digital Signal Processing and Telemedicine.

**2006 IEEE International Conference on Power Electronic, Drives and Energy Systems**

# Discrimination between Inrush current and Internal Faults using Pattern Recognition Approach

B. K. Panigrahi, *Member, IEEE*, S. R. Samantaray, P. K. Dash, *Senior Member, IEEE*, and G. Panda, *Senior Member, IEEE*

*Abstract*- **This paper presents a new approach to distinguish between inrush current and internal faults of power transformer using pattern recognition approach. The HS-transform (Hyperbolic S-transform) is used to extract patterns of inrush current and internal faults from the captured transformer current. HS-transform is a very powerful tool for non-stationary signal analysis giving the information of transient currents both in time and frequency domain. The spectral energy and standard deviation are calculated to distinguish between inrush current and internal fault. Classification of internal faults and inrush current is done through Fuzzy C-means clustering.**

*Index Terms*-- **HS – Transform, Pattern recognition, Transformer inrush**

## I. Introduction

DISCRIMINATION between inrush current and internal faults has been recognized as a very challenging power transformer protection problem. The inrush current contains a large second harmonic component in comparison to a fault. Sometimes also the second harmonic may be generated in case of internal faults in power transformer. This may be due to CT saturation and distributive capacitance in long transmission line to which the power transformer is connected. Also sometimes the magnitude of second harmonic in internal fault current can be close to that present in the inrush current. Also the inrush current magnitude is relatively less in modern power transformer due to design improvements. Therefore the traditionally provided protection system with harmonic restraint will not solve the problem. Here most important requirement is to extract features from the non-stationary signals, as both inrush current and internal faults are non-stationary signal. For feature extraction or pattern recognition

P.K.Dash is director college of Enginering, Bhubaneswar, India. (E-mail: pkdash_india@yahoo.com).

S.R.Samantaray is a Ph..D. scholar in the department of Electronics and Communications Engineering ,NIT Rourkela, India. (E-mail:sbh_samant@ rediffmail.com).

G.Panda is Professor in the department of Electronics and Communications Engg NIT, Rourkela,India. (E-mail: gpanda@nitrkl.ac.in).

Dr. B.K.Panigrahi is Asst. Prof., Dept. of Electrical Engg, IIT Delhi, India. (E-mail: bkpanigrahi@ee.iitd.ac.in ).

from non-stationary signal STFT (Short Time Fourier Transform), DWT(Discrete Wavelet Transform) are used[1,2]. Here in this paper, a new approach for patterns recognition using multi resolution HS-transform with varying window of varying shape is proposed. The S-transform is an invertible time-frequency spectral localization technique that combines elements of wavelet transforms and short-time Fourier transform.

The S-transform uses an analysis window whose width is decreasing with frequency providing a frequency dependent resolution. S-Transform is continuous wavelet transform with a phase correction. It produces a constant relative bandwidth analysis like wavelets while it maintains a direct link with Fourier spectrum. The S-transform has an advantage in that it provides multi resolution analysis while retaining the absolute phase of each frequency. This has led to its application for detection and interpretation of non-stationary signal like power system disturbance signal [5] and fault analysis. The inrush current and fault current are tuned through S-transform to get the patterns of inrush current and internal faults. Then the spectral energy and the standard deviation of inrush current and fault current are computed. The level of energy content and standard deviation gives the discrimination between inrush and fault current; accordingly the relay restrains or operates. Also time frequency contours in both fault and inrush are presented to distinguish the both events. The classification is done by fuzzy c-means clustering.

## II. Hyperbolic S-transform and Pattern Recognition

The original S-transform [6] is defined as

$$S(\tau,f) = \int_{-\infty}^{\infty} h(t)\left\{\frac{|f|}{\sqrt{2\pi}}\exp\left\{-f^2(\tau-t)/2\right\}\exp(-2\pi ft)\right\}dt \quad (1)$$

Where S denotes the S-transform of h (*t*), which is the actual current signal varying with time, frequency is denoted by f, and the quantity $\tau$ is a parameter which controls the position of gaussian window on the time-axis. A small modification of the gaussian window has been suggested [7] for better performance.

$$W_{gs}(\tau-t,f,\alpha_{gs}) = \frac{|f|}{\sqrt{2\pi}\alpha_{gs}}\exp\left[\frac{-f^2(\tau-t)}{2\alpha_{gs}^2}\right] \quad (2)$$

and the S-transform with this window is given by

$$S(\tau, F, \alpha_{gs}) = \int_{-\infty}^{\infty} h(t) W_{gs}(\tau - t, f, \alpha_{gs}) \cdot \exp(-2\pi i f t) dt \qquad (3)$$

where $\alpha_{gs}$ is to be chosen for providing suitable time and frequency resolution.

In applications, which require simultaneous identification of time-frequency signatures of different disturbance in power system like voltage sag, voltage swell, multiple notch, multiple spike, oscillatory transient, chip etc. and it may be advantageous to use a window having frequency dependent asymmetry. Thus, at high frequencies where the window is narrowed and time resolution is good, a more symmetrical window needs to be chosen. On the other hand, at low frequencies where a window is wider and frequency resolution is less critical, a more asymmetrical window may be used to prevent the event from appearing too far ahead on the S-transform. Thus a hyperbolic window of the form given below is used.

$$W_{hy} = \frac{2|f|}{\sqrt{2\pi(\alpha_{hy} + \beta_{hy})}} \cdot \exp\left\{ \frac{-f^2 X^2}{2} \right\} \qquad (4)$$

Where

$$X = \frac{\alpha_{hy} + \beta_{hy}}{2\alpha_{hy}\beta_{hy}} (\tau - t - \xi) + \frac{\alpha_{hy} - \beta_{hy}}{2\alpha_{hy}\beta_{hy}} \sqrt{(\tau - t - \xi)^2 + \lambda_{hy}^2} \qquad (5)$$

In the above expression $0 \langle \alpha_{hy} \langle \beta_{hy}$ and $\xi$ is defined as

$$\xi = \frac{\sqrt{(\beta_{hy} - \alpha_{hy})^2 \lambda_{hy}^2}}{4\alpha_{hy}\beta_{hy}} \qquad (6)$$

The translation by $\xi$ ensures that the peak $W_{hy}$ occurs at $\tau - t = 0$.

At $f = 0$, $W_{hy}$ is very asymmetrical, but when $f$ increases, the shape of $W_{hy}$ converges towards that of $W_{gs}$, the symmetrical gaussian window given in (2). For different values of $\alpha_{hy}$ and $\beta_{hy}$ and with $\lambda_{hy}^2 = 1$. Fig.1 shows the nature of the window as the function of time $\tau - t$. As seen from the figure the change in the shape from an asymmetrical window to a symmetrical one occurs more rapidly with increasing $f$.

The discrete version of the Hyperbolic S-transform of the internal faults and inrush current signal samples is calculated as $S[n, j] = \sum_{m=0}^{N-1} H[m + n] G(m, n) \exp(i2\pi mj)$, Where $N$ is the total number of samples and the indices n, m, j are $n = 0, 1 .... N - 1, m = 0, 1 ..... N - 1$, and $j = 0, 1 ....... N - 1$. The

$G(m, n)$ denotes the Fourier transform of the Hyperbolic window and is given by

$$G(m, n) = \frac{2|f|}{\sqrt{2\pi(\alpha_{hy} + \beta_{hy})}} \exp(\frac{-f^2 X^2}{2n^2}) \qquad (7)$$

and

$$X = \frac{(\alpha_{hy} + \beta_{hy})}{2\alpha_{hy}\beta_{hy}} t + \frac{\beta_{hy} - \alpha_{hy}}{2\alpha_{hy}\beta_{hy}} (\sqrt{t^2 + \lambda_{hy}}) \qquad (8)$$

and $H(m, n)$ is the frequency shifted discrete Fourier transform $H[m]$, where

$$H(m) = \frac{1}{N} \sum_{m=0}^{N-1} h(k) \exp(-i2\pi nk) \qquad (9)$$

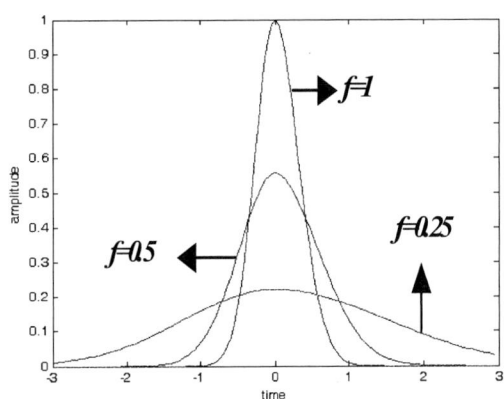

Fig. 1.  Varying window $W_{hy}$ at $f=1, f=0.5$ and $f=0.25$.

## III. SIMULATION STUDY

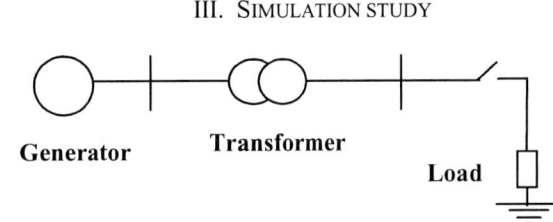

Fig. 2.  System model.

The simulation study has been done on the system shown in the fig.2. 1000MVA generator and 450 MVA transformers with 15Kv/220Kv. The study has been made for inrush current and various internal fault conditions like winding–ground, winding-winding and winding-winding-ground without load and with load. The sampling rate is 15.36 kHz. Half cycle data has been processed through HS-transform to give the energy and standard deviation. The simulation model is developed using matlab-simulink. 200 cases (examples) for internal faults and inrush current at various conditions were simulated and tested using the proposed method.

1065

## IV. RESULTS AND DISCUSSION

### A. Feature extraction using S-transform

From the simulation model data for inrush current and internal faults at different bus with and without load are generated. The HS-transform of the half cycle data from the inception of inrush and faults is computed. The normalized frequency contours are obtained as shown in Fig. 3(a) through Fig. 3(h). It is clear from the normalized frequency contours that in case of inrush current the normalized frequency contours are interrupted in nature compared to internal faults. In case of fault conditions, the normalized frequency contours are regular throughout the time series.

Apart from the normalized frequency contour for inrush current and fault current, the spectral energy and the standard deviation of the HS-transform of the signal are found out. The spectral energy and standard deviation for inrush current and fault current at various conditions are depicted in Table-I through Table-II. It is clearly seen from tables that the spectral energy of the HS-transform of the inrush current is much less compared to the spectral energy of the HS-transform of the internal fault current signal. From the spectral energy and standard deviation, the classification of inrush and internal fault is done using fuzzy C-means clustering technique.

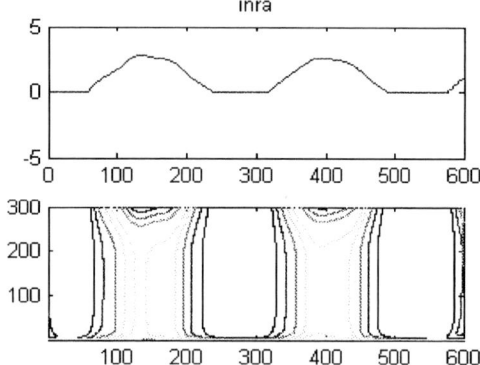

Fig. 3(a). Normalized frequency contours for inrush current of a-phase.

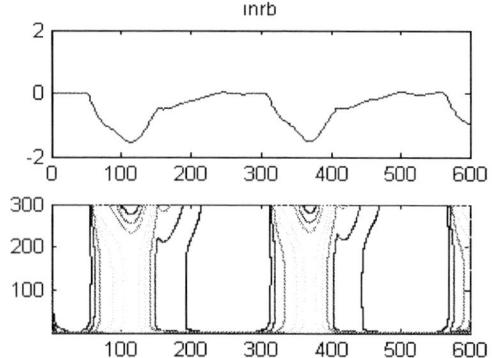

Fig. 3(b). Normalized frequency contours for Inrush current of b-phase.

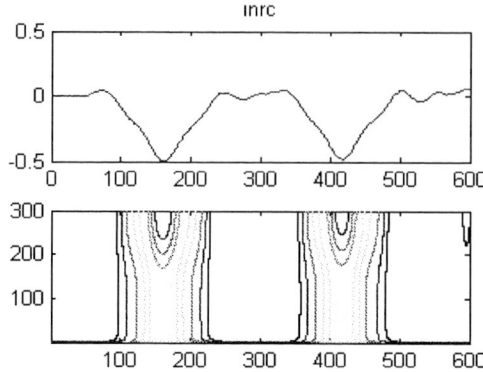

Fig. 3(c). Normalized frequency contours for Inrush current of c-phase.

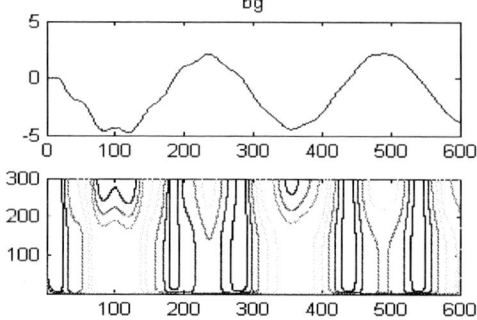

Fig. 3(d). Normalized frequency contours for Winding to ground fault of b-phase.

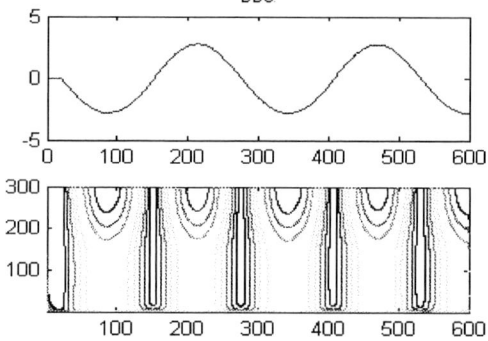

Fig. 3(e). Normalized frequency contours for Winding to winding (b-c) fault of b-phase.

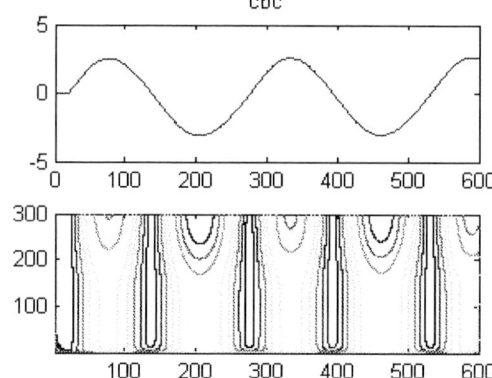

Fig. 3(f). Normalized frequency contours for Winding to winding(b-c) fault of c-phase.

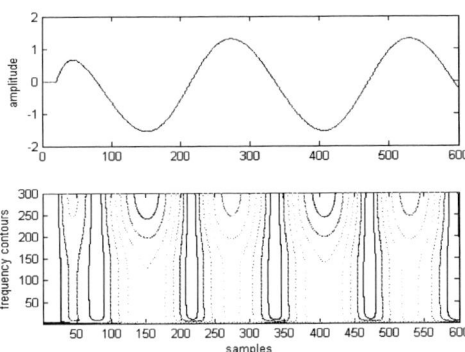

Fig.3(g). Normalized frequency contours for Winding to winding (a-c) fault of a-phase.

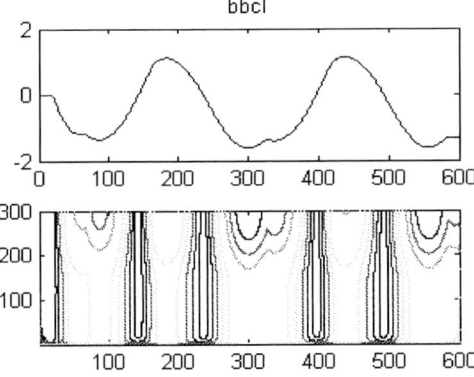

Fig.3(h). Normalized frequency contours for Winding to winding (b-c) fault of b-phase with load.

TABLE I
SPECTRAL ENERGY AND STANDARD DEVIATION FOR
INRUSH CURRENT AND FAULT WITHOUT LOAD

| Inrush/Fault (without load) | Energy | Std |
|---|---|---|
| inr-a | 461.1 | 0.3666 |
| inr-b | 212.9 | 0.2040 |
| inr-c | 65.2 | 0.0663 |
| ag | 1028.0 | 0.7995 |
| bg | 821.3 | 0.6974 |
| cg | 797.4 | 0.6562 |
| aab | 875.7 | 0.6687 |
| bab | 792.6 | 0.6184 |
| aca | 882.2 | 0.6815 |
| cac | 773.3 | 0.6211 |
| bbc | 627.0 | 0.5211 |
| cbc | 625.5 | 0.4603 |

TABLE II
SPECTRAL ENERGY AND STANDARD DEVIATION FOR
INRUSH CURRENT AND FAULT WITH LOAD

| Inrush/Fault (with load) | Energy | Std |
|---|---|---|
| inrl-a | 425.2 | 0.3538 |
| inrl-b | 184.6 | 0.2131 |
| inrl-c | 86.0 | 0.1109 |
| agl | 975.9 | 0.7632 |
| bgl | 779.9 | 0.6795 |
| cgl | 770.1 | 0.6007 |
| aabl | 879.4 | 0.6873 |
| babl | 789.4 | 0.6462 |
| acal | 887.5 | 0.7030 |
| cacl | 774.9 | 0.6600 |
| bbcl | 632.8 | 0.4464 |
| cbcl | 622.5 | 0.4460 |

## B. Classification using Fuzzy C-means clustering

After calculating the spectral energy and standard deviation from the HS-transform of the inrush and internal fault current, classification of inrush current and internal fault is done by applying clustering technique. Here in this paper Fuzzy C-means clustering is used to generate clusters to discriminate between inrush and internal fault. The data generated from HS-transform, the spectral energy and the standard deviation are used as 2-D data for Fuzzy C-means clustering as shown in Fig.4, which clearly distinguish between inrush current and internal faults.

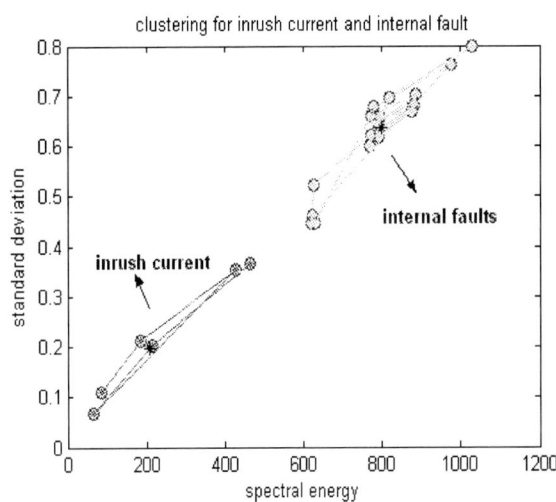

Fig. 4. Fuzzy C-means clustering to discriminate inrush current and internal fault.

## V. CONCLUSION

This paper presents a new approach for discrimination between inrush current and internal faults in power transformer by pattern recognition technique using HS-transform. The HS-transform gives the normalized frequency contours for inrush current and internal fault very distinctly as shown in the figures where second harmonic is pronounced in case of inrush current compared to faults. Also the spectral energy and standard deviation are calculated and Fuzzy C-means clustering is used to distinguish the inrush current from internal faults. As HS-transform is less prone to noise compared to Wavelet transform, it gives the effective protection for large power transformers.

## VI. REFERENCES

[1] P L.Mao and R.K. Agggarwal, "A novel approach to the classification of the transient phenomena in power transformers using combined wavelet transform and neural network", vol.16, no.4, pp.654-660, October 2001.

[2] S.N. Hon, W. Qin," A wavelet based method to Discriminate between inrush and internal faults",pp.927-931, IEEE, 2000.

[3] F. Jiang, ZQ. Bo, P.S.M Chin, M.A.Redfern, Z. chen, "Power transformer protection based on transient detection using discrete wavelet transform (DWT)",IEEE, 2000.

[4] L.G. Perez etc. al, "Training an artificial net- work to descriminate between magnetising inrush and internal faults", IEEE Trans., Power Delivery,,vol.9,No.1,pp.434-44, 1994.

[5] P.K.Dash, B.K.Panigrahi, G Panda, "Power quality analysis using S-transform", IEEE Trans. On Power Delovery,2002.

[6] Stockwell, R.G., Mansinha ,L., and Lowe, R.P., "Localization of the complex spectrum :The S-transform" , IEEE Trans. On Signal Process, vol.44, No.4pp.998-1001,April-1996.

[7] Pinnegar, C.R and Mansinha, Lalu., "The S-transform with windows of arbitrary and varying window", Geophysics,vol-68,No-1,pp.381-385, 2003.

[8] Mansinha, L., Stockwell., R..G. and Lowe, R.P., "Pattern analysis with two dimensional spectral localization: Application of two dimensional S-transforms", Physica A,239, pp.286-295, 1997.

## VII. BIOGRAPHIES

**B. K. Panigrahi** is working as an Assistant Professor in the Department of Electrical Engineering, IIT, New Delhi, India. Prior to joining IIT Delhi, he was working as Lecturer at University College of Engineering, Burla, Sambalpur, Orissa for about 13 years. The research interests of Dr. Panigrahi are in the areas of Intelligent control of FACTS devices, Application of advanced DSP techniques for Power Quality assessment and soft computing application to power system optimization

*S.R. Samantaray* is pursuing his Ph.D. in the Department of Electronics and Communication Engineering, National Institute of Technology, Rourkela, in the area of power system protection. His major interest includes intelligent protection, digital signal processing, soft computing, FACTs. Graduated in electrical engineering (Hons.), he worked in TATA group in the area of power system, industrial automation and instrumentation.

*P.K. Dash* (SM' 1990) is working as Director College of Engineering, Bhubaneswar, India. Earlier he was a professor in the Faculty of Engineering, Multimedia University, and Cyberjaya, Malaysia. He also served as a professor of electrical engineering and chairman, center for intelligent systems, National Institute of Technology, Rourkela, India for more than 25 years. Dr. Dash holds D.Sc., Ph.D., M.E., and B.E. degrees in electrical engineering and had his post-doctoral education at the University of Calgary, Canada. His research interests are in the area of power quality, FACTS, soft computing, deregulation and energy

markets, signal processing, and data mining and control. He had several visiting appointments in Canada, USA, Switzerland, and Singapore. To his credit he has published more than 150 International Journal papers and nearly 100 in International conferences. Prof. Dash is a fellow of the Indian National Academy of Engineering and senior member of the IEEE, and fellow of Institution of Engineers, India.

*Ganapati Panda* received Ph.D. degree in digital signal processing from IIT, Kharagpur, India, and post doctorate from University of Edinburgh, UK, in 1982 and 1984–1986, respectively. He has published more than 160 papers in referred research journals and conferences. He carries out research work in the field of DSP, soft-computing and digital communication. He is a fellow of INAE and National Academic of Science, India. Presently he is working as a professor in the Department of Electronics & Instrumentation Engineering at NIT, Rourkela, India.

**2006 IEEE International Conference on Power Electronic, Drives and Energy Systems**

# Stepwise Restoration of Power Distribution Network under Cold Load Pickup

Vishal Kumar, *Student Member, IEEE*, Rohith Kumar H.C., I. Gupta, and H.O. Gupta, *Senior Member, IEEE*

*Abstract*— This paper presents a scheme for step-by-step restoration of a power distribution network under Cold Load Pickup condition. The proposed scheme uses the optimal load shedding for each step of the restoration process, and determines the optimal switching sequence for the restoration process. The optimization is performed to achieve multiple objectives of minimum load curtailment and minimum switching operations while satisfying all the system constraints. Genetic Algorithm has been utilized to search the optimal solution. The scheme has been illustrated on a standard 33-bus radial distribution network.

*Index Terms*—Cold Load Pickup, Genetic Algorithms, Optimal Load Shedding, Power Distribution System, Restoration.

## Nomenclature

| | |
|---|---|
| $\alpha$ | Rate of decay of load under CLPU condition |
| $\Delta T$ | Un-diversified load duration |
| $Ge_{MAX}$ | Maximum number of generations |
| $I_V$ | Sum of the ratios of line-current violations to the thermal limit |
| $K_{TM}$ | Transformer maximum loading factor |
| $N_{br}$ | Total no of branches in the network |
| $N_{bu}$ | Total no of buses in the network |
| $P_c$ | Probability of crossover |
| $P_i(k)$ | $k^{th}$ chromosome of $i^{th}$ generation |
| $P_m$ | Probability of mutation |
| $P_{SIZE}$ | Population Size for genetic algorithm |
| $S(t)$ | Load Demand with respect to time for $t \geq T_0$ |
| $S_D$ | Diversified Load value in p.u. |
| $S_{Supplied}$ | Load supplied during CLPU in MVA |
| $S_T$ | Substation transformer rating in MVA |
| $S_{Total}$ | Total connected load in MVA |
| $S_{TV}$ | Transformer loading-limit violation |
| $S_U$ | Un-diversified load in p.u. |
| $(S_U^j)_i$ | Un-diversified load value of $j^{th}$ bus during $i^{th}$ generation number for a load-point |
| $SW$ | Number of operated switches |
| $T_0$ | Initiation of restoration process |
| $T_1$ | Initiation of decay towards $S_D$ |
| $u(t)$ | Unit step function |

The Authors are with EED, IIT Roorkee, Roorkee, India. Ph. no.: 01332-286294, Email: vish1dee@iitr.ernet.in

| | |
|---|---|
| $V_V$ | Sum of the ratios of bus-voltage violations to the limiting value |
| $W_{IV}$ | Weight for $I_V$ |
| $W_{Load}$ | Weight for ratio of lost load over total load |
| $W_{SW}$ | Weight for switching operation |
| $W_{TV}$ | Weight for $S_{TV}$ |
| $W_{VV}$ | Weight for $V_V$ |

## I. Introduction

THERMOSATICALLY controlled electrical loads act as a major portion of the total load demand in today's power system. During the normal operation, these loads maintain their diversity hence the total actual load remains less than the connected loads. However, on the occurrence of a prolonged outage this diversity is lost. The loss of diversity causes the post-outage load-demand up to 2 to 5 times the diversified load, this high load condition is known as 'Cold Load Pickup' (CLPU) [1]. CLPU condition leads to extensive loading on the distribution network elements, thus causing violation of voltage and current limits. Duration of outage, type of loads, weather conditions, habits of the user, and thermal characteristics of the building affect the magnitude and the duration of CLPU condition. In the literature [2], the behavior of CLPU current is mainly divided in to the following four phases: inrush, motor starting current, motor accelerating current and enduring current. The first three of the phases are transient in nature and die within few seconds, whereas the enduring phase can last for much longer durations. Therefore, the enduring phase restricts the restoration of the complete network in a single step. However, the loads are restored in steps, as the aggregated load tends to decrease with time. Such kind of restoration in stages is referred as step-by-step/stepwise restoration [3]. In the stepwise restoration, whole network is sub-divided into several sections using sectionalizing switches, and utilities operate them as to restore the network optimally.

Non-prominent load behavior had forced the researchers to neglect CLPU problem in initial stages. During the initial stage, researchers suggested the change of relay settings, and usage of very inverse characteristic relay to resolve the CLPU inrush phase problem [4]-[5]. However, significant rise in the amount of thermostatically controlled loads in the recent years has adverse impact on the power system elements during the restoration. In 1979, the peak

0-7803-9771-1/06/$25.00 ©2006 IEEE

load following an extended outage during cold weather conditions for electrically heated homes was predicted [1]. During 1985, Wilde [6] investigated the effects of CLPU condition on the substation transformer and protective devices. A similar study was also conducted by Aubin et al. for the distribution transformer [7]. Literature reveals that several efforts have been made to model the CLPU condition, the models ranging from simple piece-wise linear model to complex regression models are available [8]-[14]. However, a delayed exponent model suits for analysis purposes and is widely preferred by the researchers in the studies [3], [14]-[17].

The main objective of the utility is to restore the network quickly, using optimal sequence of restoring the loads to reduce the total restoration period. Pahwa has made a noticeable contribution, in the area of restoration under CLPU. Ucak and Pahwa minimized the total restoration time and customer interruption duration considering the substation transformer temperature and its loading limits, and also minimized the average customer interruption duration with adjacent pair wise interchange method [3], [15]. Walkilesh and Pahwa [16] performed a cost optimization in order to minimize the total annual cost, which comprised of cost of transformer, sectionalizing switches, energy interruption and transformer over-loadings. The same authors enhanced the work by utilizing Genetic Algorithms (GA) for optimization on an actual feeder. The feeder was compensated to as to meet the steady state operational voltage constraint [17]. Several evolutionary techniques such as GA [18], and ant algorithms [19] have been used to obtain the optimal switching sequence.

It is clear from the above-mentioned references that substantial work has been done on transformer over-loadings and cost minimization for CLPU. However, operational constraints such as voltage and current limits seem to have been neglected to an extent. Further for the operation of sectionalizing switches, precedence constraint has been used during the optimization, and this reduces the total amount of load being restored owing to the subsequent loading effect, which causes deferral in the restoration of loads present at the tail end of the feeders.

In the present paper an approach has been suggested that maximizes the utilization of the existing network capacity under CLPU condition trough load-switches, along-with their minimum number of operations. The optimal sequence of restoration of loads through the load-switches has been determined. The GA is effectively utilized to search the optimal solution of this multi-objective optimization problem, and the results obtained are demonstrated on a 33 Bus, 12.66 kV test system [20].

## II. AIM AND APPROACH

The main aim in the proposed work is to optimally utilize the existing system during the restoration of the network. The loss of diversity among the connected loads during CLPU condition causes excessive loading on the substation transformer, current violation in the feeders and voltage violations at the buses. In order to restore the network, loads are switched 'ON' stepwise such that all the operational constraints are met. The approach would be to find the load locations that need to be shed in order to operate the system optimally. The optimal locations for the switching of loads are determined by objective function given by (1).

### A. Problem Formulation

The main objective is to curtail minimum load while restoring the network, along with the minimum switching operations, subjected to the satisfaction of all the operational constraints. This multi-objective optimization problem is converted to a single objective problem with the help of suitable weights and the formulation is done as,

$$Min \; f = W_{Load} \frac{\left[ S_{Total} - S_{Supplied} \right]}{S_{Total}} \\ + W_{IV}I_V + W_{VV}V_V + W_{TV}S_{TV} + W_{SW}SW \qquad (1)$$

The objective function (1) is subjected to the following system operational constraints.
- All the network power flow equations must be satisfied
- Voltage at all the buses should be within the acceptable range (Utility's standard ANSI Std. C84.1-1989) i.e. within permissible limit (±5%).

$$V_{min} \leq V_i \leq V_{max} \qquad \forall i \in \left\{ N_{bu} \right\}$$

- Current in a feeder or conductor, must be less than the thermal capacity of the conductor.

$$I_i \leq I_i^{Rated} \qquad \forall i \in \{ N_{br} \}$$

Here, $I_i^{Rated}$ is current permissible for $i^{th}$ branch within safe limit of temperature.

- The total power supplied to the network must be within the substation's transformer capacity limit. In the proposed approach for network restoration, the maximum loading of the transformer is considered as the loading limit of the transformer for no additional loss of life.

i.e. Maximum transformer loading $< K_{TM} * S_T$

The maximum un-diversified load for different diversified loads and outage duration is reported, and the maximum supplying limits for one-hour outage for different pre and post outage load conditions have also been determined [3], [21]. In the current problem, $K_{TM} = 1.50$ has been used with no additional loss of life of transformer.

In objective function, the weights reflect the relative significance of each of the term present in the objective function. The main objective is to minimize the curtailment of load hence high penalties in terms of weights are imposed for the first term in (1). The second and the third term in the objective function are steady state security constraints and any violation of these constraints is also meant for a heavy penalty. Fourth term is of slightly lesser priority as the distribution transformers are rated for a short period of overloading.

### B. CLPU Model

The total demand during the restoration, following a prolonged outage is much higher due to the loss of diversity among the loads, mainly caused by thermostatically controlled loads. However, this total demand decays over time as the diversity is regained by the loads. The rate at which the network regains its diversity is dependent on the type of loads connected, weather conditions and living habits of the customers [2]. A number of CLPU models with diverse basis are observed in the literature, and are being used for study purpose. Few are empirical models [8], some are physical based models [9]-[12], and regression based models [13]. Model used for the purpose of study needs to be mathematically simple and also easier for the analytical computations. Therefore, in the present work a delayed exponential model [14], shown in Fig. 1, has been used. Mathematically the model can be synthesized as follows.

$$S(t) = \begin{Bmatrix} [S_D + (S_U - S_D)e^{-\alpha(t-T_1)}]u(t-T_1) \\ + S_U[1 - u(t-T_1)]u(t-T_0) \end{Bmatrix} \quad (2)$$

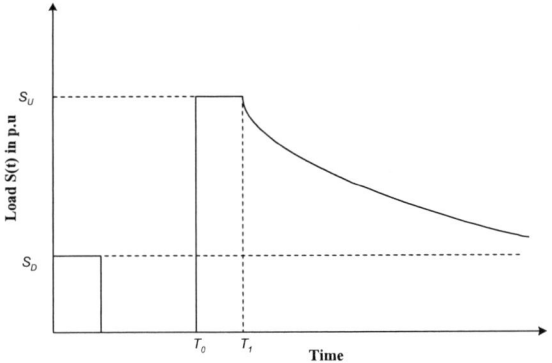

Fig. 1. CLPU model as a delayed exponent.

### C. Algorithm

Major portion of the total connected load gets restored during the initiation of the restoration process. Therefore, in the proposed approach several step/load-points are considered on the load profile of the loads switched 'ON' at initiation of the restoration. Corresponding to each load-point, locations for the optimal load shedding are determined using GA and thus determining the optimal switching sequence for entire network restoration. It is assumed that all the loads follow the delayed exponent model of CLPU [14]. Computational steps involved for the GA corresponding to a load-point for determining the optimal load shedding, are as follows.

**Input:** $P_{SIZE}$, $Ge_{MAX}$, $P_c$, $P_m$, $N_{bu}$

**Steps:**

1. Create random population $P_i$ of $P_{SIZE}$ and set $i=1$;

2. Check whether $i \leq Ge_{MAX}$. If 'Yes' then set $k = 1$; go to next step else go to step_11.

3. Check whether $k \leq P_{SIZE}$. If 'Yes' then set $j = 1$; go to next step else go to step_10.

4. Check whether $j \leq N_{bu}$. If 'Yes' then go to next step else go to step_8.

5. Update $(S_U^{\ j})_i = S(t)$.

6. Set $j=j+1$; and go to step_4.

7. Evaluate $P_i(k)$ using AC load flow analysis.

8. Set $k=k+1$; and go to step_3.

9. Apply Genetic operators and store the best of $P_i$

10. Set $i=i+1$; and go to step_2.

11. Optimal switching locations corresponding to current load-point.

## III. CASE STUDY

The proposed step-by-step restoration approach is illustrated on a standard 33-bus, 12.66 kV radial distribution network [4], with an assumed substation transformer capacity of 5MVA. The nominal supply voltage at the substation is considered as 1.045 p.u. to utilize the permitted voltage range, as the network buses face severe voltage drop during CLPU condition. The typical values of weight used are, 50.0 for the lost load, 500.0 for the current violation, 1000.0 for the voltage violation, 10.0 for the transformer violation and 10.0 for number of switching operations.

The proposed approach utilizes a delayed exponential model with parameters as: $\alpha = 1$ $hr^{-1}$, $\Delta T = T_1 - T_0 = 30$ minutes, $S_U = 2.5$ p.u., and $S_D = 1.0$ p.u. The optimal load shedding is determined at several discrete load-points considered on load profile, shown in Table-1.

TABLE I
VALUES OF TIMES ELAPSED FOR DIFFERENT LOAD-POINTS

| Step/Load-point | Time Elapsed in Minutes | Description of step/load-point |
|---|---|---|
| -1 | -120 | Outage duration |
| 0 | 0 | 2.50 x $S_D$ |
| 1 | 54.328 | 2.00 x $S_D$ |
| 2 | 95.917 | 1.50 x $S_D$ |
| 3 | 137.506 | 1.25 x $S_D$ |
| 4 | 192.480 | 1.10 x $S_D$ |

## IV. RESULTS AND DISCUSSION

The restoration problem is a case of NP-hard nonlinear combinatorial problem and GA has been used in order to search for the optimal solution. An elite preserving GA along-with roulette wheel type selection, two-point crossover and swap type mutation has been employed. GA is applied with different population sizes of 60, 80 and 100 and the best result obtained has been presented in this paper. The following are parameters used for GA

| | |
|---|---|
| Population size: | 80 |
| Crossover rate: | 0.8 |
| Mutation rate: | 0.05 |
| Maximum no of generations: | 150 |

The stepwise restoration of the network in the present work is performed for five distinct load-points considered on the load profile of the loads switched 'ON' at the initiation of the restoration under CLPU condition. The assessment of intricacies of simultaneous and stepwise restoration under CLPU condition is made, and is presented Table-II. It is observed that, buses 6 to 18 and 26 to 33 violate the under- voltage acceptable limit during CLPU condition, during simultaneous restoration leading to high system losses. Therefore single-step restoration of the network is not possible. Conversely, 70.36% of the total load is restored under CLPU condition in stepwise approach at initial step of restoration process. Moreover, significant improvement of voltage profile and reduction of network losses is obtained by optimally shedding the load trough the proposed approach. The combined GA convergence curve with different population sizes for the load-point '0' is presented in Fig.2.

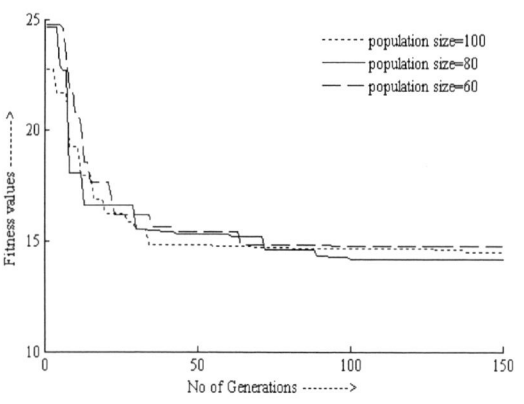

Fig. 2. GA convergence curve for various population size at load-point 0.

TABLE II
COMPARISON OF SIMULTANEOUS AND STEPWISE RESTORATION

| Restoration Technique | Restored Load in MVA | Minimum bus voltage in p.u. | Network Losses in MVA |
|---|---|---|---|
| Simultaneous | 10.1321 | 0.83651 | 1.3878 |
| Stepwise at load-point '0' | 7.1294 | 0.95310 | 0.4076 |

Further, the network regains diversity as the time elapses and the remaining loads are restored in the future load-points. The results obtained for each step/load-point is presented in Table-III, and the graphical illustration of the proposed stepwise approach for restoration presented

in Fig.3. The network is completely restored at load-point '4' with a total restoration period 192.48 minutes (i.e. ≈ 3 hr 13 minutes).

TABLE III
RESULTS FOR DISCRETE STEP/LOAD-POINTS DURING RESTORATION

| Step/Load -point | Total Load served in MVA | Load nos. to be disconnected | No. of switching operations |
|---|---|---|---|
| 0 | 7.1294 | 7-8,12-13, 17-18,31-33 | 9 |
| 1 | 6.6940 | 12-13,17-18, 31-33 | 7 |
| 2 | 5.7207 | 17-18, 32 | 3 |
| 3 | 4.8092 | 17, 32 | 2 |
| 4 | 4.6994 | -Nil- | -Nil- |

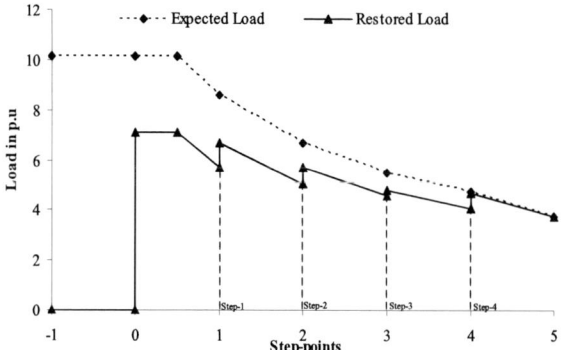

Fig. 3. Step-by-step restoration of the network under CLPU condition.

## V. CONCLUSION

Simultaneous restoration of the power distribution network under CLPU condition creates severe problems. In the present work a novel approach has been presented for optimal restoration of the network under CLPU. The approach determines the optimal sequence of restoration of the loads in step-by-step manner. The steps are considered in terms of distinct loads, relaxing the precedence constraint used in case of section-wise step-by-step restoration; it also increments the amount of the load restored in each step. The proposed approach is illustrated on a 33-bus system.

The conceptual framework of this paper in-accordance with the distribution automation will allow distribution utilities for faster restoration with maximum utilization of existing system capacities.

## REFERENCES

[1] J.E. Mcdonald, A.M. Bruning, and W.R. Mahieu, "Cold Load Pickup", *IEEE Transaction on Power Apparatus and Systems*, vol. 98, pp. 1384-1386, July/Aug.1979.

[2] V. Kumar, I. Gupta, and H. O. Gupta, "An Overview of Cold Load Pickup Issues in Power Distribution Systems," *Electric Power Components & Systems*. Vol. 34, No.6, pp. 639-651, June 2006.

[3] C. Ucak, A. Pahwa, "An analytical approach for step-by-step restoration of distribution systems following extended outages," *IEEE Transactions on Power Delivery*, vol. 9, no. 3, pp. 1717 – 1723, July 1994.

[4] O. Ramsaur, "A new approach to cold load restoration," *Electrical World*, vol. 138, pp. 101–103, October 6, 1952.

[5] R. S. Smithley, "Normal relay settings handle cold load," *Electrical World*, vol. 152, pp. 52–54, June 15, 1959.

[6] R. L. Wilde, "Effects of cold load pickup at the distribution substation transformer," *IEEE Transactions on Power Apparatus and Systems*, vol. PAS-104, no. 3, pp. 704–710, March 1985.

[7] J. Aubin, R. Bergeron, and R. Morin, "Distribution transformer overloading capability under cold-load pickup conditions," *IEEE Transactions on Power Delivery*, vol. 5, pp. 1883–1891, November 1990.

[8] J. Law, D. Minford, L. Elliott, and M. Storms, "Measured and predicted cold load pick up and feeder parameter determination using the harmonic model algorithm," *IEEE Transactions on Power Systems*, vol. 10, Issue 4, pp. 1756 – 1764, Nov. 1995.

[9] M. H. Nehrir, P. S. Dolan, V. Gerez, and W. J. Jameson, "Development and validation of a physically-based computer model for predicting winter electric heating loads," *IEEE Transactions on Power Systems*, Vol. 10, Issue: 1, pp. 266 – 272, Feb.1995.

[10] J. C. Laurent and R. P. Malhame, "A physically-based computer model of aggregate electric water heating loads", *IEEE Transactions on Power Systems*, Vol. 9, Issue: 3, pp. 1209 – 1217, Aug.1994.

[11] D. Athow and J. Law, "Development and applications of a random variable model for cold load pickup," *IEEE Transactions on Power Delivery*, Vol. 9, Issue: 3, pp. 1647 – 1653, July 1994.

[12] J. Aubin, R. Bergeron, and R. Morin, "Distribution transformer overloading capability under cold-load pickup conditions," *IEEE Transactions on Power Delivery*, Vol. 5, Issue: 4, pp. 1883 – 1891, Oct. 1990.

[13] R. C. Leou, Z. L. Gaing, C. N. Lu, B. S. Chang, and C. L. Cheng, "Distribution system feeder cold load pickup model," *Electric Power Systems Research*, Vol. 36, Issue 3, pp. 163-168, March 1996.

[14] W.W. Lang, M. D. Anderson and D.R. Fannin, "An Analytical method for quantifying the electrical space heating component of cold load pickup," *IEEE Transactions on Power Apparatus and Systems*, vol. PAS-101, no. 4, pp. 924-932, April 1982.

[15] C. Uçak and A. Pahwa, "Optimal step-by-step restoration of distribution systems during excessive loads due to cold load pickup", *Electric Power Systems Research*, Vol. 32, no. 2, pp. 121-128, February 1995.

[16] J.J. Wakileh and A. Pahwa, "Distribution system design optimization for cold load pickup," *IEEE Transactions on Power Systems*, vol. 11, no. 4, pp. 1879-1884, Nov. 1996.

[17] J.J. Wakileh and A. Pahwa, "Optimization of distribution system design to accommodate cold load pickup," *IEEE Transactions on Power Delivery*, vol.12, no.1, pp. 339-345, Jan. 1997.

[18] S. Chavali, A. Pahwa and S. Das, "A genetic algorithm approach for optimal distribution feeder restoration during cold load pickup," *Proceedings of CEC '02, Congress on Evolutionary Computation*, vol. 2, pp. 1816-1819, 12-17 May 2002.

[19] I. Mohanty, J. Kalita, S. Das, A. Pahwa, and E. Buehler, "Ant algorithms for the optimal restoration of distribution feeders during cold load pickup," *Proceedings of the Swarm Intelligence Symposium 2003*, SIS'03, pp. 132-137, 24-26 April 2003.

[20] Mesut E. Baran and Felix F. Wu. , "Network Reconfiguration in Distribution System for Loss Reduction and Load Balancing", *IEEE Transaction on Power Delivery*, Vol.4, No.2, April 1989.

[21] ANSI/IEEE C57.92-1981, Guide for loading mineral-oil-immersed power transformer up to and including 100 MVA with 55°C or 65°C winding rise, The Institute of Electrical and Electronics Engineers, Inc.,1981,Available:http://ieeexplore.ieee.org/xpl/standardstoc.jsp?isnumber=1103&isYear=1981

**Vishal Kumar** graduated in electrical engineering and post-graduated in engineering systems from DEI, Agra, India in 1995 and 1997, respectively. He is a faculty member in the Department of Electrical Engineering, Institute of Engineering & Technology, Lucknow, India. Presently, he is pursuing research work for his doctoral degree at Indian Institute of Technology Roorkee (IITR), Roorkee, India. His research interests include power distribution system operation and protection, and digital design and verification.

**Rohith Kumar.H.C** was born in Hassan, Karnataka, India on March 24, 1981. Currently, he is a post-graduate student in electrical engineering department at IITR. He graduated in electrical and electronics engineering from Visvesvaraya Technological University, Belgaum. His current research interests include power system operation and control, evolutionary computing, and artificial intelligence application to power system.

**Indra Gupta** received her B.Tech. degree in electrical engineering from HBTI, Kanpur, India, in 1984. She completed her ME and Ph.D degrees from University of Roorkee, Roorkee, in 1986 and 1996 respectively. At present, she is working as Assistant Professor in the Department of Electrical Engineering, IITR. Her research areas of interest include power distribution system automation, demand side management, and advance microprocessor architecture.

**H O Gupta** (Sr. M'2004) was born in Agra, India. He obtained his BE in Electrical Engineering from Government Engineering College, Jabalpur. He received ME in system Engineering and operation Research, and Ph.D from University of Roorkee, Roorkee, in 1975 and 1980 respectively. At present he is working as a Professor in the Department of Electrical Engineering, IITR. He visited McMaster University, Hamilton, Canada, from 1981 to 1983 as a post doctorate fellow. His research interests are in the area of computer-aided design, reliability engineering, power transformers, and power network optimization.

# Power Sector Reforms in India

Harbans L. Bajaj, *Fellow, IEEE,* and Deepak Sharma

*Abstract--* **Power sector reforms in India were initiated in the face of mounting commercial losses due to poor fiscal health of State Utilities, endemic capacity and energy shortages and increasing subsidy burden on the states. Investment in the sector was falling far short of demand in power supply. The Government of India, in 1991 embarked upon an ambitious program for reforming the sector with the prime objective of transforming the electricity industry into an efficient enterprise. This paper analyses the pre-reform era and identifies the key concerns which led to the initiation of the reform process. The paper delineates the major policy and regulatory initiatives that have been undertaken since 1991 including the provisions of the new enactment which has come into force in the form of the Electricity Act, 2003 and seek a paradigm shift. Several key elements of the reform program have been implemented by the various states, with varying degree of success, during the period 1991-2005. This paper reviews the performance of the Indian power sector in the last one and half decade since the reforms had been initiated, while undergoing restructuring. The study also examines how far the reform process during this period has been effective in realizing its set objectives and benefited the development of the power sector and the nation at large. The paper briefly discusses the issue involved in introduction of competition in the power sector primarily through development of a market for bulk power.**

*IndexTerms--***Distribution, Generation, Indian power sector, Open Access, Power markets, Power sector Reforms, Regulatory Commission, Transmission.**

## I. INTRODUCTION

UNDER India's federal constitution, the power sector is a concurrent responsibility of the central and state governments and development of this sector took place essentially through various public sector utilities – some under the central government and the majority under the state governments.

Until 1991, the sector in the state was managed by one large, vertically integrated entity that generated, transmitted and

---

Harbans L. Bajaj is pursuing doctoral programme through distance mode in Faculty of Engineering at University of Technology, Sydney, NSW 2007, Australia and is presently Member Technical, Appellate Tribunal for Electricity, Government of India, New Delhi, India (e-mail: harbans.bajaj@student.uts.edu.au).

Deepak Sharma is Associate Professor and Director, Energy Planning and Policy Program, Graduate School of Engineering, University of Technology, Sydney, City Campus, NSW 2007, Australia (e-mail: deepak.sharma@uts.edu.au).

distributed power, under the respective Ministries of Power. The rationale for state-owned integrated electric utilities reflected the internationally common view that electricity sector was a natural monopoly (with the exception of US and Japan) [1]. However, recent Reform programs in the UK, Chile, Brazil and Argentina, with vertically integrated large and inefficient public sector enterprises, have yielded significant gains in economies [2]. Gains from competition in developing countries shall largely be dependent on the market reforms and the modernization of the economic system in general.

As Hunt and Shuttleworth [3] state: "the big idea that underlies the new world of competition and choice in electricity is that it is possible and desirable to separate the transportation from the thing transported. That is, electric energy as a product can be separated commercially from transmission as a service."

The first step in any reform typically involves separation of the electricity sector from government control. Reforms package being adopted by the various countries essentially include the following elements:

- Unbundling
- Independent and Effective regulation
- Privatization
- Competition – both wholesale as well as retail

The Indian power sector has witnessed significant changes since early 1990s. Endemic power shortages, poor operational performance and precarious financial condition of State Electricity Boards (SEBs) prompted a number of policy and regulatory changes – focusing mainly on better management and control of the power industry and for attracting investment in the sector. Such policy changes of institutional and regulatory nature are expected to address technical and financial challenges facing the Indian power sector. This paper takes brief account of the state of Indian power sector and of concerns that have prompted various steps by the central and the state governments for the development of the sector. Major policy and regulatory initiatives undertaken in the power sector since 1991 are also dealt with. The paper also analyses some of the major provisions of the Electricity Act 2003 and its implications on future prospects in Indian power sector. It also puts these in perspective towards understanding the evolving market structure in the generation, T&D segments. Finally, the paper briefly discusses the issues involved in introducing competition into the power sector primarily through development of a market for bulk power.

## II. The Indian Electricity Sector

Electricity is lifeline for socio-economic development. Efficient provision of energy not only contributes to economic well being indirectly through economic development, but being, central to the basic human needs of health and education, electricity access has a direct bearing on the living standards as well as alleviating poverty. Though the Indian power sector has achieved substantial growth during the post-independence era, the sector has been ailing from serious functional problems during the past few decades. The power sector annually avails a substantial share of the outlay of the national economic plan (about 13–18%) [4], but most of the State Electricity Boards (SEBs) in India have been striving under resource crunch and operating at huge commercial losses. Consequently, the electricity services provided to the consumers by these SEBs—both in terms of quality and quantity—are 'poor'. The supply–demand gap of electricity in India has been consistently widening over years and most of the States in India are facing heavy electricity shortage. The energy deficit increased to 11.5% and peak deficit to 18% by 1990/91 [5]. The annual commercial losses also showed a spiraling trend increasing from over Rs. 15,000 million in 1985 to over Rs. 40,000 million in 1991 [6]. The gap between the average cost of supply and average tariff increased from 25 paise/kWh in 1985-86 to 110 paise/kWh in 2002-03 thereby entailing huge losses to SEBs. Per capita consumption of electricity in India increased from 178 kWh in 1985-86 to 338 kWh in 1996-97 [6] and to 592 kWh in 2003-04 [7], is less than 1/20 of that prevailing in the US and less than half that in China [8]. The generating capacity additions were totally inadequate to meet the burgeoning demand and consequently the deficits in electrical energy and peak power requirement became the order of the day. The financial as well operational performance of the sector further deteriorated with commercial losses (without subsidy) increasing to over Rs.292,520 million in the year 2001-02 [9] T&D losses reaching up to 33.98 % and energy and peak shortages continued being 7.1% and 11.8% respectively. The un-remunerative tariff and poor techno-commercial management were identified to be the major reasons for this loss [10]. Functional inefficiency of Government owned SEBs, non-rationality in all facets of electricity generation and use (including tariff setting, metering, billing and revenue collection), inability of the utilities to mobilize adequate resources for capacity expansion and modernization, etc.; have been cited as the major reasons for the crisis [10-12]. Huge percentage of unused supply due to poor operating load factor of thermal stations (about 50%), deficient energy accounting, high level of theft and pilferage of electricity, subsidy to agriculture sector, cross subsidy to domestic sector, etc., have been cited as the key factors of the inefficient functioning of the SEBs in India during this period [13-14]. Thus there was a general approved consensus that maintaining this status quo will be detrimental to the nation and may harm the sector itself. It was in this situation that in 1991 Government of India decided to restructure the power sector radically through a set of comprehensive reforms.

## III. Power Sector Reforms

The efforts to restructure the power sector in India formally commenced in 1991. The prime reasons which prompted the Government to initiate such a reform process were: (i) the ever-widening gap between the demand and availability of electricity, (ii) the poor technical and financial performance of the State Electricity Boards and (iii) inability of the Central and State Governments to finance and mobilize resources for generation capacity expansion projects, making third party investment in power sector imperative. The initial step in this direction has been the amendment of legislation governing the electricity sector in 1991. The Indian Electricity Act, 1910 and the Electricity (Supply) Act, 1948 were amended to attract private investment in power generation. This facilitated the tapping of domestic and foreign capital markets, provided assured returns on investment and reduced legal hassles to allow the private investors to set-up generation capacities or operate as licensee in distribution segments, which were hitherto a monopoly of the SEBs. Private power initiative in generation banked on long-term power purchase agreements.

In 1995, these measures were further strengthened by a Mega Power Policy, whereby plants above 1000MW capacity would receive additional incentives in the form of a 10-year tax holiday, exemption of customs duty for imports, reduced hassles for clearances, etc. This also provided for the setting up of Power Trading Corporation (PTC) to act as an intermediary between the private developers of mega projects and the SEBs. Though independent power producers (IPPs) evinced interest for adding generation capacity for about 95,000MW, only 6500MWwas added during the eighth and ninth five-year plans (1992–2002) [15]. Further, out of a targeted capacity addition of 17,588MW from the private sector during the ninth Five-year plan (1997–2002), a mere 5061MW only materialized [16].

The National Development Council set up in 1993 was the first official body to steer the reform process. This was followed by various national level conferences of State Chief Ministers during the years 1996, 1998, 2000 and 2001.

Having experienced success in restructuring the electricity industry in the Latin American countries, World Bank put forth power sector reforms as a necessary condition for future assistance to power sector in the recipient countries [17]. Therefore, at the urging of the World Bank, Orissa was the first state to enact, in 1995, comprehensive power sector reform act involving (1) an independent regulatory commission, (2) unbundling of the State Electricity Board (SEB) into separate generation, transmission and distribution entities, and (3) eventual privatization, particularly of distribution. Andhra Pradesh, Haryana, Rajasthan, Uttar Pradesh and Karnataka have since followed this pattern of power sector reform.

The Ministry of Power organized a discussion between the center and the states in October and December 1996, from which emerged the ''Common Minimum National Action Plan for Power'' (CMNAP). The CMNAP recommended:

- That the SEBs be corporatized, initially within the existing framework of public ownership followed by gradual privatization;
- That the SEBs focuses on improving efficiency in both generation and distribution via reorganization, efficient metering and energy audits;
- The creation of independent state electricity regulatory commissions (SERCs), answerable only to the state High Court;
- That tariffs be set—"with immediate effect"—to earn a return on capital employed of at least 3%;
- Cross-subsidization be continued provided that no user pays less than 50% of its average costs. A 3-year phase-in was allowed for farmers only, who would immediately pay at least Rs 0.50/kWh;
- Simplification of procedures, including that adjustments for changes in fuel charges be "automatically incorporated" in the tariff structure as a pass-through cost. This concept was incorporated in the June 1997 guidelines for private sector participation in generation.

The CMNAP formed the basis for the June, 1997 guidelines on generation in power sector. However, the 1996/97 reforms were not comprehensive as it had altogether neglected the reforms in the distribution sector which were essential to improve the fiscal health of the SEBs

In the developed world context, where there is generally surplus power availability the emphasis of Reforms is towards competition in view of the high cost of electricity [18-20]. However, in the Indian context where power deficits is the norm, Reforms inter alia means that there would be increased availability of power, better quality of power, enhanced investment in power sector, healthy competition amongst the constituents, improved efficiency in generation, transmission and distribution, cost recovery and better training and redeployment of human resources. Comprehensive Reforms of legislations including Regulatory Commission Act (1998) and Electricity Bill (2001) which culminated it the Electricity Act, 2003 also followed. Central Electricity Regulatory Commission (CERC, formed on 26 April 1999) as well as State Electricity Regulatory Commissions (SERCs) subsequently set up in 25 States is already functioning. Most of the States have initiated reform process and some have made substantial progress in restructuring of the power sector.

The main functions of CERC include regulating tariffs of generating companies, owned or controlled by the Government of India and any other generating company catering to more than one state, and also tariffs for the inter-state transmission of electricity. Apart from this, significant steps taken by CERC include introduction of Availability Based Tariff (ABT), Indian Electricity Grid Code, and Guidelines for transmission licensing, open access Regulations, Trading Regulations and fixing of trading margins. ABT has been instrumental in bringing discipline to the grid by providing frequency linked incentives and disincentives. In the ABT, a two-part tariff is supplemented with a charge for Unscheduled Interchange (UI) for the supply and consumption of energy in variation from the pre-committed daily schedule and depending on grid frequency at that point of time. The regulatory changes have brought transparency to the tariff-making process. They have

also led to the rationalization of distribution tariffs, thereby arresting increases of cross-subsidy in the system. Public hearings have been able to give voice to consumers in raising their concerns and contribute constructively to the regulatory process. In order to address the consumer complaints, SERCs have come up with a complaint-handling system.

Due to poor capacity addition by the IPPs, the need for distribution reforms was recognized. It was towards this effort that in the Meeting of the Chief Ministers on Power Sector reforms was held in March 2001where some level of political convergence concerning the reforms emerged. The most important step to improve the bottomline of the sector is effective and creative management to reduce technical and commercial losses and increase revenues. The resulting revenues along with performance-tied grants from government and multilateral and bilateral agencies can be used to improve technical performance involving reduction of T&D losses and improvement of power quality (frequency, voltage, continuity). Towards this effort, the Accelerated Power Development & Reform Program (APDRP) was launched in February 2001 by the Union government to promote distribution reforms and provide transitional finance for the SEBs undertaking reforms. The main objectives pertaining to distribution reform are aimed at achieving 100% metering, conducting energy audits, improving HT/LT ratio, replacement of distribution transformers and the use of IT solutions to ensure accountability. The APDRP aims at reduction of AT&C losses, bring about Commercial viability, reduce outages & interruptions and increase consumer satisfaction

This program has two components namely the *Investment Component* which covers strengthening and up gradation of sub-transmission & distribution and the *Incentive Component* which is a grant for states/ Utilities towards reduction of cash losses with 2000-01 as the base year. Eight (8) states namely Andhra Pradesh, Gujarat, Haryana, Kerala, Maharashtra, Punjab, Rajasthan and West Bengal have been the recipient of such incentive totaling Rs. 17233 million.

Most of the SEBs was on the verge of financial collapse. Large amount of SEBs debts to Central Public Sector Utilities (CPSUs) and the railways cast a shadow on their balance sheet. One-time settlement of SEB debts was initiated and came into effect from 17 April 2002 as a tripartite agreement between the respective state government, the Ministry of Finance and the Reserve Bank of India (RBI) [21]. As per this scheme, 60% of the surcharge and interest on delayed payment as of 30 September 2001 is waived. The scheme securities the remaining surcharge and interest and the full principal amount through tax free bonds to be issued through RBI by the respective state governments. It also provides for recovery of future defaults exceeding 90 days from the funds due to the state.

Recognizing the need for the Reform process, a comprehensive Electricity Bill was drafted in 2000 following a wide consultative process. After a number of amendments, the bill finally sailed through the legislative process and was enacted on 10 June 2003. It replaces the three existing legislations governing the power sector, namely Indian Electricity Act, 1910, the Electricity (Supply) Act, 1948 and the Electricity Regulatory Commissions Act, 1998. The

Electricity Act, 2003 mandates that Regulatory Commissions shall regulate tariff and issue of licenses and that State Electricity Boards (SEBs) will no longer exist in the existing form and will be restructured into separate generation, transmission and distribution entities. Regulatory function has been taken away from the purview of the government. The Electricity Act, 2003 mandated licensee-free thermal generation, non-discriminatory open access of the transmission system and gradual implementation of open access in the distribution system will pave way for creation of power market in India. The main additional provisions of the act are:

- De-licensing of thermal generation and captive generation.
- Provision for license-free generation and distribution in rural areas and provision for management of rural distribution by Panchayats, Cooperative Societies, non-government organizations, franchisees, etc.
- Non-discriminatory open access in transmission.
- Multiple licensing in distribution.
- Mandatory metering of all electricity supplies.
- Adoption of multi-year tariff principles.
- Open access in distribution to be introduced in phases.
- Provision for cross-subsidy surcharge on direct sale to consumers until cross subsidies are gradually phased out.
- Power Trading recognized as a distinct activity with ceilings on trading margins to be fixed by the Regulatory Commissions.
- Provision for payment of subsidies through State budget.
- Setting up of an Appellate Tribunal to hear appeals against the decisions of the CERC and the SERCs.

The Act is aimed at providing an investor friendly environment for potential developers in the power sector by removing administrative hurdles in the development of power projects and shall provide impetus to distribution reform to be undertaken in India. Provisions like de-licensing of thermal generation, open access and multiple licensing; no surcharge for captive generation shall be the basis for a competitive environment in the Indian power sector. Provisions of open access would be instrumental in the development of competitive power markets, and multi year tariffs shall bring in necessary incentives for performance improvement and to reduce regulatory risk.

## IV. EVALUATION OF THE REFORM PROCESS

Power sector reform is a long process and its impacts would be known after a long time. Though it is difficult to predict the outcome of the reform process, a mid course review of developments could, however, help learn from the past mistakes and take some mid-course corrective measures. Indeed, advocates for the power sector reforms argue that the proposed changes will bring better quality of power at lower rates, with positive ripple effects through economies and societies. In India the initial impetus of the reform was on the generation side rather than the distribution side where the actual problem lay. Distribution reform was given thrust by incentive based schemes like the Accelerated Power Development & Reform Program (APDRP). Thirteen (13) SEBs and Electricity Departments have unbundled and

corporative and another nine (9) are expected shortly, political compulsions to go ahead with privatization has forced many state governments to repeatedly postpone the same in their respective states.

Bacon and Besant–Jones advocate relying on privatization of the distribution and supply functions first [22]. This facilitates the entry of potential investors in the generation by improving the creditworthiness of buyers of power from the generators. The same model had been adopted for Orissa reforms, however, it has not yielded the desired results and it is ranked 21st amongst 29 states by ICRA/CRISIL, 2006 [23]. In the case of Orissa, post privatization, the operations of distribution companies are not viable at the tariffs fixed by the regulator, which were fixed on the basis of a normative T & D loss of 35% against the actual losses that were estimated to be around 45–47% [2]. Revenues from sale of the government stake in Orissa were not ploughed back into the sector which limited the government's ability to influence the future developments in the sector. However, Orissa provided a powerful demonstration effect for other states to follow and learn from its experiences.

The World Bank's staff appraisal report on Orissa, before commencing of the reforms, has targeted a reduction of T&D losses to the level of 25% by the year 2001 but this has not been realized at all [24]. The picture of Haryana, UP, AP and Karnataka also are not different. In the case of these States, the losses showed an increasing trend and are currently above 30%. In AP and Karnataka the percentage of T&D losses have shown about two-fold increase while undergoing restructuring. Regarding the growth in per capita consumption of electricity, the rates in these States except that of AP have not indicated any substantial improvement above the national average growth rate of 3.5%.

In the 'Delhi model' some of these lacunae were taken care of to some extent. Involvement of Regulators at all stages of privatization and long-term tariff profile led to the reduction of regulatory uncertainty. However, it continues with the single-model buyer approach as in the case of Orissa and other restructured state utilities. In the Delhi model, phased reduction in T&D losses, revenue collections and transitional financial assistance (subsidy) are the three critical areas. The Orissa model enabled others states to learn from their experiences.

## V. POWER MARKETS

A market must have the following elements to be effective and competitive [3]

- Many buyers and many sellers
- Buyers and sellers should be responsive to price
- Liquid and efficient market places
- Equal non-discriminatory access to essential facilities
- Treatment of subsidies and environmental controls so that they do not interfere with the working of market

The first serious attempt to form a liberalized electricity market was launched in Chile in 1982. Such markets became operational in England and Wales in 1990, Nordic countries in 1991, Australia in 1994, New Zealand in 1996, US in late

1990s, Spain and Netherlands in 1998 and in Texas and Canada in 2001 [25].

With a view to develop power markets in India, PTC was set up under the joint ownership of the central power producers, Powergrid and various financial institutions. New players in the trading besides PTC, like NTPC Vidyut Vyapar Nigam Limited (NVVN), Adani, Reliance have also come on to the scene after the enactment of the Electricity Act, 2003 wherein trading has been recognized a distinct activity. This has facilitated power exchange between the power producers and the state utilities. Such trade has grown from about 44 million Kwh in 2000-01 to over 14 billion Kwh in 2005-06 [26]. Due to various enabling provisions under the Act, there is for development considerable scope for bulk power markets in India. CERC have recently issued a public notice inviting comments/suggestions from the various stakeholders on paper entitled "Developing a Common Platform for Electricity Trading" in order to create a road map for developing markets for electricity in India. Power generators, traders, captive generators and emerging co-operative producers would have an opportunity to participate in such market mechanism.

## VI. CONCLUSIONS

Ongoing reforms in the power sector are expected to ensure that the burgeoning demand of electricity matching with the targeted 9%-10% growth of the annual gross domestic product (GDP) is met and that complete rural electrification shall be achieved by the end of Tenth Five Year Plan. While the main objective of power sector reforms in the developed world is to enhance competition in the sector, the need to improve the fiscal health of the State Electricity Boards (SEBs)/Utilities as well as attract private investment in the ailing power sector have been the main driving force to undertake reforms in India. Regulatory changes have been able to bring about tariff rationalization and transparency and consumer protection in the regulatory process. Benefits of the reform Programme has started bearing fruits some of which are evident in the form of reduction in the AT&C losses as well as the reduction in the commercial losses of the SEBs/utilities.

An emphasis on demand management (peak reduction, load-curve smoothing, end-use efficiency improvement) as well as energy audit is also required. Enactment of the Electricity Act, 2003 holds many promises for the sector. Licensee-free thermal generation, non-discriminatory open access of the transmission System and gradual implementation of open access in the distribution system, phased reduction of cross subsidy will pave way for creation of power market in India. Upgradation and improvement of the distribution segments is the key to long-term sustained growth of the power sector. Efforts must be channelised to address the issues that pertain to the distribution segment, especially for bulk power. Further improvement of the system could be achieved by introduction of competition. These longer-term challenges may require additional elements of reform/restructuring similar to those proved in the industrialized countries. Phasing out cross-subsidy in a gradual manner, prices reflective of costs, improving the technical and commercial efficiency which requires political will and support would go a long way in building a strong and vibrant power sector having positive net worth.

## VII. REFERENCES

[1] Dossani, "Reorganization of Power Distribution sector in India," Energy Policy, vol. 32, pp.1277-1289, 2004.

[2] Anoop Singh, "Power Sector Reforms in India: current issues and prospects" Energy policy, vol.34, no.16, pp.2480-2490, Nov 2006.

[3] S Hunt and G Shuttleworth, "Competition and Choice in Electricity," New York, Wiley, 1996, pp.1.

[4] Working of State Electricity Boards and Electricity Departments Annual Report Planning Commission, (Power & Energy Division), Government of India, pp.2, May 2002.

[5] Working of State Electricity Boards and Electricity Departments, Annual Report Planning Commission, (Power & Energy Division), Government of India, pp.77, April 1999.

[6] Working of State Electricity Boards and Electricity Departments, Annual Report Planning Commission, (Power & Energy Division), Government of India, pp.5-6, April 1999.

[7] All India Electricity Statistics: General Review, Central Electricity Authority, Ministry of Power, Government of India, New Delhi, pp.16, January, 2005.

[8] Based on data obtained from International Energy Agency website. Available: http://www.iea.org.

[9] Report on the performance of the state power utilities for the years 2002-03 to 2004-05, in Power Finance Corporation Ltd., Delhi, India, 2006.

[10] Approach Paper to the tenth five year plan (2002-07), Planning Commission, Government of India, September 2001.

[11] P. Baijal, "Restructuring power sector in India" Economic and Political Weekly, pp. 2795-2804, Sept 1999.

[12] K.S Parikh, R Radhakrishna, "India Development Report," New Delhi, Oxford University Press, 2002.

[13] K.P. Kannan and N.V Pillai, "Plight of the power sector in India: SEBs and their saga of inefficiency," Paper no. 308, Center of Development Studies, Trivandrum, India, November 2000.

[14] India: Environmental issues in the power sector national synthesis (Draft report) of World Bank, April1998.

[15] India Energy Market Reforms: WEC Report, World Energy Council, 2002.

[16] Tenth Five Year Plan, Planning Commission, Government of India, New Delhi, 2002.

[17] A.T Rajan, "Power sector reform in Orissa : an ex-post analysis of the causal factors" Energy policy, vol. 28, no. 10, pp. 657-669, 2000.

[18] R.J. Gilbert, E.P. Kahn, "Competition in institutional change in US electric power regulation". In Gilbert, R. J. Kahn, E.P. (Eds.), International Comparisons of Electricity Regulations, Cambridge University Press, Cambridge, 1996.

[19] D.M. Newbery, and R Green., "Regulation, public ownership and privatization of the English electricity industry" In Gilbert, R. J. Kahn, E.P. (Eds.), International Comparisons of Electricity Regulations, Cambridge University Press, Cambridge,1996.

[20] P.L. Joskow "Restructuring, competition and regulatory reform in the US electricity sector" Journal of Economic Perspectives vol. 11, no.3, pp. 119-138, 1997.

[21] S.M, .Ahluwalia, , "Power sector reforms: a review of the process and an evaluation of the outcome" National Council of Applied Economic Research, Delhi, March 2000.

[22] R.W Bacon. and J. Besant-Jones, "Global Electric Power Reform, Privatization and Liberalization of the Electric Power Industry in Developing Countries, Energy and Mining Sector Board". The World Bank, Washington, D.C. 2001.

[23] Report on Rating of State Power Sectors, ICRA/CRISIL, submitted to Ministry of Power, Government of India, New Delhi, 2006.

[24] Orissa power sector restructuring project, no. 14298-IN, Energy and infrastructure operations Division, World Bank, 1996.

[25] Developing a common platform for Electricity Trading, Central Electricity Regulatory Commission, Delhi, India, pp. 30, July 2006.

[26] Developing a common platform for Electricity Trading, Central Electricity Regulatory Commission, Delhi, India, pp.10, July 2006.

**2006 IEEE International Conference on Power Electronic, Drives and Energy Systems**

# A New Structure for Electricity Market Scheduling

S. Soleymani, A. M. Ranjbar, and A. R. Shirani

*Abstract--* **On pool market structure, the generating units are typically dispatched in order of lowest to highest bid as needed to meet demand requirement as well as considering network constraints. Generally, no account is taken of generating unit's reliability when scheduling them. This paper proposes a new structure for electricity market to consider generating unit's reliability in the scheduling problem. GAMS (General Algebraic Modeling System) language has been used to solve the social welfare maximization problem using CPLEX optimization software with mixed integer programming.**

*Index Terms--* **Electricity Market, Generating Units, Power Generation Scheduling, Reliability.**

## I. INTRODUCTION

RECENT changes in the electricity industry in several countries have led to a less regulated and more competitive energy market. At this condition, generating companies (Gencos) have to sell their outputs by presenting bids on the market. So instead of cost reduction, profit maximization will be the dominant bidding strategy of Gencos and hence they should consider other factors such as market structure and their interrelationship, demand forecast level and competitor behavior

In power systems world-wide, generating units are traditionally scheduled at least cost subject to operational constraints and security criteria such as provision of adequate spinning reserve levels. In a competitive electricity market, the security of the system is influenced by the level of Ancillary Services (AS), that are procured.

Generally, no account is taken of generator reliability when scheduling units. Recent studies show that, there are good reasons for considering reliability in the scheduling process. When a generating unit is forced out, the system dynamics are aggravated which may cause load shedding or put the system in a state where load shedding is more likely to occur. Also, the replacement energy and reserve that must be procured are costly. If the unit's forced outage probabilities (FOP) are

---

S. Soleymani, A. M. Ranjbar, are with the Electrical Engineering Department, Sharif University of Technology, Tehran, Iran (e-mail: ssoleymani@mehr.sharif.edu, ssoleymani@nri.ac.ir aranjbar@sharif.edu, )

A. R. Shirani is with Niroo Research Institute, Tehran, Iran (e-mail: ashirani@nri.ac.ir)

considered in the scheduling process, then the expected impact of contingencies can be lowered thus enhancing the security of the system.

A generating unit's reliability is improved through regular maintenance and good operating practice. In competitive electricity markets, generation owners are responsible for maintaining their units. This can mean that units are not maintained often enough because of cost and lost revenue. A system whereby reliable units are rewarded can be used to compensate generators for maintaining units. The use of fines when units are forced out can provide a further incentive to maintain units.

This paper proposes a new market structure, which not only considers the price and quantity bidding of Gencos, but also regards the Gencos' reliability in selecting Gencos for power generation scheduling. In the proposed market structure the bidding strategies of Gencos are assumed to be known.
The paper is organized as follows: The generating unit's bid is described in section II. The market clearing model and the proposed market structure are shown in sections III and IV. Vection V gives illustrative example with three units; section VI provides the conclusion.

## II. GENERATING UNIT'S BID

In a power market, generators or units may prepare their strategic bids according to the four known models in imperfect competition. These models are Bertrand, Cournot, Stackelberg and SFE where Stackelberg model is similar to the Cournot model [3].

In the Bertrand model, generating units compete each other using prices as strategy choices and they bid at their marginal cost at Nash equilibrium point.

In the classic model of Cournot, units compete against each other using quantities as strategy choices. unit's products are assumed to be homogenous, Demand is price-responsive, and market clearing price is the intersection of aggregated supply and market demand curves. Stackelberg model is similar to the Cournot model. However, the competitors don't offer their output quantities simultaneously. The so-called " leader" will make the first move, which is followed by that of followers who take into account the leader's action [5].

In the SFE model, Gencos compete with each other through the simultaneous choice of supply functions. Klemperer and Meyer developed SFE in order to model competition in the presence of demand uncertainty. The SFE model was used by Green and Newbery for analyzing the

---

0-7803-9771-1/06/$25.00 ©2006 IEEE

competitive strategic bidding in electricity markets [3].

Fig.1 illustrates where the intensity of competition predicted by the basic formulation of each of the models places them along the competitive spectrum.

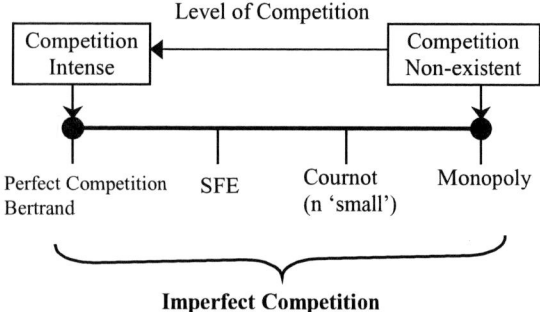

Fig. 1. Equilibrium Models and Predicted Degree of Competition [3]

Among these models, it is only the SFE model which enables a unit to link its bidding price with the bidding quantity of its product and only this model is the closest to the actual behavior of players in the actual power market.

This paper supposes that, units are requested to submit a piece-wise quantity-price curve like the one shown in Fig.2 to the ISO.

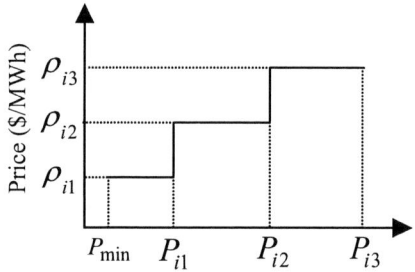

Fig. 2. Bid curve for unit i

## III. MARKET CLEARING MODEL

In the electricity markets, Gencos and Distributed Companies (Discos) will submit bid curves to the ISO, then ISO clears market after collecting bids. In this model, ISO maximizes social welfare subject to the bids. The social welfare is the sum of consumer and producer surplus. Consumer surplus is the utility or benefit the consumer gains from a given equilibrium point. Similarly, producer surplus is the benefit or profit the producer gains from a given equilibrium point. An example of supply, demand, equilibrium point and producer and consumer surplus is shown in Fig.3.

The Fig.3 represents an elastic demand case, but in some cases demand is not elastic. In this condition the energy market clearing model in the ISO's point of view for a time period of $T$ is expressed as follows:

$$\min \quad \sum_{t=1}^{T}\sum_{i=1}^{N} \rho_i^t . P_i^t . STD_i^t \tag{1}$$

subject to:

*Capacity Constraint:*
$$P_i^{\min} \leq P_i^t \leq P_i^{\max} \tag{2}$$

*Up and Down ramp rate constraints:*
$$P_i^{t+1} - P_i^t \leq Iramp$$
$$P_i^t - P_i^{t+1} \leq Dramp \tag{3}$$

*Minimum up and down time constraints:*
$$SU_i^t . \sum_{j=0}^{T_i^{up}} STD_i^{t+j} = SU_i^t . T_i^{up} \tag{4}$$

$$Sh_i^t . \sum_{j=0}^{T_i^{down}} STD_i^{t+j} = 0$$

*The state transition constraint:*
$$STD_i^{t-1} - STD_i^t + SU_i^t - Sh_i^t = 0 \tag{5}$$

*DC power flow equation:*
$$B.\theta^t = P_G^t - P_D^t \tag{6}$$

*Transmission line constraint:*
$$F_{\min l} \leq F_l^t \leq F_{\max l} (l=1,2,...,L) \tag{7}$$
where:

$T$: the analyzing period of market

$N$: the number of generating units

$\rho_i^t$ : the bidding price of unit $i$ at time $t$

$P_i^t$ : the output power of unit $i$ at time $t$

$STD_i^t$ : the status of unit $i$ at time $t$ and equals 1 if the unit is on and 0 if the unit is off

$P_i^{\min}$, $P_i^{\max}$ : the minimum and maximum generation levels of unit $i$

$Iramp$, $Dramp$: the increasing and decreasing ramp rate of unit $i$

$T_i^{up}$, $T_i^{Down}$ : the minimum Up and Down time of unit $i$

$SU_i^t$, $Sh_i^t$ : the start up and shut down variables of unit $i$ in the ISO's point of view at time $t$

$B$ : the network susceptance matrix

$\theta^t$ : vector of bus voltage angles at time $t$

$P_G^t$ : vector of bus generation at time t

$P_D^t$ : constant vector of bus loads at time t

$F_l^t$ : power flow on line $l$ at time $t$

$F_l^{\max}$, $F_l^{\min}$ : upper and lower flow limits on line $l$

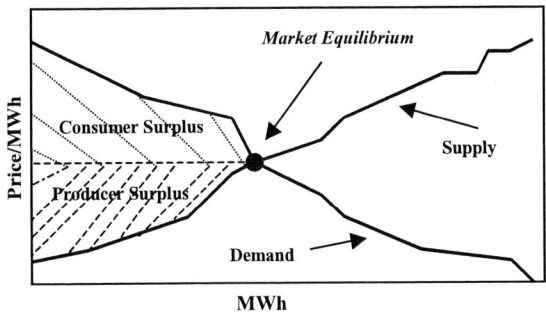

Fig. 3. Supply, demand, price equilibrium and producer and consumer surplus

Except the MW dispatched of units, Market Clearing Price (MCP) is the other result of this optimization problem. Once the energy market is cleared, each generating unit will be paid according to uniform or pay-as-bid pricing. Under the uniform pricing structure, the bid price of the last unit dispatched sets the market clearing price, then all units dispatched receive the same MCP. Under the pay-as-bid pricing structure, every winning generating unit gets its bid price as its income.

When market structure is based on uniform clearing price, the hourly MCP is calculated as follows:

$$MCP(t) = \max(i, \rho_i^t . STD_i^t) \qquad (8)$$

As shown in equations 1 to 7, only the bidding price and quantity of generating units affect on the selecting them for power generation.

## IV. THE PROPOSED MARKET STRUCTURE

In the proposed method, the reliability of the generating units is considered in generation scheduling of competitive electricity markets.

The generating units have an estimate of their Forced Outage Probability (FOP), so if we consider the FOP of each unit, a measure of the risk of capacity loss over all time periods is:

$$\sum_{t=1}^{T}\sum_{i=1}^{N} P_i^t . STD_i^t . FOP_i^t \qquad (9)$$

Thus, reliability may be considered in the scheduling problem by adding the following term to (1) to form the objective function.

$$\sum_{t=1}^{T}\sum_{i=1}^{N} VOLU_t . P_i^t . FOP_i^t \qquad (10)$$

where $VOLU_t$ is a scaling factor called the Value of Lost Units at time $t$. This scaling factor represents the cost to the system incurred by a unit being forced out and it may be different at different times of the day. VOLU will account for the cost of activating the reserve and procuring replacement reserve. Equation (10) will associate different costs with units based on their loading levels and FOP. Hence minimizing (1)+(10) will reduce the loading levels and the number of start-ups of unreliable units.

Given that unreliable units are costly to the system, the long-term expected production cost will be lowered and units with low FOPs will be rewarded by being scheduled for more MWhrs than those with high FOPs.

In order to ensure that generators provide accurate information, it is proposed that a generator will pay a penalty if and where their unit is forced out, regardless of whether or not a curtailment occurs. For unit i at hour t this penalty could take the form:

$$Penalty\ Weight \times P_i^t \times STD_i^t \times (1 - FOP_i^t)^n \qquad (11)$$

If a generator overestimates its FOP, then it will have a lower income but its penalty, if it is forced out, will be smaller. If the generator underestimates its FOP, it will be brought on-line more often but will pay larger penalties for contingencies. The historical data on operation and maintenance information can be used by generators to estimate their FOP.

## V. SIMULATION RESULTS

The 6-bus system is used in Fig.4 to illustrate the results of application of the proposed method for a daily energy market. There are three units in the system and unit1 is considered as a slack bus. The information about the hourly load service entities (LSEs) and the network are shown in Fig.5 and Table I, respectively.

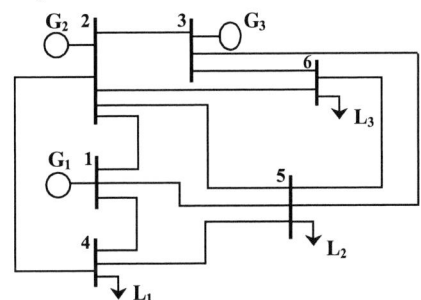

Fig. 4. Six- bus system

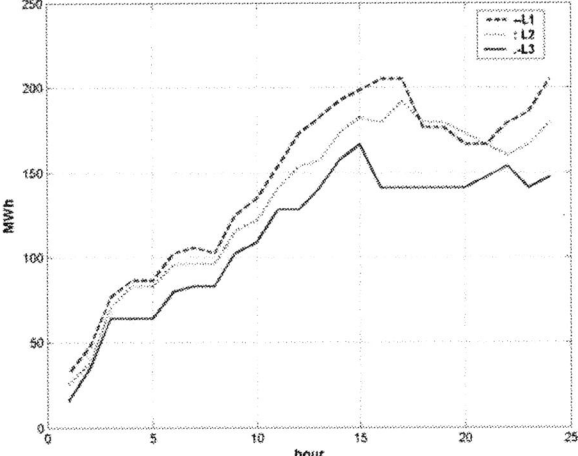

Fig. 5. Hourly LSE's Information

## TABLE I
### TRANSMISSION NETWORK DATA

| Line | From Bus | To Bus | R (pu) | X (Pu) | Limit (MW) |
|------|----------|--------|--------|--------|------------|
| 1 | 1 | 2 | 0.1 | 0.2 | 17.78 |
| 2 | 1 | 4 | 0.05 | 0.2 | 42.52 |
| 3 | 1 | 5 | 0.08 | 0.3 | 45.57 |
| 4 | 2 | 3 | 0.05 | 0.25 | 19.13 |
| 5 | 2 | 4 | 0.05 | 0.1 | 52.62 |
| 6 | 2 | 5 | 0.1 | 0.3 | 35.46 |
| 7 | 2 | 6 | 0.07 | 0.2 | 48.2 |
| 8 | 3 | 5 | 0.12 | 0.26 | 32.12 |
| 9 | 3 | 6 | 0.02 | 0.1 | 66.63 |
| 10 | 4 | 5 | 0.2 | 0.4 | 10.81 |
| 11 | 5 | 6 | 0.1 | 0.3 | 8.42 |

The a, b, and c coefficients of generating cost function (which is expressed as: $C(P) = aP^2 + bP + c$) and other required data about generating units are tabulated in Tables II and III.

## TABLE II
### COST COEFFICIENTS OF GENERATING UNITS

| Unit | a $\$/(MWh)^2$ | b $\$/MWh$ | c $\$$ |
|------|----------------|------------|--------|
| 1 | 0.00045 | 15.5 | 1078.8 |
| 2 | 0.00031 | 16 | 969.8 |
| 3 | 0.00041 | 13 | 600.9 |

## TABLE III
### GNERAL INFORMATION OF GENERATING UNITS

| Pmin MW | Pmax MW | FOP |
|---------|---------|-----|
| 0 | 600 | 0.05 |
| 30 | 455 | 0.04 |
| 30 | 230 | 0.03 |

The model described in this paper has been written in GAMS [6] language. GAMS is a high-level modeling system for mathematical programming and optimization. It consists of a language compiler and a stable of integrated high-performance solvers. In this paper, GAMS has been used to solve the ISO's objective function module or market scheduling problem using the CPLEX optimization software with Mixed Integer Programming (MIP).

*Case 1--* In this case, it is supposed that all units bid at their marginal cost. The power system scheduling problem is performed in two cases: without considering the reliability of units and with regard to their reliability. The profile of hourly MW dispatched of units and the expected payoff and market share for three units in two cases are shown in Fig.7, Fig.8, and Tables IV, and V.

The results show that the risk of capacity loss (obtained from Eq.2 ) in the first case will be more than in second case (risk value in two cases are 395.18, and 248.1). It may be seemed that the ISO's objective function (the value of eq.1 and 3) in the second case should be more greater than in the first case but the results show that it is only slightly greater

than in the first case (the ISO's objective function in the first and second cases are 1720.6 and 2170 respectively).

## TABLE IV
### PAYOFF AND MARKET SHARE OF UNITS IN CASE I
### WHITHOUT CONSIDERING THE RELIABILITY OF UNITS

| | Unit1 | Unit2 | Unit3 |
|---|-------|-------|-------|
| The expected pay off (\$) | -25353 | 0 | -735.3 |
| The market share (MW) | 4735.6 | 0 | 5280 |

## TABLE V
### PAYOFF AND MARKET SHARE OF UNITS IN CASE I
### WITH CONSIDERING THE RELIABILITY OF UNITS

| | Unit1 | Unit2 | Unit3 |
|---|-------|-------|-------|
| The expected pay off (\$) | -1078 | -21935 | 1564.6 |
| The market share (MW) | 21 | 4714.6 | 5280 |

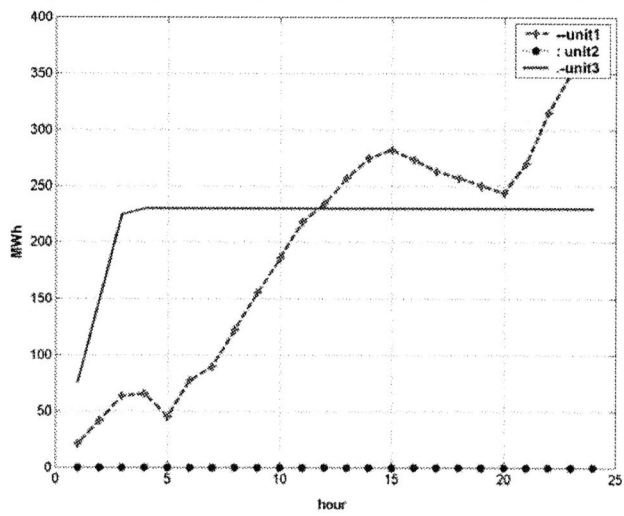

Fig. 6. The hourly MW dispatched of units when they bid at their marginal costs and without considering their FOPs

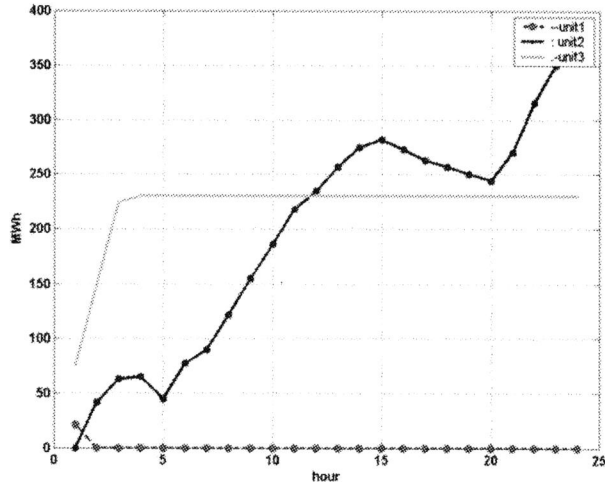

Fig. 7. The hourly MW dispatched of units when they bid at their marginal costs and with considering their FOPs

Case 2—In this case, it is assumed that generating units bid, strategically, where they devise its own bid segments according to the linear supply function equilibrium, which is expressed as follows:

$$\rho_i^t = k_i.MC_i^t = k_i.(2a_iP_i^t + b_i) \qquad (12)$$

where:

$k_i$ : bidding strategy of unit i

$MC_i^t$ : the marginal cost of unit i at time t with second-order generating (operating) cost function as :

$$C_i^t = C(P_i^t) = a_i(P_i^t)^2 + b_iP_i^t + c_i$$

$a_i$, $b_i$, and $c_i$ : generation cost coefficients.

In this case, the $k_i$ s of generating units are known as shown in Table VI.

TABLE VI
BIDDING STRATEGY OF GENERATING UNITS

|        | Unit1 | Unit2 | Unit3 |
|--------|-------|-------|-------|
| $k_i$  | 1.8   | 2.5   | 3.1   |

The expected payoff and market share for three units in this case when reliability is considered in market structure is shown in Table VII.

TABLE VII
PAYOFF AND MARKET SHARE OF UNITS IN CASE I
WITH CONSIDERING THE RELIABILITY OF UNITS

|                          | Unit1 | Unit2 | Unit3 |
|--------------------------|-------|-------|-------|
| The expected pay off ($) | 125   | 1008  | 3055  |
| The market share (MW)    | 825   | 3075  | 4100  |

In the first view of this case it is expected that the unit with lowest price bid should has the largest market share. But the simulation results show that considering reliability in the scheduling process of units has a major impact on selecting the generators.

## VI. CONCLUSIONS

Recent changes in the electricity industry have led to a less regulated and more competitive energy market. In these conditions the generating units are typically dispatched in order of lowest to highest bid as needed to meet demand and no account is taken of generating unit's reliability.

In this paper a method for considering generating unit reliability in the scheduling process, which rewards units with lower forced outage probabilities has been proposed. The results shows that, it is not only useful for ISO to decrease its loss of load, but also it may be has profit for it. The proposed structure for power system scheduling, encourages units to improve their reliability. The 6-bus system is employed to illustrate the proposed method.

## VII. REFERENCES

[1] [4] M. Shahidehpour, H. Yamin, and Z.Li,," Market operation in electric power systems", New York: Wiley,2002.

[2] Mohammad E. Elhawary," Electrical power systems", IEEE Press Power Systems Engineering Series, 1995.

[3] S. Soleymani, A. R. Shirani, A. M. Ranjbar, " Nash Equilibrium Point Computation in Electricity Market", Journal of Electrical Science and Technology, Vol. 17, Serial No.42, 2005, PP.1-11.

[4] S. Soleymani, A. M. Ranjbar, A. R. Shirani, " Optimal Bidding Strategic of GENCOs for Competition in Day- ahead Energy Market", 20th International Power System Conference, 14-16 Nov 2005, Tehran, Iran.

[5] Michael Blake," Game theory and electricity markets", Drayton Analytics research paper series, Feb. 2003.

[6] GAMS Release 2.50," A User's Guide", GAMS Development Corporation, 1999 available: www.gams.com

[7] M. Flynn, W. P. Sheridan, J. D. Dillon, and M. J. O' Malley, " Reliability and Reserve in Competitive Electricity Market Scheduling", IEEE Transactions on Power Systems, Vol.16, No.1, February 2001.

## VIII. BIOGRAPHIES

**Soodabeh Soleymani** received her B.S and M.S degrees in electrical engineering from the Sharif University of Technology. Since 2004, She is pursuing her PhD program at the same university. Her area of research includes power market simulation, and market power monitoring in deregulated power systems.

**Ali Mohammad Ranjbar** received the M.S. and Ph.D. degrees in Electrical Engineering from Tehran University in 1967 and Imperial College of London University in 1975, respectively. Since then, he has been at Sharif University of Technology, Department of Electrical Engineering, where he is currently a full professor. Dr. Ranjbar is the Editor-in-Chief of Journal of Electrical Science and Technology since 1989 and the Director of Niroo Research Institute since 1996 too. His main research interests are in the areas of electric power systems protection and operation, and electrical machine.

**Ali Reza Shirani** received the B.S. degree in the Electrical Engineering Department of Sharif University of Technology, Tehran, Iran. He is the Vice President of Niroo Research Institute since 1997. His main research interests are in the areas of electric power systems protection and operation, and electrical machine.

**2006 IEEE International Conference on Power Electronic, Drives and Energy Systems**

# Modelling of STATCOM Based Voltage Regulator for Self-Excited Induction Generator with Dynamic Loads

Bhim Singh, *Senior Member, IEEE*, S. S. Murthy, *Senior Member, IEEE*, and Sushma Gupta

*Abstract--* **This paper presents an analysis of three-phase self-excited induction generator (SEIG) with static compensator (STATCOM) as a voltage regulator. Current controlled voltage source inverter is used as STATCOM, which provides fast dynamic response to maintain constant voltage at SEIG terminals during severe load perturbation and acts as a source or sink of reactive power. The dynamic model of SEIG is developed in d-q stationary reference frame. Transient analysis of the SEIG-STATCOM system is carried out for voltage build-up, switching in STATCOM and feeding an induction motor as a dynamic load. Star-delta starter is used to reduce inrush current during starting of motors. Low valued capacitors connected in parallel with induction motors reduce the burden on STATCOM.**

*Index Terms*—**Self-excited induction generator, STATCOM, Voltage Regulator, Dynamic load, staring.**

## I. INTRODUCTION

$D$ECENTRALISED power generation has been recently considered as a viable option to grid supply for isolated regions due to considerable transmission losses and poor reliability of the latter. With recent concerns for environmental issues, exploitation of renewable energy sources such as wind, small hydro and biomass often abundantly available in such locations has received greater attention. In these applications, Self-Excited Induction Generators (SEIG) are found to be viable options due to their ruggedness, simplicity, brushless squirrel cage rotor, ease of manufacture, low maintenance and absence of DC source [1,2]. However, SEIG requires suitable controller to regulate the voltage and frequency at varying loads and prime mover input. It has to be tailor made to the type of prime mover and load. Engine driven SEIG using oil and biomass as fuel is a strong contender in this race as the engine effects energy balance for varying loads by adjusting the fuel intake and regulating the speed. But this has to compete with widely used synchronous generators in all its performance as per standards. A major demand on such generators is to satisfactorily start and run induction motors used for commercial or agricultural loads, which is considered stringent. Reactive power control is the central requirement to

such a strategy. .After estimating the reactive power requirement [2-6], one has to develop controllers to effect this variation to maintain constant terminal voltage. Several reported reactive power control schemes [3-6] have been reported but these schemes have several limitations, hence there is a need for further research efforts. Recent development of viable static compensator (STATCOM) with flexible control [6] may result in a more acceptable reactive power control scheme for SEIG, both for static and dynamic loads. This is an area not investigated in the literature and therefore chosen as the theme of this paper.

In this paper, the model of a STATCOM, connected to the terminals of a SEIG supplying an induction motor load is developed. The model is used to simulate the system's response to the following: SEIG voltage build up; direct-on-line starting of an induction motor; star-delta starting of the induction motor; and step change in mechanical load on the shaft of the induction motor.

## II. SYSTEM CONFIGURATION AND CONTROL SCHEME

The schematic diagram of SEIG with excitation capacitor, STATCOM, load and control scheme is shown in Fig.1. Excitation capacitors are selected to generate rated voltage of SEIG at no load. The additional demand of reactive power is fulfilled by the STATCOM under varying loads. The STATCOM acts as a source of lagging or leading current to maintain the terminal voltage constant with variation in load. The STATCOM consists of a three-phase insulated gate bipolar transistor (IGBT) based current controlled voltage source inverter (CC-VSI), DC bus capacitor and AC inductors. The AC output of the inverter is connected through the AC filtering inductors to the SEIG terminals. The DC bus capacitor is used as an energy storage device and provides self-supporting DC bus of VSI.

The control scheme to regulate the terminal voltage of SEIG is based on the generation of source currents having two components, in-phase and quadrature with AC voltages. The unit amplitude templates ($u_a$, $u_b$ and $u_c$) are three-phase sinusoidal functions, computed by dividing the AC voltages $v_a$, $v_b$ and $v_c$ by their amplitude $V_t$. Another set of quadrature unit amplitude templates ($w_a$, $w_b$ and $w_c$) are sinusoidal functions obtained from in-phase vectors ($u_a$, $u_b$ and $u_c$). The sensed AC terminal voltage ($V_t$) is compared with the reference voltage. The voltage error is processed in the PI (proportional-integral) controller. The output of the PI controller ($I^*_{smq}$ for AC voltage control loop decides the amplitude of reactive current

---

Bhim Singh and S.S. Murthy are with the Department of Electrical Engineering, Indian Institute of technology Delhi, New Delhi, India, and Sushma Gupta with the Deptt of Electronic and Comm. RGPV Bhopal (e-mail:bsingh@ee.iitd.ac.in@ee.iitd.ac.in, ssmurthy@yahoo.com).

.

0-7803-9771-1/06/$25.00 ©2006 IEEE      1084

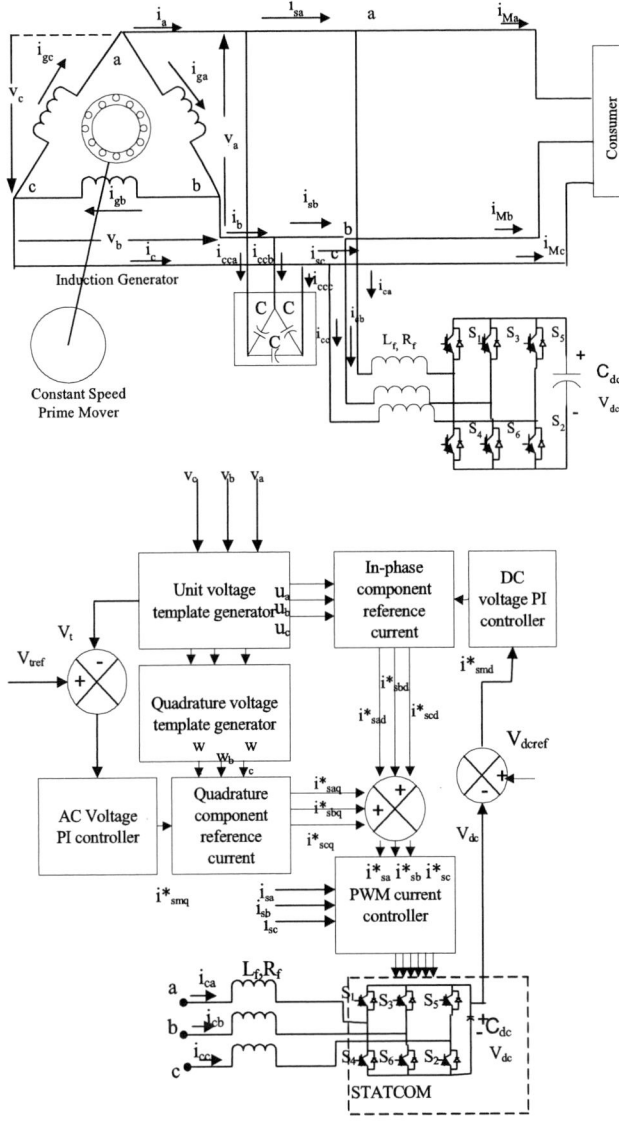

Fig. 1. Schematic diagrams of SEIG-STATCOM system.

to be generated by the STATCOM. Multiplication of quadrature amplitude templates ($w_a$, $w_b$ and $w_c$) with the AC PI voltage controller output ($I^*_{smq}$) yields the quadrature component of the reference source currents. The DC bus voltage is sensed and compared with DC reference voltage. The error voltage is processed in another PI controller. The output of the PI controller ($I_{smd}^*$) decides the amplitude of active current. Multiplication of in-phase unit amplitude templates ($u_a$, $u_b$ and $u_c$) with ($I_{smd}^*$) yields the in-phase component of the reference source currents. The sum of quadrature components ($i^*_{saq}$, $i^*_{sbq}$, $i^*_{scq}$) and in-phase components ($i^*_{sad}$, $i^*_{sbd}$, $i^*_{scd}$) of source currents is the net reference source currents, which are compared with the sensed source currents ($i_{sa}$, $i_{sb}$ and $i_{sc}$) in PWM current controller to generate gating signals for IGBTs of STATCOM.

If AC terminal voltage is greater than reference voltage ($V_{tref}$), quadrature component of reference source current lags the terminal voltage. The switching function derived from the PWM controller results in decreasing the fundamental

component of AC output voltage of inverter below the AC terminal voltage so that reactive current flows from SEIG to STATCOM and the STATCOM absorbs the reactive (inductive) power. If AC terminal voltage is less than reference voltage ($V_{tref}$), quadrature component of reference source current leads the terminal voltage. The switching function derived from the PWM controller causes the increased fundamental component of inverter output voltage above the AC terminal voltage so that current flows from STATCOM to SEIG and the STATCOM supplies the reactive (capacitive) power to the SEIG to maintain constant AC terminal voltage. When the AC terminal voltage is equal to the reference voltage, the reactive current exchange will remain at constant level.

### III. MODELLING OF SEIG-STATCOM SYSTEM

Mathematical model of SEIG-STATCOM system consists of the modelling of SEIG and STATCOM as detailed below.

*A. Modelling of control scheme of STATCOM*

Three-phase SEIG voltages ($v_a$, $v_b$ and $v_c$) are considered sinusoidal and hence their amplitude is computed as:

$$V_t = \{(2/3)\,(v_a^2 + v_b^2 + v_c^2)\}^{1/2} \qquad (1)$$

The unit amplitude phase templates in phase with $v_a$, $v_b$ and $v_c$ are:

$$u_a = v_a/V_t; \quad u_b = v_b/V_t; \quad u_c = v_c/V_t \qquad (2)$$

The unit amplitude templates in quadrature with $v_a$, $v_b$ and $v_c$ may be derived by taking a quadrature transformation of the in-phase unit amplitude templates $u_a$, $u_b$ and $u_c$ [6] as:

$$w_a = -u_b / \sqrt{3} + u_c / \sqrt{3} \qquad (3)$$
$$w_b = \sqrt{3}\,u_a / 2 + (u_b - u_c) / 2\sqrt{3} \qquad (4)$$
$$w_c = -\sqrt{3}\,u_a / 2 + (u_b - u_c) / 2\sqrt{3} \qquad (5)$$

*1) Quadrature component of reference source currents*

The AC voltage error $V_{er(n)}$ at the $n^{th}$ sampling instant is:

$$V_{er(n)} = V_{tref(n)} - V_{t(n)} \qquad (6)$$

where $V_{tref(n)}$ is the amplitude of reference AC terminal voltage and $V_{t(n)}$ is the amplitude of the three-phase AC voltage at the SEIG terminals at $n^{th}$ instant. The output of the PI controller ($I^*_{smq(n)}$) for maintaining AC terminal voltage constant at the $n^{th}$ sampling instant is expressed as:

$$I^*_{smq(n)} = I^*_{smq(n-1)} + K_{pa}\{V_{er(n)} - V_{er(n-1)}\} + K_{ia}\,V_{er(n)} \qquad (7)$$

where $K_{pa}$ and $K_{ia}$ are the proportional and integral gain constants of the PI controller, $V_{er(n)}$ and $V_{er(n-1)}$ are the voltage errors in $n^{th}$ and $(n-1)^{th}$ instant and $I^*_{smq(n-1)}$ is the amplitude of quadrature component of the reference source current at $(n-1)^{th}$ instant. The quadrature components of the reference source currents are estimated as:

$$i^*_{saq} = I^*_{smq}\,w_a; \quad i^*_{sbq} = I^*_{smq}\,w_b; \quad i^*_{scq} = I^*_{smq}\,w_c \qquad (8)$$

*2) In-phase component of reference source currents*

The DC bus voltage error $V_{dcer(n)}$ at $n^{th}$ sampling instant is as:

$$V_{dcer(n)} = V_{dcref(n)} - V_{dc(n)} \qquad (9)$$

where $V_{dcref(n)}$ is the reference DC voltage and $V_{dc(n)}$ is the instantaneous DC link voltage. The output of the PI controller for maintaining DC bus voltage of the STATCOM at the $n^{th}$ sampling instant, is expressed as:

$$I^*_{smd(n)} = I^*_{smd(n-1)} + K_{pd}\{V_{dcer(n)} - V_{dcer(n-1)}\} + K_{id}\,V_{dcer(n)} \qquad (10)$$

$I^*_{smd(n)}$ is considered as the amplitude of in phase component of the source current. $K_{pd}$ and $K_{id}$ are the proportional and integral gain constants of the DC bus PI controller. In-phase components of reference source currents are estimated as:

1085

$$i*_{sad} = I*_{smd} u_a; \quad i*_{sbd} = I*_{smd} u_b; \quad i*_{scd} = I*_{smd} u_c \quad (11)$$

### 3) Reference source currents

Total reference source currents are sum of in-phase and quadrature components of reference source currents as:

$$i*_{sa} = i*_{saq} + i*_{sad} \quad (12)$$
$$i*_{sb} = i*_{sbq} + i*_{sbd} \quad (13)$$
$$i*_{sc} = i*_{scq} + i*_{scd} \quad (14)$$

### 4) PWM current controller

Reference source currents ($i*_{sa}$, $i*_{sb}$ and $i*_{sc}$) are compared with the sensed source currents ($i_{sa}$, $i_{sb}$ and $i_{sc}$). The ON/OFF switching patterns of the gate drive signals to the IGBT of the STATCOM are generated from the PWM current controller. The current errors are computed as:

$$i_{saerr} = i*_{sa} - i_{sa} \quad (15)$$
$$i_{sberr} = i*_{sb} - i_{sb} \quad (16)$$
$$i_{scerr} = i*_{sc} - i_{sc} \quad (17)$$

These current error signals are amplified and compared with the triangular carrier wave. If the amplified current error signal of phase 'a' is greater than the triangular wave signal switch $S_1$ (upper device) is ON and switch $S_4$ (lower device) is OFF, and the value of switching function SA is set to 1. If the amplified current error signal corresponding to $i_{saerr}$ is less than the triangular wave signal switch $S_1$ is OFF and switch $S_4$ is ON, and the value of SA is set to 0. Similar logic applies to other phases.

### B. Modelling of STATCOM

The STATCOM is a current controlled voltage source inverter and modeled as follows:

The derivative of its DC bus voltage is defined as:

$$pv_{dc} = (i_{ca} SA + i_{cb} SB + i_{cc} SC) / C_{dc} \quad (18)$$

where SA, SB and SC are switching functions for the ON/OFF positions of voltage source inverter switches $S_1$-$S_6$.

The DC bus voltage reflects at the AC output of the inverter in the form of three-phase PWM line AC voltage $e_a$, $e_b$ and $e_c$. These voltages may be expressed as:

$$e_a = v_{dc} (SA - SB) \quad (19)$$
$$e_b = v_{dc} (SB - SC) \quad (20)$$
$$e_c = v_{dc} (SC - SA) \quad (21)$$

The volt-current equations of the output of voltage source inverter are:

$$v_a = R_f i_{ca} + L_f p i_{ca} + e_a - R_f i_{cb} - L_f p i_{cb} \quad (22)$$
$$v_b = R_f i_{cb} + L_f p i_{cb} + e_b - R_f i_{cc} - L_f p i_{cc} \quad (23)$$
$$i_{ca} + i_{cb} + i_{cc} = 0 \quad (24)$$

Value of $i_{cc}$ from eqn (24) is substituted in eqn. (23):

$$v_b = R_f i_{cb} + L_f p i_{cb} + e_b + r_f i_{ca} + L_f p i_{ca} + R_f i_{cb} + L_f p i_{cb} \quad (25)$$

Rearranging and simplifying eqns. (22) and (25) the STATCOM current derivatives are obtained as:

$$pi_{ca} = \{( v_b - e_b) + 2 (v_a - e_a) - 3 R_f i_{ca}\}/(3L_f) \quad (26)$$
$$pi_{cb} = \{(v_b - e_b) - (v_a - e_a) - 3 R_f i_{cb}\}/(3L_f) \quad (27)$$

### C. Modelling of SEIG

The dynamic model of three-phase SEIG is developed using d-q axes stationary references frame [6]. The relevant voltage-current equations in matrix form with notations are:

$$[v] = [R] [i] + [L] p [i] + w_g [G] [i] \quad (28)$$

from which, current derivatives can be obtained as:

$$p[i] = [L]^{-1} \{ [v] - [R] [i] - w_g [G] [i] \} \quad (29)$$

where $[v] = [v_{ds} \, v_{qs} \, v_{dr} \, v_{qr}]^T$; $[i] = [i_{ds} \, i_{qs} \, i_{dr} \, i_{qr}]^T$
$[R] = diag [ R_s \, R_s \, R_r \, R_r]$

$$[L] = \begin{bmatrix} L_s + L_m & 0 & L_m & 0 \\ 0 & L_s + L_m & 0 & L_m \\ L_m & 0 & L_r + L_m & 0 \\ 0 & L_m & 0 & L_r + L_m \end{bmatrix}; [G] = \begin{bmatrix} 0 & 0 & 0 & 0 \\ 0 & 0 & 0 & 0 \\ 0 & -L_m & 0 & L_r + L_m \\ L_m & 0 & L_r + L_m & 0 \end{bmatrix} \quad (30)$$

The electromagnetic torque balance equation of SEIG is as:

$$T_{shaft} = T_e + J (2/P) p w_g \quad (31)$$

The derivative of rotor speed of SEIG from eqn. (31) is:

$$P w_g = \{P/(2J)\} (T_{shaft} - T_e) \quad (32)$$

The developed electromagnetic torque ($T_e$) of the SEIG can be expressed as [6]:

$$T_e = (3P/4) L_m (i_{qs} i_{dr} - i_{ds} i_{qr}) \quad (33)$$

The operation of SEIG relies on magnetic saturation to maintain a steady voltage. The magnetizing inductance $L_m$ is therefore a non-linear function of the magnetizing current $I_m$ and has to be chosen appropriate dependent on the operating condition. In the dynamic modelling this feature needs to be incorporated. From the d-q axes currents, $I_m$ is computed as:

$$I_m = \{\surd (i_{ds} + i_{dr})^2 + (i_{qs} + i_{qr})^2 \}/\surd 2 \quad (34)$$

From this magnetising current ($I_m$), the magnetizing inductance ($L_m$) is computed from the magnetization characteristic relating $L_m$ and $I_m$. Relation between $L_m$ and $I_m$ is obtained by synchronous speed test [6] and which can be modelled as:

$$L_m = K_1 + K_2 I_m + K_3 I_m^2 + K_4 I_m^3 + K_5 I_m^4 \quad (35)$$

where $K_1$ to $K_5$ are constant for different machines and are given in Appendix.

It must be noted that the above modelling approach has several approximations. The effect of saturation on leakage inductance and rotor skin effects are not accommodated in rotor resistance.

### D. AC Line Tterminal Voltage at the Common Coupling

Direct and quadrature axis currents ($i_{ds}$ and $i_{qs}$) of SEIG are converted to three-phase stator currents ($i_{ga}$, $i_{gb}$ and $i_{gc}$). The line currents of SEIG may be obtained from these phase currents ($i_a = i_{ga-gb}$; $i_b = i_{gb}-i_{gc}$; $i_c = i_{gc}-i_{ga}$). The derivative of AC terminal voltage of SEIG are derived as:

$$p \, v_a = \{(i_a - i_{ca} - i_{Ma}) - (i_b - i_{cb} - i_{Mb})\} / (3 \, C) \quad (36)$$
$$p \, v_b = \{(i_a - i_{ca} - i_{Ma}) + 2 (i_b - i_{cb} - i_{Mb})\} / (3 \, C) \quad (37)$$
$$v_a + v_b + v_c = 0. \quad (38)$$

where $i_{Ma}$, $i_{Mb}$ and $i_{Mc}$ are motor (load) currents and $i_{ca}$, $i_{cb}$ and $i_{cc}$ are STATCOM currents. C is per phase excitation capacitor connected across SEIG.

### E. Modelling of the Dynamic Load

The dynamic model of three-phase squirrel cage induction motor as dynamic load is similar to that of the induction generator but the parameters concerned is related to the motor. The volt-current equations of a three-phase induction motor in the current derivative can be expressed as:

$$p [i_M] = [L_M]^{-1} \{ [v_M] - [R_M] [i_M] - w_m[G_M] [i_M] \} \quad (39)$$

where $[v_M]$, $[i_M]$, $[R_M]$, $[L_M]$ and $[G_M]$ can be defined similar to those given in eqn. (30).

The developed electromagnetic torque of the I. M. is as:

$$T_{eM} = \{3P_M/4\} L_{mM} (i_{qsM} i_{drM} - i_{dsM} i_{qrM}) \quad (40)$$

The derivative of electromechanical torque of motor speed can be expressed as:

$$p \, w_m = \{P_M/(2J_M)\} (T_L - T_{eM}) \quad (41)$$

1086

where $T_L$, $J_M$ and $P_M$ denote load torque, moment of inertia and number of poles of the motor respectively.

## IV. RESULTS AND DISCUSSION

The self-excited induction generator system with STATCOM and an induction motor load is simulated under various loading conditions and obtained results are shown in Figs. 2-8. For this study, a 15 kW, 415V, 30A, 4-pole machine has been used as a generator and 3.7 kW and 15 kW squirrel cage motors are used as dynamic loads. Parameters of the generator and motors are given in Appendix.

### A. Voltage Build-up and Switching on of STATCOM

Fig. 2 shows the transient waveforms under voltage build up and thereafter switching on the STATCOM. Response from top respectively relates to 3-phase AC terminal voltages ($v_{abc}$), generator currents ($i_{abc}$), STATCOM currents ($i_{cabc}$), amplitude of AC terminal voltage and its reference value ($V_t$, $V_{tref}$), DC

out within a few cycles. Moreover, there is a small change in speed of SEIG (1.5%), which recovers quickly to normal value due to STATCOM.

### B. Direct on line starting of an induction motor load fed from SEIG-STATCOM system

The SEIG-STATCOM system can start the induction motor direct 'on line'. Figs. 3 and 6 show dynamic response of three-phase generator voltages ($v_{abc}$) and currents ($i_{abc}$), STATCOM currents ($i_{cabc}$), motor (load) currents ($i_{mabc}$), amplitude of AC terminal voltage and its reference value ($v_t$, $v_{tref}$), DC bus voltage and its reference value ($v_{dc}$, $v_{dcref}$), generator speed ($w_g$) and motor speed ($w_m$) for 3.7 kW and 15 kW motors respectively. Induction motors of rating of 3.7 kW and 15 kW settle down in 300 ms and 600 ms respectively. At starting, the induction motor draws heavy current and hence both generator and STATCOM currents increase. As dynamic load is

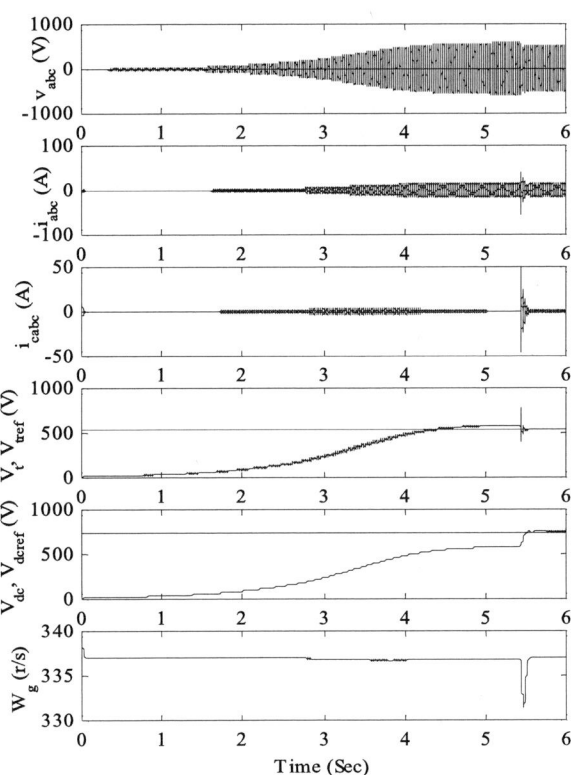

Fig. 2. Transient waveforms of voltage build-up and switch in STATCOM on three-phase SEIG.

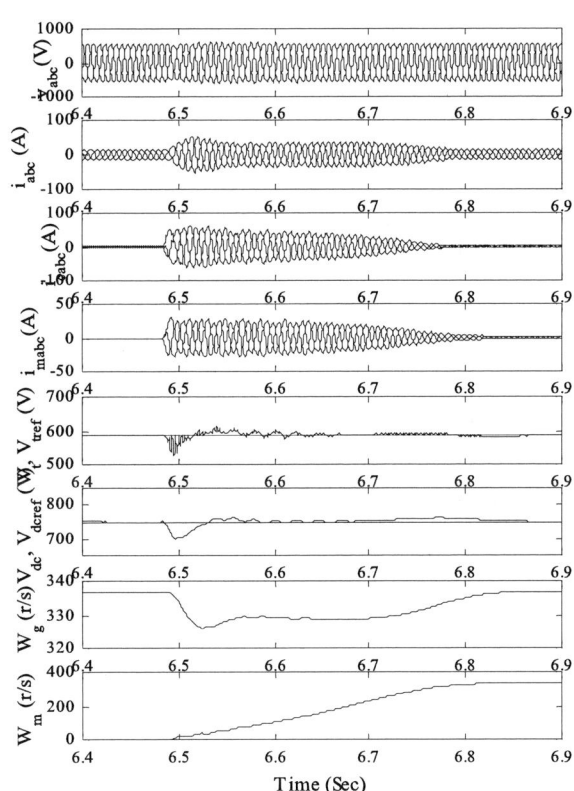

Fig. 3. Dynamic response of direct on line starting of an induction motor (3.7 kW) with SEIG-STATCOM system.

bus voltage and its reference value ($V_{dc}$, $V_{dcref}$) and generator speed ($w_g$). To generate rated voltage of 415 V rms (586 V peak) at no-load, 57 µF capacitor per phase is connected across the terminals of SEIG. Initially DC bus capacitor of current -controlled voltage source inverter charges to 586 V (peak of AC voltage) during voltage building up through the anti parallel diodes of inverter (without control action). At 5.5 sec. gate pulses are given to the insulated gate bipolar transistors and control action of voltage source inverter is activated. STATCOM behaves as a source of reactive power and draws small active power from the generator to charge its DC bus capacitor at reference voltage (750 V). There is a small oscillation at the switching in STATCOM but it damps

switched on, a small dip in the DC bus voltage is observed which shows the instantaneous energy transfer from DC bus capacitor to SEIG to maintain AC terminal voltage constant. The generator speed also drops down during starting of an induction motor load and recovers back as the motor attains rated speed due to dynamics of mechanical system.

### C. Star-delta starting of an induction motor load

Star-delta starter is preferred for the starting of an induction motor, which reduces its starting current to one-third, compared to direct on line starting. Figs. 4 and 7 show the dynamic response of generator, motor and STATCOM under

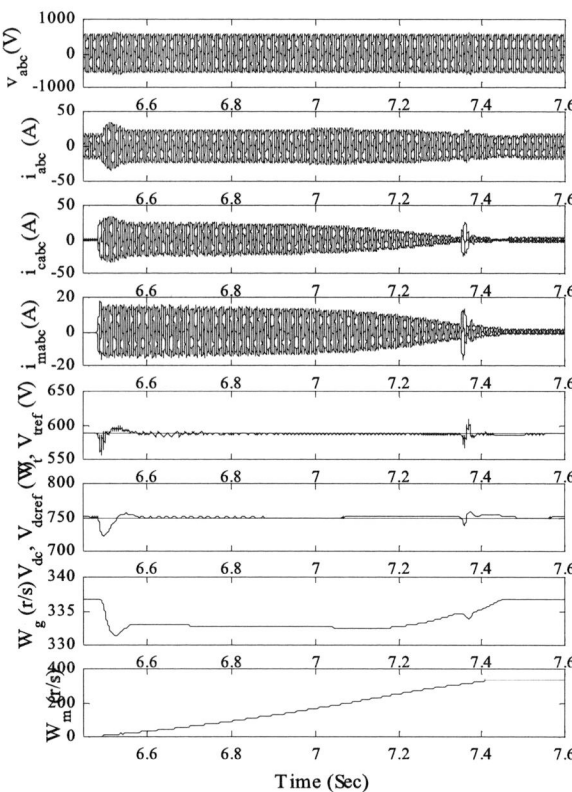

Fig. 4. Dynamic response of star-delta starting of an induction motor (3.7 kW) with SEIG-STATCOM system.

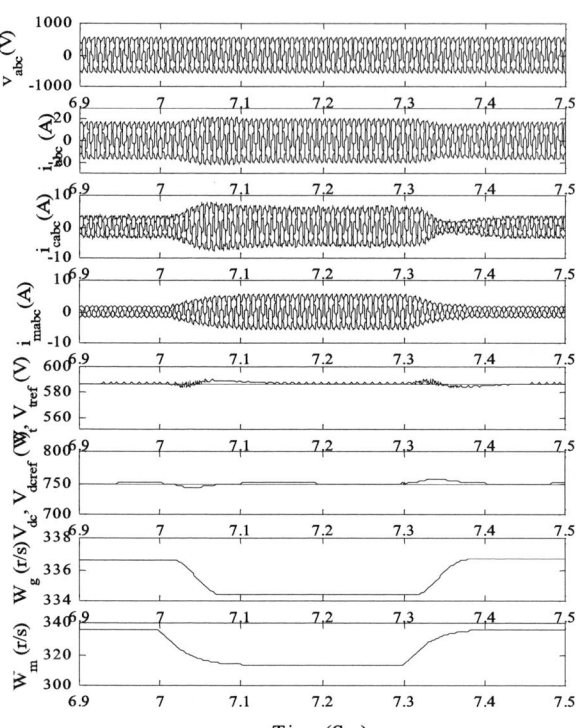

Fig. 5. Dynamic response of SEIG-STATCOM and induction motor load (3.7 kW) with application and removal of rated mechanical load (25 Nm) on IM.

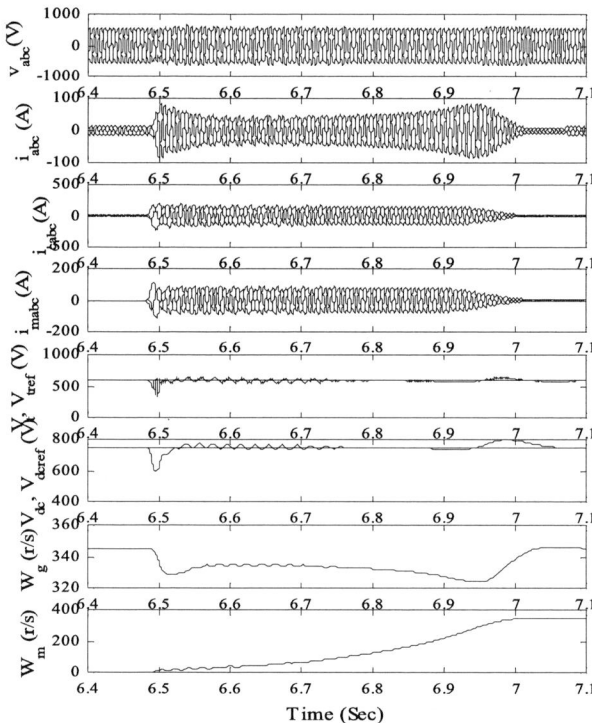

Fig. 6. Dynamic response of direct on line starting of an induction motor (15 kW) with SEIG-STATCOM system.

star-delta starting of 3.7 kW and 15 kW motors respectively. In star-delta starter, initially motor phases experience reduced voltage and it takes longer time (800 ms and 1500 ms) for the 2 motors to reach rated speed. With star-delta starter, DC bus capacitor of STATCOM experiences less burden and the motor is smoothly switched on. Starter changes the motor winding from star to delta connection at 80% of its rated speed, which can be observed in Figs 4 and 7. In star-delta starter, STATCOM currents reduce to around 50% as compared to the direct on line starter. Therefore, if star-delta starter is used, STATCOM rating reduces to 50%. A 4% dip in DC bus voltage is observed compared to 8% dip in direct on line starter. The generator speed drops by around 2% as compared to 3% in direct on line starter. This indicates better response of SEIG-STATCOM system in case of star-delta starter.

*D. Mechanical Loading on the Motor*

Fig. 5 shows the dynamic response of the SEIG-STATCOM-IM system under application and removal of rated mechanical load (25 N-m) on shaft of the 3.7 kW motor. Response quantities settle down quickly to the new operating conditions, which show the versatility of the controller. Motor and generator currents increase under mechanical loading to provide active power to the load (motor) and STATCOM currents increase to provide reactive power to the load (motor) and SEIG. Small oscillation in the terminal voltage is observed, which damps out within few cycles. An under-shoot and an over-shoot in voltages are observed on DC bus capacitor at the application and removal of mechanical load respectively. Under mechanical loading of the motor, generator speed drops down slightly and speed of the motor

also drops down. Fig 8 shows the dynamic performance of the SEIG system for 15 kW motor loads for rated load application and removal, the corresponding torque being 100 N-m.

## V. CONCLUSIONS

The proposed STATCOM based voltage regulator for the SEIG has shown promising results in maintaining constant voltage even under severe dynamic loads of an induction motor. It exhibits fast dynamic response compared to conventional reactive power controllers even during starting of an induction motor. It has been observed that the SEIG-STATCOM system is able to start the induction motor of same rating as that of SEIG without losing self-excitation. The

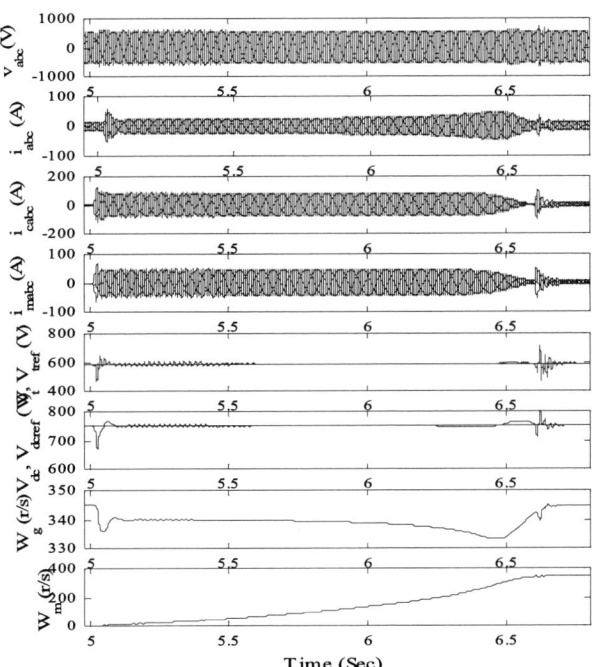

Fig. 7. Dynamic response of star-delta starting of an induction motor (15 kW) with SEIG-STATCOM system.

STATCOM rating can be reduced considerably through star-delta starter of the motor. The STATCOM has the advantage of harmonic free sinusoidal voltage at the terminals of SEIG and hence the consumer gets good quality supply.

## VI. APPENDICES

*A. STATCOM control parameters*

$L_f$= 1.2 mH, $R_f$ = 0.045 $\Omega$ and $C_{dc}$ = 4000$\mu$F.

AC voltage PI controller: $K_{pa}$ =0.05, $K_{ia}$ = 0.04.

DC bus voltage PI controller $K_{pd}$ = 0.44, $K_{id}$ =0.08

*Selection of DC bus capacitor for VSI*

The value of $C_{dc}$ is chosen on permissible dip in DC voltage at application of load and rise in DC bus voltage at sudden removal of load. For the considered voltage dip in DC bus voltage of 2% using energy conservation principle:

$1/2\ C_{dc}\ \{(v_{dc})^2 - (v_{dcl})^2\} = 3\ V\ (a\ I)\ t$

Where $v_{dc}$ = 750 V, $v_{dcl}$ = 735 V, V = 415$\sqrt{2}$ V, I = 30$\sqrt{2}$ A, t = 350 $\mu$Sec and a =1.2 (during transient current rating is likely to vary from 120% to 180% of the steady-state condition). DC bus capacitor is:

$C_{dc}$ = 2816 $\mu$F. Hence commercial value of $C_{dc}$ is taken as 4000 $\mu$F.

*Selection of DC bus voltage for VSI*

The value of DC bus voltage is choosen based on instantanous energy available to the controller and it is trade off between voltage dip and rise during sudden application of load and sudden removal of load. For the VSI DC bus voltage is defined as:

$V_{dc}$ = 2$\sqrt{2}$(V/$\sqrt{3}$)/$m_a$  where $m_a$ = modulation index (=1) and V = AC line voltage (= 415 V). The DC bus voltage is as:

$V_{dc}$ = 677.7 V  by considering the safety factor, $V_{dc}$ is taken 750 V.

*Selection of AC inductance for VSI*

The selection of AC inductance ($L_f$) depends on the current ripple $i_{cr(p-p)}$ allowed through the inductance. If 5% current ripple is allowed through the AC inductance than inductance is calculated as:

$L_f$ = $\{(\sqrt{3}/2)\ m_a\ V_{dc}\}/\ \{6.a.f_s.\ i_{cr(p-p)}\}$  = 1.06 mH where $f_s$ = 20 KHz. By considering safety factor, $L_f$ is taken 1.2 mH.

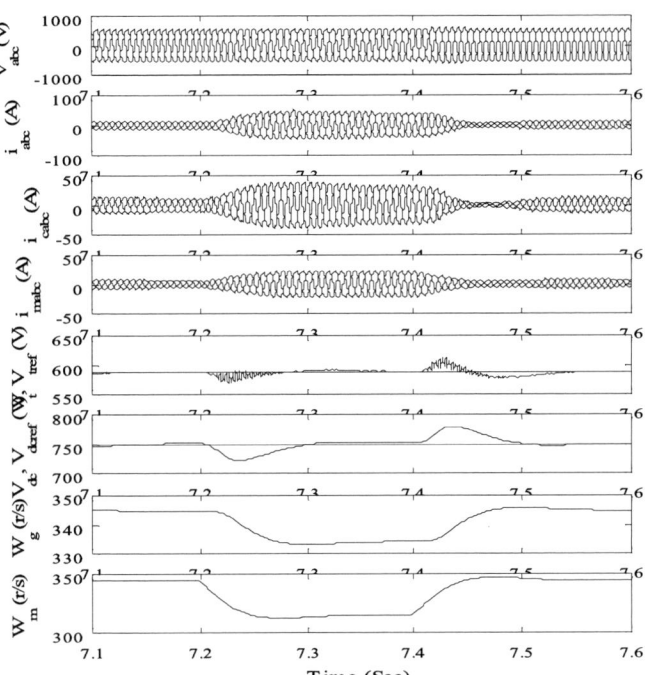

Fig. 8. Dynamic response of SEIG-STATCOM and induction motor load (15 kW) with application and removal of rated mechanical load (100 N m) on IM.

*B. Parameters of three-phase squirrel cage Motors*

All the machines are rated at 415 V, 4 pole, 50 Hz and $\Delta$ connected stator windings with squirrel cage rotors. The parameters of the induction machines are as follows:

3.7 kW Machine: $R_s$ = 5.53 $\Omega$, $R_r$ = 5.86 $\Omega$, $X_{ls}$ = $X_{lr}$ =9.6 $\Omega$, $J_m$ = 0.0842 kg/m$^2$ , $L_m$ = 0.89 - 0.0039 $I_m$ - 0.107 $I_m^2$+ 0.0245 $I_m^3$-0.0012 $I_m^4$

15 kW Machine: $R_s$ = 0.696 $\Omega$, $R_r$ = 0.743 $\Omega$, $X_{ls}$ = $X_{lr}$ = 3.49 $\Omega$, $J_m$ = 0.2 kg/m$^2$ , $L_m$ =0.205+0.0053 $I_m$-0.0023 $I_m^2$+0.0001 $I_m^3$

## VII. REFERENCES

[1] S. S. Murthy, O. P. Malik and A. K. Tandon, "Analysis of self-excited induction generator," *IEE Proc.*, Vol. 129, Pt.C, No. 5, pp. 260-265, November 1982.

[2] Bhim Singh, L. Shridhar and C. S. Jha, "Transient analysis of self-excited induction generator supplying dynamic load," *Electrical Machines and Power Systems*, Vol. 27, pp. 941-954, 1999.

[3] L. Wang and C. H. Lee, "Long-shunt and short-shunt connections on dynamic performance of a SEIG feeding an induction motor load," *IEEE Trans. on Energy Conversion*, Vol. 15, No. 1, pp. 1-7, March 2000.

[4] O. P. Malik, D. Diwan, S. S. Murthy, T. Grant, P. Walsh, "A solid state voltage regulator for self-excited induction generators," Proc. on IEEE Industrial and Commercial Power Systems Conf., Milwaukee, p. 25, May 1983.

[5] R. K. Mishra, Bhim Singh and M. K. Vasantha, "Voltage regulator for an isolated self-excited cage induction generator," *Electric Power System Research,* Vol. 24, No. 1, pp. 75-83, July 1992.

[6] Bhim Singh and L. B. Shilpakar, "Analysis of a novel solid state voltage regulator for a self-excited induction generator," *IEE Proc. Gener. Transm. Distrib.*, Vol. 145, No. 6, pp. 647-655, November 1998.

**2006 IEEE International Conference on Power Electronic, Drives and Energy Systems**

# Optimum Design of UPFC Controllers Using GEA: Decoupled Real & Reactive Power Flow and Damping Controllers

N. Ray Chaudhuri, and M. L. Kothari, *Senior Member, IEEE*

*Abstract—* **This paper presents a comprehensive approach for designing UPFC controllers (Decoupled Real & Reactive power flow controllers, dc link and ac voltage regulators & UPFC based damping controller). For designing the controllers, the detailed model of the system with UPFC has been considered. Investigations clearly show that the real and reactive power flows on the line can be effectively regulated independently using decoupled controller. Studies also reveal that a significant improvement in damping of power system oscillations can be achieved with UPFC based damping controller.**

*Index Terms--* **Damping Controller, Dynamic stability, FACTS controllers, Genetic and evolutionary algorithms, UPFC.**

## I. INTRODUCTION

THE unified power flow controller (UPFC) is a FACTS device which is capable of controlling power system parameters such as terminal voltage, line impedance and phase angle. The primary function of UPFC is to control power flow on a given line and voltage at the UPFC bus. The UPFC can also be utilized for damping power system oscillations by judiciously incorporating a damping controller.

Steady state and dynamic models of UPFC have been developed by many researchers [1-3]. Schauder and Mehta [1] have presented Vector Analysis and Control of the advanced static var compensators (ASVCs). They have developed the concept of decoupled control of real & reactive power flows on a line. Wang [2,3] has developed linear dynamic models known as modified Heffron-Phillips models for a power system including UPFC. He has also proposed an approach for designing a robust UPFC based damping controller. Tambey and Kothari [4] have proposed a comprehensive approach for designing UPFC controllers using modified Heffron-Phillips model. Papic *et al* [5] have presented a modified controller structure for UPFC with a predictive control loop & pre-control signal for dc voltage controller.

---

N. Ray Chaudhuri is with John F. Welch Technology Center, GE, Bangalore 560066 India (e-mail: nilanjan.chaudhuri@ge.com).

M. L. Kothari is with the Department of Electrical Engineering, IIT Delhi, Delhi, New Delhi 110016 India (e-mail: mohankothari@hotmail.com) .

The main objectives of the research work presented in this paper are:

(1) To present a comprehensive dynamic model of a power system including detailed model of synchronous generator, steam turbine, governor, PSS & UPFC controllers.

(2) To present a systematic approach for optimum design of PSS & UPFC controllers (i.e. decoupled real and reactive power flow controllers, DC & AC voltage regulators and damping controller) using Genetic and Evolutionary Algorithm (GEA).

(3) To investigate an interaction between PSS & UPFC controllers by examining the dynamic performance of the system with optimum PSS and UPFC controllers.

## II. SYSTEM INVESTIGATED

A single machine- infinite bus (SMIB) system with UPFC installed in one of the two parallel-connected transmission lines is considered (Fig.1). The UPFC based on 48-pulse GTO converter is considered. The static excitation system (model type IEEE-ST1A) along with delta-omega PSS has been considered. A four-stage steam turbine with appropriate governor model is included. The nominal operating condition & system parameters are given in the Appendix.

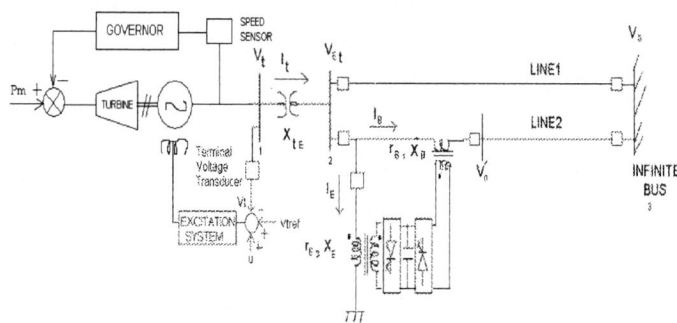

Fig. 1. Schematic diagram of a single machine infinite bus (SMIB) system with UPFC in one of the lines.

## III. DECOUPLED REAL AND REACTIVE POWER FLOW CONTROLLER

Fig. 2 shows the simplified circuit diagram of the shunt connected 3-phase voltage source converter (STATCOM). The R and L represent the resistance and inductance of transformer respectively through which VSC is connected to the 3-phase power system.

---

0-7803-9771-1/06/$25.00 ©2006 IEEE

The AC side equation of the circuit in terms of instantaneous voltage and currents is given in equation (1).

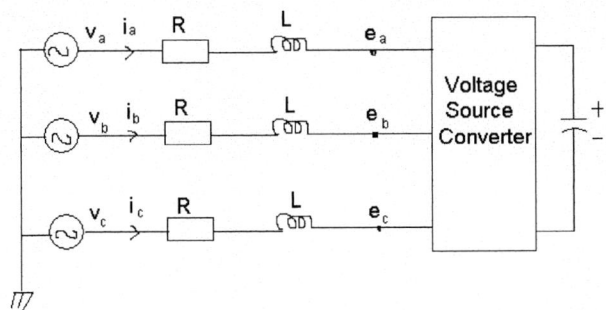

Fig. 2. Equivalent circuit of a 3-phase voltage source converter connected to the system.

$$p\begin{bmatrix} i_a \\ i_b \\ i_c \end{bmatrix} = \begin{bmatrix} -\dfrac{R\omega_B}{L} & 0 & 0 \\ 0 & -\dfrac{R\omega_B}{L} & 0 \\ 0 & 0 & -\dfrac{R\omega_B}{L} \end{bmatrix} \begin{bmatrix} i_a \\ i_b \\ i_c \end{bmatrix} + \dfrac{\omega_B}{L}\begin{bmatrix} v_a - e_a \\ v_b - e_b \\ v_c - e_c \end{bmatrix} \quad (1)$$

All quantities are expressed in per unit [1]. Transforming phase variables to d and q-axes coordinates (the d-q axes coordinate system is defined such that the d-axis coincides with instantaneous voltage vector) i.e. $v_d = |v|$ and $v_q = 0$, we get,

$$p\begin{bmatrix} i_d \\ i_q \end{bmatrix} = \begin{bmatrix} -\dfrac{R\omega_B}{L} & \omega \\ -\omega & -\dfrac{R\omega_B}{L} \end{bmatrix} \begin{bmatrix} i_d \\ i_q \end{bmatrix} + \dfrac{\omega_B}{L}\begin{bmatrix} |v| - e_d \\ -e_q \end{bmatrix} \quad (2)$$

The equation (2) can be rewritten as,

$$p\begin{bmatrix} i_d \\ i_q \end{bmatrix} = \begin{bmatrix} -\dfrac{R\omega_B}{L} & 0 \\ 0 & -\dfrac{R\omega_B}{L} \end{bmatrix} \begin{bmatrix} i_d \\ i_q \end{bmatrix} + \begin{bmatrix} 0 & \omega \\ -\omega & 0 \end{bmatrix}\begin{bmatrix} i_d \\ i_q \end{bmatrix} + \dfrac{\omega_B}{L}\begin{bmatrix} v_d - e_d \\ -e_q \end{bmatrix}$$

Where,

$$x_1 = \frac{\omega_B}{L}(|v| - e_d) \qquad x_2 = -\frac{\omega_B}{L}e_q \quad (3)$$

$$p\begin{bmatrix} i_d \\ i_q \end{bmatrix} = \begin{bmatrix} -\dfrac{R\omega_B}{L} & 0 \\ 0 & -\dfrac{R\omega_B}{L} \end{bmatrix} \begin{bmatrix} i_d \\ i_q \end{bmatrix} + \begin{bmatrix} 0 & \omega \\ -\omega & 0 \end{bmatrix}\begin{bmatrix} i_d \\ i_q \end{bmatrix} + \begin{bmatrix} x_1 \\ x_2 \end{bmatrix} \quad (4)$$

The instantaneous real power p(t) and reactive power q(t) in terms of d and q axes voltage and current components are:

$$p(t) = \frac{3}{2}v_d i_d \quad \text{and} \quad q(t) = \frac{3}{2}v_d i_q \quad (5)$$

It may be noted that the d-axis current component contributes to the instantaneous active power p(t) and the q-axis current component contributes to instantaneous reactive power q(t).

In order to develop decoupled P-Q flow control algorithm, the variables $x_1$ and $x_2$ defined above are redefined in an alternative form as follows:

$$x_1 = (K_{ps} + \frac{K_{is}}{s})(i_d^* - i_d) - \omega i_q \quad (6)$$

$$x_2 = (K_{ps} + \frac{K_{is}}{s})(i_q^* - i_q) + \omega i_d \quad (7)$$

Where, $K_{ps}$ and $K_{is}$ are the proportional and integral gain settings of the PI controllers for regulating $i_d$ and $i_q$. $i_d^*$ and $i_q^*$ are the desired values of $i_d$ and $i_q$.

The desired values $i_d^*$ and $i_q^*$ are computed from equation (5), i.e.

$$i_d^* = \frac{2p^*(t)}{3v_d} \quad \text{and} \quad i_q^* = \frac{2q^*(t)}{3v_d} \quad (8)$$

Here, p*(t) & q*(t) are the reference instantaneous real & reactive powers. The transfer functions $\dfrac{i_d(s)}{i_d^*(s)}$ and $\dfrac{i_q(s)}{i_q^*(s)}$ derived using equations 4, 6 & 7 are identical [5] and are as follows:

$$F(s) = \frac{i_d(s)}{i_d^*(s)} = \frac{i_q(s)}{i_q^*(s)} = \frac{K_{is} + sK_{ps}}{K_{is} + s(\dfrac{R\omega_B}{L} + K_{ps}) + s^2} \quad (9)$$

The equation (9) clearly shows that the current components $i_d$ and $i_q$ can be regulated independent of each other [i.e. the de-coupled control of p(t) and q(t)].

The modulating index m of the VSC and phase angle $\delta$ can be computed from $e_d$ and $e_q$ as follows:

$$m = \frac{\sqrt{e_d^2 + e_q^2}}{v_{dc}} \quad \text{and} \quad \delta = \tan^{-1}\left(\frac{e_q}{e_d}\right) \quad (10)$$

The decoupled real and reactive power flow control technique developed above considering STATCOM can be adapted to series connected VSC of the UPFC easily. For series converter, the line currents are measured & then decomposed into d & q components ($i_d$ & $i_q$) in synchronously rotating reference frame. Here the d-axis is made to coincide with the instantaneous line voltage vector at the UPFC location where the decoupled control of line real & reactive power is intended.

## IV. MODIFIED HEFFRON-PHILLIPS MODEL

A linearised dynamic model is obtained by linearising the non-linear dynamic model of the system around an operating condition [2]. The linearised dynamic model is given below:

$$\overset{\bullet}{\Delta\omega} = \frac{(\Delta P_m - \Delta P_e - D\Delta\omega)}{M} \quad (11)$$

$$\overset{\bullet}{\Delta\delta} = \omega_B\Delta\omega \quad (12)$$

$$\Delta \dot{E}_q^{'} = \frac{\left(-\Delta E_q + \Delta E_{fd}\right)}{T_{do}^{'}} \qquad (13)$$

$$\Delta \dot{E}_{fd} = \frac{-\Delta E_{fd} + K_a(\Delta V_{ref} - \Delta V_t)}{T_a} \qquad (14)$$

$$\Delta \dot{V}_{dc} = K_7 \Delta \delta + K_8 \Delta E_q^{'} - K_9 \Delta V_{dc} + K_{ce} \Delta m_E +$$
$$K_{c\delta e} \Delta \delta_E + K_{cb} \Delta m_B + K_{c\delta b} \Delta \delta_B \qquad (15)$$

Where,

$$\Delta Pe = K_1 \Delta \delta + K_2 \Delta E_q^{'} + K_{pe} \Delta m_E + K_{p\delta e} \Delta \delta_E$$
$$+ K_{pb} \Delta m_B + K_{p\delta b} \Delta \delta_B + K_{pd} \Delta V_{dc}$$

$$\Delta E_q = K_4 \Delta \delta + K_3 \Delta E_q^{'} + K_{qe} \Delta m_E + K_{q\delta e} \Delta \delta_E$$
$$+ K_{qb} \Delta m_B + K_{q\delta b} \Delta \delta_B + K_{qd} \Delta V_{dc}$$

$$\Delta V_t = K_5 \Delta \delta + K_6 \Delta E_q^{'} + K_{ve} \Delta m_E + K_{v\delta e} \Delta \delta_E$$
$$+ K_{vb} \Delta m_B + K_{v\delta b} \Delta \delta_B + K_{vd} \Delta V_{dc}$$

Fig. 3 shows the modified Heffron-Phillips transfer function model of the system including UPFC. The control vector **[u]** is defined as follows:

$$[\mathbf{u}] = [\,\Delta m_E \quad \Delta \delta_E \quad \Delta m_B \quad \Delta \delta_B\,]^T \qquad (16)$$

Here,

$$\mathbf{K_{pu}} = [K_{pe} \quad K_{p\delta e} \quad K_{pb} \quad K_{p\delta b}]$$

$$\mathbf{K_{qu}} = [K_{qe} \quad K_{q\delta e} \quad K_{qb} \quad K_{q\delta b}]$$

$$\mathbf{K_{vu}} = [K_{ve} \quad K_{v\delta e} \quad K_{vb} \quad K_{v\delta b}]$$

$$\mathbf{K_{cu}} = [K_{ce} \quad K_{c\delta e} \quad K_{cb} \quad K_{c\delta b}]$$

At this stage it is extremely important to mention that in the model derived in section III, the d-axis of the synchronously rotating reference frame coincides with infinite bus voltage vector $V_b$. However, the modified Heffron –Phillips model is derived considering the q- axis coinciding with $E_q^{'}$. An appropriate coordinate transformation is incorporated to link the modified Heffron-Phillips model with Decoupled P-Q controller. The modulating indices $m_B$ and $m_E$ are for series and shunt VSCs respectively.

## V. UPFC CONTROLLERS

The UPFC controllers are:
1. Decoupled P-Q controller
2. DC link & AC terminal voltage regulators
3. Power system oscillation damping controller

### A. Decoupled P-Q controller

The small perturbation model of the decoupled P-Q controllers is obtained by linearising the model around the nominal operating condition. Fig. 4 shows the transfer function block diagram of decoupled P-Q controller of UPFC (series converter). In this model the d and q-axes components of the current in line 2 are denoted as $\Delta I_{Bd}$ & $\Delta I_{Bq}$.

$\Delta I_{Bd}^{/}$ & $\Delta I_{Bq}^{/}$ are obtained from $\Delta I_{Bd}$ & $\Delta I_{Bq}$ after proper coordinate transformation. The reference settings of these

components are $\Delta I_{Bdref}$ & $\Delta I_{Bqref}$. Identical P-I controllers are chosen for regulating d and q axes components of line currents.

### B. DC link voltage & terminal voltage controller

Figs. 5 & 6 show the transfer function block diagrams of the Proportional-Integral type AC terminal voltage & DC link voltage regulators respectively. The DC link voltage is regulated by regulating phase angle $\delta_E$ of the shunt converter and the AC terminal voltage is regulated by regulating the modulating index $m_E$ of the shunt converter.

### C. UPFC based damping controller

Fig. 7 shows the transfer function block-diagram of the UPFC based damping controller. It consists of two cascade-connected blocks. Block 1 is provided to derive a speed –deviation signal. The real power output Pe of the generator is compared with the set point Pm (mechanical power). The error is integrated and multiplied with 1/M to derive a speed deviation signal. It may be noted that the derived speed deviation signal is used instead of the measured speed deviation signal, since the measured speed-deviation signal, in general may not be available at the UPFC location and may also contain signals corresponding to the torsional oscillations. The second block comprises gain ($K_D$), washout block and a pair of identical phase lead compensators. The output of the damping controller is used to modulate $\Delta P_{ref}$ of the decoupled P-Q flow controller (Fig 4).

## VI. DESIGN OF PSS AND UPFC CONTROLLERS

### A. Computation of the constants of the transfer- function model (Fig. 3)

The parameters of the modified Heffron- Phillips model are computed for the nominal operating condition & system parameters. These parameters are given below:

$K_1$ = 0.586,$K_2$= 0.928,$K_3$=1.883,$K_4$= 0.505,$K_5$ = -0.123
$K_6$= 0.572,$K_7$=-0.069,$K_8$= 0.045,$K_9$= 0.004,$K_{pe}$= 0.118
$K_{qb}$=0.231,$K_{qe}$=-0.255,$K_{vb}$=-0.069,$K_{ve}$= 0.134,$K_{cb}$= 0.015
$K_{ce}$=-0.006,$K_{p\delta b}$=-0.030, $K_{p\delta e}$=0.397,$K_{q\delta b}$=0.01,$K_{q\delta e}$=0.136
$K_{v\delta b}$=-0.009,$K_{v\delta e}$=0.008,$K_{c\delta b}$=0.011,$K_{c\delta e}$= 0.164,$K_{pd}$ = 0.056
$K_{pb}$ = 0.191,$K_{qd}$= -0.116,$K_{vd}$=0.061

### B. Optimization of PSS parameters

Fig. 8 shows the transfer function block diagram of the Delta-Omega PSS. The PSS comprises of a gain block, washout transfer function & two identical phase-lead networks (i.e. T11=T13 & T12=T14). Washout time constant $T_w$=10s is assumed. The optimum PSS parameters are computed applying GEA considering objective function J= $\int_0^\infty (\Delta \omega)^2 \, dt$ .

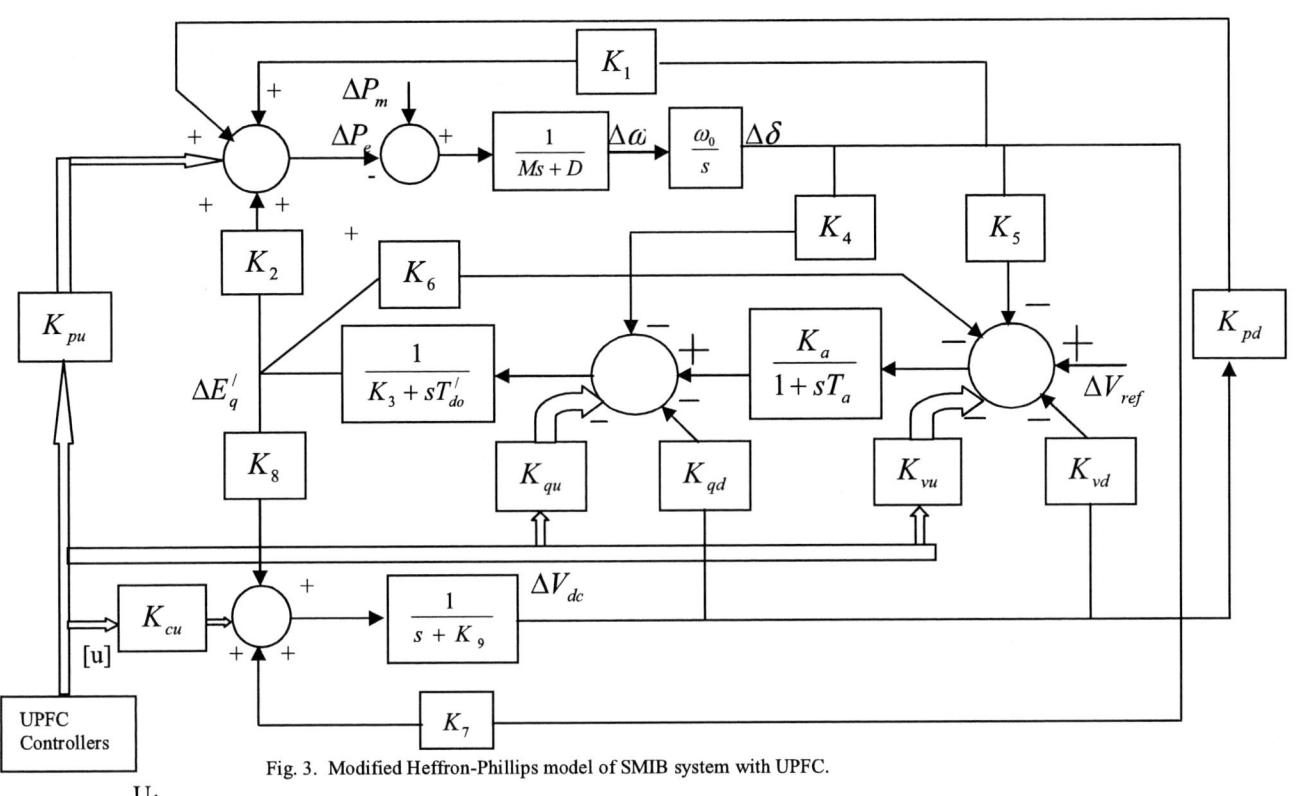

Fig. 3. Modified Heffron-Phillips model of SMIB system with UPFC.

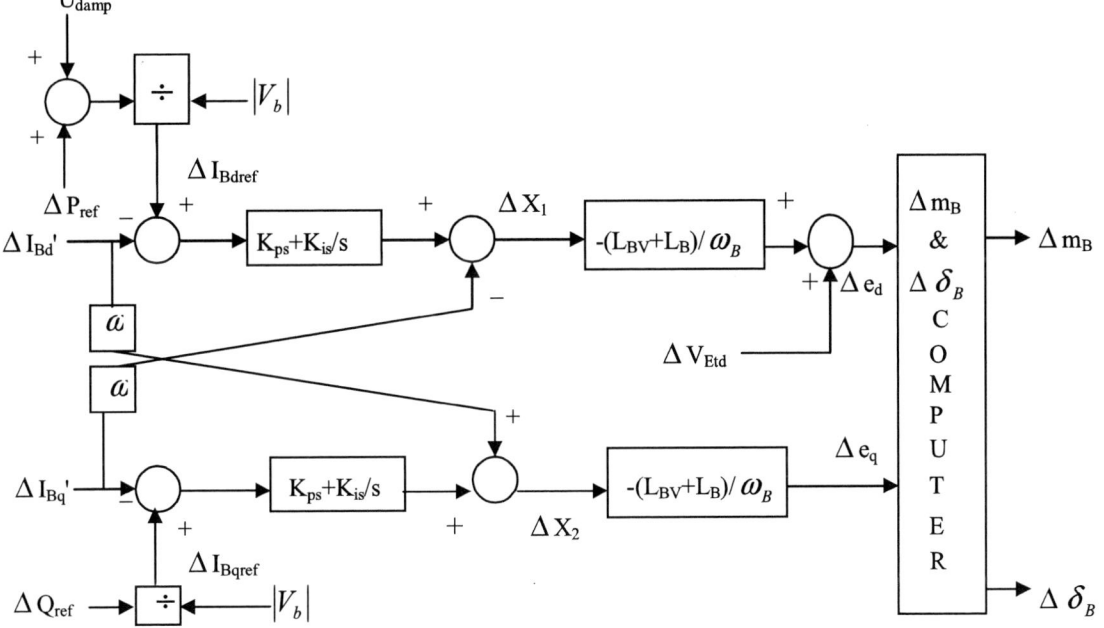

Fig. 4. Transfer function block diagram of the decoupled P-Q controller.

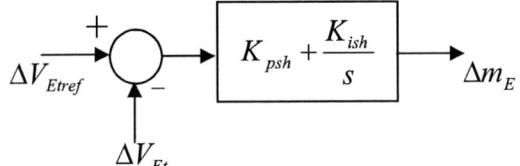

Fig. 5. Transfer function block diagram of AC terminal voltage regulator.      Fig. 6. Transfer function block diagram of DC link voltage regulator.

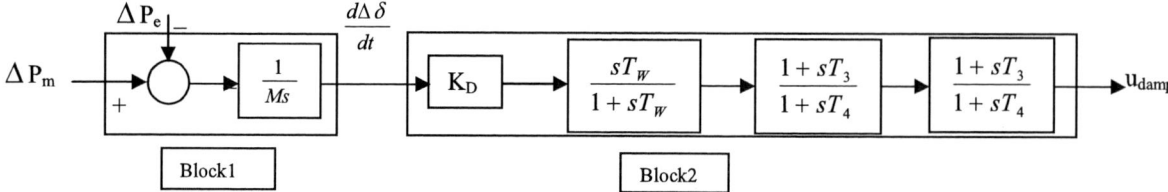

Fig. 7. Transfer function block diagram of UPFC based damping controller

The optimum PSS parameters obtained are: Ks=40; T11=T13=0.57s; T12=T14=0.32s.

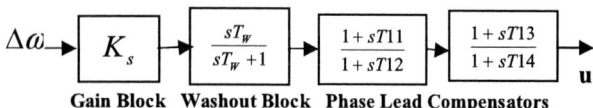

**Gain Block  Washout Block  Phase Lead Compensators**

Fig. 8. Transfer function block diagram of Delta- omega PSS.

### C. Optimization of UPFC controllers

The UPFC controllers are optimized simultaneously using GEA considering the objective function as following:

$$J= \int_0^{\infty}[(\Delta P)^2 + (\Delta Q)^2 + 100(\Delta Vdc)^2 + (\Delta V_{Et})^2]dt \quad (17)$$

While optimizing the UPFC controllers the optimum PSS settings as obtained in section *B* are considered.

The parameters of the UPFC controllers to be optimized are given in Table I. This Table also shows the range of the parameters in which GEA search is confined.

TABLE I
UPFC CONTROLLER PARAMETERS

| Controller | Parameter | Range of parameters considered |
|---|---|---|
| Decoupled P-Q flow | Kps Kis | $2000 \leq$ Kps $\leq 4000$ $4000 \leq$ Kis $\leq 8000$ |
| DC voltage Regulator | Kpsh Kish | $10 \leq$ Kpsh $\leq 20$ $50 \leq$ Kish $\leq 200$ |
| AC voltage Regular | Kpsh Kish | $10 \leq$ Kpsh $\leq 20$ $50 \leq$ Kish $\leq 200$ |

Fig. 9 shows the plot of objective function J vs generation number. The optimum parameters of the Decoupled P-Q controller, DC and terminal voltage regulators are:
Kps=4000; Kis=8000; Kpsh=19.983; Kish=106.61.

## VII. DYNAMIC PERFORMANCE OF THE SYSTEM WITH OPTIMUM UPFC CONTROLLERS

The dynamic performance of the system is obtained considering optimum UPFC controller for the nominal power flow on line2 without UPFC (P=0.3075pu, Q=0.1116pu).

The simulation studies are carried out using MATLAB POWER SYSTEM BLOCKSET. In all simulation studies the UPFC is inserted at 0.06s.

Fig 10 shows the dynamic responses for P, Q, V$_{dc}$ and sending end voltage for a 20% step increase in Pref at 0.3s followed by a step decrease in Pref by 40% at t=0.45s.

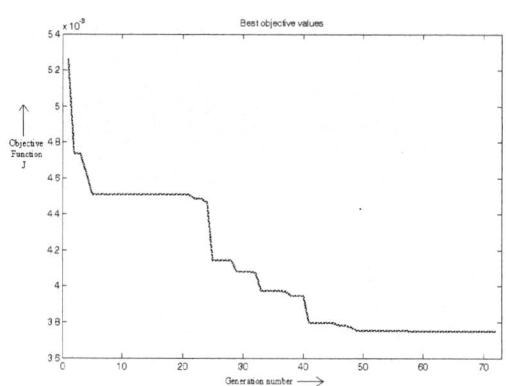

Fig. 9. Objective function J vs generation number (Optimization of UPFC controllers.)

Fig. 10 clearly shows that power flow on line 2 is regulated to the desired value. Further it can be seen that reactive power flow remains at the set value. The dc link voltage is regulated quickly to the set value. It may be noted that the control action is quite fast and smooth.

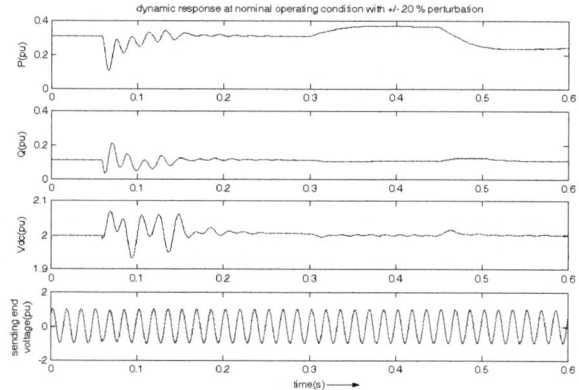

Fig. 10. Dynamic responses for line 2 real and reactive powers, DC link voltage and sending end voltage.

Fig. 11 shows the dynamic responses for P, Q, V$_{dc}$ and sending end voltage considering Qref setting reduced to zero from initial value=0.1116pu at 0.3s. It can be clearly seen that Q is regulated to the desired value without affecting the P.

## VIII. DESIGN OF UPFC BASED DAMPING CONTROLLER

The UPFC based damping controller (Fig. 7) is now designed using GEA considering system operating at nominal loading condition with optimum UPFC controllers and PSS designed in previous sections. The parameters to be optimized are given in Table II. This Table also shows the range in which GEA search is confined.

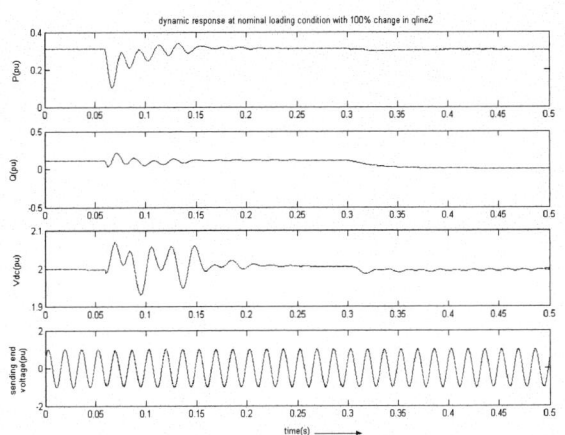

Fig. 11. Dynamic responses for line2 real and reactive powers, DC link voltage and sending end voltage.

TABLE II
UPFC BASED DAMPING CONTROLLER PARAMETERS

| Parameters | Range of parameters over which Search is confined |
|---|---|
| $K_D$ | $50 \le K_D \le 100$ |
| T3 | $0.4 \le T3 \le 0.6s$ |
| T4 | $0.2 \le T4 \le 0.5s$ |

The objective function $J = \int_0^\infty (\Delta\omega)^2 dt$ is considered.

Optimum parameters of the damping controller obtained are:
$K_D$ =99.976; T3=0.584s; T4=0.298s;

## IX. DYNAMIC PERFORMANCE OF THE SYSTEM WITH UPFC CONTROLLERS INCLUDING DAMPING CONTROLLER

Fig 12 shows the dynamic responses of system at nominal loading condition with
(a) Optimum UPFC controllers & PSS.
(b) Optimum UPFC controllers, PSS & UPFC based damping controller

A step increase in $P_{ref}$ ($\Delta P_{ref}$ =0.02 pu) is considered.

Fig 12 clearly shows that the UPFC based damping controller significantly improves the damping of the power system oscillations.

## X. CONCLUSIONS

A comprehensive approach for designing UPFC controllers (decoupled P-Q flow controller, DC & AC voltage regulators and UPFC based damping controller) has been presented using GEA. Studies clearly show that the real and reactive power flows on the line are controlled independently using UPFC decoupled controller. Investigations also reveal that the UPFC based damping controller is quite effective in enhancing damping of the power system oscillations.

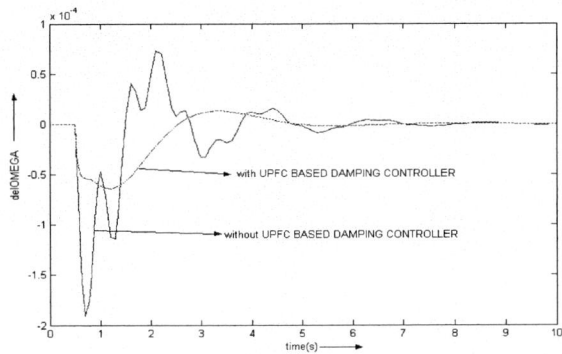

Fig. 12. Dynamic responses for $\Delta\omega$.

## APPENDIX

The nominal parameters & operating condition of the system:

Generator: M=8.0 MJ/MVA, $T_{do}^{/}$=5.044s,$X_d$=1.0pu,$X_q$=0.6 pu, $X_d^{/}$= 0.3 pu. Steam turbine & governor:  Governor Droop $R_p$=5%. Excitation system: Ka=50;Ta= 0.05s, $T_{LPfilter}$ = 20ms. Transformers: $X_{tE}$=0.1pu, $X_E$ =1.2pu,$X_B$=0.1pu, $r_E$ = 0.12 pu, $r_B$ =  0.01 pu. Transmission lines: $X_T$ =1.0pu,$X_{BV}$=1.3pu, $r_T$=0.1 pu, $r_{BV}$=0.13pu. Nominal operating condition : Pe=0.9114pu,Qe= 0.2765pu, Vt=1.032 pu, Vb=1.0pu,f= 60Hz UPFC parameters: $m_{E0}$=1.0, $m_{B0}$=0.1, $\delta_{E0}$=28.1°, $\delta_{B0}$=-21.1°, $V_{dc}$ = 2 pu,  $C_{dc}$ = 3 pu.

## REFERENCES

[1] C. Schauder and H. Mehta, "Vector Analysis and Control of Advanced Static Var Compensator," IEE proc.-Gener. Transm. Distrib., Vol. 140, No. 4, July 1993, pp. 299-306.
[2] H.F. Wang, "Damping function of Unified Power Flow Controller," IEE Proc. C. Genr. Transm. Distrib. 1999, 146,(1),pp.81-87.
[3] H.F. Wang, "A unified model for the analysis of FACTS devices in damping power system oscillations part III: unified power flow controller," IEEE Trans. Power Deliv, 2000, 15, (3), pp. 978-983.
[4] N. Tambey and M.L. Kothari, "Damping of power system oscillations with unified power flow controller (UPFC)," IEE proc.-Gener. Transm. Distrib. , Vol. 150, No. 2, March 2003, pp. 129-140.
[5] I. Papic, P. Zunko, D. Povh and M. Weinhold, "Basic control of Unified Power Flow controller," IEEE Trans. on Power Systems, Vol. 12, N0. 4, November 1997, pp. 1734-1739.

**Nilanjan Ray Chaudhuri** was born in Calcutta, W.B., India on 10th April 1981. He recieved B.E in electrical engineering from Jalpaiguri Govt. Engg. College, Jalpaiguri, W.B, India in 2003. He recieved M.Tech degree in power systems from IIT Delhi, New Delhi, India in 2005.

He is currently working as Edison engineer with GE in John F. Welch Technology Center, Bangalore, India.

**M.L.Kothari** (SM'92 ), received the B.E. degree in Electrical Engineering from University of Jodhpur in 1964 , M.E. degree in power systems from University of Rajasthan in 1970 and Ph.D. degree in 1981 from Indian Institute of Technology, New Delhi. At present he is serving as Emeritus Fellow at I.I.T. Delhi. Prof Kothari is Fellow of Indian National Academy of Engineering.

His research interests include FACTS, Automatic Generation Control, Power System Stabilizer, Dynamic security analysis, Computer Relaying and application of Artificial Neural Network and Fuzzy logic control to power systems.

**2006 IEEE International Conference on Power Electronic, Drives and Energy Systems**

# Application of Static Synchronous Series Compensator to Damp Sub-Synchronous Resonance

Hassan Barati, Afshin Lashkar Ara,
M. Ehsan, M. Fotuhi-Firuzabad, *Senior Member, IEEE*, and S. M. T. Bathaee

*Abstract*--In this paper, a Static Synchronous Series Compensator (SSSC) is used to avoid torsional mode instability in a series compensated transmission system. The power system used in this study is the IEEE first benchmark model for Sub-Synchronous Resonance (SSR) analysis. Simulation results in time domain using PSCAD/EMTDC presented and discussed.

*Index Terms*--SSR Phenomenon, SSSC, PSCAD/EMTDC.

## I. INTRODUCTION

SERIES capacitors have been extensively used as a very effective means of increasing power transfer capability of transmission system, and improving transient and steady state stability limits of a power system. This is due to partially compensating the reactance of the transmission lines. However, the application of series capacitors may lead to the phenomenon of SSR. Under a disturbance, series capacitors may excite sub-synchronous oscillations, when electrical resonant frequency of the network is close to natural torsional mode frequency of turbine-generator shaft. Under such circumstance the shaft will oscillate at this natural frequency. This oscillation might grow to endurance limit in seconds resulting in shaft fatigue and possibly damage and failure [1],[2].

Fast going development of power electronic technology have brought about wonderful possibilities for advanced new equipment in order to better exploitation form now systems.

A long last decade many controlled equipment under name of flexible AC transmission systems (FACTS) technology have been designed and completed. These devices placed in transmission lines as series, parallel and series-parallel, which

control exploitation parameters of transmission systems in steady state and also dynamic conduct of system in transient state. FACTS equipment can be used effectively for power flow, load divider between parallel corridors, voltage adjustment, increasing transient stability and reducing of system oscillations [3].

One of the FACTS devices is static synchronous series compensator (SSSC) which by using a voltage source converter connected to transformer, it is serially placed in transmission line, and injects a voltage with controlled amplitude and angle to transmission line, any time the injected voltage in relation to line current is perpendicular, the injecting voltage looks an impedance to line and it is completely the same as reactive compensator (inductive or capacitive). This mode can be choice for series capacitor compensation in system. If the angle between line current and injecting voltage SSSC is less than 90, the functional mode of SSSC will be as capacitive-ohmic and the equivalent series connection is a capacitor and a resistor. With suitable controlling strategy and depends on amplitude oscillations related to sub-harmonic mode, the value of equivalent capacitive-resistance reactance suitable to transmission line is injected and it is caused to attenuate and eliminate the negative resistance effect, which is encounter from resonance in compensated transmission lines with series capacitor and will consequently modify the SSR phenomenon [4]-[6].

The paper is organized as follows, with brief consideration of torsional oscillation in rotor turbine-generator in thermal generating units, the structure and functional modes of SSSC will be explained, and for purpose of SSR phenomena appearing simulation and method of damping that the PSCAD/EMTDC software have been used. For this purpose, the IEEE first benchmark model, that is submitted for studying SSR phenomenon in three cases have been used:

1. Compensation with capacitor only,
2. Compensation by capacitor in combination with SSSC,
3. Compensation by SSSC only,

With one symmetrical three phase short circuit.

## II. TURBINE-GENERATOR TORSIONAL CHARACTERISTICS[2]

### A. Multi-Mass Totsional Model

Fig 1 shows a typical multi-mass torsional model and simplified structure of shaft in a thermal generation unit, which

---

This work is sponsored by Islamic Azad University–Science & Research Branch, Tehran, Iran.

H. Barati is with the Department of Engineering, Islamic Azad University - Science & Research Branch, Tehran, Iran. (e-mail: barati216@gmail.com).

A. Lashkar Ara is with the Department of Engineering, Islamic Azad University-Dezful Branch, Dezful, Iran. (e-mail: a_lashkarara@hotmail.com).

M. Ehsan is with the Dept. of Electrical Engineering, Sharif University of Technology, Tehran, Iran. (e-mail: ehsan@sharif.edu).

M. Fotuhi-Firuzabad is with the Dept. of Electrical Engineering, Sharif University of Technology, Tehran, Iran. (e-mail: fotuhi@sharif.edu).

S. M. T. Bathaee is with the Dept. of Electrical Engineering, K. N. Toosi University, Tehran-Iran. (e-mail: bathaee@kntu.ac.ir)

consists of rotors torsion mass of generator, its exciting system, two sections of low pressure turbine section (LP), intermediate pressure turbine section (IP) and high pressure turbine section (HP). Dynamic characteristics of shaft system is determined by three sets of parameters: Inertia constant (H) related to individual masses, shaft spring constant (K) related to parts of shaft which connects adjacent masses together, Damping factor (D) for masses.

In transient case, the torque of generator for air gap is determined with dynamic of generator and power system, which is connected to it. Created torque by individual parts of turbine $(T_{LP_A}, T_{LP_B}, T_{IP}, T_{HP})$ depends on dynamic of steam turbine and its control system.

### B. Natural and Torsion Frequencies and Various Modes

If complete equations of shaft system written by state space form, then:

$$\dot{X} = A.X \qquad (1)$$

State variables are: velocity changes $\Delta\omega_i$, rotor angle changes $\Delta\delta_i (i = 1,2,3,4,5)$ and for a differential changes, the torque of generator air gap is as:

$$T_e = K_S . \Delta\delta_1 \qquad (2)$$

In which $K_S$ is stand for synchronized torque factor. The state of matrix $A$ for rotor system depends on inertia factor, shaft spring constant of any masses and synchronized torque factor of generator.

The eigenvalues of $A$ yield natural frequencies of shaft system and related eigen vectors "mode shapes". In general, a rotor with $N$ masses have $N-1$ torsional mode. Fig. 2 shows rotor natural frequencies and curves of different modes in a $555^{MV_2}$, $3600^{rpm}$ steam turbine-generator.

### III. SUB-SYNCHRONOUS RESONANCE PHENOMENON [1], [2]

Mainly the sub-synchronous resonance phenomenon (SSR) in compensated transmission system occurs with series capacitor. In the years 1970 to 1971 AD, the first problem of SSR come up in Mohave power plant in United States of America which caused turbine–generator shaft destruction, and after that SSR have become a considerable subject in power industry.

### A. Features of Compensated Transmission Systems by Series Capacitor

Fig. 3 shows a series compensated radial system, in which series capacitor with line inductance makes a series resonance circuit and its natural frequency is as bellow:

$$f_n = f.\sqrt{X_C/X_L} \qquad (3)$$

Where $X_C$ is capacitor reactance of each phase and $X_L$ is the total line reactance in base frequency. Since the ratio of compensation $(X_C/X_L)$ is usually between 25 to 70 percents $f_n$ is usually less than main frequency, therefore the system has sub-harmonic resonance or mode.

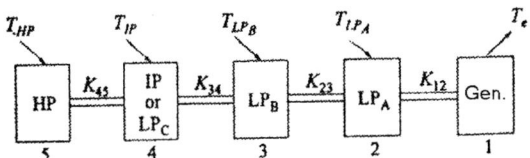

Fig. 1. Structure of a typical multi-mass model for turbine-generator shaft.

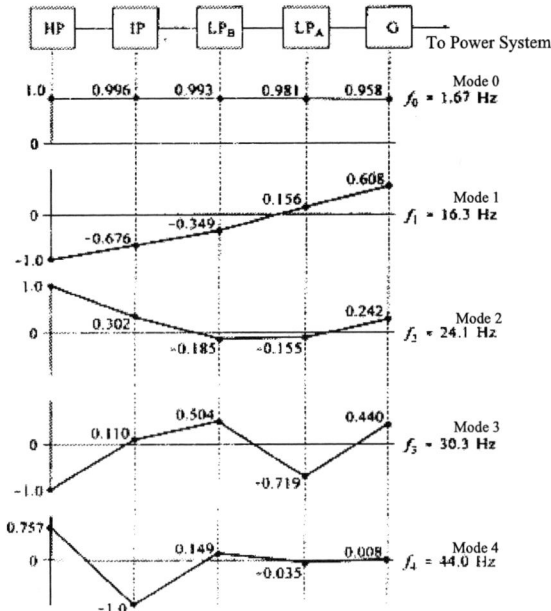

Fig. 2. Torsional natural frequencies and shapes of modes for a turbine-generator.

Fig. 3. Series compensated radial system.

During any disturbance, transient currents in sub-harmonic resonance frequency $f_n$ are excited and these currents are added to current at base frequency. The components with frequency $f_n$, which is related to stator generator, induce currents (consequently torque) in rotor with slip frequency $(f_o - f_n)$, therefore the compensated transmission networks by series capacitor can caused steady sub-synchronous oscillation or negative damped, which involve two mechanism: Self exciting from induction generator effect, Interference with torsional oscillations.

### B. Self-Exciting Encountered from Induction Generator Effect

Fig. 4 shows the simplified equivalent circuit of synchronous machine used for sub-synchronous effect consideration. In this model the effect of salient pole has not been considered, and machine has been shown with the same circuit used for induction motors in which the quantities of main frequency have been eliminated and only the effect of sub-synchronous frequency has been considered.

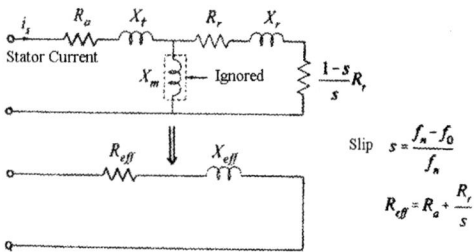

Fig. 4. The simplified equivalent circuit of synchronous machine.

Since $f_n < f_o$ the slip $S$ is negative and rotor is very similar to an induction machine in operational above synchronous speed. Depends on $f_n$, the effective resistance may be negative and in high ranges of compensation this apparent negative resistance may be exceeded the network resistance and operate like RLC circuit with negative resistance, this situation caused self exciting and in finally will be ended in unacceptable electrical oscillations. This form of self exciting is completely electrical phenomenon, and depends on shaft torsional characteristics.

*C. Interference torsional to result in SSR*

If $(f_o - f_n)$ be closed to one of torsional frequencies of turbine-generator shaft system, the torsional frequencies may be excited which it is called sub-synchronous resonance (SSR). Under such a condition the small induction voltage created from rotor oscillation may be converted to a large sub-synchronous currents, this current creates an oscillation component in rotor torque and its phase is such that to increase more rotor oscillation. When this torque is larger than the torque of mechanical damping, the electromechanical system attached to it shows the increasing oscillations, and sub-synchronous resonance occurrences may be dangerous. If torsional oscillations happens, the turbine-generator shaft will break with fearful occurrences.

IV. STATIC SYNCHRONOUS SERIES COMPENSATOR [4], [5]

The SSSC is settled in transmission line in series and injects a voltage with controlled magnitude and angle into it. This injected voltage is, directly or indirectly, always used to control the flowing power on the line. However, this injected voltage is dependent on the operating mode selected for the SSSC to control power flow. The principal operating modes are as follows:

- **Line impedance compensation mode:** When the injected voltage is kept in quadrature with respect to the line current, so that the series insertion emulates an impedance when viewed from the line, to emulate purely reactive (inductive or capacitive) compensation. This mode can be selected to match existing series capacitive line

compensation in the system.
- **Automatic power flow control mode:** The magnitude and angle of the injected voltage is controlled so as to force such a line current that results in the desired active and reactive power flow in the line. In automatic power flow control mode, the series injected voltage is determined automatically and continuously by a closed-loop control system to ensure that the desired active and reactive power flow are maintained despite power system changes.

Fig. 5 shows the effect of reactive compensation on normalized power flow of transmission line and normalized effective reactance of transmission line, in which $X_q$ is equivalent reactance of SSSC related voltage into transmission line.

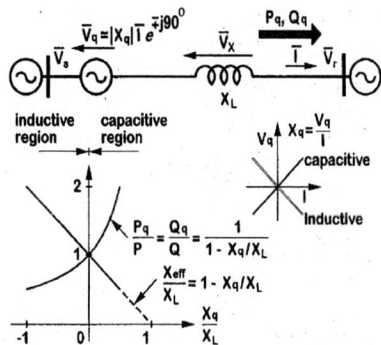

Fig. 5. The effect of reactance compensator on power flow and effective reactance.

When equivalent reactance is inductive, power flow $(P_q, Q_q)$ decreases and effective reactance $(X_{eff})$ also compensated reactance $(-X_q / X_L)$ increases.

Fig. 6 shows the $V-I$ characteristics obtainable from operating SSSC is terms of line current for series compensating of transmission line when it operate on voltage control mode (Fig. (6-a1)) and reactance-control mode (Fig. (6-b1)). Fig. (6-(a2, b2)) shows the percentage of parallel loss in terms of line current for above operating modes.

In many practical applications in which economical problems must be considered, combination of a constant capacitor and SSSC are used and called "hybrid series compensator" for series compensation as shown in Fig. 7. $V-I$ Characteristic obtained in voltage-control and reactance-control modes are shown in Fig. 8 in comparison with operating SSSC lonely.

Controlling injected voltage of SSSC to transmission line continuously, make it possible, that in capacitive mode the angle between current of line and injected voltage of less than 90 degree, in the case, SSSC is equal to a connection of a capacitor and a resistor in series. In other words, circuit works in capacitive-ohmic mode and can eliminate or attenuate the effect of negative ohmic resistance obtained from sub-harmonic resonance and finally SSR phenomena.

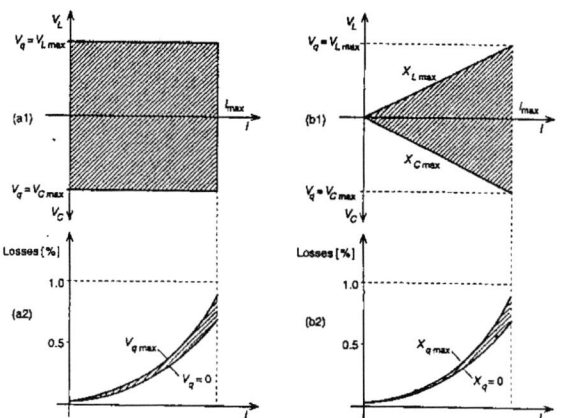

Fig. 6 V-I characteristic and losses(%) of SSSC in term of line current: a- voltage control mode, b- reactance-control mode.

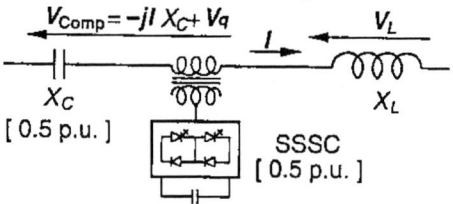

Fig. 7. Hybrid series compensation scheme.

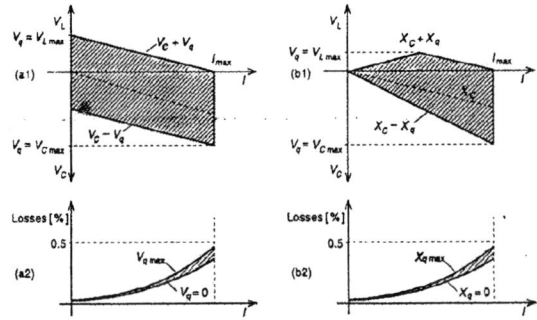

Fig. 8. V-I characteristic and losses (%) in hybrid series compensator operated in term of line current: a-Voltage-control mode, b-Reactance control mode.

## V. COMPUTER SIMULATION OF SSR PHENOMENA APPEARING AND IMPROVEMENT [8],[7]

The PSCAD/EMTDC software is a specific professional simulation tools for analyzing power systems. Also this software is the most suitable device for simulation of instantaneous time domain response and electromagnetic transients of electrical systems.

In this article, using software and the IEEE first benchmark model (which is for studying SSR phenomena), simulation have done for three cases:

1. Series compensation with capacitor only,
2. Compensation with hybrid series compensator,
3. Series compensation with SSSC only,

and with making a symmetrical three phase short circuit to ground the SSR phenomena leakage in case 1 have been shown, and by installing a SSSC with a suitable controlling device the SSR effect in 2 and 3 cases have been subsided.

### 1. Series Compensation with Fixed Capacitor

The IEEE first benchmark model has been used for SSR phenomena studies in PSCAD/EMTDC software, which includes generator-turbine, transformer, series compensated transmission line and infinite bus. With performing a computer program and making a three phase symmetrical short circuit in specific time, the SSR phenomena occurs and torsional oscillations have been excited and unstable oscillation have been appeared in electrical and mechanical parameter and if this oscillations do not damp. And attenuate, it will lead to break turbine-generator shaft.

Fig. 9 shows generator output voltage, the voltage across capacitor, Generator output current, Generator electromagnetic torque, $LP_A$ to $LP_B$ torque, and generator to exciter torque respectively.

### 2. Hybrid Series Compensation:

To prevent SSR phenomena, a SSSC installation have been used in transmission line. Switching signals of GTO's for SSSC converter are generator by SPWM technique or designing a suitable PI controller and injecting measured signals to controller input (feedback signals), which caused to inject a suitable voltage in transmission line by SSSC and under operating fixed capacitor unstable oscillations damped and severely attenuated and role of SSSC in elimination of sub-synchronous resonance effect and prevention of breaking shaft is properly obvious from computer results.

Fig. A1 in appendix shows a typical power system with hybrid series compensatory control circuit and *PI* controller.

Fig. 10 shows generator output voltage, voltage across capacitor, generator out put current, generator electromagnetic torque, $LP_A$ to $LP_B$ torque, and generator to exciter torque, related to hybrid series compensatory.

### 3. Series Compensation by SSSC

Taking in consideration the operating modes of SSSC and control ability of amplitude and angle of its injected voltage independent from line current make if possible for series compensation to give profit lonely. In this section with the same condition used for hybrid soirees compensation, But by eliminating fixed capacitor, simulation have been done and obtaining results have been indicated in Fig. 11 which in comparison with Fig. 9 and 10.

### VI. CONCLUSION

In this paper, while considering the way of appearing SSR phenomena in systems of series compensated with a fixed capacitor modes of function for static synchronous series compensatory (SSSC) also were explained. The need of installing a SSSC with a suitable Controller for continuously controlling amplitude and angle of SSSC injected voltage to transmission line, for improvement of SSR phenomena were described. By using a powerful software PSCAD/EMTDC on the IEEE first benchmark model, simulation occurred and the results indicates SSSC success in eliminating SSR phenomena and preventing turbine-generator shaft breakdown.

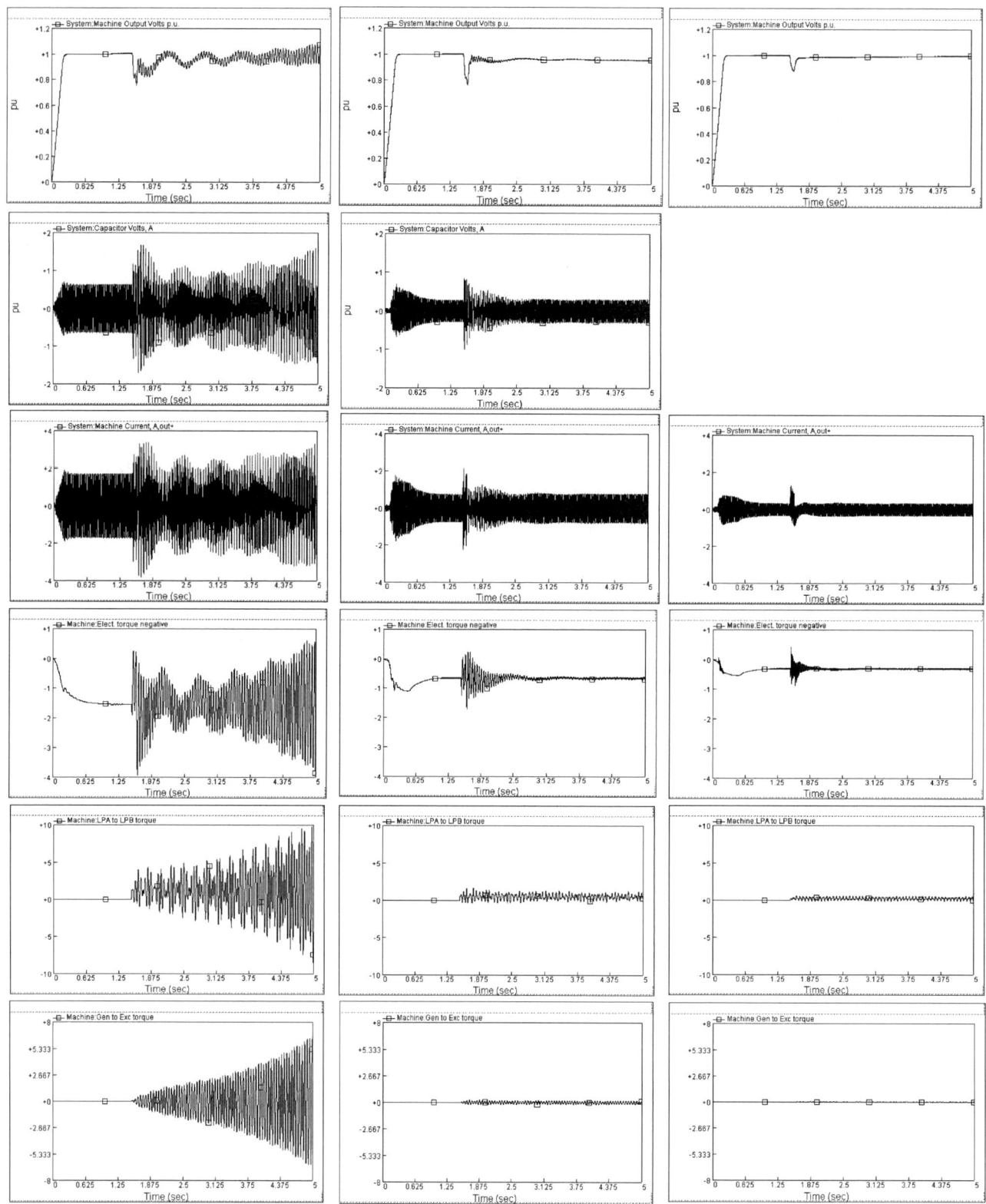

Fig. 9. Results obtained from series compensation by fixed capacitor.

Fig. 10. Results obtained from hybrid series compensatory.

Fig. 11. Results obtained from series compensation with SSSC.

## VII. REFERENCE

[1] T. J. E. Miller,"Reactive Power Control in Electric System", John Wily & Sons Inc.,1984.

[2] P. Kundur, "Power System Stability and Control", New York, NY:Mc-Graw-Hill, 1994.

[3] Yong H.Song, Allan T. Johans, "Flexible AC Transmission Systems (FACTS)", IEE Inc., 1999, ISBN 964-6426-41-7.

[4] Narain G. Hingorani, Laszlo Gyugyi, "Understanding FACTS: Concepts and Technology of Flexible AC Transmission Systems", by the IEEE, Inc., 2000.

[5] A. Lashkar Ara, S.A. Nabavi Niaki, "Comparison of the FACTS Equipment Operation in Transmission and Distribution Systems", CIRED conference, Barcelona 2003.

[6] Kalyan K. Sen, "SSSC–Static Synchronous Series Compensator: Modeling and Applications", IEEE Transactions on Power Delivery , Vol.13, No. 1, Jan. 1998.

[7] L. Sunil Kumar, Arindam Ghosh,"Modeling and Control Design of a Static Synchronous Series Compensator", IEEE Trans. on Power Delivery, Vol. 14, No. 4, October 1999.

[8] G. N. Pillai, Arindam Ghosh, A. Joshi, "Robust Control of SSSC to Improve Torsional Damping", Power Engineering Society Winter Meeting, 2001. IEEE, Vol.: 3, 28 Jan. - 1 Feb. 2001, pp 1115 -1120.

## VIII. BIOGRAPHIES

**Hassan Barati** received the B.Sc. degree in Electronic Engineering from Isfahan University of Technology, Iran, in 1992. He obtained his MSc degree in Electrical Power Engineering from Tabriz University, Iran, in 1996. He is now the Ph.D. student in Islamic Azad University – Science & Research Branch, Tehran, Iran. His research interest is application of FACTS devices for congestion management in restructured power systems and power quality.

**Afshin Lashkar Ara** received the BSc degree in Electrical Power Engineering from Islamic Azad University – Dezful Branch, Dezful, IRAN, in 1995. He obtained his MSc degree in Electrical Power Engineering from Mazandaran University, Iran, in 2001. He is now the Ph.D. student in Iran University of Science &Technology (IUST), Tehran, IRAN. His research interest is power quality, FACTS devices, electric market, restructured power system, fuzzy systems

**M. Ehsan** received BSc & MSc in Electrical Engineering from Technical College of Tehran University 1963 and DIC & Ph.D. from Imperial College, University of London in 1977. He is currently a professor in Electrical Engineering department of Sharif University of Technology. His research interests include power system dynamics, application of expert systems in operation, control and economic aspects.

**M. Fotuhi-Firuzabad** received BSc & MSc degrees in Electrical Engineering from Sharif University of Technology and Tehran University in 1986 and 1989 respectively and MSc and Ph.D. degrees in Electrical Engineering from the University of Saskatchewan in 1993 and 1997. He is currently Associate Professor and head of the Department of Electrical Engineering in Sharif University of Technology, Tehran, Iran. His research interests include Power System Reliability, Power System Analysis, Unit Commitment and Power System Operation, Computer Application in Power Systems, Renewable Energy.

**S. M. T. Bathaee** received the B.Sc. degree in math-computer science and electrical engineering from Tehran University & K.N. Toosi University of technology respectively in Iran in 1976. He joined the George Washington Univ., USA in 1977 and awarded the M.Sc degree of energy conversion in electrical engineering in 1979, and his PhD degree from Amir-Kabir Univ. of technology (Tehran-Iran) in 1995. Since 1978, he has been academic staff and assistant Prof. of Electrical Engineering Department of K.N.T. University of technology. His current research interest includes power system analysis, design and control of hybrid systems and cogeneration.

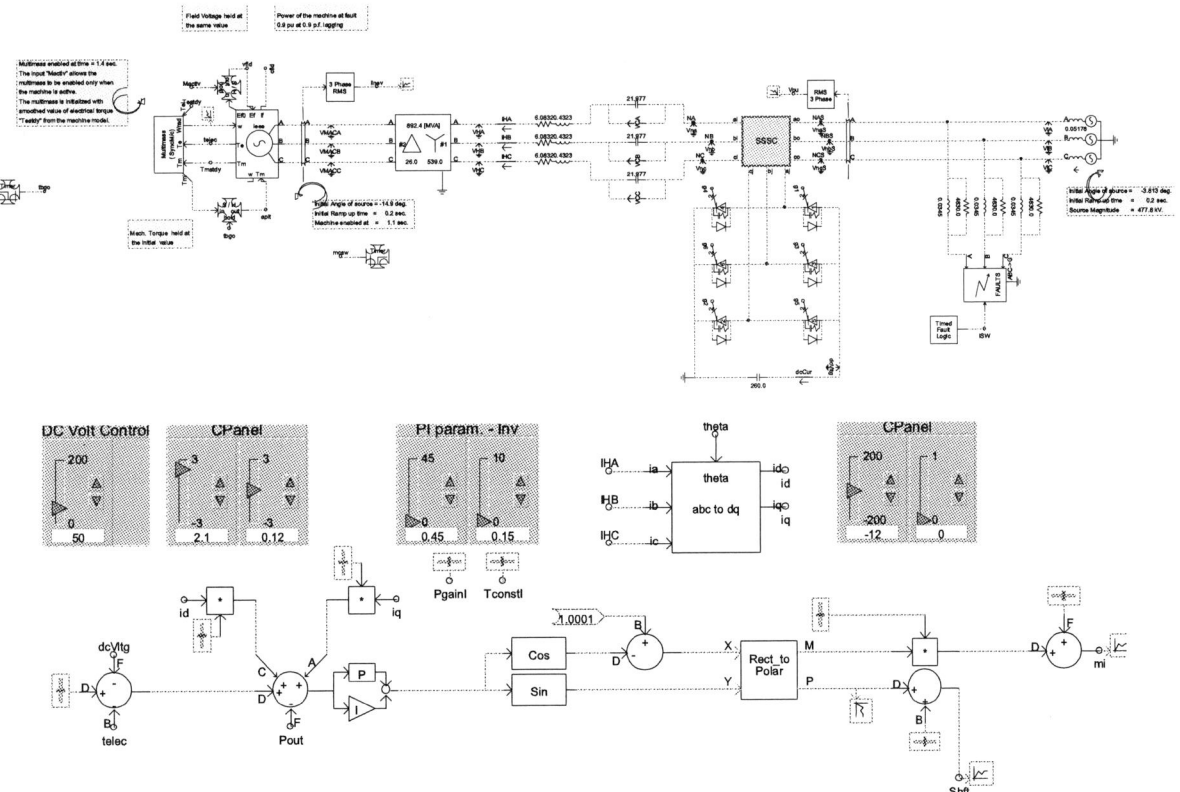

Fig. A1. The IEEE first benchmark model with hybrid series compensation and PI controller.

**2006 IEEE International Conference on Power Electronic, Drives and Energy Systems**

# A New 24-Pulse STATCOM for Voltage Regulation

Bhim Singh, *Senior Member, IEEE*, and R. Saha, *Senior Member, IEEE*

*Abstract-* **This paper is focused on design and modeling of a new fundamental frequency switching based 24-pulse 2-level $\pm$ 100MVAR STATCOM in MATLAB platform for high power applications. Only four 6-pulse elementary Gate Turn off voltage source converters (GTO-VSC) along with single stage magnetics meeting dual objectives of magnetic summing circuit and coupling transformer at PCC, and PI-controllers employing principle of phase angle control, are modeled to achieve an improved performance to regulate load voltage in an ac network with THD values within limits.**

*Index Terms-* Fast Fourier Transformation, Gate-Turn off Thyristor, Magnetics, STATCOM, Total Harmonic Distortion, Voltage Source Converter.

## I. INTRODUCTION

SELF-commutating GTO devices in VSC technology are widely used as main controllable switching element for high power rating compensators, and operated either in square wave or quasi square wave mode by means of GTO triggering once per cycle of fundamental power frequency. In the state-of-the art dynamic reactive power compensator, use of multi-pulse topology along with fundamental frequency switching control of GTO-VSC is a mature technique employed to achieve a close to sinusoidal AC output voltage from GTO-VSCs. In this topology, a number (P) of elementary six-pulse converters waveforms are electro-magnetically added to produce a multi-pulse waveform which contains harmonics in the order of $6NP \pm 1$ ,where N=1, 2, 3, 4,...and P=1, 2, 3, 4...number of six pulse VSC. For example, in a 4x6-pulse STATCOM, AC voltage output waveform will contain harmonics of the order of $23^{rd}$ , $25^{th}$, $47^{th}$, $49^{th}$,....etc.

In this paper, a 24-pulse, 2-level, $\pm$ 100MVAR STATCOM model employing 4 elementary GTO-VSCs with a design of single stage interfacing magnetics is simulated in MATLAB environment to control voltage of an inductive load in AC network [1-7]. A controllable three-phase harmonic optimized 24-pulse voltage waveform with a phase displacement of 15° is obtained after adding electro-magnetically the AC output voltages of two pairs of GTO-VSC operated at displacement angles of (37.5°, 7.5°) and (22.5°, –7.5°). As control methodology, principle of phase angle control ($\alpha$) across the reactance of the coupling transformer is adopted using standard PI-controllers to be used for voltage and current control in d-q rotating frame [2,4,6,7]. The proposed model is simulated in MATLAB/SimPowerSystems environment and results show an acceptable operating performance

---

Bhim Singh is with the Department of Electrical Engineering, Indian Institute of Technology, New Delhi 110 016, India (e-mail: bsingh@ee.iitd.ac.in, bhimsinghr@gmail.com)

R. Saha is with the Central Electricity Authority, Sewa Bhawan, R. K. Puram, New Delhi 110 066, India (e-mail: rshahacno@yahoo.com)

characteristics and reasonably low THD values [8] of supply voltage and current.

## II. WORKING PRINCIPLE OF STATCOM

The main objective of STATCOM is to control reactive current by generation and absorption of controllable reactive power. The essential components in a GTO-VSC based STATCOM are GTO-VSC bridge(s), DC capacitor (C) working as an energy storage device, interfacing magnetics forming the electrical coupling between the VSC bridge circuits, AC mains system, and controllers generating gating signals.

A controllable three-phase 24-pulse AC output voltage waveform is obtained at the point of common coupling (PCC). AC output voltage of VSC bridge circuits ($V_c$) is governed by DC capacitor voltage ($V_{dc}$), which is controlled by varying phase difference between $V_c$ and $V_s$ (system voltage at PCC). An almost sinusoidal current in quadrature ($I_q$) with the line voltage is injected into electrical system emulating an inductive or a capacitive reactance at the point of common coupling (PCC). The magnitude and phase difference of $I_q$ determine the magnitude and phase difference between $V_c$ and $V_s$ across the transformer leakage inductance, which in turn controls reactive power flow. Fig. 1 and Fig. 2 show the basic architecture of a STATCOM and its phasor diagrams for various operating modes respectively. When $V_c > V_s$, the STATCOM is considered to be operating in a capacitive mode and when $V_c < V_s$, it is operating in an inductive mode and for $V_c = V_s$, no reactive power exchange takes place and STATCOM is said to be operating in floating mode.

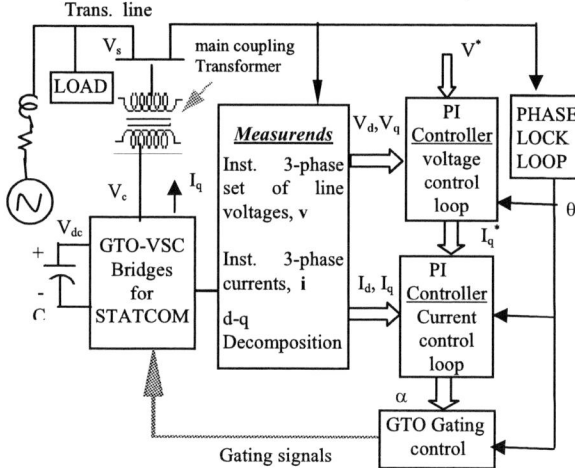

Fig. 1. GTO-VSC based STATCOM Architecture.

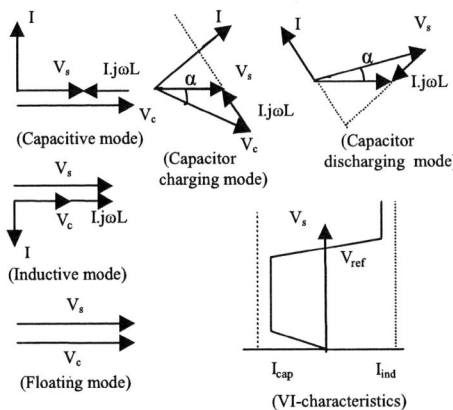

Fig. 2. Operating modes of STATCOM and phasor diagrams.

### III. MODEL OF STATCOM

Fig. 3 shows the circuit layout of the proposed 24-pulse STATCOM configuration. The 24-pulse model is achieved by 4x6-pulse GTO-VSCs which are connected in parallel on the DC side with an energy storing DC capacitor (15000μF) and operated in fundamental frequency switching mode at displacement angle of 37.5°, 7.5°, 22.5° and -7.5°. De-coupled AC output voltage waveforms of the four elementary VSCs are electro-magnetically added by two multi-winding transformers to obtain a 24-pulse waveform with a step size of 15° across the primary terminals of the transformer configuration. The AC system is represented by Thevenin equivalent voltage source with a short circuit level of about 3000MVA and X/R ratio equal to 10. The PI-controllers are employed in two control loops e.g. outer voltage control loop and inner current loop. The voltage loop determines reference reactive current ($I_q$*) based on system voltage, which is used to the inner current controller for phase angle control (α).

#### A. Inter-facing Magnetics

Multi-winding linear transformer model with a configuration of two secondary windings and single primary winding has been employed per phase. Three such multi-winding transformers have been connected to obtain a 3-phase transformer in which 6-terminals of secondary constitute Δ-Y connection and three terminals of primary constitute open Y-connection; corresponding MATLAB model of the transformer windings is shown in Fig. 4. This 3-windings circuit arrangement is used as a 3-phase transformer unit with a configuration of Y/Δ-Y and rating of 50MVA, 50Hz, 66/5.1/5.1kV, X (=0.08pu) and two such units are employed. Secondary sides (5.1kV) of a 3-phase transformer unit are connected to the output terminals of a pair of 6-pulse GTO-VSCs (out of 4x6 pulse VSC) triggered at a displacement angle of 37.5° and 7.5°. The secondary sides (5.1kV) of the second 3-phase transformer unit are connected to another pair of VSC triggered at a displacement angle of 22.5° and –7.5°. This Y-Δ secondary connection providing a phase displacement of 30° enables to eliminate 11th, 13th, 23rd, 25th (i.e. 12N±1) harmonics from the inverter AC outputs. Primary sides (open Y-windings) of the two 3-phase transformer units are connected in series to obtain a Y connected configuration, which enables to add electro-magnetically the four AC output voltage waveforms of the 4x6 pulse VSC to provide a 24-pulse waveform across the terminals with a phase displacement of 15°. The resulting 24-pulse AC voltage waveform would have 24N±1 harmonics. This Y connected primary winding is coupled with the AC supply feeding a reactive load. The turn ratio of Δ and Y secondary windings is made √3 to maintain equal line voltage across their terminals.

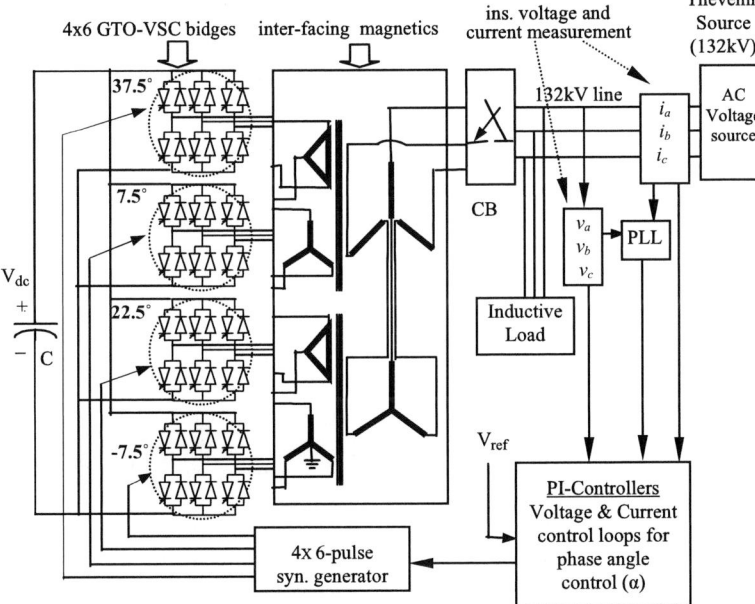

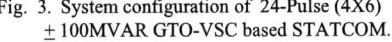

Fig. 3. System configuration of 24-Pulse (4X6) ± 100MVAR GTO-VSC based STATCOM.

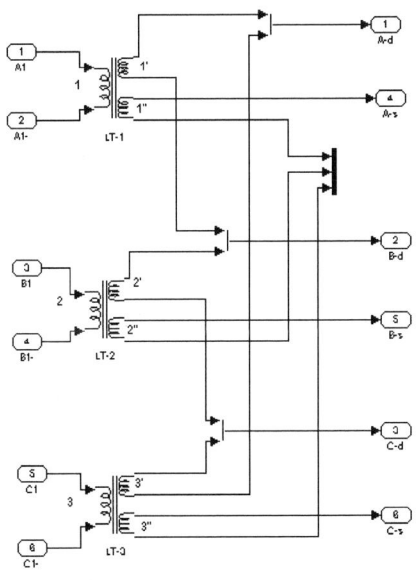

Fig. 4. Transformer configuration in MATLAB.

## IV. MATLAB MODEL OF STATCOM

Fig. 4a shows the MATLAB model diagram of the proposed STATCOM. The inner current control loop and outer voltage control loop circuits of PI-controllers are shown in Figs. 4b-4c respectively. As shown in the basic STATCOM model (Fig. 1), the AC voltage and current signals (instantaneous values) are sensed in time domain using proper sensors and synthesized/decomposed by d-q synchronous rotating axis transformation. Phase Lock Loop (PLL) is employed to calculate phase and frequency information of the fundamental positive sequence component of system voltage, which synchronizes VSC AC output voltage. The AC system is represented by Thevenin equivalent voltage source.

The PI-current control loop produces the phase angle difference ($\alpha$) of the inverter voltage ($V_c$) at PCC relative to the system line voltage ($V_s$) and thus, enables to inject an almost sinusoidal current in quadrature with the line voltage emulating an inductive or a capacitive reactance at the point of common coupling (PCC) with the electrical system. The outer PI-voltage control loop employed to regulate system voltage determines the reference reactive current ($I_q{}^*$) for the inner current loop.

## V. RESULTS AND DISCUSSION

The proposed 24-pulse STATCOM model has been simulated as a dynamic reactive power compensator in MATLAB/Simulink environment for voltage regulation.

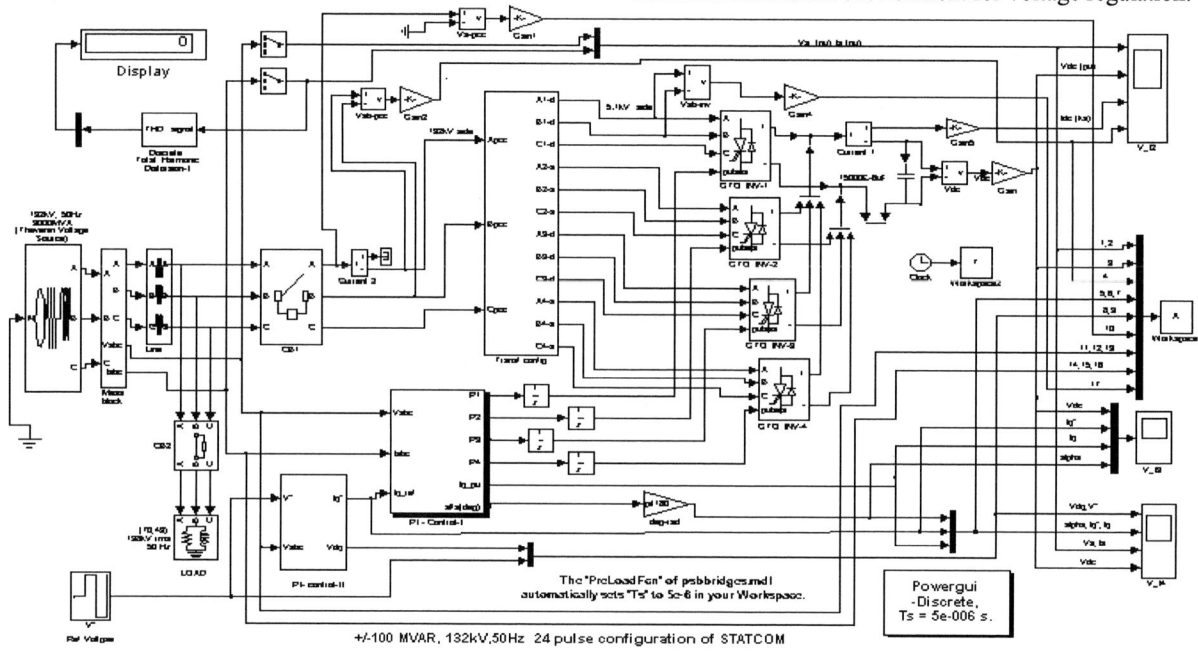

Fig. 4a. MATLAB circuit layout of the 2-level 24-pulse STATCOM.

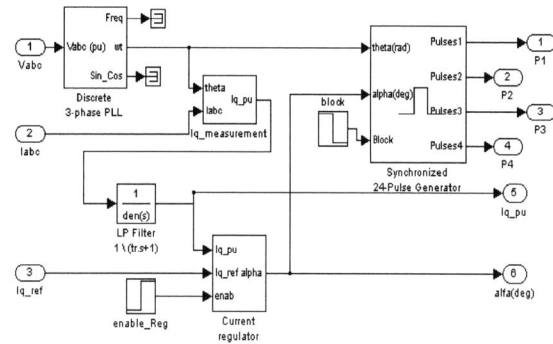

Fig. 4b. Inner current control loop layout.

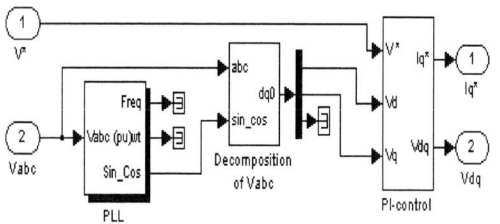

Fig. 4c. Outer voltage control loop layout.

Fig. 5a shows the AC output voltage waveform across the terminals (open) of the proposed 24-pulse compensator. The operating performance characteristics have been explained for steady state and dynamic operating conditions. During steady state operating conditions, analysis of voltage and current harmonic spectra has been carried out using FFT tools in MATLAB for determining THD levels under various operating conditions and the results show reasonably low harmonics content.

### A. Steady State Operation

Reference line voltage $V^*$ is set at 1.0pu, 1.03pu and 0.97pu corresponding to the intervals of (0s-0.22s), (0.22s-0.42s) and (0.42s-0.62s) respectively. The value of the capacitive reactive current limit is set to $I_q{}^*$=1.2pu and inductive reactive current limit at $I_q{}^*$= -1.2pu in the voltage control loop presuming that STATCOM would be operated as a voltage regulator. With the DC capacitor (C) pre-charged and total simulation time set at 0.62sec, the performance of the compensator corresponding to a load of 70MW 0.85pf lagging has been studied. Phase voltages ($v_a$, $v_b$, $v_c$) and supply current characteristics ($i_a$, $i_b$, $i_c$) corresponding to the specific time intervals are shown in Fig. 5b. The operating characteristics of other system variables e.g. $v_{a\text{-}pcc}$, $(v_a, i_a)$, $(V^*, V_{dq})$, $V_{dc}$ and $(\alpha, I_q^*, I_q)$ in respect of time (sec) are shown in Fig. 6.

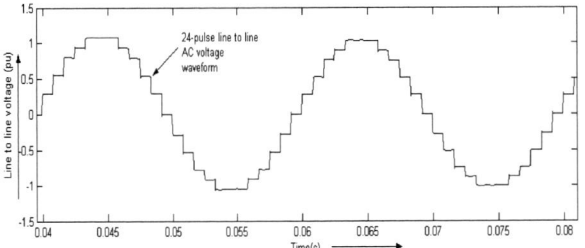

Fig. 5a.  STATCOM  open terminal  voltage waveform.

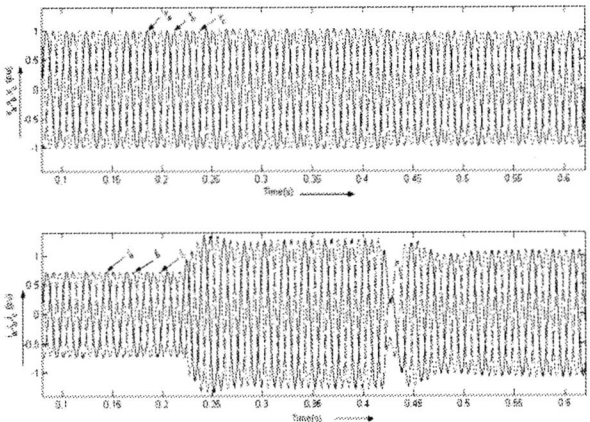

Fig. 5b.  Three-phase inst. supply voltage $v_a$, $v_b$ and $v_c$ and current $i_a$, $i_b$ and $i_c$ when $V^*$ set to 1pu, 1.03pu and 0.97pu.

### B.  Voltage Regulation at a Load of 70MW 0.85PF Lagging

Fig. 6 also shows that phase angle ($\alpha$) control employing PI-controller enables to regulate DC voltage ($V_{dc}$) across the capacitor (C), which in turn provides smooth and rapid control of load voltage ($v_a$) during the specified time intervals.

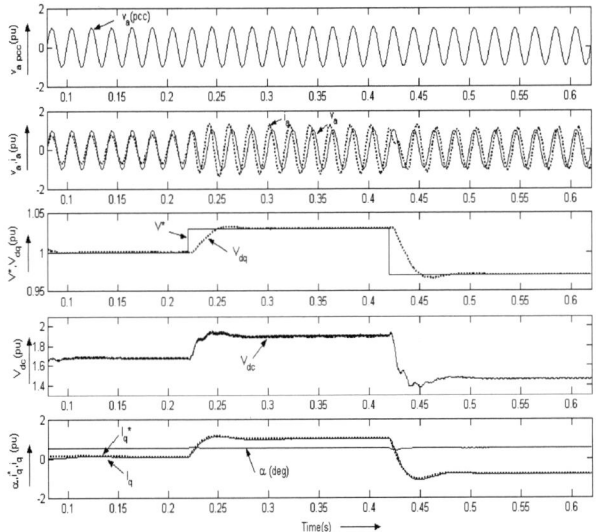

Fig. 6. Operating characteristics of the  STATCOM at 70MW 0.85pf  lagging load condition in voltage control mode.

### C.  Dynamic Characteristics

Fig. 6 shows that while reference voltage $V^*$ is dynamically changed from 1.0 pu to 1.03 pu and from 1.03pu to 0.97pu at the instant of 0.22s and 0.42s, the compensator responds within couple of cycles during both the loading conditions and it is found that the controller provides necessary damping to rapidly settle to steady states enabling smooth operation of the system. No major overshoots or undershoots in voltage and current transients have been observed from the operating characteristics.

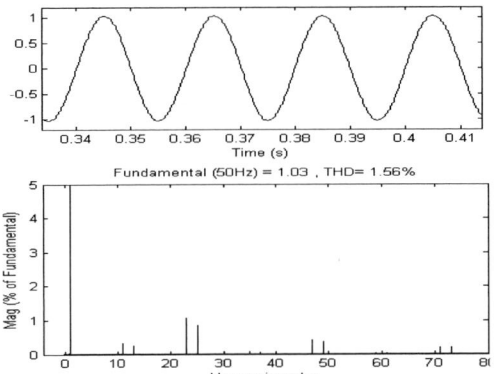

Fig. 7a.  Harmonic spectrum of phase-a voltage ($v_a$) in capacitive mode for voltage control at 70MW 0.85pf (lag).

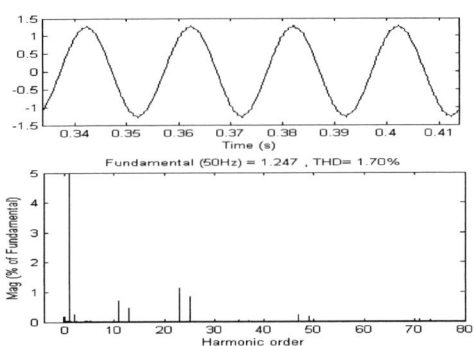

Fig. 7b.  Harmonic spectrum of phase-a current ($i_a$) in capacitive mode for voltage control at 70MW 0.85pf (lag).

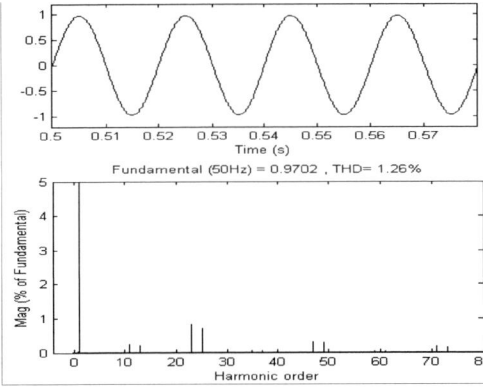

Fig. 7c.  Harmonic spectrum of phase-a voltage ($v_a$) in inductive mode for voltage control at 70MW 0.85pf (lag).

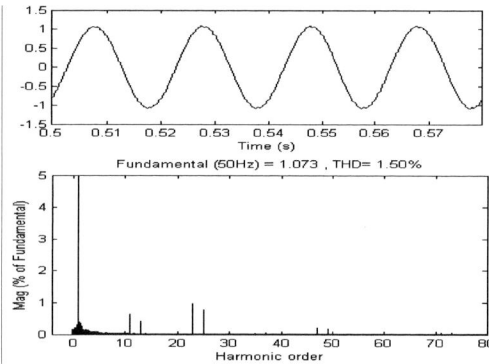

Fig. 7d. Harmonic spectrum of phase-a current (i$_a$)
in inductive mode for voltage control at 70MW 0.85pf (lag).

### E. Power Quality Aspects

The characteristics of the 24-pulse STATCOM model designed with typical magnetics have shown improvement in reducing higher order harmonics. Figs. 7a-7d show the voltage and current harmonic spectra along with FFT analysis determining THDs corresponding to capacitive and inductive modes of operation of STATCOM for a 70MW 0.85pf lagging load. It is seen that THDs in line voltage and current are reasonably low. An analysis on the voltage and current THDs for capacitive and inductive mode of operation of the proposed STATCOM has been shown in Table-I. The THD values have been found to be within IEEE-519 standard operating limits.

TABLE I
THD SUMMARY DURING CAPACITIVE AND INDUCTIVE
OPERATING MODES OF THE PROPOSED STATCOM

| Use of STATCOM | Inductive Load | Mode of Operation | %THD voltage (v$_a$) | %THD current (i$_a$) |
|---|---|---|---|---|
| Voltage Regulation | 70MW 0.85pf | Capacitive | 1.56 | 1.70 |
| | | Inductive | 1.26 | 1.50 |

### VI. CONCLUSION

The proposed 4x6-pulse 2-level GTO-VSC based ± 100MVAR STATCOM model has been designed and simulated with a single stage magnetics in place of standard two stages of magnetics topology being widely used in industries/utilities, and thus, it requires less number of transformers and simple control circuits. The magnetics is designed to meet dual objectives viz. adding AC voltage output of four VSC bridges and acting as main coupling transformer at PCC. Voltage regulation at desired set values in an inductive network has been achieved by means of phase angle control (α) across the leakage reactance of the transformer circuits and for that standard PI-control methodology and fundamental frequency switching control of GTO devices have been employed. The proposed multi-pulse topology based 24-pulse 2-level compensator model has enabled to operate it within permissible THD limits [8]. The operating characteristic of the compensator during steady state capacitive and inductive modes of operation and dynamic system conditions has been reasonably acceptable and

competitive for a design of an economical dynamic static compensator.

### VIII. REFERENCES

[1] C. W. Edwards, P. R. Nannery, K. E. Mattern and J. Gubernick, "Advanced Static VAR Generator Employing GTO Thyristors", *IEEE Trans. on Power Delivery*, Vol. 3, No.4, Oct, 1988, pp. 1622-1627.

[2] Eric J. Stacey, "Simplified Quasi-Harmonic Neutralized High Power Inverters", U.S.Patent 4 870 557, Sept. 26, 1989.

[3] Shosuke Mori, Katsuhiko Matsuno, Taizo Hasegawa, Shuichi Ohuichi, Masatoshi Takeda, Makoto Seto, Shotaro Murakami, and Fuijio Ishiguro, "Development of a Large Static VAR Generator using self-commutated inverters for improving power system", *IEEE Trans. on Power Systems*, Vol. 8, No.1, February 1993, pp. 371-377.

[4] Colin D. Schauder, "Advanced Static VAR Compensator Control System", U.S.Patent 5 329 221, Jul. 12, 1994.

[5] C. Schauder, M Gernhardt, E. Stacey, T. Lemak, L. Gyugyi, T.W.Cease and A. Edris, "Development of a ±100 MVAr Static Condenser for Voltage control of transmission systems", *IEEE Trans. on Power Delivery*, Vol. 10, No.3, July 1995, pp. 1486 – 1496.

[6] K. K Sen, "Statcom - Static Synchronous Compensator: Theory, Modeling, And Applications," *IEEE PES* WM, 1999,Vol. 2, pp. 1177 –1183.

[7] Bhim Singh and R. Saha, "A Harmonics Optimized 12-Pulse STATCOM for Power System Applications," *IEEE Power India Conf., New Delhi, April 10-11, 2006.*

[8] IEEE Std 519-1992, IEEE Recommended Practices and Requirements for Harmonic Control in Electric Power Systems, 1992.

### VII. BIOGRAPHIES

**Bhim Singh** (SM'99) was born in Rahamapur, U. P., India in 1956. He received B. E. (Electrical) degree from the University of Roorkee, India in 1977 and M. Tech. and Ph. D. degrees from the Indian Institute of technology (IIT), New Delhi, in 1979 and 1983, respectively. In 1983, he joined as a Lecturer and in 1988 became a Reader in the Department of Electrical Engineering, University of Roorkee. In December 1990, he joined as an Assistant Professor, became an Associate Professor in 1994 and Professor in 1997 at the Department of Electrical Engineering, IIT Delhi. His field of interest includes power electronics, electrical machines and drives, active filters, static VAR compensator, analysis and digital control of electrical machines. Prof. Singh is a Fellow of Indian National Academy of Engineers (INAE), Institution of Engineers (IE), India and Institution of Electronics and Telecommunication Engineers (IETE), a Life Member of Indian Society for Technical Education (ISTE), System Society of India (SSI) and National Institution of Quality and Reliability (NIQR) and Senior Member, Institute of Electrical and Electronics Engineers (IEEE).

**R Saha** (SM'06) received the B.E.E.(Hons) and M.E.E. in Electrical Engineering from the Jadavpur University, Kolkata, India in 1980 and 1982 respectively. He worked as software engineer in R&D wing of MMC Digital System Division, India from 1982 to 1983. He joined Central Electricity Authority, Govt. of India in Nov'83 through Central Power Engineering Service (Gr-A). He has been associated with Planning of National Transmission Grid in India, Power System Studies and integrated Grid Operation Management & Control. He is presently pursuing research work in the Indian Institute of Technology, Delhi. His field of interest includes Power System Planning, National Grid Development Operation and Control, FACTS technology and applications. He is a Senior Member of the Institute of Electrical and Electronics Engineers (IEEE).

**2006 IEEE International Conference on Power Electronic, Drives and Energy Systems**

# A Nonlinear Fuzzy PID Controller for CSI-STATCOM

A. Kazemi, A. Tofighi, and B.Mahdian

*Abstract*-- **This paper presents a naval approach for the control of a current source inverter (CSI) based STATCOM. The dq - frame model of the CSI STATCOM is proposed as a basis for the control design. Also steady-state characteristic of the CSI-STATCOM is proposed. A Nonlinear Fuzzy PID Controller and multivariable full state feedback has been used. The proposed STATCOM has been simulated using MATLAB/SIMULINK package. The simulation results show that no oscillatory dynamics of the ac current without overshoot or steady-state error exists.**

*Index Terms*-- **Current source inverter, (CSI), FACTS, Non-linear Fuzzy PID, STATCOM.**

## I. INTRODUCTION

RECENTLY, Flexible Alternative Current Transmission System (FACTS) controllers have been proposed to enhance the transient or dynamic stability of power systems [1]. As an important member of the FACTS controllers family, Static Synchronous Compensator has been at the center of attention and the subject of active research for many years. STATCOM is a shunt-connected device that is used to provide reactive power compensation to transmission lines [2]. Static synchronous compensator plays much more important role in reactive power compensation and voltage support because of its attractive steady state performance and operating characteristics. Through regulation of the line voltage at the point of connection, STATCOM can enhance the power transmission capability and thus extend the steady-state stability limit. STATCOM can also be used to introduce damping during power system transients and thus extend the transient stability margin. There are actually two different kinds of STATCOMs, classified by their inverter configuration: using voltage source inverters (VSIs) and current source inverters (CSIs). The VSI is the dominant topology in reactive power control, with several VSI-based STATCOMs now operating in transmission systems [3]. Most of the literature focuses on the VSI; the CSI is less well discussed. Compared with the VSI, the CSI topology offers a number of inherent advantages, including: 1) directly

A. Kazemi, and A. Tofighi, with the center of power excellent in power system and automation, Department of Electrical Engineering, Iran Instt. Of Science and Technology.
B.Mahdian with the Elec. Eng. Department, Azad university of Khorramabad, Iran.

controlling the output current of inverter; 2) implicit short-circuit protection, the output current being limited by the dc inductor; 3) high converter reliability, due to the unidirectional nature of the switches and the inherent short circuit protection. In addition, unlike the VSI STATCOM, the CSI STATCOM injects no harmonics into the ac network when it is operating at zero reactive current [4]. The research on the CSI topology and its applications in power systems has been an on-going process [5], when applied to STATCOM. CSI topology offers a distinct advantage over VSI topology. The direct output of a CSI is a controllable ac current, whereas that of a VSI is a controllable ac voltage. In most transmission systems, under normal operating conditions, the current injected by STATCOM is a small percentage of the line current. Thus, when CSI is used, the current harmonics are also small. But, when VSI is used, for a small injected current, the output voltage of VSI is large and very close to the system voltage. This results in large voltage harmonics, leading to current harmonics that are larger than those generated by CSI, and thus more costly to filters [6]. The nonlinear robust control approach is adapted to STATCOM control [7]. Design of function based variable structure fuzzy controllers for flexible AC transmission systems is presented to improve transient stability performance [8]. A nonlinear controller for GTO based static compensators (STATCOM) is presented. This controller permits to achieve high transient performances for this kind of compensators [9]. A multivariable controller based feedback linearization approach is designed for CSI based STATCOM, installed in the transmission system. Furthermore a fuzzy controller together with a feedback linearizing controller has been implemented [10, 11]. The emphasis of the paper is on the new control approach for CSI-STATCOM. State feedback and nonlinear variable PID controller have been designed for elimination of steady-state errors. The performance of the STATCOM during the steady-state and in response to step changes in the reference values of the system voltage and dc-side current are evaluated using the simulation results from MATLAB-SIMULINK software.

## II. CSI-BASED STATCOM MODELING

The schematic diagram of a CSI-based STATCOM is shown in Fig. 1. The STATCOM is connected to the transmission line through a shunt transformer. It can be explained by considering a controllable current source connected to a main ac system. The controllable current source

0-7803-9771-1/06/$25.00 ©2006 IEEE     1107

generates a three-phase sinusoidal current waveform, leading or lagging by $90^o$ with respect to the corresponding phase voltage.

The control objectives of the STATCOM are to regulate the dc-side current and give the required reactive power compensation to the transmission line.

In Fig. 1, the transformer T is modeled as a combination of an ideal transformer and a series $R$-$L$ impedance. The turns ratio of the transformer is n:1, with the transmission line on the primary-side and the converter on the secondary-side.

$C$ is the capacitance of the filter capacitors. $L_{dc}$ is the smoothing inductor, and the resistor $R_{dc}$ represents the converter switching and conduction losses. $I_{dc}$ is the dc-side current.

$[e]=[e_a\ e_b\ e_c]^T$, $[i]=[i_a\ i_b\ i_c]^T$, $[v]=[v_a\ v_b\ v_c]^T$ and $[i_i]=[i_{ia}\ i_{ib}\ i_{ic}]^T$ denote the vectors of line voltages, secondary-side currents of the transformer, voltages across the filter capacitors, and currents at the terminals of the CSI, respectively. The dc voltage $(v_{dc})$ and ac STATCOM currents are given by:

$$v_{dc}=[m]^T[v_c] \qquad (1)$$

$$[i_i]=[m]i_{dc} \qquad (2)$$

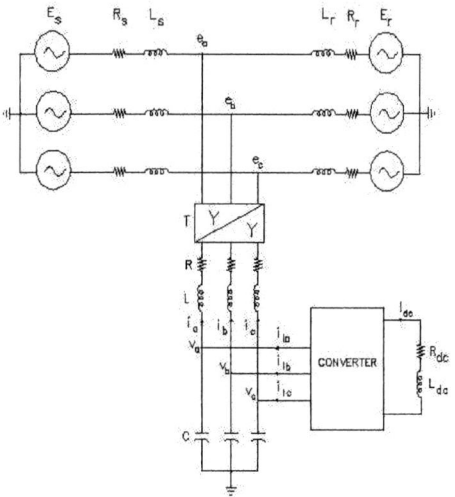

Fig. 1. CSI-based STATCOM.

Each phase of the circuit is modeled by using the state variable approach and, thereby, the model of the system shown in Fig. 2 in the $abc$ frame becomes

$$\frac{d}{dt}[i_s]=-\frac{1}{L}[v]+\frac{1}{L}[e]-\frac{R}{L}[i_s] \qquad (3)$$

$$\frac{d}{dt}[v]=\frac{1}{C}[i]-\frac{1}{C}[m]i_{dc} \qquad (4)$$

$$\frac{d}{dt}i_{dc}=\frac{1}{L_{dc}}[m]^T[v]-\frac{R_{dc}}{L_{dc}}i_{dc} \qquad (5)$$

In order to simplify the analysis, a balanced ac voltage supply is assumed. After applying Park's transformation, [e] chosen as the reference voltage vector, the above current and voltage vectors become $[E]=[E_d\ 0]^T$, $[I]=[I_d\ I_q]^T$, $[V]=[V_d\ V_q]^T$ and $[I_i]=[I_{iq}\ I_{id}]^T$, respectively. The state variables only have $d$ and $q$ components, the zero-sequence component being equal to zero. Thus, the $dq$ transformation of the model in $abc$ coordinates (3)–(5) yields:

$$\frac{d}{dt}\begin{bmatrix}I_d\\I_q\\V_d\\V_q\\I_{dc}\end{bmatrix}=\begin{bmatrix}-\dfrac{R}{L}&\omega&\dfrac{1}{L}&0&0\\-\omega&-\dfrac{R}{L}&0&\dfrac{1}{L}&0\\-\dfrac{1}{C}&0&0&\omega&\dfrac{I_{std}}{C}\\0&-\dfrac{1}{C}&-\omega&0&\dfrac{I_{stq}}{C}\\0&0&-\dfrac{I_{std}}{L_{dc}}&-\dfrac{I_{stq}}{L_{dc}}&-\dfrac{R_{dc}}{L_{dc}}\end{bmatrix}\times\begin{bmatrix}I_d\\I_q\\V_d\\V_q\\I_{dc}\end{bmatrix}+\begin{bmatrix}-V_{sd}\\-V_{sq}\\0\\0\\0\end{bmatrix} \qquad (6)$$

where $\omega$ is the rotation frequency of the $dq$ frame and is equal to the nominal frequency of the system voltage.

$I_{std}$ and $I_{stq}$ are the control signals of CSI and can be expressed by:

$$\begin{bmatrix}I_{std}\\I_{stq}\end{bmatrix}=M\times\begin{bmatrix}\cos\delta\\\sin\delta\end{bmatrix} \qquad (7)$$

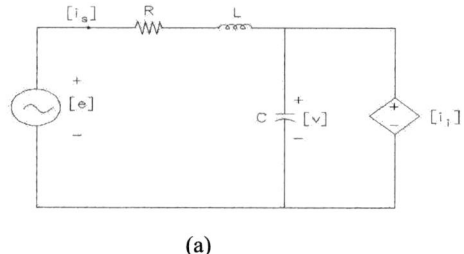

(a)

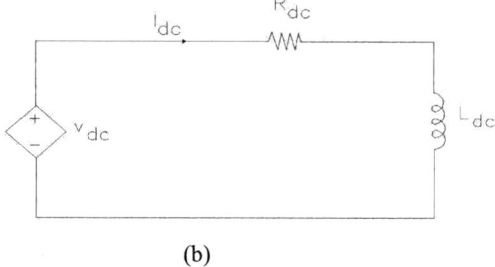

(b)

Fig. 2. CSI-STATCOM in the abc frame:(a) AC side; (b) DC side.

where M is the modulation index and $\delta$ is the phase angle of CSI output current with respect to the system phase voltage, $V_s$. In order to simplify the analysis, let the $d$ axis of the $dq$ frame coincide with the space vector of the system voltage, i.e.

$$\begin{bmatrix} V_{sd} \\ V_{sq} \end{bmatrix} = \begin{bmatrix} E_d \\ 0 \end{bmatrix} \qquad (8)$$

In addition, the equivalent resistor $R_{dc}$ on the dc side of the inverter is neglected here.

### III. NONLINEAR FUZZY PID CONTROLLER

Conventional PID controllers can not provide good performance if the controlled process is nonlinear or has a significant dead time [11].

In this paper a new algorithm is used for design of a nonlinear variable gain PID controller. This algorithm combines linear PI and PD controllers to produce a nonlinear controller. The essential idea is to combine linear controllers using fuzzy reasoning to produce a nonlinear PID controller; this makes a very fast response without any overshoot.

#### A. Basic Elements

In linear control theory the basic PD and PI controllers are expressed mathematically as:

$$u^{PD} = k_1 e + k_2 \dot{e} \qquad (9)$$

$$u^{PI} = k_3 e + k_4 \int e\,dt \qquad (10)$$

where the error $e = x - x_{ref}$

Consider the normalized absolute error $e$ defined as:

$$e^* = G|e| \qquad (11)$$

where the scaling factor G is chosen such that $e^* \in [0,1]$.

#### B. Controller Synthesis Using Fuzzy Logic

The fuzzy controller consists of the following two rules:

If $e^*$ is small, then u is $u^{PI}$

If $e^*$ is large, then u is $u^{PD}$

where "small" and "large" are two fuzzy sets defined on the normalized error domain $e^*$, the corresponding membership functions are defined as:

$$\mu_S(e^*) = 1 - e^{-\left\{\frac{\alpha}{e^*}\right\}^{2.5}} \qquad (12)$$

$$\mu_L(e^*) = 1 - e^{-\left\{\frac{\beta}{1-e^*}\right\}^{2.5}} \qquad (13)$$

where the scalars $\alpha$ and $\beta$ are real positive constants. For $\alpha = \beta$ the membership functions are symmetrical. Using the centroid defuzzification to produce the output of the controller, it can be written:

$$u = \frac{\mu_S . u^{PI} + \mu_S . u^{PD}}{\mu_S + \mu_L} \qquad (14)$$

Substituting Eqs. (9), and (10) in Eq. (14), and performing some algebraic manipulations, the control signal is given by:

$$u = k_P e + k_D \dot{e} + k_i \int e\,dt \qquad (15)$$

where the proportional gain $(k_p)$, the derivative gain $(k_D)$ and the integral gain $(k_i)$ are given by:

$$k_P = \frac{\mu_S . k_3 + \mu_L . k_1}{\mu_S + \mu_L} \qquad (16)$$

$$k_D = \frac{\mu_L . k_2}{\mu_S + \mu_L} \qquad (17)$$

$$k_i = \frac{\mu_S . k_4}{\mu_S + \mu_L} \qquad (18)$$

Substituting Eqs. (12) and (13) in Eqs. (16) and (18), the controller parameters are given by:

$$k_P = \frac{k_3(1 - e^{-\{\frac{\alpha}{e^*}\}^{2.5}}) + k_1(1 - e^{-\{\frac{\beta}{1-e^*}\}^{2.5}})}{2 - e^{-\{\frac{\alpha}{e^*}\}^{2.5}} - e^{-\{\frac{\beta}{1-e^*}\}^{2.5}}} \qquad (19)$$

$$k_D = \frac{k_2(1 - e^{-\{\frac{\alpha}{e^*}\}^{2.5}})}{2 - e^{-\{\frac{\alpha}{e^*}\}^{2.5}} - e^{-\{\frac{\beta}{1-e^*}\}^{2.5}}} \qquad (20)$$

$$k_i = \frac{k_4(1 - e^{-\{\frac{\beta}{1-e^*}\}^{2.5}})}{2 - e^{-\{\frac{\alpha}{e^*}\}^{2.5}} - e^{-\{\frac{\beta}{1-e^*}\}^{2.5}}} \qquad (21)$$

It is clear that $k_i$, $k_D$ and $k_p$ are nonlinear functions of the normalized error $e^*$.

The parameters of the Nonlinear Fuzzy-PID are chosen as:

$k_1 = 3.4$, $k_2 = 1.5$, $k_3 = 0.06$, $k_4 = 0.1$, and $\alpha = \beta = 0.8$.

### IV. CONTROL SYSTEM DESIGN FOR SCI-STATCOM

#### A. AC Side

The development of the controller for CSI STATCOM is broken into two stages. The first stage deals with achieving good ac side closed loop behavior through a combination of full state feedback to assign fast dynamics and integral control to eliminate steady-state errors. Here it is assumed to be a constant. The second stage of the development is to construct a dc current control loop to ensure constant dc link current operation.

First, a state feedback for AC side is designed, the first four equations of (6) are considered. In order to simplify further

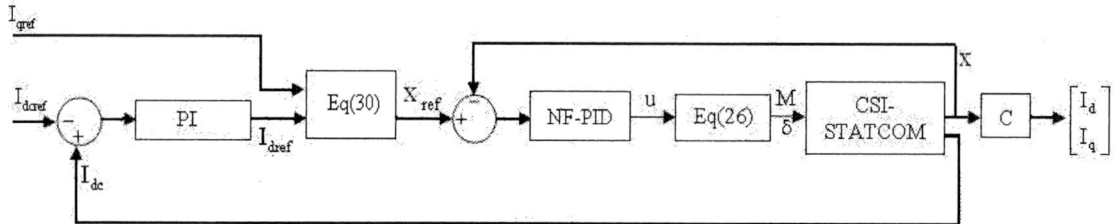

Fig. 3. Block diagram of the complete CSI STATCOM controller.

analysis, the reduced state and disturbance vectors $x = [I_d \quad I_q \quad V_d \quad V_q]^T$, and $w = [V_{sd} \quad V_{sq}]^T$ are defined and the equations are written in following standard format.

$$\dot{x} = Ax + Bu + Gw \qquad (22)$$

The output of the system is $y = Cx$ where from (6)

$$A = \begin{bmatrix} -\dfrac{R}{L} & \omega & \dfrac{1}{L} & 0 \\ -\omega & -\dfrac{R}{L} & 0 & \dfrac{1}{L} \\ \dfrac{1}{C} & 0 & 0 & \omega \\ 0 & \dfrac{1}{C} & -\omega & 0 \end{bmatrix}, \quad B = \begin{bmatrix} 0 & 0 \\ 0 & 0 \\ \dfrac{1}{C} & 0 \\ 0 & \dfrac{1}{C} \end{bmatrix} \qquad (23)$$

$$G = \begin{bmatrix} -\dfrac{1}{L} & 0 \\ 0 & -\dfrac{1}{L} \\ 0 & 0 \\ 0 & 0 \end{bmatrix}, \quad C = \begin{bmatrix} 1 & 0 & 0 & 0 \\ 0 & 1 & 0 & 0 \end{bmatrix} \qquad (24)$$

And the input vector is defined as $u = [I_{std}I_{dc} \quad I_{stq}I_{dc}]^T$.

This reduced set of system equations is linear time invariant and thus may be regulated by a linear controller. All the subsequent discussion will deal with the input vector, although the final CSI system inputs will be derived from:

$$[I_{std} \quad I_{stq}]^T = u / I_{dc} \qquad (25)$$

$$M * e^{j\delta} = I_{std} + jI_{stq} \qquad (26)$$

The dynamics of (22) can be assigned through using a full state feedback controller. The input required to assign the close loop dynamics is then of (15), this loop controls the ac current:

$$e = x - x_{ref} \qquad (27)$$

$$u = K_P e + K_D sIe + K_I \frac{1}{s} Ie \qquad (28)$$

According to Eq. (27):

$$x = (SI - A)^{-1} Bu + (SI - A)^{-1} Gw \qquad (29)$$

$Y = Cx$

to a given reference:

$$X_{ref} = \begin{bmatrix} I_{dref} & I_{qref} \end{bmatrix}^T \qquad (30)$$

where:

$x_{ref}$ : the reference input

$K_p, K_D, K_I$: $2 \times 4$ constant state-feedback gain matrix for state-feedback

To find $K_p, K_D, K_I$ we use pole placement method.

$$K_I = \begin{bmatrix} k_{11} & k_{12} & k_{13} & k_{14} \\ k_{21} & k_{22} & k_{23} & k_{24} \end{bmatrix} \quad J = P, K, I \qquad (31)$$

*B. DC Side*

For dc side, relation between $I_{dc}$ and $I_d$ from [11] is:

$$\frac{d}{dt} I_{dc} = \frac{1}{i_{dc}} \frac{1}{L_{dc}} (V_{sd} I_d + V_{sq} I_q) - \frac{R_{dc}}{L_{dc}} i_{dc} \qquad (32)$$

The output of the dc current controller is the d-axis reference current given to the ac current control loop.

The closed loop transfer functions from $\begin{bmatrix} i_{dref} & i_{qref} \end{bmatrix}^T$ to $\begin{bmatrix} i_d & i_q \end{bmatrix}^T$ is:

$$G(s) = -\frac{s_1 s_2 s_3}{(s - s_1)(s - s_2)(s - s_3)} \qquad (33)$$

The poles are selected such as to obtain a fast, smooth, non oscillatory transient response without an overshoot. For *DC* side a *PI* controller is designed. Complete block diagram of the CSI-STATCOM controller is presented in Fig. 3. Fig. 4 shows the block diagram of the *DC* side STATCOM controller.

*C. Reduced-Order State Estimator*

In section 5.A a suitable controller design is needed to all the state variables, and the disturbance input must be available. The state variables include the injected currents $i_a, i_b, i_c$, the connecting-point voltages $e_a, e_b, e_c$, and the voltages across the filter capacitors $v_a, v_b, v_c$.

1110

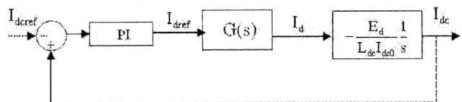

Fig. 4. Block diagram of dc current control loop design.

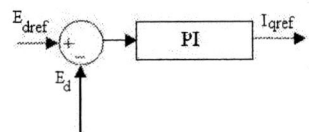

Fig. 5. Finding $I_{qref}$ using a PI controller.

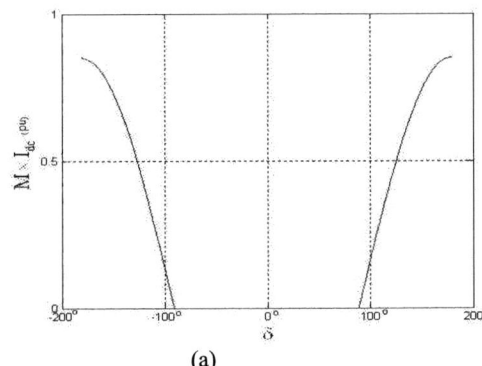

(a)

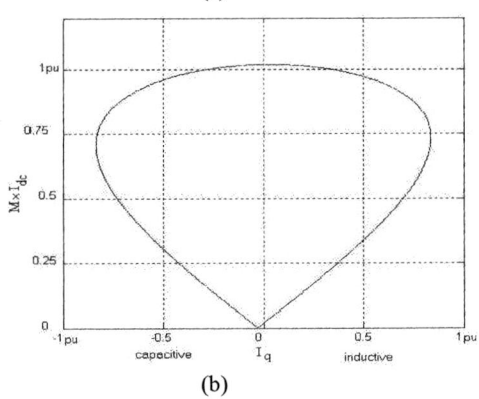

(b)

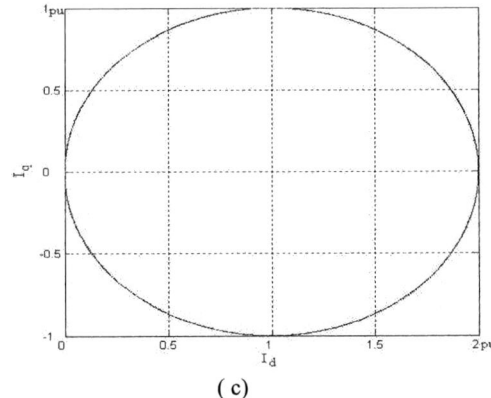

( c )

Fig. 6. The steady-state characteristics of CSI STATCOM.
(a) $M \times I_{dc}$ vs. $\delta$; (b) $M \times I_{dc}$ vs. $I_q$; (c) $I_q$ vs. $I_d$

Only input, output, and disturbance variables should be measured. These variables include $I_d, I_q$ and $E_d$. A reduced-order state estimator can be designed to eliminate the sensors for the capacitor voltages [6].

The state variable vector x is assumed to consist of two parts, i.e., $x = [x_1 \quad x_2]^T$. $x_1 = [I_d \quad I_q]^T$ is the vector of the measured state variables and $x_2 = [V_d \quad V_q]^T$ is the vector of the variables to be estimated. Thus, (27) can be re-written as:

$$\begin{bmatrix} \dot{x}_1 \\ \dot{x}_2 \end{bmatrix} = \begin{bmatrix} A_{11} & A_{12} \\ A_{21} & A_{22} \end{bmatrix} \begin{bmatrix} x_1 \\ x_2 \end{bmatrix} + \begin{bmatrix} B_1 \\ B_2 \end{bmatrix} u + \begin{bmatrix} G_1 \\ G_2 \end{bmatrix} w \quad (34)$$

where:

$A_{11}$, $A_{12}$ : the upper-left , right $2 \times 2$ sub-matrix in $A$,

$A_{21}$, $A_{22}$ : the lower-left , right $2 \times 2$ sub-matrix in $A$,

$B_1$, $B_2$    : the upper and lower 2 rows in $B$,

$G_1$ , $G_2$    : the upper and lower 2 rows in $G$, and

The estimation vector is

$$\begin{aligned} \hat{x}_2 &= Hx_1 + \int [(A_{22} - HA_{12}) \hat{x}_2 \\ &+ (A_{21} - HA_{11})x_1 \\ &+ (B_2 - HB_1) + (G_2 - HG_1)w] dt \end{aligned} \quad (35)$$

where, $H$ is a $2 \times 2$ constant matrix.

In this estimator, $H$ is designed to make $(A_{22} - HA_{12})$ a diagonal matrix.

Input reference value, $E_{dref}$ is given directly. Fig. 5 shows the schematic diagram of the method of finding $I_{qref}$ from the difference between $E_{dref}$ and the measured value $E_d$, then a PI controller can be used to obtain $I_{qref}$.

## V. SIMULATION RESULTS

### A. Steady-State Characteristic of CSI-STATCOM

From the above equations, the steady-state operation point of the CSI-STATCOM is derived [4]:

$$I_d = -\frac{E_d}{R}\cos^2 \delta \qquad I_q = -\frac{E_d}{2R}\sin 2\delta \quad (36)$$

$$V_d = E_d \sin \delta (\sin \delta + \frac{X_1}{R}\cos \delta) \quad (37)$$

$$V_q = -E_d \cos \delta (\sin \delta + \frac{X_1}{R}\cos \delta) \quad (38)$$

$$I_{dc} = \frac{1}{M}[\frac{E_d}{X_c}\sin \delta - \frac{E_d}{R}\cos \delta (1 - \frac{X_1}{X_c})] \quad (39)$$

From Eq. (36):

$$\left(I_d - \frac{V_s}{2R}\right)^2 + I_q{}^2 = \left(\frac{V_s}{2R}\right)^2 \qquad (40)$$

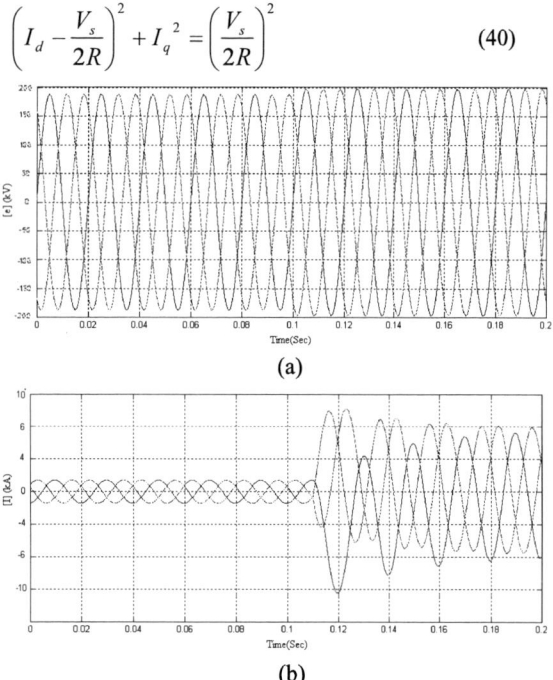

(a)

(b)

Fig. 7. Waveforms of: (a) the system phase-to-ground voltage (e); (b) The injected current (I).

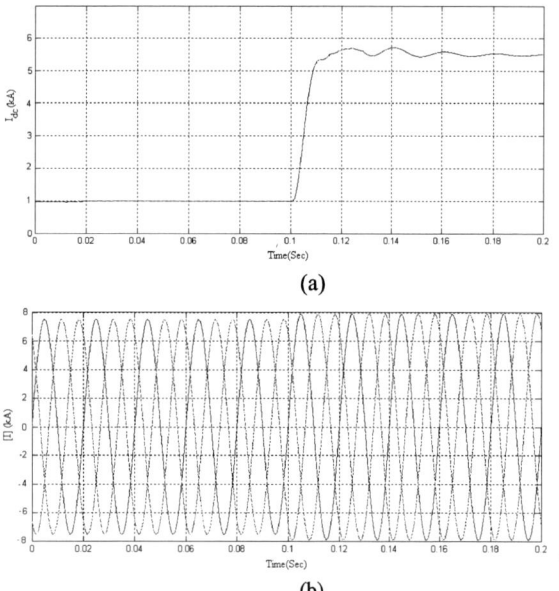

(a)

(b)

Fig. 8. Waveforms of: (a) The dc current; (b) The injected current (I).

E.q 40) is a circle with center $(\frac{V_s}{2R},0)$ and radius $\frac{V_s}{2R}$, Fig. 6(a) shows $M \times I_{dc}$ for wide ranges versus of $\delta$, Fig. 6(b) obviously illustrates $M \times I_{dc}$ is in capacitive or inductive

regain. Fig.6(c), where $X_1$ and $X_c$ are the reactance values of output filter $L$ and $C$, $X_1 = \omega_0 L, X_c = \frac{1}{\omega_0 C}$.

In steady-state operation, the output reactive current is only dependent upon $\delta$, and modulation index only influences the dc side current.

Fig. 6 shows the curves of the CSI STATCOM under steady-state conditions.

### A. Power System Simulation Using SIMULINK

The simulated power system is composed of two voltage sources at the sending- and receiving-ends of a transmission line. The base values for the voltage and the power are taken to be 230 kV and 100 MVA, respectively. The voltages at both ends are 1.0 pu. The sending-end voltage is leading the receiving-end voltage by 30 degrees.

The line impedance is $0.01j + 0.1 \text{pu}$, which is split equally on the left-and right-hand sides of the connecting point of the STATCOM. A transformer is connected in the middle of the line. The leakage reactance and resistance of the transformer, referred to the secondary side, are 0.1 pu. and 0.01 pu, respectively. The capacitance of the filter capacitors is 330 F, which places the resonant frequency of the L-C circuit at 270 Hz. The dc inductor ($L_{dc}$) is sized at 40 mH; a $0.1\,\Omega$ resistor is connected in series with the dc inductor to represent the internal losses of the converter. The switching frequency is set to 900 Hz, i.e., 18 times higher than the fundamental frequency. The simulation results are shown in Figs. 7 and 8. Fig. 7 shows how the system voltage and the dc current can be regulated to their reference values. At first, $I_{dc}$ is maintained at 1 kA. The injected current, $i_a$ is small. At 100 ms, the reference value of the line voltage is set to be 1.02 p.u. As seen in Fig. 7(a), the line voltage responds to the change in the reference and settles at the new steady-state value within two cycles of the fundamental frequency. DC current remains almost constant during the transient period of the line voltage regulation. The peak value of the injected current, as shown in Fig. 7(b), is about 6 kA, i.e. at t=100 ms, the dc current reference, $I_{dcref}$, is set to be 6 kA.

As seen in Fig. 8(a), $I_{dc}$ follows the change in the reference and reaches the new set-point within a half cycle. The system voltage is not influenced during the transient period of $I_{dc}$.

## VI. CONCLUSION

In this paper, a CSI-based STATCOM is proposed, and its $dq$-frame model is studied. The state-feedback control and nonlinear fuzzy PID is applied to the CSI-STATCOM. The performances of the CSI-STATCOM at steady-state and in response to step changes in the reference values of the system voltage and the dc-side current are evaluated using the simulation results by MATLAB/SIMULINK package. Thus, the proposed control scheme, which is a combination of full state feedback and fuzzy PID controller, gives the CSI-STATCOM a good dynamic response and steady-state tracking ability. The simulation results indicate that the CSI-based STATCOM can fulfill all the objectives of a STATCOM.

## VII. REFERENCES

[1] Y. Ni, L. Jiao S. Chen B. Zhang, "Application of a nonlinear PID controller on STATCOM with differential tracker", *1998 IEEE*

[2] Boniface H.K. Chia, Stella Moms, and P.K. Dash, "Multivariable Nonlinear Control of Current Source Inverter-based STATCQM for Synchronous Generator Stabilization,"

[3] Schauder, M. Gernhardt, and E. Stacey et al., "Operation of $\pm$ 100MVAR TVA STATCOM," *IEEE Trans. Power Delivery, vol. 12, pp. 1805–1811, Oct. 1997.*

[4] Dong Shen and P. W. Lehn, "Modeling, Analysis, and Control of a Current Source Inverter-Based STATCOM", *IEEE Trans. Power Del., vol. 17, no. 1, Jun. 2002*

[5] J. Espinoza and G. Joos, "State variable decoupling and power flow control in PWM current-source rectifiers," *IEEE Trans. Ind. Electron., vol. 45, no. 1, pp. 78–87, Feb. 1998.*

[6] Yang Ye, Mehrdad Kazerani, and Victor H. Quintana, "Current-Source Converter Based STATCOM: Modeling and Control", *IEEE Trans. Power Del., vol. 20, no. 2, April. 2005.*

[7] Qiang Lu, Feng Liu, Shengwei Mei, Masuo Goto, "Nonlinear Disturbance Attenuation Control for STATCOM," *2001 IEEE.*

[8] [8]. Stella Morris, P.K. Dash,, and K.P. Basu, "A Fuzzy Variable Structure Current Controller For Flexible AC Transmission Systems, " *2002 IEEE.*

[9] Z. Yao, Putra Kesimpar, V. Donescu, Nicolas Lechevin and V. Rajagopalan, "Nonlinear Control for STATCOM Based on Differential Algebra," *1998 IEEE.*

[10] Boniface H.K. Chia, Stella Moms, and P.K. Dash, "A Fuzzy-Feedback Linearizing Nonlinear Control of CSI based STATCOM for Synchronous Generator Stabilization", *Proceedings of the 2004 IEEE International Conference on Control Applications Taipei, Taiwan, September 2-4, 2004.*

[11] Abdel-Azim S. Ibrahim, "Nonlinear PID Controller Design Using Fuzzy Logic*", IEEE, Melecon, May 7-9, Egypt.*

[12] Jose R. Espinoza, and Geza Joos, "State Variable Decoupling and Power Flow Control in PWM Current-Source Rectifiers", *IEEE Trans. Ind. Elec., vol. 45, no. 1, Feb. 1998.*

1113

**2006 IEEE International Conference on Power Electronic, Drives and Energy Systems**

# Distance Relay Tripping Characteristic in Presence of UPFC

S. Jamali, A. Kazemi, and H. Shateri

*Abstract*--Distance relays are widely used as main and back-up protection for transmission lines. Their operation is based on impedance measurement at the relaying point, which can be affected by several factors, including pre-fault line loading and short circuit levels at the line ends. The problem of growing loads and limited paths for new transmission lines, has been somehow overcome by introduction of Flexible AC Transmission System (FACTS) devices into the network, which imposes further changes on the measured impedance at the relaying point. This paper presents the effects of Unified Power Flow Controller (UPFC), one of FACTS devices, on the measured impedance at the relaying point. In addition to UPFC controlling parameters, its installation point also affects the measured impedance. The variation of the tripping characteristic for different installation points and controlling parameters of UPFC will also be studied.

*Index Terms*--Distance protection, Fault resistance, FACTS devices, Tripping characteristic, UPFC.

## I. INTRODUCTION

THE measured impedance at the relaying point is the basis of the distance protection operation. There are several factors affecting the measured impedance at the relaying point. Some of these factors are related to the power system parameters prior to the fault instance, which can be categorized into two groups [1]-[4]. First group are the structural conditions, represented by the short circuit levels at the transmission line ends, whereas the second group are the operational conditions, represented by the line load angle and the voltage magnitude ratio at the line ends. In addition to the power system parameters, the fault resistance could greatly influence the measured impedance, in such a way that for zero fault resistance, the power system parameters do not affect the measured impedance. In other words, power system parameters affect the measured impedance only in the presence of the fault resistance, and as the fault resistance increases, the impact of power system parameters becomes more severe.

In the recent years FACTS devices are introduced to the power systems to increase the transmitting capacity of the lines

and provide the optimum utilization of the system capability. This is done by pushing the power systems to their limits. It is well documented in the literature that the introduction of FACTS devices in a power system has a great influence on its dynamics. As power system dynamics changes, many sub-systems are affected, including the protective systems. Therefore, it is essential to study effects of FACTS devices on the protective systems, especially the distance protection, which is the main protective device at EHV and HV levels.

Unlike power system parameters, the controlling parameters of FACTS devices could affect the measured impedance even in the case of zero fault resistance. In the presence of FACTS devices, the conventional distance characteristic such as Mho and Quadrilateral are greatly subjected to mal-operation in the both form of over-reaching and under-reaching the fault point. Therefore, the conventional characteristics might not be useful in the presence of FACTS devices.

The impact of STATCOM on the measured impedance has been discussed in [4], by assuming the instantaneous operation of its control system. The effects of series connected FACTS devices on the measured impedance at the relaying point have been presented in [5] and more detailed studies for UPFC have been presented in [6], where it has been assumed that the protective system operate before the control system of FACTS devices.

This paper presents the effects of the installation of UPFC on a transmission line, in the form of the measured impedance at the relaying point and the tripping characteristic of distance relay. For three installation points of UPFC, i.e. at near end, mid-point, and far end, the measured impedance at the relaying point and its tripping characteristic will be presented.

## II. UPFC AND ITS MODELING

Unified Power Flow Controller, UPFC, consists of two converters, series and shunt connected converters to the transmission line, which have a common dc link via a dc storage capacitor [6]-[7], see Fig. 1. Converter 1 is the shunt connected converter and could inject or absorb the reactive power and provide the required active power of the series converter. Converter 2 is the series connected converter and injects a variable voltage, in the form of both variable magnitude and phase angle. These converters are operating independently from the reactive power point of view, but the required active power of Converter 2, and the losses of both converters and storage capacitor, is provided via Converter 1.

---

The authors are with the Center of Excellence for Power Systems Automation and Operation, Department of Electrical Engineering, Iran University of Science and Technology (IUST), Narmak 16846, Tehran, Iran, (e-mails: sjamali@iust.ac.ir, kazemi@iust.ac.ir, and shateri@iust.ac.ir).

0-7803-9771-1/06/$25.00 ©2006 IEEE      1114

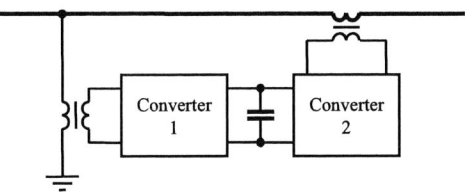

Fig. 1. Circuit arrangement of UPFC.

The equivalent circuit of UPFC is shown in Fig. 2, which consists of two branches, a shunt and a series branch. The shunt branch is related to the Converter 1, which is presented by the impedance $Z_{Sh}$ and the voltage source $E_{Sh}$. The series branch is corresponding to the Converter 2, which is represented by the impedance $Z_{Se}$ and the voltage source $re^{j\theta}V_i$.

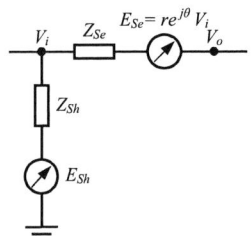

Fig. 2. Equivalent circuit of UPFC

### III. MEASURED IMPEDANCE AT RELAYING POINT

Distance relays operate based on the measured impedance at the relaying point. In the absence of UPFC and for zero fault resistance, the measured impedance by a distance relay only depends on the length of the line section located between the fault and the relaying points. In Fig. 3 this impedance is equal to $pZ_{1L}$, where $p$ is per unit length of the line section lied between the fault and the relaying points, and $Z_{1L}$ is the line positive sequence impedance in ohms.

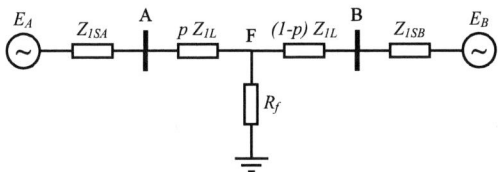

Fig. 3. Equivalent circuit for single phase to ground fault.

In the case of a non-zero fault resistance, the measured impedance is not equal to the impedance of the line section located between the relaying and fault points. In this case, the structural and operational conditions of the power system affect the measured impedance. The operational conditions prior to the fault instance can be represented by the load angle of the line and the ratio of the voltage magnitude at the line ends. The structural conditions are evaluated by the short circuit levels at the line ends. In the absence of UPFC and with respect to Fig. 3 and Fig. 4, the apparent impedance measured by the distance relay can be expressed by the following equations. More detailed calculations can be found in [2].

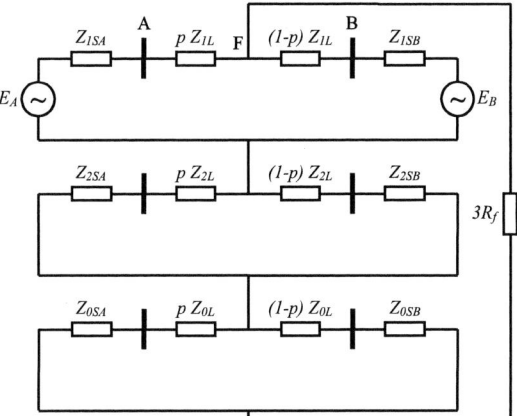

Fig. 4. Equivalent circuit of phase A to ground fault.

$$Z_{1A} = Z_{1SA} + pZ_{1L} \tag{1}$$

$$Z_{1B} = Z_{1SB} + (1-p)Z_{1L} \tag{2}$$

$$Z_{0A} = Z_{0SA} + pZ_{0L} \tag{3}$$

$$Z_{0B} = Z_{0SB} + (1-p)Z_{0L} \tag{4}$$

$$Z_{\Sigma} = 2\frac{Z_{1A}Z_{1B}}{Z_{1A}+Z_{1B}} + \frac{Z_{0A}Z_{0B}}{Z_{0A}+Z_{0B}} \tag{5}$$

$$C_1 = \frac{Z_{1B}}{Z_{1A}+Z_{1B}} \tag{6}$$

$$C_0 = \frac{Z_{0B}}{Z_{0A}+Z_{0B}} \tag{7}$$

$$K_{0L} = \frac{Z_{0L}-Z_{1L}}{3Z_{1L}} \tag{8}$$

$$K_{ld} = \frac{1-he^{-j\delta}}{Z_{1B}+Z_{1A}he^{-j\delta}} \tag{9}$$

$$C_{ld} = (Z_{\Sigma}+3R_f)K_{ld} \tag{10}$$

$$Z_A = pZ_{1L} + \frac{3R_f}{C_{ld}+2C_1+C_0(1+3K_{0L})} \tag{11}$$

It can be seen when the fault resistance is equal to zero, the measured impedance at the relaying point is equal to the impedance of the line section lied between the relaying and the fault points. The power system conditions only affect the measured impedance in the presence of the fault resistance.

Depending on whether UPFC is present in the fault loop or not, the measured impedance at the relaying point would change.

#### A. UPFC Including Fault Loop

When UPFC is present in the fault loop, (1), (3), (9) and (10) are changed and some new equations are introduced as following:

$$Z_{1AF} = Z_{1SA} + Z_{Se} + pZ_{1L} \tag{12}$$

$$Z_{1BF} = Z_{1SB} + (1-p)Z_{1L} \tag{13}$$

$$Z_{1A} = Z_{1FI} + \frac{Z_{Sh}Z_{1AI}}{Z_{Sh}+Z_{1AI}} \tag{14}$$

1115

$$Z_{0A} = Z_{0FI} + \frac{Z_{Sh}Z_{0AI}}{Z_{Sh}+Z_{0AI}} \tag{15}$$

$$C_{1A} = \frac{Z_{Sh}}{Z_{1SA}+Z_{Sh}} \tag{16}$$

$$C_{0A} = \frac{Z_{Sh}}{Z_{0SA}+Z_{Sh}} \tag{17}$$

$$C_{Sh} = Z_{1FI}\left[\begin{array}{l} C_{ld_\Delta}+2C_1(1-C_{1A})+ \\ C_0(1-C_{0A})(1+3K_{0L}) \end{array}\right] \tag{18}$$

$$C_{Z_{Se}} = 3Z_{Se}C_0K_{0L} \tag{19}$$

$$Den = Z_{1AI}[Z_{1FI}he^{-j\delta}+Z_{1BF}(1+re^{j\theta})E_{Sh}]$$
$$+Z_{Sh}\left[\begin{array}{l}[Z_{1AI}(1+re^{j\theta})+Z_{1FI}]he^{-j\delta} \\ +Z_{1BF}(1+re^{j\theta}) \end{array}\right] \tag{20}$$

$$K_{ld} = \frac{Z_{1BI}[1-E_{sh}]+Z_{Sh}[1+re^{j\theta}-he^{-j\delta}]}{Den} \tag{21}$$

$$K_{ld_\Delta} = \frac{Z_{1AI}[(1+re^{j\theta})E_{Sh}-he^{-j\delta}]-Z_{1BI}[1-E_{sh}]}{Den} \tag{22}$$

$$C_{ld_\Delta} = (Z_\Sigma+3R_f)K_{ld_\Delta} \tag{23}$$

$$K_{V_{Se}} = \frac{Z_{1AI}Z_{1BI}E_{sh}+Z_{Sh}[Z_{1AI}he^{-j\delta}+Z_{1BI}]}{Den}re^{j\theta} \tag{24}$$

$$C_{V_{Se}} = (Z_\Sigma+3R_f)K_{V_{Se}} \tag{25}$$

$$Z_A = Z_{Se}+pZ_{1L}+\frac{C_{Sh}-C_{Z_{Se}}-C_{V_{Se}}+3R_f}{C_{ld}+2C_1C_{1A}+C_0C_{0A}(1+3K_{0L})} \tag{26}$$

where:

$Z_{1AI}$: Positive sequence impedance between UPFC and A

$Z_{0AI}$: Negative sequence impedance between UPFC and A

$Z_{1BI}$: Positive sequence impedance between UPFC and B

$Z_{1FI}$: Positive sequence impedance between UPFC and F

$Z_{0FI}$: Negative sequence impedance between UPFC and F

It can be seen that in the absence of fault resistance, the measured impedance is not equal to the actual impedance up to the fault point.

*B. UPFC Excluding Fault Loop*

When UPFC is not present in the fault loop, (2), (4), and (9) are changed:

$$Z_{1AF} = Z_{1SA}+pZ_{1L} \tag{27}$$

$$Z_{1BF} = Z_{1SB}+Z_{Se}+(1-p)Z_{1L} \tag{28}$$

$$Z_{1B} = Z_{1FI}+\frac{Z_{Sh}Z_{1BI}}{Z_{Sh}+Z_{1BI}} \tag{29}$$

$$Z_{0B} = Z_{0FI}+\frac{Z_{Sh}Z_{0BI}}{Z_{Sh}+Z_{0BI}} \tag{30}$$

$$Den = Z_{1BI}[Z_{1AF}E_{Sh}+Z_{1FI}]$$
$$+Z_{Sh}[Z_{1AF}he^{-j\delta}+Z_{1FI}(1+re^{j\theta})+Z_{1BI}] \tag{31}$$

$$K_{ld} = \frac{Z_{1BI}[1-E_{sh}]+Z_{Sh}[1+re^{j\theta}-he^{-j\delta}]}{Den} \tag{32}$$

It can be seen when UPFC is not present in the fault loop, in spite of its presence on the line, in the absence of the fault resistance, the measured impedance is not affected by the power system conditions, nor by controlling parameters of UPFC.

## IV. EFFECTS OF UPFC ON DISTANCE RELAY TRIPPING CHARACTERISTIC

The impacts of the presence of UPFC on a transmission line have been tested for a practical system. A 400kV Iranian transmission line with the length of 300 km has been used for this study. The structure of this line is shown in [8]. By utilizing the Electro-Magnetic Transient Program (EMTP) [9] various sequence impedances of the line are evaluated according to its physical dimensions. The calculated impedances and the other parameters of the system are as following:

$R_{1L} = 0.01133$    $\Omega/km$
$X_{1L} = 0.3037$    $\Omega/km$
$R_{0L} = 0.1535$    $\Omega/km$
$X_{0L} = 1.1478$    $\Omega/km$
$Z_{1SA} = 8\angle 85°$    $\Omega$
$Z_{0SA} = 12\angle 75°$    $\Omega$
$Z_{1SB} = 16\angle 85°$    $\Omega$
$Z_{0SB} = 24\angle 85°$    $\Omega$
$h = 0.96$
$\delta = 16°$

In the absence of UPFC, Fig. 5 shows the tripping characteristic of the distance relay, which is the measured impedance at the relaying point by distance relay as the fault resistance varies from 0 to 200 ohms, while the fault location changes from the relaying point up to the far end of the transmission line.

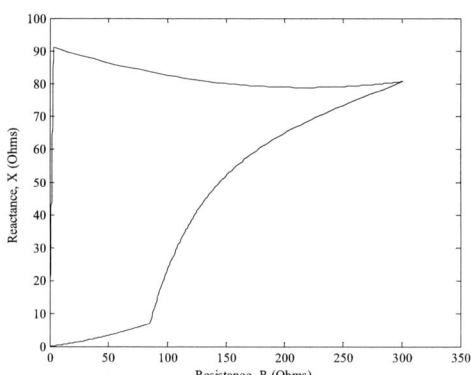

Fig. 5. Distance relay tripping characteristic, without UPFC.

Normally UPFC controls its connecting point voltage and its flowing current according to its controlling strategy. Therefore, UPFC controlling parameters vary by the variation of the loading conditions of the power system. But in this study the operational conditions of the power system are assumed to be constant and it is assumed these conditions are achieved by the different UPFC controlling parameters. Here, $r$, $\theta$, and $|E_{Sh}|$ could be controlled.

1116

## A. UPFC at Near End

In this case, UPFC is always present in the fault loop. Fig. 6 shows the tripping characteristic for zero $r$, i.e. the series converter is inactive, while $|E_{Sh}|$ takes the values of 0.90, 0.97, 1.03, and 1.10. For comparison, the tripping characteristic without UPFC is also shown in the dotted form.

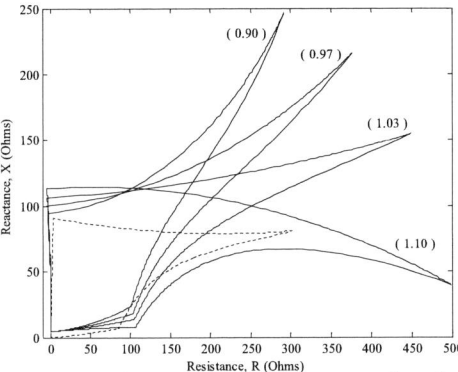

Fig. 6. Tripping characteristic: UPFC at near end; $r = 0$.

It can be seen that as the magnitude of $E_{sh}$ increases, for the low fault resistances, the measured reactance increases, while in the case of higher fault resistances it decreases. The measured resistance increases considerably, as the result of increase in the magnitude of $E_{Sh}$.

Fig. 7 shows the effect of $\theta$ variation on the tripping characteristic for $r = 0.1$ and $|E_{Sh}| = 0.98$. Here, $\theta$ takes the values of 0°, 45°, 90°, and 135°. The tripping characteristic without UPFC is also shown in the dotted form.

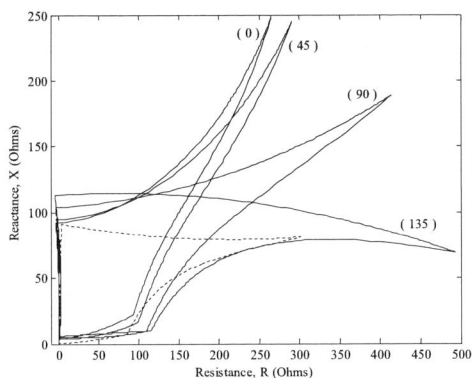

Fig. 7. Tripping characteristic: UPFC at near end; $|E_{Sh}| = 0.98$; $r = 0.1$.

It can be seen that as $\theta$ increases, the measured reactance decreases in the case of low fault resistances, while it increases for the higher fault resistances. The measured resistance increases as $\theta$ increases.

Fig. 8 shows the effect of changes in the magnitude of the series injected voltage, $r$, on the tripping characteristic in the case of $\theta = 0°$ and $|E_{Sh}| = 0.98$. Here, $r$ takes the values of 0.00, 0.05, 0.10, and 0.15. The tripping characteristic without UPFC is also shown in the dotted form.

It can be seen that as $r$ increases, the measured reactance decreases in the case of low fault resistances, while it increases

for the higher fault resistances. The measured resistance decreases by increase in $r$.

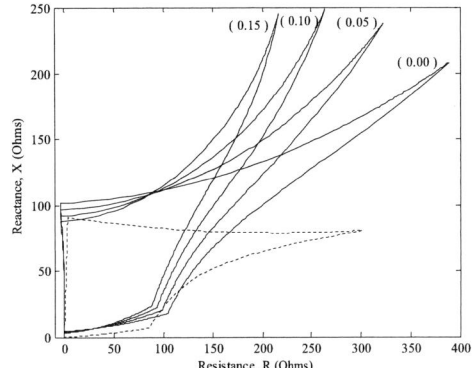

Fig. 8. Tripping characteristic: UPFC at near end; $\theta = 0°$; $|E_{Sh}| = 0.98$.

## B. UPFC at Mid-Point

In this case, UPFC is not present in the fault loop for the faults at the near half on the line, while it is included in the fault loop in the case of faults at the far half. Fig. 9 shows the tripping characteristic for zero $r$, i.e. inactive series converter, while $|E_{Sh}|$ takes the values of 0.90, 0.97, 1.03, and 1.10. The tripping characteristic without UPFC is also shown in the dotted form.

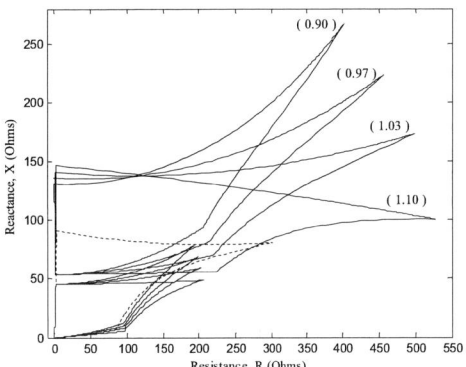

Fig. 9. Tripping characteristic: UPFC at mid-point; $r = 0$.

It can be seen that, in this case the tripping characteristic composed of two separate parts. The lower part is related to the near half of the line, while the upper part is corresponded to the far half. It can be seen that as the magnitude of $E_{sh}$ increases, in the case of the low fault resistances, the measured reactance increases, while for the higher fault resistances it decreases. The measured resistance increases, as the result of increase in the magnitude of $E_{Sh}$.

Fig. 10 shows the effect of $\theta$ variation on the tripping characteristic in the case of $r = 0.1$ and $|E_{Sh}| = 0.98$. Here, $\theta$ takes the values of 0°, 45°, 90°, and 135°. The tripping characteristic without UPFC is also shown in the dotted form.

It can be seen that as $\theta$ increases, the measured reactance decreases for low fault resistances, while it increases for the higher fault resistances, and the measured resistance increases.

1117

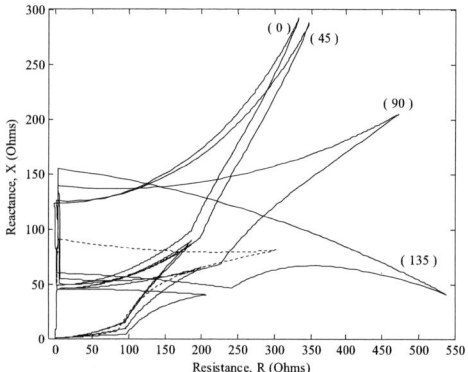

Fig. 10. Tripping characteristic: UPFC at mid-point; $|E_{Sh}| = 0.98$; $r = 0.1$.

Fig. 11 shows the effect of changes in the magnitude of the series injected voltage, $r$, on the tripping characteristic in the case of $\theta = 0°$ and $|E_{Sh}| = 0.98$. Here, $r$ takes the values of 0.00, 0.05, 0.10, and 0.15. The tripping characteristic without UPFC is also shown in the dotted form.

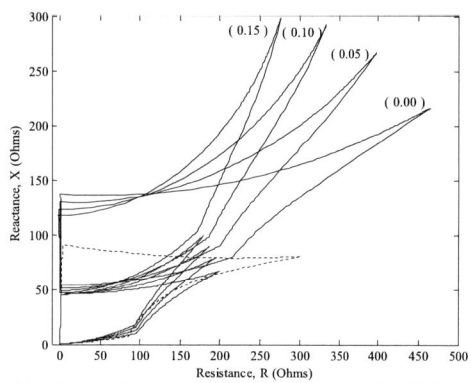

Fig. 11. Tripping characteristic: UPFC at mid-point; $\theta = 0°$; $|E_{Sh}| = 0.98$.

It can be seen that as $r$ increases, the measured reactance decreases in the case of the low fault resistances, while it increases for the higher fault resistances. The measured resistance decreases as $r$ increases.

### C. UPFC at Far End

In this case, UPFC is never present in the fault loop. Fig. 6 shows the tripping characteristic for zero $r$, while $|E_{Sh}|$ takes

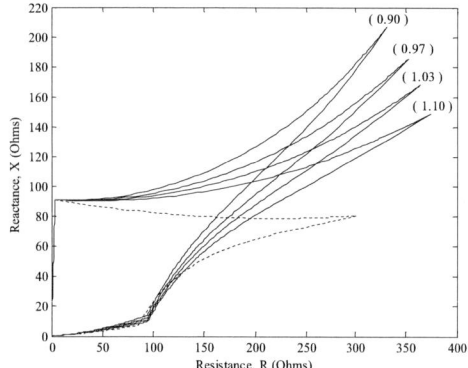

Fig. 12. Tripping characteristic: UPFC at far end; $r = 0$.

the values of 0.90, 0.97, 1.03, and 1.10. The tripping characteristic without UPFC is also shown in the dotted form.

It can be seen that as the magnitude of $E_{sh}$ increases, the measured reactance decreases, while the measured resistance increases slightly. In the case of zero fault resistance, the measured impedance is the actual impedance between the relaying and the fault points.

Fig. 13 shows the effect of variation of the injected voltage angle, $\theta$, on the tripping characteristic in the case of $r = 0.1$ and $|E_{Sh}| = 0.98$. Here, $\theta$ takes the values of 0° and 135°. The tripping characteristic without UPFC is also shown in the dotted form.

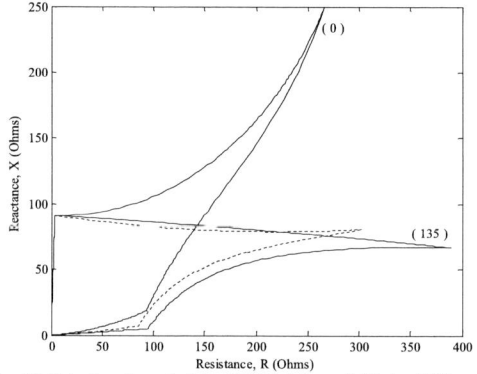

Fig. 13. Tripping characteristic: UPFC at far end; $|E_{Sh}| = 0.98$; $r = 0.1$.

It can be seen that as $\theta$ increases, the measured reactance decreases, while the measured resistance increases. For zero fault resistance, the measured impedance is the actual impedance up to the fault point.

Fig. 14 shows the effect of changes in the magnitude of the series injected voltage, $r$, on the tripping characteristic in the case of $\theta = 0°$ and $|E_{Sh}| = 0.98$. Here, $r$ takes the values of 0.00, 0.05, 0.10, and 0.15. The tripping characteristic without UPFC is also shown in the dotted form.

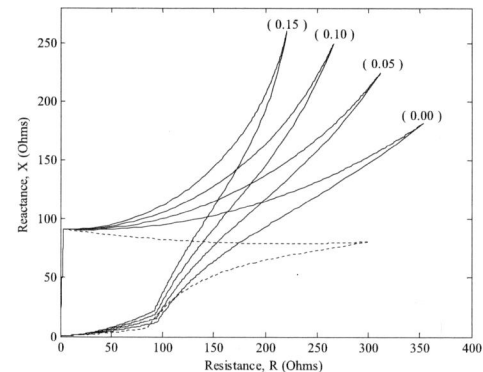

Fig. 14. Tripping characteristic: UPFC at far end; $\theta = 0°$; $|E_{Sh}| = 0.98$.

It can be seen that as $r$ increases, the measured reactance increases, while the measured resistance decreases. In the case of zero fault resistance, the measured impedance is the actual impedance between the relaying and the fault points, the same as two pervious cases.

## V. CONCLUSIONS

The measured impedance at the relaying point is affected by the power system conditions, including operational and structural conditions. On the other hand, FACTS devices have a great influence on the power system dynamics. In this study, at first, the tripping characteristic without UPFC is introduced and then the effects of variations in the controlling parameters and installation point of UPFC are studied. Here, the operational and the structural conditions of the power system are assumed to be constant. The combination of the variation of the system conditions and UPFC controlling parameters and installation locations has a complicated influence on the measured impedance at the relaying point. Therefore, with the exception of the case of installation of UPFC at the far end, once UPFC is installed on a transmission line, adaptive distance protection methods is required to overcome the measured impedance deviation problem.

## VI. REFERENCES

[1] Zhang Zhizhe, C. Deshu, "An adaptive approach in digital distance protection", *IEEE Trans. Power Delivery*, vol. 6, no. 1, pp. 135-142, Jan. 1991.

[2] Y. Q. Xia, K. K. Li, A. K. David, "Adaptive relay setting for stand-alone digital distance protection", *IEEE Trans. Power Delivery*, vol. 9, no. 1, pp. 480-491, Jan. 1994.

[3] S. Jamali, "A fast adaptive digital distance protection", in *Proc. 2001 IEE 7th International Conference on Developments in Power System Protection, DPSP2001*, pp. 149-152.

[4] Khalil El-Arroudi, Geza Joos, and Donald. T. McGillis, "Operation of impedance protection relays with the STATCOM", *IEEE Trans. Power Delivery*, vol. 17, no. 2, pp. 381-387, April 2002.

[5] P. K. dash, A. K. Pradhan, G. Panda, A. C. Liew, "Digital protection of power transmission lines in the presence of series connected FACTS devices", *IEEE Trans. Power Delivery*, vol. 15, no. 1, pp. 38-43, Jan. 2000.

[6] P. K. dash, A. K. Pradhan, G. Panda, A. C. Liew, "Adaptive relay setting for flexible AC transmission systems (FACTS)", in *Proc. 2000 IEEE Power Engineering Society Winter Meeting*, vol. 3, pp. 1967-1972.

[7] A. T. Johns, A. Ter-Gazarian, D. F. Warne, *Flexible ac transmission systems (FACTS)*, Padstow, Cornwall: TJ International Ltd., 1999.

[8] S. Jamali, H. Shateri, "Robustness of distance relay with quadrilateral characteristic against fault resistance", in *Proc. 2004 IEEE International Conference on Power System Technology, PowerCon2004*.

[9] H. W. Dommel, "EMPT reference manual", Microtran Power System Analysis Corporation, Vancouver, British Columbia, Canada, August 1997.

## VII. BIOGRAPHIES

**Sadegh Jamali**, was born in 1956 in Tehran, Iran. He received his BSc from Sharif university of Technology in Tehran in 1979, MSc from UMIST, Manchester, UK in 1986 and PhD from City University, London, UK in 1990, all in Electrical Engineering. Dr. Jamali is currently an Associate Professor in the Department of Electrical Engineering at Iran University of Science and Technology in Tehran. Dr. Jamali is a Fellow of the Institution of Electrical Engineers (IEE) and the IEE Council Representative in Iran. His field of interest includes Power System Protection and Distribution Systems.

**Ahad Kazemi**, was born in Tehran, Iran, in 1952. He received his MSc degree in electrical engineering from Oklahoma Statasze University, U.S.A in 1979. He is currently an associate professor in electrical engineering department of Iran University of Science and Technology, Tehran, Iran. His research interests are reactive power control, power system dynamics, stability and control and FACTS devices.

**Hossein Shateri**, was born in 1979 in Karaj, Iran. He received his BSc and MSc from Iran University of Science and Technology in Tehran in 2001 and 2003, respectively all in Electrical Engineering. He is currently working towards a Phd degree in the Department of Electrical Engineering at Iran University of Science and Technology (IUST) in Tehran, Iran since Sep. 2004 and the research assistant of Digital Power System Protection Lab. H. Shateri is a Member of the Institution of Electrical Engineers (IEE). His field of interest includes Power System Protection, and Distribution Systems Protection and Automation.

**2006 IEEE International Conference on Power Electronic, Drives and Energy Systems**

# Investigations on Boundaries of Controllable Power Flow with Unified Power Flow Controller

S. Srividhya, C. Nagamani, and A. Karthikeyan

*Abstract*--The limits of controllable power flow in the transmission line with the UPFC are determined by the equipment ratings and the line limits. In this paper, boundaries of controllable power flow in a simple two machine power system with UPFC based on a geometrical approach [3] are investigated. The main advantage of this geometrical approach is that even with a minimum computational effort, it gives a clear insight regarding the operational area of the power system with UPFC. However, certain subtle aspects of this geometrical method have not been addressed. A clear and comprehensive explanation of the mathematical basis is presented in this paper for the geometrical approach [3] for evaluating the controllable power flow. Although in a general sense, the location of UPFC within a line is not a crucial parameter affecting the overall performance, results of the geometric approach indicate discontinuities in the P- δ curves specifically in certain ranges of δ. Overall, the results indicate that there is no significant effect of the location of UPFC on the minimum and maximum real power transmitted at a given power angle within the boundaries of controllable power flow. However the corresponding reactive power and currents are affected by location of UPFC. Further, according to the geometrical approach, the feasible range of δ itself depends upon the location of UPFC. Also, the investigations reveal the restricted scope of the geometrical approach. Overall, it is observed that the geometrical approach [3] is appropriate and useful only for some ranges of power angle and location of UPFC.

*Index Terms*-- FACTS, P- δ curves, Q- δ curves, UPFC.

## I. NOMENCLATURE

$V_S$ : Sending end voltage, (p.u.)
$V_R$ : Receiving end voltage, (p.u.)
$V_B$ : Series voltage source, (p.u.)
$I_E$ : Shunt current source, (p.u.)
$I_S$ : Sending end current, (p.u.)
$I_R$ : Receiving end current, (p.u.)
$I_O$ : Line current without compensation, (p.u.)
δ : angle between the sending and receiving end voltages, (°)
α : angle between X-Y and x-y plane, (°)
k : Factor defining electrical distance between sending end of the line and UPFC, (p.u.)

## II. INTRODUCTION

THE Unified Power Flow Controller (UPFC) enables independent and simultaneous control of a transmission line voltage, impedance, and phase angle.

The authors are with the Department of Electrical and Electronics Engineering, National Institute of Technology, Tiruchirappalli, India. (email: s_srivi1@yahoo.co.in ; cnmani@nitt.edu ;akarthik@nitt.edu)

This controller offers substantial advantages for the static and dynamic operation of power system, but it also brings with it major challenges in power electronics and power system design. Several time domain simulation techniques which generate exhaustive amount of data with a number of trial runs are the only tools available for analysis of a general UPFC in the presence of practical equipment and system limits. A few approaches have been earlier reported for determining the operating limits of controllable power flow with UPFC in the power transmission line [1]-[4]. Most commonly a pair of ideal voltage or current sources is used to model the UPFC. In other studies the two voltage source converters with a common dc link (supported by a capacitor) are used. In steady state operation the active power exchanged between the series converter and the line must be handled by the shunt converter. Thus the UPFC as a whole exchanges zero real power with the transmission line.

In a recent study [3] a geometrical approach has been reported to evaluate the boundaries of the power flow in a simple two machine power system with UPFC. The main advantage of this geometrical approach is that even with a minimum computational effort, it gives a straight forward and clear insight regarding the operational area of the power system with UPFC. Moreover the inclusion as well as the interpretation of each additional constraint is easier with such an approach compared to conventional computing methods which involve handling of large amount of data.

However, even in this paper [3] certain subtle aspects of the geometrical approach such as the effect of UPFC location on the distribution of shunt injection current and consequently on the P- δ curve, have not been explained. In the present study, a clear and comprehensive explanation of the mathematical basis including the subtle aspects is presented for the geometrical approach [3] for evaluating the controllable power flow.

The reference paper considered only one location of UPFC (k) and hence the overall picture of the study is incomplete. Although in a general sense, the location of UPFC within a line is not a crucial parameter affecting the overall performance, results of the geometric approach indicate the presence of discontinuities in the family of P- δ curves in specific ranges of k and δ. This is due to the sudden and drastic change(s) in the orientation of the major axis of ellipse representing the distribution of shunt injection current in the line. Also certain non-feasible ranges of P-δ curves with different limits which were not reported earlier are discussed in this paper. Overall, it is observed that the geometrical approach is appropriate only for some ranges of k and δ.

0-7803-9771-1/06/$25.00 ©2006 IEEE

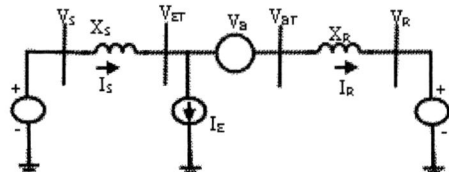

Fig. 1. Representation of line with UPFC.

The present study firstly explains the mathematical and geometrical implications of varying UPFC location and then the reachable operating ranges of the UPFC with various constraints imposed based on the same geometrical approach [3]. Finally the anomalies in the family of P- $\delta$ curves in certain specific ranges of k and $\delta$ are explained with reference to the discontinuities in the orientation of the major axis of ellipse representing the distribution of shunt injection current in the line.

### III. GEOMETRICAL APPROACH

Fig. 1 shows the per phase equivalent circuit of a simple two machine power system in which a UPFC is installed at an arbitrary point in the transmission line. For ease of comparison, the same set of symbols as in [3] is considered in this paper. The UPFC is represented by the combination of a shunt current source ($I_E$) and the series voltage source ($V_B$). The equivalent reactance between the sources $V_S$ and $V_B$ and $V_B$ and $V_R$ are denoted as $X_S$ and $X_R$ respectively. All the line losses and converter losses are neglected.

The basic explanation for constant power lines, locus of equal constant power lines and the definition of d-axis is given in the Appendix A. Since the prime objective of the transmission system is to maximize the power transmitted through the line, it is imperative to examine the controllable ranges of power transmitted subject to the practical constraints. The constraints imposed in the operation of power system with UPFC, are as follows:
- Voltage limit of series converter or $|V_B| \leq V_{Bmax}$
- Current limit of shunt converter or $|I_E| \leq I_{Emax}$

#### A. Geometrical Interpretation of series injection voltage

Fig. 2 shows the phasor diagram of the system. The current $I_A$ is composed of $I_O$ and $I_B$ and is defined as,

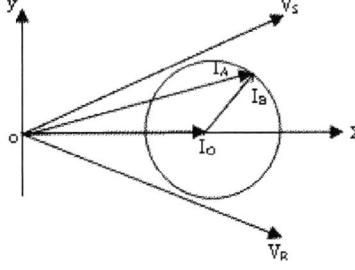

Fig. 2. Reachable set of line current without shunt compensation.

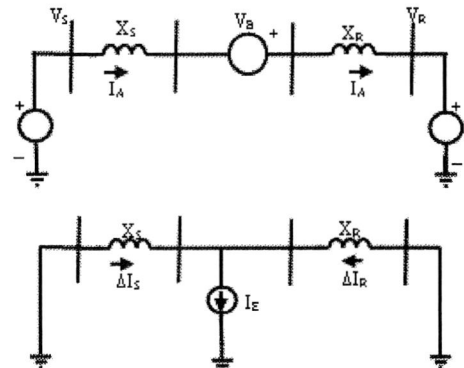

Fig. 3 (a) & (b). Decomposition of current components.

$$I_A = \frac{V_S - V_R}{jX_L} + \frac{V_B}{jX_L} \qquad (1)$$

Where $I_0 = \frac{V_S - V_R}{jX_L}$ and $I_B = \frac{V_B}{jX_L}$ $\qquad (2)$

The vector $I_A$ resides within the circle centered at $I_O$ and with a radius specified by the voltage limit of the series converter. This circle is called as $I_A$ circle and it represents the series injection voltage constraint.

#### B. Geometrical interpretation of series and shunt injection component currents

Using the principle of superposition, the circuit can be decomposed as shown in Fig. 3 (a) and (b). This allows an independent analysis of influence of series and shunt converters on the currents $I_S$ and $I_R$. Comparing the Fig.s 3(a) and (b), the various currents can be expressed as

$$I_S = I_A + \Delta I_S \quad \text{and} \quad I_R = I_A - \Delta I_R \qquad (3)$$

Since $I_E = I_S - I_R : I_E = \Delta I_S + \Delta I_R$ $\qquad (4)$

and due to the fact that

$X_S = kX_L$ and $X_R = (1-k)X_L$, we can write

$$\Delta I_S = (1-k)I_E \quad \text{and} \quad \Delta I_R = kI_E \qquad (5)$$

Fig. 4 shows the vectorial composition of the currents $I_S$ and $I_R$ in relation to that of $I_E$, $\Delta I_S$ and $\Delta I_R$. In steady state operation UPFC exchanges zero real power with the line. This implies that for the lossless line, $P_S$ and $P_R$ are equal. Hence the tips of current vectors of $I_S$ and $I_R$ should lie on a pair of constant power lines. Thus, for a given power transfer $P_1$, the tip of $I_E$ must lie on the constant power line $P_S = P_1$, and its tail on the line $P_R = P_1$. (refer Fig. 15)

Since the current $I_E$ which is the difference of currents $I_S$ and $I_R$ is also the sum of $\Delta I_S$ and $\Delta I_R$, a solution pair consisting

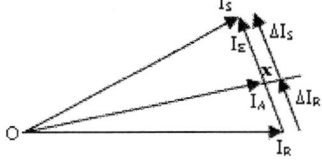

Fig. 4. Vectorial composition of $I_S$ and $I_R$.

1121

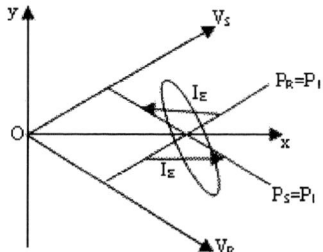

Fig. 5. Family of solutions for $I_E$ and $I_A$.

of $I_A$ and $I_E$ vectors exists where the tip of the $I_A$ lies on $I_E$. This "point of contact" is denoted as "x" in Fig. 4. Location of point "x" on $I_E$ is uniquely determined by the factor k since at this point, the tip of $I_A$ divides the $I_E$ vector in the ratio k: (1-k) according to (5). This is an additional constraint on $I_A$. Thus for a given amplitude of $I_E$, a family of possible solution pairs exist, as shown in Fig. 5. The locus of $I_A$ tips (corresponding to the given power transfer $P_1$ and a given amplitude of the vector $I_E$) is an ellipse as shown in Fig. 5. This ellipse is called the $I_A$ ellipse.

The real power balance of the whole system with UPFC implies that there should be an overlapping of the circular area representing the series voltage injection limit ($I_A$ circle) and the elliptical area representing the shunt current injection limit ($I_A$ ellipse).

*C. Effect of UPFC location and the orientation of ellipse major or minor axis on the operating point*

As explained earlier, the area common to both $I_A$ circle and the $I_A$ ellipse defines the operational area of the system with UPFC due to the implication of power balance. Consider the system of Fig.1 with known voltages at the sending and receiving ends and the line reactance. With a given series voltage injection, the $I_A$ circle is fixed with its centre at the tip of current vector $I_0$ (Fig. 2.). Considering a fixed value of k and $I_E$, for a given power to be transmitted, the $I_A$ ellipse can be drawn as shown in Fig. 5. The equation of $I_A$ ellipse for the present case from (B3) is in the form

$$Ax^2 + 2Hxy + By^2 = C \quad \text{this may be written as}$$

$A'X^2 + B'Y^2 = C'$ by defining a new X-Y plane which makes an angle $\alpha$ with the original x-y plane (refer Fig. 18)

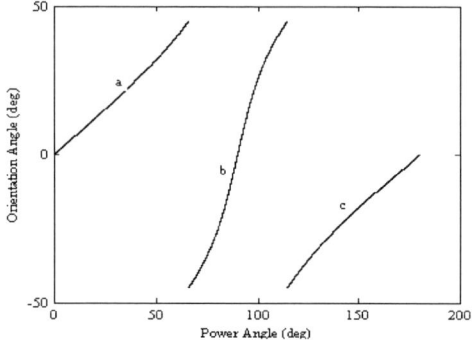

Fig. 6. Variation of $\alpha$ with $\delta$ for k=0.6 .

TABLE I
VARIATION OF $\alpha$ WITH $\delta$ FOR K=0.6

| K=0.6 | $\delta$ range in ° | $\alpha$ range in ° |
|---|---|---|
| Segment a | 0.0 to 65.90515 | 0 to 45 |
| Segment b | 65.90516 to 114.09484 | -45 to 45 |
| Segment c | 114.09485 to 180.0 | -45 to 0 |
| K=0.9 | $\delta$ range in ° | $\alpha$ range in ° |
| Segment a | 0.0 to 48.18968 | 0 to 45 |
| Segment b | 48.18969 to 131.81031 | -45 to 45 |
| Segment c | 131.81032 to 180.0 | -45 to 0 |

The plane transformation is given by

$$x = X \cos \alpha - Y \sin \alpha \tag{6}$$
$$y = X \sin \alpha + Y \cos \alpha$$

Where $\alpha = \frac{1}{2} \tan^{-1}\left( \frac{k \sin 2\delta}{1 - 2k \sin^2 \delta} \right)$ (7)

Hence $\alpha$ is the angle of inclination between the ellipse major or minor axis and the original x-axis. Since $\alpha$ is the inverse tan function of both k and $\delta$, discontinuities in $\alpha$ value exist in certain ranges of k and $\delta$. From (7) it can be anticipated that when a discontinuity occurs the magnitude as well as the direction (clockwise or anti-clockwise) of the angle $\alpha$ i.e., the orientation of $I_A$ ellipse will change drastically. The variation of $\alpha$ with $\delta$ for k=0.6 is illustrated in Fig. 6. Also Table 1 indicates that the change in $\alpha$ is sudden and drastic from the segment 'a' to 'b' and then to segment 'c'. From the Table 1 it is clear that for k=0.6, from $\delta$=0° to 65.90515°, the orientation of the ellipse i.e., $\alpha$ varies from 0 to 45° in the anti-clockwise direction. With a further minute increment in $\delta$, $\alpha$ changes abruptly from 45° to -45°. Similarly at $\delta$=114.09°, $\alpha$ changes again from 45° to -45° suddenly. Such discontinuity in the system variable $\alpha$ is bound to result in similar changes in the other performance variables of the system. Also for k=0.9, the variation of $\alpha$ with $\delta$ is shown to contain discontinuities. The effect of these discontinuities on the P-$\delta$, Q-$\delta$ and I-$\delta$ curves is investigated and the results are presented in the next section.

## IV. RESULTS AND DISCUSSIONS

Based on the geometrical approach, the P-$\delta$ curves, Q-$\delta$ curves and I-$\delta$ curves are plotted. The following constraints are applied:

- $V_{Bmax}$ = 0.4 p.u.
- $I_{Emax}$ = 0.379 p.u.

Within the above constraints, for the sake of illustration, $V_B$=0.4 p.u. and $I_E$=0.379 p.u. for two cases of k=0.6 and k=0.9 are considered. The results are shown in Figs. 7-9. Fig. 7 illustrates the variation of minimum and maximum controllable real powers $P_{min}$ and $P_{max}$ with $\delta$ for k=0.6 and 0.9. It can be observed that for k=0.6, the variation of $P_{min}$ and $P_{max}$ with $\delta$ follow the pattern of a typical P-$\delta$ curve for UPFC in the $\delta$ range from 0° to 65.90515°. However, with a further

1122

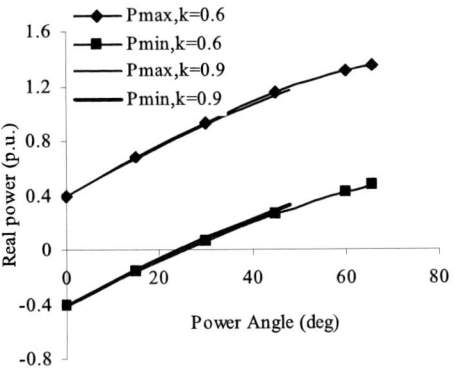

Fig. 7. Real power Pmax and Pmin Vs. δ for k=0.6 and 0.9 .

increment in δ, there is no operating point satisfying the given $I_E$ limit either for the $P_{max}$ curve or for the $P_{min}$ curve. It can be readily noted from Fig. 6 and Table. 1 that the discontinuity in the ellipse orientation angle α occurs for the same set of k and δ values. A similar feature is observed in the $P_{max}$ and $P_{min}$ vs. δ curves for k=0.9 wherein the discontinuities occur at δ=48.18969° (shown in the same Fig. 7). Fig. 8 illustrates the Q-δ curves and Fig. 9 illustrates the I-δ curves for k=0.6 and 0.9. There is no significant effect of k on the real power while there is some change in the reactive power and currents. However, the operating range of $\delta$ is affected noticeably by k.

The effect of discontinuity in α at certain k and δ values on the system operating region is illustrated through the Figs. 10 and 11 for k=0.6 and δ=65.9° and Figs. 12 and 13 for k=0.9 and δ=48.1°. Fig. 10 shows the phasor diagram for the case k=0.6 and δ=65.9° wherein there is only one point P common to both $I_A$ circle and the $I_A$ ellipse. The line segment AB drawn through P between the $P_S$ and $P_R$ lines measures 0.379 p.u (as stipulated by the constraint on $I_E$). Also the point P divides the distance AB between $P_S$ and $P_R$ lines in the ratio (1-k) and k. This operating point yields a feasible $P_{max}$ for the given k and δ. However, when a small increment in δ is

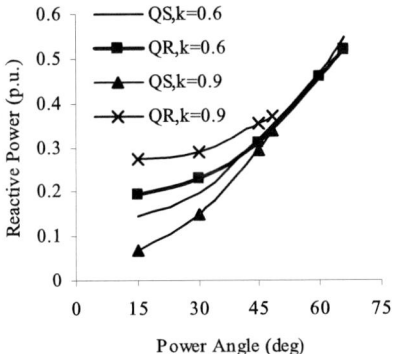

Fig. 8. Reactive power $Q_S$ and $Q_R$ Vs. δ for k=0.6 & 0.9 corresponding to $P_{min}$.

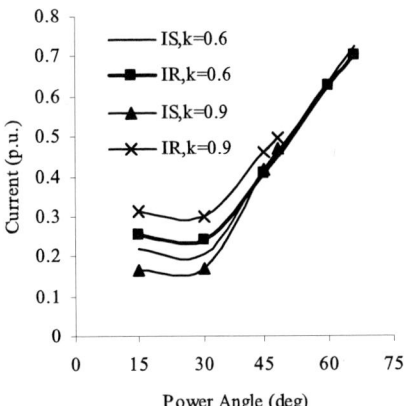

Fig. 9. Currents $I_S$ and $I_R$ Vs. δ for k=0.6 & 0.9 corresponding to $P_{min}$ .

considered i.e., for the same case at δ=65.91° the situation is totally different, as shown in Fig. 11. As before, the $I_A$ circle and $I_A$ ellipse are drawn and here they have only a single point P in common. However, through the point P, it is not possible to draw any straight line segment touching $P_R$ and $P_S$ lines which fulfils the two conditions that i) the segment measures $I_E$ =0.379 p.u. and ii) that the point P divides the line segment in the ratio k and (1-k). This drastic difference in the operating diagram is brought about by the discontinuity in $\alpha$ at the particular k and δ. Thus it can be inferred that, based on the geometrical approach, the operating region for k=0.6, with the two constraints $V_B$ = 0.4 p.u and $I_E$ = 0.379 p.u is restricted for δ=0° to 65.90515° only as against δ =0° to 180° as reported in [3].

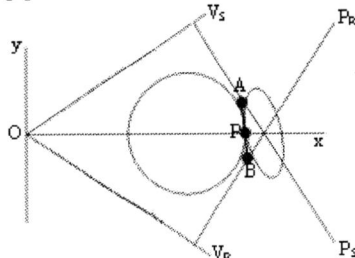

Fig. 10. To find Pmax for k=0.6 when δ=65.90515 (before discontinuity).

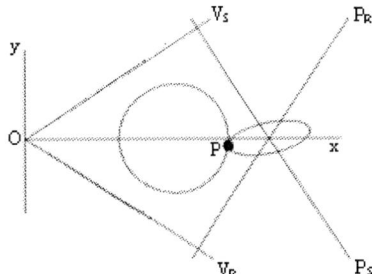

Fig. 11. To find Pmax for k=0.6 when δ=65.90516 (after discontinuity).

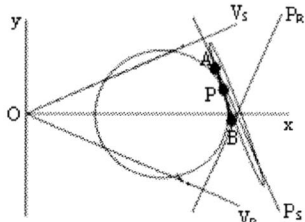

Fig.12. To find Pmax for k=0.9 when δ=48.18968 (before discontinuity).

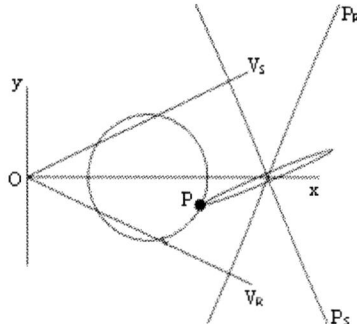

Fig. 13. To find Pmax for k=0.9 when δ=48.18969 (after discontinuity).

Another illustration, for k=0.9, establishing the effect of discontinuity in $\alpha$ at δ= 48.189° on the system operating region is shown in Figs. 12 and 13. It should be noted that even with only two of the total five constraints considered in [3], the operating region is restricted to much less than δ=180° as claimed.

Thus a close examination of the geometrical approach and the results reveal that the said [3] approach is valid only for some ranges of k and δ. In general, the scope of the geometrical approach can be defined to be corresponding to the range of α where α is positive and increasing. In other words, α (angle of orientation of the $I_A$ ellipse) should be in this range $0 \leq 45$ ° for feasible operation.

## V. CONCLUSIONS

A clear and comprehensive explanation is presented for the basis of a geometrical approach [3] (reported earlier) for finding the boundaries of controllable power flow in a simple two machine power system with UPFC. Overall, the results indicate that there is no significant effect of the location of UPFC on the minimum and maximum real power transmitted at a given δ value within the boundaries of controllable power flow. However the corresponding reactive power and currents are affected by k. Further, according to the geometrical approach, the feasible range of δ itself, does depend upon the location of UPFC. Also, the investigations reveal the restricted scope of the geometrical approach. Further investigations are on-going to verify these findings using alternative computational methods.

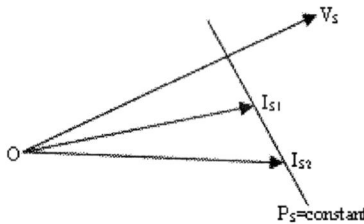

Fig. 14. Constant power line.

## VI. APPENDIX - A

Consider the sending end power. The active power is proportional to the projection of the sending end current vector onto the sending end voltage as shown in Fig. 14. The current vectors $I_{S1}$ and $I_{S2}$ transfer the same sending end power, as would any other current vector that has its tip on the same line perpendicular to $V_S$. This line will be called a "constant power line $P_S$."

Fig. 15. shows constant power lines for both the ends corresponding to the same power flow for two different cases. Thus, if $I_S$ lies anywhere on the line $P_S = P_1$, then power

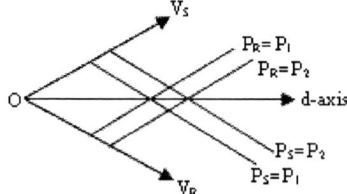

Fig. 15. Lines of sending and receiving end power lines.

balance requires $I_R$ to lie somewhere on the $P_R = P_1$ line. When multiple pairs of constant power lines are drawn, the locus of their intersection points defines a straight line, which is denoted as the d-axis. Since for increasing magnitudes of $P_S$ and $P_R$, the point of intersection moves along the positive d-axis, the d-axis co-ordinate is proportional to the power transmitted through the line.

## VII. APPENDIX- B

Ellipse Equation

In this section, the parametric equation of the ellipse is presented. In general, if the straight line $A(x_1,y_1)$ $B(x_2,y_2)$ (Fig. 16) is divided in the ratio m and n, then the co-ordinates of dividing point C (x,y) are given by $\frac{mx_2 + nx_1}{m+n}, \frac{my_2 + ny_1}{m+n}$

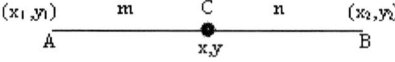

Fig. 16. Determination of points (x,y).

In the present problem of power flow the coordinates of C (x,y) are given as follows (refer Fig. 17):

$$x = (1-k)u + kv\cos\delta \quad : \quad y = kv\sin\delta \qquad (8)$$

1124

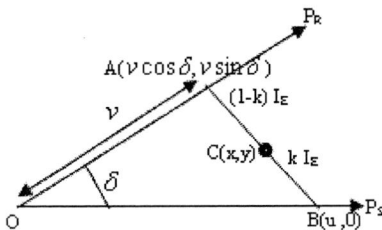

Fig. 17. Determination of co-ordinates C (x,y).

The distance between the points ($v\cos\delta, v\sin\delta$) and (u, 0) is $I_E$ and is expressed as

$$(v\cos\delta - u)^2 + (v\sin\delta)^2 = I_E^2 \qquad (9)$$

On subtracting $v\cos\delta$ on both sides in (B1), we get

$$x - v\cos\delta = (1-k)u + v\cos\delta(k-1)$$

i.e., $\left(\dfrac{x - v\cos\delta}{1-k}\right)^2 = I_E^2 - (v\sin\delta)^2$    from (9).

Also $\left(\dfrac{x - v\cos\delta}{1-k}\right)^2 + \left(\dfrac{y}{k}\right)^2 = I_E^2$    from (8).

Simplifying, $k^2\left(x - v\cos\delta\right)^2 + \left(1-k\right)^2 y^2 = I_E^2 k^2 \left(1-k\right)^2$

$$k^2\left(x - \frac{y}{k\sin\delta}\cos\delta\right)^2 + \left(1-k\right)^2 y^2 = I_E^2 k^2 \left(1-k\right)^2 \quad \text{from (8).}$$

Thus, $\left[k^2 x^2 - 2xyk\cot\delta + y^2\left[\left(1-k\right)^2 + \cot^2\delta\right]\right]/k^2\left(1-k\right)^2 = I_E^2$    (10)

Equation (10) is of the form $Ax^2 + Hxy + By^2 = C$ wherein it can be easily verified that this represents an ellipse except in the case δ=90° and k=0.5 in which case it is a circle.

If X,Y are the new axes obtained by rotating x-y axis through an angle α then the co-ordinates of the points of the two system are given by,

$$x = X\cos\alpha - Y\sin\alpha \quad \& \quad y = X\sin\alpha + Y\cos\alpha \qquad (11)$$

In Fig. 19. $\theta$ is the angle from $P_R$ line to the new axis X and the angle between the original x-axis and the $P_R$ line is $\pi/2 - \delta_R$.

$$\theta = \alpha - \pi/2 + \delta_R$$

In order to have the co-ordinate axes as the major and minor axes of the ellipse, we must choose α in such a way that there is no xy term. Therefore equating the co-efficient of xy term to zero and substituting (11) in (10) we get,

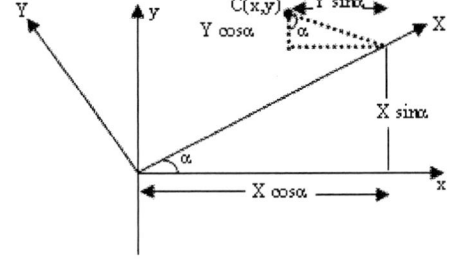

Fig. 18. New X-Y plane.

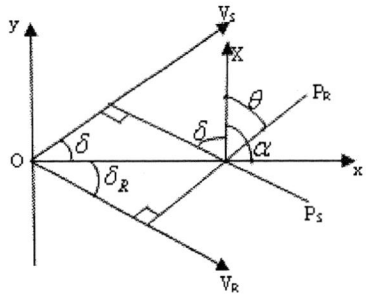

Fig. 19. Determination of angle.

$$\tan 2\alpha = \frac{2H}{A - B} \qquad (12)$$

On substituting the values of A,B,C and H in (12 ) we get,

$$\tan 2\alpha = \frac{k\sin 2\delta}{1 - 2k\sin^2\delta}$$

The length of major axis and minor axis are given by respectively,

$$2r_1 = I_E\left(\cos ec\,\delta + \sqrt{\cos ec^2\delta - 4k(1-k)}\right) \qquad (13)$$

$$2r_2 = I_E\left(\cos ec\,\delta - \sqrt{\cos ec^2\delta - 4k(1-k)}\right) \qquad (14)$$

Therefore the ellipse that corresponds to $P_1=0$ can be described by the parametric equation,

$$\begin{pmatrix} x_d \\ x_q \end{pmatrix} = R(\theta)\begin{bmatrix} r_1 & 0 \\ 0 & r_2 \end{bmatrix}\begin{bmatrix} \cos(\xi) \\ \sin(\xi) \end{bmatrix}$$

where ξ is the parameter taking values in the set [0 2π] and R(θ) is the rotation matrix defined as

$$R(\theta) = \begin{bmatrix} \cos(\theta) & -\sin(\theta) \\ \sin(\theta) & \cos(\theta) \end{bmatrix} \qquad (15)$$

## VIII. ACKNOWLEDGEMENTS

The authors gratefully acknowledge the valuable discussions they had with Dr. C. S. Karuppan Chetty, Professor and Head of Department of Mathematics, National Institute of Technology, Tiruchirappalli, India.

## IX. REFERENCES

[1] L.Gyugyi, "Unified power-flow control concept for flexibleAC transmission systems," *IEE Proceedings- C*, vol.139, no. 4, pp. 323–331, July 1992.

[2] J. Bian, D. G. Ramey, R. J. Nelson, and A. Edris, "A study of equipment sizes and constraints for a unified power flow controller," *IEEE Trans.Power Delivery*, vol. 12, pp. 1385–1391, July 1997.

[3] J. Z. Bebic, P. W. Lehn, and M. R. Iravani," P– δ Characteristics for the Unified Power Flow Controller— Analysis Inclusive of Equipment Ratings and Line Limits," *IEEE Trans.Power Delivery*, vol. 18, , No. 3, pp 1066-1072, July 2003.

[4] Nabavi-Niaki and M. R. Iravani, "Steady-state and dynamic models of unified power flow controller (UPFC) for power system studies," *IEEE Trans. Power Syst.*, vol. 11, pp. 1937–1943, Nov. 1996.

[5] H.K. Dass, *Engineering Mathematics*, Part-I, S.Chand & company Ltd, 1994.

## X. BIOGRAPHIES

**S.Srividhya** had undergone her B.E. degree in Electrical & Electronics Engineering in Annamalai University, Chidambaram, India during the year 2001-2005. Subsequently she joined for M.Tech. in 2005 in Power Systems at the E.E.E. Department, National Institute of Technology, Tiruchirapalli, India. This work has been carried out by her while working towards her project in M.Tech. degree. Her areas of interest include FACTS controllers and HVDC.

**A.Karthikeyan** received his B.E. degree in Electrical & Electronics Engineering from Bharathidasan University, Tiruchirappalli in 2002. He is presently working as a Project Assistant in the MHRD sponsored project in the E.E.E. Department, National Institute of Technology, Tiruchirapalli, India. He is also pursuing his M.S. (by research) degree in the same department. His current interests include A.I. applications in FACTS, Power System Optimization studies.

**C.Nagamani** received her M.Tech and Ph.D. degrees from I.I.T., Kanpur and University of Technology, Sydney, respectively. From 1985 to 1991 she was with the Central Power Research Institute, Bangalore, India. Subsequently she joined the E.E.E. Department, National Institute of Technology (then known as Regional Engineering College), Tiruchirapalli, India as a lecturer. At present, she is an Assistant Professor in the same department. Her areas of interest include Power electronics and Drives, Renewable Energy Systems and FACTS controllers.

**2006 IEEE International Conference on Power Electronic, Drives and Energy Systems**

# VSC Based HVDC System for Passive Network with Fuzzy Controller

A. K. Moharana , Ms. K. Panigrahi, B. K. Panigrahi, *Member, IEEE,*
and P. K. Dash, *Senior Member, IEEE*

*Abstract*--**For stable operation of a power system reliable control of active & reactive power is necessary. Voltage source converter based HVDC system bears the advantages of being able to change the control strategy immediately with respect to the active and reactive power changes. In this paper a function based fuzzy control system is proposed for the control of active power, DC voltage at converter station and voltage at inverter station. Basically two input one output methodology based fuzzy logic controller has been proposed. The basic features of proposed system used in VSC based HVDC systems are: 1) only two rule base system is used.2) all controllers are not self correcting. At both stations error and change in error have been taken as input to the controller. Design of high pass filter with FFT analysis has also been proposed for a better dynamic performance of the function based fuzzy controller. Computer simulation using SIMULINK gives clear result of the performance of the proposed controller.**

*Index Terms*-- **VSC, HVDC, AC filter, FLC, PI.**

## I. INTRODUCTION

TILL date thyristor based converters are widely used for HVDC transmission system. Since for a long period it is working, so many researches have been done for its control and this system is now well established. Though this system is working very successfully it has certain remarkable limitations. As a result VSC based HVDC light system using IGBT technology has attracted the attention of many researchers in this field [1]. Here the control strategy discussed is slightly different from the conventional PI controller.

In this paper a new control strategy of VSC HVDC has been developed. The controller is of function based fuzzy controller [2]. This paper describes the advantages of function based fuzzy controller over conventional PI controller. The validity of all the controllers and filters have been tested by computer based simulation using MATLAB/ SIMULINK. The possibilities and promising future to maintain the stability of VSC based HVDC system has been tested.

---

This work is supported in part by the Department of Science and Technology , Government of India. (Project ID: SR/S3/EECE/02/2005-2005-Engg dated 12th Aug 2005)

A. K. Moharana is with Electrical Engineering Cell, Steel Authority of India Limited (e-mail: mrakshaya@yahoo.co.in)

Ms. K. Panigrahi is with Electrical Science Research center, College of Engineering Bhubaneswar. (e-mail: kabitanjali@gmail.com)

B. K. Panigrahi is with Department of Electrical engineering, Indian Institute of Technology Delhi (e-mail: bkpanigrahi@ee.iitd.ac.in)

P. K. Dash is with Electrical Science Research center, College of Engineering Bhubaneswar. (e-mail: pkdash_india@yahoo.com)

## II. MATHEMATICAL MODEL OF VSC BASED HVDC SYSTEM

In this paper mathematical model of the whole system has been derived in d-q frame. The whole model is based upon the state space control theory ($\dot{x} = A.x + B.u$). An input - output linearization approach is used to derive a small signal model for the system. Then the controllers have been designed according to the system state. The controller performance is studied under different operating conditions and results are recorded. Performance of the proposed controller is compared with the classical PI controller. The whole model is simulated in the MATLAB/ SIMULINK environment. The model studied in this paper is shown below.

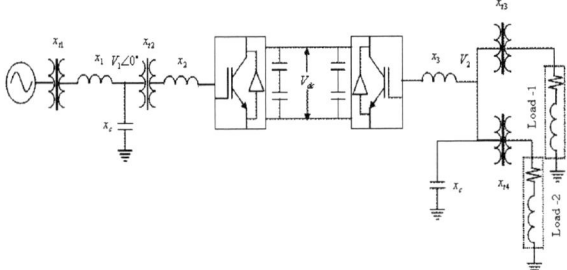

Fig. 1. VSC HVDC system.

Here VSC HVDC system feeds a passive network. There are two loads in the passive network. $x_{t1}, x_{t2}$, $x_{t3}$ and $x_{t4}$ are the transformer reactance. $x_1$ is the line reactance and $x_2, x_3$ are the reactance of the Rectifier and Inverter respectively. $P_{L1}, Q_{L1}$ and $P_{L2}, Q_{L2}$ are the load in the passive network. Dc line resistance is taken as the $R_{dc}$. $C_1$ and $C_2$ are the DC link capacitors. $V_1$ is the bus voltage. $V_R$ is the rectifier bus voltage. $V_I$ is the inverter bus voltage. $V_2$ is the voltage at the passive network bus which is also the load voltage.

The differential equations of the generator are as follows

$$\dot{\delta} = \omega_1 - \omega_0 \tag{1}$$

$$\dot{\omega} = \frac{Pm - Pe}{IC} \tag{2}$$

$$\dot{E}'_q = \frac{1}{T'_0}\left\{E'_{q0} + E'_f - E'_d - (x'_d - x'_{dd}).I_{1d}\right\} \tag{3}$$

---

0-7803-9771-1/06/$25.00 ©2006 IEEE     1127

$$\dot{E}'_f = \frac{1}{T'_e}\left\{-E'_f + k_e\left(V_{1ref} - V\right)\right\} \qquad (4)$$

Where

$\delta$ is the power angle of the generator

$\omega$ is the angular speed of the generator.

$E'_f$ is Field voltage of the generator

$E'_q$ is Quadrature axis voltage of the generator

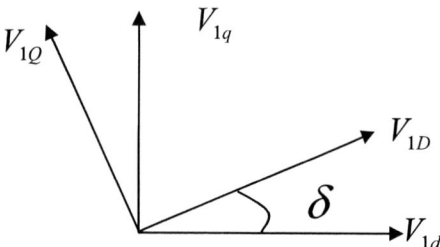

Fig. 2. The phasor of Network and Machine frame voltages.

Transforming the bus voltage and current from machine frame to Network frame using the above Fig. 2,

$$\begin{bmatrix} V_{1D} \\ V_{1Q} \end{bmatrix} = \begin{bmatrix} \cos\delta & \sin\delta \\ -\sin\delta & \cos\delta \end{bmatrix} \cdot \begin{bmatrix} V_{1d} \\ V_{1q} \end{bmatrix} \qquad (5)$$

$$\begin{bmatrix} I_{1D} \\ I_{1Q} \end{bmatrix} = \begin{bmatrix} \cos\delta & \sin\delta \\ -\sin\delta & \cos\delta \end{bmatrix} \cdot \begin{bmatrix} I_{1d} \\ I_{1q} \end{bmatrix} \qquad (6)$$

The convert terminal voltage in network frame

$$\begin{bmatrix} V_{rD} \\ V_{rQ} \end{bmatrix} = \begin{bmatrix} 0 & (x_2+x_{t2}) \\ -(x_2+x_{t2}) & 0 \end{bmatrix} \begin{bmatrix} I_{rd} \\ I_{rq} \end{bmatrix} + \begin{bmatrix} 1 & 0 \\ 0 & 1 \end{bmatrix} \begin{bmatrix} V_{1D} \\ V_{1Q} \end{bmatrix}$$

Now following the equation to calculate the $m_1$ and $\alpha_1$

$$V_{rD} = m_1 . V_{dc1} . \sin\alpha_1 \qquad (7)$$

$$V_{rQ} = m_1 . V_{dc1} . \cos\alpha_1 \qquad (8)$$

Now writing the dynamic equations for the rectifier [4]

$$\dot{I}_{rD} = \frac{V_{1D} - V_{rD}}{L_3} + \omega . I_{rQ} \qquad (9)$$

$$\dot{I}_{rQ} = \frac{V_{1Q} - V_{rQ}}{L_3} - \omega . I_{rD} \qquad (10)$$

$$\dot{V}_{dc1} = \frac{1}{C}\left\{\frac{P_{dc1}}{V_{dc1}} - \left(\frac{V_{dc1} - V_{dc2}}{r_{dc}}\right)\right\} \qquad (11)$$

we can replace the $P_{dc1}$ by the equation

The dynamic equations of the Inverter are as follows

$$\dot{I}_{iD} = \frac{V_{iD} - V_{2D}}{L_4} + \omega . I_{iQ} \qquad (12)$$

$$\dot{I}_{iQ} = \frac{V_{iQ} - V_{2Q}}{L_3} - \omega . I_{iD} \qquad (13)$$

Since our objective is to control the AC voltage and DC voltage at the rectifier side and to control the AC voltage at the inverter side. Output (y) can be chosen as follows

$$[y] = \begin{bmatrix} I_{rQ} \\ V_{dc1} \\ I_{iD} \\ I_{iQ} \end{bmatrix}$$

And the control variables are as follows:

$$[u] = \begin{bmatrix} m_1 \\ \alpha_1 \\ m_2 \\ \alpha_2 \end{bmatrix}$$

## III. Control Strategy for HVDC Light System

The conventional control strategy for a HVDC light is concerned with the control of AC voltage and DC voltage. Using different types of controllers, active and reactive power is controlled effectively at each station. Normally conventional PI controller is used for this purpose. In conventional PI controller direct axis and quadrature axis current is controlled that changes $m$ and $\delta1$. Choosing the right value for $Kp$ and $Ki$ of a PI controller is really a difficult job.

But most of the times a non linear fuzzy PI controller is selected for the control of the HVDC light system. Generally non linear fuzzy PI controller improves the dynamic performance of the total HVDC system.

### A. Two Rule Based PI Like FLC

To overcome the tuning problem of conventional PI controller, here a two rule based PI like Fuzzy logic controller has been proposed. This section describes the main features of the proposed two ruled base two input one output FLC [2].

PI like FLC is based upon Metarule as follows

"*If e is not self correcting, then control action du is not zero and depends upon the sign and magnitude of e and de*". Rule base is applied to control the voltage and power in VSC based HVDC system. Here two inputs are considered as one is error (e) and another is change in error (de).

When a statement like '*If sign (e) and sign (de) are not zero and sign (e) = sign (de)*' is true then error is certainly not self correcting. Control action is required at this stage that is sign (u) is not zero. So we have assumed that under this situation sign (u) can be related to sign (e). This is achieved by defining the relationship as sign (u) = sign (e). This is quite non complete strategy. It can also be stated as a single Metarule.

"*If sign (e) = sign (de) ≠ 0, Then sign (u) = sign (e)*"

So here we have got three linguistic variables $E$ is error, $DE$ is change in error and $U$ is output of the controller. These variables have two linguistic values $P$ (positive) and $N$ (negative). Using the above defined variables two rules are stated as follows.

"*If E is $\tilde{E}P$ and $\tilde{DE}$ is $\tilde{DEP}$ then U is $\tilde{U}P$*"

"*If E is $\tilde{N}E$ and DE is $\tilde{N}P$ then U is $\tilde{U}N$*"

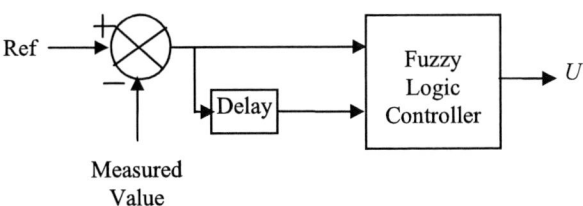

Fig. 3. Generalized Fuzzy Like PI Controller.

*B. Comparison of PI Like FLC with conventional PI controller*

General Equations of a PI controller taking two input $e$ and $de$ are as follows which are not linear.

$$du = k_1 e + k_2 de \qquad (14)$$

$$k_1 = \left(\frac{2.G}{3}\right)\lambda_\mu K_e e \qquad (15)$$

$$k_2 = \left(\frac{2.G}{3}\right)\mu(1-\lambda)K_{de}de \qquad (16)$$

Where $\lambda \, \varepsilon \, [0,1]$, $\mu$ is parameter used for weightage and G is the gain.

After linearization

$$u = k_1 e + k_2 de \qquad (17)$$

$$k_1 = \left(\frac{2.G}{3}\right)\lambda_\mu K_e e \qquad (18)$$

$$k_2 = \left(\frac{2.G}{3}\right)\mu(1-\lambda)K_{de}de \qquad (19)$$

In this design one point is remarkable that is magnitude of the output of the controller '$u$' is more than in case of conventional PI controller. This concept has been applied to first order non linear plant by Ying [6]. The relation followed by the model was as follows.

$$\frac{dy}{dt} = -y + \alpha.y^\beta + u \qquad (20)$$

Where $\alpha$ and $\beta$ are positive parameters.

But we have applied the concept to control the AC voltage at the inverter station and to control active power and DC voltage at the converter station. For the control the equation used is

$$u(t) = u(t-1) + \frac{2}{3}G\tanh(2(K_1 e(t) + K_2 de(t))) \qquad (21)$$

Where $K_1$ acts as integral constant and $K_2$ as proportional constant. Keeping $K_1=0$, G and $K_2$ are tuned. Then $K_1$ is gradually increased to achieve the required out put. In this way all the gains are tuned at one operating condition but for different operating conditions the gains are optimized using the Genetic Algorithm program.

## IV. OPTIMIZATION OF G, K₁ AND K₂

Parameters *G, K₁ and K₂* are optimized using genetic algorithm. GA parameters are as follows.

Population Size: 50,

Crossover Fraction: 0.8,

Mutation probability 0.05,

Maximum number of generations: 100.

## V. AC FILTER DESIGN

Due to the power electronic converters in the VSC HVDC system a large no of harmonics are generated, which are injected to the power system. Due to these distorted voltage and current wave form power looses in the network increases. There fore shunt filters are used to bypass the harmonic wave forms.

Here in this paper PWM triggering scheme is used so harmonics produced are centered at the switching frequency. The order of harmonics produced during the switching of the Converter and Inverter can be formulated as $f_h = f_s \pm 2nf_0$.

Where $f_s$ is the switching frequency, $f_0$ is the fundamental frequency and n is the order of harmonic. In this paper $f_s$ is taken as 1350 Hz and $f_0$ is taken as 50 Hz. Since $m_f$ is more than 21 Asynchronous triggering scheme is adopted to minimize the generation of sub-harmonics. Where $m_f = \dfrac{f_s}{f_0}$.

Depending upon the harmonics produced in this paper a high pass C-Type filter has been proposed in the Rectifier side and a double tuned filter has been designed in the Inverter side. The transformers chosen are $\Delta Y$ in the rectifier side and $Y\Delta$ in the inverter side.

## VI. SIMULATION RESULTS

*A. Specification of the system studied*

Generator: 25 KV, 600 MVA

Transformer 1: 25/400 KV, 600 MVA

Transformer 2: 132/66 KV, 300 MVA

Transformer 3: 132/11 KV, 200 MVA

Transformer 4: 66/11 KV, 100 MVA

Rdc=0.1 Ω, Ldc=1 mH; C=800uf.

Load 1: 80 MW, 20 MVAR

Load 2; 160 MW, 35 MVAR

*B. Filter Specifications:*

| Parameter | High Pass Filter | Double Tuned Filter |
|---|---|---|
| Operating Voltage | 132 KV | 66 KV |
| Cut Off Frequency | 1050 Hz | 150 & 250 Hz |
| Reactive Power | 40 MVAR | 20 MVAR |

*C. GAINS of the Controllers*

*Rectifier Station:*

For $M_1$, $K1 = 0.008, K2 = 0.003, G = 120$

For $\alpha_1$, $K1 = 0.008, K2 = 0.003, G = 1$

$T1 = 7 \times 10^{-6}$   $T2 = 25 \times 10^{-6}$

*Inverter Station:*

For $M_2$, $K1 = 0.003$, $K2 = 0.001$, $G = 77$

For $\alpha_2$, $K1 = 0.001$, $K2 = 0.05$, $G = 1.5$

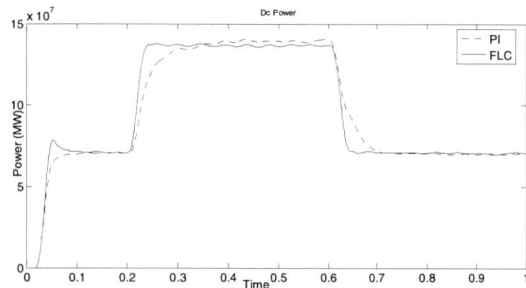

Fig. 4. Power flow in the DC line with sudden change in demand.

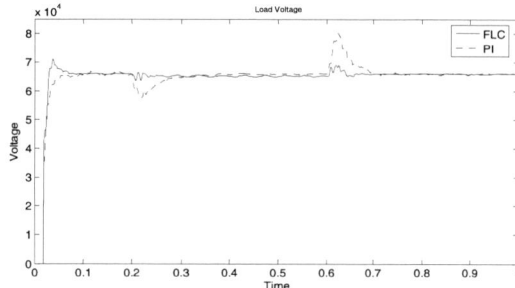

Fig. 5. Load Voltage with sudden change in demand.

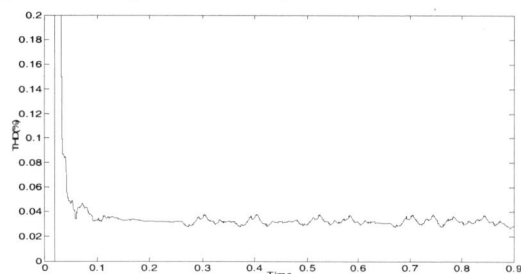

Fig. 6. Inverter Side THD.

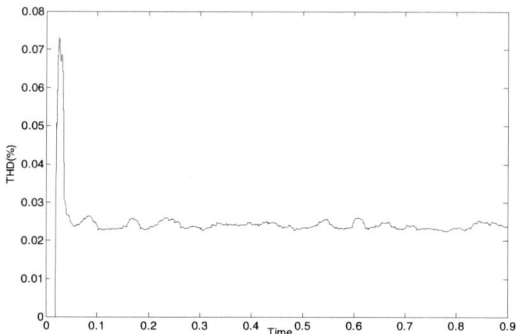

Fig. 7. Converter side THD.

## VII. CONCLUSION

The VSC based HVDC light system with function based fuzzy controller has been proposed and designed in this paper. Here for the control of voltage and power proposed controller is used which controls the system by taking two input as *error* and *change in error*. The methodology used to design the PI like FLC for the HVDC light system works effectively. Under normal and fault condition the controller is tested which gives the satisfactory result. Simulation result shows that with proposed control strategy, quick response and dynamic stability have been achieved for any kind of changes and high level control accuracy is attained at different operating condition.

## VIII. ACKNOWLEDGEMENT

The project is supported by department of science and technology. Government of India. (Project ID: SR/S3/EECE/ 02/2005-SERC-Engg, Dated 12th Aug 2005)

## IX. REFERENCES

[1] Stefan G Johansson, G Asplund, E Jansson & Roberto Rudervall, "Power System Stability Benefits with VSC DC-Transmission System" ; *Cigre conference in Paris, France 2004.*

[2] Christian Melin and Boris Vidolov, "Two Rule Based Linguistic Fuzzy Controllers", *IEEE transaction on fuzzy system, VOL 11, No 1,pp 79-87, February 2003.*

[3] Guibin Zhang, Zheng Xu, Ye Cai, "An Equivalent Model for Simulating VSC based HVDC system" *IEEE transaction on power system. March 2003.*

[4] Guibin Zhang, Zheng Xu, Hongtao Liu, "Supply Passive Networks with VDC-HVDC", *IEEE transaction on power system, March 2001*

[5] Stella Morris, P.K.Dash, "Function Based Hybrid Fuzzy Genetic Controller For VSI Based STATCOM: Single Machine Infinite System".

[6] H Ying, W. Siler and J.J. Buckley, "Fuzzy Control Theory: A Nonlinear Case", *Automatica, Vol.26, No.3, 1990*

**2006 IEEE International Conference on Power Electronic, Drives and Energy Systems**

# Voltage Regulation and Power Flow Control of VSC Based HVDC System

Bhim Singh, *Senior Member, IEEE,* B. K. Panigrahi, *Member, IEEE*, and D. Madhan Mohan

**Abstract--This paper presents an approach to control the power flow and regulate the DC voltage at the DC link of Voltage Source Converter (VSC) based High Voltage DC (HVDC) Transmission system. Two identical converters used at both rectifier and inverter ends. The switching of the converter is carried out at fundamental frequency switching. The power flow between the two stations is controlled by controlling the phase shift between the two AC voltages, and harmonic mitigation is done using 12-pulse configuration. Simulation results are presented the validate the control of the HVDC system.**

*Index Terms--* **Voltage Source Converter, fundamental switching, HVDC.**

## I. INTRODUCTION

HVDC transmission systems providing economic solutions for special kind of transmission like bulk, long distance and underwater transmission lines. The voltage source converter technology made it more suitable for such kind of applications with a number of benefits and improved performances [1-7]. This type of voltage source converter provides the reactive power compensation in FACTS applications [8]. The benefits of VSC can also be used in the HVDC transmission technology by which the important drawbacks of conventional HVDC are overcome. i.e. the conventional HVDC demands lagging reactive power. VSC converters used for power transmission or voltage support combined with an energy storage source offer continuous and independent control of real and reactive power. This reactive power control is independent of other terminal also. VSC based HVDC system is normally used in two configurations. One is point-to-point, where the two converter stations are connected by a long transmission and the second is back-to-back, in which transmission line of zero distance is used. The VSC is used in HVDC systems for both high and medium power range. IGBT and GTO are used with PWM (pulse width modulation) technology, for medium and high power respectively. In case of high power system the PWM may not be suitable as it increases the switching losses. Normally PWM is used for harmonics mitigation at the same time it increases the loss proportional to the power rating. Hence it is not advisable to use PWM control for the high power system. Here the control of converter is carried at the fundamental frequency switching

Bhim Singh, B. K. Panigrahi and D. Madhan Mohan are with the Department of Electrical Engineering, Indian Institue of Technology Delhi, Hauz Khas, New Delhi-110016,India(e-mail: bsingh@ee.iitd.ac.in, bkpanigrahi@ee.iitd.ac.in, dmadhanmohan@gmail.com)

in which each device is switched one time per cycle. In this paper the HVDC model with fundamental switching of the converter is used for power flow control. The IGBT based HVDC system for medium power applications with PWM control are called as HVDC Light, and there are some installations around the world of high power HVDC systems using 9 times of fundamental switching.

## II. SYSTEM CONFIGURATION

The converter used here is a two level converter. Fig. 1 shows HVDC system based on voltage source converter.

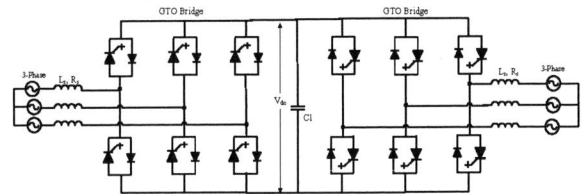

Fig. 1. Basic VSC-HVDC System.

The converter is a GTO (Gate Turn Off thyristor) based in which each device has an anti parallel diode. It is a 6-pulse 2-level bridge connected in back to back with zero distance DC link. AC systems are connected to the bridge through transformer or AC inductor. Two units are identical. A capacitor is used at DC side to maintain the DC voltage. The voltage conversion from AC to DC side is given by the model equation. Transformer separates the output voltage from the AC grid voltage in the inverter and rectifier unit from the input AC voltages. The converter cell consists of six GTO thyristors and diodes, forming a three phase voltage source converter.

Fig. 2 shows the phasor diagram of the converter and inverter voltages. Here P and Q are real and reactive power absorbed by the VSC. $V_s$ is the supply voltage, $V_c$ is converter voltage, and $\gamma$ is angle difference between $V_s$ and $V_c$. Angle of $V_s$ is same as the supply voltage angle and is fixed, and the angle of the converter voltage $V_c$ is decided by the switching angle of the converter with respect to supply voltage angle. The nature of the converter voltage is a square for fundamental frequency switching of GTOs. DC voltage at the output of the converter is decided by converter co-efficient K. The DC voltage is given by the formula. $V_{DC} = K \cdot V_c$. where $V_{DC}$ is DC output voltage of the rectifier. K is converter co-efficient or constant. The AC side equations are given by

0-7803-9771-1/06/$25.00 ©2006 IEEE     1131

$$\begin{Bmatrix} V_{sa} \\ V_{sb} \\ V_{sc} \end{Bmatrix} \cong R \cdot L \frac{di}{dt} \begin{Bmatrix} i_a \\ i_b \\ i_c \end{Bmatrix} \cdot \begin{Bmatrix} V_{ia} \\ V_{ib} \\ V_{ic} \end{Bmatrix} \tag{1}$$

and

$$P_{ac} \cong v_{ia}.i_a \cdot v_{ib}.i_b \cdot v_{ib}.i_b \tag{2}$$

The equation on DC side is given by

$$P_{dc} \cong V_{dc}.I_{dc} \tag{3}$$

as per the energy conversion process, total AC and DC power must be equal as

$$P_{dc} \cong P_{ac} \tag{4}$$

Where Real power and reactive power is given by

$$P \cong \frac{V_s V_c}{X} \sin \gamma$$

$$Q \cong \frac{V_s (V_s - V_c \cos \gamma)}{X} \tag{5}$$

Where $V_s$ is the supply voltage, $V_c$ is converter voltage; X is AC link interface reactance, and $\delta$ the angle difference between $V_s$ and $V_c$. The steady state performance of the VSC based HVDC system is studied in detail. The control scheme evaluates the phase shift required between two voltages to control the power flow. Gate signals are derived according to the phase shift requirement. Similar control is applied to the inverter unit. The important feature of VSC is that it is operated at fundamental frequency switching.

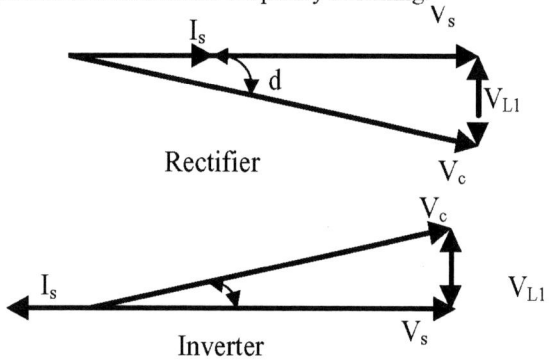

Fig. 2. Phasor diagram of Rectifier and Inverter.

Fig. 3 shows a 12-pulse GTO based voltage source converter based HVDC system. Two 6-pulse converters are connected in parallel to realize a 12-pulse converter. Two sets of cr qyhuhu#wdqvir up hu#r i#\ 2, #dqg# 2, #duh#xvhg#iro qh# for converter side and the other for inverter side. The primary of the transformers are connected in series and secondary windings of \ #dqg#l #duh#fr qqhfwhg#w #khh#w r #eulgj hv##Nese \ #dqg#l #z lqglqj s provide 30″ phase shift between two 6-pulse bridges. The same configuration is used in the inverter terminal. A DC capacitor is used in the DC side to hold the DC voltage.

## III. CONTROL SCHEME

Here the real and reactive power is controlled independently by controlling the amplitude of q and d axis

voltage vector respectively. The d-axis vector $V_d$ is parallel to the supply voltage $V_s$ and it controls the real power. Either lagging or leading power can be drawn from the system. The q-axis vector is perpendicular to the supply voltage which controls the reactive power. The phase difference between the voltages $V_s$ and $V_1$ is $\gamma$, and adjustment of this $\gamma$ enables to control of phase angle of $V_1$ with respect to the d-axis and control the power flow. The d-q transformation is used to convert the 3-phase quantities into d-q quantities. The voltage current equations at the rectifier and inverter end in terms of d-q values is given by

$$\begin{Bmatrix} v_{sa} \\ v_{sb} \\ v_{sc} \end{Bmatrix} \cong R \cdot L_1 \frac{di}{dt} \begin{Bmatrix} i_{1a} \\ i_{1b} \\ i_{1c} \end{Bmatrix} \cdot \begin{Bmatrix} v_{1a} \\ v_{1b} \\ v_{1c} \end{Bmatrix} \tag{6}$$

$$\begin{Bmatrix} v_{sa} \\ v_{sb} \\ v_{sc} \end{Bmatrix} \{0 \begin{Bmatrix} v_{1a} \\ v_{1b} \\ v_{1c} \end{Bmatrix} \cong R \cdot L_1 \frac{di}{dt} \begin{Bmatrix} i_{1a} \\ i_{1b} \\ i_{1c} \end{Bmatrix} \tag{7}$$

where $v_{sa}$, $v_{sb}$, $v_{sc}$, $i_a$, $i_b$, $i_c$, 3-phase supply voltage and current. $v_{1a}$, $v_{1b}$, $v_{1c}$, $i_{1a}$, $i_{1b}$, $i_{1c}$ converter voltage and current.

Representing the above eqn. in d-q form,

$$\begin{bmatrix} R \cdot L_1 \frac{d}{dt} & \zeta_1 L_1 \\ 0\zeta_1 L_1 & R \cdot L_1 \frac{d}{dt} \end{bmatrix} \begin{Bmatrix} i_{1d} \\ i_{1q} \end{Bmatrix} \cong \begin{Bmatrix} v_{sd} \, 0 \, v_{1d} \\ v_{sq} \, 0 \, v_{1q} \end{Bmatrix} \tag{8}$$

in the above equation $v_{sd}$, $v_{sq}$ are the d and q axis component of supply voltage and $v_{ad}$, $v_{aq}$ are the d and q axis component of converter voltage. Similarly the current components are given. R and $L_1$ is the resistance and reactance of the AC inductor used in between the supply and converter station. The active and reactive power drawn from the utility is given by

$$p_1 \cong v_{sd}.i_{1d} + v_{sq}.i_q$$

$$q_1 \cong v_{sd}.i_{1q} + v_{sq}.i_{1d} \tag{9}$$

$i_d*$ and $i_q*$ are the reference currents in the d and q axes, respectively. Here the q is considered as zero, as the real power flow is controlled here. When the $i_q$ is zero from (9), $p_1$ is depends upon $i_d$ value and $i_d$ is measured from the supply current $i_{abc}$. DC voltage is measured from the DC link and it is compared with the reference DC voltage in the voltage regulator.

The error signal is processed through a PI (Proportional plus Integral controller which gives the output as $i_d*$, and it is used as a reference current for current controller. Converted $i_d$ value from $i_{abc}$ is compared with the reference $i_d*$, and passed through another PI controller, and the output of the controller is taken as , or del ($\gamma$) the angle required between the supply voltage and converter voltage. This value is used in the pulse generator to introduce phase shift in the firing of the converter. The pulse generator produces pulses at fundamental frequency with zero phase shift. Required phase shift is introduced in the pulse generation by this value. Another phase shift required between the two bridges for the 12-pulse converter operation is introduced in the pulse generator.

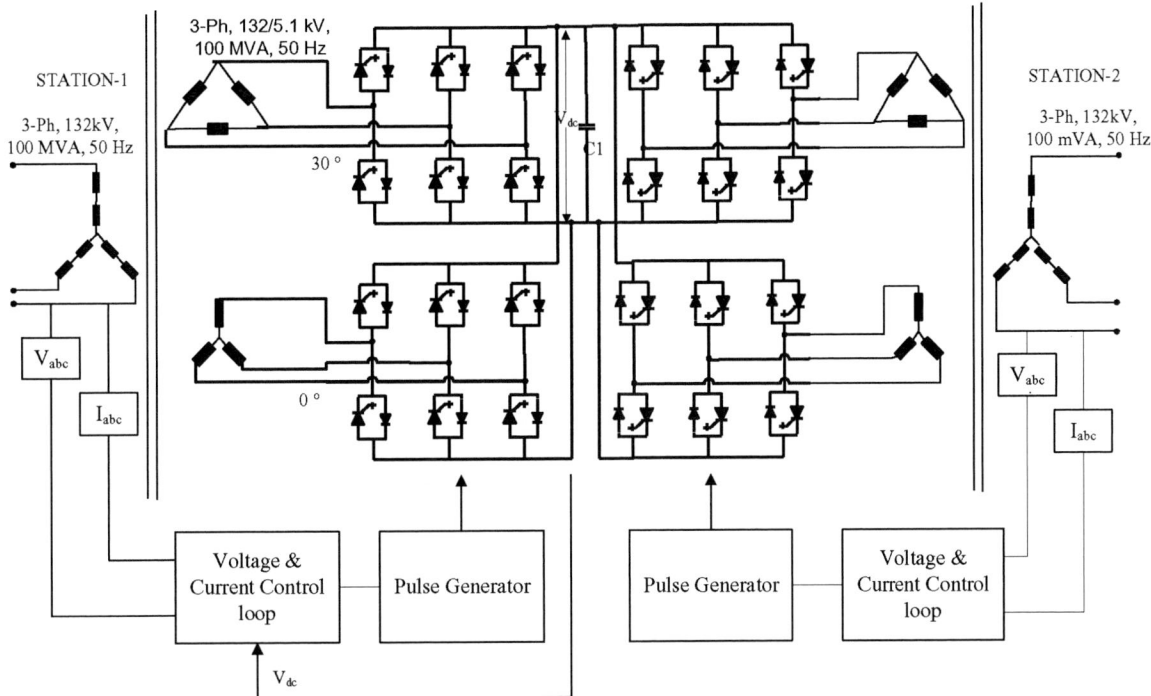

Fig. 3. A 12-pulse 100MVA GTO based VSC- Based HVDC System.

## A. Fixed Voltage Control

The main aim of the voltage control is to regulate it at the reference value and it outputs the required reactive power to AC grid. At the DC side, the DC capacitance holds out the DC bus voltage and the DC cable or line is the channel to flow active power. The energy stored in the capacitance will reduce or increase if the active power is not balanced between the two sides of the VSC stations, thereby, the DC bus voltage fluctuates. Fixed DC voltage control holds out the DC voltage by changing the active power exchanged between the VSC and the AC grid. The VSC has an inner current controller in rotation co-ordinates and an outer voltage loop.

The DC link voltage is sensed and compared to the voltage reference. The AC active power current, used for controlling the balance of DC power, consists of two parts: one is the steady portion; the other is variational current portion to compensate the DC voltage fluctuation. This, of course, yields a stationary error in the DC bus voltage, Proportional plus integral gain (PI) controller are employed to control the AC side currents, and generate references for the AC active power currents in the synchronous rotation (dq) frames.

In these control schemes, the output DC voltage is controlled by outside voltage loop, the inside feed forward decoupled PI current regulators ensure that the input AC currents track these references. This fixed DC voltage control also provides the exact reactive power to AC grid as its needed. The block diagram of DC bus voltage controller is shown in Fig. 4.

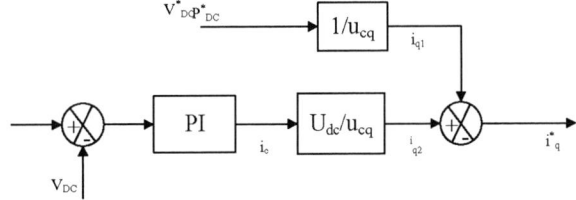

Fig. 4. DC bus Voltage Controller.

## IV. MODELING AND SIMULATION

The model of the 12-pulse GTO based VSC based HVDC system is implemented in MATLAB environment using SIMULINK and PSB block set tool boxes. Converter bridges and transformer configurations are implemented using the universal bridge and transformer blocks. The Total Harmonics Distortion (THD) during the steady state condition has been estimated using FFT tools in MATLAB. The connection diagram of the transformer and bridges to realize a 12-pulse operation is shown in Fig. 5, The MATLAB model is

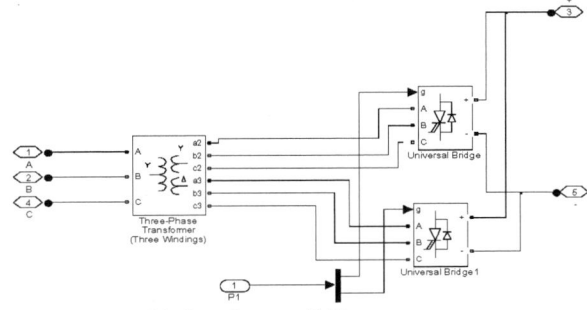

Fig.5. MATLAB model of transformer and bridge arrangement.

1133

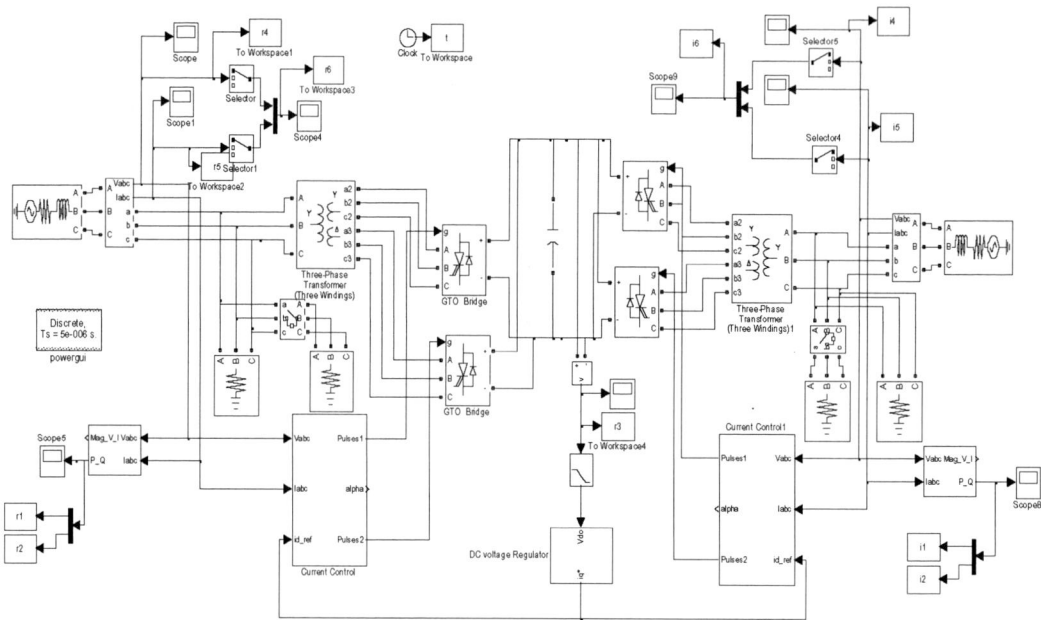

Fig. 6. MATLAB model of 12-pulse GTO VSC-HVDC System.

shown in Fig. 6. The THD in the voltage and current are determined using power GUI block, which gives the harmonics spectrum on both rectifier and inverter side. The transformer and bridge connection shown in Fig. 6 is used in both converter stations.

## V. RESULTS AND DISCUSSION

The 12-pulse GTO based voltage source converter based HVDC system is simulated in MATLAB environment for

voltage regulation and power flow control and harmonics mitigation. The obtained results show the performance of the system in respect of harmonics mitigation and voltage regulation and power flow control of both the converter stations. The performance of the converter station-1 is shown in Fig. 7. real power flow, reactive power, DC link voltage, supply voltage, supply current and phase voltage and phase current are shown. The simulation is done of 1 sec. Upto 0.5 sec load is applied on both side of the system. The load is

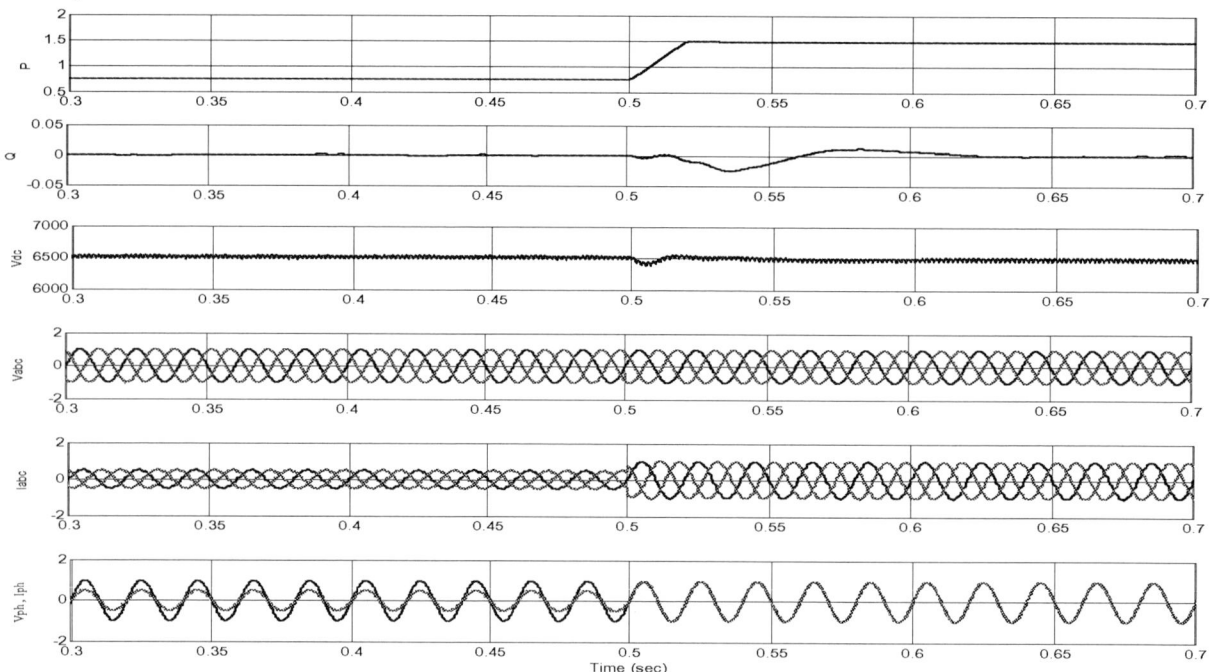

Fig. 7. Performance of Converter-1.

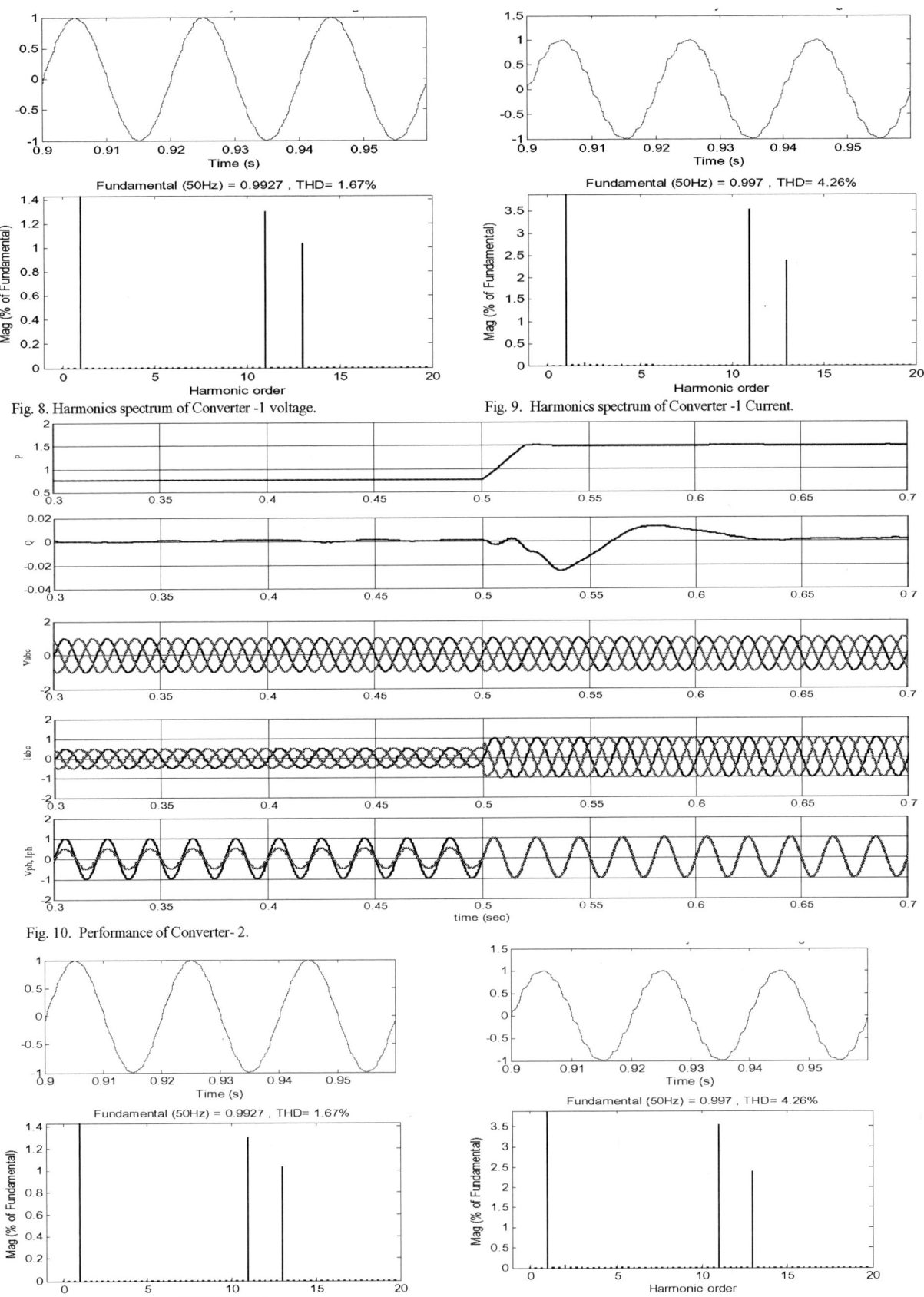

Fig. 8. Harmonics spectrum of Converter -1 voltage.

Fig. 9. Harmonics spectrum of Converter -1 Current.

Fig. 10. Performance of Converter- 2.

Fig. 11. Harmonics spectrum of Converter -2 voltage.

Fig. 12. Harmonics spectrum of Converter -2 current.

equally shared by two stations. At 0.5 sec load is increased on both sides which is again equally shared by two stations. The DC link voltage is disturbed when load changes, but again brought back to the reference value by the control action. Reactive power variation is shown in Fig.7 for a load change. The harmonics spectrum of supply voltage and current of converter- 1 station is shown in Figs. 8-9. The voltage THD is reported as 1.67% and current THD is as 4.26%. The performance of the inverter station is shown in Fig. 10, which gives real power flow, reactive power, supply voltage, supply current and phase voltage and phase current. The THD is reported for the inverter station is shown in Figs. 11-12.

The harmonics spectra are shown in Figs. 8-9 for the voltage and current of the converter-1 station. The voltage harmonics is reported as 1.67% and current harmonics is reported as 4.26 % which well with in the acceptable level. Still the harmonics level can be reduced by increasing the pulse number or other reinjection circuits with out increasing the number of devices.

## VI. CONCLUSIONS

The power flow control and the voltage regulation at the DC link terminal have been achieved for the VSC based HVDC system using GTO thyristors at fundamental switching frequency. The harmonics injection due to the converter in the AC system is reduced by using a 12-pulse converter configuration, along with the DC voltage regulation and power factor at the supply side is maintained close to unity. The results have shown the power conversion in HVDC system can be achieved effectively at low switching losses with better control.

## VII. APPENDIX

*Parameters used for the simulation:*
Converter type of VSC, with GTO thyrisotrs. No. of Pulse 12. Converter AC voltage of 5.1 kV, and dc voltage of 6.5kV, control at fundamental frequency (50Hz) switching. DC capacitor of 75000 µF,
Source of 132kV (rms), 50Hz, with short circuit level of 3000MVA, and X/R=7.
Converter transformer: 2 X 3-phse, transformer of Y/Y and \ 2 /#r qilj xudvlr q/#z lvk#s rlrp duhv#r qqhg#lq#vhulhv#
PI controller gain: current controller $K_p$ = 29, $K_i$ = 2000
Voltage controller $K_p$ = 250, $K_i$ = 1600.

## VIII. REFERENCES

[1] Mojtaba Noroozian, Abdel-Aty Edris, David Kidd, "Vkh#Sr vldqvldo#Xse of Voltage-Sourced Converter-Based Back-to-Back Tie in Load Uhvwrudwlr qv"# *IEEE Trans. on Industry Applications,* vol. 18, no. 4. pp-1416-1412, Oct-2003.

[2] Gunnar Dvs o xqg/#Nhoa#Hulnvvr q/#Nhoa#Vyhqvvr q/#"BGF #Vudqvp lvvlr q#edvhg# on Voltage Source Converter," in *Proc. CIGRE SC 14 Colloquium in South Africa 1997.*

[3] D. A. Paice, Power Electronic Converter Harmonics- Multipulse Methods for clean power. New York: *IEEE Press,* 1996.

[4] Makoto Hagiwara, Hideaki Fujita, Hirofumi Akagi, "Shuirup dqfh#r i#a Self-Commutated BTB HVDC Link System under a Single-Line-to-

Ground Fault Condition", *IEEE Trans. on Power Electronics,* vol. 18, no.1, pp- 278-285, Jan-2003.

[5] Makoto Hagiwara, Hirofumi Akagi, "Dq#Dssur dfk#wr#Uhj xodwlqj #kh#GF -Link Voltage of a Voltage Source BTB system during power flow line idxowv"# *IEEE Trans. on Industry Applications,* vol. 41, no. 5, pp-1263-1271, Sep/Oct-2005.

[6] Uxlkxd/#Vr qj #Fkdr /#] khqj #Uxrp hl/#Ol/#[ l}r {lq/#] kr x/#"EYVF #edvhg# KYGF #dqg#lw#fr qwr oVvudwhj | Š/#in *Proc. 2005 IEEE/PES Transmission and Distribution Conference Exhibition: Asia and Pacific, China.*

[7] J xlelq#] kdq #/#] khqj #[ x/#h#F dl/#Dq#Ht xlydchqwP r ghofir ir u#Mp x odVqj # YVF # Edvhg# KYGF Š/# in *Proc. IEEE/PES Trantn. and Distrn.* - 2001.pp.20-24.

[8] Eklp #Mqj k/#U#Vdkd/#D#Kdrmonics Optimized 12-pulse STATCOM for Srz hu#V\ wp #Dssdfdwr qv"Š/#*In Proc. IEEE Power India Conference, 2006,* 10-12 April 2006. pp.1-7.

[9] Nhqqhvk#Olsp dq/#"Kdup r qlf#Uhgxfwlr q#r u#P xov-Eulgj h#Fr qyhuwuv"Š# U.S. Patent 4 975 822, Dec. 4, 1990.

## IX. BIOGRAPHIES

**Bhim Singh** (SM'99) was born in Rahamapur, U. P., India in 1956. He received B. E. (Electrical) degree from University of Roorkee, India in 1977 and M. Tech. and Ph. D. degrees from Indian Institute of technology (IIT), New Delhi, in 1979 and 1983, respectively. In 1983, he joined as a Lecturer and in 1988 became a Reader in the Department of Electrical Engineering, University of Roorkee. In December1990, he joined as an Assistant Professor, became an Associate Professor in 1994 and Professor in 1997 at the Department of Electrical Engineering, IIT Delhi. His field of interest includes power electronics, electrical machines and drives, active filters, static VAR compensator, analysis and digital control of electrical machines. Prof. Singh is a Fellow of Indian National Academy of Engineering (INAE), Institution of Engineers (India) (IE (I)) and Institution of Electronics and Telecommunication Engineers (IETE), a Life Member of Indian Society for Technical Education (ISTE), System Society of India (SSI) and National Institution of Quality and Reliability (NIQR) and Senior Member of IEEE (Institute of Electrical and Electronics Engineers).

**B.K.Panigrahi** is presently working as an Assistant Professor in the Department of Electrical Engineering, IIT, New Delhi. Prior to joining IIT Delhi, he was working as Lecturer at University College of Engineering, Burla, Sambalpur, and Orissa for about 13 years. His field of interest includes the areas of Intelligent control of FACTS devices, Application of advanced DSP techniques for Power Quality assessment.

**D. Madhan Mohan** was born in Kancheepuram, Tamil Nadu, India in 1980. He received his diploma from Bhakvatsalam Polytechnic, kanceepuram in 1998, B. E in Electrical from Madras University in 2001, and M.E in Power Systems from Anna University in 2004. Then he worked as a Lecturer in RMD Engg College, Chennai. Presently he is pursuing his Ph.D from Department of Electrical Engg, Indian Institute of Technology and Delhi. His field of interest includes power electronics, power quality, HVDC and FACTS.

**2006 IEEE International Conference on Power Electronic, Drives and Energy Systems**

# Modeling and Simulation of Electromagnetic Conducted Emission Due to Power Electronics Converters

A. Farhadi, and A. Jalilian

*Abstract* - **Electromagnetic Interference (EMI) refers to the undesired generation of radiated or conducted energy in electrical systems. High-speed semiconductors are applied in power electronics converters to improve efficiency. But high frequency switching leads to generation of interference over a wide range of frequency. EMI is an inevitable problem in modern power electronic circuits. Electromagnetic compatibility (EMC), which has recently gained a high importance, is the solution against electromagnetic interference. The first step of EMC evaluation is modeling and simulation of EMI to help power electronics designers to have an estimation of EMC status in their designs. Modeling and simulation of different typical samples of power electronics converters are studied in this paper in point of view of EMC. Simulation results demonstrated noncompliance behavior of most common power electronic converters in terms of EMC.**

*Index Terms* - **Converter, EMC, EMI, Power Electronics,**

## I. INTRODUCTION

FAST semiconductor devices make it possible to have high speed and high frequency switching in power electronics converters [1]. High speed switching helps to reduce weight and volumes of equipment; however, it causes some unwanted effects such as radio frequency interference (RFI) emission [2]. Compliance with electromagnetic compatibility (EMC) regulations is a problem for producers to present their products cost effective to the markets. Post development modifications would be too costly; therefore it is important to take EMC aspects already in design phase [3]. Modeling and simulation is the most cost effective tool to analyze EMC consideration before developing the products. Most of the previous studies concerned the low frequency analysis of power electronics components [4], [5]. However, different types of power electronics converters are capable to be considered as source of EMI. They could propagate the EMI in both radiated and conducted forms. Line Impedance Stabilization Network (LISN) is required for measurement and calculation of interference level [6]. Interference spectrum measurement at the output of LISN will be introduced as the

criteria of EMC evaluation [7], [8].

National or international regulations are the references for the evaluation of equipment in point of view of EMC [7], [8]. This paper studies interference spectrum in radio frequencies range in conducted. Simulation of four different typical converters is carried out by Pspice/Orcad 9.2 conventional software [9]. The simulation results have been compared with the regulation limitation, which shows noncompliance behavior of power electronic converters in EMC point of view.

## II. SOURCE, PATH AND VICTIM OF EMI

Undesired voltage or current is called interference and their cause is called interference source. In this paper high-speed converters are the source of interference. Interference propagated by radiation in area around of an interference source or by conduction through common cabling or wiring connections. In this study power electronics conducted emission is considered only. Equipment such as computers, receivers, amplifiers, industrial controllers, etc that are exposed to interference corruption are called victims. The common connections of elements, source lines and cabling provide paths for conducted noise or interference. Electromagnetic conducted interference has two components as differential mode and common mode [10].

### A. *Differential mode conducted interference*

This mode is related to the noise that is imposed between different lines of a test circuit by a noise source. Deduced current path is shown in Fig. 1 [10]. The interference source, path impedances, differential mode current and load impedance are also shown in Fig 1.

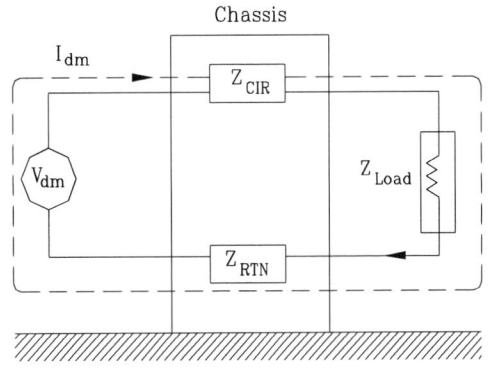

Fig. 1. Differential mode conducted interference path.

---

Amir Farhadi is Msc. student of Electrical Engineering in Iran University of Science & Technology (IUST). Meanwhile he is with Niroo Research Institute (N.R.I) as research engineer (e-mail: afarhadi@nri.ac.ir).

Alireza Jalilian is an academic member of Iran University of Science & Technology (IUST). He is assistant professor of Electrical Engineering department (e-mail: Jalilian@iust.ac.ir).

0-7803-9771-1/06/$25.00 ©2006 IEEE

*B. Common mode conducted interference*

Noise or interference could appear and impose between the lines, cables or connections and common ground that is called common mode interference. Any leakage current between load and common ground could be modeled by interference voltage source. Fig. 2 demonstrates the common mode interference source, currents, $I_{cm1}$ and $I_{cm2}$ and the related current paths [10].

The power electronics converters perform as noise source between lines of the supply network. In this study differential mode of conducted interference is particularly important and discussion will be continued considering this mode only.

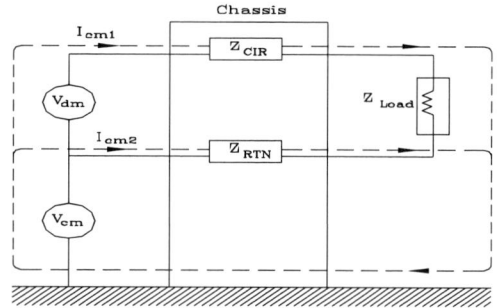

Fig. 2. Common mode conducted interference paths.

## III. ELECTROMAGNETIC COMPATIBILITY REGULATION

Application of electrical equipment especially static power electronic converters in different equipment is increasing more and more. As mentioned before, power electronics converters are considered as an important source of electromagnetic interference and have corrupting effects on the electric networks [2]. High level of pollution reduces the quality of power resulting from various disturbances in electric networks. On the other side some residential, commercial and especially medical consumers are so sensitive to power system disturbances including voltage and frequency variations. The best solution to reduce corruption and improve power quality is complying national or international EMC regulations. CISPR, IEC, FCC and VDE are among the most famous organizations from Europe, USA and Germany who are responsible for determining and publishing the most important EMC regulations. IEC and VDE requirement and limitation on conducted emission are shown in Fig. 3. and Fig. 4 [7], [10].

For different groups of consumers different classes of regulations could be complied. Class A for common consumers and class B with more hard limitations for special consumers are separated in Fig. 3. and Fig. 4. Frequency range of limitation is different for IEC and VDE that are 150 kHz up to 30 MHz and 10 kHz up to 30 MHz respectively. Compliance of regulations is evaluated by comparison of measured or calculated conducted interference level in the mentioned frequency range with the stated requirements in regulations. In united European community compliance of regulation is mandatory and products must have certified label to show covering of requirements [8].

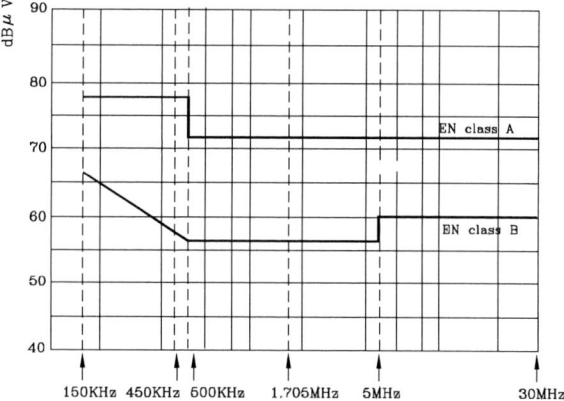

Fig. 3. IEC conducted emission limits.

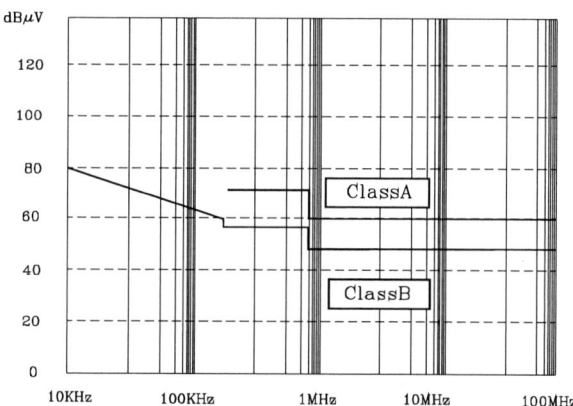

Fig. 4. VDE conducted emission limits.

## IV. ELECTROMAGNETIC CONDUCTED INTERFERENCE MEASUREMENT

*A. Line Impedance Stabilization Network (LISN)*

LISN is an industrial element offered by standards to place between the supply and power electronics converter including load as an interface to make it possible measuring the conducted interference [7]. The stated situation is shown in Fig. 5 [6].

LISN should have the following characteristics to satisfy measurement conditions [6].

1-Providing a low impedance path to transfer power from source to power electronics converter and load.

2-Providing a low impedance path from interference source, here power electronics converter, to measurement port.

1138

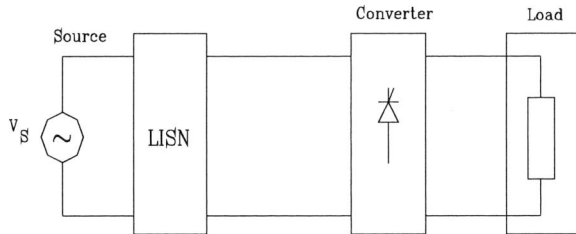

Fig. 5. LISN placement to measure conducted interference.

### B. LISN topology

The common topology for LISN is shown in Fig. 6 [7]. LISN elements quantity based on the common topology are classified as shown in Table I [7].

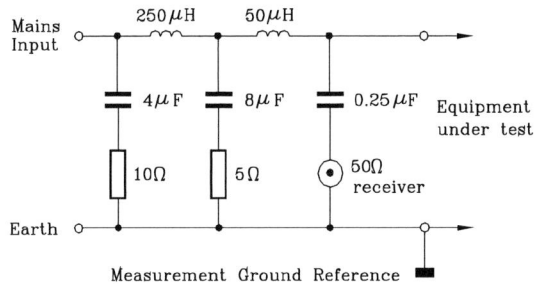

Fig. 6. LISN common topology.

TABLE I
LISN ELEMENTS QUANTITY

| L1 | C1 | R1 | L2 |
|----|----|----|----|
| 250µH | 4µF | 10Ω | 50µH |
| C2 | R2 | C3 | R3 |
| 8µF | 5Ω | 250nF | 50Ω |

Variation of LISN impedance versus frequency with the mentioned topology and quantity for elements is presented in Fig. 7, which shows, somehow the stabilized impedance [7].

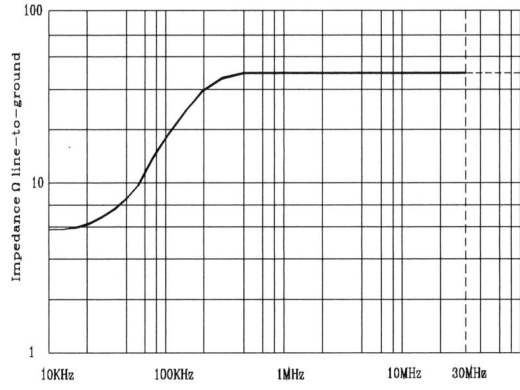

Fig. 7. LISN impedance versus frequency.

Variation of level of signal at the output of LISN versus frequency is the spectrum of interference. The electromagnetic compatibility of a device can be evaluated by comparison of its interference spectrum with the standard limitations. The level of signal at the output of LISN in frequency range 10 kHz up to 30 MHz or 150 kHz up to 30 MHz is criteria of compatibility and should be under the standard limitations. In practical situations, the LISN output is connected to a spectrum analyzer and interference measurement is carried out. But for modeling and simulation purposes, the LISN output spectrum must be calculated using appropriate software.

## V. SIMULATION OF EMI DUE TO POWER ELECTRONIC CONVERTERS

Pspice/Orcad 9.2 is a verified and conventional software for electrical and electronic circuits simulations [9]. This software is used to analyze and calculate the conducted interference spectrum in this study. Nonideal behavior of resistances, capacitors and inductors are taken into account in the simulations [11]. The simulation results are presented in the following sections.

### A. AC/DC Converter EMI simulation

Three-phase thyristor controlled bridge rectifiers have wide industrial applications [12]. Sample of such equipment including source, LISN, converter and load in accordance with the block placement of Fig. 5. is shown in Fig. 8. in the last page of this paper. Common and typical parameters of the simulated circuit are presented in appendix. Fig. 9 shows the simulation results, which is the variation of V(A5) at LISN output versus frequency.

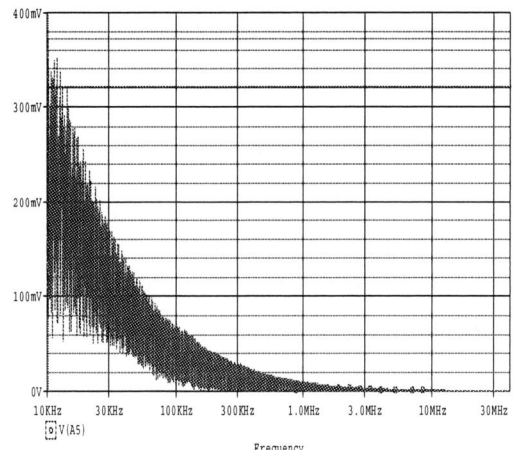

Fig. 9. EMI simulation results of a controlled AC/DC converter.

Converting the results to dBµv make it possible to compare them with standard requirements. It is seen that for this sample the level of conducted interference due to converter is not tolerable by the regulations in Fig. 3. and Fig 4. As a consequence this converter with the mentioned parameters does not comply the regulations. It is possible to repeat the simulation with other parameters, in rare of them the performance may improve and in most of the others the results may become worse.

## B. AC/AC Converter EMI simulation

The second category of power electronic converters considered for EMI simulation is three phase thyristor controlled AC/AC converters. Typical sample of this converter is shown in Fig. 10 [12]. The parameters are presented in appendix. Favorite result is the spectrum of V(A5) as the conducted interference at the output of LISN. Fig. 11. shows the favorite results. Comparison the calculated or simulated results for this special sample with limitations of standards stated noncompliance.

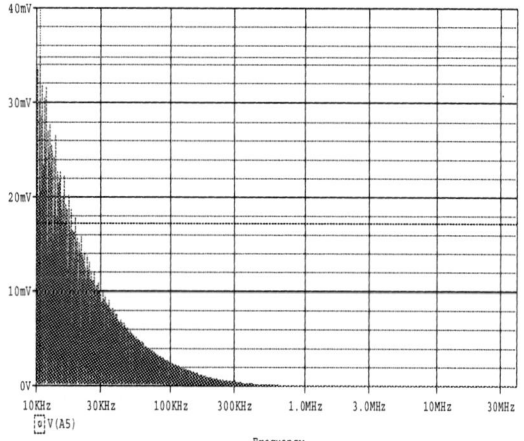

Fig. 11. EMI simulation of AC/AC converter.

## C. DC/DC Converter EMI simulation

In this study conducted electromagnetic interference due to the operation of a fixed frequency buck DC/DC converter is simulated [12]. The applied method could be generalized to other type of DC/DC converters with any switching controllers. Typical diagram of Fig. 12. shows the converter and the required parameters are given in appendix. Fig. 13. shows LISN spectrum output.

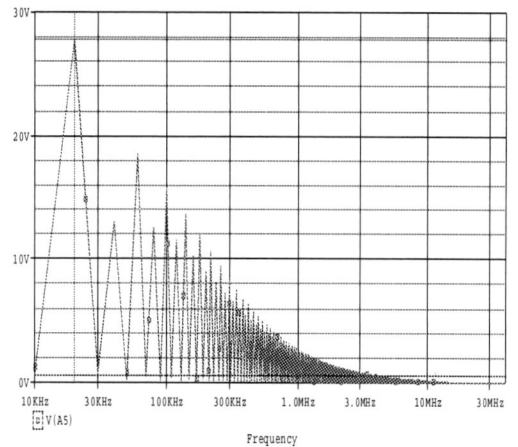

Fig. 13. EMI simulation results of DC/DC buck converter.

Though high frequency switching improves the efficiency of the converter [12] but from Fig. 13. destructive effect on EMC is considerable. The level of interference is more than standard limitations. Simulation results states not only noncompliance but also inconsistency with regulations.

## D. DC/AC Converter EMI simulation

It is not necessary to explain about wide application of DC/AC converters in different industries [12]. Simulation of three-phase pulse width modulated (PWM) DC/AC converter is a time consuming process by Pspice/Orcad 9.2 [9] and needs a huge memory, so single phase of this type of converter has been simulated based on the circuit diagram of Fig. 14. and related parameters in appendix. Fig. 15. shows the spectrum of LISN output.

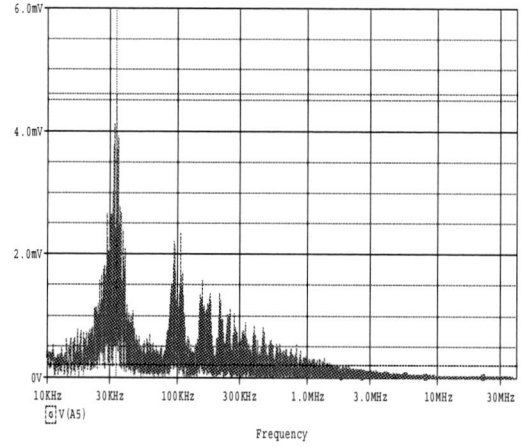

Fig. 15. EMI simulation results of DC/AC single phase PWM converter.

The noncompliance is obvious. The simulation could be easily generalized to three phase DC/AC converters with any controlling strategy.

## VI. CONCLUSION

Appearance of electromagnetic interference due to the operation of fast semiconductor devices in power electronics converter is introduced in this paper. Radiated and conducted interference coupling are introduced as two major types of electromagnetic interference where conducted type is studied in this paper. Compatibility regulations and techniques of conducted interference measurement were explained. LISN as an important part of measuring process besides its topology, parameters and impedance were described. Sample of four different types of common power electronic converters were considered and their EMI were simulated using common verified and conventional Pspice/Orcad 9.2 software. The most important point of this study is that none of the mentioned converters comply with the EMC standard regulations. It is necessary to present hardware and software mechanisms to reduce the level of interference to the standard level.

## VII. REFRENCES

[1] Mohan, Undeland, and Robbins, "Power Electronics Converters, Applications and Design" 3rd edition, John Wiley & Sons, 2003.

[2] P. Moy, "EMC Related Issues for Power Electronics", IEEE, Automotive Power Electronics, 1989, 28-29 Aug. 1989 pp. 46 – 53.

[3] M. J. Nave, "Prediction of Conducted Interference in Switched Mode Power Supplies", Session 3B, IEEE International Symp. on EMC, 1986.

[4] Henderson, R. D. and Rose, P. J., "Harmonics and their Effects on Power Quality and Transformers", IEEE Trans. On Ind. App., 1994, pp. 528-532.

[5] I. Kasikci, "A New Method for Power Factor Correction and Harmonic Elimination in Power System", Proceedings of IEEE Ninth International Conference on Harmonics and Quality of Power, Volume 3, pp. 810 – 815, Oct. 2000.

[6] M. J. Nave, "Line Impedance Stabilization Networks: Theory and Applications", RFI/EMI Corner, April 1985, pp. 54-56.

[7] T. Williams, "EMC for Product Designers" 3rd edition 2001 Newnes.

[8] B. Keisier, "Principles of Electromagnetic Compatibility", 3rd edition ARTECH HOUSE 1987.

[9] Pspice/Orcad 9.2 User's Guide.

[10] J. C. Fluke, "Controlling Conducted Emission by Design", Vanhostrand Reinhold 1991.

[11] L. Kenneth Kaiser, "Electromagnetic Compatibility Handbook" CRC Press 2005.

[12] M. Rashid, "Power Electronics Circuits, Devices and Application", Printice Hall 1993.

## VIII. APPENDIX

### A. Simulated AC/DC converter parameters:

| | |
|---|---|
| RMS input voltage | 220V |
| Frequency | 50 Hz |
| Load impedance | 1.5 $\Omega$ |
| Load angle | 30° |
| Firing angle | 60° |
| RsourceA, B, C | 0.1m $\Omega$ |
| LsourceA, B, C | 100nH |
| CparA, B, C | 2pF |
| Cpar$_L$ | 2pF |
| R$_{floating}$ | 1E12$\Omega$ |

### B. Simulated AC/AC converter parameters:

| | |
|---|---|
| RMS input voltage | 220V |
| Frequency | 50 Hz |
| Load Impedance | 1.5 $\Omega$ |
| Load Angle | 30° |
| Firing Angle | 45° |
| RsourceA, B, C | 0.1m $\Omega$ |
| LsourceA, B, C | 100nH |
| CparA, B, C | 2pF |
| Cpar$_{LA, B, C}$ | 2pF |
| R$_{floating}$ | 10E9$\Omega$ |

### C. Simulated DC/DC converter parameters:

| | |
|---|---|
| Vs | 110V |
| Rsource | 0.1 $\Omega$ |
| Lsource | 10nH |
| Cpars | 2pF |
| Rload | 6 $\Omega$ |
| Lload | 50μH |
| CparL | 2pF |
| Rfloating | 10E12 |

### D. Simulated DC/AC converter parameters:

| | |
|---|---|
| Vs | 110V |
| Rsource | 0.1m $\Omega$ |
| Lsource | 10nH |
| Cpars | 2pF |
| Rload | 0.5 $\Omega$ |
| Rg1, 2,3,4 | 100 $\Omega$ |

## IX. BIOGRAPHIES

**Amir Farhadi** was born in Tehran, Iran on first of January 1966. He received his B.Sc degree in Electrical Engineering from SHARIF University of Technology, Iran in 1989. He joined Niroo Research Institute as a research engineer. He has started for M.Sc of Electrical Engineering in Iran University of Science and Technology since 2004. His special field of interest is Power Quality.

**Alireza Jalilian** was horn in Yazd, Iran in 1961. He received his BSc degree in Electrical Engineering from Mazandran University, Iran in 1989 and his ME (Hons) and PhD degree in Electrical Engineering from University of Wollongong, Australia in 1992 and 1997 respectively. Dr Jalilian joined the power engineering group of the Department of Electrical Engineering of Iran University of Science and Technology (IUST) in 1998 as an academic member. Dr Jalilian's research interests are Power Quality causes, effects and mitigations.

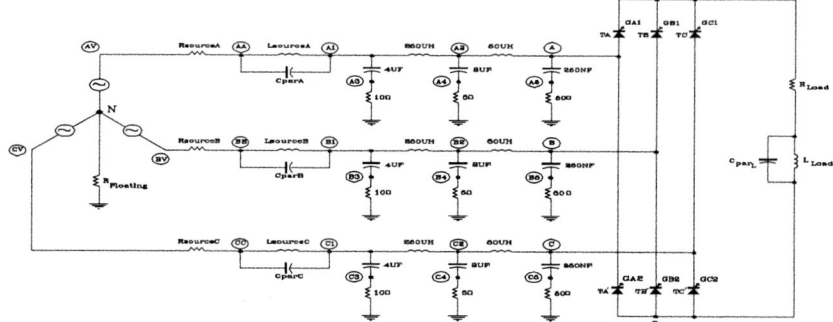

Fig. 8. Simulation diagram of controlled AC/DC converter.

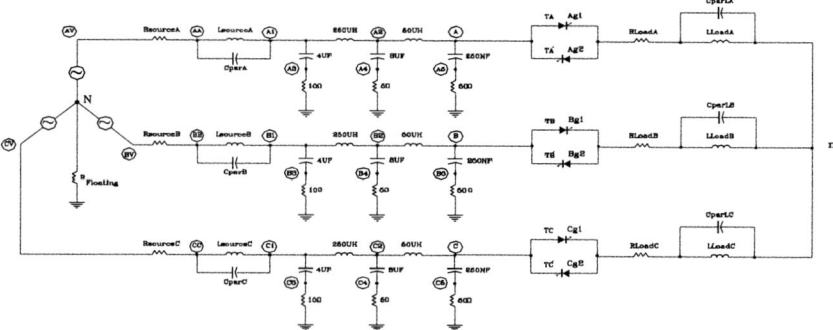

Fig. 10. Simulation diagram of AC/AC converter.

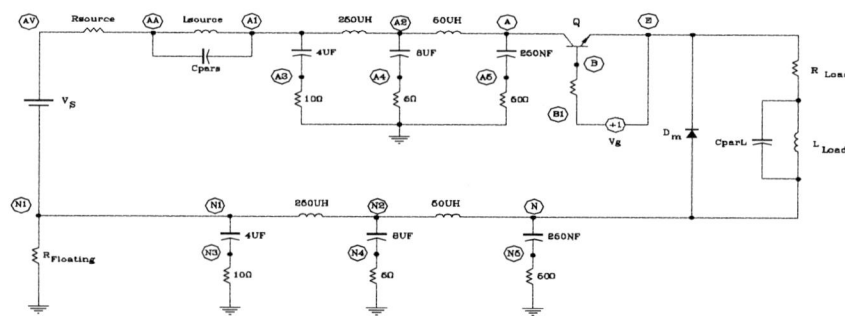

Fig. 12. Simulation diagram of DC/DC Buck converter.

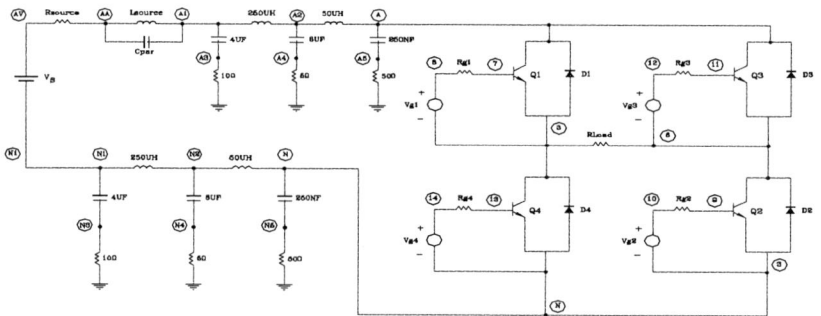

Fig. 14. Simulation diagram of PWM DC/AC converter.

**2006 IEEE International Conference on Power Electronic, Drives and Energy Systems**

# Evaluation of Operational Characteristics Of Electronic Ballasts For Metal-Halide HID Lamps

Ahteshamul Haque, and M.S.Jamil Asghar, *Member,* IEEE

*Abstract--*Electronic ballasts are needed to shape the voltage and current waveforms which suit the needs of the metal-halide, high-intensity discharge (HID) lamps. This paper deals with the evaluation of the low frequency and the high frequency operations of electronic ballasts. The results are analyzed in terms of lamps operating characteristics. A boost converter works as CCM PFC in both the operations. A dc/dc buck and a full-bridge inverters are used for low frequency operations. An LCC resonant inverter is used for high frequency operations. Simulation results and actual test bench results are shown for both 100W and 400W metal-halide HID lamps. It is concluded that the choice is between cost and reliability corresponding to high frequency and low frequency operations, respectively.

*Index Term--* dc/dc Converter, Electronic Ballasts, High Intensity Discharge Metal Halide Lamps, Power factor correction (PFC), Resonant Inverters.

## I. INTRODUCTION

THE concept of energy saving has achieved great importance and various areas are being explored in this regards. Artificial lighting is one of the area, where energy consumption is of great significance. Tremendous efforts are being made to get highly efficient lamps, e.g. fluorescents, metal-halide high intensity discharge (HID), LED etc.

Metal-halide, HID lamps has tremendous market acceptance for lighting. It has long dominated commercial and industrial applications such as high bay, industrial and retail outdoor lighting due to its inherently high lumen output, high efficacy and superior quality white light.

Alternative light sources for these applications were not as viable. The conventional fluorescent lightings had insufficient lumen output while the incandescent lighting has poor efficacy.

The electronic ballasts have now been developed for metal halide, establishing new benchmarks in performance and energy savings, and reasserting metal halide value.

The electronic ballasts function to shape the voltage and current waveforms to best meet the needs of HID lamps, operating the lamp at higher and lower frequencies. However due to certain characteristics of HID lamps, a band of operating frequency is to be avoided i.e. in normal running condition.

---

Ahteshamul Haque is doing Ph.D. from Department of Electrical Engineering, Aligarh Muslim University, Aligarh- India. (email: ahtshm@gmail.com)
M.S.Jamil Asghar is with Department of Electrical Engineering, Aligarh Muslim University, Aligarh- India. (email: msjasghar@gmail.com)

Because of the compact shape and size of the arc tubes used in HID lamps, these operating frequencies can create acoustic resonance–a phenomenon that can destroy lamps immediately. Various topologies has been proposed in the past for driving metal-halide HID lamps [2,3,4, 6,10,11,12].

It is important for power electronics engineer to understand and analyze the electronic ballast operation with these lamps, under low and high frequency operating conditions, in accordance with lamp operating characteristics.

This understanding becomes more important, particularly when no lamp model is available. This paper is an attempt to provide this understanding.

## II. LAMP OPERATIONS

It is important for power electronics engineer to understand the operating sequence of metal-halide HID lamps, before designing electronic control for it [1,2,5,13]. The operations of HID lamps can be classified in three phases:

1.  Starting
2.  Warm up
3.  Rated power.

*1) Starting Phase:*
Lamp starting phase can be further classified into three phases.

*A) Breakdown (Ignition)*
An adequate high ignition voltage is required to the electrical breakdown of the starting gases within the arc tube. The breakdown depends upon fill pressure of the gas, choice of gas, electrode construction and geometry of the discharge tube.

*B) Glow Discharge:*
The initial breakdown causes a glow discharge to take place. During this phase, little light is emitted and the lamp presents a positive impedance characteristic.

*C) Glow to Arc Transition:*
During the glow to arc transition, due to continuous ionization, the condition of glow discharge (thermo-ionic arc) changes from high-voltage and low-current to low-voltage and high-current.

*2) . Warm up Phase:*
In this phase, when the electron collision maintains the rise of temperature of arc tube, it vaporizes the metal and increases the pressure. During this phase, which may last many seconds, the voltage across the lamp increases and therefore current must be limited.

0-7803-9771-1/06/$25.00 ©2006 IEEE

The period of the warm up phase starts from the thermo-ionic emission of electrodes, until the lamp reaches to its nominal operating temperature.

The lamp maintenance can be varied and affected, if the time of the electrode-heating phase is decreased, resulting in less electrode sputtering.

Increasing the lamp current during warm up phase decreases the duration of the electrode-heating phase, with the possibility of improved maintenance [1]. Therefore, the electronic ballasts, should have the capability of higher warm up current, while at the same time keeping low crest-factor of the warm up current.

*3) . Rated Power or Normal Operation:*

When the lamp power reaches its rated value, the controller acts to compensate both, the increasing voltage of the lamp and the utility line voltage variation.

The desired lamp current (I) and lamp power (P), versus lamp voltage characteristics, for an electronic control gear [2], is illustrated in Fig.1.

## III. ACOUSTIC RESONANCE PHENOMENON

Acoustic waves in a discharge tube can be generated by periodic power, which will periodically heat the gas in the discharge tube, resulting in pressure oscillations (superimposed on an average pressure). These waves reflect against the discharge tube, and at certain frequencies standing pressure waves called acoustic resonance occur. Acoustic resonance may cause strong arc instabilities, or even extinction of the arc. It may change the lamp properties. As a result of the acoustic resonance, the discharge path may increase in length, lamp voltage may rise and the arc may oscillate, resulting in flickering of arc (Fig. 2).

The acoustic resonance depends on the lamp tube geometry and dimensions, gas compositions and thermodynamic conditions of the gas.

In other words, the periodic input power and the subsequent energy exchange by elastic collisions between charge particles and neutral gas are the source of pressure perturbations. As the input frequency is increased and an eigen frequency is approached, a pressure wave mode become propagational which, in turn, perturbs the discharge path.

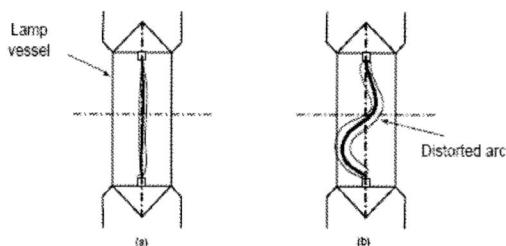

Fig. 2. Arc discharge: (a) normal operation and (b) deformation due to acoustic resonance.

Various methods had been proposed for this investigation [7,8,9,13,14]. The lamp properties that determine the eigen frequencies are known to vary with manufacturing tolerances and by lamp age.

It is important for power electronic engineer to know the frequency range, in which this acoustic resonance phenomenon can occur, for designing high frequency electronic ballasts.

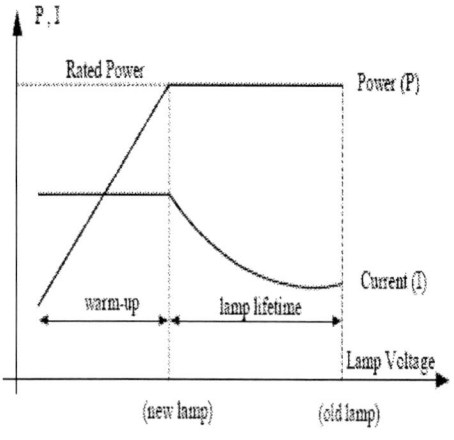

Fig. 1. The variation of current, (I) and power, (P) with respect to lamp voltage.

## IV- TOPOLOGIES

*A) - For Low Frequency Operation:*

As shown in Fig. 3, the output low frequency topology comprises of EMI filter, used to meet EMC-IEC/FCC Standard, full-bridge rectifier, boost PFC section (Fig. 4), followed by a buck dc/dc converter (Fig. 5) and low frequency full-bridge with a lamp (Fig. 6).

Since operation is complex, and three converters i.e. two dc/dc and one dc/ac converters are used, an intelligent control is required to perform the operation, as per the lamp characteristics, discussed in Section II. To perform this intelligent control, a micro-controller (ATMEL 90PWM) is used along with the driver ICs, whose block diagram is shown in Fig. 7. These low frequency topologies are discussed in [2,3,10,11].

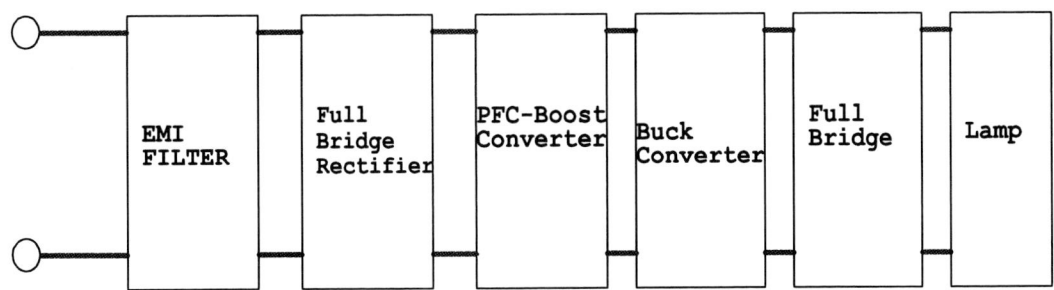

Fig. 3. Block diagram of low frequency topology used.

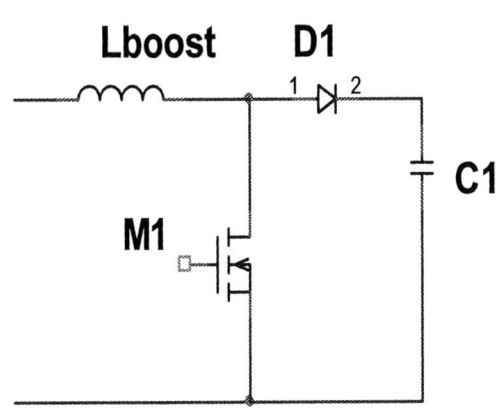

Fig. 4. PFC boost converter topology.

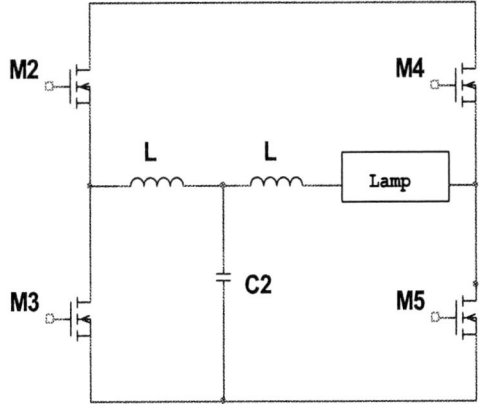

Fig. 6. Full-bridge topology with lamp schematic.

A boost section is used for power factor correction. A buck converter is used for controlling the lamp current and the full-bridge at the output is used at normal condition for low frequency operations. The transformer, L is used to provide the ignition voltage at starting or breakdown and subsequently forms series circuit with the lamp.

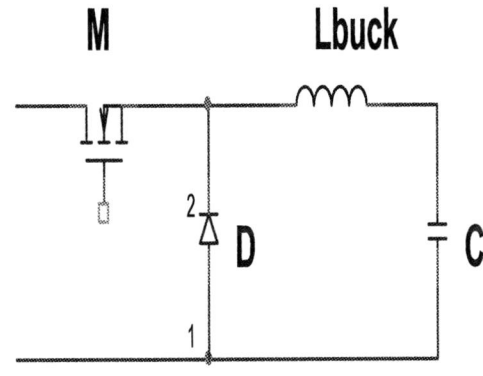

Fig. 5. Buck converter topology.

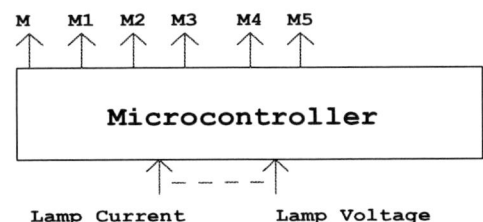

Fig. 7. Block diagram of a micro-controller.

1145

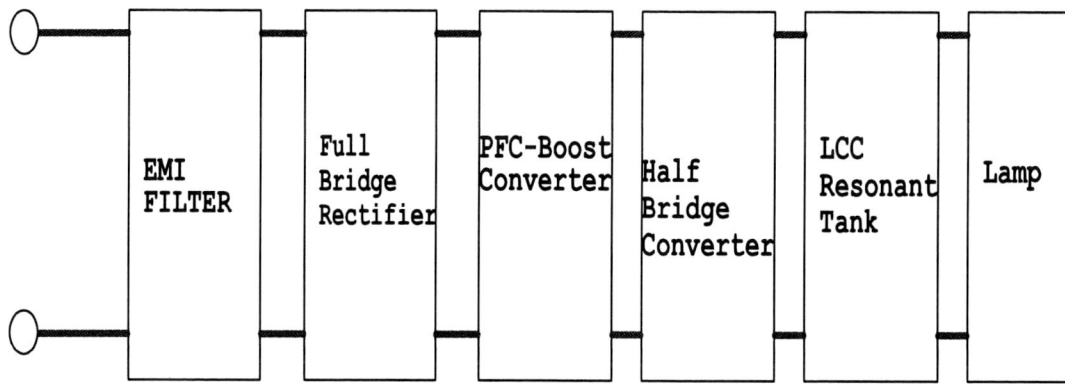

Fig. 8. Block Diagram of High Frequency Topology used.

*B) For High Frequency Operations:*

Figure 8 shows the block diagram of the high frequency topology used for evaluation of metal- halide HID lamps. Up to PFC boost section, it is same as the low frequency full-bridge topology. After PFC section, a half-bridge, LCC resonant inverter is used (Fig. 9). Both switches M7 and M8 are controlled by the micro-controller. The tank values i.e. $L_{res}$, $C_s$, $C_p$ is calculated for a 400W lamp load [6].

Both the capacitors, $C_s$ and $C_p$, play an important role during ignition phase. However, during normal running condition, $C_p$ branch impedance is higher than the impedance of the branch consisting of the lamp and dc blocker capacitor, C. It is one of the main reasons for having almost no circulating current in the resonant tank during normal running condition. It makes the system more efficient.

To achieve the desired lamp current/lamp power the frequency is swept from the no-load resonant frequency of LCC resonant tank, till the normal condition is achieved. This topology is also controlled by the micro-controller.

## V. SIMULATION AND EXPERIMENTAL RESULTS

A prototype single board is assembled for evaluation of both high frequency and low frequency topologies.
Following is the ballast specifications: -

- Rms AC Mains Voltage-230V, 50 Hz
- DC Bus Voltage – 465V
- Output Power – 100W for Low Frequency
-                       - 400W for High Frequency

INVERTER SWITCHING FREQUENCY-

- For Low Frequency  - 100Hz
- For High Frequency – 108kHz

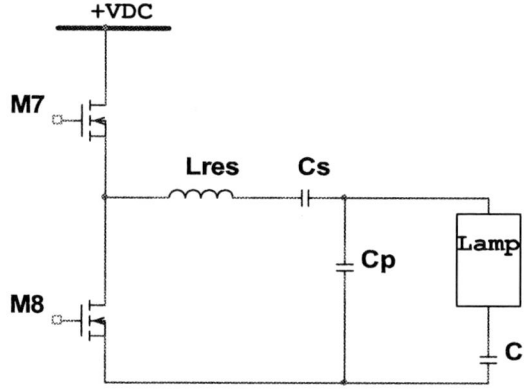

Fig. 9. Half-bridge, high frequency LCC resonant tank.

The results shown in this section is recorded for both low frequency and high frequency operations.

Figure 10 shows the simulation results of ignition voltage of a 100W, metal-halide HID lamp (make: Osram) in low frequency operations. The actual waveforms of the experimental setup are shown in Figs. 11 and 12. The simulated and the actual waveforms of the gate signals are shown in Figs. 13 and 14. Figure 15 shows the experimental waveform of current and voltage in normal condition for low frequency operations.

As per the characteristics of metal-halide, HID lamps discussed in section 2, the ignition voltage is applied across the lamps for starting of lamp operation (Figs. 10, 11 and 12). A dc/dc buck regulator is provided to control the lamp current in warm up as well as normal condition. The lamp power during the normal operating condition is controlled as per the strategy discussed in section II (Fig. 1). The output lamp current and the lamp voltage at low frequency are shown in Fig. 15.

1146

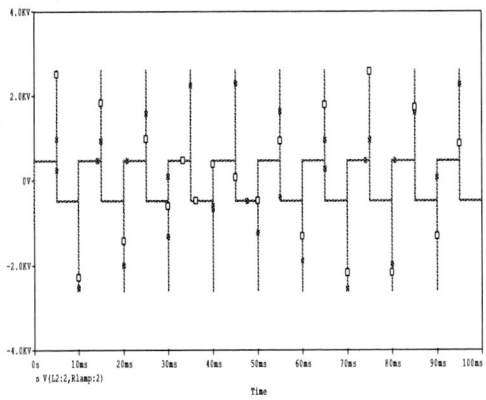

Fig. 10. Simulation result- lamp ignition voltage.

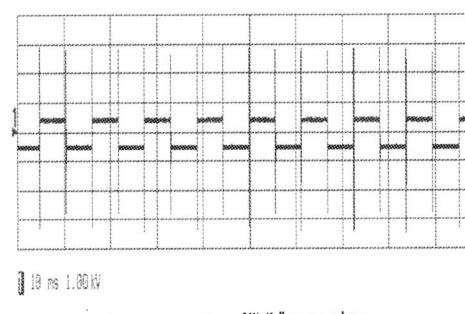

Fig. 11. CRO waveforms: Lamp voltage ignition phase.

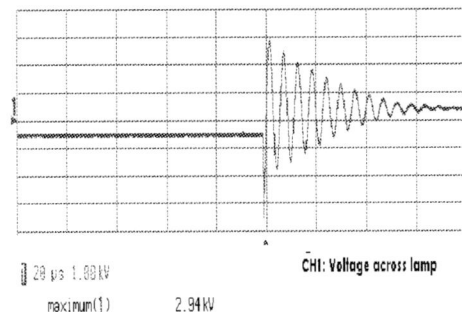

Fig. 12. CRO waveforms: zoom of lamp ignition voltage.

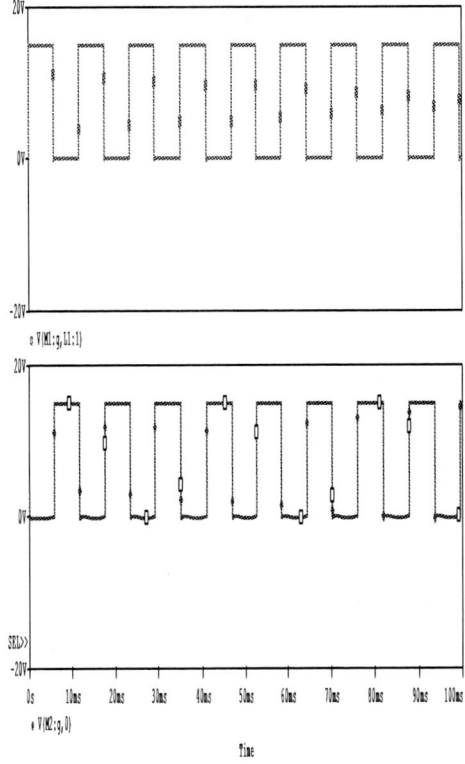

Fig. 13. Simulation results: Mosfet gate signal M2, M3.

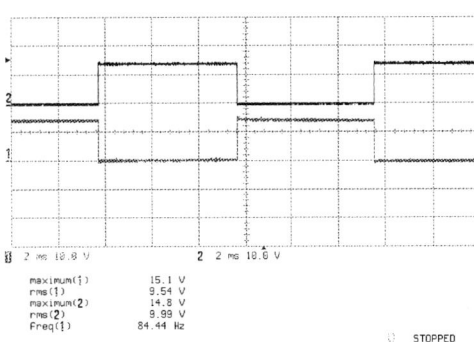

Fig. 14. CRO waveforms: M2, M3 gate signal.

1147

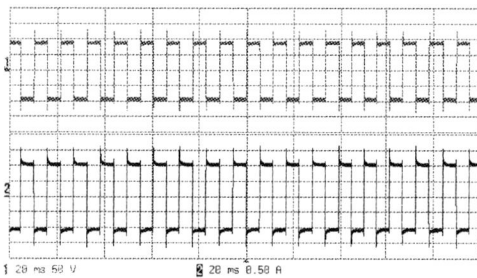

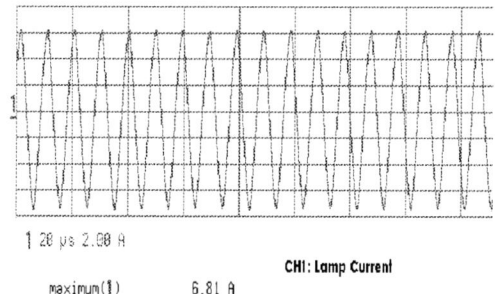

Fig. 17. CRO waveforms: lamp current–warm up phase.

Fig. 15. CRO waveforms: lamp voltage and current in normal operating condition.

The biggest advantage of low frequency topology is to minimize the occurrence of acoustic resonance phenomenon in these lamps, as discussed in section III. On the other hand, the ballast is having a complex operation due to the addition of three converter stages.

In high frequency topology evaluation (Fig. 16), the breakdown voltage is applied across the lamp. Figure 17 shows the lamp current during the warm up phase, which is higher than the normal condition (Fig. 18).

From power electronics design point of view, it is always preferable to go with high frequency operation, as it will minimize the size of the system. But the main drawback of this topology is from lamp characteristics point of view i.e. acoustic resonance phenomenon. The acoustic resonance phenomenon occurs always at high frequencies [7,8,9]. Moreover, this high frequency range, is not fixed. It varies with the age of lamp, and with other thermodynamic properties. This range may also vary from one vendor to another vendor of metal- halide, HID lamps.

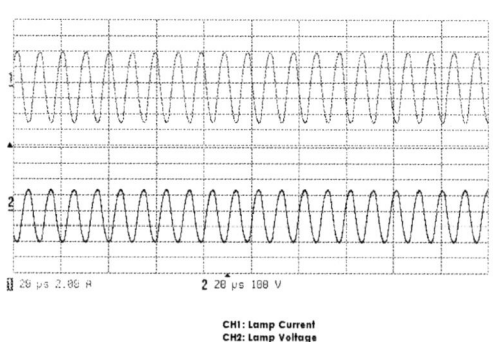

Fig. 18. CRO waveforms: lamp current & voltage –normal operation.

## VI. CONCLUSION

The operational characteristics of electronic ballasts for metal-halide, HID lamps are discussed. It is evaluated by simulation as well as experimentation. It is found that the design of electronic ballasts at higher and lower frequencies have certain advantages as well as disadvantages. The cost of low frequency ballast is at the higher side, while the risk of lamp damaging is present with high frequency operations. Therefore, it is concluded that the choice is between cost and reliability.

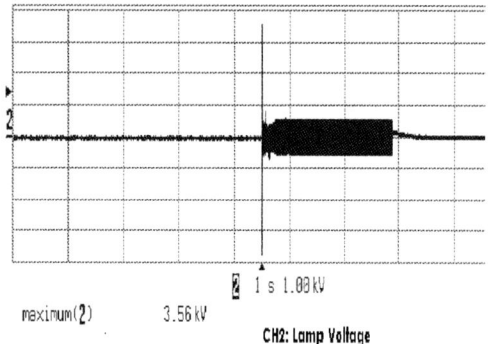

Fig. 16. CRO waveforms: ignition phase of 400W lamp.

## VII. REFERENCES

[1]  M. W. Fellows, "A study of high intensity discharge lamp-Electronic Ballast Interface," in 38th IAS Annual Meeting, Conference Record vol 2,pp.1043-1048 Oct.2003.

[2]  M.A.CO, "Micro-controlled Electronic gear for low wattage Metal Halide (MH) and High Pressure Sodium (HPS) lamps," IEEE Trans. Industry Application, vol 3, pp.1863-1868,Oct.2002.

[3]  M. Shen , Z. Qian and F. Z. Peng, "Design of a two-stage low-frequency square-wave electronic ballast for HID lamps," IEEE Trans. Industry Applications, vol,39,pp.424-430,April.2003.

[4]  M.A. Jo , M. Brumatti, D.S.L. Simonetti and J.F.L. Vieira, "Single stage electronic ballast for HID lamps," 38th IAS Annual Meeting, Conference Record, vol 1, pp. 339 – 344, Oct.2003.

[5]  Advance Transformer inc, Literature," ABC's of High intensity Discharge (HID) Ballasts.". Available: http://www.Advancetransformer.com

[6] D. Zhang; W. Zhang, Y. Liu and X. Zhao, "Design of LCC resonant inverter for metal halide lamp ballast," The 4th International Power Electronics and Motion Control Conference, IPEMC 2004, vol 3, pp. 1558-1562, Aug 2003.

[7] J.Zhou, L.Ma, and Z.Qian, " A novel method for testing acoustic resonance of HID lamps." Fourteenth Annual Applied Power Electronics Conference and Exposition, APEC'99, vol 1, pp. 480-485, March 1999.

[8] J. Olsen and W. P. Moskowitz, "Time resolved measurement of HID lamp Acoustic Frequency Spectra," Thirty-Third IAS Annual Meeting, vol 3, pp. 2111-2116, Oct. 1998.

[9] W. Yan, Y.K.E. HO, and S.Y.R. Hui, "Investigation on methods of eliminating Acoustic Resonance in small wattage High Intensity Discharge lamp", in Proc. of the IEEE-IAS Annual Meeting, 2000.

[10] H Li, M. Shen, Y. Jiang and Z. Qian, "A novel low-frequency electronic ballast for HID lamps," IEEE Trans. Industry Applications, vol. 41, pp. 1401-1408 Sept.-Oct. 2005.

[11] M. Shen, Z. Qian and F. Z. Peng, "Design of a two-stage low-frequency square-wave electronic ballast for HID lamps," IEEE Trans. on Industry Applications, vol 39, pp.424-430, March-April 2003.

[12] Cardesin.R.J, Garcia.J, Dalla-Costa.J, Alonso.J.M,"Electronic ballast for metal halide lamps based on a class E resonant inverter operating at 1 MHz." Twentieth Annual IEEE conference-Power electronics conference and Exposition-APEC 2005, vol 1,pp.600-604, March 2005.

[13] Sheng Y. Tang, Chin.S. Moo,Ching R Lee," High-Frequency operating characteristics of metal halide lamps." International Conference on Power Electronics and Drives Systems- PEDS 2005, vol.1,pp.667-671, Jan 2006.

[14] Moo.C.S, Huang.C.K, Hsiao.Y.N,"High- frequency electronic ballast with auto-tracking control for metal halide lamps." 38th IAS Annual Meeting, Conference Record, vol 2,pp. 1025-1029, Oct.2003.

## VIII. BIOGRAPHIES

**Ahteshamul Haque** was born in Varanasi, India. He obtained B.Tech degree in Electrical Engineering , from Zakir Husain College of Engineering and

Technology, Aligarh Muslim University, Aligarh-India, in Year    1999, and received M.Tech degree in Power Systems, from the Department of Electrical Engineering, Indian Institute of Technology (I.I.T) Delhi –India, in year 2000.He has the working experience of five years duration in Lighting Industry for designing Electronic Ballast for various kind of lamps, including T8/T5/T5PHO/HID/LED of international standard for various markets. Currently he is Pursuing Ph.D. from Department of Electrical Engineering, Zakir Husain college of Engineering and Technology, Aligarh Muslim University, Aligarh, India. He was member of IEEE from year 2000 to 2004.

**M.Syed. Jamil Asghar** was born in Patna, India. He obtained B.Sc. Engg. (Electrical), M.Sc. Engg (Power Systems), and Ph.D. (Power Electronics) degrees from Aligarh Muslim University, Aligarh (India) in 1978, 1982 and 1995, respectively. He Joined the Department of Electrical Engineering of the same University in 1983, where he is presently working as a Professor. He has over 20 years of teaching and research experience in power electronics. He has puiblished several dozen papers in international journals and proceedings, including several single-authored papers in IEEE Transactions. He holds several patents. He has contributed a chapter (Gate drive circuits) to Power Electronics Handbook, edited by Mohammad H. Rashid, Academic Press/Elsevier Science, California, 2001. He also authored a book, Power Electronics (Prentice-Hall of India, 2004). His research and teaching interests include power electronics, renewable energy systems and electrical machines. He is a member of IEEE and a Fellow of IETE India.

**2006 IEEE International Conference on Power Electronic, Drives and Energy Systems**

# Active Power Filter Control Algorithm using Wavelets

### Karunesh K Gupta, Rajneesh Kumar, and H. V. Manjunath

*Abstract*--This paper presents a Wavelet Transform(WT) based technique to extract fundamental frequency component from a nonsinusoidal and unbalanced load current in a three phase system. The fundamental frequency component is extracted using Multiresolution analysis (MRA). The remaining harmonics can be used by the active filter for compensation. Simulation result obtained for a rectifier load current shows the usefulness of the proposed method.

*Index Terms*-- Active Power Filter, Harmonics, Multiresolution Analysis, THD, Wavelet Transform.

## I. INTRODUCTION

THE Nonlinear devices, like power electronics converters, inject harmonic currents in the AC system and increase overall reactive power demanded by the equivalent load. Also, the number of sensitive loads that require ideal sinusoidal supply voltage for their proper operation has increased. In order to keep power quality under limits proposed by standards, different compensating techniques using series and shunt active filters have been proposed [1]. Active power Filter (APF) characteristics depends on the accuracy of reference/extracted signal and its speed of computation. Different schemes such as Fast Fourier Transform (FFT), Kalman filter, Instantaneous Reactive Power (IRP) and Synchronous Reference Frame (SRF) theory, for the extraction of the desired signal have been proposed [2]-[4]. The extraction by FFT leads to inaccurate results if the signal is contaminated by noise. Although high frequency noise can be eliminated through low pass filtering before applying the Fourier transform, it is still troublesome to eliminate possible sub-harmonics that exist in the distorted waveform. The main problem of the Fourier transform is the number of points in the observation window, which should be a multiple of the numbers of samples per period. When the fundamental's frequency varies around the 50Hz value, this corresponds to a modification in the number of samples per period. Thus, the number of points in the observation window is not a multiple of the number of sample per period. As a result, the accuracy

of the extraction is reduced. The Wavelet Transform based technique is investigated in this paper to extract fundamental frequency component from nonsinusoidal current. This technique can eliminate the above mentioned drawbacks up to certain extent.

This paper presented in four parts. Starting with an introduction, the subsequent sections cover operating principle of active power filters, wavelet transform based controller and simulation results.

## II. ACTIVE POWER FILTERS

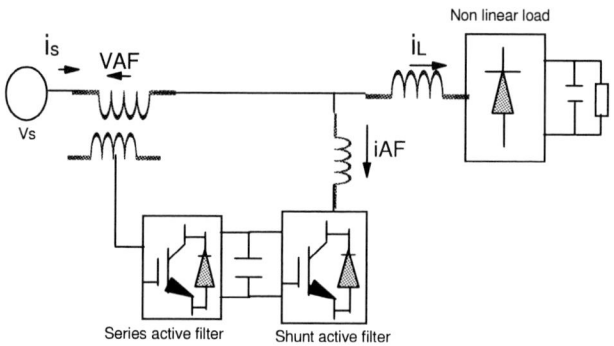

Fig. 1. Single Phase or three phase active filter.

The system configuration of a combined series and shunt active filter is shown in Fig.1. The operating principle of series and shunt active filter is described as followed [5]. The series active filter for harmonic voltage filtering is controlled on the following feedback manner

- The controller detects the instantaneous supply current $i_S$
- It extracts the harmonic current $i_{Sh}$ from the detected supply current by means of digital signal processing
- The active filter applies the compensating voltage $v_{AF}$ ($=-K\ i_{Sh}$) across the primary of the transformer. This results in significantly reducing the supply harmonic current $i_{Sh}$ when feedback gain K is set to be high enough.

The shunt active filter is for harmonic current filtering is controlled on the basis of the following feedforward manner

- The controller detects the instantaneous load current $i_L$.

---

The authors are with Electrical and Electronics Engineering group BITS, Pilani-Rajasthan-333031, India.
(e-mail: kgupta@bits-pilani.ac.in, rajneesh@bits-pilani.ac.in, hvmanju@bits-pilani.ac.in )

0-7803-9771-1/06/$25.00 ©2006 IEEE

- It extracts the harmonic current $i_{Lh}$ from the detected load current by means of digital signal processing.
- The active filter draws the compensating current $i_{AF}(=-i_{Lh})$ from the utility supply voltage $V_s$, so as to cancel out the harmonic current $i_{Lh}$.

The effectiveness of active power filter depends on accurate extraction of fundamental component of current waveform and fastness of control strategy. The wavelet transform based control strategy is described below.

### III. WT Based Controller

The block diagram of proposed WT based controller of APF is shown in Fig. 2. The three phase load currents $i_r$, $i_y$, and $i_b$, are first transformed from three phase system to two phase system currents $i_\alpha$, $i_\beta$ and $i_0$ using d-q model. The alpha and beta axis currents do not contain zero sequence current. Hence, the fundamental components of $i_\alpha$ and $i_\beta$ are extracted by proposed technique. The two phase harmonic currents $i_{\alpha h}$ and $i_{\beta h}$ are then obtained by subtracting the extracted fundamental currents $i_{\alpha f}$ and $i_{\beta f}$ from the two phase source currents $i_\alpha$ and $i_\beta$, respectively. Finally, these two-phase harmonic components and zero sequence current are transformed into three phases system using d-q model. These three phase currents obtained after transformation are then used as reference signals in hysteresis current controller for the generation of switching signals for IGBTs in VSI.

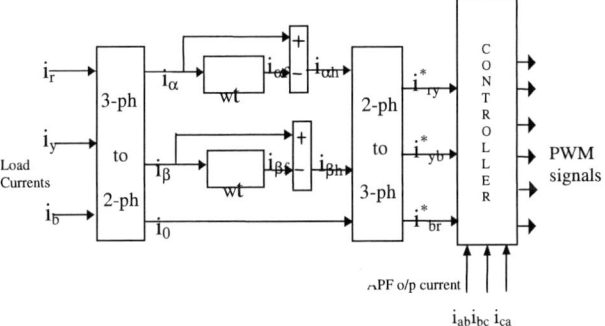

Fig. 2. Block diagram of WT based controller.

Fig.3 shows the architecture of proposed WT technique. Discrete Wavelet Transform (DWT) is an orthogonal function, which is applied to finite group of data. Multiresolution analysis [6] [7] decomposes a signal into multiple levels of details. Through a repeated change of basis, the original data is mapped into a new representation, comprising one scaling function and a sequence of wavelets. The coefficient of the scaling function refers to the coarse value, the coefficients of the wavelets, the successive details.

The framework of multiresolution can be stated as follows: Let $L_2(\Omega)$ denote the collection of all functions with finite energy over $\Omega$. The existence of a nested set of subspaces of $L_2(\Omega)$,

$$V_n \subset V_{n+1}, \text{ with } \bigcup V_n \text{ dense in } \Omega \tag{1}$$

guarantees that for any function $f$ in $L_2(\Omega)$ the successive

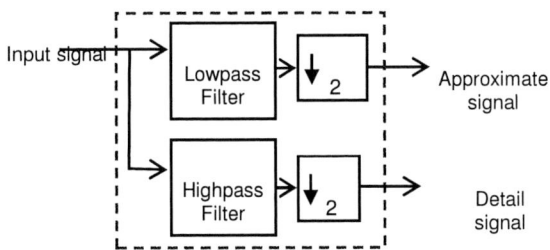

Fig. 3. Single level decomposition.

approximations $f_n$ in $V_n$ converge to $f$.
Successive approximations $f_n$ is express as

$$f_{n+1} = f_n + g_n \tag{2}$$

where $g_n$ is some function in $W_n$ which is the orthogonal complement of $V_n$ in $V_{n+1}$. This means that

$$V_{n+1} = V_n \oplus W_n \tag{3}$$

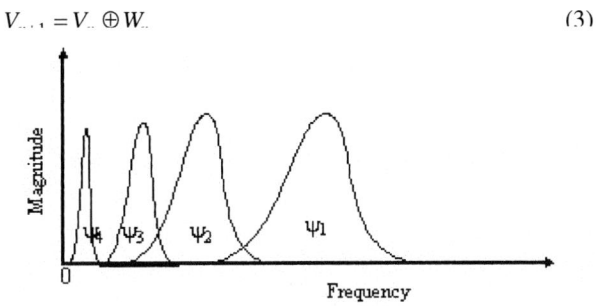

Fig. 4. Spectrums of Wavelet Filter Banks.

Therefore $f_{n+1}$ is always a better approximation than $f_n$. The ability to actually compute the approximations $f_n$ relies on finding basis functions in $V_n$ (the so-called *scaling functions*) and basis functions in $W_n$, the *wavelets*. Over the domain of definition, the first set is generated by scaling and shifting a function which satisfies a two-scale equation:

$$\varphi(x) = \sqrt{2} \sum_n h_n^{low} \varphi(2x - n) \tag{4}$$

The mother wavelet satisfies the wavelet equation:

$$\psi(x) = \sqrt{2} \sum_n h_n^{high} \varphi(2x - n) \tag{5}$$

The coefficients $h_n^{low}$ and $h_n^{high}$ are the coefficients of the corresponding filter bank. The frequency spectrums of wavelet filter banks are shown in Fig. 4. Mallat's algorithm provides a scheme to evaluate the coefficients $a_{00}$ and $b_{jk}$ in the final new representation. The signal is separated in an approximation (lowpass) of the signal and details (highpass) by the pyramid algorithm. For more than one level decomposition the pyramid algorithm is applied to the approximation, and so on [5]. The initial signal can be recomposed using a sum of signal's approximations and details. The proposed method consists in the applying the Mallat's pyramid algorithm to the given waveform The threshold is applied to the detail coefficients, the signal will be reconstructed without high frequency information contained in these coefficients, which is related to harmonic components. Thus the harmonics within the signal is eliminated. For sub-harmonics the same method can be applied to the approximation coefficients.

A signal can be fully decomposed into n levels, given by $N=2^n$, where N is the total number of data points. Each of these wavelet levels corresponds to a frequency band given by

$$f = 2^v \left( \frac{f_s}{N} \right) \qquad (6)$$

where

f    higher frequency limit of the frequency band represented by the level v

$f_s$    sampling frequency;

N    number of data points in the original input signal.

The maximum frequency that can be measured is given by the Nyquist theory as

$$f_{max} = \frac{f_s}{2} \qquad (7)$$

where $f_s$ is sampling frequency.

Table I gives the frequency band information for the different levels of the wavelet decomposition.

TABLE I
DIFFERENT LEVELS OF WAVELETS DECOMPOSITION FOR $F_s$ = 10240 Hz

| S.N. | Wavelet Level | Frequency band Hz | Center frequency Hz |
|------|---------------|-------------------|---------------------|
| 1 | $0(a_{10})$ | DC-5 | 2.5 |
| 2 | $1(d_{10})$ | 5-10 | 7.5 |
| 3 | $2(d_9)$ | 10-20 | 15 |
| 4 | $3(d_8)$ | 20-40 | 30 |
| 5 | $4(d_7)$ | 40-80 | 60 |
| 6 | $5(d_6)$ | 80-160 | 120 |
| 7 | $6(d_5)$ | 160-320 | 240 |
| 8 | $7(d_4)$ | 320-640 | 480 |
| 9 | $8(d_3)$ | 640-1280 | 960 |
| 10 | $9(d_2)$ | 1280-2560 | 1920 |
| 11 | $10(d_1)$ | 2560-5120 | 3840 |

## IV. SIMULATION RESULTS

Some simulations results are presented. The signal is generated by nonlinear load simulation in Simulink. The signal has a length of 879 samples during the observation period T=60ms. The sampling frequency is 14650Hz. The approximate and detail signals are shown in Fig. 5 and 6. The mother wavelet db4 and db20 is used for decomposition of load current. The db20 gives better results because it is smoother than db4 wavelet. The drawback of db20 or higher order wavelets is its length. The frequency contents are shown in Fig. 7 and 8. As we observed from Fig.8, the harmonic contents are drastically reduced.

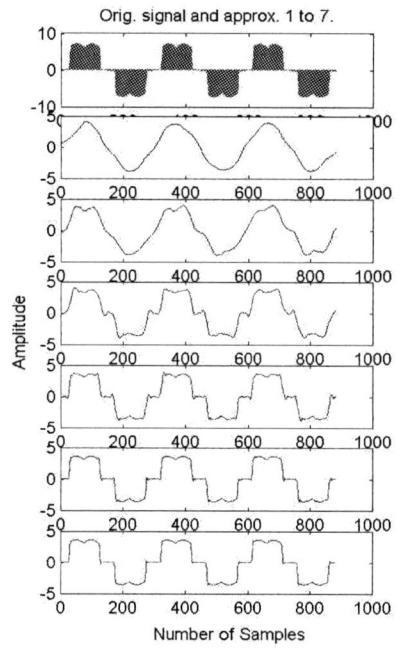
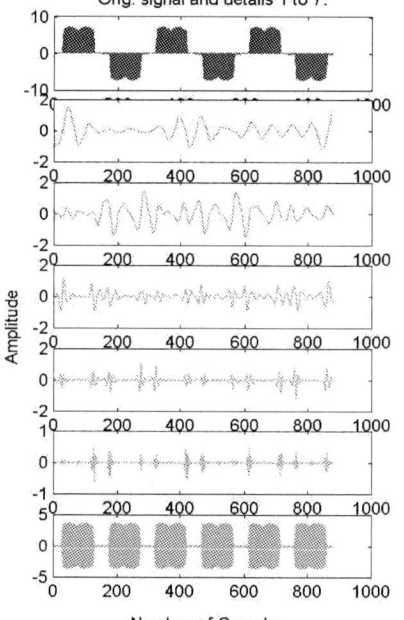
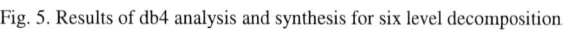

Fig. 5. Results of db4 analysis and synthesis for six level decomposition.

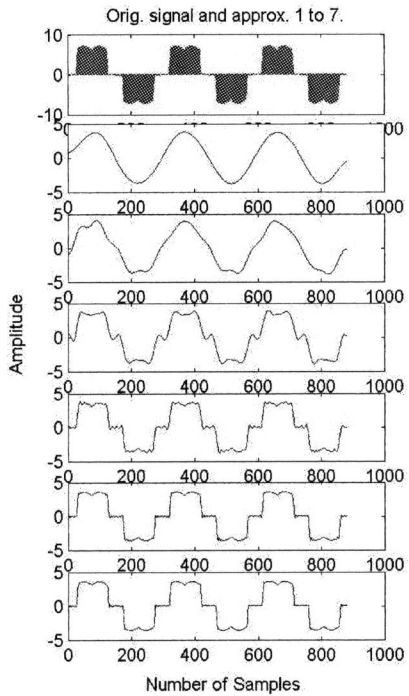
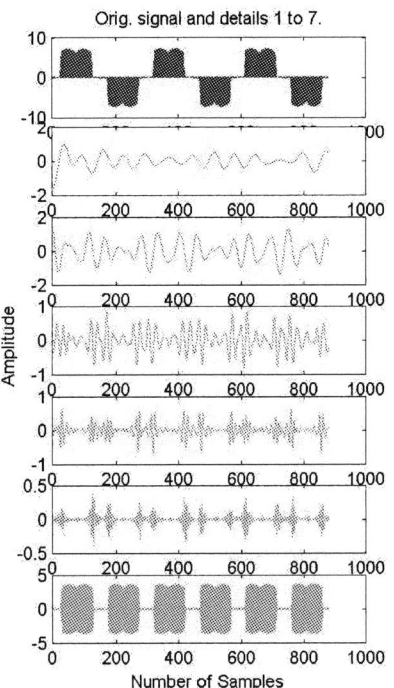

Fig. 6. Results of db20 analysis and synthesis for six level decomposition.

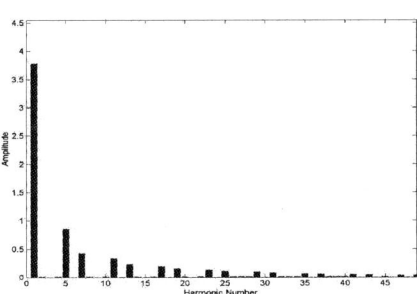

Fig. 7. Amplitude of Frequency content of load current (db4)

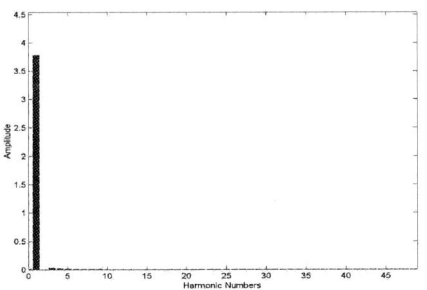

Fig. 8. Amplitude of frequency content of the extracted waveform (db4).

## V. CONCLUSIONS

In this paper WT-based technique has been presented. The simulated results demonstrate their effectiveness under various operating conditions. Some of the advantages of the proposed technique are:

- Insensitive to small shift in frequency.
- It can be used to eliminate subharmonics and interharmonics.
- Simple architecture and hence easy for implementation.

## REFERENCES

[1]  Singh, B., Al-Haddad, K., Chandra, A. "A review of active filters for power quality improvement", IEEE Trans. Industrial Electronics, vol. 46, Issue: 5, pp. 960 –971, Oct. 1999.

[2]  El-Habrouk, M., Darwish, M.K., Mehta, P., "Active power filters: a review", pp. 403 –413, IEE Proc. Electric Power Applications, vol.147, Issue:5, Sept. 2000.

[3]  M. Suman, D. W. P. Thomas, P. Zanchetta, "Power System Impedance Estimation for Improved Active Filter Control, using Continuous Wavelet Transforms," PES TD 2005/2006 May 21-24, 2006 pp. 653- 658.

[4]  M. Marinelli, L. Salvatore, "EKF-and wavelet-based algorithms applied to harmonic detection for active shunt filters," 11th International Conference on Harmonics and Quality of Power 2004, Sept. 12-15, 2004, pp.721-727.

[5]  Hirofumi Akagi, "Active Harmonic Filters," Proceedings of the IEEE,Vol. 93, No.12, pp.2128- 2141, Dec. 2005.

[6]  S.Mallat and S.Zhong, " Characterisation of Signals from Multiscale edges", IEEE Trans.on Pattern Analysis and Machine Intelligence, 14(7): 710-732, July 1992

[7]  S.Mallat, "A theory for multiresolution signal decomposition: the wavelet representation, IEEE Trans.on Pattern Analysis and Machine Intelligence, vol.11, no.7, pp.674-693, July 1989.

**2006 IEEE International Conference on Power Electronic, Drives and Energy Systems**

# Effects of Power Lines on Performance of Home Control System

V. Chunduru, *Student Member, IEEE*, and N. Subramanian, *Member, IEEE*

*Abstract*--Home Control System (HCS) helps to monitor and control the home appliances as well as security aspects of the digital home that is expected to be the standard for the future home. Chiefly, HCS is an integration of Home Appliance Control Systems (HACS) and Home Security Systems (HSS). HACS enables the home-owner to control appliances such as stove, refrigerator, air-conditioner, and the like, remotely, while the HSS helps to monitor the status of various networked security devices in the home and control certain aspects of the devices. Monitoring and control may be done by a personal digital device such as a laptop, PDA, telephone, or even a cell phone. One of the technologies widely used by HCS to connect the home controller with the appliances, equipments, and devices, is the X10 protocol that uses power lines for data transmission. In this paper we analyze the performance of power lines for HCS and suggest recommendations that will help increase the performance of HCS.

*Index Terms*--Home Control Systems, Home Appliance Control Systems, Home Security Systems, Performance, Power Lines, X10.

## I. INTRODUCTION

HOME Control System (HCS) helps to monitor and control the home appliances as well as security aspects of the digital home and digital homes are expected to be the standard for future homes [1]. Chiefly, HCS is an integration of Home Appliance Control Systems (HACS) [2] and Home Security Systems (HSS) [3]. The HCS system helps the home owner in a situation where she is miles away and recalls that she hasn't closed the stove or some other appliance, at which point she could reliably perform the desired functions remotely which helps her save money , time, and her property from potential fire, theft and public disturbances. For example, as our survey of homeowners has indicated, one of the frequently occurring problems has been the failure to close the garage door when leaving the home - open garage doors can be detected and closed remotely using the HCS: the status of the garage door can be observed on a personal digital device such as a laptop, personal digital assistant (PDA), a telephone, or even a cell phone, and, if necessary, the garage door can be remotely closed using the same digital device.

For this to work there is at least one home controller in the house that is connected to other appliances and equipments that need to be controlled. One of the technologies widely used by HCS to interconnect the controller to the appliances and equipments is the X10 protocol [4] that uses the existing power lines for data transmission. The main advantage with this technique is that the infrastructure for connecting the controller to the appliances is readily available and therefore can potentially reduce costs for the installation of the HCS.

Performance is an important aspect of HCS – high speed of information delivery and reduced time to wait for system response are important characteristics for technology acceptance [5, 6]. Performance has been defined [7] as the accomplishment of system functions within the constraints of speed, accuracy, and memory usage. For HCS, time to respond to the user, bandwidth, cost, ease of use, and accessibility, are some of the most important aspects related to performance. We analyzed the performance of HCS using X10 protocol and we found that the HCS performance is significantly impacted using the X10 protocol on current electrical wiring in the house. In this paper we present the results of our experiments with the X10 protocol and recommend techniques to improve the performance of HCS using home electrical wiring.

Our literature survey demonstrated that very little work has been done in analyzing performance of X10 for HCS – manufacturers of X10 equipments (such as www.smarthome.com) merely list the advantages. Our previous study [3] focused on performance of different types of HCS whereas in this study we focused on the suitability of X10 technology for HCS. The paper is organized as follows: in Section 2 we discuss the HCS especially with respect to distinguishing characteristics of the HCS domain; Section 3 discusses the X10 protocol and its use for HCS; Section 4 analyses the performance of X10 technology for HCS; and Section 6 discusses the conclusions and future work.

## II. HOME CONTROL SYSTEM

A Home Control System is an integration of HACS and HSS.HCS is an integration of the following technologies: home networking, smart appliances, the internet, and mobile

---

V. Chunduru is with Department of Computer Science, University of Texas at Tyler, Tyler, TX 75799 USA (e-mail: vishnuchunduru@gmail.com).

N. Subramanian is with Department of Computer Science, University of Texas at Tyler, Tyler, TX 75799 USA (e-mail: nsubramanian@uttyler.edu)

0-7803-9771-1/06/$25.00 ©2006 IEEE

wireless access. Home networking is the collection of elements that process, manage transport and store information enabling the connection and integration of multiple computing, control, monitoring and communication devices in the home. Home networking, in turn, has been enabled by the emergence of new trends such as broadband access, telecommuting, multi-PC households, remote home security services, remote home energy services, and even remote assistive solutions for disabled people (for example, Sensor Information Systems for Assisted Living or SISAL [8]). Home networking led to the development of the residential gateway (RG) that interfaces the home with the outside world, and the home network controller that provides the interface between the devices at home and RG.

Smart appliances is a relatively newer development and several major appliance manufacturers (Toshiba [9], Samsung [10], LG [11], and Carrier [12]) are developing internet-ready appliances such as stoves, refrigerators, washers, dryers, and the microwave, so that these smart appliances may be directly plugged-into the home network. Once these smart appliances are plugged-in, they become another element in the home network and may be controlled via the controller, either from outside or the inside of the home.

Internet has really helped propel the ability of remote control facility of the HCS. With several hand-held wireless devices being internet accessible, for example, the laptop, personal digital assistants (PDA's, such as for example, Blackberry [13]) and cell phones, the RG is now accessible via the internet from any of these hand-held devices and as such permits the access and control of devices at home from these devices remotely – while home networking enabled control from a short distance, the internet has enabled control of the home from a very large distance possibly hundreds or even thousands of miles away. Therefore, HCS provides unprecedented level of control to the home owner and as a result may increase the quality of her life.

The distinguishing features of the data sent over the HCS system are the following:

1. short bursts of control commands from the controller
2. short bursts of response commands from the appliance or equipment
3. typically several nodes connected to the system, where the node refers to a controller, appliance, or equipment
4. typically long average distance usually measured in tens of feet
5. occasionally large data transmissions
6. Repetitive use of the technology due to habit.

### III. X10 FOR HOME CONTROL SYSTEMS

When different networks are joined a gate way must perform the functions of media translation, address translation, authentication/filtering and system management. The Residential Gateway (RG) performs these functions for the home [14].

One of the technologies widely used for HCS is X10 protocol which is used for data transmission. The X10 protocol [15, 16] is perhaps the oldest standard for home networking. It was introduced in 1978 for the Sears home control system and the Radio Shack plug'n'power system [17]. X10 communicates between transmitter and receiver by sending and receiving signals over the power line wiring. These signals involve short RF bursts which represent digital information. This protocol has be used as it has many advantages including being inexpensive, no new wiring required, simple to install, compatible with many products, controls up to 256 devices.

The X10 home automation system [23] provides a convenient means for interfacing the appliances in the home with the help of the RG. Household electrical wiring is used to send digital data between X10 devices. This digital data is encoded onto a 120 kHz carrier which is transmitted as bursts during the relatively quiet zero crossings of the 50 or 60 Hz AC alternating current waveform. One bit is transmitted at each zero crossing. The digital data consists of an address and a command sent from a controller to a controlled device. Controllers query equally advanced devices to respond with their status. This status may be as simple as "off" or "on", or the current dimming level, or even the temperature or other sensor reading. Devices usually plug into the wall where a lamp, television, or other household appliance plugs in; however some built-in controllers are also available for wall switches and ceiling fixtures.

The relatively high-frequency carrier frequency carrying the signal cannot pass through a power transformer or across the phases of a multiphase system. In addition, because the signals are timed to coincide with the zero crossings of the voltage waveform, they would not be timed correctly to be coupled from phase-to-phase in a three-phase power system.

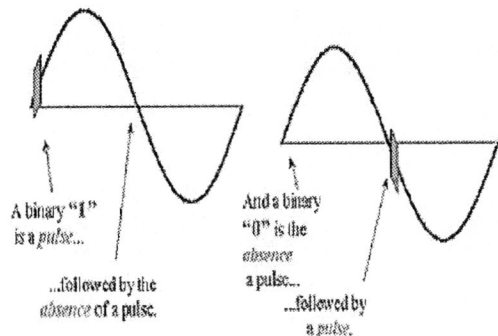

Fig. 1. Transmissions synchronized to zero crossing.

For split phase systems, the signal can be passively coupled from phase-to-phase using a passive capacitor, but for three phase systems or where the capacitor provides insufficient coupling, an active X10 repeater is sometimes used. It may also be desirable to block X10 signals from leaving the local area so, for example, the X10 controls in one house don't interfere with the X10 controls in a neighboring house. In this situation, inductive filters can be used to attenuate the X10 signals coming into or going out of the local area.

The X10 system was chosen over others primarily consists of a number of individual nodes that receive the signals transmitted by the controller. The X10 transmissions are synchronized to the zero crossing point of the AC power line "Fig. 1".Every node has a zero crossing detector that is used to detect signals transmitted by the controller. The receivers look for the signal within 0.6 microseconds of the zero crossing point.

**Initiation Signal**

| 1 | 1 | 1 | 0 | 0 | 1 | 1 | 0 | 1 | 1 | 1 | 0 | 0 |
|---|---|---|---|---|---|---|---|---|---|---|---|---|
| Start Signal | | | House Code "A" | | | | Device Code "2" | | | | | |

**Command Codes**

| On = 00101 | All Lights On = 00011 | Bright = 01011 |
|---|---|---|
| Off = 00111 | All Units Off = 00001 | Dim = 01001 |

Fig. 2. X10 Transmission Command Codes.

The controller transmits signals to the nodes using existing electrical wiring it initiates communication by sending a start packet followed by the house code and the device code. This initiation signal is followed by a command signal that is used to control the specified node. The command codes "Fig. 2" include options for toggling nodes on/off and for dimming/brightening lamp modules.

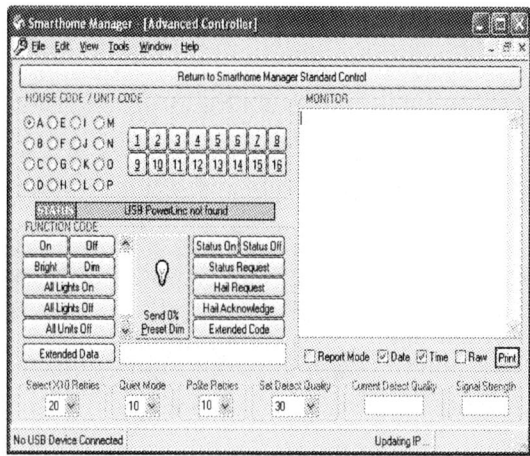

Fig. 3. X10 Advanced Integrated Controller.

By installing a series of nodes any home can be automated to a reasonable degree. The installation would only require setting a house and device code on a node and attaching the X10 node between a device and the wall outlet. The screen "Fig. 3" below shows the snapshot of the commercially available X10 systems advanced integrated controller.

## IV.  PERFORMANCE ANALYSIS OF X10

Using internal grants, several equipments were purchased to evaluate the performance of commercial home control systems: the web server, X10 home controller, signal analyzers, X10 motion sensors, X10 telephone controller switches and X10 door sensors were purchased from Smarthome (www.smarthome.com), the cameras were purchased from Toshiba (www.toshiba.com), the laptop and the PDA were purchased from Dell (www.dell.com), while the cell phone used was Sony Ericsson's with service from T-Mobile. "Fig. 4' shows the basic configuration and the equipments that are a part of the HCS used for performance analysis.

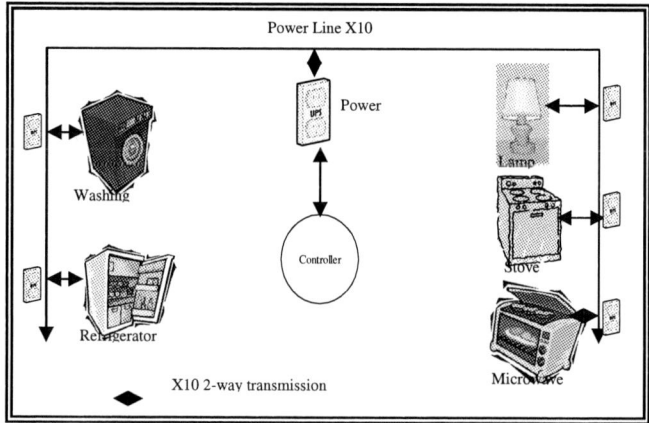

Fig. 4. Basic Configuration HCS interconnected with X10.

In the above configuration all the appliances are interconnected to the X10 advance controller with the help of the existing power line. Appliances such as washing machine, refrigerator, stove, microwave oven are connected to the appliance module where as the lamp is connected to the lamp module, both the appliance module and lamp module are bi-directional i.e. they provide two way data transmission of X10 signals.

Like regular receivers and transmitters, they can communicate on all 256 addresses. 2-Way products are helpful for status reporting and triggering other receivers to turn on, off or even run a macro event (a multi-step event run by an intelligent controller).

The homeowner can remotely control any appliance connected in this configuration using the X10 advanced controller by providing the specific house code and unit code. The user has the option to turn on or off all or specific appliances and can even check the status of the appliances. Table I highlights the results of our experiments with X10 based HCS. While speed of switching is quick for short

TABLE I
PERFORMANCE ANALYSIS RESULTS OF USING X10 FOR HCS

| Row No. | Parameter Measured | Time taken for the system to react | Comments |
|---|---|---|---|
| 1 | Time taken for Telephone Controller to control | < 1 sec | Almost immediate and accurate for short distances (< 5 feet) |
| 2 | Time taken for appliance control | < 1 sec | Almost immediate and accurate for short distances (< 5 feet) |
| 3 | Bandwidth of computer accessing internet over network | 12000 bps | Without X10 it is 20kbps-So sufficient redirection in bandwidth |
| 4 | Bandwidth of computer accessing another element in network | 12000 bps | Without X10 (for example, using wireless) it is much faster |
| 5 | Connecting appliance more than 20 feet from controller | Very poor (does not work at all) | Data signal attenuation seems to be the issue |
| 6 | Propagation of X10 signals in case of 240 volt devices | No reliable path(does not work at all) | Low-impedance bridge between two phase wires seems to be the issue |
| 7 | Propagation of X10 signals incase of low power devices below 50 watts | May not work well | Minimum resistive loads is must for accurate signal transmission |
| 8 | Transmission of signals | One at a time | Signals can only be transmitted one after the other for proper transmission and to avoid collision |

distances (less than 5 feet) as shown in rows 1 and 2 of Table I, however, over longer distances (> 20 feet) there is no signal received at all (row 5 of Table I). Moreover, bandwidth using X10 is usually considerably less than other alternatives – in fact it is only about 65% of dial-up phone connections (rows 3 and 4 of Table I). Then there is the problem of unreliable signal propagation (rows 6 and 7 of Table I) and of half-duplex transmission (row 8 of Table I). Home Control System is an integration of HACS and HSS.

As seen in table I the performance, power consumption of X10 depends on several factors apart from its advantages including being inexpensive, no new wiring required, simple to install, compatible with many products, controls up to 256 devices it has many drawbacks as well. The drawbacks of X10 are signals from a transmitter in one live conductor may not propagate through the high impedance of the distributed transformer winding to the other live conductor. Often, there's simply no reliable path to allow the X10 signals to propagate from one phase wire to the other; this failure may come and go as large 240 volt devices such as stoves or dryers are turned on and off. (When turned on, such devices provide a low-impedance bridge for the X10 signals between the two phase wires.) This problem can be permanently overcome by installing a capacitor between the phase wires as a path for the X10 signals; the manufacturers commonly sell signal couplers that plug into 240 volt sockets that perform this function. More sophisticated installations install an active repeater device between the phases, while others combine signal amplifiers with a coupling device. A repeater is also needed for inter-phase communication in homes with three-phase electric power.

Some X10 controllers may not work well or at all with low power devices (below 50 watts) or devices like fluorescent bulbs that do not present resistive loads. Use of an appliance module rather than a lamp module may resolve this problem.X10 signals can only be transmitted one command at a time. If two X10 signals are transmitted at the same time, they will collide and the receivers will not be able to decode the signal commands.

The X10 protocol is also slow. It takes roughly three quarters of a second to transmit a device address and a command. While generally not noticeable when using a tabletop controller, it becomes a noticeable problem when using 2-way switches or when utilizing some sort of computerized controller. The apparent delay can be lessened somewhat through the use of scenes and by using slower device dim rates.

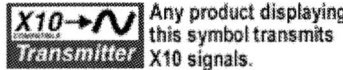 Any product displaying this symbol transmits X10 signals.

These transmitters send a specially coded low-voltage signal that is super-imposed over the 120 volts on the home's electrical wires. A transmitter is usually capable of sending up to 256 different addresses on the AC line. Multiple transmitters can send signals to the same module.

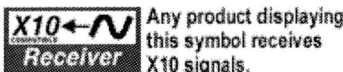 Any product displaying this symbol receives X10 signals.

Devices with this symbol receive the special signals sent by the transmitters. Once a matching signal comes in, the device responds and turns ON or OFF or dims or brightens. Receivers generally have "code dials" that are adjusted by the user to set the address. Multiple devices with the same address can co-exist in the same home.

**X10↔∿ 2-Way** Any product displaying this symbol transmits & receives X10 signals. These devices both send and receive X10 signals. Like regular receivers and transmitters, they can communicate on all 256 addresses. 2-Way products are helpful for status reporting and triggering other receivers to turn on, off or even run a macro event (a multi-step event run by an intelligent controller).

Based on this analysis we feel that X10 negatively impacts the performance of HCS due to the following reasons:

1. Time to respond to the user is fast for short distances – however, over the typical average long distances between the controller and the appliance encountered at home, the time to respond increases rapidly; the signal is sometimes so weak at the receiver that the receiver is not able to detect the signal at all.
2. Bandwidth is significantly reduced with X10 mainly due to the overhead involved – one may argue that since the average data lengths are short whether bandwidth is an issue: this advantage is quickly overcome by the fact that the communication is half duplex and that when several nodes try to communicate the bandwidth becomes a liability.
3. Cost is usually low since no connectors need be installed between the controller and the appliance – however, amplifiers and noise filters are needed to send signals properly to distant nodes on the network and this quickly drives up the cost.
4. Ease of use may initially appear to be a strong point of the X10 technology – however, if the signal attenuation due to distance and line quality is significant then the system may well become unusable.
5. This technology is only accessible from home and not from outside – some form of converter will need to be used between outside connections and internal wiring; for example, if one requires that an appliance respond to commands from laptop in the homeowner's office, then the commands will have to be received over DSL or cable modem at the home and converted into X10 signals for transmission over home wiring which not only adds to the cost but could also result in undesirable delays.
6. Presence of noise and other disturbances on the power lines significantly impact the performance of HCS negatively: X10 devices such as lights are triggered randomly without any control command being sent to them. Heating pads and fluorescent lights also seem to affect X10 devices. An experiment that used a USB connection between the computer and the power line also did not alleviate the problems due to power line disturbances.

*A. Suggested Improvements to HCS*

Based on our experiments we would like to suggest the following improvements to the technology of using power lines as connectors for HCS:

1. Use of different protocol such as PL-201 [34] or BPL (Broadband over Power Line) [36] for HCS – this could significantly improve bandwidth and access of the line for the nodes.
2. Change of the protocol frame structure so that overhead is reduced and thereby possibly increasing throughput.
3. Possibly adopting different wiring standards (similar to the Home plug initiative [36]) for homes in the future so that the capabilities of the power line are increased for broadband internet access as well.
4. Use power line noise eliminators for reducing effects of power line disturbances on X10 devices. Ground fault circuit detectors may also help to reduce this problem.

V. CONCLUSION AND FUTURE WORK

Home Control Systems (HCS) are expected to be increasingly prevalent in future homes. HCS is an integrated system that includes HACS and HSS. HCS is expected to improve quality of life for the people living in the future home by giving them unparalleled ability to remotely control their way of life at home. Performance related measurements of HCS are important to the industry and consumers so that concerned stakeholders may make the appropriate choice of HCS for their homes.

We have procured some of the commercial HCS equipments and evaluated their performance aspects. A commonly used technology in HCS for communicating with household appliances and equipments is the power line infrastructure at home. The X10 protocol is used to communicate between the controller and the appliance over the power line. While this technology is cost effective and relatively easy to use, the performance of this technology has considerable scope for improvement. During our experiments we observed the problem is weakening of X10 signal due to distance from the transmitter, reduction in bandwidth compared to even analog telephone line for data transmission, and inability to easily switch from one phase to another. Our recommendation is that HCS may need to use another type of protocol such as the PL-201 [35, 36] to the receiver, possible change of protocol frame so that overhead is reduced and possible adoptions of better wiring standards for the home. Noise and other disturbances in power lines negatively affect X10 devices by spuriously triggering the devices.

For the future we plan to study the application of the improvements suggested above for practical HCS including techniques to mitigate the effects of power line disturbances.

Also integrated approaches that use different types of technologies for HACS and HSS may need to be used so that case one system fails the others may step in (in the spirit of the view-based approach [2]) – this is similar in concept to the INSTEON technology [37]. Finally, we would like to adopt the HCS technologies to other larger networks such as the LAN, MAN, WAN, and finally the Internet itself.

## VI. ACKNOWLEDGEMENT

We would like to thank Dr.Larry Manning, Adjunct Professor, Department of Computer Science, University of Texas at Tyler, a keen X10 hobbyist, for his valuable suggestions and comments.

## VII. *REFERENCES*

[1]. "This New House", *Fortune Magazine* Special Issue supplement on How the World Will Work The Next 75 Years, September 19, 2005.

[2]. V. Chunduru and N. Subramanian, "A View-Based Approach to Improve Reliability of Home Appliance Control Systems", *Proceedings of IEEE Region 5 Conference*, San Antonio, May, 2006, to be published.

[3]. V. Chunduru and N. Subramanian, "Performance Analysis of Home Control Systems", presented at the *IEEE Emerging Technology Conference*, September, 2006, Richardson Texas, unpublished.

[4]. www.x10.com

[5]. S. E. Chang and M. S. H. Heng, "An Empirical Study on Voice-Enabled Web Applications", *IEEE Pervasive Computing Journal*, Vol. 5, No. 3, July-September 2006, pp. 76-81

[6]. C. Frankish, R. Hull, and P. Morgan, "Recognition Accuracy and User Acceptance of Pen Interfaces", available at http://sigchi.org/chi95/proceedings/papers/crf_bdy.htm

[7]. IEEE Std 610.12-1990, *IEEE Standard Glossary of Software Engineering Terminology*, IEEE, 1990.

[8]. J. A. Stankovic, I. Lee, A. Mok, and R. Rajkumar, "Opportunities and Obligations for Physical Computing Systems", IEEE Computer, November 2005, Volume 38, No. 11, pp. 23-31.

[9]. http://feminity.toshiba.co.jp/feminity/feminity_eng/about/index.html

[10]. www.samsung.com/homenetwork

[11]. www.lgeus.com

[12]. www.carrier.com

[13]. www.blackberry.com/solutions/home_personal

[14]. www.wipro.com/homenet

[15]. X10 Communications Protocol available at www.homecontrols.com.

[16]. www.en.wikipedia.org.

[17]. www.act-solutions.com/kingery13.htm

[18]. *X10, Bluetooth, Ethernet FAQ's* from www.smarthomeforum.com.

[19]. J. L. Camp, *Performance Analysis of a Secure IEEE 802.11B Wireless Network Incorporating Personal Digital Assistants*, Master's Thesis, School of Engineering and Management, Air Force Institute of Technology, Wright-Patterson AFB, Ohio, Report No. A164904, June 2002, Pentagon Reports.

[20]. I. Potamitis, K. Georgila, N. Fakotakis, and G. Kokkinakis, "An Integrated System for Smart-Home Control of Appliances Based on Remote Speech Interaction", *8th European Conference on Speech Communication and Technology*, pp. 2197-2200, Geneva, Switzerland, Sept. 1-4, 2003.

[21]. http://www2.sims.berkeley.edu/academics/courses/is213/s05/projects/thermostat/a3_analysis.php

[22]. www.adt.com/adt

[23]. http://www.windowschallenge.com/Final%20Reports/2006/WiNCE_Final.pdf#search=%22wince_final%22

[24]. S. Gupta, "Residential gateway and protocol from Home Networking" available at www.wipro.com.

[25]. V. C. Zandy and B. P. Miller, "Reliable Network Connections", *Proceedings of 8th ACM International Conference on Mobile Computing and Networking*, 2002, Atlanta, USA, pp. 95 – 106

[26]. L. Chung, B. A. Nixon, E. Yu, and J. Mylopoulos, *Non-Functional Requirements in Software Engineering*, Kluwer Academic Publishers, Boston, 2000.

[27]. N. Subramanian, R. Puerzer, and L.Chung, "A comparative Evaluation of Maintainability: A study of Engineering Department's Website Maintainability", *Proceedings of the International Conference on Software Maintenance (ICSM)*, September 2005, pp. 669 – 672.

[28]. *Wired and wireless networking* from http://compnetworking.about.com/cs/homenetworking/a/homewiredless.htm

[29]. T. Jorgensen and N. Johansen, "Control the home with a wireless network" available at www.wirelessnetdesignline.com

[30]. www.washington.edu/admin/hr/ocpsp/ps.research/comp.glossary.html.

[31]. www.sei.cmu.edu/opensystems/glossary.html

[32]. Mark Reis, CIO, Trane Company, Tyler, email communication dated 21st November, 2005.

[33]. http://en.wikipedia.org/wiki/X10_%28industry_standard%29

[34]. http://www.geocities.com/ido_bartana/

[35]. http://www.planet.com.tw

[36]. http://www.qrpis.org/~k3ng/bpl.html

[37]. http://www.insteon.com

## VIII. BIOGRAPHIES

**Nary Subramanian** (M' 1990) is currently serving as the Assistant Professor of Computer Science at the University of Texas at Tyler, Tyler, Texas. Earlier he served as the Assistant Professor of Computer Engineering in the Department of Engineering at Hofstra University, New York. Dr. Subramanian received his Ph.D. in Computer Science from the University of Texas at Dallas, an MSEE from Louisiana State University, Baton Rouge, and another MSEE from Delhi University, Delhi, India. Dr. Subramanian has about 15 years' experience in the industry in engineering, sales, and management. He has been a co-chair of the International Workshop on System/Software Architectures for the past five years, has been a guest-editor for conference proceedings and special journal issues that dealt with System and Software Architectures, and has served on the Program Committees of several international conferences and workshops. His research interests include software architecture, software engineering, software metrics, software security, non-functional requirements, expert systems, computational biology, home appliance control systems, information systems, and legal systems. He has published almost 20 papers on software architectures alone and he has several more papers to his credit. Dr. Subramanian has served as the judge for several high-school science fairs. He has received awards from both the industry and the academia. He is a member of IEEE.

**Vishnu Chunduru** (SM '2005) is a graduate student in the Department of Computer Science at the University of Texas at Tyler, research interests include Home Appliance control systems, Software Security, Web Designing. He has published and presented his work in conferences. He is a student member of IEEE.

# Author Index

## A

Abiri, E. .................................................298
Abrishamifar, A. ......................................298
Acarnley, P. P .......................................1020
Achari, V. T. Sadasivan .........................122
Adachi, Shun-ichi ...................................310
Adhinarayanan, T. ..................................942
Adya, A .................................................892
Afjei, E ..................................................151
Agarwal, P. ...........................................274
Agarwal, Pramod ....................86, 531, 673
Agarwal, Vivek ...............................443, 460
Agrawal, Pramod .............................467, 473
Ahmad, Mukhtar .....................................179
Ahmadian, J. .........................................678
Ahsan, Faisal M. ...................................602
Aktarujjaman, M. ...................................753
Alam, Shahabur .....................................729
Ali, Iqbal .............................................1041
Amaresh, K. ..........................................662
Amarnath , J. .........................................424
Amarnath, J. ..........................................511
Ambusaidi, K. A. ...................................389
Anand, Keerthi ......................................401
Anuradha, K. .........................................718
Arvindan, A.N. .......................................315
Asghar, Ali ............................................46
Asghar, M. S. Jamil ..............................1143
Aware, M.V. .........................517, 782, 395

## B

B, Isha T ...............................................824
Babu, K. Hari .........................................267
Bajaj, Harbans L. .................................1074
Balasubramanian, R. ...............................771
Banerjee, S. ...........................................20
Barati, Hassan .......................................959
Basavaraja, B. .......................................219
Basu, K. P. ...........................................46
Bates, Ian .............................................871
Bathaee, S. M. T. .................................1096
Behera, Ranjan K. ..................................430
Beig, A. R. .............................................7
Benedict, Eric ........................................740
Bhalodi, Kalpesh H. ...............................467
Bhargava, Annapurna ..............................572
Bhat, A. H. ............................................274
Bhudamani, RM .......................................555
Bhuvaneswari, G. ...........39, 560, 647, 652
Bodkhe, S. B. ........................................395

## C

C, Sreekumar .........................................443
C., Rohith Kumar H. .............................1069
Chanana, Saurabh ..................................1013
Chandra, A. ...........................................608
Chandra, Ambrish ...................................583
Chandra, D. ............................................370
Chatterjee, Dheeman ...............................818
Chatterjee, J.K. ...............................602, 812
Chatterjee, Kishore .................................460

Chaturvedi, Ganesh Dutt ..........................876
Chaturvedi, P. K. ...................................473
Chaudhari, M. A. ...................................348
Chaudhuri, N. Ray .................................1090
Chen, Po-Jia .........................................199
Chiang, Sheng-Feng ...............................199
Choudhuri, S. Ghatak .............................204
Chowdhury, S. .........................................90
Chowdhury, S. P. .....................................90
Chunduru, V. .......................................1154

## D

Dahiya, Surender ....................................620
Daigavane, M. B. ...................................506
Dalvand, H. ...........................................577
Dananjayan, P. ......................................331
Das, Anandarup .....................................602
Das, Biswarup ........................................572
Das, G.Tulasi Ram ..................................449
Das, Shyama P. ................................360, 430
Dash, P. K. ....................................1064, 1127
Dash, Subhransu Sekhar ..........................966
Dehkordi, B. Mirzaeian .............................54
Dehkordi, Behzad Mirzaeian ...............545, 101
Devadoss, Surendran ...............................871
Devasahayam, Robert .......................122, 131
Dixit, T. V. ............................................430
Dobariya, C. V. ......................................792
Dong-Fang, Zhou ...............................11, 522
Donyavi, F. ...........................................657
Dwivedi, Sanjeet ....................................342

## E

Easwarlal, C. .........................................15
Ehsan, M. ...........................................1096
Ehsan, Mehdi .........................................959
Ekram, S. ..............................................171
Elangovan, S. ........................................683
Ethny, S. A. ........................................1020
Etingov, P. V. ........................................909

## F

Farhadi, A. .........................................1137
Fazil, M. ...............................................50
Feng, Bai ......................................11, 522
Fernandez, E. ........................................898
Fotuhi-Firuzabad , Mahmud ......................959
Fotuhi-Firuzabad, M. ............................1096

## G

G, Subhash Joshi T ................................636
Gairola, Sanjay ...............................336, 701
Garg, Vipin ....................................647, 652
Gaur, Prerna ...........................................96
Gautam, Ashutosh .................................1036
Gayathri, M. S. L. ..................................842
Geethalakshmi, B. .................................331
Ghawghawe, N. D. .................................985
Gholami, A. .........................................1053
Ghose, T ...............................................888
Ghosh, Arindam .....................................181

# Author Index

Giaouris, D. .................................................... 20, 1020
Gilreath, Phil ................................................... 1024
Goedtel, A. ....................................................... 918, 926
Goel, Ankur ...................................................... 735
Goel, Manish ..................................................... 836
Gopakumar, K. ................................................... 256
Gopila, M. ........................................................ 15
Gounden, N. Ammasai .......................................... 854
Goyal, Devendra ................................................. 135
Goyal, Himani .................................................... 902
Gujarathi, P. K. .................................................. 517
Gupta , Sushma .................................................. 1084
Gupta, Ajai ...................................................... 801
Gupta, H.O. ...................................................... 673, 1069
Gupta, I. .......................................................... 1069
Gupta, J.R.P ...................................................... 892
Gupta, Karunesh K .............................................. 1150
Gupta, R. A. ...................................................... 210
Gupta, Ranjan K. ................................................ 707
Gupta, S P ........................................................ 1031, 1059
Gupta, Swapnil ................................................... 888
Gusev, B.A. ...................................................... 383

## H

Hangal, A. ........................................................ 50
Hanmandlu, M. ................................................... 902
Haque, Ahteshamul .............................................. 1143
Hasani, S. ........................................................ 657
Hidayat, Nabil M ................................................ 310, 307
Hojabri, H. ....................................................... 631
Huang, Rongjun .................................................. 860

## I

Iqbal, Atif ....................................................... 179

## J

Jack, Alan G. .................................................... 500
Jadid, Shahram ................................................... 970
Jain, D.K. ........................................................ 620
Jain, Shailendra K ............................................... 473
Jaiswal, V. ....................................................... 50
Jalalifar, M. ...................................................... 59, 193
Jalilian, A. ....................................................... 25, 678, 1137
Jamali, S. ......................................................... 1114
Janakiraman, P. A. .............................................. 641, 490, 485
Jaswant, L ........................................................ 712
Jaswanti, .......................................................... 992
Jeyabharath, R. .................................................. 117
Jha, A.N. ......................................................... 735
Jha, Aman Kumar ................................................ 267
Jhang, Jyun-Jhong ............................................... 199
Jie, Shi Yu- ...................................................... 11, 522
John, Vinod ...................................................... 495, 740
Joseph, Aby ...................................................... 636
Joseph, Achari, ................................................... 131
Joseph, C. C. ..................................................... 122
Joshi, R R ........................................................ 303
Joshi, R. R. ...................................................... 210

## K

K, Aroul. ......................................................... 378

K, Kaushik ....................................................... 1007
K, Unnikrishnan A. .............................................. 636
K, Vadirajacharya ............................................... 673
Kadwane, S.G. ................................................... 888
Kanabar, M. G. .................................................. 792
Kangsanant, Theo ................................................ 871
Kank, Amogh ..................................................... 975
Kapoor, Rajiv .................................................... 620
Karan, B. M. ..................................................... 267, 888
Karthikeyan, A. .................................................. 1120
Karthikeyan, K. .................................................. 537
Kasal, Gaurav Kumar ............................................ 64, 847
Kashem, M. A. ................................................... 830, 753
Kastha, D. ........................................................ 824
KATO, Yoshito ................................................... 307, 310
Kazemi ,A. ....................................................... 970, 1053, 1107, 1114
Khadkikar, V. .................................................... 608
Khan, M. Ahfaz .................................................. 1047
Khan, M.Rizwan .................................................. 179
Khan, Md. Haseeb ................................................ 511
Khan, Z. J. ....................................................... 506
Khaparde, S. A. .................................................. 792
Khatre, M. ....................................................... 500
Khincha, H. P. .................................................... 936, 996
Kiranmai, K. S. Phani ........................................... 354
Kothari, D. P. .................................................... 902, 882
Kotharl, M. L. ................................................... 1090
Kottayil, Sasi K .................................................. 747
Krishnaswami, Hariharan ........................................ 707
Kulkarni, S. V. .................................................. 1007
Kulkarni, S. V. .................................................. 975
Kumar, A. D. Raj ................................................ 718
Kumar, Amit ..................................................... 888
kumar, Anil ...................................................... 370
Kumar, Arun Shailendra .......................................... 526
Kumar, Ashok .................................................... 620
Kumar, Ashwani .................................................. 1013
Kumar, CH.Siva .................................................. 759
Kumar, M. Vijaya ................................................ 262
Kumar, Manish ................................................... 620
Kumar, Mukesh ................................................... 107, 126
Kumar, Parveen ................................................... 70
Kumar, Rajneesh ................................................. 1150
Kumar, Vinod ..................................................... 303, 1031, 1059, 1069
Kumbhar, G. B. ................................................... 975

## L

Lakshminarasamma, N. ........................................... 455
Lakshminarayanan, Sanjay ....................................... 256
Lavanya, K. ...................................................... 463
Lavanya, V. ...................................................... 854
Le, An D.T. ...................................................... 830
Ledwich , G. ..................................................... 753
Ledwich, G. ...................................................... 830
Lee, Fu-Shin ..................................................... 199
Lei, Yung-Tsung .................................................. 199
Lone, Shameem Ahmad ........................................... 865

## M

M, Veerachary .................................................... 354, 413, 479, 526
Ma, D.D. ......................................................... 20
Mabel, M. Carolin ............................................... 898
Madhusudan, ...................................................... 836

# Author Index

Mahajan, D. .................................... 171
Mahato, S. N. ...................................80
Mahdian , B. ................................1107
Mallesham, Gaddam ......................401
Manjunath, H. V. ..........................1150
Manna, Manpreet Singh ................948
Marwaha, Anupma ...................30, 948
Marwaha, Sanjay .....................30, 948
Marzband, M. ...........................1, 625
Masoudi, M. ..................................657
Masoum, M.A.S. ..........678, 970, 1053
Maswood, Ali. I. ...........................366
Mazumder, Sudip K. ......................860
Meleshin, V.I. ...............................383
Mirzaeian, B. ................................193
Mishra, Mahesh K. ........................537
Mishra, S. ..............................735, 979
Mittal, A. P. ............................892, 96
Moallem, Peyman ..........................101
Modi, P. K. ....................................473
Mohan, D. Madhan .......................1131
Mohan, Ned .................250, 707, 765
Mohapatra, K.K. ...........................765
Mohapatra, Krushna K ...................250
Moharana , A. K. .........................1127
Mohod , S. W. ........................395, 782
Mokhtari, H. ..............596, 631, 657
Mondal, Gopal ..............................256
Moradi, H. ....................................151
Morris, Stella .................................46
Mufti, Mairaj-ud-Din .....................865
Muni, B.P. .....................................718
Muni, Bishnu P. ......................111, 667
Murthy, K. V. S. Ramachandra .......550
Murthy, S. S. ............7, 39, 842, 1084, 836
Muthuselvan, N.B. .........................966

## N

Nabav, S.M.H ......................970, 1053
Nagamani, C. ...............................1120
Naidu, Kiran ..........................39, 842
Nair, Manjula G. ...........................560
NAKAMURA, Masaaki ...................307
Nakamura, Masaaki .......................310
Narayanan, G. ...............................231
Natarajan, S.P. ..............................373
Negnevitsky, M. .....................753, 830
nezhad, S. M. Saghaeian ................193
Nirody, J. S. ...................................75

## O

Ovchinnikov, D.A. .........................383

## P

P, Vinodh Kumar ...........................747
Pai, M. A. .....................................818
Palanisamy, V. ................................15
Panda, A. K. .................................378
Panda, G. ....................................1064
Panda, R. C. .................................463
Pandi, V. Ravikumar ......................695
Panigrahi, B. K. .....695, 724, 1064, 1127, 1131

Panigrahi, Ms. K. ........................1127
Pant, Vinay ...................................572
Parmar, K.P.Singh .........................882
Parthiban, P. .................................531
Patel , Rajesh ..............................1031
Payam, A. Farrokh ......54, 59, 193, 215
Perumal, B.Venkatesa .....................812
Peterson, Maryclaire ...............280, 1024
Phattanasak, M. ............................407
Pickert, V. ..............................20, 389
Pinto, A. J. P. ...................................7
Poshtan, Majid .............................565
Pradhan, A K ..................................70
Prakash, Anupama .......................1036
Prasad, P.V.N. ...............175, 225, 759

## R

Rahimi, M. ....................................596
Rahmati, A. ...................................298
Raj, C. Thanga ................................86
Rajagopal, K. R. ...146, 157, 163, 168, 182, 187
Rajaram, M. ..................................117
Ramakrishnan, K. ..........................244
Ramalingam, BMSM ......................555
Ramanarayanan, V. ........................455
Ramesh, L. .....................................90
Ramrathnam, ................................842
Ranjbar, A. M. ......................931, 1079
Ranjbar, M. ...................................931
Rao, E.V. Chandra Sekhara ............175
Rao, G. Govinda ...........................550
Rao, M.V. Ramana ........................225
Rao, Polimera Malleswara ..............854
Rao, S. Eswar ...............................667
Rao, T.K. Nagaraja ........................787
Rastgoufard, Parviz .......................565
Ravi, Jally ....................................237
Ravi, N. ...............................50, 171
Ravichandran, M. H. ...............122, 131
Ravikumar, B. ...............................936
Ravindranath, G. ...........................175
Reddy, K. V. V. ............................1013
Reddy, Sathish Kumar ....................560
Reddy, T. Brahmananda ......262, 424, 511
Reddy, Y.V. Siva ...........................262
Rlavi, Jally ...................................141

## S

S, Meera K ...................................747
Sadati, N. .....................................931
Sadhukhan, Gautam ........................70
Sagar, Prem .................................1002
Saha, A. K. .....................................90
Saha, R. .......................................1102
Saini, R P .............................801, 776
Samantaray, S. R. .........................1064
Sanavullah, M.Y. .............................15
Sanglikar, Amit .............................495
Sanjeevikumar, P. ..........................331
Sankar, V. ....................................662
Saritha, B. ..............................485, 490
Sarkar, Arghya ..............................689

# Author Index

Sarma, A.V.R.S. ....................................... 759
Sarma, D.V.S.S.Siva ............................... 219
Sathaiah, Chippa ................................... 157
Satish, T. ................................................ 765
Saxena, Rakesh .................................... 1047
Sekhar, K.Chandra ................................ 449
Selvajyothi, K. ...................................... 641
Sengupta, S. .......................................... 689
Seni, ...................................................... 830
Serni, P. J. A. ............................... 918, 926
Seshachalam, D. .................................... 370
Shaikholeslami, A. ............................ 1, 625
Shakarami, M. R. ..................................... 25
Sharma , Deepak .................................. 1074
Sharma ,V.K. ......................................... 315
Sharma, A.K. ........................................ 1047
Sharma, Deepen .................................... 413
Sharma, K. Manjunatha ......................... 787
Sharma, M P .................................. 801, 80
Sharma, V. K. ....................................... 436
Sharma, Vishnu K ................................. 460
Shateri, H. ............................................ 1114
Shazreen, .............................................. 740
Shenoy, T. P. .......................................... 75
Shet, Vinayak N. ............................ 286, 292
Sheth, Nimit K. .................... 182, 146, 187
Shirani, A. R. ....................................... 1079
Shyamala, P. ......................................... 996
Silva, I. N. da ................................. 918, 926
Singh , Ravindra Kumar .................. 321, 417
Singh, B. P. ...................... 107, 126, 342
Singh, Bhim ........39, 64, 96, 107, 126, 204, 237, 336, 342
Singh, Brij N. .................... 280, 565, 1024
Singh, Chanan ...................................... 806
Singh, G. K. ......................................... 776
Singh, S. P. ............................................ 80
SinghT , Fhim ....................................... 141
Singla, Bhoj Raj ..................................... 30
Sinha, S. K. ........................................... 724
Siva, U. ................................................. 842
Siva, Uddanti .......................................... 39
Sivakumaran, T.S. .................................. 373
Sivanagaraju, S. .................................... 662
Sivanandakumar, D. ............................... 244
Skandari, R. ........................................... 298
Slabharwal, Slatish Chander ................... 35
Solanki, Jitendra ................................... 614
Soleymani, S. ................................ 931, 1079
Somasundaram, P. ................................. 966
Song, Y. H. ............................................. 90
Soni, K. M. .......................................... 436
Sood, Vijay K. ...................................... 729
Srinivasu, B. ......................................... 225
Srivastava, S. P. ............................... 86, 531
Srividhya, S. ........................................ 1120
Subbarayudu, D. ............................ 424, 511
Subramanian, N. ................................... 1154
Sudhakar, Singamaneni Bala ................. 479
Suryawanshi, H. M. ........................ 348, 506
Sydulu, M. ............................................ 942

## T

TAKAHASHI, Nobuo ............................. 307

Takahashi, Nobuo ................................. 310
Tandon, A. K. ....................................... 836
Tekwani, P.N ........................................ 256
Tewolde, Meharegzi ............................. 360
Thakre, K. L. ........................................ 985
Thakur, T. ..................................... 712, 992
Thakur, Tripta ...................................... 797
Thomas, Mini S. ......................... 1036, 1041
Thukaram, D. ..................... 936, 953, 996
Tofighi, A. ........................................... 1107
Toliyat, H. ............................................ 151
Tripathi, R. K. ...................................... 370
Tripathy, M. ......................................... 979
Tseng, Shao-Chun ................................ 199

## U

Umamaheswari, B. ................................ 463
Upadhyay, Parag ........................... 163, 168

## V

Vaitheeswaran, N. ................................ 771
Vasudevan, JM ..................................... 555
Veena, P. .............................................. 117
Venugopal, S. ....................................... 231
Verma, Vishal ...................................... 589
Virulkar , V. B. .................................... 395
Vithal, JVR .......................................... 667
Vittal, K.P. ........................................... 787
Voropai , N. I. ...................................... 909
Vyjayanthi, C. ...................................... 953

## W

Wadhwani , Sulochana ......................... 1059
Wadhwani, A. K. .......................... 210, 1059
Wang, Lingfeng .................................... 806
Wani, M. G. .......................................... 436

## Y

Yadav, K. B. ......................................... 776
Yesuratnam, G. .................................... 953
Yokozeki , Ichiro ................................. 310

## Z

Zahawi, B. ...................................... 20, 389, 1020
Zhong-Xia, Niu ............................... 11, 522
Zué, Aslain Ovono ............................... 583

CURRAN ASSOCIATES INC.
proceedings
.com

9780780397712